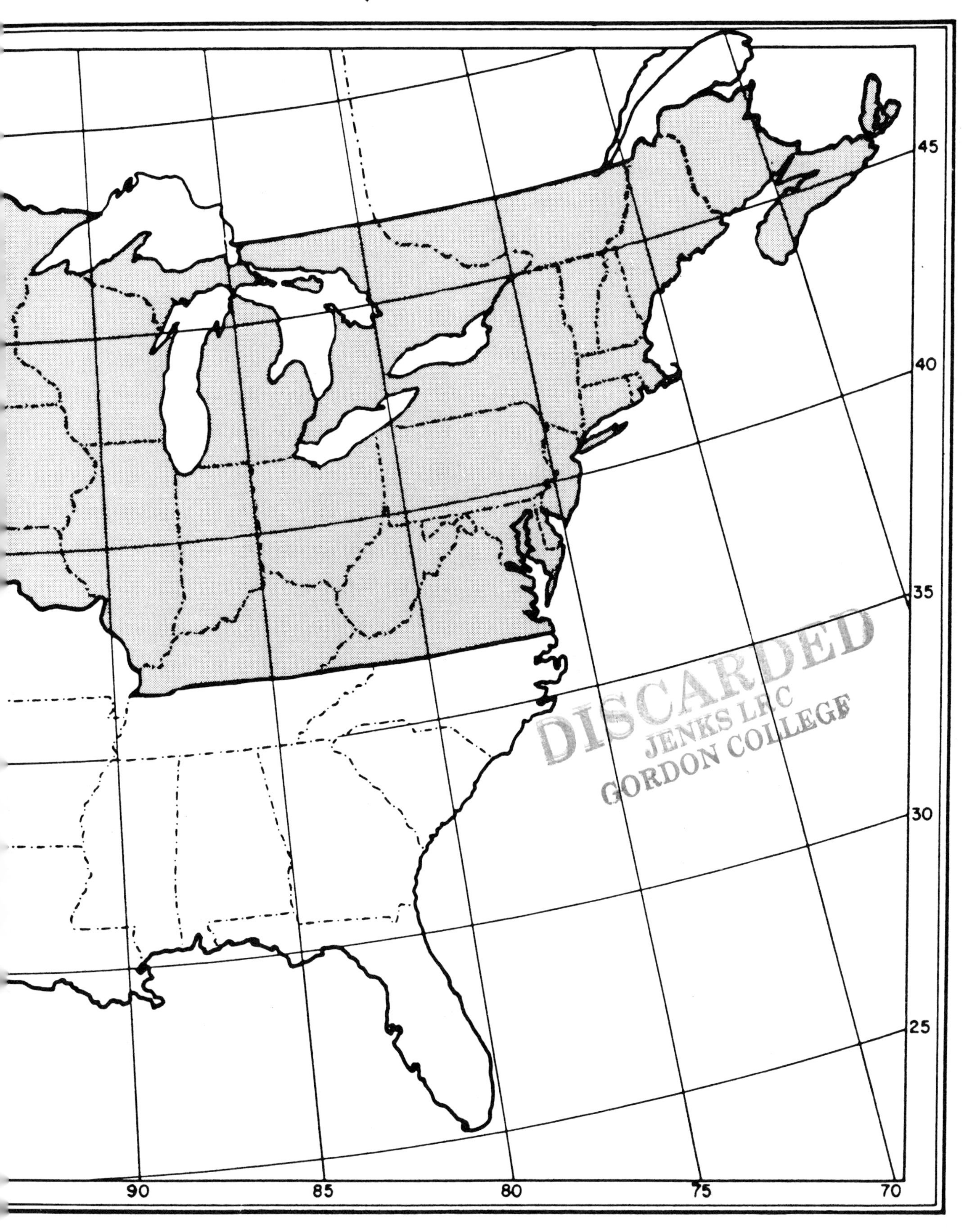

Illustrated Companion to Gleason and Cronquist's Manual

Illustrations of the Vascular Plants of Northeastern United States and Adjacent Canada

Noel H. Holmgren

with the artistic and editorial assistance of
Patricia K. Holmgren
Robin A. Jess
Kathleen M. McCauley
Laura Vogel

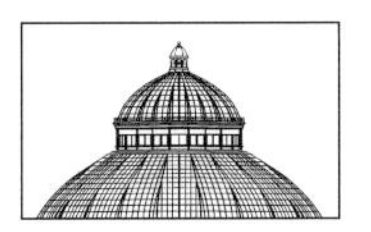

THE NEW YORK BOTANICAL GARDEN
1998

Second Printing, 2004
Third Printing, 2005
Fourth Printing, 2006

Cover illustration © 2003 Bobbi Angell: tulip-tree, *Liriodendron tulipifera* L. (Magnoliaceae)

The paper used in this publication meets the requirements of the
American National Standard for Information Sciences—Permanence of Paper for
Publications and Documents in Libraries and Archives, ANSI/NISO (Z39.48–1992).

Printed in the United States of America by the Maple-Vail Book Manufacturing Group
using soy-based ink on recycled paper.

MetLife Foundation is a Leadership Funder of The New York Botanical Garden Press.

Library of Congress Cataloging-in-Publication Data

Holmgren, Noel H.
     Illustrated companion to Gleason & Cronquist's manual:
illustrations of the vascular plants of northeastern United
States and adjacent Canada / Noel H. Holmgren; with the artistic
and editorial assistance of Patricia K. Holmgren . . . [et al.].
       p. cm
     Includes bibliographical references and indexes.
     ISBN 0-89327-399-6
     1. Botany — Northeastern States. 2. Botany — Canada, Eastern.
3. Botany — Northeastern States — Pictorial works. 4. Botany — Canada,
Eastern — Pictorial works. 5. Plants — Identification. I. Holmgren, Noel H.
II. Holmgren, Patricia K. III. Gleason, Henry A. (Henry Allan), 1882–1975,
Manual of vascular plants of northeastern United States and adjacent Canada.
III. Title
QK117.H65 1998
581.974—dc21
                                    97-47032

This book is dedicated to the illustrators whose skillful drawings brought vivid life first to the text of Gleason's *New Britton and Brown Illustrated Flora,* subsequently to many other publications over the past 45 years, and now to the Gleason and Cronquist *Manual.*

Walter Lincoln Graham
Eduardo Salgado
Anne Rogelberg
Lucille E. Kopp
Mary Content Easton
N. Davis

The Scientific Publications Program at The New York Botanical Garden thanks the following donors for their generous support of this publication:

The Andrew W. Mellon Foundation
The Davis Conservation Foundation
Mary Flagler Cary Charitable Trust
Metropolitan Life Foundation

CONTENTS

Foreword	vii	Urticaceae	71
Preface	ix	Juglandaceae	72
Acknowledgments	xi	Myricaceae	75
Introduction	xiii	Fagaceae	75
Illustrations:		Betulaceae	82
Lycopodiaceae	1	Phytolaccaceae	84
Selaginellaceae	4	Nyctaginaceae	85
Isoetaceae	4	Aizoaceae	85
Equisetaceae	5	Cactaceae	86
Ophioglossaceae	8	Chenopodiaceae	86
Osmundaceae	10	Amaranthaceae	94
Schizaeaceae	11	Portulacaceae	98
Hymenophyllaceae	11	Molluginaceae	100
Polypodiaceae	12	Caryophyllaceae	100
Dennstaedtiaceae	12	Polygonaceae	112
Adiantaceae	14	Plumbaginaceae	125
Aspleniaceae	17	Theaceae	125
Onocleaceae	29	Elatinaceae	125
Blechnaceae	30	Clusiaceae	126
Marsileaceae	30	Tiliaceae	131
Salviniaceae	30	Sterculiaceae	131
Taxaceae	31	Malvaceae	131
Pinaceae	31	Sarraceniaceae	138
Taxodiaceae	34	Droseraceae	138
Cupressaceae	35	Cistaceae	139
Magnoliaceae	36	Violaceae	142
Annonaceae	38	Tamaricaceae	149
Calycanthaceae	38	Passifloraceae	149
Lauraceae	38	Cucurbitaceae	150
Saururaceae	39	Loasaceae	151
Aristolochiaceae	40	Salicaceae	152
Nelumbonaceae	42	Capparaceae	161
Nymphaeaceae	42	Brassicaceae	162
Cabombaceae	44	Resedaceae	182
Ceratophyllaceae	44	Cyrillaceae	183
Ranunculaceae	45	Clethraceae	183
Berberidaceae	62	Empetraceae	183
Lardizabalaceae	63	Ericaceae	184
Menispermaceae	63	Pyrolaceae	195
Papaveraceae	64	Monotropaceae	196
Fumariaceae	66	Diapensiaceae	197
Platanaceae	68	Sapotaceae	198
Hamamelidaceae	68	Ebenaceae	198
Ulmaceae	68	Styracaceae	198
Cannabaceae	70	Symplocaceae	199
Moraceae	70	Primulaceae	199

Hydrangeaceae	204
Grossulariaceae	205
Crassulaceae	208
Saxifragaceae	211
Rosaceae	217
Mimosaceae	255
Caesalpiniaceae	256
Fabaceae	258
Elaeagnaceae	288
Podostemaceae	289
Haloragaceae	289
Lythraceae	291
Thymelaeaceae	293
Trapaceae	293
Onagraceae	294
Melastomataceae	303
Cornaceae	304
Santalaceae	306
Viscaceae	307
Celastraceae	308
Aquifoliaceae	309
Buxaceae	311
Euphorbiaceae	311
Rhamnaceae	321
Vitaceae	322
Linaceae	326
Polygalaceae	328
Staphyleaceae	330
Sapindaceae	330
Hippocastanaceae	330
Aceraceae	331
Anacardiaceae	334
Simaroubaceae	335
Meliaceae	335
Rutaceae	336
Zygophyllaceae	337
Oxalidaceae	337
Geraniaceae	338
Limnanthaceae	341
Balsaminaceae	341
Araliaceae	342
Apiaceae	344
Loganiaceae	361
Gentianaceae	362
Apocynaceae	368
Asclepiadaceae	370
Solanaceae	376
Convolvulaceae	384
Cuscutaceae	386
Menyanthaceae	388
Polemoniaceae	389
Hydrophyllaceae	392
Boraginaceae	395
Verbenaceae	403
Lamiaceae	408
Hippuridaceae	431
Callitrichaceae	431
Plantaginaceae	431
Buddlejaceae	434
Oleaceae	435
Scrophulariaceae	438
Orobanchaceae	458
Acanthaceae	459
Pedaliaceae	460
Bignoniaceae	461
Lentibulariaceae	462
Campanulaceae	464
Rubiaceae	469
Caprifoliaceae	478
Adoxaceae	485
Valerianaceae	486
Dipsacaceae	487
Asteraceae	489
Butomaceae	602
Alismataceae	602
Hydrocharitaceae	606
Scheuchzeriaceae	607
Juncaginaceae	608
Potamogetonaceae	608
Ruppiaceae	613
Najadaceae	614
Zannichelliaceae	614
Zosteraceae	614
Acoraceae	615
Araceae	615
Lemnaceae	616
Xyridaceae	617
Commelinaceae	619
Eriocaulaceae	621
Juncaceae	622
Cyperaceae	631
Poaceae	690
Sparganiaceae	776
Typhaceae	778
Bromeliaceae	779
Pontederiaceae	779
Liliaceae	780
Haemodoraceae	781
Agavaceae	804
Smilacaceae	804
Dioscoreaceae	806
Iridaceae	807
Burmanniaceae	812
Orchidaceae	812
Index to Common Names	829
Index to Scientific Names	891

FOREWORD

The second edition of Gleason and Cronquist's *Manual of the Vascular Plants of Northeastern United States and Adjacent Canada*, on which the *Illustrated Companion* is based, has gone through several reprintings since its original publication in 1991. Some of the most conspicuous errors were corrected in the second printing in 1993, and subsequent errors brought to our attention by many helpful users of the *Manual* have been incorporated in a corrected edition of the *Manual*, which is scheduled for publication in early 2004. Considerable care was given to re-paging the *Manual* so that the cross-referencing with the *Illustrated Companion* could be maintained. The name changes in the corrected edition of the *Manual* have been incorporated into this reprinting of the *Illustrated Companion*.

The following list includes the 20 name changes and 13 orthographic corrections that have been made on the plates.

Noel H. Holmgren
Mary Flagler Cary Curator
The New York Botanical Garden
December 2003

Page	Old Name	New Name
148	23. *Viola adunca* var. *minor*	23. *Viola adunca* var. *glabra*
218	*Spiraea vanhouttei*	*Spiraea ×vanhouttei*
322	5. *Rhamnus citrifolia*	5. *Rhamnus davurica*
334	3. *Rhus copallinum*	3. *Rhus copallina*
347	4. *Osmorhiza chilensis*	4. *Osmorhiza berterii*
366	2. *Gentianopsis procera*	2. *Gentianopsis virgata*
419	1. *Orthodon dianthera*	1. *Mosla dianthera*
514	4. *Iva xanthifolia*	4. *Iva xanthiifolia*
608	1. *Triglochin maritimum*	1. *Triglochin maritima*
608	2. *Triglochin palustre*	2. *Triglochin palustris*
608	3. *Triglochin striatum*	3. *Triglochin striata*
616	1. *Wolffiella floridana*	1. *Wolffiella gladiata*
616	1. *Wolffia papulifera*	1. *Wolffia brasiliensis*
616	2. *Wolffia punctata*	2. *Wolffia borealis*
638	8. *Eriophorum polystachion*	8. *Eriophorum angustifolium*
689	1. *Cymophyllus fraseri*	1. *Cymophyllus fraserianus*
692	1. *Brachyelytrum erectum* var. *septentrionale*	1. *Brachyelytrum erectum* var. *glabratum*
695	2. *Festuca elatior*	2. *Festuca arundinacea*
721	2. *Bromus willdenowii*	2. *Bromus catharticus*
721	7. *Bromus altissimus*	7. *Bromus latiglumis*
761	1. *Amphicarpum purshii*	1. *Amphicarpum amphicarpon*
766	1. *Setaria geniculata*	1. *Setaria parviflora*
778	9. *Sparganium minimum*	9. *Sparganium natans*
784	2. *Melanthium hybridum*	2. *Melanthium latifolium*
786	*Scilla nonscripta*	*Scilla non-scripta*
799	2. *Streptopus roseus*	2. *Streptopus lanceolatus*
804	3. *Smilax ecirrata*	3. *Smilax ecirrhata*
807	3. *Dioscorea batatas*	3. *Dioscorea polystachya*
826	4. *Corallorrhiza wisteriana*	4. *Corallorhiza wisteriana*
827	1. *Corallorrhiza maculata*	1. *Corallorhiza maculata*
827	2. *Corallorrhiza trifida*	2. *Corallorhiza trifida*
827	3. *Corallorrhiza striata*	3. *Corallorhiza striata*
827	5. *Corallorrhiza odontorhiza*	5. *Corallorhiza odontorhiza*

PREFACE

The veracity of the adage "a picture is worth a thousand words" is probably never better borne out than when it comes to well-prepared botanical illustrations. No matter how detailed and thoughtfully rendered written descriptions of plant species may be, there is no substitute for a good line drawing to confirm or reject a plant identification. An illustration is the next best thing to having an actual, authoritatively annotated herbarium specimen at hand for comparison. And yes, even professional botanists rely on illustrations to confirm the conclusions they reach initially through the use of technical keys.

This illustrated companion volume to the second edition of Gleason and Cronquist's 1991 *Manual of Vascular Plants of Northeastern United States and Adjacent Canada* serves to make whole the utility of the initial undertaking. Noel H. Holmgren, with editorial assistance from Patricia K. Holmgren and artistic and editorial assistance from Laura Vogel, Kathleen M. McCauley, and Robin A. Jess, has planned and coordinated the preparation of this *Illustrated Companion to Gleason and Cronquist's Manual.* The result is a book that will be of tremendous value to seasoned users of the original Gleason and Cronquist as well as to individuals who found the non-illustrated Gleason and Cronquist too inaccessible.

Preparation of the present volume was much more than a mere cut-and-paste job from Henry A. Gleason's *New Britton and Brown Illustrated Flora.* Decisions were made on which drawings best represented the species in those cases where names had been misapplied or where two or more species have been placed in synonymy. Noel Holmgren and his associates added many new details, and specimens to best represent the taxon were selected for the artist. Gleason and Cronquist's *Manual* of 1991 contains many additional species that were not treated in Gleason's *Illustrated Flora* of 1952. New illustrations were made for all of these species.

Noel Holmgren and his collaborators are to be congratulated for the great service they have rendered to all students of the flora of northeastern North America by making available again the illustrations from Gleason's *New Britton and Brown Illustrated Flora.* This book follows in a proud tradition of scholarly publishing by The New York Botanical Garden in North American floristic studies.

Brian M. Boom
Vice President for Botanical Science and Pfizer Curator of Botany
The New York Botanical Garden
October 1997

ACKNOWLEDGMENTS

I am grateful to many people who have been involved in one way or another during the four years it has taken to complete the *Illustrated Companion*. Of these, six are singled out for special acknowledgment. Lucille (Kopp) Blum, one of the original artists of the *New Britton and Brown Illustrated Flora*, helped immeasurably by reconstructing the history of the project during her years of involvement. Sandi Frank, then Director of the Garden's Scientific Publications Department, had the courage to take responsibility for arranging all funds necessary for carrying out the project. She administered the disbursement of funds and dealt with making photostat copies of each of the existing 1055 original plates (19 were missing). Patricia K. Holmgren edited the plates for spelling, magnifications, and page references. Robin A. Jess compiled a computer checklist of all the taxa recognized in Gleason and Cronquist's *Manual*, did most of the plate composition (cut-and-paste), retraced the illustrations from 19 missing original plates plus some blurred photostats, and drew some of the new artwork. Kathleen M. McCauley labeled most of the plates, composed several plates, did the research on the name of the artist for each species, checked the labeled drawings against the originals in the *Illustrated Flora*, and added the common names from Gleason and Cronquist's *Manual* to the index for the *Illustrated Companion*. Laura Vogel provided most of the new artwork and composed and labeled many plates.

Others have helped bring this project to completion. Bobbi Angell offered valuable advice and encouragement throughout the course of the project. Brian M. Boom, the Garden's Vice President for Botanical Science, played an important role in the early planning stages. Susan Fraser, the Garden's Archivist, has allowed access to the original plates and has helped locate historical information. C. Eric Hellquist did the research on the dates of employment of the original artists. Jacquelyn A. Kallunki, Assistant Director of the Garden's Herbarium, constructively reviewed the manuscript. The recollections of Celia K. Maguire, who was on the Garden's staff during the latter years of production of Gleason's *New Britton and Brown Illustrated Flora*, have helped us understand and appreciate the history. Scott Mori, Director of the Garden's Institute of Systematic Botany, gave enthusiastic encouragement. Melissa A. Rossow, Herbarium Intern, spent one hectic day-before-deadline helping place protective covers on each of the 827 plates. Joy E. Runyon, Project Editor, and Susan Frayman, Director of the Garden's Scientific Publications Department, have overseen the technical aspects of publication, enabling us to proceed without giving this aspect any consideration. Dennis W. Stevenson helped convert the Apple Computer taxon checklist tables compiled by Robin Jess into a format recognizable by my database management system.

I gratefully acknowledge the support of the *Illustrated Companion* project by the Mary Flagler Cary Charitable Trust, The Andrew W. Mellon Foundation, The Davis Conservation Foundation, the Metropolitan Life Foundation, and The New York Botanical Garden.

Noel H. Holmgren, *Mary Flagler Cary Curator*
The New York Botanical Garden
October 1997, revised December 2003

INTRODUCTION

The collection of illustrations in the *Illustrated Companion to Gleason and Cronquist's Manual* (*Illustrated Companion*) is based on the second edition of Henry A. Gleason and Arthur Cronquist's (1991) *Manual of Vascular Plants of Northeastern United States and Adjacent Canada* (*G&C Manual*). The *G&C Manual* was published in 1991 as a substantially revised version of the 1963 first edition.

The first edition was based on the 1952 *New Britton and Brown Illustrated Flora of the Northeastern United States and Adjacent Canada* (*Illustrated Flora*) by Gleason and collaborators. The *G&C Manual* correctly bore Gleason's name as senior author, even though he was not directly involved in its production. Gleason envisioned the book as a field manual to accompany the *Illustrated Flora*, using Britton's 1901 *Manual of the Flora of the Northern States and Canada* as a model. To this end, Cronquist followed the three-volume work as closely as possible, complete with references to volume and page number, even to the extent of retaining the outmoded Englerian sequence of families. Most of the first edition was a direct copy of Gleason's descriptions and keys in the *Illustrated Flora*. The changes were few, mainly incorporating results of taxonomic revisions published during the interim. The book was designed to be small enough for carrying into the field, and the cross-references allowed the user to refer to the *Illustrated Flora* for its illustrations.

In preparing the second edition, Cronquist used Gleason's wording in keys and descriptions wherever possible, but he made substantial taxonomic changes. The family sequence follows his own widely used system (Cronquist, 1981). The second edition was intended to stand on its own, with no attempt to cross-reference it to the *Illustrated Flora*.

Users of the second edition of the *G&C Manual* who have access to the *Illustrated Flora* have learned that the differences between the two are great, making it difficult and time consuming to track down the appropriate illustrations. Moreover, Gleason's *Illustrated Flora* has been out of print for some time now. The time had come for this compilation of illustrations.

For the *Illustrated Companion*, photostats were made from the original artwork used in the *Illustrated Flora*. The photostats were cut, and the pieces were rearranged into the sequence of the second edition of the *G&C Manual*. New illustrations of the several species not treated in the earlier work were prepared, and many useful diagnostic details were added to old illustrations.

New York Botanical Garden's Succession of Three-Volume Illustrated Floras

Since 1898, The New York Botanical Garden has provided professional and amateur botanists with a fully illustrated account of the vascular flora of northeastern

United States and southeastern Canada, beginning with the completion of the first edition of Nathaniel Lord Britton and Addison Brown's *An Illustrated Flora of the Northern United States, Canada and the British Possessions.* The three volumes appeared in sequence in 1896, 1897, and 1898. Britton and Brown published a second edition in 1913, and this work remained the only illustrated resource for the region until 1952, with the appearance of Gleason's three-volume *Illustrated Flora.* In the earlier Britton and Brown editions, each species illustration was restricted to a 2 × 2½ inch rectangle, a confinement that allowed for simplified habit and diagnostic details. The habit drawings resemble the flattened and often folded herbarium specimens from which they were drawn. Britton justified this format for reasons of economy: less page space and less costly to produce. The 1913 edition of the *Illustrated Flora* was reprinted in 1970 in paperback by Dover Publications, with a slight alteration in the title and slight reduction in page size.

In 1939, the Director of The New York Botanical Garden, William J. Robbins, directed Gleason to prepare a completely revised edition of the *Illustrated Flora,* to fill the growing need for a modern treatment to replace Britton and Brown's aging 1913 edition. Gleason completed the project within 13 years of its inception, which is a remarkable achievement. He was able to do it so expeditiously by his own efficient work habits and by delegating many aspects of the production to others. He enlisted ten colleagues to contribute critical groups, turned over supervision of the artists to members of his staff, and received considerable editorial assistance from other Garden staff. Also, because it was a New York Botanical Garden project, with the charge from the Director and encouragement of the Board of Directors, he did not have to concern himself with finding sources of financial support.

The *New Britton and Brown Illustrated Flora* was given a completely new look, with a new format design, new text, and new and much improved artwork. The geographical region treated was trimmed down in size to exclude many western and subarctic North American species (Fig. 1). The artwork was much improved in style and execution. The artists, liberated from the constraints of the small rectangles, had the freedom to use the full 6-inch width of the page and a height ranging from 2½ inches to 8½ inches, and they were encouraged to make the drawings appear as three-dimensional as possible. Without increasing the number of pages, more space for the illustrations was made available by eliminating the nomenclatural bibliographic citations and by treating fewer species through reduction in regional coverage and with more modern species concepts that resulted in synonymizing many of the taxa recognized in the earlier work.

Illustrators

The artists of the *New Britton and Brown Illustrated Flora* were not acknowledged by name, nor did they initial their work. The only illustrator mentioned by name in the introductory text, Henry C. Creutzburg, was not even directly involved. His work was cited because it was the only non-original artwork; parts of his illustrations of *Carex* from Kenneth K. Mackenzie's *North American Cariceae* (1940) were used. Luckily, most of the original plates were signed and dated on the back of the boards as they were completed, and so we have been able to identify the respective artists for most of the illustrations.

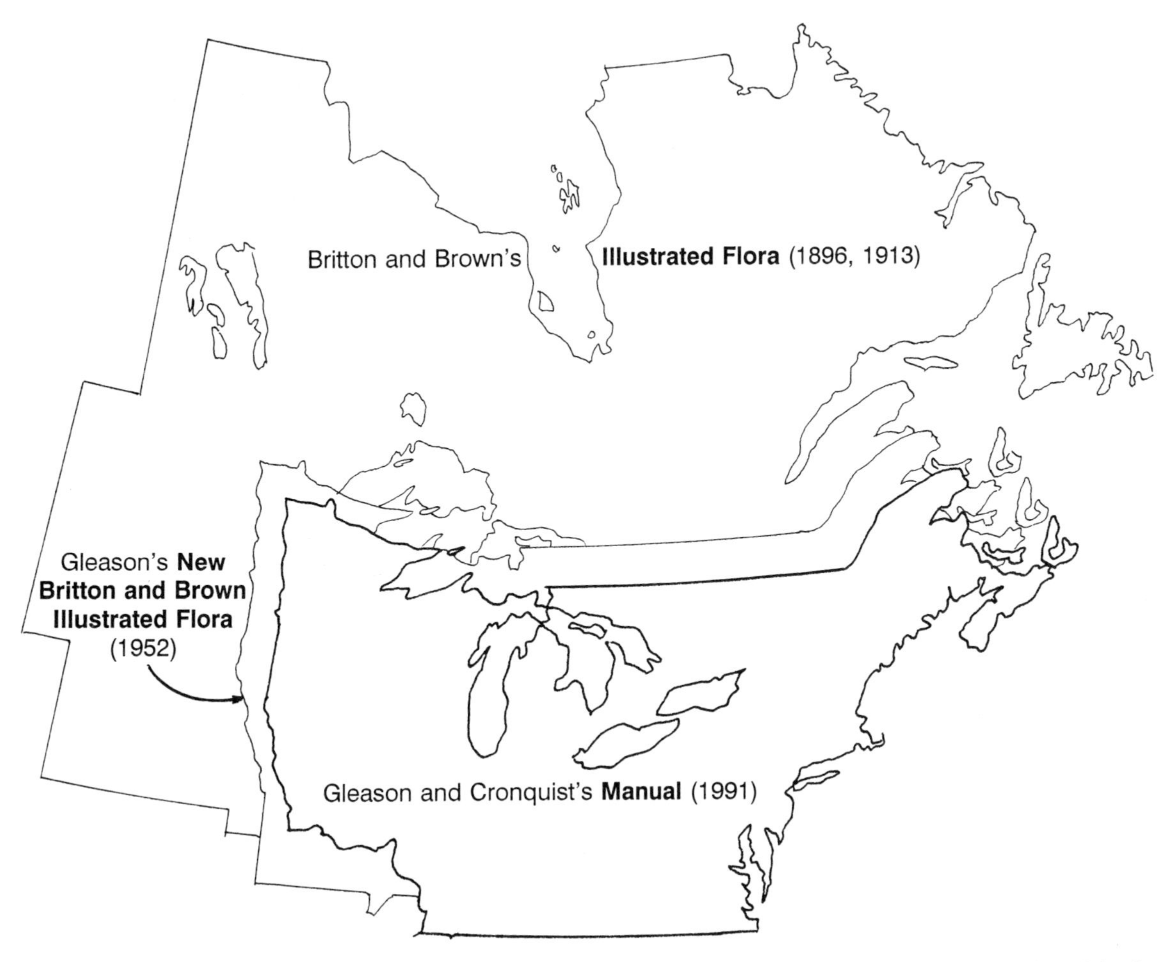

Figure 1. Northeastern United States and eastern Canada showing coverage of the floras mentioned in the text.

While Gleason was writing descriptions and keys, making field excursions to critical areas of the region, and coordinating the work of other contributors, the supervision of the artists was in the hands of Frances Elizabeth Wynne, Harold Norman Moldenke, and Edward Johnston Alexander. Wynne was Assistant Curator at the Garden from 1943 to 1946, during which time she also wrote the treatment of Droseraceae for the *Illustrated Flora*. Moldenke was on the staff from 1934 to 1951, rising through the ranks from Assistant Curator to Curator, and he also provided a treatment of his specialty, the Verbenaceae. Wynne and Moldenke would select the appropriate specimens from The New York Botanical Garden's Herbarium and give them to the artists along with the relevant manuscript and instructions. Alexander was acknowledged as one of the supervisors of the artists, but I found no information indicating direct involvement. Because he was so well acquainted with the Eastern U.S. flora, it is likely that his role was checking the finished illustrations for accuracy. Alexander was at the Garden from 1926 to 1963, joining the staff as John Kunkel Small's research assistant and retiring at the rank of Associate Curator. He contributed the families Iridaceae and Violaceae to the *Illustrated Flora*.

The artists were, in descending order of the size of their contribution, Walter L. Graham, Eduardo Salgado, Anne Rogelberg, Lucille E. Kopp, Mary C. Easton, and N. Davis. The artwork of four others was used. As already mentioned, parts of

Harry C. Creutzburg's published *Carex* illustrations were reused, and some of Mary Emily Eaton's and Margaret Sorenson's paintings from the journal *Addisonia* were copied. Six plates in the Rubiaceae were prepared by an unidentified artist.

Most of the following information on the artists was provided by Lucille (Kopp) Blum. Lucille E. Kopp joined the project in the winter of 1946, following graduation from Hunter College. At that time, Harold Moldenke was supervising Walter Graham and Anne Rogelberg. Mary Content Easton and N. Davis, who had been supervised by Frances Wynne, were no longer on the project; very little is known about them. Easton started in 1943 and Davis in 1945. Davis illustrated only two complete plates and small parts of four others.

Lucille recalled that in the summer of 1948, a student of Alma L. Moldenke (Harold's wife), from the Evander Childs High School, did some of the illustrations. We have not come across his name, but perhaps the six unsigned plates in the Rubiaceae, which are in a style different from any of the others, are in fact his.

On some of the original plates, additions and corrections have been patched over the original. Many of these do not appear in any of the reprinted versions of the *Illustrated Flora* in The New York Botanical Garden Library, including the last printing. Their origin or reason remains a mystery to us. However, as a consequence of their having been done, they are used in the *Illustrated Companion.*

The page layout of the *Illustrated Flora* was carefully done with each species illustration located conveniently near its description. To accomplish this, the size and content of each plate was predetermined. Lucille recollected that when Moldenke brought herbarium specimens and manuscript of the text to her, he would give her instructions on which species to include on each plate, the details to depict, and exactly how much vertical space to use. How Moldenke arrived at these calculations for size and inclusion is not clear, unless he and the editorial staff had typeset galley proofs in hand and were concurrently composing a dummy copy of the single- and double-column text. With the specimens in front of her, Lucille would have to carefully design the plate to include all the necessary details and achieve an overall aesthetic balance. She had to keep in mind the rule that parts of the illustrations should touch all four sides of the given space. Lucille's preliminary drawing was worked out on tracing paper and transferred to a four-ply bristol board by tracing over a light box (which had to be strongly lit to give such a heavy board sufficient transparency); the resulting sketch was then enhanced and inked. In some instances she would cover the backside of the tracing paper with overlapping pencil strokes and, positioning the sketch over the bristol board, would retrace it, producing a "carbon copy." In making corrections, Lucille would draw the corrected figure on a four-ply board, peel off the top two plies bearing the new rendition, and paste it over the incorrect depiction. The labeling was done by someone other than the artists.

The project came to an end for Lucille in 1950 as the project generally was winding up. She then enrolled in the graduate program at Columbia University, assisted in botany labs, and made illustrations for various teaching purposes. She also found part-time work at The New York Botanical Garden in various capacities ranging from illustrating for the Horticulture Department to working in the Native Plant Garden. She earned her Ph.D. in 1968, producing a revision of the genus *Persea* (Lauraceae) in the Western Hemisphere. She then became a wife and mother,

but worked part-time in various horticultural and illustration jobs on Long Island, closer to her home.

Walter Lincoln Graham was originally hired by the Garden as an artist under the Federal Art Project program of the Work Projects Administration (WPA) in November 1942. He was a bachelor who lived in the Bronx and retired from the Garden in 1950 in poor health.

Anne Rogelberg was a Hunter College graduate with a major in geology. She was hired sometime in late 1945 or early 1946. She married James T. Stromquist and in early 1949 moved to Chicago. She was a snazzy dresser, wearing clothes that she designed and made.

Eduardo Salgado, who was hired in early 1947, was retained until 1951 to finish illustrating the remaining few species and to redraw some required corrections. Beyond his employment at the Garden, he illustrated for several horticultural books and did commercial artwork in Manhattan. An accomplished painter in oils and watercolors, he had several one-man exhibitions of his work. Salgado was born in the Philippines, where he received his early training in art at the School of Fine Arts, University of the Philippines. Before coming to the Garden, he did postgraduate work in art at the University of Michigan.

Explanation of the Plates

The numbers (or number) in each genus-name header in the upper right or upper left corner of each plate refer to the pages where the descriptions for each illustrated species appear in the *G&C Manual*. The numbers preceding the names of the species match those used in the *G&C Manual*. Unnumbered species are those mentioned only incidentally in the *Manual*. The other numbers represent magnification and reduction sizes of the drawings. The illustrator of each species is indicated by initials. The initials of the artists of the *New Britton and Brown Illustrated Flora* are typeset in a serif font. In a few instances initials appear in parentheses. These refer to the original artist whose work first appeared in *Addisonia* and whose work was copied by the artists of the *Illustrated Flora*. The new illustrations by Laura Vogel and Robin Jess are initialed by them.

Five species and two hybrids were not illustrated because of the lack of plant specimens in the New York Botanical Garden Herbarium. Missing are (numbers refer to pages in the *G&C Manual*): *Cimicifuga rubifolia* (Ranunculaceae, 49), *Corispermum orientale* (Chenopodiaceae, 100), *Cerastium diffusum* (Caryophyllaceae, 117), *Populus ×canadensis* (Salicaceae, 168), *Chrysanthemum lacustre* (Asteraceae, 552), *Sphenopholis ×pallens* (Poaceae, 762), and *Habenaria correlliana* (Orchidaceae, 858).

We have used the second printing of the *G&C Manual* (1993), which contains some corrections in nomenclature and spelling of epithets. Our labeling reflects some further corrections in spelling, mainly involving orthographic changes, such as changes in gender or changing "ae" to "i."

Literature Cited

Britton, N. L. 1901. Manual of the flora of the northern States and Canada. Henry Holt, New York.

————— & A. Brown. 1896–1898. An illustrated flora of the northern United States, Canada and the British possessions. 3 vols. Charles Scribner's Sons, New York.

————— & —————. 1913. An illustrated flora of the northern United States, Canada and the British possessions. Ed. 2. 3 vols. Charles Scribner's Sons, New York.

————— & —————. 1970 [reprint of the 1913 edition]. An illustrated flora of the northern United States and Canada. 3 vols. Dover Publications, New York.

Cronquist, A. 1981. An integrated system of classification of flowering plants. Columbia Univ. Press, New York.

Gleason, H. A. 1952. The new Britton and Brown illustrated flora of the northeastern United States and adjacent Canada. 3 vols. Hafner Press, New York.

————— & A. Cronquist. 1963. Manual of vascular plants of northeastern United States and adjacent Canada. Van Nostrand, Princeton, New Jersey.

————— & —————. 1991 [reprinted in 1993 and 1995 with minor revisions]. Manual of vascular plants of northeastern United States and adjacent Canada. Ed. 2. New York Botanical Garden, Bronx.

Kopp, L. E. 1968. A taxonomic revision of the genus *Persea* in the Western Hemisphere (Lauraceae). Ph.D. thesis, Columbia University, New York.

Mackenzie, K. K. 1940. North American Cariceae. 2 vols. New York Botanical Garden, Bronx.

Illustrated Companion to Gleason and Cronquist's Manual

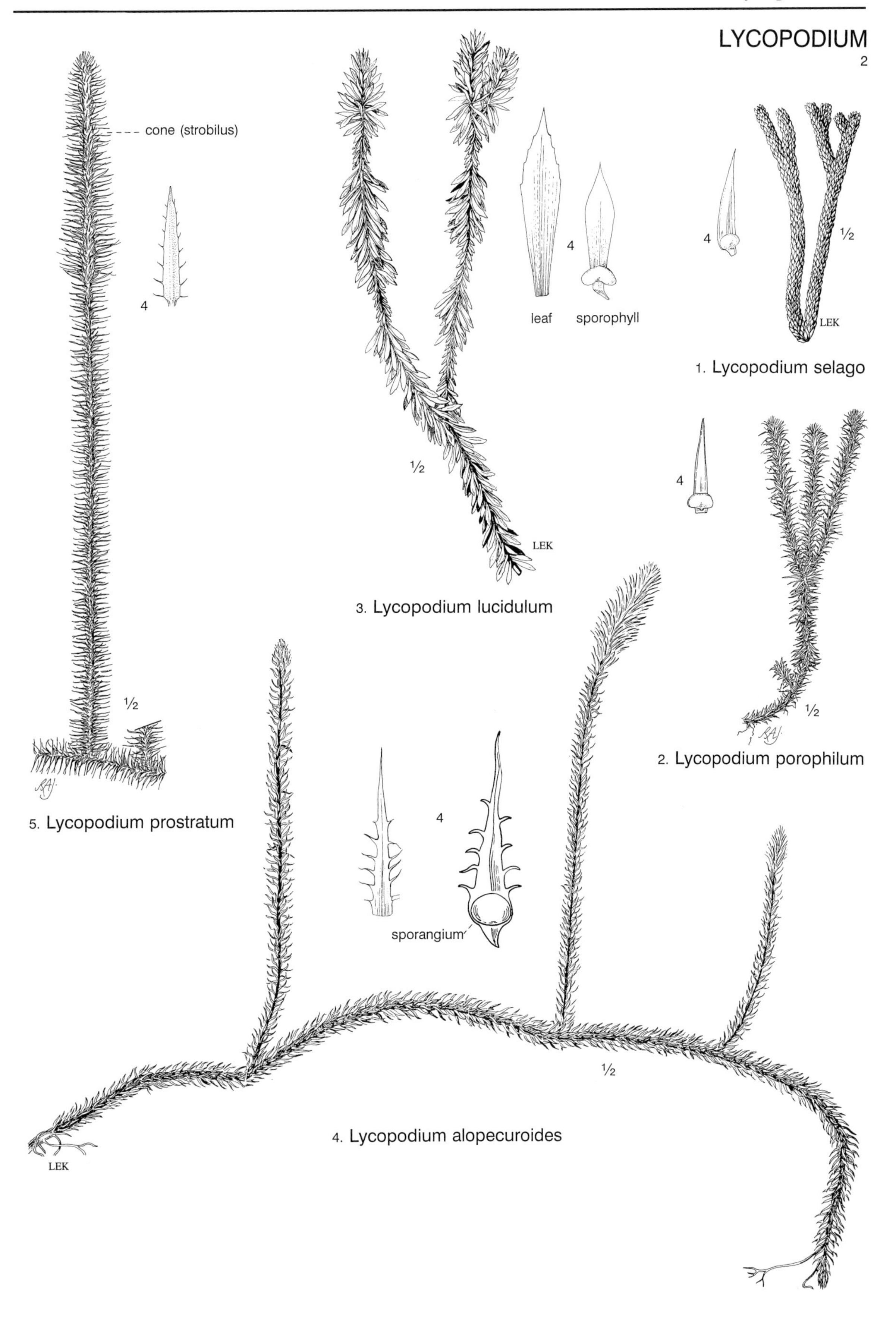

cone (strobilus)

4

leaf   sporophyll

4

4   ½

LEK

1. Lycopodium selago

½

LEK

3. Lycopodium lucidulum

4

4

½

2. Lycopodium porophilum

½

LEK

5. Lycopodium prostratum

4

sporangium

½

4. Lycopodium alopecuroides

LEK

LYCOPODIUM
2, 3

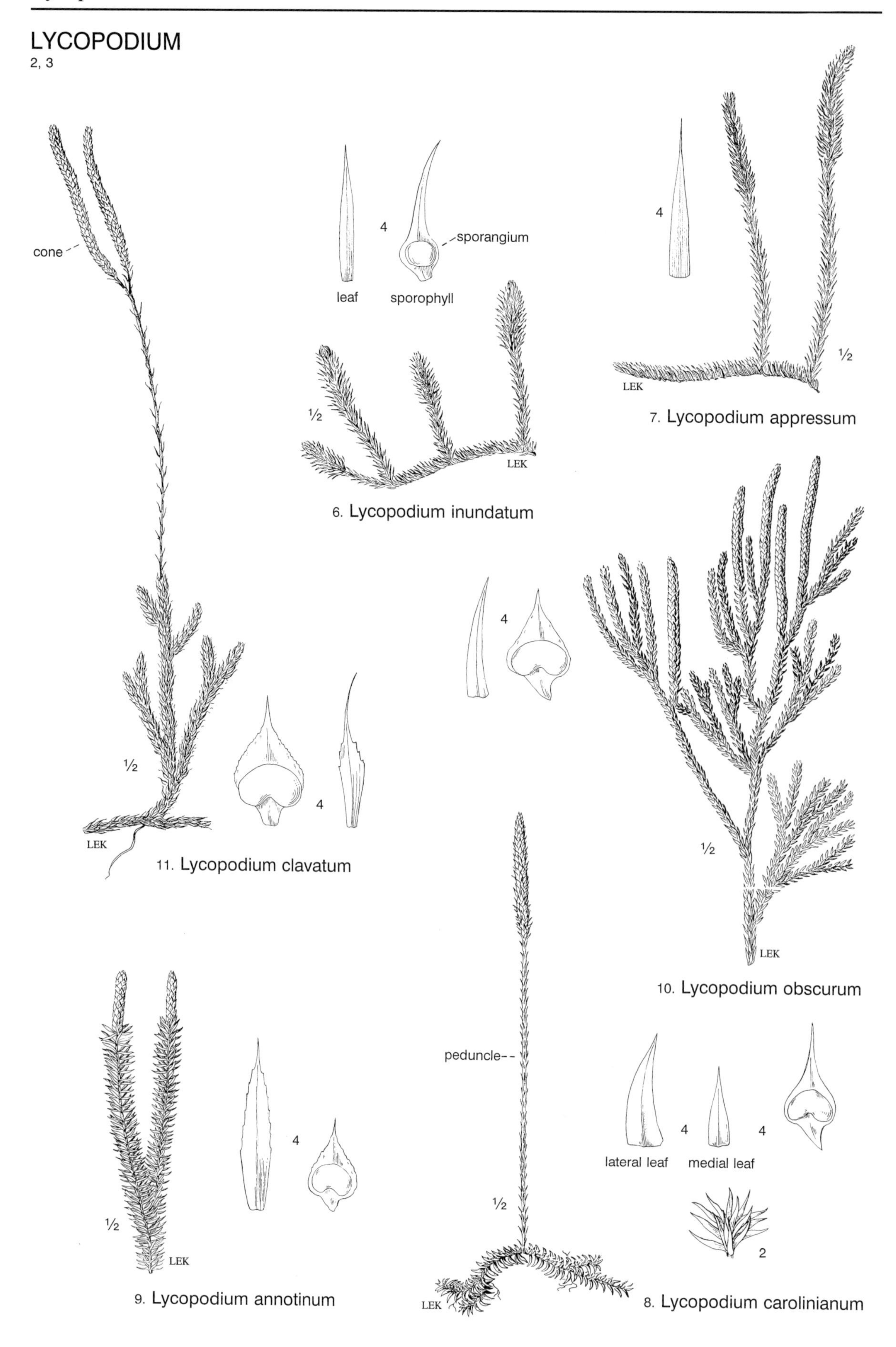

cone

leaf   sporophyll   4   sporangium

½   LEK

6. Lycopodium inundatum

4

½   LEK

7. Lycopodium appressum

4

½   LEK

11. Lycopodium clavatum   4

4

½   LEK

10. Lycopodium obscurum

peduncle

½   LEK

4   4

lateral leaf   medial leaf

2

½   LEK

9. Lycopodium annotinum

½   LEK

8. Lycopodium carolinianum

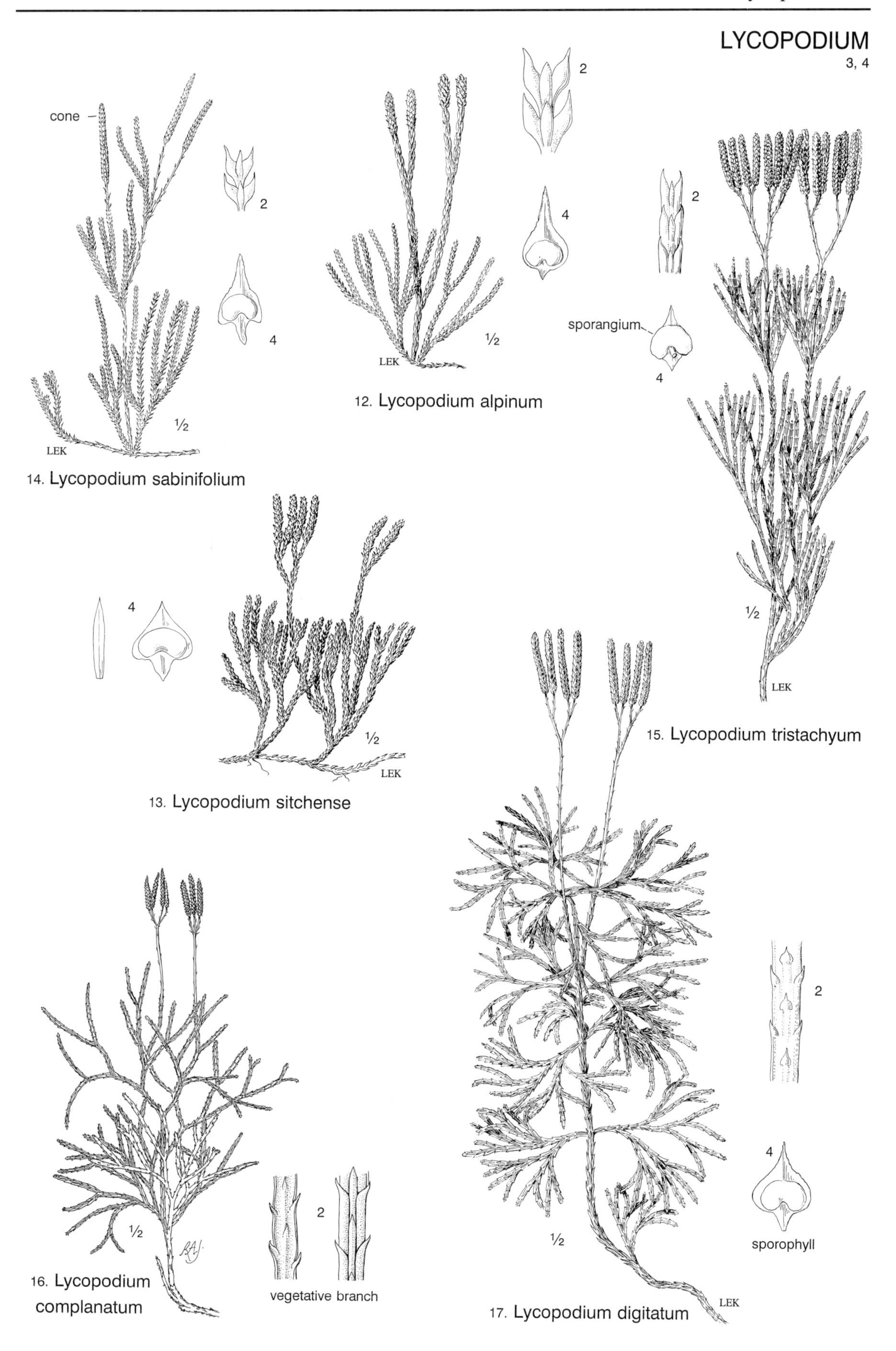

cone

2

4

LEK

½

14. Lycopodium sabinifolium

2

4

½

LEK

12. Lycopodium alpinum

2

sporangium

4

sporangium

LEK

½

LEK

15. Lycopodium tristachyum

4

½

LEK

13. Lycopodium sitchense

2

4

sporophyll

2

½

vegetative branch

16. Lycopodium complanatum

½

LEK

17. Lycopodium digitatum

SELAGINELLA
ISOETES
5, 6

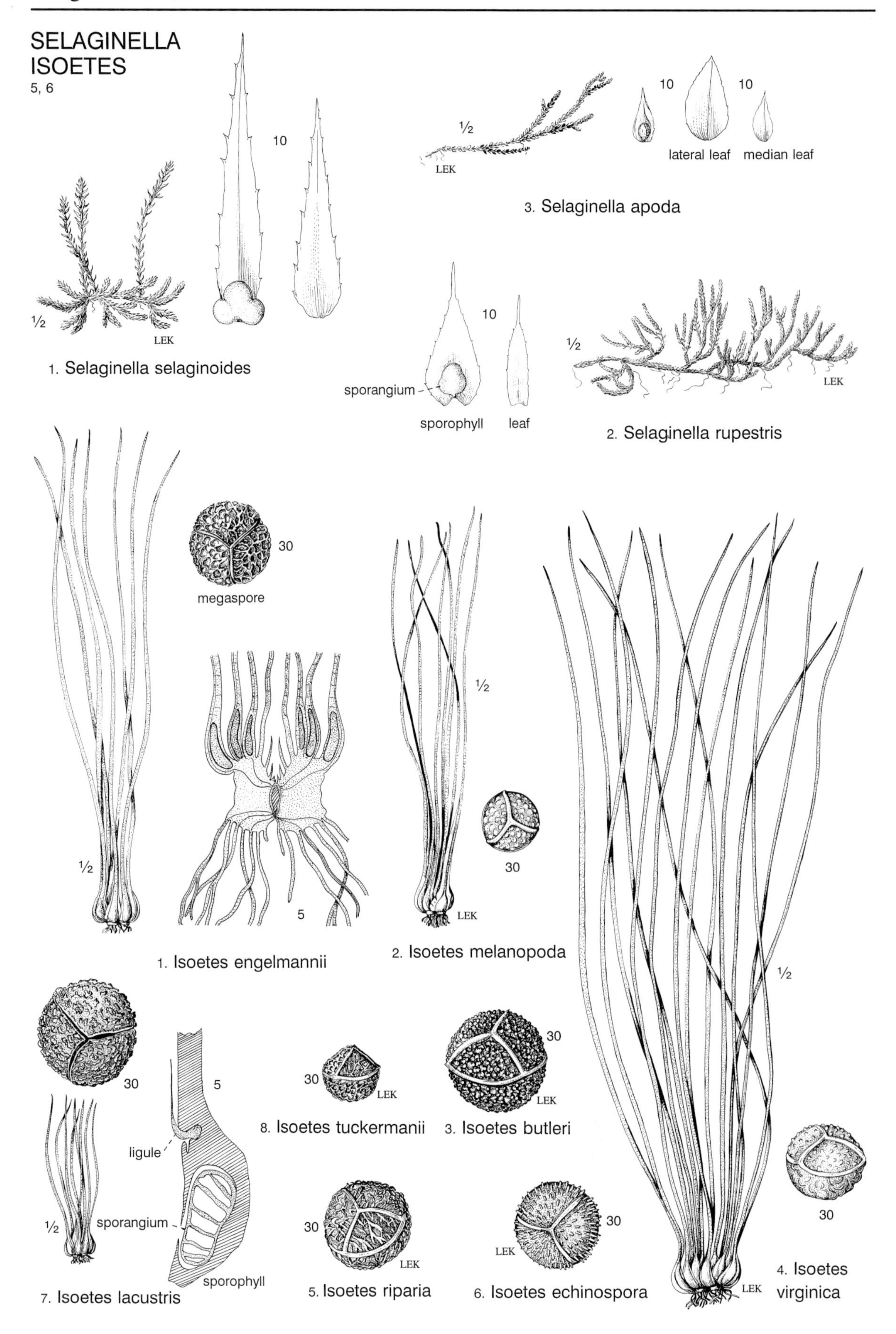

10

½ LEK

1. Selaginella selaginoides

10

10 lateral leaf median leaf

3. Selaginella apoda

10

sporangium

sporophyll   leaf

½ LEK

2. Selaginella rupestris

30
megaspore

½

5

1. Isoetes engelmannii

½

30

2. Isoetes melanopoda

30

5

ligule

sporangium

sporophyll

½

7. Isoetes lacustris

30 LEK

8. Isoetes tuckermanii

30 LEK

3. Isoetes butleri

30 LEK

5. Isoetes riparia

30 LEK

6. Isoetes echinospora

½

30

4. Isoetes virginica

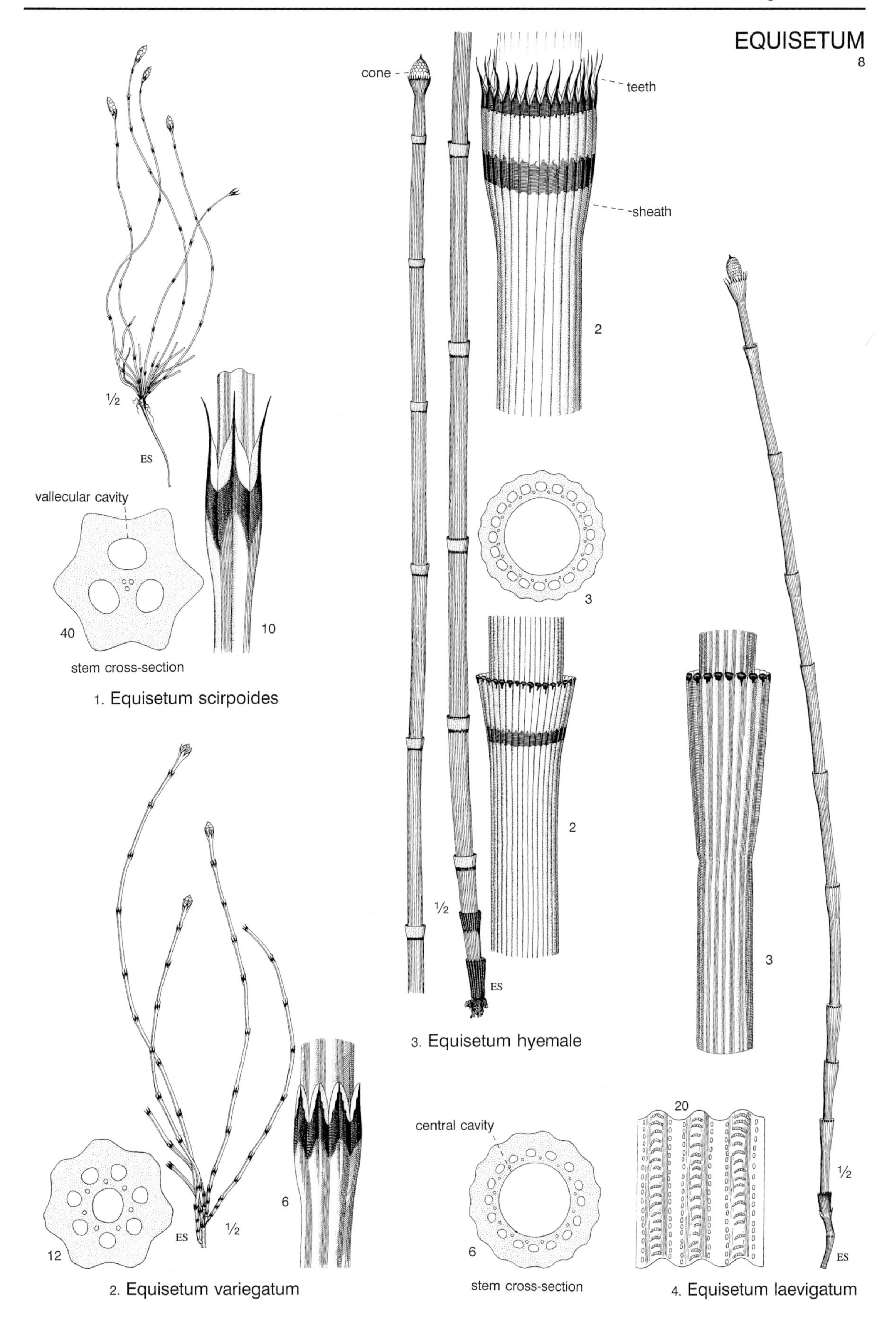

cone

teeth

sheath

2

vallecular cavity

40

stem cross-section

1. Equisetum scirpoides

ES

10

3

2

½

ES

3. Equisetum hyemale

central cavity

6

stem cross-section

3

20

½

ES

4. Equisetum laevigatum

12

ES

½

6

2. Equisetum variegatum

EQUISETUM
8

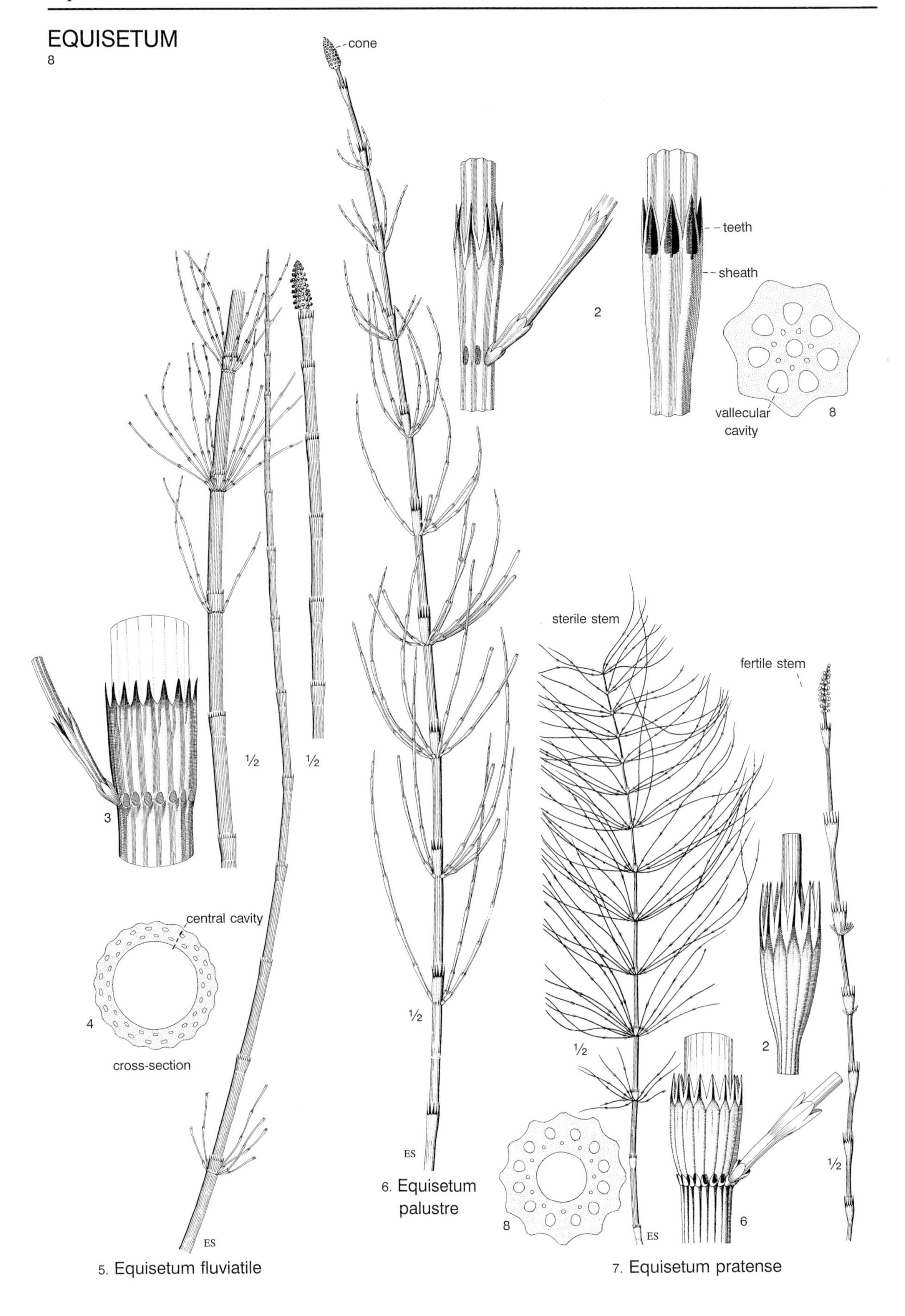

cone

teeth

sheath

vallecular
cavity

8

2

½

½

3

central cavity

4

cross-section

ES

5. Equisetum fluviatile

½

ES

6. Equisetum
palustre

sterile stem

fertile stem

½

2

8

½

6

ES

7. Equisetum pratense

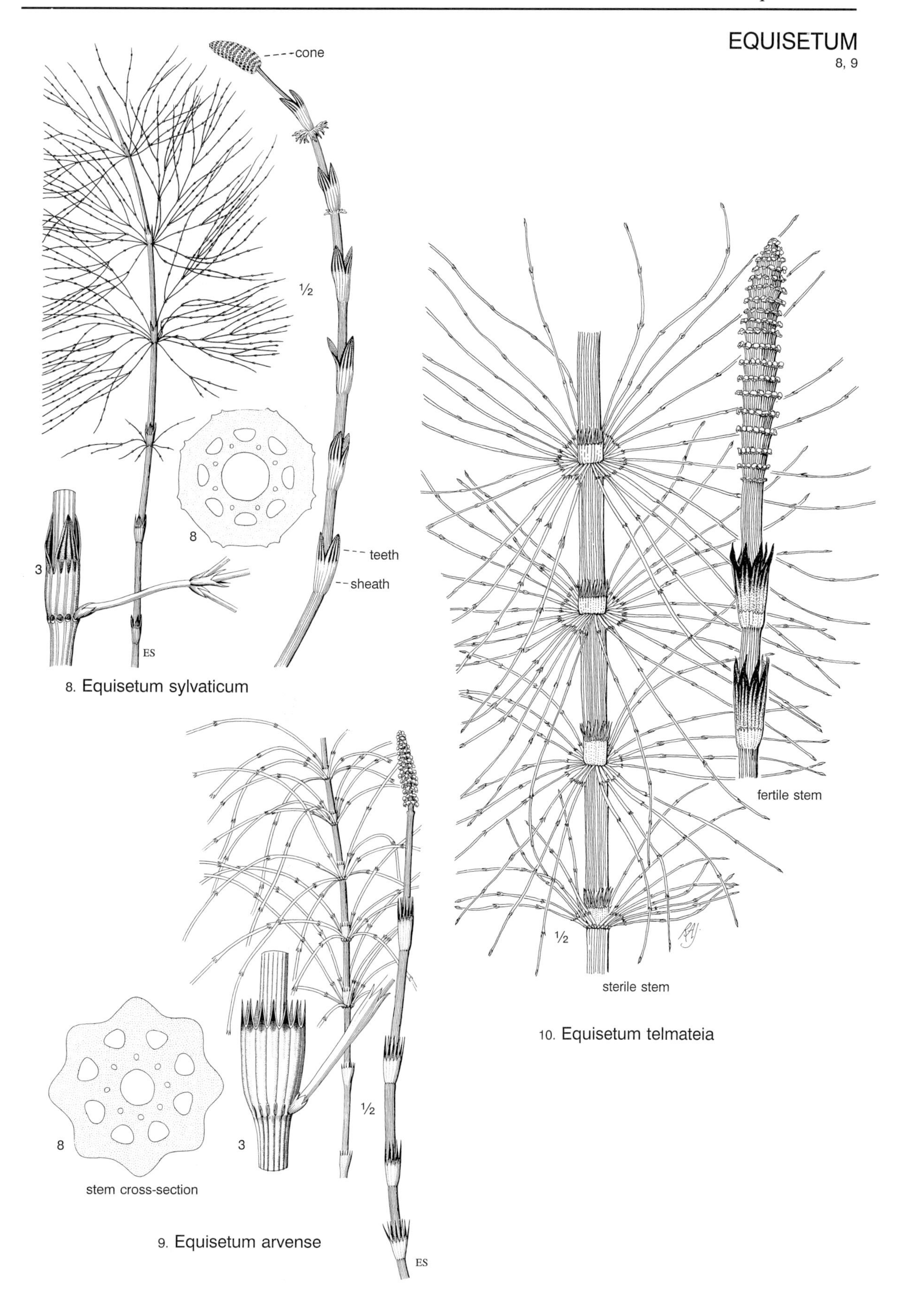

cone

½

8

3

ES

teeth

sheath

8. Equisetum sylvaticum

8

3

½

ES

stem cross-section

9. Equisetum arvense

fertile stem

½

sterile stem

10. Equisetum telmateia

BOTRYCHIUM
10, 11

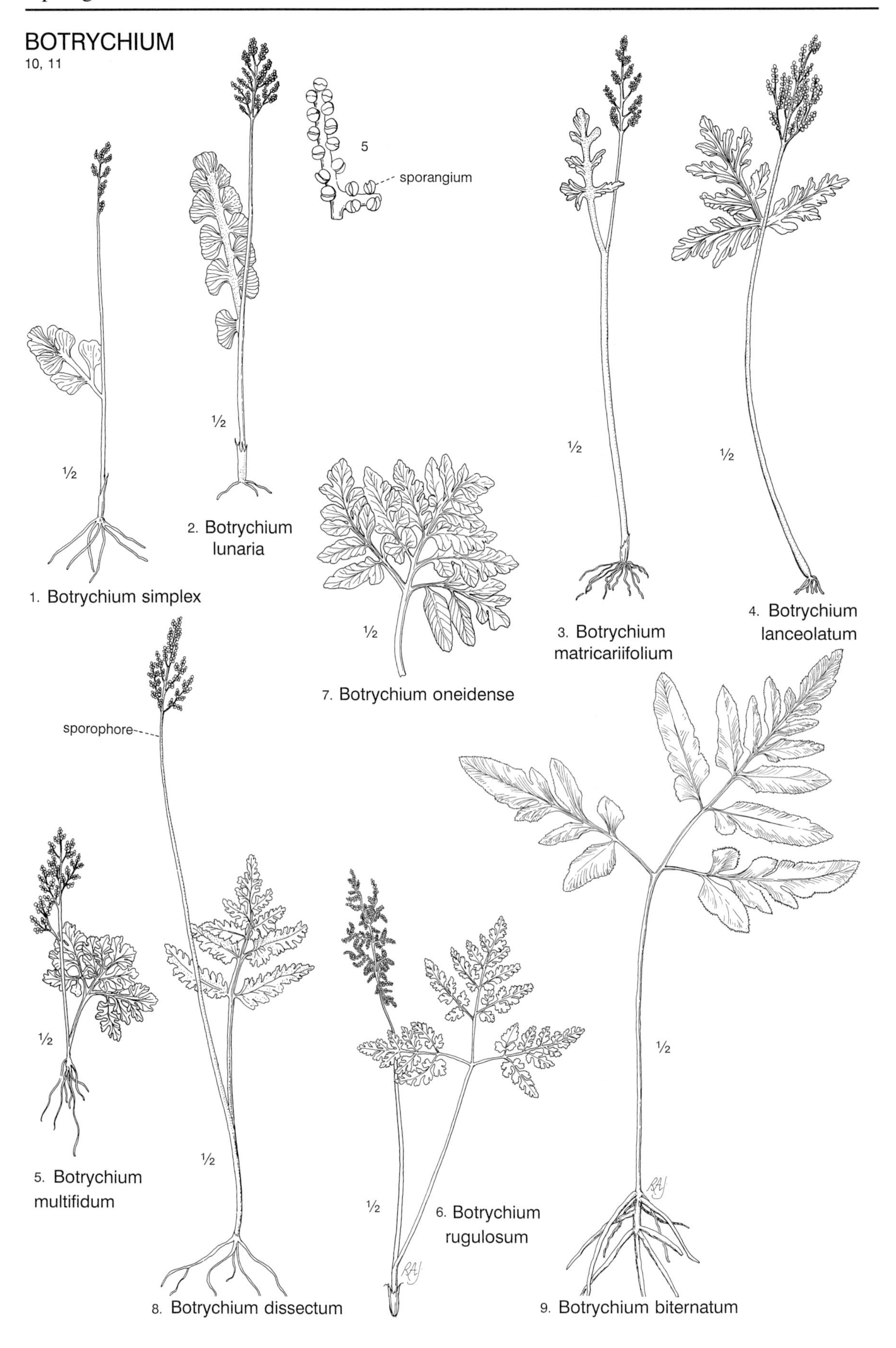

5

— sporangium

½

½

½

2. Botrychium
lunaria

1. Botrychium simplex

½

7. Botrychium oneidense

½

½

3. Botrychium
matricariifolium

4. Botrychium
lanceolatum

sporophore----

½

½

½

½

5. Botrychium
multifidum

8. Botrychium dissectum

6. Botrychium
rugulosum

9. Botrychium biternatum

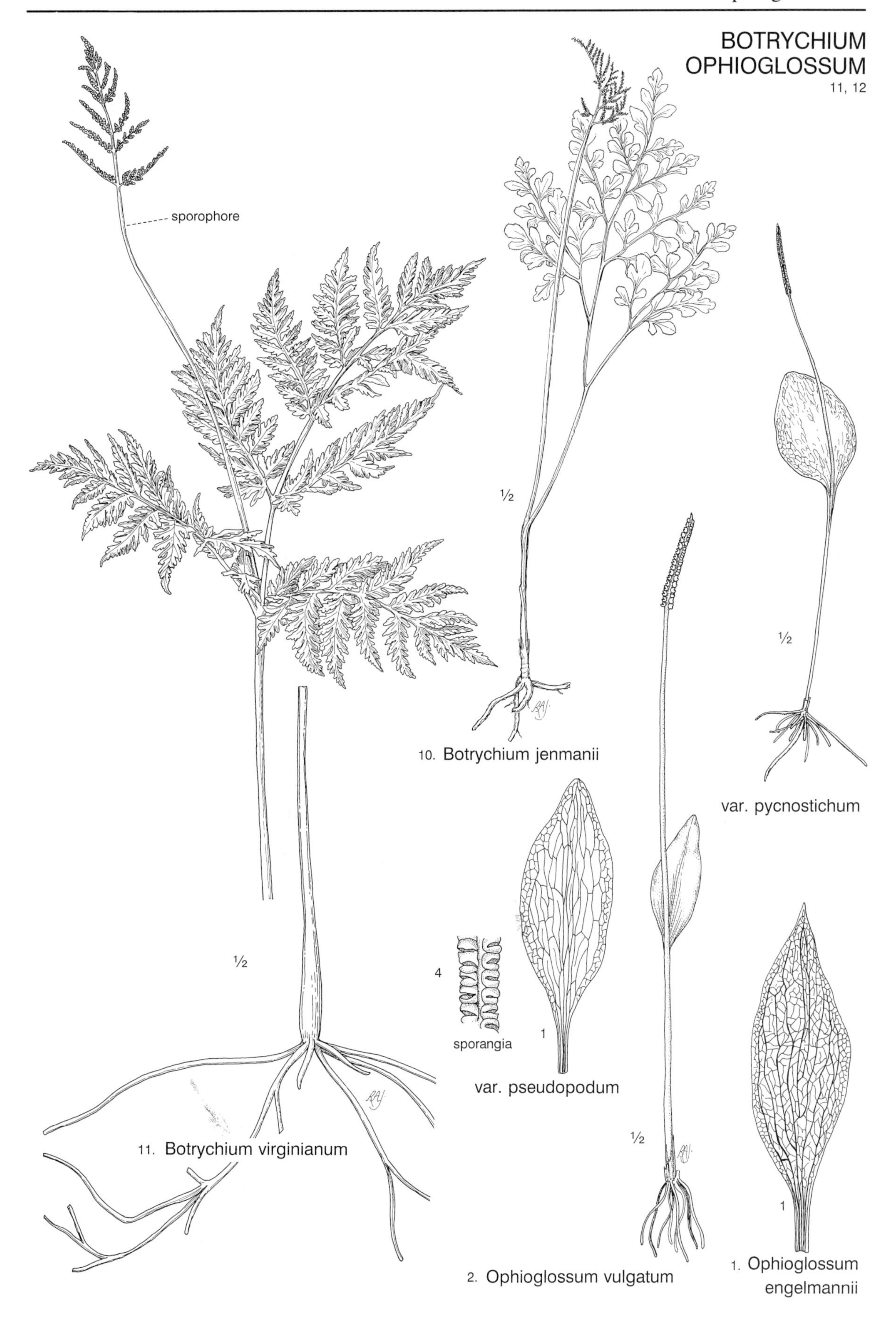

sporophore

½

10. Botrychium jenmanii

var. pycnostichum

½

11. Botrychium virginianum

4

sporangia

1

var. pseudopodum

½

2. Ophioglossum vulgatum

1

1. Ophioglossum
engelmannii

OSMUNDA
12

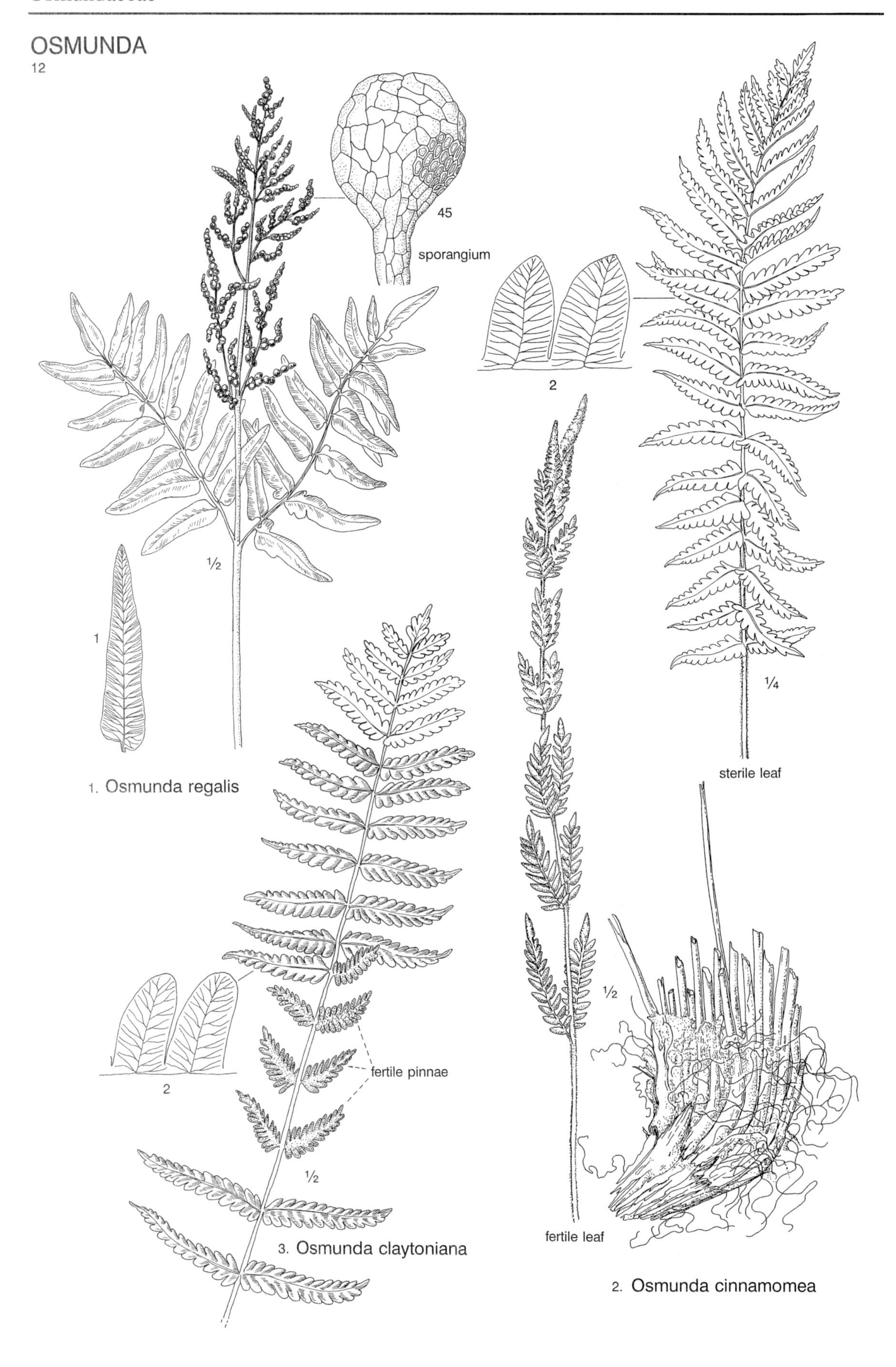

45
sporangium

2

½

1

1. Osmunda regalis

¼

sterile leaf

2

fertile pinnae

½

3. Osmunda claytoniana

½

fertile leaf

2. Osmunda cinnamomea

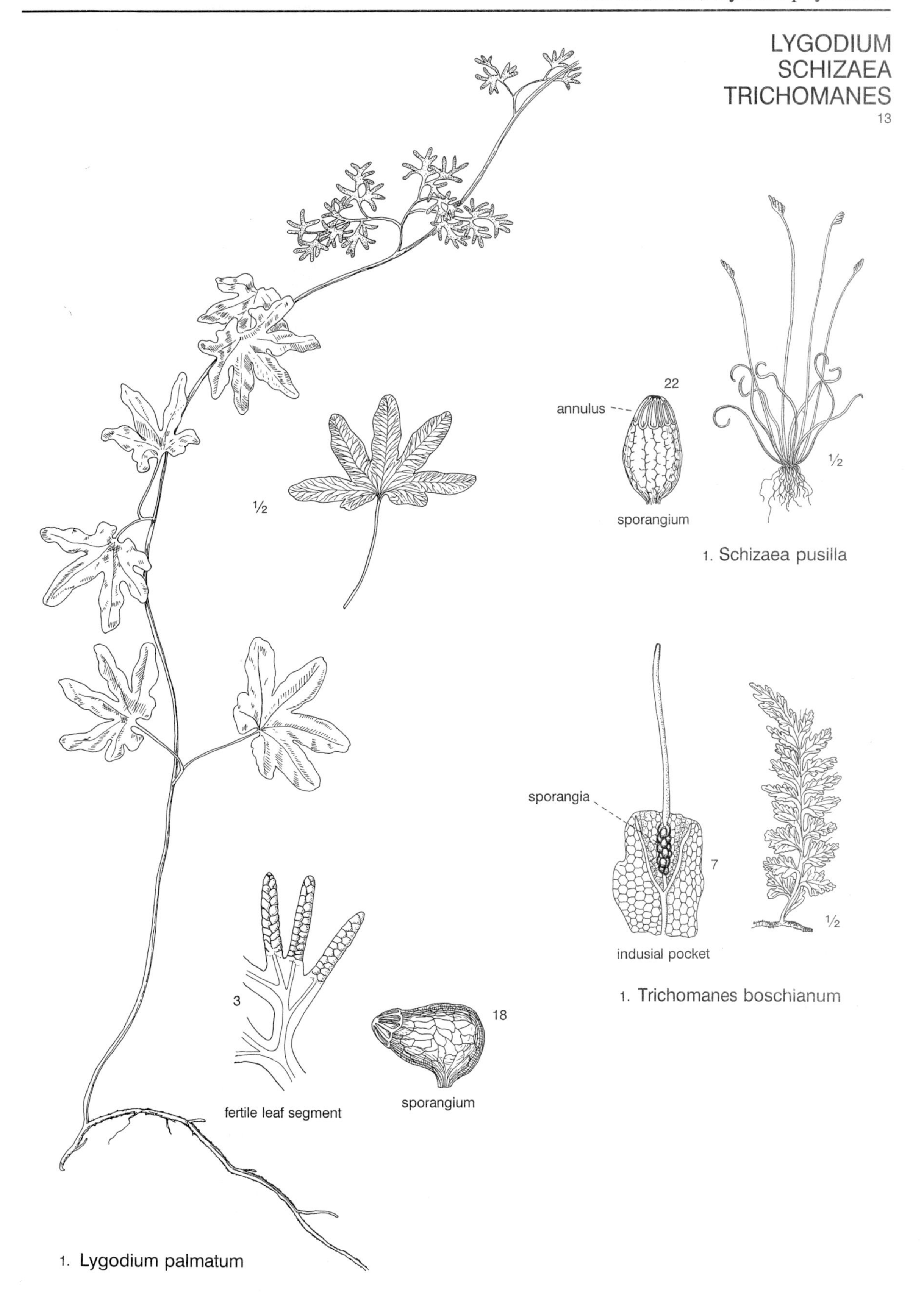

½

½

1. Schizaea pusilla

annulus

22

sporangium

sporangia

7

indusial pocket

½

1. Trichomanes boschianum

3

fertile leaf segment

18

sporangium

1. Lygodium palmatum

POLYPODIUM
DENNSTAEDTIA
14, 15

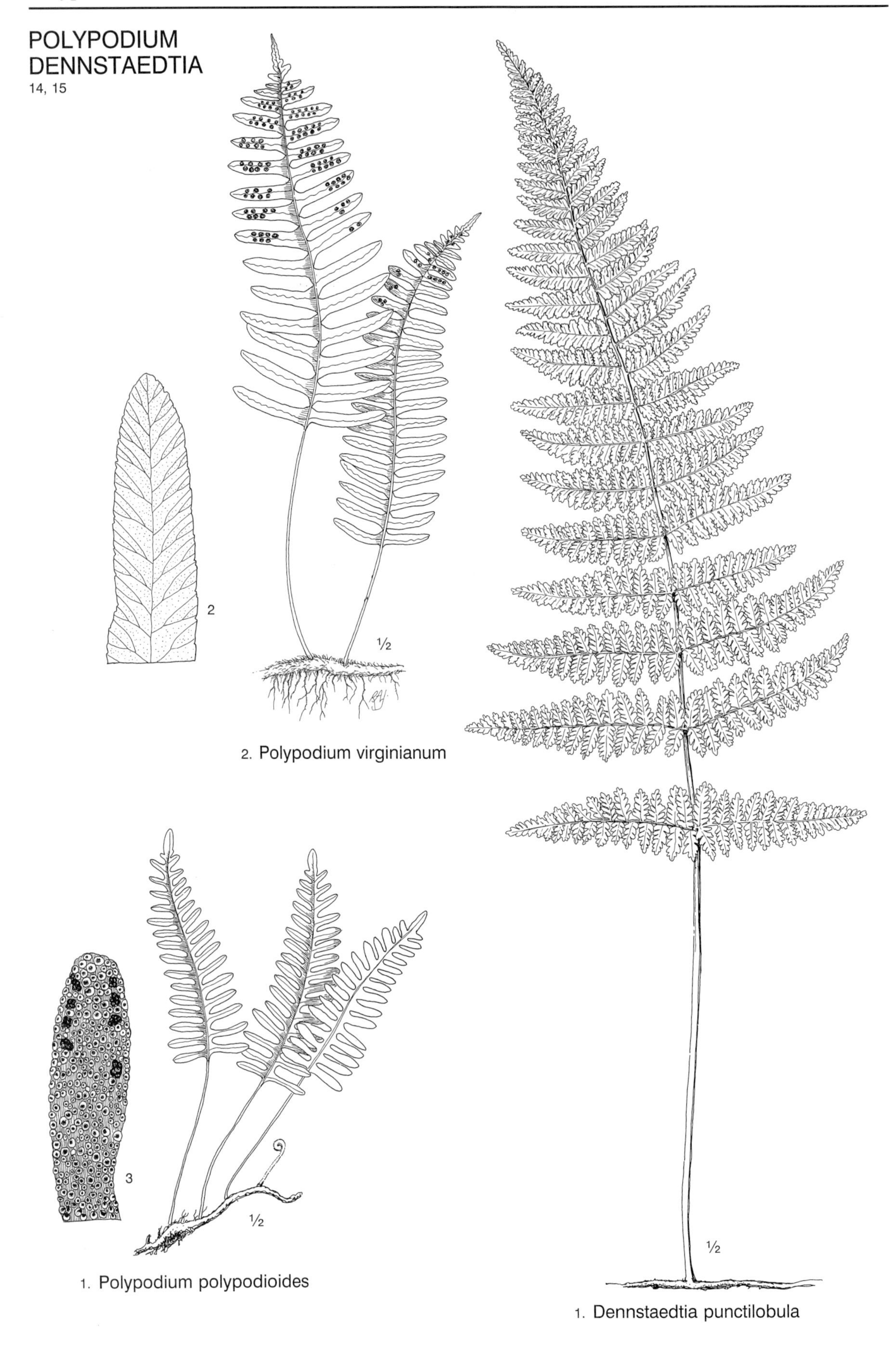

2. Polypodium virginianum

1. Polypodium polypodioides

1. Dennstaedtia punctilobula

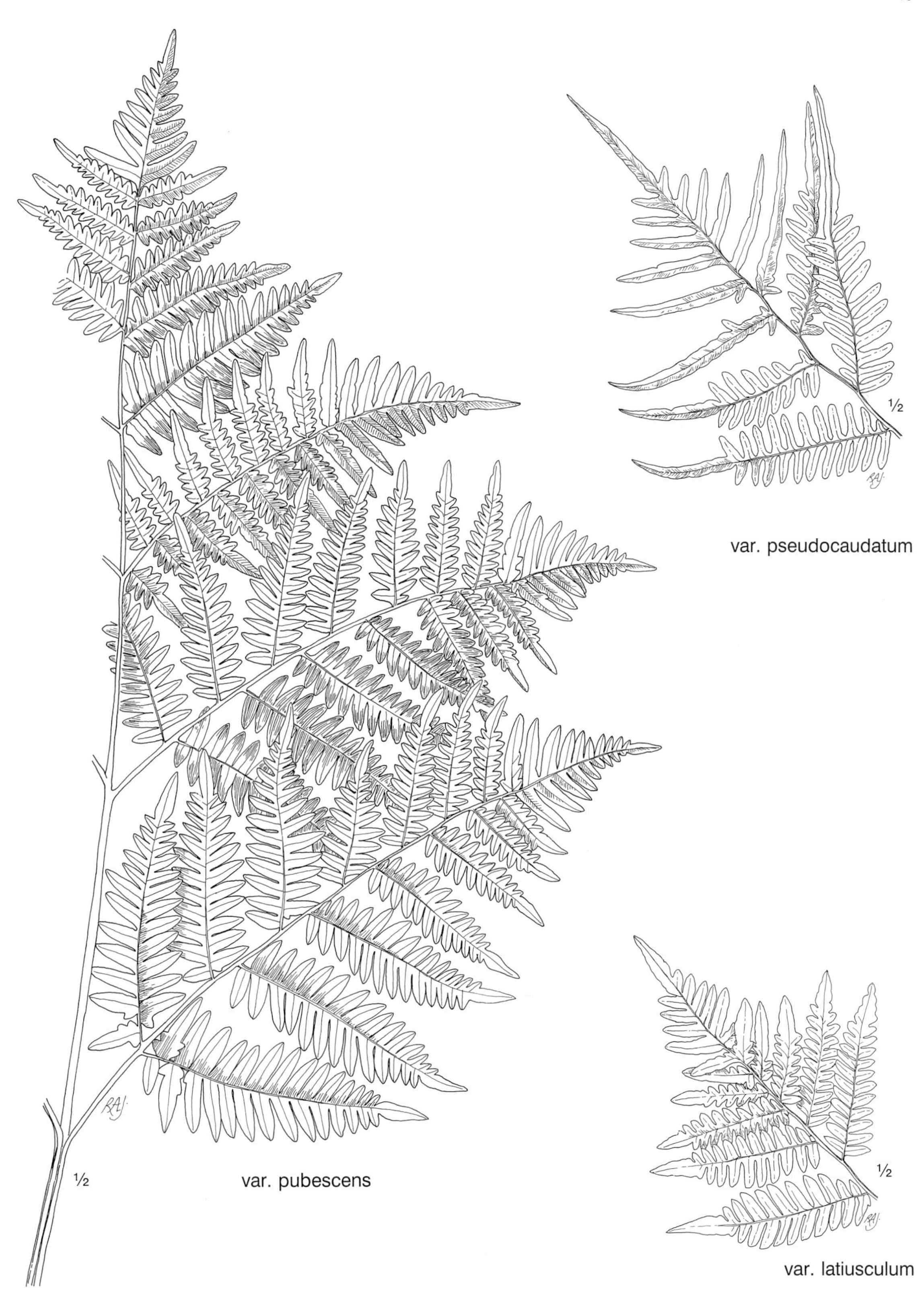

var. pseudocaudatum

var. latiusculum

½   var. pubescens

1. Pteridium aquilinum

ADIANTUM
CRYPTOGRAMMA

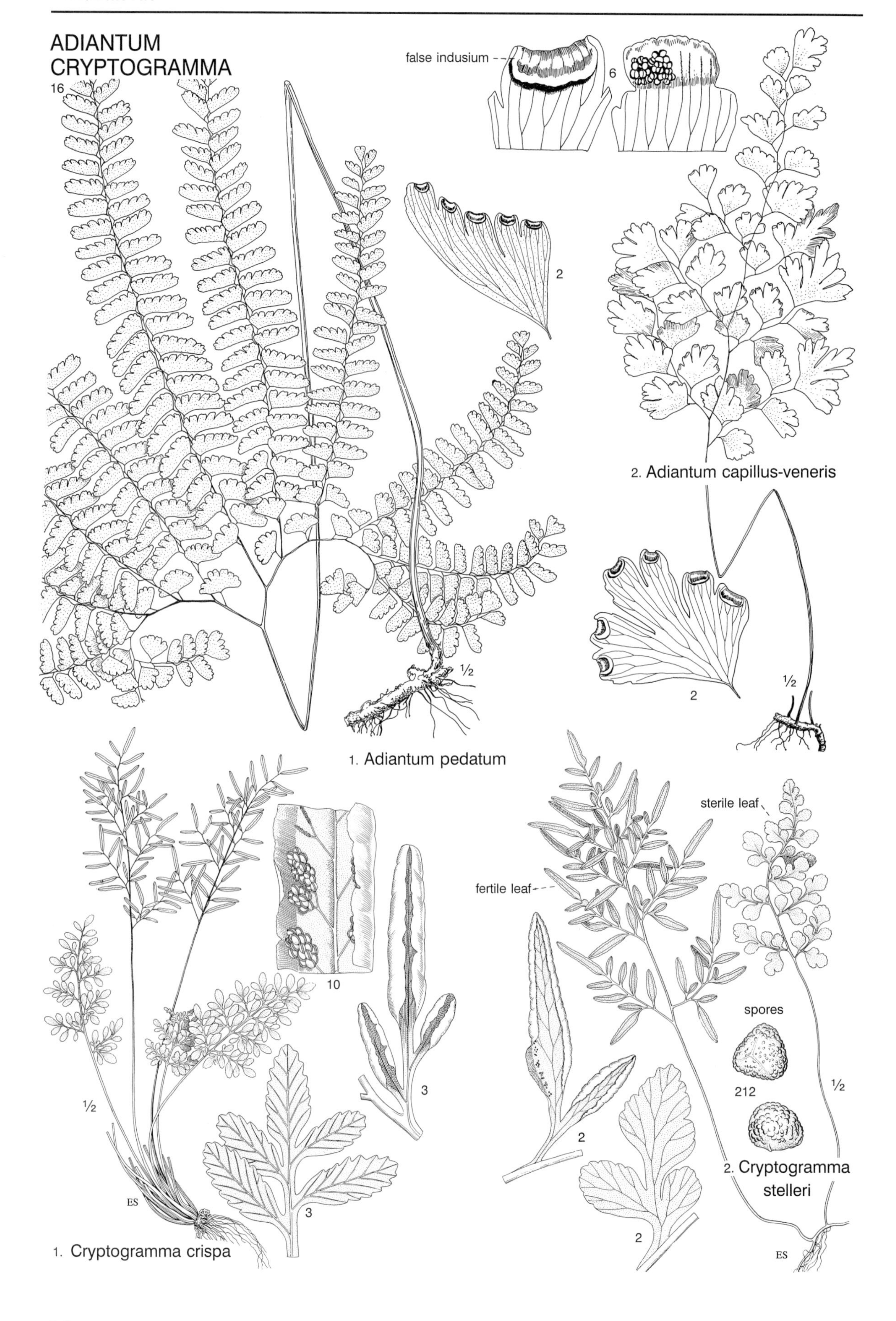

false indusium

16

2

6

2. Adiantum capillus-veneris

1. Adiantum pedatum

2

10

3

fertile leaf

sterile leaf

spores

212

3

2

1. Cryptogramma crispa

ES

2

2

ES

2. Cryptogramma
stelleri

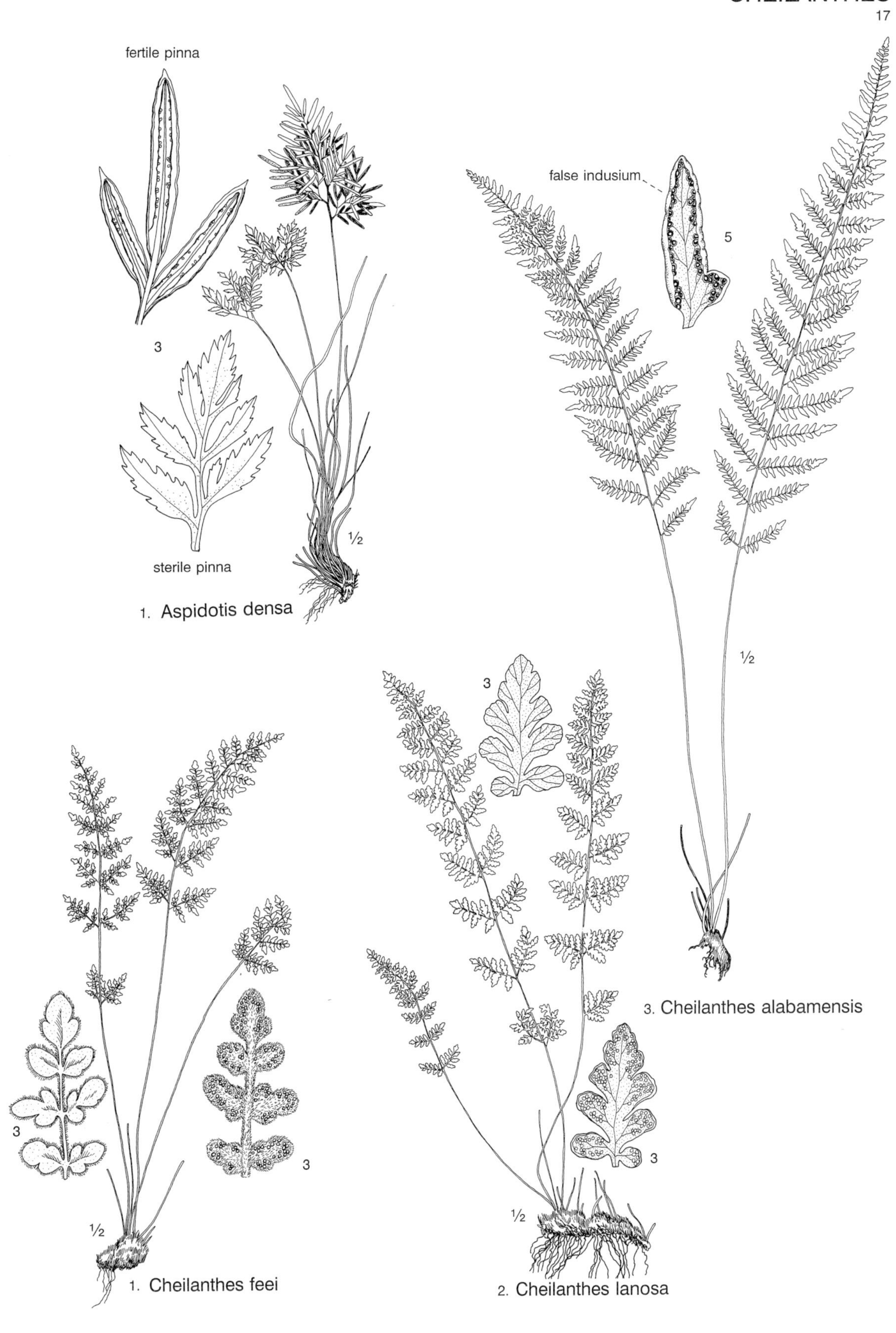

fertile pinna

3

sterile pinna

1. Aspidotis densa

½

false indusium

5

½

3. Cheilanthes alabamensis

3

3

½

1. Cheilanthes feei

3

3

½

2. Cheilanthes lanosa

CHEILANTHES
NOTHOLAENA
PELLAEA
17, 18

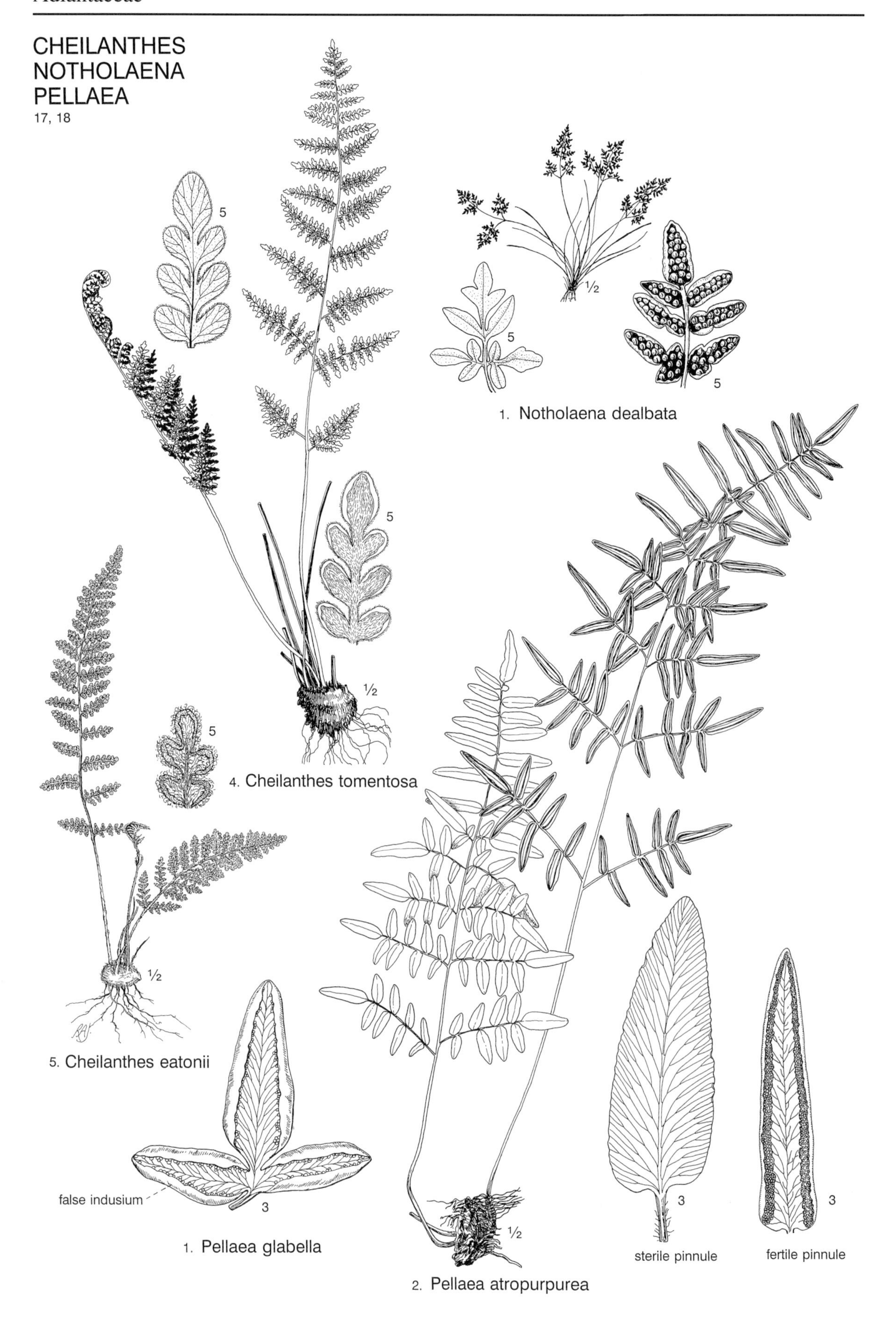

1. Notholaena dealbata

4. Cheilanthes tomentosa

5. Cheilanthes eatonii

false indusium

1. Pellaea glabella

2. Pellaea atropurpurea

sterile pinnule

fertile pinnule

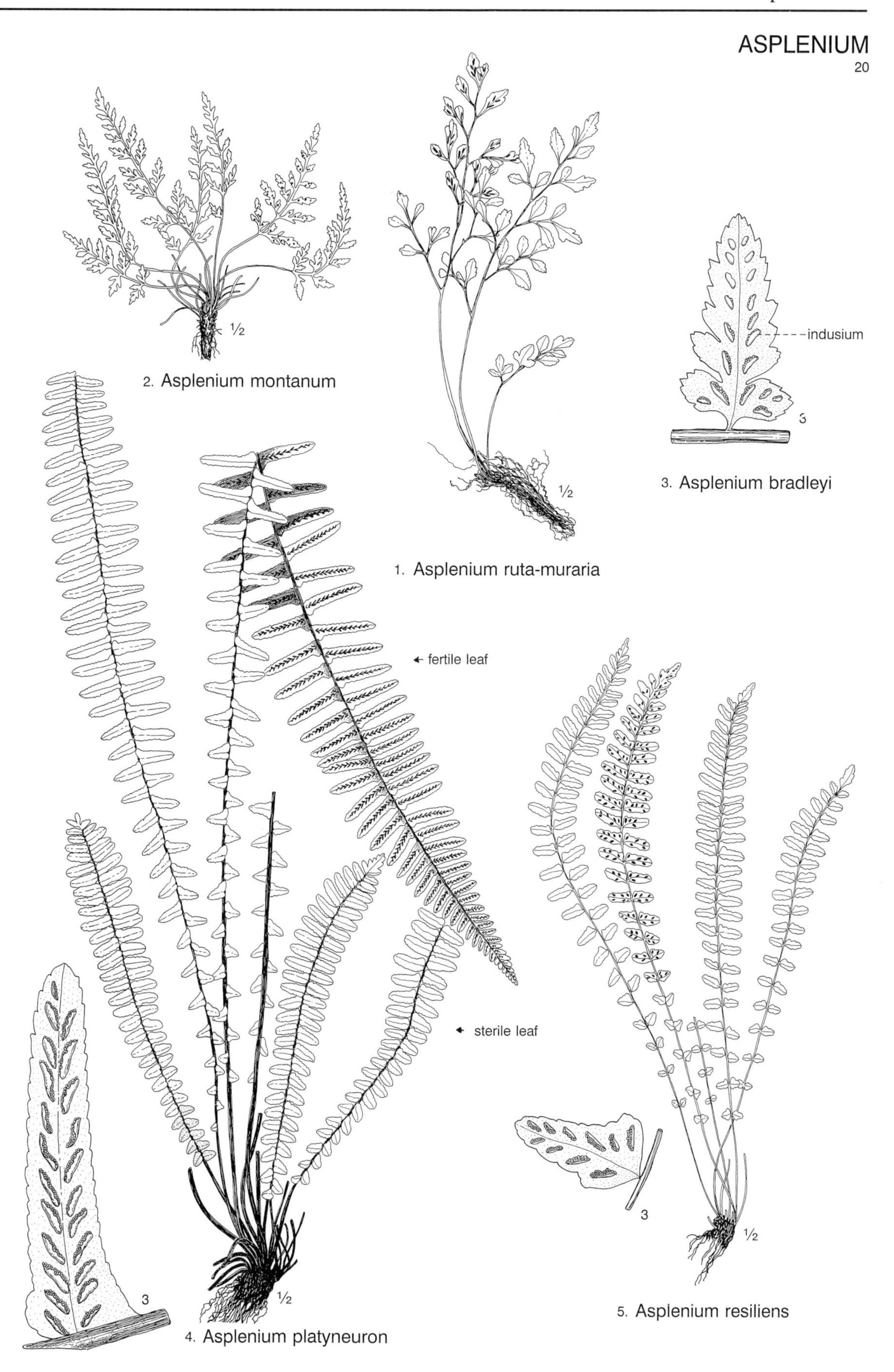

2. Asplenium montanum

3. Asplenium bradleyi

---indusium

1. Asplenium ruta-muraria

← fertile leaf

← sterile leaf

4. Asplenium platyneuron

5. Asplenium resiliens

ASPLENIUM
PHYLLITIS
20, 21

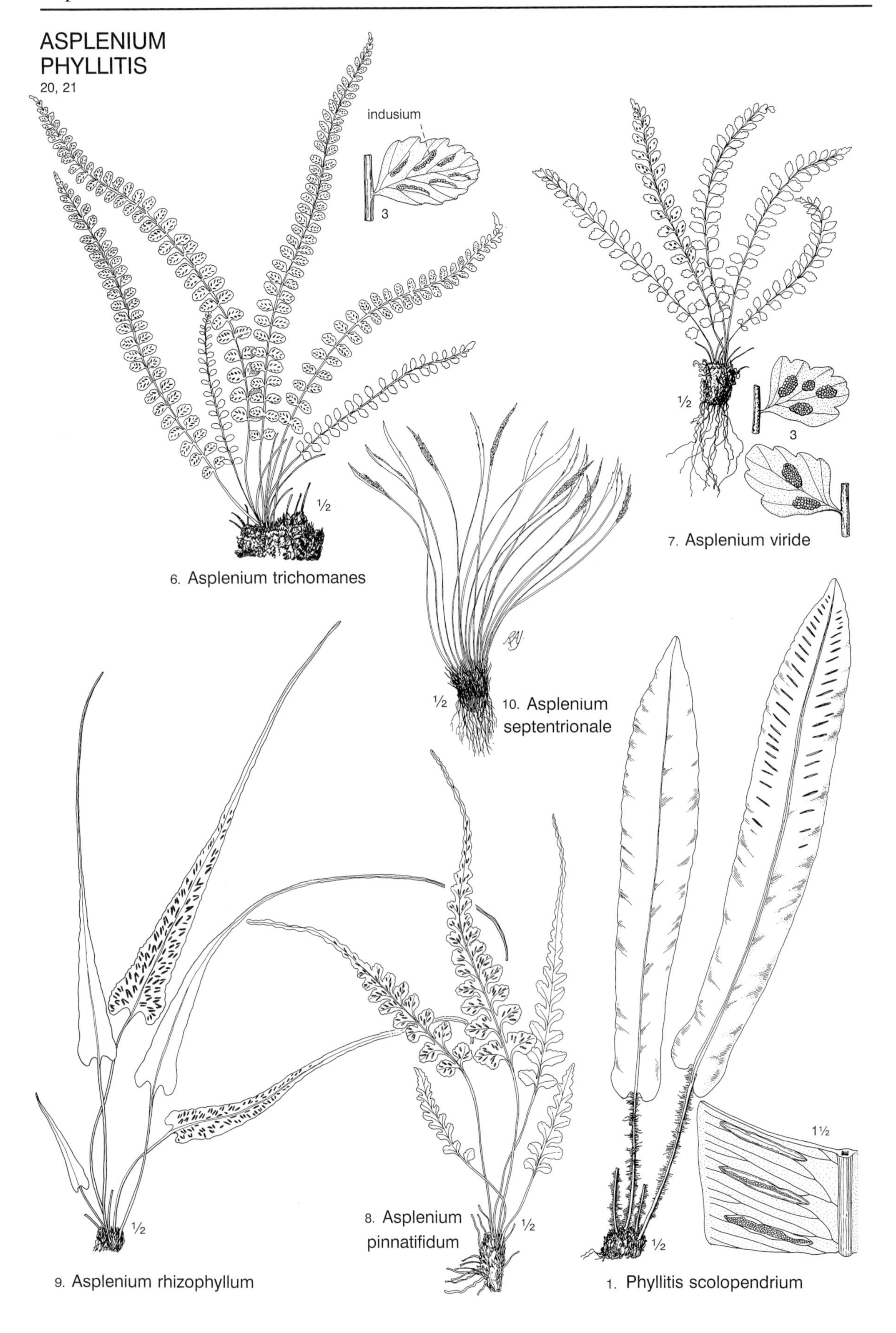

indusium

3

6. Asplenium trichomanes

7. Asplenium viride

10. Asplenium septentrionale

9. Asplenium rhizophyllum

8. Asplenium pinnatifidum

1. Phyllitis scolopendrium

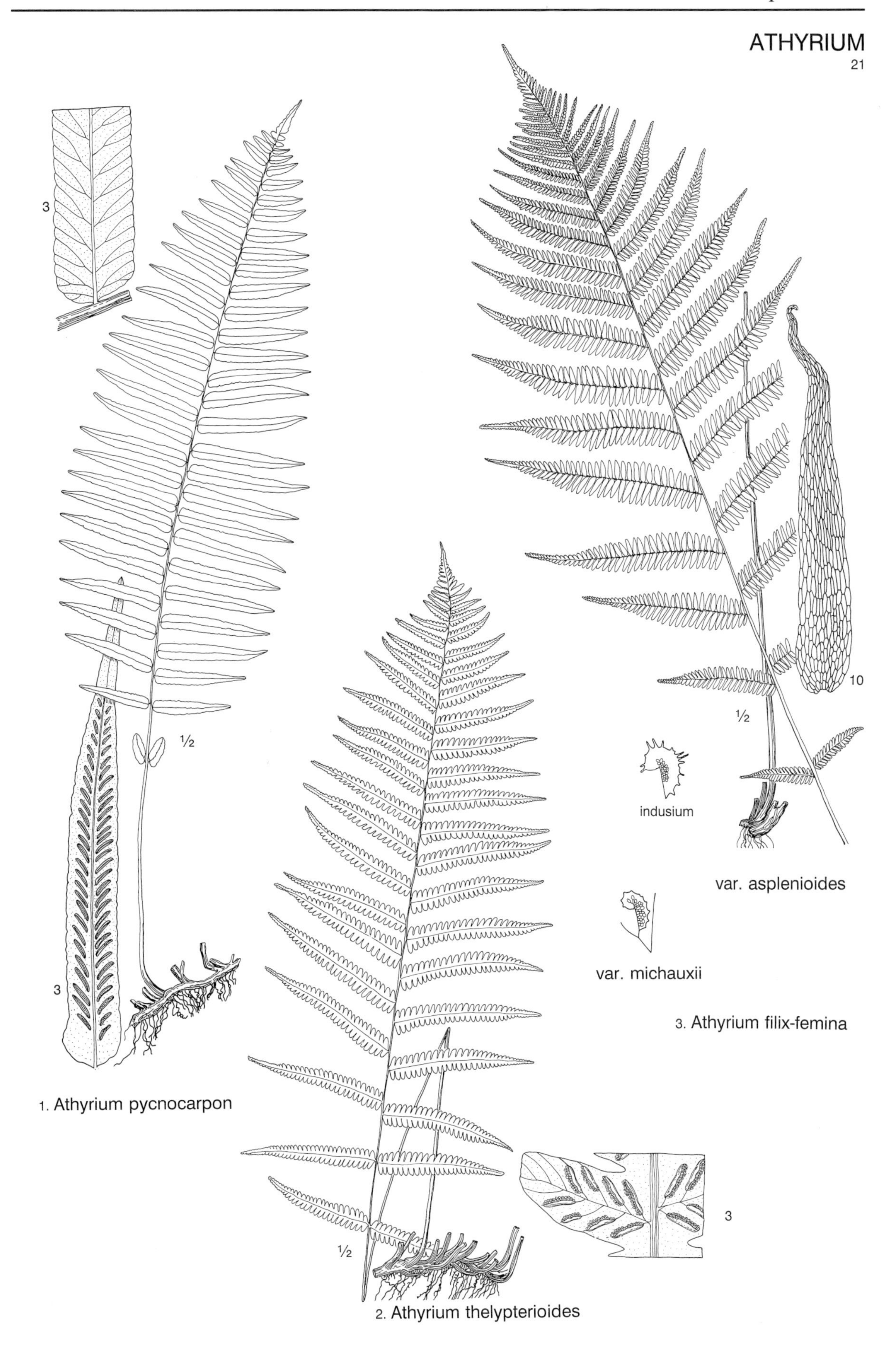

3

1. Athyrium pycnocarpon

½

2. Athyrium thelypterioides

indusium

var. michauxii

var. asplenioides

10

3. Athyrium filix-femina

CYSTOPTERIS
WOODSIA
22, 23

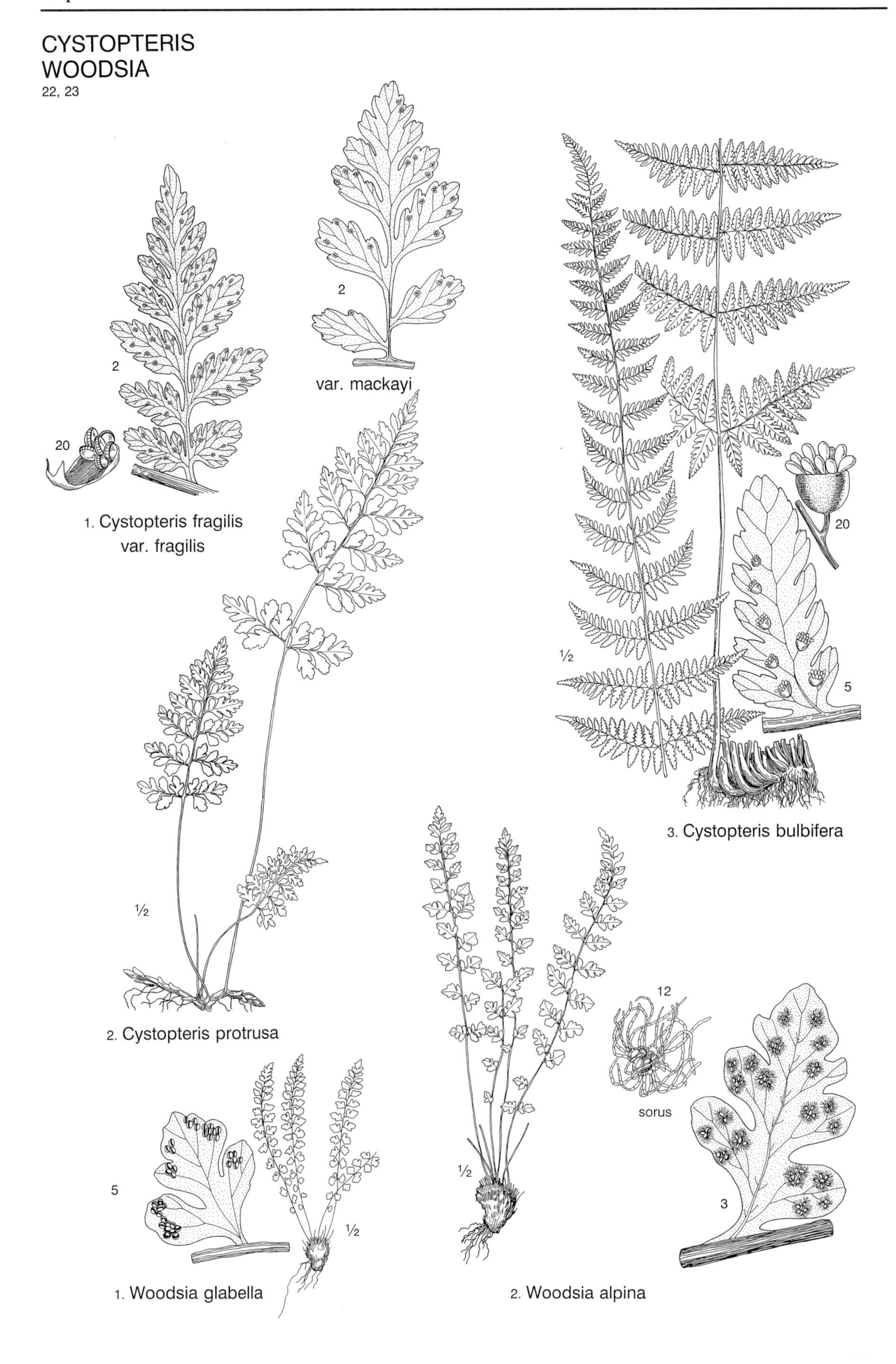

var. mackayi

20

1. Cystopteris fragilis
var. fragilis

½

2. Cystopteris protrusa

½

20

5

3. Cystopteris bulbifera

12

sorus

3

5

½

1. Woodsia glabella

½

2. Woodsia alpina

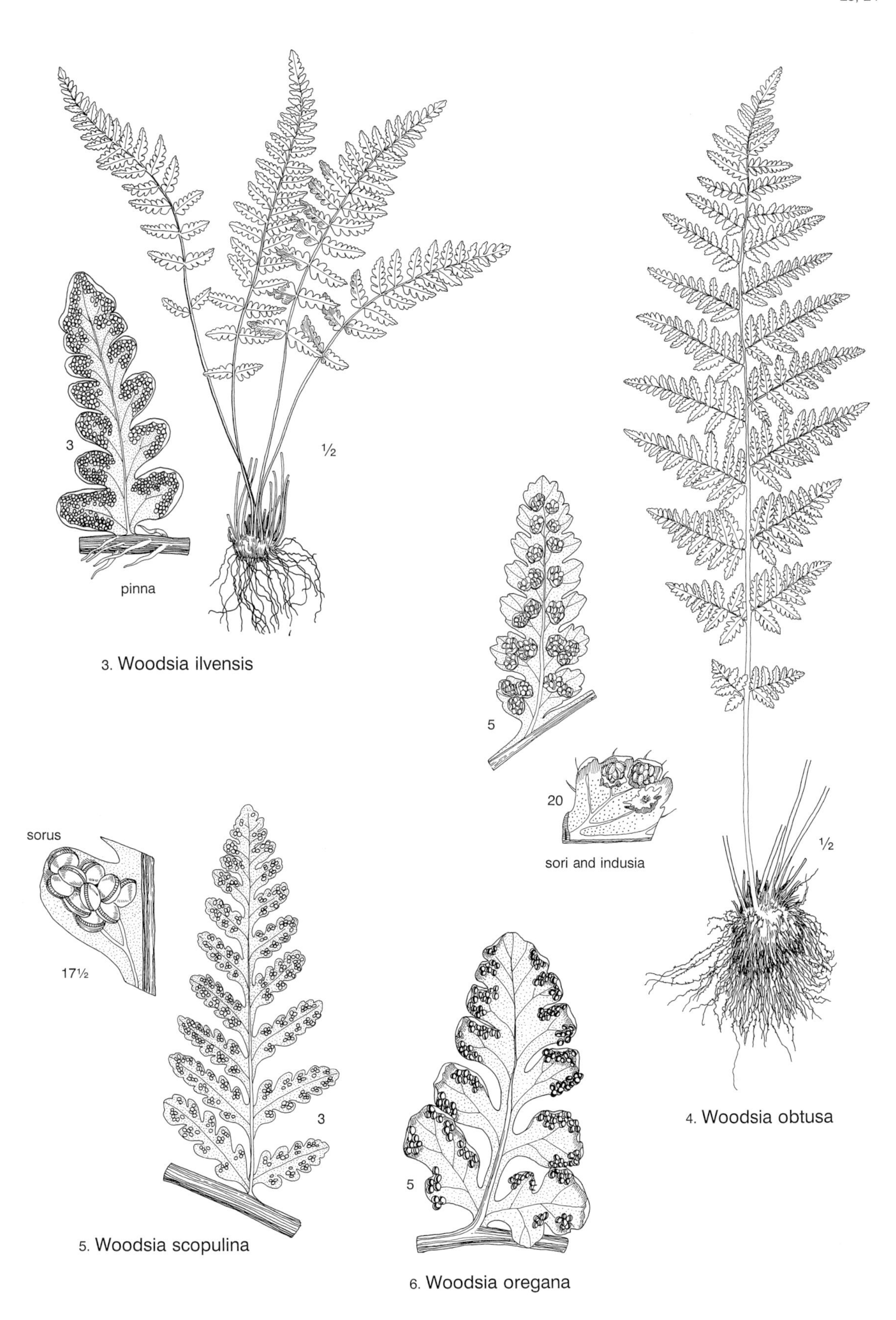

3

pinna

½

3. Woodsia ilvensis

sorus

17½

5

3

5. Woodsia scopulina

5

20

sori and indusia

5

6. Woodsia oregana

½

4. Woodsia obtusa

GYMNOCARPIUM
THELYPTERIS
24, 25

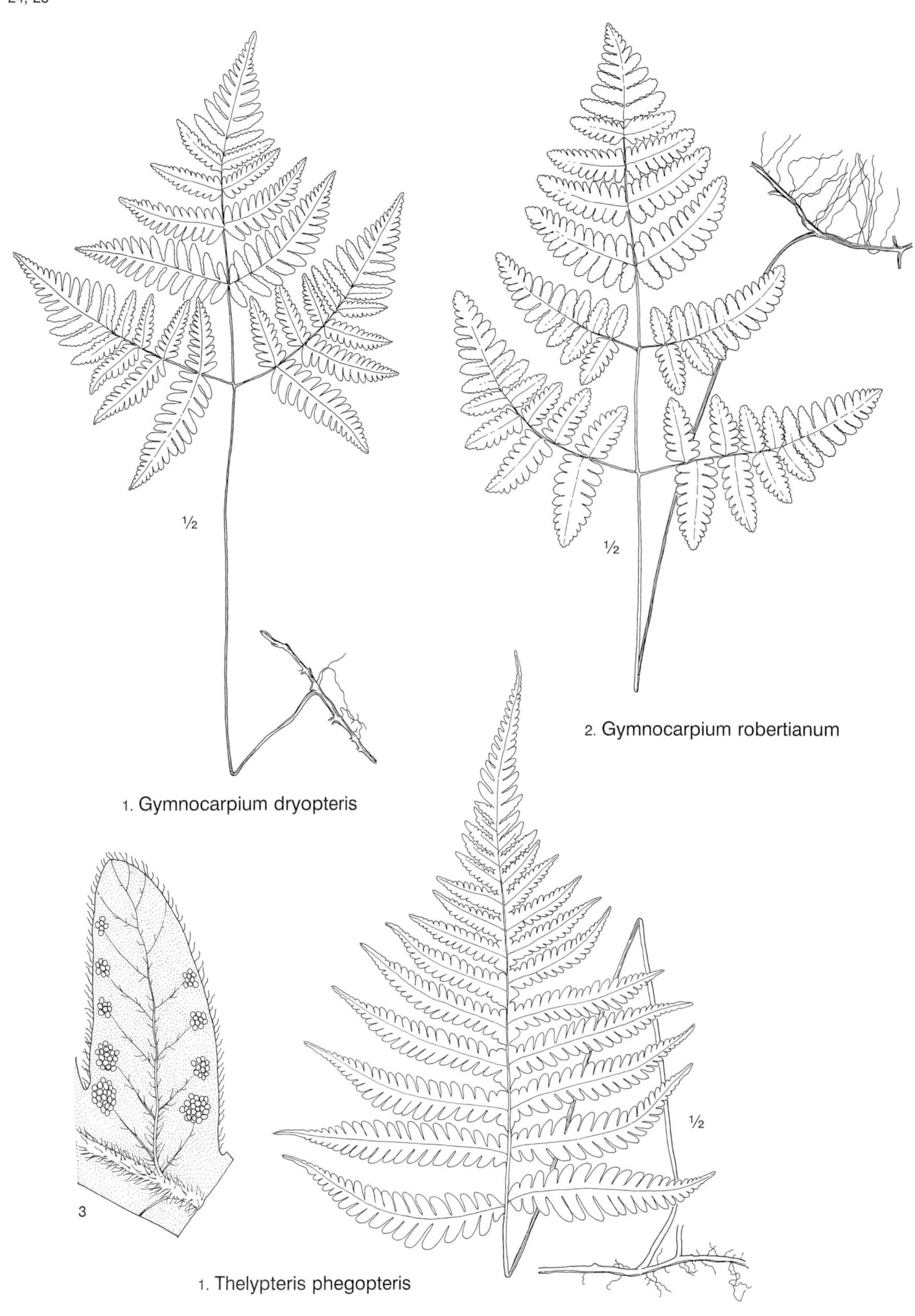

½

1. Gymnocarpium dryopteris

2. Gymnocarpium robertianum

½

3

½

1. Thelypteris phegopteris

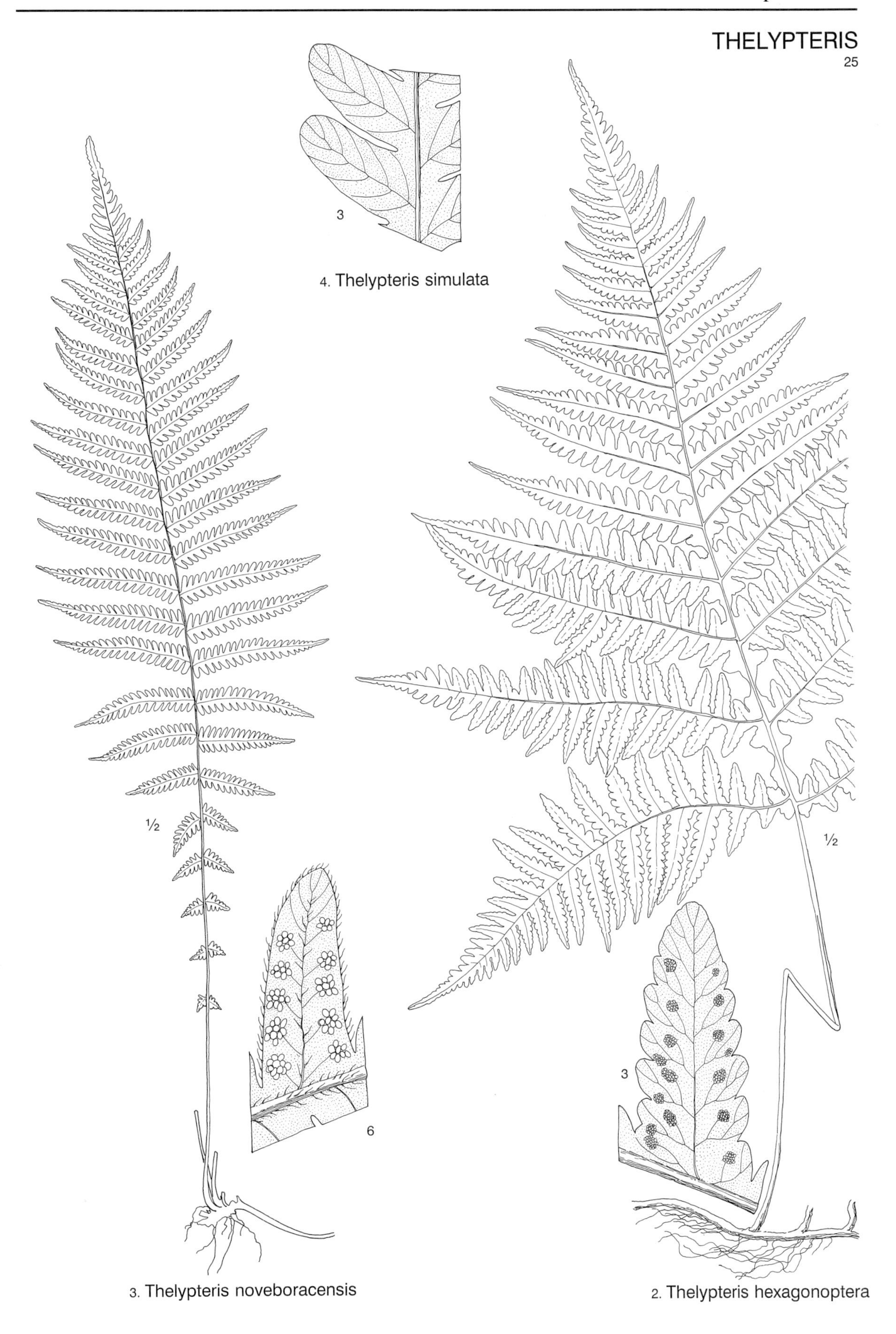

4. Thelypteris simulata

3

½

6

3. Thelypteris noveboracensis

½

3

2. Thelypteris hexagonoptera

THELYPTERIS
DRYOPTERIS
25, 26

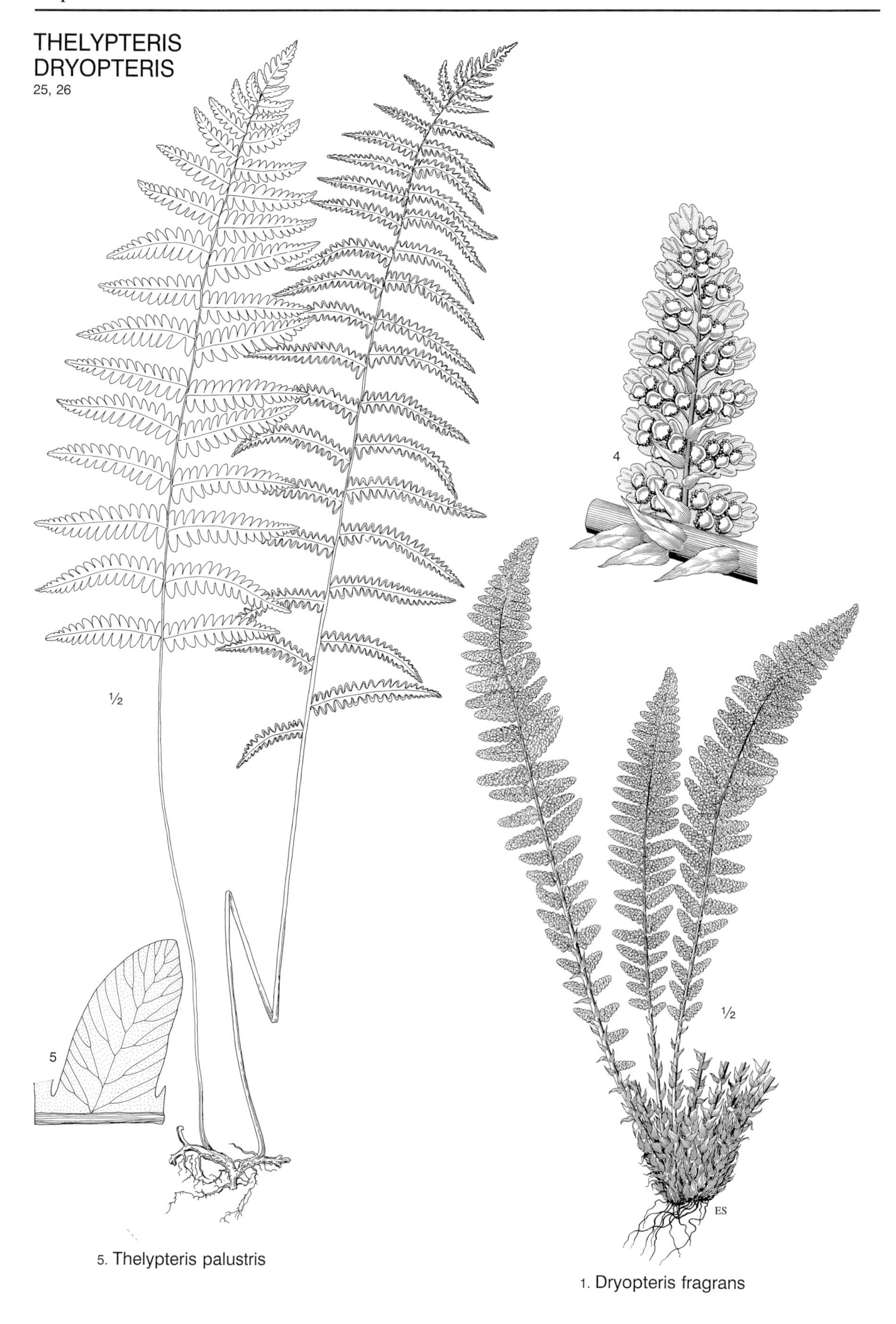

½

5

5. Thelypteris palustris

4

½

ES

1. Dryopteris fragrans

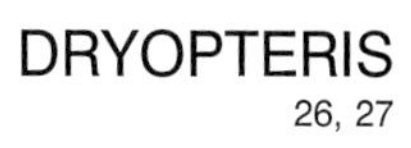

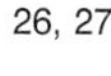

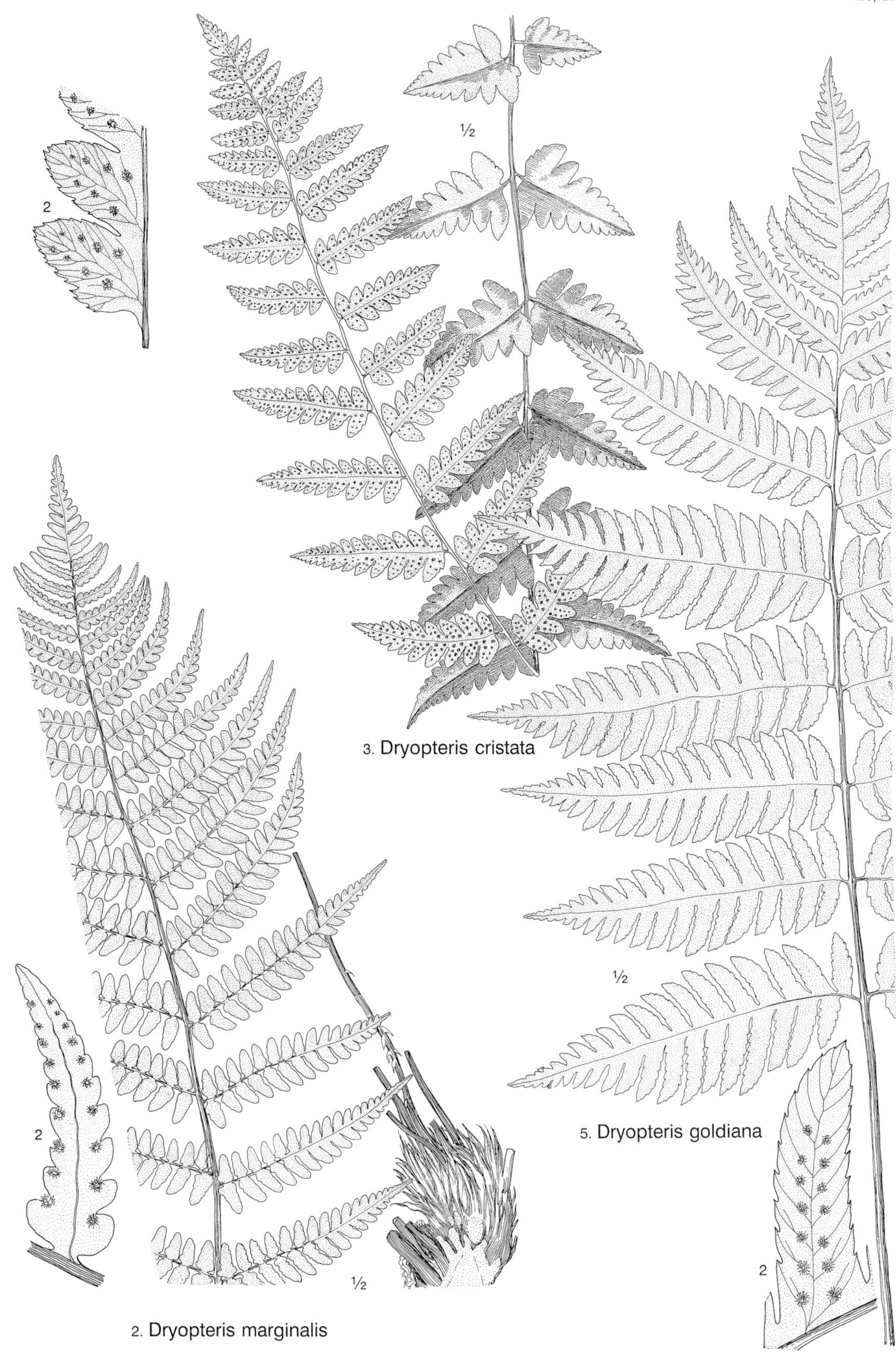

3. Dryopteris cristata

5. Dryopteris goldiana

2. Dryopteris marginalis

DRYOPTERIS
26, 27

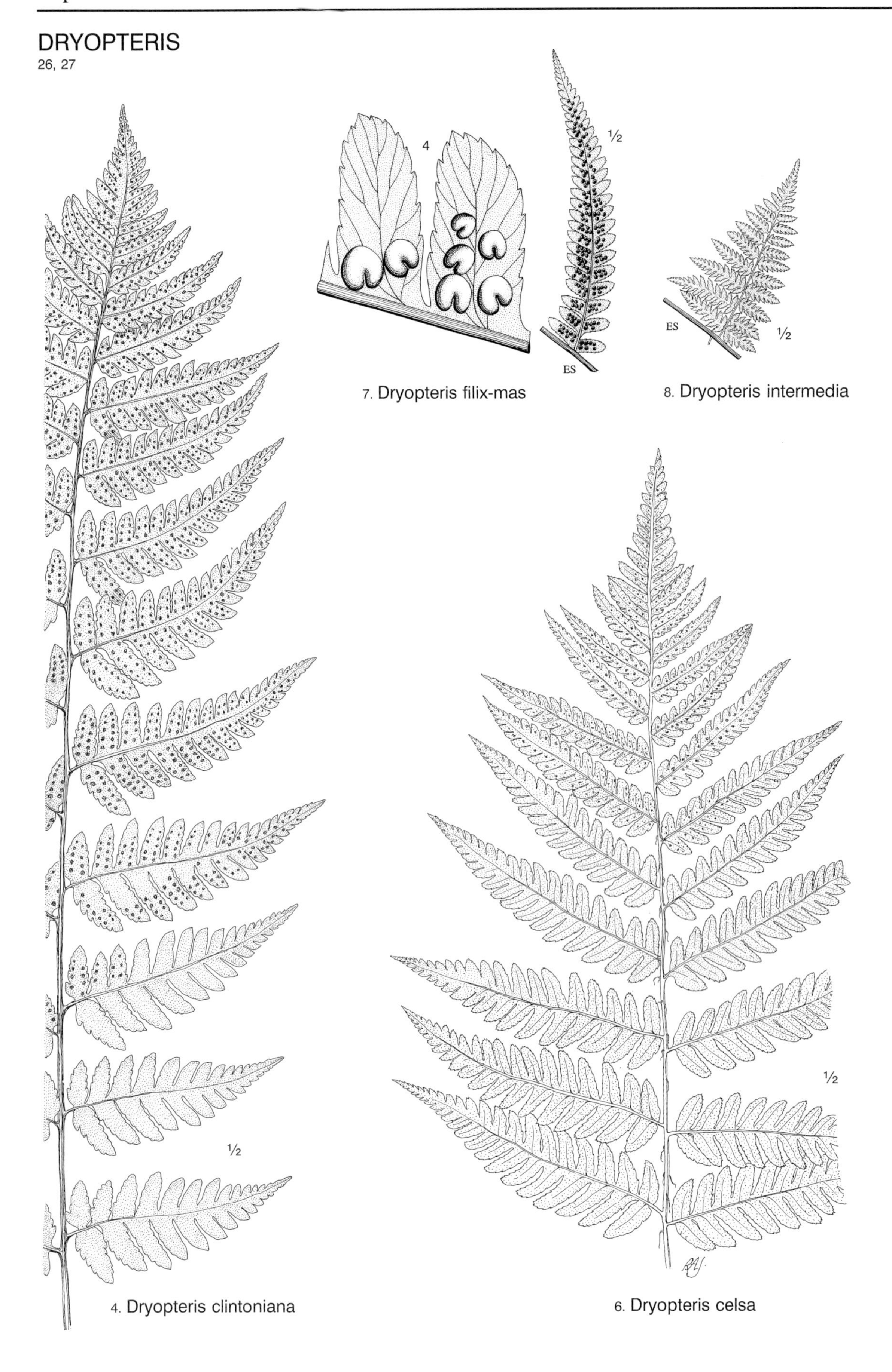

7. Dryopteris filix-mas

8. Dryopteris intermedia

4. Dryopteris clintoniana

6. Dryopteris celsa

ILLUSTRATED COMPANION TO

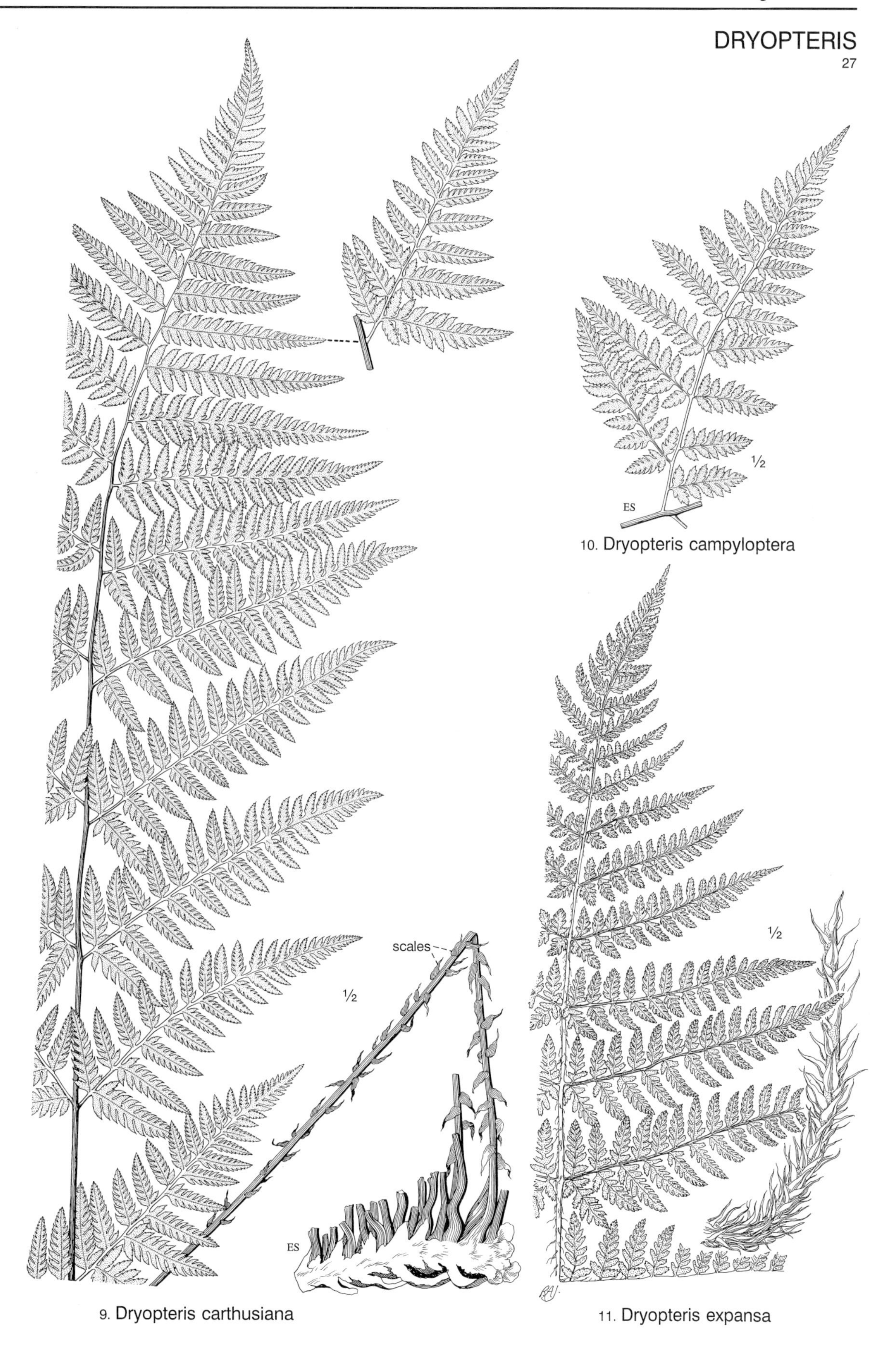

9. Dryopteris carthusiana

scales

½

ES

10. Dryopteris campyloptera

½

ES

11. Dryopteris expansa

½

POLYSTICHUM
28

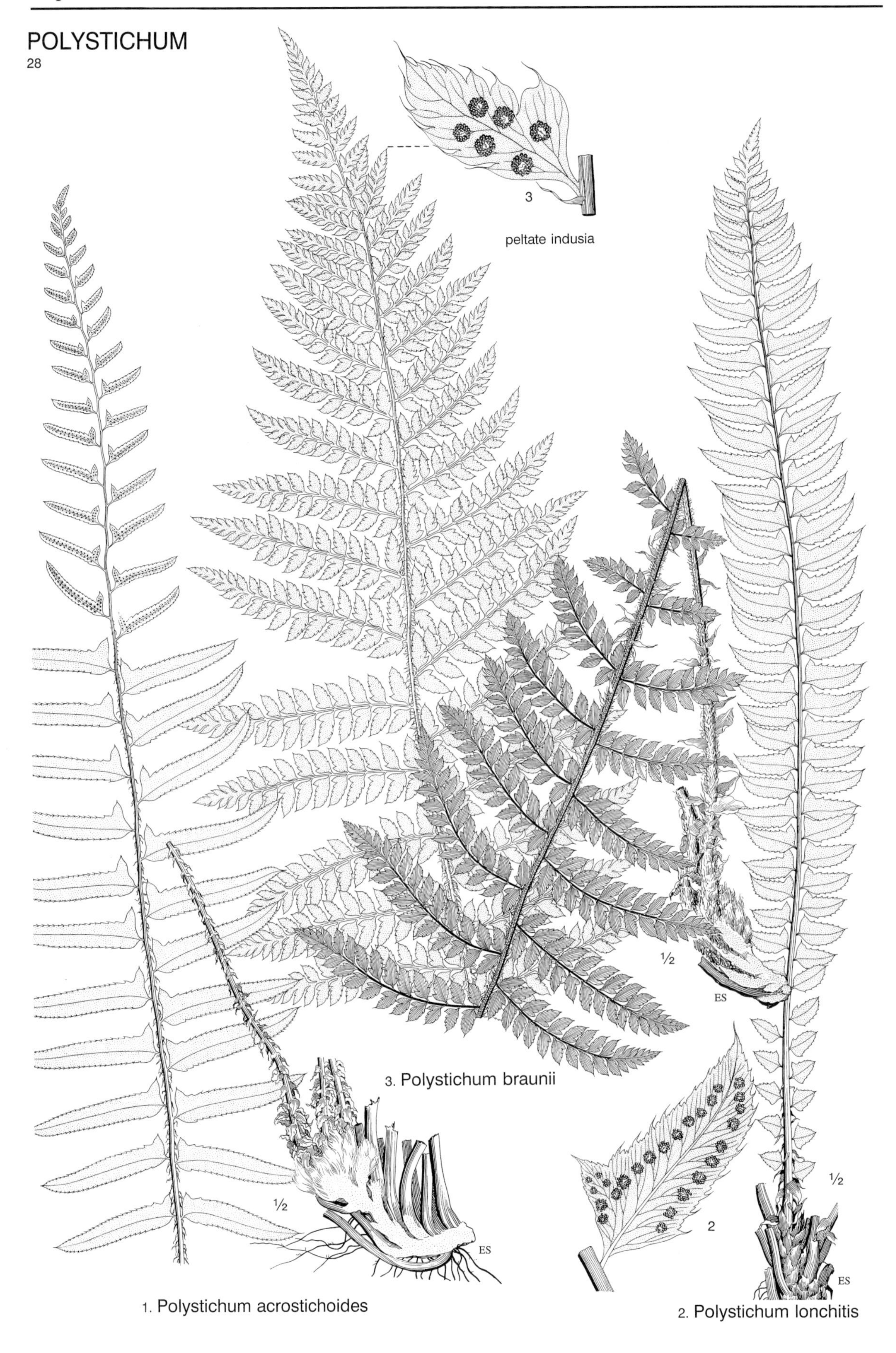

3

peltate indusia

3. Polystichum braunii

1. Polystichum acrostichoides

2. Polystichum lonchitis

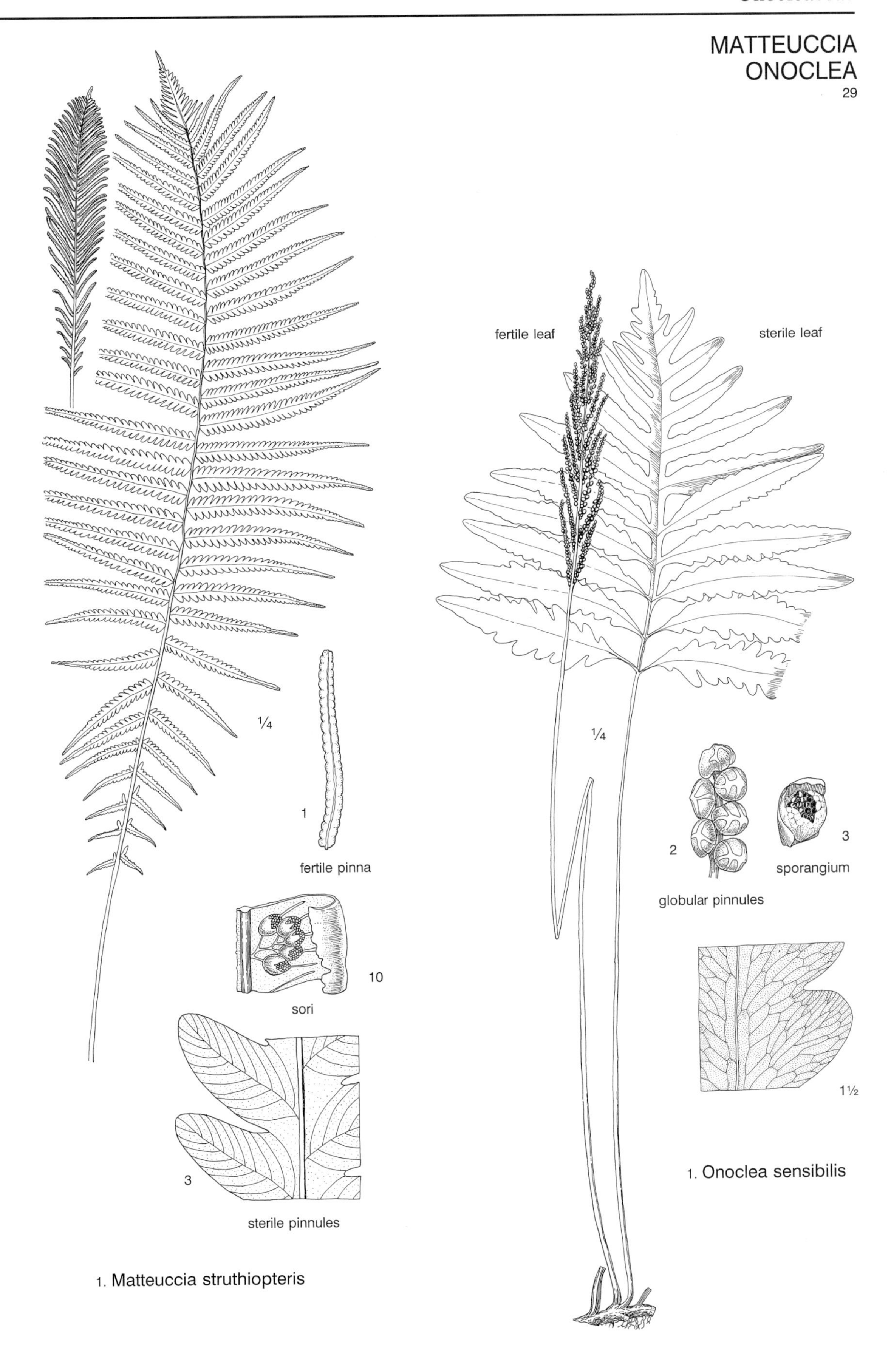

fertile leaf

sterile leaf

1/4

1

fertile pinna

10

sori

3

sterile pinnules

1/4

2

3

sporangium

globular pinnules

1½

1. Onoclea sensibilis

1. Matteuccia struthiopteris

WOODWARDIA
MARSILEA
AZOLLA
29–31

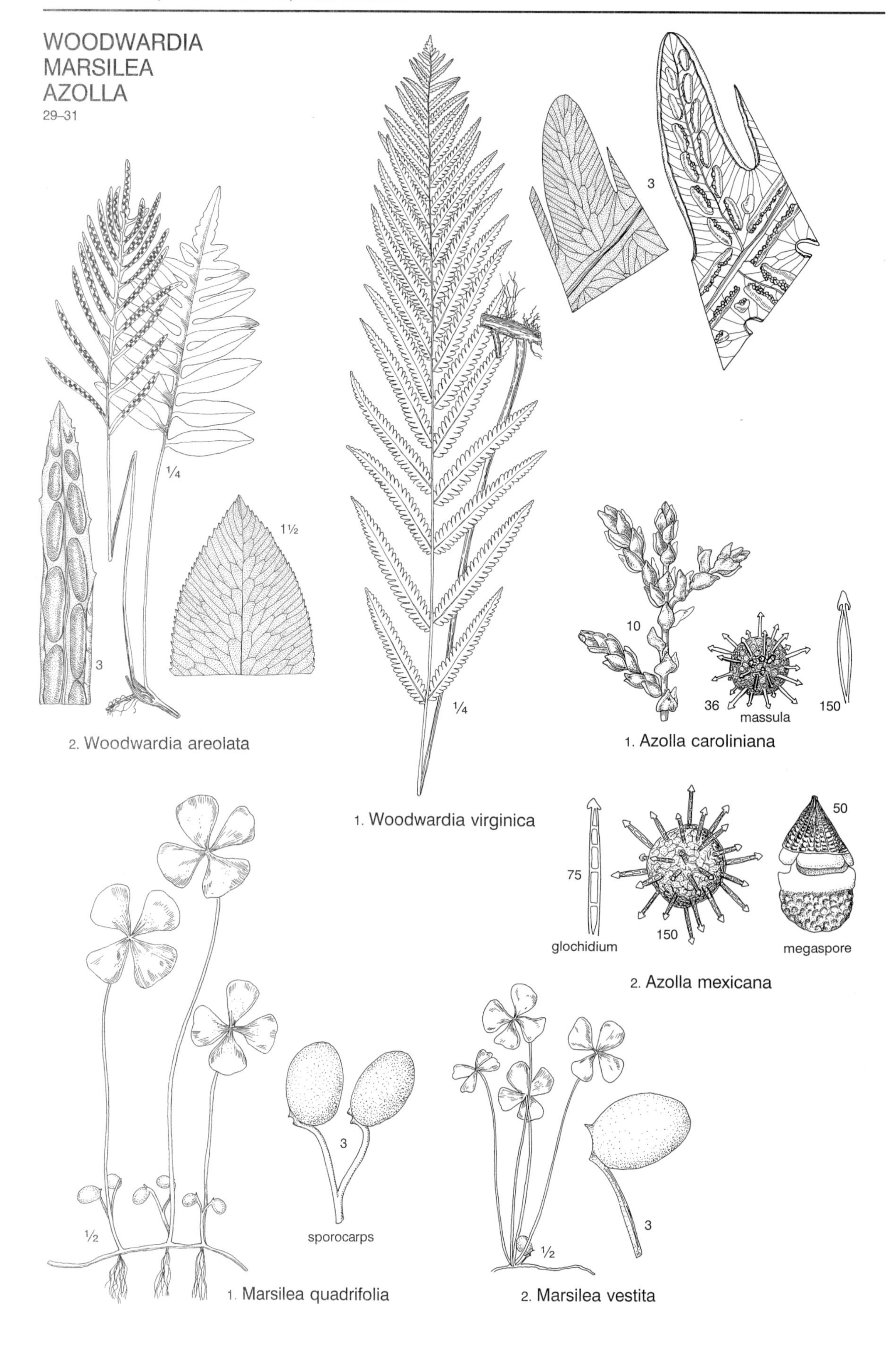

2. Woodwardia areolata

1. Woodwardia virginica

1. Azolla caroliniana

massula

2. Azolla mexicana

glochidium megaspore

1. Marsilea quadrifolia

sporocarps

2. Marsilea vestita

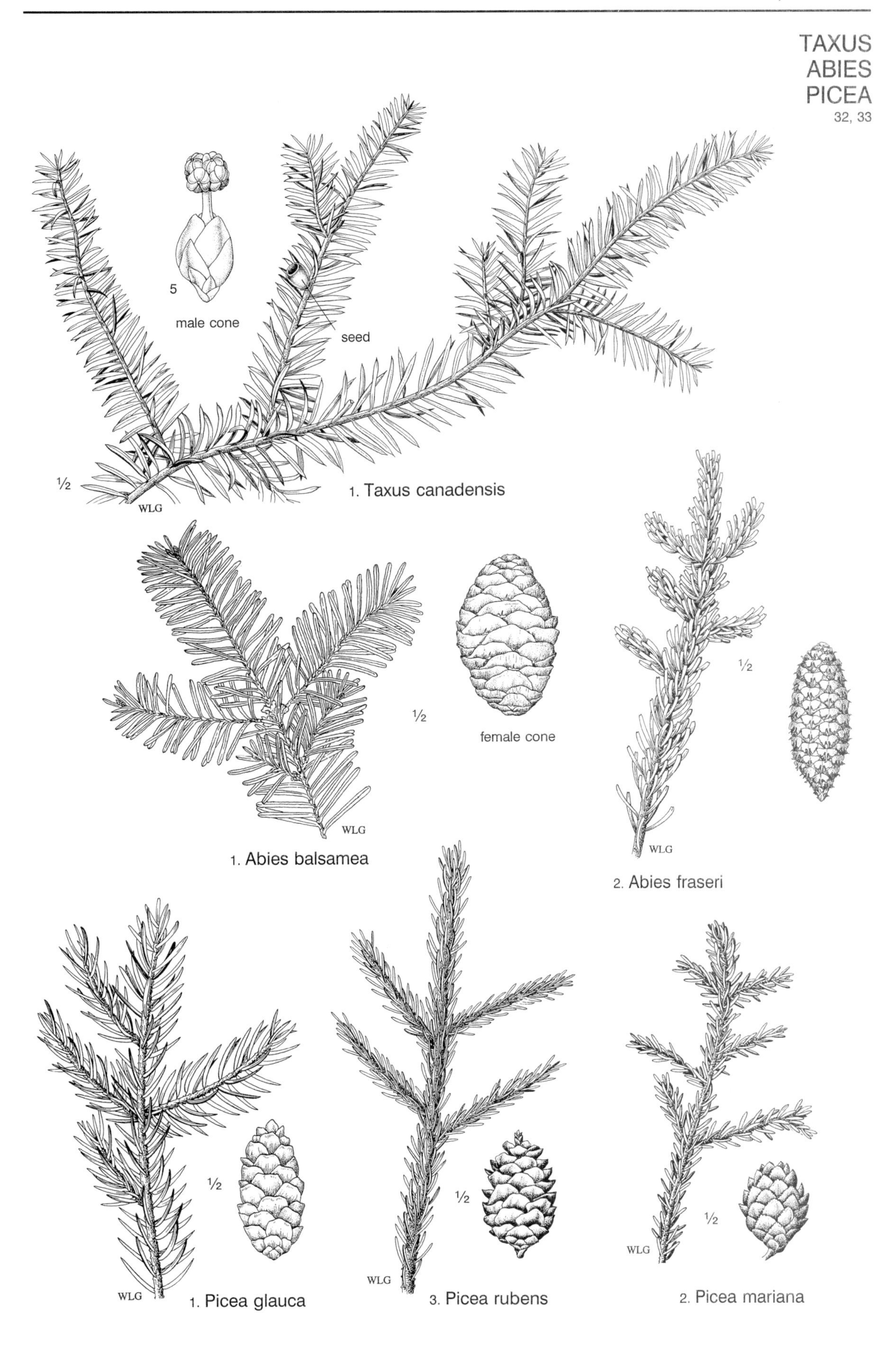

5
male cone

seed

½
WLG

1. Taxus canadensis

½
female cone

½

WLG

1. Abies balsamea

½

WLG

2. Abies fraseri

½

WLG

1. Picea glauca

½

WLG

3. Picea rubens

½

WLG

2. Picea mariana

TSUGA
LARIX
PINUS
34, 35

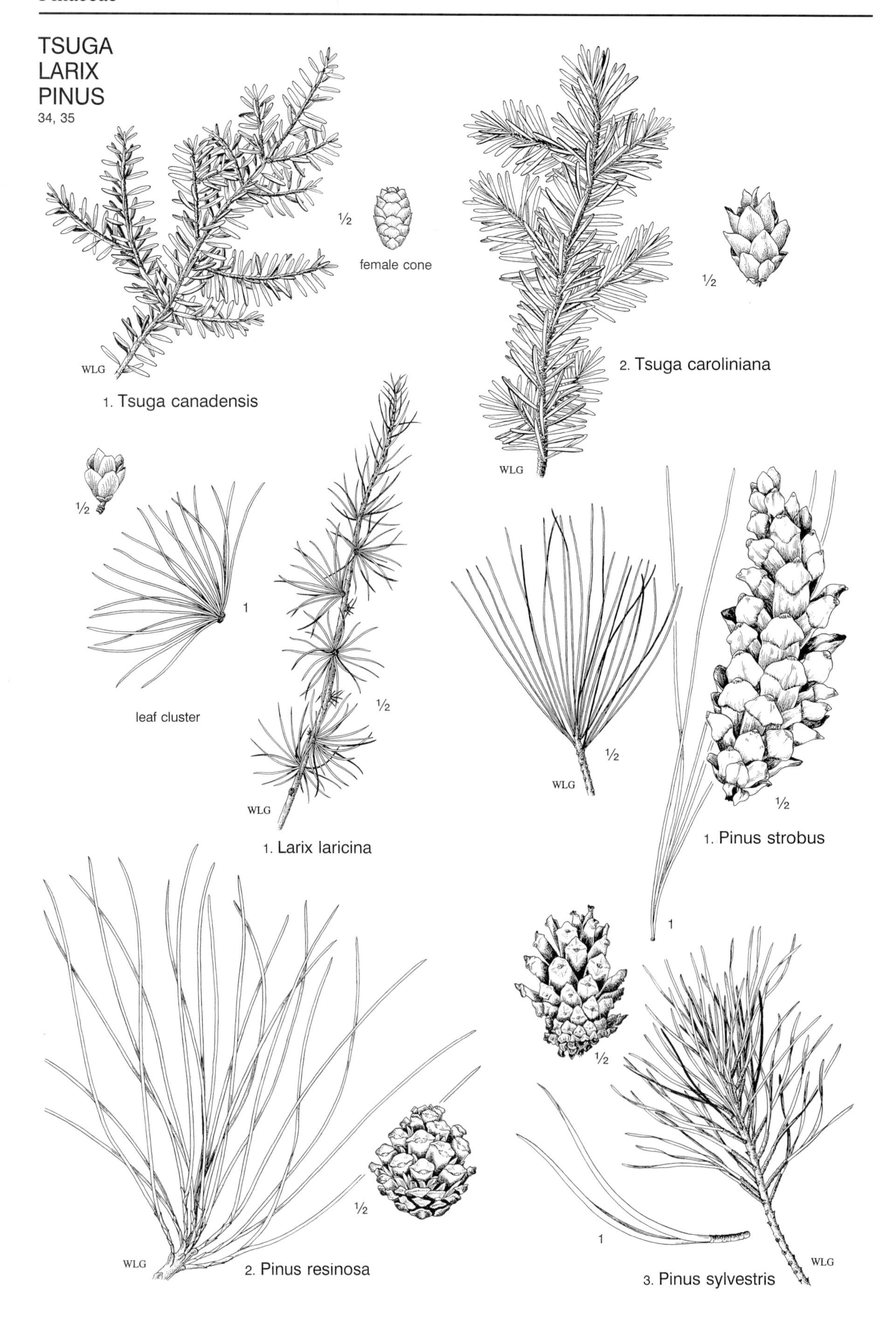

½ female cone

1. Tsuga canadensis

½

2. Tsuga caroliniana

WLG

½

leaf cluster

1

½

WLG

1. Larix laricina

½

WLG

½

WLG

1

1. Pinus strobus

½

1

1

½

WLG

2. Pinus resinosa

3. Pinus sylvestris

WLG

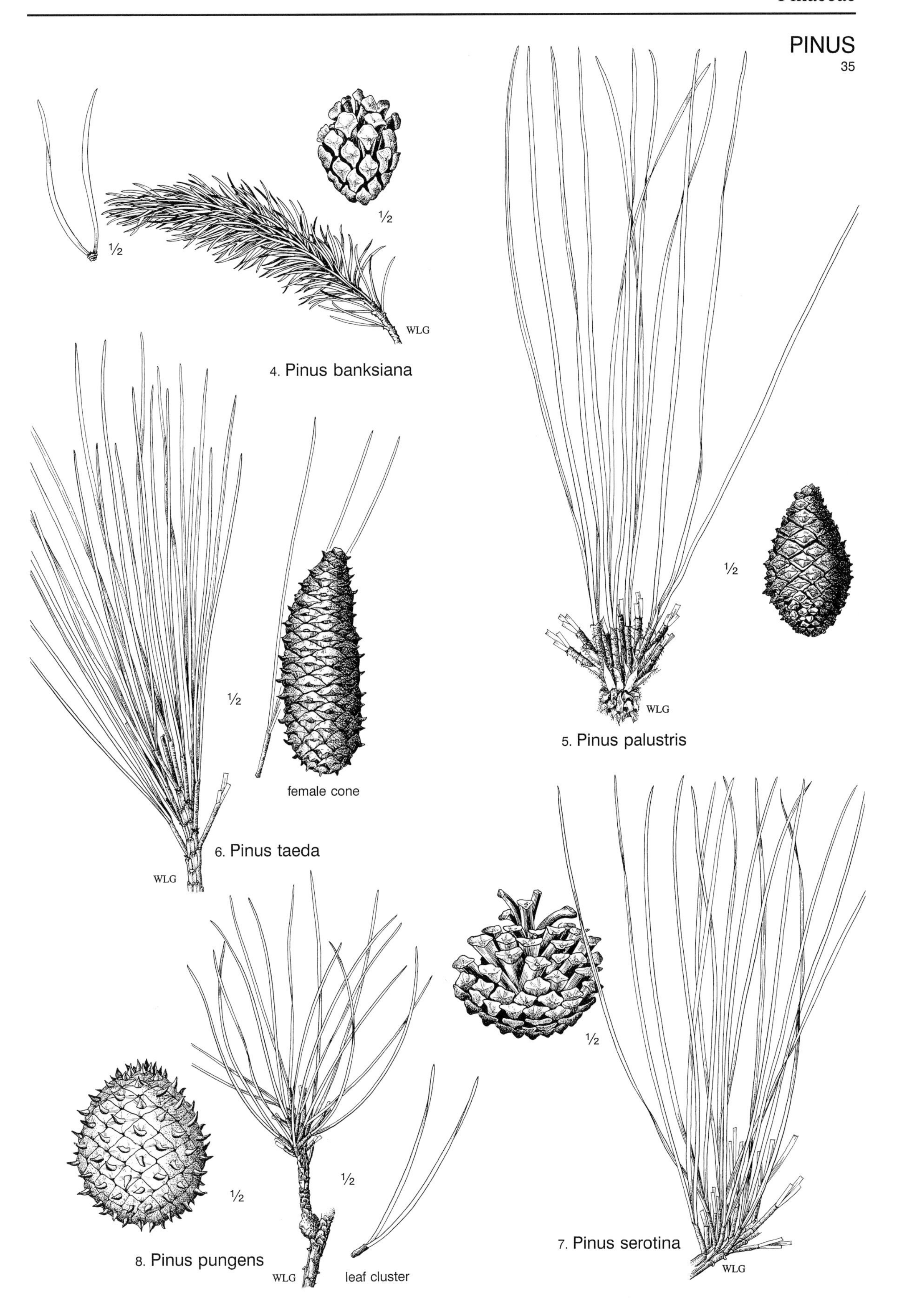

4. Pinus banksiana

5. Pinus palustris

½

female cone

6. Pinus taeda

8. Pinus pungens

leaf cluster

7. Pinus serotina

PINUS
TAXODIUM
35, 36

½

WLG

female cone

9. Pinus rigida

leaf cluster

WLG

½

11. Pinus echinata

½

WLG

10. Pinus virginiana

1

WLG   2

½

1. Taxodium distichum

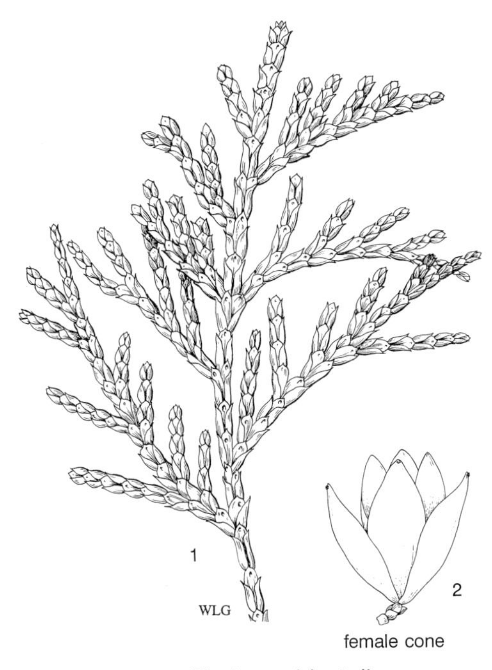

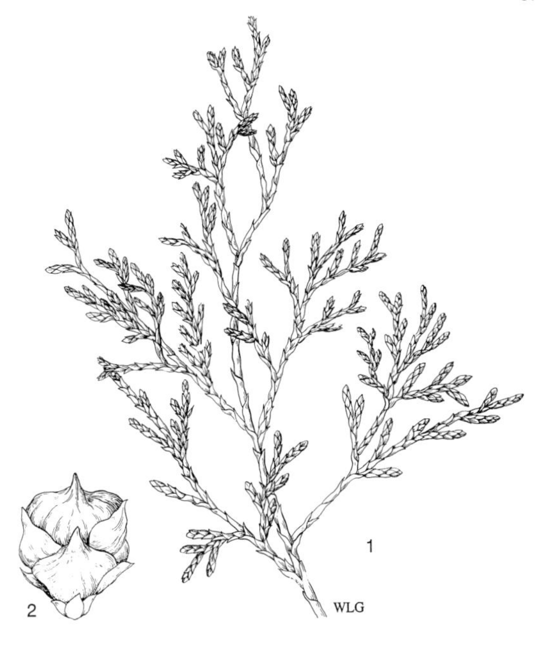

female cone

1. Thuja occidentalis

1. Chamaecyparis thyoides

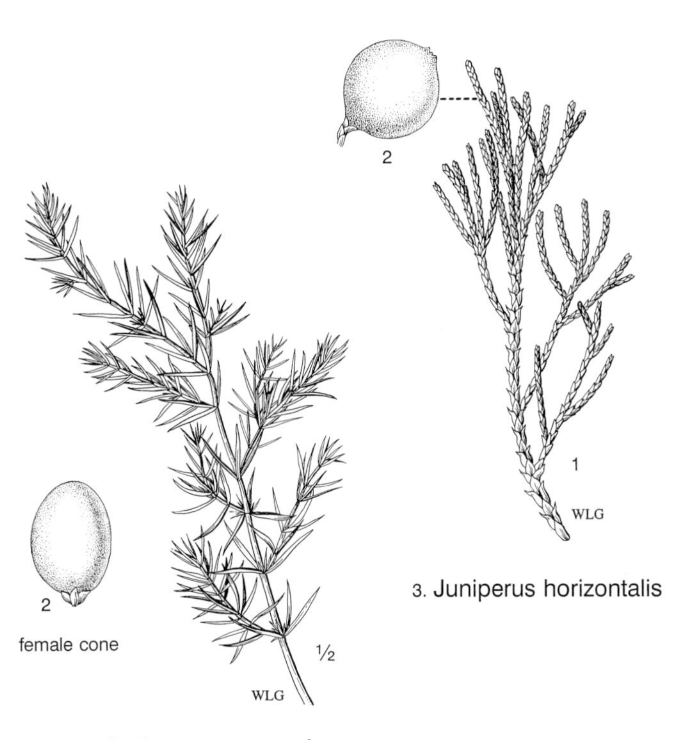

3. Juniperus horizontalis

female cone

½

1. Juniperus communis

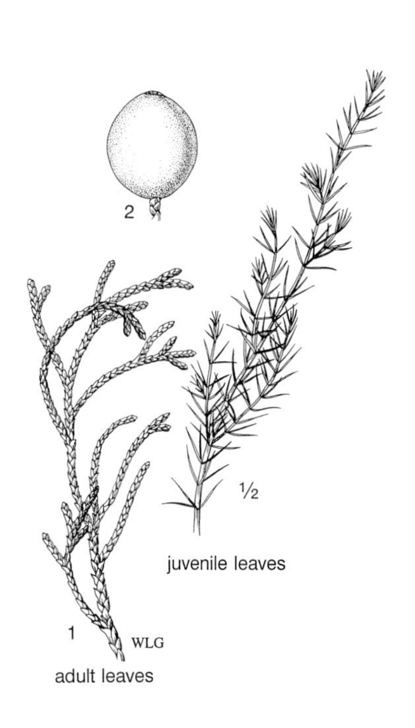

juvenile leaves

adult leaves

2. Juniperus virginiana

MAGNOLIA
39

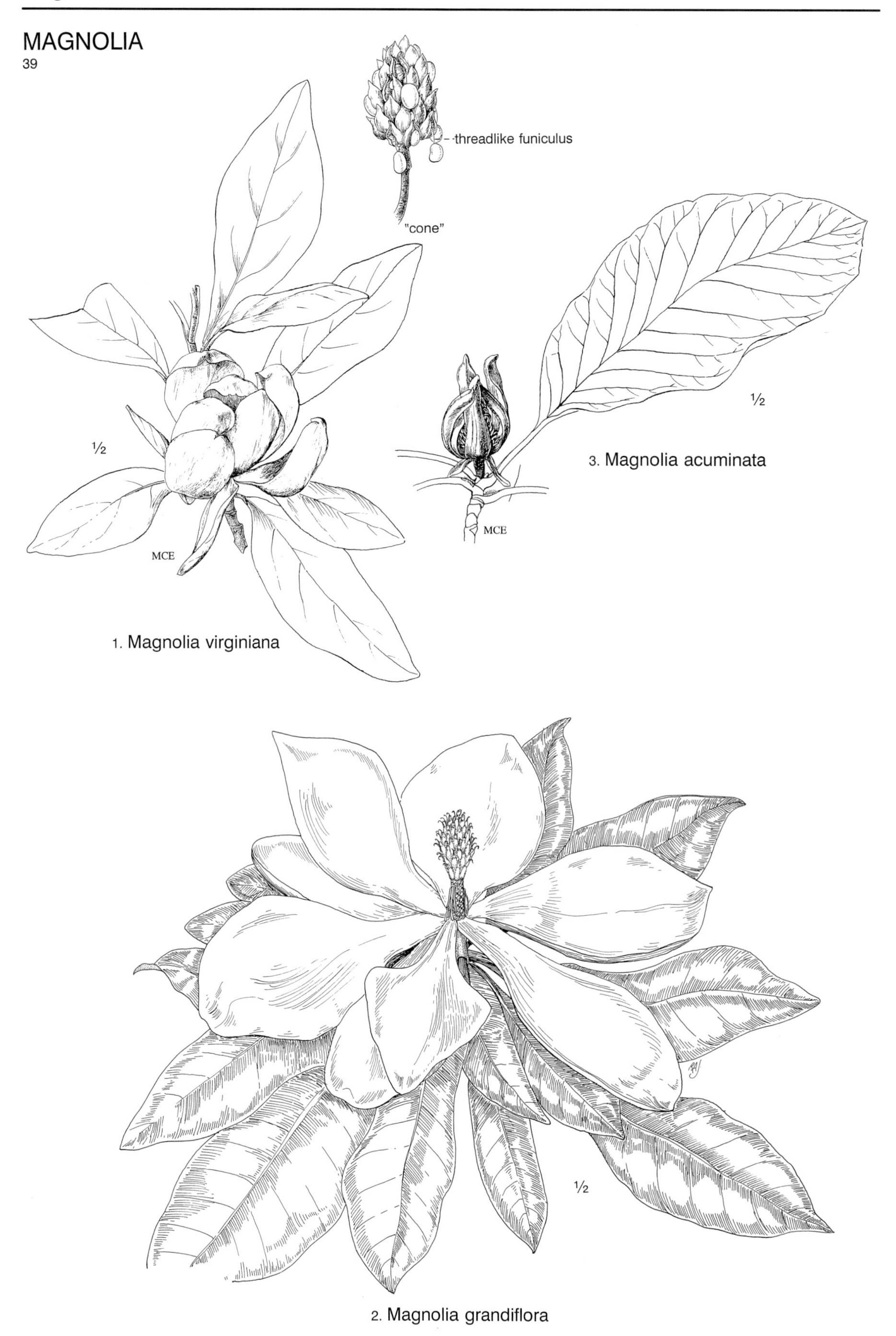

threadlike funiculus

"cone"

½

MCE

1. Magnolia virginiana

3. Magnolia acuminata

MCE

½

2. Magnolia grandiflora

½

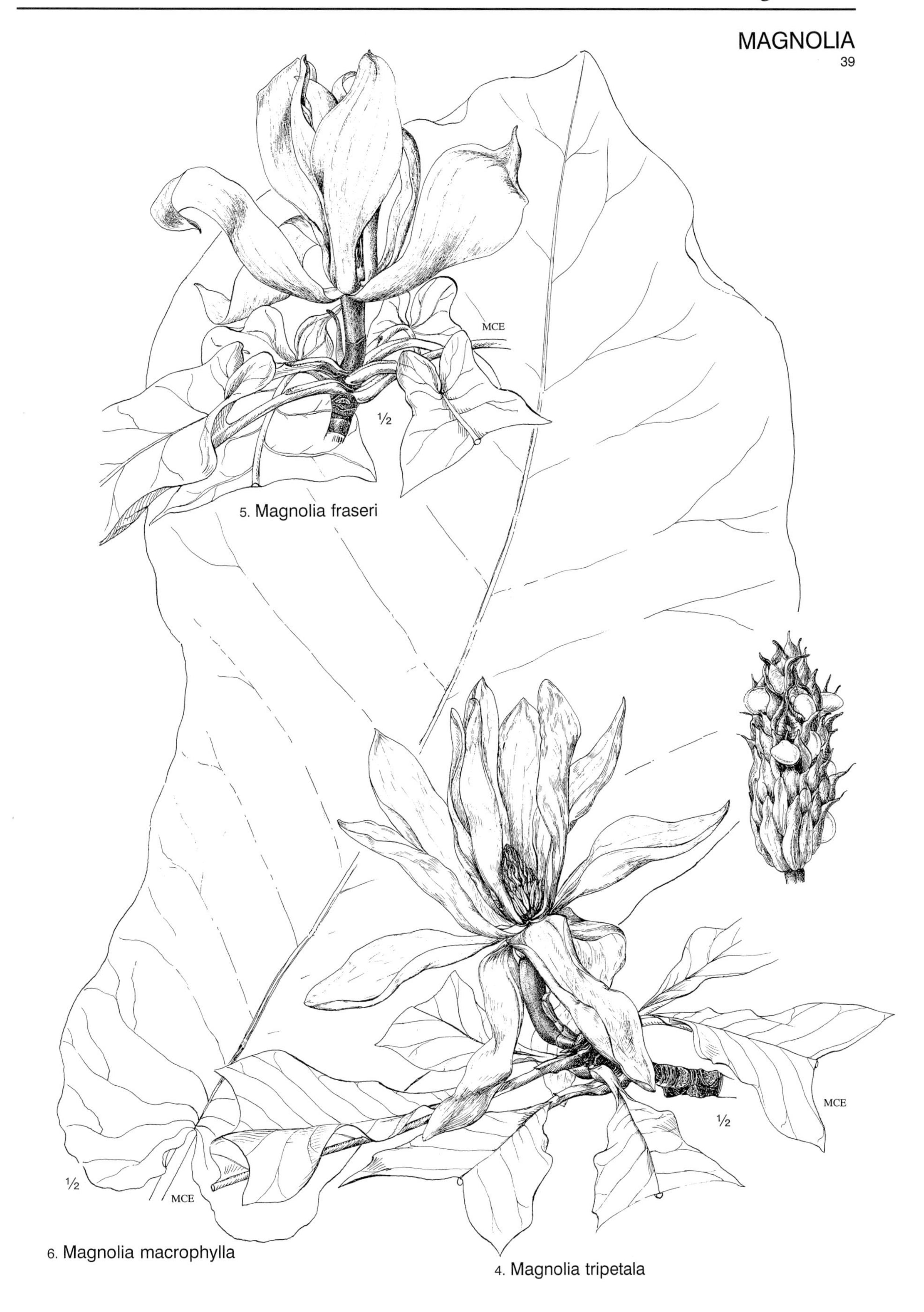

5. Magnolia fraseri

6. Magnolia macrophylla

4. Magnolia tripetala

LIRIODENDRON
ASIMINA
CALYCANTHUS
PERSEA
39–41

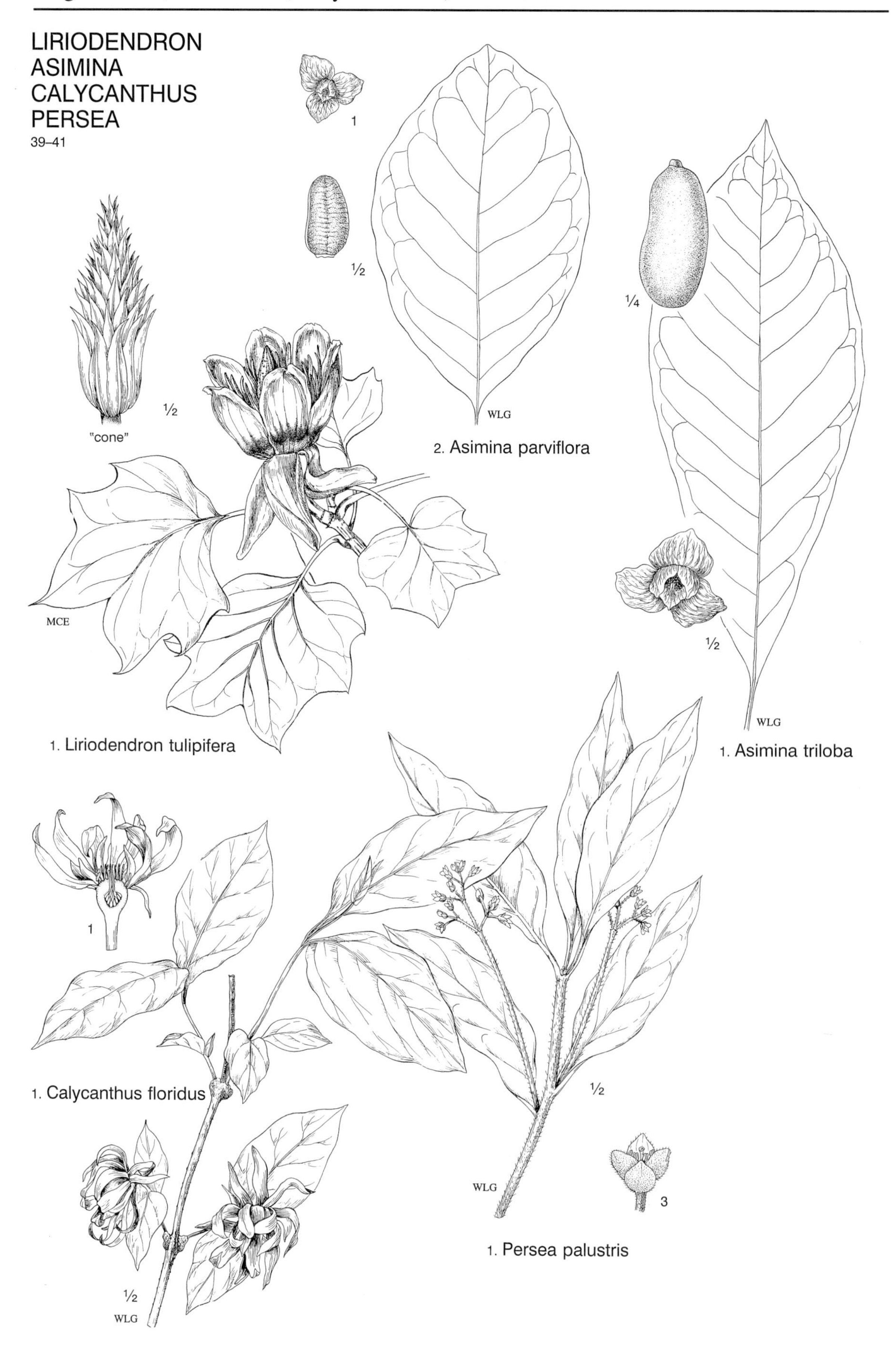

"cone"

1. Liriodendron tulipifera

MCE

2. Asimina parviflora

WLG

1. Asimina triloba

WLG

1. Calycanthus floridus

1. Persea palustris

WLG

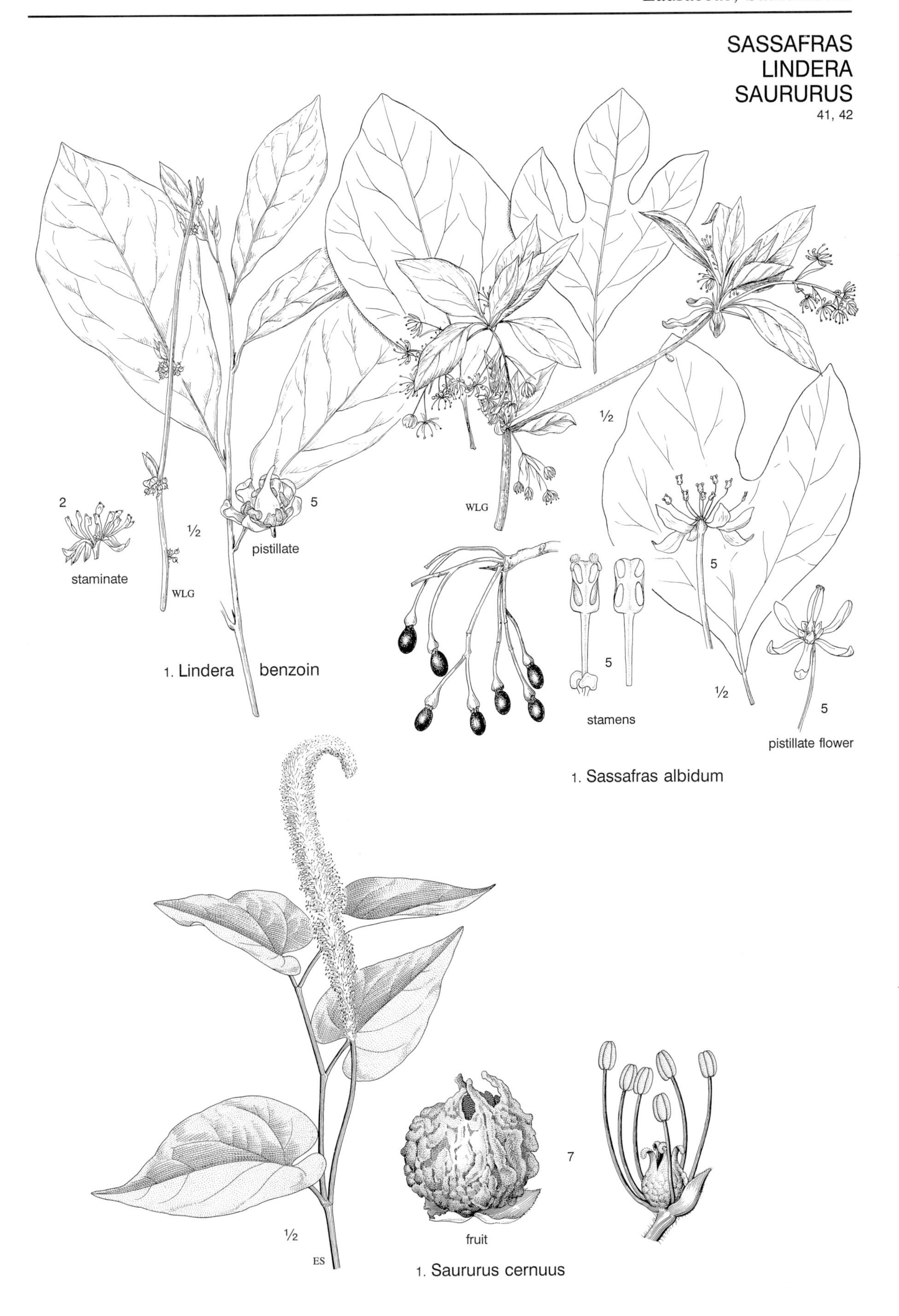

2

staminate

½

WLG

5

pistillate

1. Lindera benzoin

½

WLG

½

5

stamens

5

½

5

pistillate flower

1. Sassafras albidum

7

fruit

½

ES

1. Saururus cernuus

HEXASTYLIS
42, 43

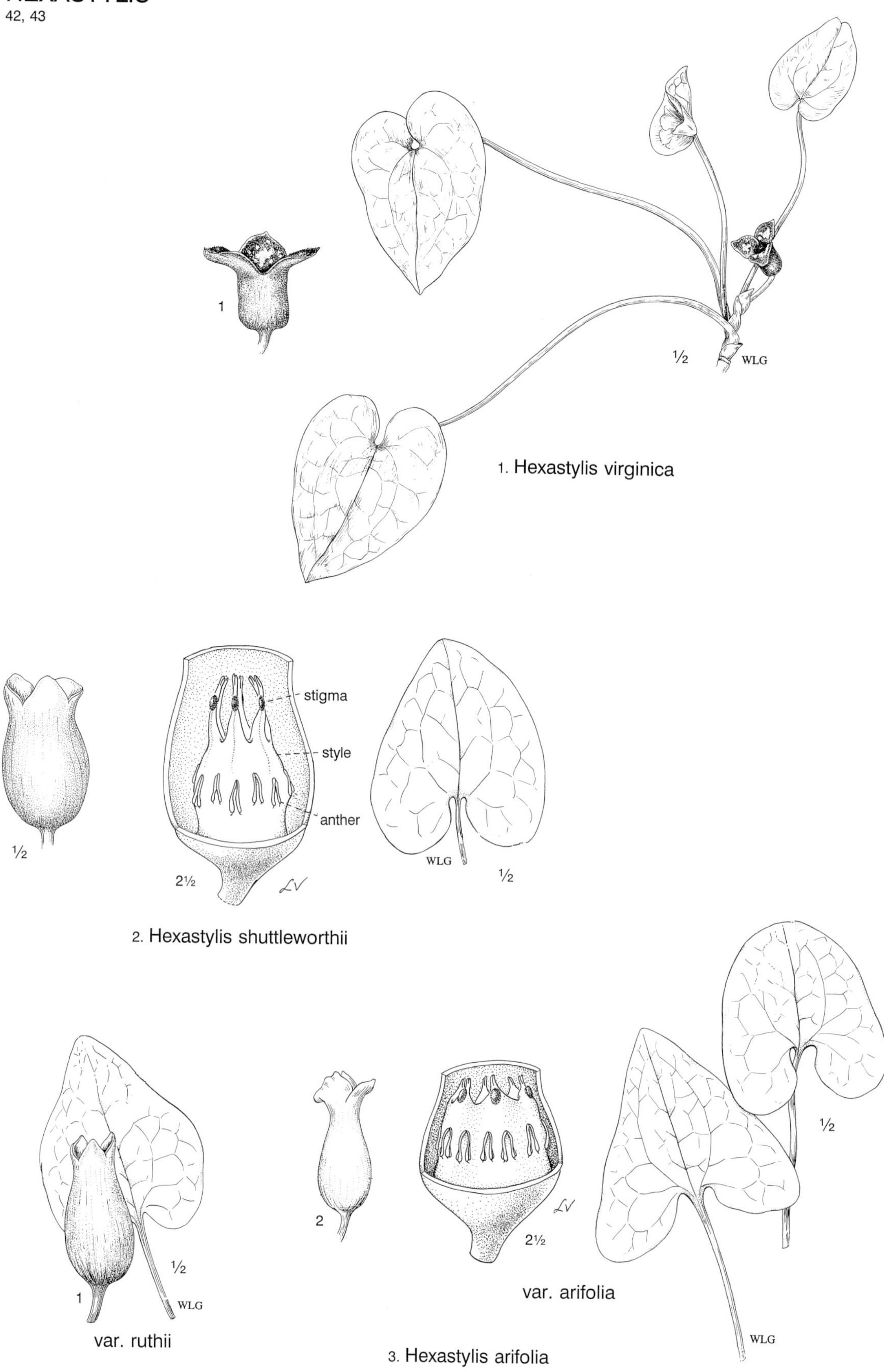

1. Hexastylis virginica

2. Hexastylis shuttleworthii

stigma

style

anther

var. ruthii

var. arifolia

3. Hexastylis arifolia

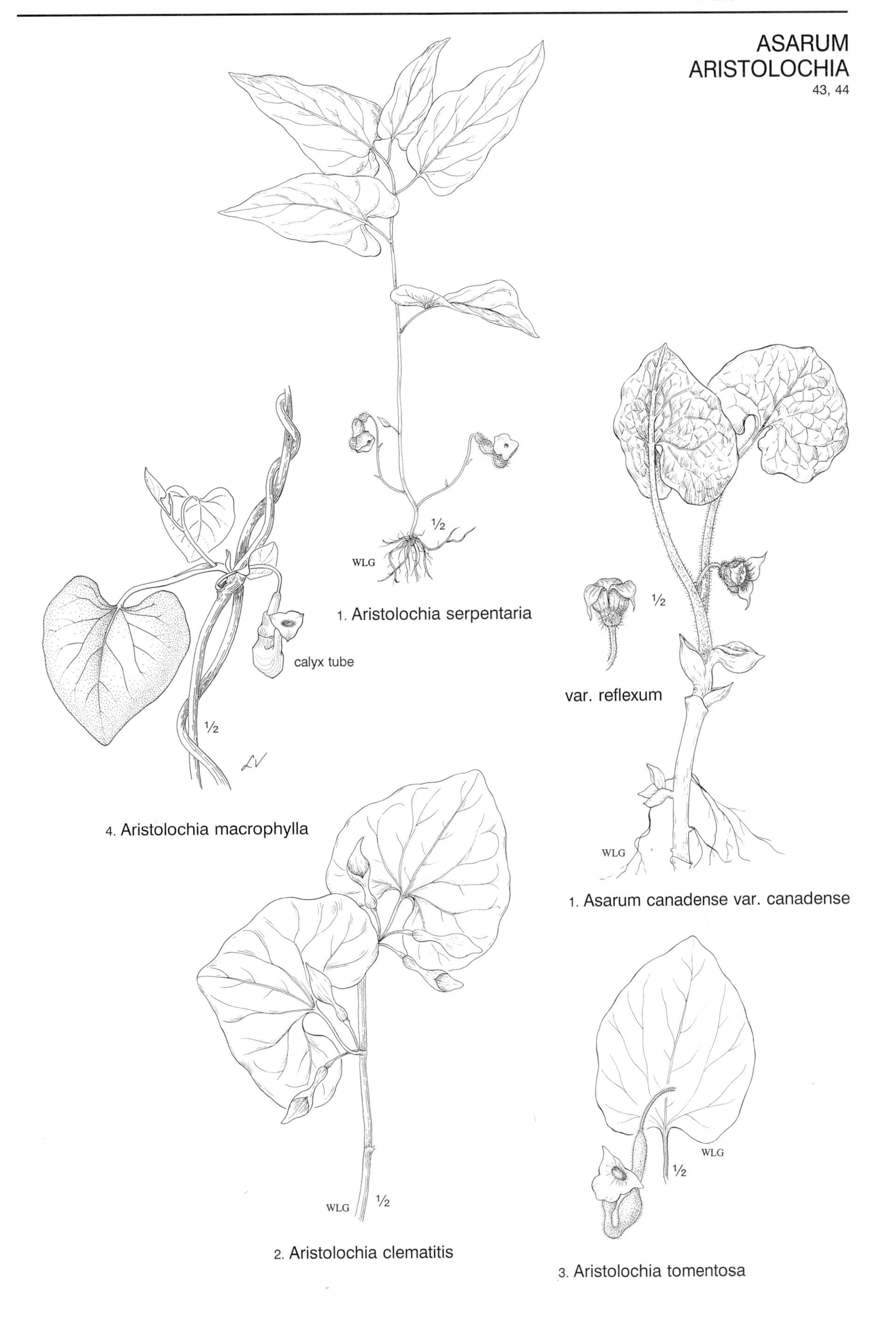

1. Aristolochia serpentaria

WLG

calyx tube

4. Aristolochia macrophylla

var. reflexum

1. Asarum canadense var. canadense

WLG

2. Aristolochia clematitis

3. Aristolochia tomentosa

NELUMBO
NUPHAR
44, 45

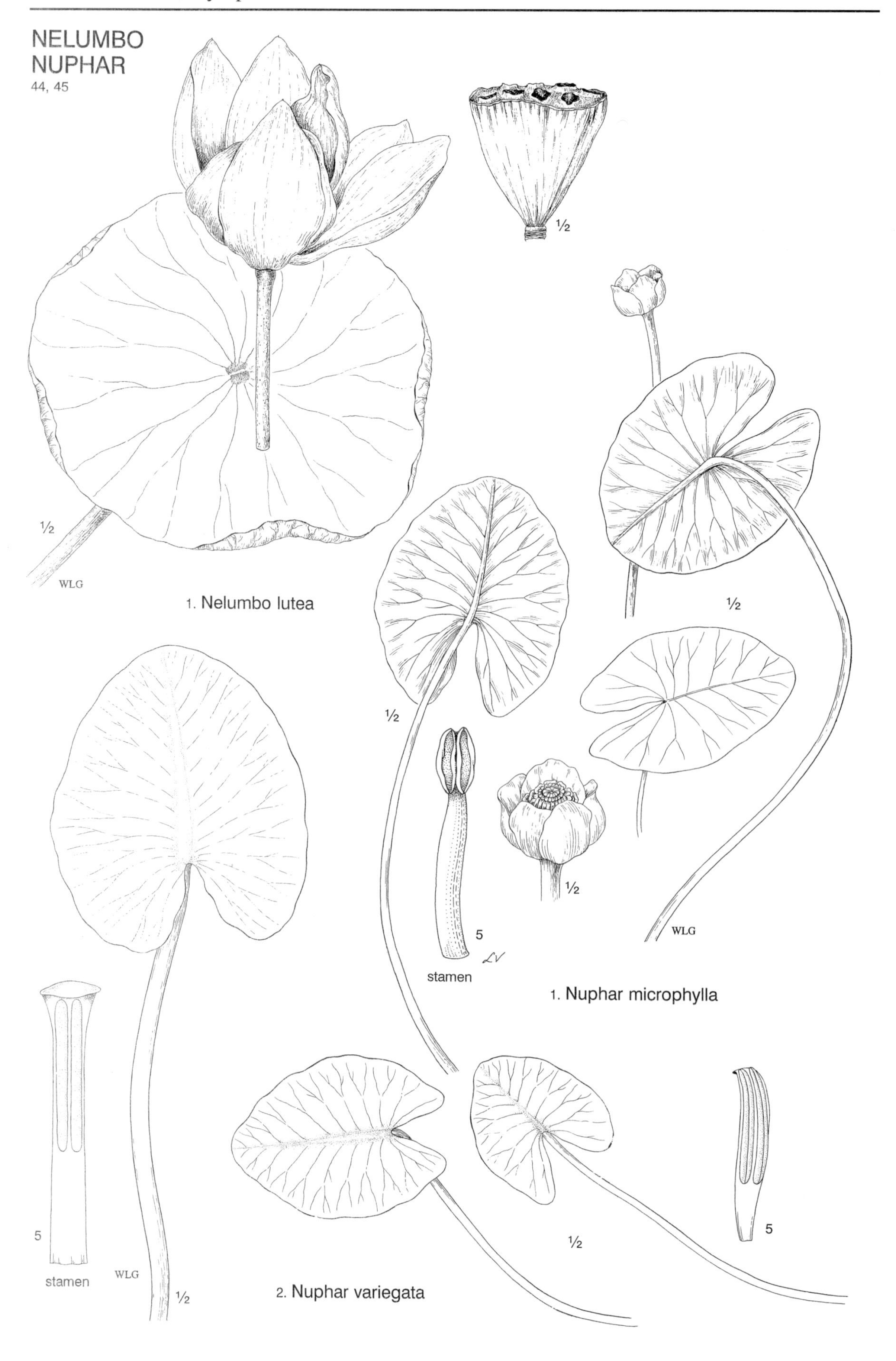

½

WLG

1. Nelumbo lutea

½

½

½

½

5

stamen

1. Nuphar microphylla

WLG

5

stamen

WLG

½

2. Nuphar variegata

½

5

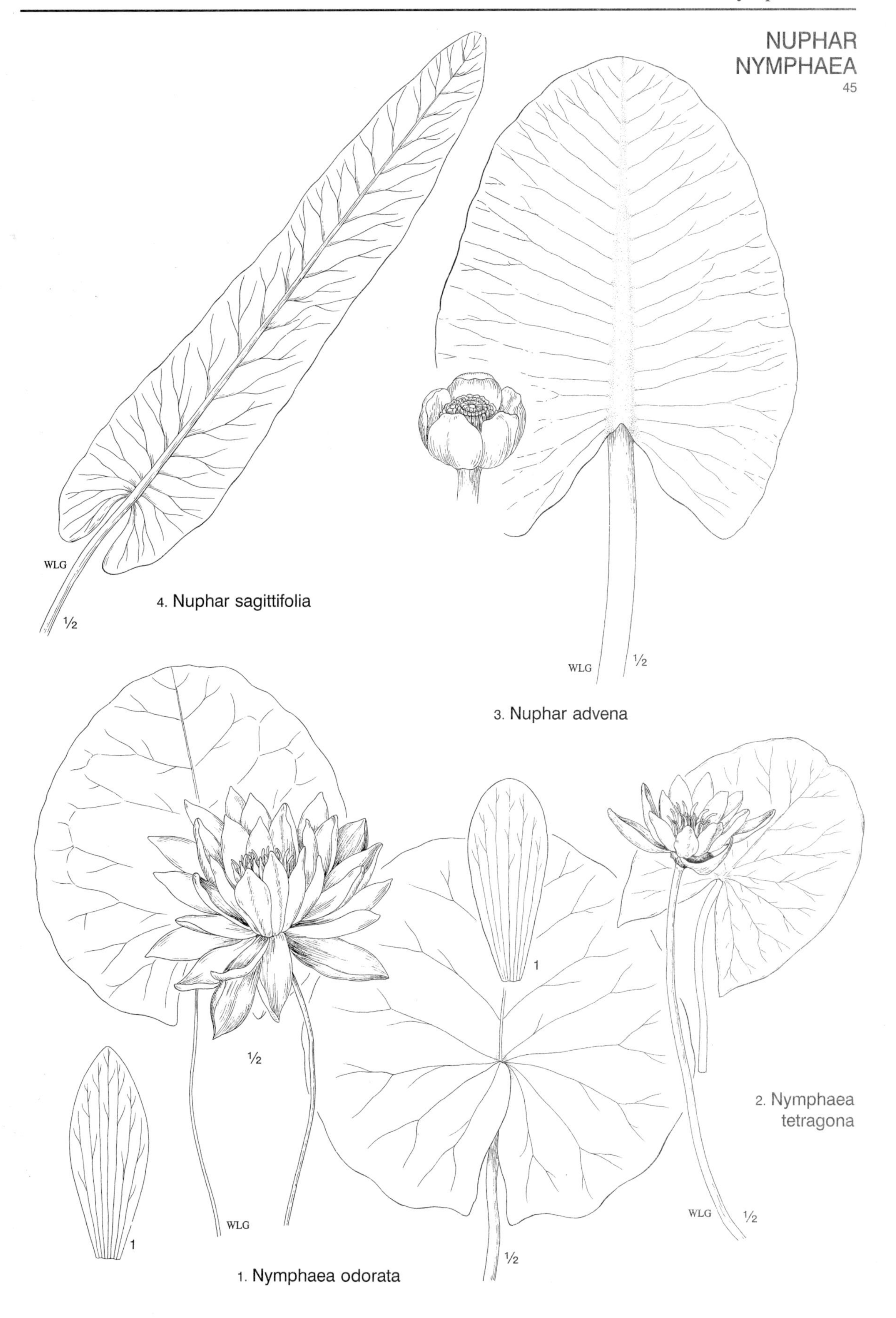

4. Nuphar sagittifolia

WLG

½

3. Nuphar advena

WLG   ½

½

WLG

1

2. Nymphaea
tetragona

WLG   ½

1. Nymphaea odorata

1

½

BRASENIA
CABOMBA
CERATOPHYLLUM
46, 47

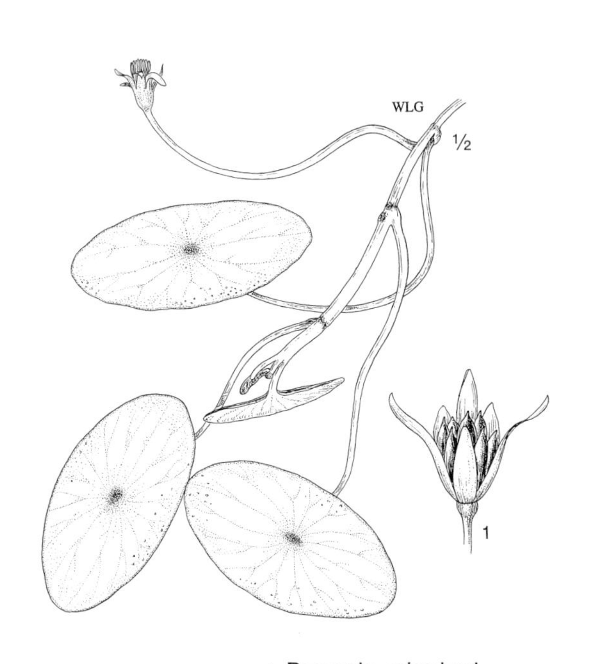

1. Brasenia schreberi

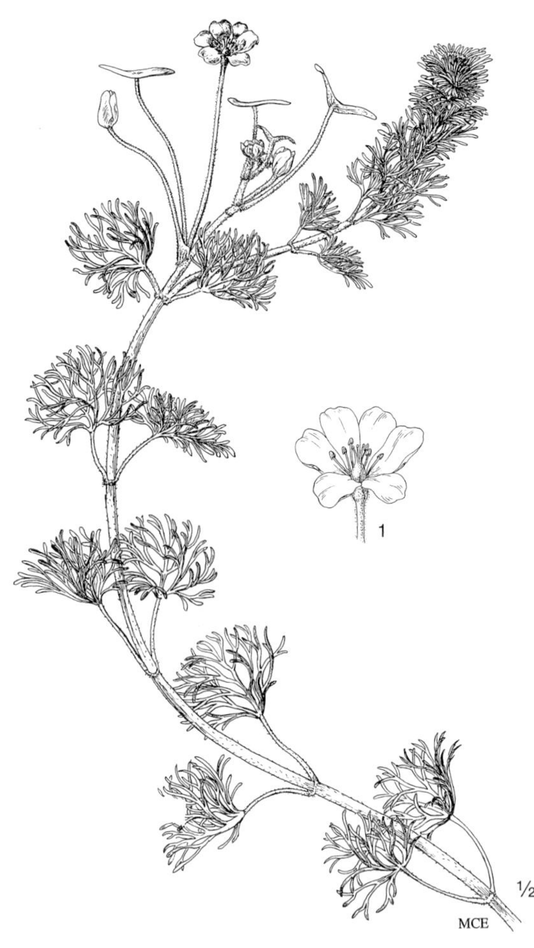

1. Cabomba caroliniana

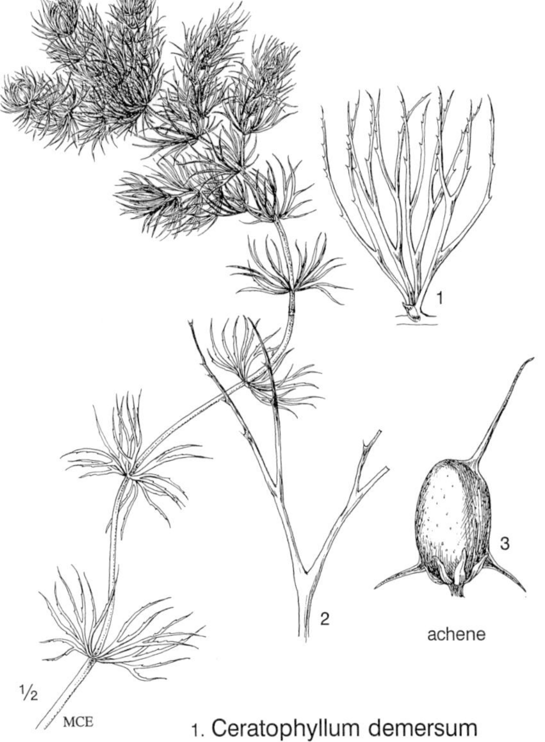

1. Ceratophyllum demersum

2. Ceratophyllum echinatum

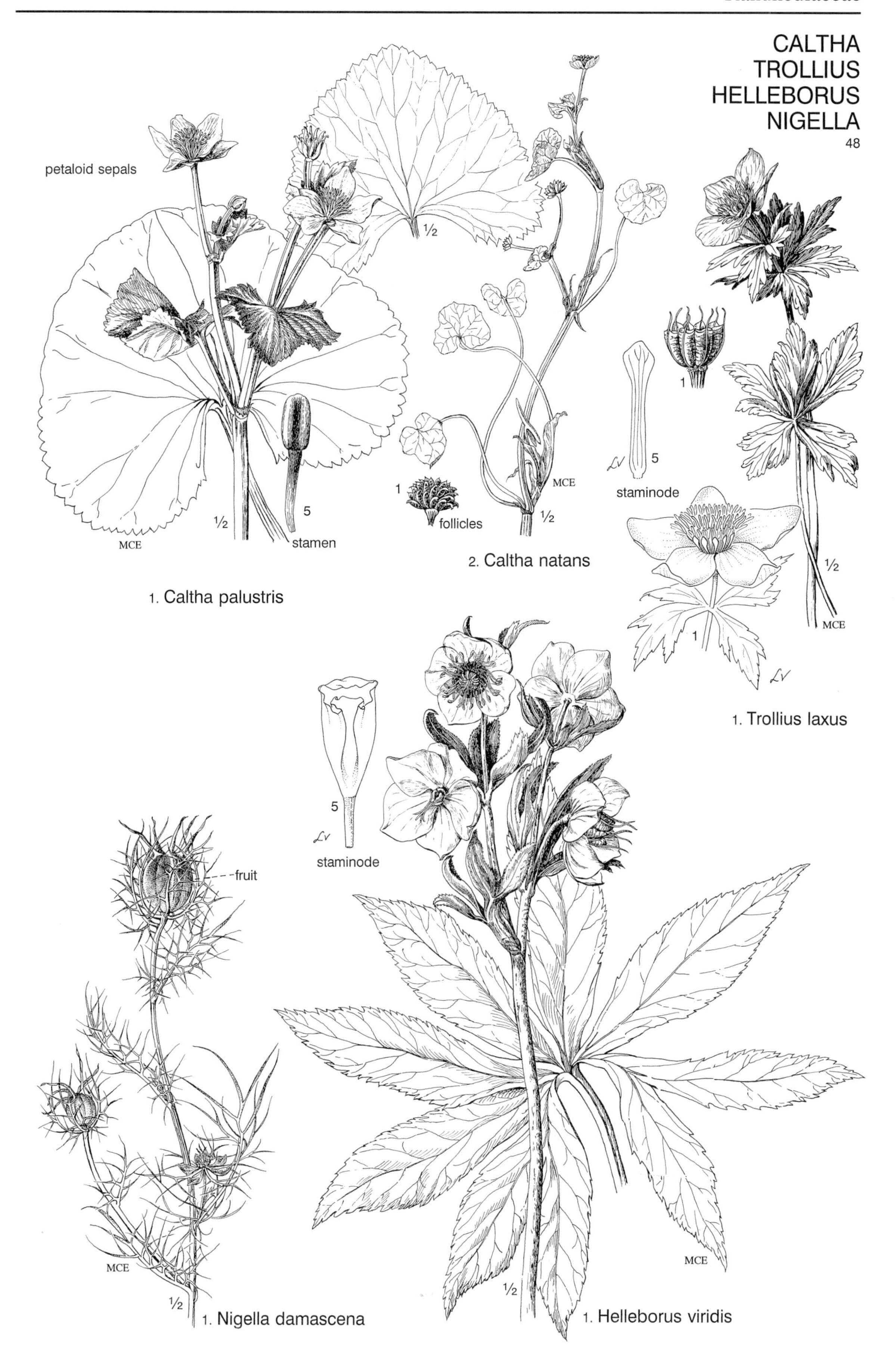

petaloid sepals

½

MCE

½

1. Caltha palustris

5

stamen

½

MCE

1

follicles

½

MCE

2. Caltha natans

5

staminode

1

staminode

1

½

MCE

1. Trollius laxus

5

staminode

fruit

½

1. Nigella damascena

½

MCE

1. Helleborus viridis

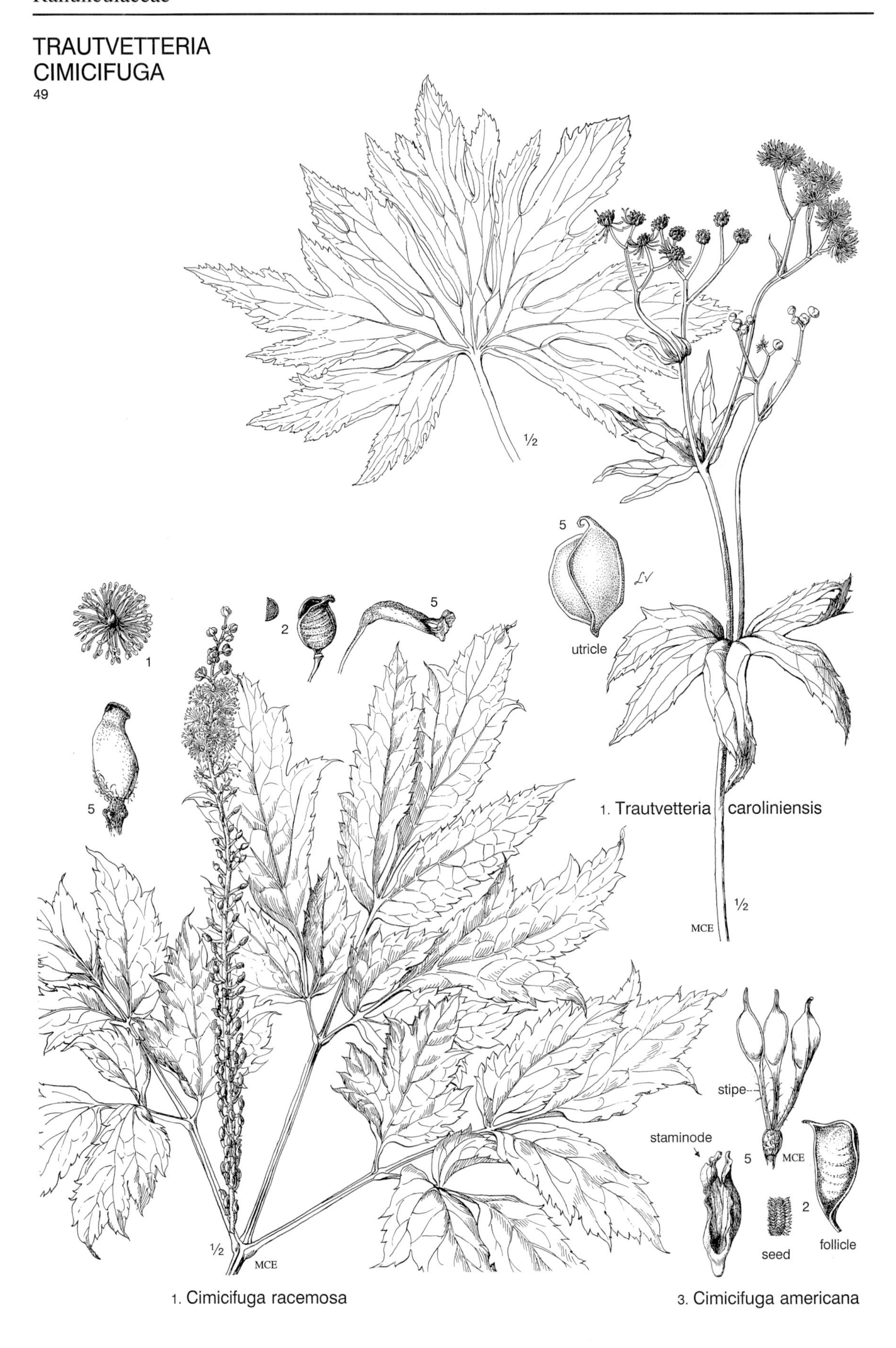

½

5

utricle

1. Trautvetteria caroliniensis

½

MCE

1

5

2

5

5

½

MCE

1. Cimicifuga racemosa

stipe

staminode

5 MCE

2

seed

follicle

3. Cimicifuga americana

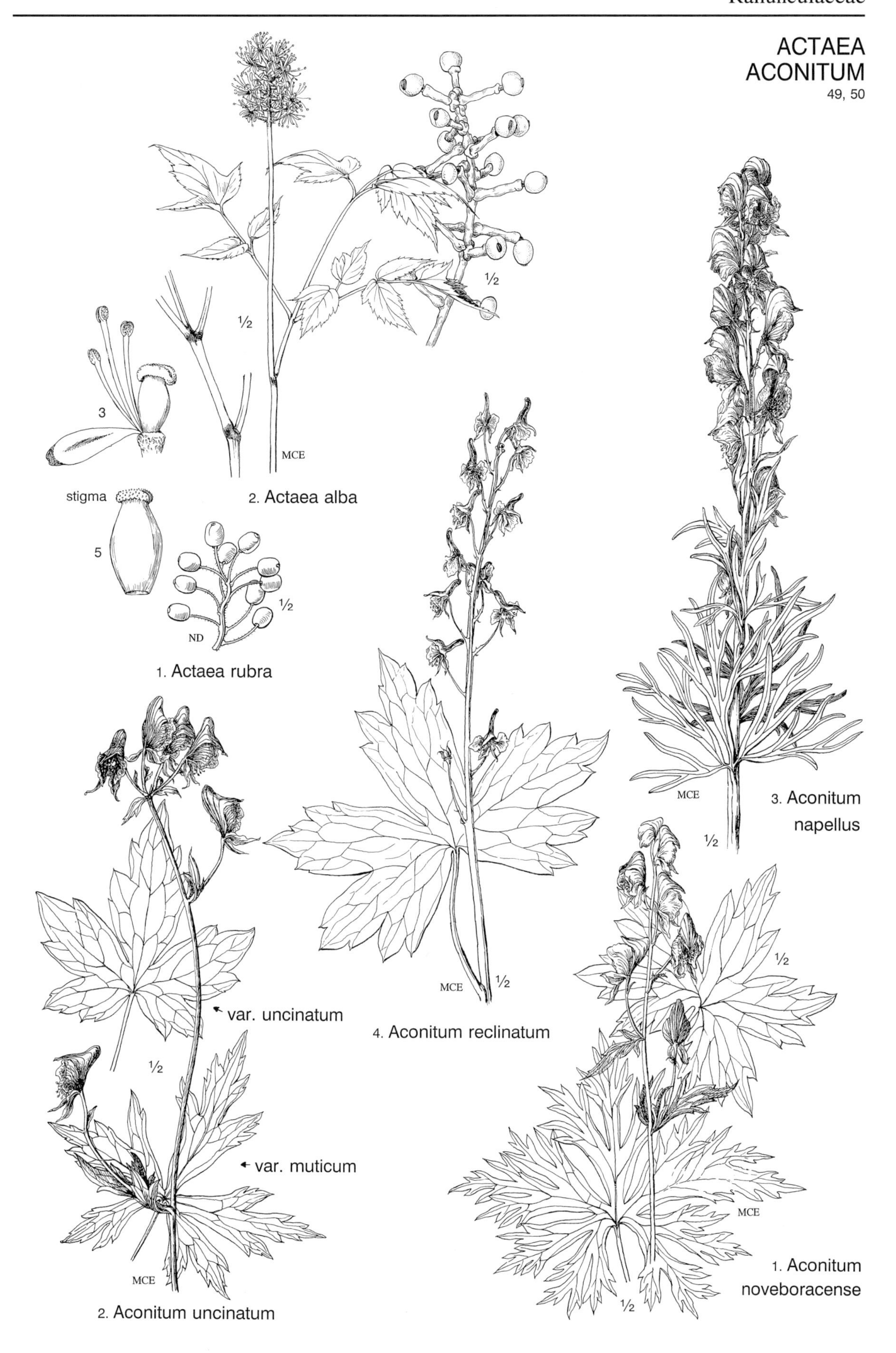

3

stigma

5

ND

1. Actaea rubra

MCE

2. Actaea alba

var. uncinatum

2. Aconitum uncinatum

var. muticum

MCE

4. Aconitum reclinatum

MCE

3. Aconitum
napellus

MCE

1. Aconitum
noveboracense

DELPHINIUM
51

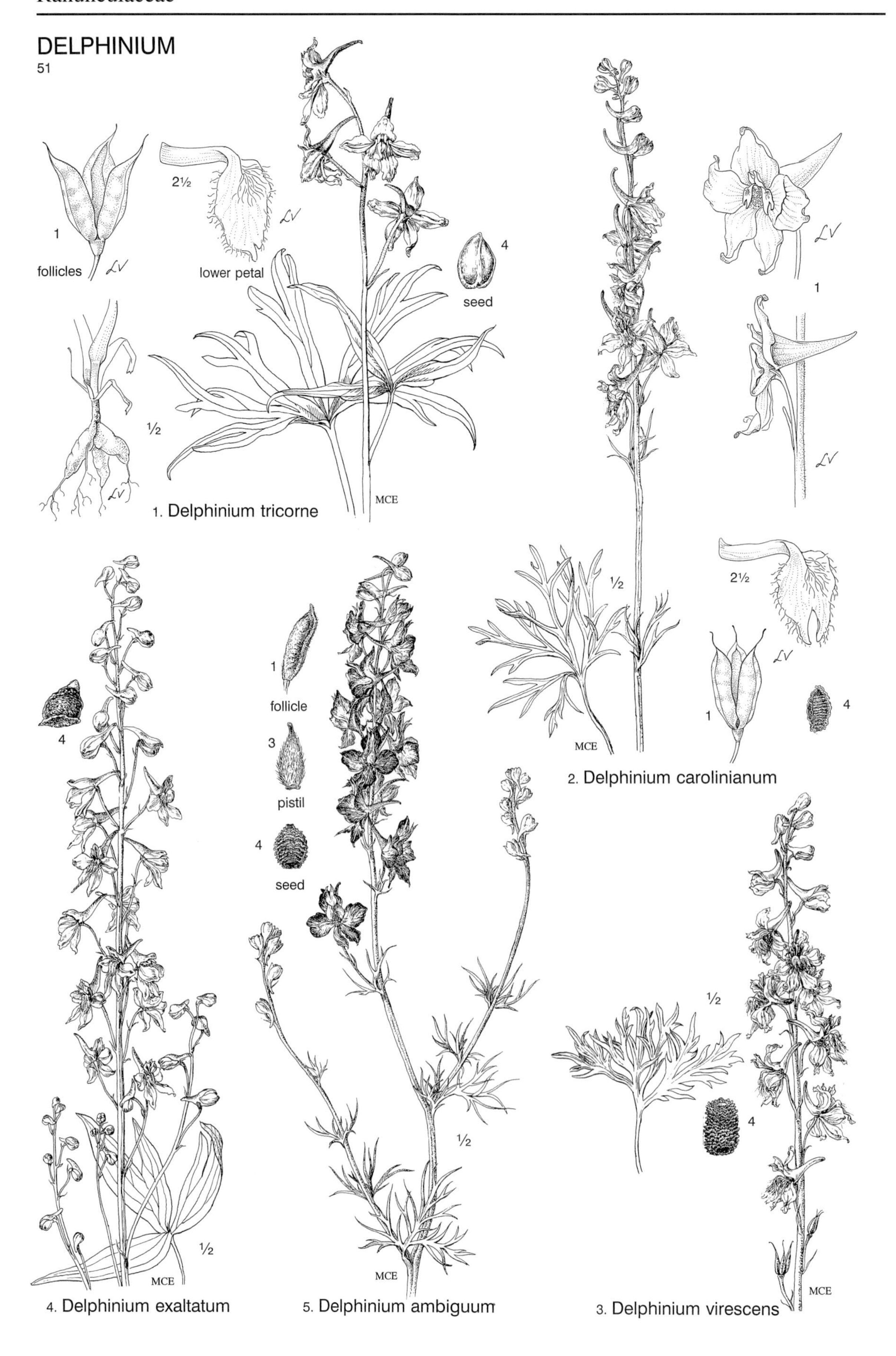

follicles

lower petal

2½

seed

4

1

½

1. Delphinium tricorne

MCE

1

4

follicle

3

pistil

4

seed

2. Delphinium carolinianum

½

MCE

2½

4

4

½

MCE

4. Delphinium exaltatum

5. Delphinium ambiguum

½

½

4

MCE

3. Delphinium virescens

MCE

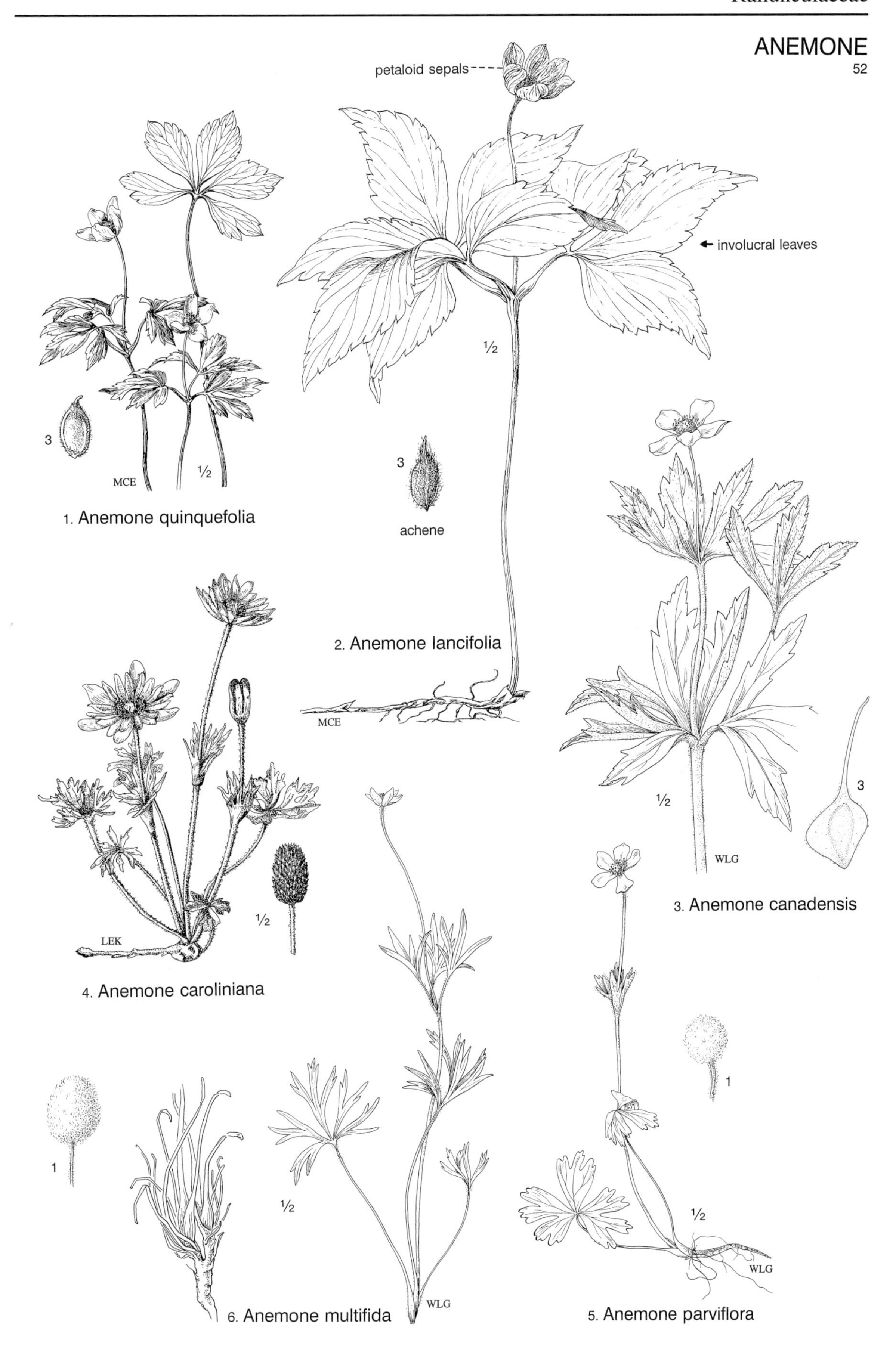

petaloid sepals

← involucral leaves

½

1. Anemone quinquefolia

3

MCE ½

3

achene

2. Anemone lancifolia

MCE

½

½

WLG

3

3. Anemone canadensis

4. Anemone caroliniana

LEK ½

1

6. Anemone multifida

WLG

½

1

5. Anemone parviflora

WLG ½

ANEMONE
HEPATICA
52, 53

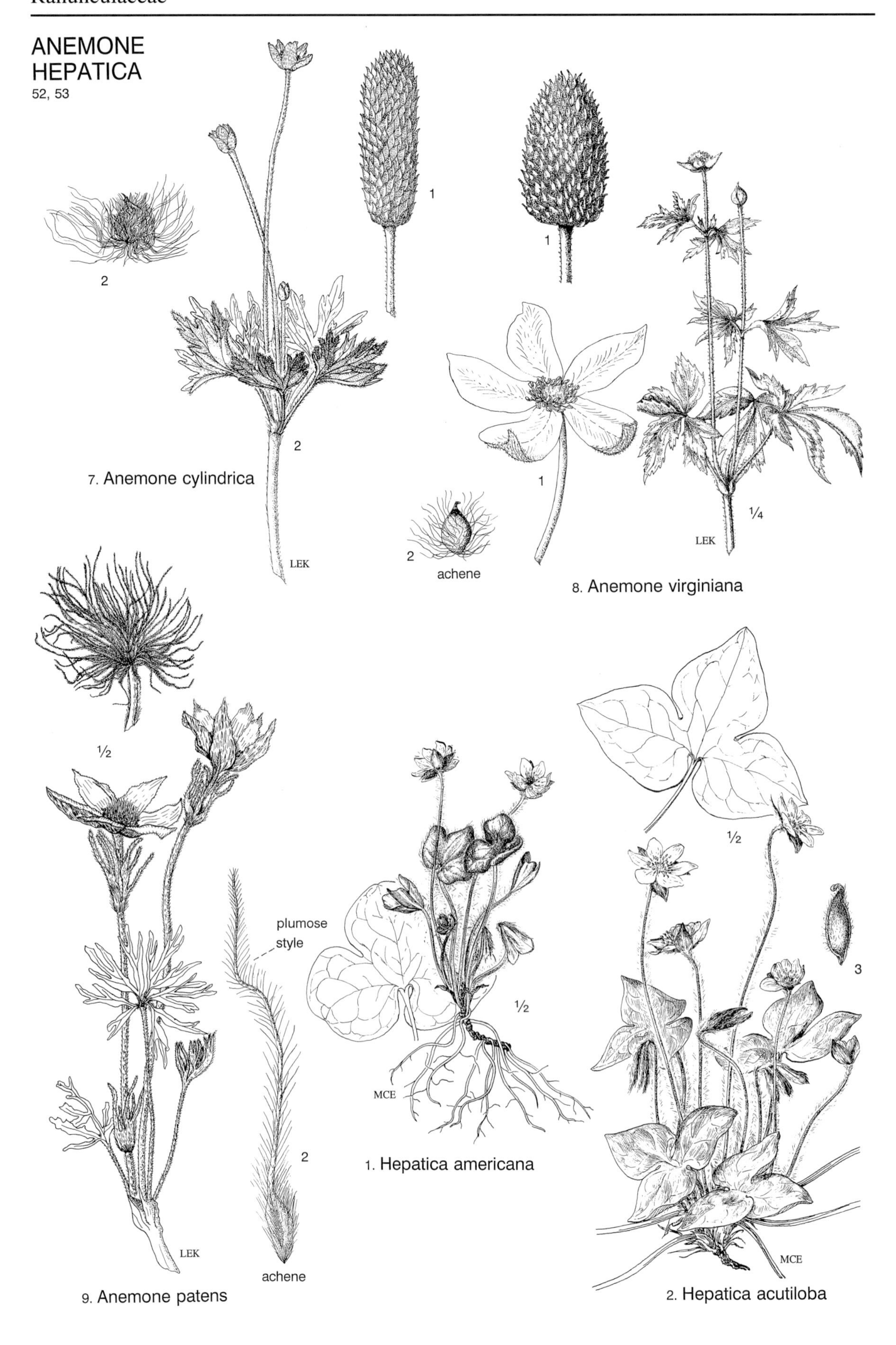

7. Anemone cylindrica

8. Anemone virginiana

achene

1. Hepatica americana

plumose
style

9. Anemone patens

achene

2. Hepatica acutiloba

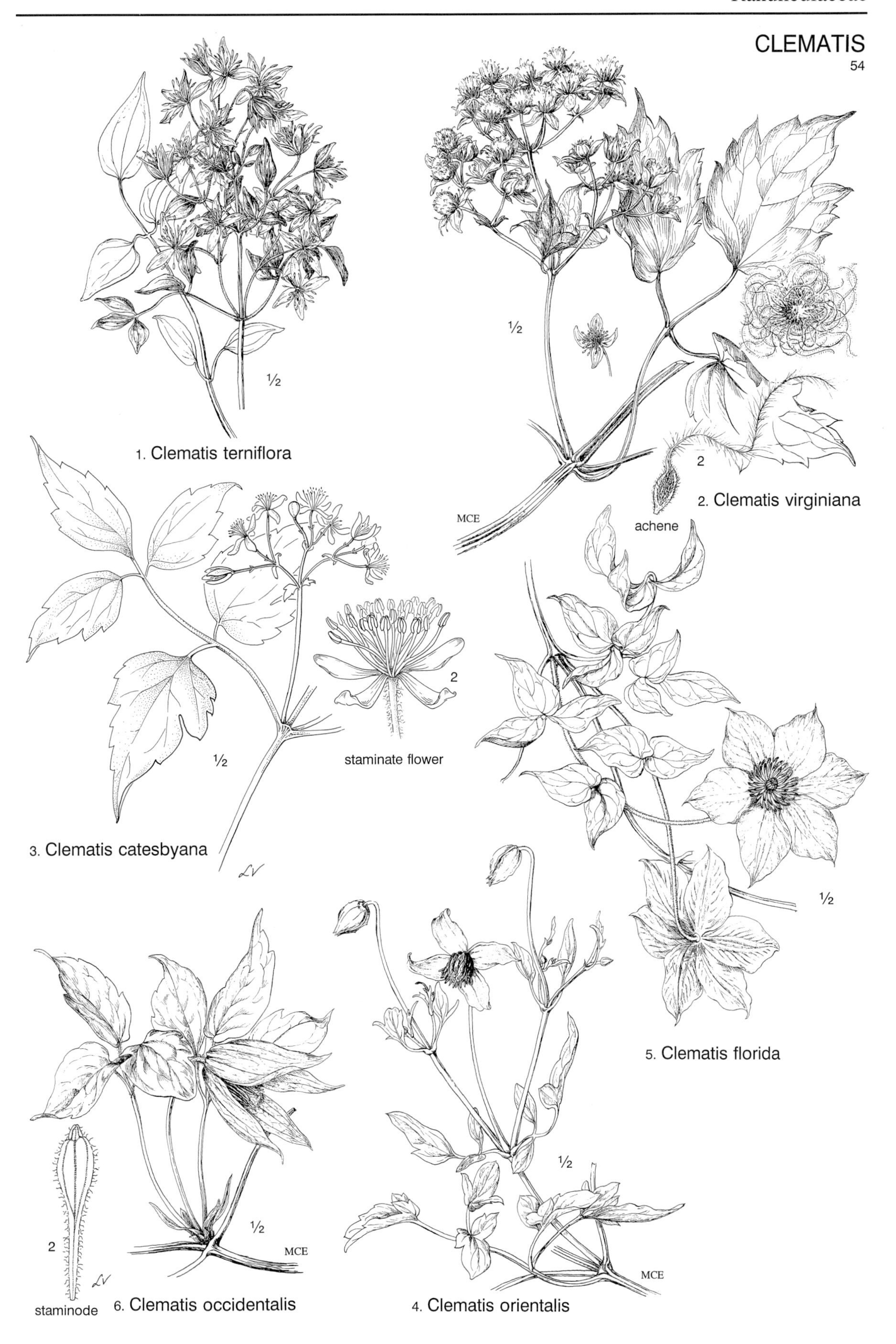

1. Clematis terniflora

2. Clematis virginiana

achene

MCE

staminate flower

3. Clematis catesbyana

5. Clematis florida

staminode   6. Clematis occidentalis

4. Clematis orientalis

MCE

CLEMATIS
54

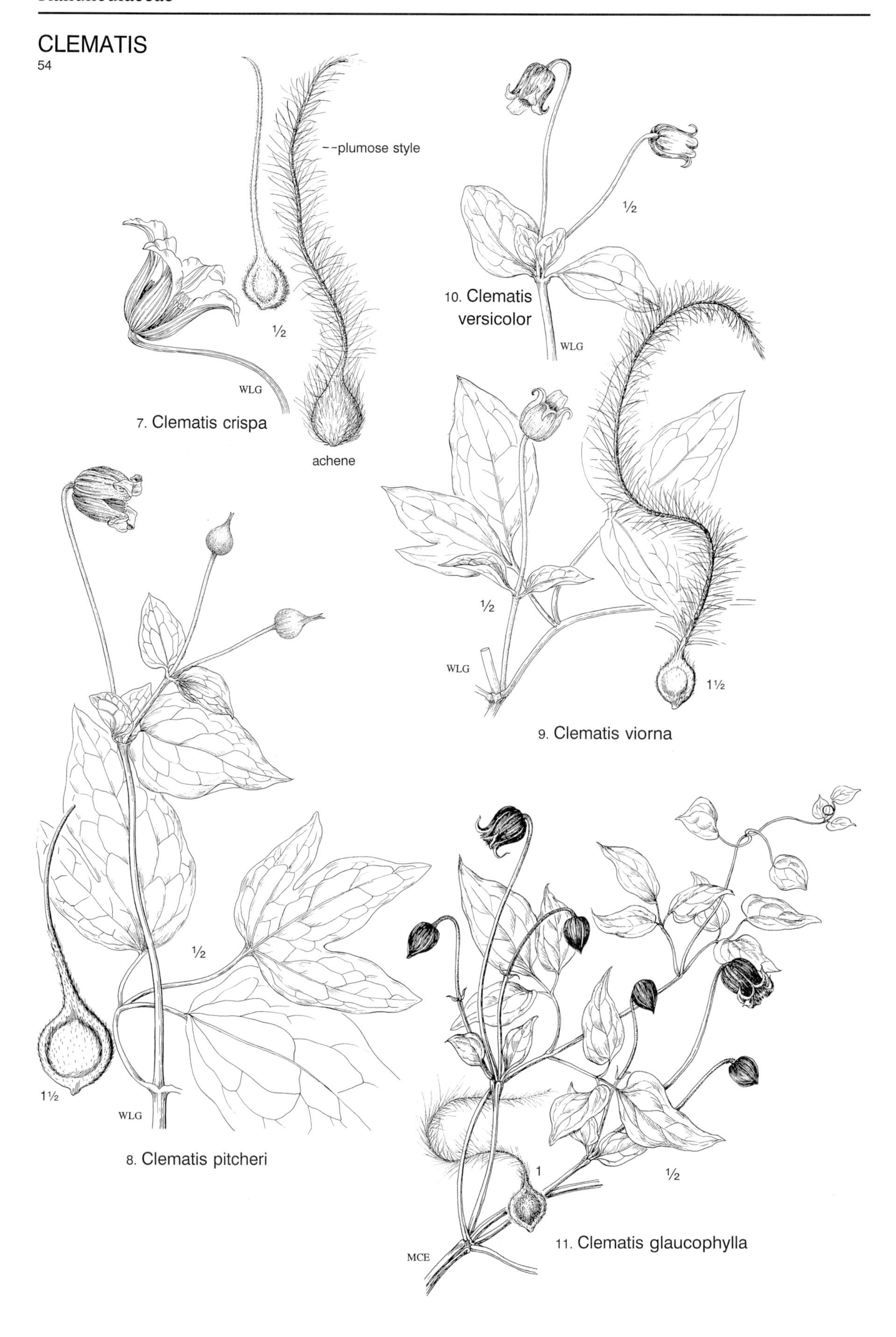

--plumose style

7. Clematis crispa

WLG

½

achene

10. Clematis versicolor

WLG

½

9. Clematis viorna

WLG

½

1½

8. Clematis pitcheri

1½

WLG

11. Clematis glaucophylla

MCE

1

½

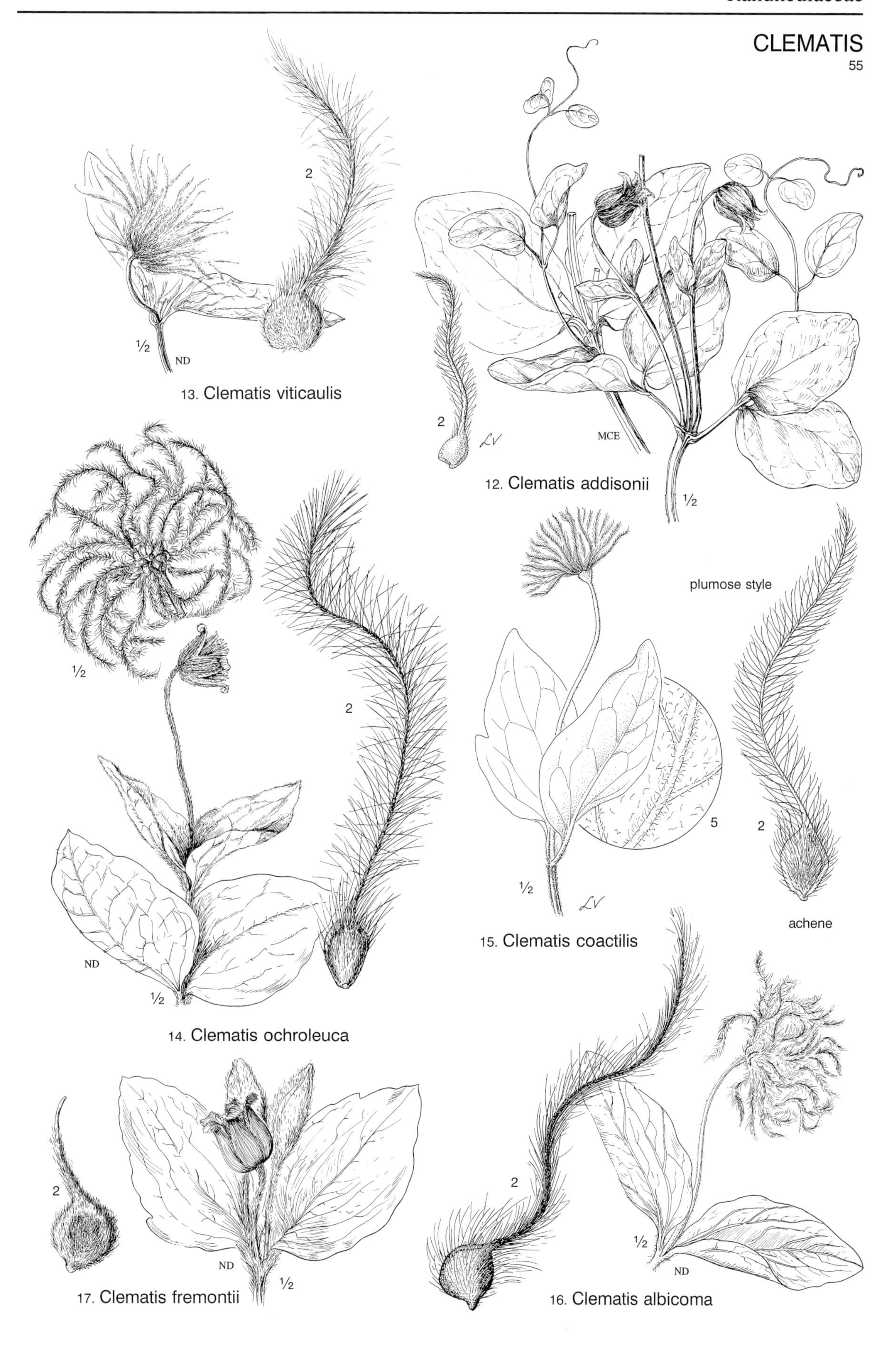

13. Clematis viticaulis

12. Clematis addisonii

plumose style

15. Clematis coactilis

achene

14. Clematis ochroleuca

17. Clematis fremontii

16. Clematis albicoma

RANUNCULUS
56, 57

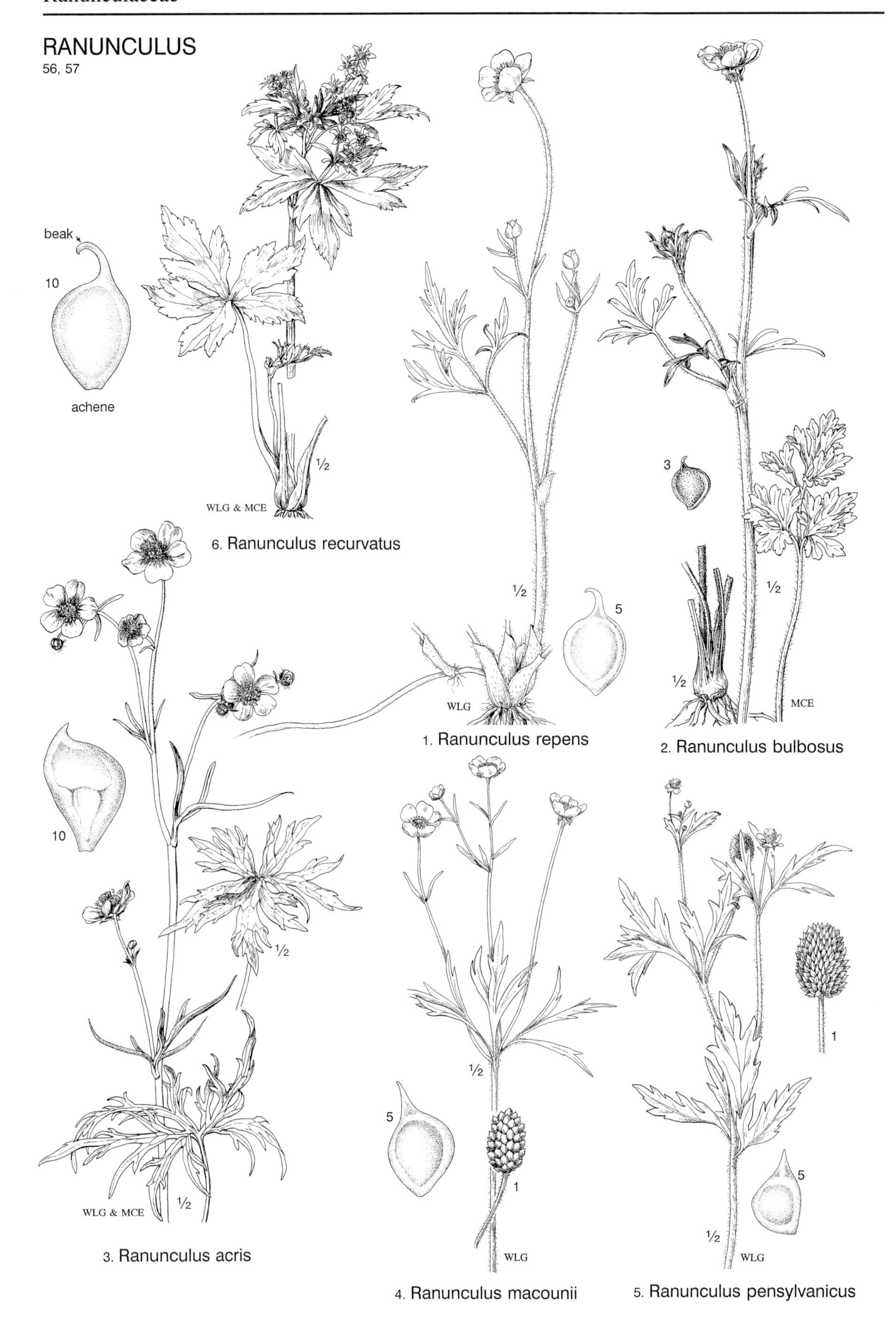

beak

10

achene

½

WLG & MCE

6. Ranunculus recurvatus

3

½

½

MCE

2. Ranunculus bulbosus

5

WLG

1. Ranunculus repens

10

½

WLG & MCE   ½

3. Ranunculus acris

5

½

1

WLG

4. Ranunculus macounii

1

5

½   WLG

5. Ranunculus pensylvanicus

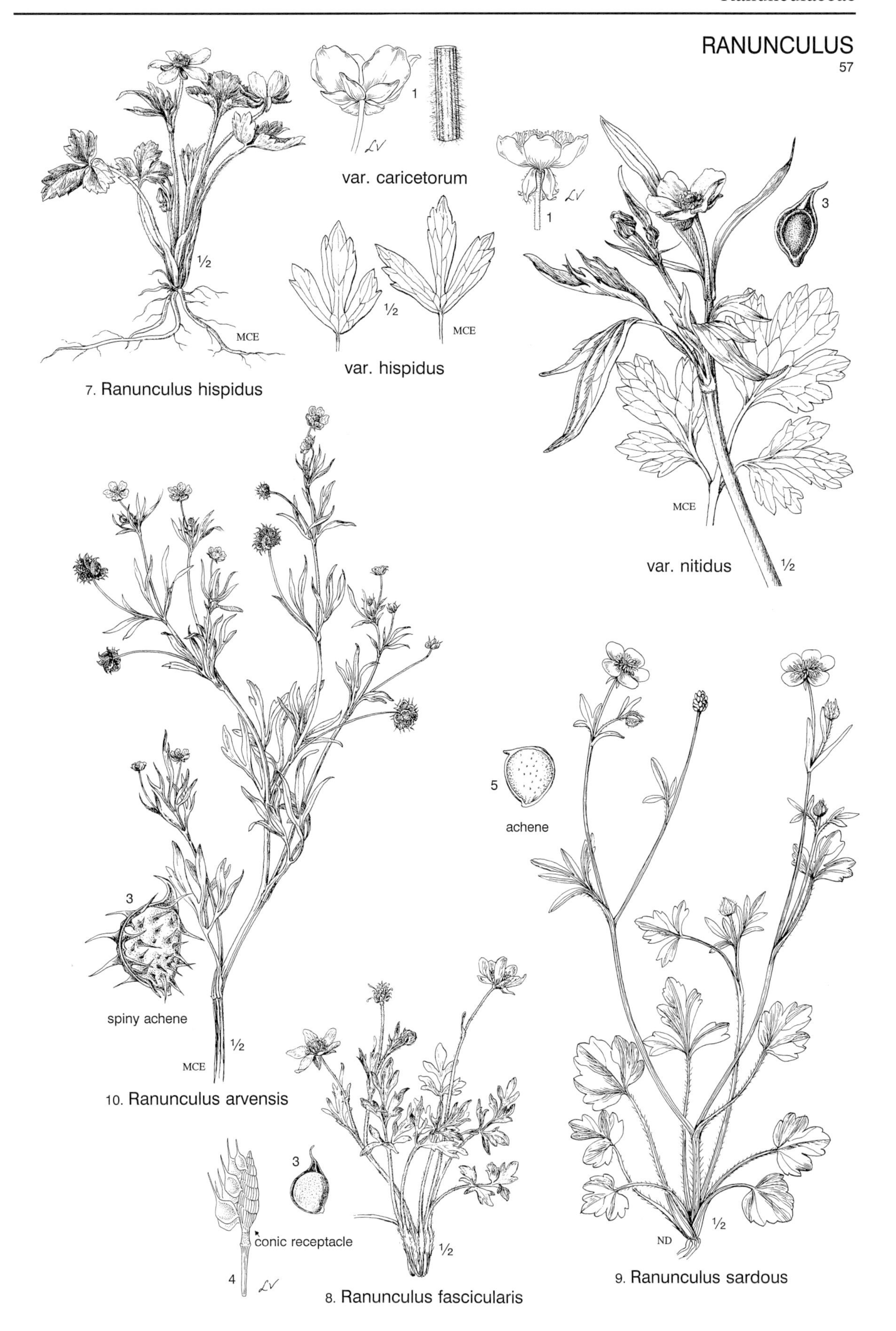

var. caricetorum

var. hispidus

7. Ranunculus hispidus

var. nitidus ½

5 achene

spiny achene

10. Ranunculus arvensis

conic receptacle

8. Ranunculus fascicularis

9. Ranunculus sardous

RANUNCULUS
57, 58

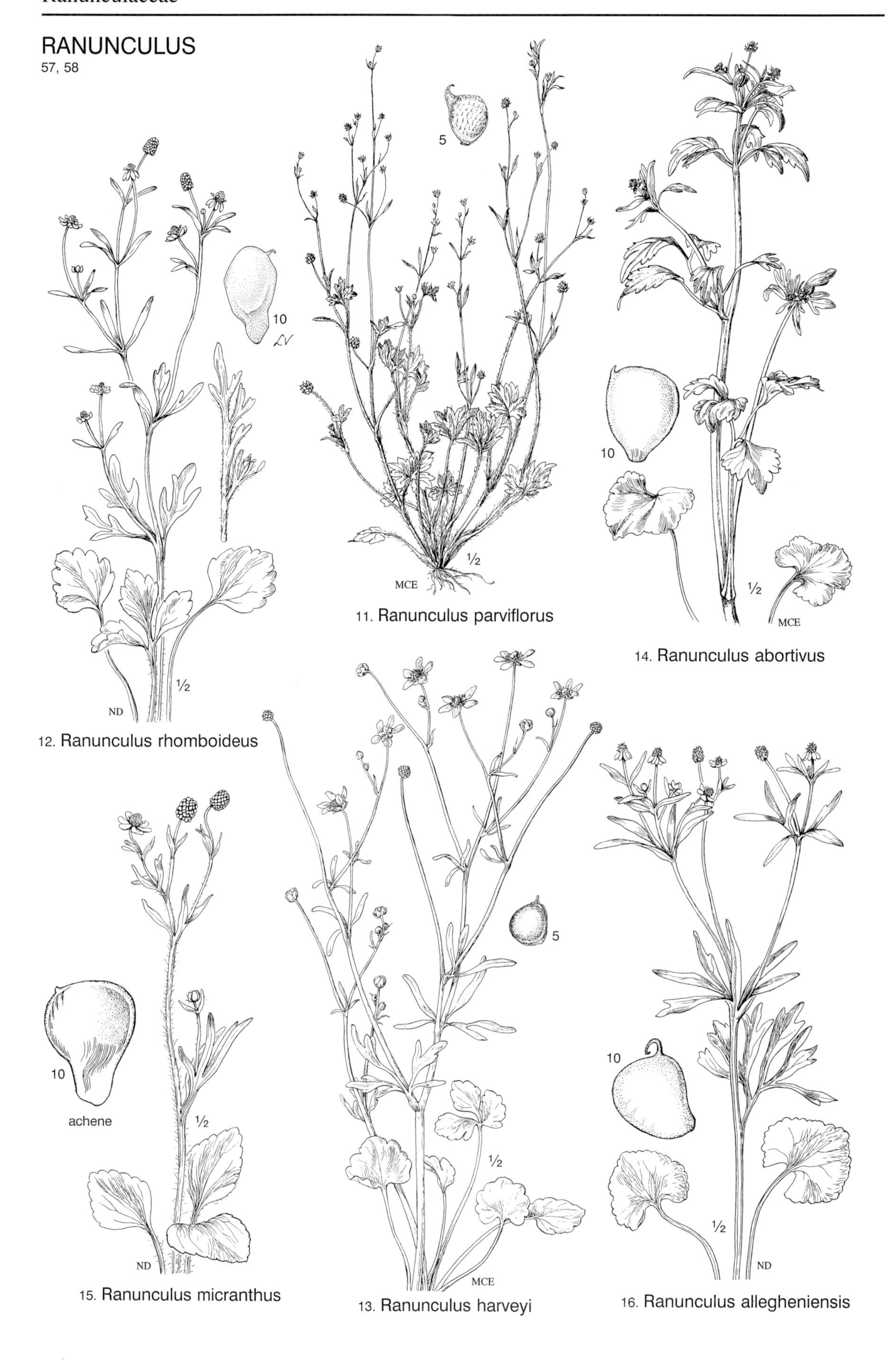

11. Ranunculus parviflorus

14. Ranunculus abortivus

12. Ranunculus rhomboideus

15. Ranunculus micranthus

13. Ranunculus harveyi

16. Ranunculus allegheniensis

RANUNCULUS

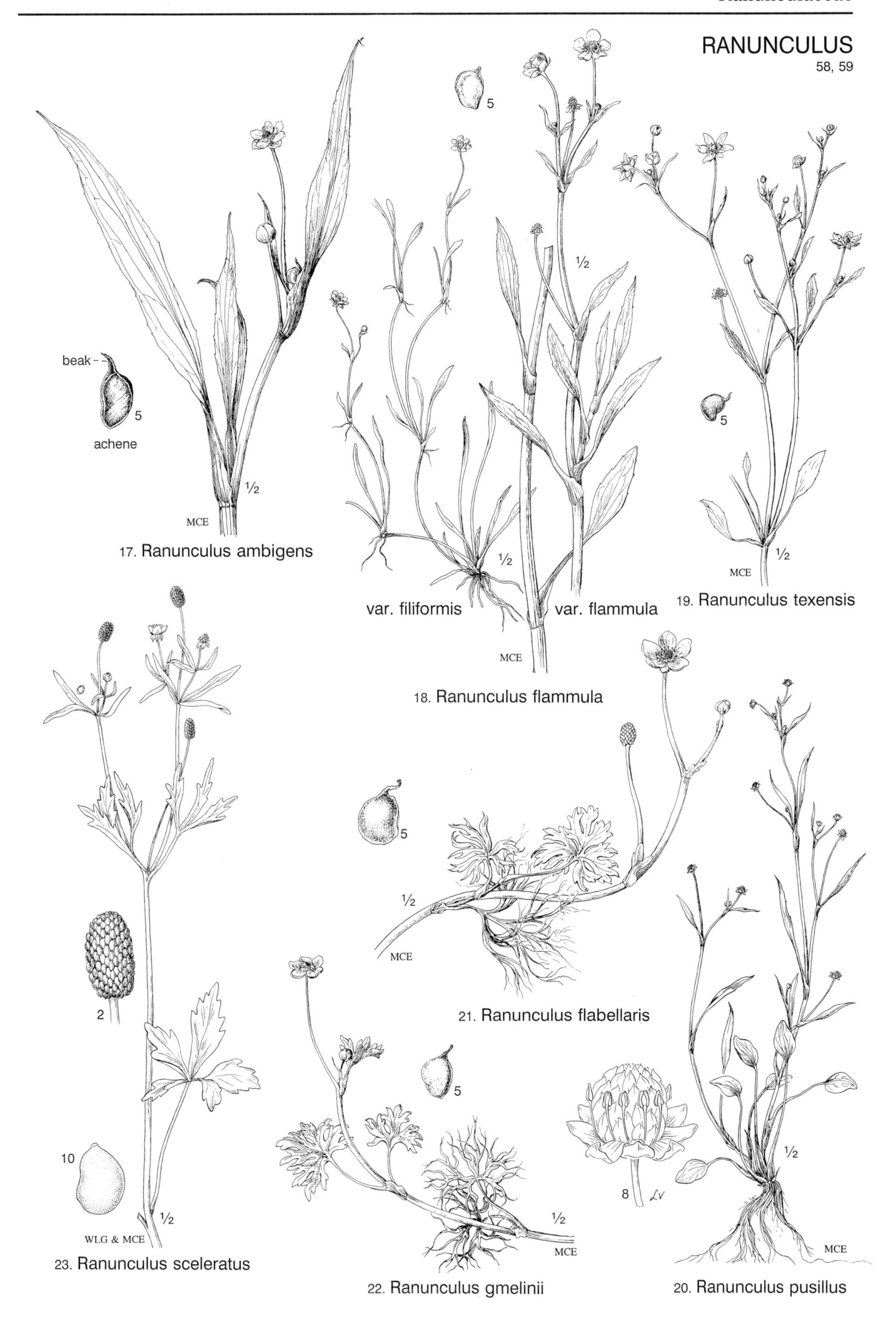

beak

5

achene

½

MCE

17. Ranunculus ambigens

5

½

var. filiformis   var. flammula

MCE

18. Ranunculus flammula

5

½

MCE

19. Ranunculus texensis

2

10

½

WLG & MCE

23. Ranunculus sceleratus

5

½

MCE

21. Ranunculus flabellaris

5

½

MCE

22. Ranunculus gmelinii

8

½

MCE

20. Ranunculus pusillus

RANUNCULUS
59

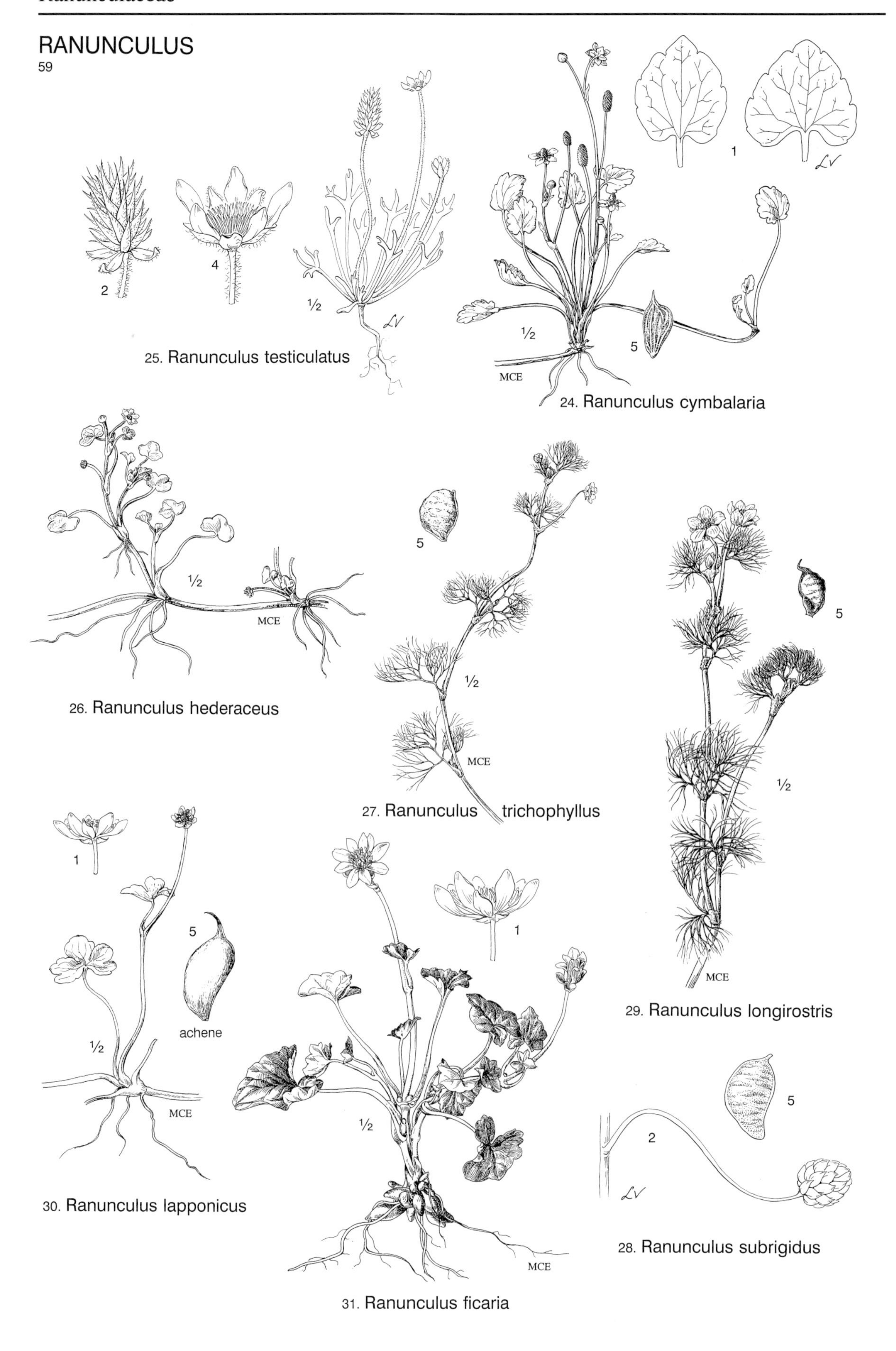

25. Ranunculus testiculatus

24. Ranunculus cymbalaria

26. Ranunculus hederaceus

27. Ranunculus trichophyllus

29. Ranunculus longirostris

30. Ranunculus lapponicus

28. Ranunculus subrigidus

31. Ranunculus ficaria

58

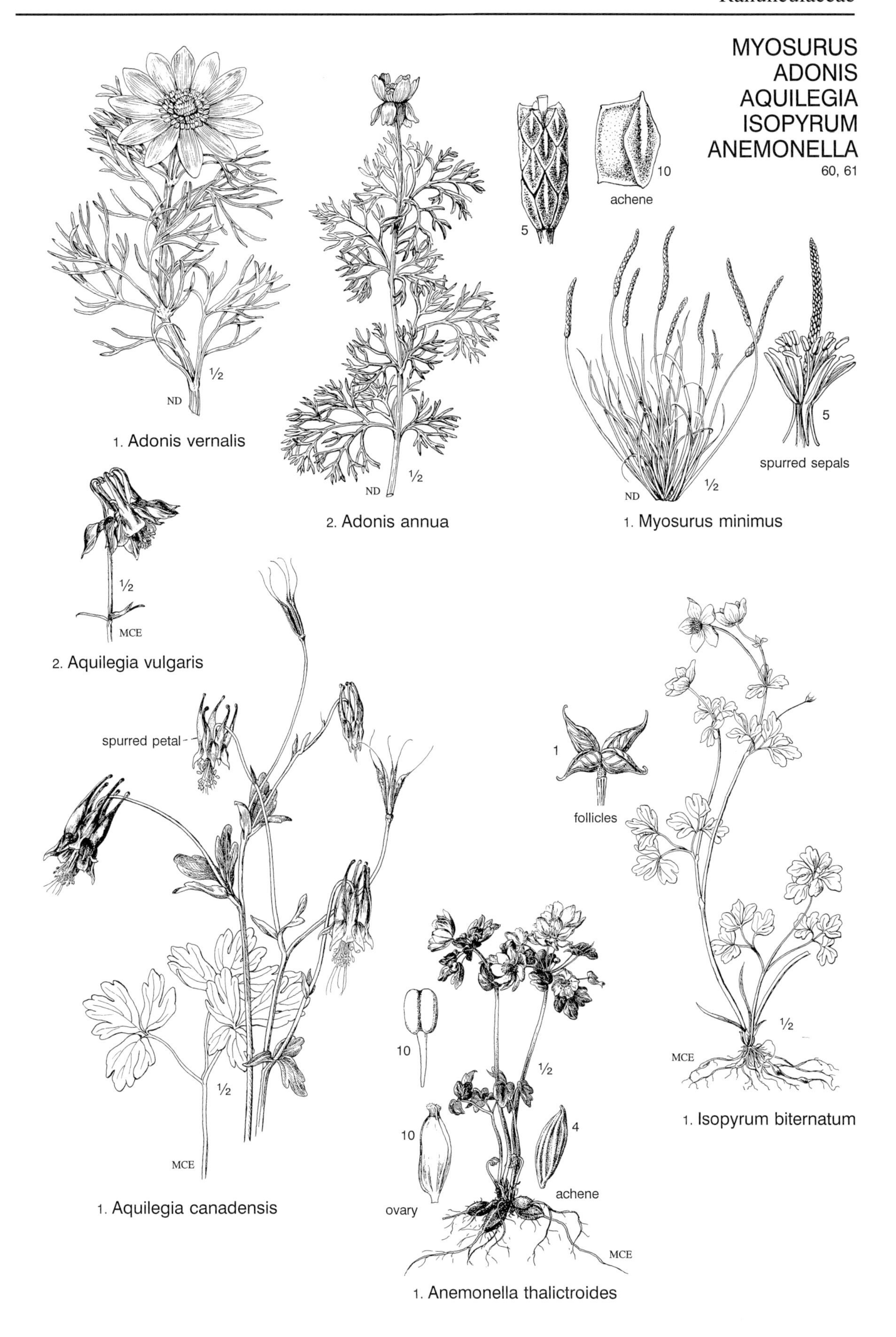

MYOSURUS
ADONIS
AQUILEGIA
ISOPYRUM
ANEMONELLA
60, 61

10
achene

5

1. Adonis vernalis
ND

2. Adonis annua
ND

spurred sepals
5

1. Myosurus minimus
ND

2. Aquilegia vulgaris
MCE

spurred petal

1
follicles

1. Aquilegia canadensis
MCE

10
10
ovary
4
achene

1. Anemonella thalictroides
MCE

1. Isopyrum biternatum
MCE

THALICTRUM
61, 62

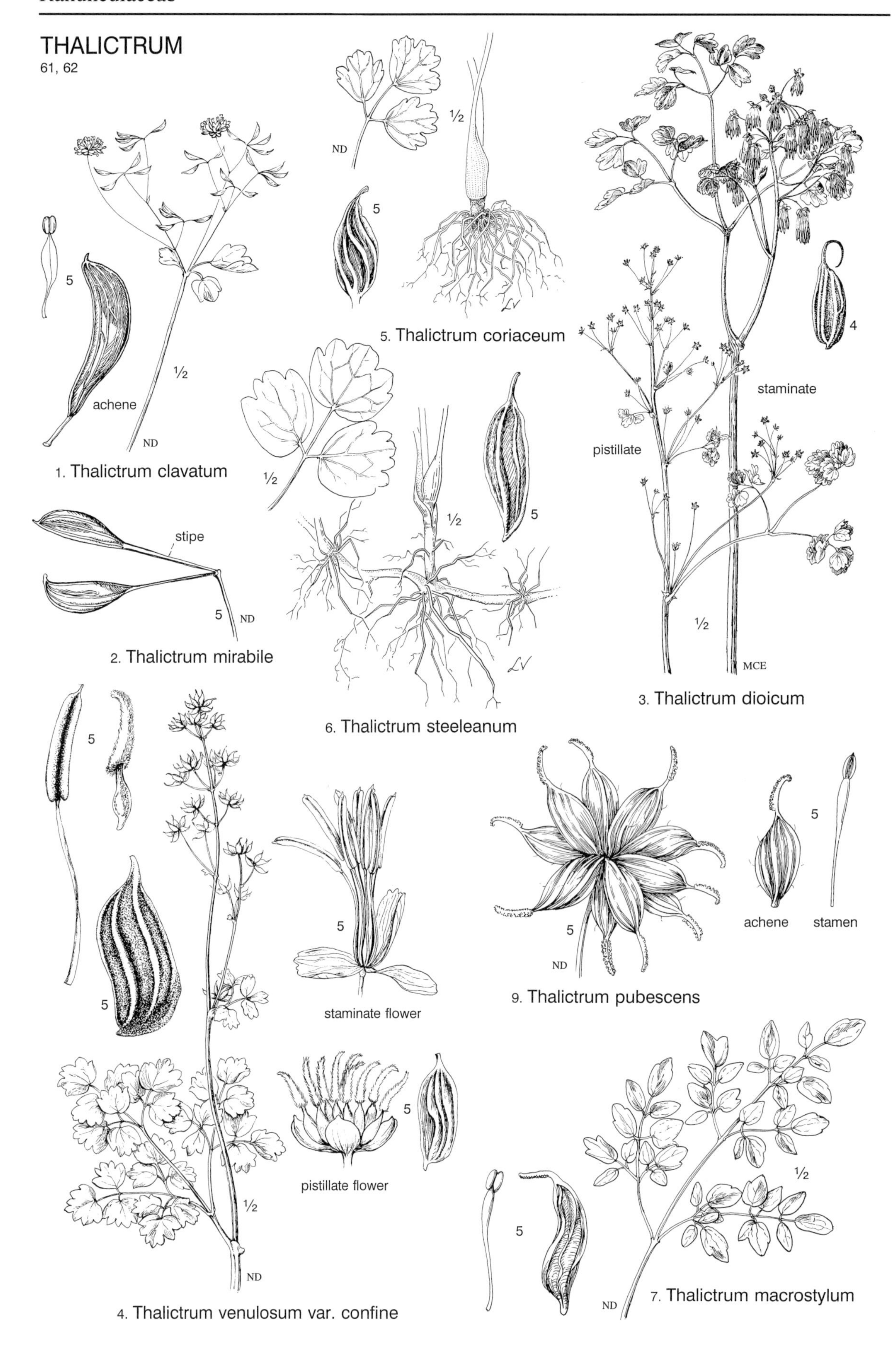

1. Thalictrum clavatum

achene

ND

2. Thalictrum mirabile

stipe

5. Thalictrum coriaceum

6. Thalictrum steeleanum

3. Thalictrum dioicum

pistillate   staminate

MCE

9. Thalictrum pubescens

achene   stamen

staminate flower

pistillate flower

4. Thalictrum venulosum var. confine

7. Thalictrum macrostylum

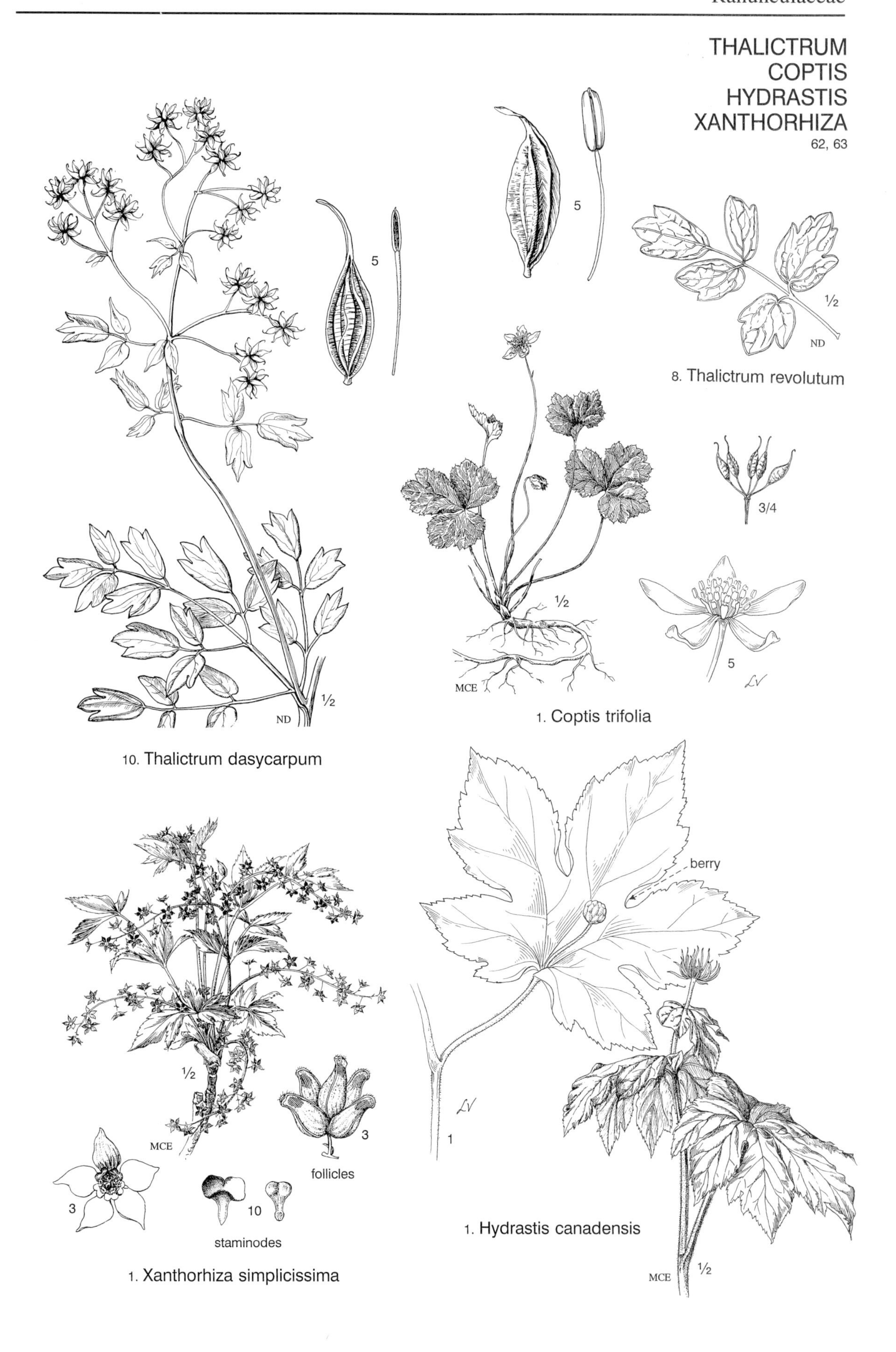

8. Thalictrum revolutum

10. Thalictrum dasycarpum

1. Coptis trifolia

follicles

staminodes

1. Xanthorhiza simplicissima

berry

1. Hydrastis canadensis

PODOPHYLLUM
DIPHYLLEIA
JEFFERSONIA
CAULOPHYLLUM
BERBERIS
63, 64

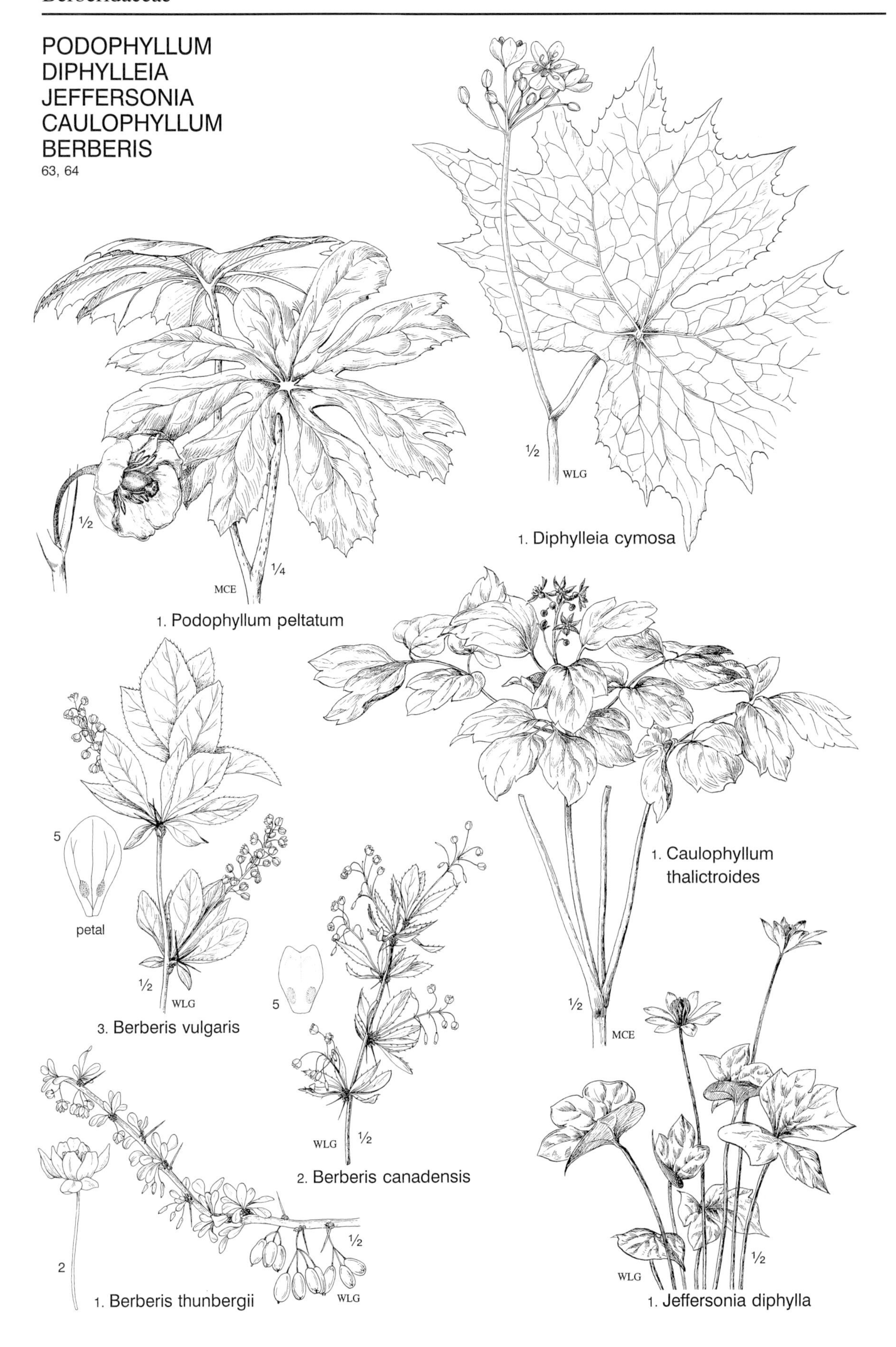

1. Diphylleia cymosa

1. Podophyllum peltatum

1. Caulophyllum thalictroides

3. Berberis vulgaris

2. Berberis canadensis

1. Berberis thunbergii

1. Jeffersonia diphylla

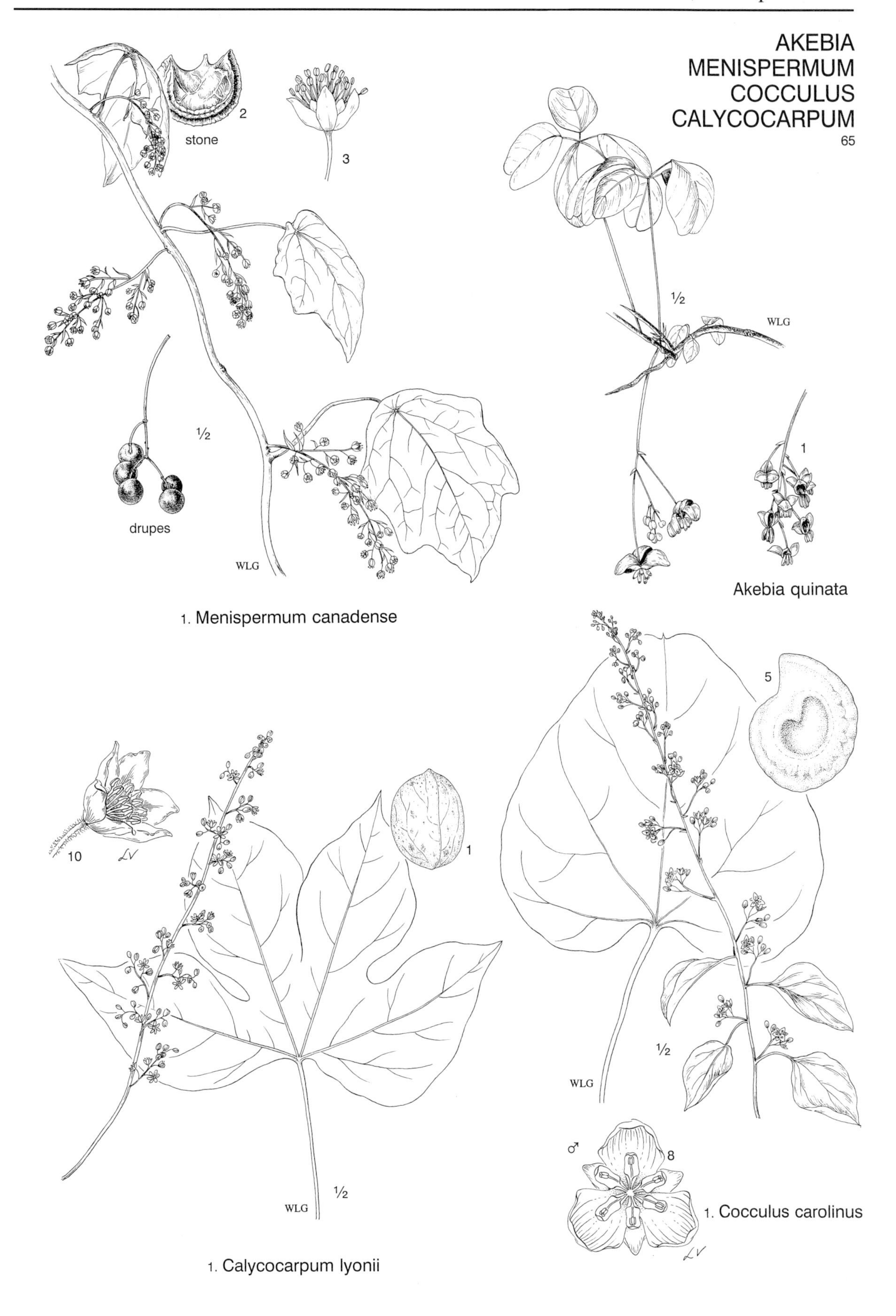

AKEBIA
MENISPERMUM
COCCULUS
CALYCOCARPUM
65

stone

2

3

½

drupes

WLG

1. Menispermum canadense

½

WLG

1

Akebia quinata

10 ∠V

1

5

½

WLG

8 ∠V

½ WLG

1. Calycocarpum lyonii

1. Cocculus carolinus

SANGUINARIA
STYLOPHORUM
CHELIDONIUM
GLAUCIUM
66, 67

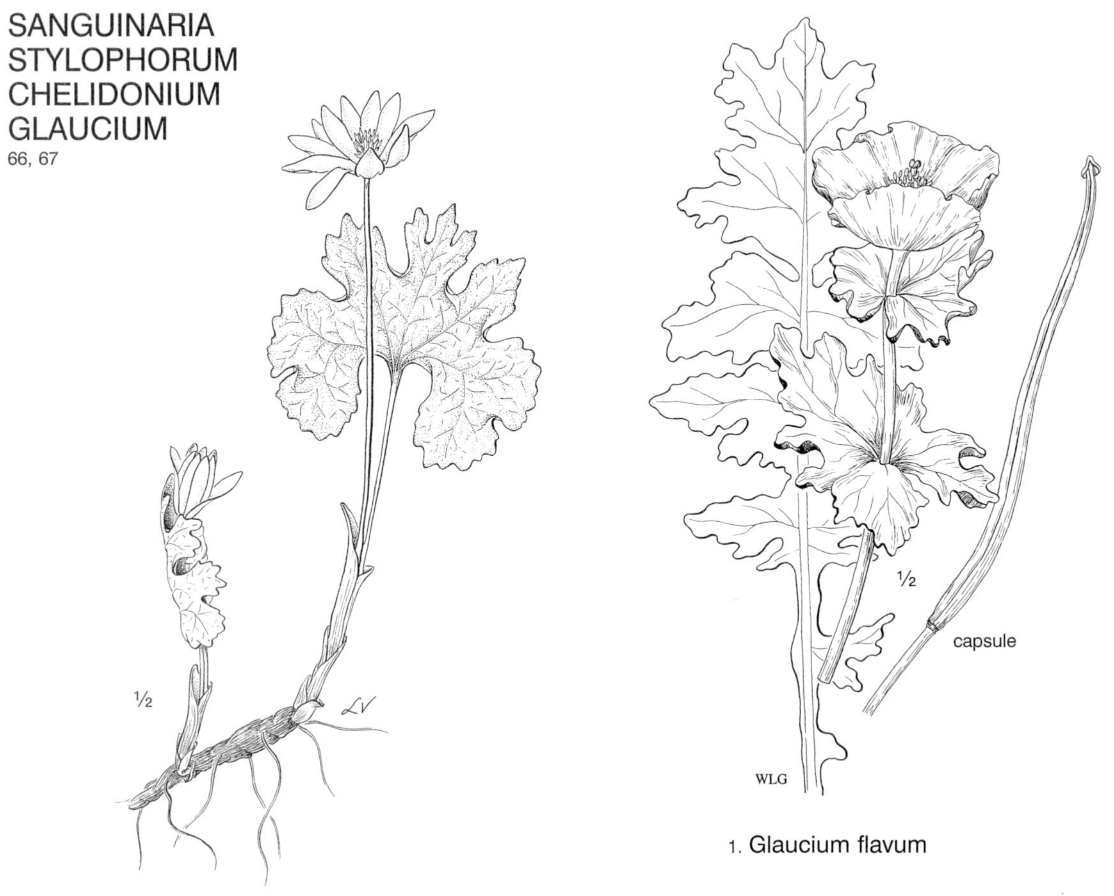

1. Sanguinaria canadensis

1. Glaucium flavum

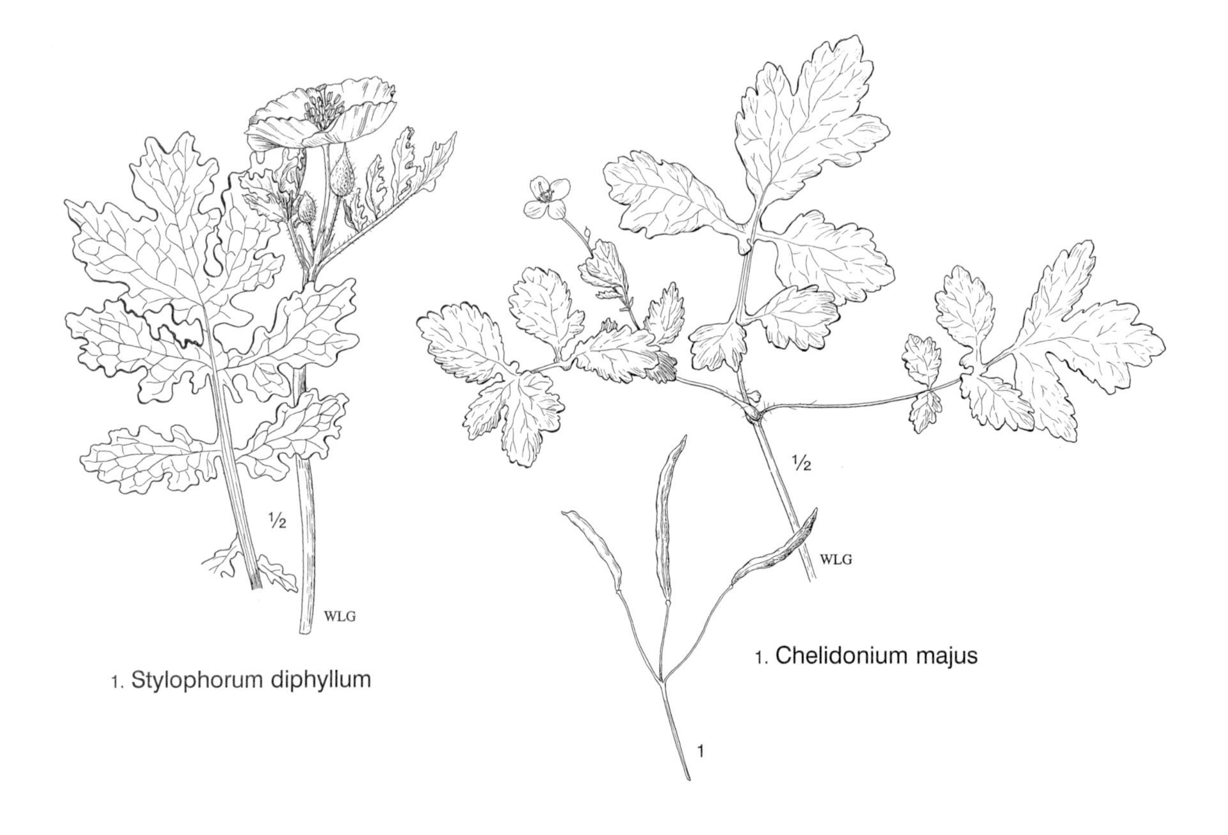

1. Stylophorum diphyllum

1. Chelidonium majus

ILLUSTRATED COMPANION TO

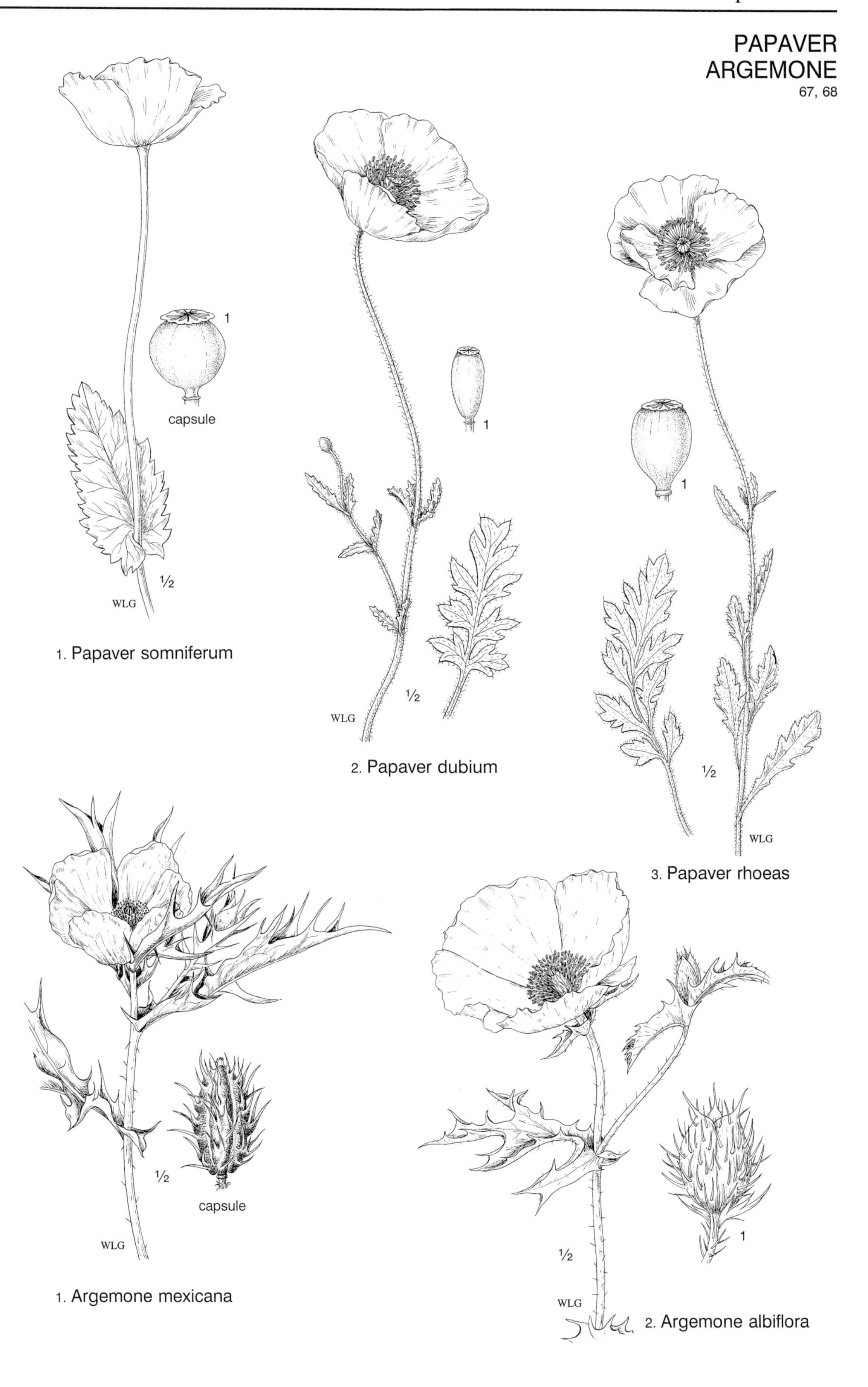

1. Papaver somniferum

2. Papaver dubium

3. Papaver rhoeas

1. Argemone mexicana

2. Argemone albiflora

DICENTRA
ADLUMIA
FUMARIA
68–70

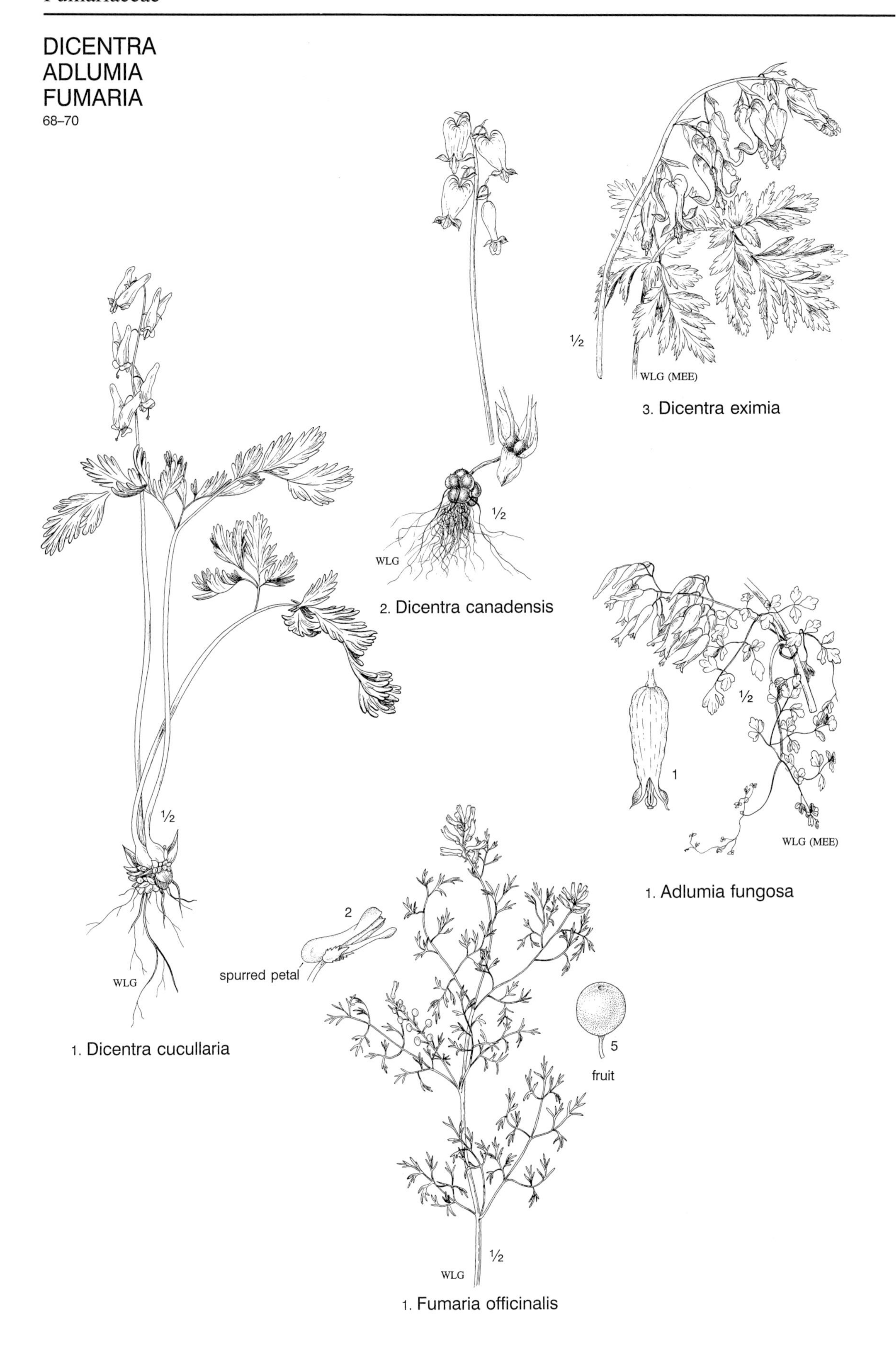

¹⁄₂

WLG (MEE)

3. Dicentra eximia

¹⁄₂

WLG

2. Dicentra canadensis

¹⁄₂

WLG

1. Dicentra cucullaria

2

spurred petal

¹⁄₂

1

WLG (MEE)

1. Adlumia fungosa

5

fruit

¹⁄₂

WLG

1. Fumaria officinalis

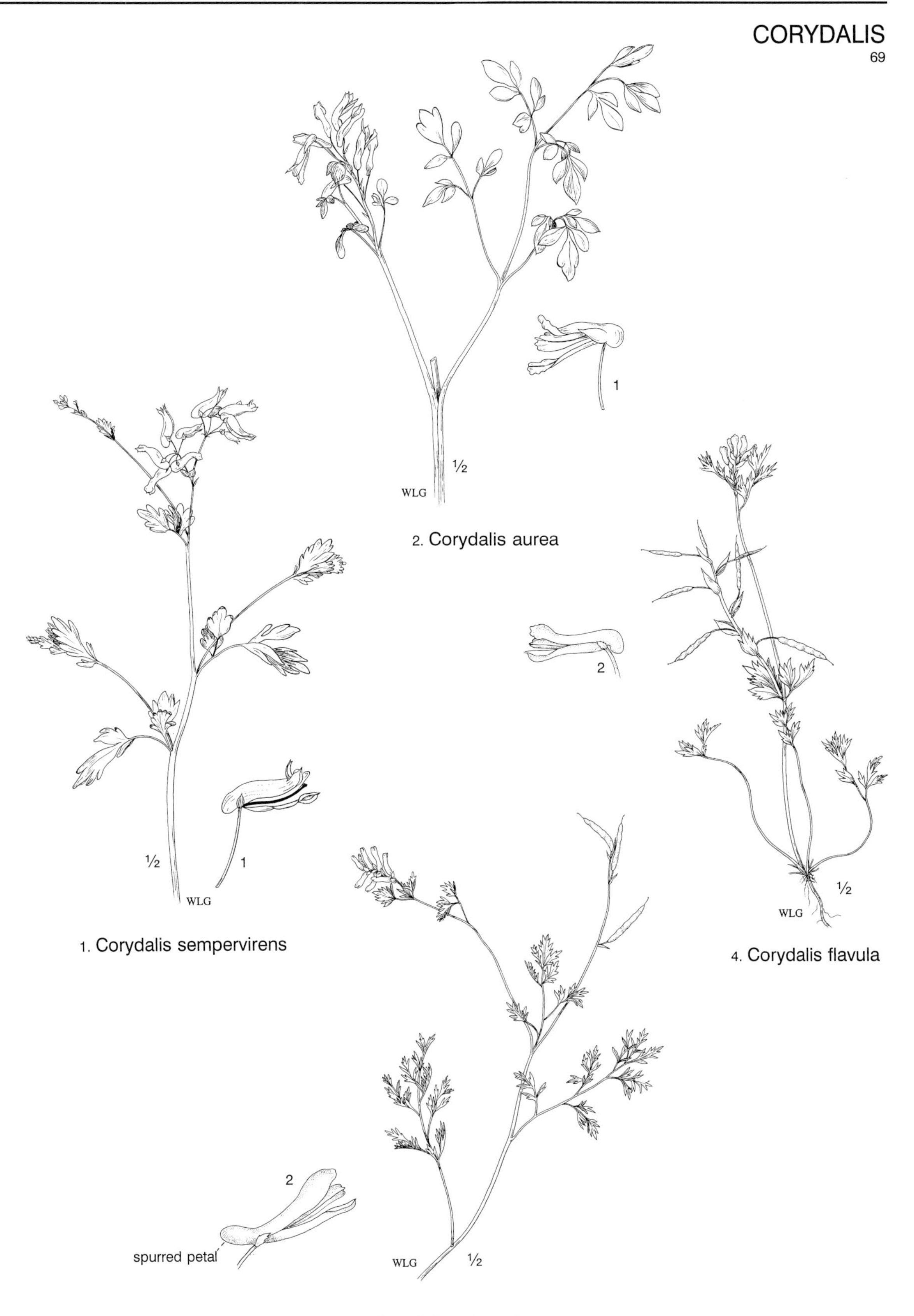

2. Corydalis aurea

1. Corydalis sempervirens

4. Corydalis flavula

spurred petal

3. Corydalis micrantha

PLATANUS
LIQUIDAMBAR
HAMAMELIS
ULMUS
70–72

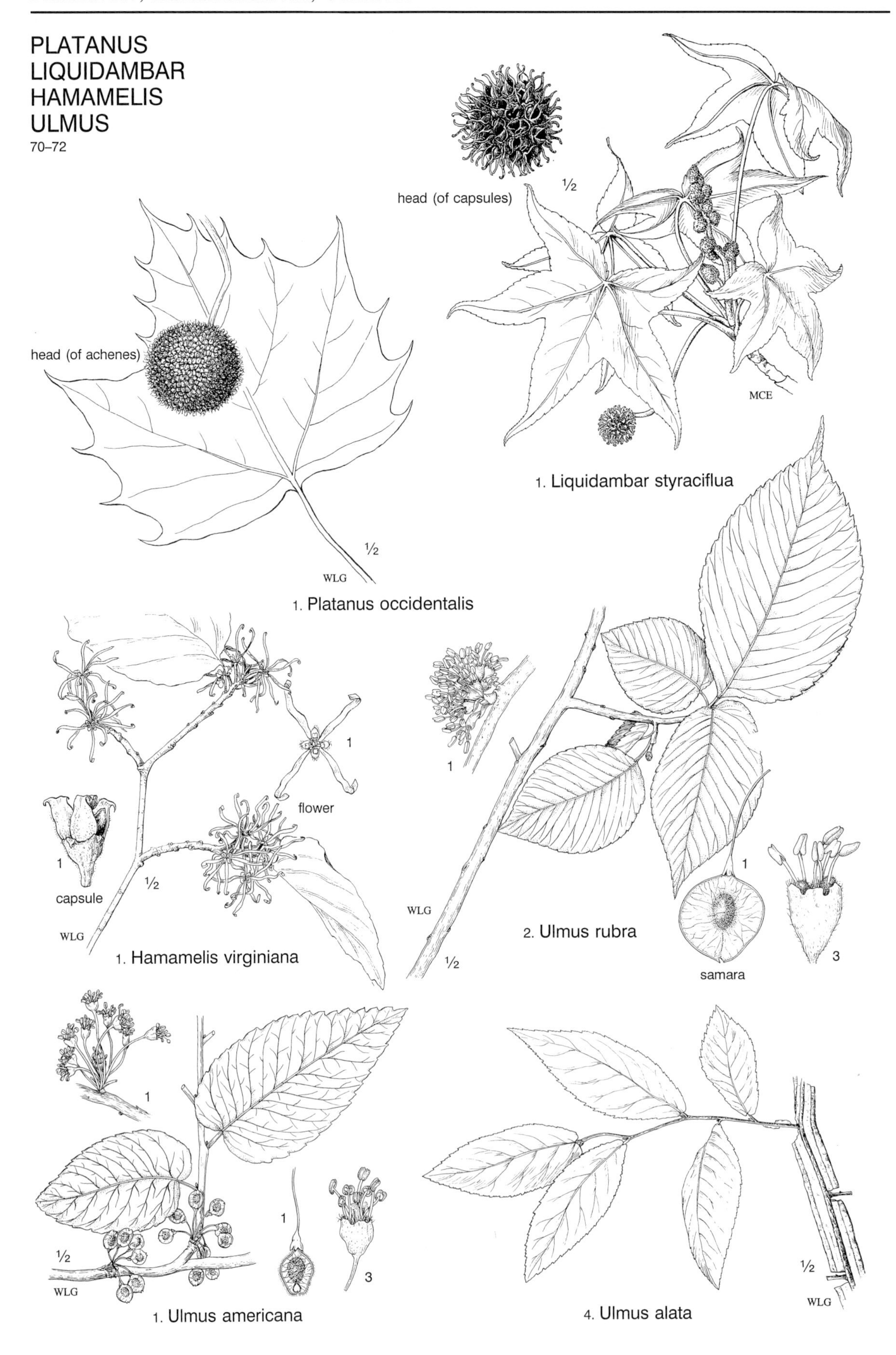

head (of capsules)

½

head (of achenes)

MCE

1. Liquidambar styraciflua

½

WLG

1. Platanus occidentalis

1

flower

1

capsule

½

WLG

1. Hamamelis virginiana

1

WLG

½

1

samara

3

2. Ulmus rubra

1

½

WLG

1

3

1. Ulmus americana

½

WLG

4. Ulmus alata

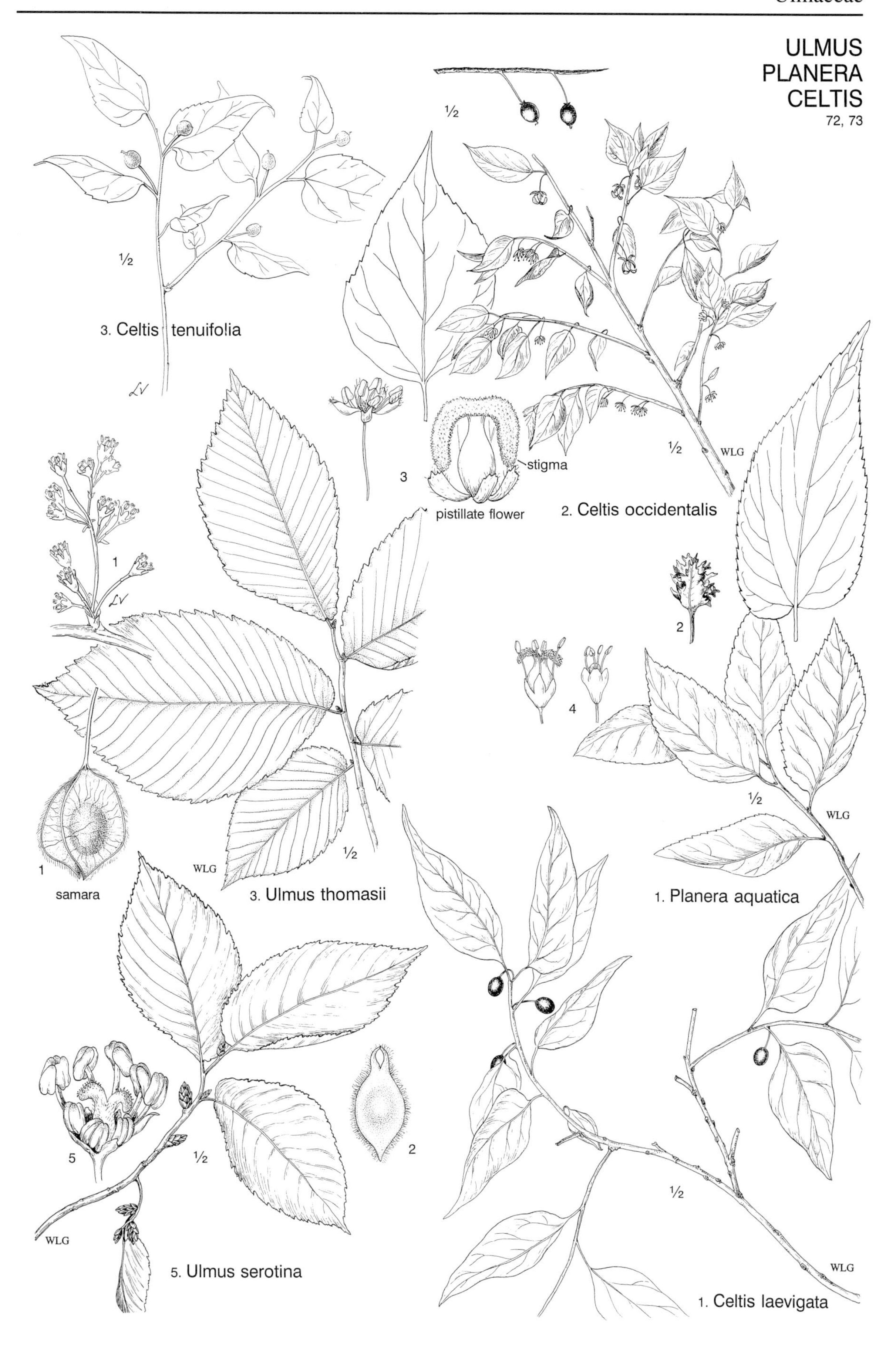

½

3. Celtis tenuifolia

½

3
pistillate flower
stigma

2. Celtis occidentalis

½ WLG

1

1
samara

3. Ulmus thomasii

½ WLG

2

4

½ WLG

1. Planera aquatica

5

2

½ WLG

5. Ulmus serotina

½

1. Celtis laevigata

½ WLG

HUMULUS
CANNABIS
MACLURA
BROUSSONETIA
MORUS
73–75

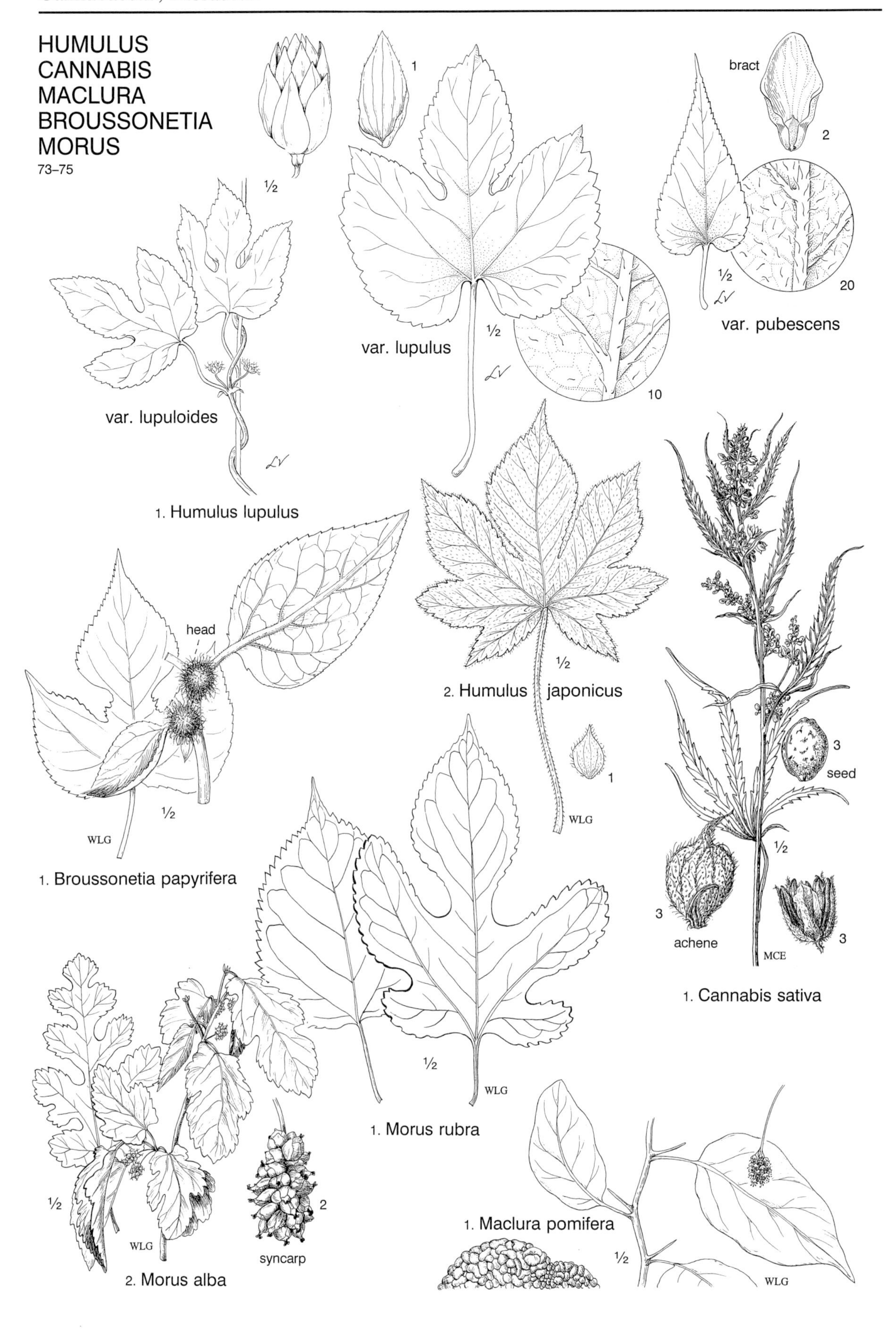

½

var. lupuloides

1. Humulus lupulus

var. lupulus

½

½

bract

2

var. pubescens

½

20

10

2. Humulus japonicus

1

WLG

head

½

WLG

1. Broussonetia papyrifera

3

seed

½

3

achene

3

MCE

1. Cannabis sativa

1. Morus rubra

½

WLG

½

WLG

2

syncarp

2. Morus alba

1. Maclura pomifera

½

WLG

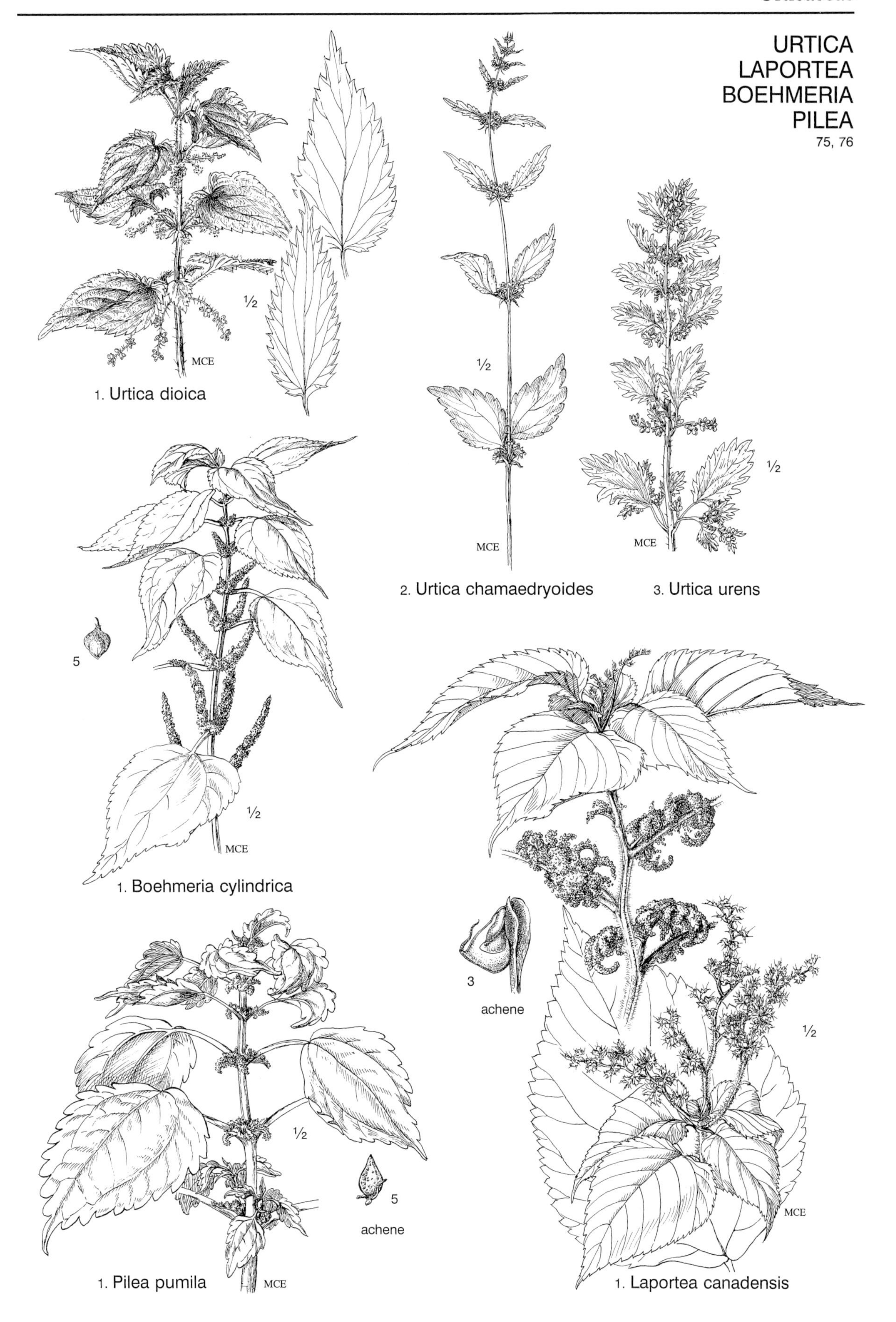

1. Urtica dioica

2. Urtica chamaedryoides

3. Urtica urens

1. Boehmeria cylindrica

3. achene

1. Pilea pumila

5. achene

1. Laportea canadensis

PILEA
PARIETARIA
JUGLANS
76–78

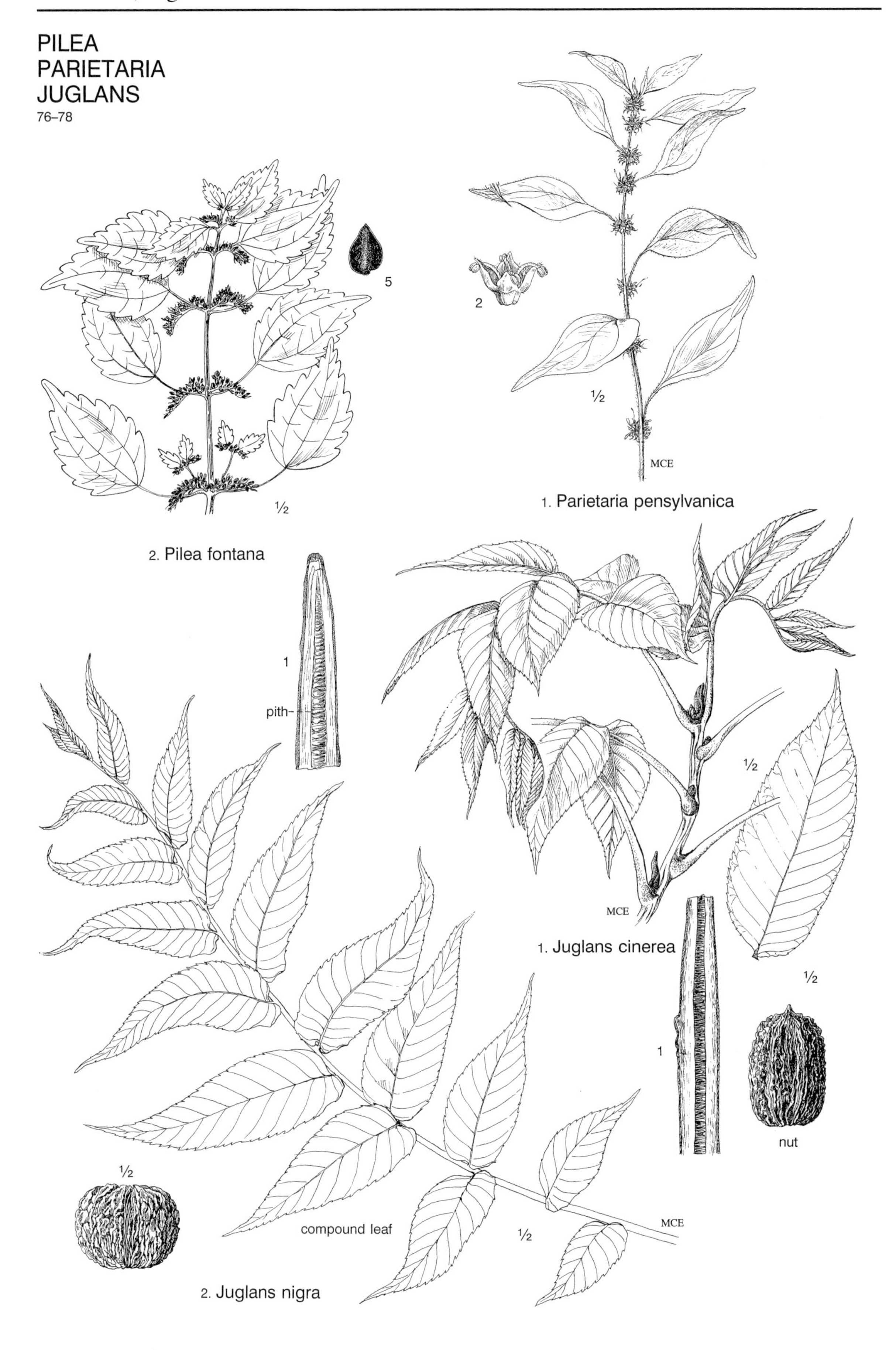

5

2

1. Parietaria pensylvanica

MCE

2. Pilea fontana

1

pith-

1. Juglans cinerea

MCE

½

1

nut

½

compound leaf

½

MCE

2. Juglans nigra

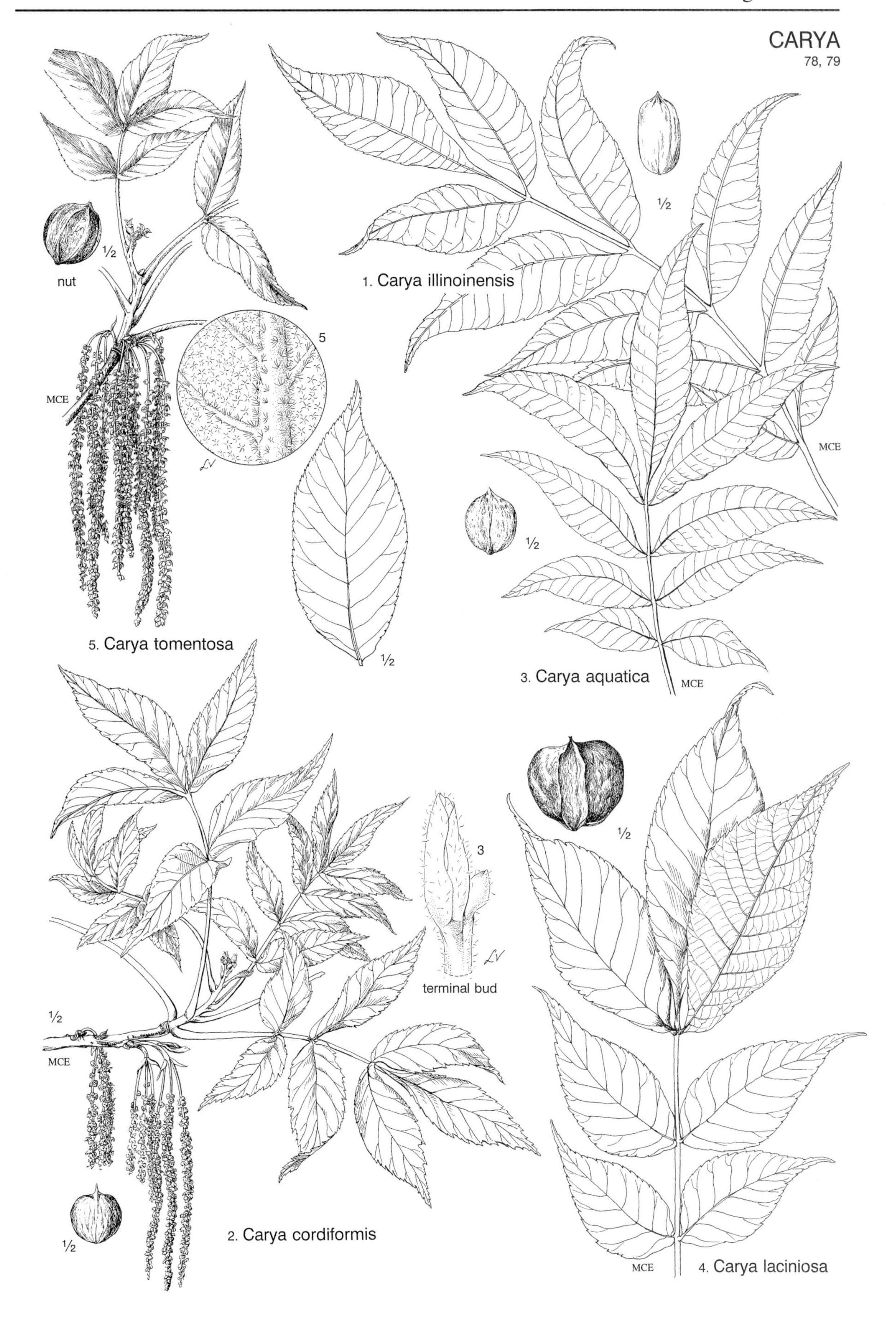

nut

½

MCE

5

1. Carya illinoinensis

½

MCE

½

3. Carya aquatica

MCE

5. Carya tomentosa

½

terminal bud

3

½

½

MCE

2. Carya cordiformis

½

MCE

4. Carya laciniosa

CARYA
79

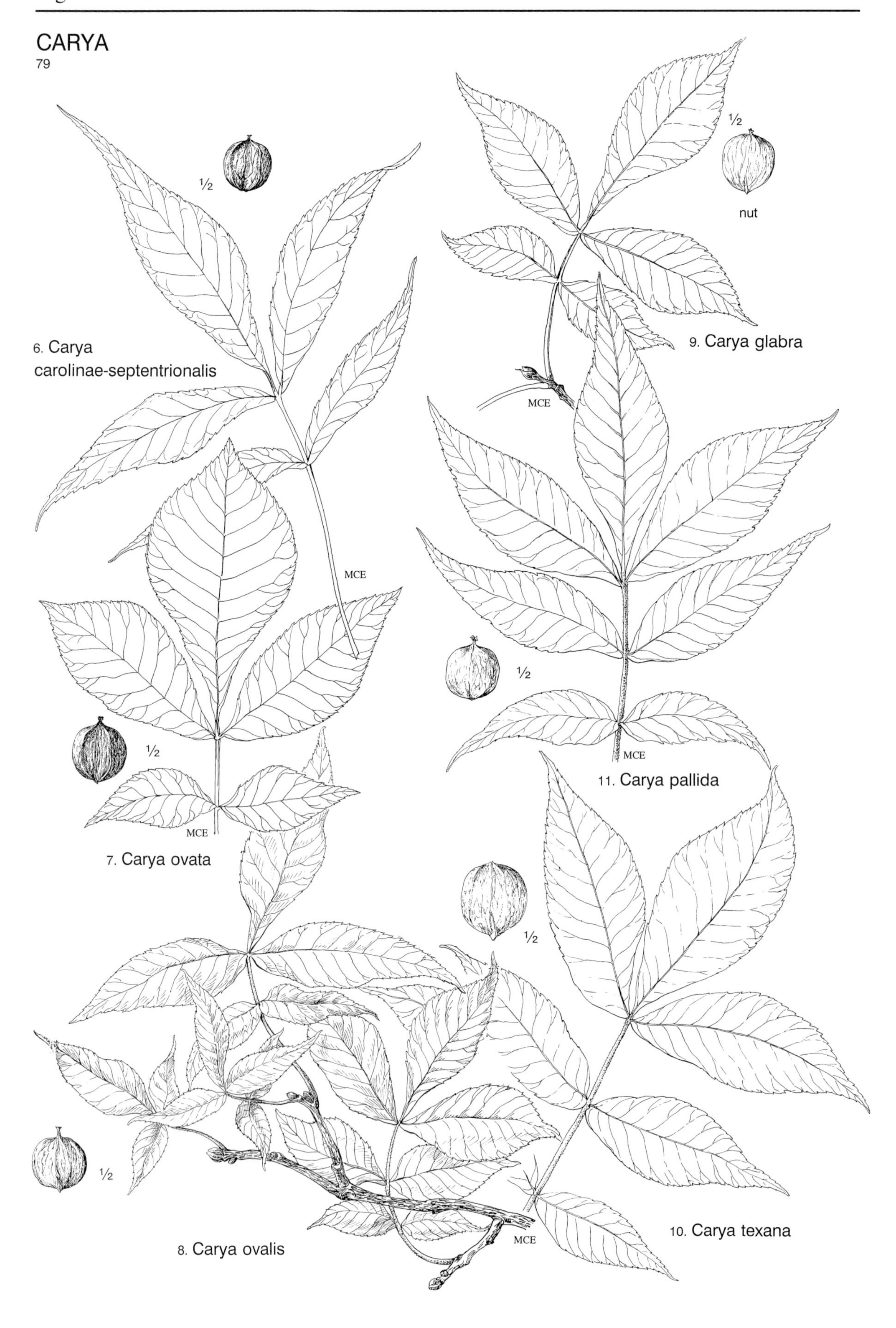

6. Carya carolinae-septentrionalis

7. Carya ovata

8. Carya ovalis

9. Carya glabra

nut

11. Carya pallida

10. Carya texana

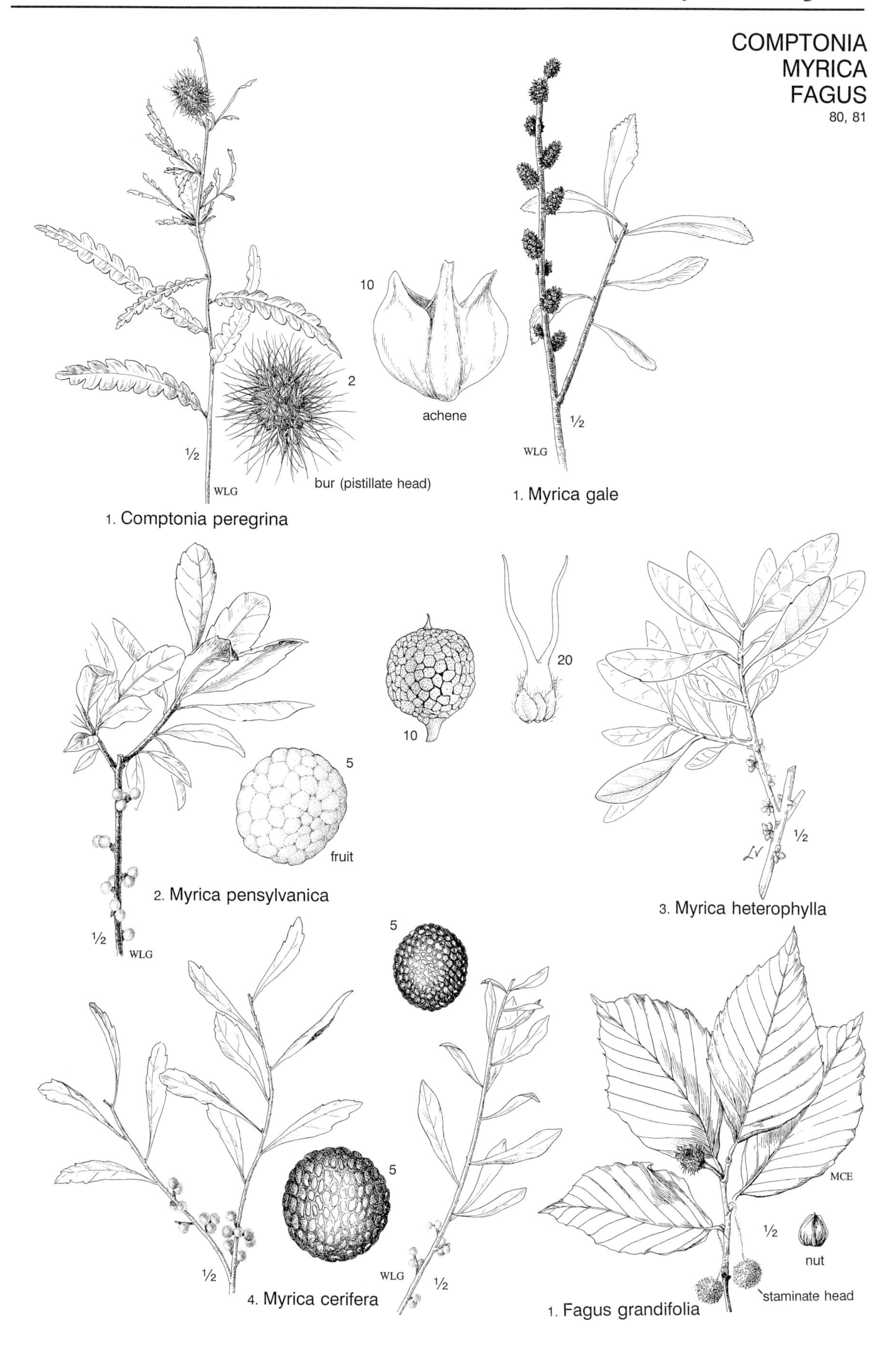

10

achene

2

bur (pistillate head)

WLG

1. Comptonia peregrina

½

WLG

½

1. Myrica gale

10

20

5

fruit

2. Myrica pensylvanica

½

WLG

3. Myrica heterophylla

½

5

5

½

WLG

½

4. Myrica cerifera

MCE

½

nut

staminate head

1. Fagus grandifolia

CASTANEA
QUERCUS
82–84

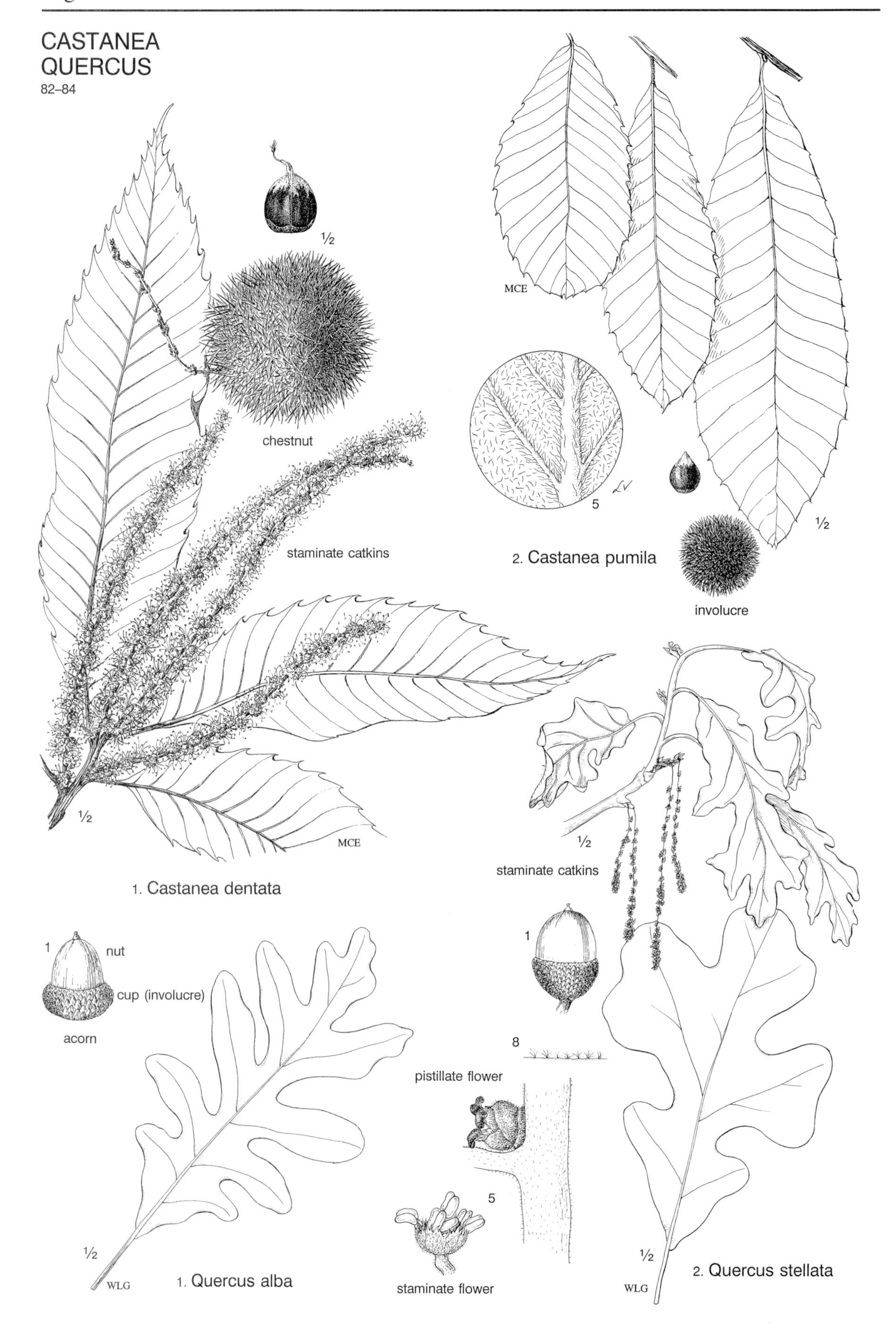

chestnut

staminate catkins

1. Castanea dentata

MCE

MCE

2. Castanea pumila

5

involucre

nut

cup (involucre)

acorn

1. Quercus alba

½

WLG

pistillate flower

8

5

staminate flower

staminate catkins

1

2. Quercus stellata

½

WLG

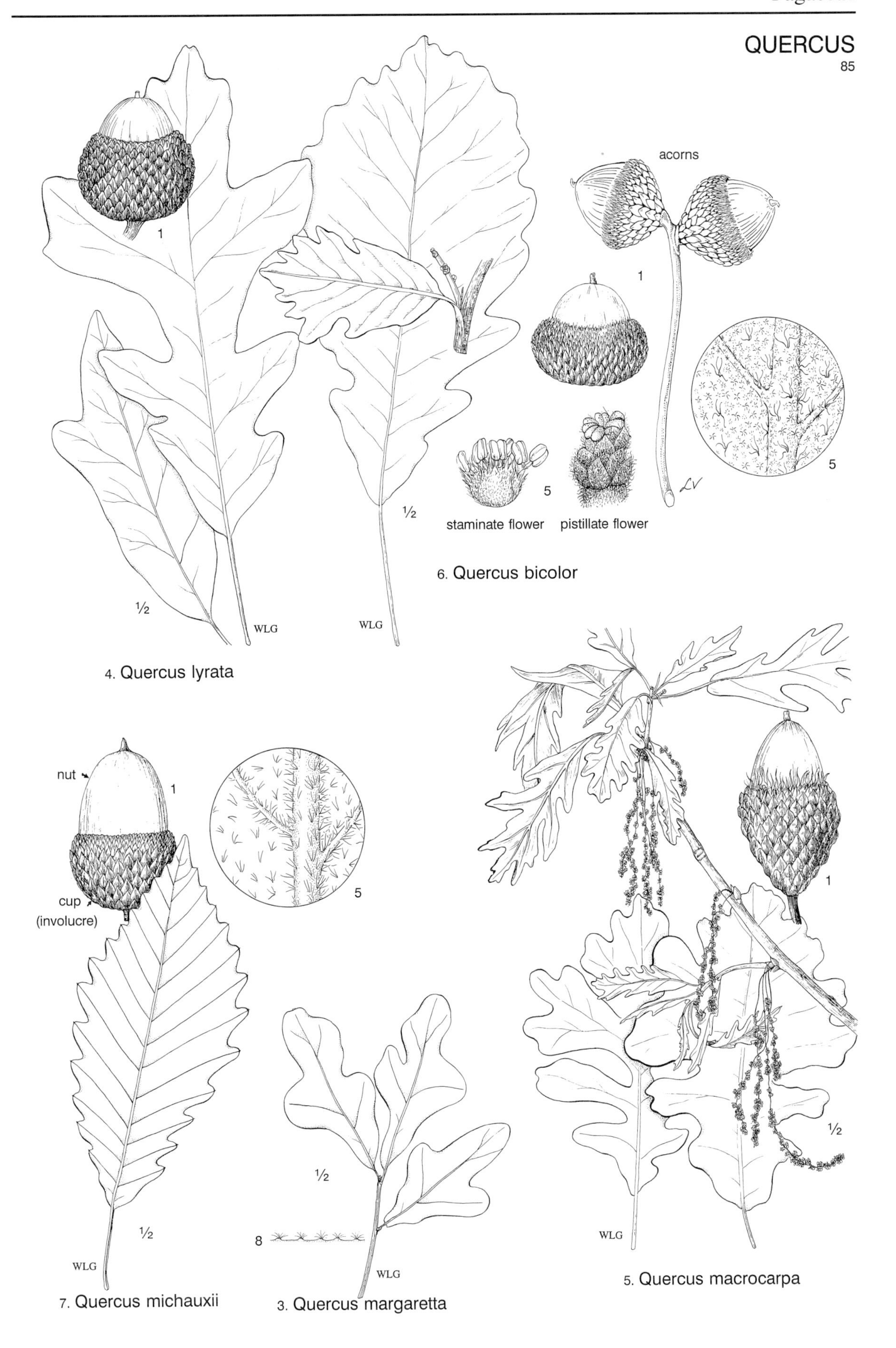

1

5

acorns

1

5

staminate flower   pistillate flower

6. Quercus bicolor

½

½

WLG

WLG

4. Quercus lyrata

nut

cup
(involucre)

1

5

½

WLG

7. Quercus michauxii

½

8

WLG

3. Quercus margaretta

1

½

WLG

5. Quercus macrocarpa

QUERCUS
85, 86

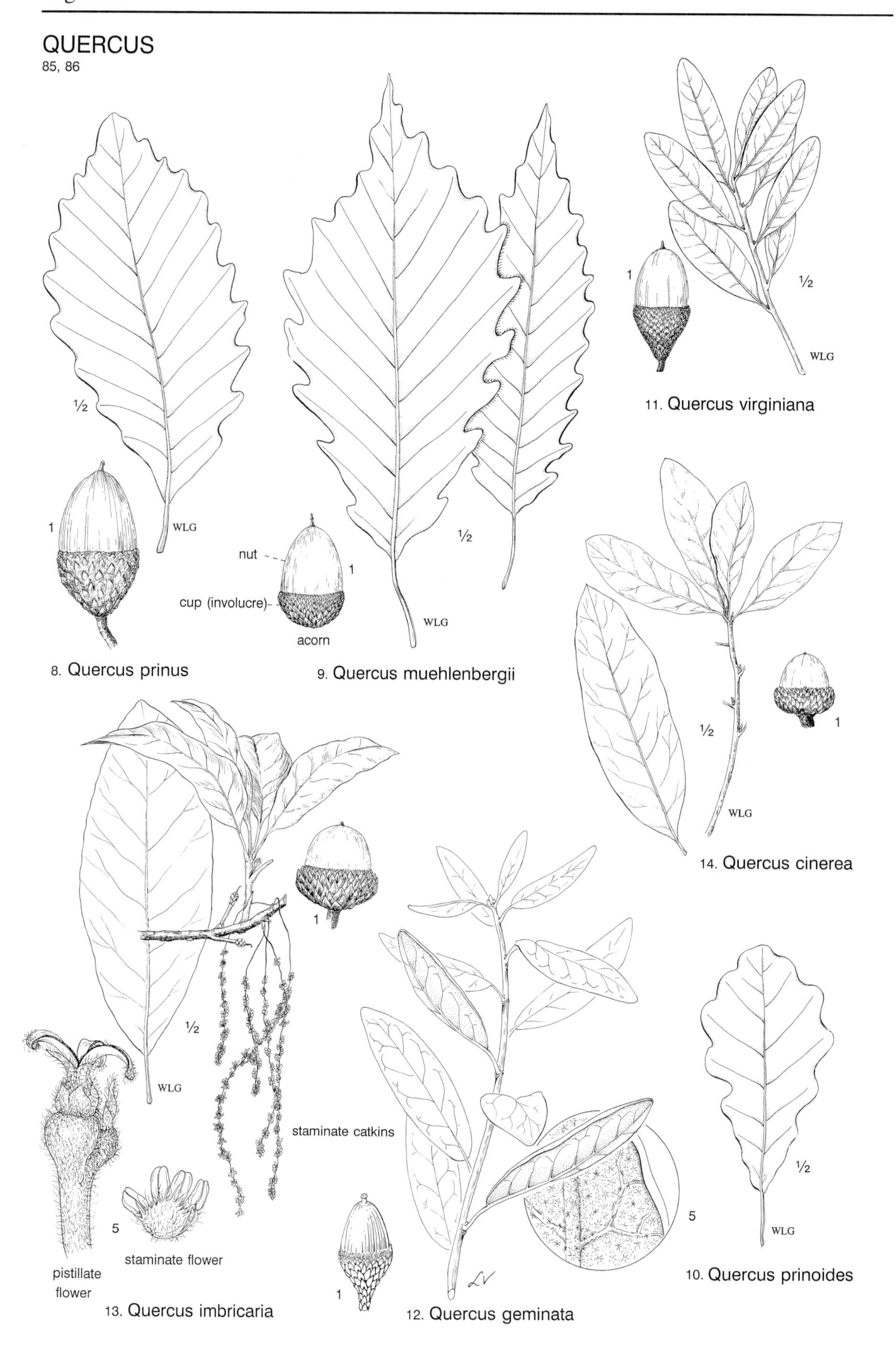

1/2

1   WLG

nut --

cup (involucre)-

acorn

1

8. Quercus prinus

9. Quercus muehlenbergii

1/2

WLG

11. Quercus virginiana

1   WLG

1/2

1/2   1

WLG

14. Quercus cinerea

1

1/2

WLG

staminate catkins

pistillate
flower

5

staminate flower

13. Quercus imbricaria

1

12. Quercus geminata

5

1/2

WLG

10. Quercus prinoides

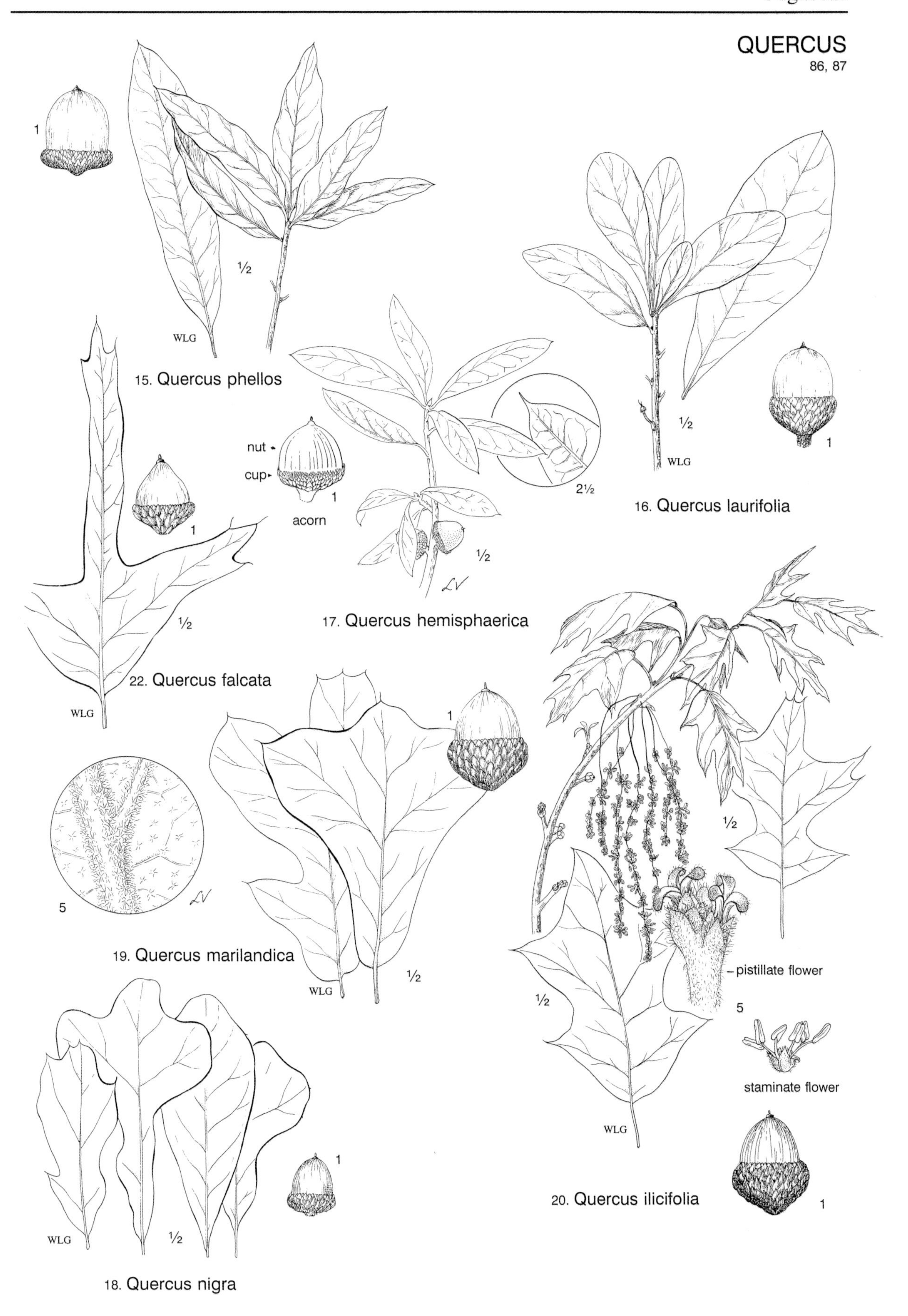

15. Quercus phellos

16. Quercus laurifolia

nut →
cup →

acorn

17. Quercus hemisphaerica

22. Quercus falcata

19. Quercus marilandica

– pistillate flower

staminate flower

20. Quercus ilicifolia

18. Quercus nigra

QUERCUS
86, 87

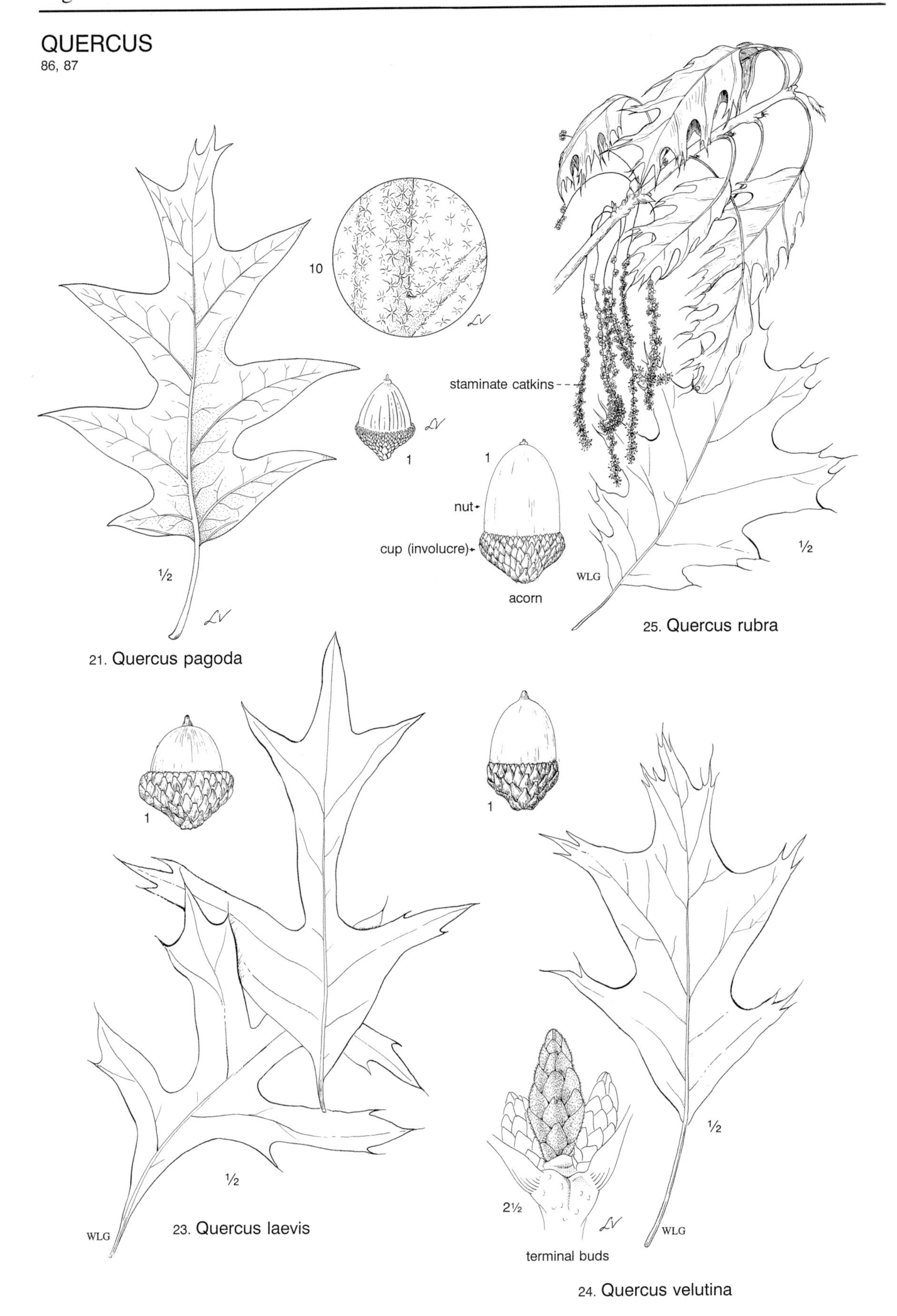

10

staminate catkins

1

1

nut

cup (involucre)

acorn

WLG

25. Quercus rubra

½

½

21. Quercus pagoda

1

1

½

½

23. Quercus laevis

2½

terminal buds

24. Quercus velutina

WLG

WLG

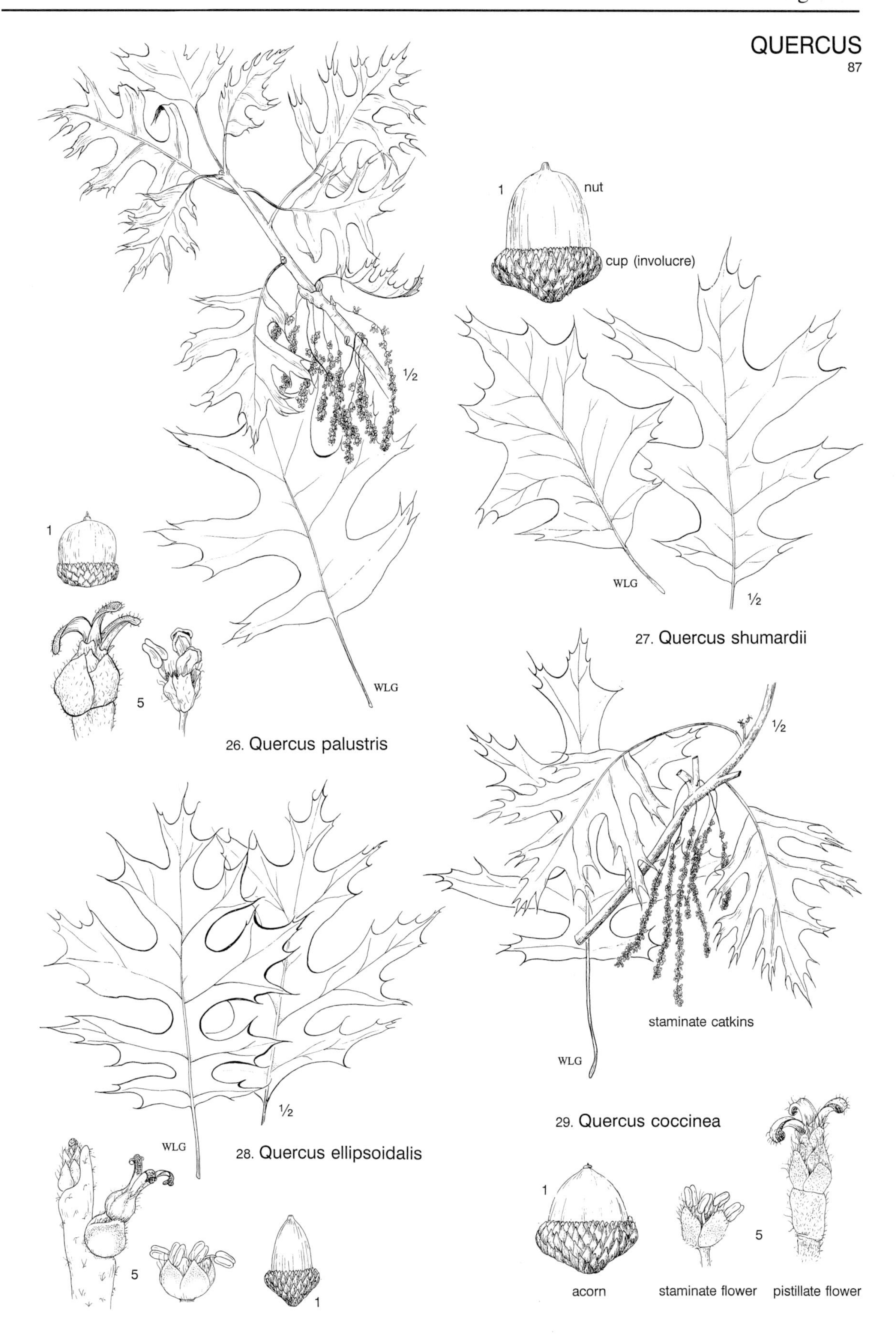

1 nut

cup (involucre)

27. **Quercus shumardii**

WLG

½

1

5

26. **Quercus palustris**

½

staminate catkins

WLG

29. **Quercus coccinea**

WLG

½

28. **Quercus ellipsoidalis**

5

1

acorn

staminate flower

pistillate flower

5

CORYLUS
OSTRYA
CARPINUS
BETULA
88–90

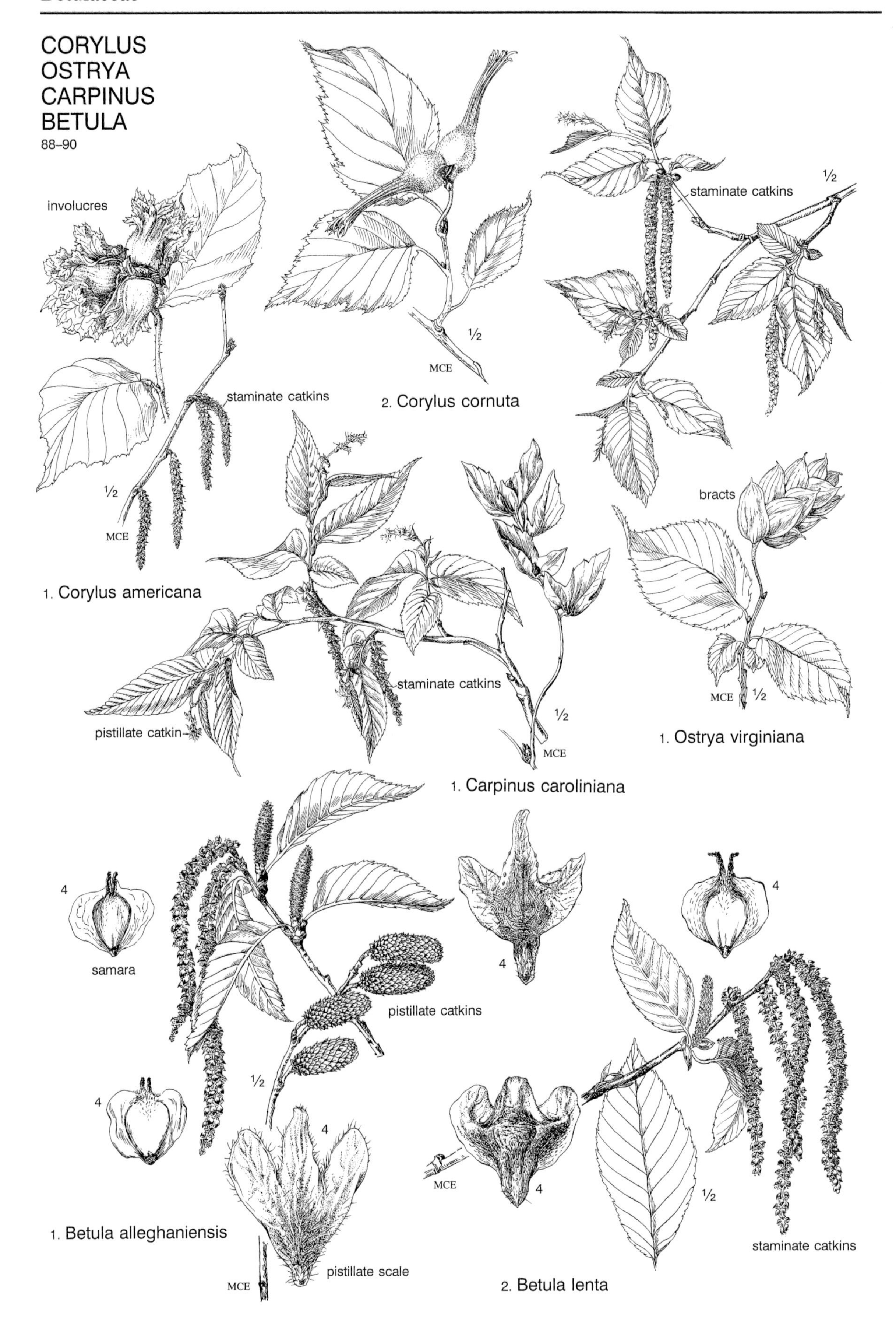

involucres

staminate catkins

½

MCE

2. Corylus cornuta

staminate catkins

½

staminate catkins

½

MCE

1. Corylus americana

bracts

MCE ½

1. Ostrya virginiana

pistillate catkin

staminate catkins

½

MCE

1. Carpinus caroliniana

4

samara

4

pistillate catkins

4

4

½

4

4

1. Betula alleghaniensis

MCE

pistillate scale

MCE

4

½

staminate catkins

2. Betula lenta

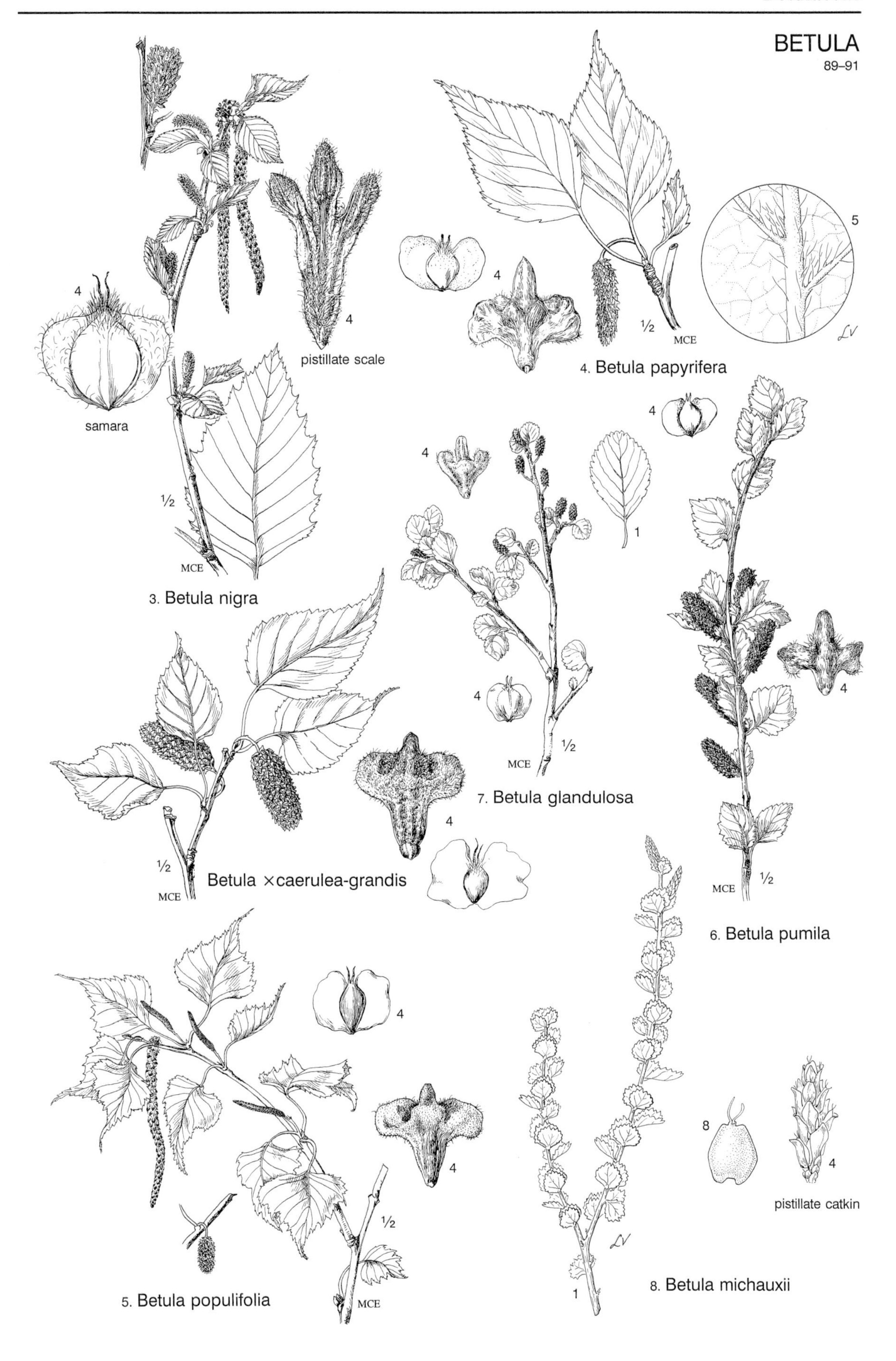

4

samara

4

pistillate scale

4

1/2
MCE

5

4. Betula papyrifera

1/2
MCE

3. Betula nigra

4

4

1

4

4

7. Betula glandulosa

1/2
MCE

Betula ×caerulea-grandis

4

4

MCE
1/2

6. Betula pumila

4

4

4

1

8

4

pistillate catkin

5. Betula populifolia

MCE

8. Betula michauxii

ALNUS
PHYTOLACCA
91, 92

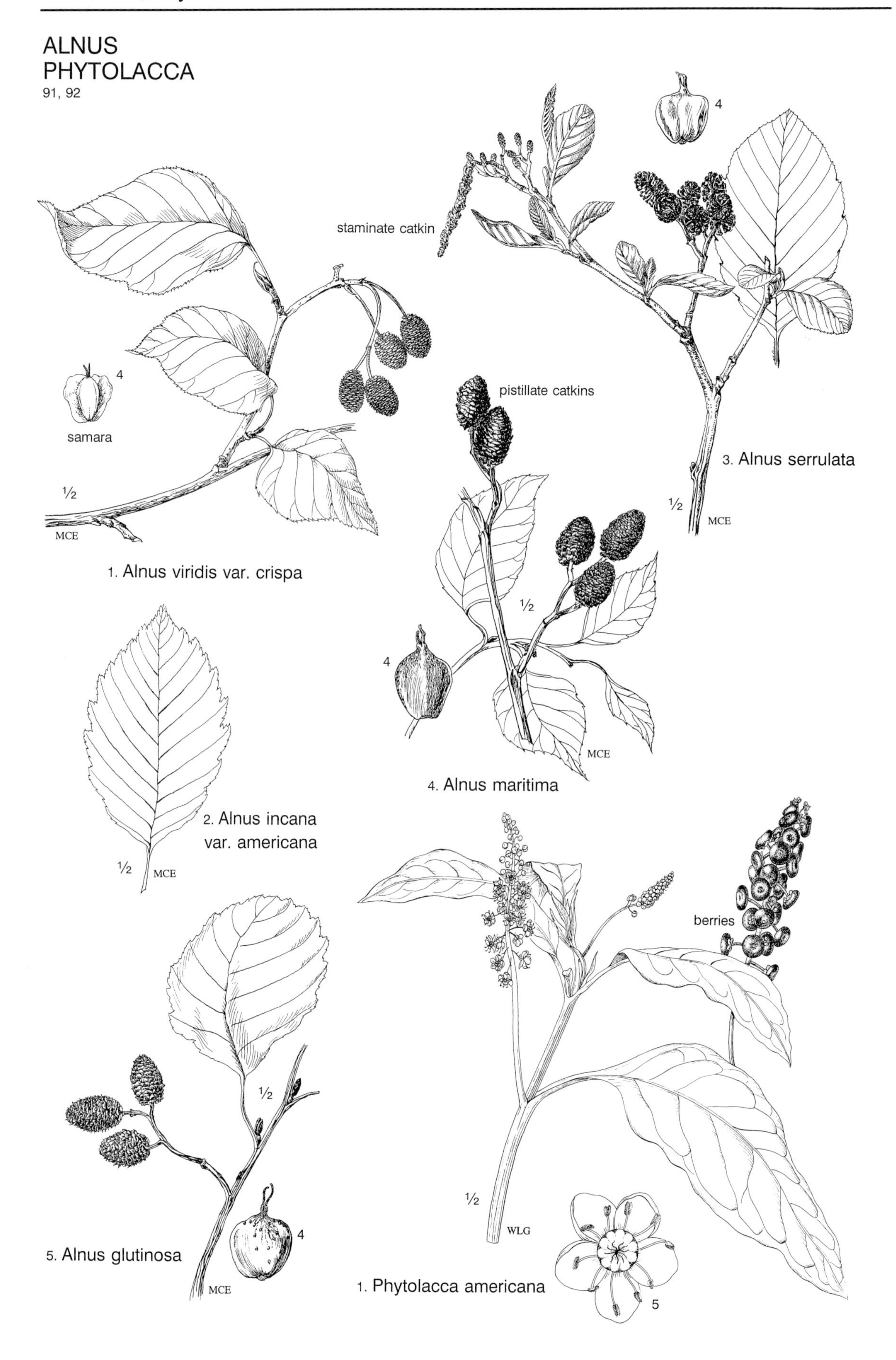

staminate catkin

samara

½

MCE

1. Alnus viridis var. crispa

2. Alnus incana
var. americana

½ MCE

pistillate catkins

½

½

4

MCE

4. Alnus maritima

3. Alnus serrulata

½ MCE

berries

5. Alnus glutinosa

½ WLG

1. Phytolacca americana

MCE

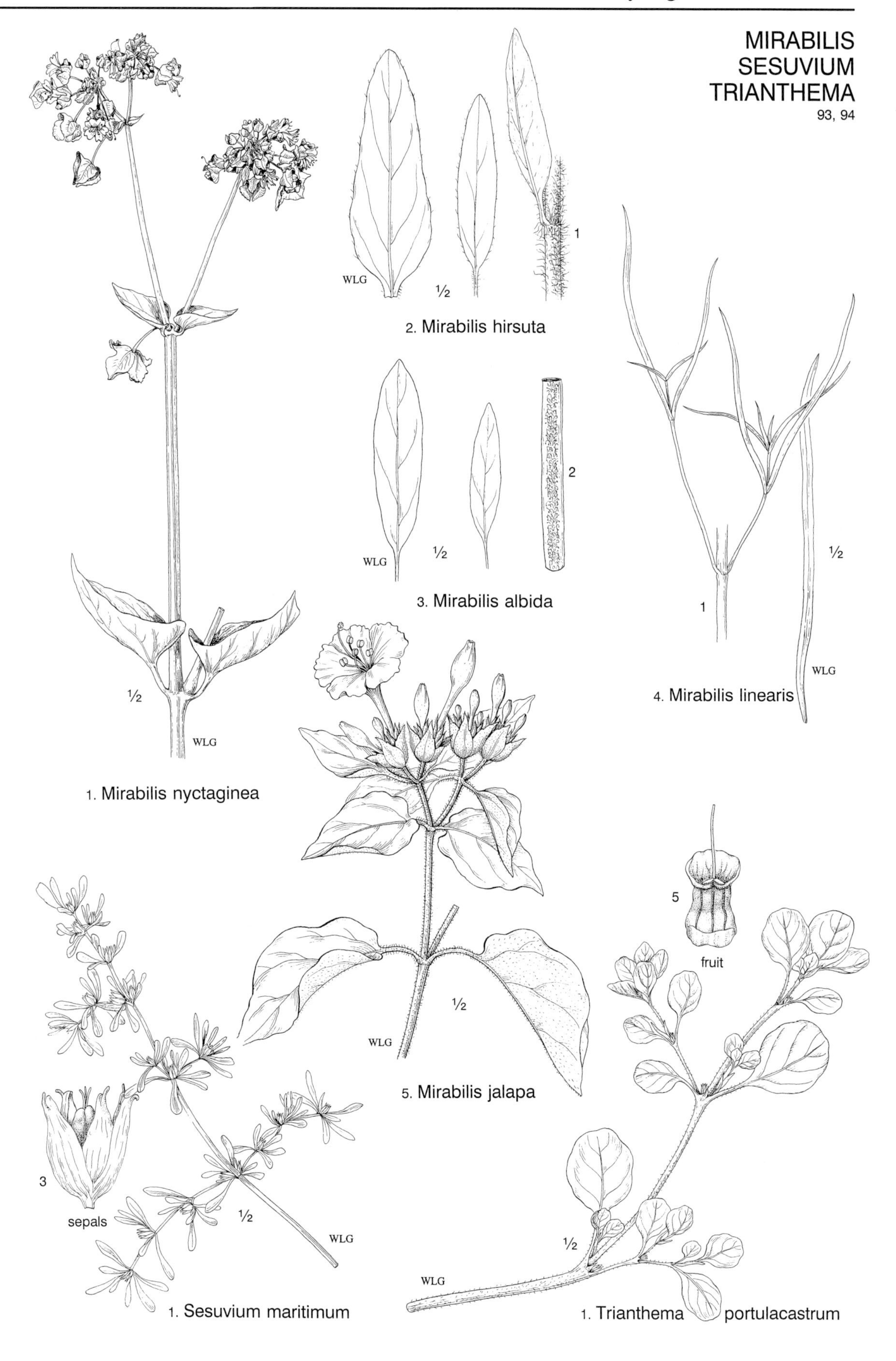

2. Mirabilis hirsuta

3. Mirabilis albida

4. Mirabilis linearis

1. Mirabilis nyctaginea

5. Mirabilis jalapa

fruit

1. Sesuvium maritimum

1. Trianthema portulacastrum

OPUNTIA
CORYPHANTHA
CHENOPODIUM
94–97

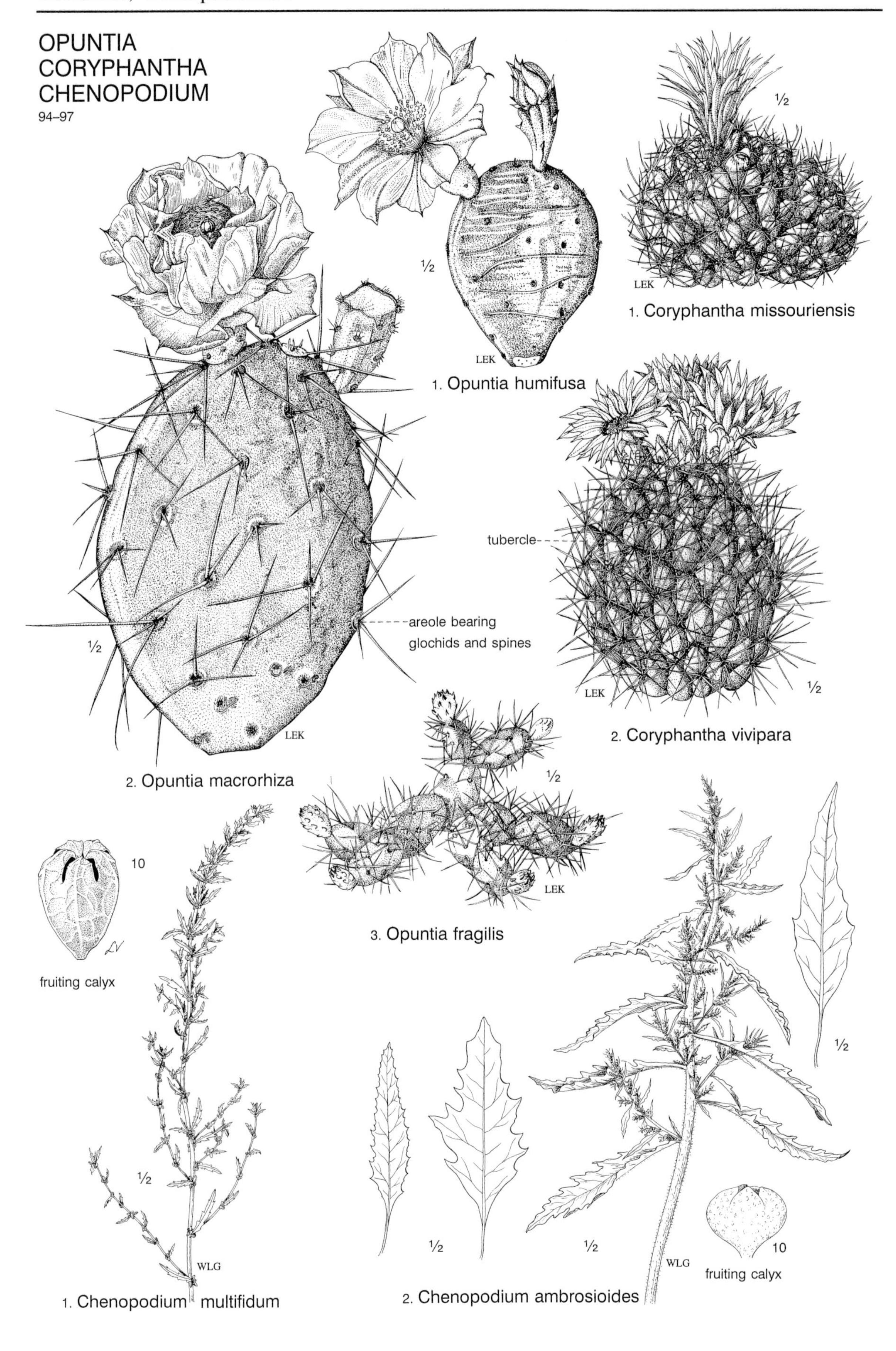

½

½

1. Coryphantha missouriensis

½

1. Opuntia humifusa

tubercle

areole bearing
glochids and spines

½

2. Coryphantha vivipara

½

2. Opuntia macrorhiza

½

10

fruiting calyx

½

3. Opuntia fragilis

½

½

½

½

10

½

WLG

fruiting calyx

1. Chenopodium multifidum

2. Chenopodium ambrosioides

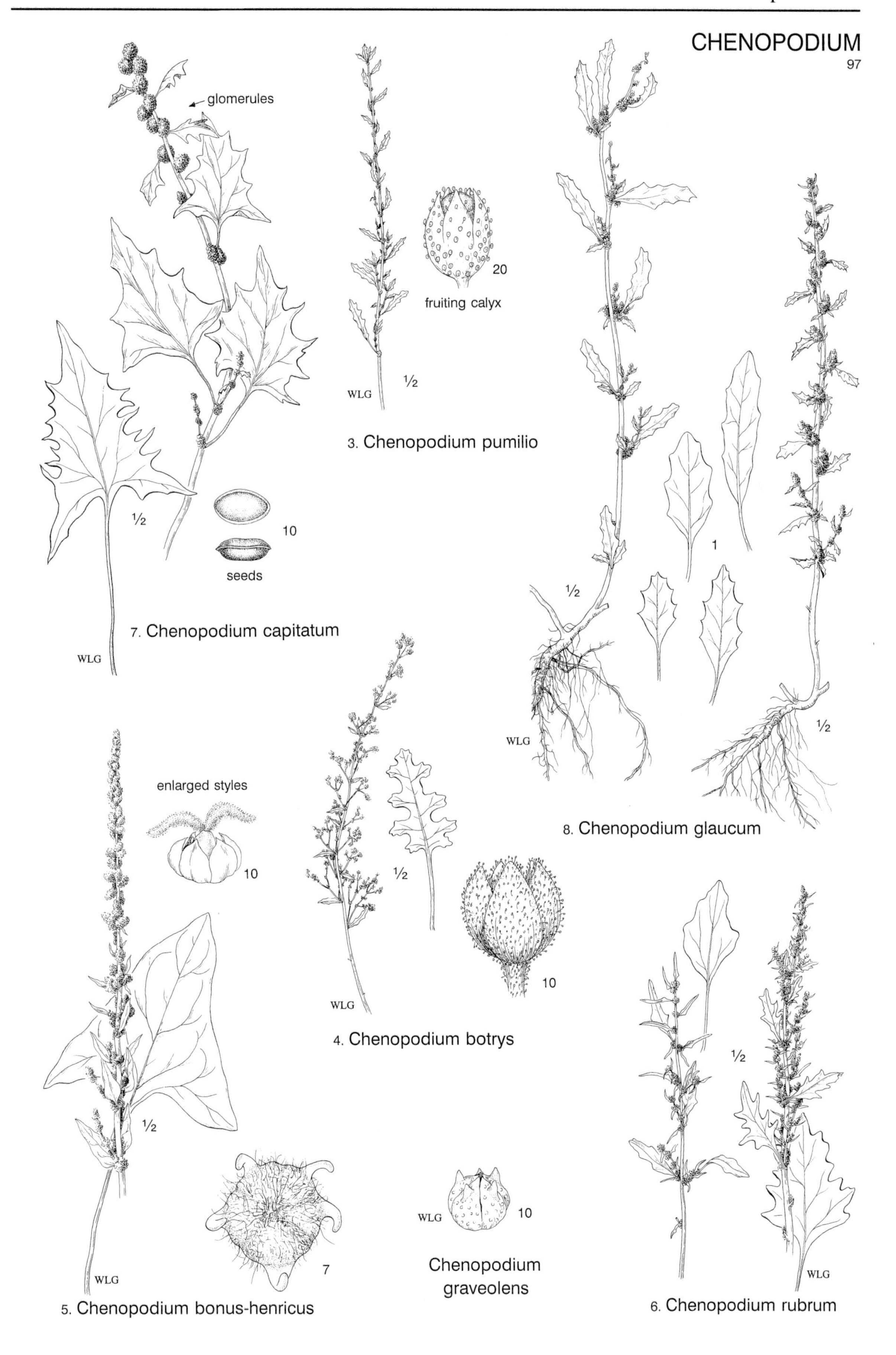

glomerules

20
fruiting calyx

½
WLG

3. Chenopodium pumilio

½

10
seeds

7. Chenopodium capitatum

WLG

1

½
WLG

½

8. Chenopodium glaucum

enlarged styles

10

½

WLG

4. Chenopodium botrys

10

5. Chenopodium bonus-henricus

WLG

7

WLG   10

Chenopodium
graveolens

½

WLG

6. Chenopodium rubrum

CHENOPODIUM
97, 98

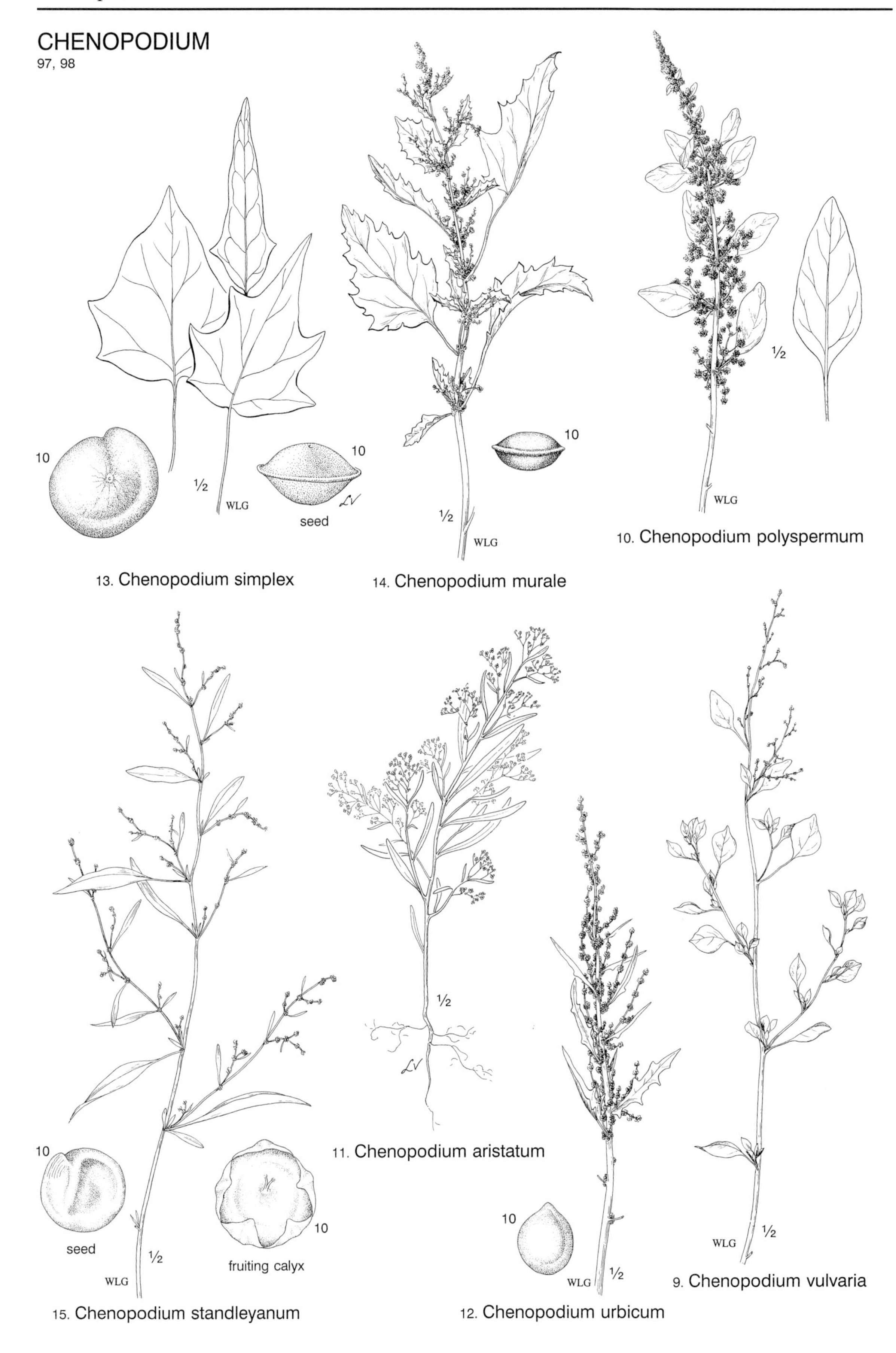

13. Chenopodium simplex

14. Chenopodium murale

10. Chenopodium polyspermum

11. Chenopodium aristatum

15. Chenopodium standleyanum

12. Chenopodium urbicum

9. Chenopodium vulvaria

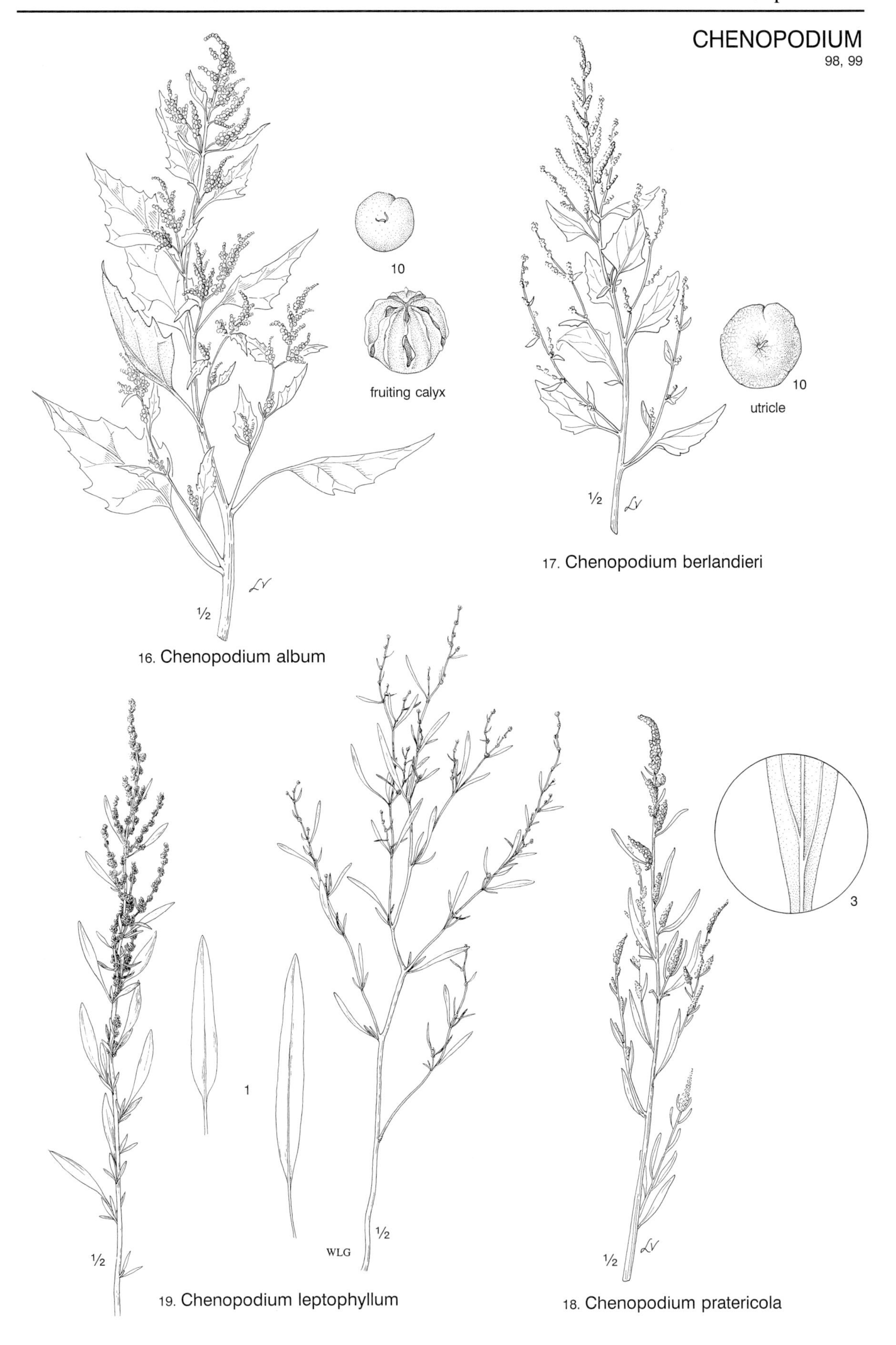

10

fruiting calyx

16. Chenopodium album

10

utricle

17. Chenopodium berlandieri

1

19. Chenopodium leptophyllum

WLG

3

18. Chenopodium pratericola

CYCLOLOMA
BASSIA
KOCHIA
AXYRIS
MONOLEPIS
CORISPERMUM
99, 100

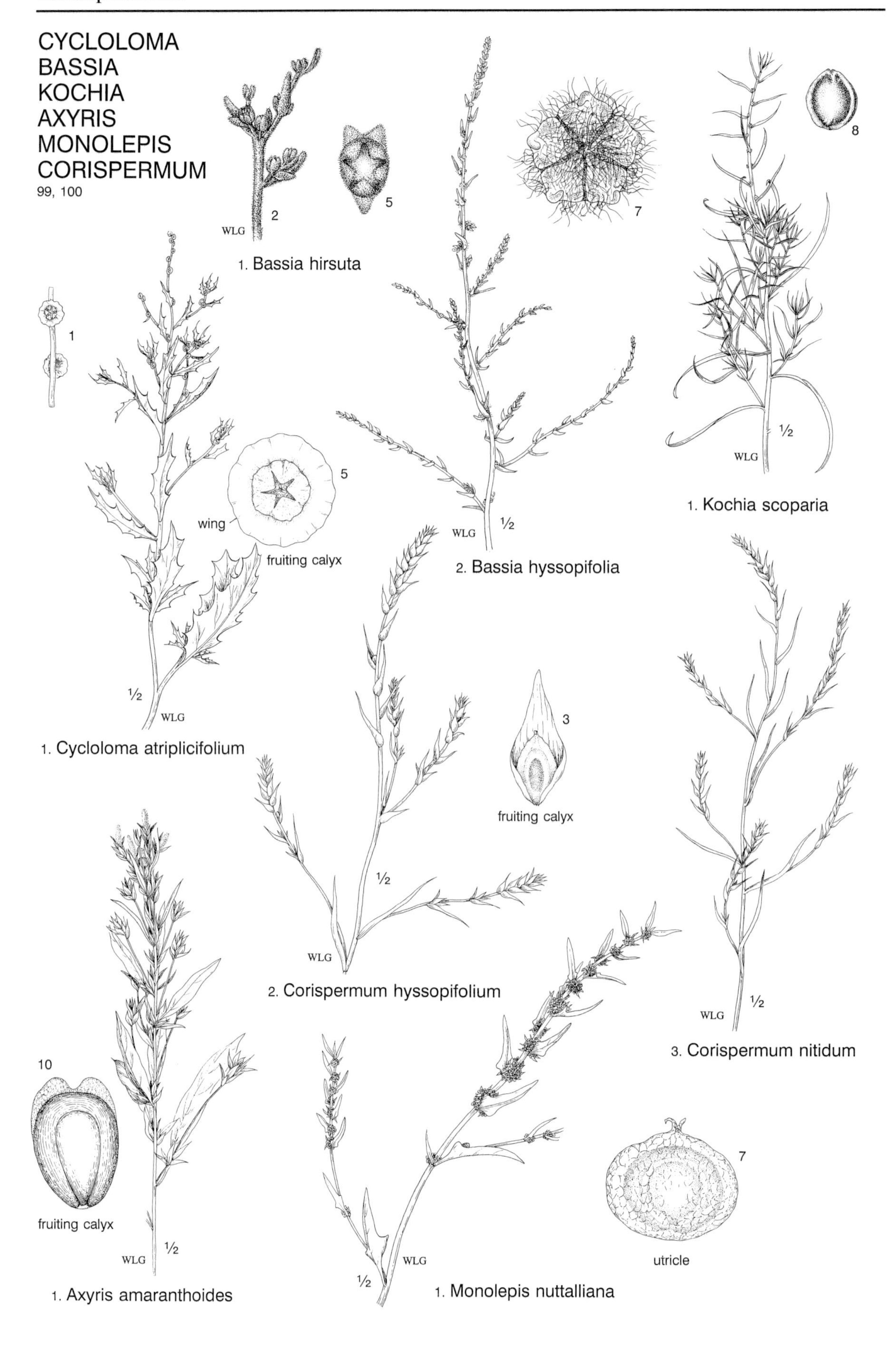

1. Bassia hirsuta

2. Bassia hyssopifolia

1. Kochia scoparia

wing

fruiting calyx

1. Cycloloma atriplicifolium

fruiting calyx

2. Corispermum hyssopifolium

3. Corispermum nitidum

fruiting calyx

1. Axyris amaranthoides

1. Monolepis nuttalliana

utricle

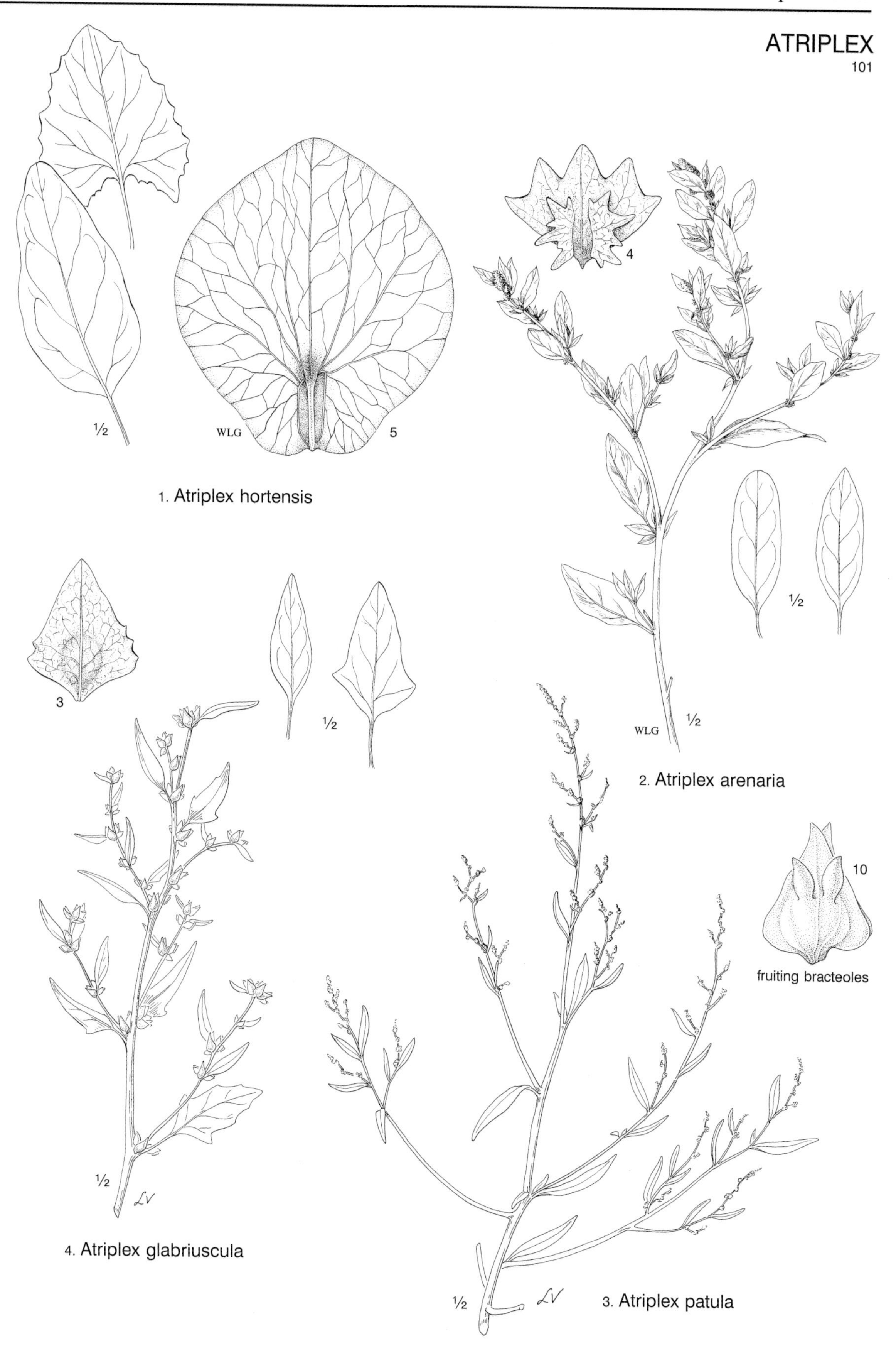

1. Atriplex hortensis

2. Atriplex arenaria

4. Atriplex glabriuscula

fruiting bracteoles

3. Atriplex patula

ATRIPLEX
SALICORNIA
102

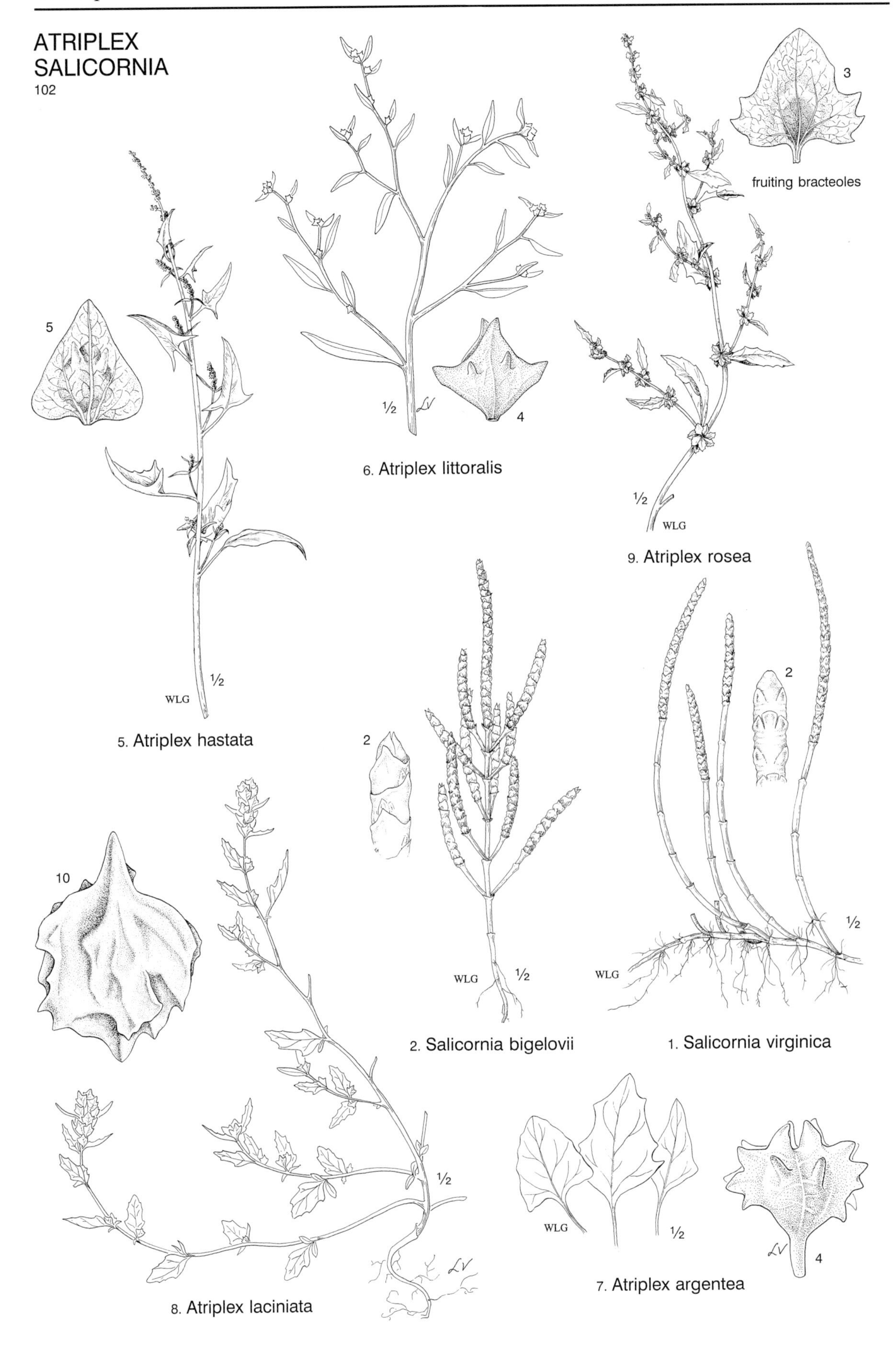

3 fruiting bracteoles

½ ½ 4

6. Atriplex littoralis

5

½ WLG

5. Atriplex hastata

½ WLG

9. Atriplex rosea

10

2

2

WLG ½

2. Salicornia bigelovii

2

½

WLG

1. Salicornia virginica

½

8. Atriplex laciniata

WLG ½

4

7. Atriplex argentea

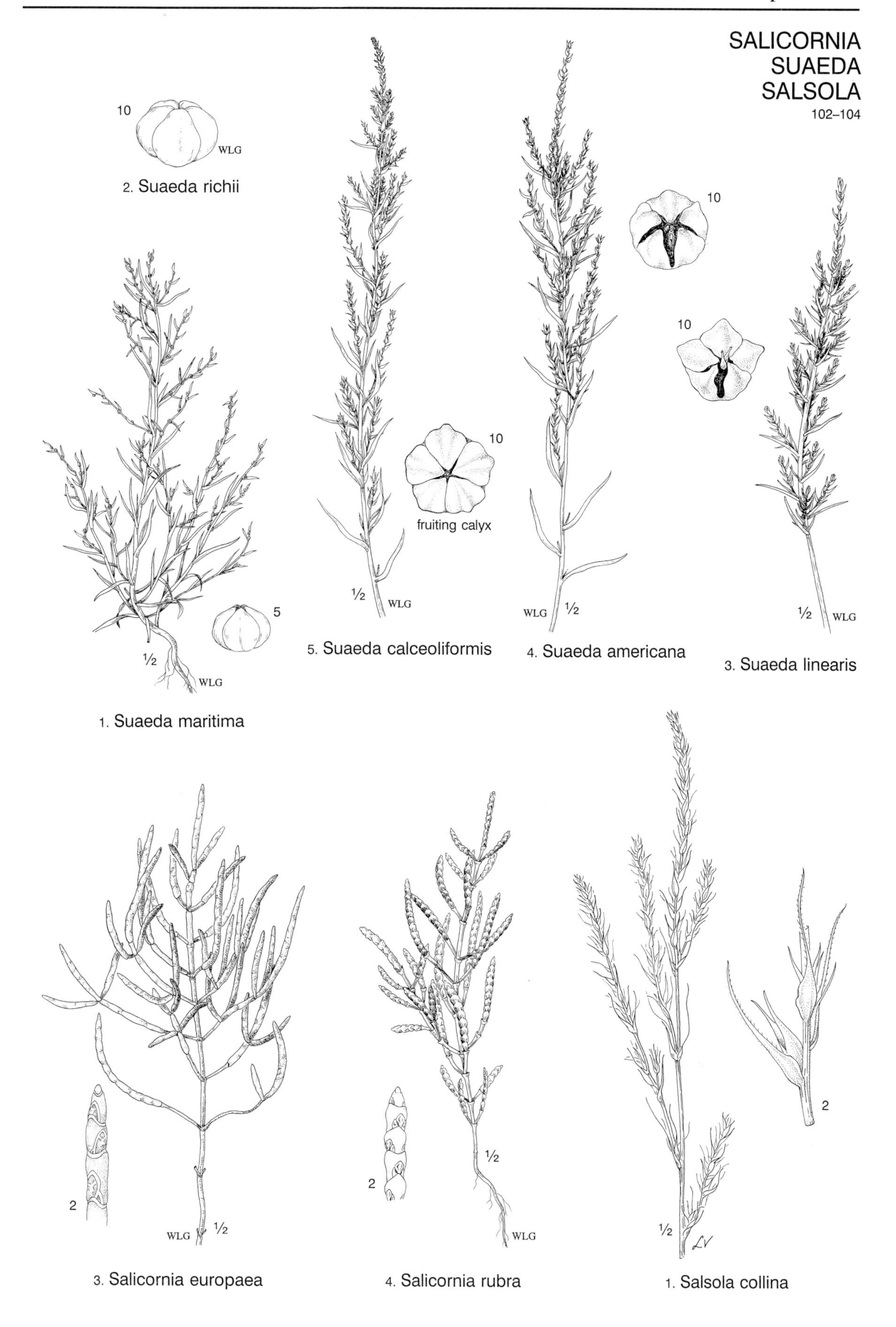

2. Suaeda richii

5. Suaeda calceoliformis

4. Suaeda americana

3. Suaeda linearis

1. Suaeda maritima

fruiting calyx

3. Salicornia europaea

4. Salicornia rubra

1. Salsola collina

SALSOLA
AMARANTHUS
104–106

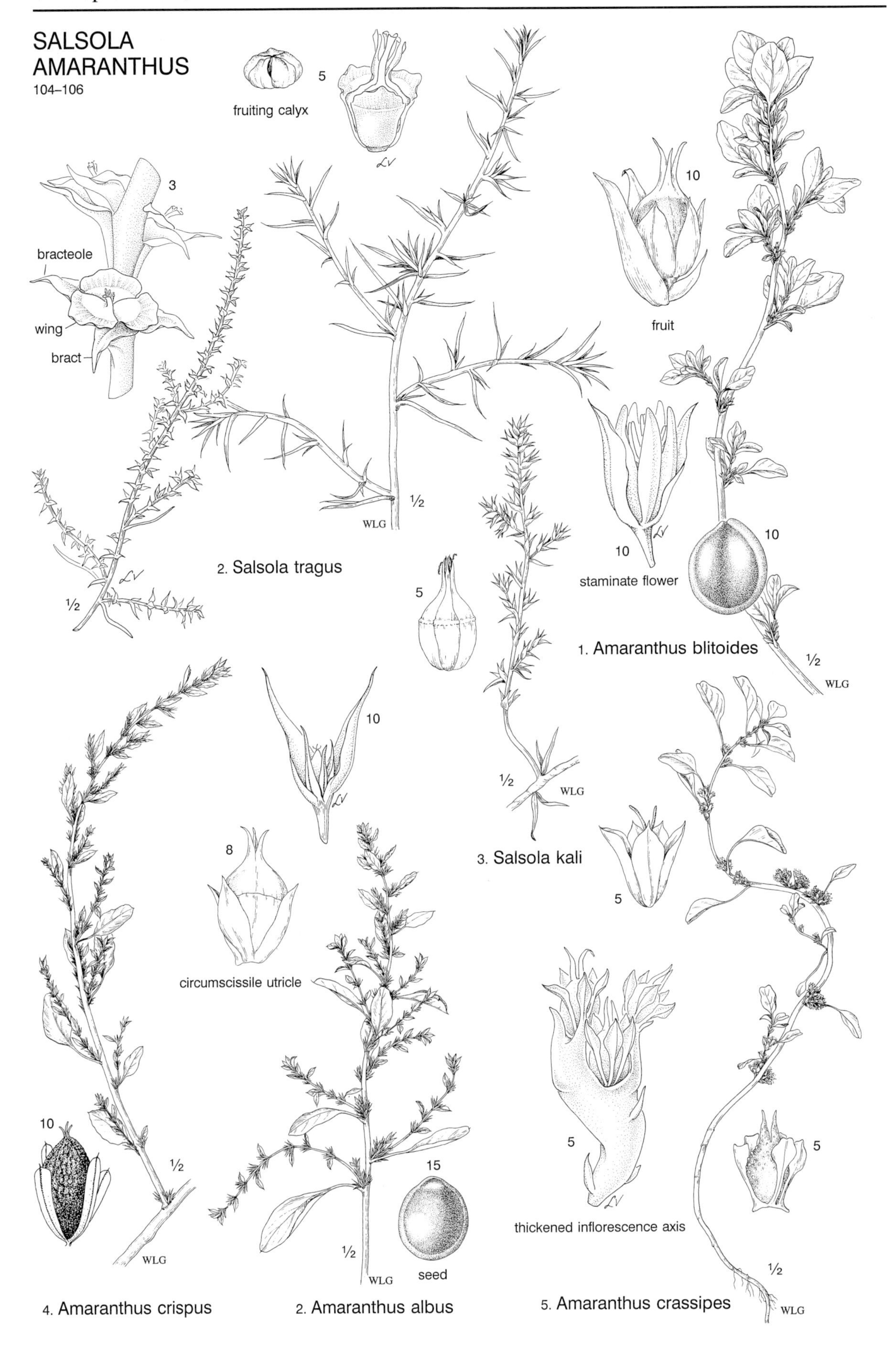

fruiting calyx 5

3

bracteole

wing

bract

½

WLG

½

2. Salsola tragus

5

5

10

10
staminate flower

fruit

10

10

1. Amaranthus blitoides

½

WLG

½

WLG

3. Salsola kali

5

10

8

circumscissile utricle

5

thickened inflorescence axis

5

5

10

½

15

½

seed

½

WLG

WLG

WLG

4. Amaranthus crispus   2. Amaranthus albus   5. Amaranthus crassipes

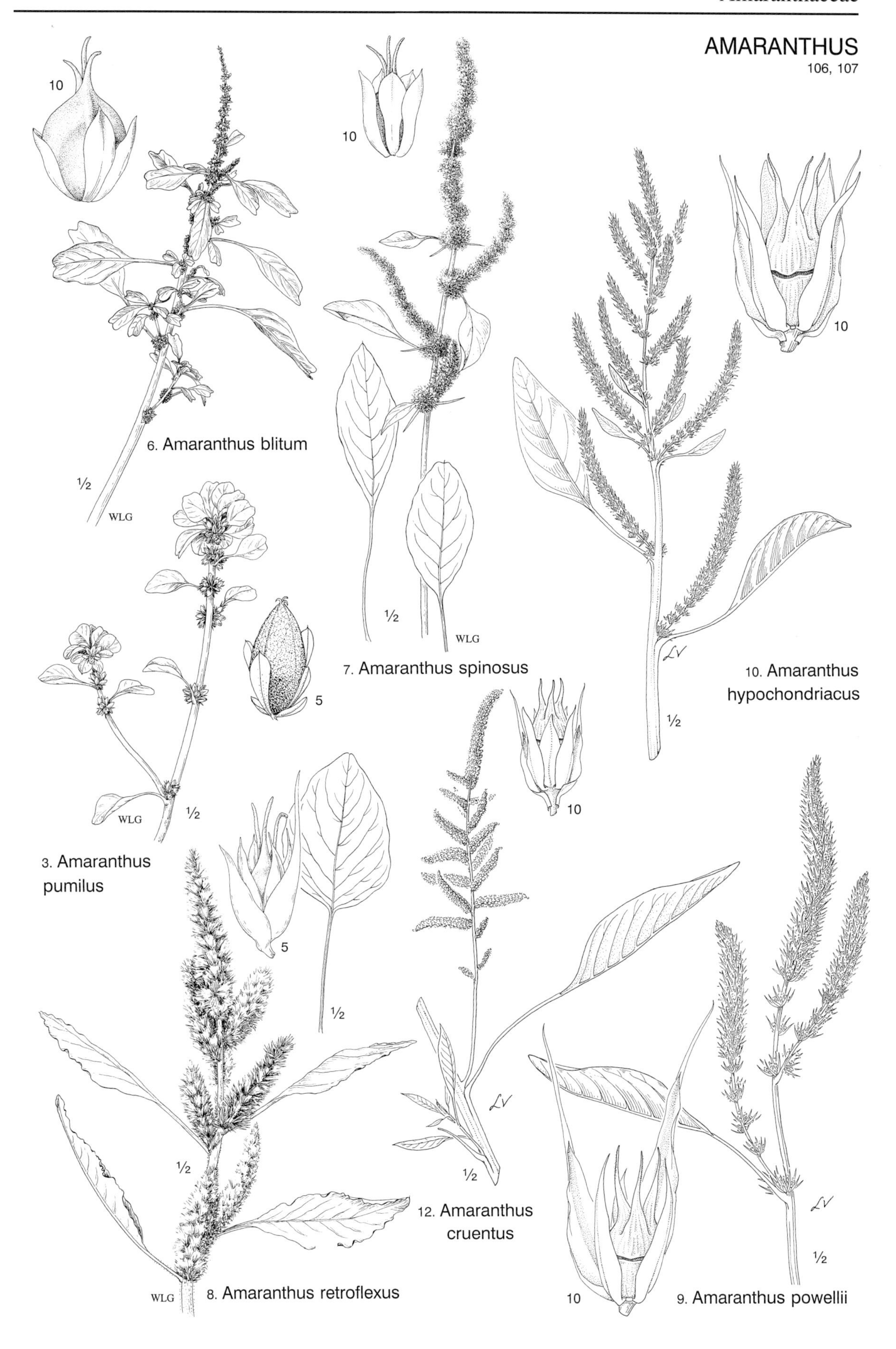

10

10

10

6. Amaranthus blitum

½

WLG

7. Amaranthus spinosus

½

WLG

½

WLG

5

3. Amaranthus
pumilus

½

WLG

5

½

10

10. Amaranthus
hypochondriacus

½

½

12. Amaranthus
cruentus

½

½

10

9. Amaranthus powellii

WLG   8. Amaranthus retroflexus

AMARANTHUS
106, 107

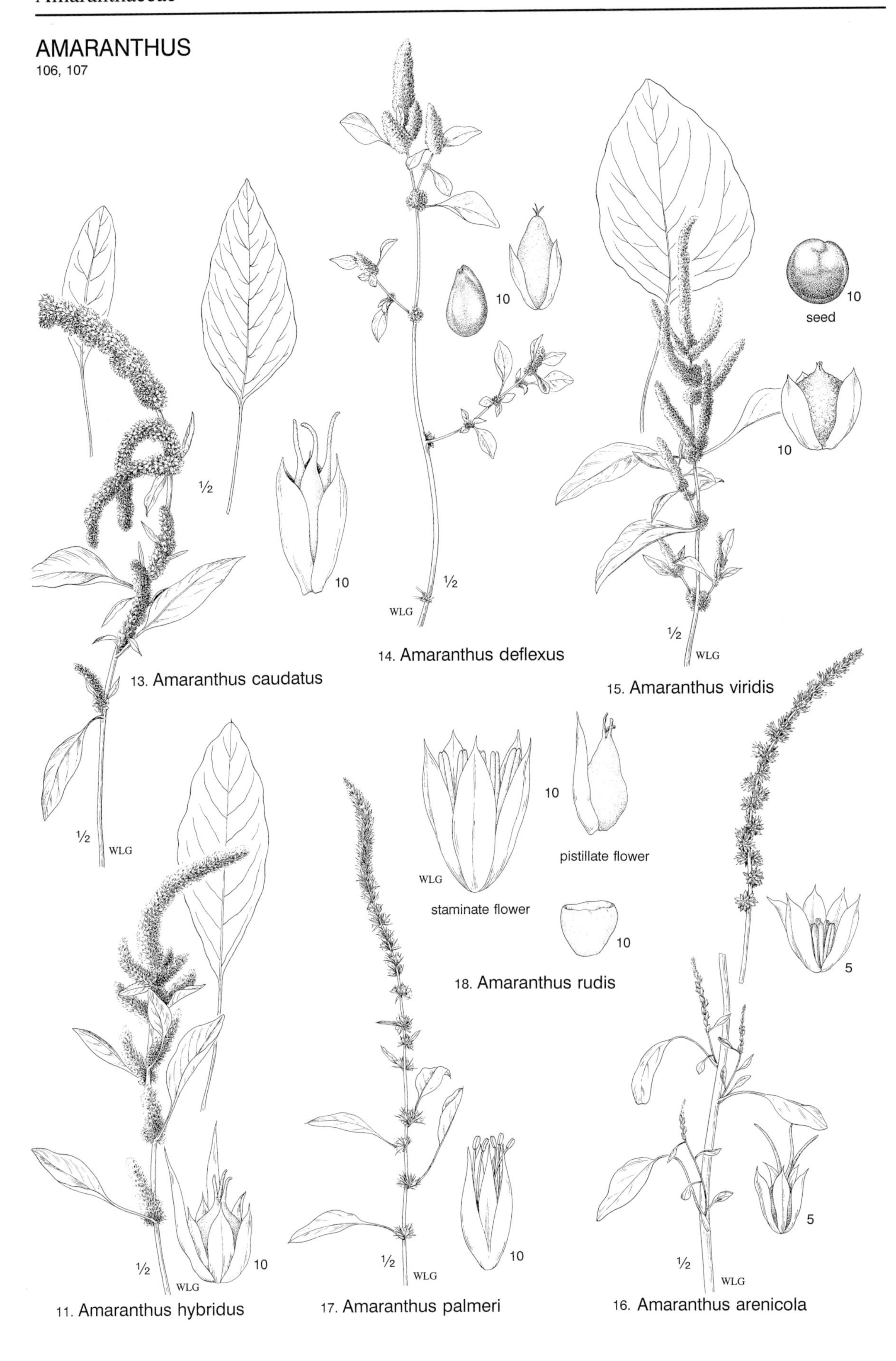

13. Amaranthus caudatus

14. Amaranthus deflexus

15. Amaranthus viridis

seed

staminate flower

pistillate flower

18. Amaranthus rudis

11. Amaranthus hybridus

17. Amaranthus palmeri

16. Amaranthus arenicola

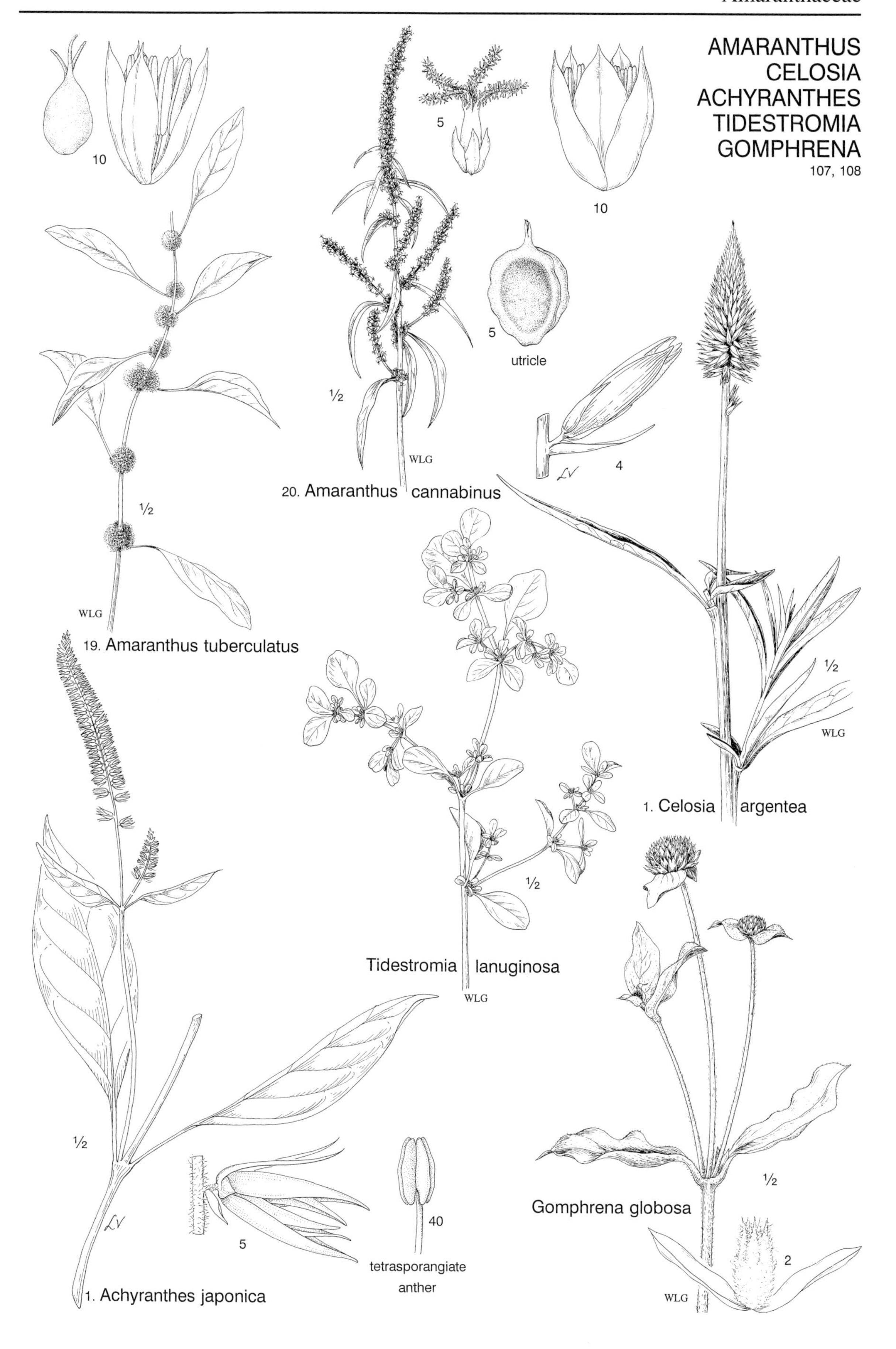

AMARANTHUS
CELOSIA
ACHYRANTHES
TIDESTROMIA
GOMPHRENA
107, 108

10

5

10

5

utricle

½

4

WLG

20. Amaranthus cannabinus

½

WLG

19. Amaranthus tuberculatus

½

WLG

1. Celosia argentea

½

Tidestromia lanuginosa

WLG

½

½

Gomphrena globosa

WLG

2

1. Achyranthes japonica

5

40

tetrasporangiate
anther

ALTERNANTHERA
FROELICHIA
IRESINE
PORTULACA
108–110

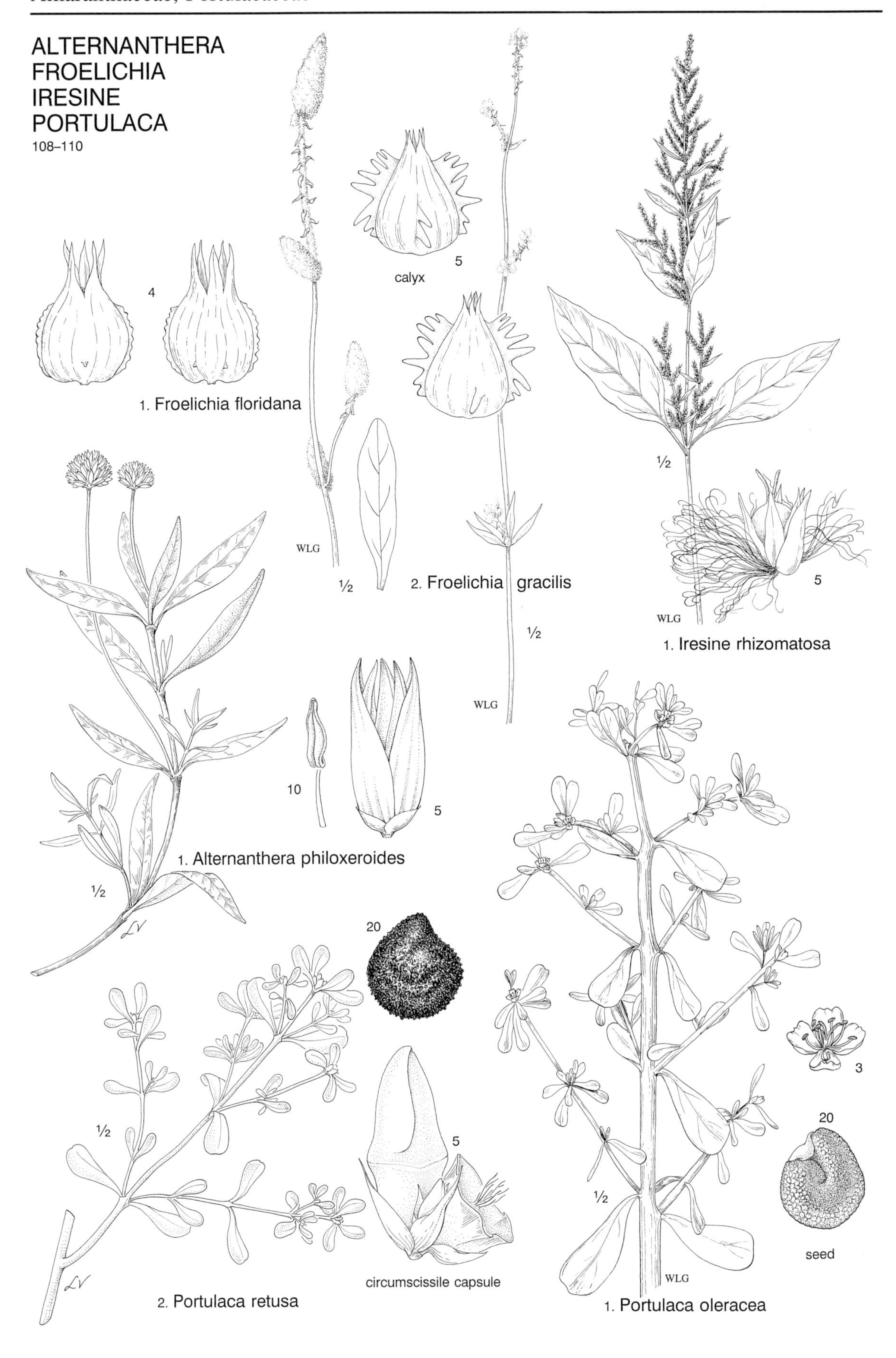

4

1. Froelichia floridana

5
calyx

2. Froelichia gracilis

½

WLG

½

½

WLG

1. Iresine rhizomatosa

5

WLG

10

5

1. Alternanthera philoxeroides

½

20

½

5

circumscissile capsule

2. Portulaca retusa

3

20

seed

½

WLG

1. Portulaca oleracea

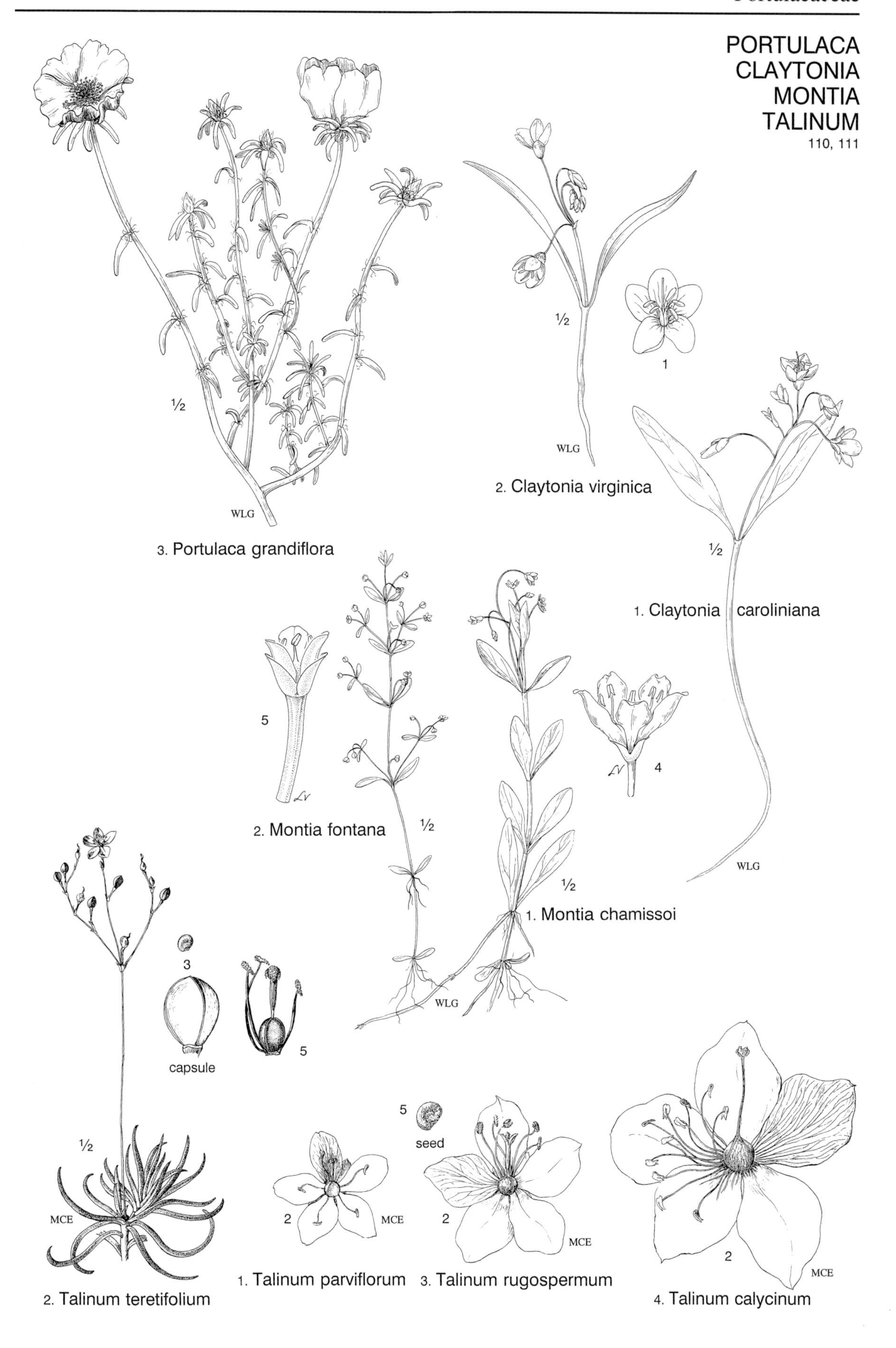

2. Claytonia virginica

WLG

1

1/2

3. Portulaca grandiflora

WLG

1/2

1. Claytonia caroliniana

1/2

WLG

5

2. Montia fontana

1/2

4

1. Montia chamissoi

1/2

WLG

3

capsule

5

1/2

MCE

2. Talinum teretifolium

5

seed

1. Talinum parviflorum

2

MCE

3. Talinum rugospermum

2

MCE

4. Talinum calycinum

2

MCE

MOLLUGO
GLINUS
SAGINA
ARENARIA
112–114

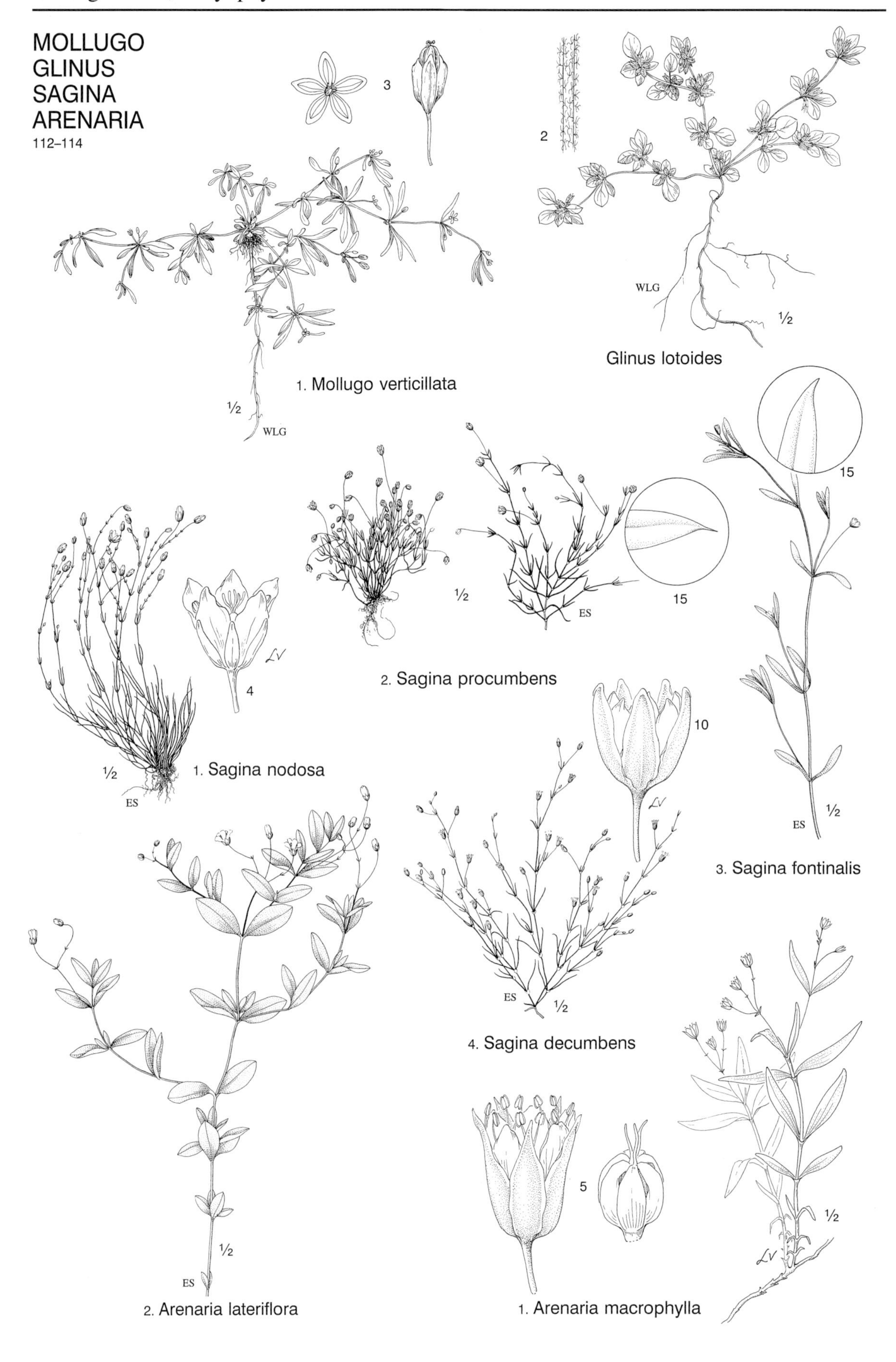

3

2

WLG   ½

Glinus lotoides

1. Mollugo verticillata

½

WLG

15

15

2. Sagina procumbens

10

4

ES

1. Sagina nodosa

½

ES

½

ES

3. Sagina fontinalis

4. Sagina decumbens

ES   ½

5

½

2. Arenaria lateriflora

1. Arenaria macrophylla

½

ES

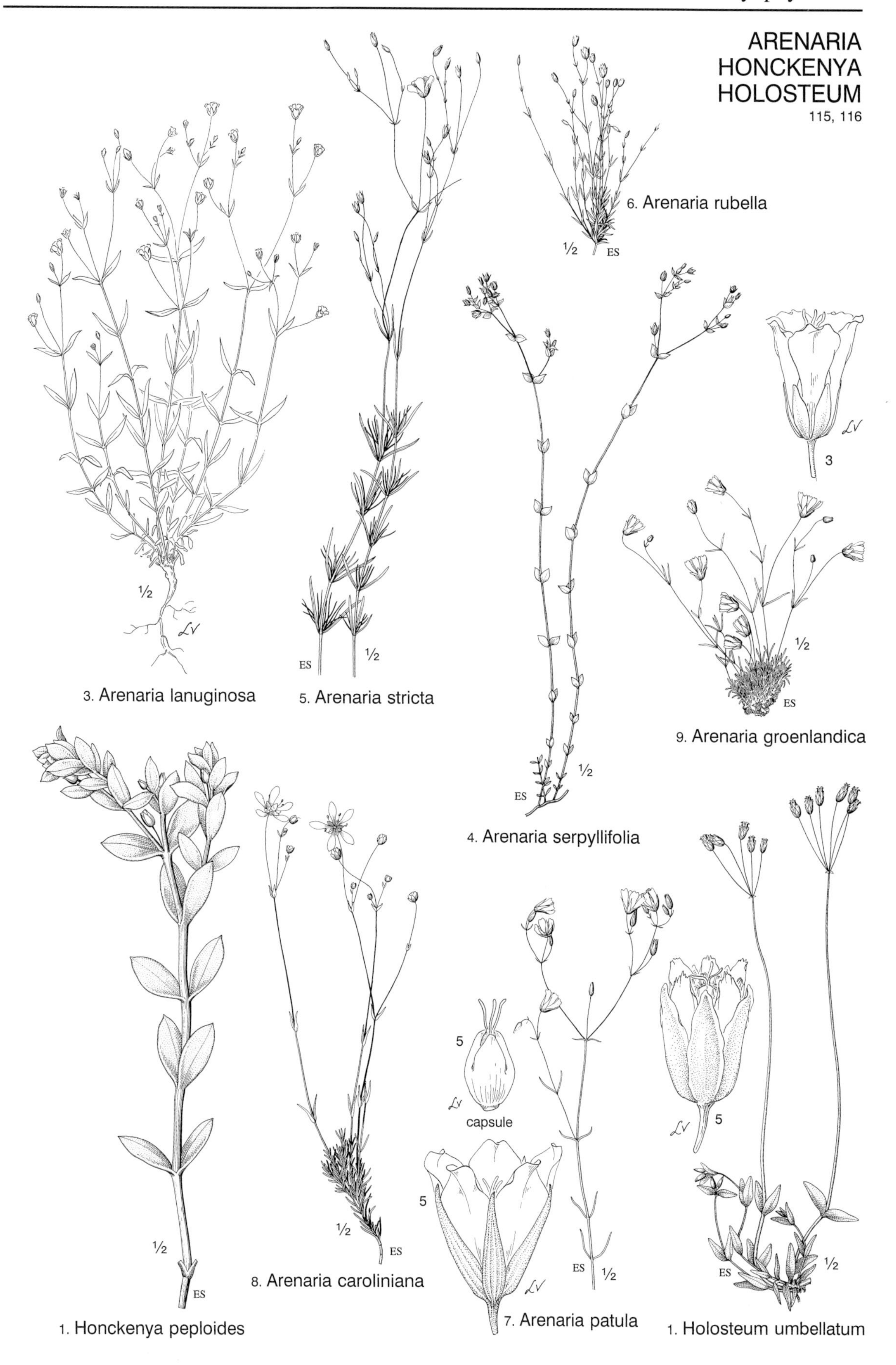

6. Arenaria rubella

3. Arenaria lanuginosa

5. Arenaria stricta

9. Arenaria groenlandica

4. Arenaria serpyllifolia

capsule

1. Honckenya peploides

8. Arenaria caroliniana

7. Arenaria patula

1. Holosteum umbellatum

STIPULICIDA
CERASTIUM
116, 117

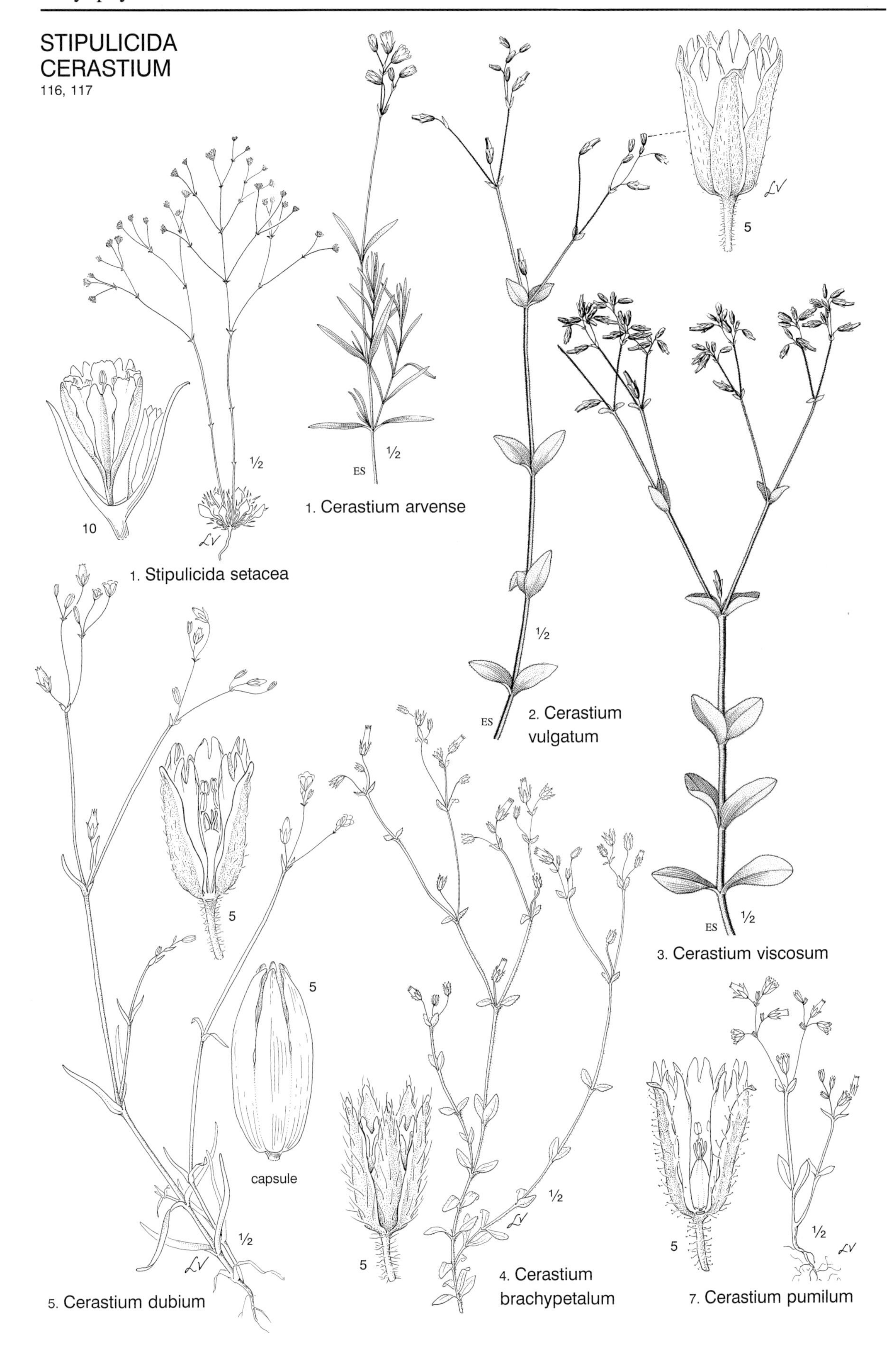

5

10

1. Stipulicida setacea

1. Cerastium arvense

ES

2. Cerastium vulgatum

ES

5

capsule

5

5

3. Cerastium viscosum

ES

5. Cerastium dubium

5

4. Cerastium brachypetalum

5

7. Cerastium pumilum

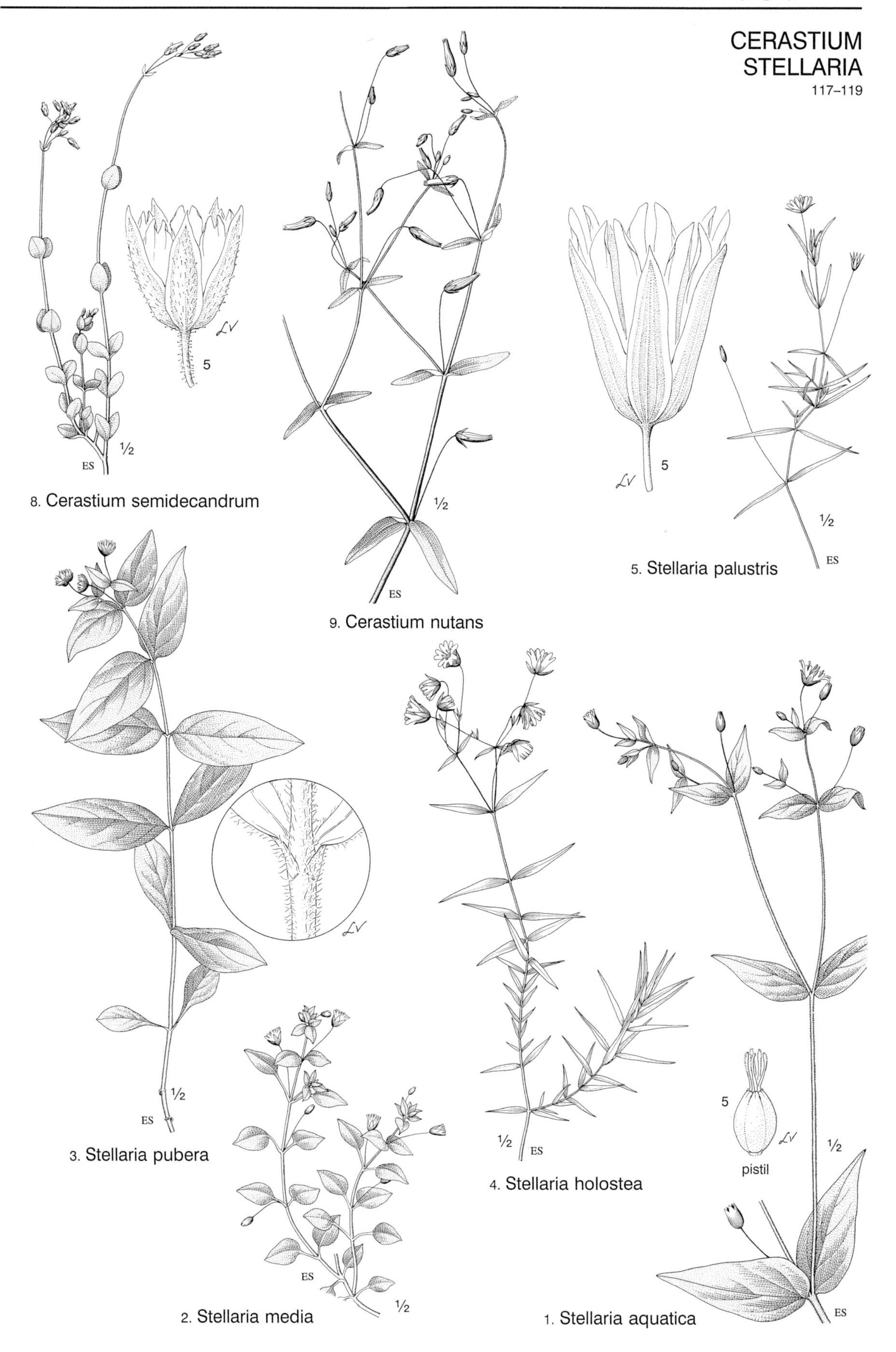

8. Cerastium semidecandrum

9. Cerastium nutans

5. Stellaria palustris

3. Stellaria pubera

2. Stellaria media

4. Stellaria holostea

pistil

1. Stellaria aquatica

STELLARIA
119, 120

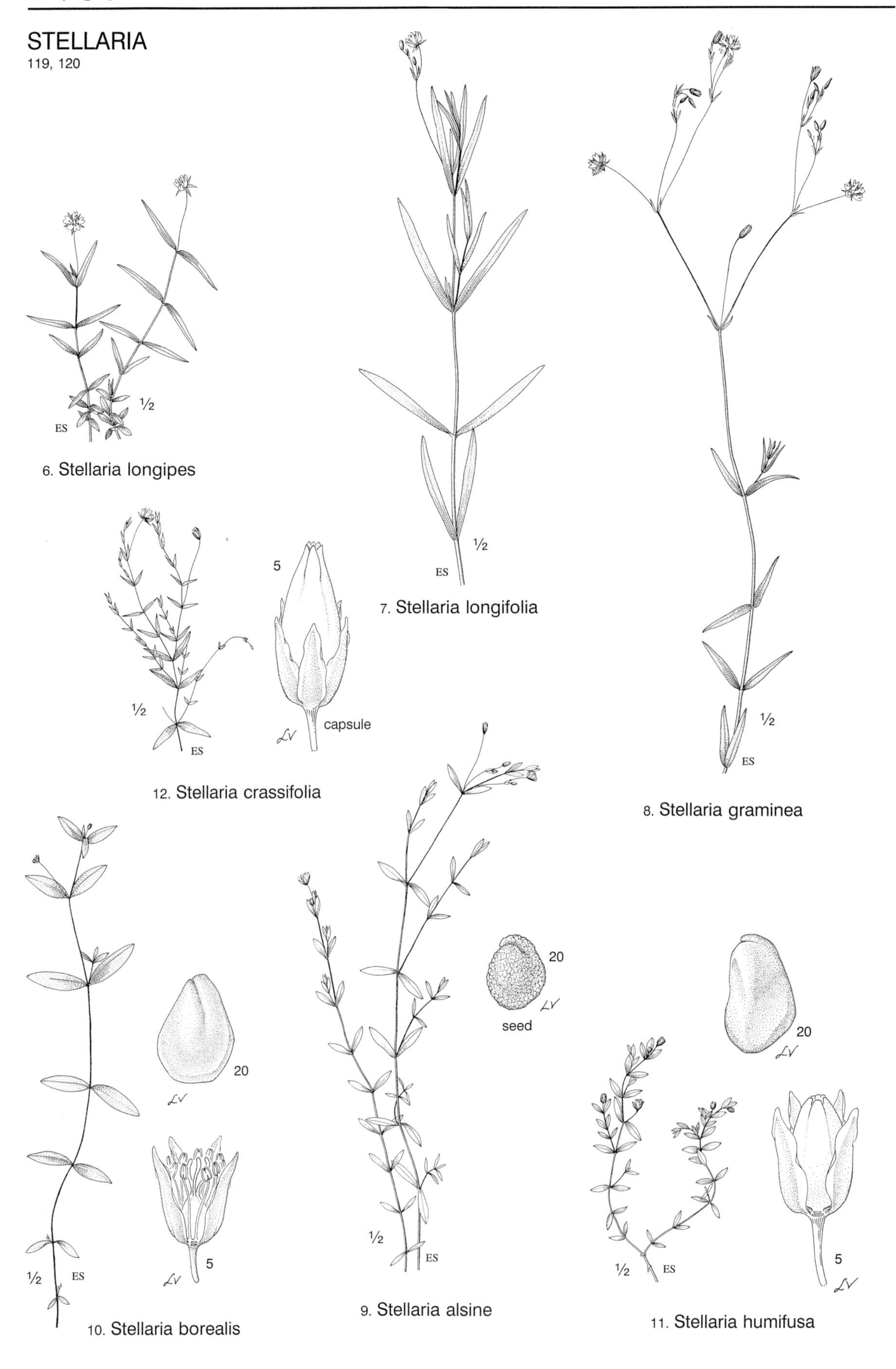

6. Stellaria longipes

7. Stellaria longifolia

12. Stellaria crassifolia

capsule

8. Stellaria graminea

5

seed 20

20

20

10. Stellaria borealis

9. Stellaria alsine

11. Stellaria humifusa

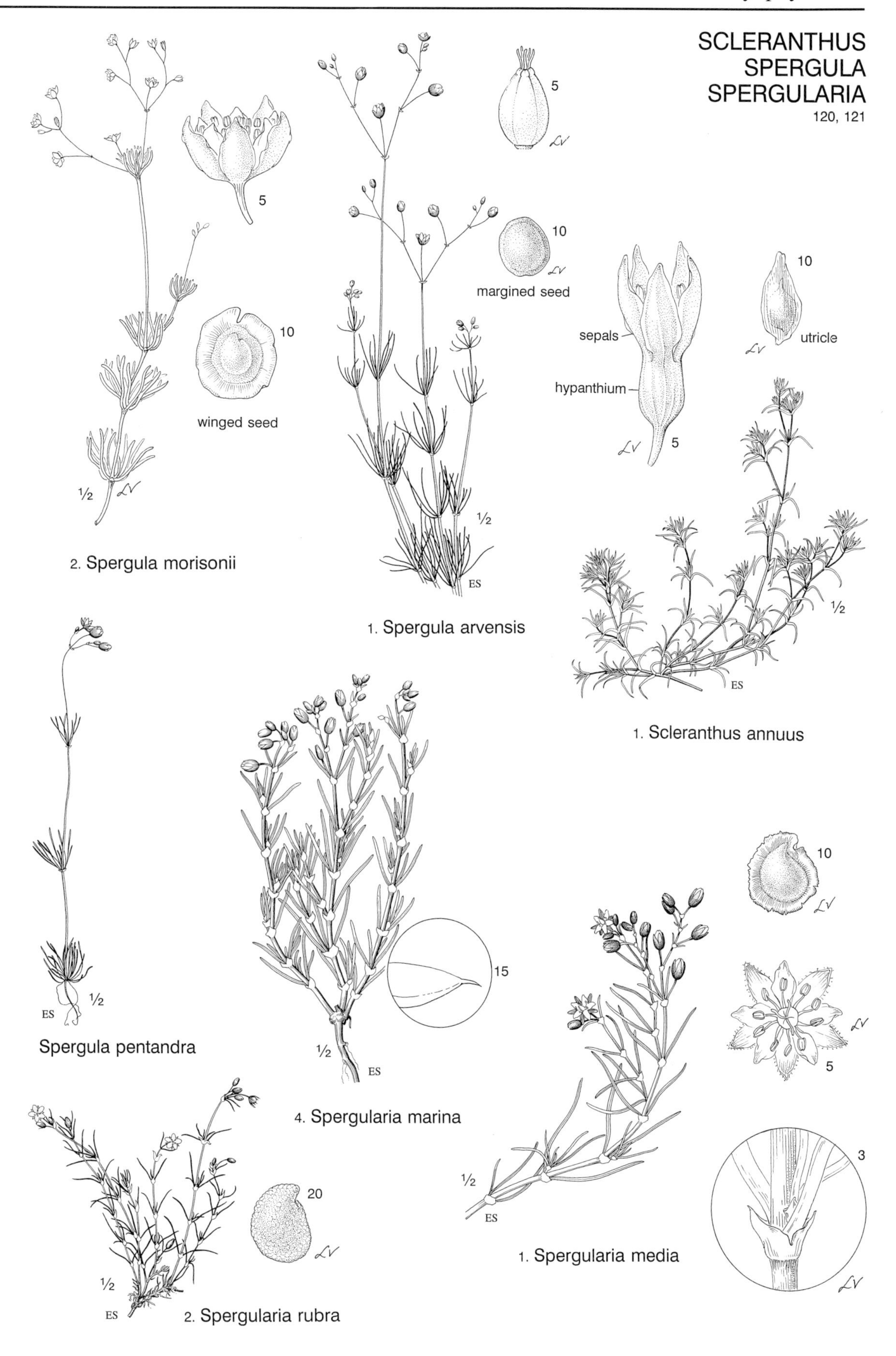

5

10

margined seed

5

winged seed

10

½

2. Spergula morisonii

½

ES

1. Spergula arvensis

sepals

hypanthium

5

10

utricle

½

ES

1. Scleranthus annuus

Spergula pentandra

ES

½

15

½

ES

4. Spergularia marina

10

5

20

½

ES

2. Spergularia rubra

½

ES

1. Spergularia media

3

SPERGULARIA
PARONYCHIA
121, 122

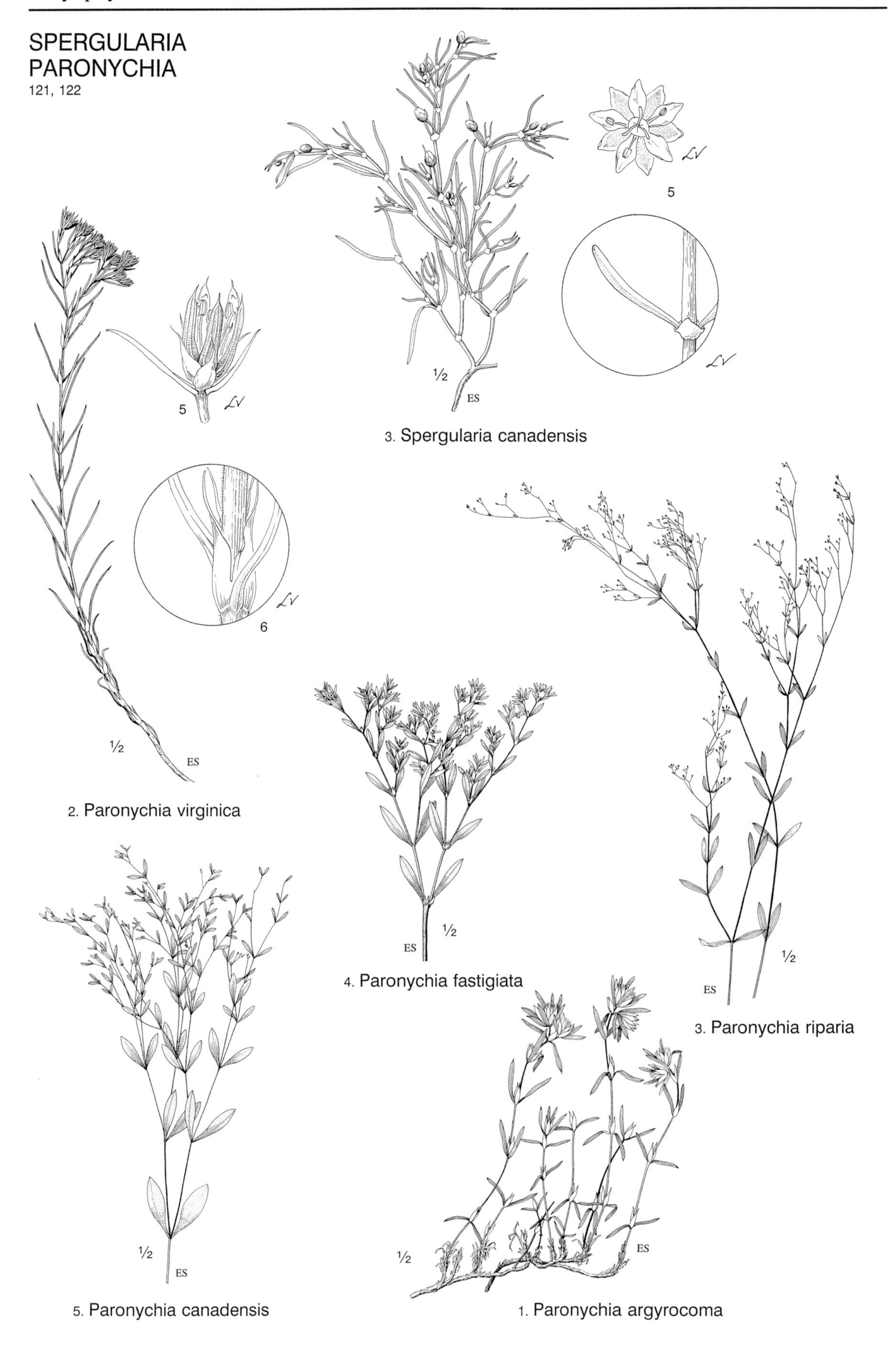

5

3. Spergularia canadensis

6

2. Paronychia virginica

4. Paronychia fastigiata

3. Paronychia riparia

5. Paronychia canadensis

1. Paronychia argyrocoma

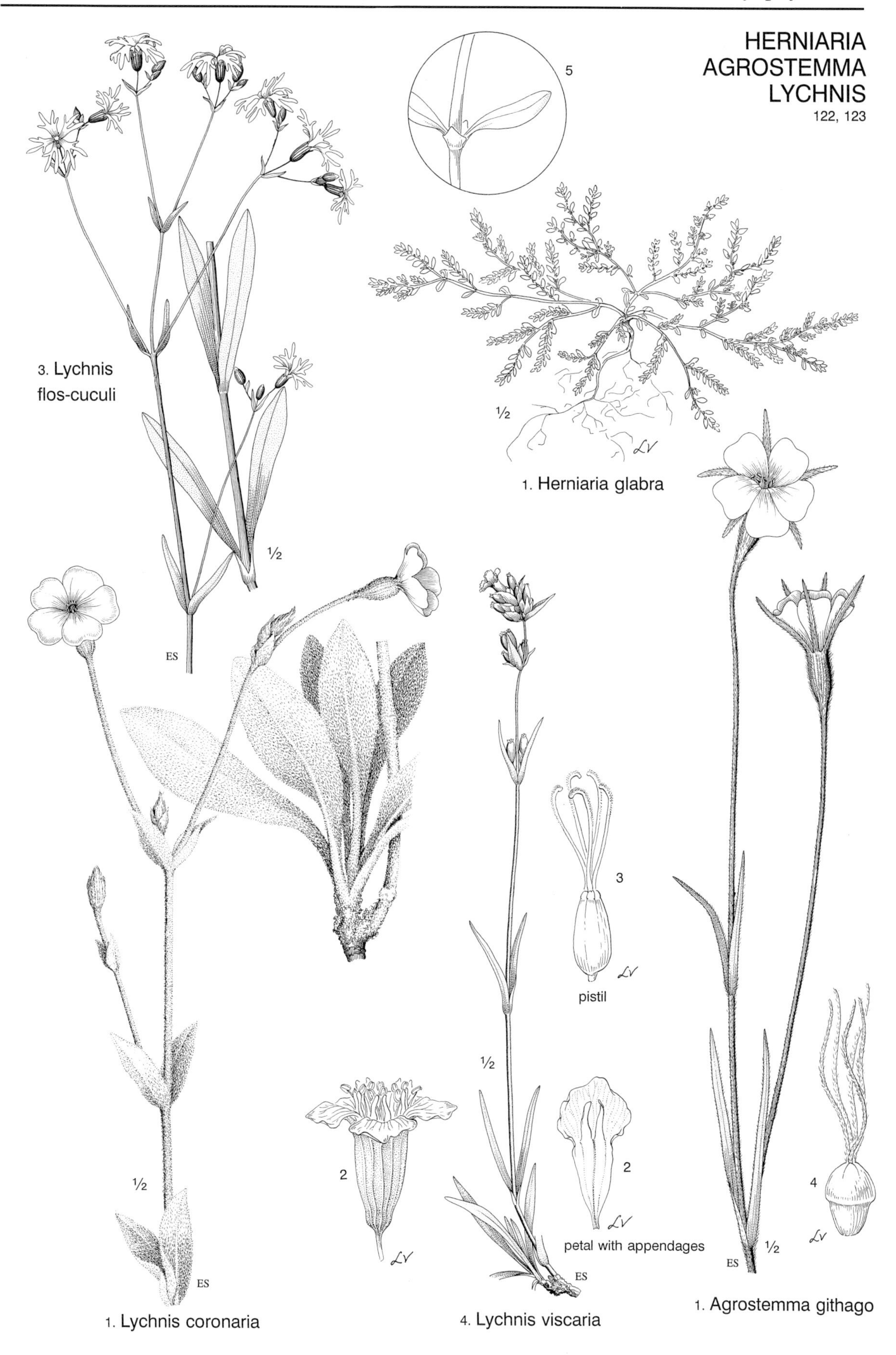

5

3. Lychnis
flos-cuculi

½

ES

½

1. Herniaria glabra

½

3

pistil

½

2

2

petal with appendages

4

1. Lychnis coronaria

ES

½

ES

4. Lychnis viscaria

ES

½

1. Agrostemma githago

LYCHNIS
SILENE
123, 124

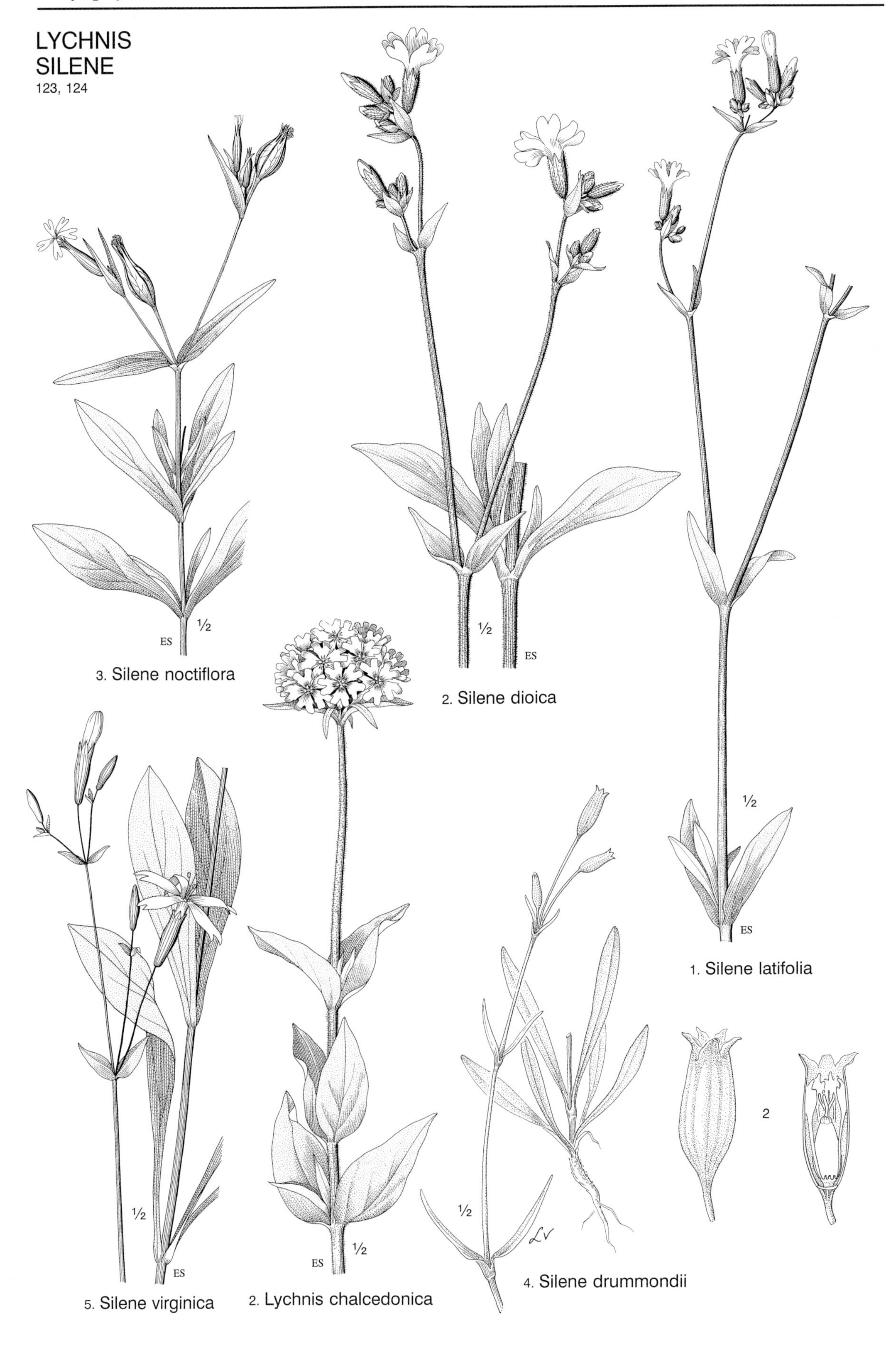

3. Silene noctiflora

2. Silene dioica

1. Silene latifolia

5. Silene virginica

2. Lychnis chalcedonica

4. Silene drummondii

2

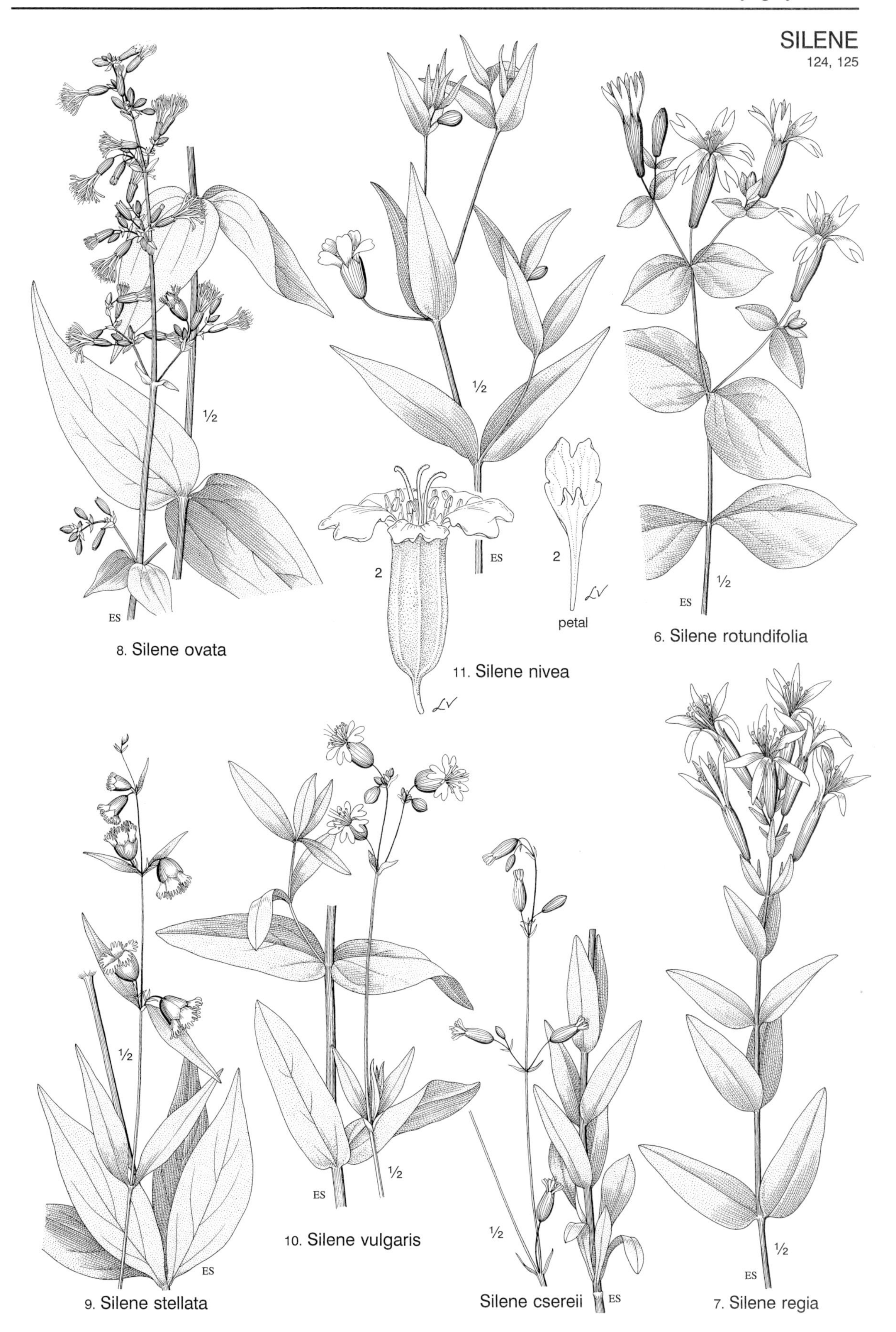

8. Silene ovata

11. Silene nivea

petal

2

6. Silene rotundifolia

9. Silene stellata

10. Silene vulgaris

Silene csereii

7. Silene regia

SILENE
125, 126

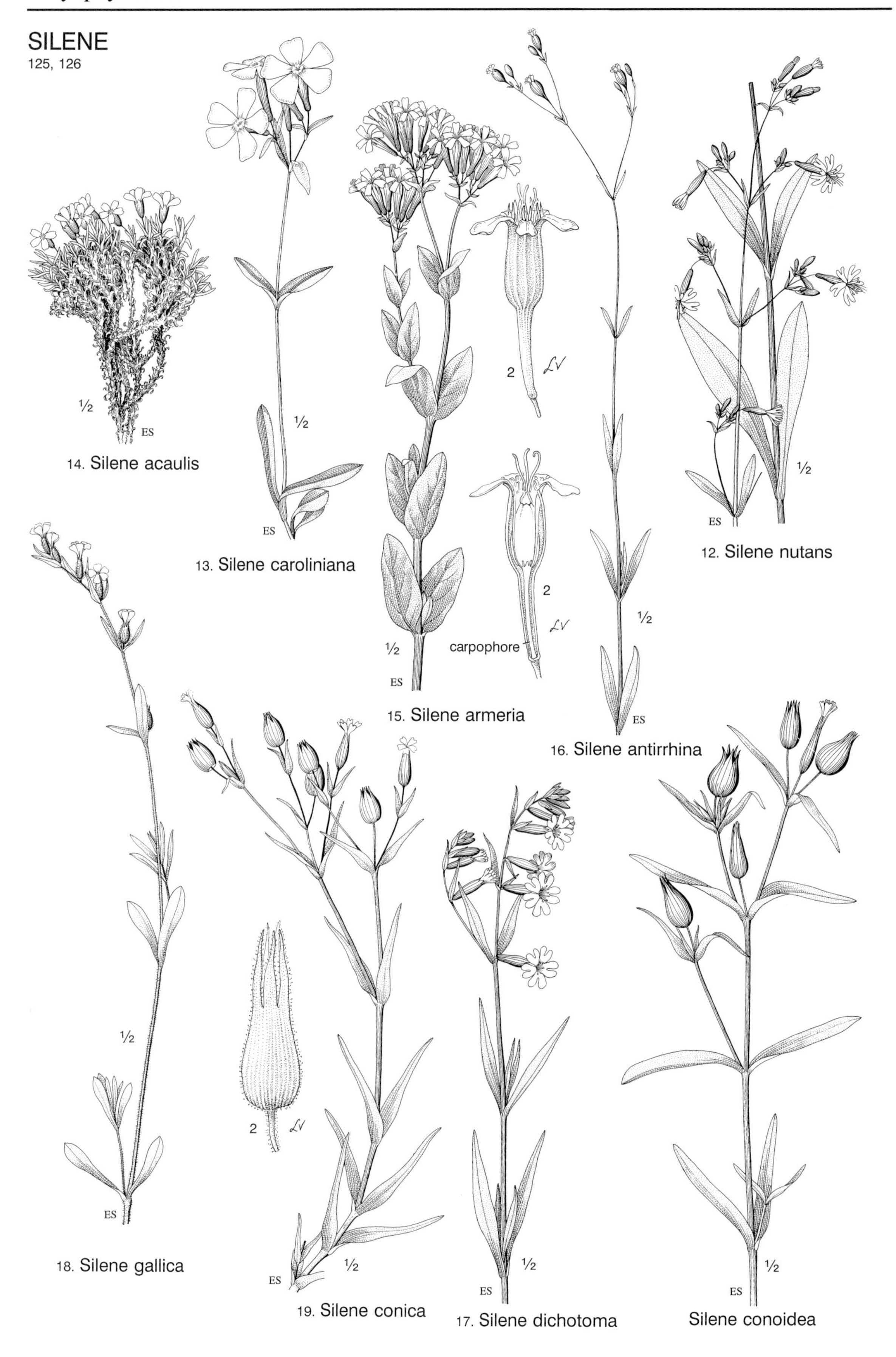

14. Silene acaulis

13. Silene caroliniana

12. Silene nutans

2

carpophore

15. Silene armeria

16. Silene antirrhina

18. Silene gallica

2

19. Silene conica

17. Silene dichotoma

Silene conoidea

SAPONARIA
VACCARIA
GYPSOPHILA
DIANTHUS
126, 127

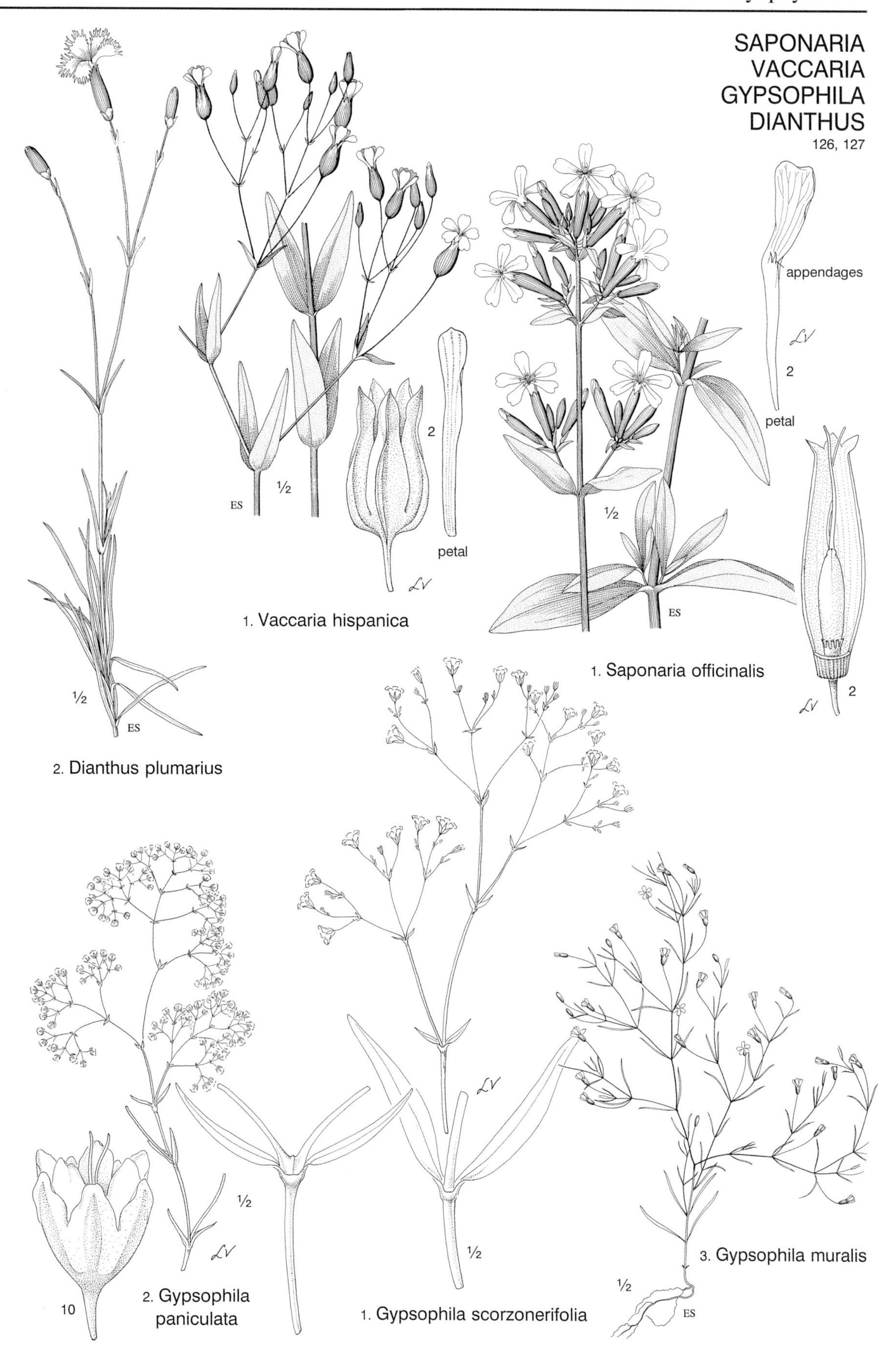

appendages

petal

2

petal

1. Vaccaria hispanica

1. Saponaria officinalis

2. Dianthus plumarius

2. Gypsophila
paniculata

1. Gypsophila scorzonerifolia

3. Gypsophila muralis

DIANTHUS
PETRORHAGIA
ERIOGONUM
127–129

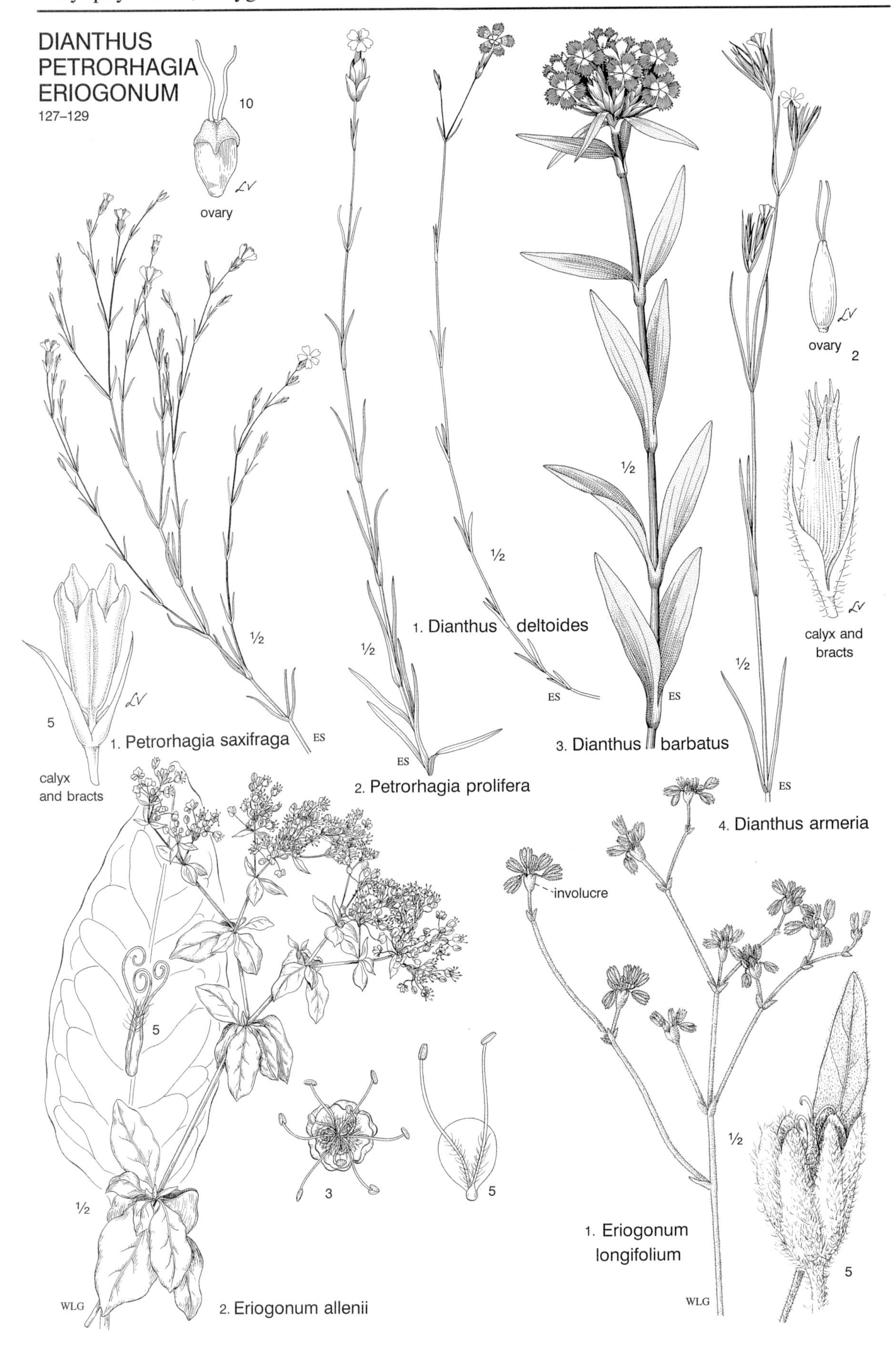

10

ovary

ovary 2

½

½

½

½

½

½

½

5

calyx and bracts

calyx and bracts

1. Petrorhagia saxifraga   ES

ES

2. Petrorhagia prolifera

1. Dianthus deltoides

ES

ES

3. Dianthus barbatus

ES

4. Dianthus armeria

involucre

5

3

5

½

½

1. Eriogonum longifolium

WLG

2. Eriogonum allenii

WLG

5

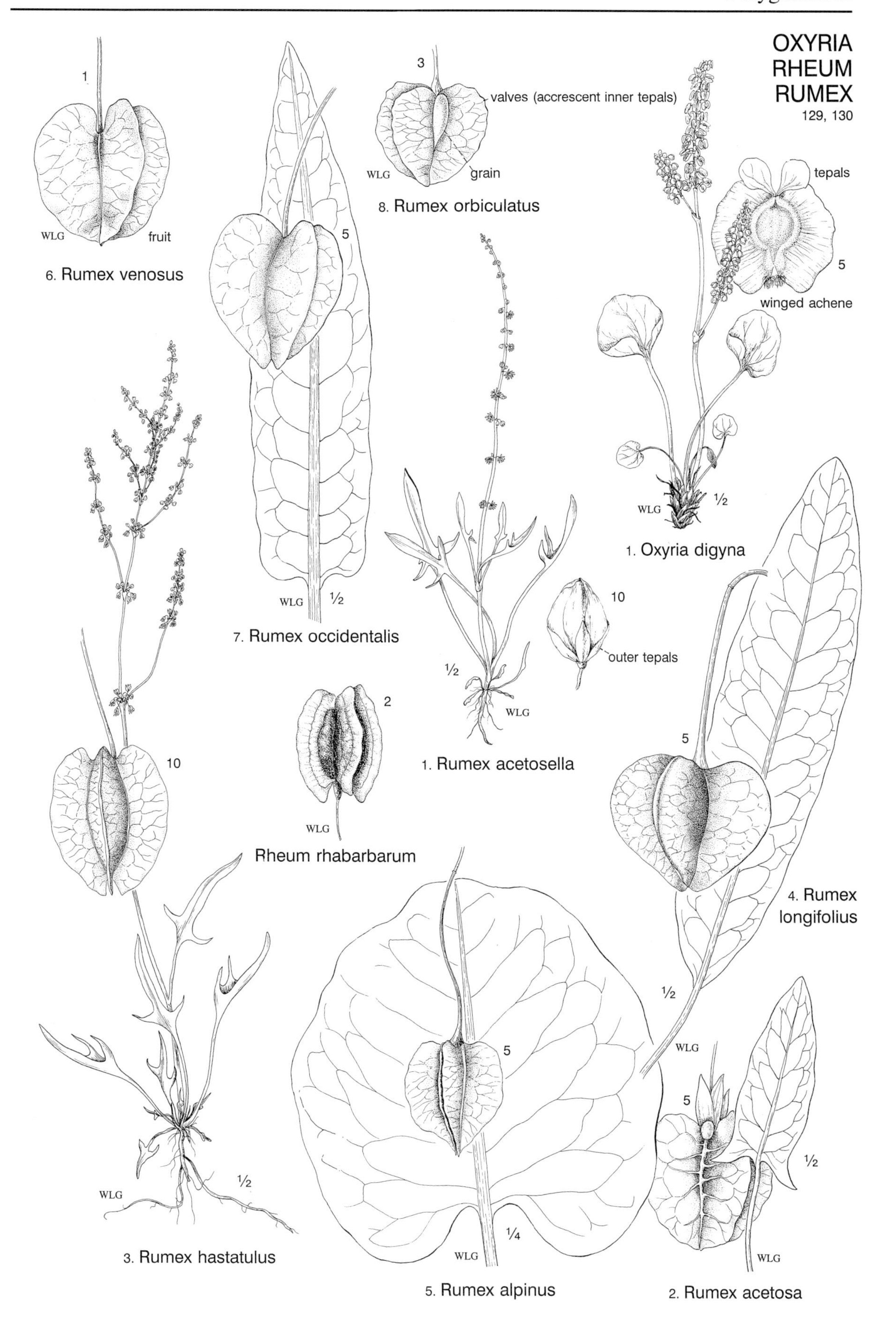

6. Rumex venosus

WLG   fruit

valves (accrescent inner tepals)

WLG   grain

8. Rumex orbiculatus

tepals

5

winged achene

1. Oxyria digyna

WLG   ½

5

WLG   ½

7. Rumex occidentalis

10

Rheum rhabarbarum

2

WLG

½

WLG

1. Rumex acetosella

10

outer tepals

5

4. Rumex
longifolius

½

WLG

3. Rumex hastatulus

WLG   ½

5

WLG   ¼

5. Rumex alpinus

5

WLG   ½

2. Rumex acetosa

RUMEX
130, 131

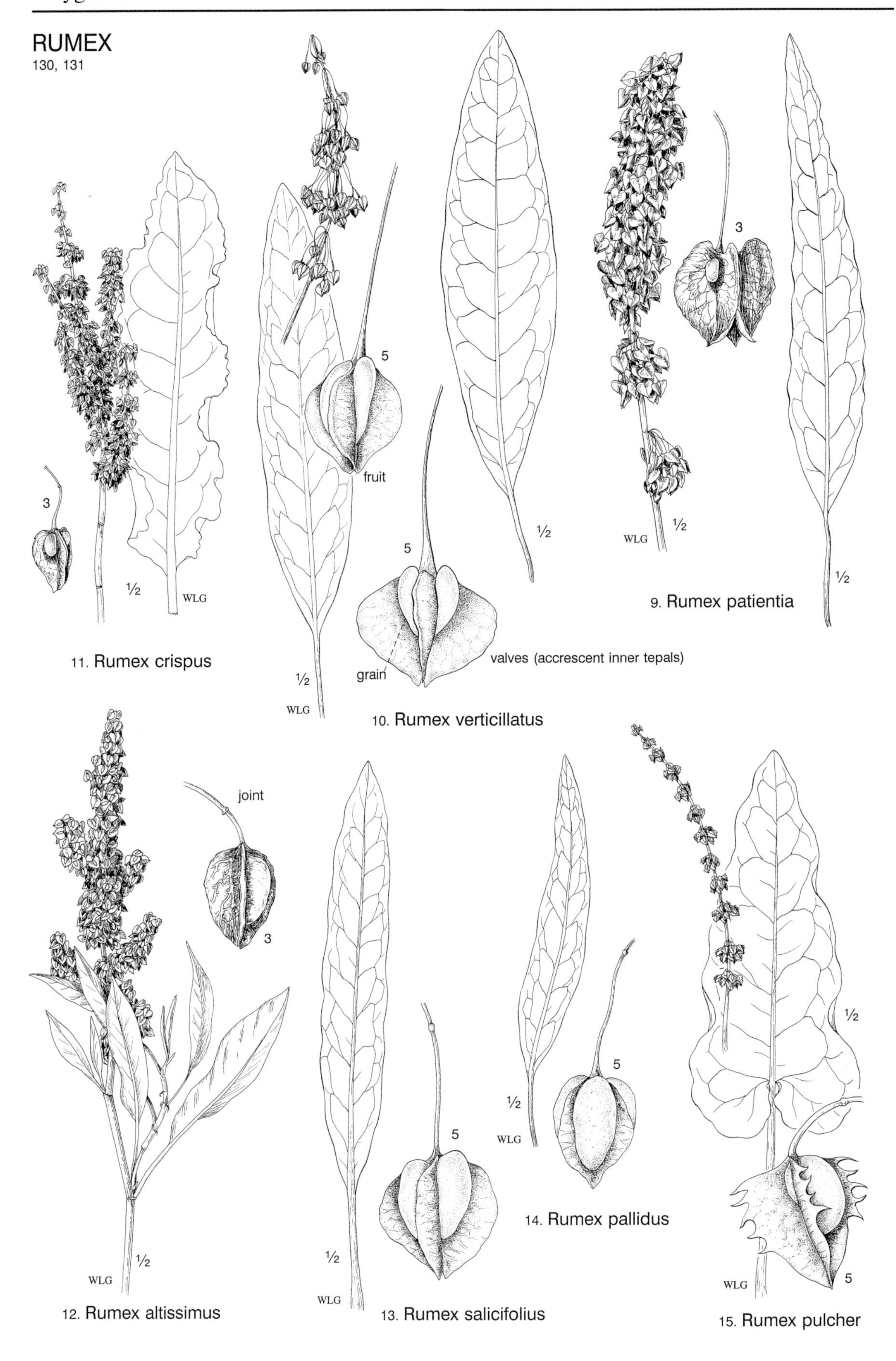

3

11. Rumex crispus

5

fruit

½

WLG

9. Rumex patientia

½

WLG

½

5

grain

valves (accrescent inner tepals)

10. Rumex verticillatus

½

WLG

joint

3

12. Rumex altissimus

½

WLG

13. Rumex salicifolius

½

WLG

5

½

WLG

14. Rumex pallidus

5

½

WLG

15. Rumex pulcher

5

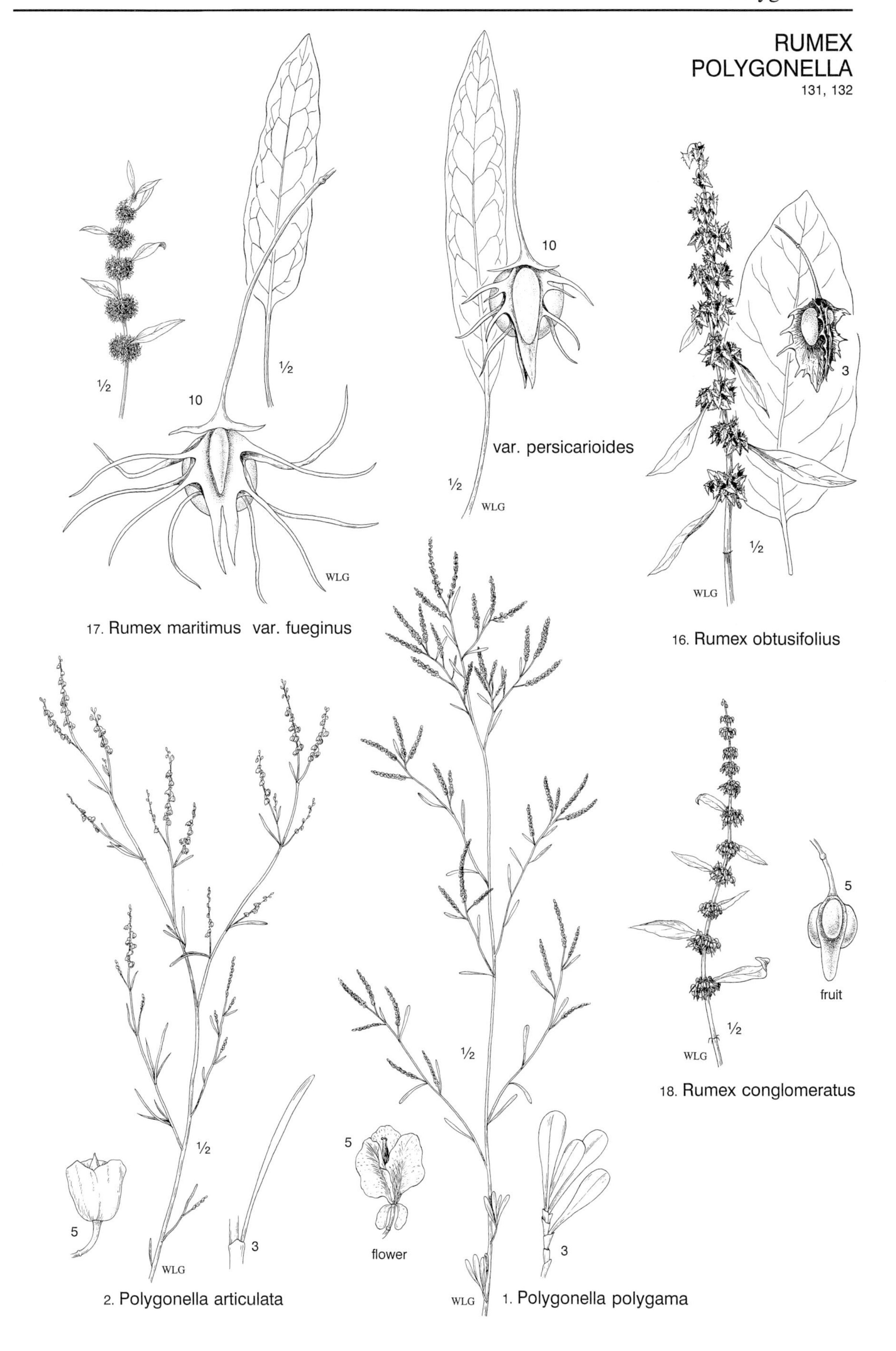

¹⁰ var. persicarioides

17. Rumex maritimus var. fueginus

16. Rumex obtusifolius

fruit

18. Rumex conglomeratus

2. Polygonella articulata

flower

1. Polygonella polygama

POLYGONUM
135

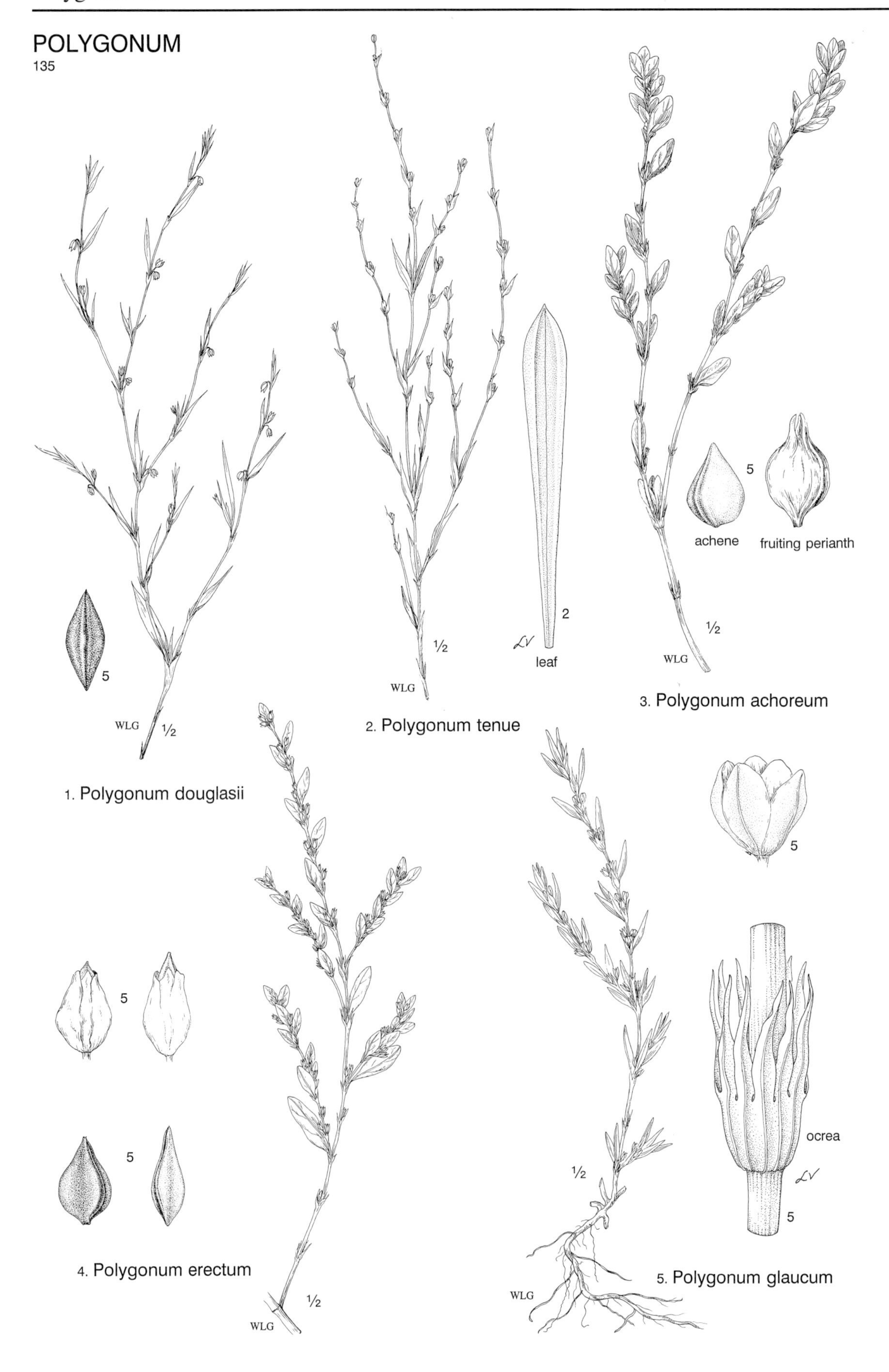

1. Polygonum douglasii

WLG   ½

5

2. Polygonum tenue

WLG   ½

2
leaf

3. Polygonum achoreum

5
achene   fruiting perianth

WLG   ½

4. Polygonum erectum

5

5

WLG   ½

5. Polygonum glaucum

5

ocrea

5

WLG   ½

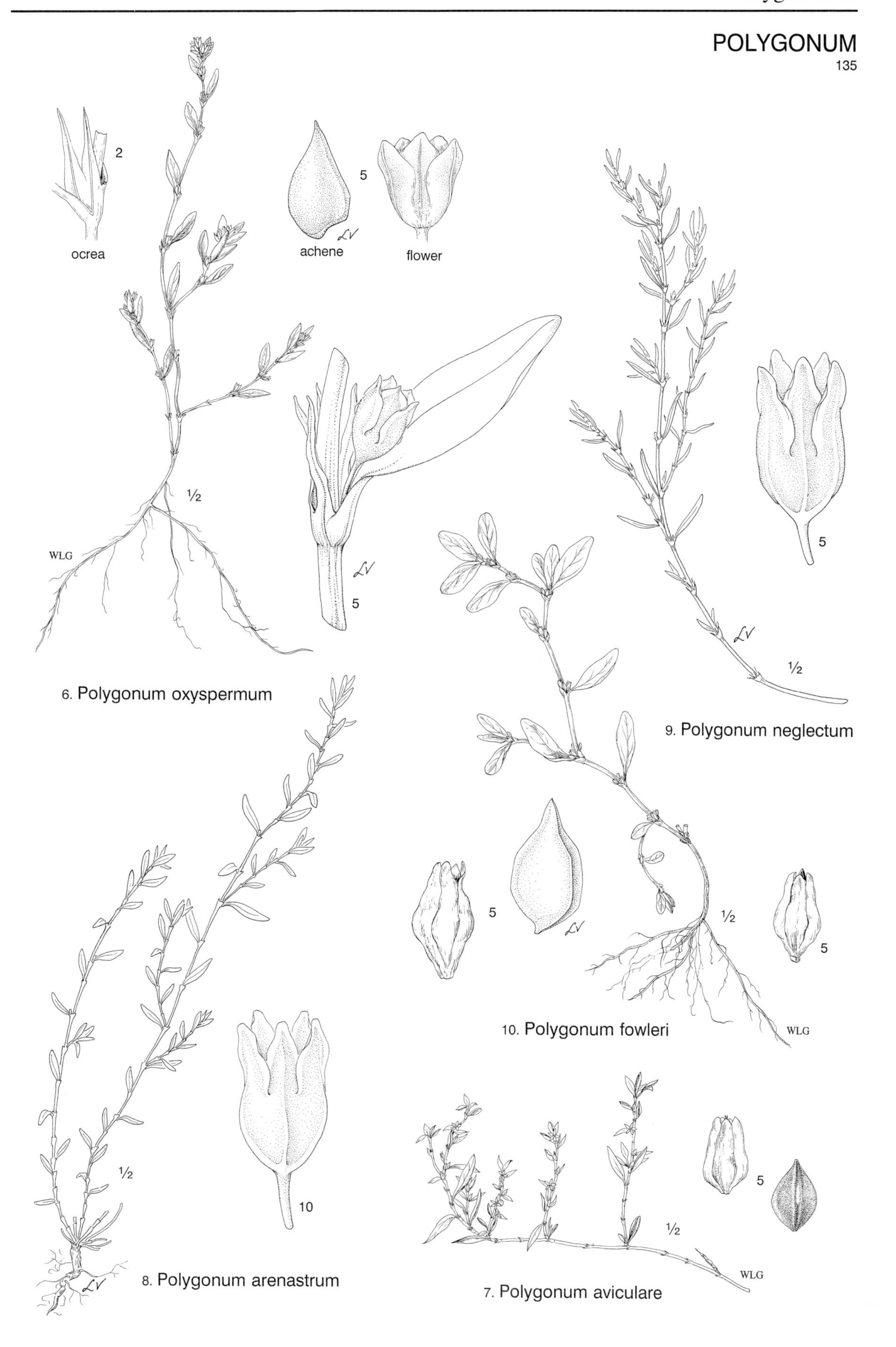

2

ocrea

5

achene

flower

½

WLG

5

6. Polygonum oxyspermum

5

9. Polygonum neglectum

½

5

10. Polygonum fowleri

WLG

½

10

8. Polygonum arenastrum

½

5

WLG

7. Polygonum aviculare

POLYGONUM
135, 136

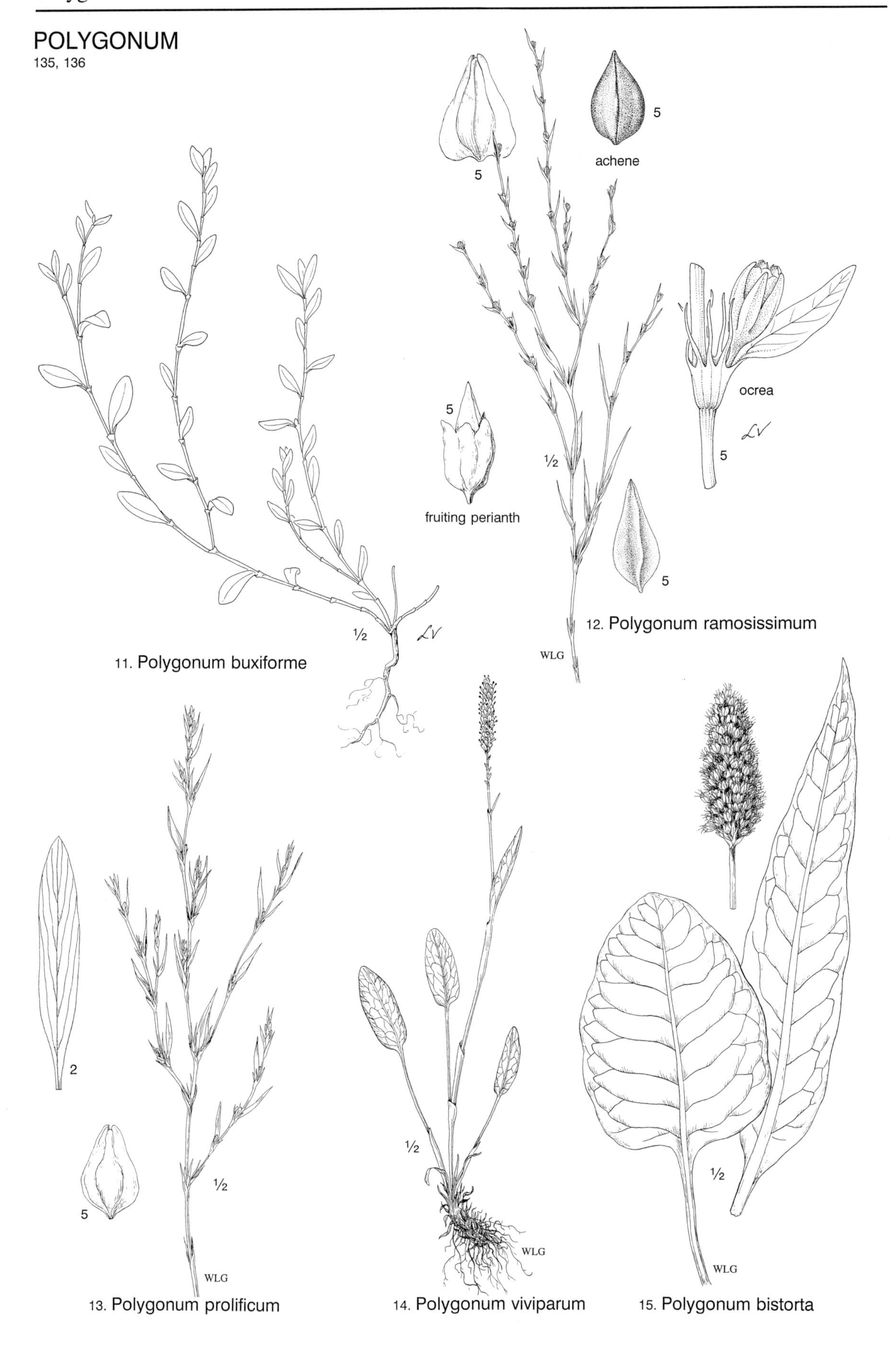

5

achene

5

5

fruiting perianth

½

ocrea

5

½

WLG

12. Polygonum ramosissimum

11. Polygonum buxiforme

½

2

5

½

WLG

13. Polygonum prolificum

½

WLG

14. Polygonum viviparum

½

WLG

15. Polygonum bistorta

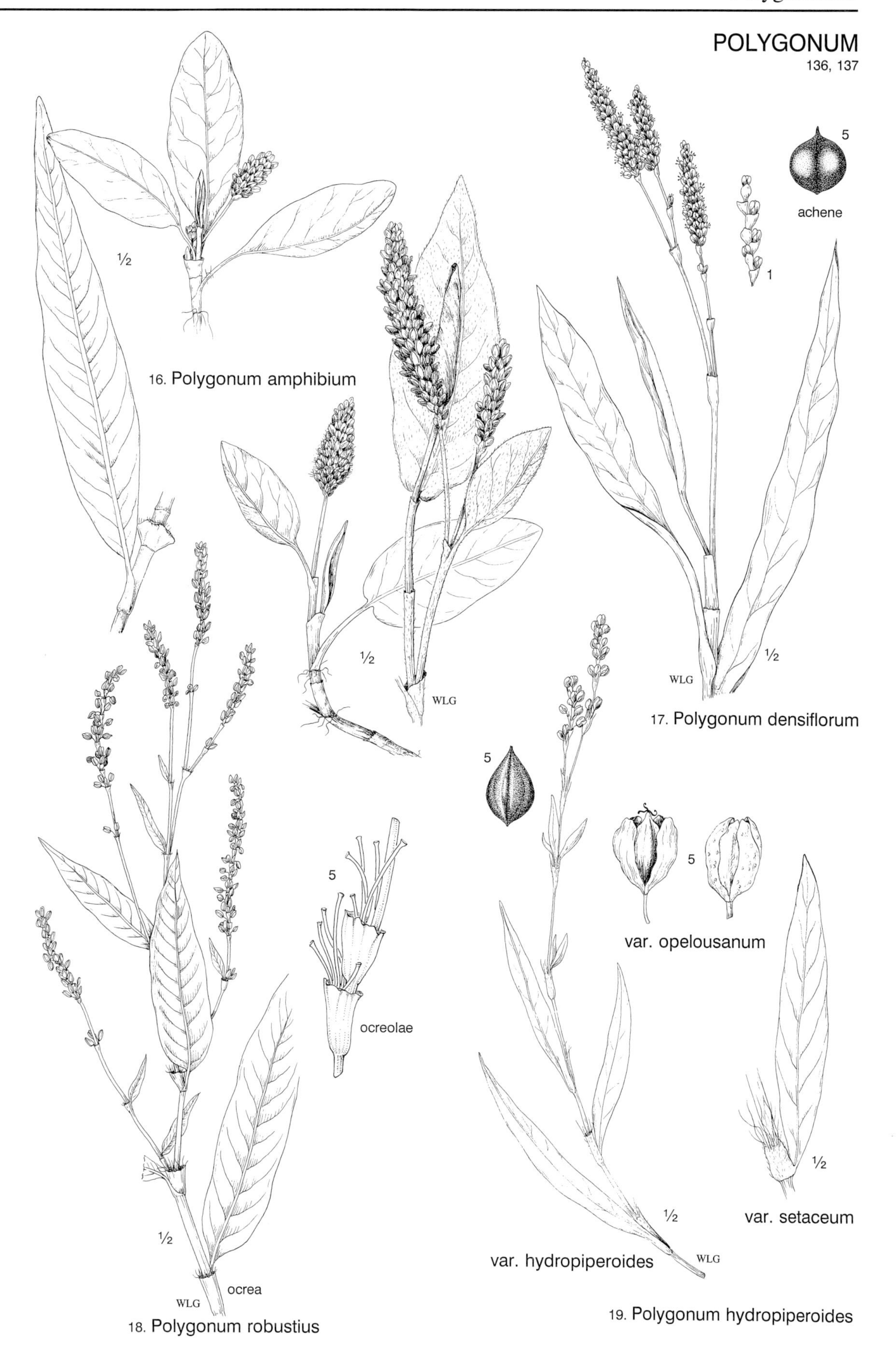

16. Polygonum amphibium

5
achene

1

17. Polygonum densiflorum

WLG

5
ocreolae

18. Polygonum robustius

ocrea
WLG

5

var. opelousanum

var. hydropiperoides

var. setaceum

19. Polygonum hydropiperoides

POLYGONUM
137

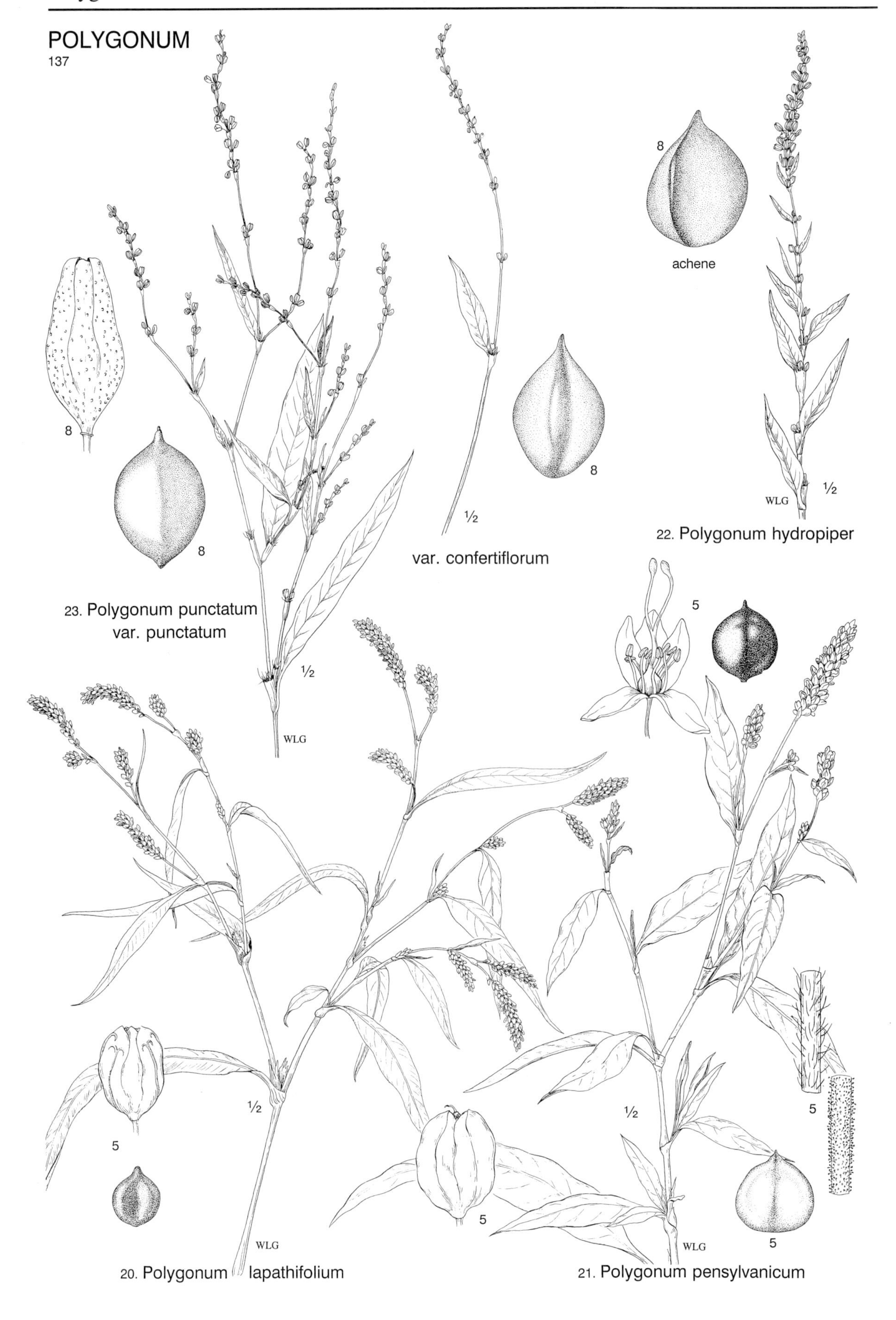

achene

8

½

var. confertiflorum

22. Polygonum hydropiper

8

23. Polygonum punctatum
var. punctatum

WLG

5

20. Polygonum lapathifolium

21. Polygonum pensylvanicum

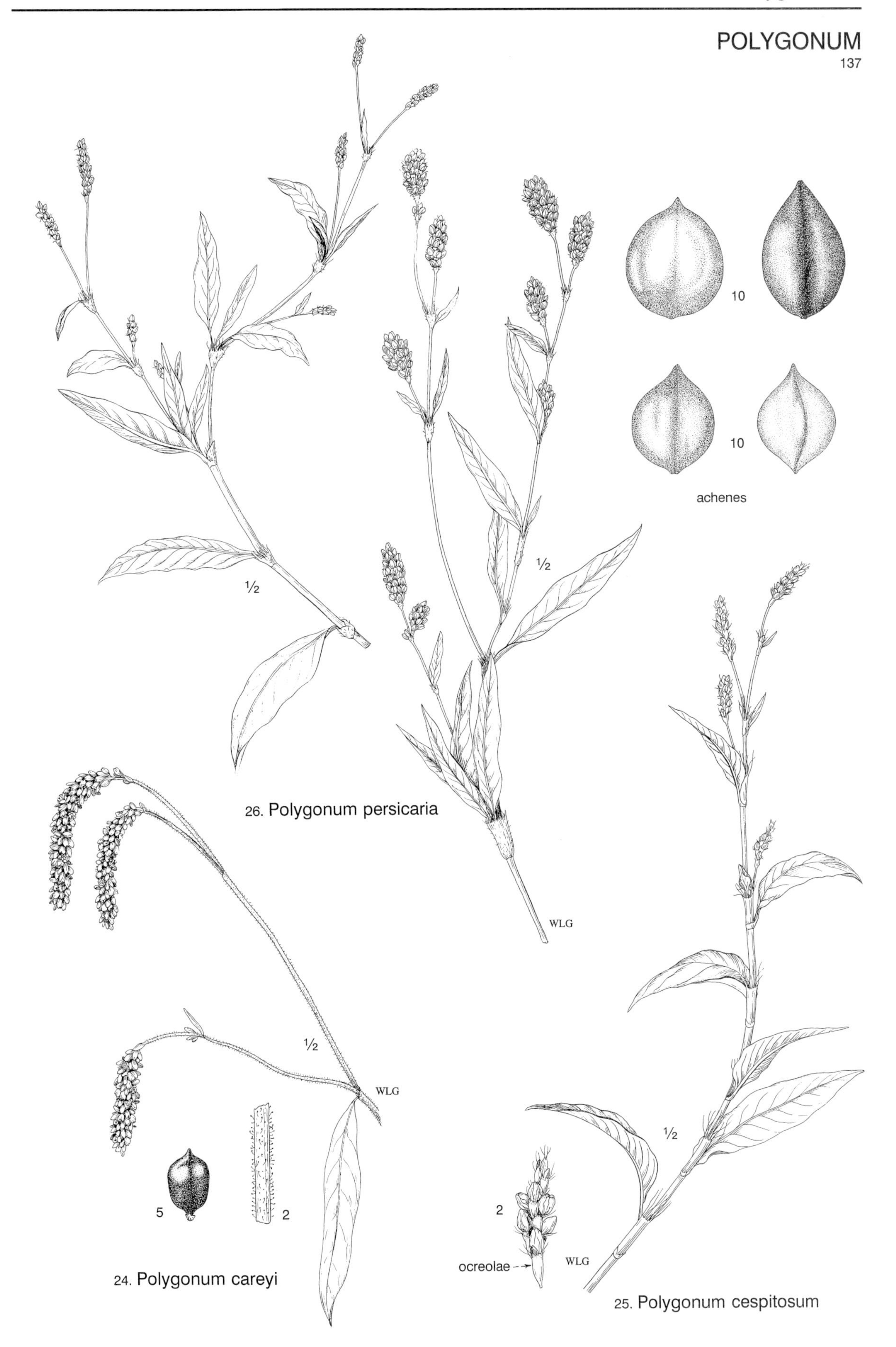

10

10

achenes

26. Polygonum persicaria

WLG

24. Polygonum careyi

5

2

½

WLG

2

ocreolae

WLG

25. Polygonum cespitosum

POLYGONUM
138

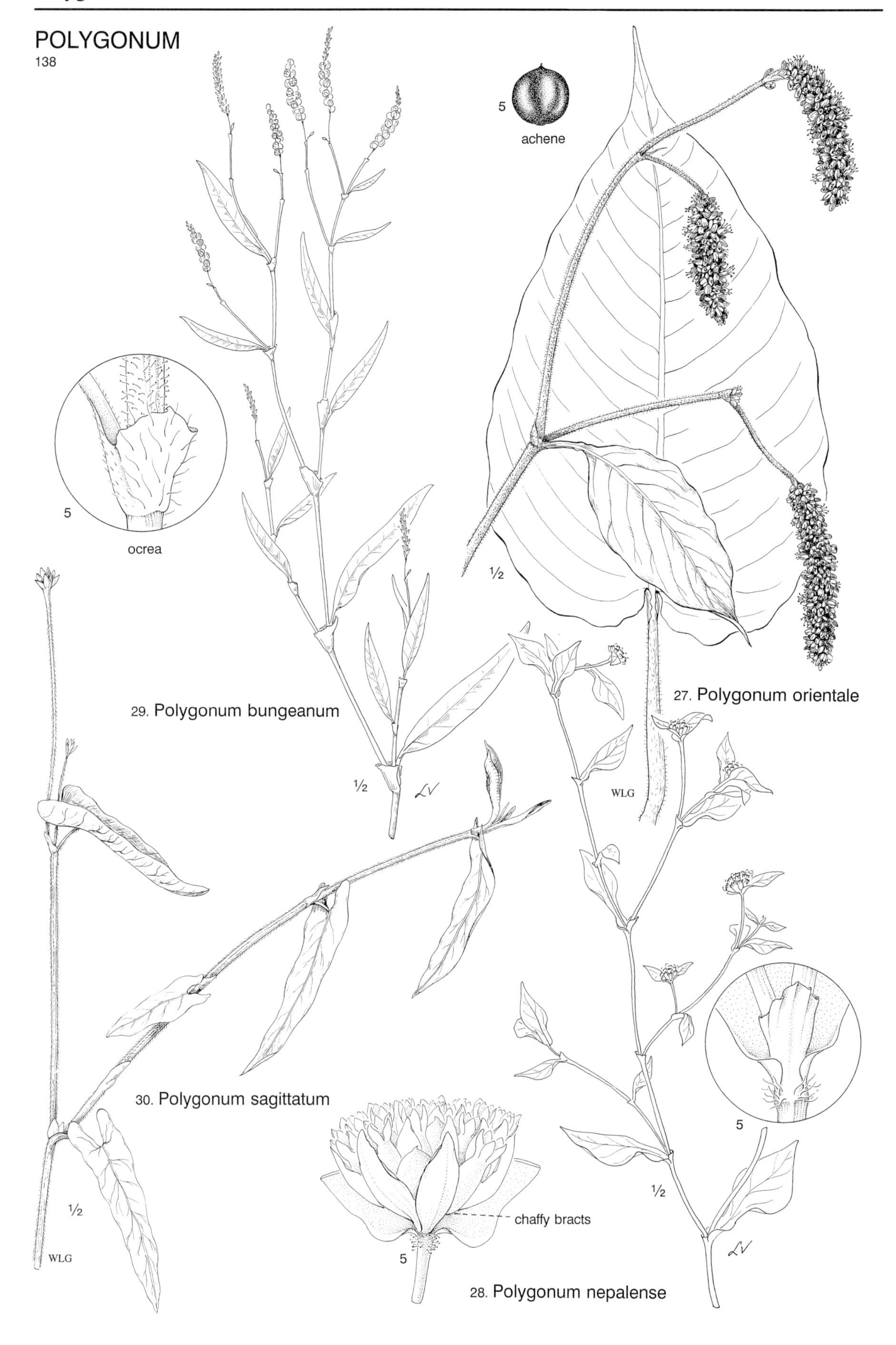

5
achene

5
ocrea

29. Polygonum bungeanum

½

27. Polygonum orientale

½

WLG

30. Polygonum sagittatum

½

WLG

chaffy bracts

5

28. Polygonum nepalense

½

5

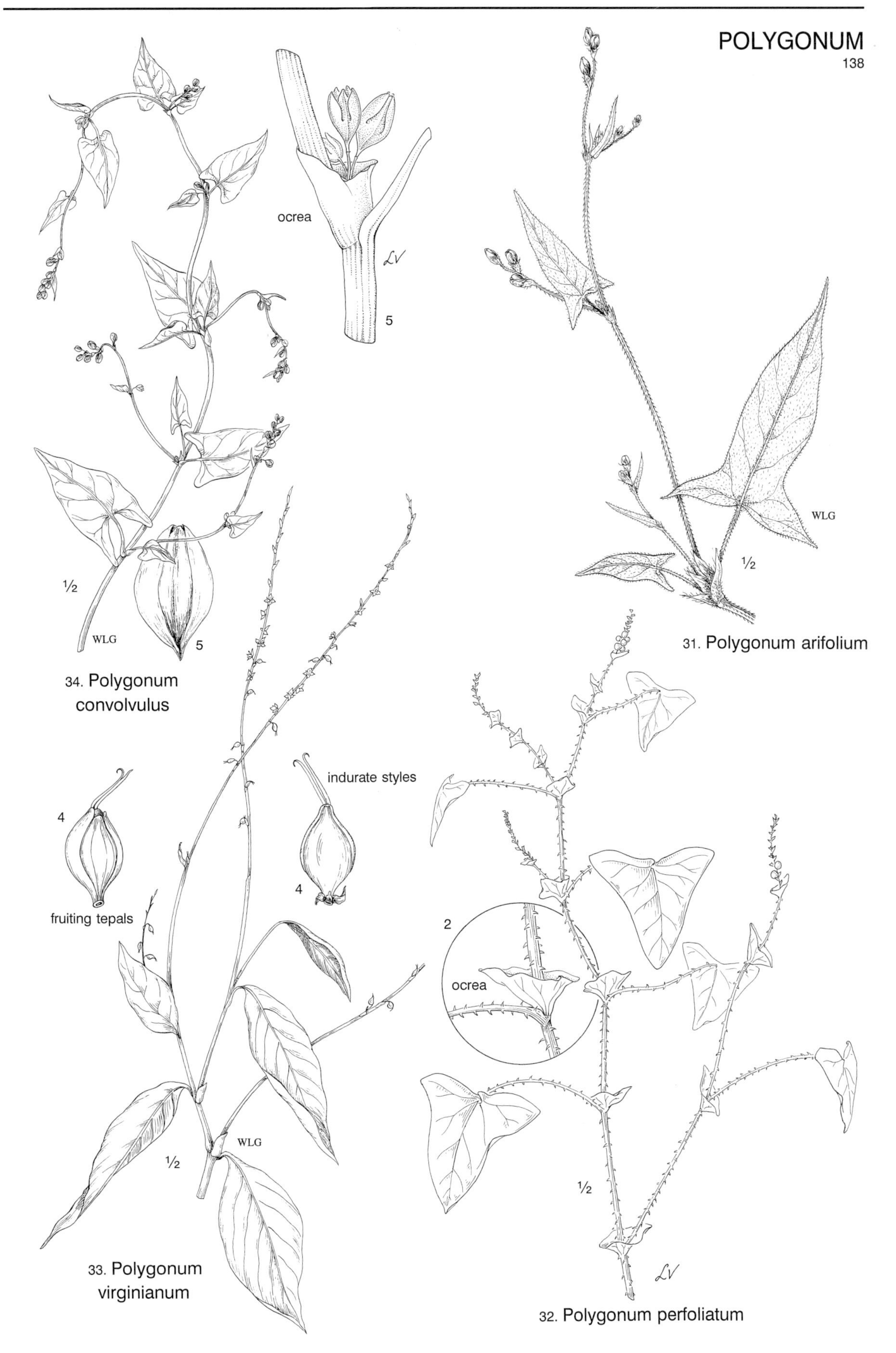

ocrea

5

34. Polygonum
convolvulus

31. Polygonum arifolium

indurate styles

4

fruiting tepals

4

2

ocrea

33. Polygonum
virginianum

32. Polygonum perfoliatum

POLYGONUM
FAGOPYRUM
BRUNNICHIA
138, 139

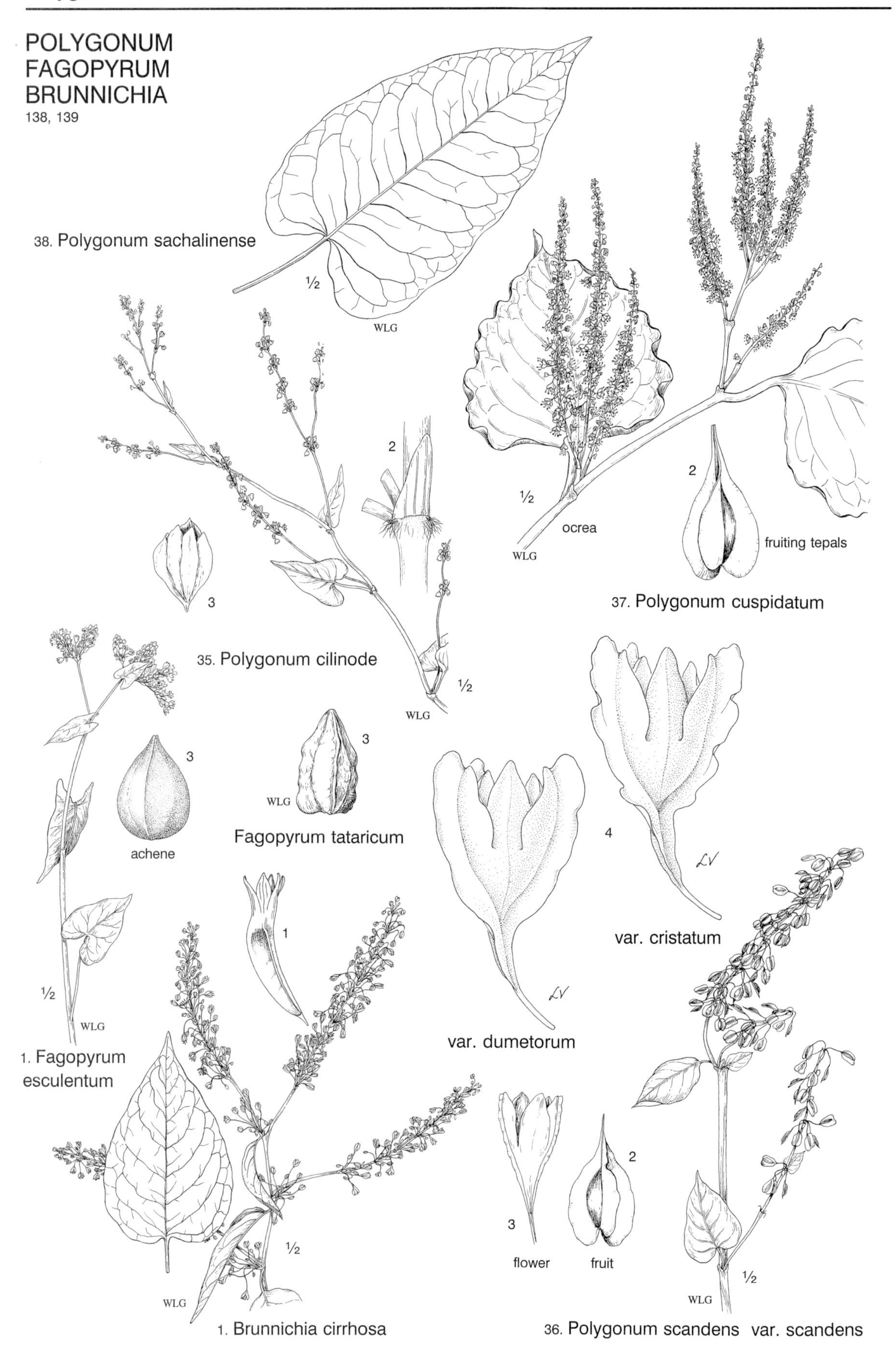

38. Polygonum sachalinense

½

WLG

37. Polygonum cuspidatum

½

WLG

ocrea

2

fruiting tepals

35. Polygonum cilinode

3

2

½

WLG

Fagopyrum tataricum

3

WLG

achene

3

var. cristatum

4

var. dumetorum

1. Fagopyrum
esculentum

½

WLG

1

flower

fruit

3

2

1. Brunnichia cirrhosa

½

WLG

36. Polygonum scandens var. scandens

½

WLG

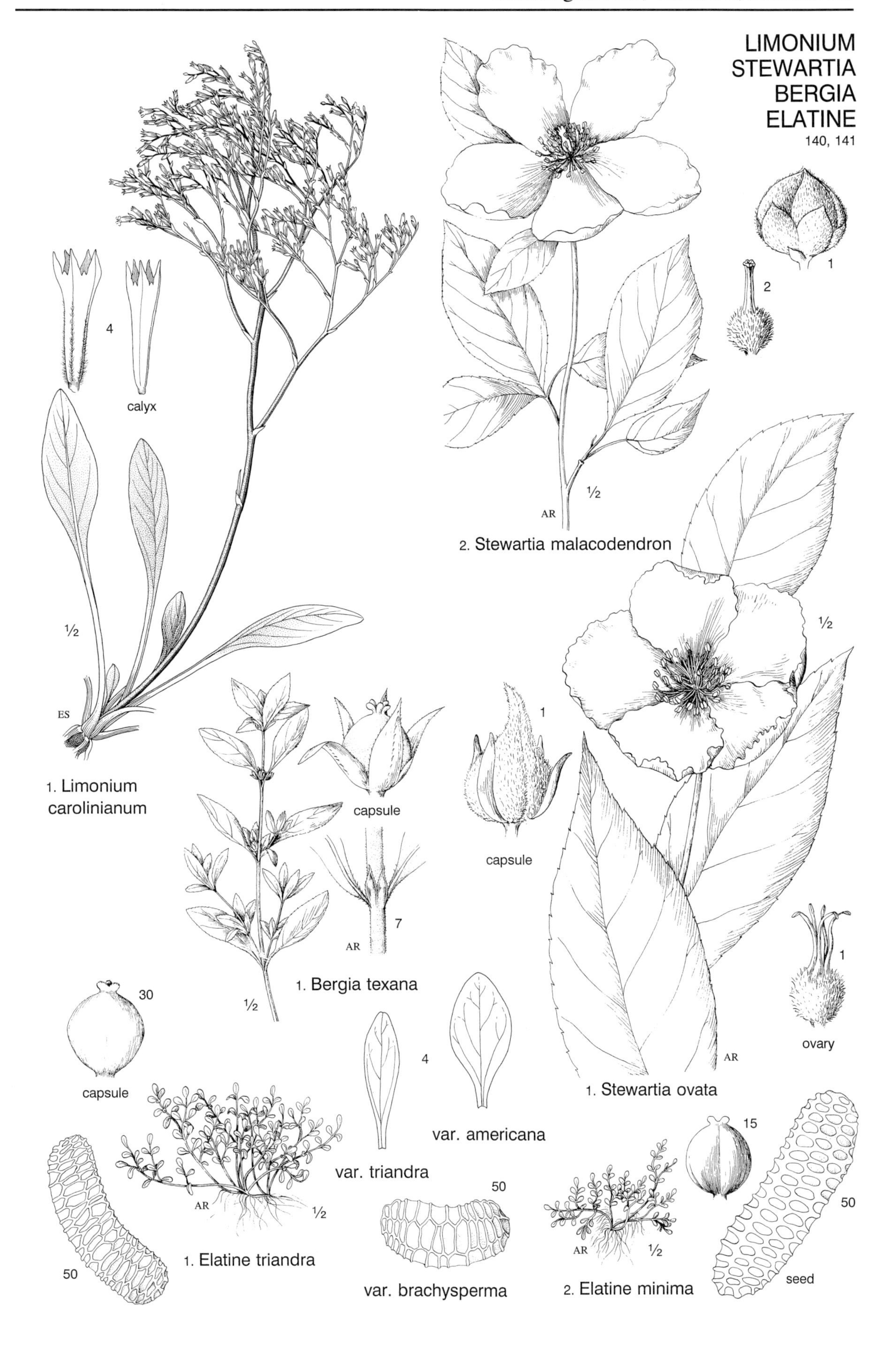

4

calyx

½

ES

1. Limonium
carolinianum

2

1

½

AR

2. Stewartia malacodendron

capsule

7

AR

1. Bergia texana

½

30

capsule

½

4

var. triandra

var. americana

1

capsule

½

1

ovary

AR

1. Stewartia ovata

15

50

50

var. brachysperma

AR

½

2. Elatine minima

50

seed

AR

½

1. Elatine triandra

50

HYPERICUM
142, 143

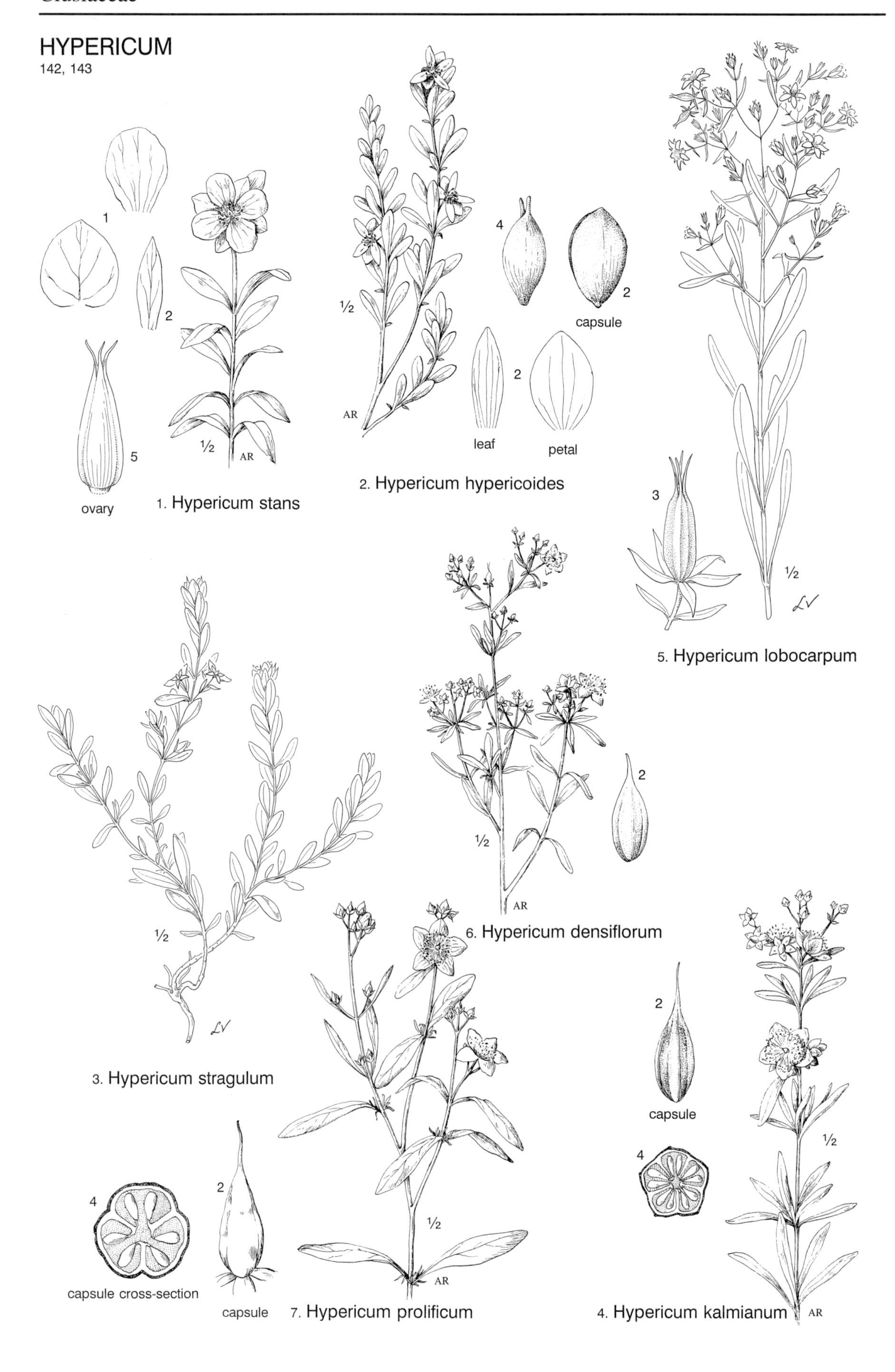

1. Hypericum stans

ovary

5 ovary

1

2

2. Hypericum hypericoides

capsule

4

2

leaf

2

petal

2

5. Hypericum lobocarpum

3

½

6. Hypericum densiflorum

2

AR

3. Hypericum stragulum

½

½

AR

4

capsule cross-section

2

capsule

7. Hypericum prolificum

½

AR

2

capsule

4

4. Hypericum kalmianum

½

AR

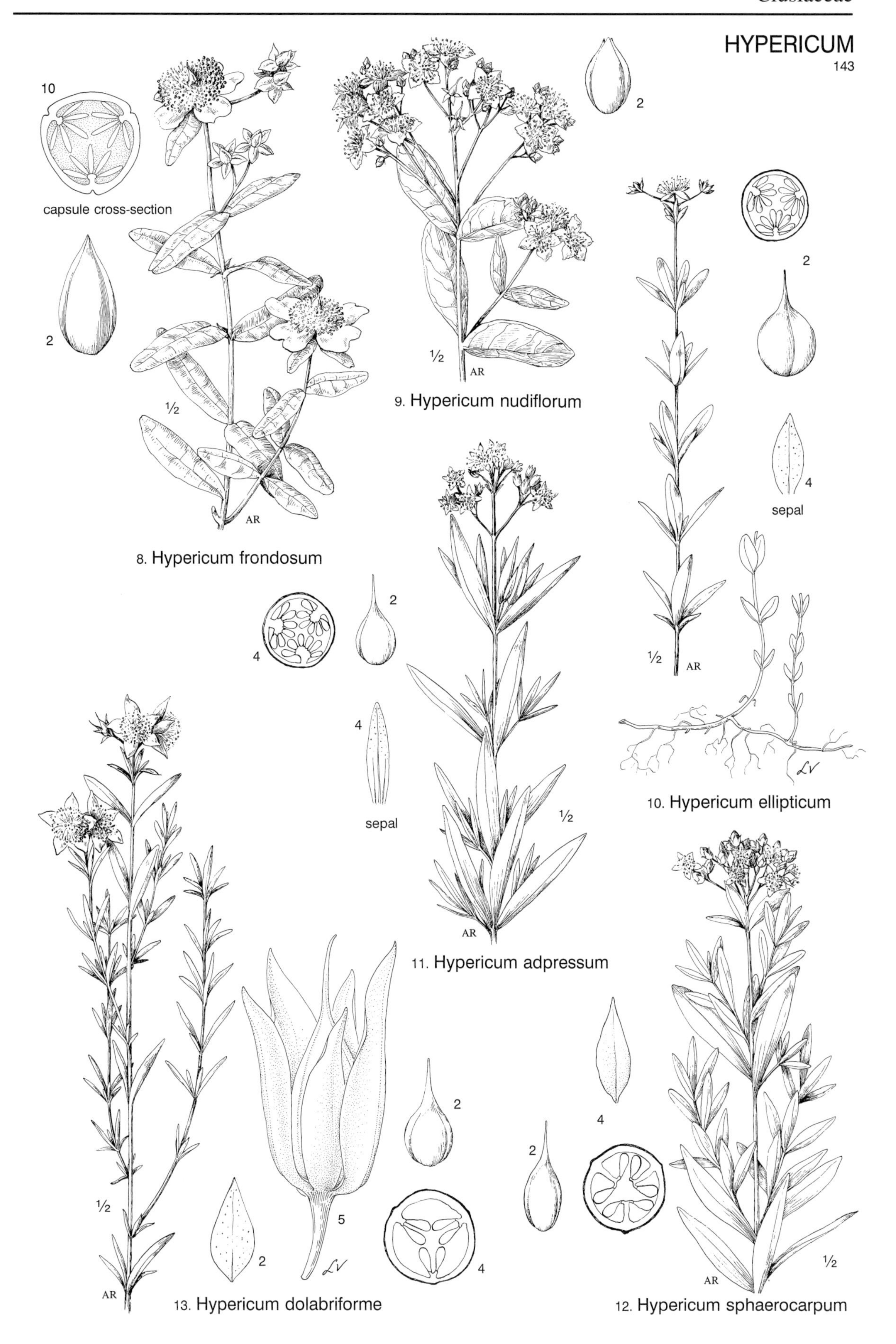

10
capsule cross-section

2

8. Hypericum frondosum

9. Hypericum nudiflorum

2

4
sepal

10. Hypericum ellipticum

4

2

4
sepal

11. Hypericum adpressum

13. Hypericum dolabriforme

12. Hypericum sphaerocarpum

HYPERICUM
144

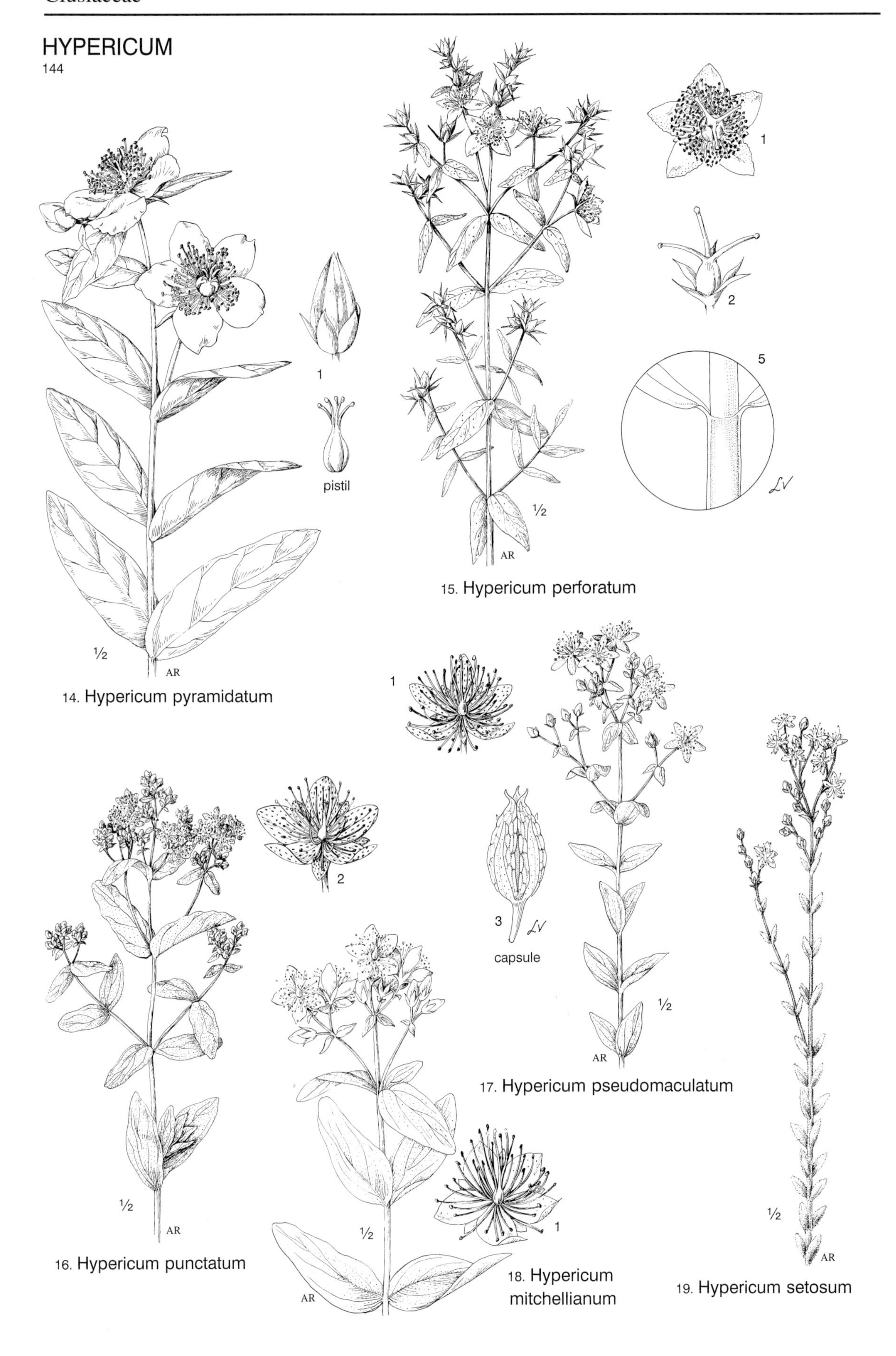

1

pistil

14. Hypericum pyramidatum

15. Hypericum perforatum

1

2

5

ℒⱱ

1

3 ℒⱱ
capsule

17. Hypericum pseudomaculatum

½

2

16. Hypericum punctatum

1

18. Hypericum mitchellianum

19. Hypericum setosum

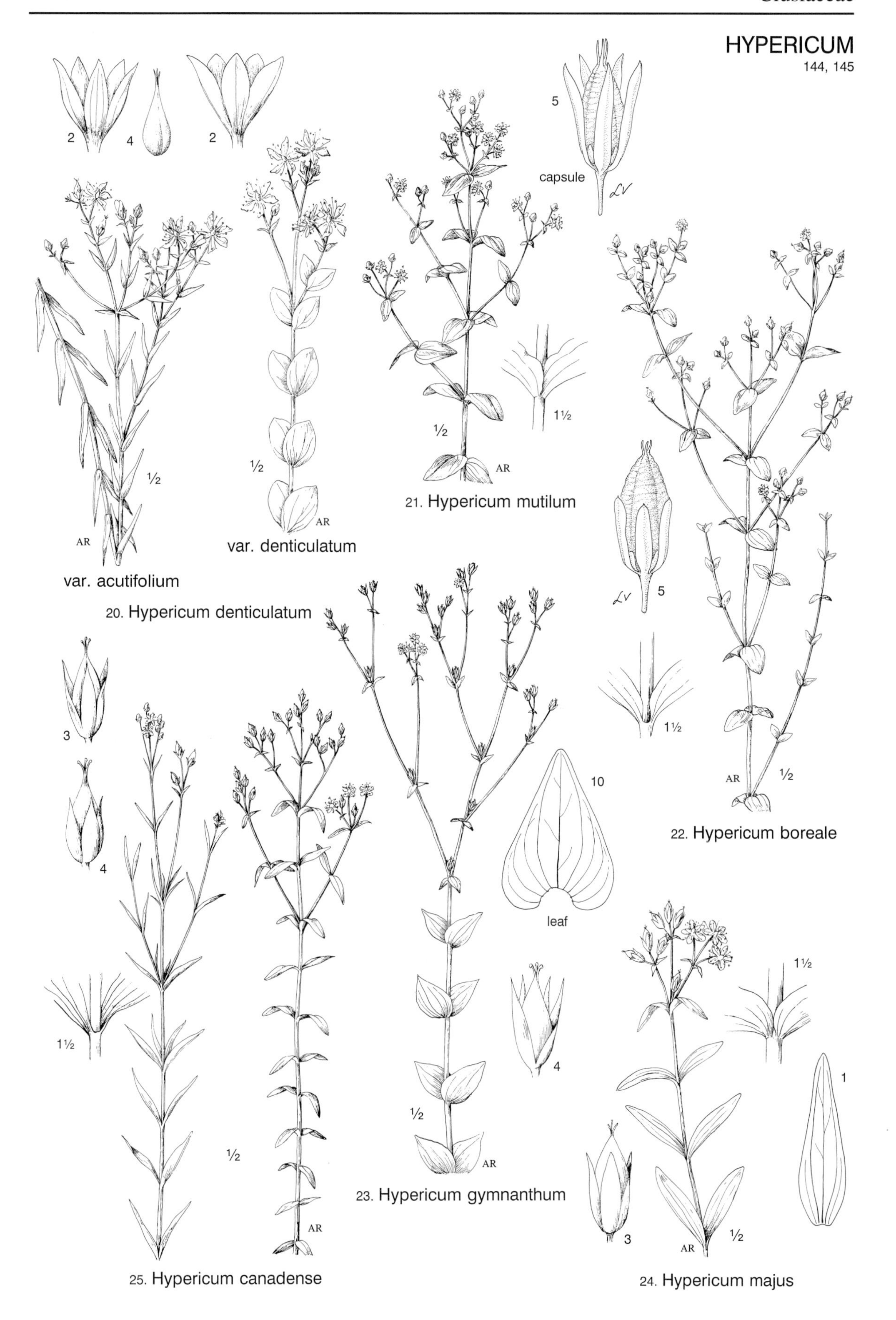

5 capsule

2   4   2

½   ½

AR

var. acutifolium

20. Hypericum denticulatum

var. denticulatum

AR

1½

AR

21. Hypericum mutilum

5

1½

AR   ½

22. Hypericum boreale

3

4

1½

½   ½

AR

25. Hypericum canadense

10

leaf

4

½

AR

23. Hypericum gymnanthum

1½

3   AR   ½

1

24. Hypericum majus

HYPERICUM
TRIADENUM
145

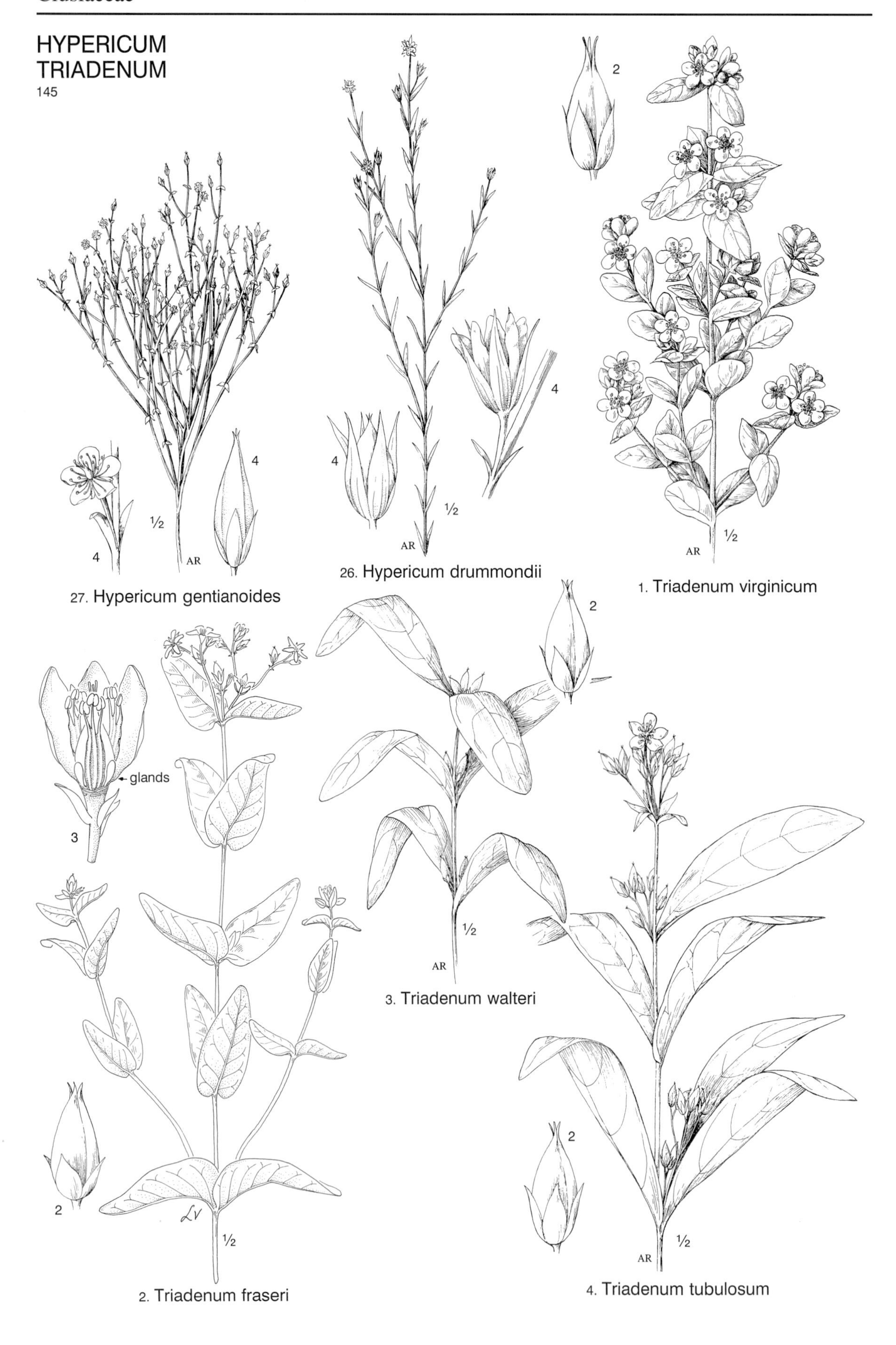

27. Hypericum gentianoides

26. Hypericum drummondii

1. Triadenum virginicum

glands

3. Triadenum walteri

2. Triadenum fraseri

4. Triadenum tubulosum

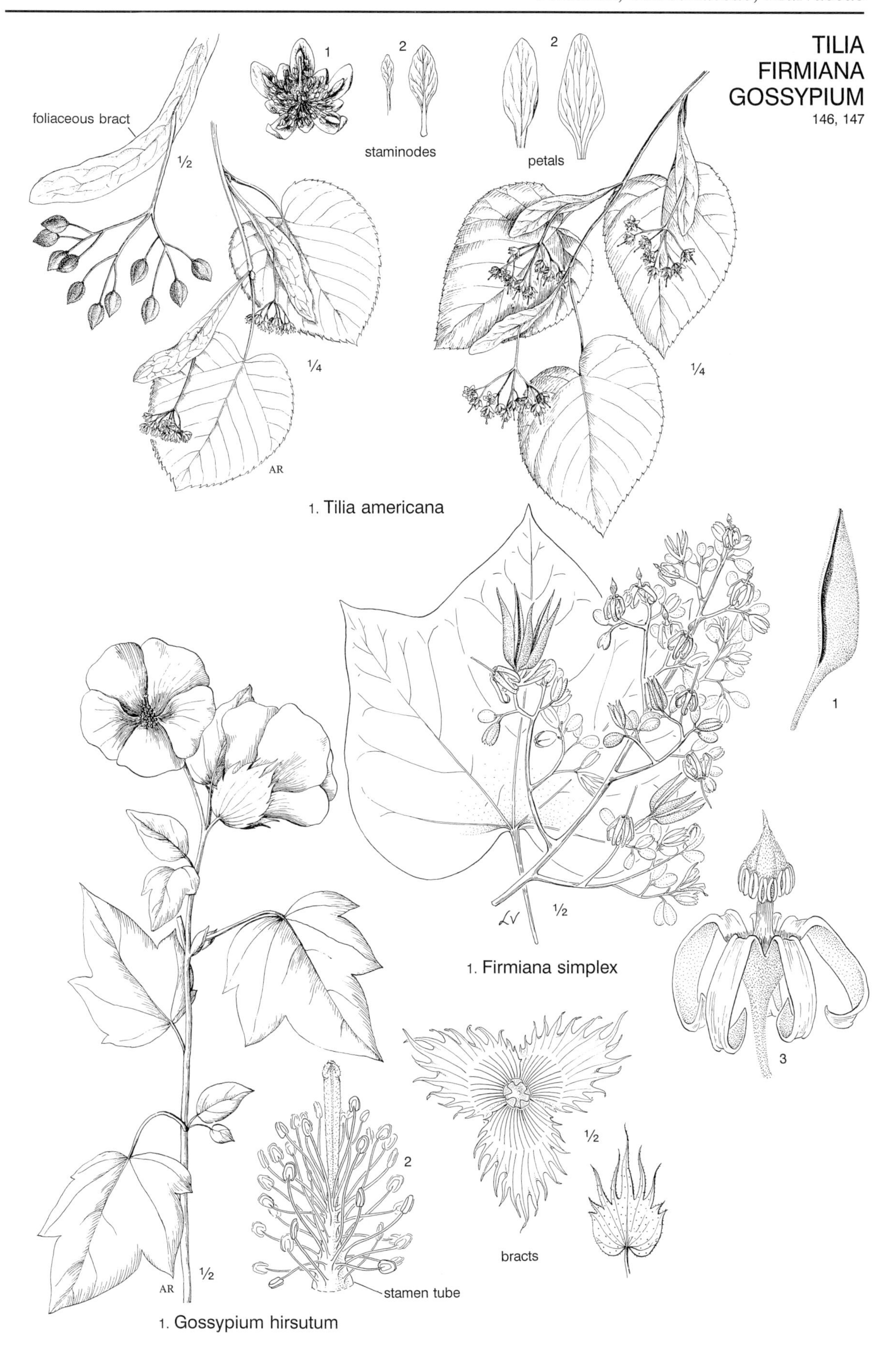

foliaceous bract

1/2

1

2

staminodes

2

petals

1/4

AR

1/4

1. Tilia americana

1/2

1

1. Firmiana simplex

3

2

stamen tube

AR

1/2

bracts

1/2

1. Gossypium hirsutum

HIBISCUS
148

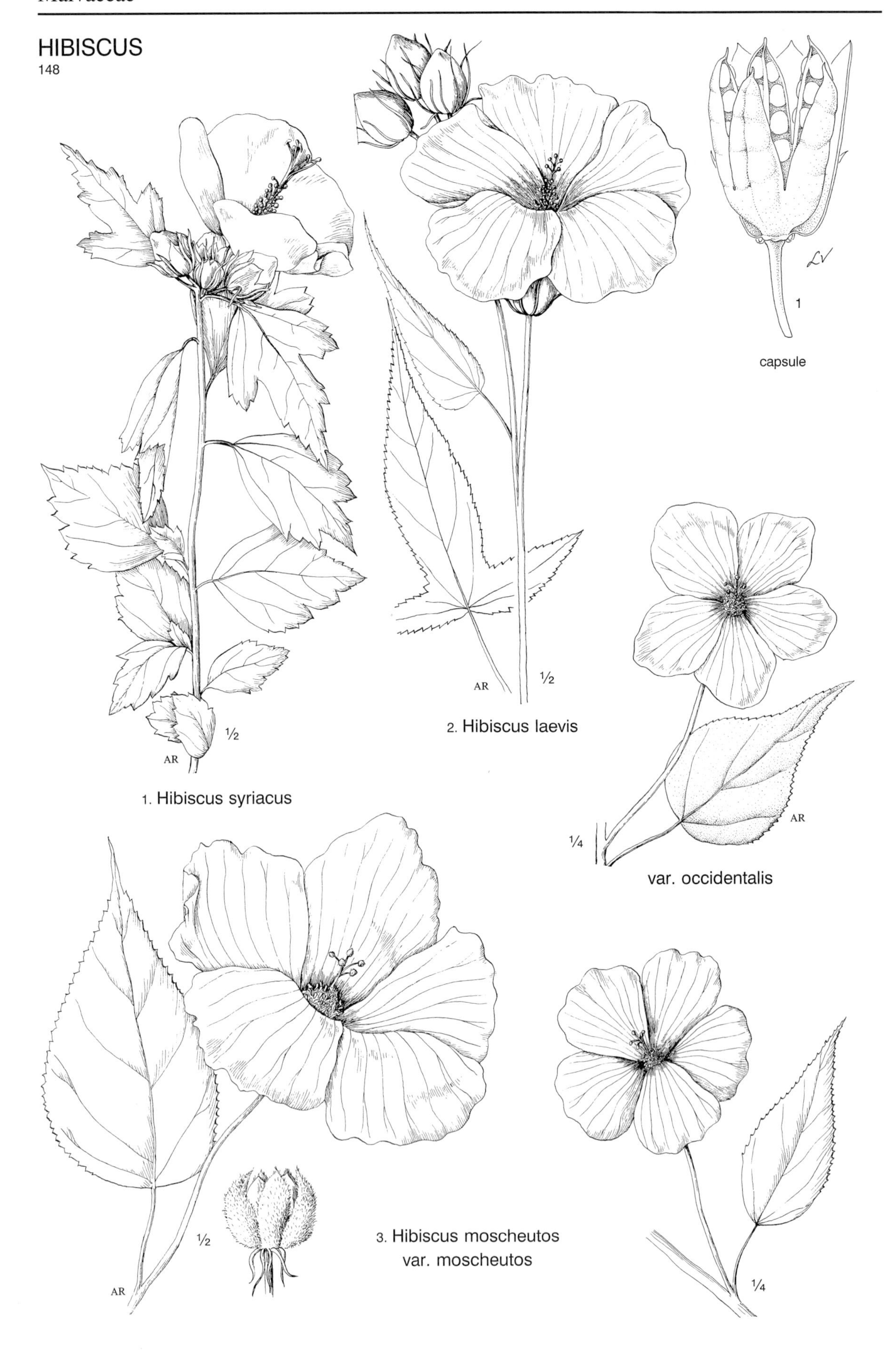

1

capsule

1. Hibiscus syriacus

2. Hibiscus laevis

var. occidentalis

3. Hibiscus moscheutos
var. moscheutos

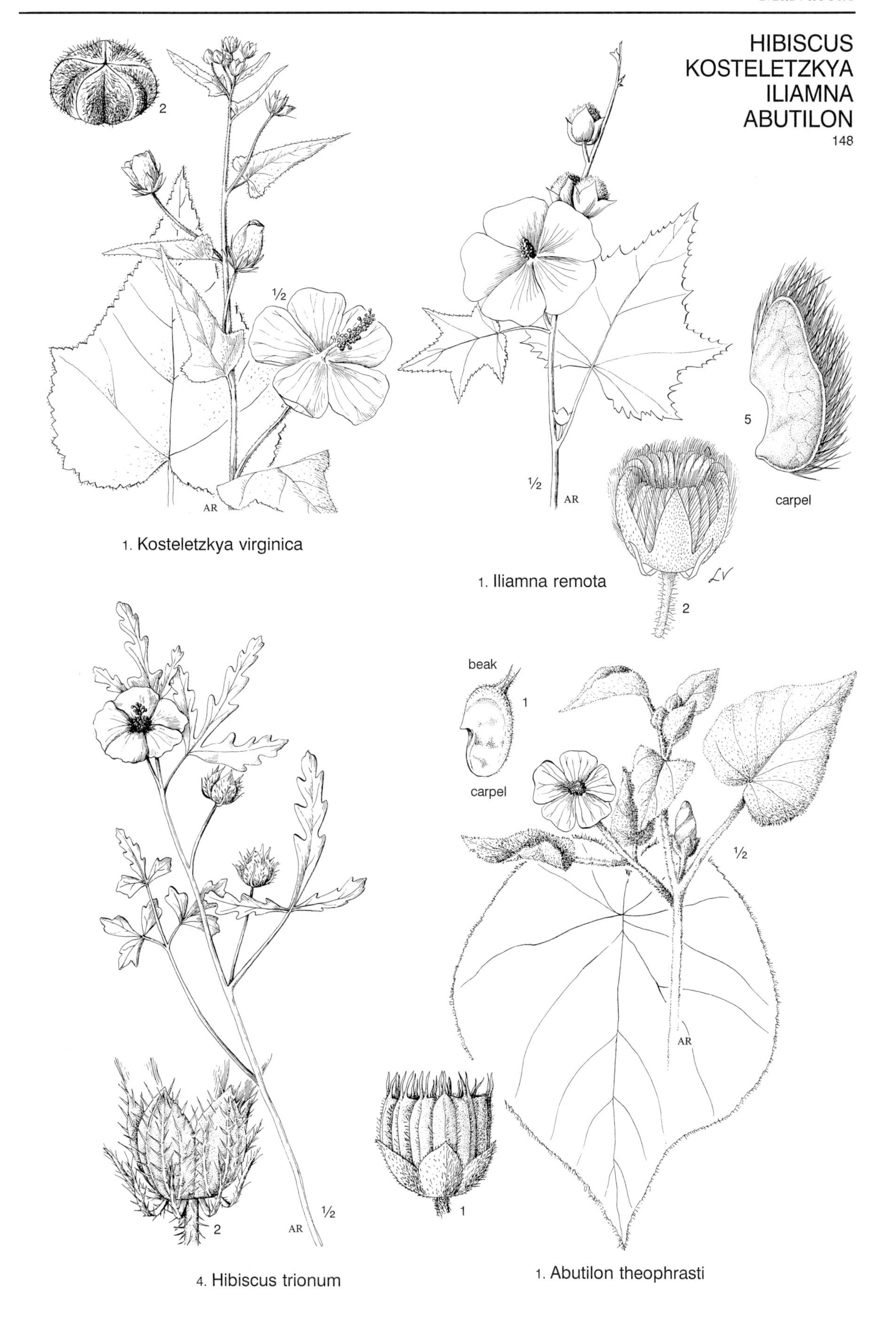

1. Kosteletzkya virginica

5 carpel

1. Iliamna remota

beak

carpel

4. Hibiscus trionum

1. Abutilon theophrasti

MODIOLA
ALTHAEA
MALVA
149

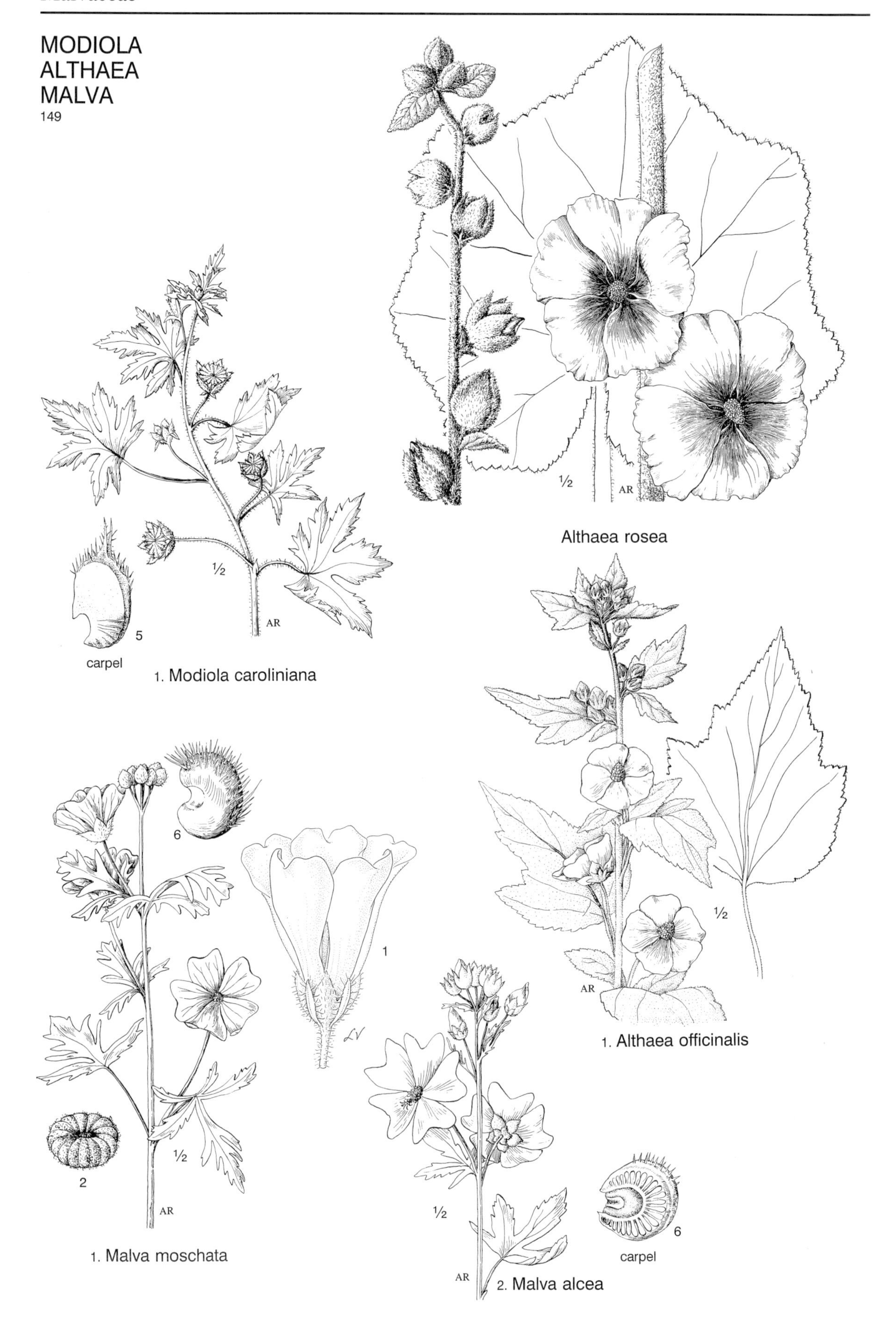

Althaea rosea

carpel

1. Modiola caroliniana

1. Althaea officinalis

1. Malva moschata

2. Malva alcea

carpel

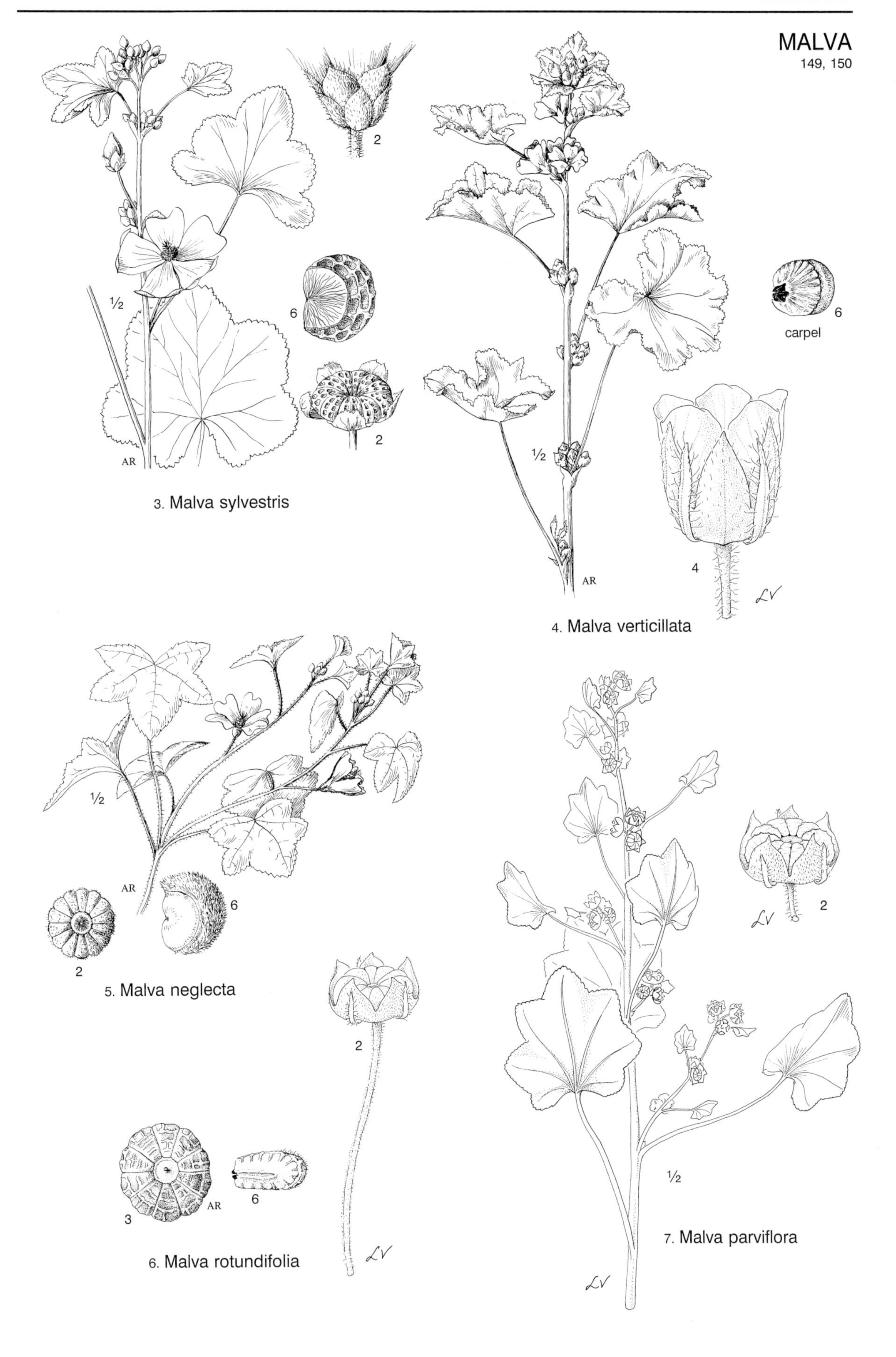

3. Malva sylvestris

carpel

4. Malva verticillata

5. Malva neglecta

6. Malva rotundifolia

7. Malva parviflora

CALLIRHOE
NAPAEA
150, 151

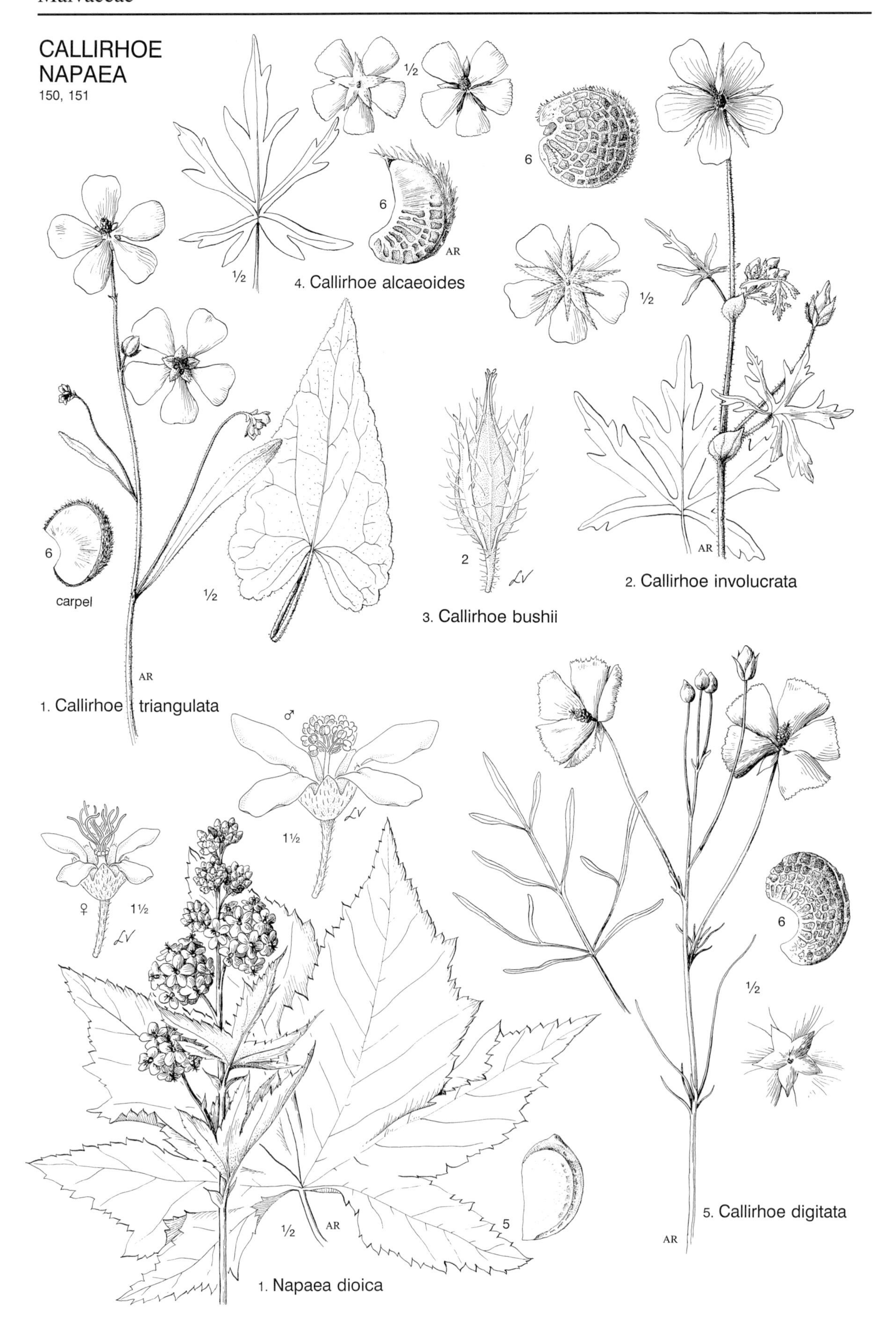

4. Callirhoe alcaeoides

2. Callirhoe involucrata

carpel

6

3. Callirhoe bushii

1. Callirhoe triangulata

5. Callirhoe digitata

1. Napaea dioica

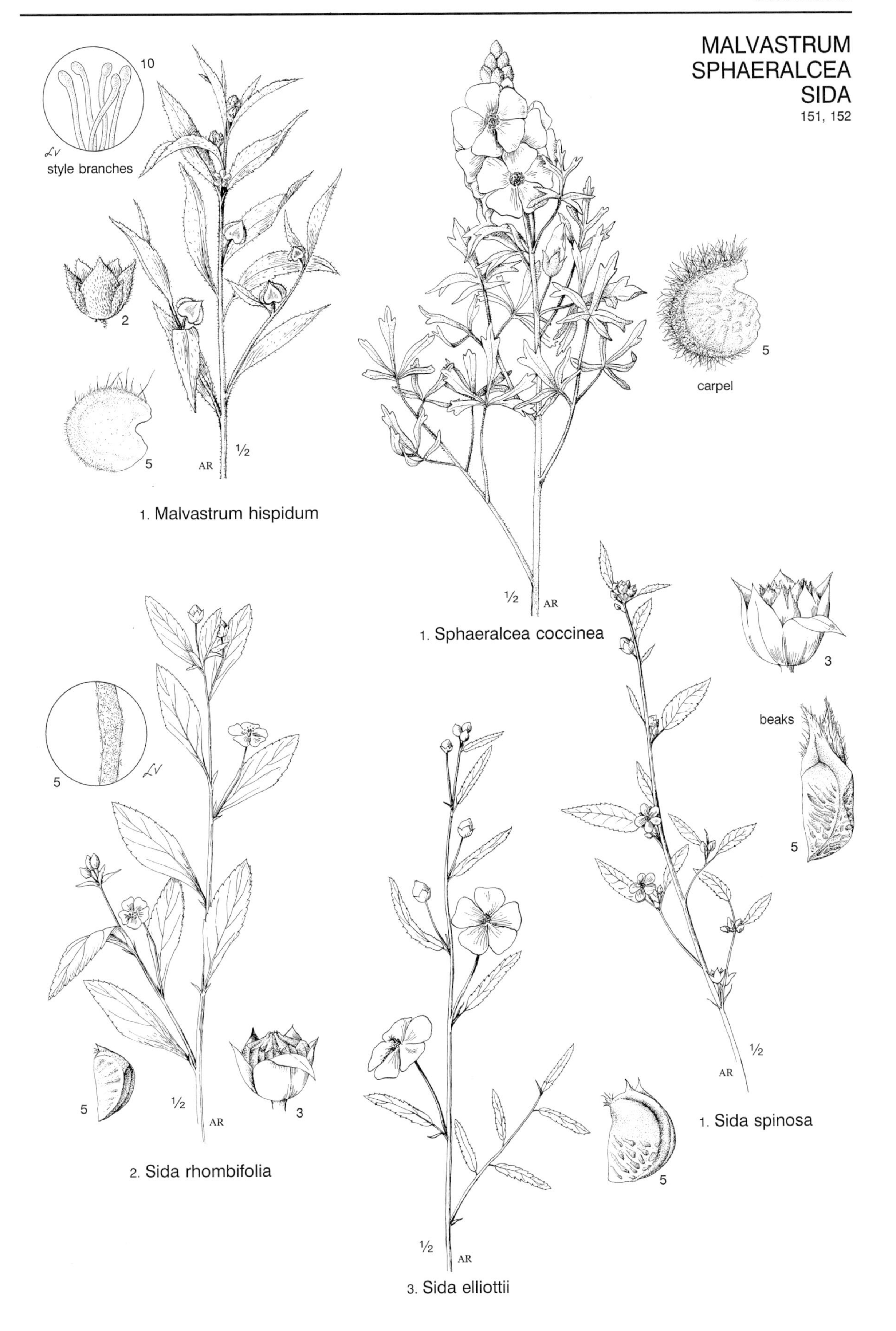

10

style branches

2

5

½

AR

1. Malvastrum hispidum

5

carpel

½ AR

1. Sphaeralcea coccinea

3

beaks

5

5

5

½

AR

1. Sida spinosa

5

½

AR

3

2. Sida rhombifolia

5

½

AR

3. Sida elliottii

SIDA
ANODA
SARRACENIA
DROSERA
152–154

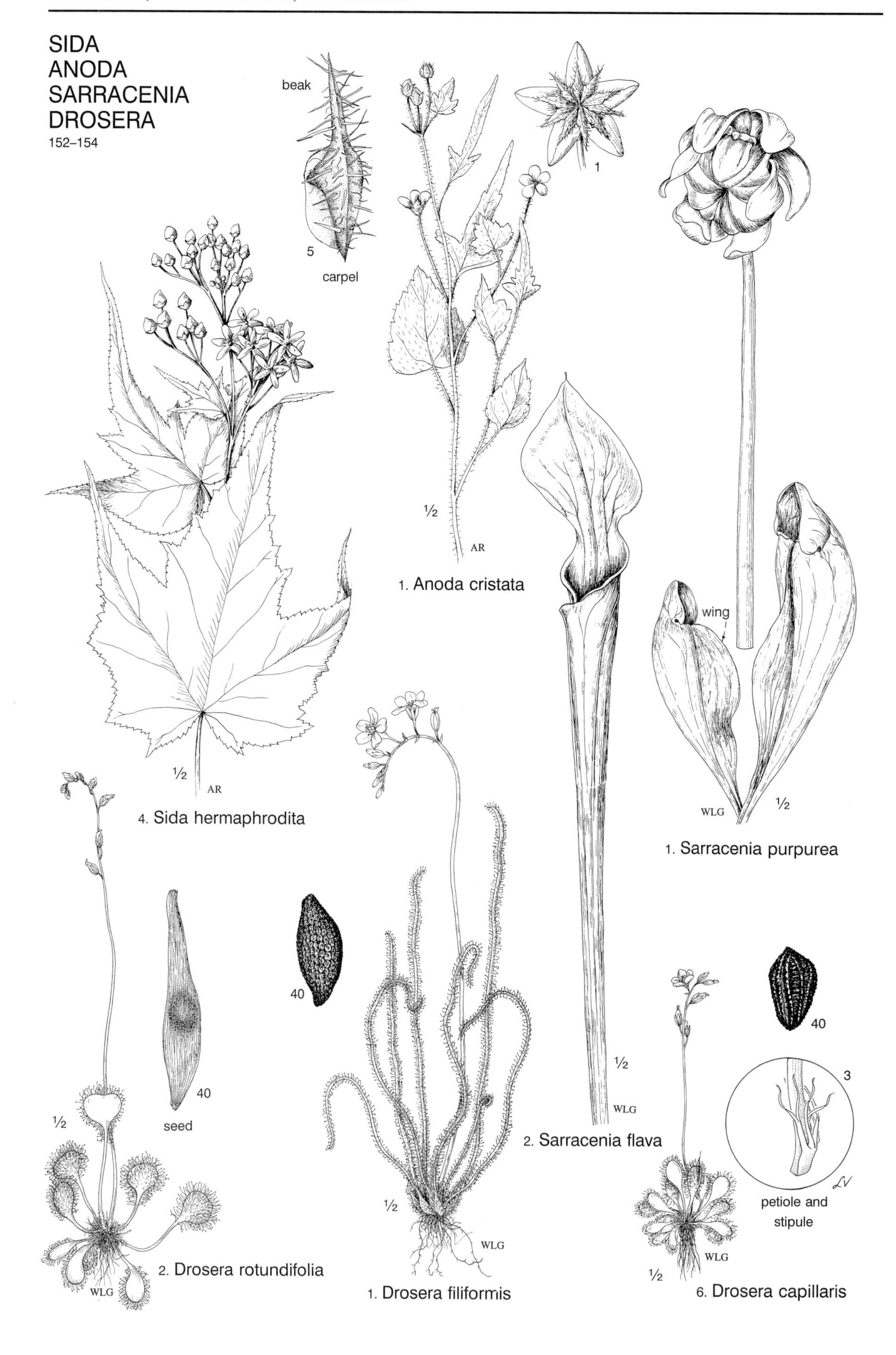

beak

5

carpel

1. Anoda cristata

½ AR

4. Sida hermaphrodita

½ AR

1. Sarracenia purpurea

wing

WLG ½

40

seed

40

½

2. Drosera rotundifolia

WLG

½

1. Drosera filiformis

WLG

½

2. Sarracenia flava

WLG

40

3

petiole and stipule

½

6. Drosera capillaris

WLG

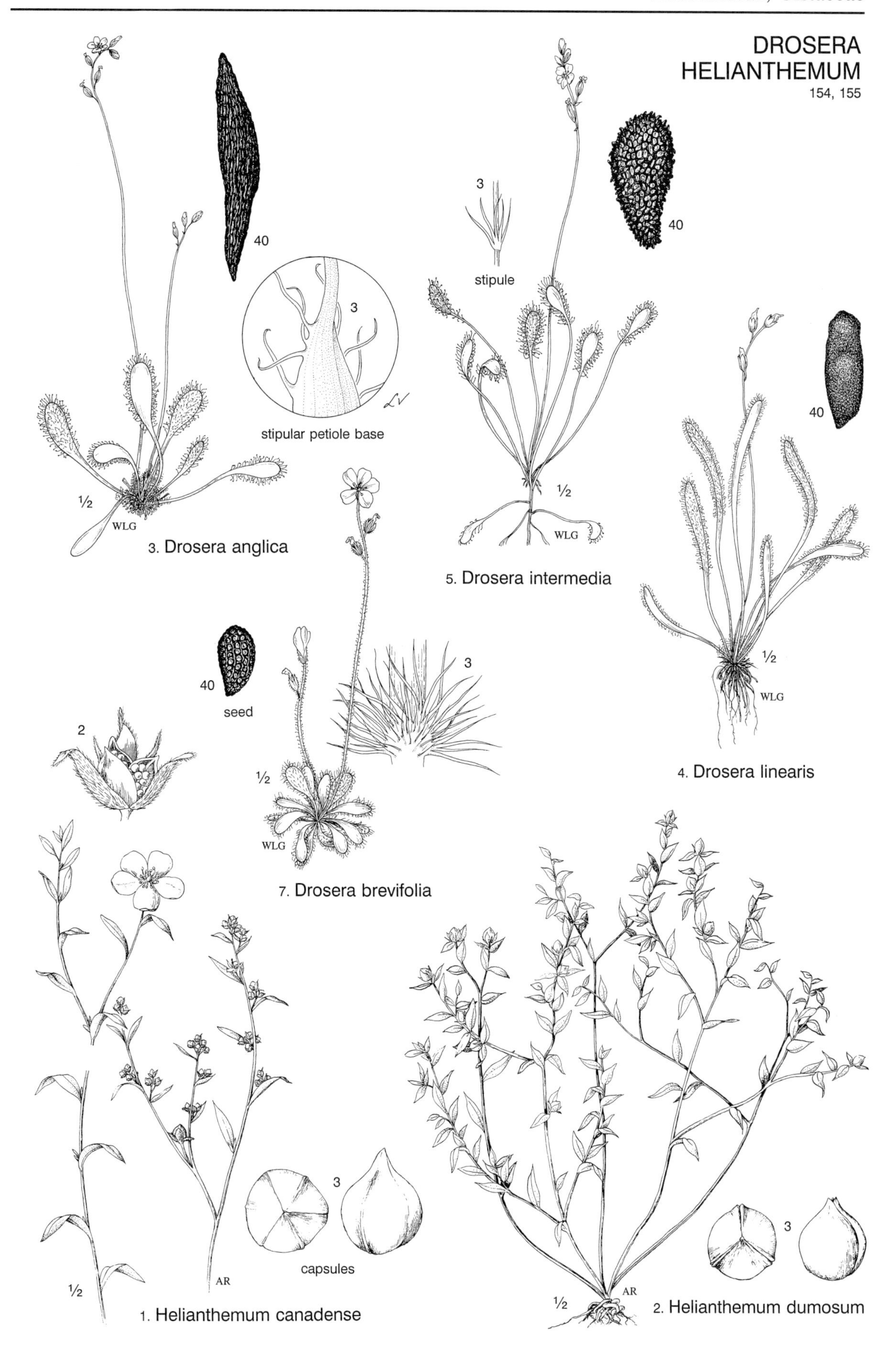

40

3

stipule

40

stipular petiole base

1/2
WLG

3. Drosera anglica

40

1/2
WLG

5. Drosera intermedia

40

1/2
WLG

4. Drosera linearis

40

seed

2

3

1/2
WLG

7. Drosera brevifolia

capsules

1/2

AR

1. Helianthemum canadense

3

1/2
AR

2. Helianthemum dumosum

HELIANTHEMUM
HUDSONIA
LECHEA
155, 156

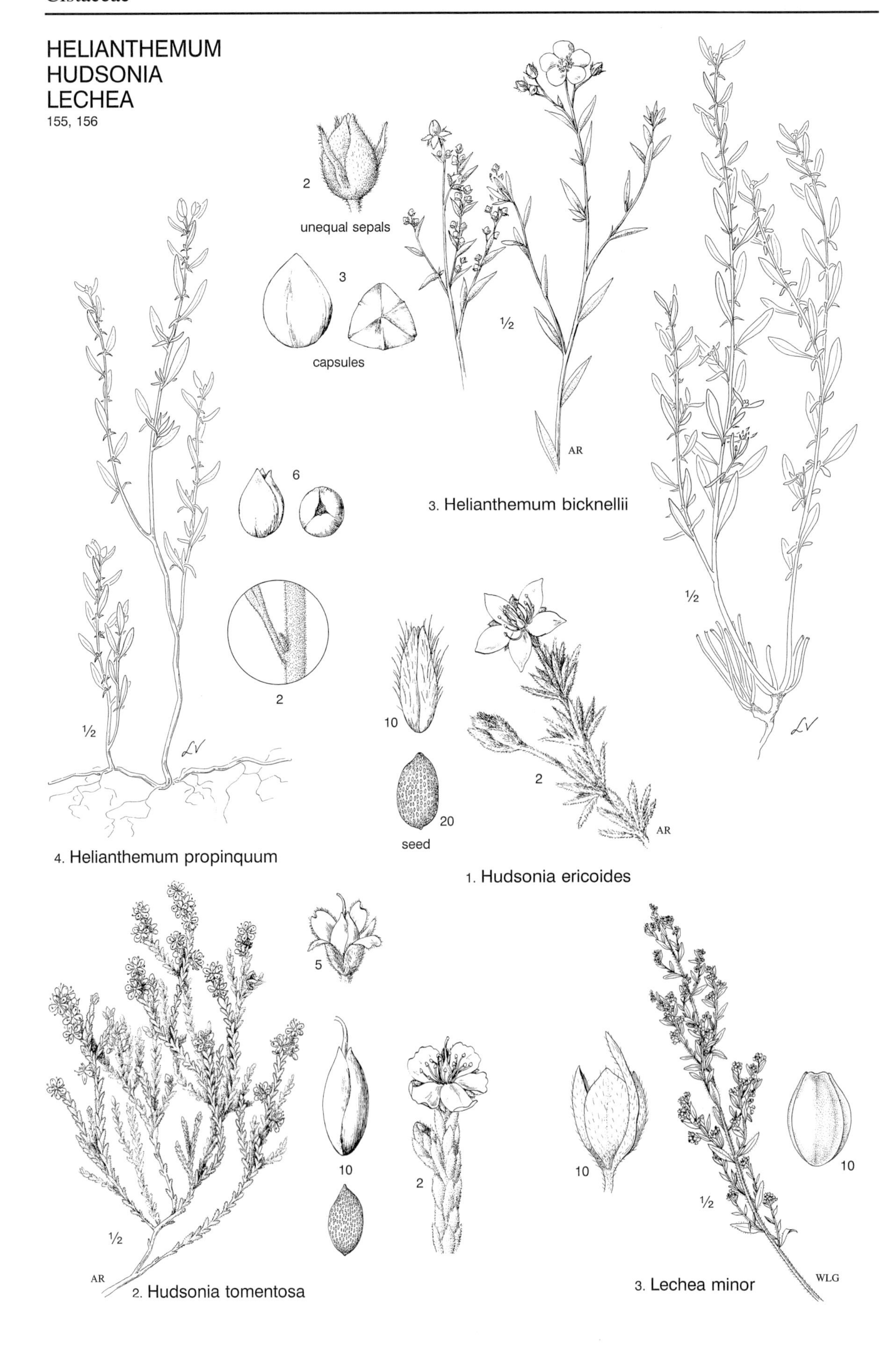

2

unequal sepals

3

capsules

½

AR

3. Helianthemum bicknellii

6

2

½

ℒⱽ

4. Helianthemum propinquum

10

20

seed

2

AR

½

ℒⱽ

1. Hudsonia ericoides

5

10

2

10

10

½

½

AR

2. Hudsonia tomentosa

3. Lechea minor

WLG

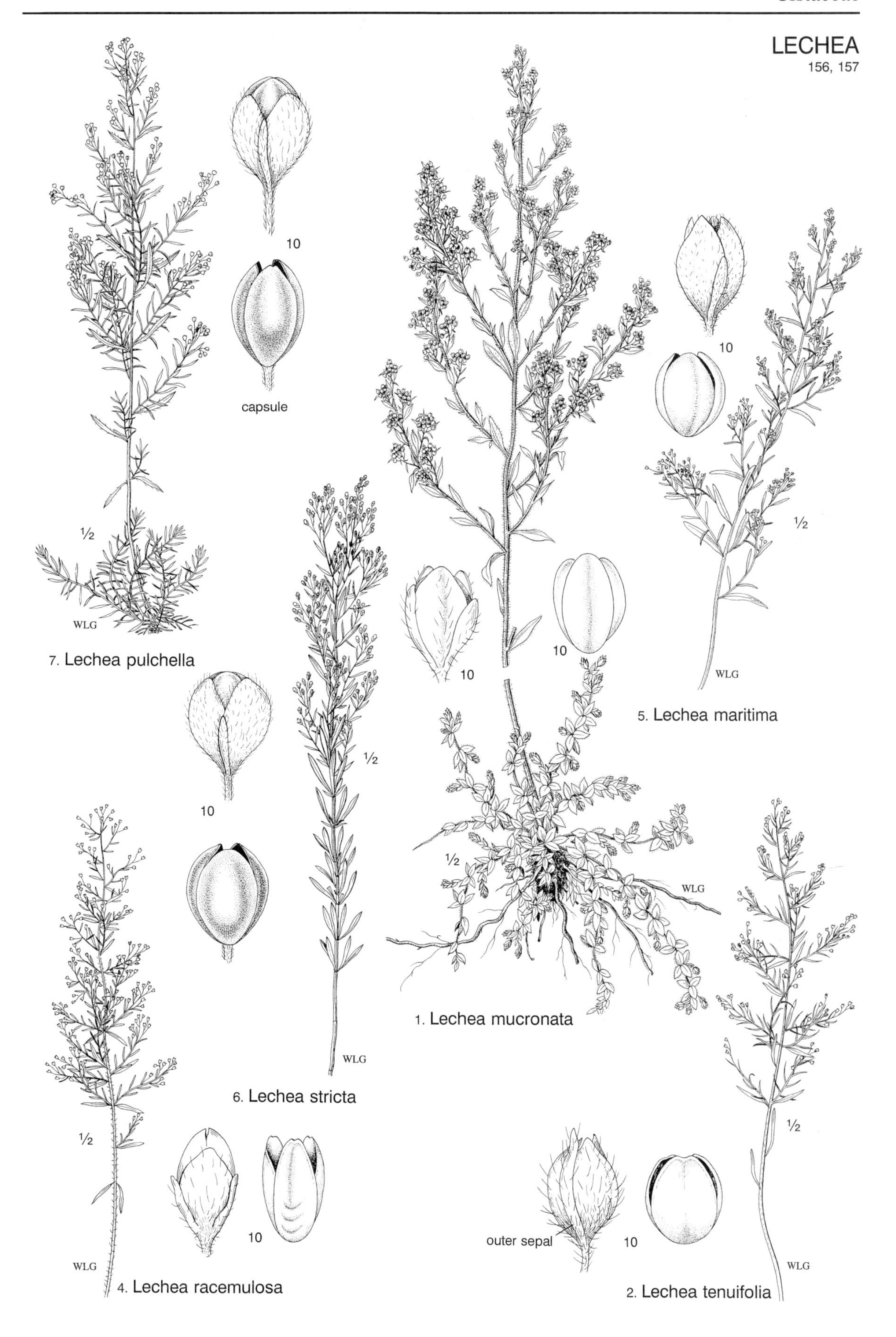

10

capsule

½

WLG

7. Lechea pulchella

½

WLG

6. Lechea stricta

10

½

WLG

4. Lechea racemulosa

10

10

½

WLG

1. Lechea mucronata

10

10

WLG

5. Lechea maritima

½

outer sepal   10

2. Lechea tenuifolia

½

WLG

LECHEA
HYBANTHUS
VIOLA
157–159

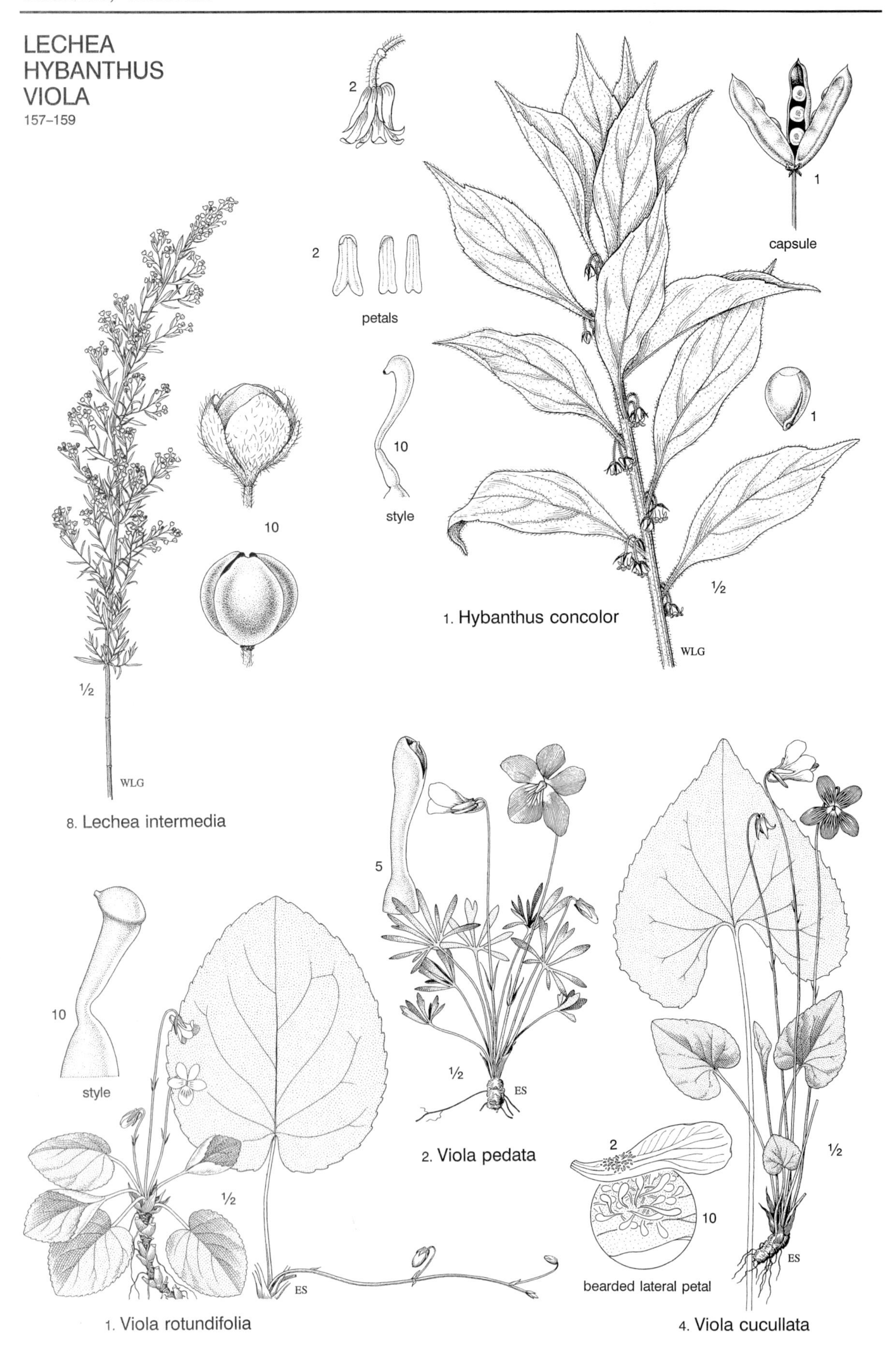

1. Hybanthus concolor

capsule

petals

style

8. Lechea intermedia

1. Viola rotundifolia

2. Viola pedata

bearded lateral petal

4. Viola cucullata

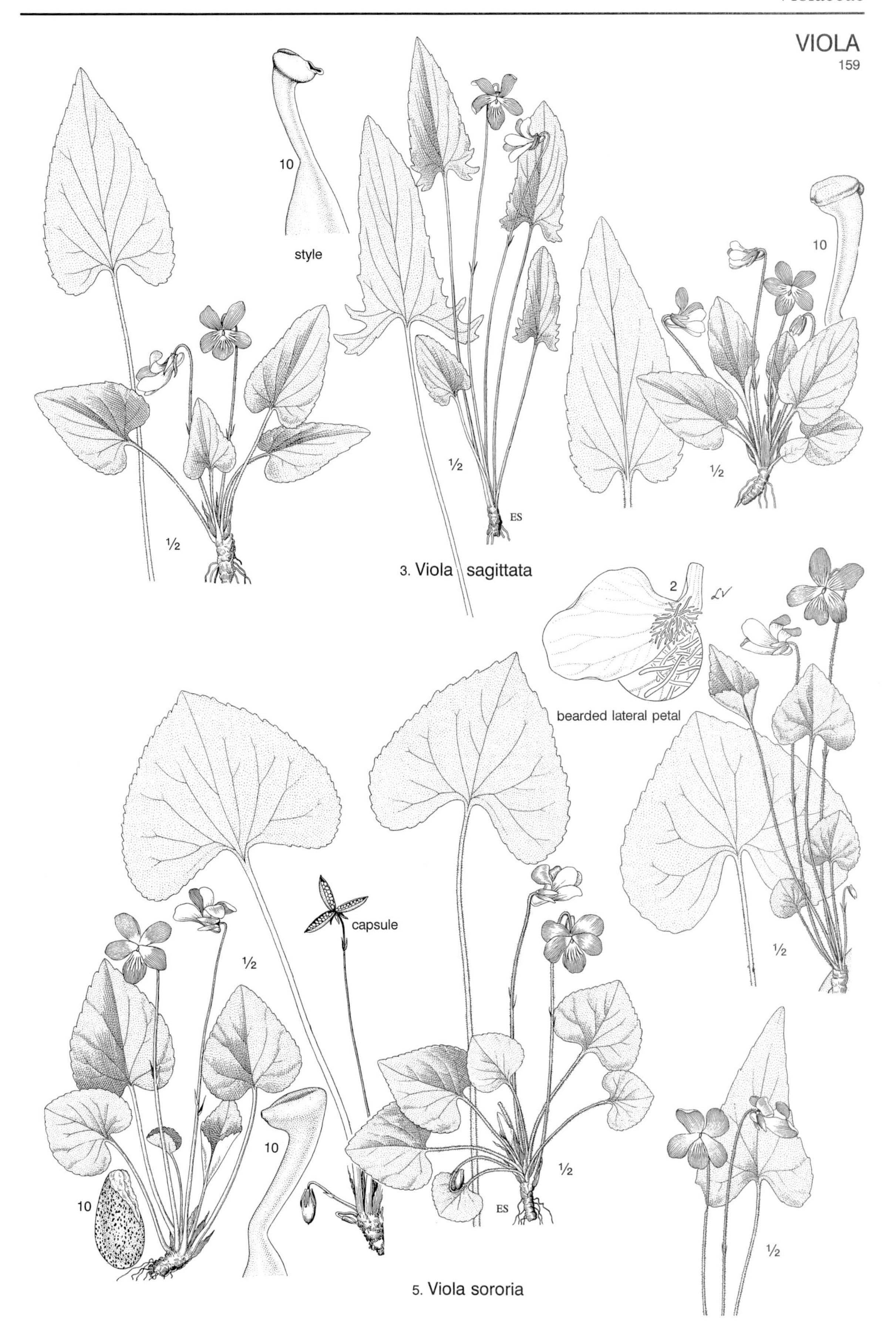

10
style

3. Viola sagittata

ES

10

bearded lateral petal

2

capsule

10

10

10

ES

½

5. Viola sororia

½

½

½

½

½

VIOLA
160

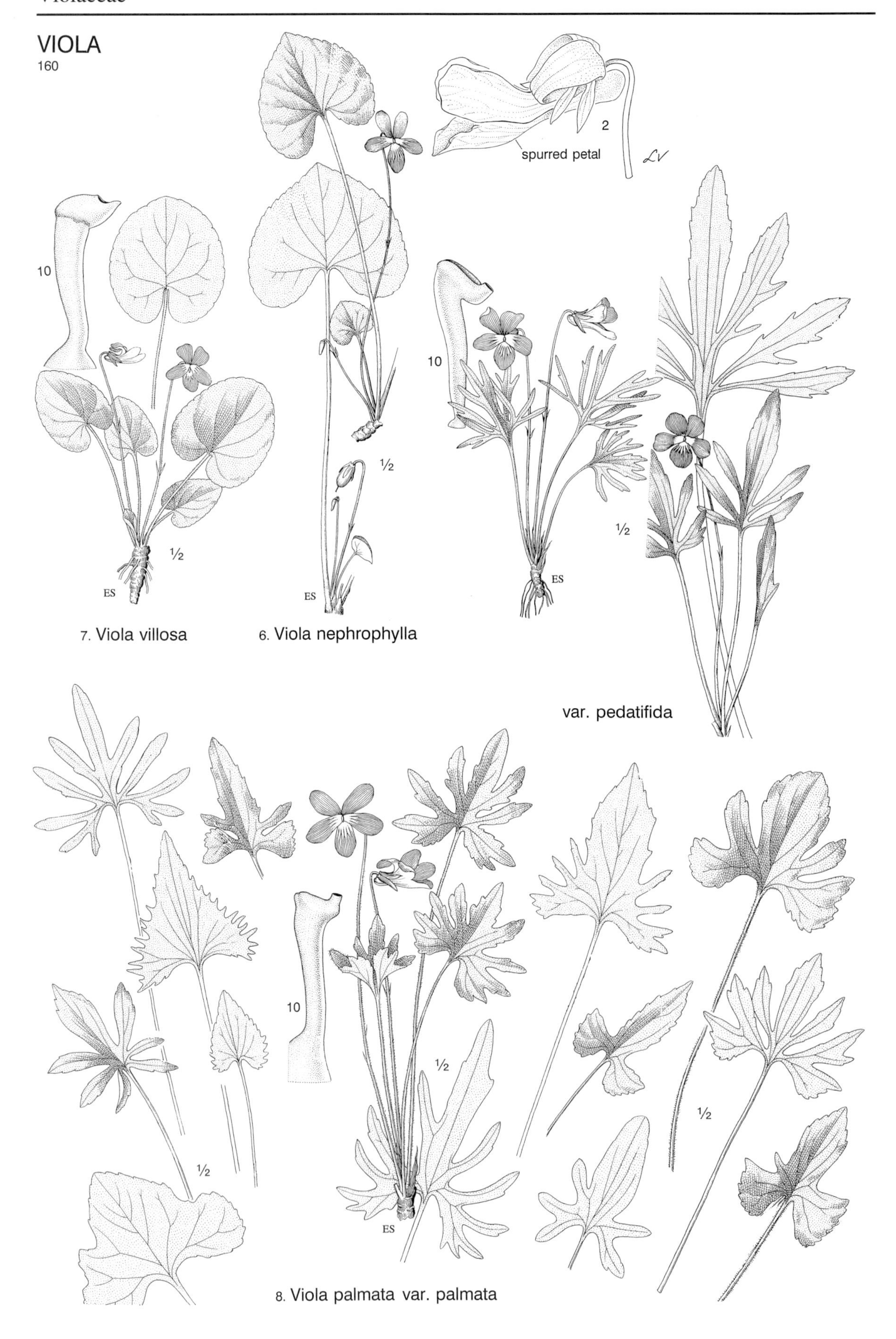

2

spurred petal

10

½

7. Viola villosa

6. Viola nephrophylla

10

½

var. pedatifida

10

½

8. Viola palmata var. palmata

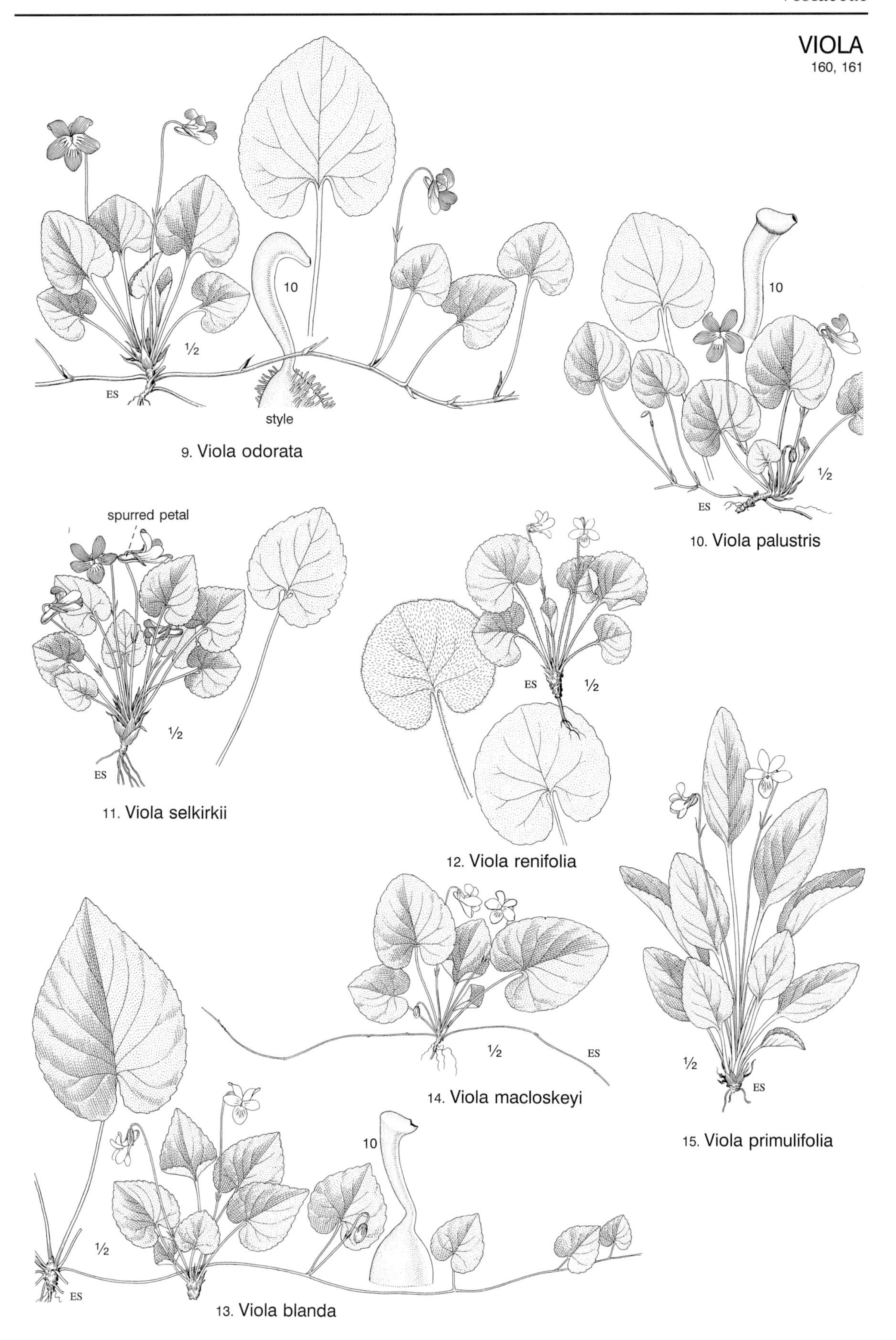

9. Viola odorata

style

10. Viola palustris

spurred petal

11. Viola selkirkii

12. Viola renifolia

14. Viola macloskeyi

15. Viola primulifolia

13. Viola blanda

VIOLA
161

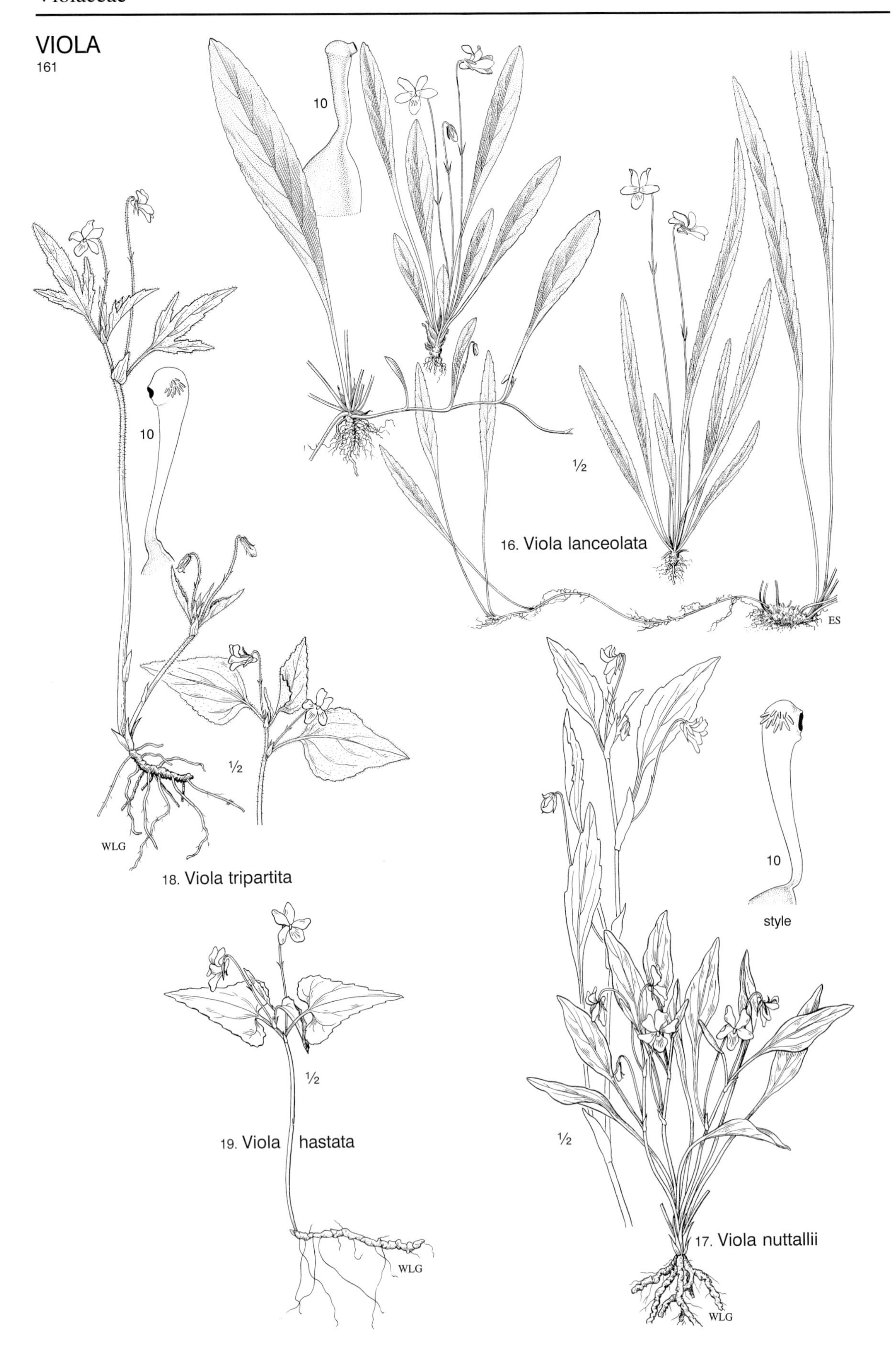

16. Viola lanceolata

18. Viola tripartita

19. Viola hastata

17. Viola nuttallii

style

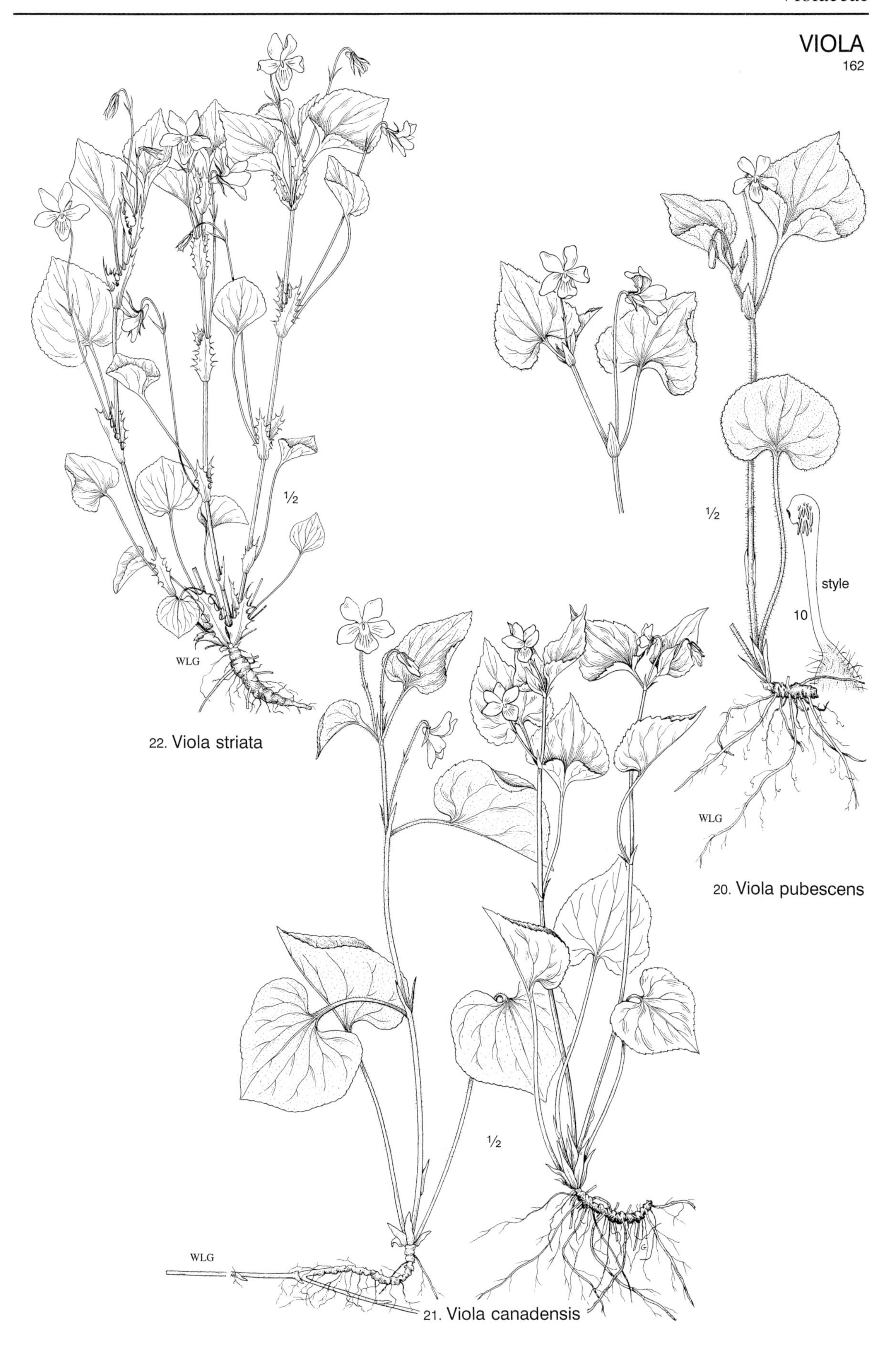

½

WLG

22. Viola striata

½

½

style

10

WLG

20. Viola pubescens

½

WLG

21. Viola canadensis

VIOLA
162, 163

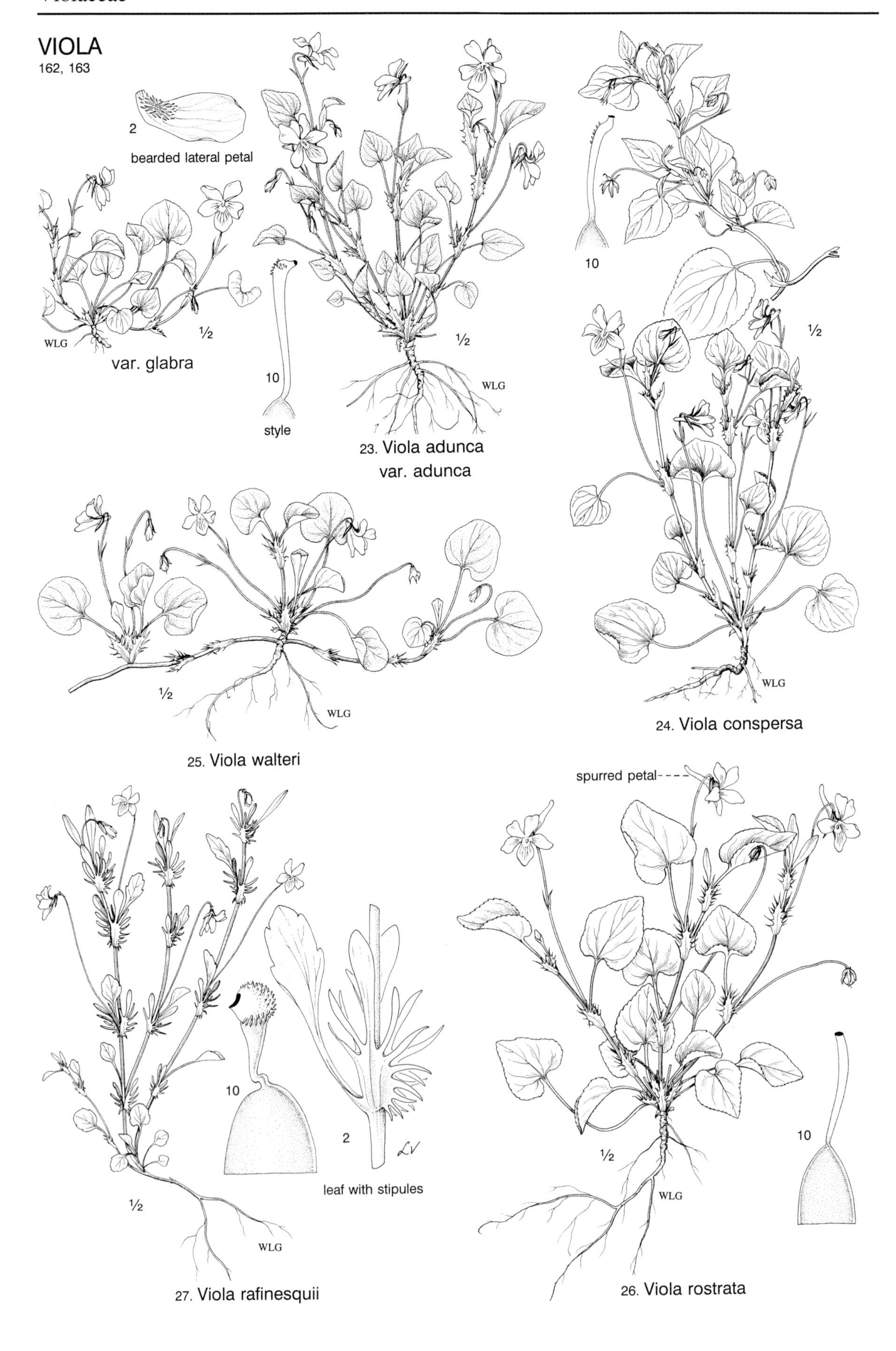

2
bearded lateral petal

var. glabra

WLG   ½

10
style

23. Viola adunca
var. adunca

WLG

½

10

½

24. Viola conspersa

WLG

½

WLG

25. Viola walteri

10

2

leaf with stipules

½

WLG

27. Viola rafinesquii

spurred petal - - - -

½

WLG

10

26. Viola rostrata

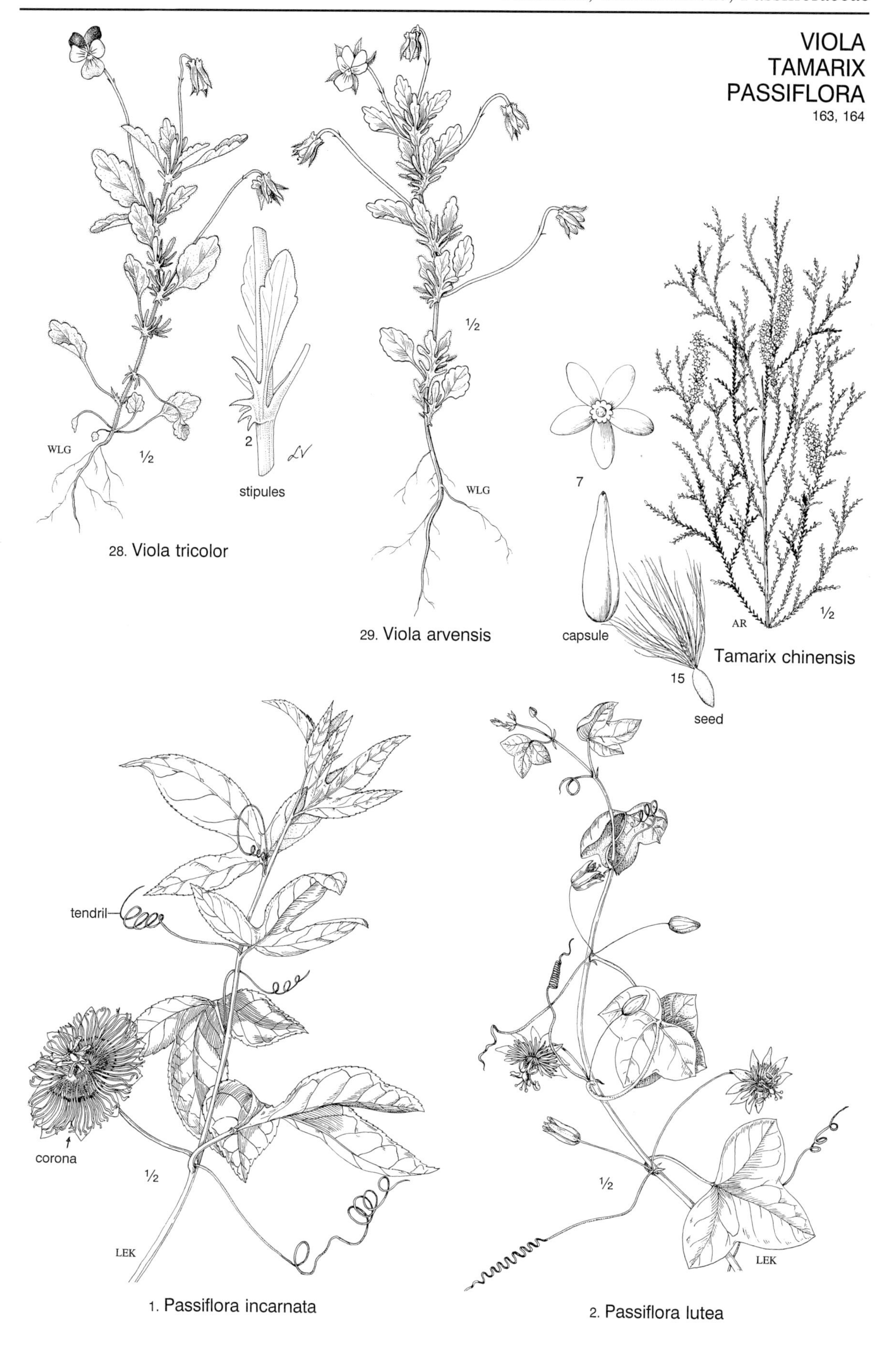

28. Viola tricolor

stipules

29. Viola arvensis

7

capsule

15

seed

Tamarix chinensis

AR

tendril

corona

1. Passiflora incarnata

LEK

2. Passiflora lutea

LEK

CUCURBITA
MELOTHRIA
ECHINOCYSTIS
164, 165

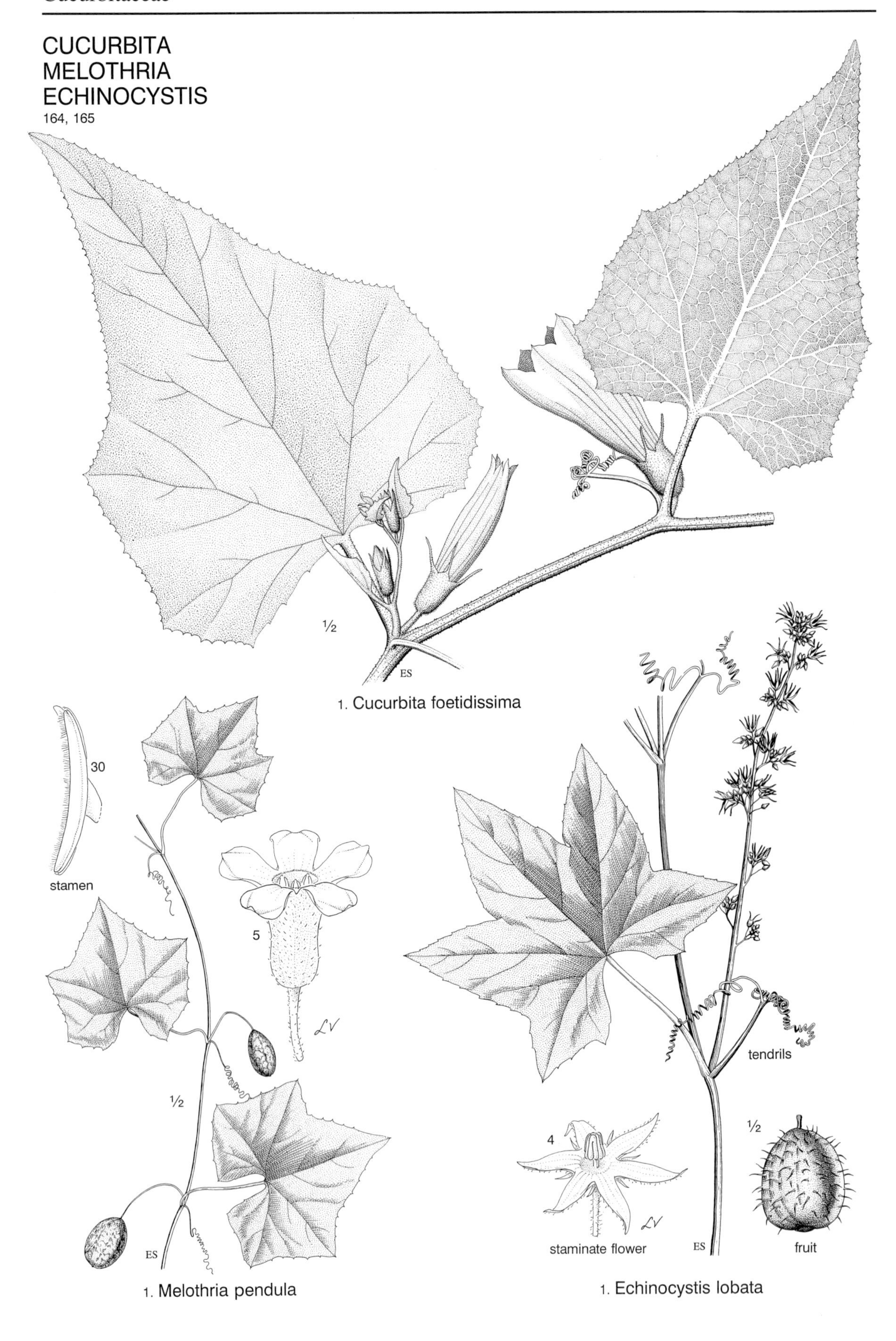

1. Cucurbita foetidissima

30

stamen

5

1. Melothria pendula

tendrils

4

staminate flower

fruit

1. Echinocystis lobata

ILLUSTRATED COMPANION TO

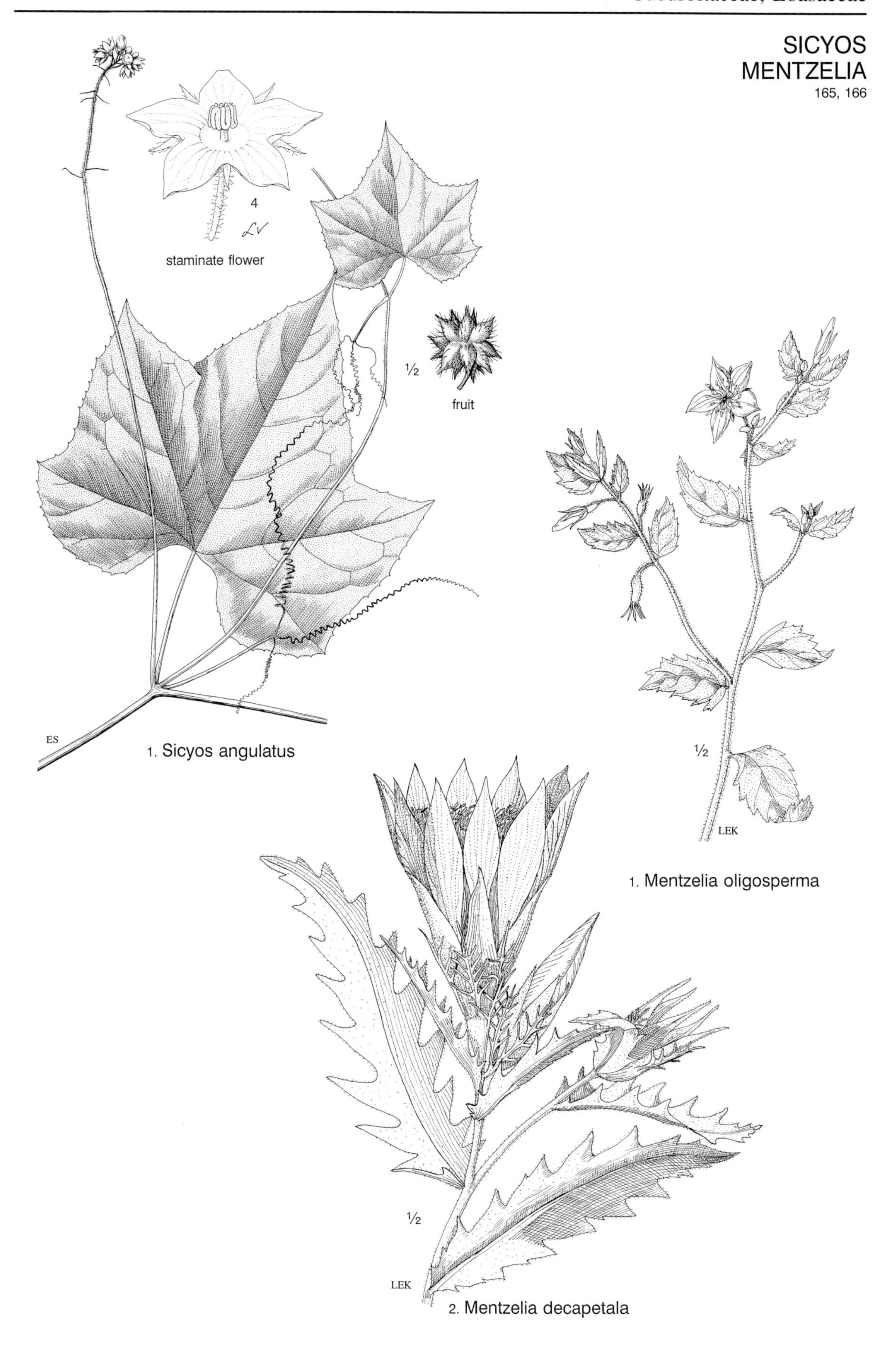

4

staminate flower

½

fruit

ES

1. Sicyos angulatus

½

LEK

1. Mentzelia oligosperma

½

LEK

2. Mentzelia decapetala

POPULUS
167, 168

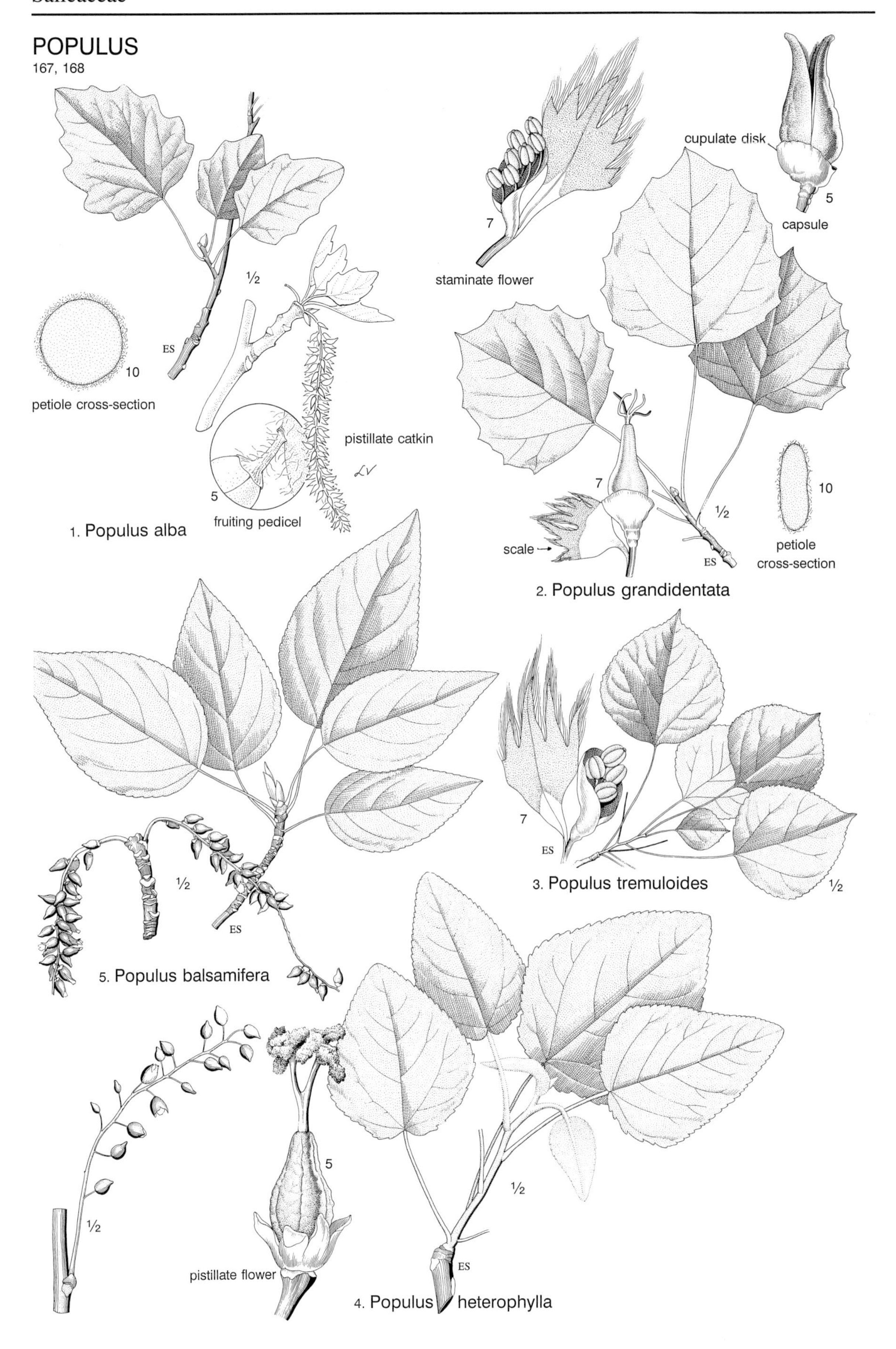

10
petiole cross-section

ES

½

1. Populus alba

5

fruiting pedicel

pistillate catkin

7

staminate flower

cupulate disk

5

capsule

7

scale →

ES

½

10

petiole cross-section

2. Populus grandidentata

½

ES

5. Populus balsamifera

7

ES

3. Populus tremuloides

½

½

5

pistillate flower

ES

½

4. Populus heterophylla

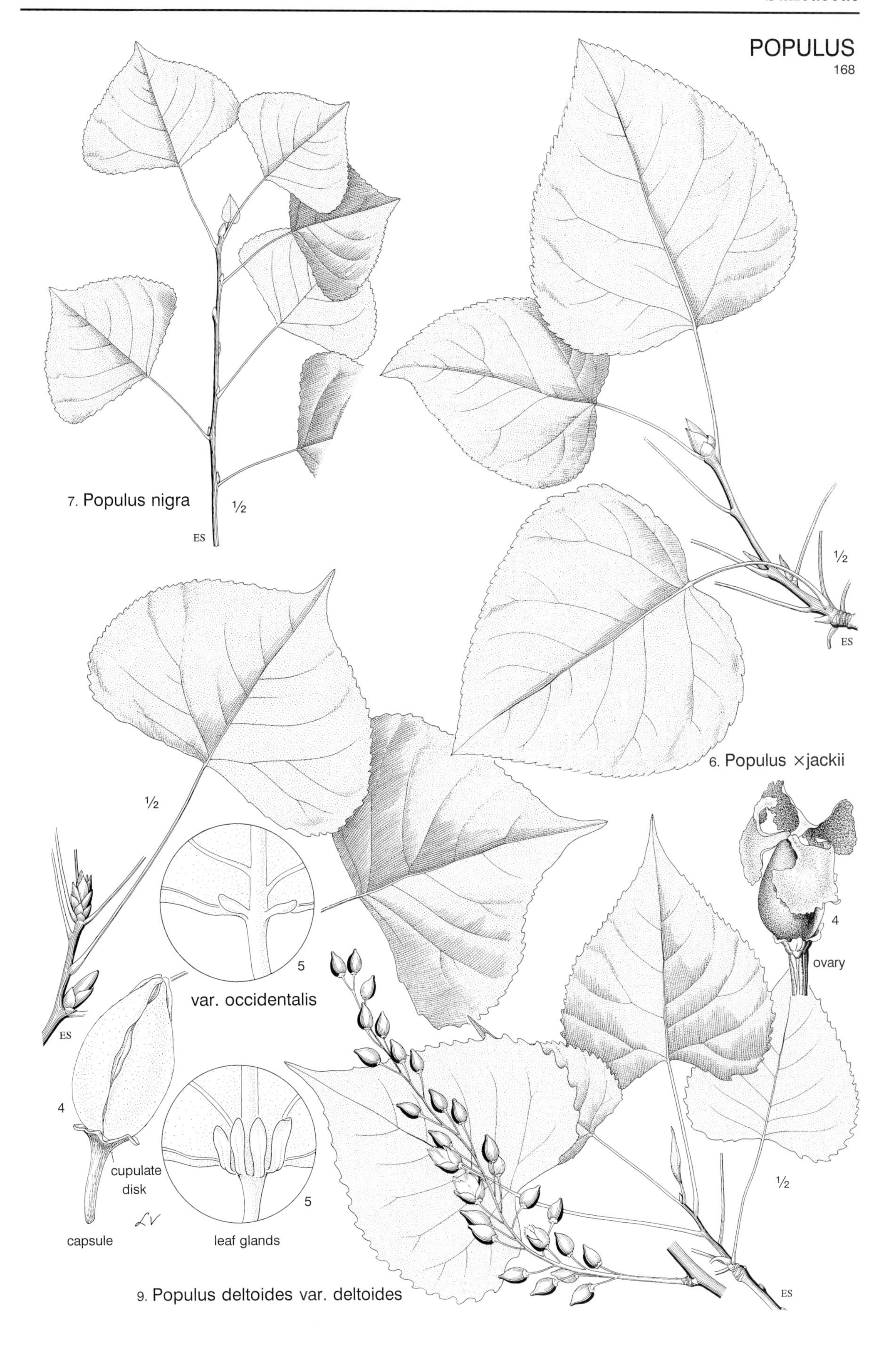

7. Populus nigra ½
ES

6. Populus ×jackii

½
ES

½

var. occidentalis
5

4
ovary

capsule

4
cupulate
disk

leaf glands
5

9. Populus deltoides var. deltoides

½
ES

ES

SALIX
169, 170

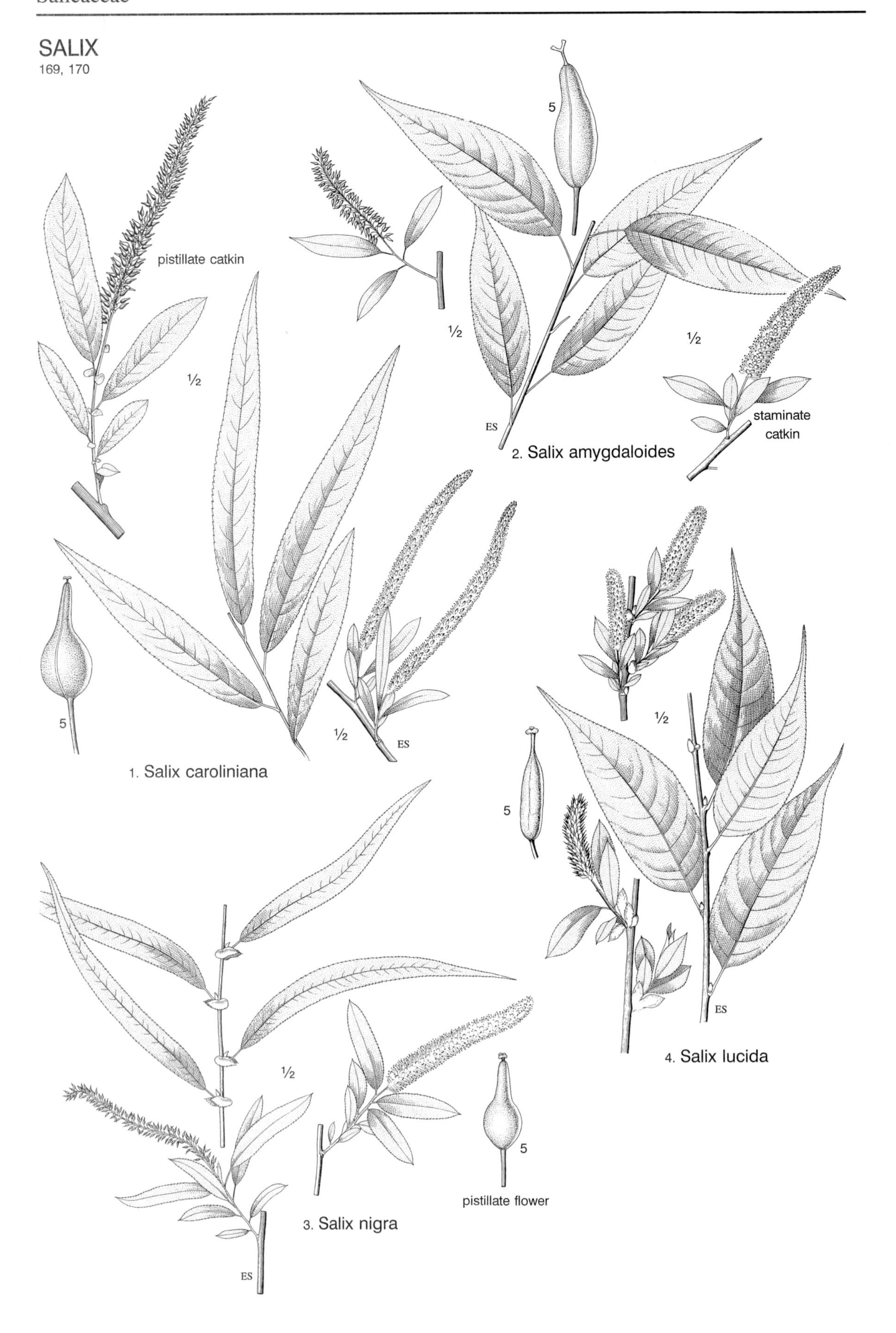

pistillate catkin

½

5

1. Salix caroliniana

½

ES

5

2. Salix amygdaloides

½

staminate catkin

½

½

5

4. Salix lucida

ES

½

½

pistillate flower

5

3. Salix nigra

ES

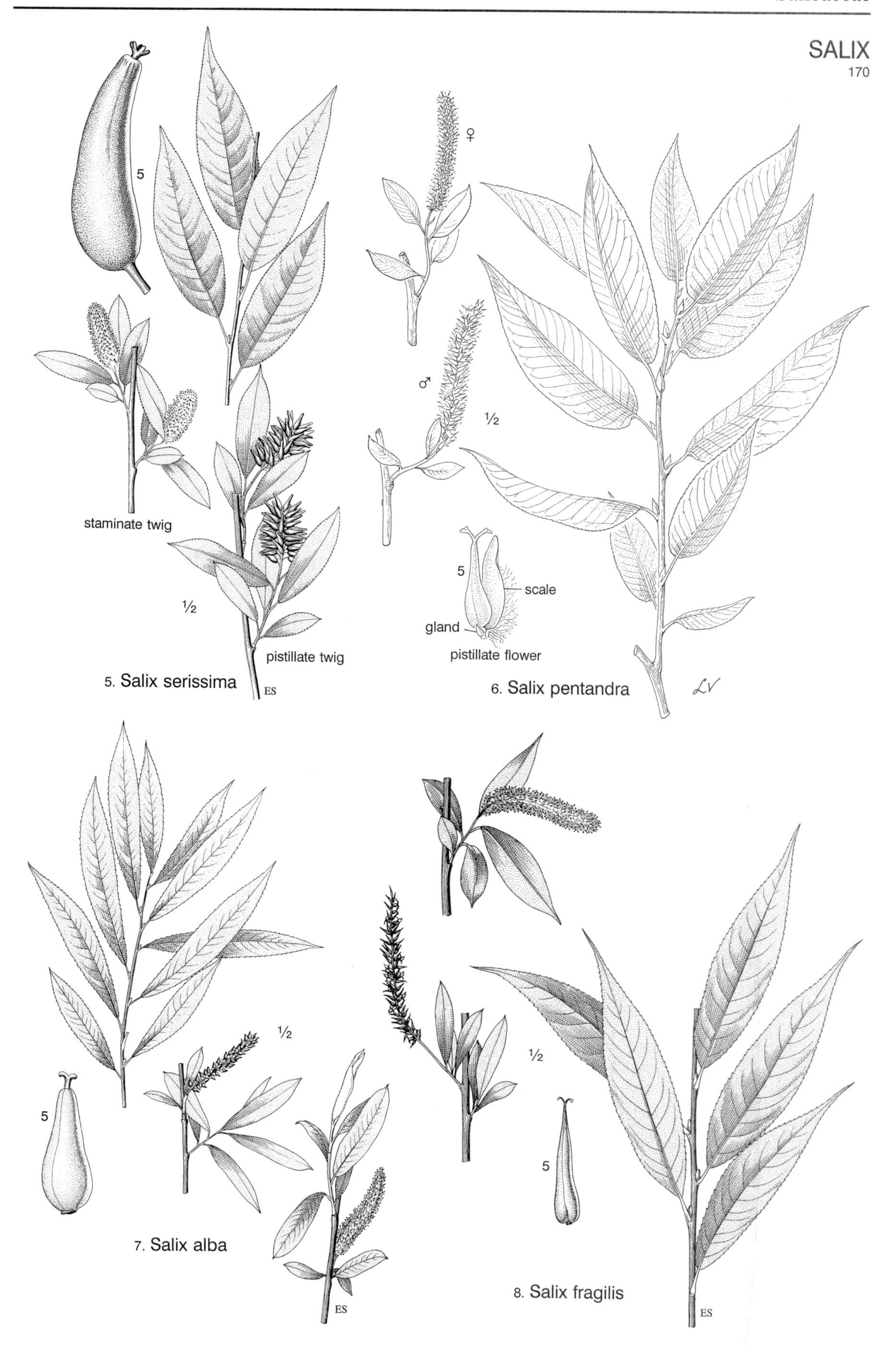

5

staminate twig

½

pistillate twig

5. Salix serissima
ES

♀

♂

½

5

scale

gland

pistillate flower

½

6. Salix pentandra

5

½

7. Salix alba

½

5

½

5

8. Salix fragilis
ES

ES

SALIX
170, 171

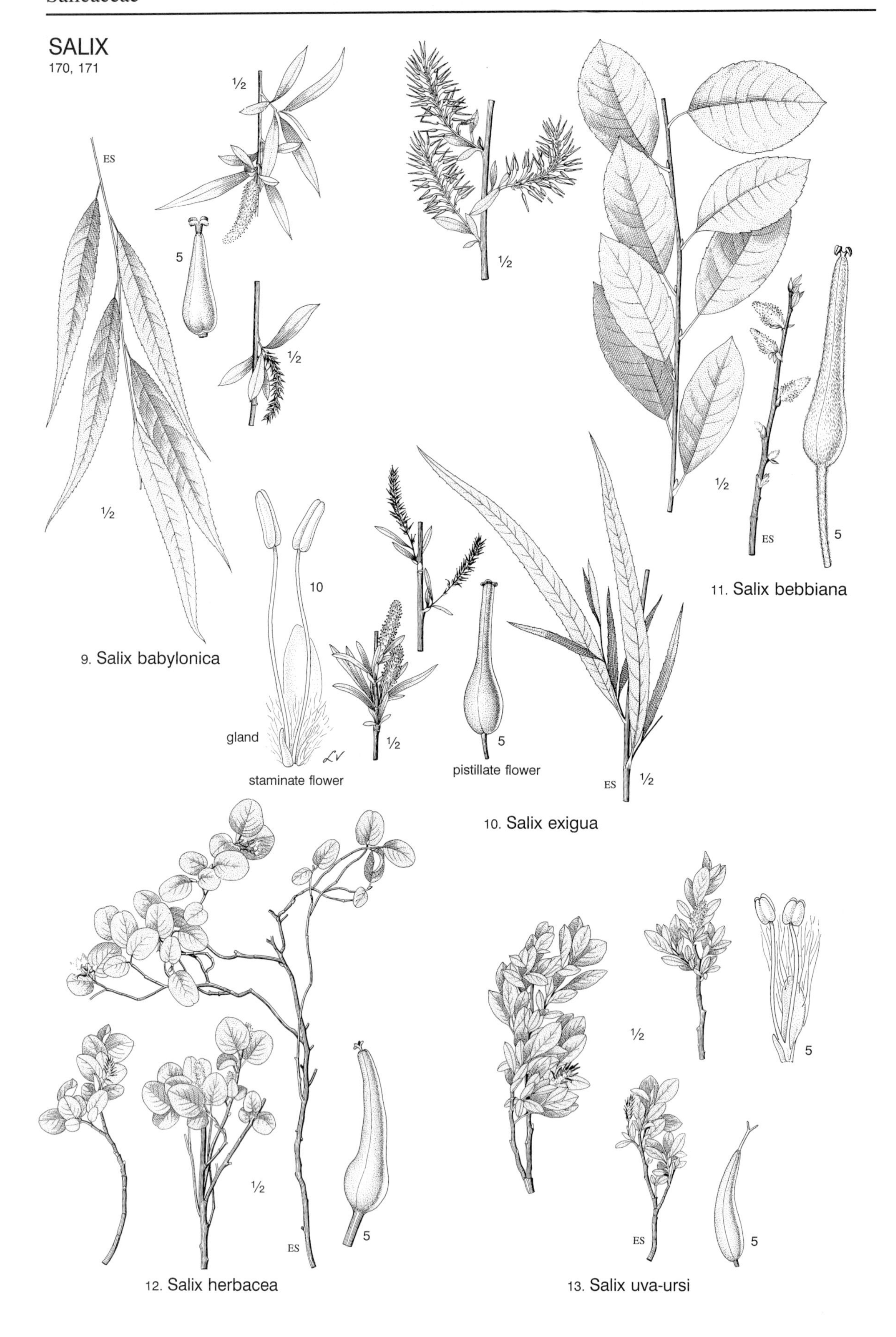

ES

½

5

½

½

9. Salix babylonica

10

gland

staminate flower

½

5

pistillate flower

½

10. Salix exigua

ES

½

½

11. Salix bebbiana

ES

5

½

5

ES

5

12. Salix herbacea

½

ES

5

½

ES

5

13. Salix uva-ursi

ILLUSTRATED COMPANION TO

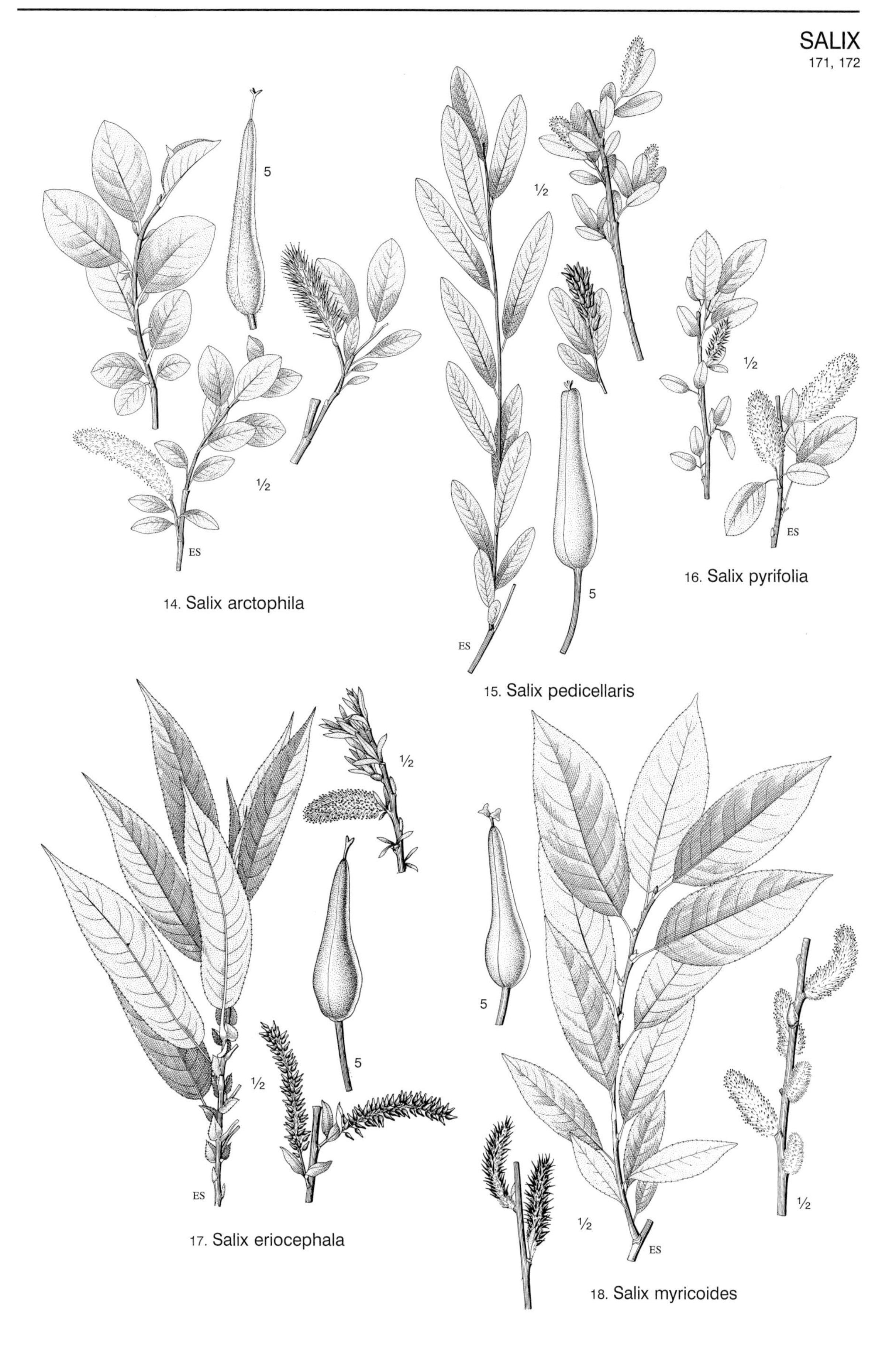

14. Salix arctophila

16. Salix pyrifolia

15. Salix pedicellaris

17. Salix eriocephala

18. Salix myricoides

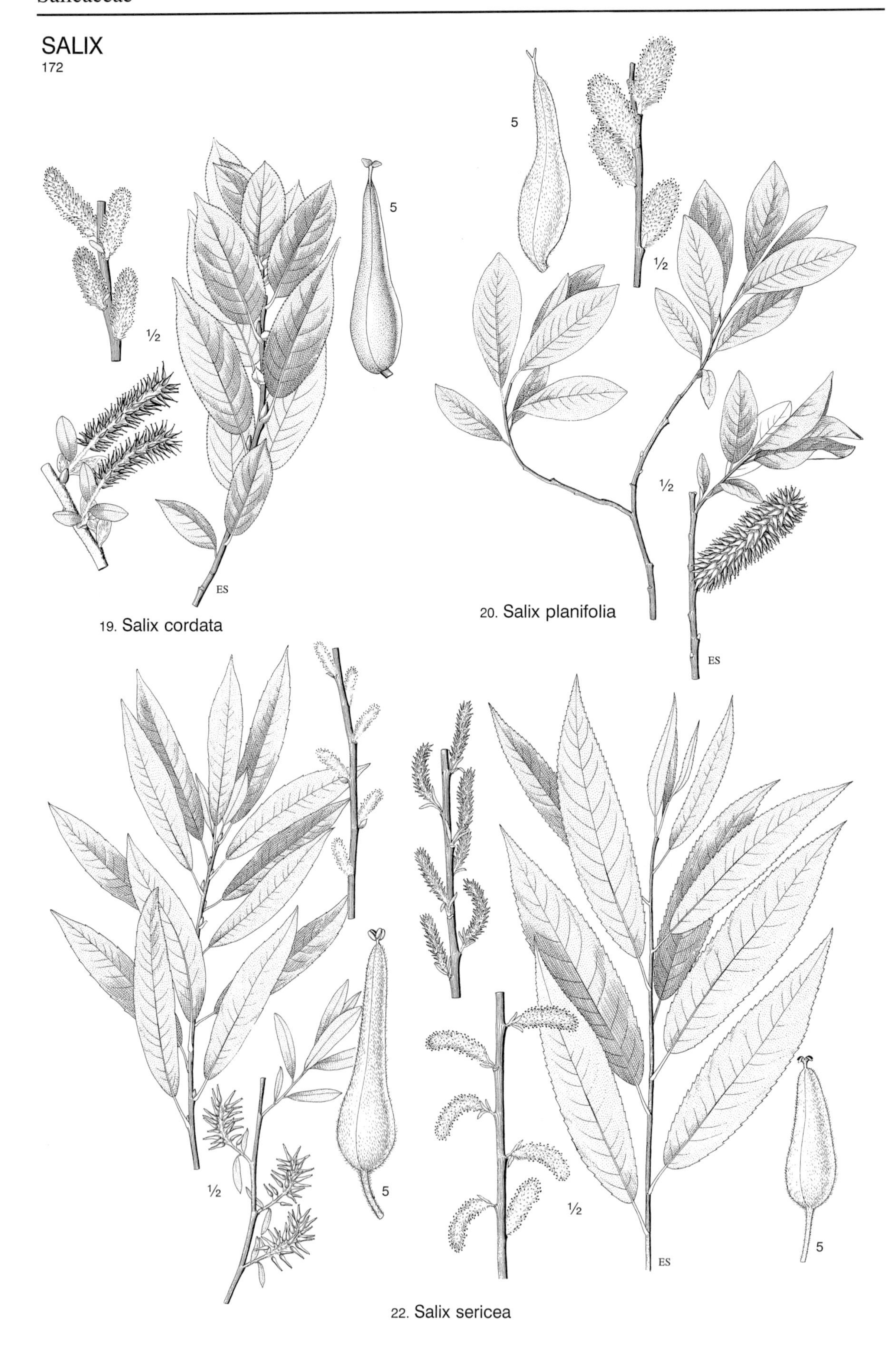

5

½

5

¼ 5

½

ES

19. Salix cordata

5

½

½

½

ES

20. Salix planifolia

½ 5

½

ES

5

22. Salix sericea

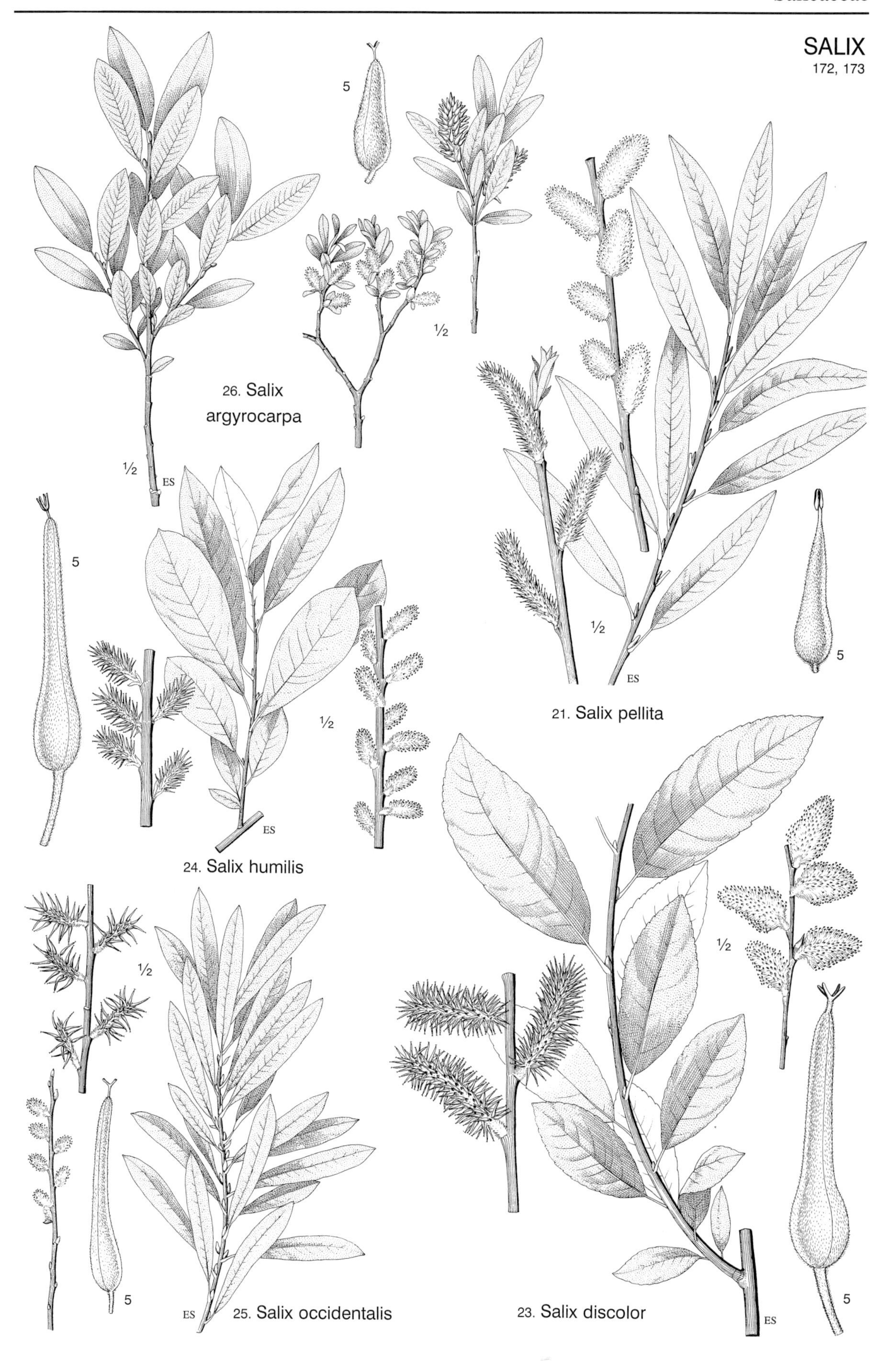

26. Salix argyrocarpa

21. Salix pellita

24. Salix humilis

25. Salix occidentalis

23. Salix discolor

SALIX
173

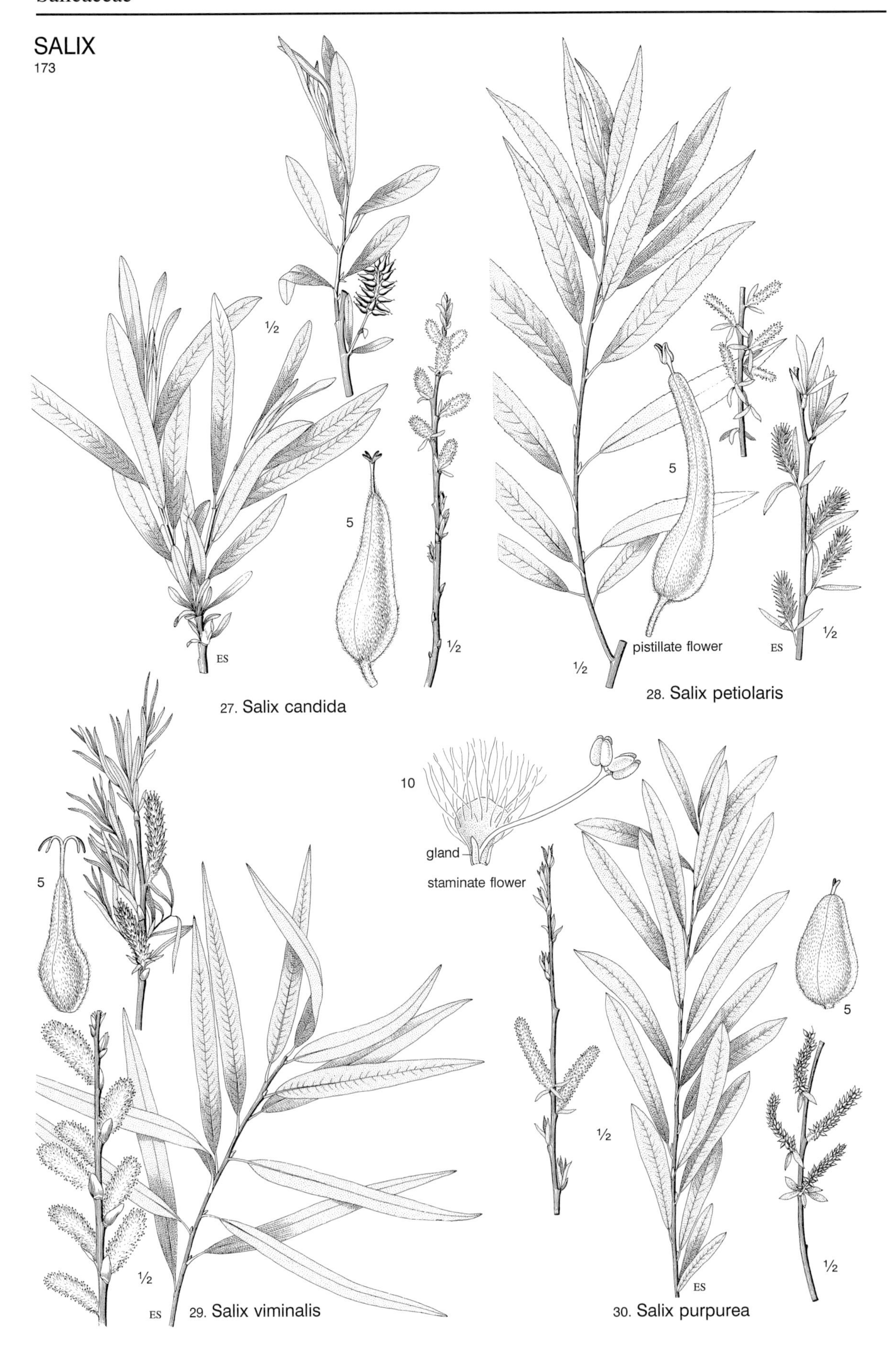

5

½

ES

5

½

27. Salix candida

5

pistillate flower

½

ES

½

28. Salix petiolaris

5

10

gland

staminate flower

½

ES

5

½

½

ES

½

29. Salix viminalis

30. Salix purpurea

CLEOME
POLANISIA
174, 175

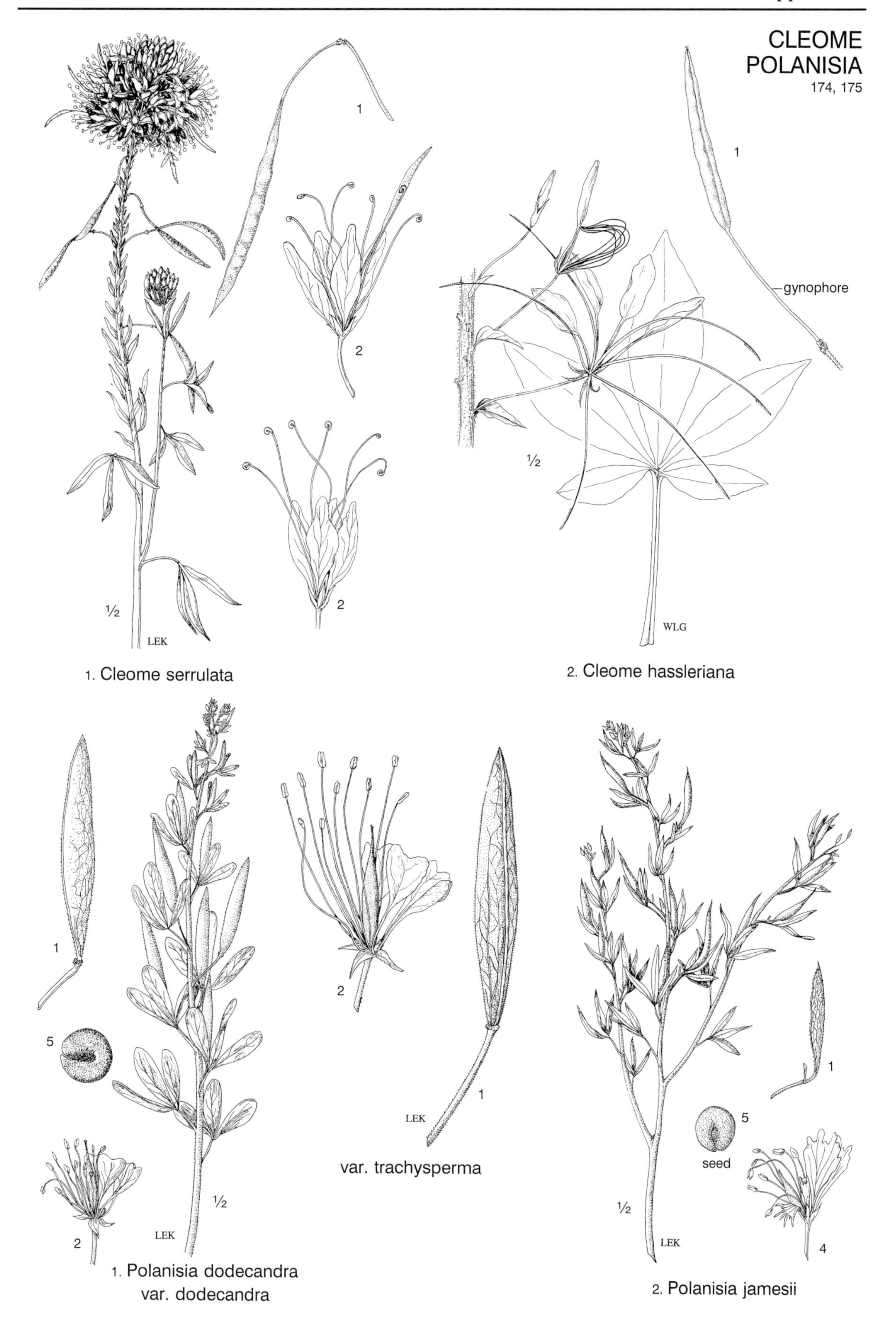

1. Cleome serrulata

2. Cleome hassleriana

gynophore

var. trachysperma

1. Polanisia dodecandra
var. dodecandra

2. Polanisia jamesii

seed

BRASSICA
SINAPIS
178

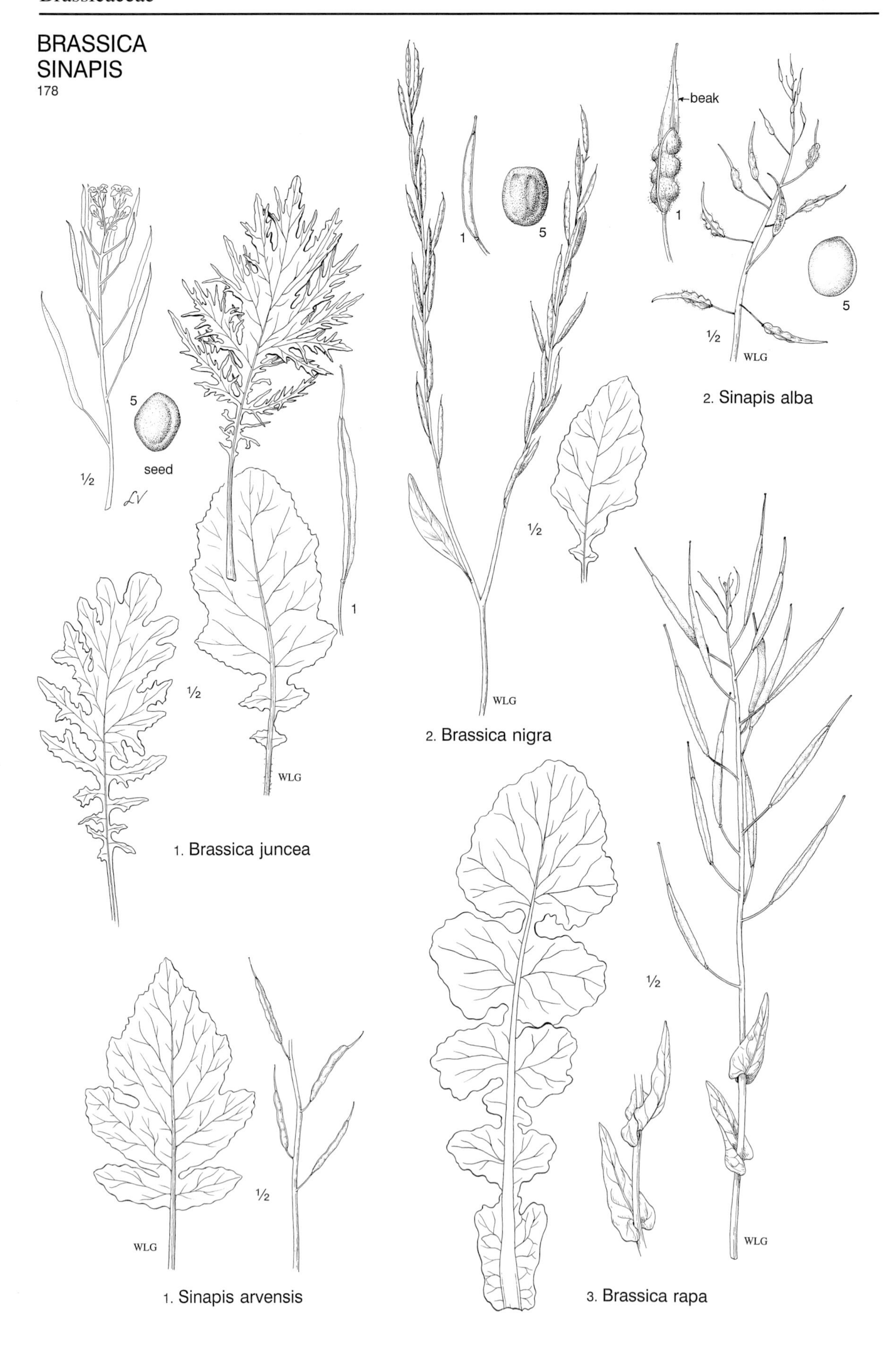

½

5 seed

½

1. Brassica juncea

½

WLG

1. Sinapis arvensis

½

WLG

1

5

½

WLG

2. Brassica nigra

½

WLG

3. Brassica rapa

←beak

1

5

½

WLG

5

2. Sinapis alba

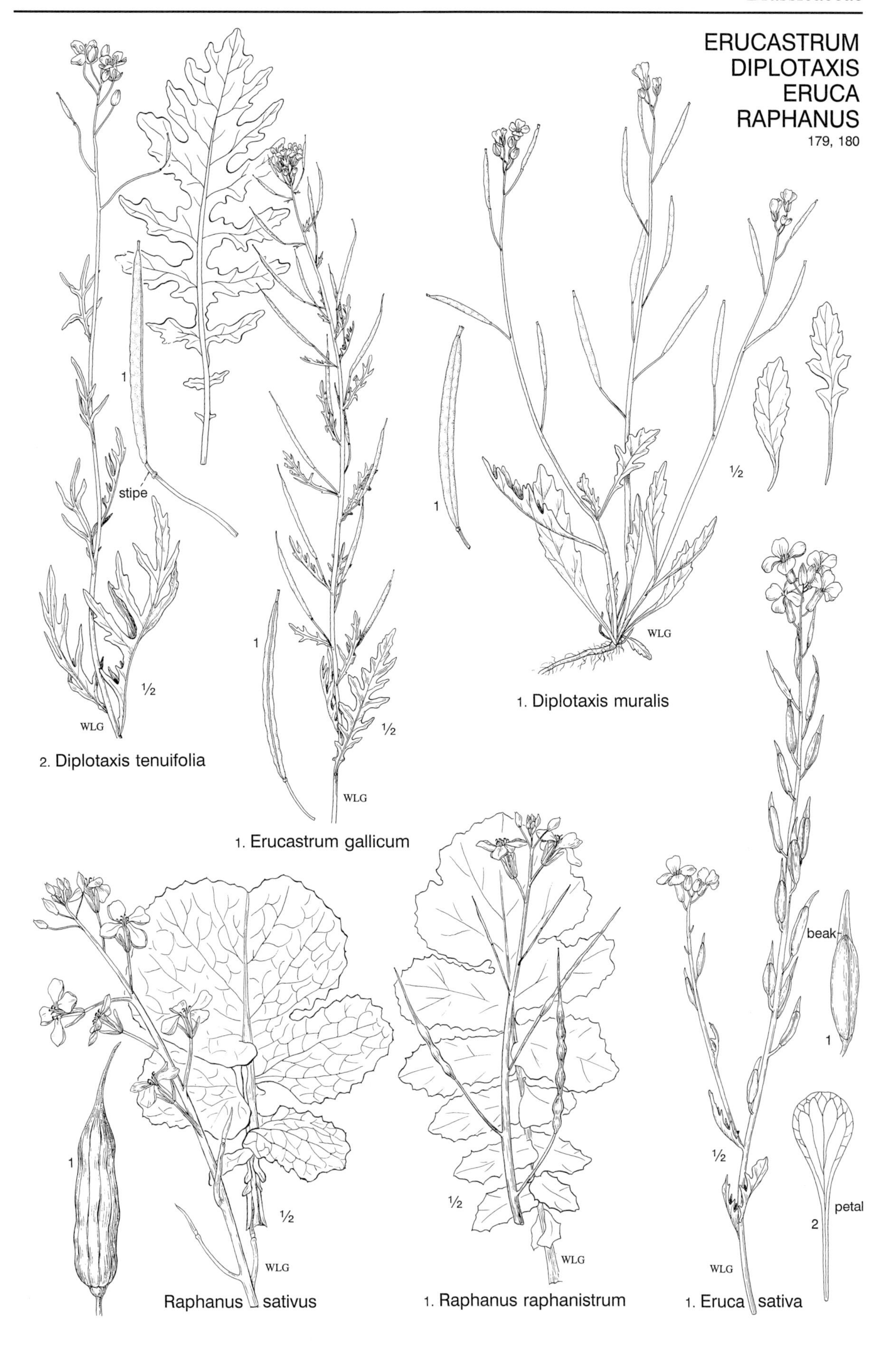

2. Diplotaxis tenuifolia

1. Erucastrum gallicum

1. Diplotaxis muralis

Raphanus sativus

1. Raphanus raphanistrum

1. Eruca sativa

RAPISTRUM
CAKILE
CONRINGIA
LEPIDIUM
180, 181

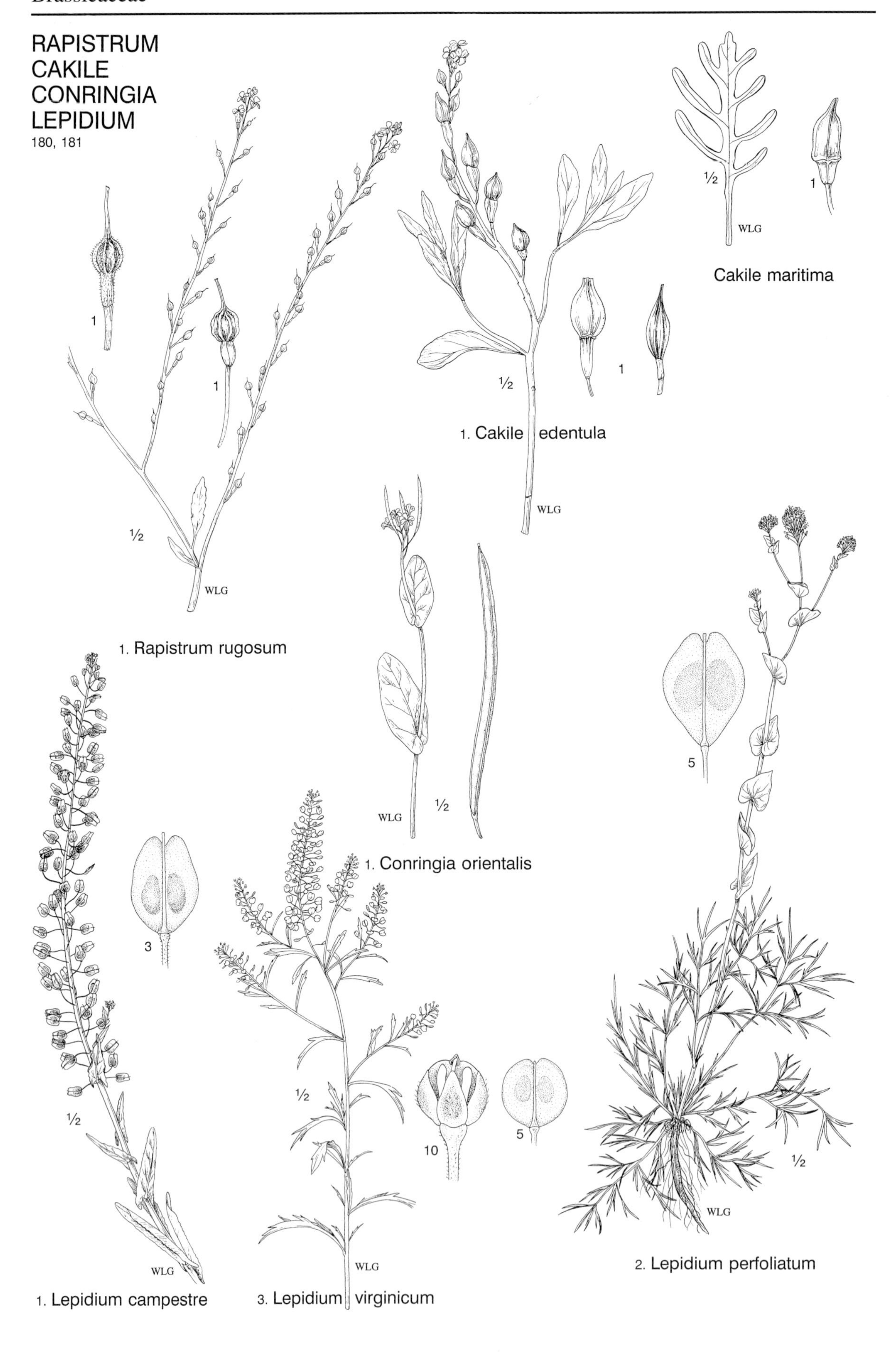

Cakile maritima

1. Cakile edentula

1. Rapistrum rugosum

1. Conringia orientalis

2. Lepidium perfoliatum

1. Lepidium campestre

3. Lepidium virginicum

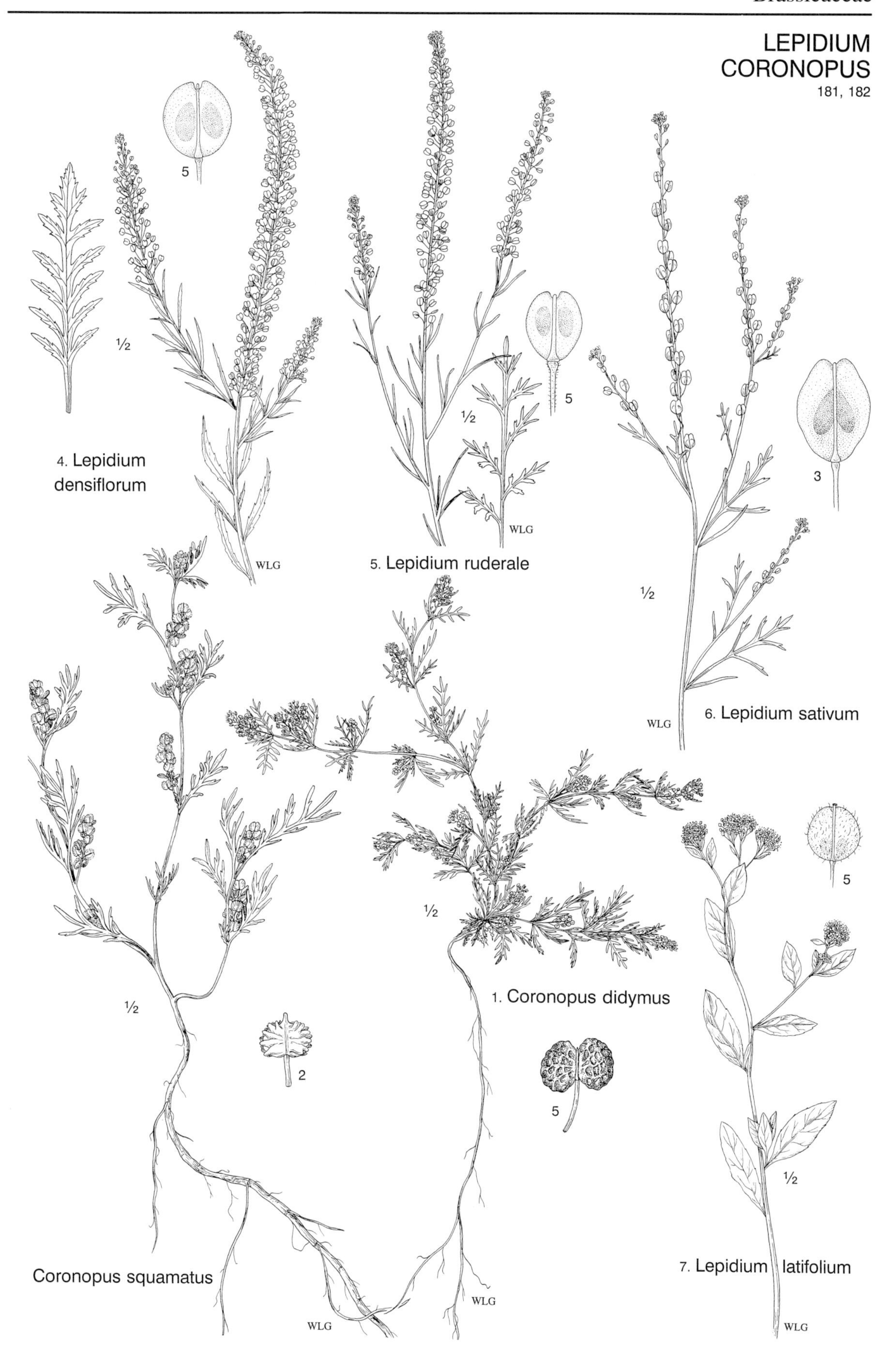

4. Lepidium
densiflorum

5. Lepidium ruderale

6. Lepidium sativum

1. Coronopus didymus

Coronopus squamatus

7. Lepidium latifolium

CARDARIA
ISATIS
THLASPI
TEESDALIA
182, 183

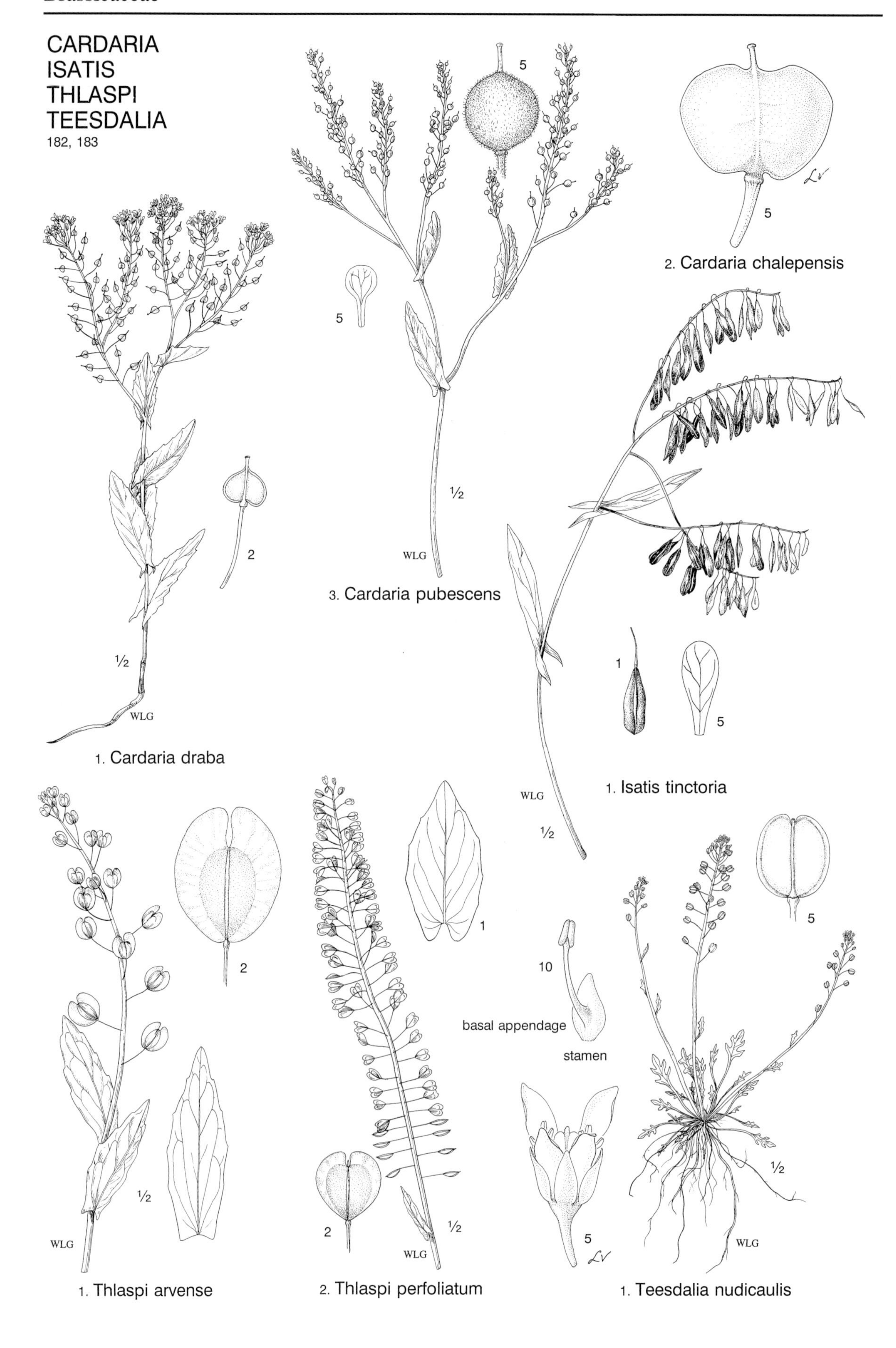

1. Cardaria draba

5

1/2

WLG

2. Cardaria chalepensis

3. Cardaria pubescens

1. Isatis tinctoria

10

basal appendage

stamen

1. Thlaspi arvense

2. Thlaspi perfoliatum

1. Teesdalia nudicaulis

IBERIS
CAPSELLA
SUBULARIA
MYAGRUM
NESLIA
BUNIAS
183, 184

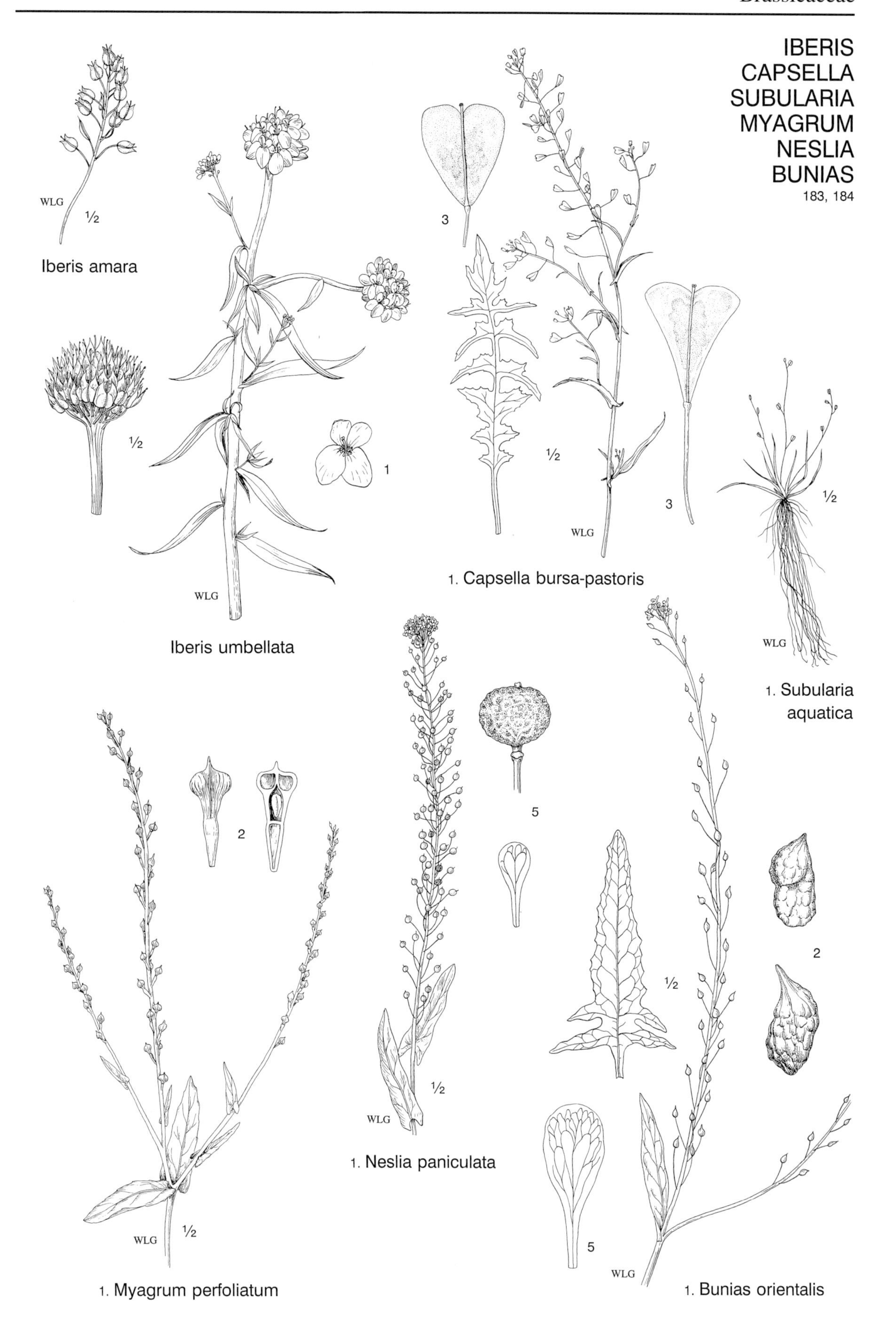

WLG ½

Iberis amara

½

½

1

WLG

Iberis umbellata

3

½

WLG

1. Capsella bursa-pastoris

3

½

WLG

1. Subularia
aquatica

2

5

½

2

2

½

5

WLG

1. Myagrum perfoliatum

1. Neslia paniculata

1. Bunias orientalis

LUNARIA
ALYSSUM
LOBULARIA
BERTEROA
DRABA
185, 186

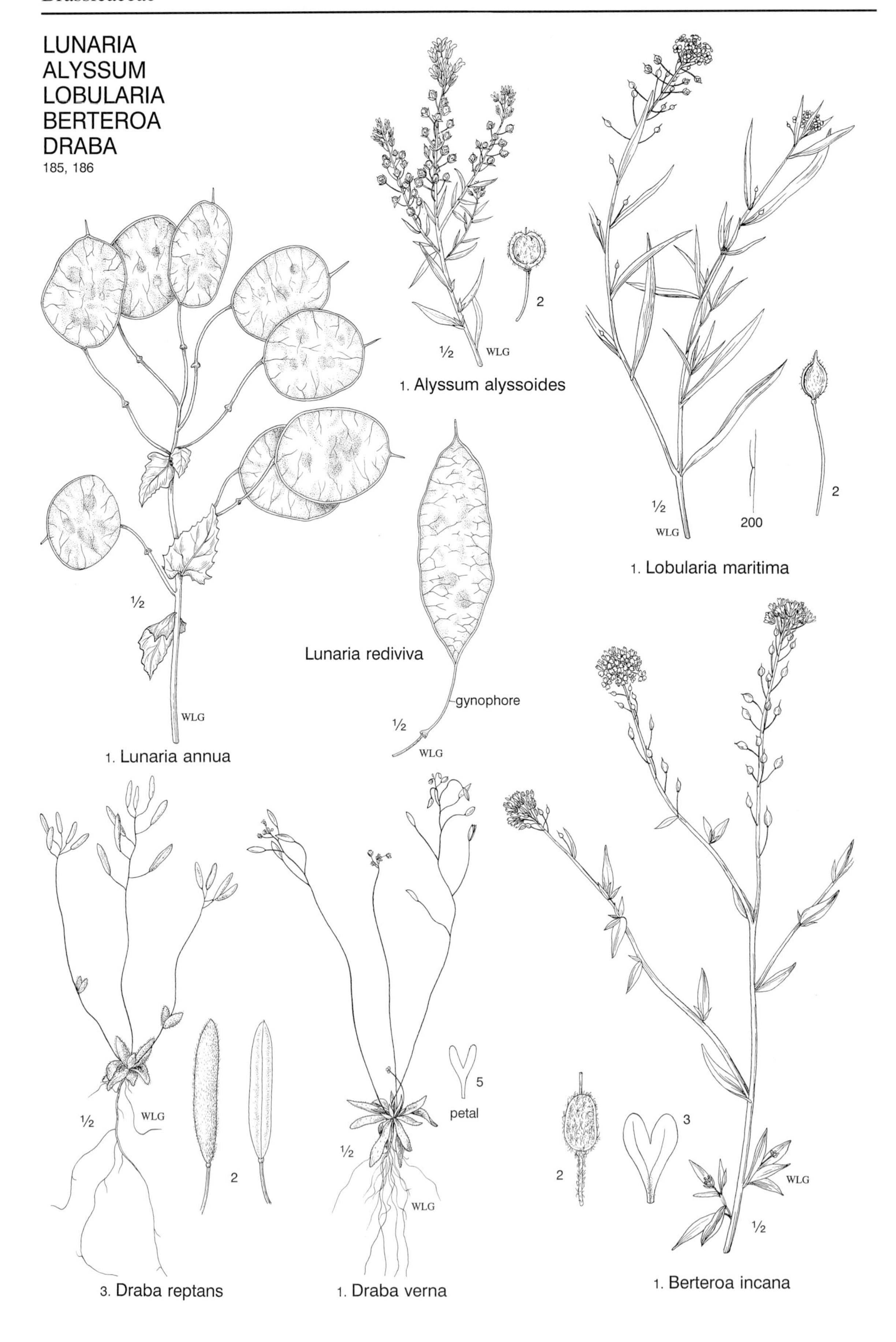

1. Alyssum alyssoides

½ WLG

1. Lobularia maritima

½ WLG 200 2

Lunaria rediviva

gynophore

½ WLG

1. Lunaria annua

WLG

3. Draba reptans

½ WLG 2

1. Draba verna

5 petal

½ WLG

1. Berteroa incana

2 3

½ WLG

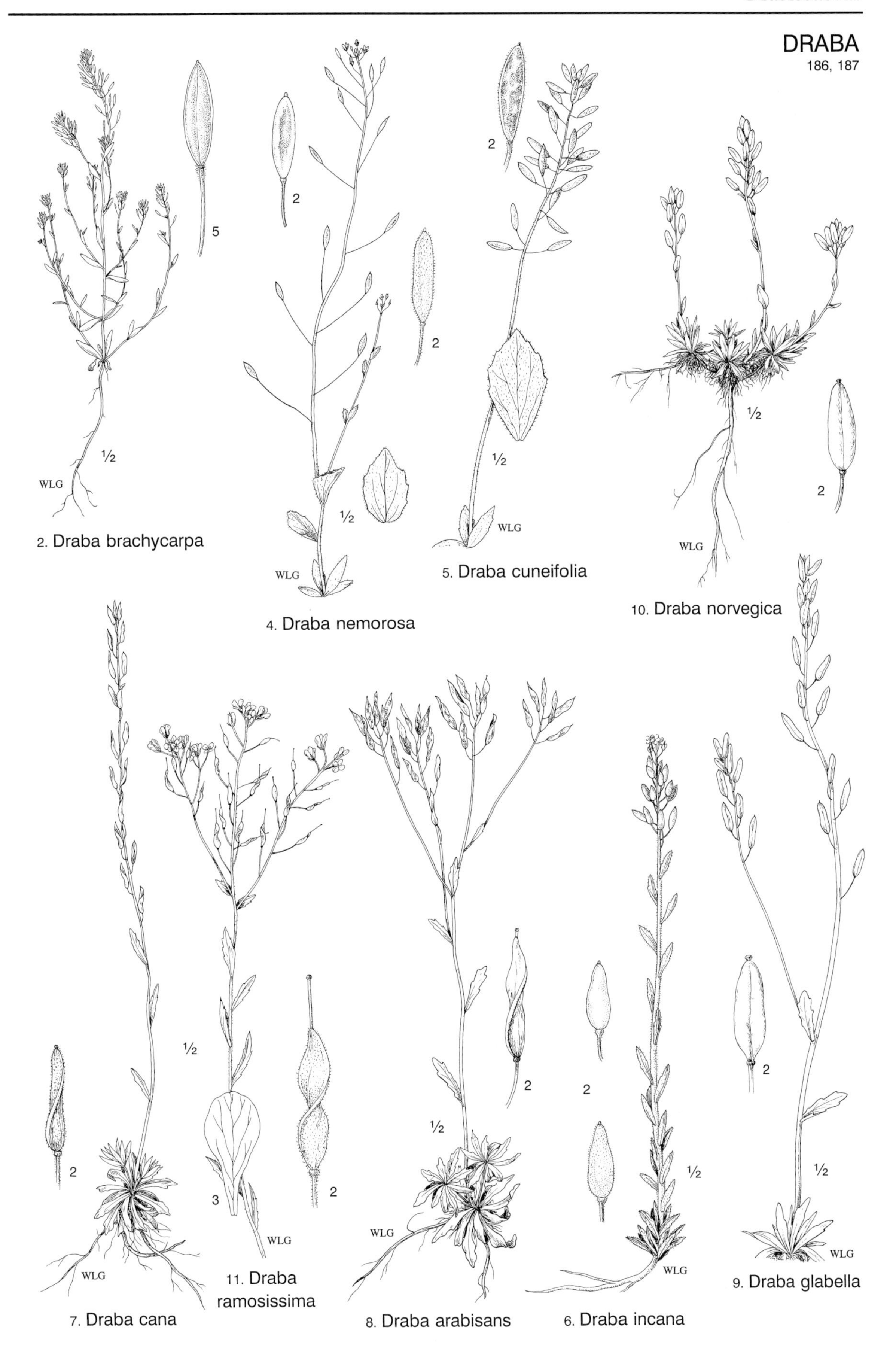

2. Draba brachycarpa

4. Draba nemorosa

5. Draba cuneifolia

10. Draba norvegica

7. Draba cana

11. Draba ramosissima

8. Draba arabisans

6. Draba incana

9. Draba glabella

LESQUERELLA
ARMORACIA
LEAVENWORTHIA
187, 188

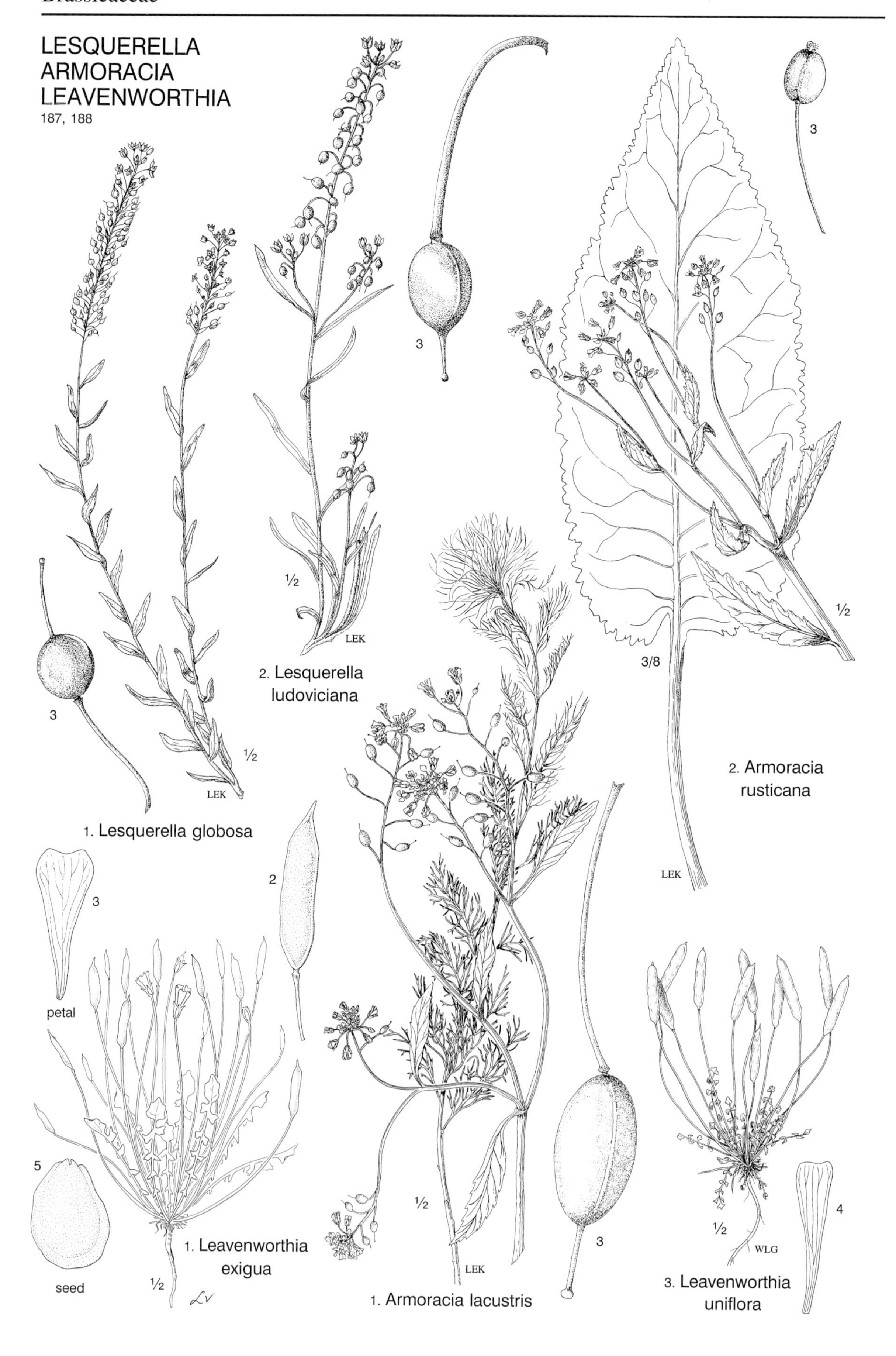

2. Lesquerella
ludoviciana

3

1. Lesquerella globosa

3

petal

5

seed   ½

1. Leavenworthia
exigua

2

1. Armoracia lacustris

3/8

2. Armoracia
rusticana

3

½   WLG

4

3. Leavenworthia
uniflora

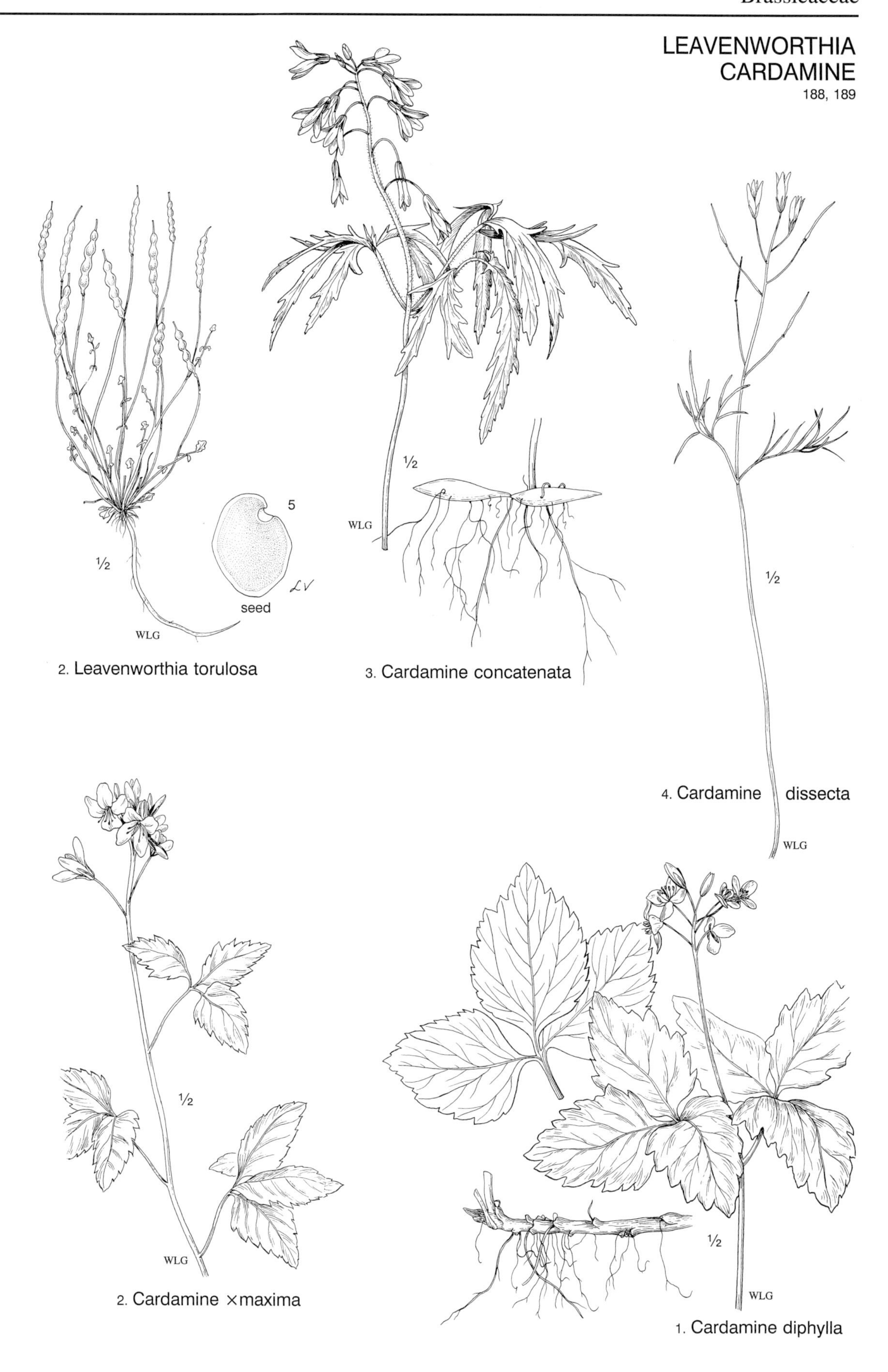

2. Leavenworthia torulosa

seed

3. Cardamine concatenata

4. Cardamine dissecta

2. Cardamine ×maxima

1. Cardamine diphylla

CARDAMINE
189, 190

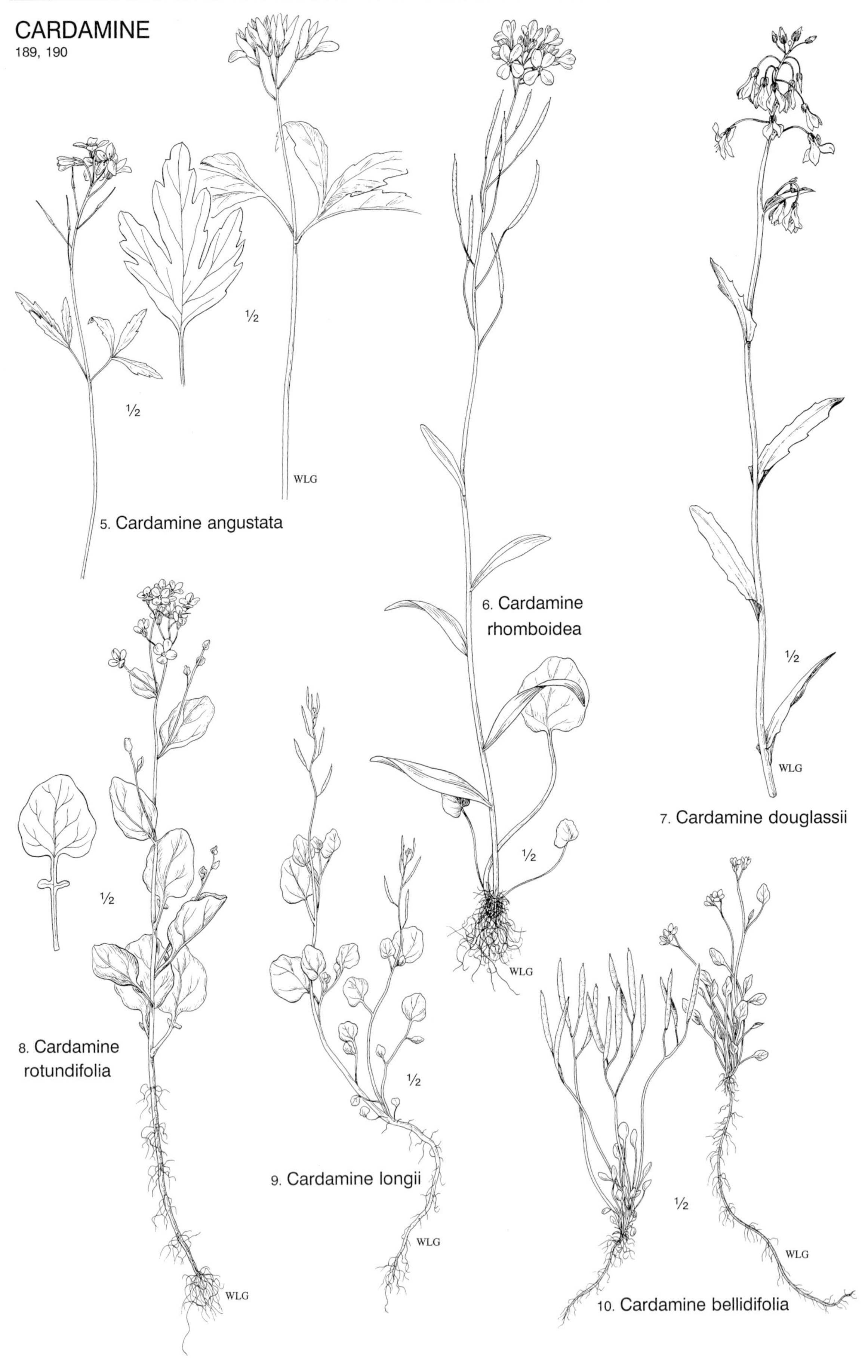

5. Cardamine angustata

½

½

WLG

6. Cardamine rhomboidea

½

½

WLG

7. Cardamine douglassii

½

WLG

8. Cardamine rotundifolia

½

WLG

9. Cardamine longii

½

WLG

10. Cardamine bellidifolia

½

WLG

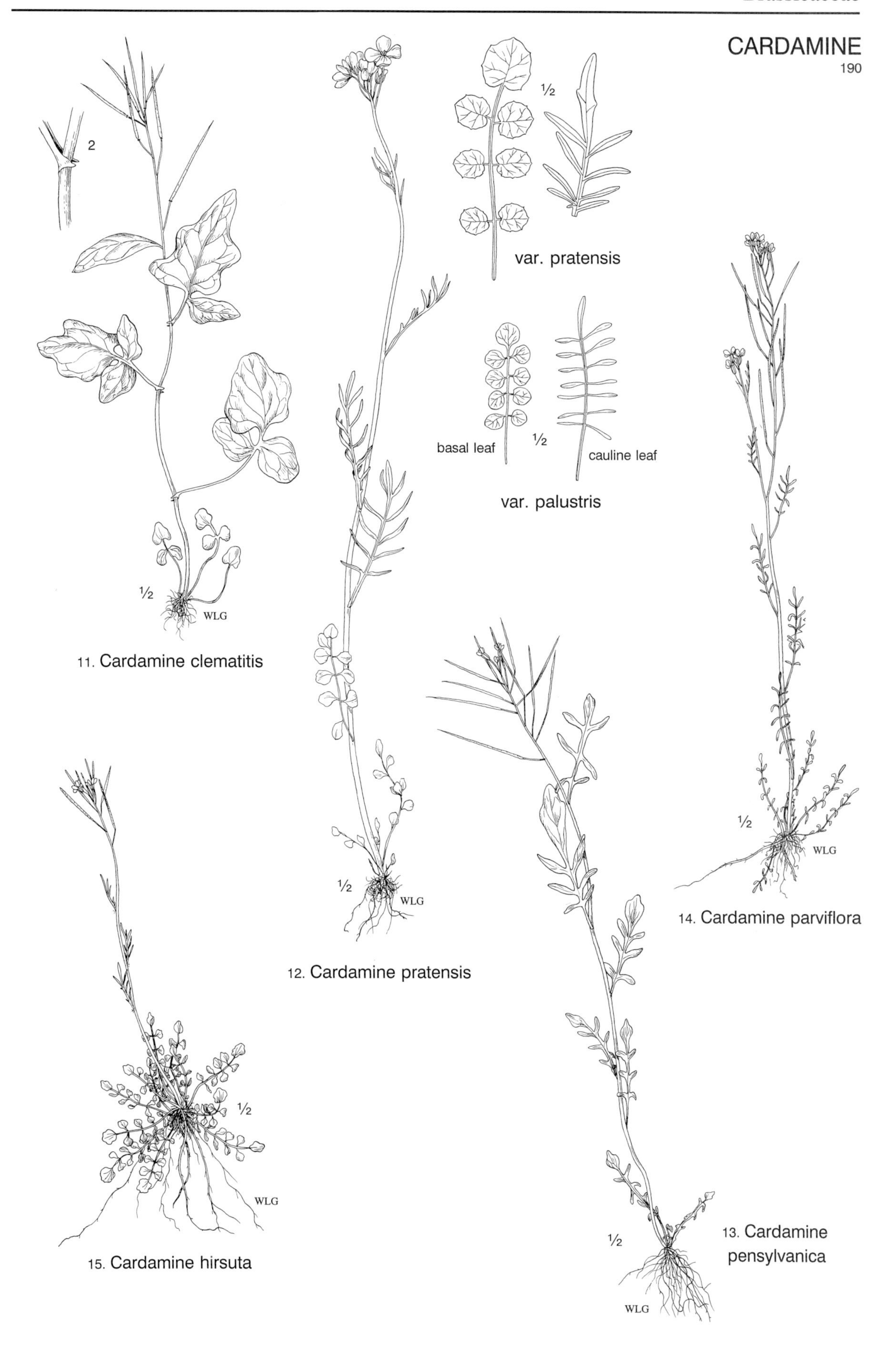

var. pratensis

basal leaf   cauline leaf

var. palustris

11. Cardamine clematitis

12. Cardamine pratensis

14. Cardamine parviflora

15. Cardamine hirsuta

13. Cardamine
pensylvanica

SIBARA
ARABIS
191, 192

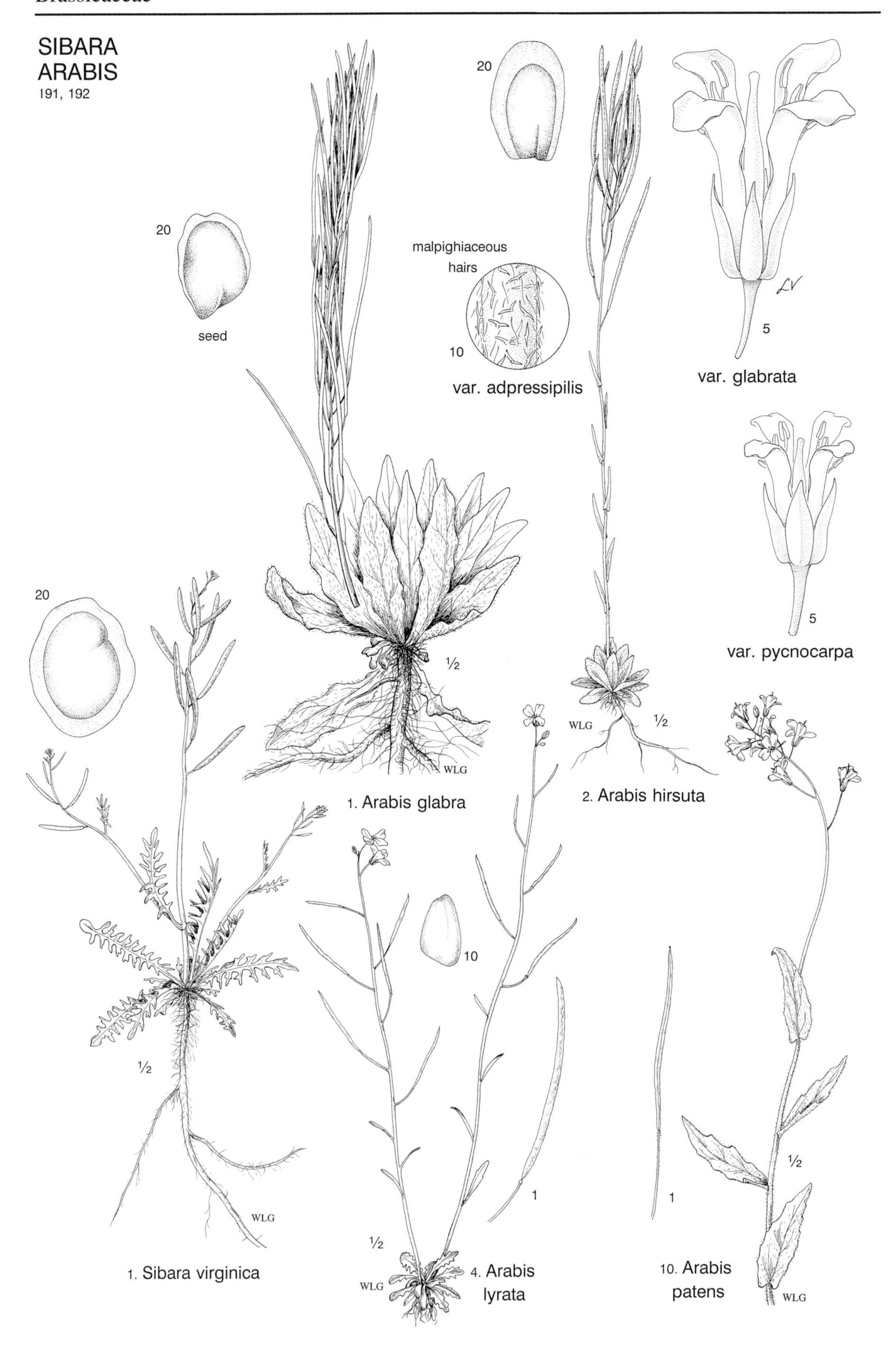

seed

malpighiaceous
hairs

10

var. adpressipilis

20

var. glabrata

5

var. pycnocarpa

5

1. Arabis glabra

2. Arabis hirsuta

20

10

1. Sibara virginica

4. Arabis lyrata

10. Arabis patens

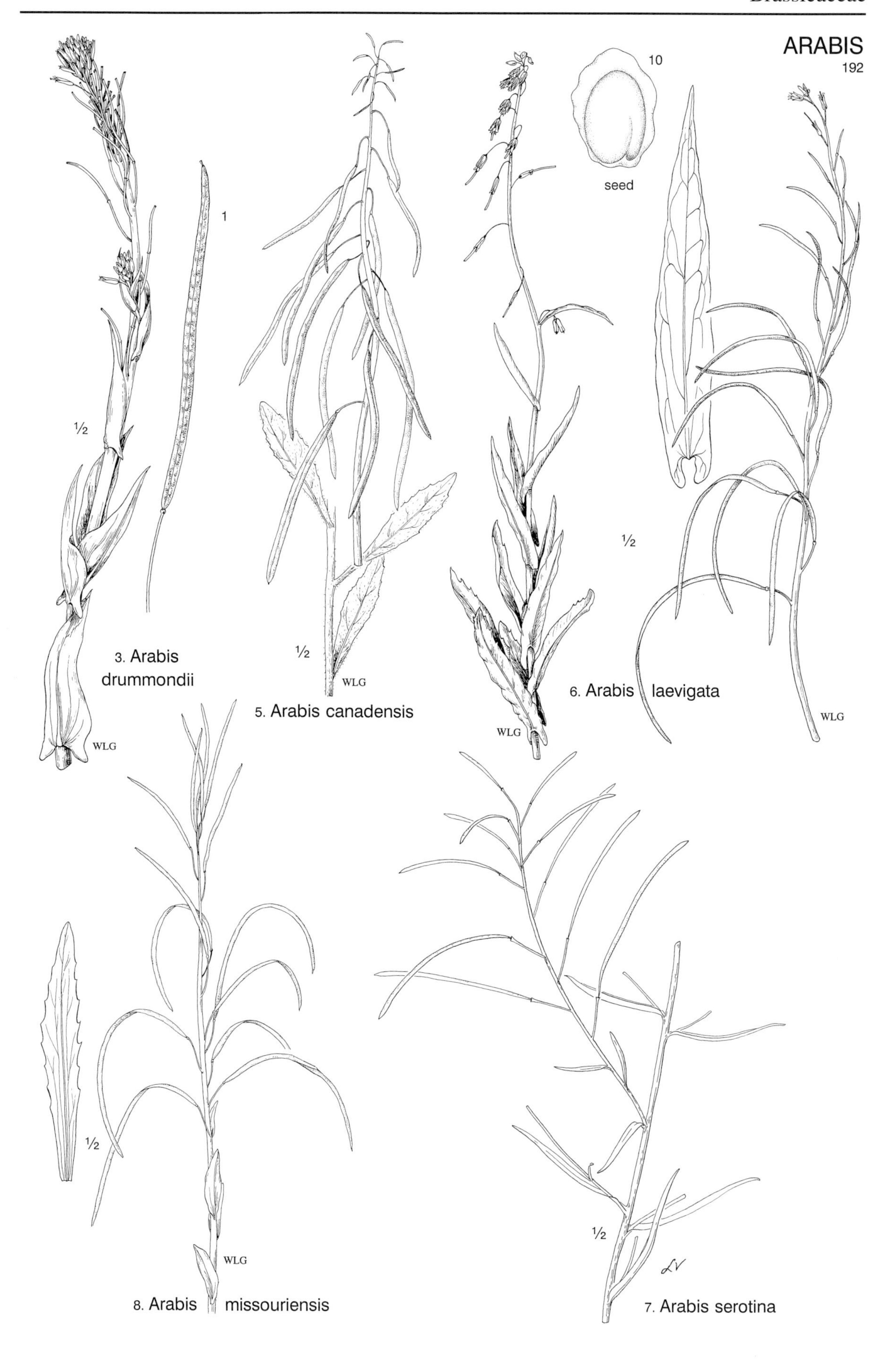

10

seed

½

1

3. Arabis
drummondii

WLG

½

WLG

5. Arabis canadensis

½

WLG

6. Arabis laevigata

½

WLG

½

WLG

8. Arabis missouriensis

½

7. Arabis serotina

ARABIS
192, 193

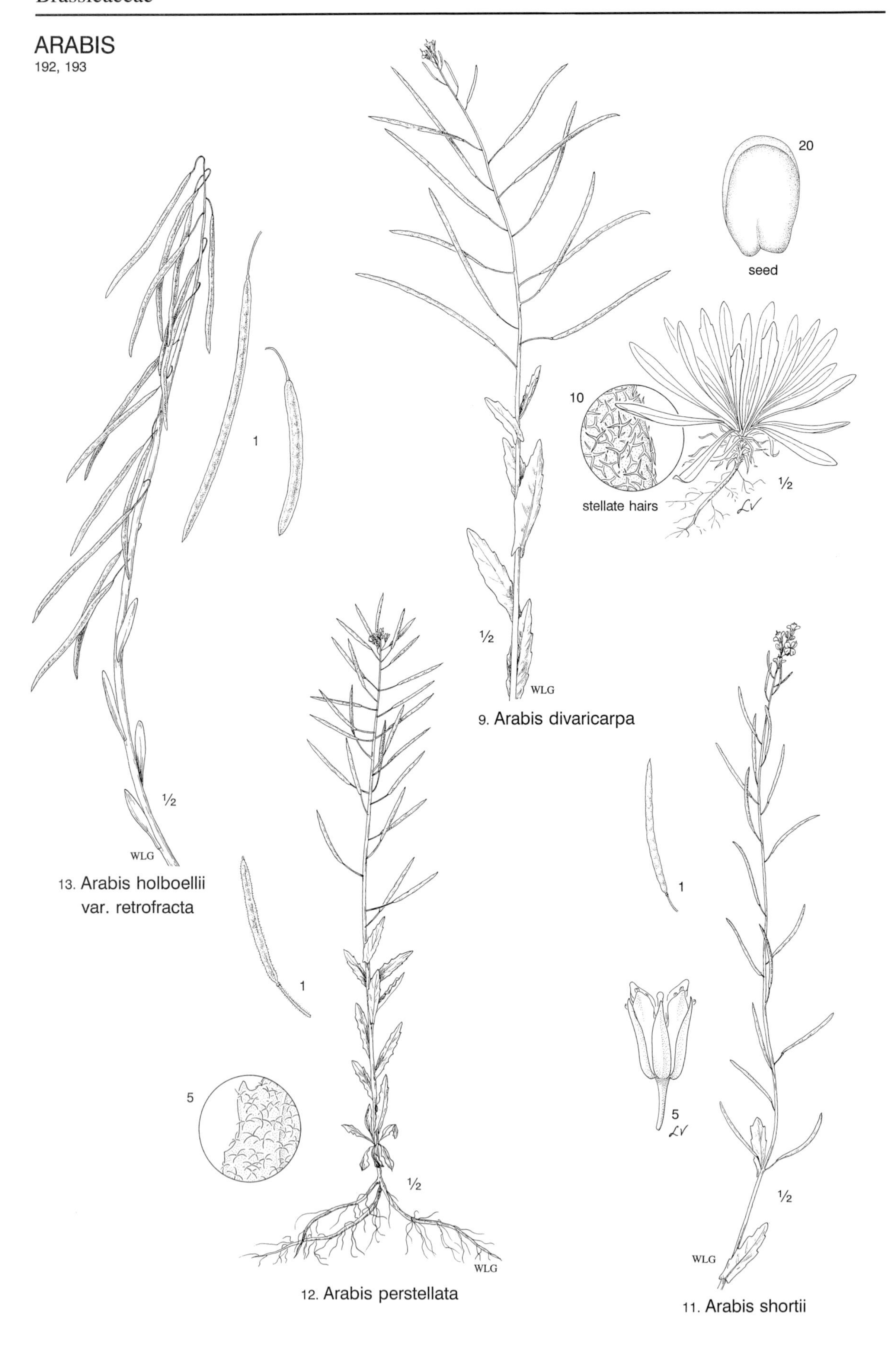

20

seed

10

stellate hairs

½

9. Arabis divaricarpa

WLG

1

½

WLG

13. Arabis holboellii
var. retrofracta

5

1

½

WLG

12. Arabis perstellata

1

5

½

WLG

11. Arabis shortii

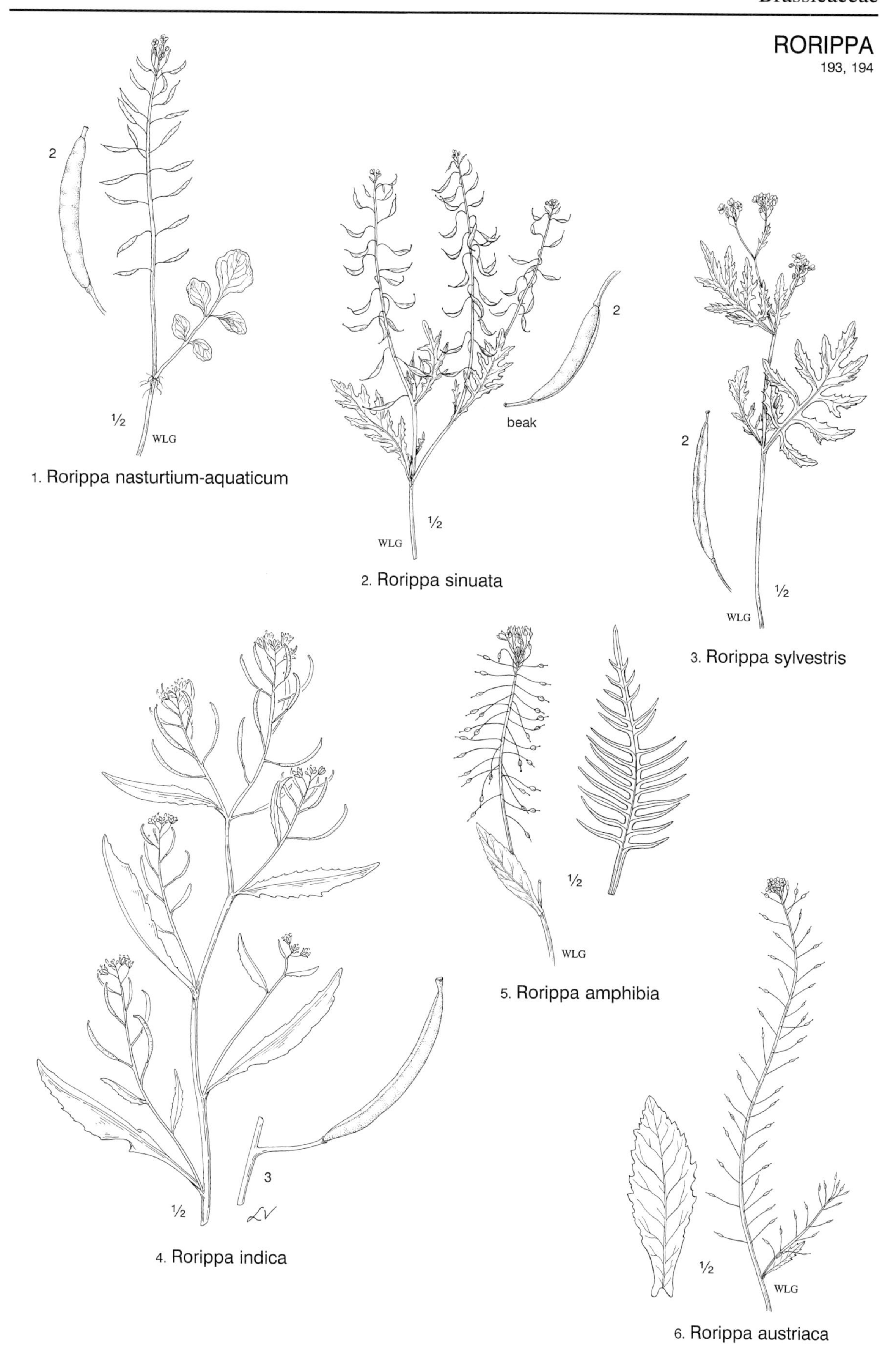

1. Rorippa nasturtium-aquaticum

2. Rorippa sinuata

3. Rorippa sylvestris

4. Rorippa indica

5. Rorippa amphibia

6. Rorippa austriaca

RORIPPA
BARBAREA
194, 195

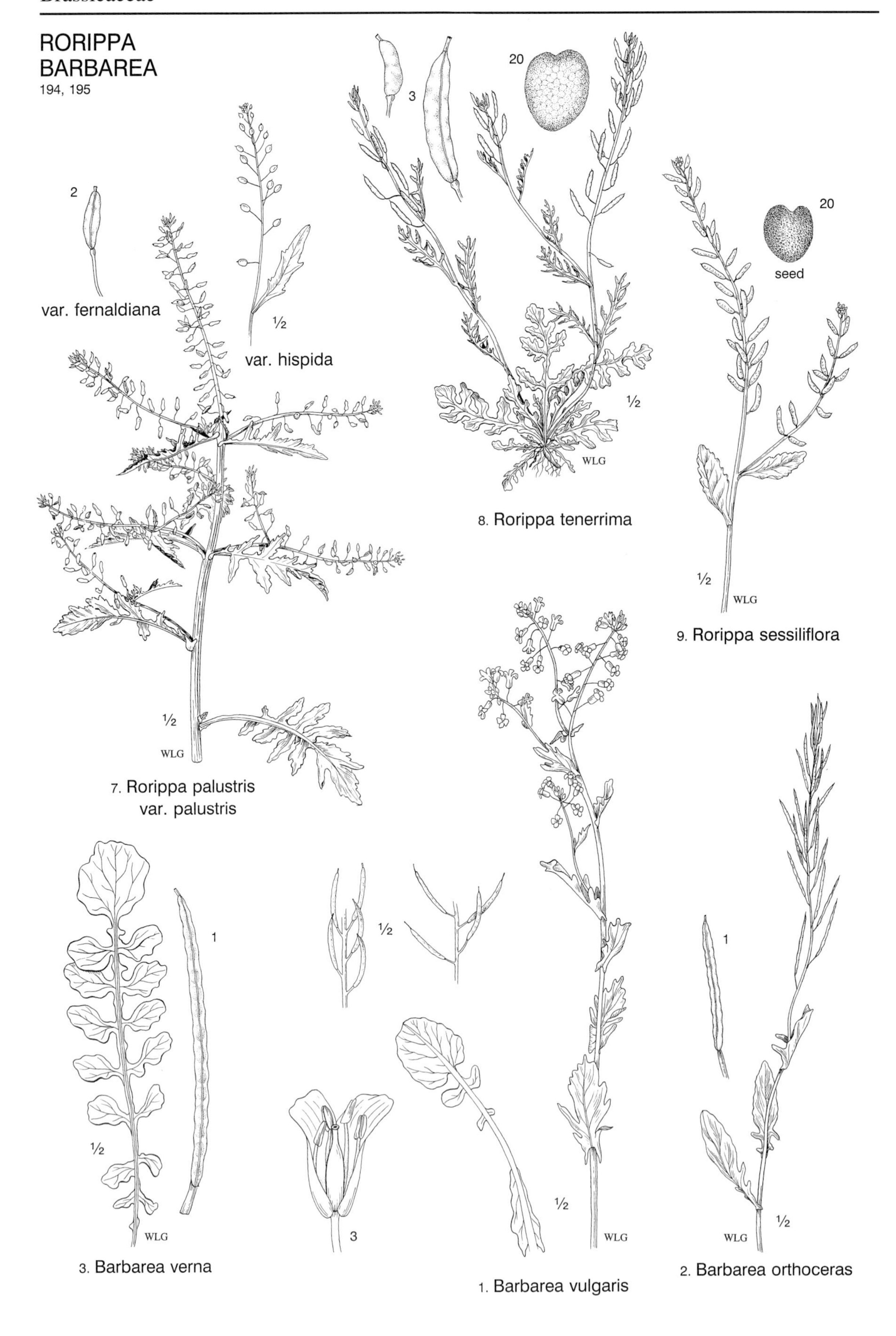

var. fernaldiana

var. hispida

20

20

seed

8. Rorippa tenerrima

9. Rorippa sessiliflora

7. Rorippa palustris
var. palustris

3. Barbarea verna

1. Barbarea vulgaris

2. Barbarea orthoceras

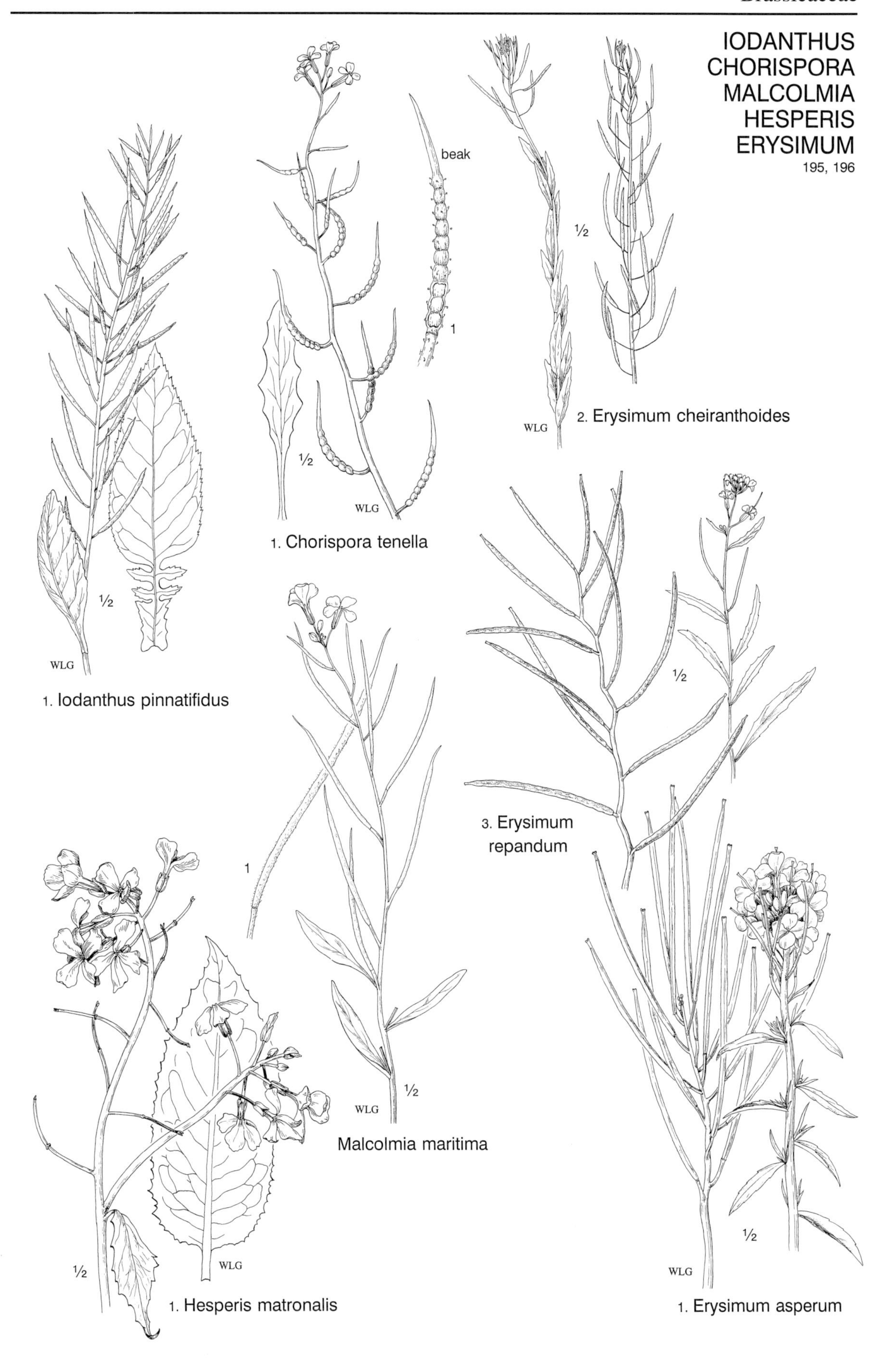

IODANTHUS
CHORISPORA
MALCOLMIA
HESPERIS
ERYSIMUM
195, 196

beak

½

WLG   2. Erysimum cheiranthoides

½

WLG

1. Chorispora tenella

½

WLG

1. Iodanthus pinnatifidus

½

3. Erysimum
repandum

1

½

WLG

Malcolmia maritima

½

½

WLG

1. Hesperis matronalis

½

WLG

1. Erysimum asperum

ERYSIMUM ALLIARIA SISYMBRIUM
196, 197

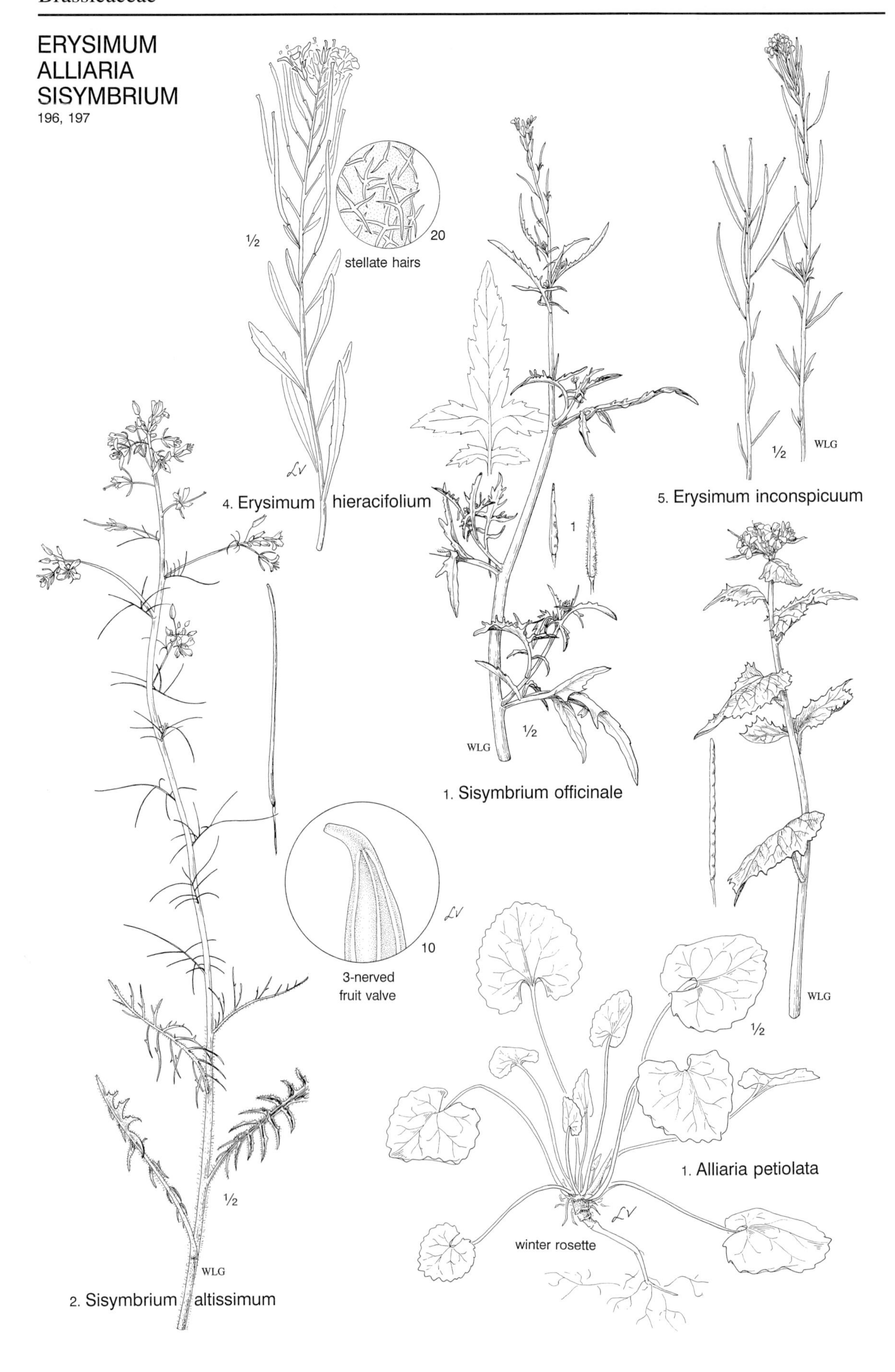

½   20
stellate hairs

4. Erysimum hieracifolium

5. Erysimum inconspicuum
½   WLG

½   1   WLG   ½

1. Sisymbrium officinale

3-nerved fruit valve   10

1. Alliaria petiolata

winter rosette

½   WLG

½   WLG

2. Sisymbrium altissimum

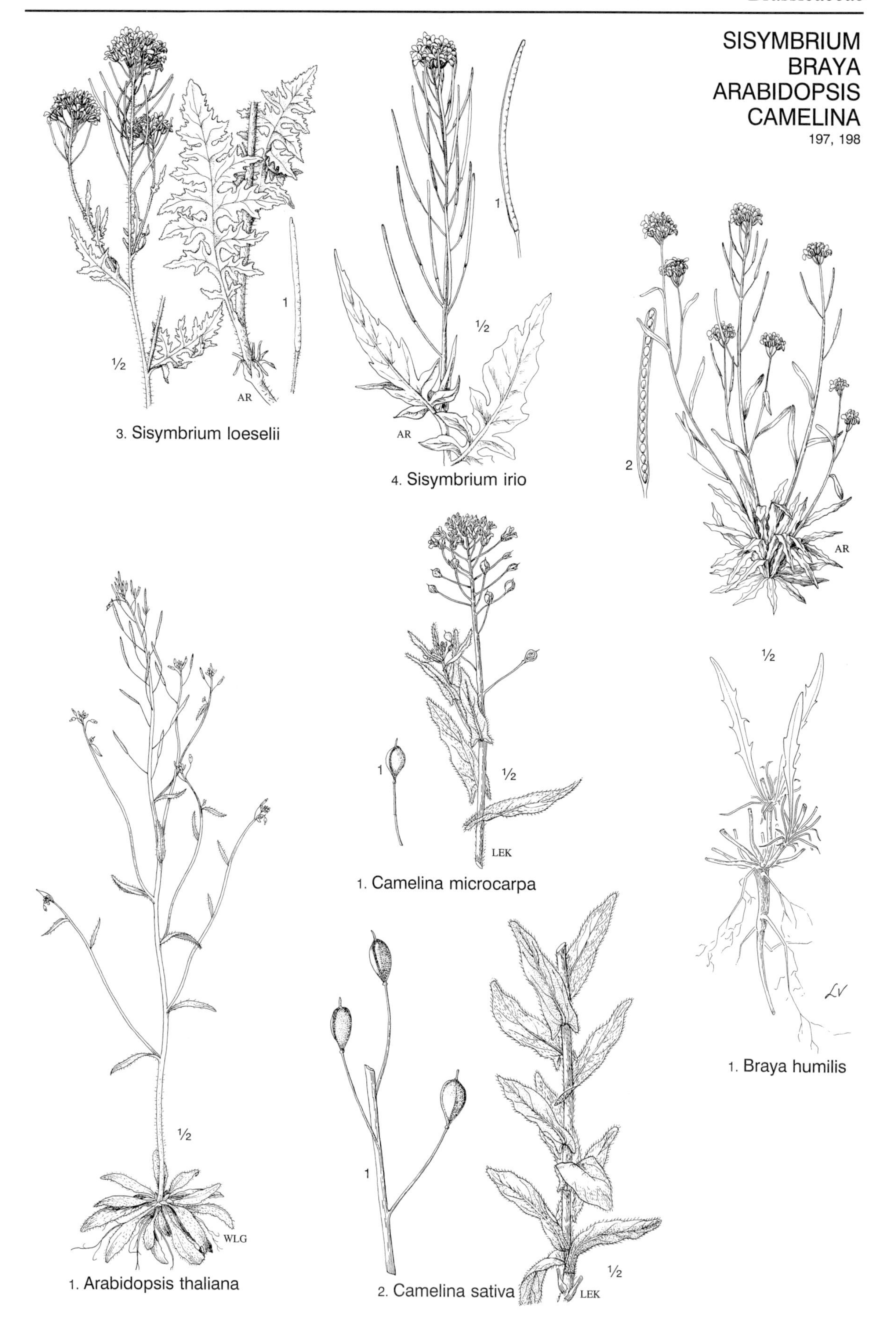

3. Sisymbrium loeselii

4. Sisymbrium irio

1. Camelina microcarpa

1. Arabidopsis thaliana

2. Camelina sativa

1. Braya humilis

DESCURAINIA
RESEDA
198, 199

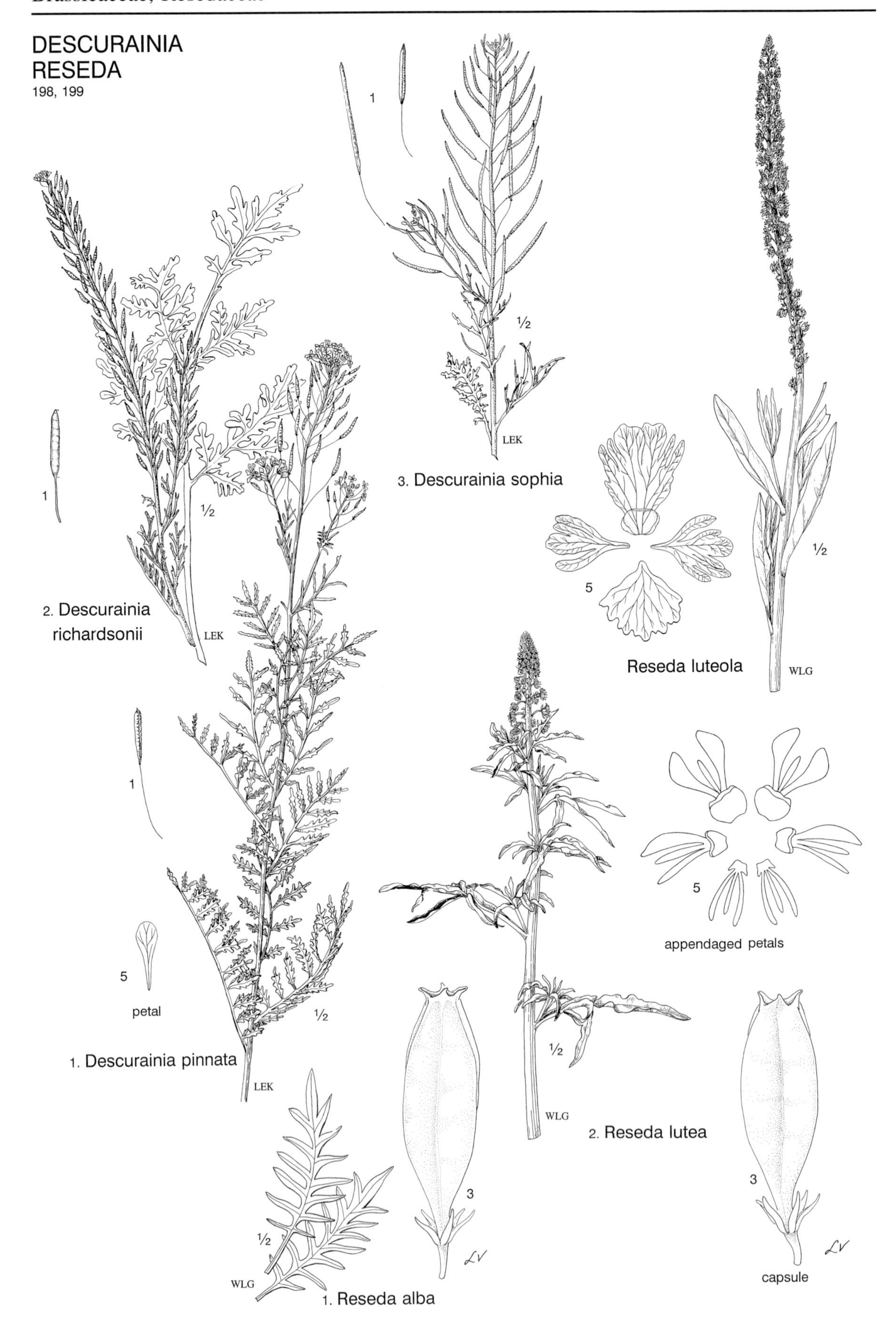

2. Descurainia richardsonii

LEK

1. Descurainia pinnata

5 petal

LEK

½

3. Descurainia sophia

LEK

Reseda luteola

WLG

appendaged petals

5

2. Reseda lutea

WLG

capsule

3

1. Reseda alba

WLG

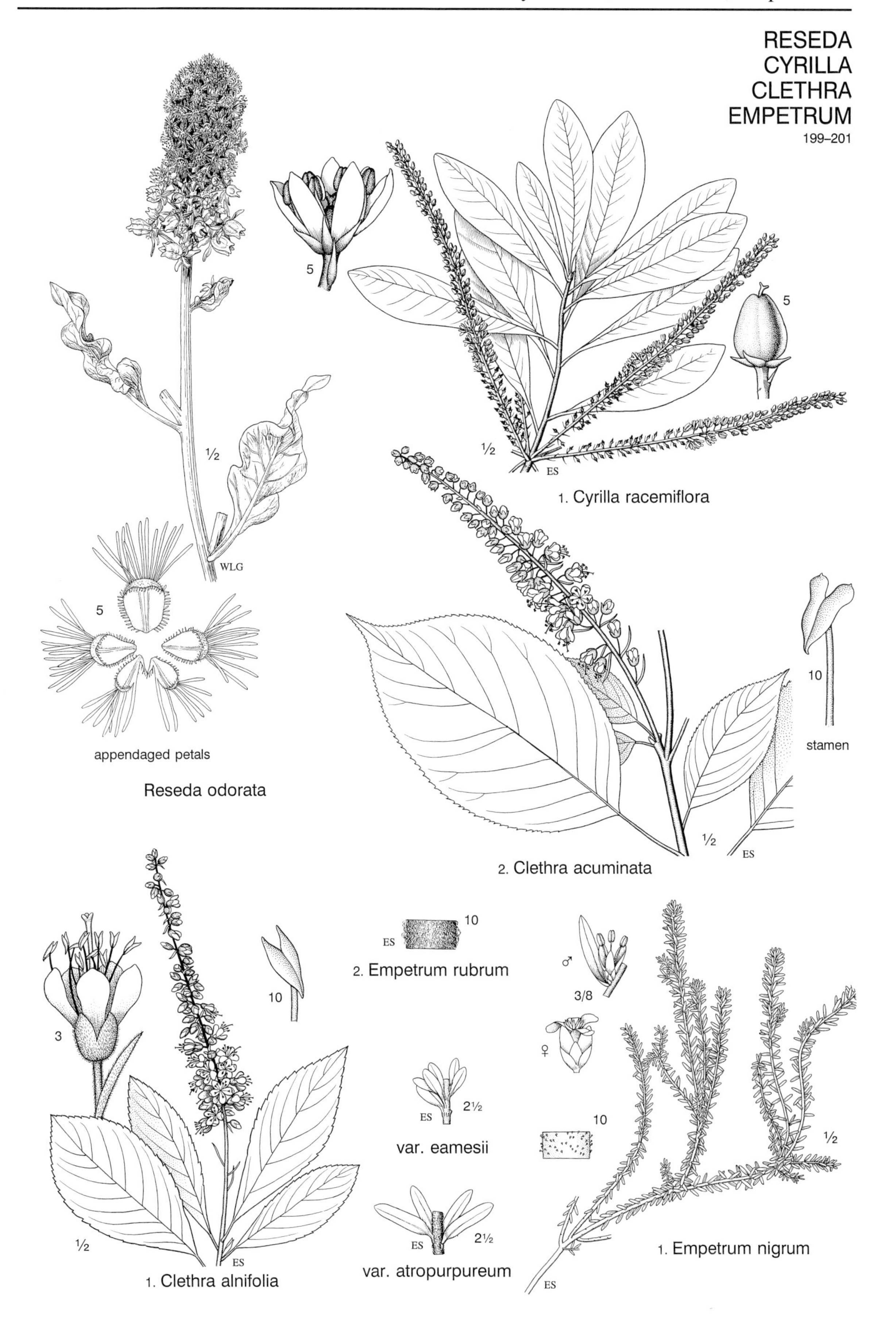

½

5

WLG

5

appendaged petals

Reseda odorata

1. Cyrilla racemiflora

½

ES

5

10

stamen

2. Clethra acuminata

½

ES

3

10

½

ES

1. Clethra alnifolia

10

ES

2. Empetrum rubrum

ES   2½

var. eamesii

ES   2½

var. atropurpureum

ES

♂

3/8

♀

10

½

1. Empetrum nigrum

COREMA
LEDUM
LEIOPHYLLUM
MENZIESIA
RHODODENDRON
201–204

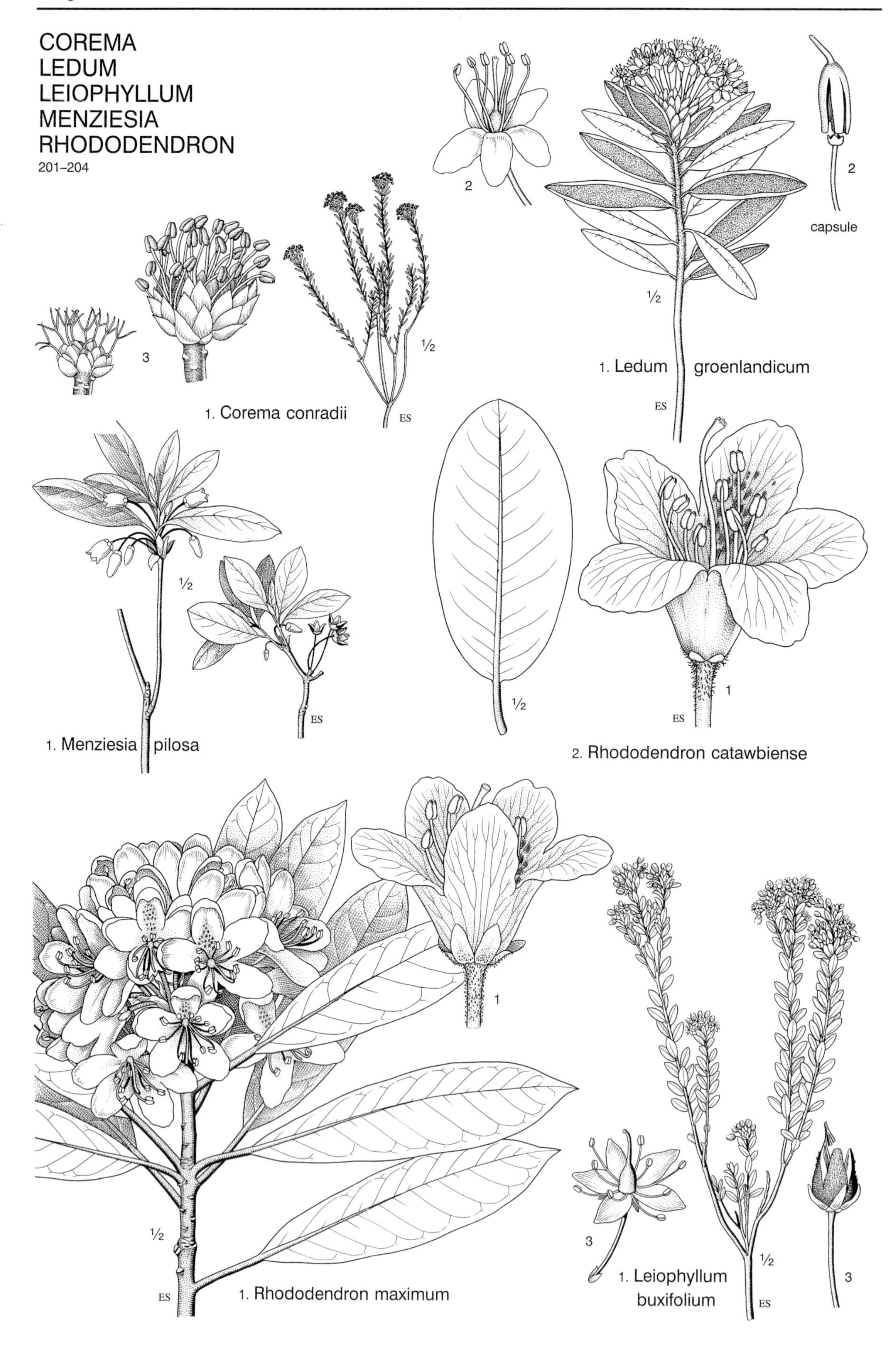

1. Corema conradii

3

1. Ledum groenlandicum

2

capsule

1. Menziesia pilosa

2. Rhododendron catawbiense

1. Rhododendron maximum

1. Leiophyllum buxifolium

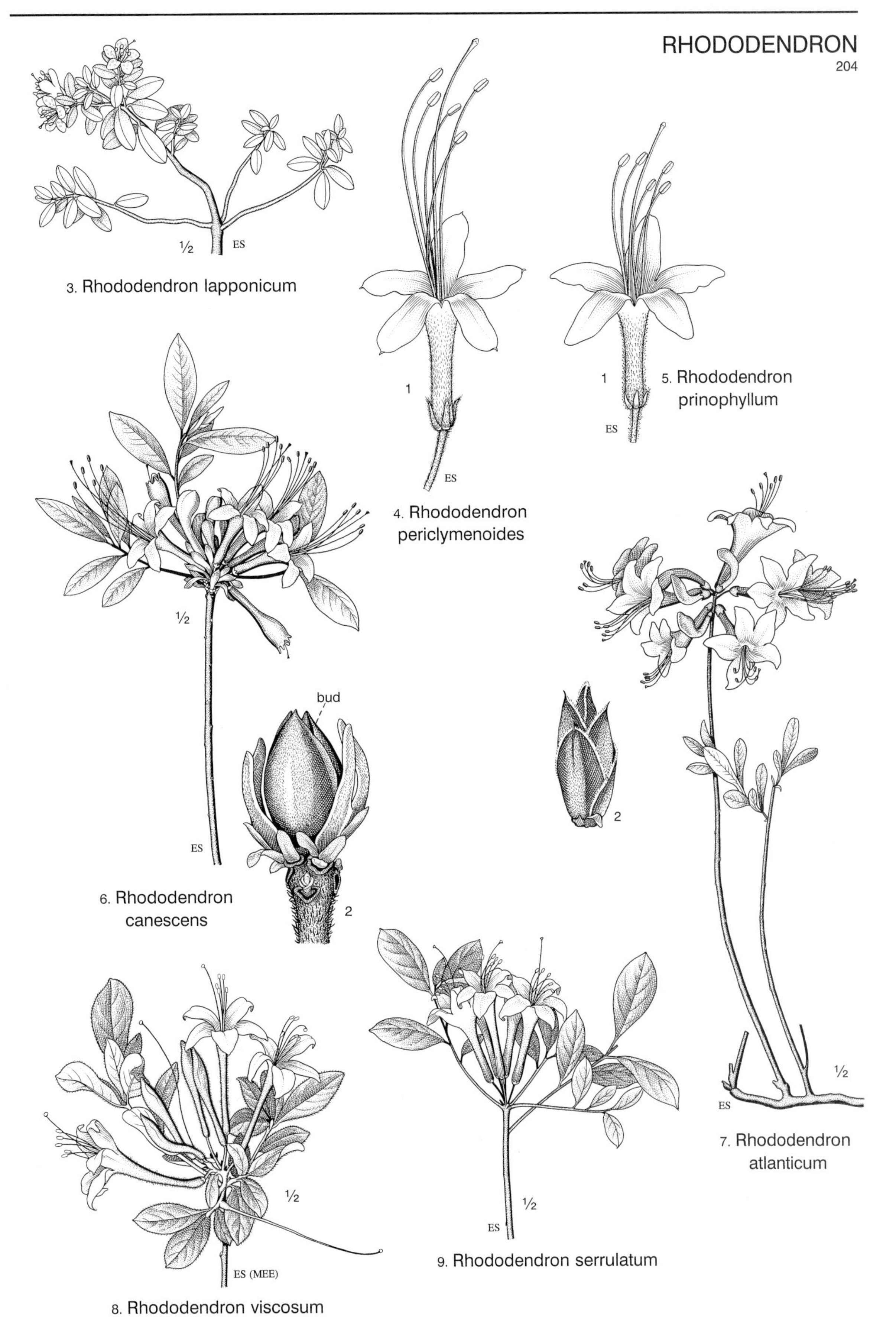

3. Rhododendron lapponicum

4. Rhododendron periclymenoides

5. Rhododendron prinophyllum

6. Rhododendron canescens

bud

7. Rhododendron atlanticum

8. Rhododendron viscosum

9. Rhododendron serrulatum

RHODODENDRON
KALMIA
204, 205

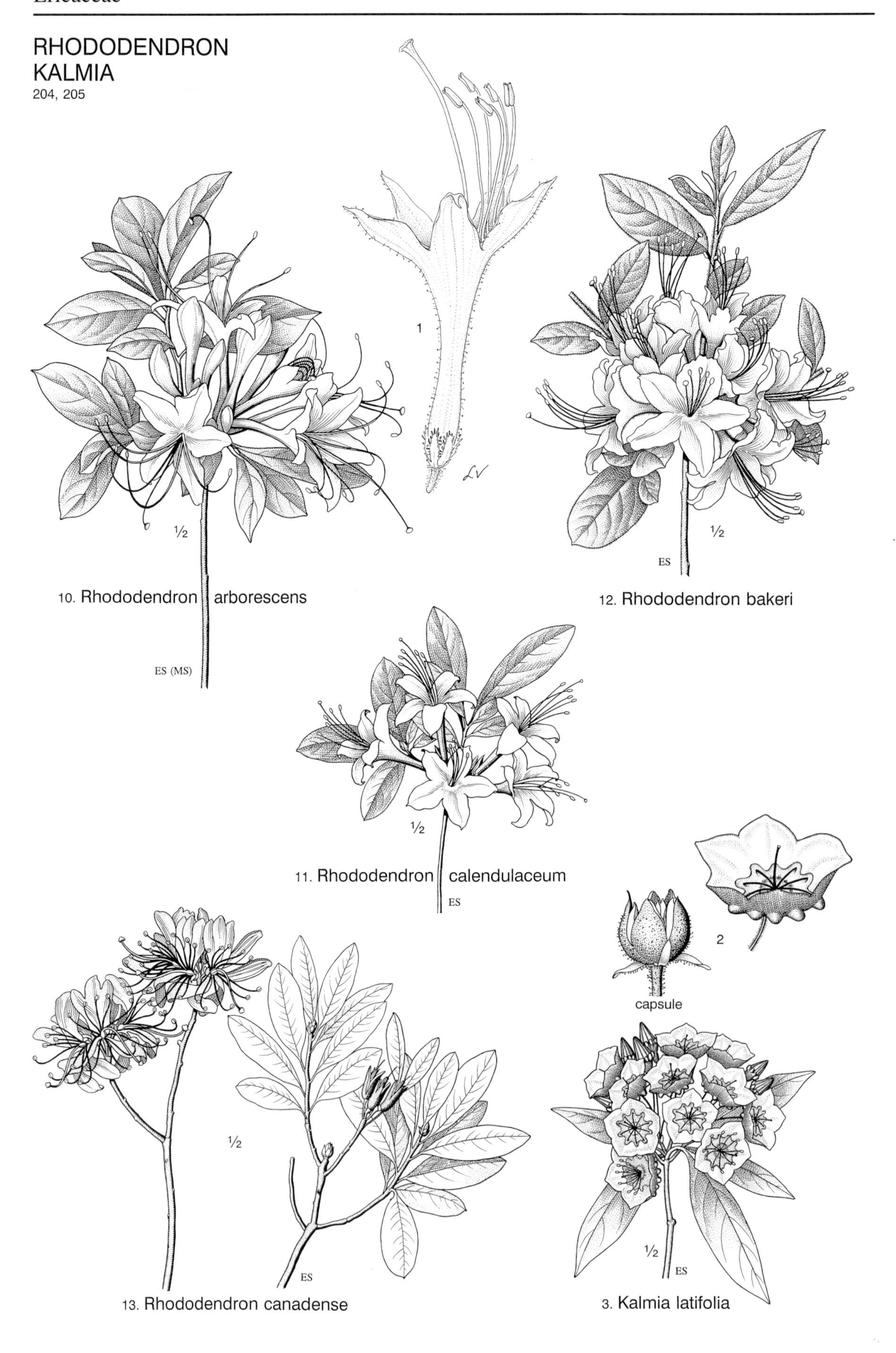

1

10. Rhododendron arborescens

ES (MS)

12. Rhododendron bakeri

ES

11. Rhododendron calendulaceum

ES

capsule

2

13. Rhododendron canadense

ES

3. Kalmia latifolia

ES

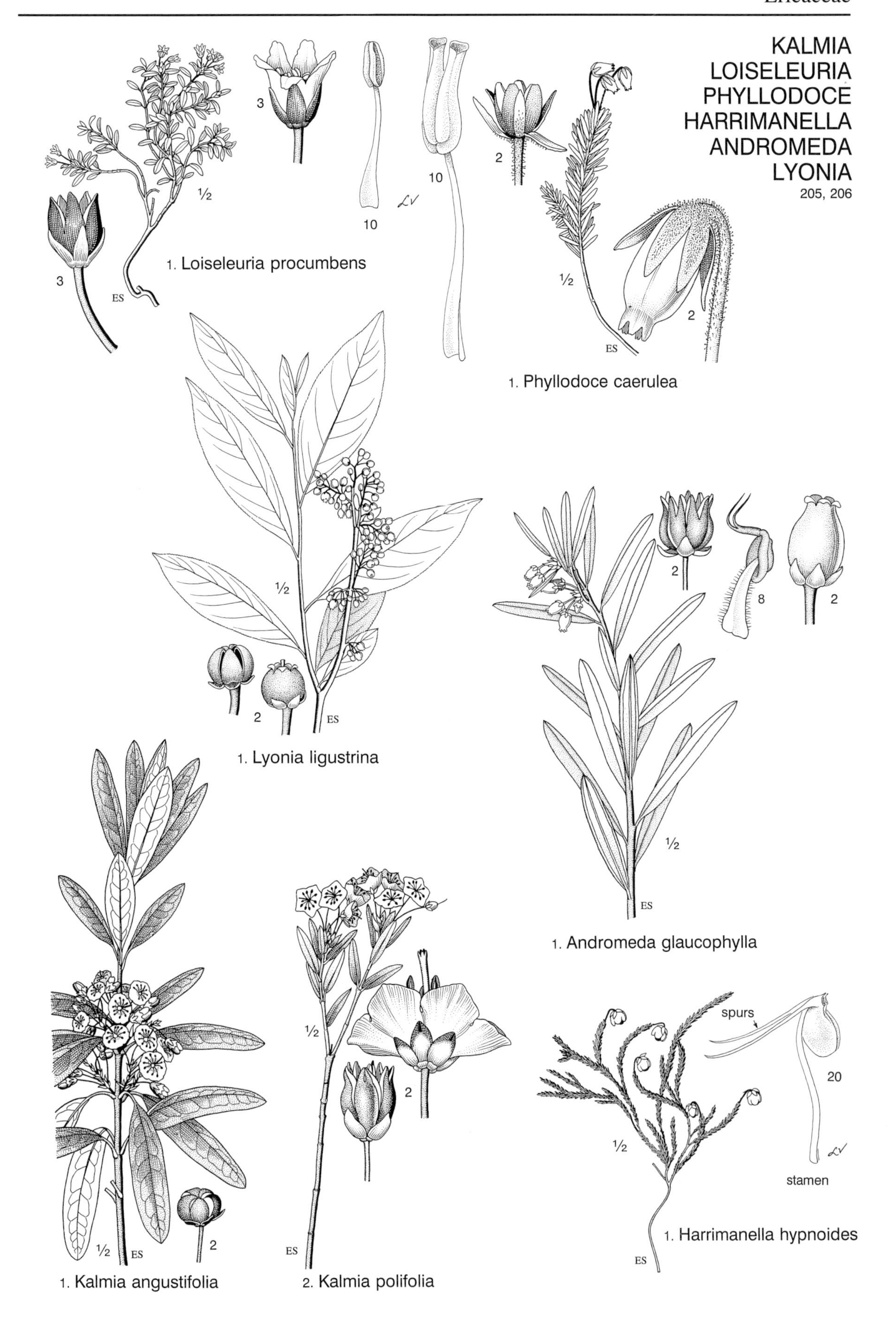

KALMIA
LOISELEURIA
PHYLLODOCE
HARRIMANELLA
ANDROMEDA
LYONIA
205, 206

1. Loiseleuria procumbens

1. Phyllodoce caerulea

1. Lyonia ligustrina

1. Andromeda glaucophylla

spurs

stamen

1. Harrimanella hypnoides

1. Kalmia angustifolia

2. Kalmia polifolia

LYONIA
LEUCOTHOE
EUBOTRYS
207

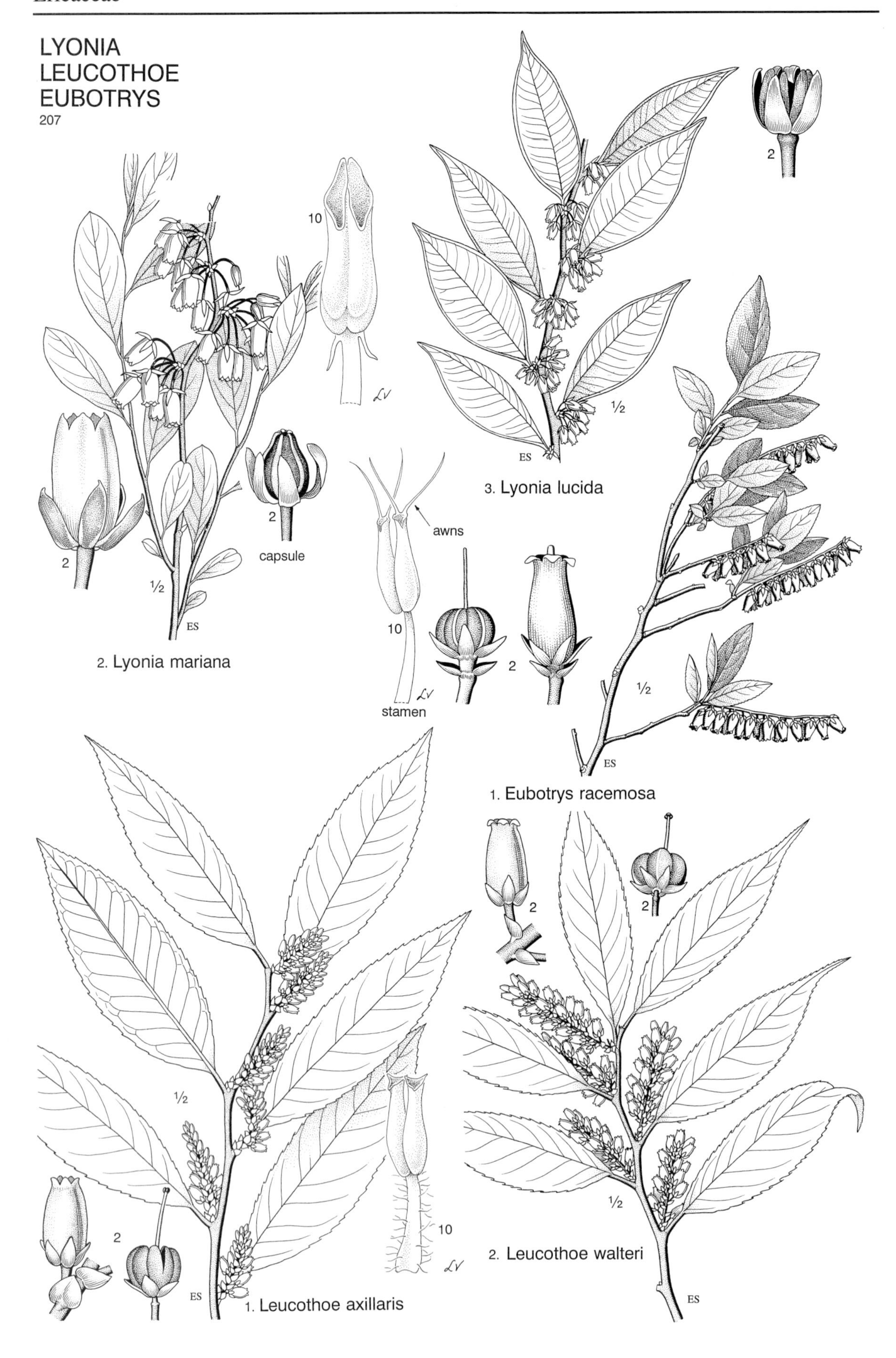

2. Lyonia mariana

capsule

10

awns

stamen

3. Lyonia lucida

1. Eubotrys racemosa

1. Leucothoe axillaris

2. Leucothoe walteri

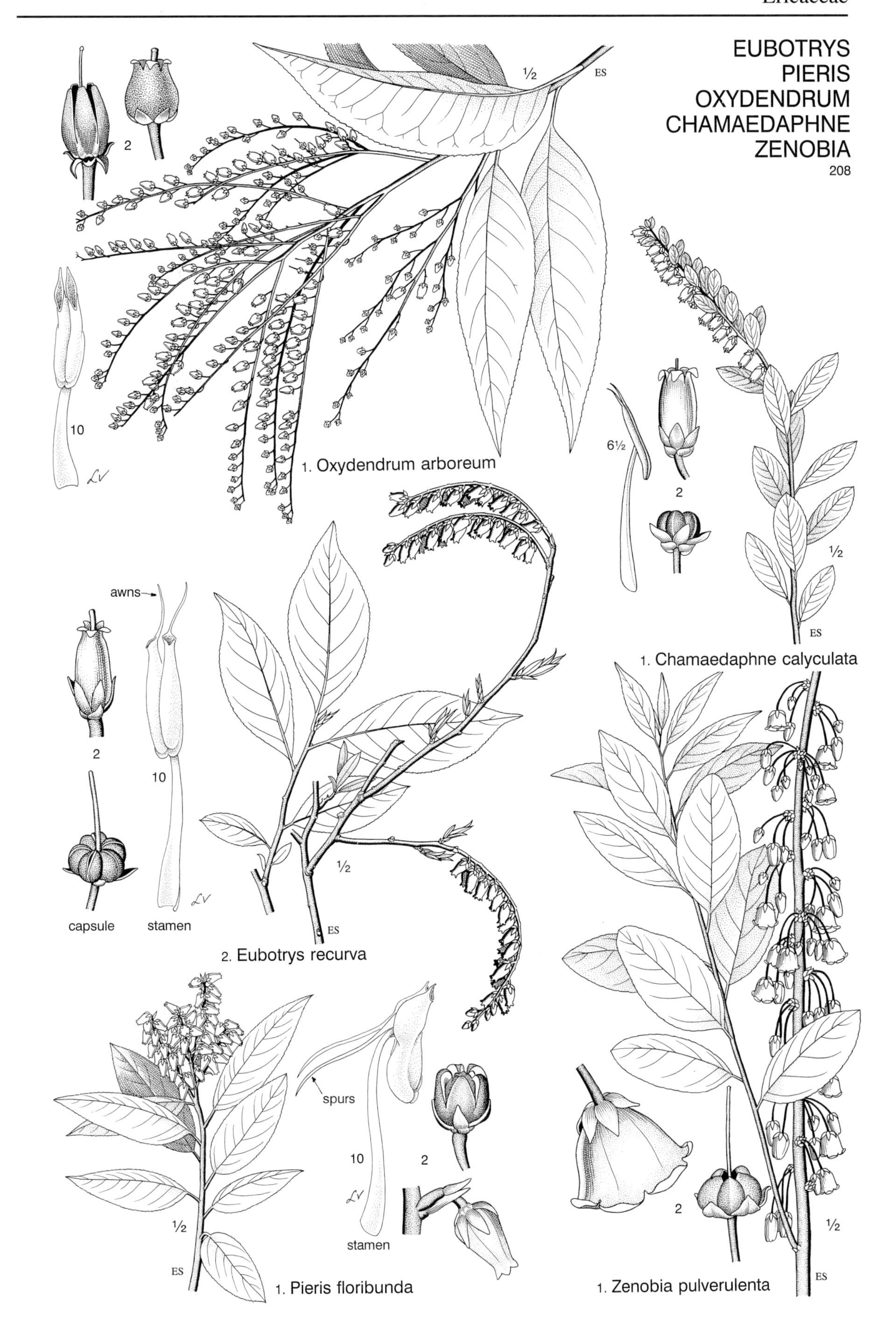

EUBOTRYS
PIERIS
OXYDENDRUM
CHAMAEDAPHNE
ZENOBIA
208

½

ES

2

10

LV

1. Oxydendrum arboreum

6½

2

½

ES

1. Chamaedaphne calyculata

awns→

2

10

LV

capsule   stamen

2. Eubotrys recurva

½

ES

spurs

10

2

LV

stamen

½

ES

1. Pieris floribunda

2

½

ES

1. Zenobia pulverulenta

GAULTHERIA
EPIGAEA
ARCTOSTAPHYLOS
CALLUNA
ERICA
209, 210

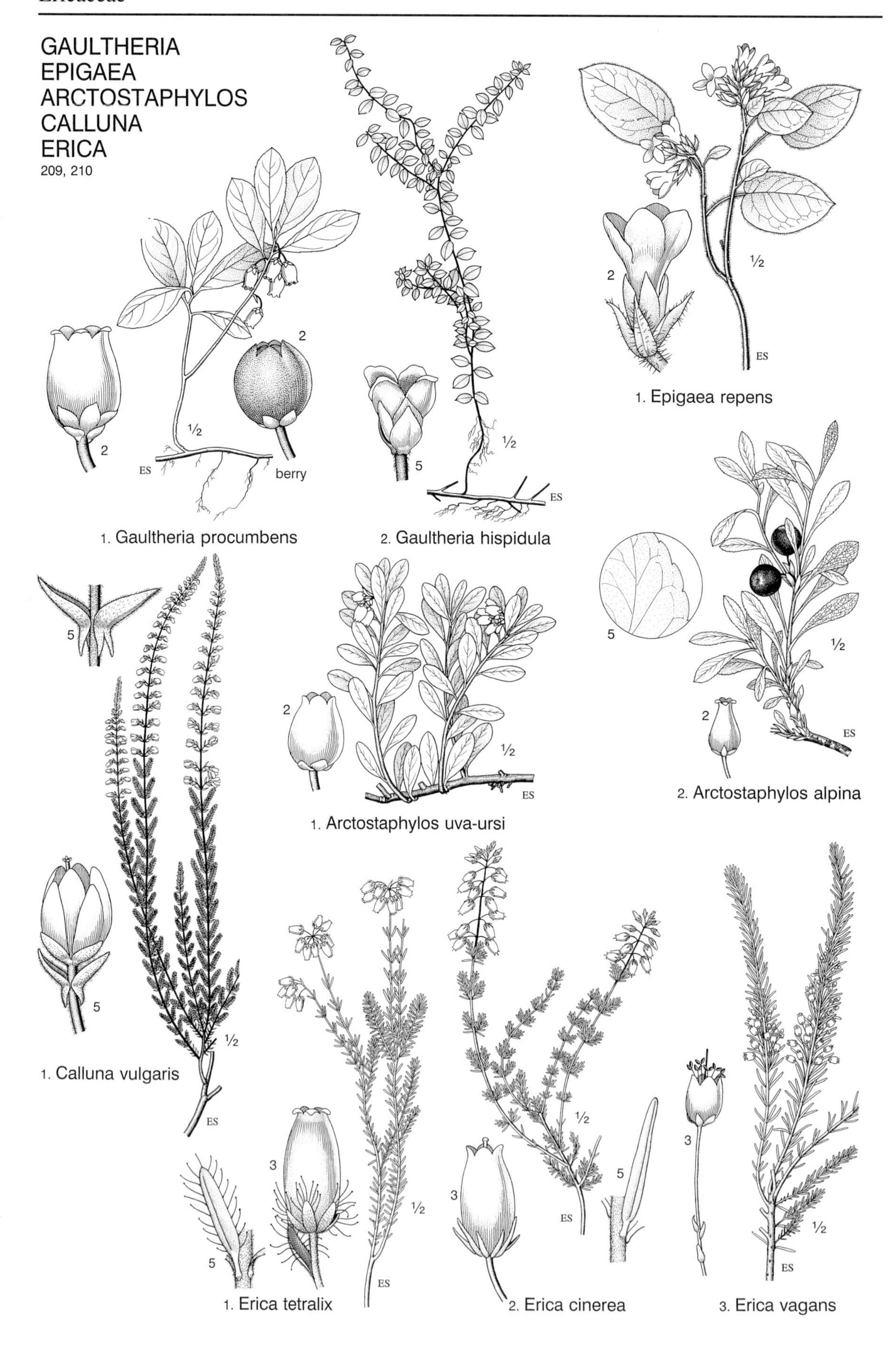

1. Gaultheria procumbens

2. Gaultheria hispidula

1. Epigaea repens

2. Arctostaphylos alpina

1. Arctostaphylos uva-ursi

1. Calluna vulgaris

1. Erica tetralix

2. Erica cinerea

3. Erica vagans

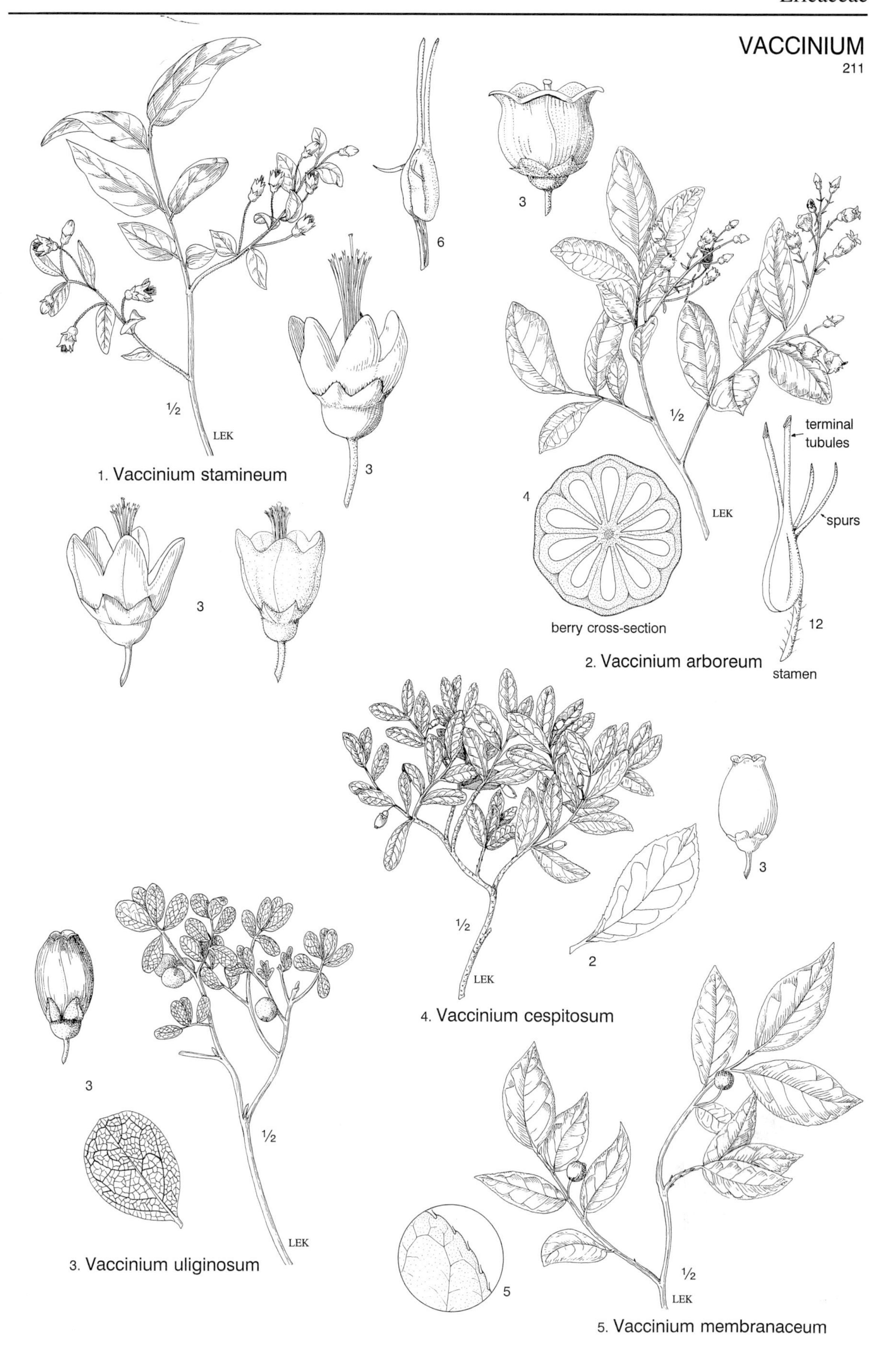

1. Vaccinium stamineum

6

3

3

3

2. Vaccinium arboreum

4 berry cross-section

terminal tubules

spurs

12 stamen

3. Vaccinium uliginosum

3

4. Vaccinium cespitosum

2

3

5. Vaccinium membranaceum

5

VACCINIUM
211, 212

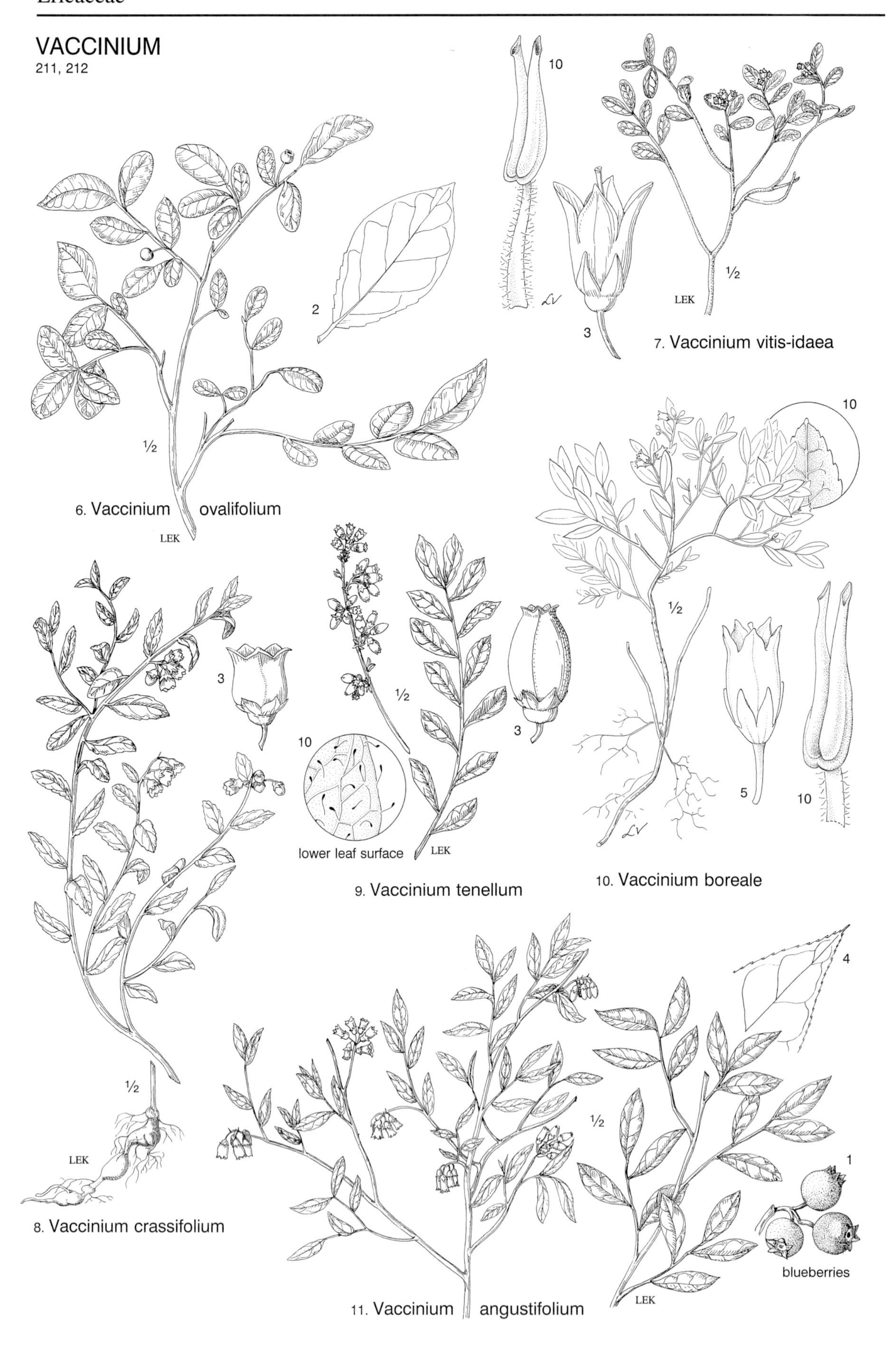

7. Vaccinium vitis-idaea

6. Vaccinium ovalifolium

9. Vaccinium tenellum

lower leaf surface

10. Vaccinium boreale

8. Vaccinium crassifolium

11. Vaccinium angustifolium

blueberries

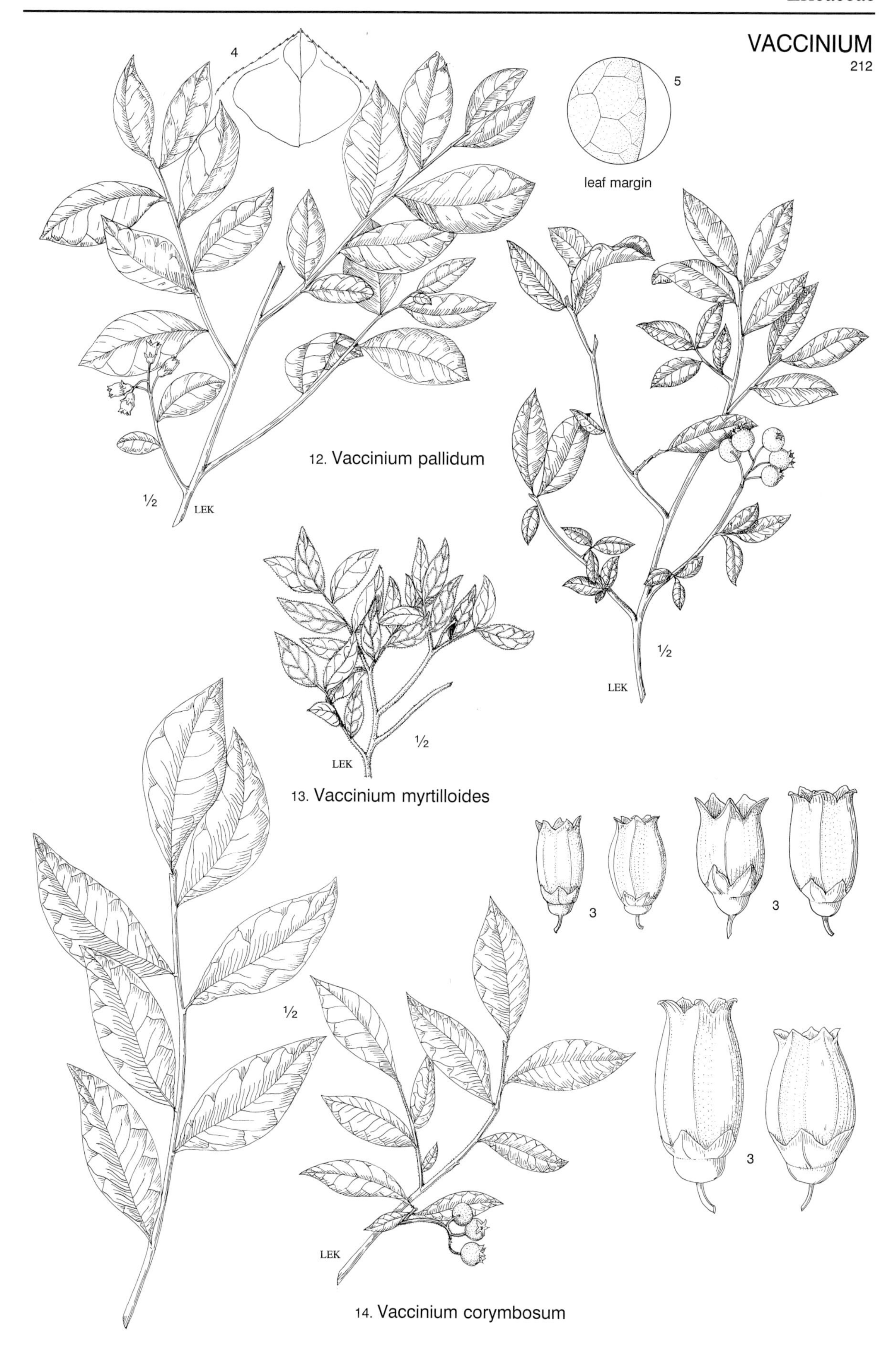

4

5

leaf margin

12. Vaccinium pallidum

½   LEK

½   LEK

13. Vaccinium myrtilloides

½   LEK

3   3

3

½

½   LEK

14. Vaccinium corymbosum

VACCINIUM
GAYLUSSACIA
212, 213

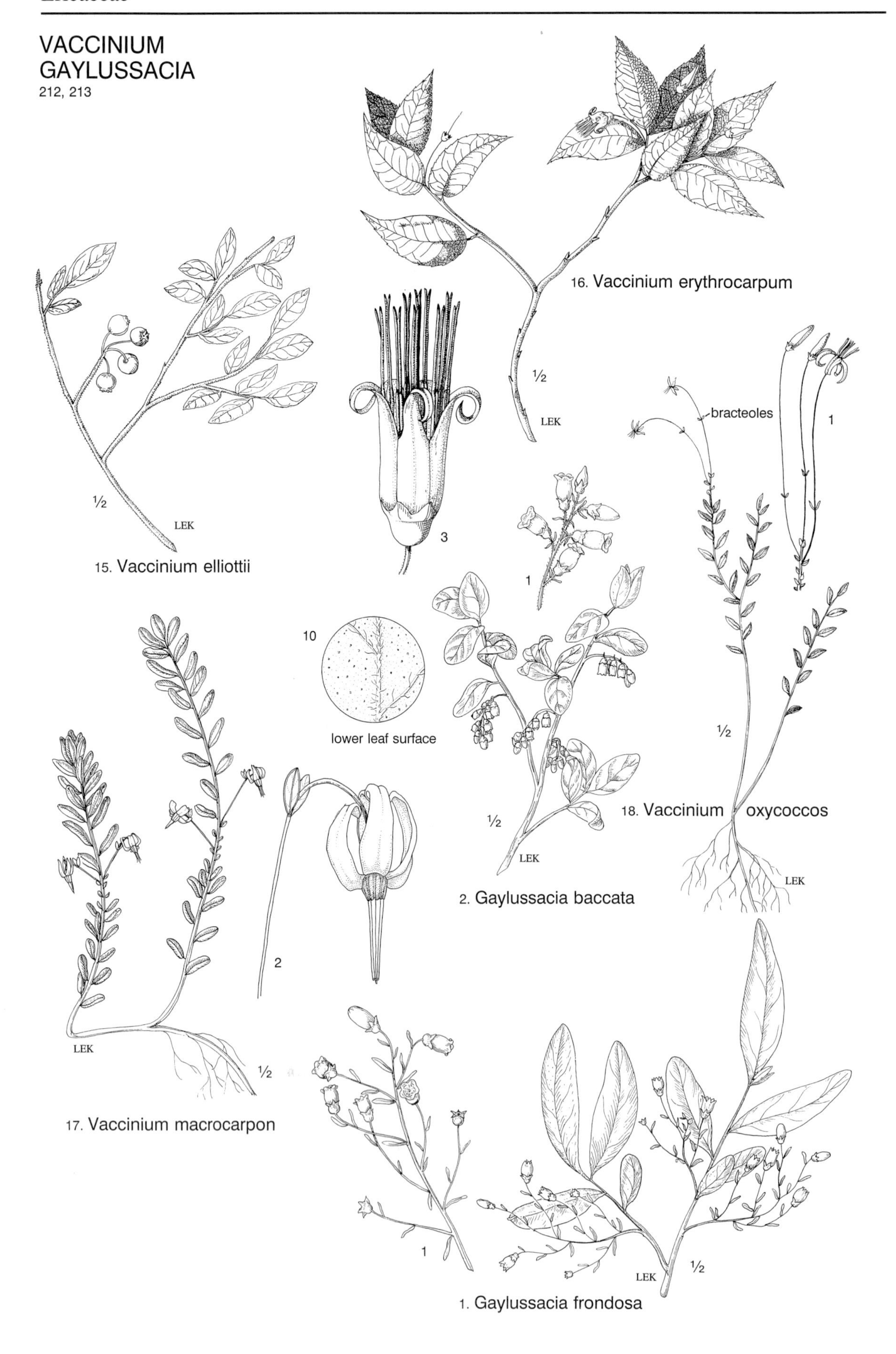

16. Vaccinium erythrocarpum

½

LEK

bracteoles

1

3

1

10

lower leaf surface

½

15. Vaccinium elliottii

½

18. Vaccinium oxycoccos

LEK

2. Gaylussacia baccata

2

½

LEK

17. Vaccinium macrocarpon

1

½

LEK

1

½

LEK

1. Gaylussacia frondosa

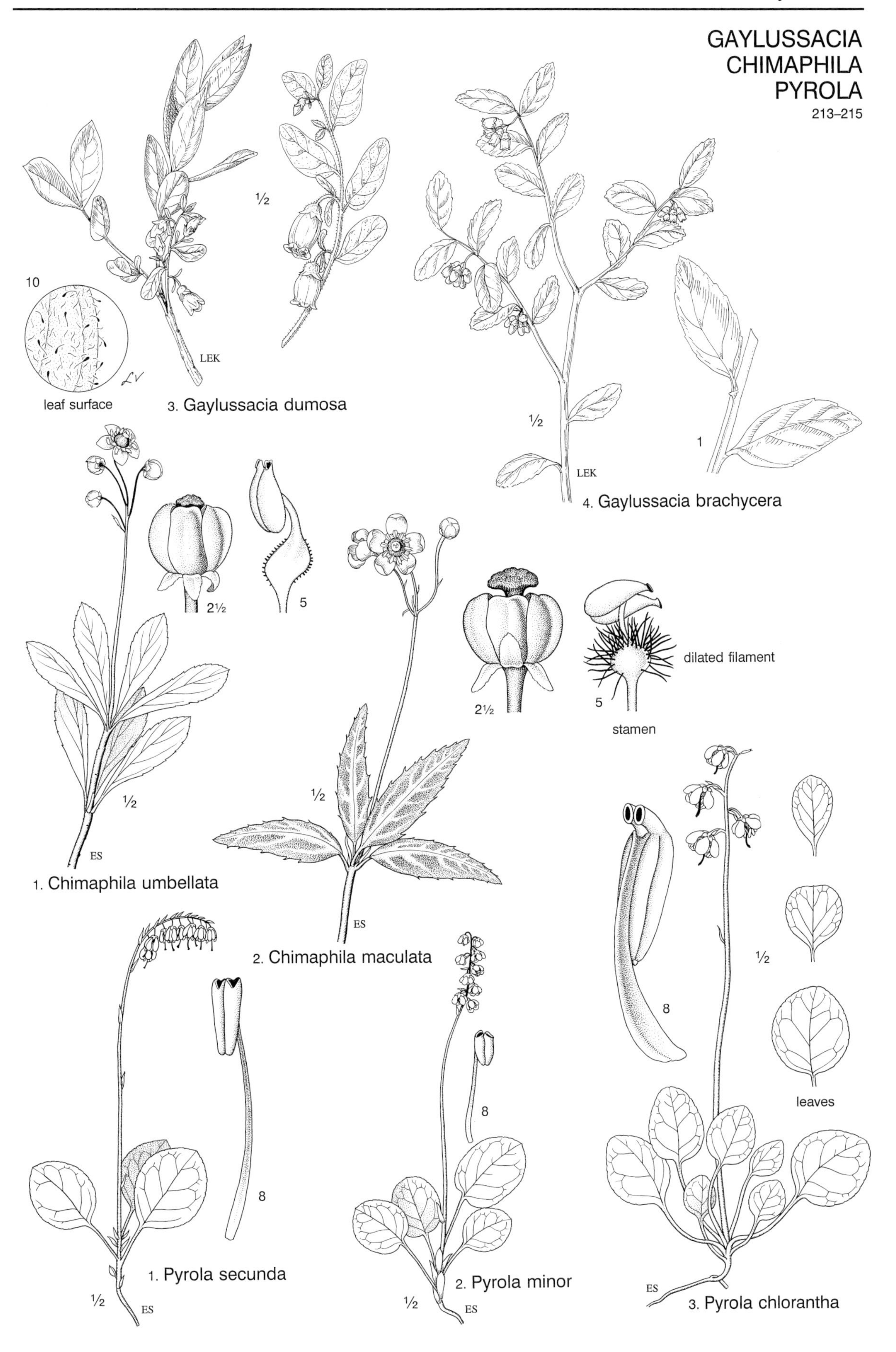

10
leaf surface

3. Gaylussacia dumosa

LEK

1

4. Gaylussacia brachycera

LEK

2½   5

1. Chimaphila umbellata

ES

2½   5
stamen

dilated filament

2. Chimaphila maculata

ES

8

leaves

1. Pyrola secunda

8

8

2. Pyrola minor

ES

3. Pyrola chlorantha

ES

PYROLA
MONESES
MONOTROPA
MONOTROPSIS
215, 216

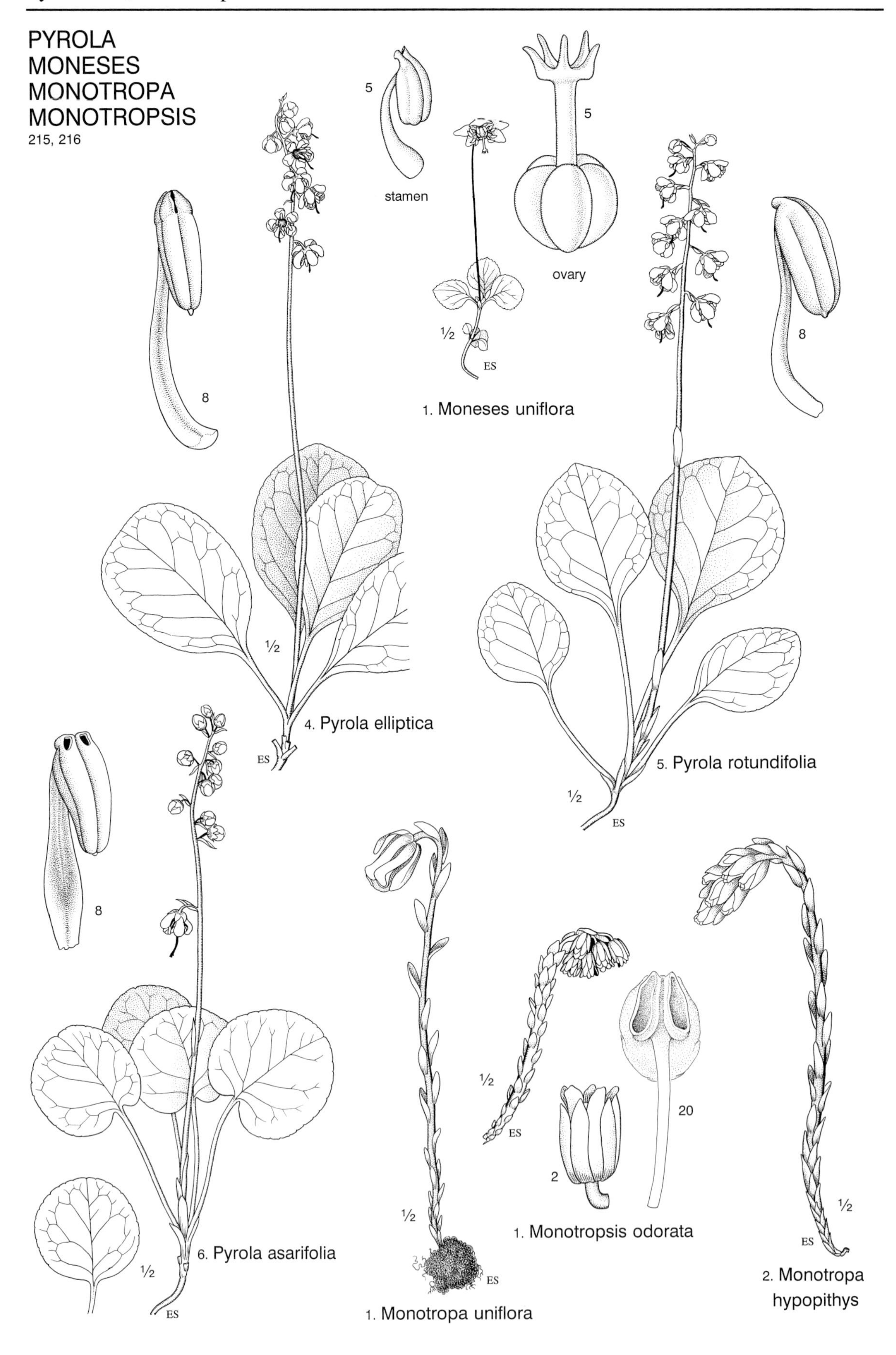

stamen

5

5

ovary

½
ES

1. Moneses uniflora

8

8

½

4. Pyrola elliptica

ES

5. Pyrola rotundifolia

½
ES

8

6. Pyrola asarifolia

½
ES

½
ES

½
ES

20

2

1. Monotropsis odorata

½
ES

2. Monotropa hypopithys

½
ES

1. Monotropa uniflora

PTEROSPORA
DIAPENSIA
PYXIDANTHERA
GALAX
SHORTIA
216–218

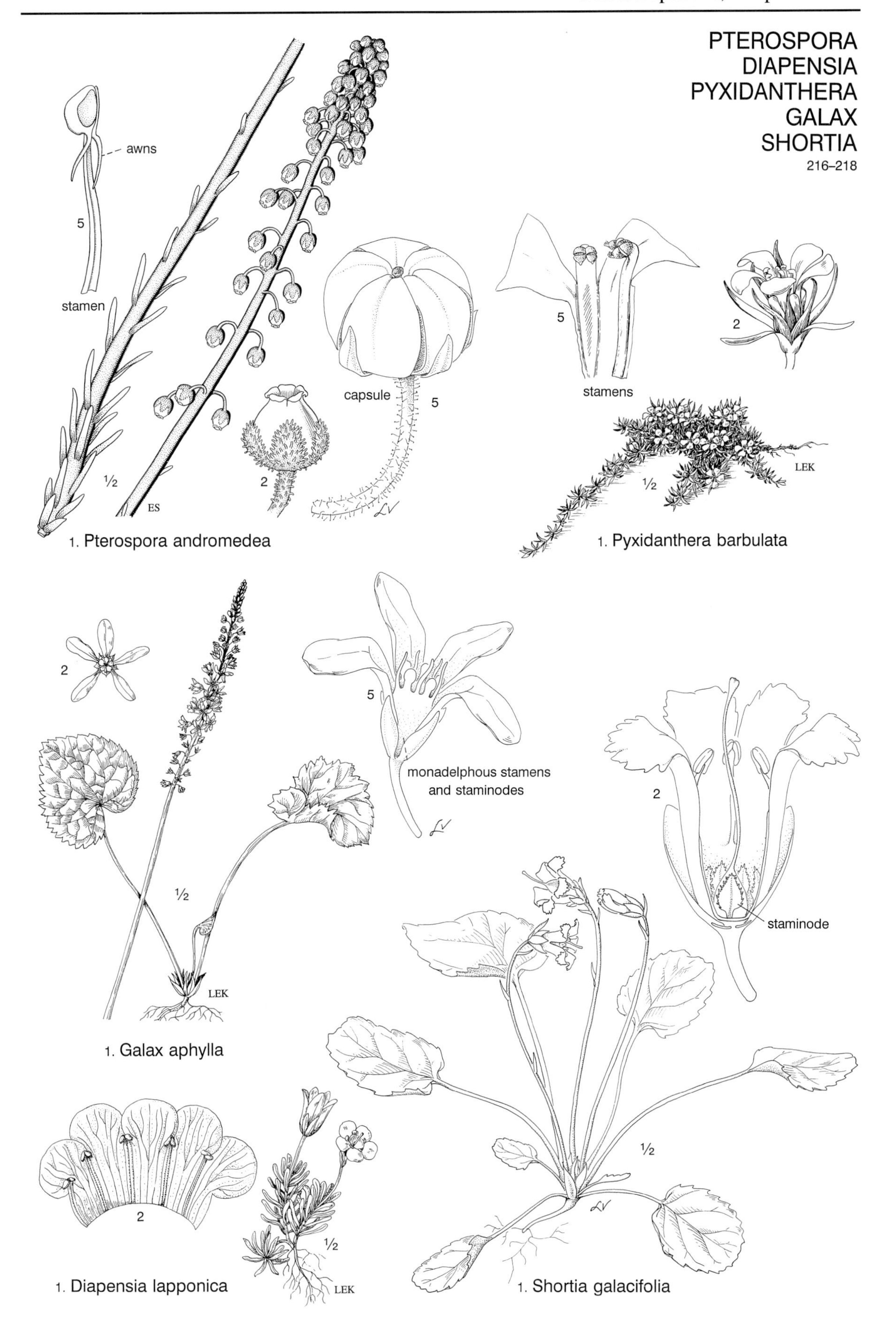

awns

5

stamen

½

ES

1. Pterospora andromedea

capsule

2

5

stamens

5

2

½

LEK

1. Pyxidanthera barbulata

2

½

LEK

1. Galax aphylla

5

monadelphous stamens
and staminodes

2

staminode

2

1. Diapensia lapponica

½

LEK

½

1. Shortia galacifolia

BUMELIA
DIOSPYROS
STYRAX
218, 219

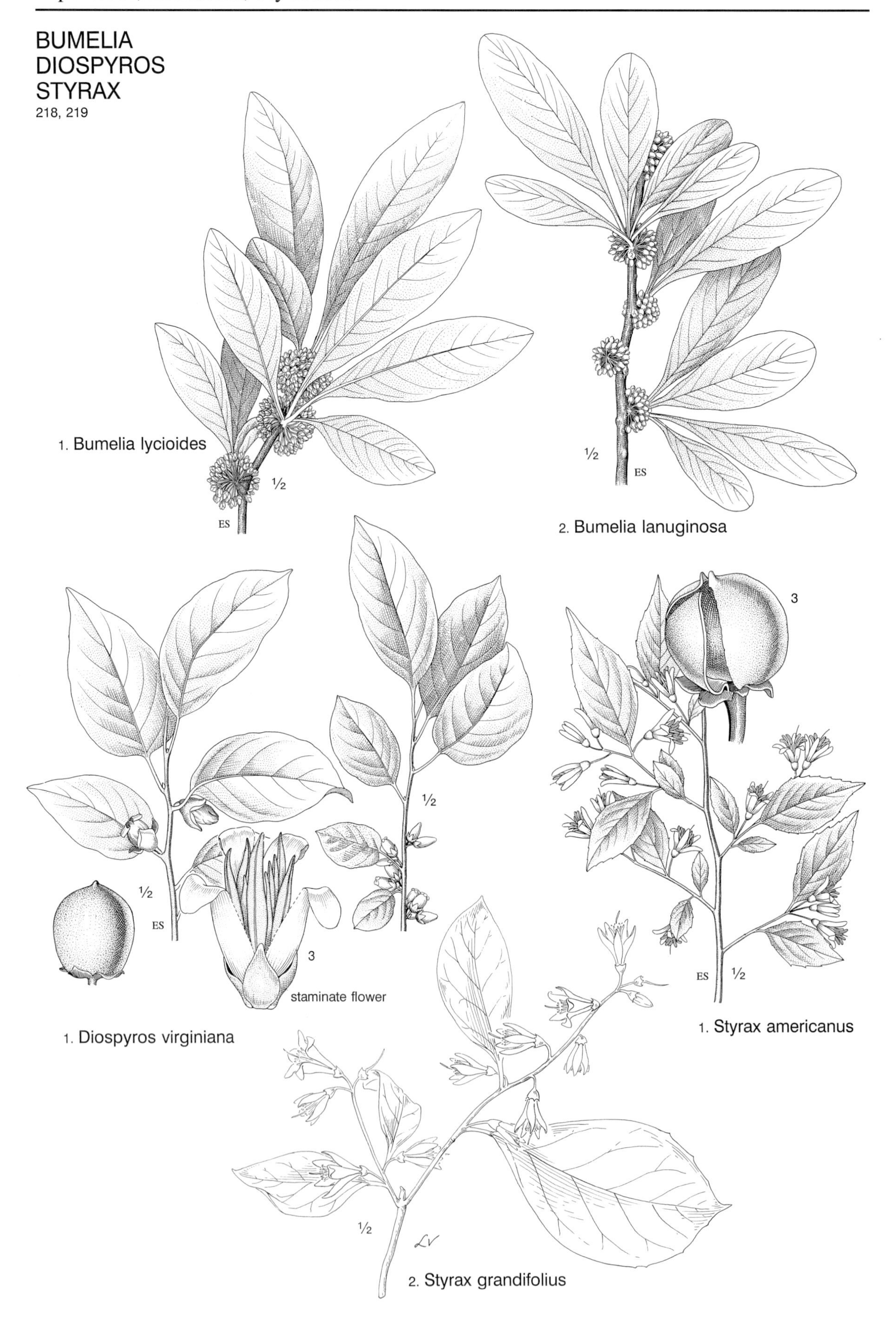

1. Bumelia lycioides

2. Bumelia lanuginosa

1. Diospyros virginiana

staminate flower

3

1. Styrax americanus

2. Styrax grandifolius

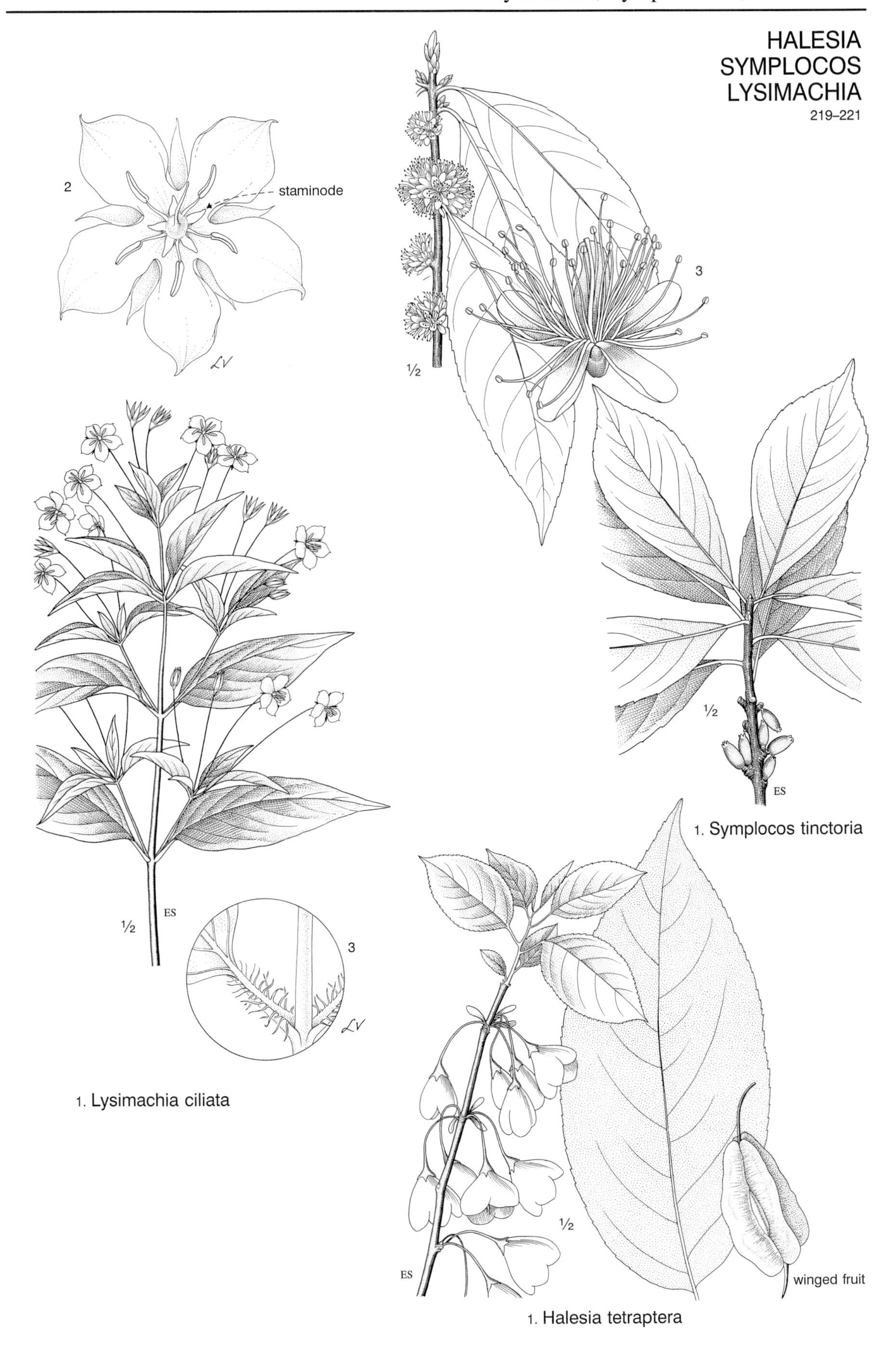

staminode

2

1. Lysimachia ciliata

3

1. Symplocos tinctoria

winged fruit

1. Halesia tetraptera

LYSIMACHIA
221

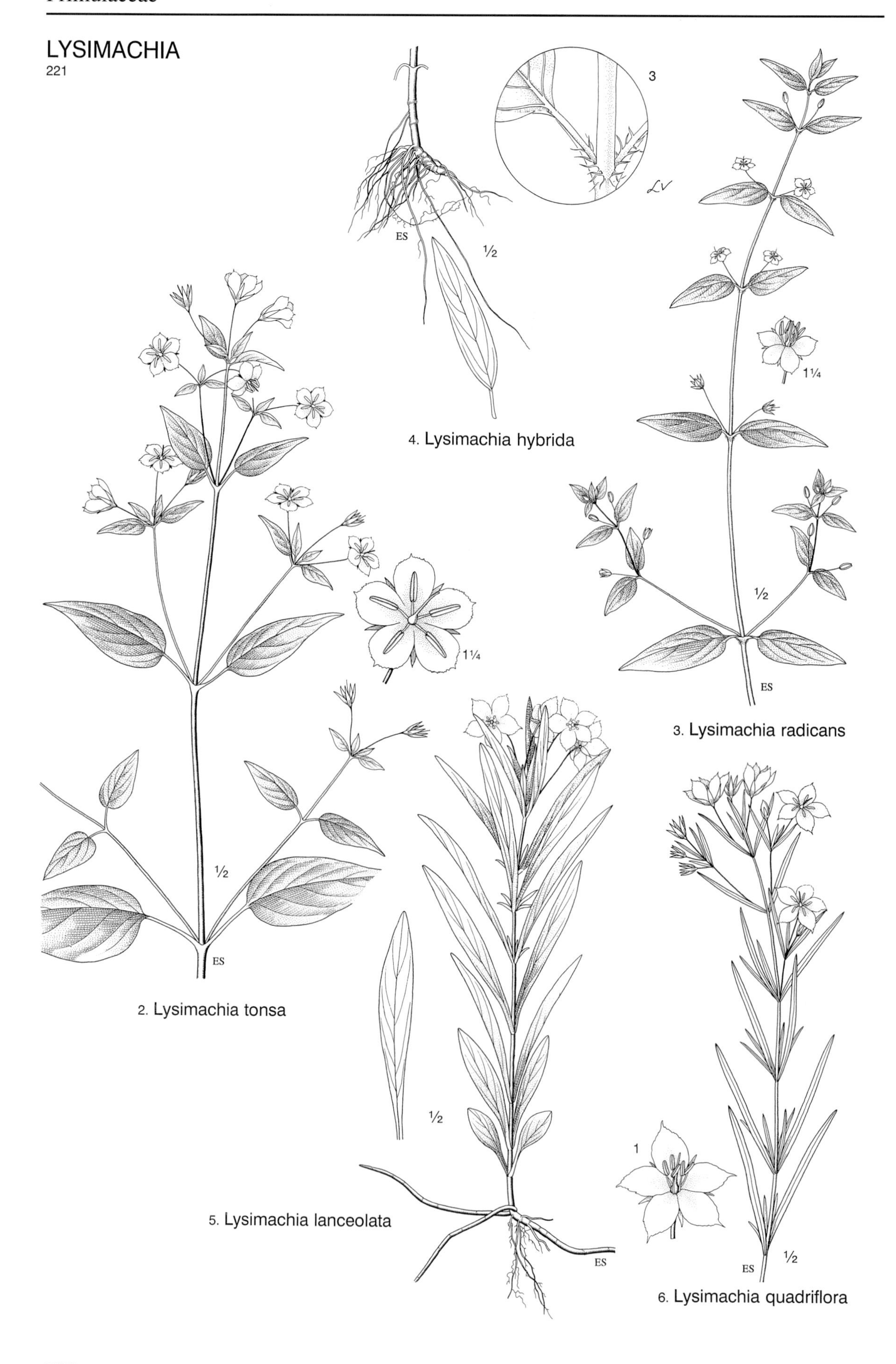

3

½

4. Lysimachia hybrida

1¼

1¼

3. Lysimachia radicans

ES

ES

2. Lysimachia tonsa

½

½

½

1

ES

ES

5. Lysimachia lanceolata

6. Lysimachia quadriflora

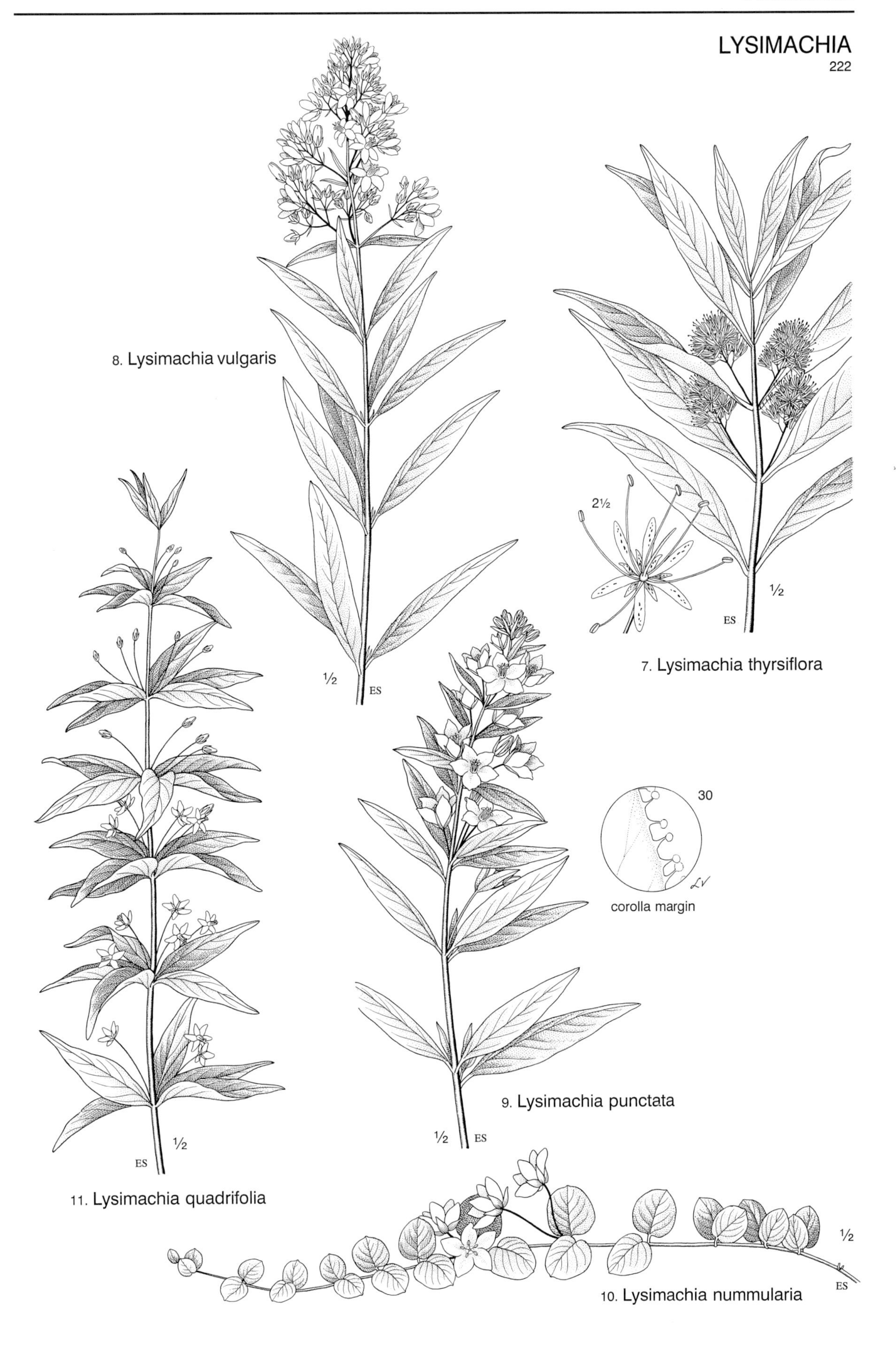

8. Lysimachia vulgaris

7. Lysimachia thyrsiflora

2½

½

ES

½

ES

30

corolla margin

9. Lysimachia punctata

½

ES

11. Lysimachia quadrifolia

½

ES

10. Lysimachia nummularia

½

ES

LYSIMACHIA
TRIENTALIS
GLAUX
ANAGALLIS
222, 223

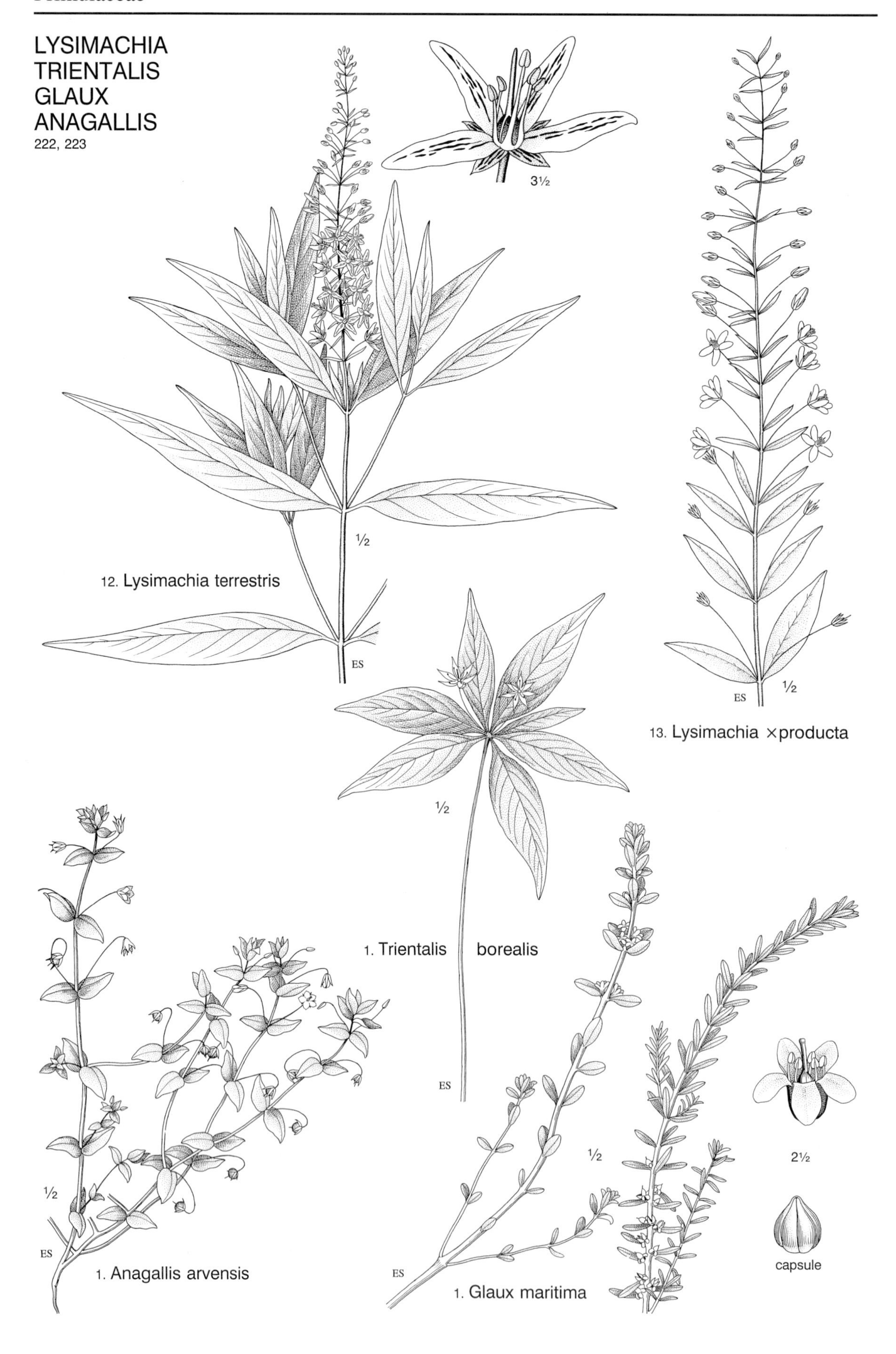

3½

12. Lysimachia terrestris

½

ES

13. Lysimachia ×producta

½

ES

½

1. Trientalis   borealis

ES

1. Anagallis arvensis

½

ES

1. Glaux maritima

½

ES

2½

capsule

ILLUSTRATED COMPANION TO

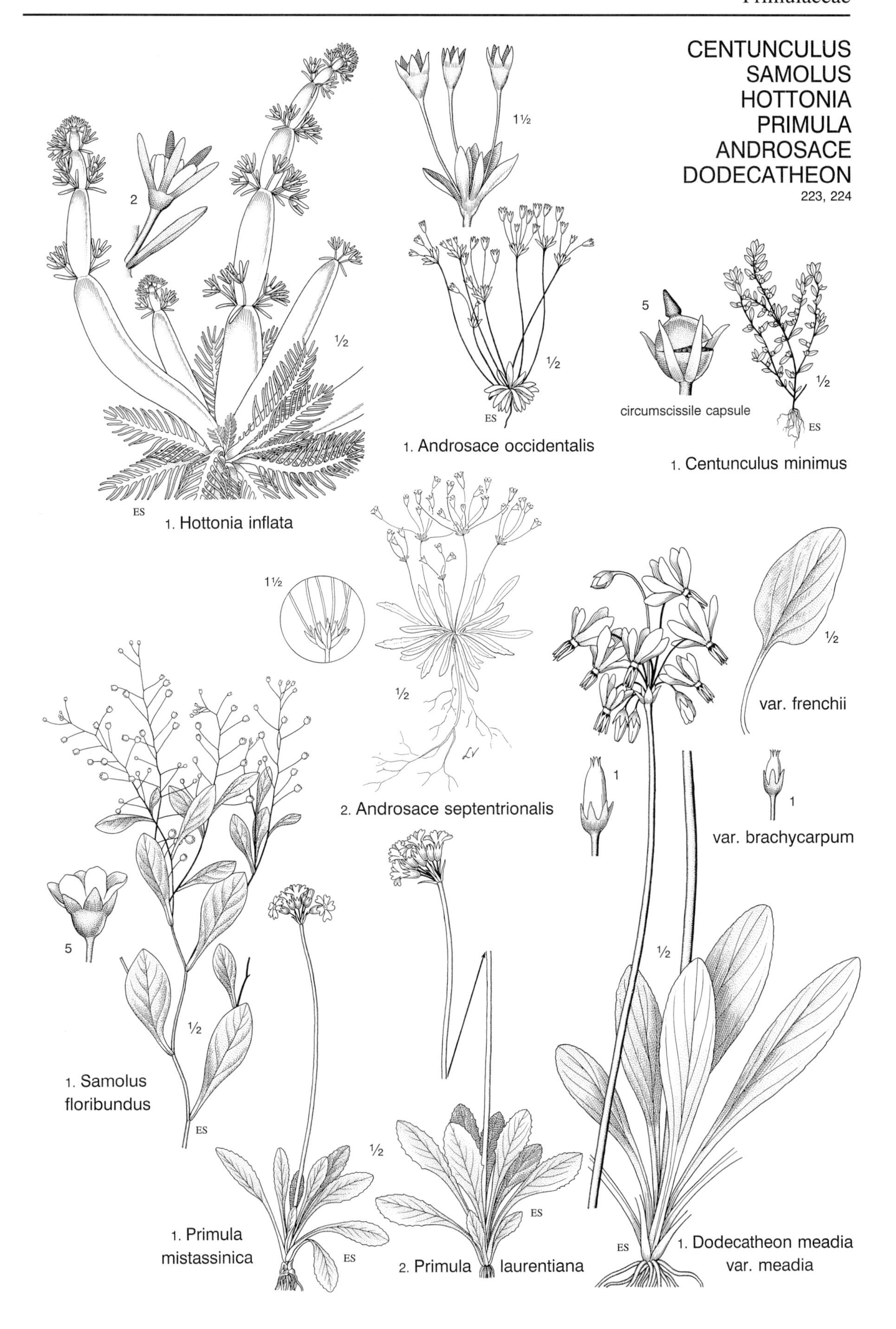

CENTUNCULUS
SAMOLUS
HOTTONIA
PRIMULA
ANDROSACE
DODECATHEON
223, 224

2

1½

½

1. Androsace occidentalis

ES

5

circumscissile capsule

½

ES

1. Centunculus minimus

ES

1. Hottonia inflata

1½

½

2. Androsace septentrionalis

var. frenchii

½

1

var. brachycarpum

1

½

5

½

1. Samolus
floribundus

ES

1. Primula
mistassinica

ES

½

ES

½

ES

2. Primula laurentiana

ES

1. Dodecatheon meadia
var. meadia

DODECATHEON
PHILADELPHUS
DECUMARIA
224–226

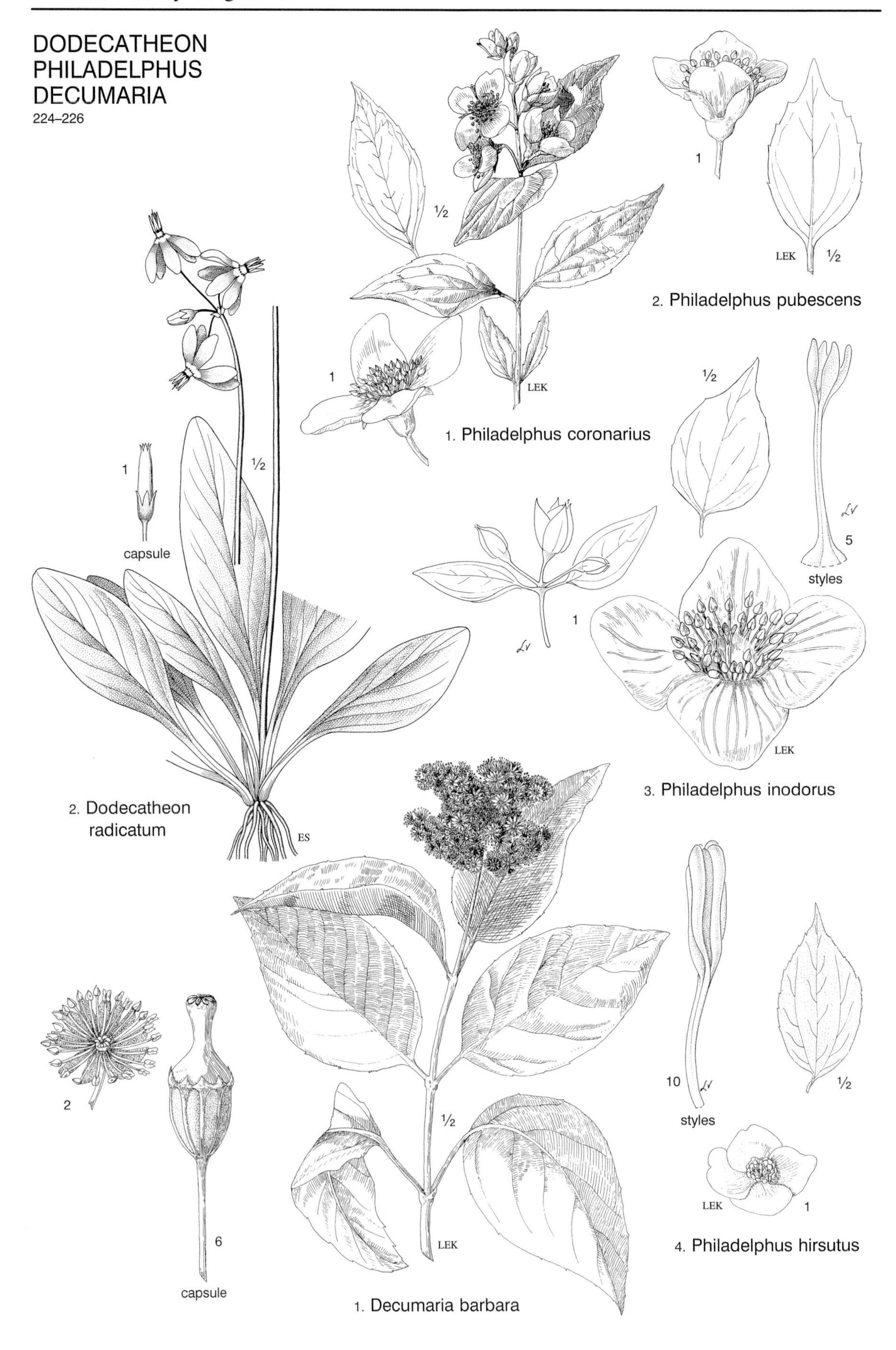

1. Philadelphus coronarius

2. Philadelphus pubescens

3. Philadelphus inodorus

4. Philadelphus hirsutus

2. Dodecatheon radicatum

1. Decumaria barbara

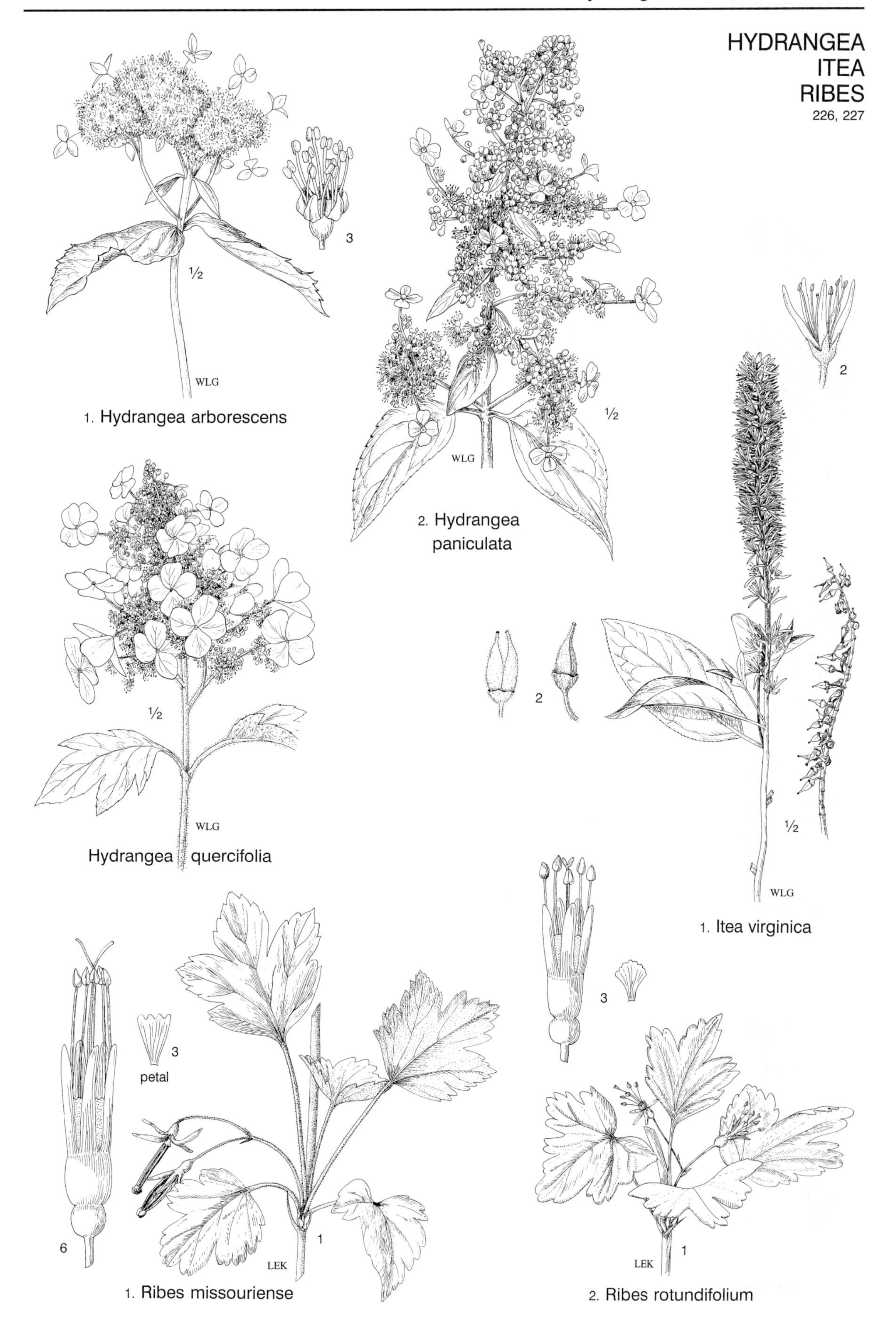

1. Hydrangea arborescens

2. Hydrangea paniculata

Hydrangea quercifolia

1. Itea virginica

petal

1. Ribes missouriense

2. Ribes rotundifolium

RIBES

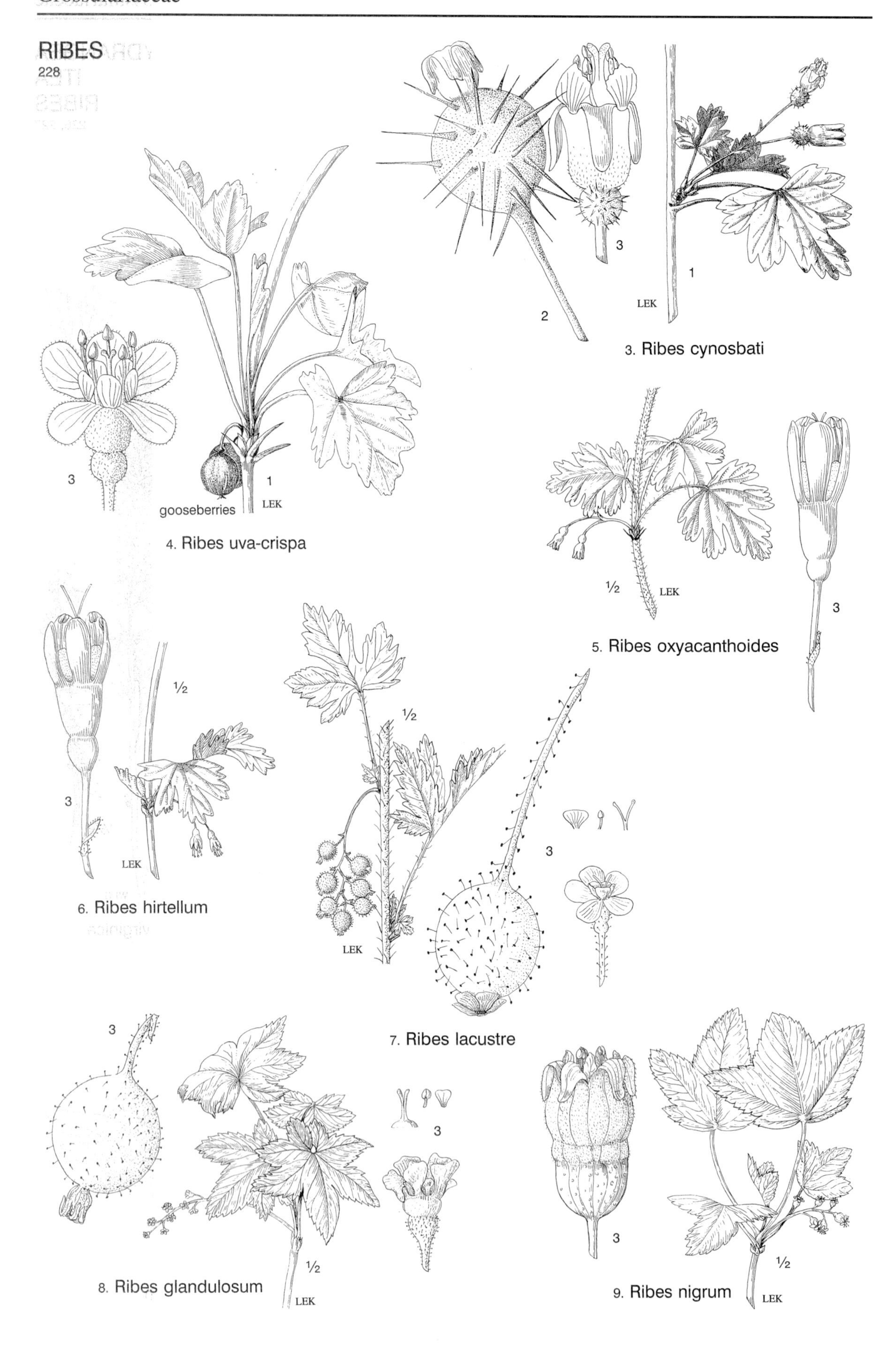

3. Ribes cynosbati

4. Ribes uva-crispa

gooseberries

5. Ribes oxyacanthoides

6. Ribes hirtellum

7. Ribes lacustre

8. Ribes glandulosum

9. Ribes nigrum

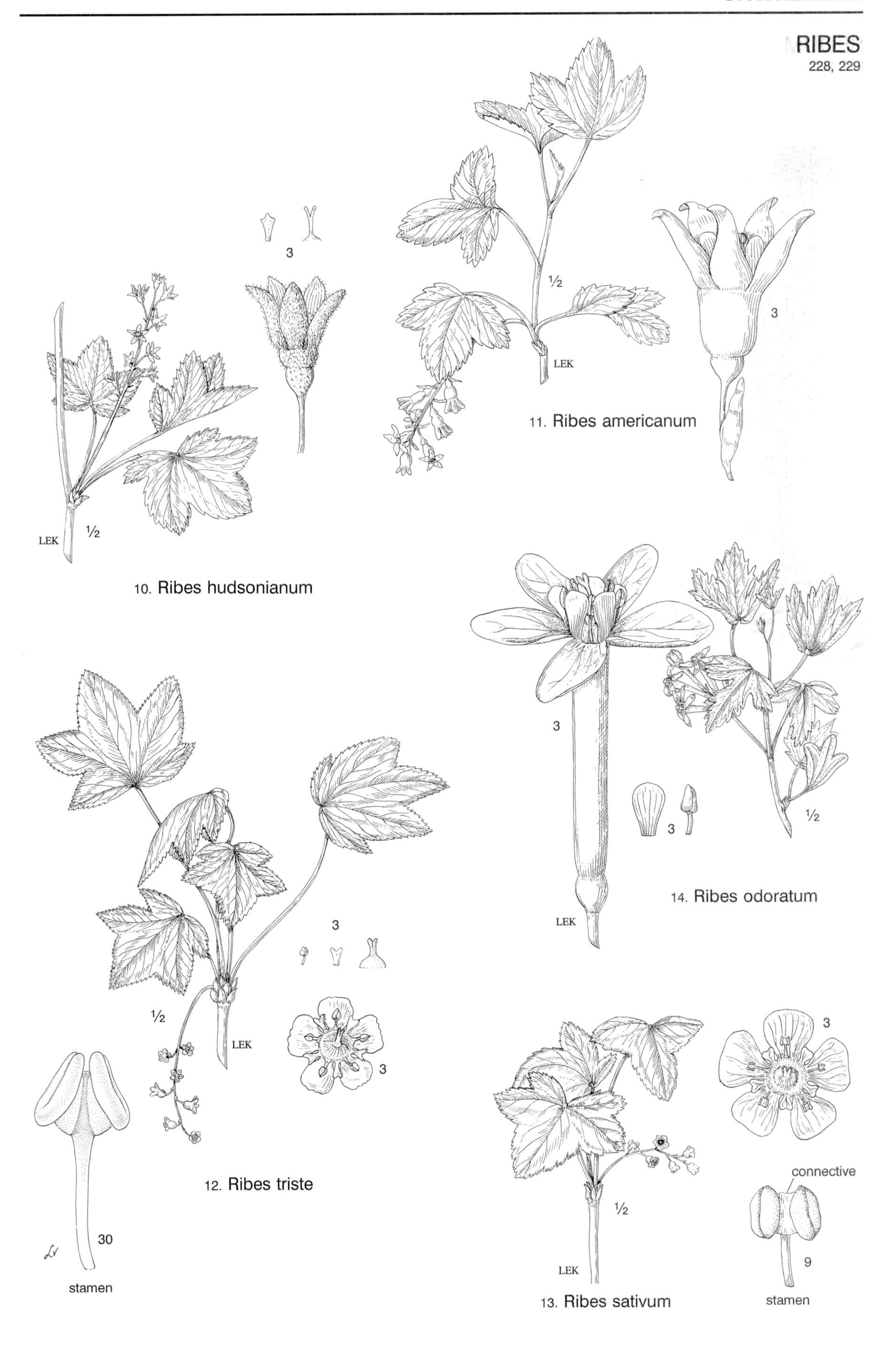

3

11. Ribes americanum

10. Ribes hudsonianum

3

14. Ribes odoratum

12. Ribes triste

30

stamen

13. Ribes sativum

connective

9

stamen

SEDUM
229, 230

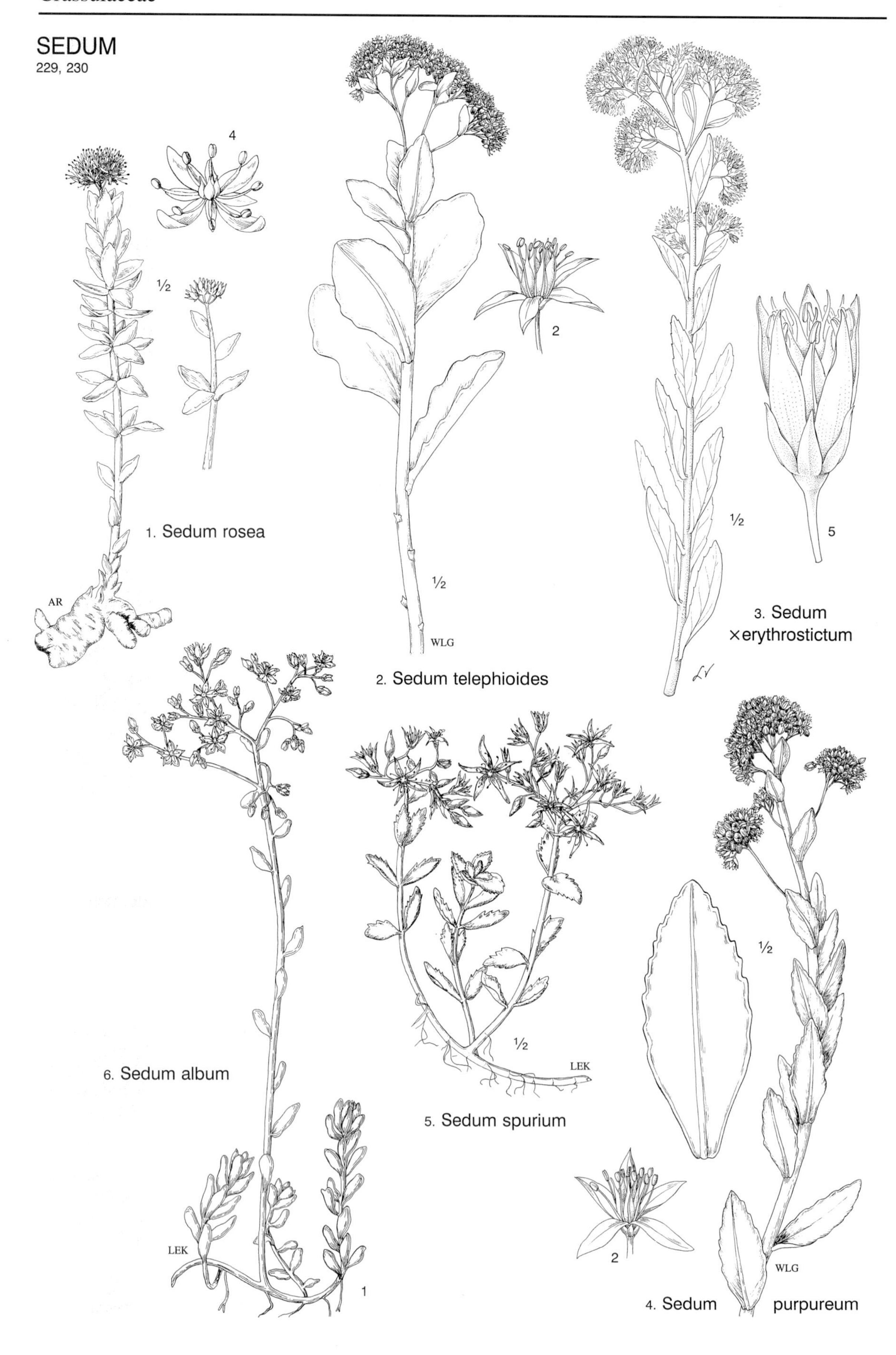

1. Sedum rosea

AR

½

4

½

2. Sedum telephioides

WLG

2

3. Sedum ×erythrostictum

½

5

6. Sedum album

LEK

½

5. Sedum spurium

LEK

1

½

2

4. Sedum purpureum

WLG

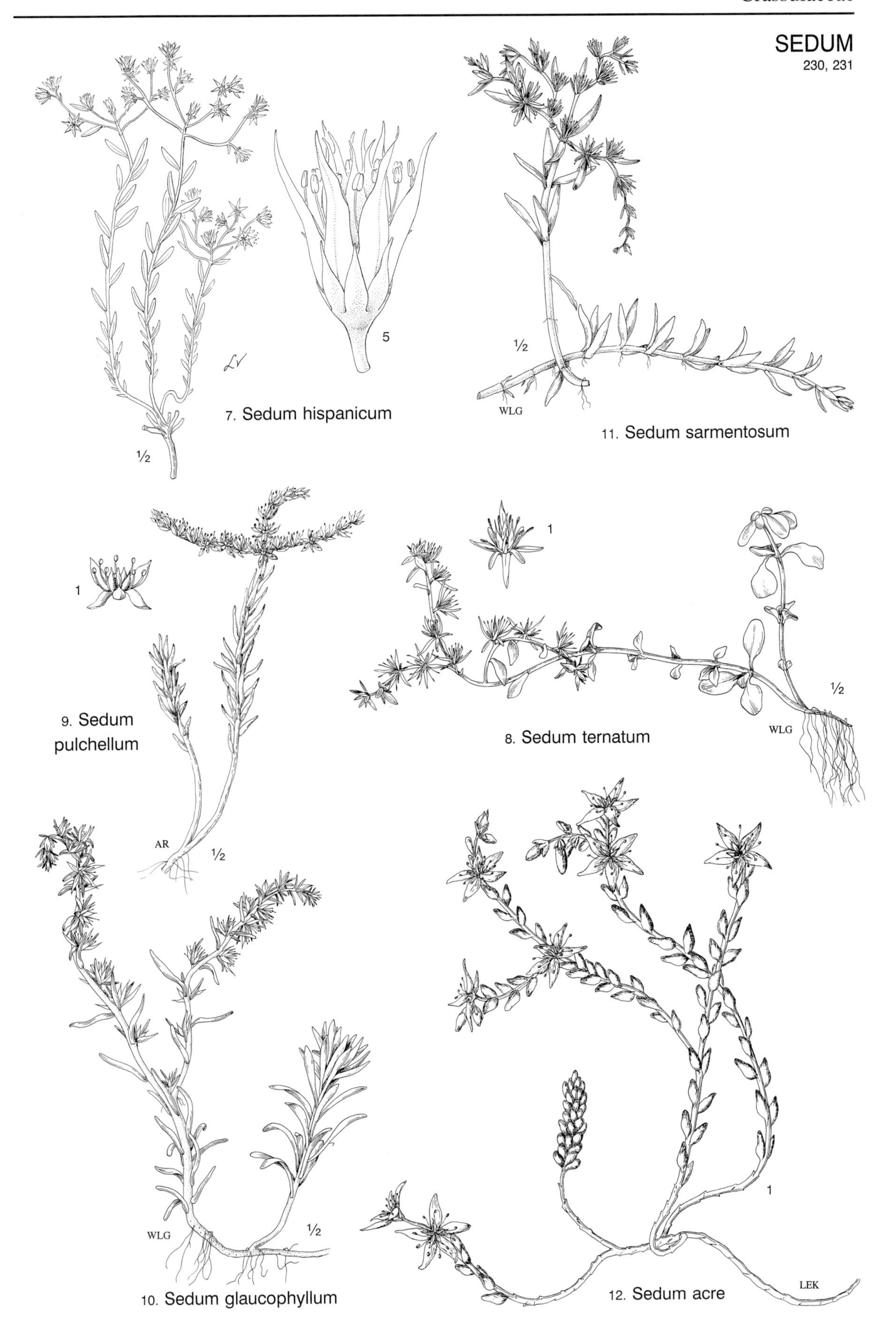

7. Sedum hispanicum

5

11. Sedum sarmentosum

WLG

9. Sedum pulchellum

1

8. Sedum ternatum

WLG

AR

10. Sedum glaucophyllum

WLG

12. Sedum acre

LEK

SEDUM
SEMPERVIVUM
CRASSULA
231

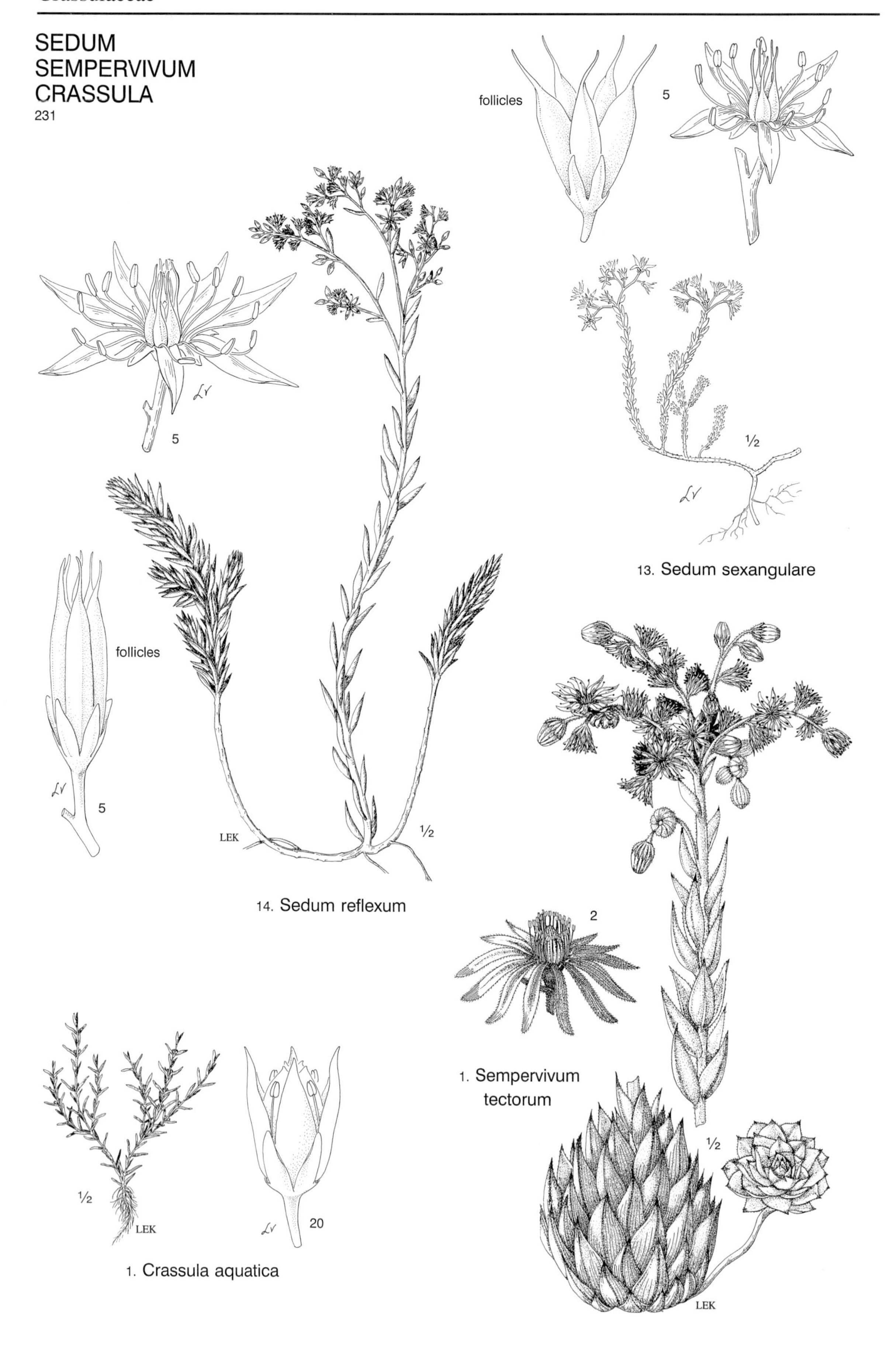

follicles 5

5

follicles

5

14. Sedum reflexum

LEK

½

13. Sedum sexangulare

½

2

1. Sempervivum
tectorum

½

LEK

½

LEK

1. Crassula aquatica

20

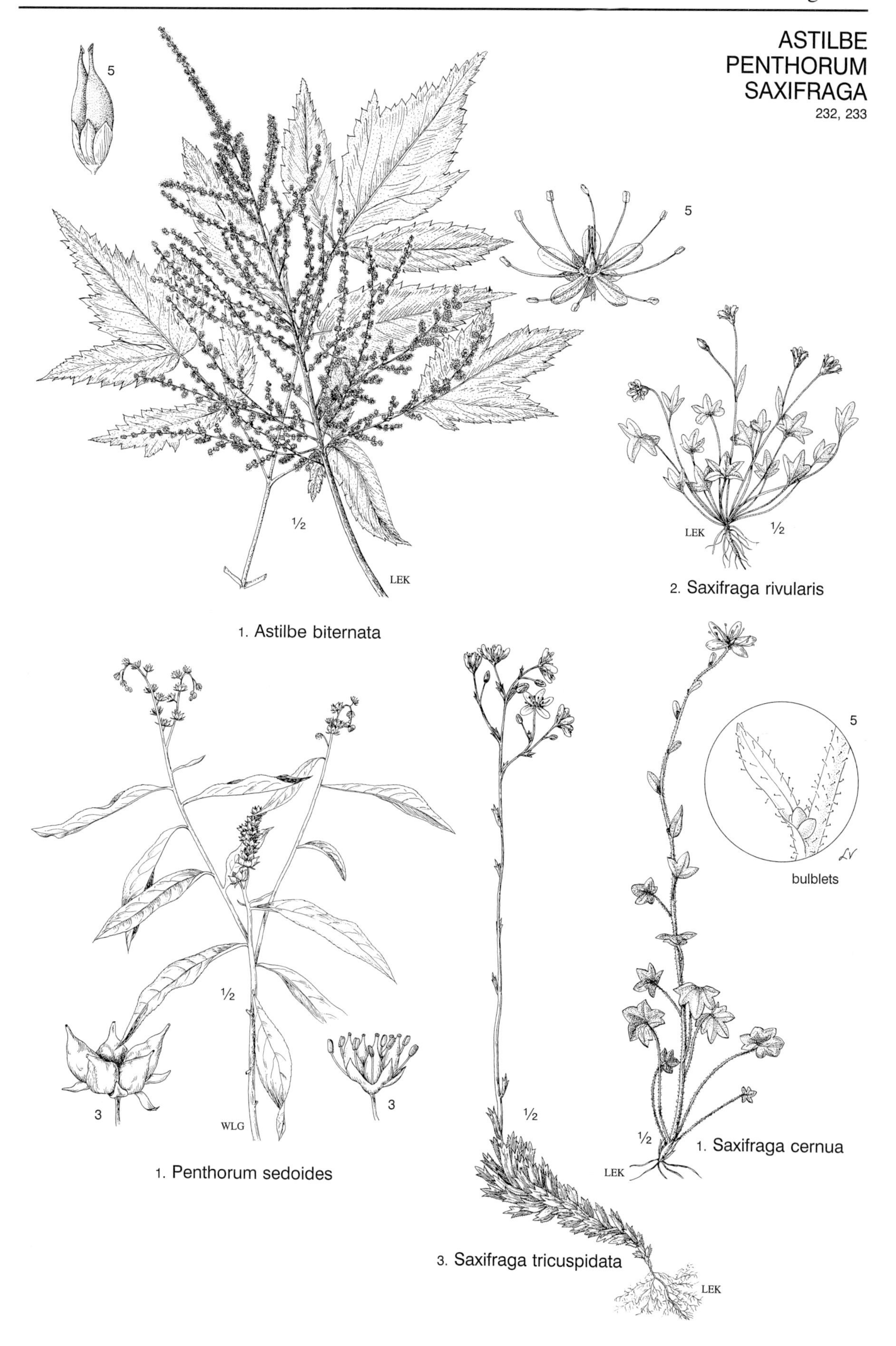

1. Astilbe biternata

2. Saxifraga rivularis

bulblets

1. Penthorum sedoides

3. Saxifraga tricuspidata

1. Saxifraga cernua

SAXIFRAGA
233

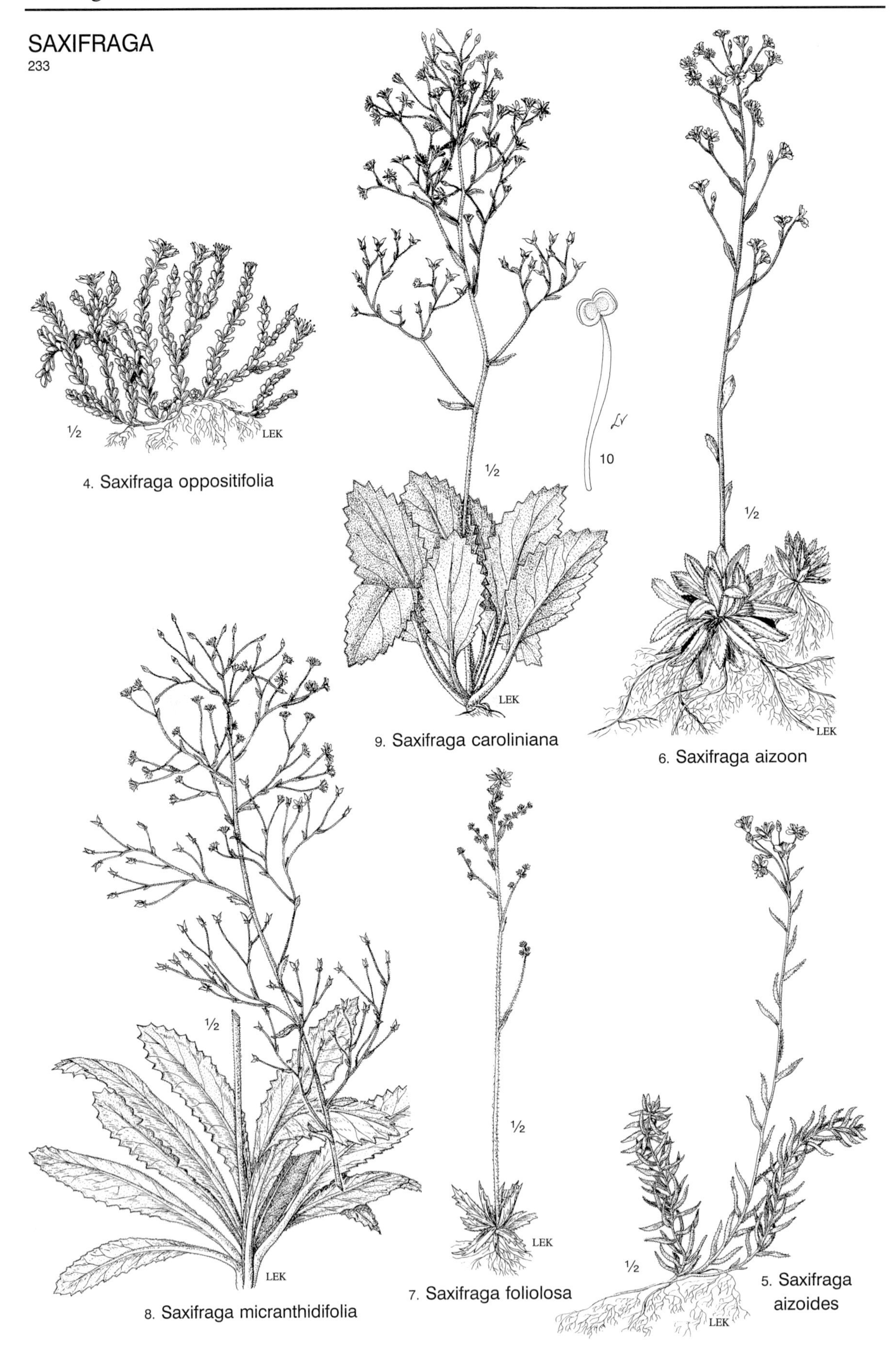

4. Saxifraga oppositifolia

10

9. Saxifraga caroliniana

6. Saxifraga aizoon

8. Saxifraga micranthidifolia

7. Saxifraga foliolosa

5. Saxifraga aizoides

ILLUSTRATED COMPANION TO

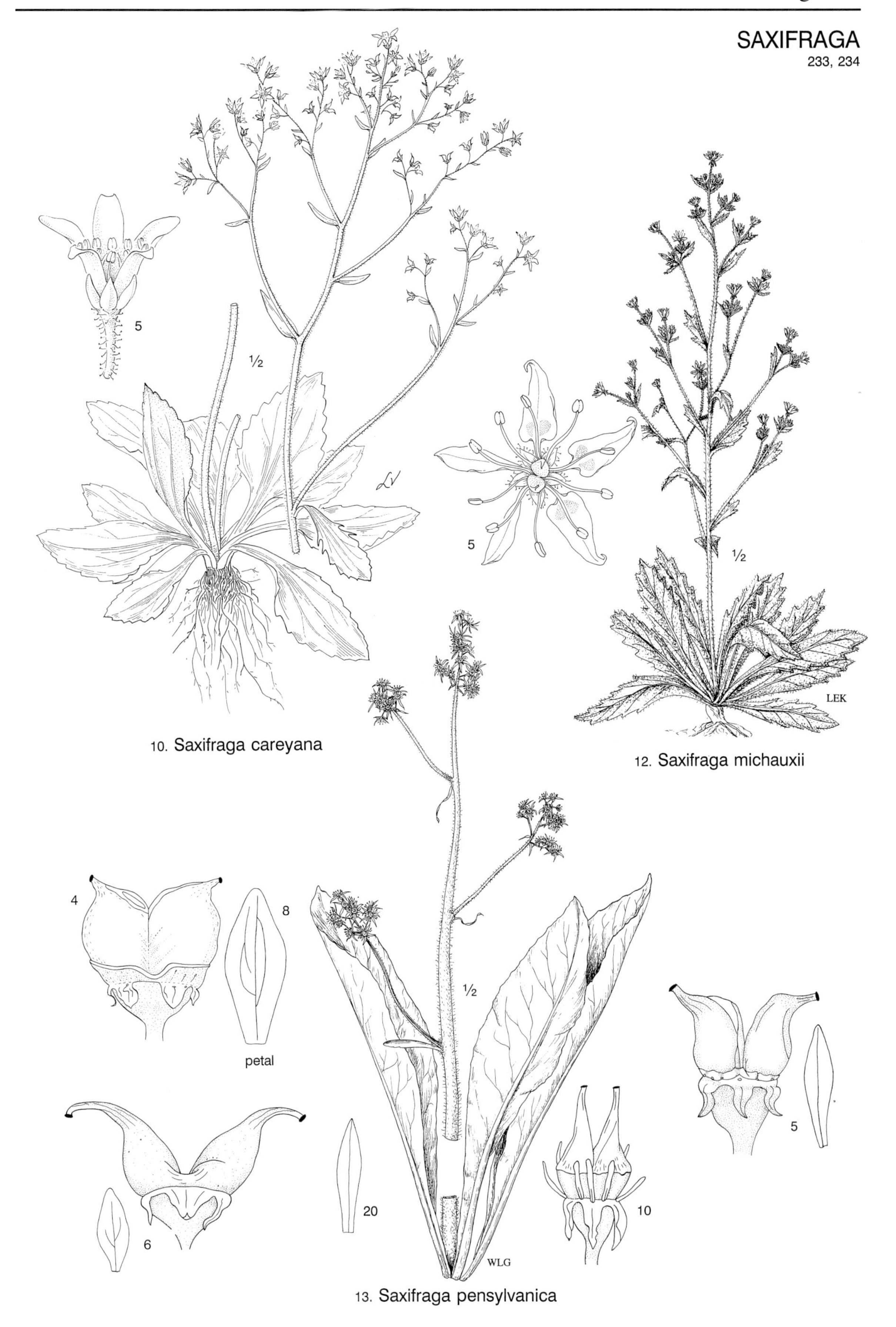

10. Saxifraga careyana

12. Saxifraga michauxii

petal

13. Saxifraga pensylvanica

SAXIFRAGA
TIARELLA
MITELLA
SULLIVANTIA
234, 235

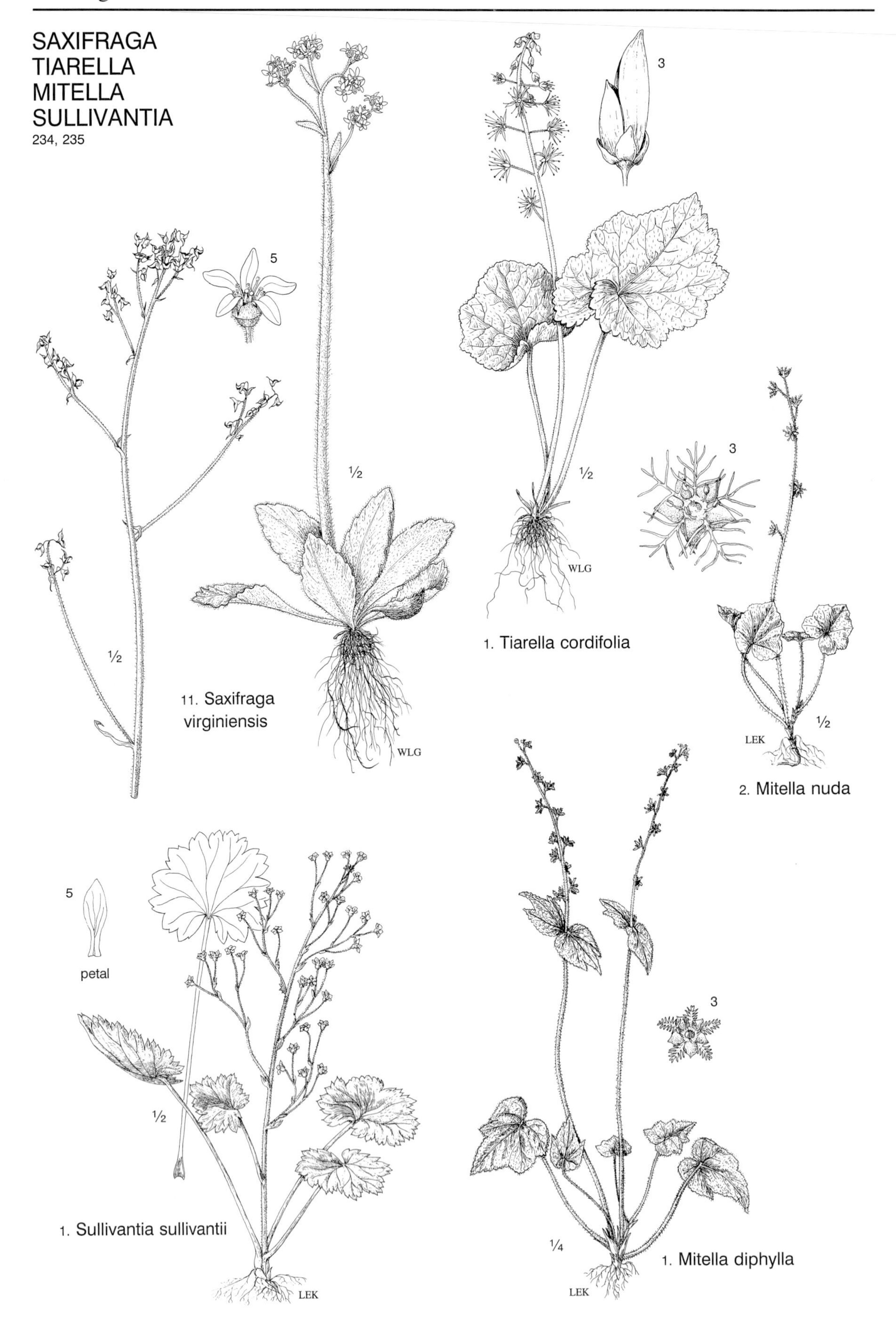

5

½

½

½

½

3

3

1. Tiarella cordifolia

WLG

11. Saxifraga
virginiensis

WLG

5

petal

½

1. Sullivantia sullivantii

LEK

2. Mitella nuda

LEK

½

3

¼

1. Mitella diphylla

LEK

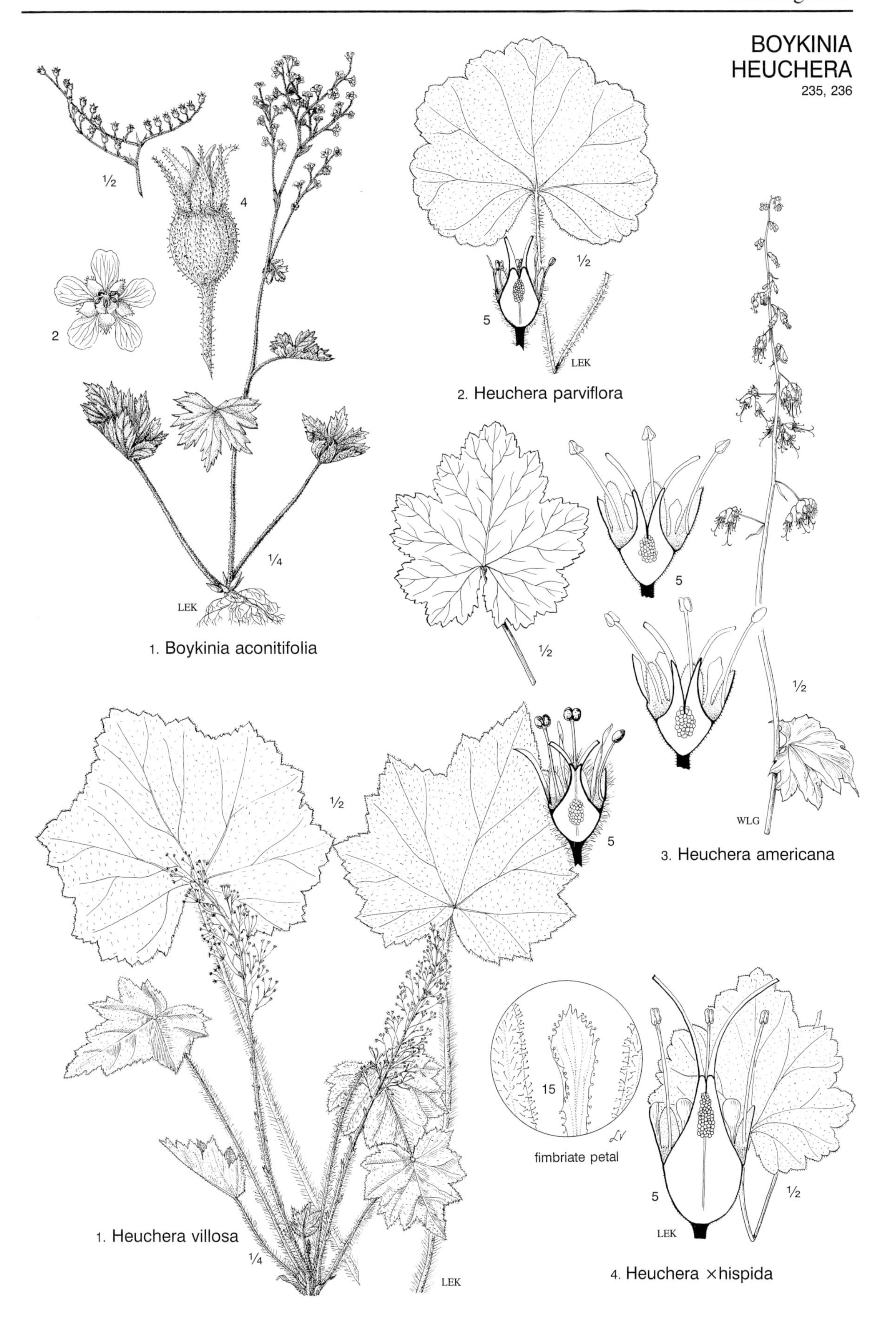

2. Heuchera parviflora

1. Boykinia aconitifolia

3. Heuchera americana

fimbriate petal

1. Heuchera villosa

4. Heuchera ×hispida

HEUCHERA
PARNASSIA
236, 237

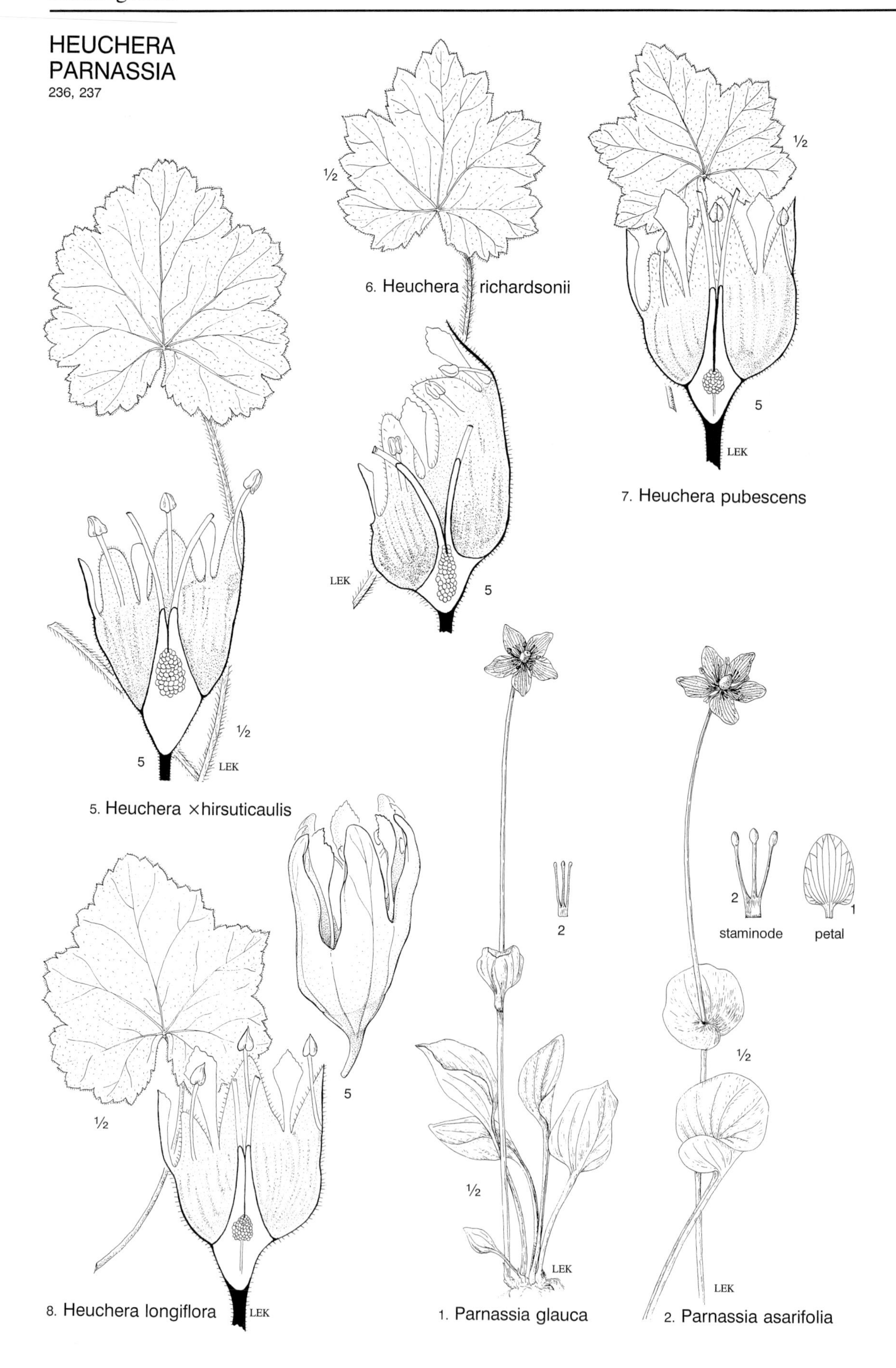

6. Heuchera richardsonii

7. Heuchera pubescens

5. Heuchera ×hirsuticaulis

staminode   petal

8. Heuchera longiflora

1. Parnassia glauca

2. Parnassia asarifolia

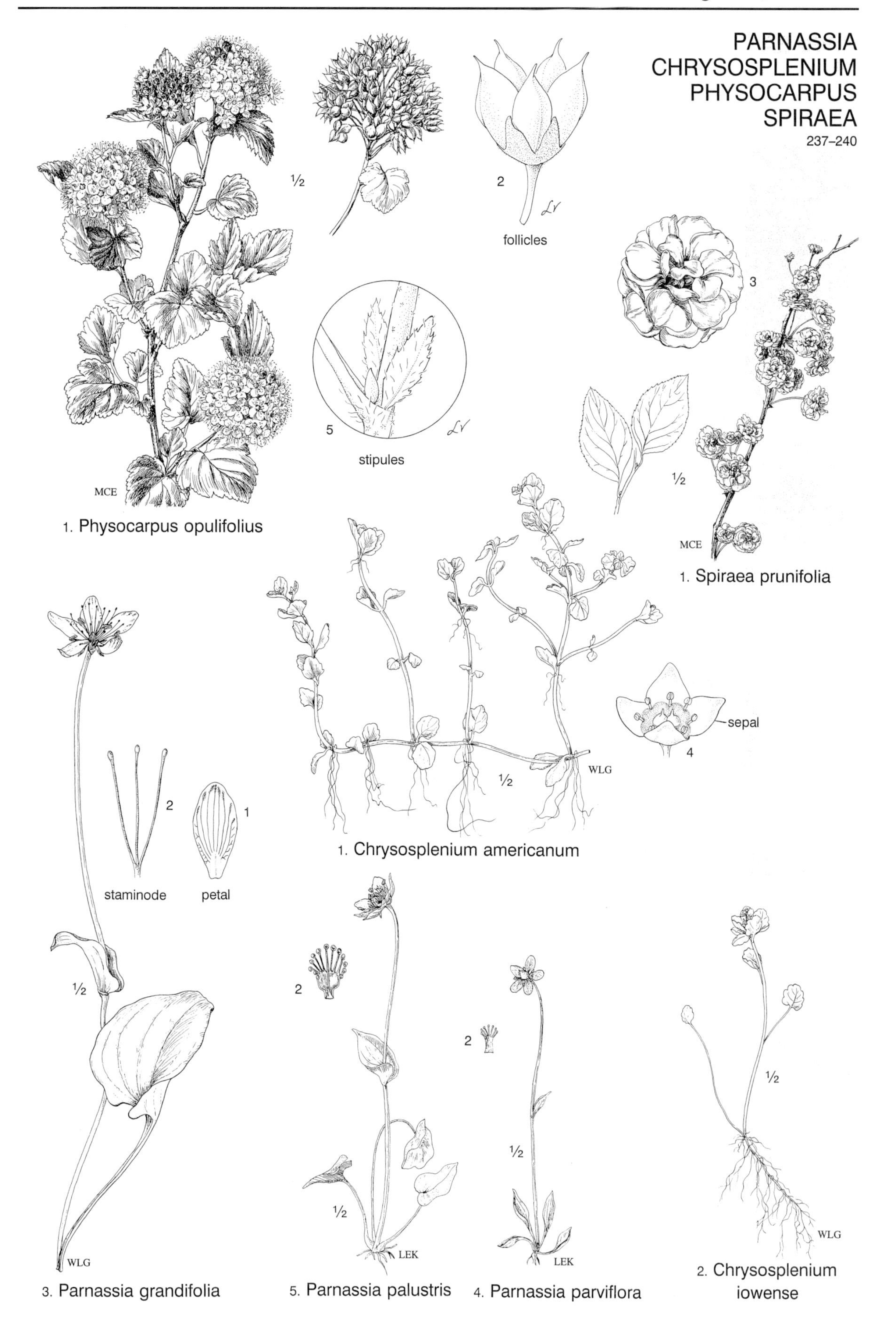

½

2

follicles

3

5

stipules

½

MCE

1. Physocarpus opulifolius

MCE

1. Spiraea prunifolia

½

WLG

1. Chrysosplenium americanum

sepal

4

2

1

staminode   petal

½

2

2

½

WLG

3. Parnassia grandifolia

2

½

LEK

5. Parnassia palustris

½

LEK

4. Parnassia parviflora

½

WLG

2. Chrysosplenium iowense

SPIRAEA

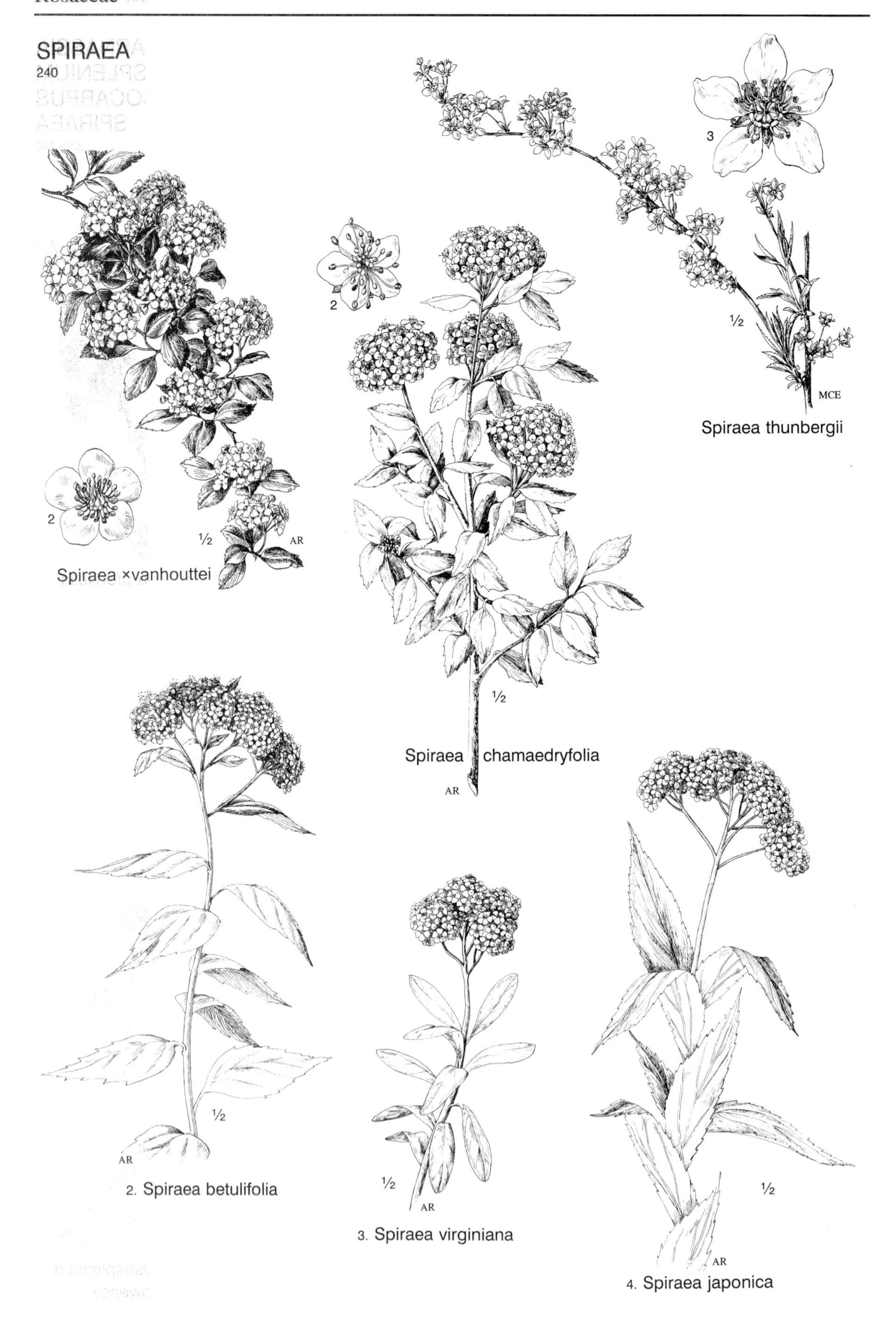

Spiraea ×vanhouttei

Spiraea thunbergii

Spiraea chamaedryfolia

2. Spiraea betulifolia

3. Spiraea virginiana

4. Spiraea japonica

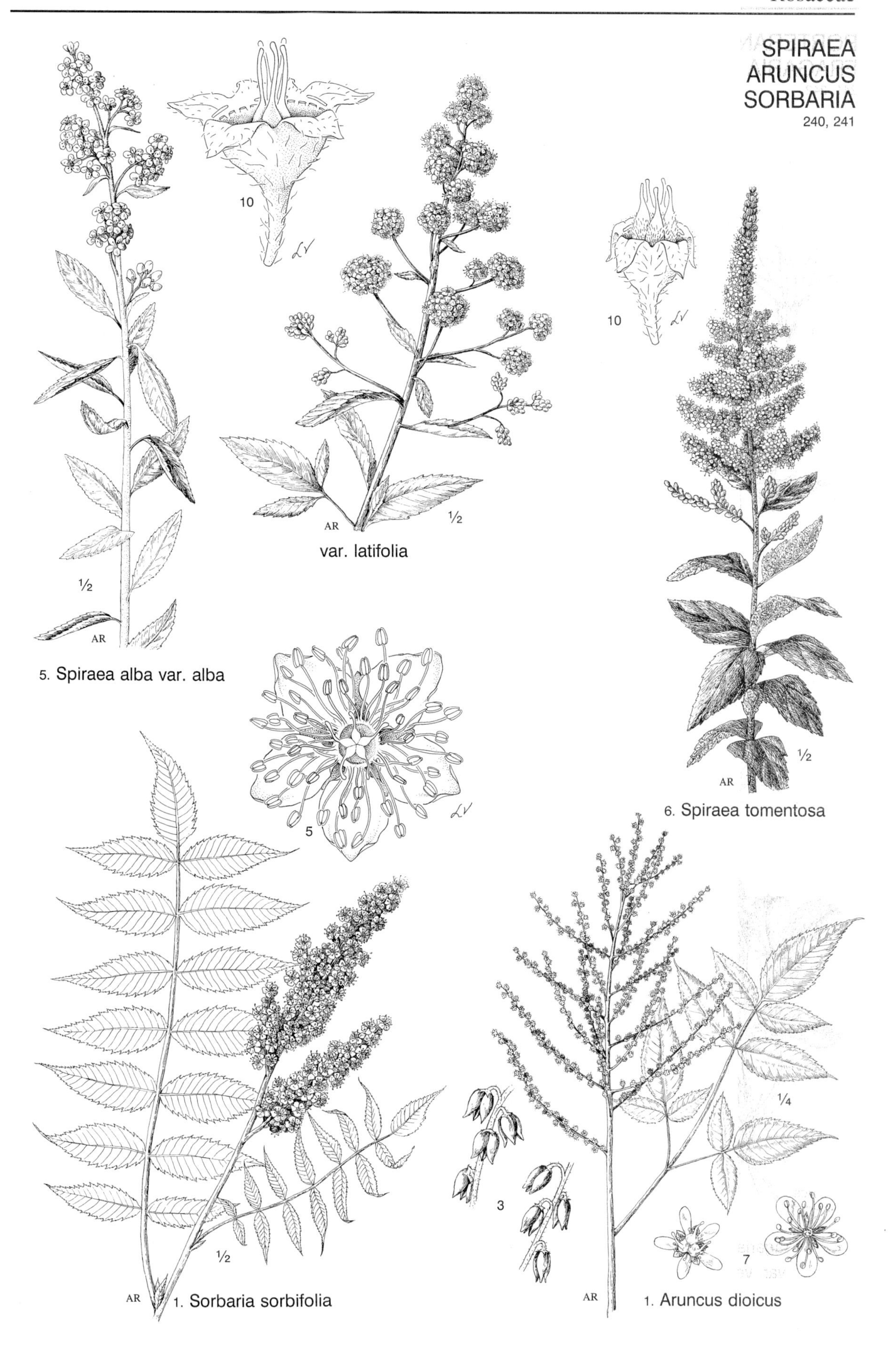

10

var. latifolia

10

5. Spiraea alba var. alba

5

6. Spiraea tomentosa

1. Sorbaria sorbifolia

3

7

1. Aruncus dioicus

PORTERANTHUS
FRAGARIA
241, 242

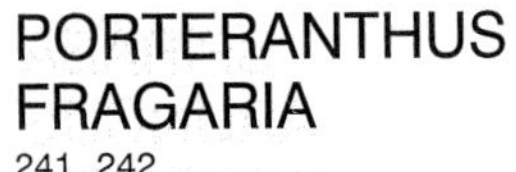

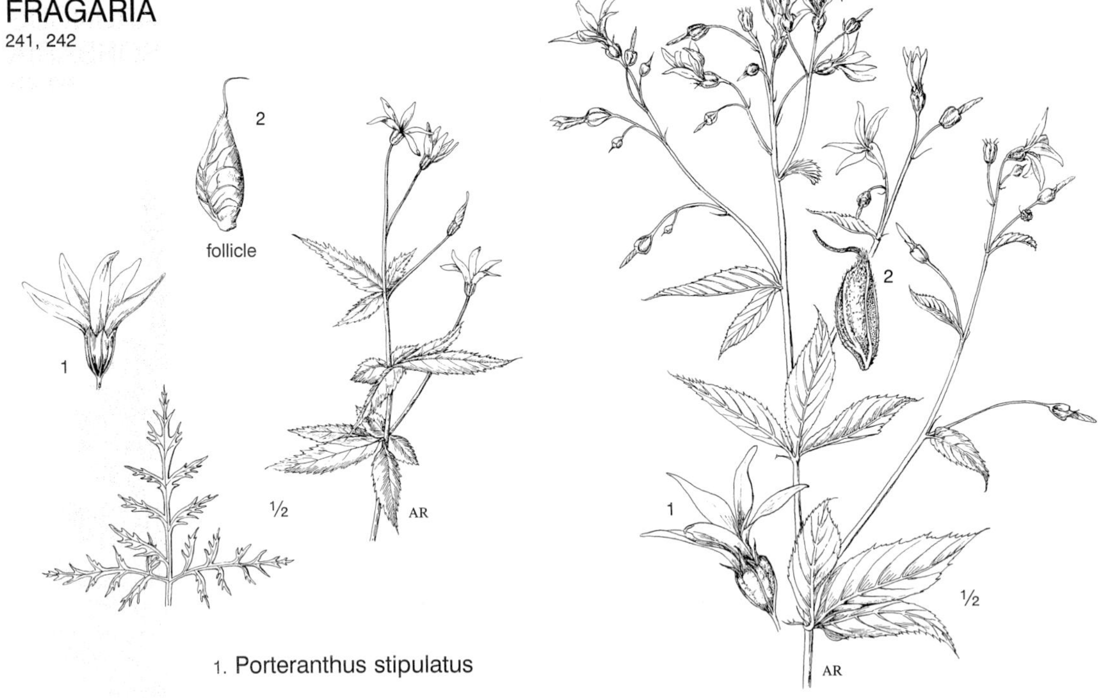

follicle

1

2

½   AR

1. Porteranthus stipulatus

1

2

½

AR

2. Porteranthus trifoliatus

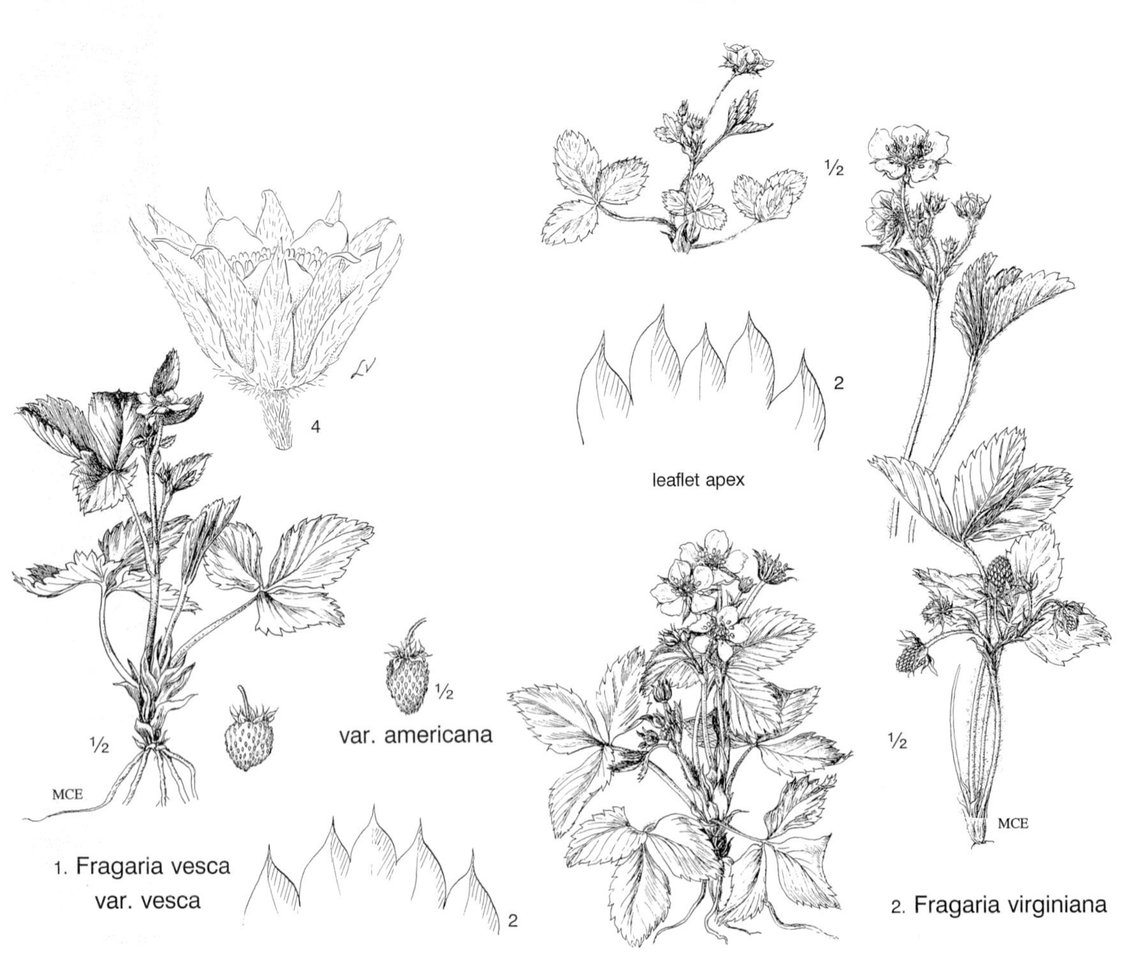

4

½

2

leaflet apex

var. americana   ½

½

MCE

1. Fragaria vesca
var. vesca

2

½

MCE

2. Fragaria virginiana

FRAGARIA
DUCHESNEA
SIBBALDIA
CHAMAERHODOS
WALDSTEINIA
242, 243

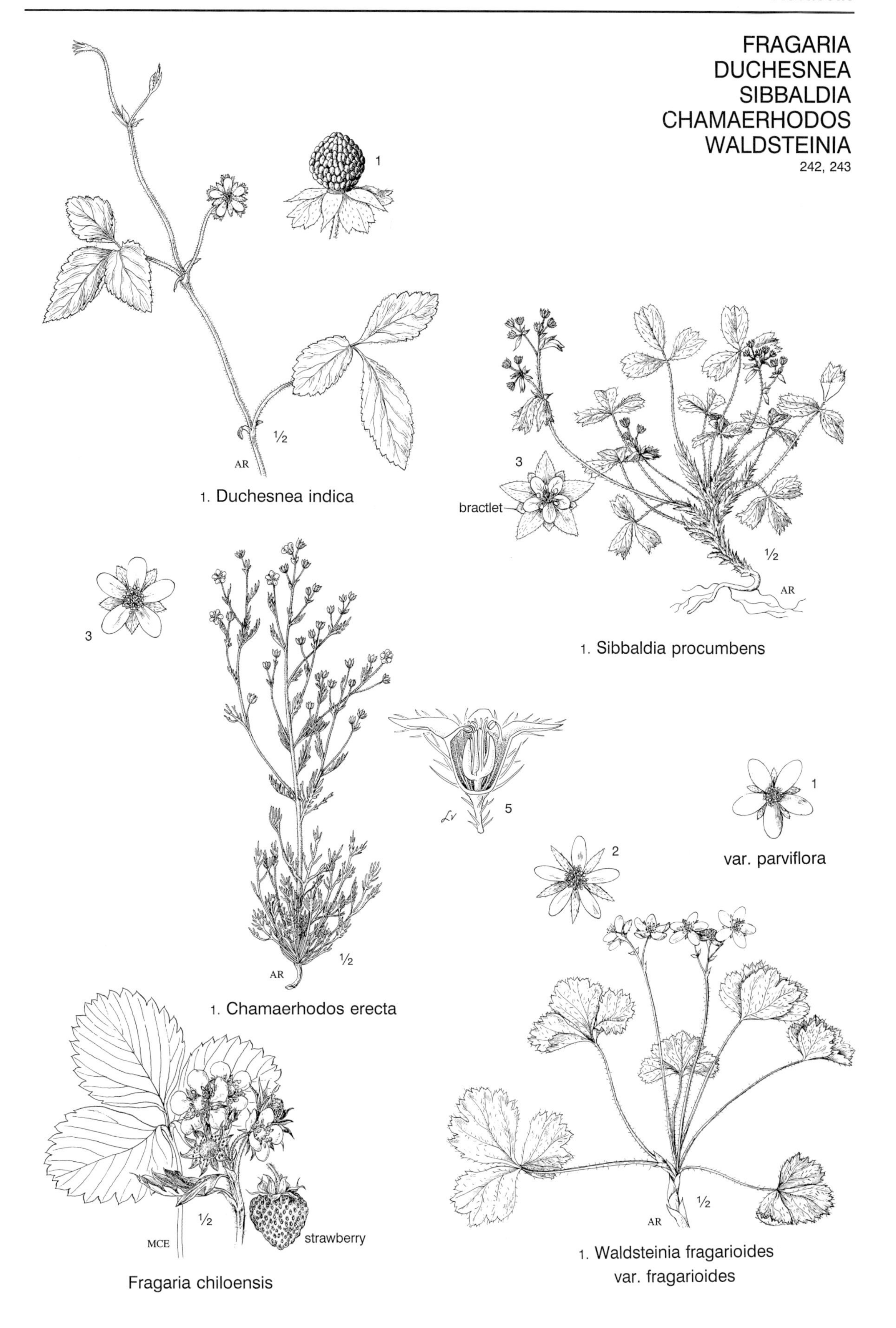

1. Duchesnea indica

1. Sibbaldia procumbens

bractlet

var. parviflora

1. Chamaerhodos erecta

Fragaria chiloensis

strawberry

MCE

1. Waldsteinia fragarioides
var. fragarioides

POTENTILLA
244

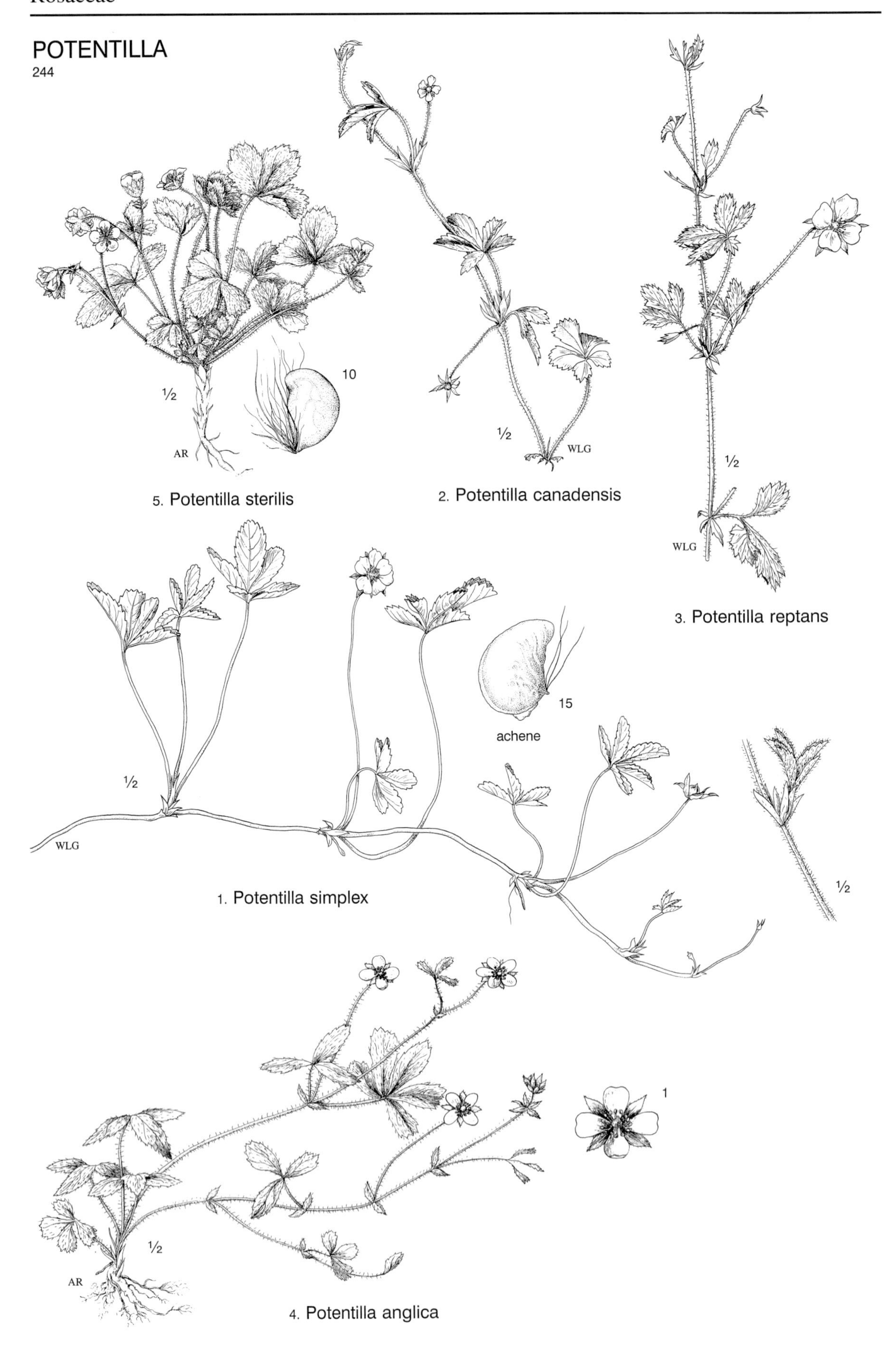

5. Potentilla sterilis

2. Potentilla canadensis

3. Potentilla reptans

10

15
achene

1. Potentilla simplex

4. Potentilla anglica

1

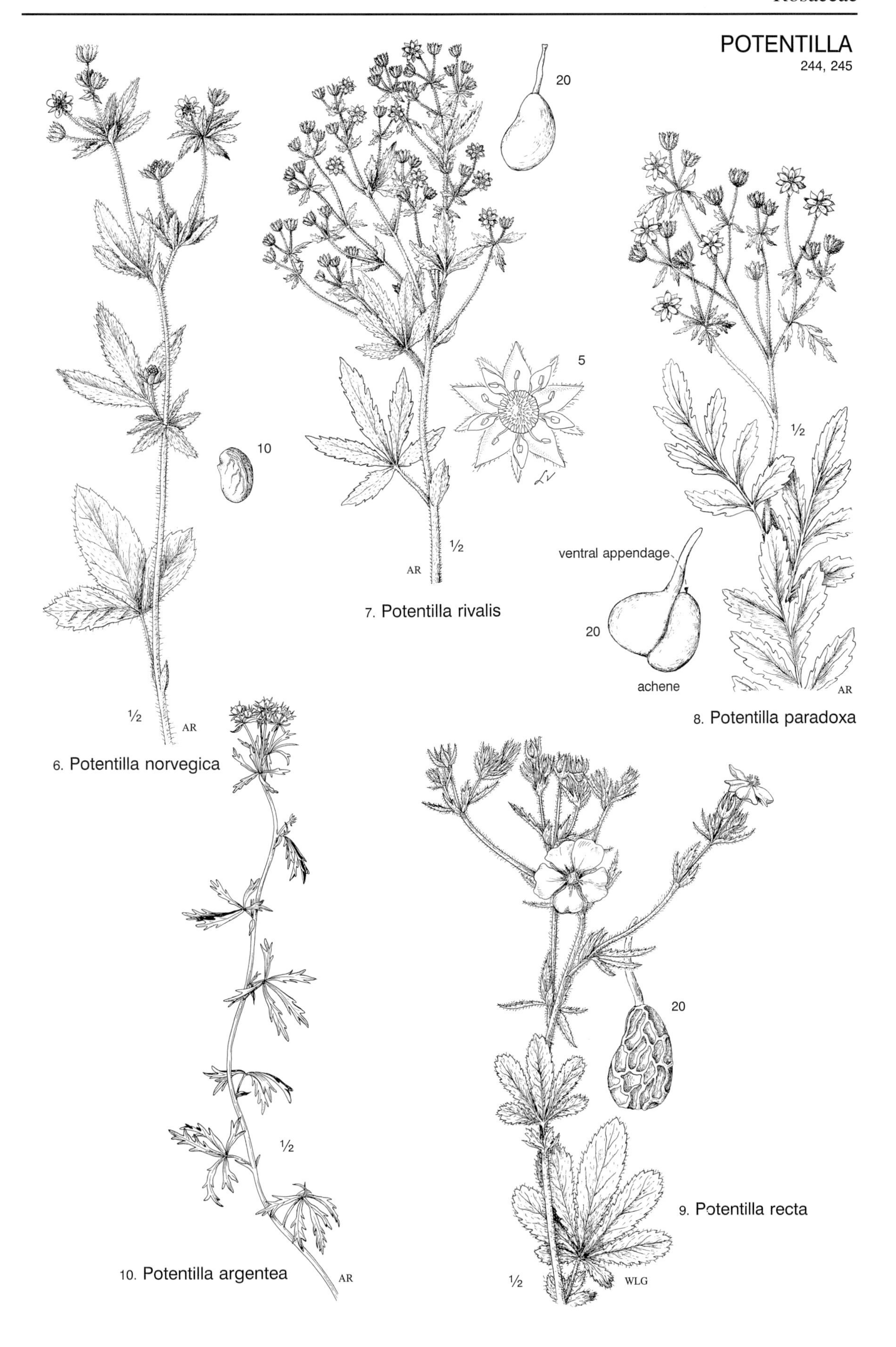

20

10

5

7. Potentilla rivalis

ventral appendage

20

achene

8. Potentilla paradoxa

6. Potentilla norvegica

½

20

9. Potentilla recta

10. Potentilla argentea

POTENTILLA
245

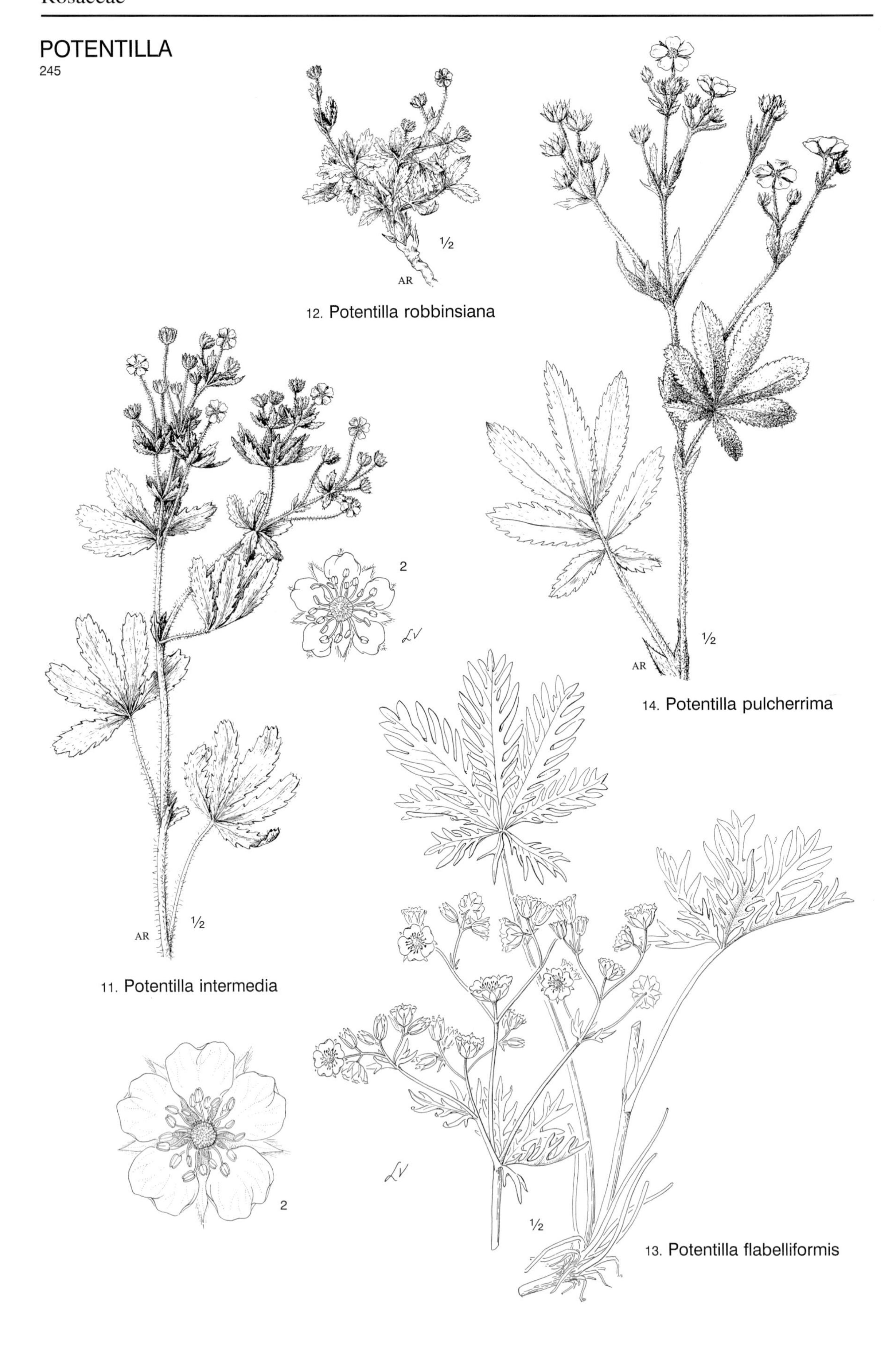

12. Potentilla robbinsiana

14. Potentilla pulcherrima

11. Potentilla intermedia

13. Potentilla flabelliformis

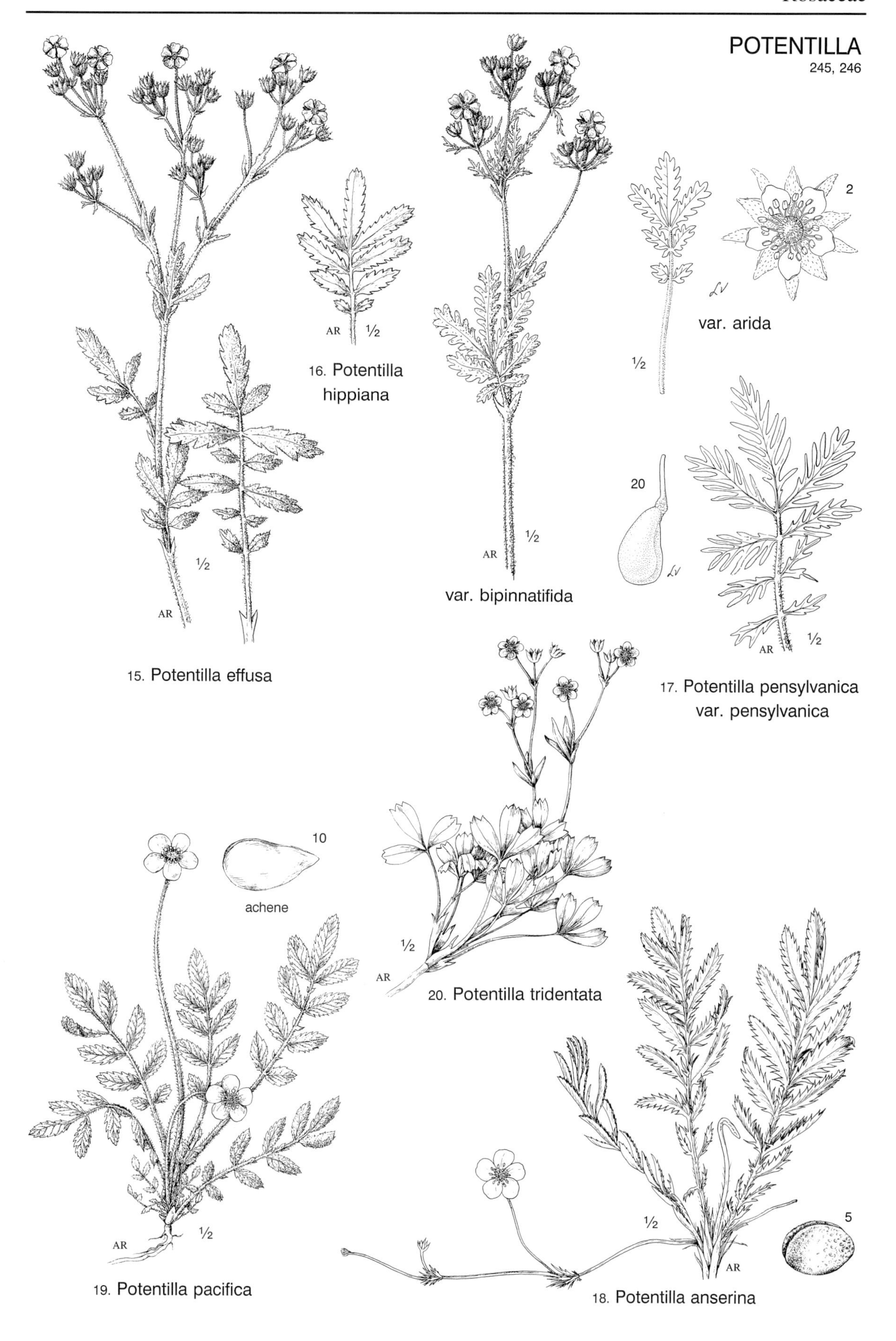

16. Potentilla
hippiana

var. arida

2

20

var. bipinnatifida

17. Potentilla pensylvanica
var. pensylvanica

15. Potentilla effusa

10

achene

20. Potentilla tridentata

19. Potentilla pacifica

5

18. Potentilla anserina

POTENTILLA
FILIPENDULA
246, 247

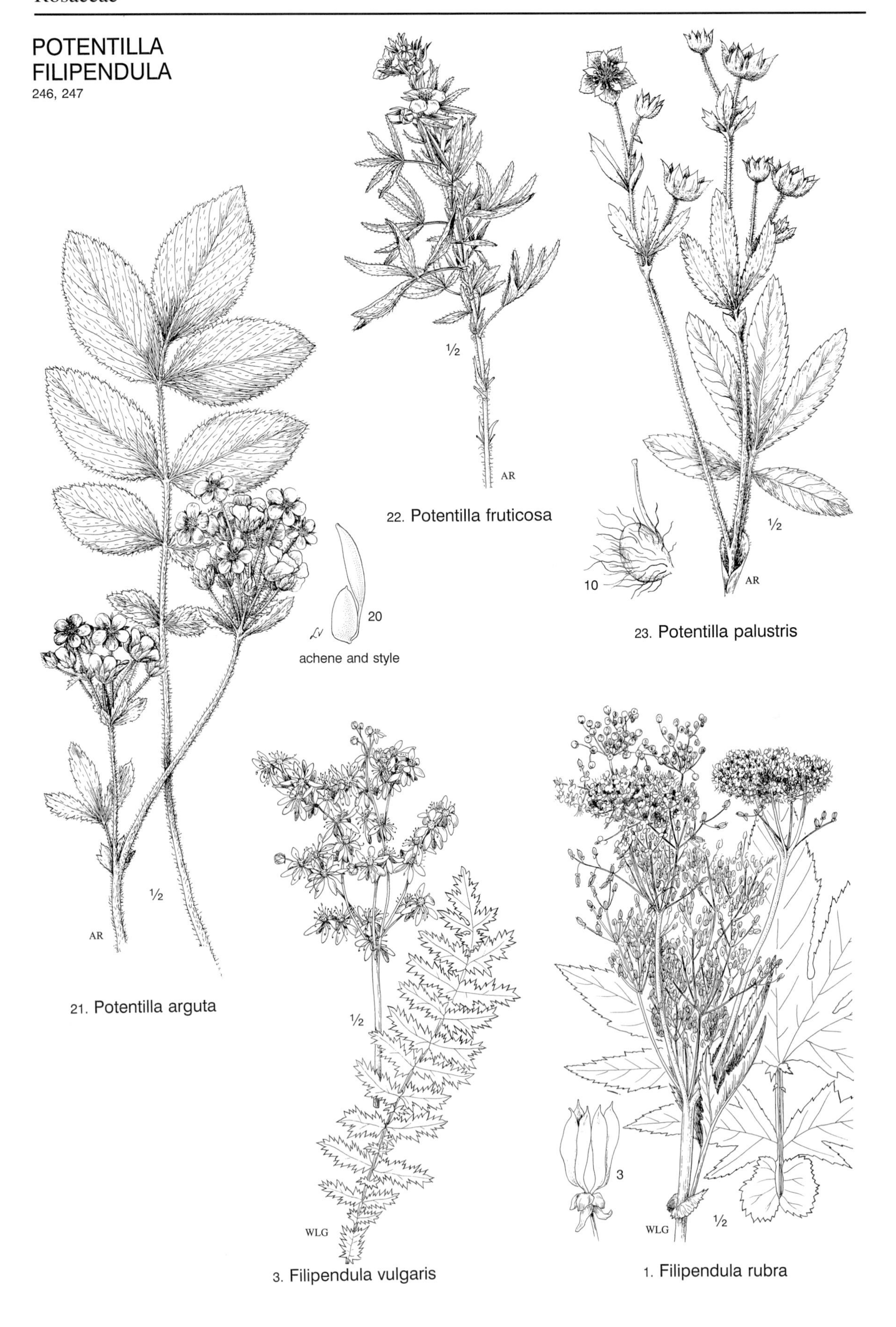

22. Potentilla fruticosa

23. Potentilla palustris

20
achene and style

21. Potentilla arguta

3. Filipendula vulgaris

1. Filipendula rubra

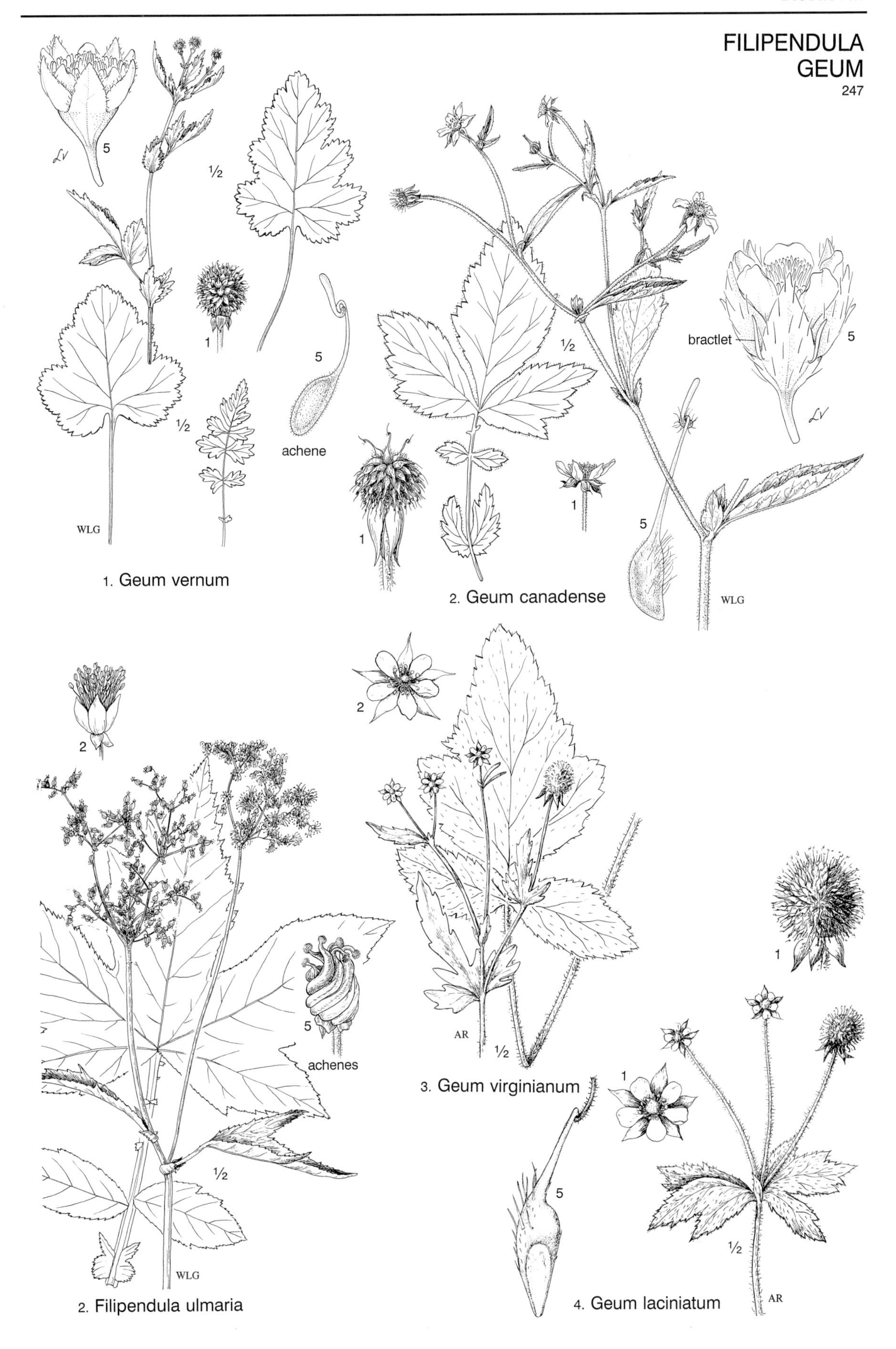

1. Geum vernum

2. Geum canadense

bractlet

achene

WLG

2. Filipendula ulmaria

3. Geum virginianum

AR

4. Geum laciniatum

achenes

GEUM
248

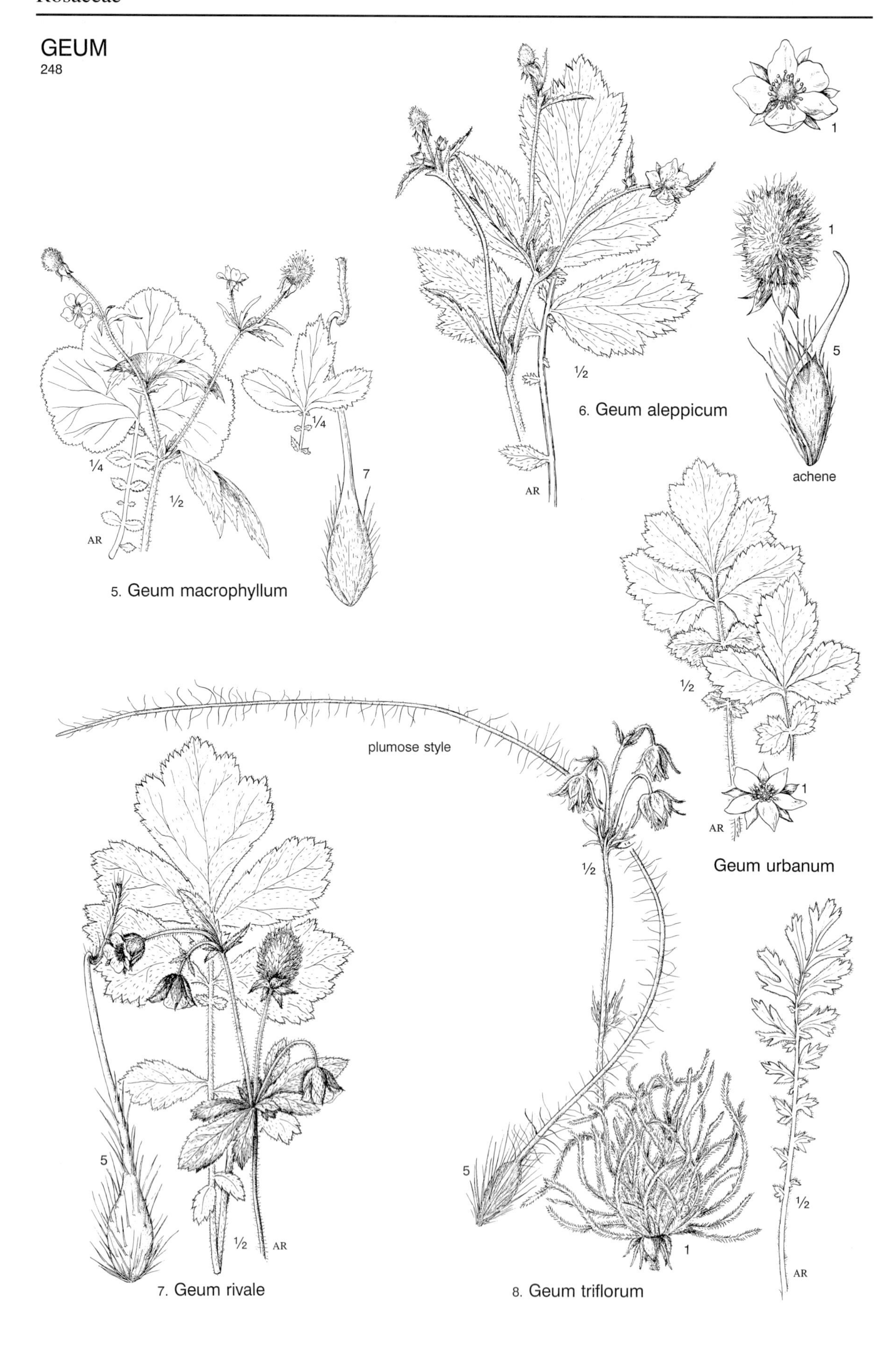

5. Geum macrophyllum

6. Geum aleppicum

achene

½

Geum urbanum

plumose style

7. Geum rivale

8. Geum triflorum

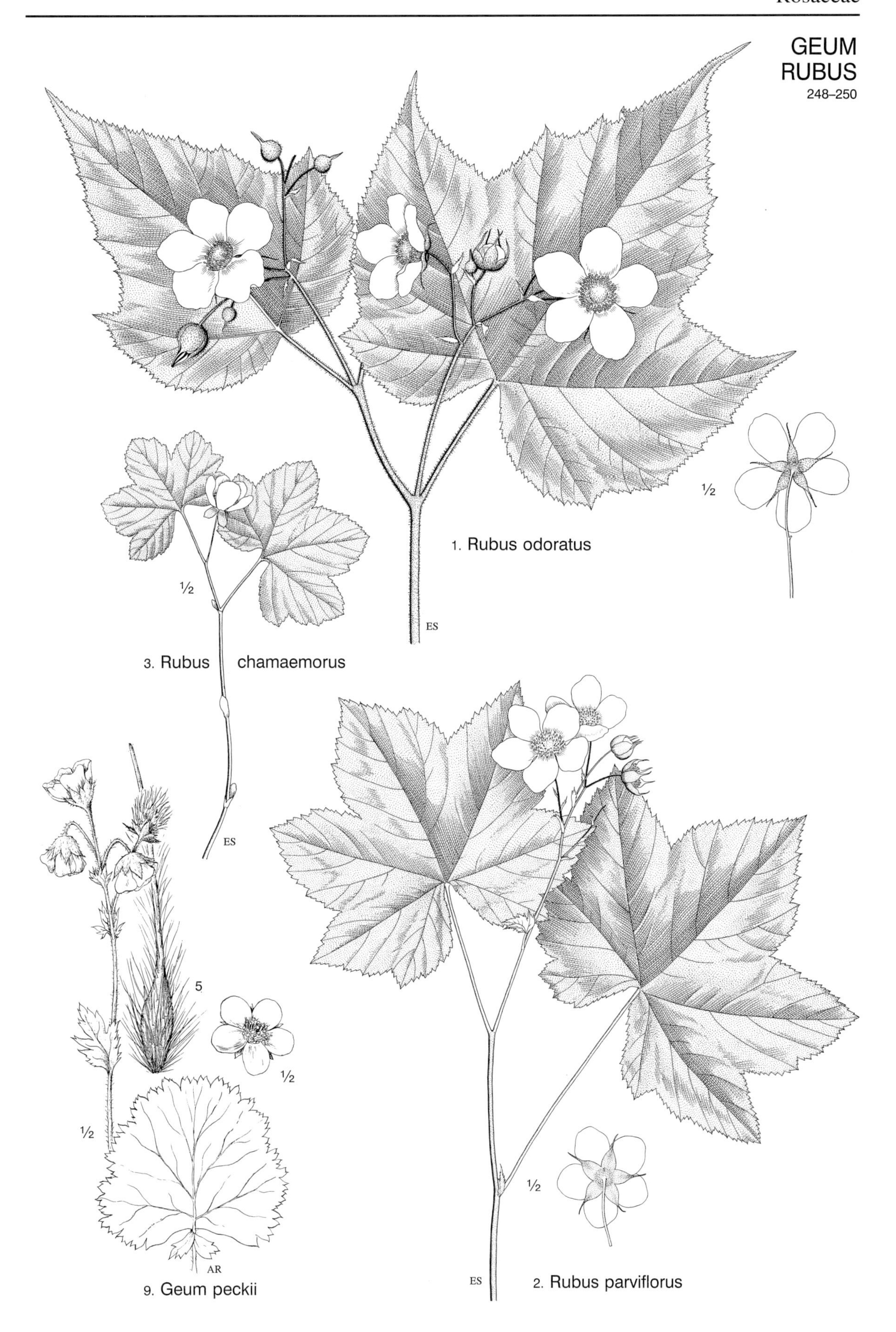

1. Rubus odoratus

3. Rubus chamaemorus

9. Geum peckii

2. Rubus parviflorus

RUBUS
250, 251

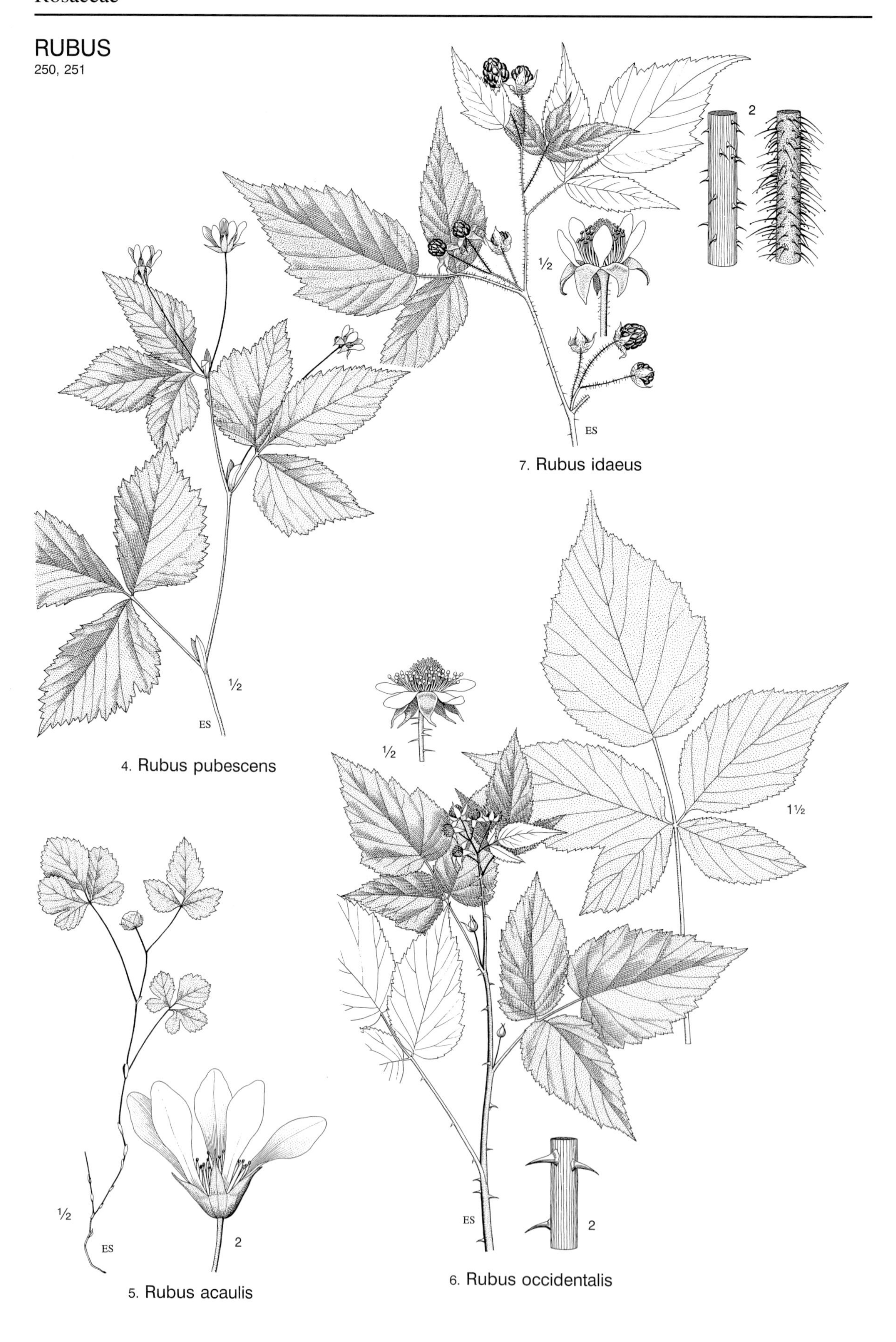

7. Rubus idaeus

4. Rubus pubescens

5. Rubus acaulis

6. Rubus occidentalis

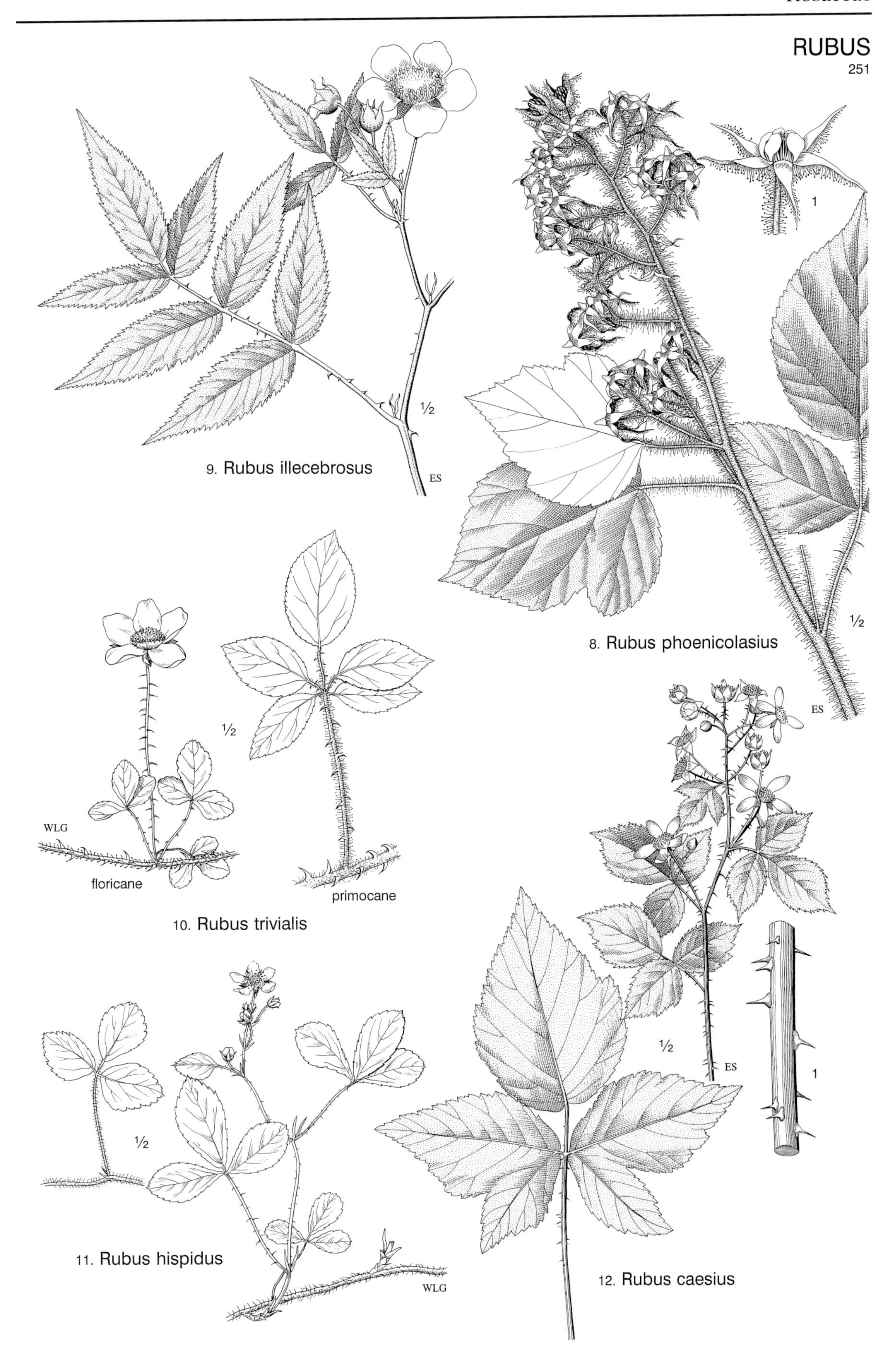

9. Rubus illecebrosus

8. Rubus phoenicolasius

10. Rubus trivialis

floricane

primocane

WLG

11. Rubus hispidus

12. Rubus caesius

WLG

RUBUS
251, 252

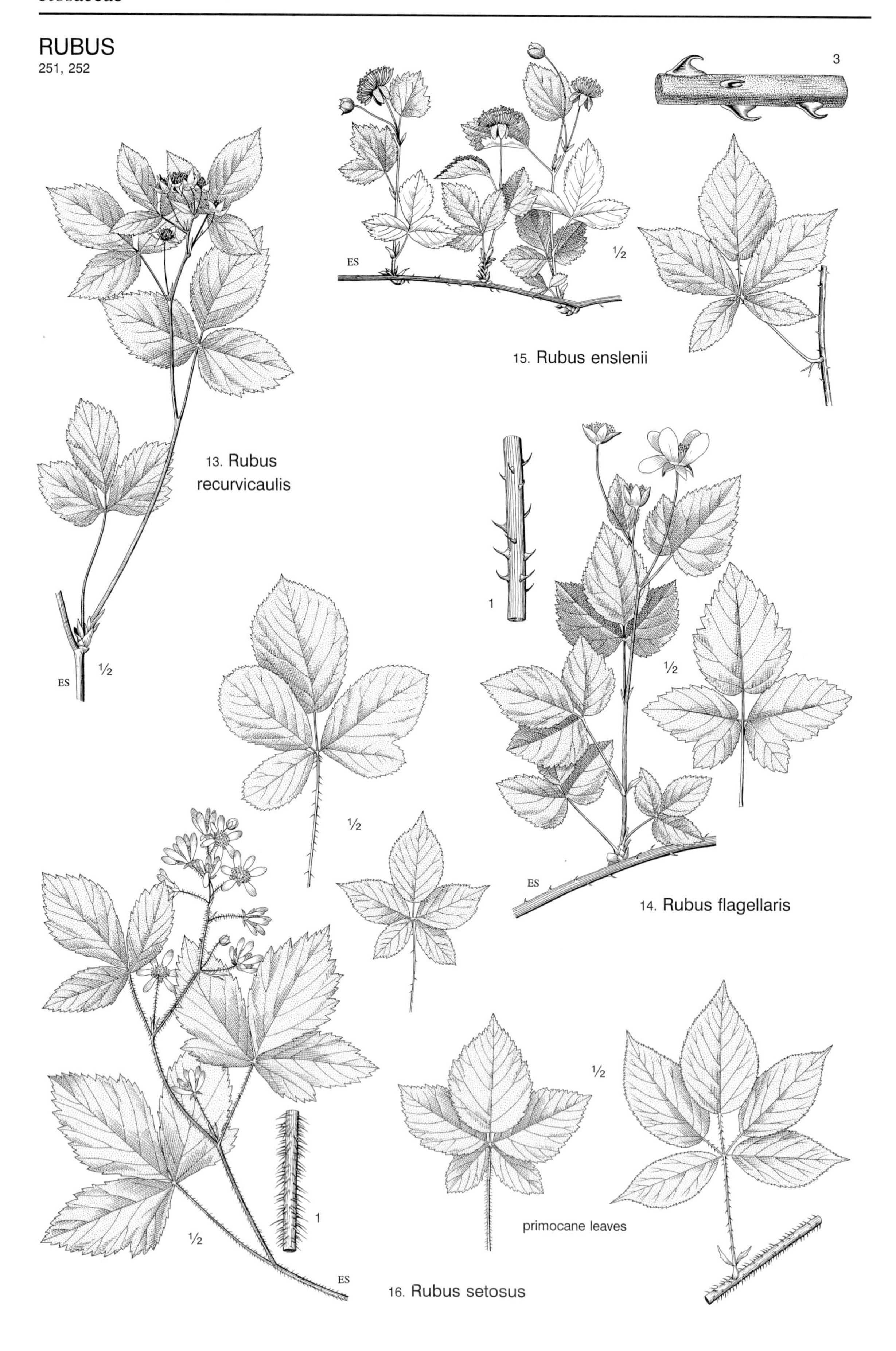

13. Rubus recurvicaulis

15. Rubus enslenii

14. Rubus flagellaris

16. Rubus setosus

primocane leaves

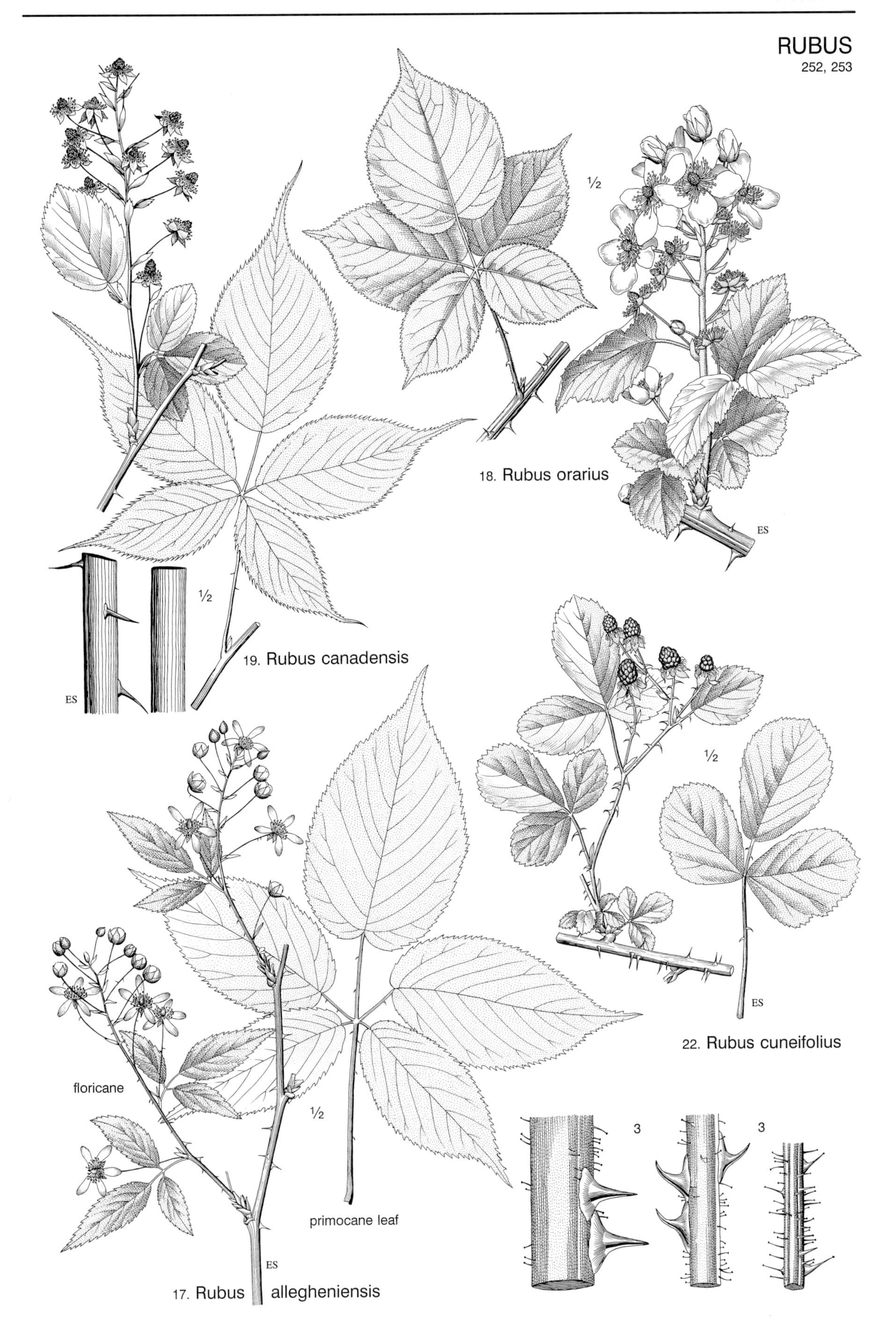

18. Rubus orarius

19. Rubus canadensis

22. Rubus cuneifolius

floricane

primocane leaf

17. Rubus allegheniensis

RUBUS
252, 253

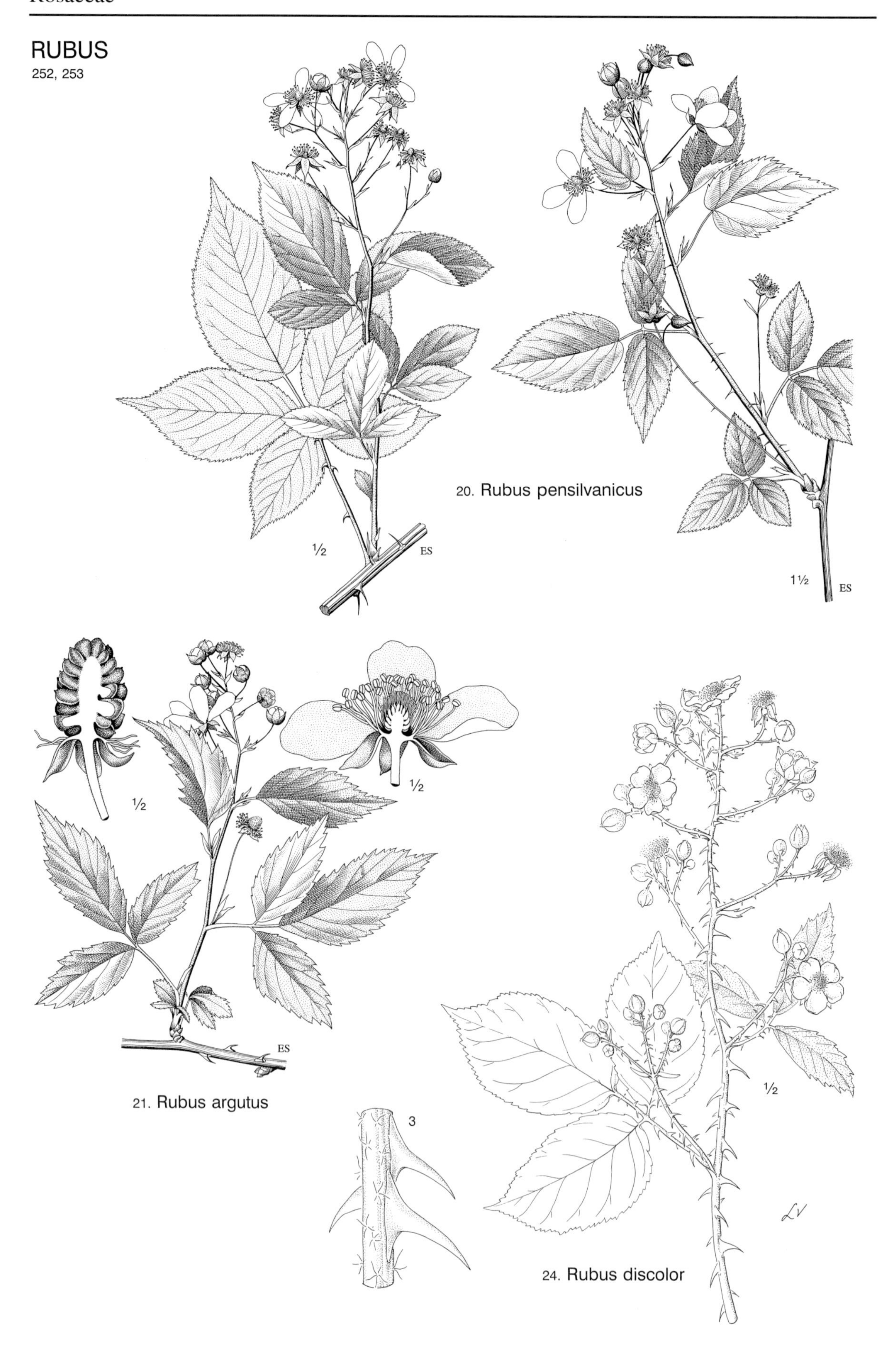

20. Rubus pensilvanicus

½ ES

1½ ES

½

½

21. Rubus argutus

ES

3

½

24. Rubus discolor

Rosaceae

RUBUS
DALIBARDA
ALCHEMILLA
APHANES
253, 254

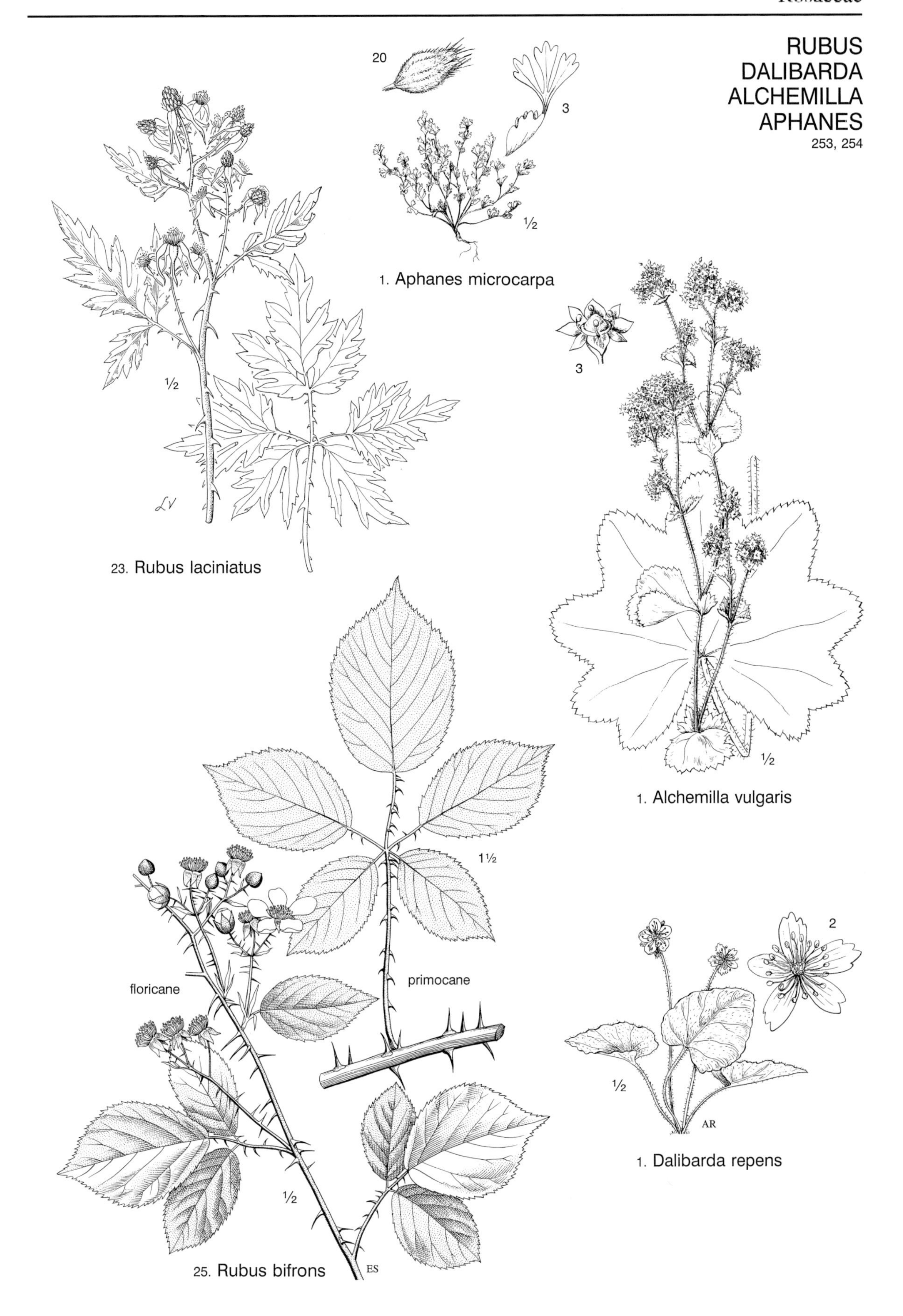

20

3

½

1. Aphanes microcarpa

½

23. Rubus laciniatus

3

½

1. Alchemilla vulgaris

1½

floricane

primocane

½

25. Rubus bifrons   ES

2

½

AR

1. Dalibarda repens

AGRIMONIA
254, 255

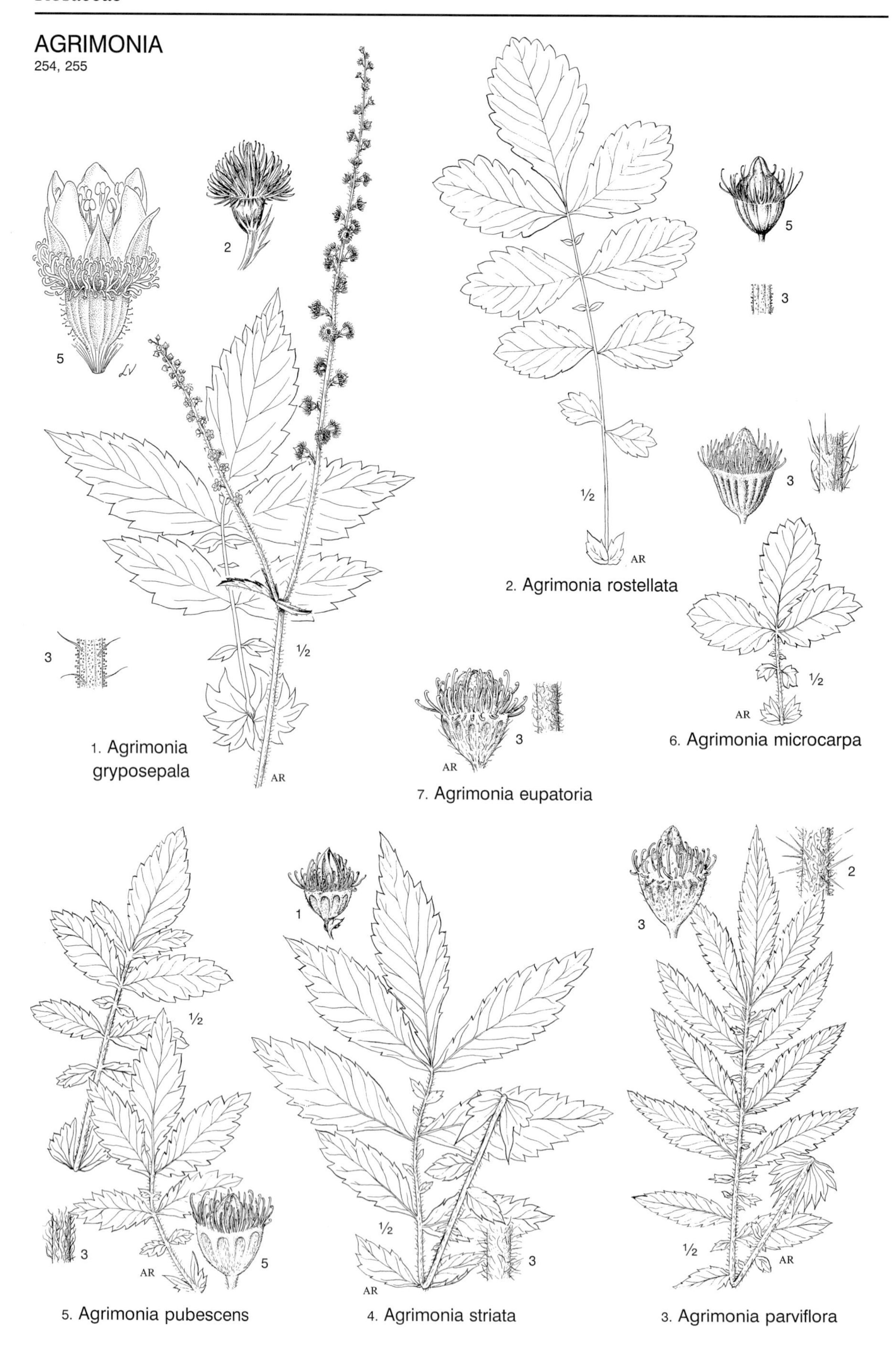

1. Agrimonia
   gryposepala

2. Agrimonia rostellata

7. Agrimonia eupatoria

6. Agrimonia microcarpa

5. Agrimonia pubescens

4. Agrimonia striata

3. Agrimonia parviflora

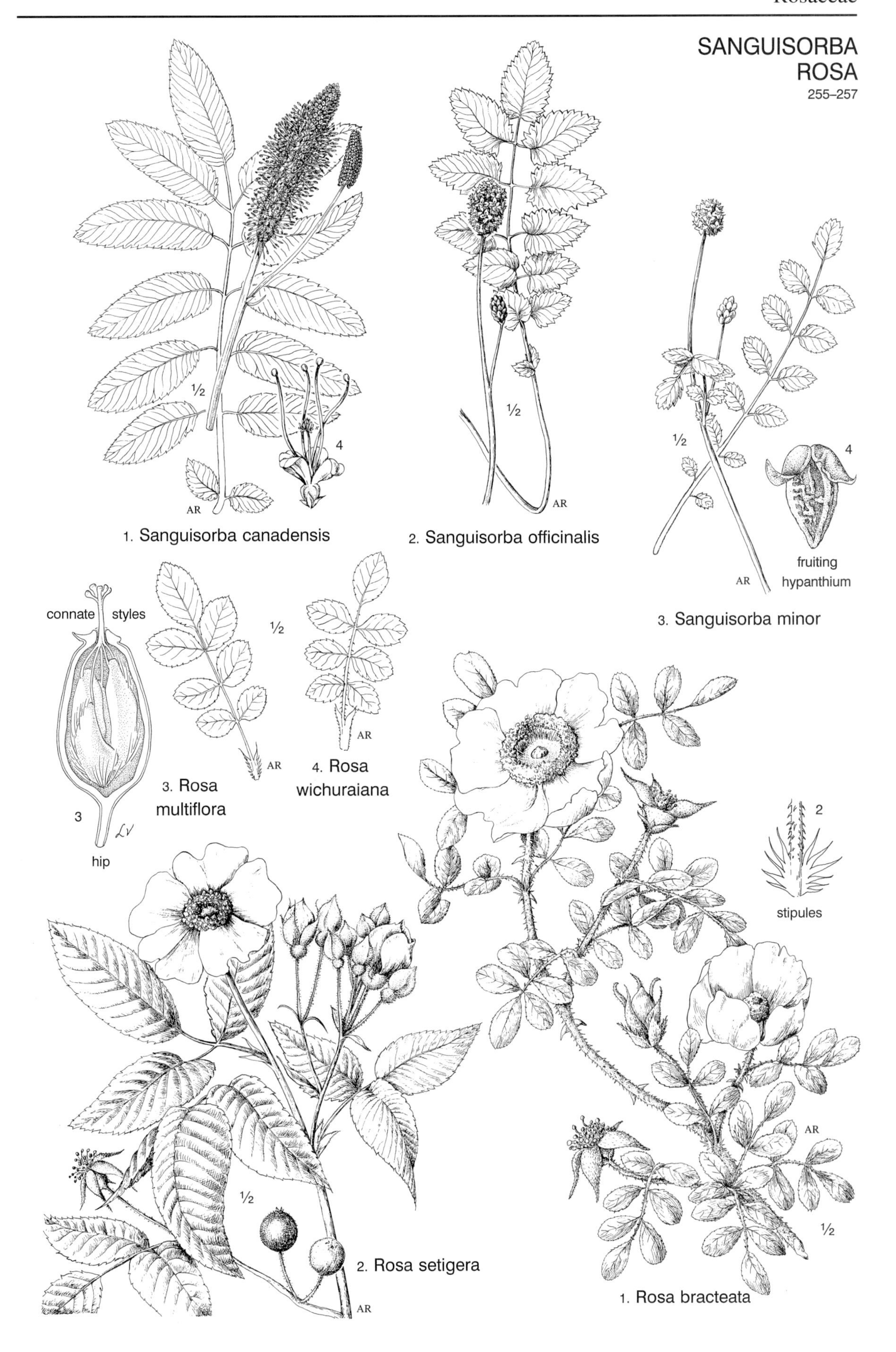

1. Sanguisorba canadensis

2. Sanguisorba officinalis

fruiting
hypanthium

3. Sanguisorba minor

connate styles

½

3. Rosa
multiflora

hip

4. Rosa
wichuraiana

2
stipules

2. Rosa setigera

1. Rosa bracteata

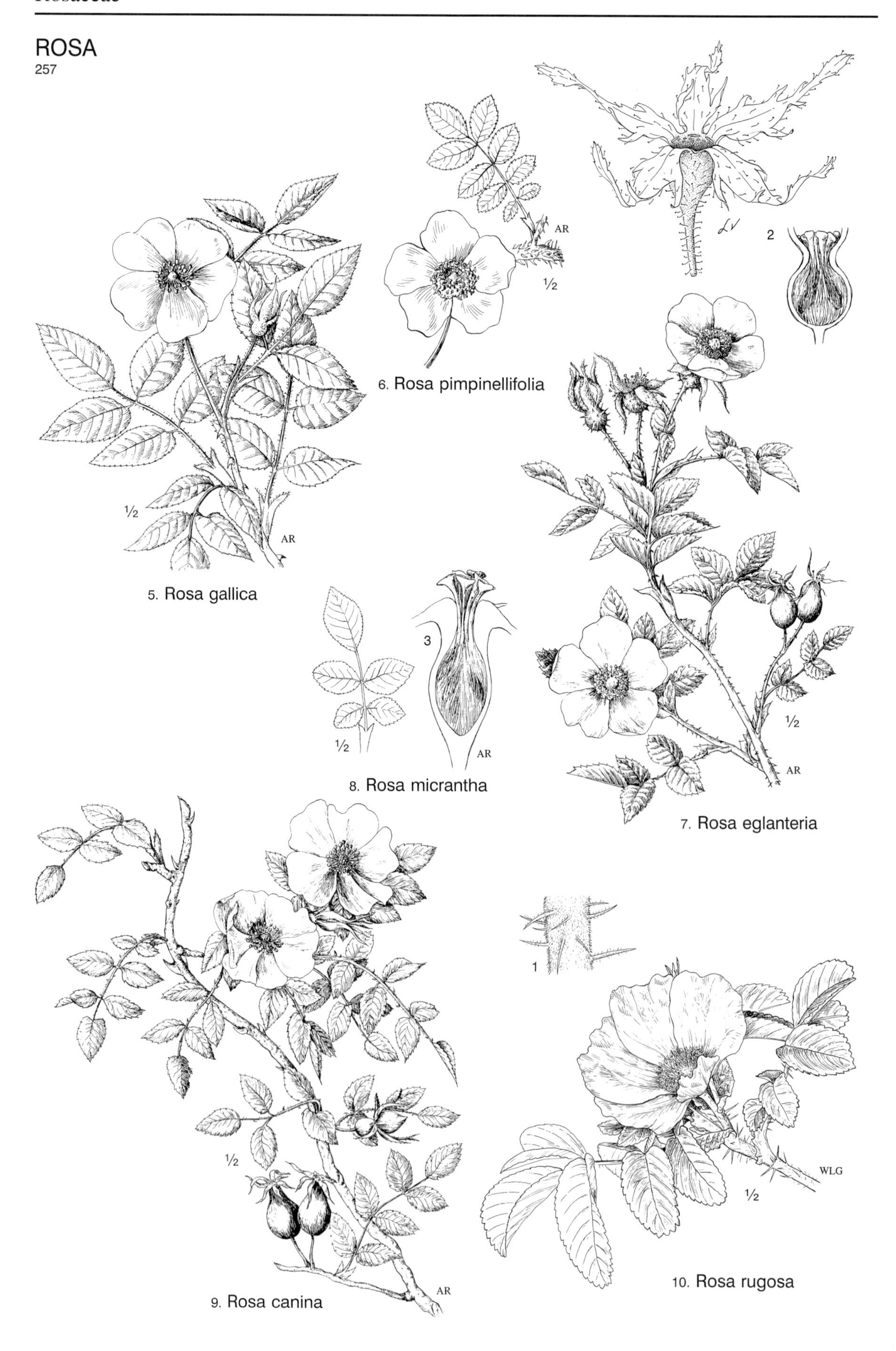

6. Rosa pimpinellifolia

5. Rosa gallica

8. Rosa micrantha

7. Rosa eglanteria

9. Rosa canina

10. Rosa rugosa

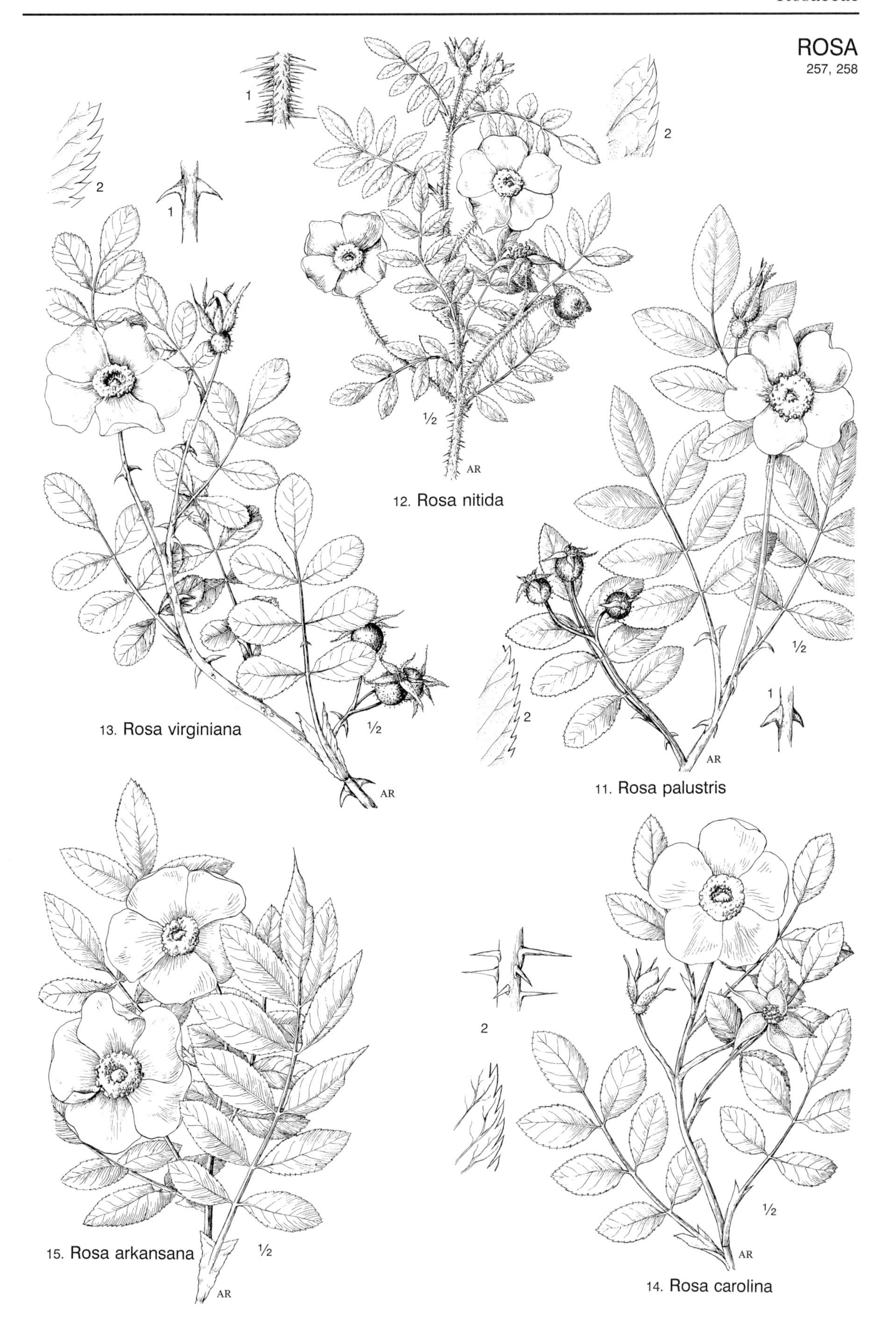

12. Rosa nitida

13. Rosa virginiana

11. Rosa palustris

15. Rosa arkansana

14. Rosa carolina

ROSA
PRUNUS
258–260

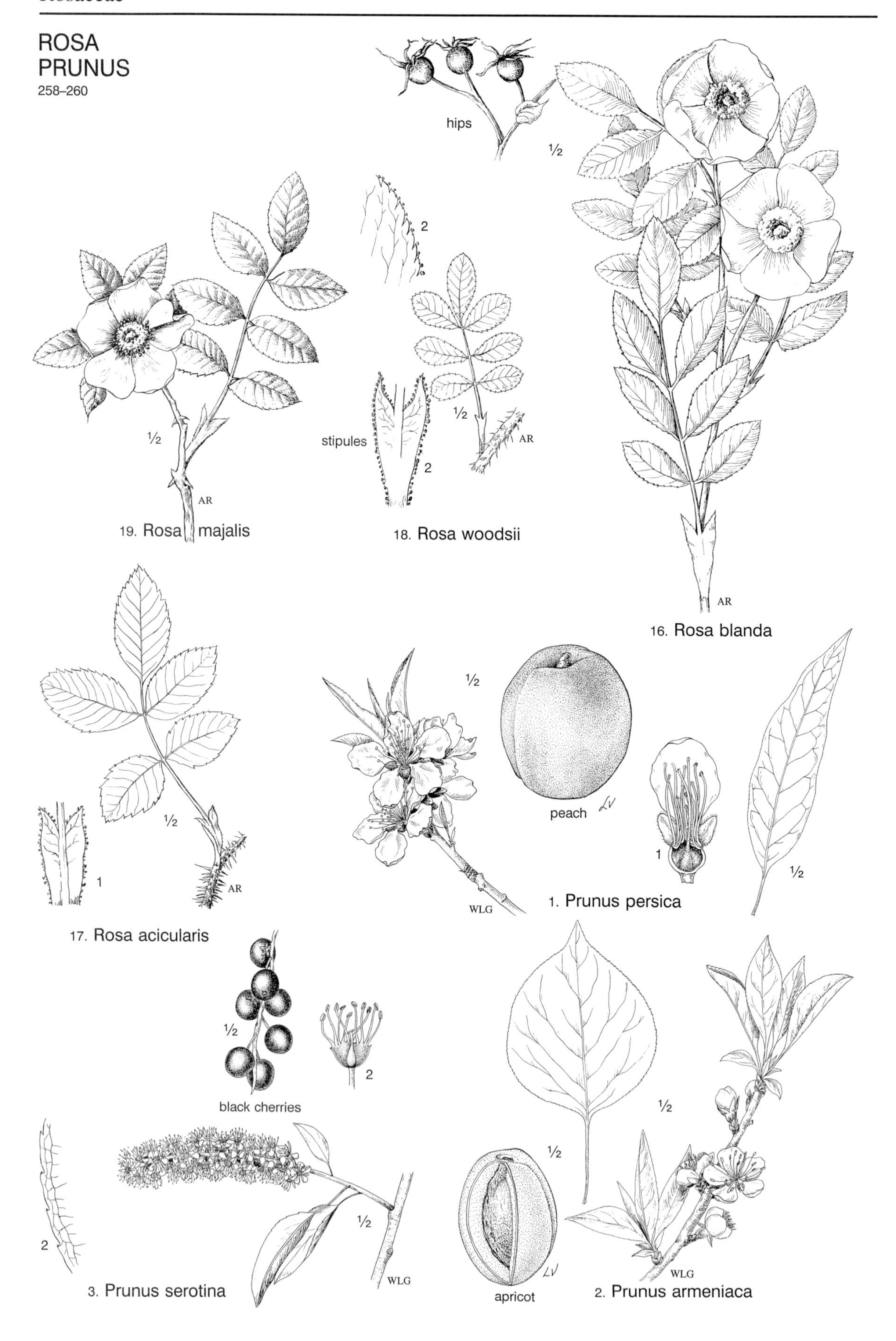

hips

½

19. Rosa majalis

2

stipules

2

½

AR

18. Rosa woodsii

½

16. Rosa blanda

AR

½

1

AR

17. Rosa acicularis

½

peach

1

½

1. Prunus persica

½

black cherries

2

½

½

½

2. Prunus armeniaca

2

3. Prunus serotina

½

WLG

apricot

WLG

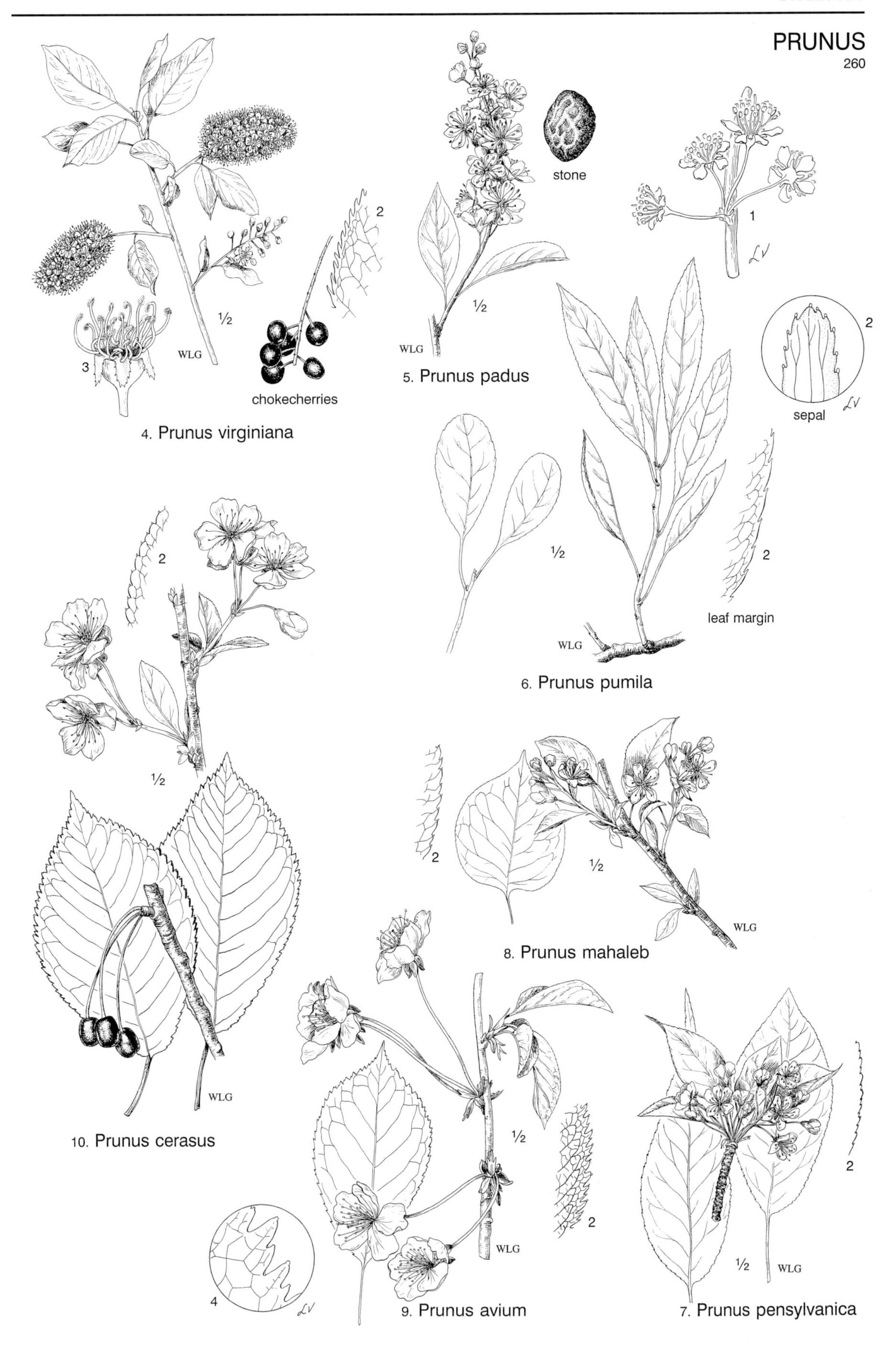

3

WLG

½

4. Prunus virginiana

2

chokecherries

5. Prunus padus

stone

1

2

sepal

½

WLG

2

½

WLG

leaf margin

2

6. Prunus pumila

2

½

10. Prunus cerasus

WLG

8. Prunus mahaleb

WLG

9. Prunus avium

WLG

2

4

7. Prunus pensylvanica

2

½

WLG

PRUNUS
261

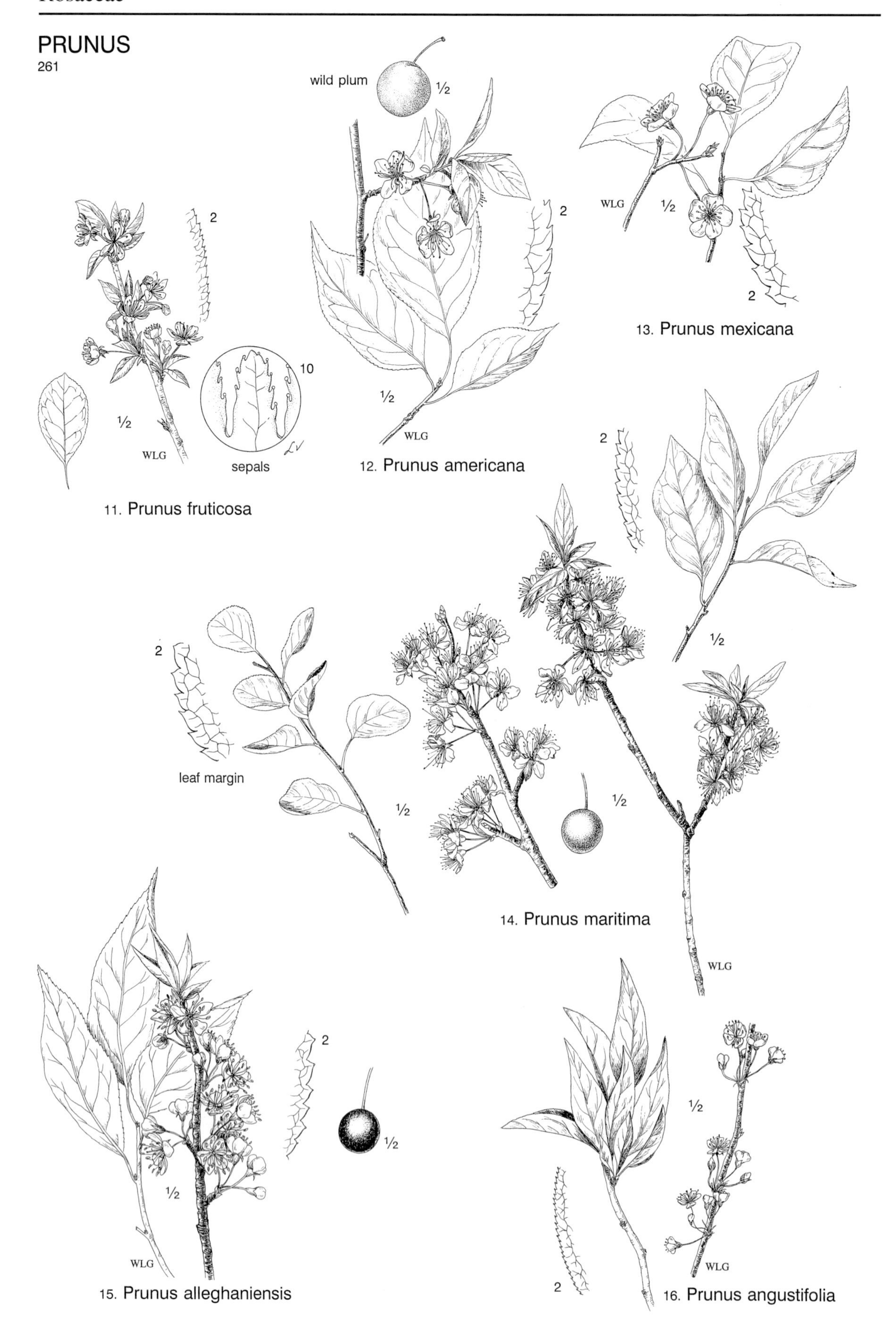

wild plum ½

2

13. Prunus mexicana

WLG ½

2

2

10

WLG

sepals

11. Prunus fruticosa

½

WLG

12. Prunus americana

½

WLG

2

½

2

leaf margin

½

14. Prunus maritima

½

½

WLG

2

½

2

15. Prunus alleghaniensis

WLG

½

½

WLG

16. Prunus angustifolia

2

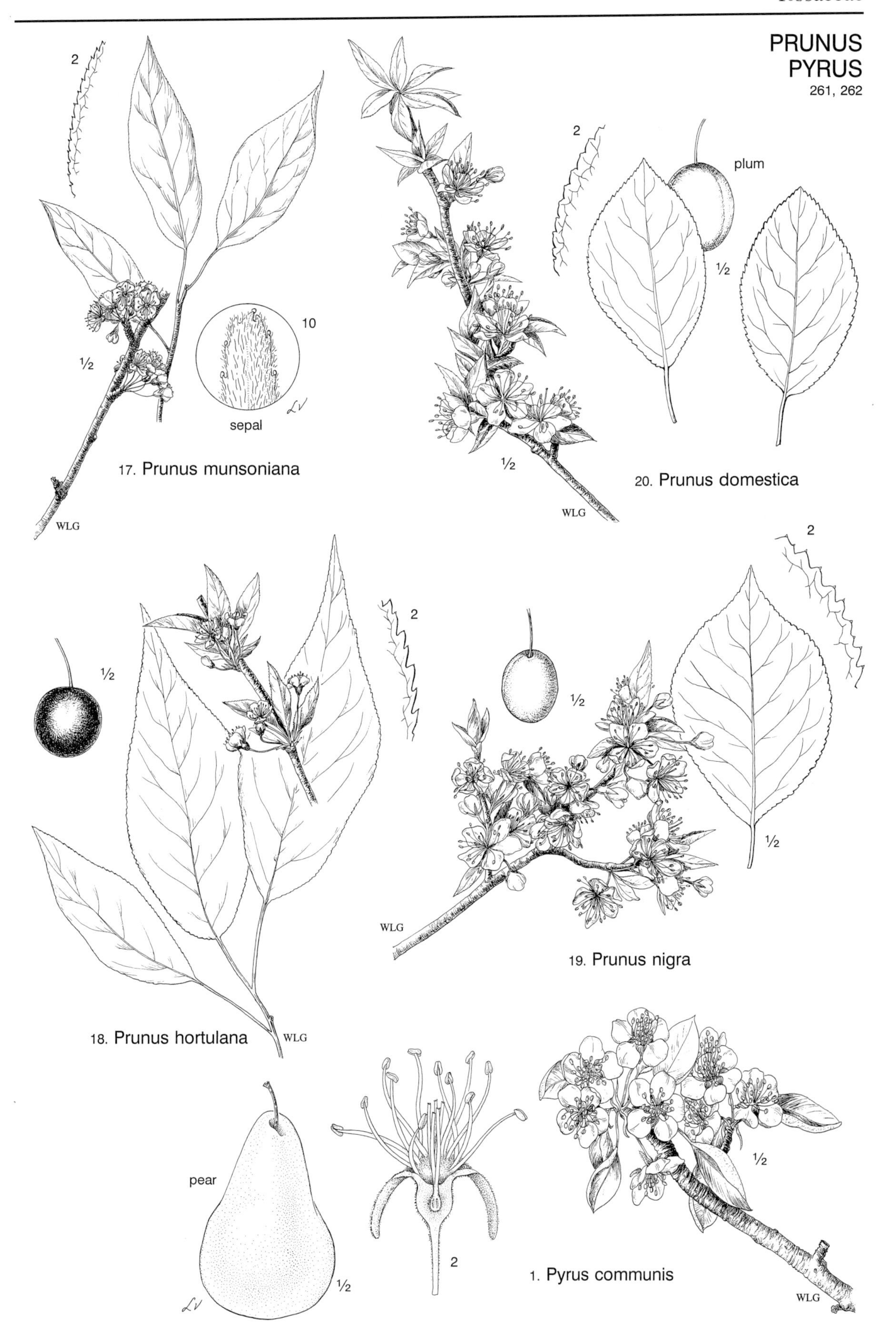

2

10

sepal

17. Prunus munsoniana

WLG

½

plum

½

2

½

20. Prunus domestica

WLG

2

½

½

½

18. Prunus hortulana   WLG

2

½

19. Prunus nigra

WLG

pear

½

2

1. Pyrus communis

½

WLG

PYRUS
262

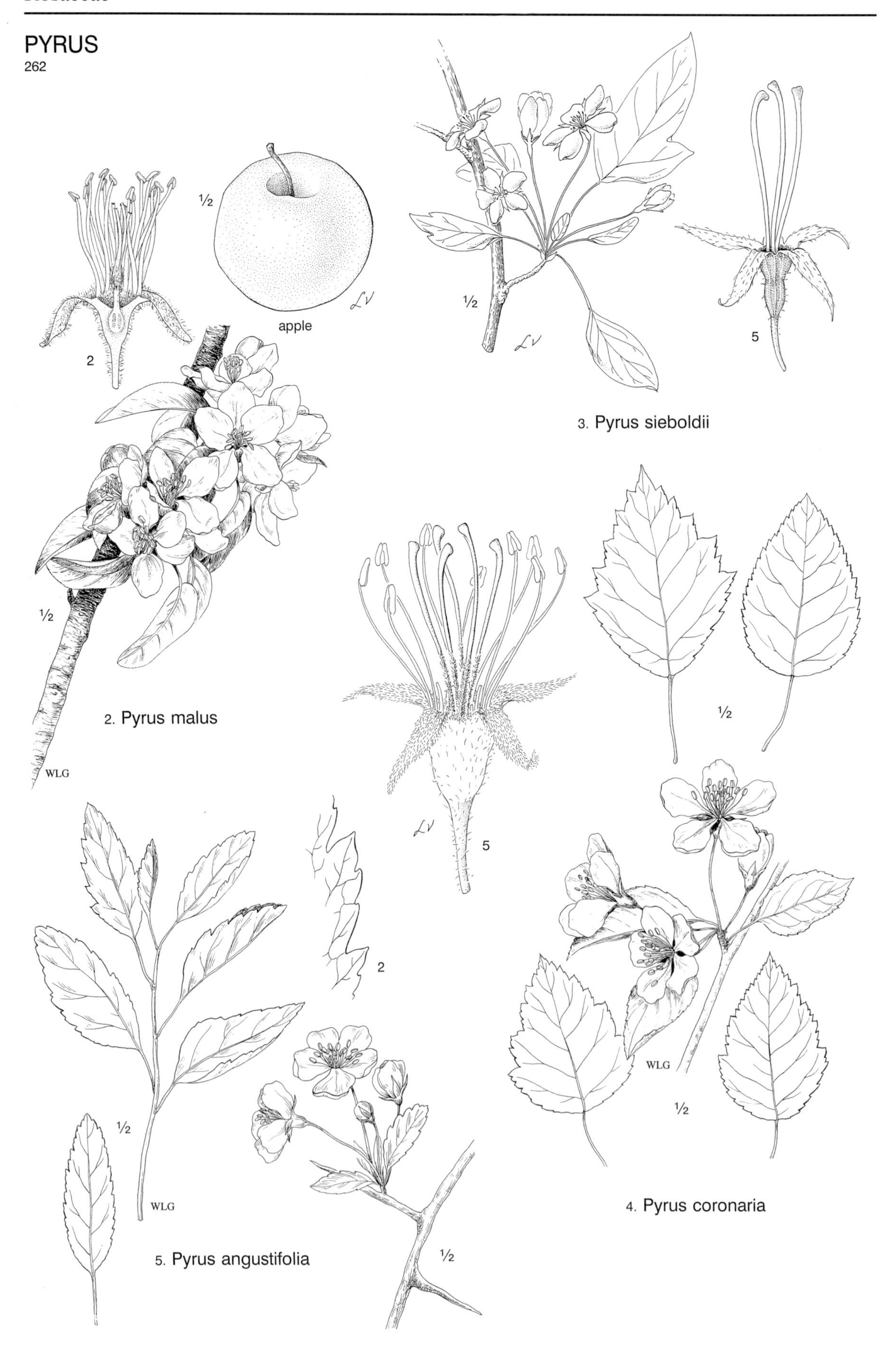

2. Pyrus malus

apple

2

½

3. Pyrus sieboldii

5

5

5

2

4. Pyrus coronaria

WLG

½

5. Pyrus angustifolia

½

WLG

½

½

WLG

½

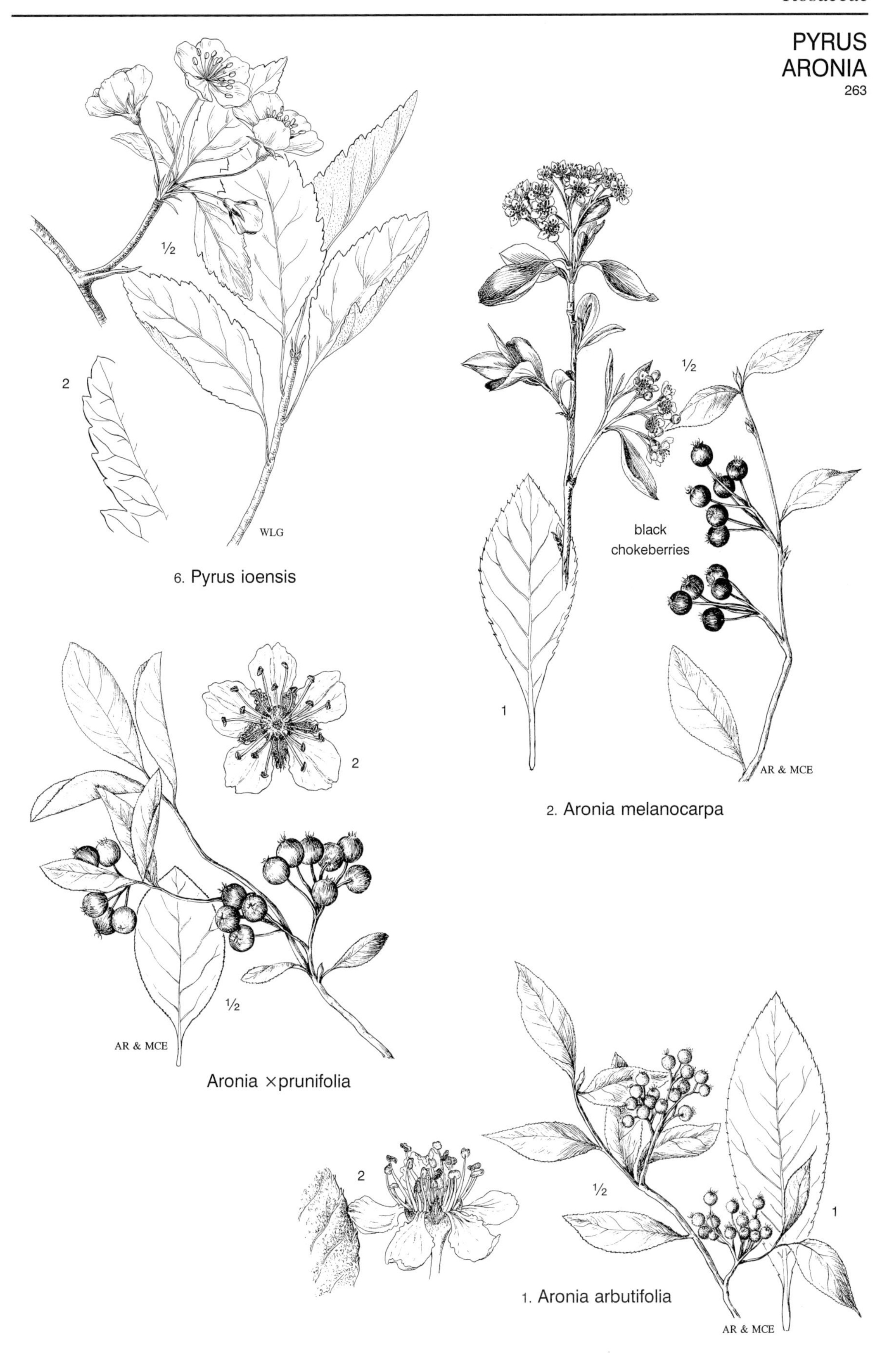

6. Pyrus ioensis

WLG

black
chokeberries

AR & MCE

2. Aronia melanocarpa

AR & MCE

Aronia ×prunifolia

1. Aronia arbutifolia

AR & MCE

SORBUS
263, 264

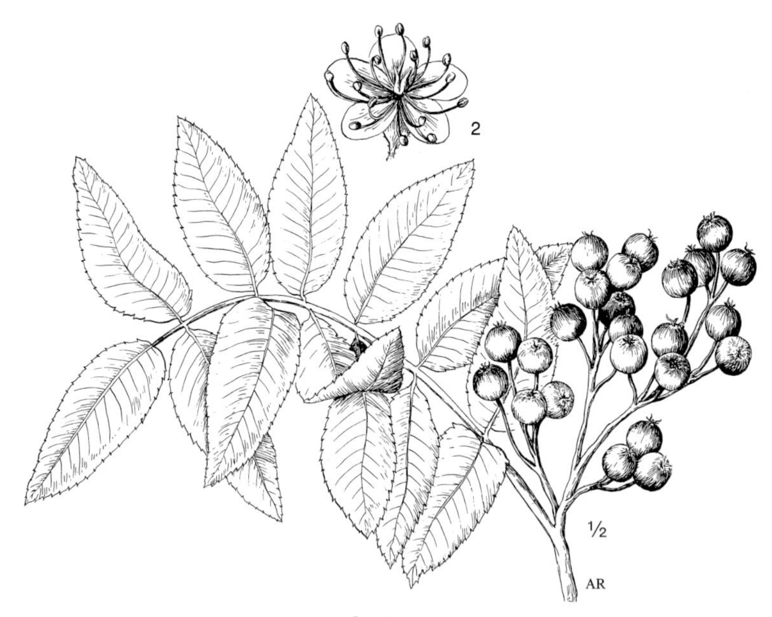

2. Sorbus decora

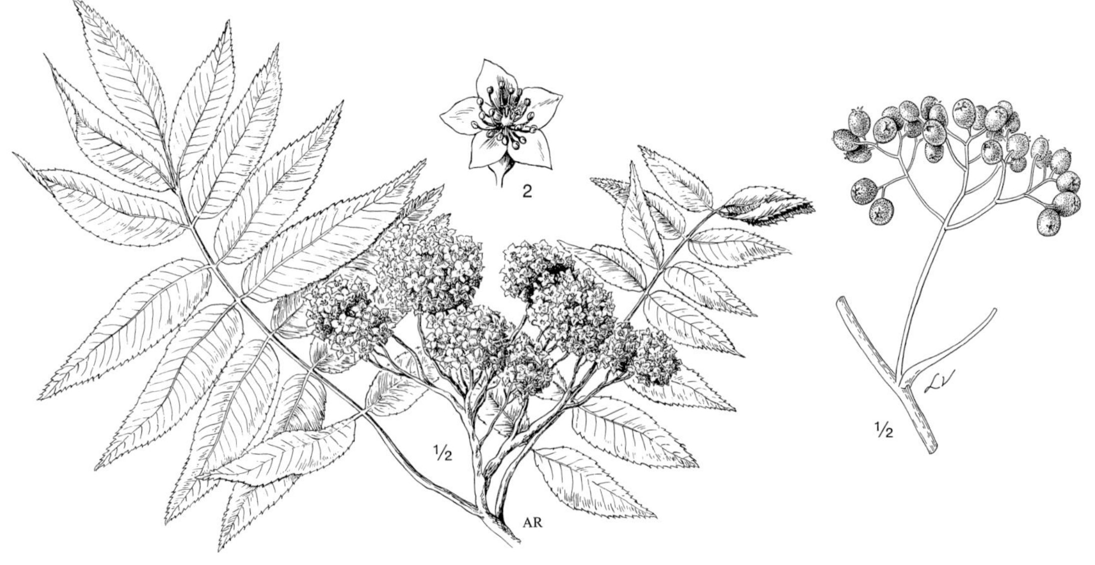

1. Sorbus americana

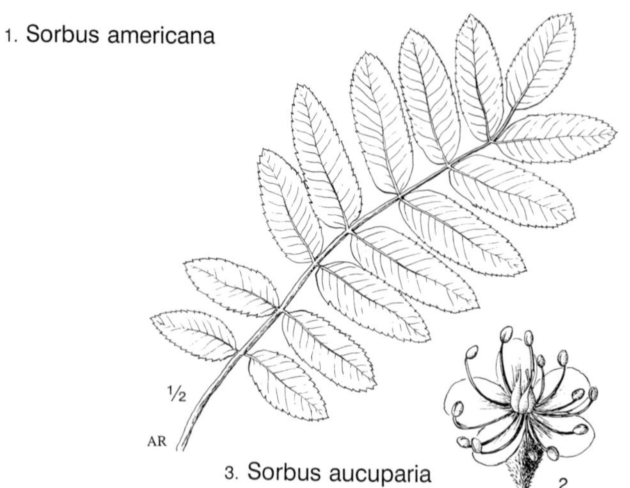

3. Sorbus aucuparia

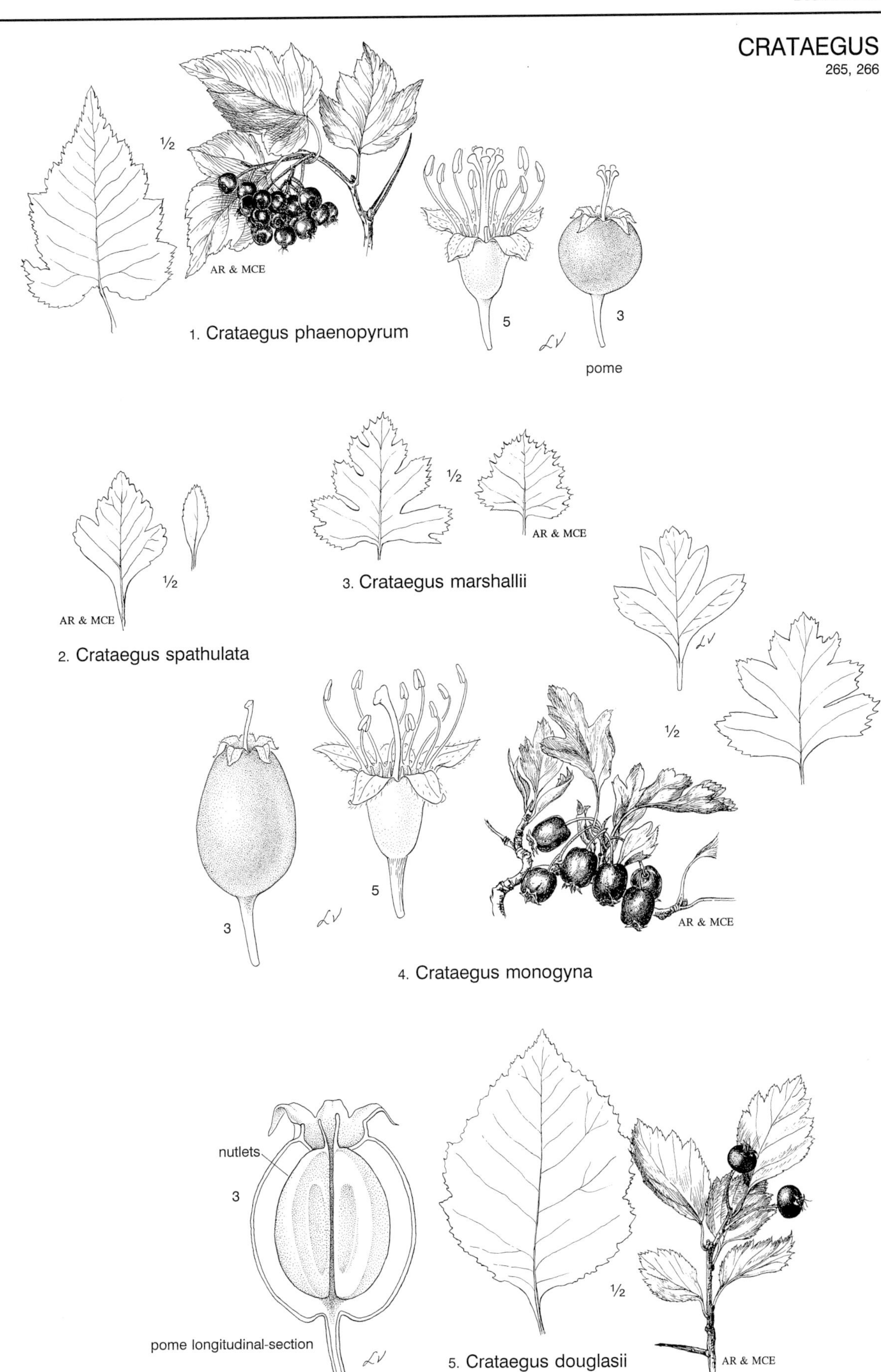

1. Crataegus phaenopyrum

5

3

pome

2. Crataegus spathulata

3. Crataegus marshallii

4. Crataegus monogyna

nutlets

3

5

pome longitudinal-section

5. Crataegus douglasii

AR & MCE

CRATAEGUS
266

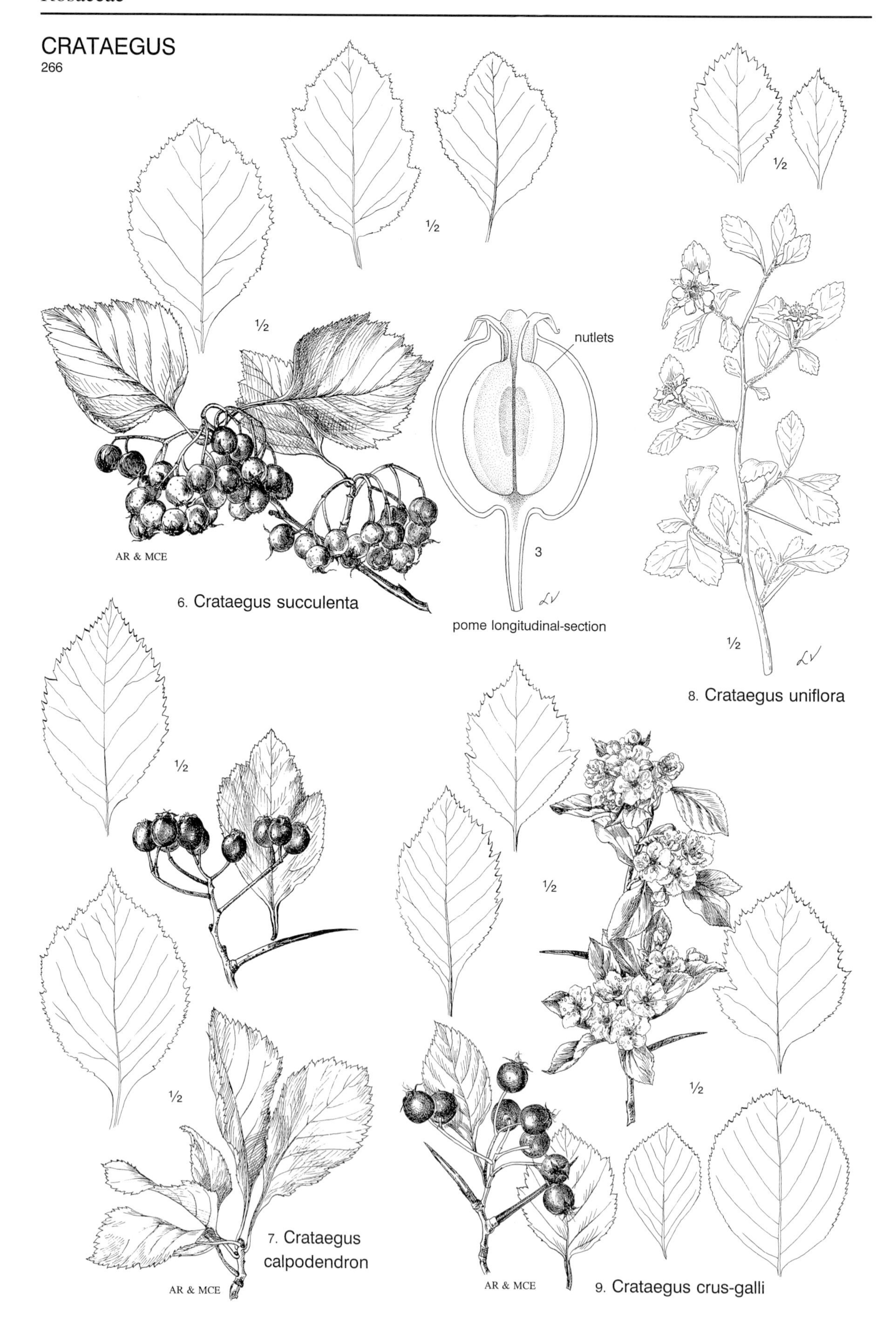

nutlets

3

pome longitudinal-section

AR & MCE

6. Crataegus succulenta

8. Crataegus uniflora

7. Crataegus calpodendron

AR & MCE

9. Crataegus crus-galli

AR & MCE

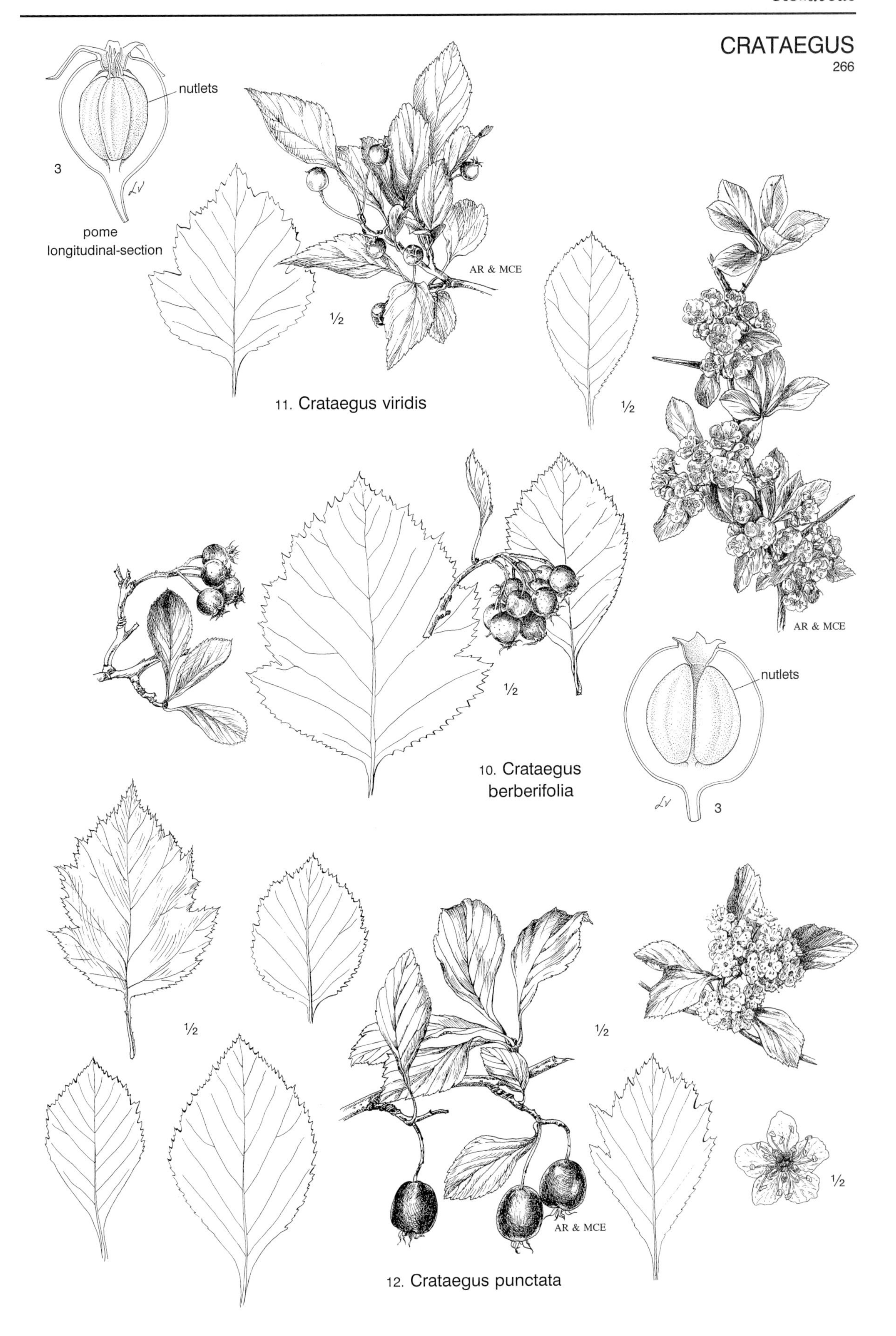

nutlets

3

pome
longitudinal-section

½

AR & MCE

11. Crataegus viridis

½

½

AR & MCE

nutlets

3

10. Crataegus
berberifolia

½

½

½

½

½

AR & MCE

12. Crataegus punctata

CRATAEGUS
267

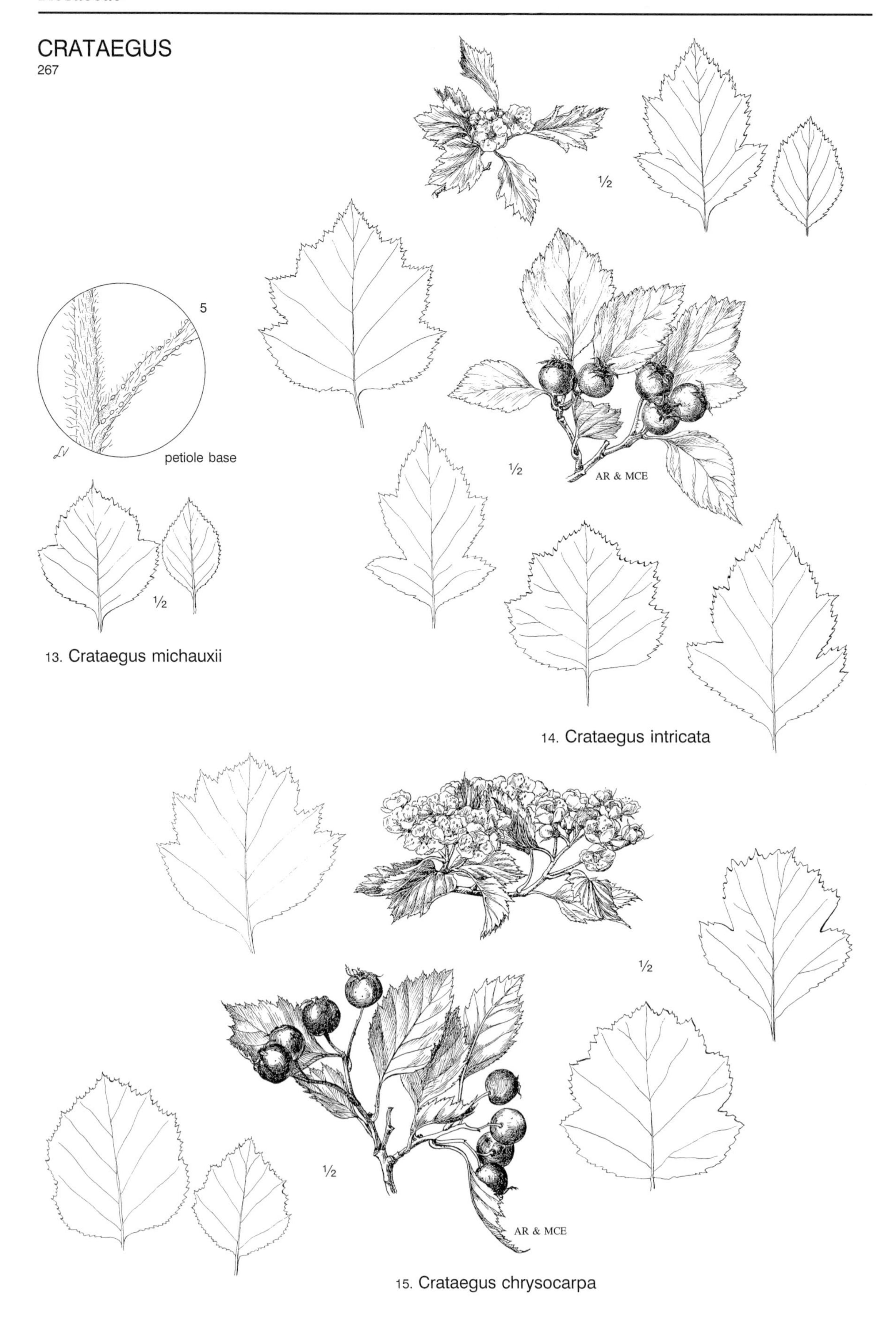

petiole base

13. Crataegus michauxii

½

AR & MCE

14. Crataegus intricata

½

AR & MCE

15. Crataegus chrysocarpa

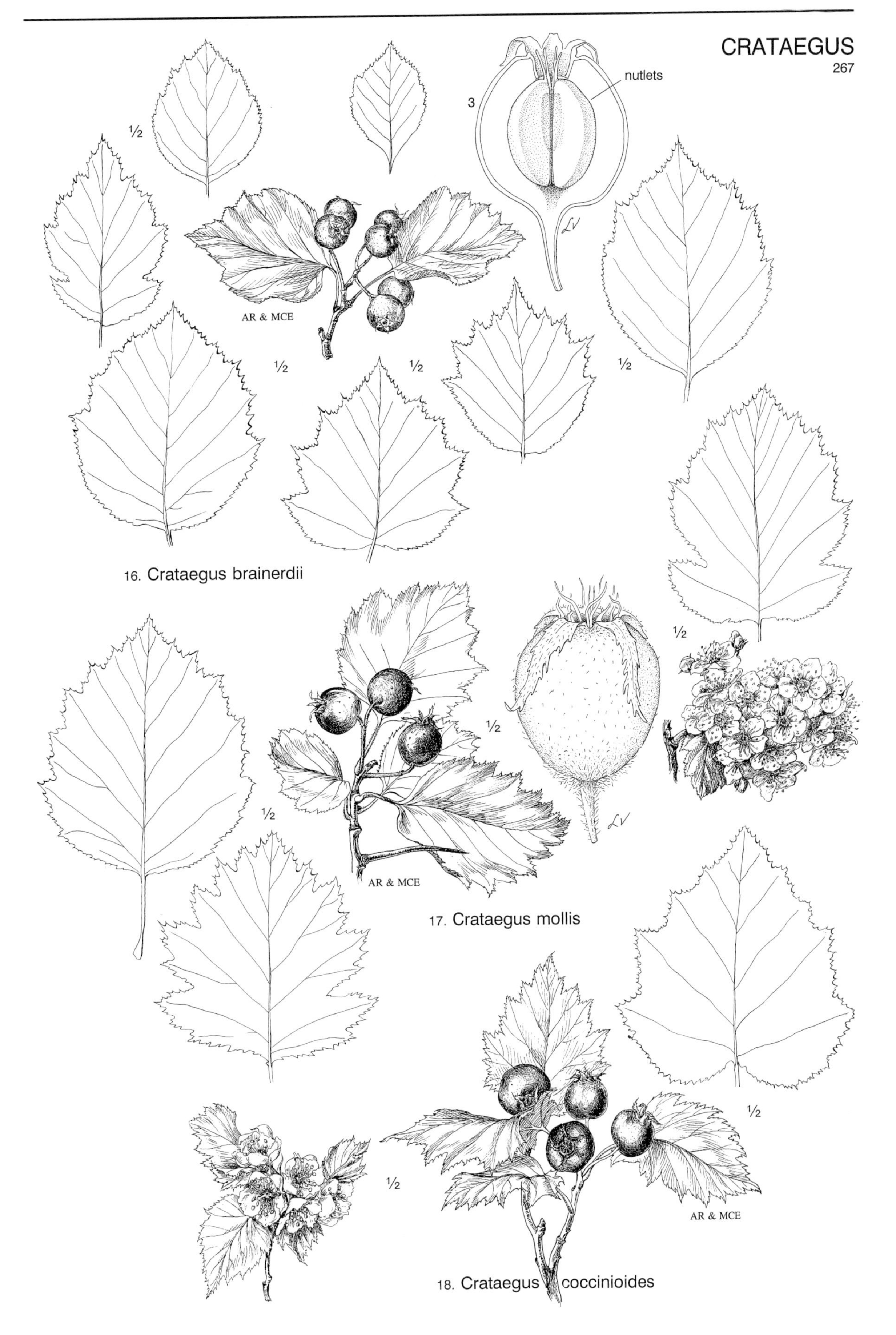

nutlets

3

AR & MCE

½   ½   ½   ½

16. Crataegus brainerdii

½

½

AR & MCE

17. Crataegus mollis

½

½

½

18. Crataegus coccinioides

AR & MCE

CRATAEGUS
267

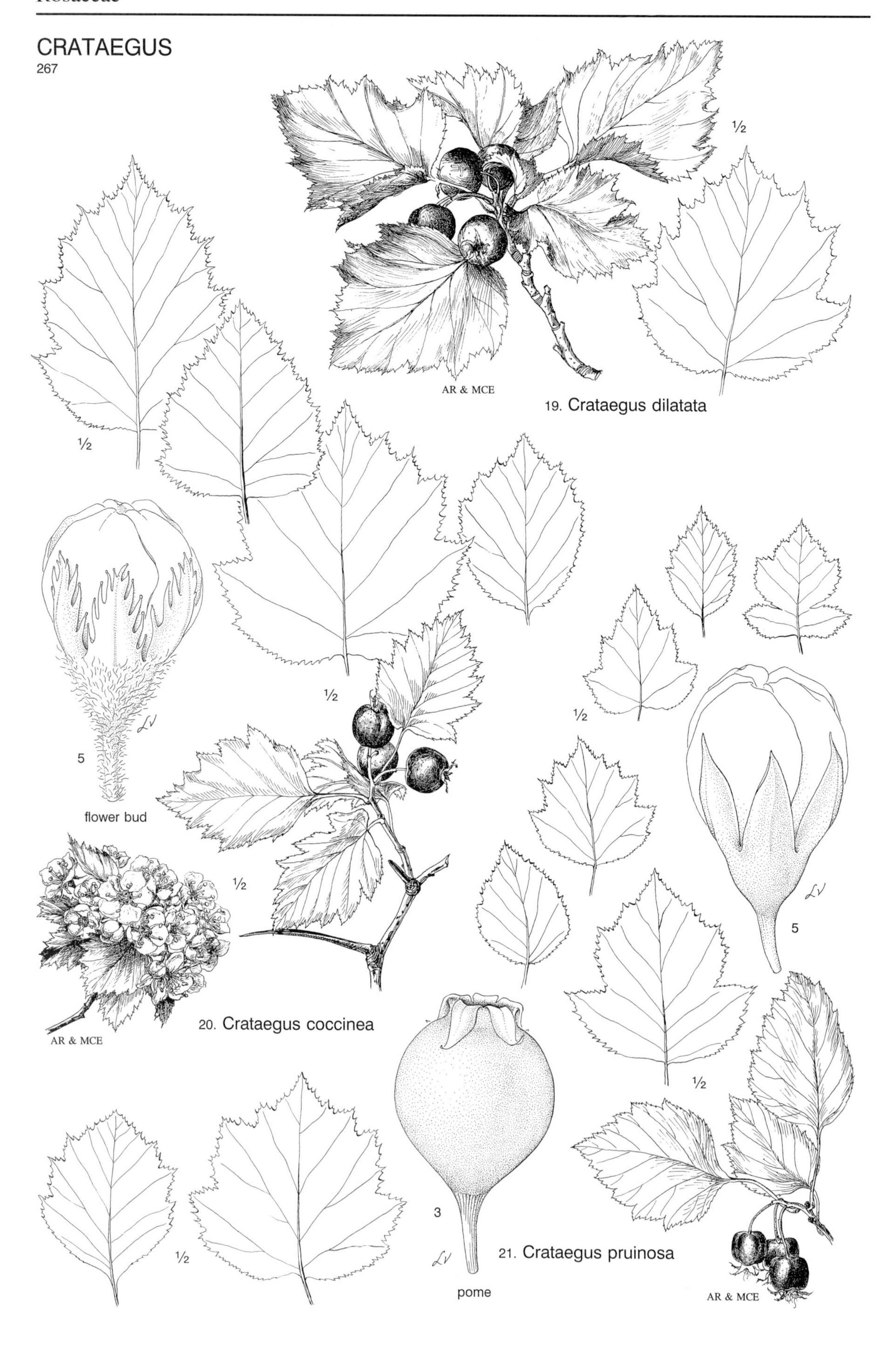

AR & MCE
19. Crataegus dilatata

½

½

5
flower bud

½

20. Crataegus coccinea
AR & MCE

½

½

½

½

5

½

3

21. Crataegus pruinosa

pome

AR & MCE

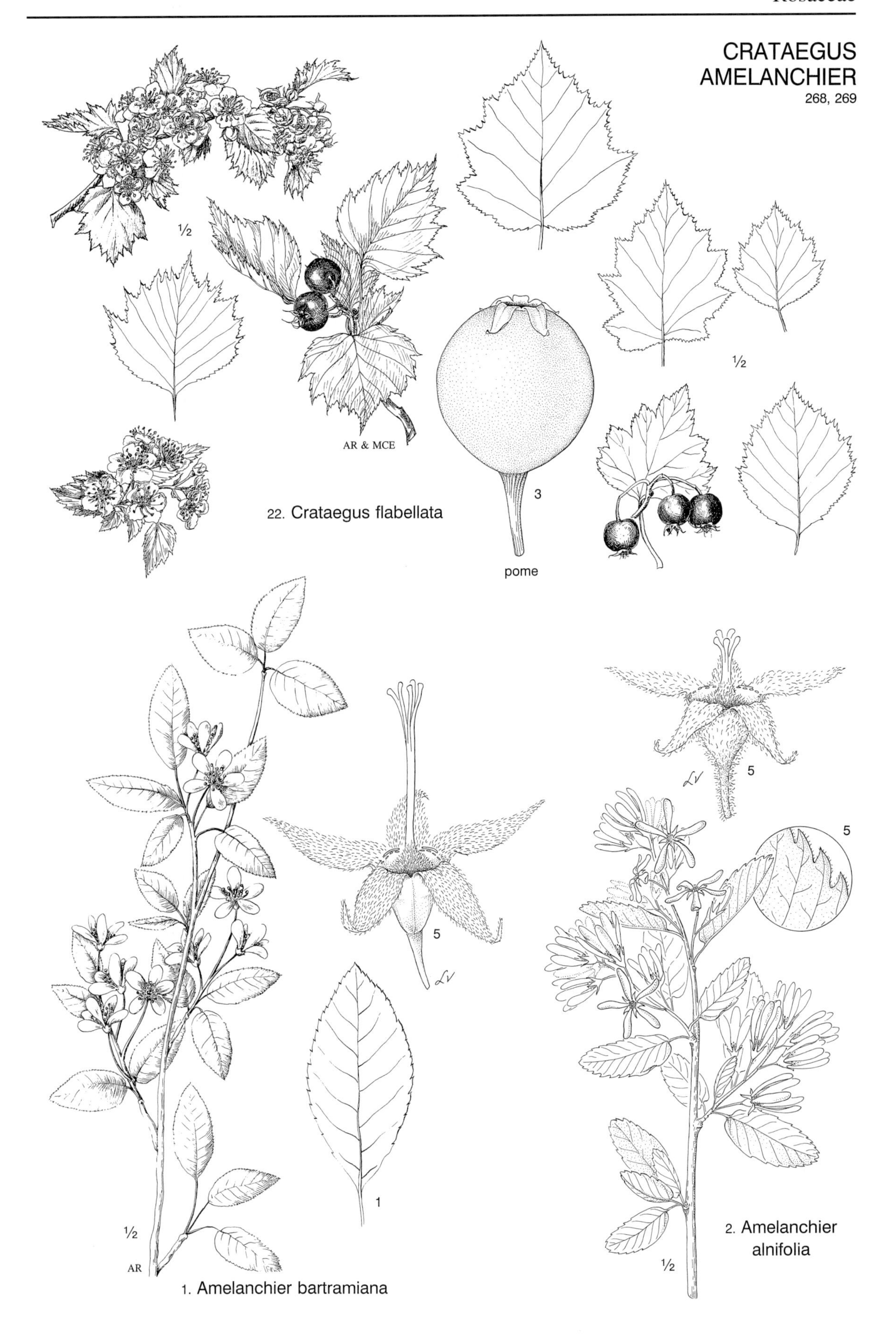

½

AR & MCE

22. Crataegus flabellata

3

pome

½

½

5

5

5

1

½

AR

1. Amelanchier bartramiana

5

2. Amelanchier
alnifolia

½

AMELANCHIER
269, 270

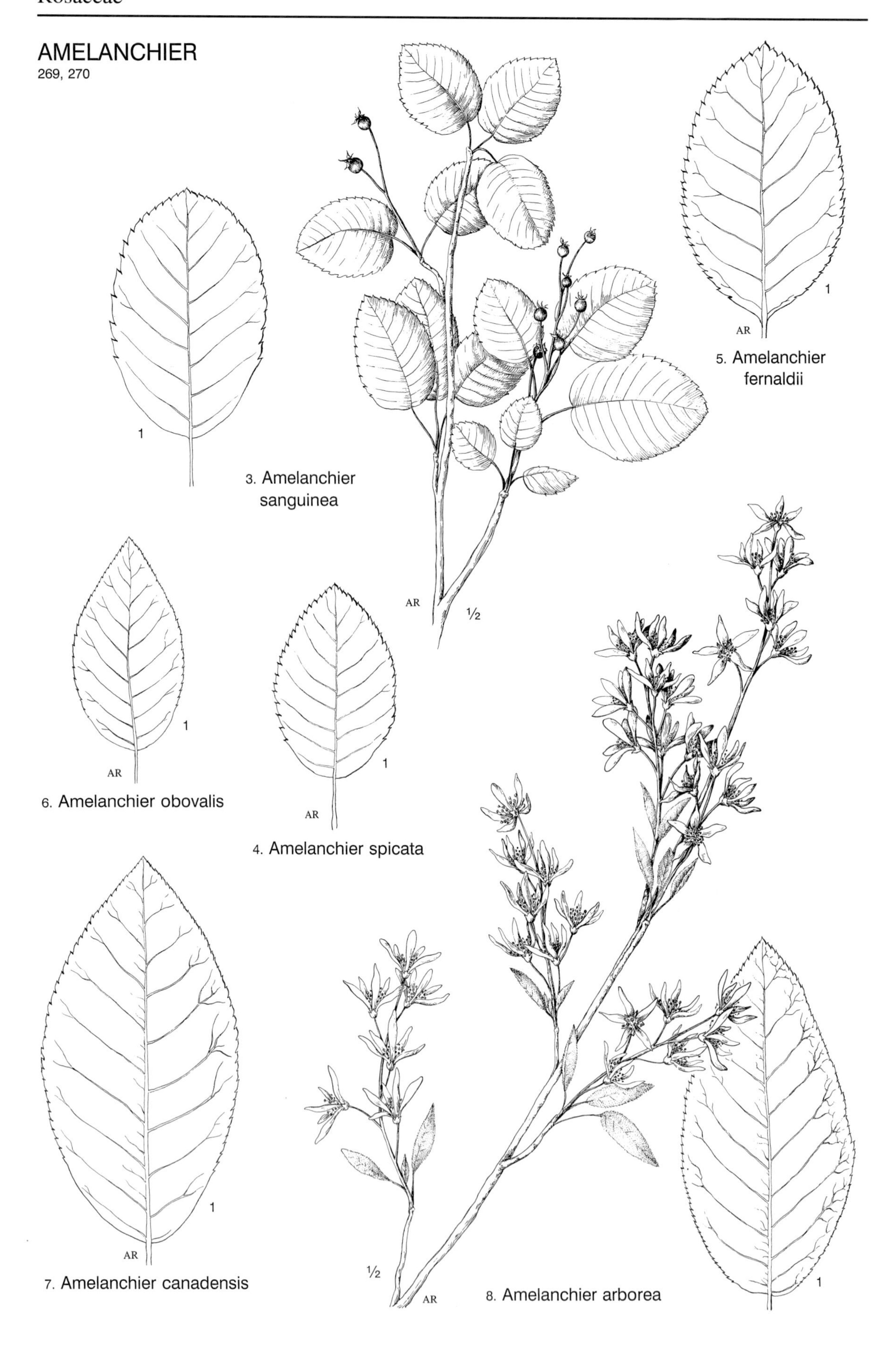

3. Amelanchier
sanguinea

5. Amelanchier
fernaldii

6. Amelanchier obovalis

4. Amelanchier spicata

7. Amelanchier canadensis

8. Amelanchier arborea

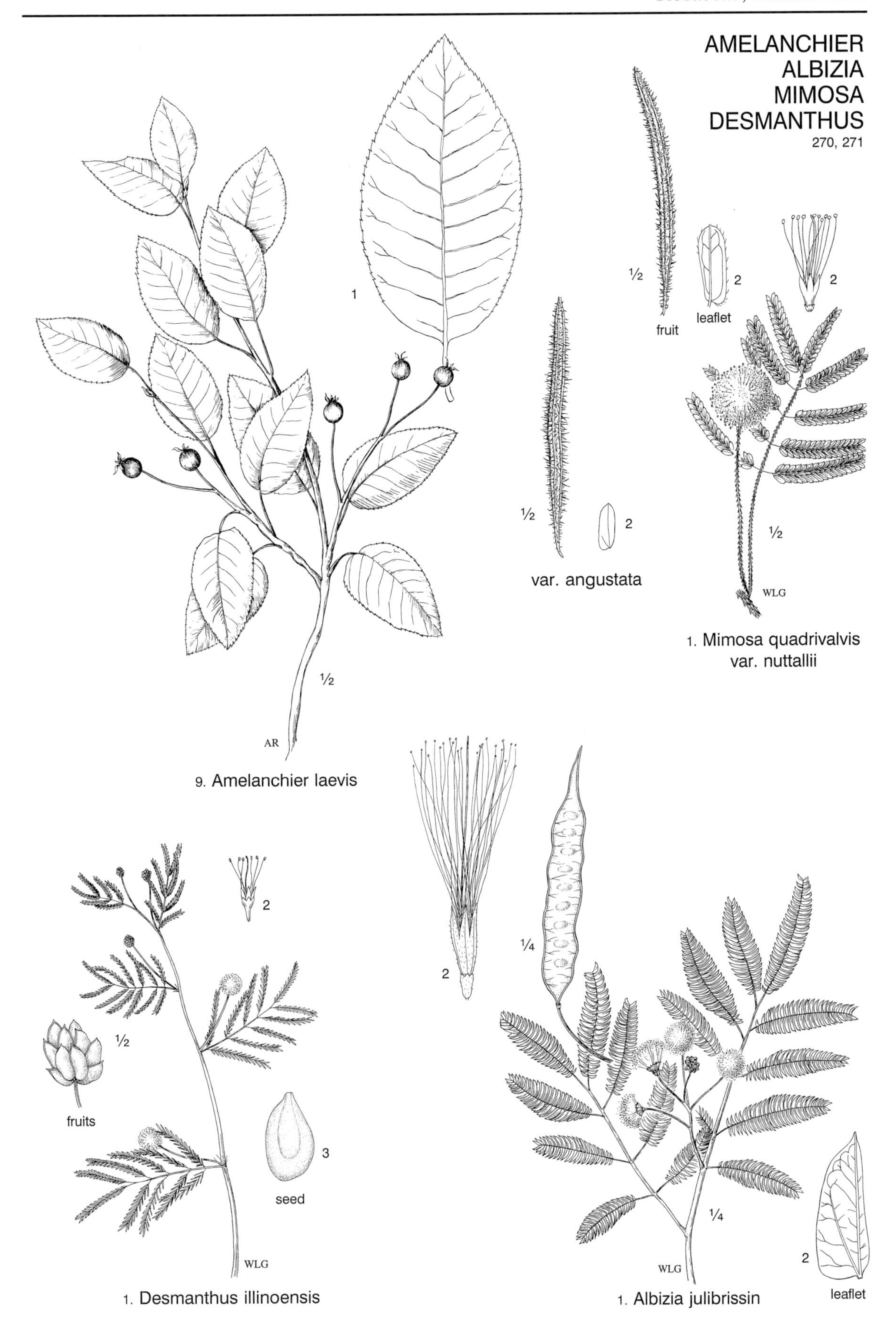

½
fruit

leaflet
2

2

var. angustata

½

2

½

WLG

1. Mimosa quadrivalvis
var. nuttallii

9. Amelanchier laevis

2

¼

2

½

fruits

3

seed

WLG

1. Desmanthus illinoensis

¼

¼

WLG

2

leaflet

1. Albizia julibrissin

CERCIS
GLEDITSIA
GYMNOCLADUS
CHAMAECRISTA
271, 272

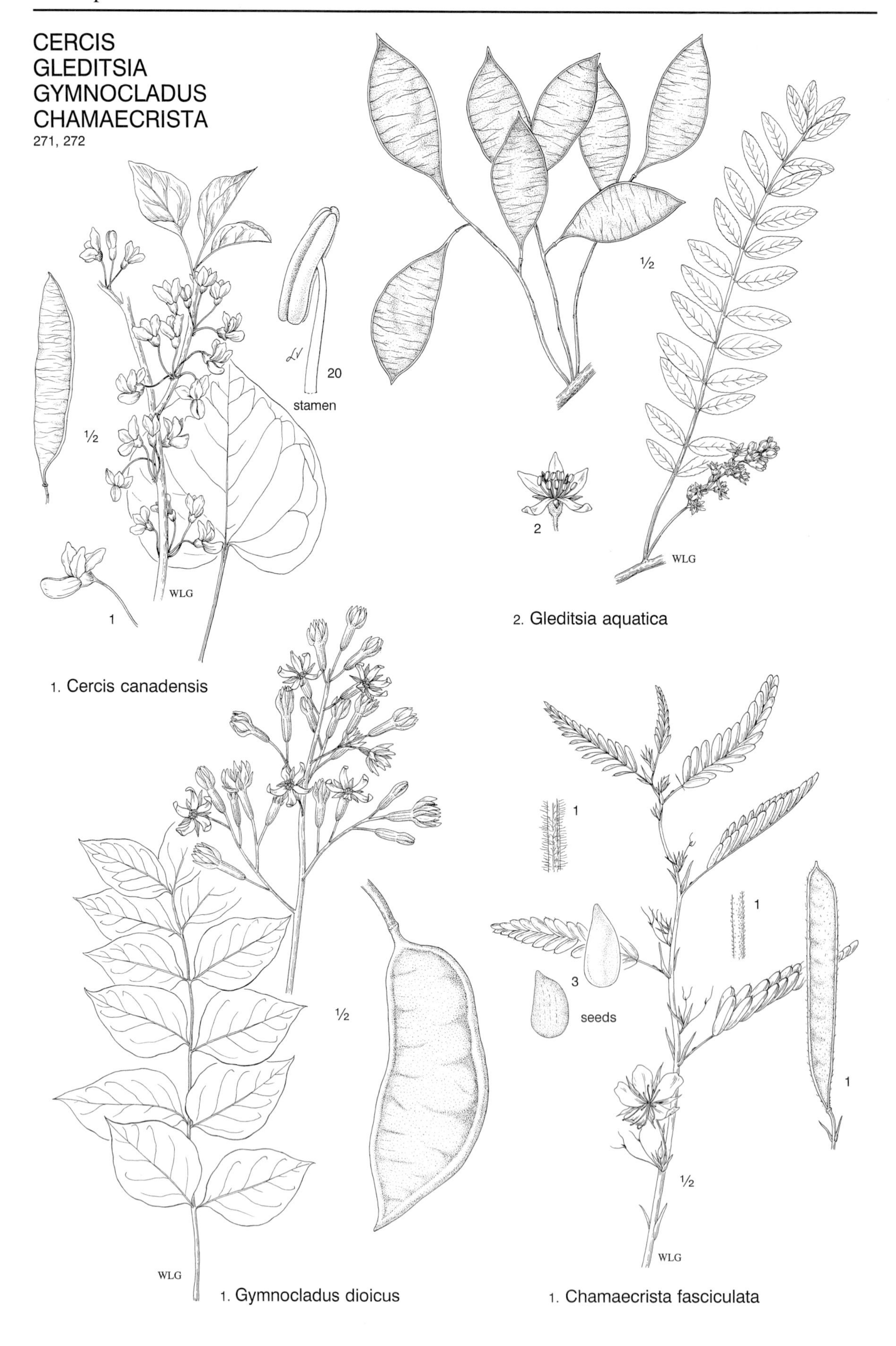

20
stamen

1. Cercis canadensis

2. Gleditsia aquatica

½

seeds

1. Gymnocladus dioicus

1. Chamaecrista fasciculata

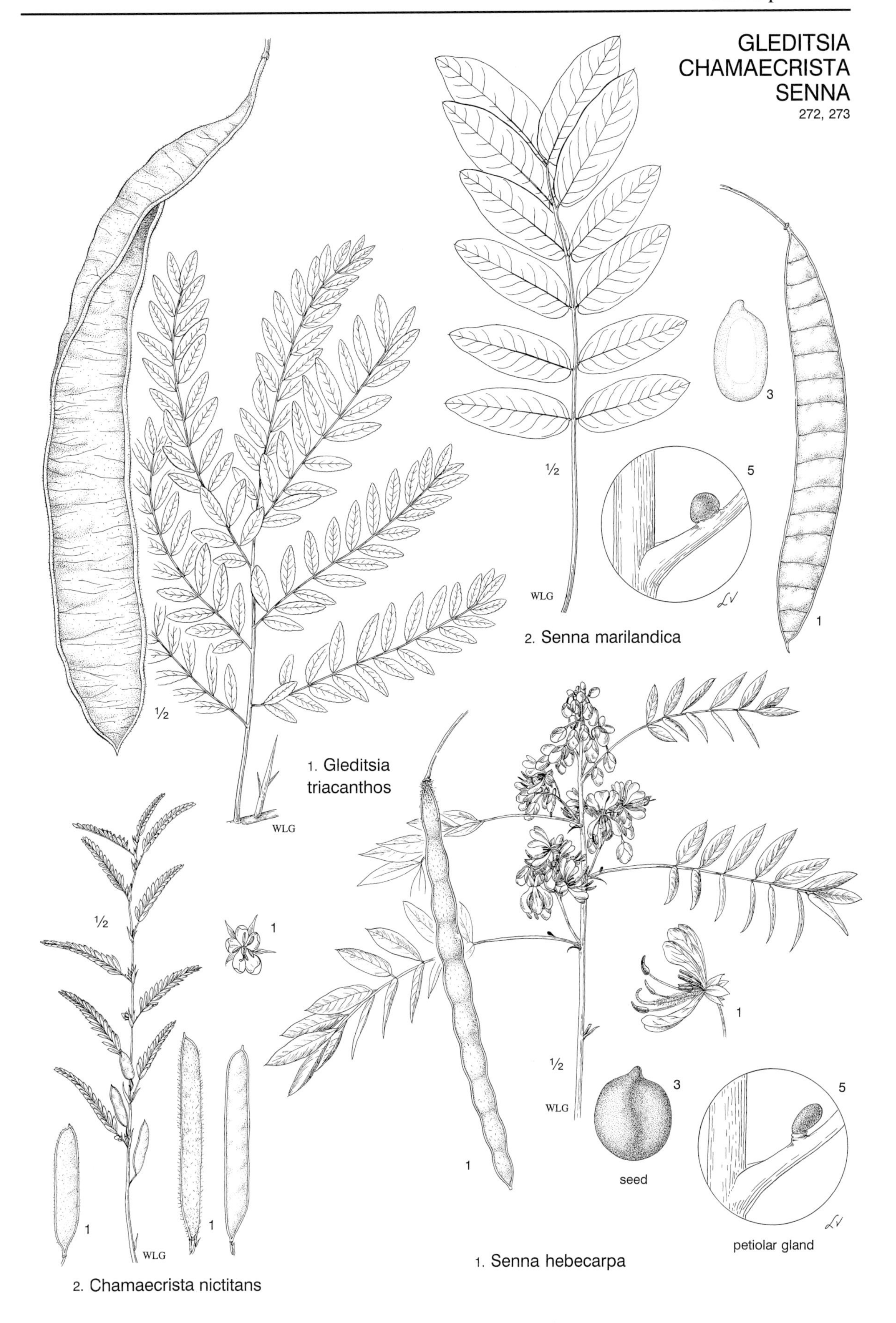

2. Senna marilandica

WLG

3

5

½

1

1. Gleditsia
triacanthos

WLG

½

½

WLG

1

1. Senna hebecarpa

½

WLG

3

seed

5

petiolar gland

1

1

2. Chamaecrista nictitans

WLG

1

1

SENNA
CLADRASTIS
BAPTISIA
273–276

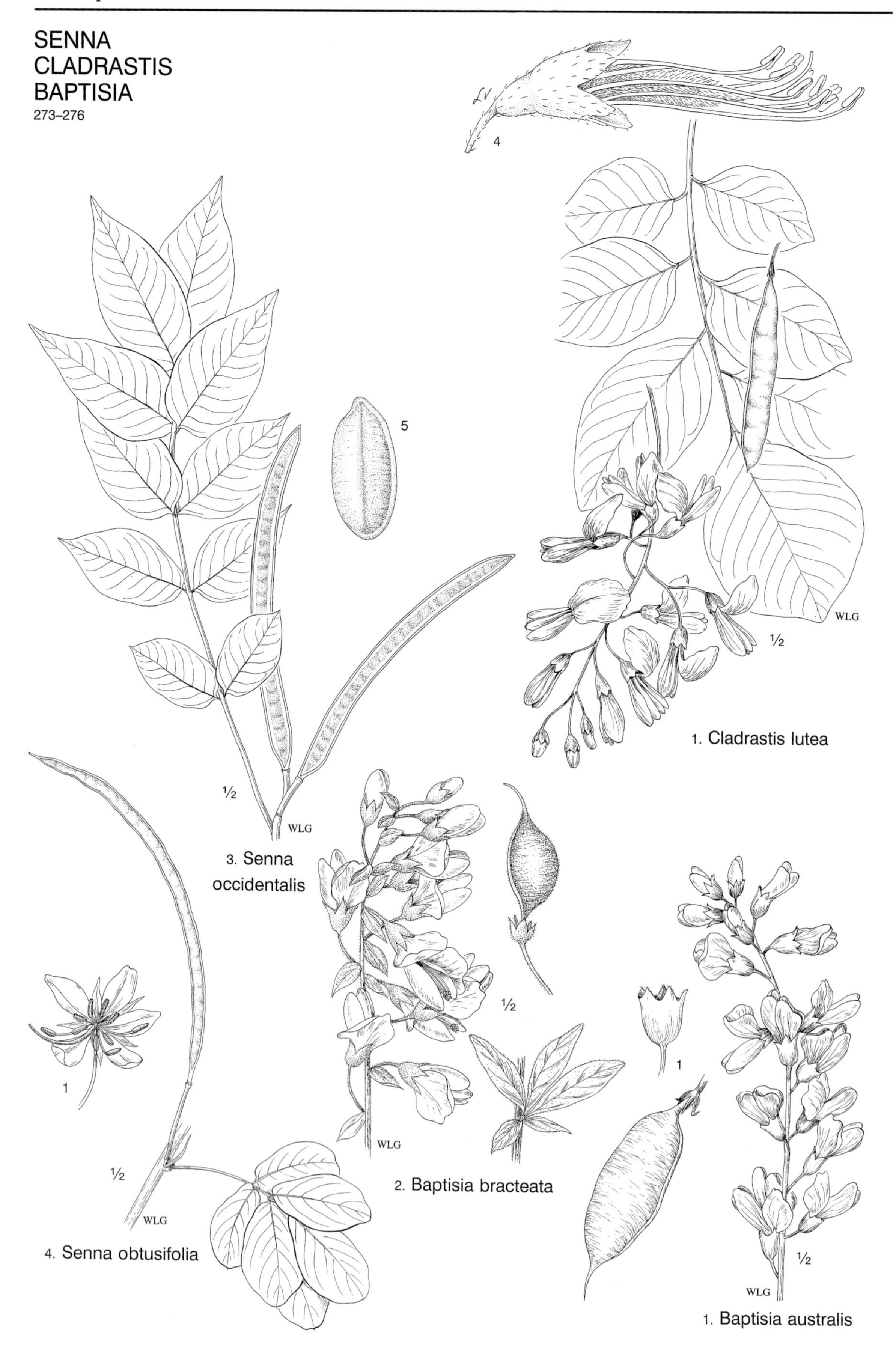

4

5

1. Cladrastis lutea

3. Senna occidentalis

2. Baptisia bracteata

1

4. Senna obtusifolia

1. Baptisia australis

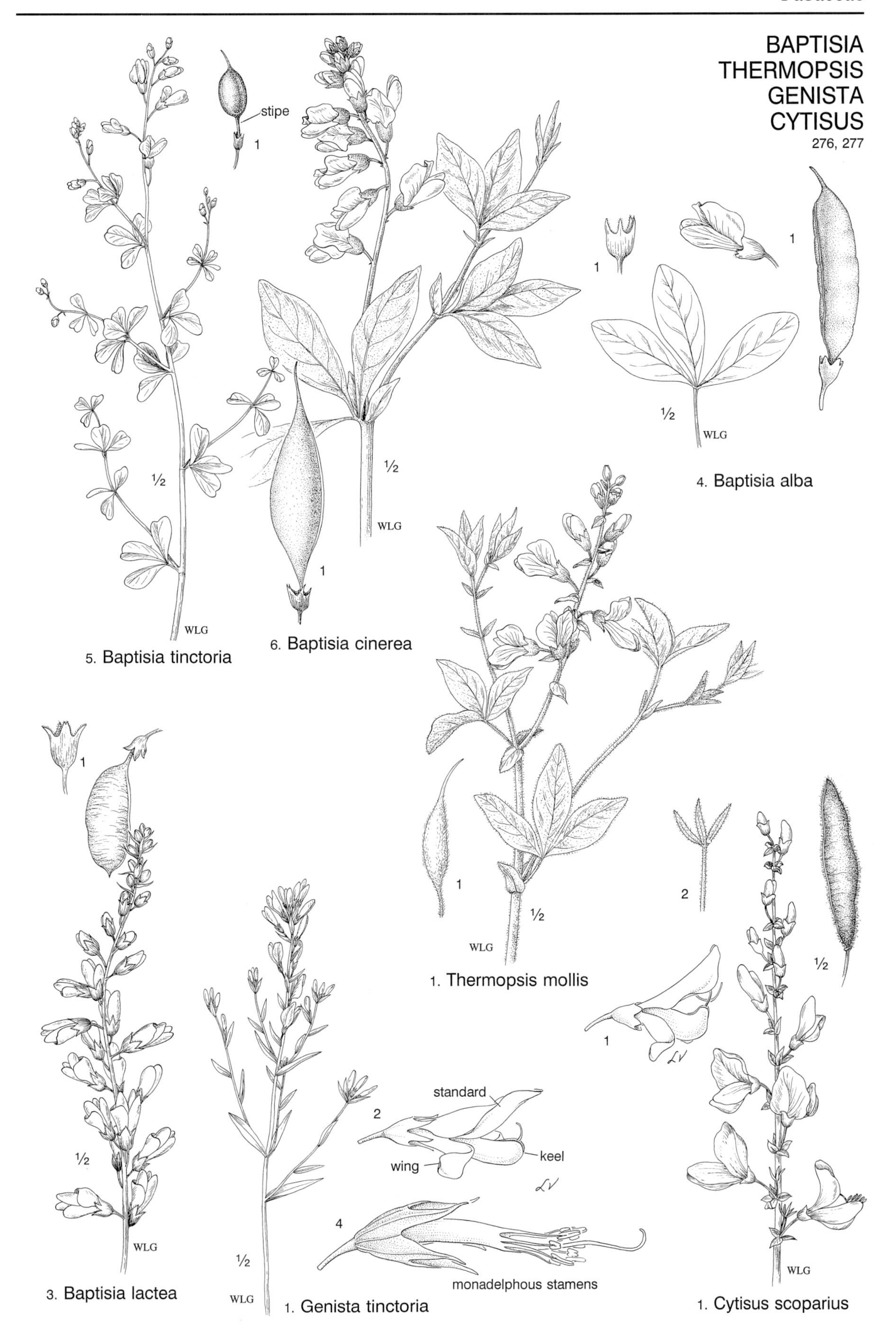

stipe

1

1

½

WLG

1

5. Baptisia tinctoria

1

6. Baptisia cinerea

½

WLG

4. Baptisia alba

½

WLG

1

½

WLG

1. Thermopsis mollis

2

½

1

½

WLG

3. Baptisia lactea

½

WLG

standard

2

wing

keel

4

monadelphous stamens

1. Genista tinctoria

1

1. Cytisus scoparius

½

WLG

ULEX
LUPINUS
CROTALARIA

278

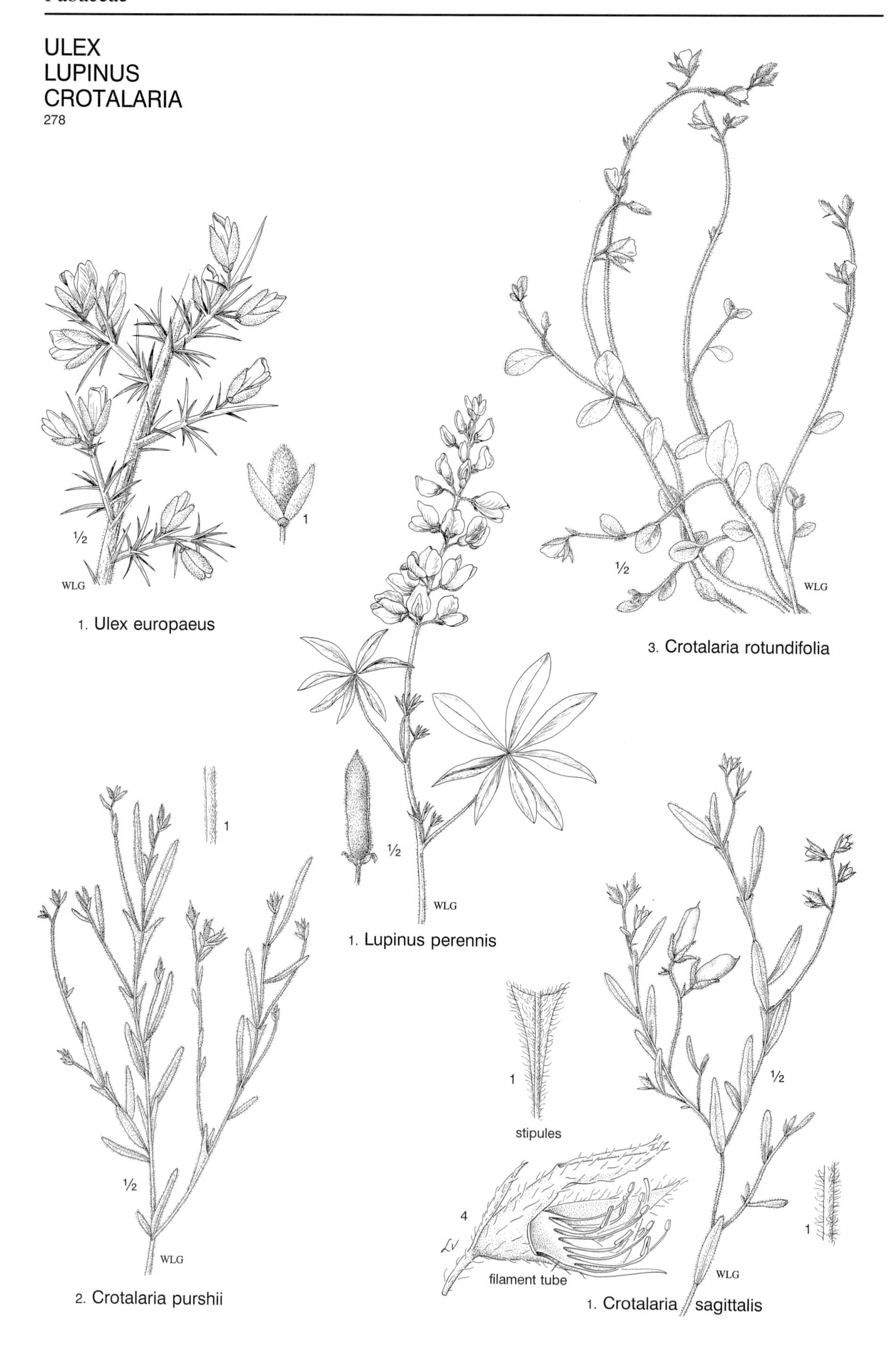

1. Ulex europaeus

3. Crotalaria rotundifolia

1. Lupinus perennis

2. Crotalaria purshii

stipules

filament tube

1. Crotalaria sagittalis

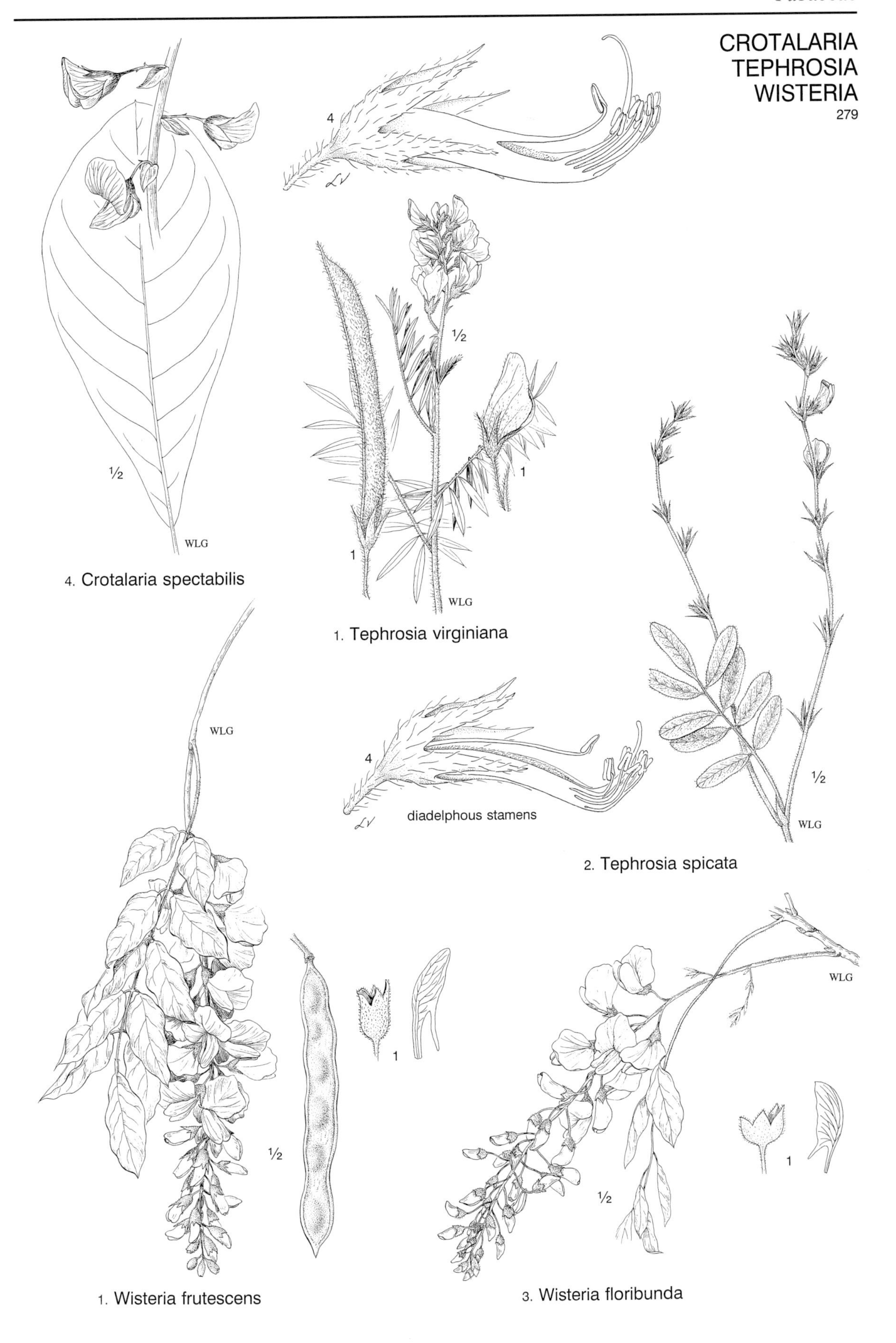

4. Crotalaria spectabilis

1. Tephrosia virginiana

diadelphous stamens

2. Tephrosia spicata

1. Wisteria frutescens

3. Wisteria floribunda

WISTERIA
ROBINIA
SESBANIA
279, 280

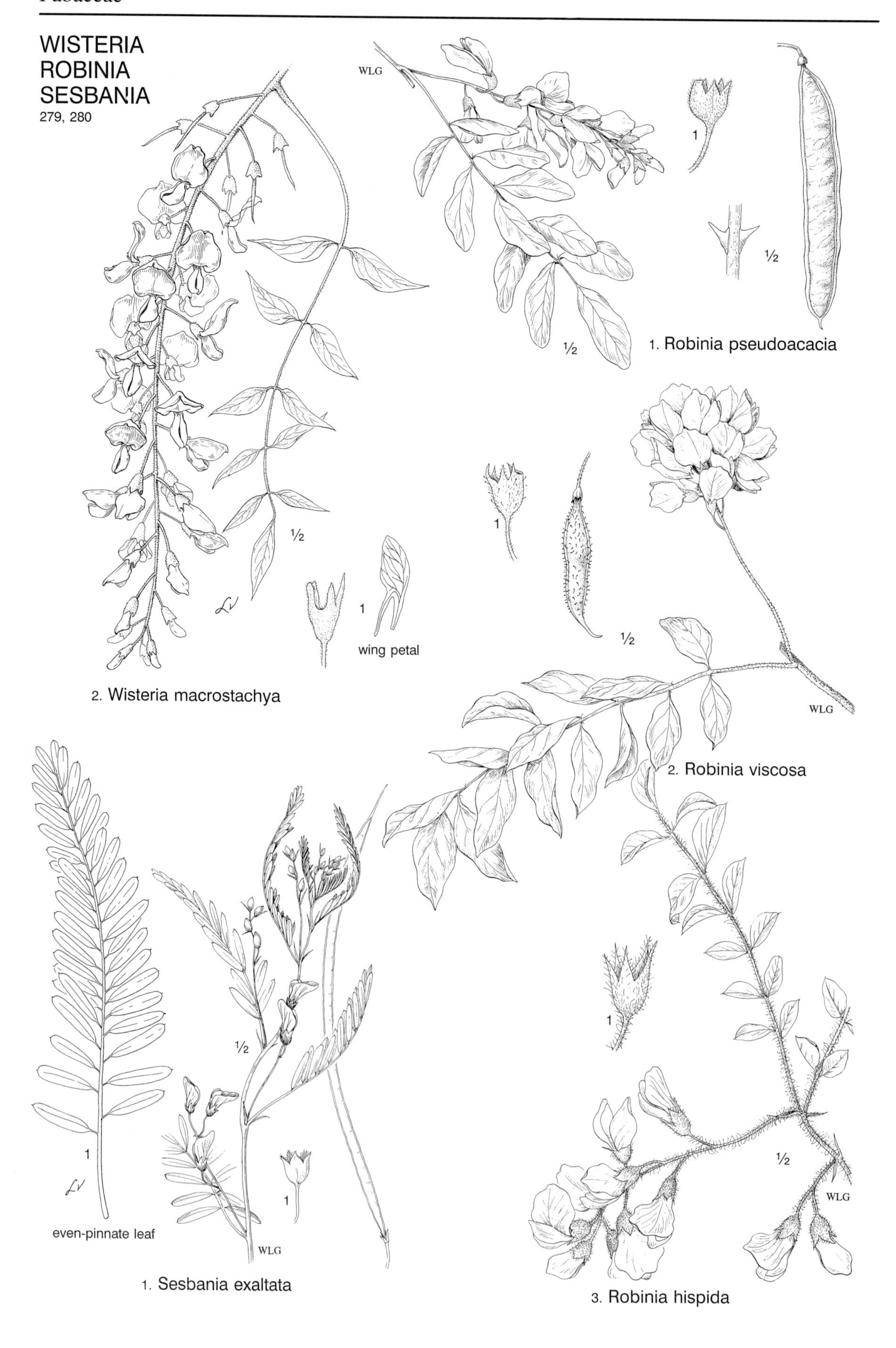

WLG

1

½

1. Robinia pseudoacacia

1

½

2. Robinia viscosa

WLG

1

wing petal

2. Wisteria macrostachya

½

even-pinnate leaf

1

WLG

1. Sesbania exaltata

1

½

WLG

3. Robinia hispida

Fabaceae

AESCHYNOMENE
STYLOSANTHES
ARACHIS
ZORNIA
281

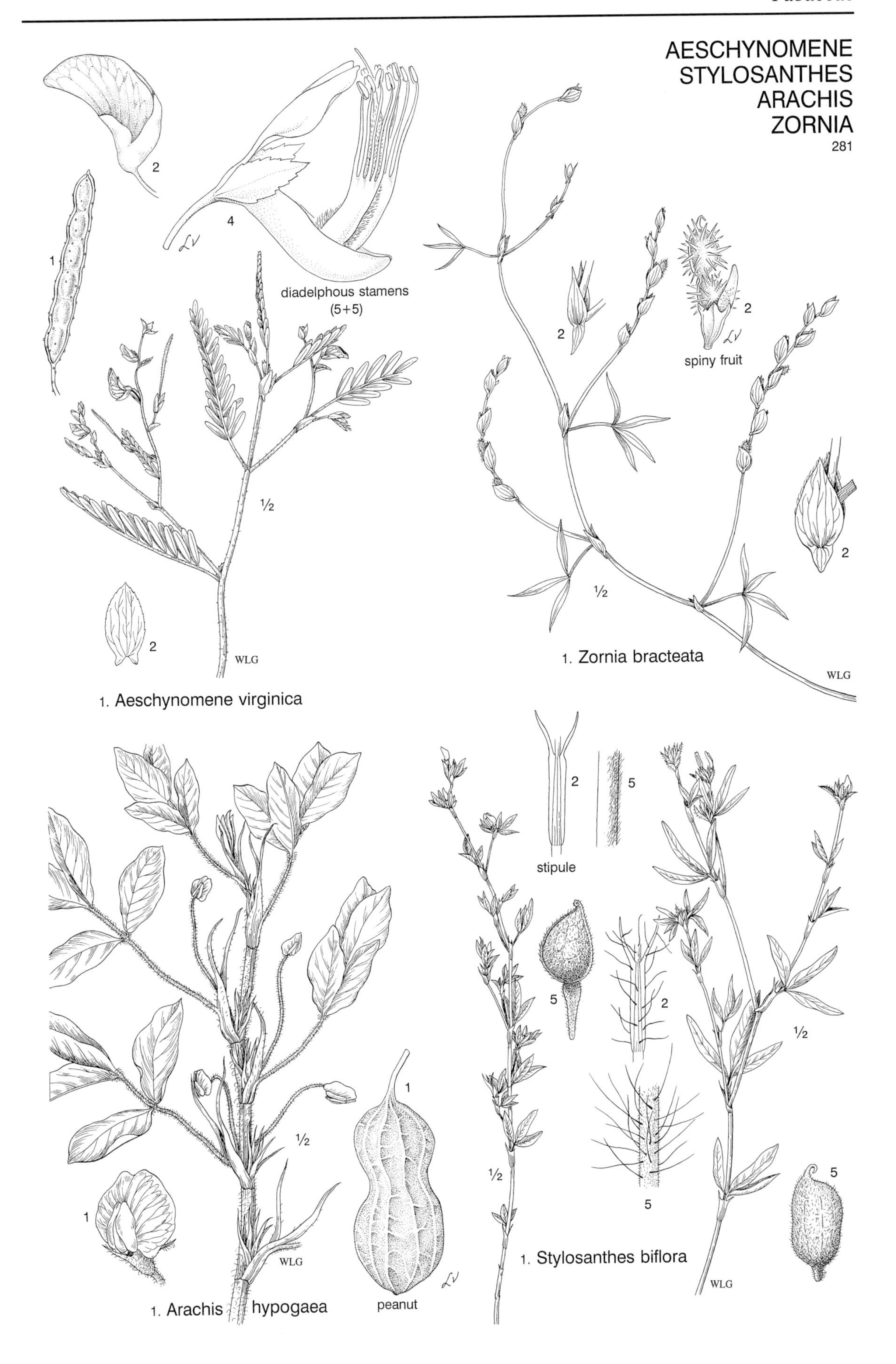

diadelphous stamens
(5+5)

1. Aeschynomene virginica

spiny fruit

1. Zornia bracteata

stipule

1. Arachis hypogaea

peanut

1. Stylosanthes biflora

COLUTEA
ASTRAGALUS
281–283

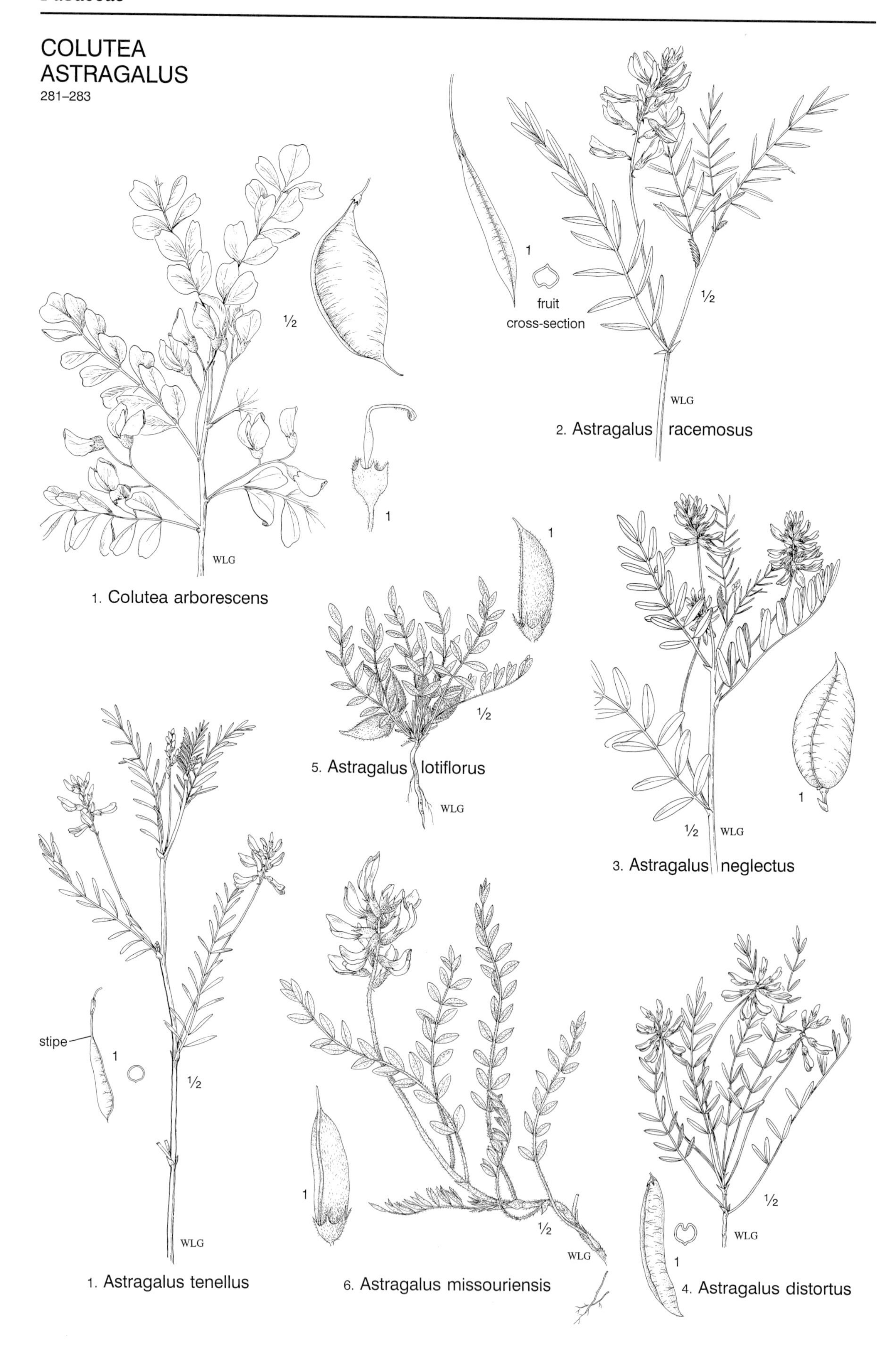

1. Colutea arborescens

2. Astragalus racemosus

fruit
cross-section

5. Astragalus lotiflorus

3. Astragalus neglectus

stipe

1. Astragalus tenellus

6. Astragalus missouriensis

4. Astragalus distortus

264

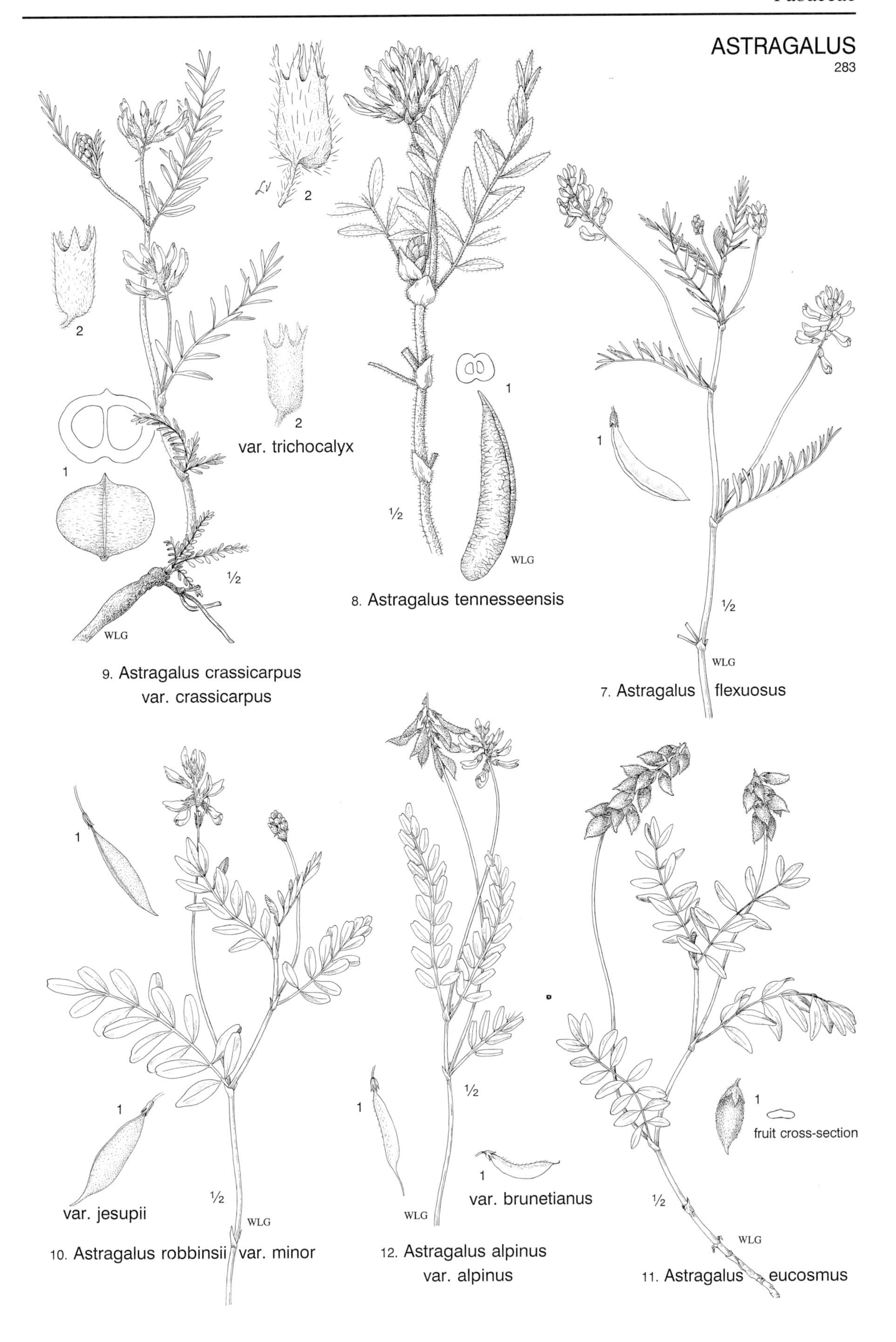

var. trichocalyx

8. Astragalus tennesseensis

9. Astragalus crassicarpus
var. crassicarpus

7. Astragalus flexuosus

var. jesupii

10. Astragalus robbinsii var. minor

var. brunetianus

12. Astragalus alpinus
var. alpinus

fruit cross-section

11. Astragalus eucosmus

ASTRAGALUS
284

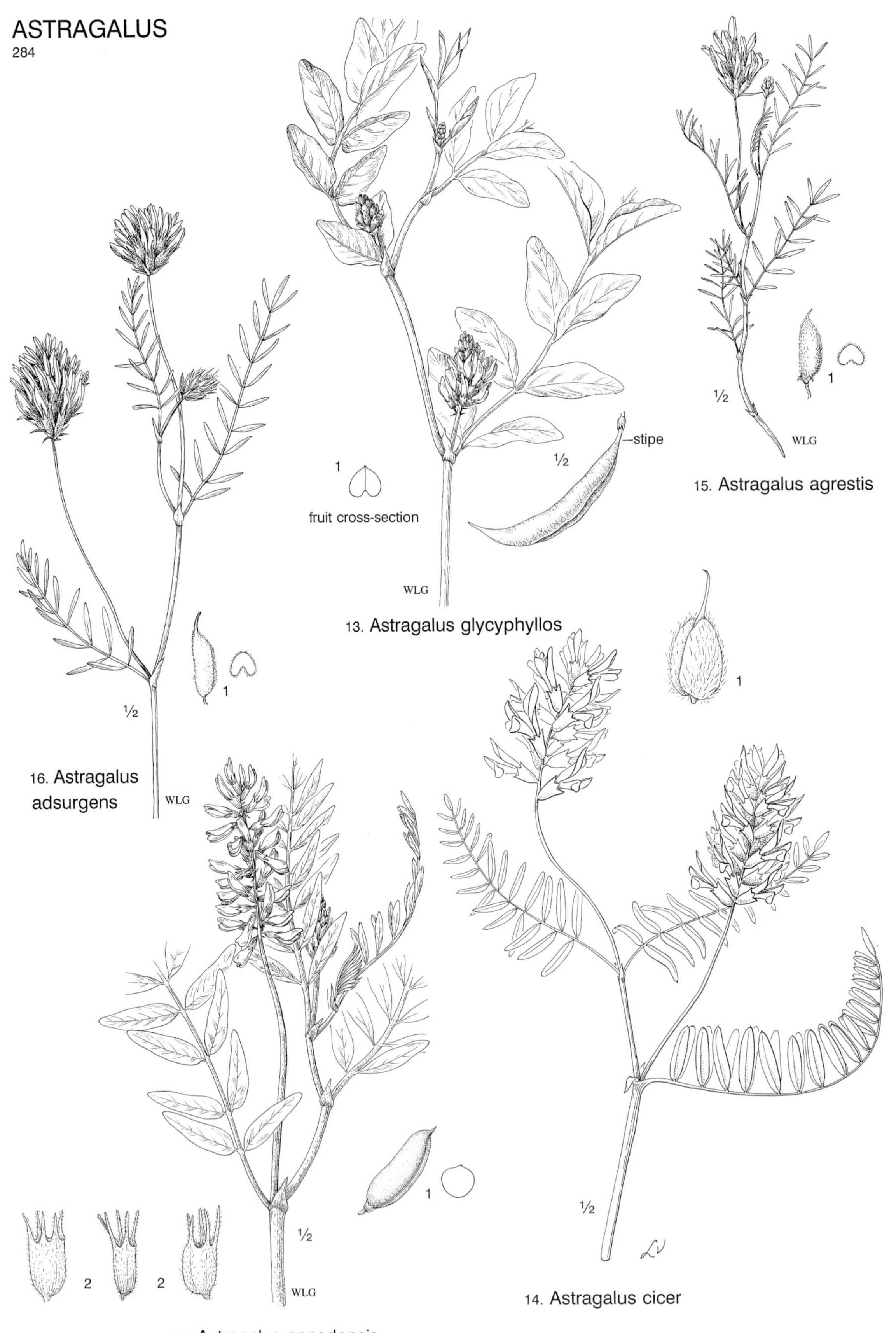

1
fruit cross-section

½ —stipe

½

15. Astragalus agrestis

WLG

WLG

13. Astragalus glycyphyllos

1

½

1

16. Astragalus adsurgens

WLG

1

2   2

½

WLG

17. Astragalus canadensis

½

14. Astragalus cicer

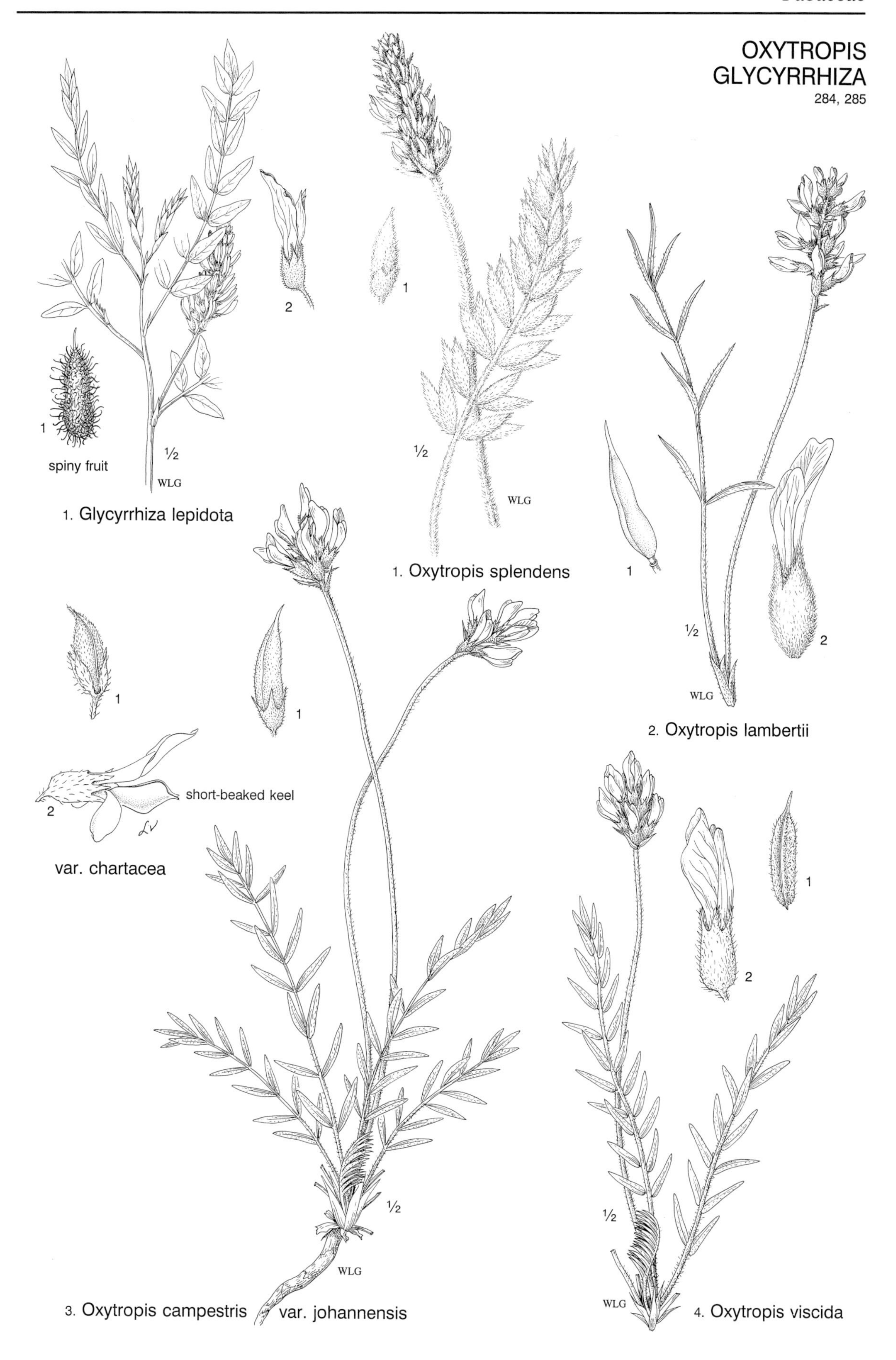

1 spiny fruit

1. Glycyrrhiza lepidota

1. Oxytropis splendens

2. Oxytropis lambertii

var. chartacea

short-beaked keel

3. Oxytropis campestris var. johannensis

4. Oxytropis viscida

LOTUS
ANTHYLLIS
CORONILLA
ORNITHOPUS
285, 286

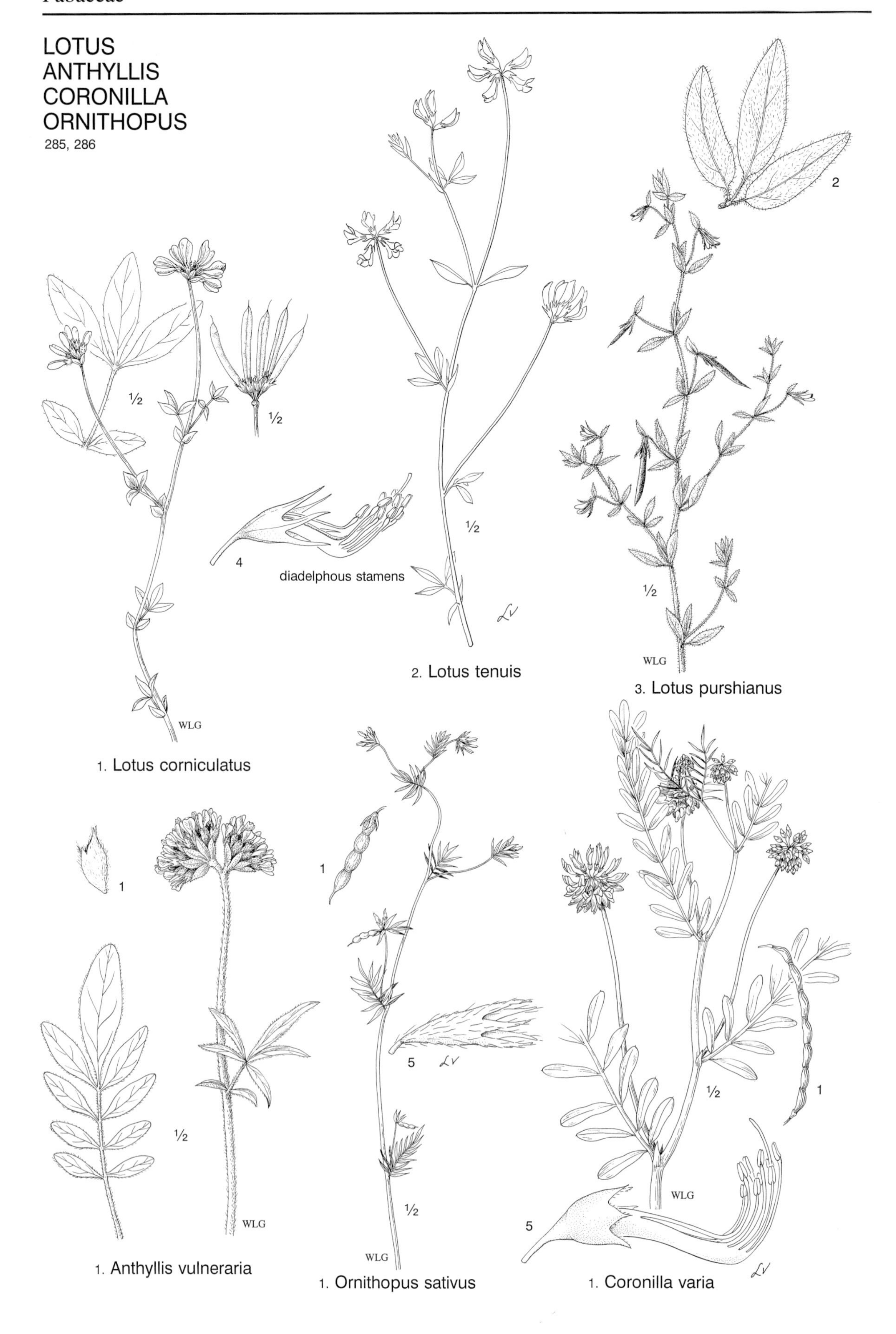

½   ½

4

diadelphous stamens

2

2. Lotus tenuis

½

WLG

1. Lotus corniculatus

3. Lotus purshianus
½
WLG

1

1

5

½

WLG

1. Anthyllis vulneraria

½

5

WLG

1

½

WLG

1. Coronilla varia

½

WLG

1. Ornithopus sativus

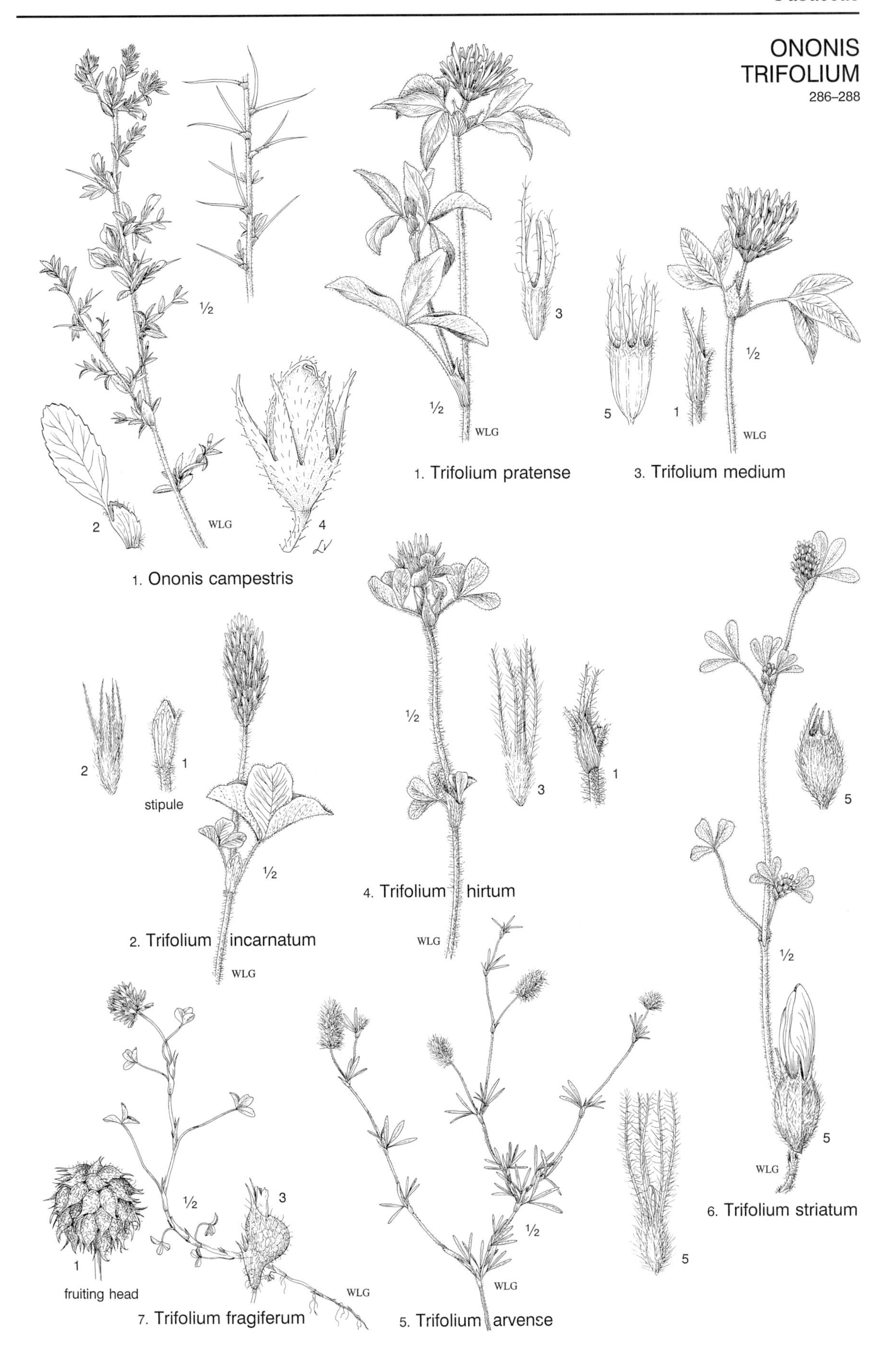

1. Ononis campestris

1. Trifolium pratense

3. Trifolium medium

2. Trifolium incarnatum

4. Trifolium hirtum

6. Trifolium striatum

7. Trifolium fragiferum

5. Trifolium arvense

stipule

fruiting head

TRIFOLIUM
288

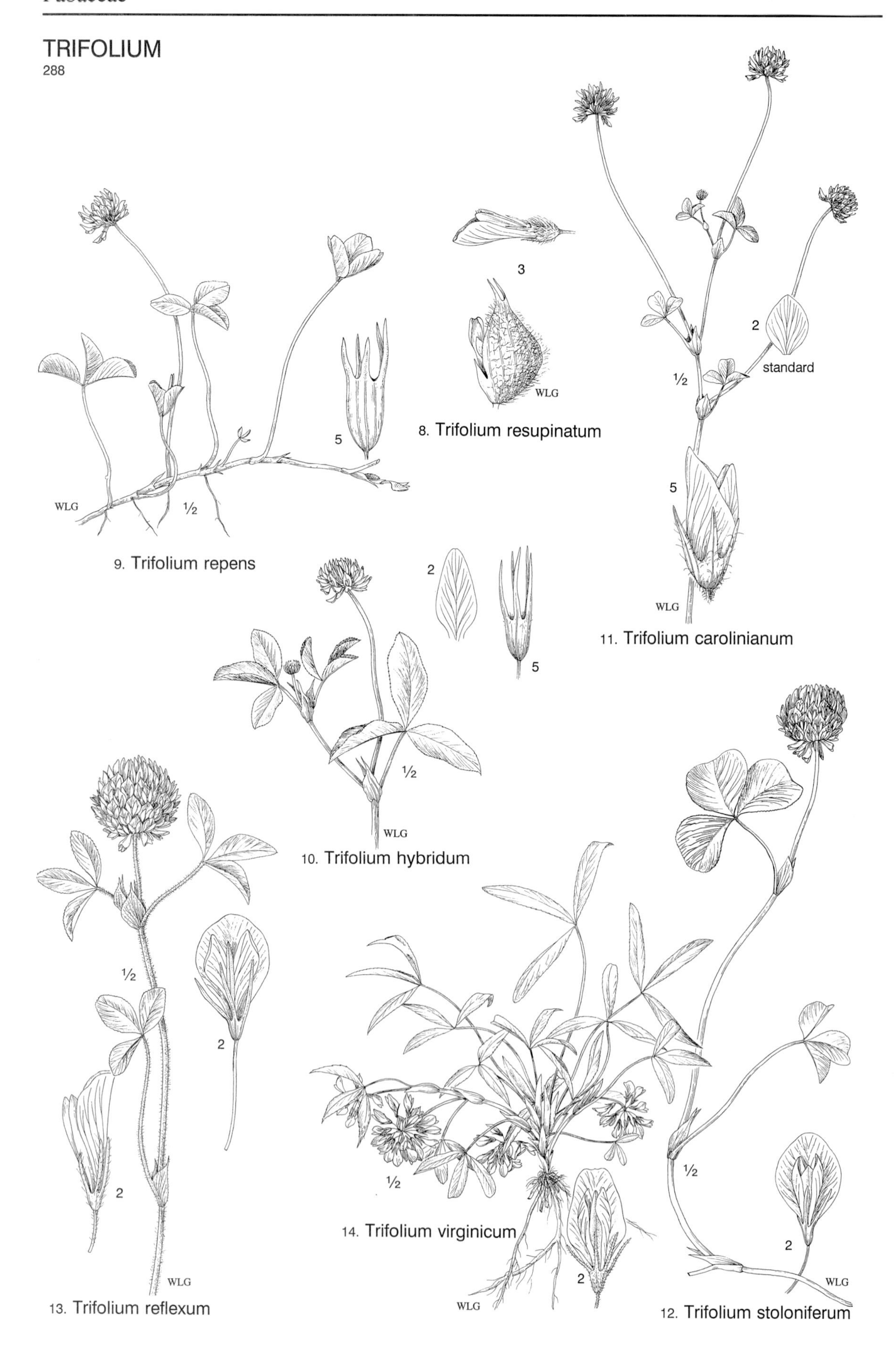

3

8. Trifolium resupinatum

5

9. Trifolium repens

2

standard

5

11. Trifolium carolinianum

2

5

10. Trifolium hybridum

2

2

13. Trifolium reflexum

14. Trifolium virginicum

2

2

12. Trifolium stoloniferum

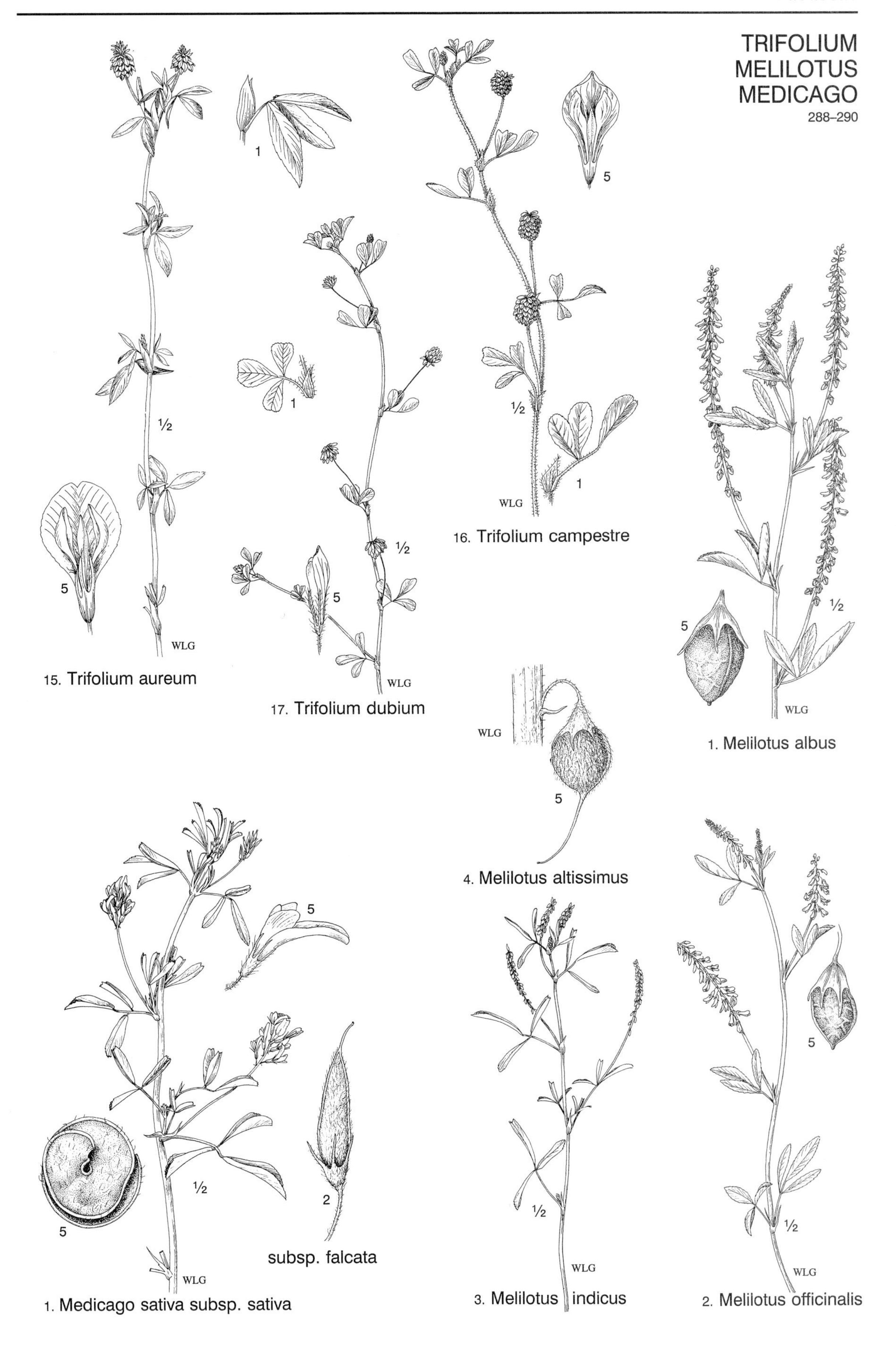

15. Trifolium aureum

17. Trifolium dubium

16. Trifolium campestre

1. Melilotus albus

4. Melilotus altissimus

subsp. falcata

1. Medicago sativa subsp. sativa

3. Melilotus indicus

2. Melilotus officinalis

MEDICAGO
VICIA
290, 291

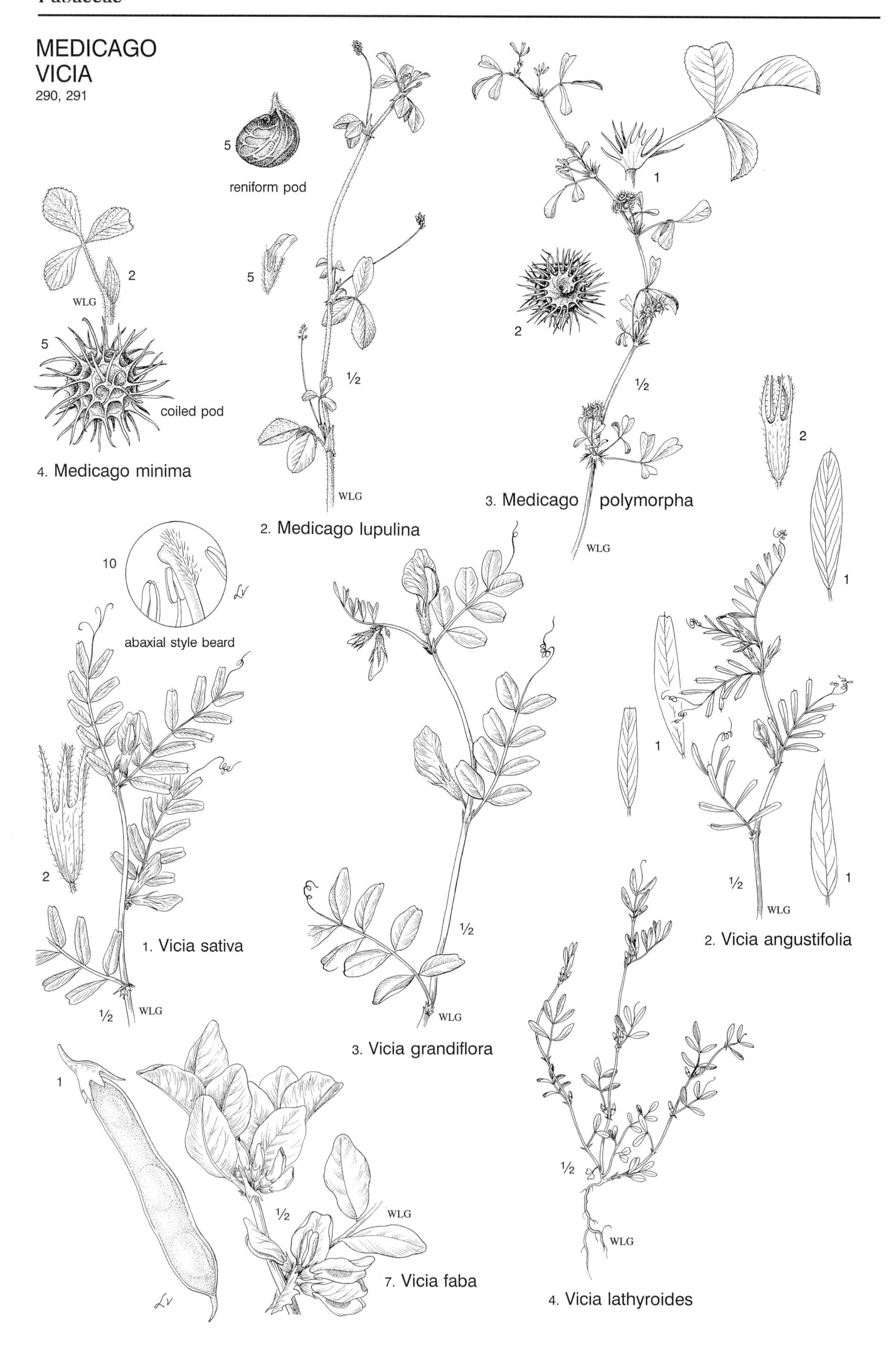

5 reniform pod

5

4. Medicago minima

5 coiled pod

2

WLG

½

WLG

2. Medicago lupulina

1

2

½

WLG

3. Medicago polymorpha

2

1

10

abaxial style beard

2

1. Vicia sativa

½   WLG

2

½   WLG

3. Vicia grandiflora

1

1

½

WLG

2. Vicia angustifolia

1

1

½

WLG

4. Vicia lathyroides

1

½

WLG

7. Vicia faba

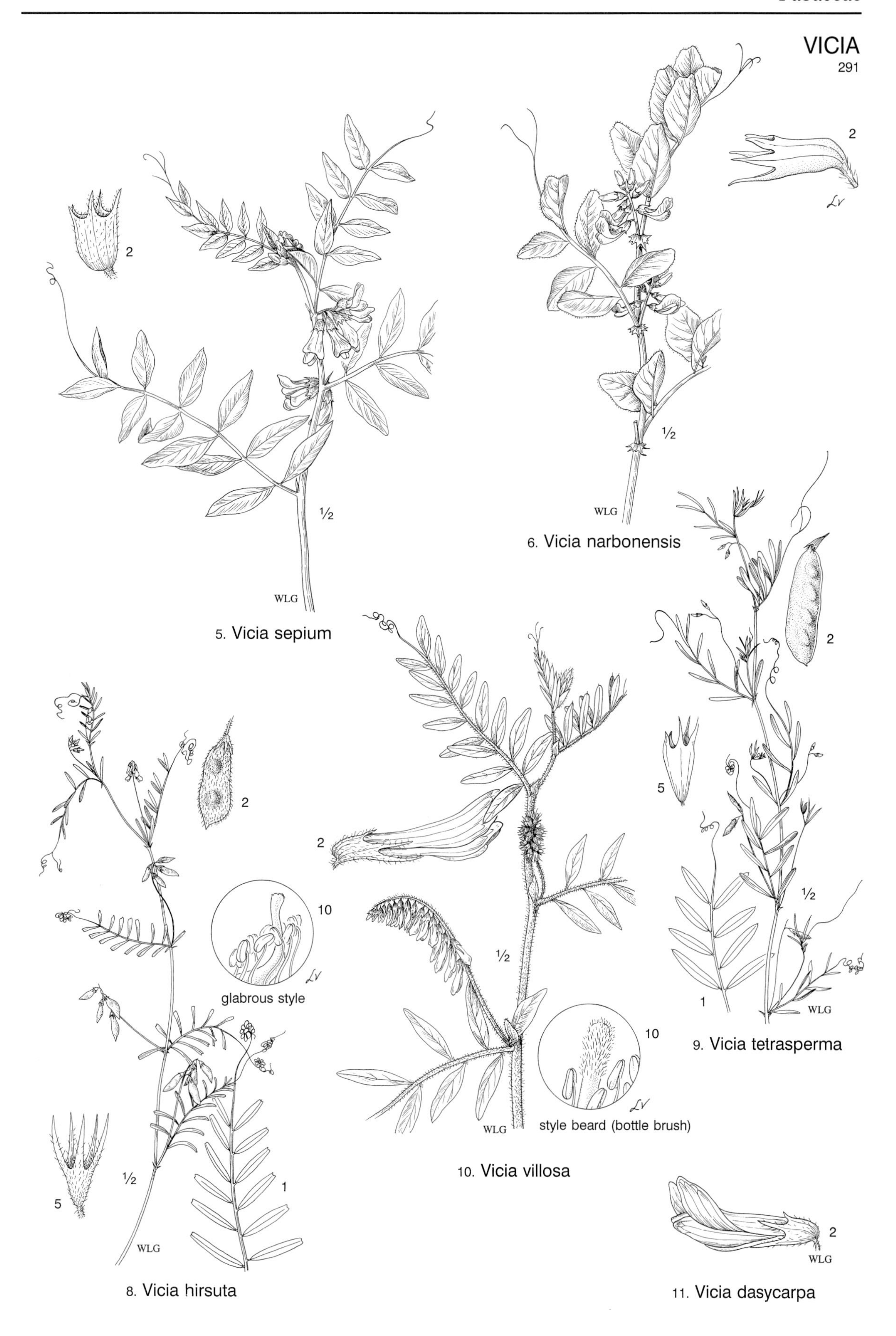

2

½

WLG

5. Vicia sepium

2

½

WLG

6. Vicia narbonensis

2

½

1

WLG

9. Vicia tetrasperma

5

2

10
glabrous style

2

½

10
style beard (bottle brush)

WLG

10. Vicia villosa

2

½

1

5

WLG

8. Vicia hirsuta

2

WLG

11. Vicia dasycarpa

VICIA
LATHYRUS
291–293

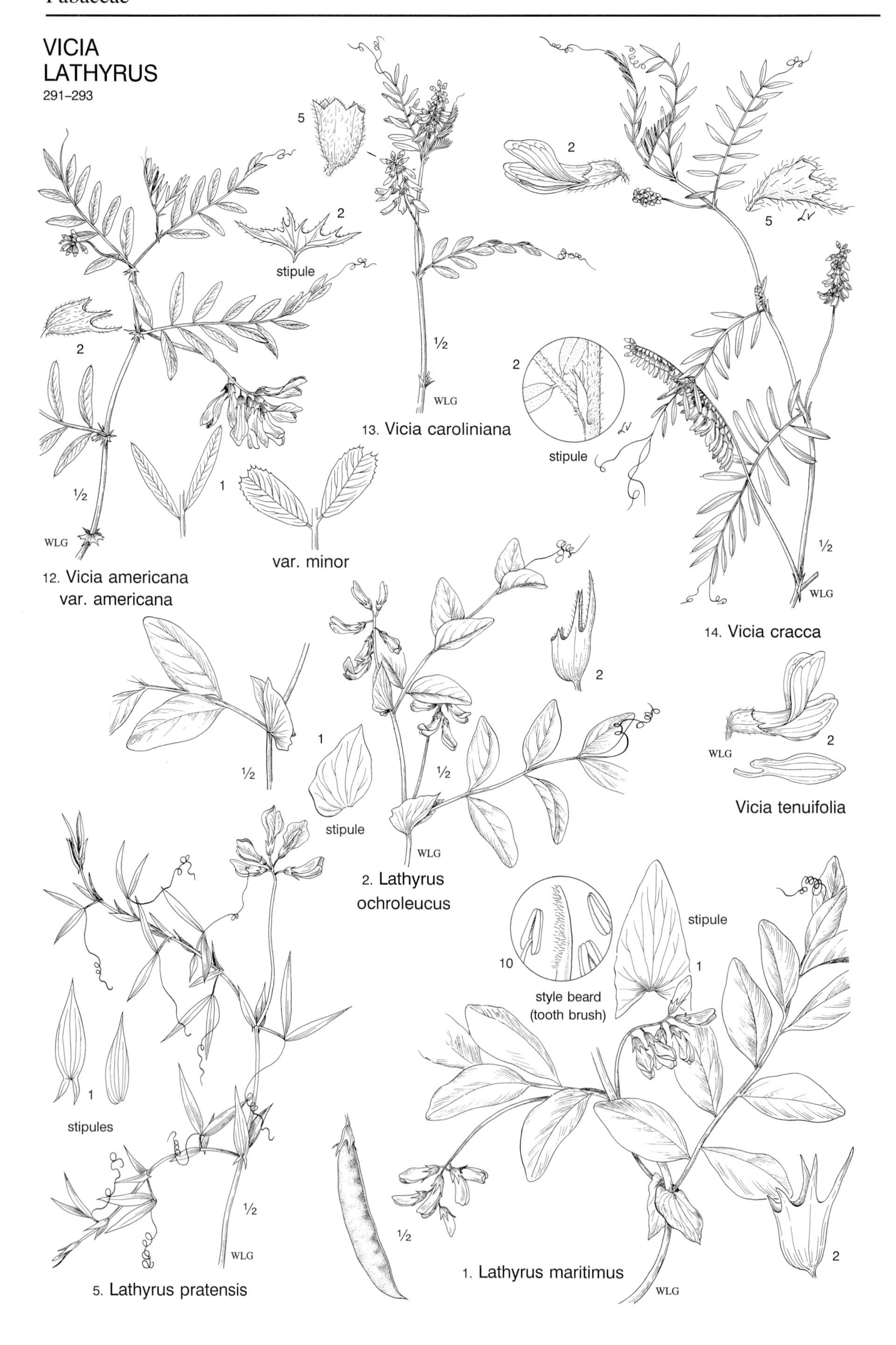

5

2
stipule

2

13. Vicia caroliniana

½
WLG

5

2
stipule

2

12. Vicia americana
var. americana

½
WLG

1
var. minor

½
WLG

14. Vicia cracca

½
WLG

2

½

1
stipule

½
WLG

2. Lathyrus
ochroleucus

2
WLG

Vicia tenuifolia

2

10
style beard
(tooth brush)

stipule

1

1
stipules

½
WLG

5. Lathyrus pratensis

½

1. Lathyrus maritimus

2

WLG

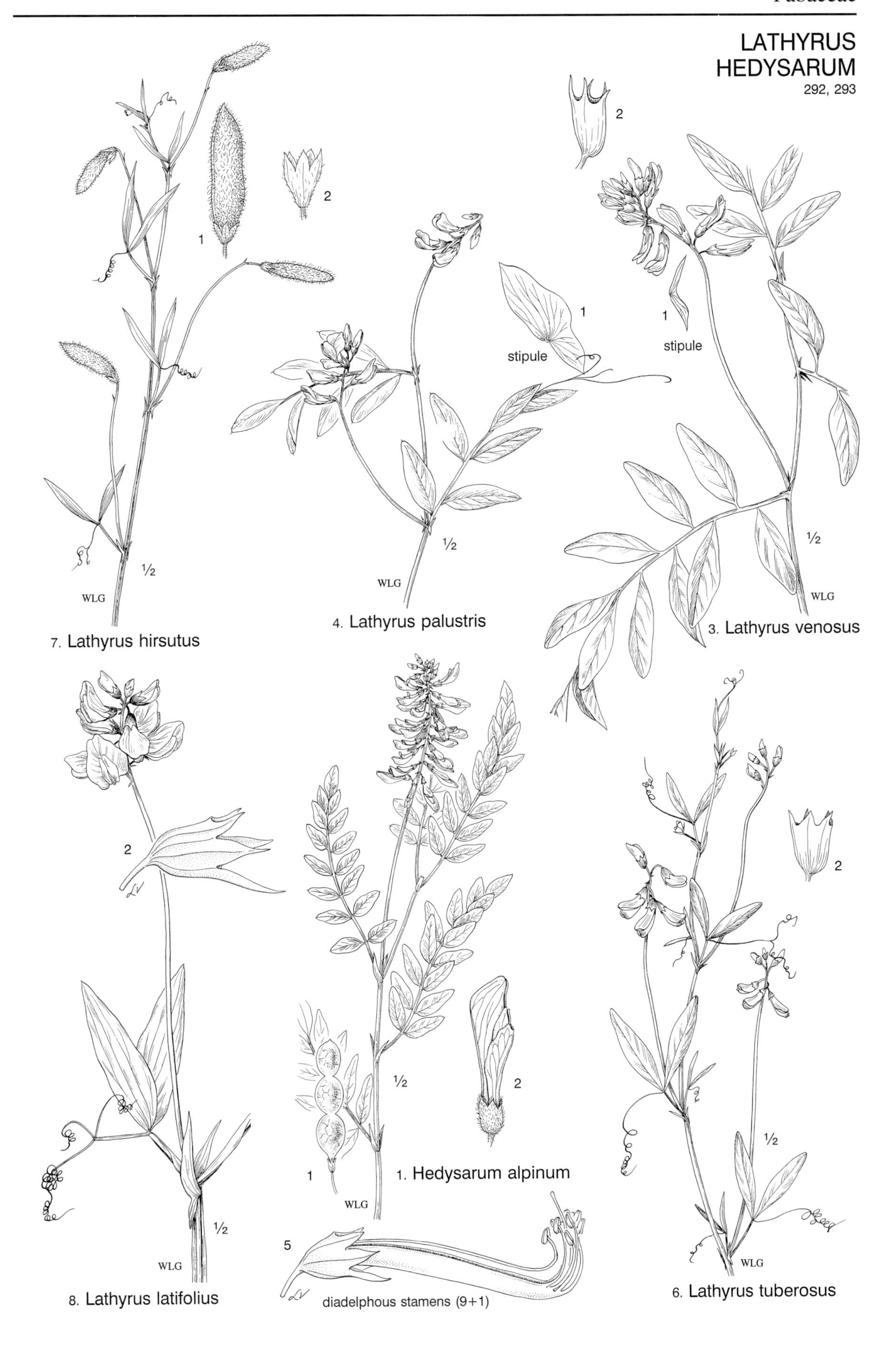

2

1

2. Lathyrus palustris — stipule

1. stipule

4. Lathyrus palustris

3. Lathyrus venosus

7. Lathyrus hirsutus

2

8. Lathyrus latifolius

2

1. 1. Hedysarum alpinum

5 diadelphous stamens (9+1)

2

6. Lathyrus tuberosus

ONOBRYCHIS
DESMODIUM
293–295

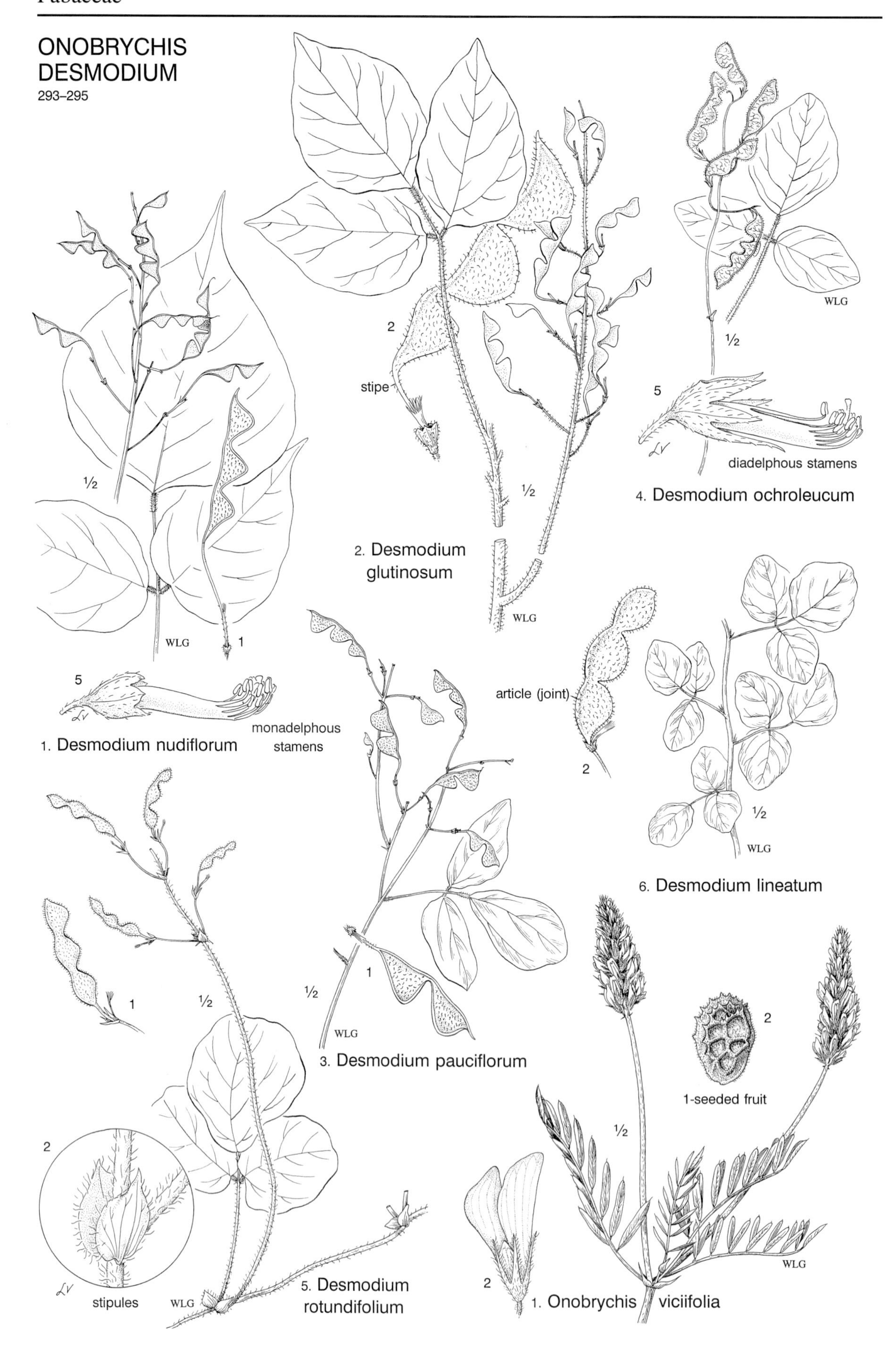

stipe

2. **Desmodium glutinosum**

WLG

4. **Desmodium ochroleucum**

diadelphous stamens

5

1. **Desmodium nudiflorum**

monadelphous stamens

article (joint)

2

6. **Desmodium lineatum**

3. **Desmodium pauciflorum**

WLG

1-seeded fruit

stipules

5. **Desmodium rotundifolium**

1. **Onobrychis viciifolia**

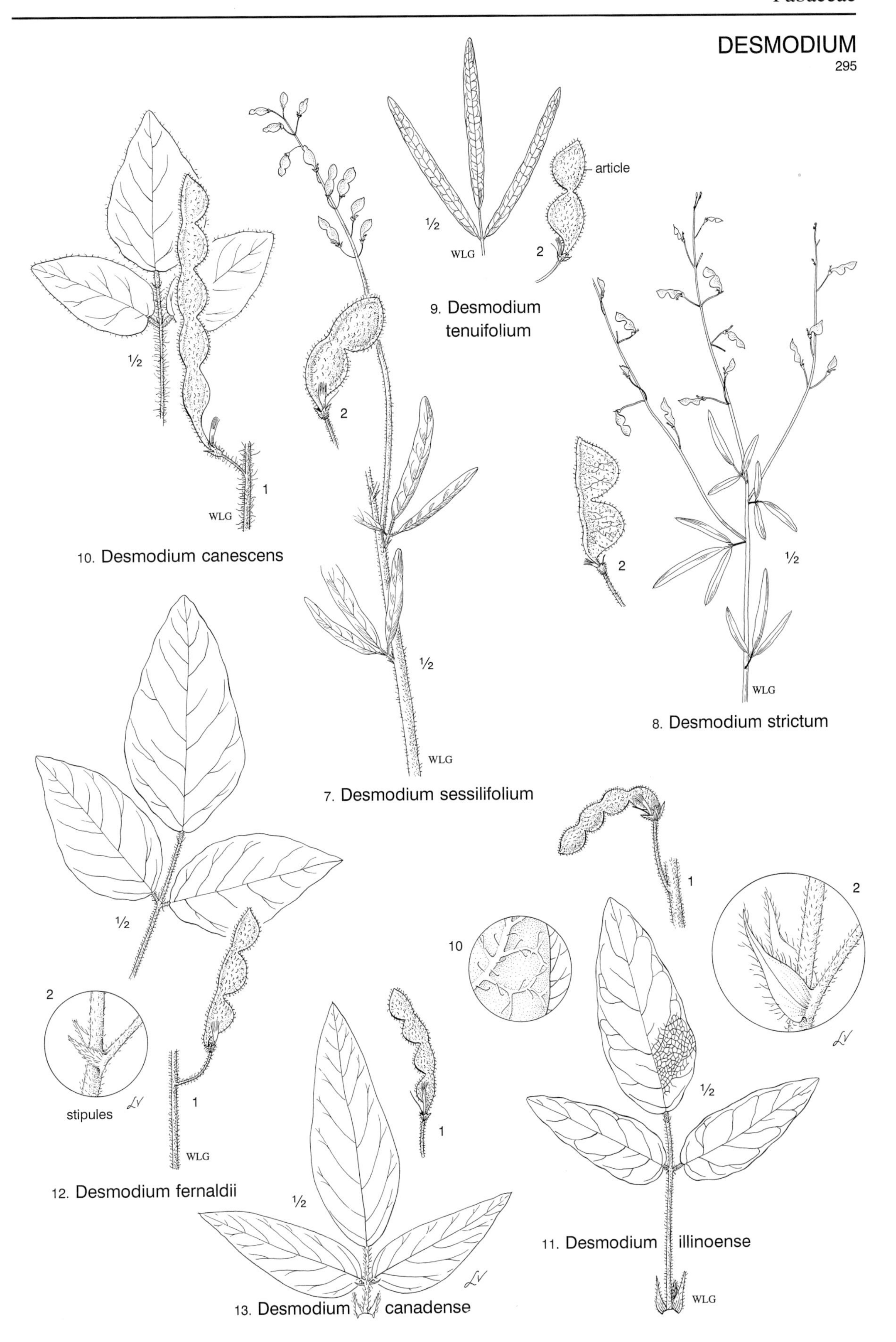

9. Desmodium
tenuifolium

article

10. Desmodium canescens

8. Desmodium strictum

7. Desmodium sessilifolium

12. Desmodium fernaldii

stipules

13. Desmodium canadense

11. Desmodium illinoense

DESMODIUM
296

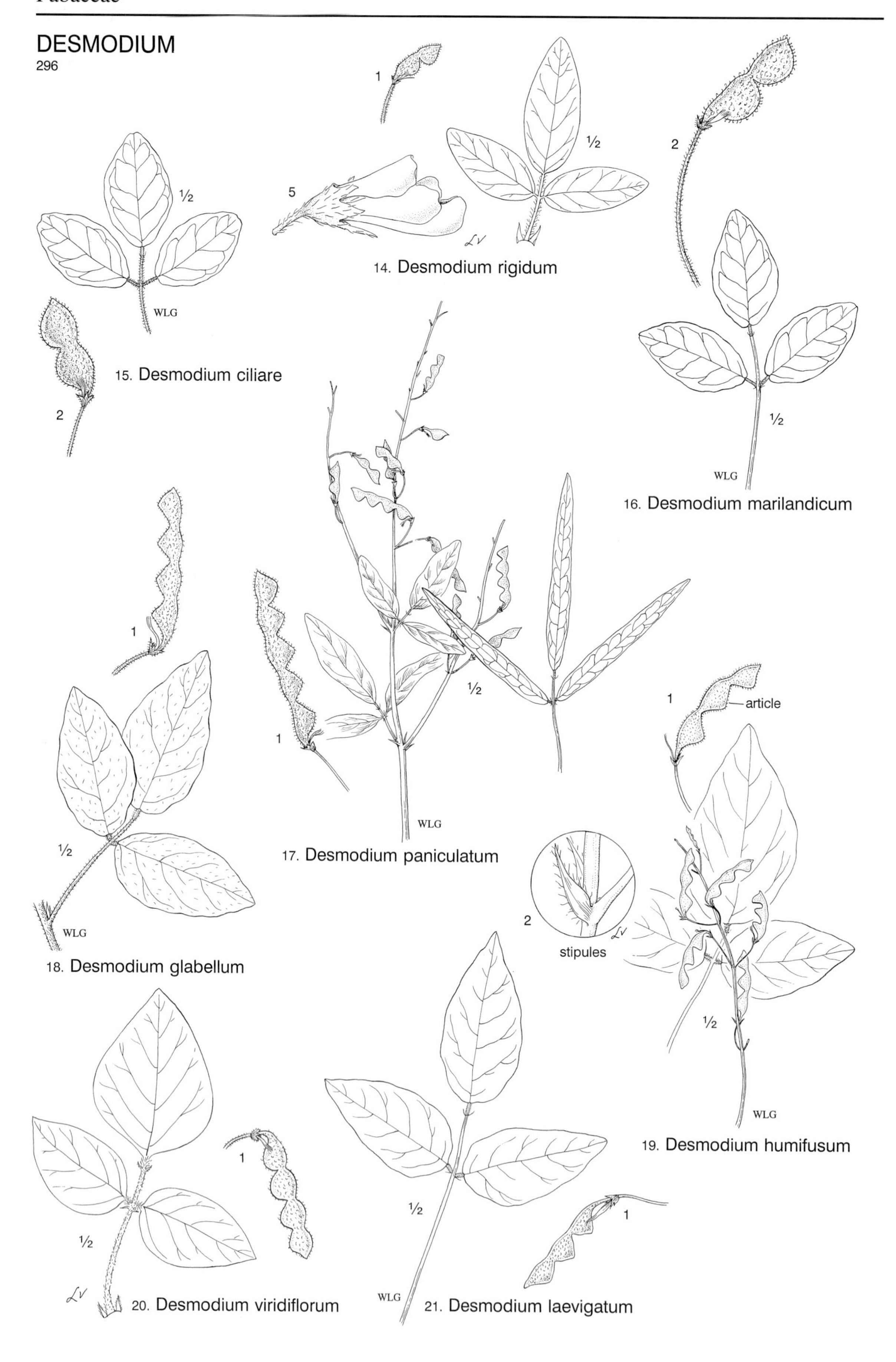

14. **Desmodium rigidum**

15. **Desmodium ciliare**

16. **Desmodium marilandicum**

17. **Desmodium paniculatum**

18. **Desmodium glabellum**

article

stipules

19. **Desmodium humifusum**

20. **Desmodium viridiflorum**

21. **Desmodium laevigatum**

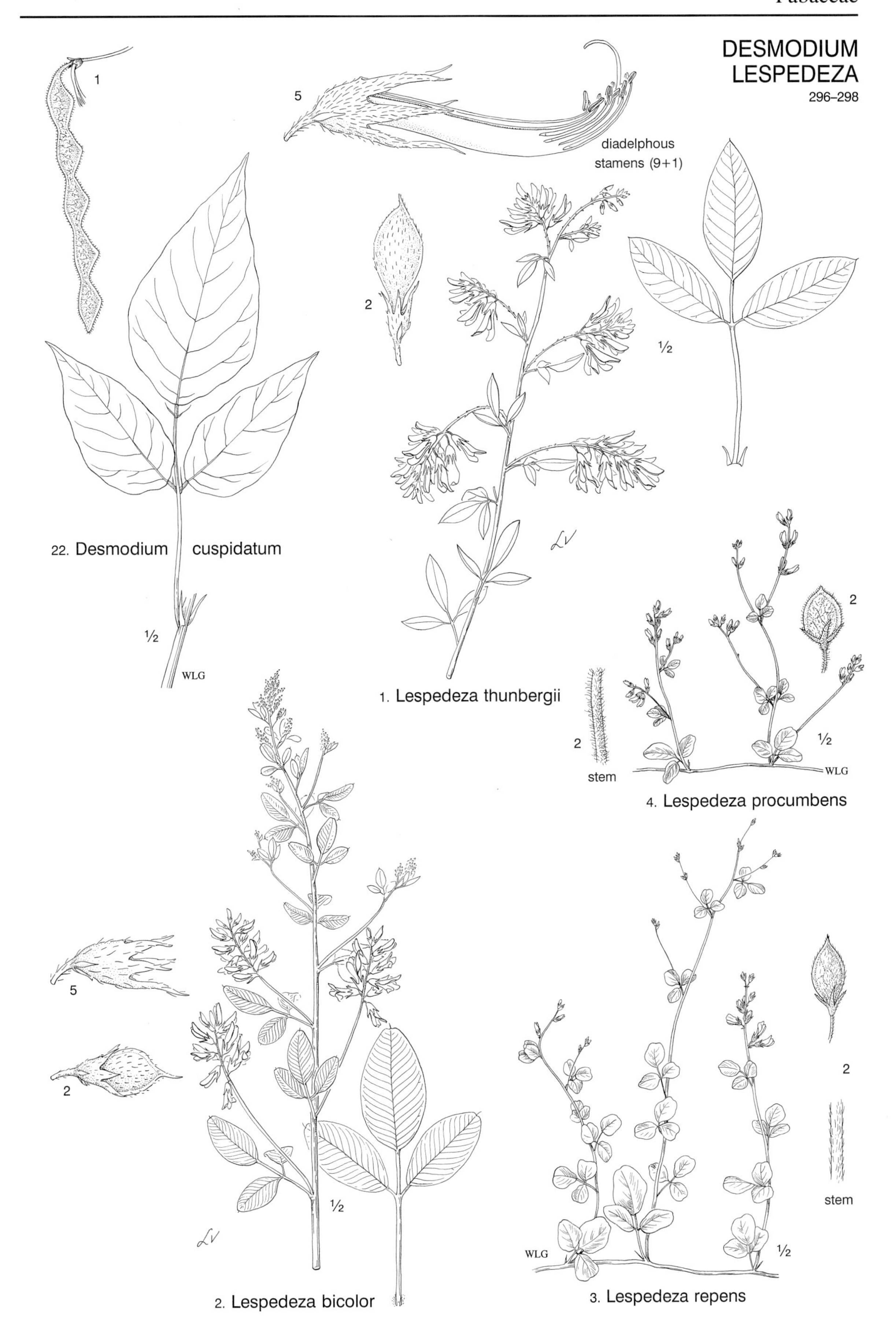

diadelphous
stamens (9+1)

22. Desmodium   cuspidatum

½

WLG

1. Lespedeza thunbergii

4. Lespedeza procumbens

stem

2. Lespedeza bicolor

3. Lespedeza repens

stem

WLG

LESPEDEZA
297, 298

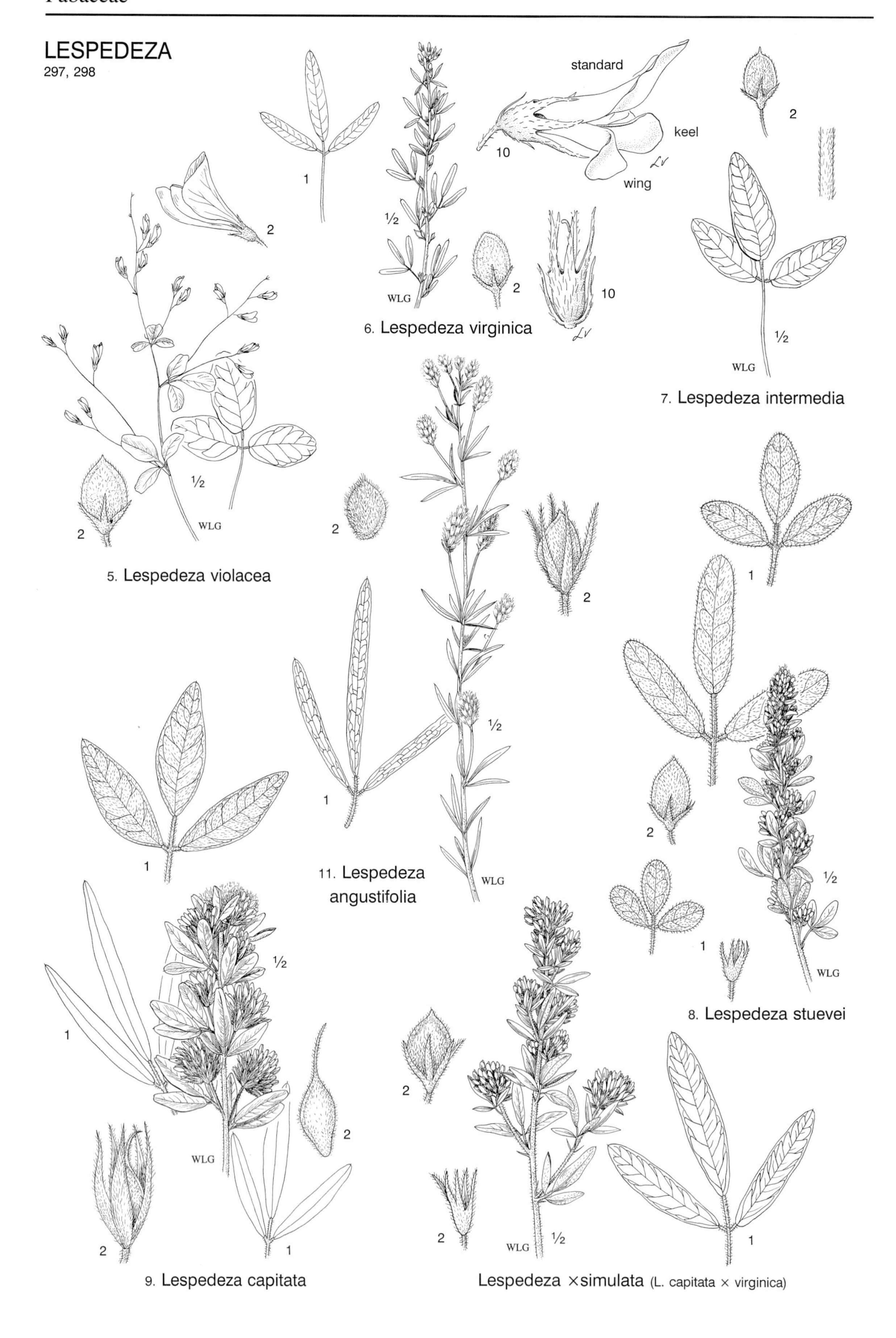

standard

keel

wing

10

6. **Lespedeza virginica**

7. **Lespedeza intermedia**

5. **Lespedeza violacea**

11. **Lespedeza angustifolia**

8. **Lespedeza stuevei**

9. **Lespedeza capitata**

Lespedeza ×simulata (L. capitata × virginica)

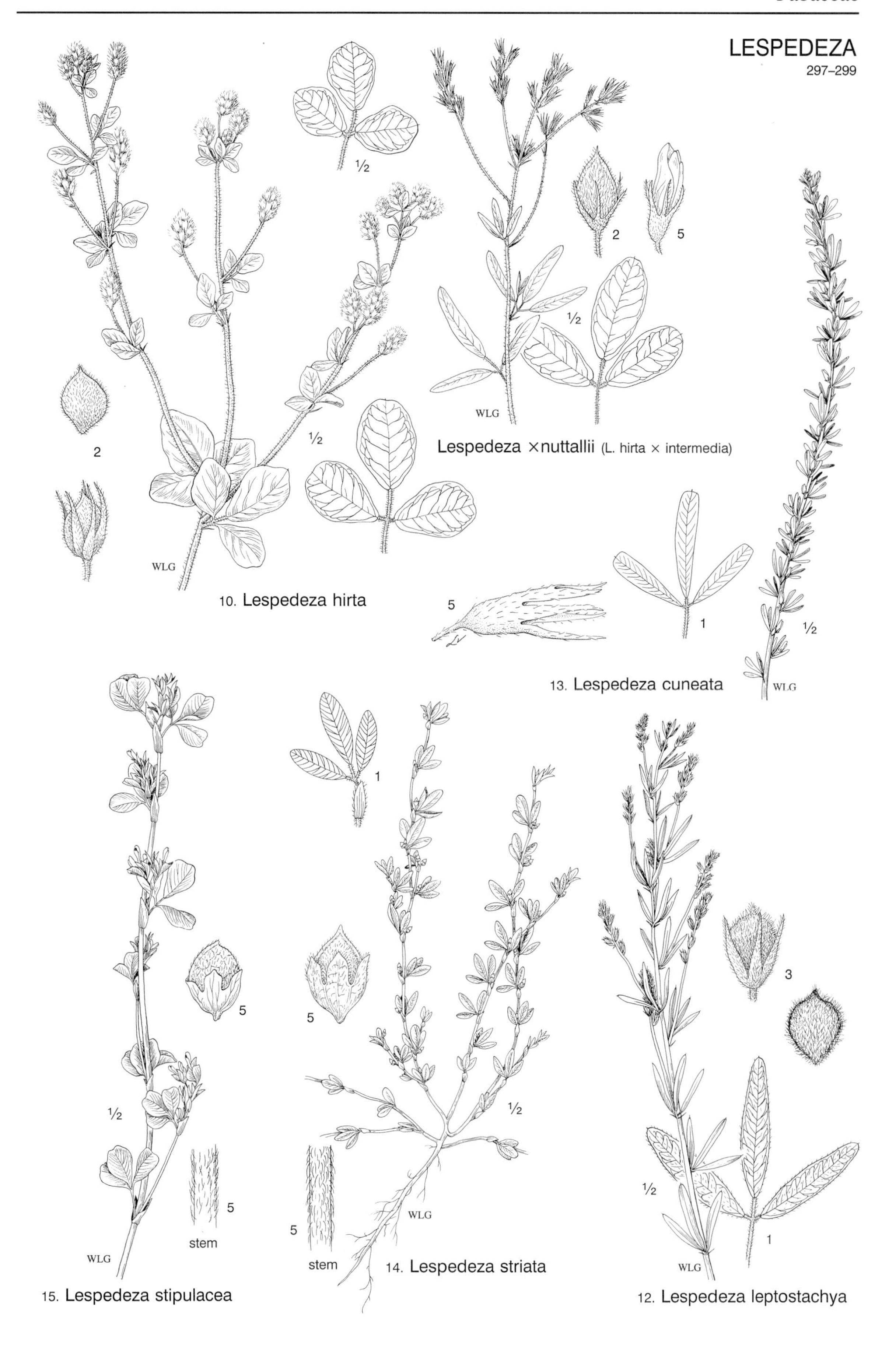

10. Lespedeza hirta

Lespedeza ×nuttallii (L. hirta × intermedia)

13. Lespedeza cuneata

15. Lespedeza stipulacea

14. Lespedeza striata

12. Lespedeza leptostachya

DALEA
AMORPHA
299, 300

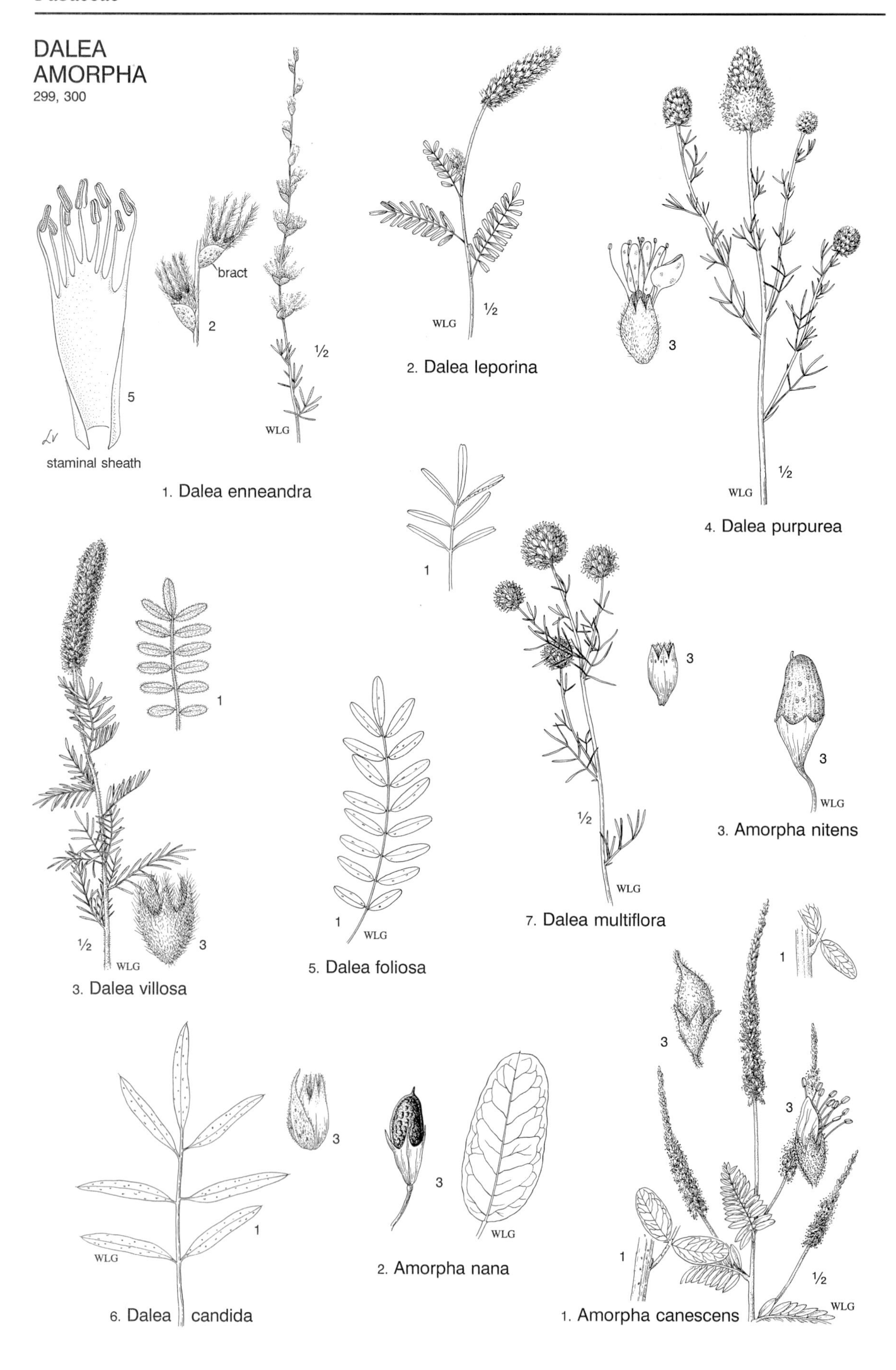

5

staminal sheath

bract

2

½

WLG

1. Dalea enneandra

½

WLG

2. Dalea leporina

3

½

WLG

4. Dalea purpurea

1

3

3

3. Amorpha nitens

1

½

WLG

3. Dalea villosa

1

WLG

5. Dalea foliosa

½

WLG

7. Dalea multiflora

3

3

1

1

½

WLG

1. Amorpha canescens

3

WLG

6. Dalea candida

3

3

WLG

2. Amorpha nana

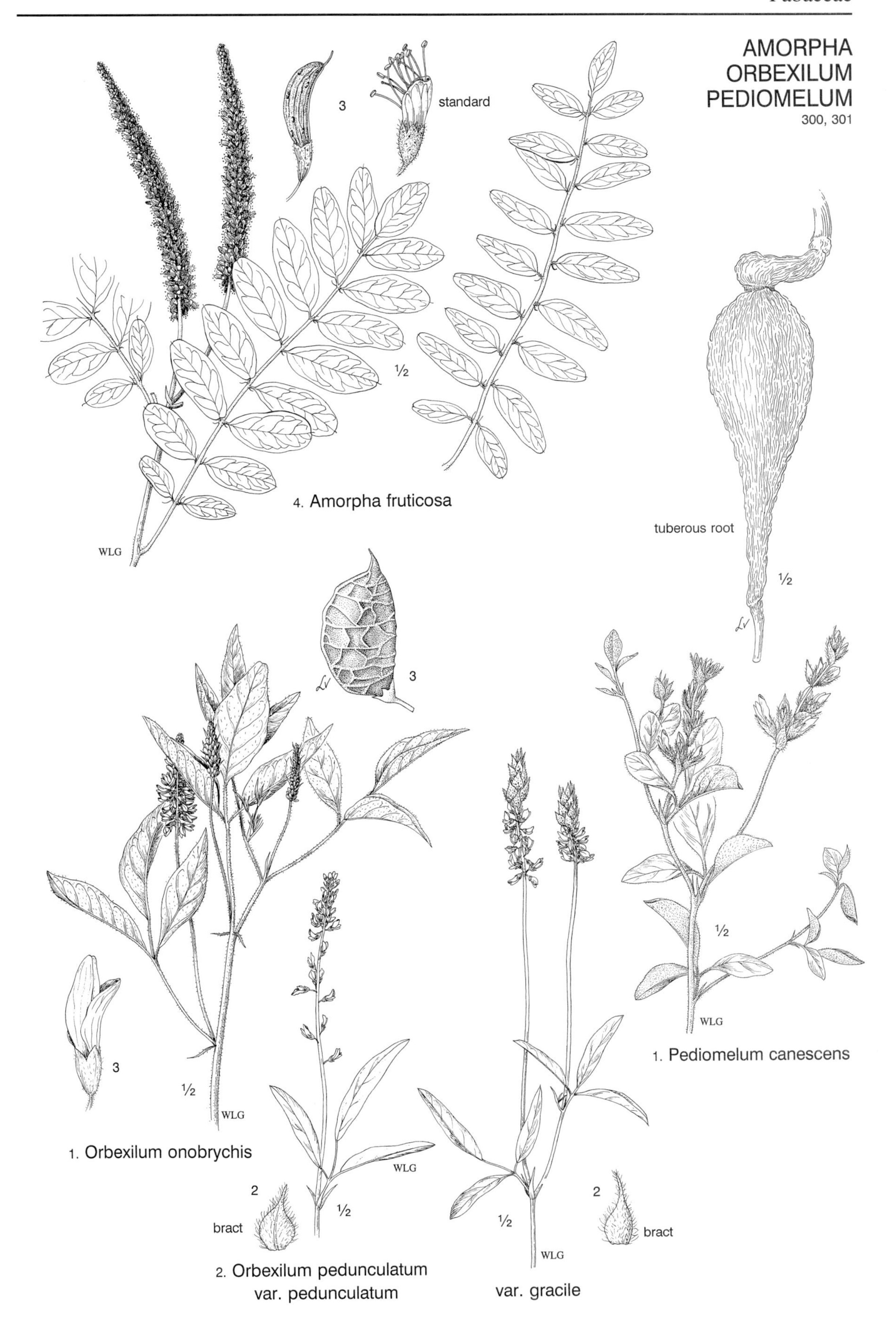

3

standard

½

4. Amorpha fruticosa

WLG

tuberous root

½

3

1. Pediomelum canescens

½

WLG

3

½

WLG

1. Orbexilum onobrychis

WLG

2

bract

2. Orbexilum pedunculatum
var. pedunculatum

½

½

WLG

var. gracile

2

bract

PEDIOMELUM
PSORALIDIUM
301, 302

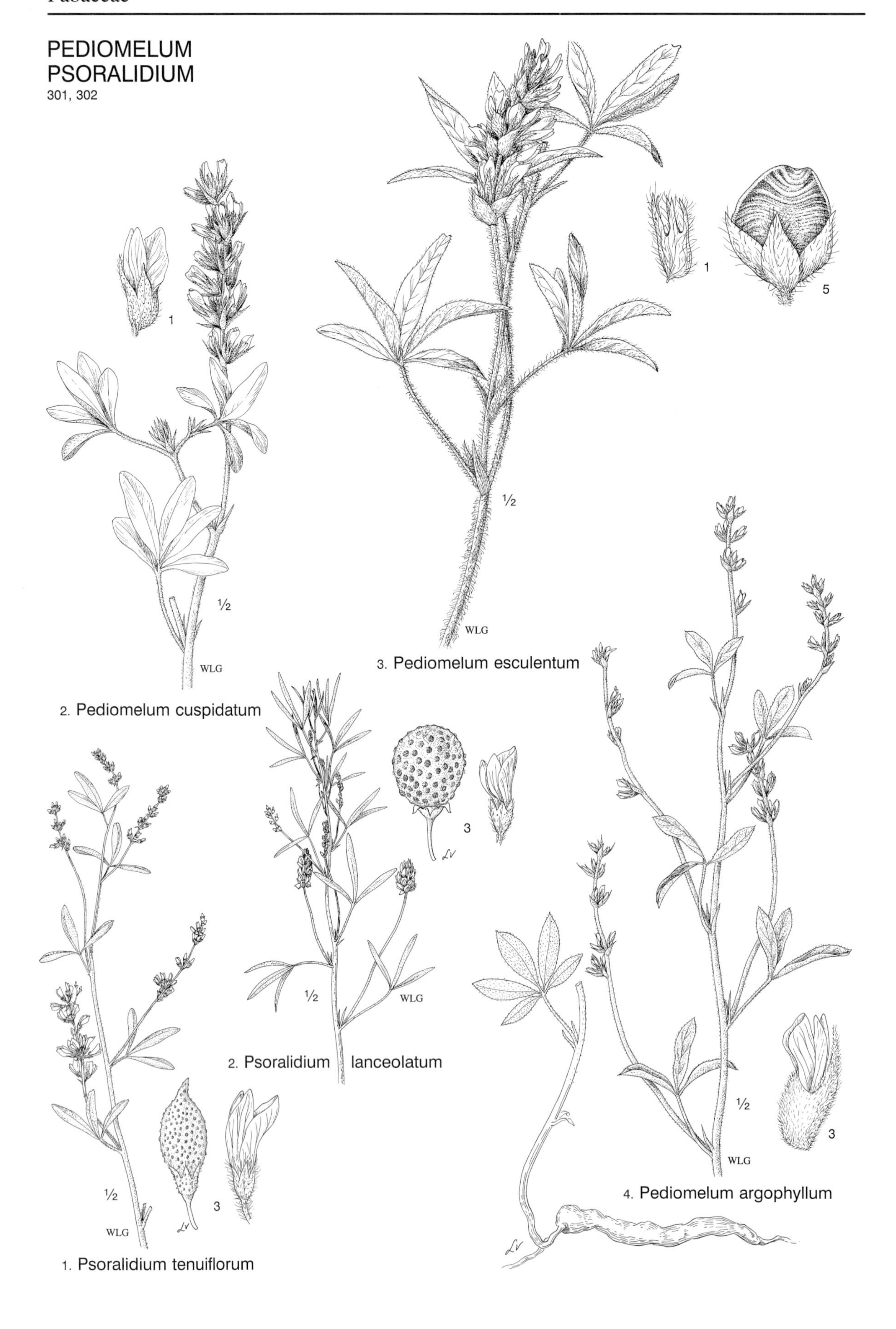

1

1. Pediomelum esculentum

5

2. Pediomelum cuspidatum

3

2. Psoralidium lanceolatum

3

1. Psoralidium tenuiflorum

3

4. Pediomelum argophyllum

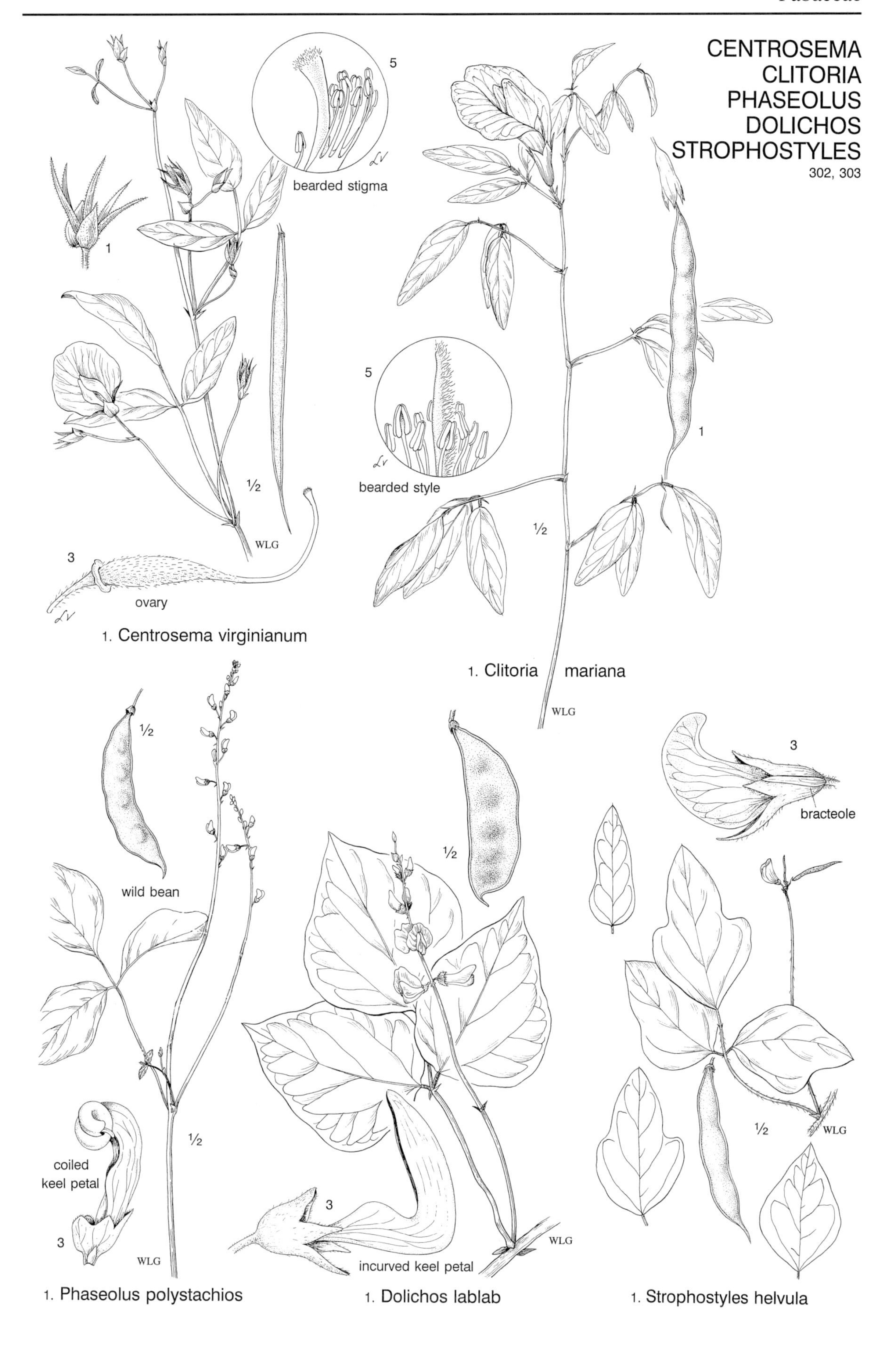

CENTROSEMA
CLITORIA
PHASEOLUS
DOLICHOS
STROPHOSTYLES
302, 303

bearded stigma

5

bearded style

5

½

WLG

3

ovary

1. Centrosema virginianum

1. Clitoria mariana

½

WLG

3

bracteole

½

wild bean

coiled
keel petal

3

WLG

3

incurved keel petal

WLG

½

WLG

1. Phaseolus polystachios

1. Dolichos lablab

1. Strophostyles helvula

STROPHOSTYLES
RHYNCHOSIA
APIOS
AMPHICARPAEA
303, 304

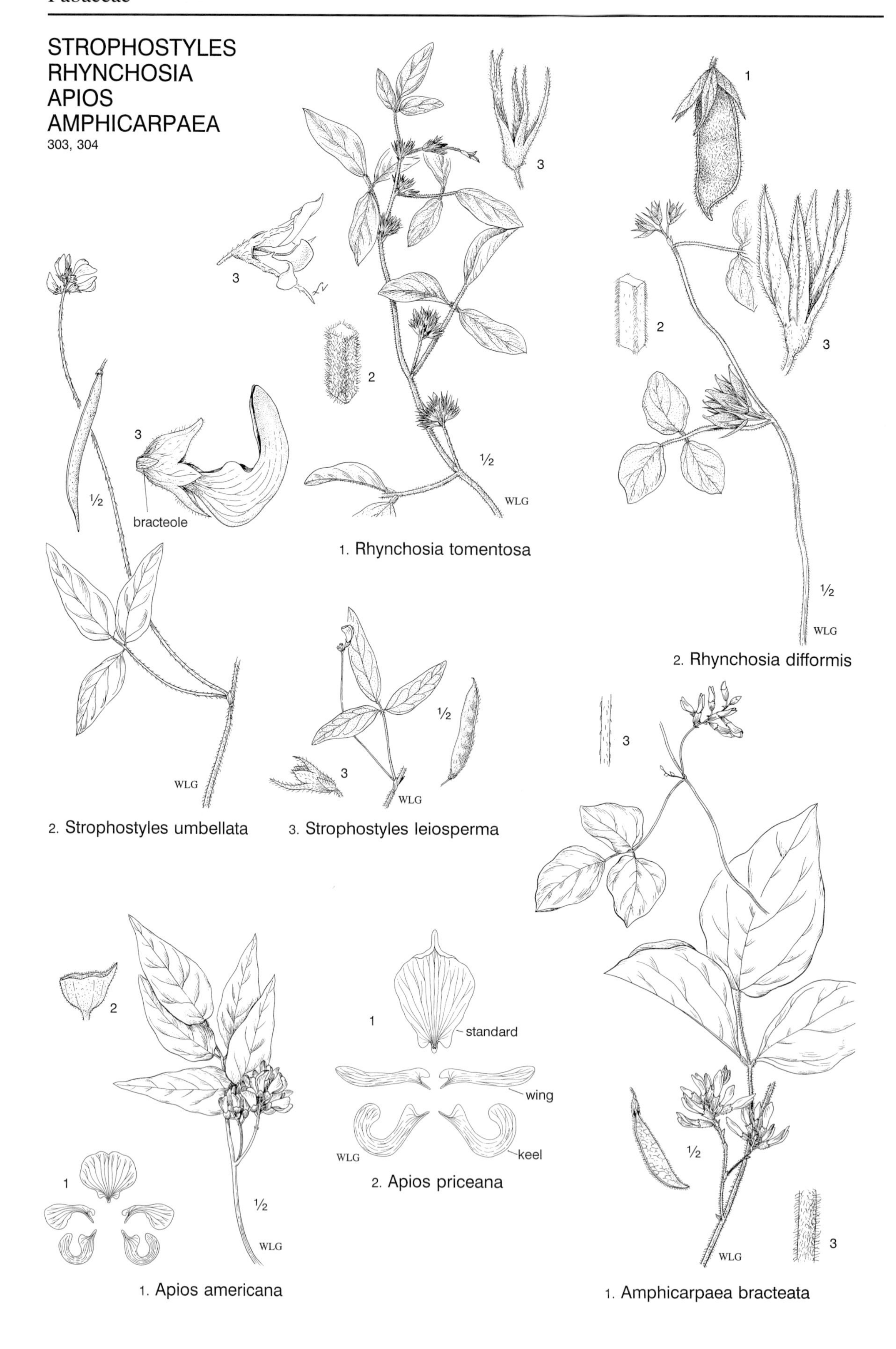

bracteole

1. Rhynchosia tomentosa

2. Rhynchosia difformis

2. Strophostyles umbellata

3. Strophostyles leiosperma

standard

wing

keel

2. Apios priceana

1. Apios americana

1. Amphicarpaea bracteata

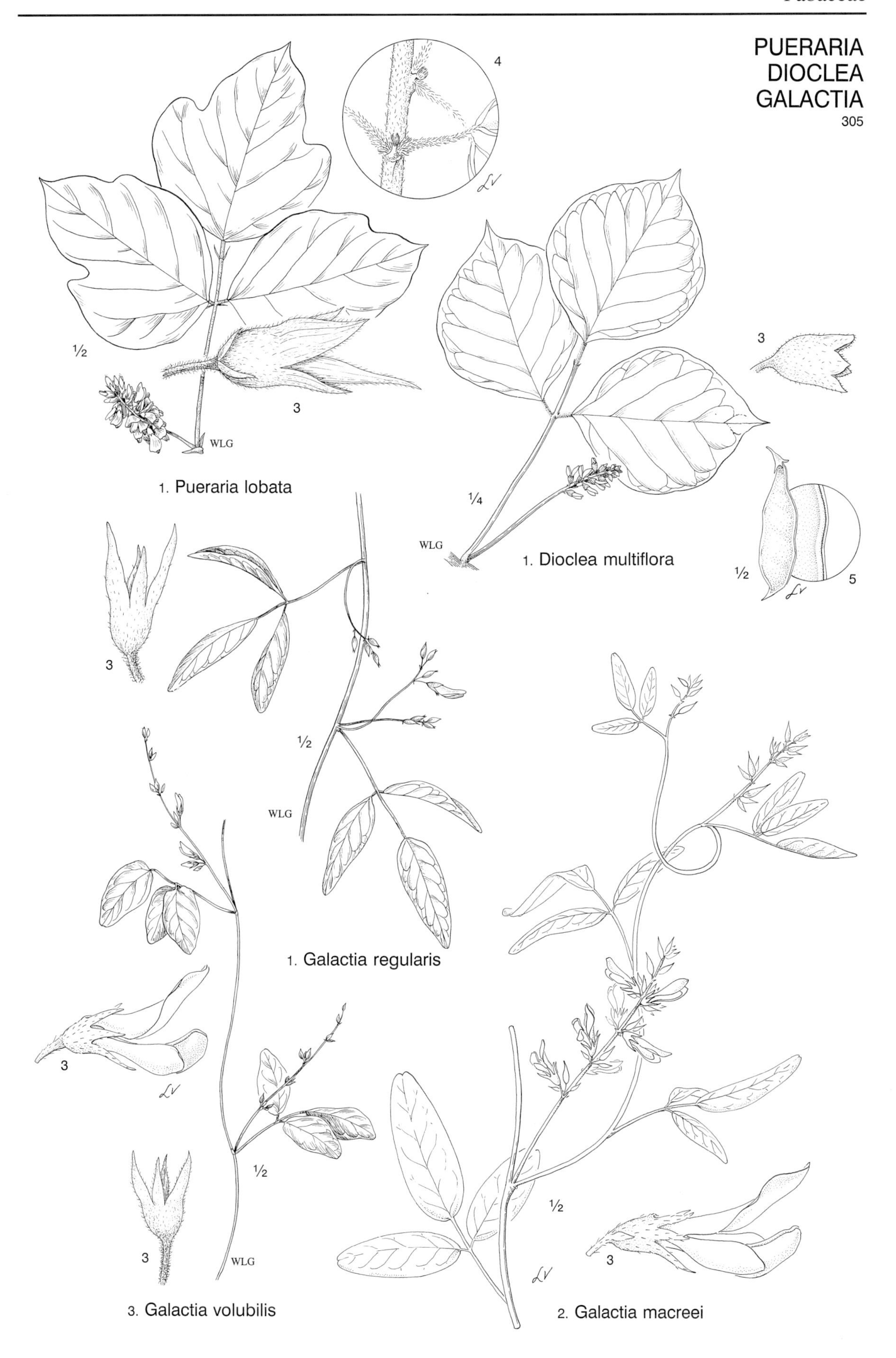

1. Pueraria lobata

1. Dioclea multiflora

1. Galactia regularis

3. Galactia volubilis

2. Galactia macreei

SHEPHERDIA
ELAEAGNUS
306, 307

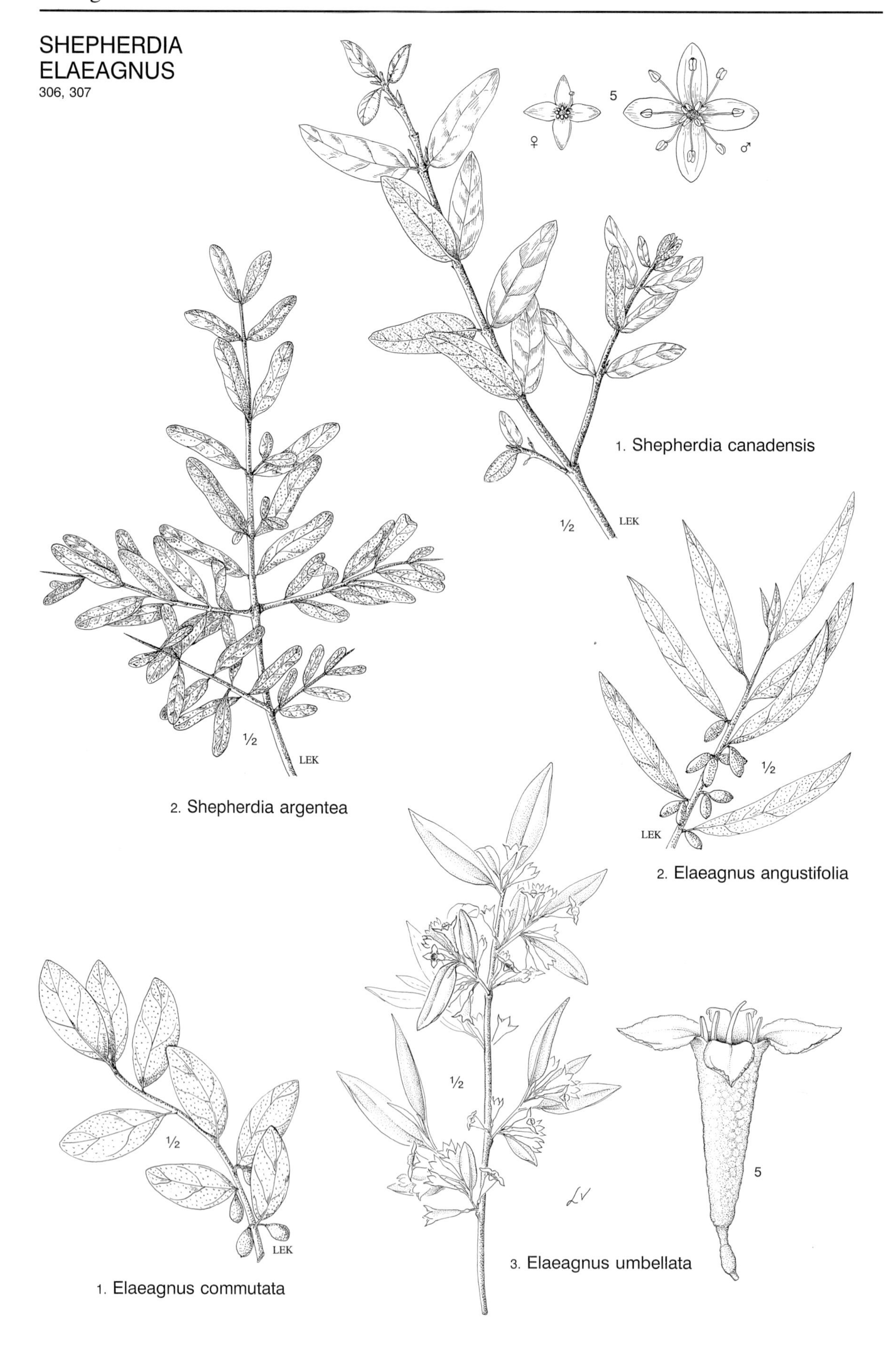

1. Shepherdia canadensis

½ LEK

2. Shepherdia argentea

½ LEK

2. Elaeagnus angustifolia

½ LEK

1. Elaeagnus commutata

½ LEK

3. Elaeagnus umbellata

½

5

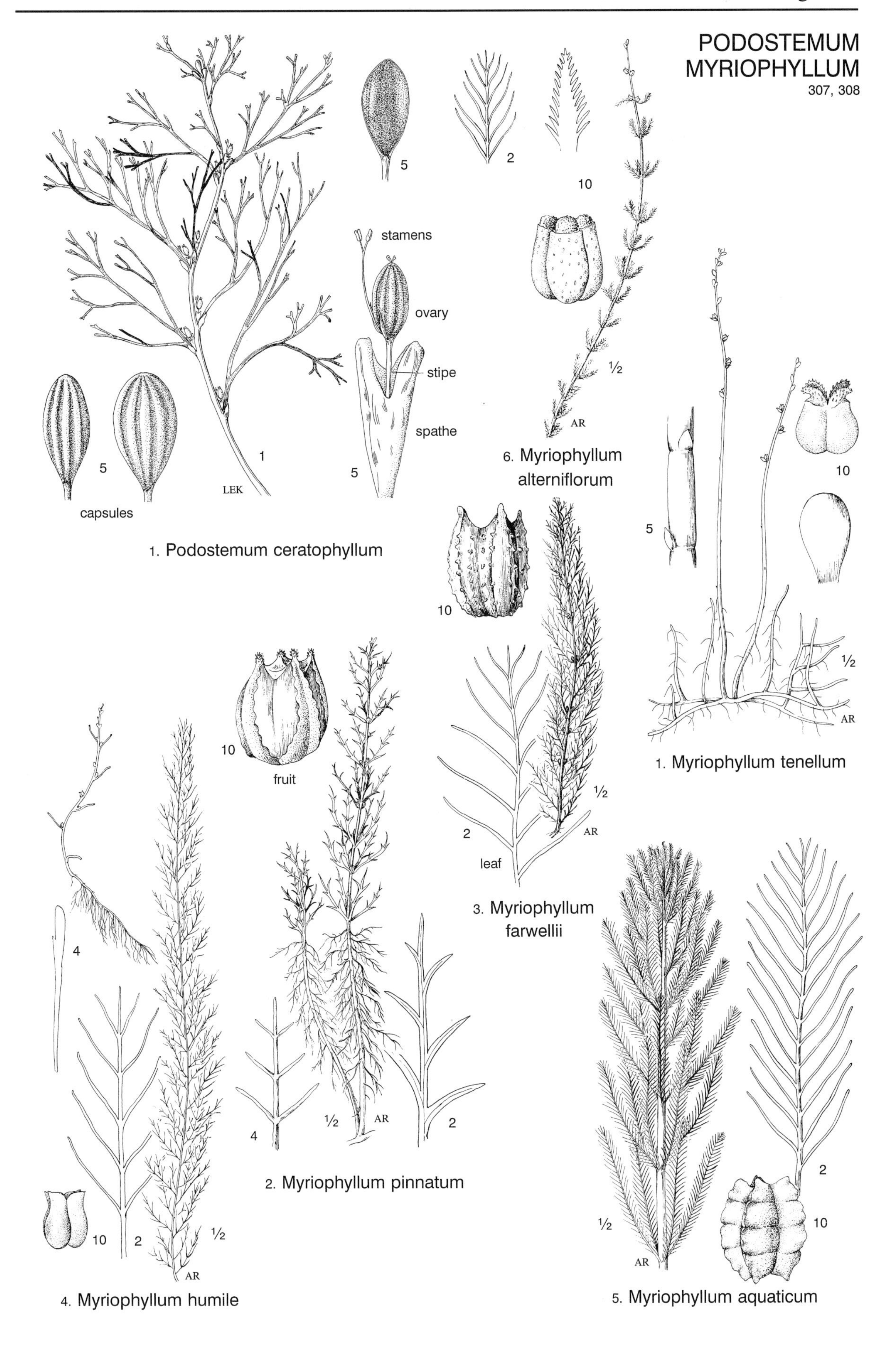

5

2

10

½
AR

stamens

ovary

stipe

spathe

5

6. Myriophyllum
alterniflorum

10

5

5

capsules

1. Podostemum ceratophyllum

10

10

5

½

10

½
AR

1. Myriophyllum tenellum

10

fruit

½
2

leaf

3. Myriophyllum
farwellii

4

4

½ AR 2

2. Myriophyllum pinnatum

2

½
AR

10

10

2

10

2

½
AR

4. Myriophyllum humile

5. Myriophyllum aquaticum

MYRIOPHYLLUM
PROSERPINACA
308, 309

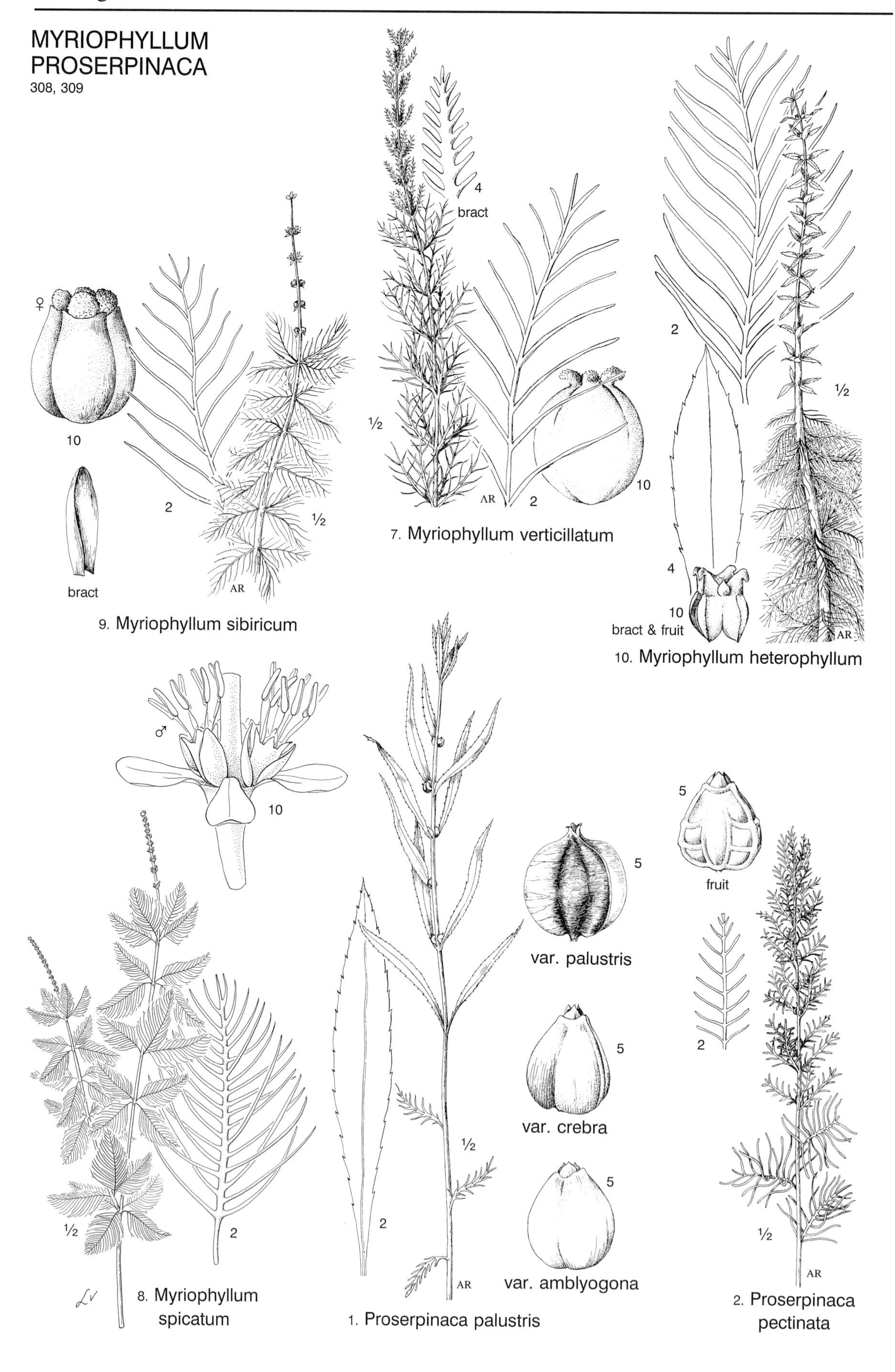

10

bract

9. Myriophyllum sibiricum

4
bract

½

2

10

7. Myriophyllum verticillatum

2

½

4

10
bract & fruit

10. Myriophyllum heterophyllum

♂

10

5

var. palustris

5
fruit

5

var. crebra

2

5

var. amblyogona

½

½

2

½

2

8. Myriophyllum
spicatum

1. Proserpinaca palustris

2. Proserpinaca
pectinata

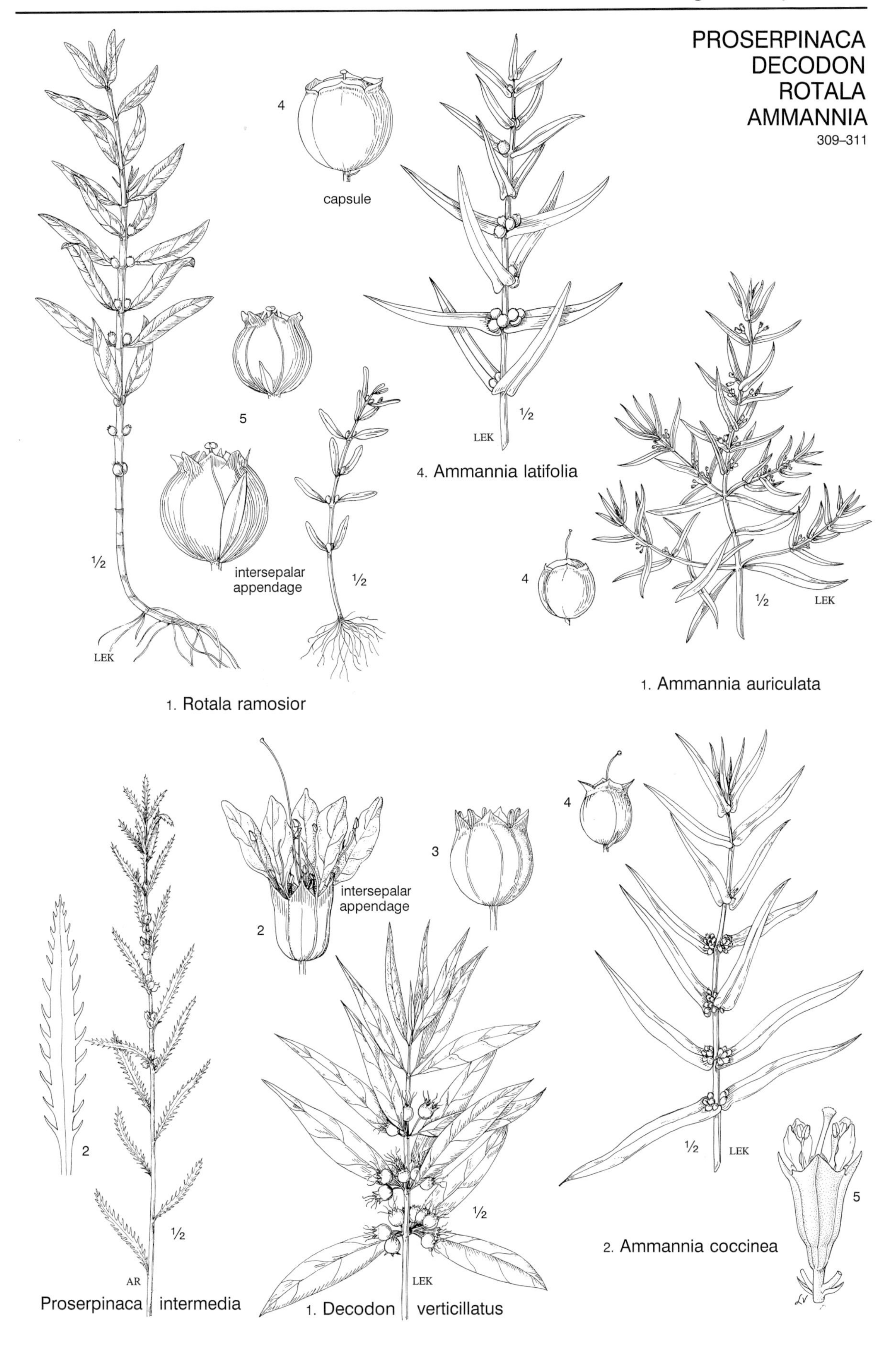

4

capsule

5

½

LEK

½

intersepalar
appendage

½

1. Rotala ramosior

½

LEK

4. Ammannia latifolia

4

½   LEK

1. Ammannia auriculata

2

intersepalar
appendage

3

4

2

½

AR

Proserpinaca intermedia

½

LEK

1. Decodon verticillatus

½   LEK

2. Ammannia coccinea

5

AMMANNIA
DIDIPLIS
LYTHRUM
310, 311

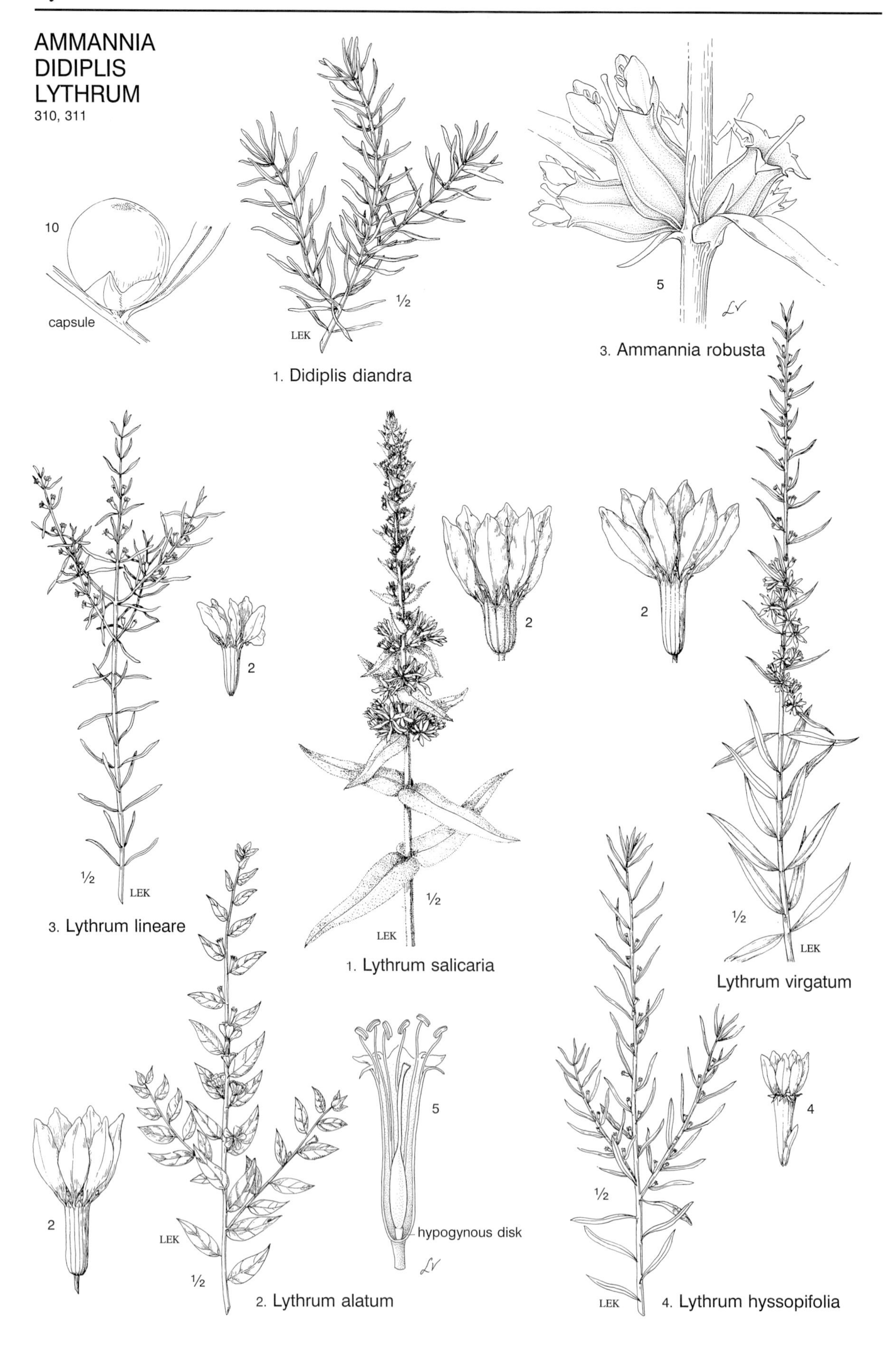

10

capsule

1. Didiplis diandra

3. Ammannia robusta

5

3. Lythrum lineare

1. Lythrum salicaria

Lythrum virgatum

2. Lythrum alatum

hypogynous disk

4. Lythrum hyssopifolia

CUPHEA
DAPHNE
DIRCA
THYMELAEA
TRAPA
312, 313

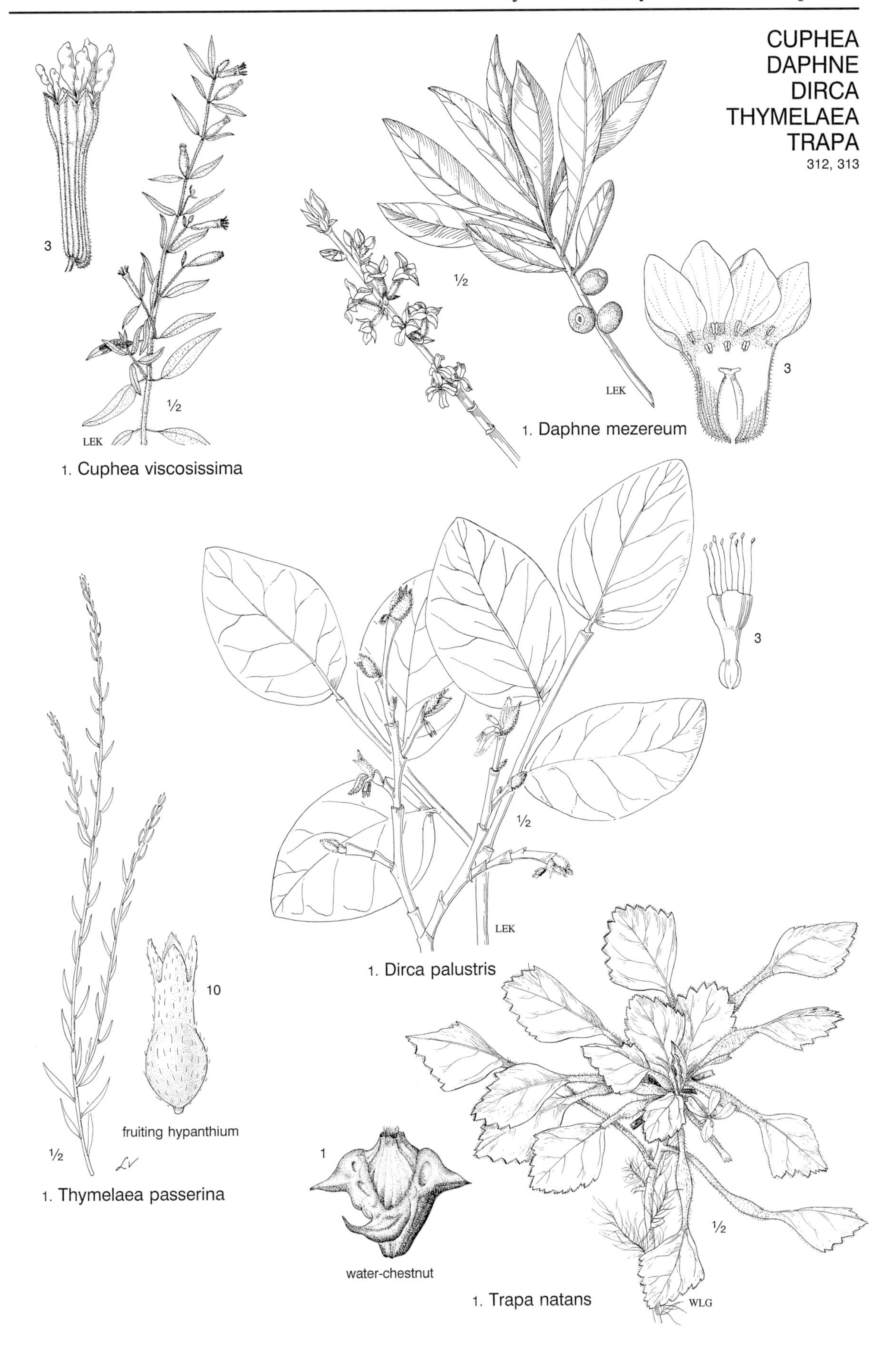

3

1. Cuphea viscosissima

1. Daphne mezereum

3

1. Dirca palustris

10

fruiting hypanthium

1. Thymelaea passerina

water-chestnut

1. Trapa natans

LUDWIGIA
314, 315

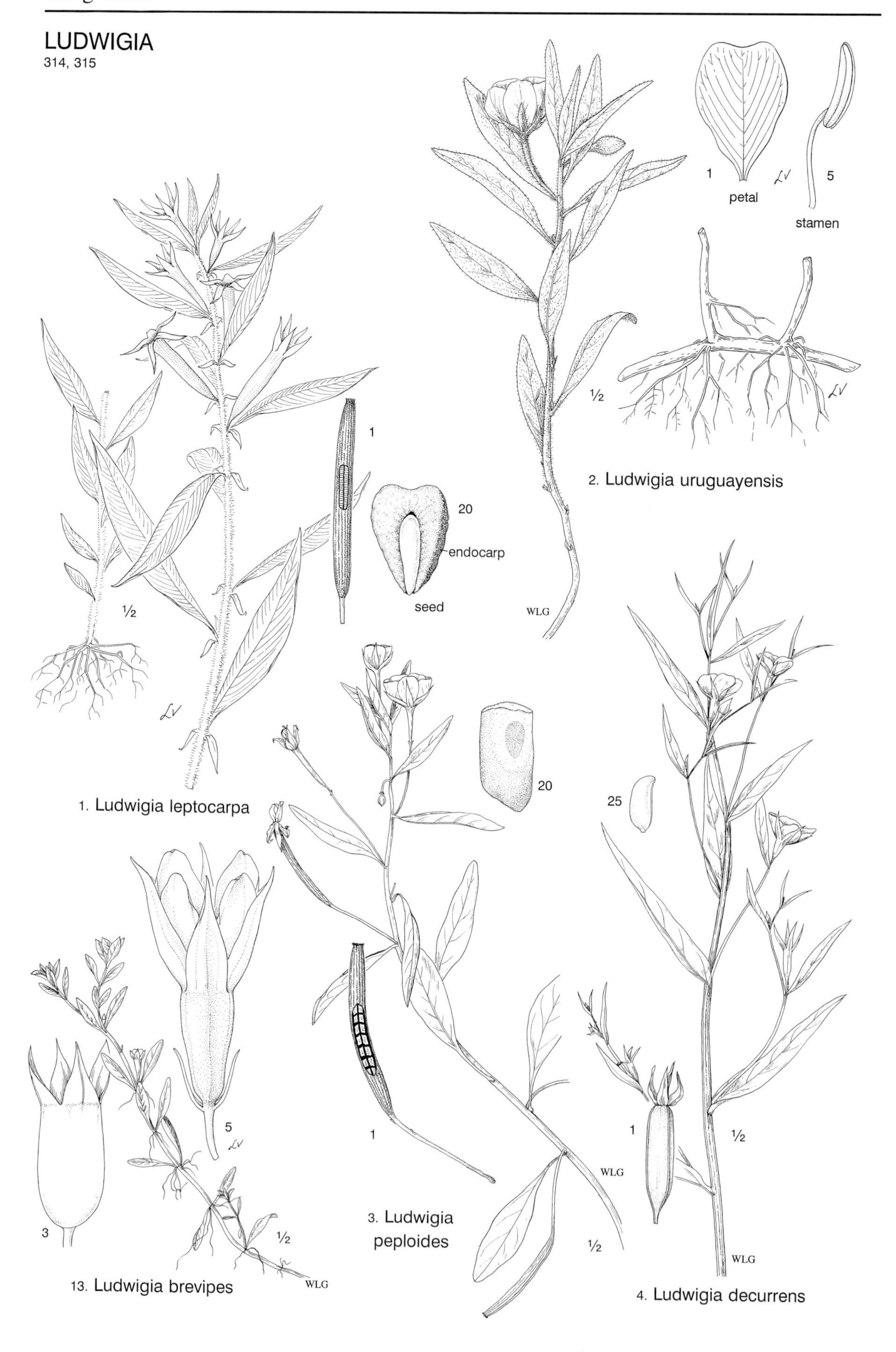

1   petal

5

stamen

1

20
endocarp

seed

2. Ludwigia uruguayensis

WLG

1. Ludwigia leptocarpa

20

25

3. Ludwigia
peploides

WLG

13. Ludwigia brevipes

5

3

½

WLG

1

½

4. Ludwigia decurrens

WLG

1

½

WLG

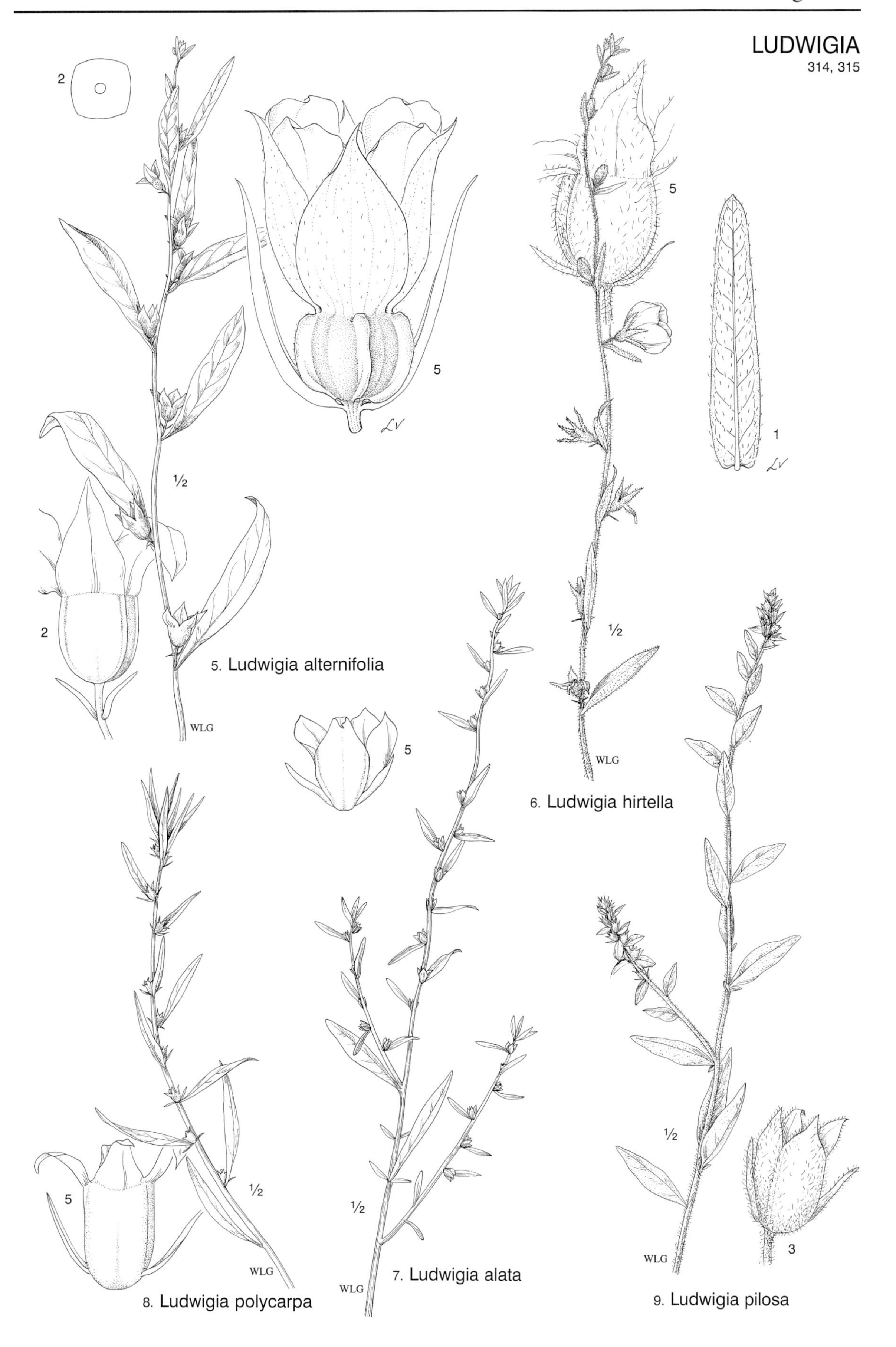

5. Ludwigia alternifolia

6. Ludwigia hirtella

8. Ludwigia polycarpa

7. Ludwigia alata

9. Ludwigia pilosa

LUDWIGIA
EPILOBIUM
315, 316

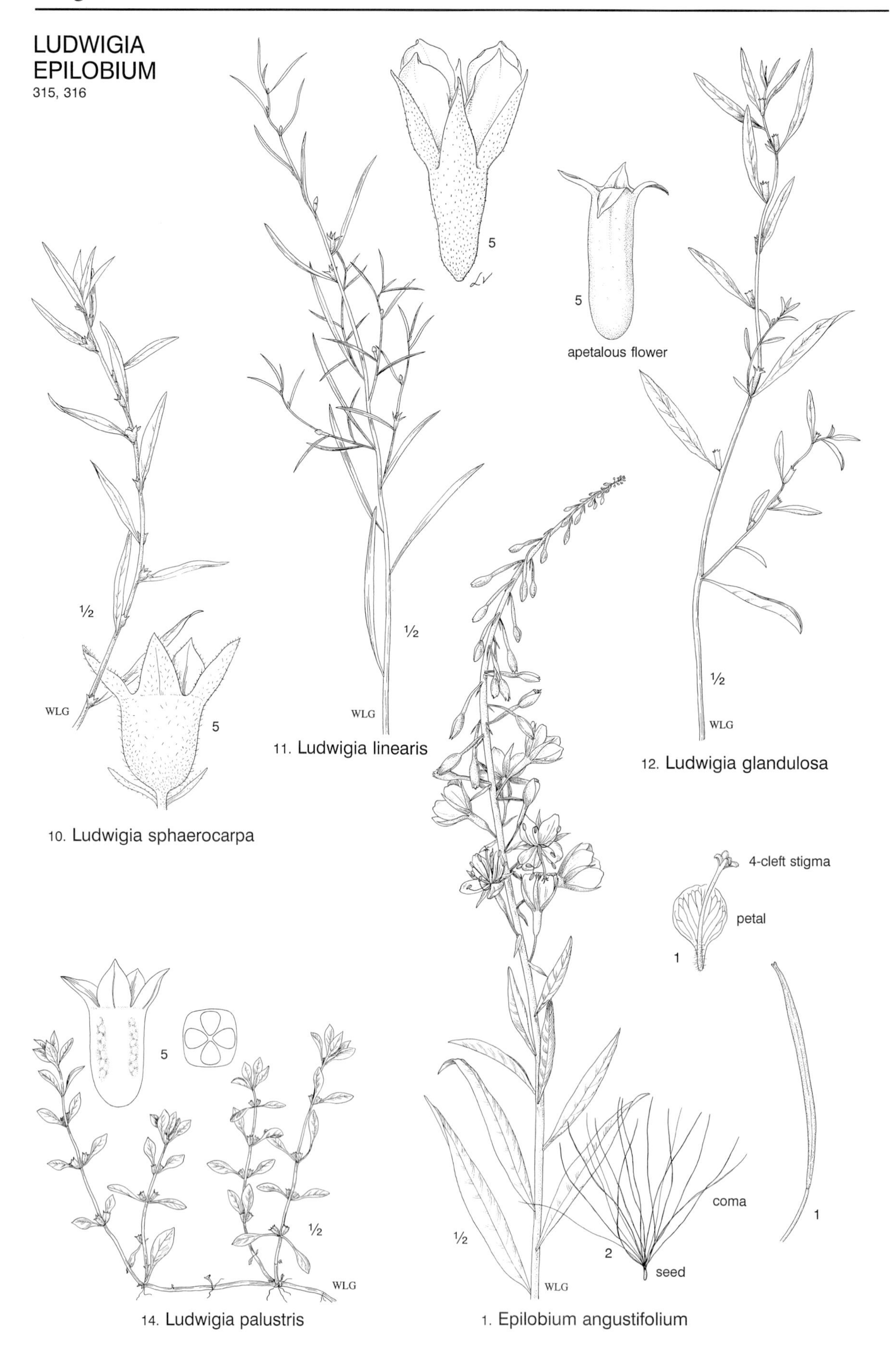

5

apetalous flower

11. **Ludwigia linearis**

12. **Ludwigia glandulosa**

10. Ludwigia sphaerocarpa

4-cleft stigma

petal

1

coma

seed

14. **Ludwigia palustris**

1. **Epilobium angustifolium**

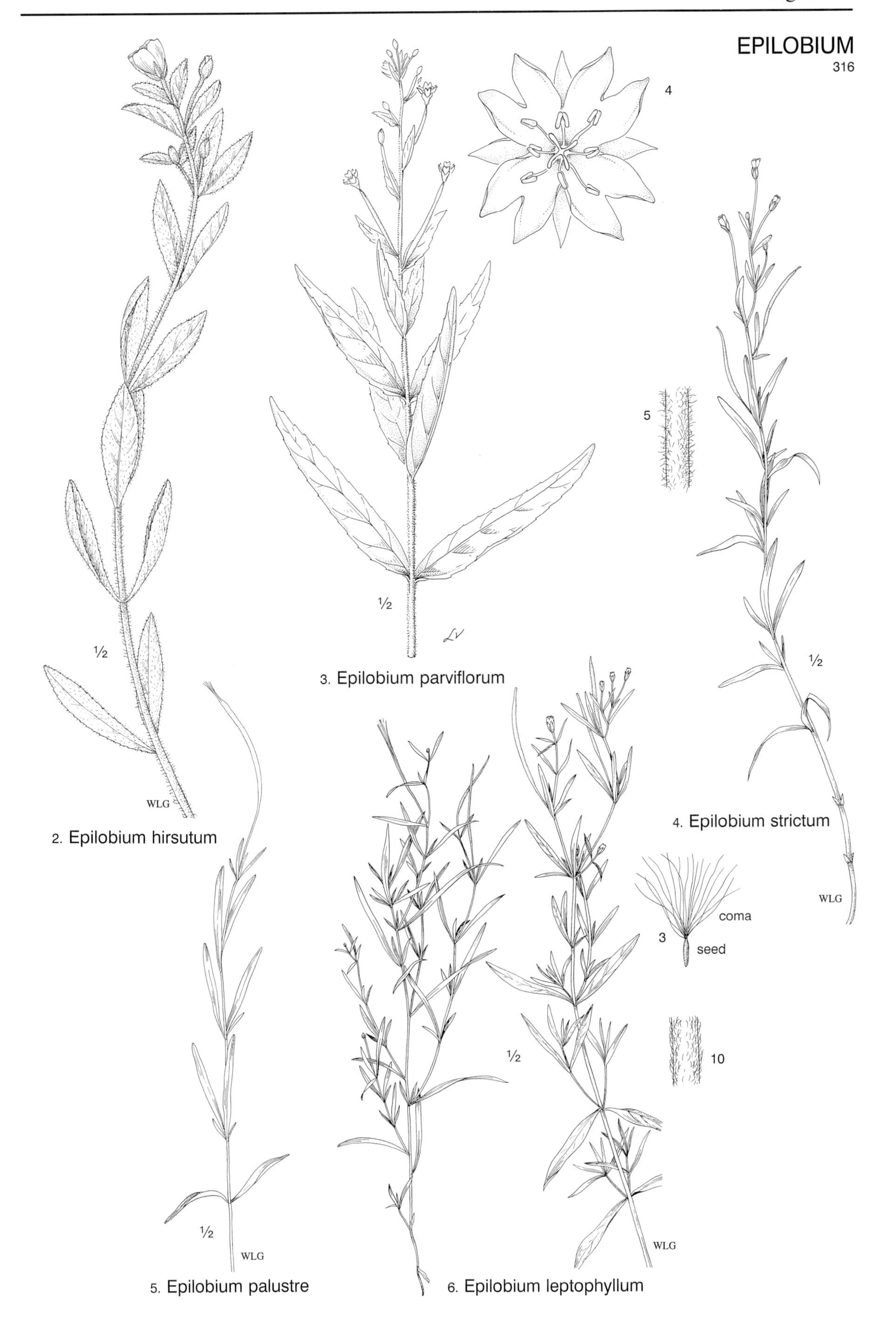

4

5

2. Epilobium hirsutum

3. Epilobium parviflorum

4. Epilobium strictum

coma

3 seed

10

5. Epilobium palustre

6. Epilobium leptophyllum

EPILOBIUM
316, 317

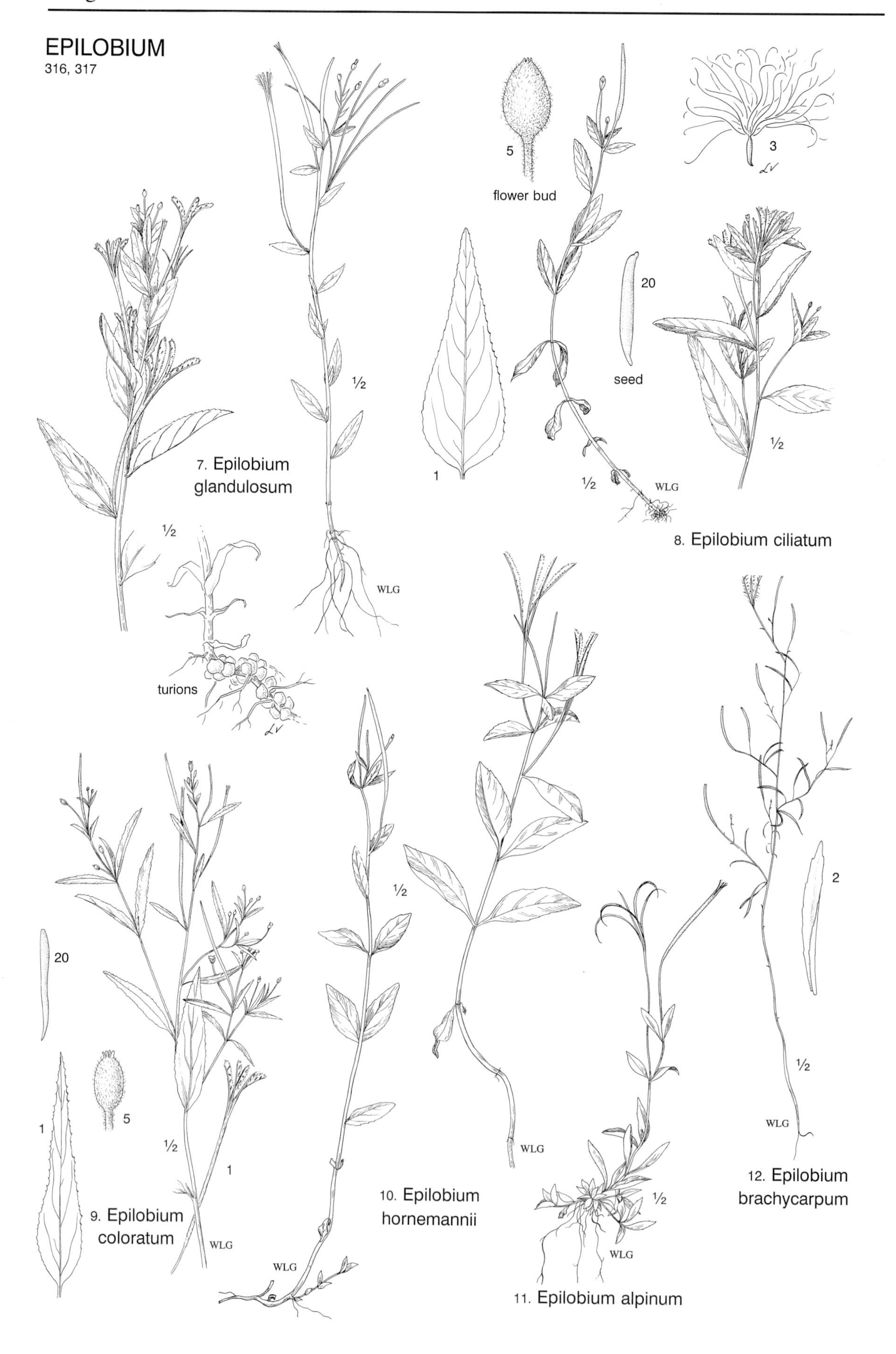

5
flower bud

3

1

20

seed

7. Epilobium
glandulosum

8. Epilobium ciliatum

WLG

WLG

turions

WLG

20

5

1

9. Epilobium
coloratum

WLG

10. Epilobium
hornemannii

WLG

2

1/2

WLG

12. Epilobium
brachycarpum

11. Epilobium alpinum

WLG

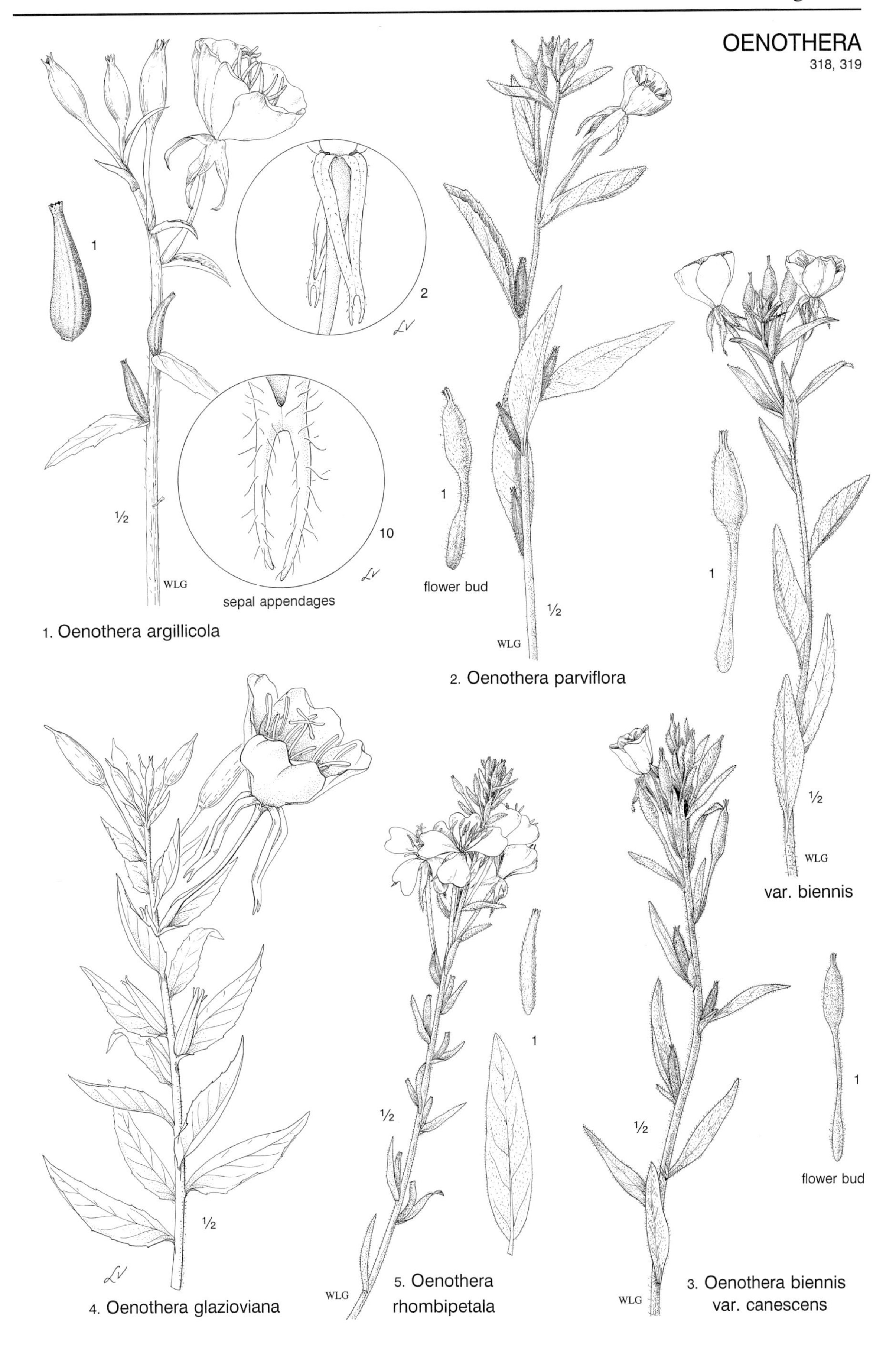

1 WLG

sepal appendages

1. Oenothera argillicola

2 LV

10 LV

½

flower bud

1

½

WLG

2. Oenothera parviflora

1

½

WLG

var. biennis

½

WLG

4. Oenothera glazioviana

½

1

WLG

5. Oenothera rhombipetala

½

1

flower bud

WLG

3. Oenothera biennis var. canescens

OENOTHERA
319

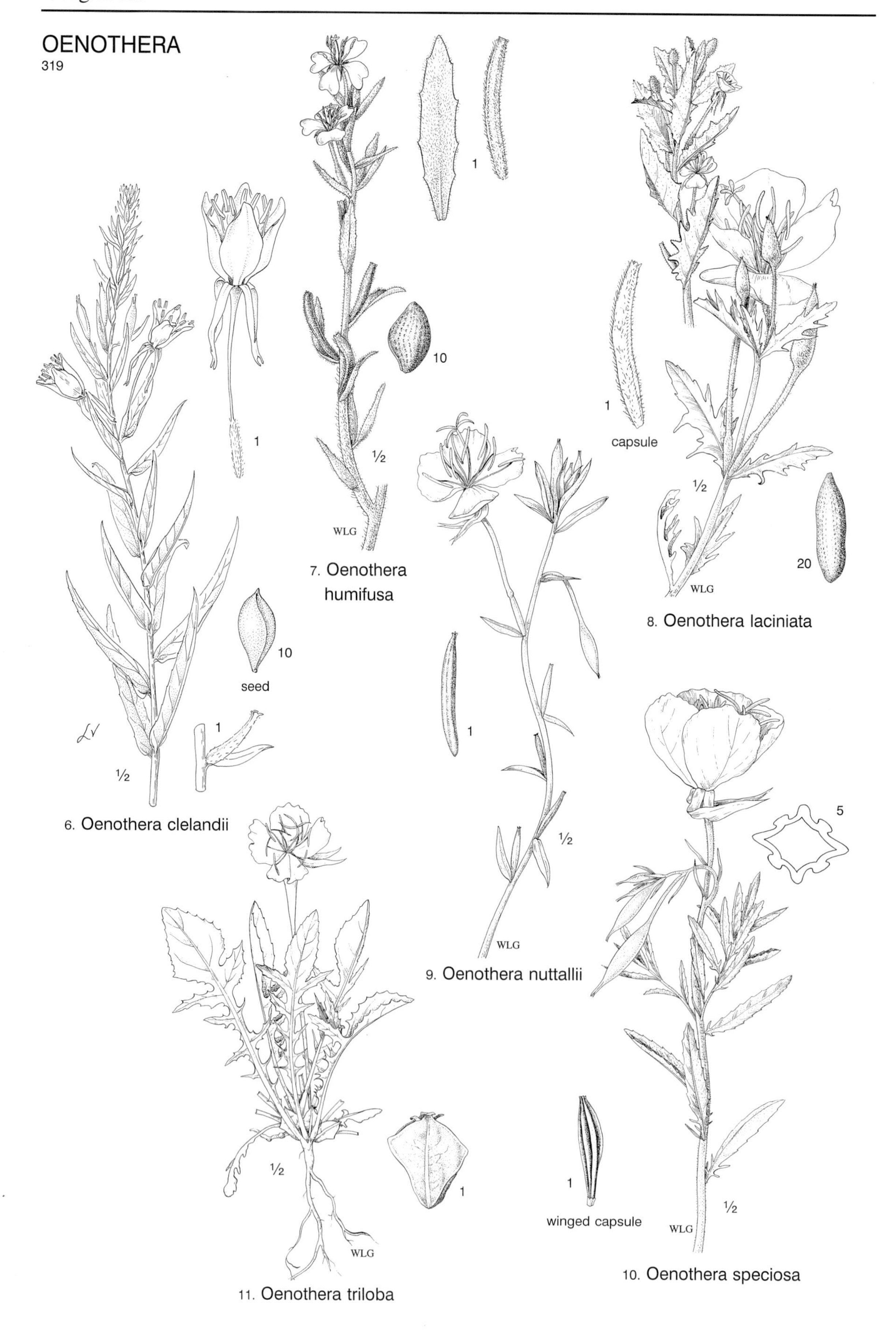

7. Oenothera
humifusa

10

seed

6. Oenothera clelandii

capsule

8. Oenothera laciniata

9. Oenothera nuttallii

winged capsule

10. Oenothera speciosa

11. Oenothera triloba

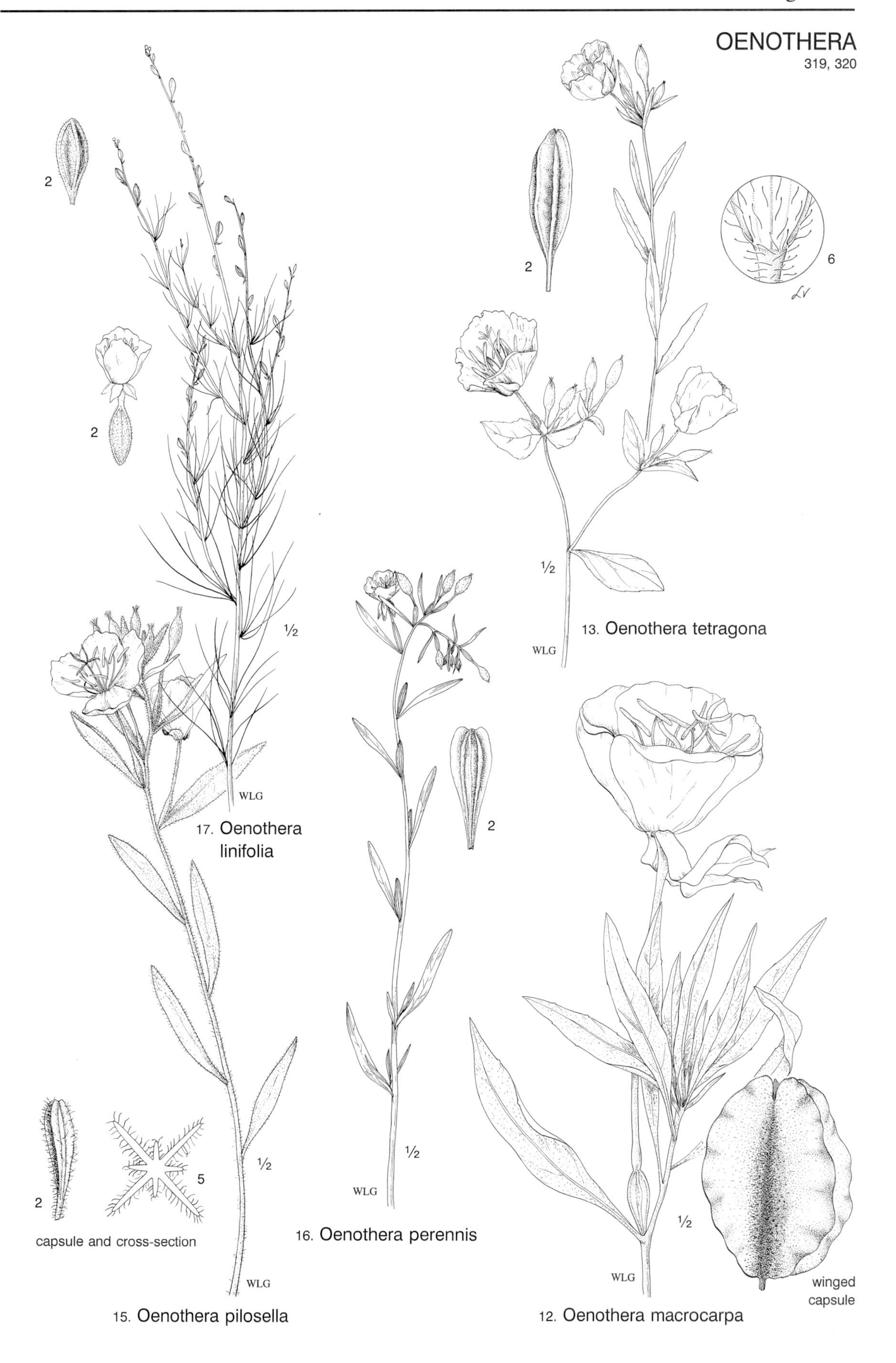

2

2

½

2

17. Oenothera
linifolia

WLG

2

5

capsule and cross-section

15. Oenothera pilosella

WLG

½

2

½

WLG

16. Oenothera perennis

2

½

WLG

13. Oenothera tetragona

6

½

WLG

12. Oenothera macrocarpa

winged
capsule

OENOTHERA
CALYLOPHUS
GAURA
319–321

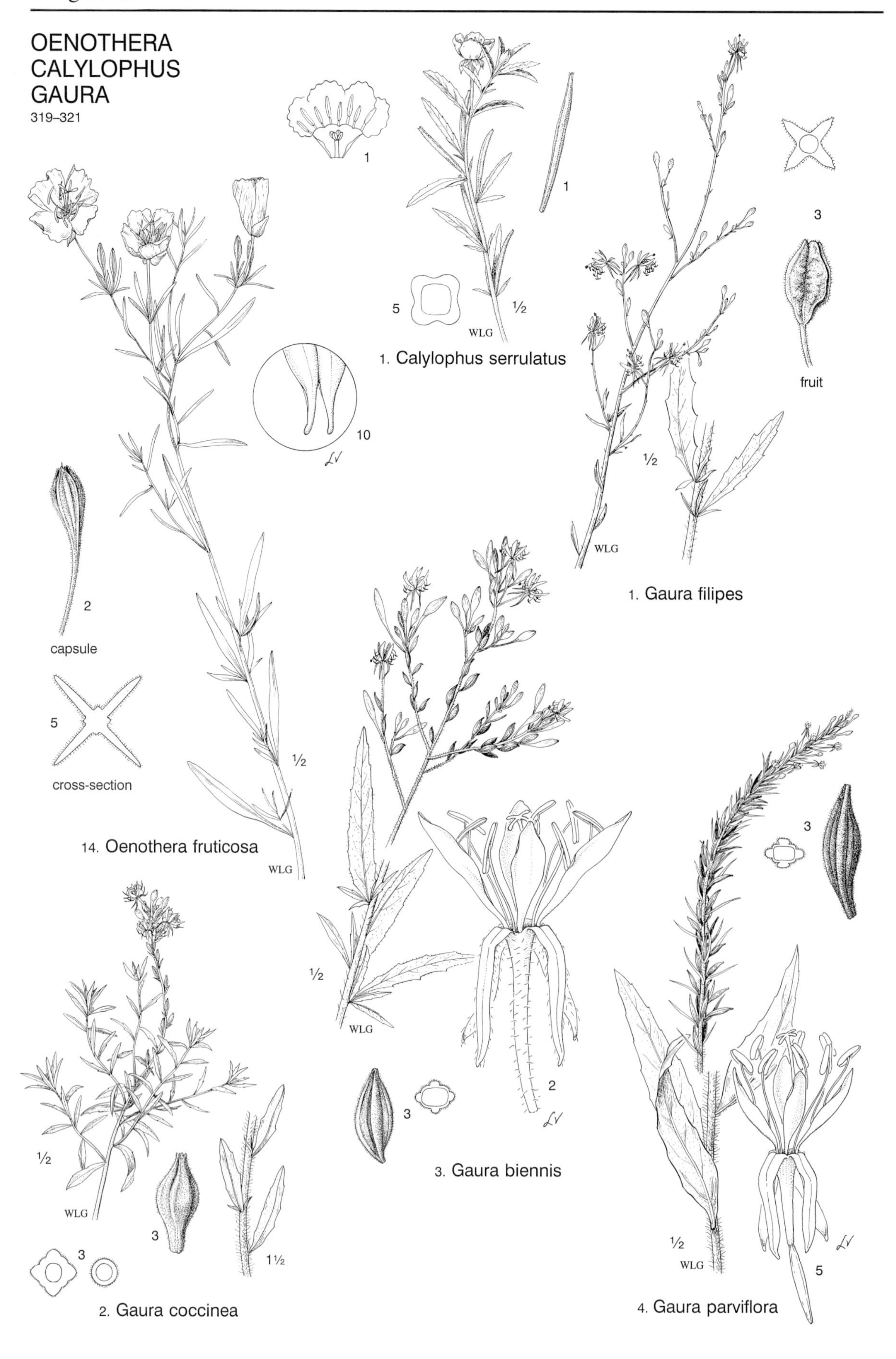

5 WLG   ½

1. Calylophus serrulatus

10

2
capsule

5
cross-section

14. Oenothera fruticosa

WLG

3

fruit

½

WLG

1. Gaura filipes

½

WLG

½

WLG

2

3

3. Gaura biennis

½

WLG

3

3

1½

2. Gaura coccinea

3

½

WLG

5

4. Gaura parviflora

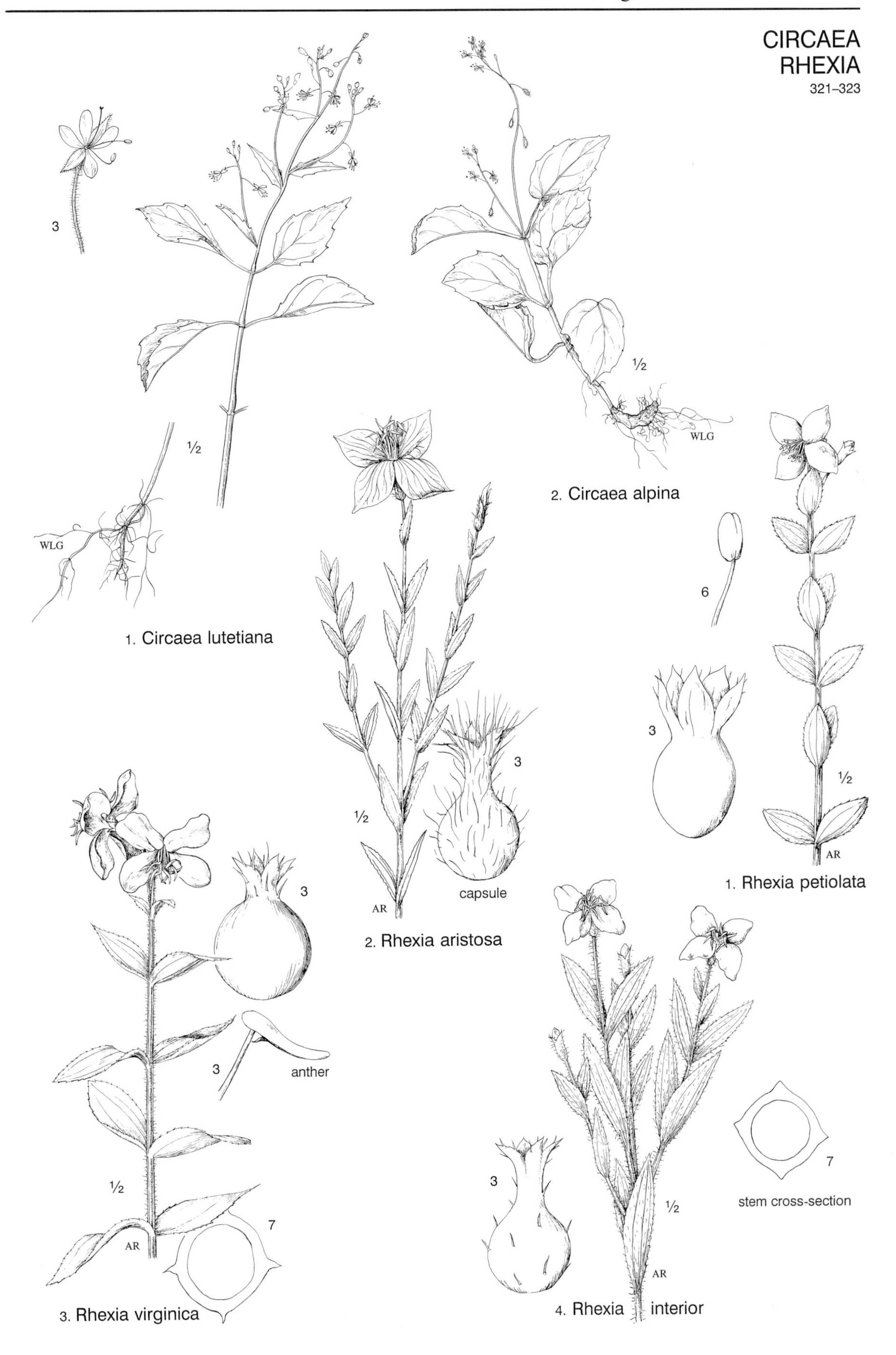

3

1/2

1/2

WLG

WLG

2. Circaea alpina

1. Circaea lutetiana

6

3

1/2

AR

1. Rhexia petiolata

1/2

3

AR

capsule

2. Rhexia aristosa

3

anther

3

7

1/2

3

AR

7

stem cross-section

1/2

AR

3. Rhexia virginica

4. Rhexia interior

RHEXIA
CORNUS
323, 324

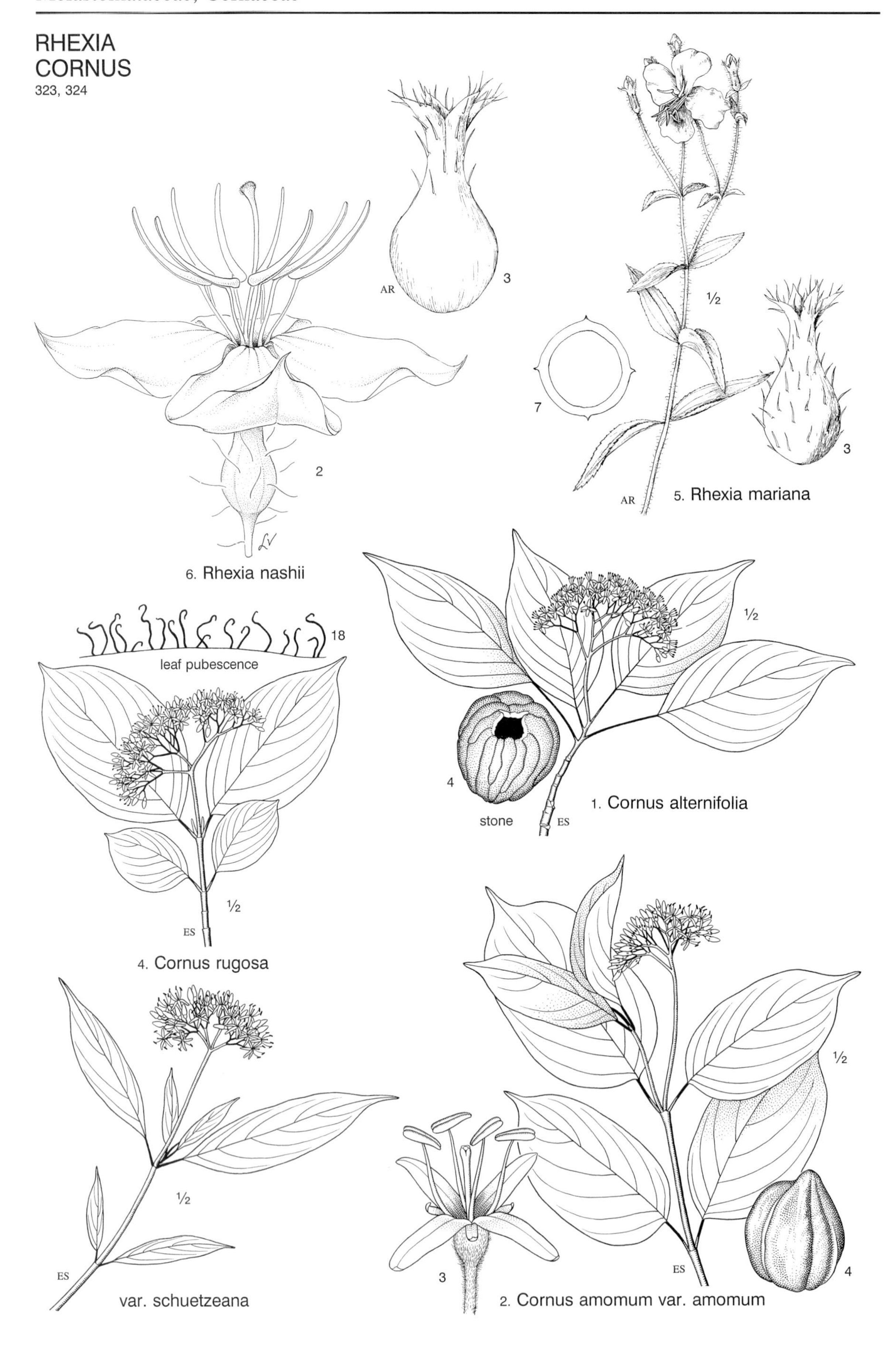

AR

3

½

7

2

AR 5. Rhexia mariana

3

6. Rhexia nashii

18
leaf pubescence

½

1. Cornus alternifolia

4

stone ES

½

ES

4. Cornus rugosa

½

ES

var. schuetzeana

3

ES

2. Cornus amomum var. amomum

4

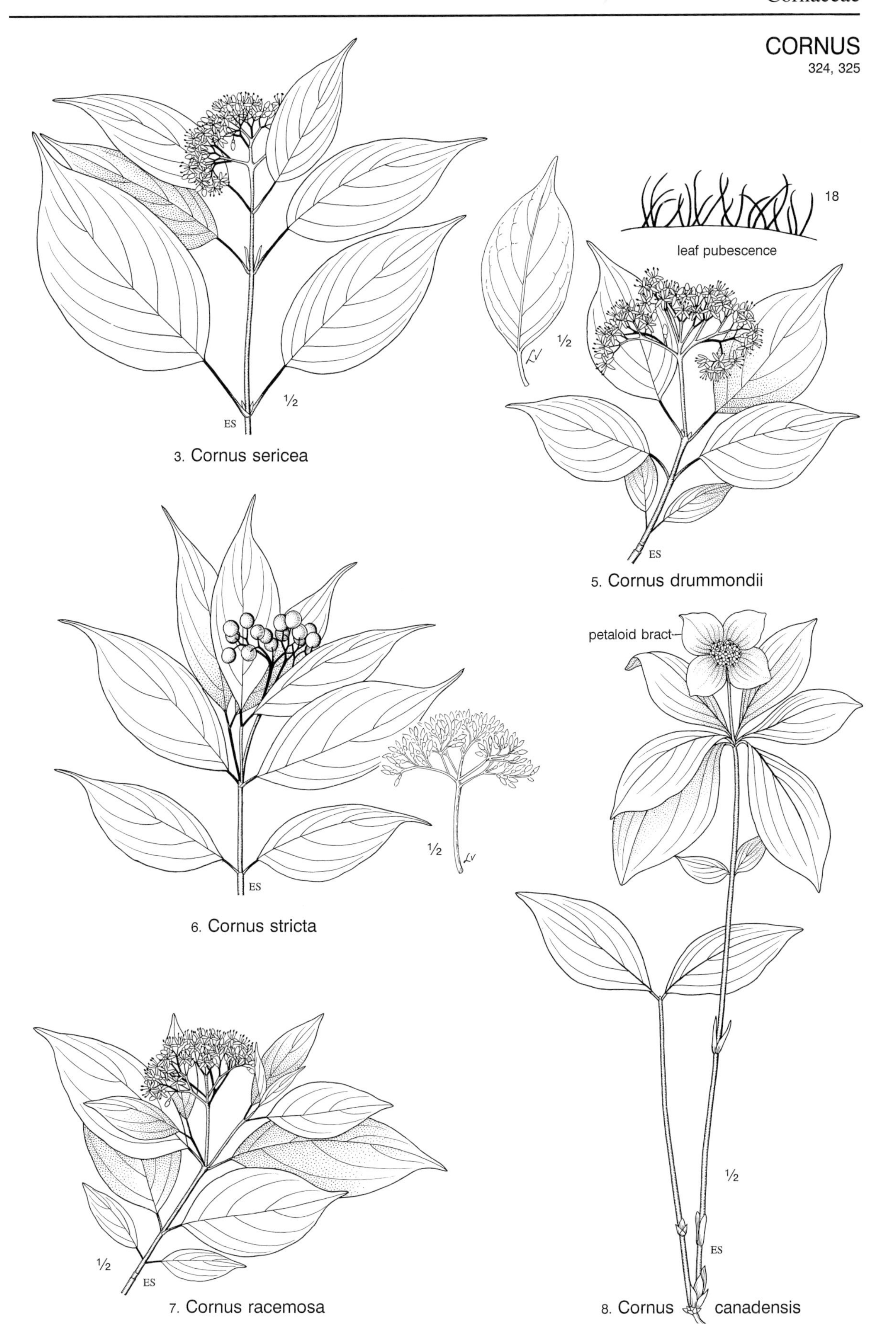

18
leaf pubescence

1/2

3. Cornus sericea

5. Cornus drummondii

1/2

6. Cornus stricta

petaloid bract—

1/2

7. Cornus racemosa

8. Cornus canadensis

CORNUS
NYSSA
COMANDRA
325, 326

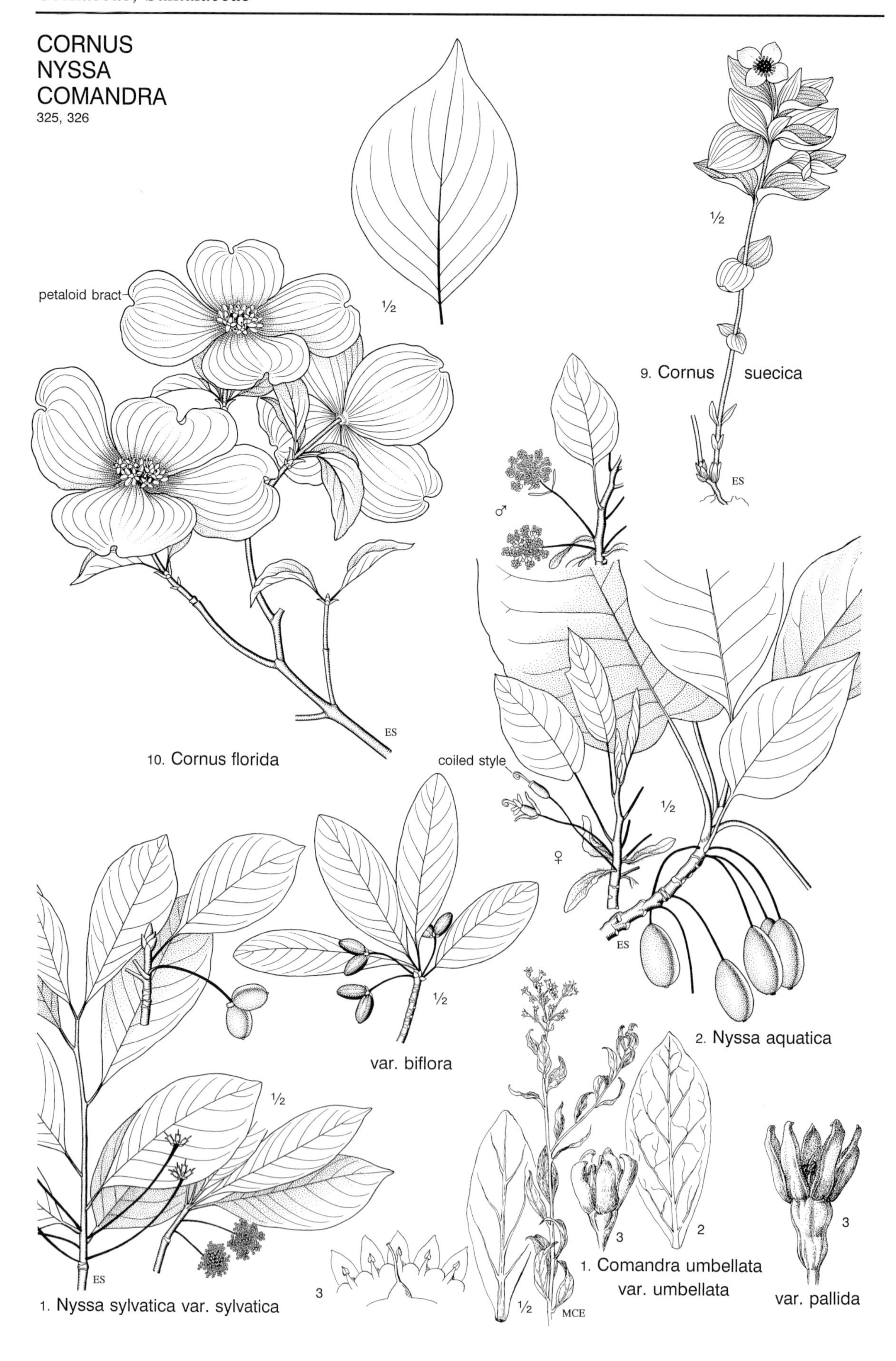

petaloid bract

½

9. Cornus suecica

ES

10. Cornus florida

coiled style

½

♂

♀

ES

2. Nyssa aquatica

var. biflora

½

½

1. Nyssa sylvatica var. sylvatica

ES

3

½

MCE

3

2

1. Comandra umbellata
   var. umbellata

3

var. pallida

GEOCAULON
PYRULARIA
NESTRONIA
BUCKLEYA
PHORADENDRON
ARCEUTHOBIUM
326–328

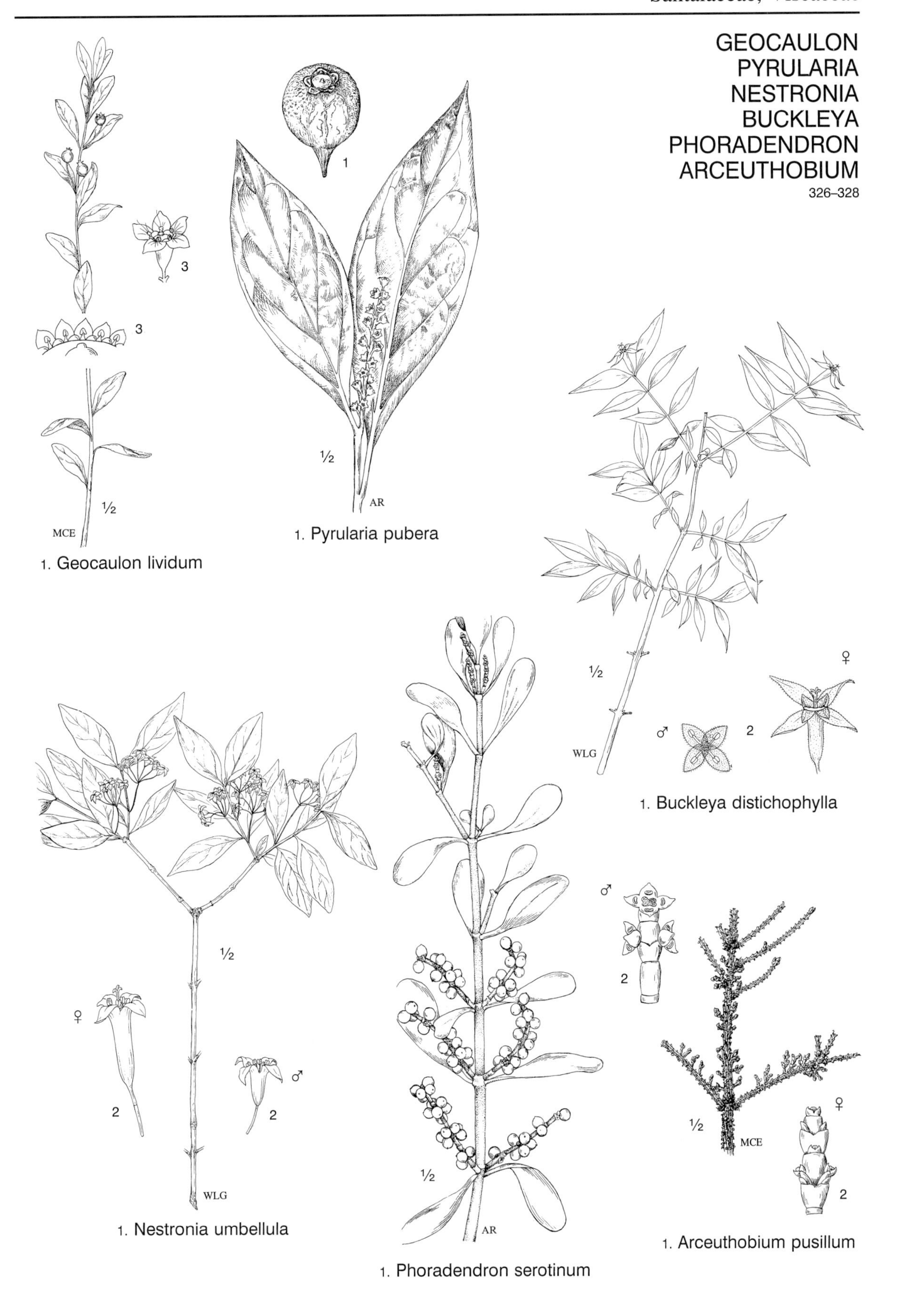

1. Pyrularia pubera

1. Geocaulon lividum

1. Buckleya distichophylla

1. Nestronia umbellula

1. Phoradendron serotinum

1. Arceuthobium pusillum

CELASTRUS
EUONYMUS
328, 329

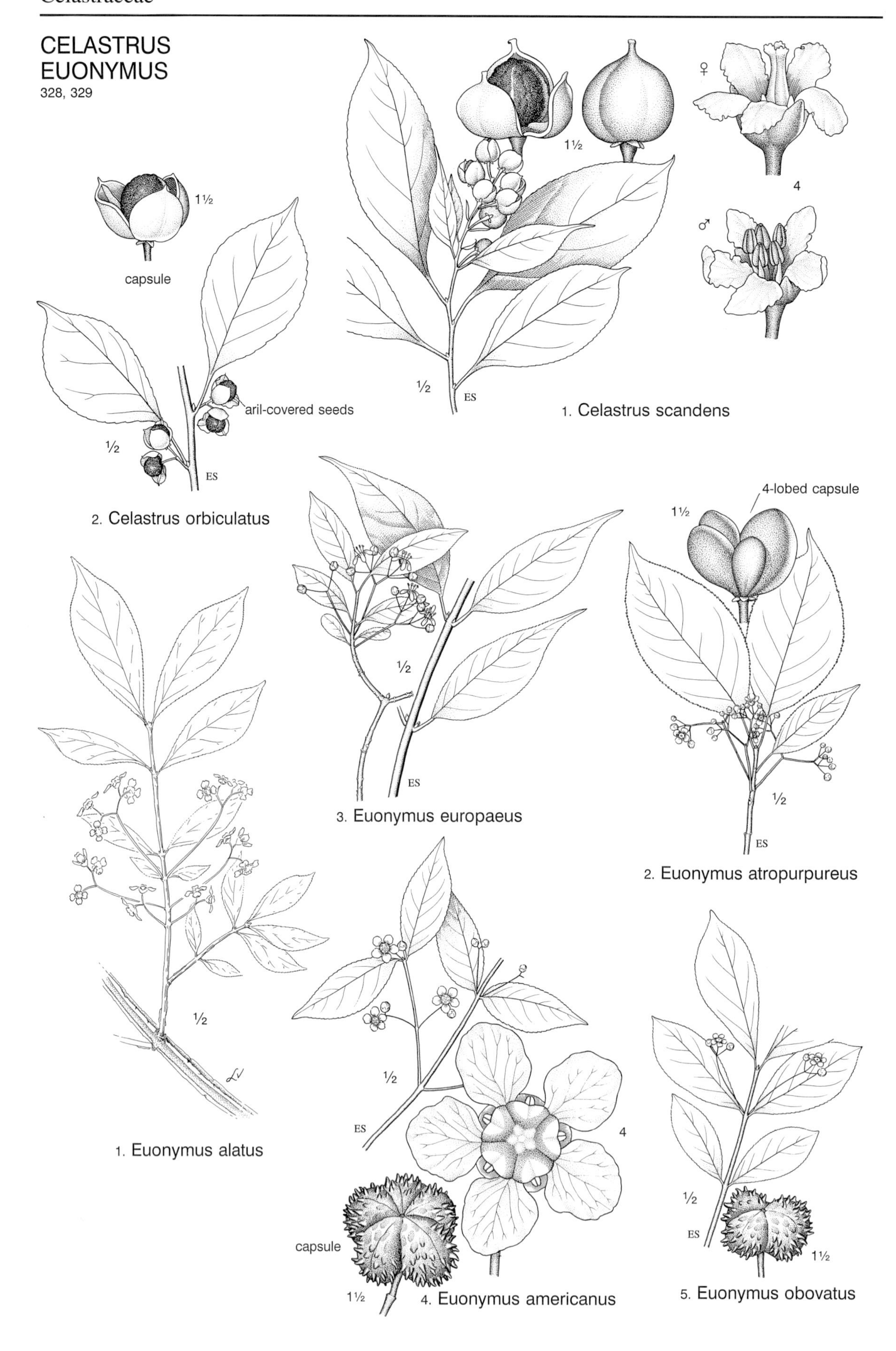

capsule 1½

aril-covered seeds

½ ES

2. Celastrus orbiculatus

1½

½ ES

♀ 4

♂

1. Celastrus scandens

½

3. Euonymus europaeus

4-lobed capsule

1½

½ ES

2. Euonymus atropurpureus

1. Euonymus alatus

½

ES

½

ES

4

capsule

1½

4. Euonymus americanus

½

ES

1½

5. Euonymus obovatus

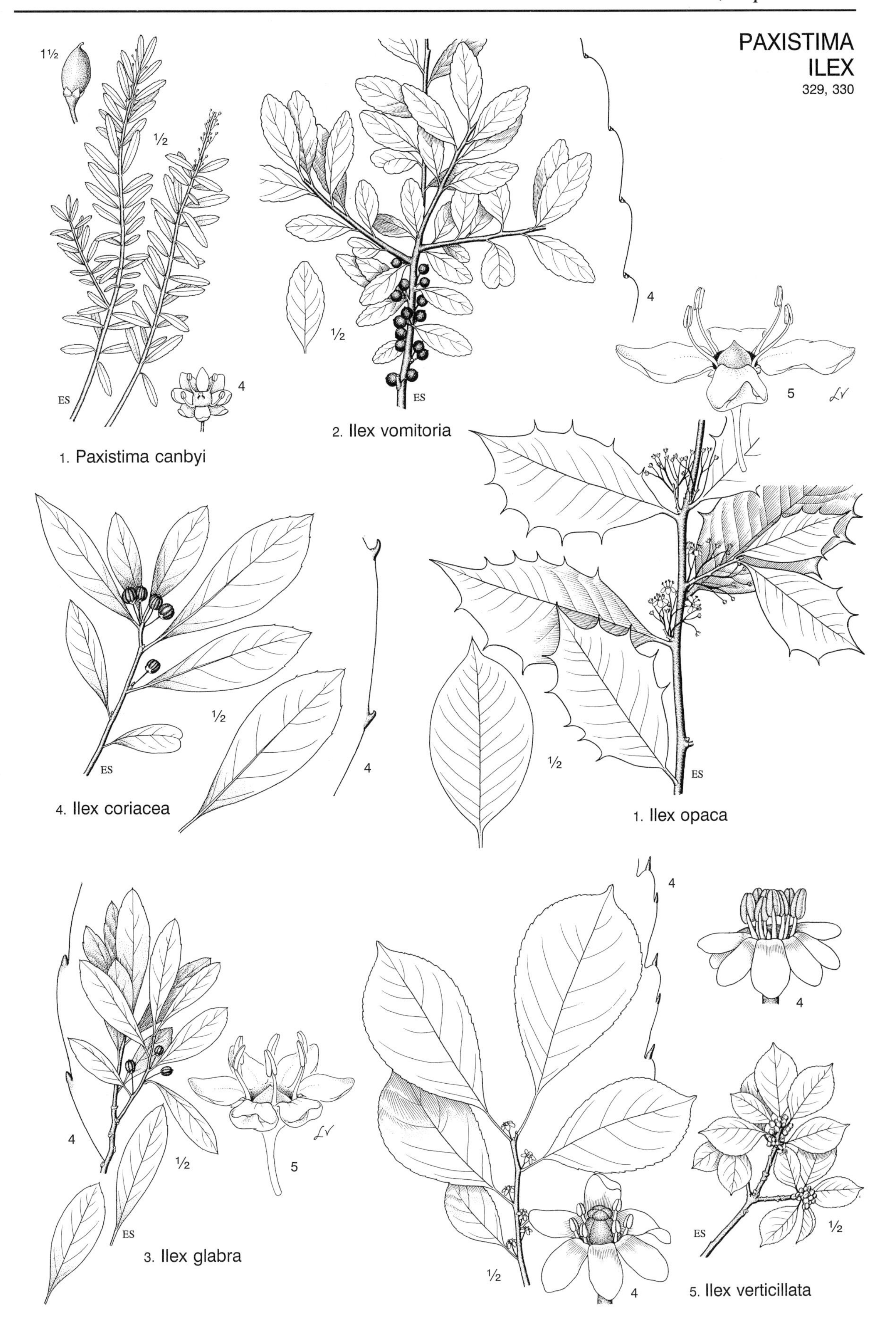

1½

½

4

ES

1. Paxistima canbyi

½

½

ES

2. Ilex vomitoria

4

5

½

ES

4. Ilex coriacea

4

½

ES

1. Ilex opaca

4

½

5

3. Ilex glabra

4

½

4

4

½

4

5. Ilex verticillata

½

ES

ILEX
NEMOPANTHUS
330, 331

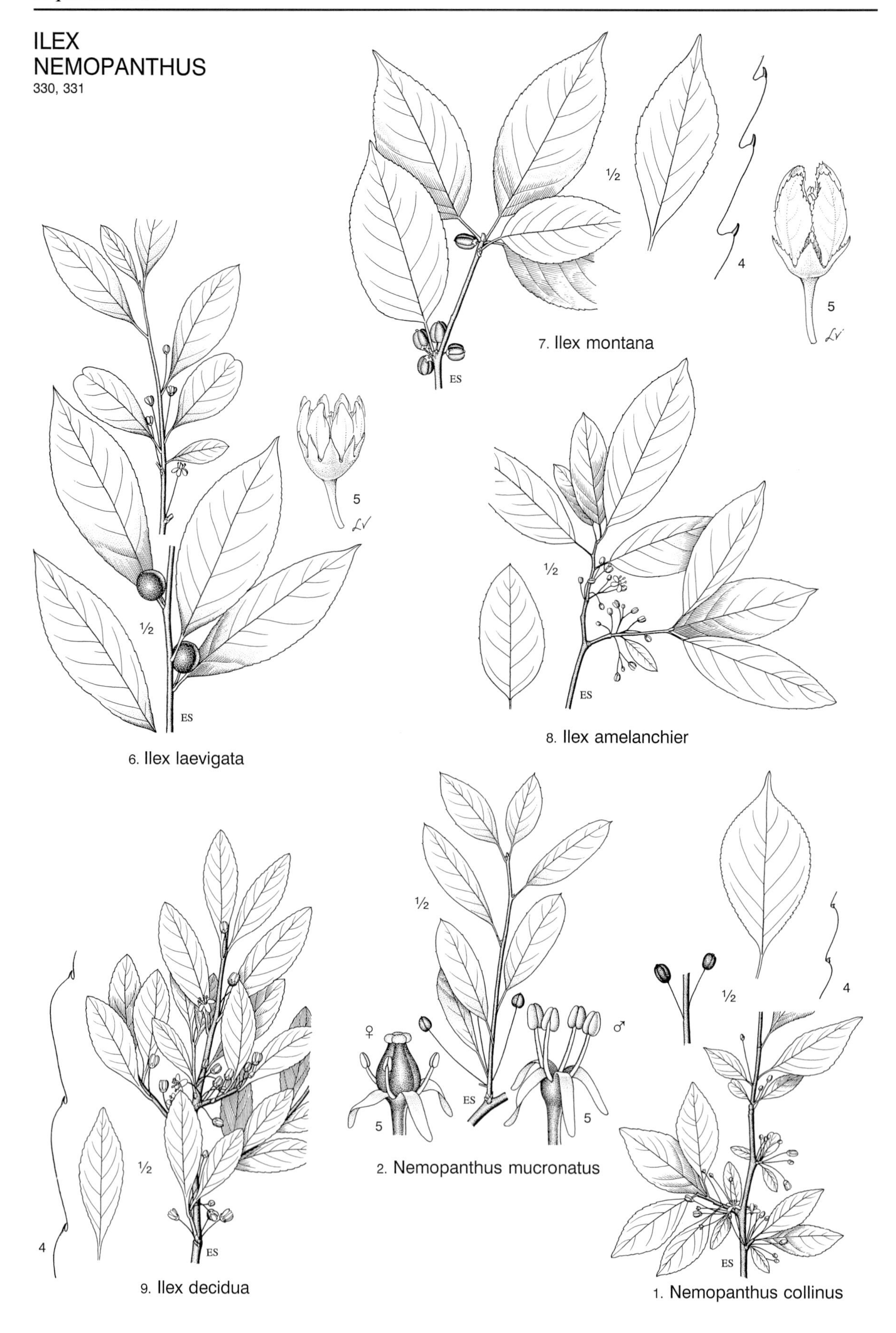

7. Ilex montana

6. Ilex laevigata

8. Ilex amelanchier

9. Ilex decidua

2. Nemopanthus mucronatus

1. Nemopanthus collinus

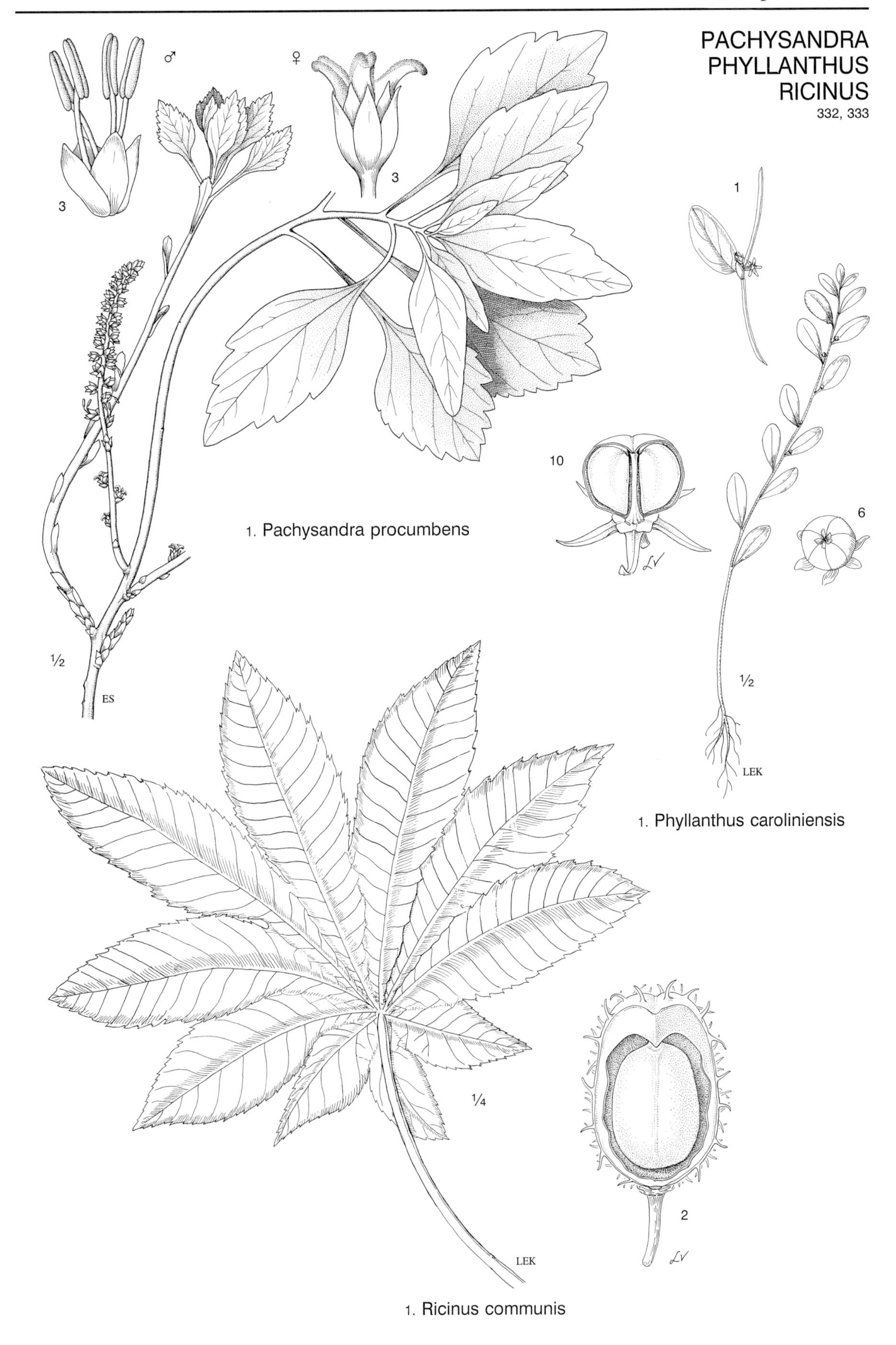

1. Pachysandra procumbens

1. Phyllanthus caroliniensis

1. Ricinus communis

CNIDOSCOLUS
CROTON
333

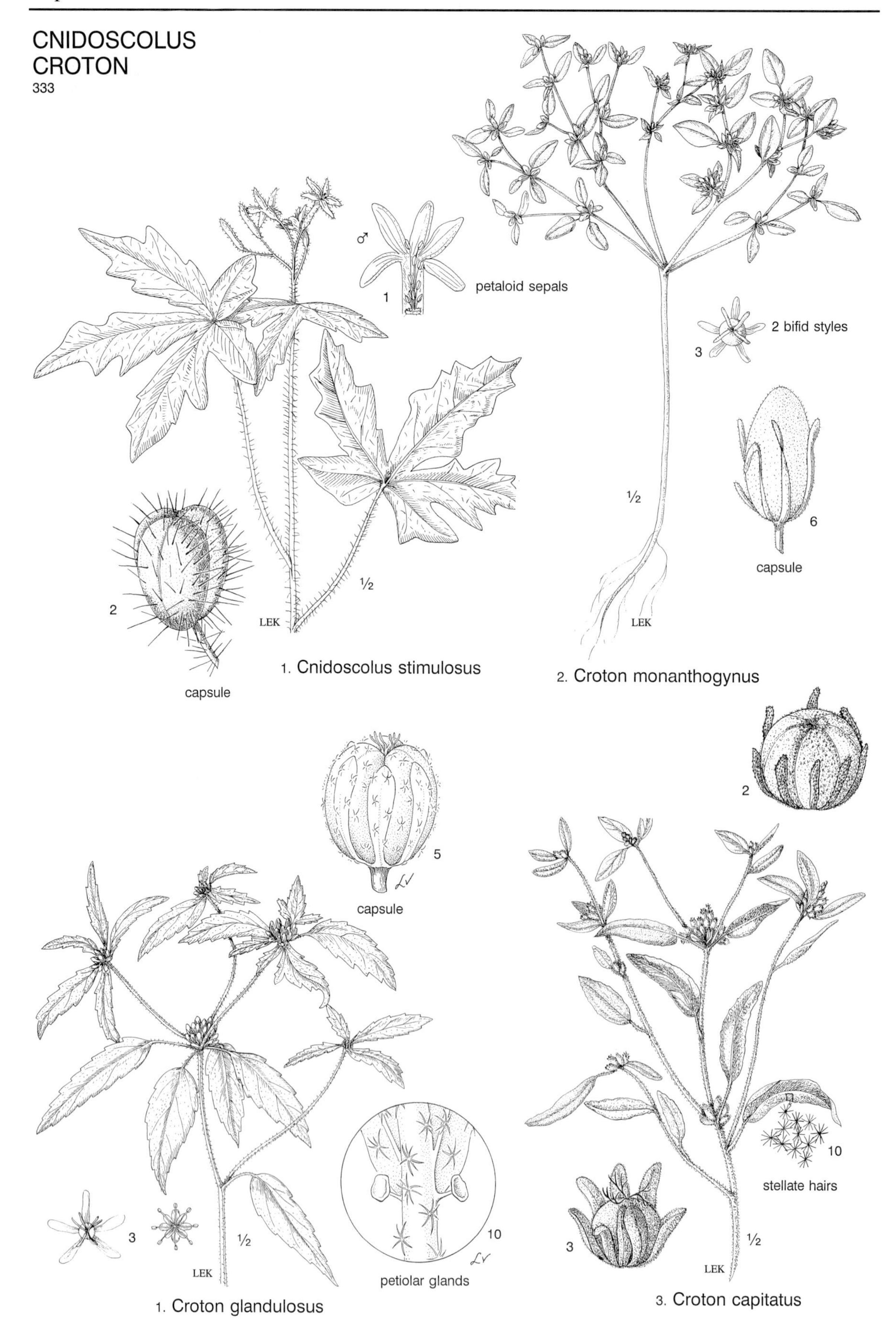

1 petaloid sepals

♂

2 bifid styles

3

6 capsule

½

1. Cnidoscolus stimulosus

2 capsule

capsule

2. Croton monanthogynus

5 capsule

2

1. Croton glandulosus

3

10 petiolar glands

10 stellate hairs

3 ½

3. Croton capitatus

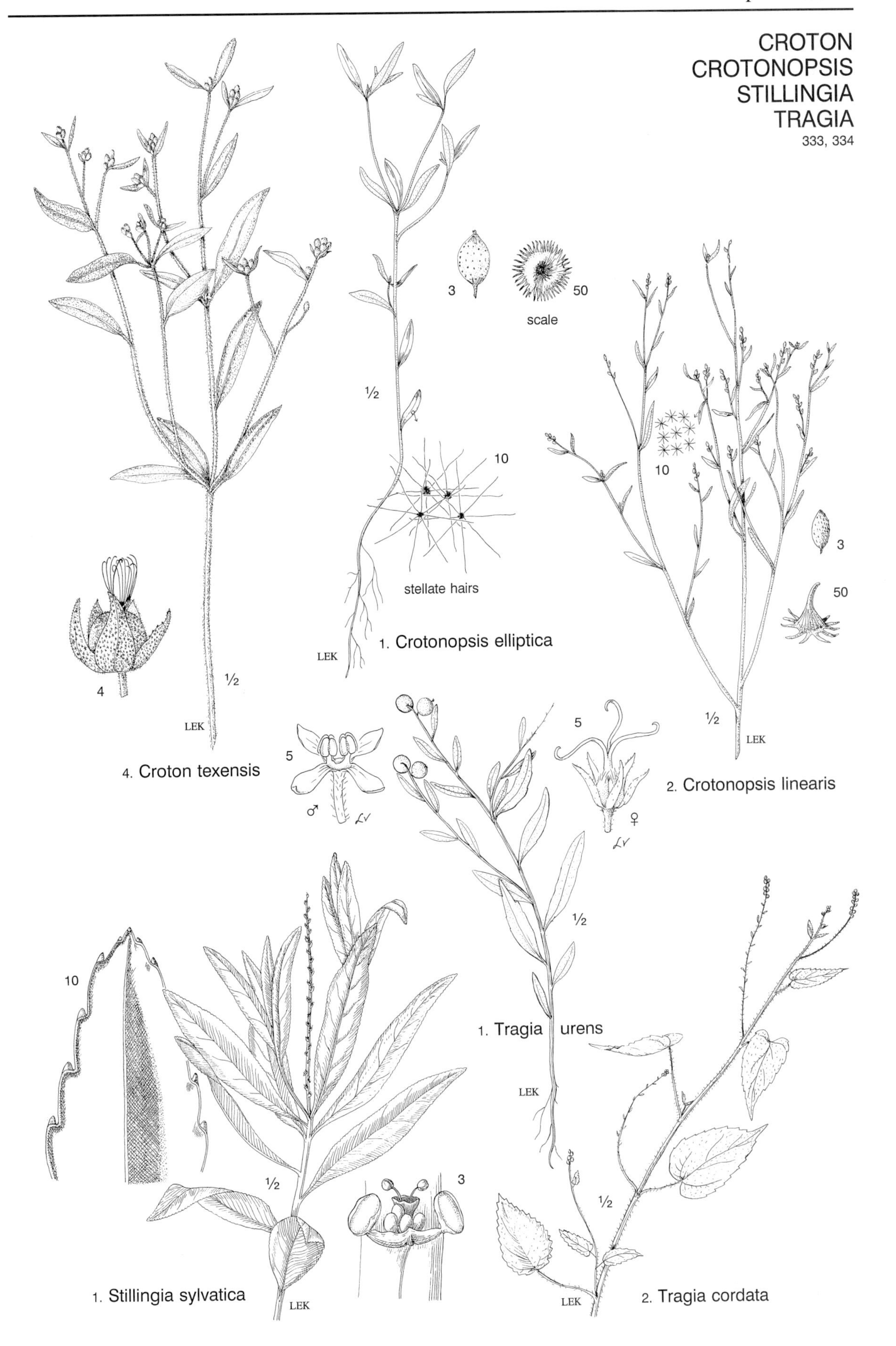

3

50

scale

1/2

10

stellate hairs

1. Crotonopsis elliptica

LEK

10

3

50

1/2

LEK

2. Crotonopsis linearis

4

LEK

1/2

LEK

4. Croton texensis

5

♂

5

♀

1/2

1. Tragia urens

LEK

10

1/2

3

1. Stillingia sylvatica

LEK

1/2

LEK

2. Tragia cordata

ACALYPHA
335

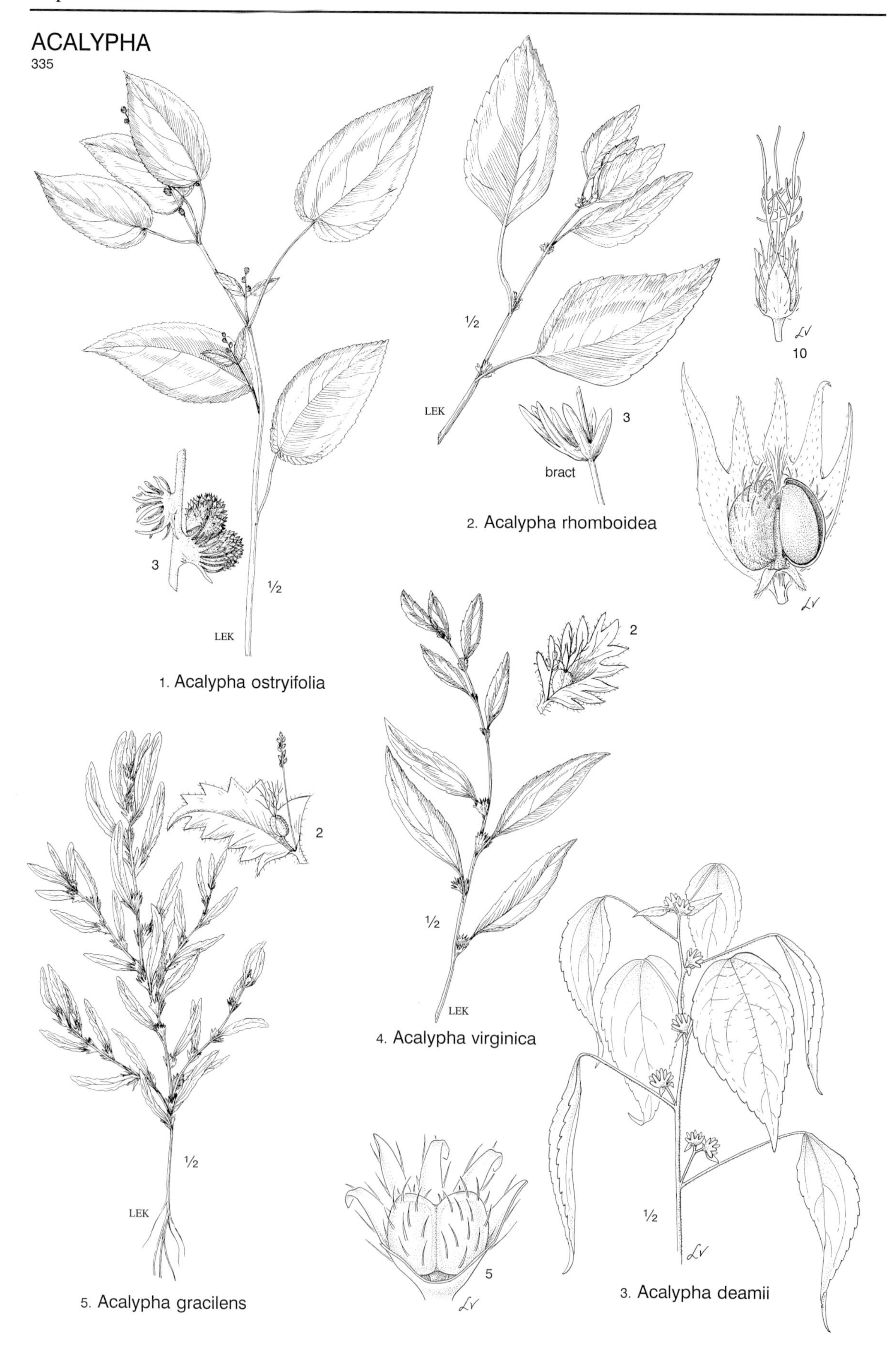

1. Acalypha ostryifolia

2. Acalypha rhomboidea

bract

10

4. Acalypha virginica

5. Acalypha gracilens

5

3. Acalypha deamii

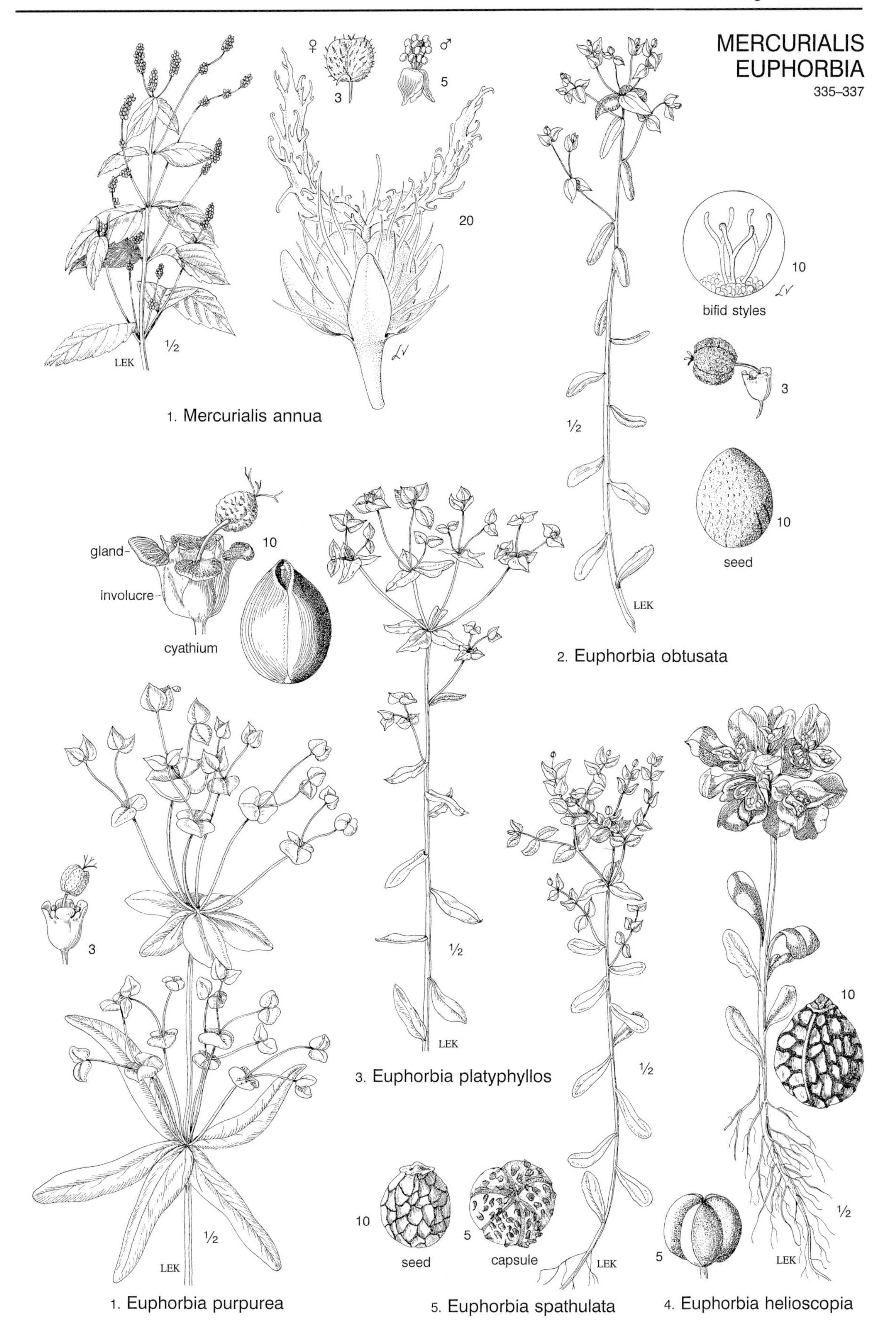

1. Mercurialis annua

bifid styles

seed

gland

involucre

cyathium

2. Euphorbia obtusata

1. Euphorbia purpurea

3. Euphorbia platyphyllos

seed

capsule

5. Euphorbia spathulata

4. Euphorbia helioscopia

EUPHORBIA
338

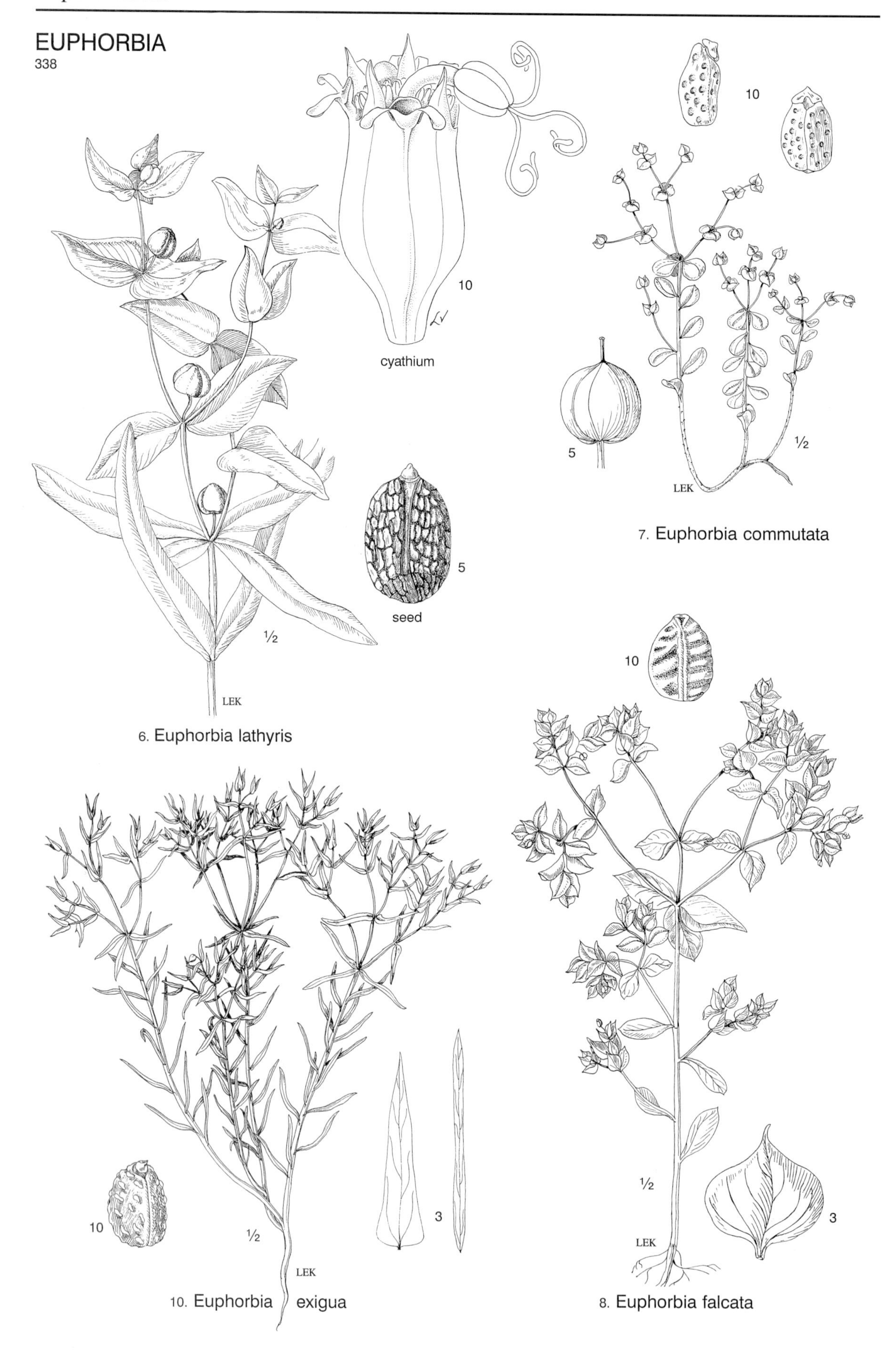

cyathium

10

5

seed

½

LEK

6. Euphorbia lathyris

10

5

½

LEK

7. Euphorbia commutata

10

10

½

LEK

10. Euphorbia exigua

3

10

½

LEK

3

8. Euphorbia falcata

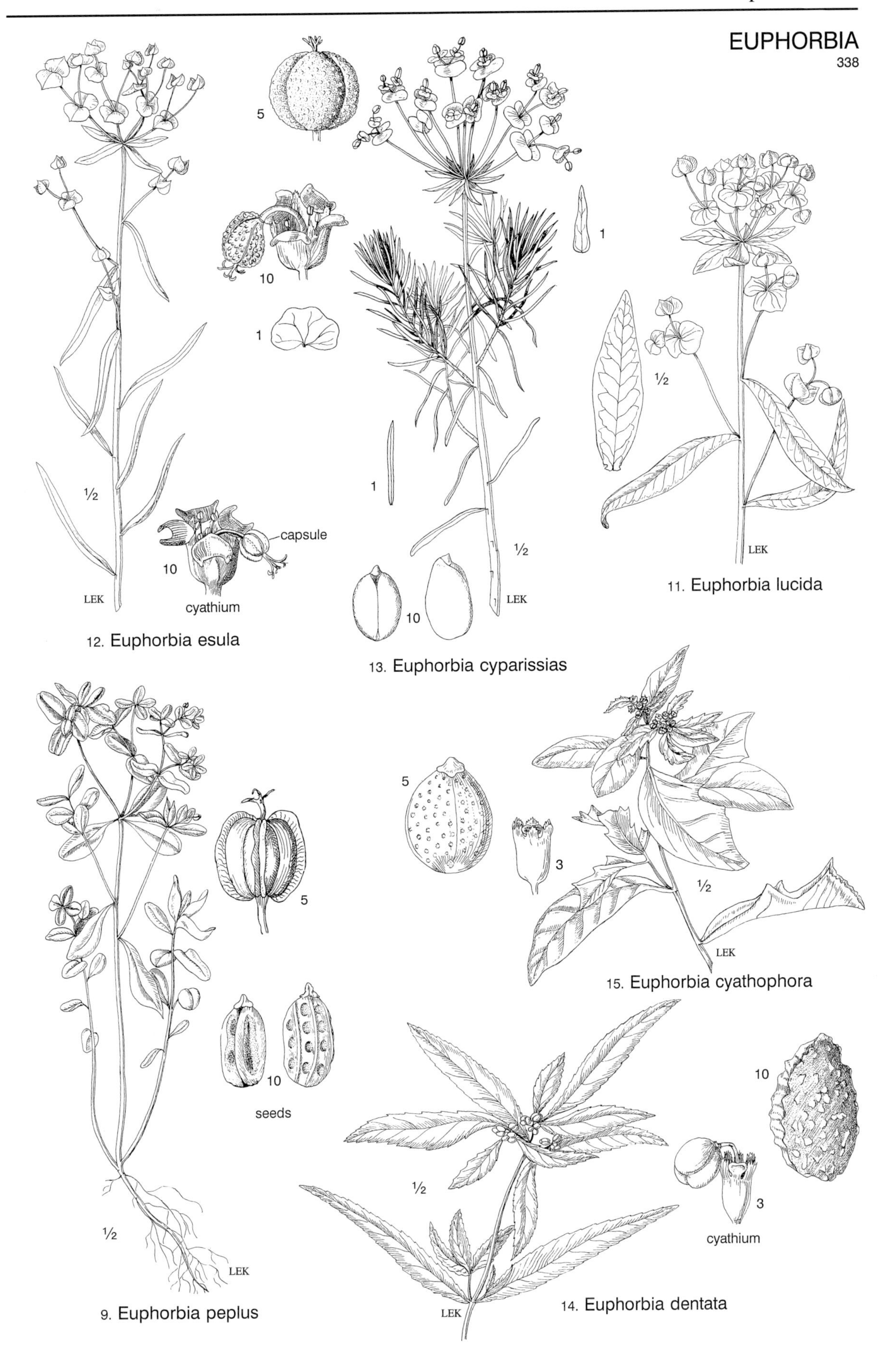

5

10

1

1

10 capsule

cyathium
LEK

12. Euphorbia esula

1

1

10

1/2
LEK

13. Euphorbia cyparissias

1

1/2
LEK

11. Euphorbia lucida

5

3

1/2
LEK

15. Euphorbia cyathophora

5

10

seeds

1/2

LEK

9. Euphorbia peplus

1/2

LEK

10

3

cyathium

14. Euphorbia dentata

EUPHORBIA
339

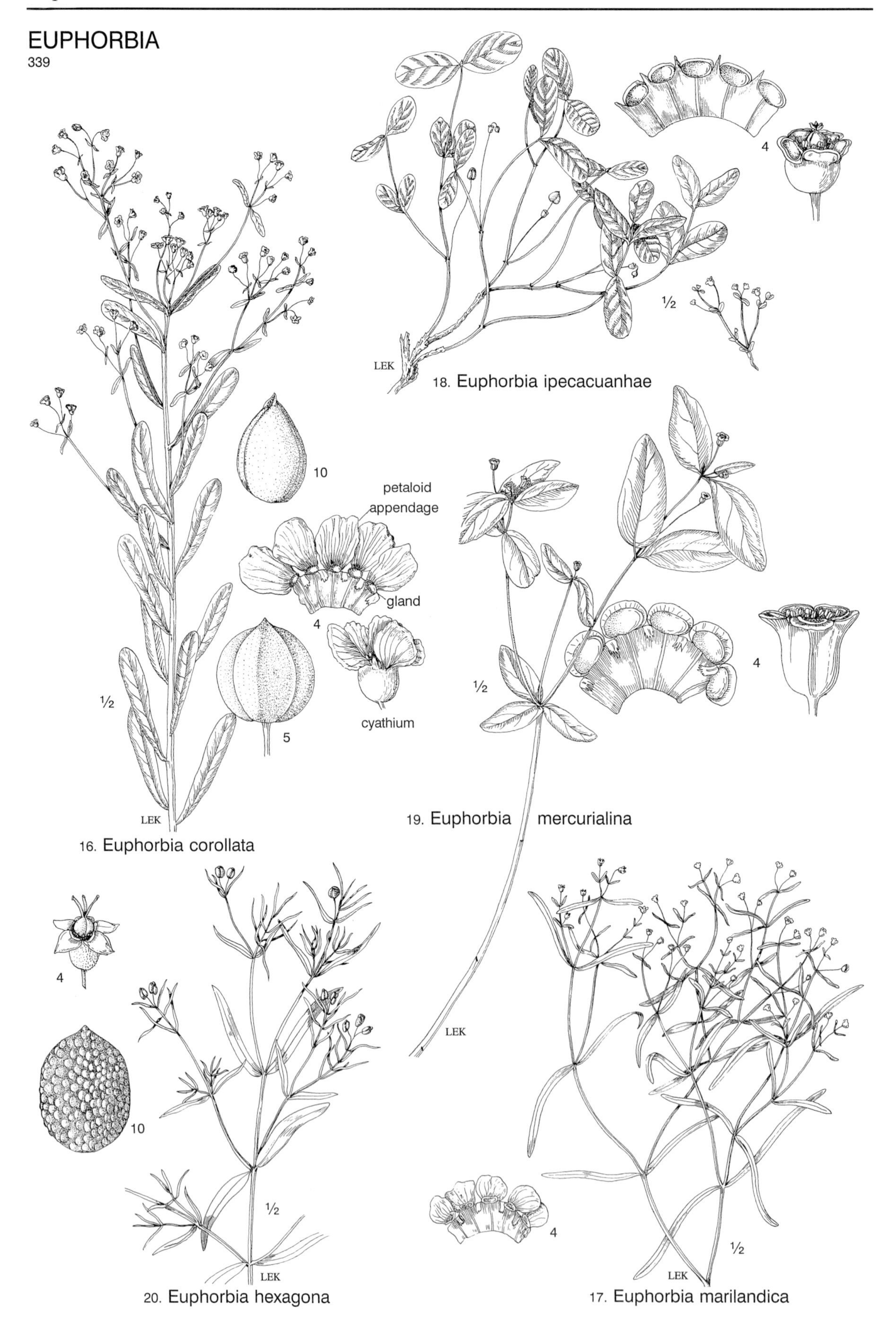

18. Euphorbia ipecacuanhae

petaloid appendage

gland

cyathium

16. Euphorbia corollata

19. Euphorbia   mercurialina

20. Euphorbia hexagona

17. Euphorbia marilandica

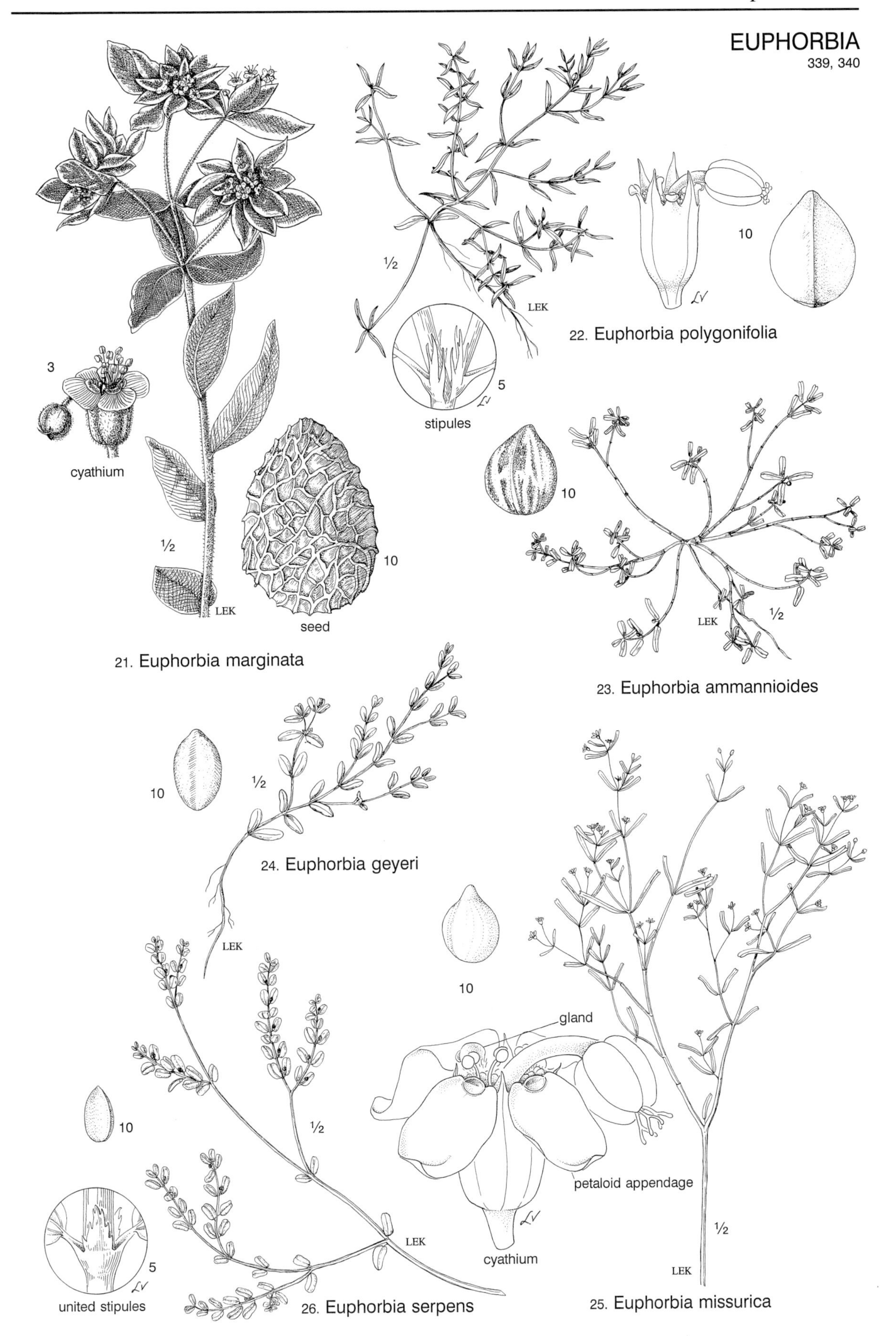

3

cyathium

½

¼

LEK

21. Euphorbia marginata

seed

10

½

LEK

stipules

5

22. Euphorbia polygonifolia

10

10

23. Euphorbia ammannioides

½

LEK

10

½

LEK

24. Euphorbia geyeri

10

½

10

gland

petaloid appendage

cyathium

25. Euphorbia missurica

½

LEK

LEK

5

united stipules

26. Euphorbia serpens

EUPHORBIA
340

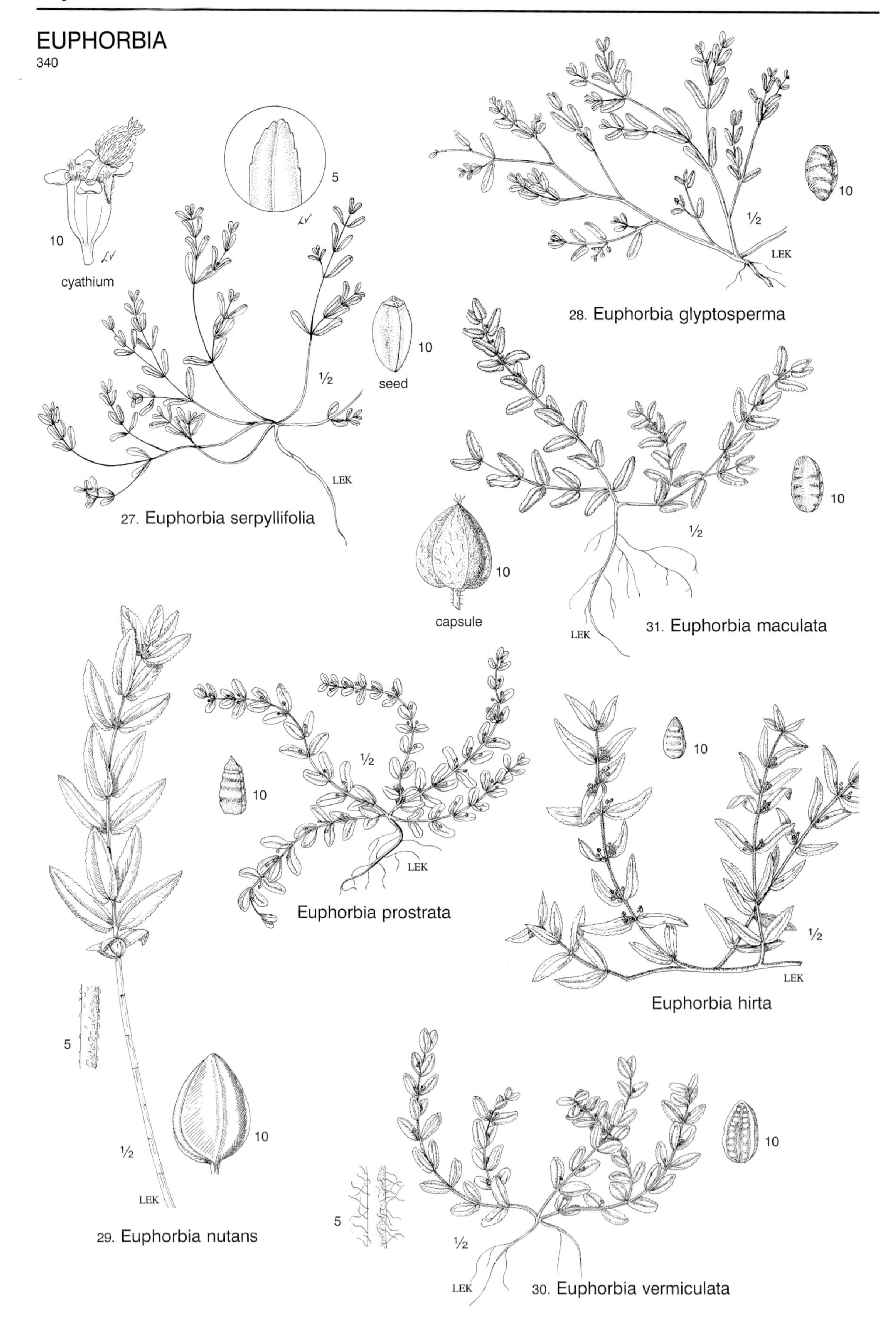

10

cyathium

5

27. Euphorbia serpyllifolia

10
seed

28. Euphorbia glyptosperma

10

10
capsule

31. Euphorbia maculata

10

Euphorbia prostrata

10

Euphorbia hirta

5

10

29. Euphorbia nutans

5

10

30. Euphorbia vermiculata

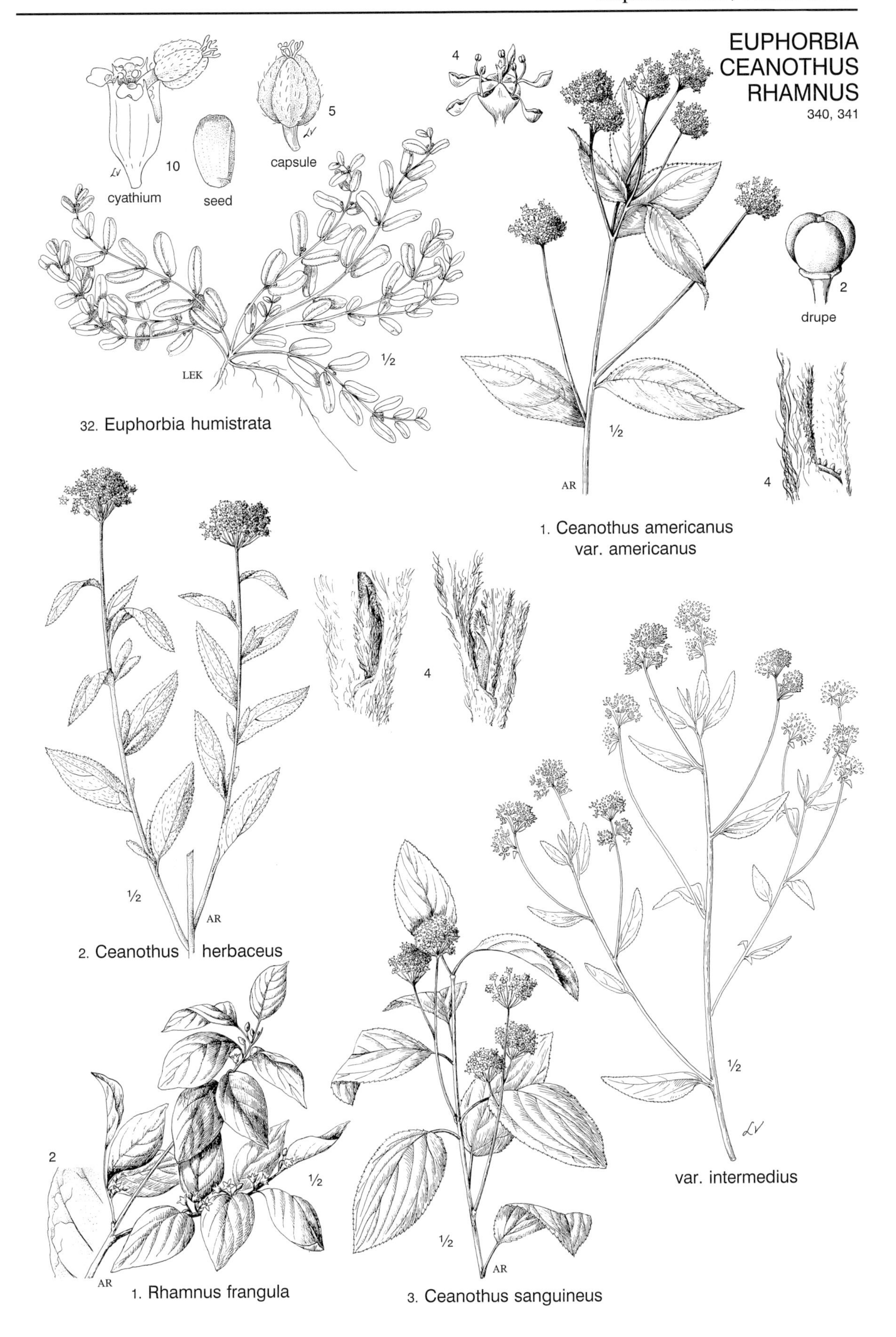

10 cyathium

seed

5 capsule

32. Euphorbia humistrata

4

2 drupe

1. Ceanothus americanus
var. americanus

4

2. Ceanothus herbaceus

var. intermedius

1. Rhamnus frangula

3. Ceanothus sanguineus

RHAMNUS
BERCHEMIA
AMPELOPSIS
342, 343

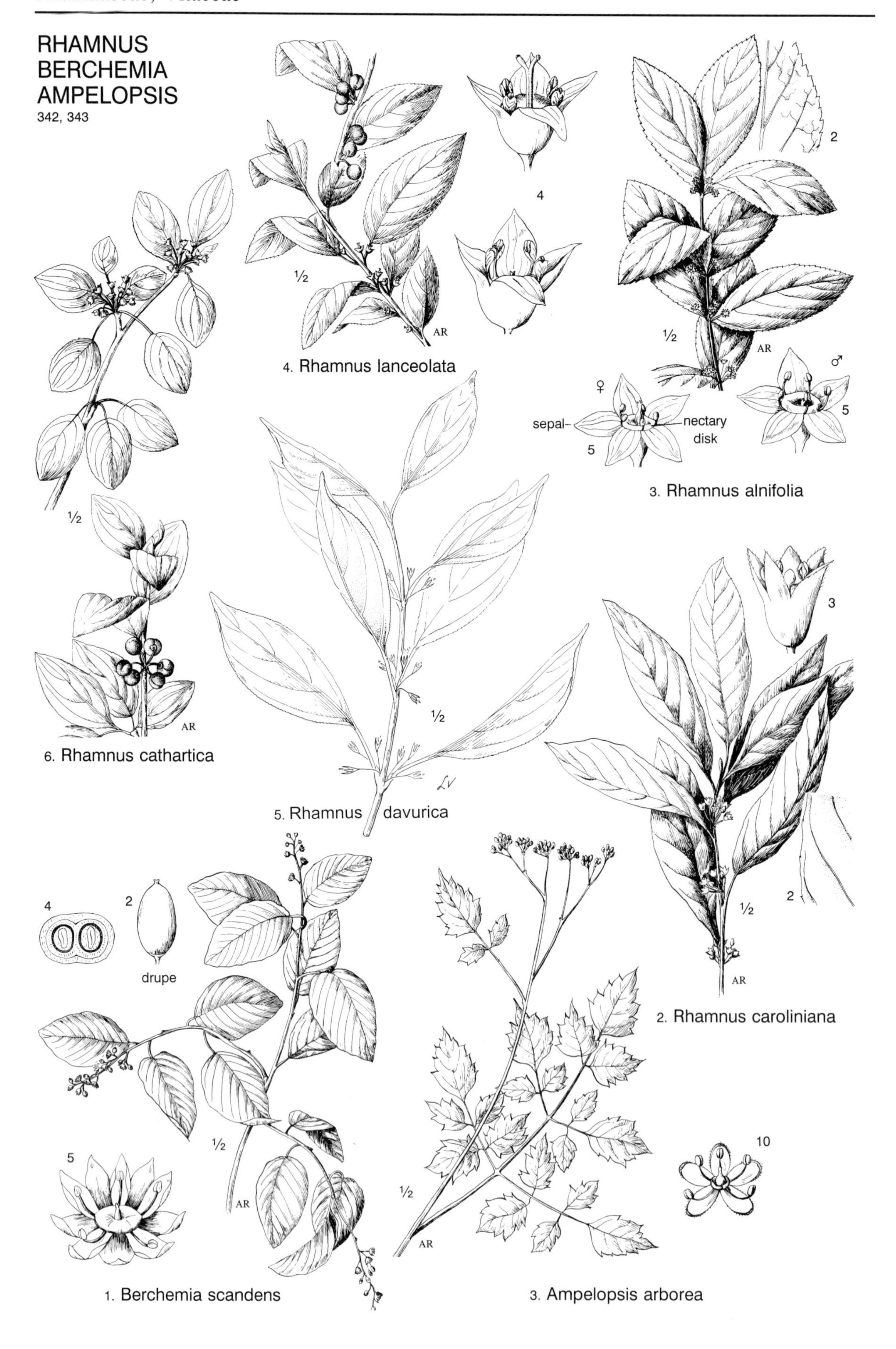

4. Rhamnus lanceolata

sepal—   —nectary
disk

3. Rhamnus alnifolia

6. Rhamnus cathartica

5. Rhamnus davurica

2. Rhamnus caroliniana

drupe

1. Berchemia scandens

3. Ampelopsis arborea

ILLUSTRATED COMPANION TO

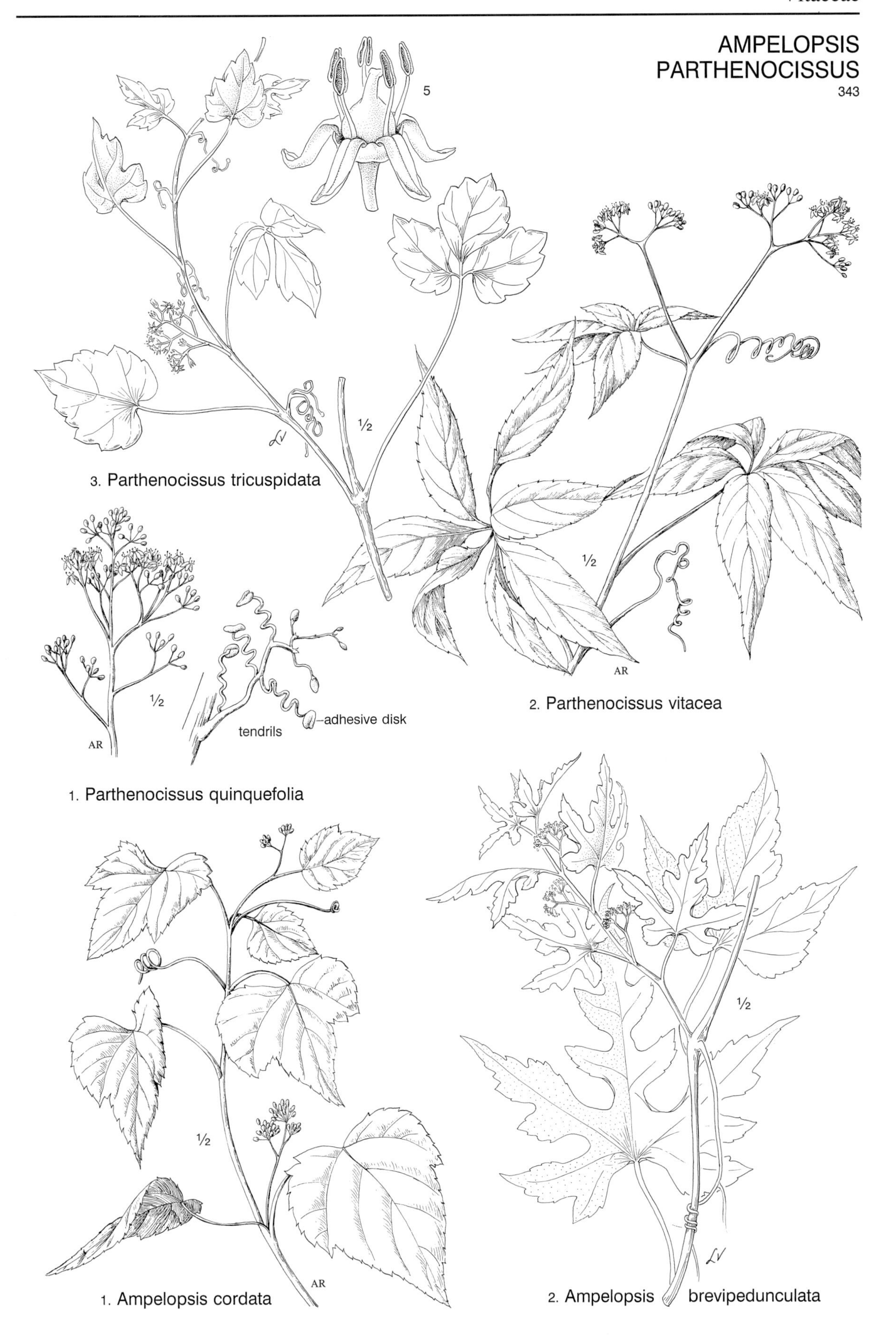

5

3. Parthenocissus tricuspidata

½

½

tendrils ─adhesive disk

½

AR

AR

1. Parthenocissus quinquefolia

2. Parthenocissus vitacea

½

1. Ampelopsis cordata

AR

2. Ampelopsis brevipedunculata

VITIS
344, 345

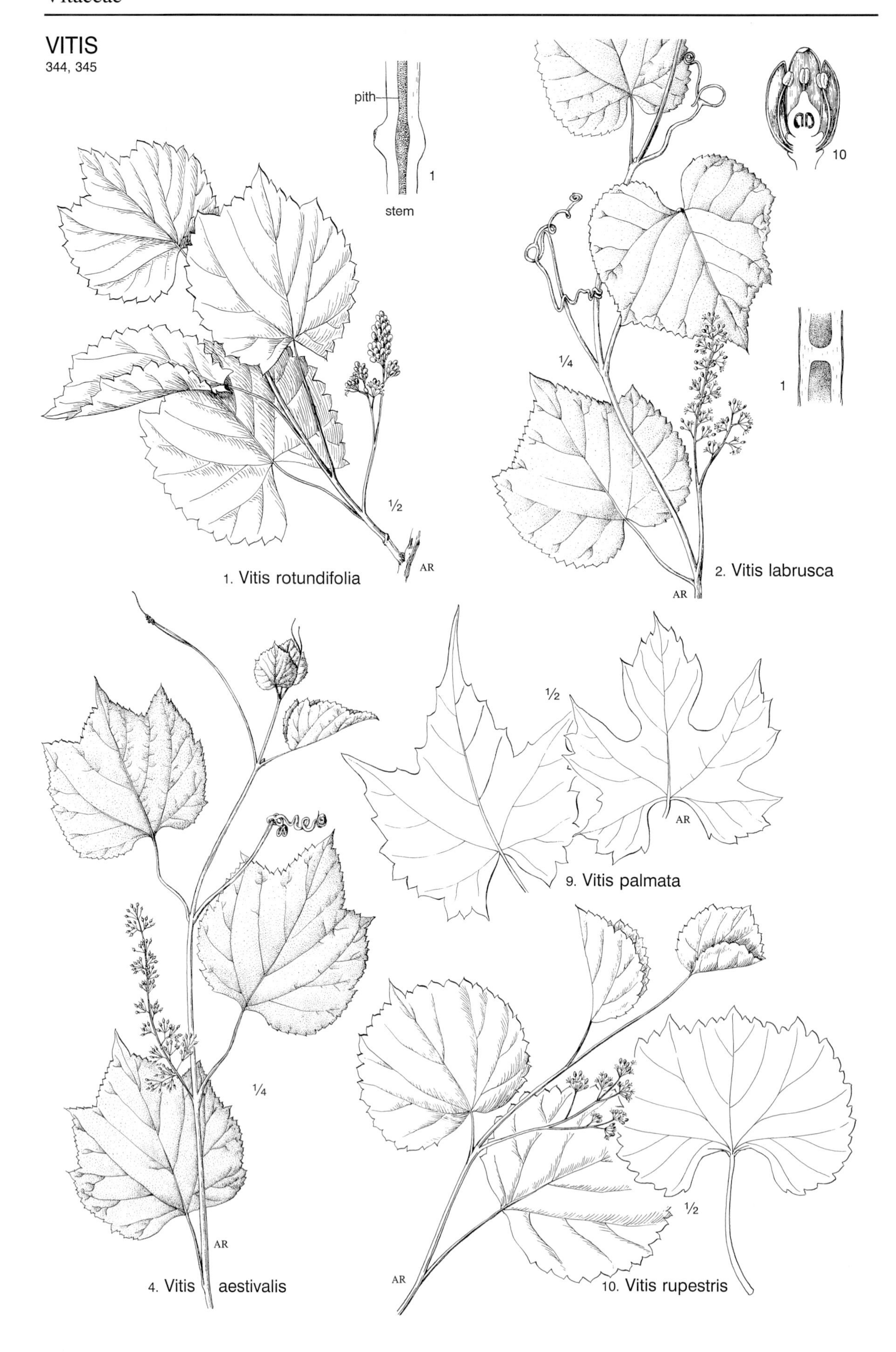

pith

1

stem

10

1

1/2

1/4

1/2

AR

1. Vitis rotundifolia

AR

2. Vitis labrusca

1/2

AR

9. Vitis palmata

1/4

AR

4. Vitis aestivalis

AR

1/2

10. Vitis rupestris

ILLUSTRATED COMPANION TO

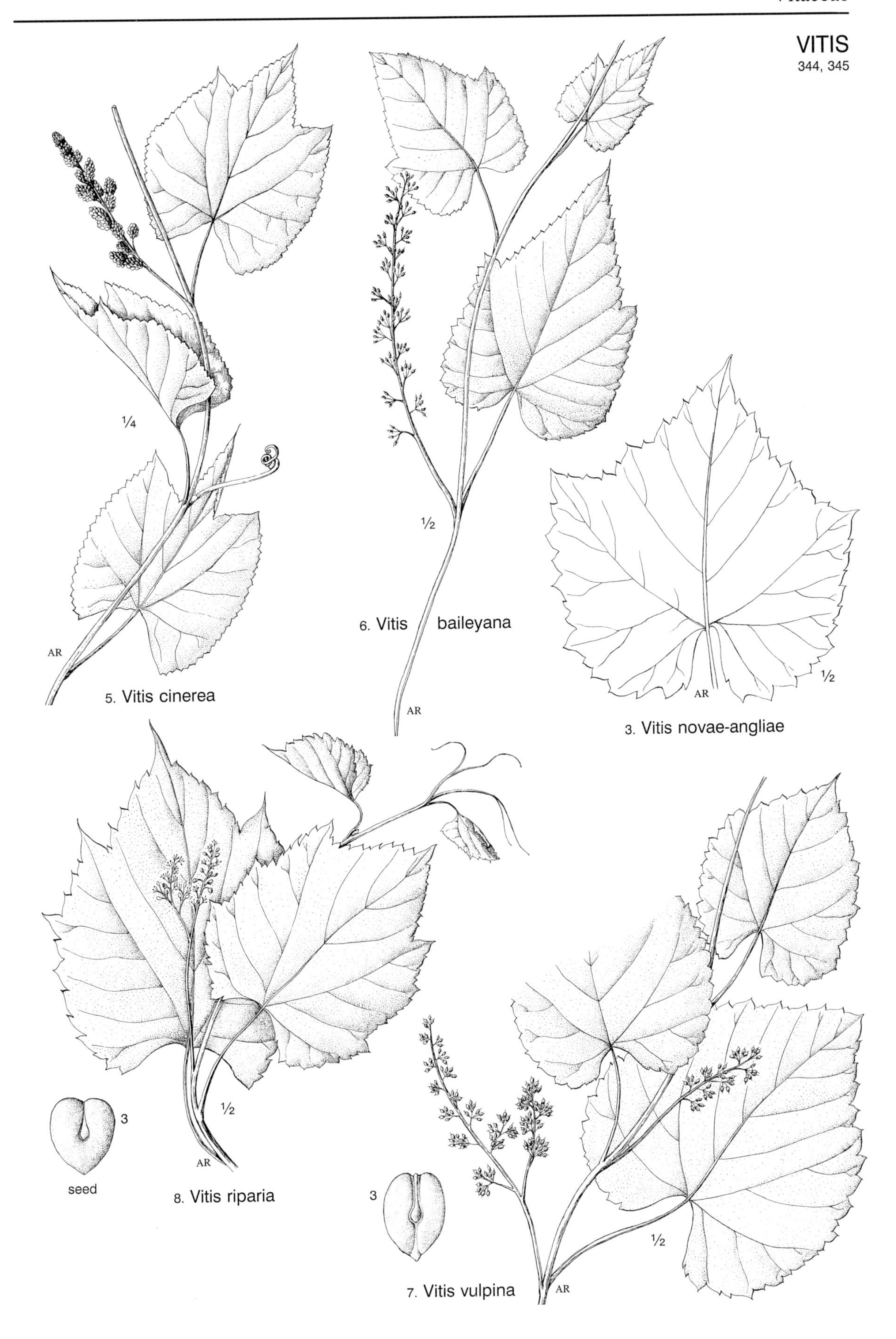

1/4

AR

5. Vitis cinerea

1/2

6. Vitis   baileyana

AR

3. Vitis novae-angliae

1/2

AR

8. Vitis riparia

3

seed

1/2

AR

3

7. Vitis vulpina

AR

1/2

LINUM
346

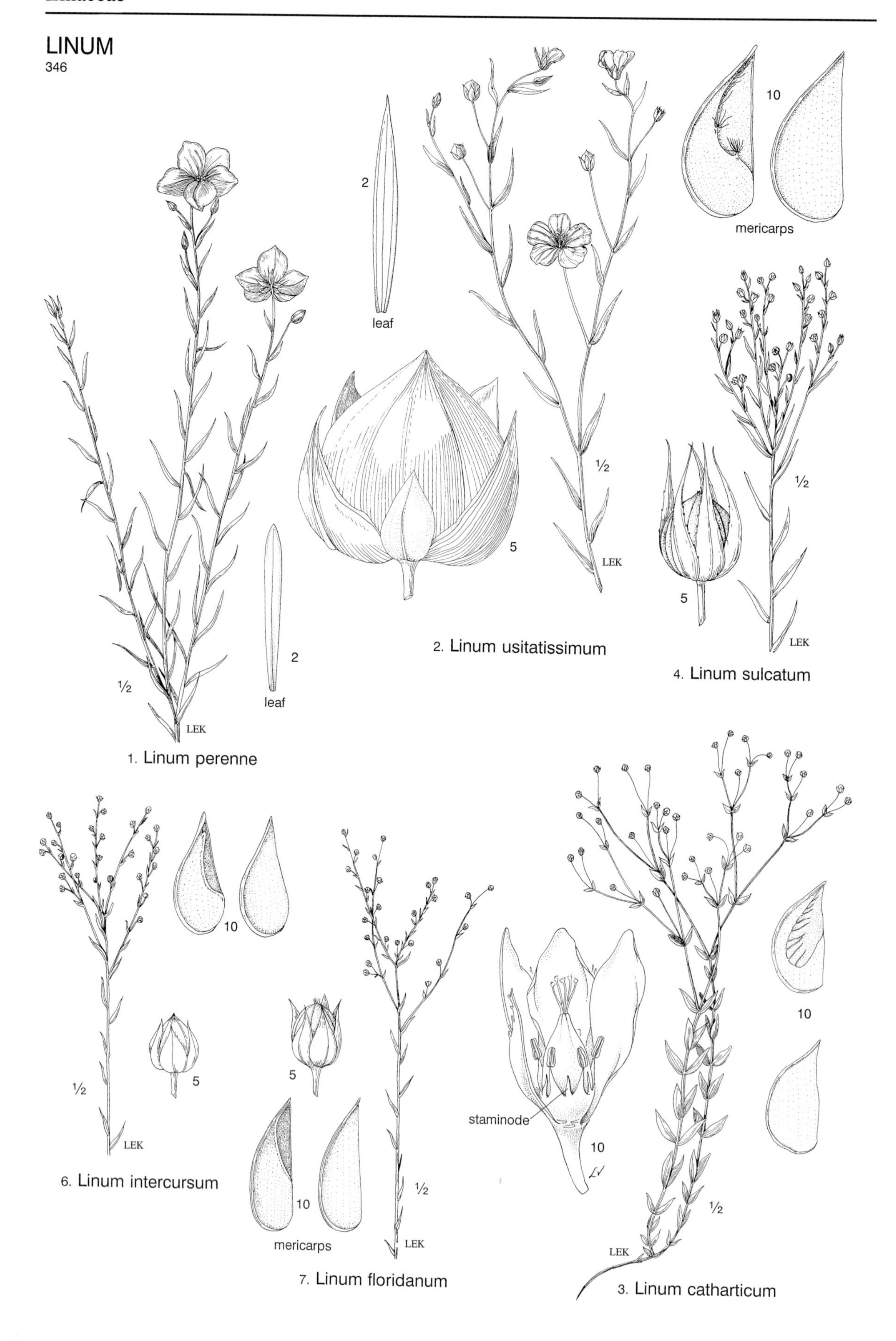

leaf

1. Linum perenne

2. Linum usitatissimum

mericarps

4. Linum sulcatum

6. Linum intercursum

7. Linum floridanum

staminode

3. Linum catharticum

mericarps

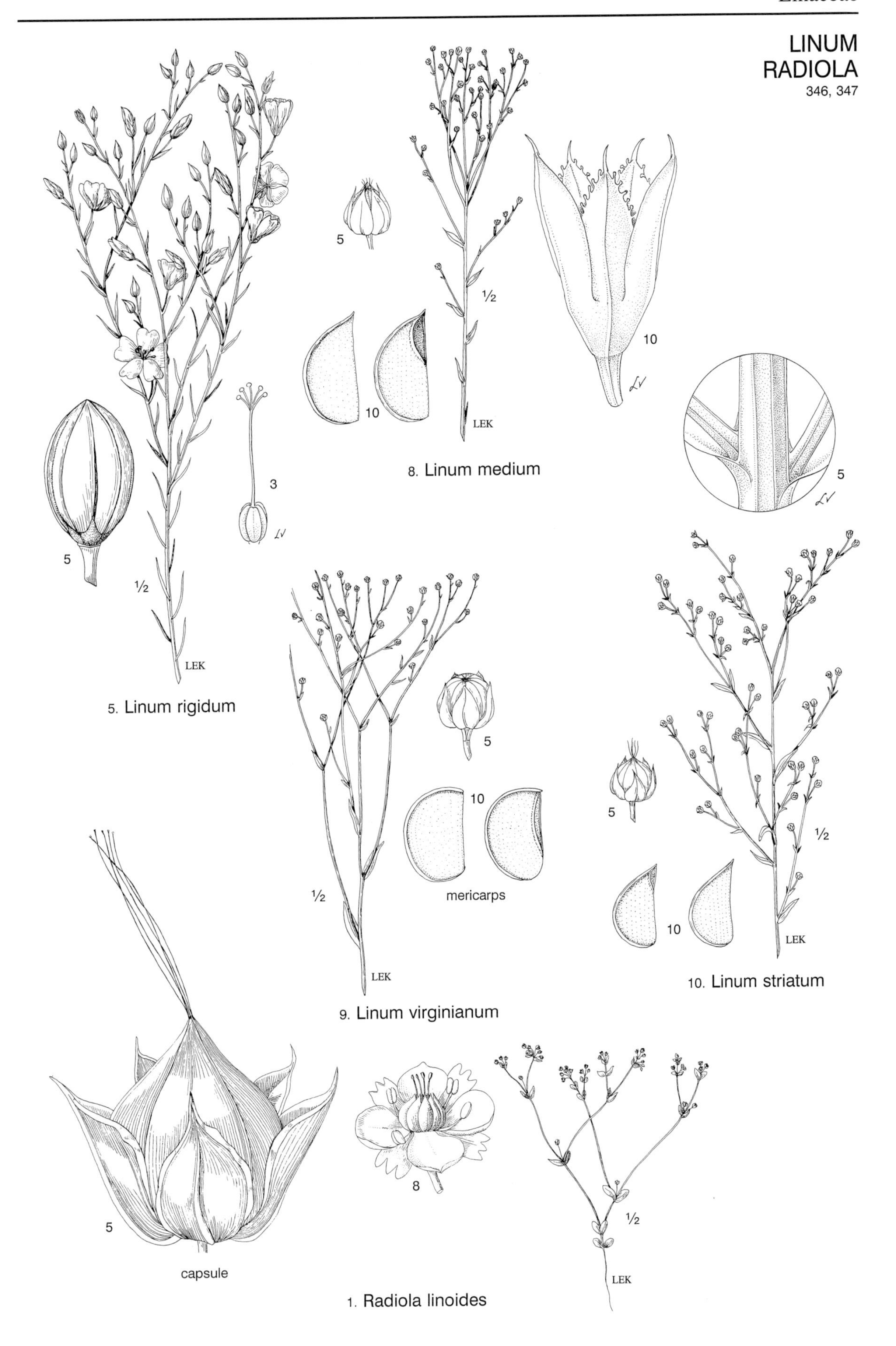

5. Linum rigidum

3

5

½

8. Linum medium

10

5

9. Linum virginianum

mericarps

10. Linum striatum

5

capsule

8

1. Radiola linoides

POLYGALA
348

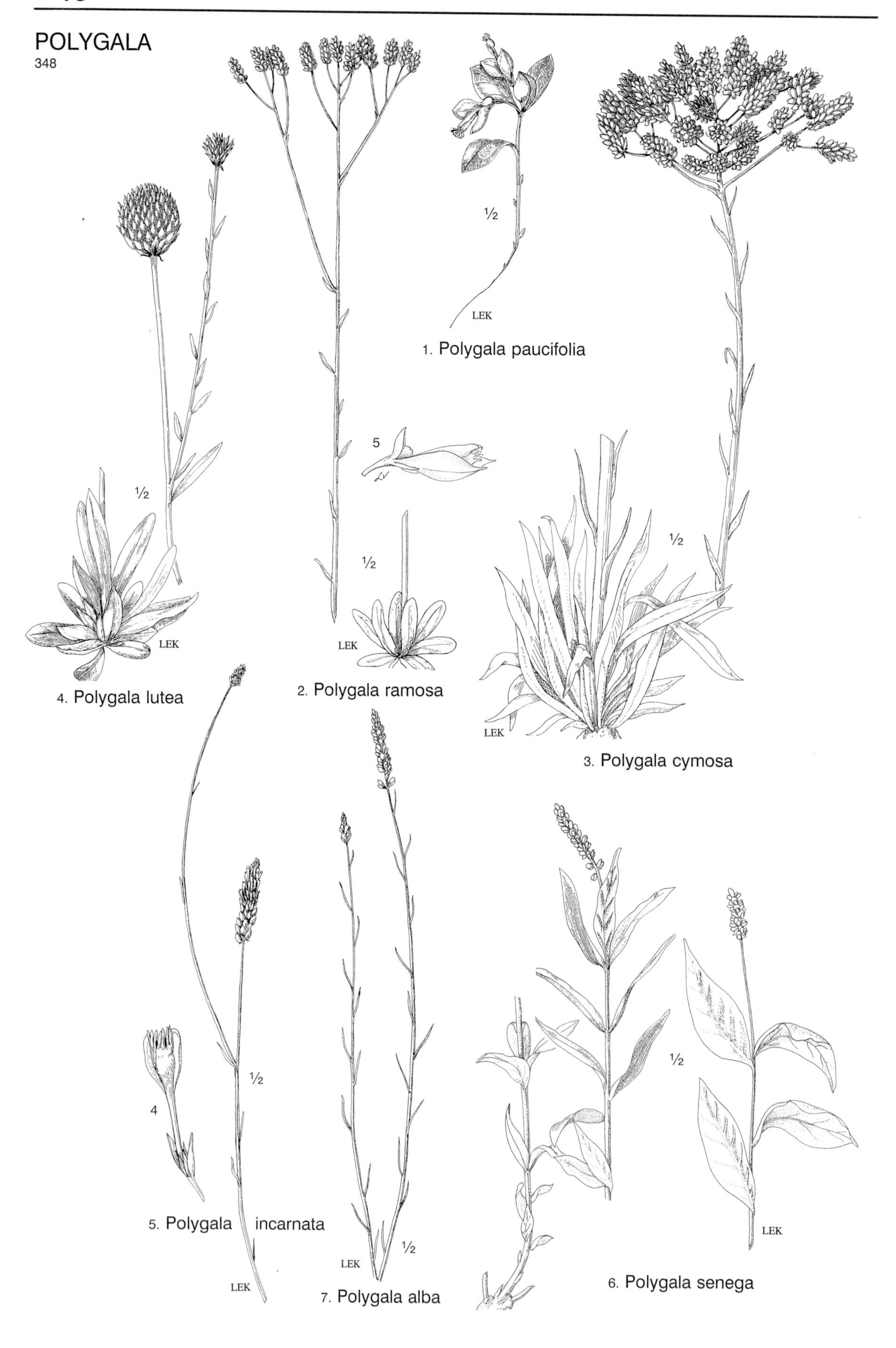

1. Polygala paucifolia

5

4. Polygala lutea

2. Polygala ramosa

3. Polygala cymosa

5. Polygala incarnata

7. Polygala alba

6. Polygala senega

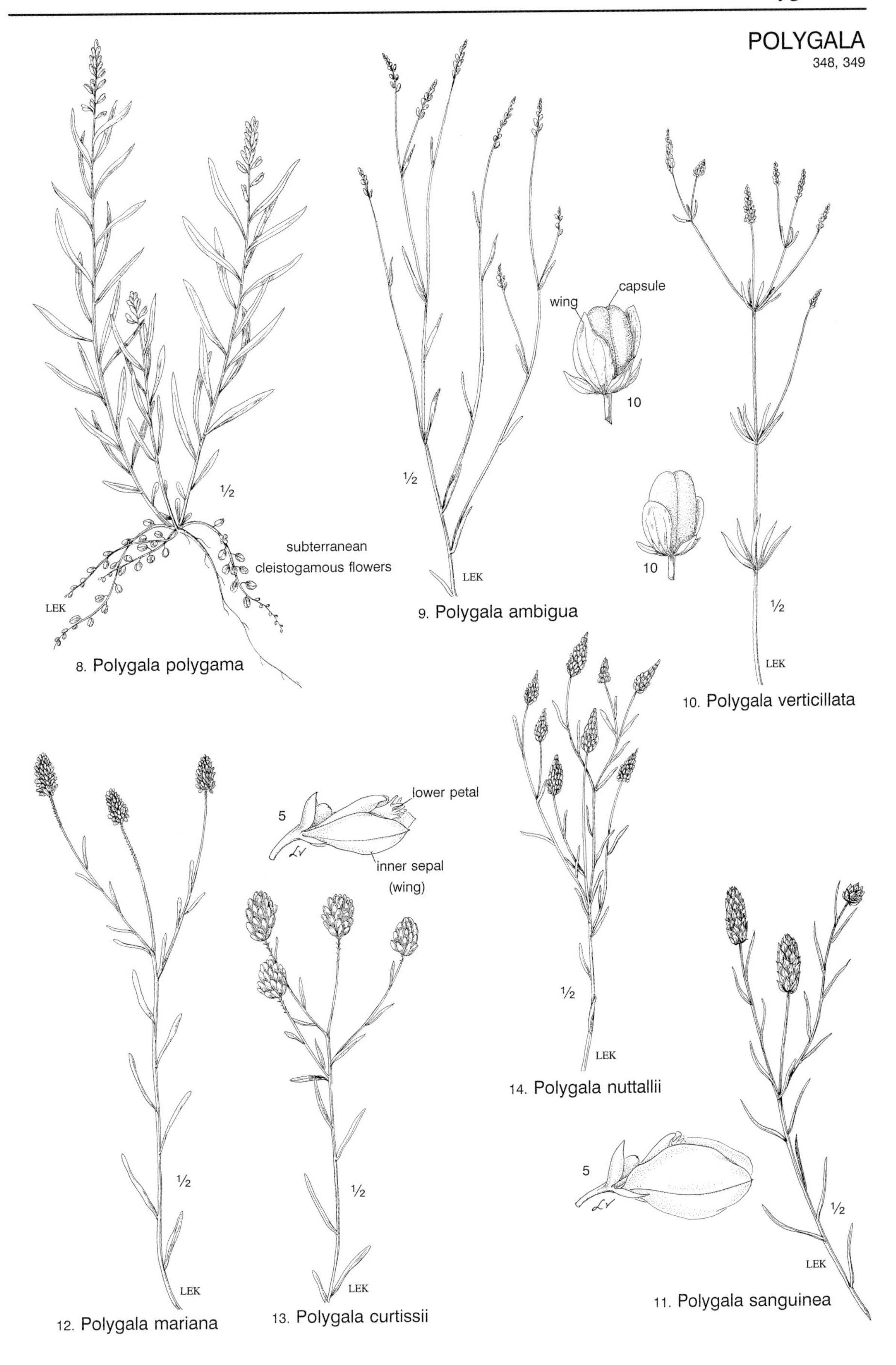

½

LEK

8. Polygala polygama

subterranean
cleistogamous flowers

½

LEK

9. Polygala ambigua

wing   capsule

10

10

½

LEK

10. Polygala verticillata

5   lower petal

inner sepal
(wing)

½

½

LEK

14. Polygala nuttallii

½

LEK

12. Polygala mariana

½

LEK

13. Polygala curtissii

5

½

LEK

11. Polygala sanguinea

POLYGALA
STAPHYLEA
CARDIOSPERMUM
KOELREUTERIA
AESCULUS
349–351

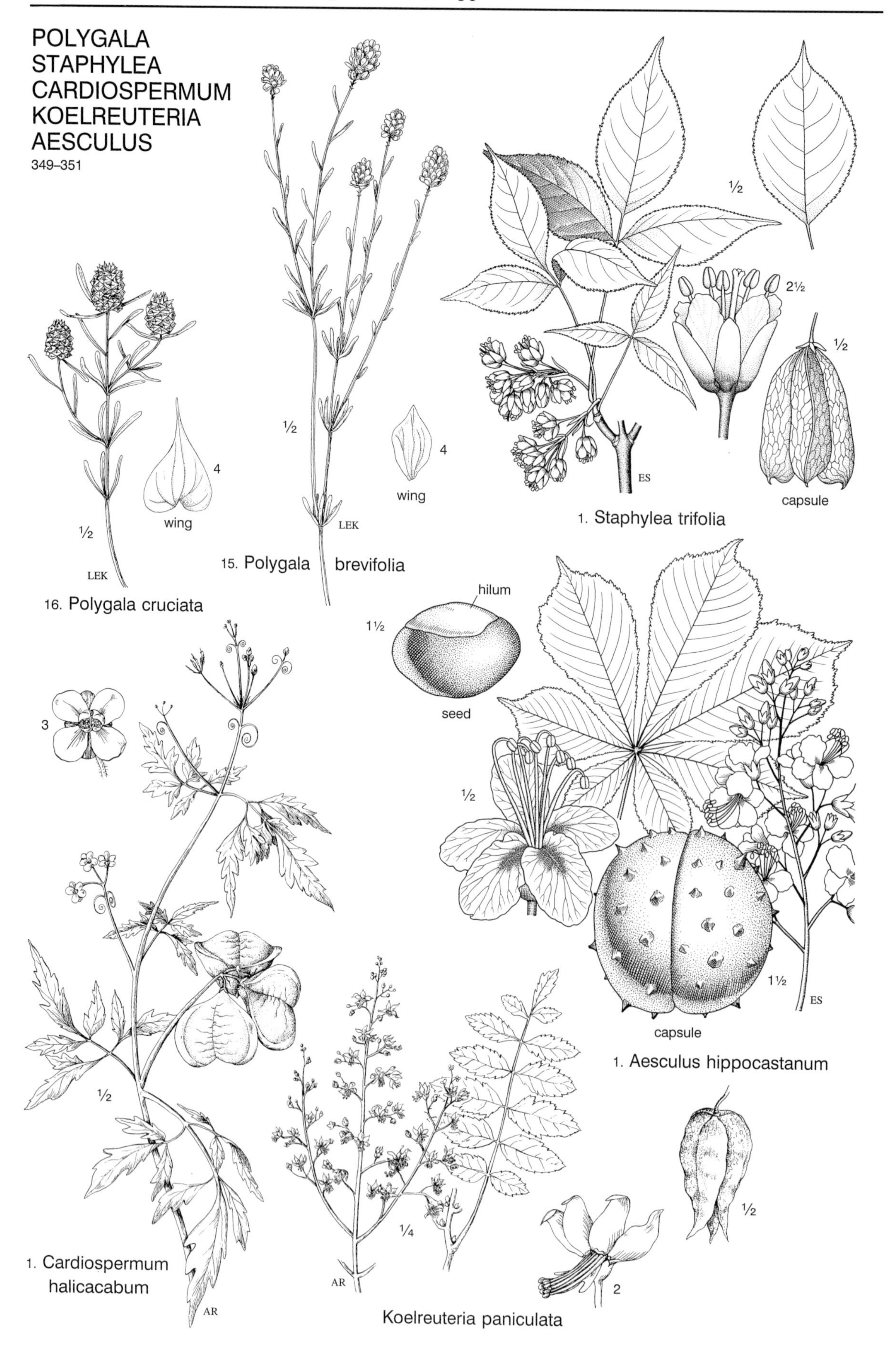

16. Polygala cruciata

15. Polygala brevifolia

1. Staphylea trifolia

capsule

hilum

seed

1. Aesculus hippocastanum

capsule

1. Cardiospermum halicacabum

Koelreuteria paniculata

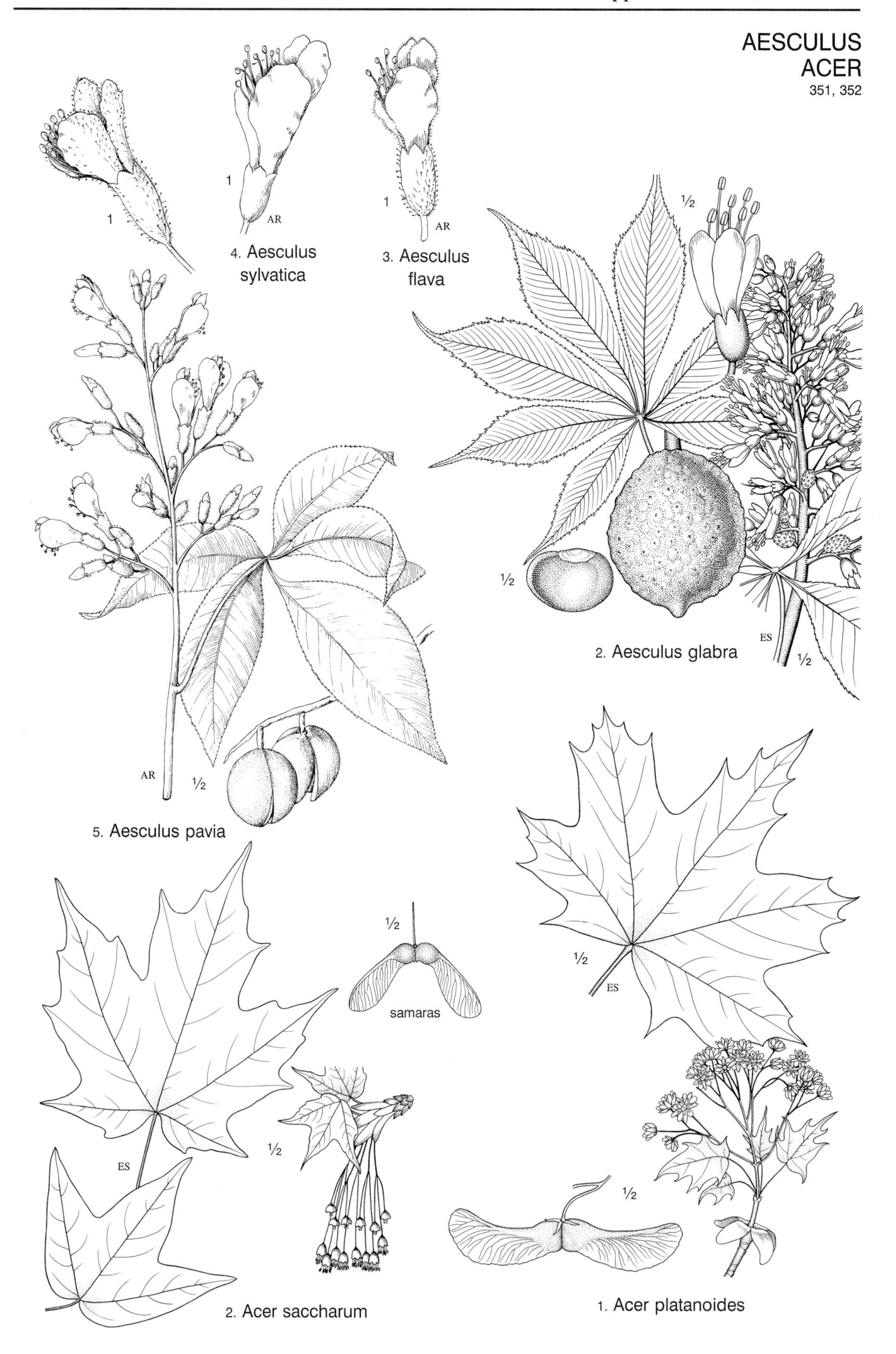

4. Aesculus
sylvatica

3. Aesculus
flava

2. Aesculus glabra

5. Aesculus pavia

samaras

2. Acer saccharum

1. Acer platanoides

ACER
352

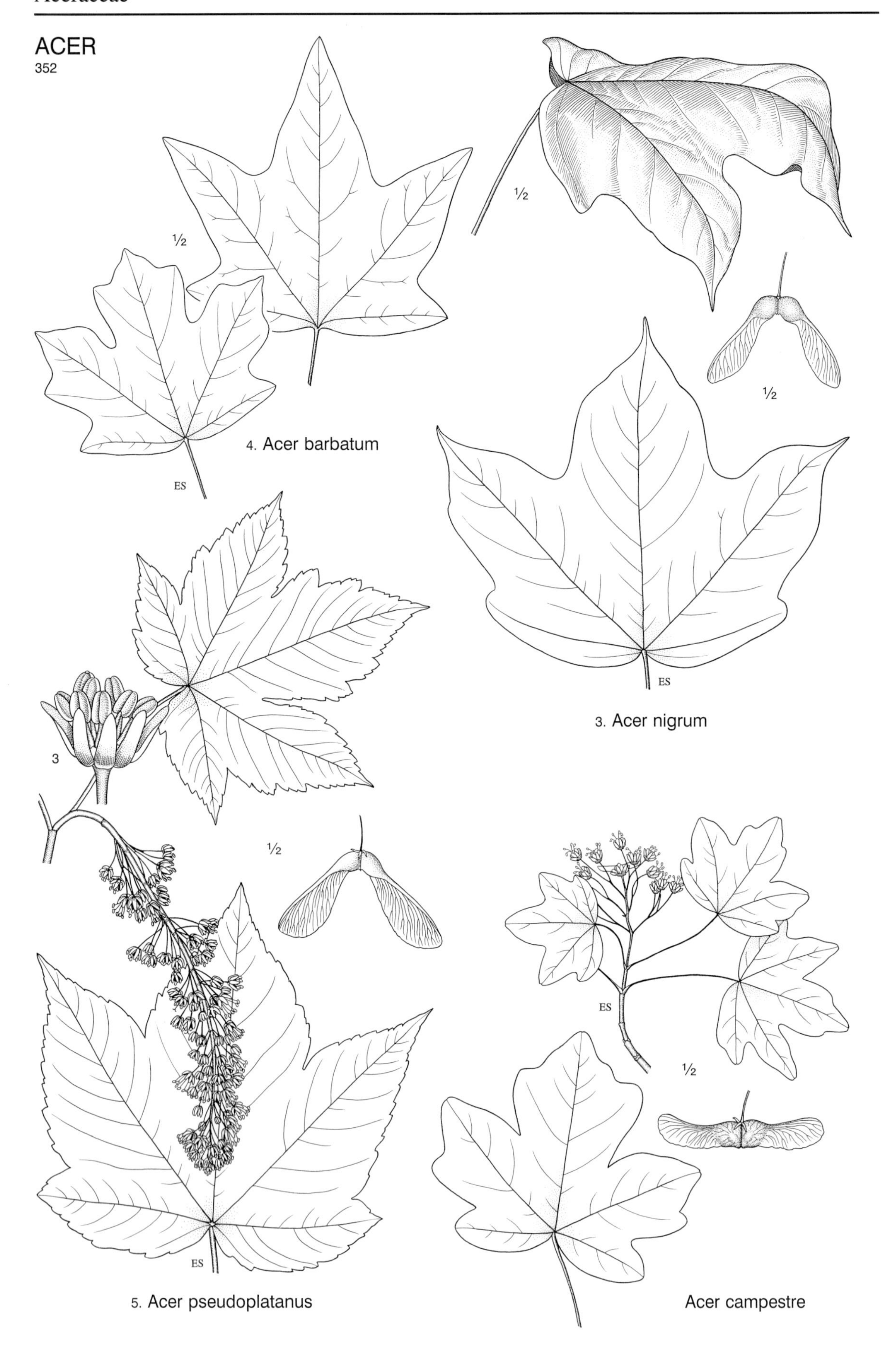

4. Acer barbatum

3. Acer nigrum

5. Acer pseudoplatanus

Acer campestre

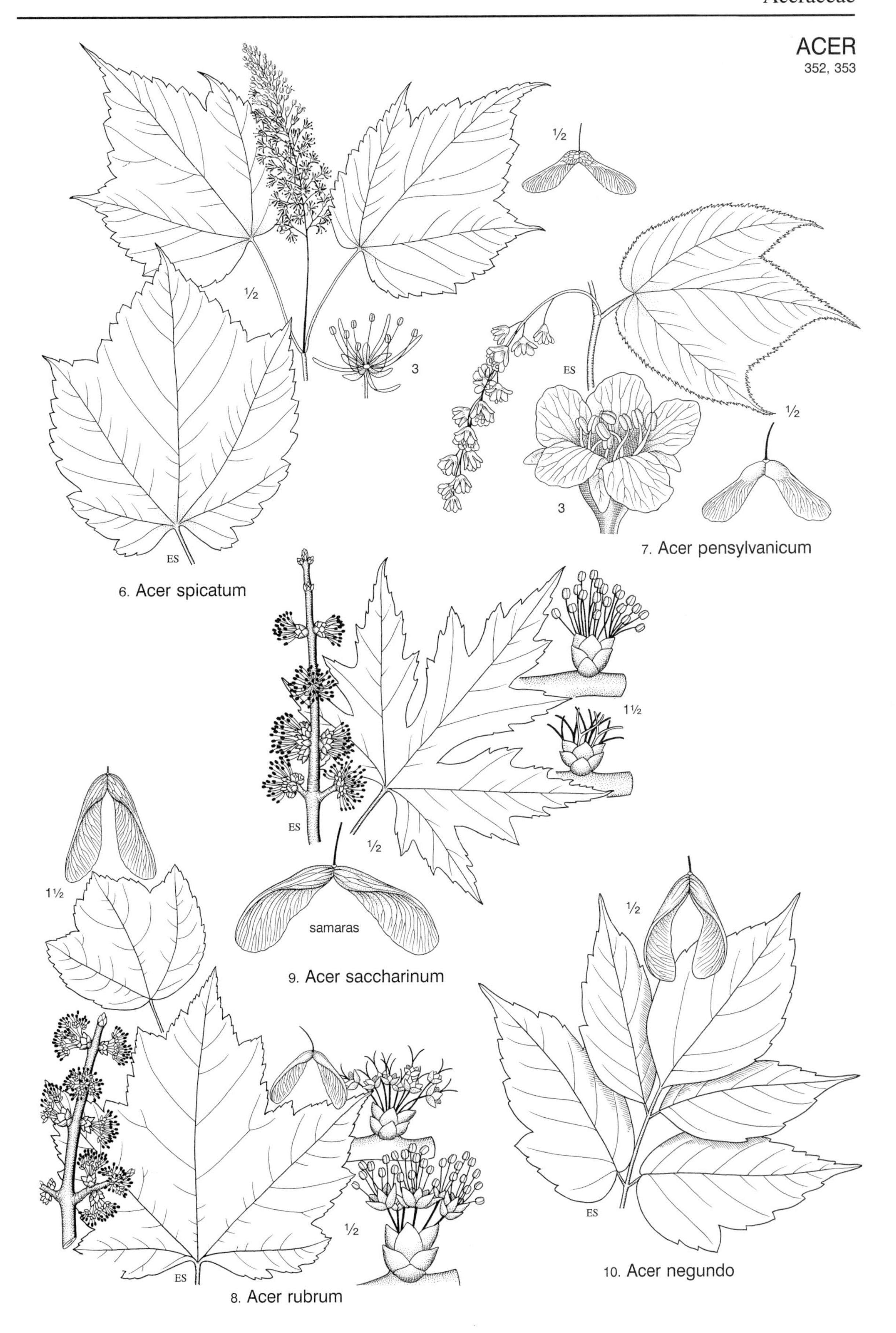

6. Acer spicatum

7. Acer pensylvanicum

9. Acer saccharinum

samaras

8. Acer rubrum

10. Acer negundo

RHUS
353, 354

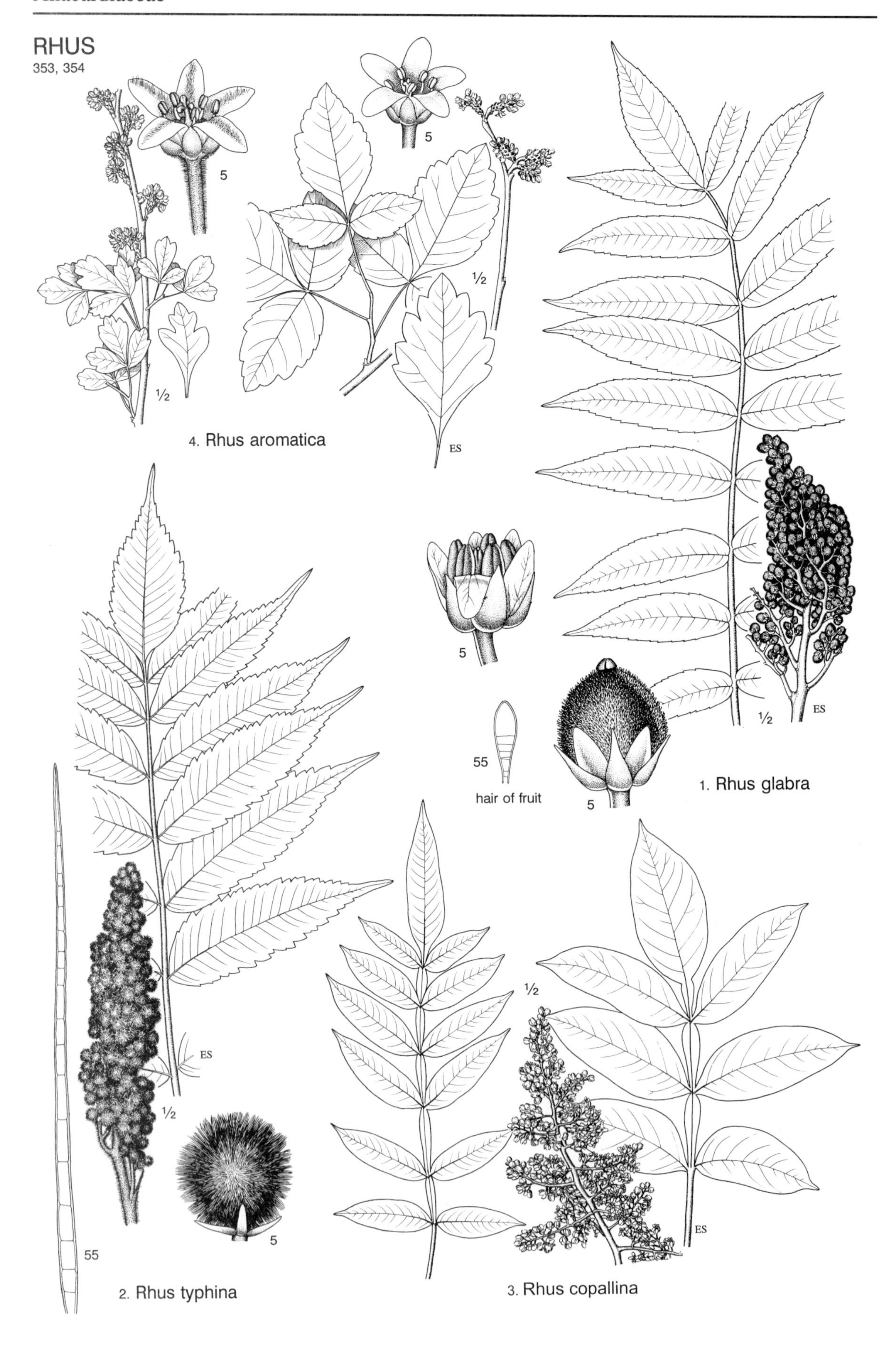

4. Rhus aromatica

5

½

ES

1. Rhus glabra

5

55
hair of fruit

5

½

ES

2. Rhus typhina

55

5

3. Rhus copallina

½

ES

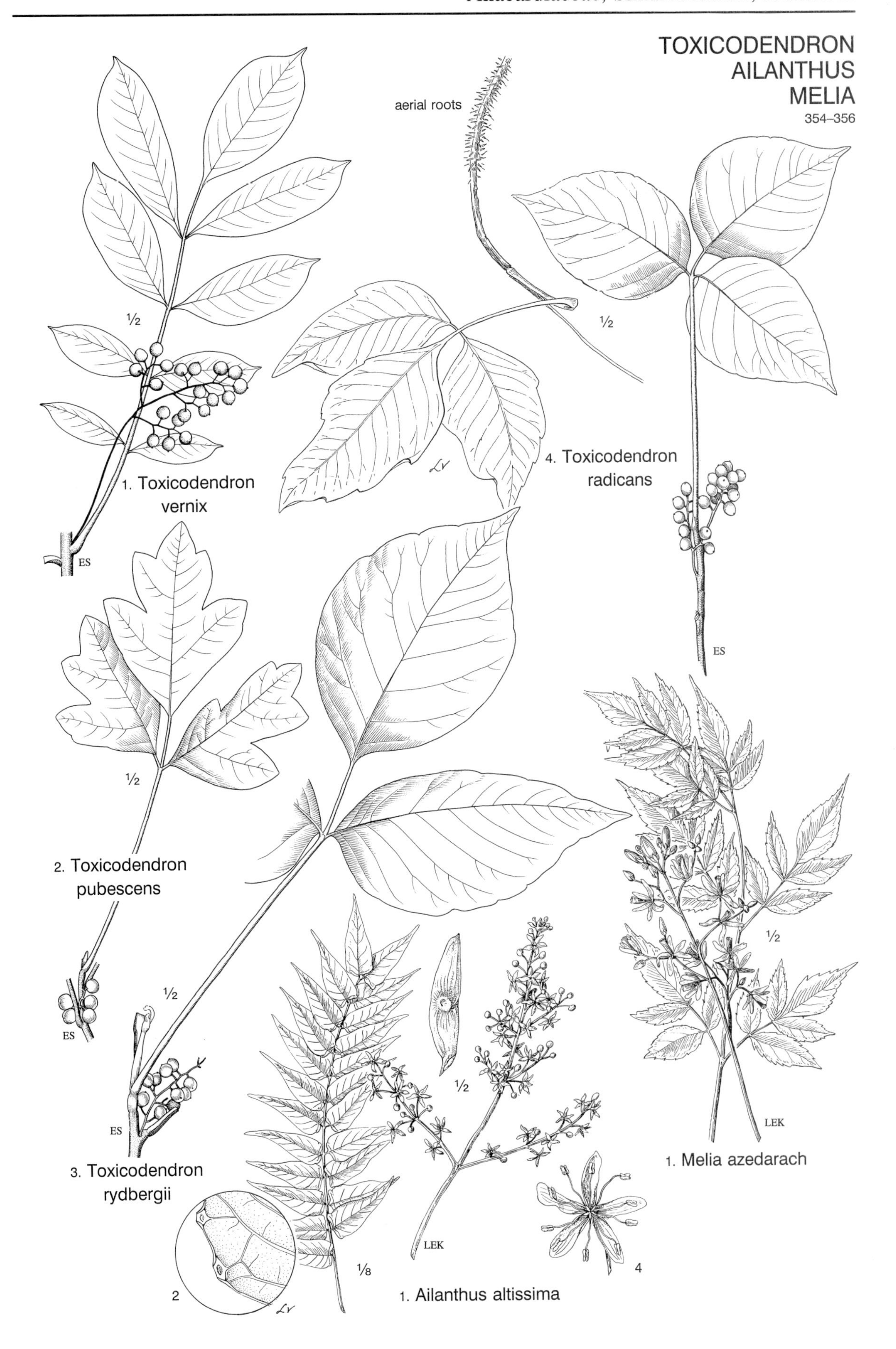

aerial roots

½

1. Toxicodendron
vernix

ES

4. Toxicodendron
radicans

ES

½

2. Toxicodendron
pubescens

ES

½

ES

3. Toxicodendron
rydbergii

2

1. Melia azedarach

½

LEK

LEK

½

4

⅛

1. Ailanthus altissima

ZANTHOXYLUM
PTELEA
RUTA
356, 357

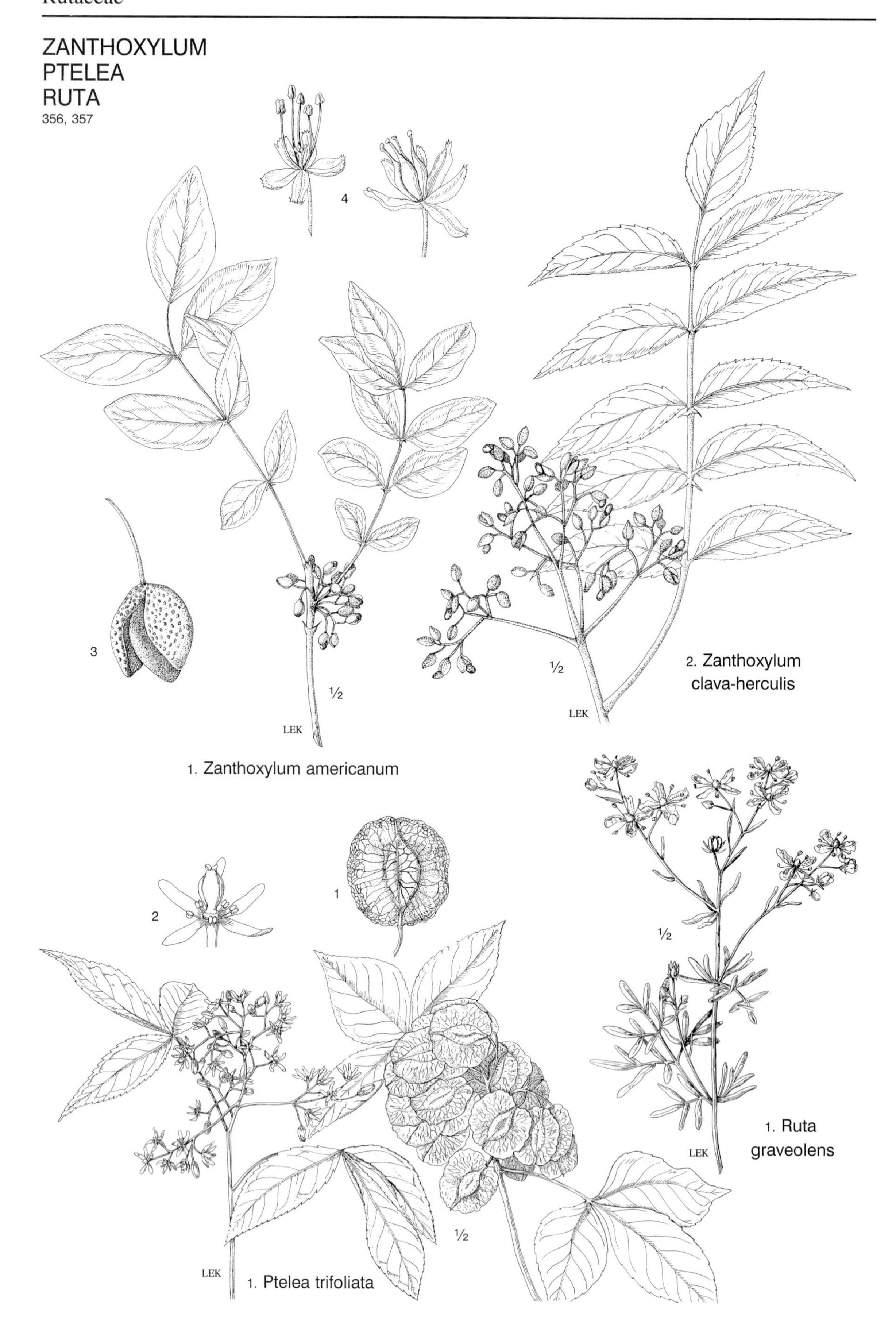

4

3

½

LEK

1. Zanthoxylum americanum

½

LEK

2. Zanthoxylum
clava-herculis

2

1

½

LEK

1. Ruta
graveolens

1. Ptelea trifoliata

½

LEK

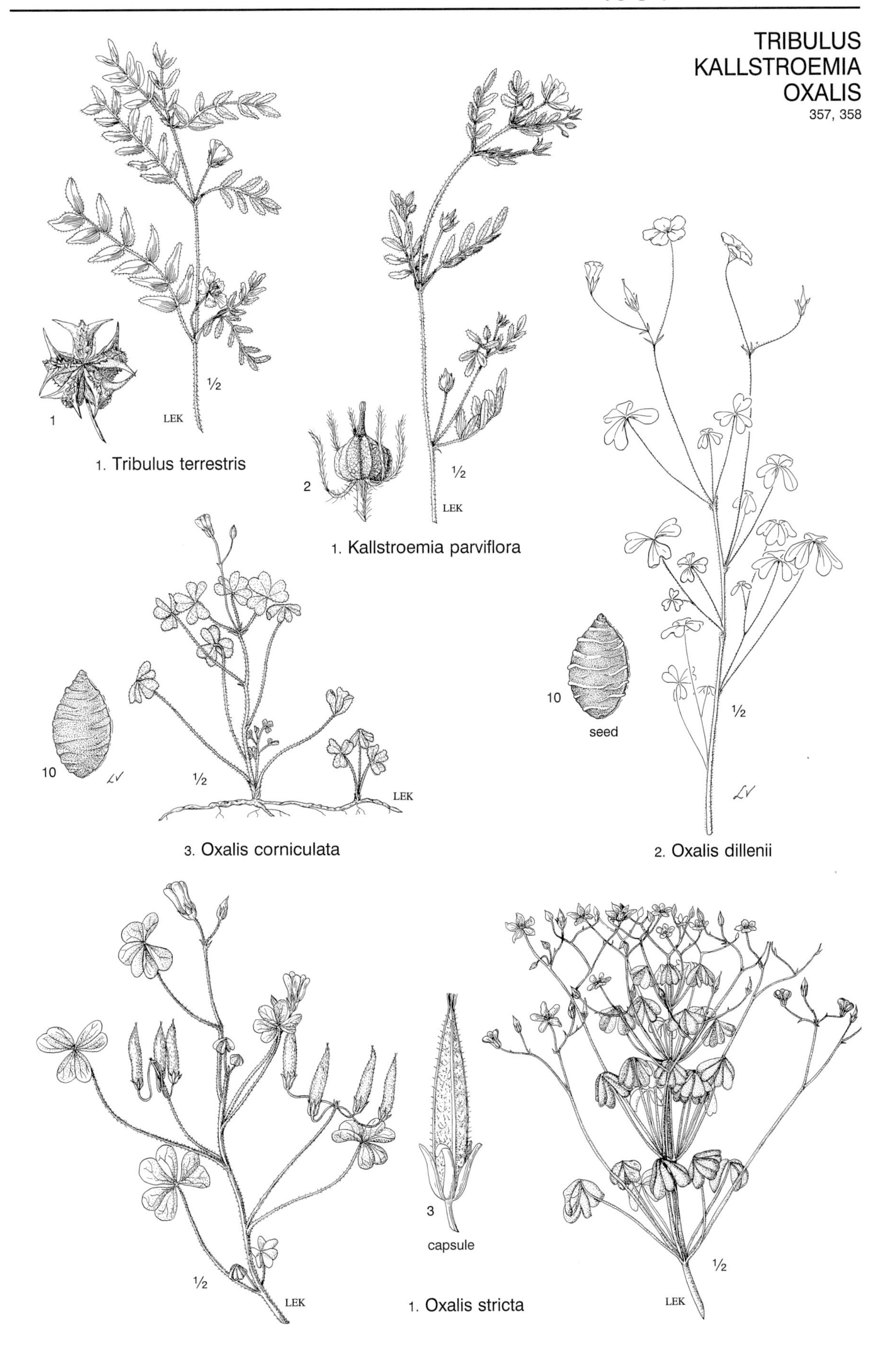

1. Tribulus terrestris

1. Kallstroemia parviflora

10 seed

2. Oxalis dillenii

3. Oxalis corniculata

3 capsule

1. Oxalis stricta

OXALIS
GERANIUM
358–360

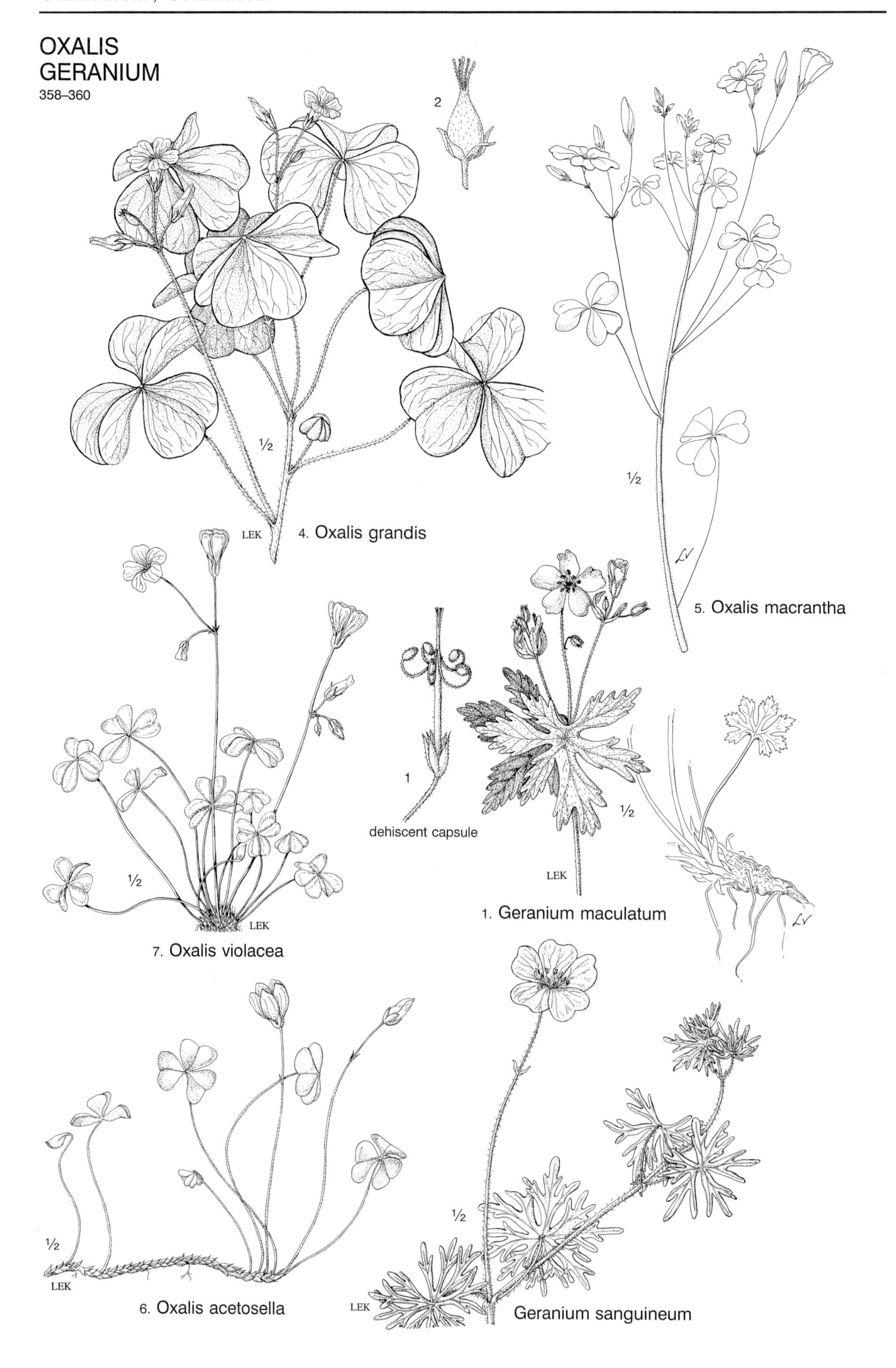

2

4. Oxalis grandis

½

LEK

5. Oxalis macrantha

½

LEK

7. Oxalis violacea

½

LEK

1

dehiscent capsule

1. Geranium maculatum

½

LEK

LEK

6. Oxalis acetosella

½

LEK

½

Geranium sanguineum

LEK

½

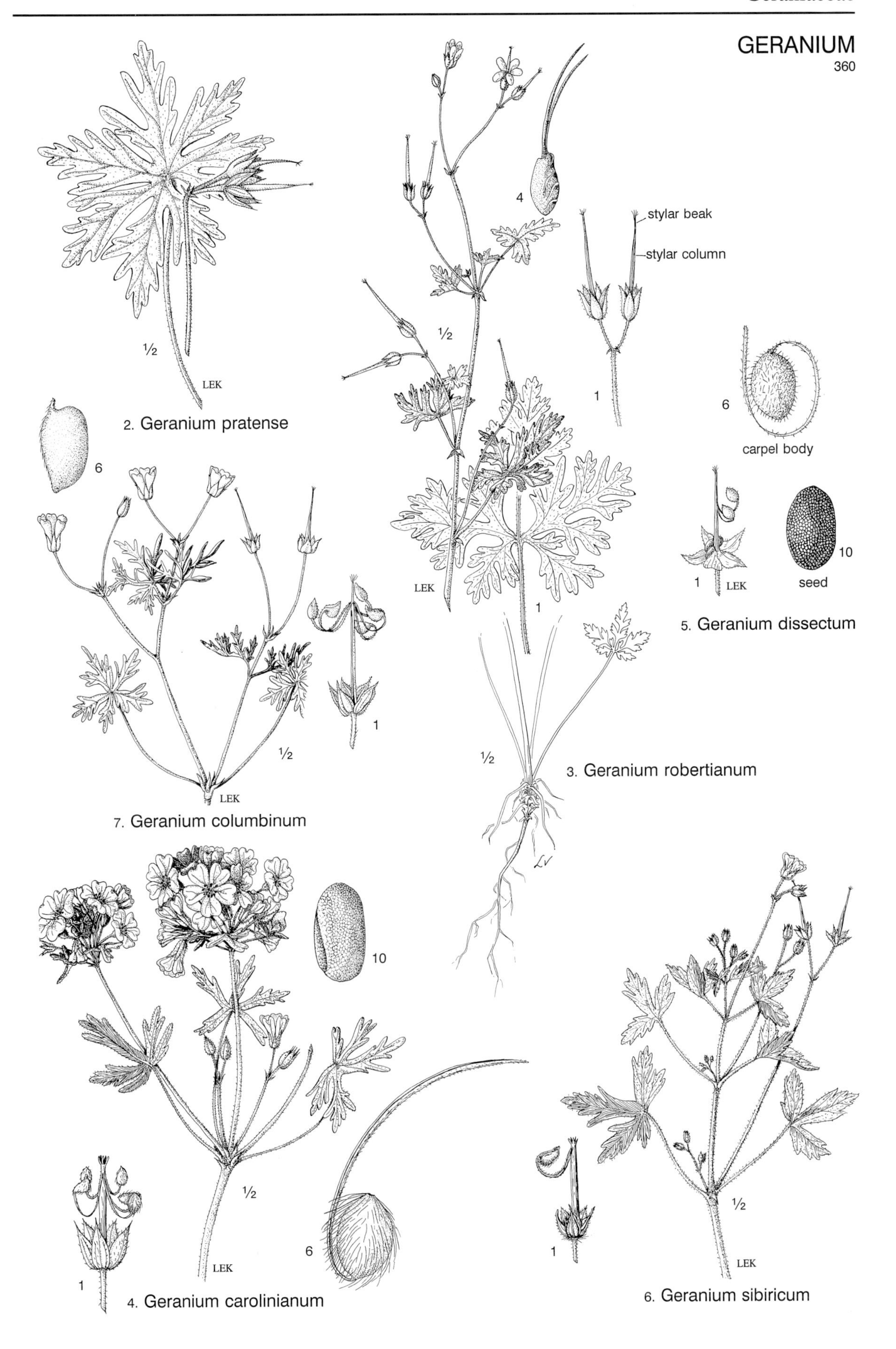

2. Geranium pratense

stylar beak

stylar column

carpel body

5. Geranium dissectum

seed

7. Geranium columbinum

3. Geranium robertianum

4. Geranium carolinianum

6. Geranium sibiricum

GERANIUM
ERODIUM
360, 361

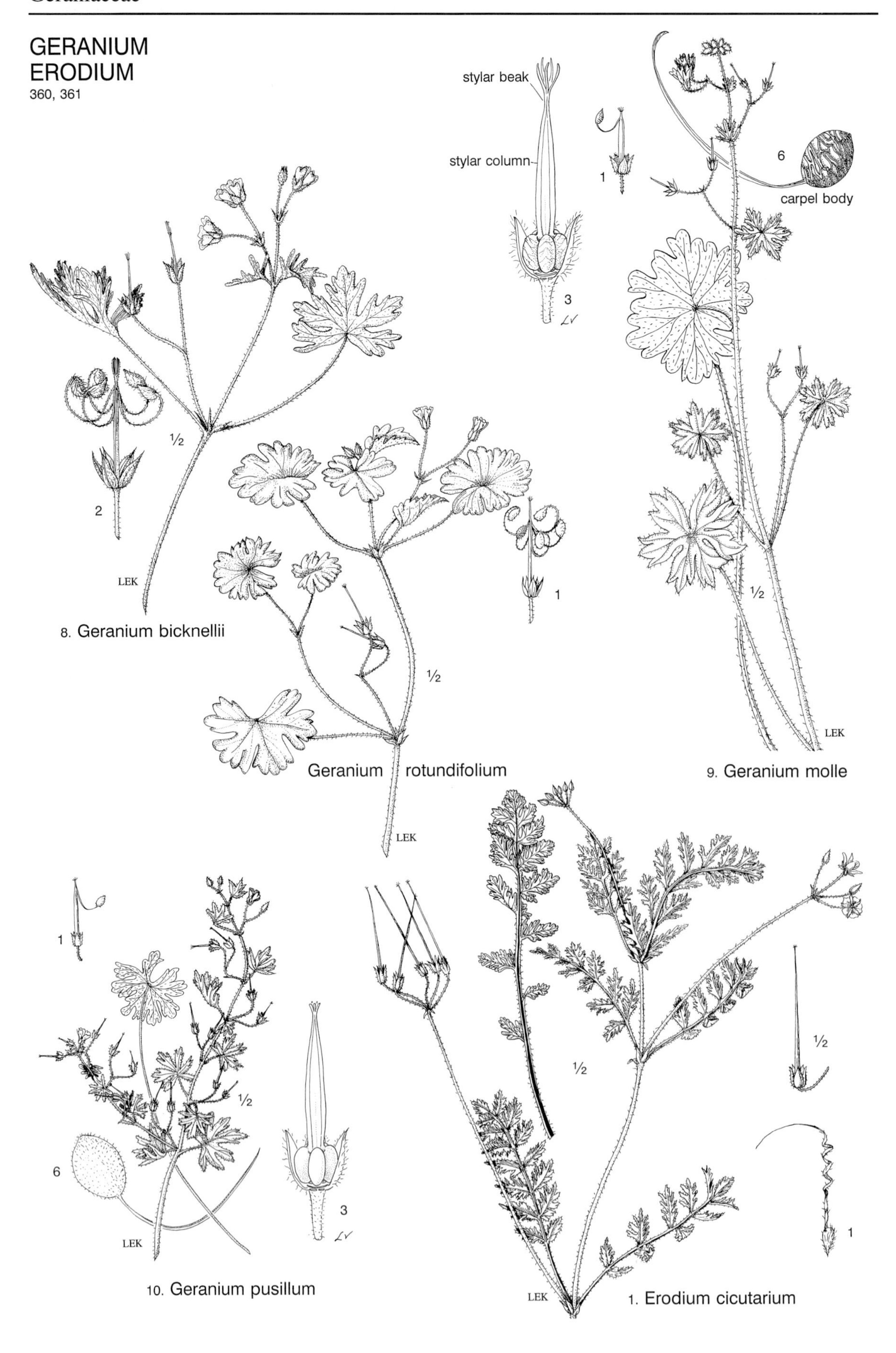

stylar beak

stylar column

carpel body

1

6

3

8. Geranium bicknellii

2

LEK

Geranium rotundifolium

LEK

1

9. Geranium molle

LEK

½

10. Geranium pusillum

6

1

3

LEK

1. Erodium cicutarium

LEK

1

½

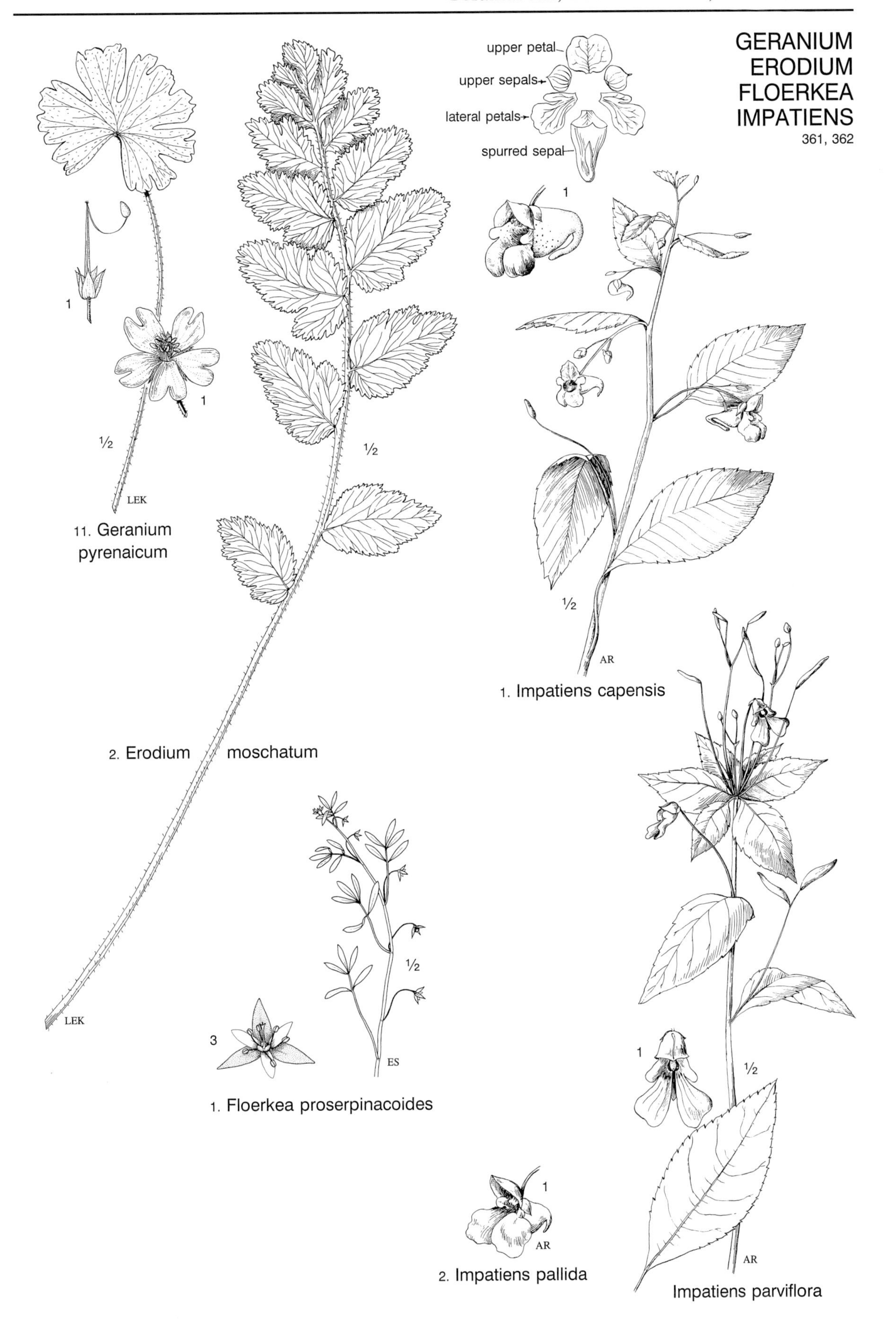

GERANIUM
ERODIUM
FLOERKEA
IMPATIENS
361, 362

upper petal
upper sepals
lateral petals
spurred sepal

1

1. Impatiens capensis

11. Geranium
pyrenaicum

2. Erodium moschatum

1. Floerkea proserpinacoides

3

2. Impatiens pallida

Impatiens parviflora

IMPATIENS
ARALIA
362, 363

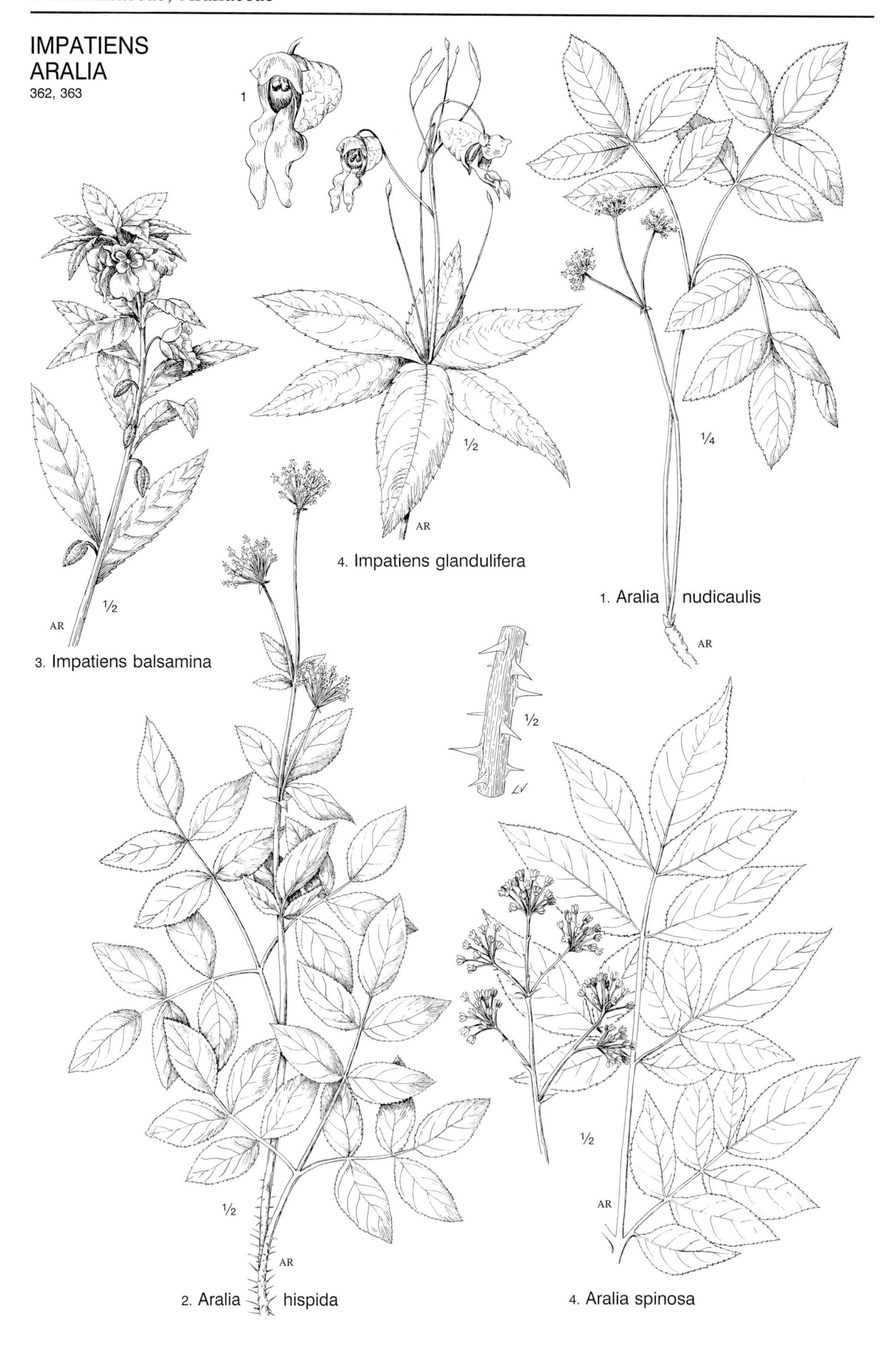

1

½

4. Impatiens glandulifera

AR

¼

1. Aralia nudicaulis

AR

3. Impatiens balsamina

½

AR

½

LV

½

AR

2. Aralia hispida

½

AR

4. Aralia spinosa

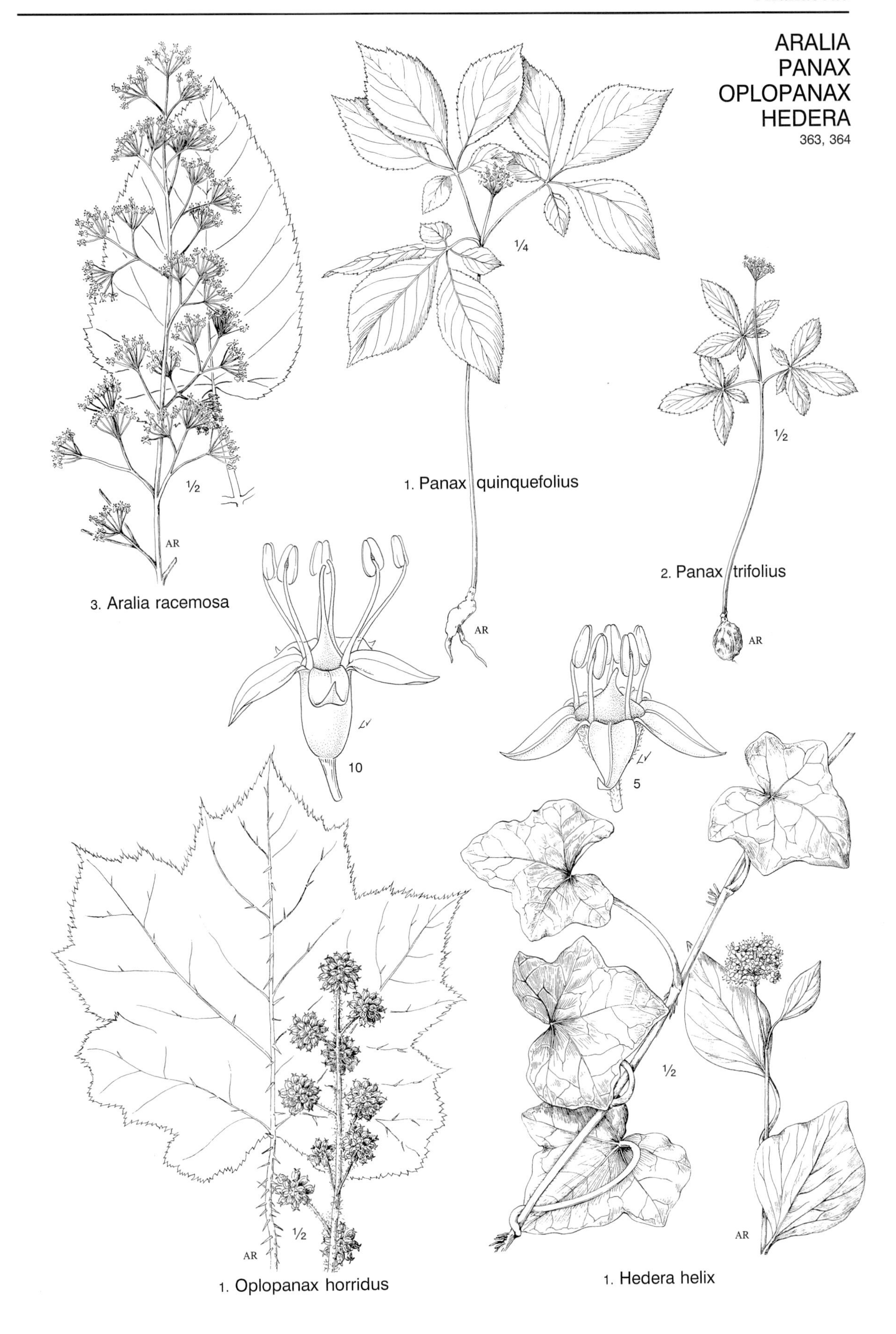

1. Panax quinquefolius

3. Aralia racemosa

2. Panax trifolius

10

5

1. Oplopanax horridus

1. Hedera helix

HYDROCOTYLE
CENTELLA
369

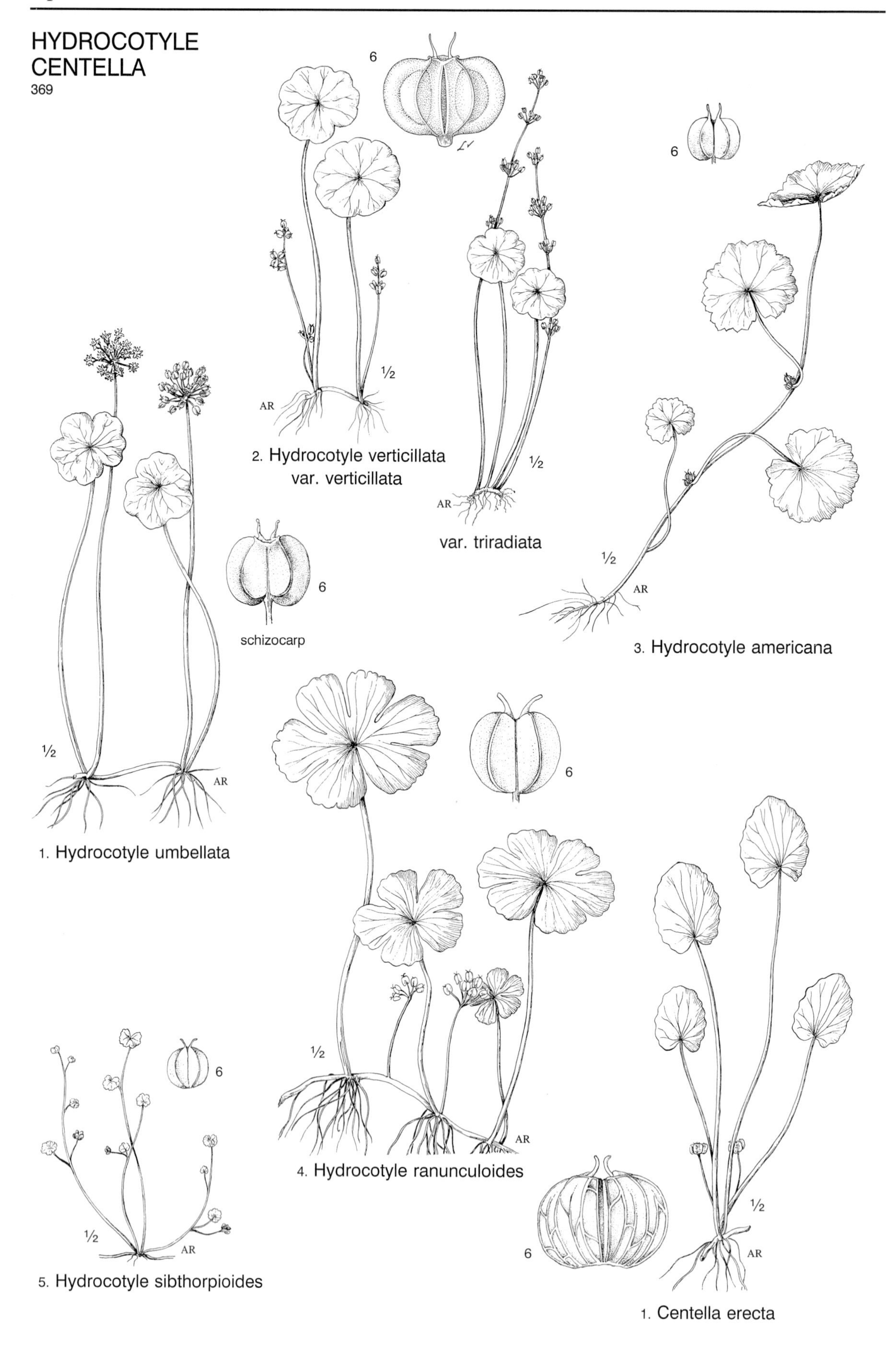

2. Hydrocotyle verticillata
var. verticillata

var. triradiata

6

schizocarp

3. Hydrocotyle americana

1. Hydrocotyle umbellata

4. Hydrocotyle ranunculoides

5. Hydrocotyle sibthorpioides

1. Centella erecta

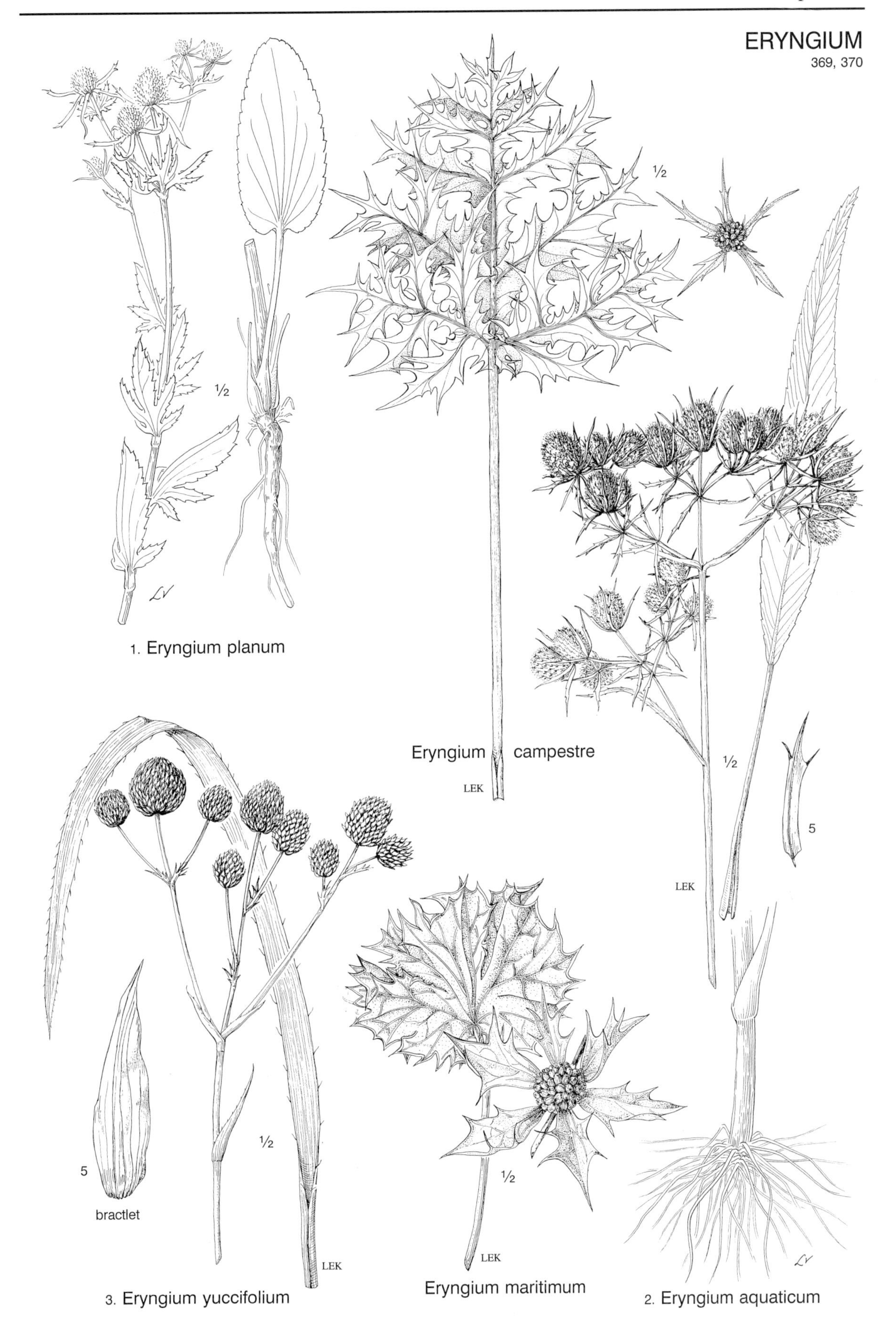

1. Eryngium planum

Eryngium   campestre

3. Eryngium yuccifolium

bractlet

Eryngium maritimum

2. Eryngium aquaticum

ERYNGIUM
SANICULA
CRYPTOTAENIA
370, 371

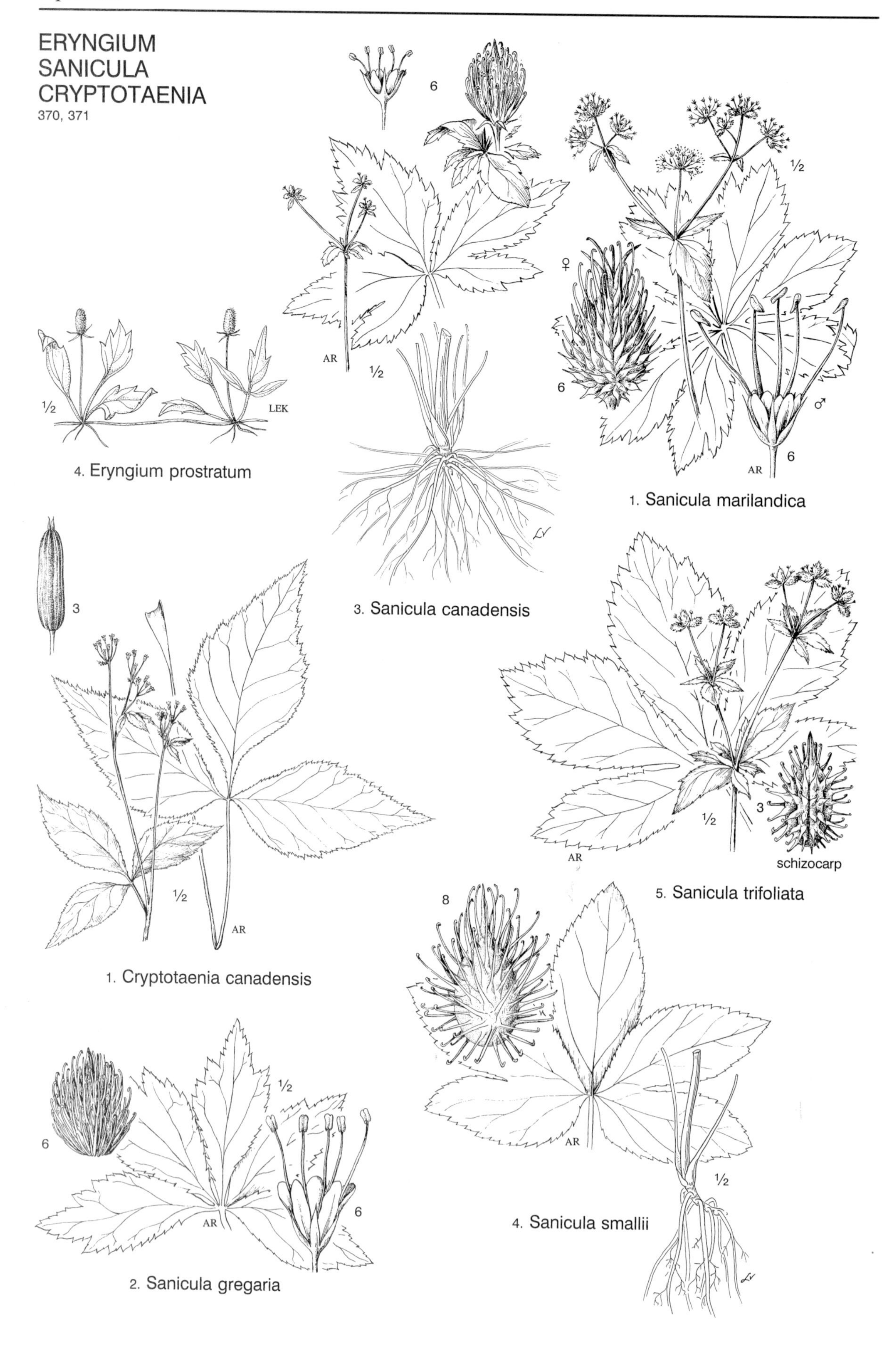

4. Eryngium prostratum

3. Sanicula canadensis

1. Sanicula marilandica

1. Cryptotaenia canadensis

5. Sanicula trifoliata

schizocarp

2. Sanicula gregaria

4. Sanicula smallii

ILLUSTRATED COMPANION TO

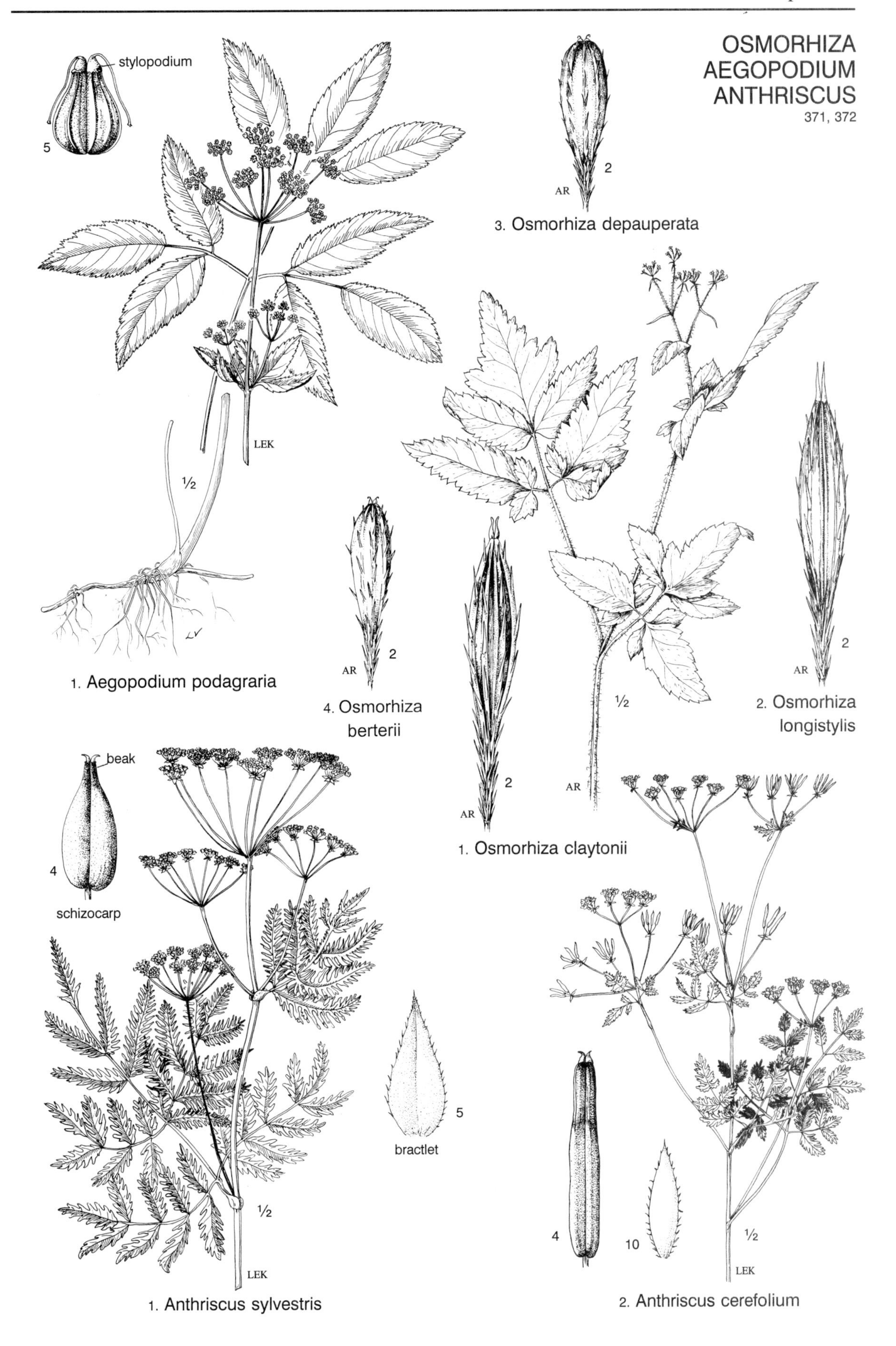

5 stylopodium

3. Osmorhiza depauperata

1. Aegopodium podagraria

4. Osmorhiza berterii

2. Osmorhiza longistylis

1. Osmorhiza claytonii

beak

4 schizocarp

5 bractlet

1. Anthriscus sylvestris

4   10   2. Anthriscus cerefolium

ANTHRISCUS
TREPOCARPUS
CARUM
FALCARIA
CHAEROPHYLLUM
372, 373

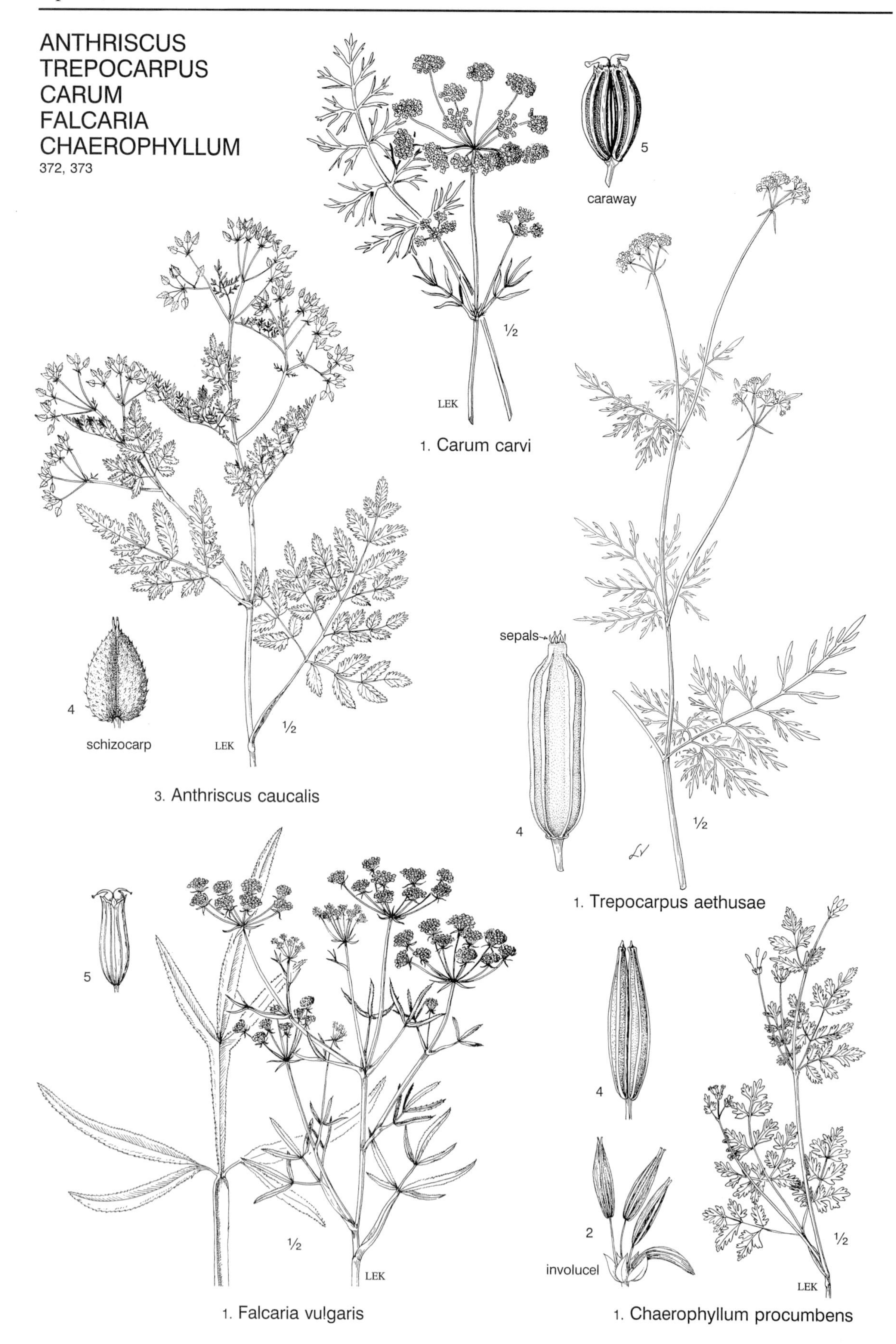

5

caraway

½

LEK

1. Carum carvi

4

schizocarp   LEK   ½

3. Anthriscus caucalis

sepals

4

1. Trepocarpus aethusae

½

5

½   LEK

1. Falcaria vulgaris

4

2

involucel

½   LEK

1. Chaerophyllum procumbens

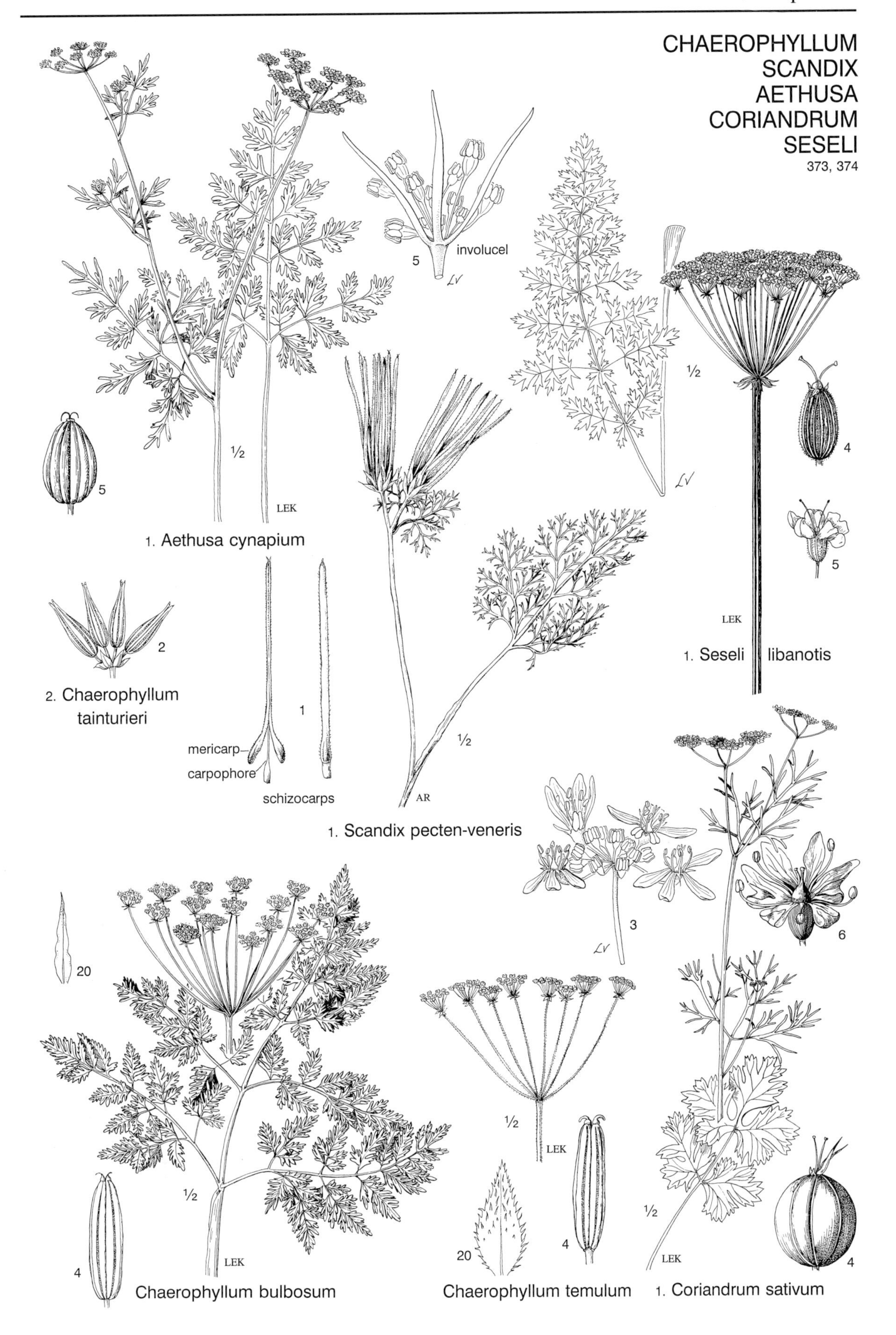

CHAEROPHYLLUM
SCANDIX
AETHUSA
CORIANDRUM
SESELI
373, 374

5 involucel

1. Aethusa cynapium

1. Seseli libanotis

2. Chaerophyllum tainturieri

mericarp
carpophore
schizocarps

1. Scandix pecten-veneris

Chaerophyllum bulbosum

Chaerophyllum temulum

1. Coriandrum sativum

SPERMOLEPIS
TORILIS
DAUCUS
374, 375

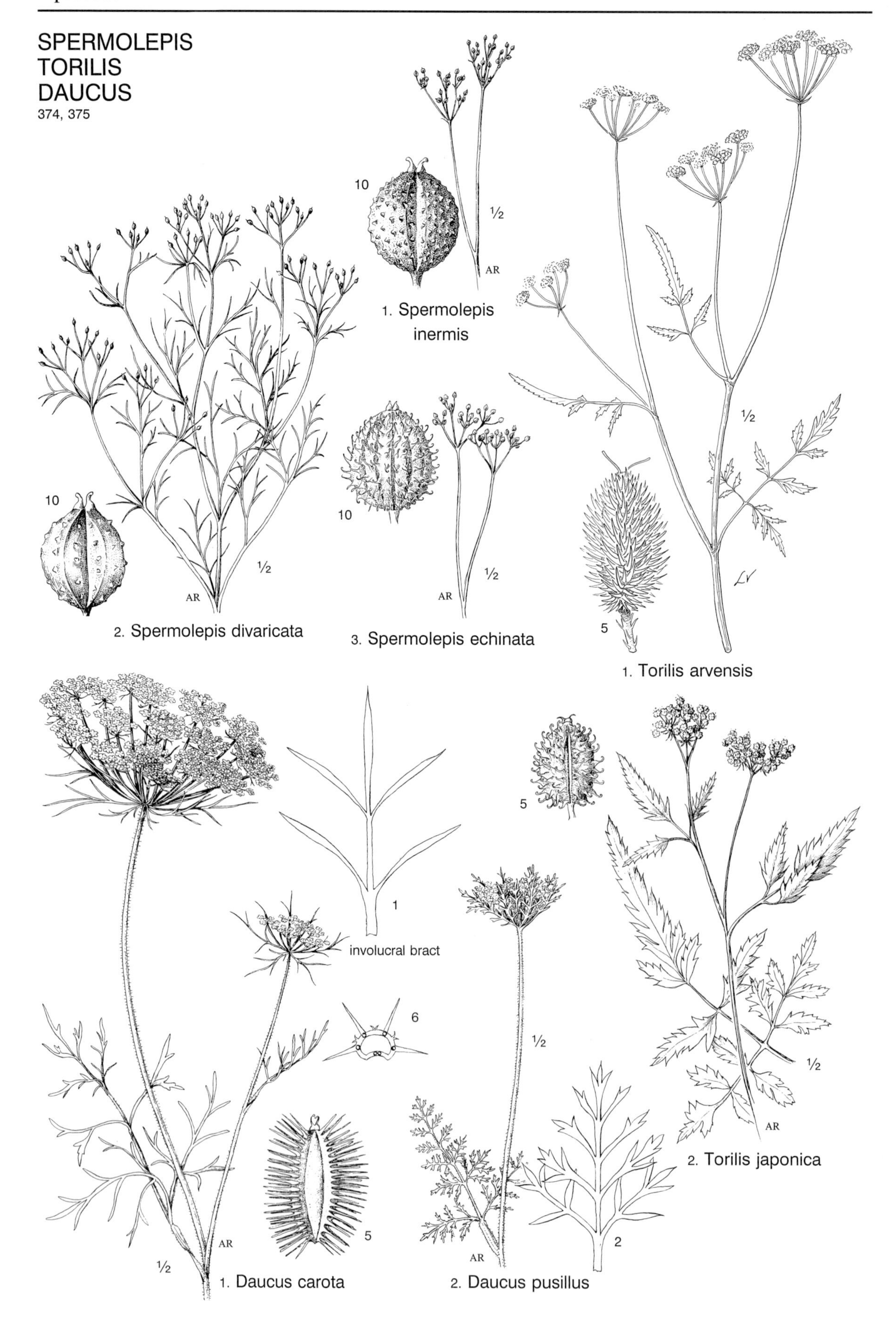

1. Spermolepis inermis

2. Spermolepis divaricata

3. Spermolepis echinata

1. Torilis arvensis

2. Torilis japonica

involucral bract

1. Daucus carota

2. Daucus pusillus

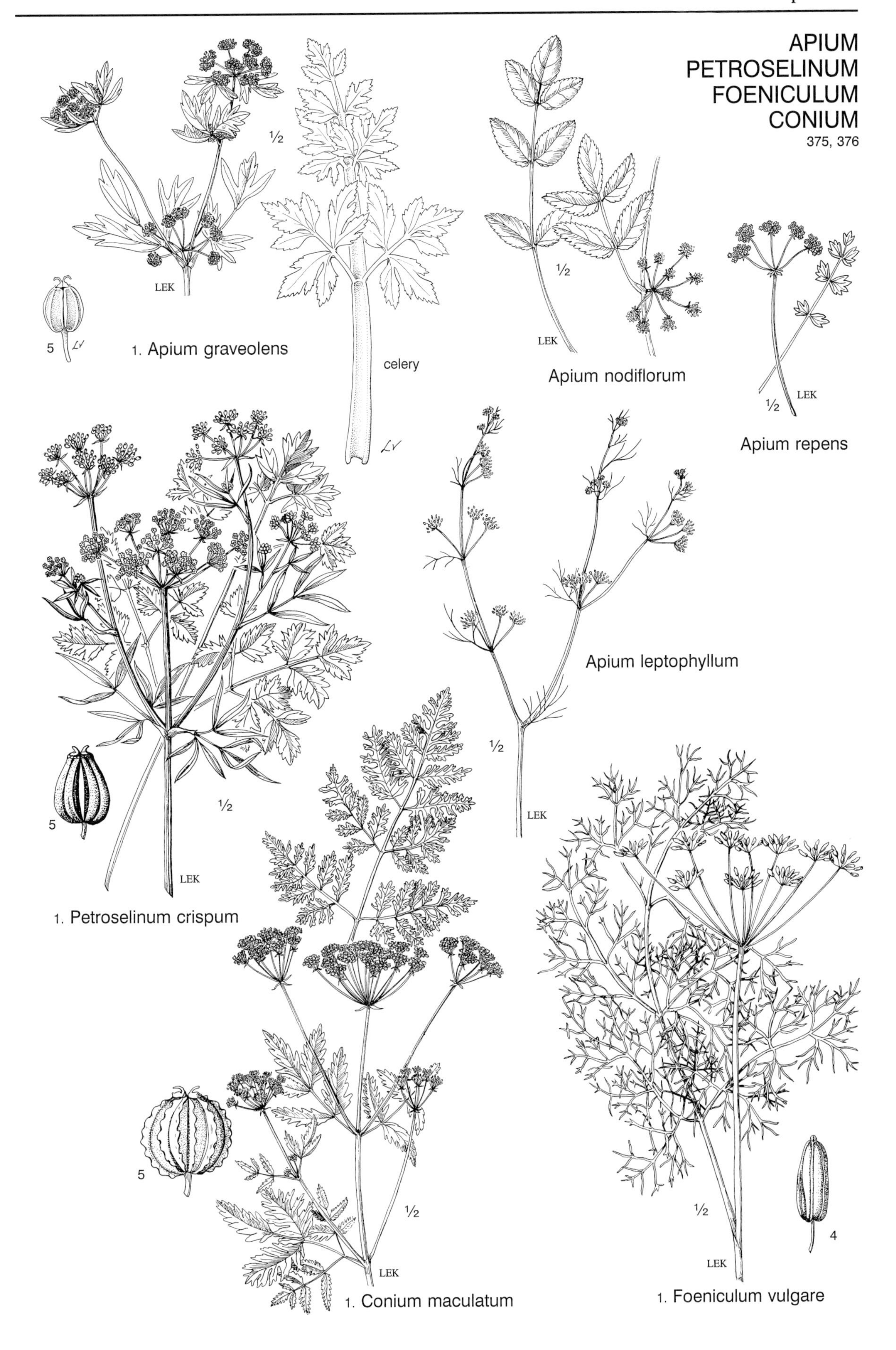

APIUM
PETROSELINUM
FOENICULUM
CONIUM
375, 376

1. Apium graveolens

5

LEK

celery

Apium nodiflorum

LEK

Apium repens

LEK

Apium leptophyllum

LEK

1. Petroselinum crispum

5

LEK

1. Conium maculatum

5

LEK

1. Foeniculum vulgare

4

LEK

ANETHUM
BUPLEURUM
TAENIDIA
376, 377

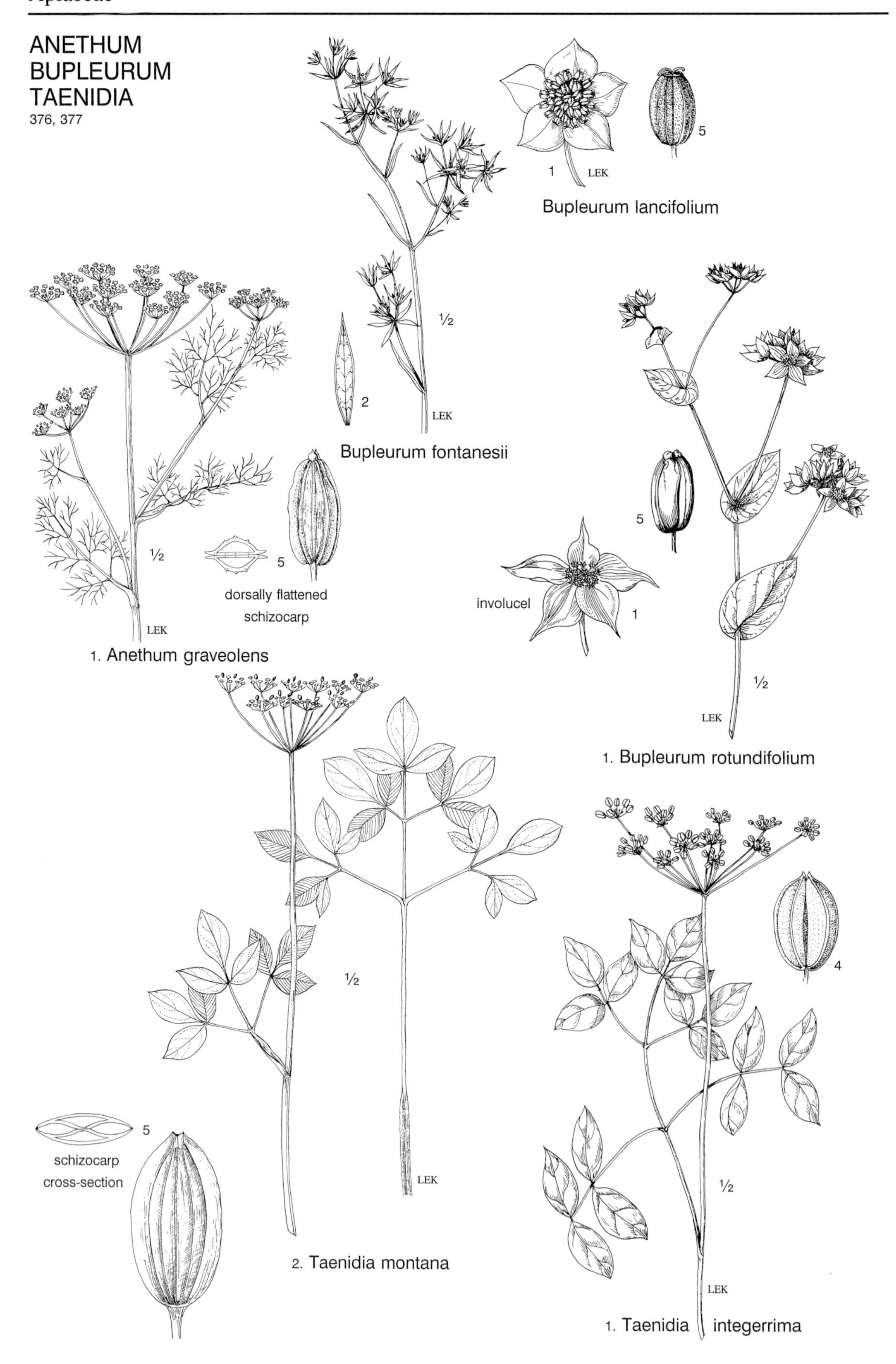

Bupleurum lancifolium

1   LEK   5

Bupleurum fontanesii

2   ½   LEK

dorsally flattened
schizocarp   5

½   LEK

1. Anethum graveolens

involucel   5   1   ½   LEK

1. Bupleurum rotundifolium

schizocarp
cross-section   5   ½   LEK

2. Taenidia montana

4   ½   LEK

1. Taenidia integerrima

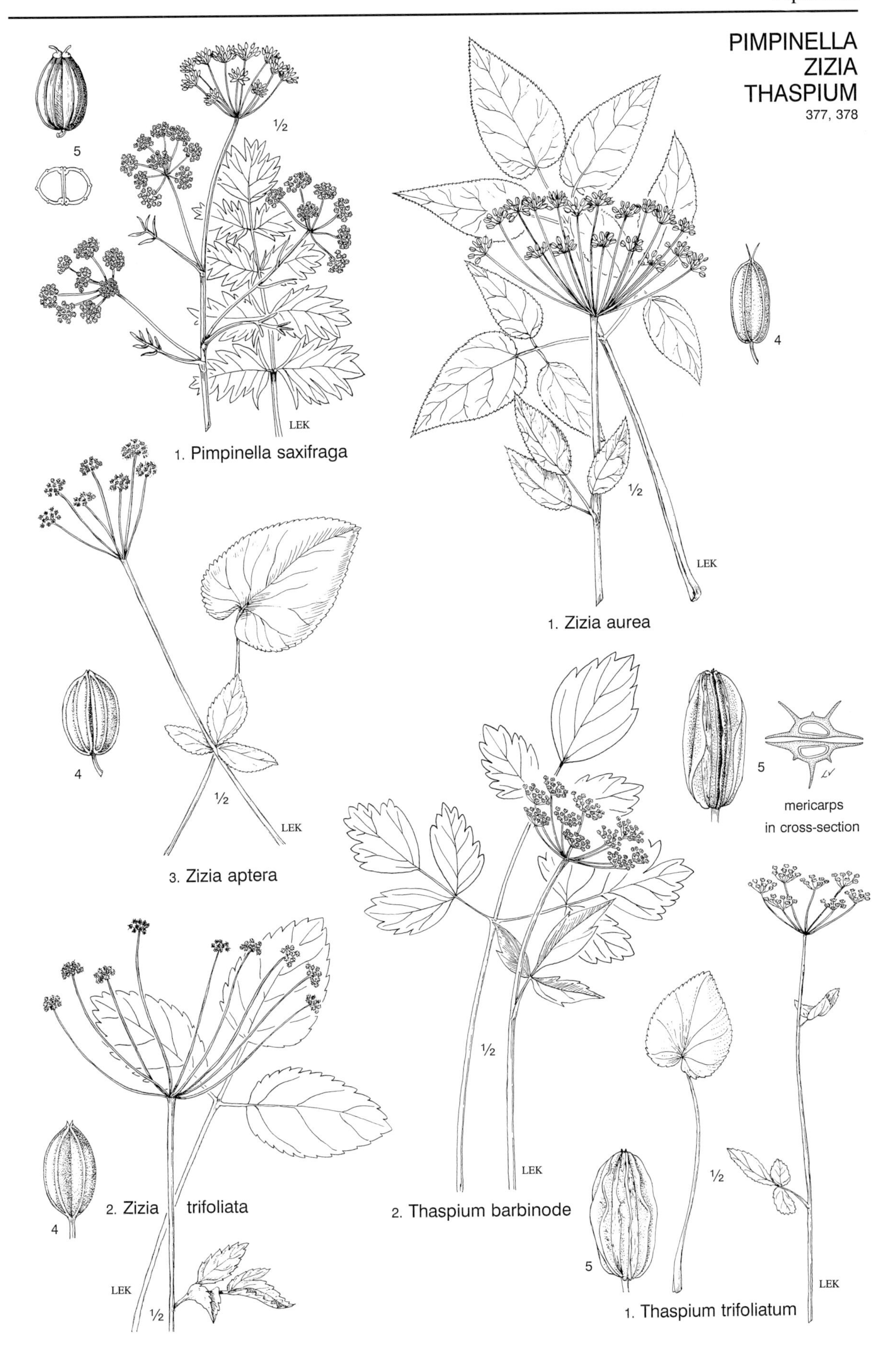

1. Pimpinella saxifraga

1. Zizia aurea

3. Zizia aptera

2. Zizia   trifoliata

5

mericarps
in cross-section

2. Thaspium barbinode

1. Thaspium trifoliatum

THASPIUM
LIGUSTICUM
PERIDERIDIA
OENANTHE
378, 379

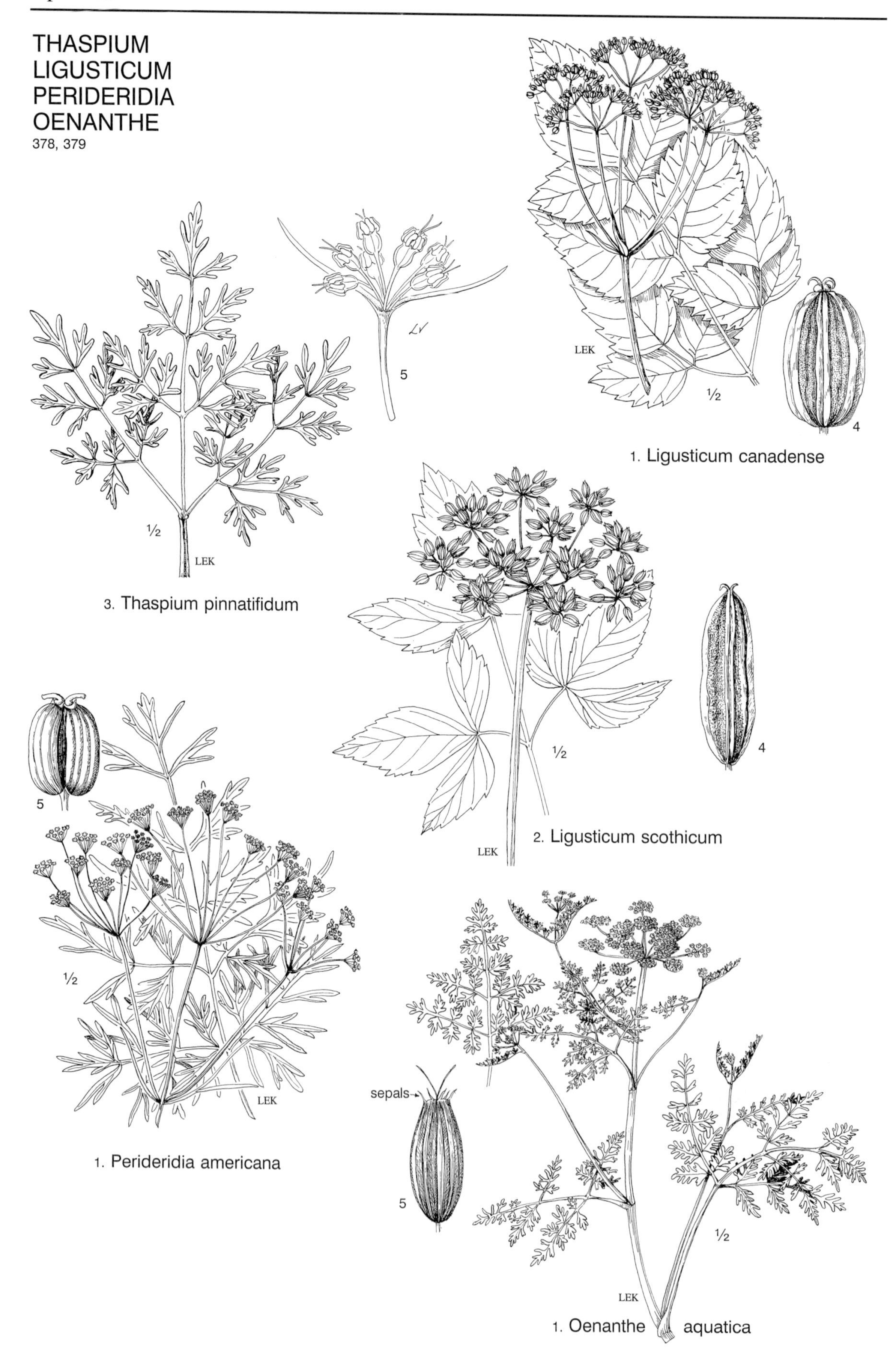

1. Ligusticum canadense

3. Thaspium pinnatifidum

2. Ligusticum scothicum

1. Perideridia americana

sepals→

1. Oenanthe aquatica

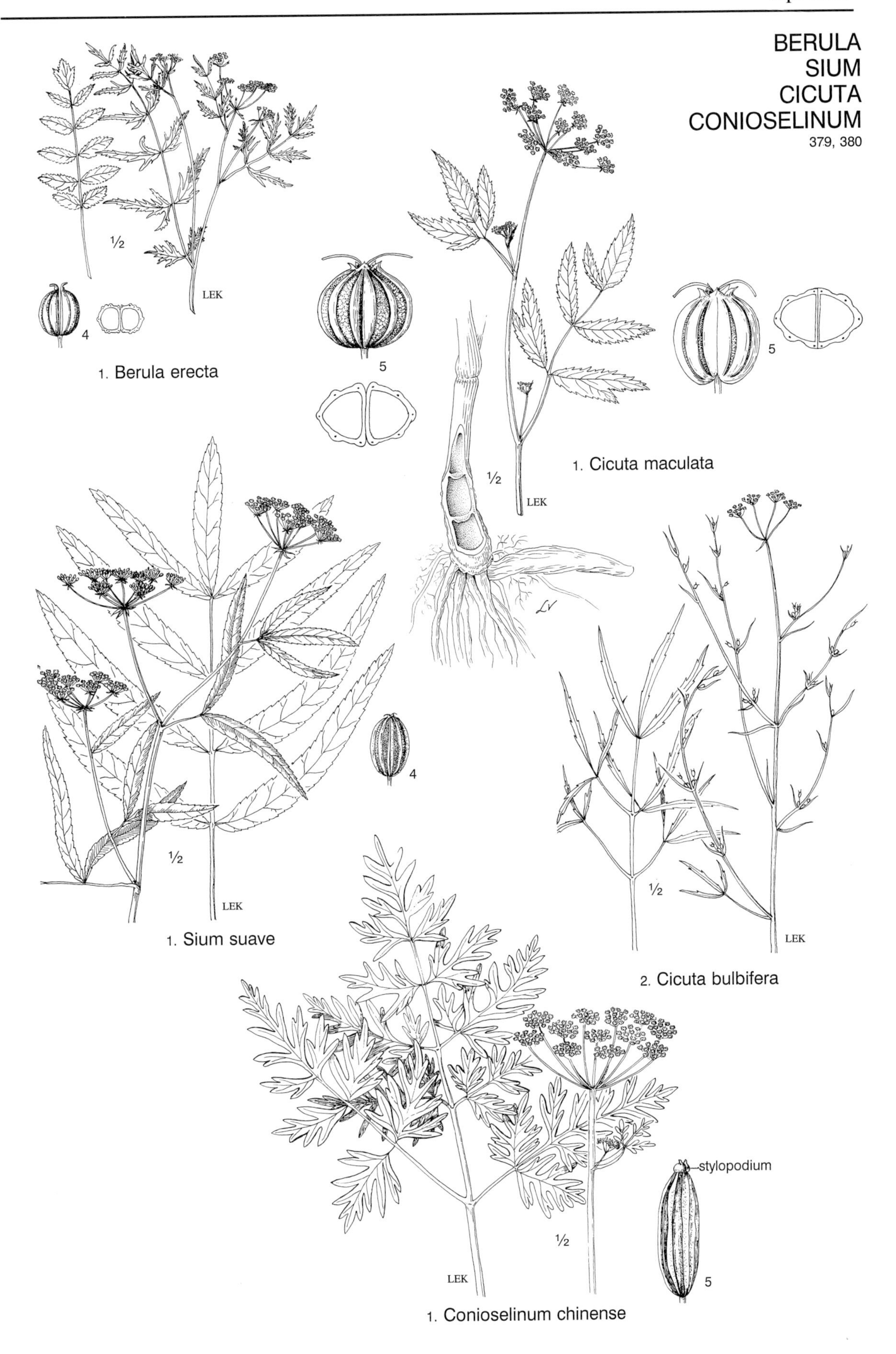

1. Berula erecta

1. Cicuta maculata

1. Sium suave

2. Cicuta bulbifera

stylopodium

1. Conioselinum chinense

OXYPOLIS
PTILIMNIUM
380, 381

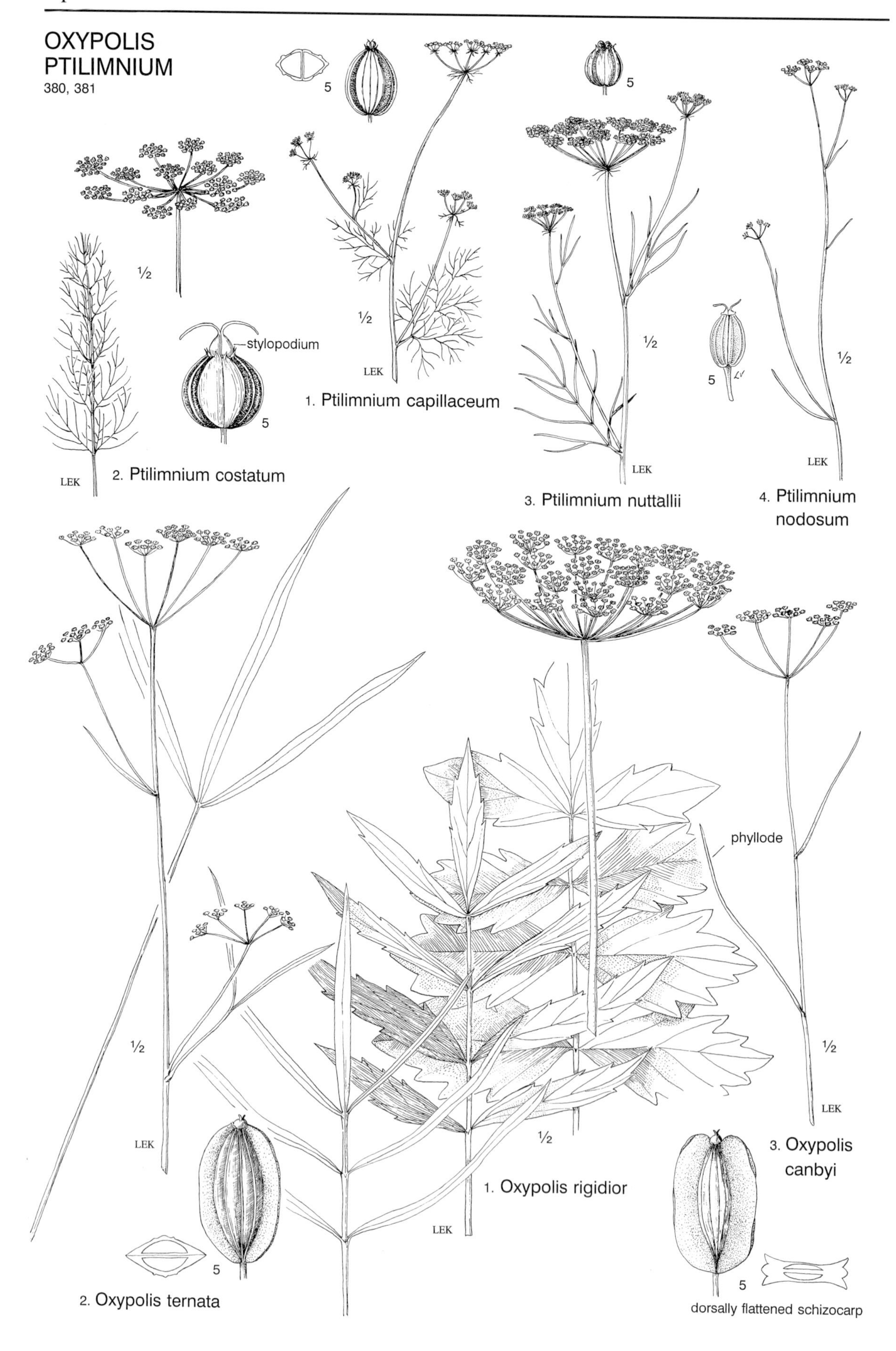

5

1. Ptilimnium capillaceum

LEK

—stylopodium

5

LEK   2. Ptilimnium costatum

3. Ptilimnium nuttallii

5

4. Ptilimnium nodosum

LEK

phyllode

1. Oxypolis rigidior

3. Oxypolis canbyi

2. Oxypolis ternata

5

dorsally flattened schizocarp

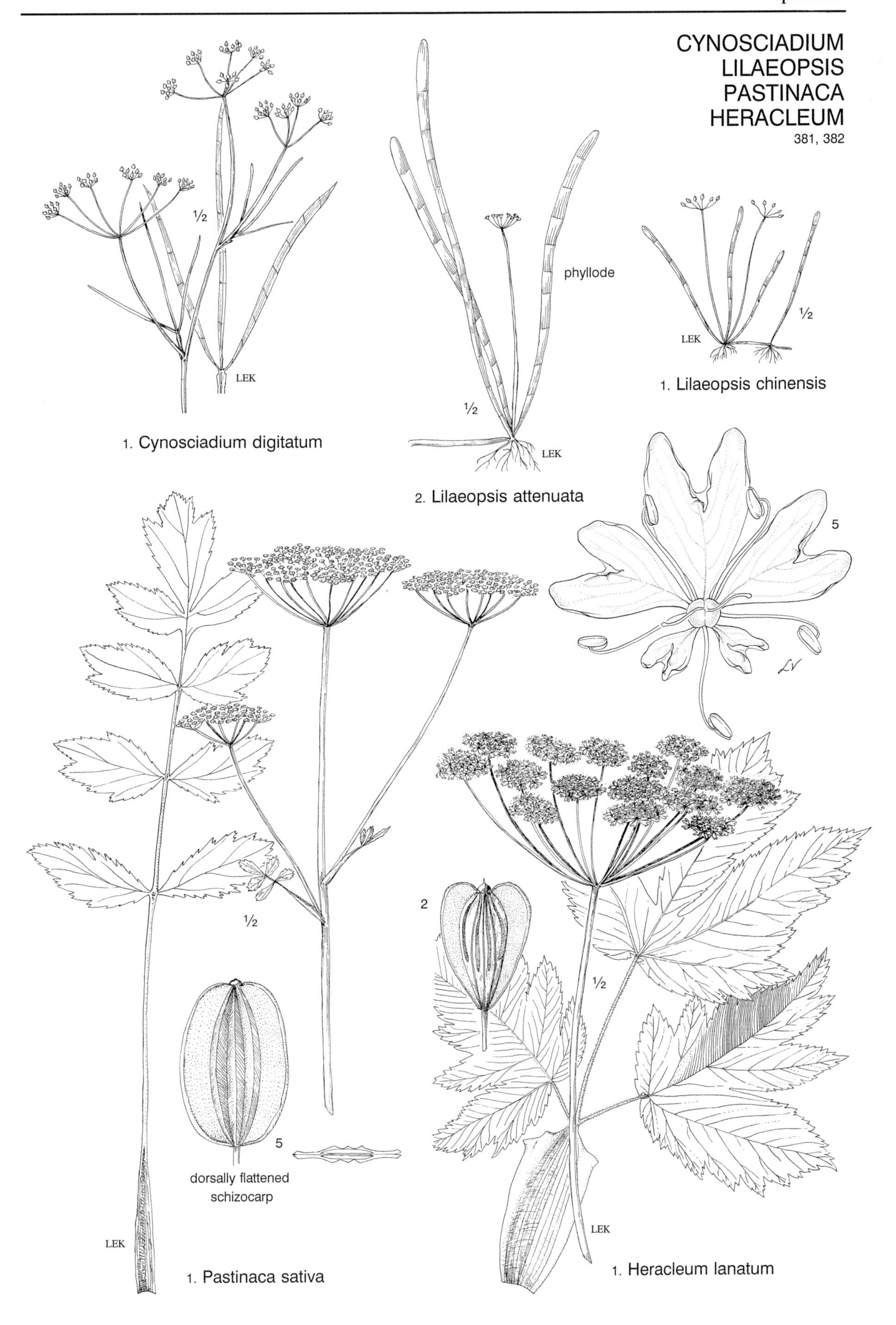

½

LEK

1. Cynosciadium digitatum

phyllode

½

LEK

2. Lilaeopsis attenuata

½

LEK

½

1. Lilaeopsis chinensis

5

½

2

½

5

dorsally flattened
schizocarp

LEK

1. Pastinaca sativa

LEK

1. Heracleum lanatum

HERACLEUM
ANGELICA
382, 383

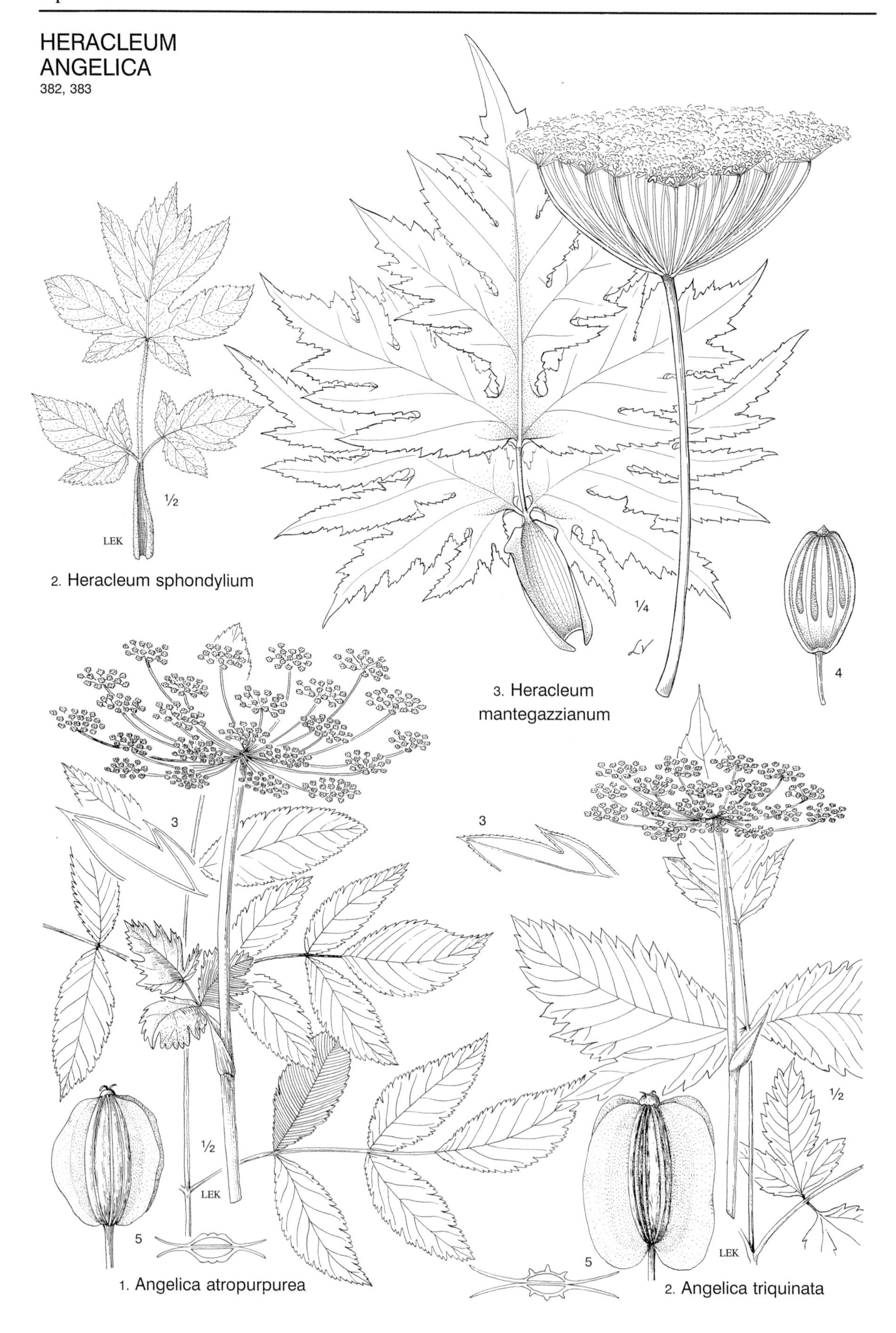

2. Heracleum sphondylium

LEK

½

¼

3. Heracleum
mantegazzianum

4

1. Angelica atropurpurea

LEK

½

5

3

2. Angelica triquinata

LEK

½

5

3

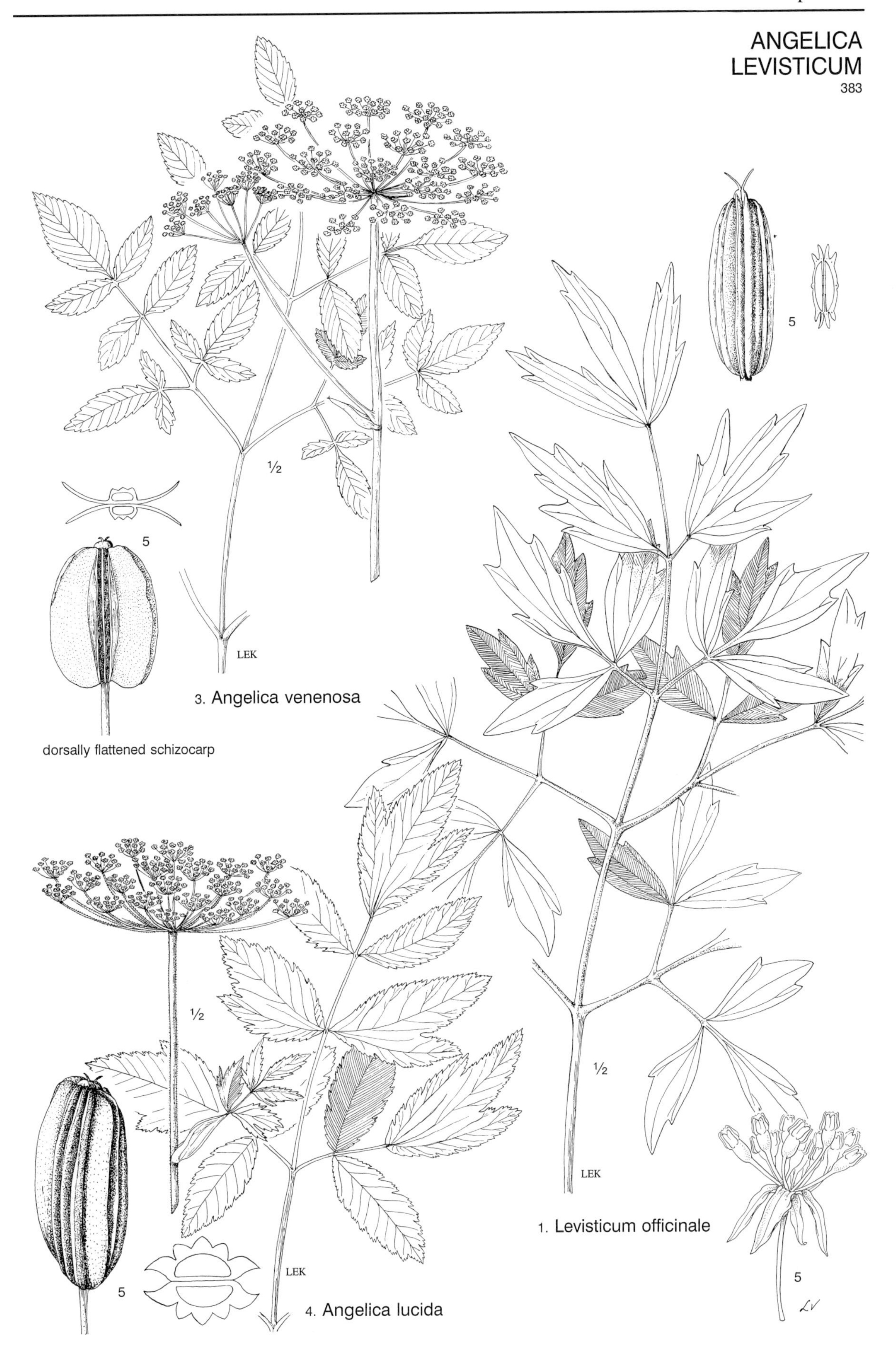

3. Angelica venenosa

dorsally flattened schizocarp

4. Angelica lucida

1. Levisticum officinale

PEUCEDANUM
POLYTAENIA
LOMATIUM
383, 384

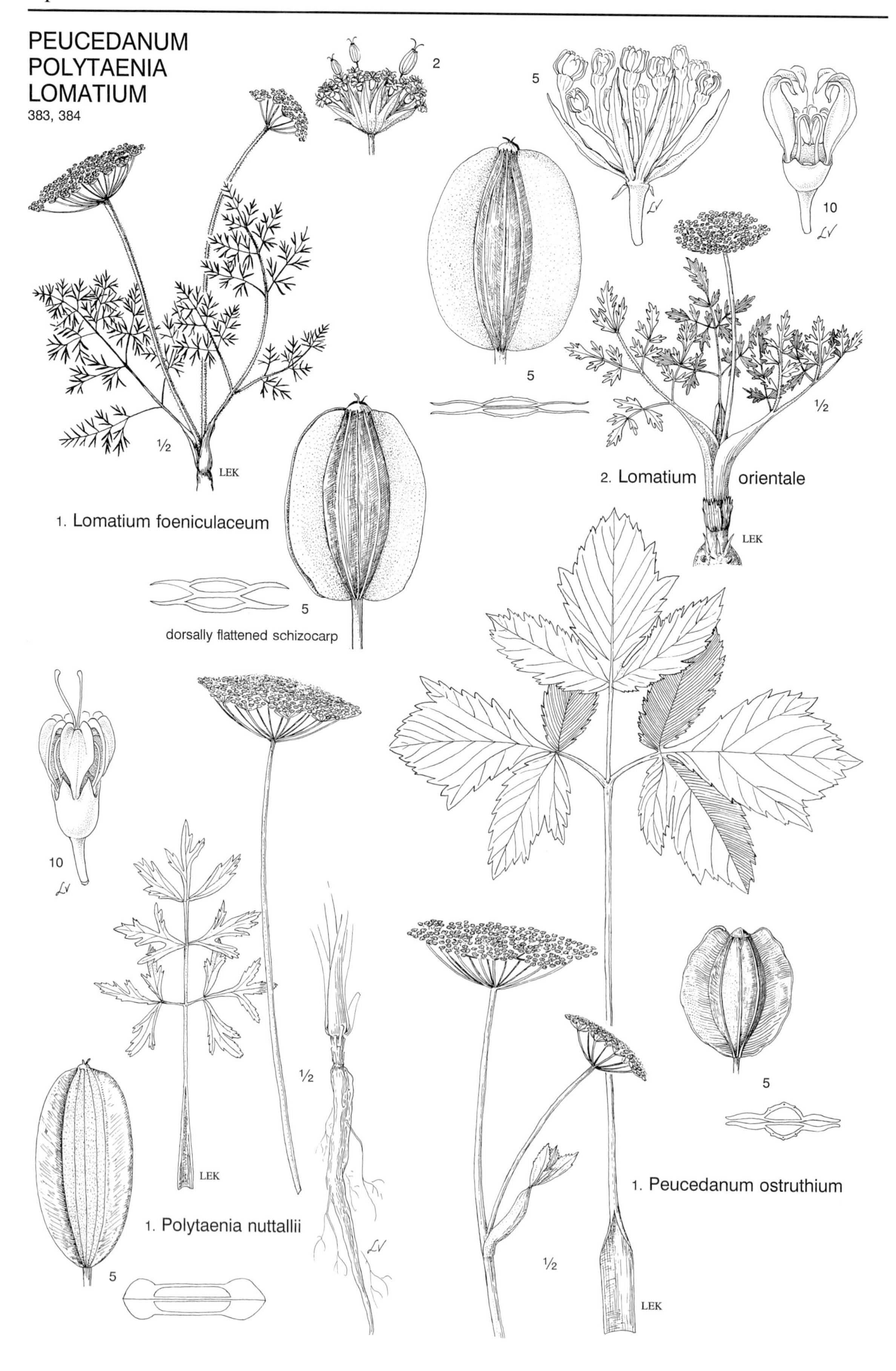

2

5

5

10

5

5

1. Lomatium foeniculaceum

2. Lomatium orientale

LEK

dorsally flattened schizocarp

10

1. Polytaenia nuttallii

1. Peucedanum ostruthium

LEK

LEK

½

½

½

½

5

5

5

5

LEK

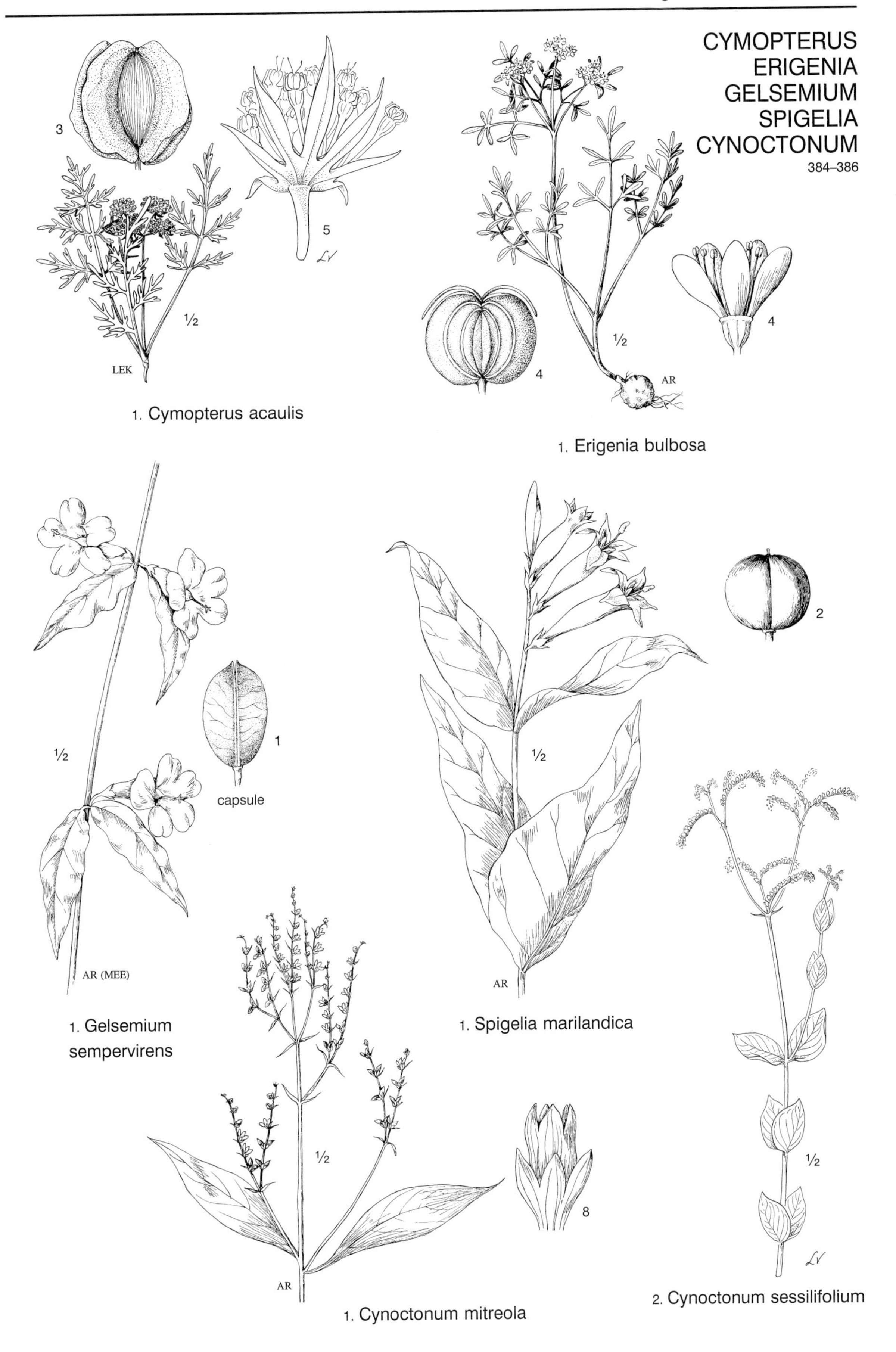

CYMOPTERUS
ERIGENIA
GELSEMIUM
SPIGELIA
CYNOCTONUM
384–386

1. Cymopterus acaulis

1. Erigenia bulbosa

1. Gelsemium
sempervirens

1. Spigelia marilandica

1. Cynoctonum mitreola

2. Cynoctonum sessilifolium

SABATIA
387

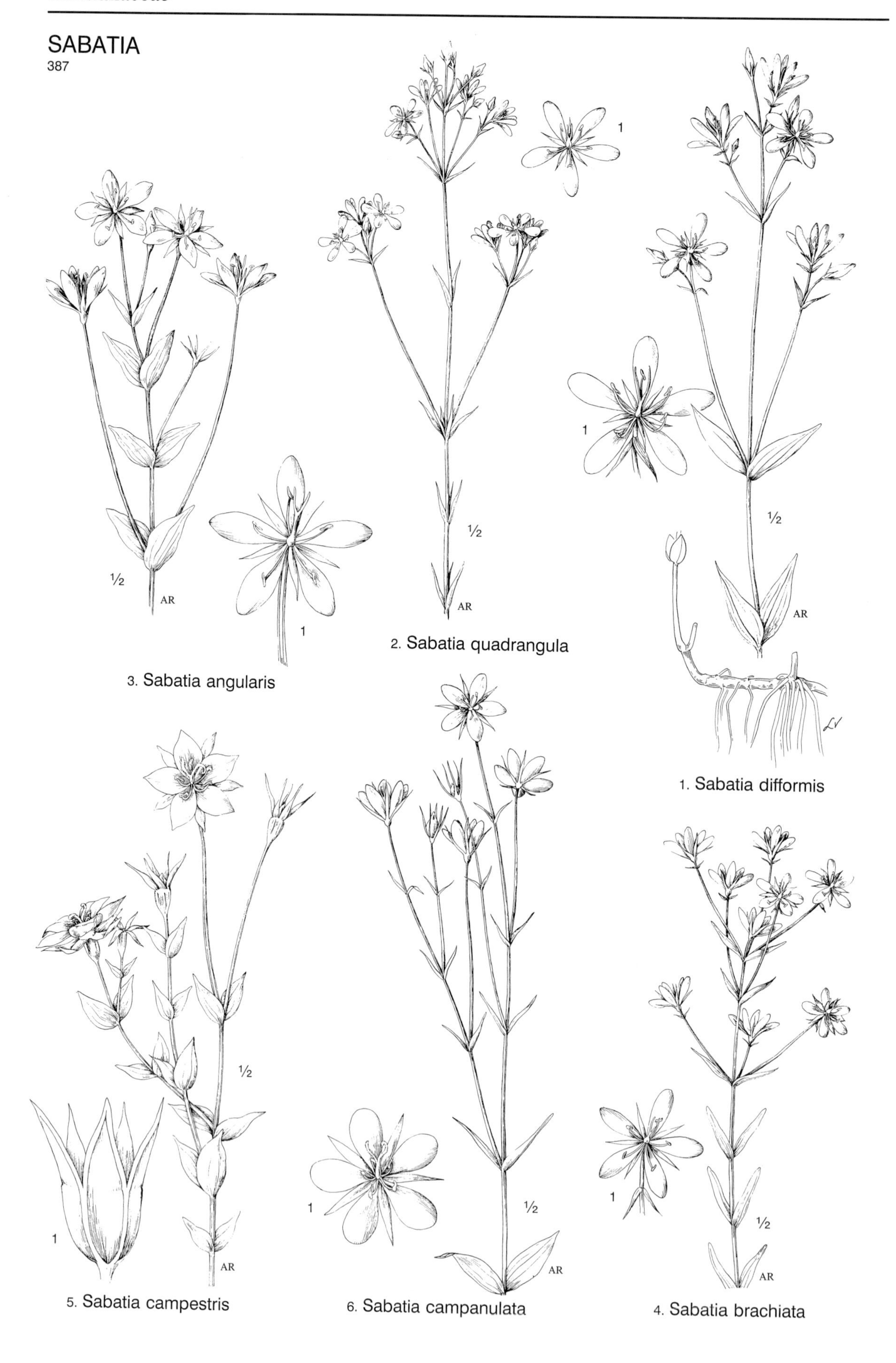

2. Sabatia quadrangula

3. Sabatia angularis

1. Sabatia difformis

5. Sabatia campestris

6. Sabatia campanulata

4. Sabatia brachiata

ILLUSTRATED COMPANION TO

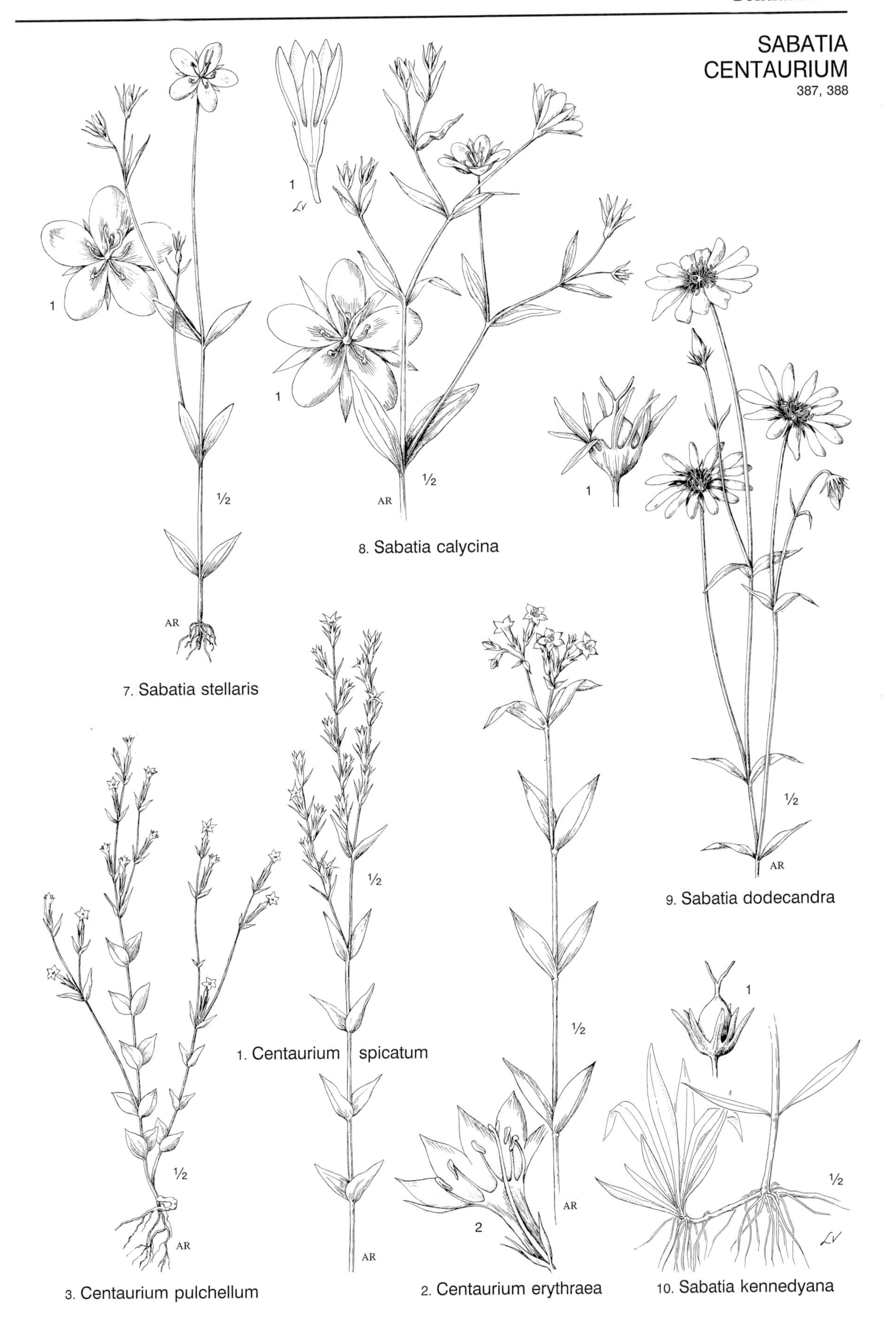

1

8. Sabatia calycina

7. Sabatia stellaris

1. Centaurium spicatum

9. Sabatia dodecandra

3. Centaurium pulchellum

2. Centaurium erythraea

10. Sabatia kennedyana

GENTIANA
389

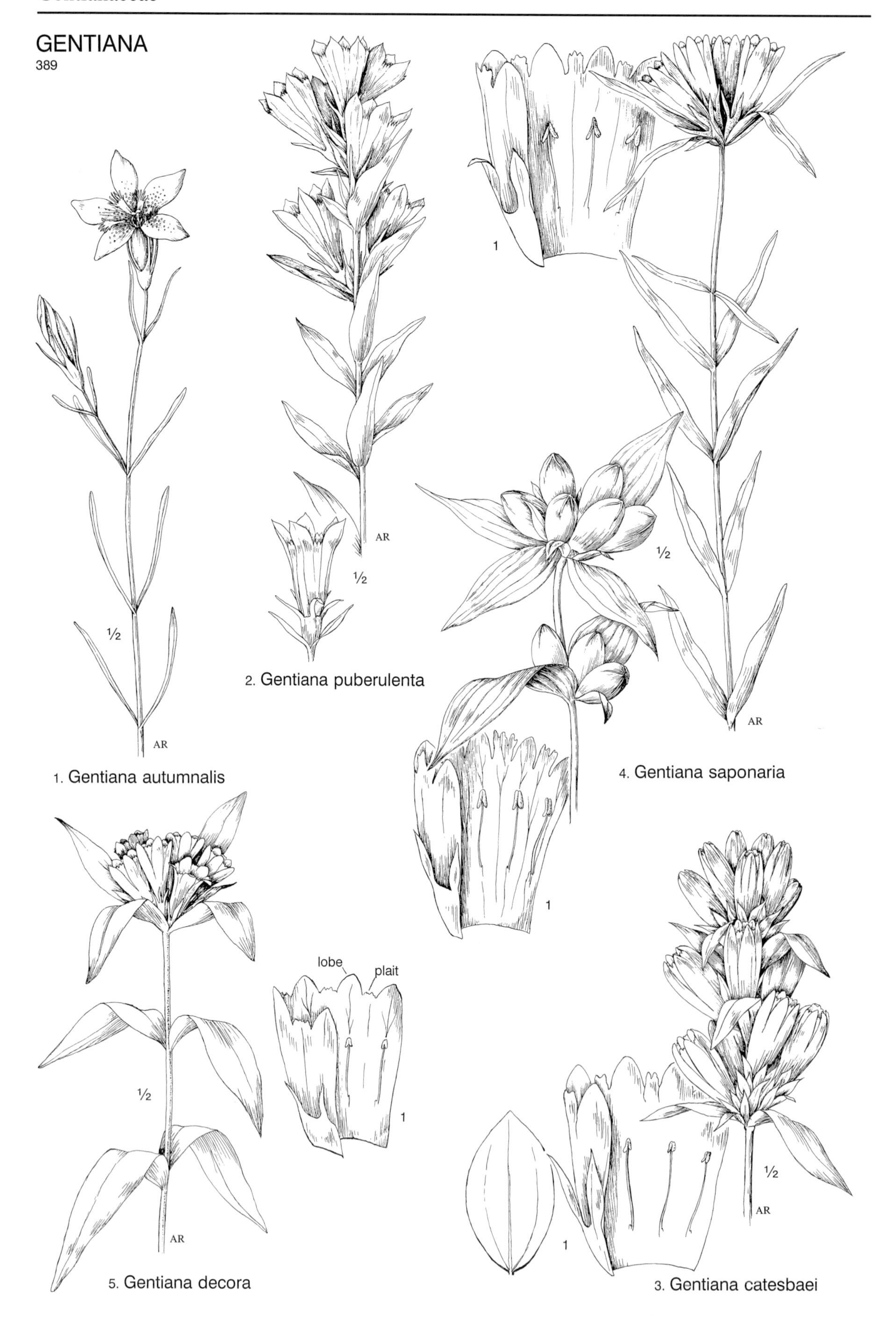

2. Gentiana puberulenta

1. Gentiana autumnalis

4. Gentiana saponaria

lobe   plait

5. Gentiana decora

3. Gentiana catesbaei

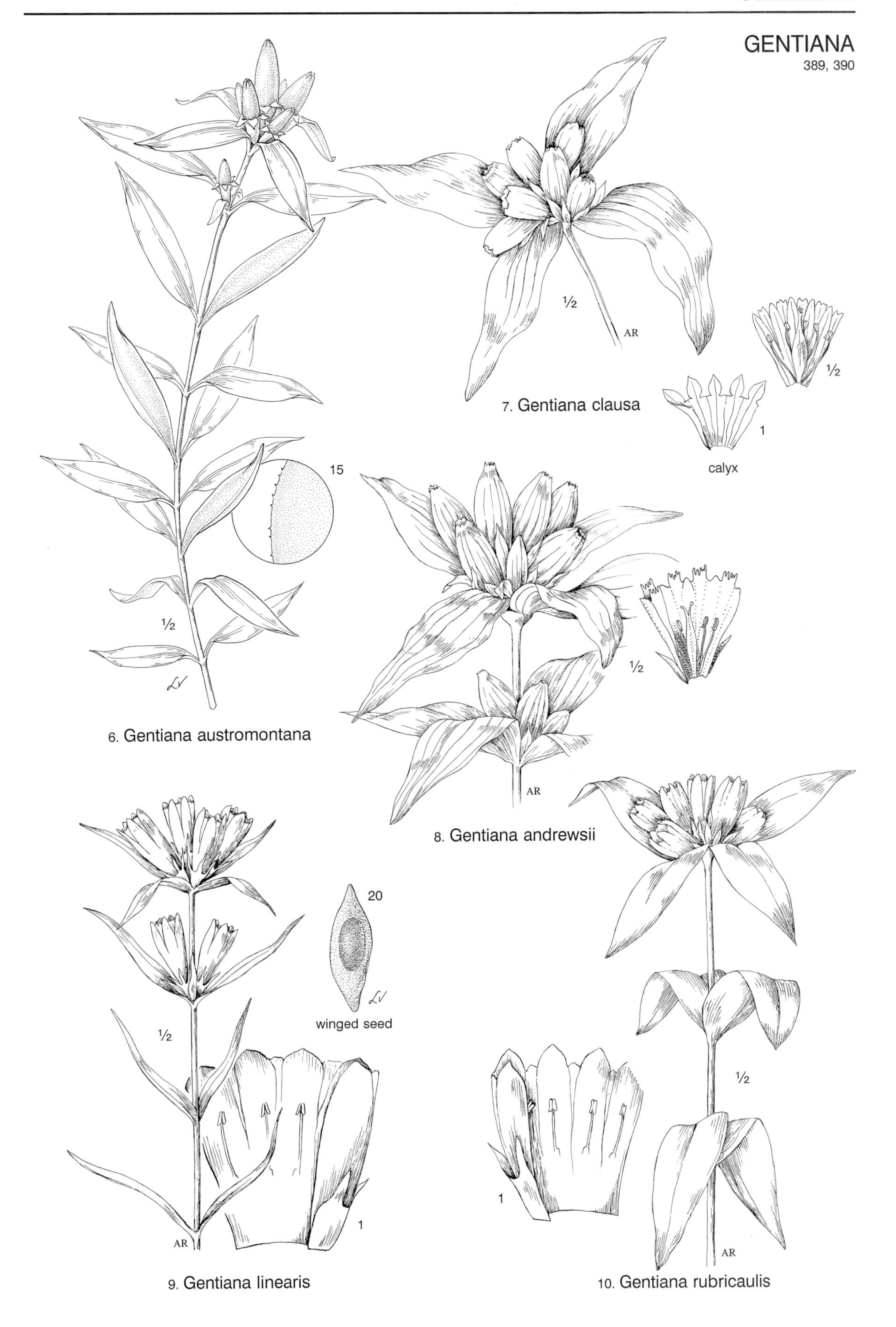

7. Gentiana clausa

½ AR

½

1

calyx

15

6. Gentiana austromontana

8. Gentiana andrewsii

½

20

winged seed

9. Gentiana linearis

½

1

AR

10. Gentiana rubricaulis

1

½

AR

GENTIANA
GENTIANOPSIS
GENTIANELLA
390, 391

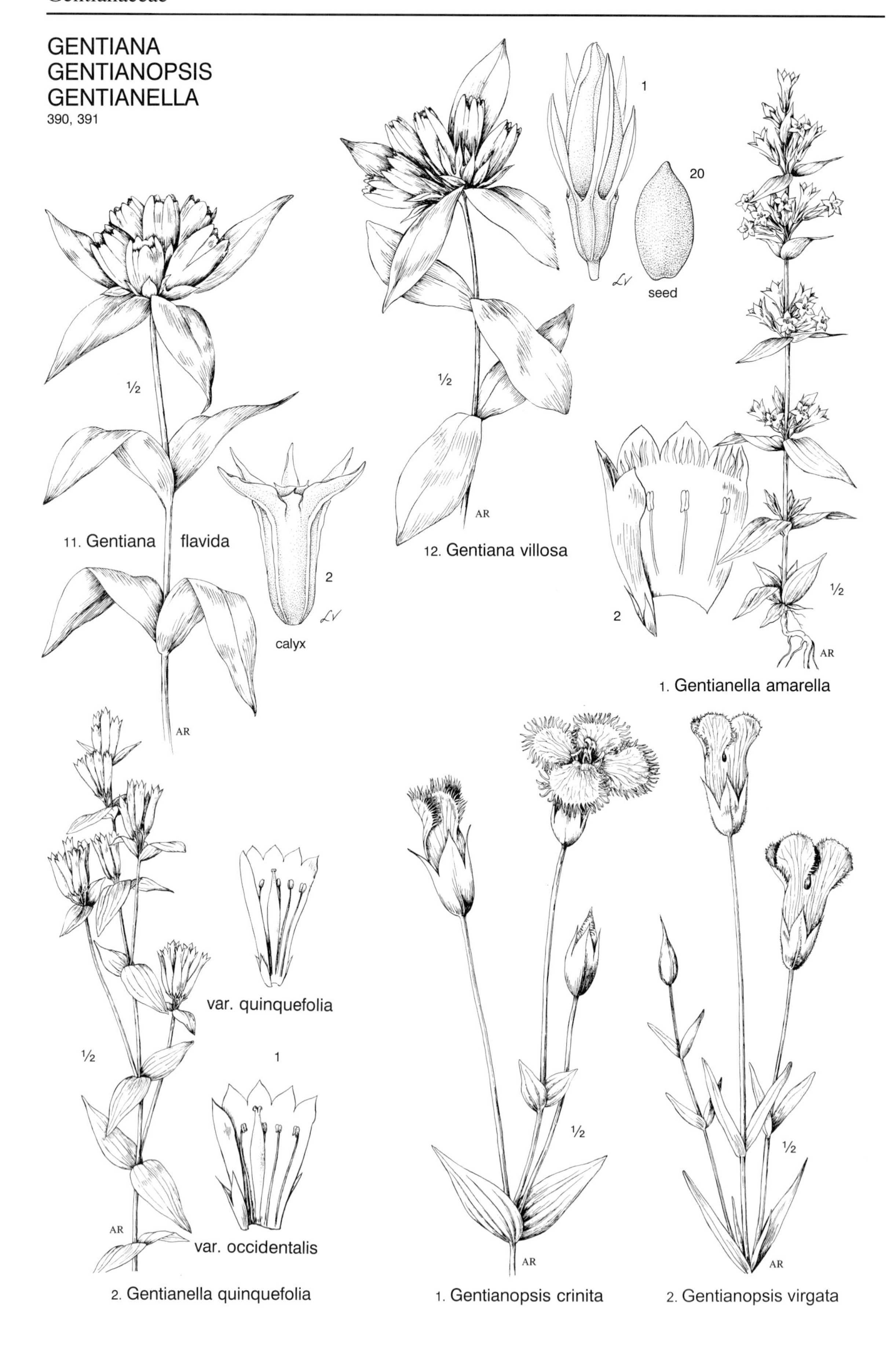

11. Gentiana flavida

calyx

12. Gentiana villosa

seed

1. Gentianella amarella

var. quinquefolia

var. occidentalis

2. Gentianella quinquefolia

1. Gentianopsis crinita

2. Gentianopsis virgata

HALENIA
LOMATOGONIUM
FRASERA
BARTONIA
OBOLARIA
391, 392

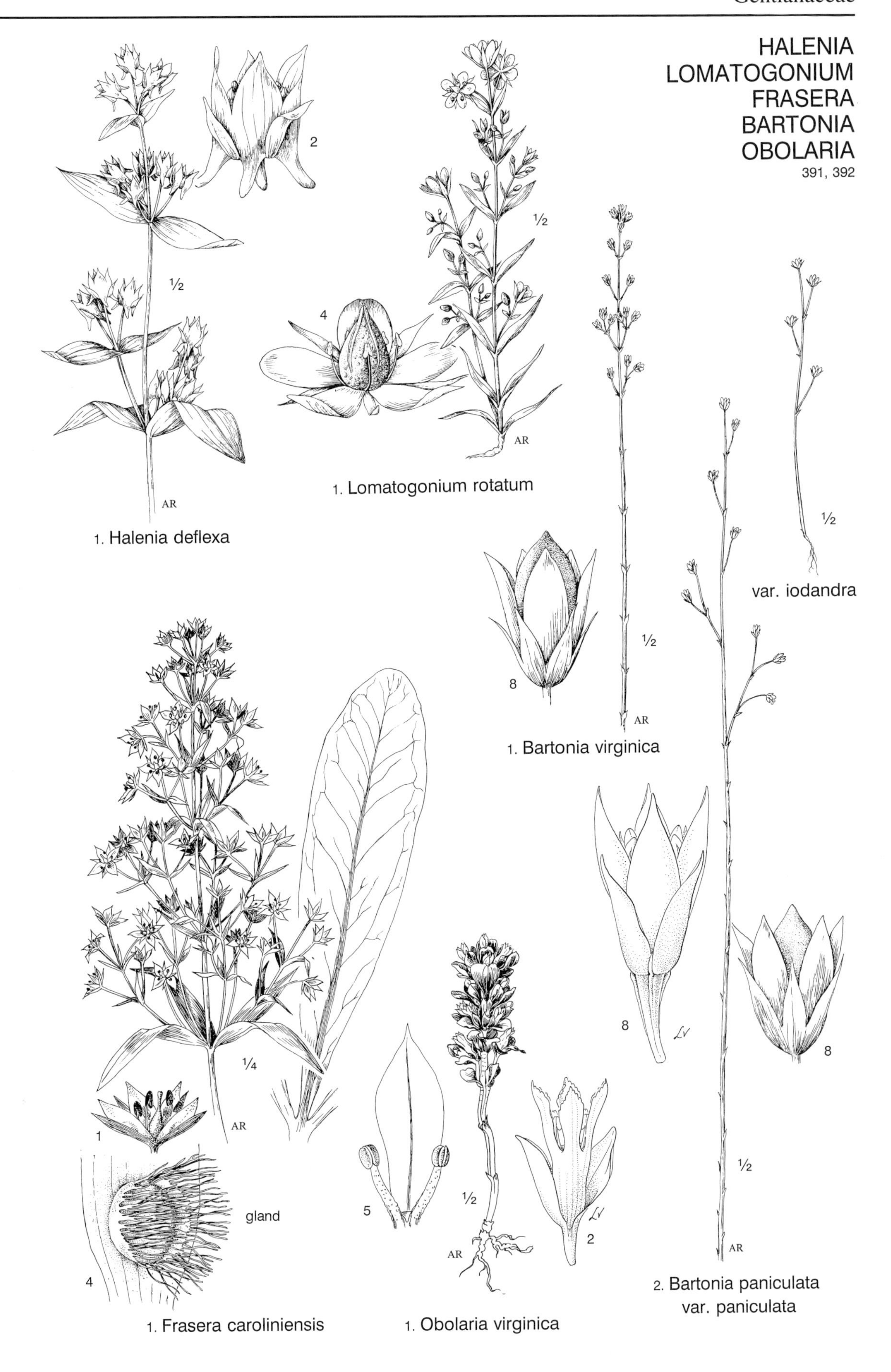

1. Halenia deflexa

1. Lomatogonium rotatum

var. iodandra

1. Bartonia virginica

1. Frasera caroliniensis

gland

1. Obolaria virginica

2. Bartonia paniculata
var. paniculata

VINCA
TRACHELOSPERMUM
AMSONIA
393, 394

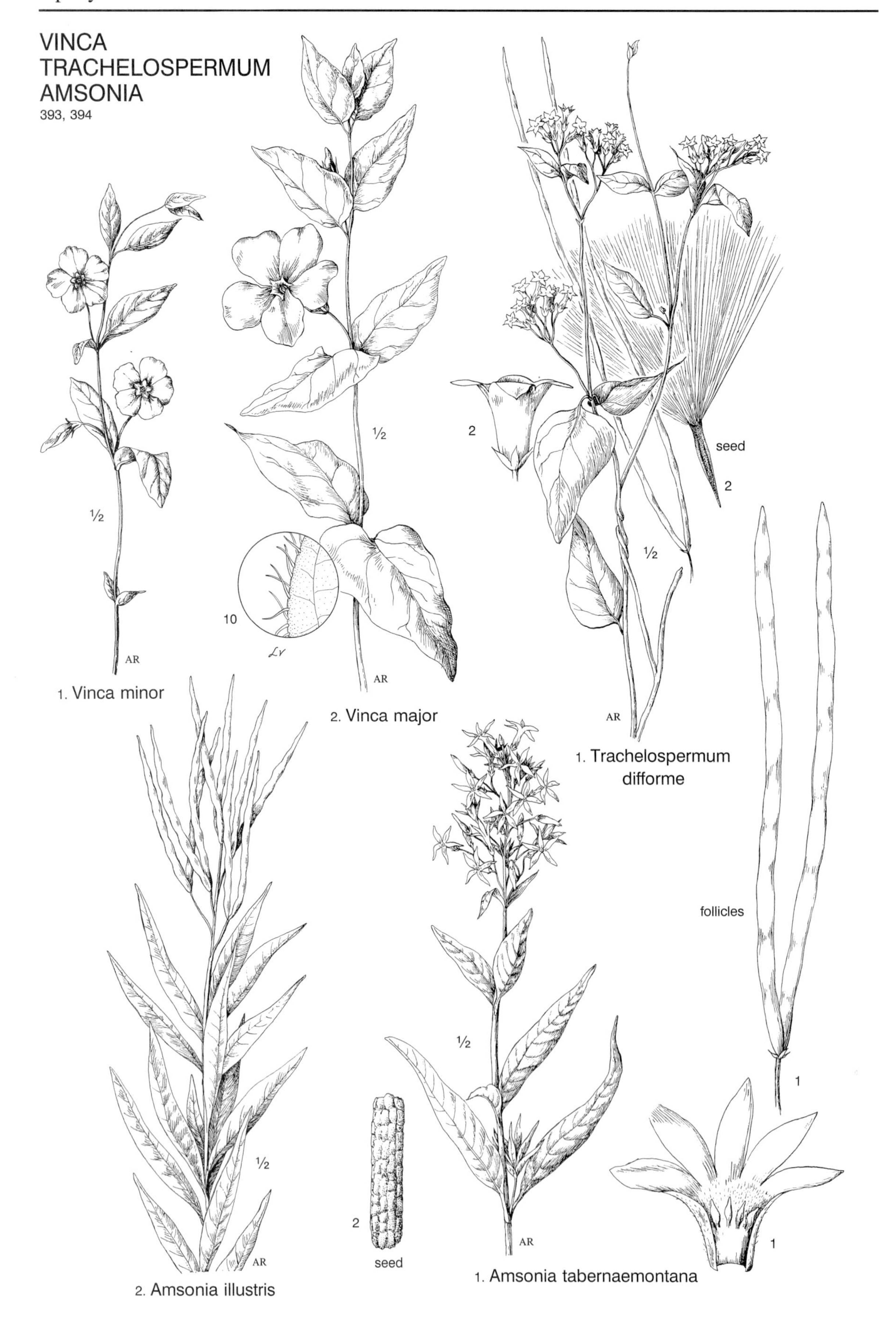

1. Vinca minor

2. Vinca major

1. Trachelospermum difforme

seed

follicles

2. Amsonia illustris

seed

1. Amsonia tabernaemontana

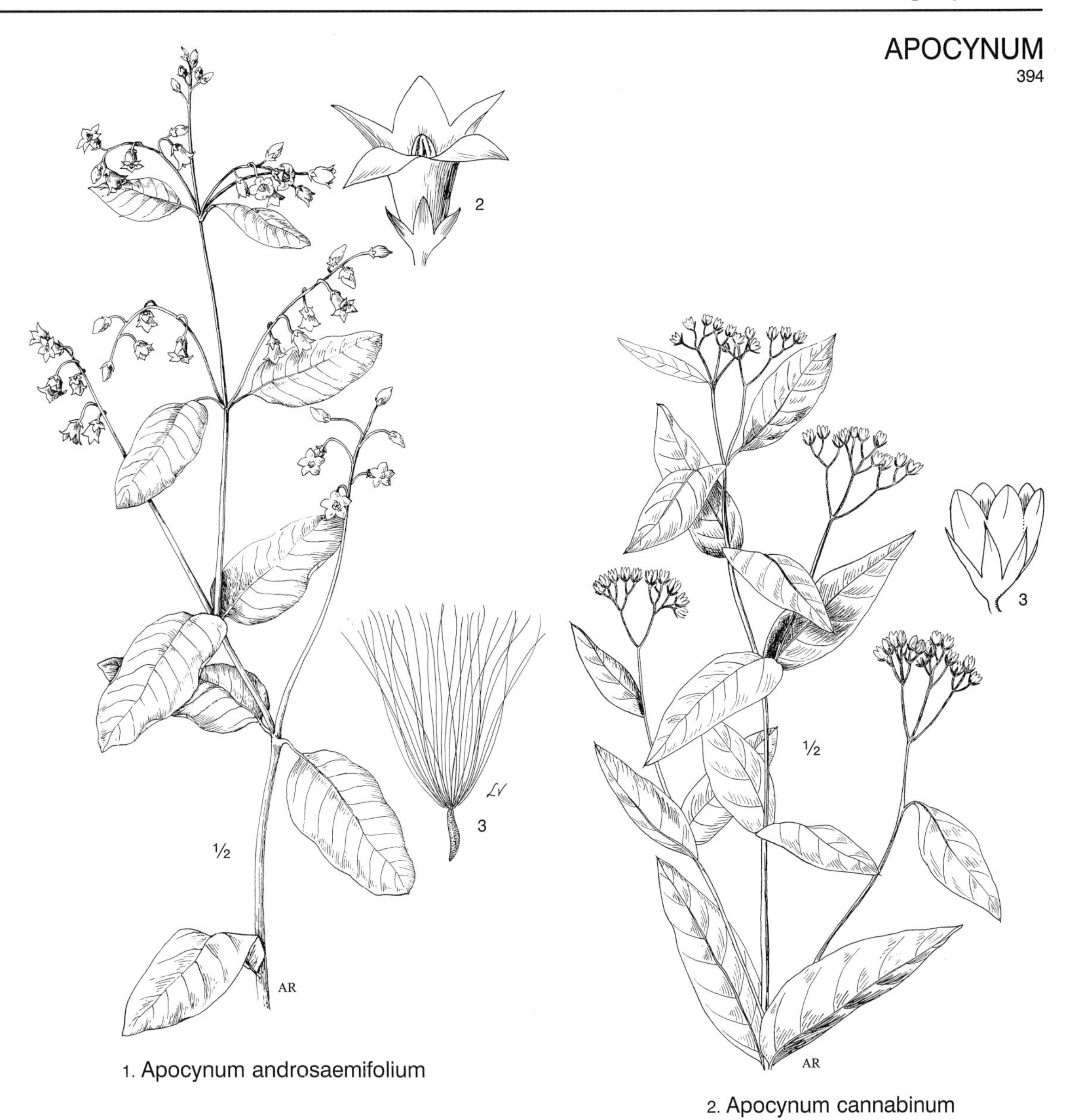

2

3

½

AR

1. Apocynum androsaemifolium

3

½

AR

2. Apocynum cannabinum

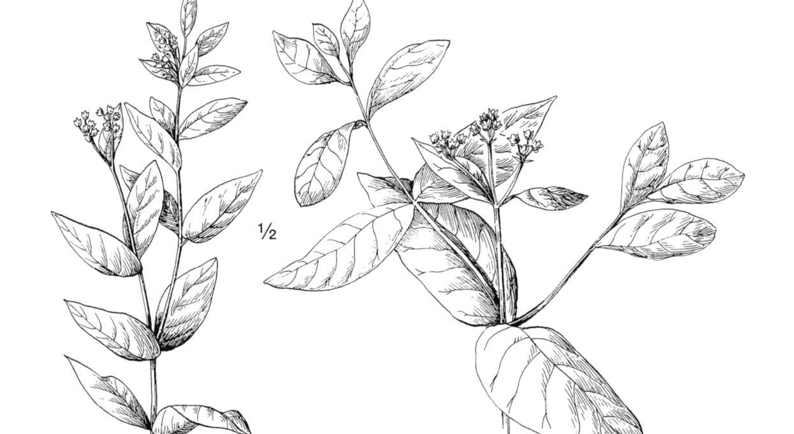

½

AR

3. Apocynum sibiricum

PERIPLOCA
ASCLEPIAS
395–397

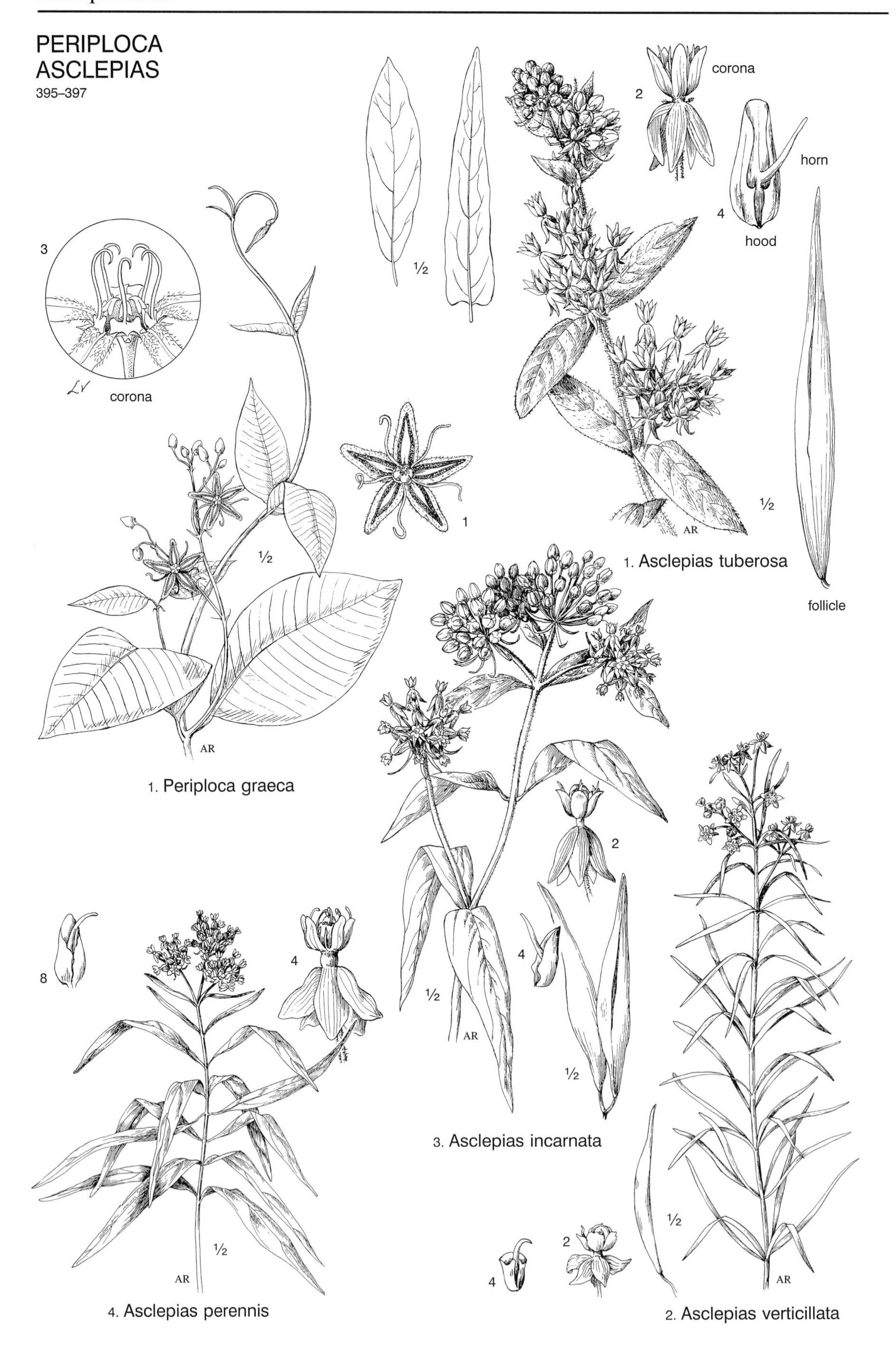

corona

2

4

horn

hood

3

corona

1

1. Asclepias tuberosa

½

follicle

1. Periploca graeca

2

4

½

4

½

3. Asclepias incarnata

8

4

½

AR

½

½

2

4

AR

4. Asclepias perennis

2. Asclepias verticillata

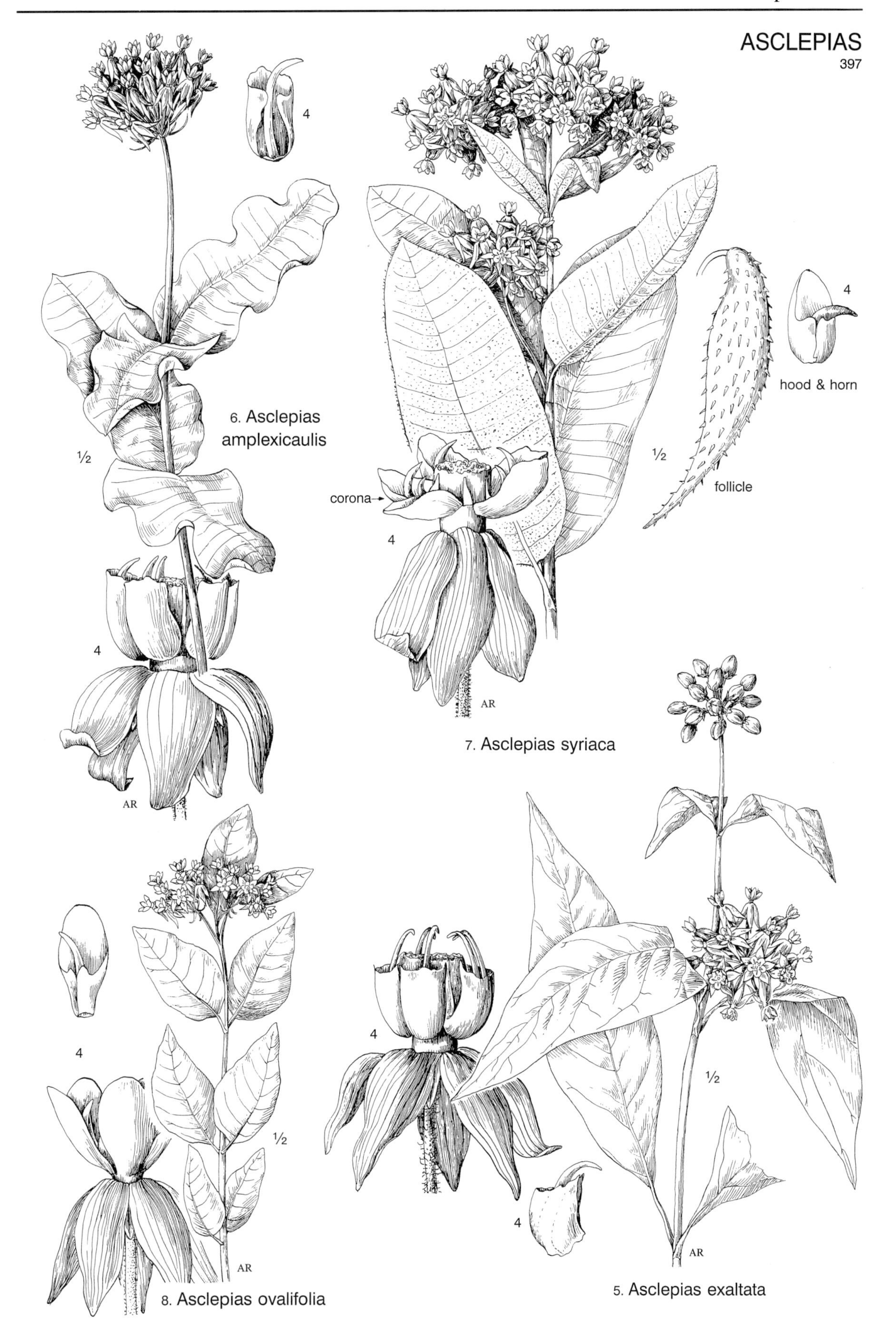

6. Asclepias amplexicaulis

½

4

4

AR

corona→

4

hood & horn

½

follicle

AR

7. Asclepias syriaca

4

½

AR

5. Asclepias exaltata

4

4

½

AR

8. Asclepias ovalifolia

ASCLEPIAS
397

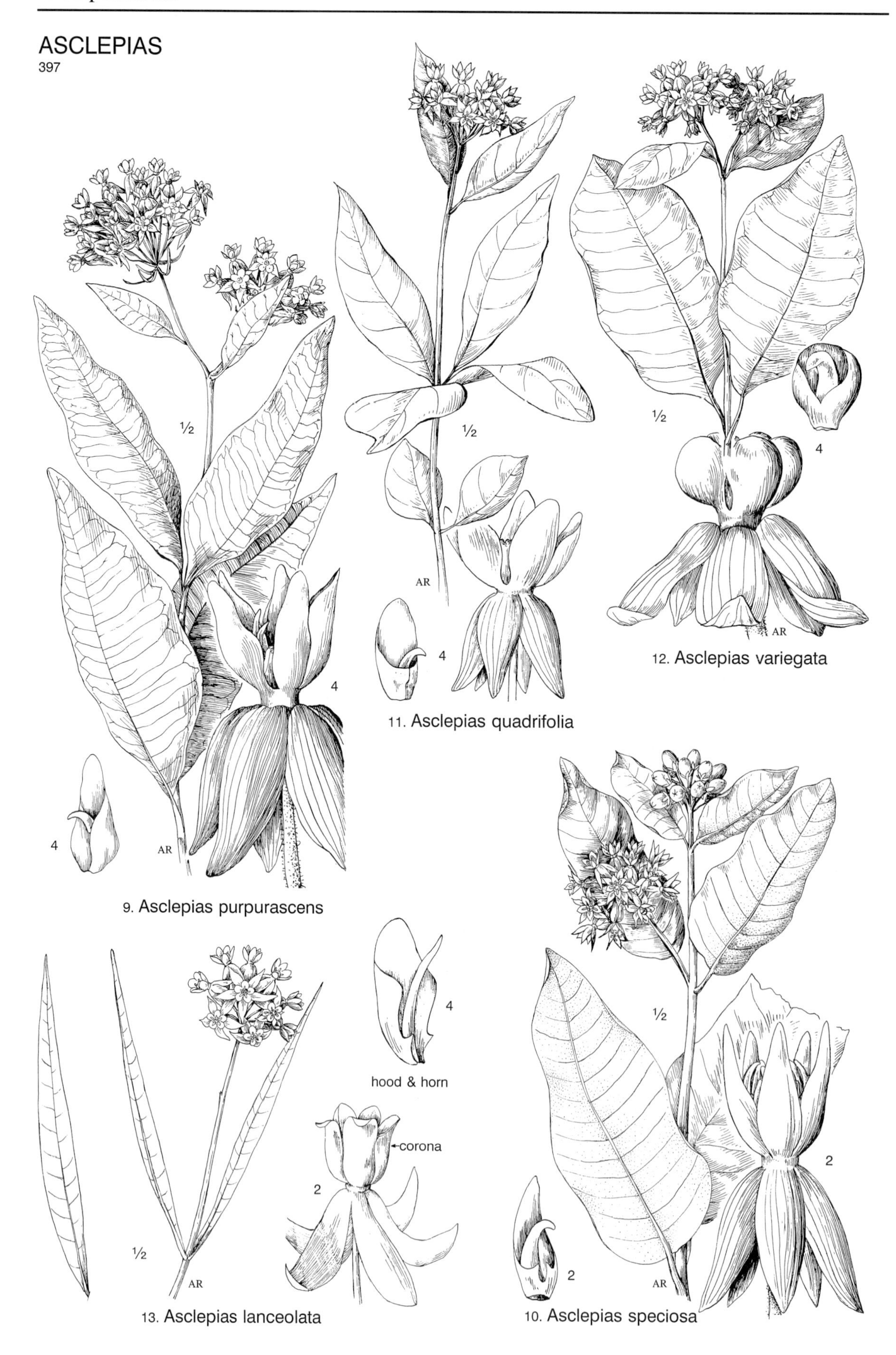

11. Asclepias quadrifolia

12. Asclepias variegata

9. Asclepias purpurascens

hood & horn

corona

13. Asclepias lanceolata

10. Asclepias speciosa

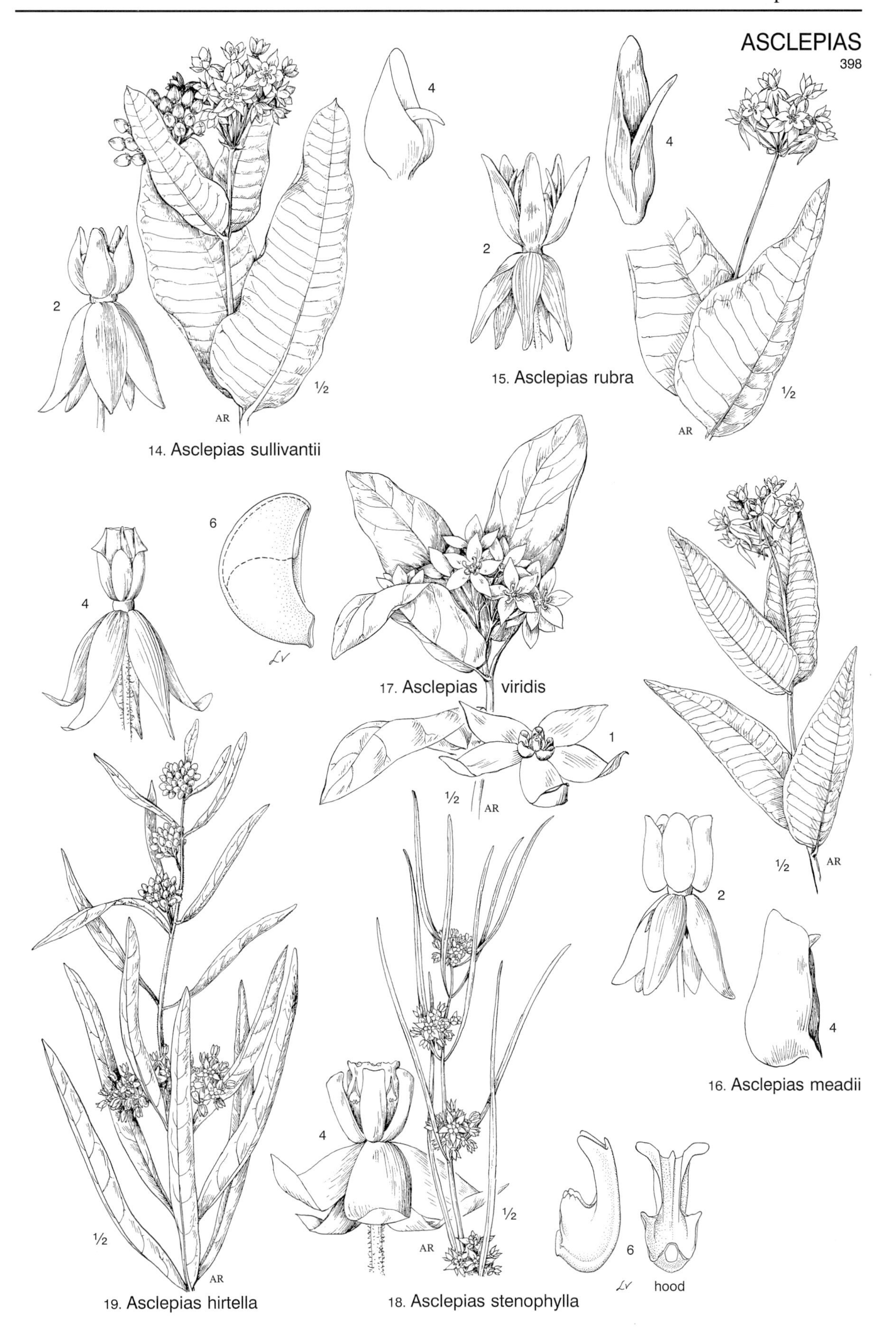

14. Asclepias sullivantii

15. Asclepias rubra

17. Asclepias viridis

16. Asclepias meadii

19. Asclepias hirtella

18. Asclepias stenophylla

hood

ASCLEPIAS
METAPLEXIS
VINCETOXICUM
398, 399

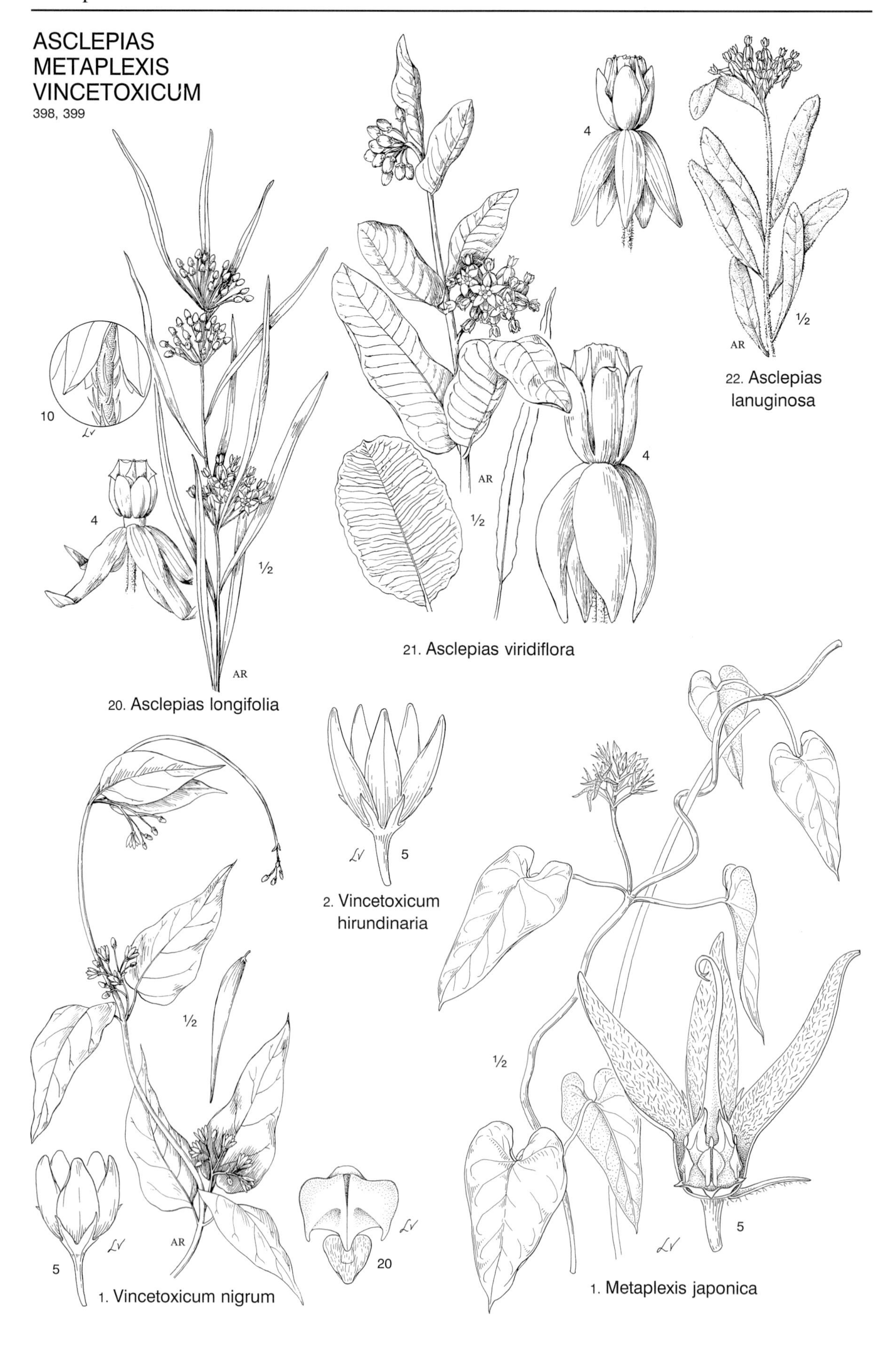

20. Asclepias longifolia

21. Asclepias viridiflora

22. Asclepias
lanuginosa

2. Vincetoxicum
hirundinaria

1. Vincetoxicum nigrum

1. Metaplexis japonica

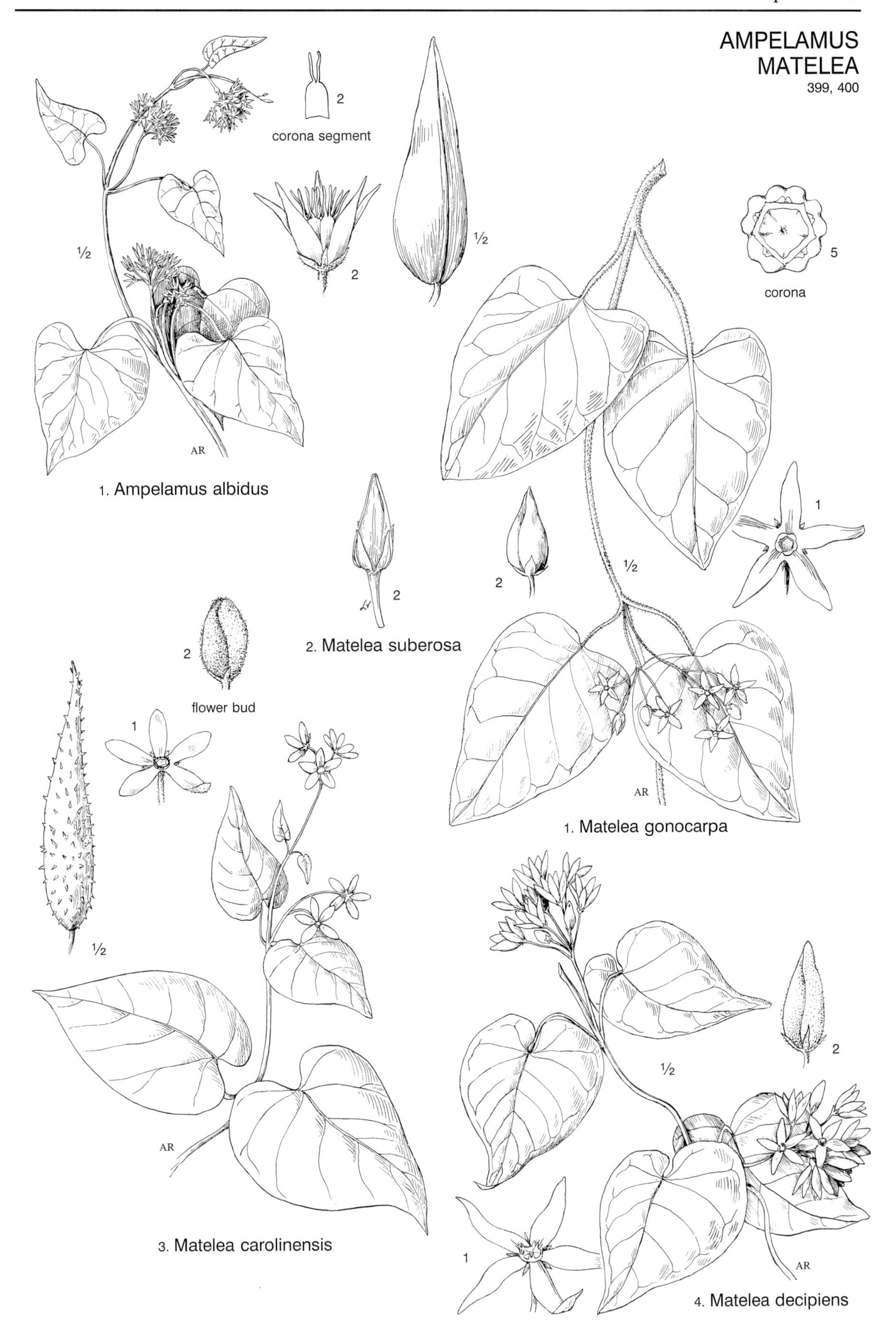

corona segment

2. corona

1. Ampelamus albidus

2. Matelea suberosa

flower bud

1. Matelea gonocarpa

3. Matelea carolinensis

4. Matelea decipiens

MATELEA
NICANDRA
LEUCOPHYSALIS
PHYSALIS
400–402

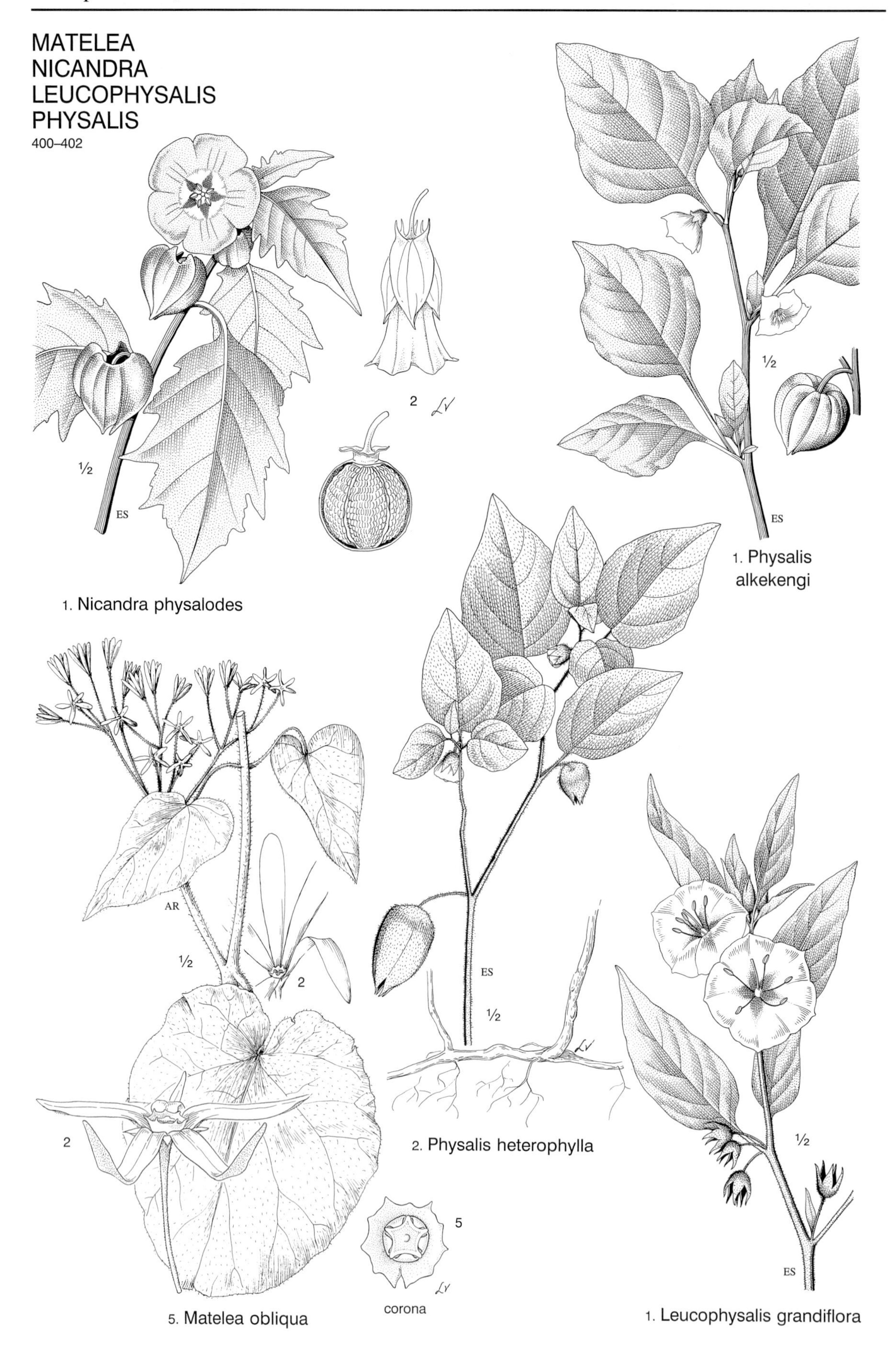

1. Nicandra physalodes

1. Physalis
alkekengi

5. Matelea obliqua

corona

2. Physalis heterophylla

1. Leucophysalis grandiflora

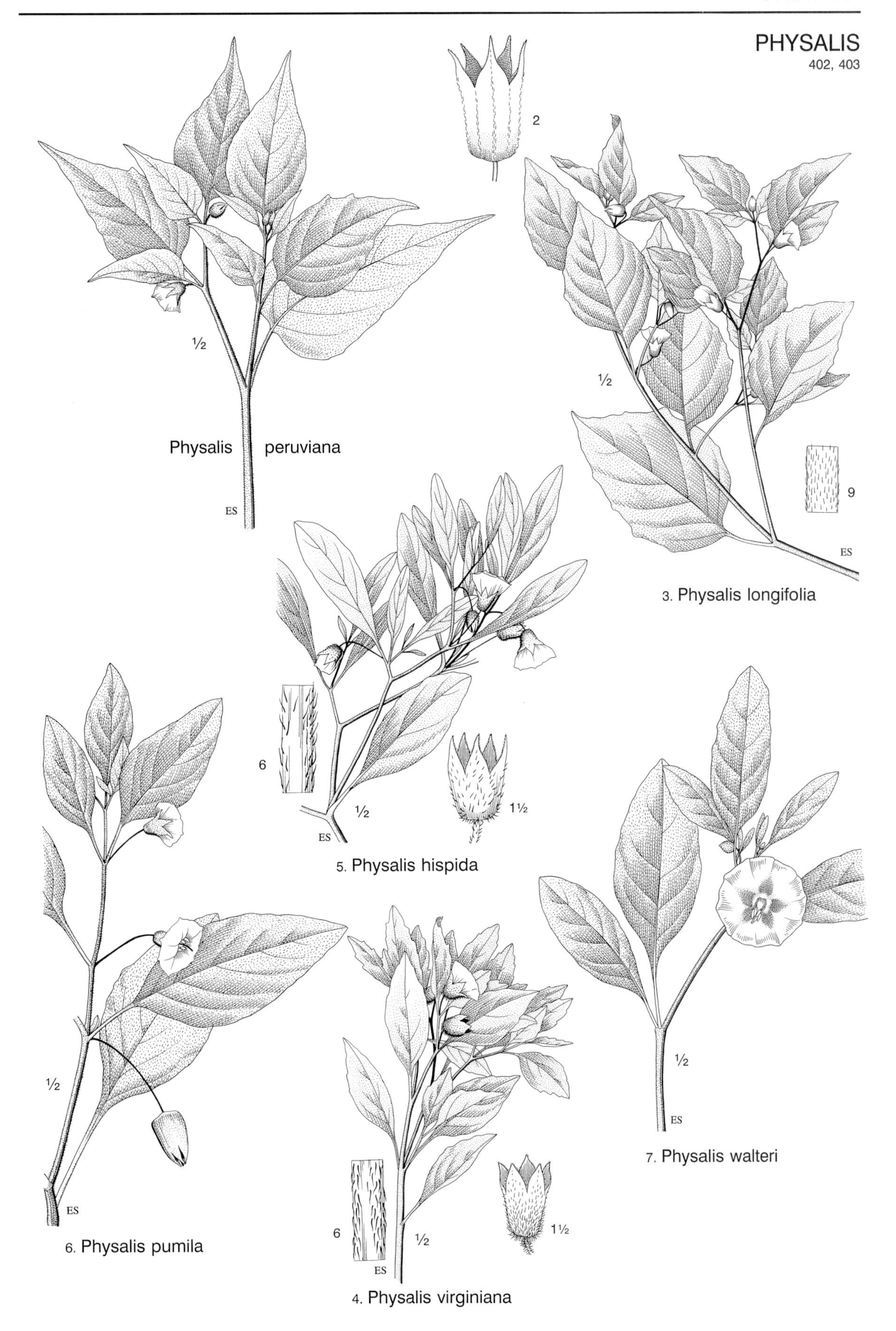

Physalis   peruviana

2

3. Physalis longifolia

9

5. Physalis hispida

6. Physalis pumila

4. Physalis virginiana

7. Physalis walteri

PHYSALIS
403

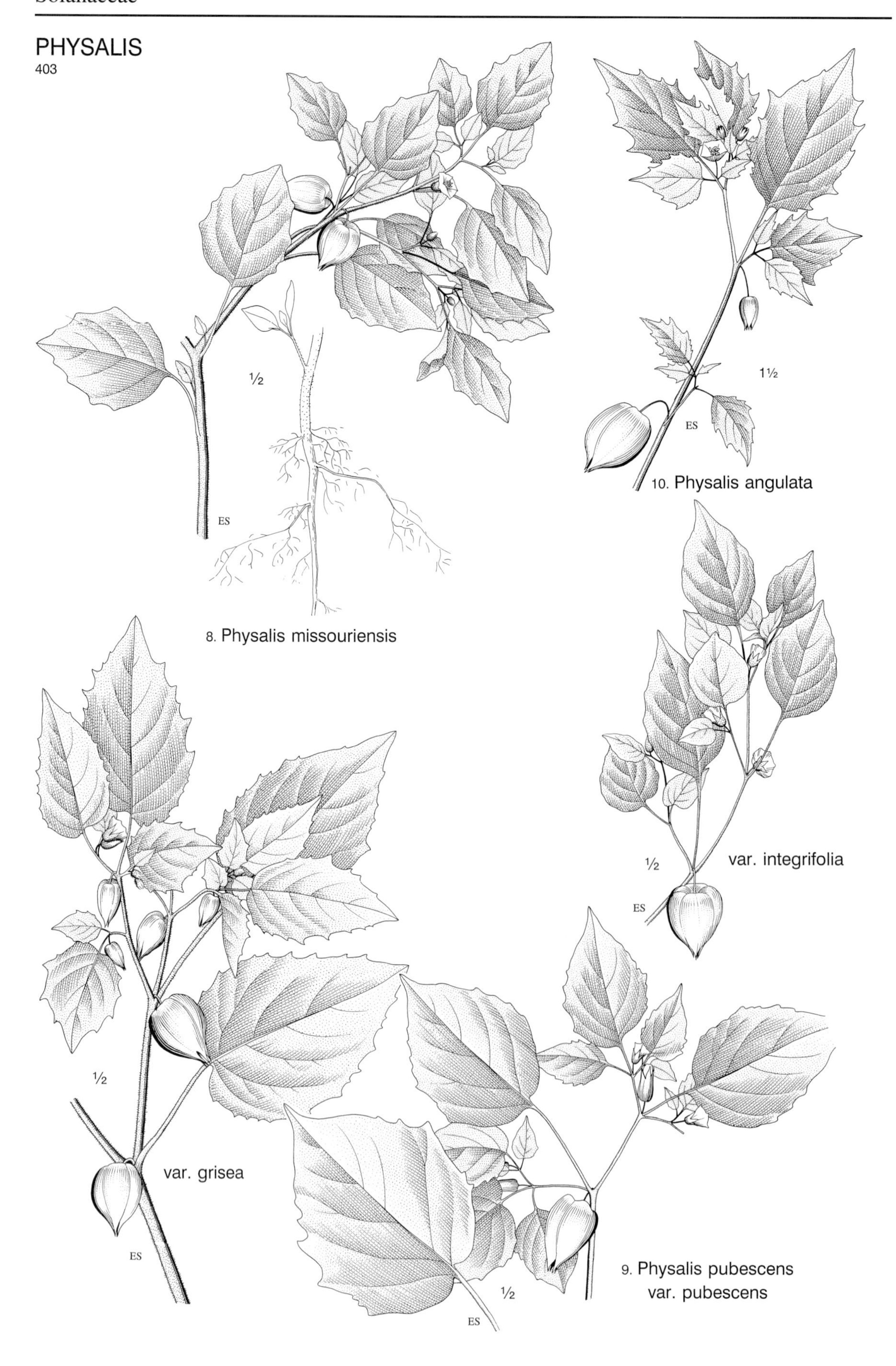

½

ES

8. Physalis missouriensis

1½

ES

10. Physalis angulata

½ var. integrifolia

ES

½ var. grisea

ES

9. Physalis pubescens
var. pubescens

½

ES

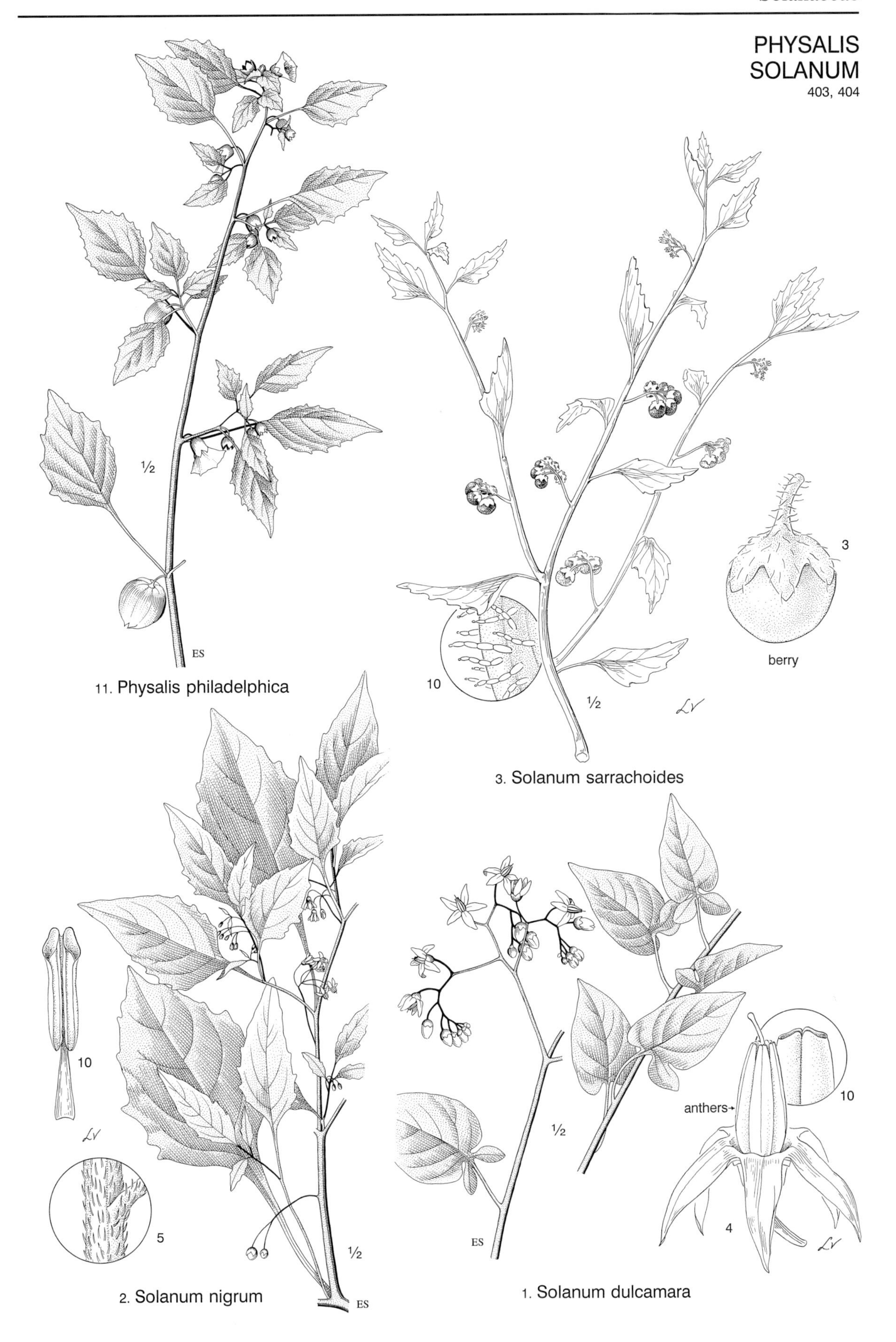

11. Physalis philadelphica

3. Solanum sarrachoides

berry

2. Solanum nigrum

1. Solanum dulcamara

anthers

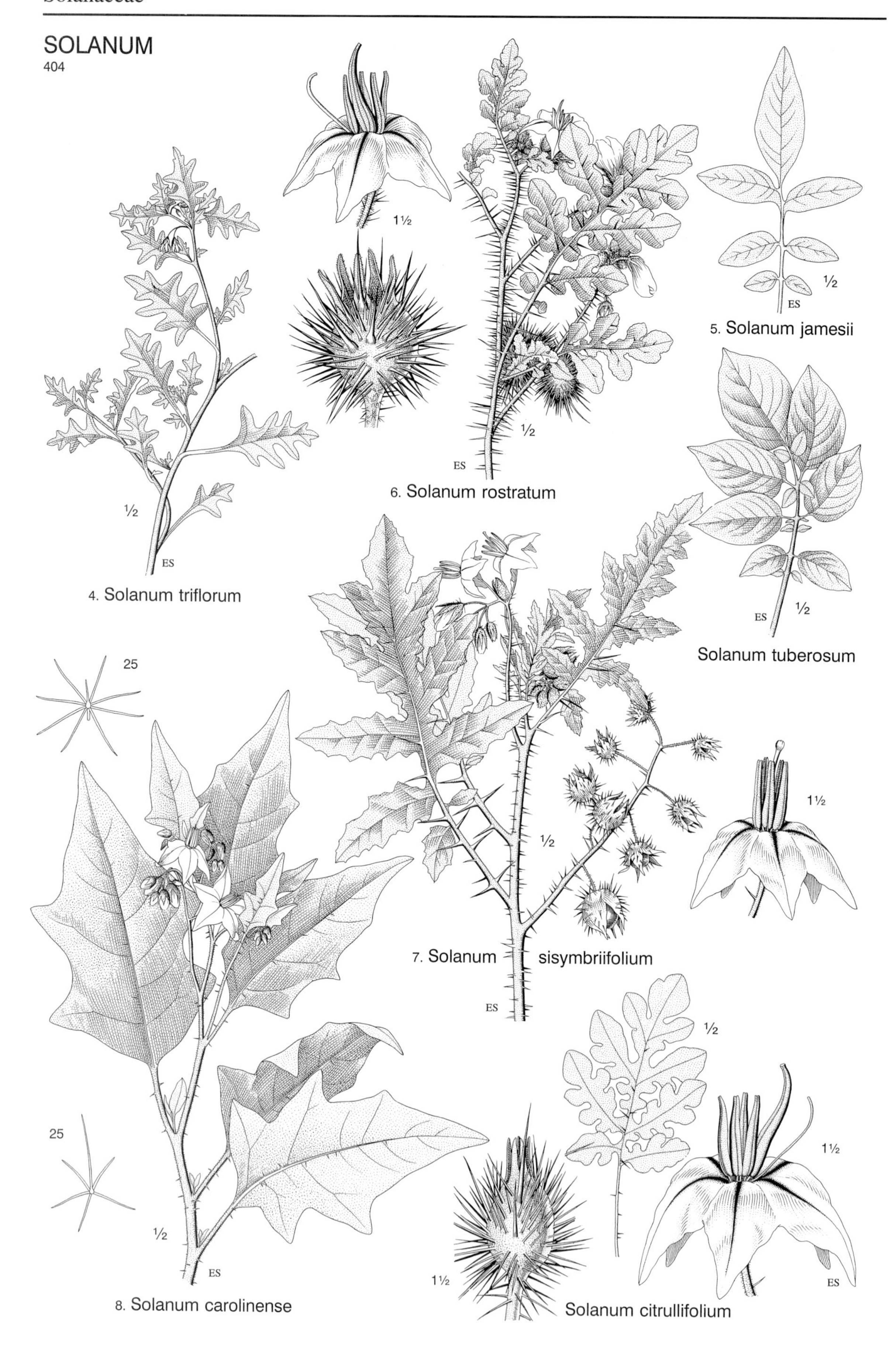

1½

ES

½

ES

4. Solanum triflorum

½

ES

6. Solanum rostratum

½

ES

5. Solanum jamesii

½

ES

Solanum tuberosum

25

25

½

ES

7. Solanum   sisymbriifolium

½

ES

1½

½

1½

8. Solanum carolinense

1½

Solanum citrullifolium

1½

ES

CHAEROPHYLLUM
SCANDIX
AETHUSA
CORIANDRUM
SESELI
373, 374

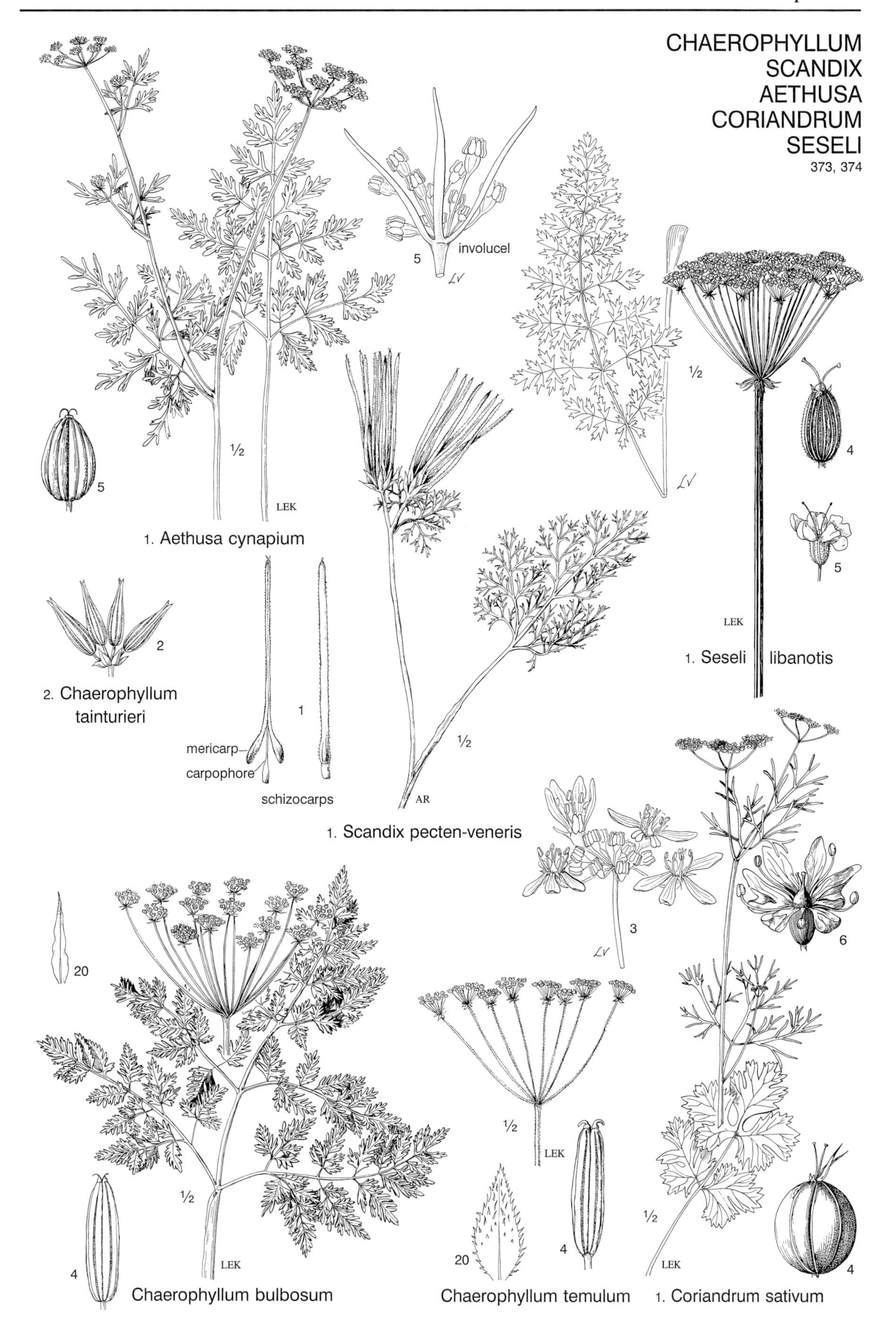

5 involucel

5

1. Aethusa cynapium

2

2. Chaerophyllum
tainturieri

mericarp
carpophore
schizocarps
1

1. Scandix pecten-veneris

½

½

½

½

1. Seseli libanotis

4

5

3

6

20

½

20

4

4

½

Chaerophyllum bulbosum

Chaerophyllum temulum

1. Coriandrum sativum

SPERMOLEPIS
TORILIS
DAUCUS
374, 375

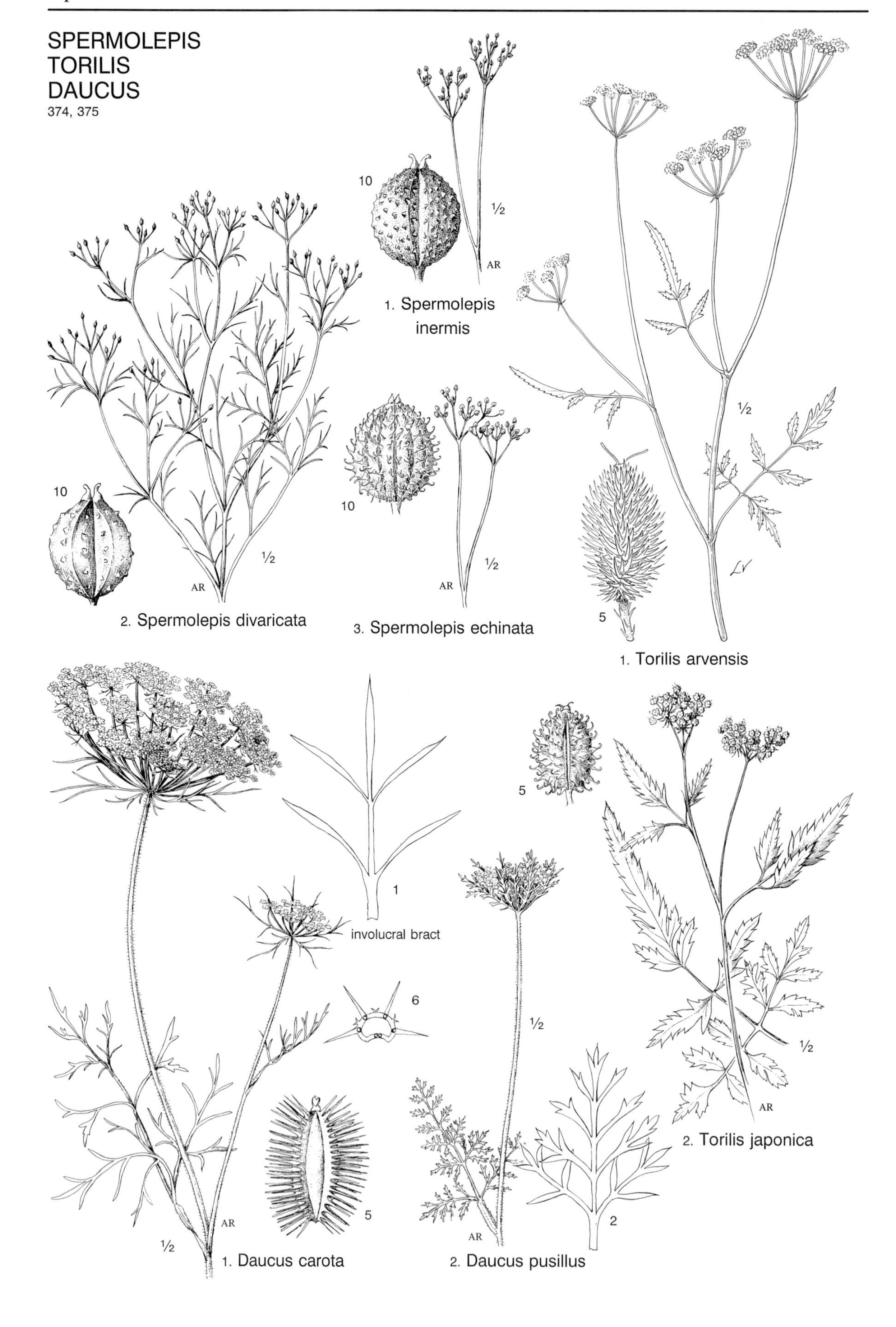

1. Spermolepis inermis

2. Spermolepis divaricata

3. Spermolepis echinata

1. Torilis arvensis

involucral bract

1. Daucus carota

2. Daucus pusillus

2. Torilis japonica

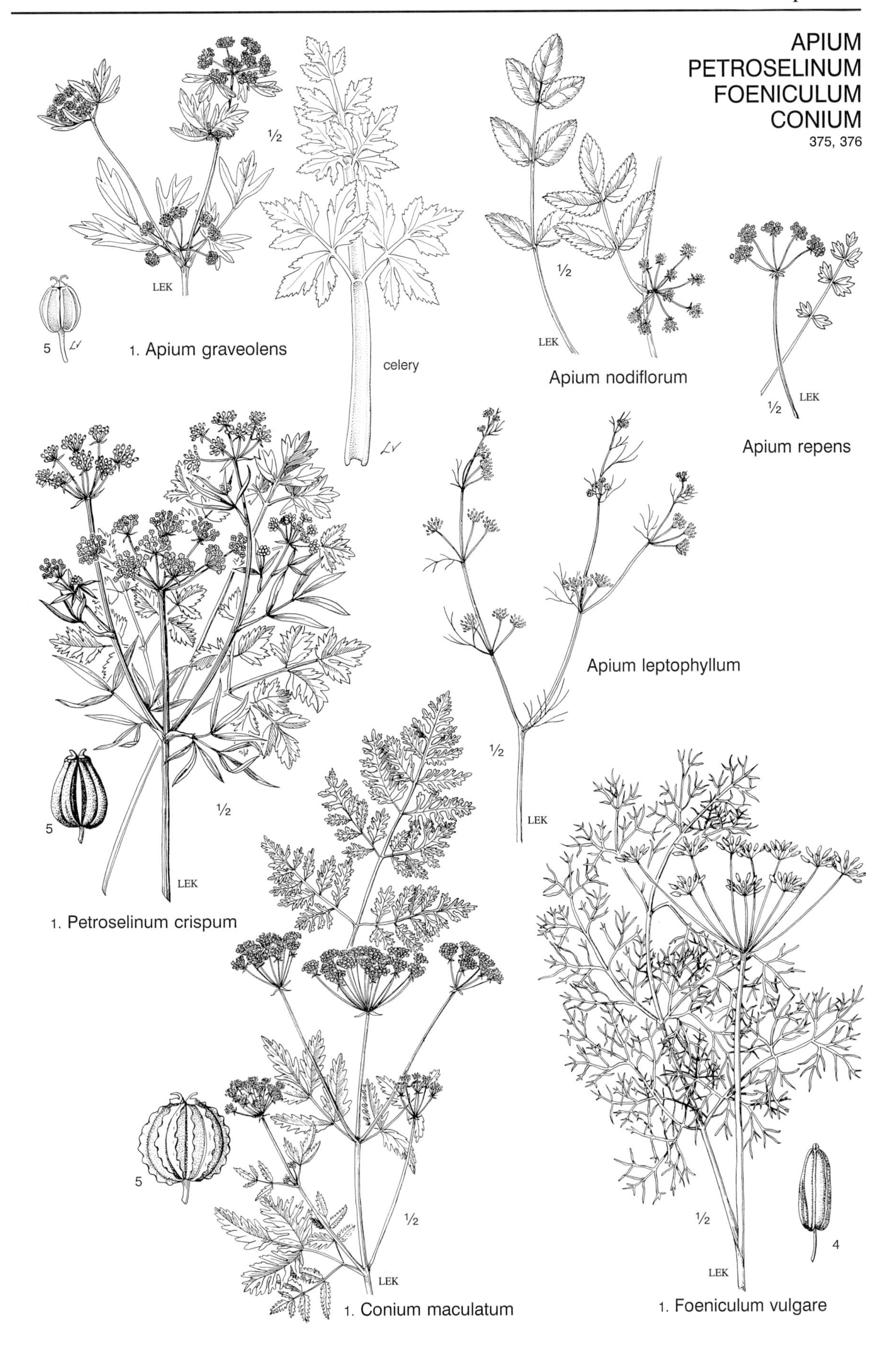

1. Apium graveolens

celery

Apium nodiflorum

Apium repens

Apium leptophyllum

1. Petroselinum crispum

1. Conium maculatum

1. Foeniculum vulgare

ANETHUM
BUPLEURUM
TAENIDIA
376, 377

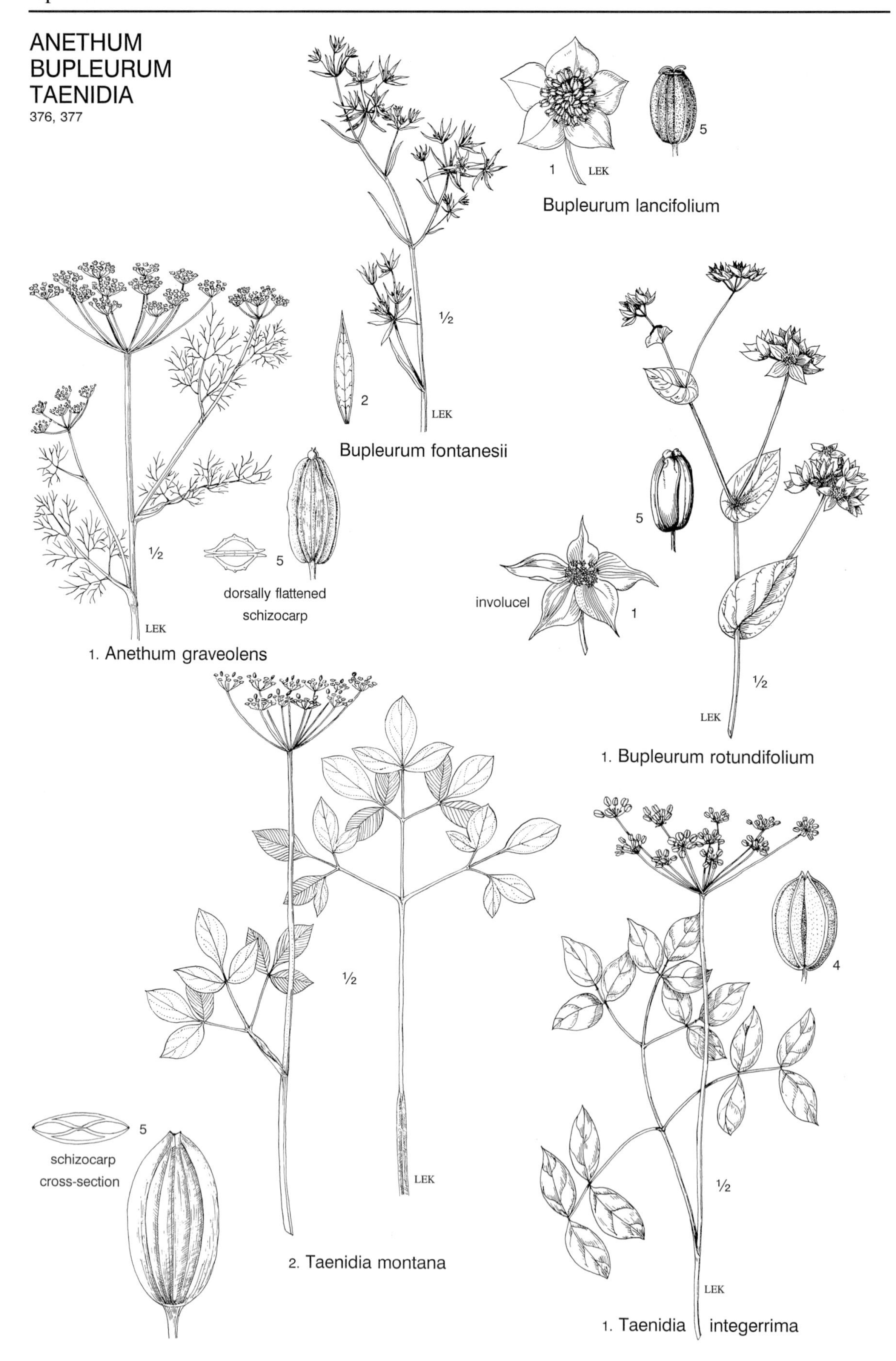

Bupleurum lancifolium

1   LEK

5

½

2

LEK

Bupleurum fontanesii

½

dorsally flattened
schizocarp

5

1. Anethum graveolens

involucel

5

1

½

LEK

1. Bupleurum rotundifolium

½

LEK

5
schizocarp
cross-section

2. Taenidia montana

4

½

LEK

1. Taenidia integerrima

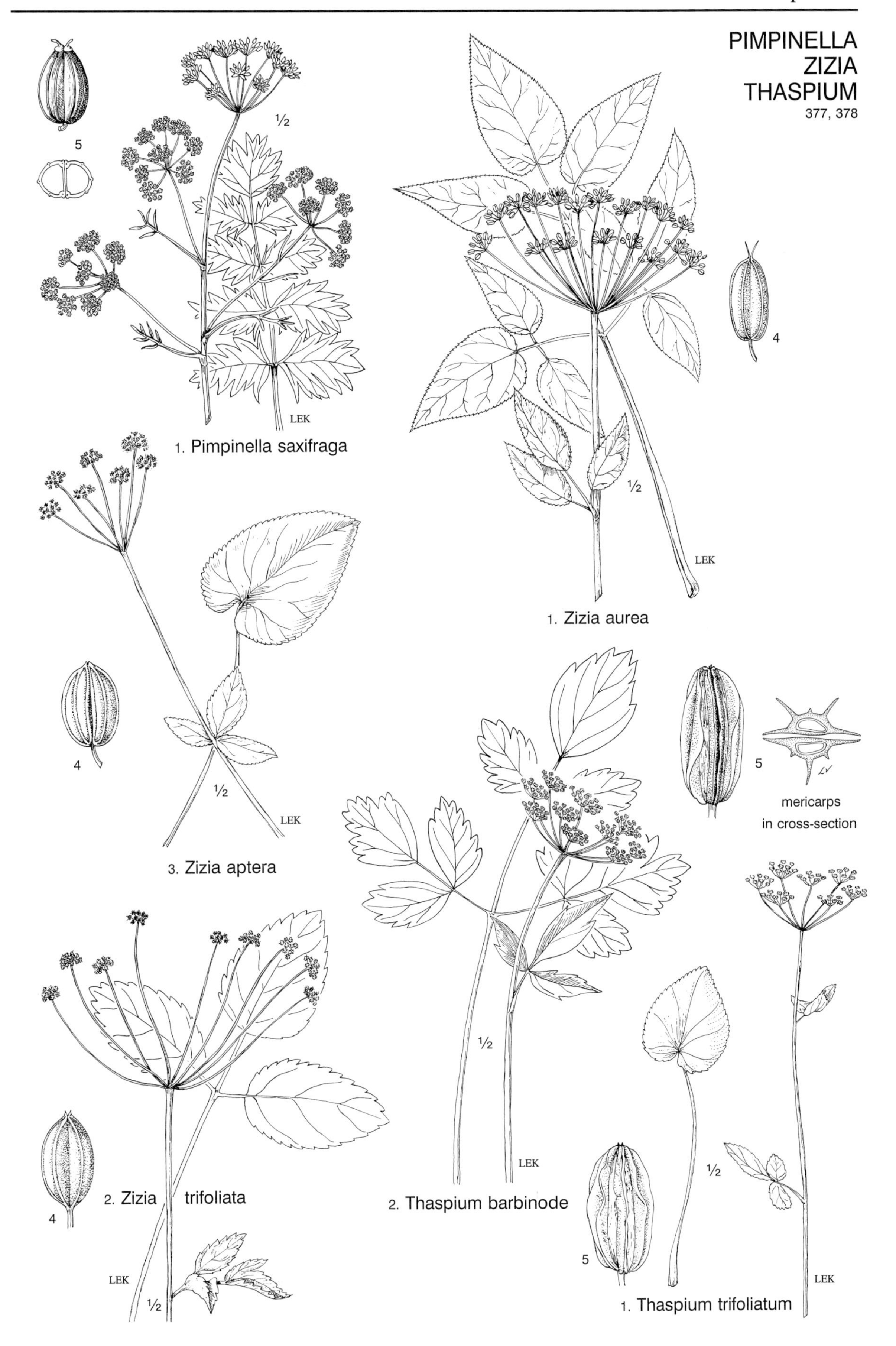

1. Pimpinella saxifraga

1. Zizia aurea

3. Zizia aptera

mericarps
in cross-section

2. Zizia   trifoliata

2. Thaspium barbinode

1. Thaspium trifoliatum

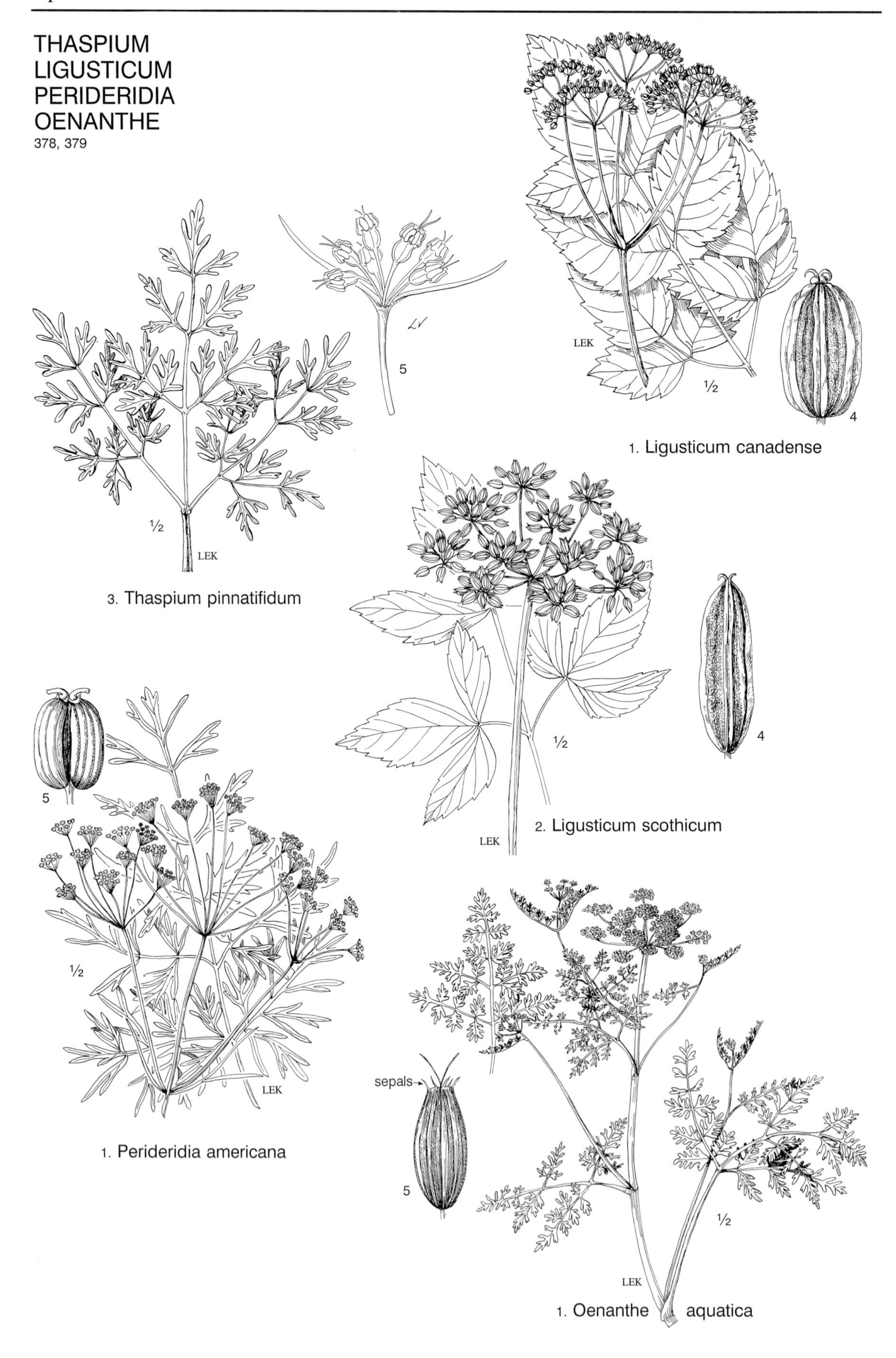

THASPIUM
LIGUSTICUM
PERIDERIDIA
OENANTHE
378, 379

5

1. Ligusticum canadense

3. Thaspium pinnatifidum

2. Ligusticum scothicum

1. Perideridia americana

sepals

1. Oenanthe aquatica

354

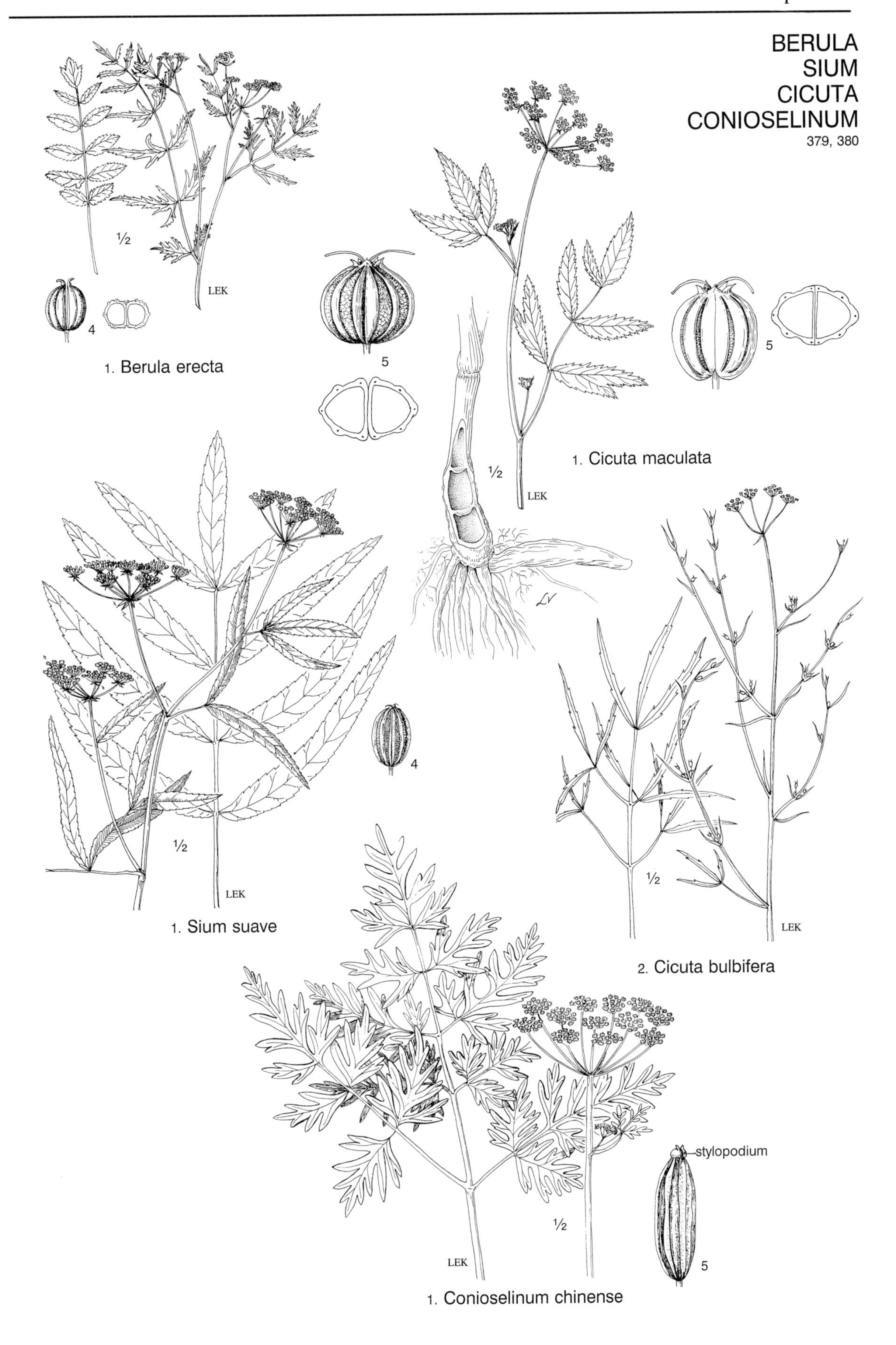

1. Berula erecta

5

1. Cicuta maculata

1. Sium suave

2. Cicuta bulbifera

stylopodium

1. Conioselinum chinense

OXYPOLIS
PTILIMNIUM
380, 381

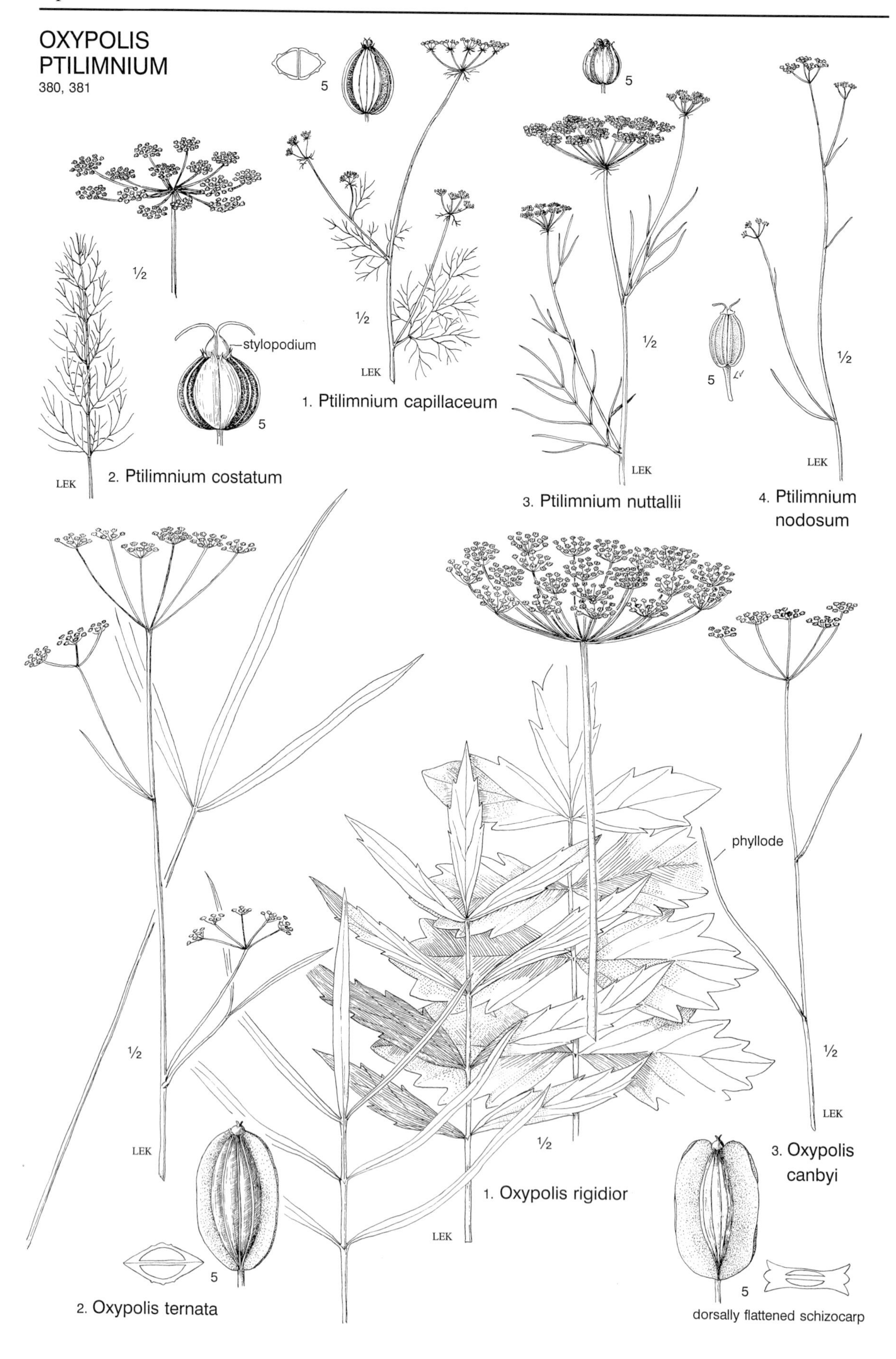

5

½

—stylopodium

5

1. Ptilimnium capillaceum

½

LEK

½

LEK

2. Ptilimnium costatum

½

LEK

3. Ptilimnium nuttallii

5

½

LEK

4. Ptilimnium nodosum

½

LEK

½

phyllode

½

LEK

3. Oxypolis canbyi

½

LEK

1. Oxypolis rigidior

LEK

5

2. Oxypolis ternata

5

dorsally flattened schizocarp

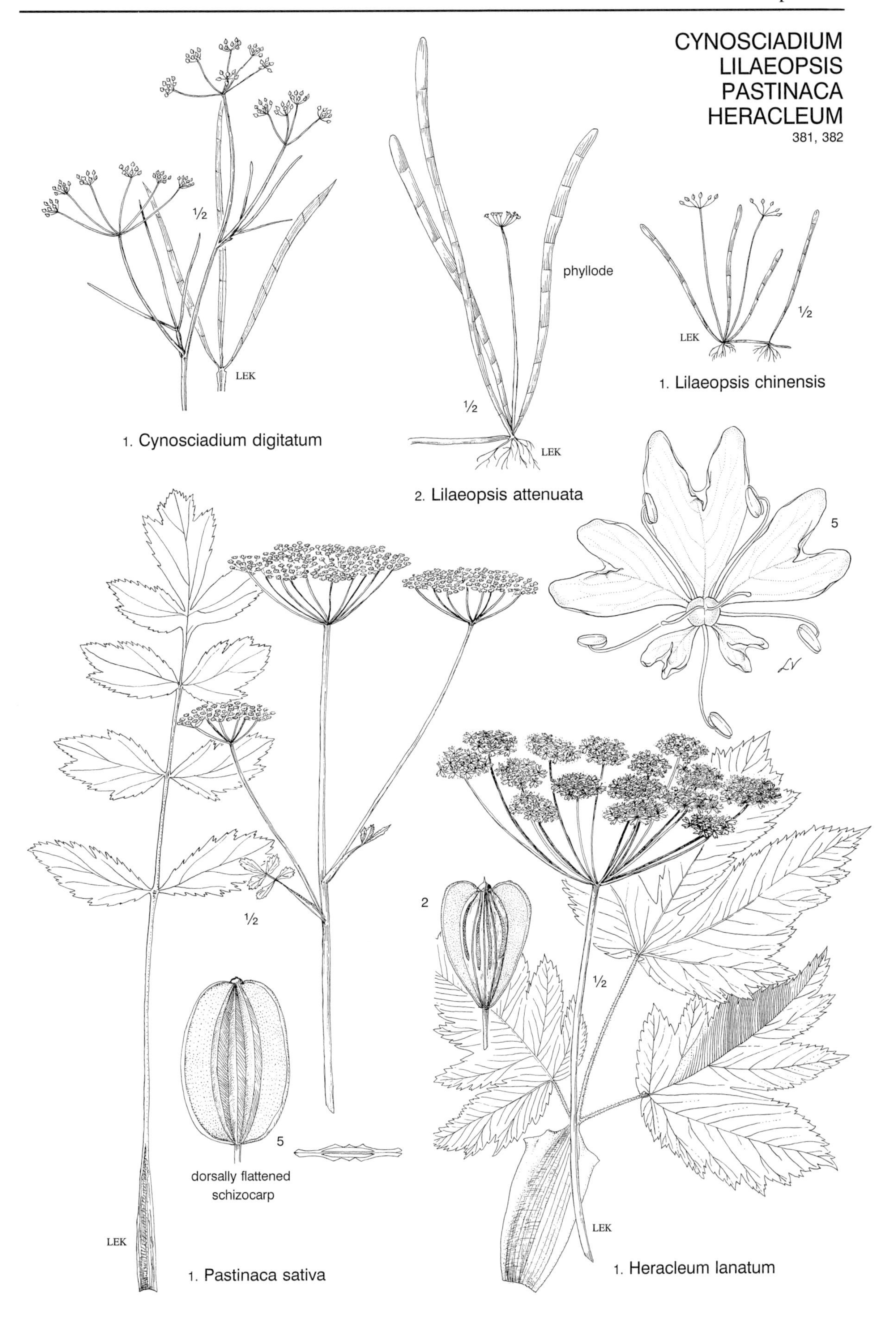

1. Cynosciadium digitatum

phyllode

2. Lilaeopsis attenuata

1. Lilaeopsis chinensis

5

2

dorsally flattened
schizocarp

1. Pastinaca sativa

1. Heracleum lanatum

HERACLEUM
ANGELICA
382, 383

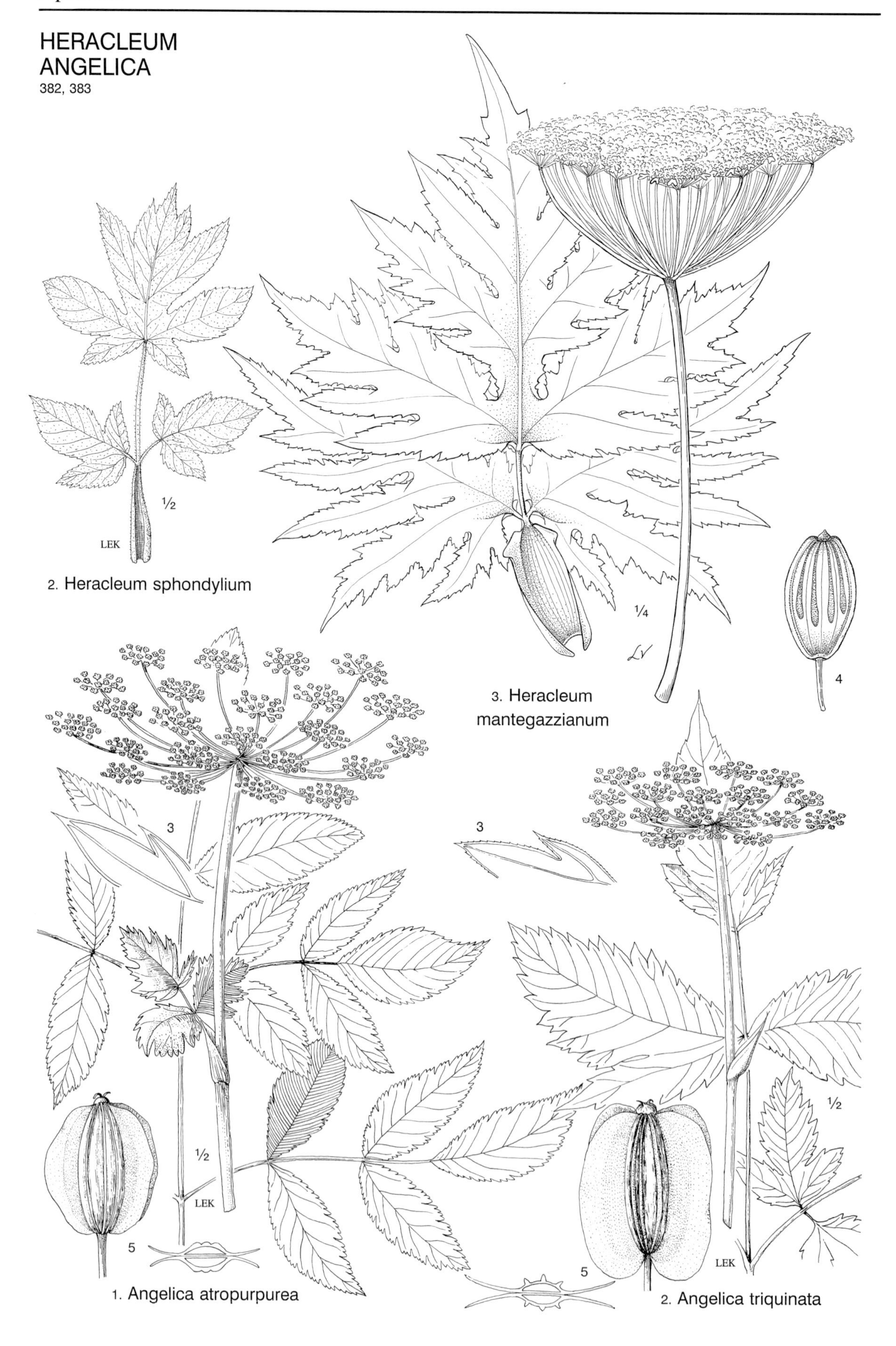

2. Heracleum sphondylium

LEK

½

3. Heracleum
mantegazzianum

¼

4

3

1. Angelica atropurpurea

½

LEK

5

3

2. Angelica triquinata

½

LEK

5

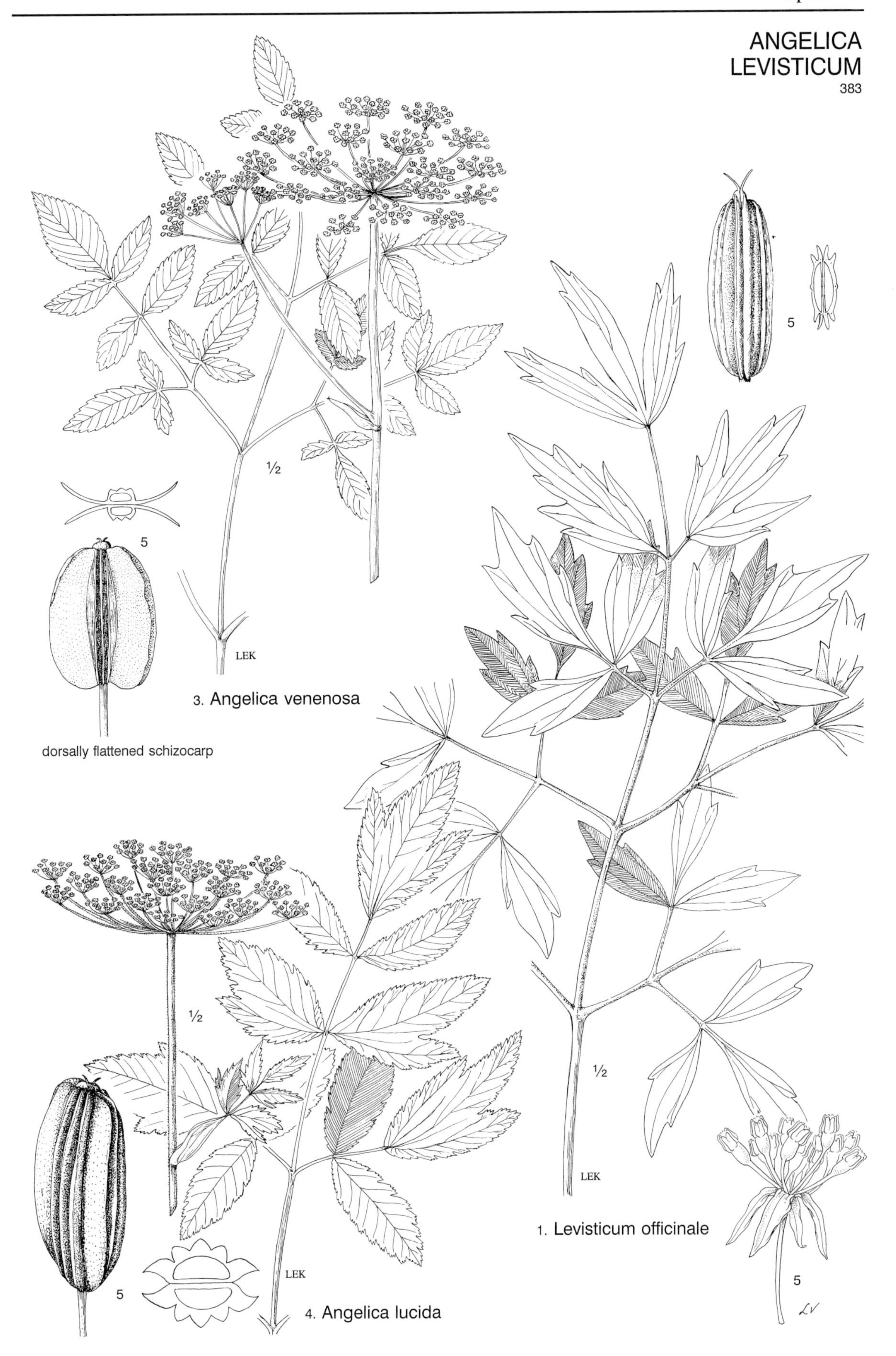

5

½

5

LEK

3. Angelica venenosa

dorsally flattened schizocarp

½

LEK

1. Levisticum officinale

5

½

5

LEK

4. Angelica lucida

PEUCEDANUM
POLYTAENIA
LOMATIUM
383, 384

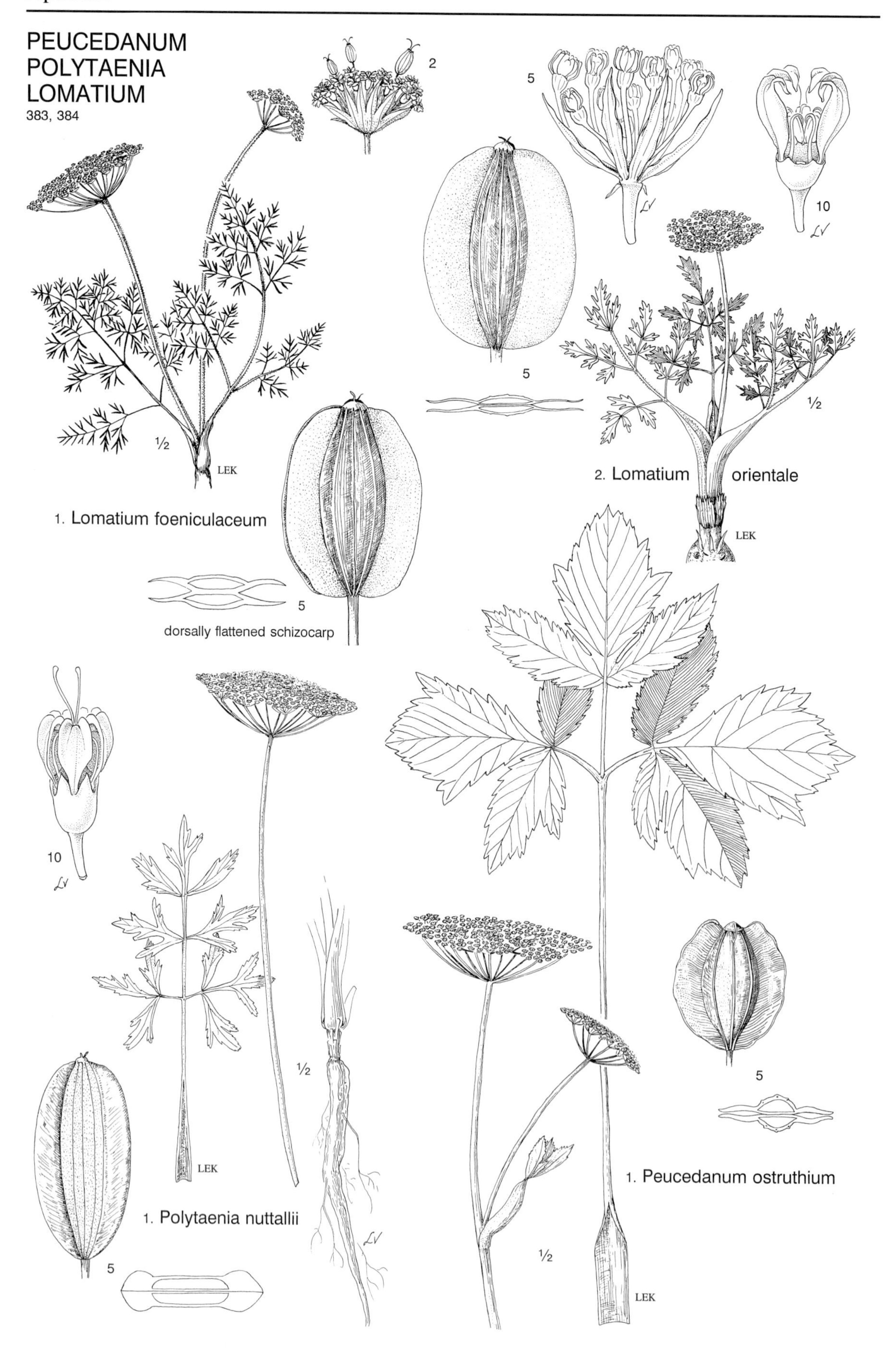

2

5

5

10

5

5

dorsally flattened schizocarp

1. Lomatium foeniculaceum

2. Lomatium orientale

LEK

10

1. Polytaenia nuttallii

1. Peucedanum ostruthium

5

LEK

5

½

LEK

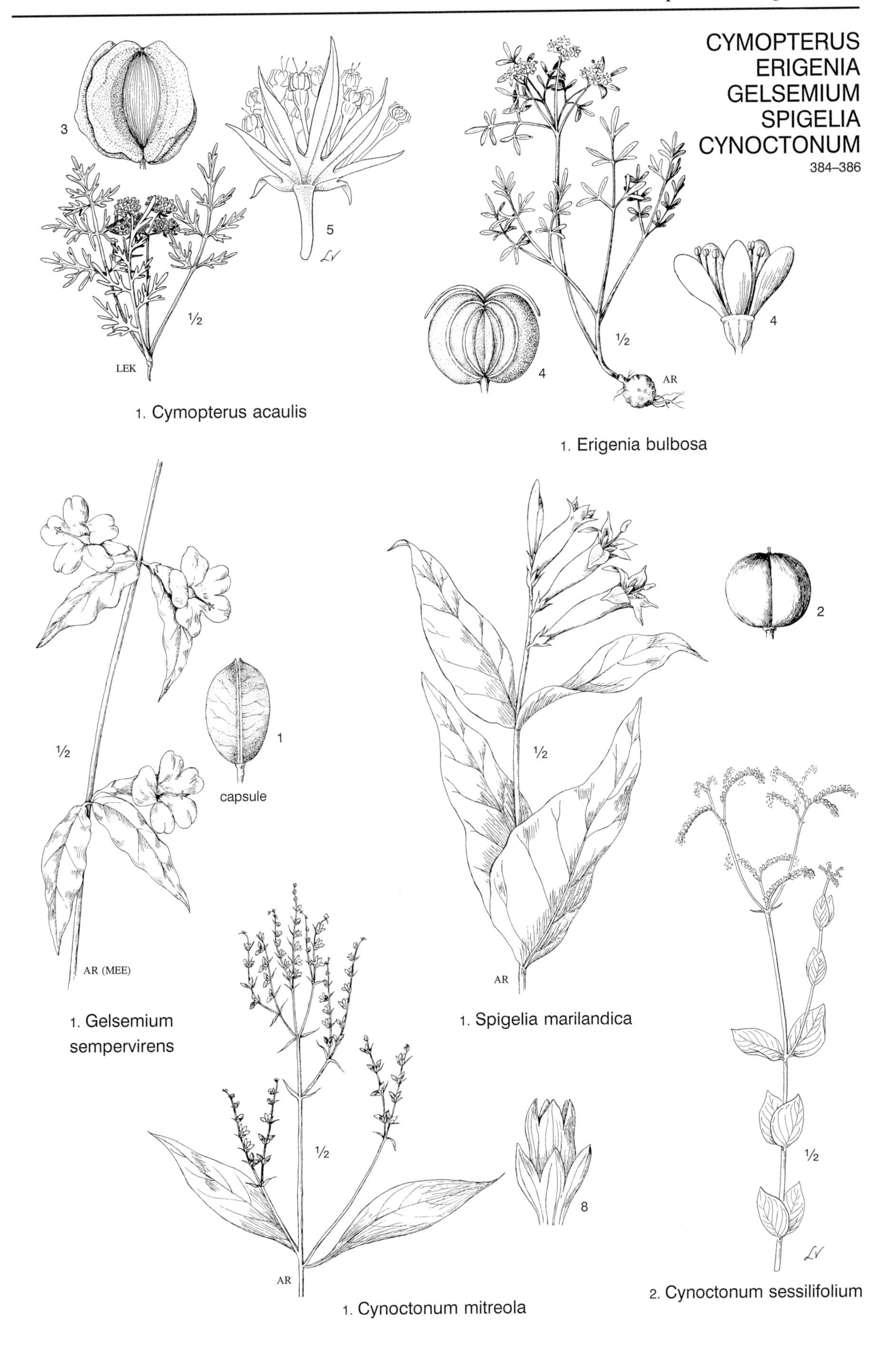

CYMOPTERUS
ERIGENIA
GELSEMIUM
SPIGELIA
CYNOCTONUM
384–386

1. Cymopterus acaulis

1. Erigenia bulbosa

1. Gelsemium sempervirens

1. Spigelia marilandica

capsule

1. Cynoctonum mitreola

2. Cynoctonum sessilifolium

SABATIA
387

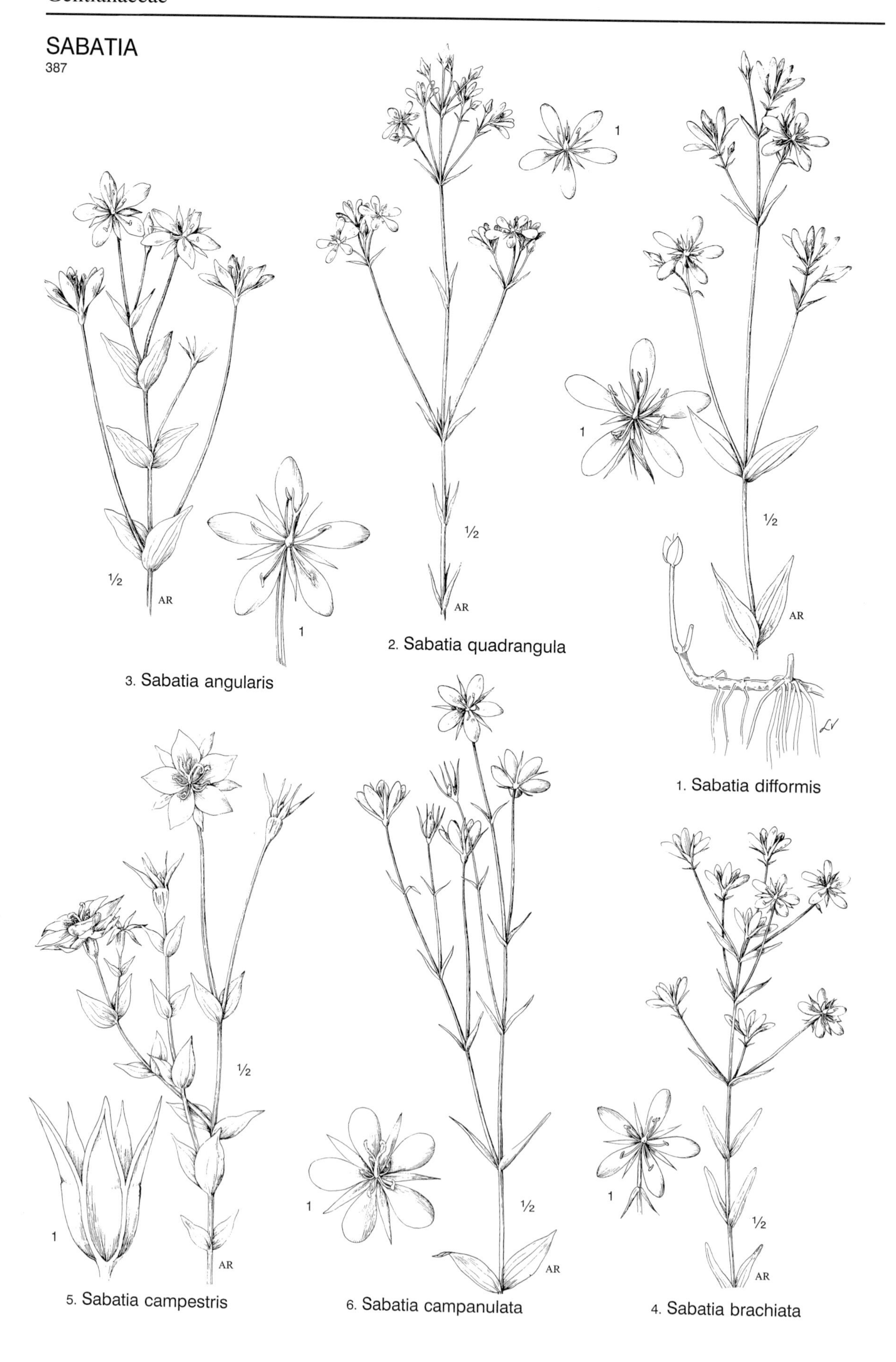

3. Sabatia angularis

2. Sabatia quadrangula

1. Sabatia difformis

5. Sabatia campestris

6. Sabatia campanulata

4. Sabatia brachiata

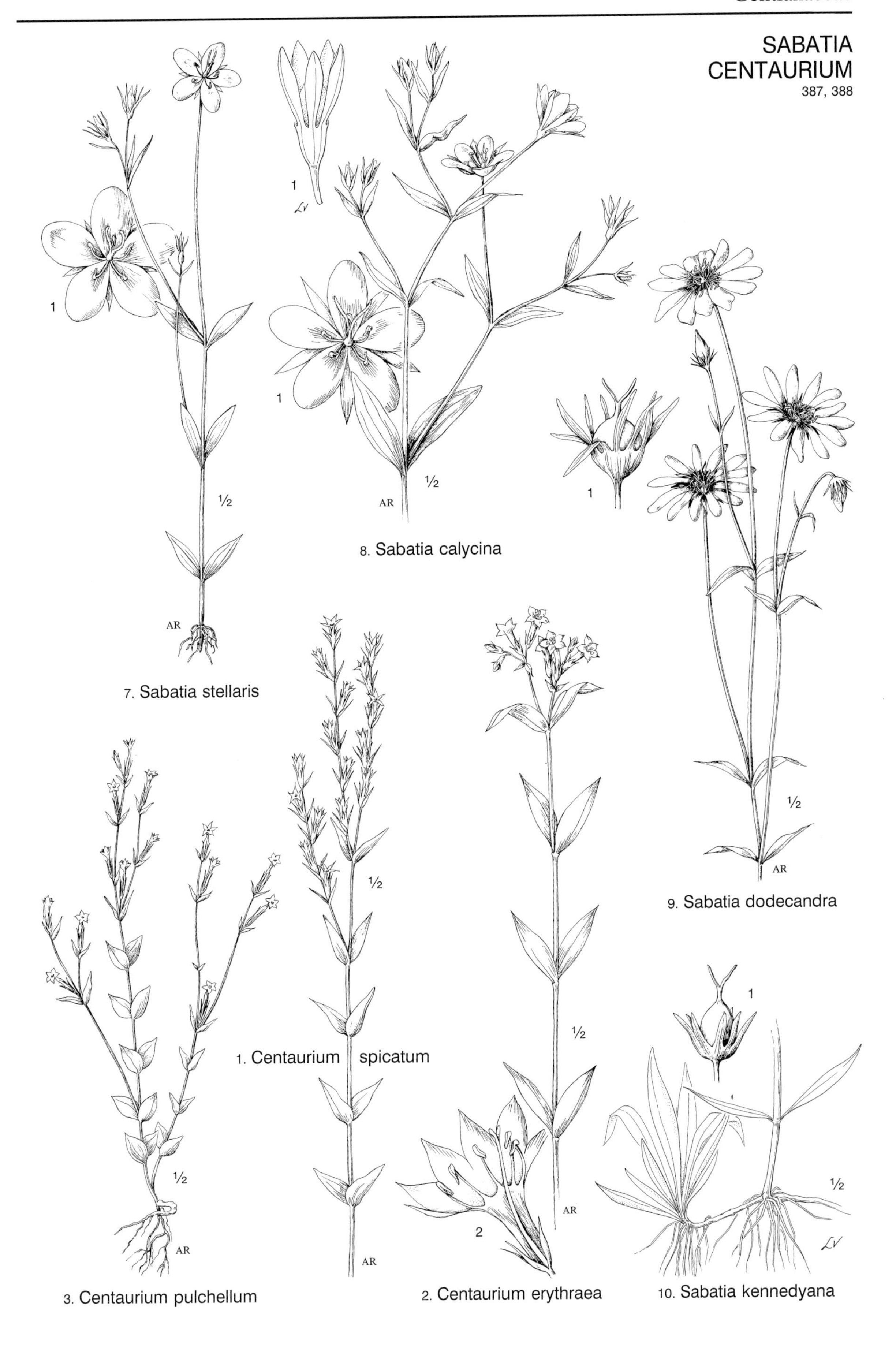

8. Sabatia calycina

7. Sabatia stellaris

1. Centaurium spicatum

9. Sabatia dodecandra

3. Centaurium pulchellum

2. Centaurium erythraea

10. Sabatia kennedyana

GENTIANA
389

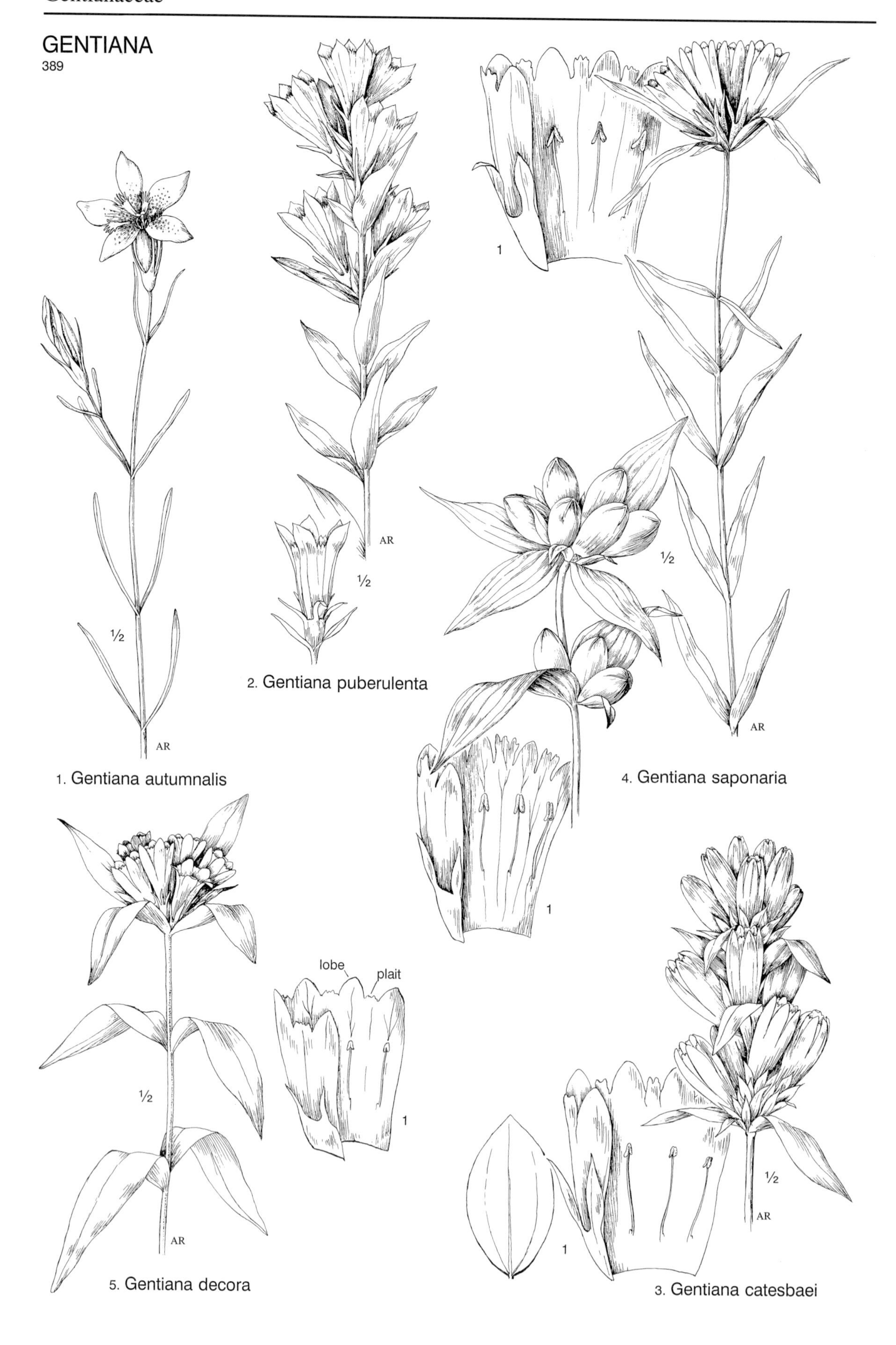

1. Gentiana autumnalis

2. Gentiana puberulenta

4. Gentiana saponaria

5. Gentiana decora

3. Gentiana catesbaei

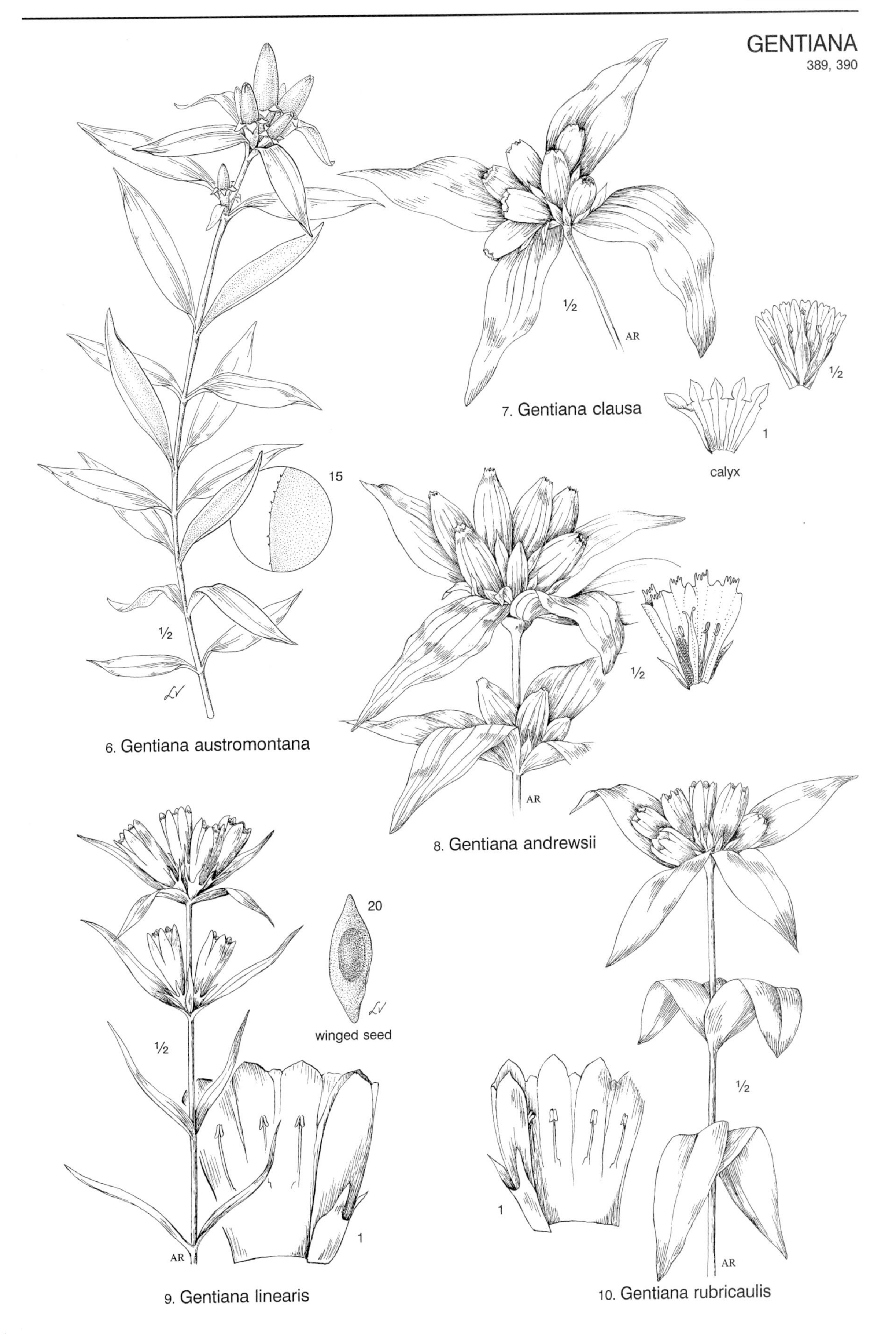

7. Gentiana clausa

calyx

6. Gentiana austromontana

8. Gentiana andrewsii

winged seed

9. Gentiana linearis

10. Gentiana rubricaulis

GENTIANA
GENTIANOPSIS
GENTIANELLA
390, 391

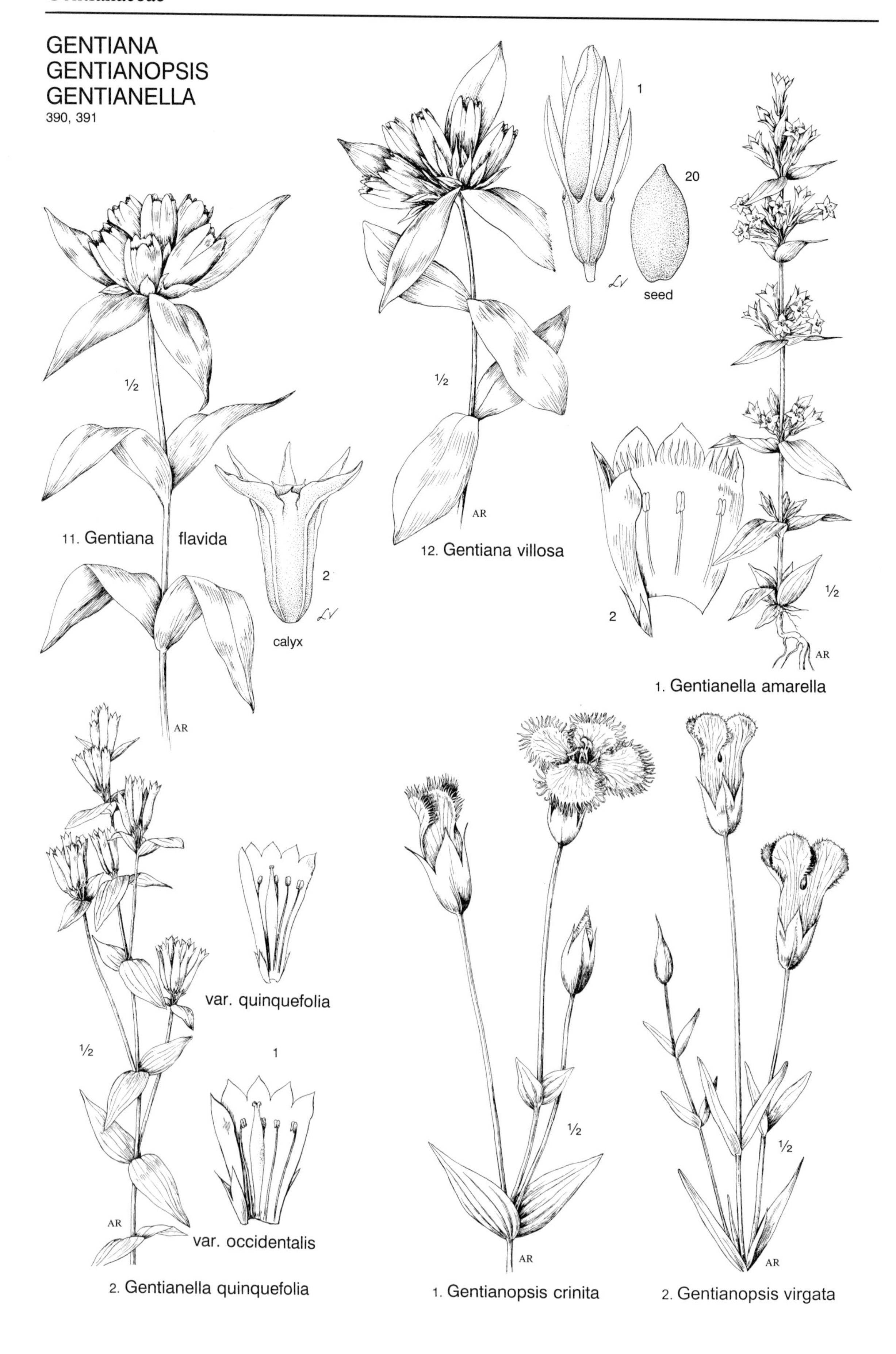

11. Gentiana flavida

calyx

12. Gentiana villosa

seed

1. Gentianella amarella

var. quinquefolia

var. occidentalis

2. Gentianella quinquefolia

1. Gentianopsis crinita

2. Gentianopsis virgata

ILLUSTRATED COMPANION TO

HALENIA
LOMATOGONIUM
FRASERA
BARTONIA
OBOLARIA
391, 392

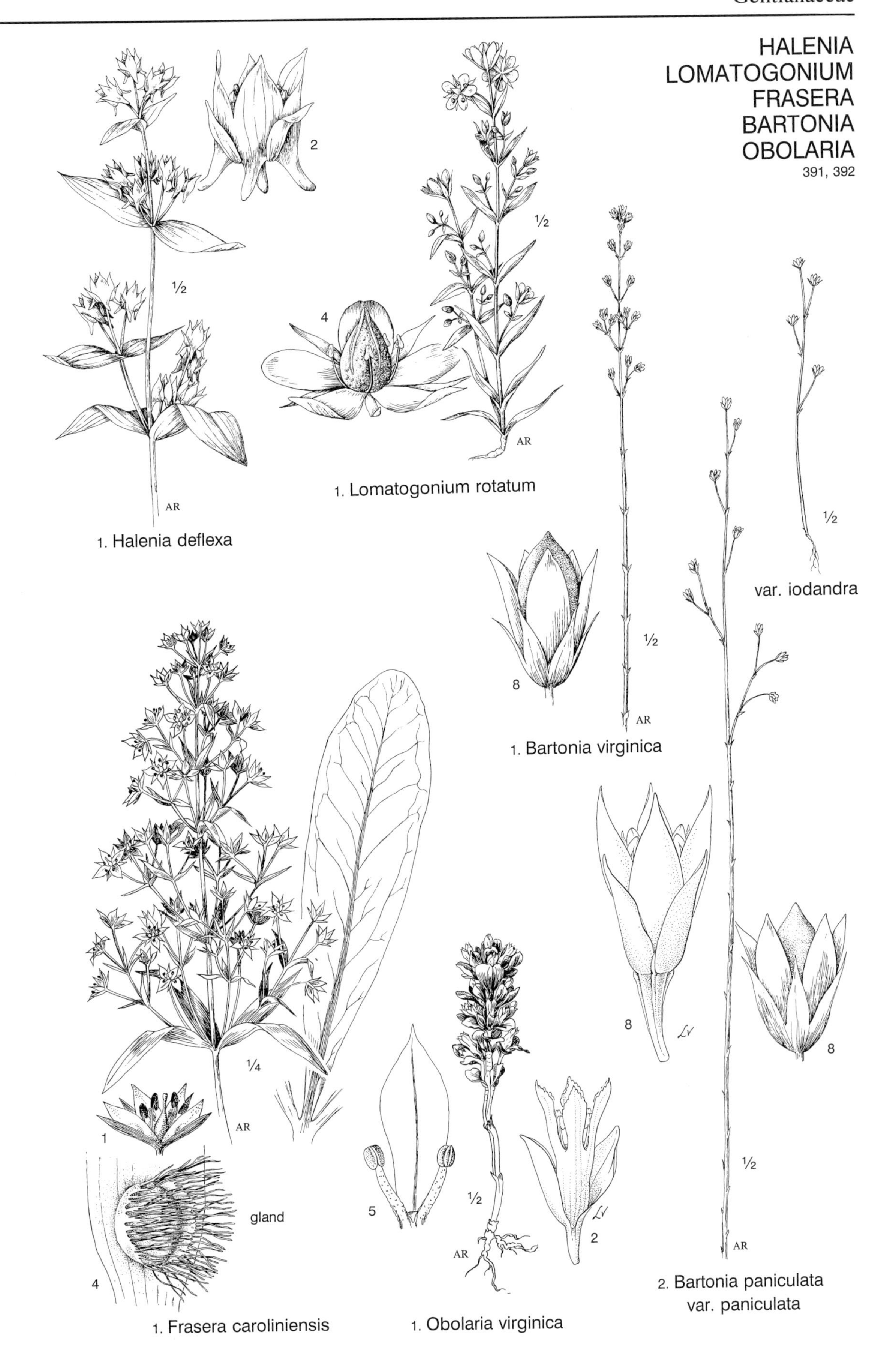

½

2

1. Halenia deflexa

½

4

AR

1. Lomatogonium rotatum

½

var. iodandra

8

½

AR

1. Bartonia virginica

¼

AR

gland

1

4

1. Frasera caroliniensis

5

½

AR

2

1. Obolaria virginica

8

8

½

AR

2. Bartonia paniculata
var. paniculata

VINCA
TRACHELOSPERMUM
AMSONIA
393, 394

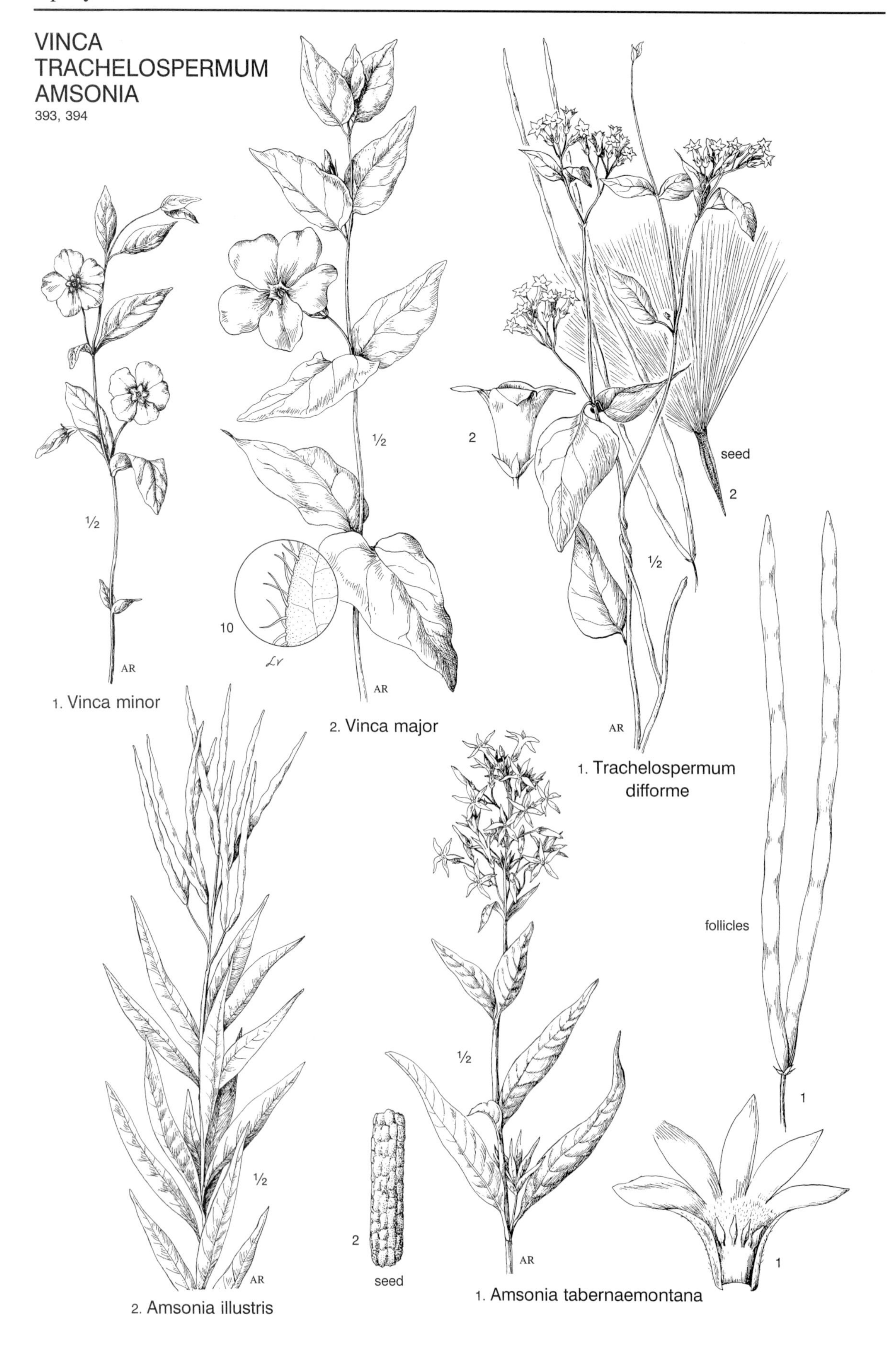

1. Vinca minor

2. Vinca major

1. Trachelospermum difforme

2. Amsonia illustris

1. Amsonia tabernaemontana

seed

follicles

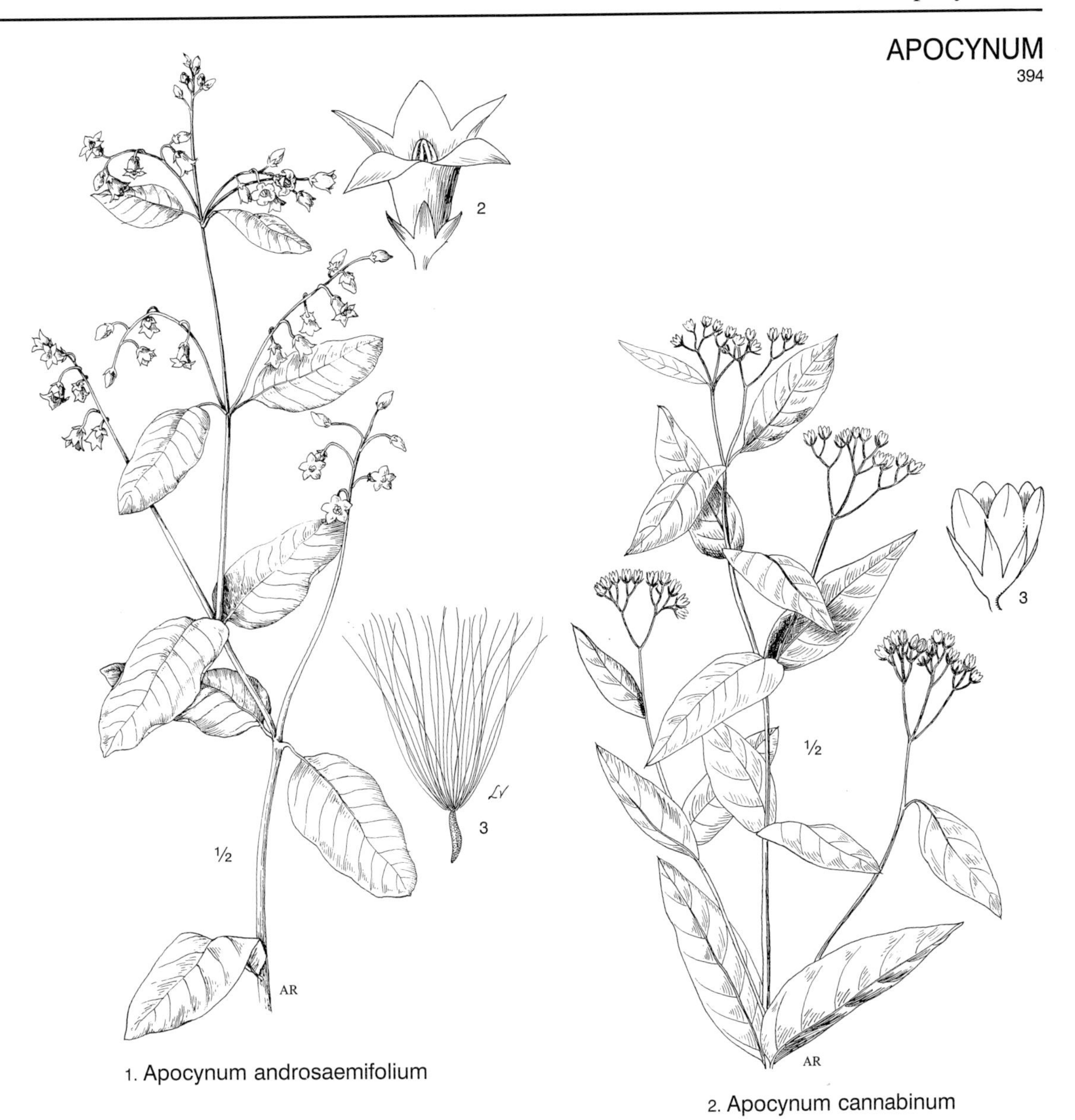

1. Apocynum androsaemifolium

2. Apocynum cannabinum

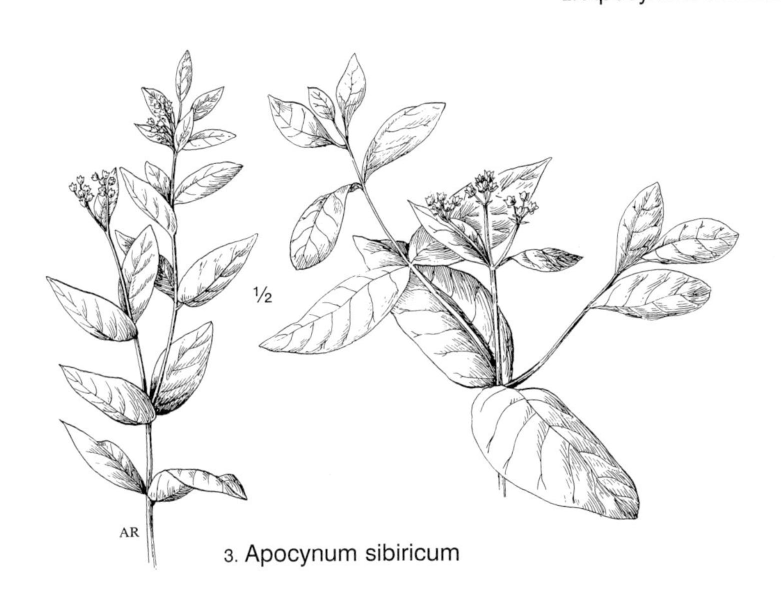

3. Apocynum sibiricum

PERIPLOCA
ASCLEPIAS
395–397

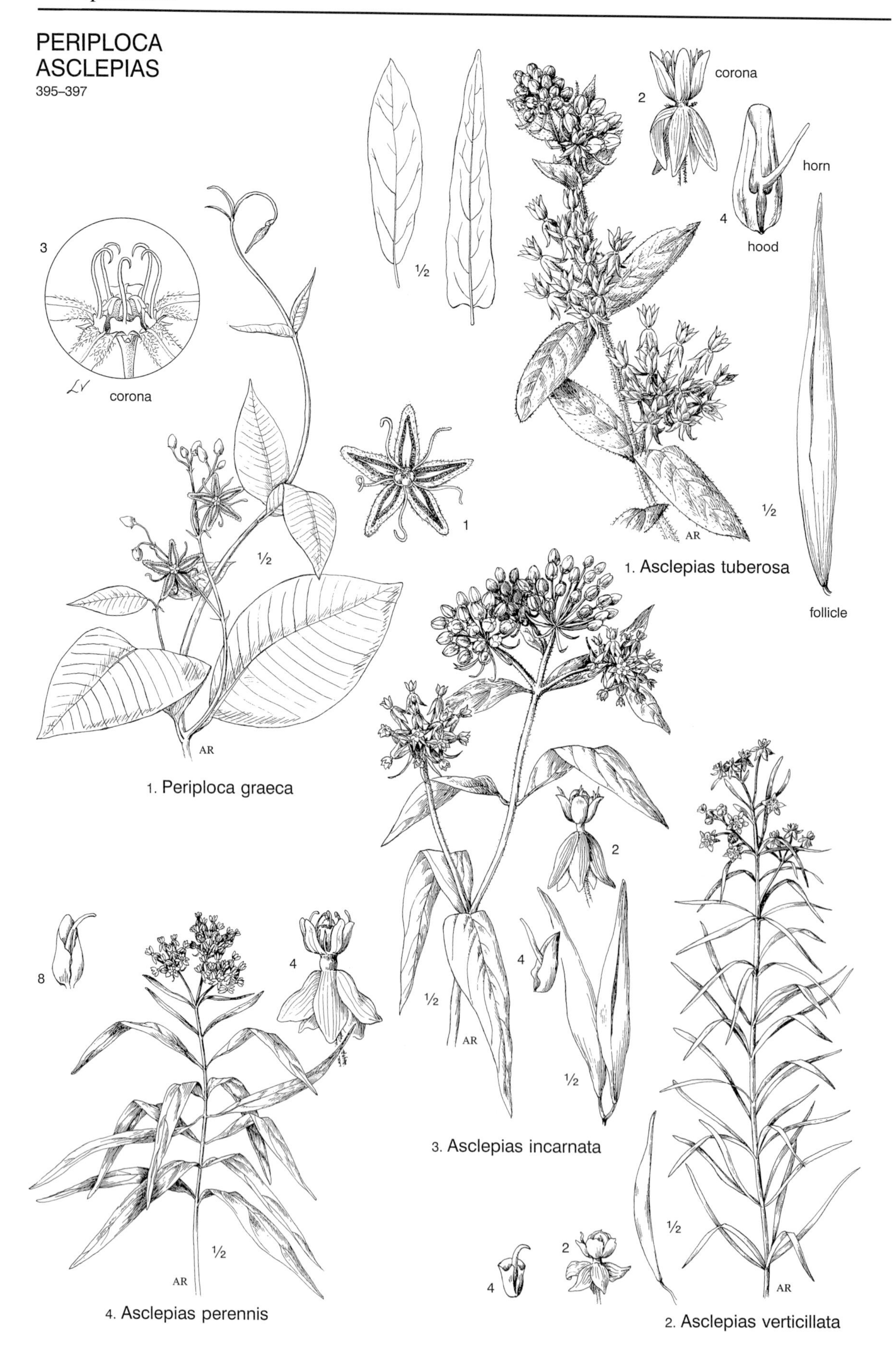

corona

3

corona

2

corona

4

horn

hood

½

½

1

1. Asclepias tuberosa

follicle

½

1. Periploca graeca

2

AR

½

4

½

3. Asclepias incarnata

8

4

½

AR

4. Asclepias perennis

2

4

½

AR

2. Asclepias verticillata

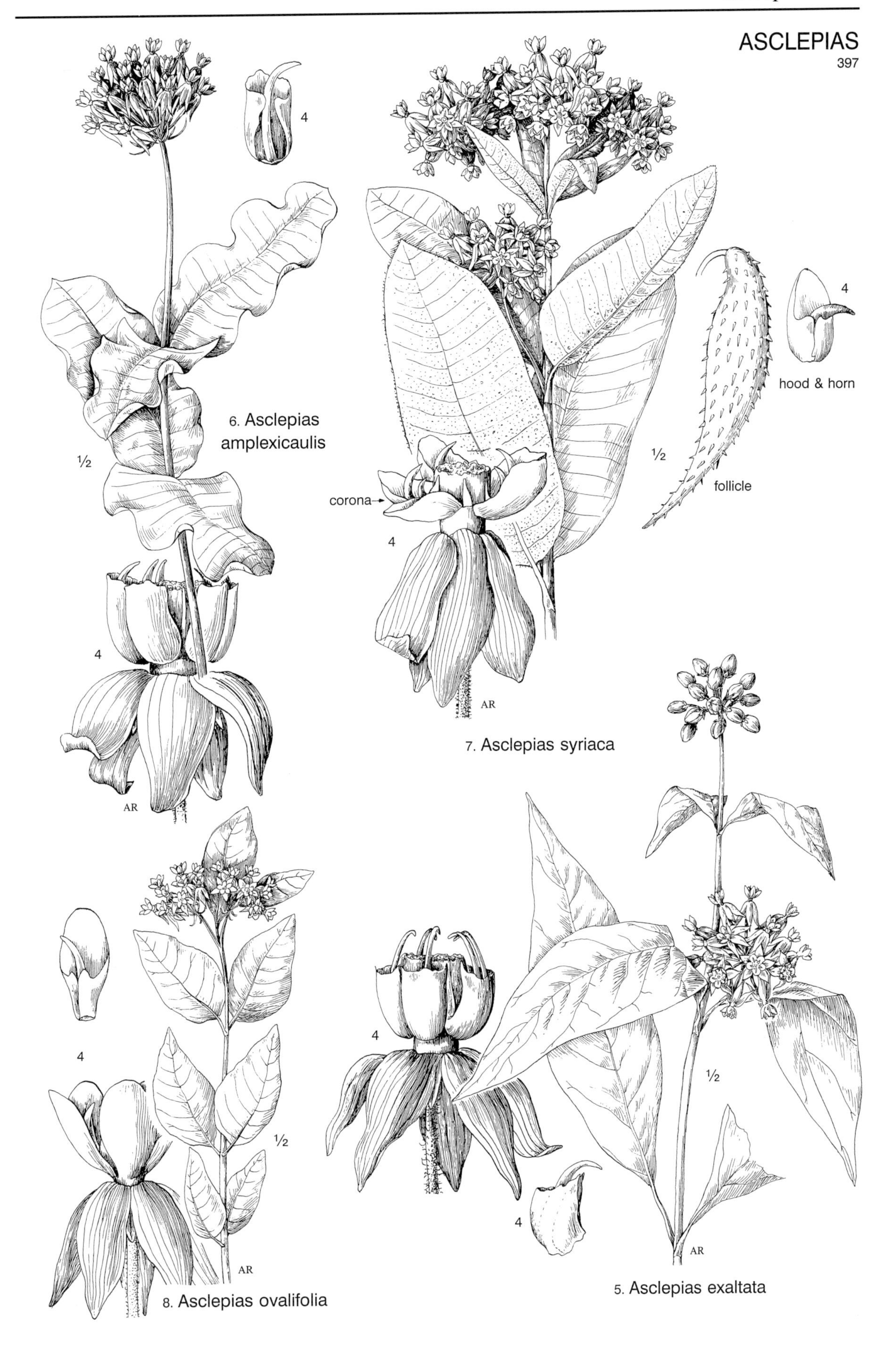

6. Asclepias
amplexicaulis

½

4

4

corona→

4

½

hood & horn

follicle

7. Asclepias syriaca

AR

4

½

4

AR

8. Asclepias ovalifolia

4

½

AR

5. Asclepias exaltata

ASCLEPIAS
397

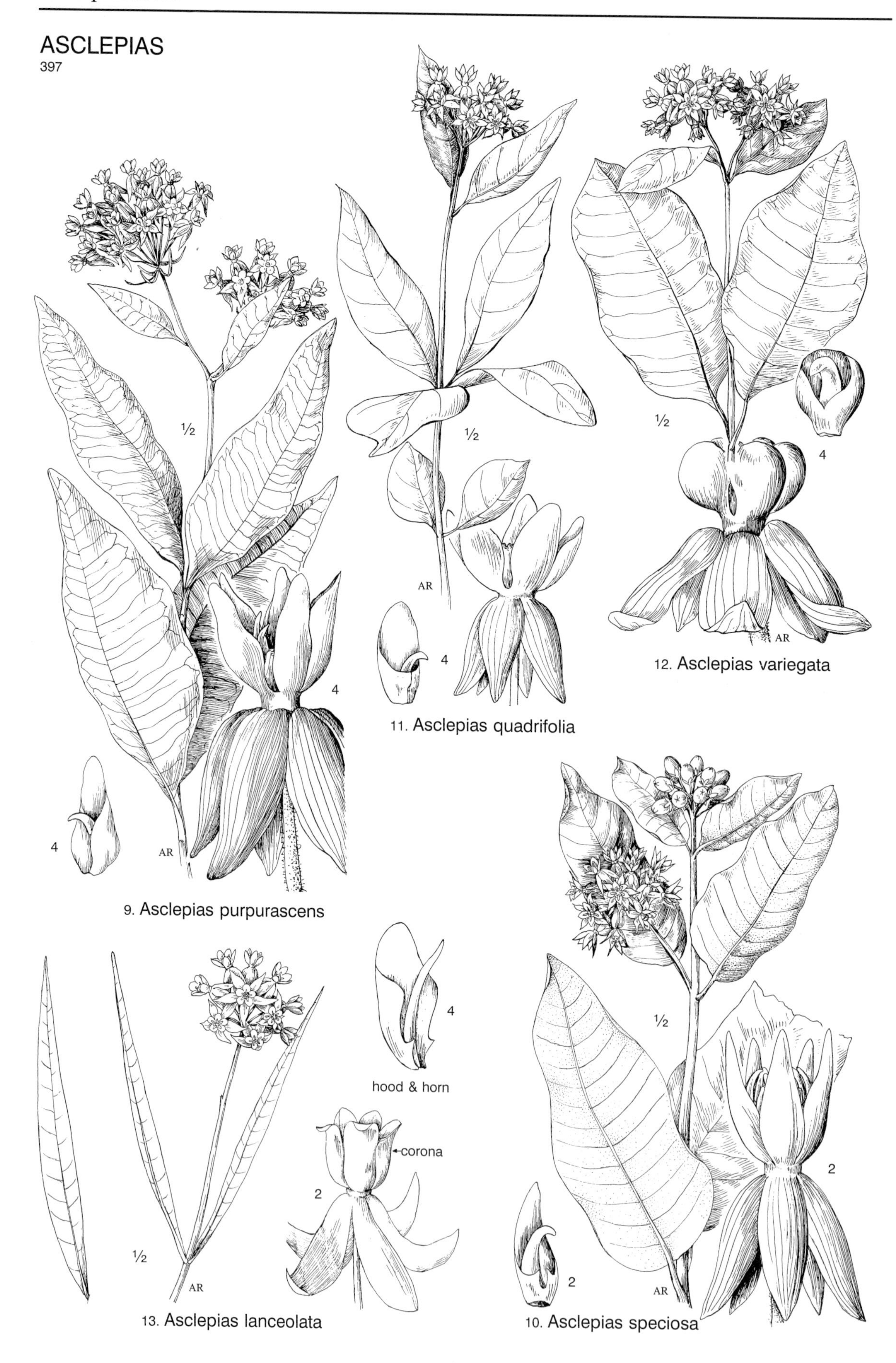

11. Asclepias quadrifolia

12. Asclepias variegata

9. Asclepias purpurascens

hood & horn

corona

13. Asclepias lanceolata

10. Asclepias speciosa

ILLUSTRATED COMPANION TO

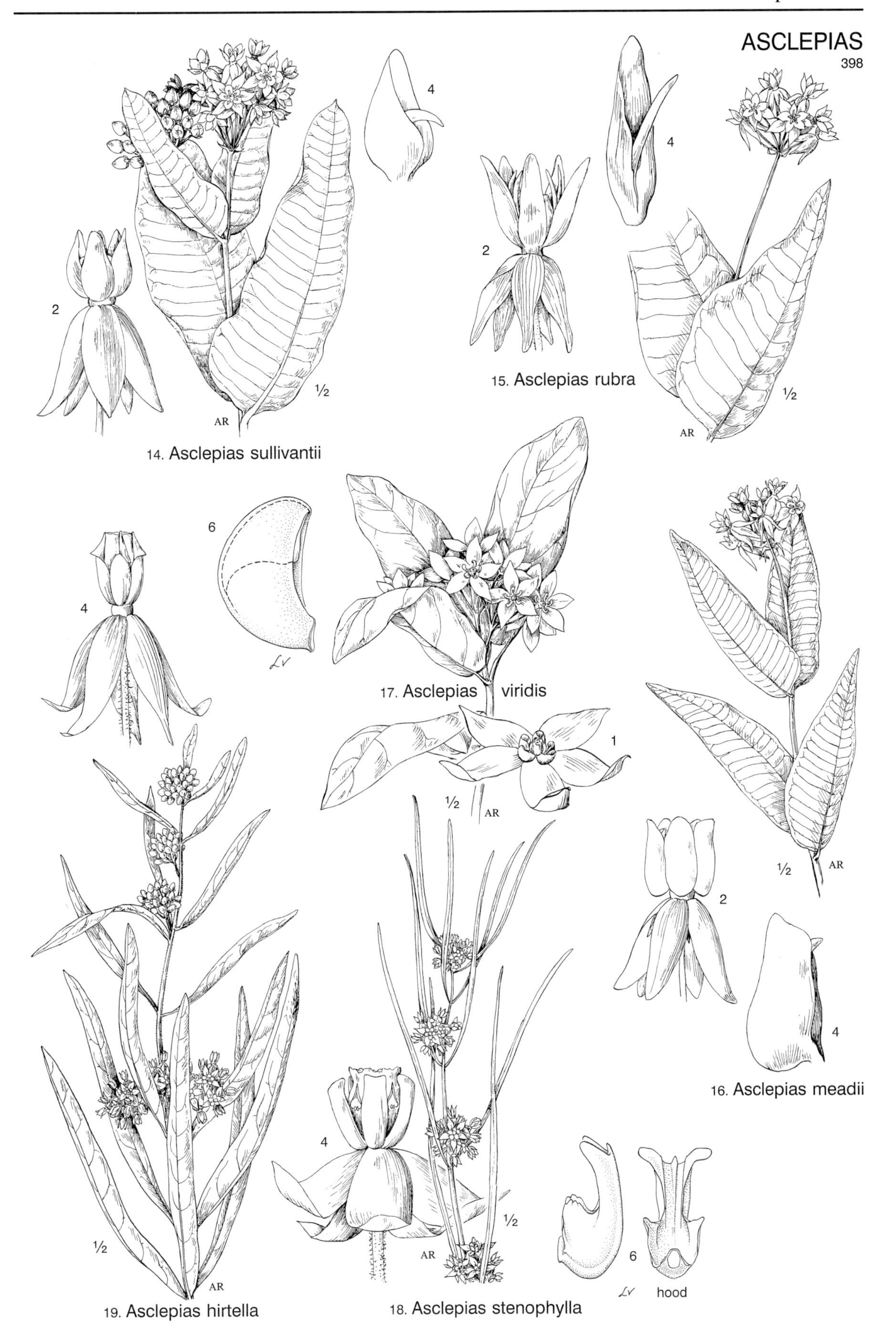

14. Asclepias sullivantii

15. Asclepias rubra

17. Asclepias viridis

16. Asclepias meadii

19. Asclepias hirtella

18. Asclepias stenophylla

hood

ASCLEPIAS
METAPLEXIS
VINCETOXICUM
398, 399

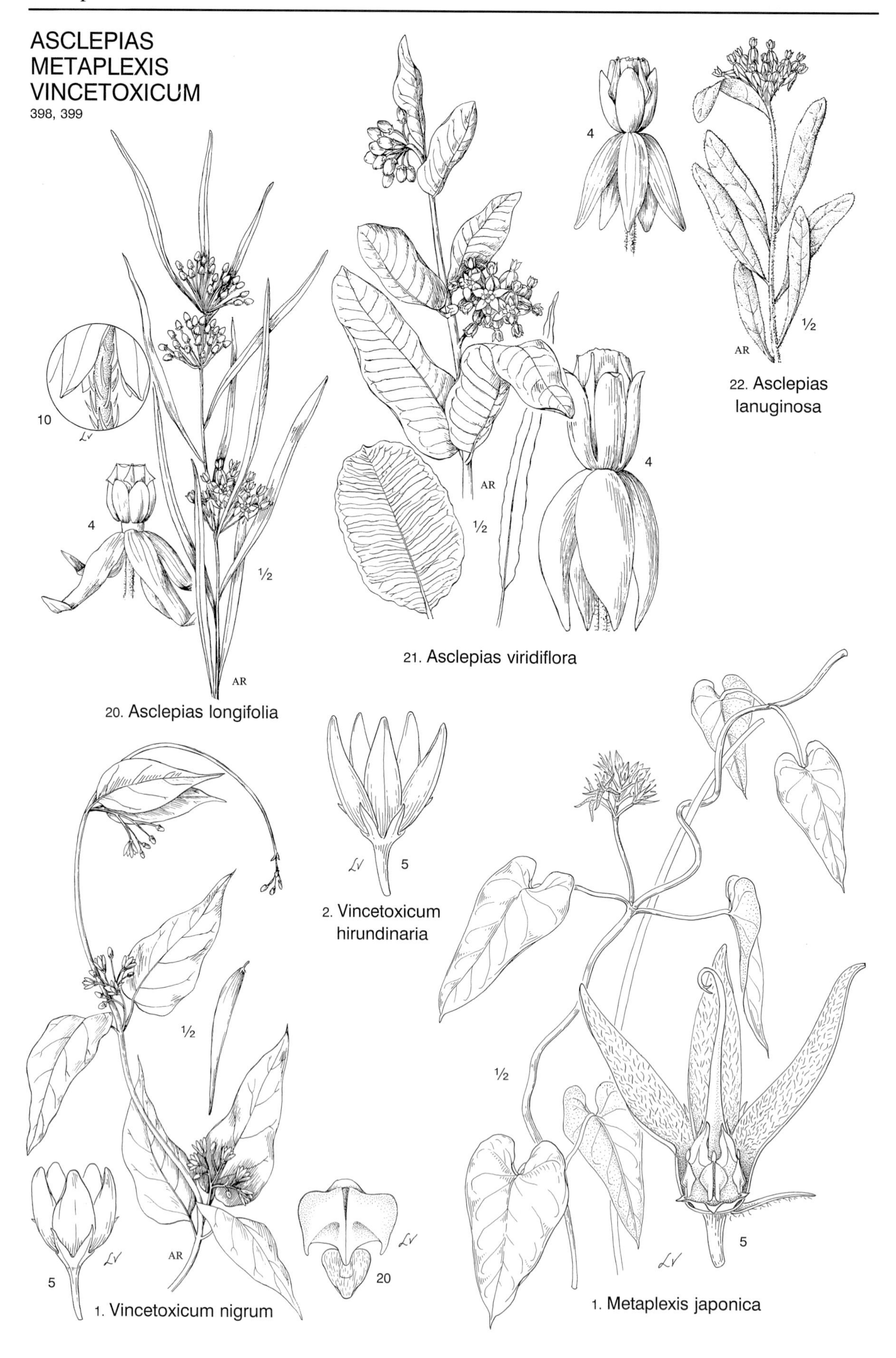

10

4

20. Asclepias longifolia

21. Asclepias viridiflora

4

22. Asclepias
lanuginosa

2. Vincetoxicum
hirundinaria

5

1. Vincetoxicum nigrum

20

1. Metaplexis japonica

5

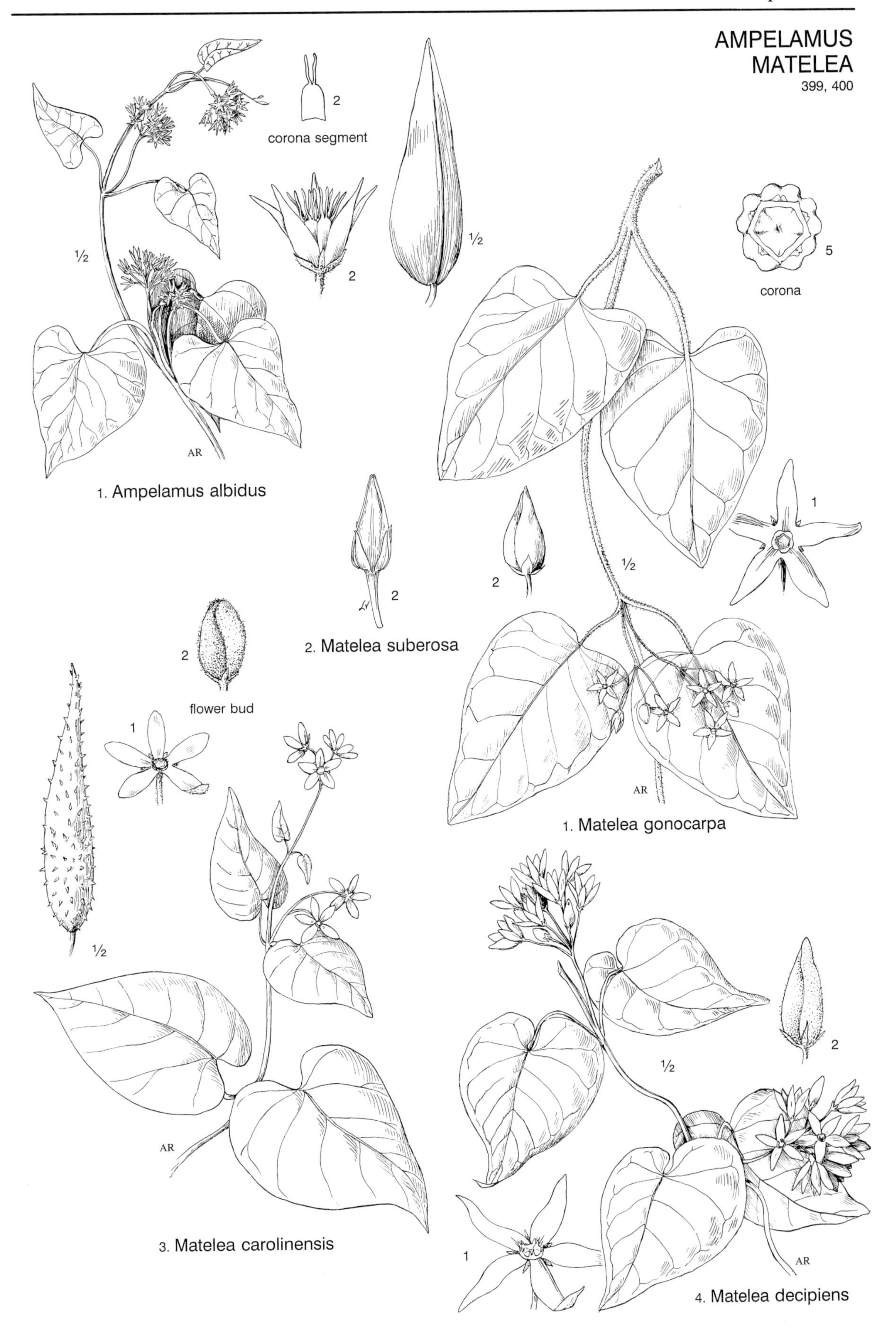

corona segment

2

corona

5

1. Ampelamus albidus

2. Matelea suberosa

flower bud

1. Matelea gonocarpa

3. Matelea carolinensis

4. Matelea decipiens

MATELEA
NICANDRA
LEUCOPHYSALIS
PHYSALIS
400–402

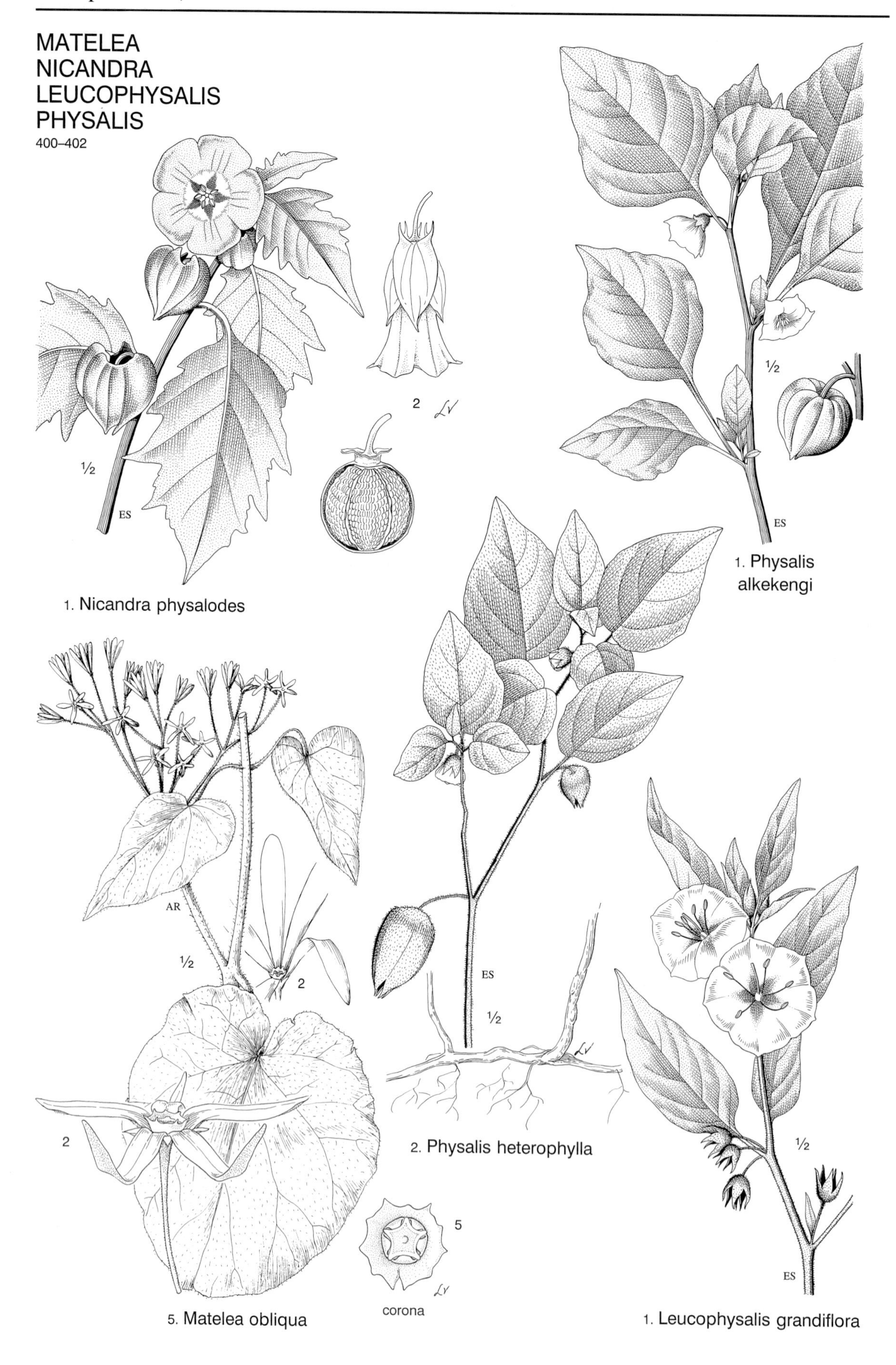

1. Nicandra physalodes

2

1. Physalis alkekengi

2. Physalis heterophylla

5. Matelea obliqua

corona

1. Leucophysalis grandiflora

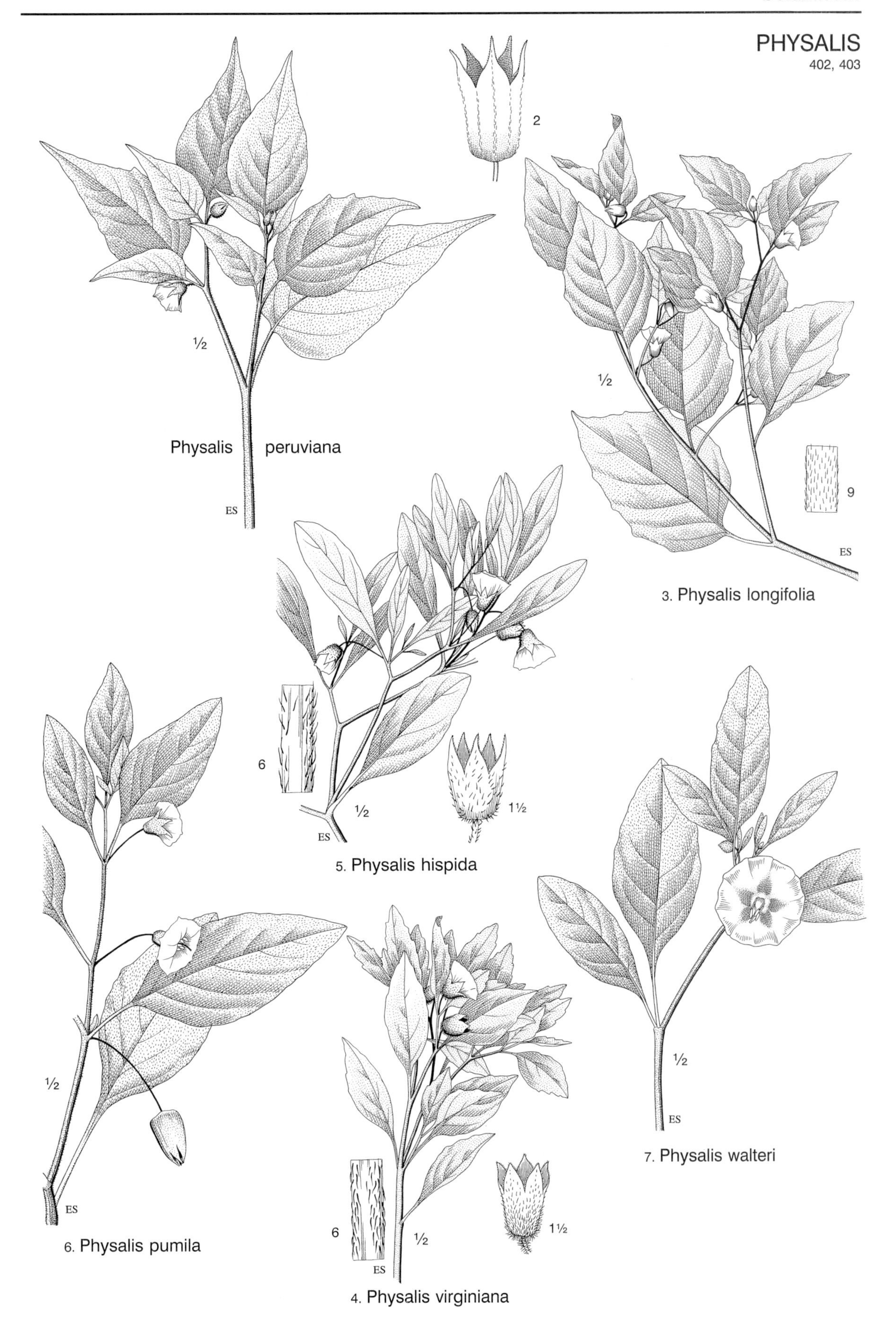

2

Physalis   peruviana

ES

3. Physalis longifolia

9

ES

6

5. Physalis hispida

½   1½

ES

6. Physalis pumila

½

ES

6

4. Physalis virginiana

½   1½

ES

7. Physalis walteri

½

ES

PHYSALIS

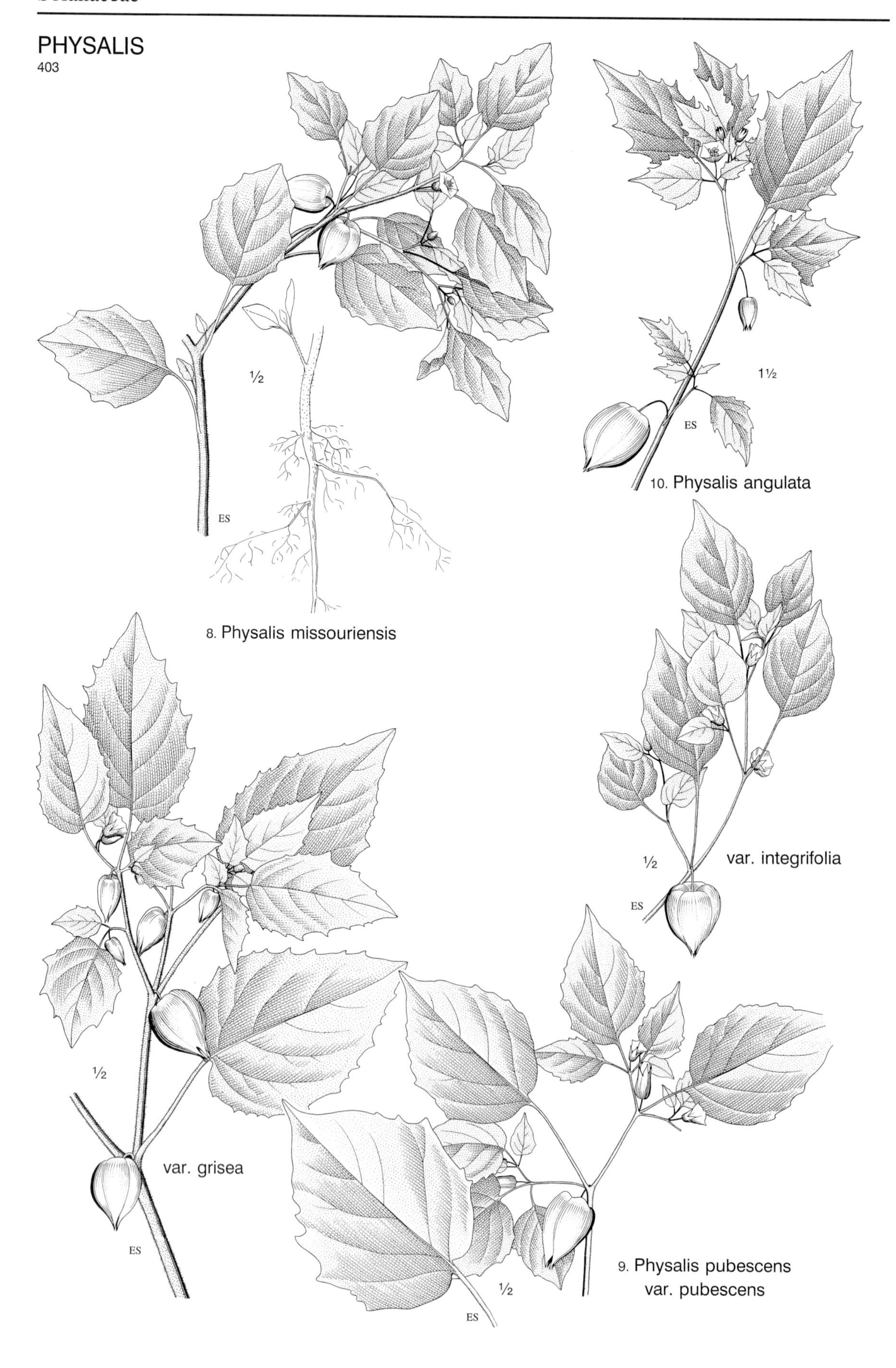

½

ES

8. Physalis missouriensis

10. Physalis angulata

1½

ES

var. integrifolia

½

ES

var. grisea

½

ES

9. Physalis pubescens
var. pubescens

½

ES

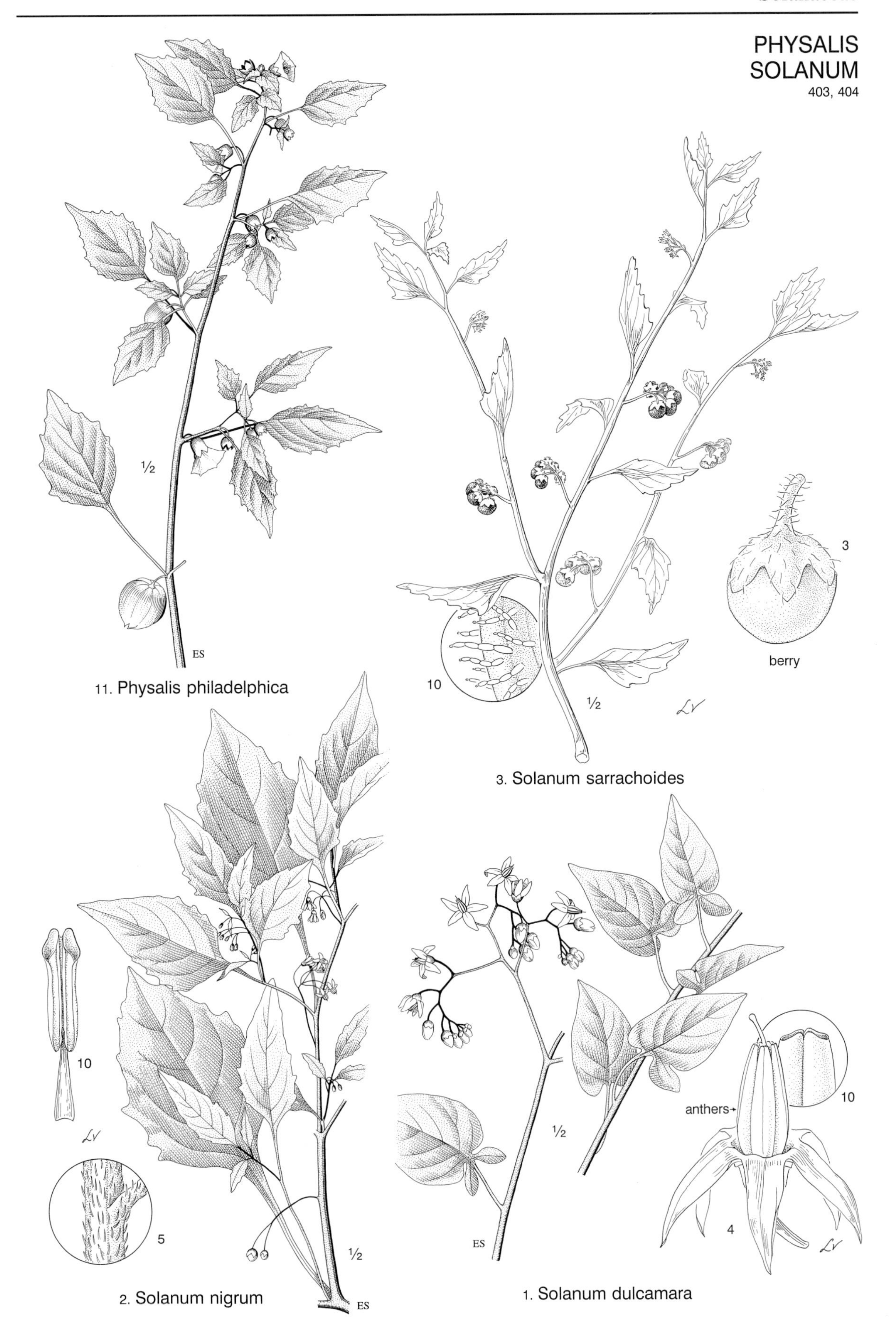

11. Physalis philadelphica

3. Solanum sarrachoides

berry

2. Solanum nigrum

1. Solanum dulcamara

anthers

SOLANUM
404

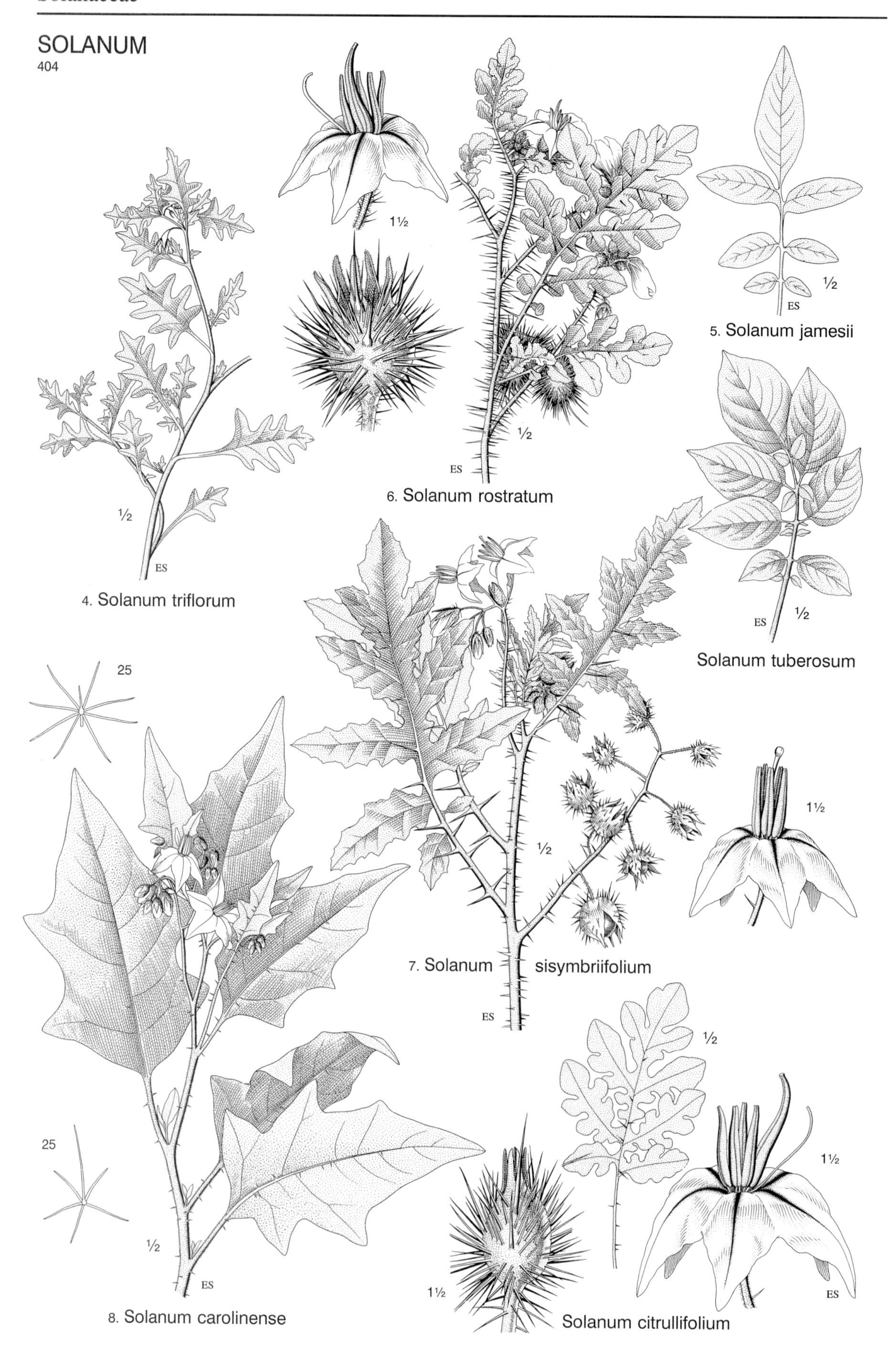

1½

½
ES

4. Solanum triflorum

½
ES

6. Solanum rostratum

½
ES

5. Solanum jamesii

½
ES

Solanum tuberosum

25

25

½

7. Solanum   sisymbriifolium

½
ES

1½

½
ES

8. Solanum carolinense

1½

½

1½
ES

Solanum citrullifolium

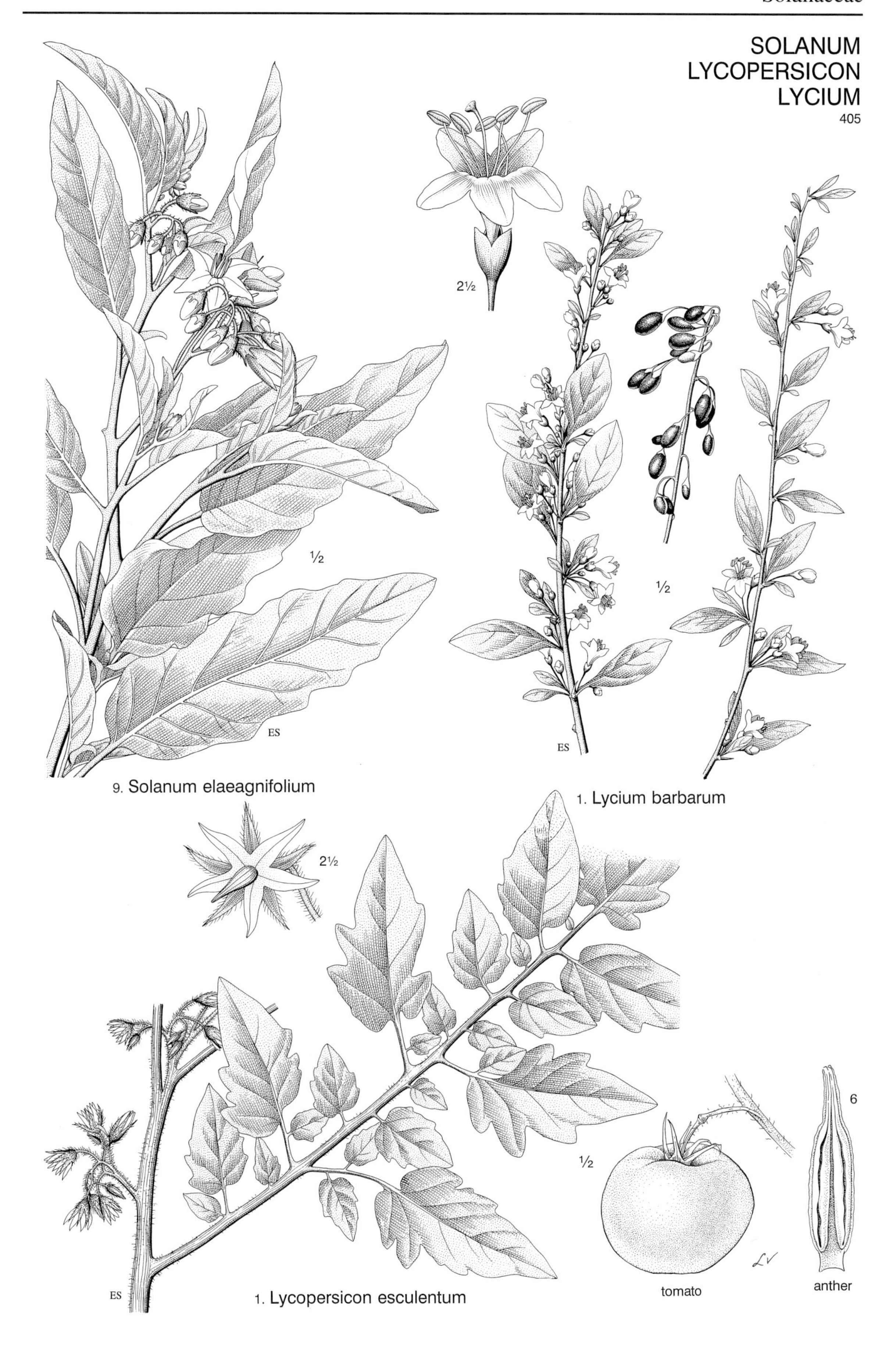

2½

½

ES

9. Solanum elaeagnifolium

½

ES

1. Lycium barbarum

2½

½

ES

1. Lycopersicon esculentum

tomato

6

anther

HYOSCYAMUS
DATURA
NICOTIANA
405, 406

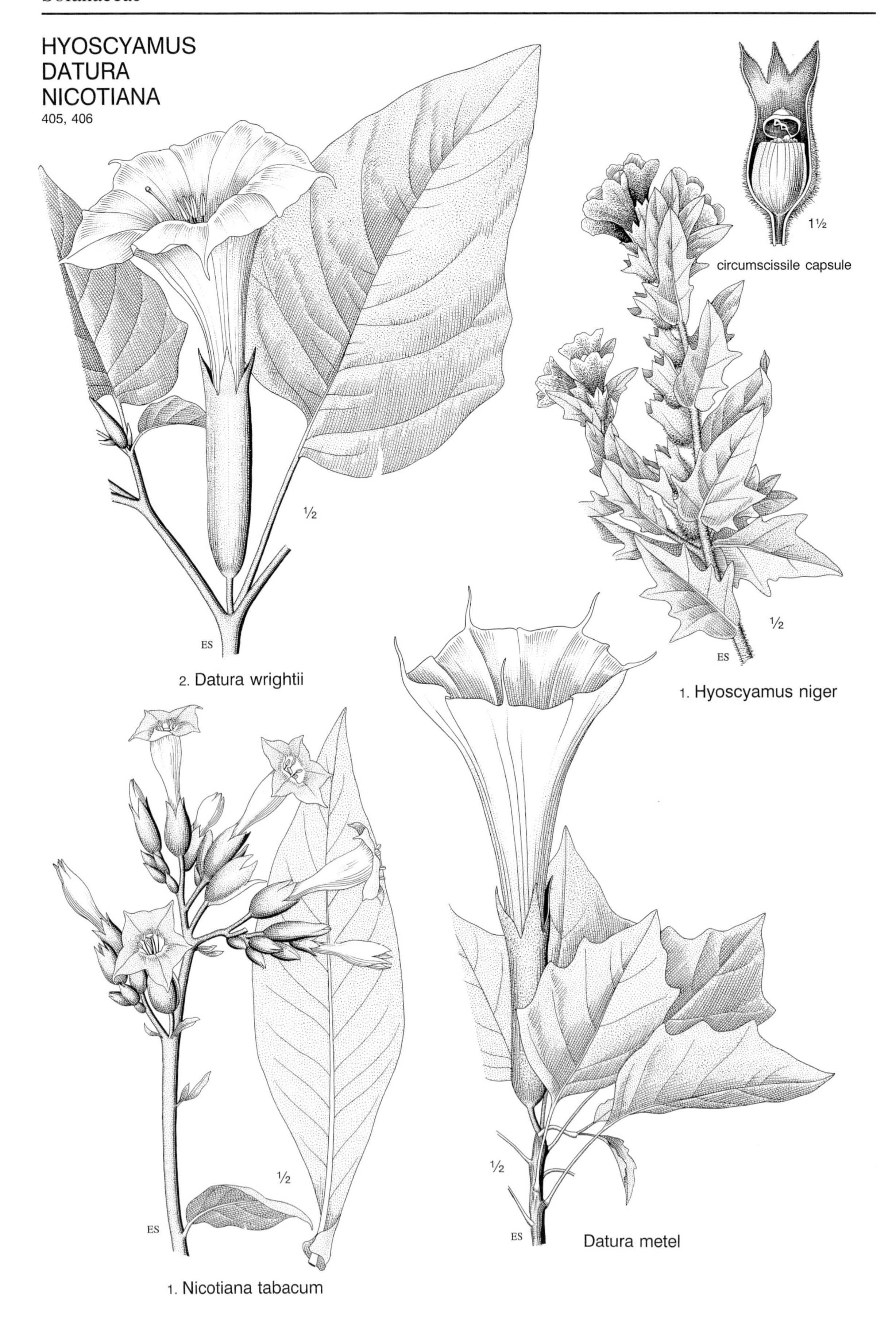

1½

circumscissile capsule

2. Datura wrightii

1. Hyoscyamus niger

1. Nicotiana tabacum

Datura metel

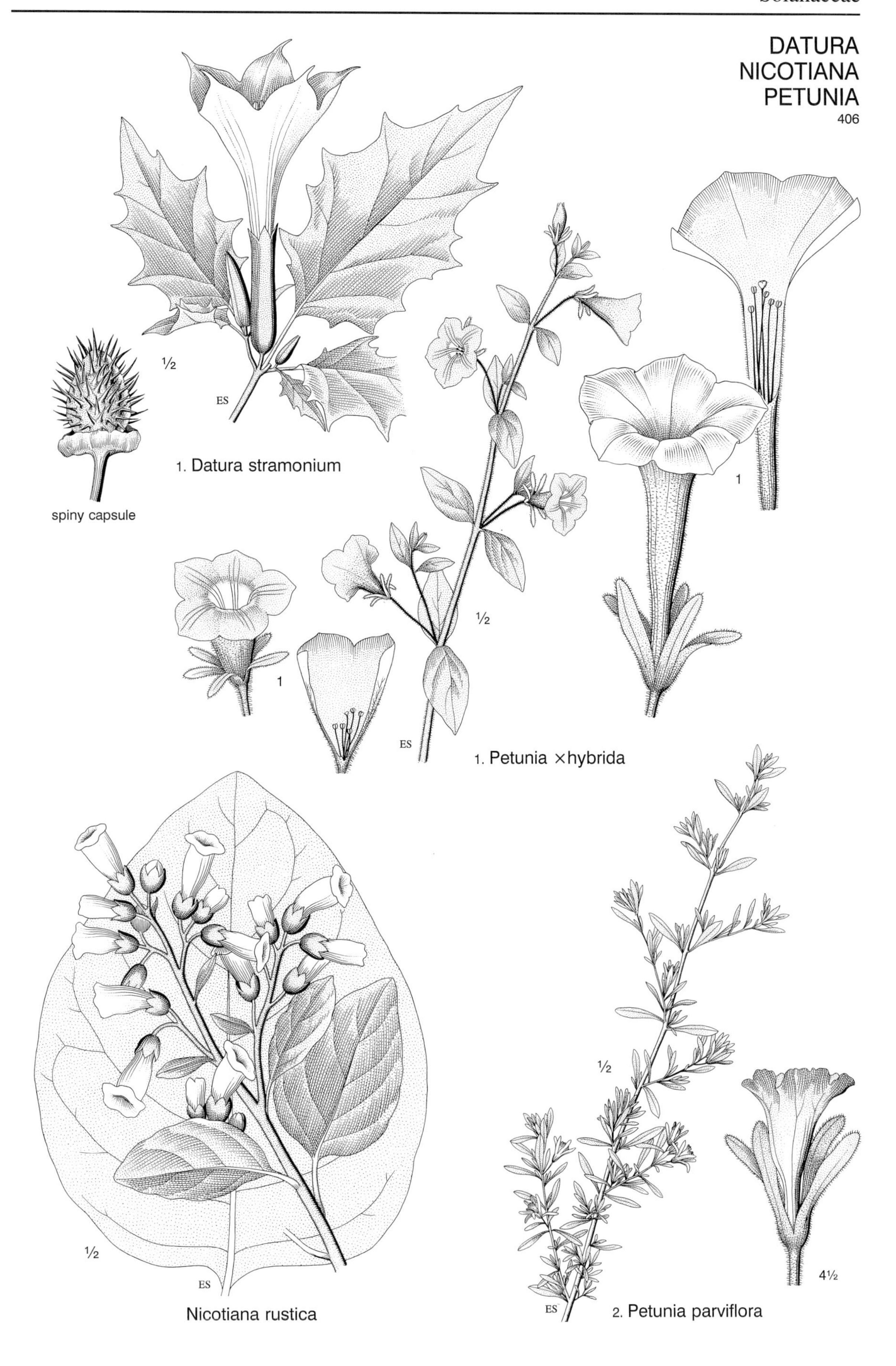

1. Datura stramonium

spiny capsule

1. Petunia ×hybrida

Nicotiana rustica

2. Petunia parviflora

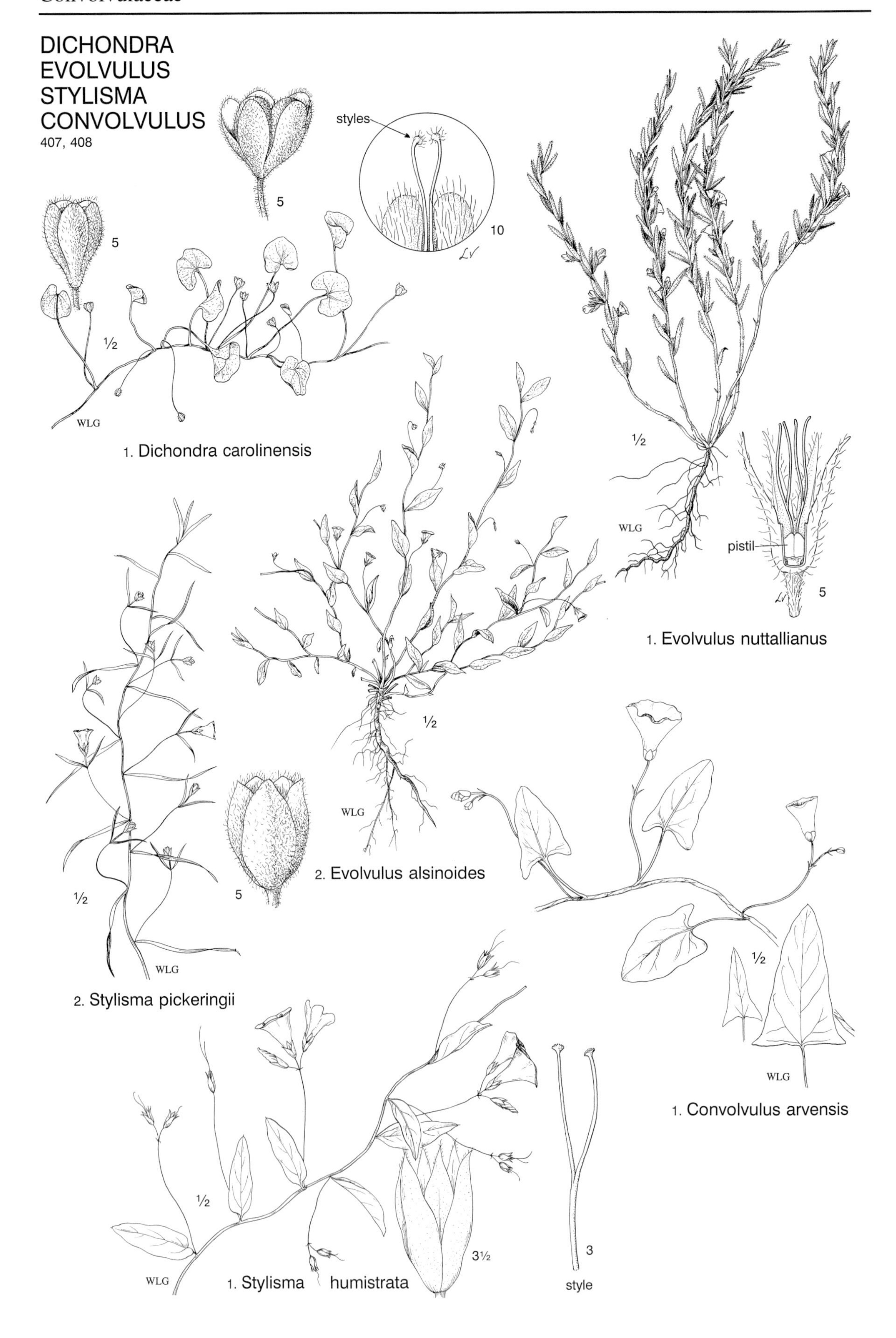

DICHONDRA
EVOLVULUS
STYLISMA
CONVOLVULUS
407, 408

styles

5

5

10

½

WLG

1. Dichondra carolinensis

½

WLG

pistil

5

1. Evolvulus nuttallianus

½

½

WLG

WLG

5

2. Evolvulus alsinoides

2. Stylisma pickeringii

½

WLG

1. Convolvulus arvensis

½

3½

3

style

1. Stylisma   humistrata

384

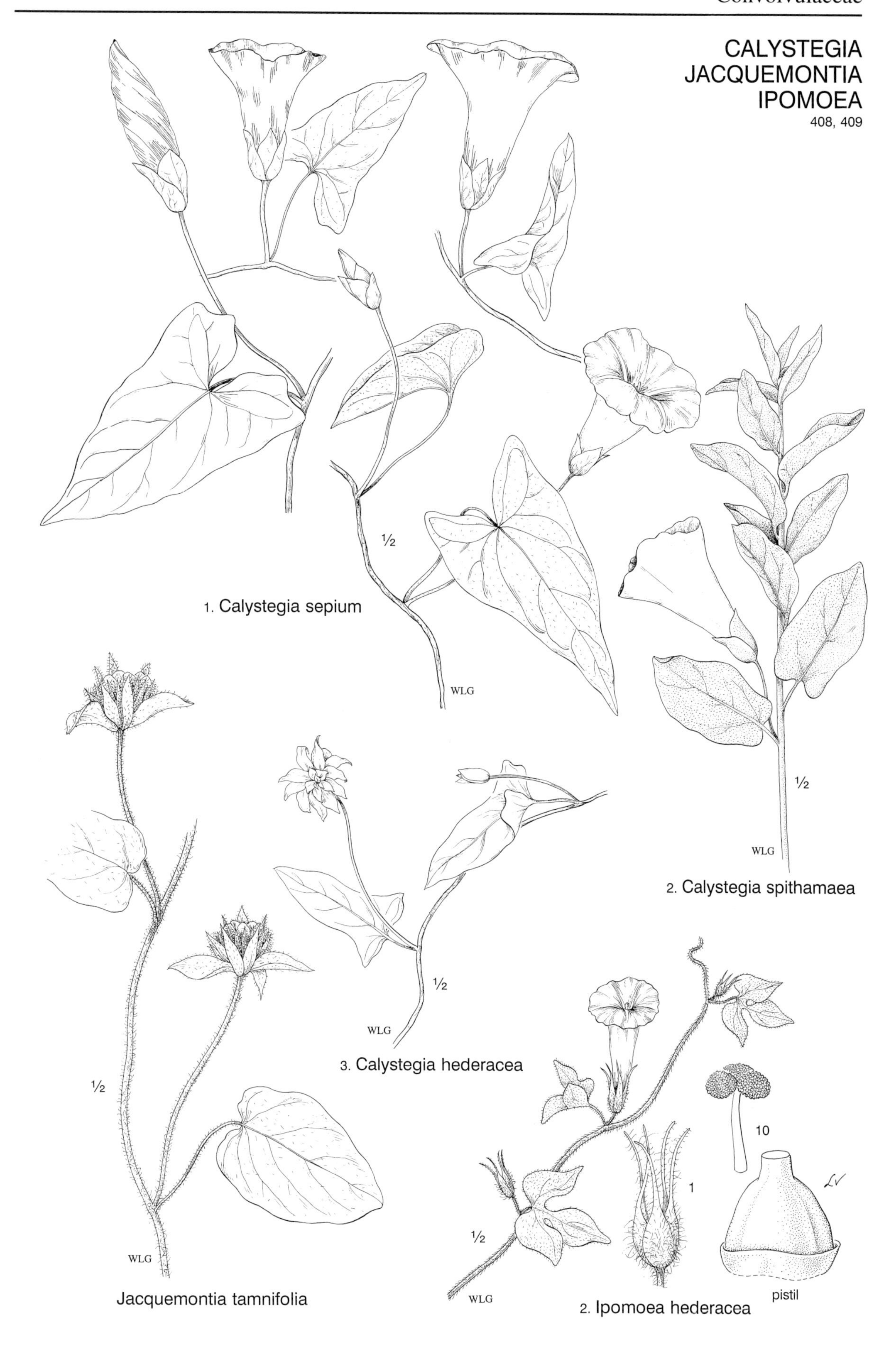

1. Calystegia sepium

½

WLG

2. Calystegia spithamaea

½

WLG

½

WLG

3. Calystegia hederacea

½

WLG

Jacquemontia tamnifolia

½

WLG

10

1

½

WLG

2. Ipomoea hederacea

pistil

IPOMOEA
CUSCUTA
409–411

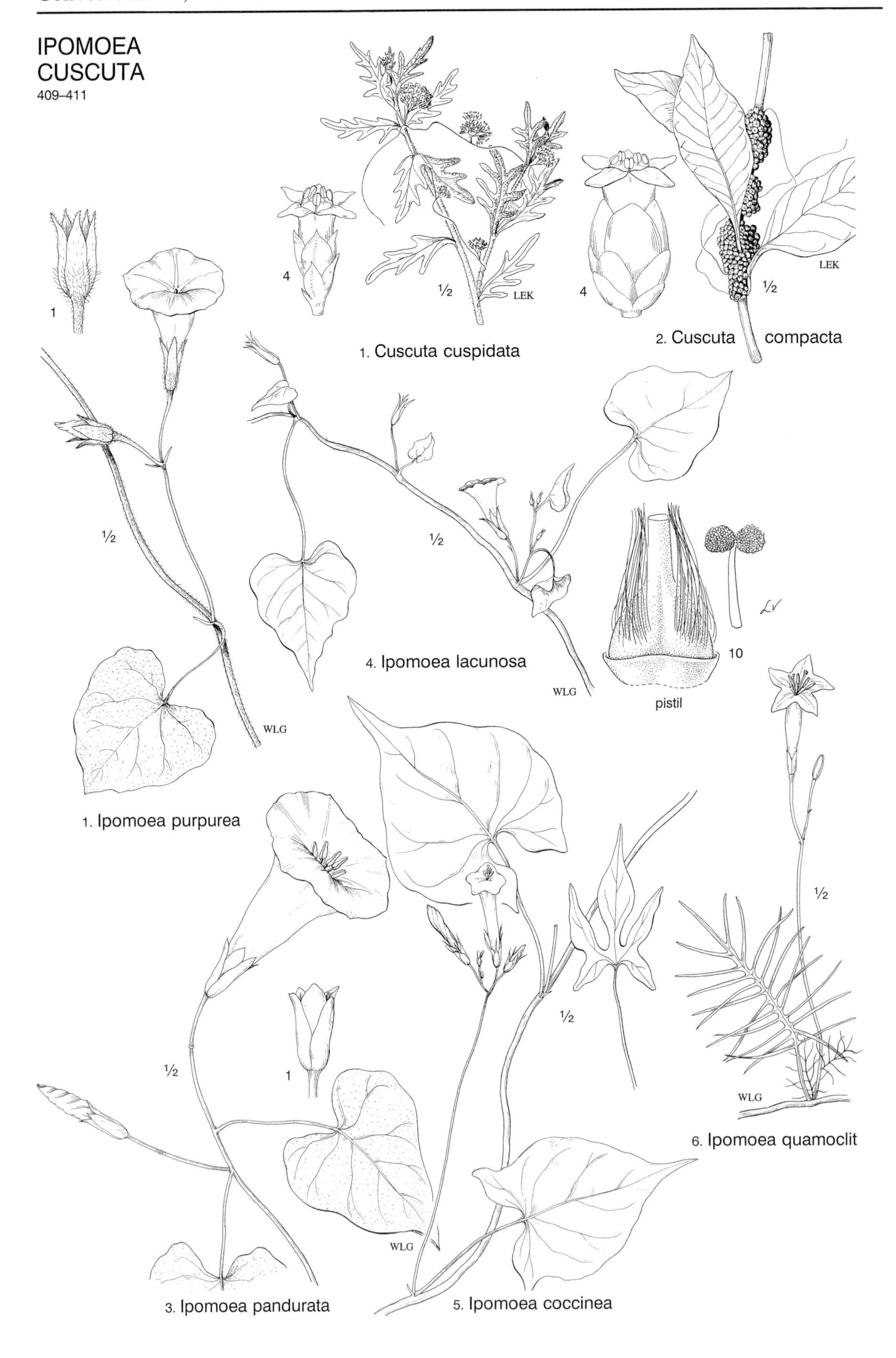

1. Cuscuta cuspidata

2. Cuscuta compacta

4. Ipomoea lacunosa

pistil

1. Ipomoea purpurea

3. Ipomoea pandurata

5. Ipomoea coccinea

6. Ipomoea quamoclit

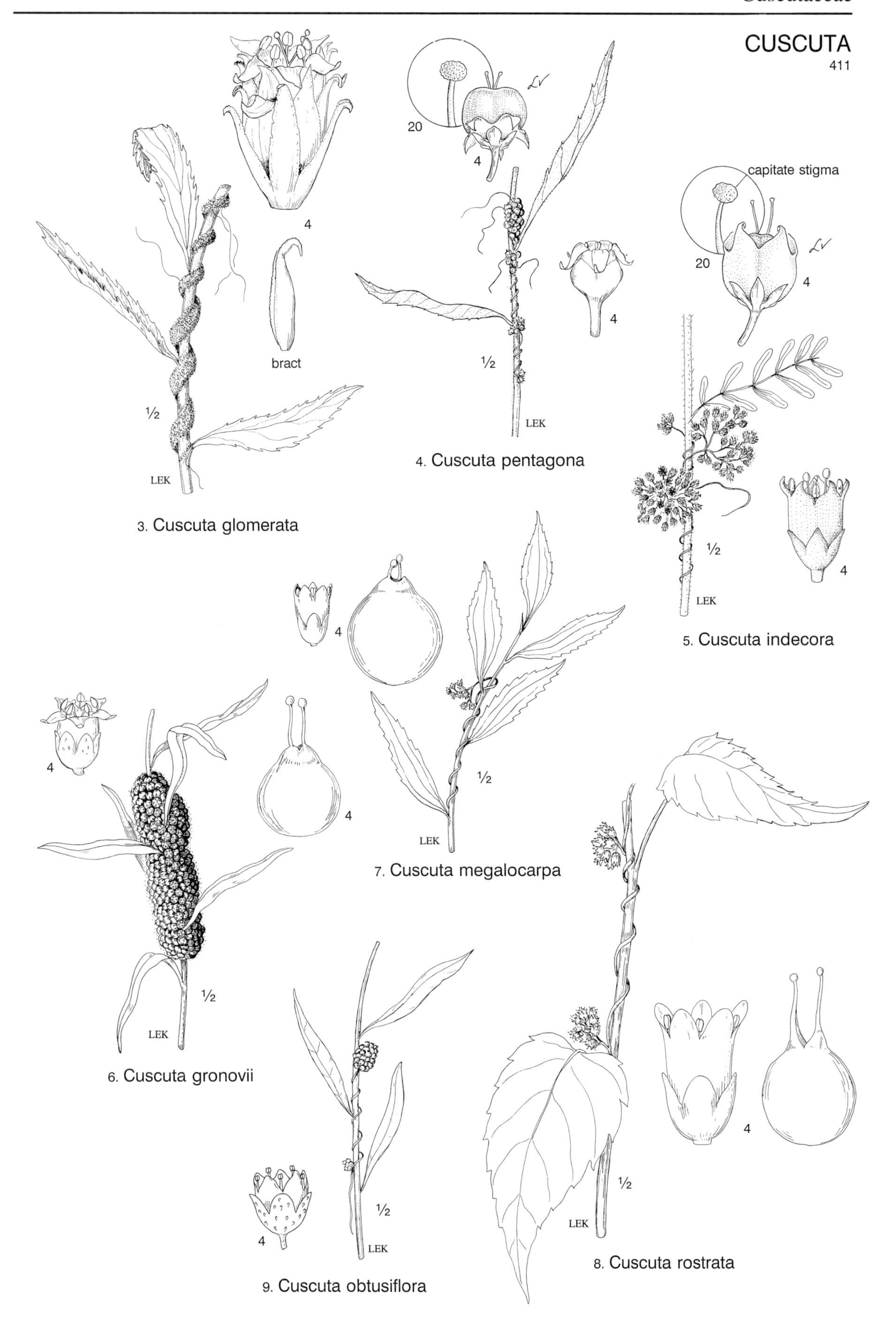

bract

4. Cuscuta pentagona

3. Cuscuta glomerata

capitate stigma

5. Cuscuta indecora

7. Cuscuta megalocarpa

6. Cuscuta gronovii

9. Cuscuta obtusiflora

8. Cuscuta rostrata

CUSCUTA
MENYANTHES
411, 412

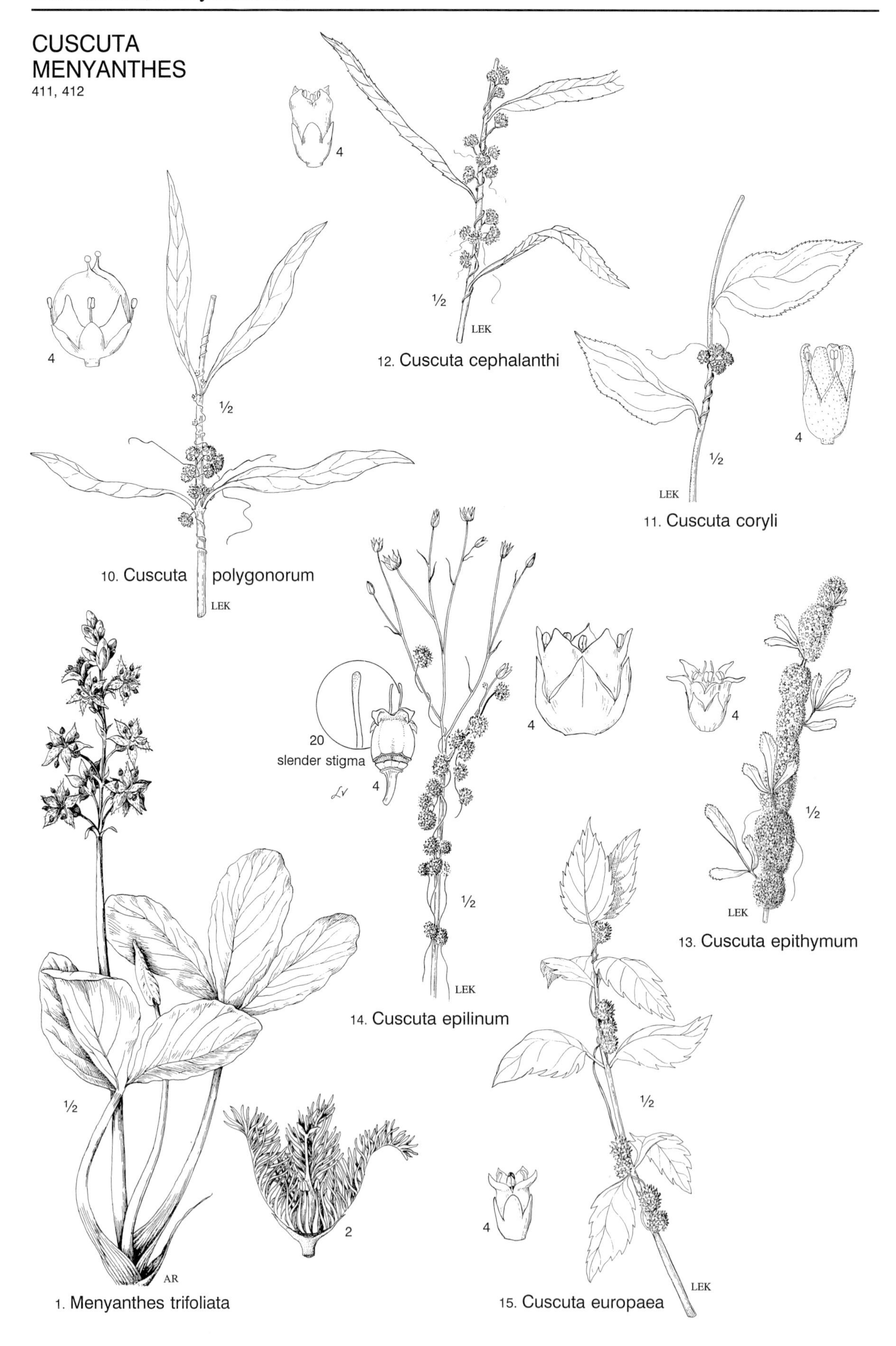

4

4

½

12. Cuscuta cephalanthi

LEK

11. Cuscuta coryli

LEK

4

½

10. Cuscuta   polygonorum

LEK

4

4

20
slender stigma

4

4

½

13. Cuscuta epithymum

LEK

½

LEK

14. Cuscuta epilinum

½

2

½

4

1. Menyanthes trifoliata

AR

15. Cuscuta europaea

LEK

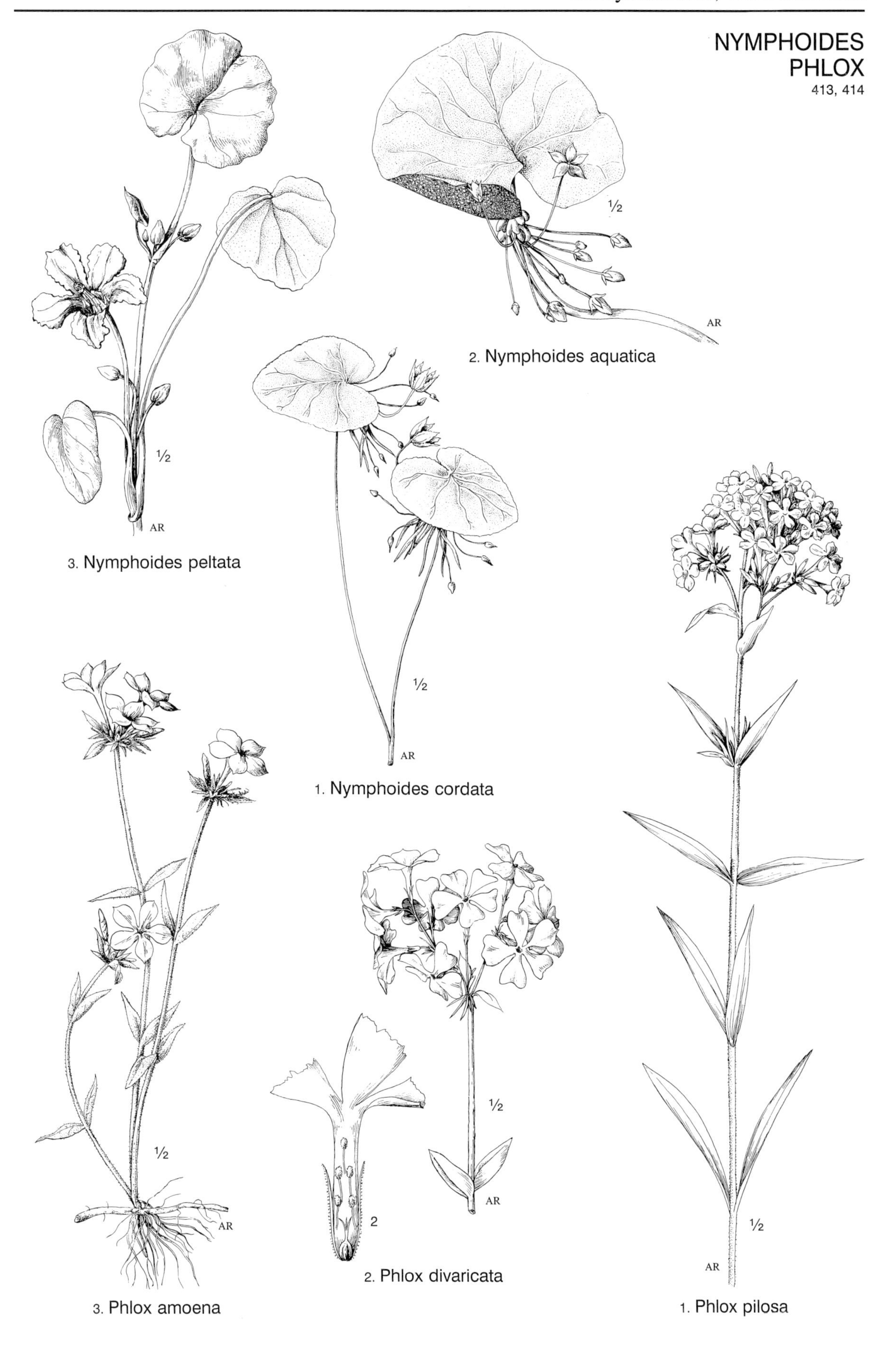

2. Nymphoides aquatica

3. Nymphoides peltata

1. Nymphoides cordata

3. Phlox amoena

2. Phlox divaricata

1. Phlox pilosa

PHLOX
414, 415

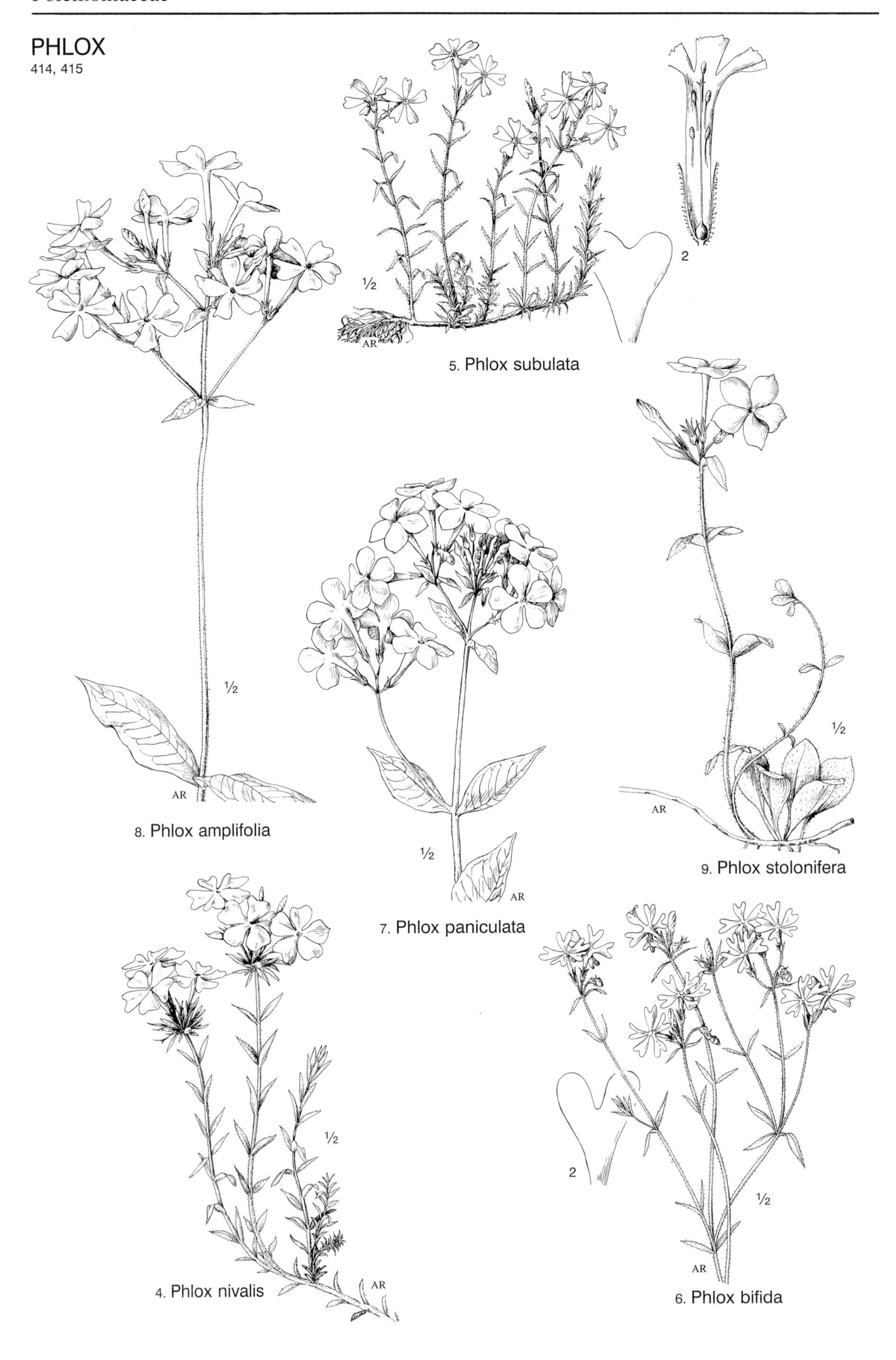

½

2

5. Phlox subulata

8. Phlox amplifolia

½

AR

7. Phlox paniculata

½

AR

½

9. Phlox stolonifera

AR

4. Phlox nivalis

½

AR

2

½

AR

6. Phlox bifida

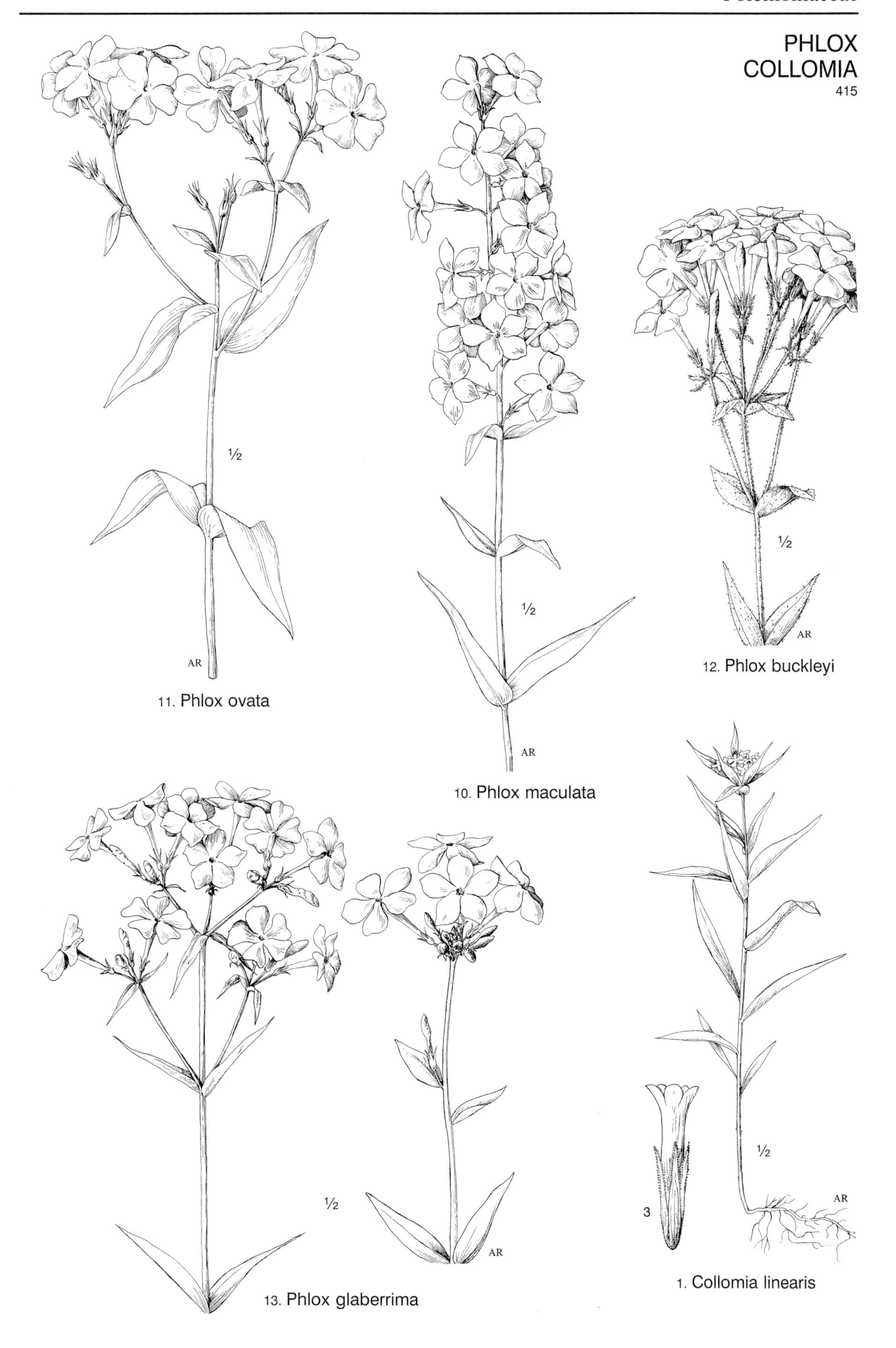

11. Phlox ovata

10. Phlox maculata

12. Phlox buckleyi

13. Phlox glaberrima

1. Collomia linearis

POLEMONIUM
GILIA
PHACELIA
416, 417

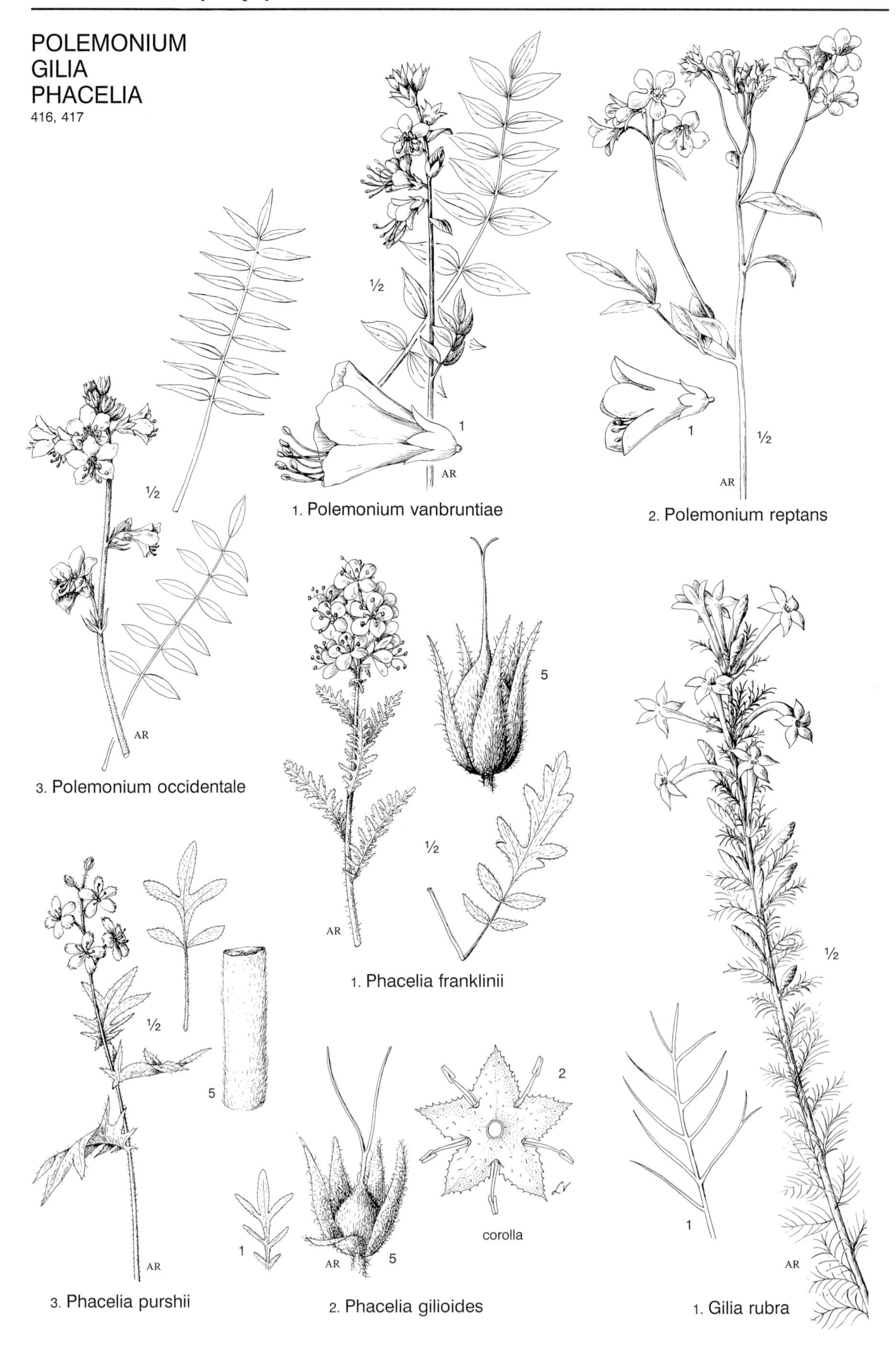

1. Polemonium vanbruntiae

2. Polemonium reptans

3. Polemonium occidentale

1. Phacelia franklinii

3. Phacelia purshii

2. Phacelia gilioides

corolla

1. Gilia rubra

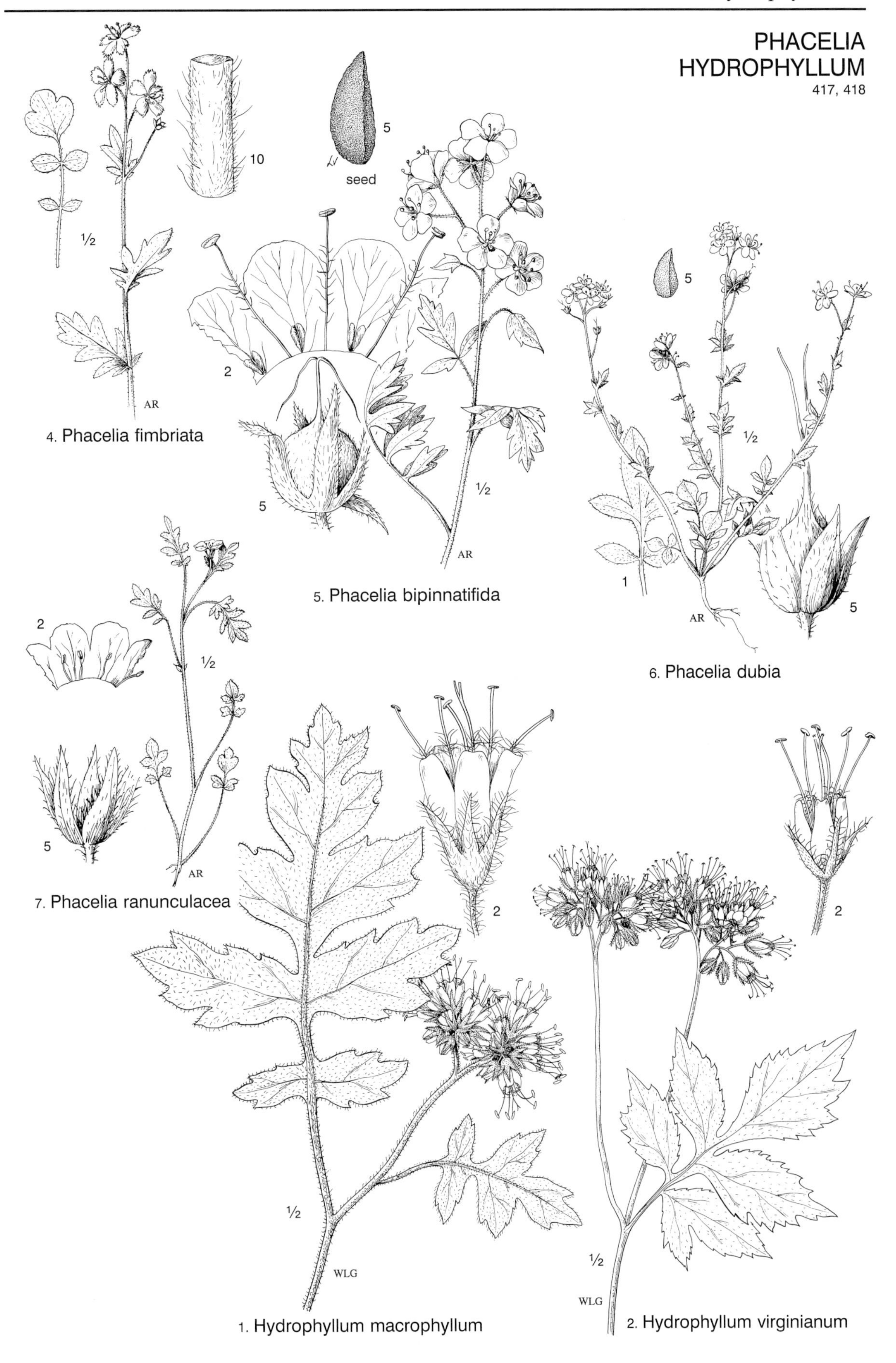

10

5
seed

4. Phacelia fimbriata

5. Phacelia bipinnatifida

5

6. Phacelia dubia

7. Phacelia ranunculacea

1. Hydrophyllum macrophyllum

2. Hydrophyllum virginianum

HYDROPHYLLUM
ELLISIA
NEMOPHILA
HYDROLEA
418, 419

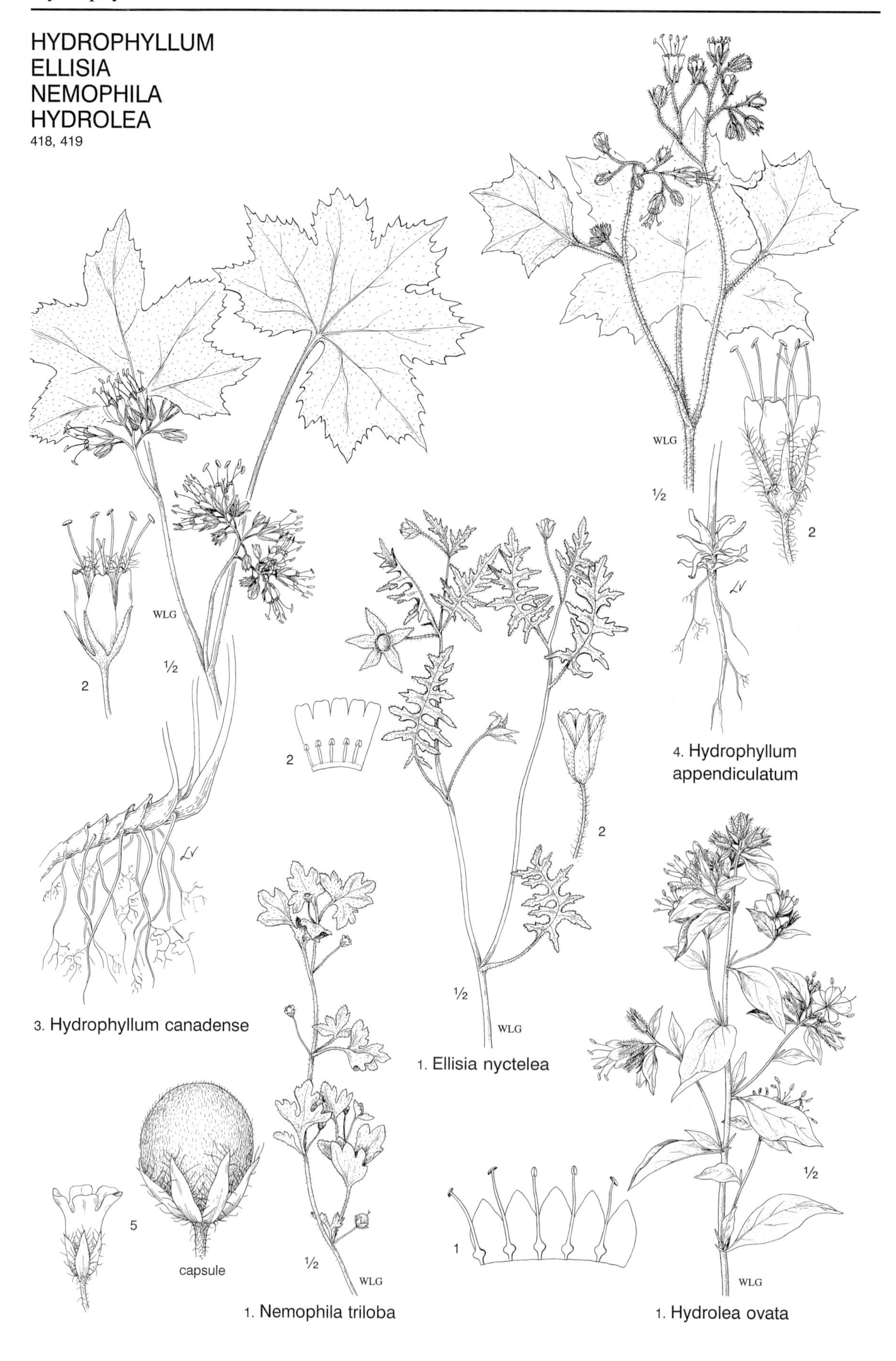

4. Hydrophyllum
appendiculatum

3. Hydrophyllum canadense

1. Ellisia nyctelea

capsule

1. Nemophila triloba

1. Hydrolea ovata

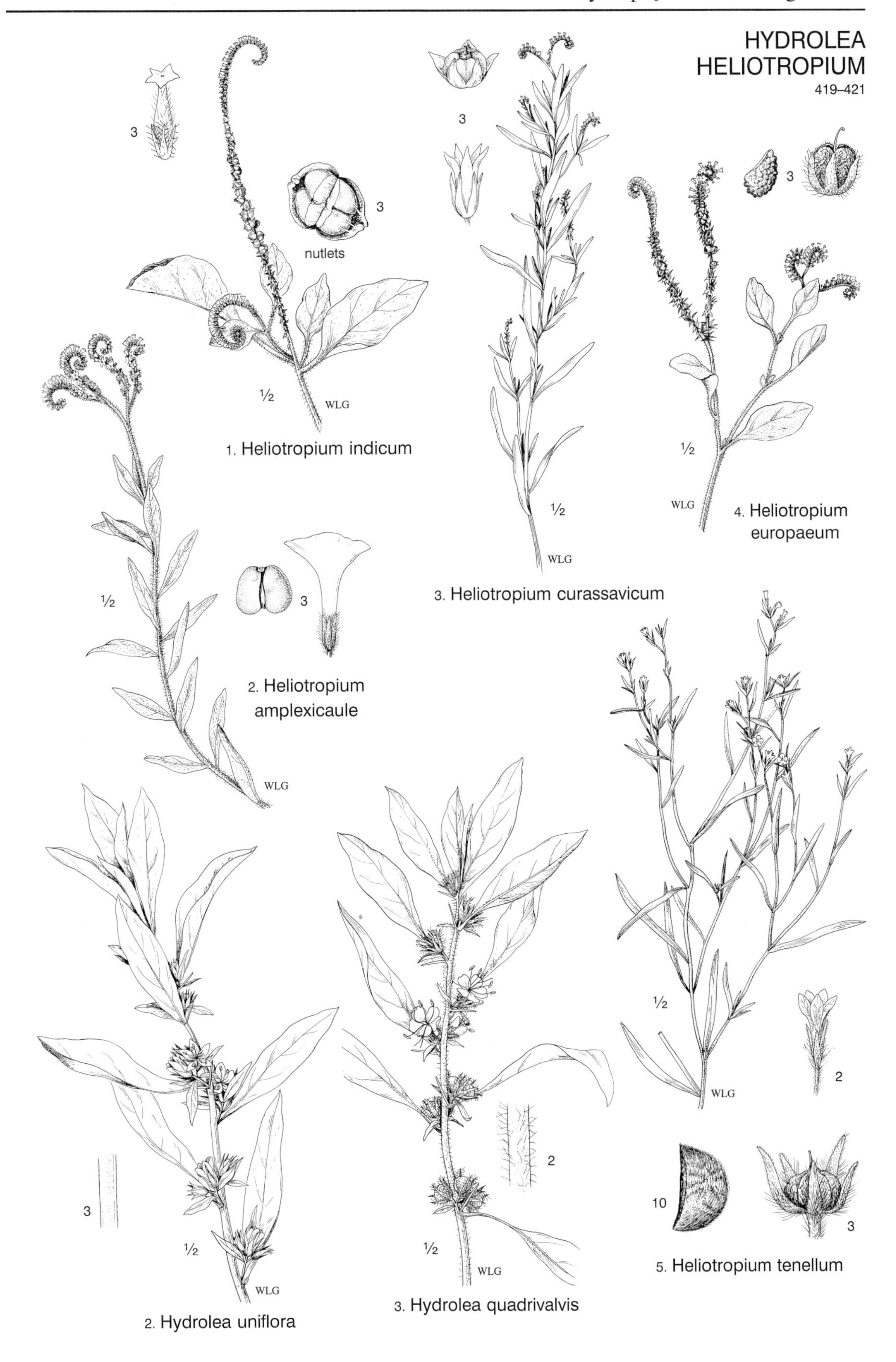

3

nutlets

1. Heliotropium indicum

3. Heliotropium curassavicum

4. Heliotropium europaeum

2. Heliotropium amplexicaule

2. Hydrolea uniflora

3. Hydrolea quadrivalvis

5. Heliotropium tenellum

MERTENSIA
ASPERUGO
MYOSOTIS
421, 422

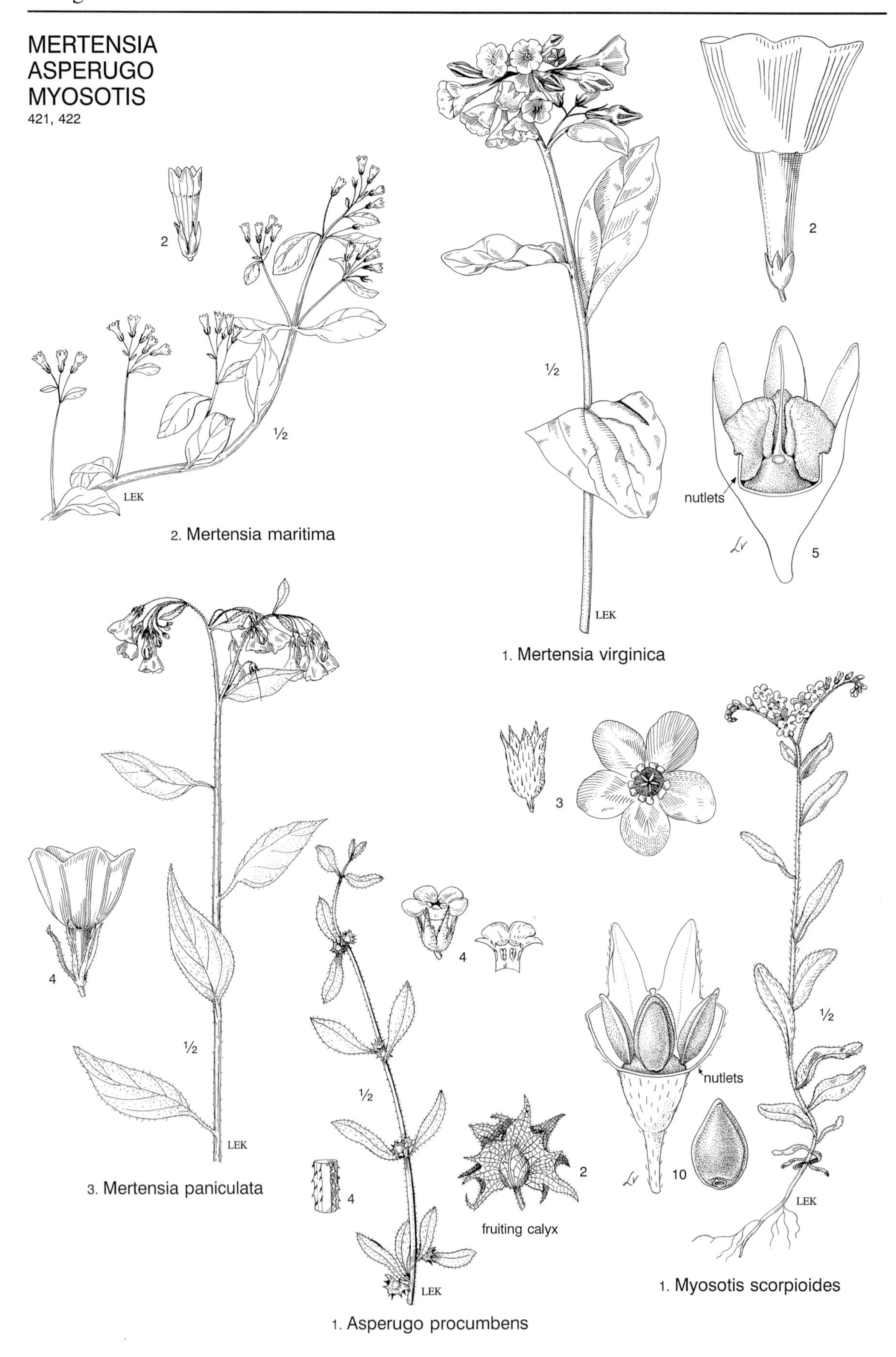

2. Mertensia maritima

1. Mertensia virginica

nutlets

3. Mertensia paniculata

fruiting calyx

nutlets

1. Asperugo procumbens

1. Myosotis scorpioides

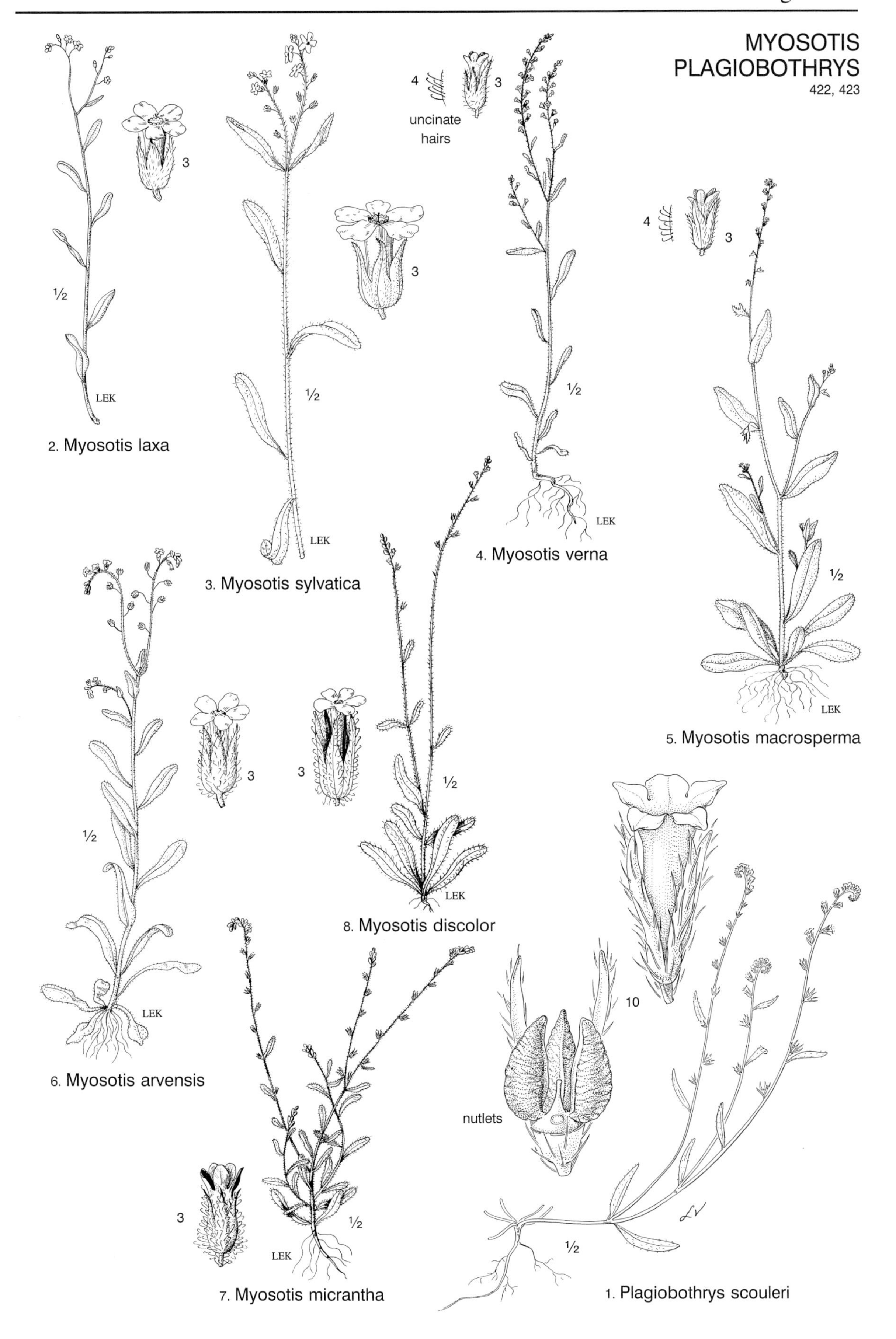

4 uncinate hairs 3

3

2. Myosotis laxa

3. Myosotis sylvatica

4. Myosotis verna

4 3

5. Myosotis macrosperma

3 3

8. Myosotis discolor

6. Myosotis arvensis

10

nutlets

7. Myosotis micrantha

1. Plagiobothrys scouleri

AMSINCKIA
HACKELIA
LAPPULA
423, 424

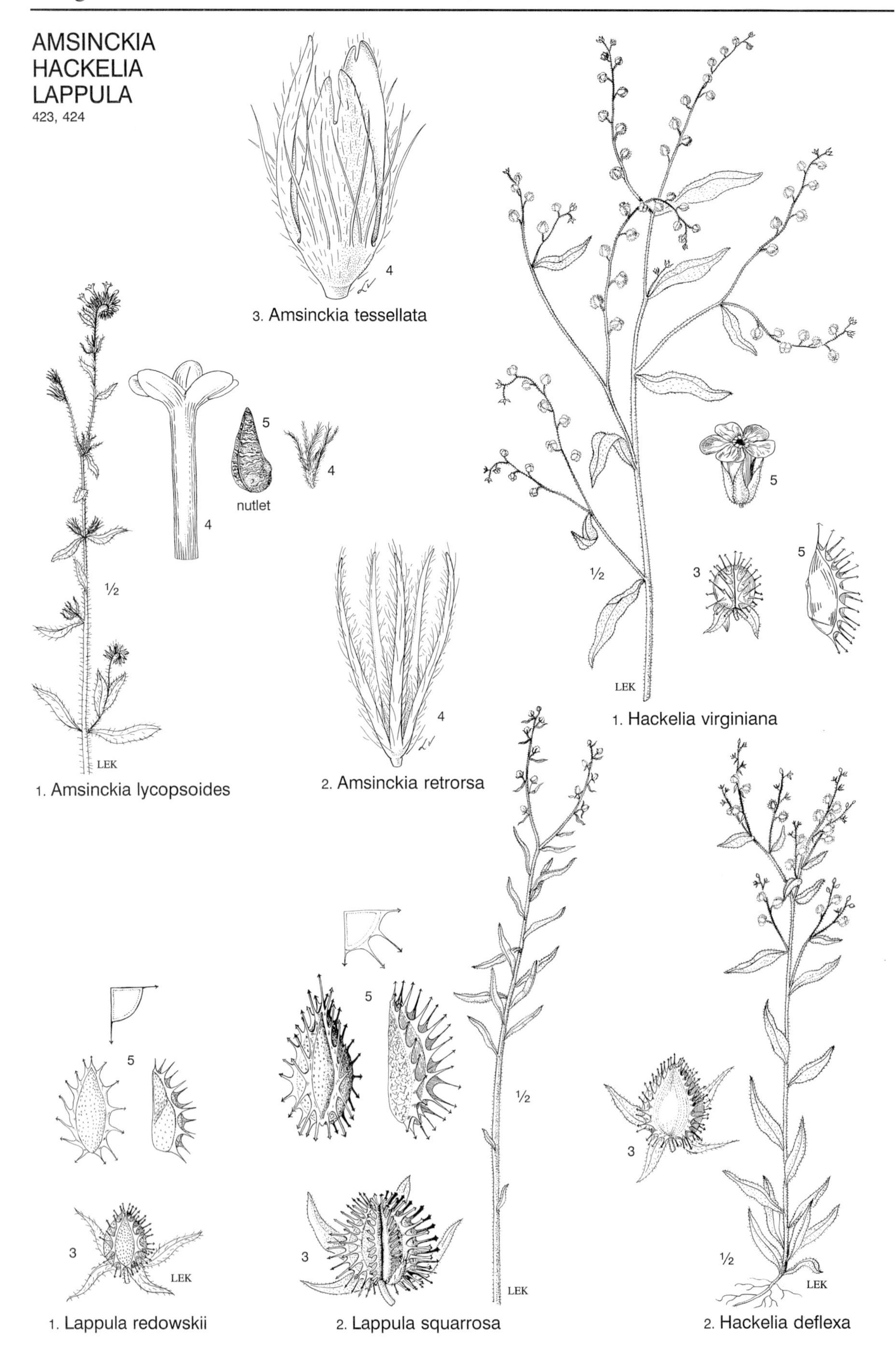

3. Amsinckia tessellata

nutlet

1. Amsinckia lycopsoides

2. Amsinckia retrorsa

1. Hackelia virginiana

1. Lappula redowskii

2. Lappula squarrosa

2. Hackelia deflexa

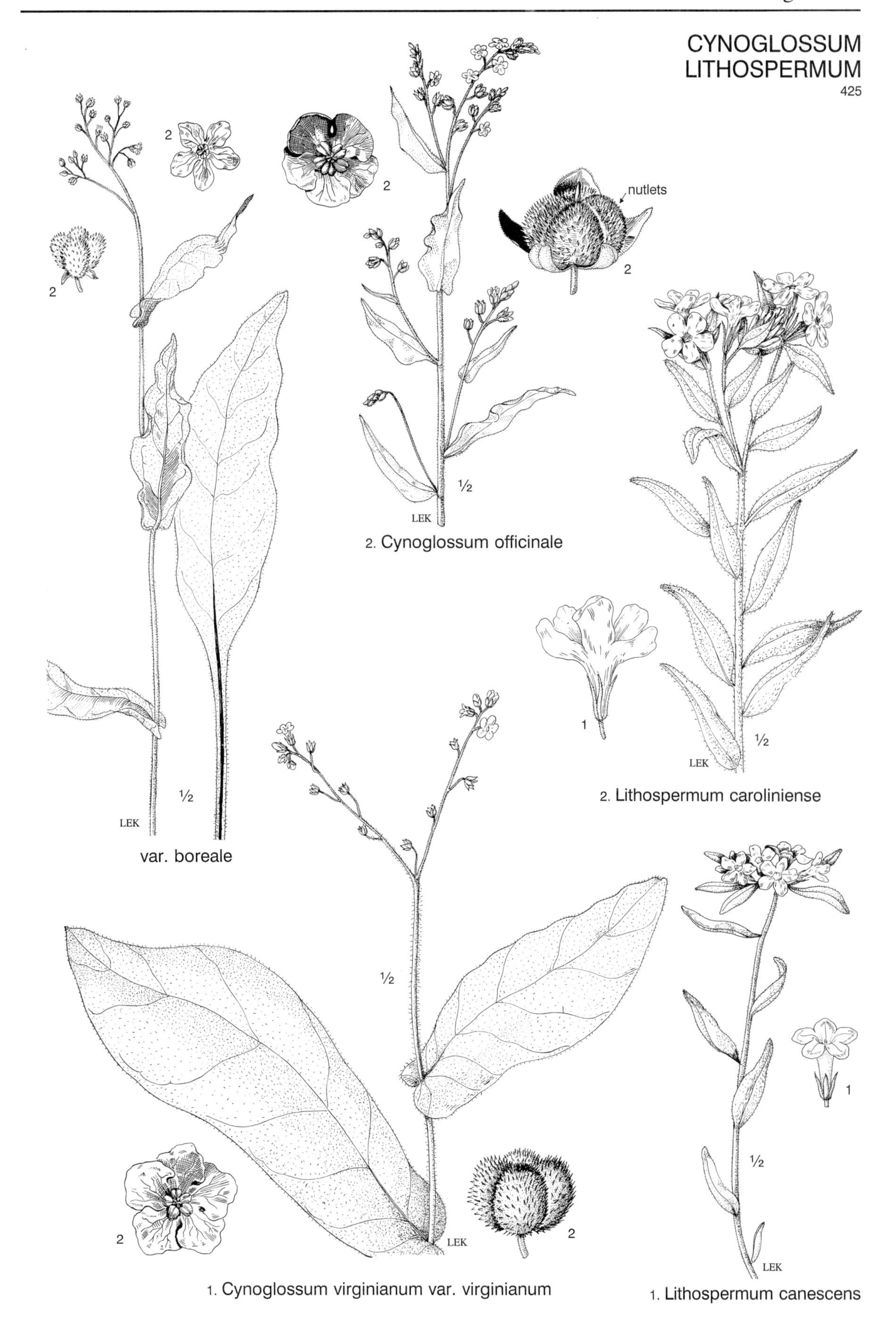

nutlets

2. Cynoglossum officinale

var. boreale

2. Lithospermum caroliniense

1. Cynoglossum virginianum var. virginianum

1. Lithospermum canescens

LITHOSPERMUM
426

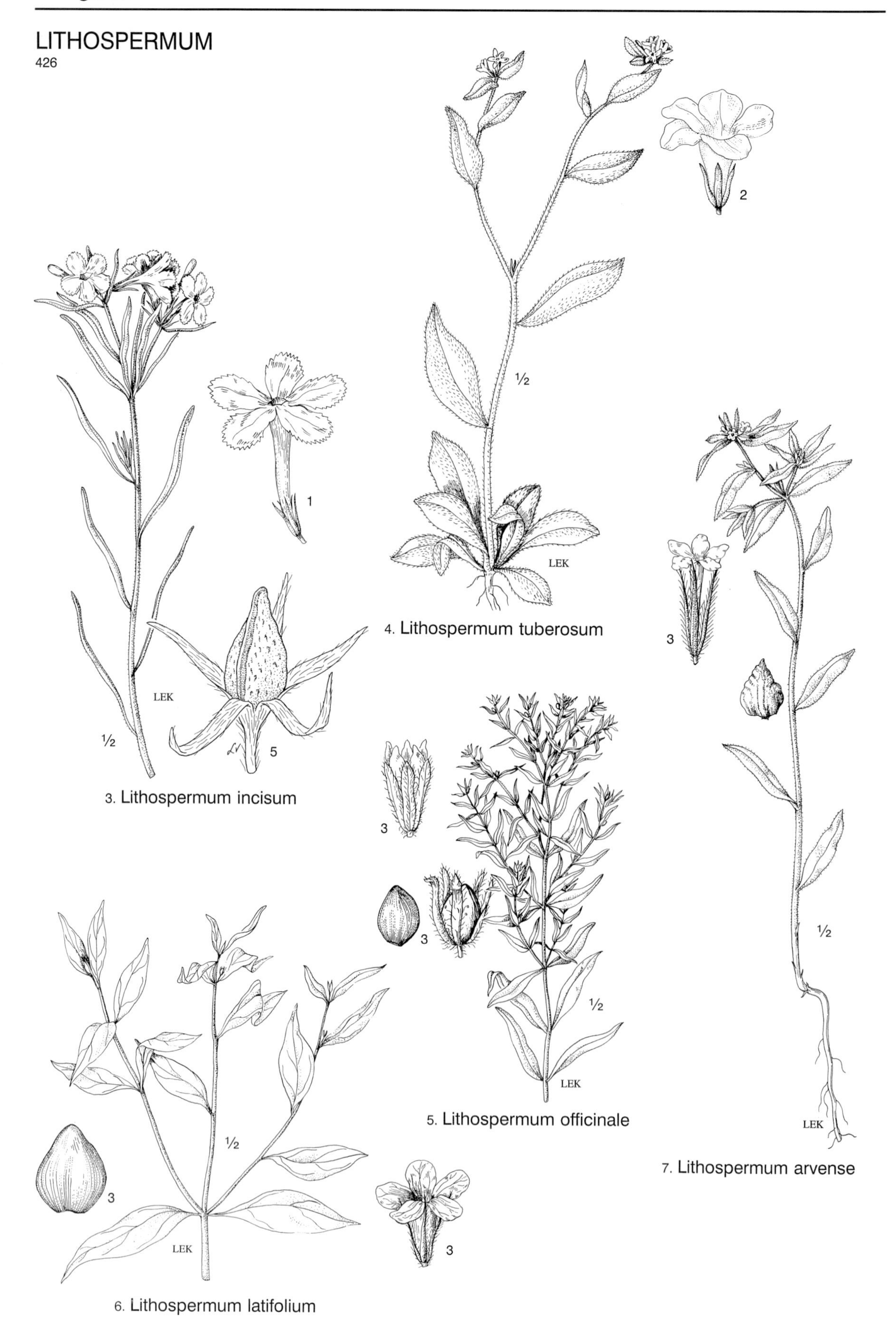

1

2

4. Lithospermum tuberosum

3

3. Lithospermum incisum

5

3

3

5. Lithospermum officinale

3

7. Lithospermum arvense

3

6. Lithospermum latifolium

3

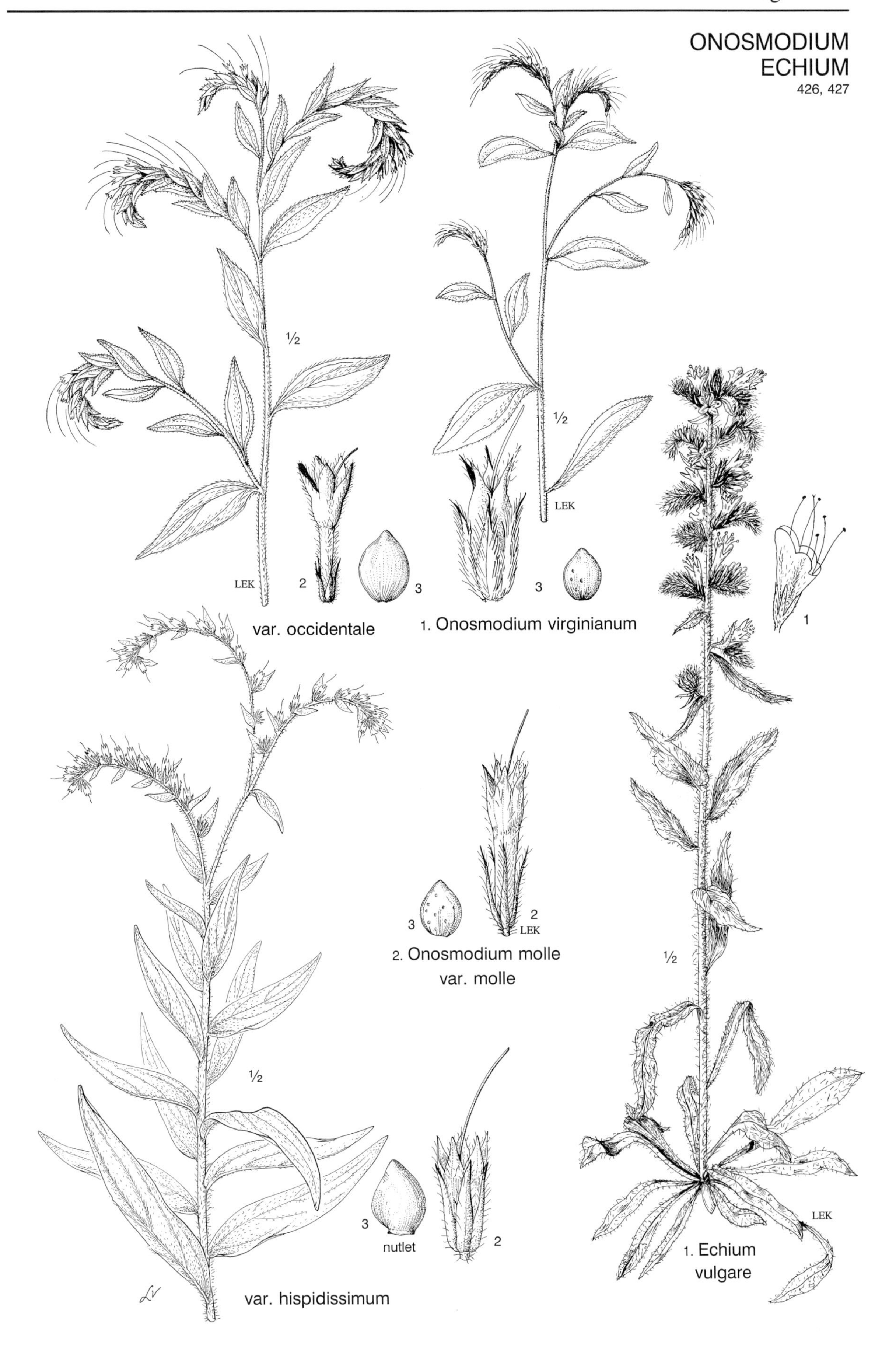

var. occidentale

1. Onosmodium virginianum

2. Onosmodium molle
var. molle

var. hispidissimum

nutlet

1. Echium
vulgare

ANCHUSA
SYMPHYTUM
BORAGO
427, 428

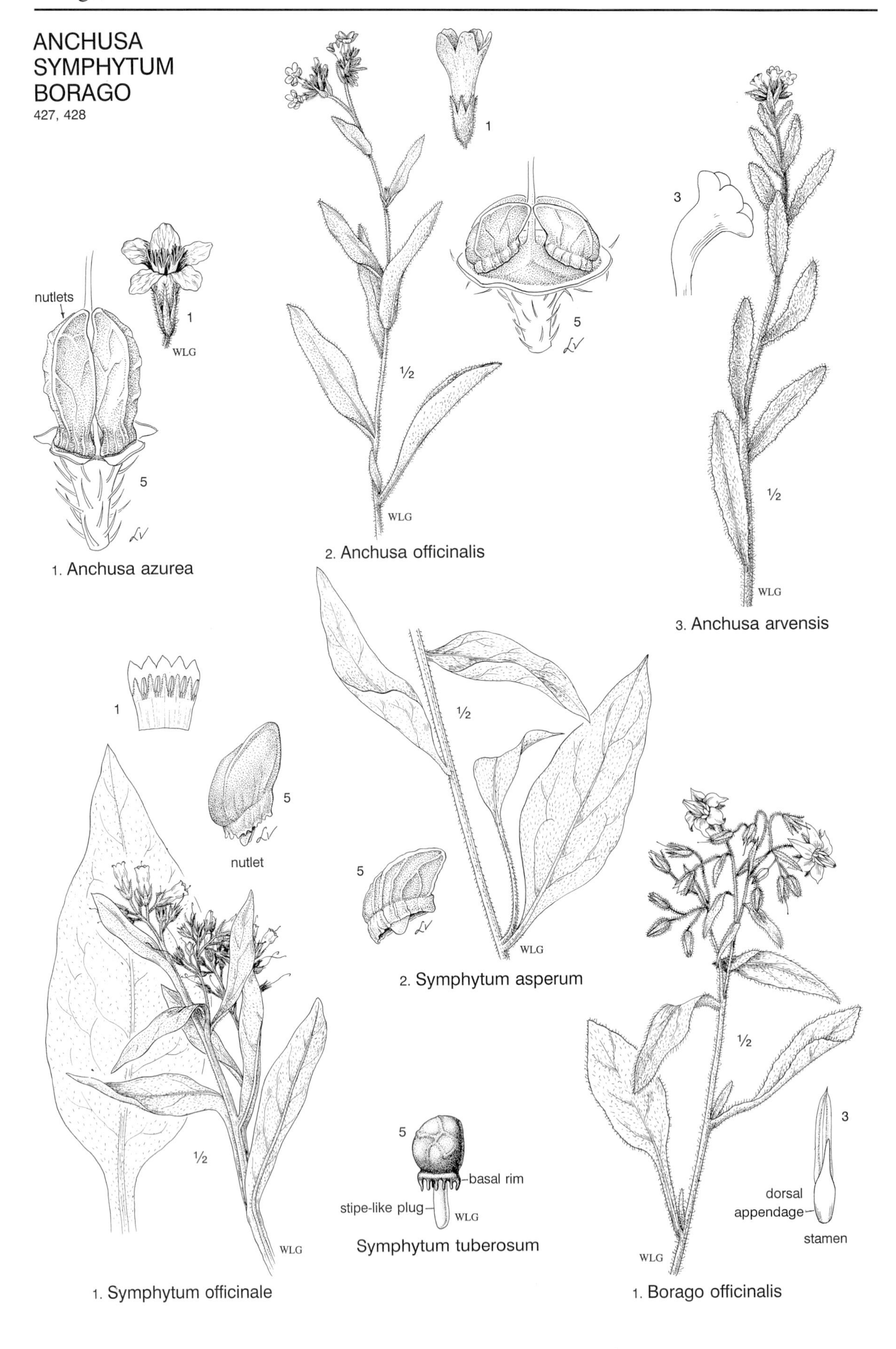

nutlets

1. Anchusa azurea

WLG

2. Anchusa officinalis

WLG

3. Anchusa arvensis

WLG

nutlet

2. Symphytum asperum

WLG

1. Symphytum officinale

WLG

basal rim

stipe-like plug

WLG

Symphytum tuberosum

dorsal
appendage

stamen

1. Borago officinalis

WLG

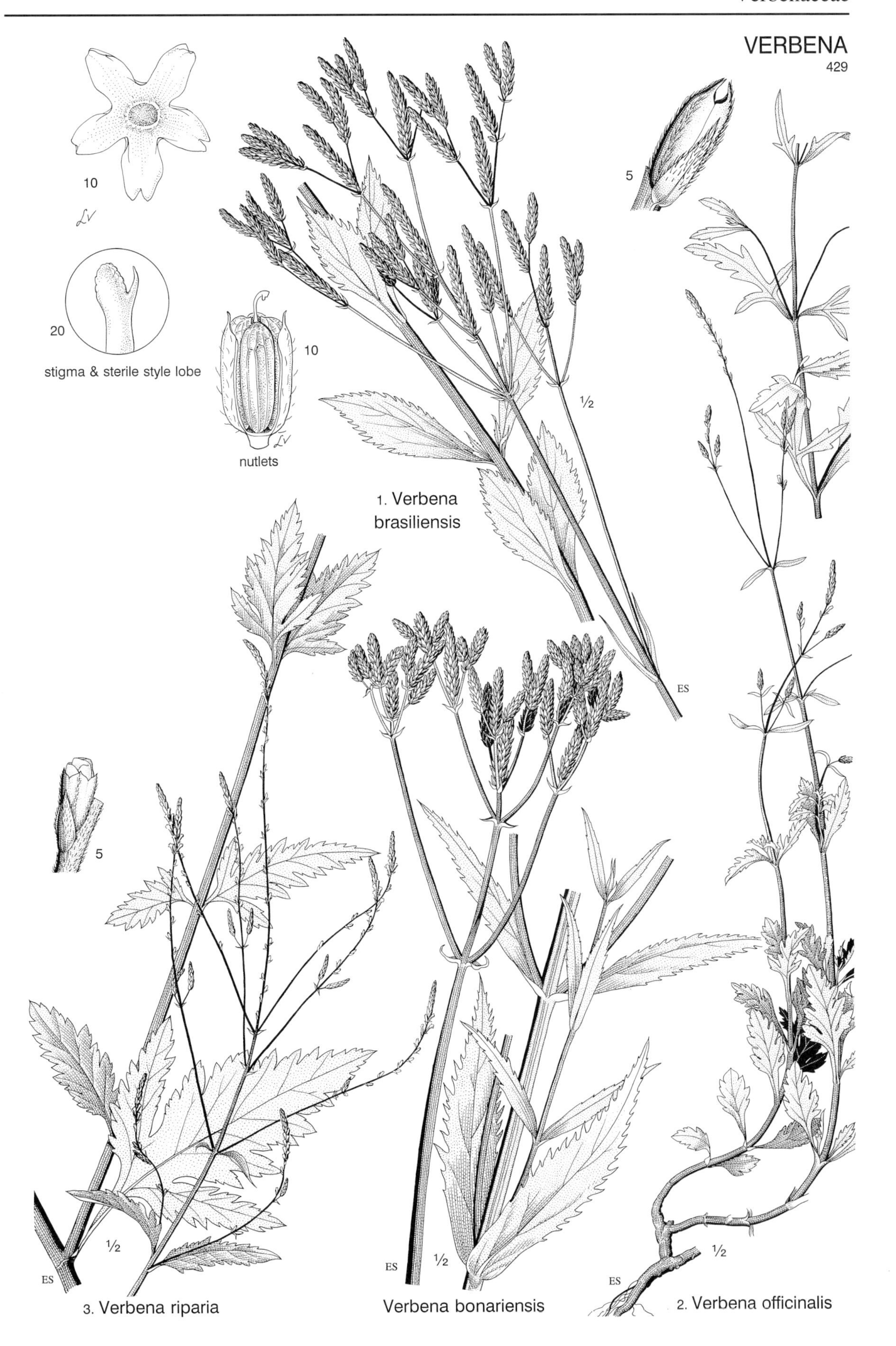

10

20

stigma & sterile style lobe

nutlets

10

5

1. Verbena
brasiliensis

5

ES

3. Verbena riparia

Verbena bonariensis

2. Verbena officinalis

VERBENA
429

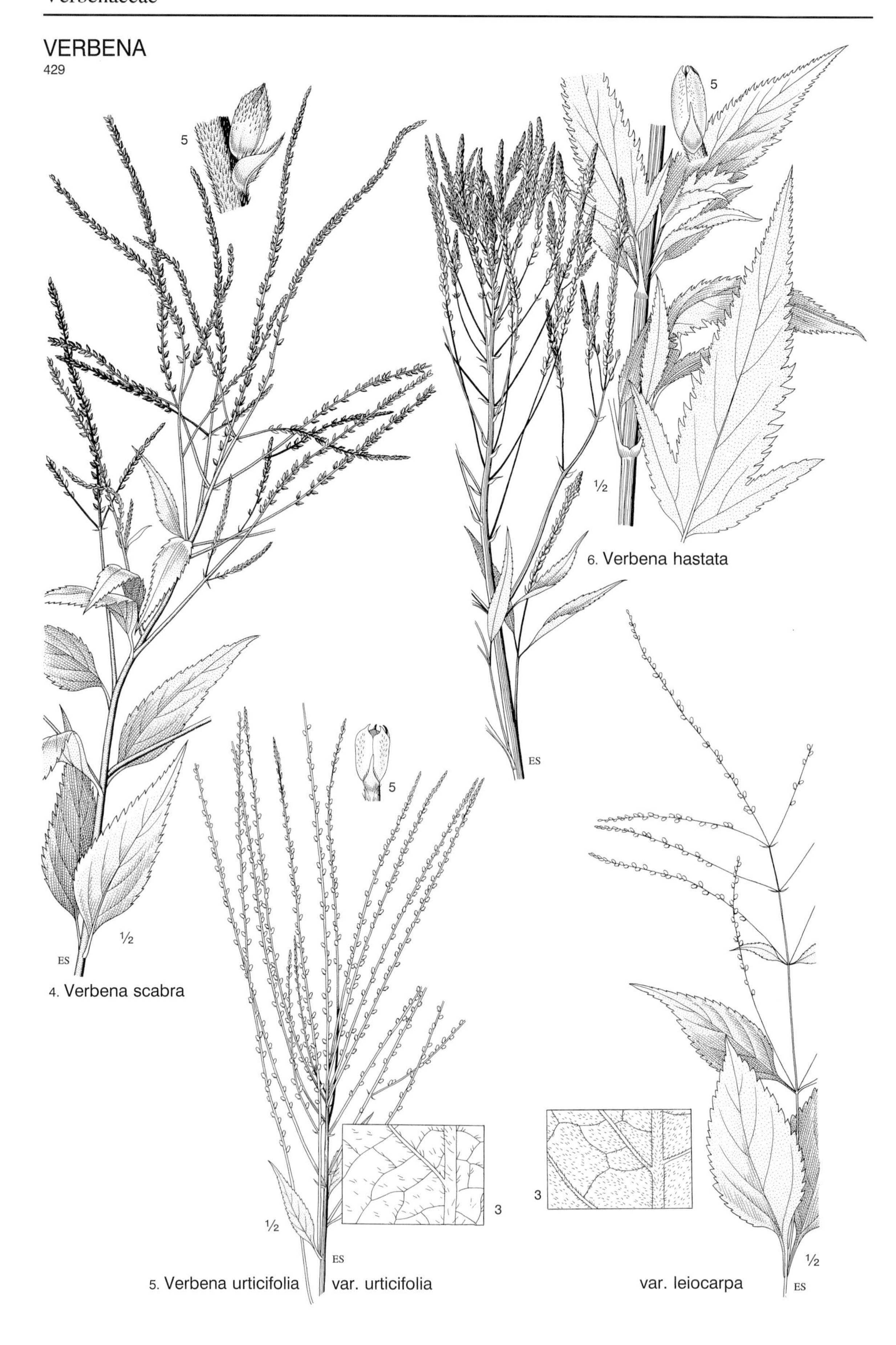

6. Verbena hastata

4. Verbena scabra

5. Verbena urticifolia   var. urticifolia

var. leiocarpa

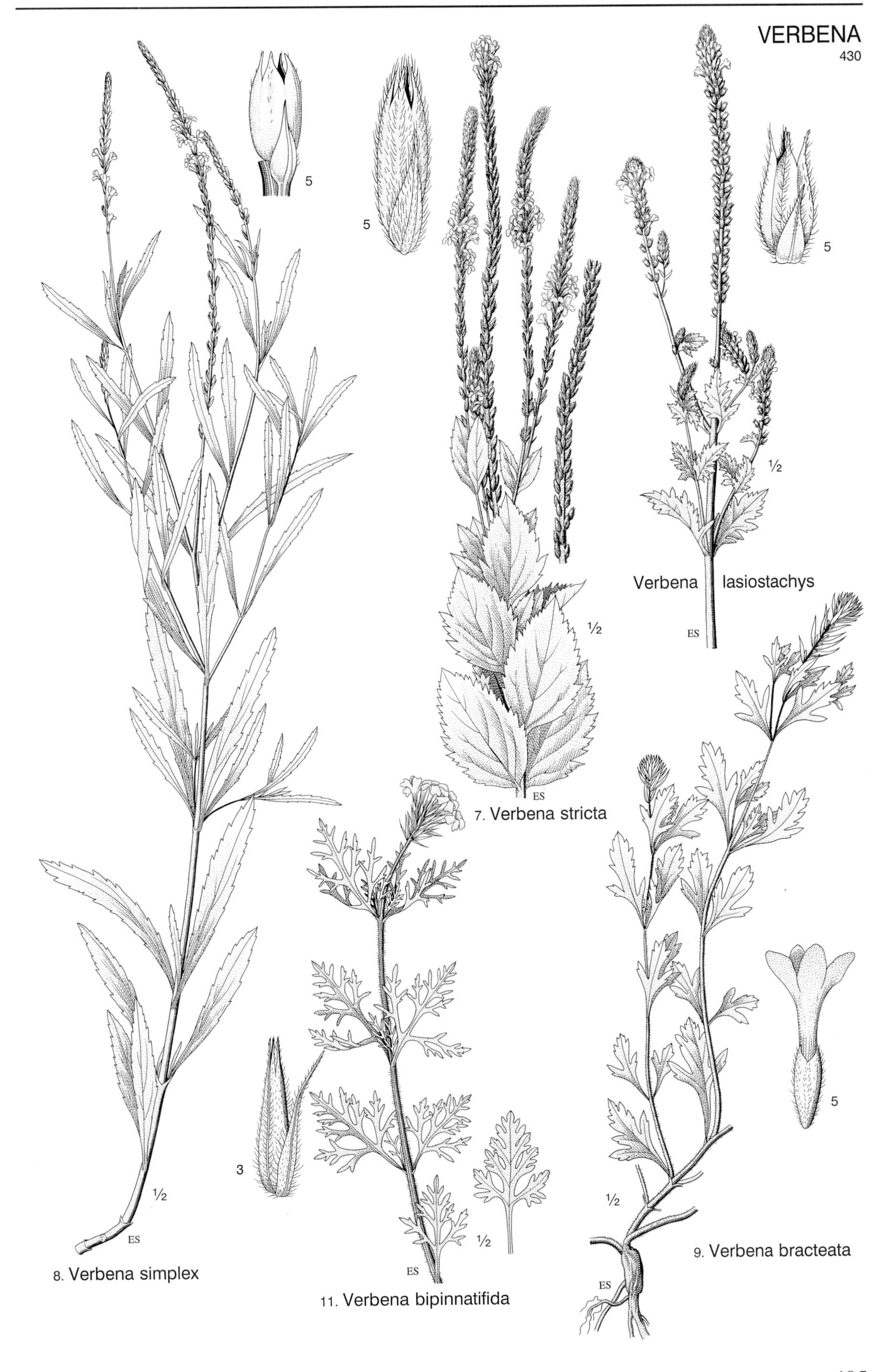

5

5

5

½

Verbena   lasiostachys

ES

7. Verbena stricta

ES

5

½

ES

8. Verbena simplex

3

½

ES

11. Verbena bipinnatifida

½

ES

9. Verbena bracteata

5

VERBENA
PHYLA
430, 431

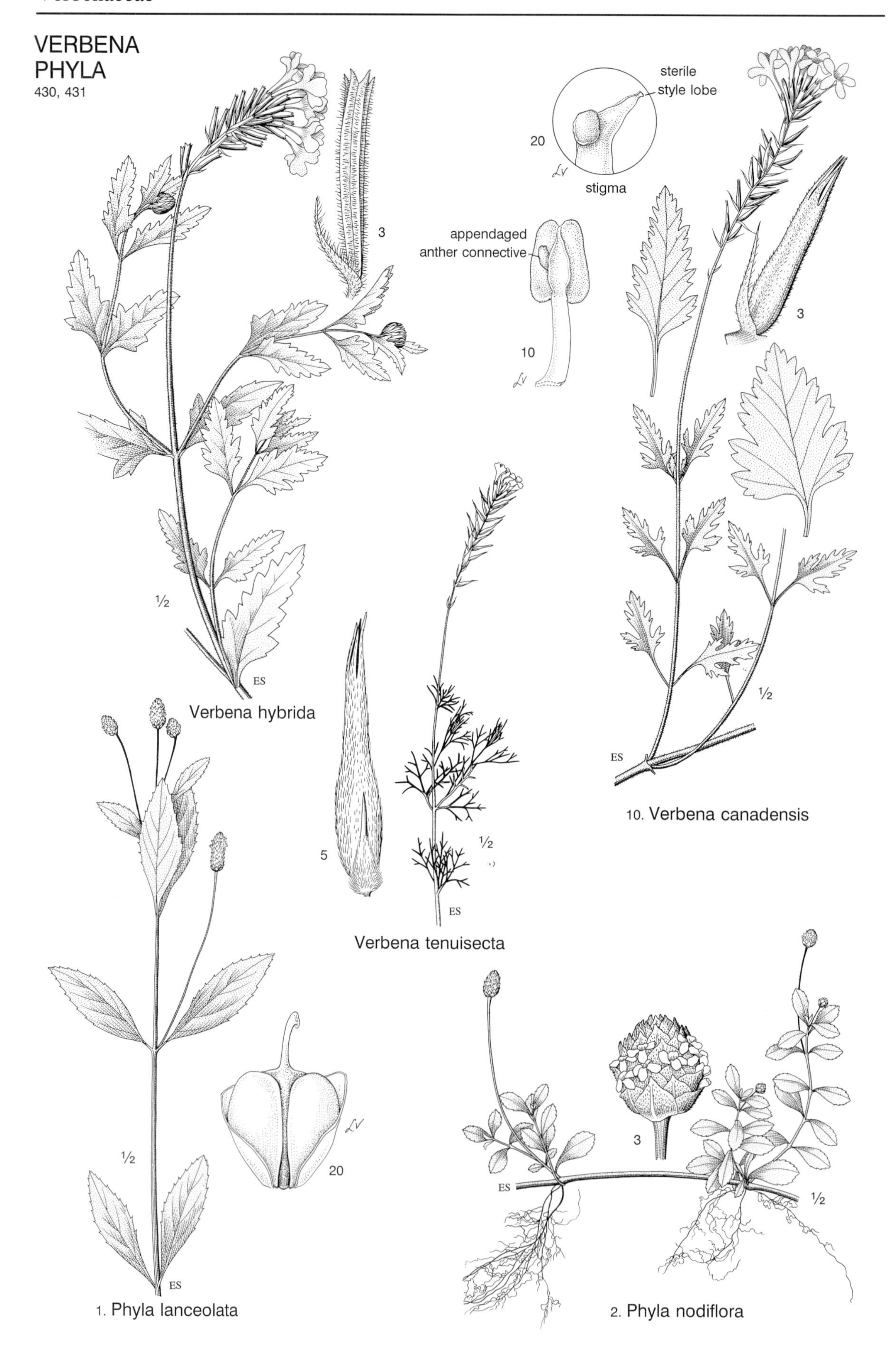

sterile
style lobe

20

stigma

appendaged
anther connective

10

3

Verbena hybrida

5

Verbena tenuisecta

10. Verbena canadensis

20

1. Phyla lanceolata

3

2. Phyla nodiflora

Verbenaceae

PHRYMA
CALLICARPA
CLERODENDRUM
VITEX
431, 432

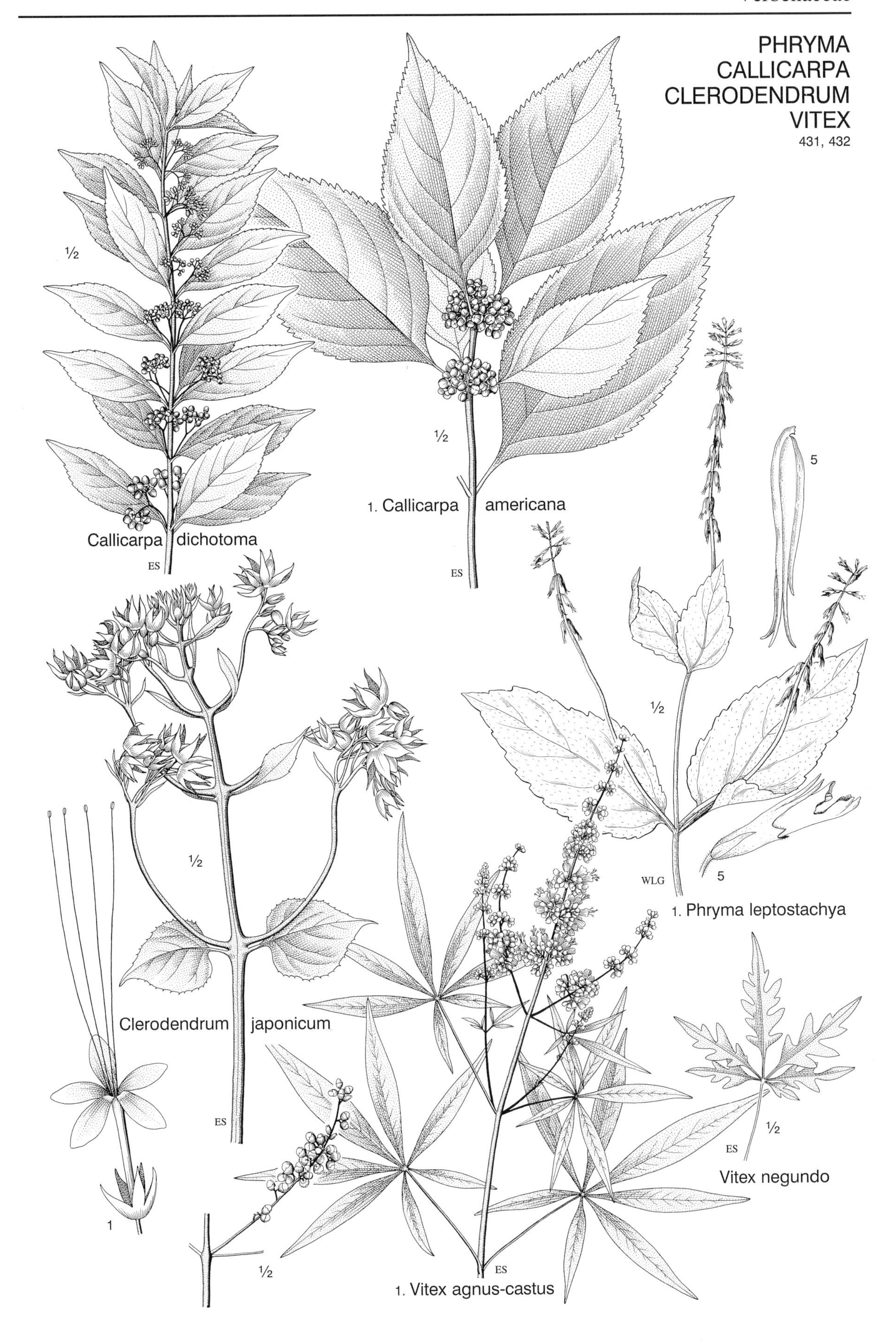

½

Callicarpa dichotoma
ES

½

1. Callicarpa americana
ES

5

½

WLG

5

1. Phryma leptostachya

½

Clerodendrum japonicum
ES

1

½

1. Vitex agnus-castus
ES

½
ES

Vitex negundo

AJUGA
TEUCRIUM
434, 435

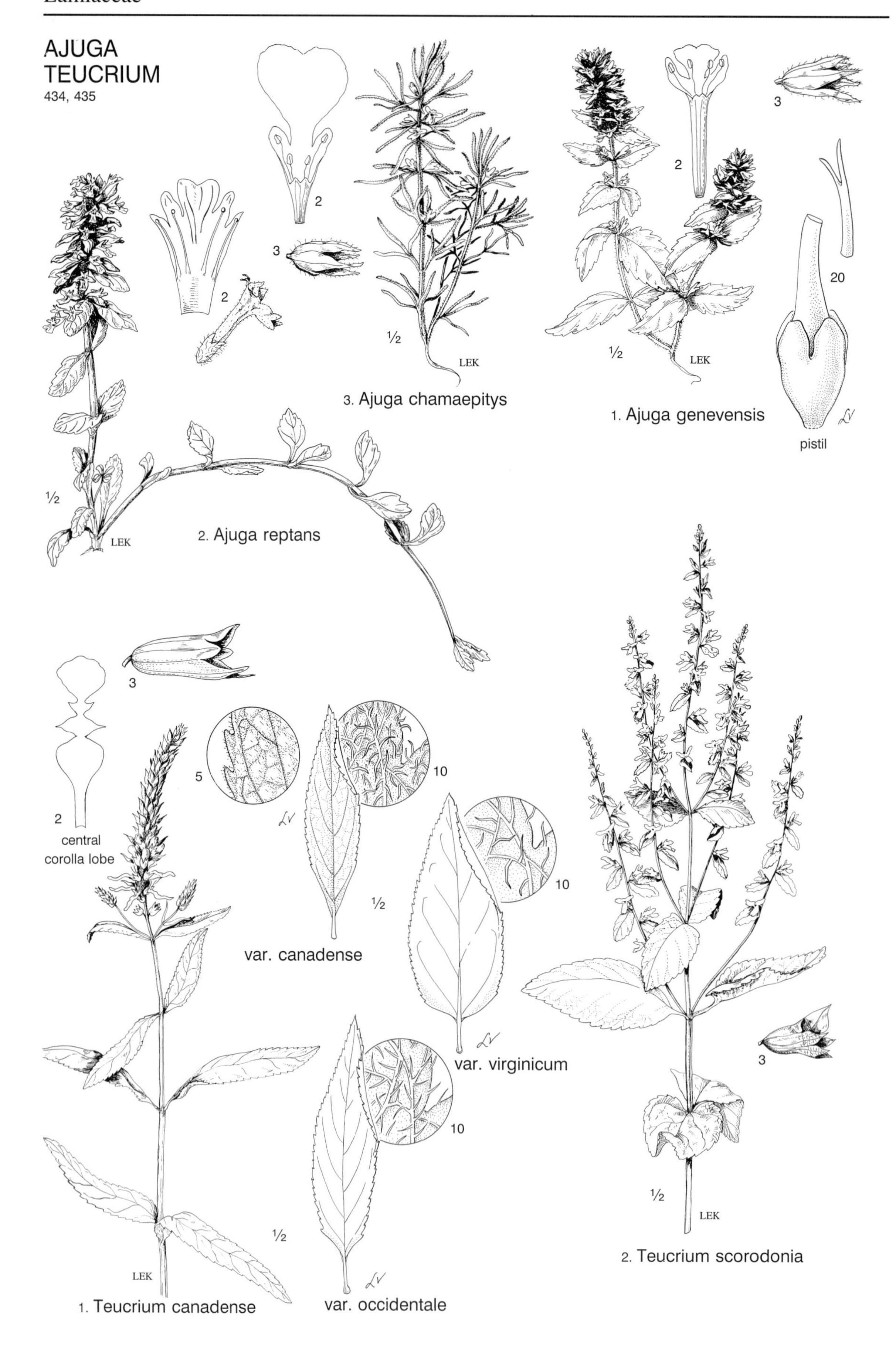

3. Ajuga chamaepitys

1. Ajuga genevensis

pistil

2. Ajuga reptans

central
corolla lobe

var. canadense

var. virginicum

var. occidentale

1. Teucrium canadense

2. Teucrium scorodonia

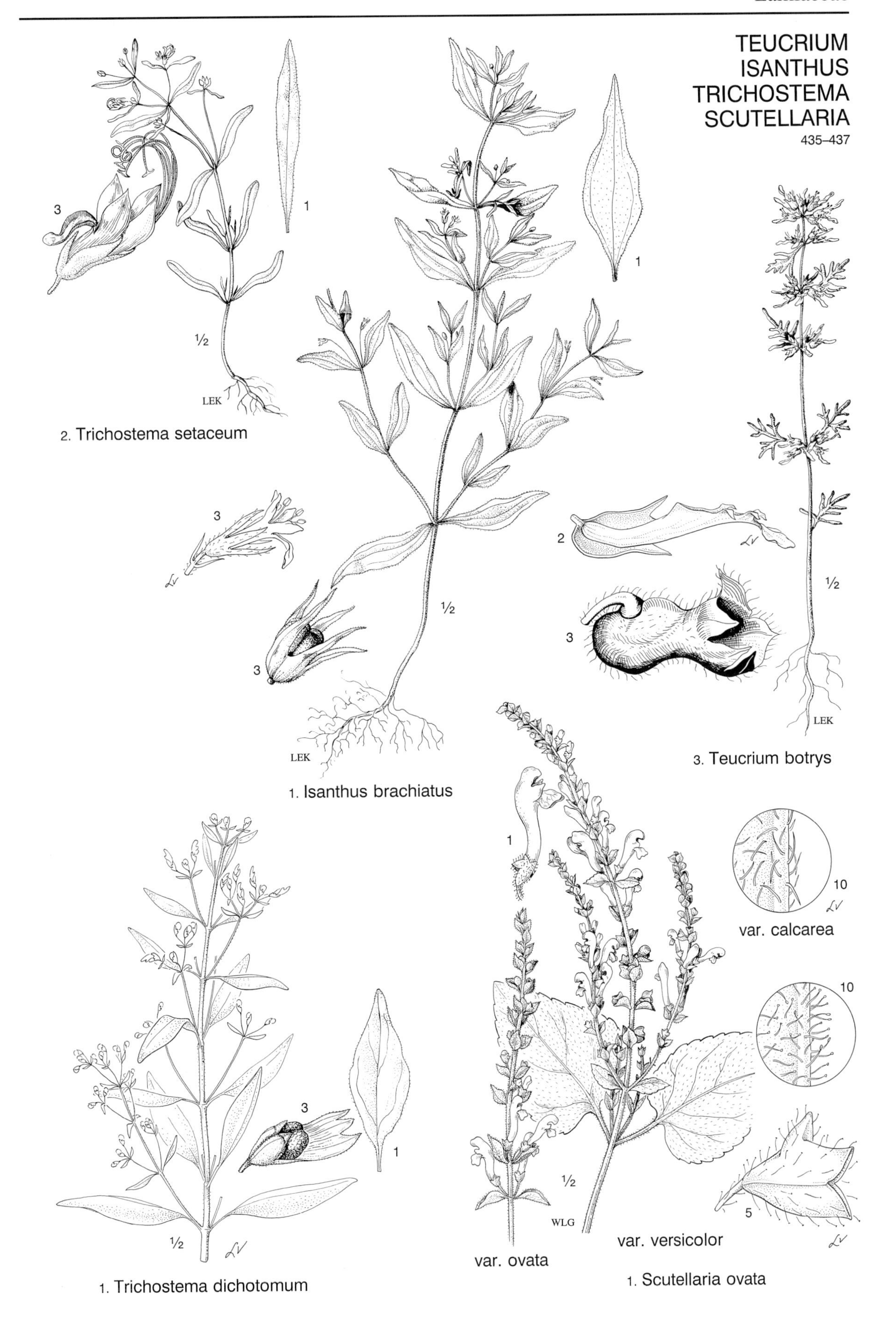

2. Trichostema setaceum

1. Isanthus brachiatus

3. Teucrium botrys

1. Trichostema dichotomum

var. calcarea

var. ovata

var. versicolor

1. Scutellaria ovata

SCUTELLARIA
437

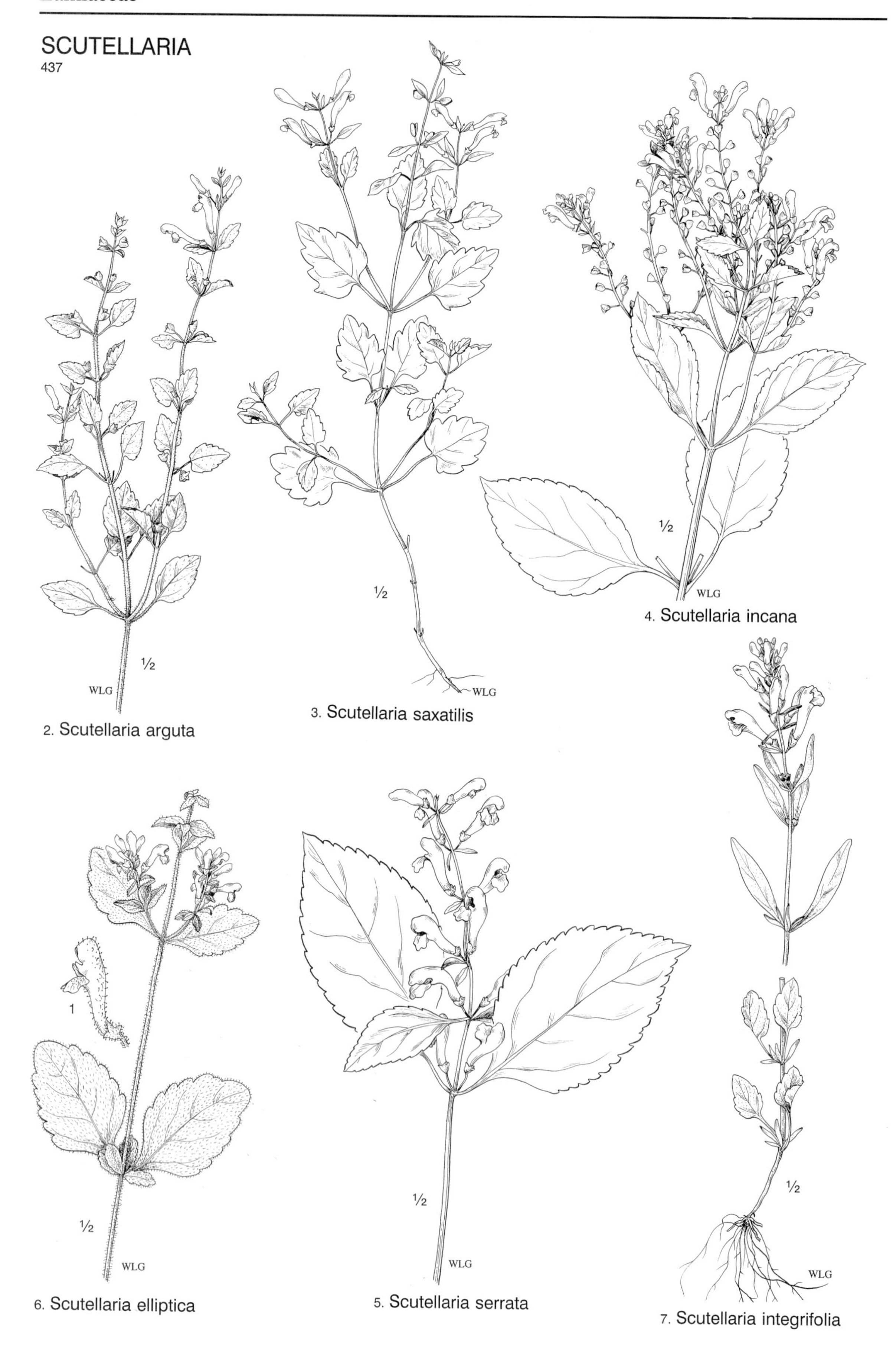

2. Scutellaria arguta

3. Scutellaria saxatilis

4. Scutellaria incana

6. Scutellaria elliptica

5. Scutellaria serrata

7. Scutellaria integrifolia

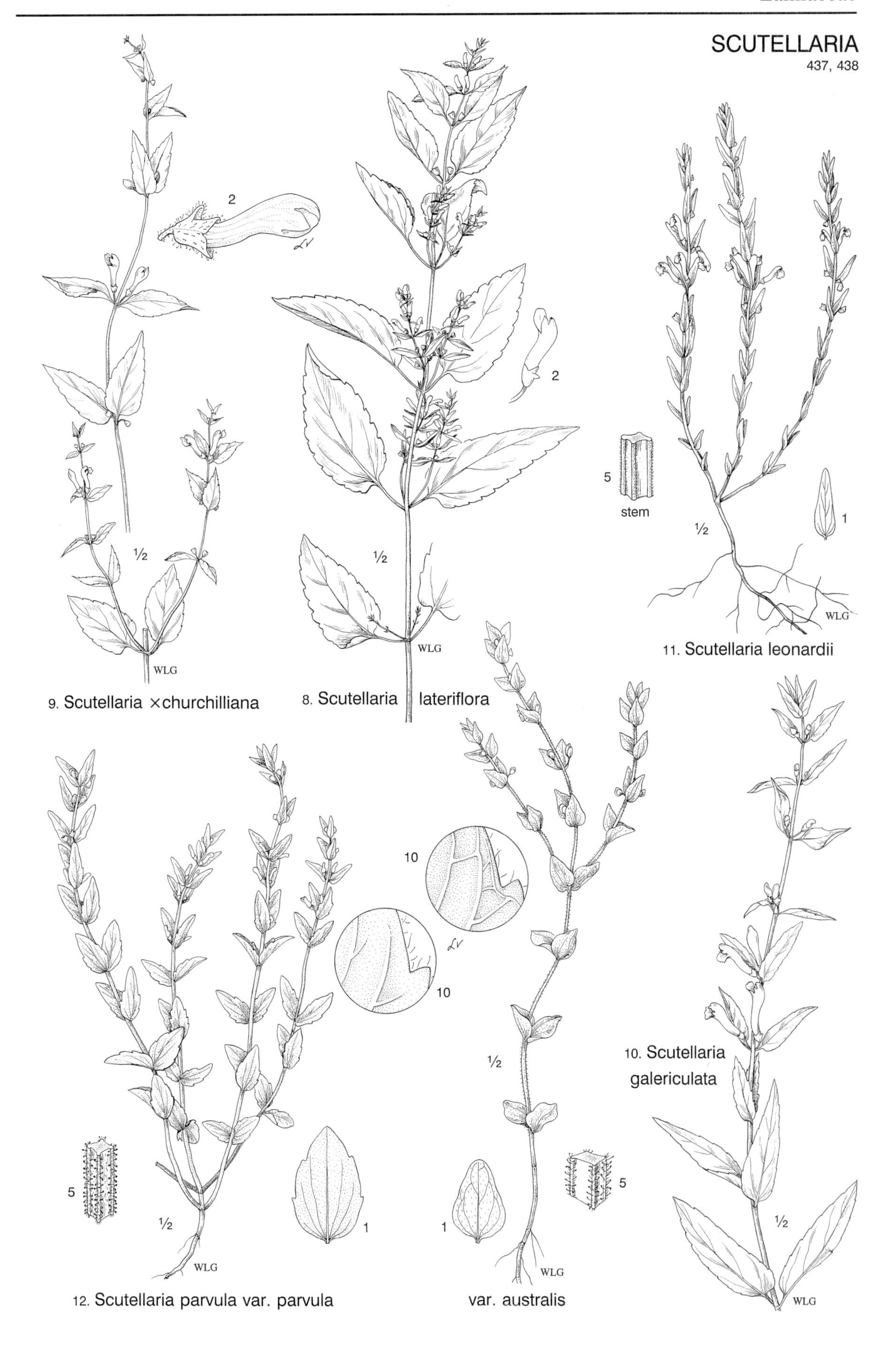

2

2

5
stem

½

1

11. Scutellaria leonardii

9. Scutellaria ×churchilliana

8. Scutellaria lateriflora

10

10

½

½

10. Scutellaria
galericulata

5

½

1

1

5

½

12. Scutellaria parvula var. parvula

var. australis

SCUTELLARIA
SIDERITIS
MARRUBIUM
HEDEOMA
MELISSA
CONRADINA
438, 439

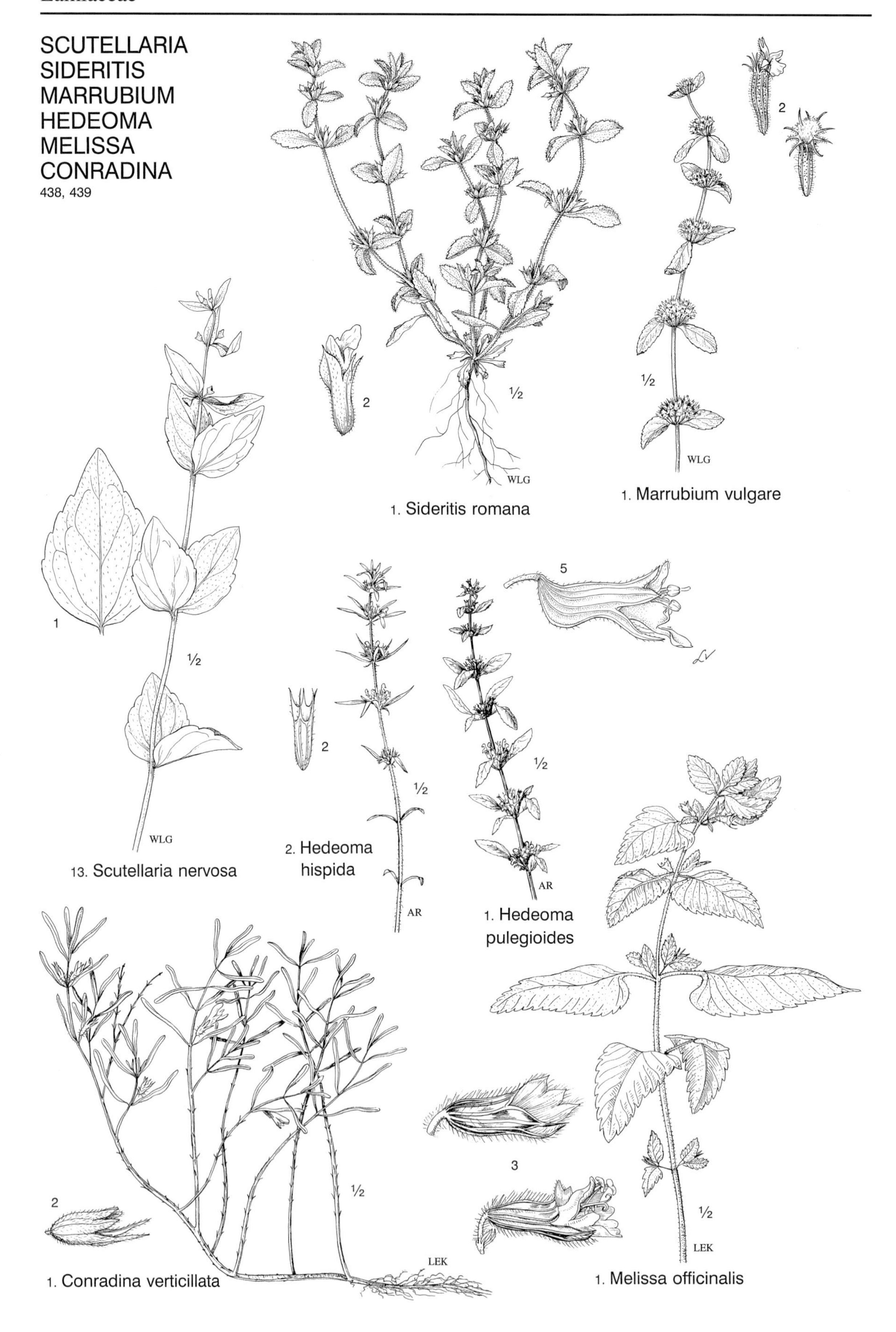

1. Sideritis romana

WLG

1. Marrubium vulgare

WLG

13. Scutellaria nervosa

WLG

2. Hedeoma
hispida

AR

1. Hedeoma
pulegioides

AR

5

1. Conradina verticillata

LEK

1. Melissa officinalis

LEK

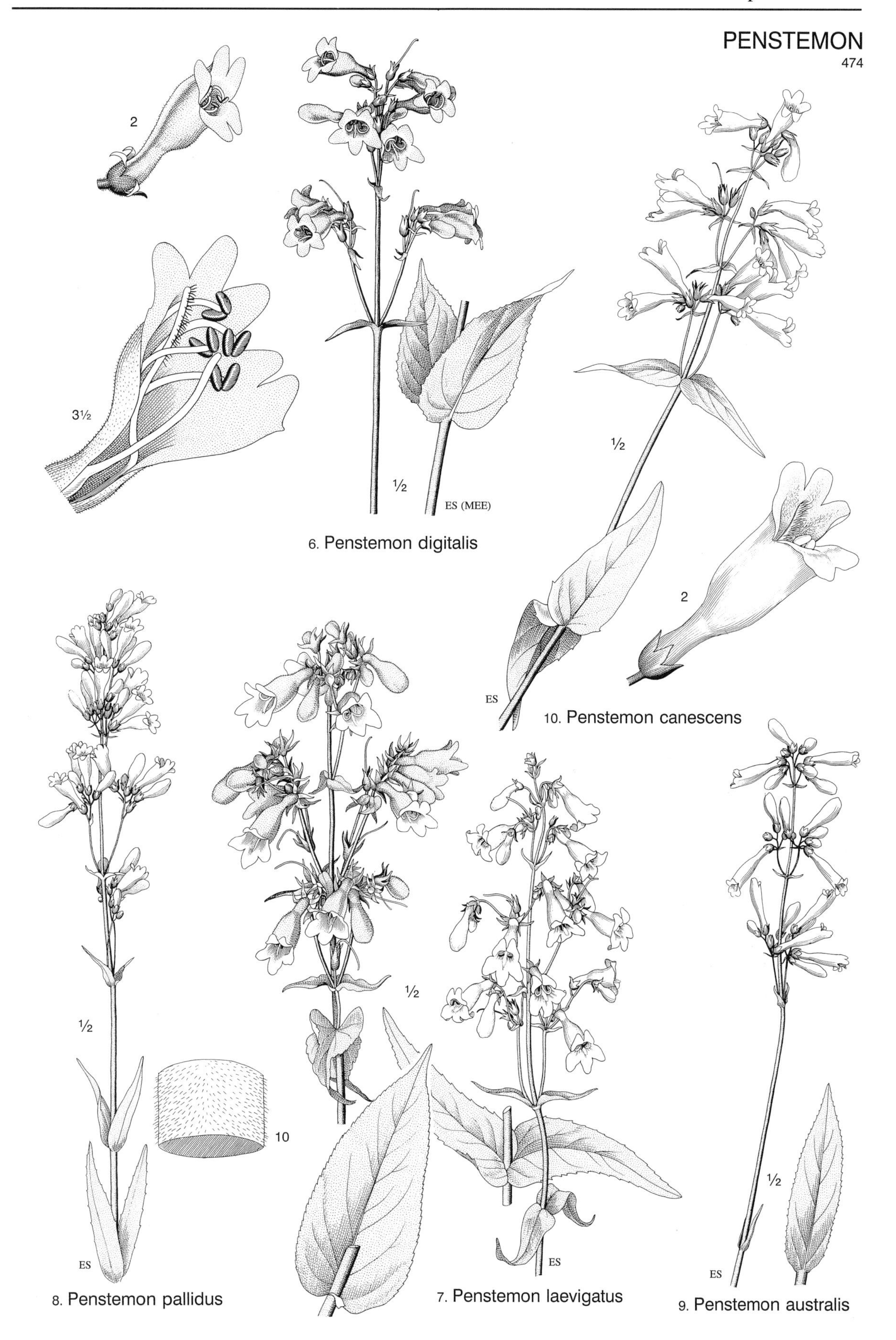

2

3½

½

½   ES (MEE)

6. Penstemon digitalis

½

2

ES

10. Penstemon canescens

½

10

½

ES

8. Penstemon pallidus

½

ES

7. Penstemon laevigatus

½

ES

9. Penstemon australis

PENSTEMON
SCROPHULARIA
COLLINSIA
474, 475

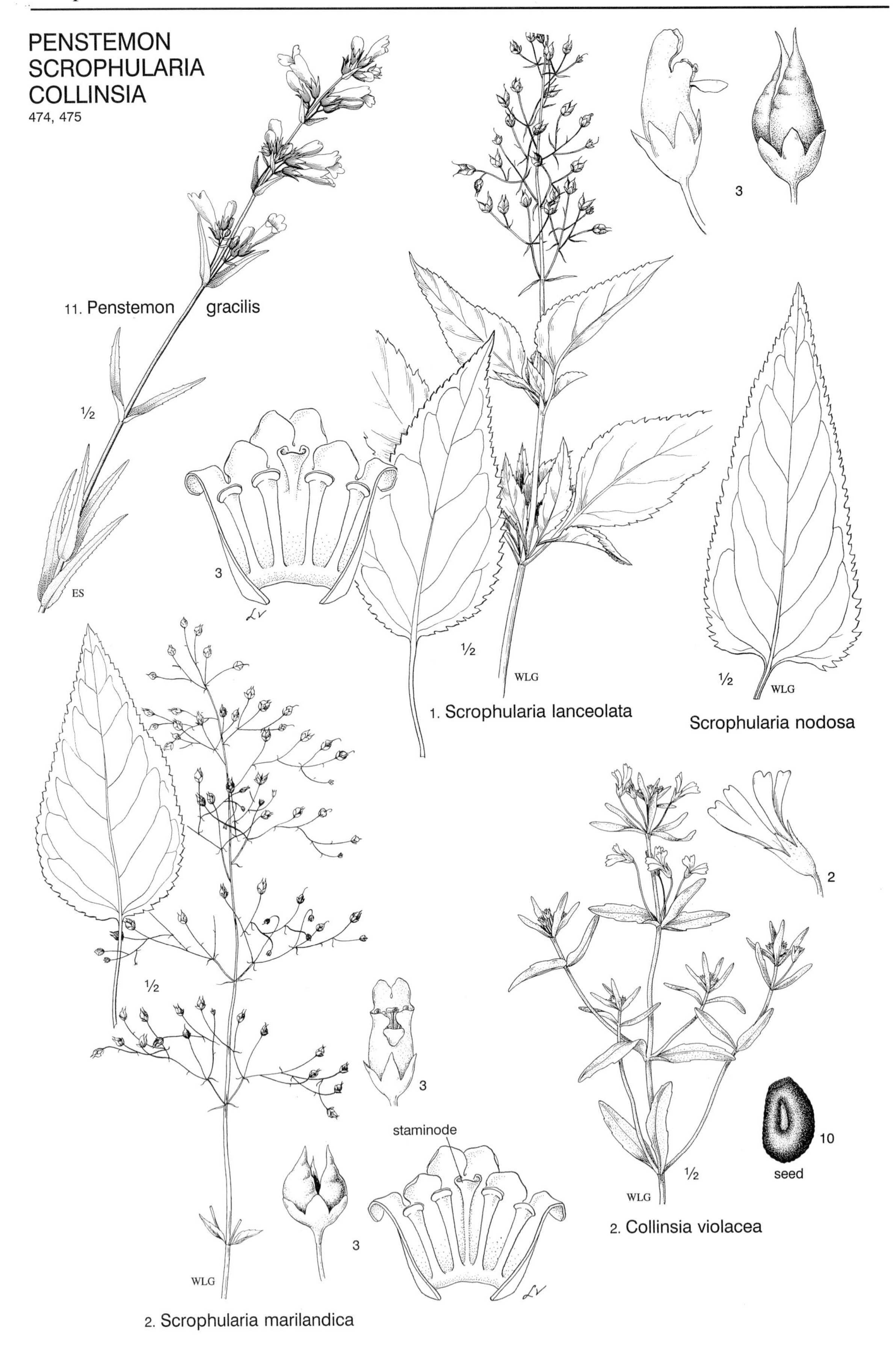

11. Penstemon gracilis

½

ES

3

½

1. Scrophularia lanceolata

WLG

½

WLG

Scrophularia nodosa

3

staminode

3

½

WLG

2. Scrophularia marilandica

3

2

½

WLG

2. Collinsia violacea

10
seed

COLLINSIA
LINARIA
475, 476

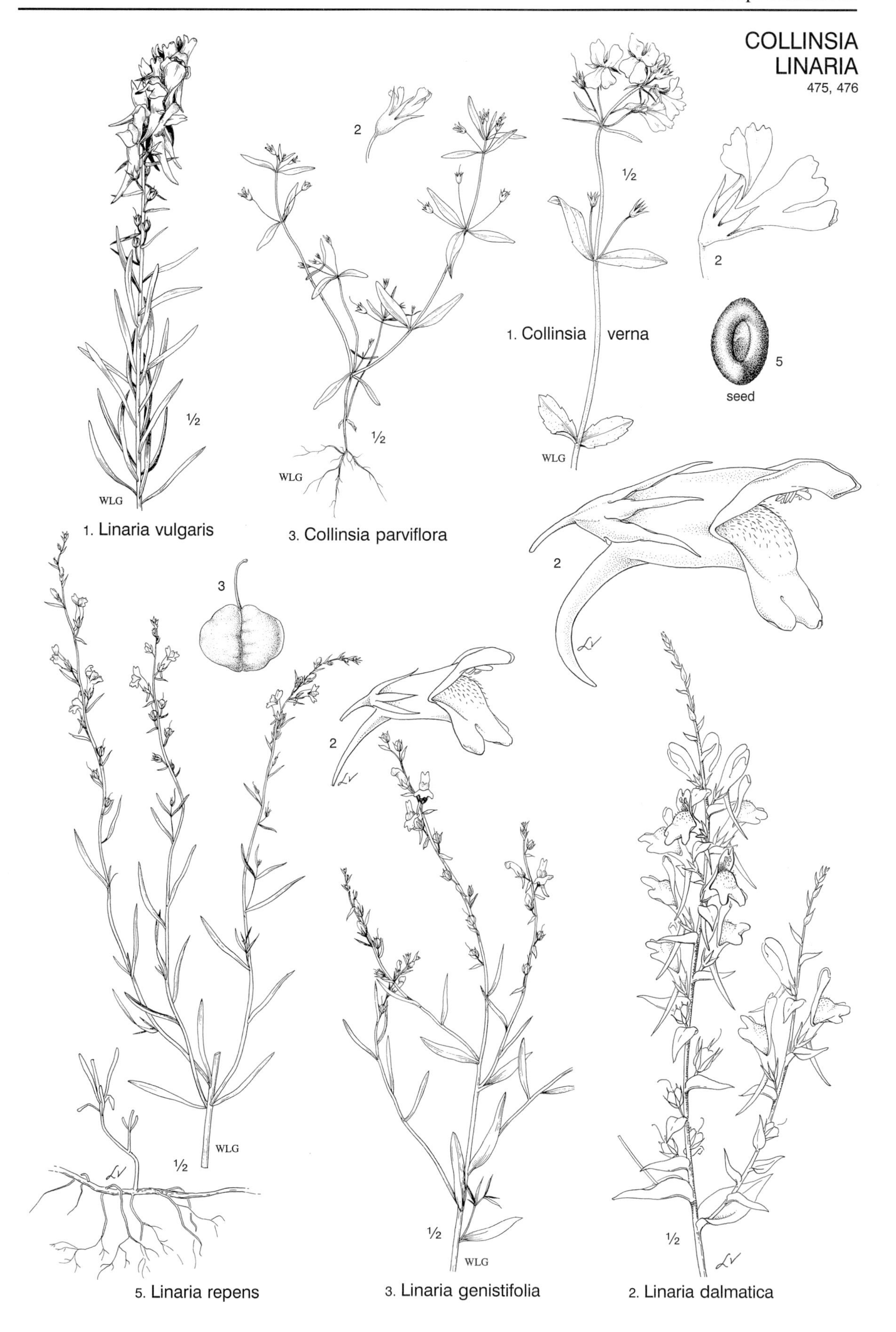

1. Linaria vulgaris

3. Collinsia parviflora

1. Collinsia verna

5 seed

2

1. Linaria vulgaris

3. Collinsia parviflora

5. Linaria repens

3. Linaria genistifolia

2. Linaria dalmatica

LINARIA
KICKXIA
CYMBALARIA
ANTIRRHINUM

476, 477

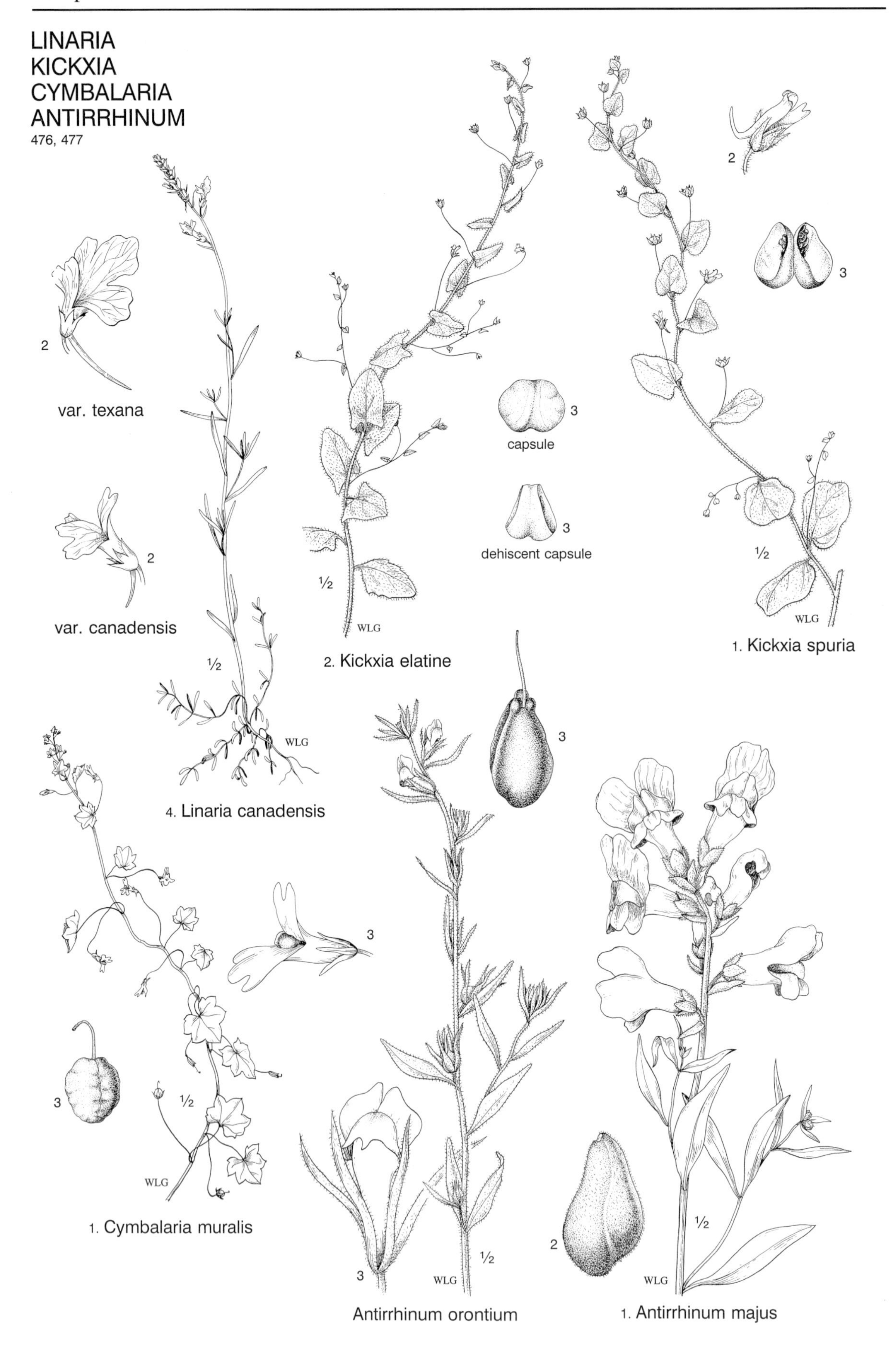

2 var. texana

2 var. canadensis

½ 4. Linaria canadensis

½ 2. Kickxia elatine

3 capsule

3 dehiscent capsule

3 ½ 1. Kickxia spuria

3 ½ 1. Cymbalaria muralis

3 ½ Antirrhinum orontium

2 ½ 1. Antirrhinum majus

CHAENORRHINUM
DIGITALIS
DASISTOMA
SEYMERIA
AUREOLARIA
477, 478

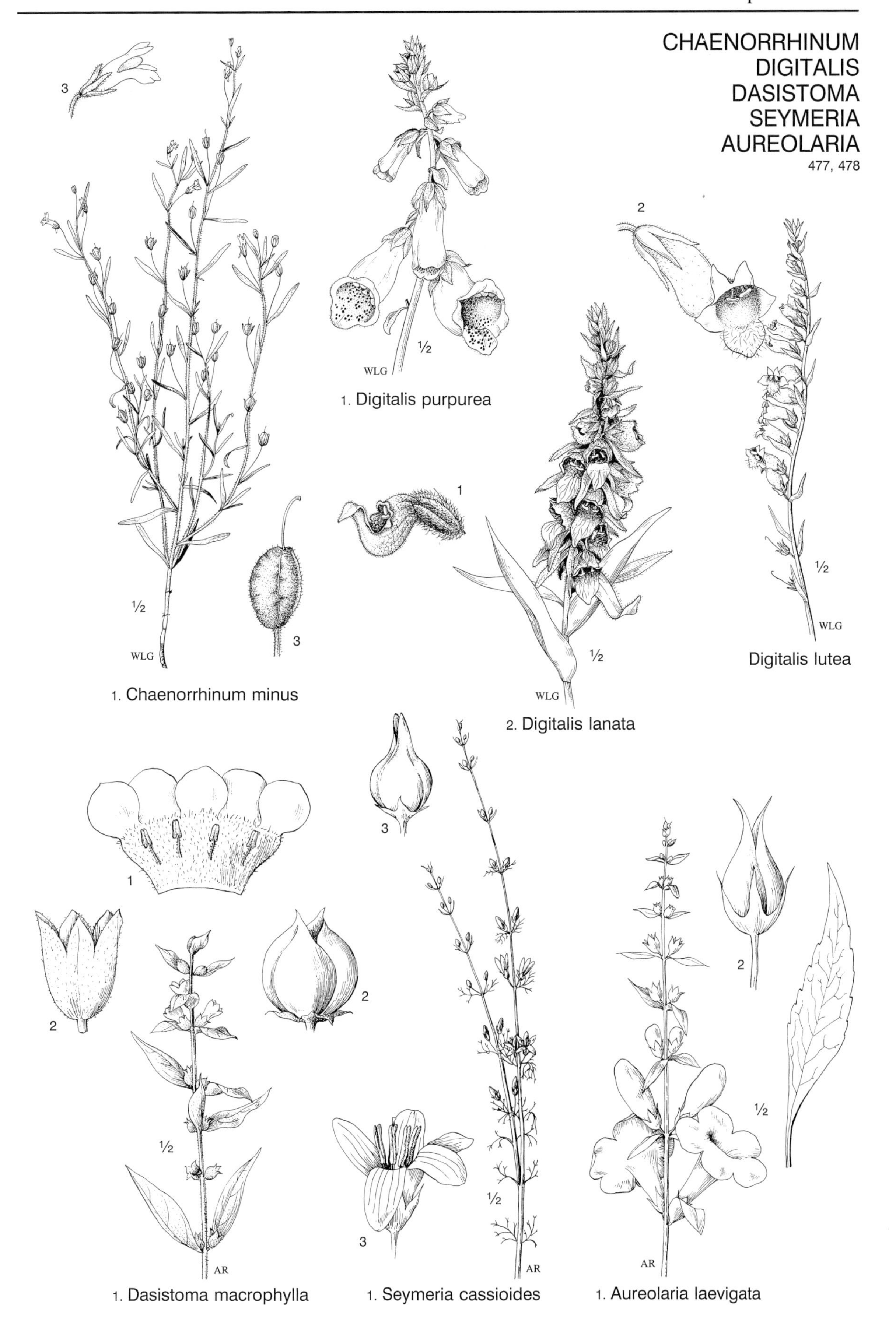

1. Digitalis purpurea

WLG

Digitalis lutea

WLG

1. Chaenorrhinum minus

2. Digitalis lanata

WLG

1. Dasistoma macrophylla

1. Seymeria cassioides

1. Aureolaria laevigata

AR

AUREOLARIA
478, 479

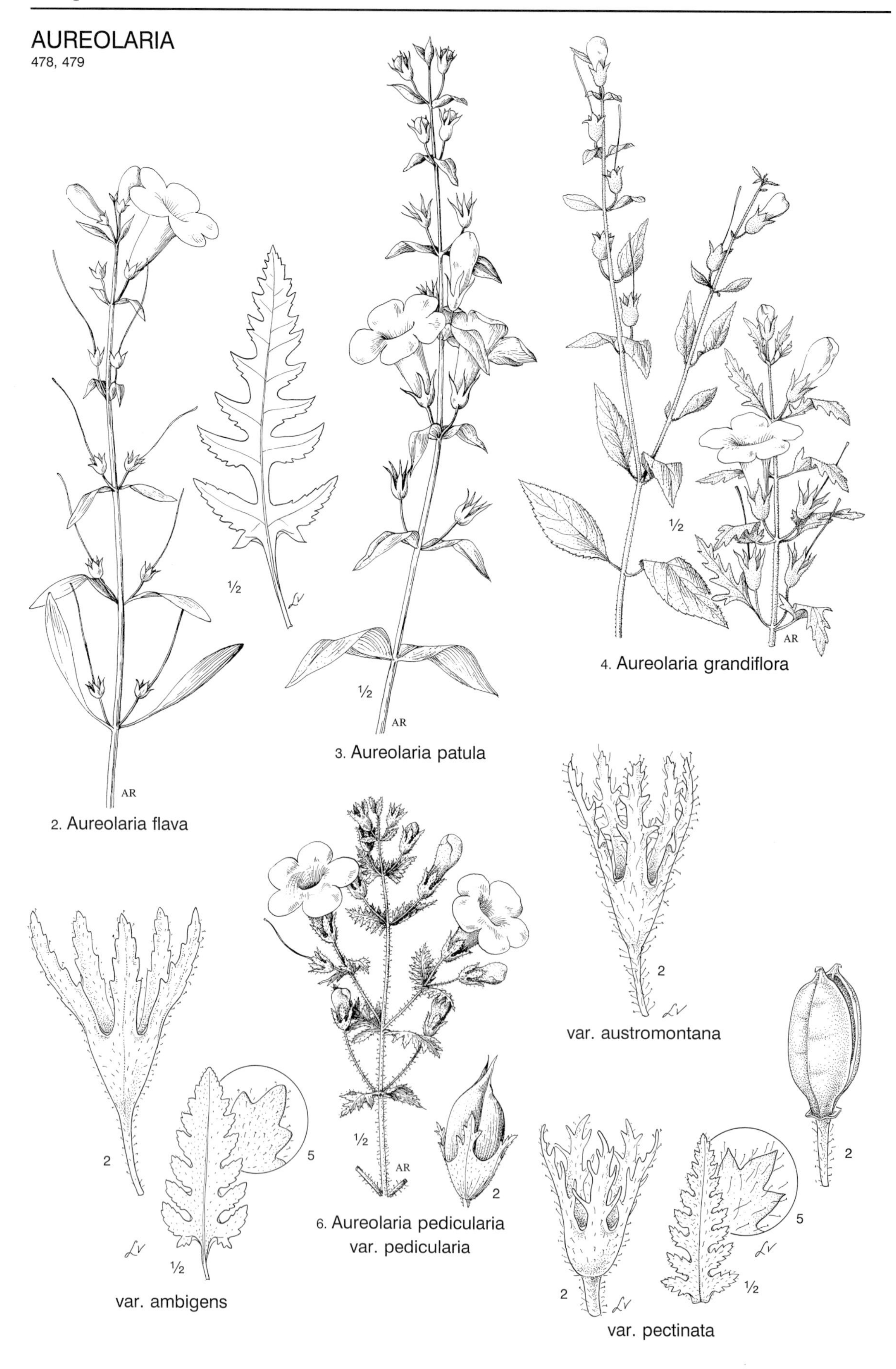

2. Aureolaria flava

3. Aureolaria patula

4. Aureolaria grandiflora

var. austromontana

var. ambigens

6. Aureolaria pedicularia
var. pedicularia

var. pectinata

ILLUSTRATED COMPANION TO

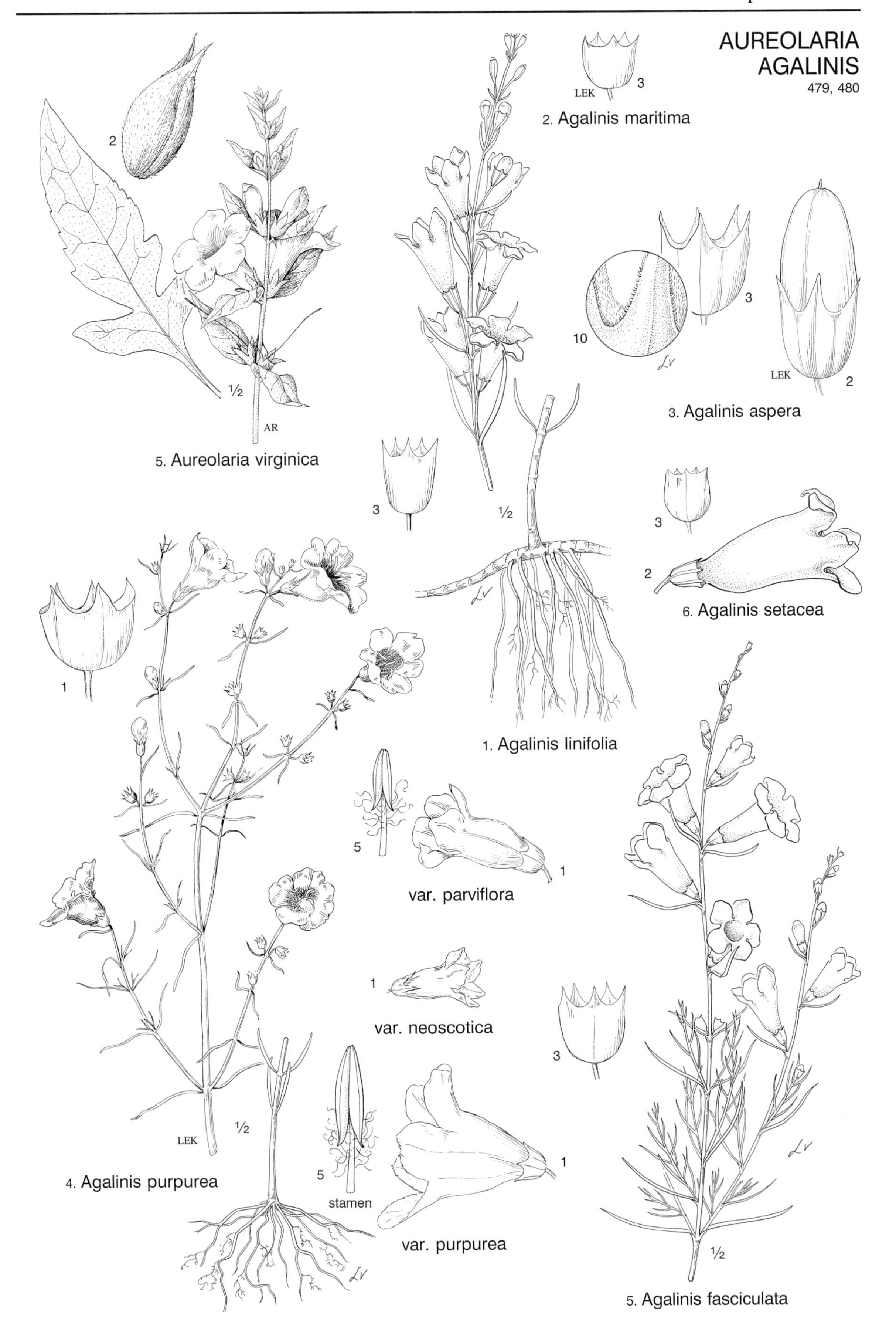

2. Agalinis maritima

3. Agalinis aspera

5. Aureolaria virginica

6. Agalinis setacea

1. Agalinis linifolia

var. parviflora

var. neoscotica

stamen

var. purpurea

4. Agalinis purpurea

5. Agalinis fasciculata

AGALINIS
BUCHNERA
480, 481

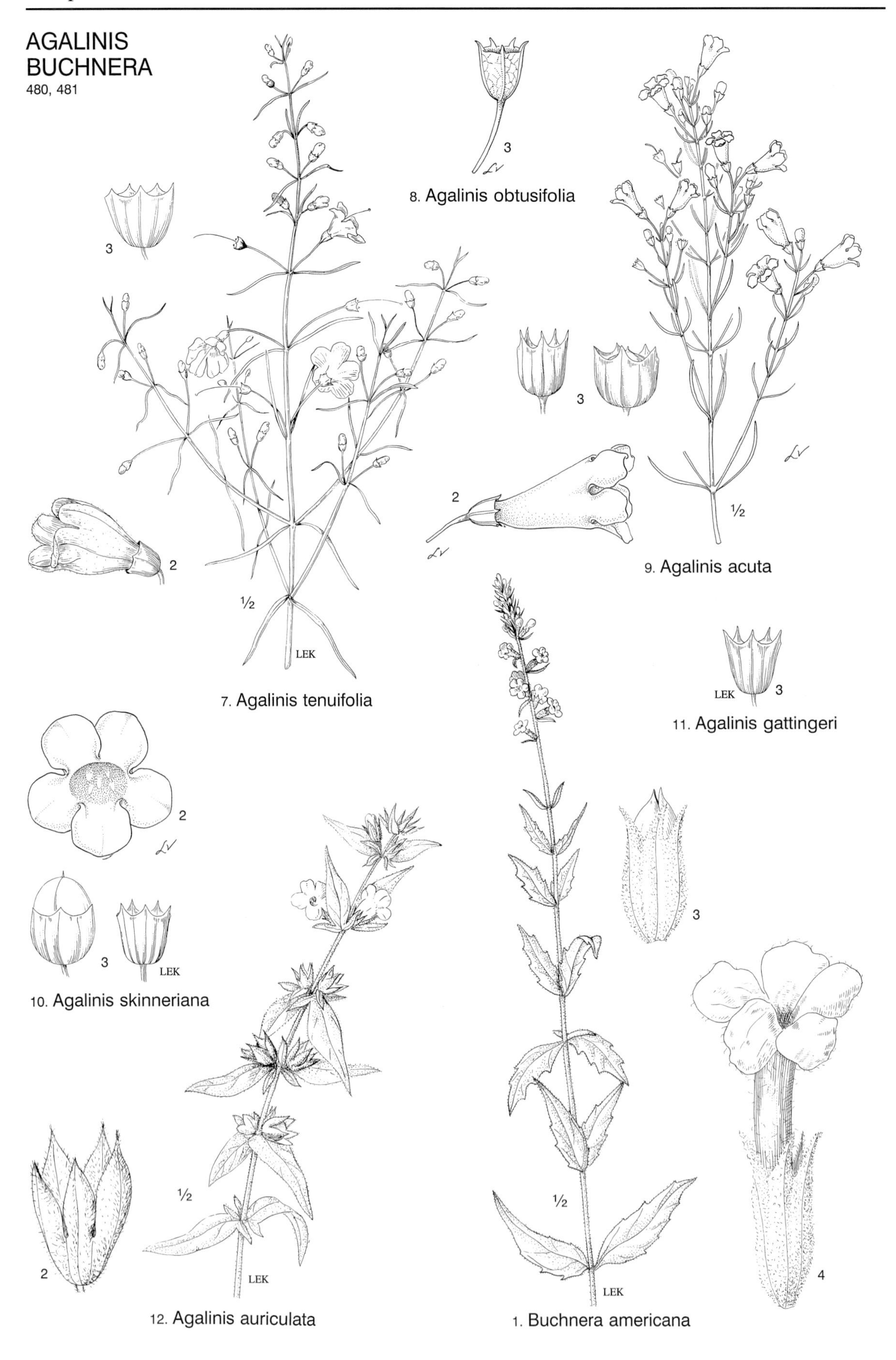

3

2

½

LEK

7. Agalinis tenuifolia

3

8. Agalinis obtusifolia

3

2

9. Agalinis acuta

LEK 3

11. Agalinis gattingeri

2

3 LEK

10. Agalinis skinneriana

2

½

LEK

12. Agalinis auriculata

3

½

LEK

1. Buchnera americana

4

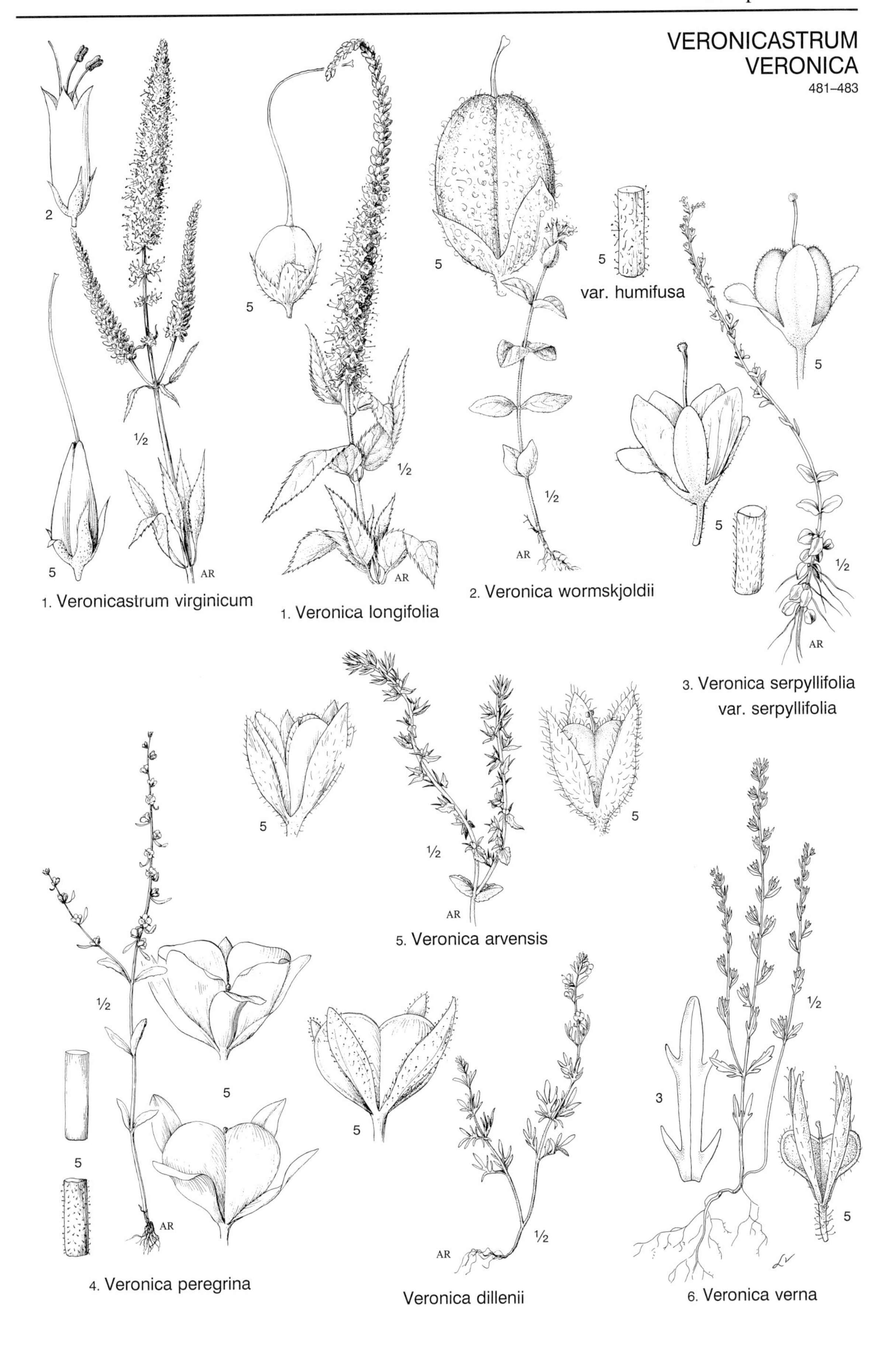

var. humifusa

1. Veronicastrum virginicum

1. Veronica longifolia

2. Veronica wormskjoldii

3. Veronica serpyllifolia
var. serpyllifolia

5. Veronica arvensis

4. Veronica peregrina

Veronica dillenii

6. Veronica verna

VERONICA
483, 484

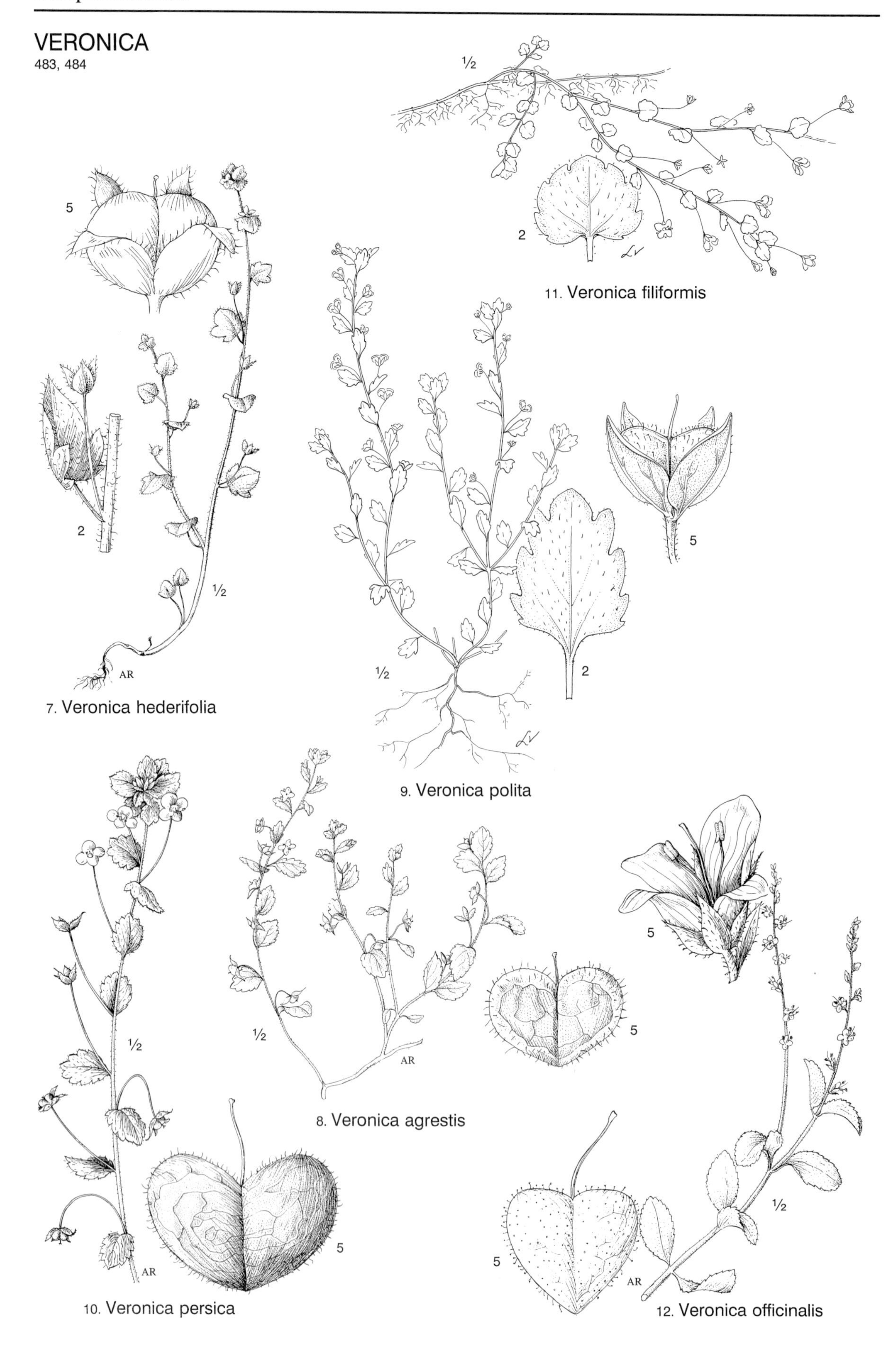

11. Veronica filiformis

7. Veronica hederifolia

9. Veronica polita

8. Veronica agrestis

10. Veronica persica

12. Veronica officinalis

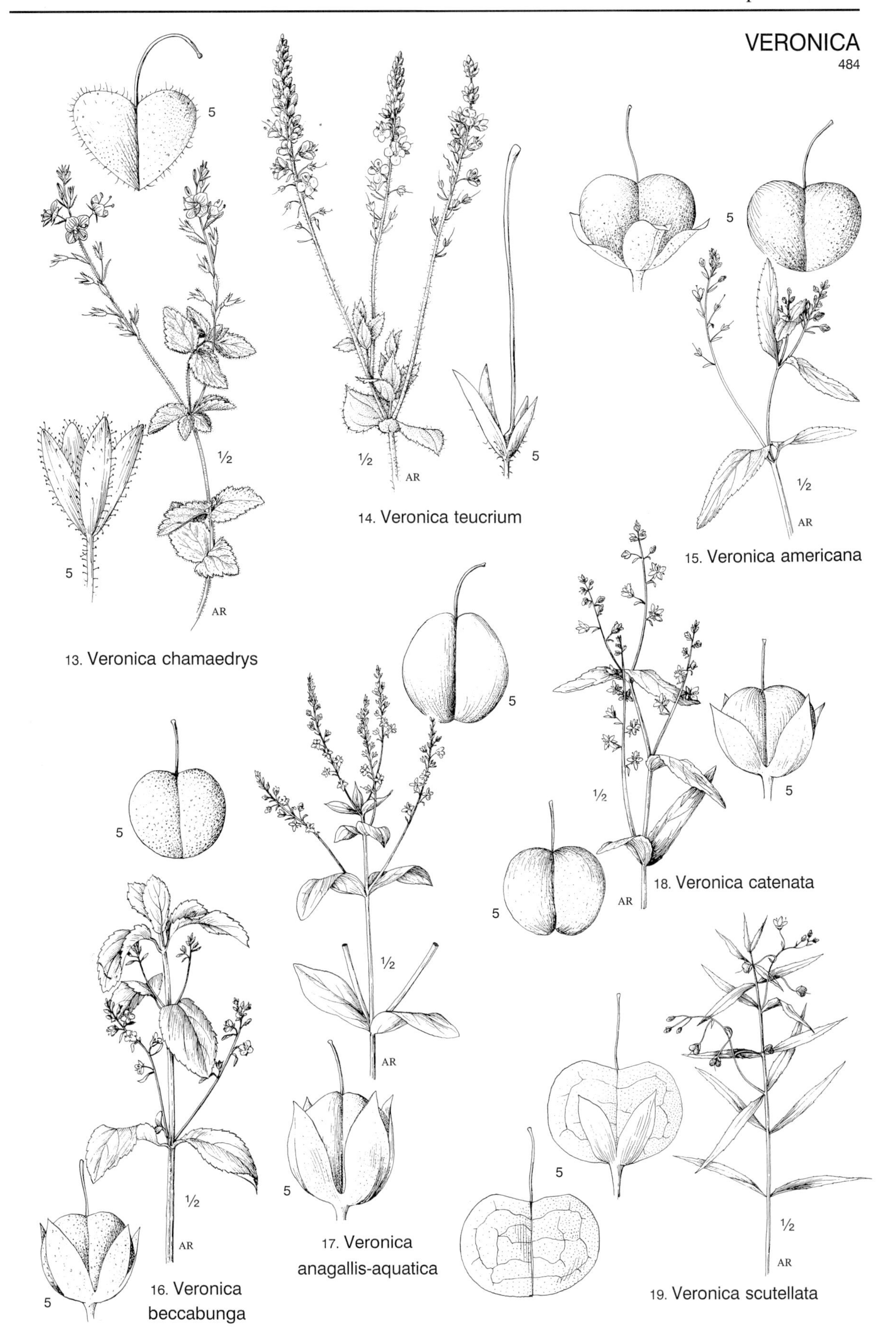

13. Veronica chamaedrys

14. Veronica teucrium

15. Veronica americana

16. Veronica
beccabunga

17. Veronica
anagallis-aquatica

18. Veronica catenata

19. Veronica scutellata

BESSEYA
SCHWALBEA
ODONTITES
EUPHRASIA
484–486

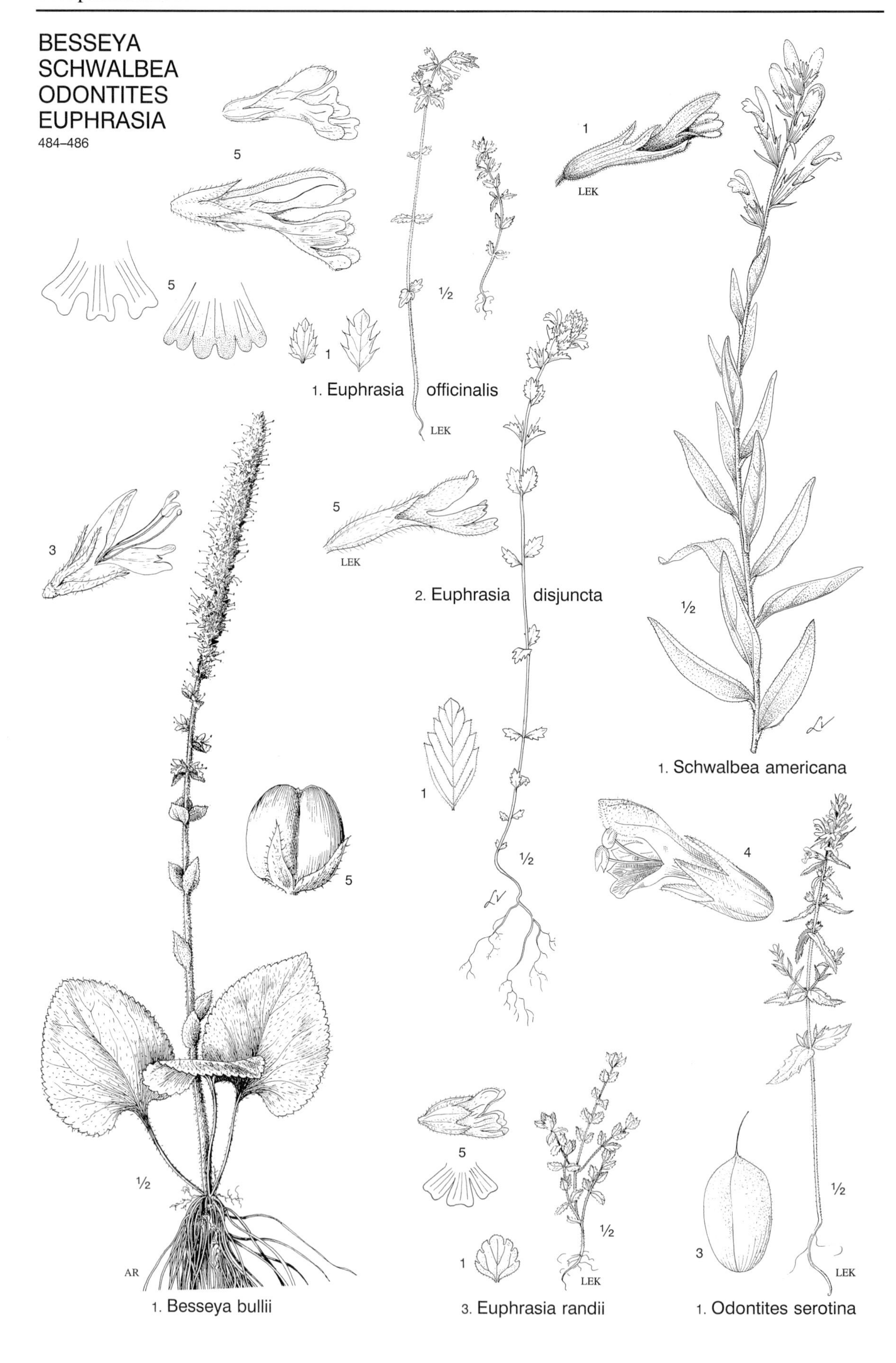

1. Euphrasia officinalis

2. Euphrasia disjuncta

1. Schwalbea americana

1. Besseya bullii

3. Euphrasia randii

1. Odontites serotina

ILLUSTRATED COMPANION TO

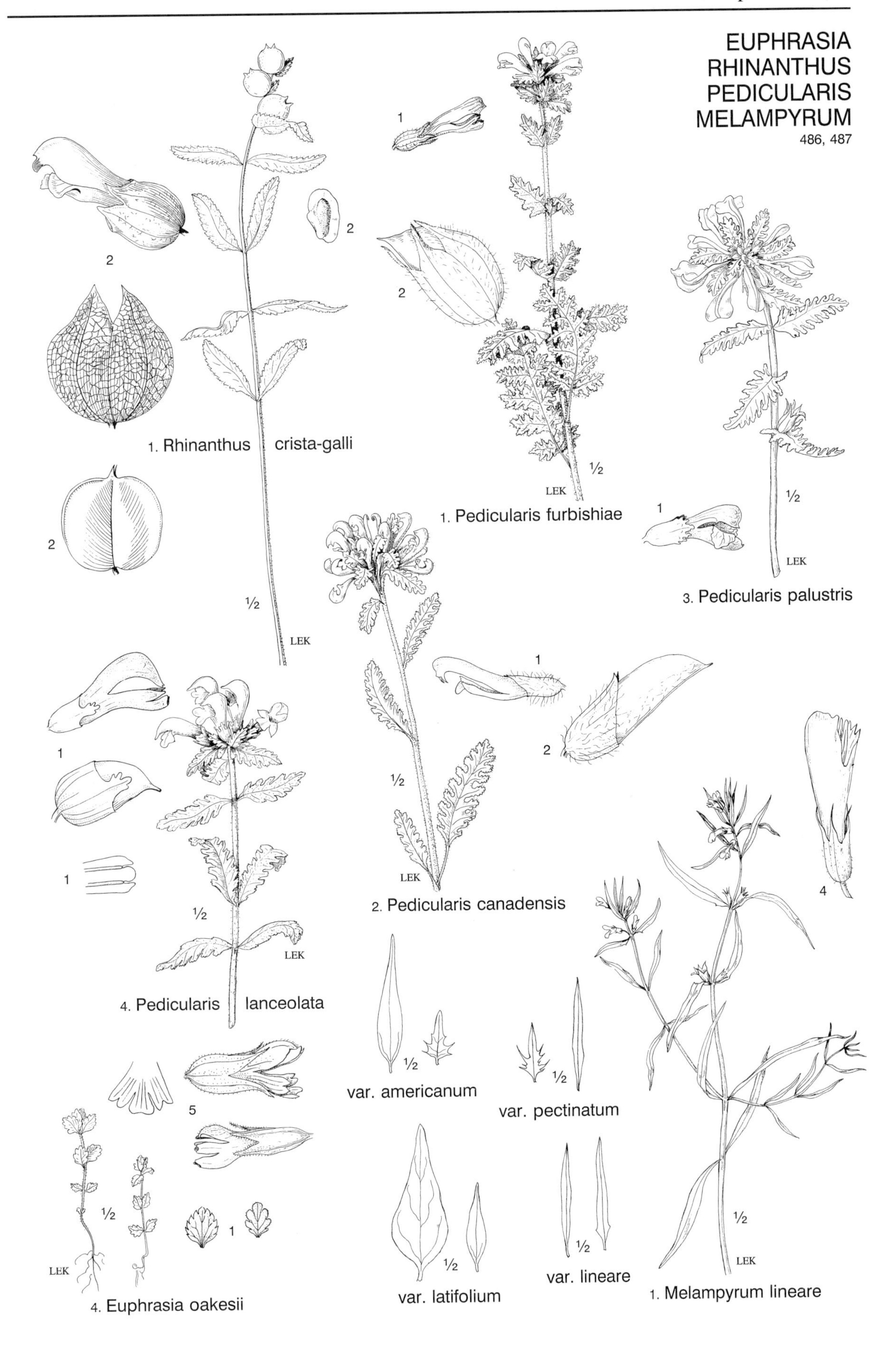

1. Rhinanthus   crista-galli

1. Pedicularis furbishiae

3. Pedicularis palustris

2. Pedicularis canadensis

4. Pedicularis   lanceolata

var. americanum

var. pectinatum

4. Euphrasia oakesii

var. latifolium

var. lineare

1. Melampyrum lineare

ORTHOCARPUS
CASTILLEJA
OROBANCHE
488, 489

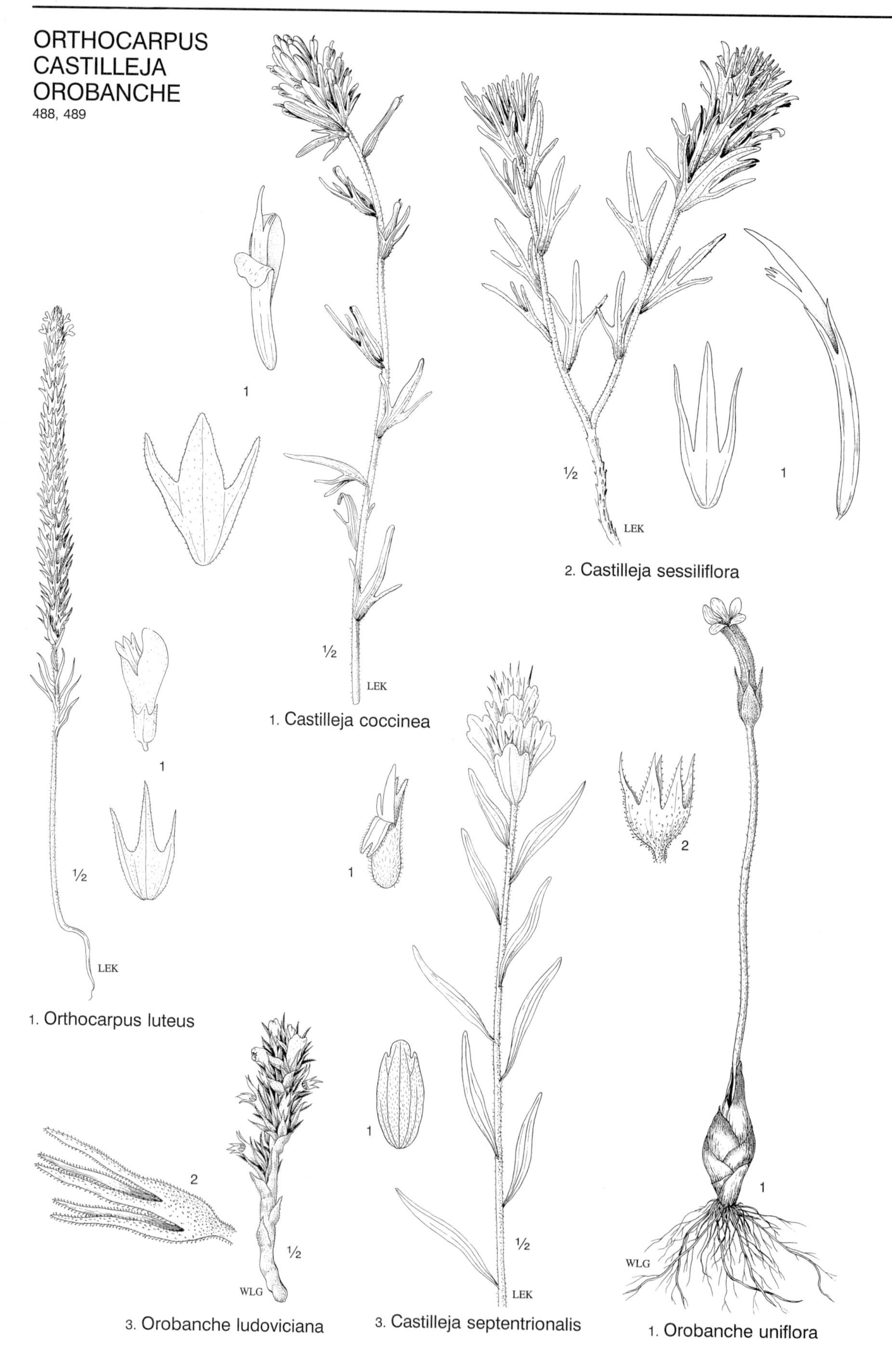

1. Castilleja coccinea

2. Castilleja sessiliflora

1. Orthocarpus luteus

3. Orobanche ludoviciana

3. Castilleja septentrionalis

1. Orobanche uniflora

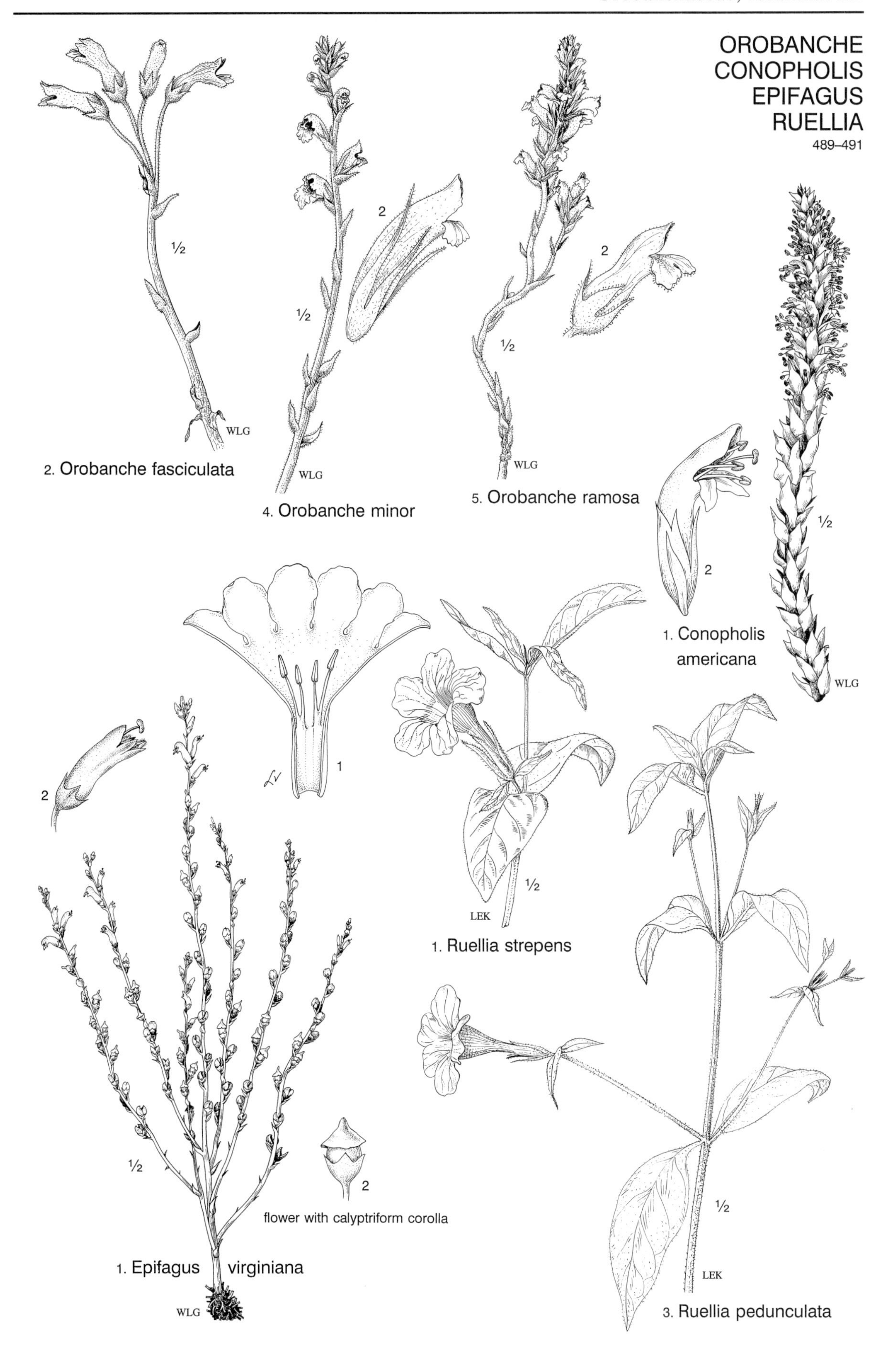

2. Orobanche fasciculata

4. Orobanche minor

5. Orobanche ramosa

1. Conopholis americana

1. Ruellia strepens

flower with calyptriform corolla

1. Epifagus virginiana

3. Ruellia pedunculata

RUELLIA
JUSTICIA
DICLIPTERA
SESAMUM
490–492

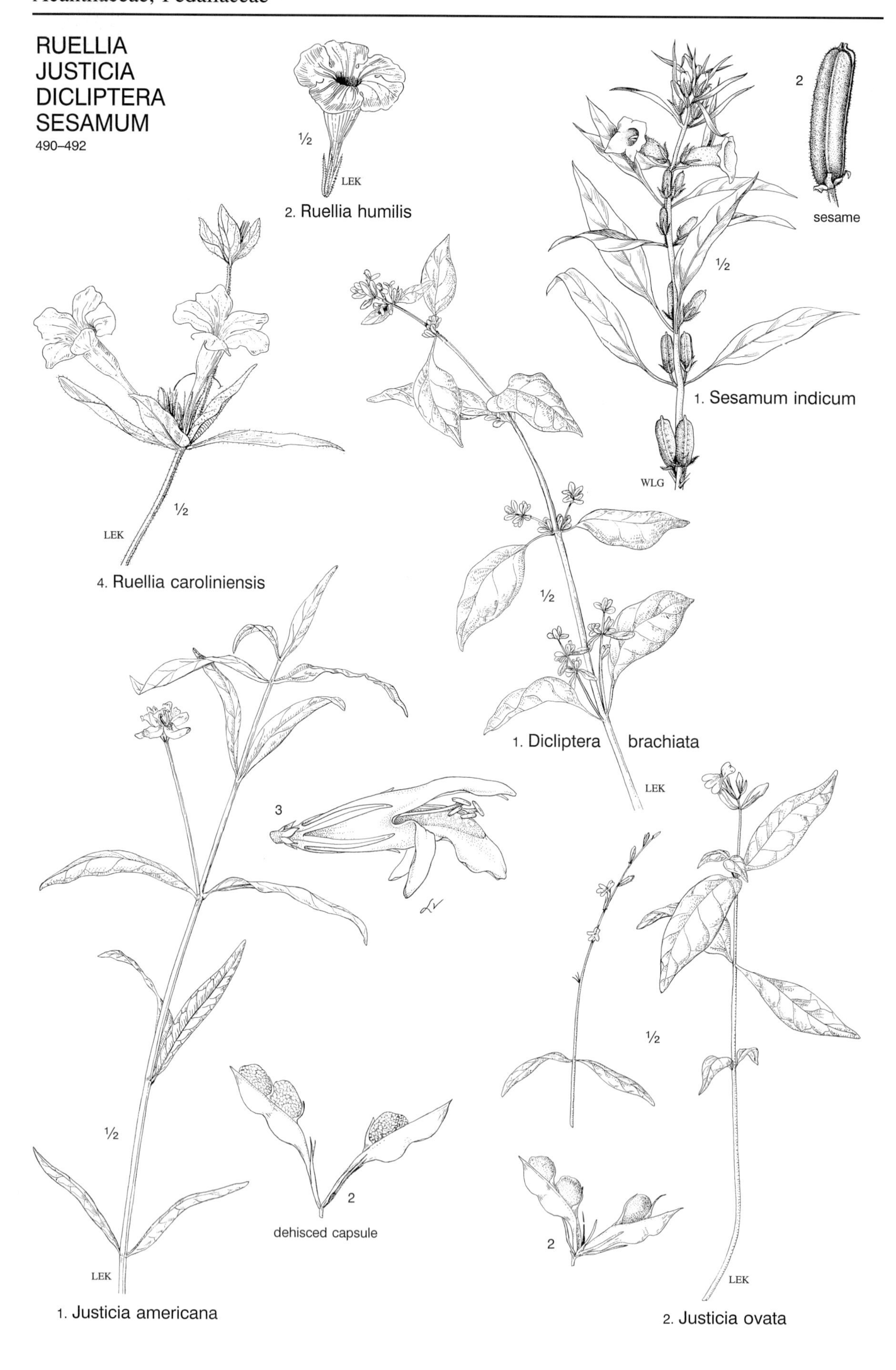

½
LEK
2. Ruellia humilis

2

sesame

½

1. Sesamum indicum
WLG

½
LEK
4. Ruellia caroliniensis

½

1. Dicliptera   brachiata
LEK

3

½
LEK
1. Justicia americana

2
dehisced capsule

½

2
LEK
2. Justicia ovata

PROBOSCIDEA
PAULOWNIA
CATALPA
492, 493

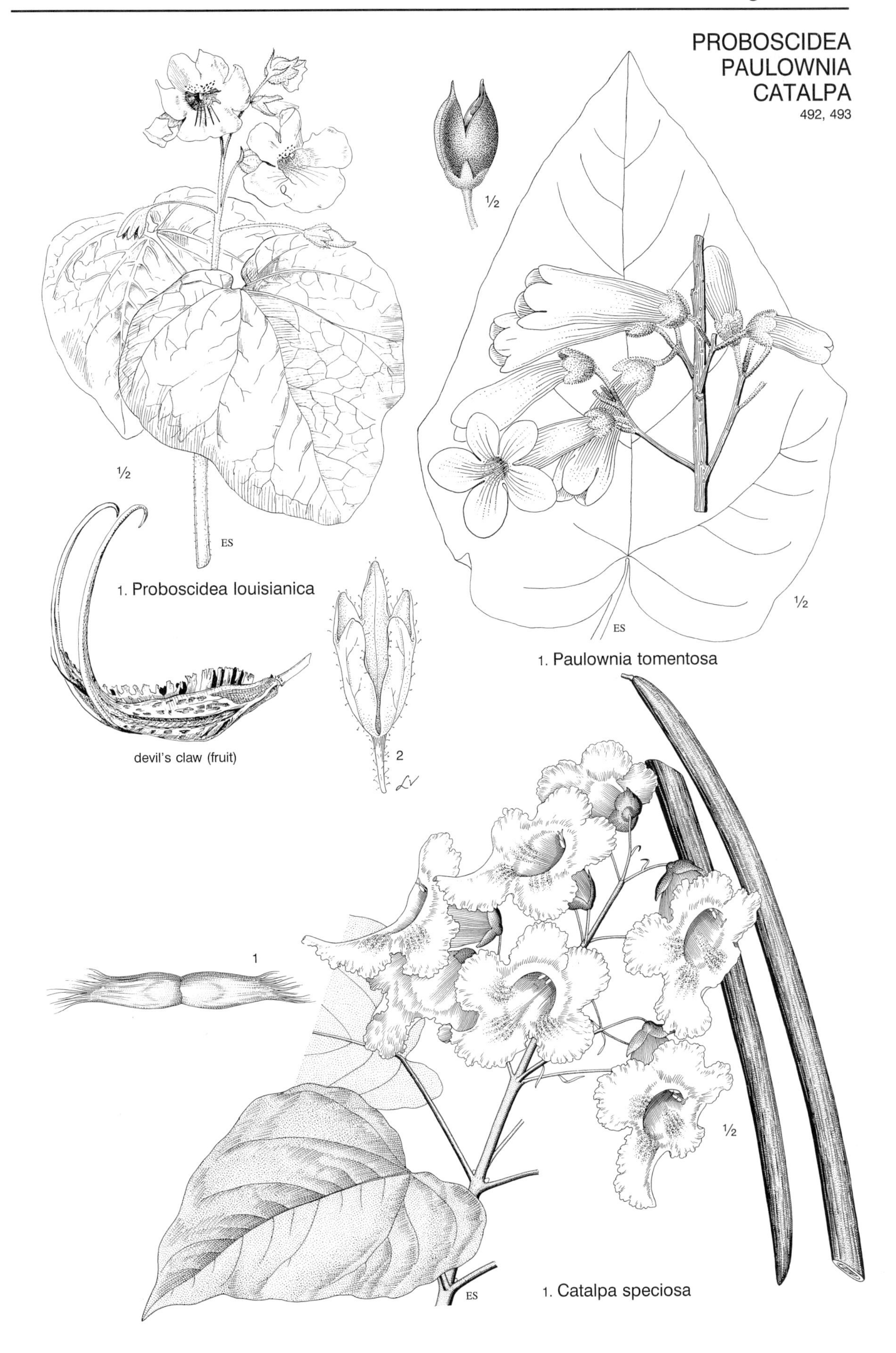

1. Proboscidea louisianica

devil's claw (fruit)

2

1. Paulownia tomentosa

1. Catalpa speciosa

CATALPA
CAMPSIS
BIGNONIA
PINGUICULA
UTRICULARIA
493–495

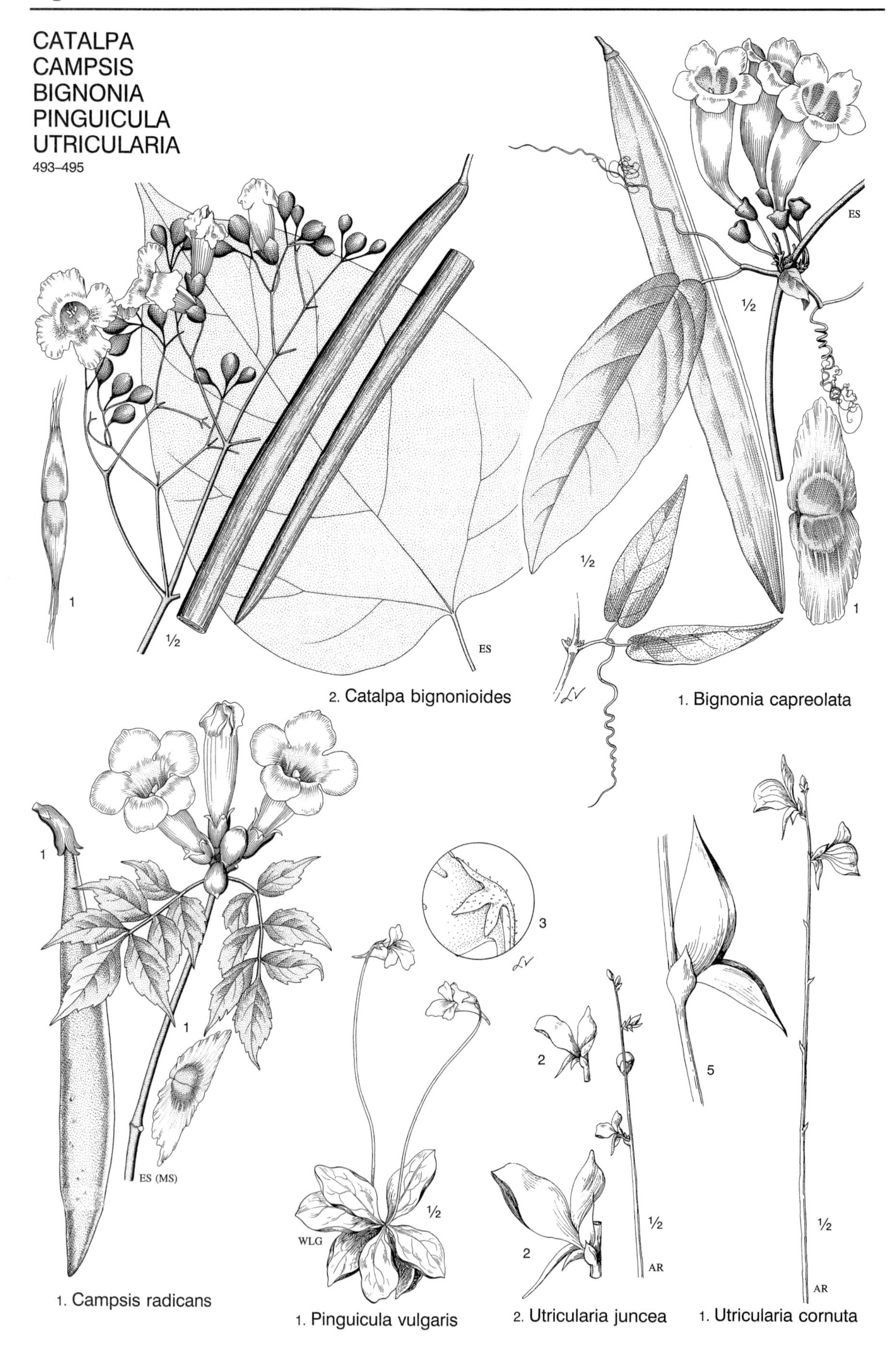

2. Catalpa bignonioides

1. Bignonia capreolata

1. Campsis radicans

1. Pinguicula vulgaris

2. Utricularia juncea

1. Utricularia cornuta

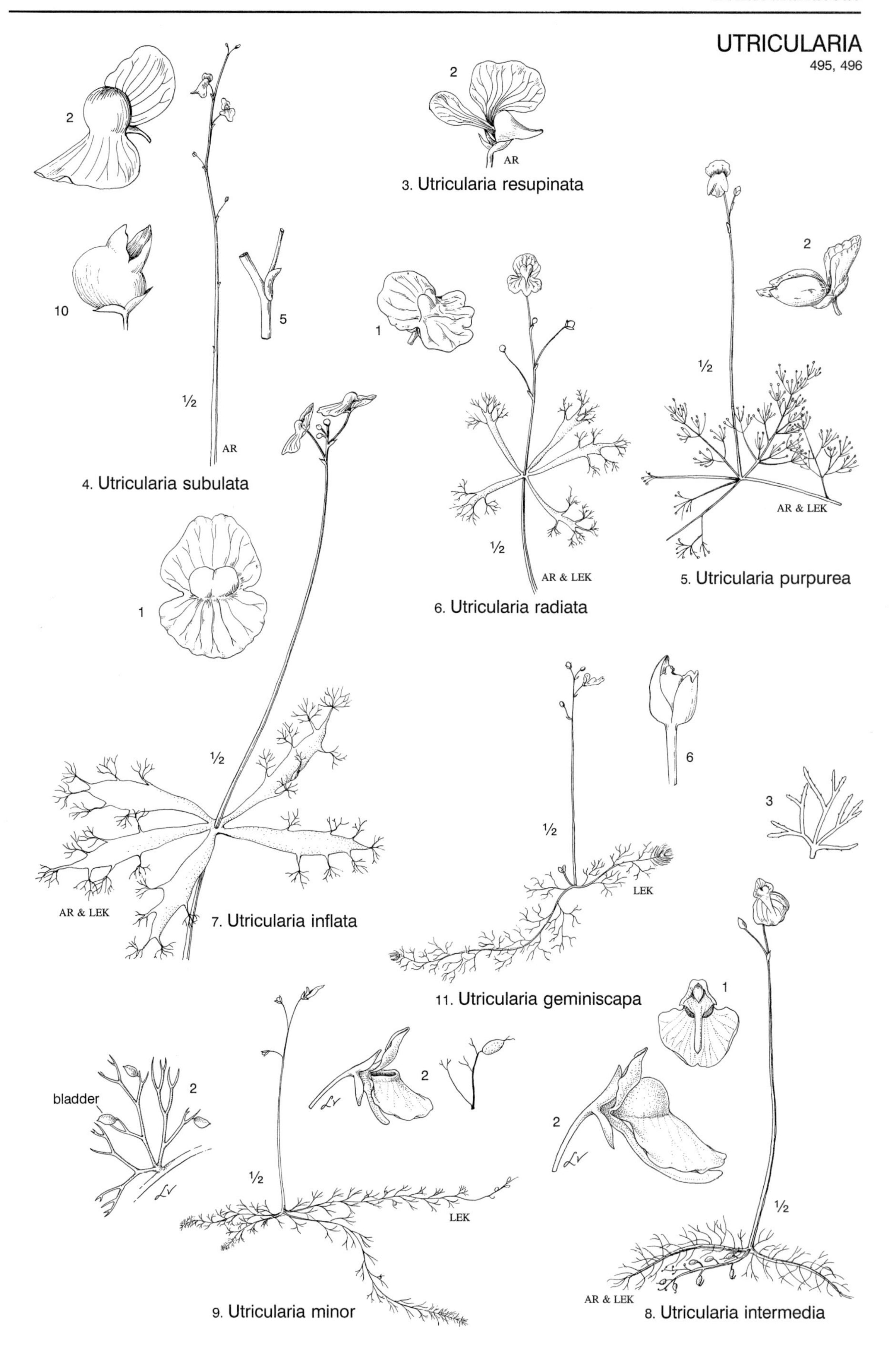

3. Utricularia resupinata

4. Utricularia subulata

5. Utricularia purpurea

6. Utricularia radiata

7. Utricularia inflata

11. Utricularia geminiscapa

9. Utricularia minor

bladder

8. Utricularia intermedia

UTRICULARIA
CAMPANULA
496, 497

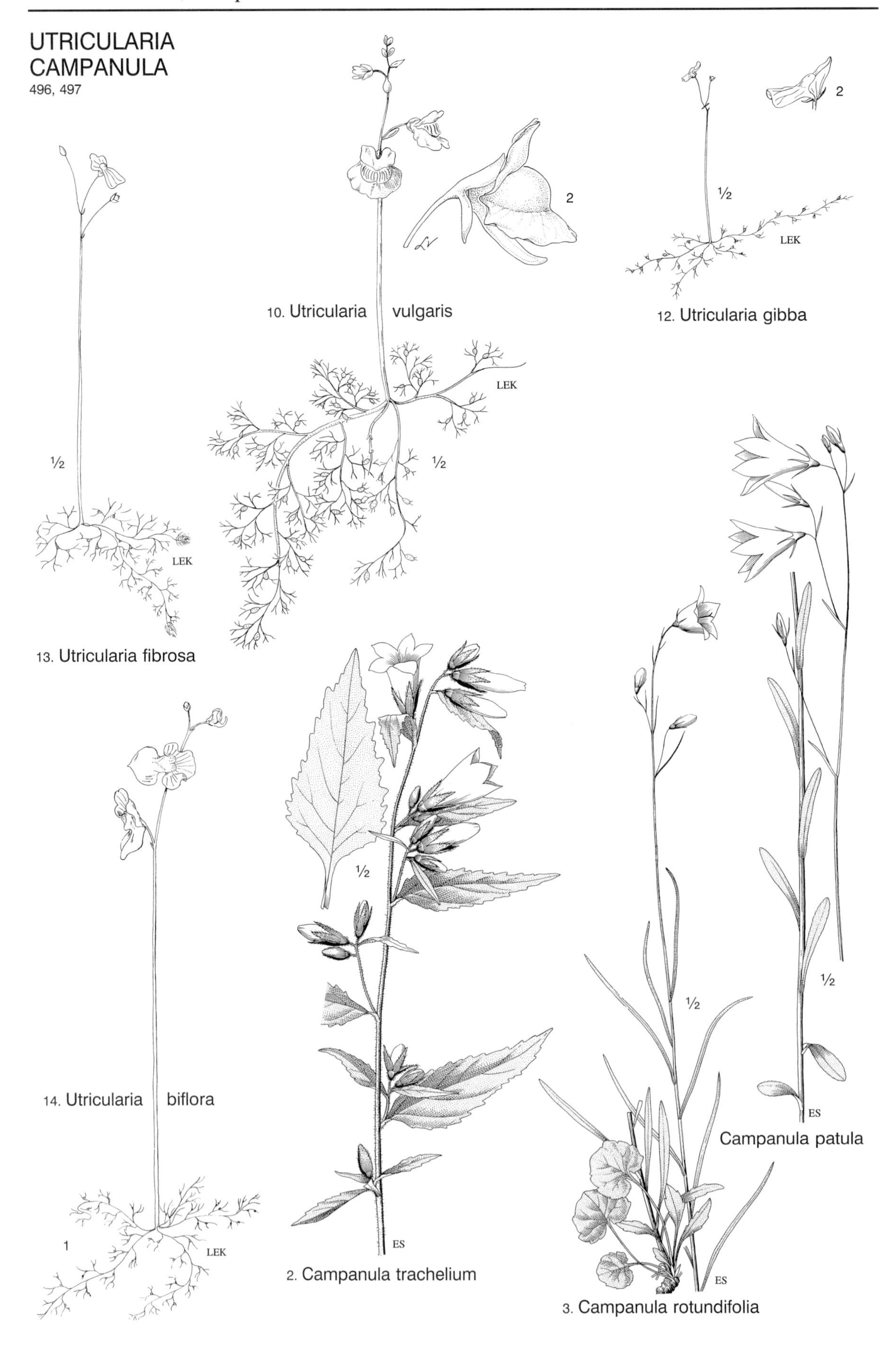

10. Utricularia   vulgaris

12. Utricularia gibba

13. Utricularia fibrosa

14. Utricularia   biflora

2. Campanula trachelium

3. Campanula rotundifolia

Campanula patula

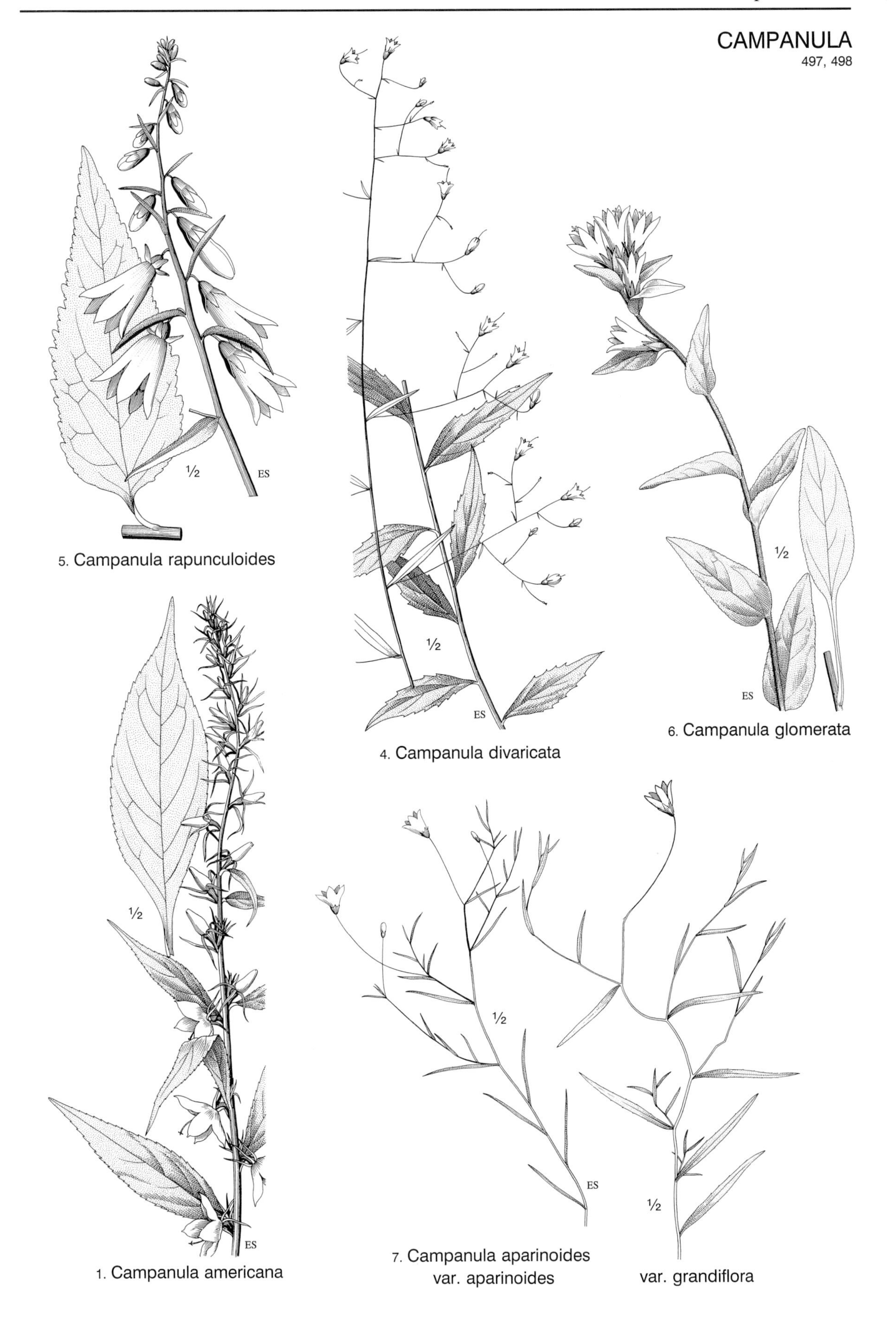

5. Campanula rapunculoides

4. Campanula divaricata

6. Campanula glomerata

1. Campanula americana

7. Campanula aparinoides
var. aparinoides

var. grandiflora

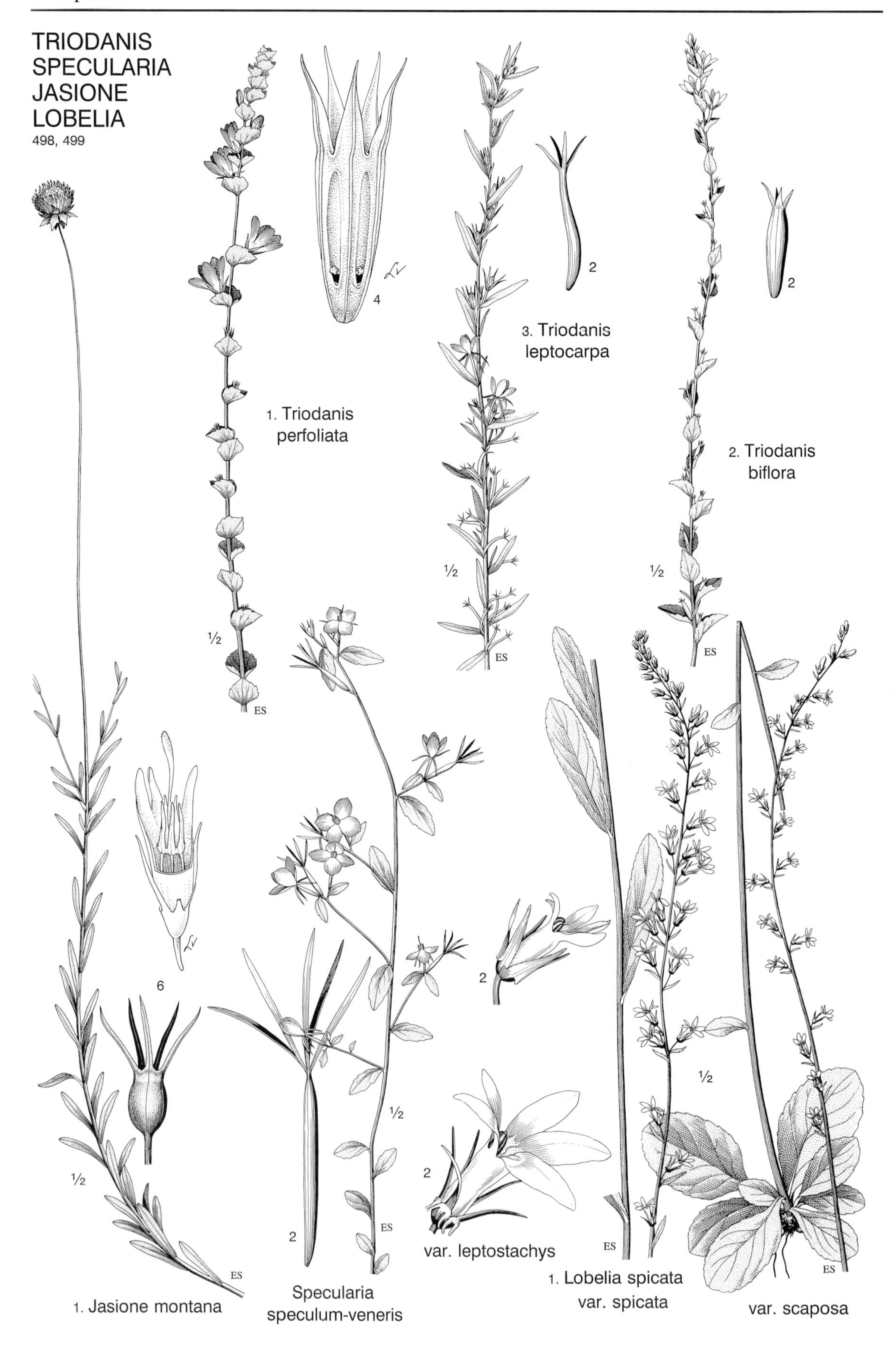

TRIODANIS
SPECULARIA
JASIONE
LOBELIA
498, 499

1. Triodanis
perfoliata

4

3. Triodanis
leptocarpa

2

2. Triodanis
biflora

2

1. Jasione montana

6

Specularia
speculum-veneris

var. leptostachys

2

1. Lobelia spicata
var. spicata

var. scaposa

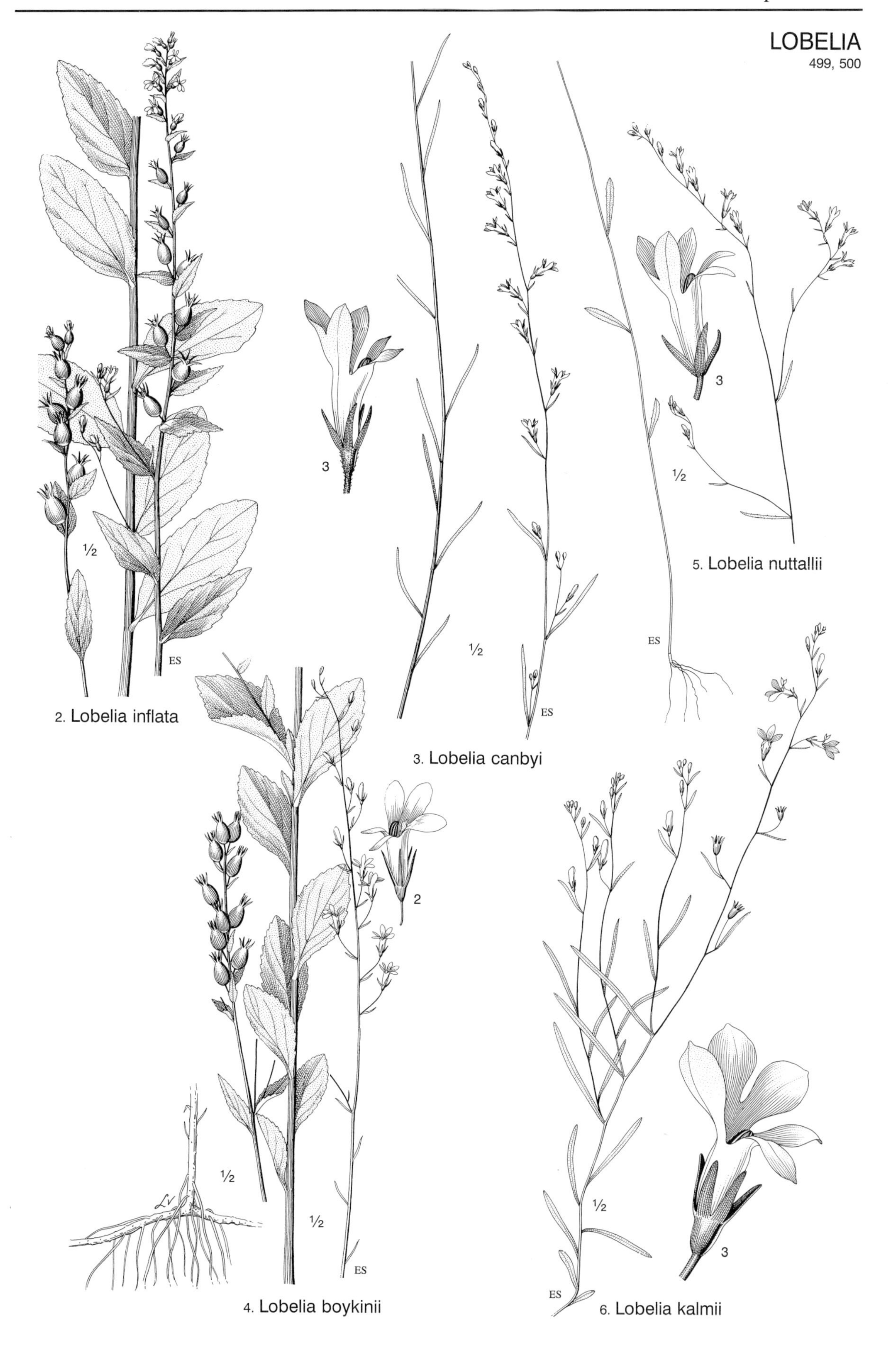

2. Lobelia inflata

3. Lobelia canbyi

5. Lobelia nuttallii

4. Lobelia boykinii

6. Lobelia kalmii

LOBELIA
500

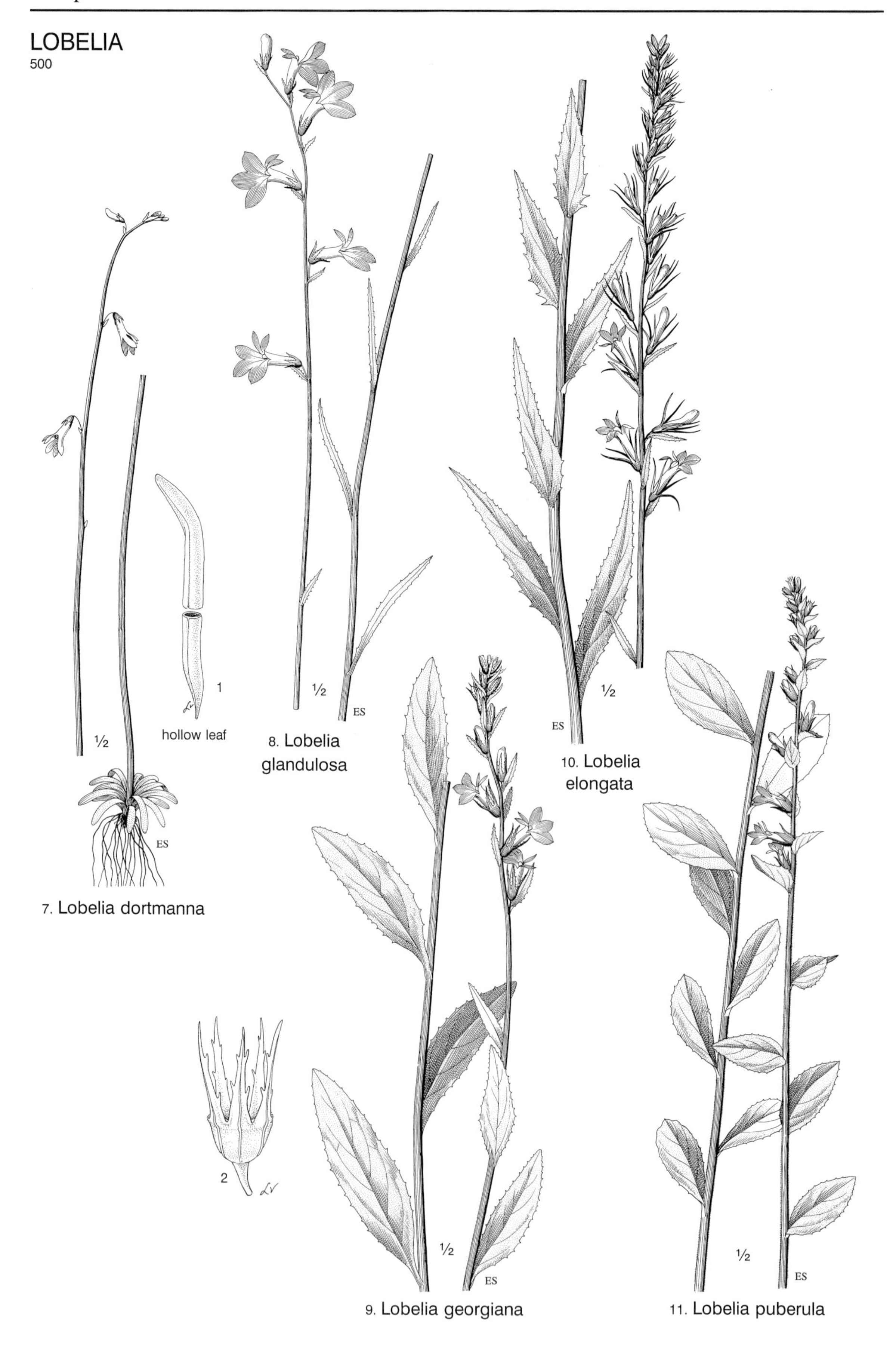

7. Lobelia dortmanna

hollow leaf

1

8. Lobelia glandulosa

2

9. Lobelia georgiana

10. Lobelia elongata

11. Lobelia puberula

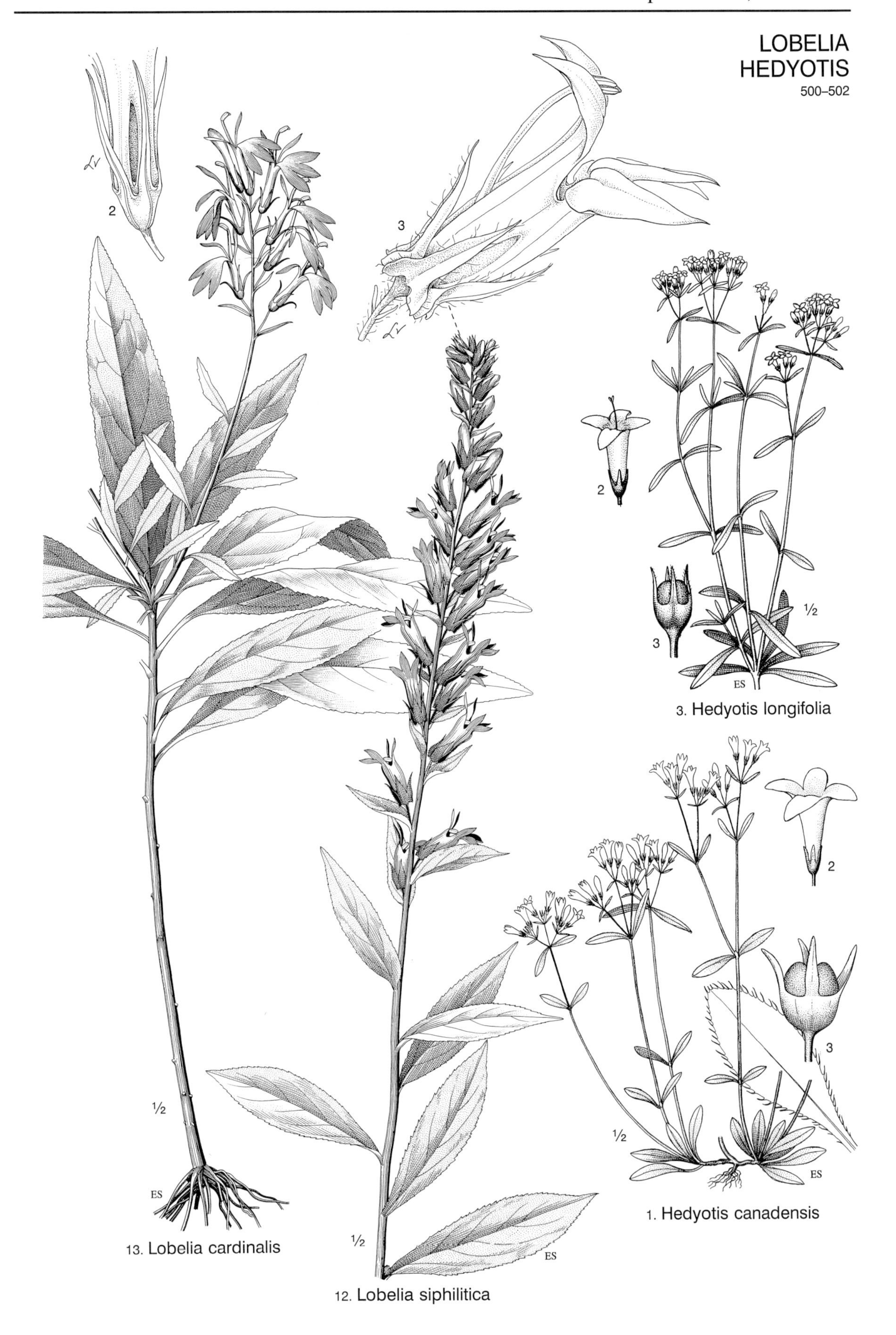

3. Hedyotis longifolia

1. Hedyotis canadensis

13. Lobelia cardinalis

12. Lobelia siphilitica

HEDYOTIS
501, 502

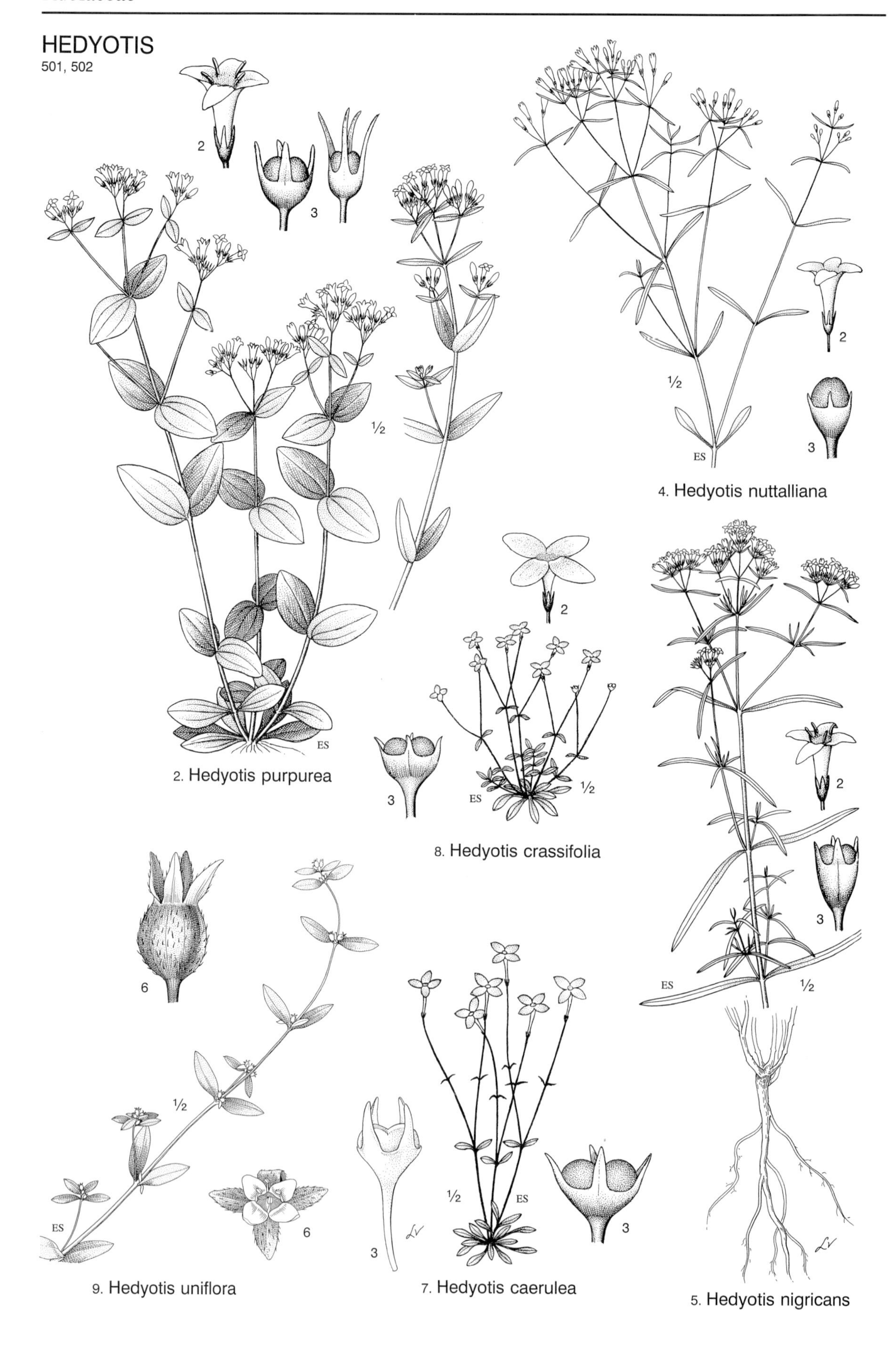

2. Hedyotis purpurea

4. Hedyotis nuttalliana

8. Hedyotis crassifolia

9. Hedyotis uniflora

7. Hedyotis caerulea

5. Hedyotis nigricans

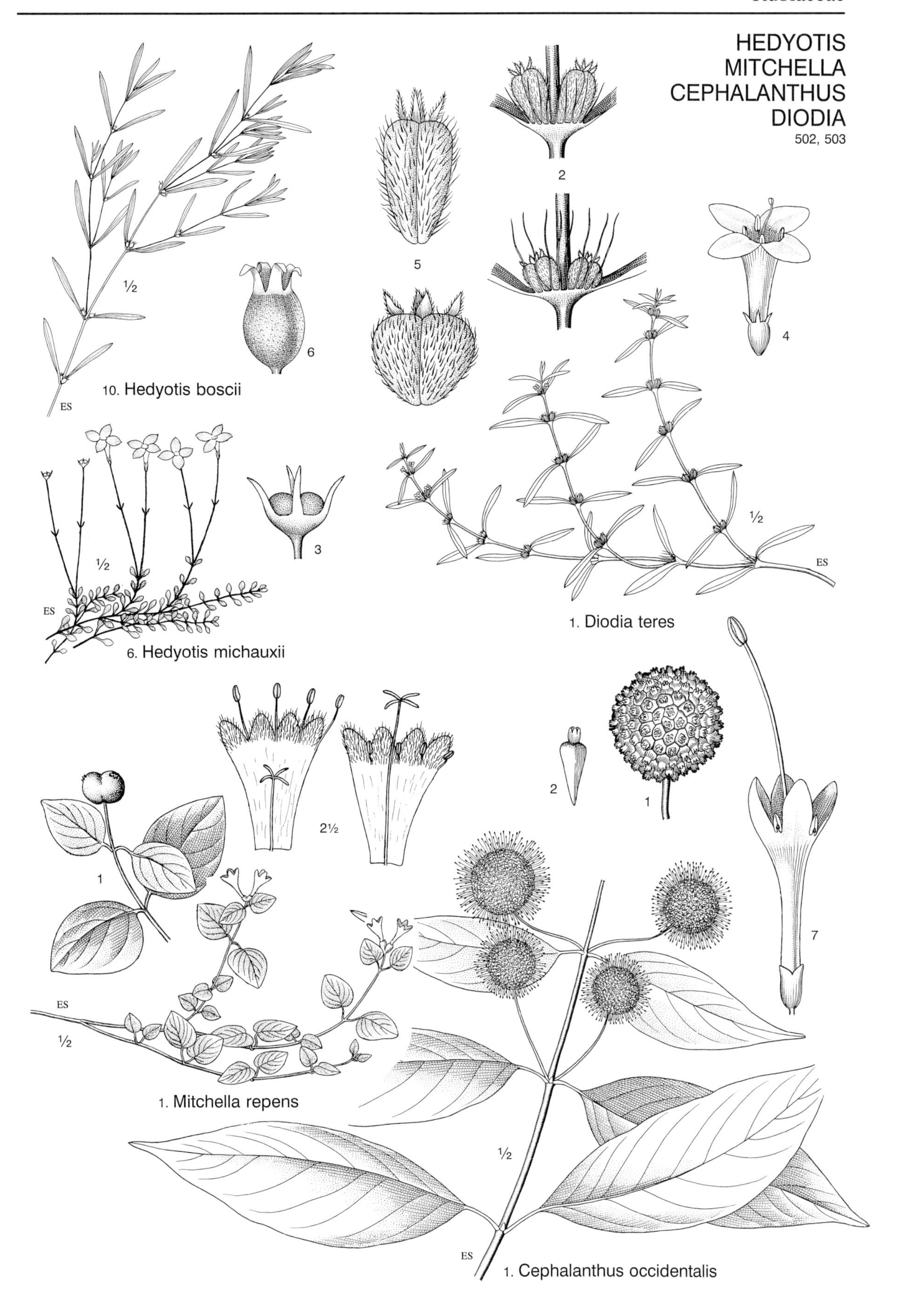

10. Hedyotis boscii

6. Hedyotis michauxii

1. Diodia teres

1. Mitchella repens

1. Cephalanthus occidentalis

DIODIA
SPERMACOCE
RICHARDIA
503, 504

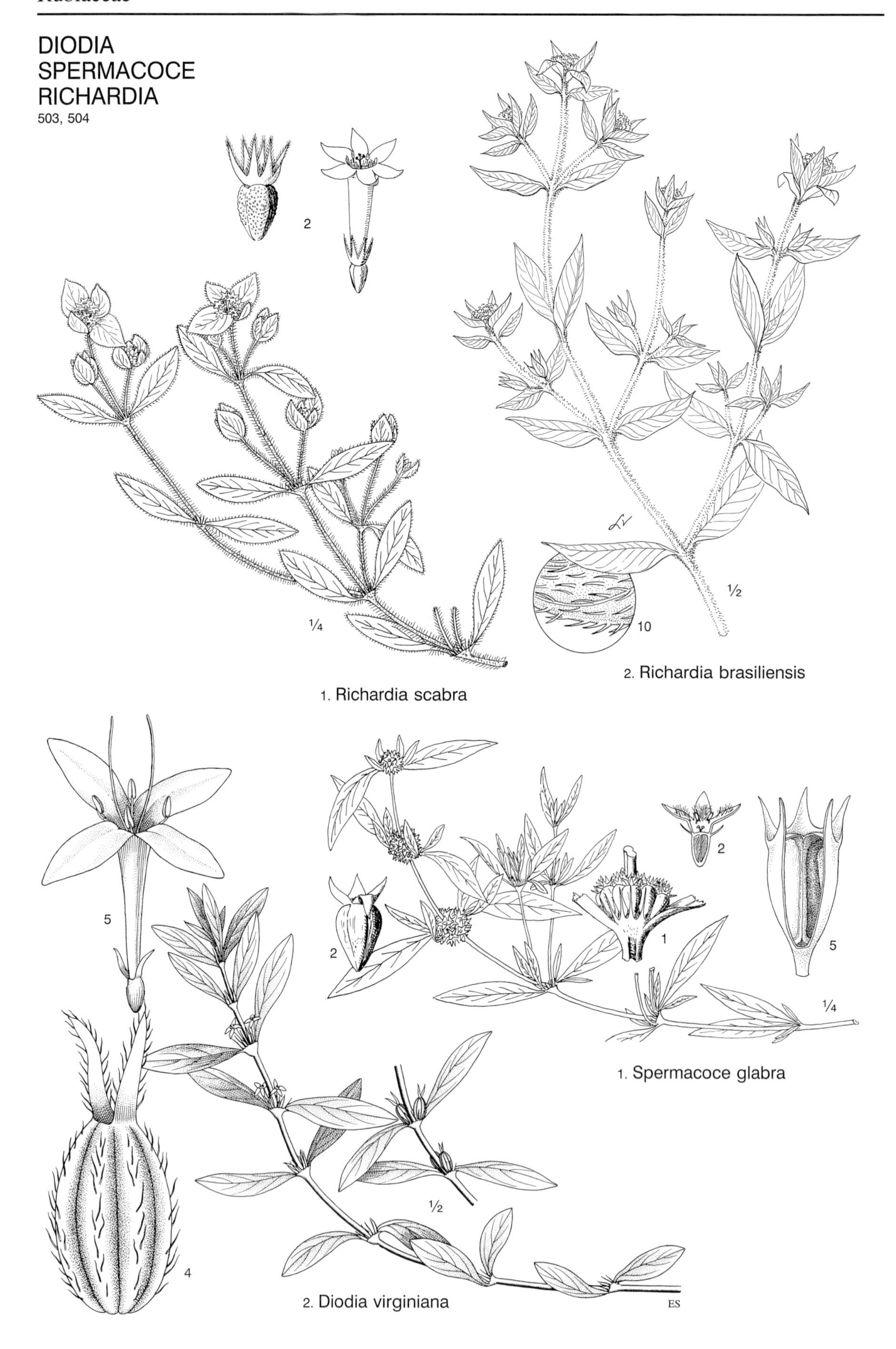

1. Richardia scabra

2. Richardia brasiliensis

1. Spermacoce glabra

2. Diodia virginiana

ES

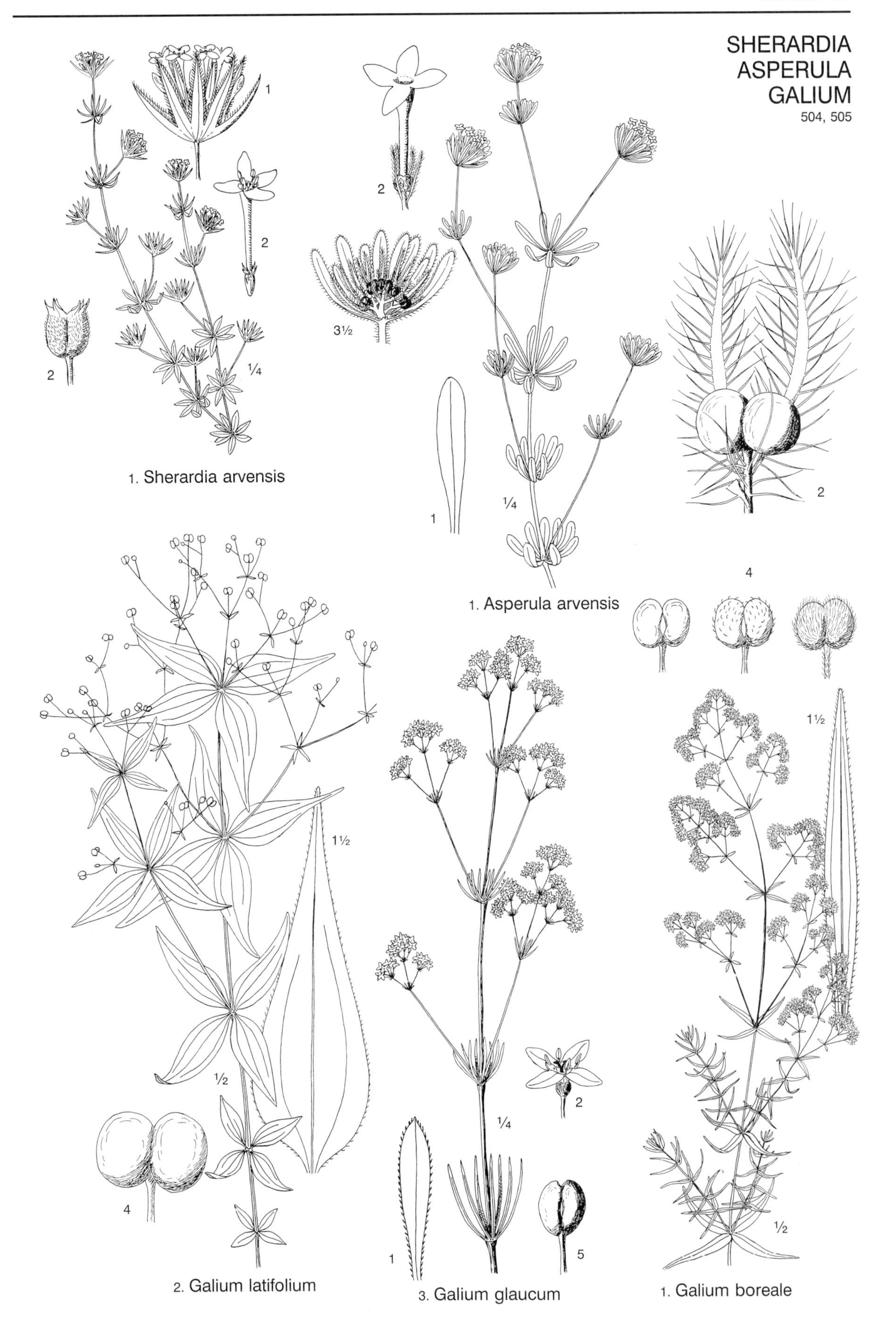

1. Sherardia arvensis

1. Asperula arvensis

2. Galium latifolium

3. Galium glaucum

1. Galium boreale

GALIUM
505, 506

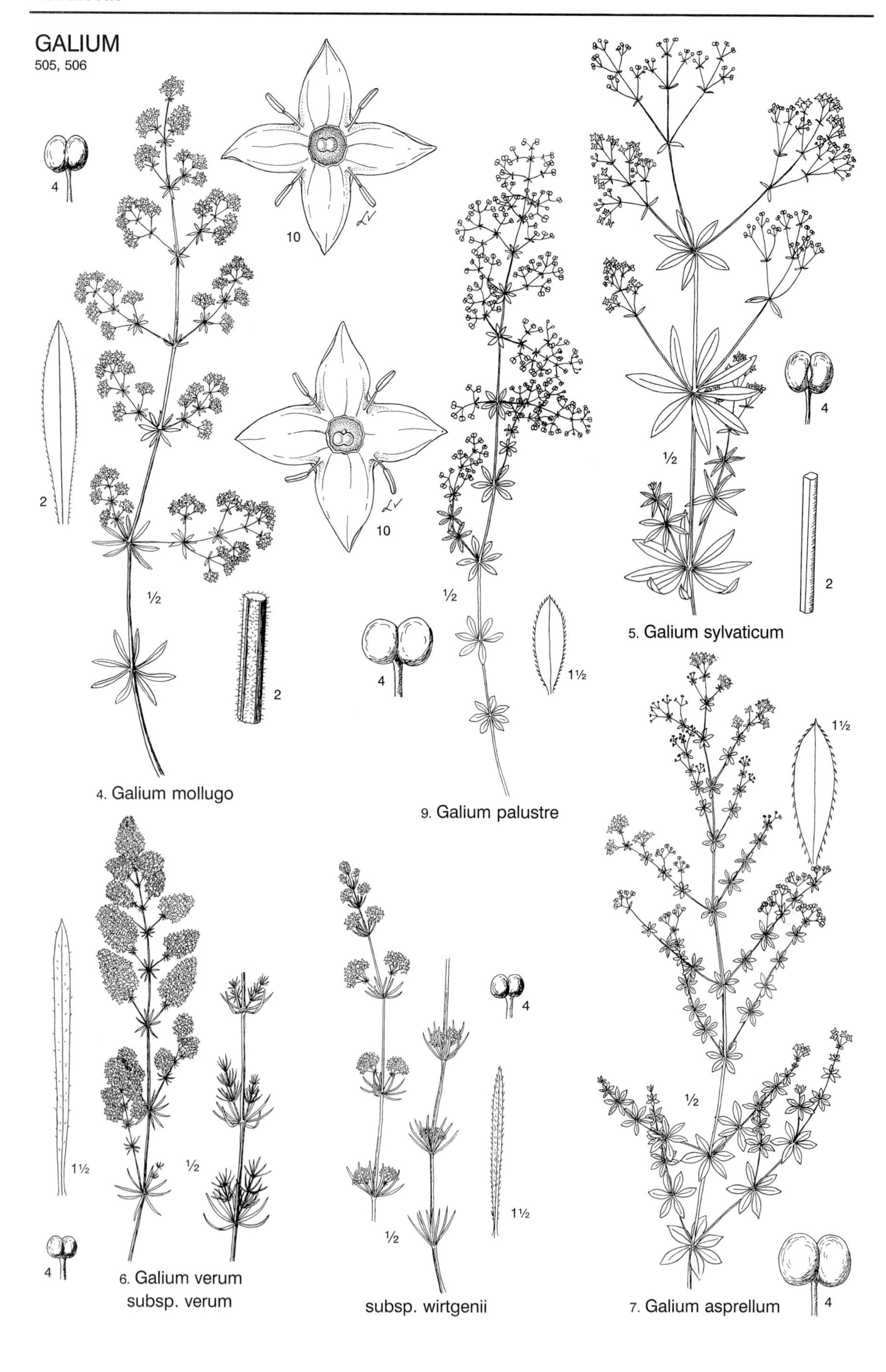

4. Galium mollugo

10

10

5. Galium sylvaticum

9. Galium palustre

6. Galium verum
subsp. verum

subsp. wirtgenii

7. Galium asprellum

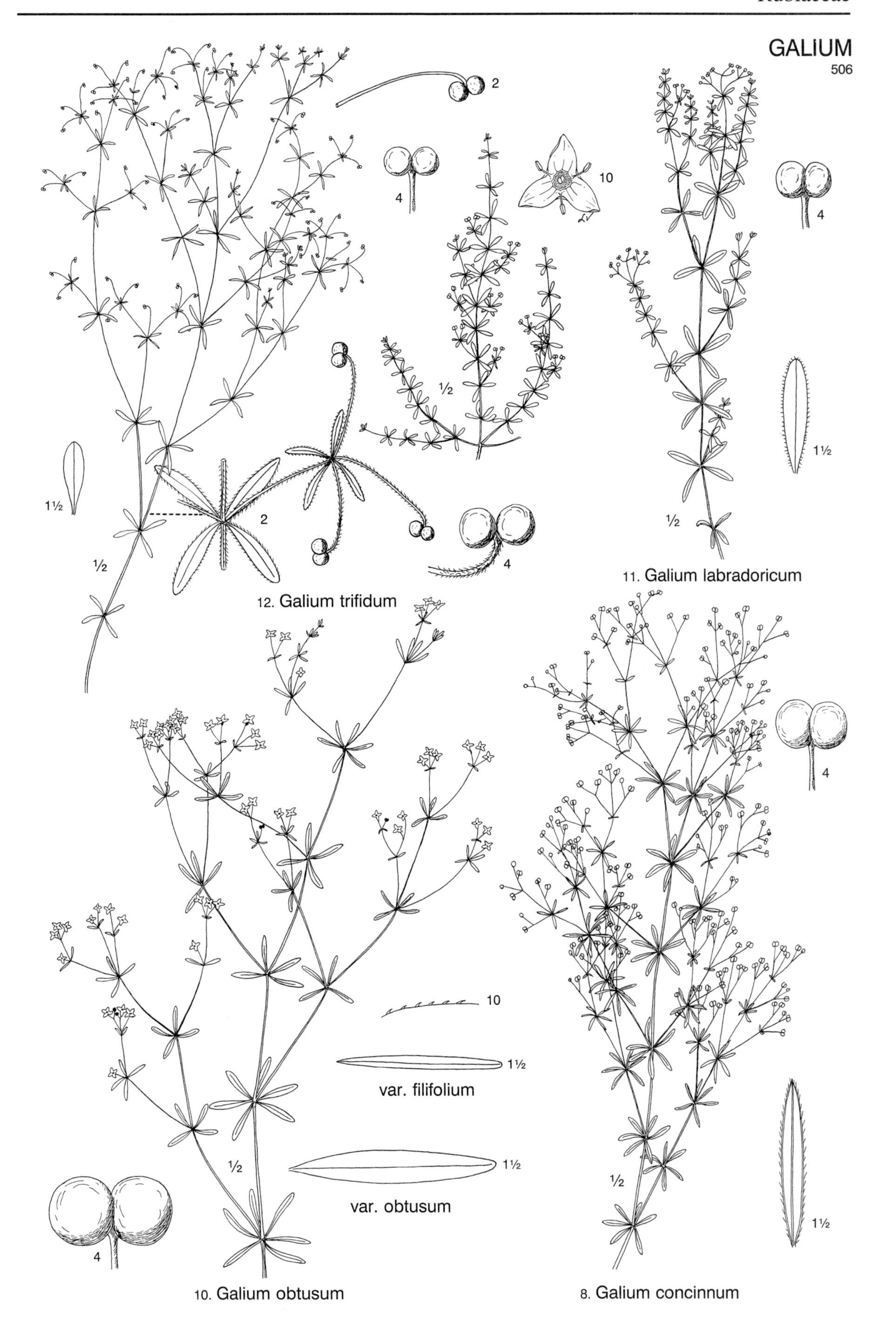

12. Galium trifidum

11. Galium labradoricum

var. filifolium

var. obtusum

10. Galium obtusum

8. Galium concinnum

GALIUM
506, 507

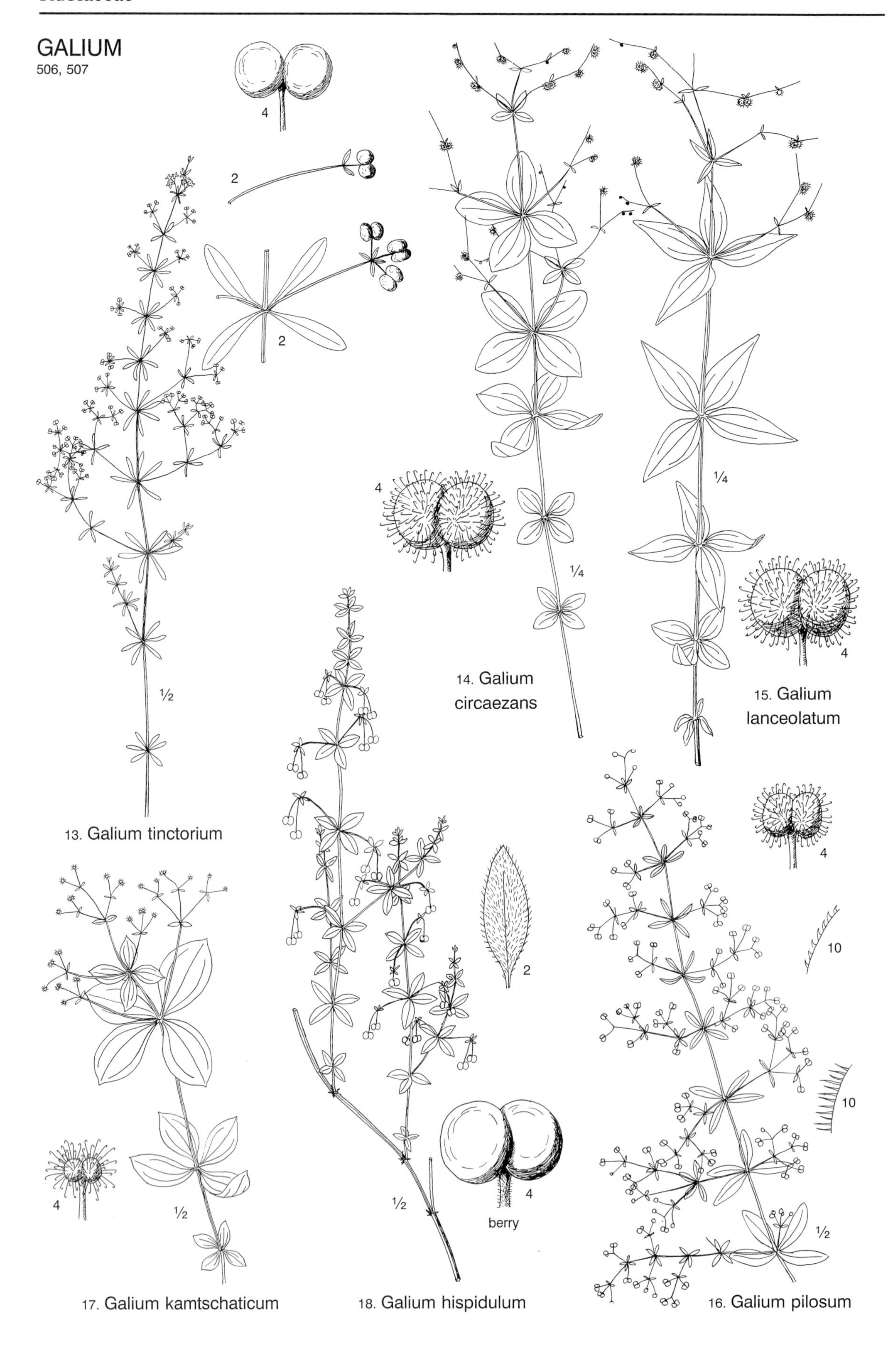

2

2

4

13. Galium tinctorium

14. Galium
circaezans

¼

¼

15. Galium
lanceolatum

4

4

2

berry

4

10

10

17. Galium kamtschaticum

18. Galium hispidulum

16. Galium pilosum

4

½

½

½

½

½

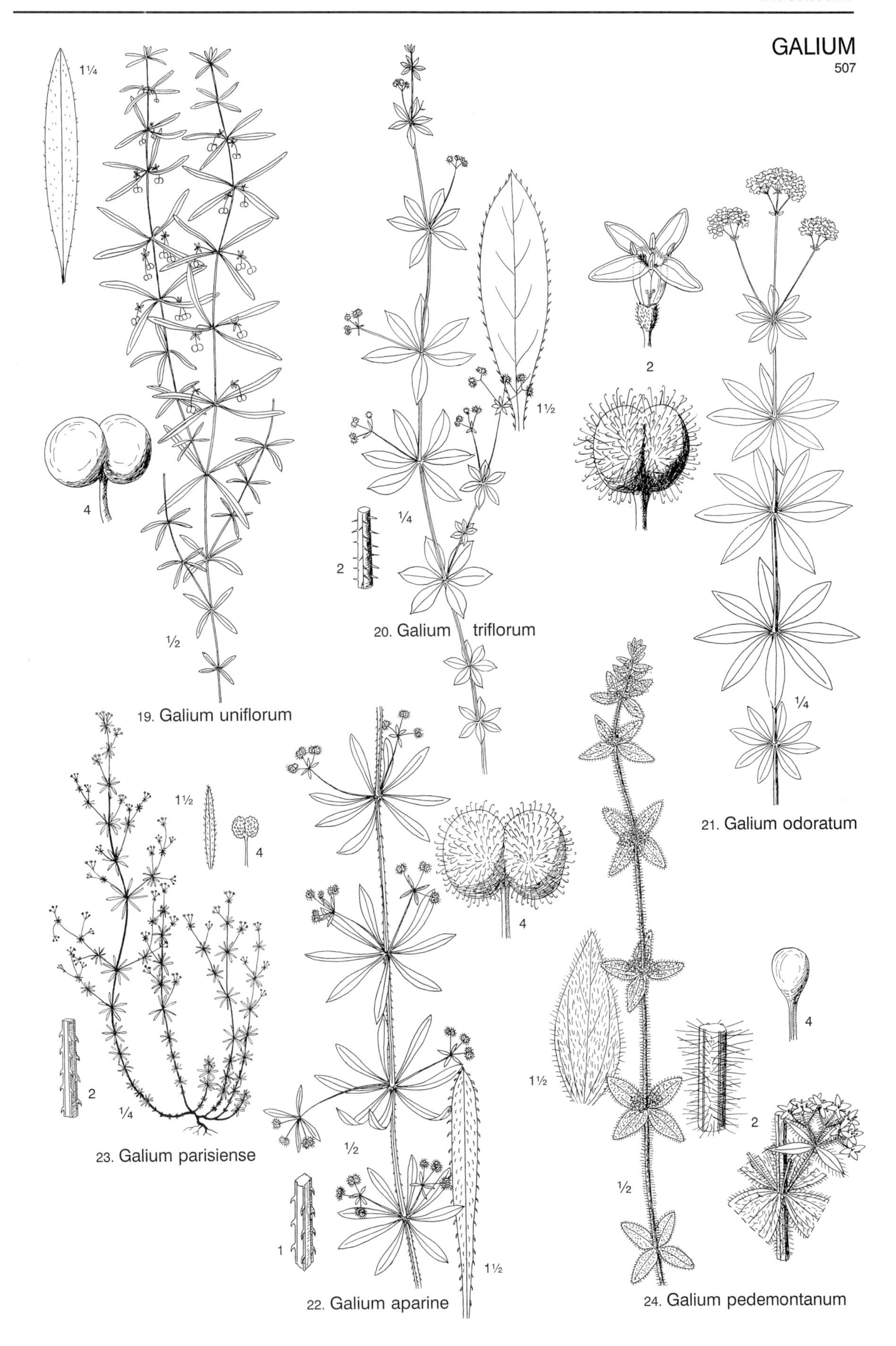

19. Galium uniflorum

20. Galium triflorum

21. Galium odoratum

23. Galium parisiense

22. Galium aparine

24. Galium pedemontanum

DIERVILLA
LONICERA
508, 509

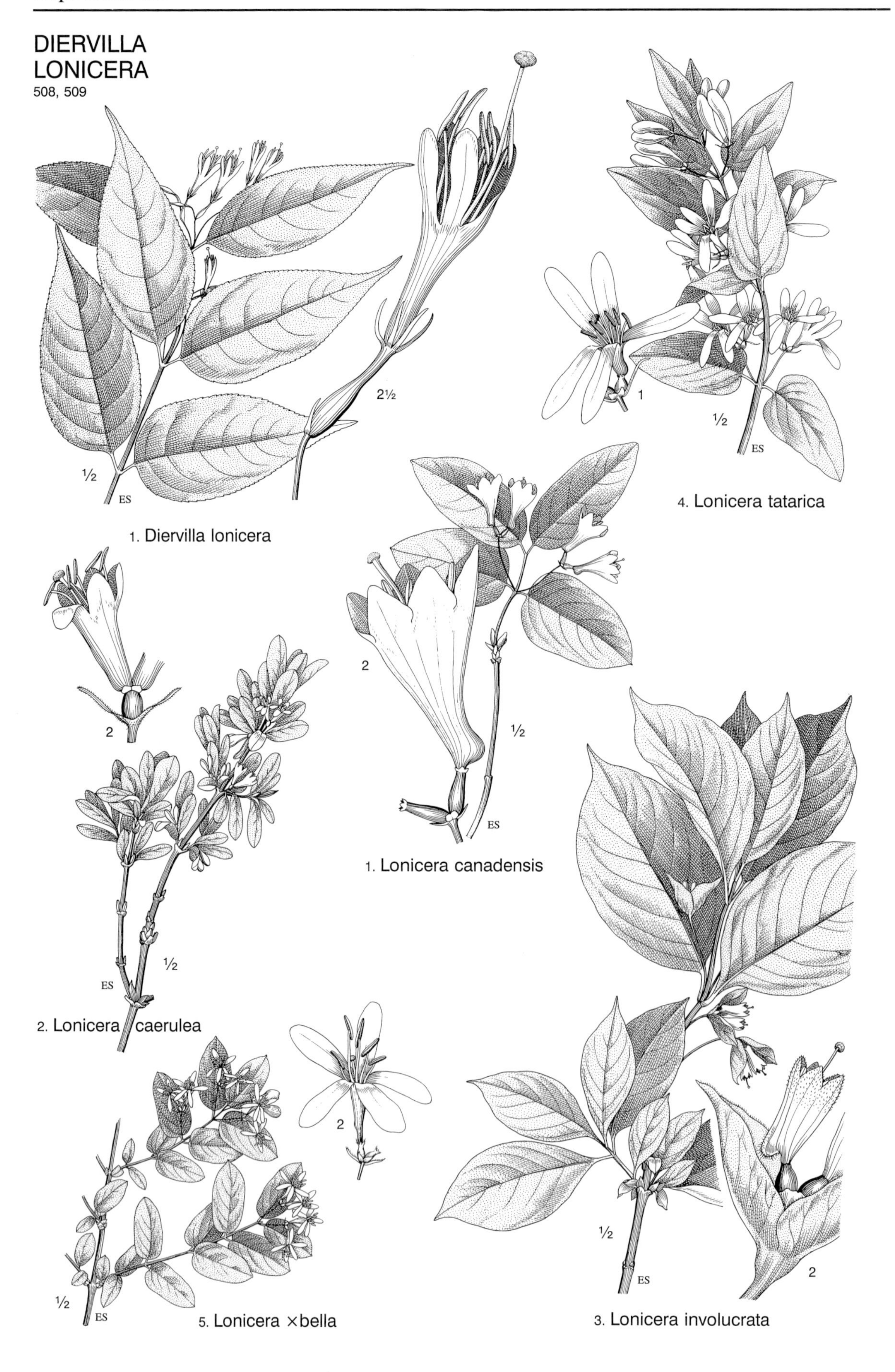

1. Diervilla lonicera

4. Lonicera tatarica

2. Lonicera caerulea

1. Lonicera canadensis

5. Lonicera ×bella

3. Lonicera involucrata

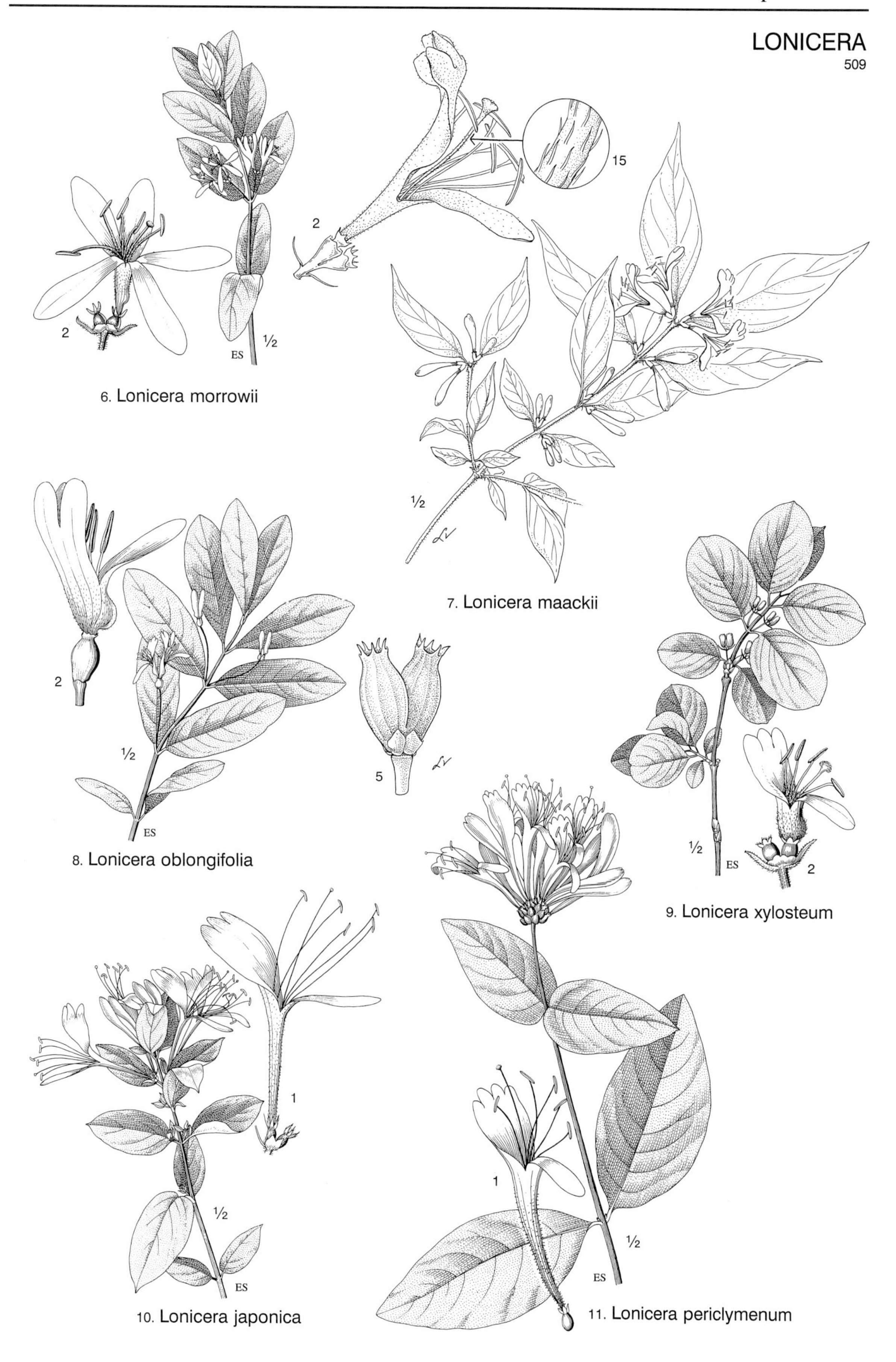

6. Lonicera morrowii

7. Lonicera maackii

8. Lonicera oblongifolia

9. Lonicera xylosteum

10. Lonicera japonica

11. Lonicera periclymenum

LONICERA
509, 510

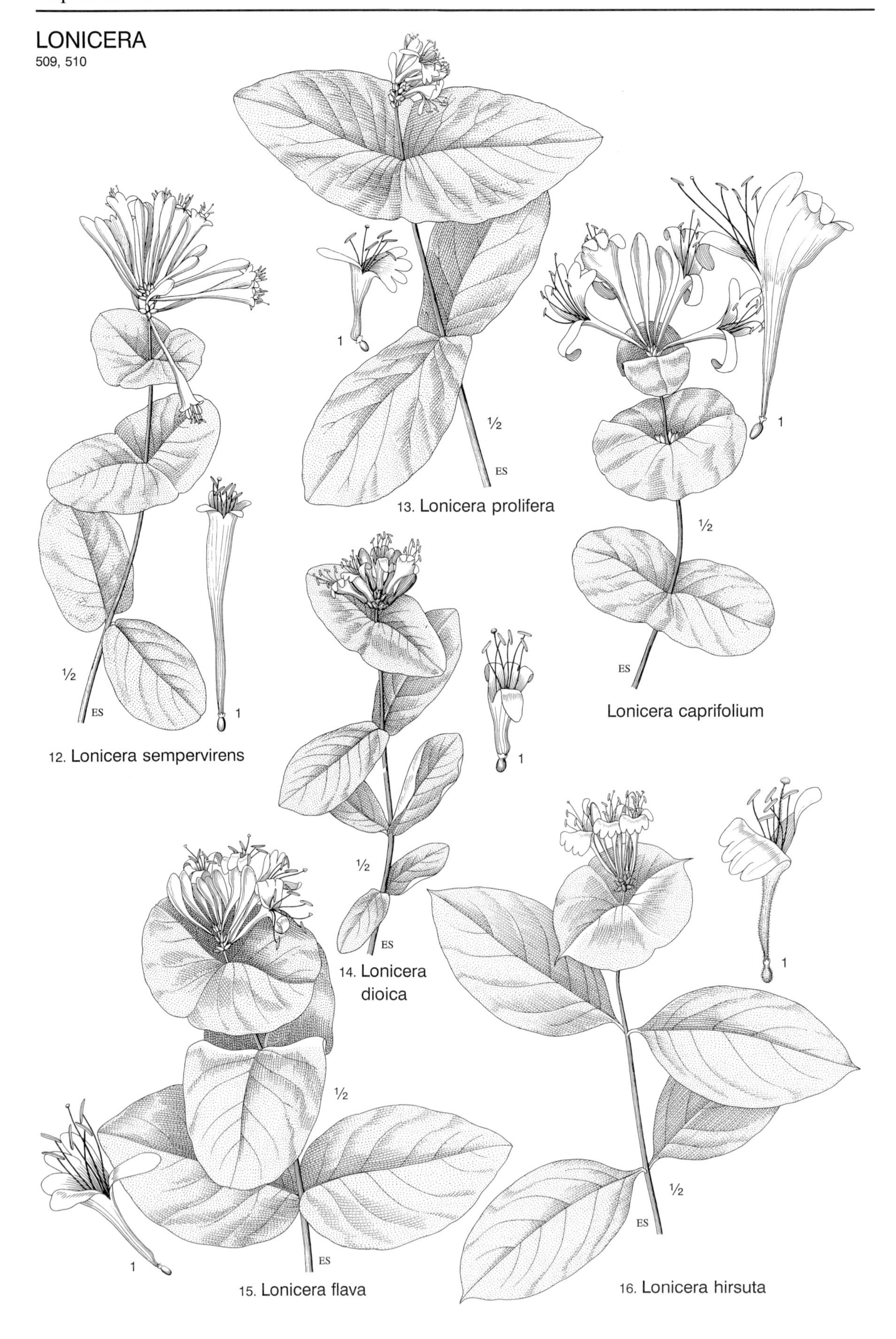

1

½

ES

13. Lonicera prolifera

½

½

ES

Lonicera caprifolium

12. Lonicera sempervirens

1

1

½

ES

14. Lonicera dioica

1

½

½

ES

15. Lonicera flava

½

ES

16. Lonicera hirsuta

1

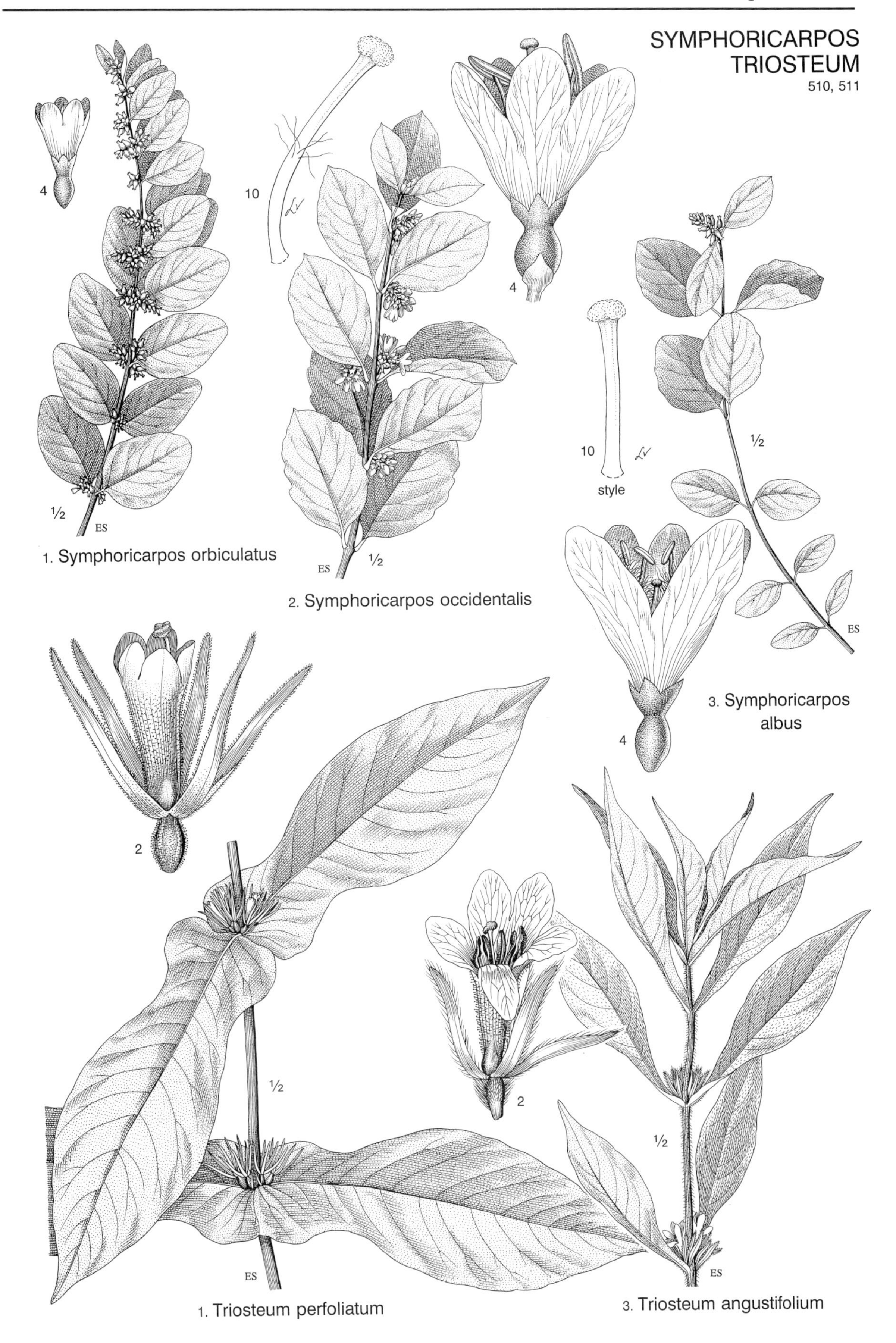

4

10

4

10

style

½

ES

1. Symphoricarpos orbiculatus

½

ES

2. Symphoricarpos occidentalis

½

ES

3. Symphoricarpos
albus

4

2

½

ES

1. Triosteum perfoliatum

2

½

ES

3. Triosteum angustifolium

TRIOSTEUM
LINNAEA
VIBURNUM
511, 512

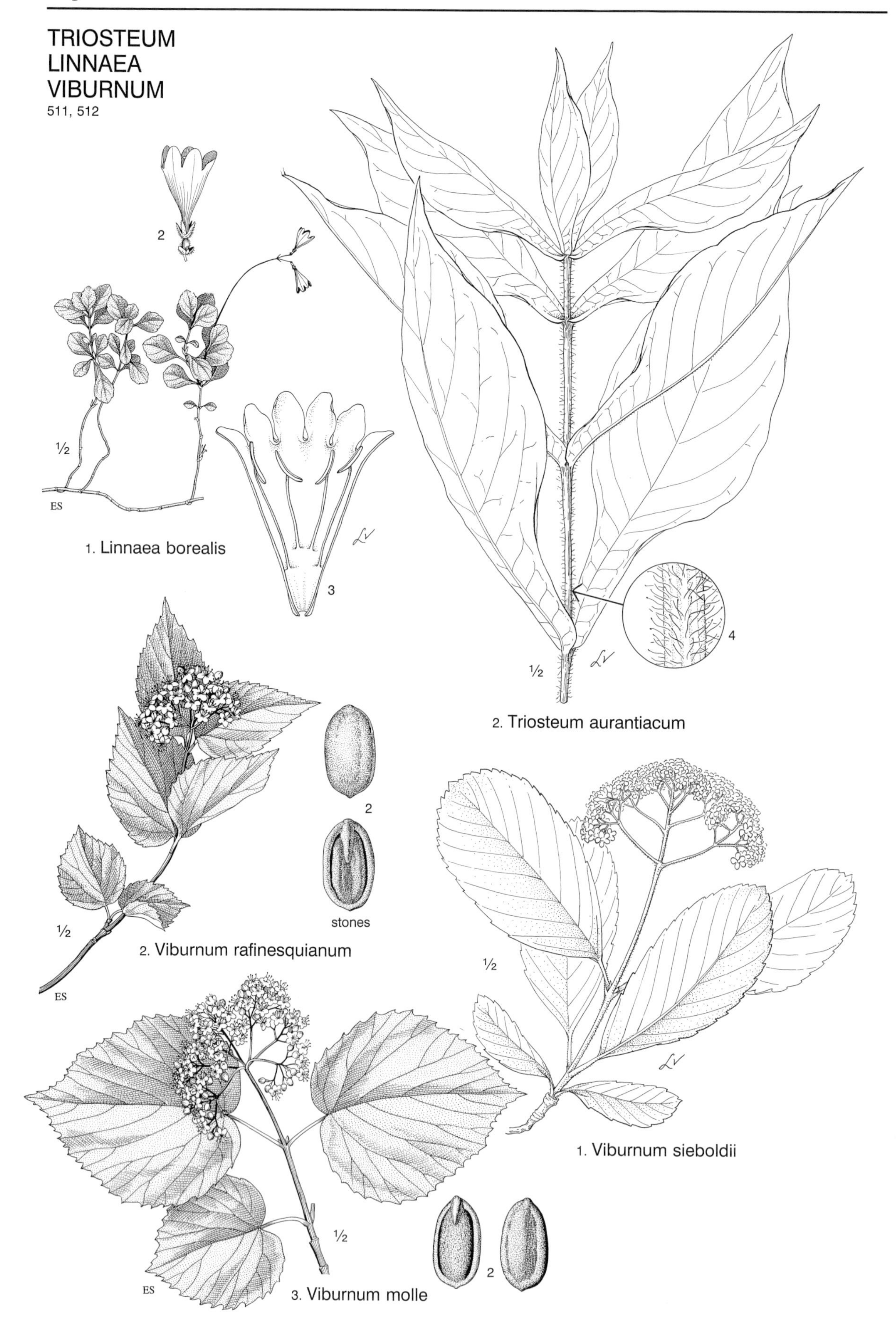

1. Linnaea borealis

2. Triosteum aurantiacum

2. Viburnum rafinesquianum

stones

1. Viburnum sieboldii

3. Viburnum molle

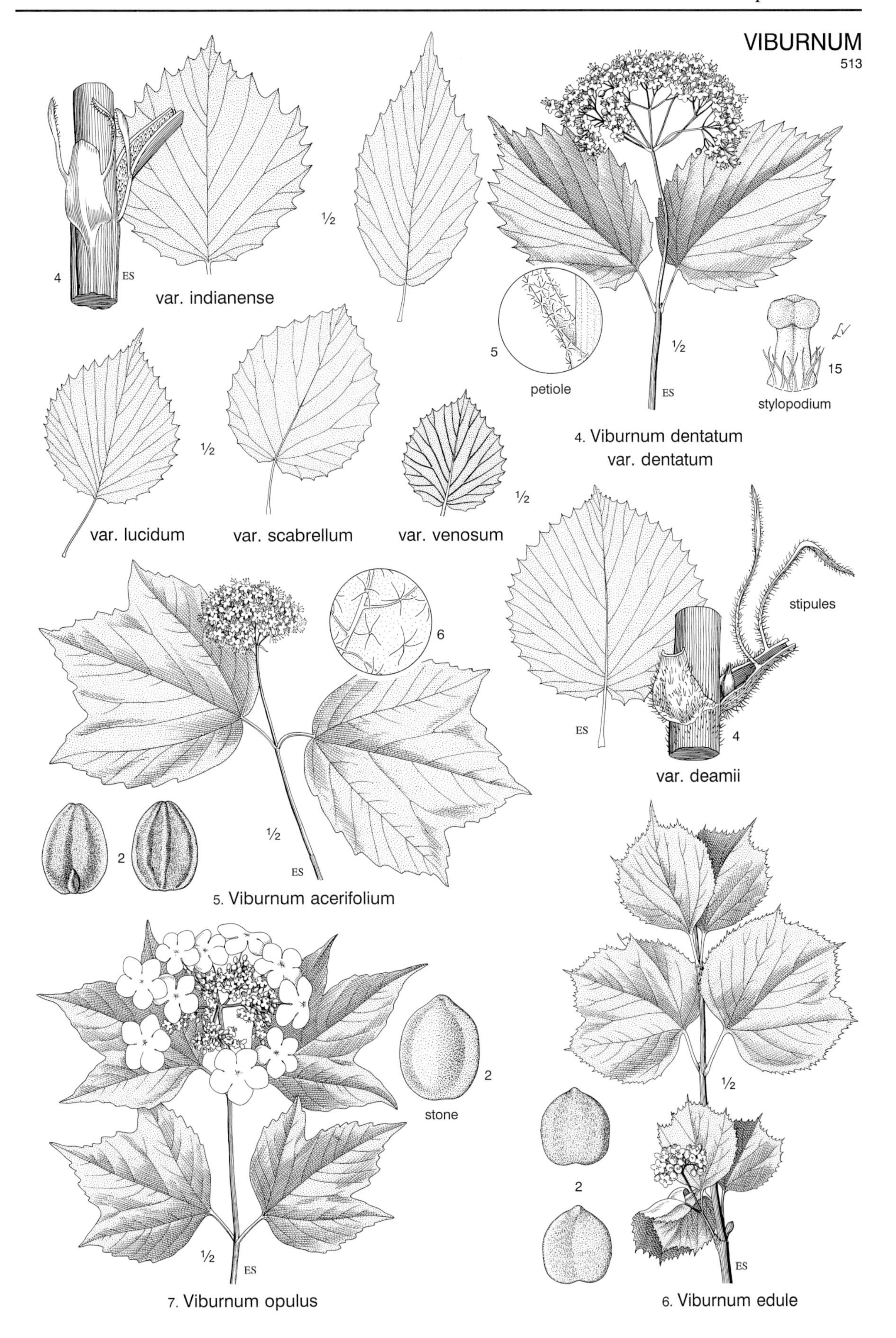

4

ES

var. indianense

½

5

petiole

½

ES

stylopodium

15

4. Viburnum dentatum
var. dentatum

var. lucidum   var. scabrellum   var. venosum

½   ½   ½

6

stipules

ES

4

var. deamii

2

½

ES

5. Viburnum acerifolium

2

stone

½

2

2

½

ES

7. Viburnum opulus

6. Viburnum edule

VIBURNUM
513, 514

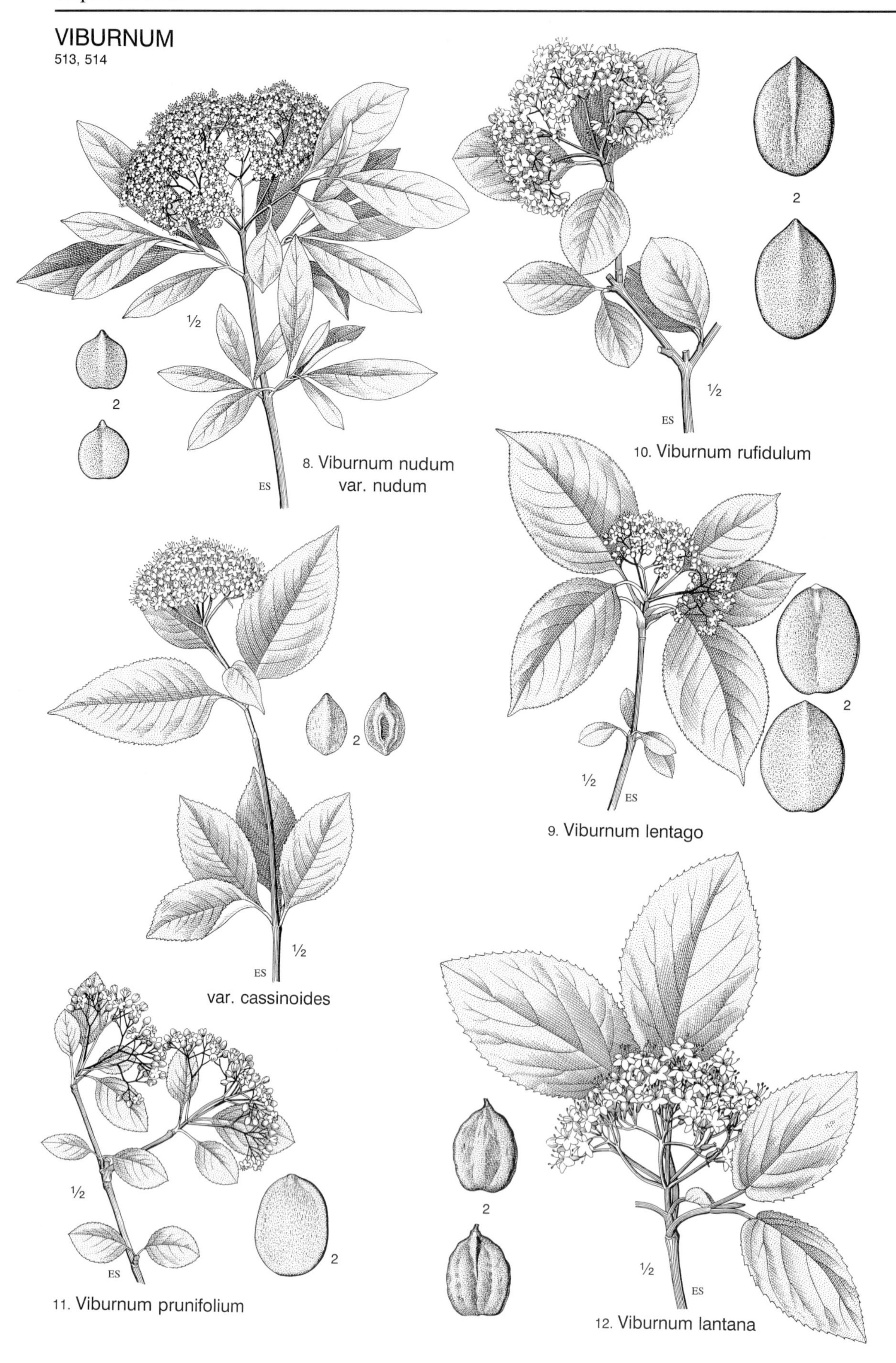

8. Viburnum nudum
var. nudum

10. Viburnum rufidulum

var. cassinoides

9. Viburnum lentago

11. Viburnum prunifolium

12. Viburnum lantana

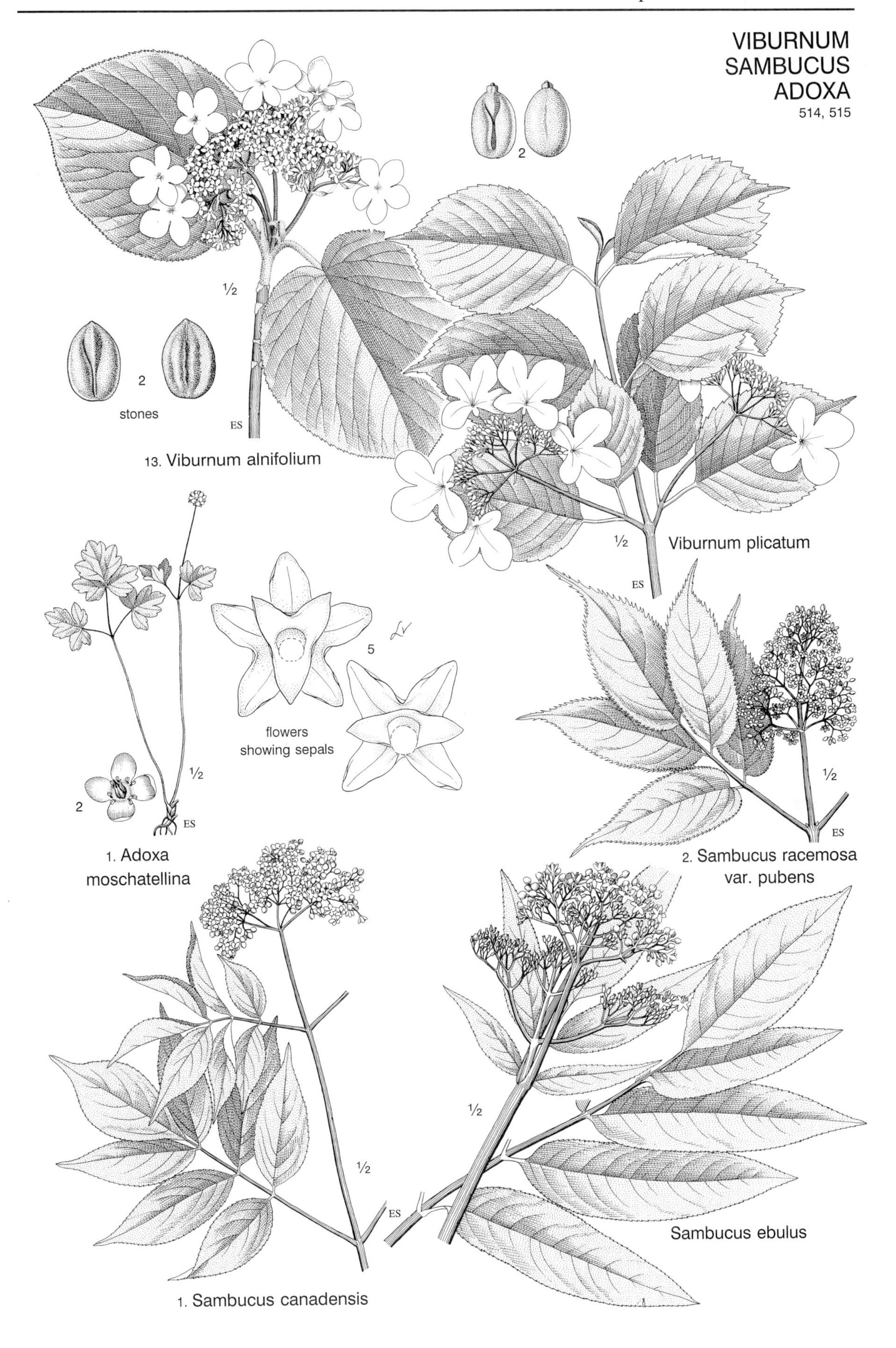

2

stones

13. Viburnum alnifolium

2

½

ES

½

Viburnum plicatum

ES

½

2

flowers
showing sepals

5

1. Adoxa
moschatellina

½

2

ES

2. Sambucus racemosa
var. pubens

½

ES

1. Sambucus canadensis

½

ES

½

Sambucus ebulus

VALERIANA
VALERIANELLA
515, 516

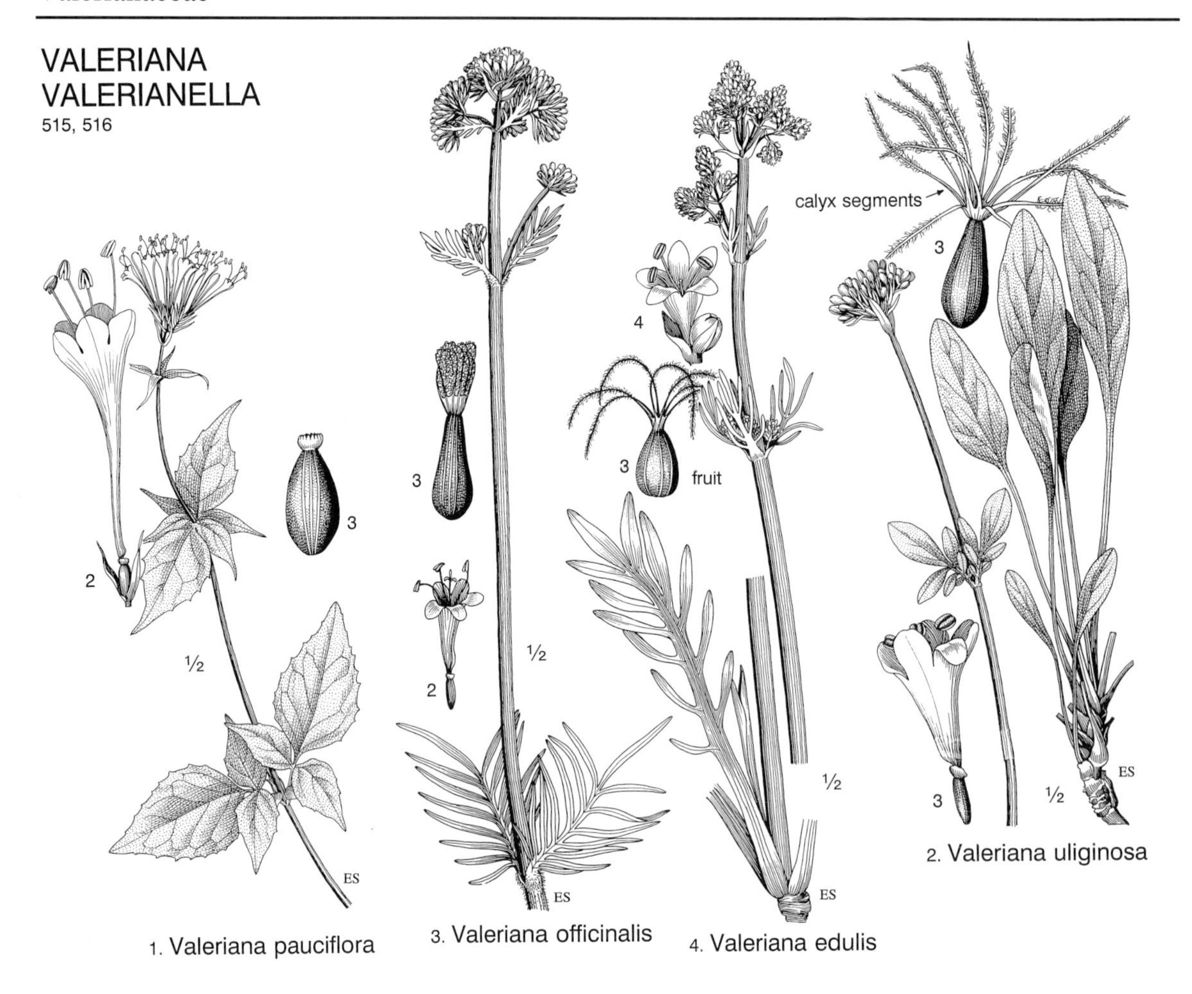

1. Valeriana pauciflora

3. Valeriana officinalis

4. Valeriana edulis

2. Valeriana uliginosa

calyx segments

fruit

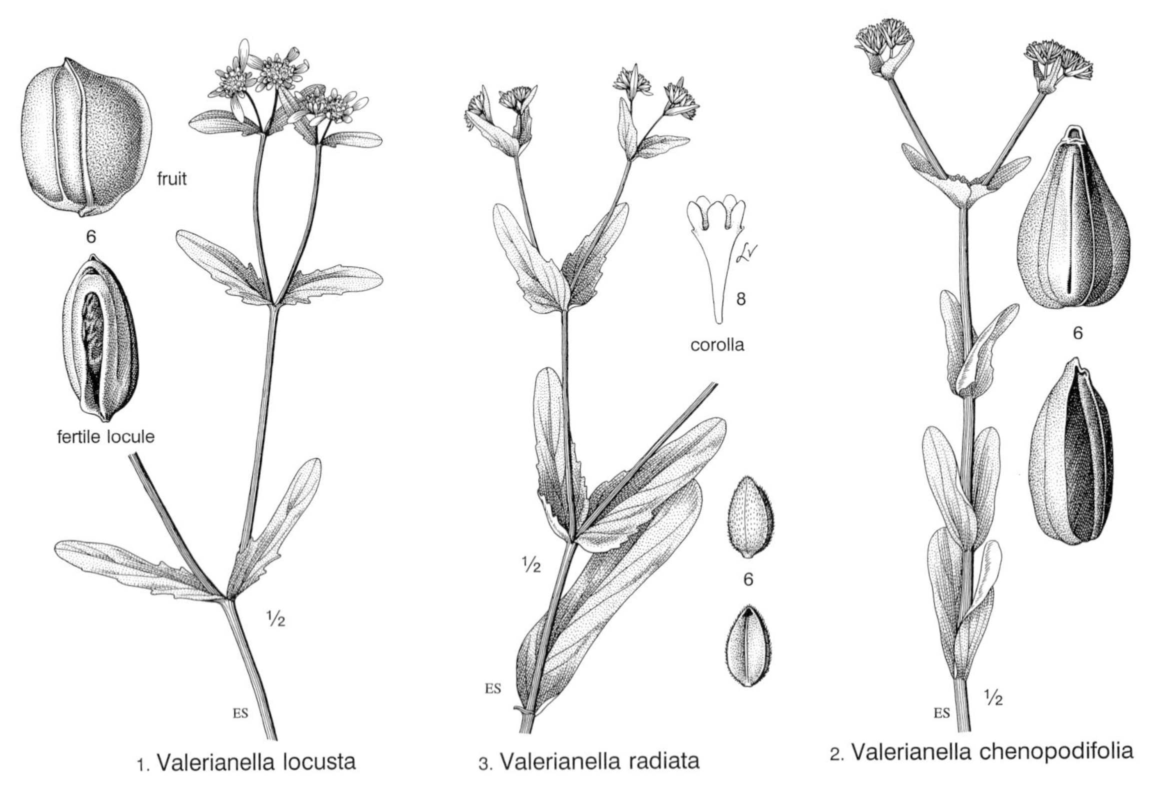

fruit

fertile locule

corolla

1. Valerianella locusta

3. Valerianella radiata

2. Valerianella chenopodifolia

ILLUSTRATED COMPANION TO

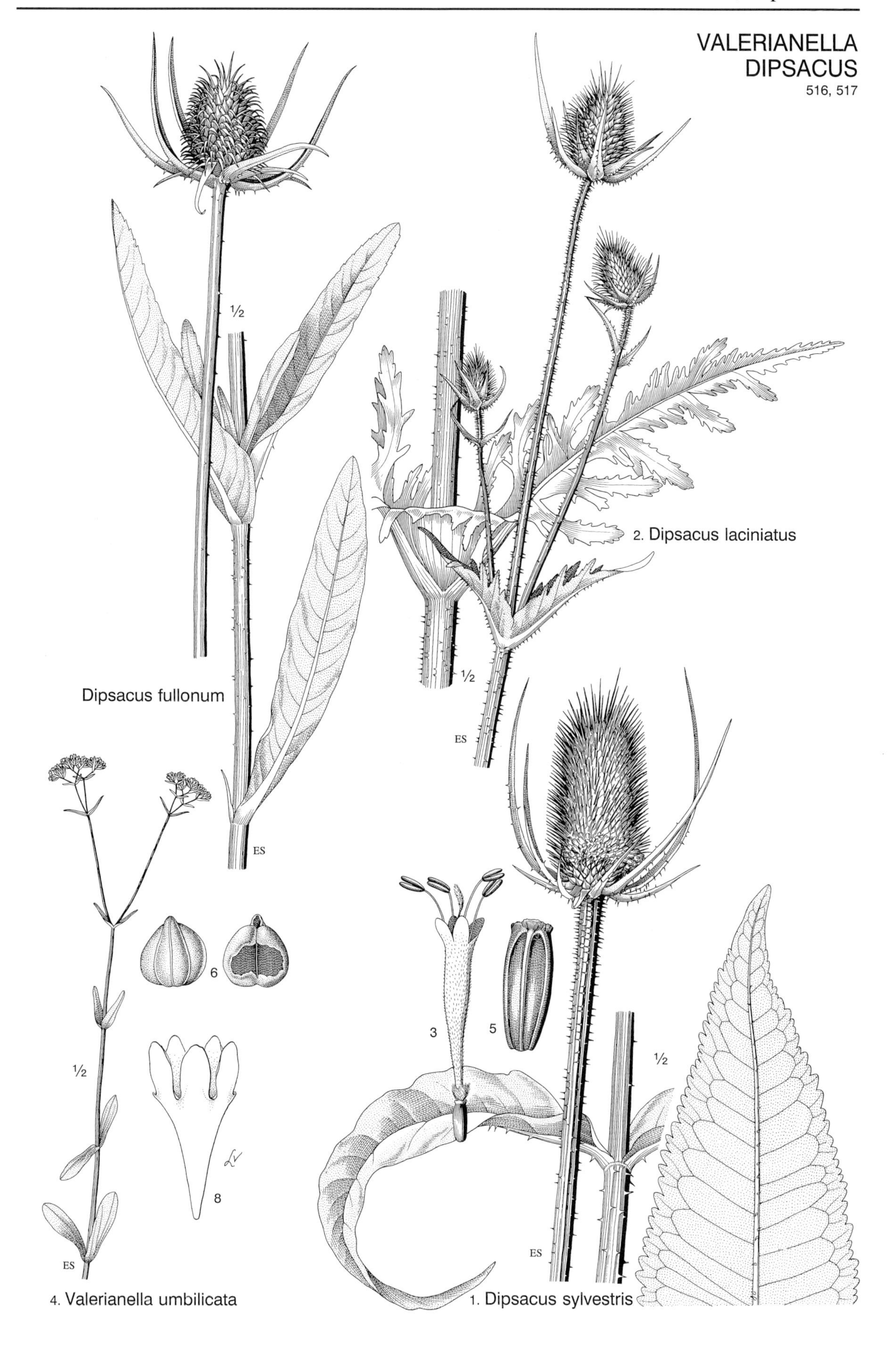

Dipsacus fullonum

2. Dipsacus laciniatus

6

3   5

8

4. Valerianella umbilicata

1. Dipsacus sylvestris

SUCCISELLA
SUCCISA
SCABIOSA
KNAUTIA
517, 518

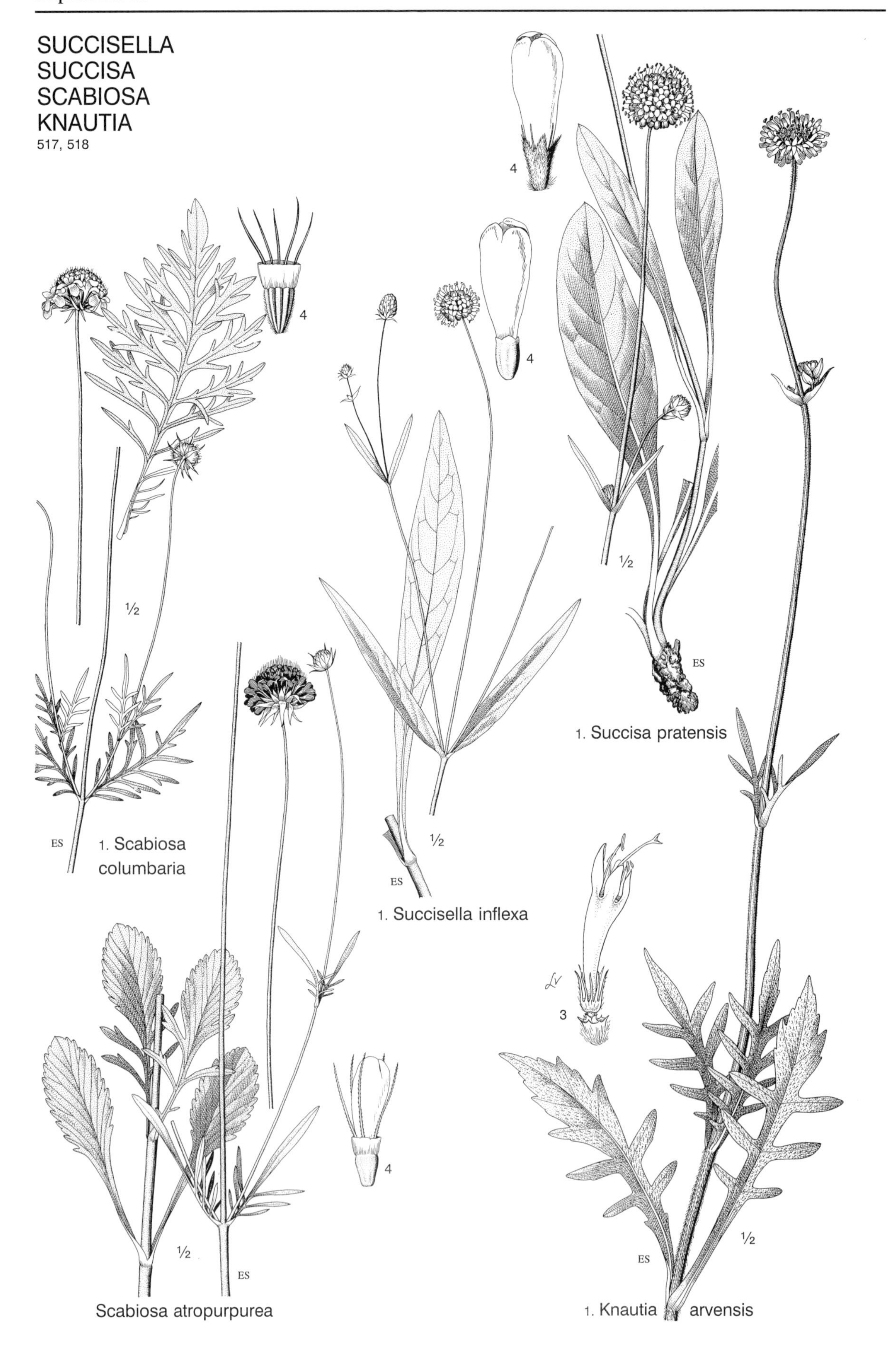

4

4

4

½

1. Succisa pratensis

ES

½

½

ES

1. Scabiosa
columbaria

ES

1. Succisella inflexa

3

4

½

ES

Scabiosa atropurpurea

½

ES

1. Knautia   arvensis

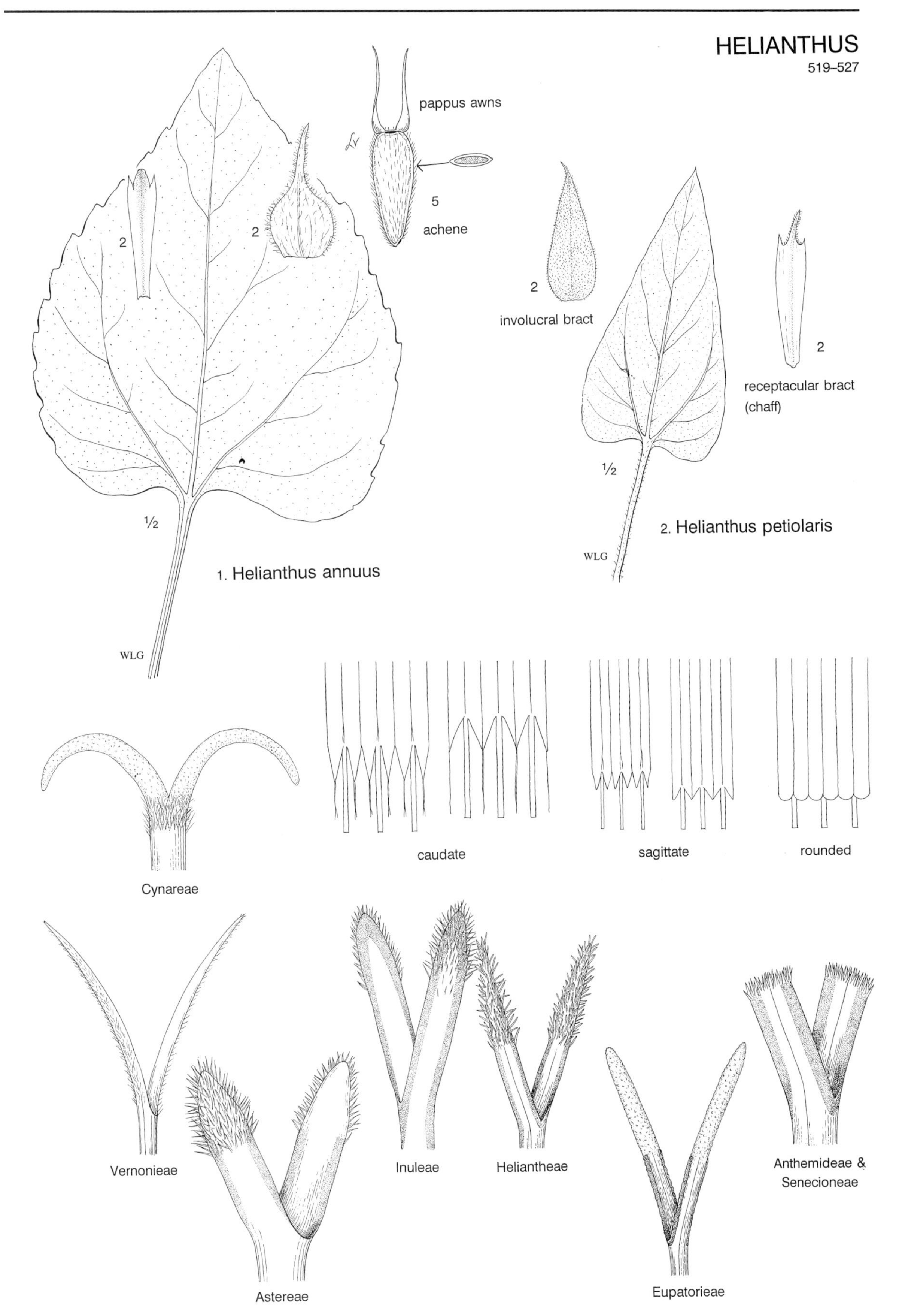

pappus awns

5

achene

2
involucral bract

2
receptacular bract
(chaff)

½

1. Helianthus annuus

½

WLG

2. Helianthus petiolaris

WLG

Cynareae

caudate

sagittate

rounded

Vernonieae

Astereae

Inuleae

Heliantheae

Eupatorieae

Anthemideae &
Senecioneae

Characteristic style branches and anther bases

HELIANTHUS
527

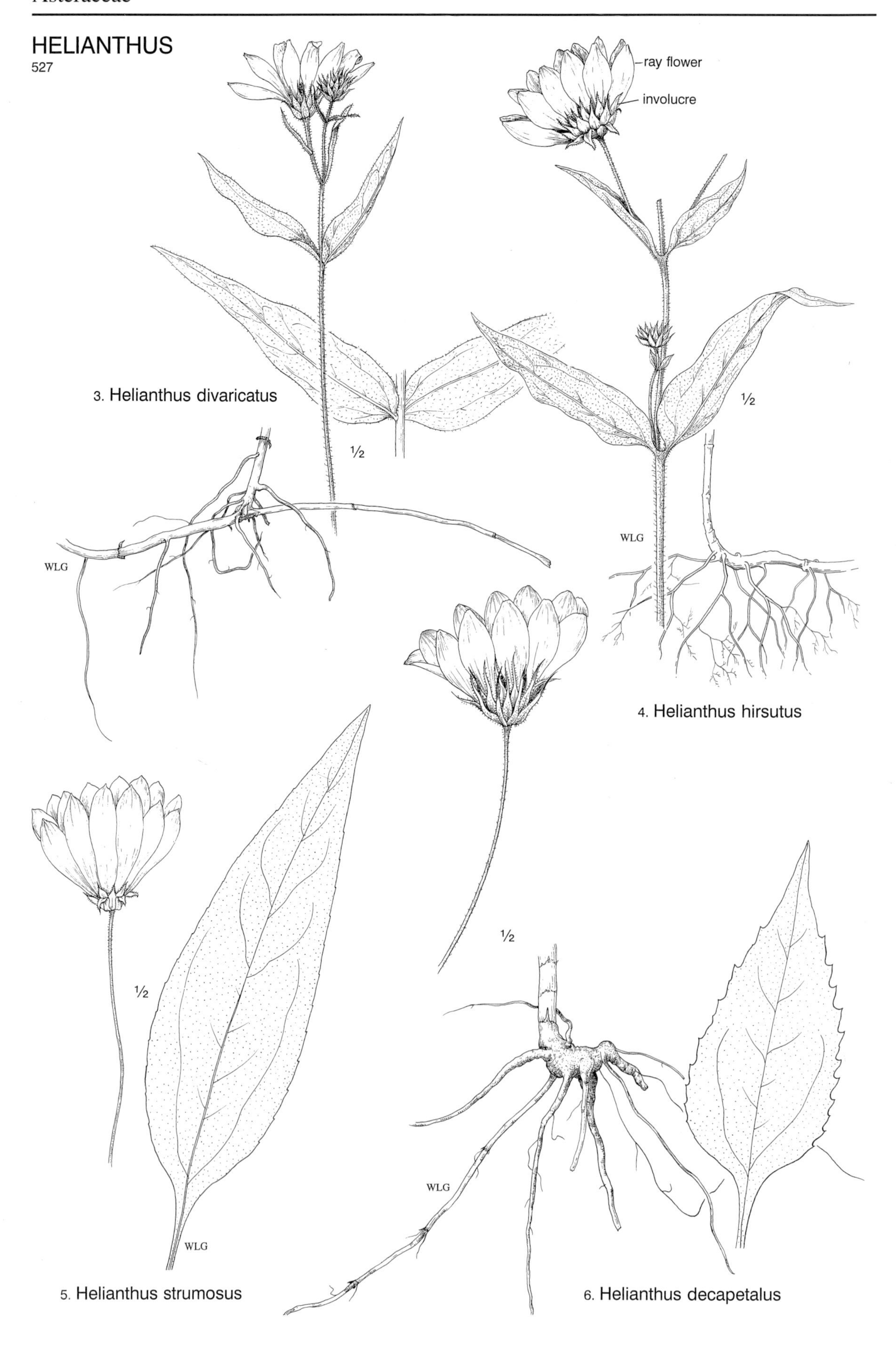

ray flower

involucre

3. Helianthus divaricatus

½

½

WLG

WLG

4. Helianthus hirsutus

½

½

WLG

WLG

5. Helianthus strumosus

6. Helianthus decapetalus

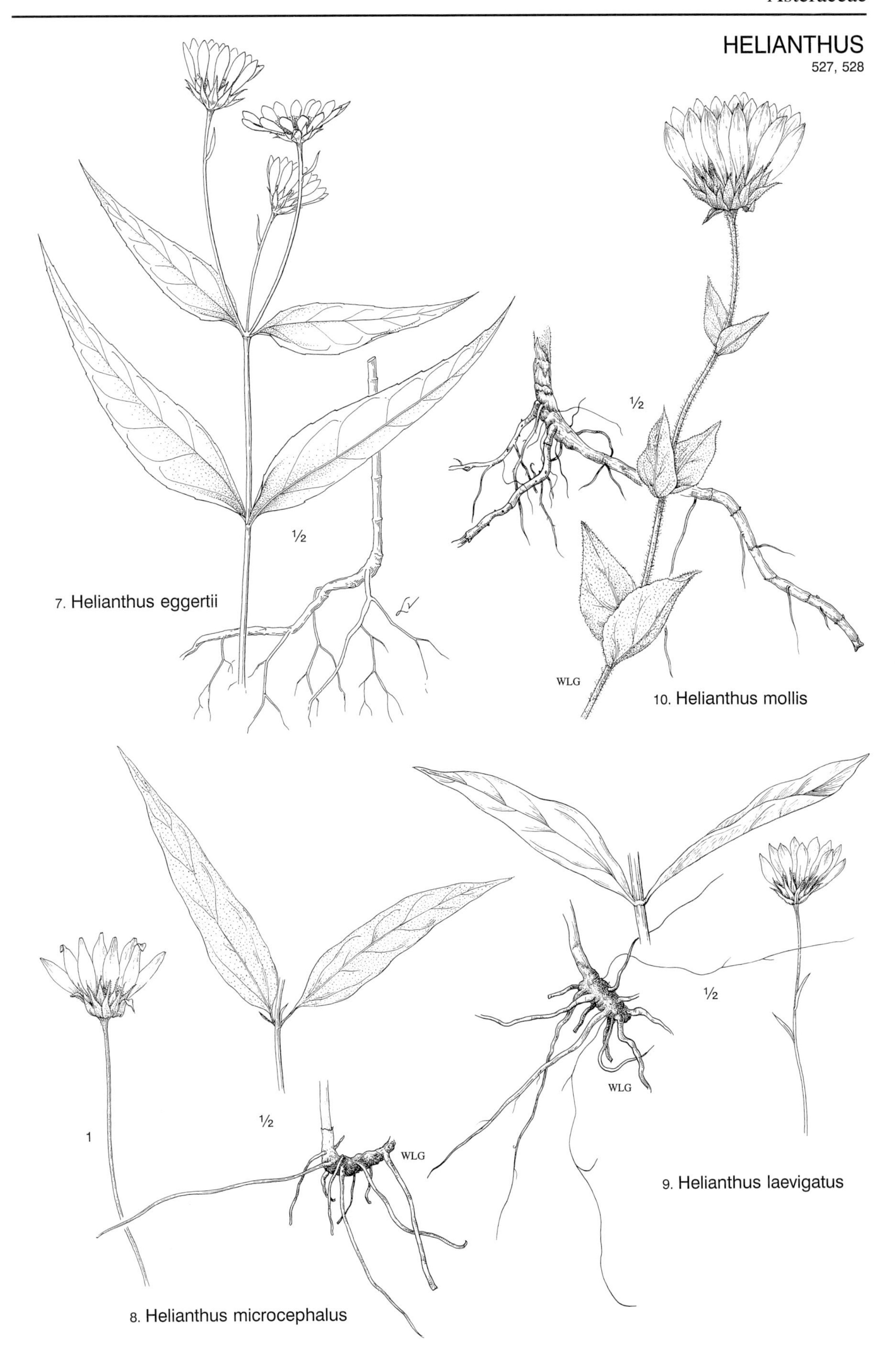

7. Helianthus eggertii

½

10. Helianthus mollis

WLG

½

8. Helianthus microcephalus

1

½

WLG

9. Helianthus laevigatus

½

WLG

HELIANTHUS
528

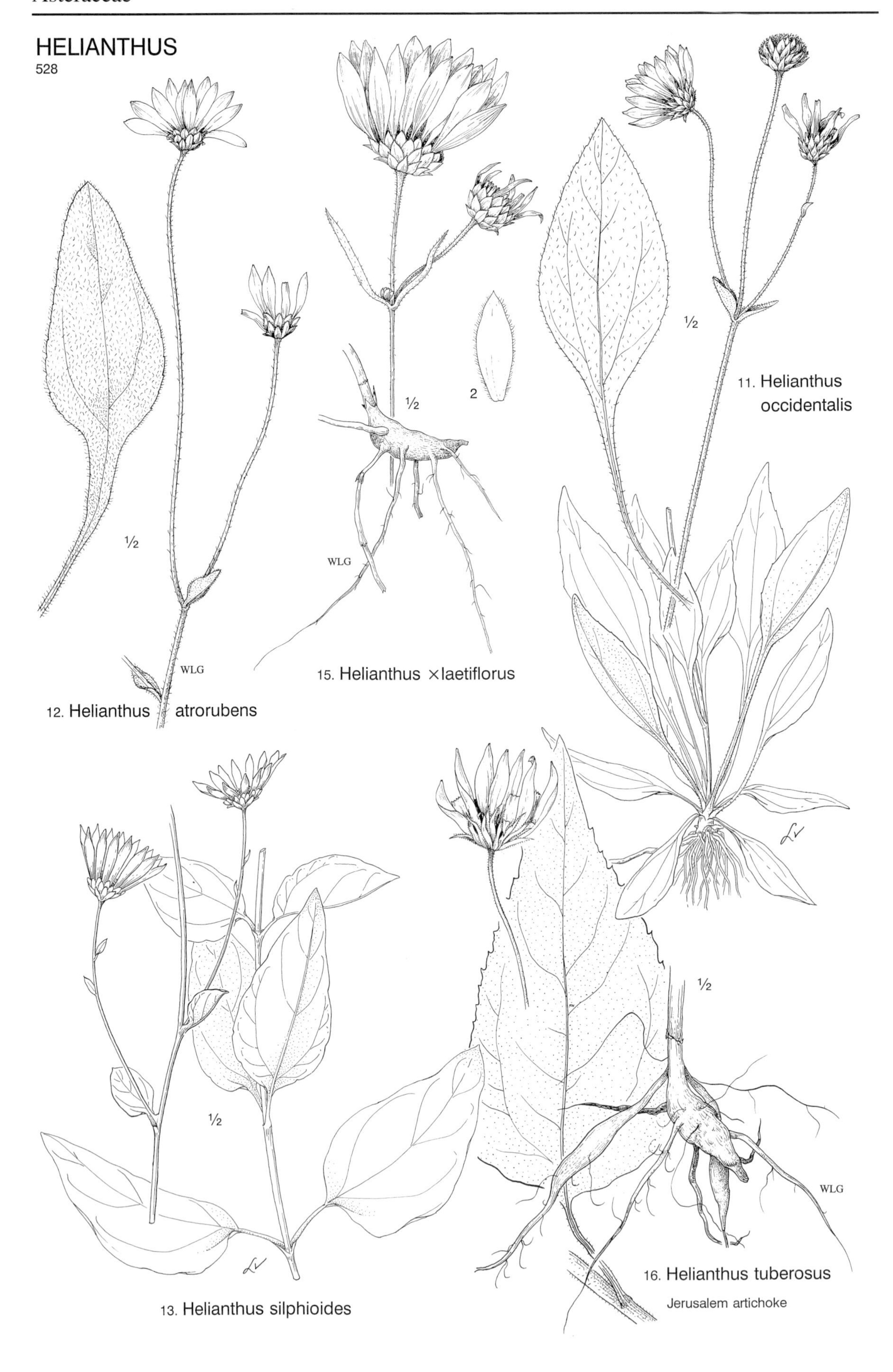

2

½

11. Helianthus occidentalis

WLG

½

12. Helianthus atrorubens

15. Helianthus ×laetiflorus

WLG

½

½

13. Helianthus silphioides

16. Helianthus tuberosus

Jerusalem artichoke

WLG

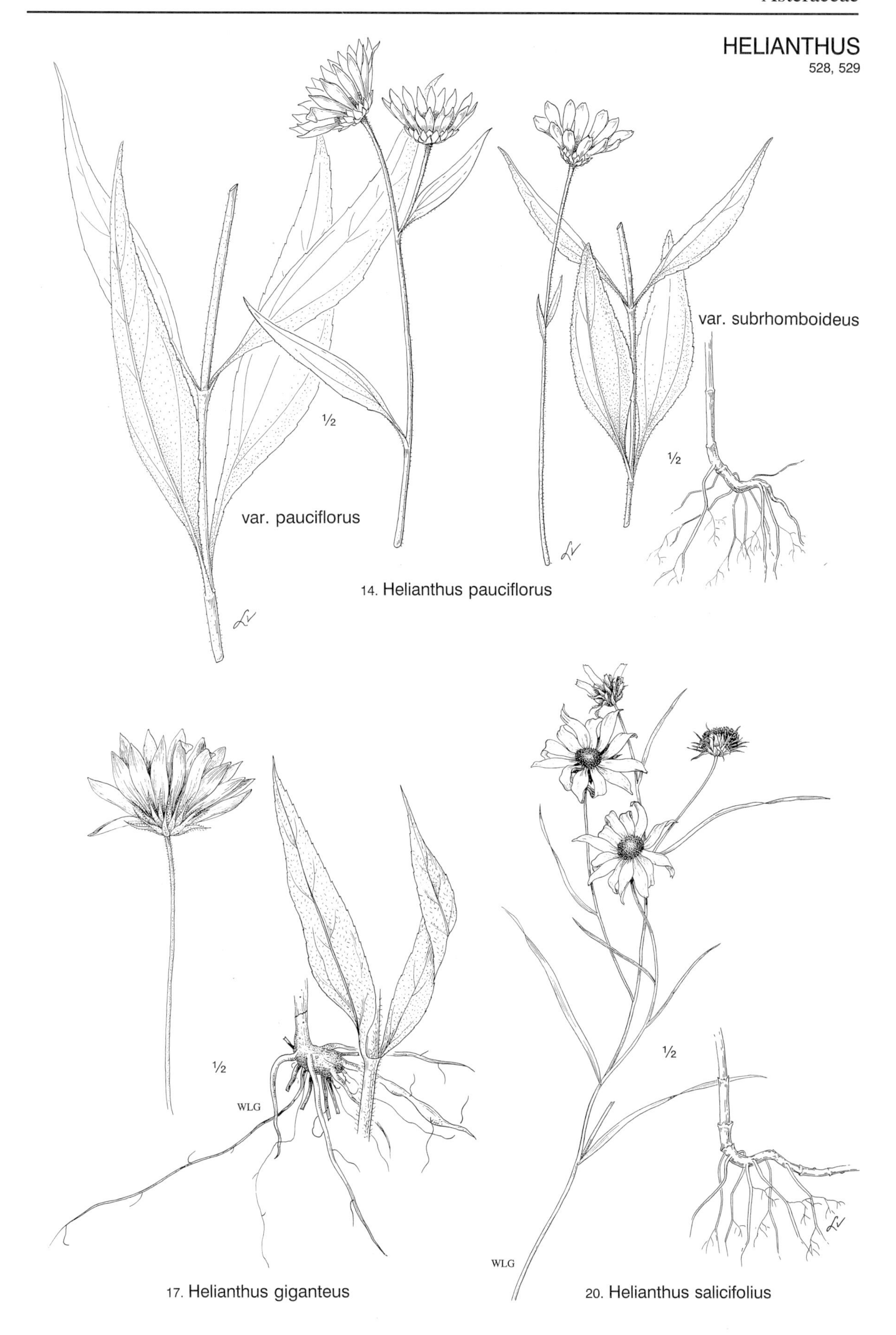

var. subrhomboideus

var. pauciflorus

14. Helianthus pauciflorus

17. Helianthus giganteus

20. Helianthus salicifolius

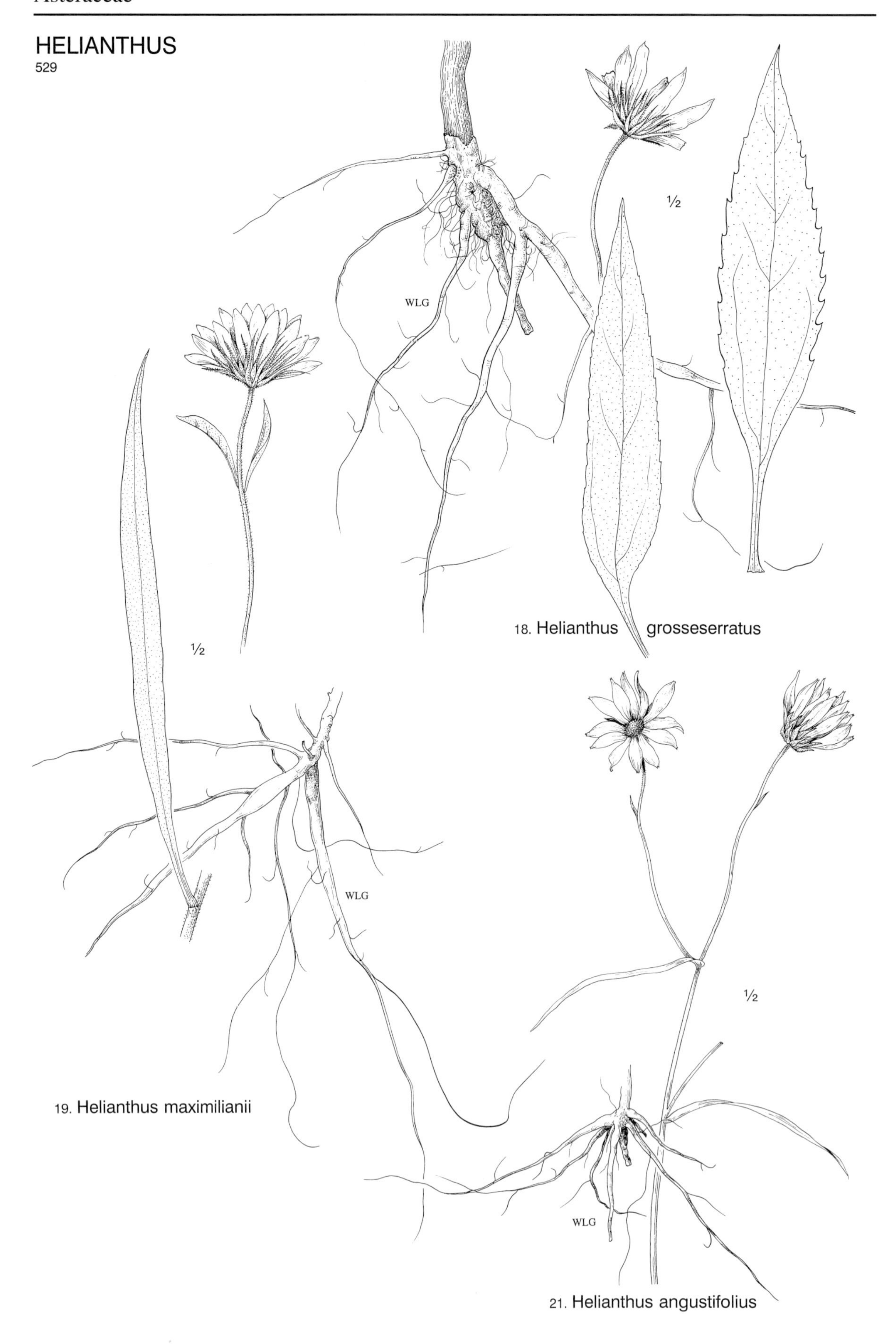

18. Helianthus grosseserratus

19. Helianthus maximilianii

21. Helianthus angustifolius

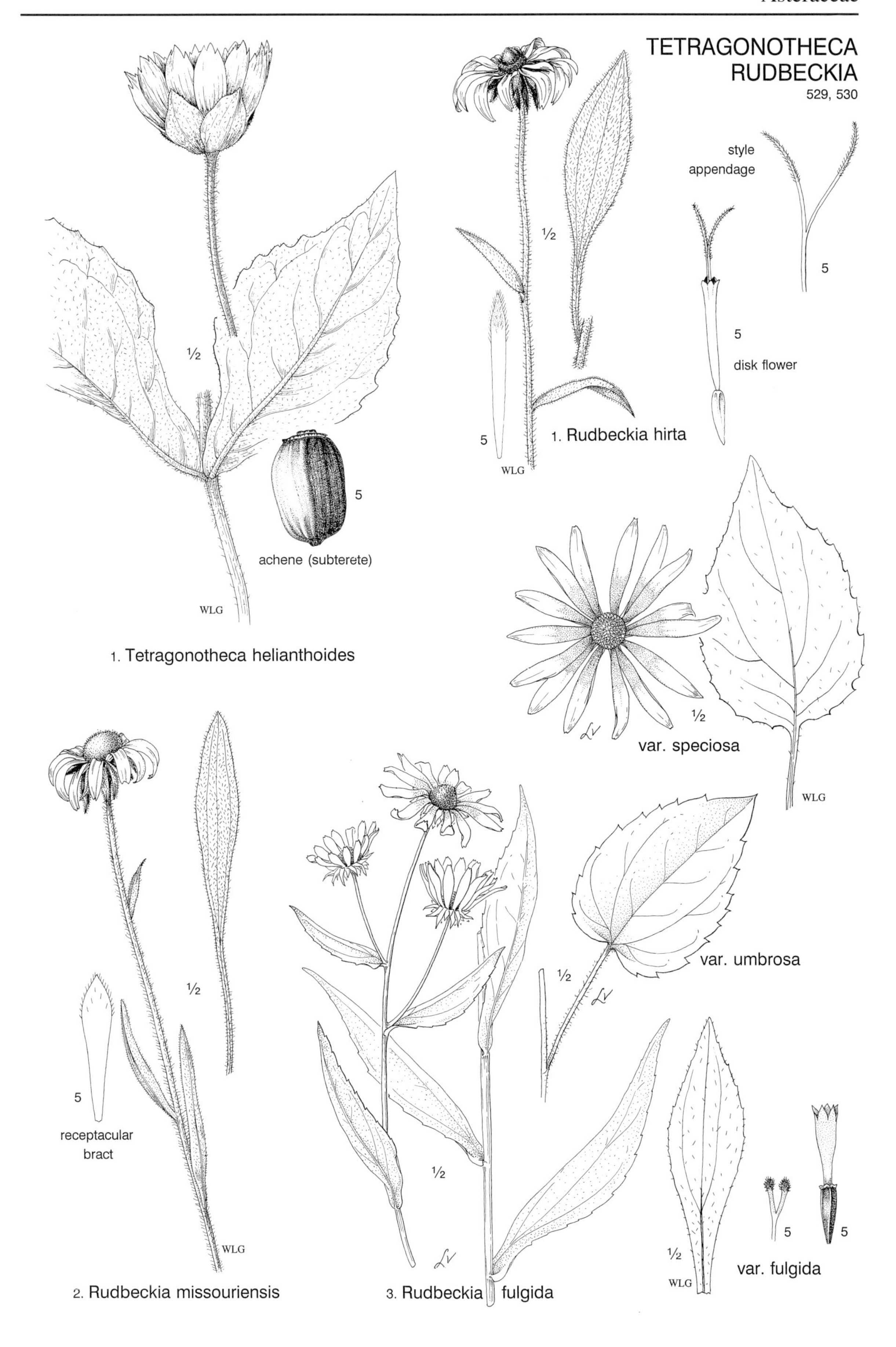

style
appendage

5

5

disk flower

1. Rudbeckia hirta

WLG

5

achene (subterete)

5

1. Tetragonotheca helianthoides

var. speciosa

½

WLG

var. umbrosa

5

receptacular
bract

½

½

½

WLG

½

WLG

5   5

var. fulgida

2. Rudbeckia missouriensis

3. Rudbeckia fulgida

RUDBECKIA
530, 531

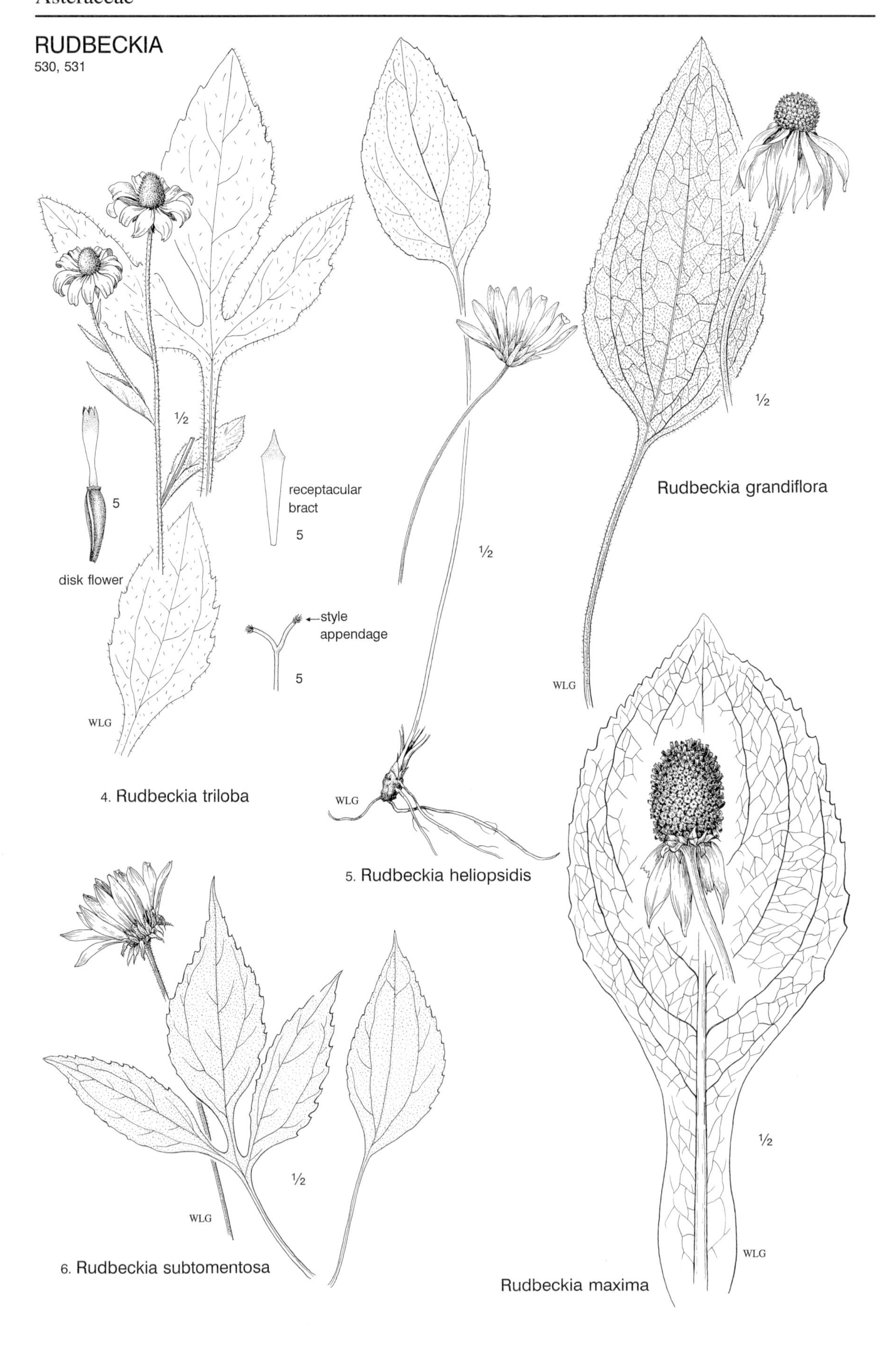

disk flower

5

receptacular
bract

5

←style
appendage

5

½

WLG

4. Rudbeckia triloba

½

WLG

5. Rudbeckia heliopsidis

½

Rudbeckia grandiflora

WLG

½

WLG

6. Rudbeckia subtomentosa

½

WLG

Rudbeckia maxima

ILLUSTRATED COMPANION TO

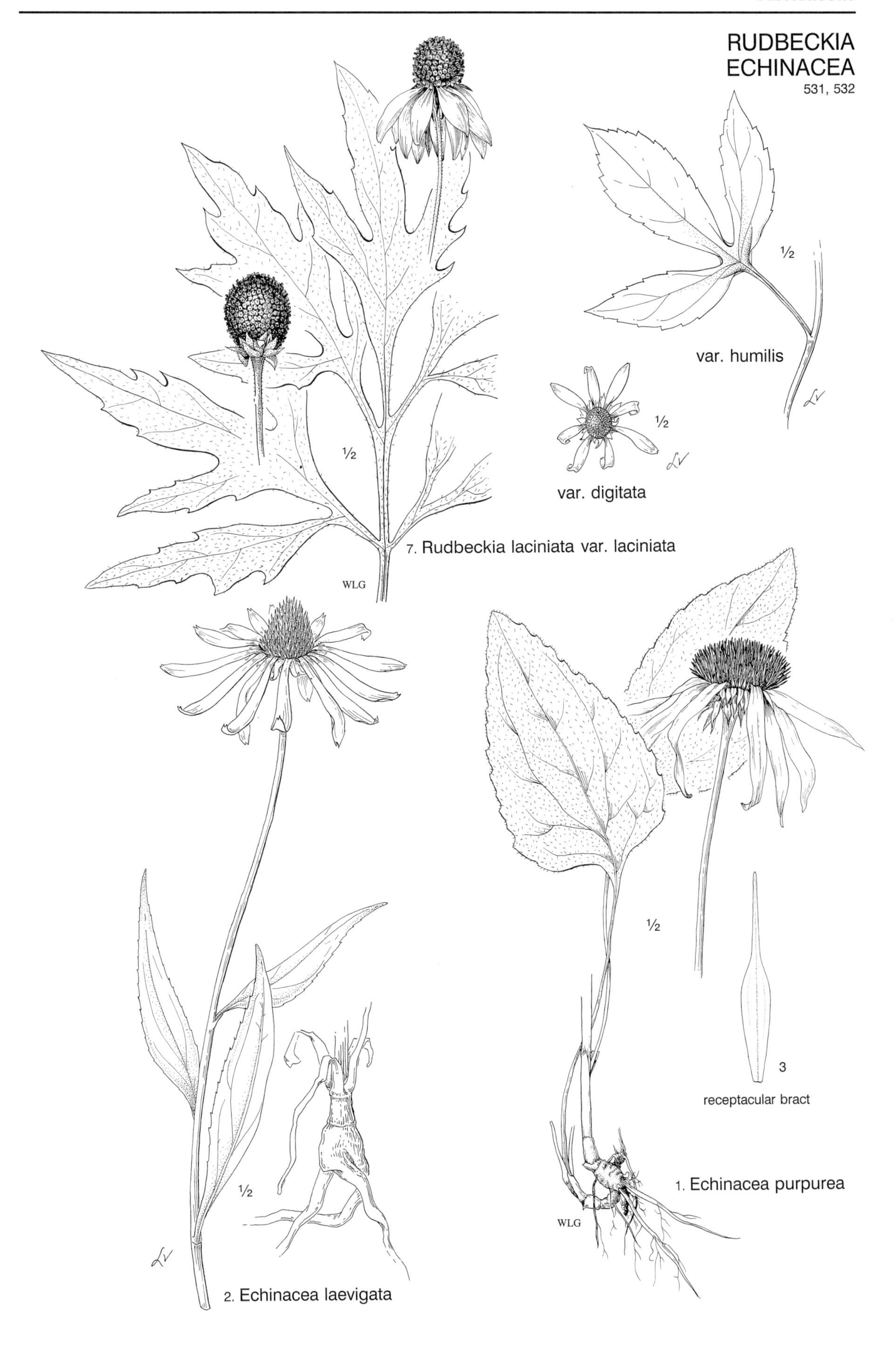

½

var. humilis

½

var. digitata

½

7. Rudbeckia laciniata var. laciniata

WLG

½

3

receptacular bract

½

1. Echinacea purpurea

WLG

½

2. Echinacea laevigata

ECHINACEA
RATIBIDA
GAILLARDIA
532, 533

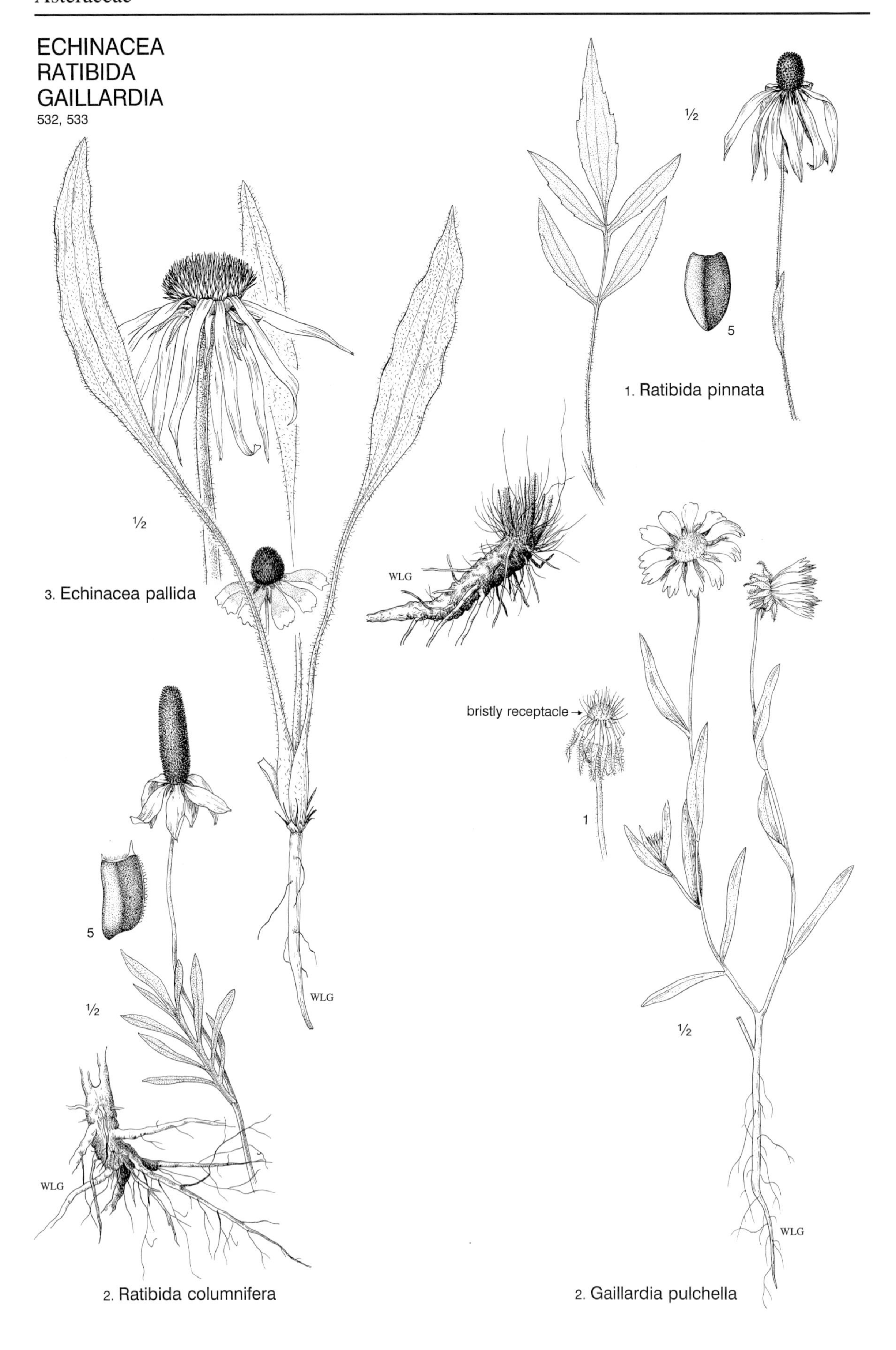

½

1. Ratibida pinnata

5

½

3. Echinacea pallida

WLG

bristly receptacle →

1

5

½

WLG

2. Ratibida columnifera

½

WLG

2. Gaillardia pulchella

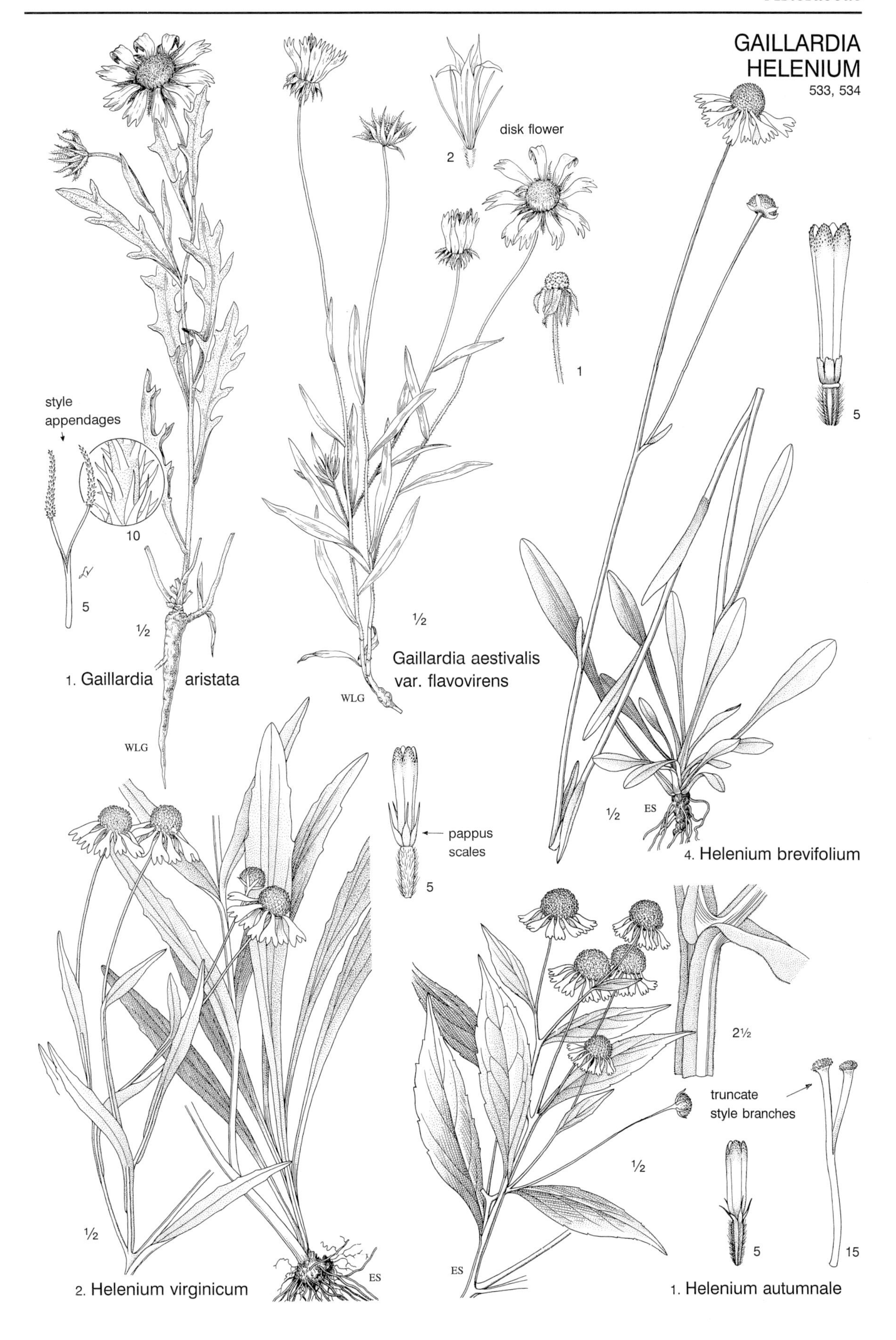

disk flower

2

1

5

style
appendages

10

5

½

1. Gaillardia aristata

WLG

½

Gaillardia aestivalis
var. flavovirens

WLG

pappus
scales

5

4. Helenium brevifolium

½ ES

2½

truncate
style branches

2. Helenium virginicum

½

ES

ES

½

5

15

1. Helenium autumnale

HELENIUM
HYMENOXYS
ARNICA
533–535

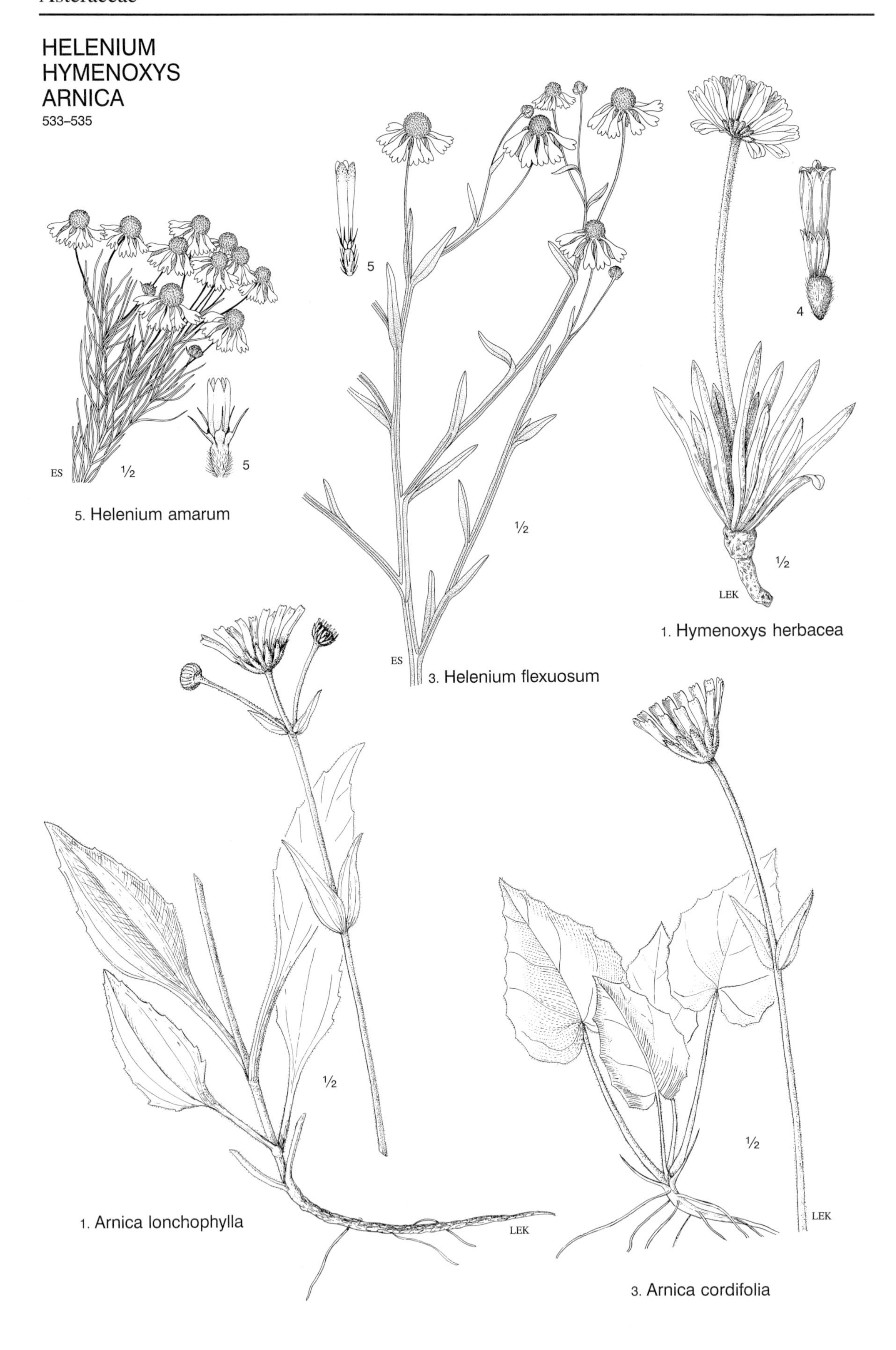

5. Helenium amarum

3. Helenium flexuosum

1. Hymenoxys herbacea

1. Arnica lonchophylla

3. Arnica cordifolia

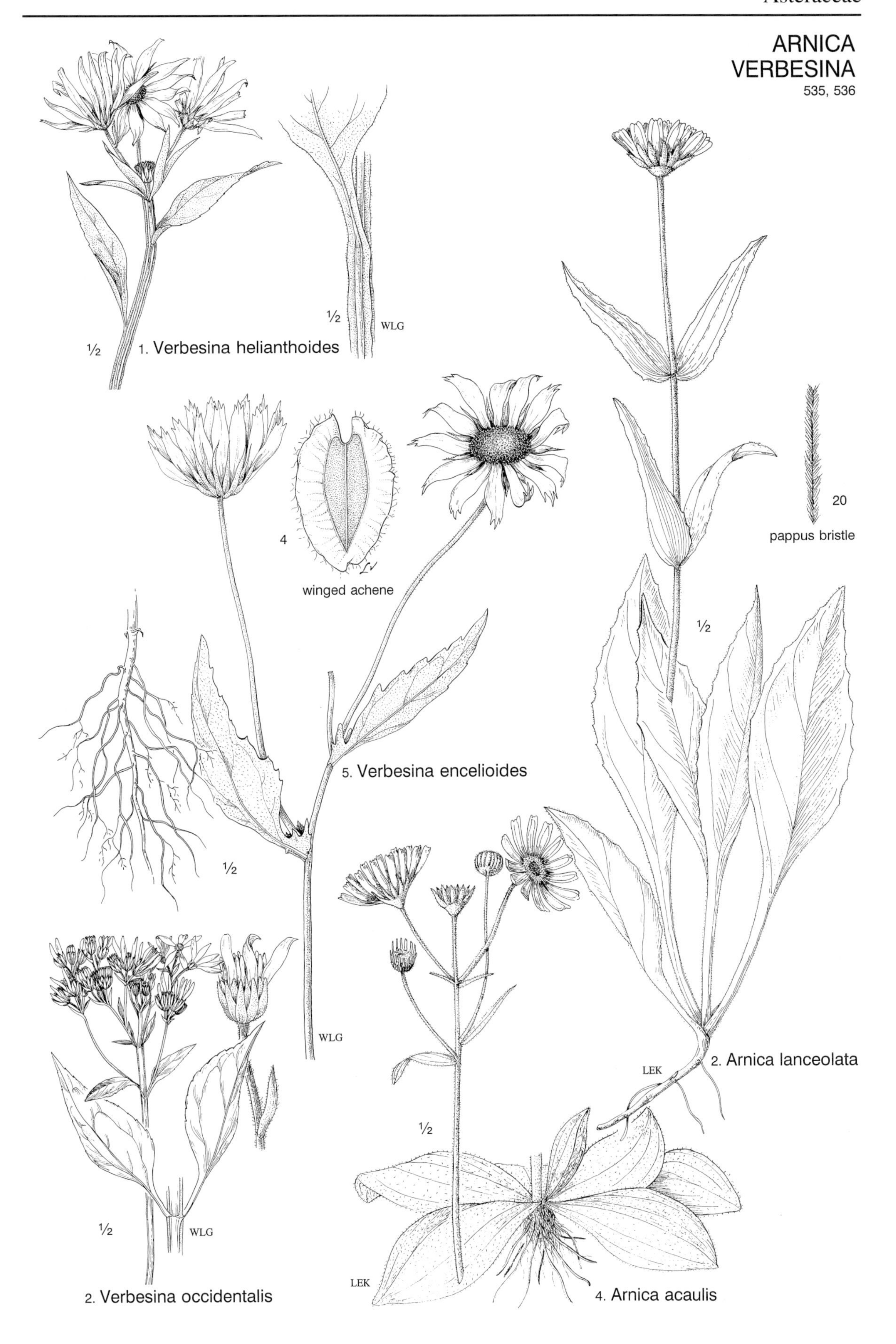

½   1. Verbesina helianthoides

½   WLG

4

winged achene

5. Verbesina encelioides

20

pappus bristle

½

½

2. Verbesina occidentalis

WLG

½   WLG

½

½   WLG

LEK

2. Arnica lanceolata

LEK   4. Arnica acaulis

VERBESINA
ECLIPTA
ZINNIA
BORRICHIA
SPILANTHES
535–537

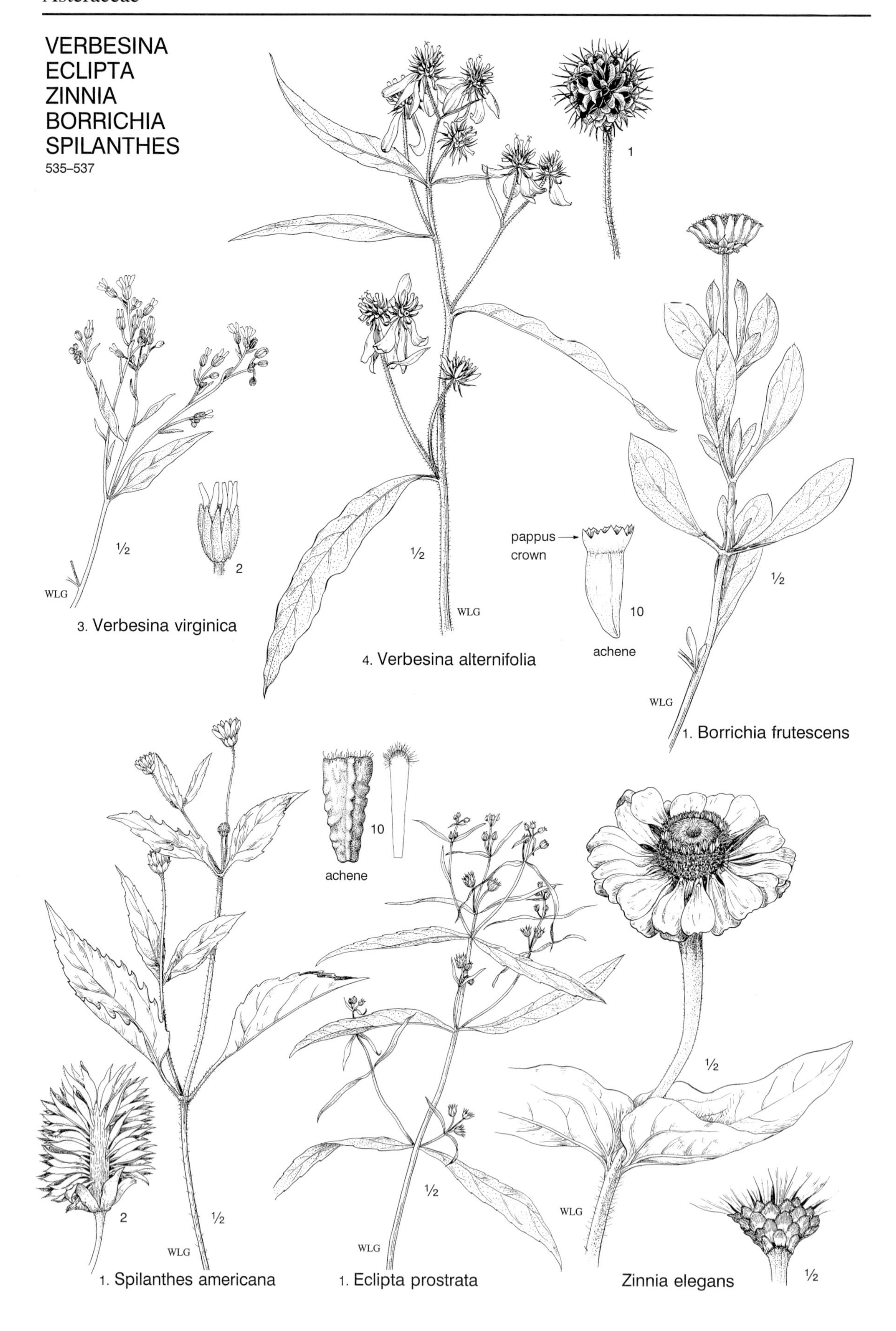

½

2

WLG

3. Verbesina virginica

½

WLG

4. Verbesina alternifolia

pappus → crown

10

achene

½

WLG

1. Borrichia frutescens

10

achene

½

2

½

WLG

1. Spilanthes americana

½

WLG

1. Eclipta prostrata

½

WLG

Zinnia elegans

½

ILLUSTRATED COMPANION TO

HELIOPSIS
GUIZOTIA
SIGESBECKIA
COREOPSIS
536–538

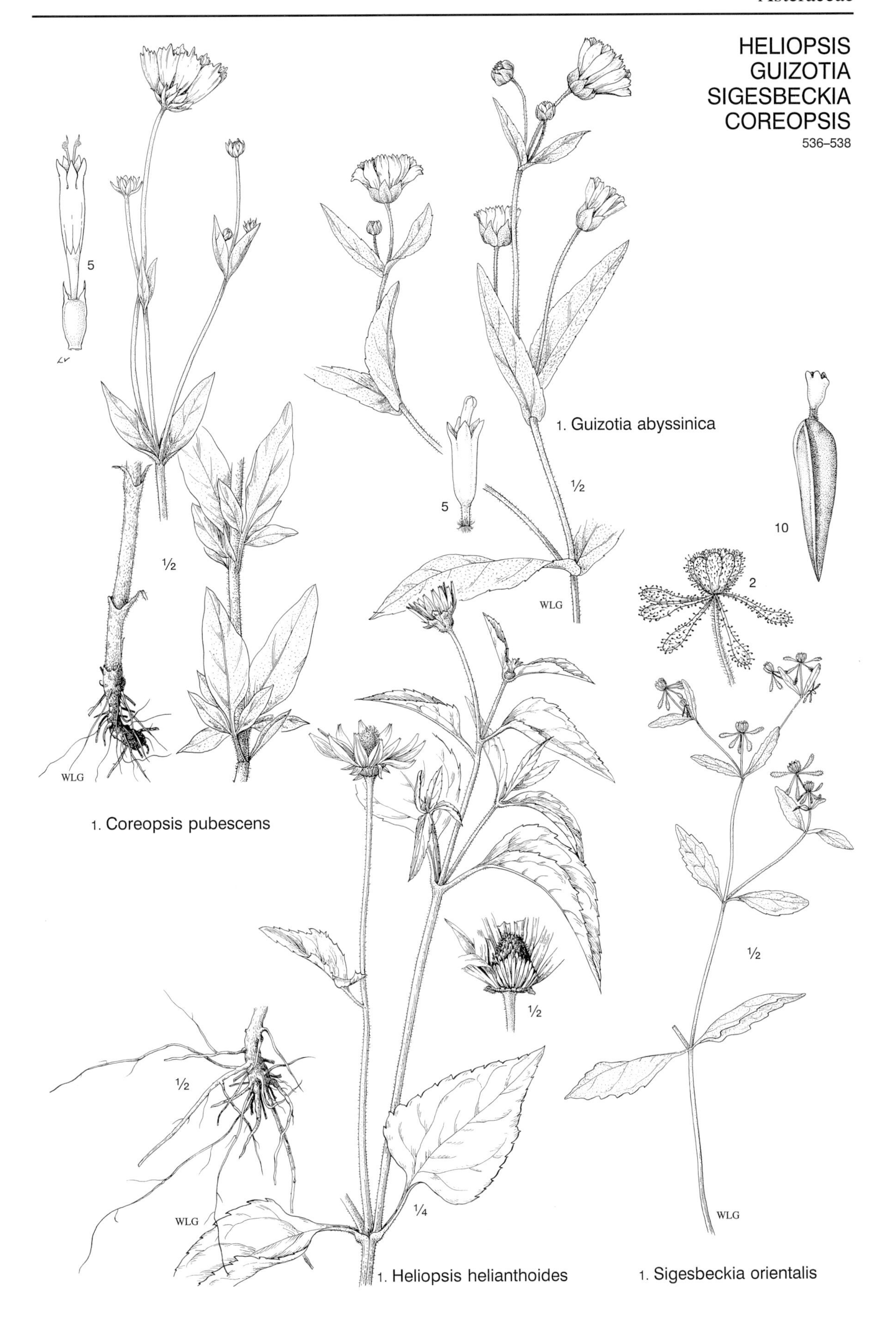

5

WLG

1. Guizotia abyssinica

½

5

10

2

½

WLG

1. Coreopsis pubescens

½

WLG

½

½

¼

WLG

½

WLG

1. Heliopsis helianthoides

1. Sigesbeckia orientalis

COREOPSIS
538

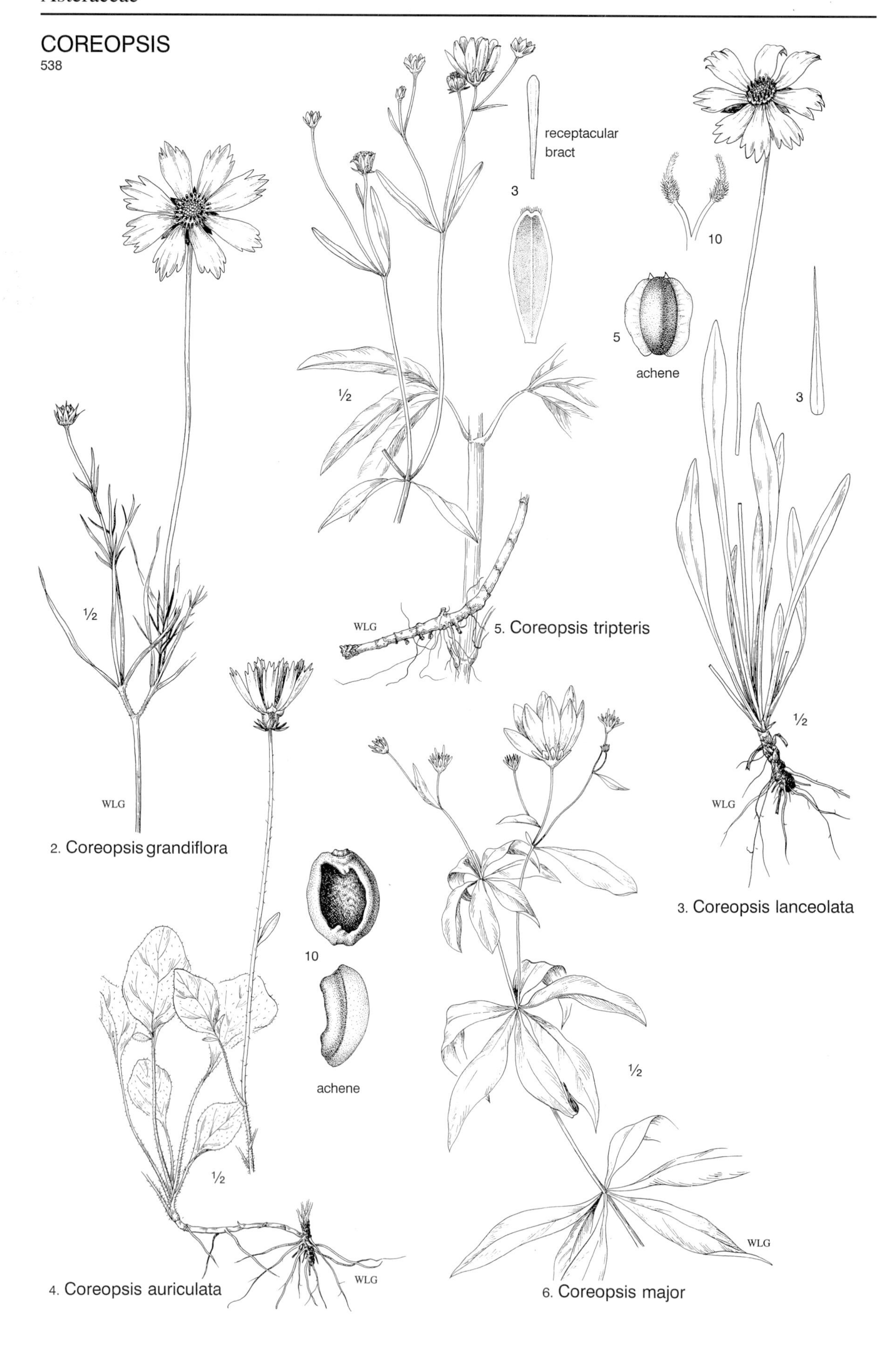

receptacular
bract

3

5 achene

10

3

1/2

WLG

5. Coreopsis tripteris

1/2

WLG

3. Coreopsis lanceolata

1/2

WLG

2. Coreopsis grandiflora

10

achene

1/2

4. Coreopsis auriculata

WLG

1/2

WLG

6. Coreopsis major

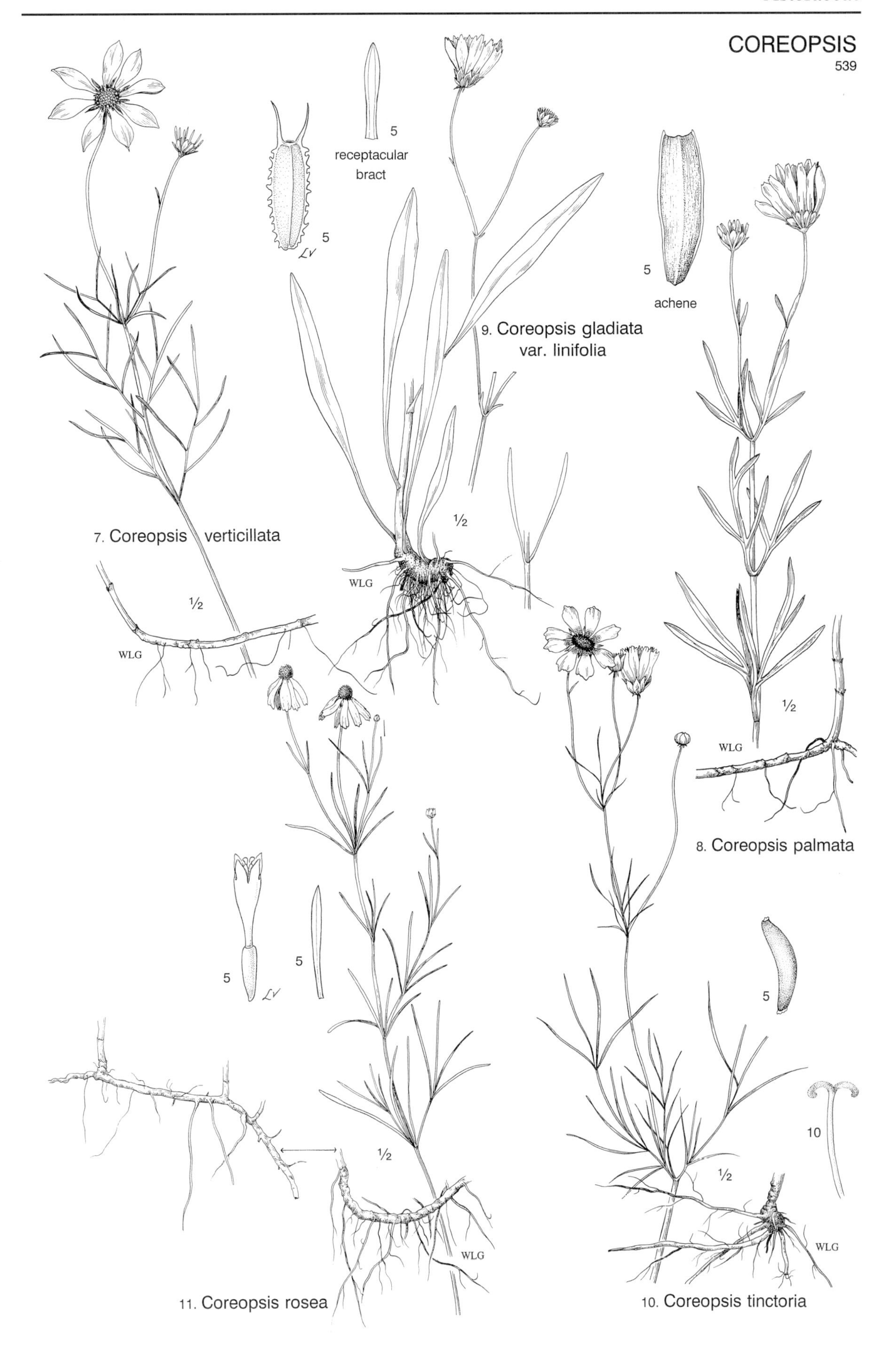

5
receptacular
bract

5
LV

9. Coreopsis gladiata
var. linifolia

5
achene

7. Coreopsis verticillata

½

½

WLG

WLG

½

WLG

8. Coreopsis palmata

5

5
LV

5

½

WLG

10

½

11. Coreopsis rosea

10. Coreopsis tinctoria

WLG

THELESPERMA
BIDENS
539, 540

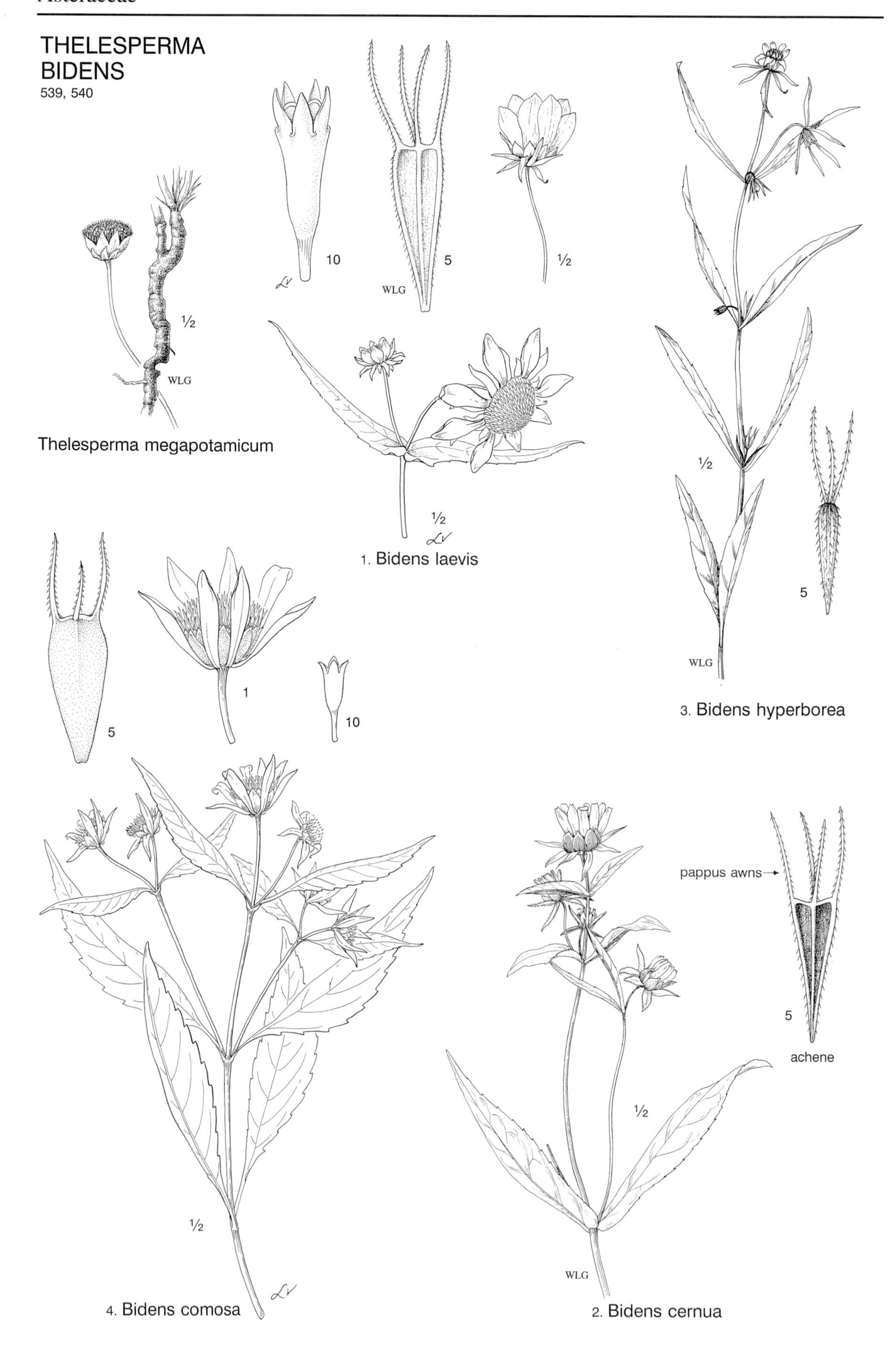

Thelesperma megapotamicum

1. Bidens laevis

3. Bidens hyperborea

pappus awns

achene

4. Bidens comosa

2. Bidens cernua

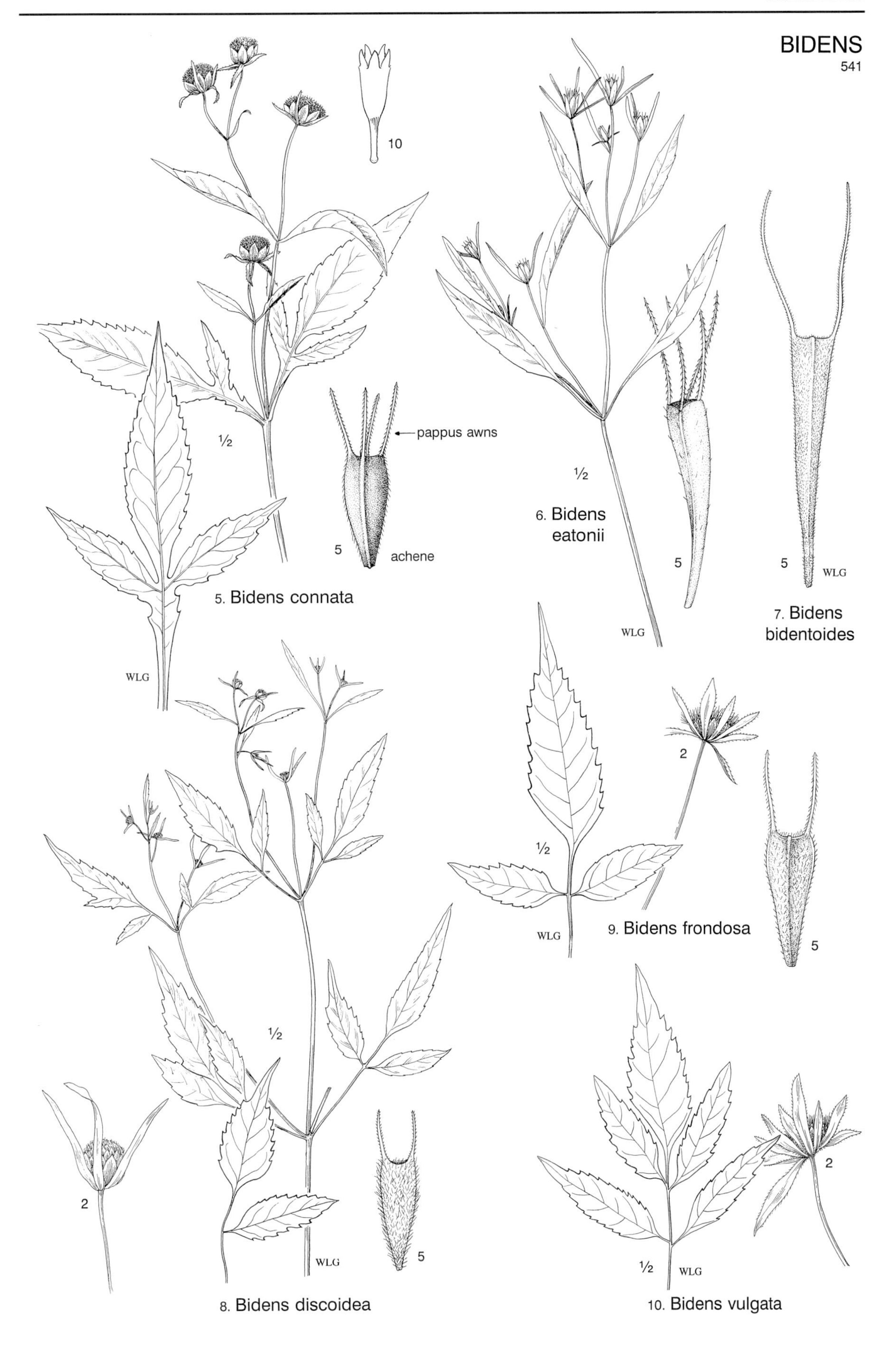

10

5. Bidens connata

pappus awns

5 achene

½

WLG

6. Bidens
eatonii

½

5

WLG

7. Bidens
bidentoides

5

5 WLG

2

9. Bidens frondosa

½

WLG

5

½

WLG

8. Bidens discoidea

2

5

2

10. Bidens vulgata

½ WLG

BIDENS
COSMOS
541, 542

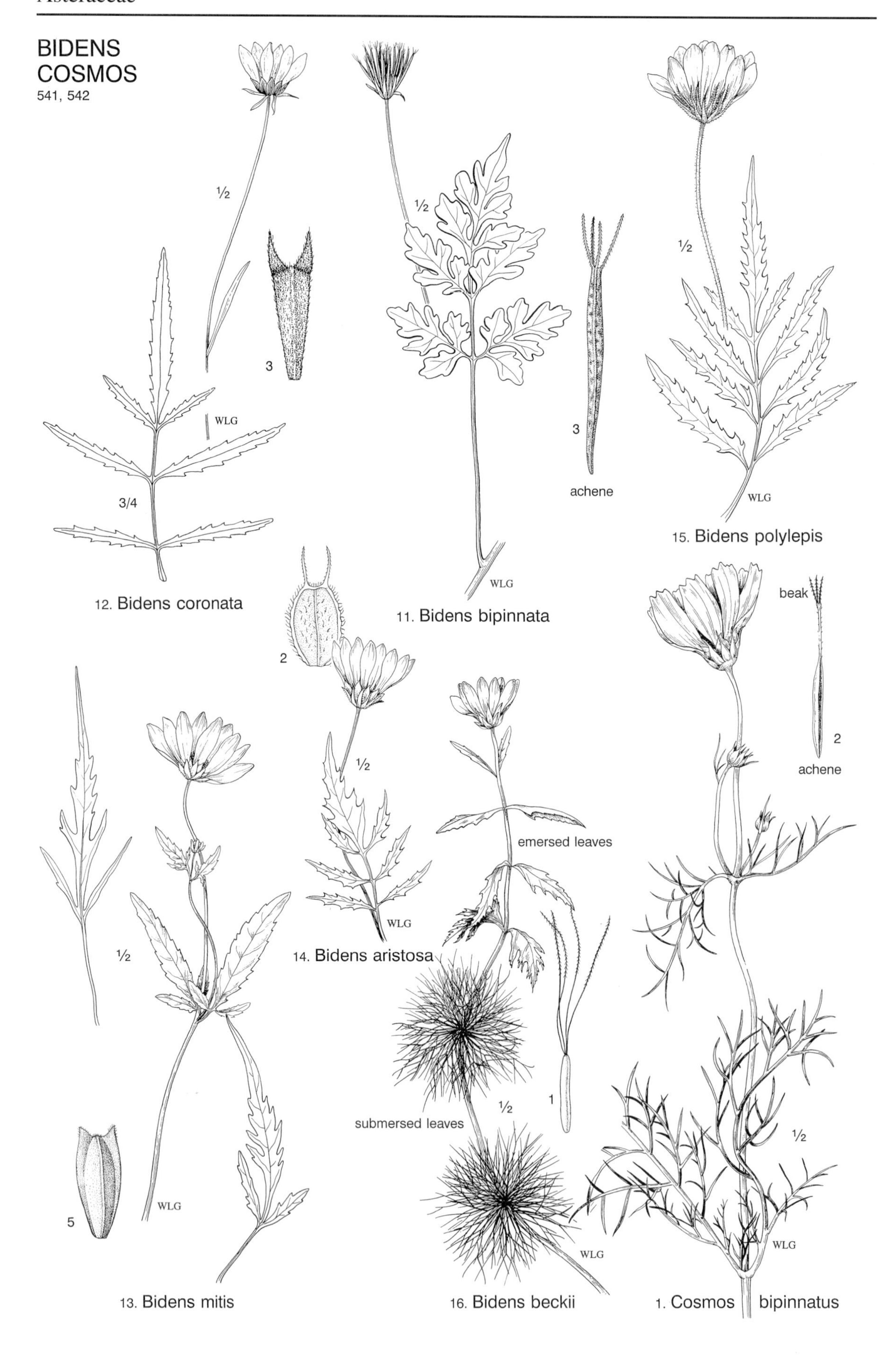

½

3

WLG

3/4

12. Bidens coronata

½

3

achene

11. Bidens bipinnata

½

WLG

½

WLG

15. Bidens polylepis

2

½

WLG

14. Bidens aristosa

½

½

emersed leaves

beak

2

achene

½

WLG

5

WLG

13. Bidens mitis

submersed leaves

1

16. Bidens beckii

½

WLG

1. Cosmos bipinnatus

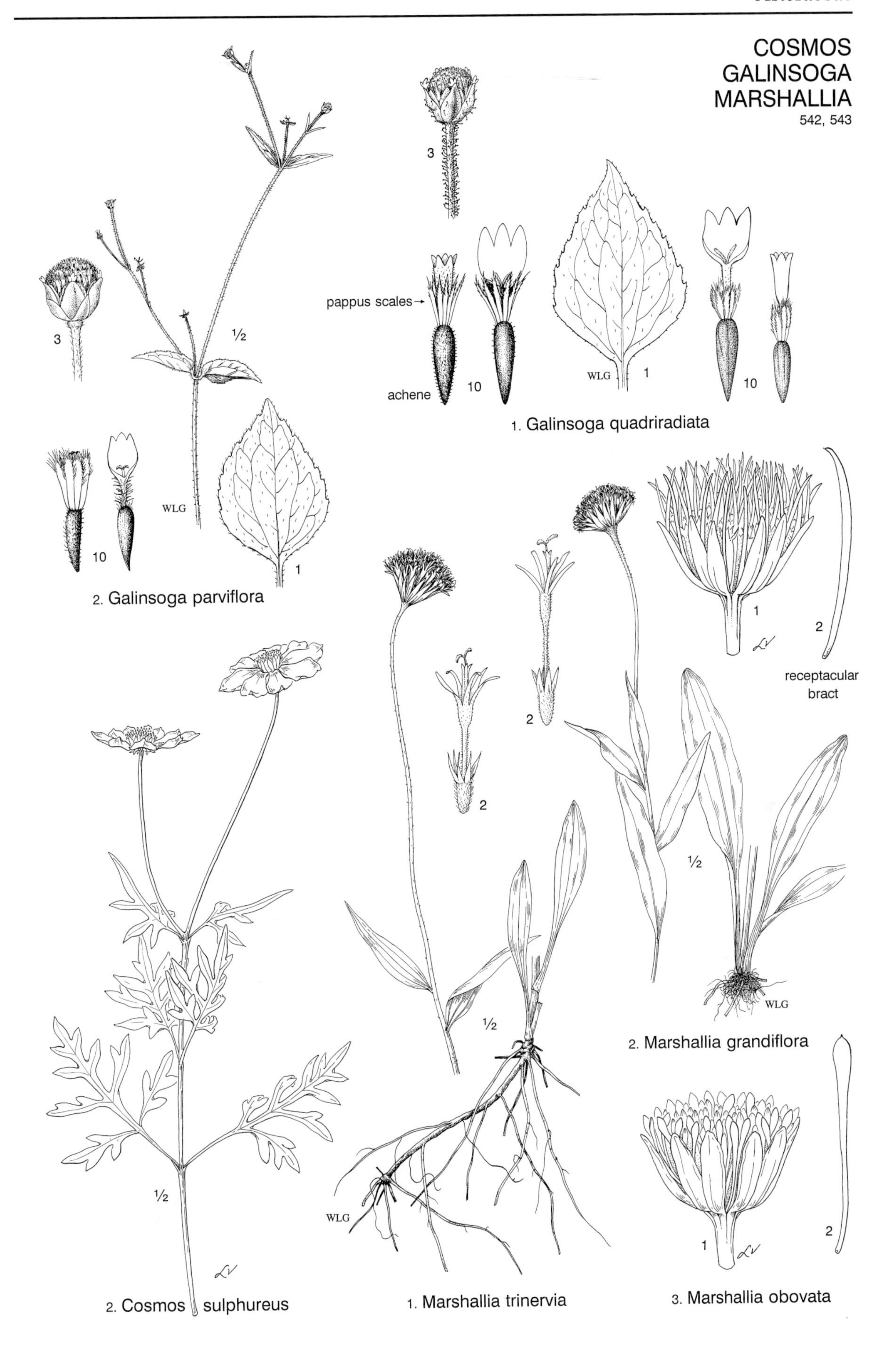

pappus scales→

achene

1. Galinsoga quadriradiata

3

10

WLG 1

10

2. Galinsoga parviflora

WLG

10

1

1

2

receptacular
bract

2

2

2

½

WLG

2. Marshallia grandiflora

½

½

WLG

1

2

2. Cosmos sulphureus

1. Marshallia trinervia

3. Marshallia obovata

MADIA
FLAVERIA
DYSSODIA
TAGETES
544

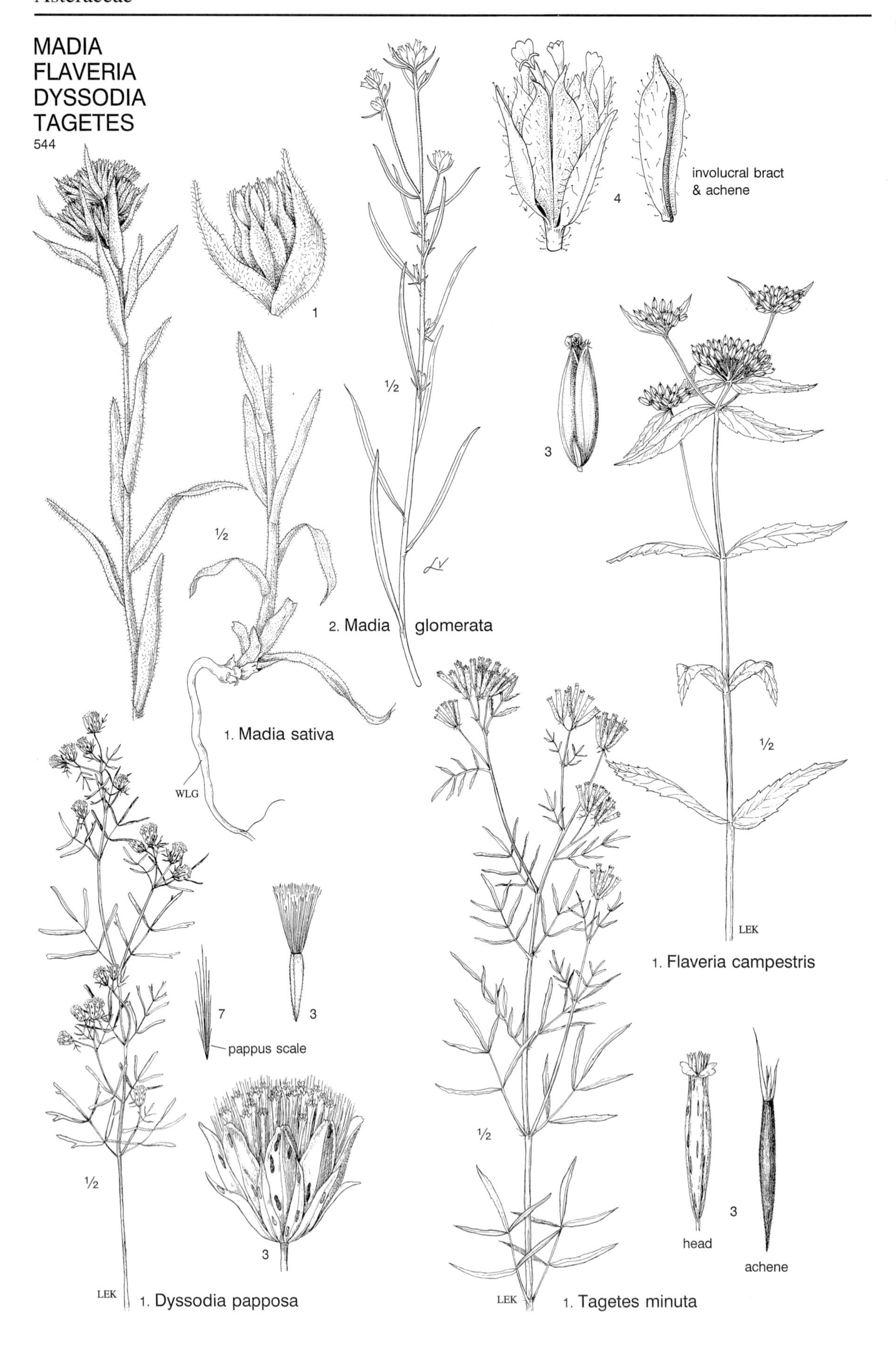

involucral bract
& achene

4

1

½

½

½

3

2. Madia glomerata

1. Madia sativa

WLG

1. Flaveria campestris

LEK

7

3

pappus scale

3

½

½

head

3

achene

LEK 1. Dyssodia papposa

LEK 1. Tagetes minuta

POLYMNIA
ACANTHOSPERMUM
SILPHIUM
545, 546

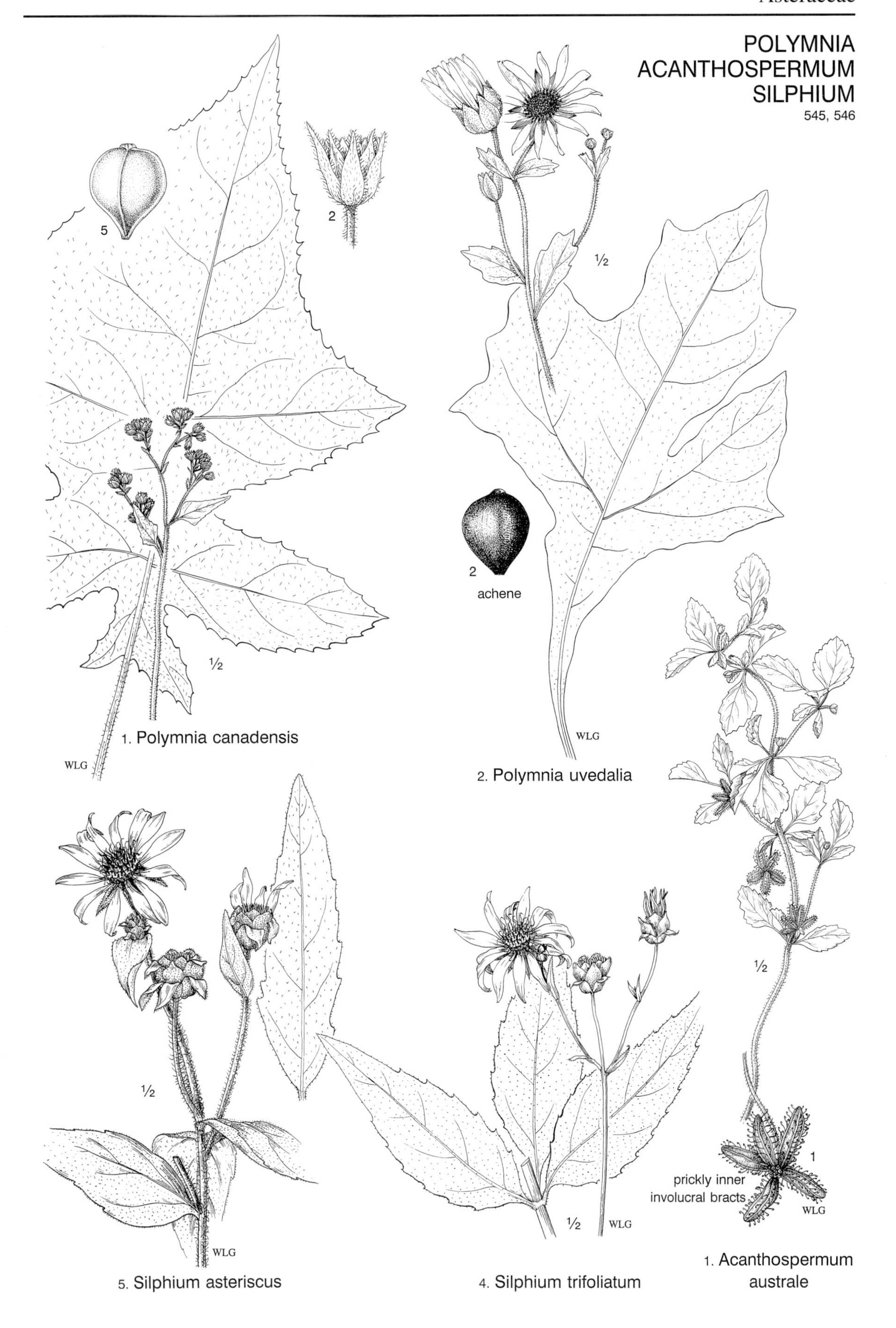

5

2

2
achene

1. Polymnia canadensis

WLG

2. Polymnia uvedalia

WLG

1

prickly inner
involucral bracts

WLG

5. Silphium asteriscus

4. Silphium trifoliatum

WLG

1. Acanthospermum
australe

SILPHIUM
546, 547

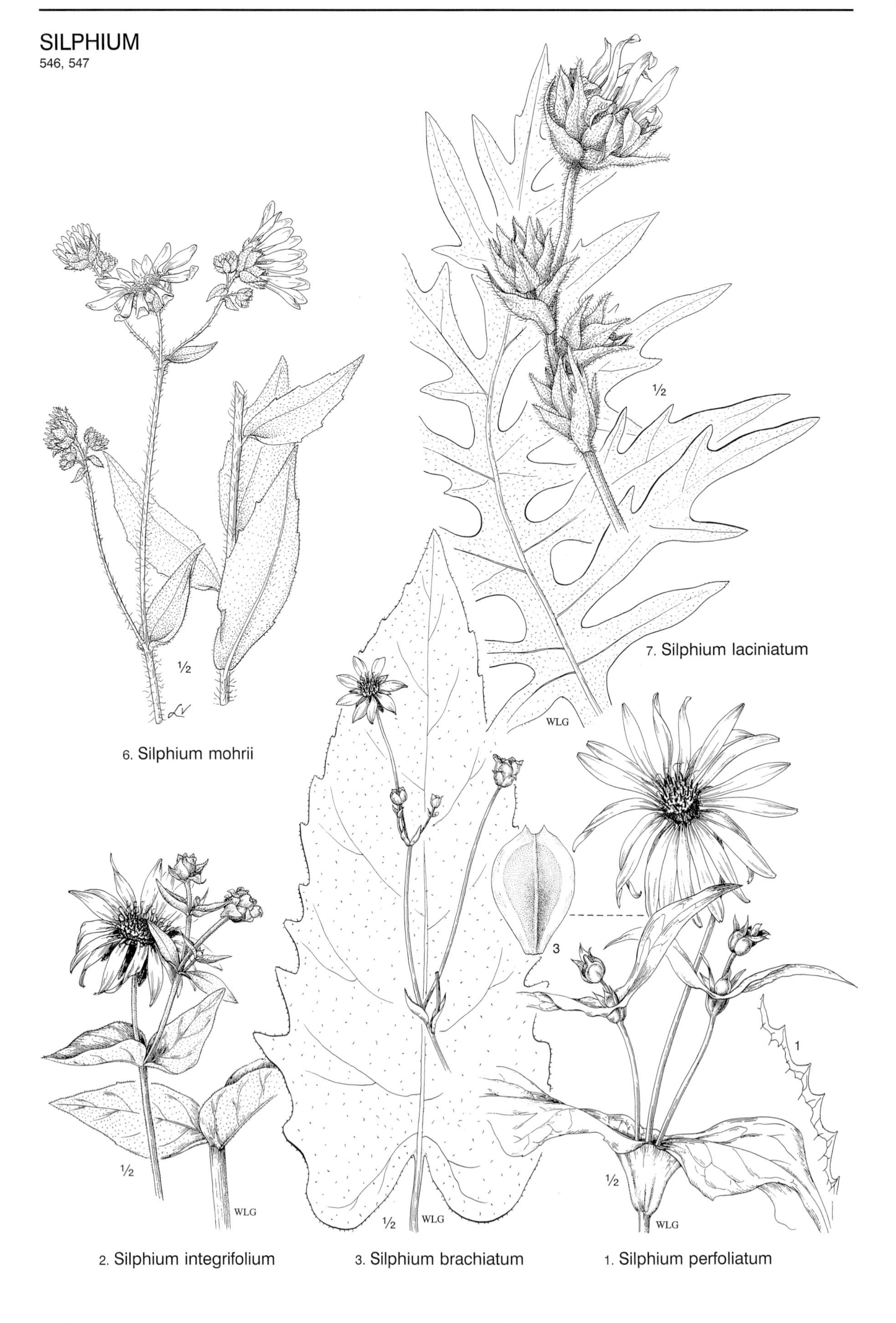

6. Silphium mohrii

7. Silphium laciniatum

2. Silphium integrifolium

3. Silphium brachiatum

1. Silphium perfoliatum

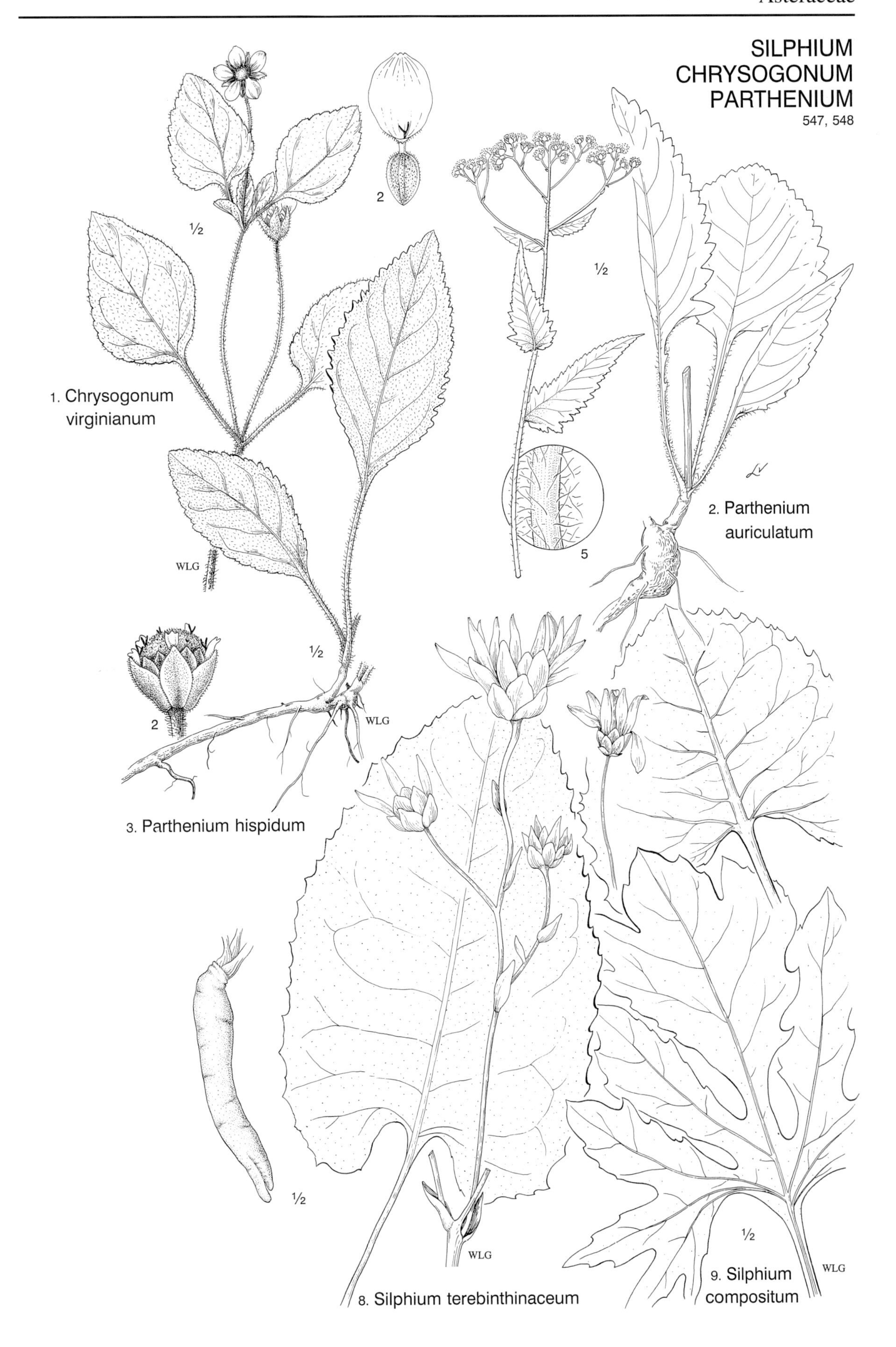

1. Chrysogonum
virginianum

2

½

2

½

WLG

2. Parthenium
auriculatum

5

2

½

WLG

3. Parthenium hispidum

½

8. Silphium terebinthinaceum

WLG

9. Silphium
compositum

½

WLG

PARTHENIUM
IVA
AMBROSIA
548, 549

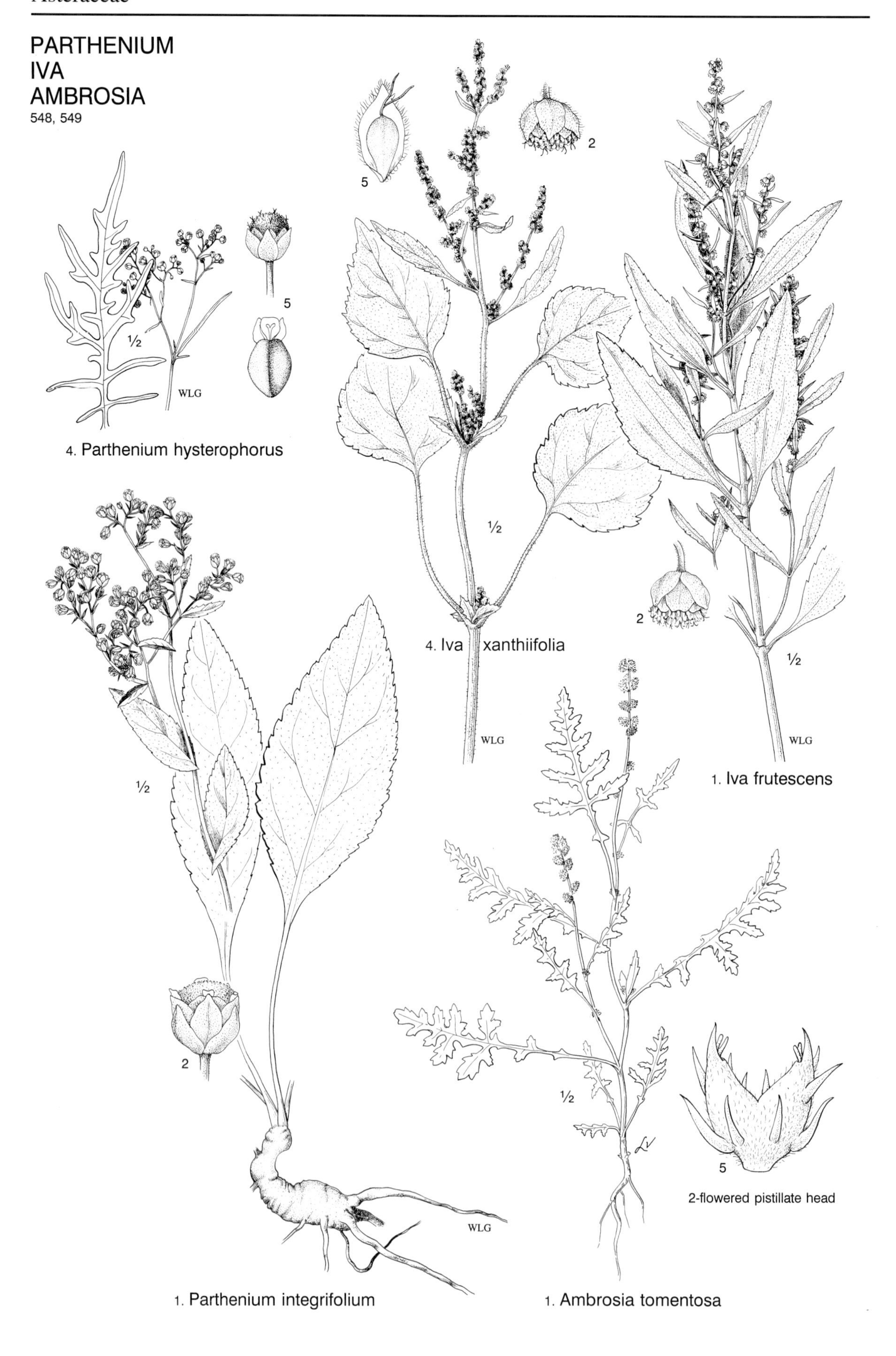

4. Parthenium hysterophorus

4. Iva xanthiifolia

5

2

1. Iva frutescens

1. Parthenium integrifolium

1. Ambrosia tomentosa

2-flowered pistillate head

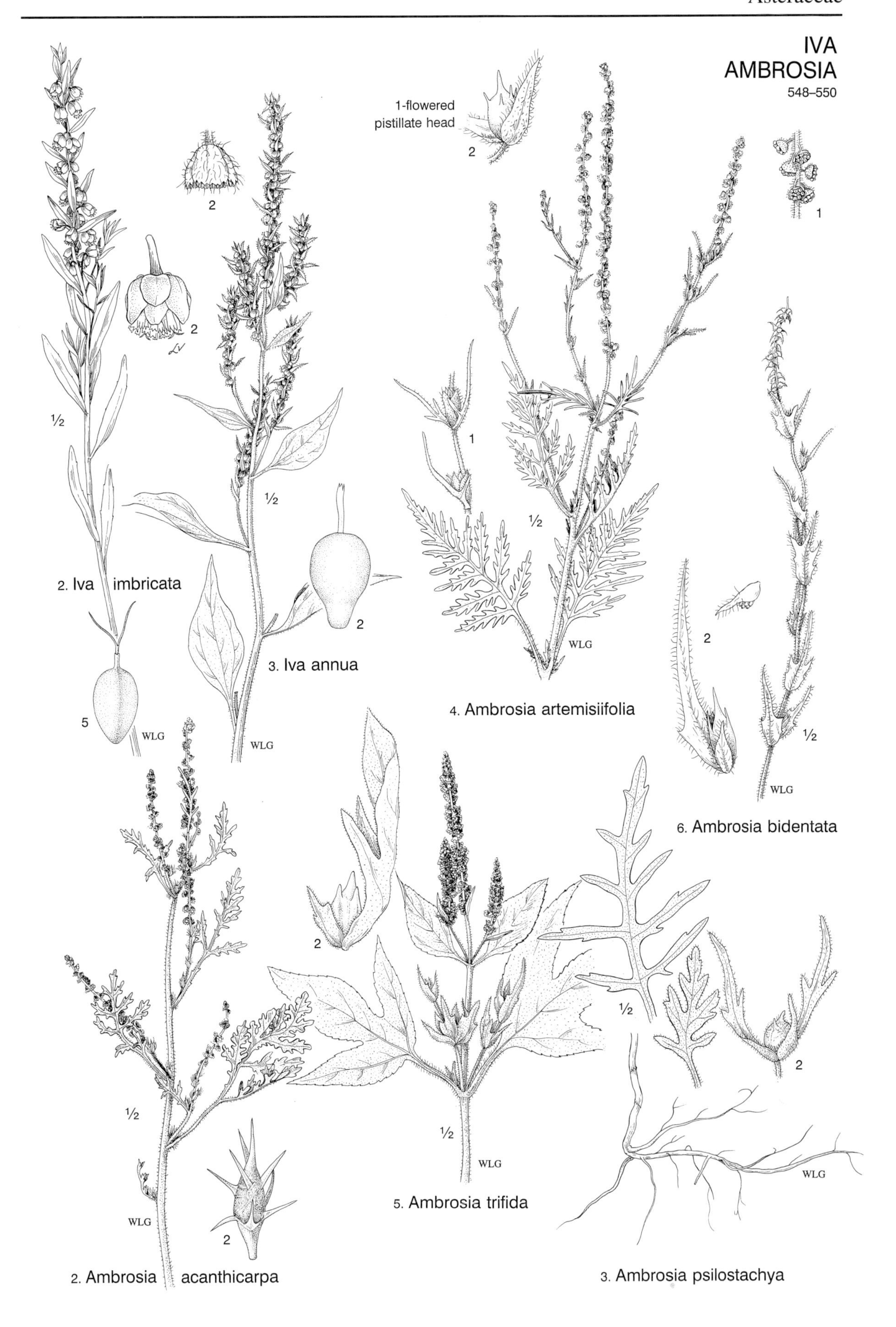

1-flowered
pistillate head

2. Iva imbricata

3. Iva annua

4. Ambrosia artemisiifolia

6. Ambrosia bidentata

5. Ambrosia trifida

2. Ambrosia acanthicarpa

3. Ambrosia psilostachya

XANTHIUM
HYMENOPAPPUS
ANTHEMIS
550, 551

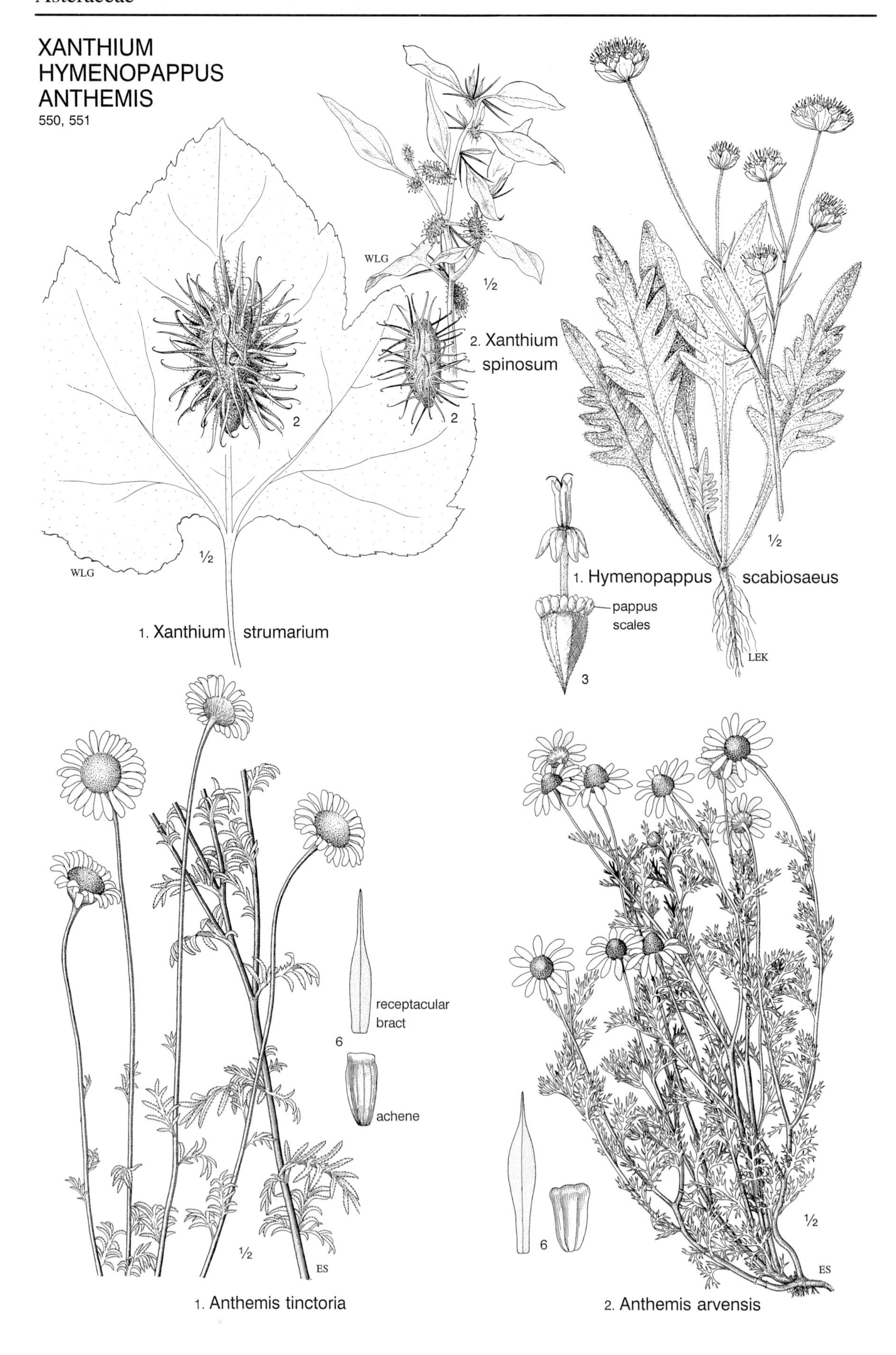

WLG

2. **Xanthium
spinosum**

WLG   ½

1. **Xanthium   strumarium**

1. **Hymenopappus   scabiosaeus**

pappus
scales

3

LEK

receptacular
bract

6

achene

6

1. **Anthemis tinctoria**

½   ES

2. **Anthemis arvensis**

½

ES

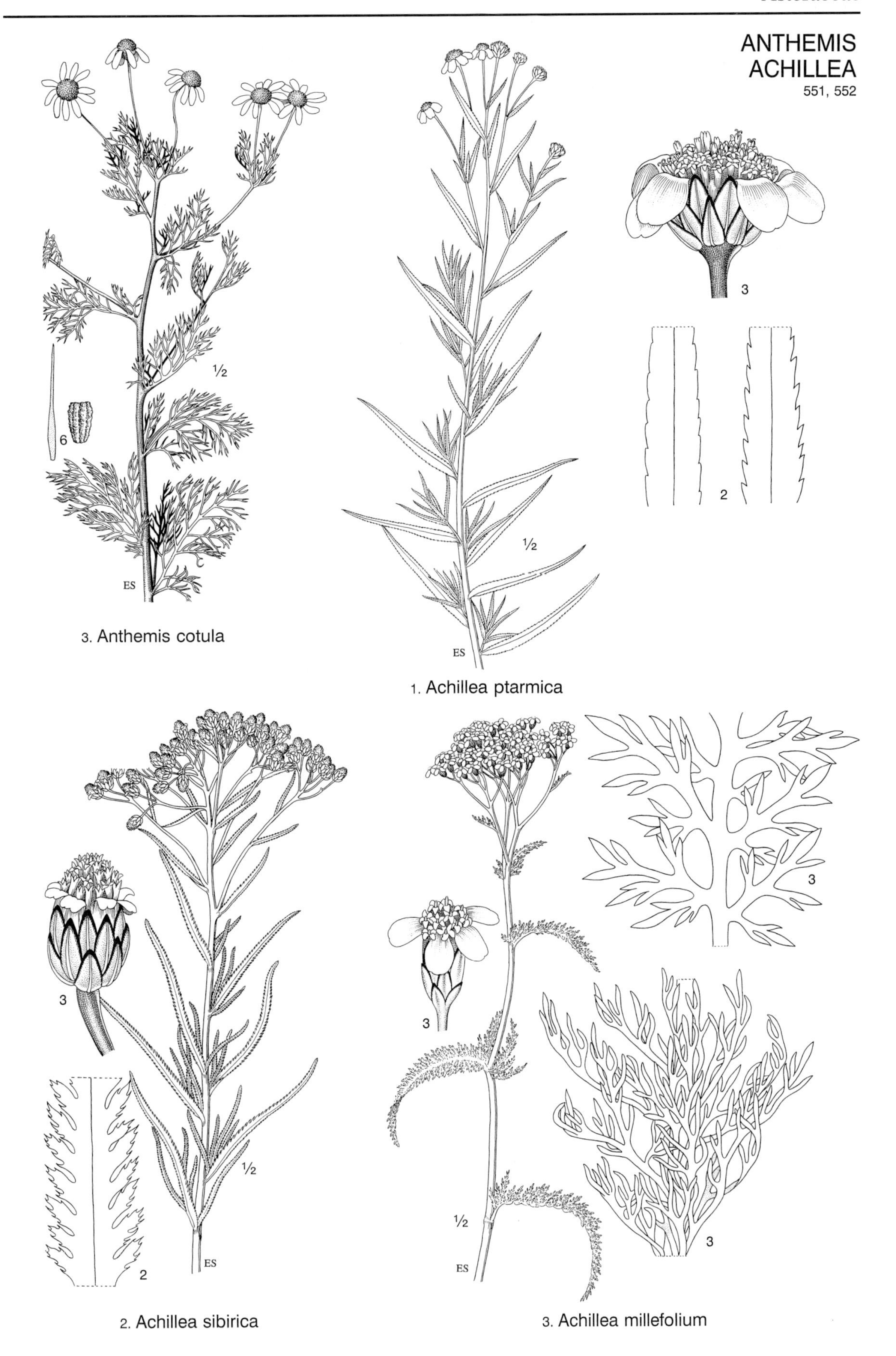

3. Anthemis cotula

1. Achillea ptarmica

2. Achillea sibirica

3. Achillea millefolium

SANTOLINA
CHRYSANTHEMUM
552, 553

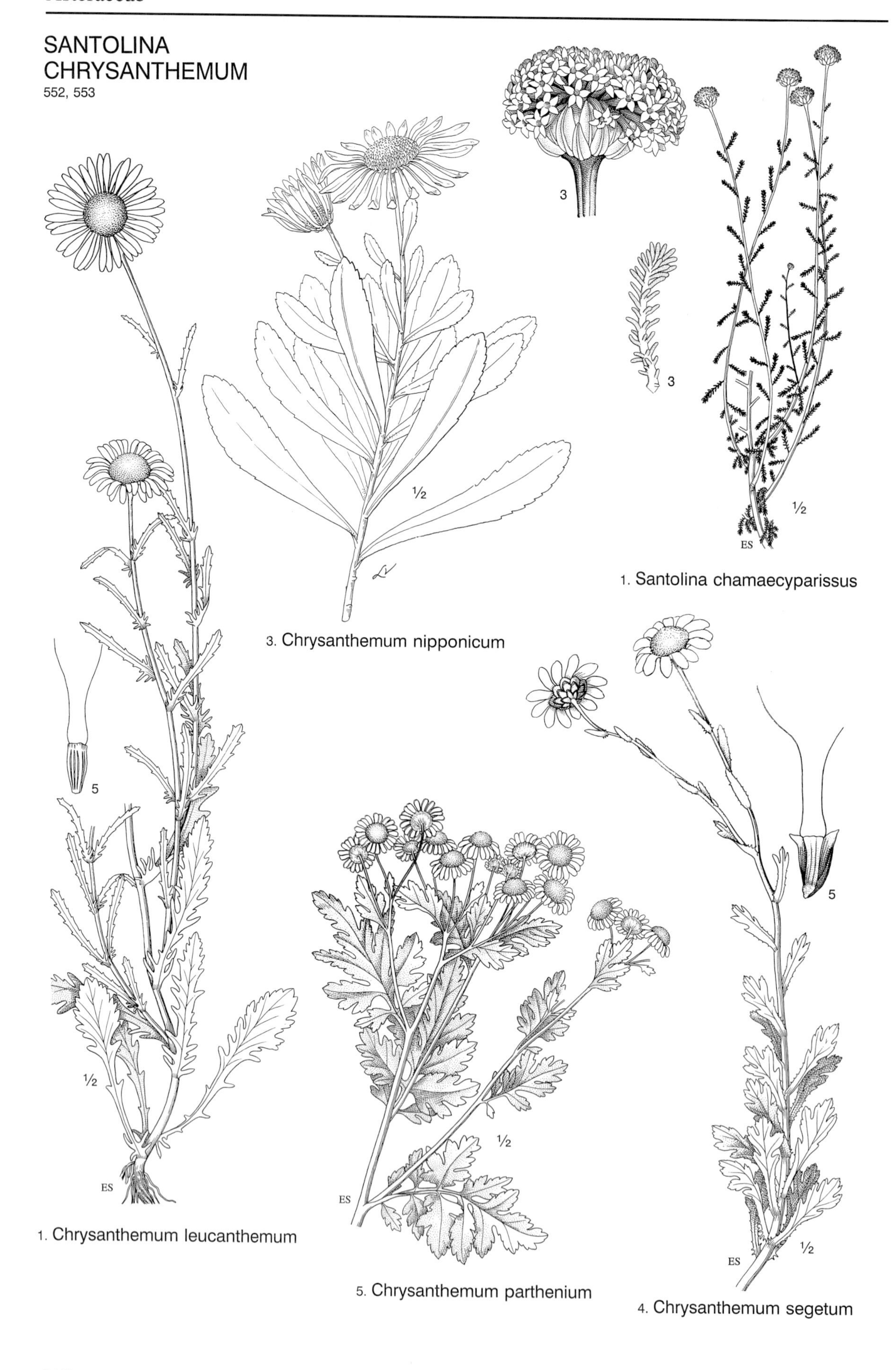

3

3. Chrysanthemum nipponicum

1. Santolina chamaecyparissus

5

1. Chrysanthemum leucanthemum

5

5. Chrysanthemum parthenium

4. Chrysanthemum segetum

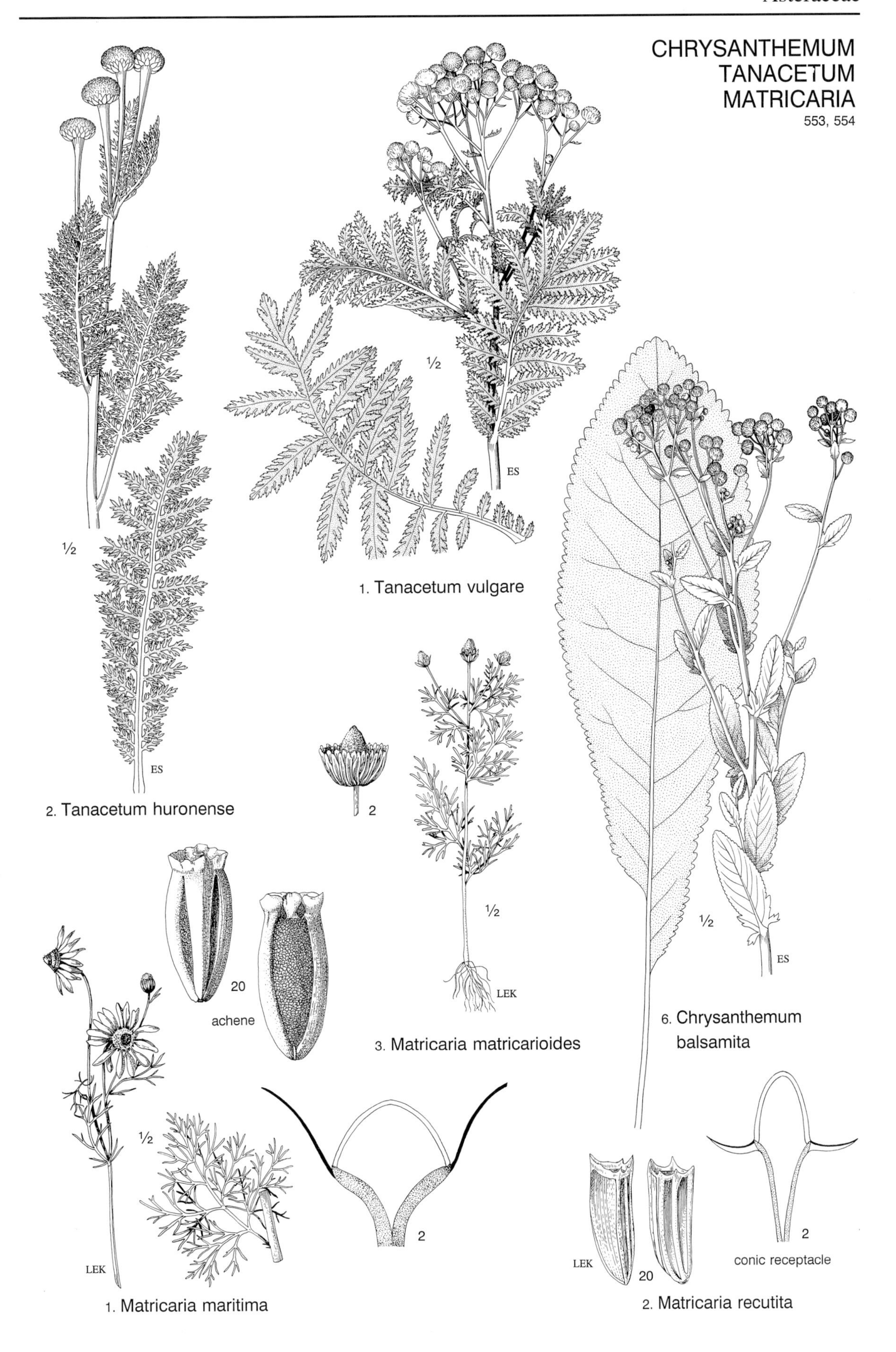

1. Tanacetum vulgare

2. Tanacetum huronense

3. Matricaria matricarioides

achene

1. Matricaria maritima

6. Chrysanthemum
balsamita

conic receptacle

2. Matricaria recutita

COTULA
ARTEMISIA
554, 555

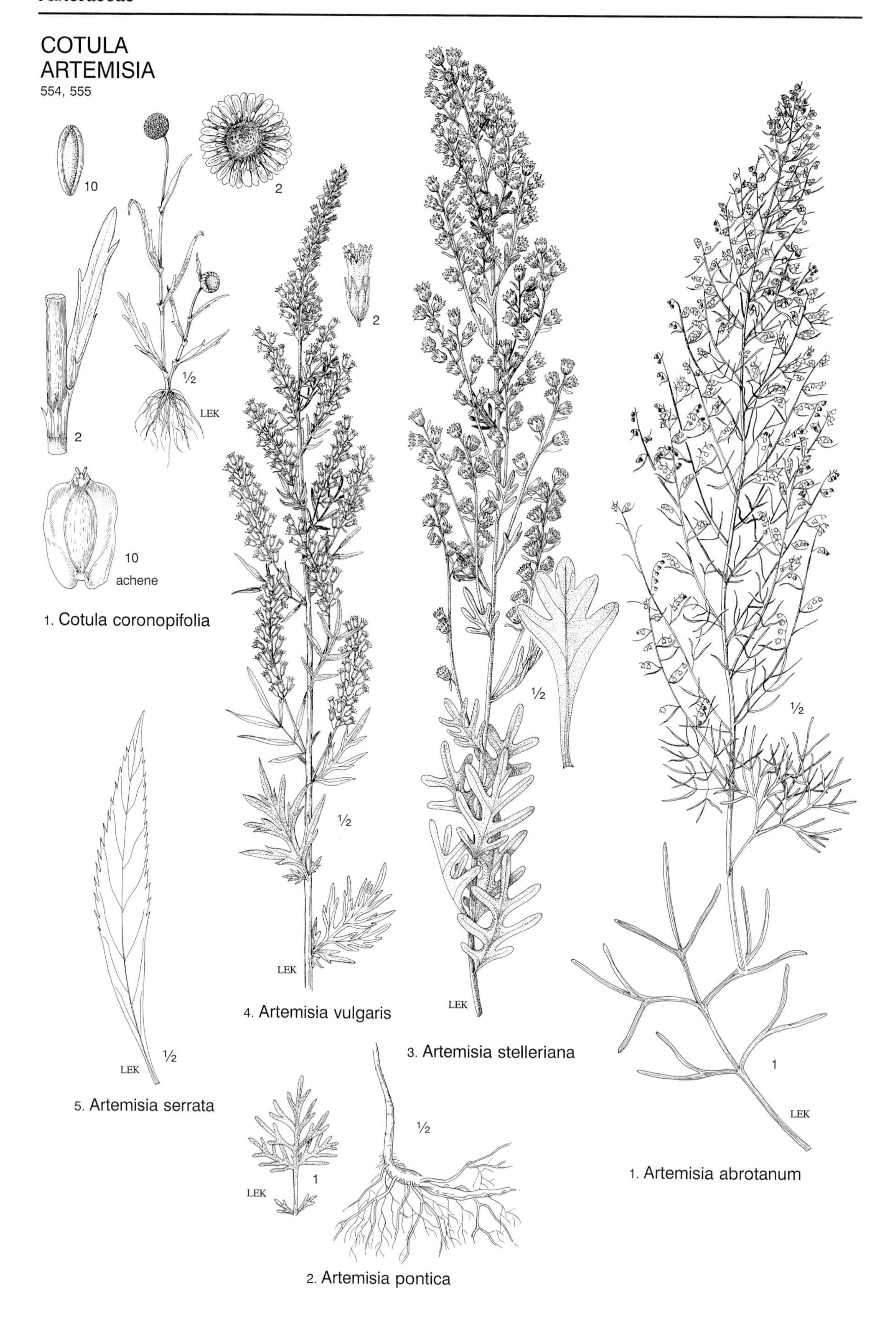

10

2

10
achene

1. Cotula coronopifolia

2

2

½
LEK

2

5. Artemisia serrata

½
LEK

4. Artemisia vulgaris

½
LEK

½

3. Artemisia stelleriana

½

1
LEK

2. Artemisia pontica

½

1. Artemisia abrotanum

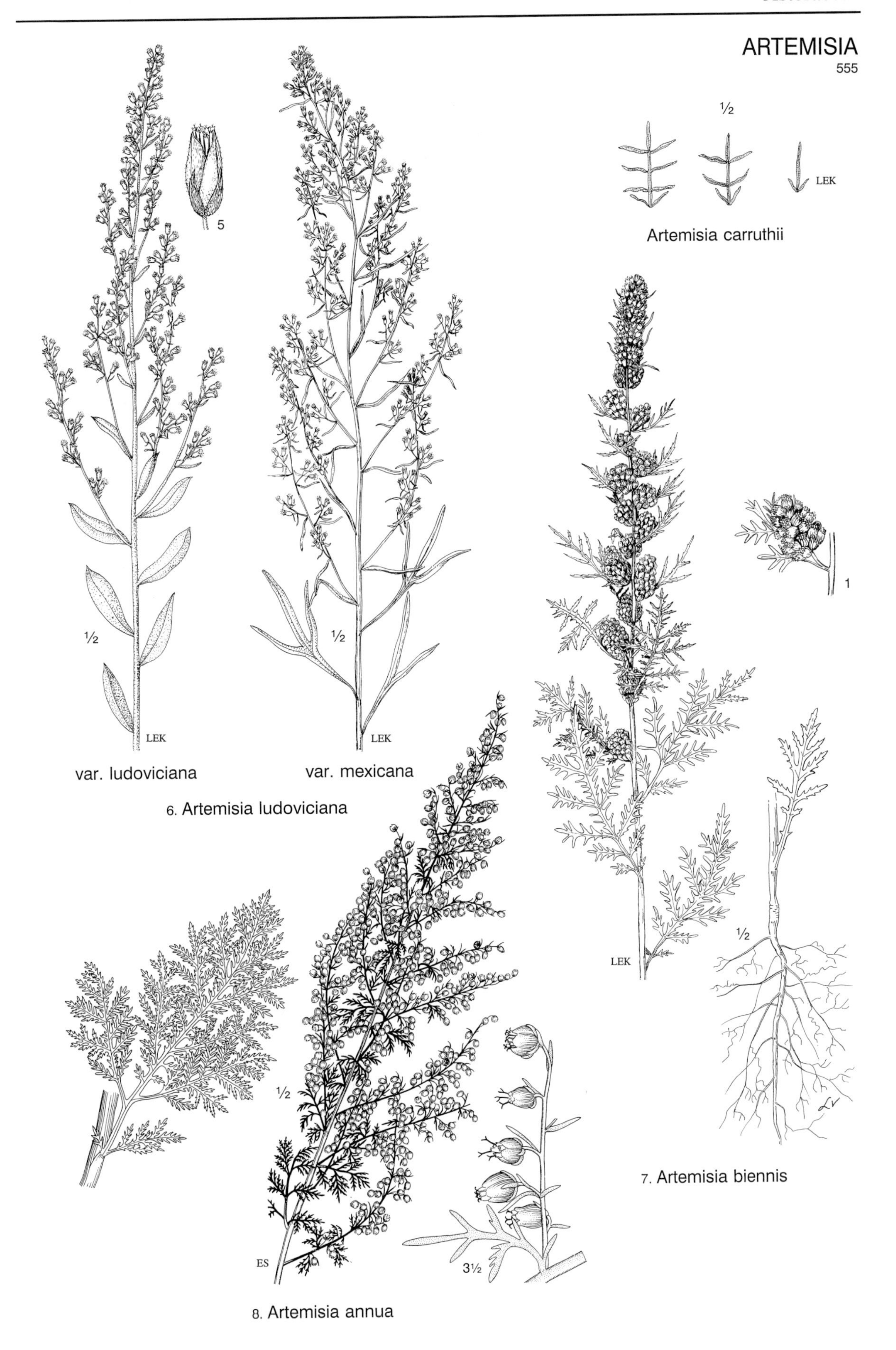

½

Artemisia carruthii

5

½

var. ludoviciana

LEK

½

var. mexicana

LEK

6. Artemisia ludoviciana

½

LEK

7. Artemisia biennis

½

ES

3½

8. Artemisia annua

ARTEMISIA
555, 556

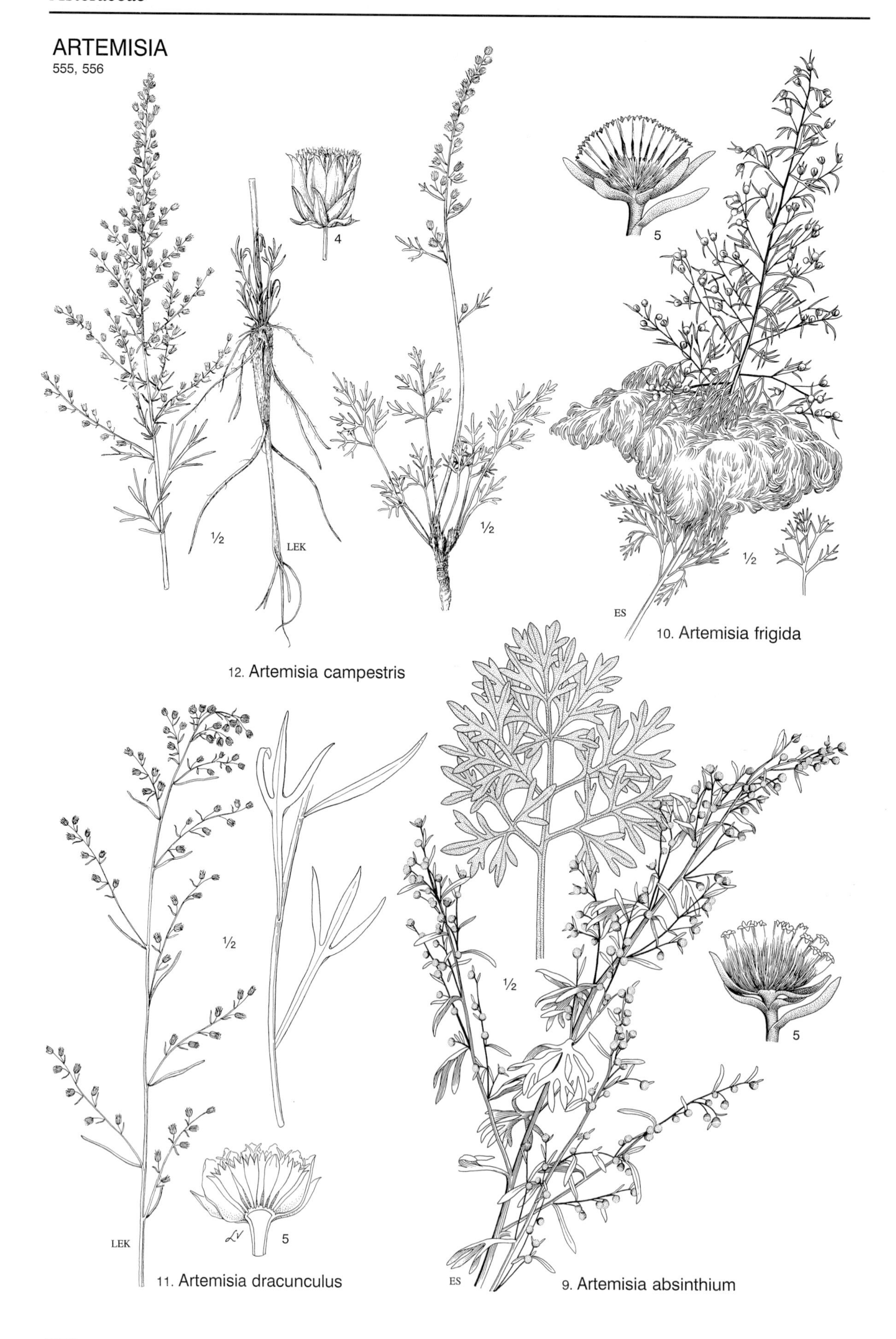

4

5

10. Artemisia frigida

ES

12. Artemisia campestris

½

LEK

½

½

½

11. Artemisia dracunculus

5

9. Artemisia absinthium

5

½

½

ES

LEK

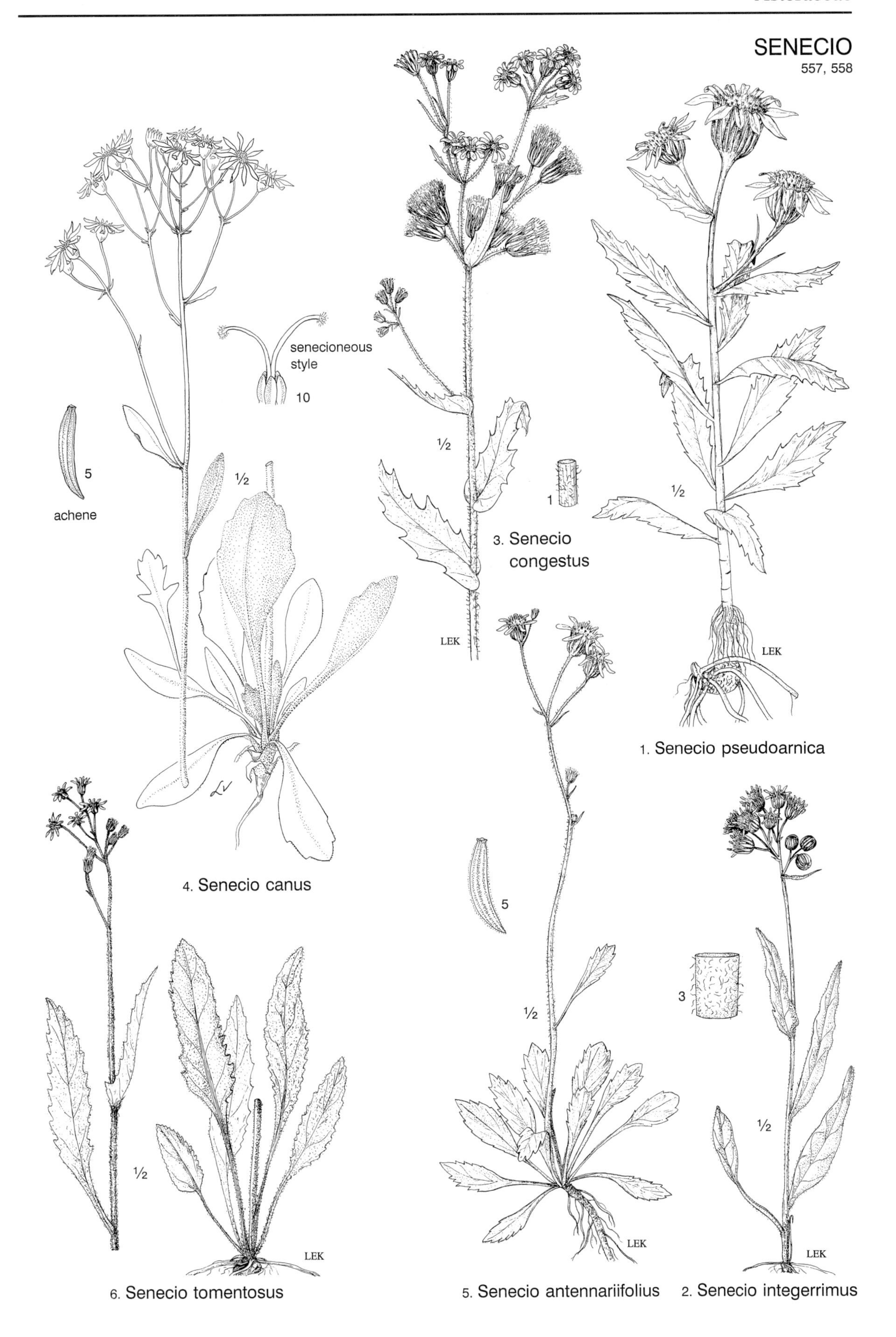

senecioneous
style
10

5
achene

3. Senecio
congestus

1

½

LEK

1. Senecio pseudoarnica

4. Senecio canus

5

3

½

½

½

6. Senecio tomentosus

LEK

5. Senecio antennariifolius

LEK

2. Senecio integerrimus

LEK

SENECIO
558

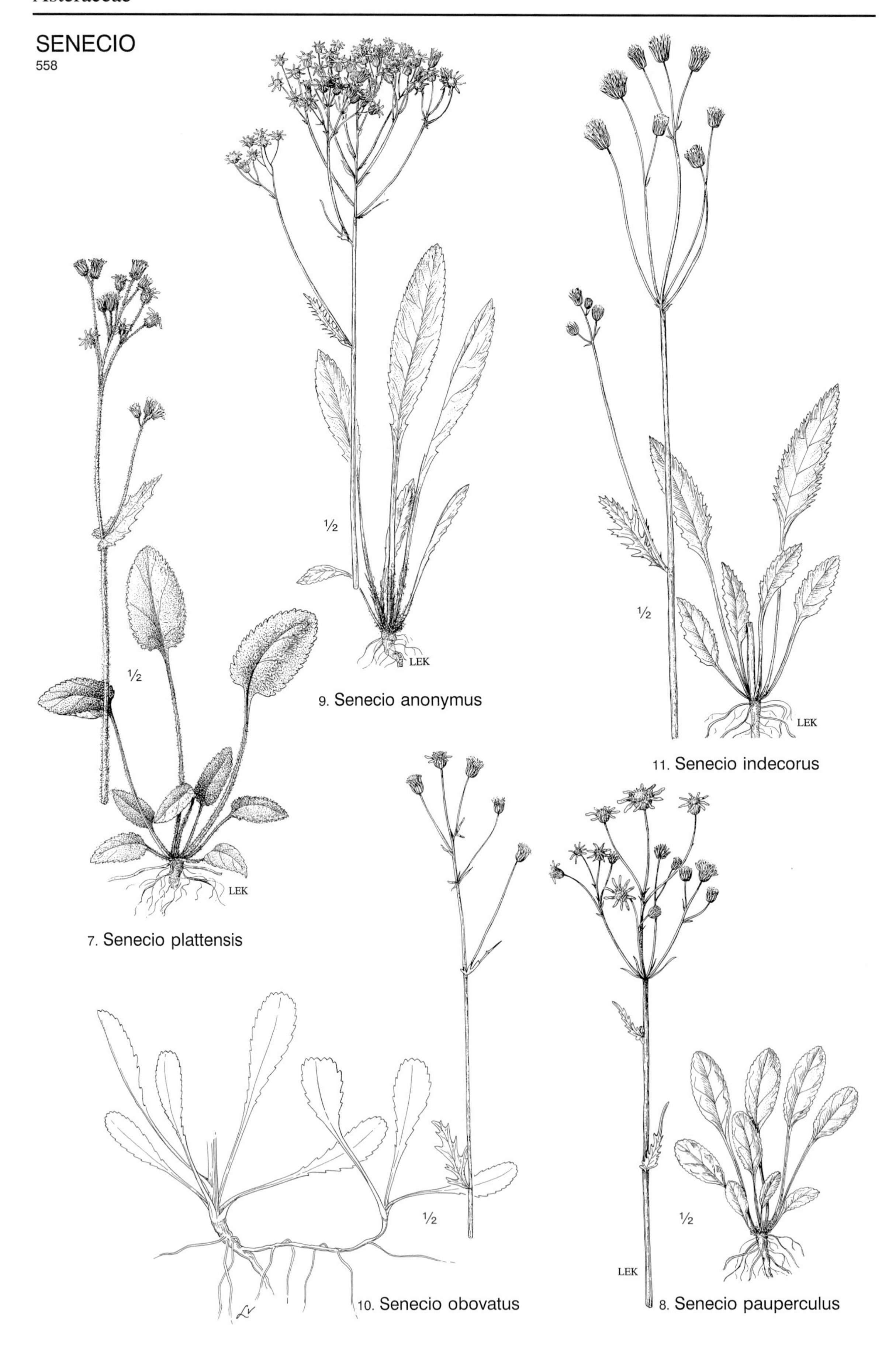

9. Senecio anonymus

7. Senecio plattensis

11. Senecio indecorus

10. Senecio obovatus

8. Senecio pauperculus

ILLUSTRATED COMPANION TO

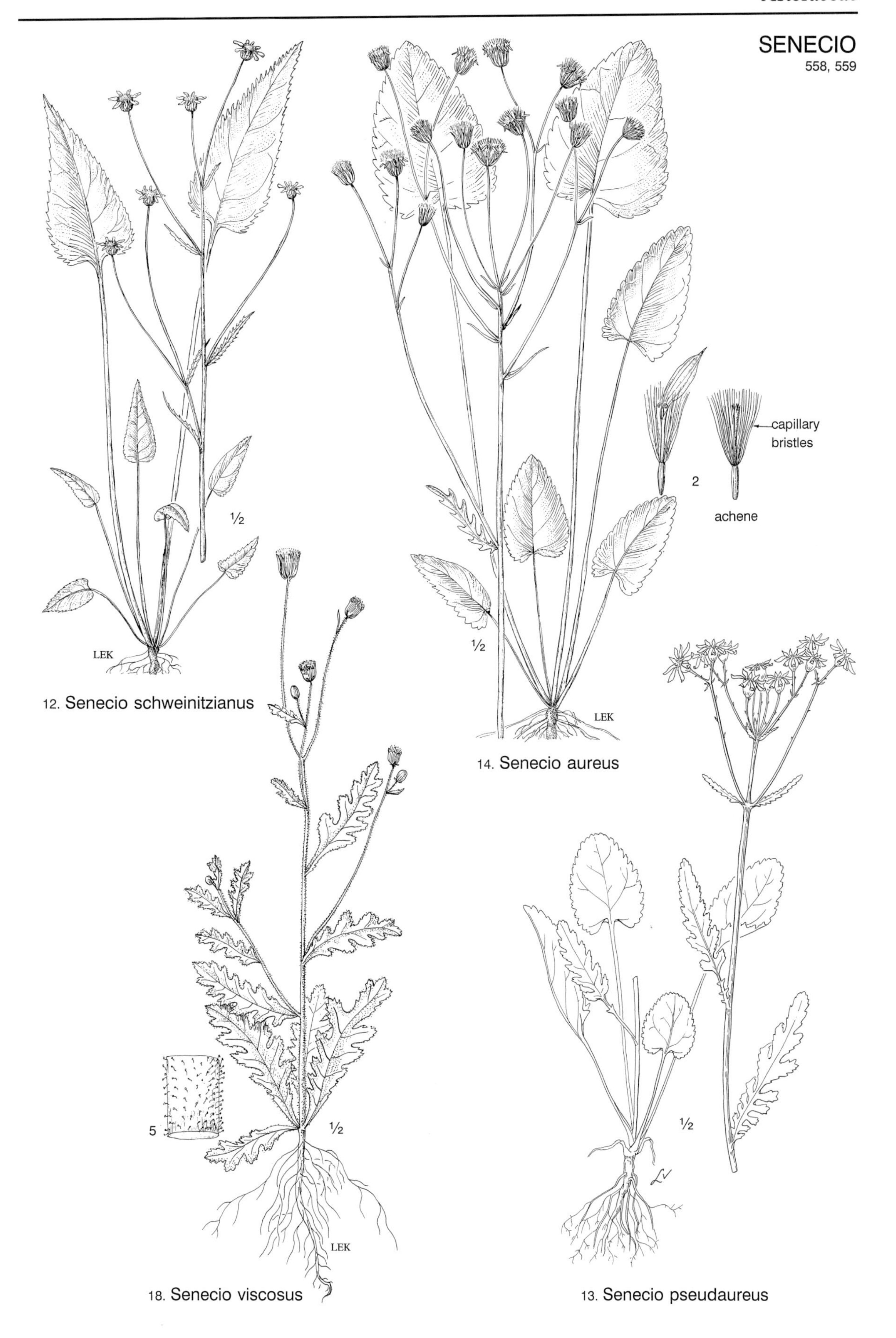

capillary
bristles

2

achene

12. Senecio schweinitzianus

14. Senecio aureus

18. Senecio viscosus

13. Senecio pseudaureus

SENECIO
559

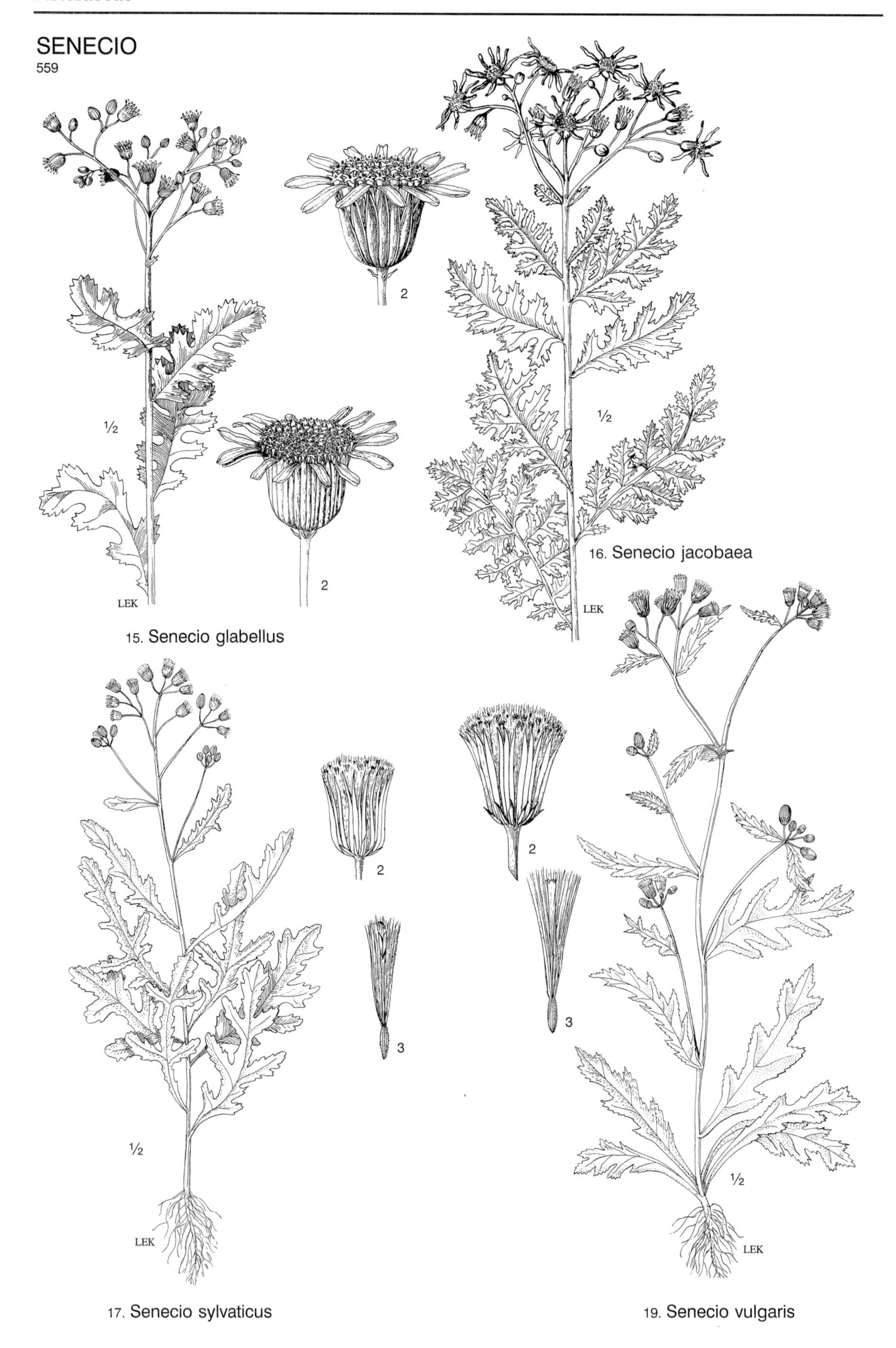

15. Senecio glabellus

16. Senecio jacobaea

17. Senecio sylvaticus

19. Senecio vulgaris

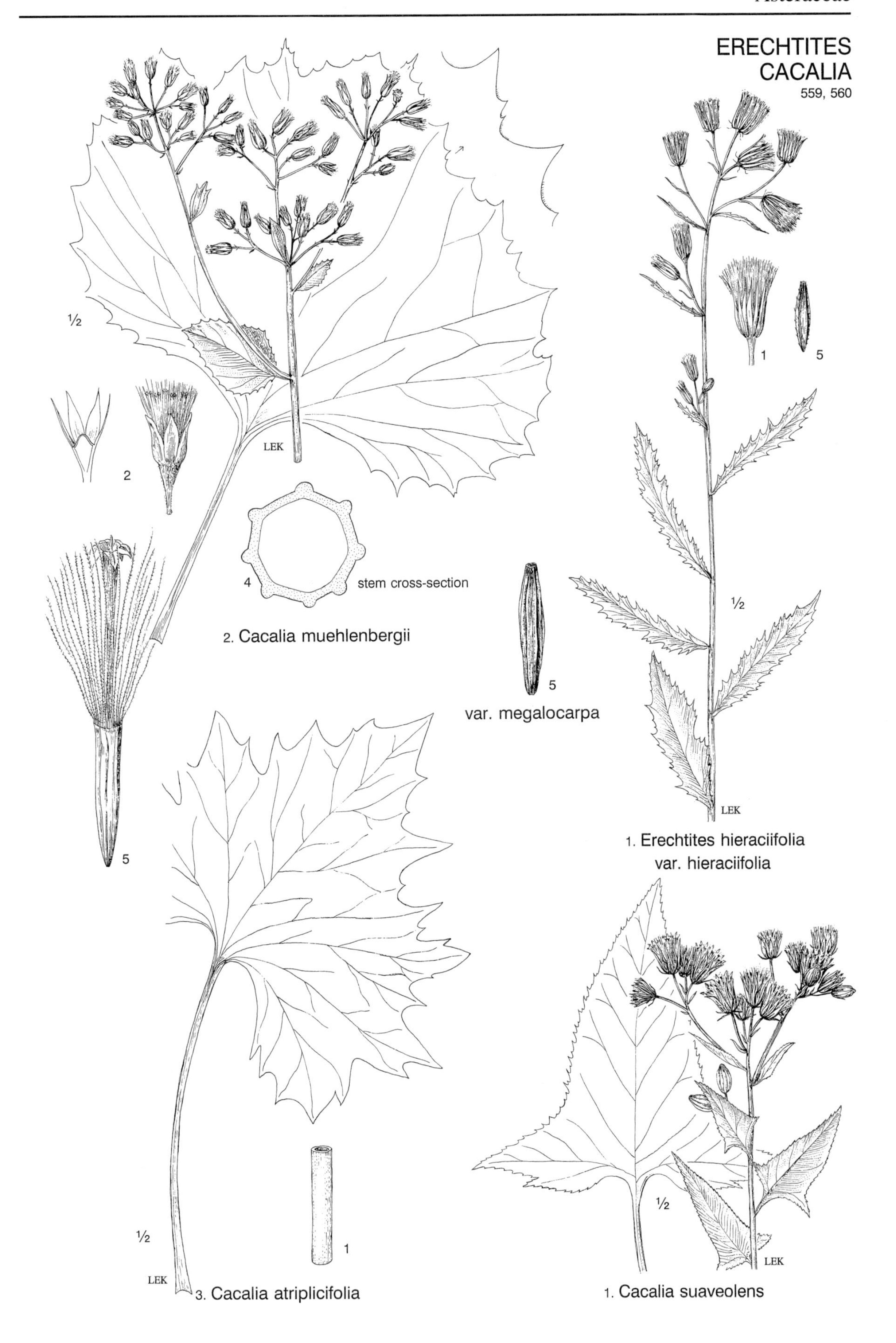

½

LEK

2

4   stem cross-section

2. Cacalia muehlenbergii

5

var. megalocarpa

1

5

1. Erechtites hieraciifolia
var. hieraciifolia

½

5

½

LEK
3. Cacalia atriplicifolia

½

LEK
1. Cacalia suaveolens

CACALIA
TUSSILAGO
PETASITES
560, 561

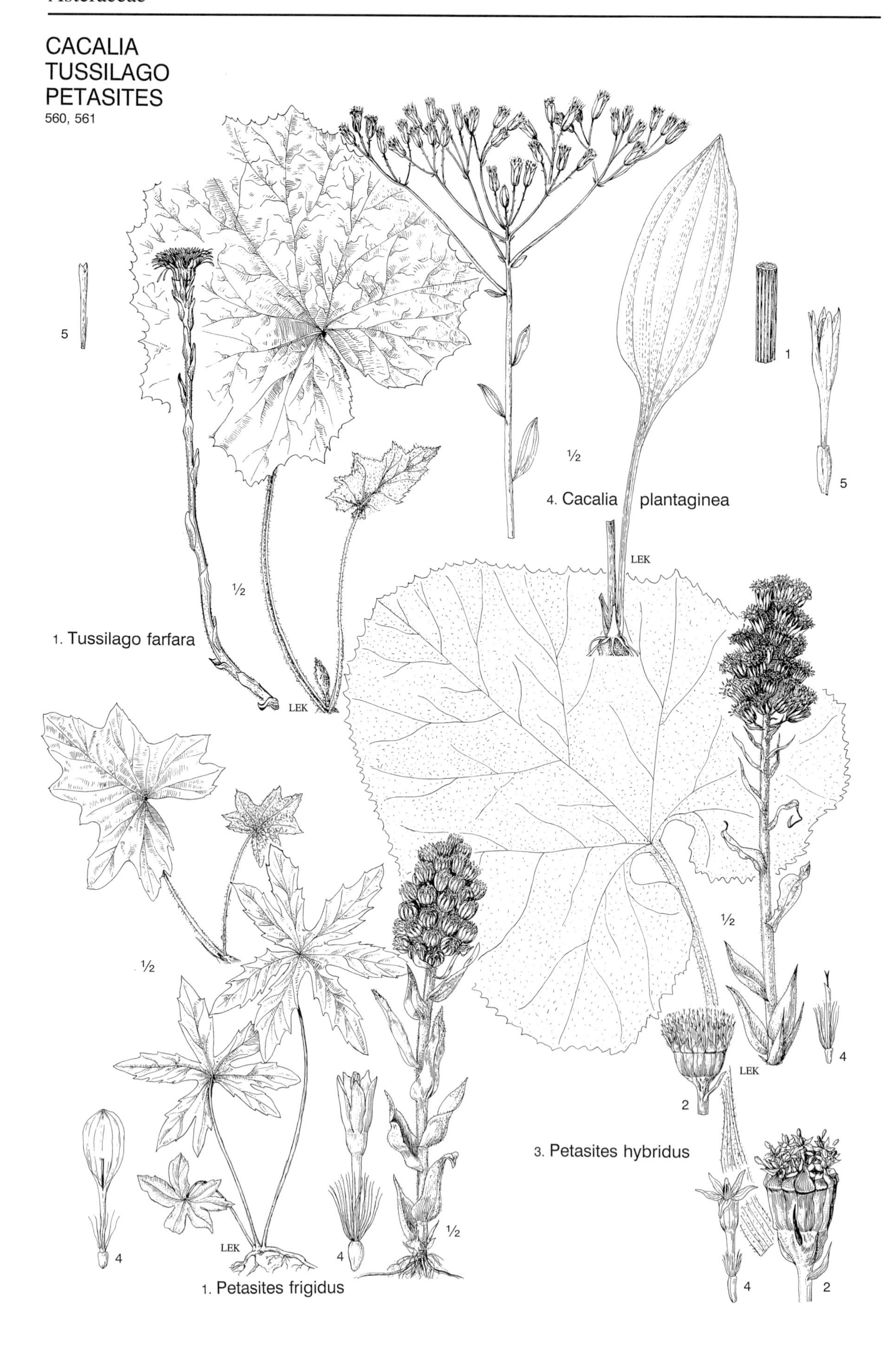

4. Cacalia plantaginea

1. Tussilago farfara

3. Petasites hybridus

1. Petasites frigidus

528

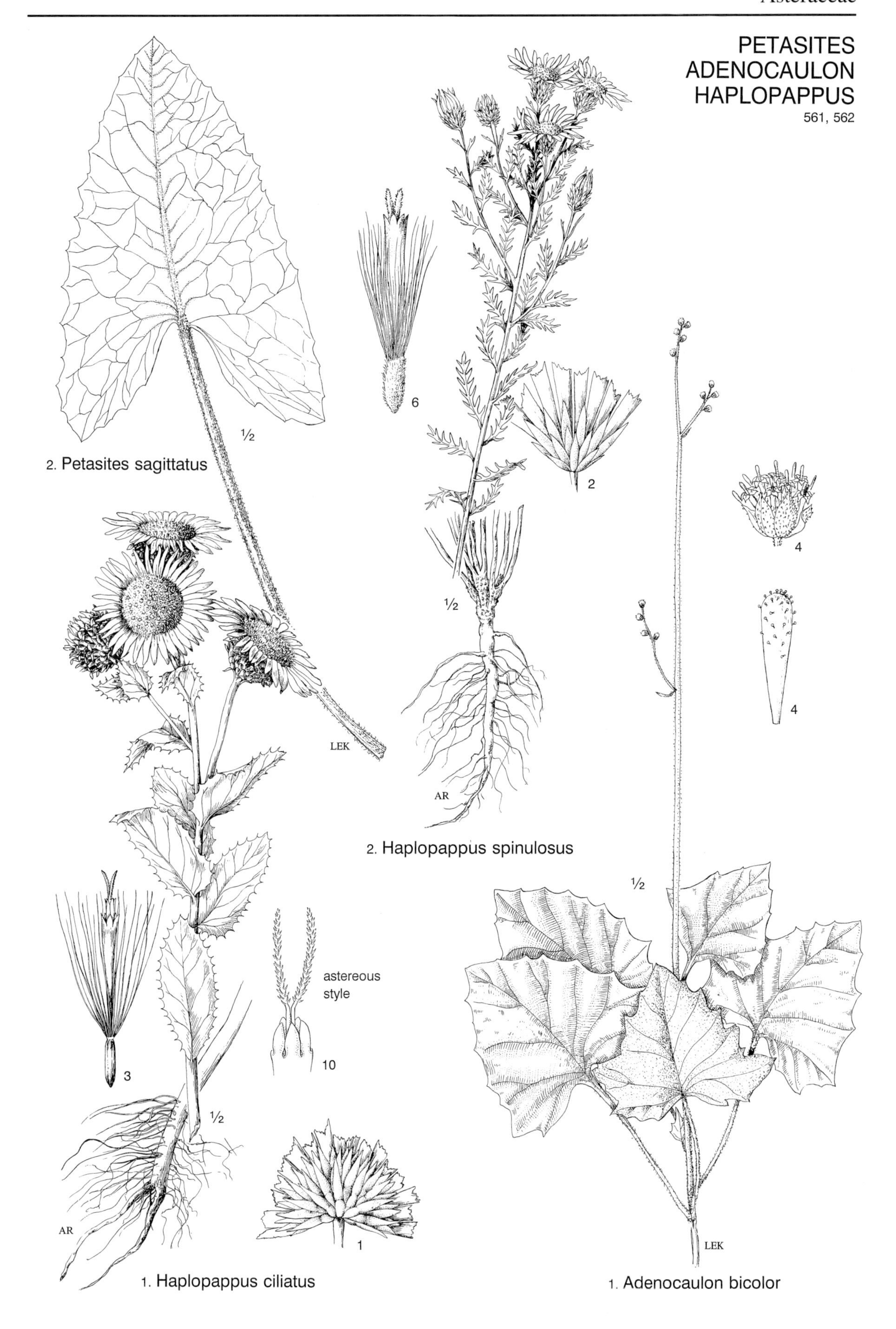

2. Petasites sagittatus

½

6

2

½

4

4

LEK

AR

2. Haplopappus spinulosus

astereous
style

10

3

½

AR

1

1. Haplopappus ciliatus

½

LEK

1. Adenocaulon bicolor

HAPLOPAPPUS
CHRYSOPSIS
562, 563

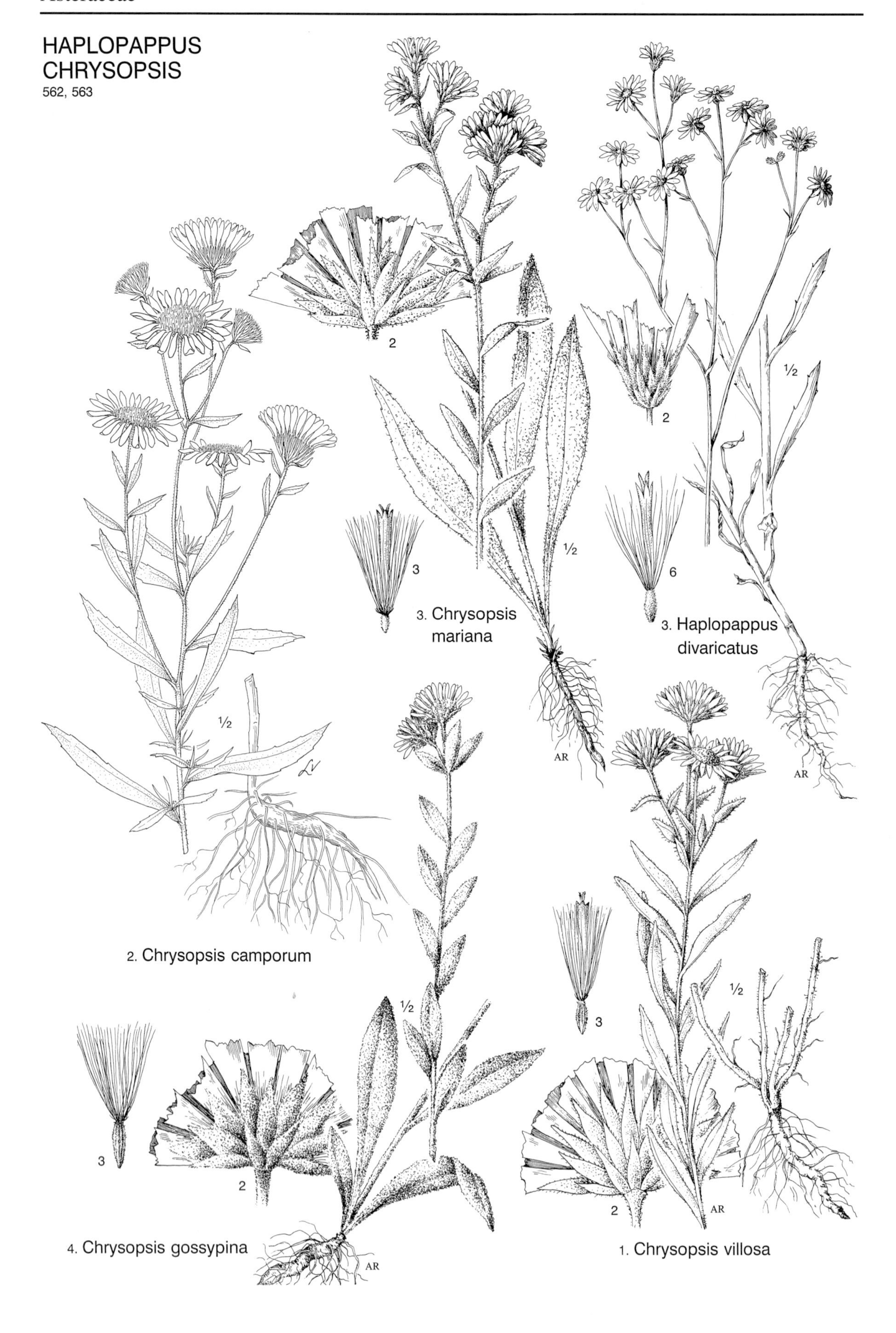

2. Chrysopsis camporum

3. Chrysopsis mariana

3. Haplopappus divaricatus

4. Chrysopsis gossypina

1. Chrysopsis villosa

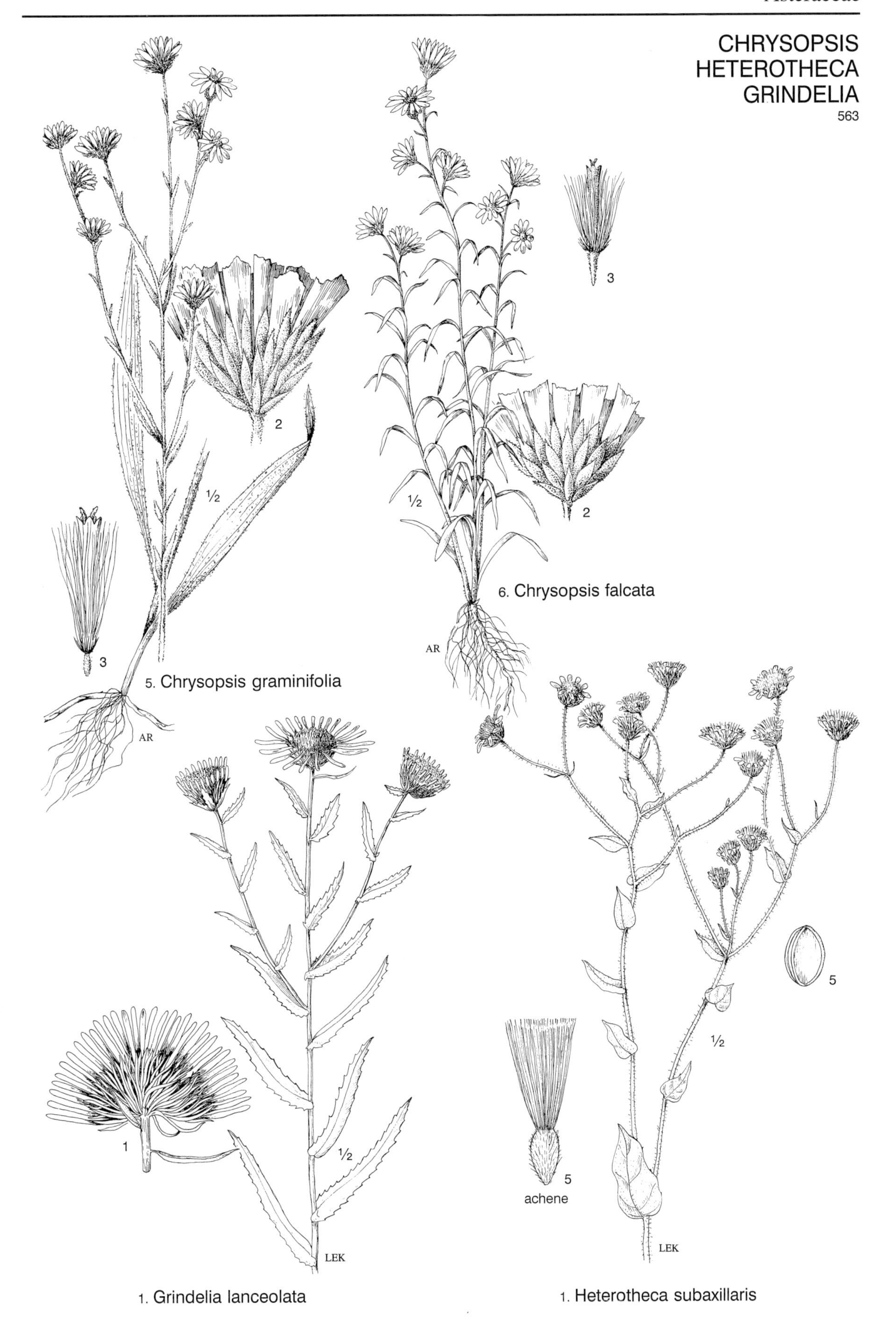

5. Chrysopsis graminifolia

6. Chrysopsis falcata

1. Grindelia lanceolata

5
achene

1. Heterotheca subaxillaris

GRINDELIA
SOLIDAGO
564–567

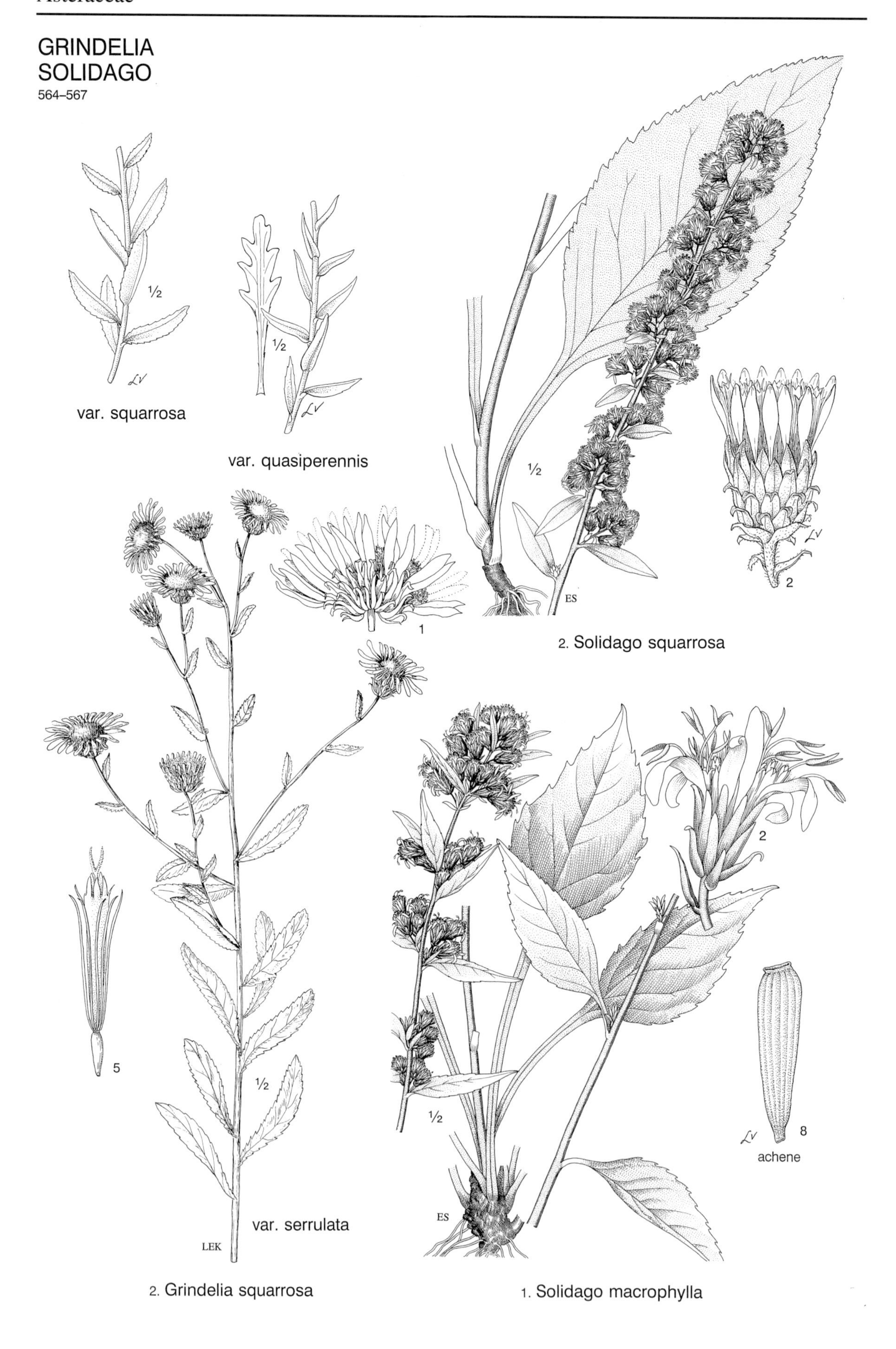

var. squarrosa

var. quasiperennis

1

2. Solidago squarrosa

2

2. Grindelia squarrosa

5

var. serrulata

LEK

ES

ES

1. Solidago macrophylla

8
achene

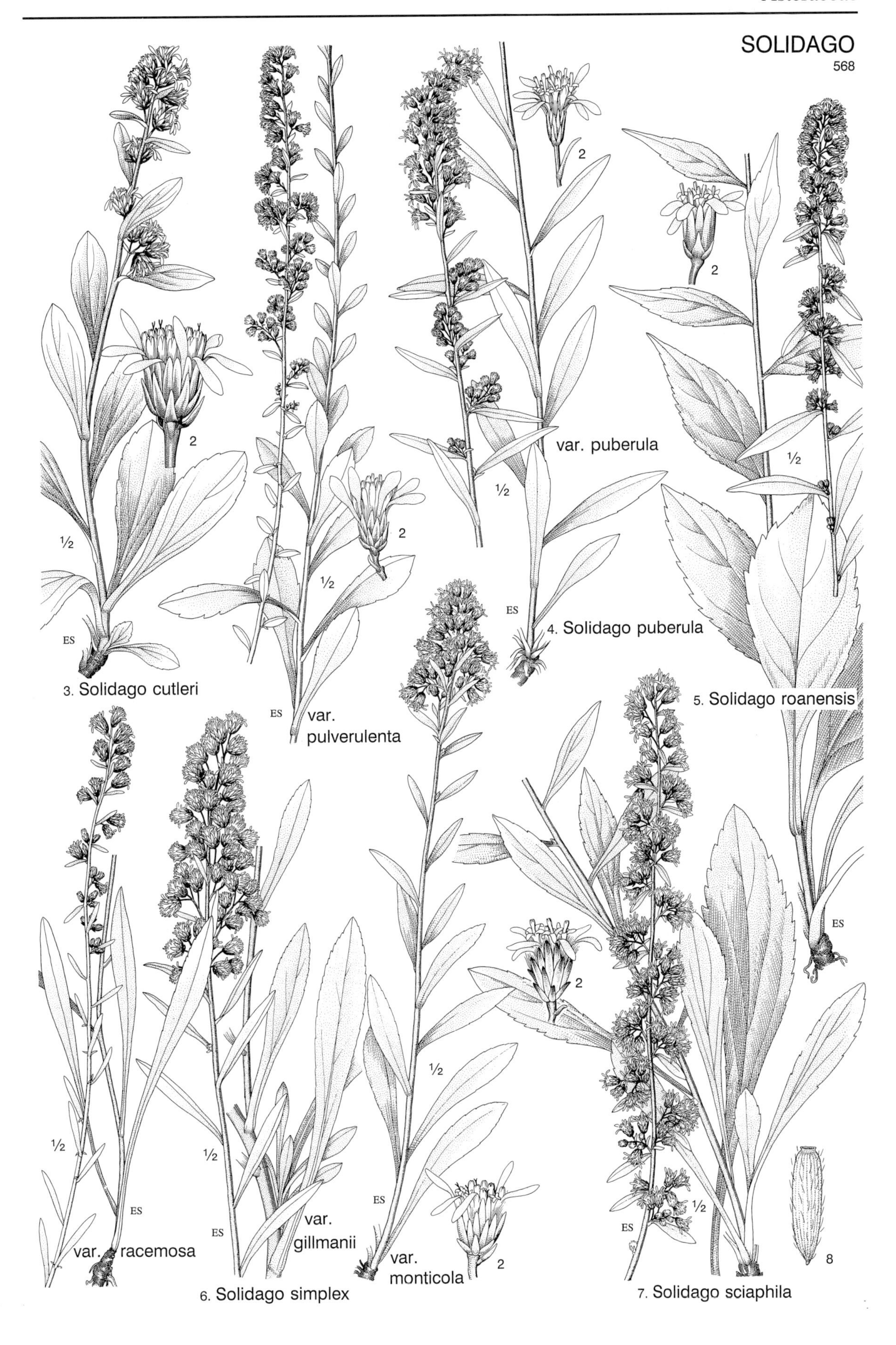

2

2

2

var. puberula

½

½

2

ES

4. Solidago puberula

3. Solidago cutleri

ES

var.
pulverulenta

5. Solidago roanensis

ES

2

½

½

2

½

var.
racemosa

ES

ES

var.
gillmanii

ES

var.
monticola

2

8

½

ES

6. Solidago simplex

7. Solidago sciaphila

SOLIDAGO
568, 569

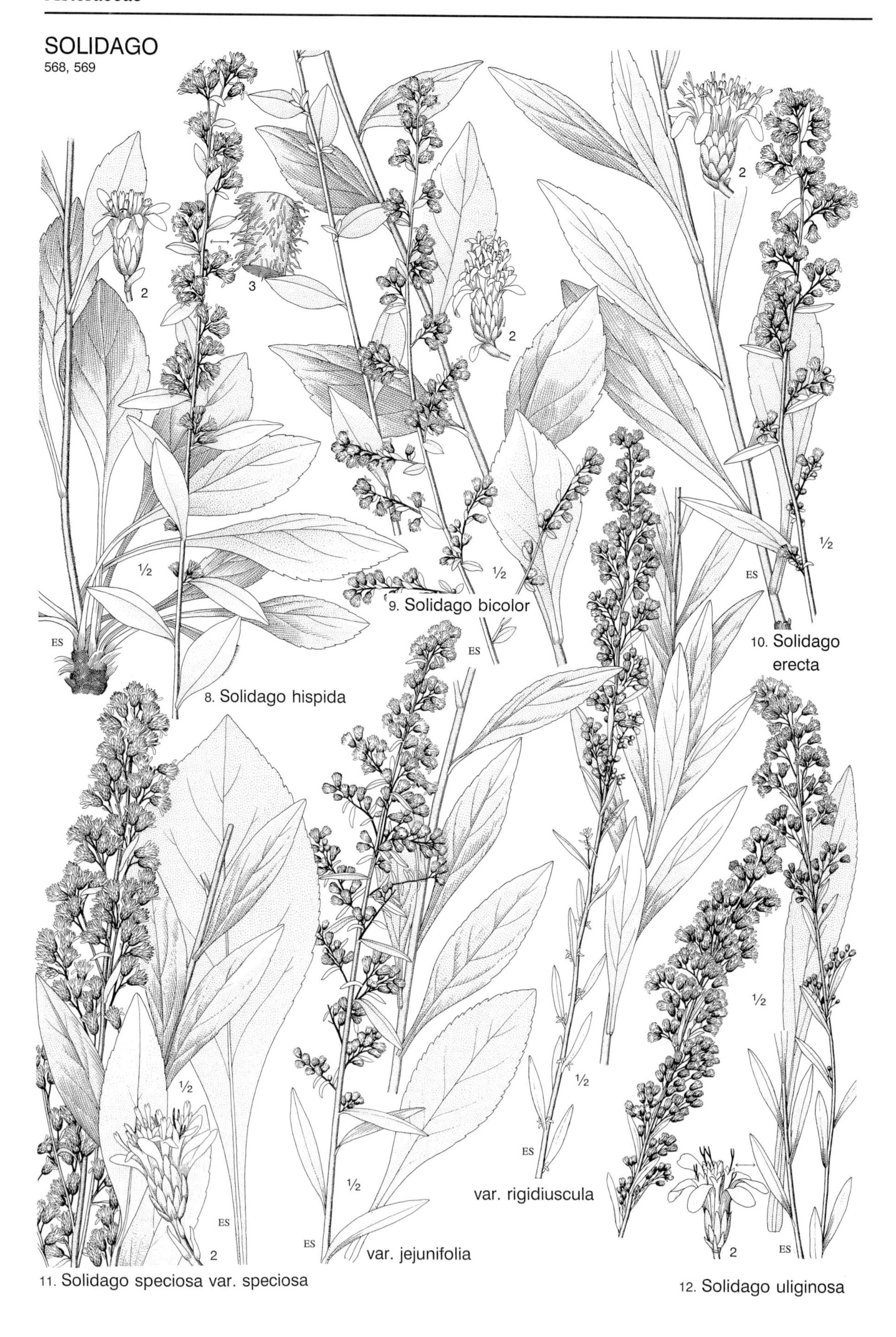

2

3

2

2

½

ES

8. Solidago hispida

9. Solidago bicolor

½

ES

10. Solidago erecta

½

ES

½

11. Solidago speciosa var. speciosa

2

ES

ES

½

½

var. jejunifolia

var. rigidiuscula

ES

½

ES

2

ES

12. Solidago uliginosa

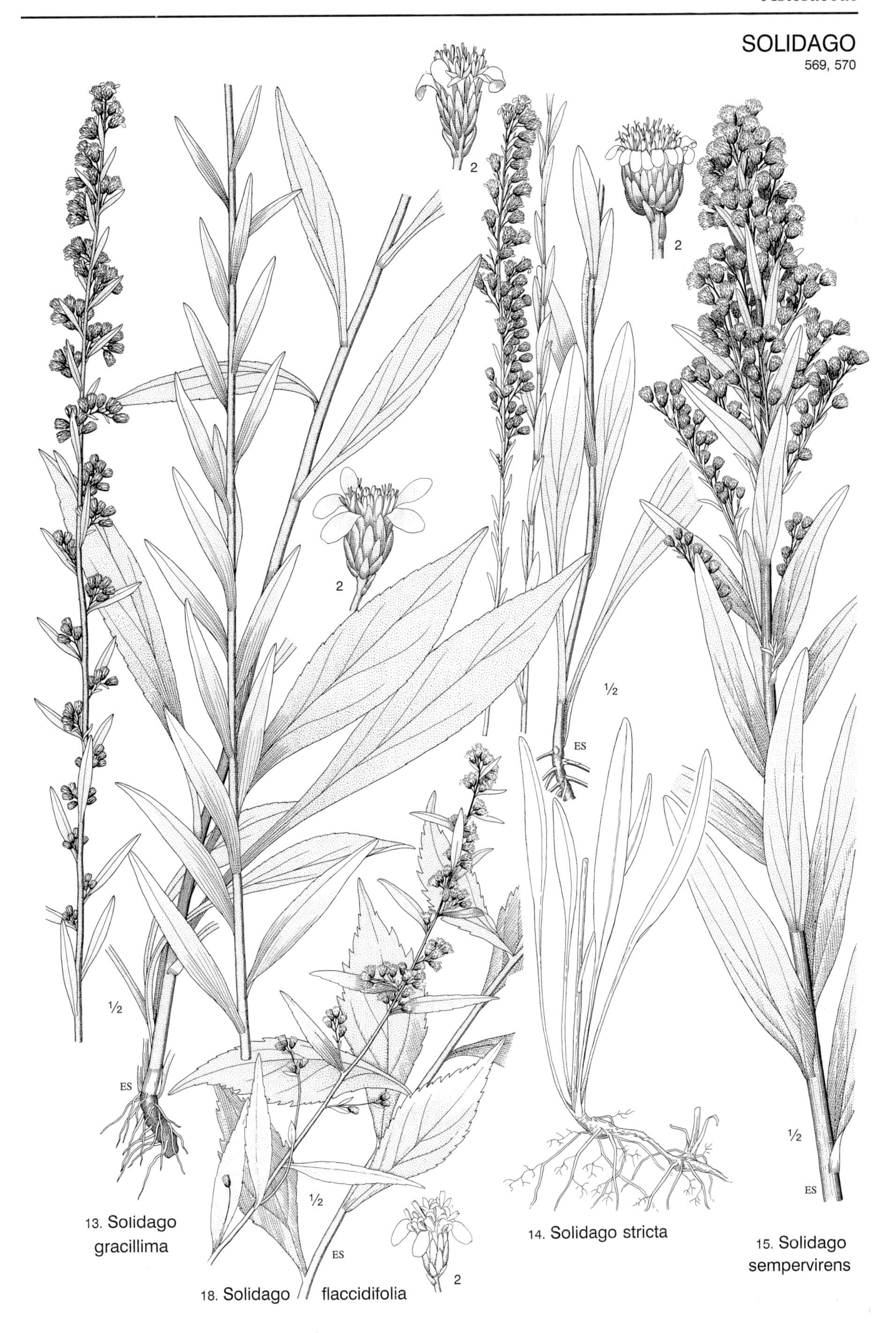

13. Solidago
gracillima

18. Solidago   flaccidifolia

2

14. Solidago stricta

15. Solidago
sempervirens

SOLIDAGO
569, 570

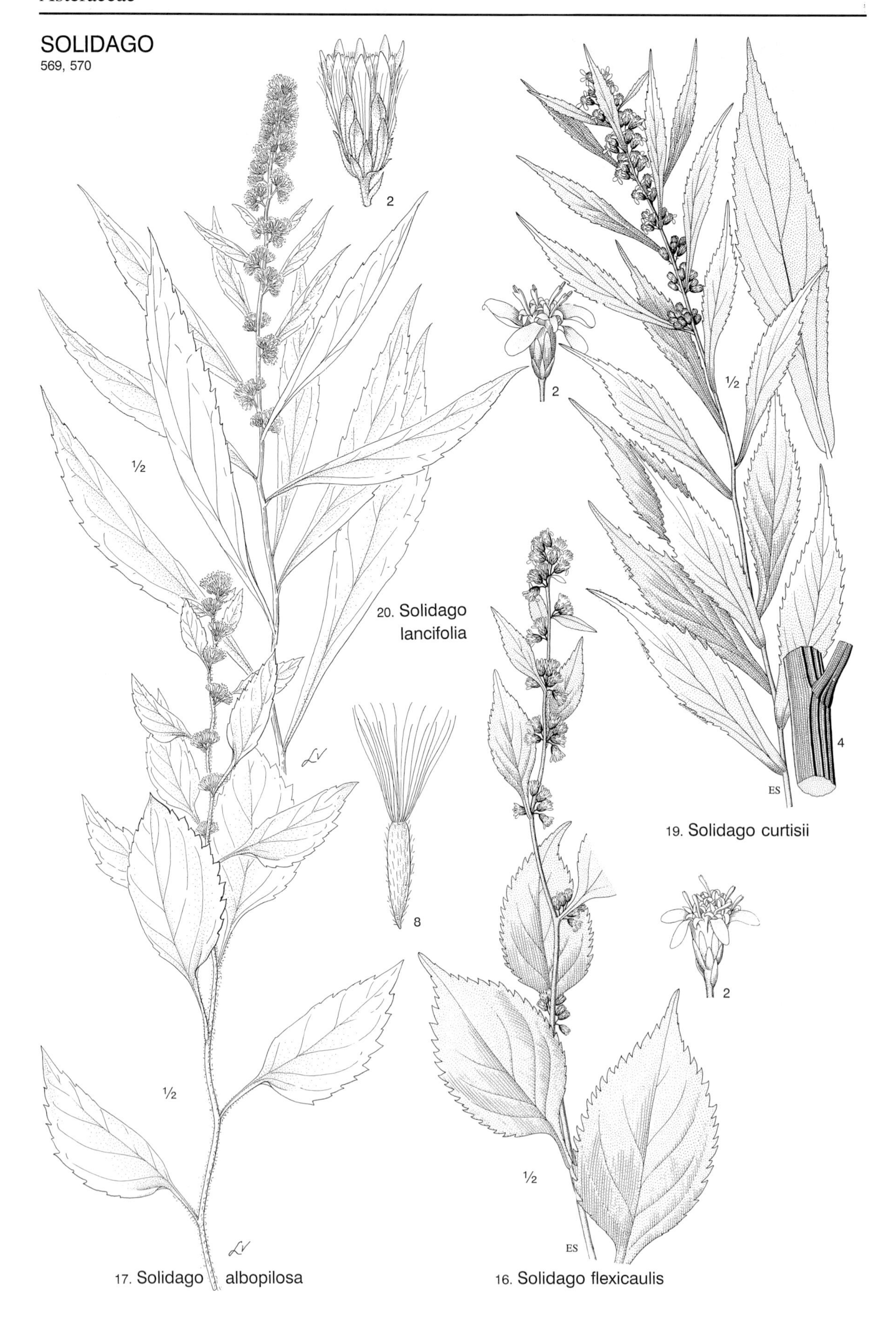

20. Solidago
lancifolia

8

19. Solidago curtisii

ES

17. Solidago albopilosa

16. Solidago flexicaulis

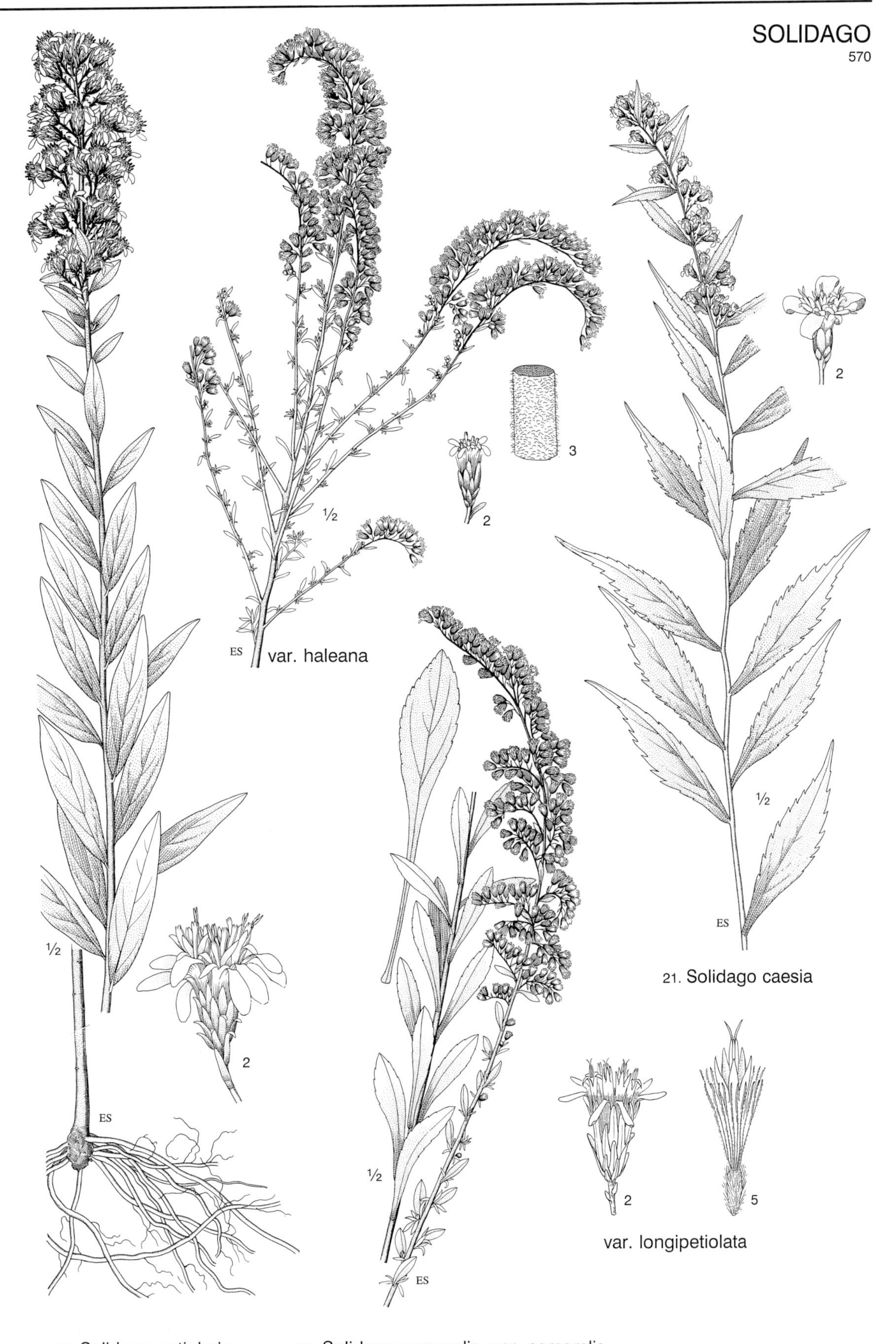

ES var. haleana

3

2

21. Solidago caesia

2

var. longipetiolata

22. Solidago petiolaris   23. Solidago nemoralis var. nemoralis

SOLIDAGO
570, 571

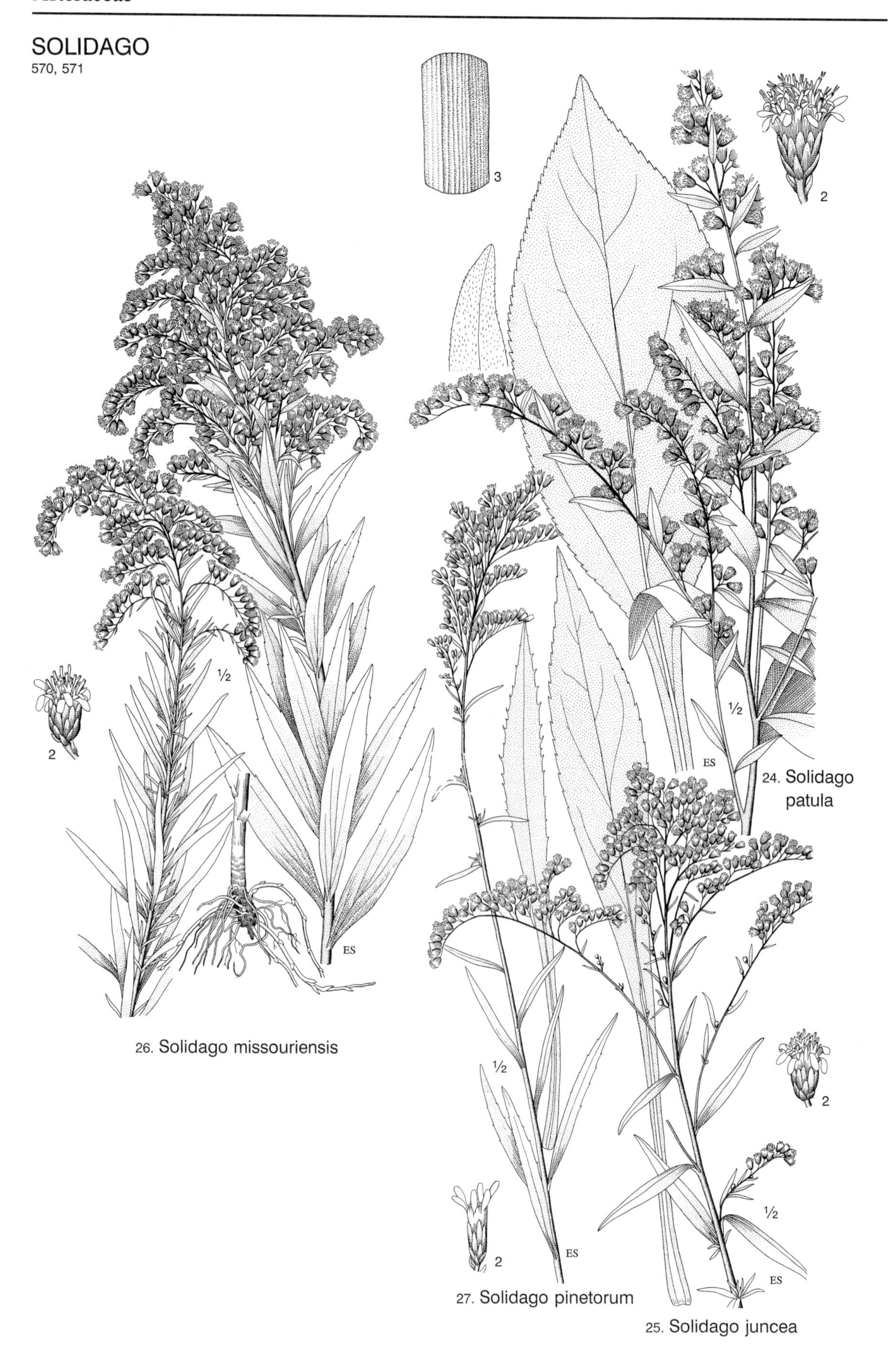

3

2

2

26. Solidago missouriensis

24. Solidago patula

27. Solidago pinetorum

2

25. Solidago juncea

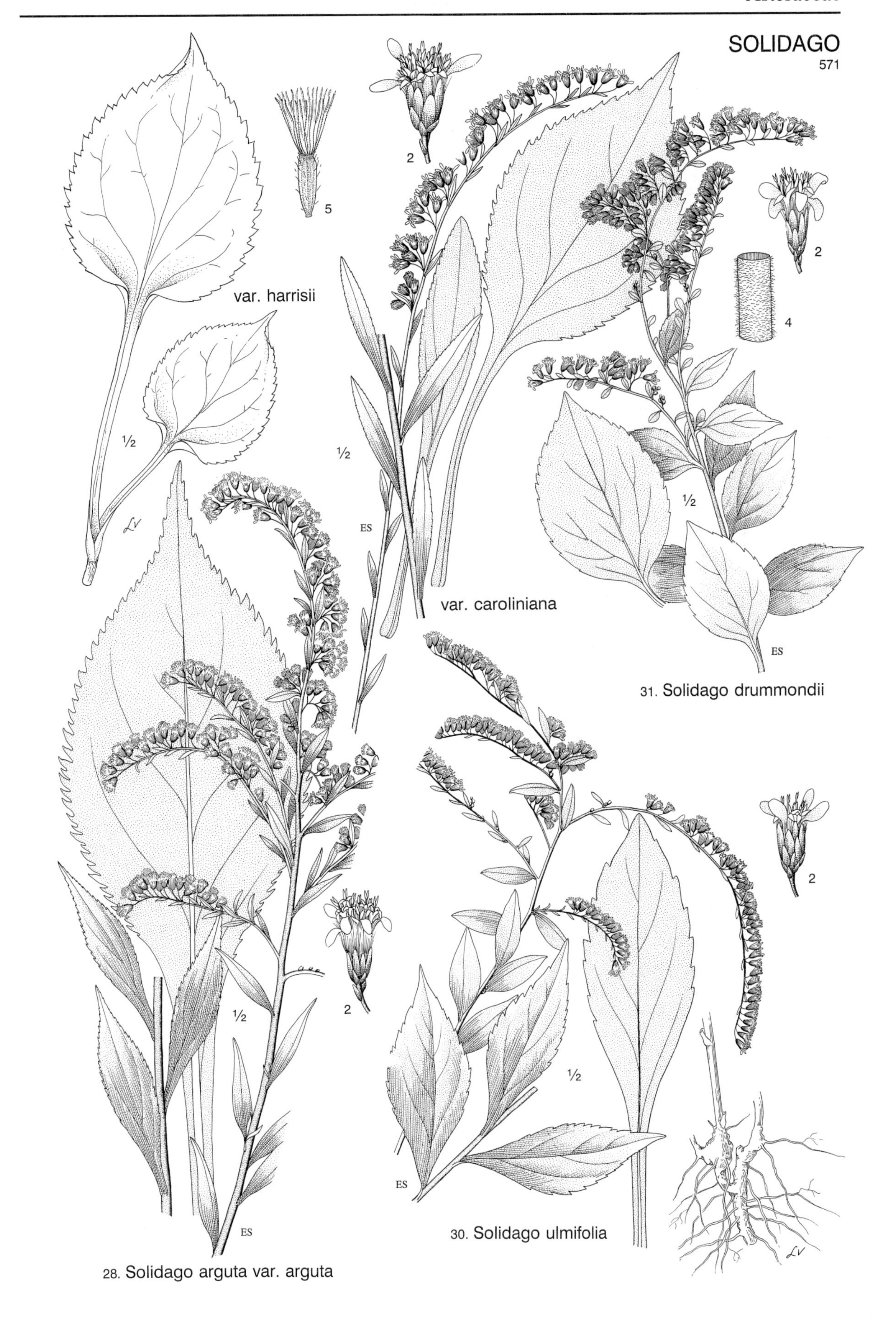

var. harrisii

½

5

2

2

4

½

½

var. caroliniana

ES

31. Solidago drummondii

ES

½

2

2

28. Solidago arguta var. arguta

30. Solidago ulmifolia

½

ES

SOLIDAGO
571, 572

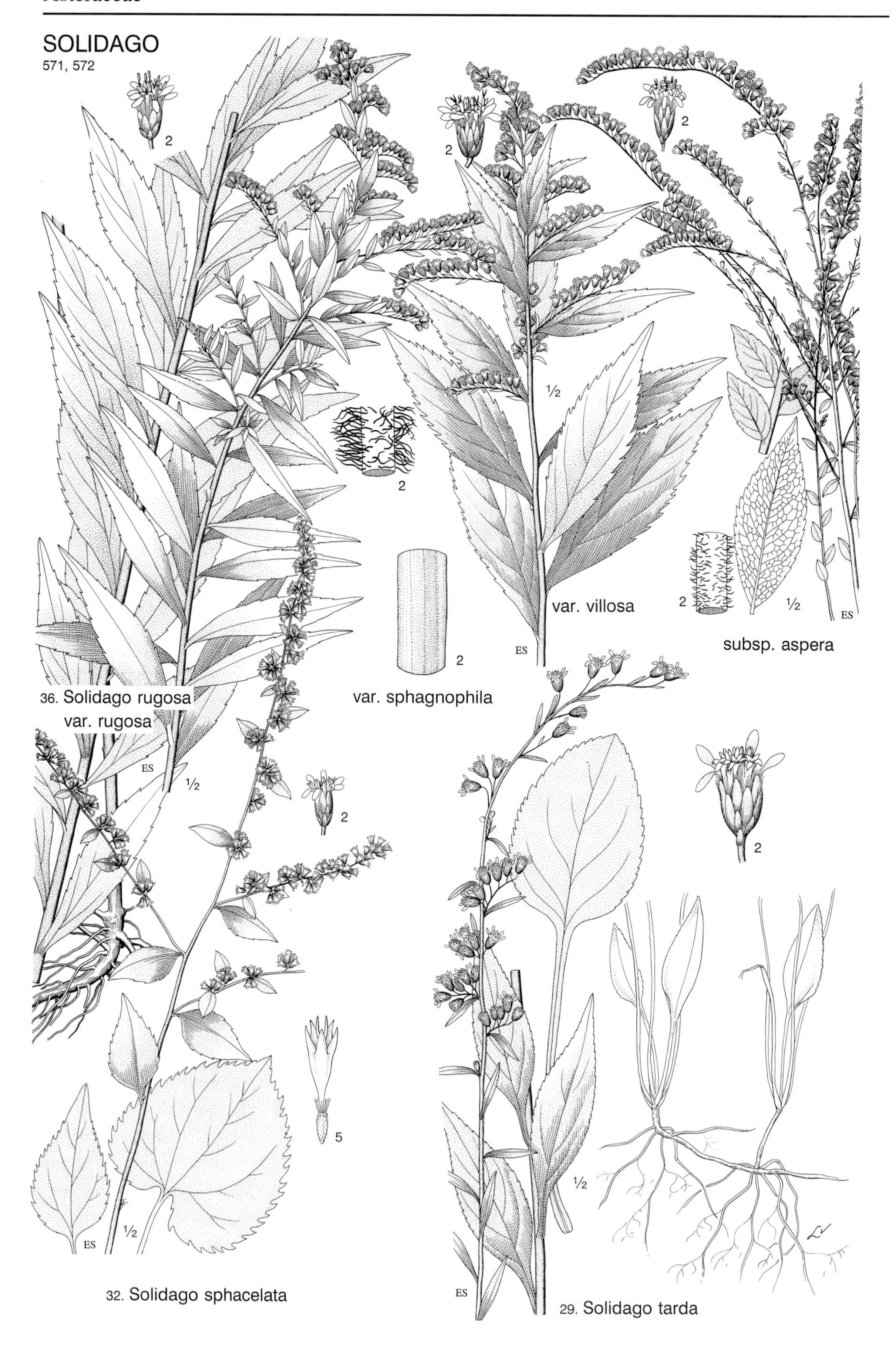

var. villosa

var. sphagnophila

subsp. aspera

36. Solidago rugosa
var. rugosa

32. Solidago sphacelata

29. Solidago tarda

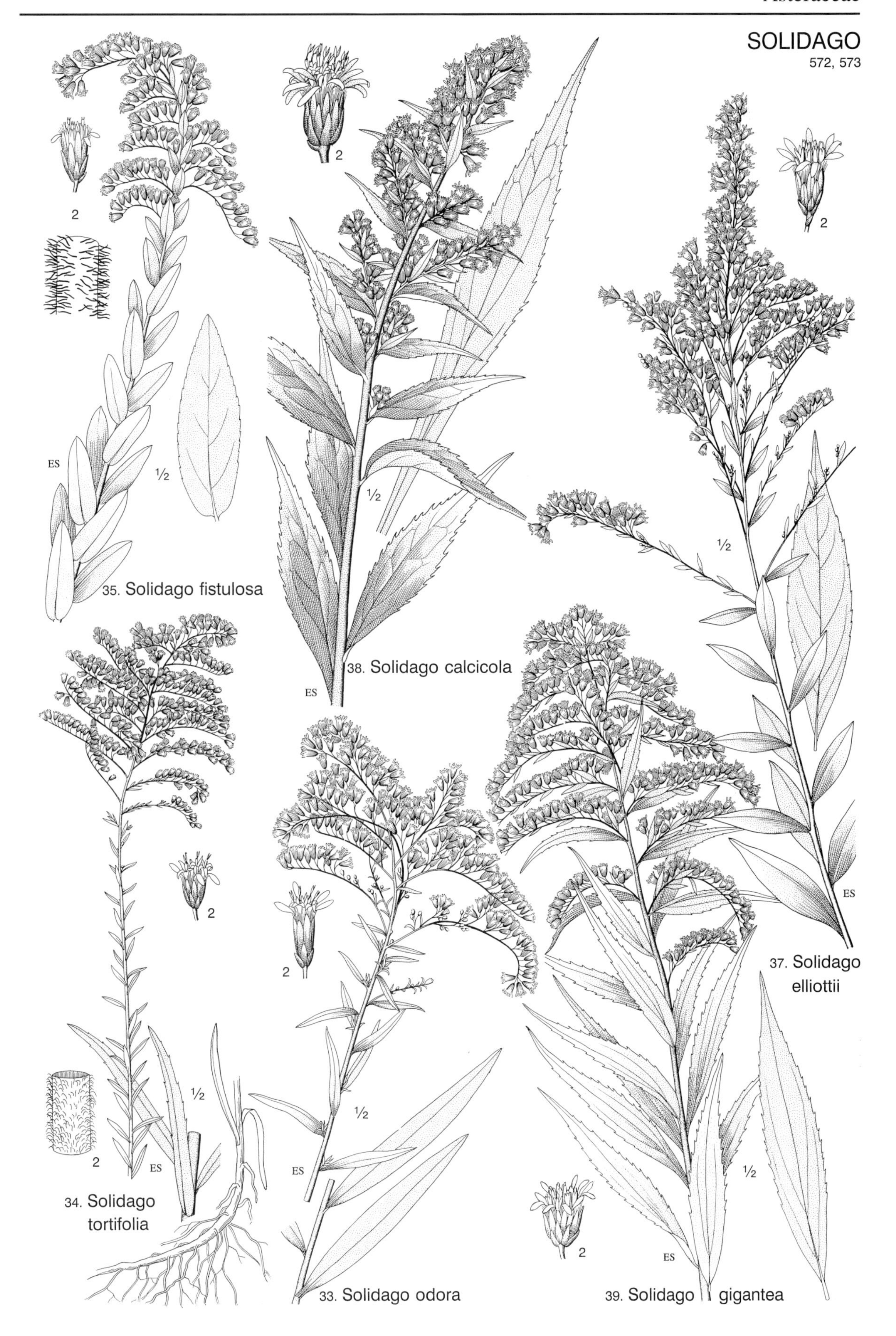

2

2

2

ES

½

35. Solidago fistulosa

38. Solidago calcicola

ES

½

2

½

34. Solidago tortifolia

2

ES

33. Solidago odora

½

ES

2

½

ES

37. Solidago elliottii

2

½

ES

39. Solidago gigantea

SOLIDAGO
573

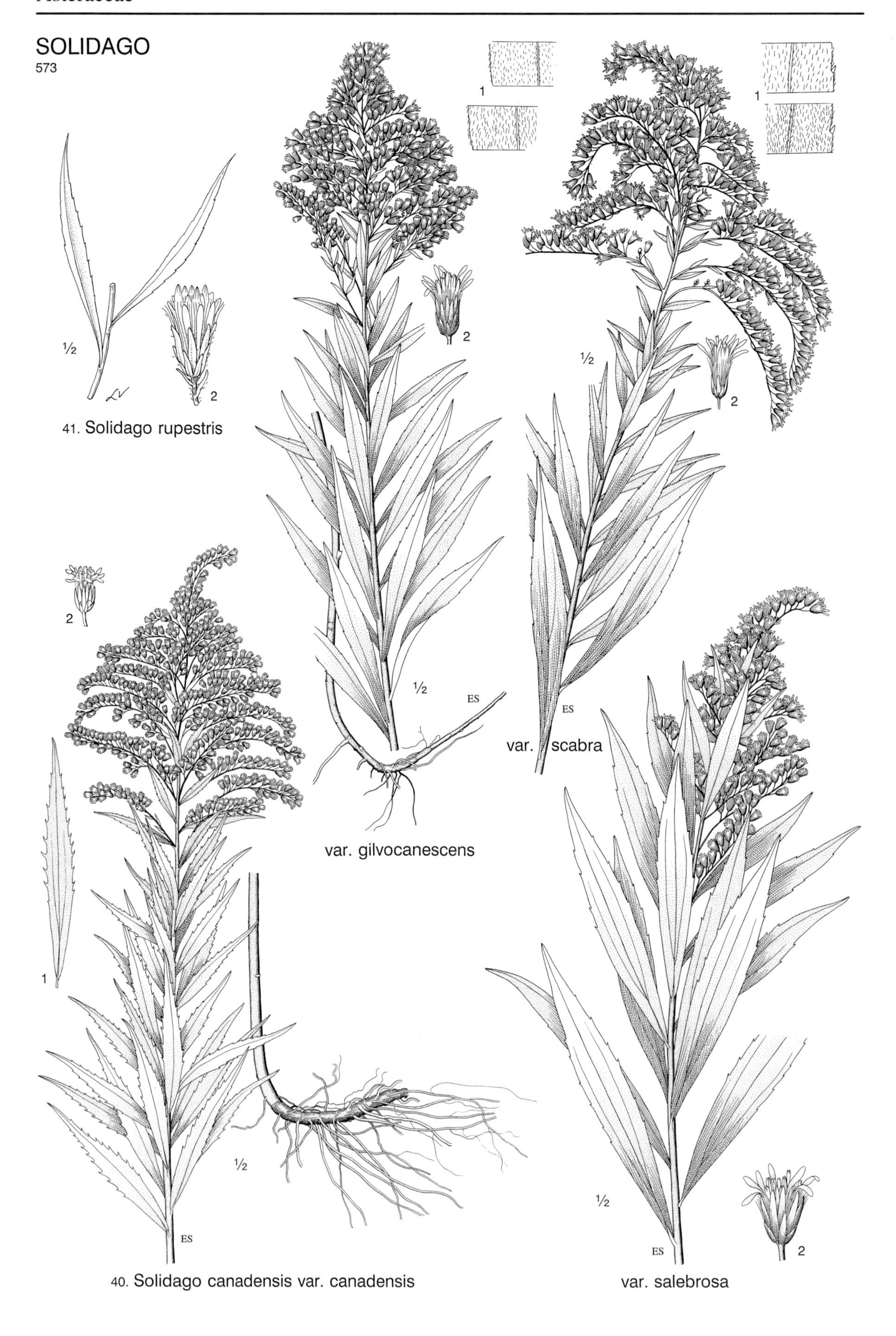

41. Solidago rupestris

var. gilvocanescens

var. scabra

40. Solidago canadensis var. canadensis

var. salebrosa

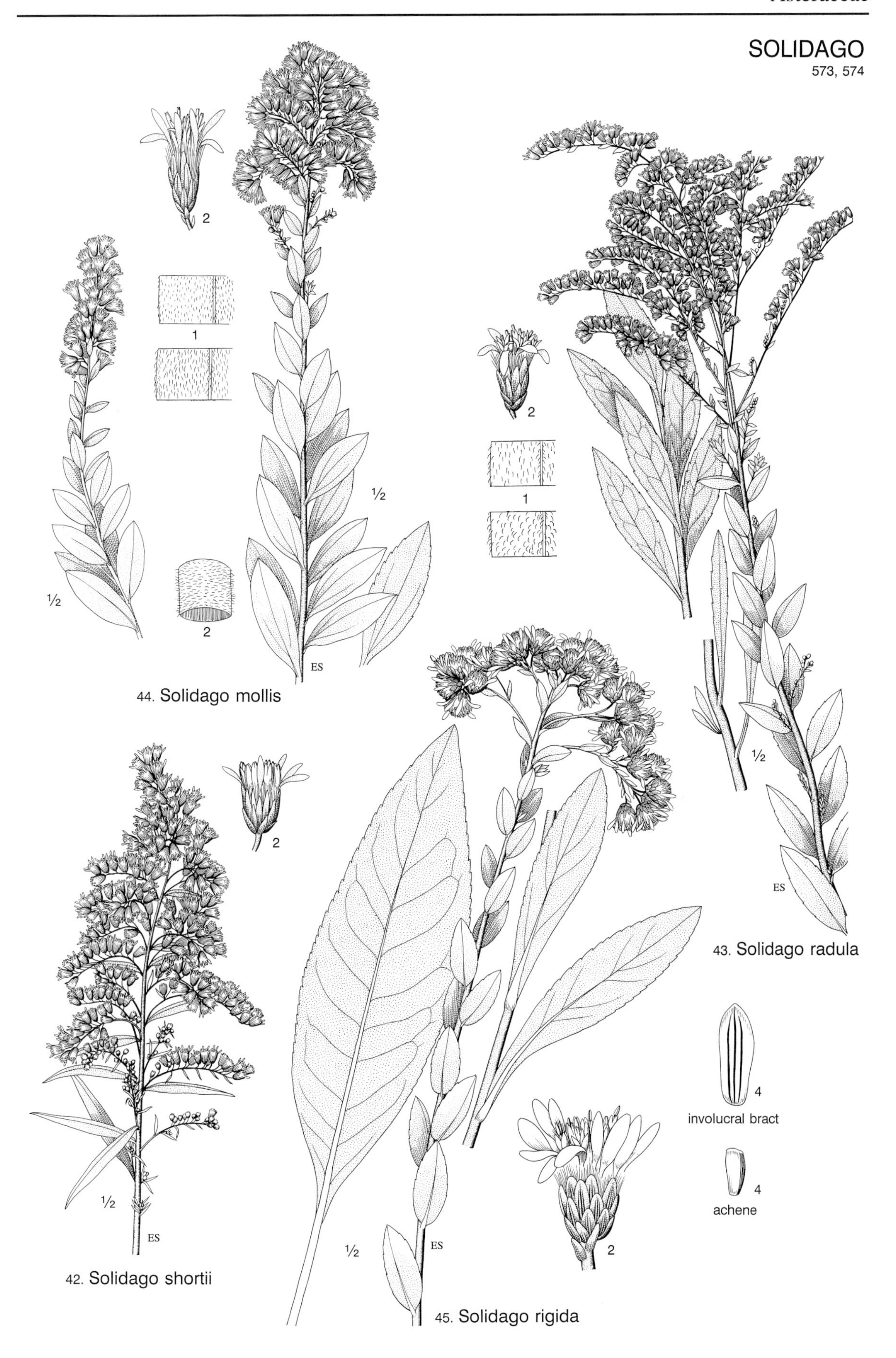

44. Solidago mollis

43. Solidago radula

42. Solidago shortii

45. Solidago rigida

involucral bract

achene

SOLIDAGO
574

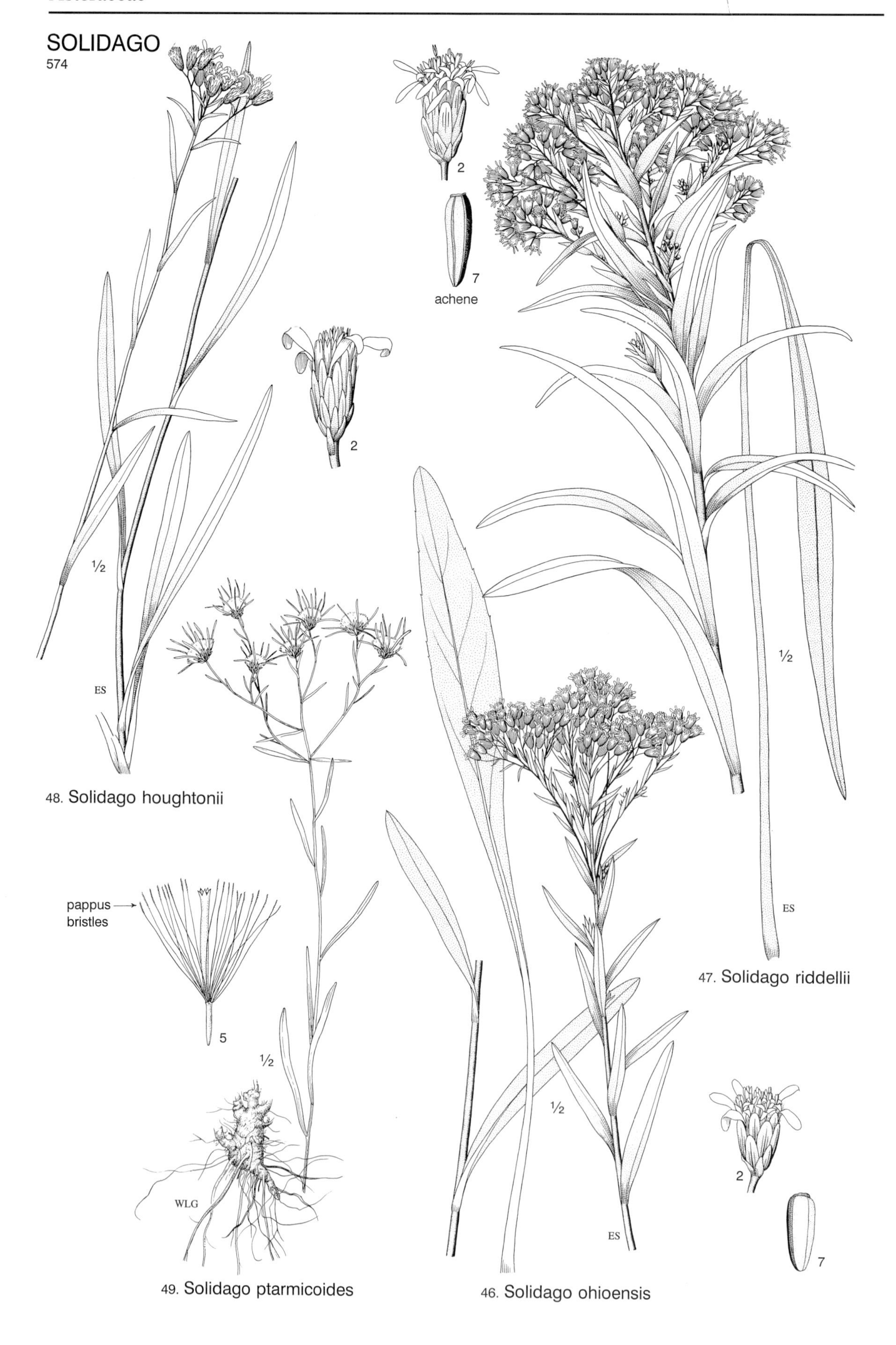

2

7
achene

2

½

ES

48. Solidago houghtonii

pappus
bristles →

5

½

WLG

49. Solidago ptarmicoides

½

ES

46. Solidago ohioensis

½

ES

47. Solidago riddellii

2

7

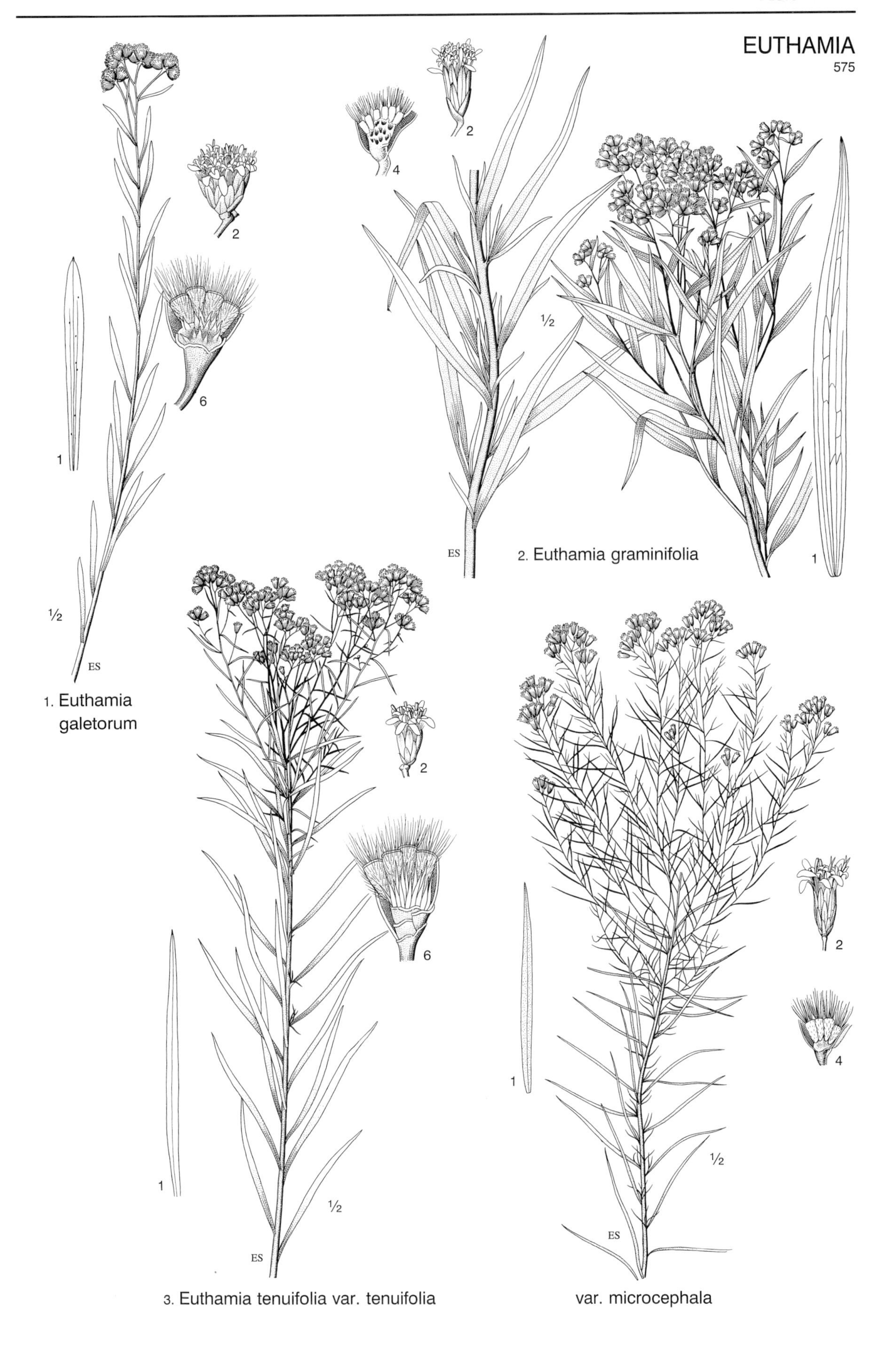

1. Euthamia
galetorum

2. Euthamia graminifolia

3. Euthamia tenuifolia var. tenuifolia

var. microcephala

EUTHAMIA
GUTIERREZIA
AMPHIACHYRIS
575, 576

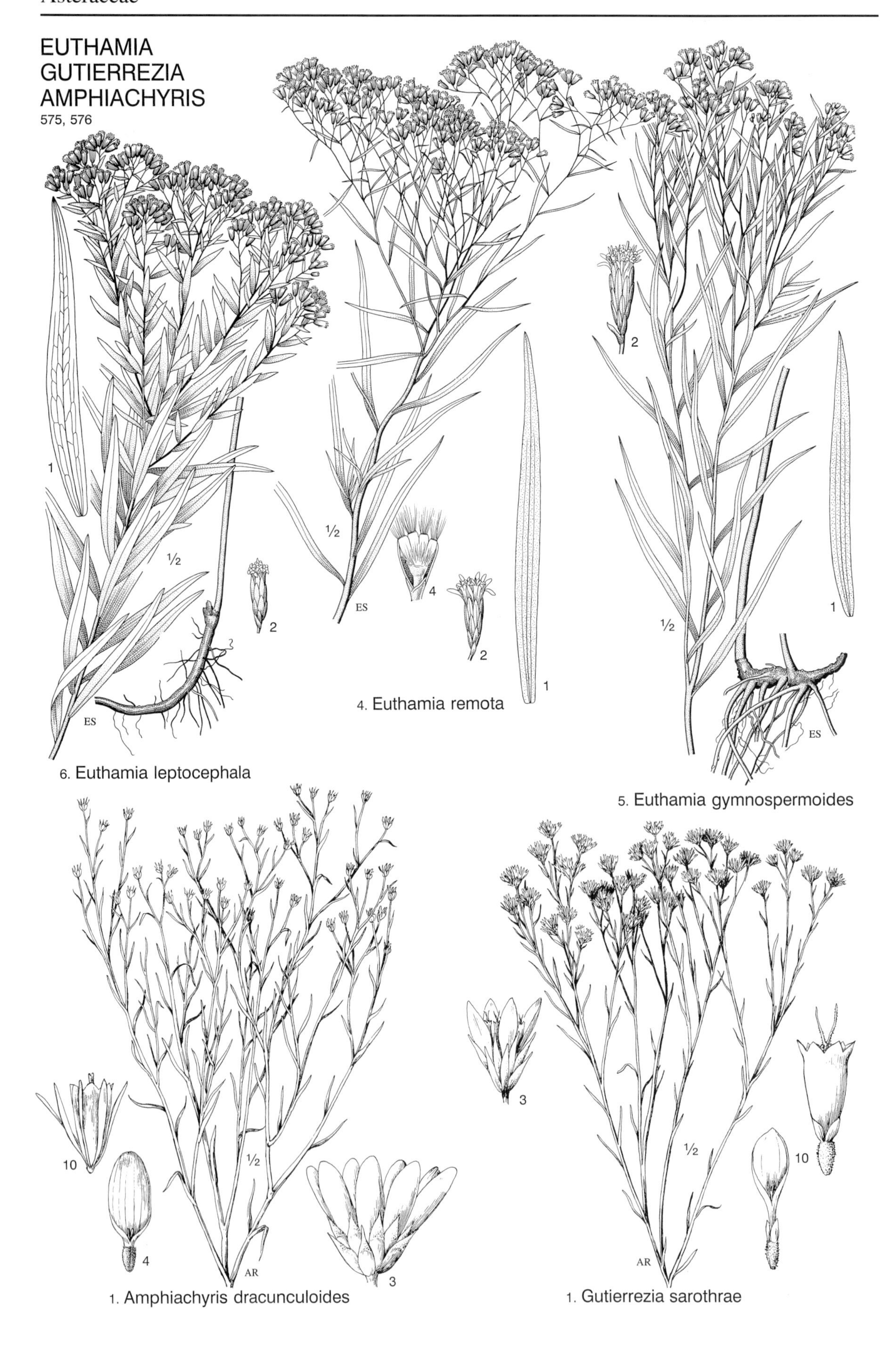

6. Euthamia leptocephala

4. Euthamia remota

5. Euthamia gymnospermoides

1. Amphiachyris dracunculoides

1. Gutierrezia sarothrae

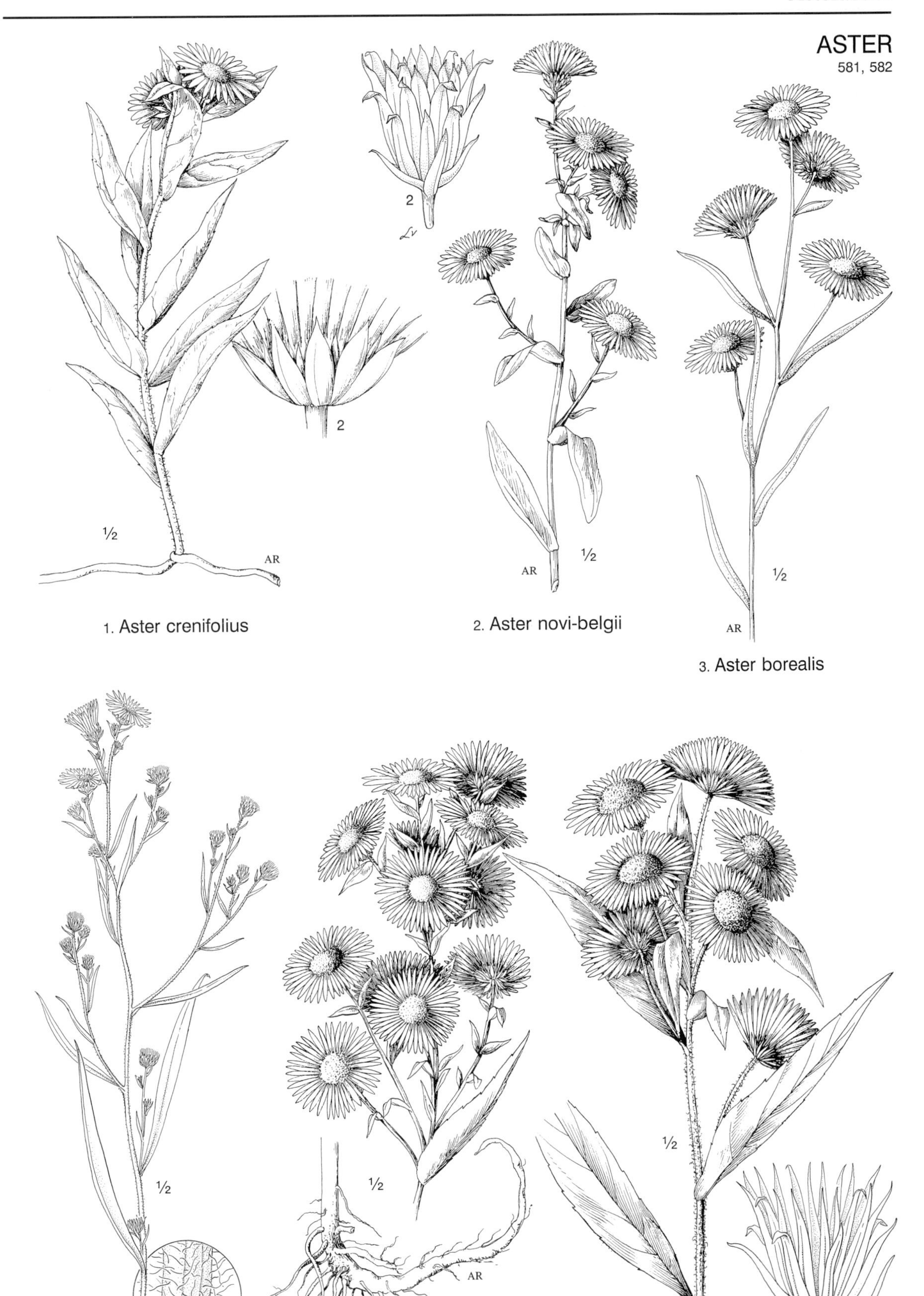

1. Aster crenifolius

2. Aster novi-belgii

3. Aster borealis

4. Aster ×longulus

6. Aster firmus

5. Aster puniceus

ASTER

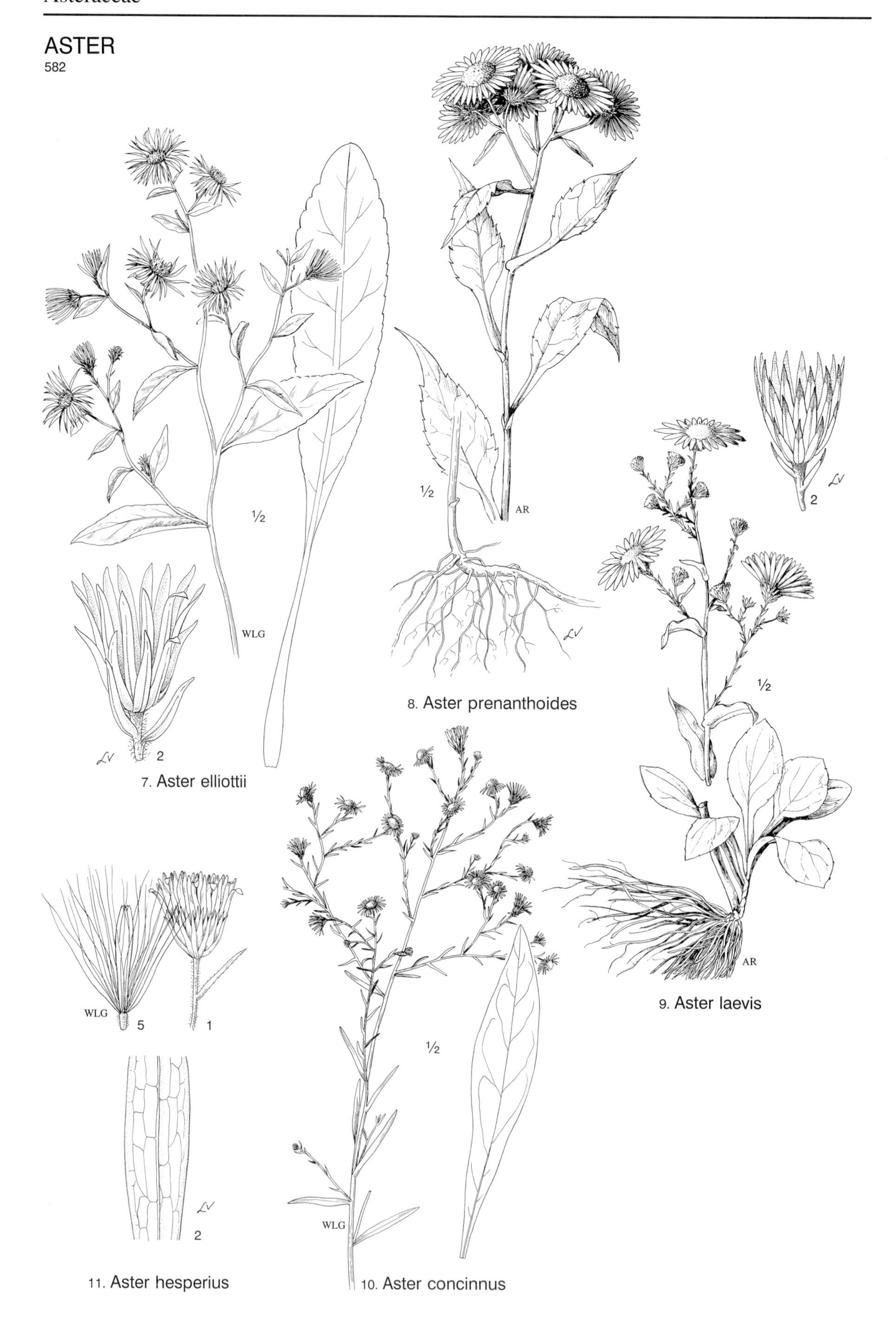

7. Aster elliottii

8. Aster prenanthoides

9. Aster laevis

11. Aster hesperius

10. Aster concinnus

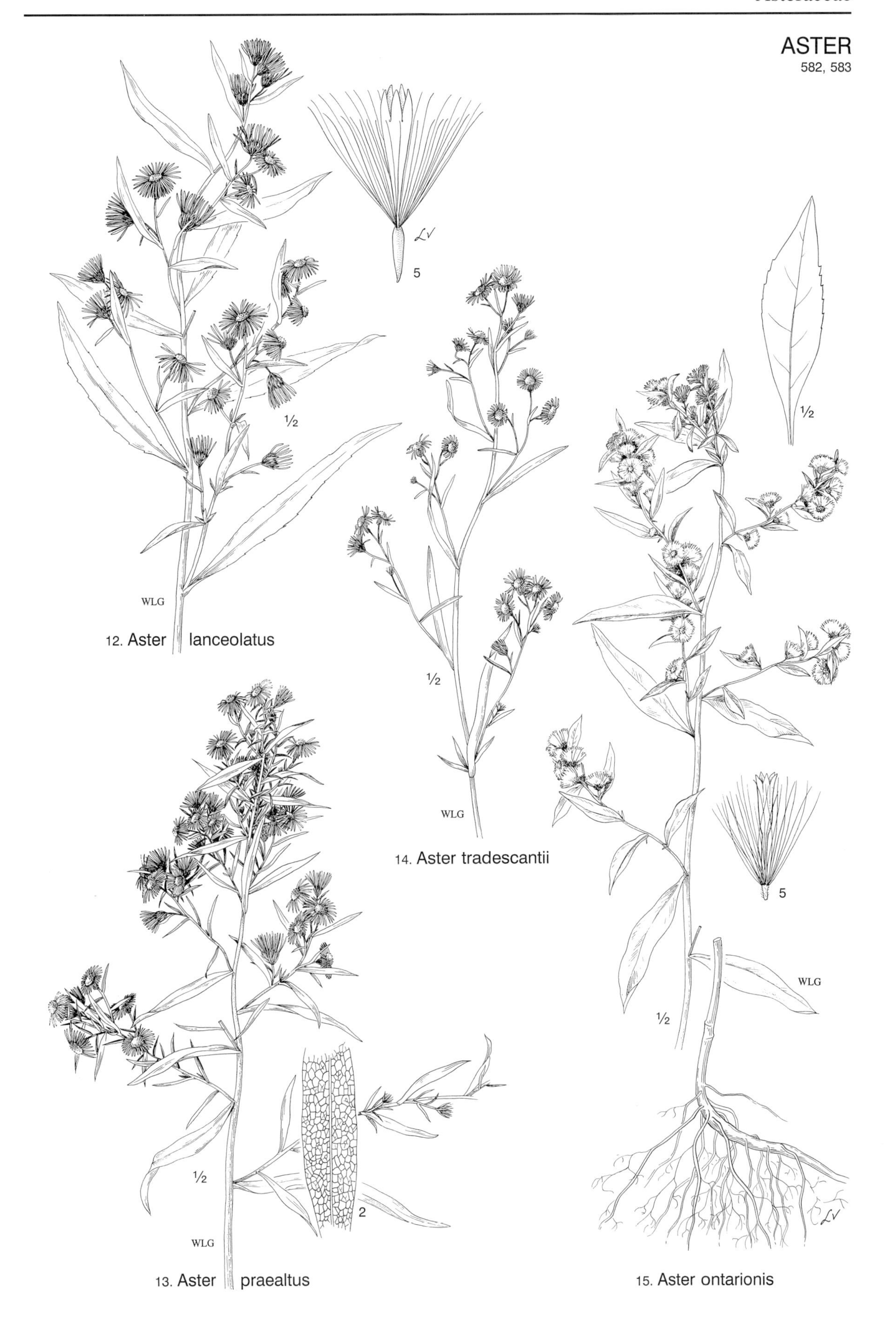

12. Aster lanceolatus

WLG

14. Aster tradescantii

WLG

13. Aster praealtus

WLG

15. Aster ontarionis

ASTER
583, 584

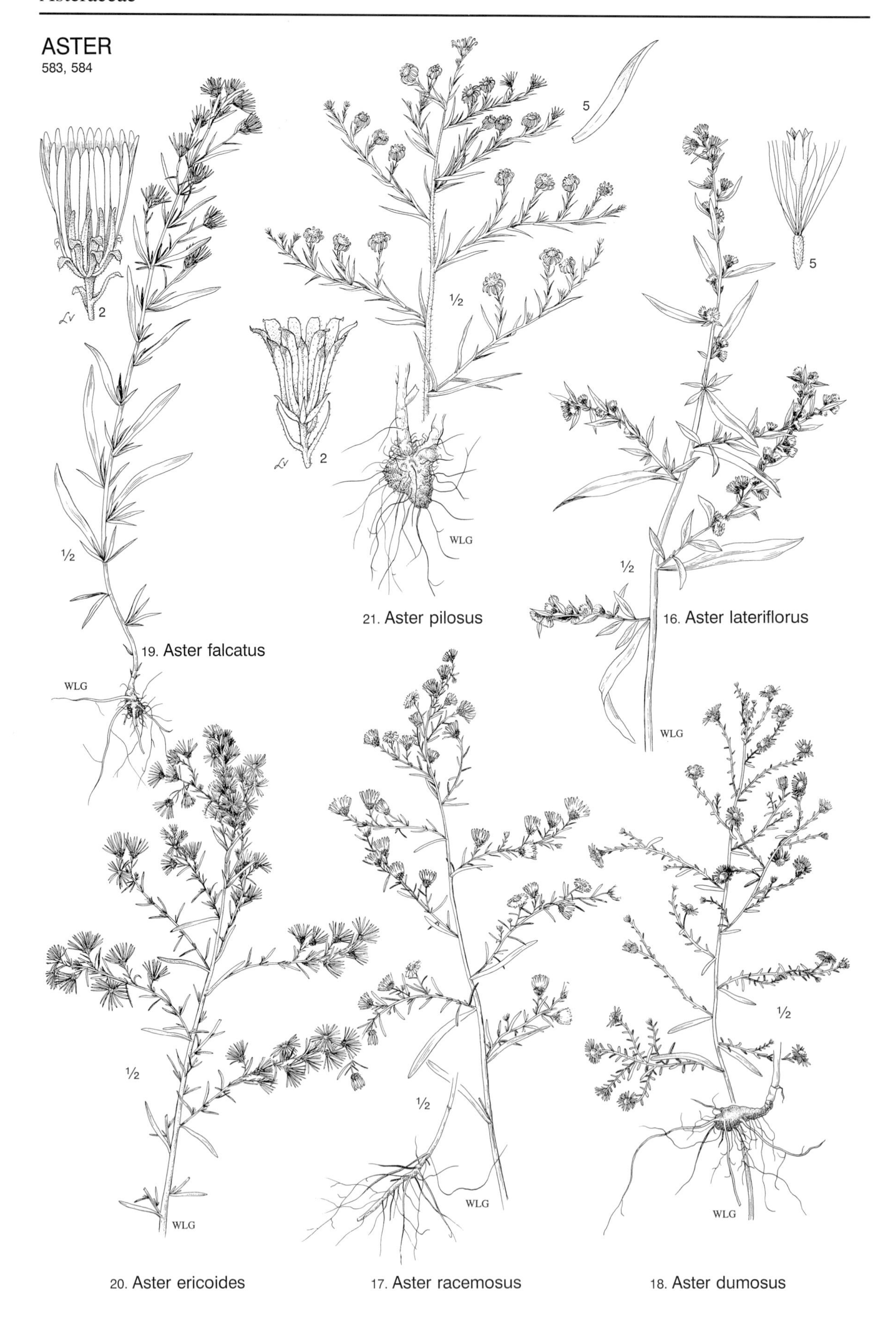

2

1/2

5

1/2

5

WLG

21. Aster pilosus

2

1/2

16. Aster lateriflorus

WLG

19. Aster falcatus

WLG

1/2

1/2

1/2

1/2

WLG

20. Aster ericoides

WLG

17. Aster racemosus

WLG

18. Aster dumosus

ILLUSTRATED COMPANION TO

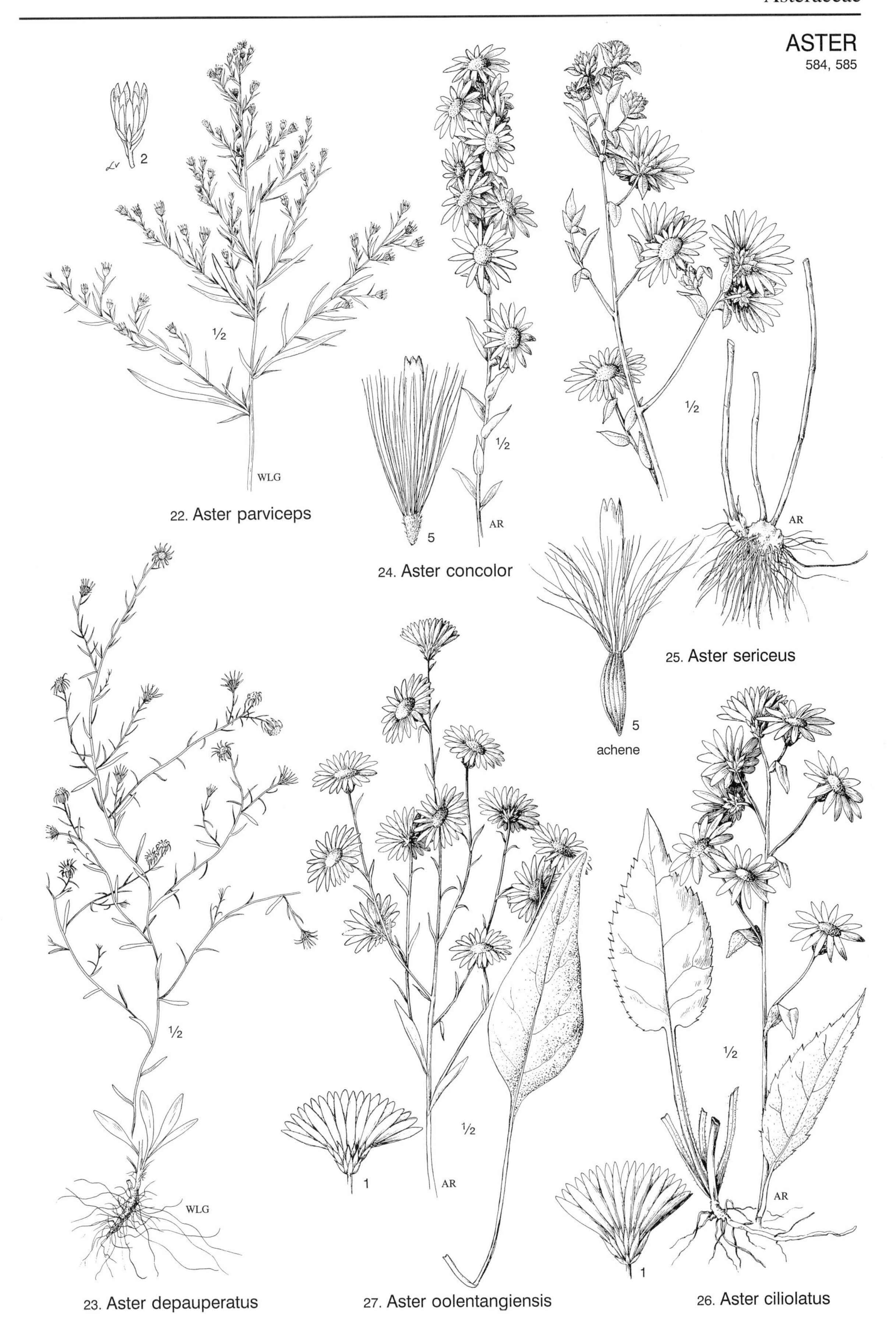

22. Aster parviceps

24. Aster concolor

25. Aster sericeus

23. Aster depauperatus

27. Aster oolentangiensis

26. Aster ciliolatus

ASTER
585

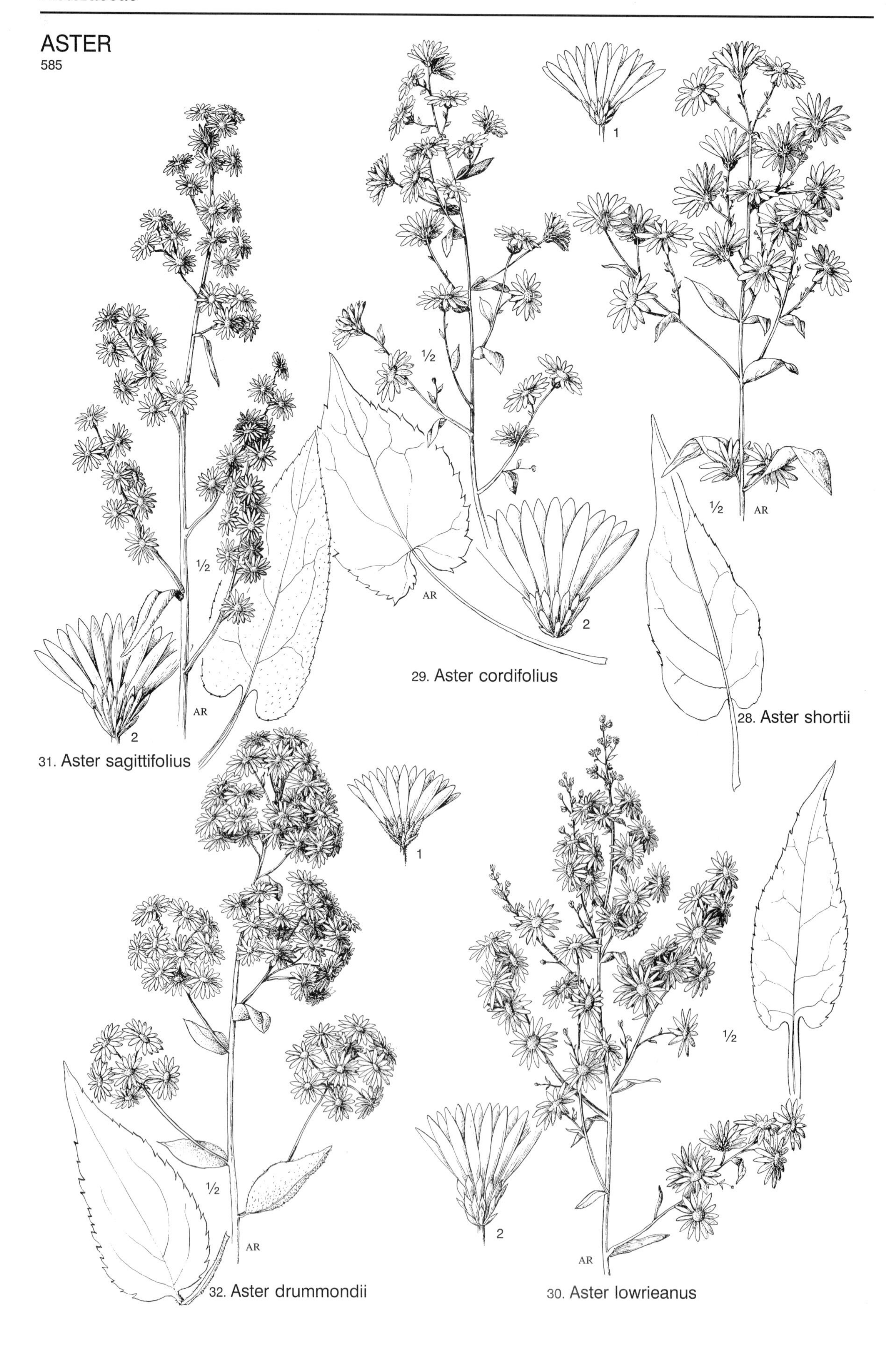

29. Aster cordifolius

28. Aster shortii

31. Aster sagittifolius

32. Aster drummondii

30. Aster lowrieanus

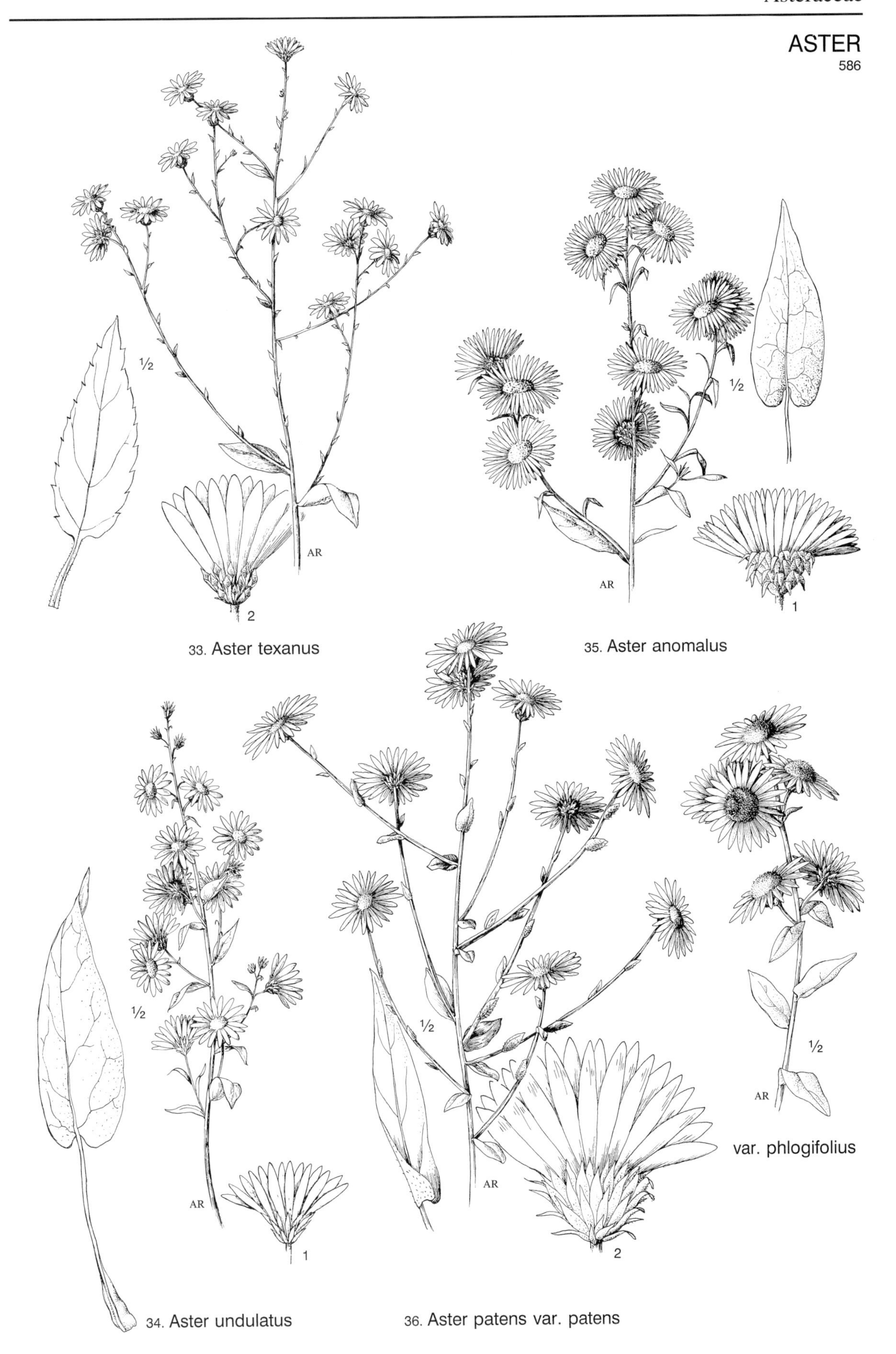

33. Aster texanus

35. Aster anomalus

34. Aster undulatus

36. Aster patens var. patens

var. phlogifolius

ASTER
586, 587

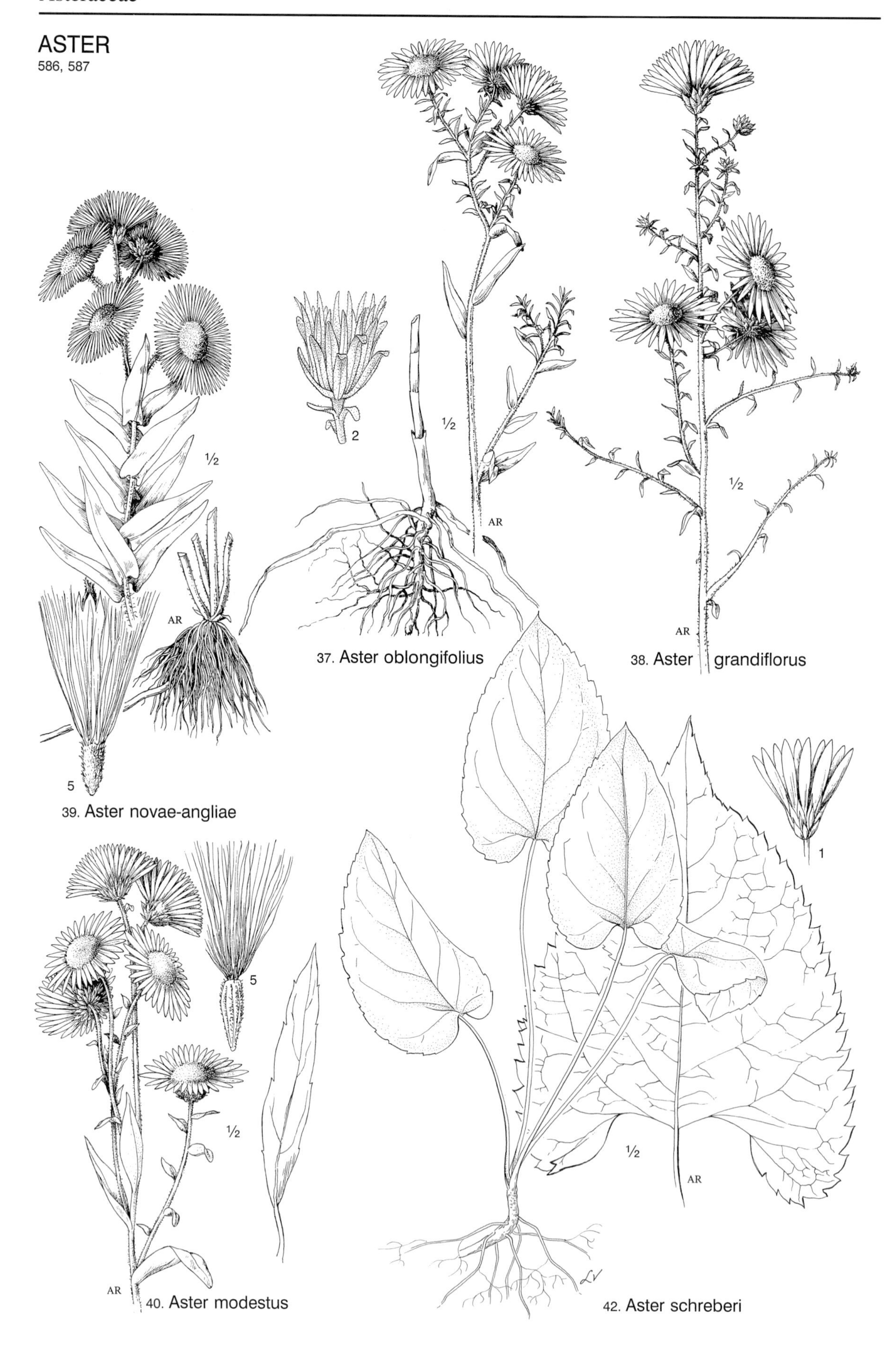

2

37. Aster oblongifolius

38. Aster grandiflorus

5

39. Aster novae-angliae

1

5

40. Aster modestus

42. Aster schreberi

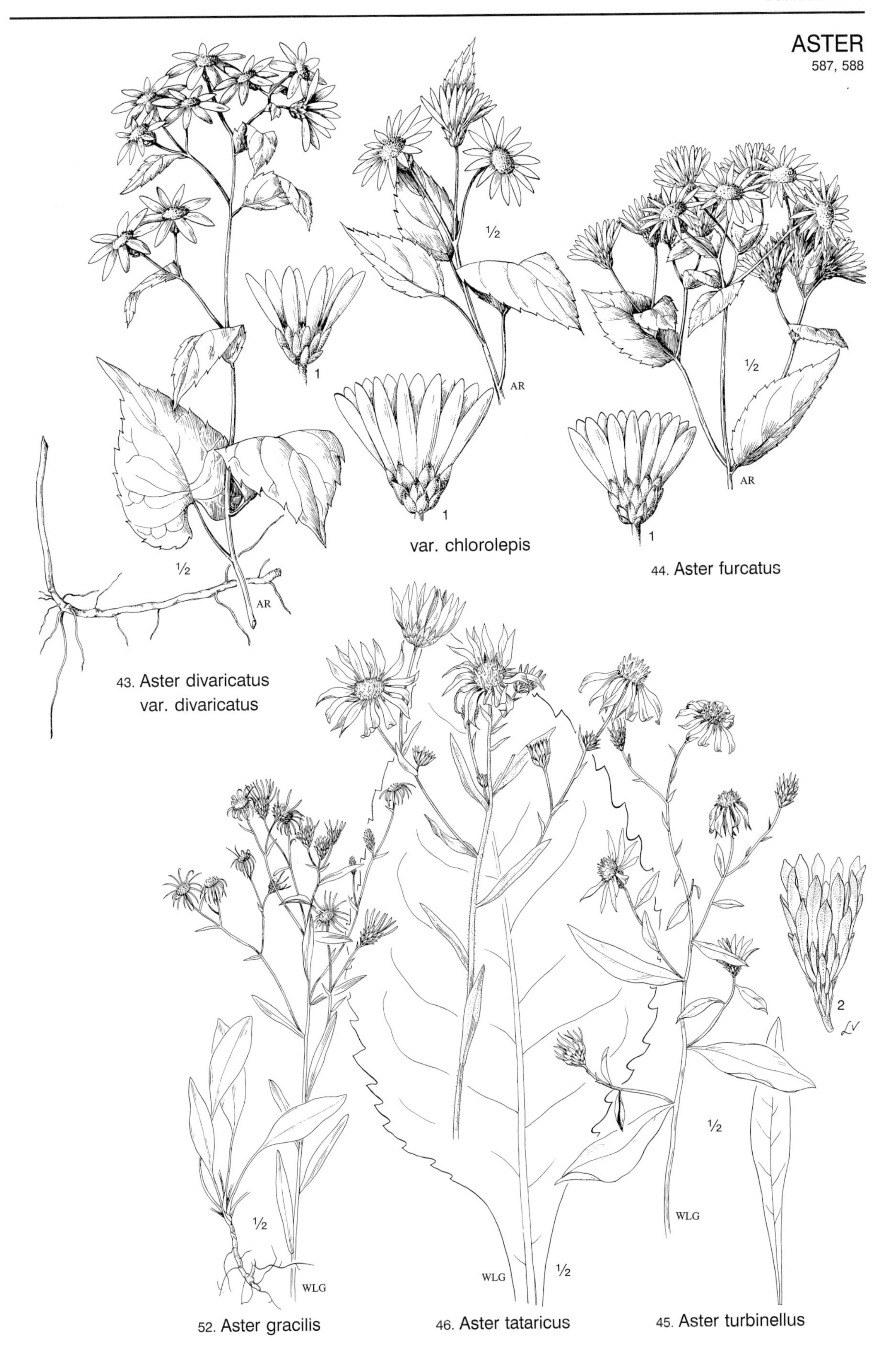

var. chlorolepis

44. Aster furcatus

43. Aster divaricatus
var. divaricatus

52. Aster gracilis

46. Aster tataricus

45. Aster turbinellus

ASTER
588

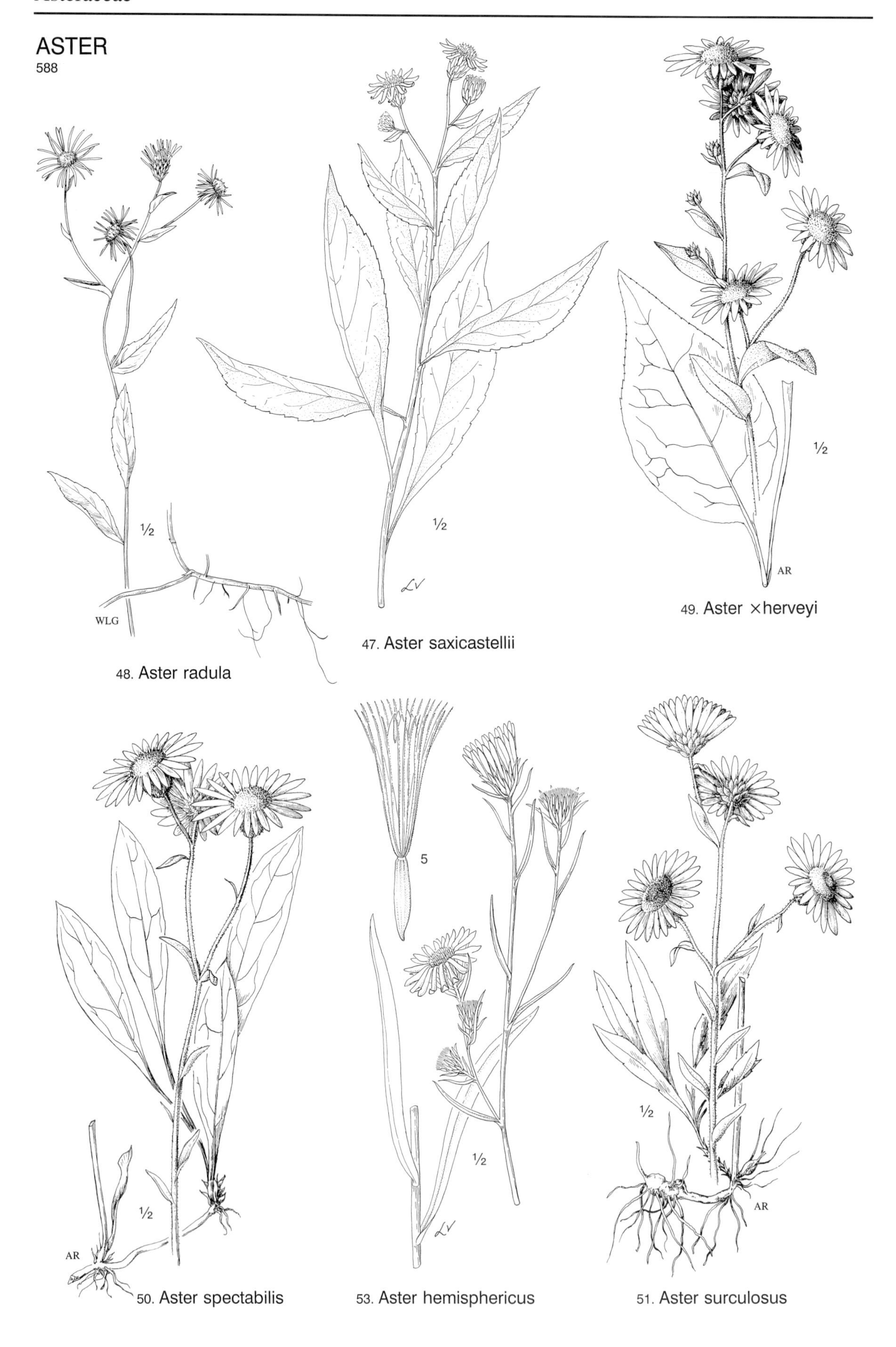

WLG

½

48. Aster radula

½

∡V

47. Aster saxicastellii

½

AR

49. Aster ×herveyi

½

AR

50. Aster spectabilis

5

½

∡V

53. Aster hemisphericus

½

AR

51. Aster surculosus

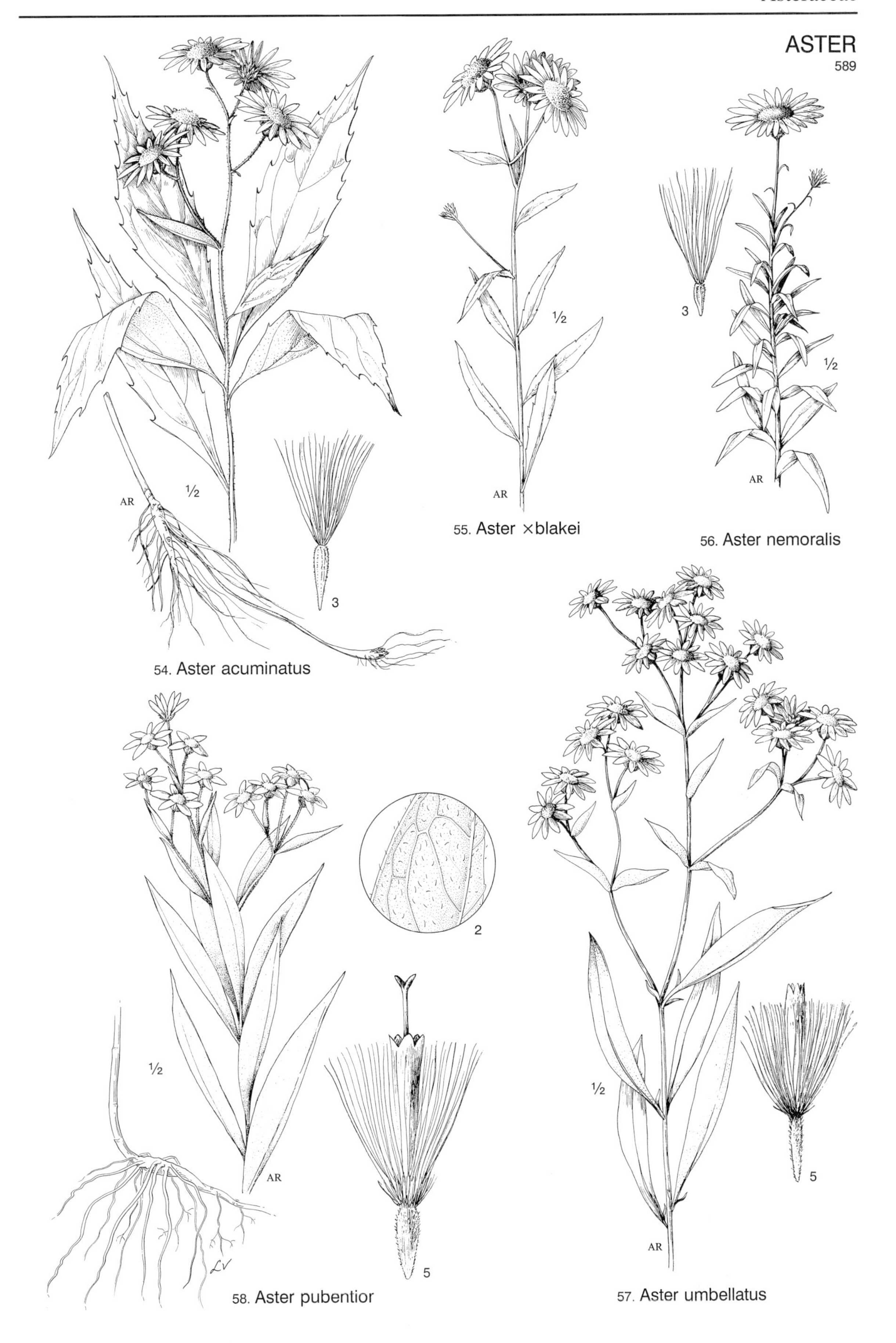

54. Aster acuminatus

55. Aster ×blakei

56. Aster nemoralis

58. Aster pubentior

57. Aster umbellatus

ASTER
589, 590

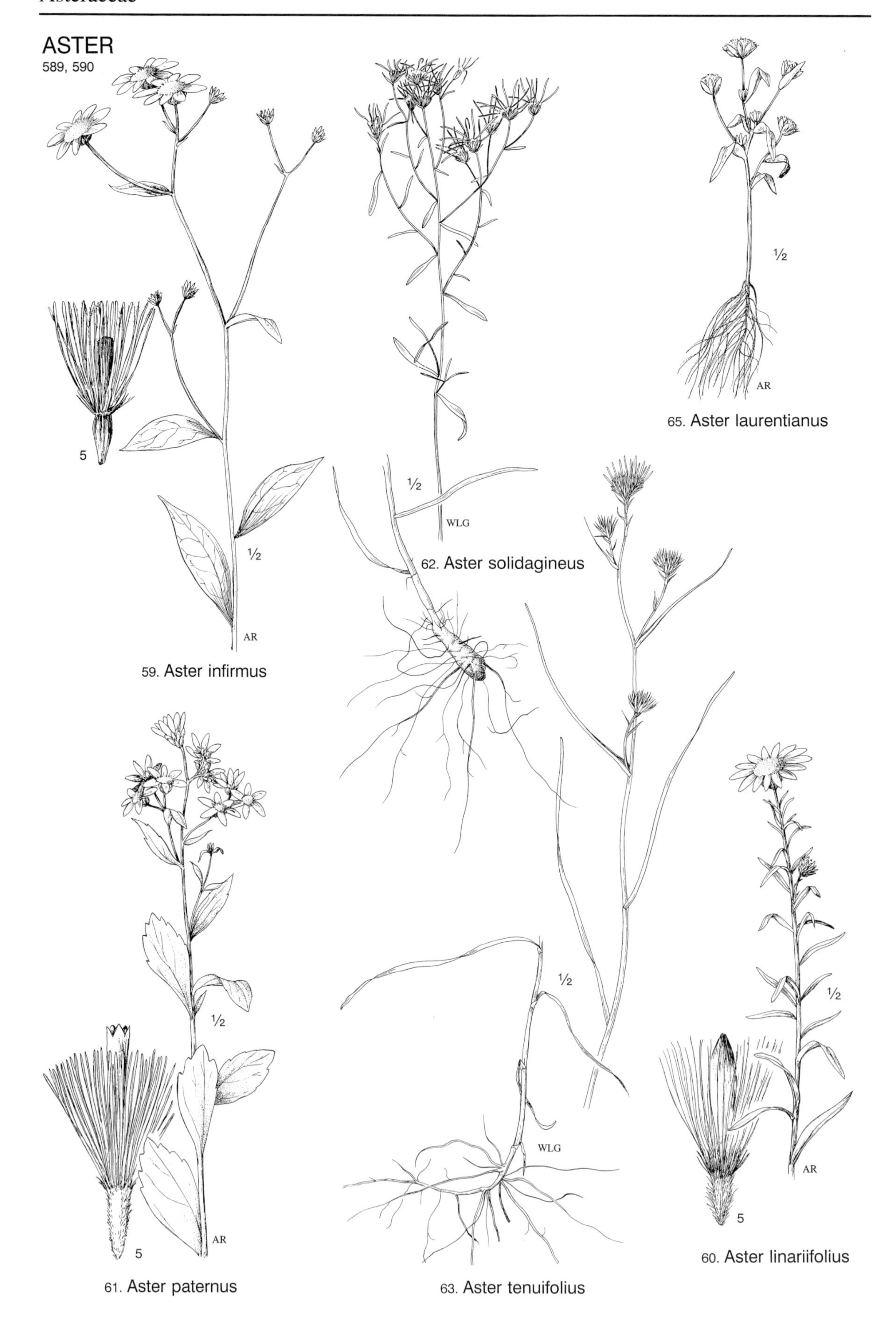

59. Aster infirmus

65. Aster laurentianus

62. Aster solidagineus

61. Aster paternus

63. Aster tenuifolius

60. Aster linariifolius

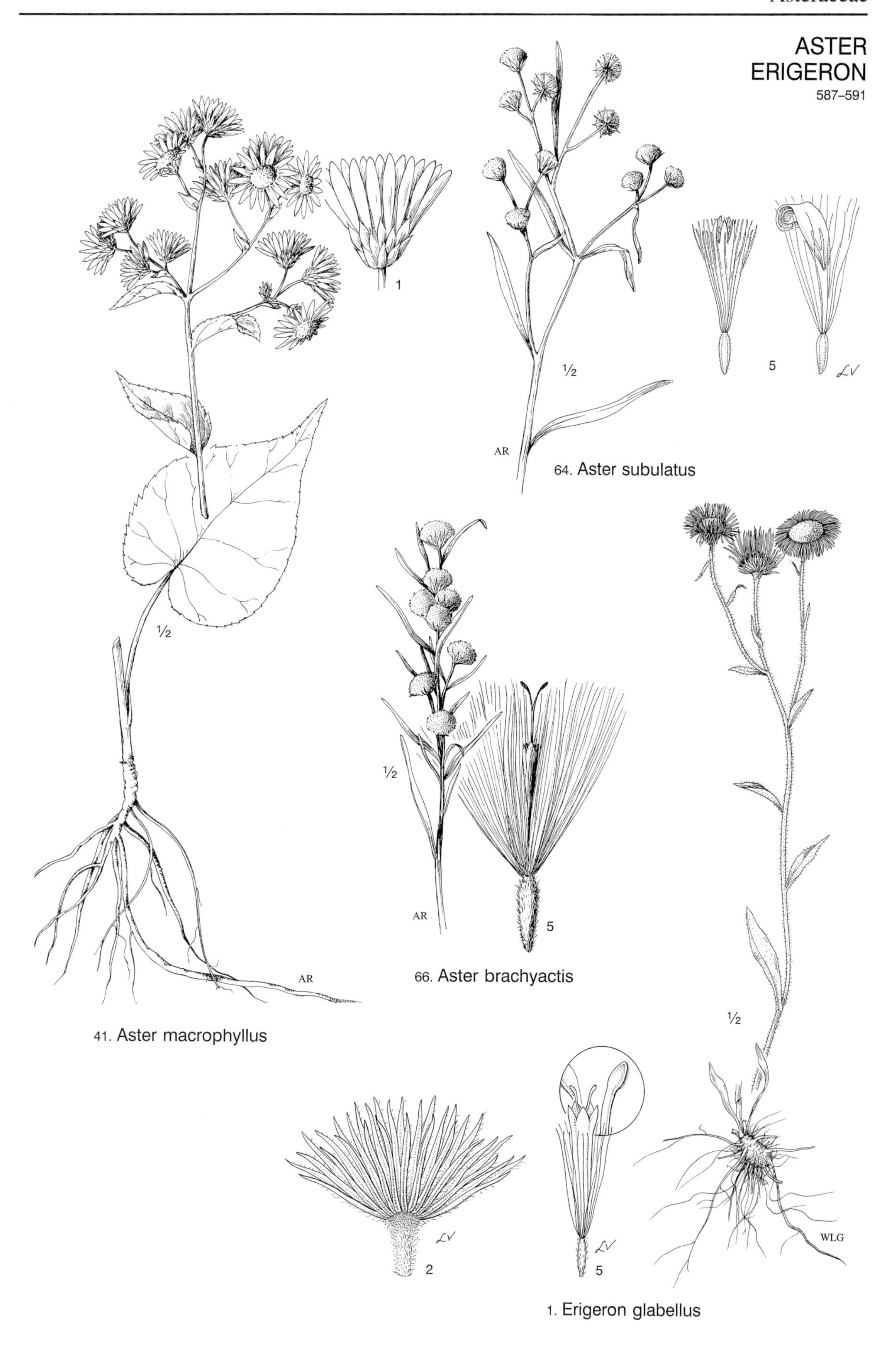

64. Aster subulatus

66. Aster brachyactis

41. Aster macrophyllus

1. Erigeron glabellus

ERIGERON
591

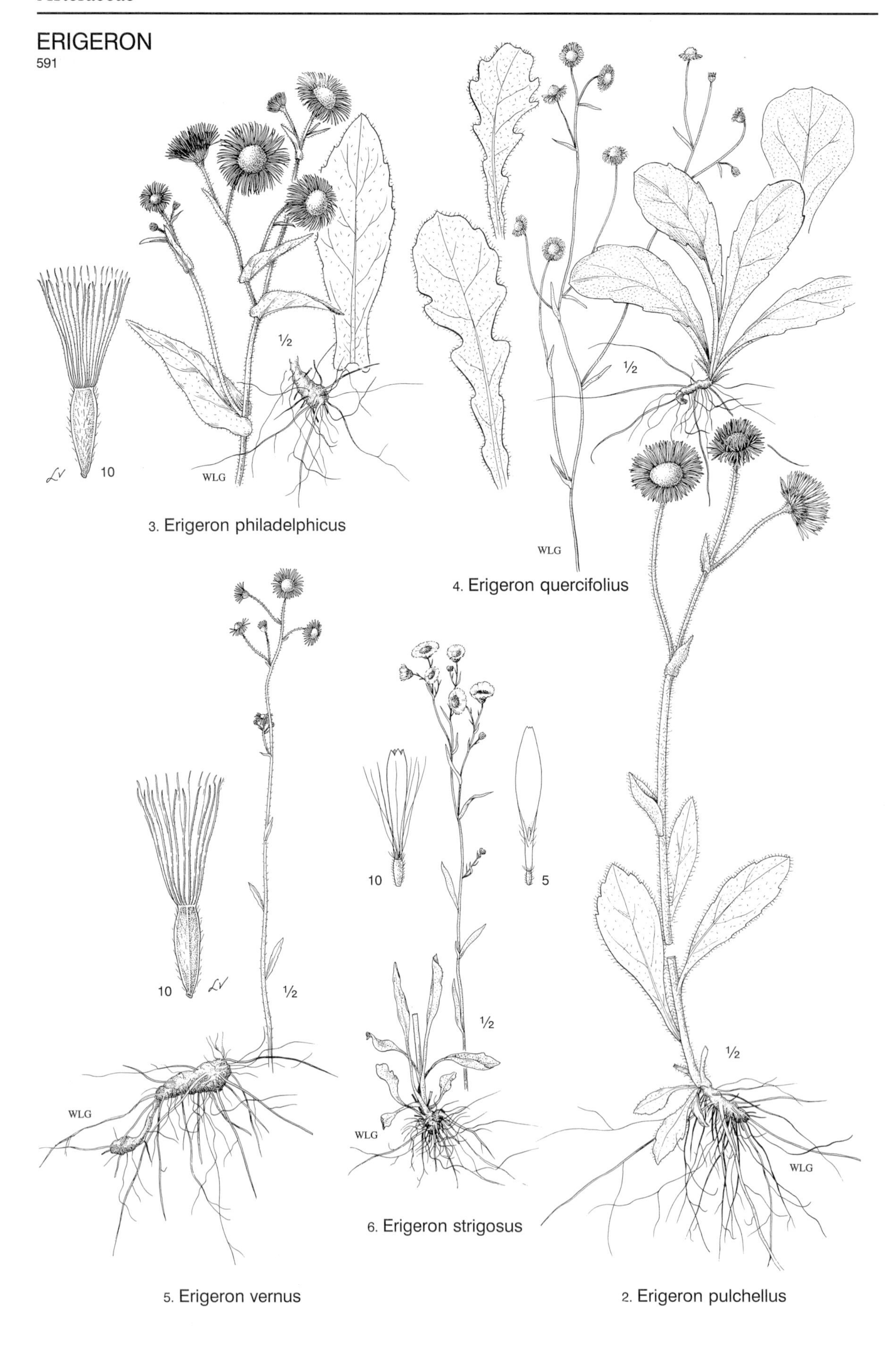

10

3. Erigeron philadelphicus

WLG

½

4. Erigeron quercifolius

10

10   ½

5. Erigeron vernus

10   5

½

6. Erigeron strigosus

½

WLG

2. Erigeron pulchellus

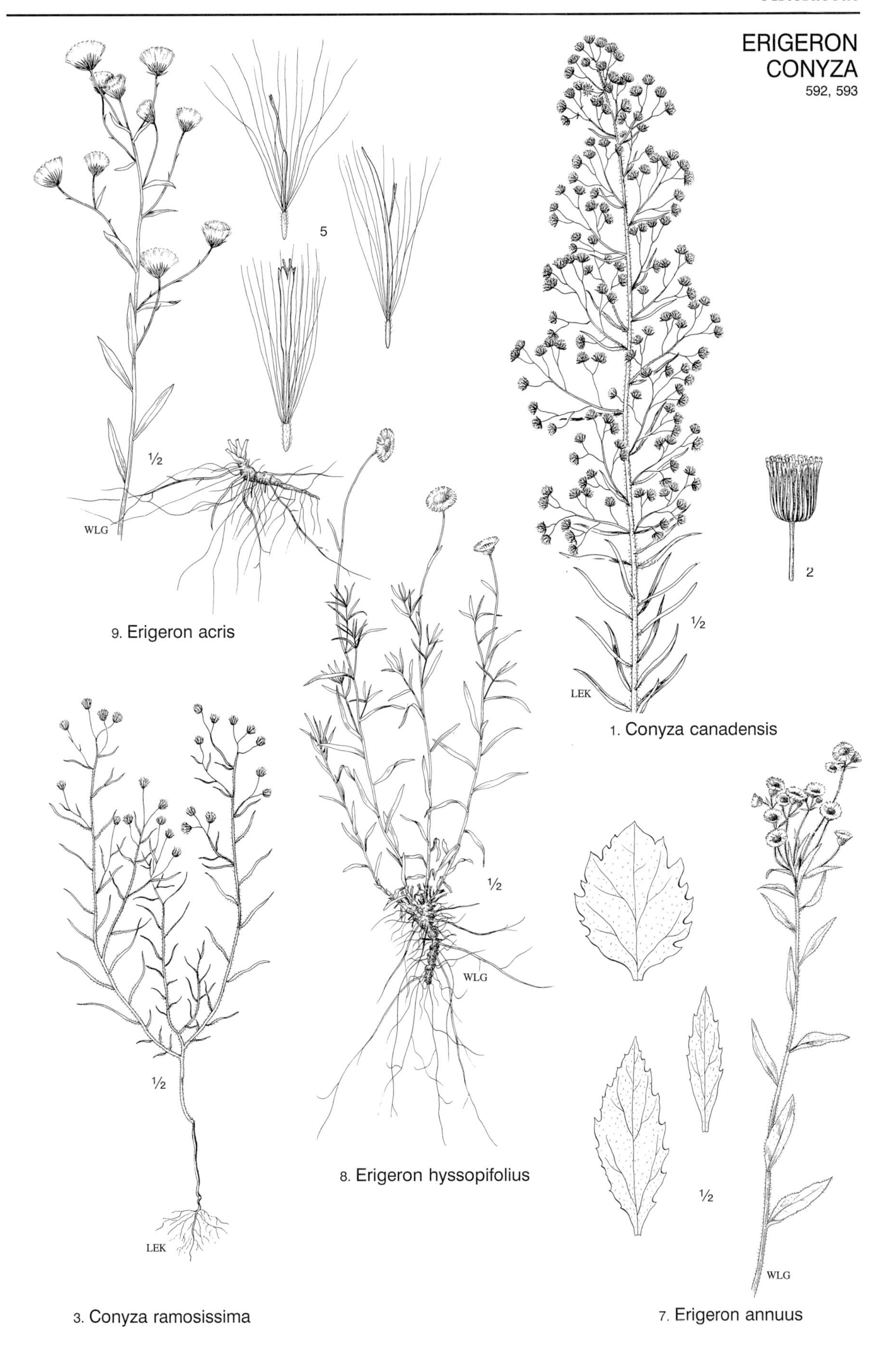

5

½

WLG

9. Erigeron acris

1. Conyza canadensis

LEK

2

½

½

3. Conyza ramosissima

LEK

8. Erigeron hyssopifolius

WLG

½

½

7. Erigeron annuus

WLG

CONYZA
BACCHARIS
BOLTONIA
592–594

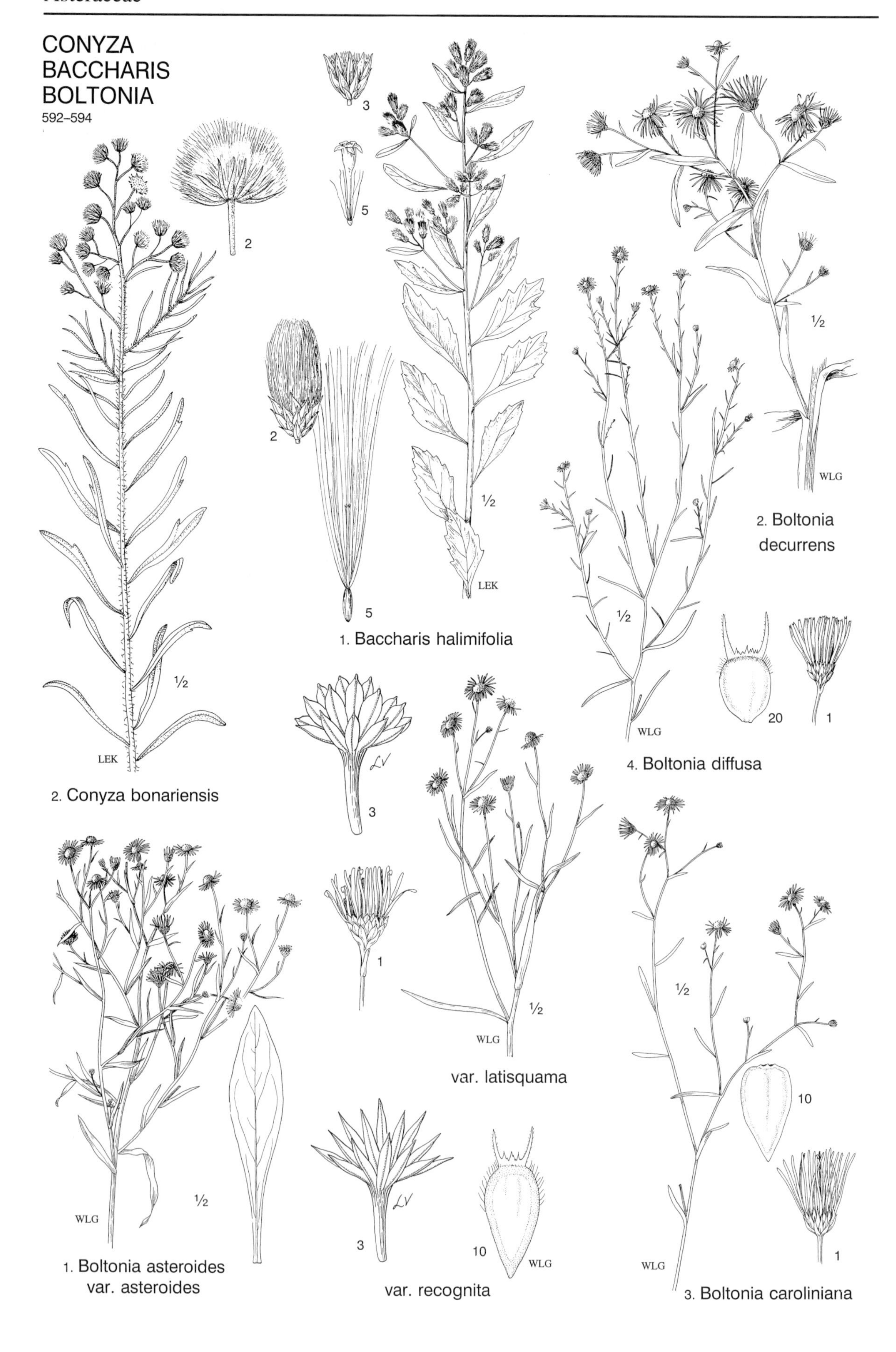

2. Conyza bonariensis

1. Baccharis halimifolia

2. Boltonia
decurrens

4. Boltonia diffusa

var. latisquama

1. Boltonia asteroides
var. asteroides

var. recognita

3. Boltonia caroliniana

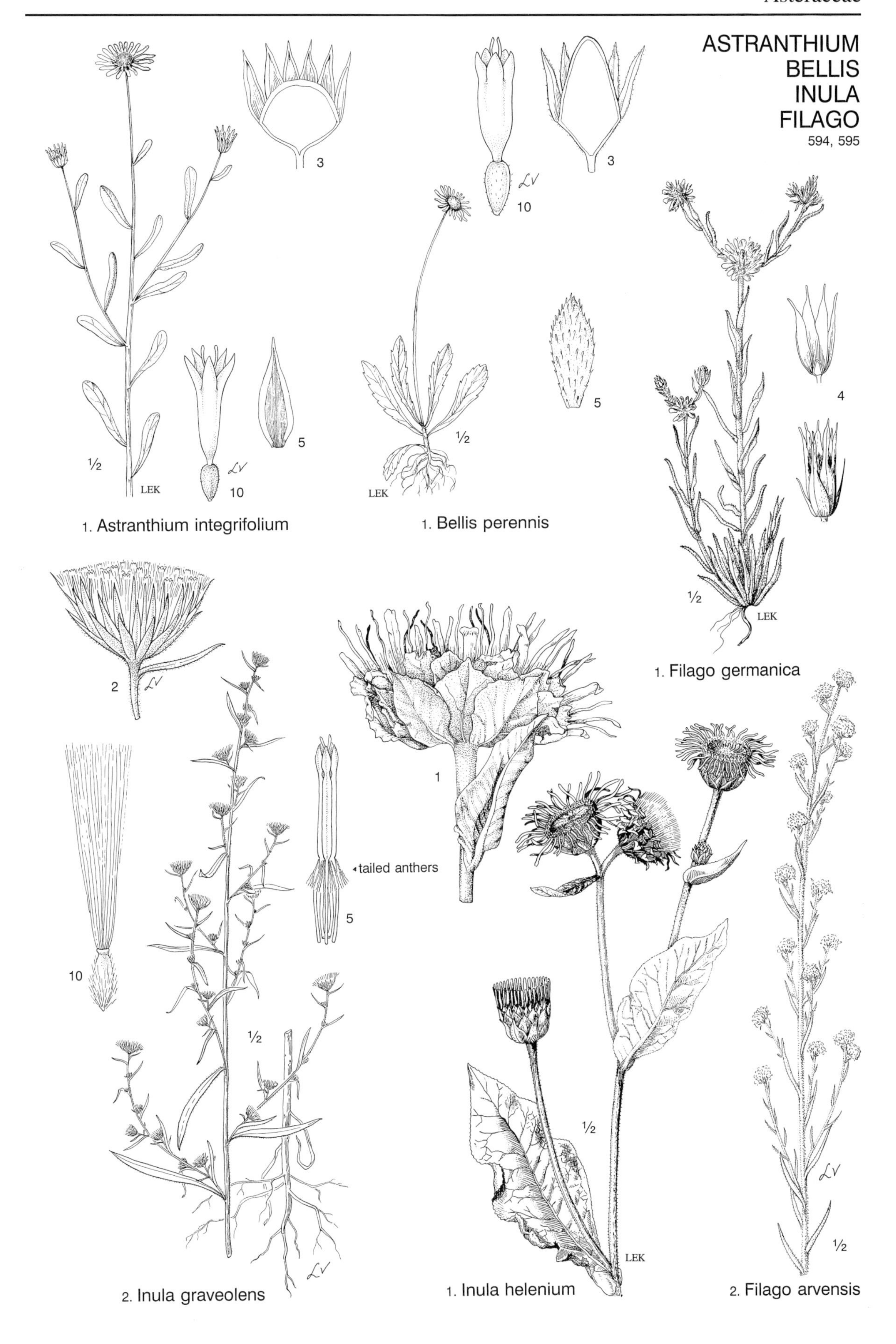

3

10

3

10

5

½

LEK

½

10

1. Astranthium integrifolium

½

LEK

1. Bellis perennis

5

4

½

LEK

1. Filago germanica

2

5

10

½

◄ tailed anthers

1

½

½

2. Inula graveolens

LEK

1. Inula helenium

½

2. Filago arvensis

PLUCHEA
GNAPHALIUM
595, 596

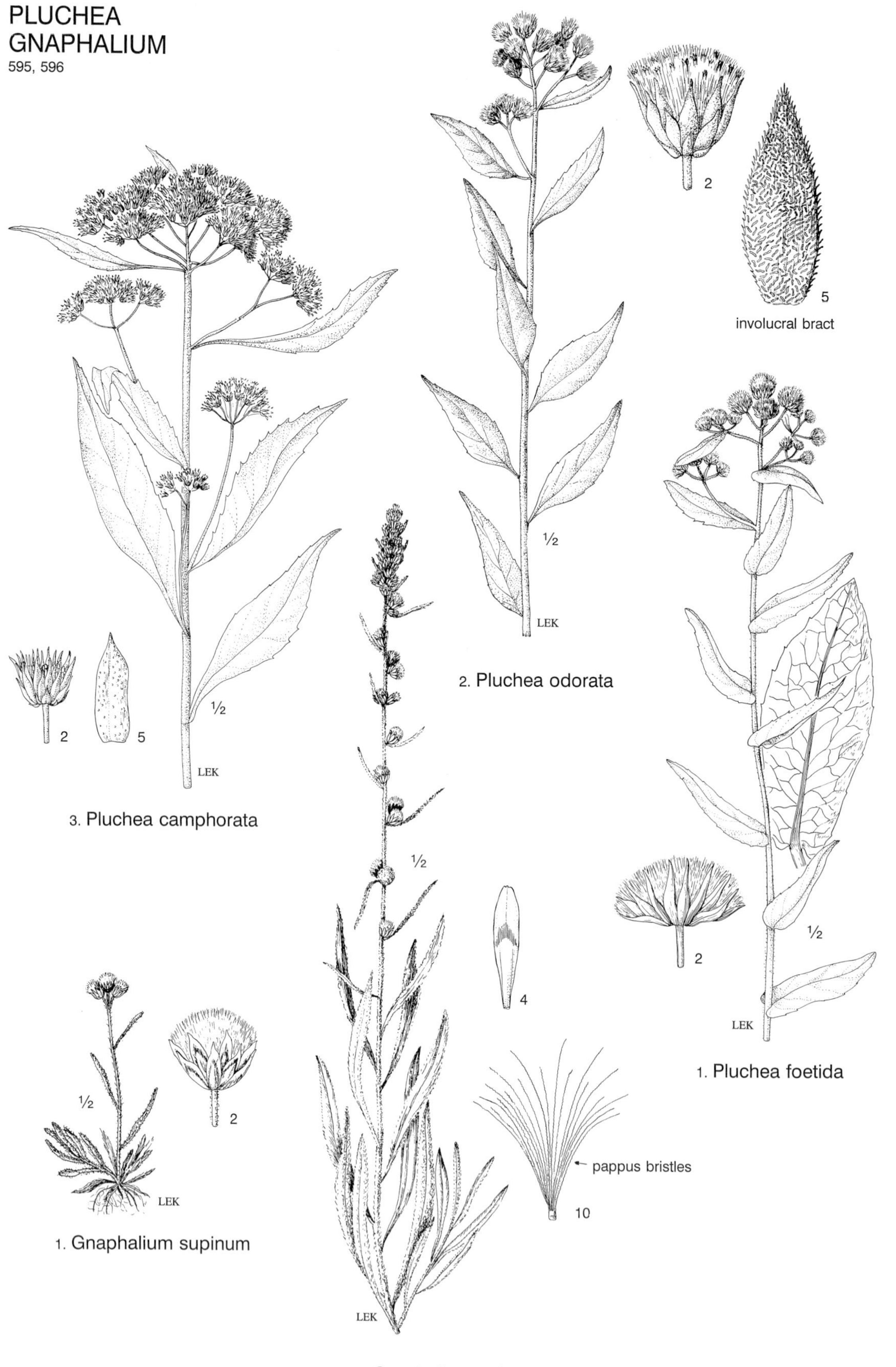

5
involucral bract

2. Pluchea odorata

2
5

3. Pluchea camphorata

2

1. Pluchea foetida

4

← pappus bristles

10

1. Gnaphalium supinum

2. Gnaphalium sylvaticum

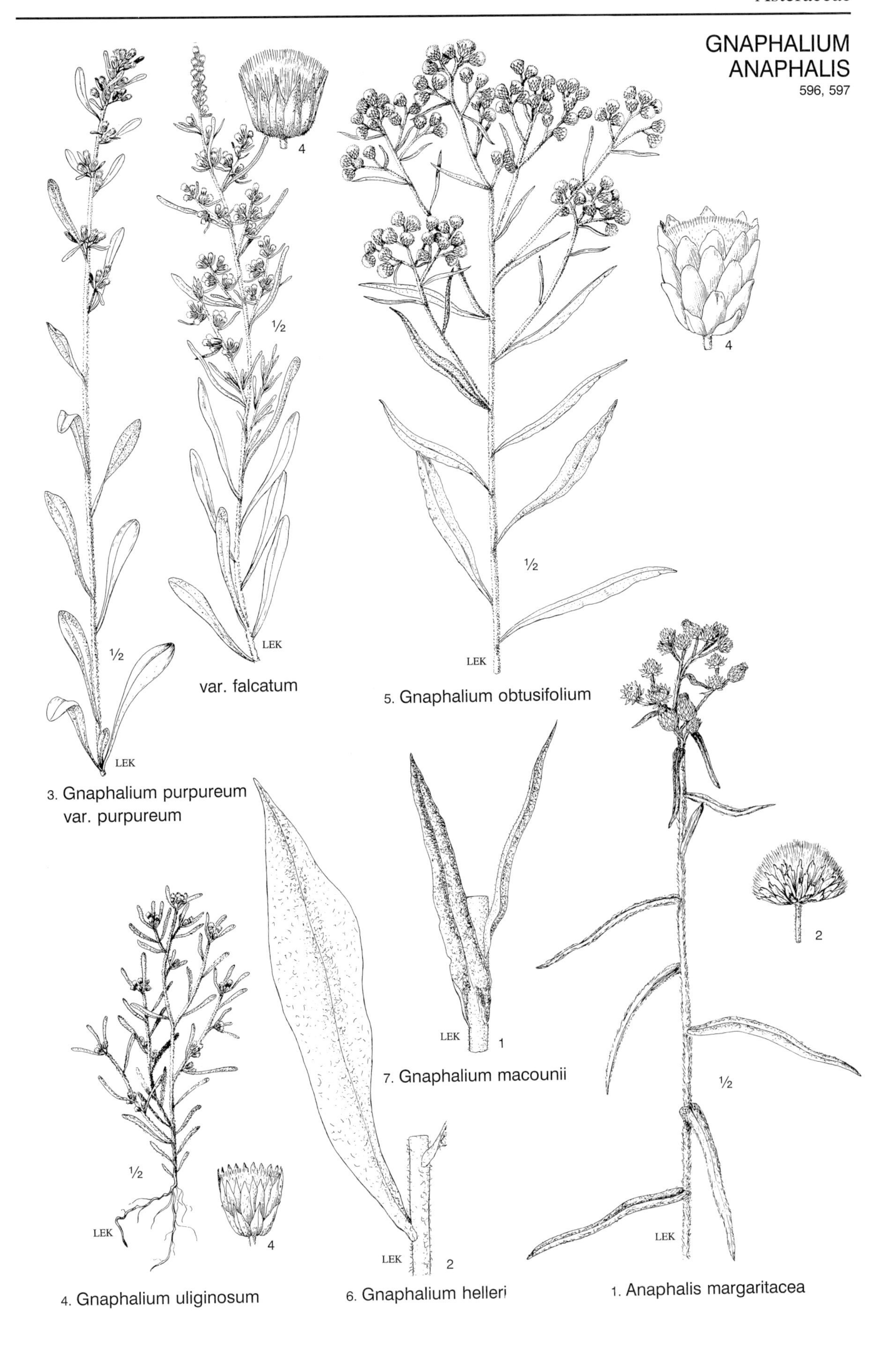

4

var. falcatum

3. Gnaphalium purpureum
var. purpureum

5. Gnaphalium obtusifolium

7. Gnaphalium macounii

4. Gnaphalium uliginosum

6. Gnaphalium helleri

1. Anaphalis margaritacea

ANTENNARIA
598

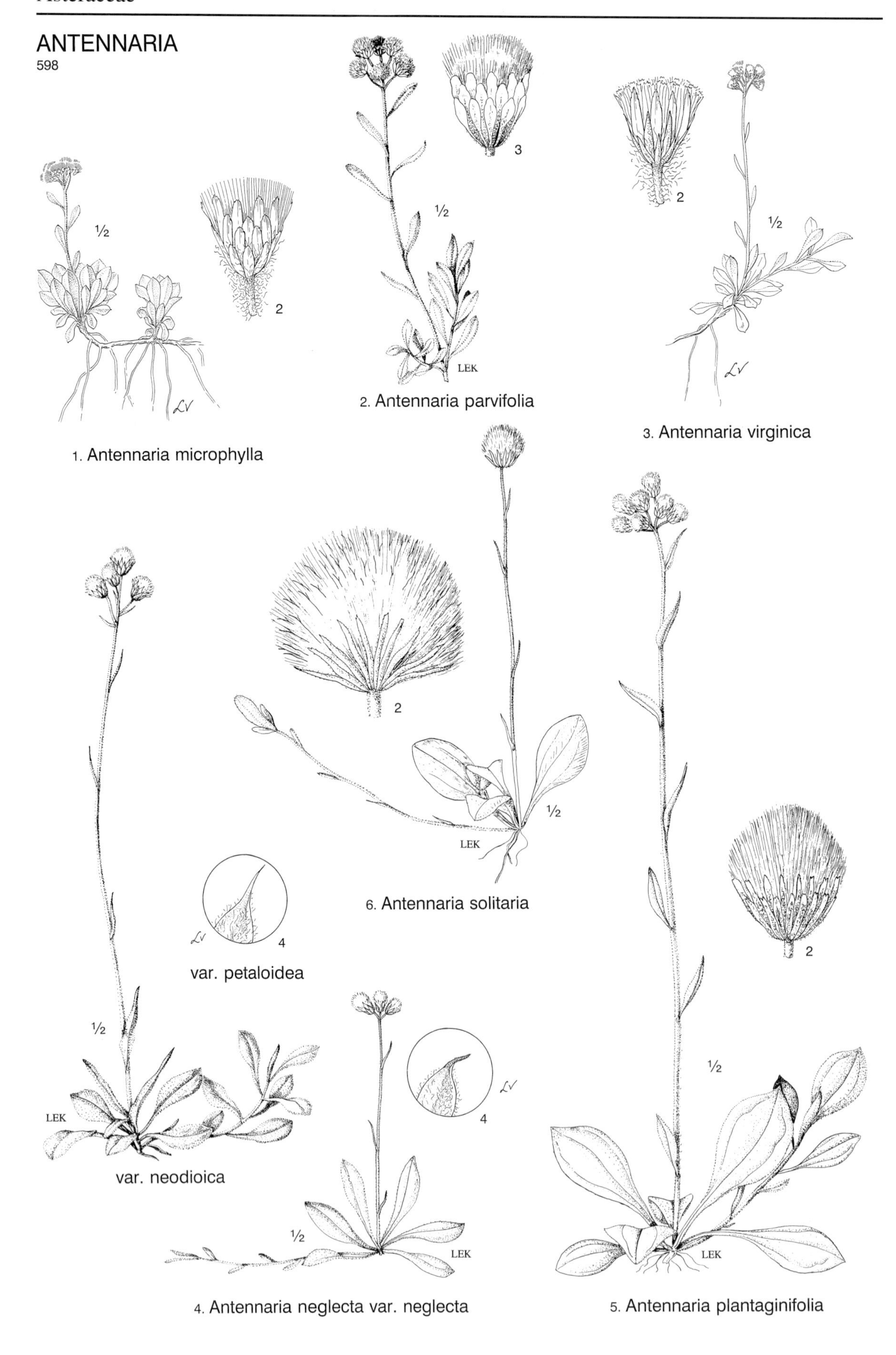

1. Antennaria microphylla

2. Antennaria parvifolia

3. Antennaria virginica

var. petaloidea

6. Antennaria solitaria

var. neodioica

4. Antennaria neglecta var. neglecta

5. Antennaria plantaginifolia

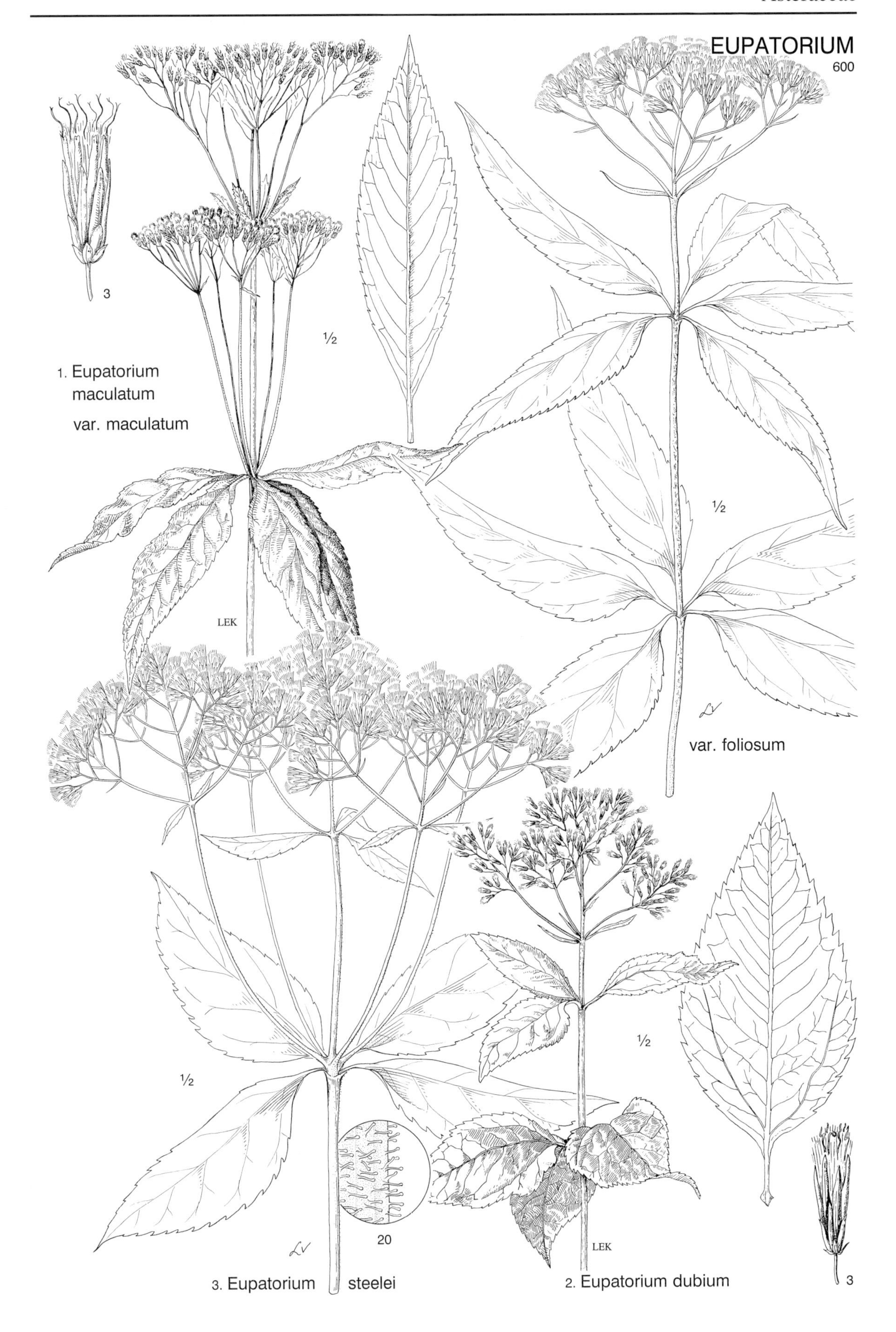

3

½

1. Eupatorium
maculatum

var. maculatum

LEK

½

var. foliosum

½

20

3. Eupatorium   steelei

½

2. Eupatorium dubium

LEK

3

EUPATORIUM
600, 601

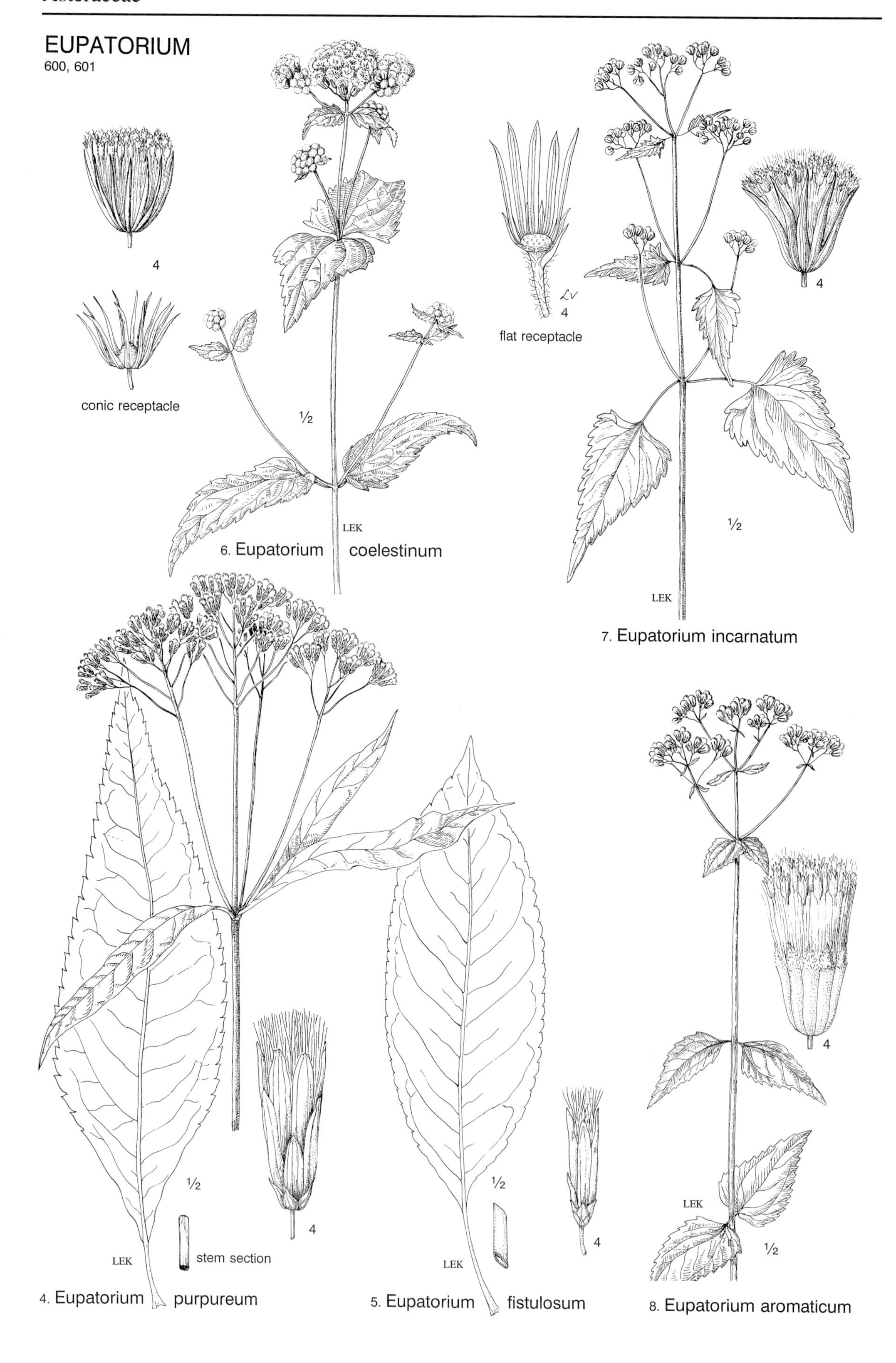

4

conic receptacle

½

6. Eupatorium   coelestinum

LEK

flat receptacle

4

4

7. Eupatorium incarnatum

½

LEK

½

4. Eupatorium   purpureum

½

LEK

stem section

4

5. Eupatorium   fistulosum

½

LEK

4

8. Eupatorium aromaticum

½

LEK

4

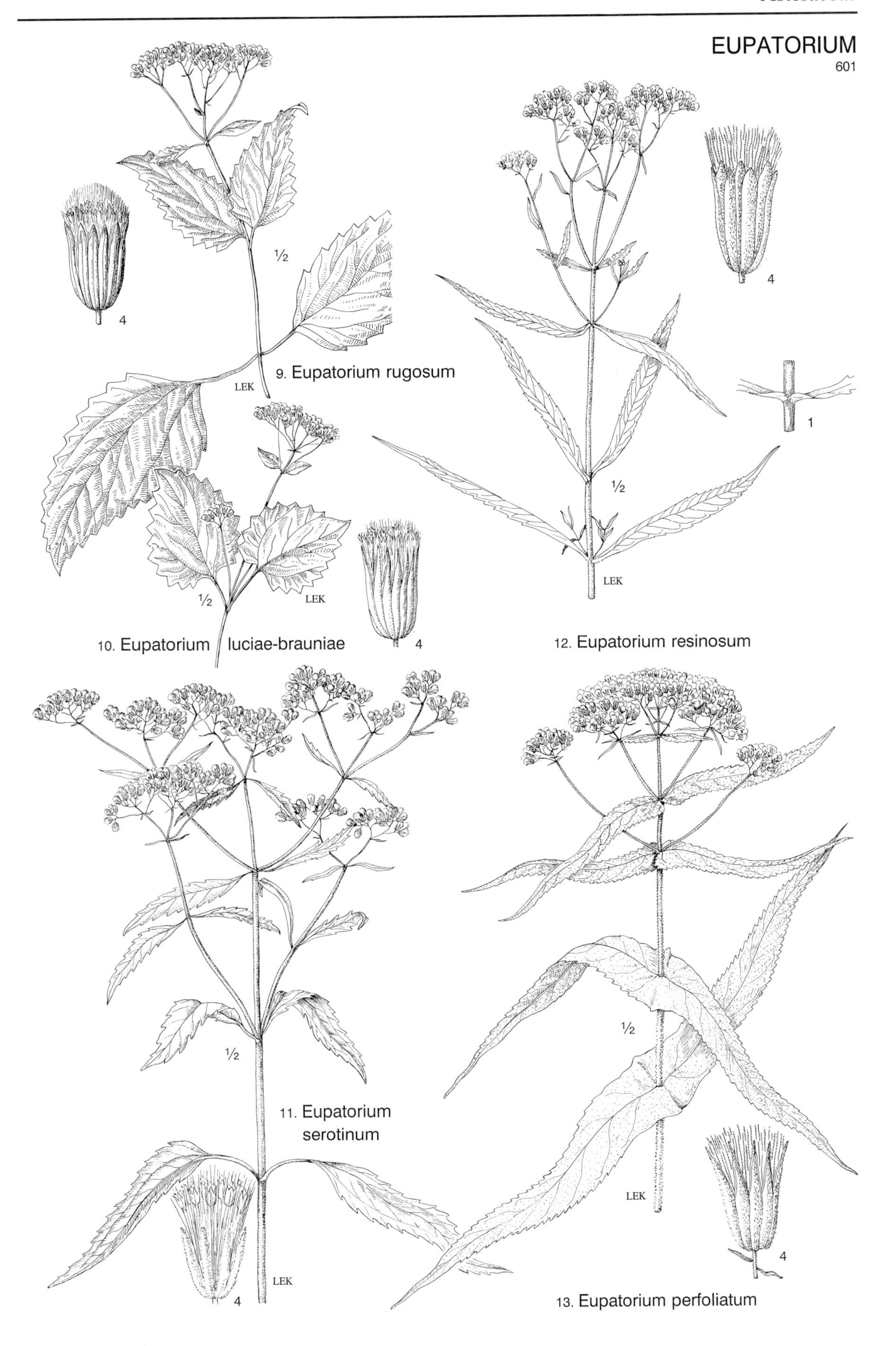

9. Eupatorium rugosum

10. Eupatorium luciae-brauniae

11. Eupatorium serotinum

12. Eupatorium resinosum

13. Eupatorium perfoliatum

EUPATORIUM
602

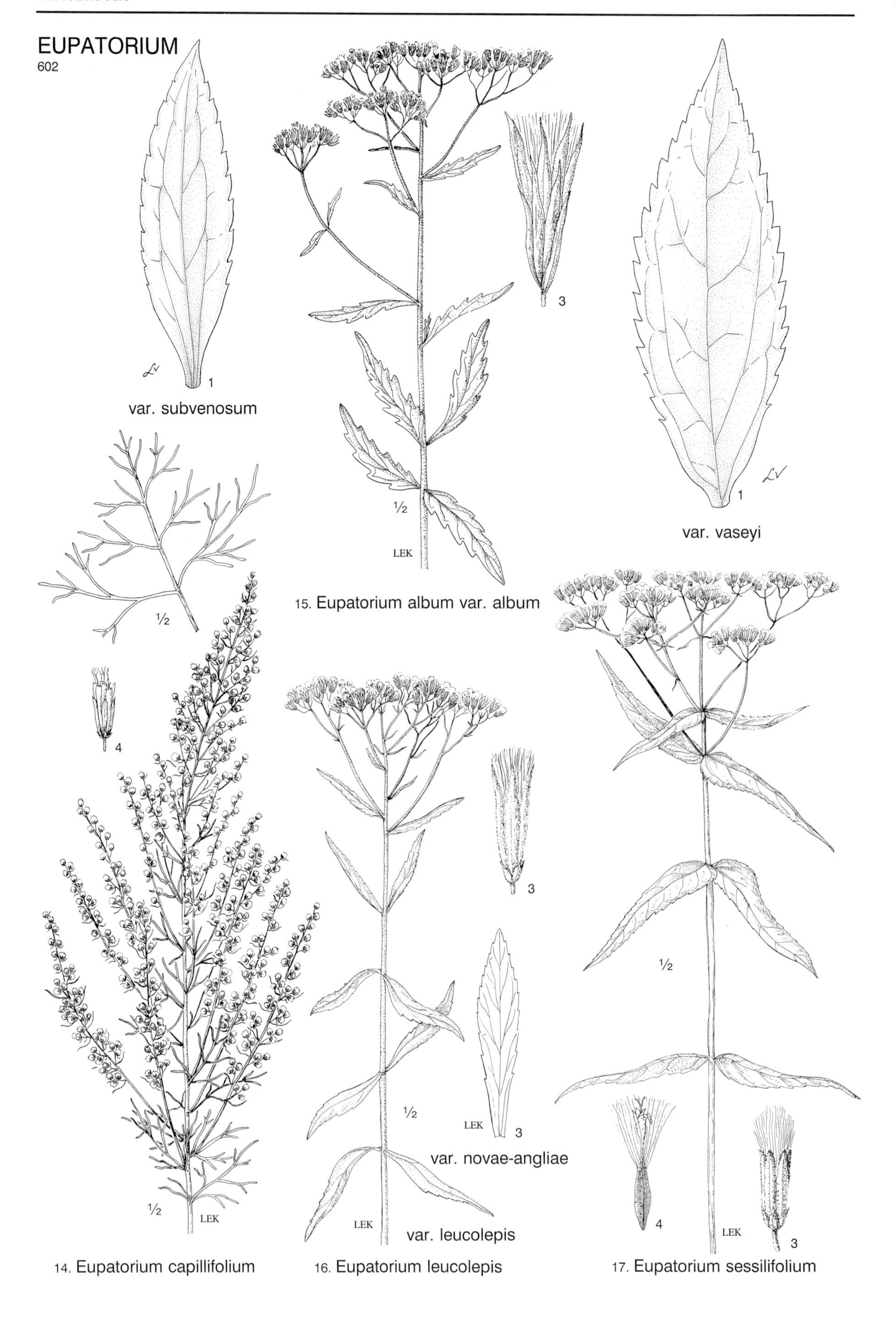

var. subvenosum

1

15. Eupatorium album var. album

var. vaseyi

1

½

4

14. Eupatorium capillifolium

½

3

var. novae-angliae

3

var. leucolepis

16. Eupatorium leucolepis

½

4

3

17. Eupatorium sessilifolium

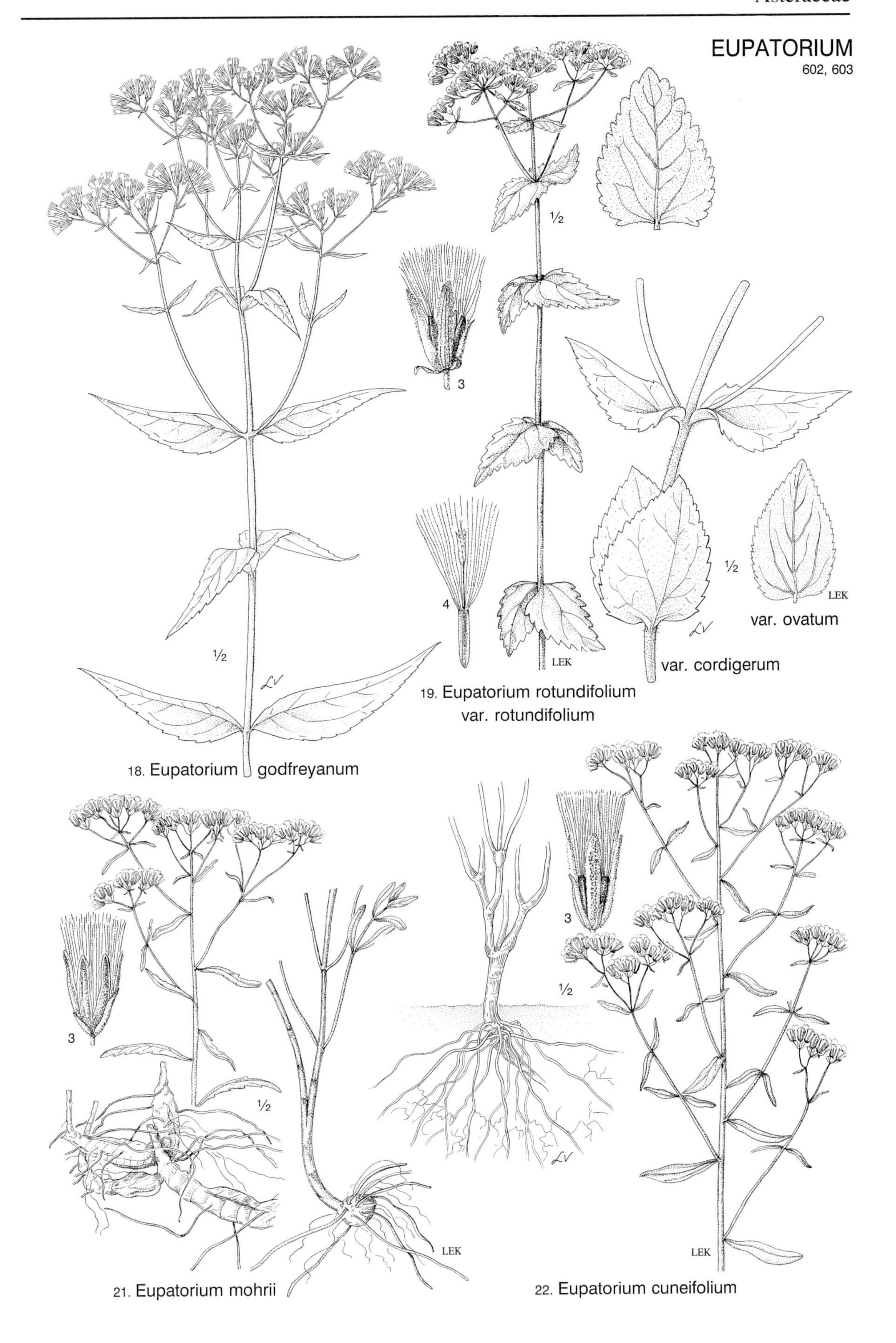

18. Eupatorium godfreyanum

19. Eupatorium rotundifolium
var. rotundifolium

var. cordigerum

var. ovatum

21. Eupatorium mohrii

22. Eupatorium cuneifolium

EUPATORIUM
603

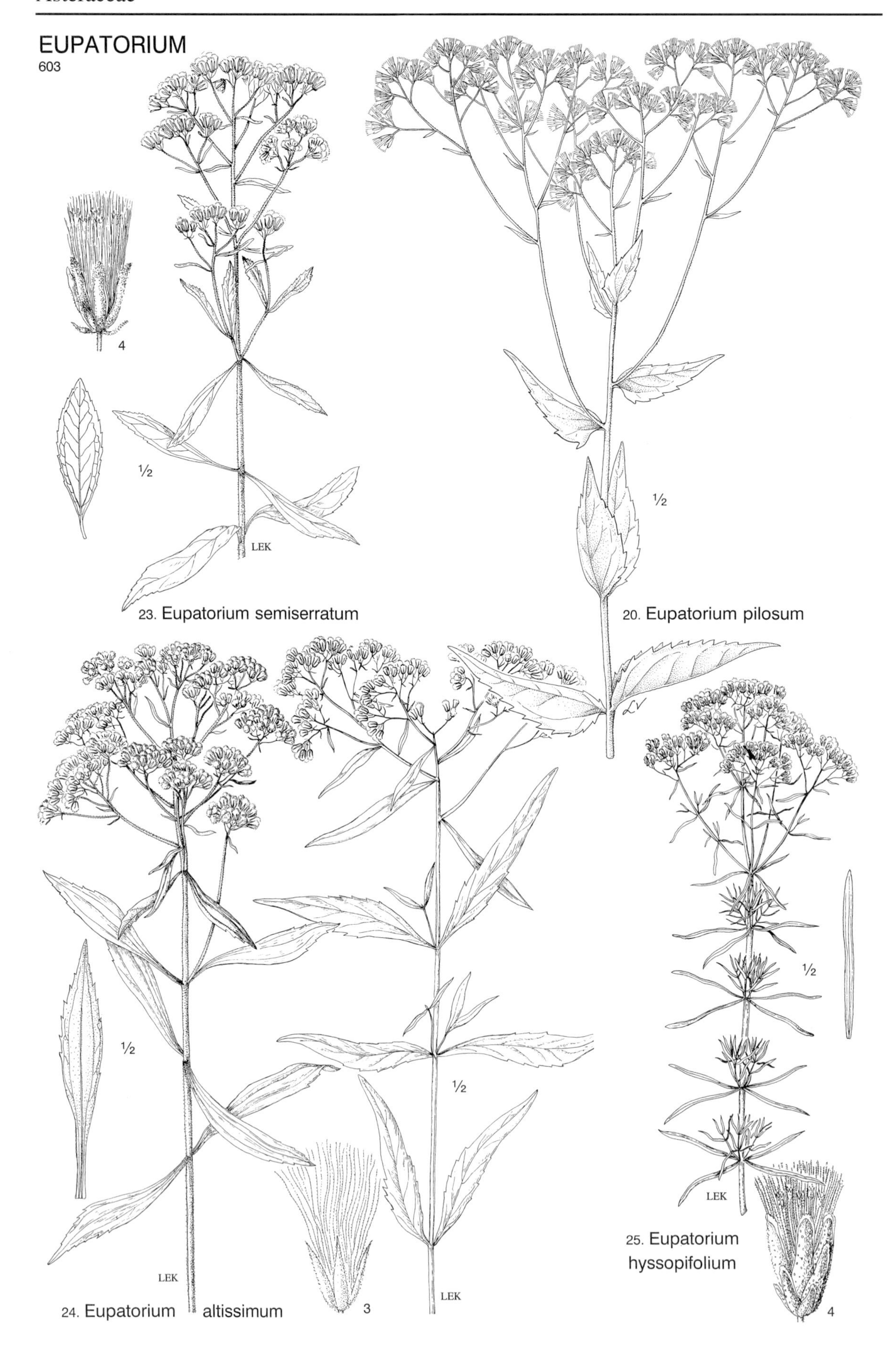

23. Eupatorium semiserratum

20. Eupatorium pilosum

24. Eupatorium altissimum

25. Eupatorium hyssopifolium

Asteraceae

MIKANIA
SCLEROLEPIS
KUHNIA
BRICKELLIA
CARPHEPHORUS
604, 605

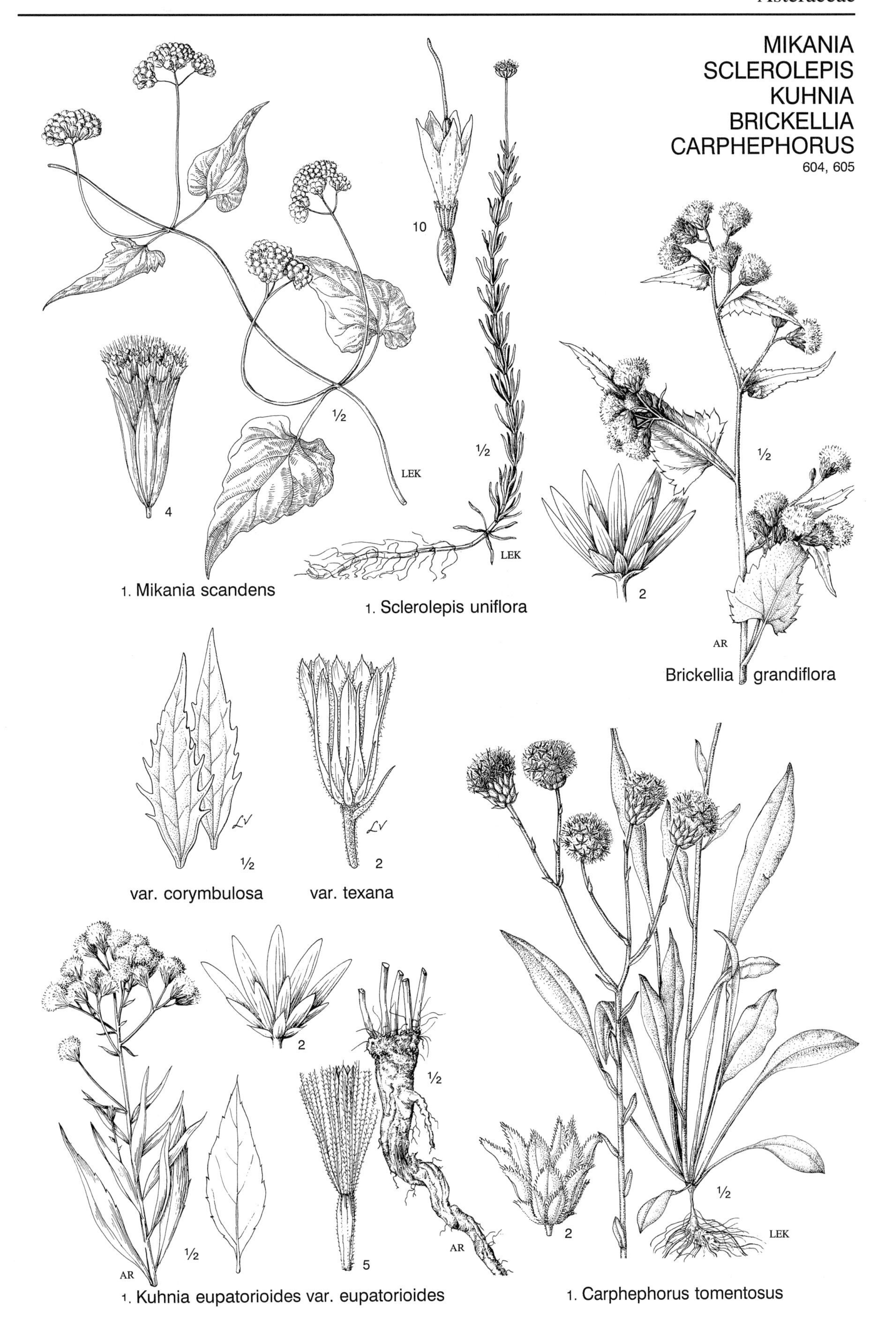

1. Mikania scandens

1. Sclerolepis uniflora

Brickellia grandiflora

var. corymbulosa

var. texana

1. Kuhnia eupatorioides var. eupatorioides

1. Carphephorus tomentosus

CARPHEPHORUS
LIATRIS
605, 606

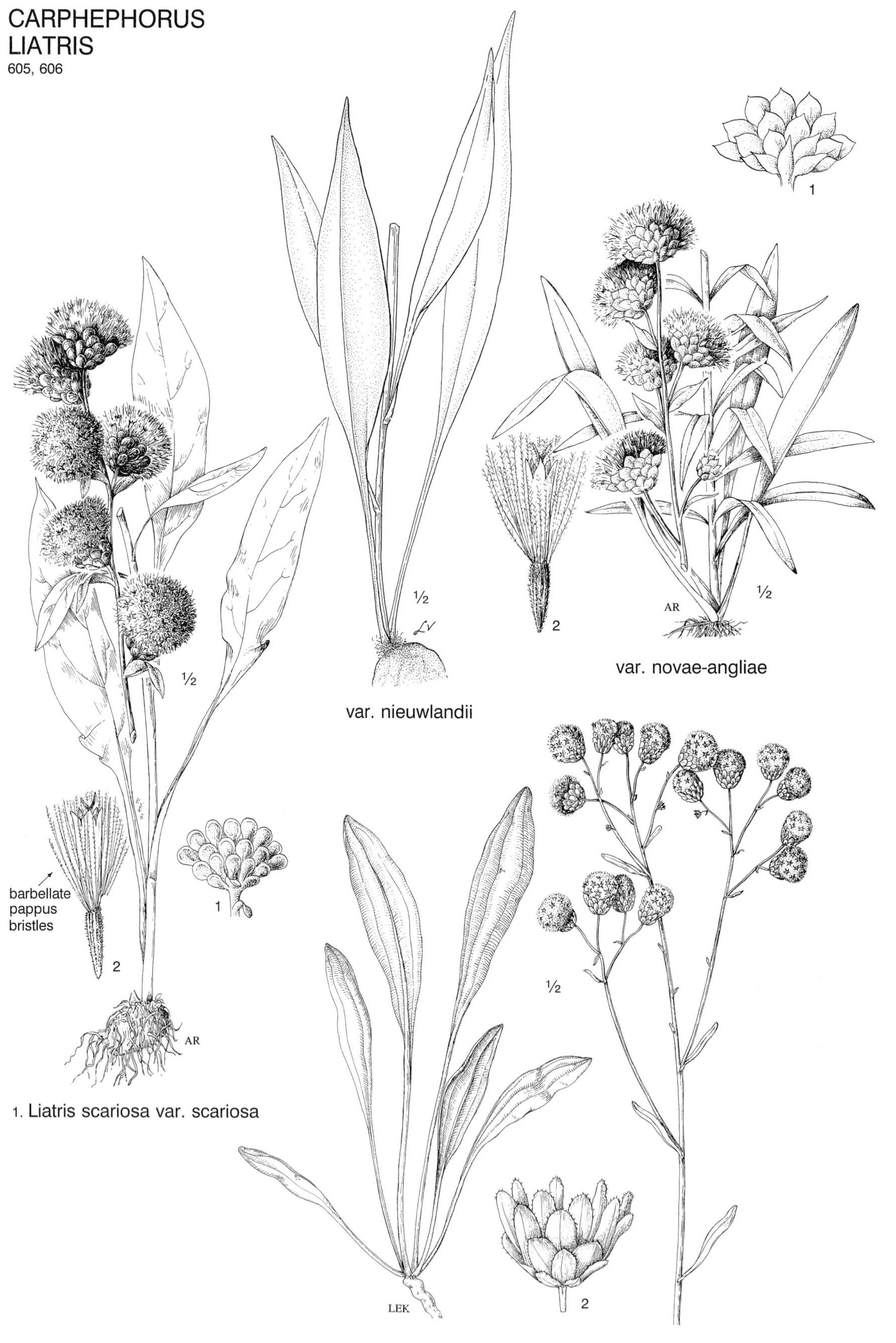

var. nieuwlandii

1

2

barbellate
pappus
bristles

2

AR

1. Liatris scariosa var. scariosa

var. novae-angliae

AR

½

½

½

½

LEK

2

2. Carphephorus bellidifolius

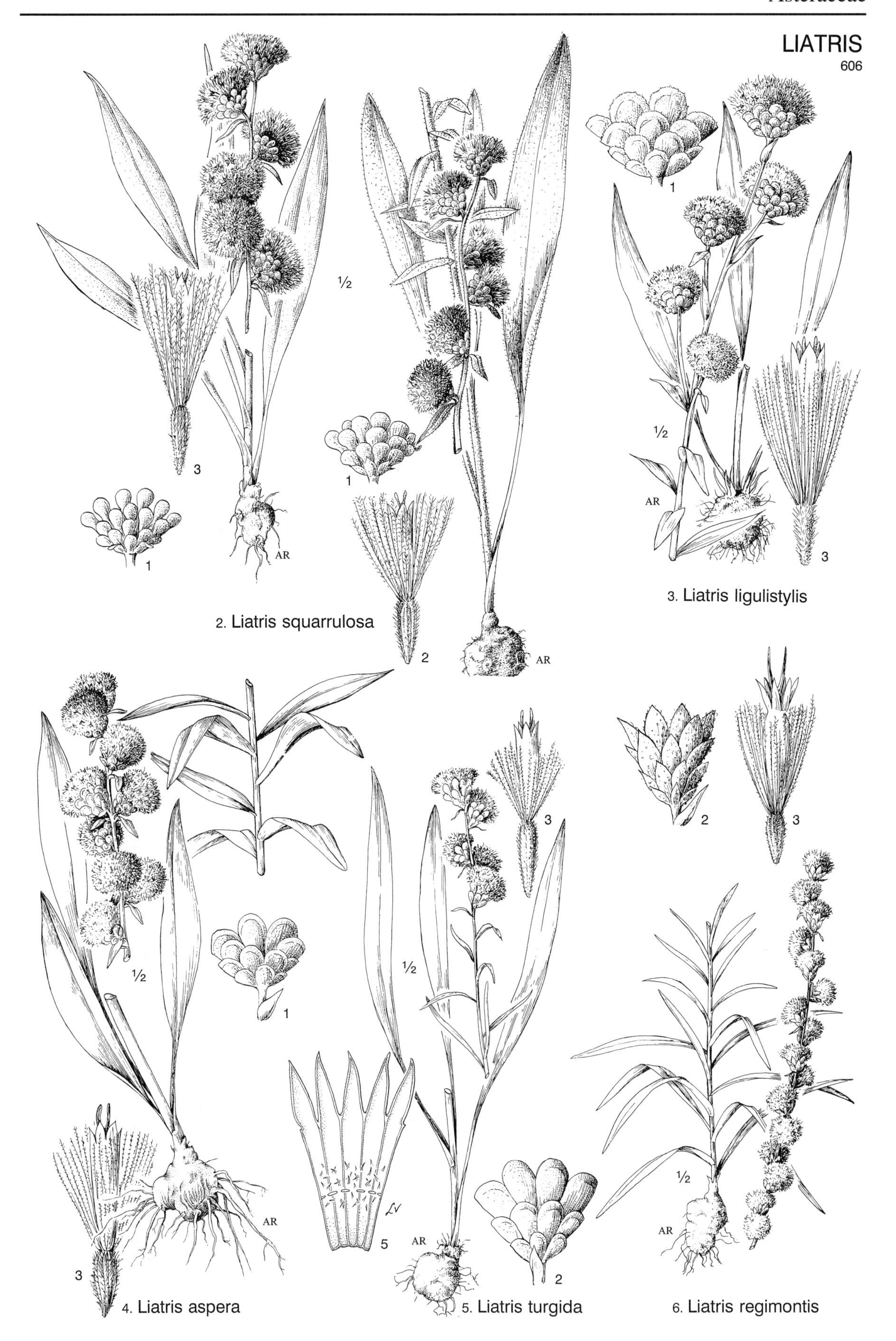

2. Liatris squarrulosa

3. Liatris ligulistylis

4. Liatris aspera

5. Liatris turgida

6. Liatris regimontis

LIATRIS
607

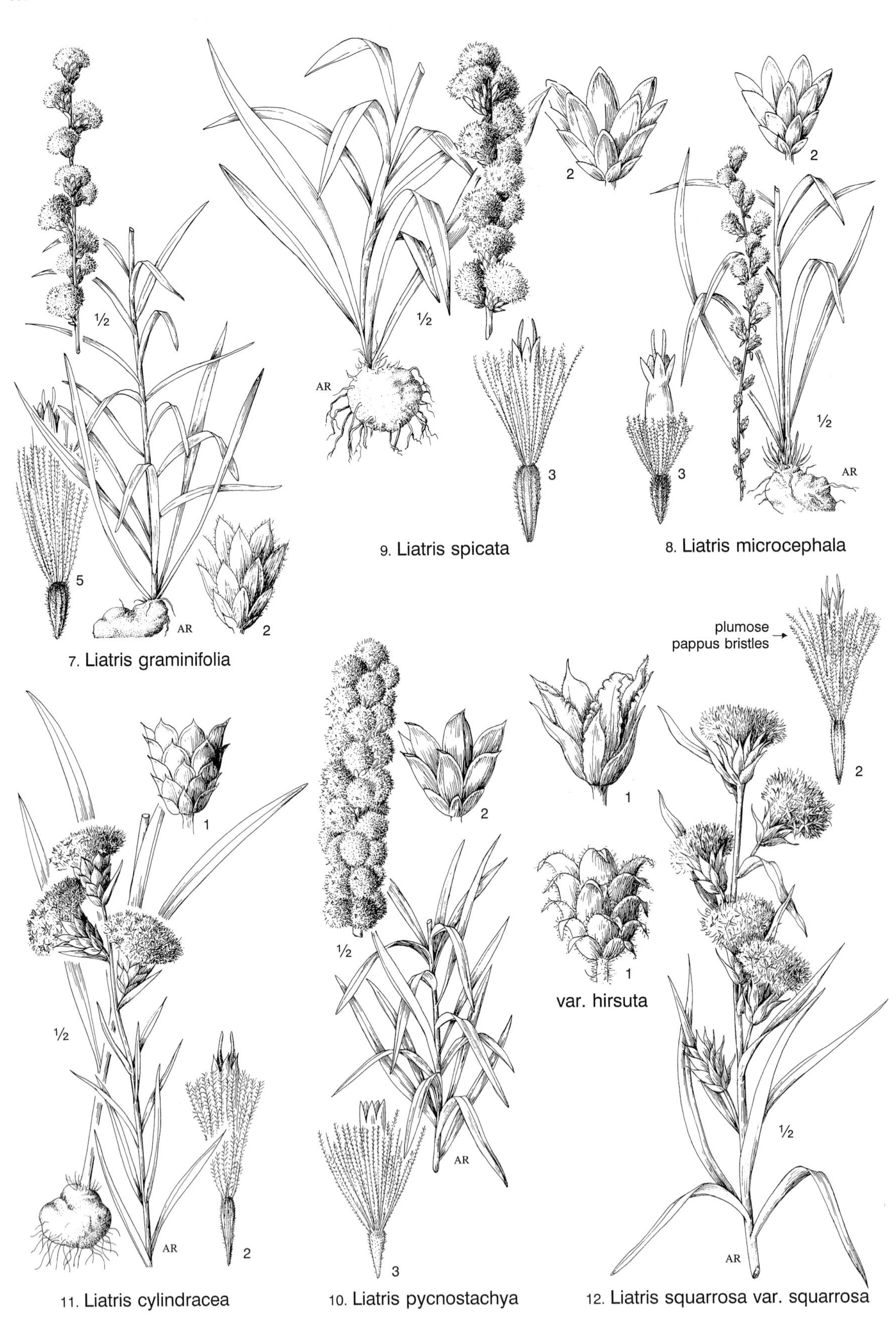

9. Liatris spicata

8. Liatris microcephala

7. Liatris graminifolia

plumose
pappus bristles

var. hirsuta

11. Liatris cylindracea

10. Liatris pycnostachya

12. Liatris squarrosa var. squarrosa

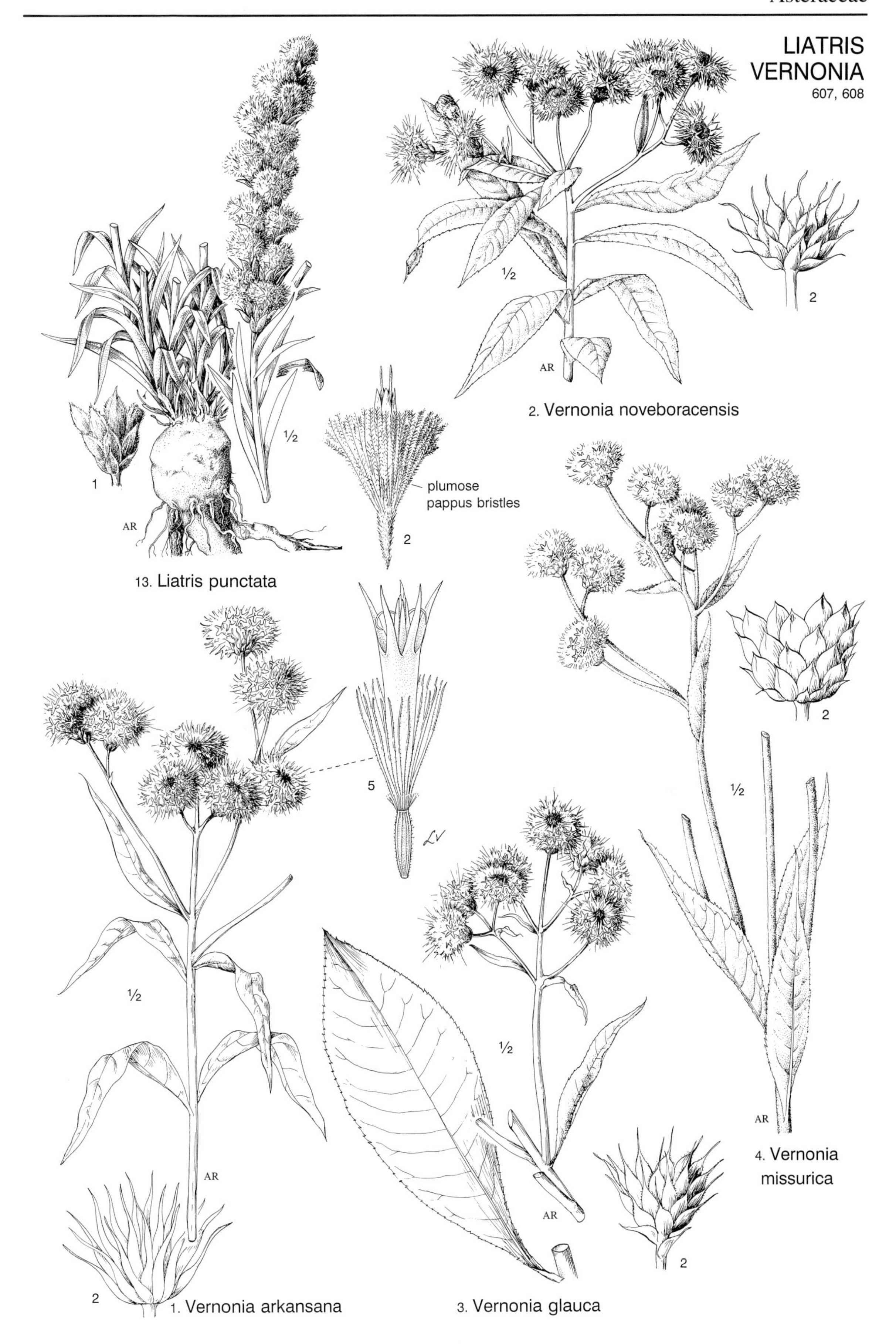

13. Liatris punctata

2. Vernonia noveboracensis

plumose
pappus bristles

1. Vernonia arkansana

3. Vernonia glauca

4. Vernonia missurica

VERNONIA
ELEPHANTOPUS
609

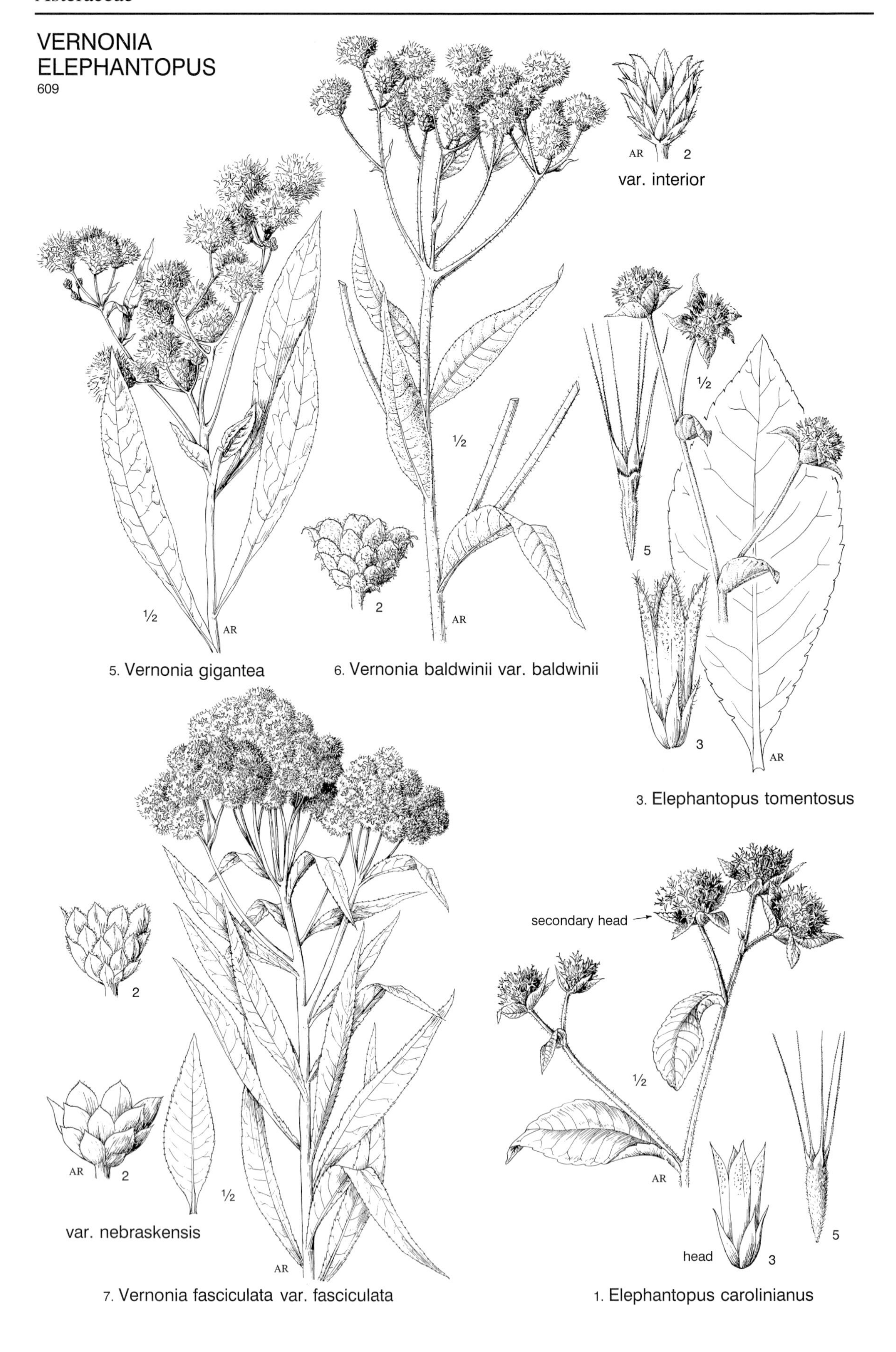

AR 2
var. interior

½

5. Vernonia gigantea

2

AR

6. Vernonia baldwinii var. baldwinii

½

5

3

AR

3. Elephantopus tomentosus

2

AR 2

½

var. nebraskensis

7. Vernonia fasciculata var. fasciculata

secondary head

½

AR

head 3

5

1. Elephantopus carolinianus

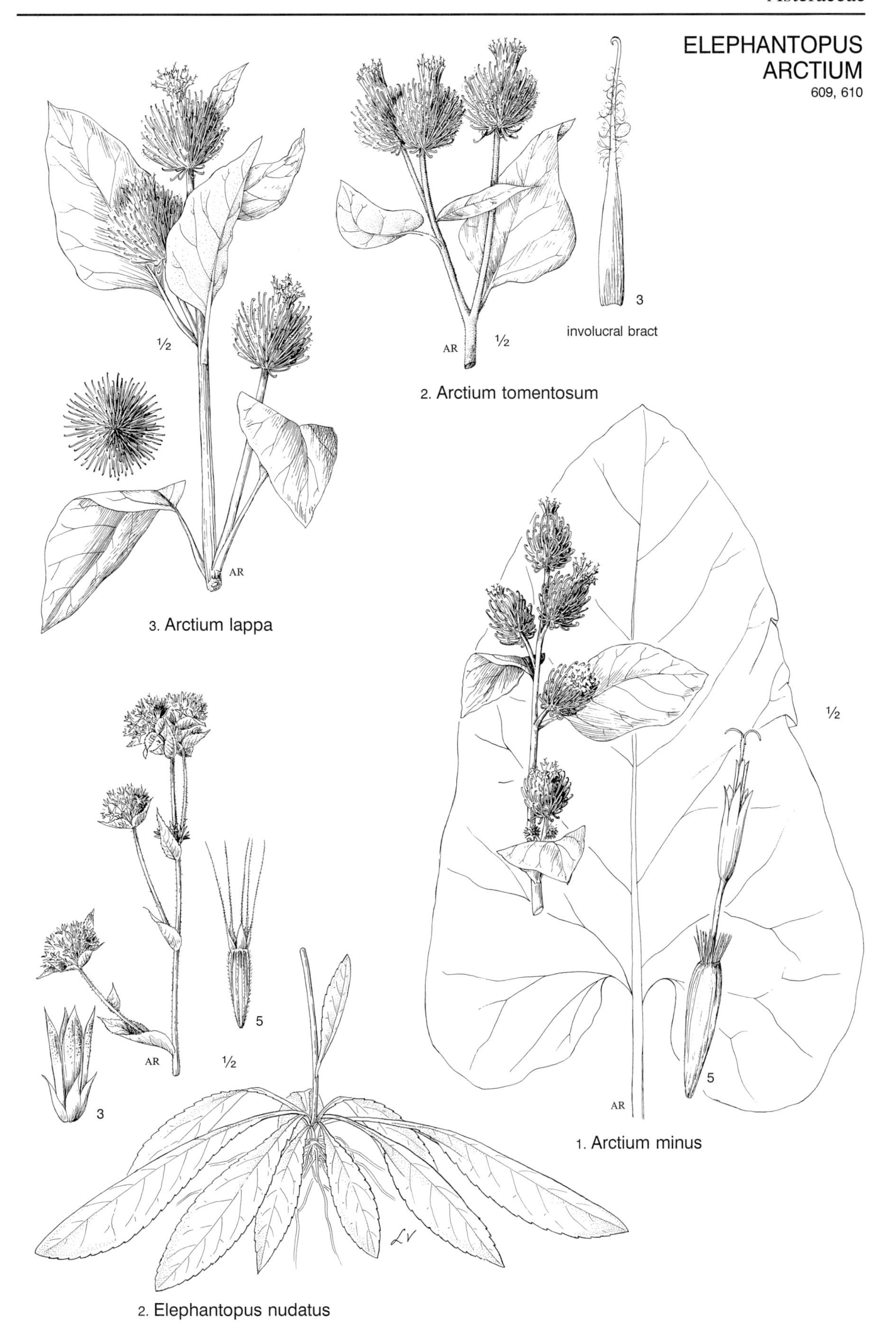

3
involucral bract

2. Arctium tomentosum

3. Arctium lappa

½

1. Arctium minus

5

2. Elephantopus nudatus

5

CARDUUS
CIRSIUM
610, 611

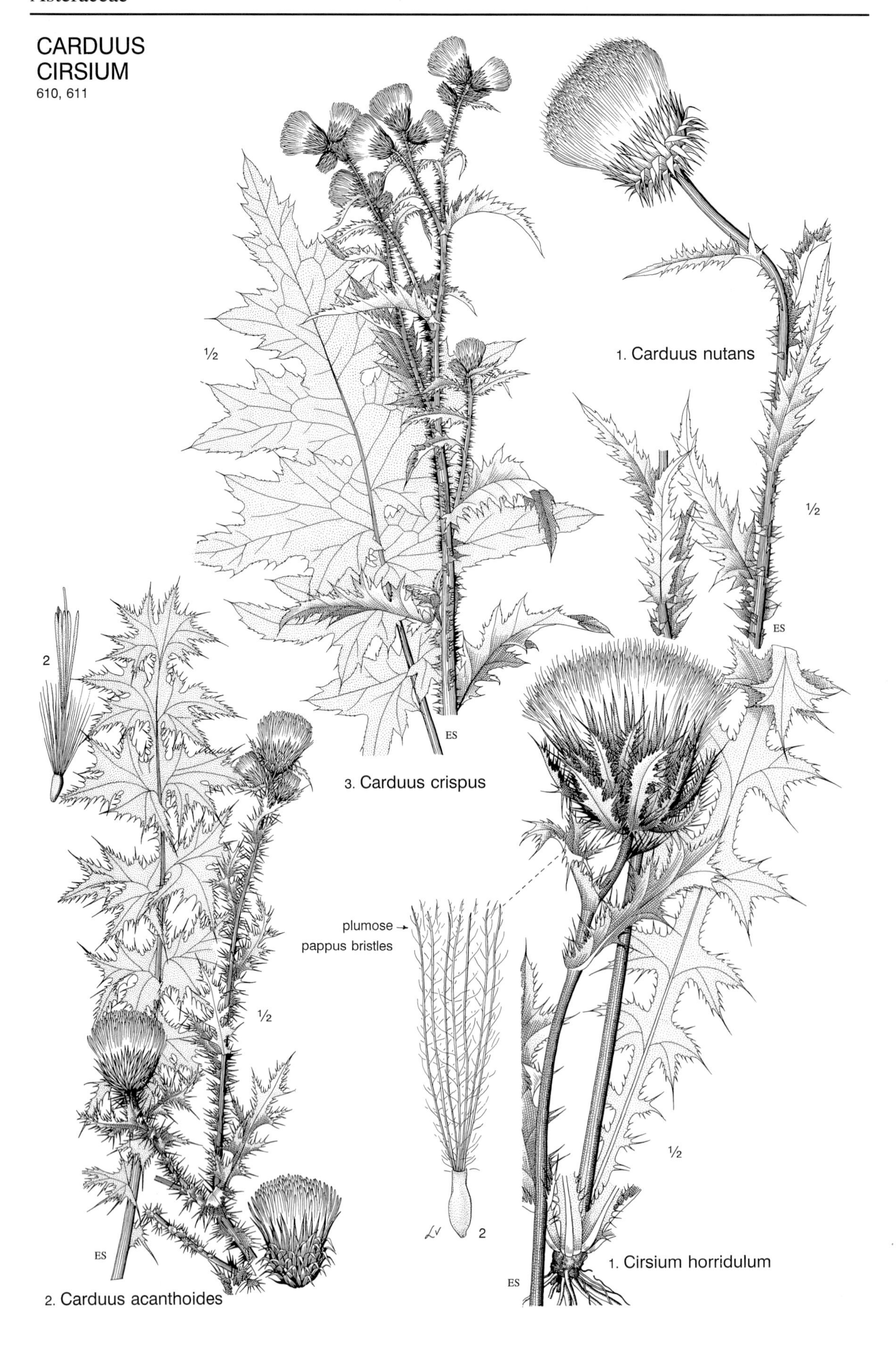

1. Carduus nutans

3. Carduus crispus

plumose →
pappus bristles

2. Carduus acanthoides

1. Cirsium horridulum

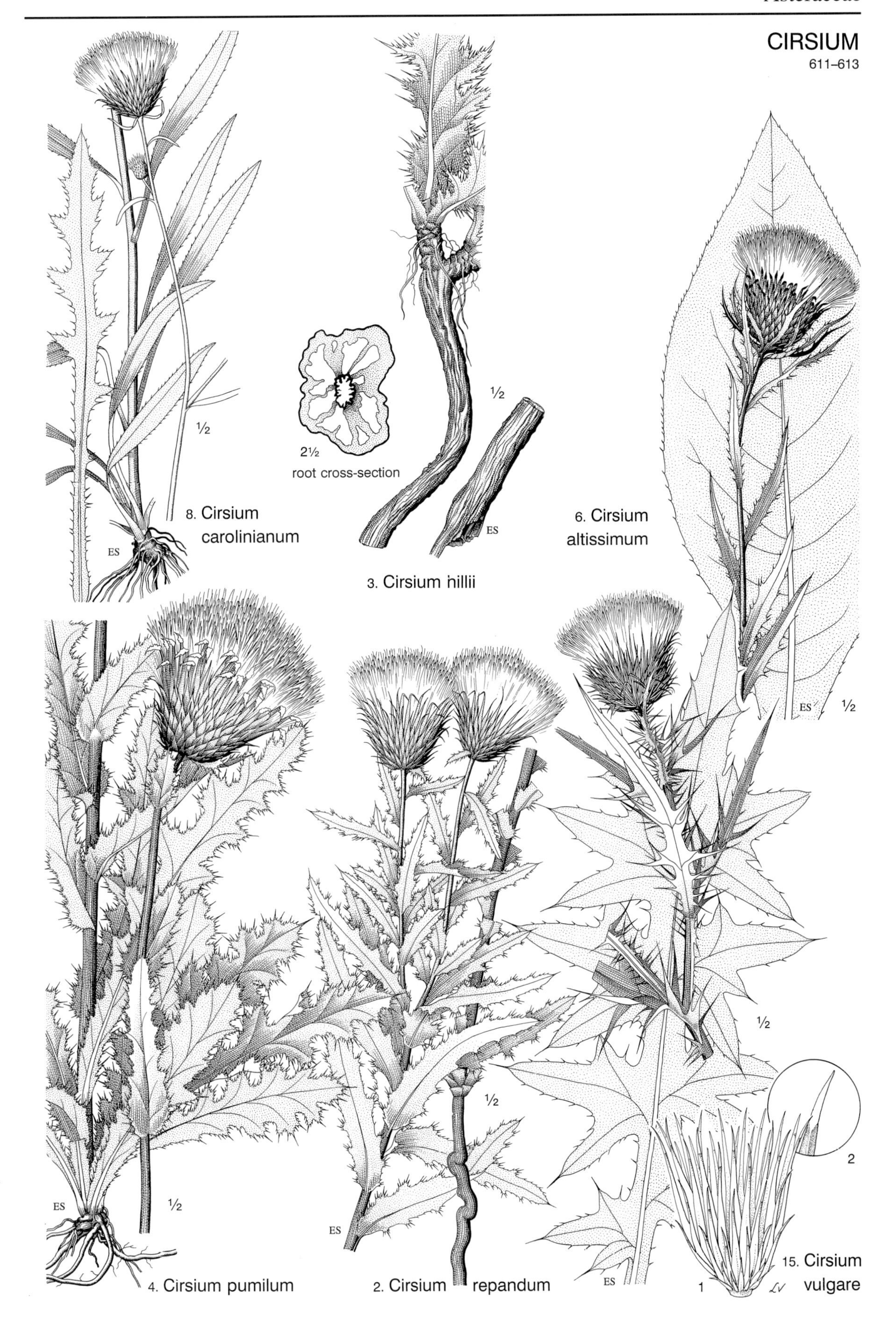

8. Cirsium carolinianum

2½ root cross-section

3. Cirsium hillii

6. Cirsium altissimum

4. Cirsium pumilum

2. Cirsium repandum

15. Cirsium vulgare

CIRSIUM
612, 613

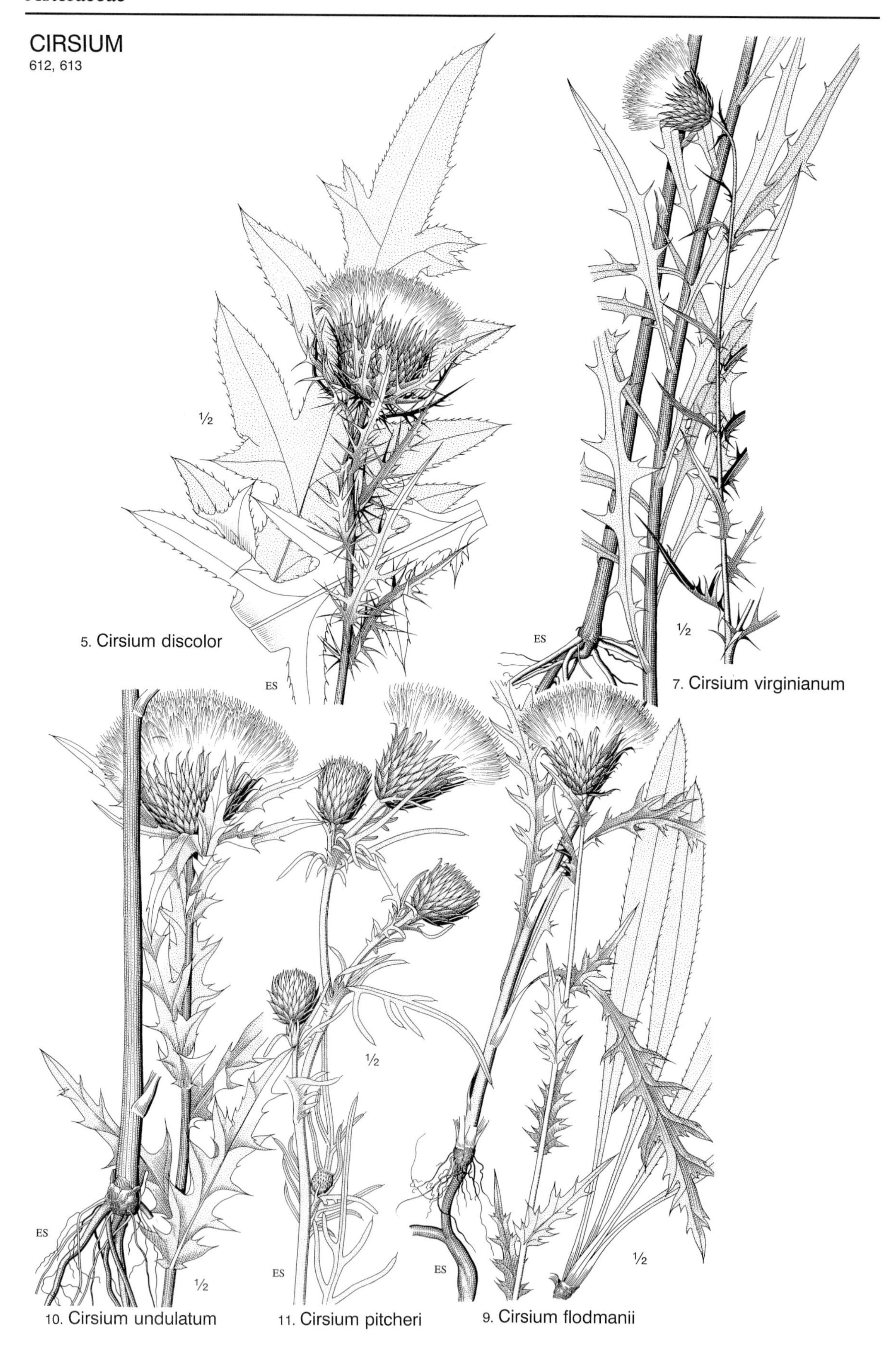

5. Cirsium discolor

7. Cirsium virginianum

10. Cirsium undulatum

11. Cirsium pitcheri

9. Cirsium flodmanii

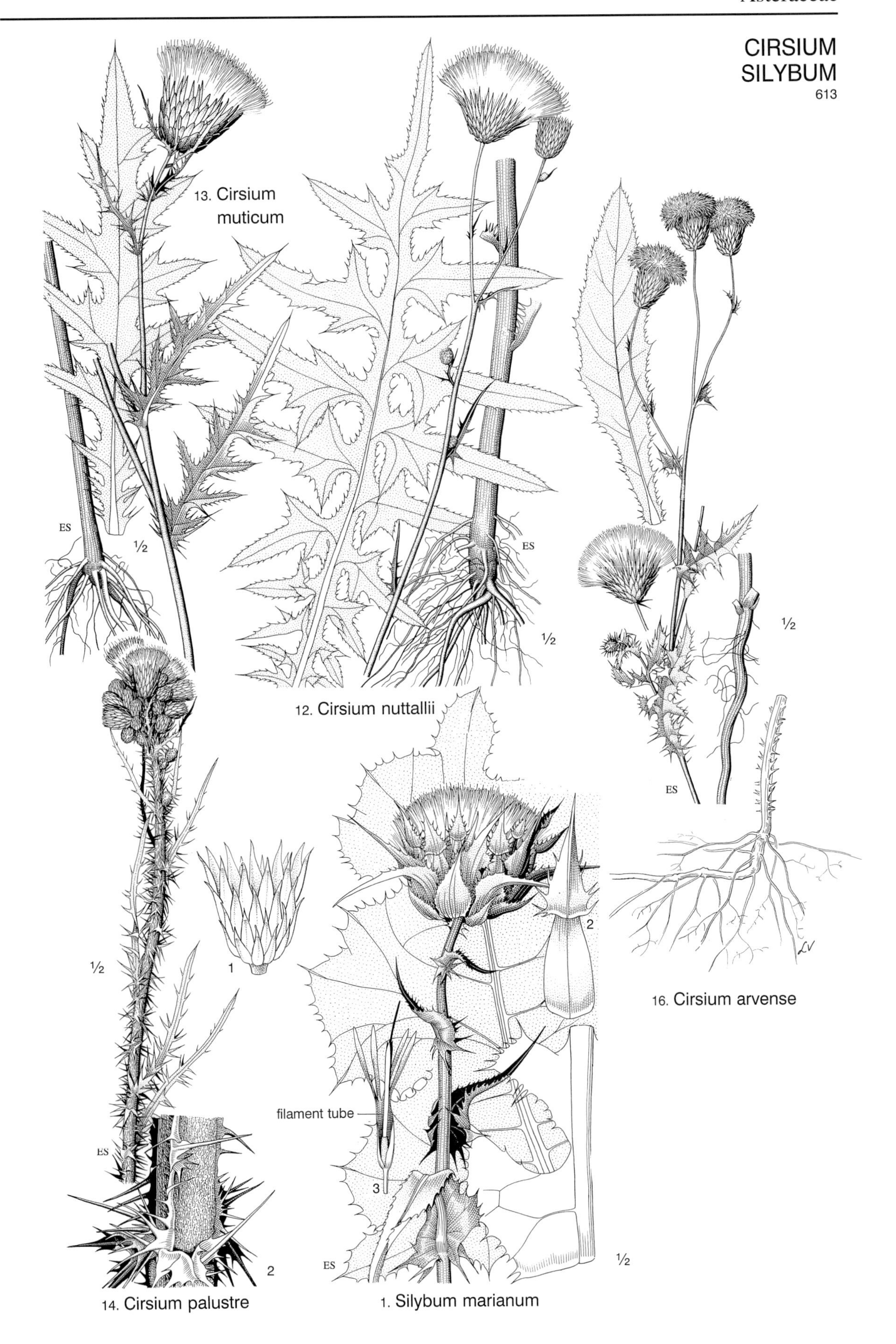

13. Cirsium muticum

12. Cirsium nuttallii

16. Cirsium arvense

14. Cirsium palustre

filament tube

1. Silybum marianum

ONOPORDUM
CARLINA
ECHINOPS
CENTAUREA
614–616

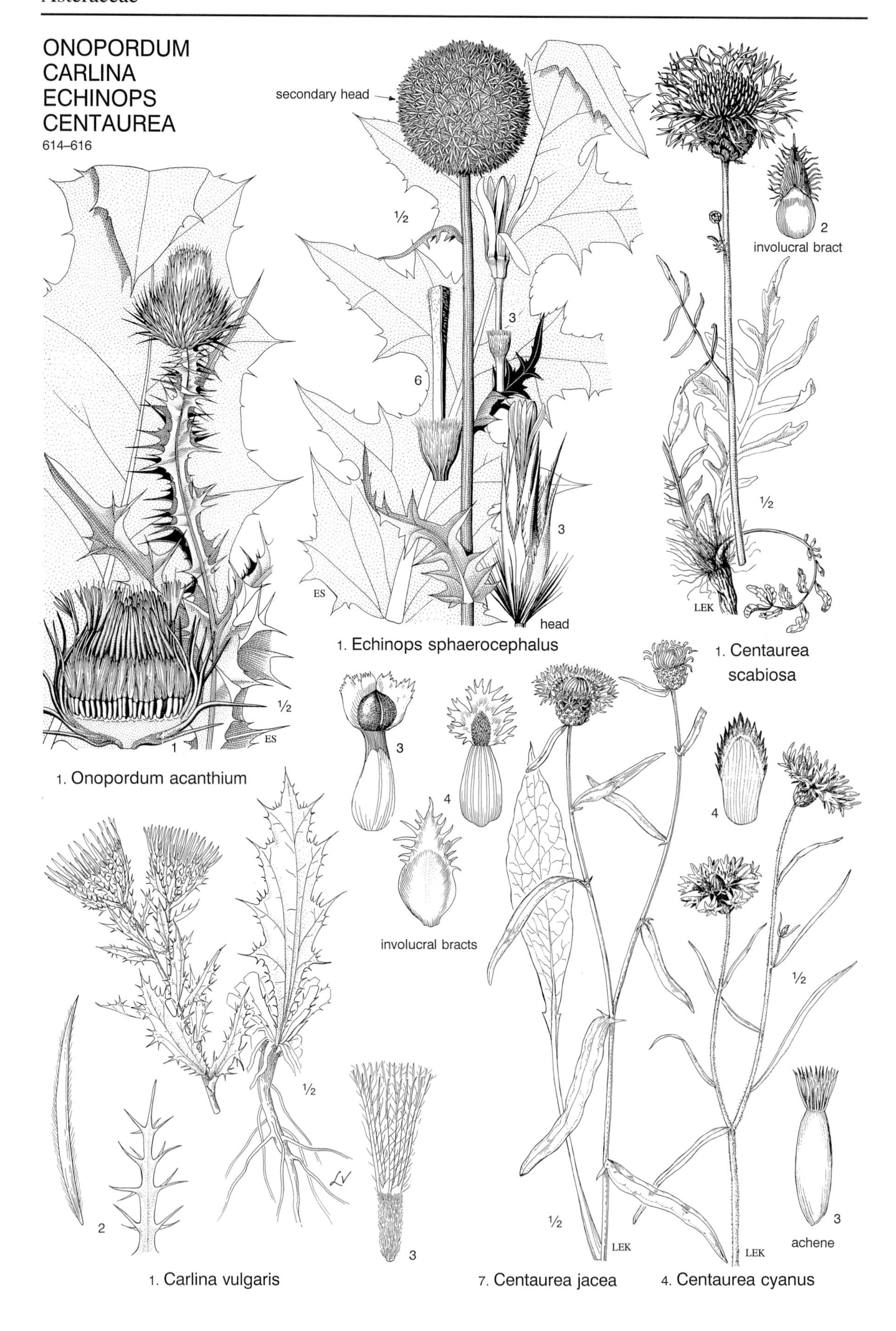

secondary head

½

3

6

ES

head

1. Echinops sphaerocephalus

2
involucral bract

½

LEK

1. Centaurea
scabiosa

½

ES

½

ES

1. Onopordum acanthium

3

4

involucral bracts

4

½

3
achene

2

½

3

1. Carlina vulgaris

½

LEK

7. Centaurea jacea

LEK

4. Centaurea cyanus

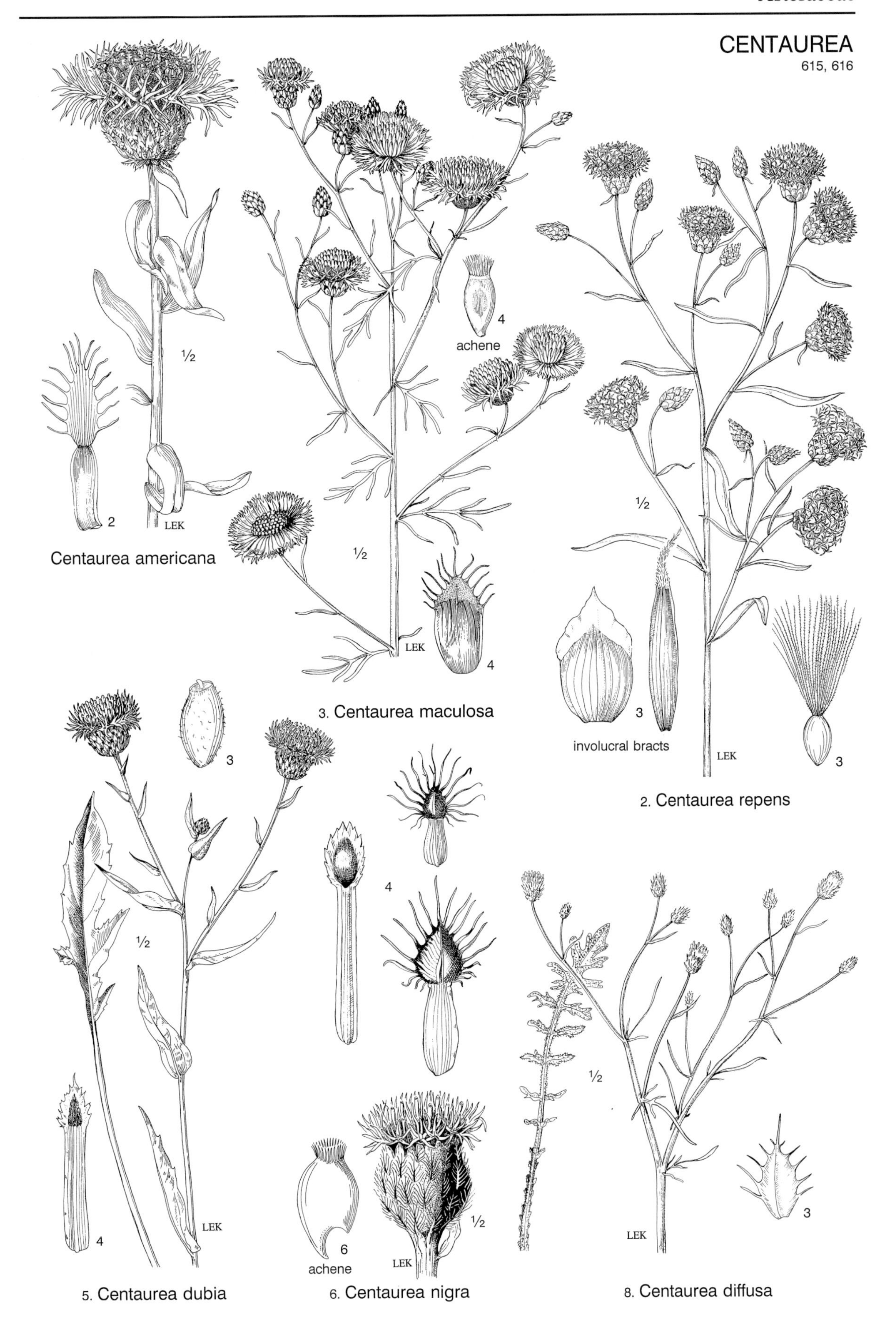

Centaurea americana

achene

3. Centaurea maculosa

involucral bracts

2. Centaurea repens

5. Centaurea dubia

achene

6. Centaurea nigra

8. Centaurea diffusa

CENTAUREA
CNICUS
PRENANTHES
616, 617

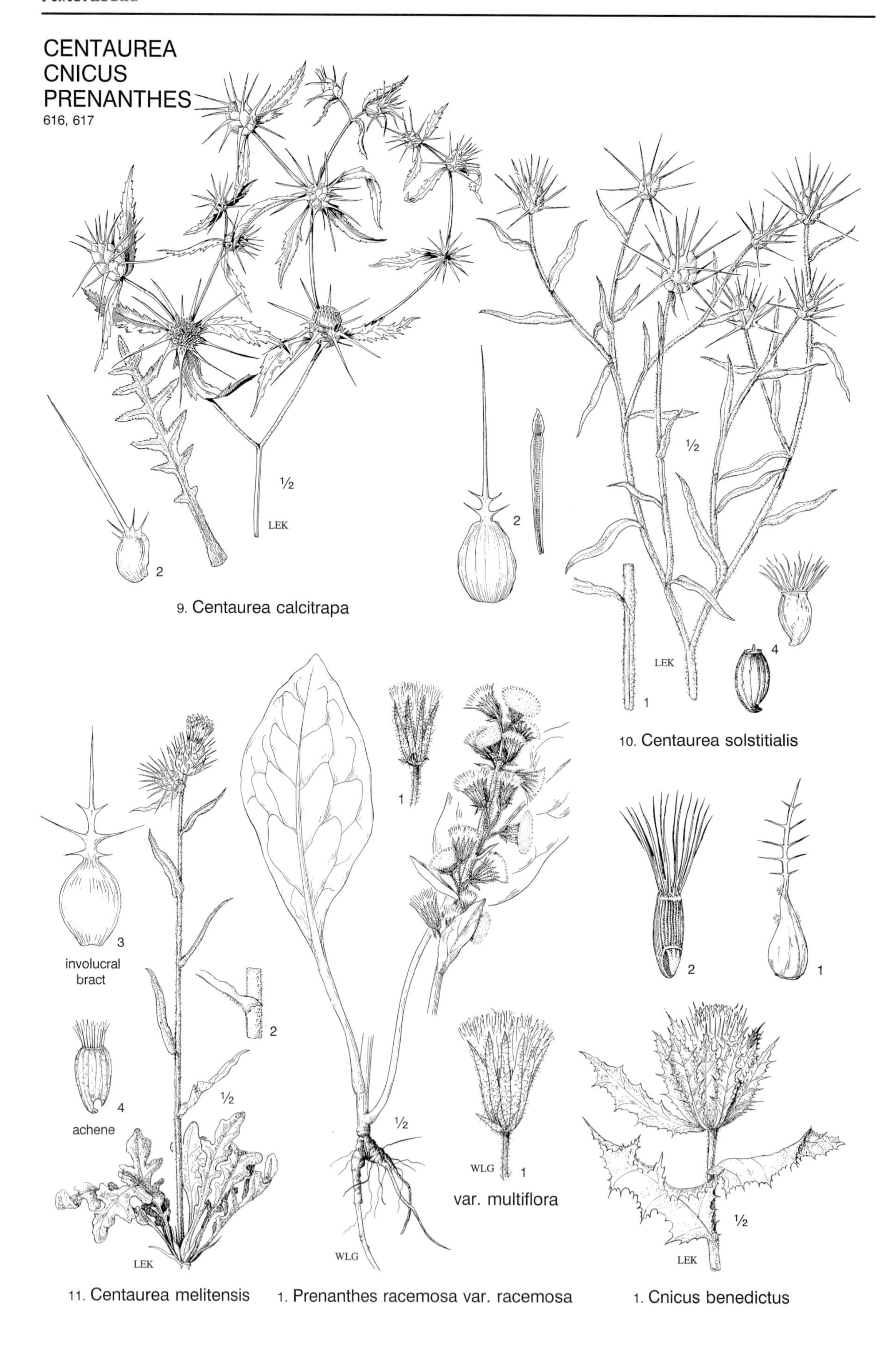

9. Centaurea calcitrapa

10. Centaurea solstitialis

involucral
bract

achene

11. Centaurea melitensis

var. multiflora

1. Prenanthes racemosa var. racemosa

1. Cnicus benedictus

586

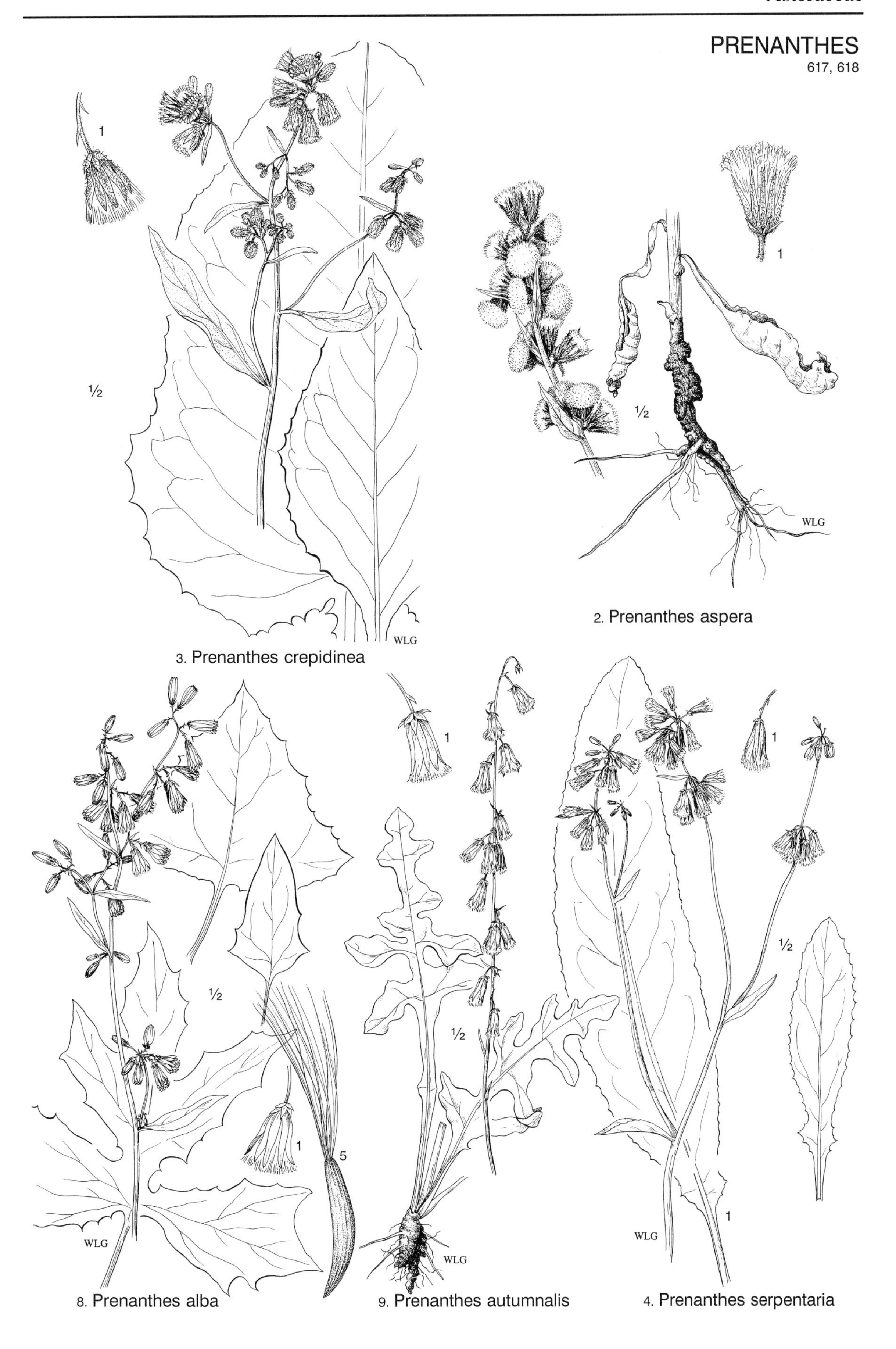

1

½

3. Prenanthes crepidinea

WLG

1

2. Prenanthes aspera

½

WLG

½

1

5

WLG

8. Prenanthes alba

1

½

WLG

9. Prenanthes autumnalis

1

½

1

WLG

4. Prenanthes serpentaria

PRENANTHES
LYGODESMIA
618, 619

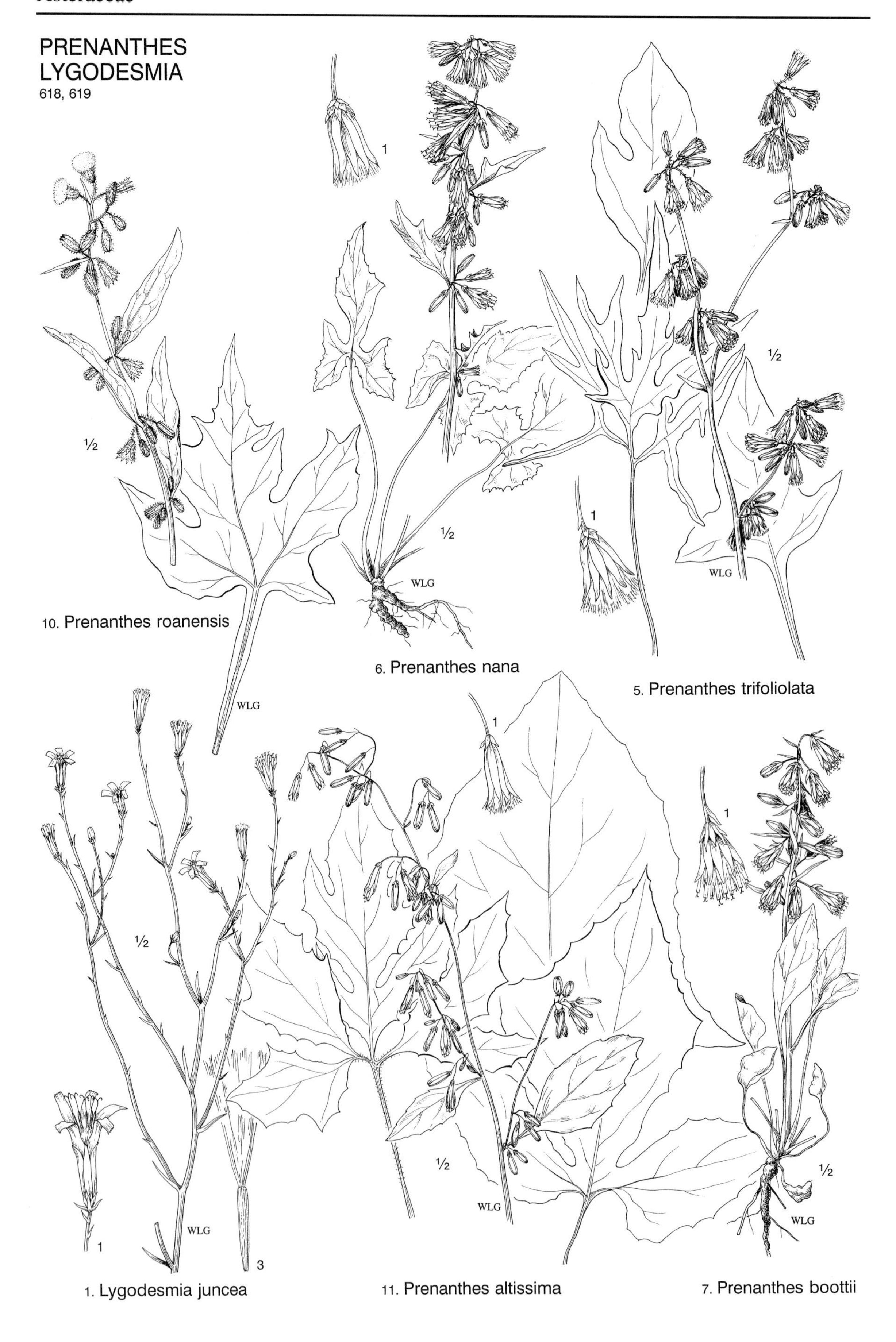

10. Prenanthes roanensis

6. Prenanthes nana

5. Prenanthes trifoliolata

1. Lygodesmia juncea

11. Prenanthes altissima

7. Prenanthes boottii

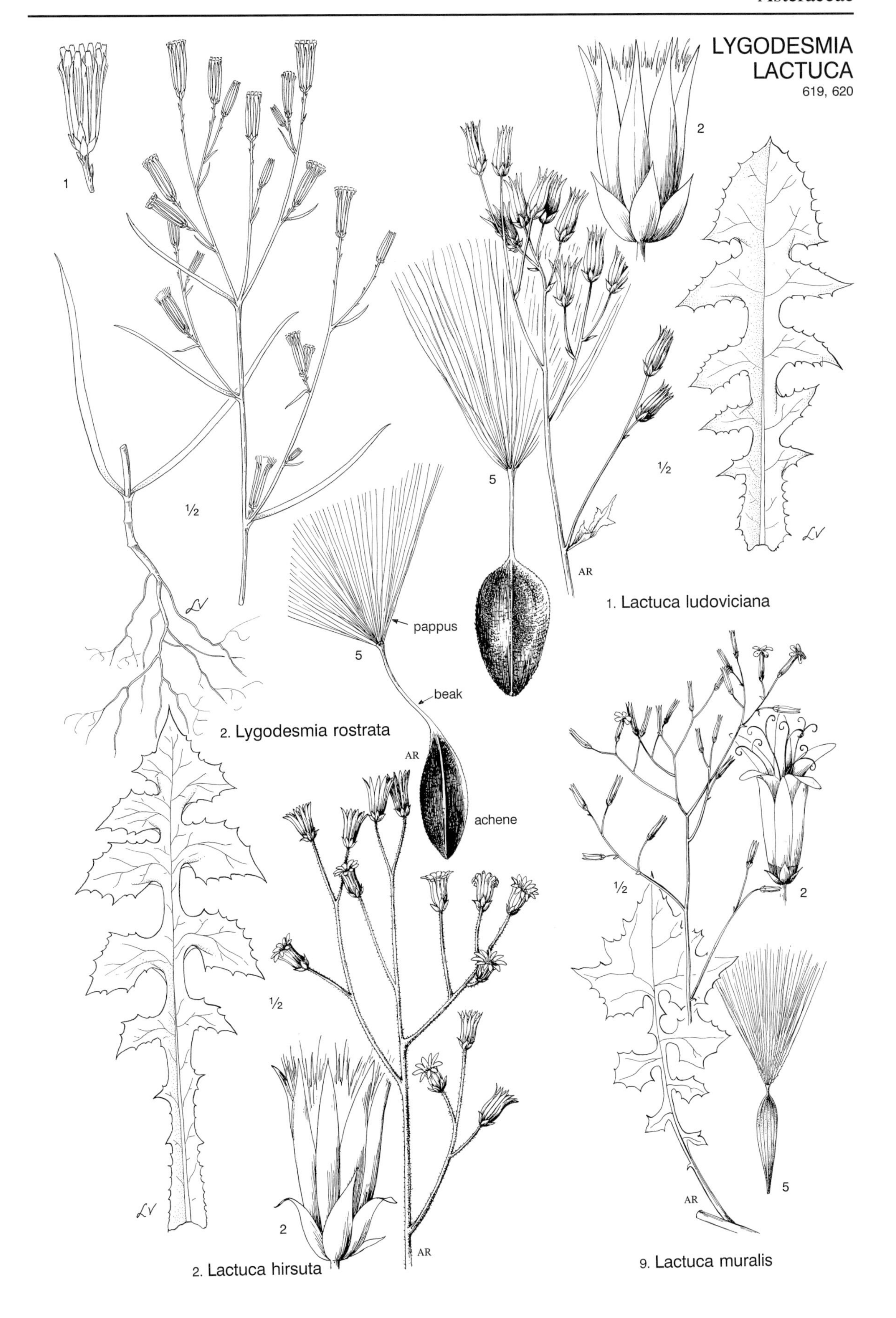

LYGODESMIA
LACTUCA
619, 620

1. Lactuca ludoviciana

pappus

beak

2. Lygodesmia rostrata

AR

achene

2. Lactuca hirsuta

9. Lactuca muralis

LACTUCA
620

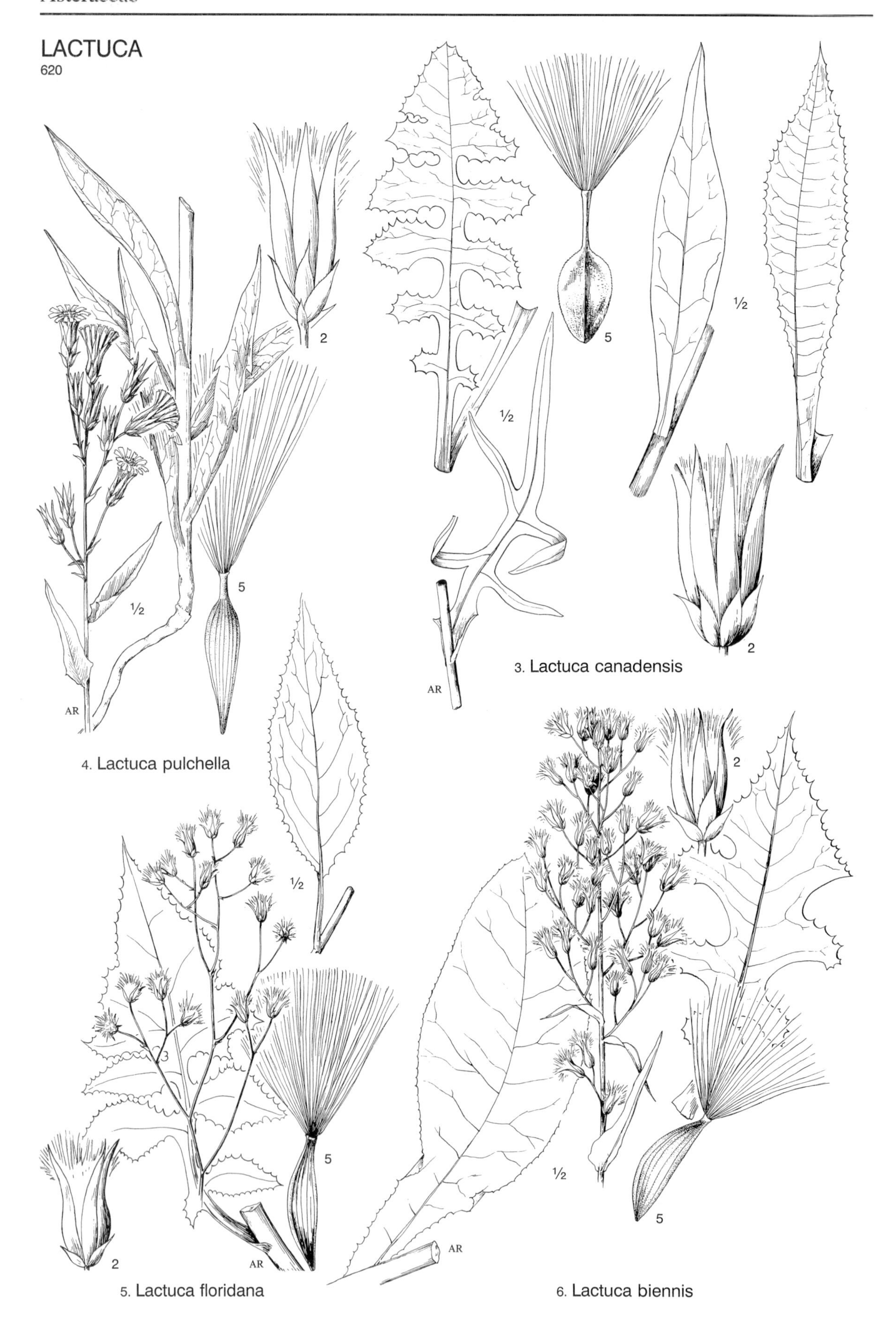

3. Lactuca canadensis

4. Lactuca pulchella

5. Lactuca floridana

6. Lactuca biennis

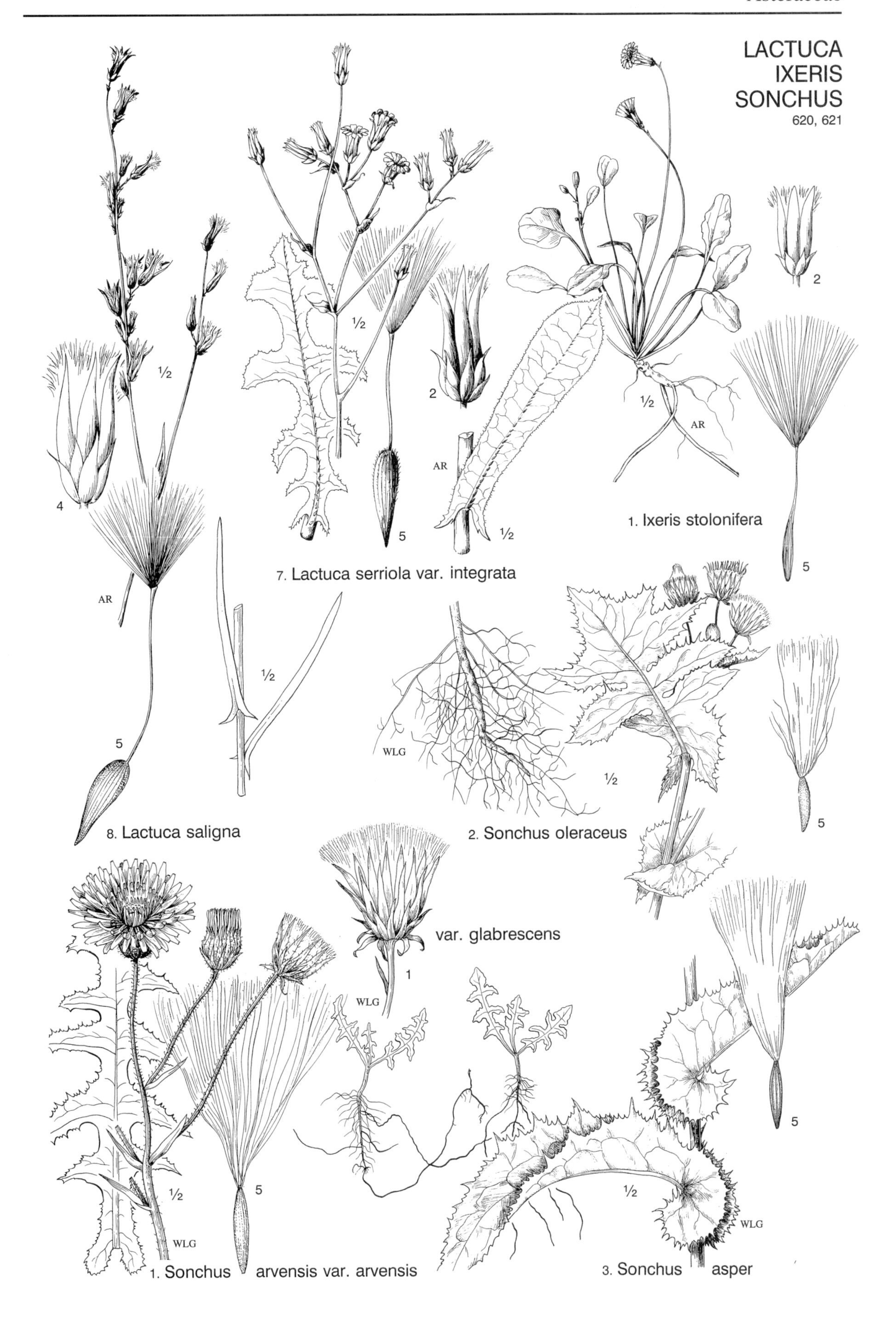

1. Ixeris stolonifera

7. Lactuca serriola var. integrata

8. Lactuca saligna

2. Sonchus oleraceus

var. glabrescens

1. Sonchus arvensis var. arvensis

3. Sonchus asper

HIERACIUM
622, 623

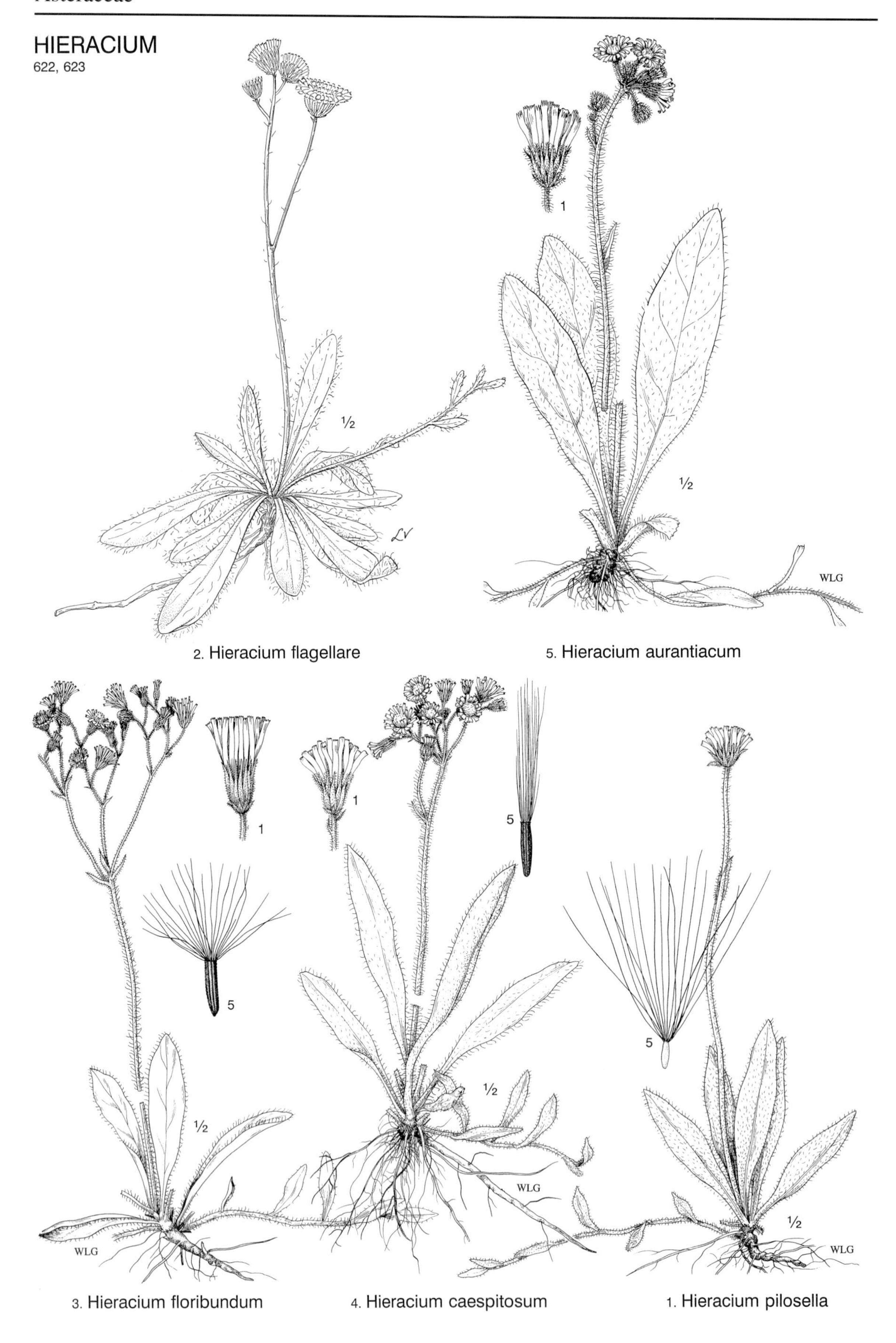

2. Hieracium flagellare

5. Hieracium aurantiacum

3. Hieracium floribundum

4. Hieracium caespitosum

1. Hieracium pilosella

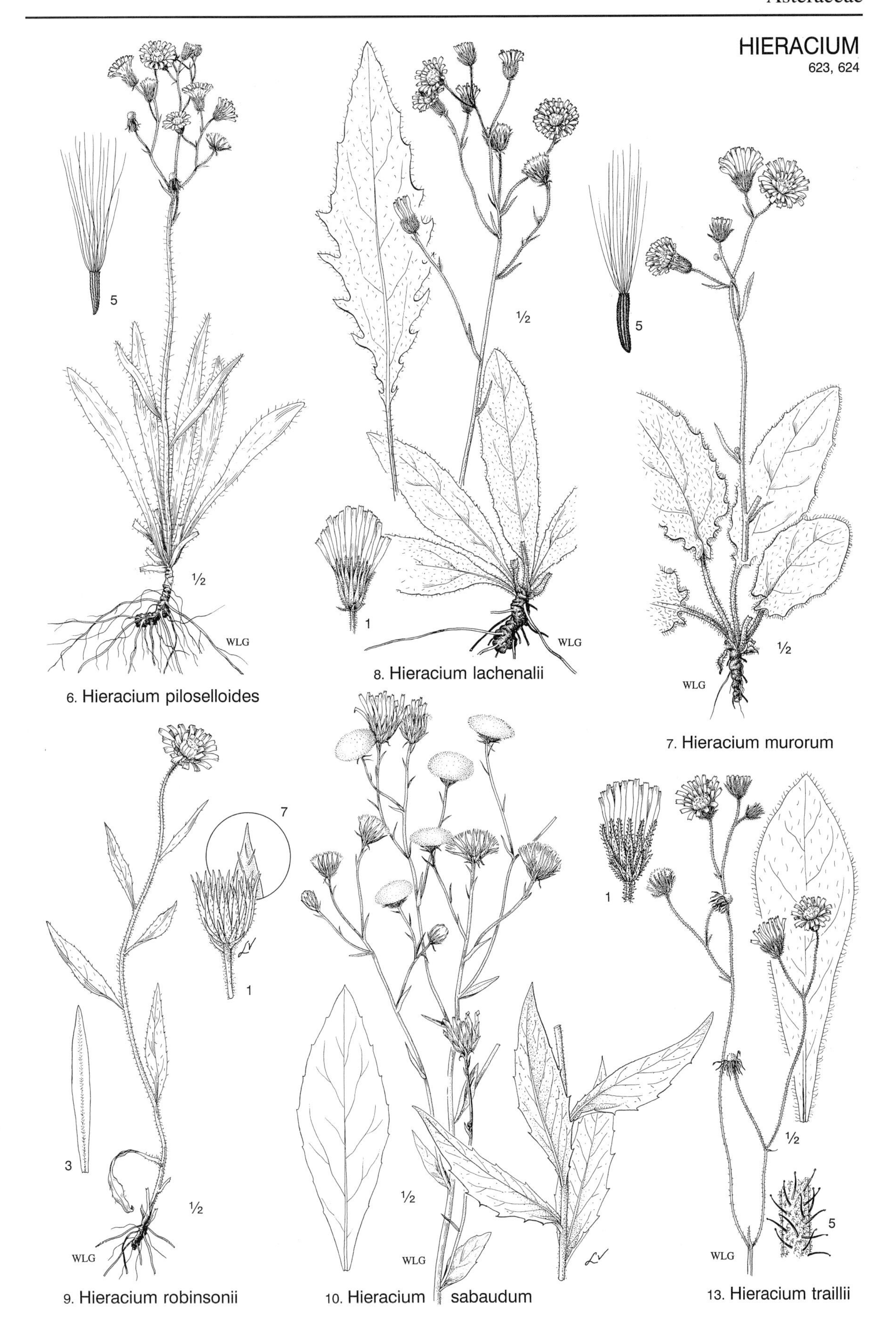

6. Hieracium piloselloides

8. Hieracium lachenalii

7. Hieracium murorum

9. Hieracium robinsonii

10. Hieracium sabaudum

13. Hieracium traillii

HIERACIUM
623, 624

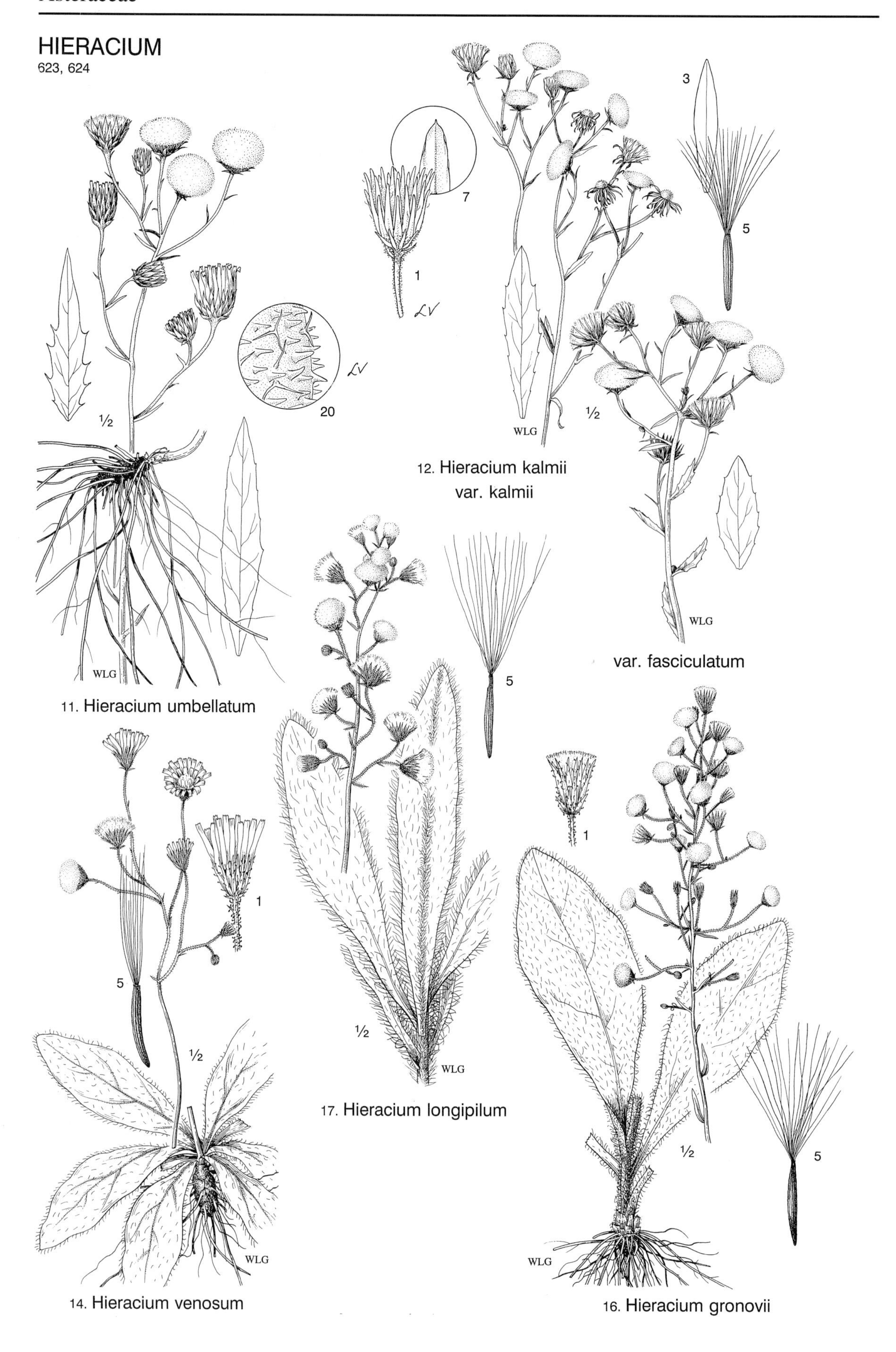

11. Hieracium umbellatum

12. Hieracium kalmii
var. kalmii

var. fasciculatum

14. Hieracium venosum

17. Hieracium longipilum

16. Hieracium gronovii

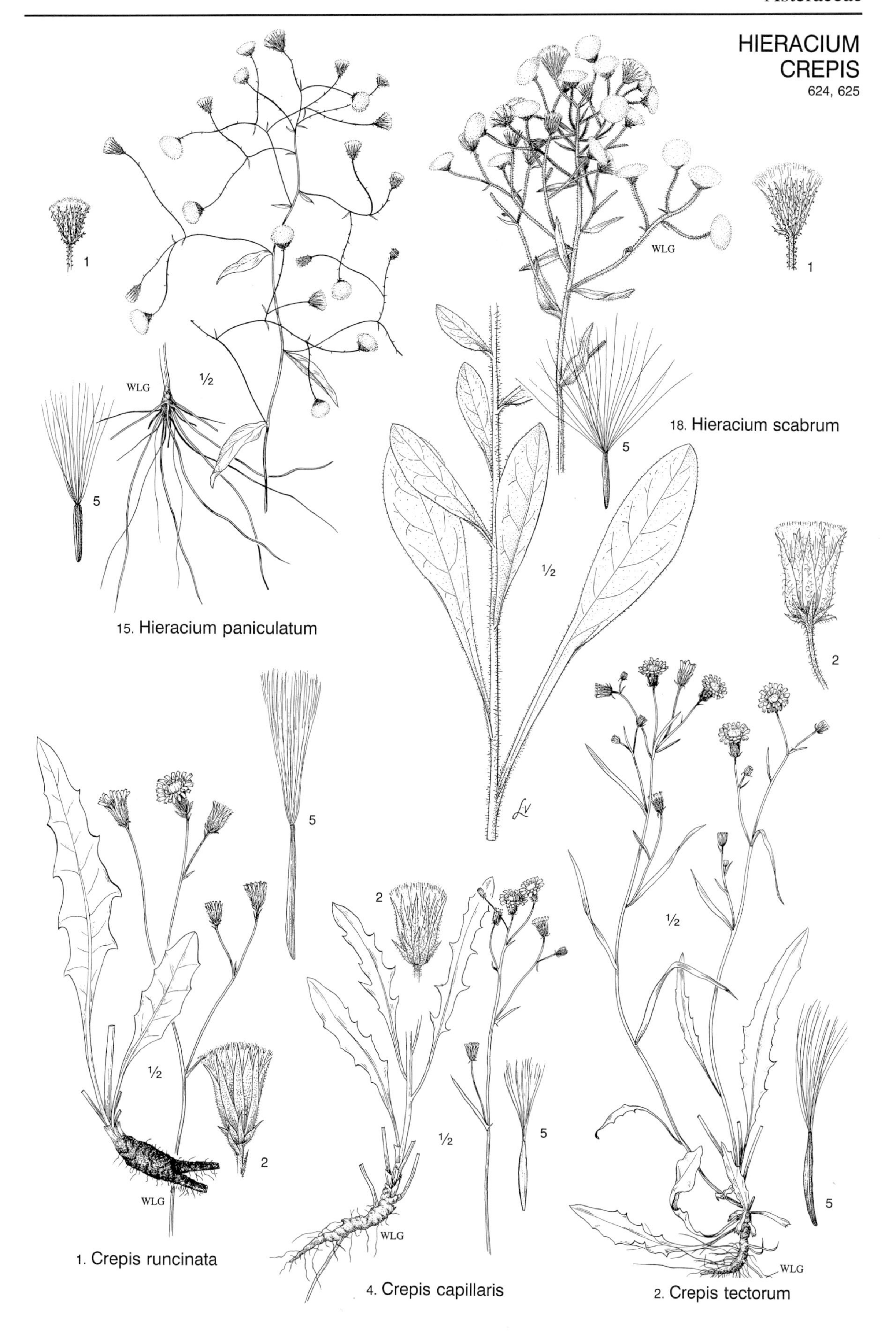

15. Hieracium paniculatum

18. Hieracium scabrum

1. Crepis runcinata

4. Crepis capillaris

2. Crepis tectorum

CREPIS
YOUNGIA
AGOSERIS
625, 626

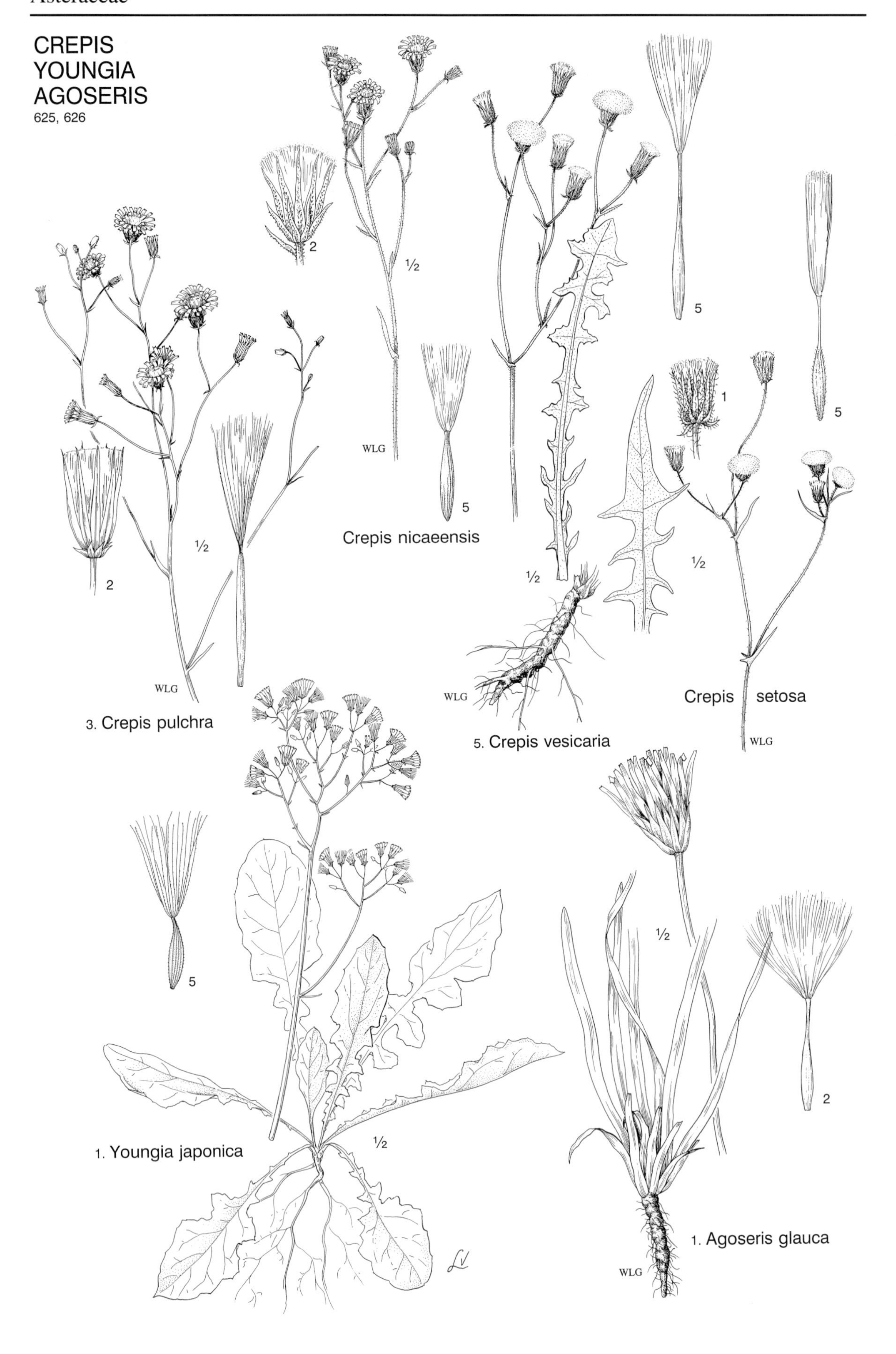

Crepis nicaeensis

3. Crepis pulchra

5. Crepis vesicaria

Crepis setosa

1. Youngia japonica

1. Agoseris glauca

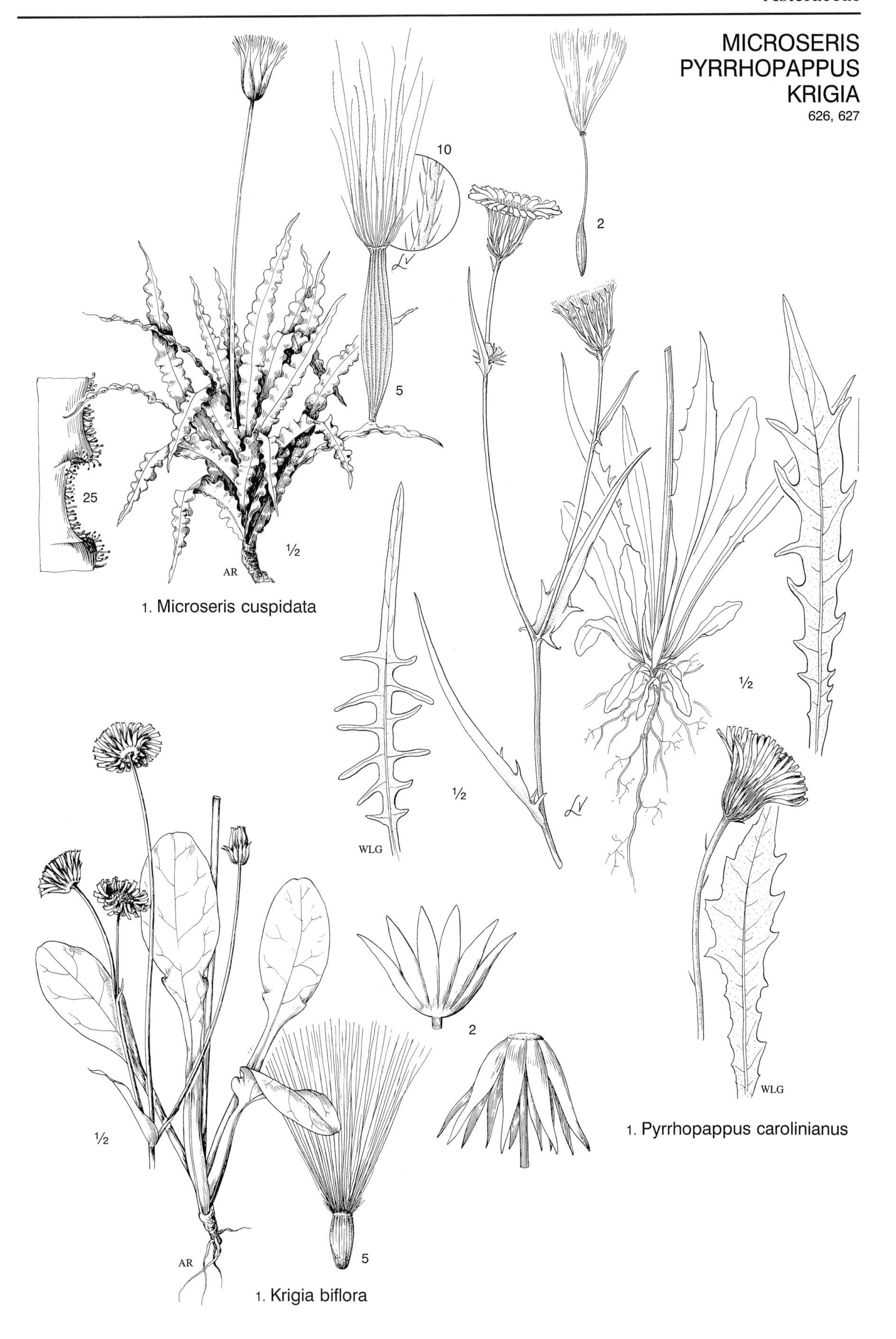

1. Microseris cuspidata

1. Pyrrhopappus carolinianus

1. Krigia biflora

KRIGIA
TARAXACUM
627, 628

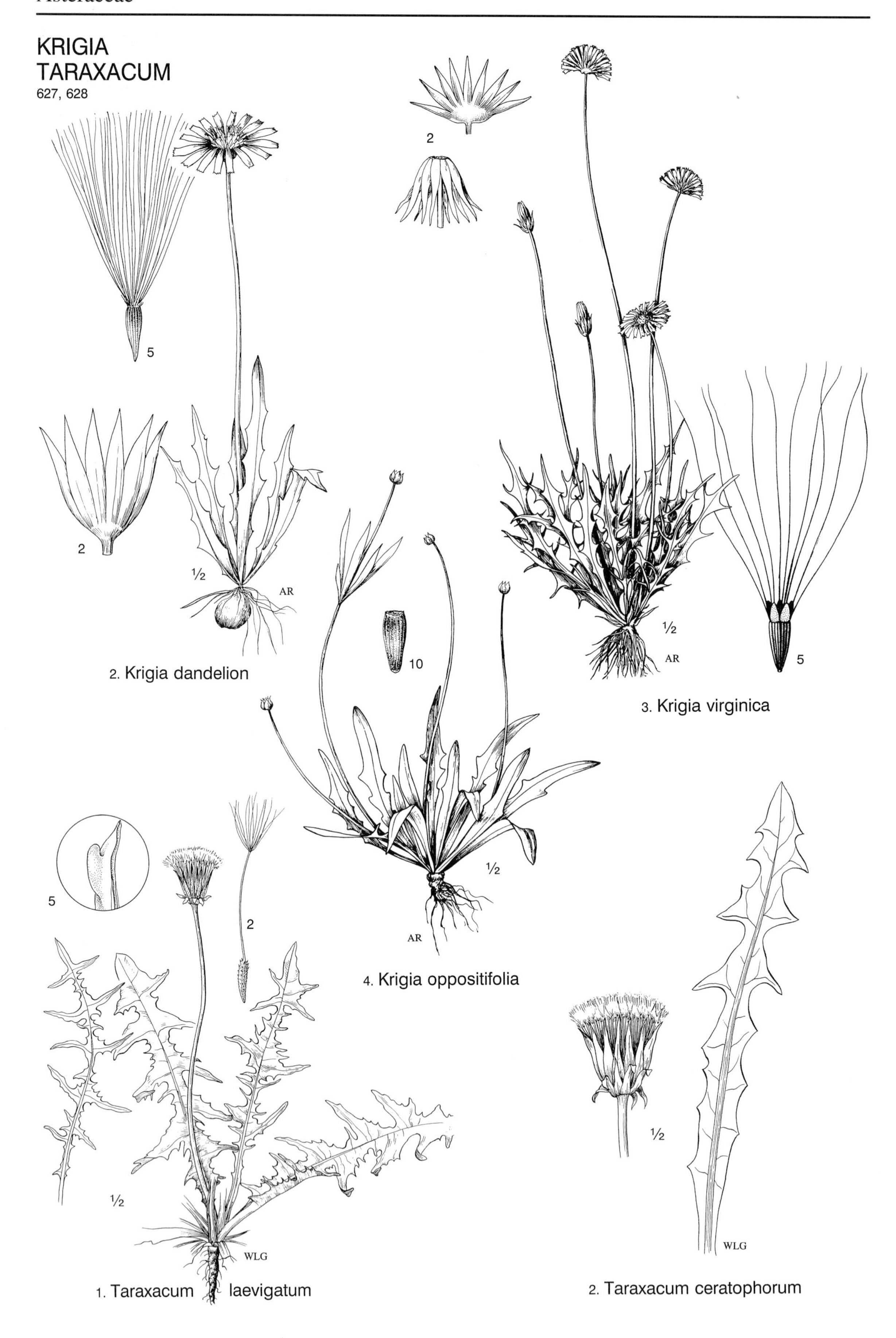

2. Krigia dandelion

3. Krigia virginica

4. Krigia oppositifolia

1. Taraxacum laevigatum

2. Taraxacum ceratophorum

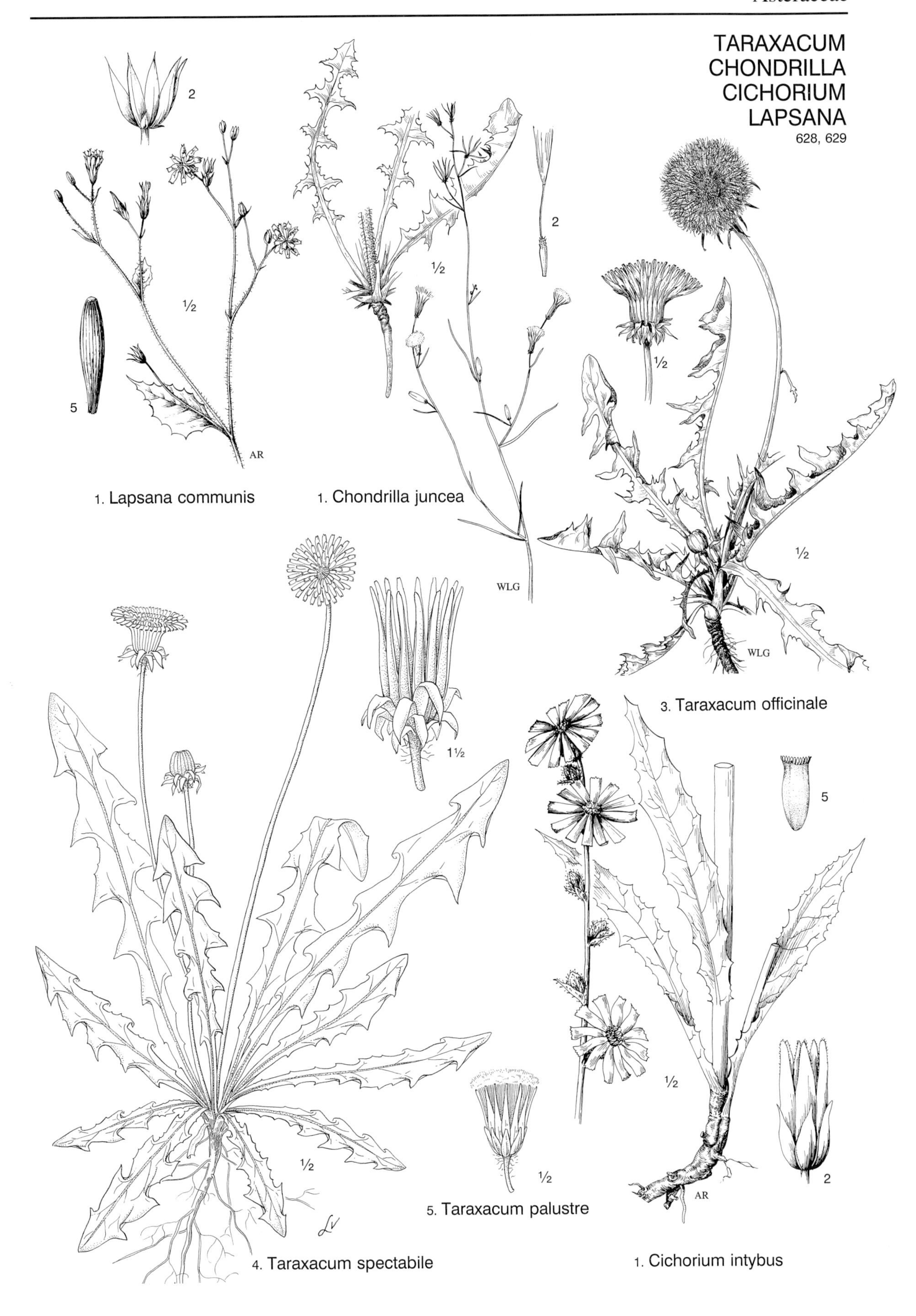

1. Lapsana communis

1. Chondrilla juncea

3. Taraxacum officinale

5. Taraxacum palustre

4. Taraxacum spectabile

1. Cichorium intybus

ARNOSERIS
PICRIS
HYPOCHAERIS
629, 630

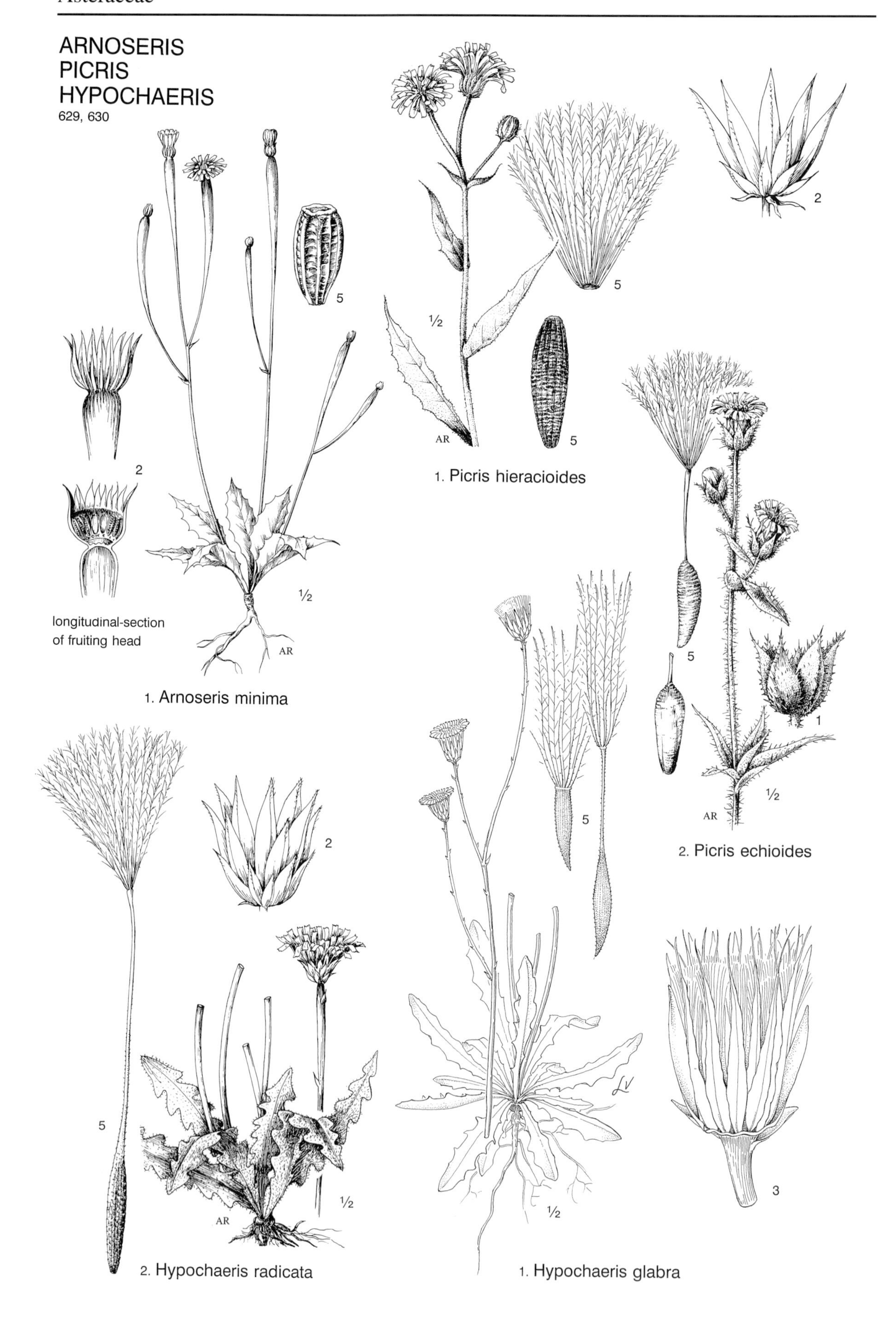

2

longitudinal-section
of fruiting head

1. Arnoseris minima

5

1. Picris hieracioides

2

5

2. Picris echioides

5

2. Hypochaeris radicata

1. Hypochaeris glabra

3

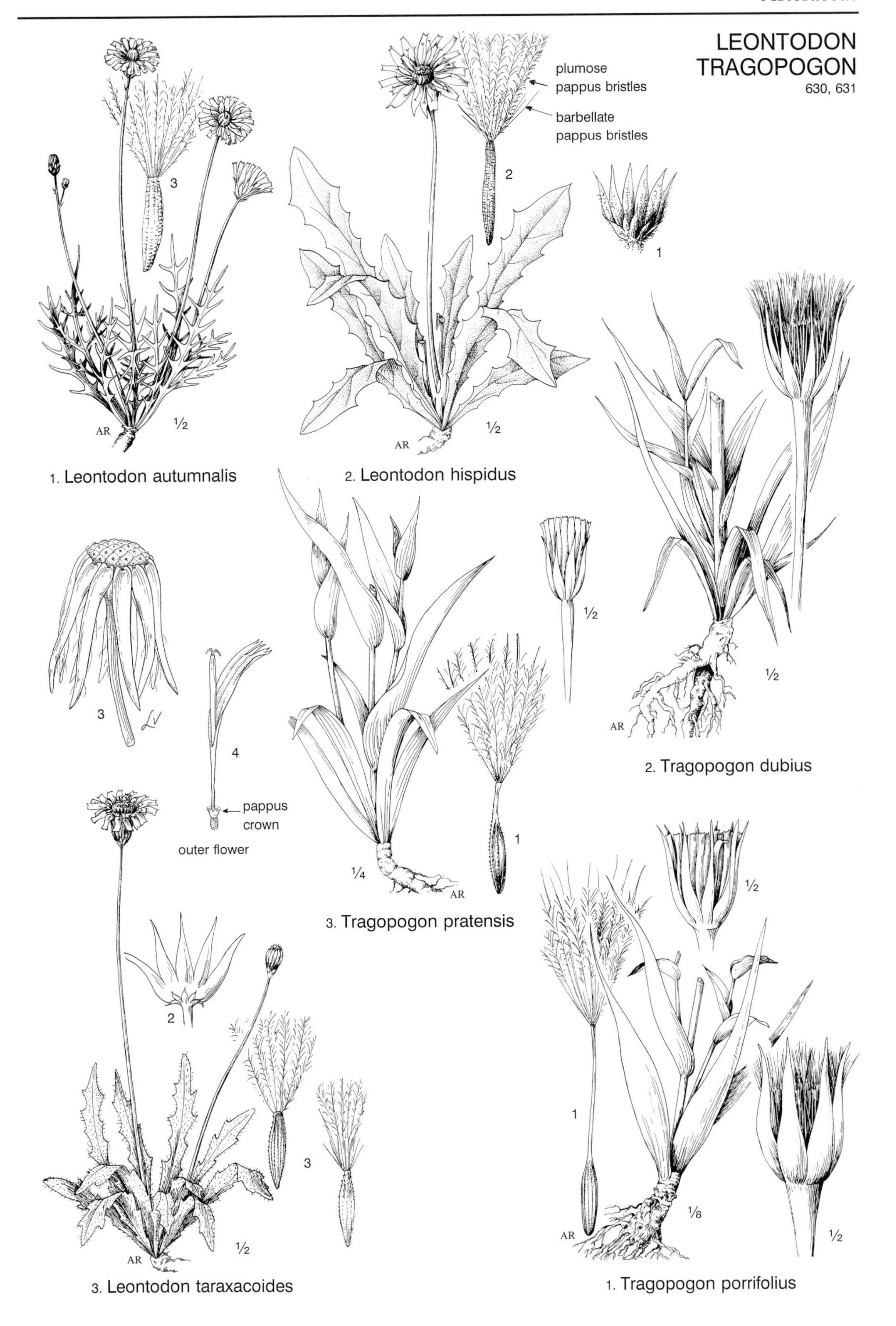

plumose
pappus bristles

barbellate
pappus bristles

1. Leontodon autumnalis

2. Leontodon hispidus

2. Tragopogon dubius

pappus
crown

outer flower

3. Tragopogon pratensis

3. Leontodon taraxacoides

1. Tragopogon porrifolius

BUTOMUS
ALISMA
SAGITTARIA
632, 633

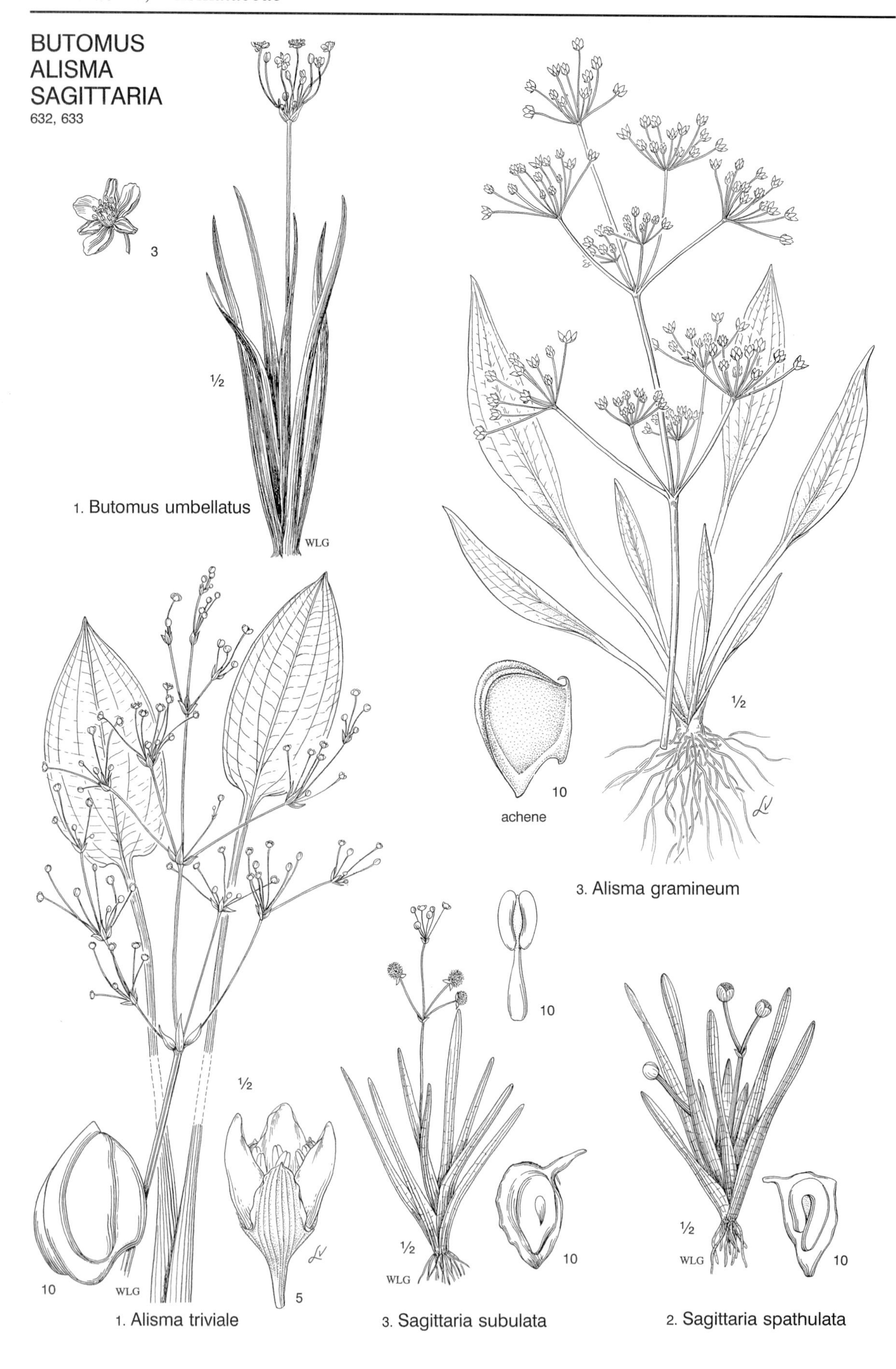

3

½

1. Butomus umbellatus

WLG

½

10
achene

3. Alisma gramineum

10

½

½

10

1. Alisma triviale

5

½

WLG

3. Sagittaria subulata

½

WLG

10

2. Sagittaria spathulata

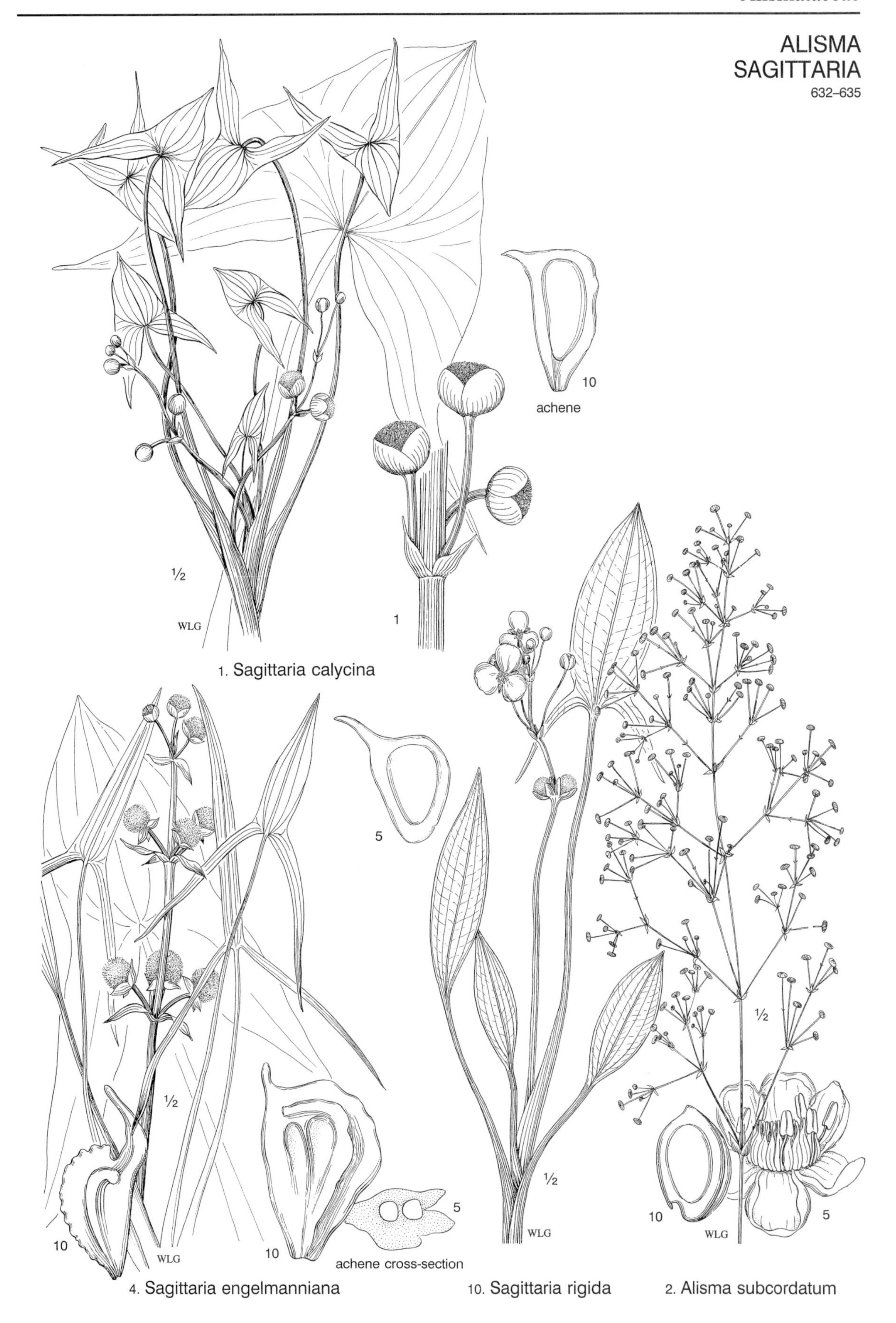

1. Sagittaria calycina

10 achene

5

4. Sagittaria engelmanniana

10 achene cross-section

10. Sagittaria rigida

2. Alisma subcordatum

SAGITTARIA
634

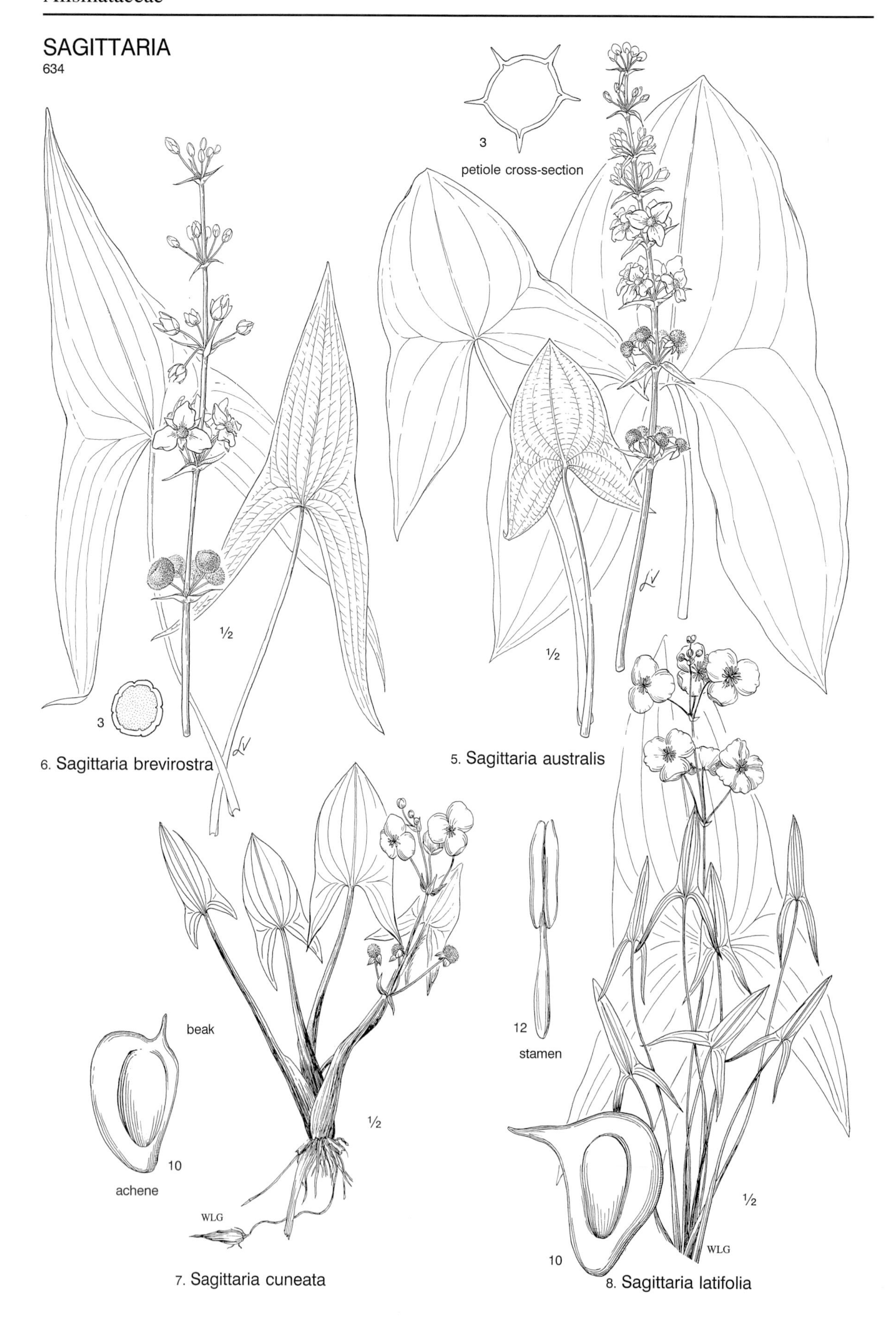

3
petiole cross-section

1/2

3

6. Sagittaria brevirostra

5. Sagittaria australis

1/2

beak

10

achene

1/2

WLG

7. Sagittaria cuneata

12
stamen

10

1/2

WLG

8. Sagittaria latifolia

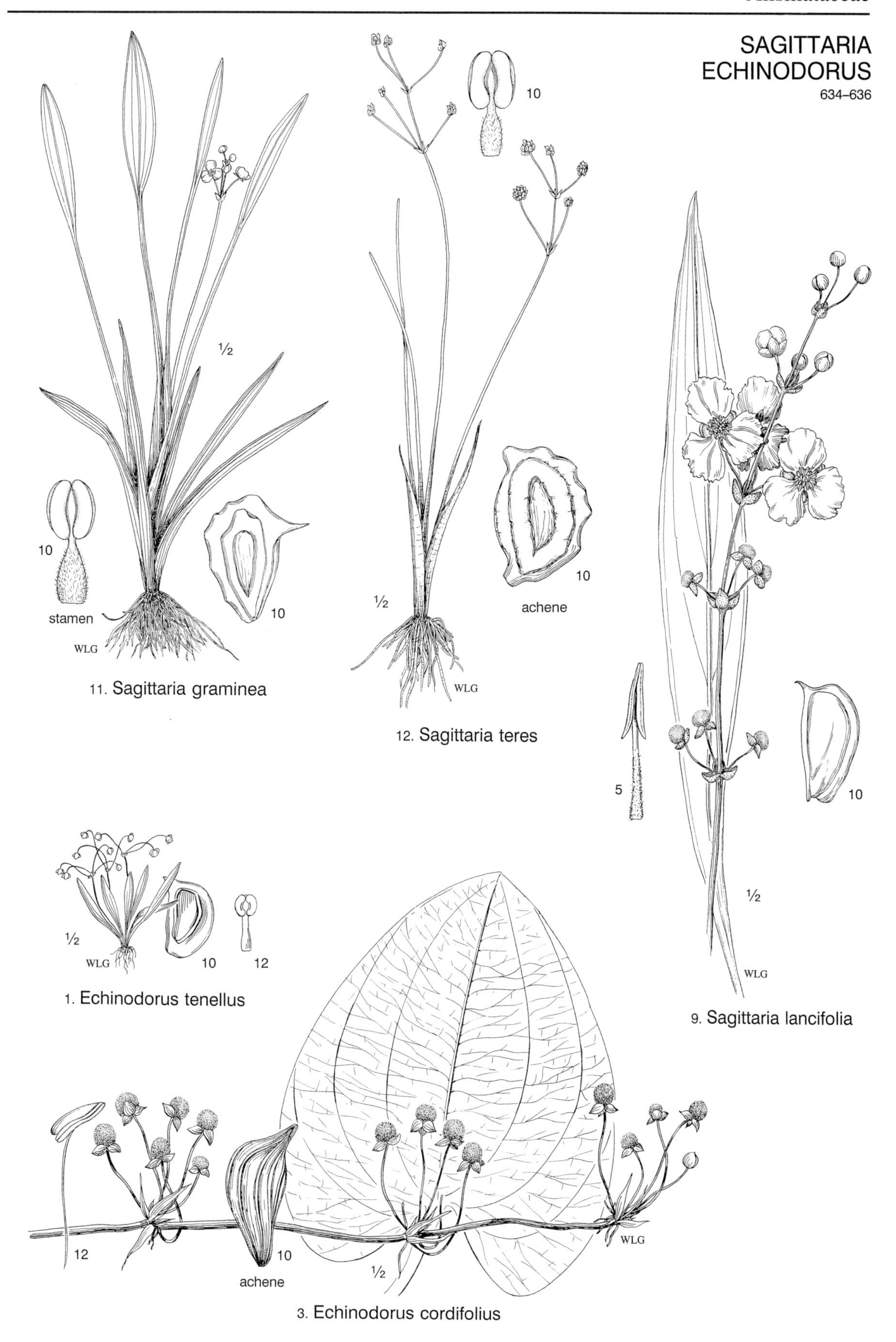

11. Sagittaria graminea

stamen

WLG

10

12. Sagittaria teres

achene

10

1. Echinodorus tenellus

WLG 10 12

9. Sagittaria lancifolia

WLG

3. Echinodorus cordifolius

achene

ECHINODORUS
EGERIA
ELODEA
HYDRILLA
HYDROCHARIS
635–638

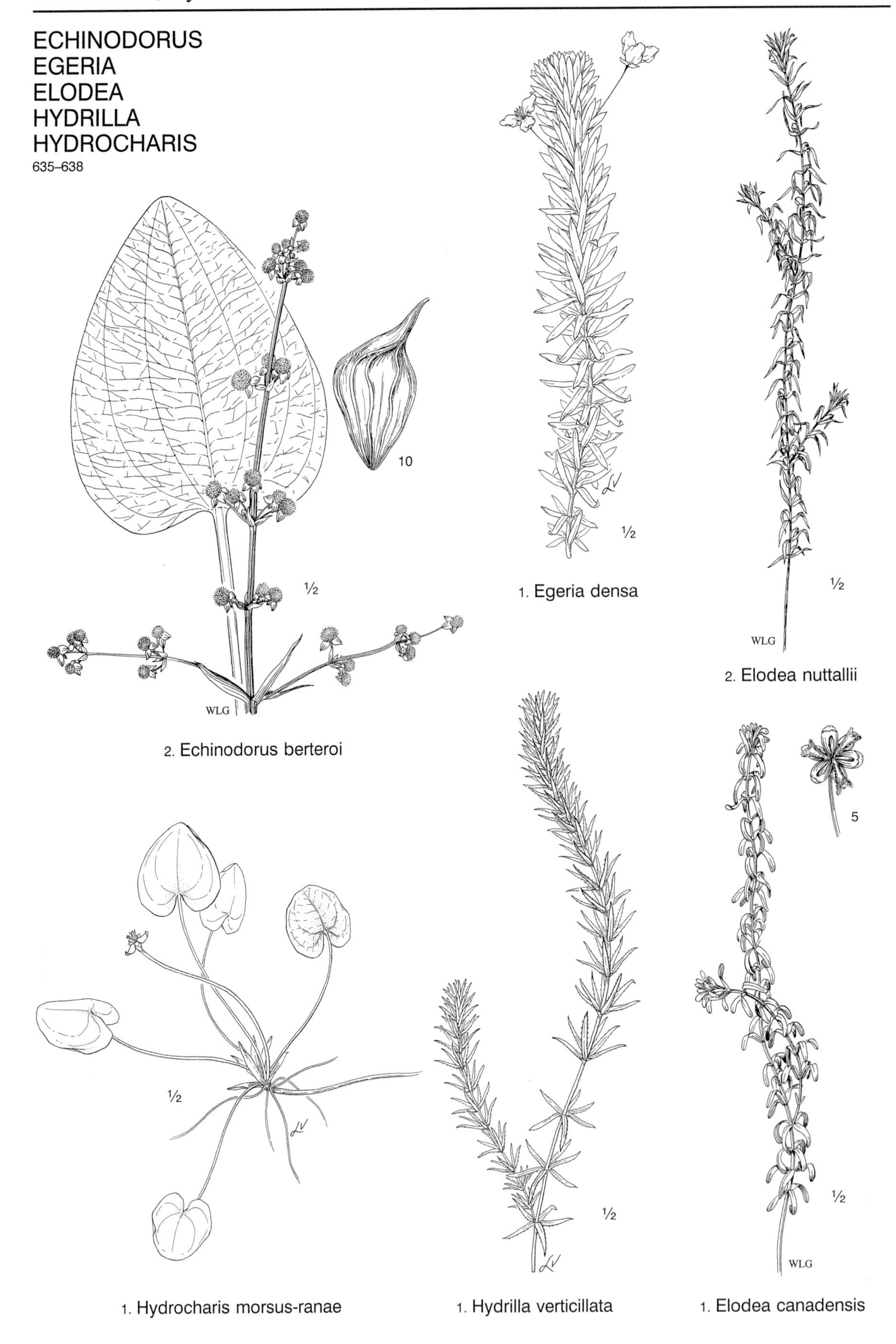

1. Egeria densa

2. Elodea nuttallii

2. Echinodorus berteroi

1. Hydrocharis morsus-ranae

1. Hydrilla verticillata

1. Elodea canadensis

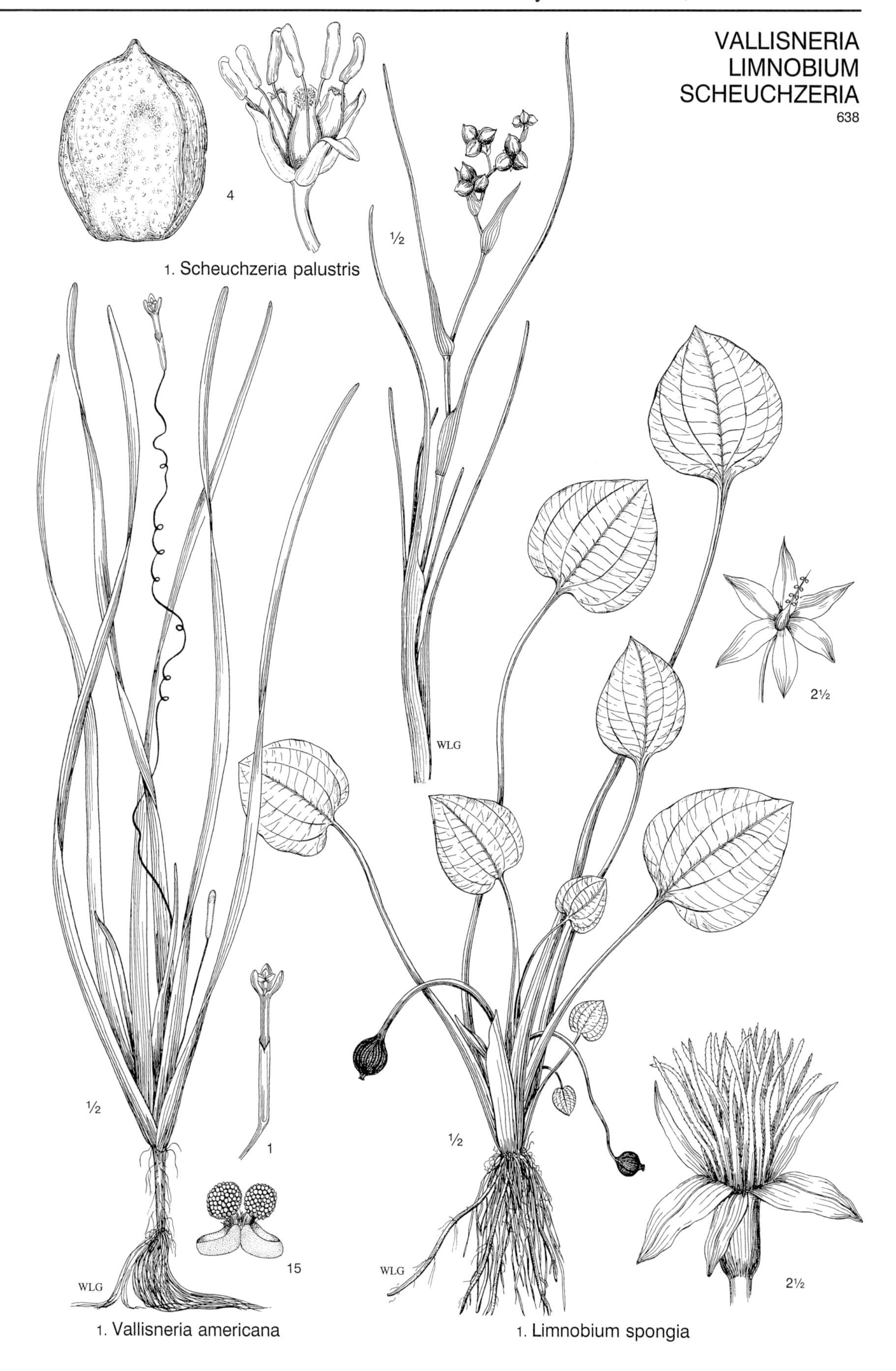

4

1. Scheuchzeria palustris

½

WLG

½

WLG

1

15

1. Vallisneria americana

½

WLG

2½

2½

1. Limnobium spongia

TRIGLOCHIN
POTAMOGETON
639–643

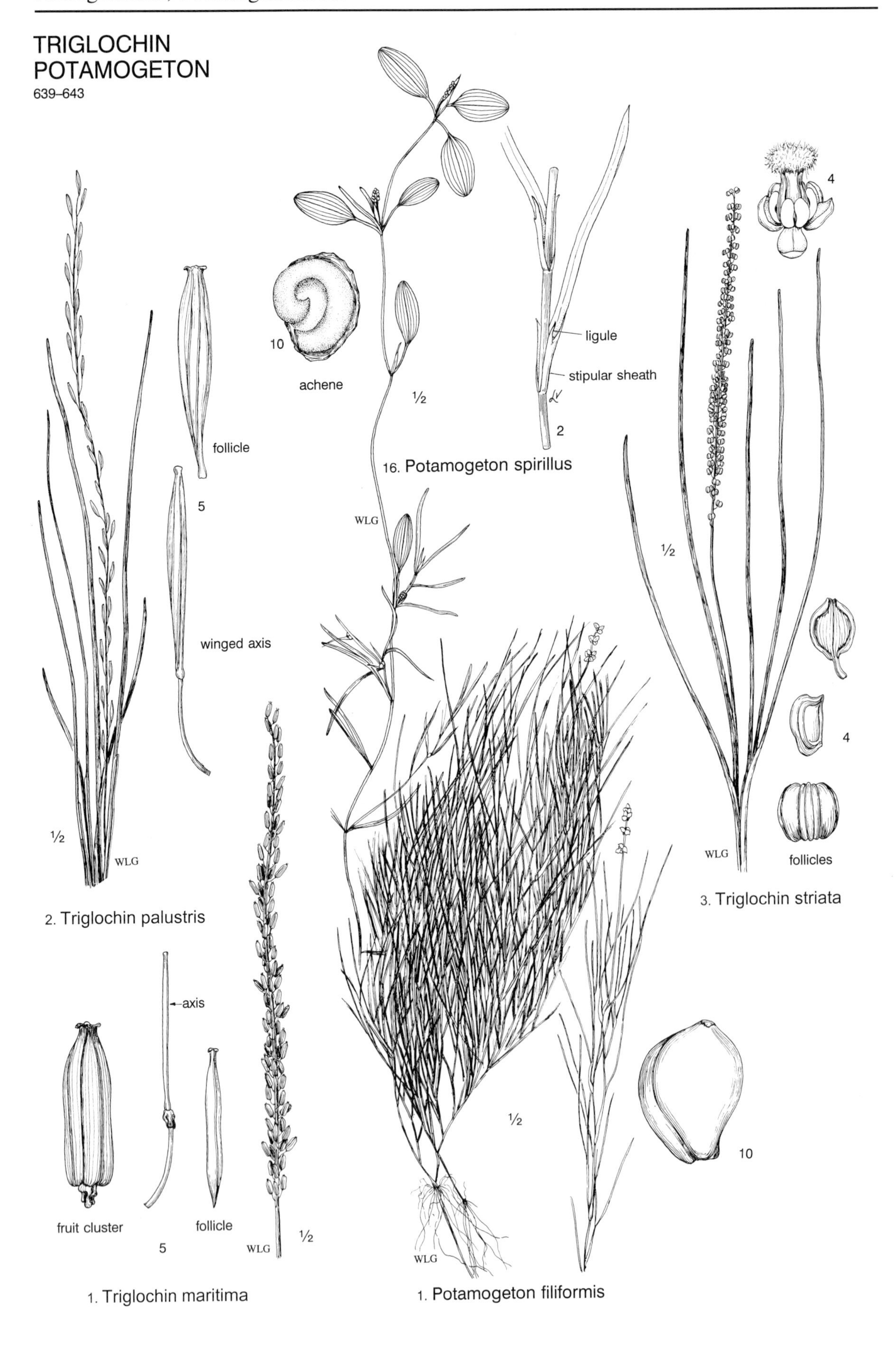

10

achene

follicle

5

winged axis

½

WLG

2. Triglochin palustris

ligule

stipular sheath

2

16. Potamogeton spirillus

½

WLG

½

½

WLG

follicles

4

3. Triglochin striata

→axis

fruit cluster   follicle

5

WLG   ½

½

10

WLG

1. Triglochin maritima

1. Potamogeton filiformis

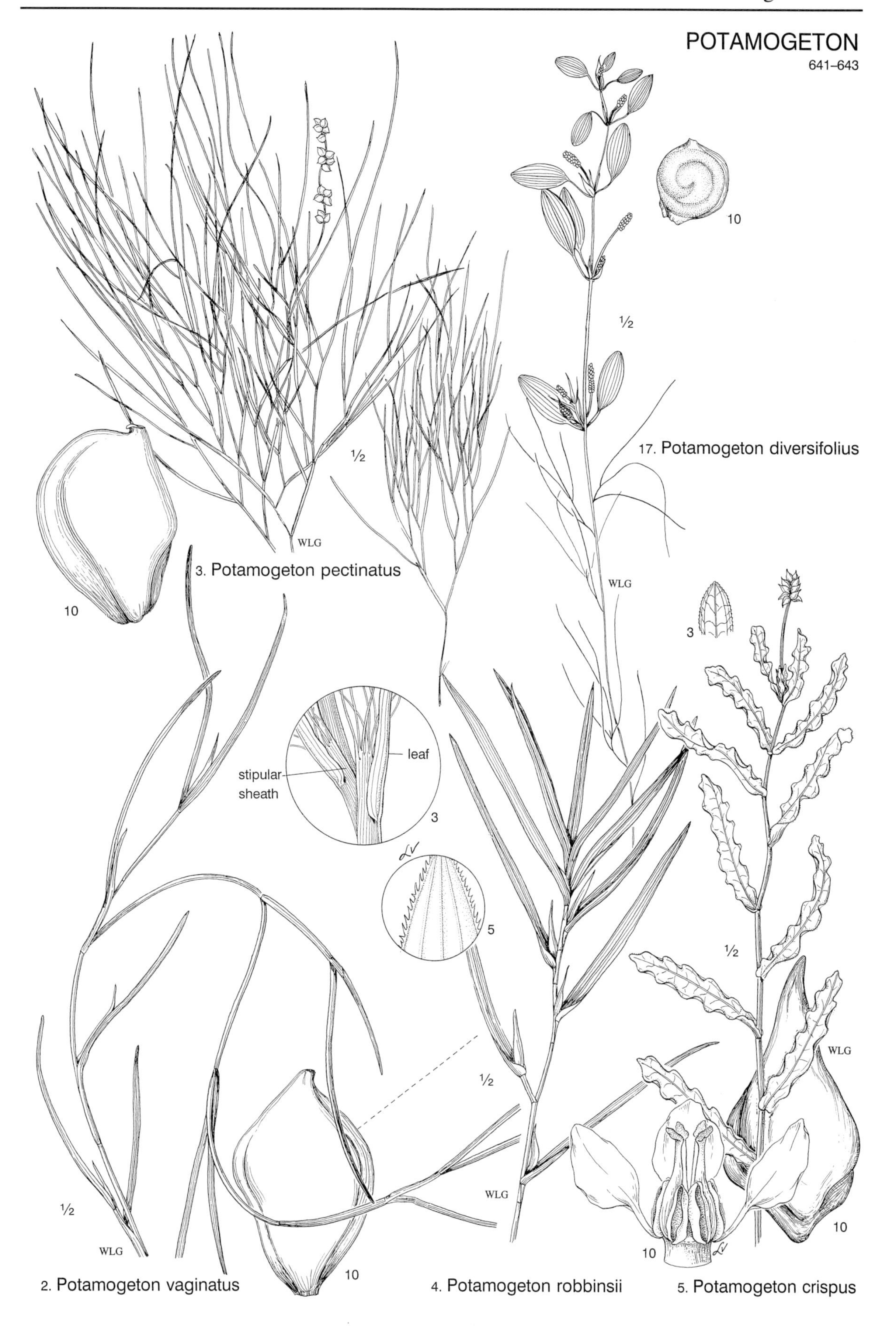

3. Potamogeton pectinatus

17. Potamogeton diversifolius

stipular-
sheath

leaf

3

5

2. Potamogeton vaginatus

4. Potamogeton robbinsii

5. Potamogeton crispus

POTAMOGETON
642, 643

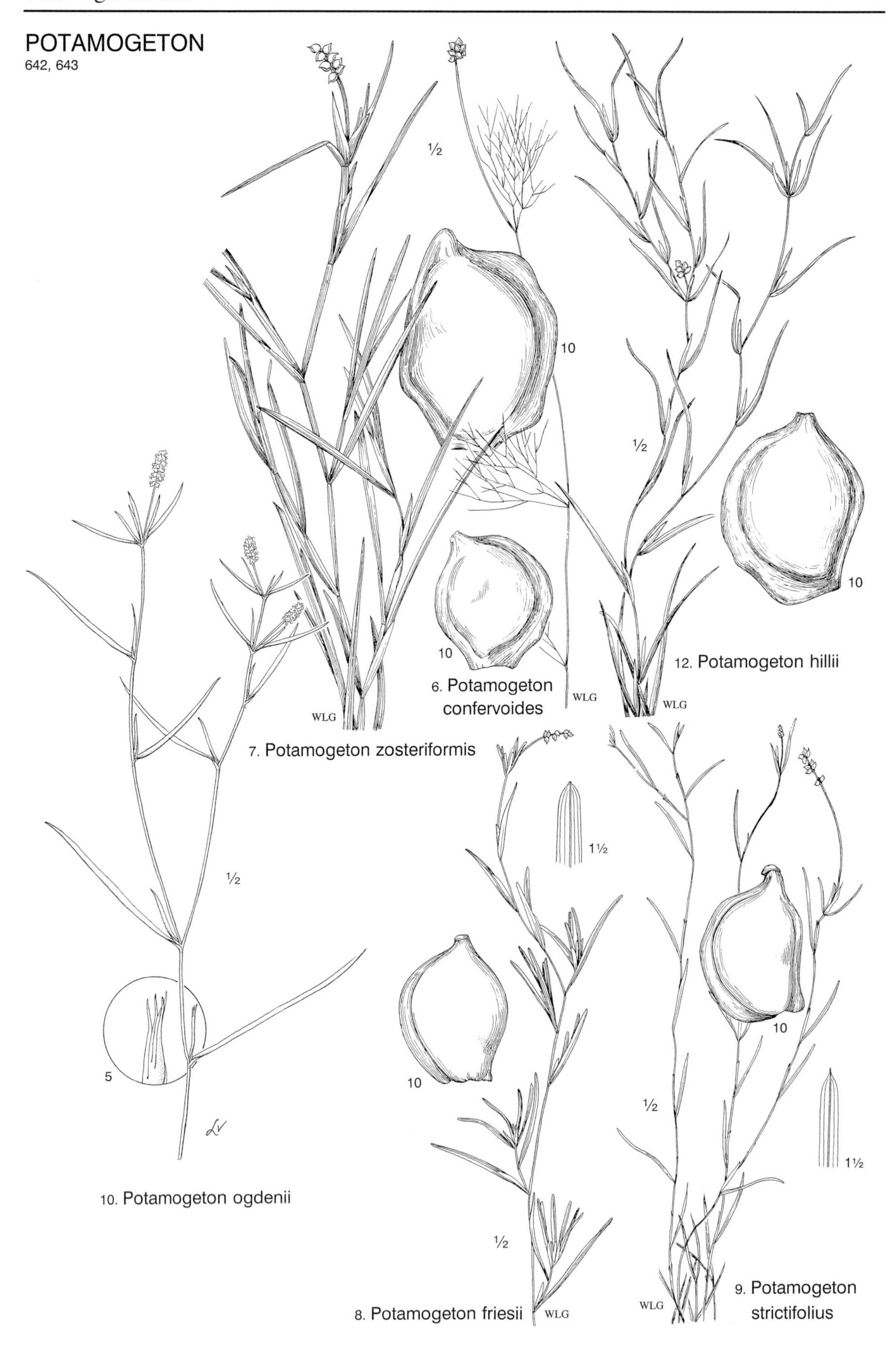

½

10

½

10

6. Potamogeton
   confervoides

WLG

7. Potamogeton zosteriformis

WLG

½

10

½

10

12. Potamogeton hillii

WLG

WLG

5

10. Potamogeton ogdenii

1½

10

½

10

½

1½

½

WLG

8. Potamogeton friesii

9. Potamogeton
   strictifolius

WLG

610

POTAMOGETON
642, 643

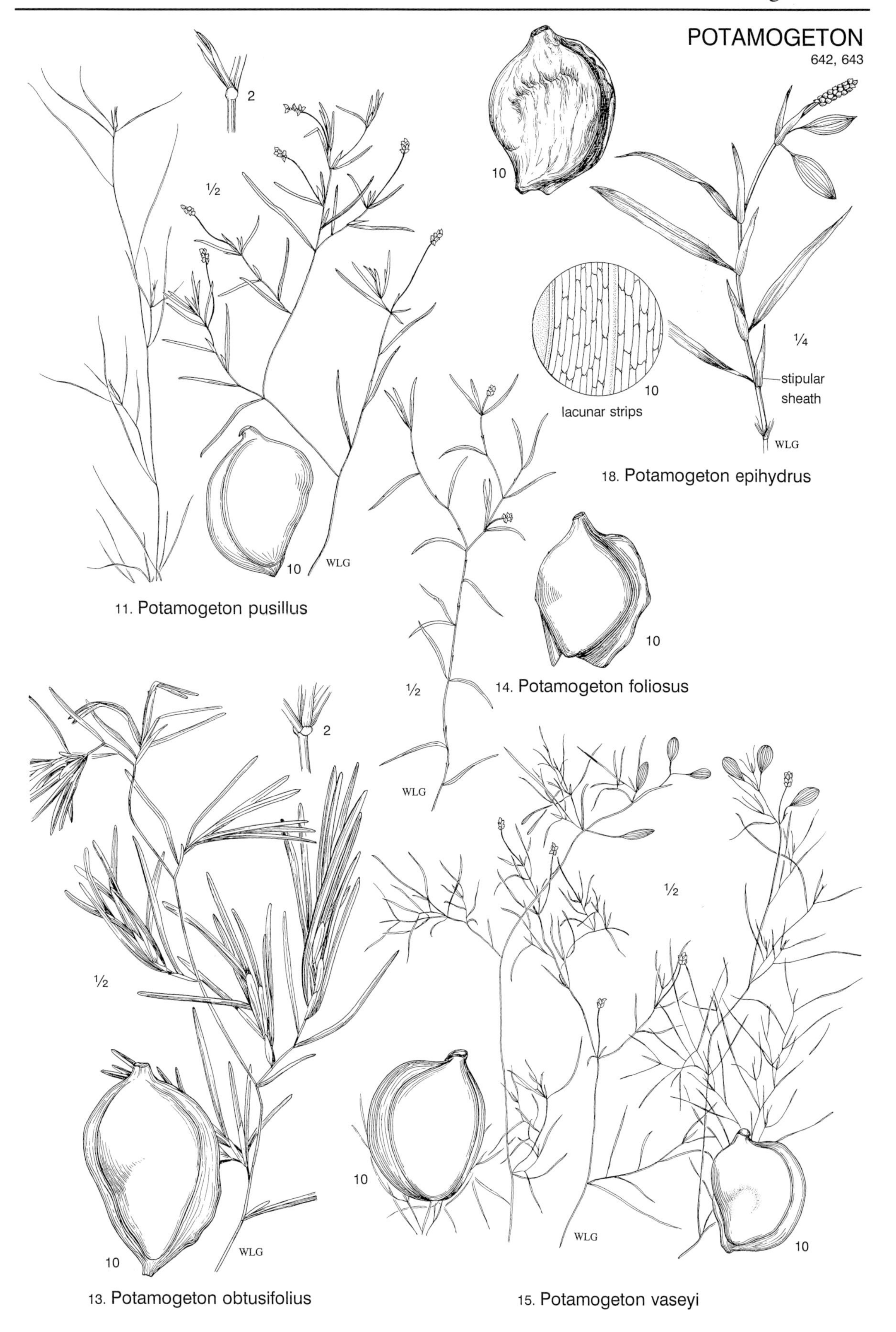

2

½

10

10
lacunar strips

¼

stipular
sheath

WLG

18. Potamogeton epihydrus

10   WLG

11. Potamogeton pusillus

½

10

14. Potamogeton foliosus

½

2

WLG

½

10

WLG

10

13. Potamogeton obtusifolius

15. Potamogeton vaseyi

POTAMOGETON
644

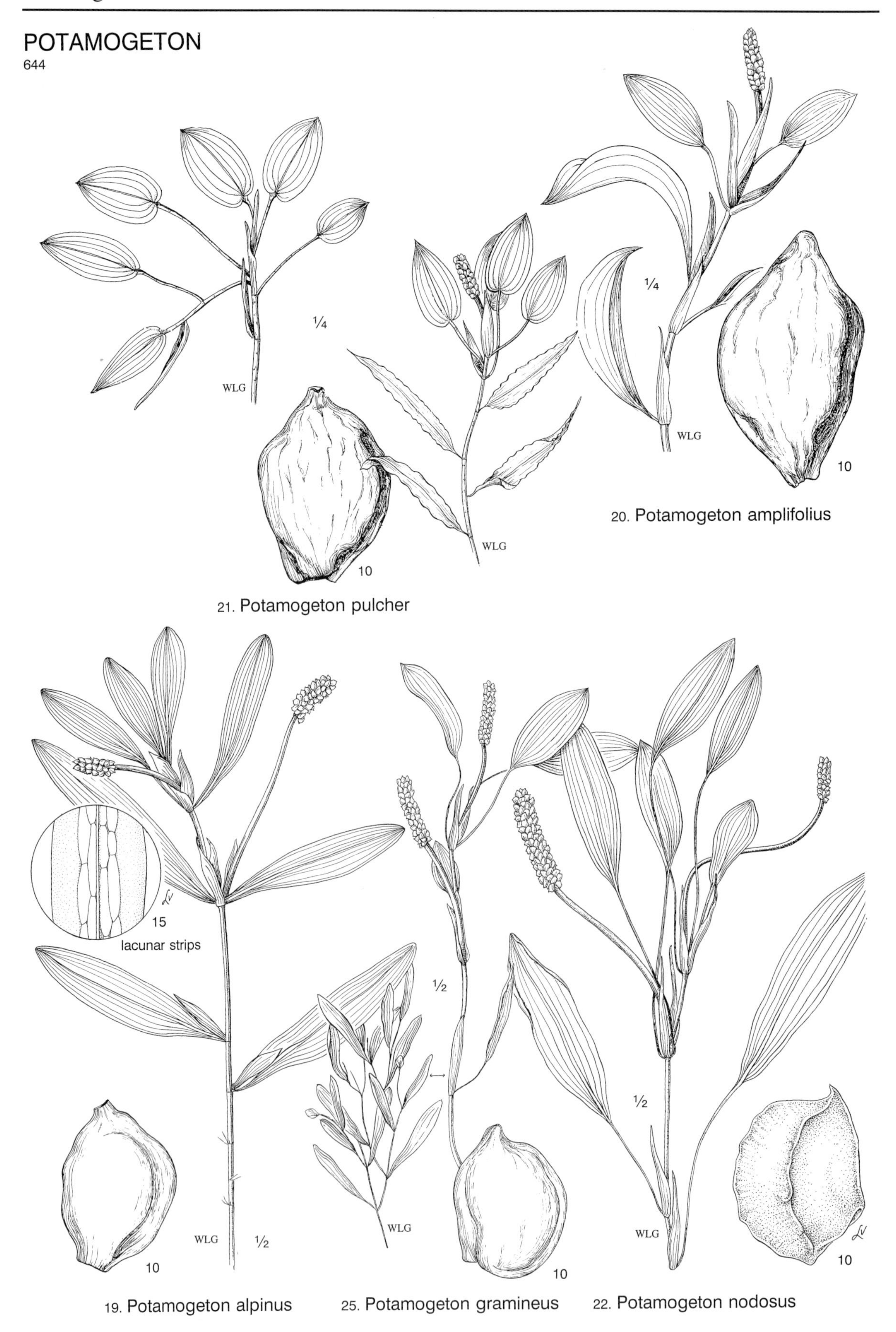

1/4

WLG

21. Potamogeton pulcher

1/4

WLG

10

20. Potamogeton amplifolius

10

WLG

15
lacunar strips

1/2

1/2

1/2

WLG   1/2

10

19. Potamogeton alpinus

WLG

10

25. Potamogeton gramineus

WLG

10

22. Potamogeton nodosus

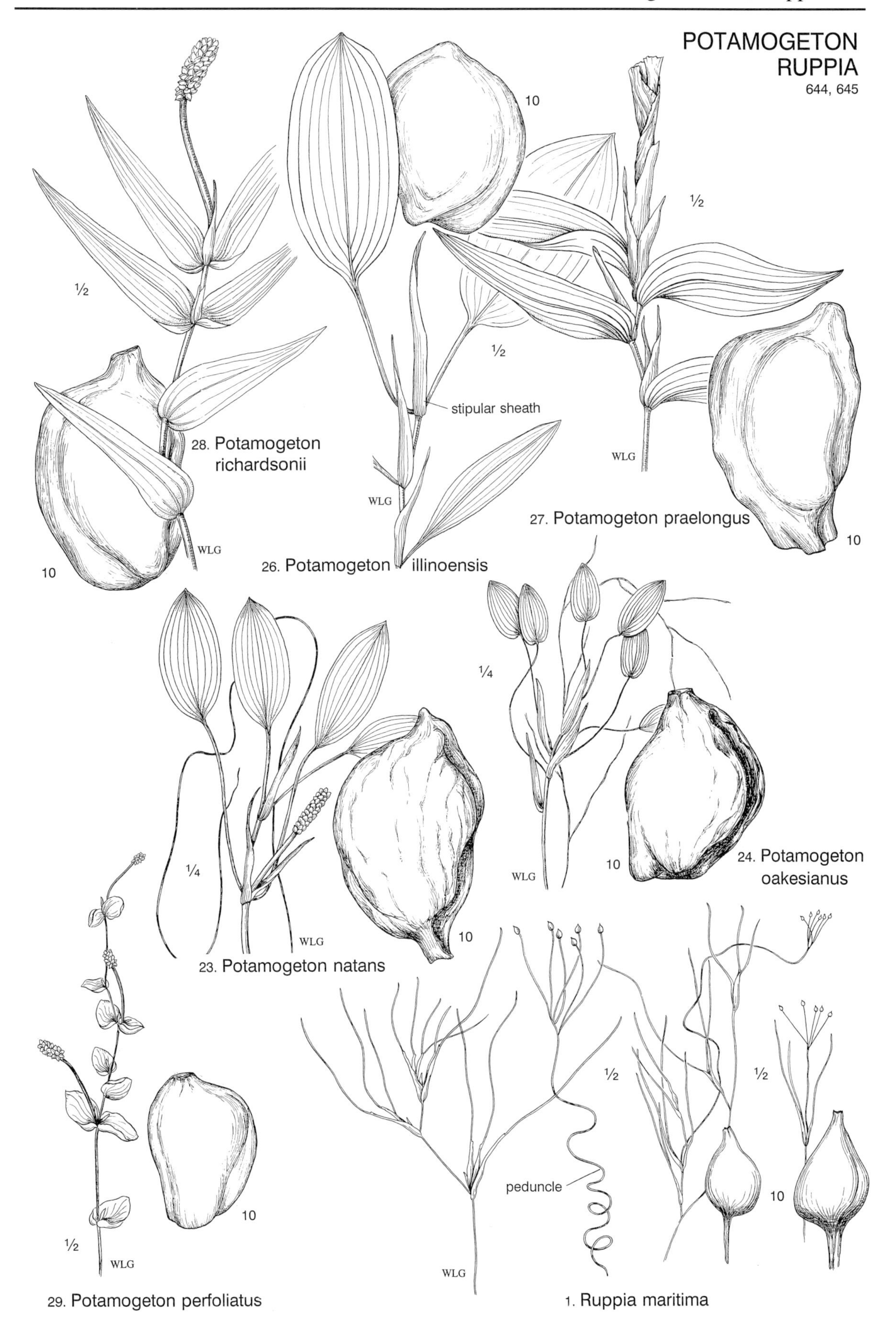

28. Potamogeton richardsonii

26. Potamogeton illinoensis

stipular sheath

27. Potamogeton praelongus

23. Potamogeton natans

24. Potamogeton oakesianus

29. Potamogeton perfoliatus

peduncle

1. Ruppia maritima

NAJAS
ZANNICHELLIA
ZOSTERA
646, 647

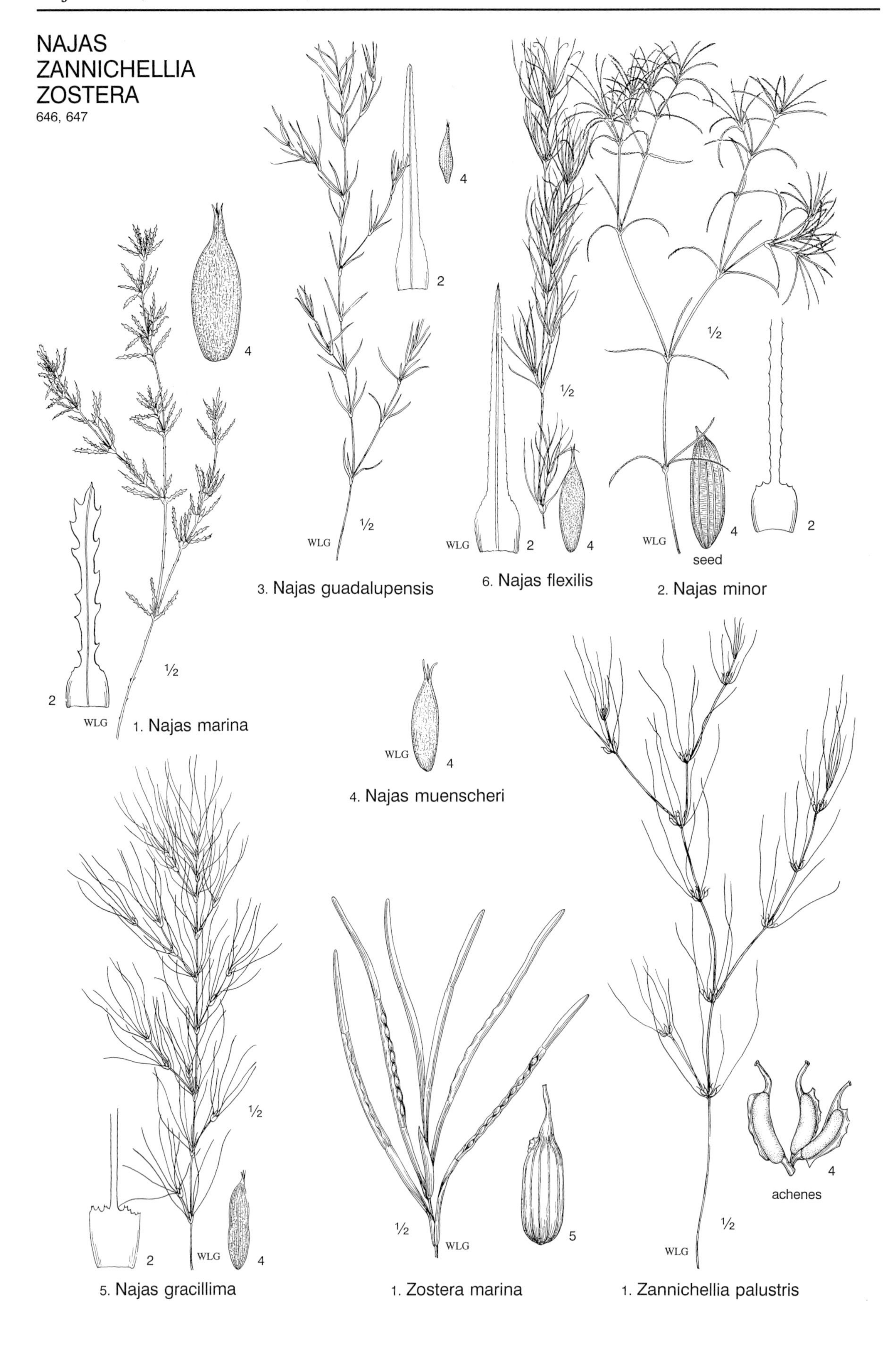

3. Najas guadalupensis

6. Najas flexilis

2. Najas minor

seed

1. Najas marina

4. Najas muenscheri

5. Najas gracillima

1. Zostera marina

1. Zannichellia palustris

achenes

ILLUSTRATED COMPANION TO

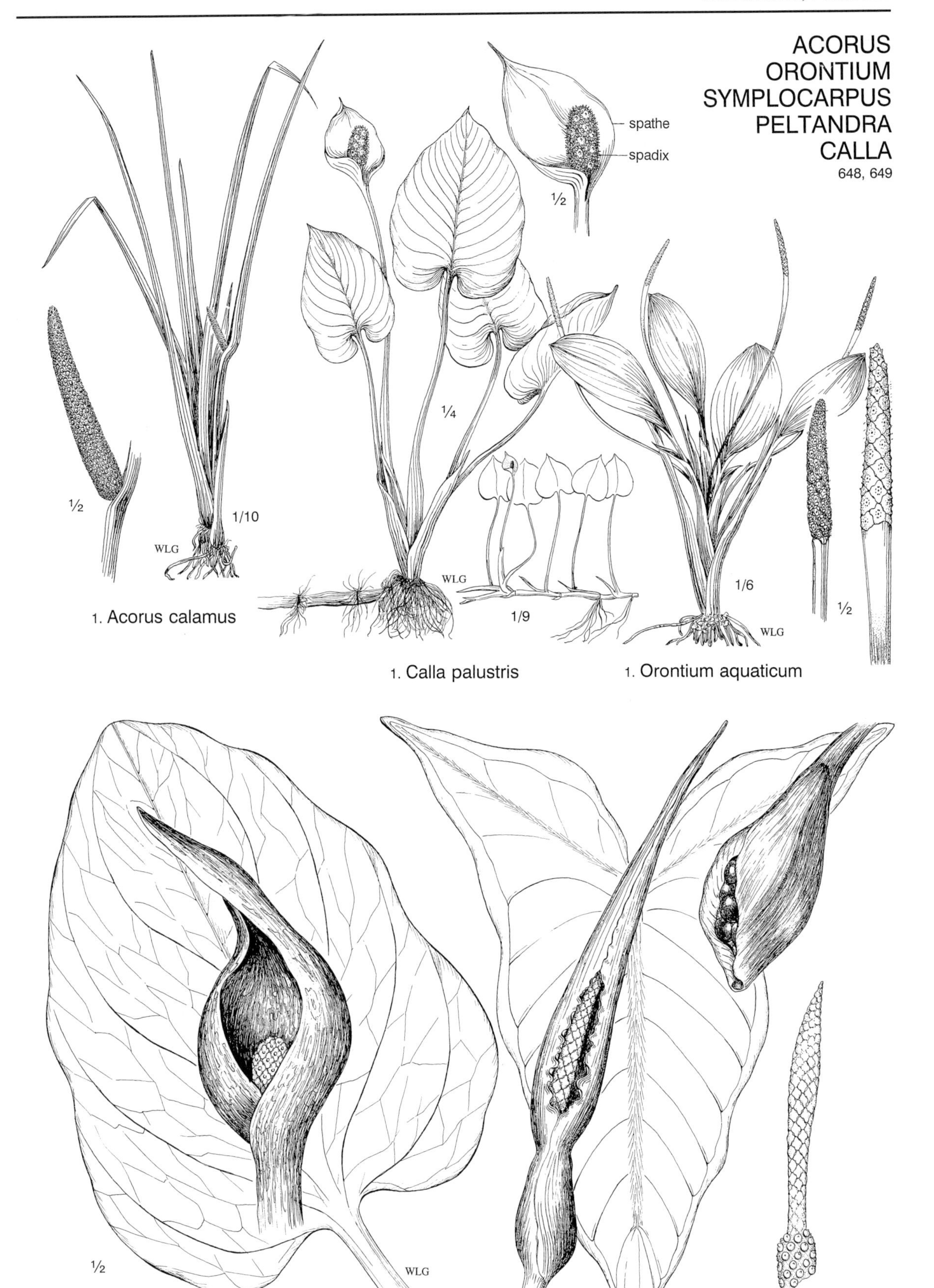

ACORUS
ORONTIUM
SYMPLOCARPUS
PELTANDRA
CALLA
648, 649

spathe

spadix

1. Acorus calamus

1. Calla palustris

1. Orontium aquaticum

1. Symplocarpus foetidus

1. Peltandra virginica

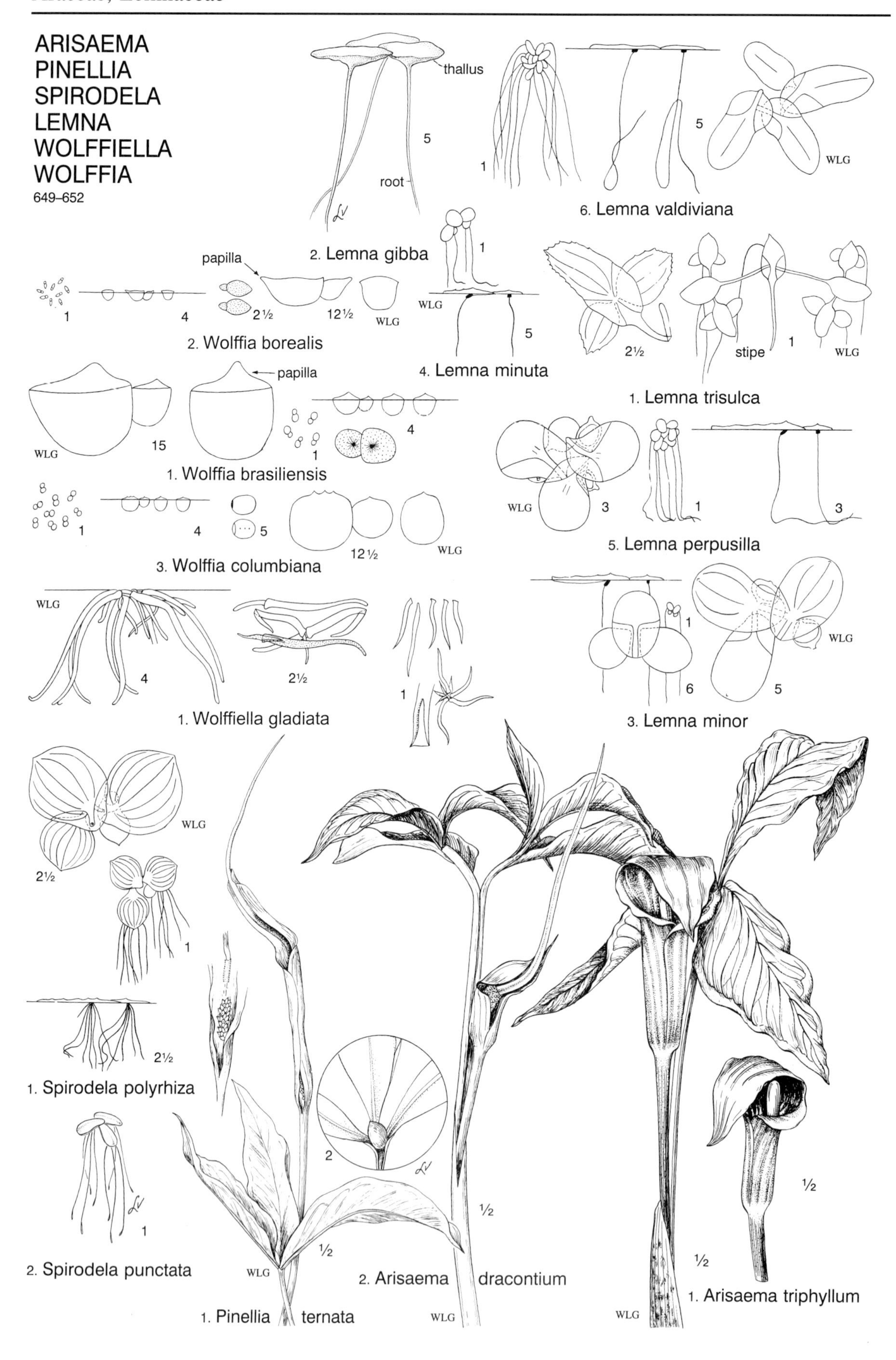

ARISAEMA
PINELLIA
SPIRODELA
LEMNA
WOLFFIELLA
WOLFFIA
649–652

thallus
root
5

1

5

WLG

6. Lemna valdiviana

2. Lemna gibba

papilla
1
4
2½
12½
WLG

2. Wolffia borealis

WLG
5

4. Lemna minuta

2½
stipe
1
WLG

1. Lemna trisulca

papilla

WLG
15

4

1

1. Wolffia brasiliensis

WLG
3
1
3

5. Lemna perpusilla

1
4
5

12½
WLG

3. Wolffia columbiana

WLG
4

2½

1

1. Wolffiella gladiata

WLG
1
6
5

3. Lemna minor

WLG

2½

1

2½

1. Spirodela polyrhiza

2

1

2. Spirodela punctata

½

1. Pinellia ternata
WLG

½
2. Arisaema dracontium
WLG

½

½

1. Arisaema triphyllum
WLG

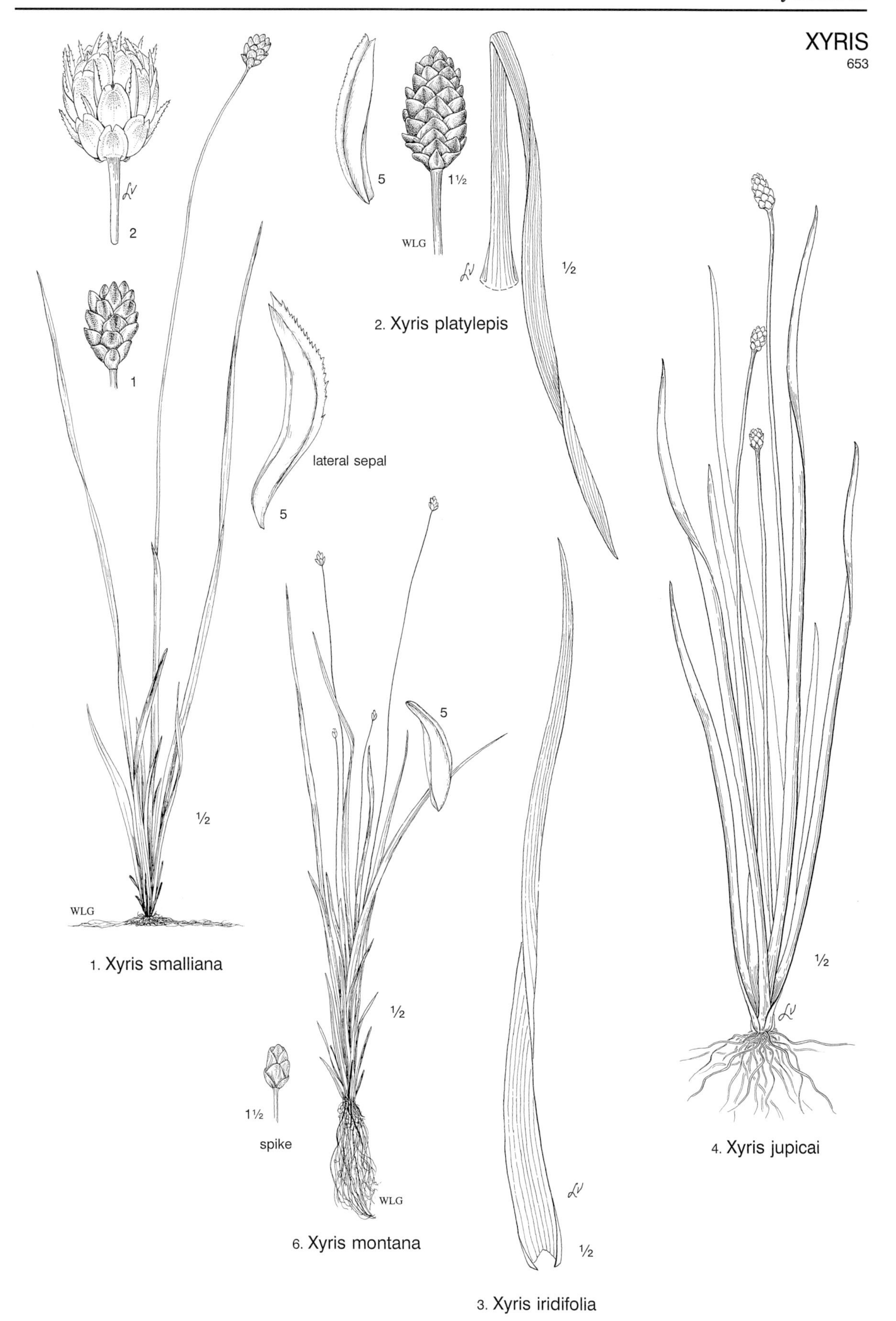

2. Xyris platylepis

lateral sepal
5

2. Xyris platylepis

1. Xyris smalliana

spike

6. Xyris montana

3. Xyris iridifolia

4. Xyris jupicai

XYRIS
653, 654

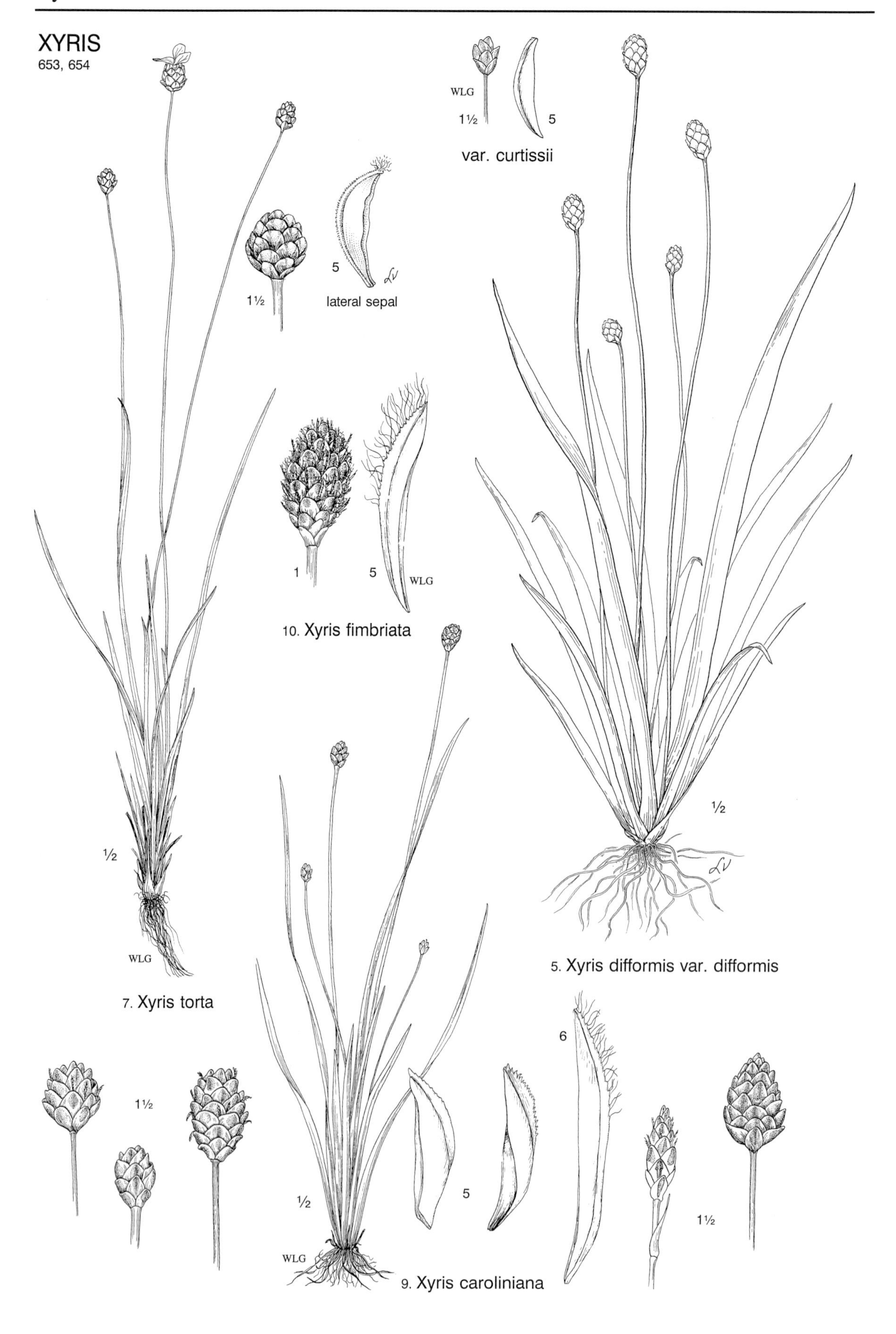

var. curtissii

5 lateral sepal

10. Xyris fimbriata

7. Xyris torta

9. Xyris caroliniana

5. Xyris difformis var. difformis

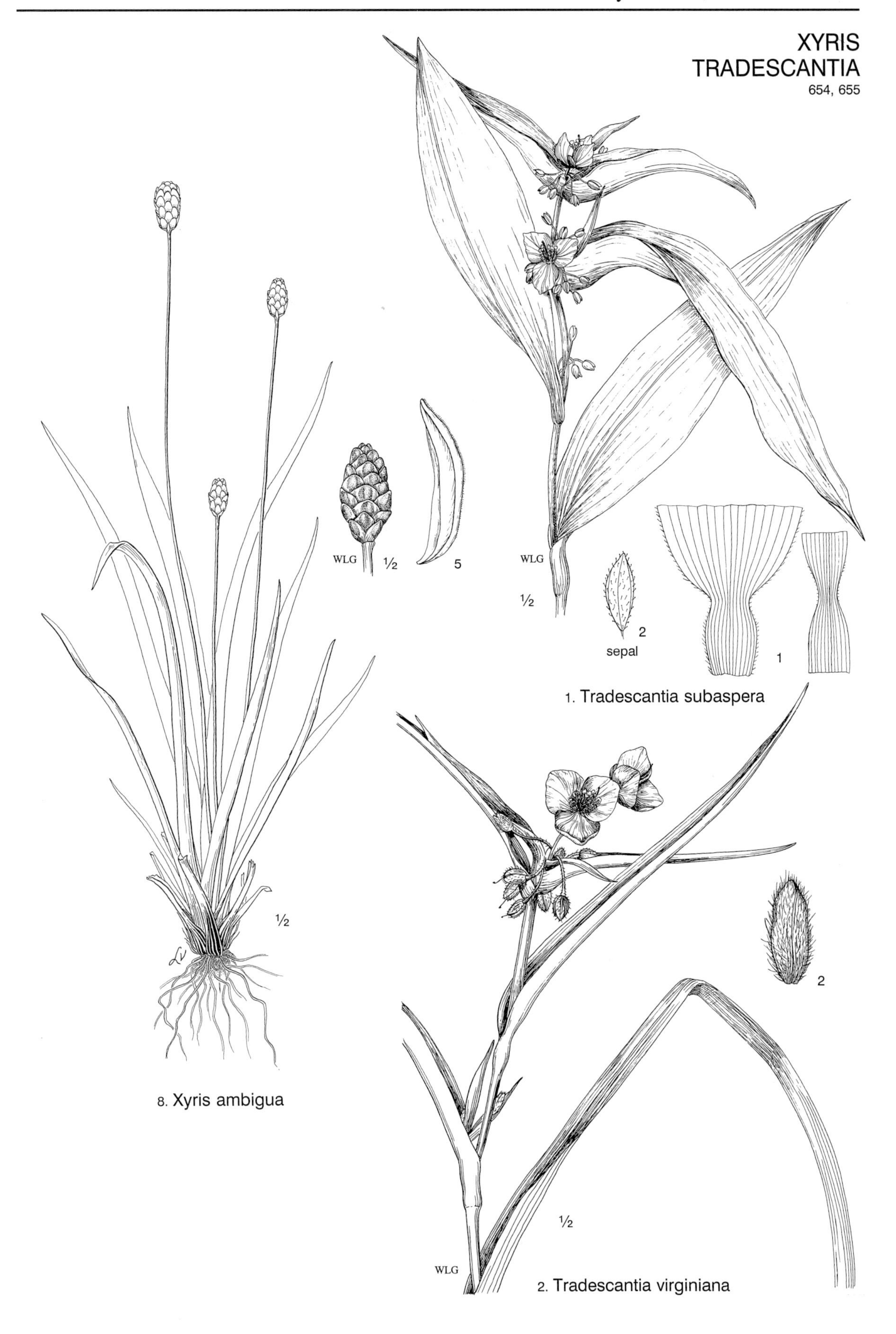

WLG ½ 5

WLG ½

2
sepal

1

1. Tradescantia subaspera

2

8. Xyris ambigua

½

WLG

2. Tradescantia virginiana

½

TRADESCANTIA
MURDANNIA
COMMELINA
655, 656

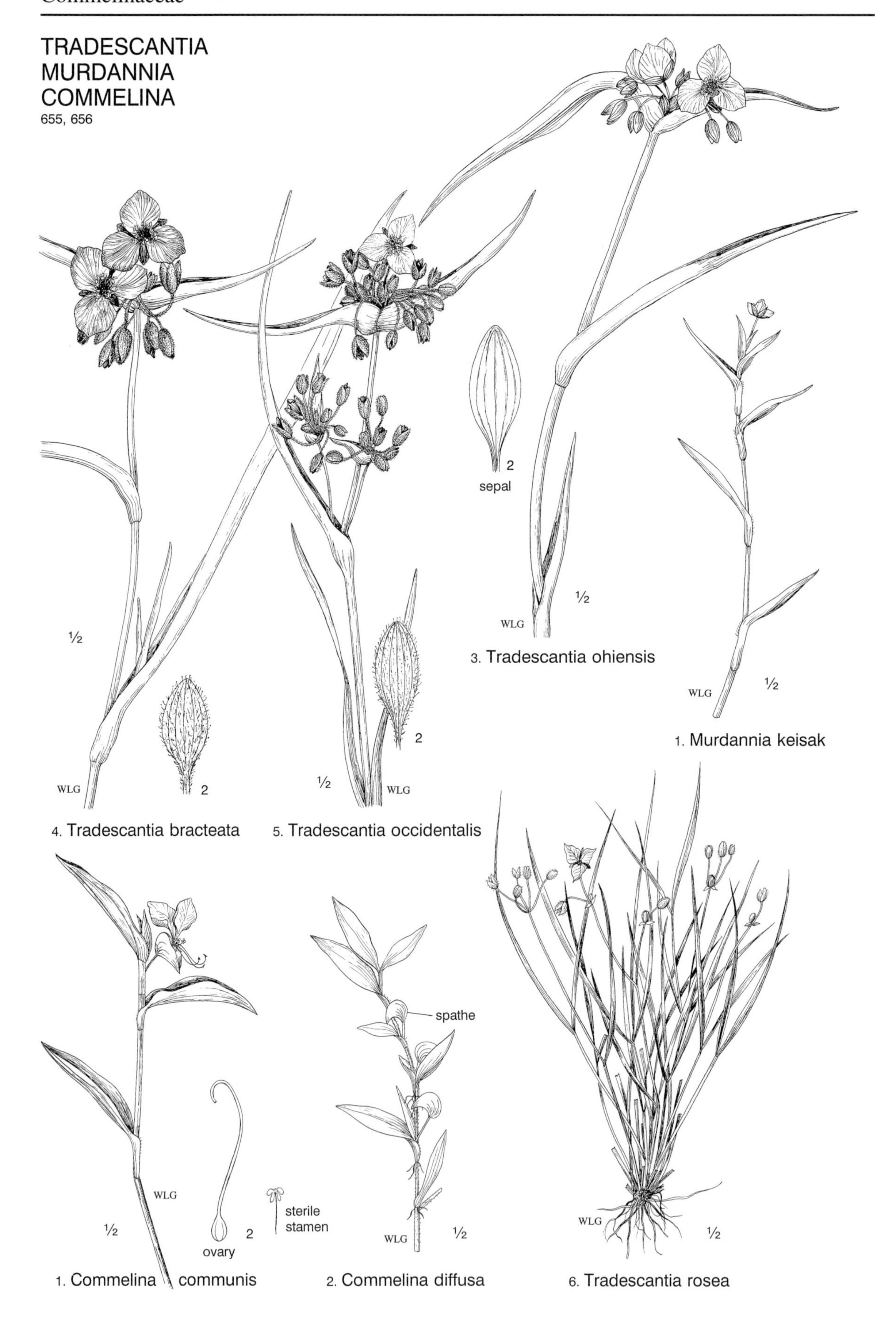

2 sepal

3. Tradescantia ohiensis

WLG ½

½

2

WLG

4. Tradescantia bracteata

½ 2

WLG

5. Tradescantia occidentalis

WLG ½

1. Murdannia keisak

spathe

WLG

½ 2
ovary

sterile stamen

1. Commelina communis

WLG

½

2. Commelina diffusa

WLG ½

6. Tradescantia rosea

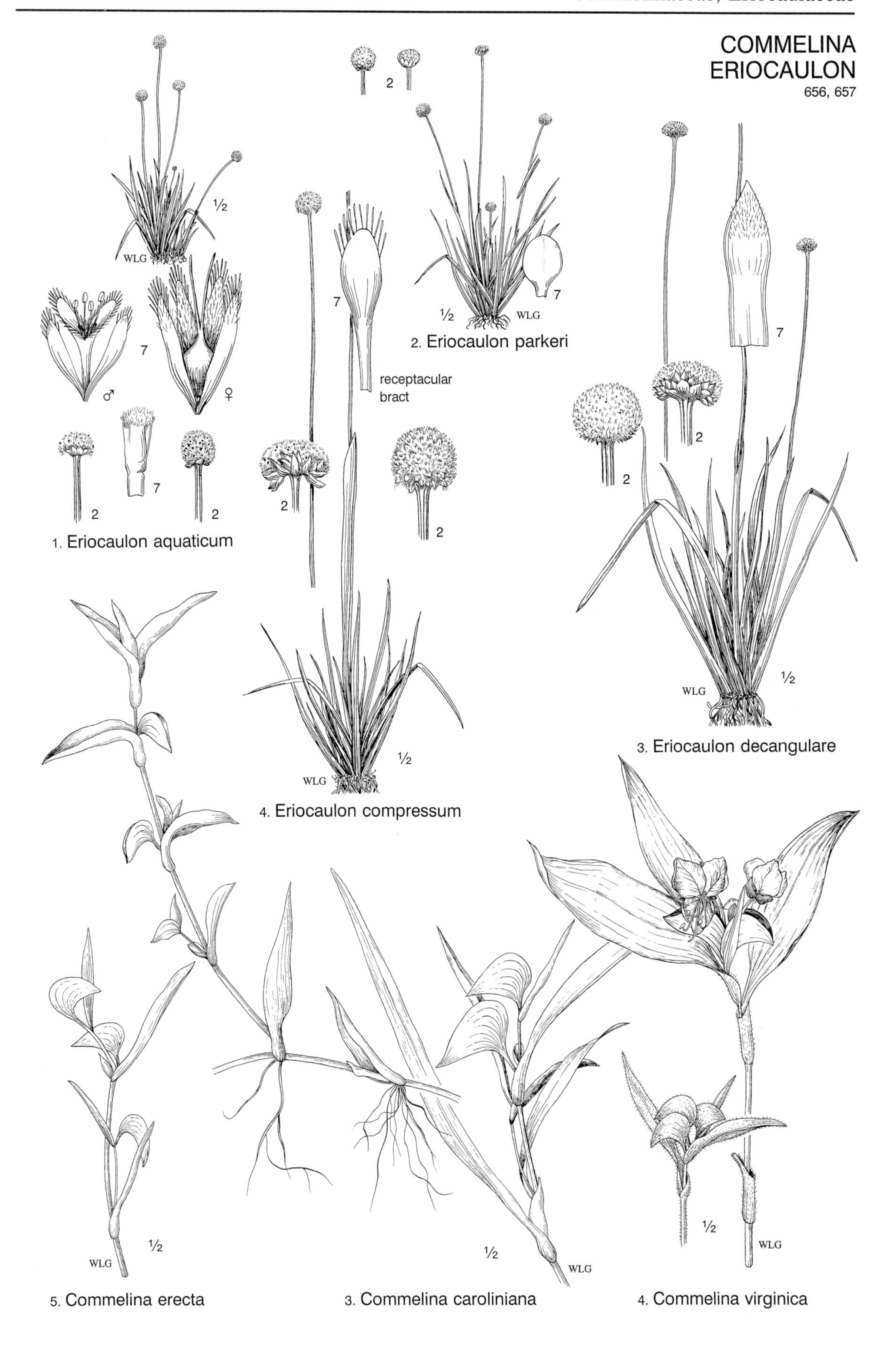

1. Eriocaulon aquaticum

receptacular
bract

2. Eriocaulon parkeri

3. Eriocaulon decangulare

4. Eriocaulon compressum

5. Commelina erecta

3. Commelina caroliniana

4. Commelina virginica

LACHNOCAULON
JUNCUS
658–661

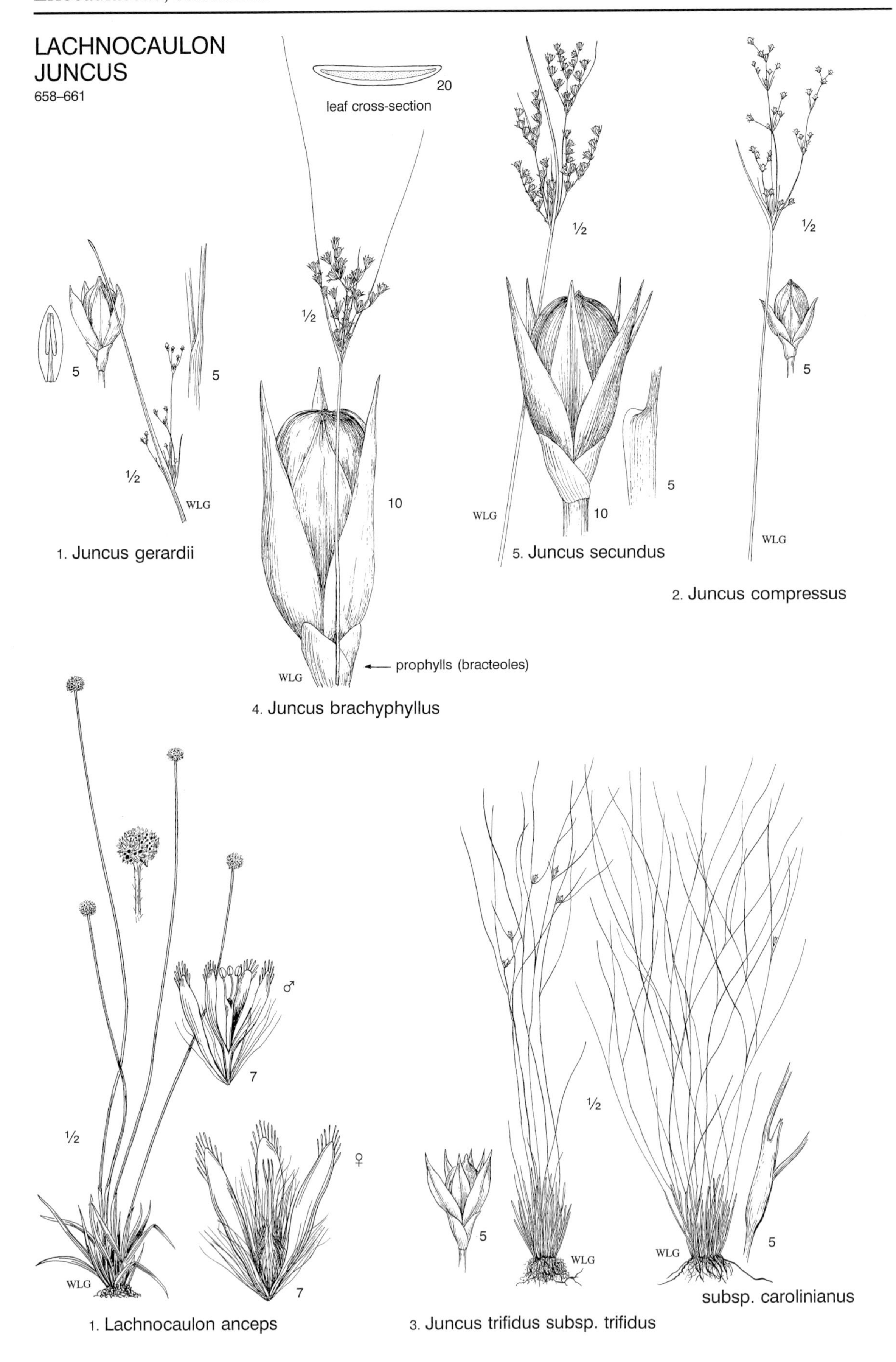

20
leaf cross-section

½

5

5

½

WLG

1. Juncus gerardii

½

5

10

WLG

← prophylls (bracteoles)

4. Juncus brachyphyllus

½

½

5

5

10

WLG

5

WLG

5. Juncus secundus

2. Juncus compressus

♂

7

♀

7

½

½

7

5

WLG

WLG

5

WLG

1. Lachnocaulon anceps

3. Juncus trifidus subsp. trifidus

subsp. carolinianus

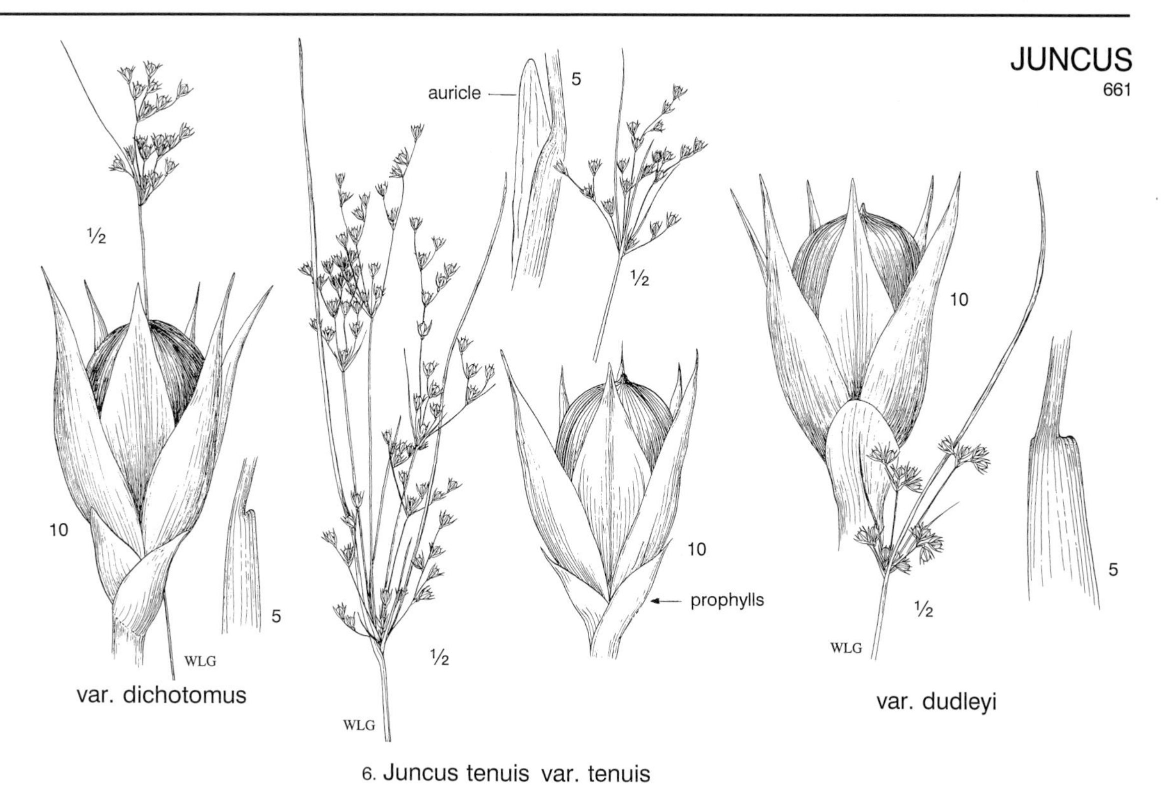

var. dichotomus

6. Juncus tenuis var. tenuis

var. dudleyi

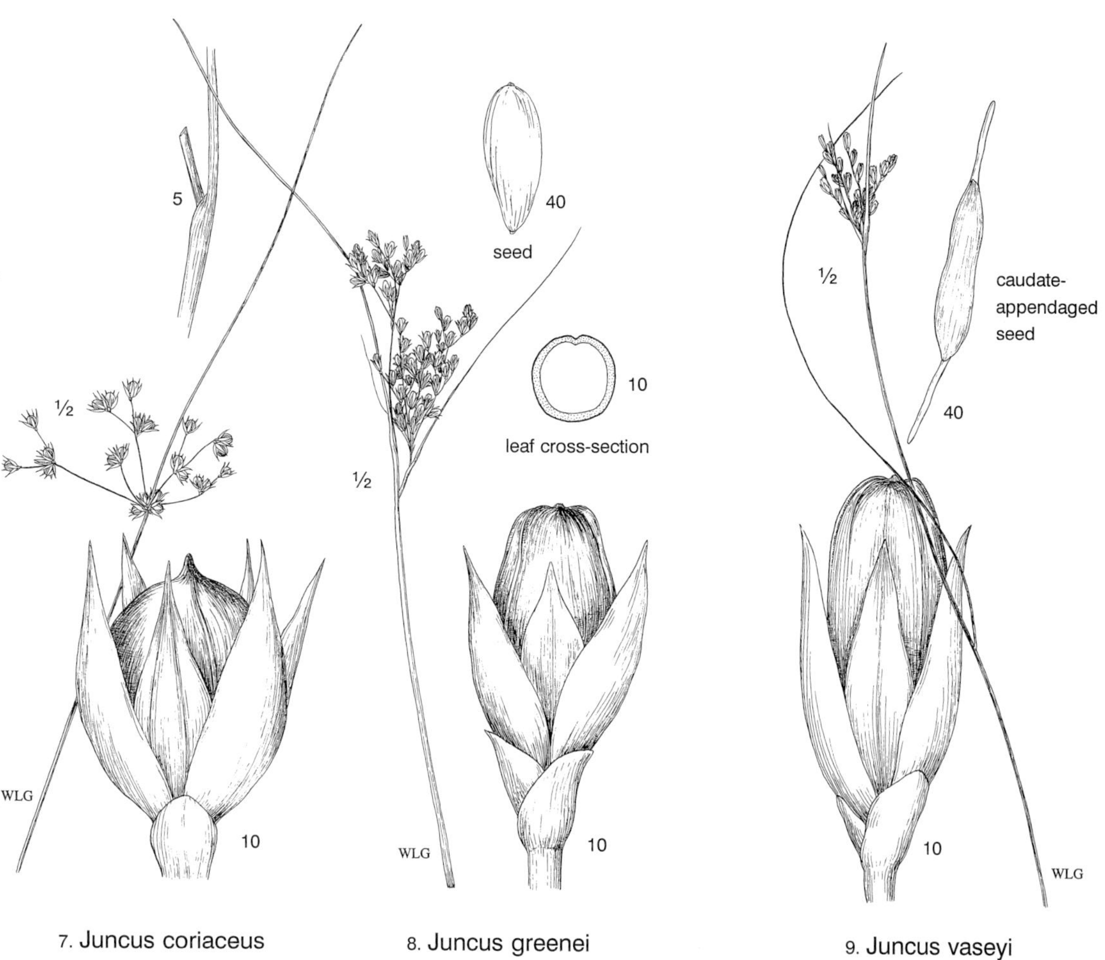

7. Juncus coriaceus

8. Juncus greenei

9. Juncus vaseyi

JUNCUS
661, 662

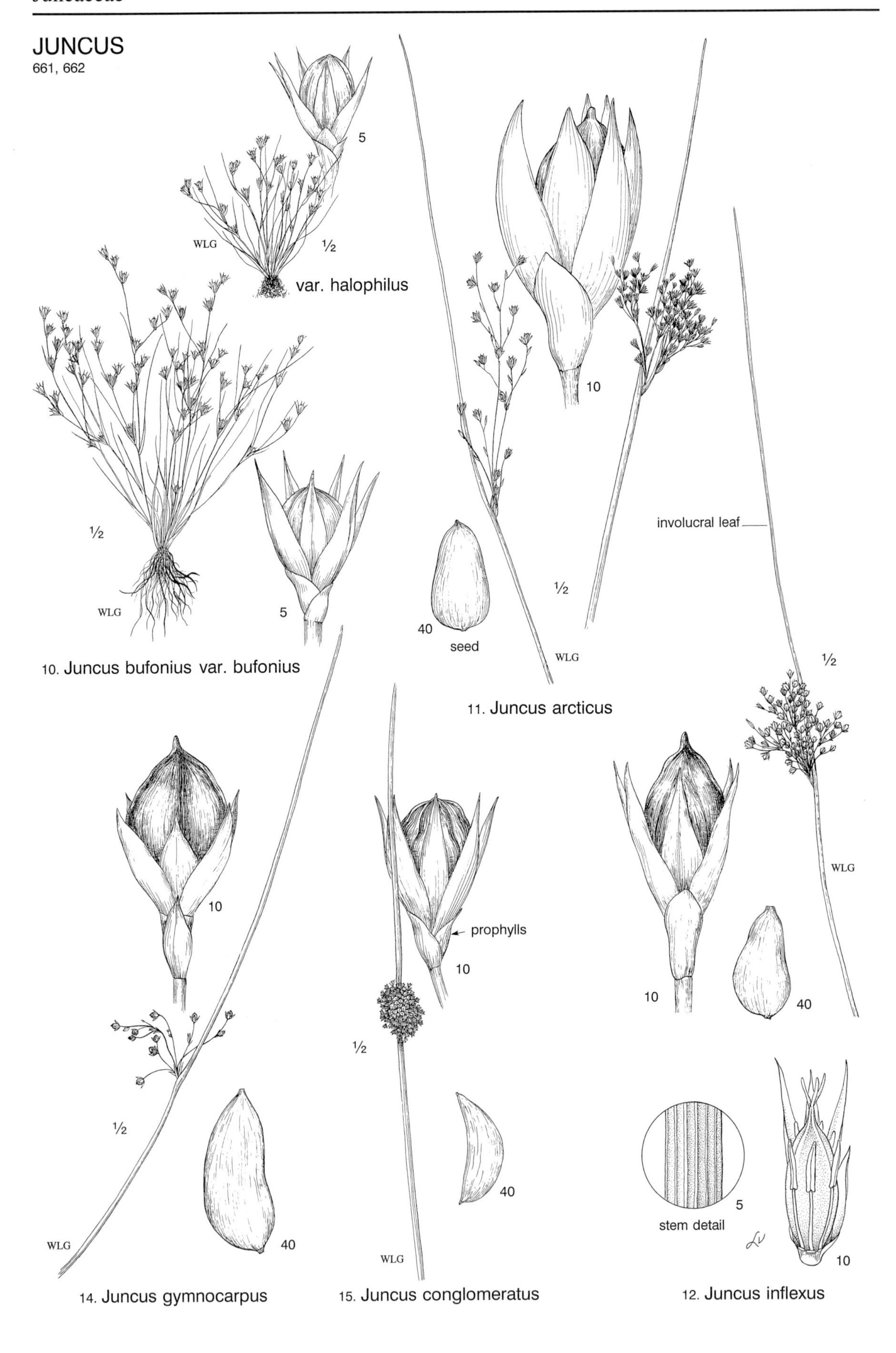

var. halophilus

10. Juncus bufonius var. bufonius

seed

11. Juncus arcticus

involucral leaf

14. Juncus gymnocarpus

prophylls

15. Juncus conglomeratus

stem detail

12. Juncus inflexus

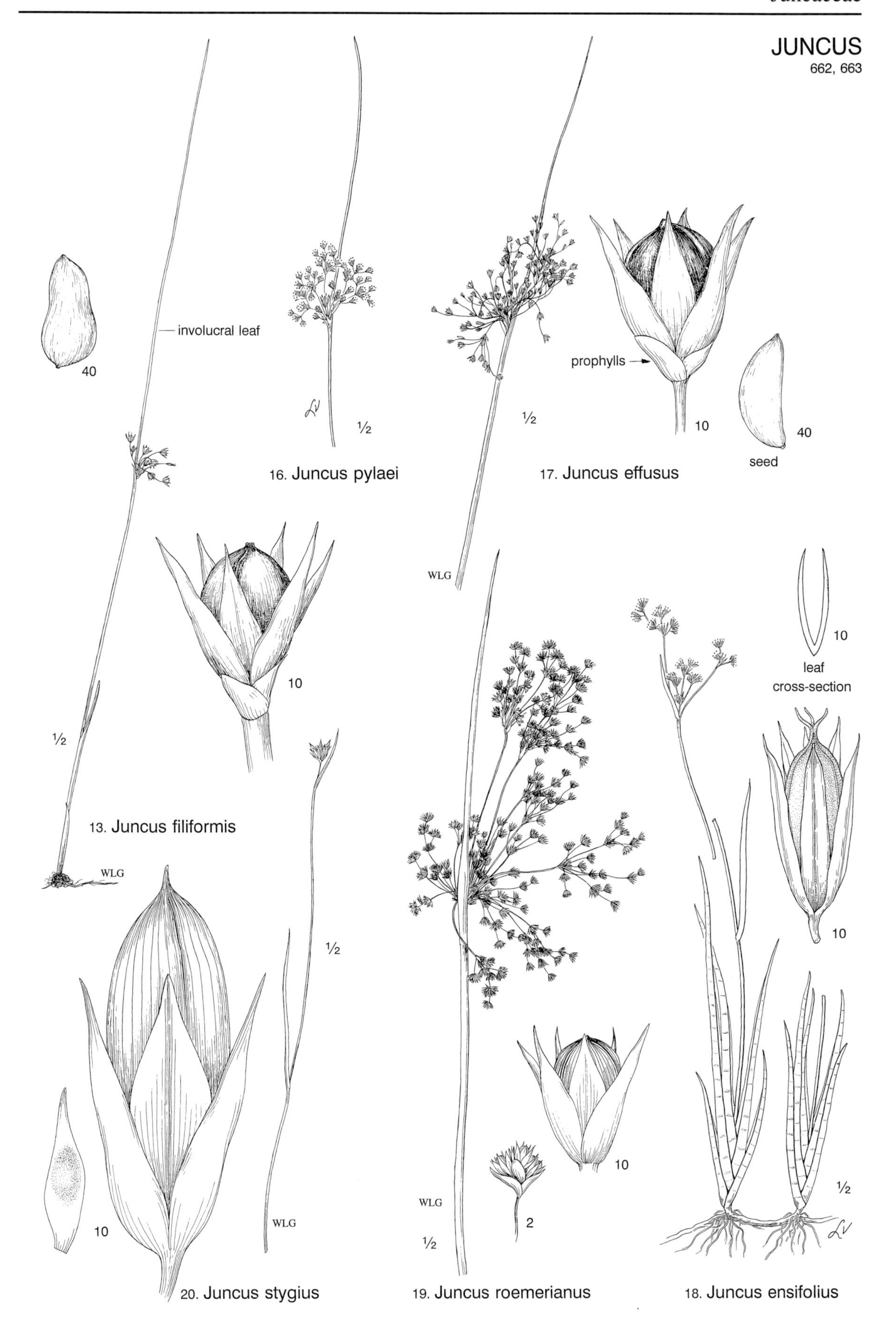

40

— involucral leaf

16. Juncus pylaei

½

½

prophylls →

10

40

seed

17. Juncus effusus

WLG

10

½

13. Juncus filiformis

WLG

½

10

leaf
cross-section

10

WLG

10

½

2

10

½

20. Juncus stygius

WLG

WLG

19. Juncus roemerianus

18. Juncus ensifolius

JUNCUS
663

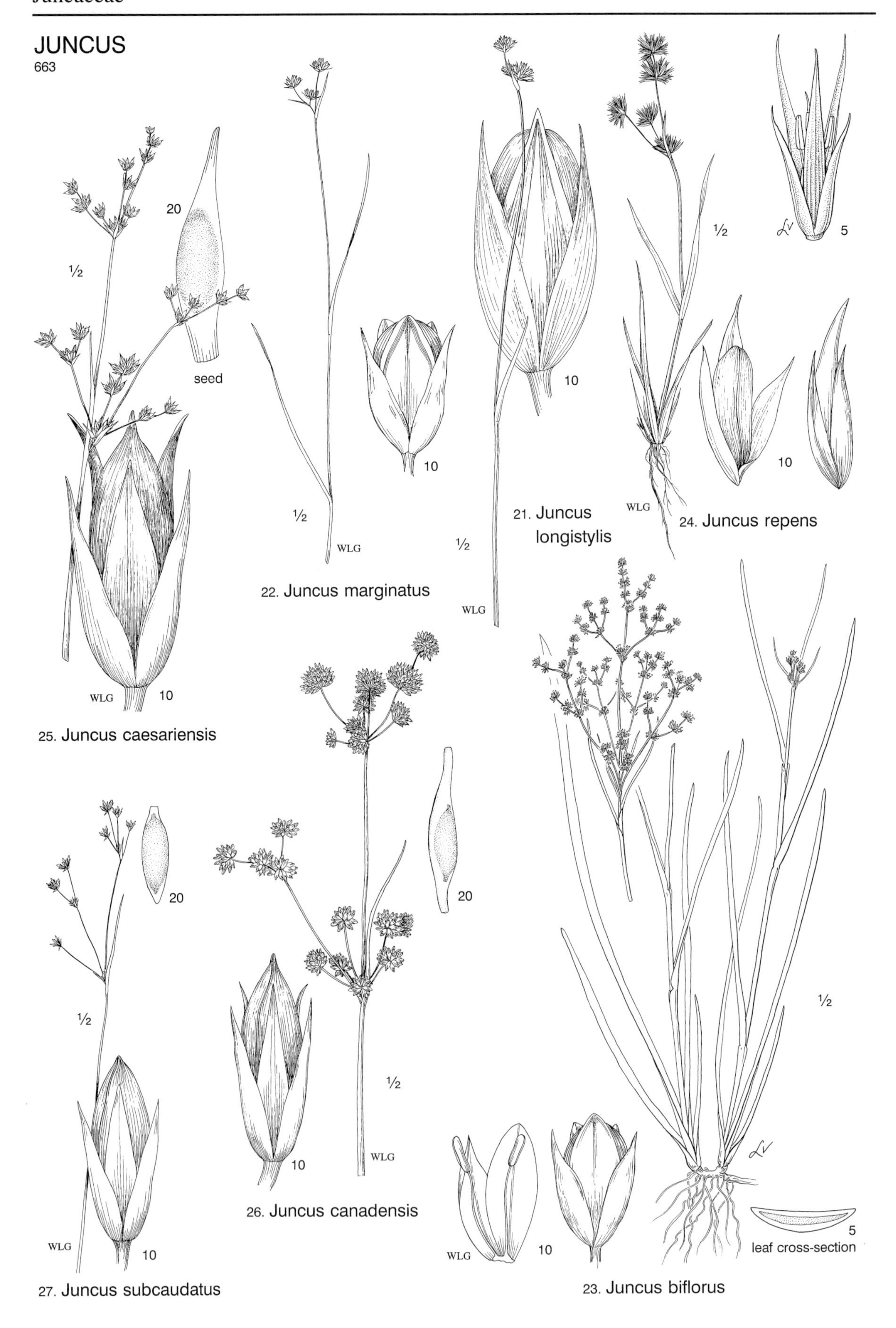

20

1/2

seed

1/2

WLG

22. Juncus marginatus

1/2

WLG

10

10

WLG

25. Juncus caesariensis

21. Juncus
longistylis

1/2

WLG

24. Juncus repens

1/2

5

10

20

1/2

WLG

10

WLG

27. Juncus subcaudatus

20

1/2

WLG

10

26. Juncus canadensis

1/2

WLG

10

1/2

5

leaf cross-section

23. Juncus biflorus

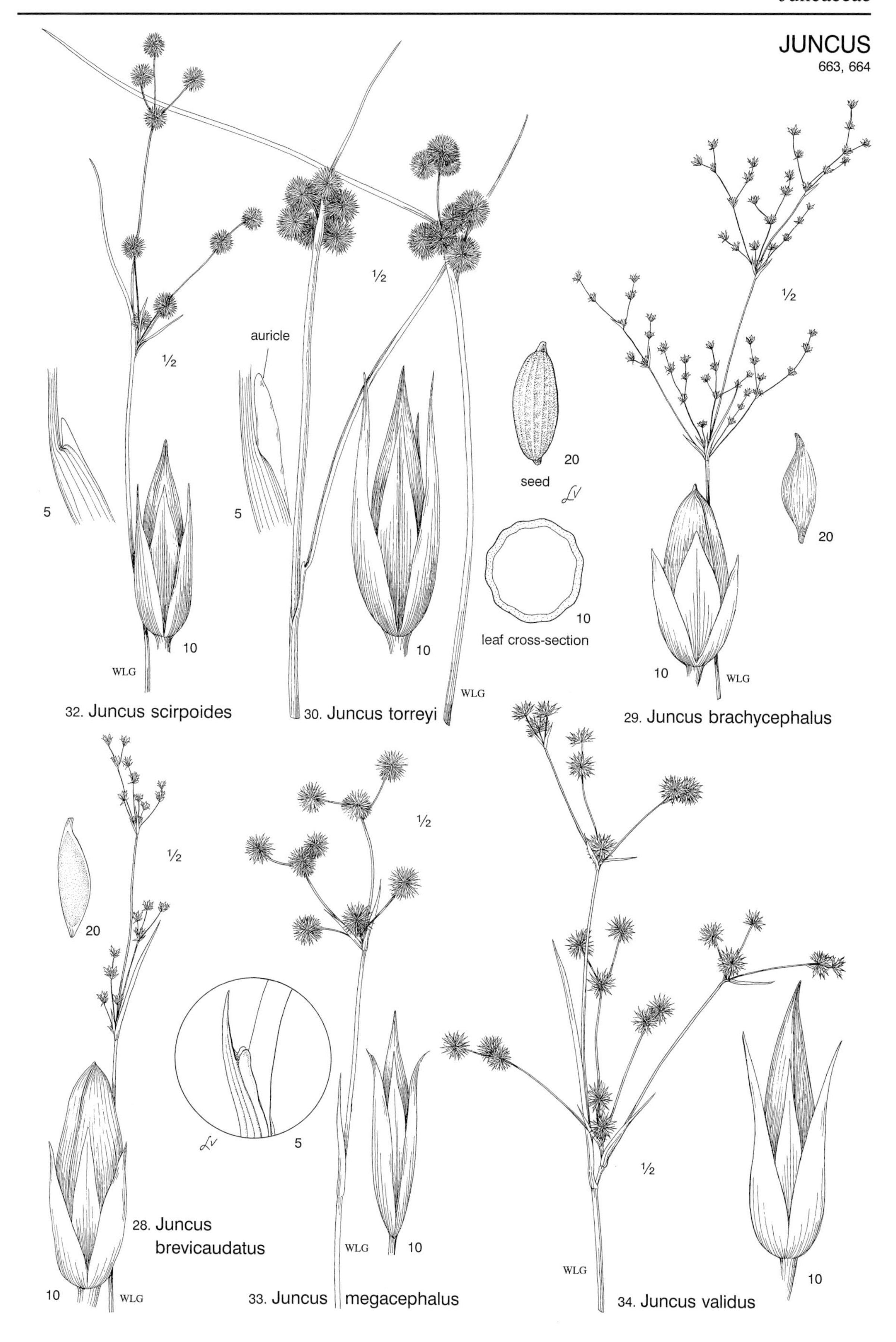

5

auricle

5

½

½

½

seed

20

_ℓν_

10

leaf cross-section

½

20

10   WLG

32. **Juncus scirpoides**

10

WLG

30. **Juncus torreyi**

WLG

10   WLG

29. **Juncus brachycephalus**

½

20

_ℓν_   5

28. **Juncus brevicaudatus**

10   WLG

½

10

WLG

33. **Juncus megacephalus**

½

WLG

10

34. **Juncus validus**

JUNCUS
664, 665

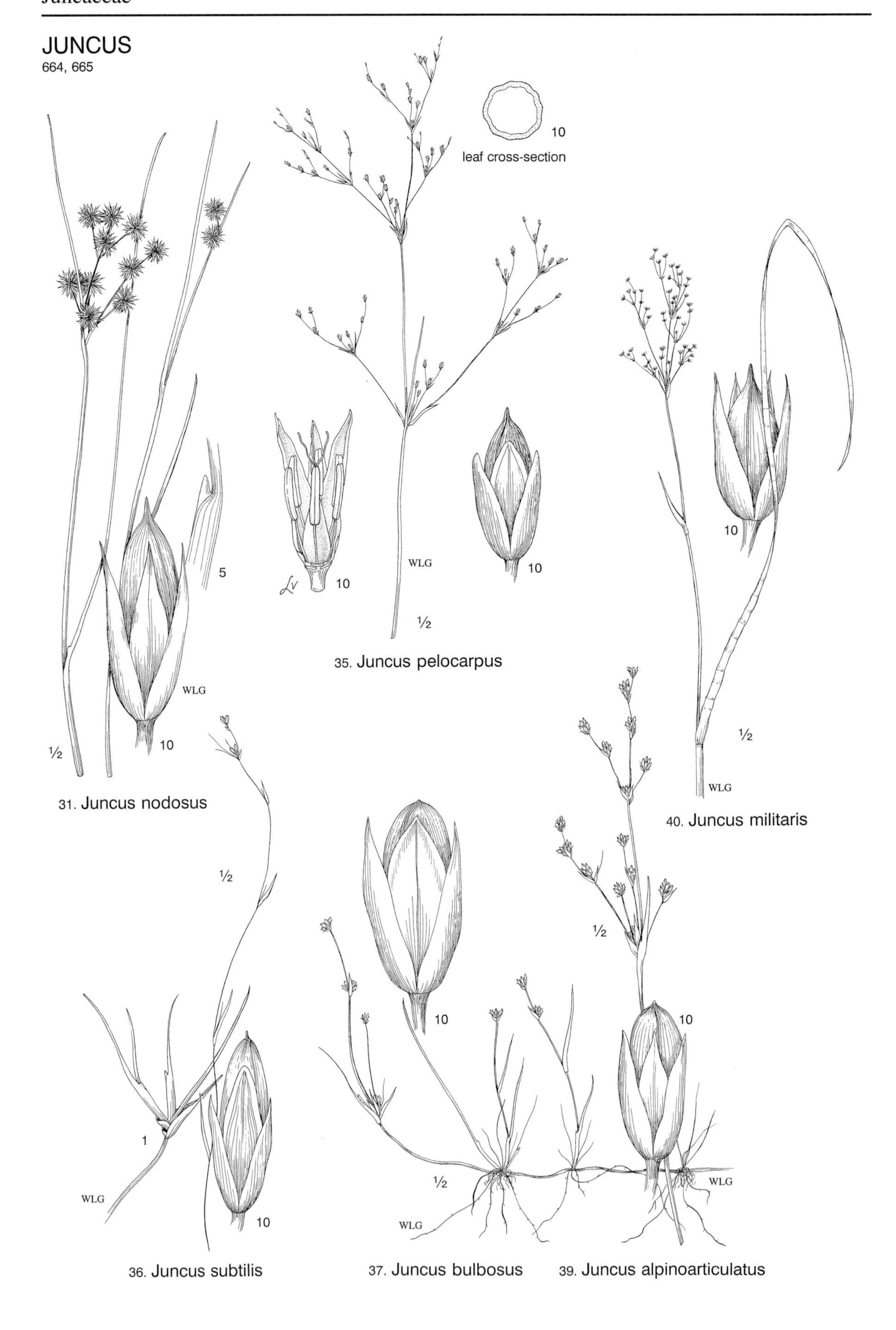

10
leaf cross-section

5

WLG

½

10

31. Juncus nodosus

WLG 10

½

WLG 10

½

10

35. Juncus pelocarpus

½

10

WLG

40. Juncus militaris

½

10

½

½

1

WLG

10

36. Juncus subtilis

10

WLG

½

WLG

37. Juncus bulbosus

10

WLG

39. Juncus alpinoarticulatus

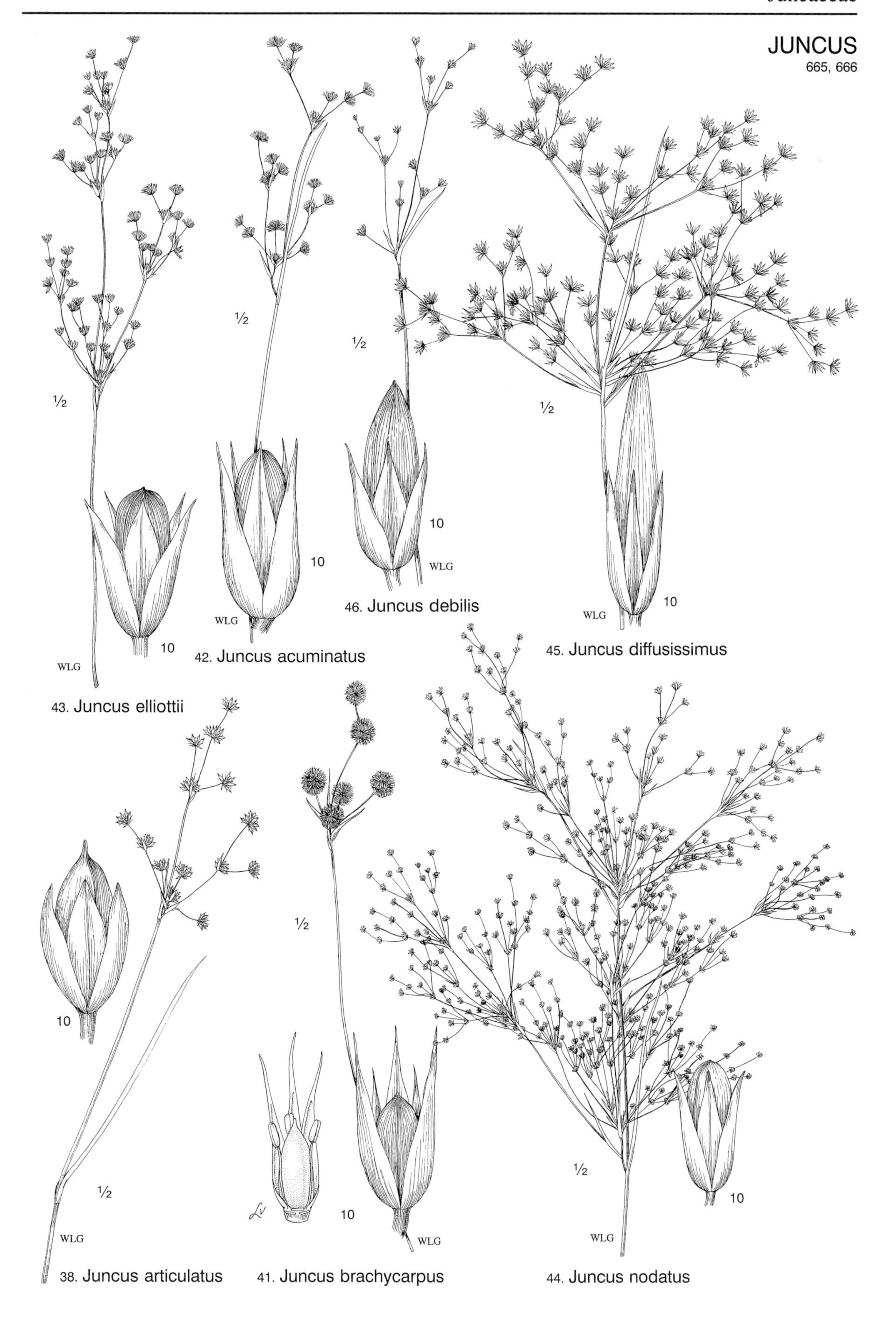

42. **Juncus acuminatus**

46. **Juncus debilis**

43. **Juncus elliottii**

45. **Juncus diffusissimus**

38. **Juncus articulatus**

41. **Juncus brachycarpus**

44. **Juncus nodatus**

LUZULA
666, 667

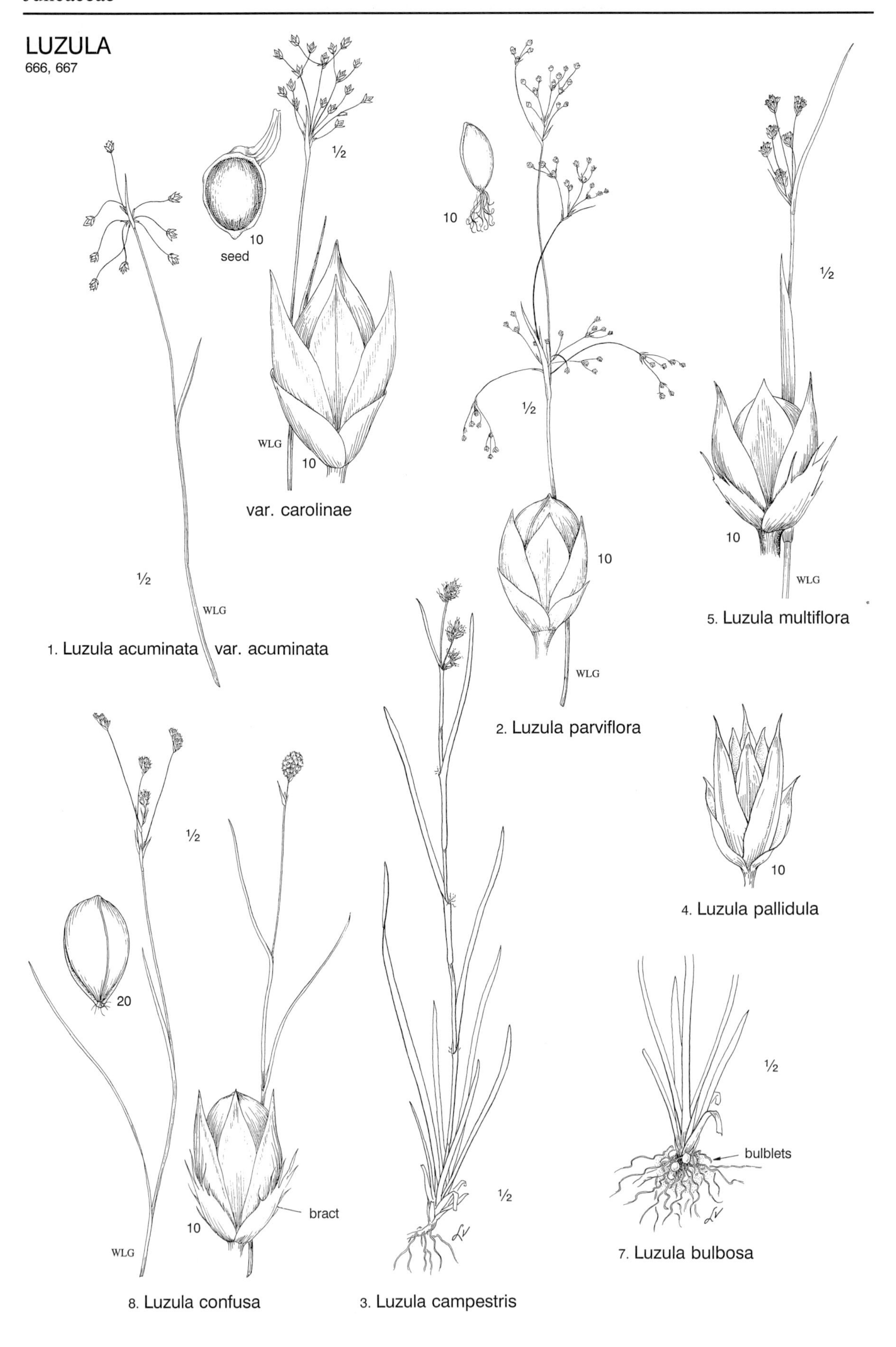

½

seed

var. carolinae

1. Luzula acuminata var. acuminata

2. Luzula parviflora

5. Luzula multiflora

4. Luzula pallidula

7. Luzula bulbosa

bract

8. Luzula confusa

3. Luzula campestris

bulblets

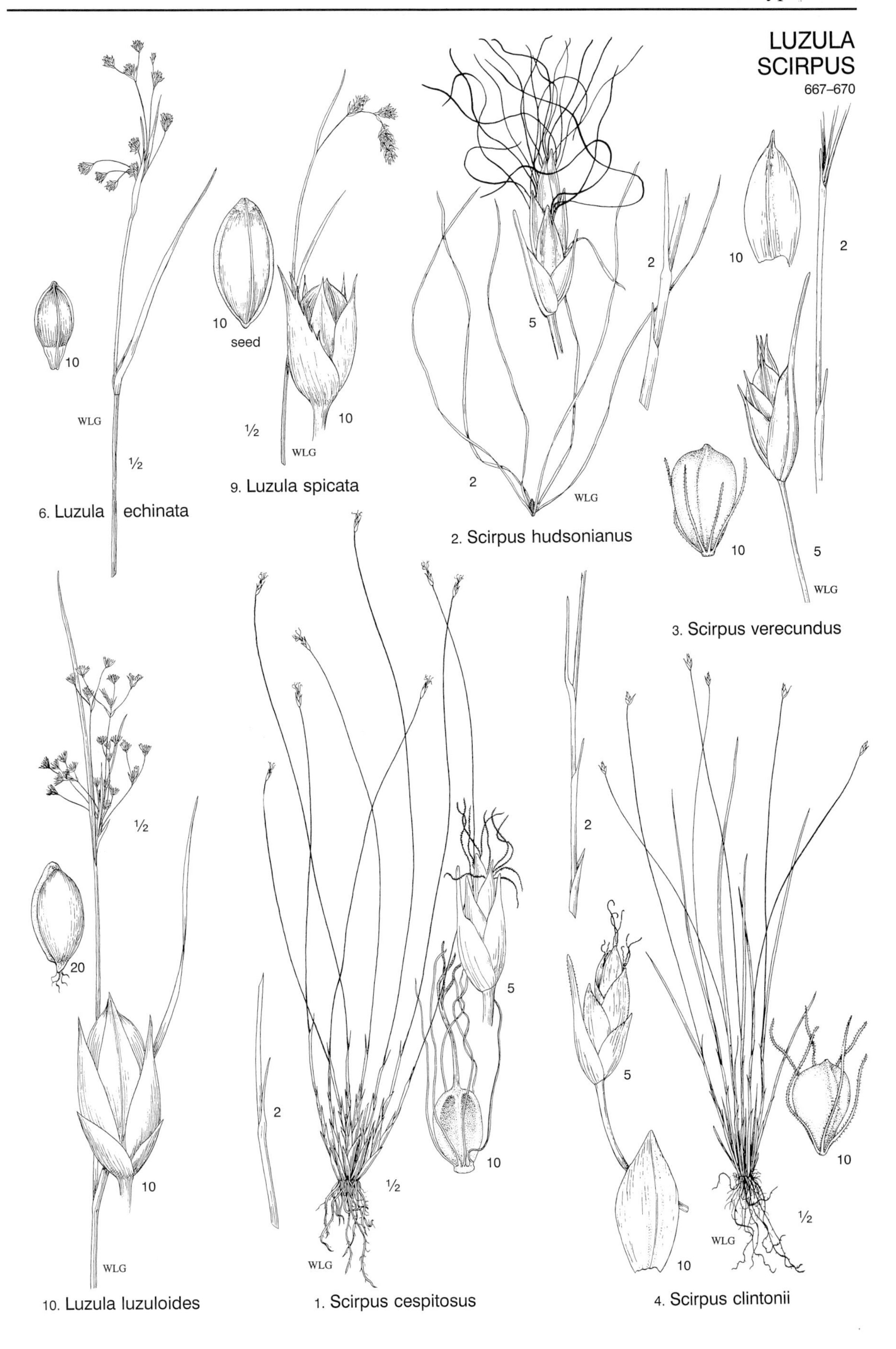

10 seed

1/2

6. Luzula   echinata

9. Luzula spicata

2. Scirpus hudsonianus

3. Scirpus verecundus

10. Luzula luzuloides

1. Scirpus cespitosus

4. Scirpus clintonii

SCIRPUS
670, 671

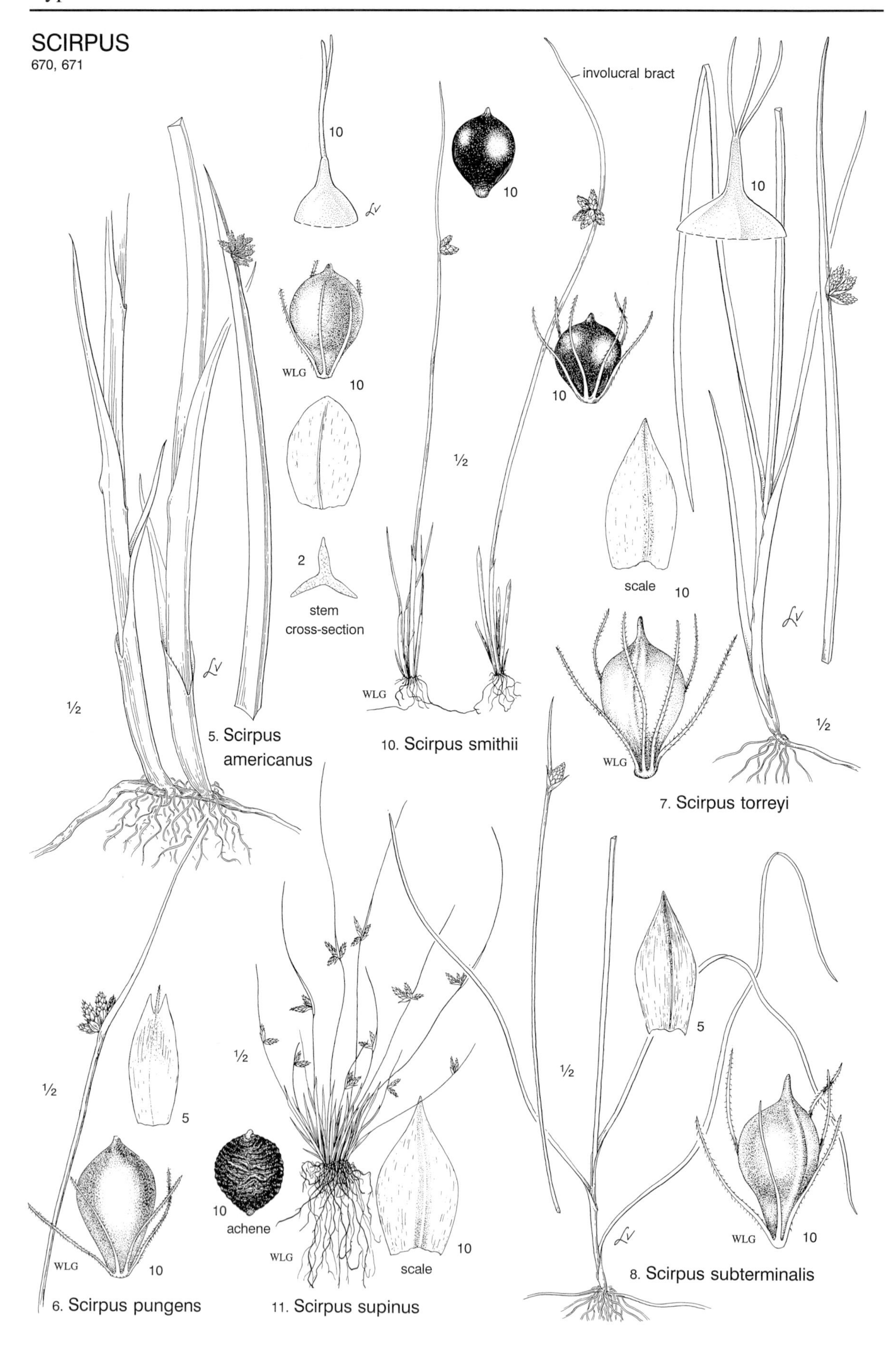

involucral bract

5. Scirpus americanus

10. Scirpus smithii

stem cross-section

scale

7. Scirpus torreyi

6. Scirpus pungens

achene

11. Scirpus supinus

scale

8. Scirpus subterminalis

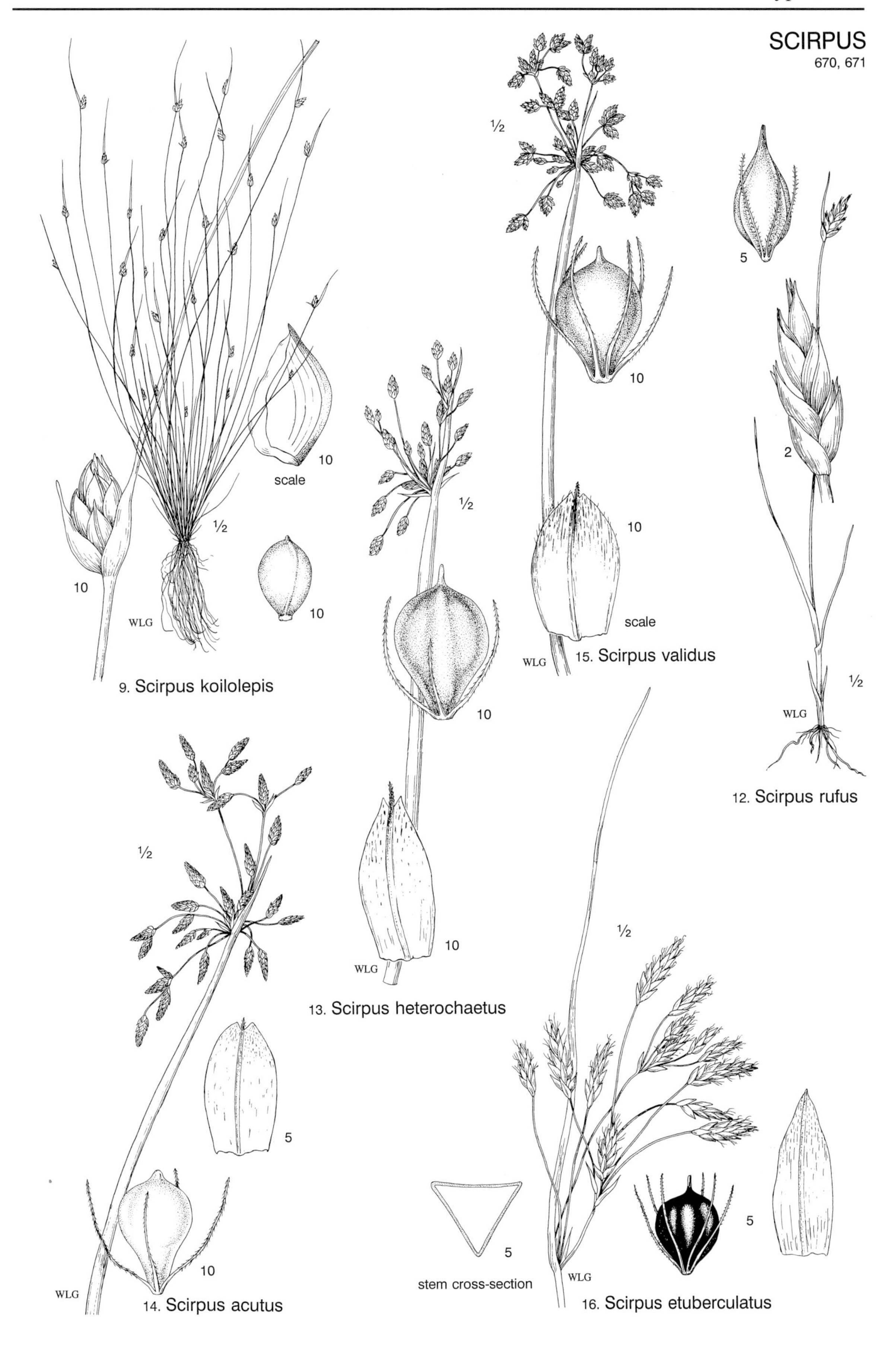

9. Scirpus koilolepis

10 scale

10

12. Scirpus rufus

13. Scirpus heterochaetus

15. Scirpus validus

14. Scirpus acutus

stem cross-section

16. Scirpus etuberculatus

SCIRPUS
671, 672

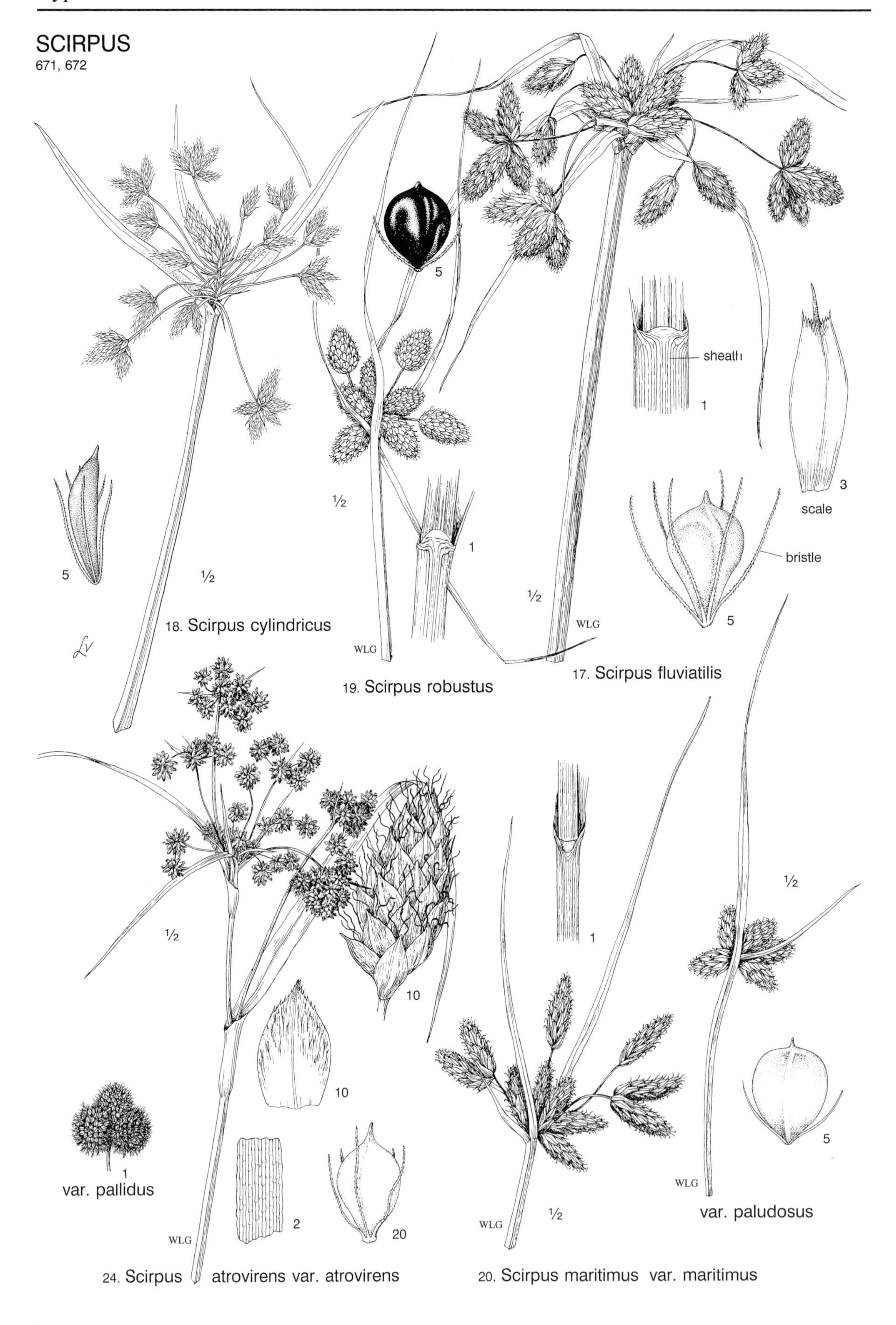

5

½

5

½

18. Scirpus cylindricus

½

1

WLG

19. Scirpus robustus

sheath

1

3

scale

bristle

5

WLG

½

WLG

17. Scirpus fluviatilis

½

10

10

2

20

var. pallidus

1

WLG

24. Scirpus atrovirens var. atrovirens

1

½

WLG

20. Scirpus maritimus var. maritimus

½

5

WLG

var. paludosus

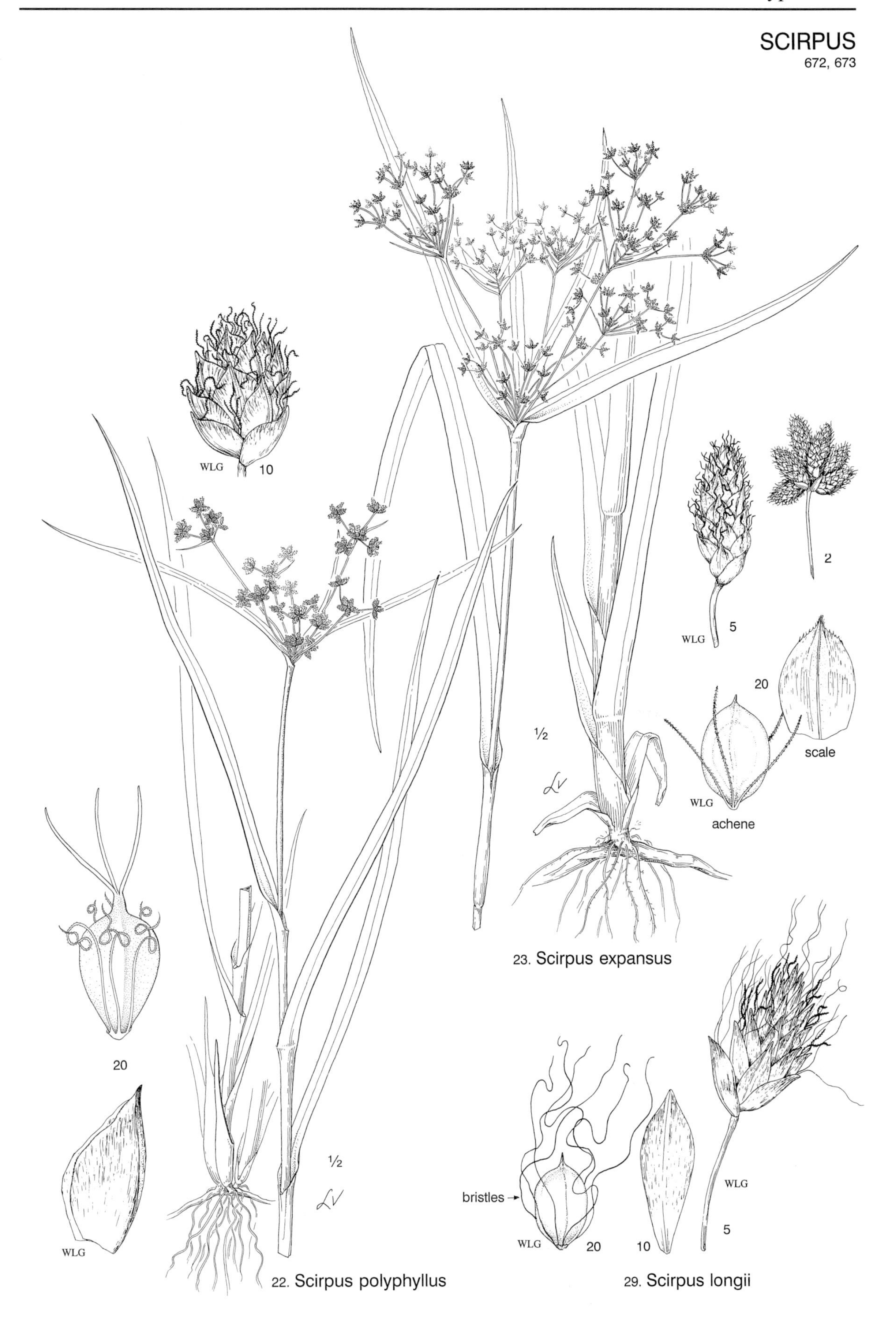

WLG 10

½

23. Scirpus expansus

WLG 5

2

20

scale

WLG

achene

20

WLG

½

22. Scirpus polyphyllus

bristles →

WLG 20

10

WLG 5

29. Scirpus longii

SCIRPUS
672, 673

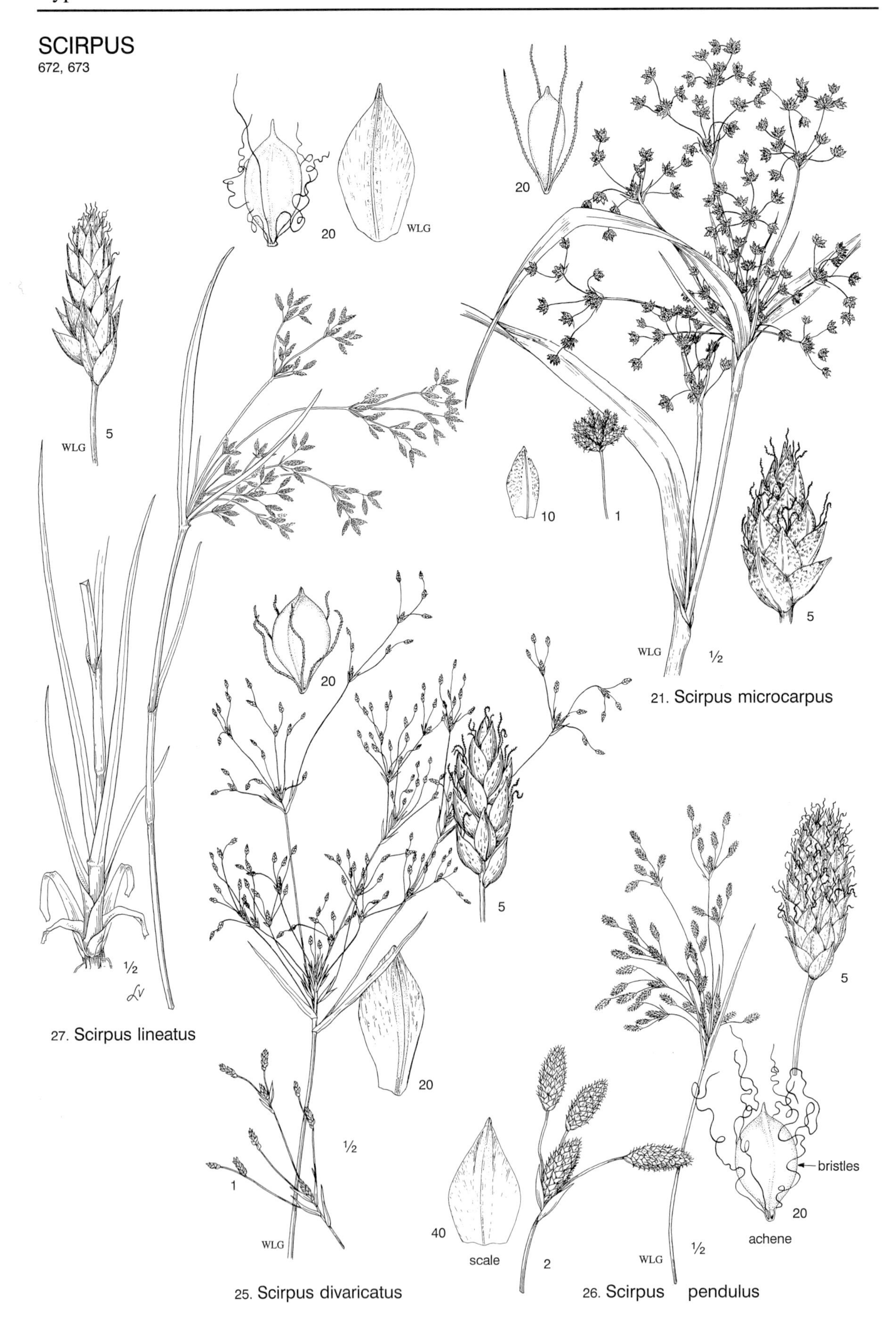

20   WLG

20

20

5

WLG   5

10   1

5

WLG   ½

21. Scirpus microcarpus

20

5

5

27. Scirpus lineatus   5

20

½

bristles

½

20

achene

40   scale   2   WLG   ½

25. Scirpus divaricatus   26. Scirpus pendulus

1   WLG

½

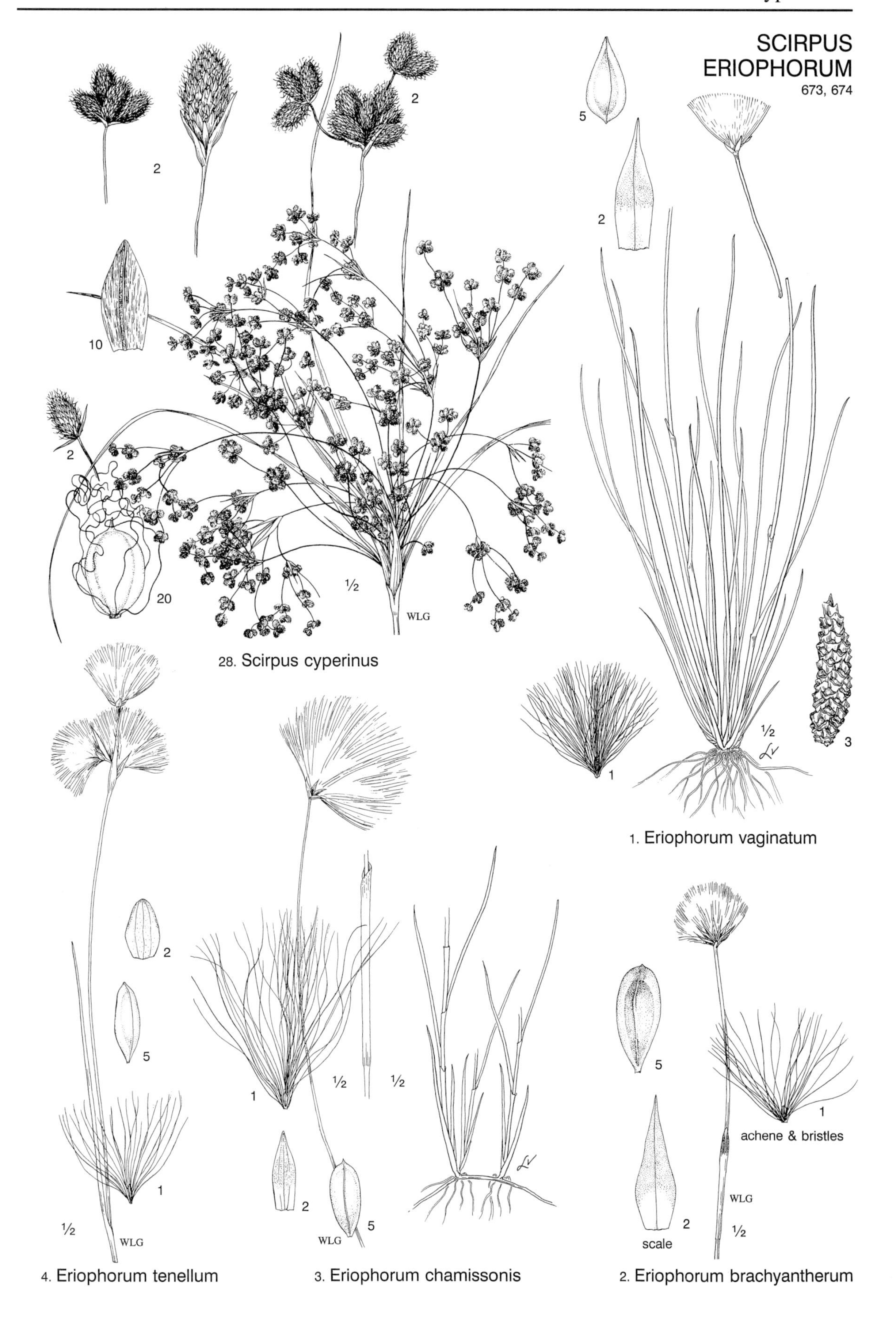

28. Scirpus cyperinus

1. Eriophorum vaginatum

4. Eriophorum tenellum

3. Eriophorum chamissonis

2. Eriophorum brachyantherum

ERIOPHORUM
ELEOCHARIS
674, 675

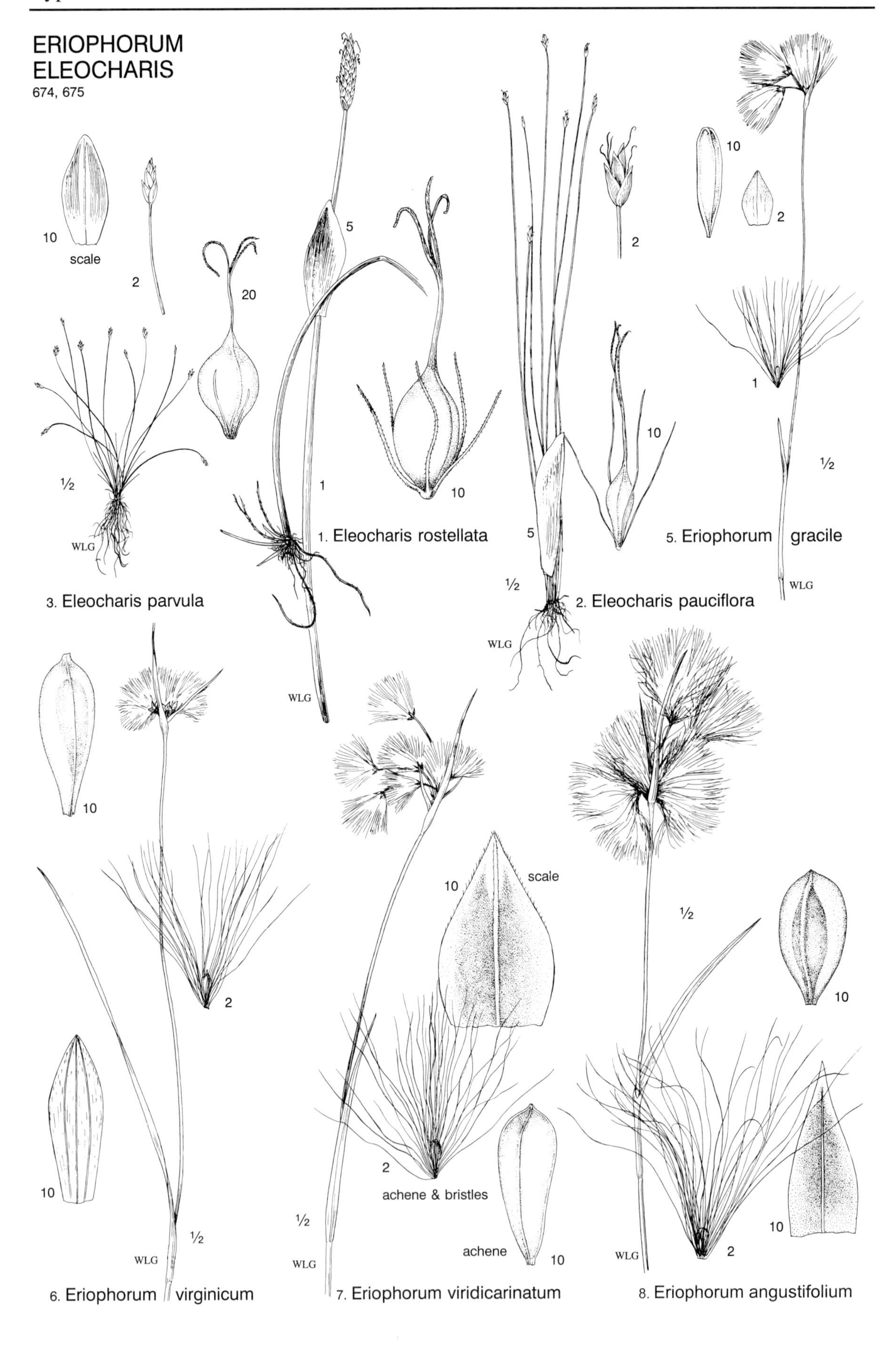

10 scale

2

20

1/2

WLG

3. Eleocharis parvula

1. Eleocharis rostellata

5

10

5

1/2

WLG

WLG

2. Eleocharis pauciflora

10

2

10

2

1

1/2

WLG

5. Eriophorum gracile

10

2

10

1/2

WLG

6. Eriophorum virginicum

10 scale

2

achene & bristles

achene

10

1/2

WLG

7. Eriophorum viridicarinatum

1/2

10

10

2

WLG

8. Eriophorum angustifolium

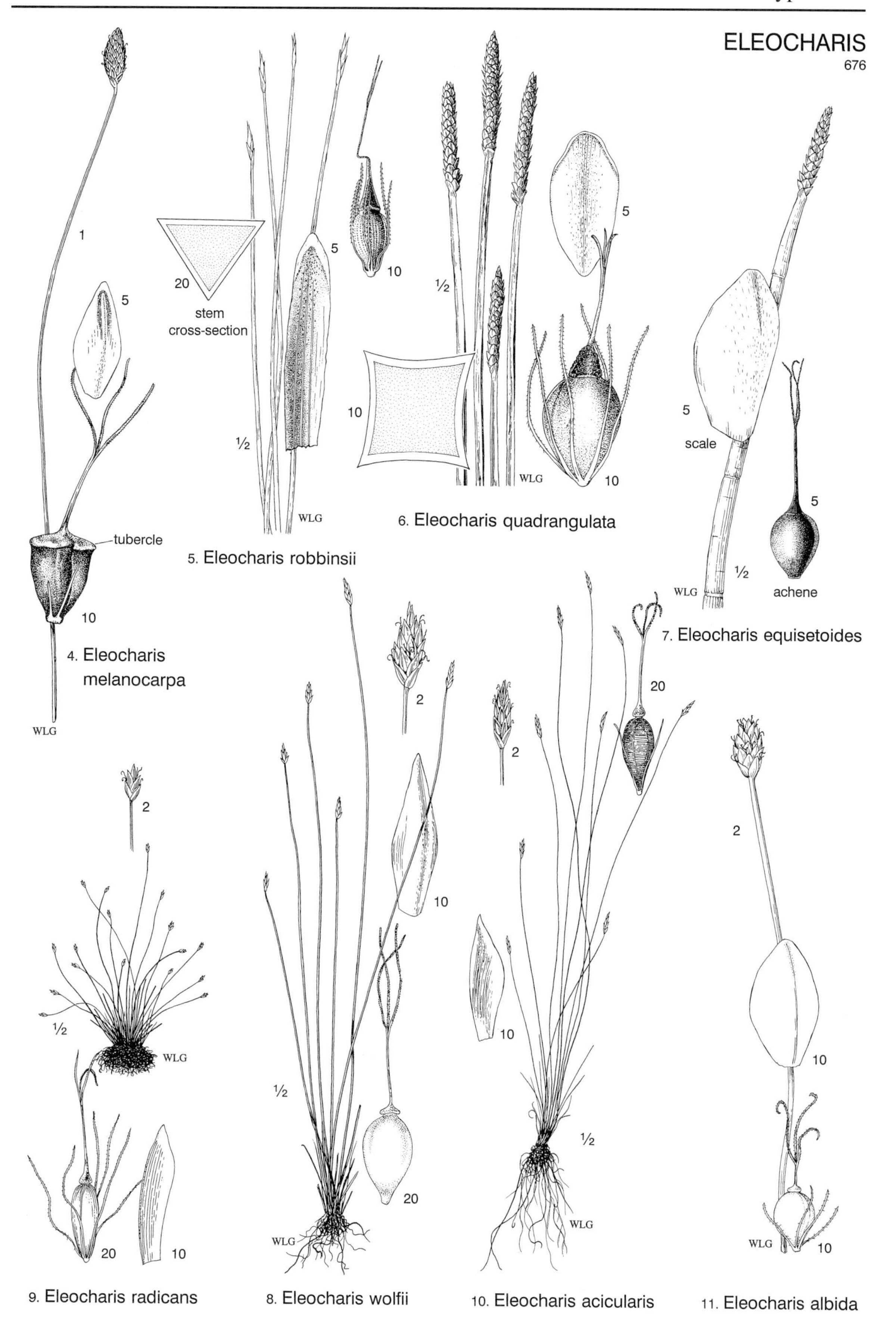

20
stem
cross-section

tubercle

10

4. Eleocharis
melanocarpa

WLG

5. Eleocharis robbinsii

6. Eleocharis quadrangulata

scale

achene

7. Eleocharis equisetoides

9. Eleocharis radicans

8. Eleocharis wolfii

10. Eleocharis acicularis

11. Eleocharis albida

ELEOCHARIS
676, 677

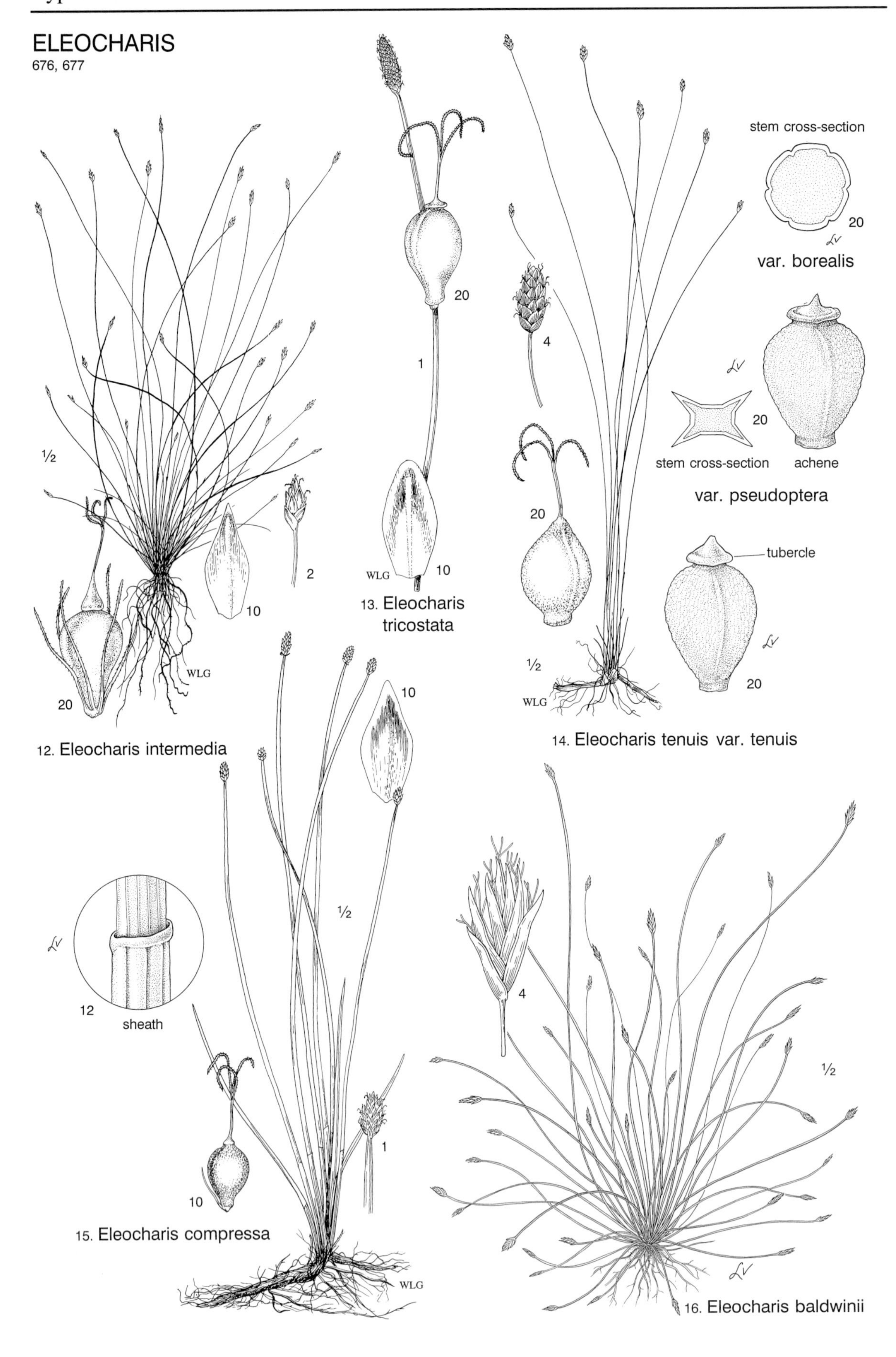

½

stem cross-section

20

var. borealis

20

4

20

var. pseudoptera

stem cross-section   achene

20

tubercle

20

½

WLG

14. Eleocharis tenuis var. tenuis

2

10

20

WLG

12. Eleocharis intermedia

20

1

WLG   10

13. Eleocharis
tricostata

10

½

12

sheath

10

1

15. Eleocharis compressa

WLG

4

½

16. Eleocharis baldwinii

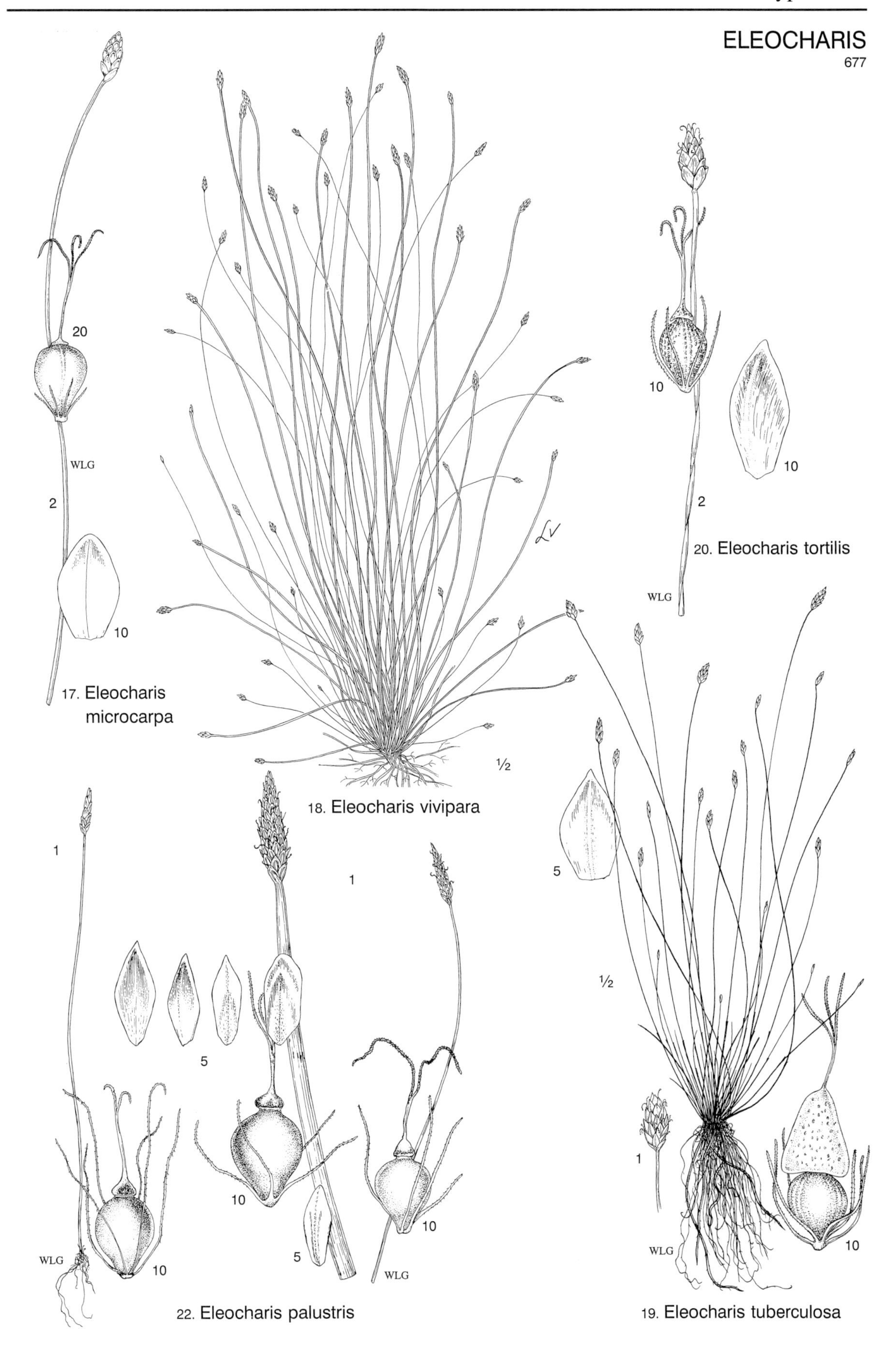

17. Eleocharis microcarpa

18. Eleocharis vivipara

20. Eleocharis tortilis

22. Eleocharis palustris

19. Eleocharis tuberculosa

ELEOCHARIS
677, 678

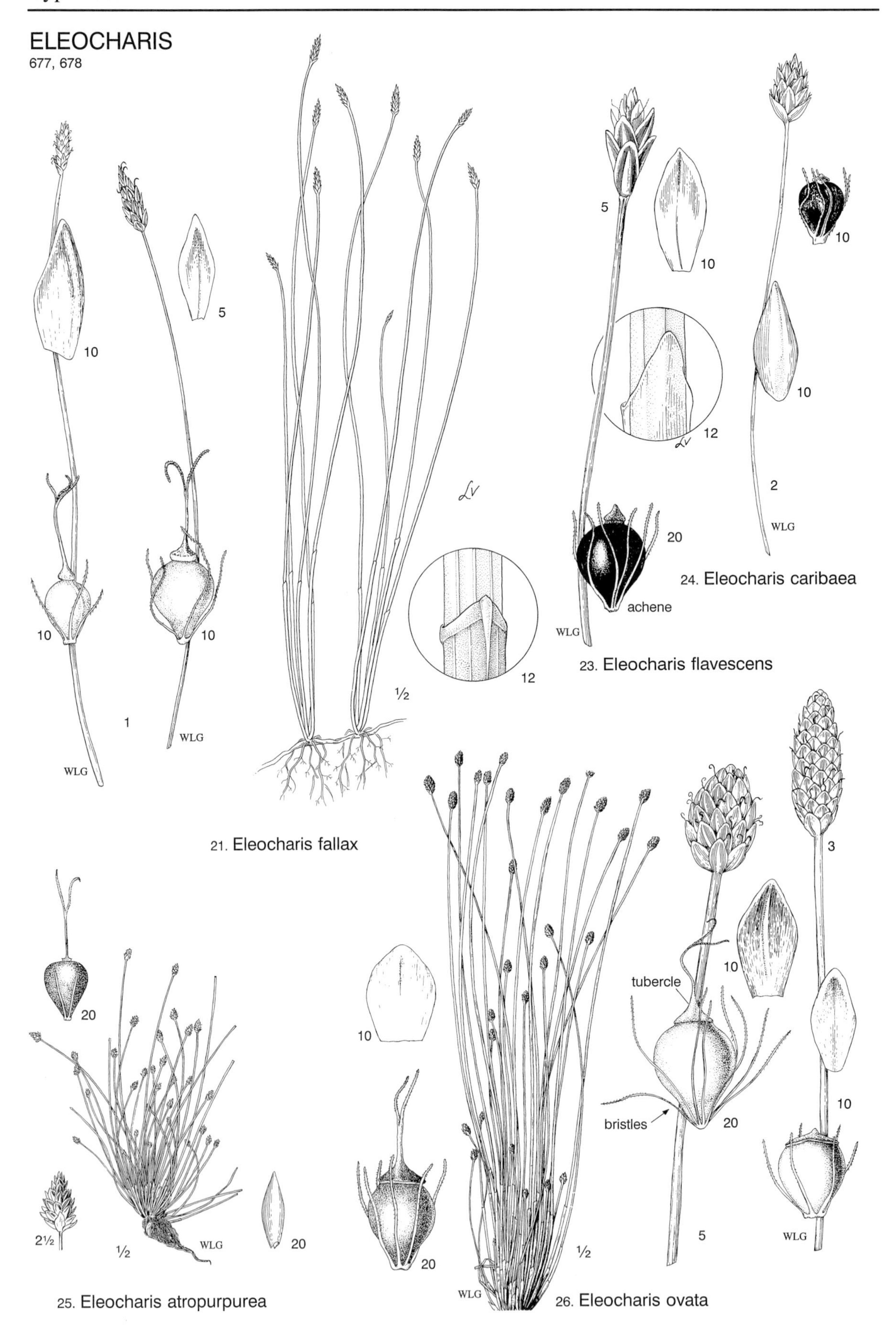

21. Eleocharis fallax

24. Eleocharis caribaea

23. Eleocharis flavescens

25. Eleocharis atropurpurea

26. Eleocharis ovata

tubercle

bristles

achene

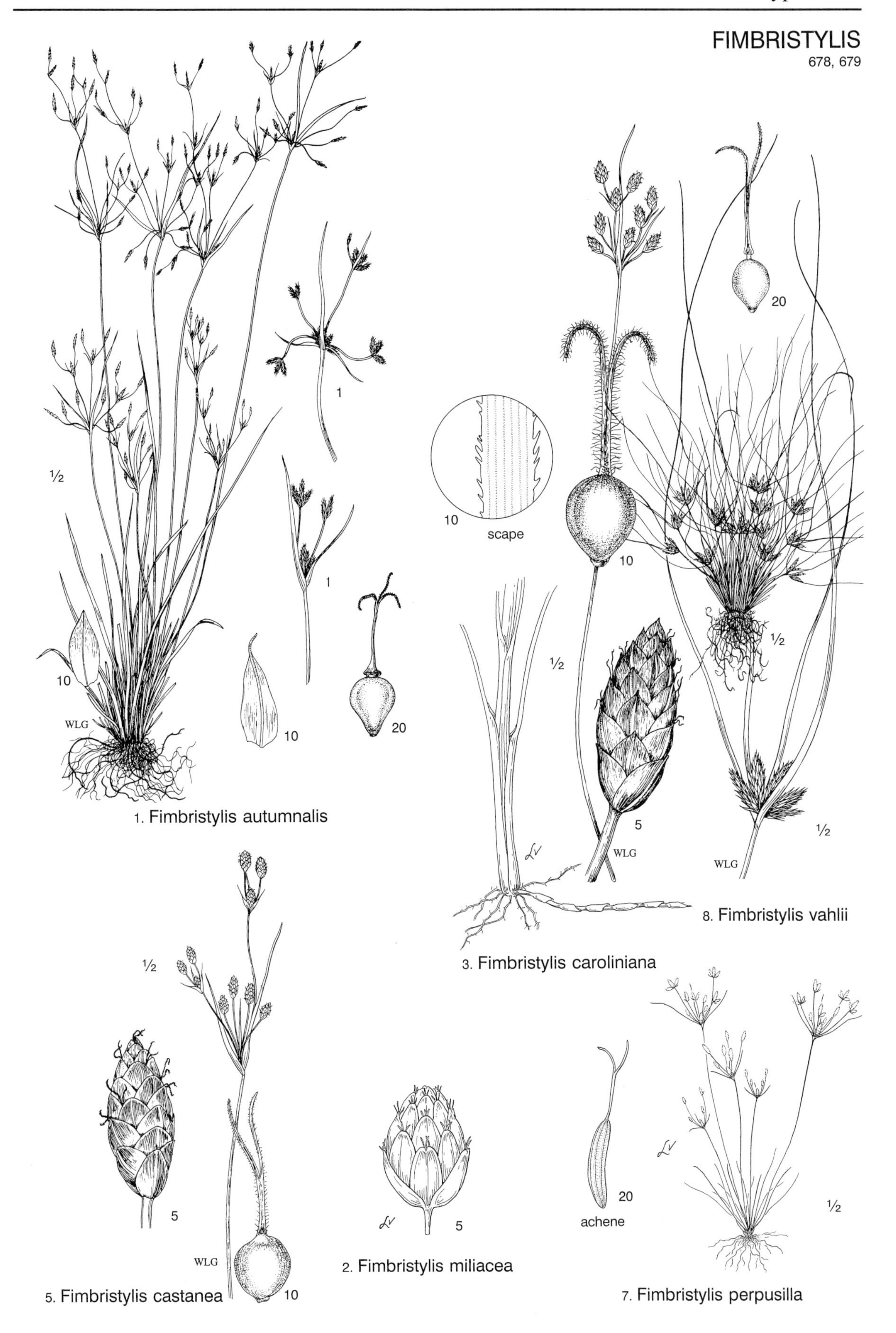

1. Fimbristylis autumnalis

scape

8. Fimbristylis vahlii

3. Fimbristylis caroliniana

5. Fimbristylis castanea

2. Fimbristylis miliacea

achene

7. Fimbristylis perpusilla

FIMBRISTYLIS
BULBOSTYLIS
679, 680

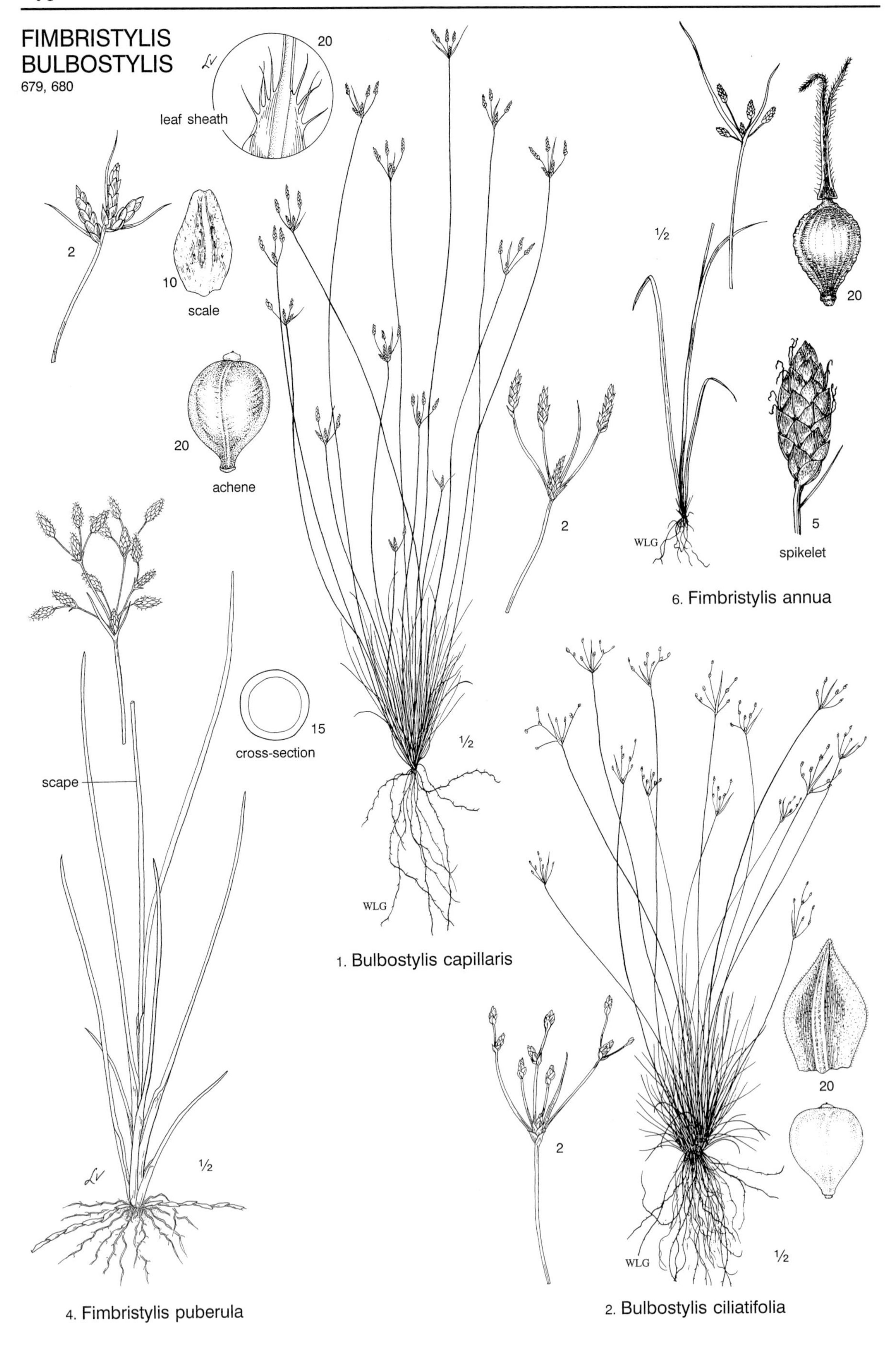

leaf sheath 20

2

10 scale

20 achene

scape

15 cross-section

1. Bulbostylis capillaris

WLG

4. Fimbristylis puberula

½

½ 20

5 spikelet

6. Fimbristylis annua

WLG

2

WLG

20

2. Bulbostylis ciliatifolia

½

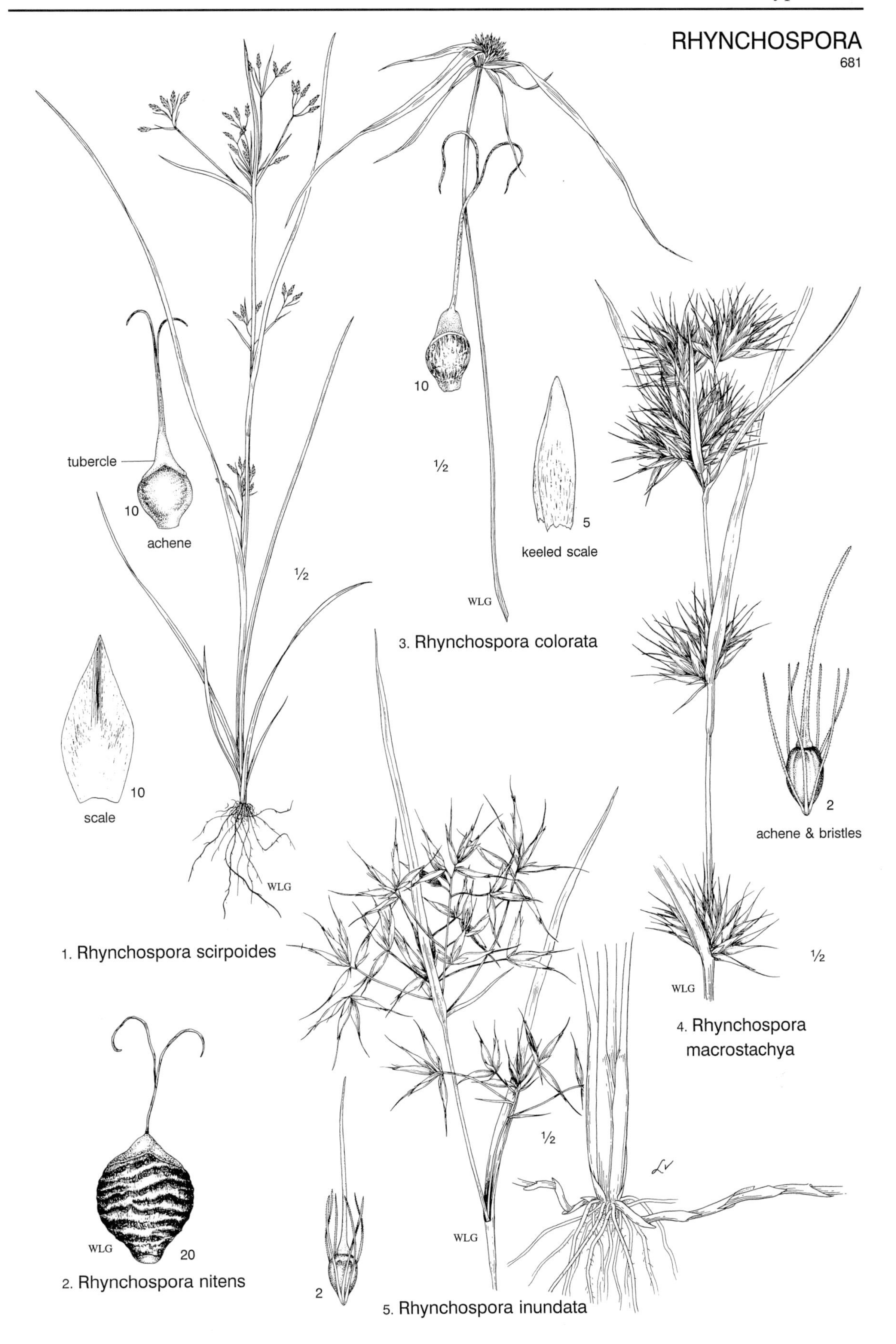

tubercle

10

achene

scale

10

½

1. Rhynchospora scirpoides

WLG

10

½

WLG

keeled scale

5

3. Rhynchospora colorata

2

achene & bristles

½

WLG

4. Rhynchospora macrostachya

WLG

20

2. Rhynchospora nitens

½

WLG

2

5. Rhynchospora inundata

RHYNCHOSPORA
681, 682

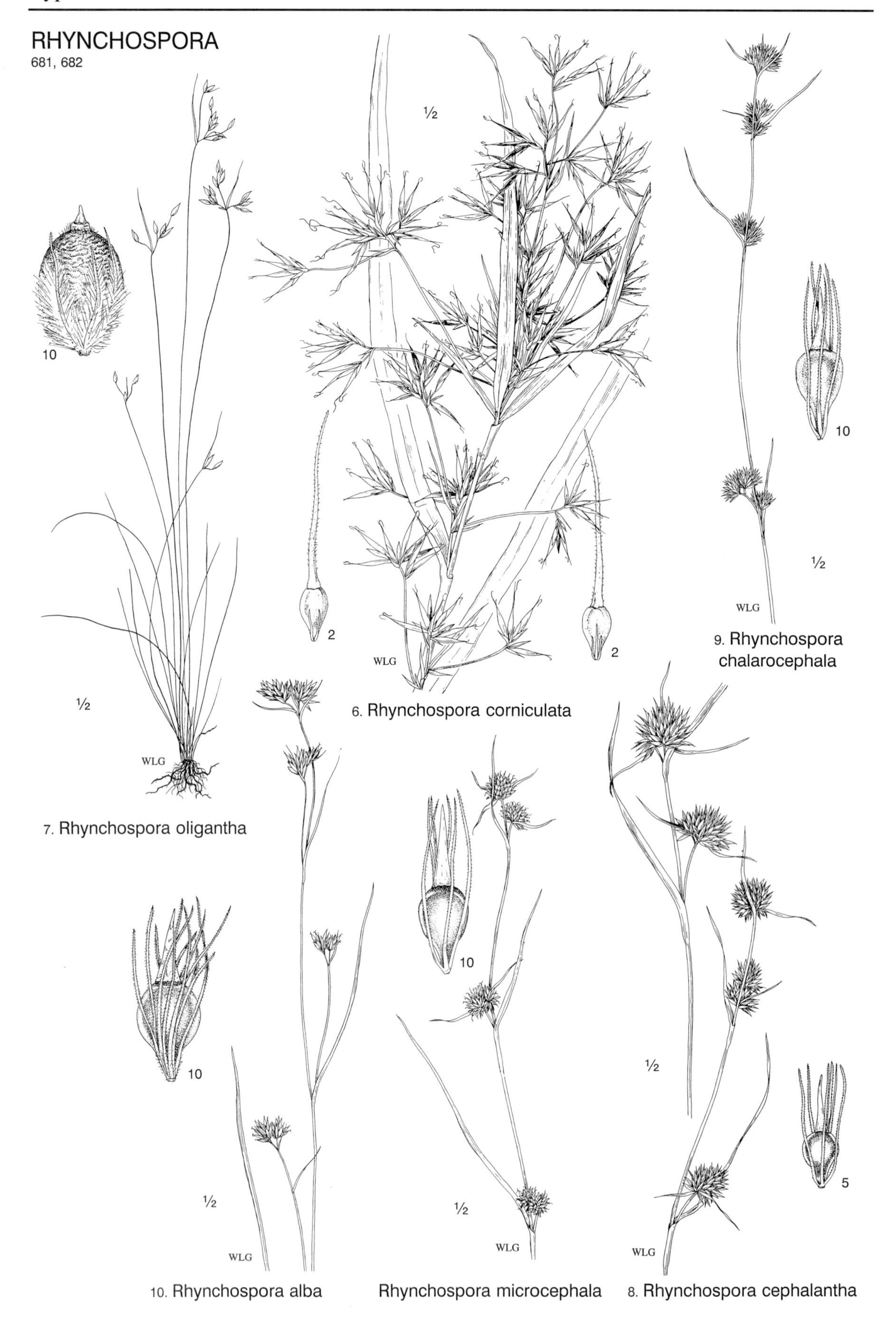

10

½

10

½

WLG

7. Rhynchospora oligantha

½

2

WLG

6. Rhynchospora corniculata

2

½

WLG

9. Rhynchospora chalarocephala

10

10

½

½

WLG

5

½

½

WLG

WLG

10. Rhynchospora alba

Rhynchospora microcephala

8. Rhynchospora cephalantha

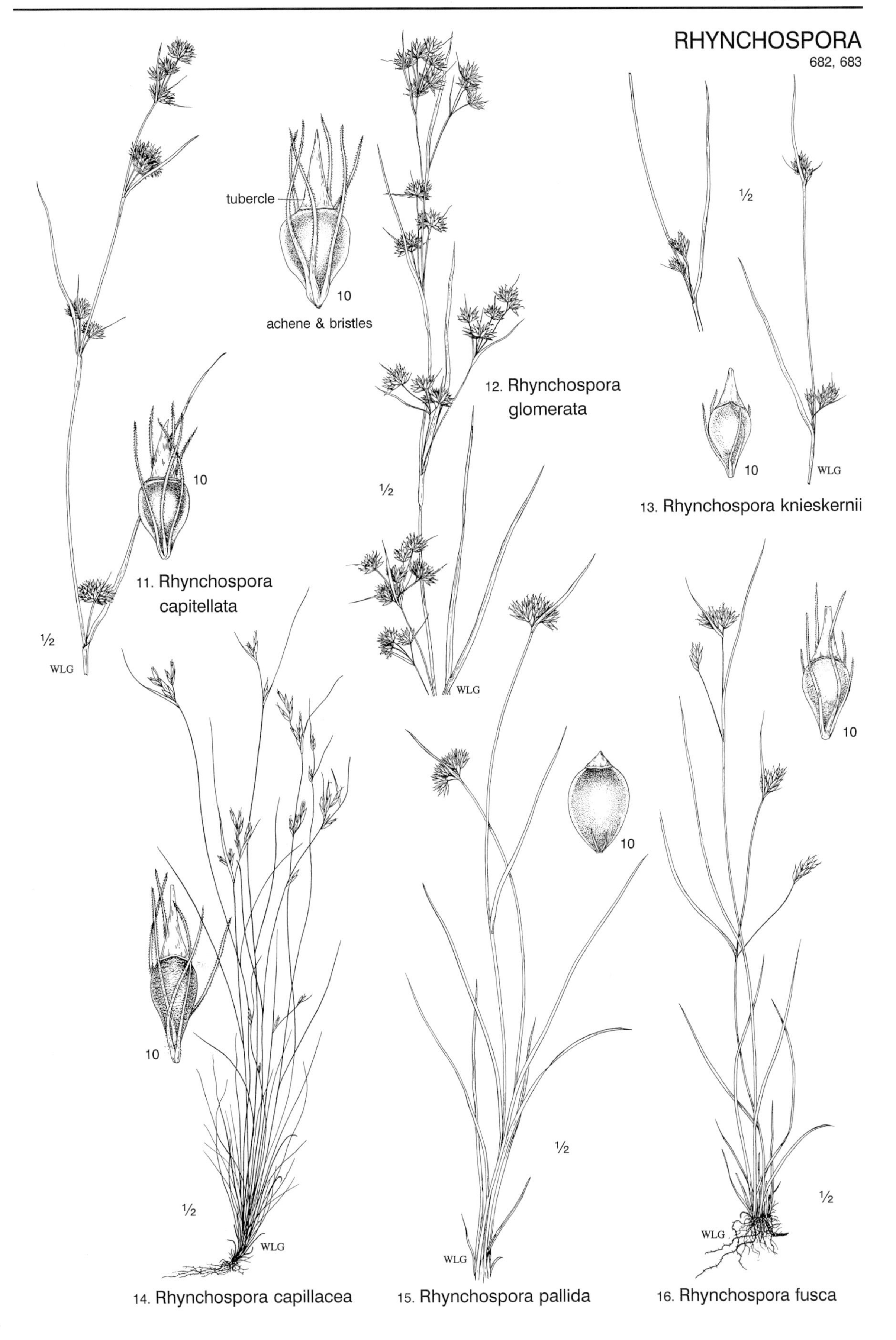

tubercle

10

achene & bristles

10

11. Rhynchospora
capitellata

½

WLG

½

12. Rhynchospora
glomerata

½

WLG

½

WLG

13. Rhynchospora knieskernii

10

10

10

10

14. Rhynchospora capillacea

½

WLG

15. Rhynchospora pallida

½

WLG

16. Rhynchospora fusca

½

WLG

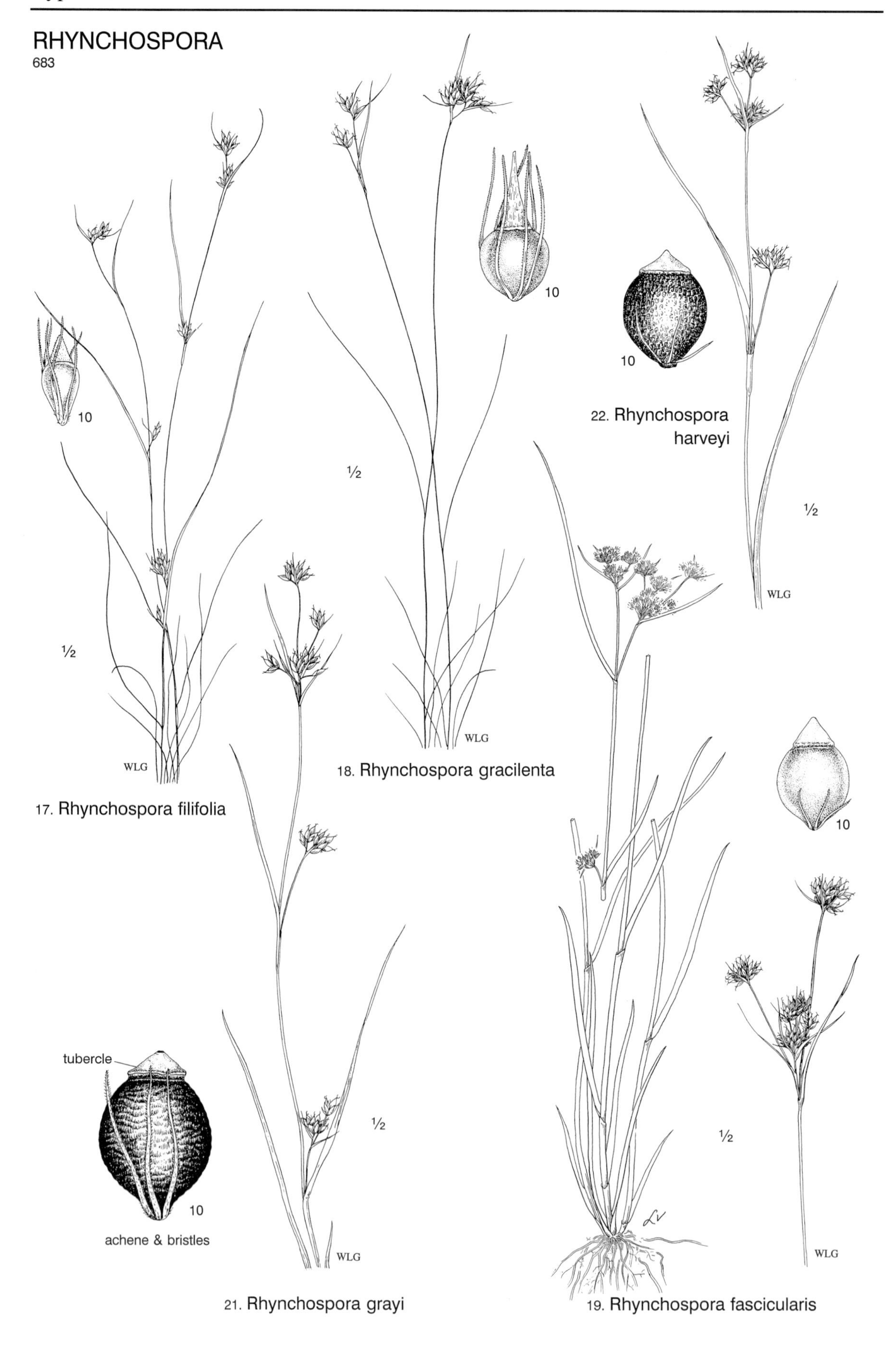

10

22. **Rhynchospora harveyi**

½

WLG

10

½

WLG

17. **Rhynchospora filifolia**

½

WLG

18. **Rhynchospora gracilenta**

10

tubercle

10

achene & bristles

½

WLG

21. **Rhynchospora grayi**

½

WLG

19. **Rhynchospora fascicularis**

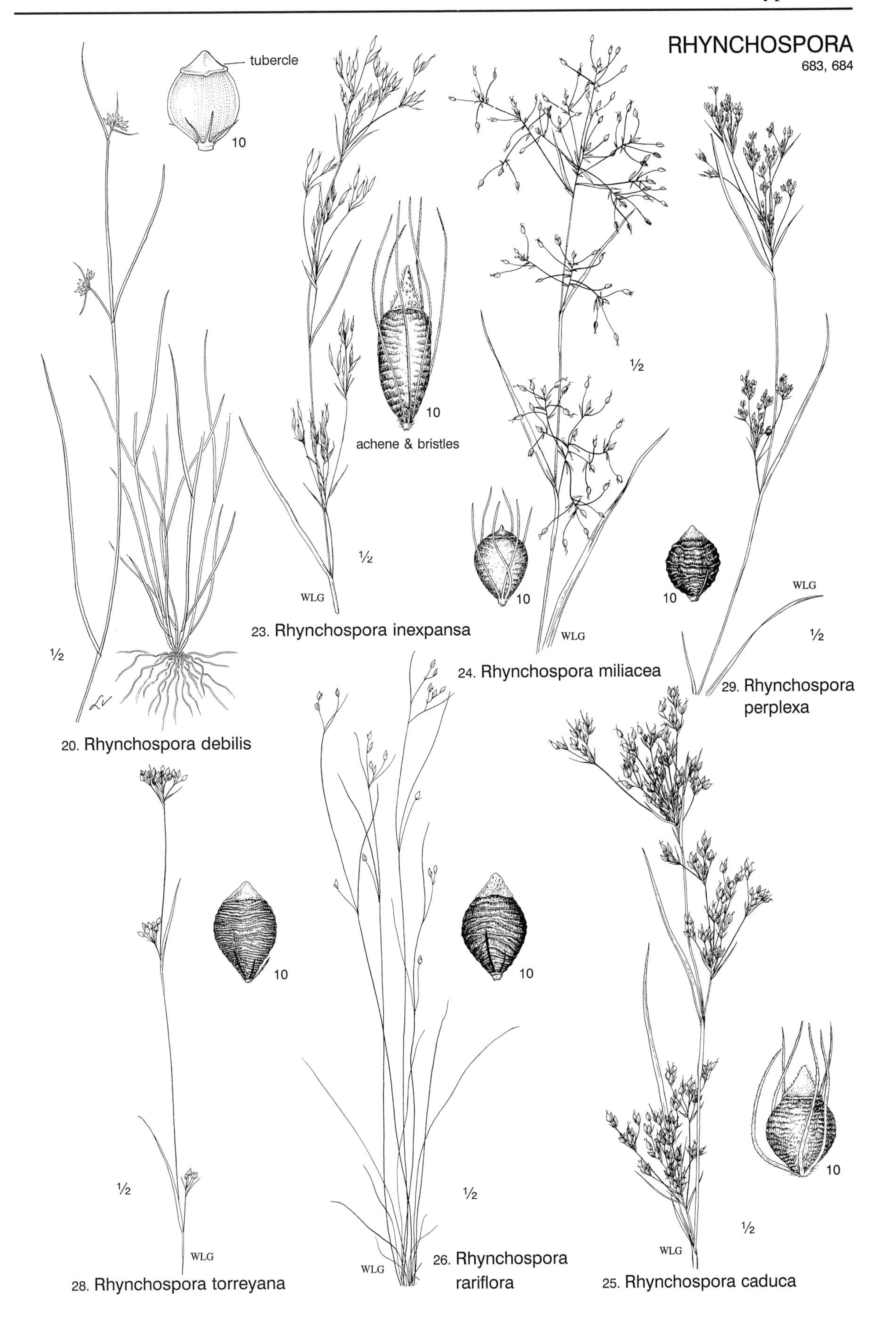

tubercle

10

achene & bristles

10

23. Rhynchospora inexpansa

10

24. Rhynchospora miliacea

10

29. Rhynchospora
perplexa

½

20. Rhynchospora debilis

10

10

½

28. Rhynchospora torreyana

½

26. Rhynchospora
rariflora

10

½

25. Rhynchospora caduca

WLG

RHYNCHOSPORA
CLADIUM
CYPERUS
684–686

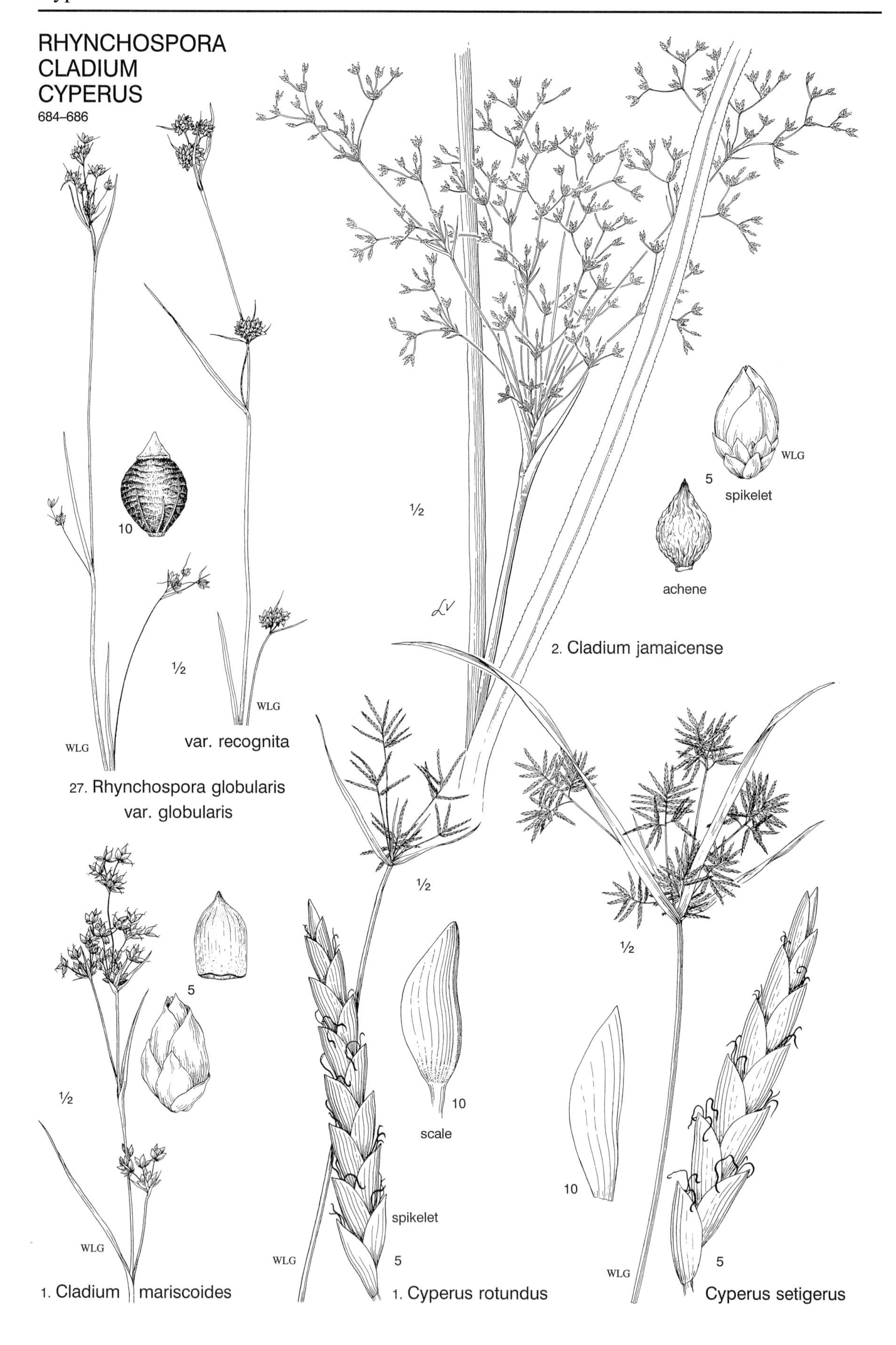

2. Cladium jamaicense

27. Rhynchospora globularis
var. globularis

var. recognita

1. Cladium mariscoides

1. Cyperus rotundus

Cyperus setigerus

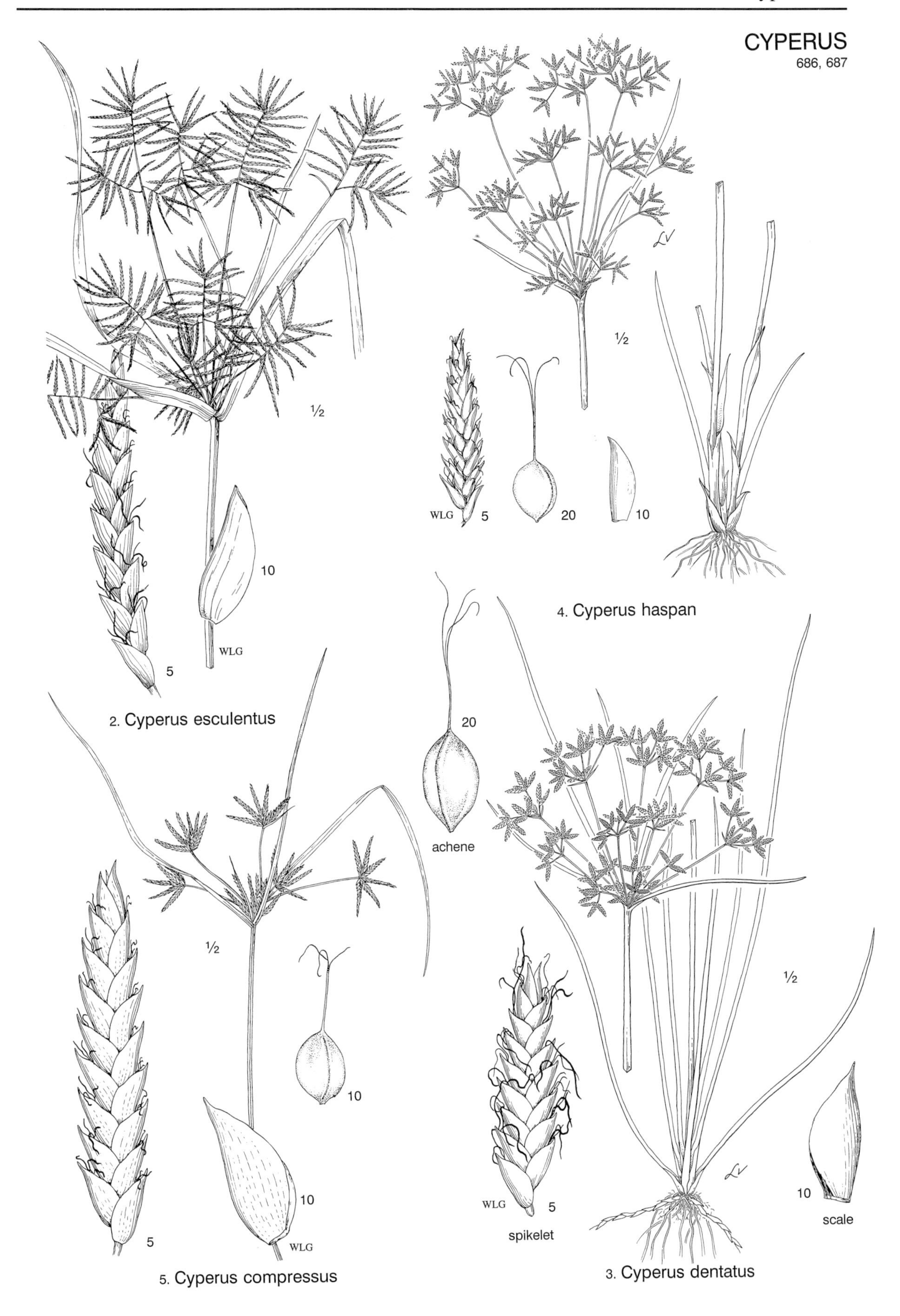

1/2

10

WLG

5

2. Cyperus esculentus

WLG 5   20   10

1/2

4. Cyperus haspan

20

achene

1/2

1/2

10

5

5. Cyperus compressus

WLG

WLG 5

spikelet

10

scale

3. Cyperus dentatus

CYPERUS
687

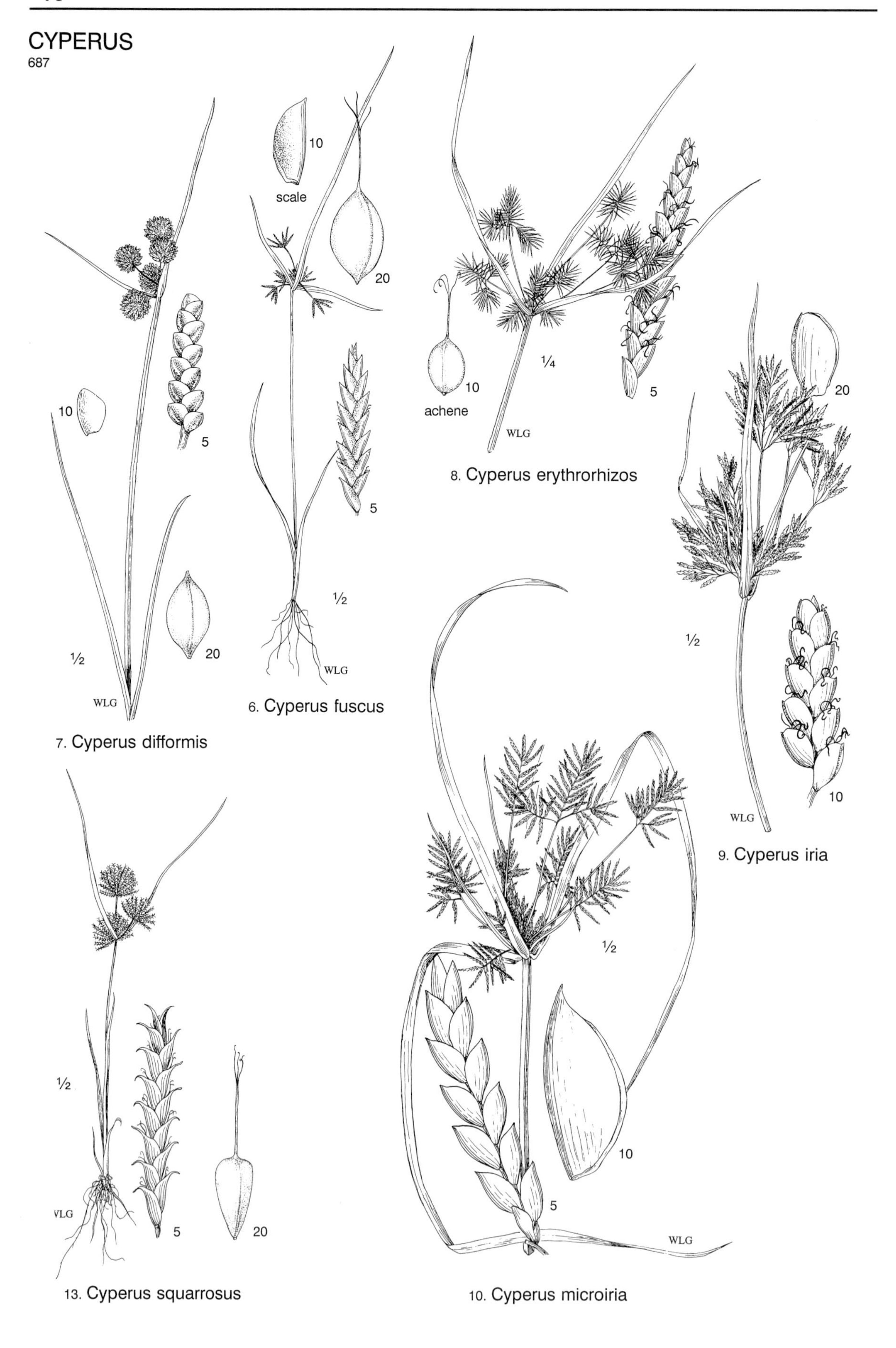

scale

10

20

10

5

10

1/2

20

WLG

7. Cyperus difformis

6. Cyperus fuscus

5

5

1/2

WLG

10

achene

1/4

5

WLG

8. Cyperus erythrorhizos

20

1/2

20

10

WLG

9. Cyperus iria

1/2

1/2

VLG

5

20

13. Cyperus squarrosus

1/2

10

5

WLG

10. Cyperus microiria

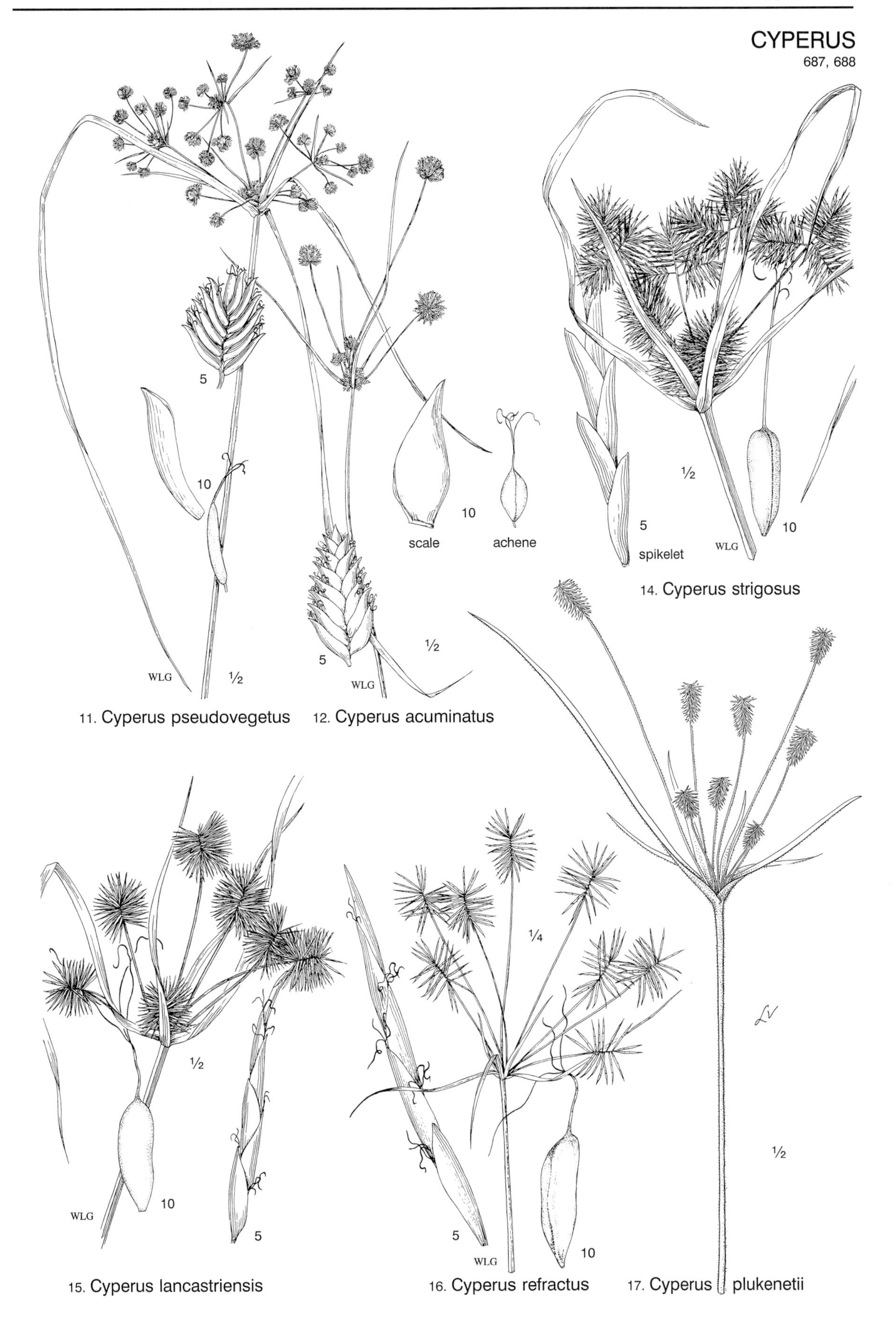

5

10

scale

10

achene

½

5

10

11. Cyperus pseudovegetus

5

½

12. Cyperus acuminatus

½

5

spikelet

10

WLG

14. Cyperus strigosus

½

10

5

15. Cyperus lancastriensis

¼

5

10

16. Cyperus refractus

½

17. Cyperus plukenetii

CYPERUS
688

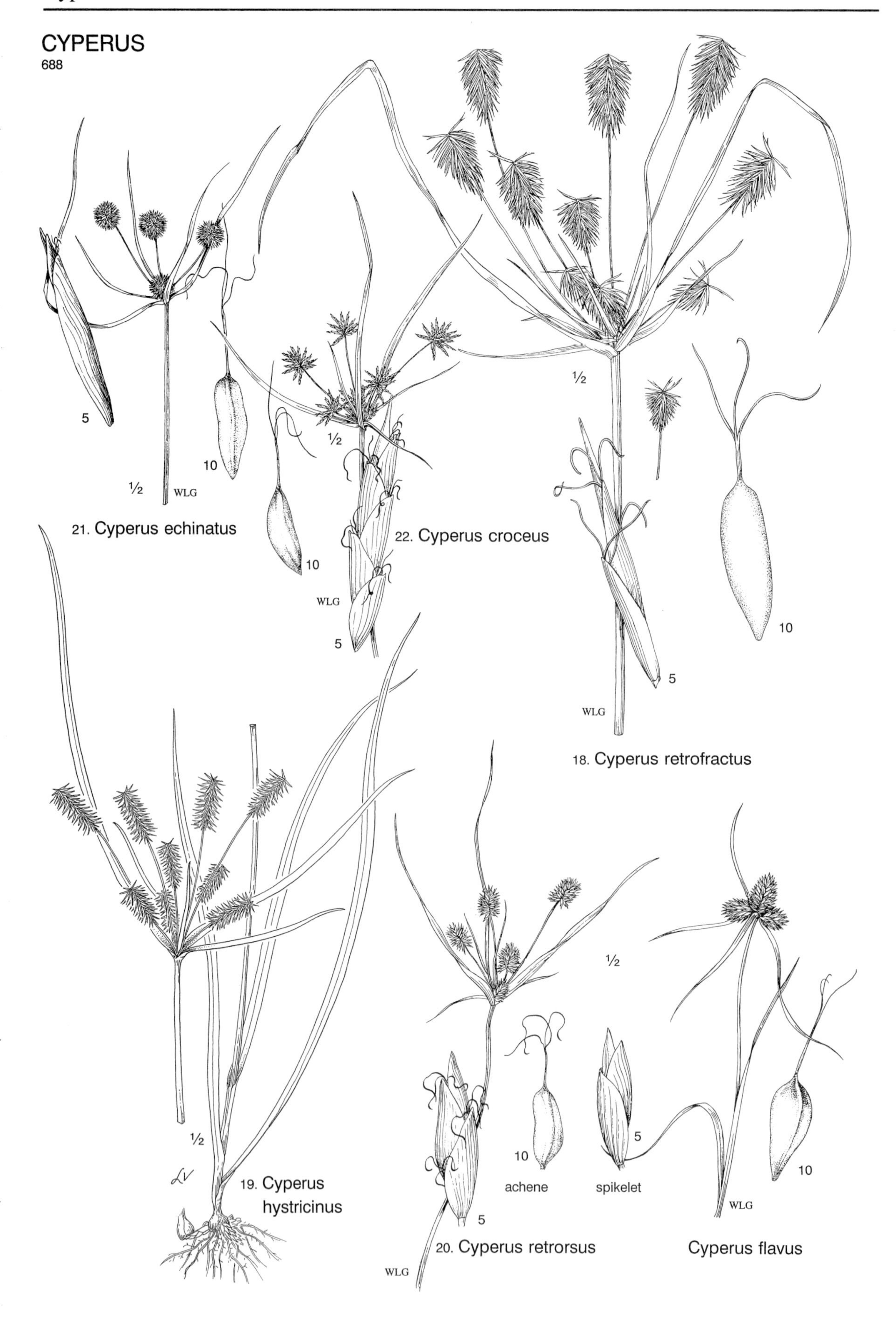

5

10

½ WLG

21. Cyperus echinatus

½

10

10

WLG

5

22. Cyperus croceus

½

5

WLG

18. Cyperus retrofractus

10

½

½

19. Cyperus
hystricinus

10
achene

5

5
spikelet

20. Cyperus retrorsus

WLG

10

WLG

Cyperus flavus

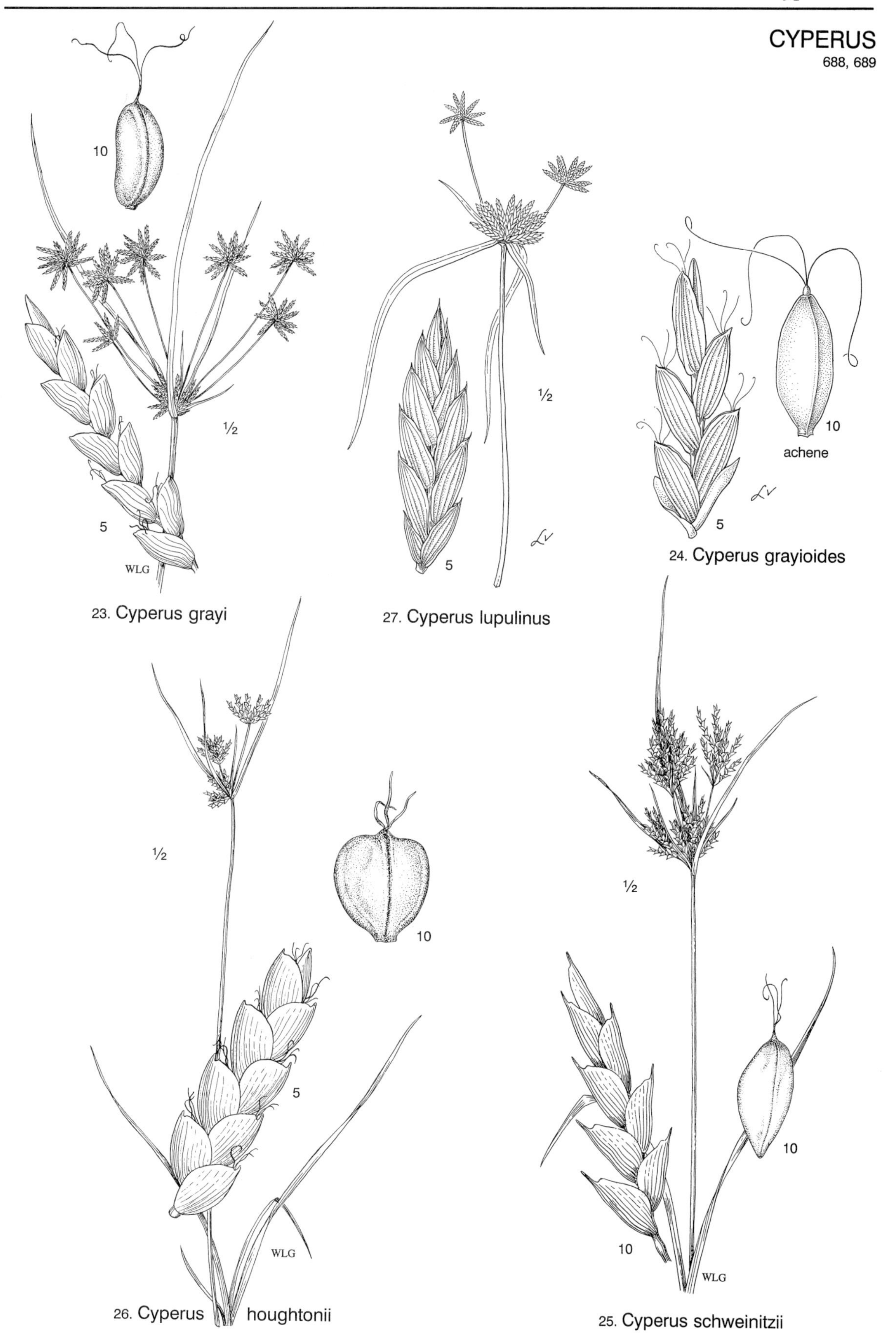

10

½

5

WLG

23. Cyperus grayi

½

5

27. Cyperus lupulinus

10

achene

5

24. Cyperus grayioides

½

5

WLG

26. Cyperus houghtonii

10

½

10

10

WLG

25. Cyperus schweinitzii

CYPERUS
689

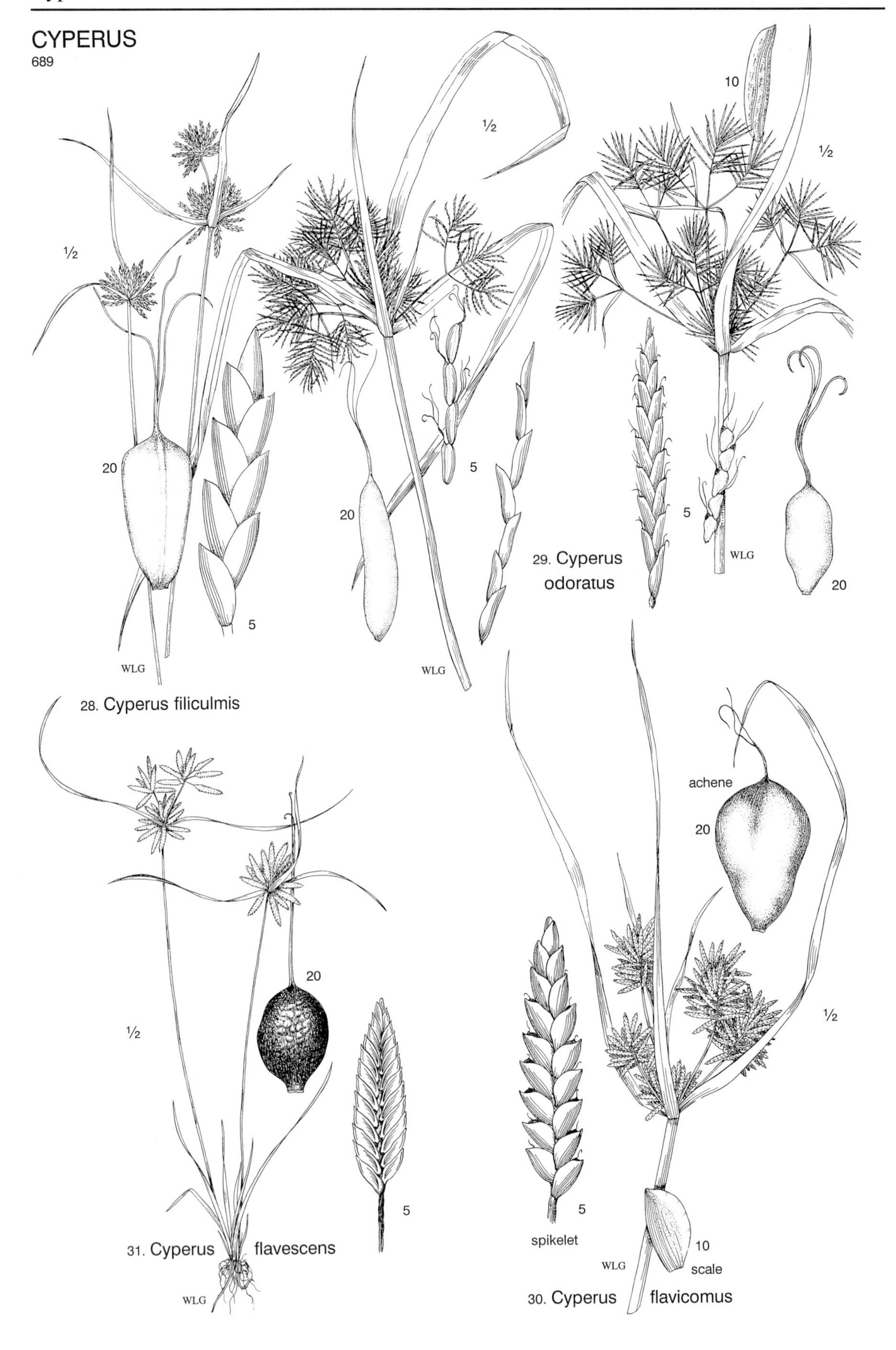

28. Cyperus filiculmis

29. Cyperus odoratus

30. Cyperus flavicomus

31. Cyperus flavescens

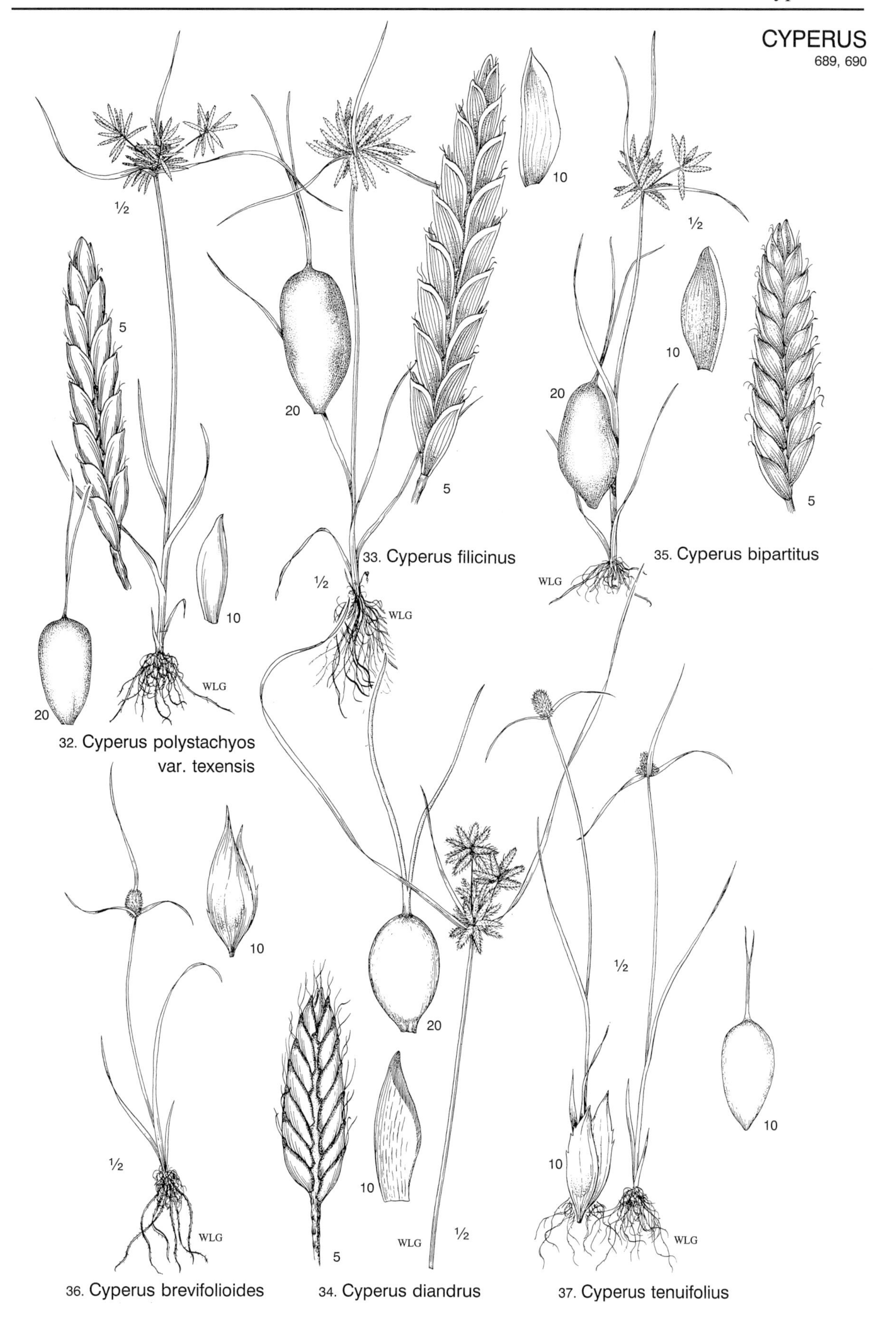

1/2

5

20

10

WLG

32. Cyperus polystachyos var. texensis

20

10

1/2

WLG

33. Cyperus filicinus

5

10

5

20

10

1/2

WLG

35. Cyperus bipartitus

5

10

1/2

WLG

36. Cyperus brevifolioides

5

10

20

10

1/2

WLG

34. Cyperus diandrus

10

1/2

10

1/2

WLG

37. Cyperus tenuifolius

DULICHIUM
HEMICARPHA
LIPOCARPHA
FUIRENA
690, 691

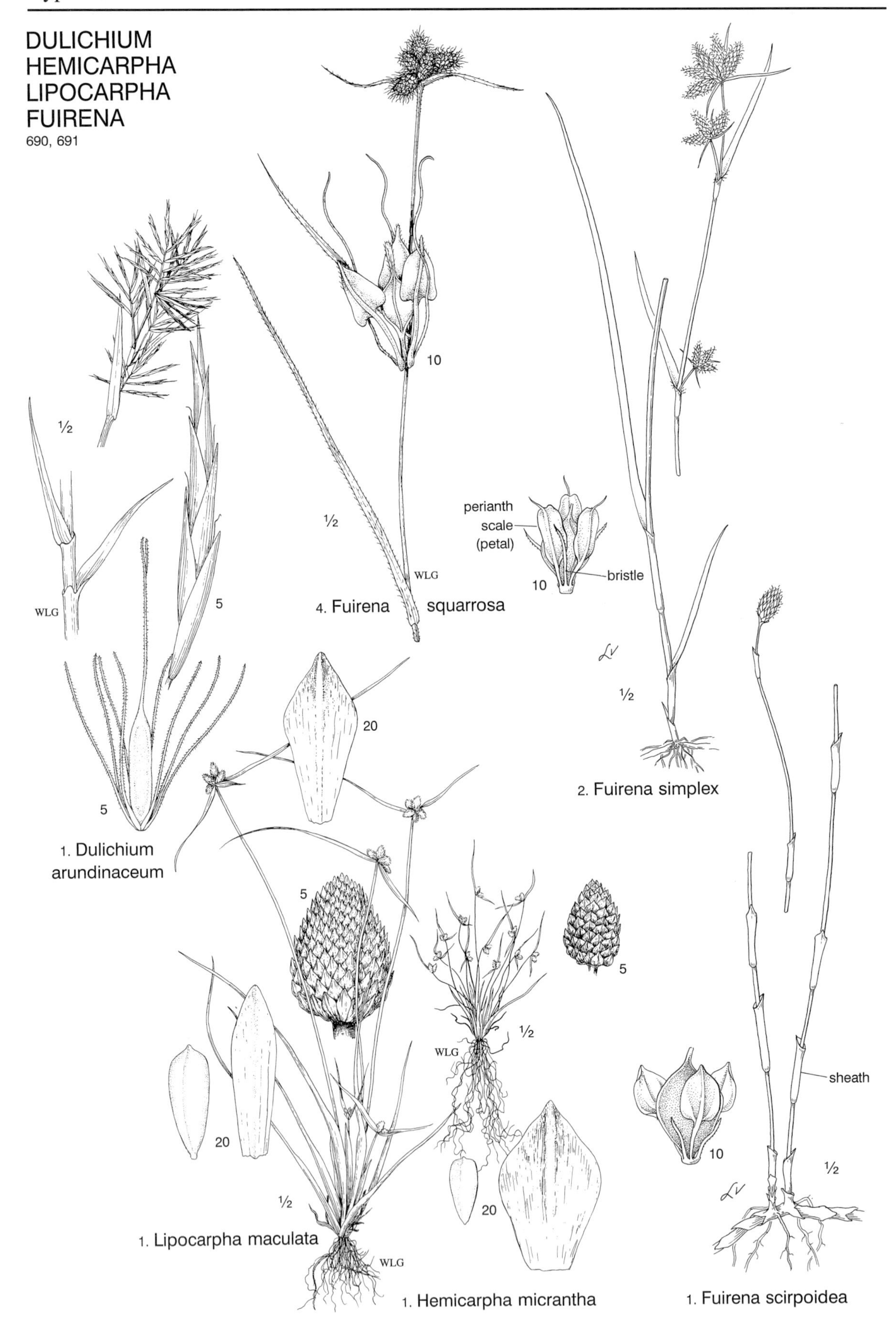

½

WLG

5

WLG

5

5

**1. Dulichium
arundinaceum**

½

WLG

½

4. **Fuirena   squarrosa**

perianth
scale
(petal)

10

bristle

10

½

2. Fuirena simplex

20

5

20

½

1. Lipocarpha maculata

WLG

20

½

WLG

20

1. Hemicarpha micrantha

sheath

10

½

1. Fuirena scirpoidea

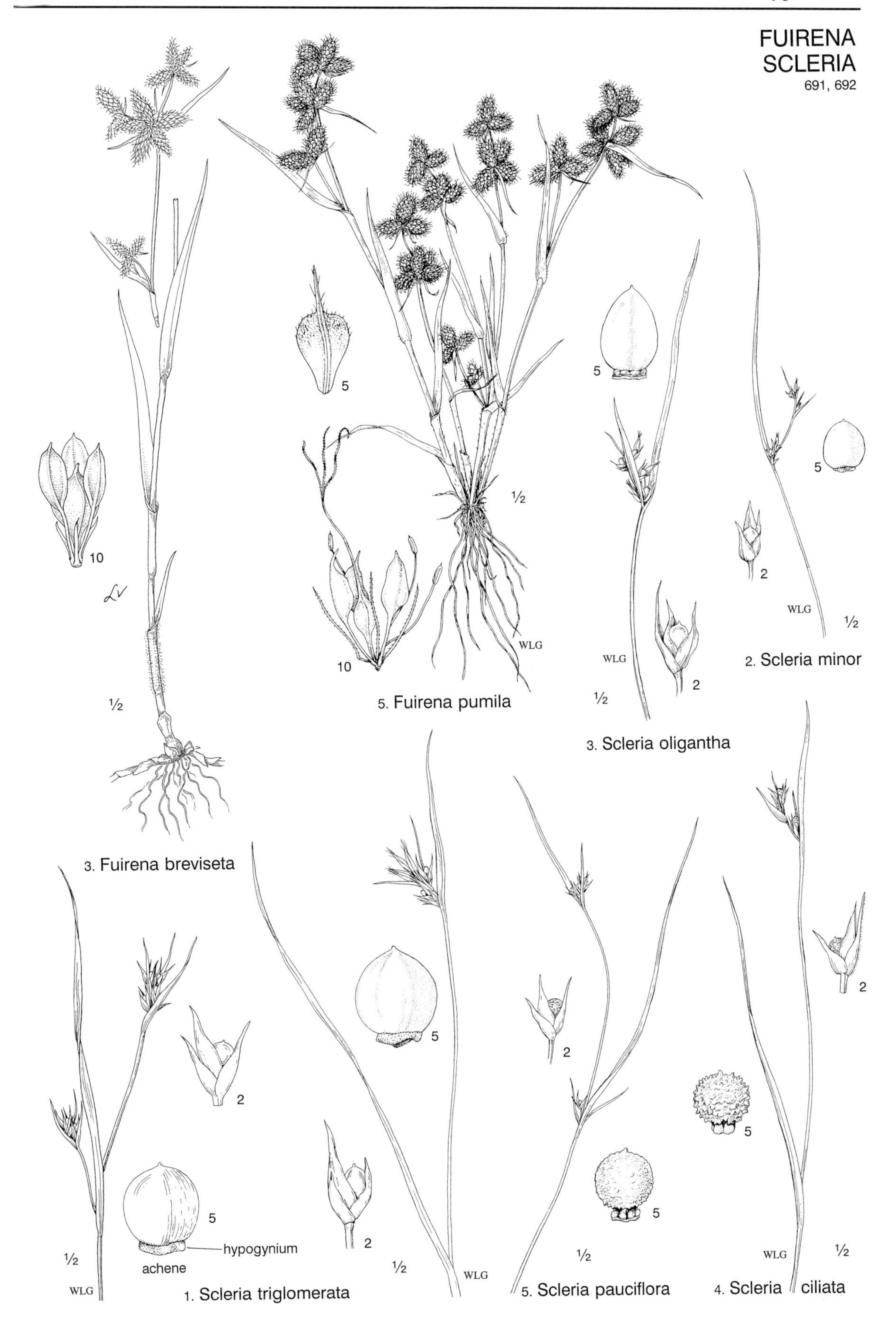

5. Fuirena pumila

10

3. Fuirena breviseta

5. Scleria minor

3. Scleria oligantha

1. Scleria triglomerata

hypogynium

achene

5. Scleria pauciflora

4. Scleria ciliata

SCLERIA
CAREX
692–708

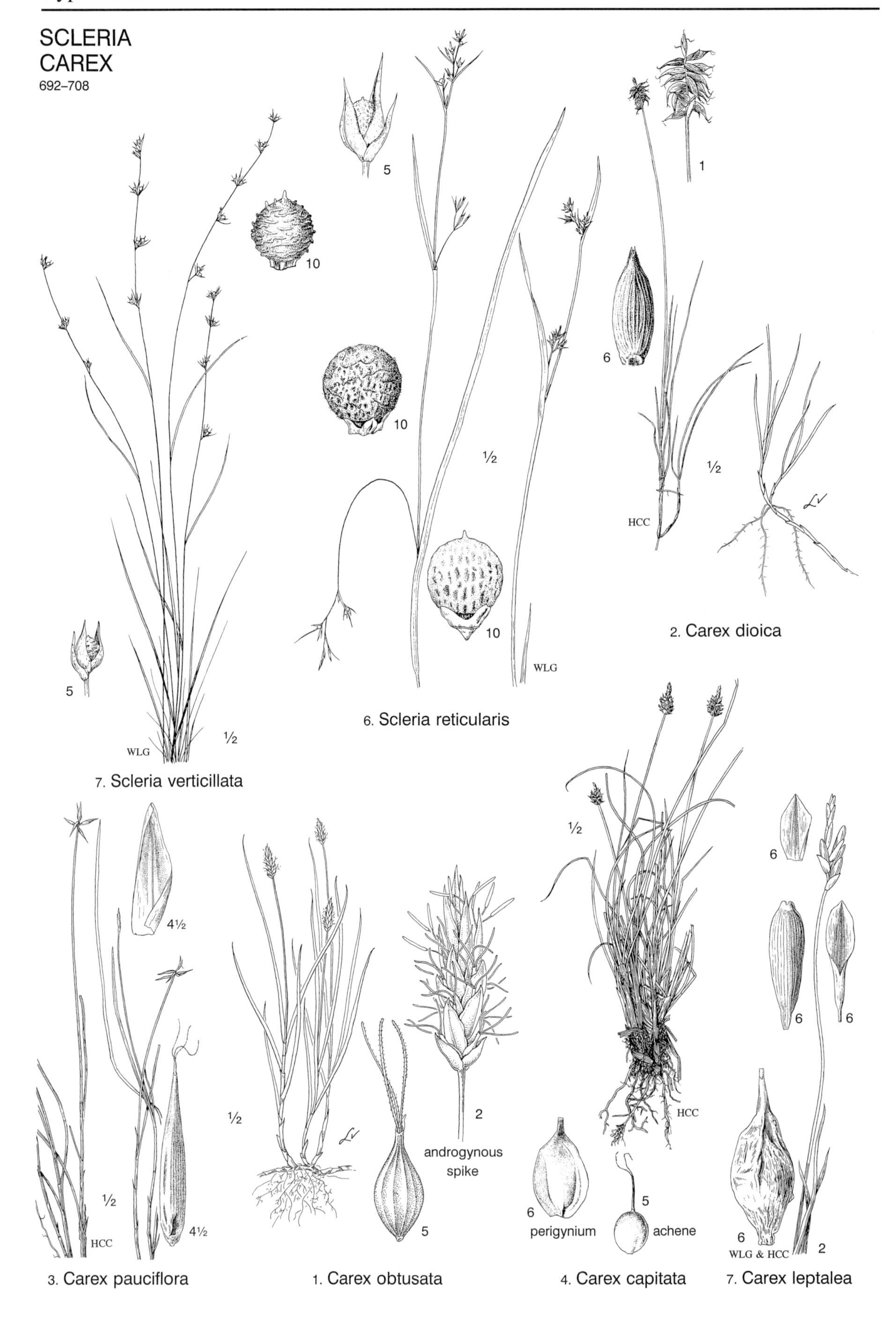

5

10

1

6

½

10

10

½

HCC

2. **Carex dioica**

½

6. **Scleria reticularis**

WLG

5

WLG

½

7. **Scleria verticillata**

½

6

4½

½

6

6

6

½

HCC

4½

androgynous
spike

2

½

HCC

6
perigynium

5
achene

6
WLG & HCC

2

5

3. **Carex pauciflora**

1. **Carex obtusata**

4. **Carex capitata**

7. **Carex leptalea**

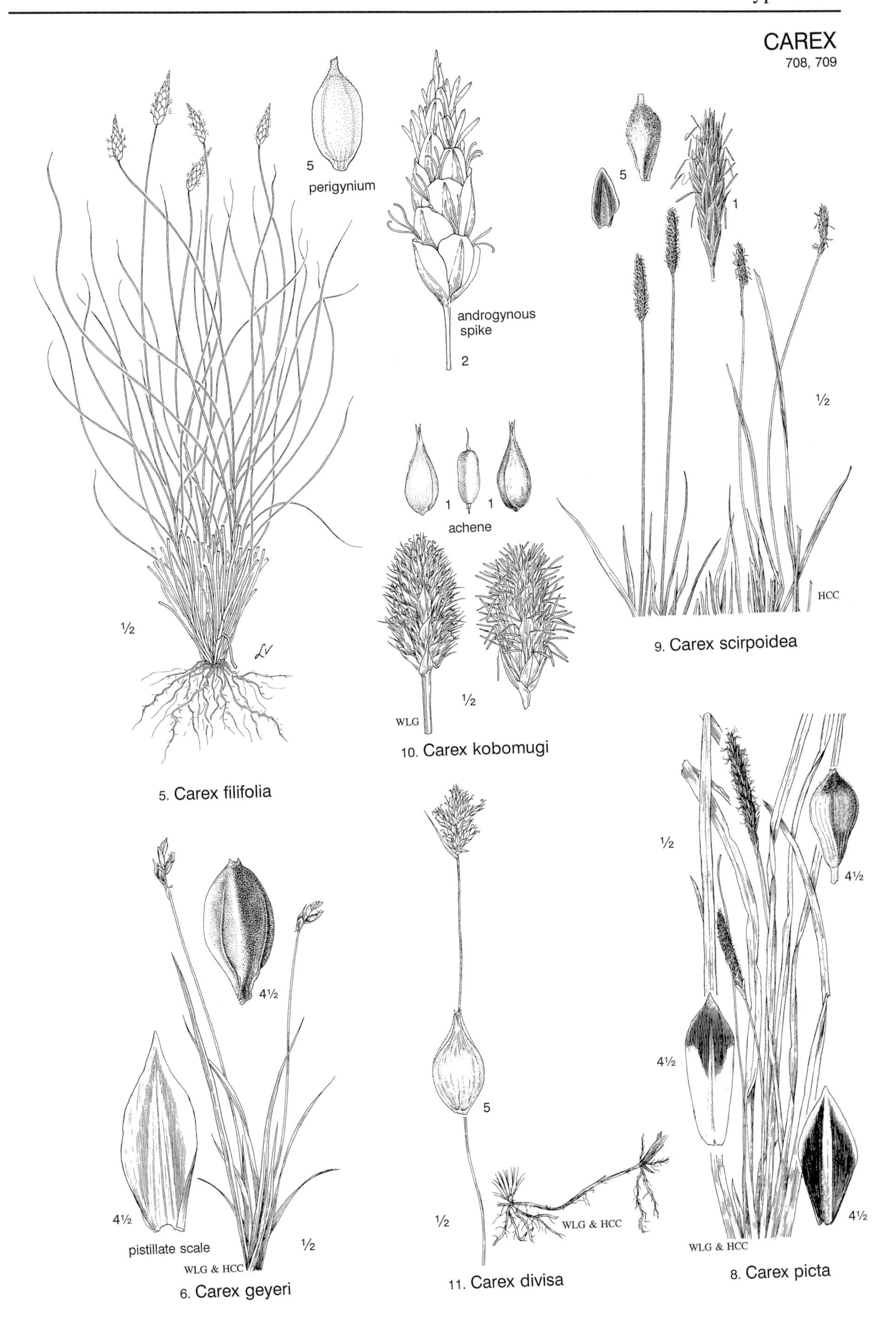

5 perigynium

androgynous spike
2

5

1

½

9. Carex scirpoidea

achene
1 1

½

10. Carex kobomugi
WLG

5. Carex filifolia

½

4½

4½

pistillate scale
WLG & HCC

½

6. Carex geyeri

5

½

11. Carex divisa
WLG & HCC

½

4½

4½

4½

8. Carex picta
WLG & HCC

CAREX
709, 710

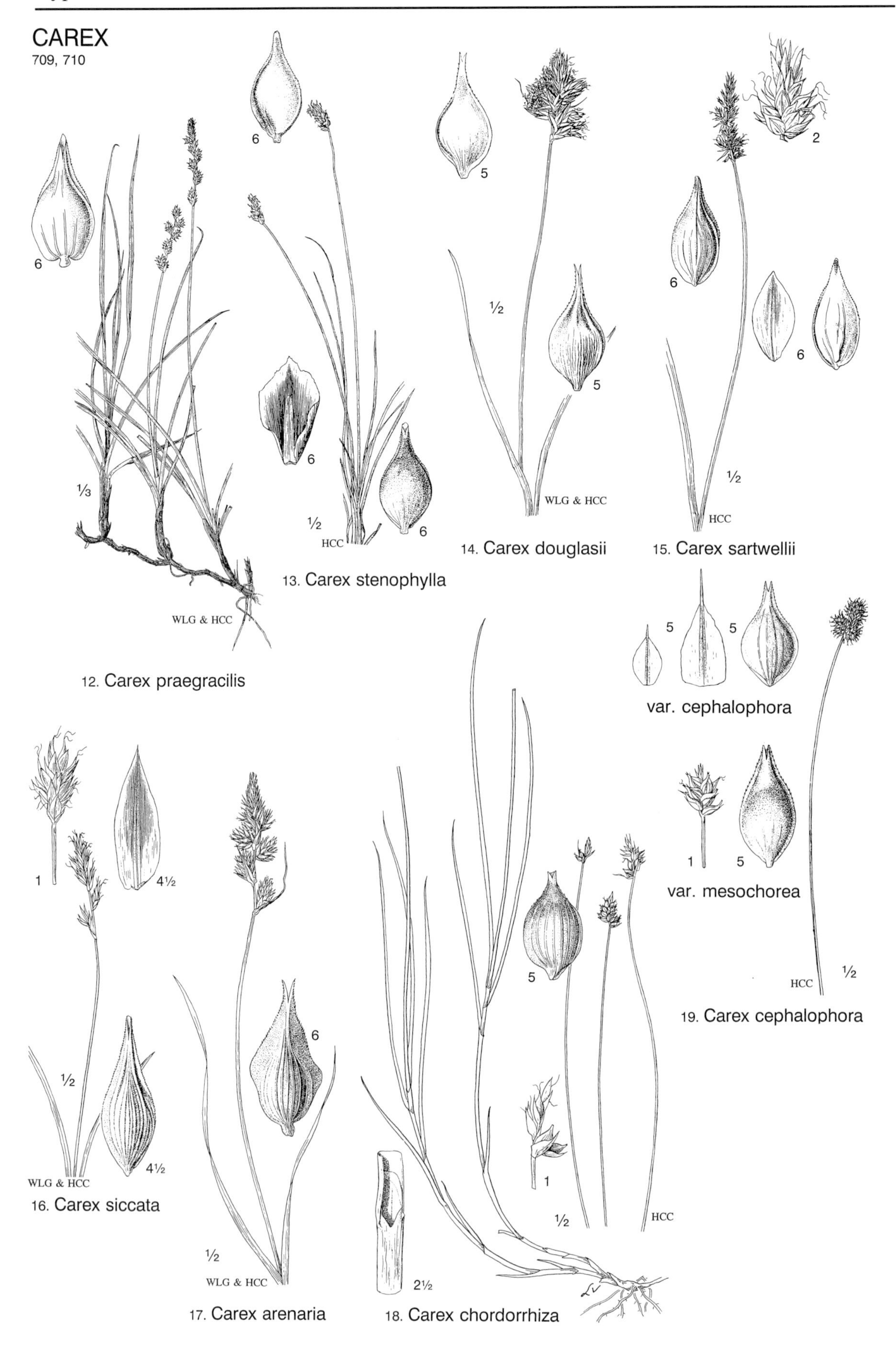

12. Carex praegracilis

13. Carex stenophylla

14. Carex douglasii

15. Carex sartwellii

var. cephalophora

var. mesochorea

19. Carex cephalophora

16. Carex siccata

17. Carex arenaria

18. Carex chordorrhiza

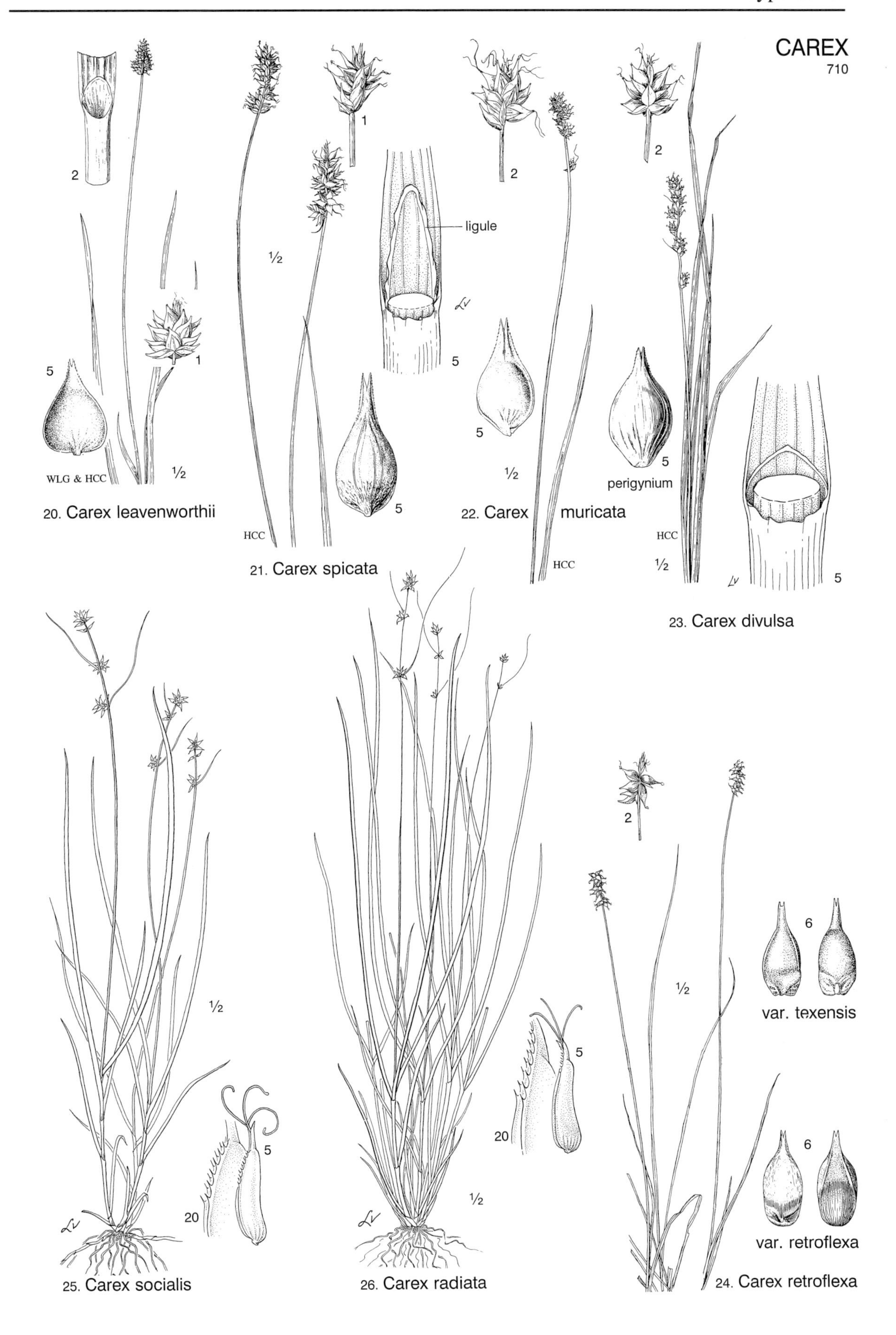

20. Carex leavenworthii

21. Carex spicata

22. Carex muricata

ligule

perigynium

23. Carex divulsa

var. texensis

var. retroflexa

24. Carex retroflexa

25. Carex socialis

26. Carex radiata

CAREX
711, 712

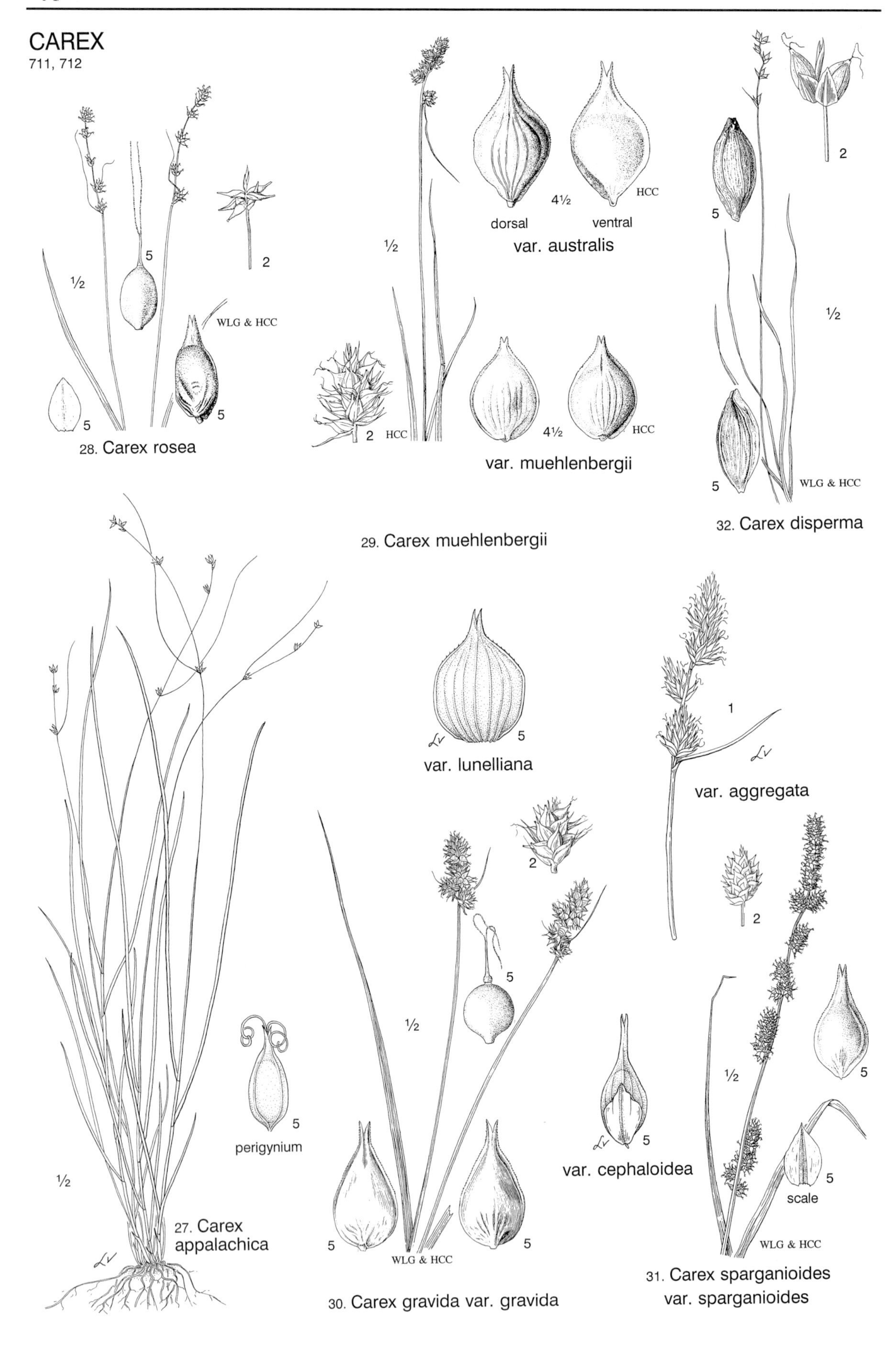

28. Carex rosea

dorsal   ventral
var. australis

var. muehlenbergii

29. Carex muehlenbergii

32. Carex disperma

27. Carex appalachica

perigynium

var. lunelliana

var. aggregata

30. Carex gravida var. gravida

var. cephaloidea

scale

31. Carex sparganioides
var. sparganioides

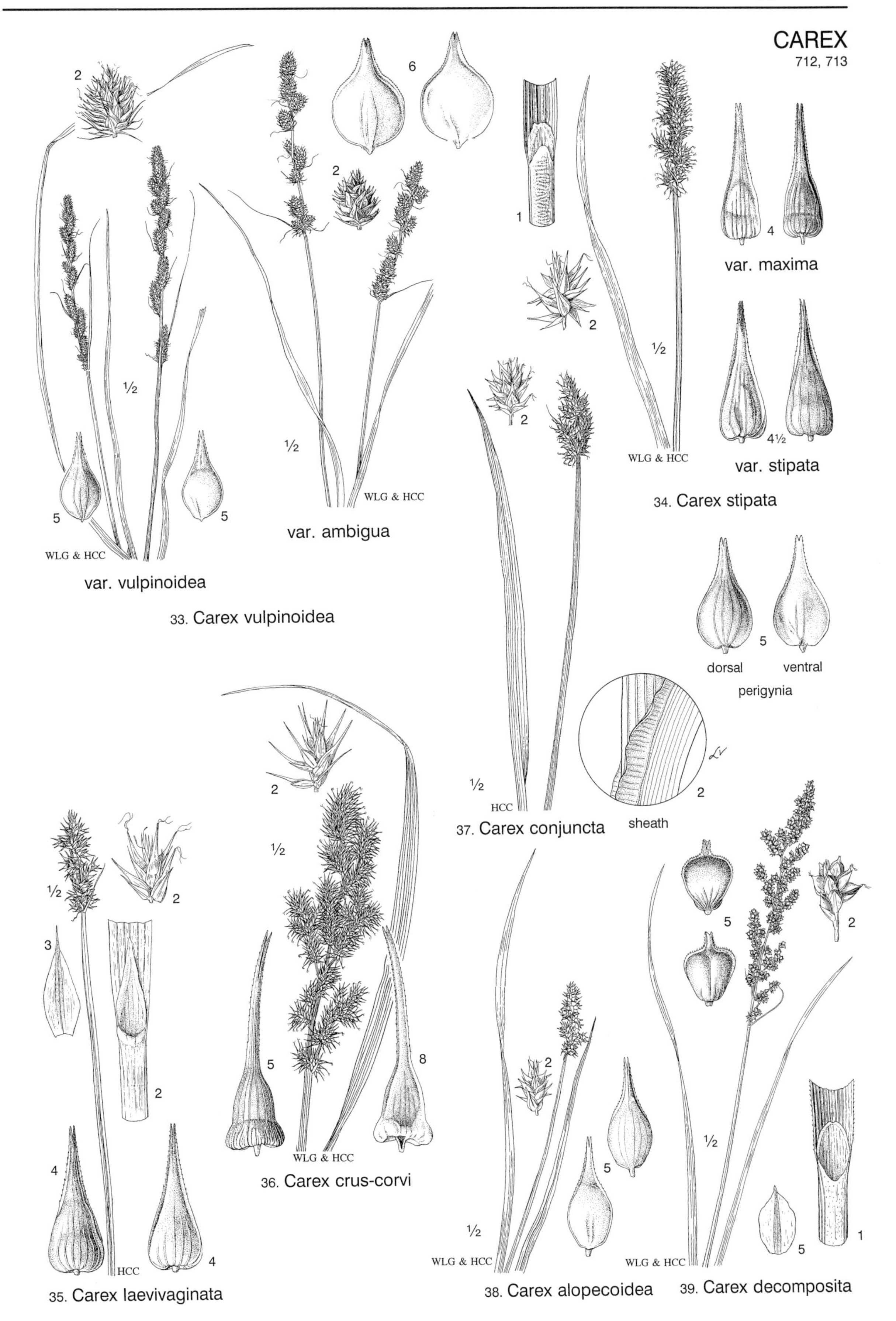

var. maxima

4

var. stipata

4½

34. Carex stipata

var. vulpinoidea

5

WLG & HCC

33. Carex vulpinoidea

var. ambigua

WLG & HCC

6

2

dorsal   ventral
perigynia

5

37. Carex conjuncta

sheath

2

HCC

36. Carex crus-corvi

WLG & HCC

35. Carex laevivaginata

HCC

38. Carex alopecoidea

WLG & HCC

39. Carex decomposita

WLG & HCC

CAREX
713, 714

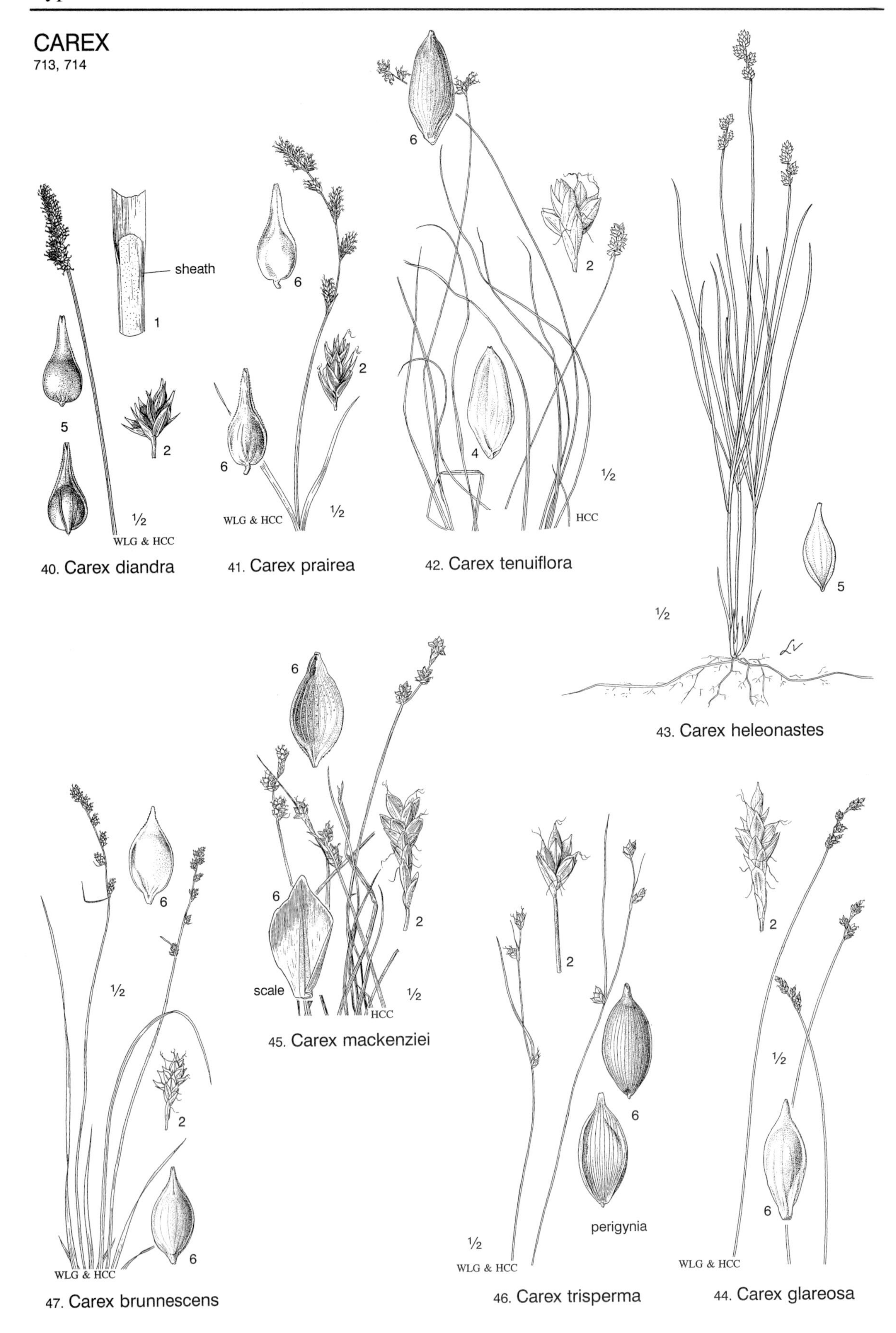

40. Carex diandra

sheath

1

41. Carex prairea

WLG & HCC

42. Carex tenuiflora

HCC

43. Carex heleonastes

45. Carex mackenziei

scale

HCC

47. Carex brunnescens

WLG & HCC

46. Carex trisperma

perigynia

WLG & HCC

44. Carex glareosa

WLG & HCC

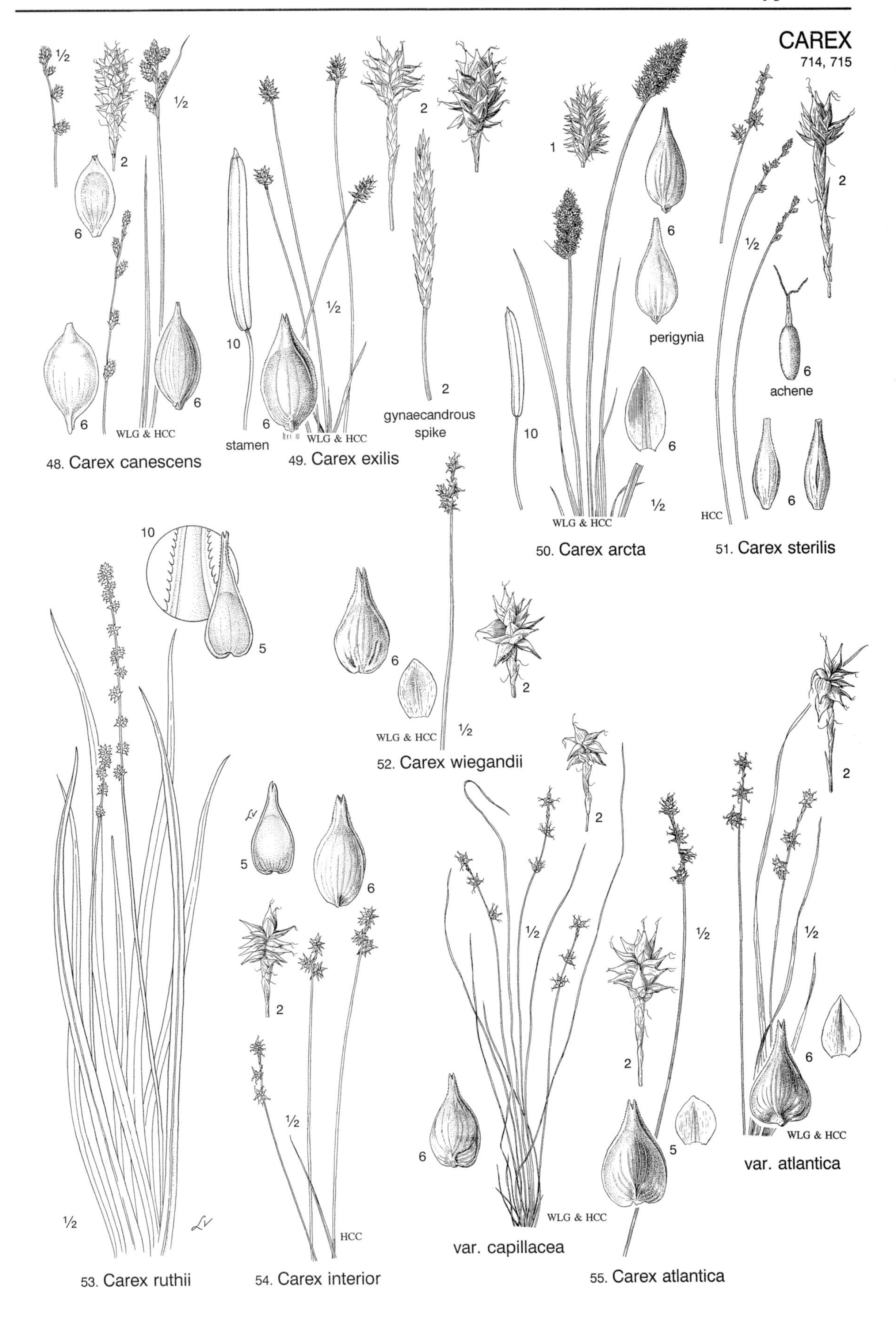

48. Carex canescens

stamen

49. Carex exilis

gynaecandrous spike

WLG & HCC

perigynia

50. Carex arcta

WLG & HCC

51. Carex sterilis

achene

HCC

52. Carex wiegandii

WLG & HCC

53. Carex ruthii

54. Carex interior

HCC

var. capillacea

WLG & HCC

55. Carex atlantica

var. atlantica

WLG & HCC

CAREX
715, 716

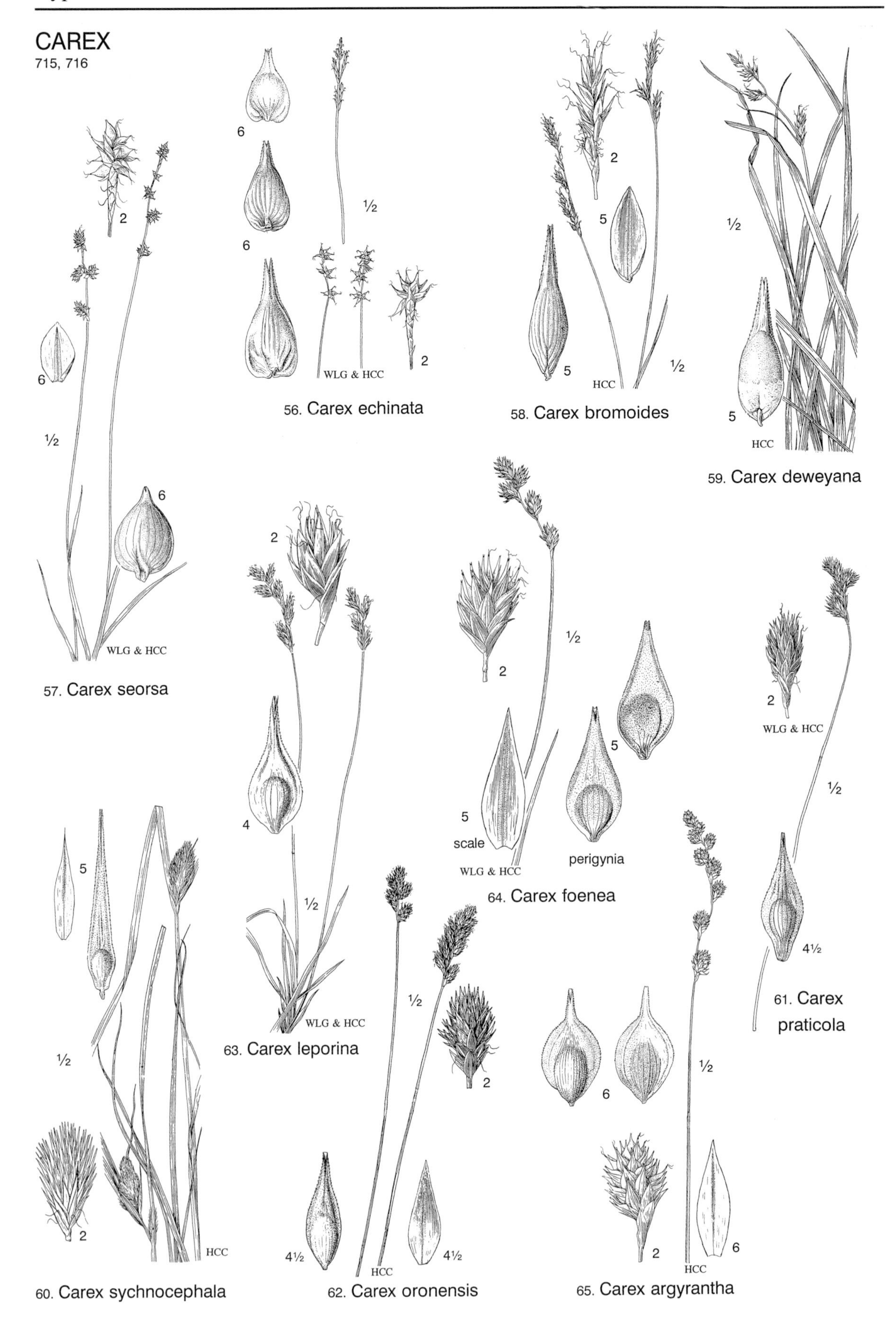

56. Carex echinata

WLG & HCC

57. Carex seorsa

WLG & HCC

58. Carex bromoides

HCC

59. Carex deweyana

HCC

60. Carex sychnocephala

HCC

61. Carex praticola

62. Carex oronensis

HCC

63. Carex leporina

WLG & HCC

64. Carex foenea

scale

perigynia

WLG & HCC

65. Carex argyrantha

HCC

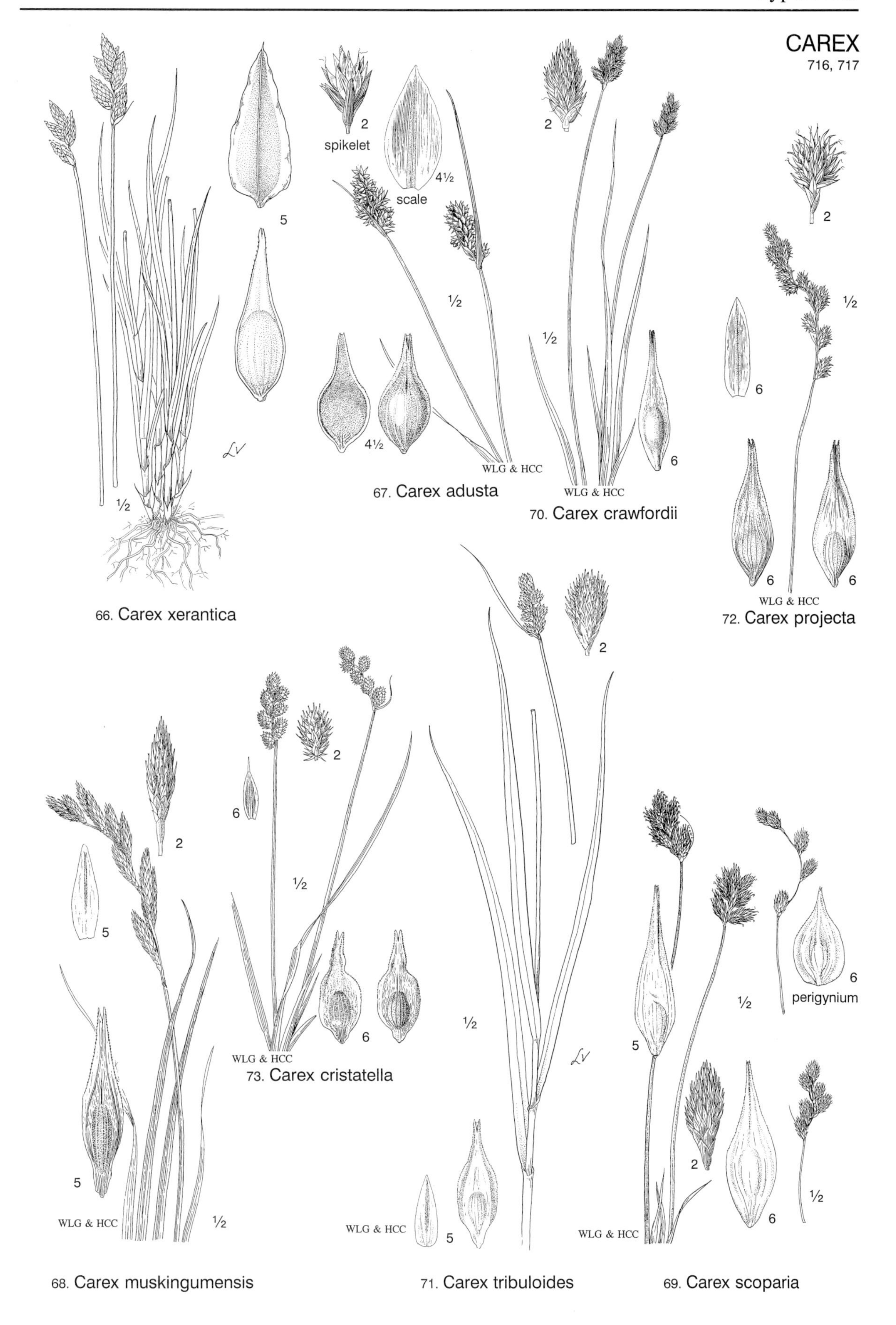

2
spikelet

4½
scale

5

67. Carex adusta

2

½

2

½

6

½

6

WLG & HCC

70. Carex crawfordii

WLG & HCC

2

½

6

6

6

WLG & HCC

72. Carex projecta

½

66. Carex xerantica

2

6

½

2

5

½

6

WLG & HCC

73. Carex cristatella

2

½

½

6

perigynium

5

5

5

2

WLG & HCC

5

½

WLG & HCC

WLG & HCC

6

½

68. Carex muskingumensis

71. Carex tribuloides

69. Carex scoparia

CAREX
717, 718

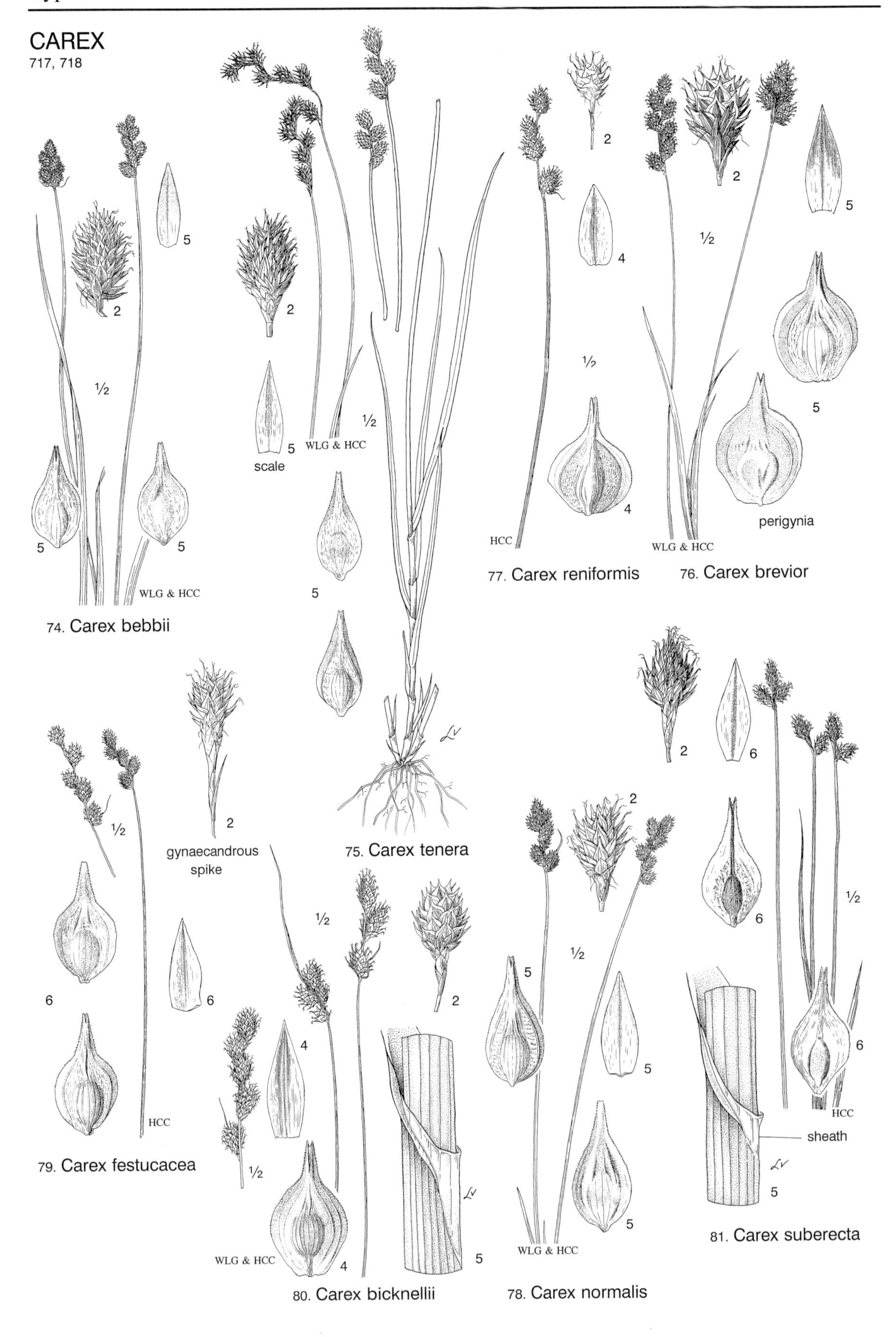

74. Carex bebbii

5 scale

WLG & HCC

75. Carex tenera

gynaecandrous spike

79. Carex festucacea

80. Carex bicknellii

78. Carex normalis

77. Carex reniformis

76. Carex brevior

perigynia

81. Carex suberecta

sheath

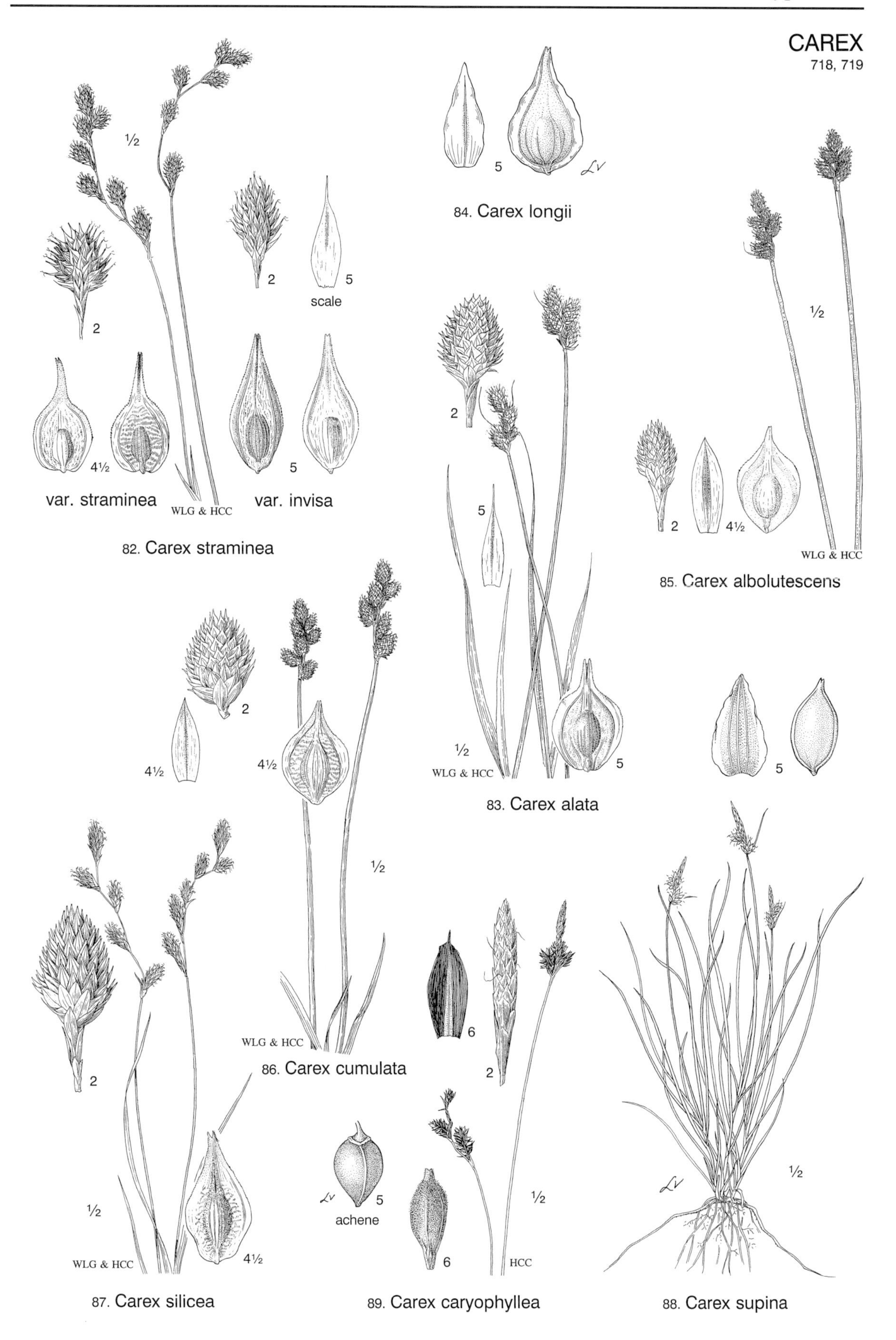

½

2

2

5
scale

2

4½

var. straminea
WLG & HCC

5

var. invisa

82. Carex straminea

5

84. Carex longii

2

5

½
WLG & HCC

5

83. Carex alata

½

2

4½

2

½
WLG & HCC

86. Carex cumulata

2

4½

½

5

½

2

4½

WLG & HCC

87. Carex silicea

½

2

4½

2

5

WLG & HCC

85. Carex albolutescens

5

6

2

5
achene

6

½

HCC

89. Carex caryophyllea

½

88. Carex supina

CAREX
719, 720

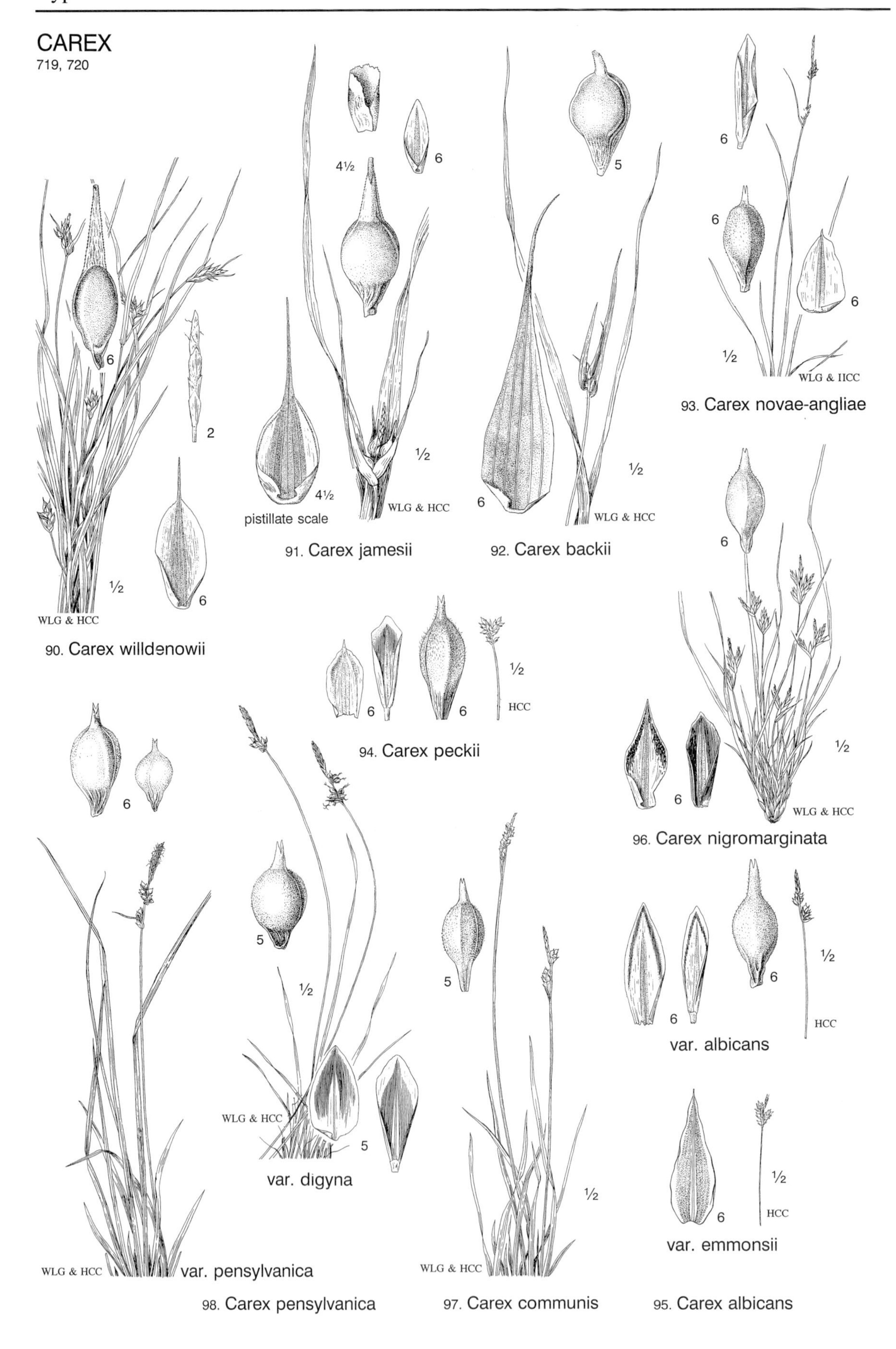

2

6

½

WLG & HCC

90. Carex willdenowii

4½

6

pistillate scale

4½

½

WLG & HCC

91. Carex jamesii

5

6

½

WLG & HCC

92. Carex backii

6

6

6

½

WLG & IICC

93. Carex novae-angliae

6

6

½

HCC

94. Carex peckii

6

6

½

WLG & HCC

96. Carex nigromarginata

6

6

6

½

HCC

var. albicans

6

½

HCC

var. emmonsii

95. Carex albicans

6

6

5

½

WLG & HCC

var. digyna

5

½

WLG & HCC

var. pensylvanica

98. Carex pensylvanica

5

5

½

WLG & HCC

97. Carex communis

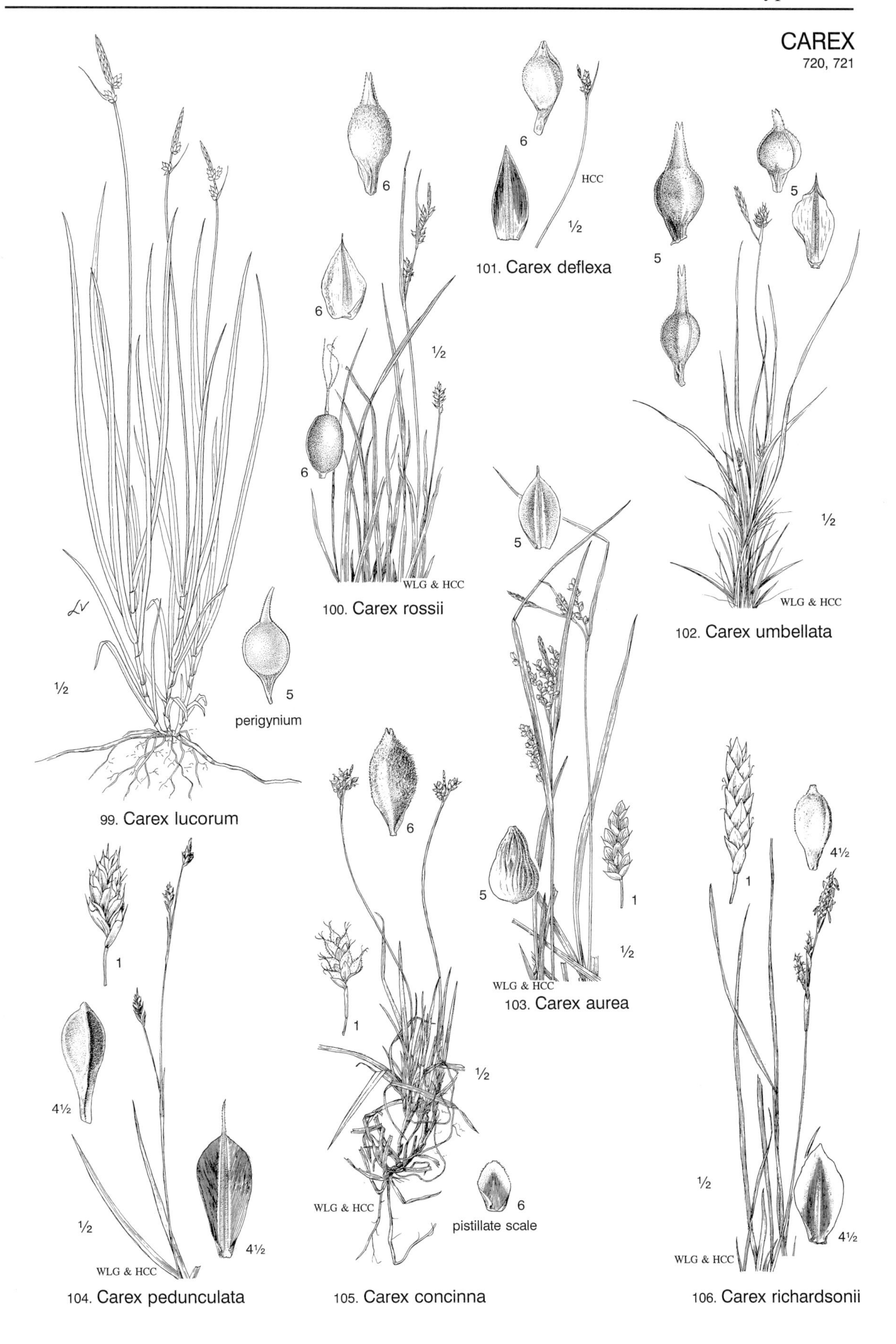

101. Carex deflexa

100. Carex rossii

102. Carex umbellata

5
perigynium

99. Carex lucorum

103. Carex aurea

104. Carex pedunculata

105. Carex concinna

6
pistillate scale

106. Carex richardsonii

CAREX
721, 722

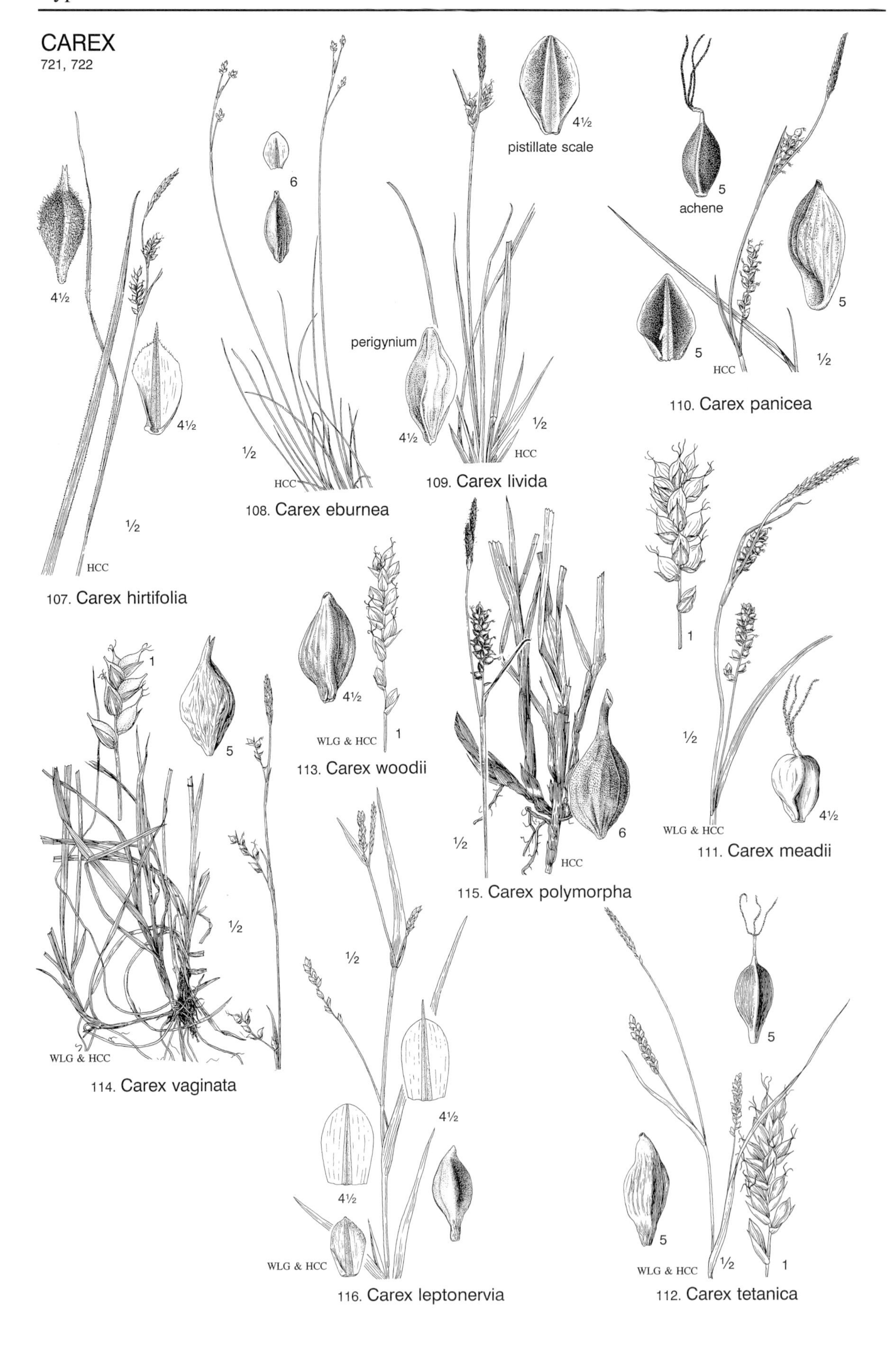

pistillate scale 4½

perigynium

6

4½

½
HCC

108. Carex eburnea

4½
4½

½
HCC

109. Carex livida

5
achene

5

5

½
HCC

110. Carex panicea

4½

½
HCC

107. Carex hirtifolia

1

5

4½

WLG & HCC   1

113. Carex woodii

1

½

4½

WLG & HCC

111. Carex meadii

1

½

1

½

6

½
HCC

115. Carex polymorpha

5

114. Carex vaginata

WLG & HCC

½

½

4½

4½

WLG & HCC

116. Carex leptonervia

5

1

5

½

WLG & HCC

112. Carex tetanica

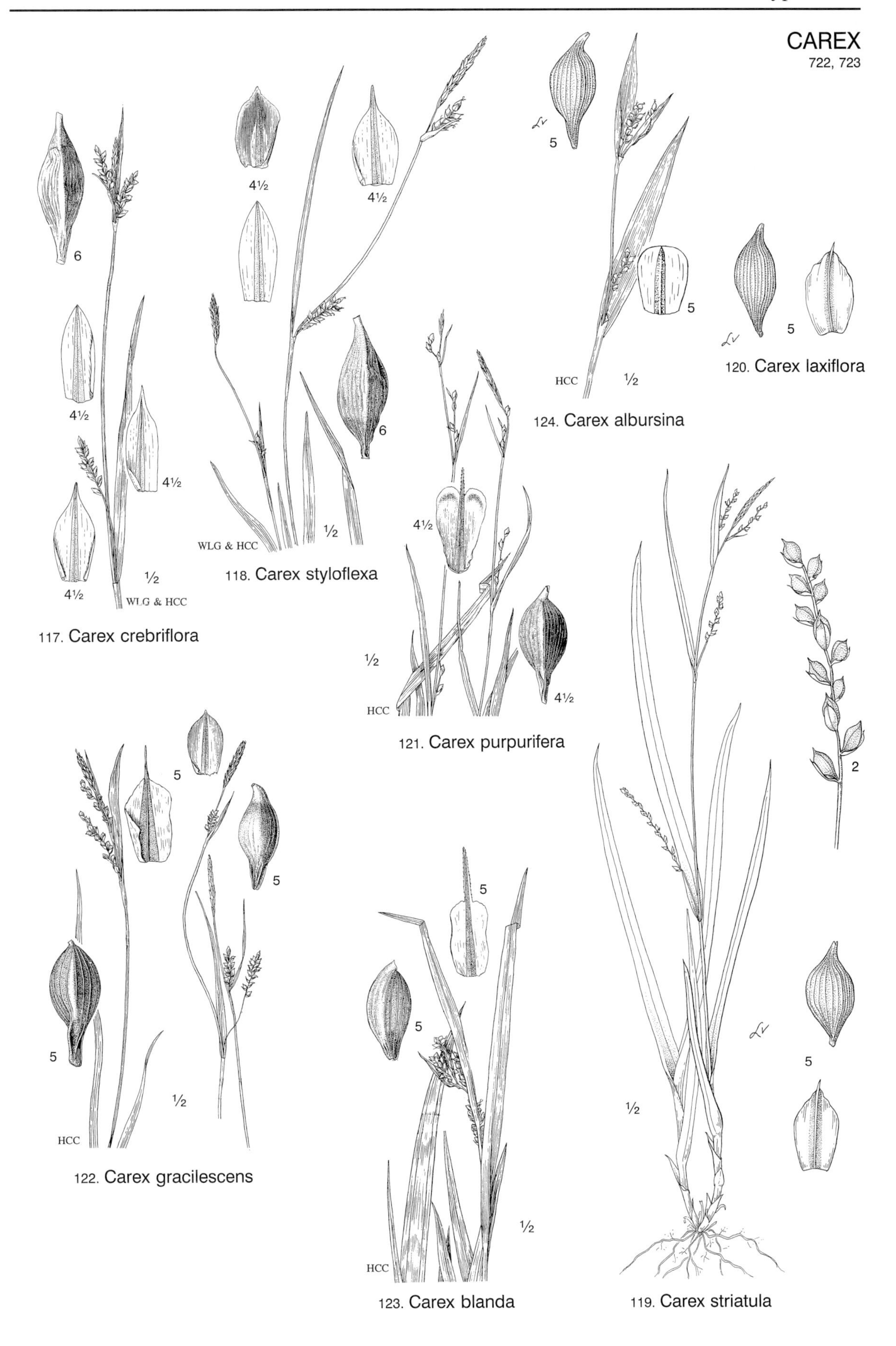

117. Carex crebriflora

118. Carex styloflexa

WLG & HCC

WLG & HCC

120. Carex laxiflora

124. Carex albursina

HCC

121. Carex purpurifera

HCC

122. Carex gracilescens

HCC

123. Carex blanda

HCC

119. Carex striatula

CAREX
723, 724

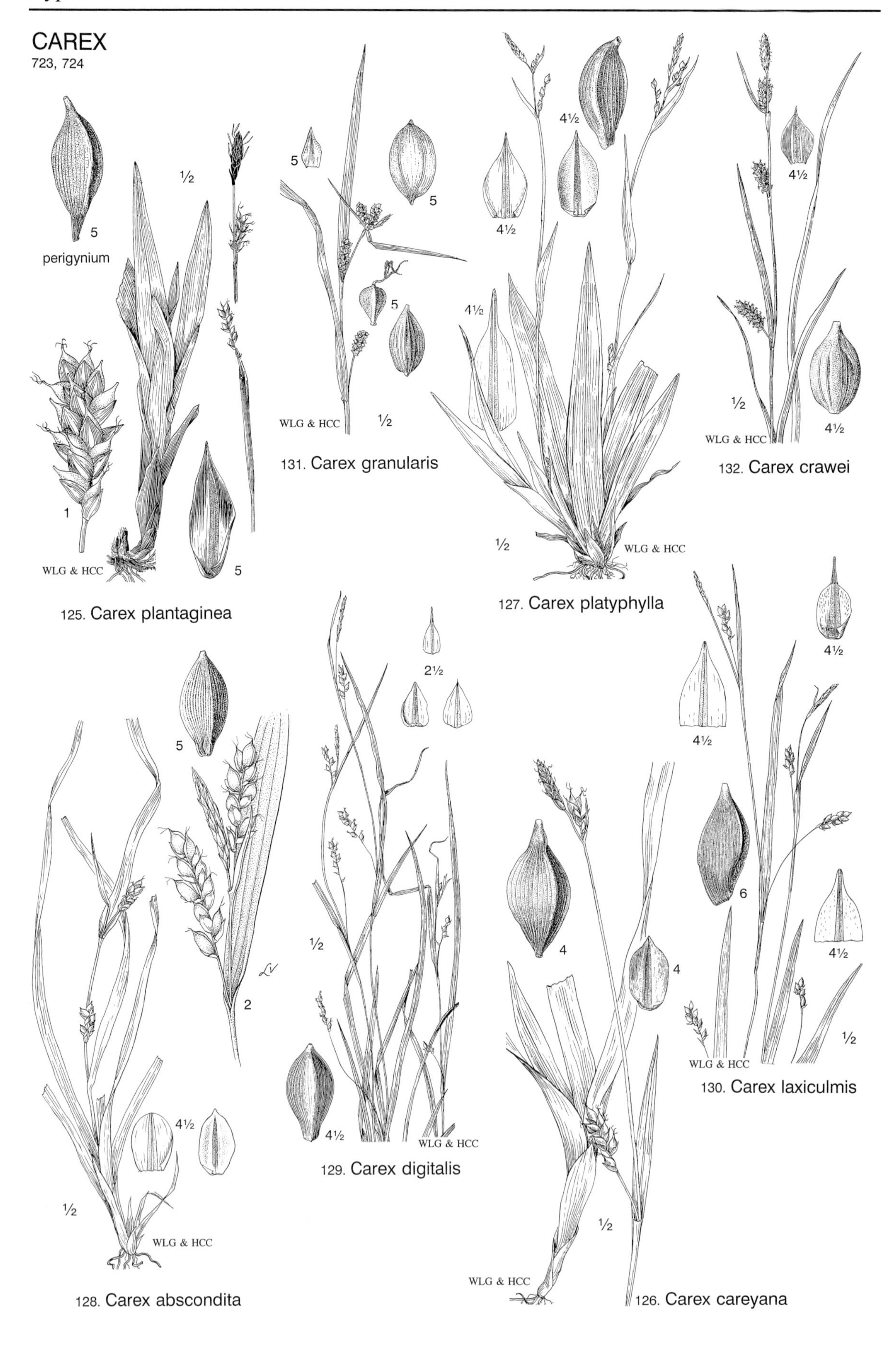

5
perigynium

1
WLG & HCC

125. **Carex plantaginea**

131. **Carex granularis**

WLG & HCC ½

127. **Carex platyphylla**

WLG & HCC ½

132. **Carex crawei**

WLG & HCC

128. **Carex abscondita**

129. **Carex digitalis**

WLG & HCC

130. **Carex laxiculmis**

WLG & HCC

126. **Carex careyana**

WLG & HCC ½

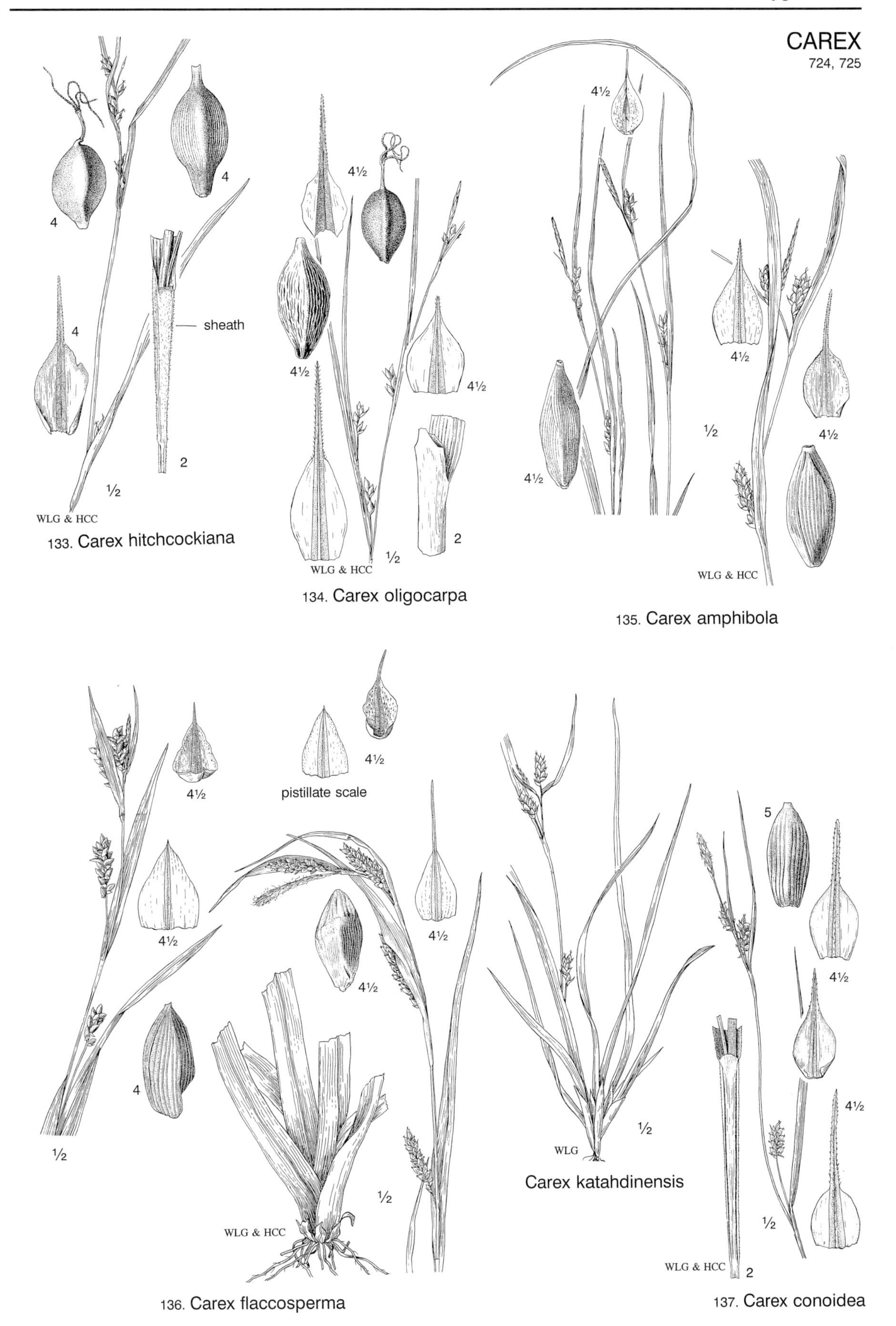

WLG & HCC
133. Carex hitchcockiana

WLG & HCC
134. Carex oligocarpa

WLG & HCC
135. Carex amphibola

pistillate scale

WLG & HCC
136. Carex flaccosperma

WLG
Carex katahdinensis

WLG & HCC
137. Carex conoidea

CAREX
725, 726

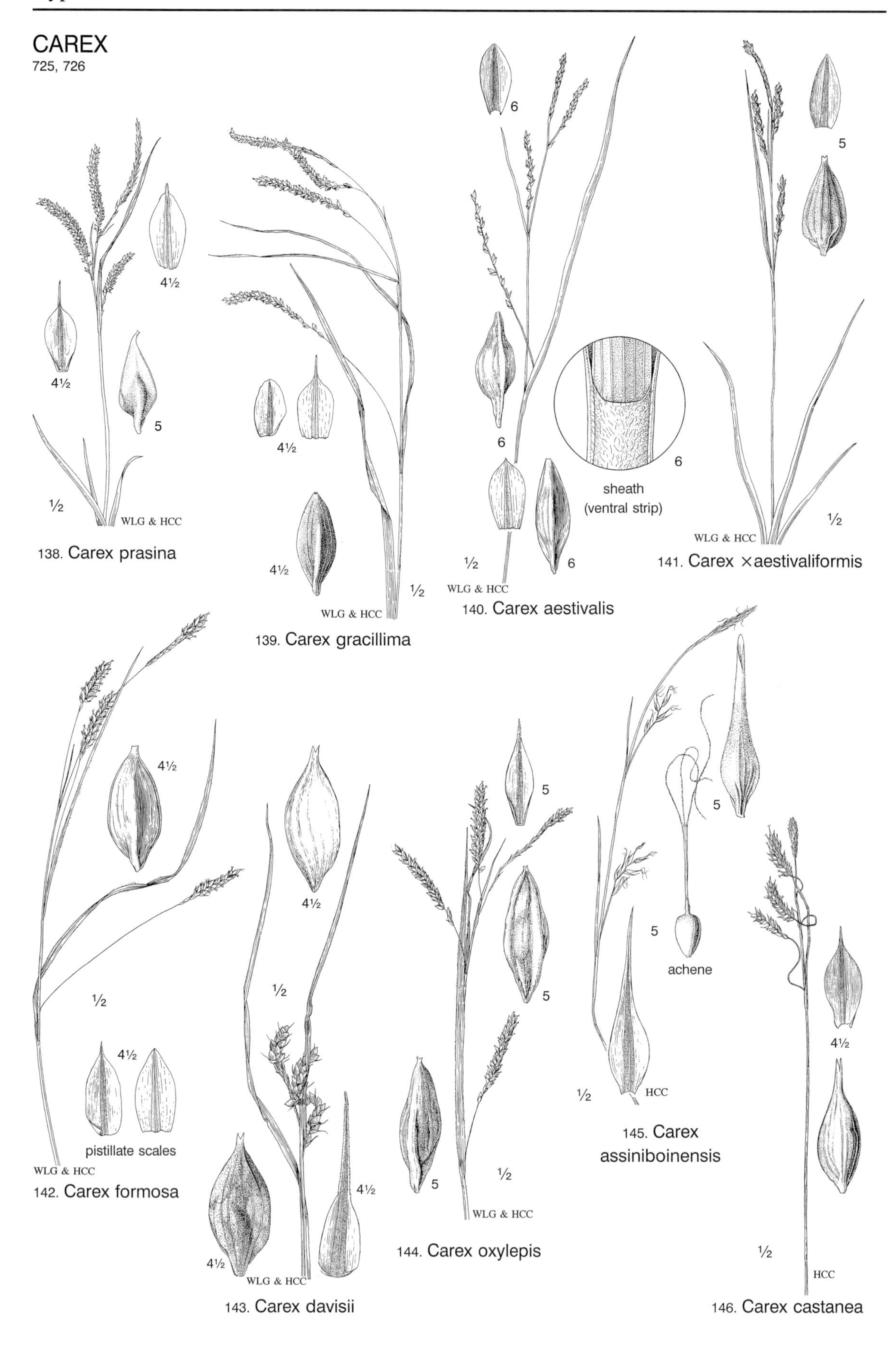

138. **Carex prasina**

139. **Carex gracillima**

140. **Carex aestivalis**

sheath
(ventral strip)

141. **Carex ×aestivaliformis**

142. **Carex formosa**

pistillate scales

143. **Carex davisii**

144. **Carex oxylepis**

achene

145. **Carex assiniboinensis**

146. **Carex castanea**

WLG & HCC

HCC

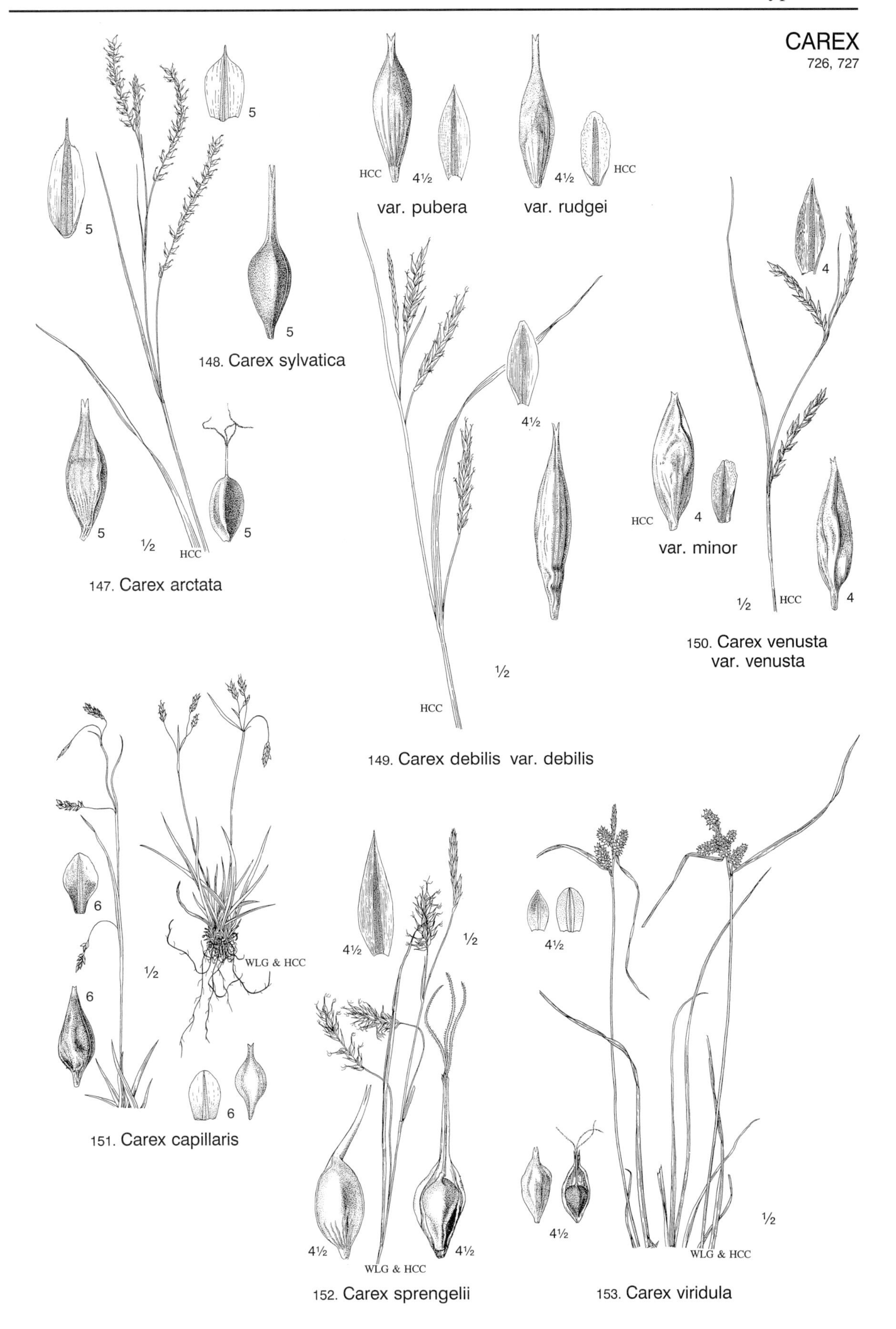

var. pubera

var. rudgei

148. Carex sylvatica

147. Carex arctata

149. Carex debilis var. debilis

var. minor

150. Carex venusta
var. venusta

151. Carex capillaris

152. Carex sprengelii

153. Carex viridula

CAREX
727, 728

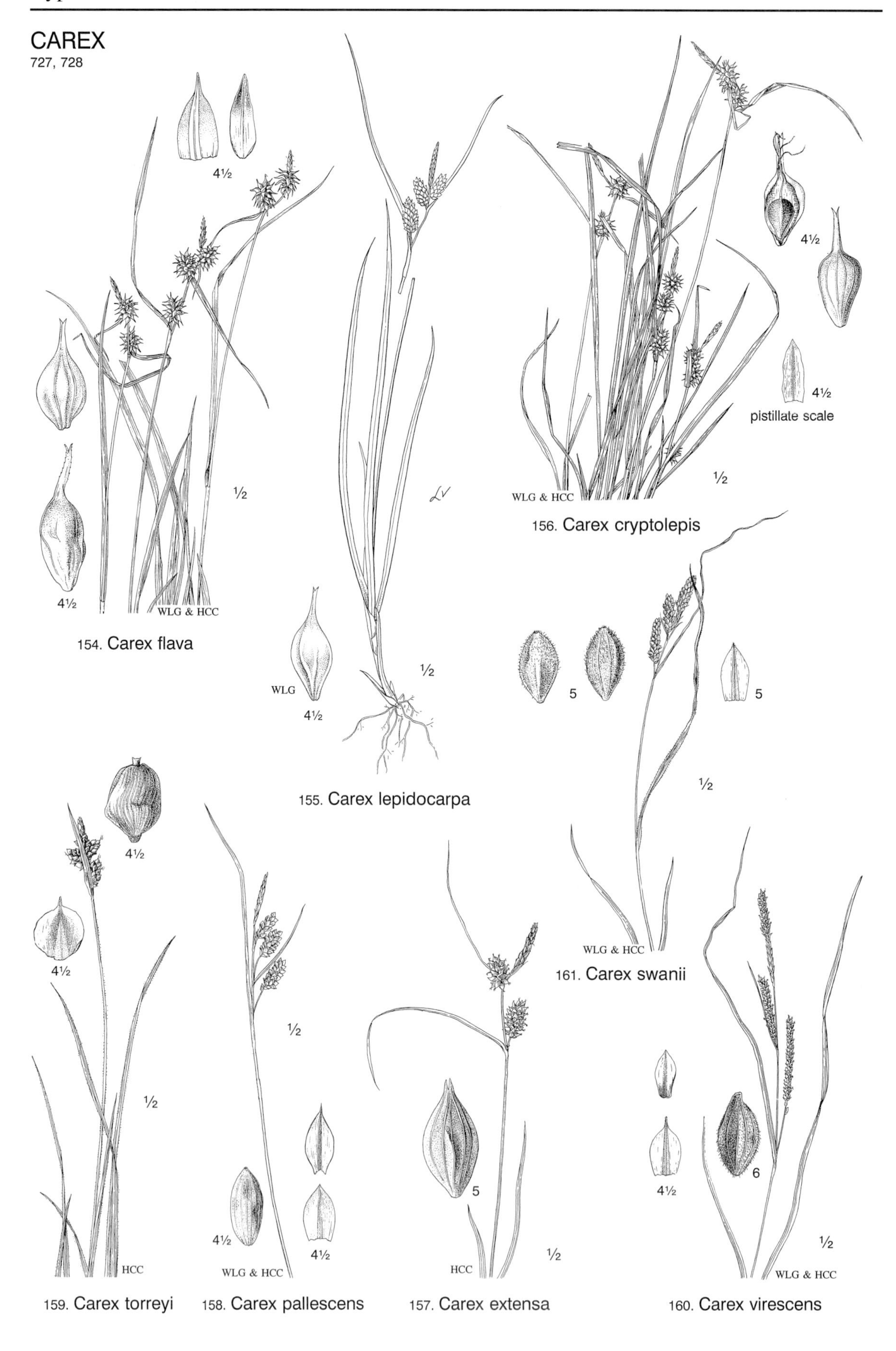

154. Carex flava

155. Carex lepidocarpa

156. Carex cryptolepis

pistillate scale

159. Carex torreyi

158. Carex pallescens

157. Carex extensa

160. Carex virescens

161. Carex swanii

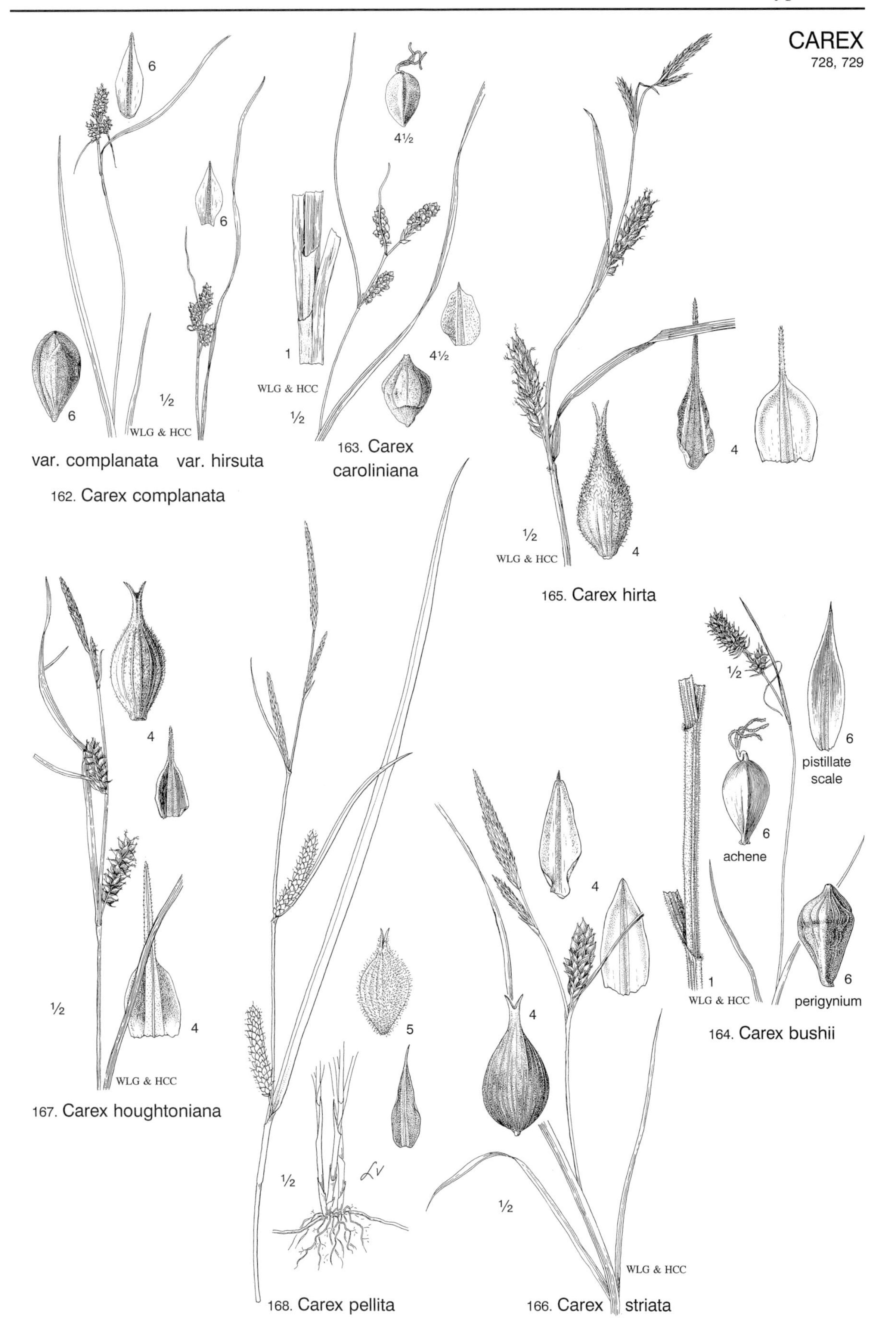

var. complanata   var. hirsuta

162. Carex complanata

163. Carex caroliniana

165. Carex hirta

167. Carex houghtoniana

168. Carex pellita

166. Carex striata

pistillate scale

achene

perigynium

164. Carex bushii

CAREX
729

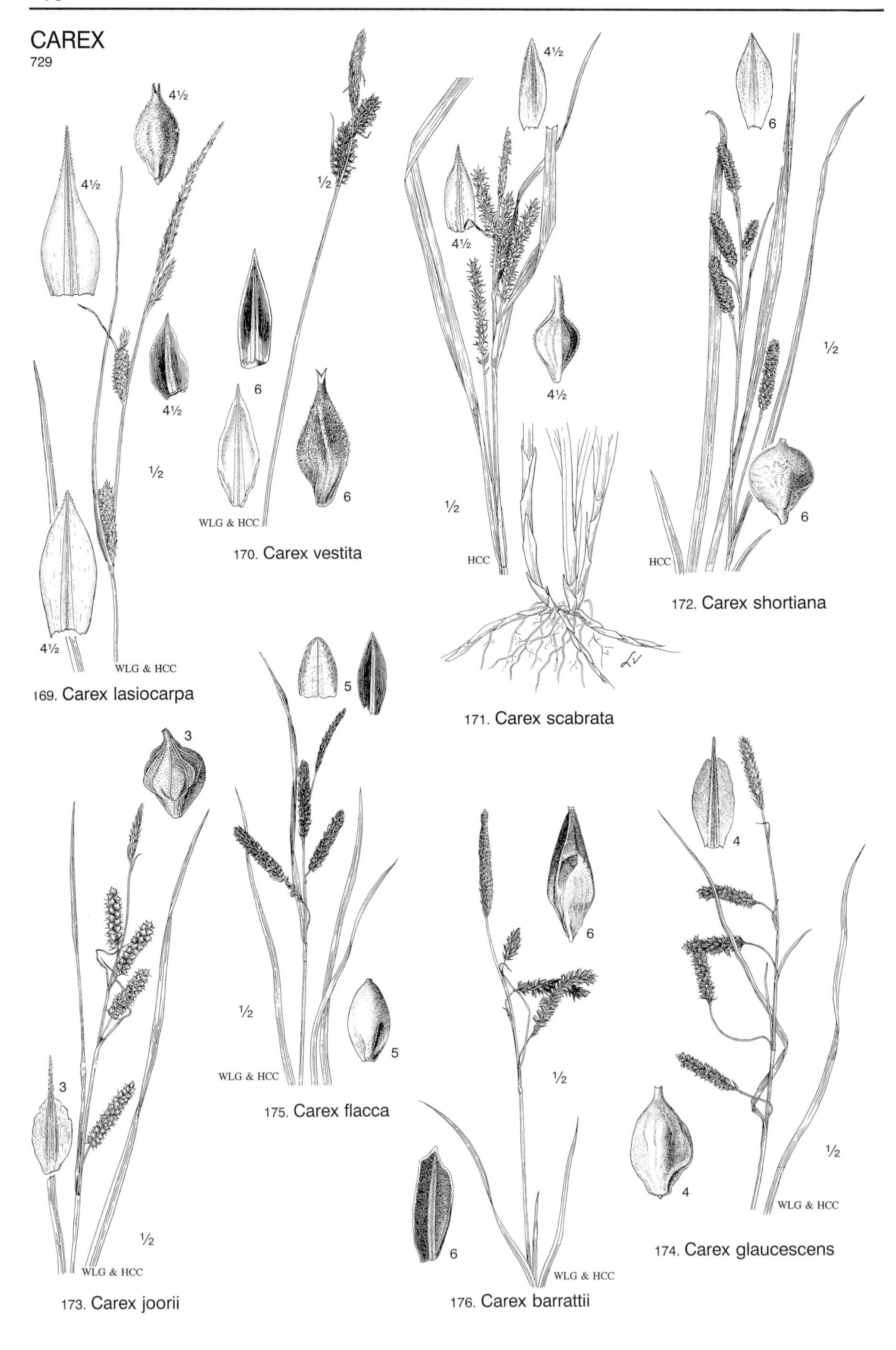

170. Carex vestita

172. Carex shortiana

169. Carex lasiocarpa

171. Carex scabrata

175. Carex flacca

173. Carex joorii

176. Carex barrattii

174. Carex glaucescens

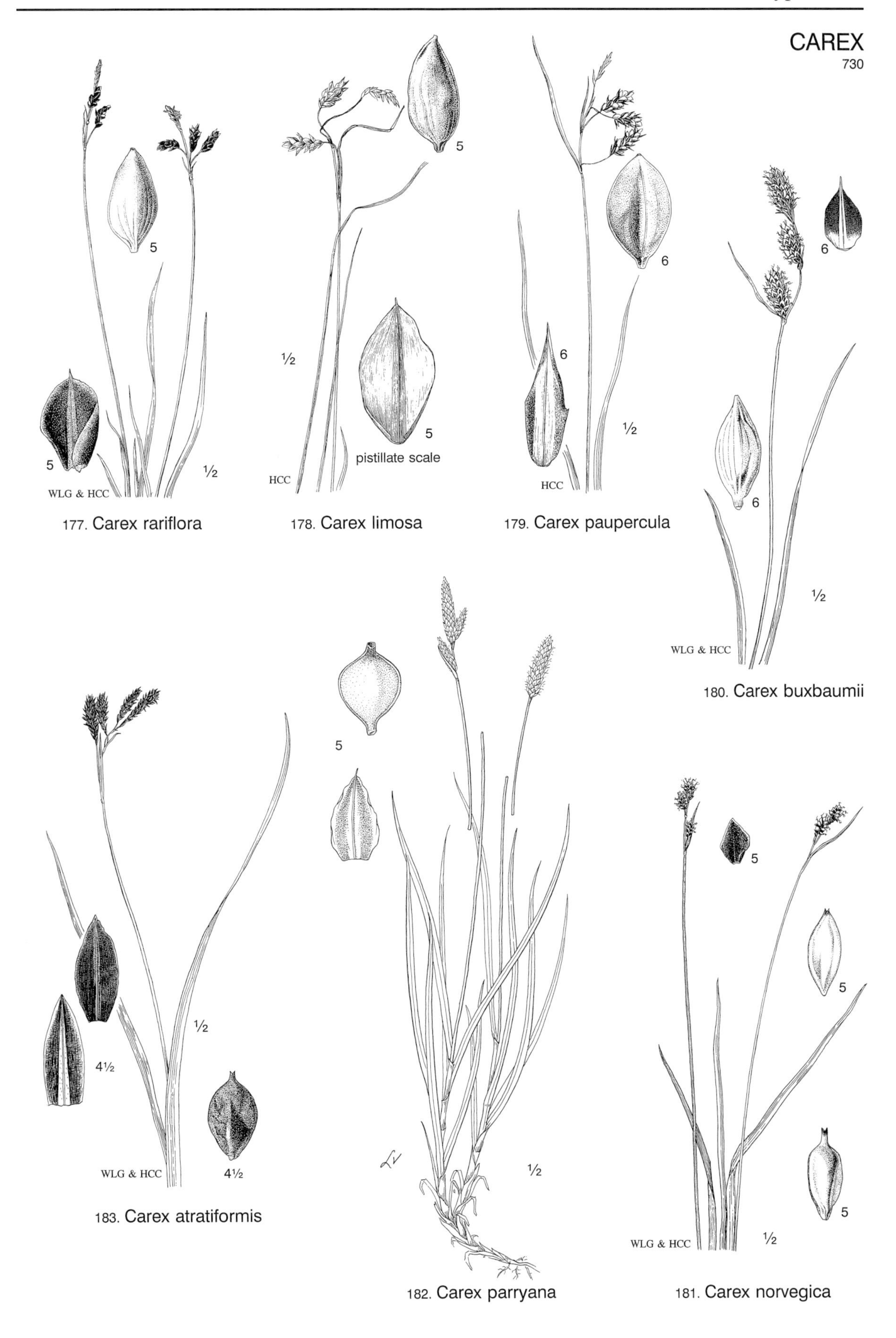

177. Carex rariflora

178. Carex limosa

pistillate scale

179. Carex paupercula

180. Carex buxbaumii

183. Carex atratiformis

182. Carex parryana

181. Carex norvegica

CAREX
731, 732

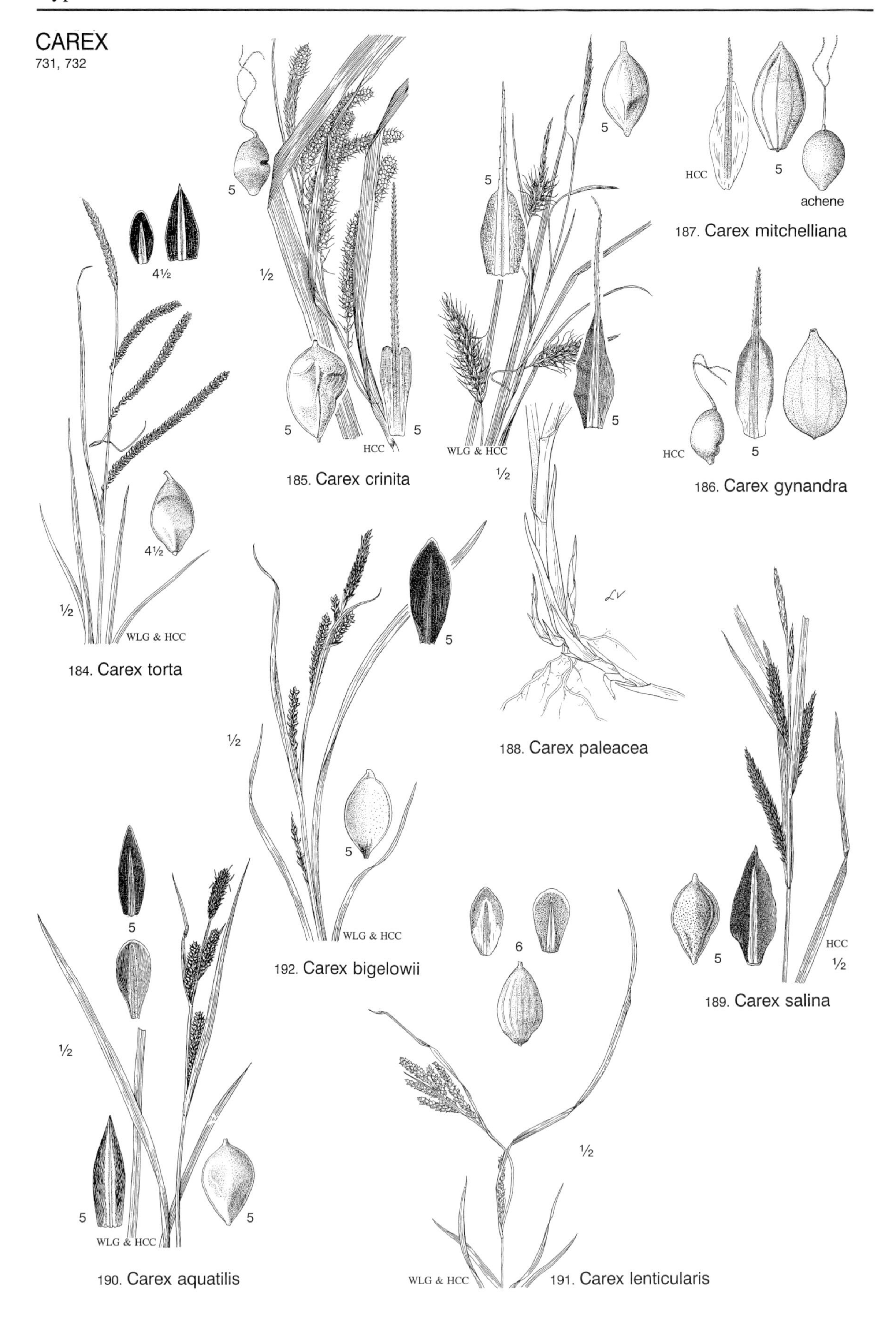

187. Carex mitchelliana

HCC   5   achene

185. Carex crinita

HCC   5   5   WLG & HCC   5   ½

186. Carex gynandra

HCC   5

184. Carex torta

4½   ½   WLG & HCC

188. Carex paleacea

192. Carex bigelowii

5   ½   WLG & HCC

189. Carex salina

5   HCC   ½

190. Carex aquatilis

5   ½   5   WLG & HCC

191. Carex lenticularis

6   ½   WLG & HCC

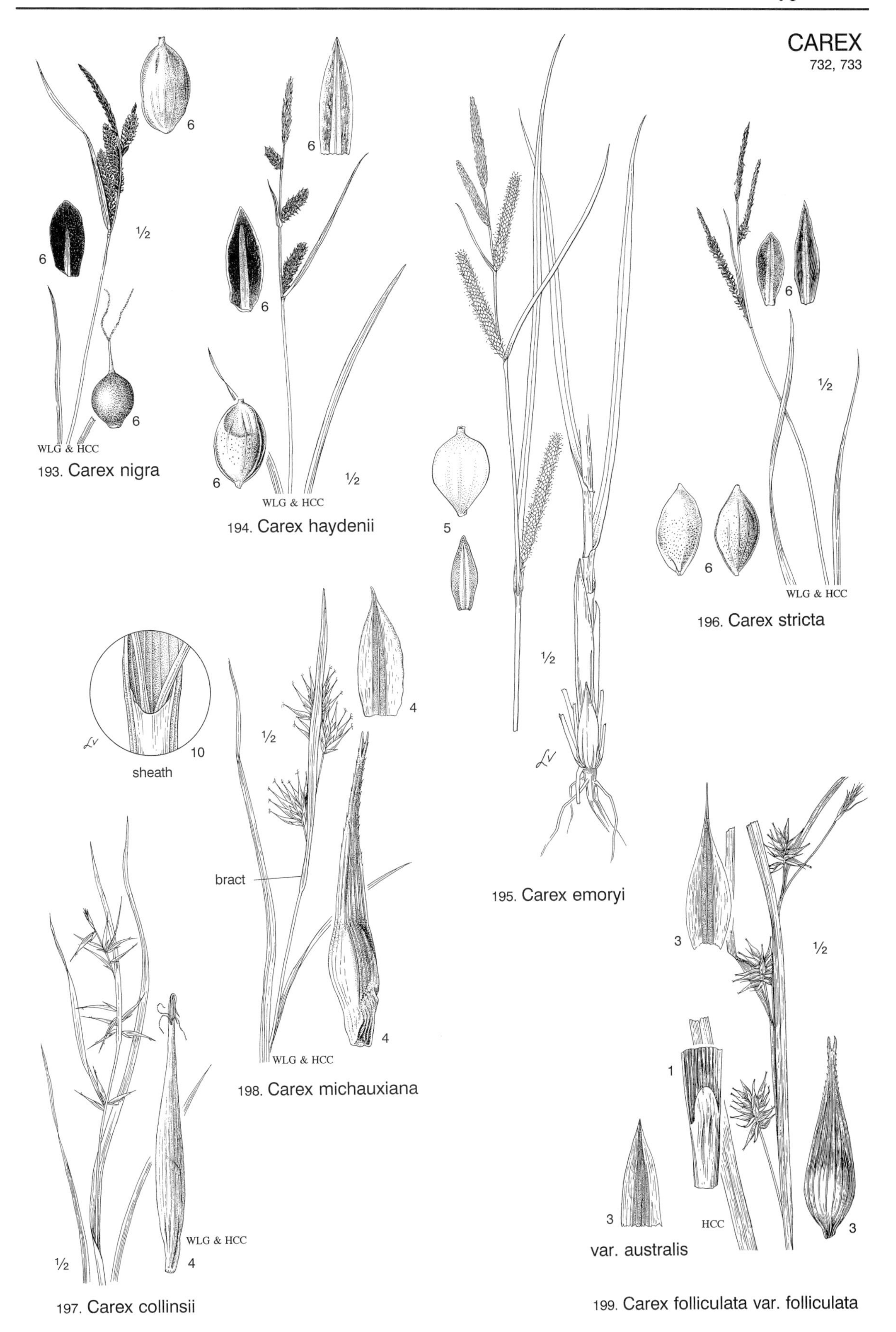

193. Carex nigra

WLG & HCC

194. Carex haydenii

WLG & HCC

195. Carex emoryi

196. Carex stricta

WLG & HCC

sheath

bract

197. Carex collinsii

WLG & HCC

198. Carex michauxiana

WLG & HCC

var. australis

HCC

199. Carex folliculata var. folliculata

CAREX
733, 734

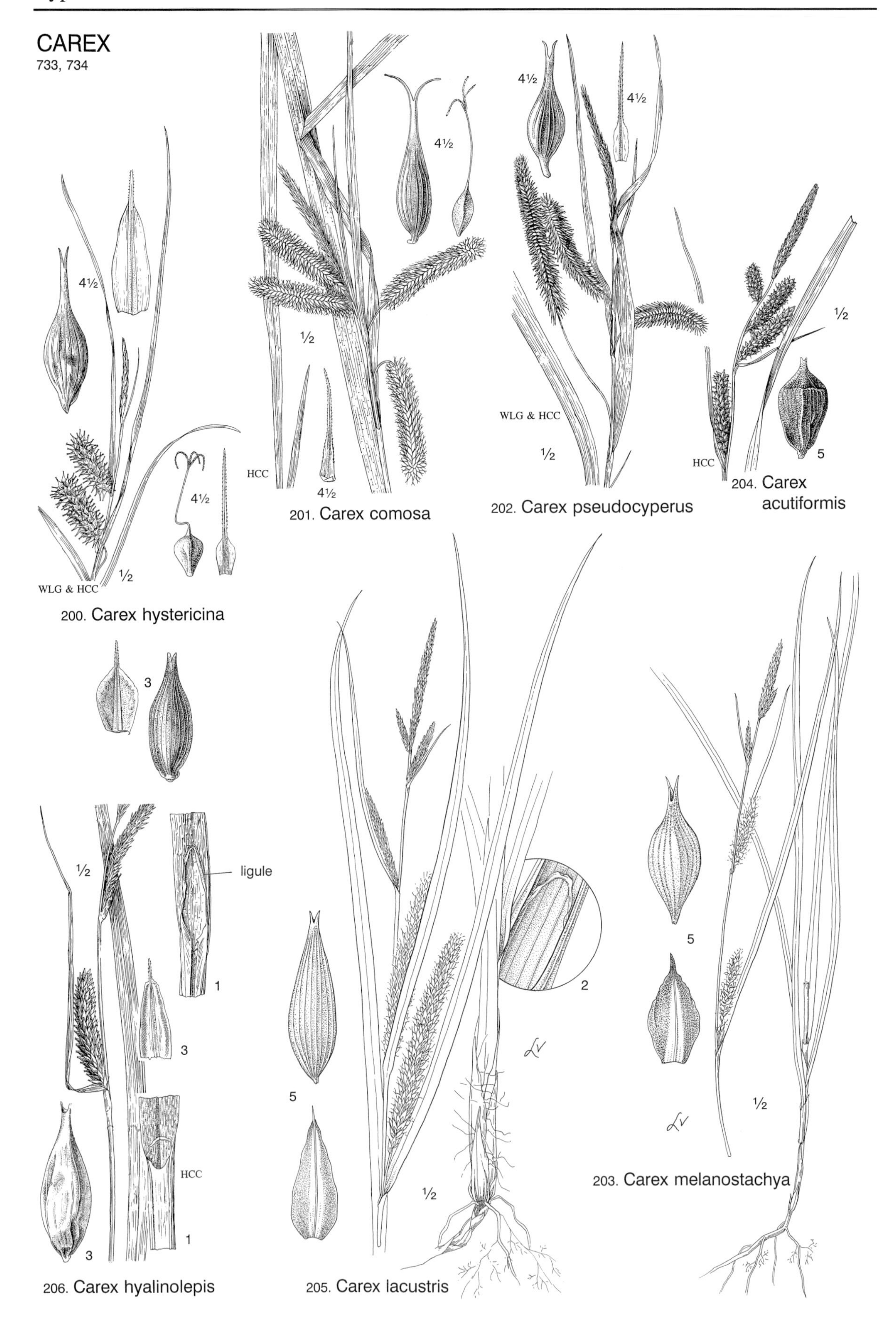

200. Carex hystericina

WLG & HCC

4½

4½

201. Carex comosa

HCC

4½

202. Carex pseudocyperus

WLG & HCC

HCC

4½

4½

204. Carex acutiformis

½

5

HCC

3

3

206. Carex hyalinolepis

½

1

ligule

3

3

HCC

1

205. Carex lacustris

5

5

½

2

5

5

½

203. Carex melanostachya

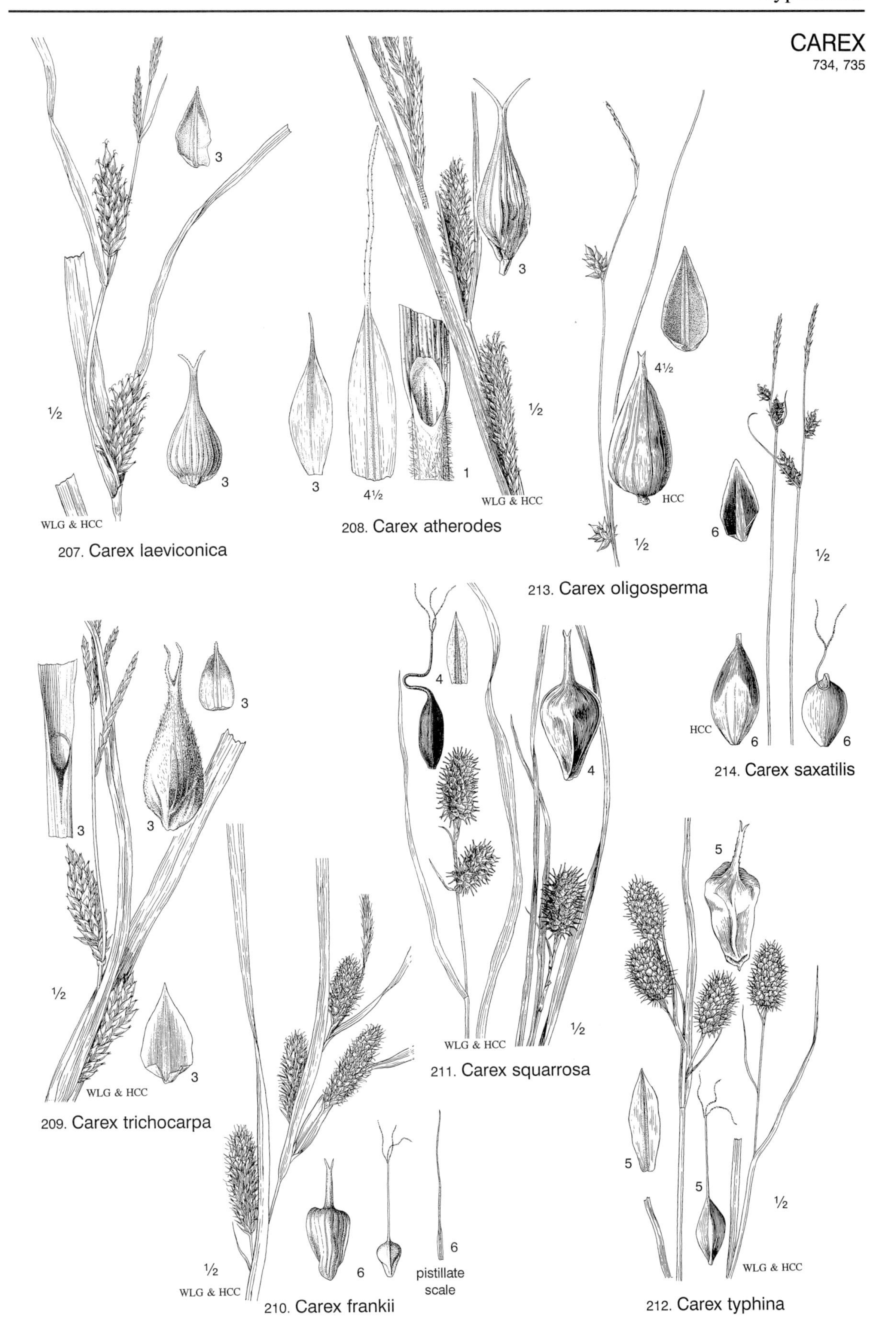

207. Carex laeviconica

208. Carex atherodes

213. Carex oligosperma

214. Carex saxatilis

209. Carex trichocarpa

211. Carex squarrosa

210. Carex frankii

pistillate scale

212. Carex typhina

CAREX
735, 736

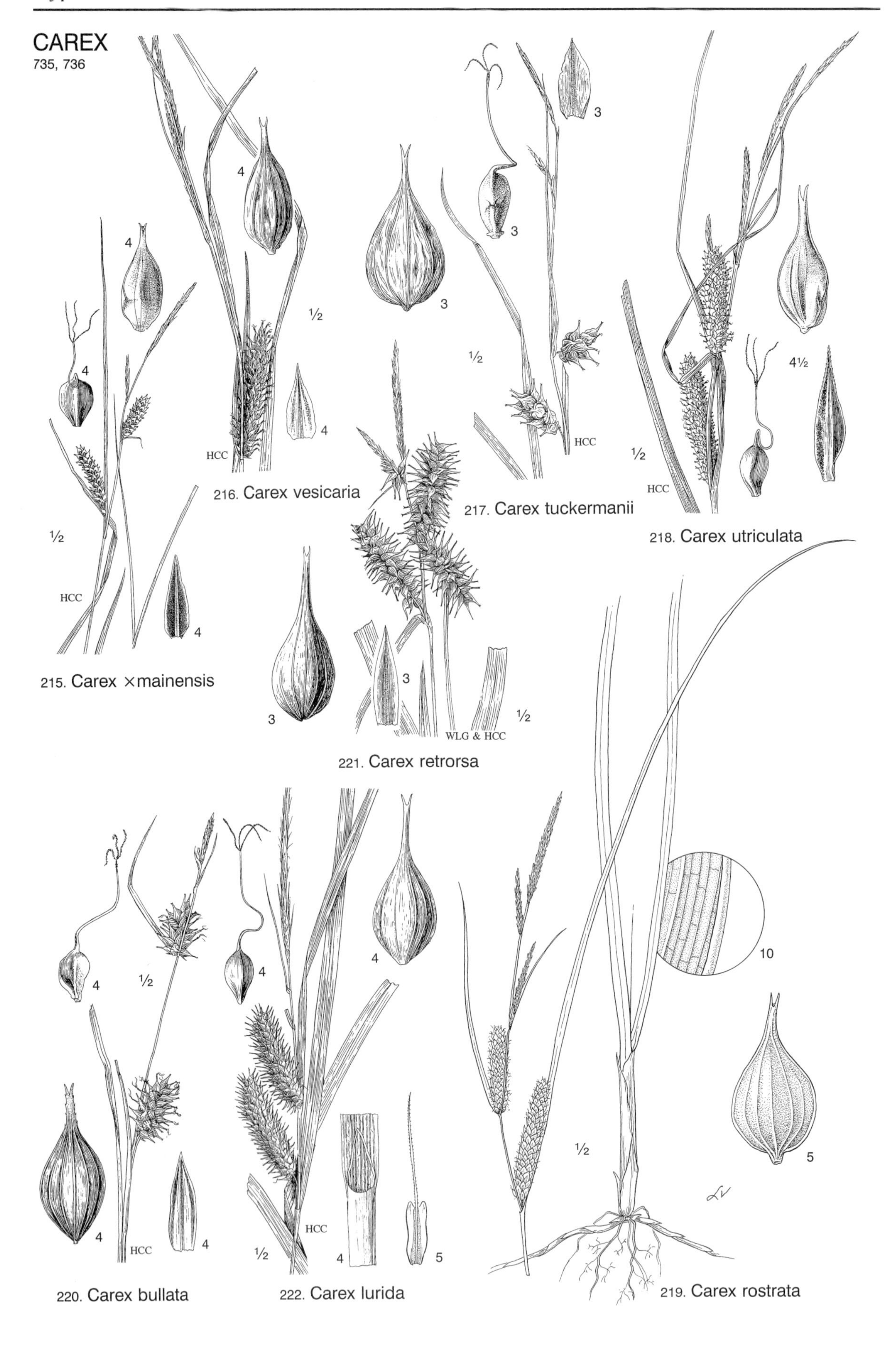

215. Carex ×mainensis

216. Carex vesicaria

217. Carex tuckermanii

218. Carex utriculata

221. Carex retrorsa

220. Carex bullata

222. Carex lurida

219. Carex rostrata

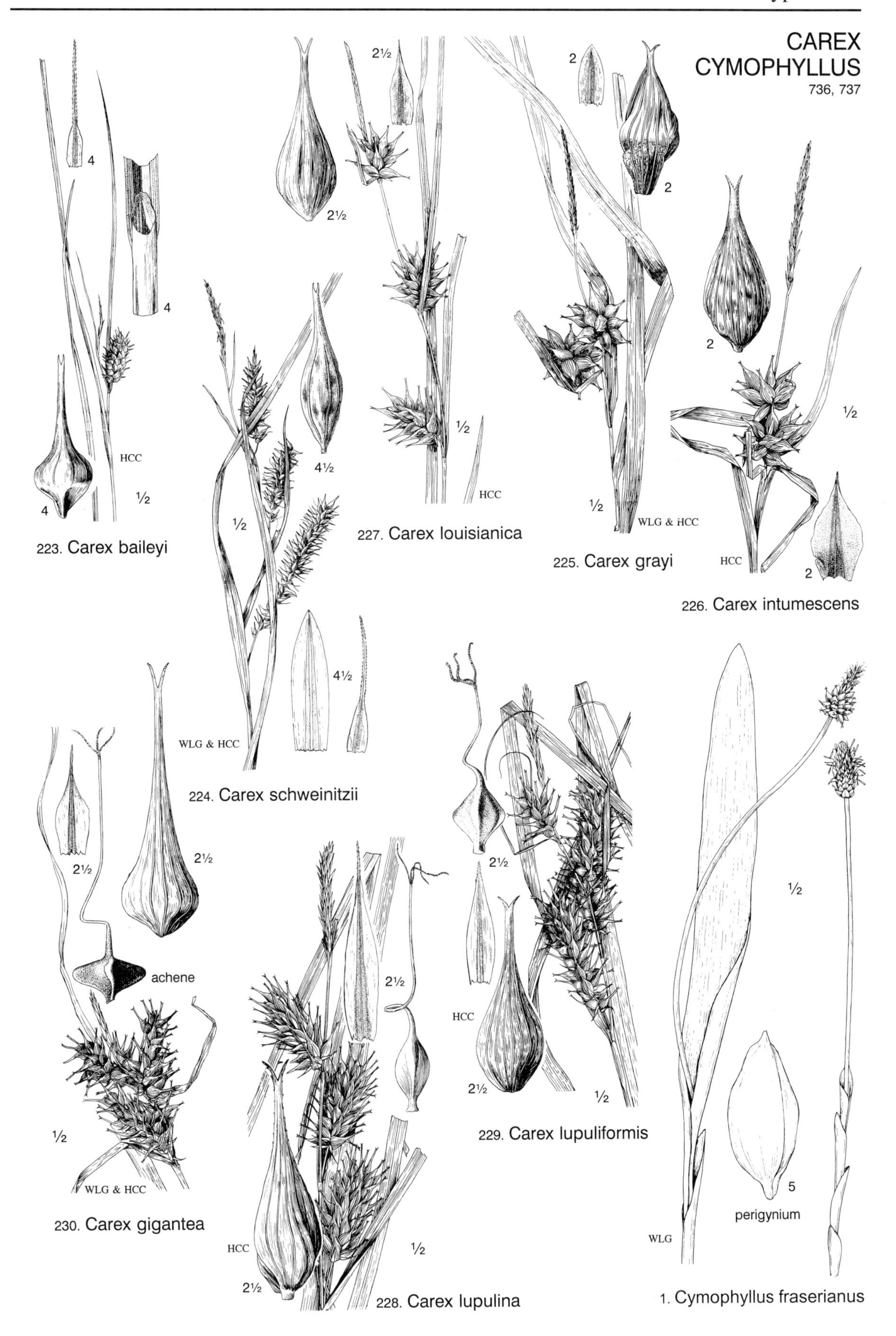

223. Carex baileyi

227. Carex louisianica

225. Carex grayi

226. Carex intumescens

224. Carex schweinitzii

230. Carex gigantea

228. Carex lupulina

229. Carex lupuliformis

1. Cymophyllus fraserianus

ARUNDINARIA
ORYZA
LEERSIA
743, 744

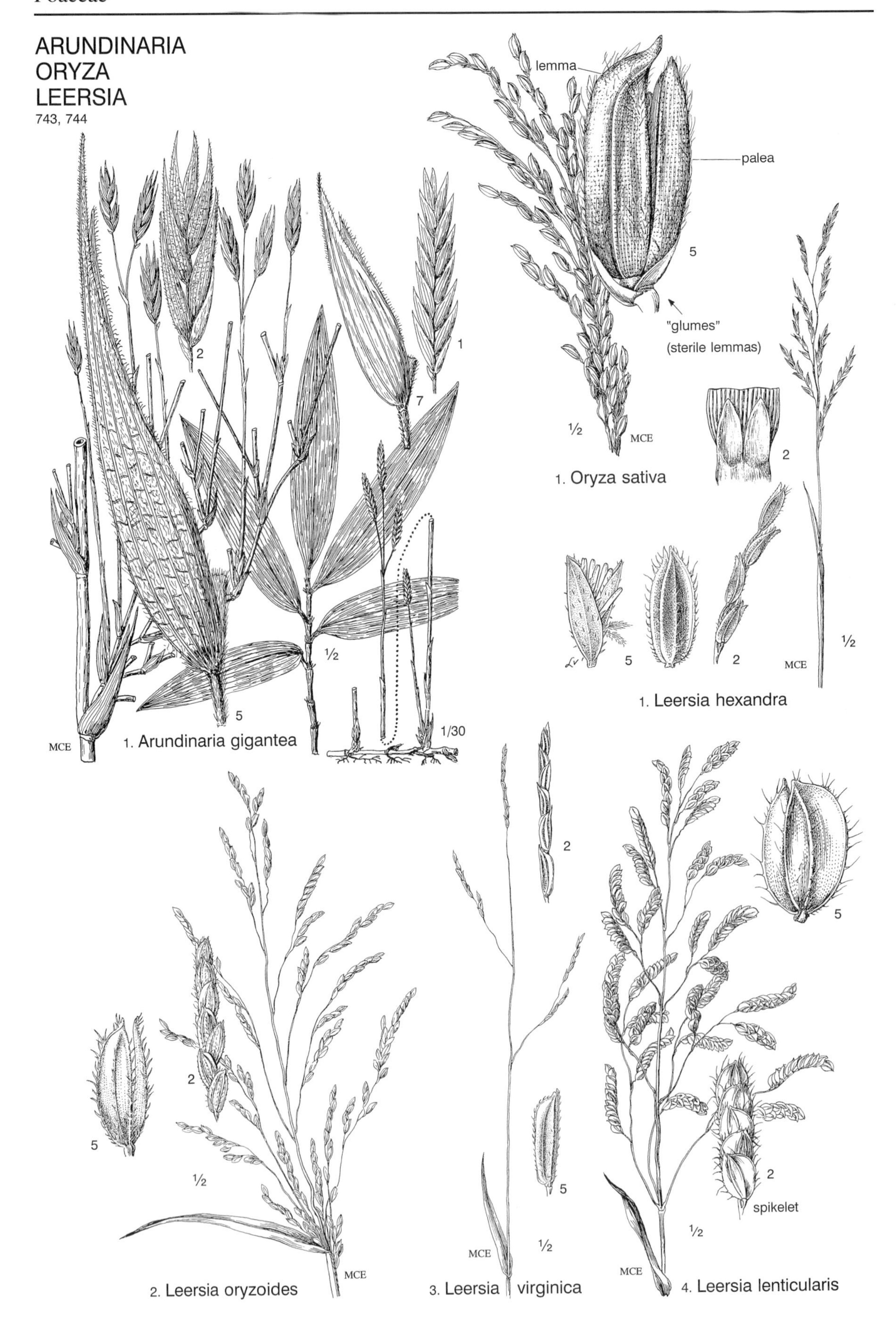

lemma

palea

5

"glumes"
(sterile lemmas)

½ MCE

2

1. Oryza sativa

5   2   MCE

½

1. Leersia hexandra

1

7

2

5

½

1/30

MCE   1. Arundinaria gigantea

2

5

½   2. Leersia oryzoides   MCE

2

5

MCE   ½   3. Leersia virginica

5

2

spikelet

½

MCE   4. Leersia lenticularis

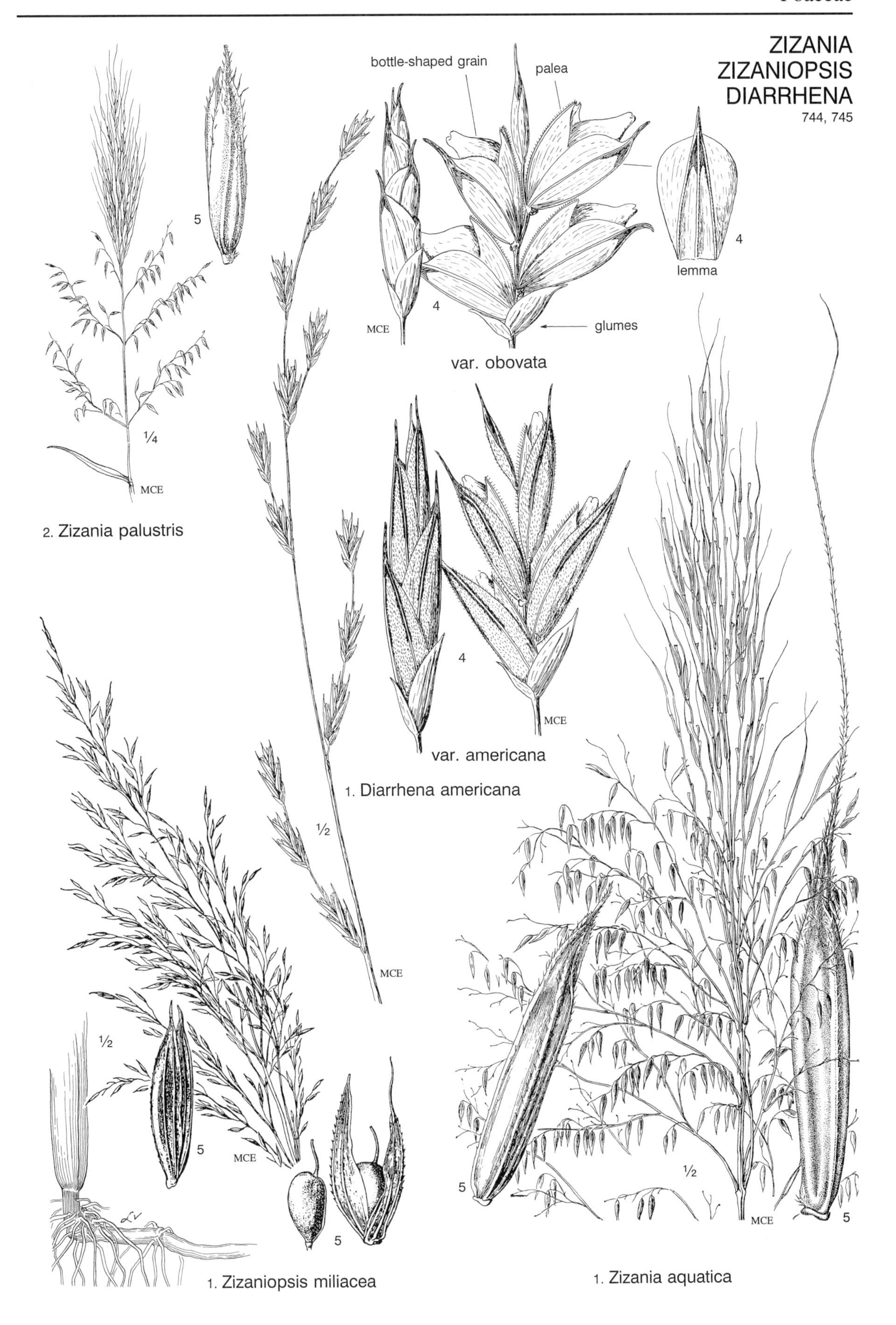

bottle-shaped grain

palea

lemma

glumes

MCE

var. obovata

4

var. americana

1. Diarrhena americana

2. Zizania palustris

1. Zizaniopsis miliacea

1. Zizania aquatica

BRACHYELYTRUM
NARDUS
ORYZOPSIS
745, 746

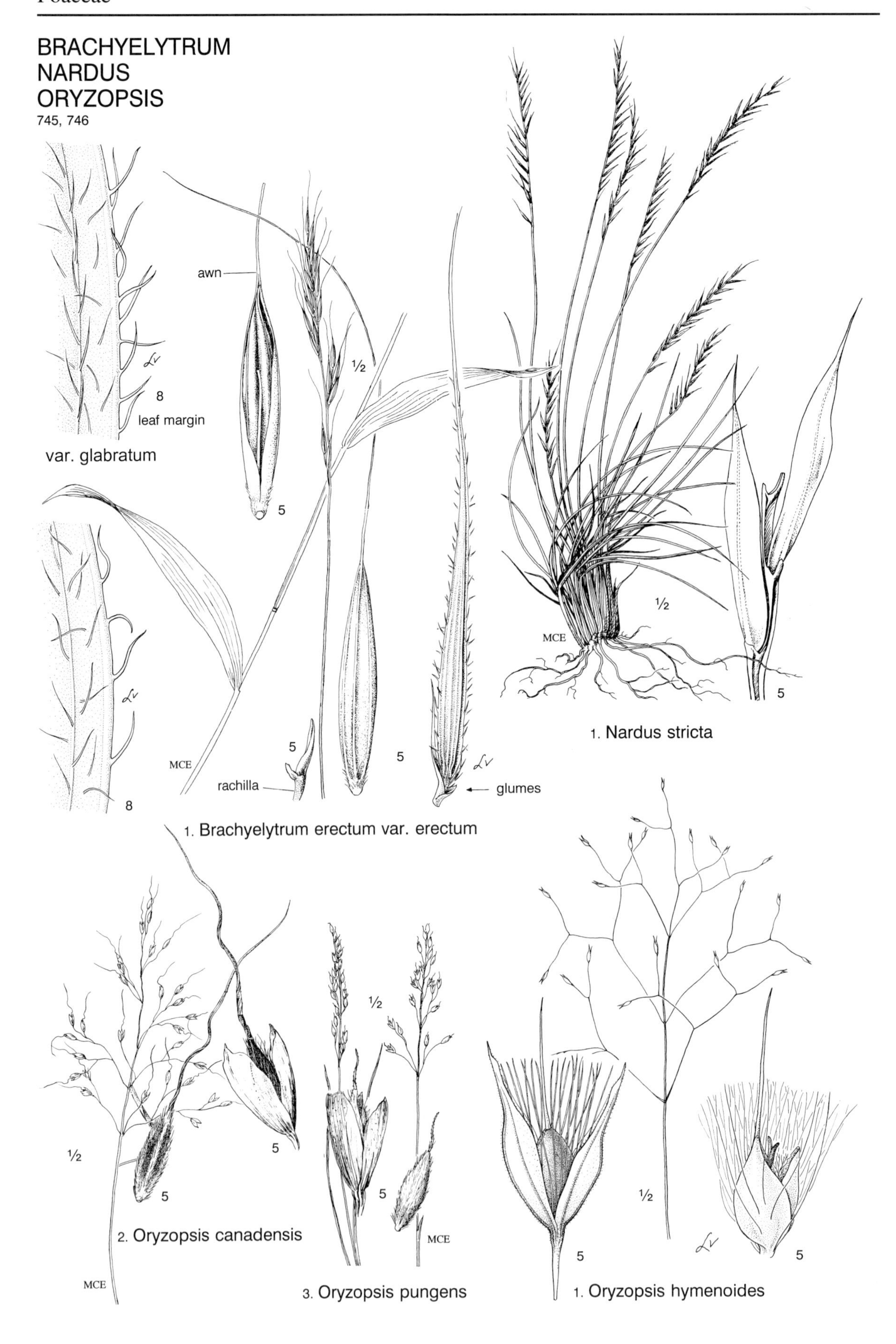

var. glabratum

awn

leaf margin

8

½

5

1. Nardus stricta

MCE

½

5

MCE

rachilla

glumes

5

5

5

8

1. Brachyelytrum erectum var. erectum

½

5

5

2. Oryzopsis canadensis

MCE

½

5

5

MCE

3. Oryzopsis pungens

½

5

5

1. Oryzopsis hymenoides

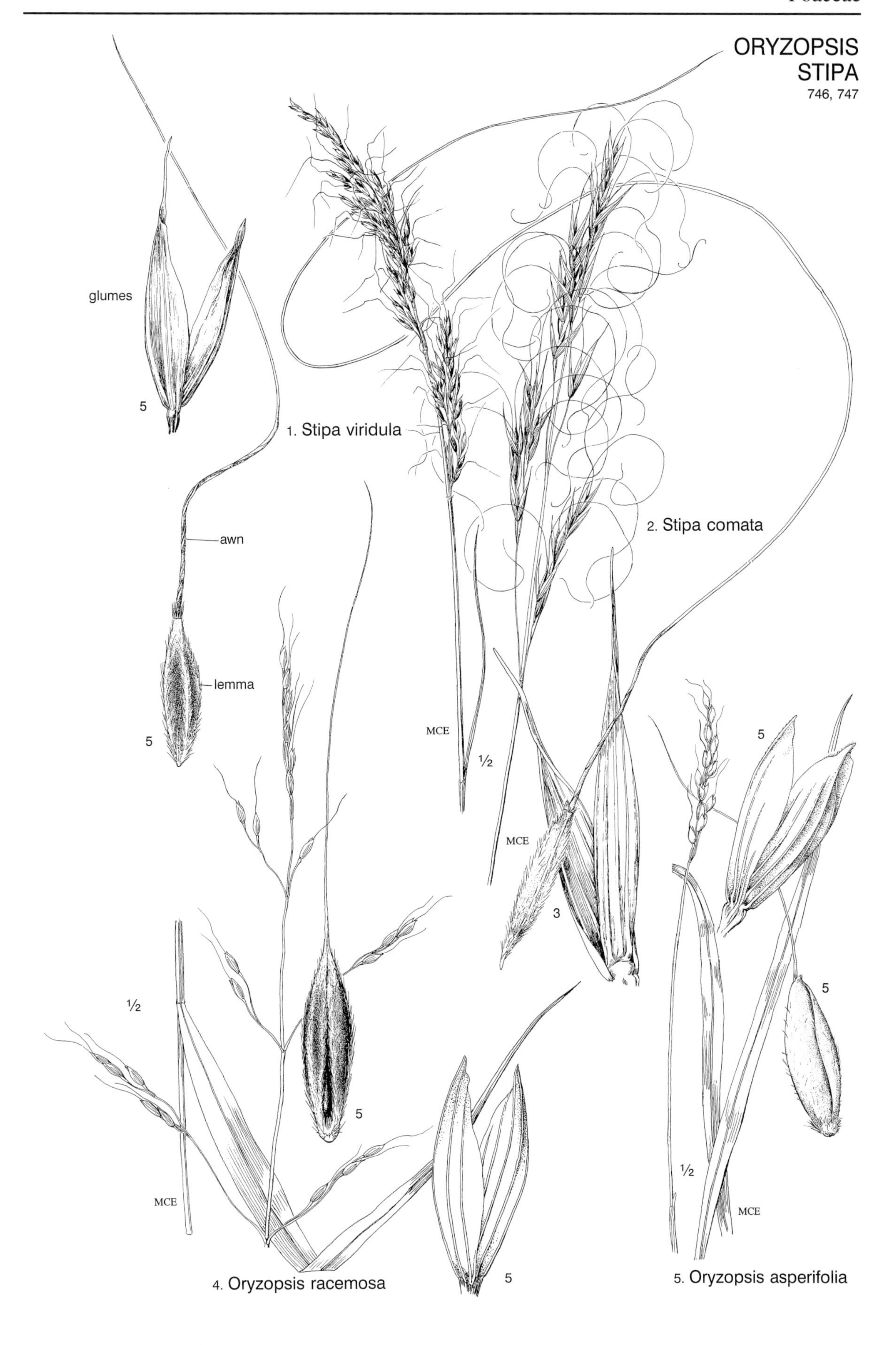

glumes

5

1. Stipa viridula

2. Stipa comata

awn

lemma

5

MCE

½

MCE

3

5

½

MCE

4. Oryzopsis racemosa

5

5. Oryzopsis asperifolia

5

5

½

MCE

STIPA
PIPTOCHAETIUM
MILIUM
747, 748

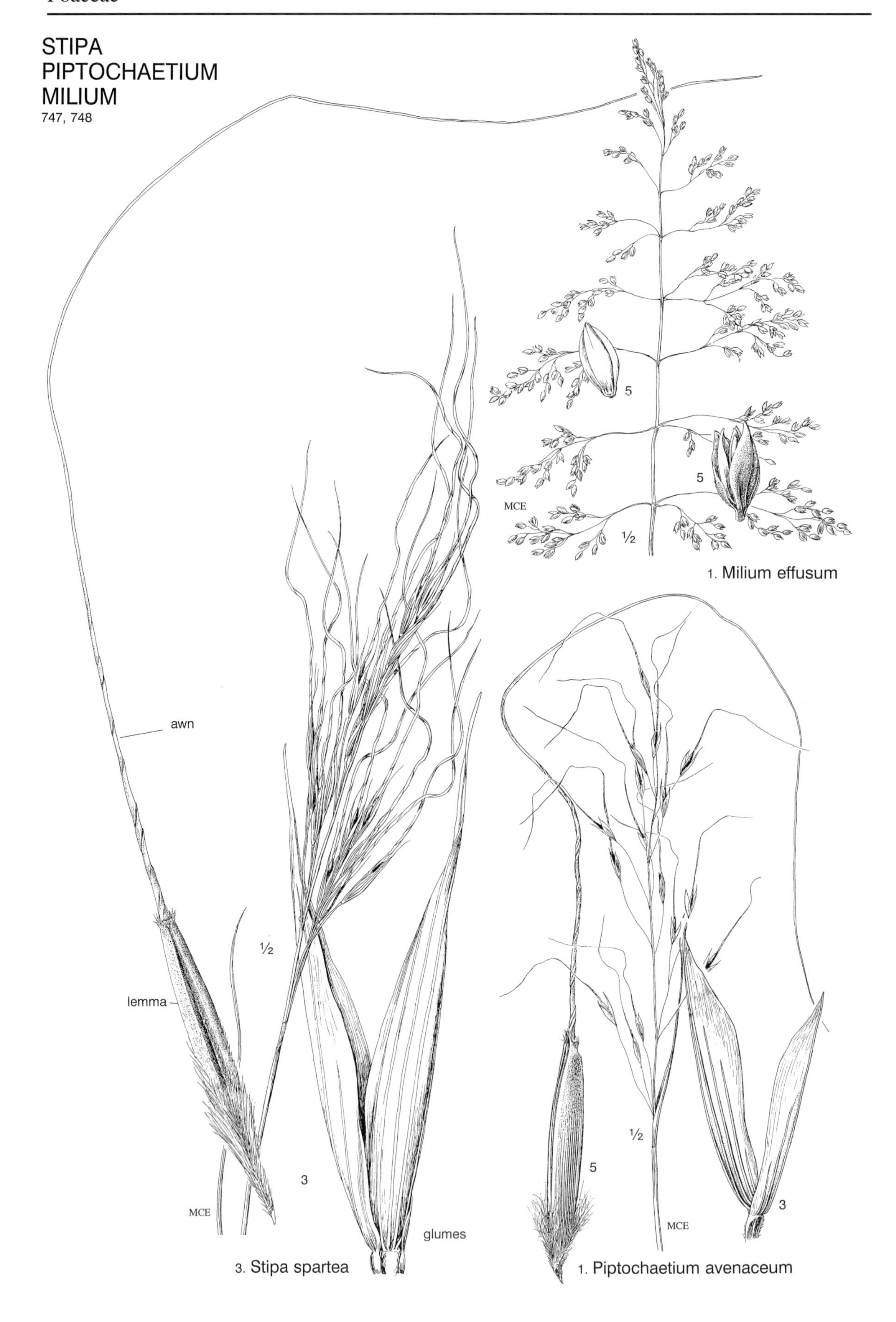

1. Milium effusum

3. Stipa spartea

awn

lemma

glumes

½

3

MCE

1. Piptochaetium avenaceum

½

5

MCE

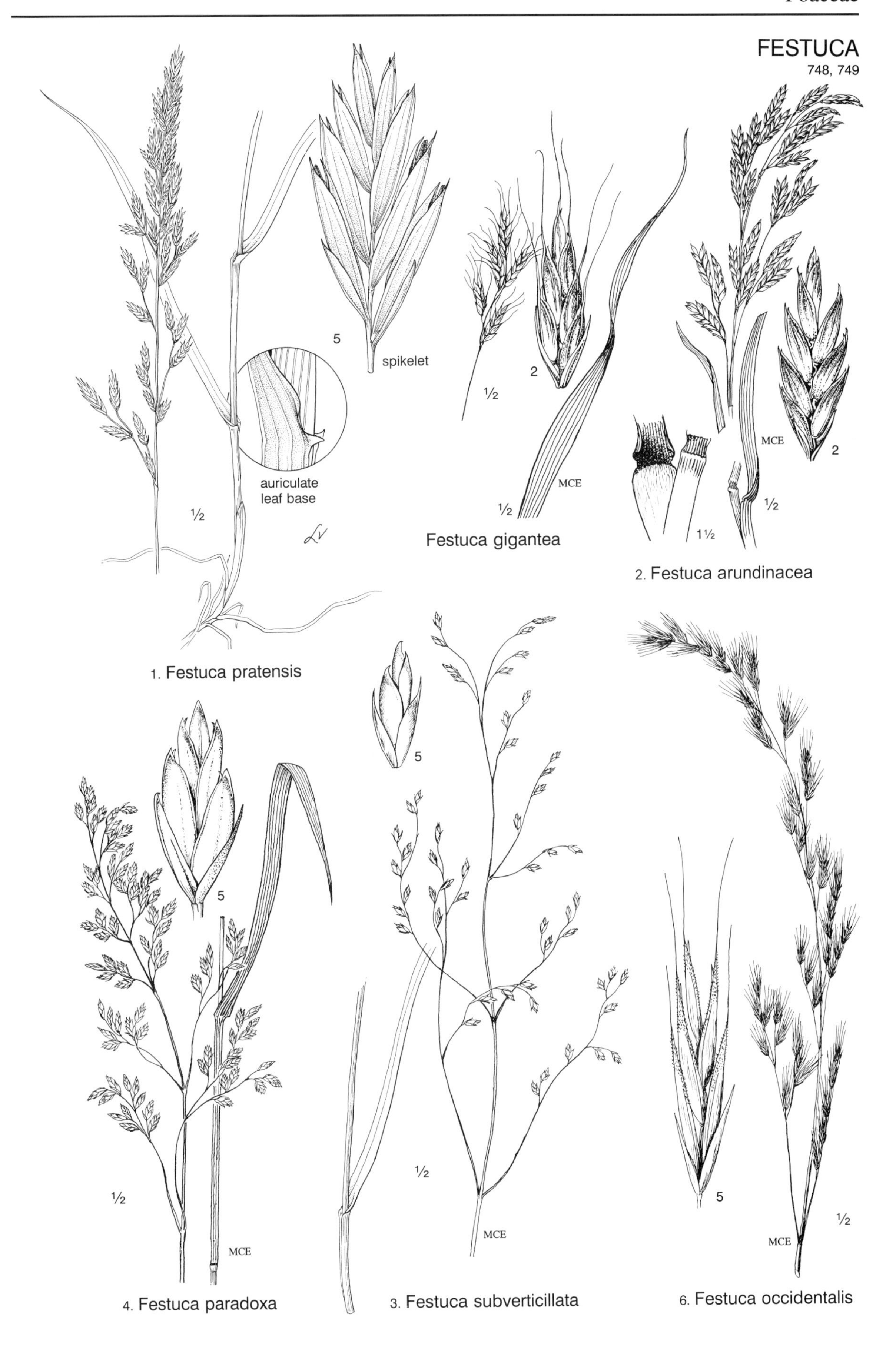

5
spikelet

auriculate
leaf base

½

1. Festuca pratensis

½

2

½

MCE

Festuca gigantea

½

MCE

2

1½

2. Festuca arundinacea

5

5

½

½

5

4. Festuca paradoxa

MCE

½

MCE

3. Festuca subverticillata

5

MCE

½

6. Festuca occidentalis

FESTUCA
749

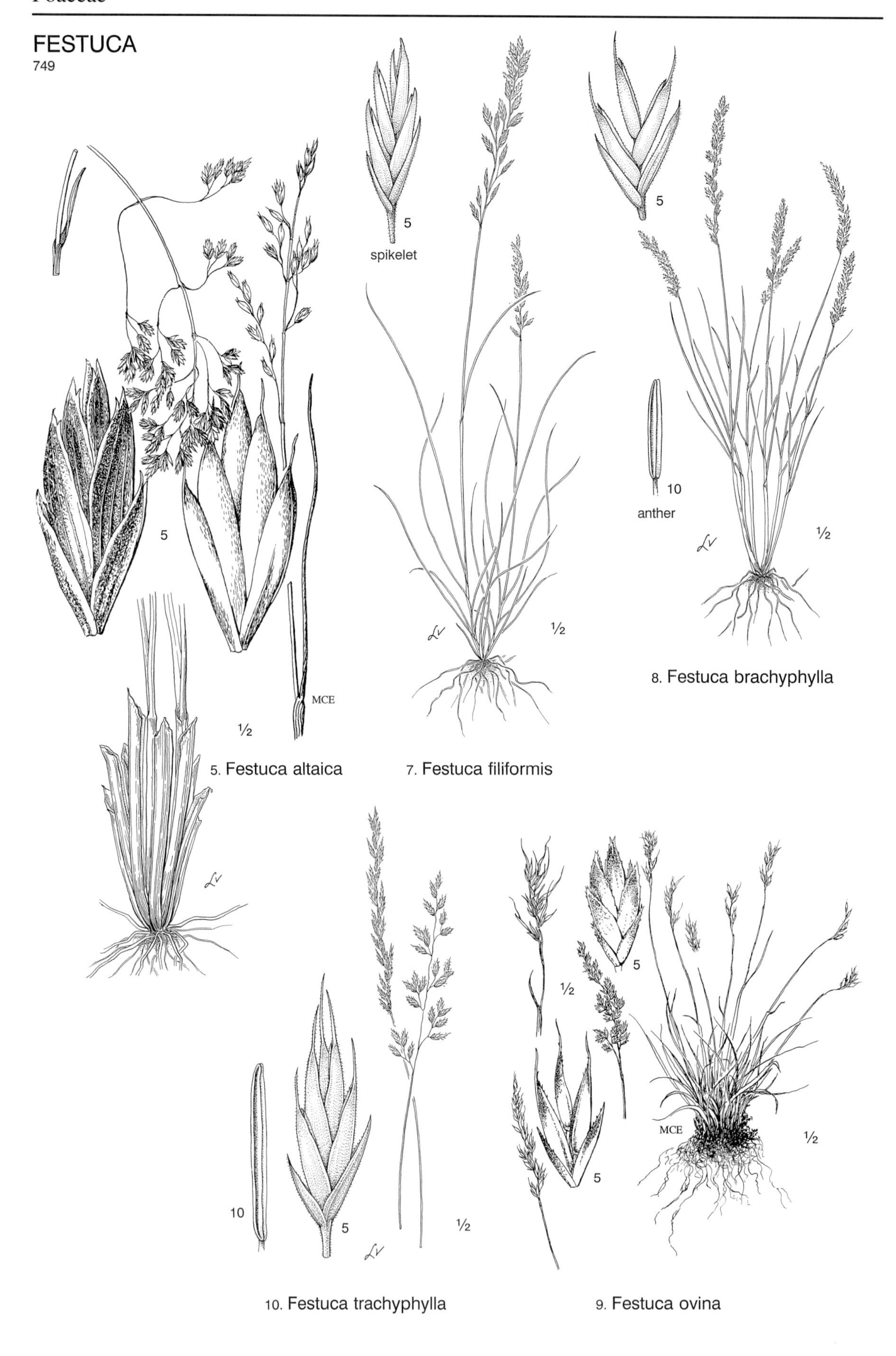

5
spikelet

5

10
anther

8. Festuca brachyphylla

5

MCE

½

5. Festuca altaica

7. Festuca filiformis

½

10

5

½

½

5

MCE

½

5

10. Festuca trachyphylla

9. Festuca ovina

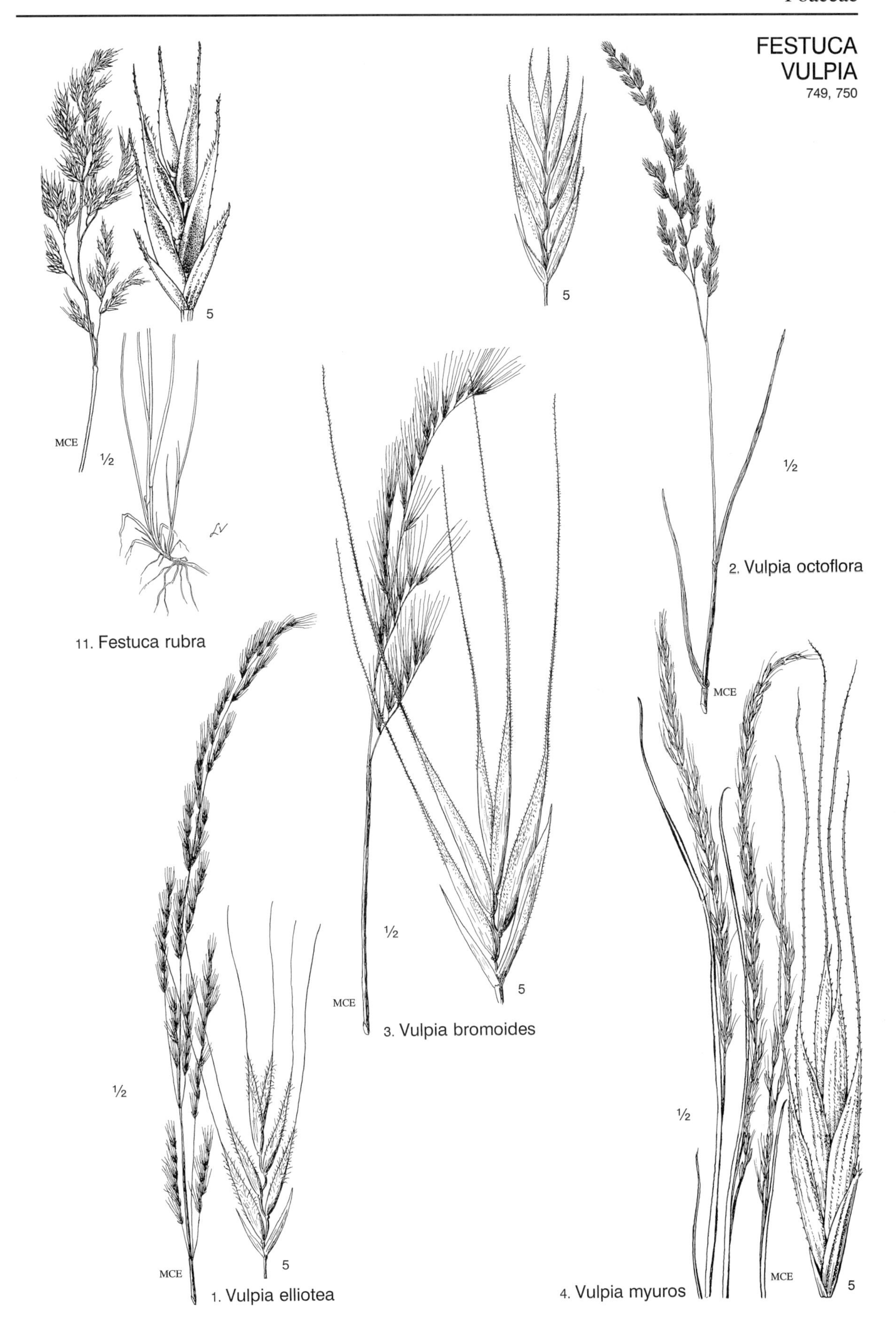

5

MCE   ½

11. Festuca rubra

5

½

2. Vulpia octoflora

MCE

½

5

3. Vulpia bromoides

MCE

½

1. Vulpia elliotea

MCE   5

4. Vulpia myuros

½

MCE   5

LOLIUM
SCOLOCHLOA
750, 751

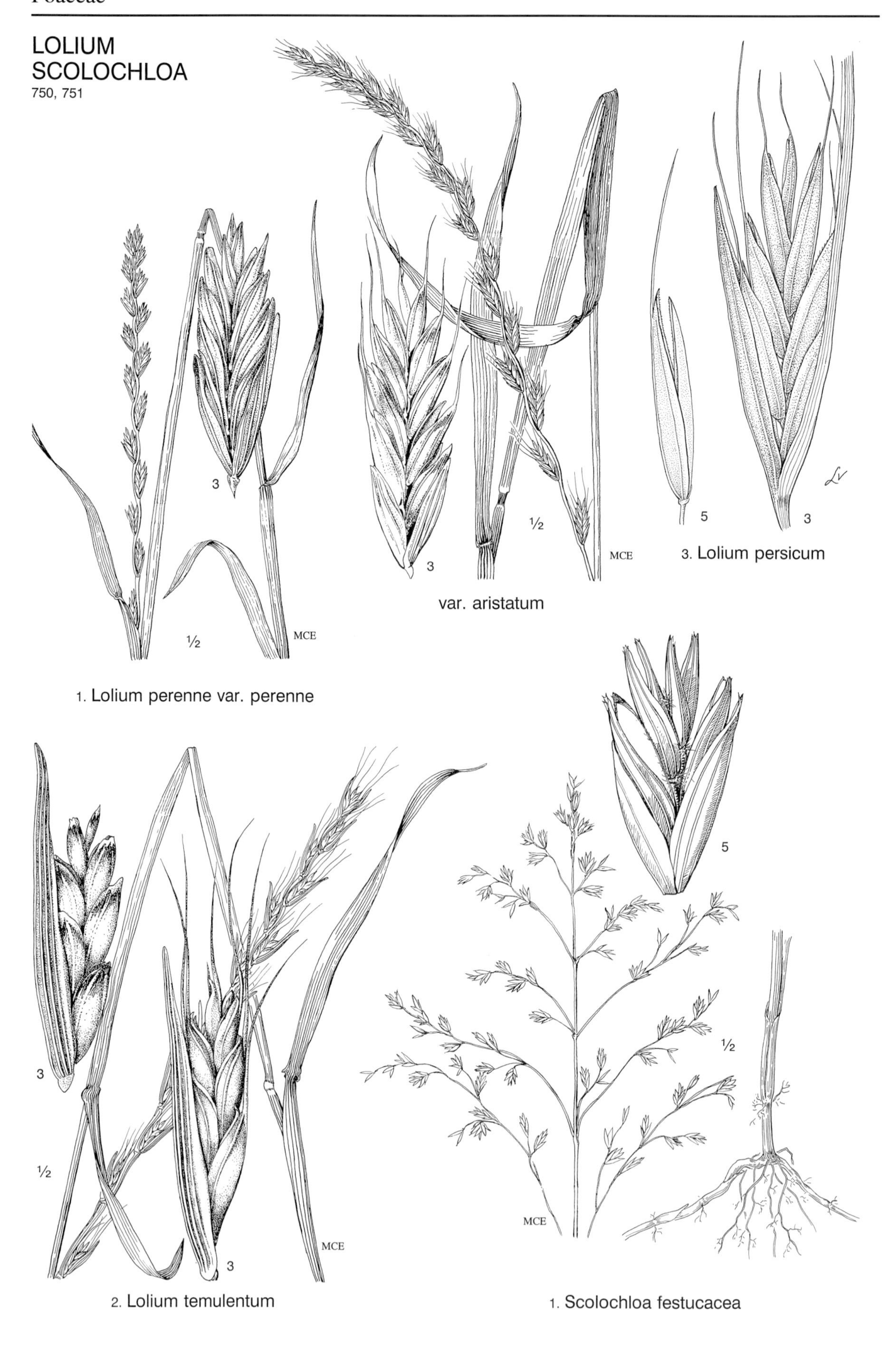

3

½

MCE

1. Lolium perenne var. perenne

3

½

MCE

var. aristatum

5

3

3. Lolium persicum

3

½

3

2. Lolium temulentum

5

½

MCE

1. Scolochloa festucacea

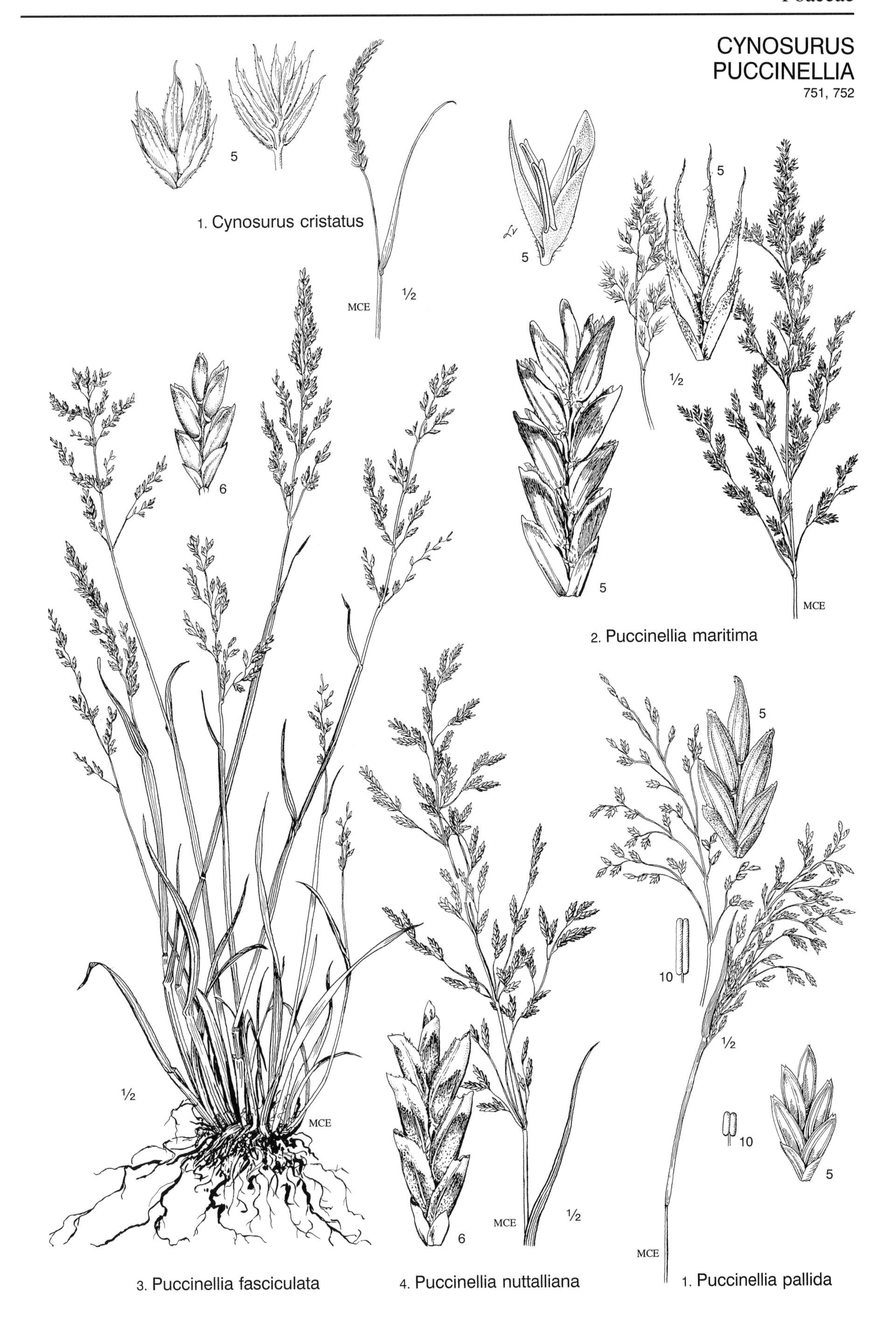

1. Cynosurus cristatus

2. Puccinellia maritima

3. Puccinellia fasciculata

4. Puccinellia nuttalliana

1. Puccinellia pallida

PUCCINELLIA
BRIZA
752, 753

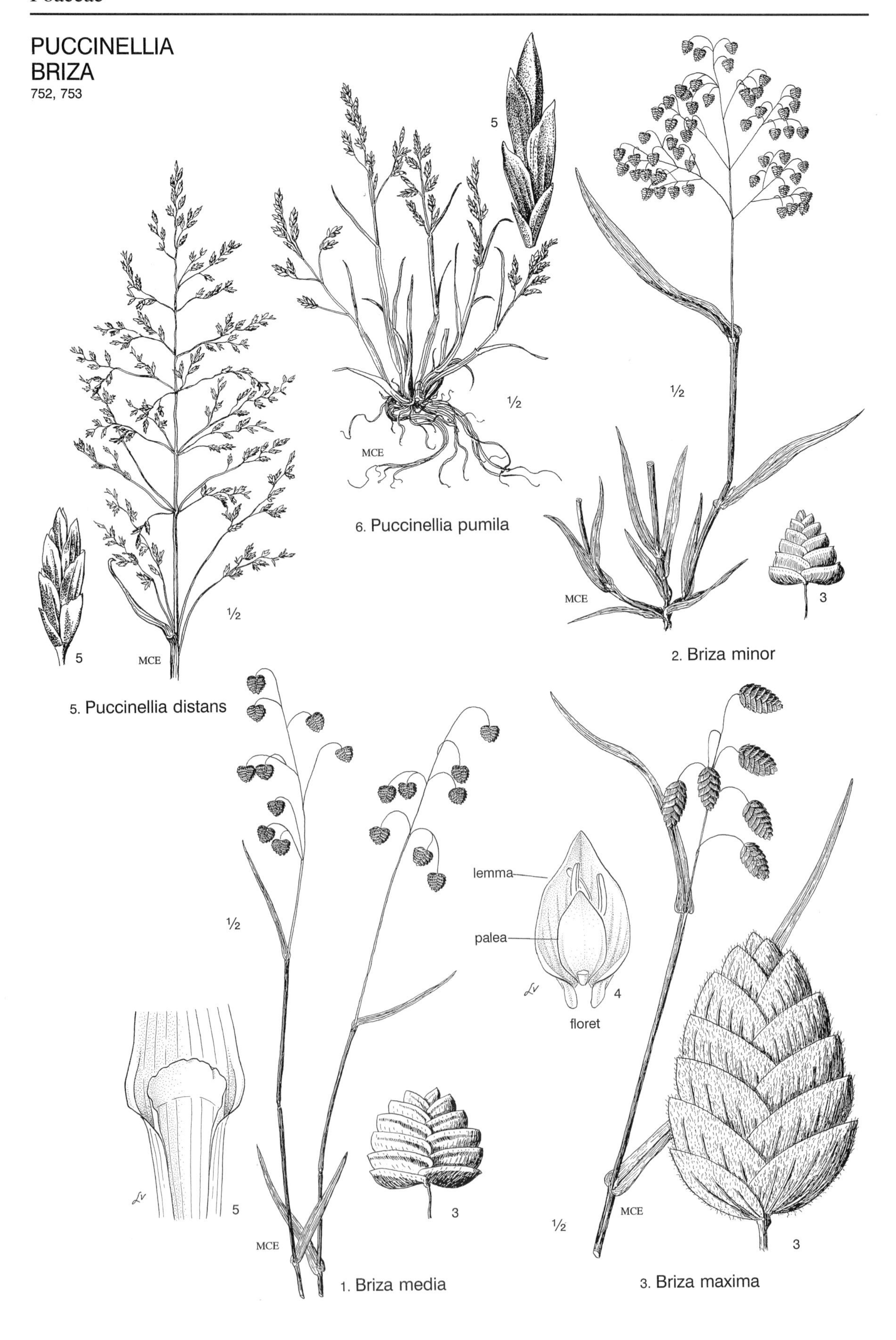

6. Puccinellia pumila

5. Puccinellia distans

2. Briza minor

lemma

palea

floret

4

1. Briza media

3. Briza maxima

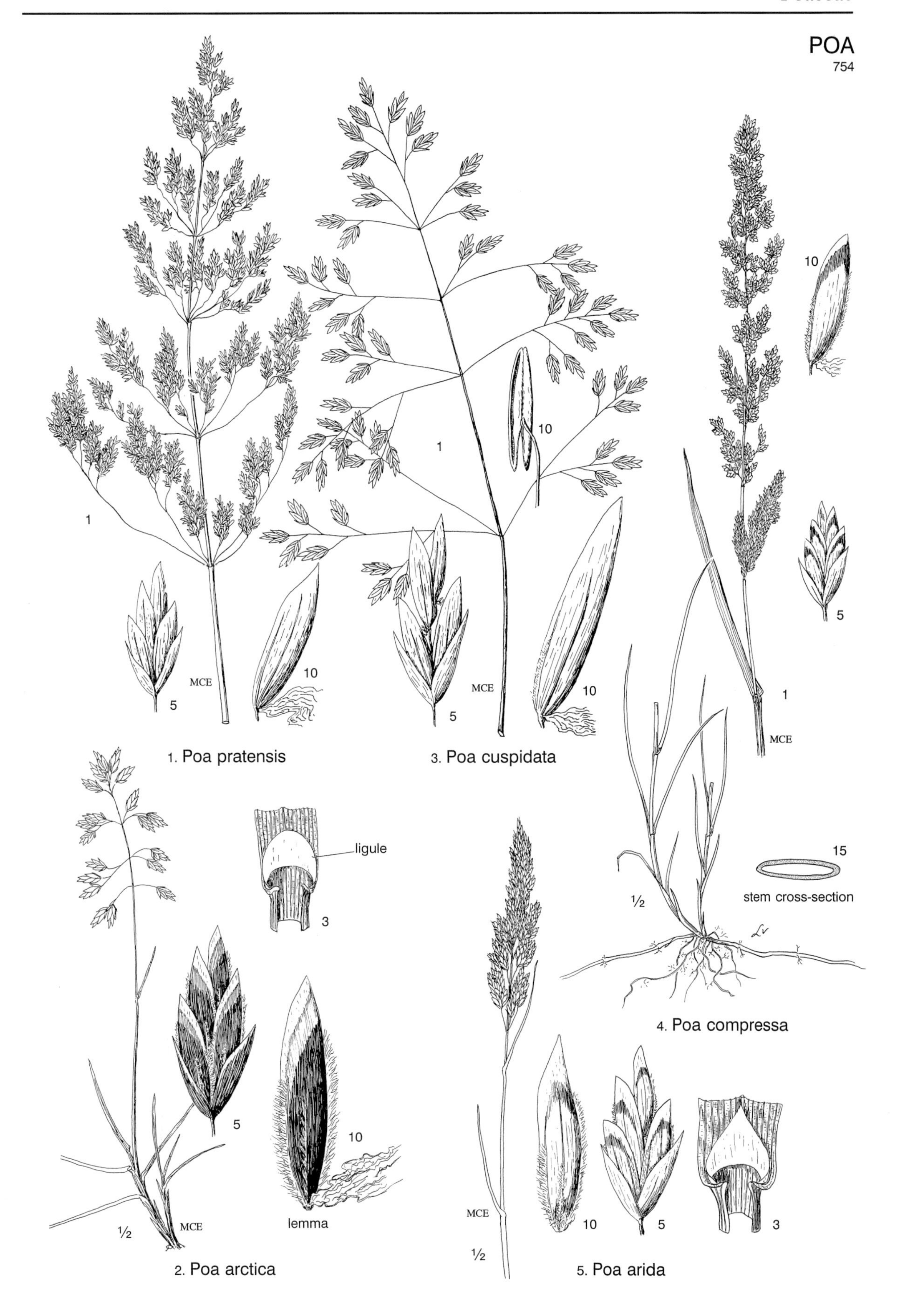

1. Poa pratensis

3. Poa cuspidata

ligule

2. Poa arctica

lemma

4. Poa compressa

stem cross-section

5. Poa arida

POA
754, 755

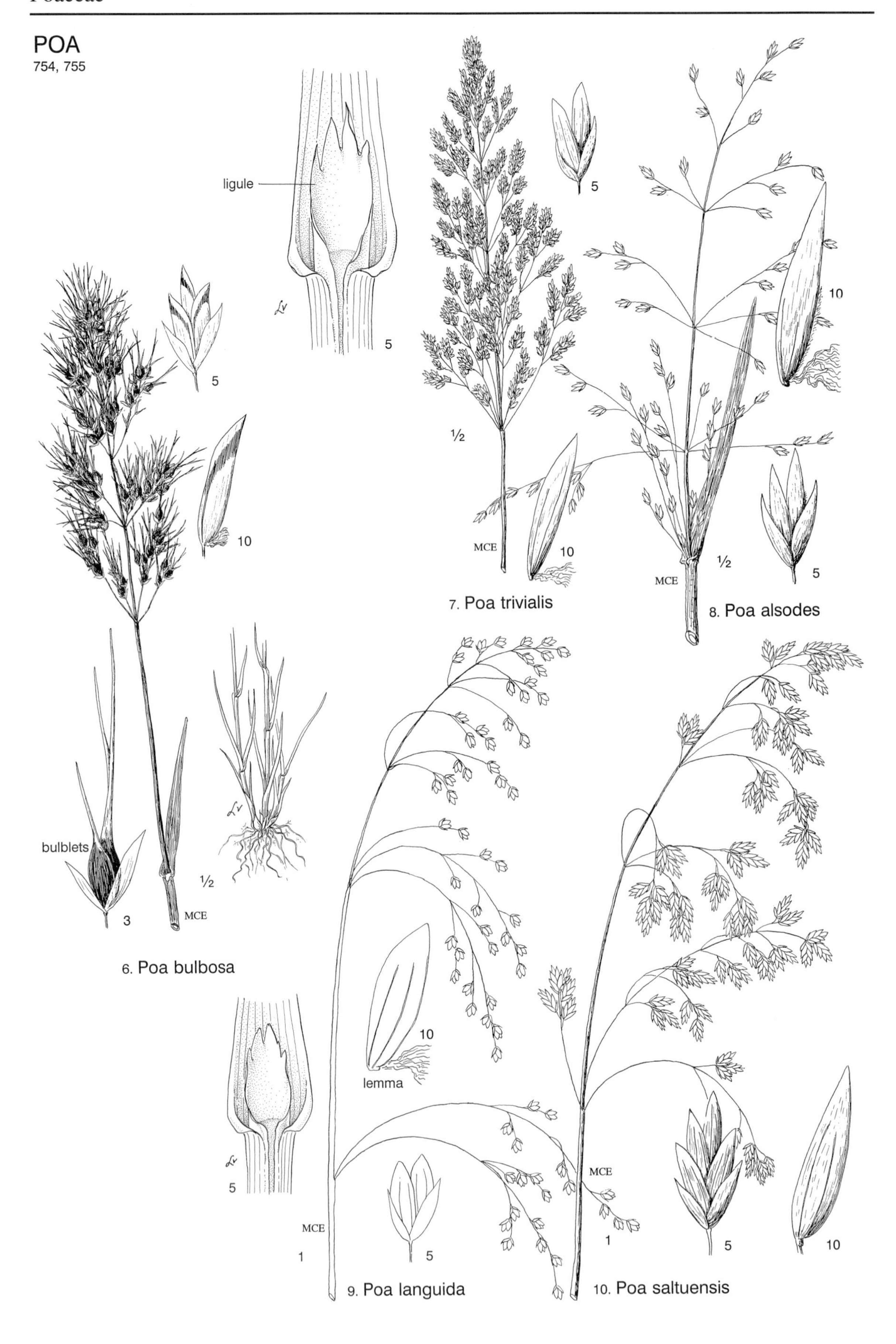

ligule

5

5

10

5

½

MCE

6. Poa bulbosa

bulblets

3

½

MCE

½

MCE

7. Poa trivialis

5

10

MCE

10

½

MCE

5

8. Poa alsodes

5

lemma

10

5

MCE

1

9. Poa languida

MCE

1

5

10

10. Poa saltuensis

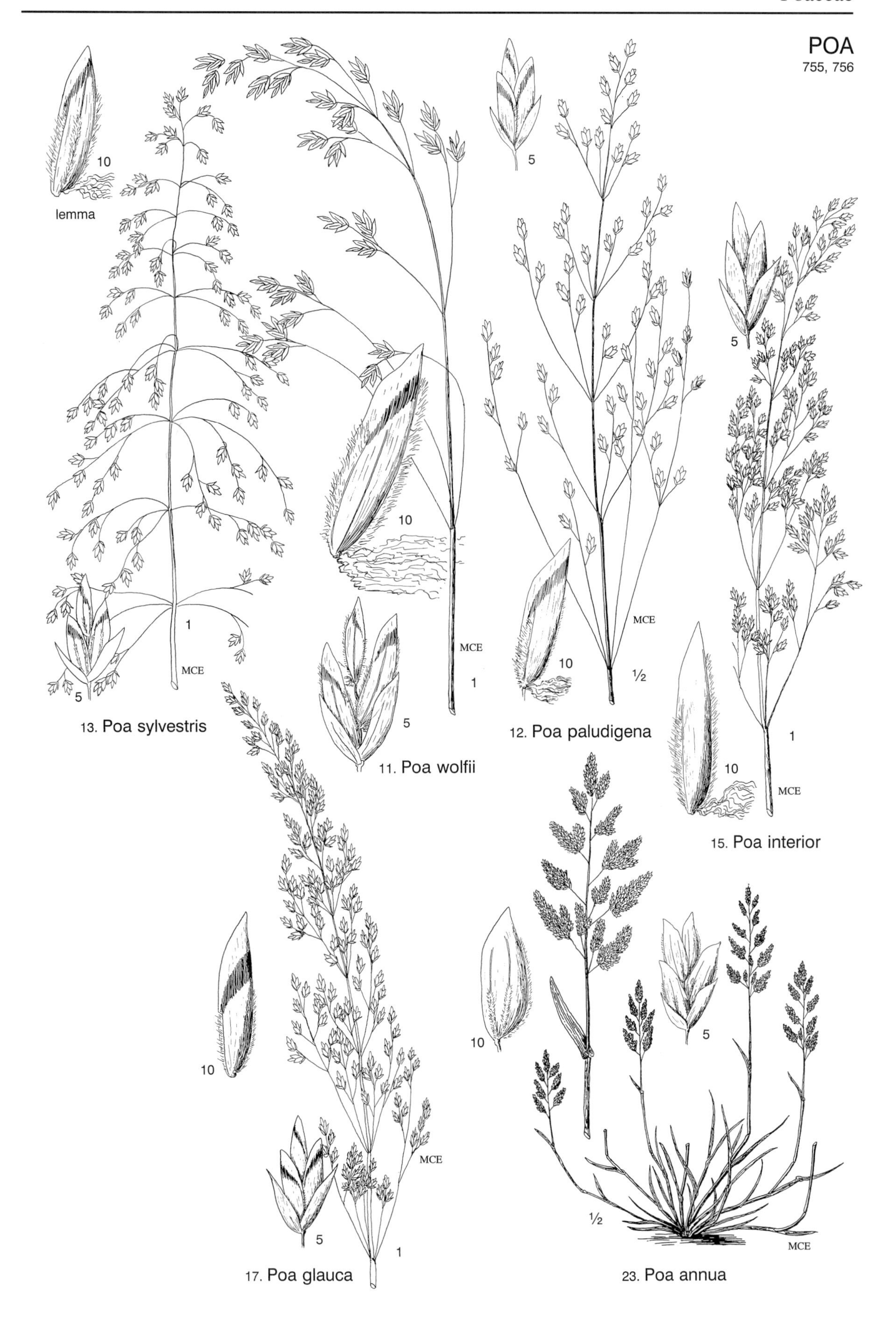

10
lemma

13. Poa sylvestris

11. Poa wolfii

12. Poa paludigena

15. Poa interior

17. Poa glauca

23. Poa annua

POA
755, 756

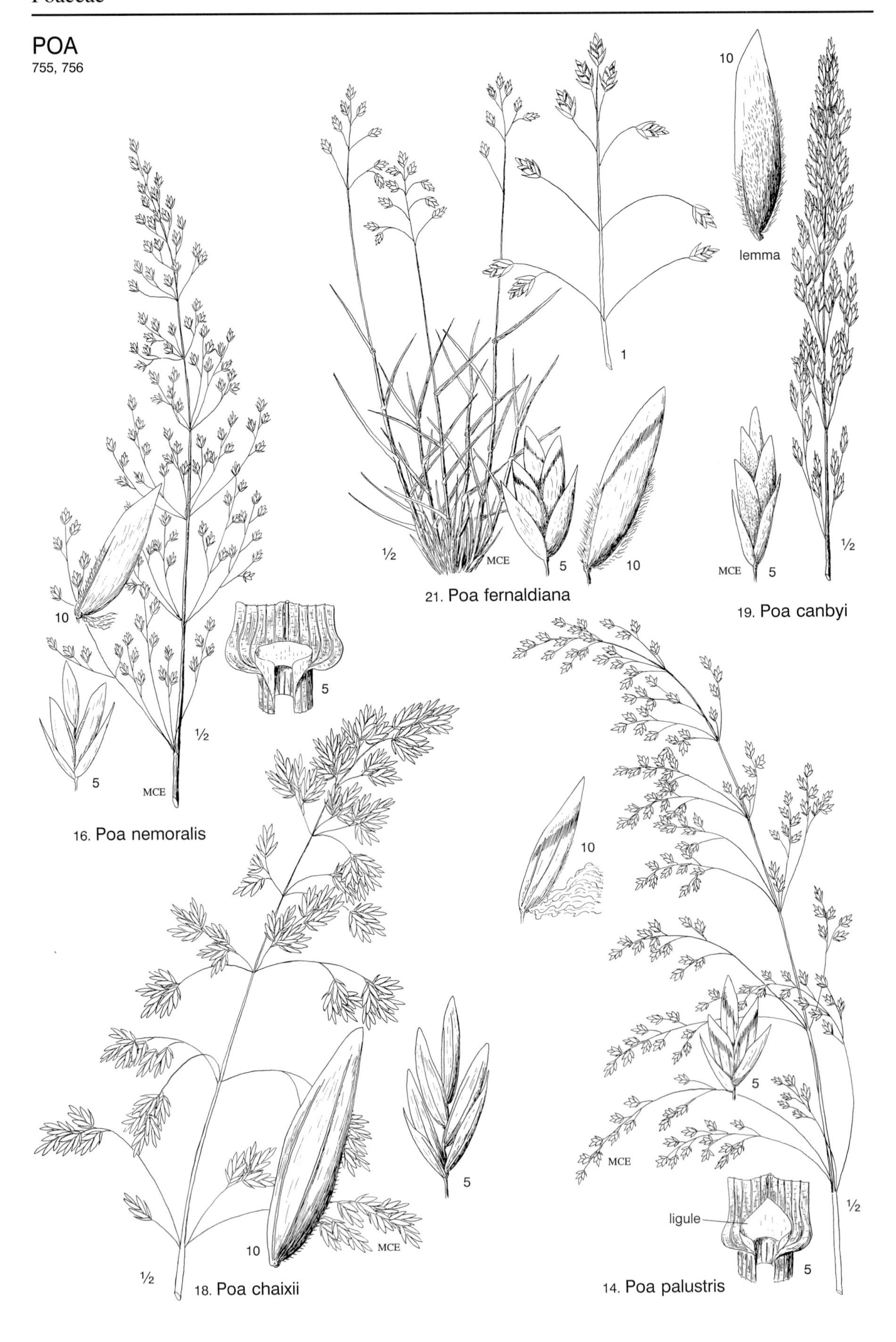

lemma

21. Poa fernaldiana

19. Poa canbyi

16. Poa nemoralis

18. Poa chaixii

ligule

14. Poa palustris

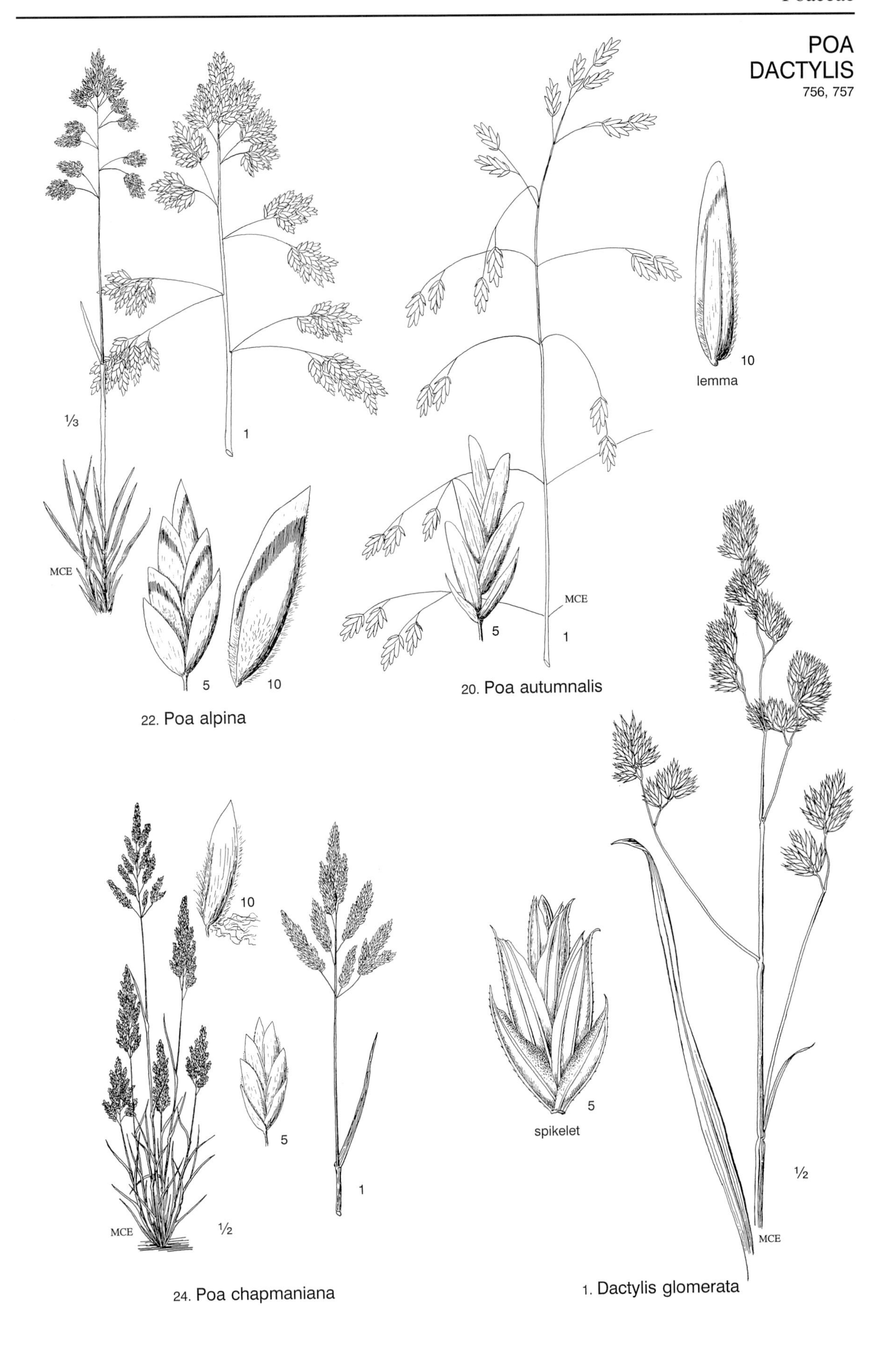

10
lemma

1/3

1

MCE

5 10

22. Poa alpina

20. Poa autumnalis

MCE

5 1

10

5

24. Poa chapmaniana

1

1/2

MCE 1/2

spikelet

5

1/2

MCE

1. Dactylis glomerata

CATABROSA
PARAPHOLIS
GLYCERIA
757, 758

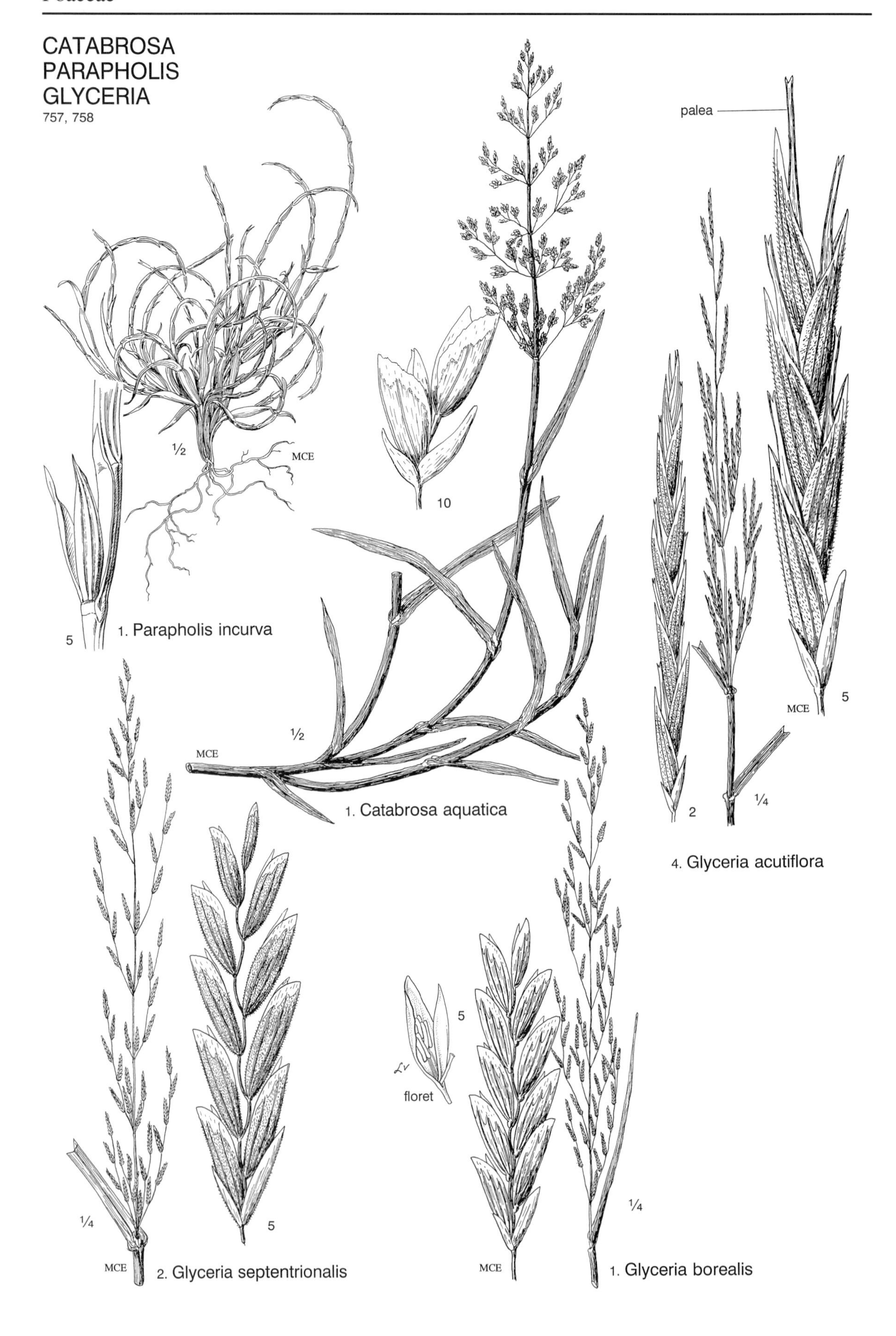

palea

½

MCE

10

5   1. Parapholis incurva

MCE   ½

1. Catabrosa aquatica

5

2   ¼

MCE

4. Glyceria acutiflora

¼

5

floret   5

MCE   2. Glyceria septentrionalis

¼

MCE   1. Glyceria borealis

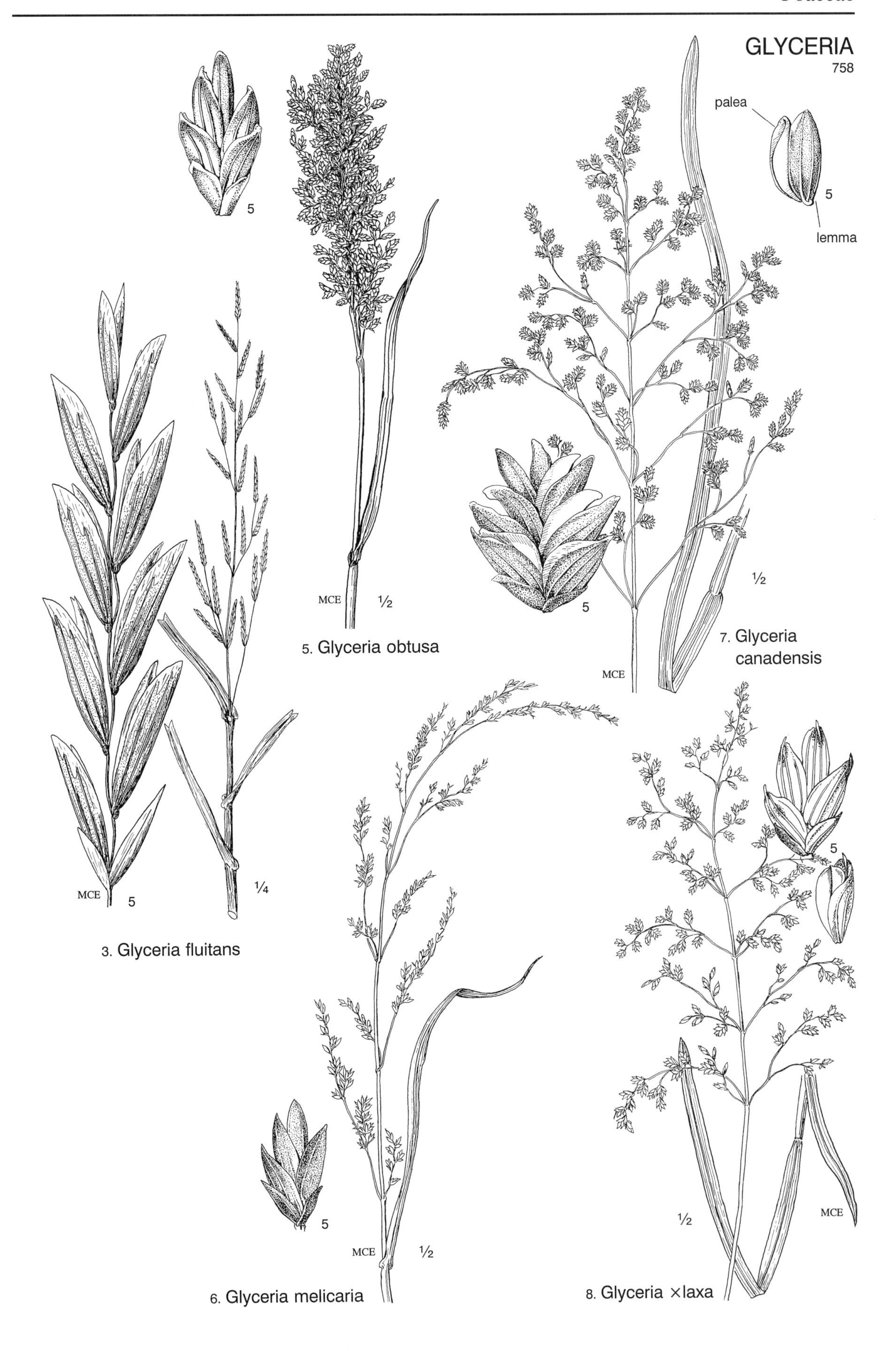

palea

5

lemma

5

MCE   ½

5. Glyceria obtusa

7. Glyceria
canadensis

MCE

MCE   5   ¼

3. Glyceria fluitans

5

MCE   ½

6. Glyceria melicaria

5

½   MCE

8. Glyceria ×laxa

GLYCERIA
MELICA
758, 759

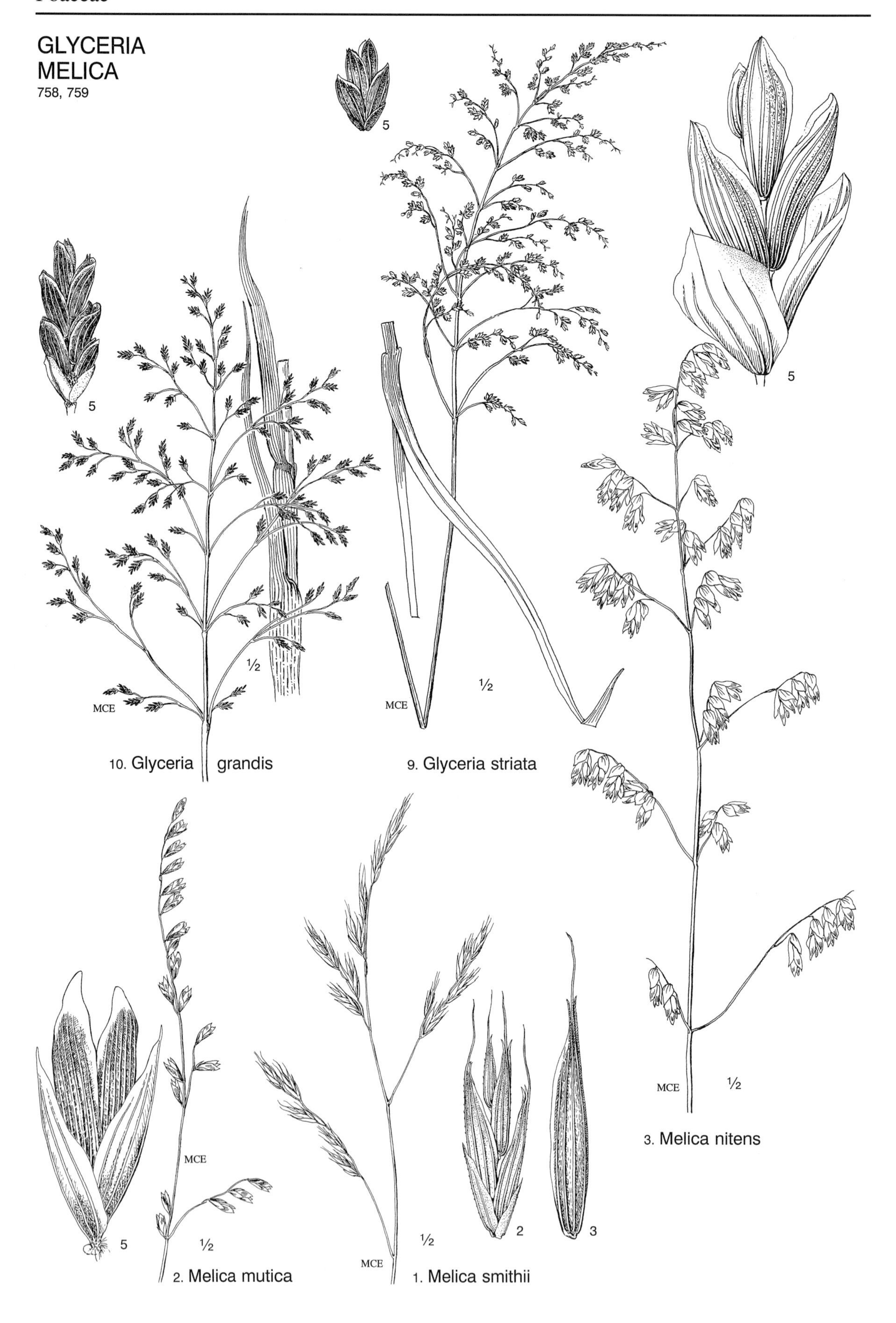

10. Glyceria | grandis

9. Glyceria striata

3. Melica nitens

2. Melica mutica

1. Melica smithii

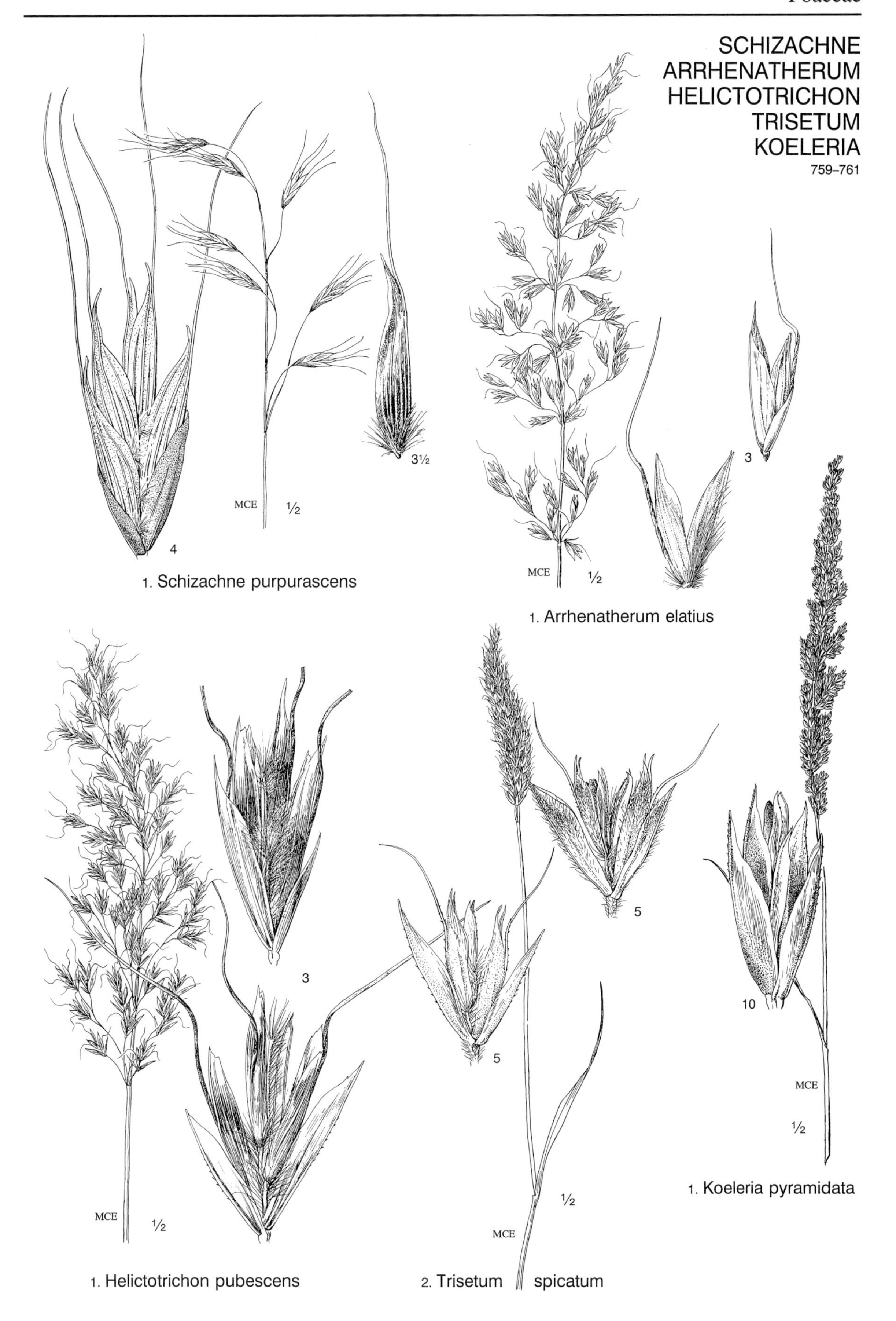

SCHIZACHNE
ARRHENATHERUM
HELICTOTRICHON
TRISETUM
KOELERIA
759–761

1. Schizachne purpurascens

1. Arrhenatherum elatius

1. Helictotrichon pubescens

2. Trisetum spicatum

1. Koeleria pyramidata

AVENA
TRISETUM
760, 761

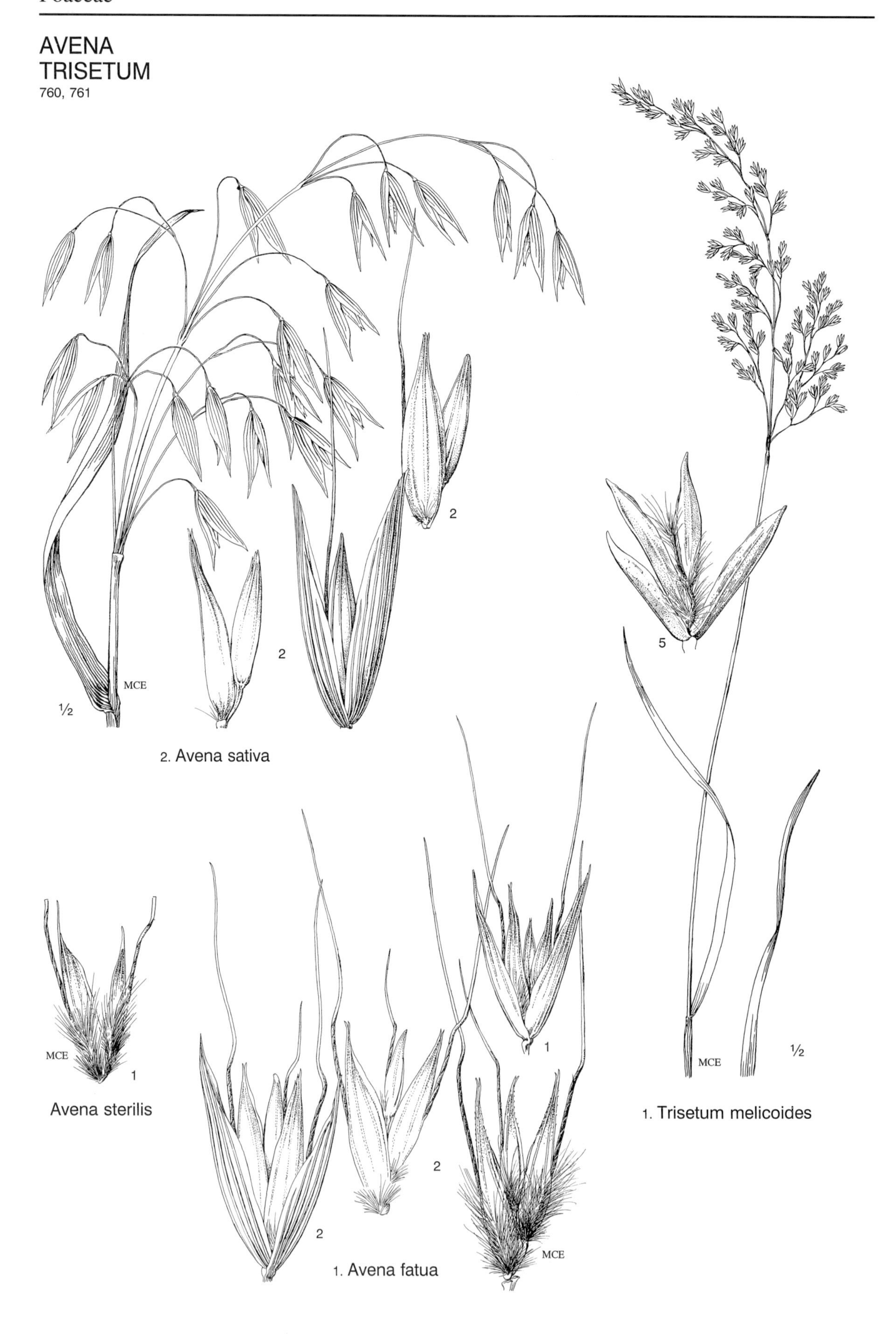

2. Avena sativa

Avena sterilis

1. Avena fatua

1. Trisetum melicoides

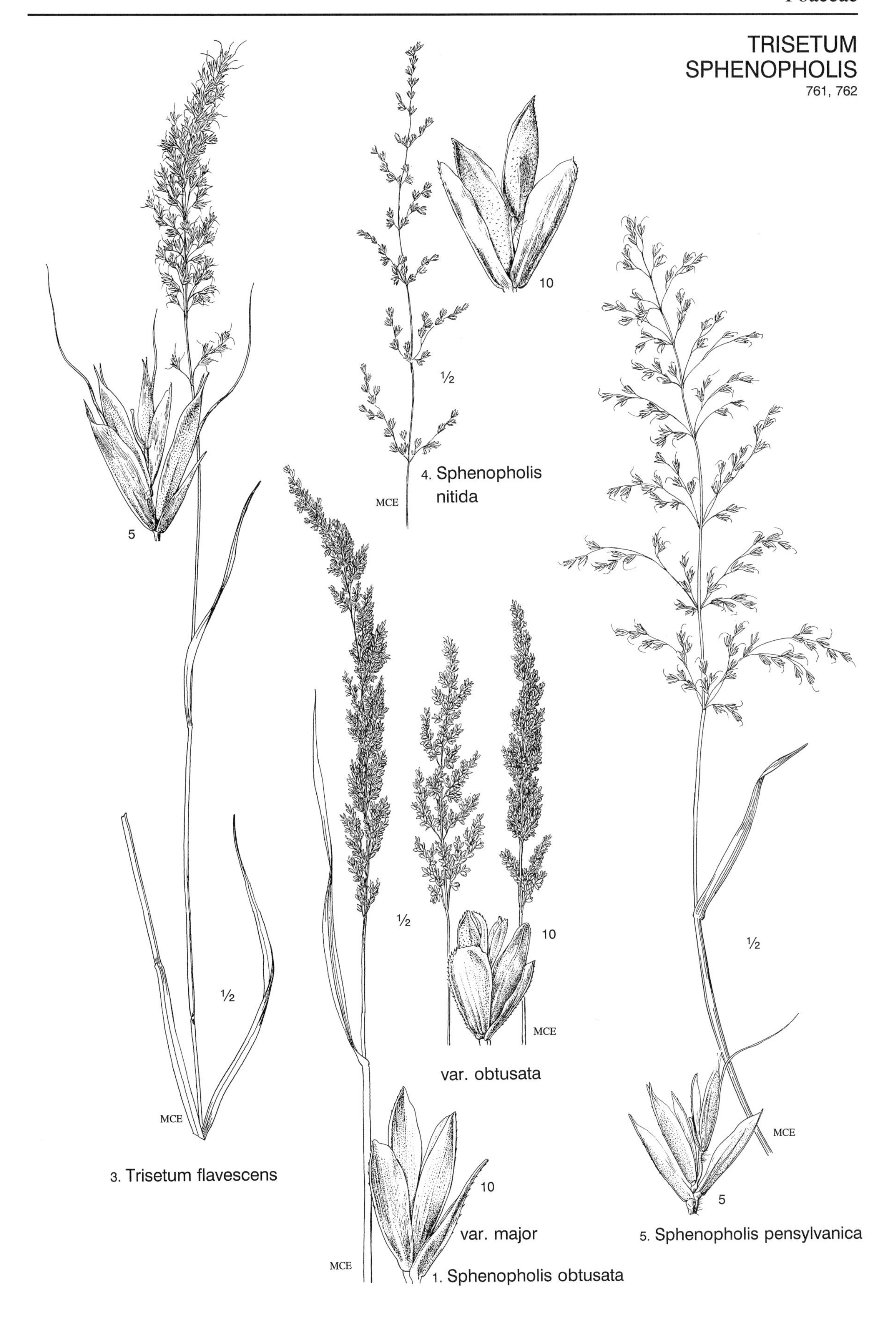

5

10

1/2

4. Sphenopholis
nitida

MCE

1/2

var. obtusata

10

MCE

3. Trisetum flavescens

1/2

MCE

var. major

10

MCE

1. Sphenopholis obtusata

1/2

5

MCE

5. Sphenopholis pensylvanica

SPHENOPHOLIS
DESCHAMPSIA
HOLCUS
762, 763

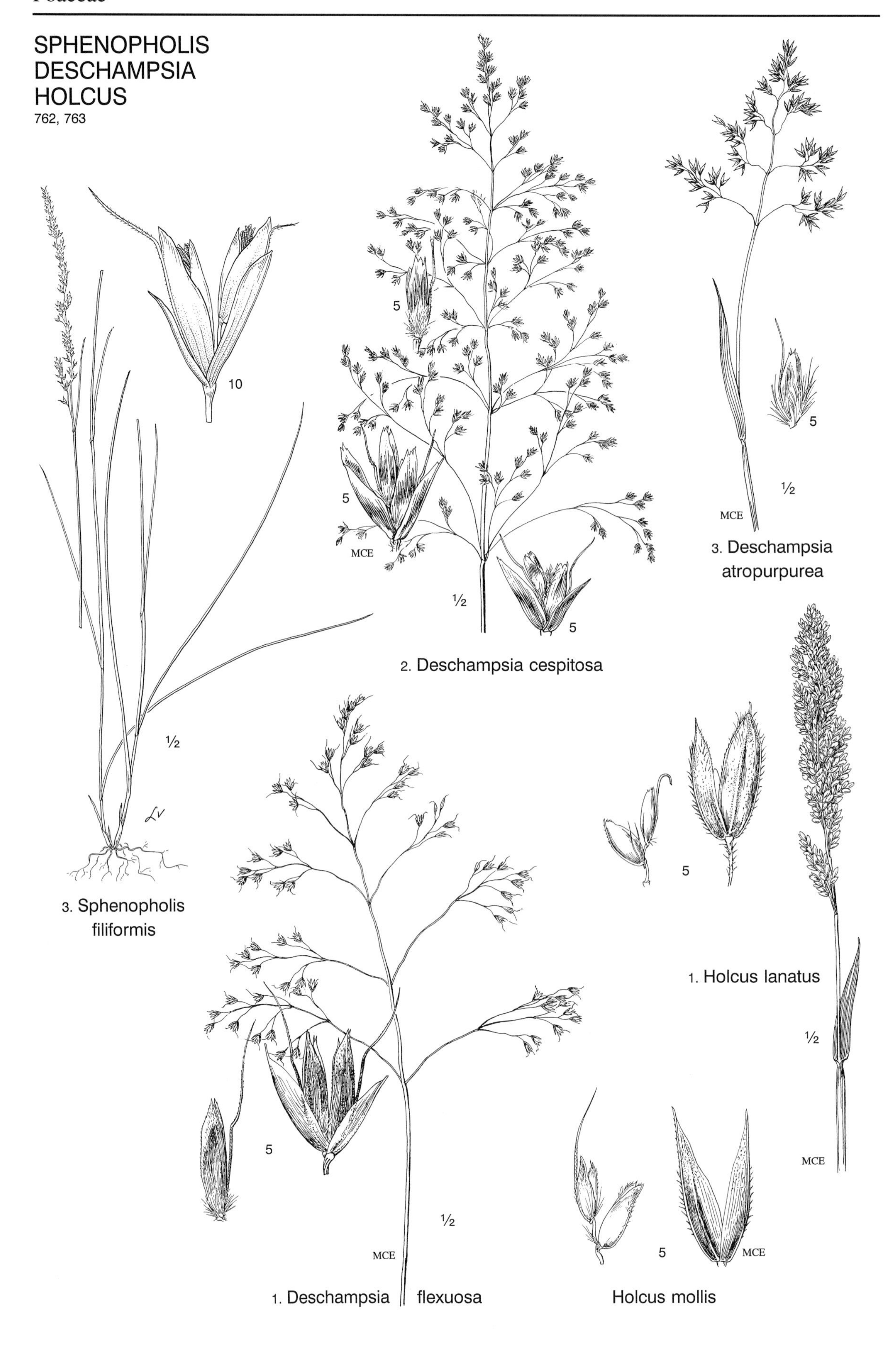

10

5

MCE

5

MCE

½

5

½

2. Deschampsia cespitosa

5

MCE

½

3. Deschampsia
atropurpurea

3. Sphenopholis
filiformis

½

5

MCE

½

5

MCE

1. Deschampsia flexuosa

5

5

MCE

Holcus mollis

5

1. Holcus lanatus

½

MCE

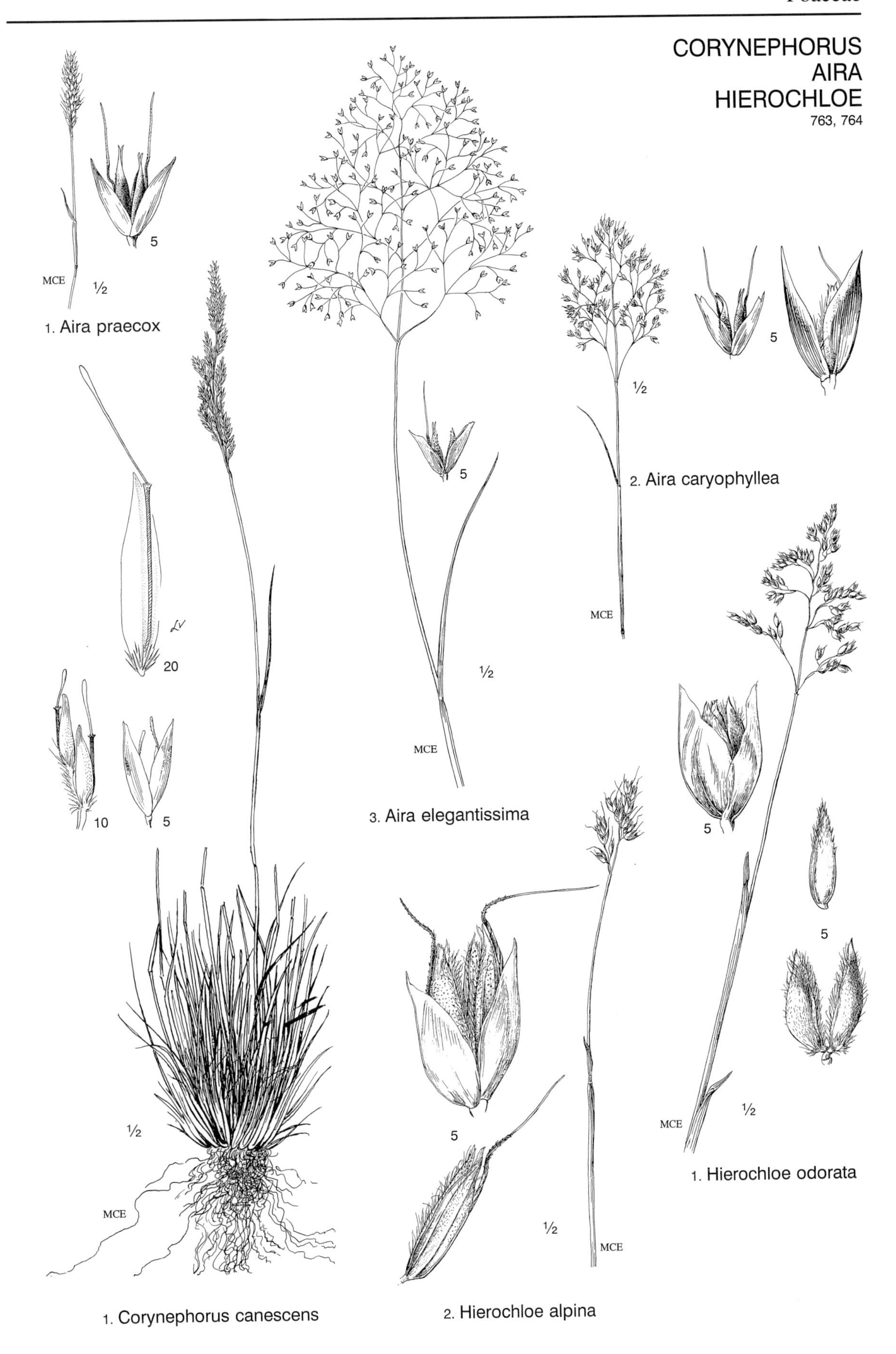

1. Aira praecox

2. Aira caryophyllea

3. Aira elegantissima

1. Hierochloe odorata

1. Corynephorus canescens

2. Hierochloe alpina

ANTHOXANTHUM
PHALARIS
CALAMAGROSTIS
764–766

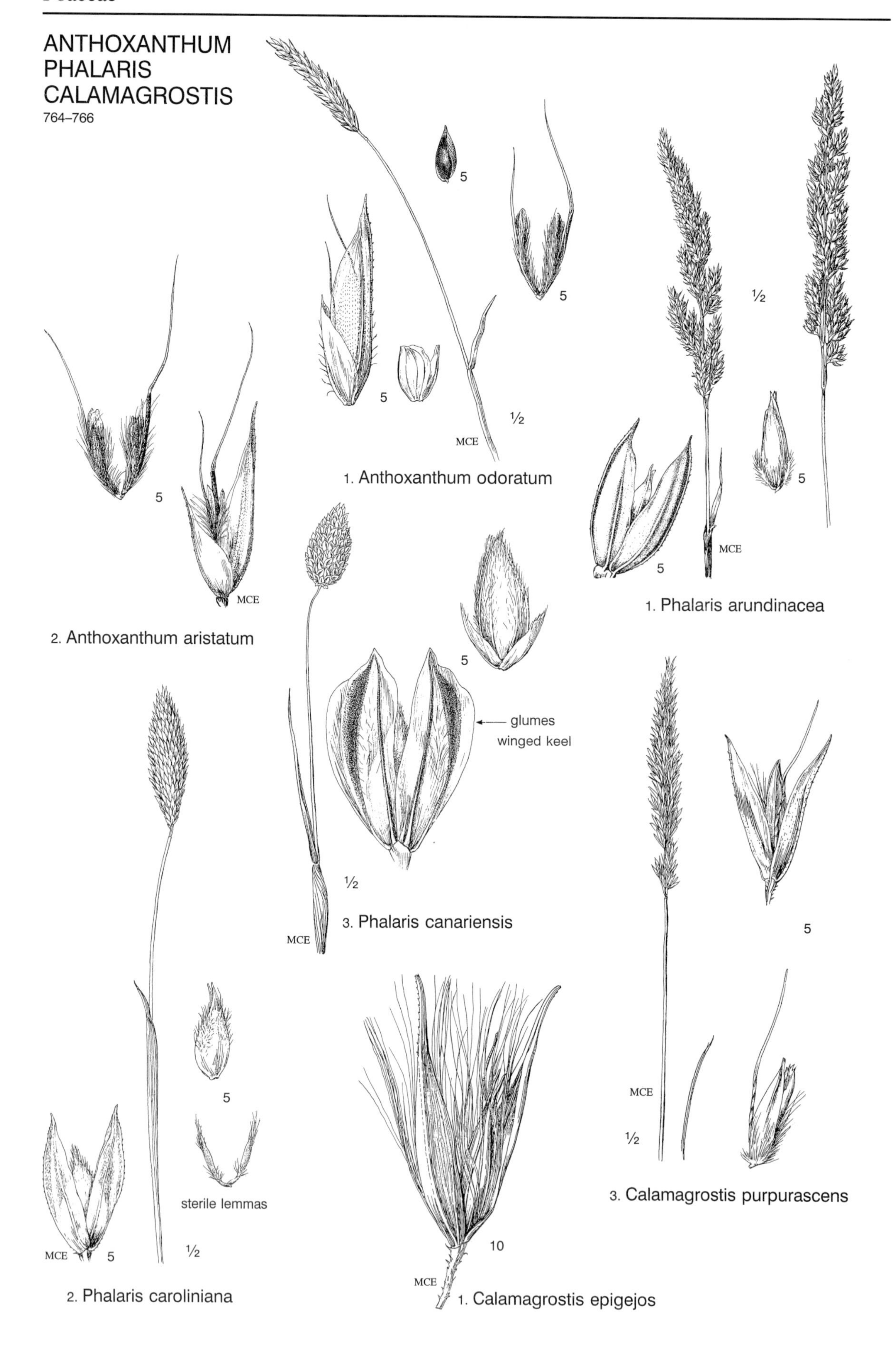

1. Anthoxanthum odoratum

2. Anthoxanthum aristatum

1. Phalaris arundinacea

glumes winged keel

3. Phalaris canariensis

3. Calamagrostis purpurascens

sterile lemmas

2. Phalaris caroliniana

1. Calamagrostis epigejos

ILLUSTRATED COMPANION TO

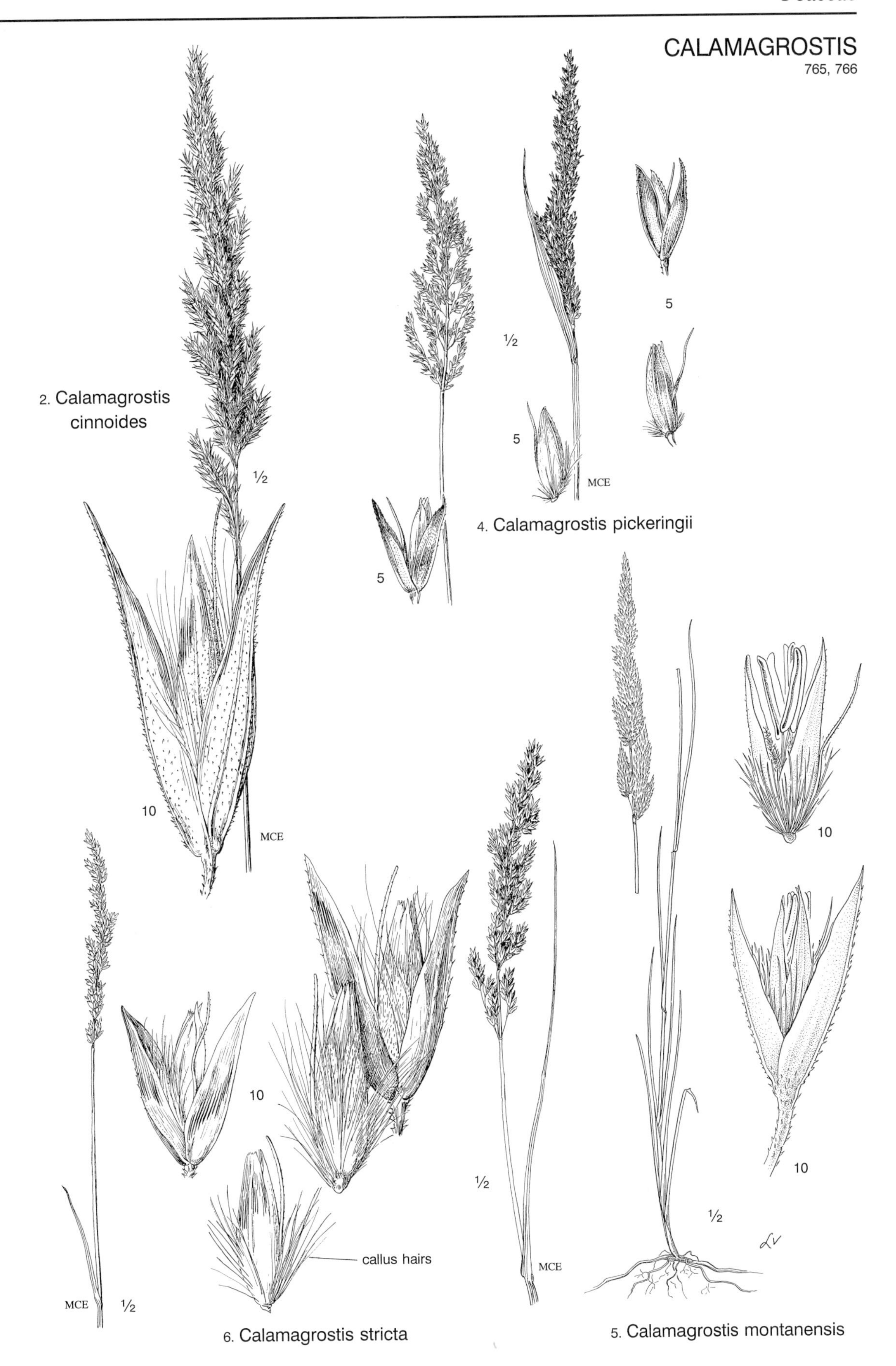

2. Calamagrostis
cinnoides

4. Calamagrostis pickeringii

6. Calamagrostis stricta

callus hairs

5. Calamagrostis montanensis

CALAMAGROSTIS
766

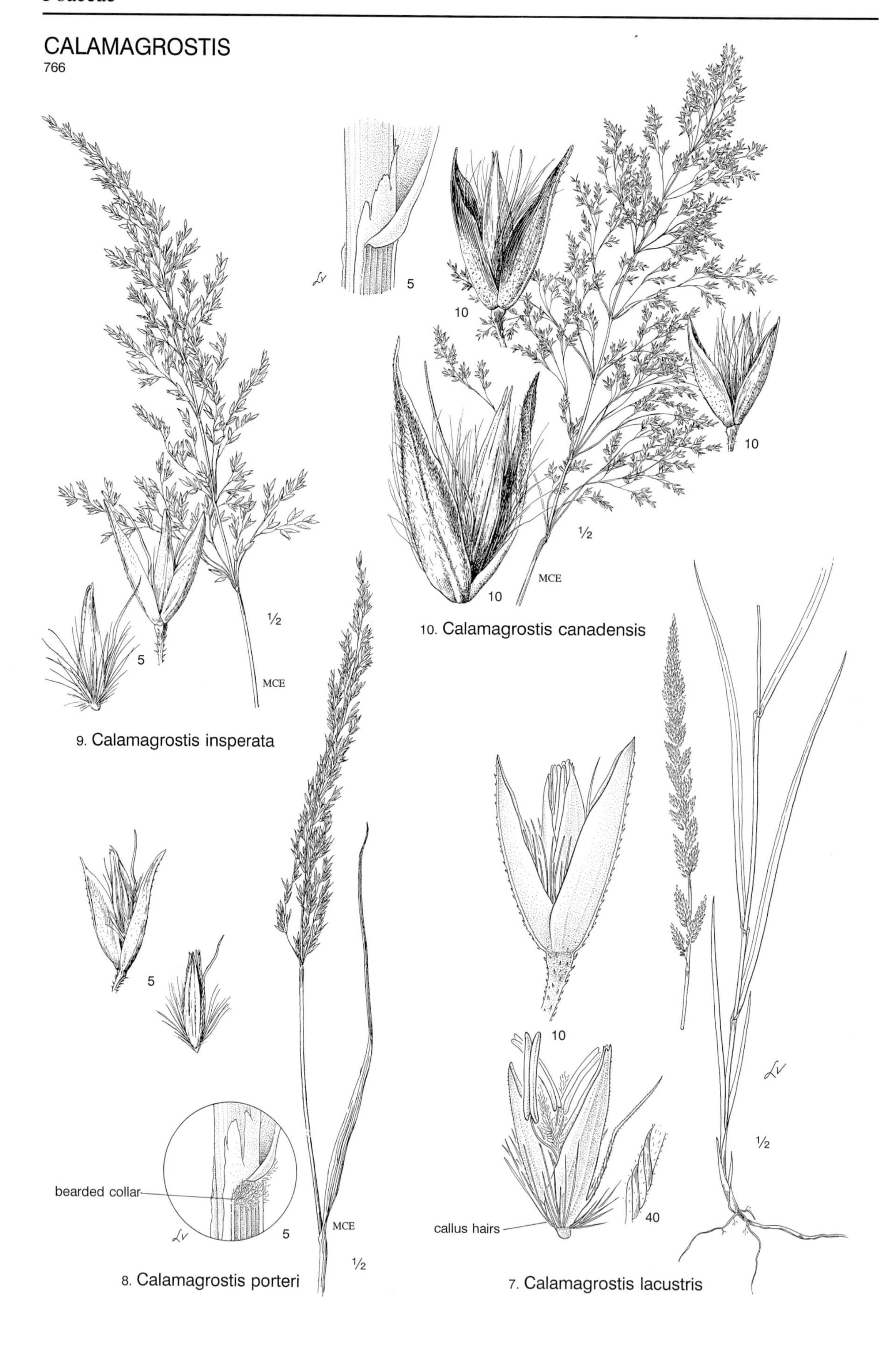

9. Calamagrostis insperata

10. Calamagrostis canadensis

8. Calamagrostis porteri

bearded collar

7. Calamagrostis lacustris

callus hairs

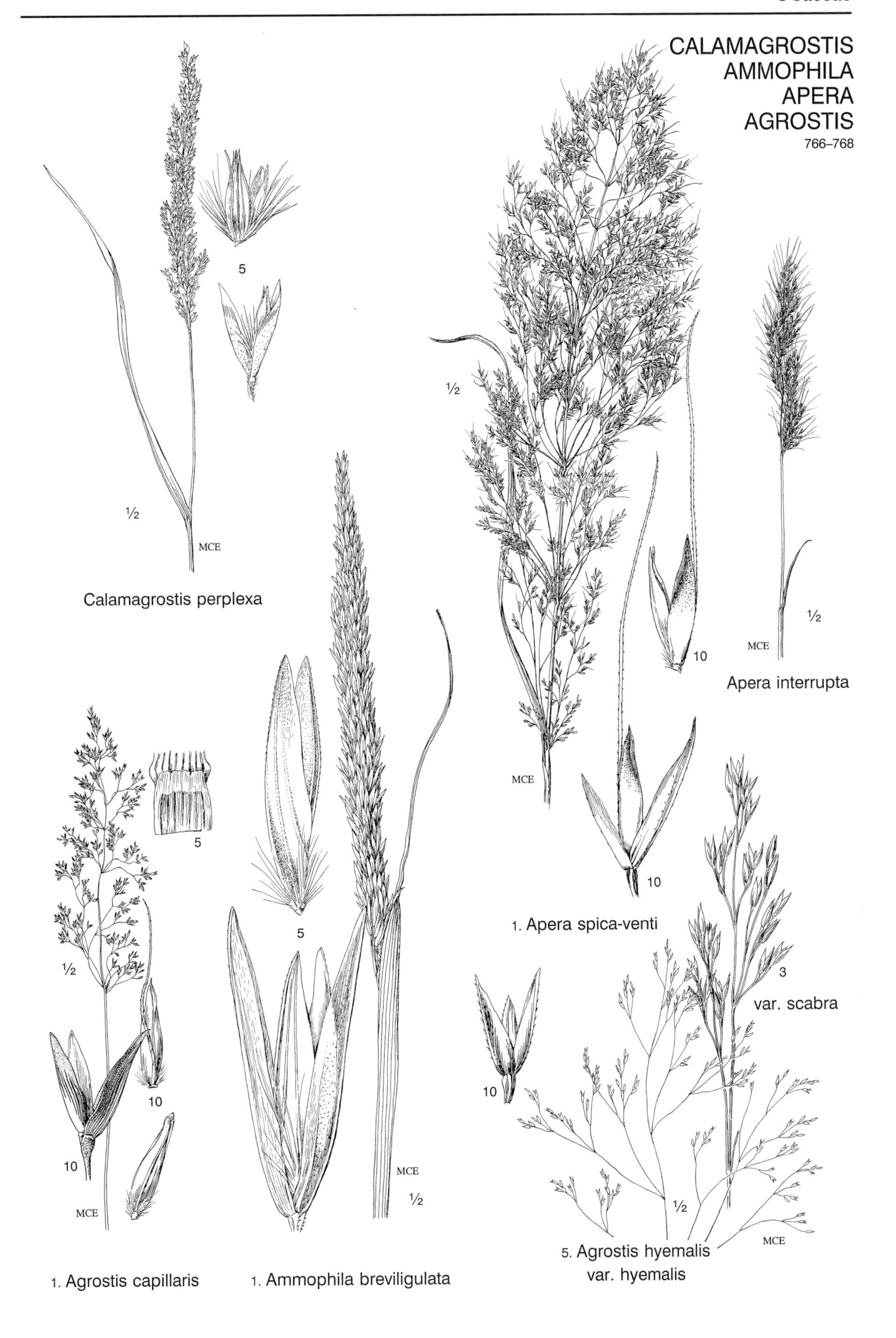

Calamagrostis perplexa

Apera interrupta

1. Apera spica-venti

var. scabra

5. Agrostis hyemalis
var. hyemalis

1. Agrostis capillaris

1. Ammophila breviligulata

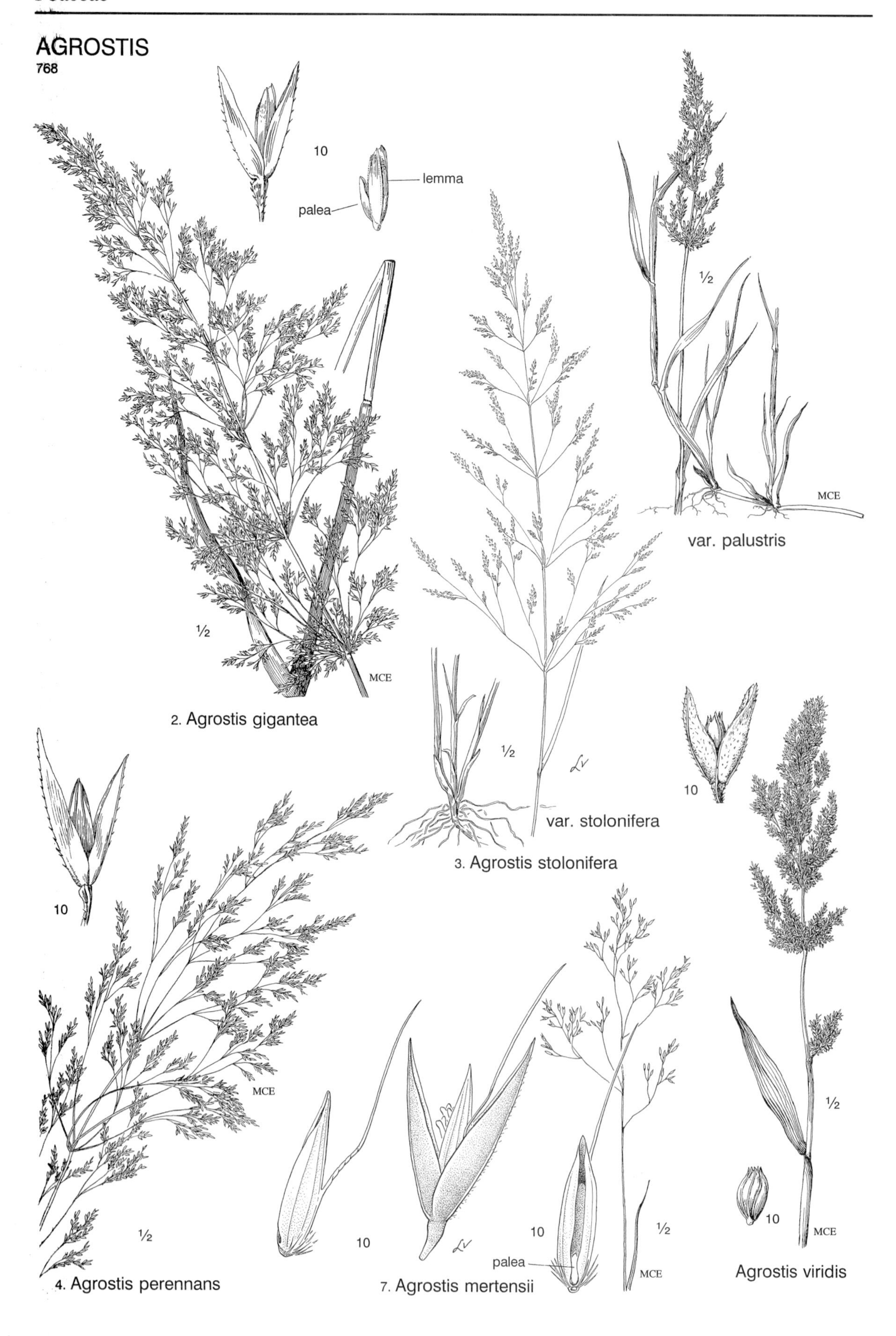

10

lemma

palea

2. Agrostis gigantea

var. palustris

MCE

var. stolonifera

3. Agrostis stolonifera

10

10

4. Agrostis perennans

10

10

palea

7. Agrostis mertensii

10

MCE

Agrostis viridis

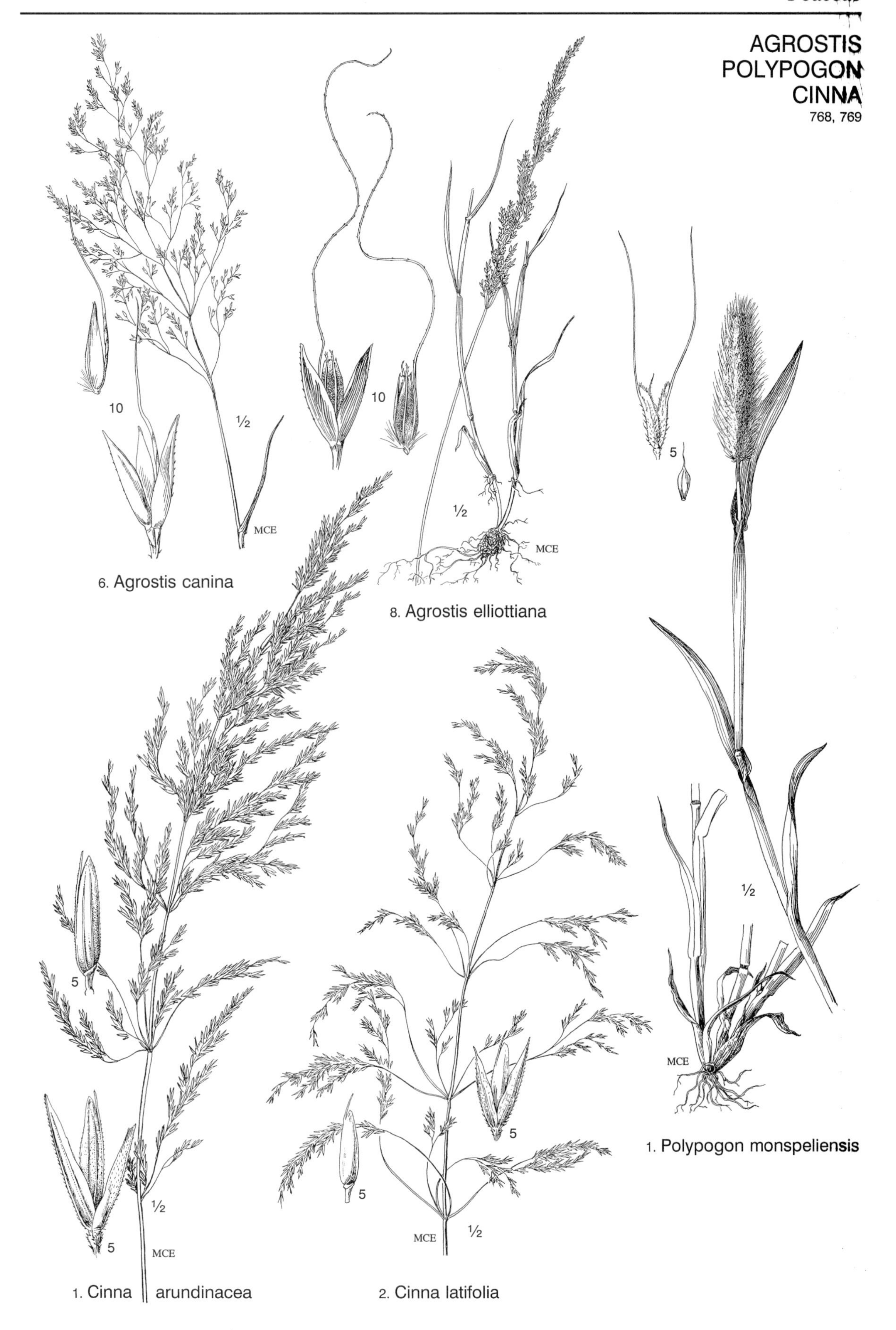

6. Agrostis canina

8. Agrostis elliottiana

1. Cinna arundinacea

2. Cinna latifolia

1. Polypogon monspeliensis

ALOPECURUS
PHLEUM
BECKMANNIA
BROMUS
769–771

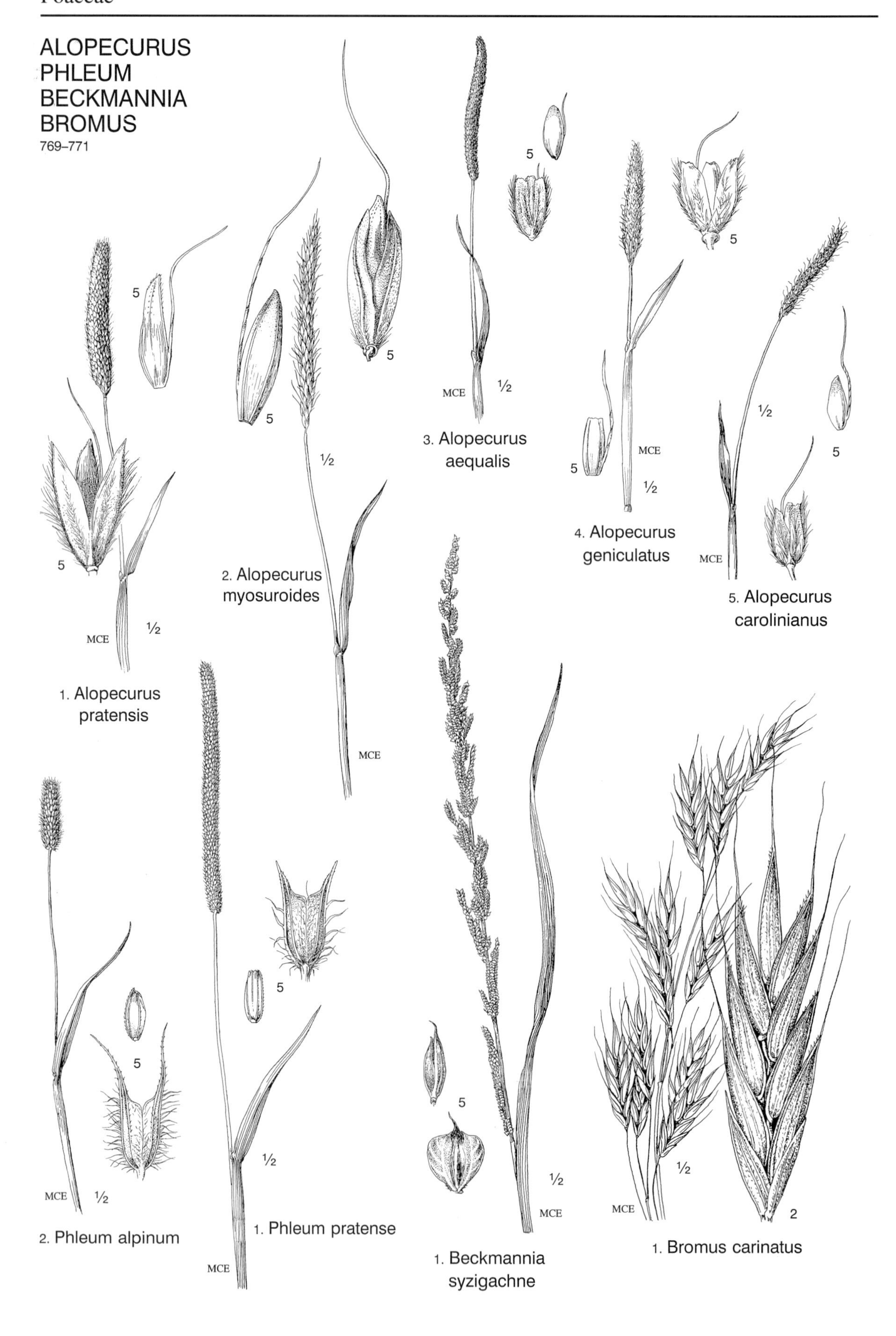

1. Alopecurus
pratensis

2. Alopecurus
myosuroides

3. Alopecurus
aequalis

4. Alopecurus
geniculatus

5. Alopecurus
carolinianus

2. Phleum alpinum

1. Phleum pratense

1. Beckmannia
syzigachne

1. Bromus carinatus

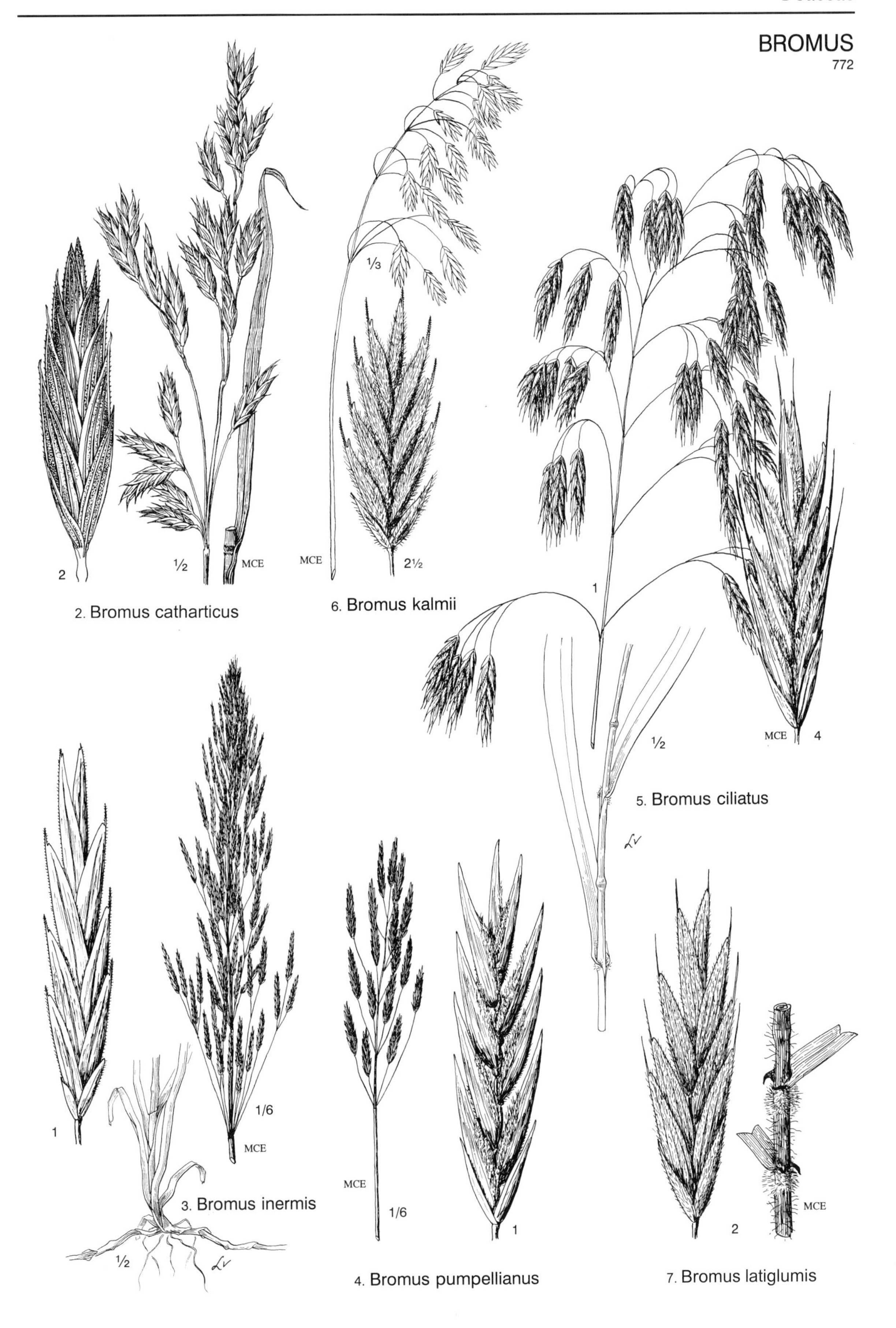

2. Bromus catharticus

6. Bromus kalmii

5. Bromus ciliatus

3. Bromus inermis

4. Bromus pumpellianus

7. Bromus latiglumis

BROMUS
772, 773

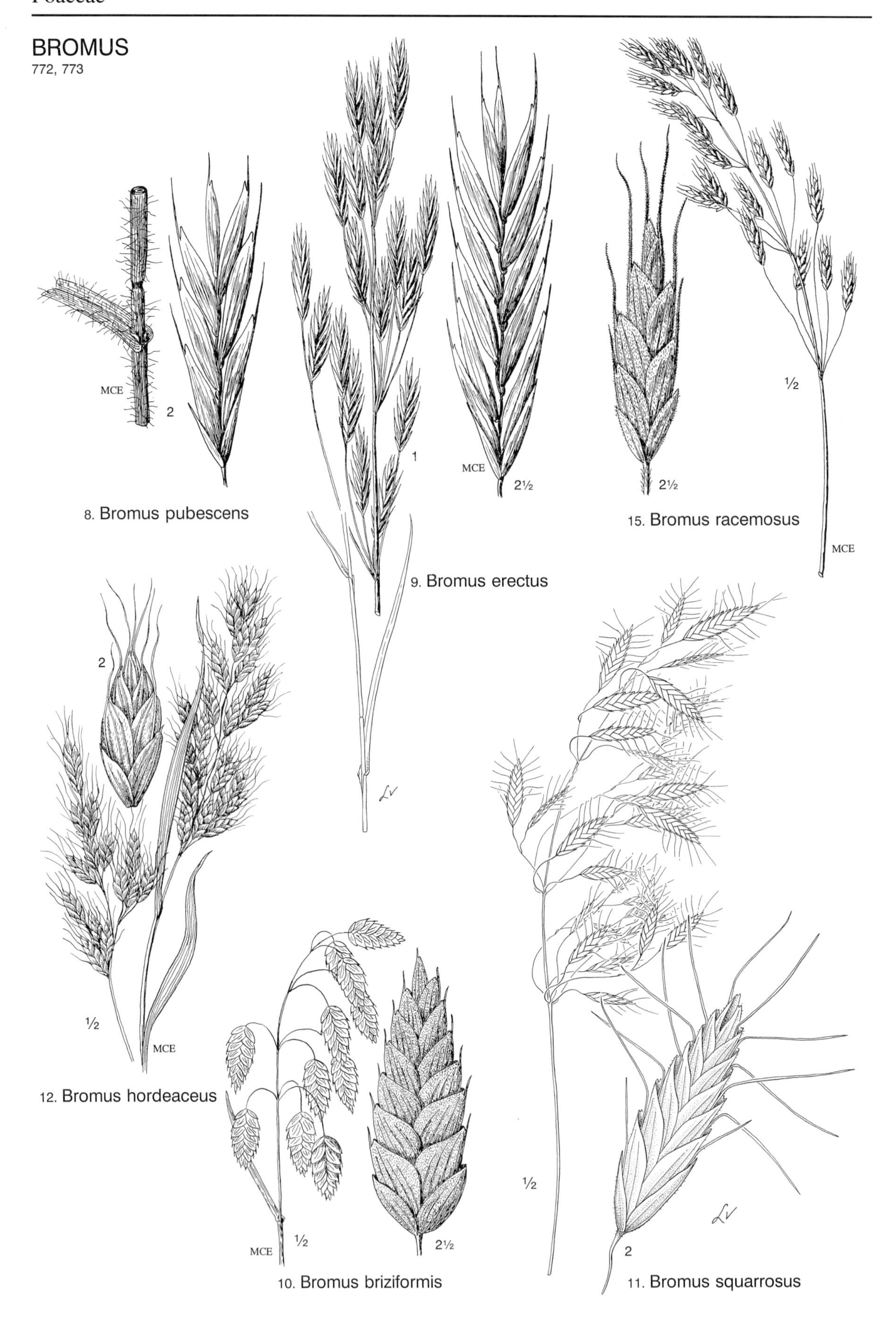

8. Bromus pubescens

9. Bromus erectus

15. Bromus racemosus

12. Bromus hordeaceus

10. Bromus briziformis

11. Bromus squarrosus

BROMUS

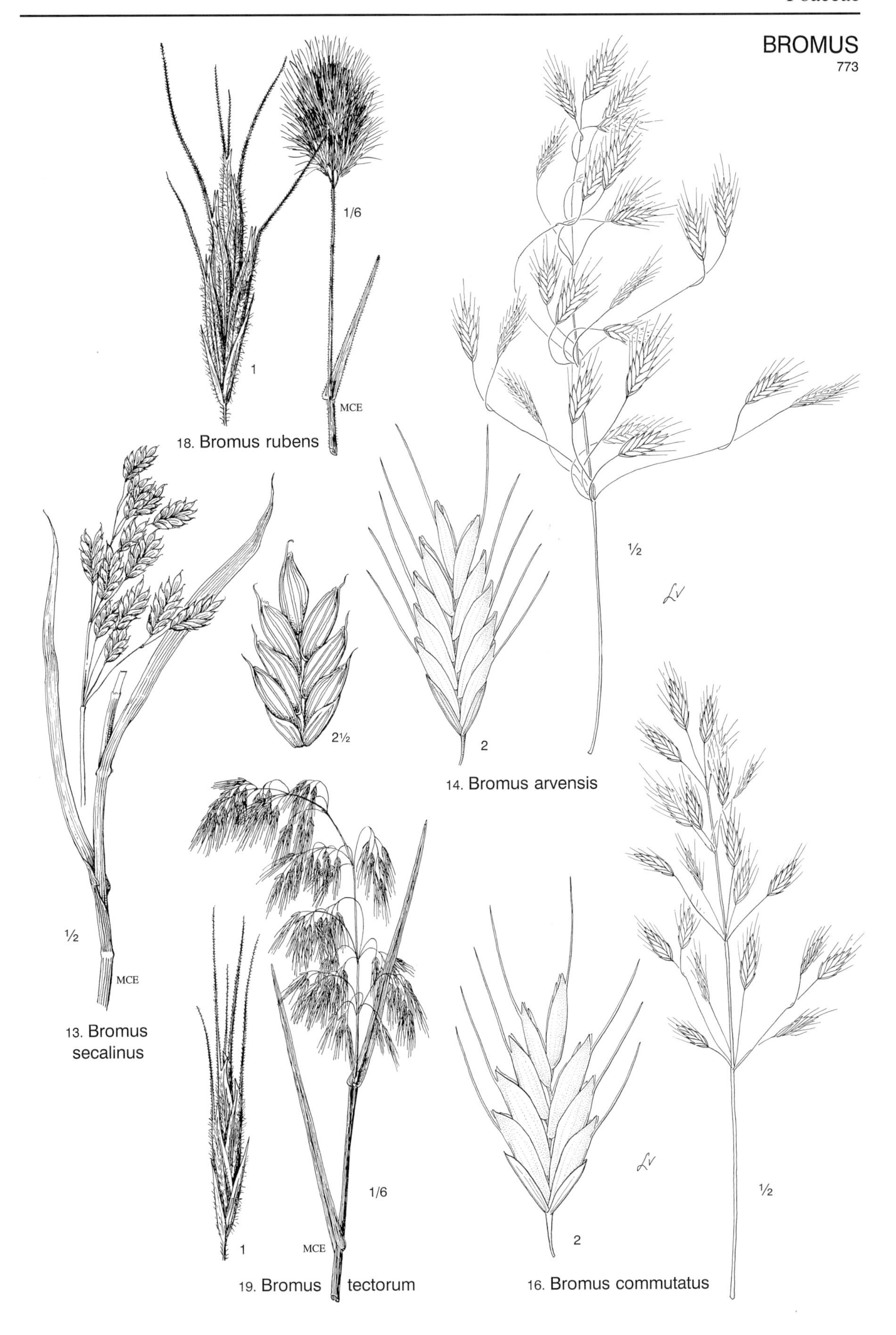

18. Bromus rubens

13. Bromus secalinus

14. Bromus arvensis

19. Bromus tectorum

16. Bromus commutatus

BROMUS
ELYMUS
773–775

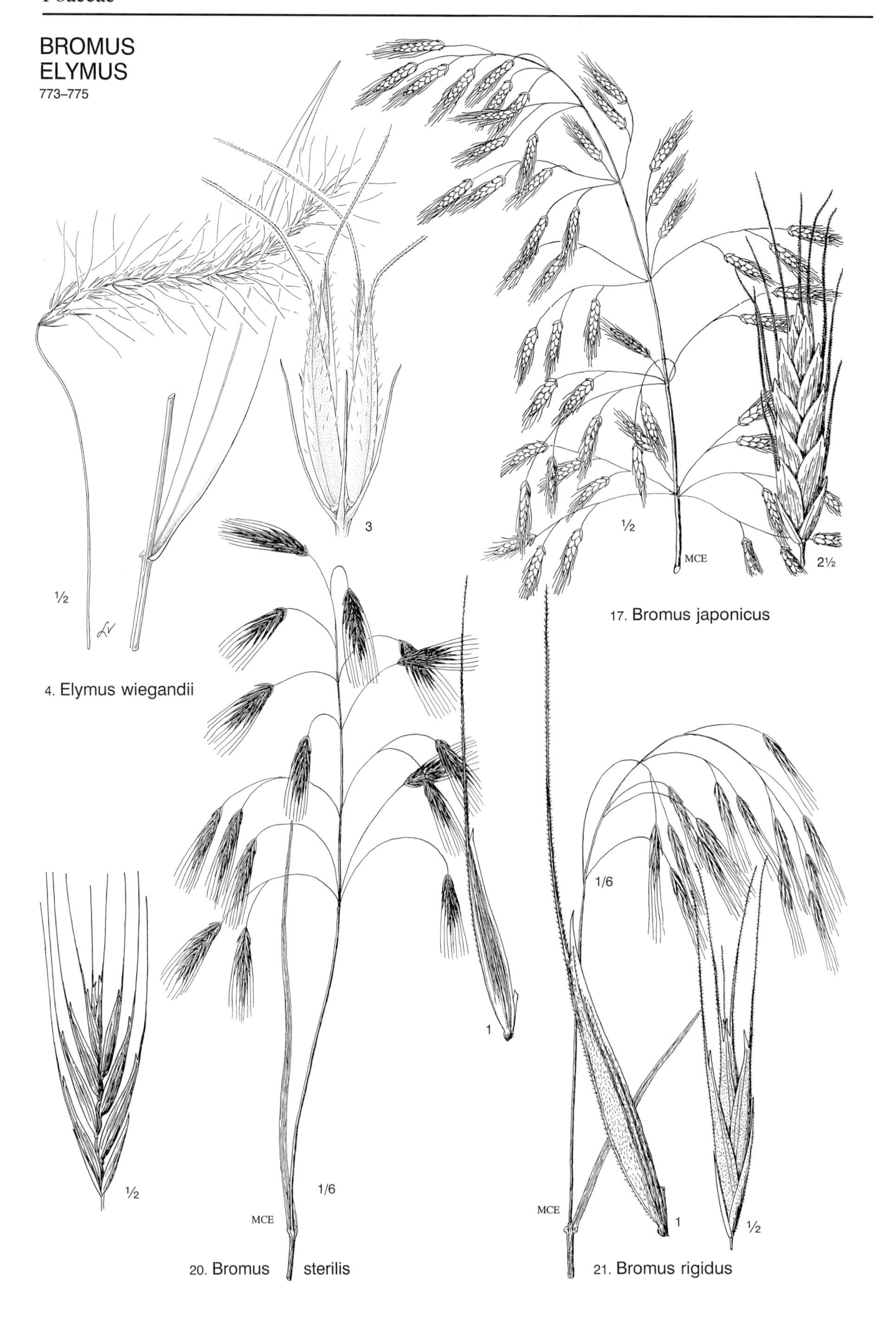

3

1/2

4. Elymus wiegandii

17. Bromus japonicus

1/2

MCE

2½

1/6

1/2

MCE

1/6

20. Bromus sterilis

1

MCE

1

1/2

21. Bromus rigidus

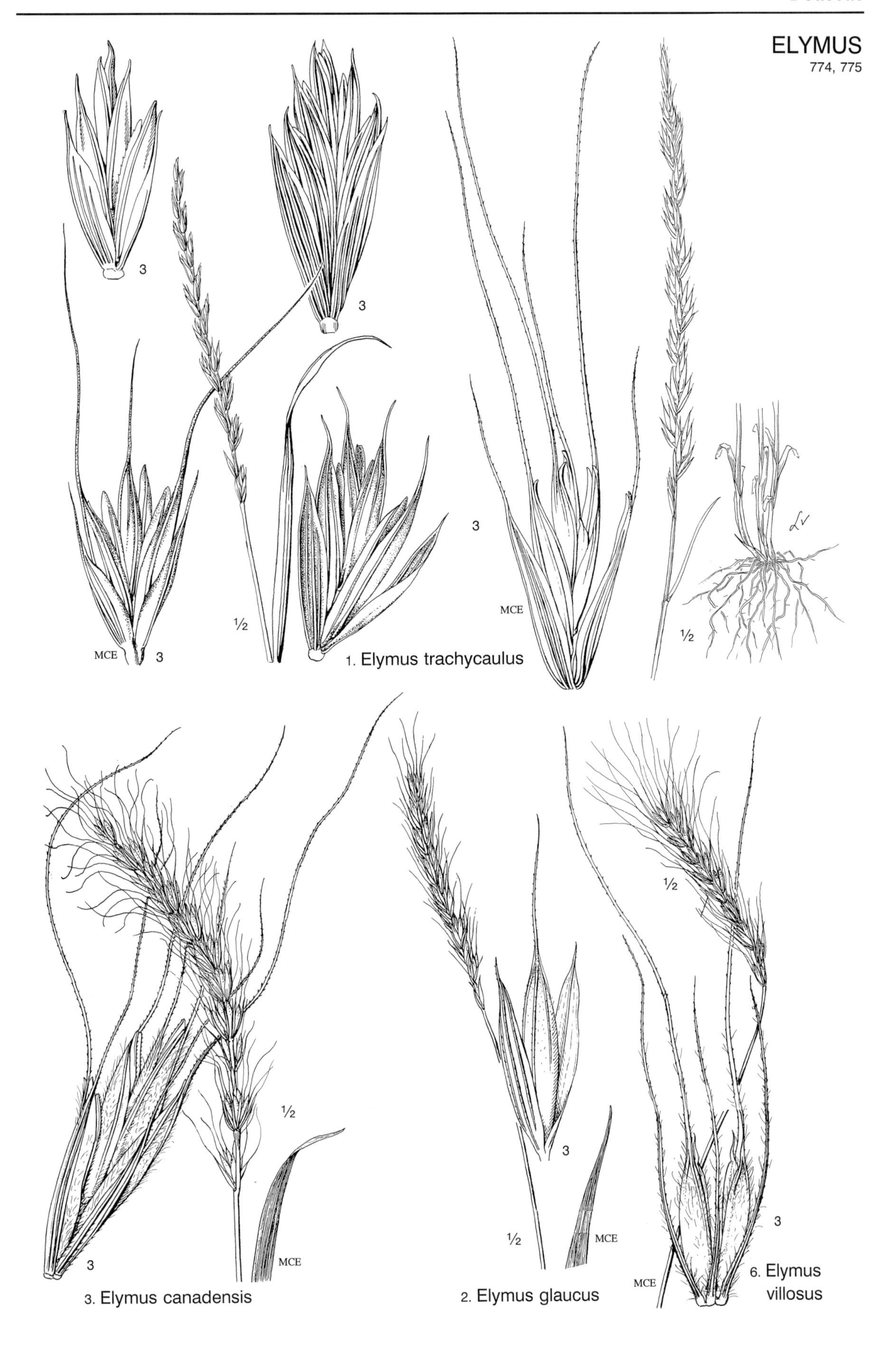

1. Elymus trachycaulus

3. Elymus canadensis

2. Elymus glaucus

6. Elymus villosus

ELYMUS

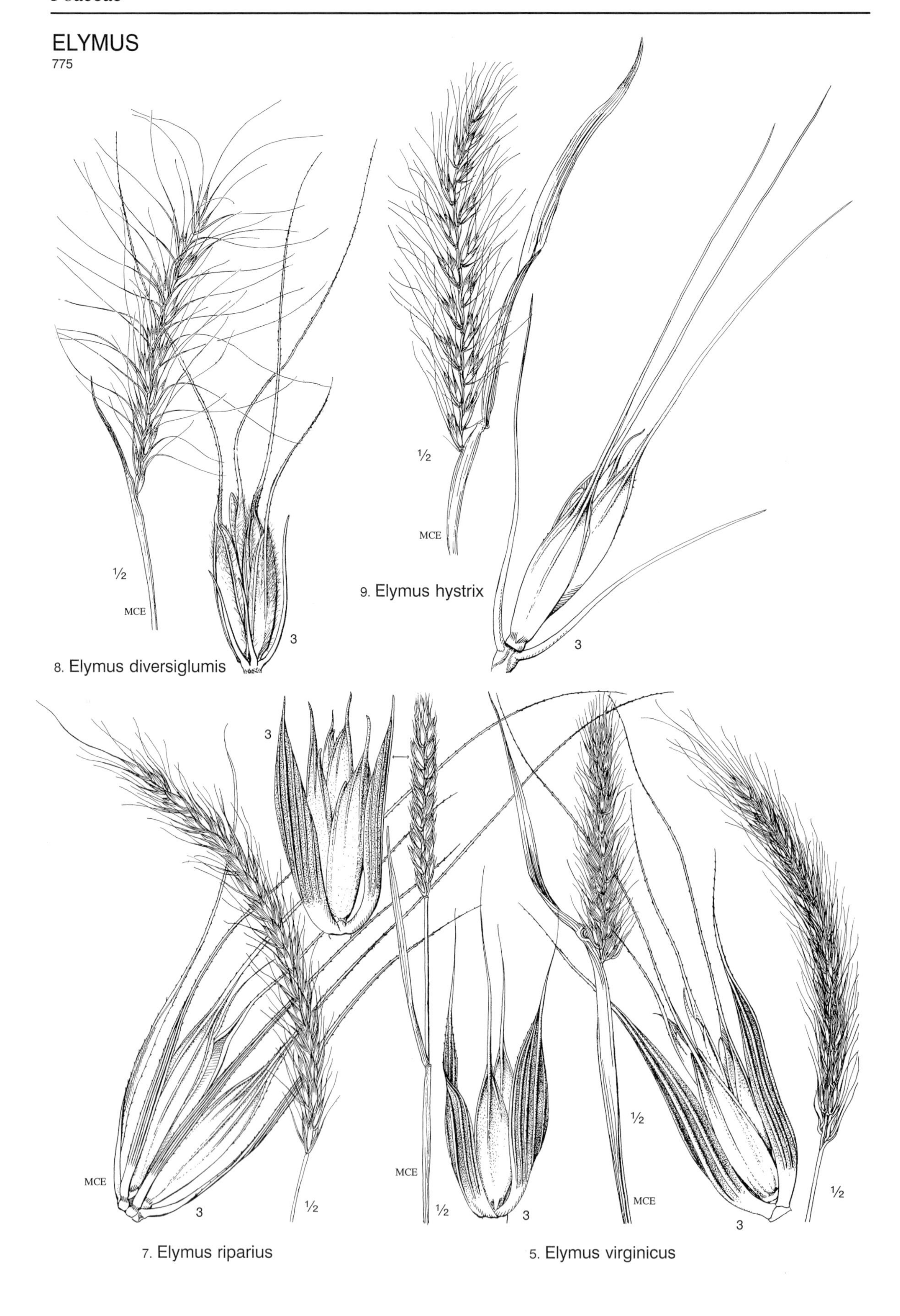

8. Elymus diversiglumis

9. Elymus hystrix

7. Elymus riparius

5. Elymus virginicus

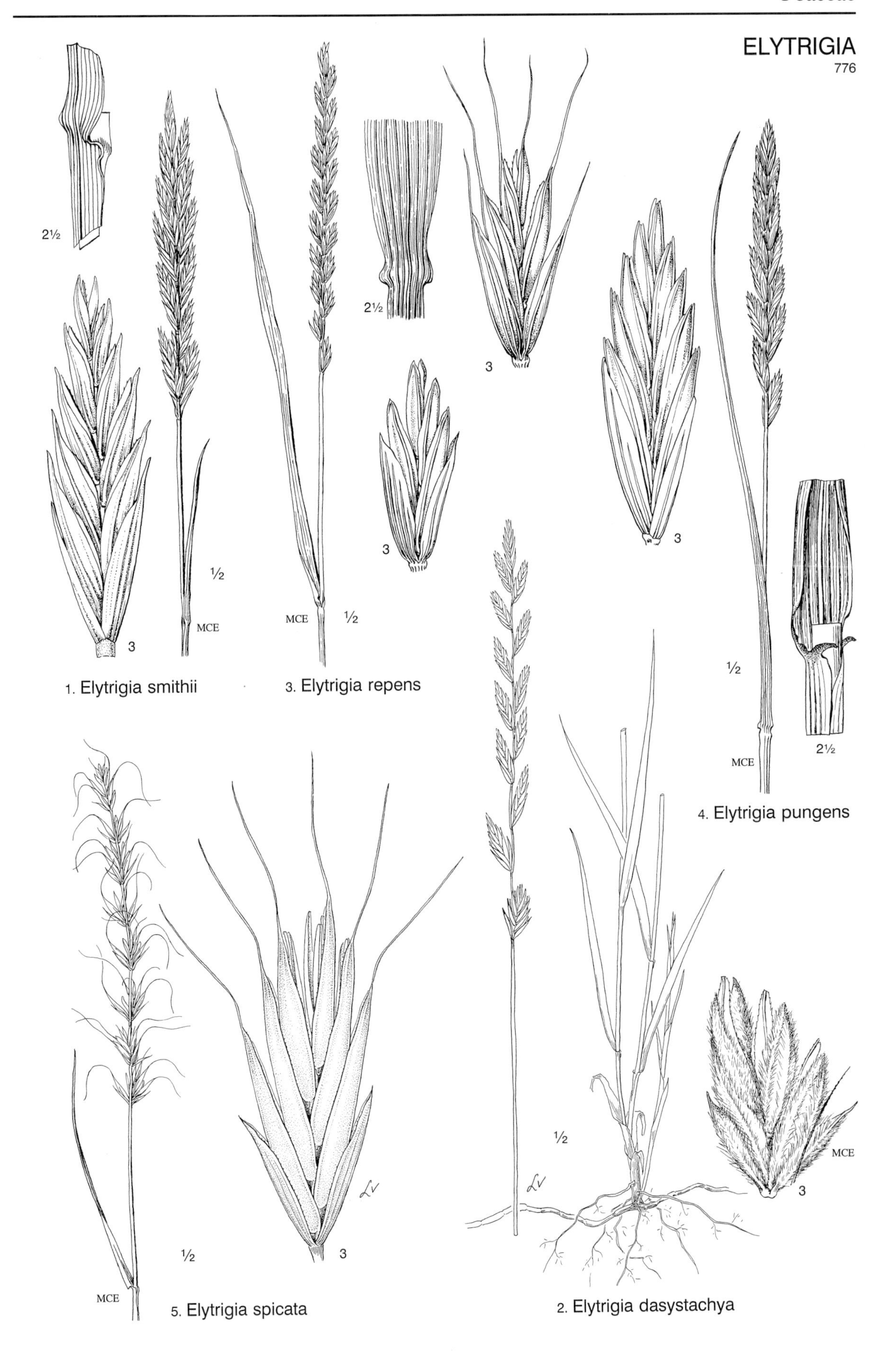

1. Elytrigia smithii

3. Elytrigia repens

4. Elytrigia pungens

5. Elytrigia spicata

2. Elytrigia dasystachya

ELYTRIGIA
LEYMUS
776, 777

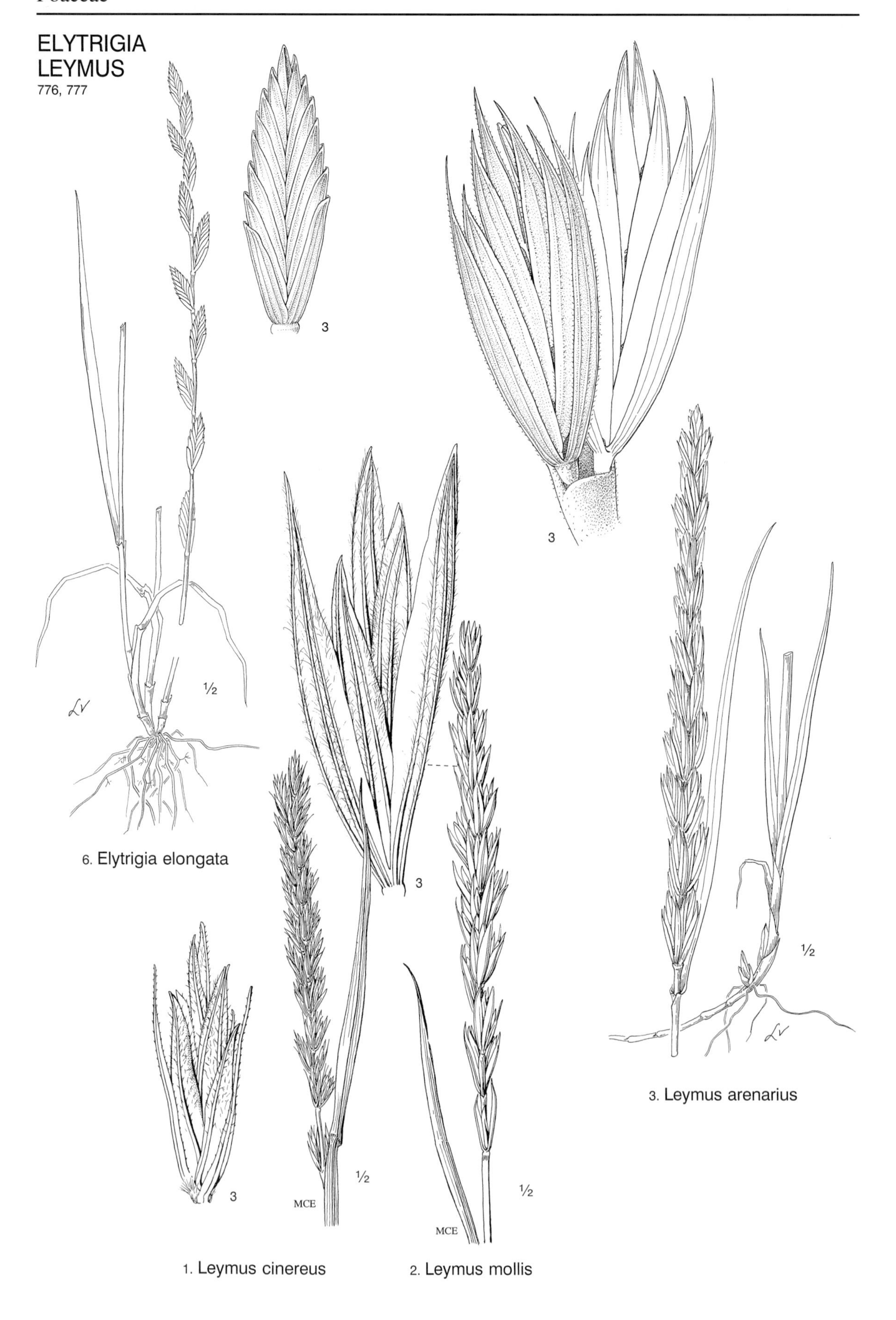

6. Elytrigia elongata

3. Leymus arenarius

1. Leymus cinereus

2. Leymus mollis

MCE

MCE

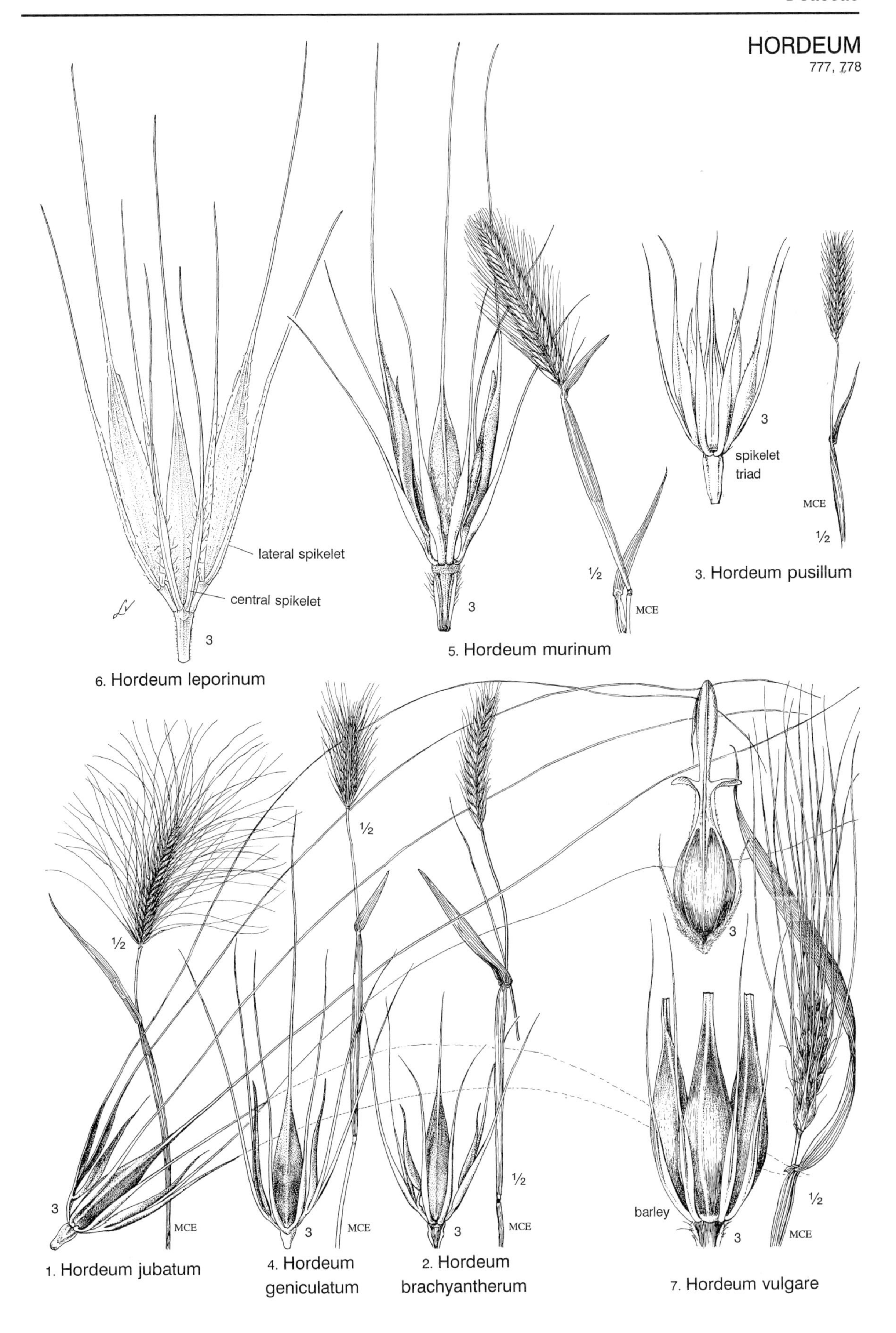

lateral spikelet

central spikelet

3

6. Hordeum leporinum

½

MCE

3

5. Hordeum murinum

3

spikelet triad

MCE

½

3. Hordeum pusillum

½

½

3

MCE

1. Hordeum jubatum

3

MCE

4. Hordeum geniculatum

3

MCE

2. Hordeum brachyantherum

½

3

½

barley

3

MCE

7. Hordeum vulgare

AGROPYRON
SECALE
TRITICUM
AEGILOPS
778, 779

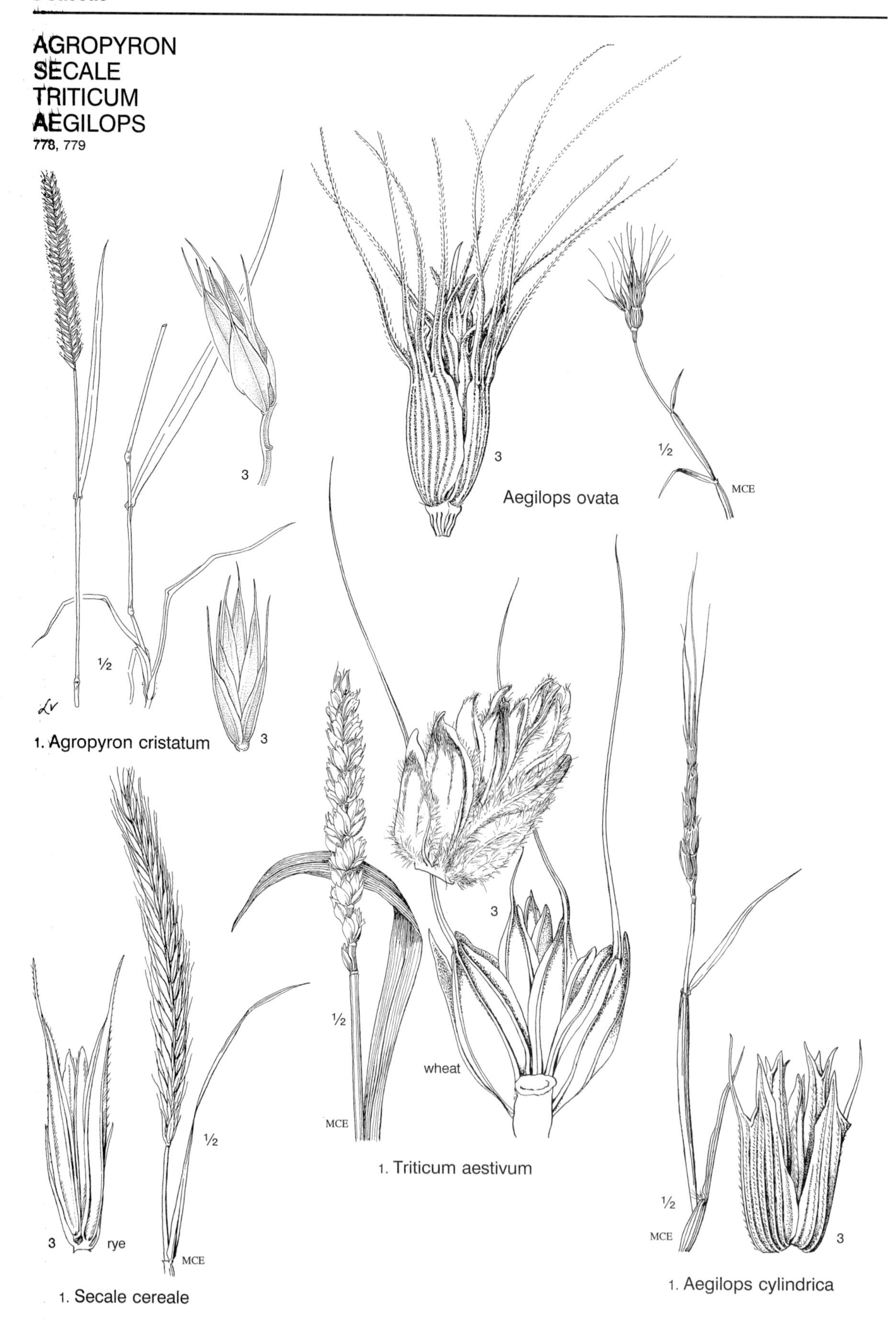

Aegilops ovata

MCE

1. Agropyron cristatum

1. Triticum aestivum

wheat

MCE

rye

MCE

1. Secale cereale

MCE

1. Aegilops cylindrica

ILLUSTRATED COMPANION TO

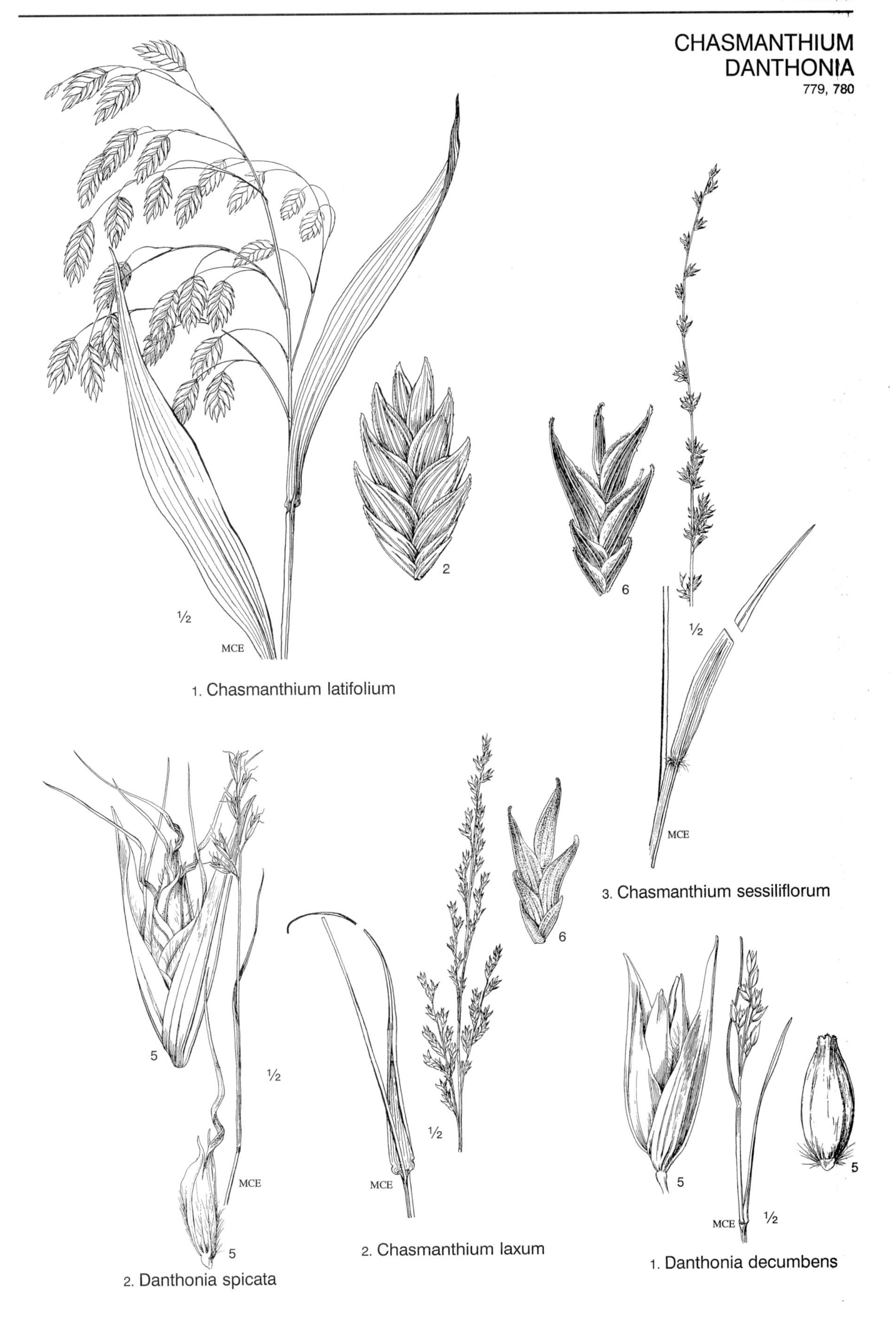

1. Chasmanthium latifolium

3. Chasmanthium sessiliflorum

2. Danthonia spicata

2. Chasmanthium laxum

1. Danthonia decumbens

DANTHONIA
MOLINIA
PHRAGMITES
780, 781

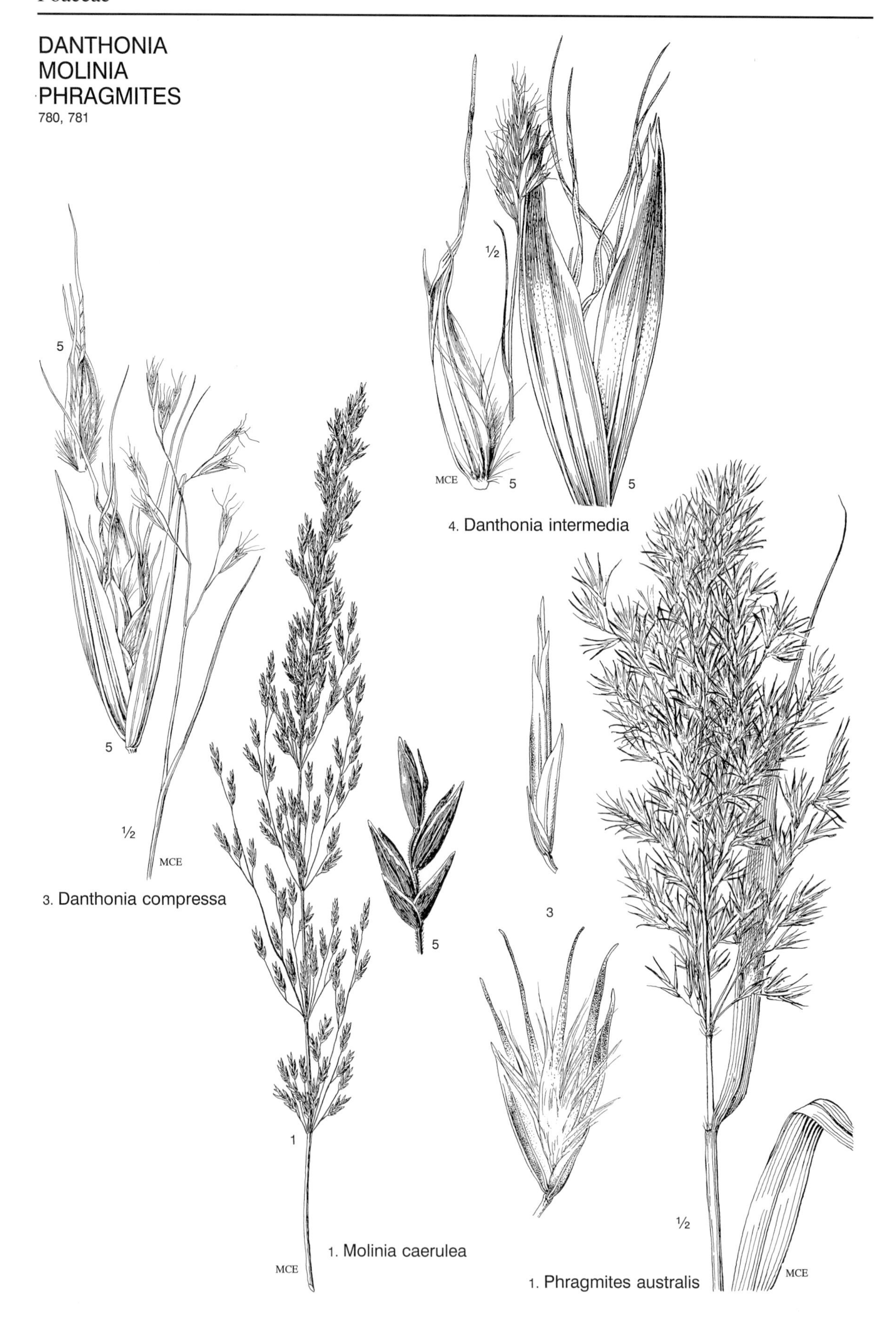

4. Danthonia intermedia

3. Danthonia compressa

½

MCE

1. Molinia caerulea

1. Phragmites australis

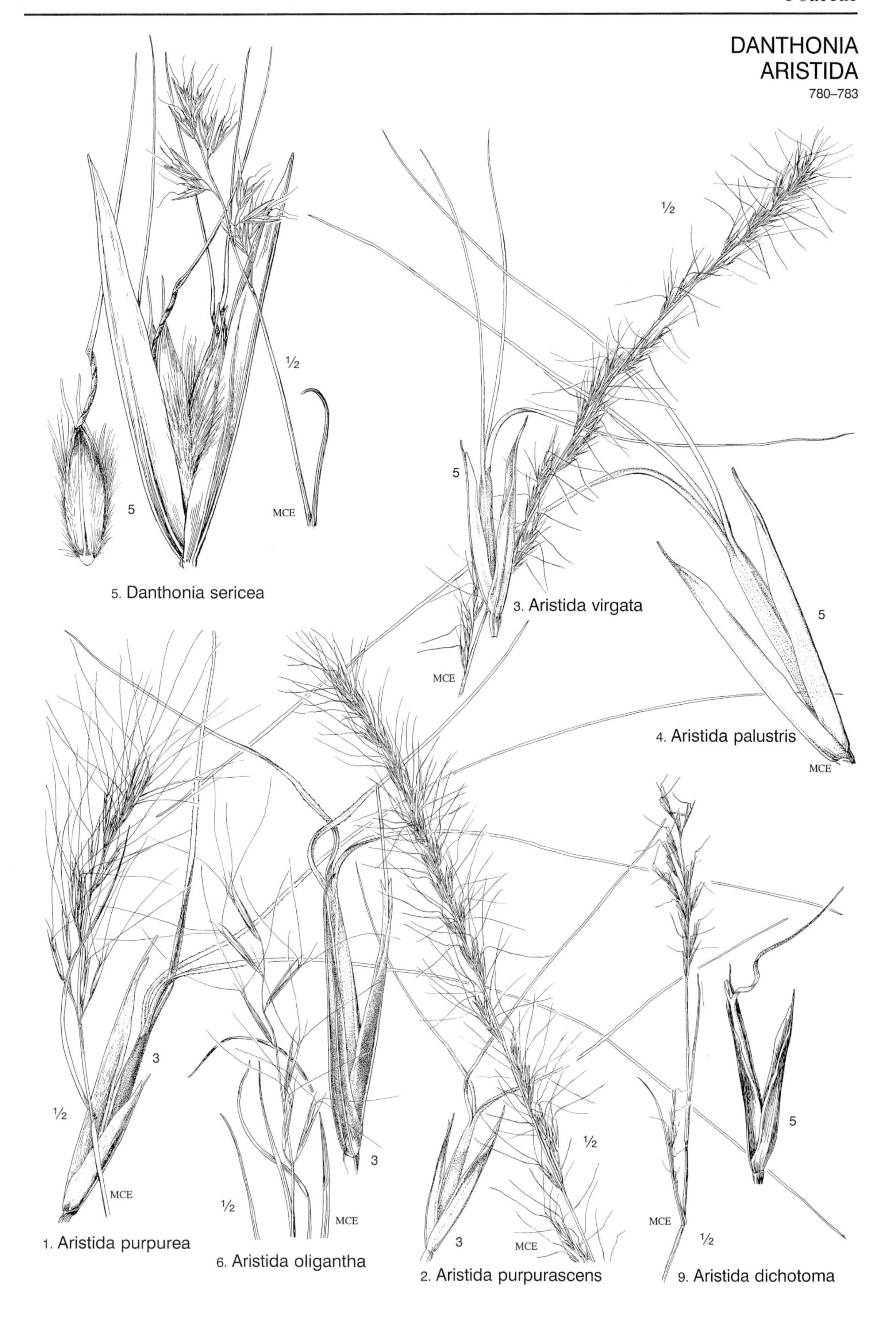

5. Danthonia sericea

3. Aristida virgata

4. Aristida palustris

1. Aristida purpurea

6. Aristida oligantha

2. Aristida purpurascens

9. Aristida dichotoma

ARISTIDA
782, 783

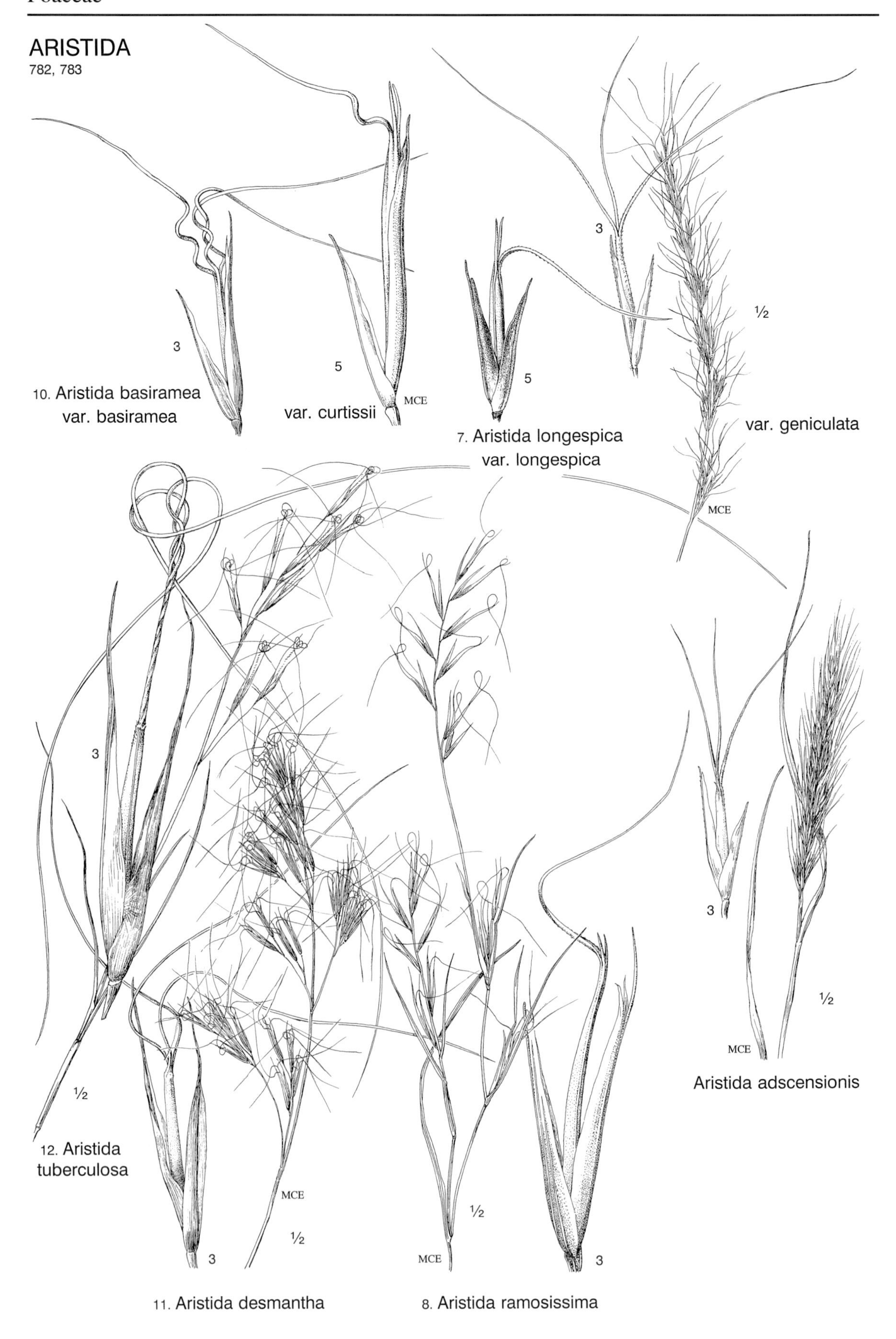

10. Aristida basiramea
var. basiramea

var. curtissii

MCE

7. Aristida longespica
var. longespica

var. geniculata

½

MCE

3

½

3

Aristida adscensionis

MCE

12. Aristida
tuberculosa

½

MCE

½

11. Aristida desmantha

½

MCE

8. Aristida ramosissima

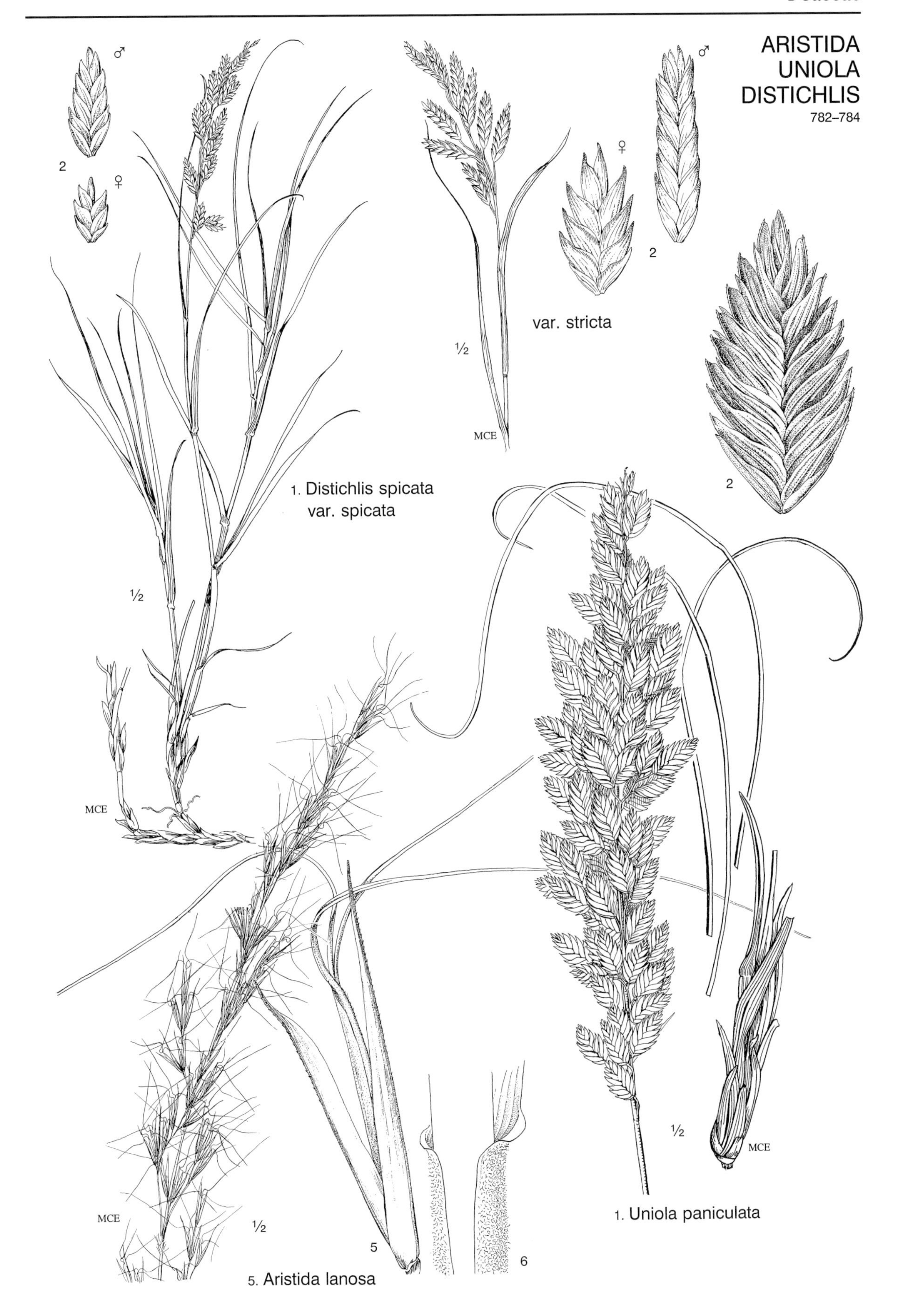

2

♂

♀

1. Distichlis spicata
var. spicata

½

½

MCE

var. stricta

♀

2

♂

2

2

½

MCE

MCE

1. Uniola paniculata

MCE

½

5

6

5. Aristida lanosa

TRIDENS
TRIPLASIS
LEPTOCHLOA
784, 785

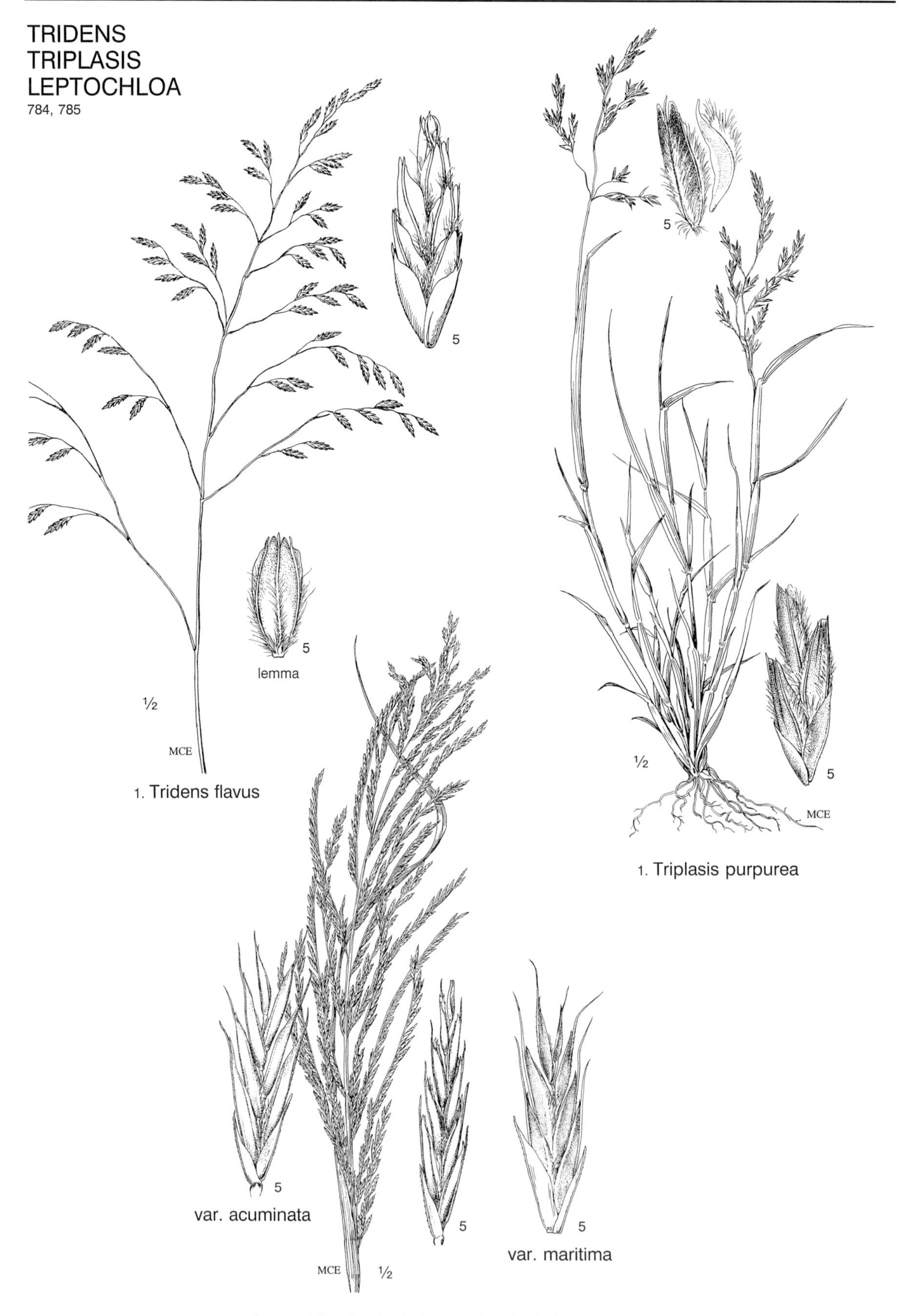

1/2

MCE

1. Tridens flavus

lemma

5

5

1. Triplasis purpurea

1/2

MCE

5

var. acuminata

5

5

var. maritima

MCE   1/2

3. Leptochloa fascicularis var. fascicularis

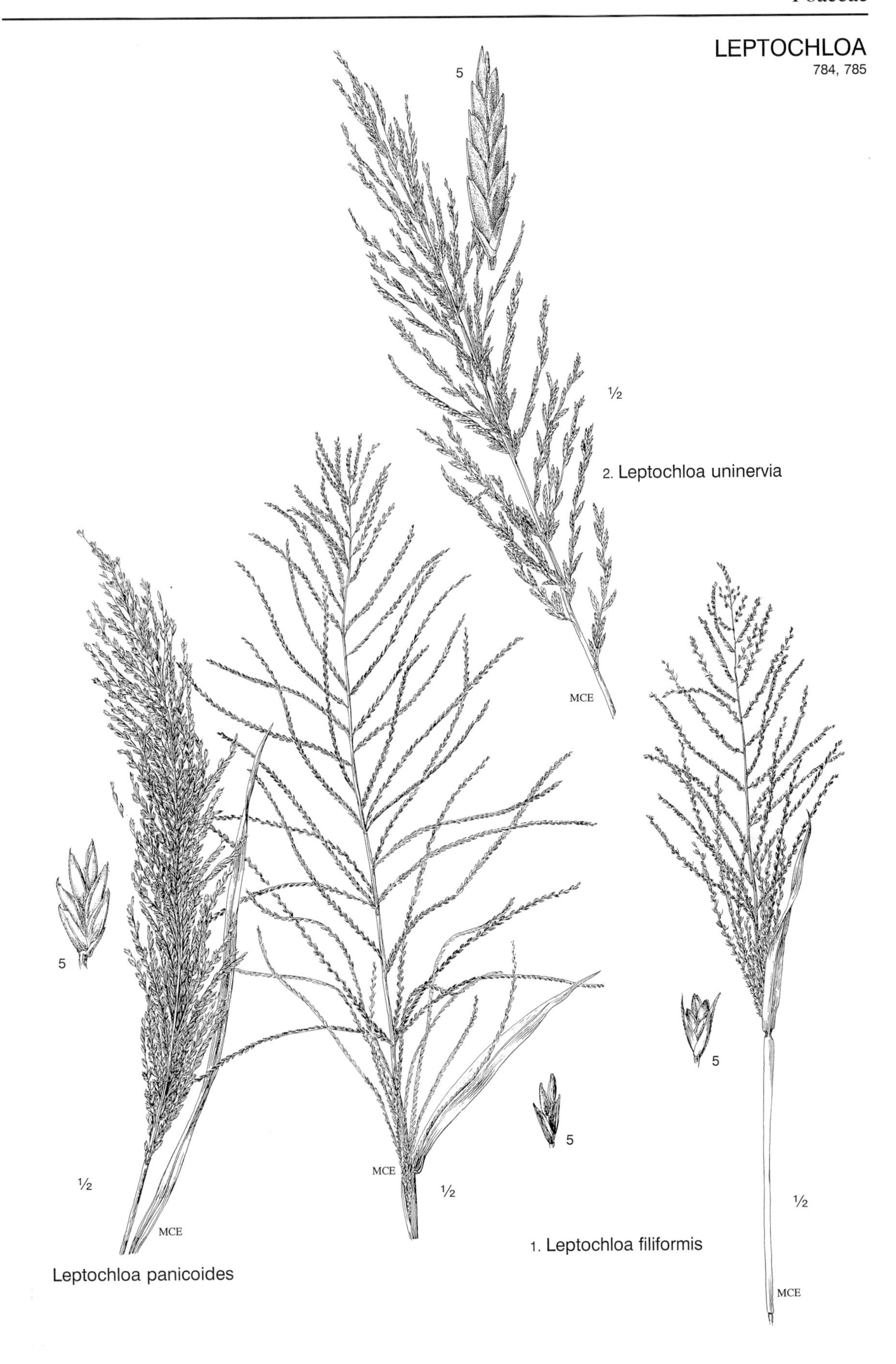

5

½

2. Leptochloa uninervia

MCE

5

½

MCE

Leptochloa panicoides

MCE

½

5

5

½

MCE

1. Leptochloa filiformis

ERAGROSTIS
786

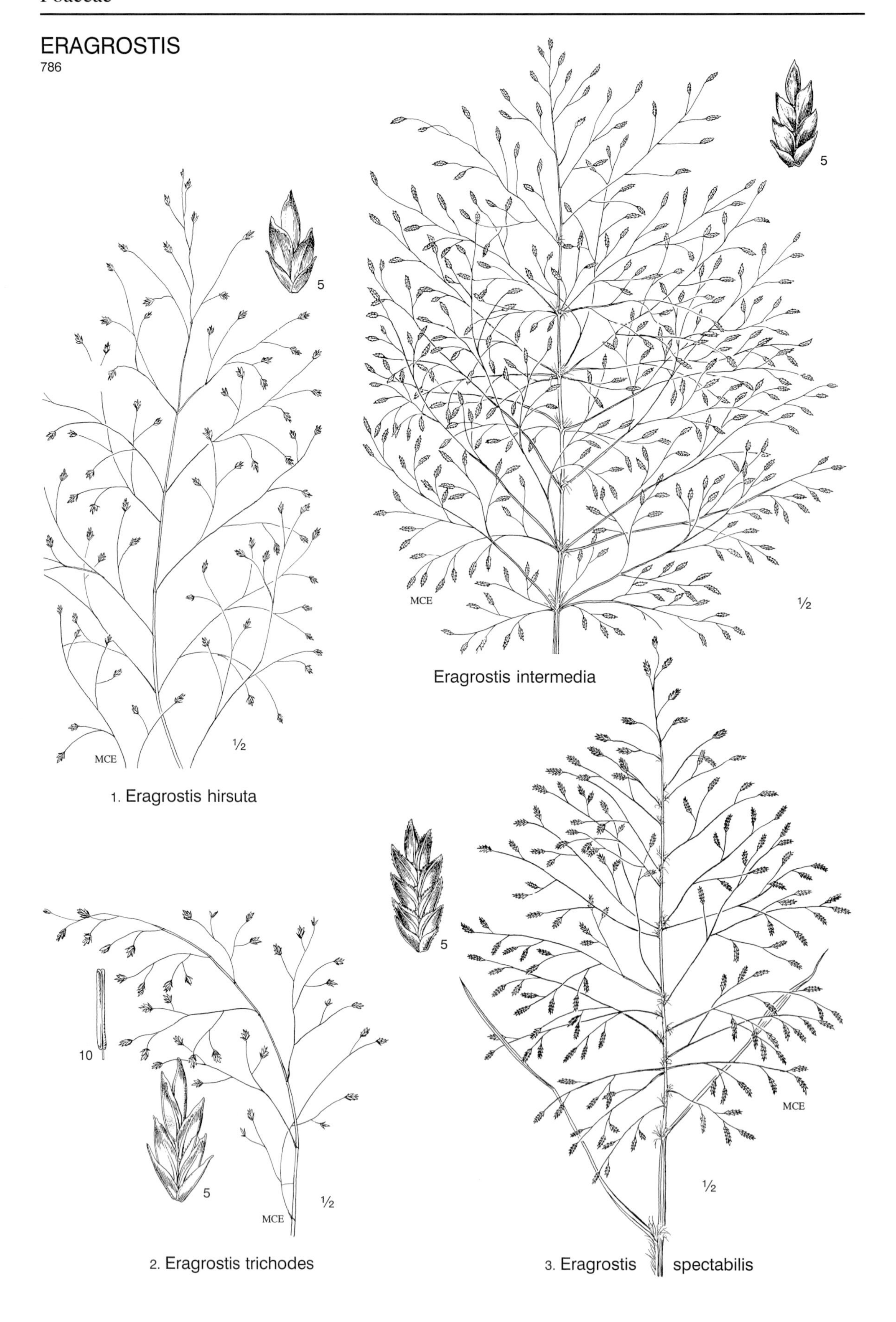

Eragrostis intermedia

1. Eragrostis hirsuta

2. Eragrostis trichodes

3. Eragrostis spectabilis

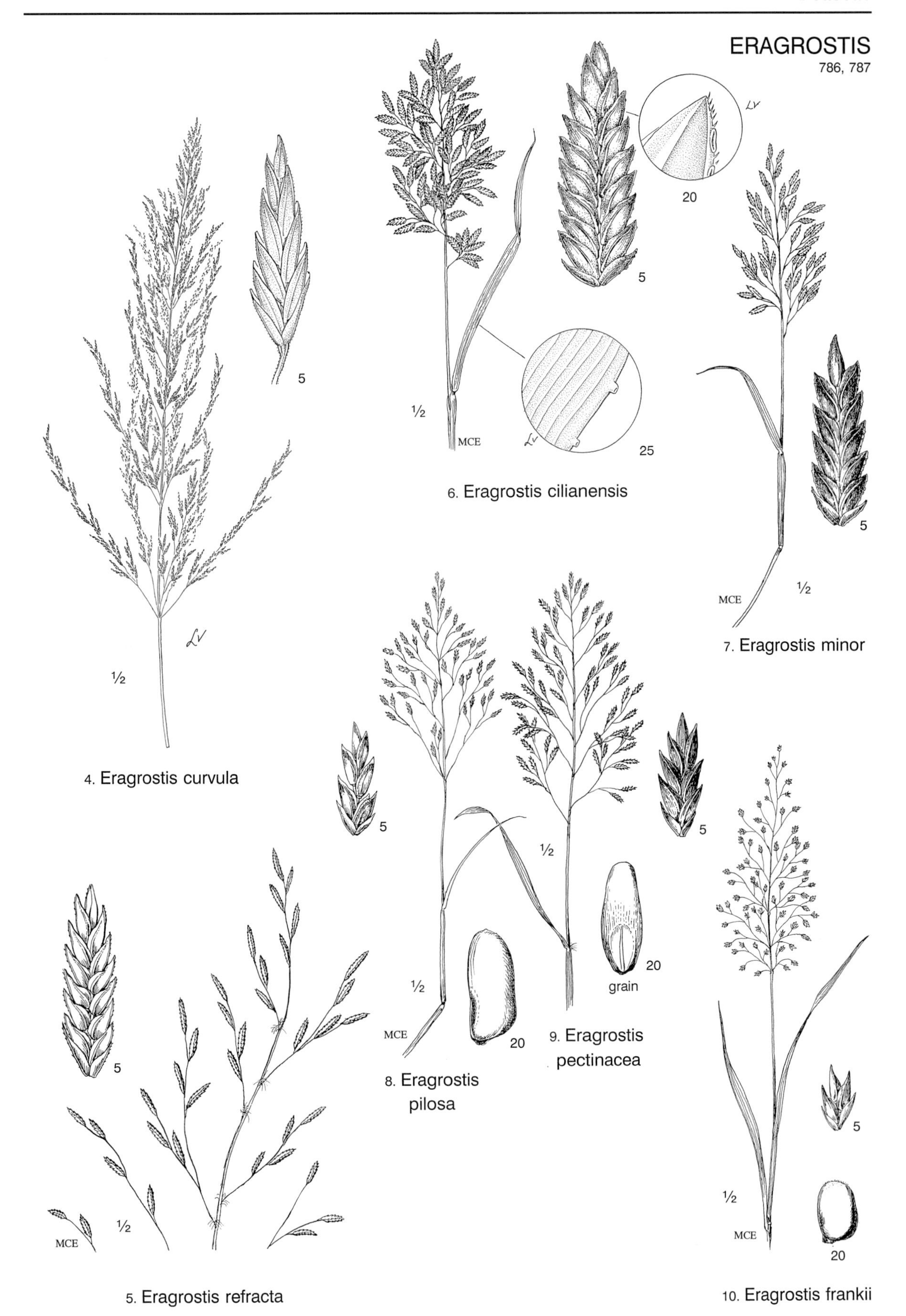

4. *Eragrostis curvula*

6. *Eragrostis cilianensis*

7. *Eragrostis minor*

8. *Eragrostis pilosa*

9. *Eragrostis pectinacea*

5. *Eragrostis refracta*

10. *Eragrostis frankii*

ERAGROSTIS
787

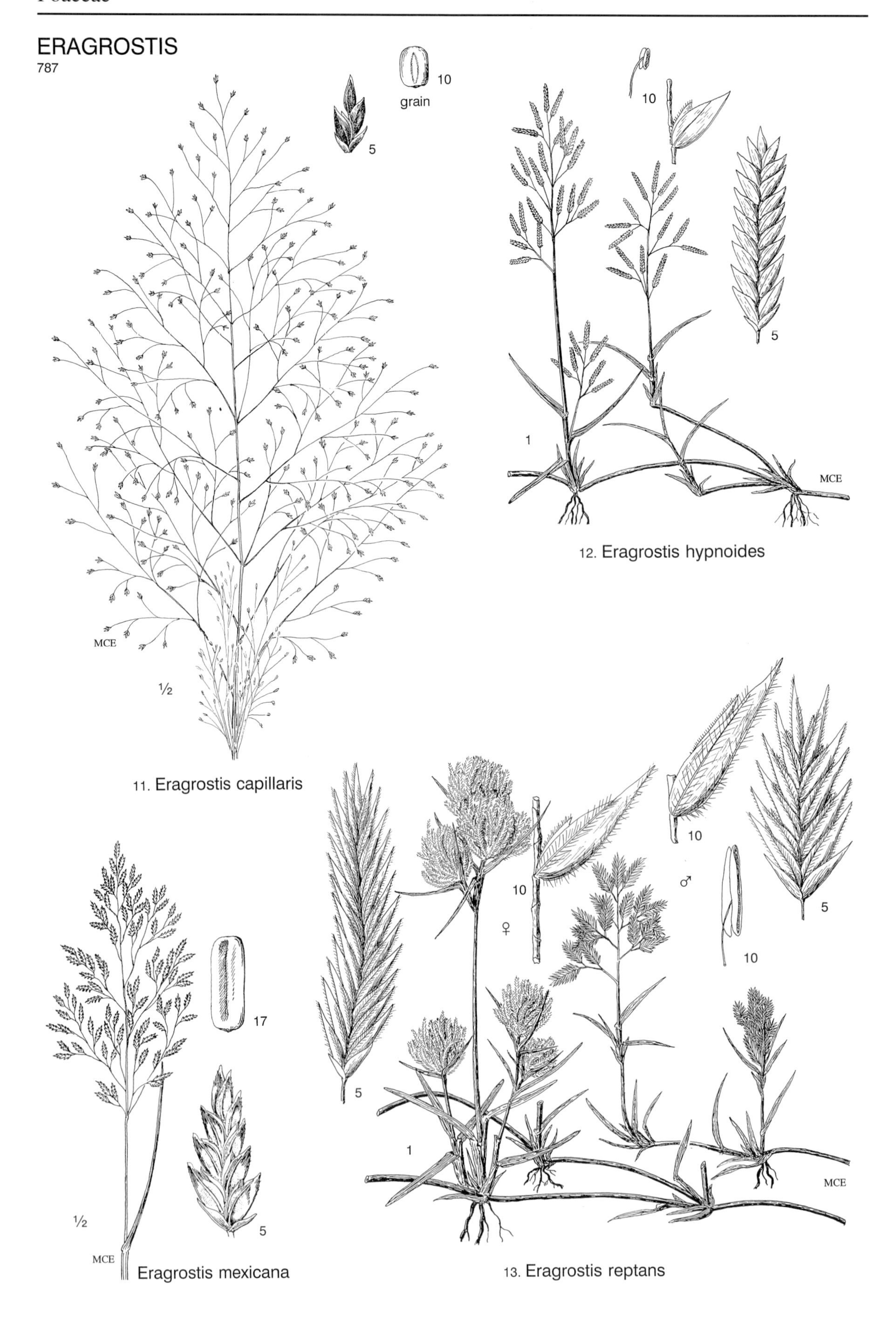

grain

10

5

11. Eragrostis capillaris

12. Eragrostis hypnoides

17

1/2

MCE

Eragrostis mexicana

13. Eragrostis reptans

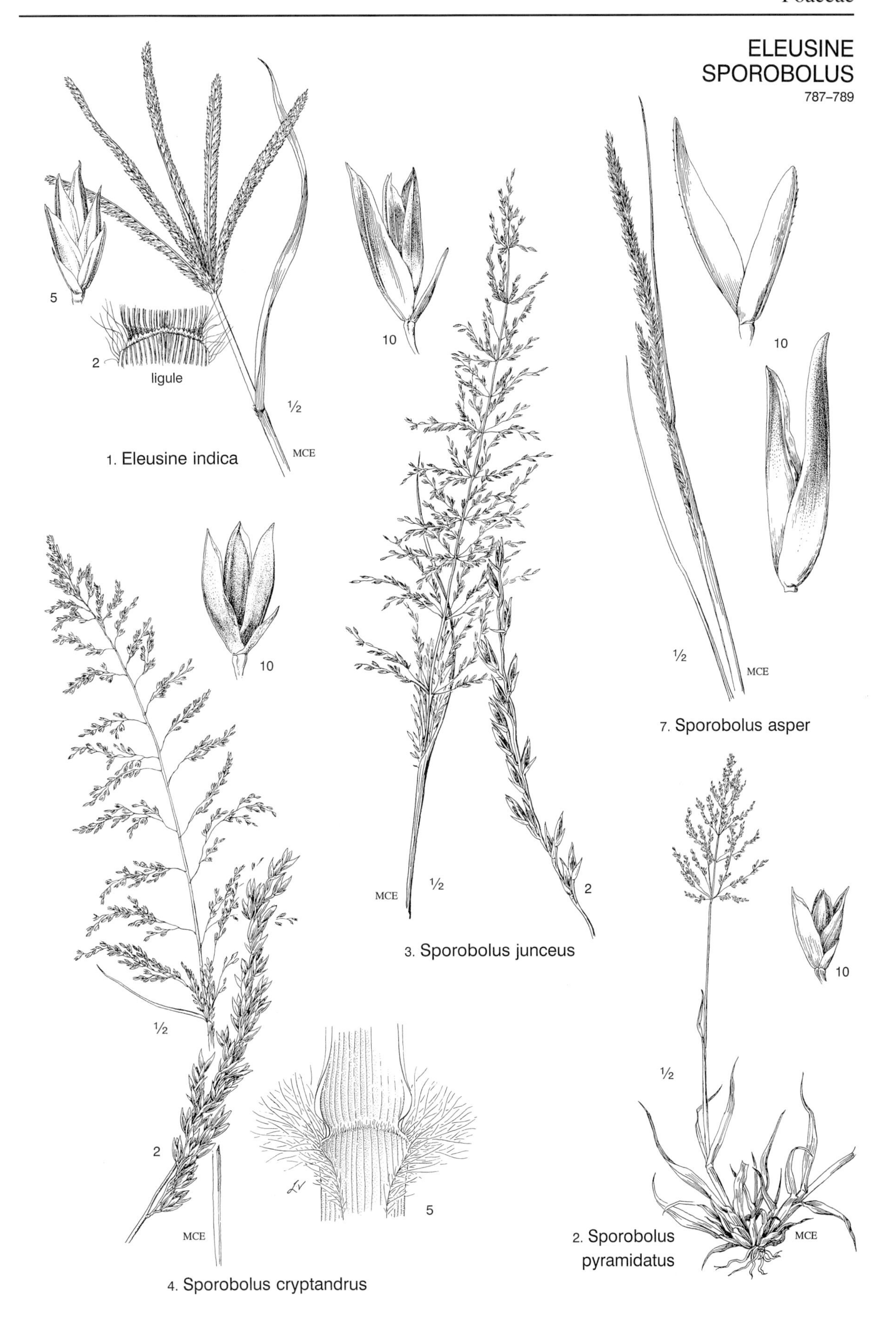

5

2

ligule

½

MCE

1. Eleusine indica

10

10

½

MCE

7. Sporobolus asper

10

½

2

MCE

½

2

3. Sporobolus junceus

10

½

2

MCE

5

MCE

4. Sporobolus cryptandrus

2. Sporobolus
pyramidatus

SPOROBOLUS
789

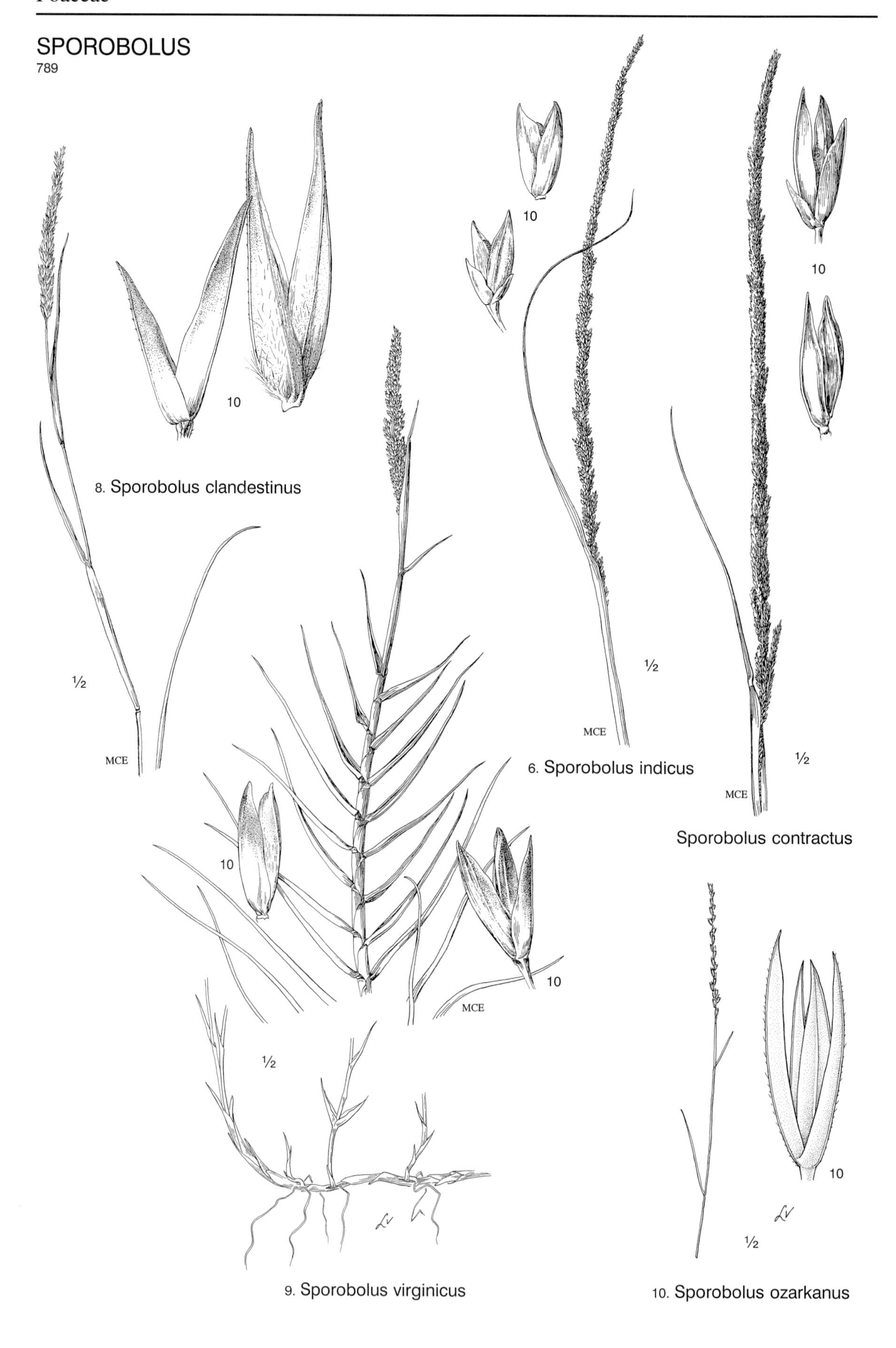

8. Sporobolus clandestinus

6. Sporobolus indicus

Sporobolus contractus

9. Sporobolus virginicus

10. Sporobolus ozarkanus

SPOROBOLUS
788, 789

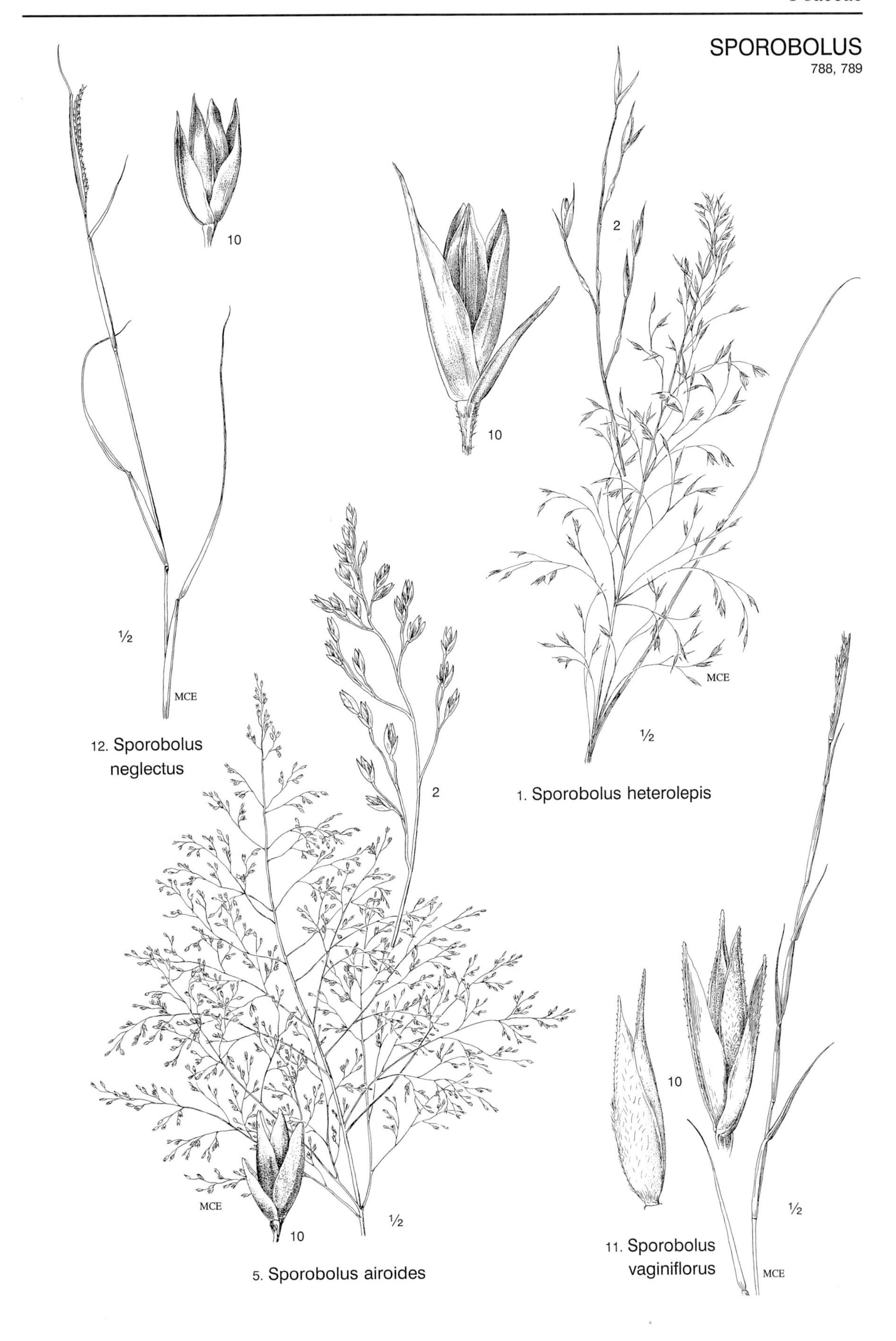

10

10

2

12. Sporobolus
neglectus

MCE

2

1. Sporobolus heterolepis

½

10

5. Sporobolus airoides

MCE

10

11. Sporobolus
vaginiflorus

½

MCE

743

CRYPSIS
MUHLENBERGIA
789–791

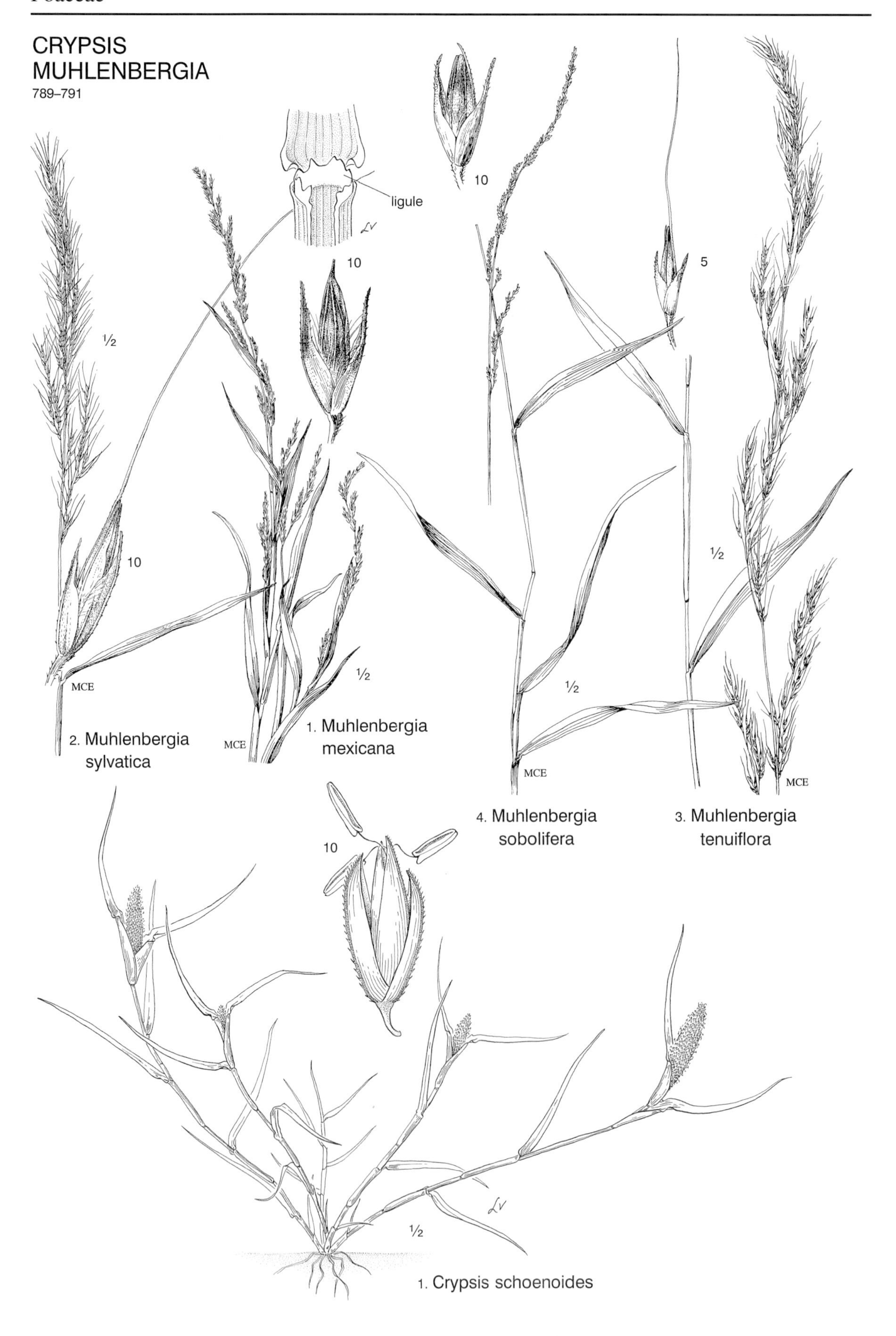

ligule

10

10

½

10

MCE

2. Muhlenbergia
sylvatica

MCE

1. Muhlenbergia
mexicana

½

5

½

½

MCE

4. Muhlenbergia
sobolifera

MCE

3. Muhlenbergia
tenuiflora

10

½

1. Crypsis schoenoides

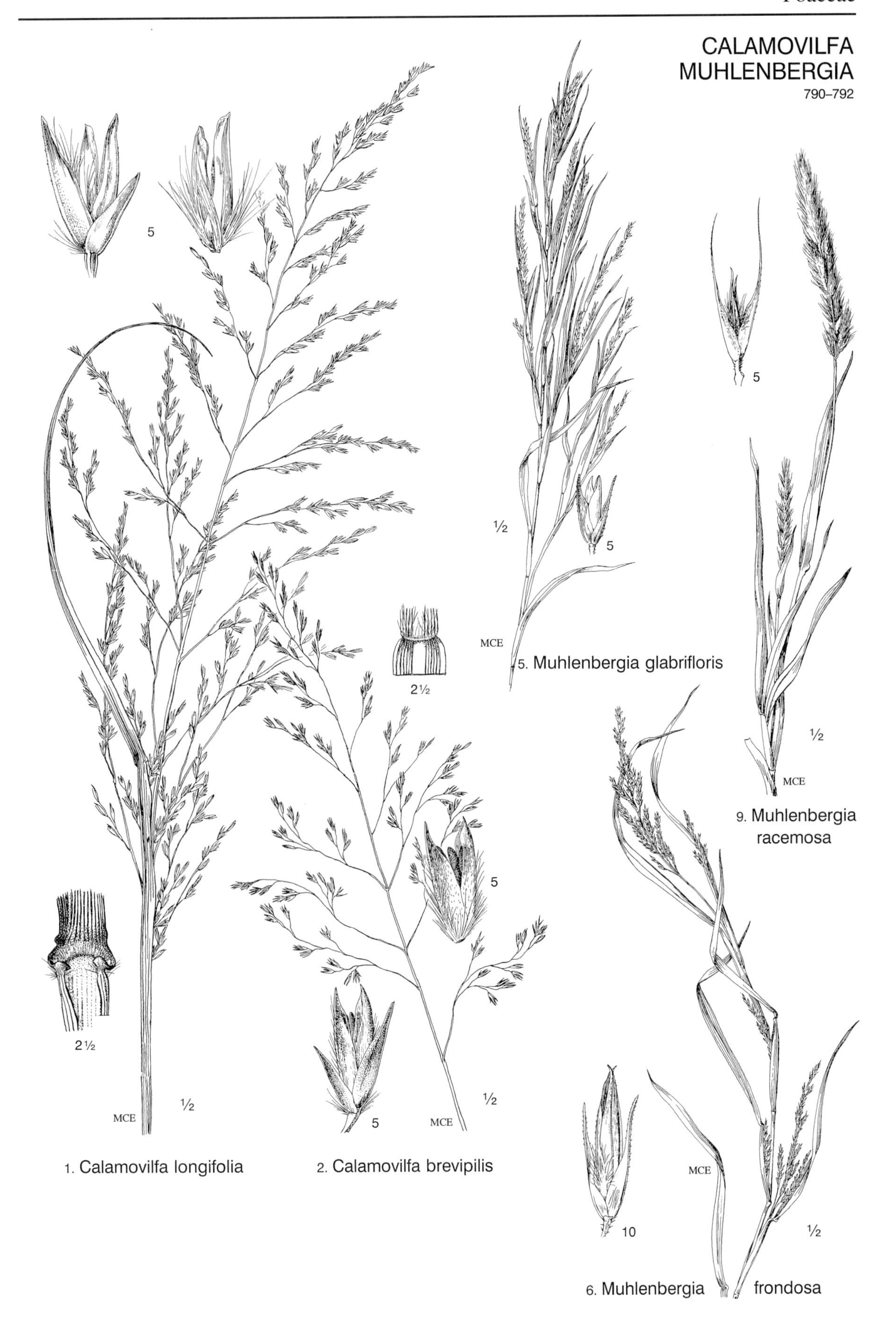

5. Muhlenbergia glabrifloris

9. Muhlenbergia
racemosa

1. Calamovilfa longifolia

2. Calamovilfa brevipilis

6. Muhlenbergia frondosa

MUHLENBERGIA
791, 792

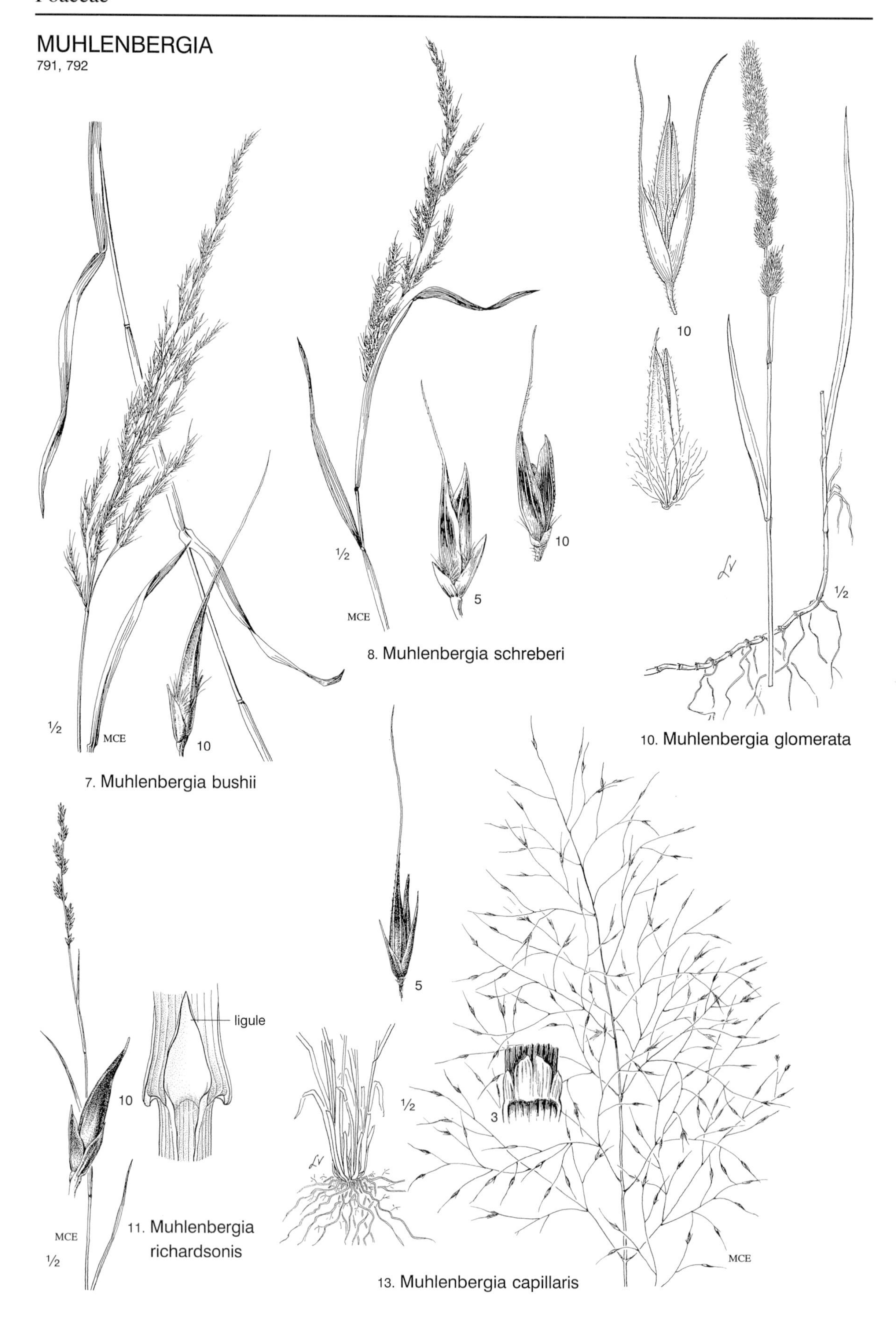

7. Muhlenbergia bushii

8. Muhlenbergia schreberi

10. Muhlenbergia glomerata

ligule

11. Muhlenbergia richardsonis

13. Muhlenbergia capillaris

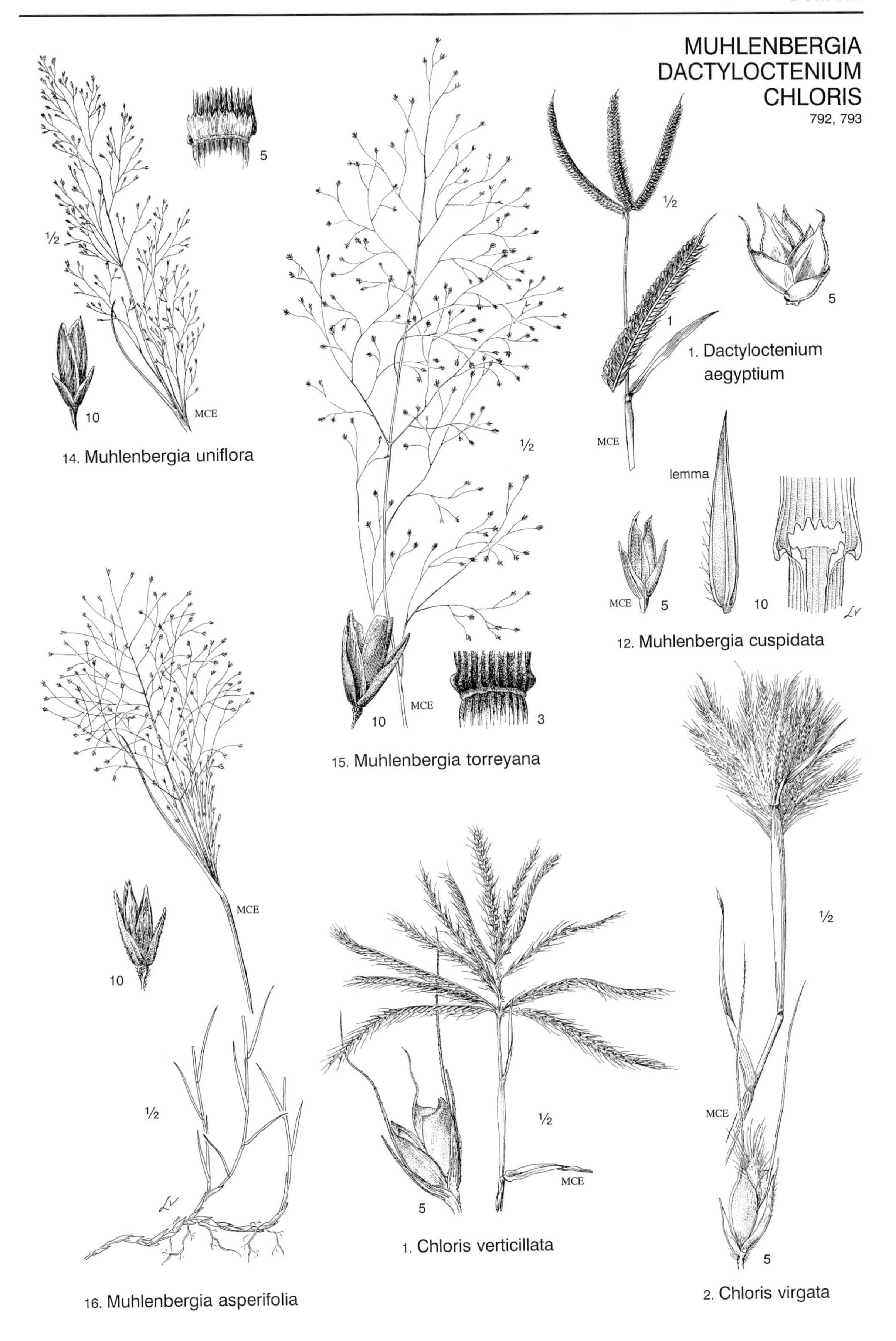

14. Muhlenbergia uniflora

15. Muhlenbergia torreyana

1. Dactyloctenium
aegyptium

lemma

12. Muhlenbergia cuspidata

16. Muhlenbergia asperifolia

1. Chloris verticillata

2. Chloris virgata

SCHEDONNARDUS
GYMNOPOGON
CTENIUM
793, 794

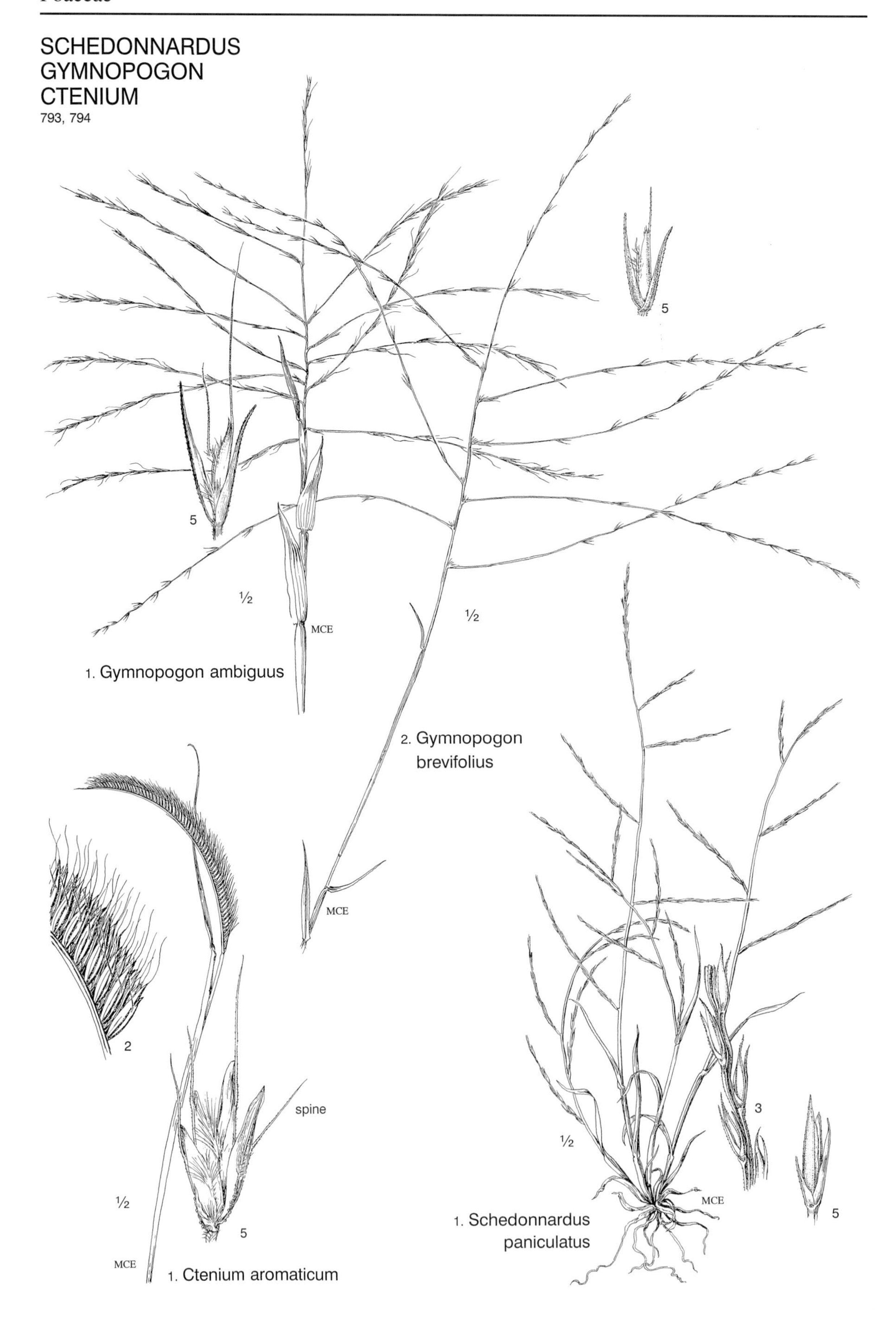

1. Gymnopogon ambiguus

½   ½

MCE

2. Gymnopogon brevifolius

MCE

2

spine

5

½

MCE

1. Ctenium aromaticum

½

MCE

3

5

1. Schedonnardus paniculatus

5

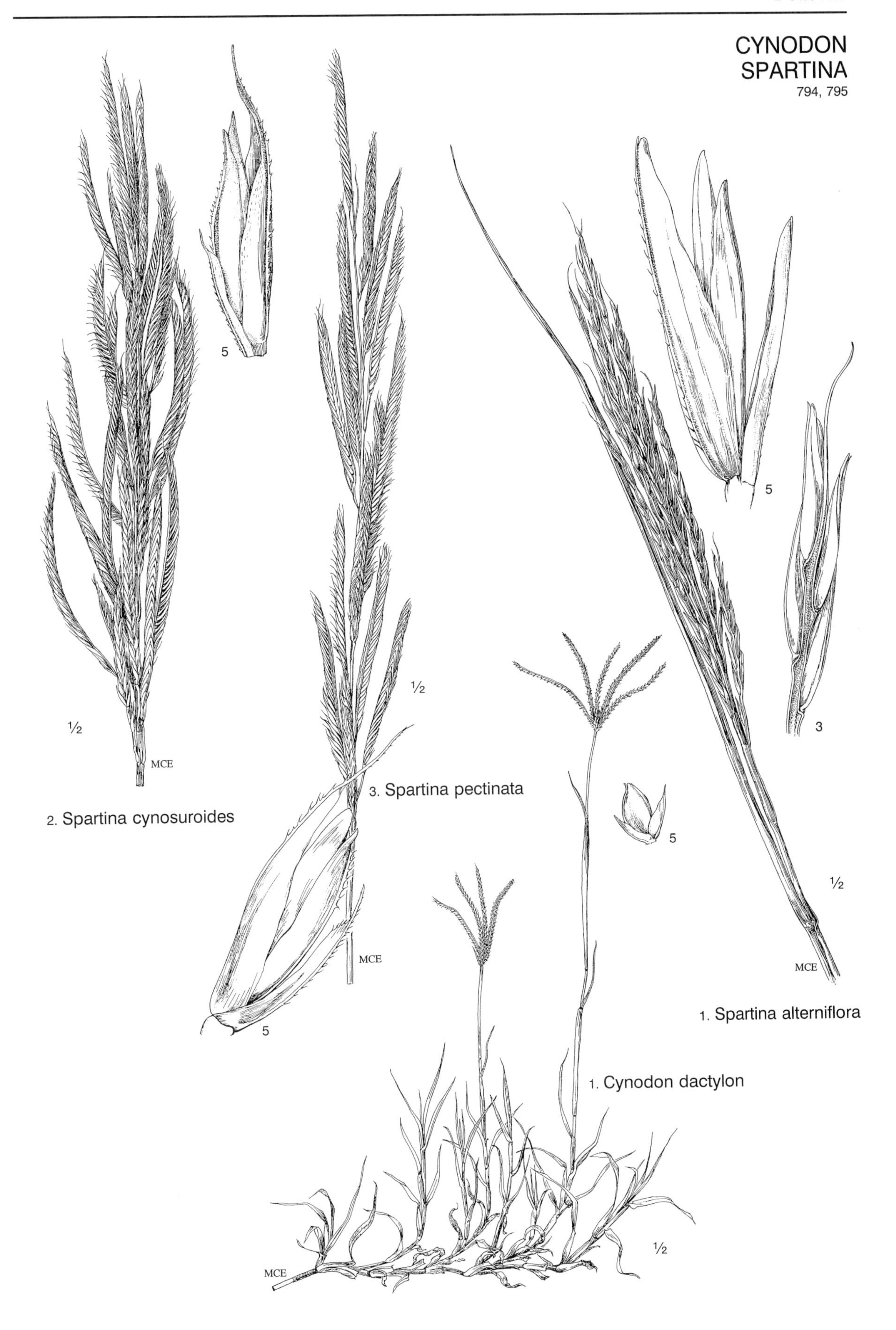

5

2. Spartina cynosuroides

3. Spartina pectinata

5

5

3

5

1. Spartina alterniflora

1. Cynodon dactylon

SPARTINA
BOUTELOUA
ZOYSIA
795, 796

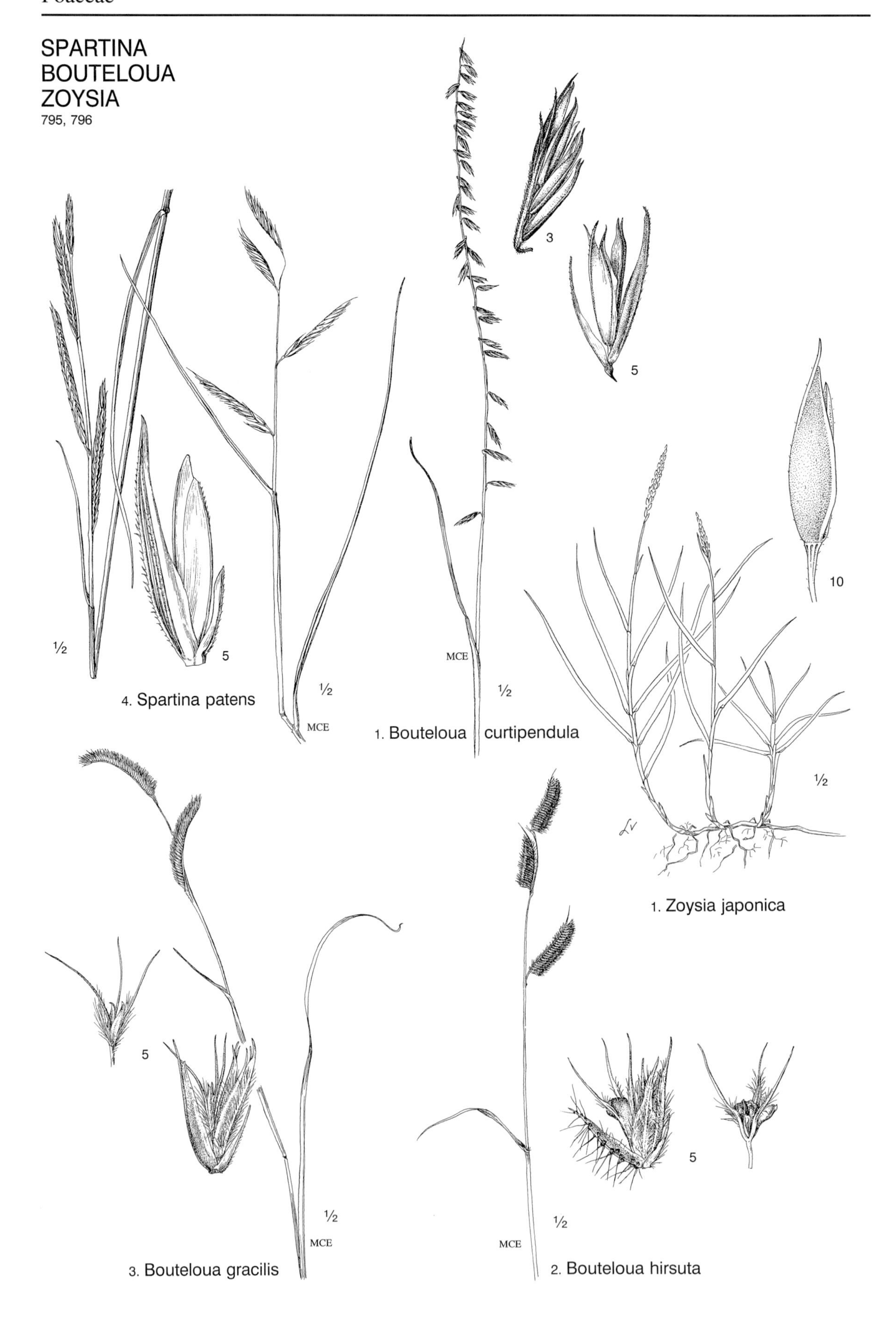

½

4. Spartina patens

5

½

MCE

MCE

½

1. Bouteloua curtipendula

3

5

10

½

1. Zoysia japonica

5

½

MCE

3. Bouteloua gracilis

½

MCE

5

½

2. Bouteloua hirsuta

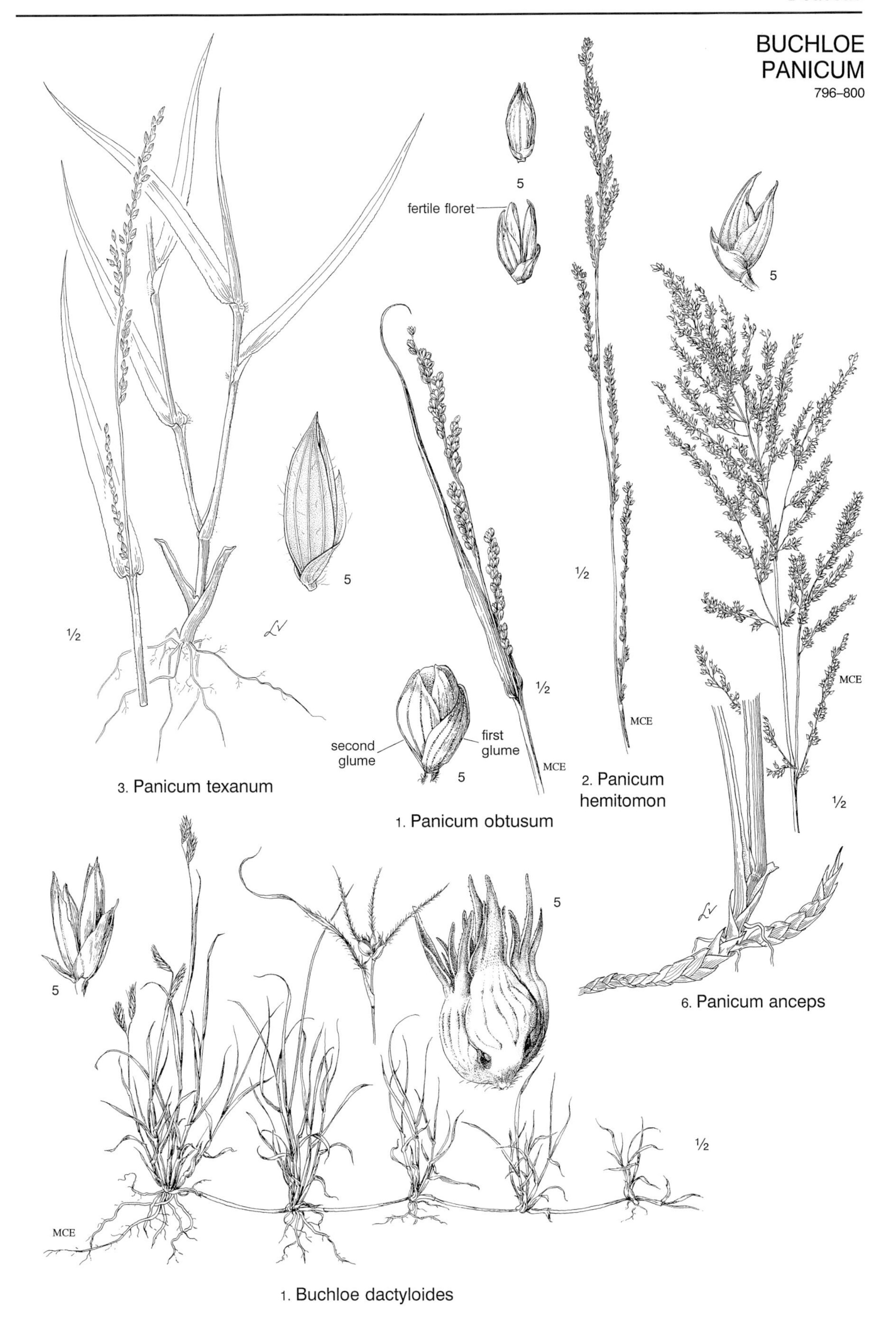

5

fertile floret

5

5

3. Panicum texanum

second
glume

first
glume

5

1. Panicum obtusum

½

MCE

2. Panicum
hemitomon

½

MCE

½

6. Panicum anceps

5

5

½

MCE

1. Buchloe dactyloides

PANICUM
799, 800

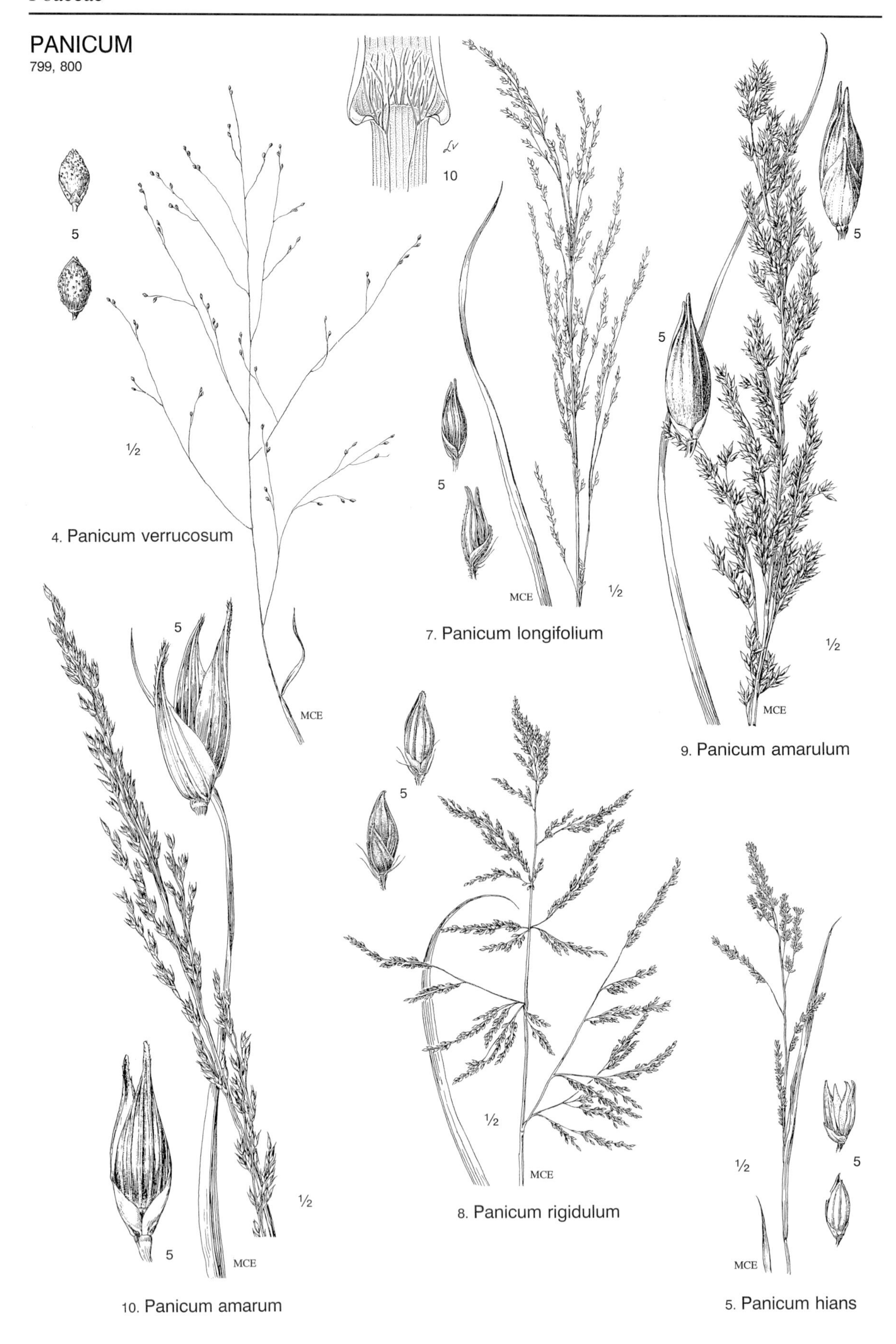

4. Panicum verrucosum

10

7. Panicum longifolium

9. Panicum amarulum

8. Panicum rigidulum

10. Panicum amarum

5. Panicum hians

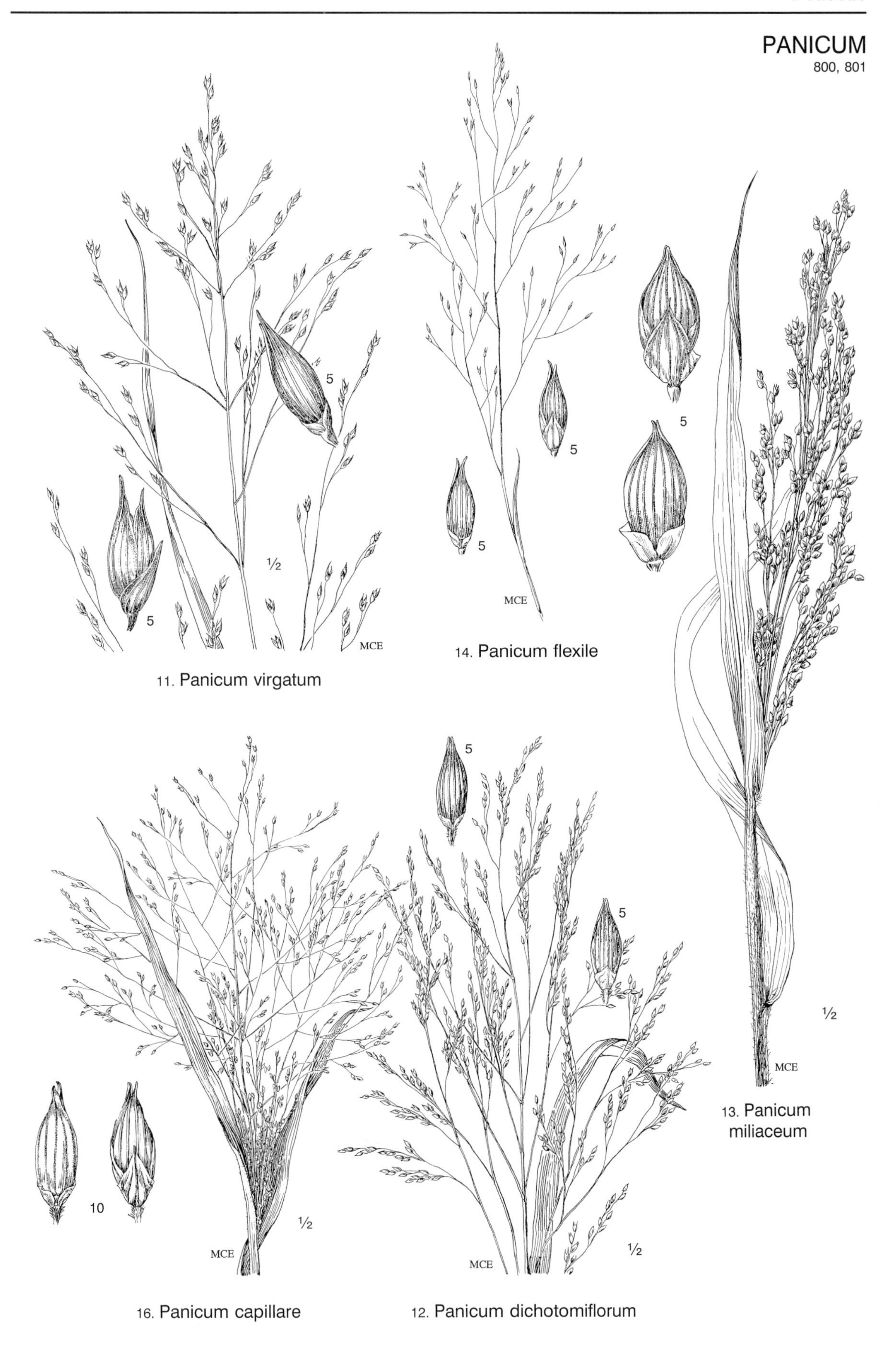

11. Panicum virgatum

14. Panicum flexile

13. Panicum miliaceum

16. Panicum capillare

12. Panicum dichotomiflorum

PANICUM
800, 801

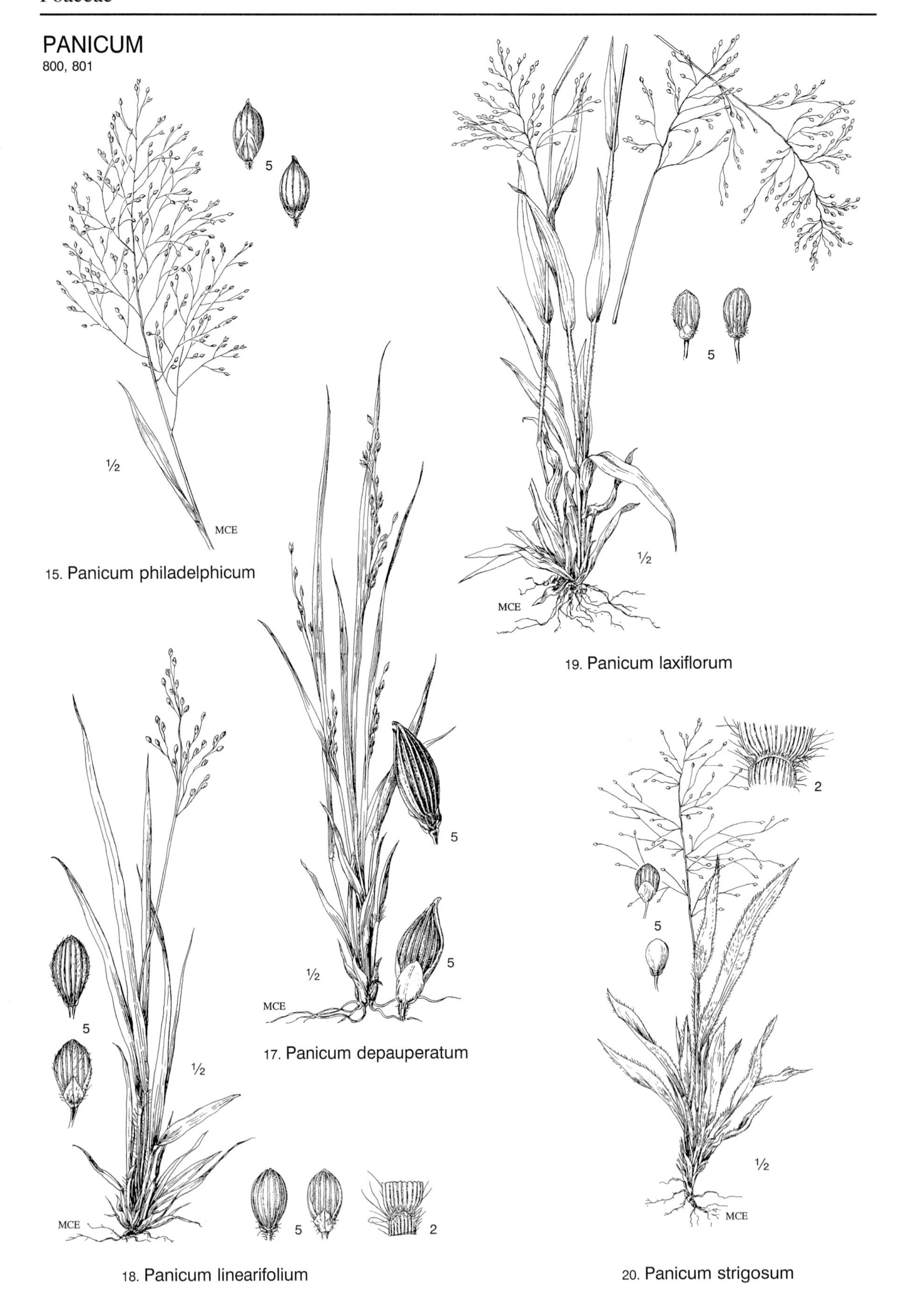

15. Panicum philadelphicum

19. Panicum laxiflorum

17. Panicum depauperatum

18. Panicum linearifolium

20. Panicum strigosum

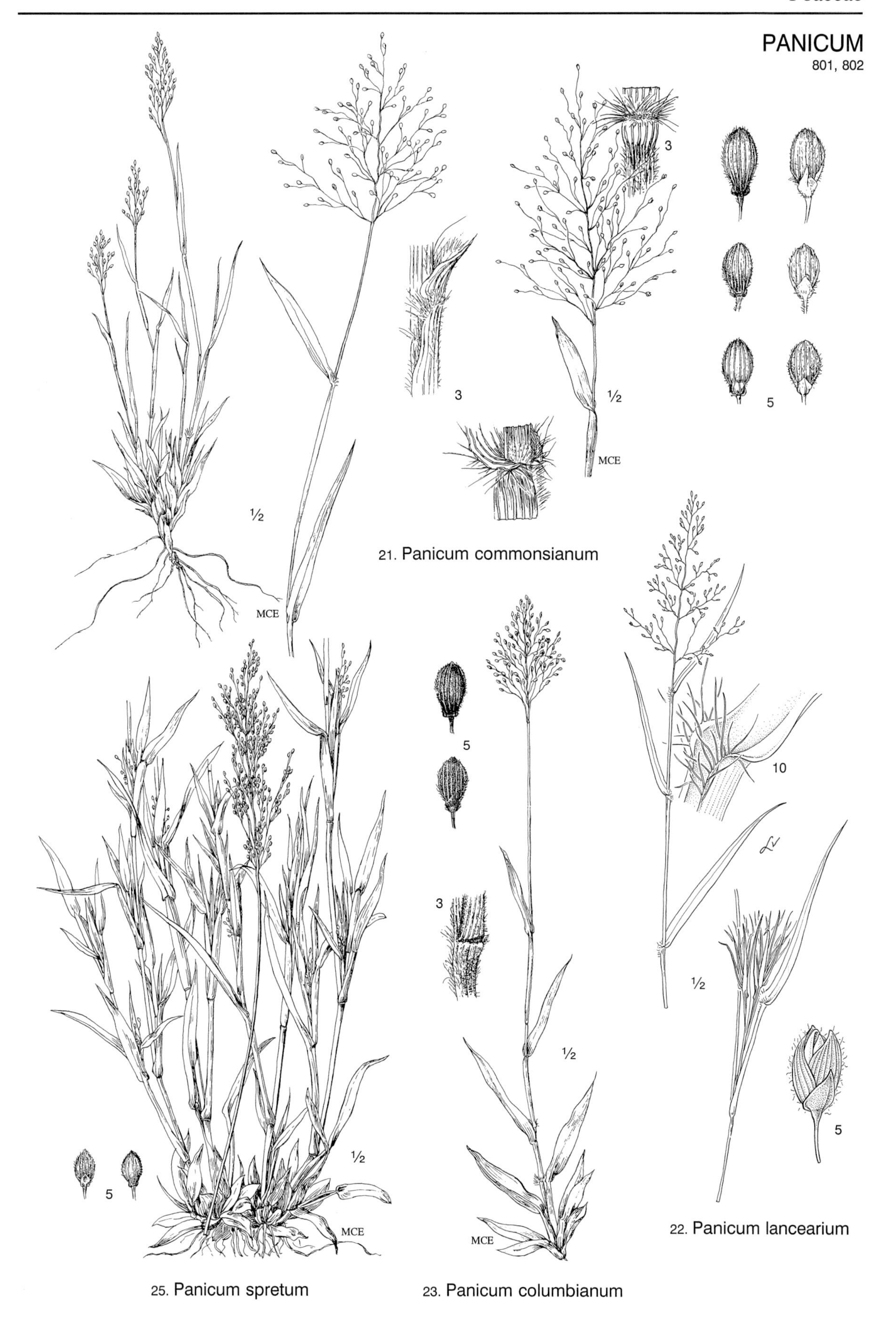

21. Panicum commonsianum

22. Panicum lancearium

25. Panicum spretum

23. Panicum columbianum

PANICUM
802, 803

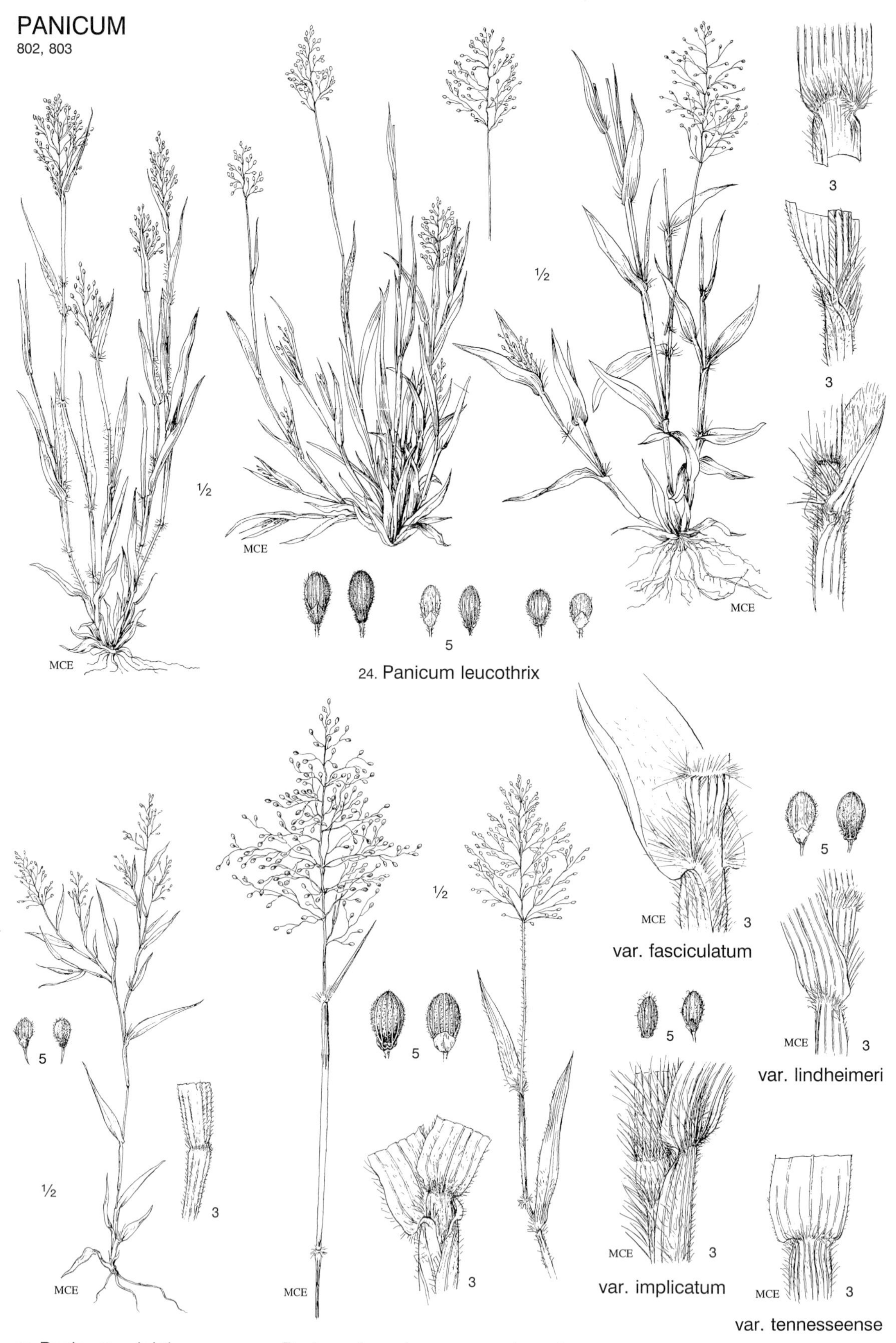

24. Panicum leucothrix

26. Panicum wrightianum   27. Panicum lanuginosum var. lanuginosum

var. fasciculatum

var. lindheimeri

var. implicatum

var. tennesseense

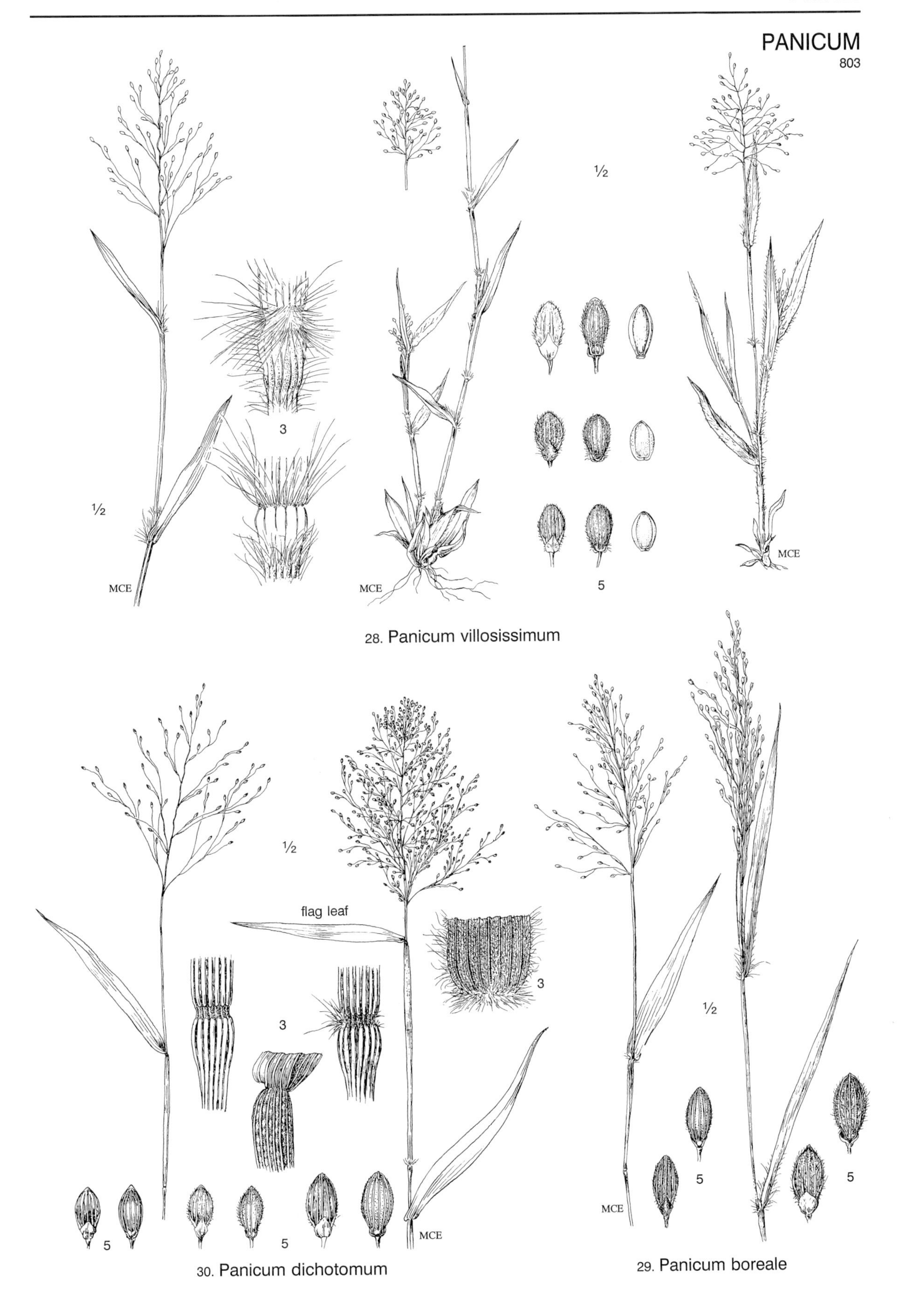

28. Panicum villosissimum

30. Panicum dichotomum

29. Panicum boreale

PANICUM
803, 804

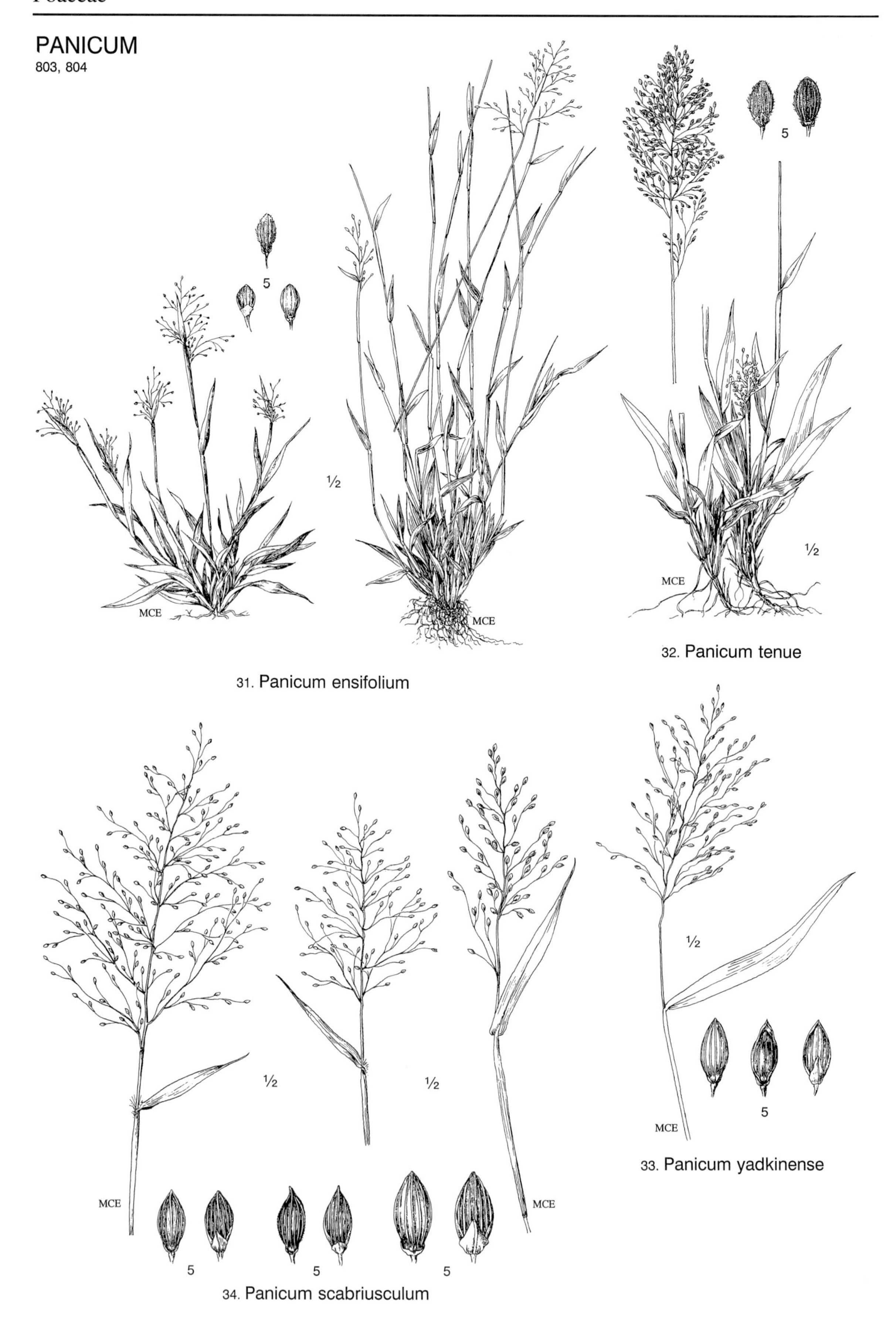

31. Panicum ensifolium

32. Panicum tenue

34. Panicum scabriusculum

33. Panicum yadkinense

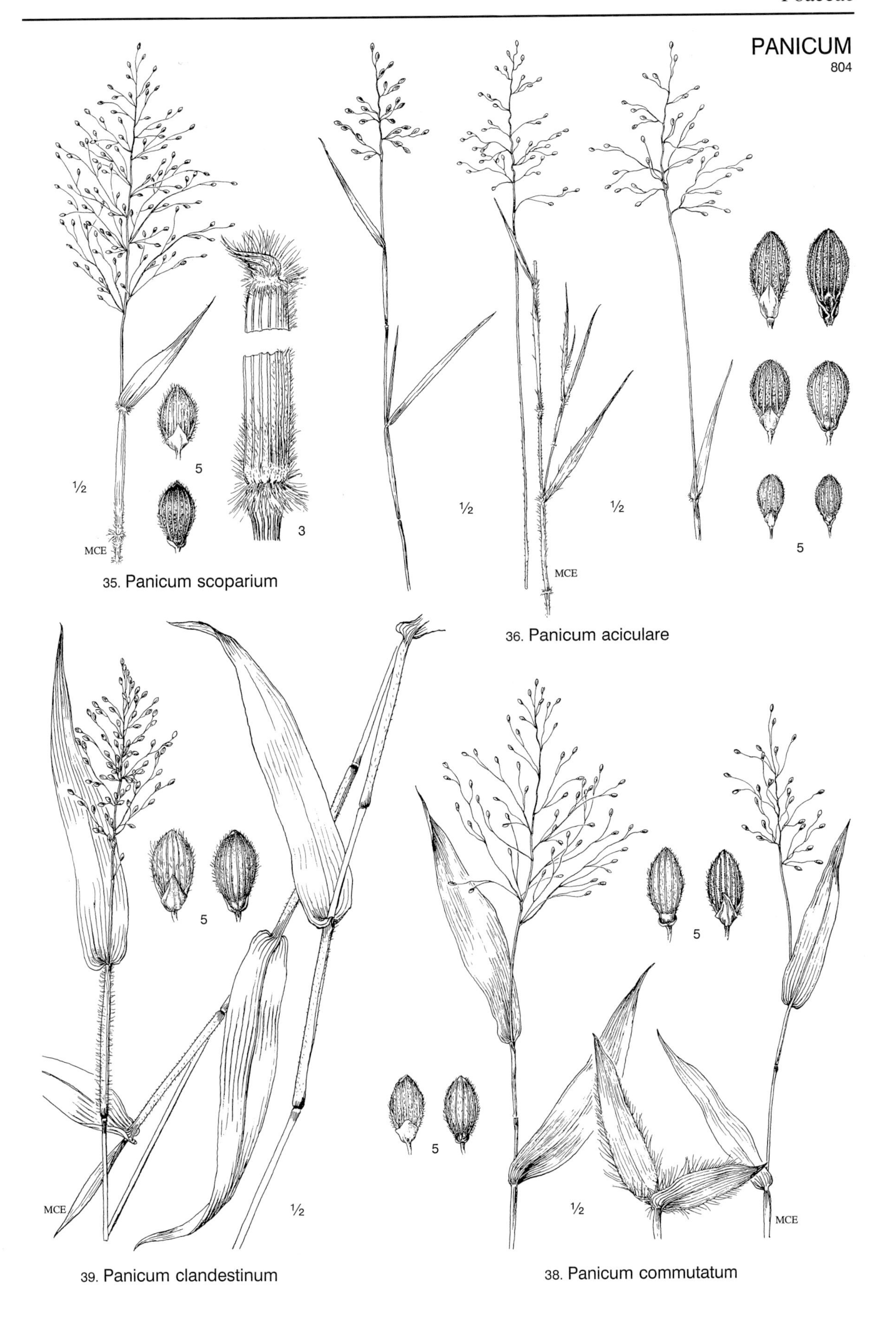

35. Panicum scoparium

36. Panicum aciculare

39. Panicum clandestinum

38. Panicum commutatum

PANICUM
804–806

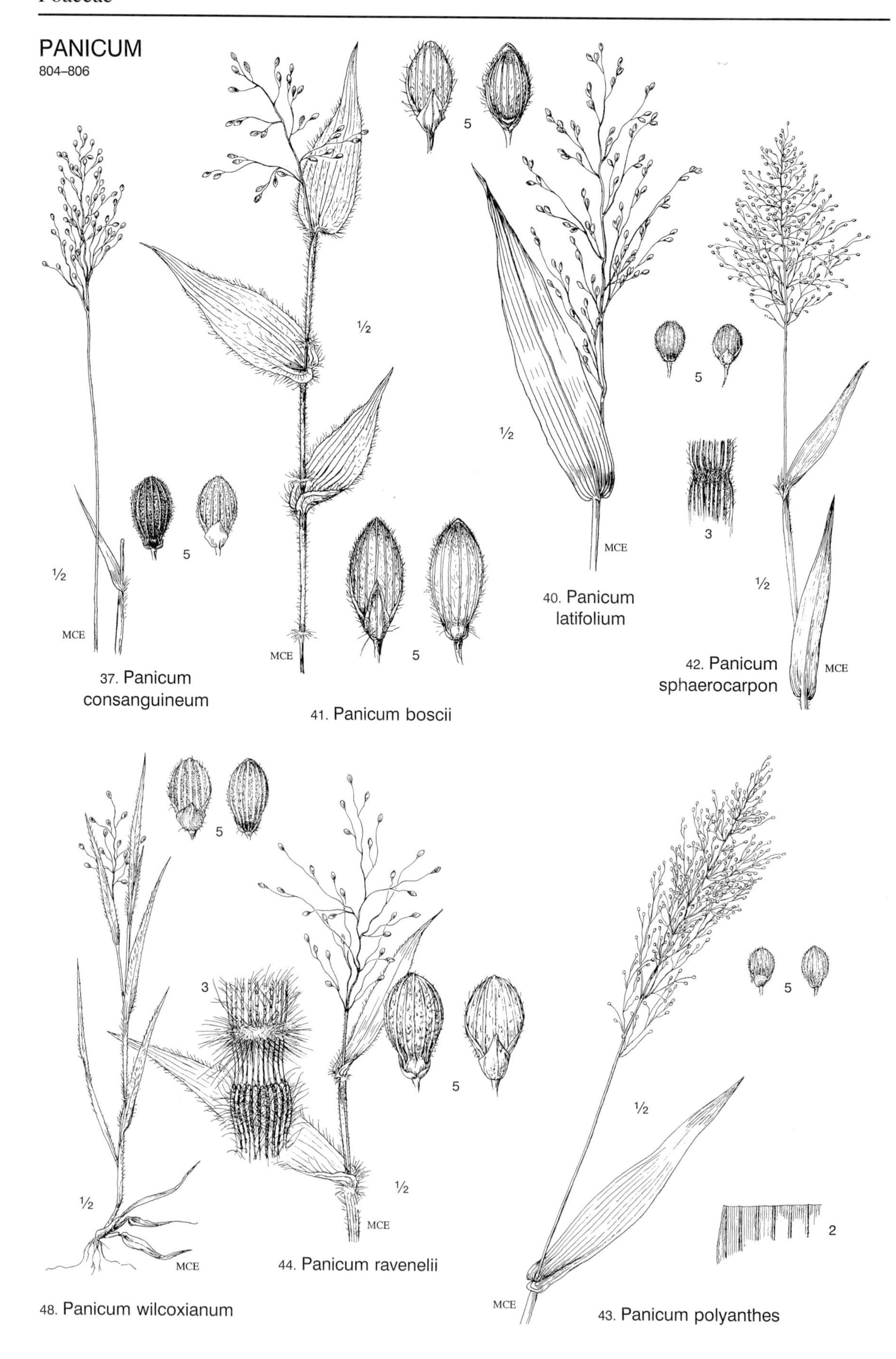

37. Panicum
consanguineum

41. Panicum boscii

40. Panicum
latifolium

42. Panicum
sphaerocarpon

48. Panicum wilcoxianum

44. Panicum ravenelii

43. Panicum polyanthes

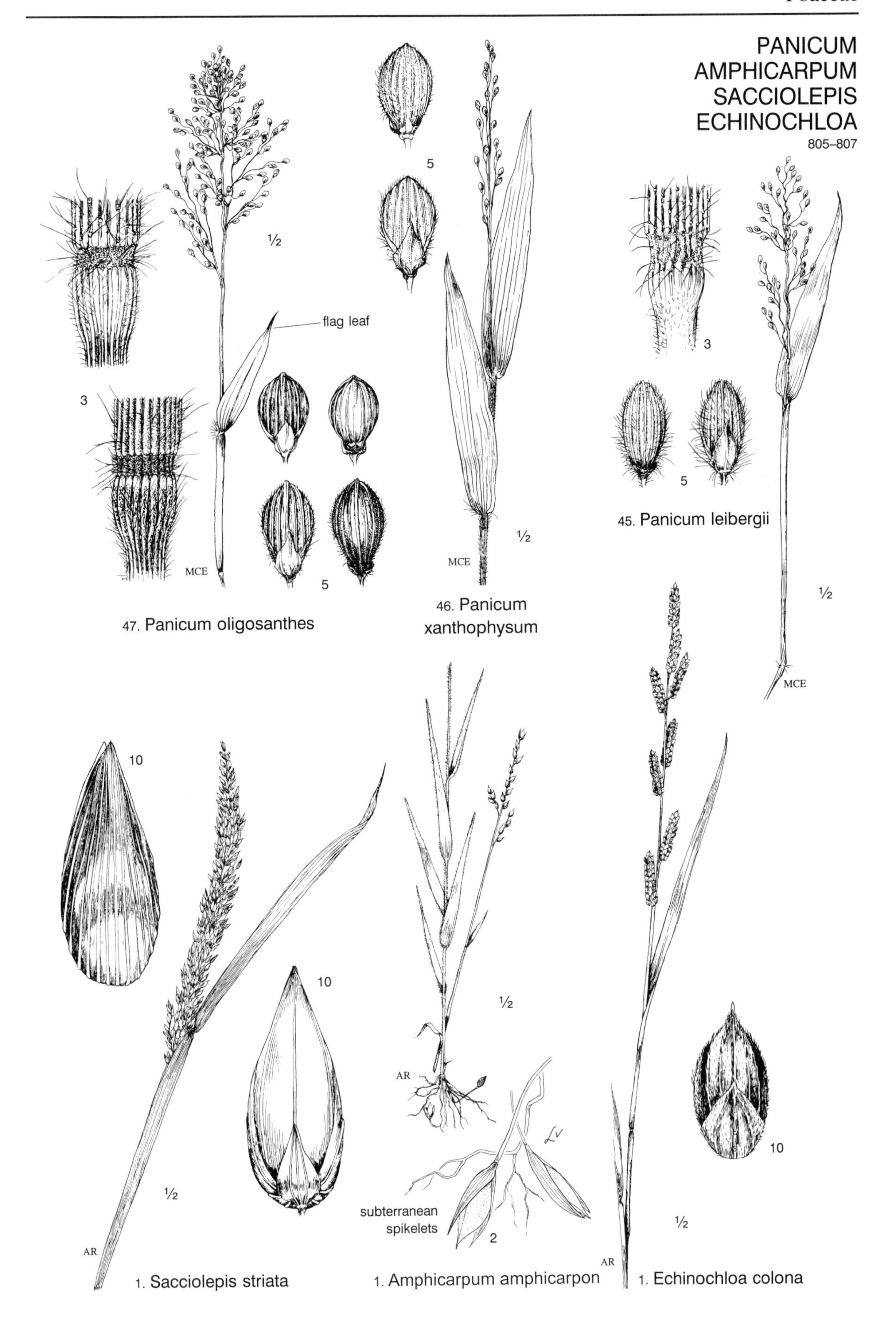

flag leaf

3

5

MCE

47. Panicum oligosanthes

5

MCE

½

46. Panicum
xanthophysum

3

5

45. Panicum leibergii

½

MCE

10

10

½

AR

1. Sacciolepis striata

½

AR

subterranean
spikelets

2

1. Amphicarpum amphicarpon

10

½

AR

1. Echinochloa colona

ECHINOCHLOA
ERIOCHLOA
807, 808

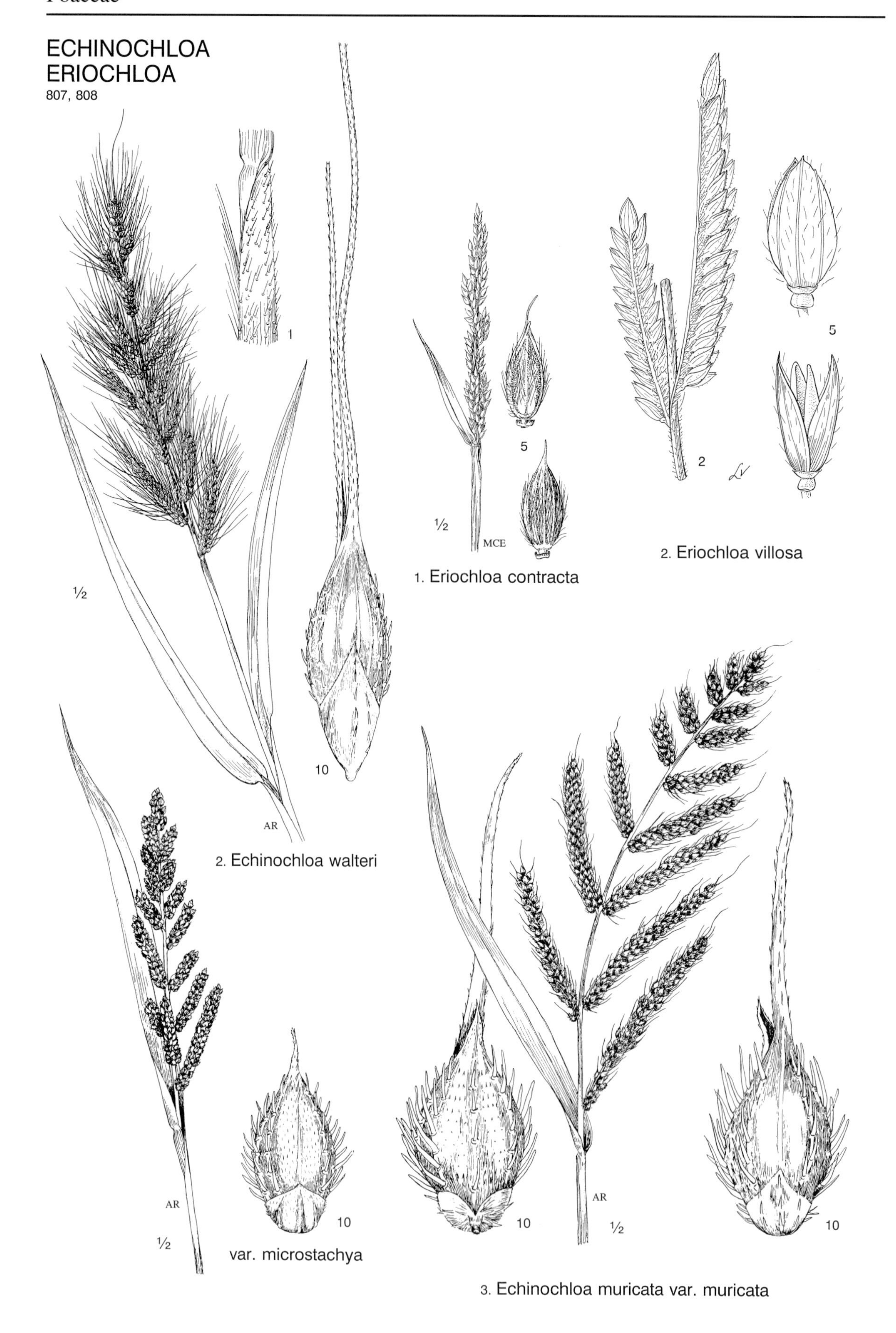

1 ½

1. Eriochloa contracta

5

MCE

2. Echinochloa walteri

2. Eriochloa villosa

5

var. microstachya

3. Echinochloa muricata var. muricata

ECHINOCHLOA
PASPALUM
807, 808

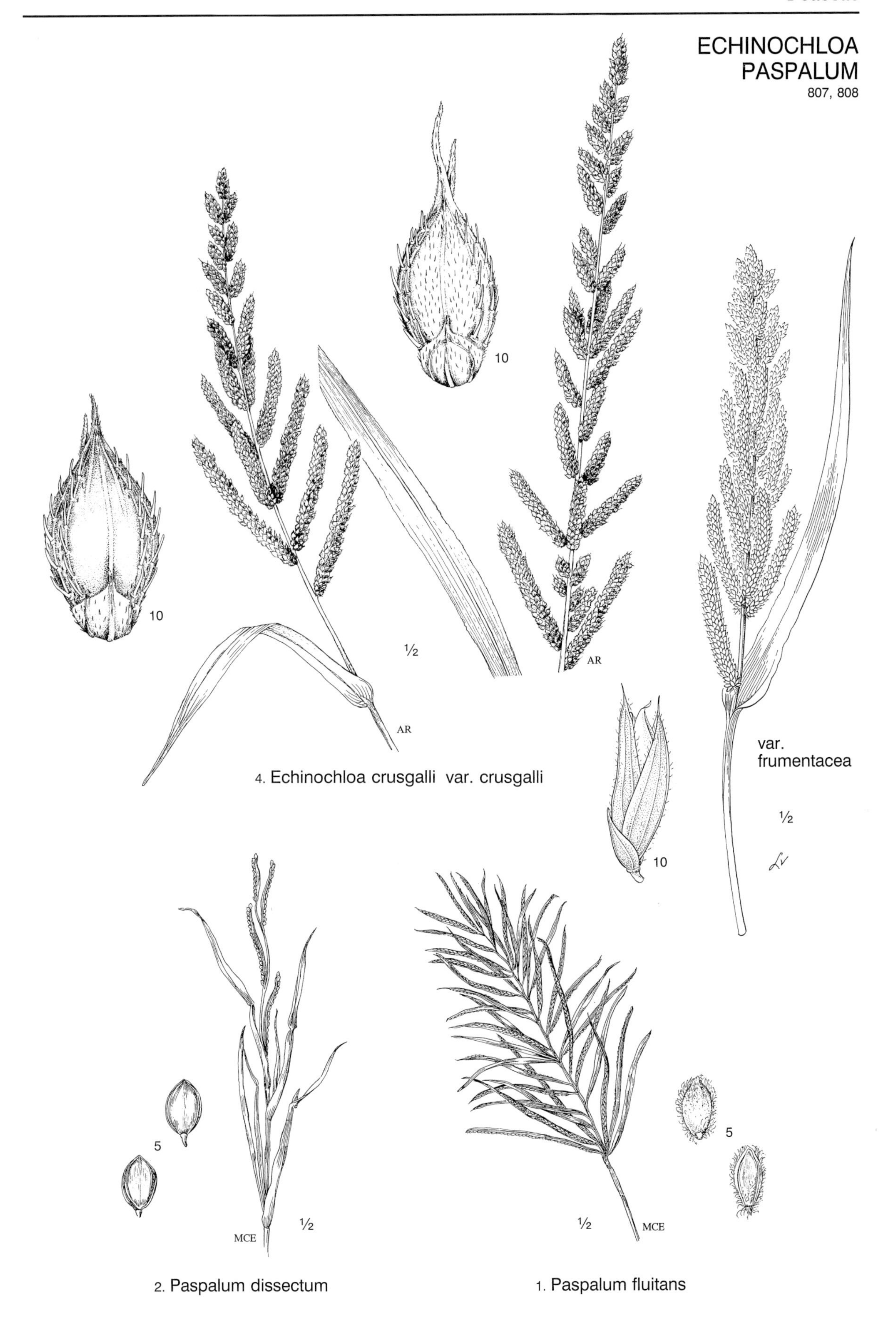

10

10

½

AR

AR

4. Echinochloa crusgalli var. crusgalli

10

var.
frumentacea

½

5

5

½

MCE

2. Paspalum dissectum

½

MCE

1. Paspalum fluitans

PASPALUM
809

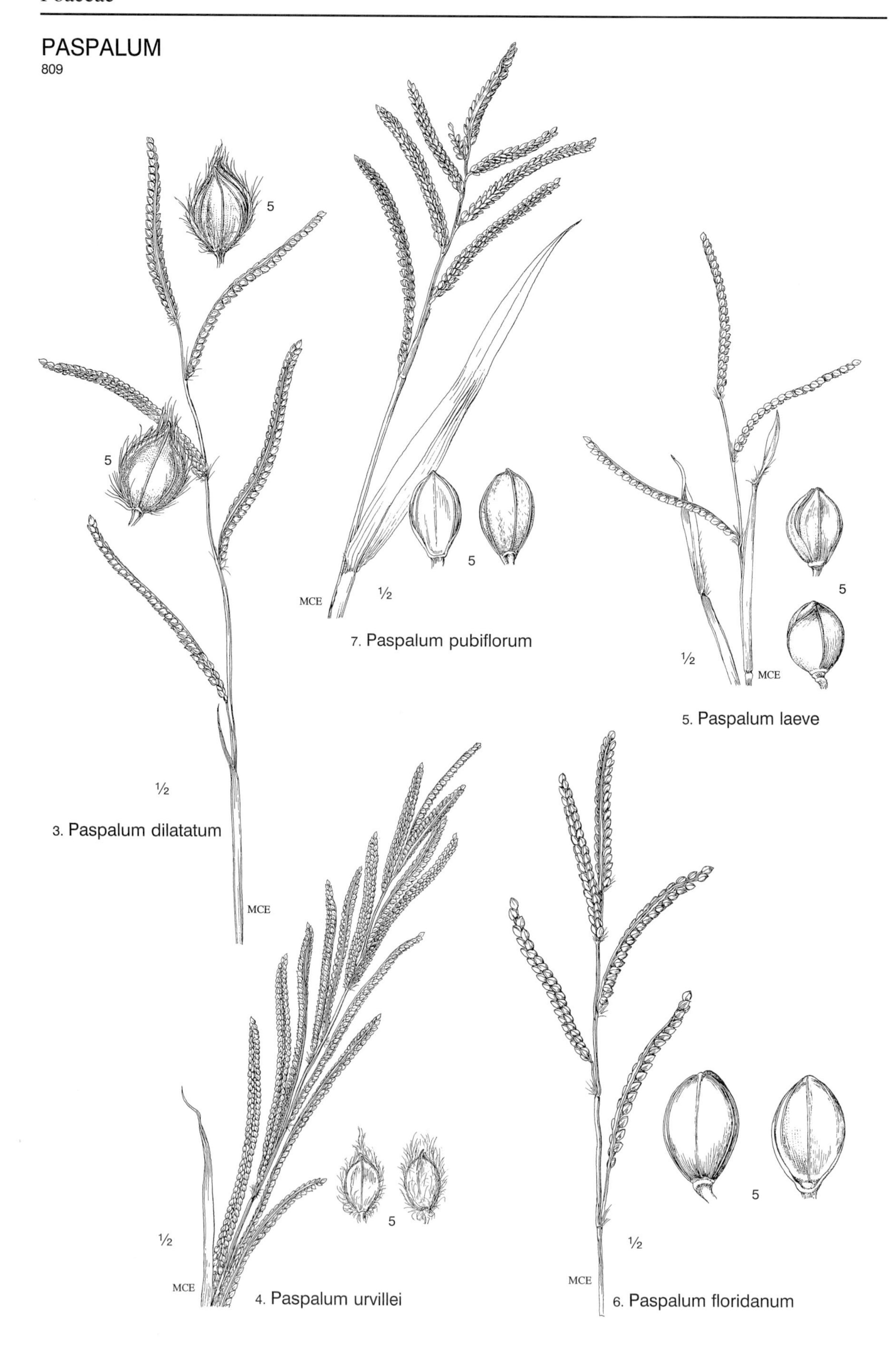

5

5

½

3. Paspalum dilatatum

MCE

MCE

½

7. Paspalum pubiflorum

5

MCE

5

½

MCE

5. Paspalum laeve

½

MCE

5

4. Paspalum urvillei

5

½

MCE

6. Paspalum floridanum

ILLUSTRATED COMPANION TO

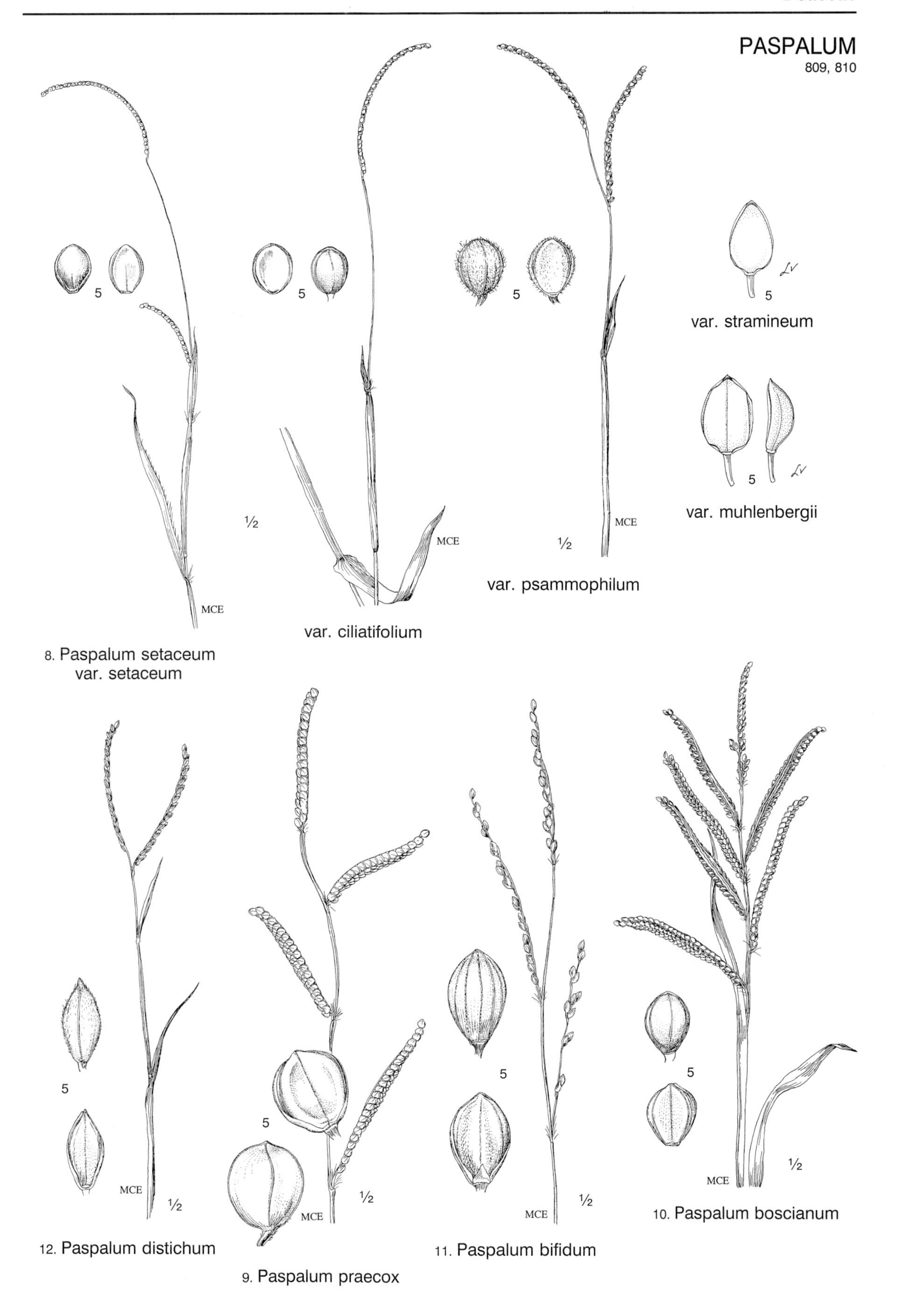

var. stramineum

var. muhlenbergii

var. psammophilum

var. ciliatifolium

8. Paspalum setaceum
var. setaceum

12. Paspalum distichum

9. Paspalum praecox

11. Paspalum bifidum

10. Paspalum boscianum

AXONOPUS
SETARIA
810, 811

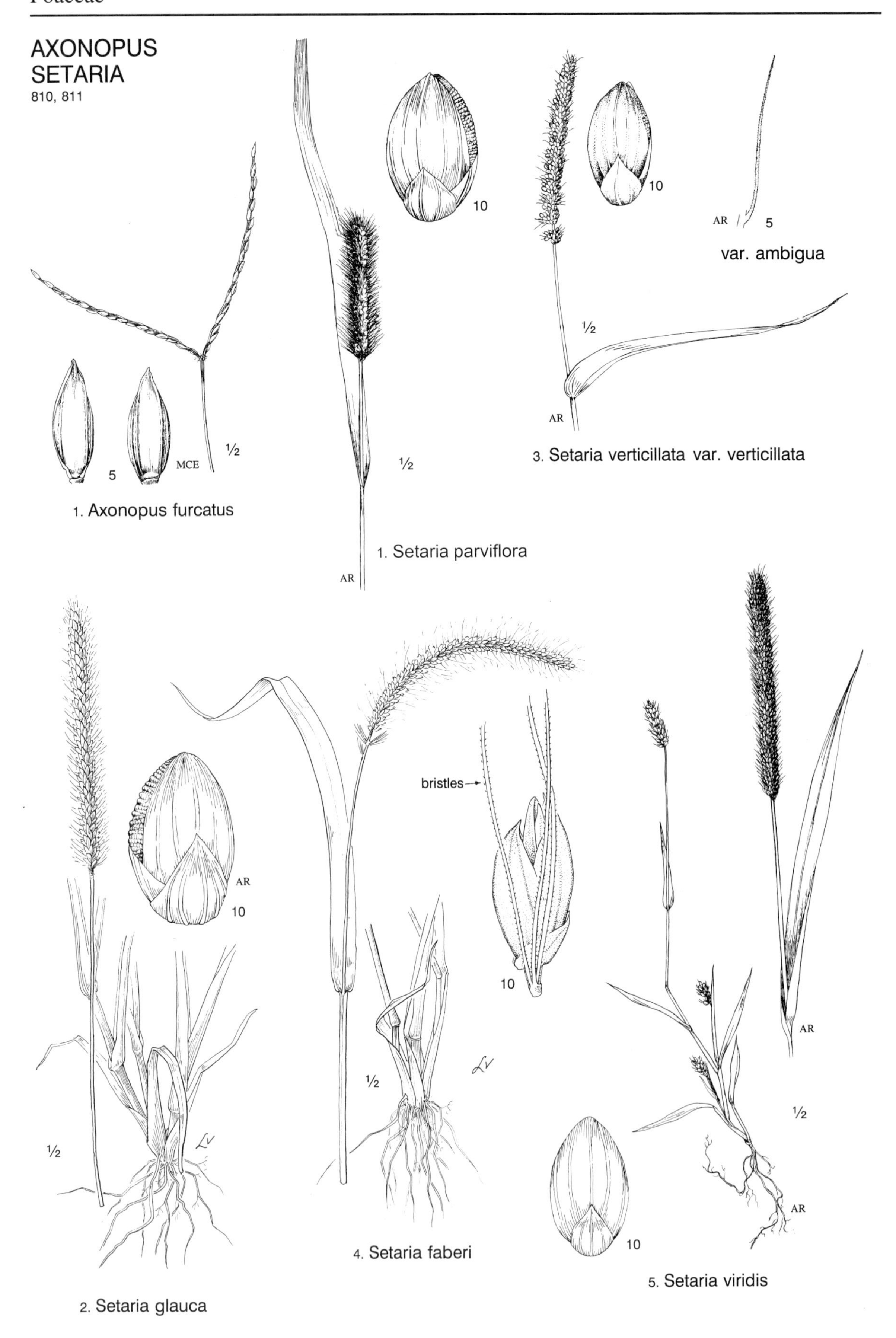

5   MCE   ½

1. Axonopus furcatus

½

AR

1. Setaria parviflora

10

10

AR   5

var. ambigua

½

AR

3. Setaria verticillata var. verticillata

AR
10

½

2. Setaria glauca

bristles→

10

½

4. Setaria faberi

10

5. Setaria viridis

AR

½

AR

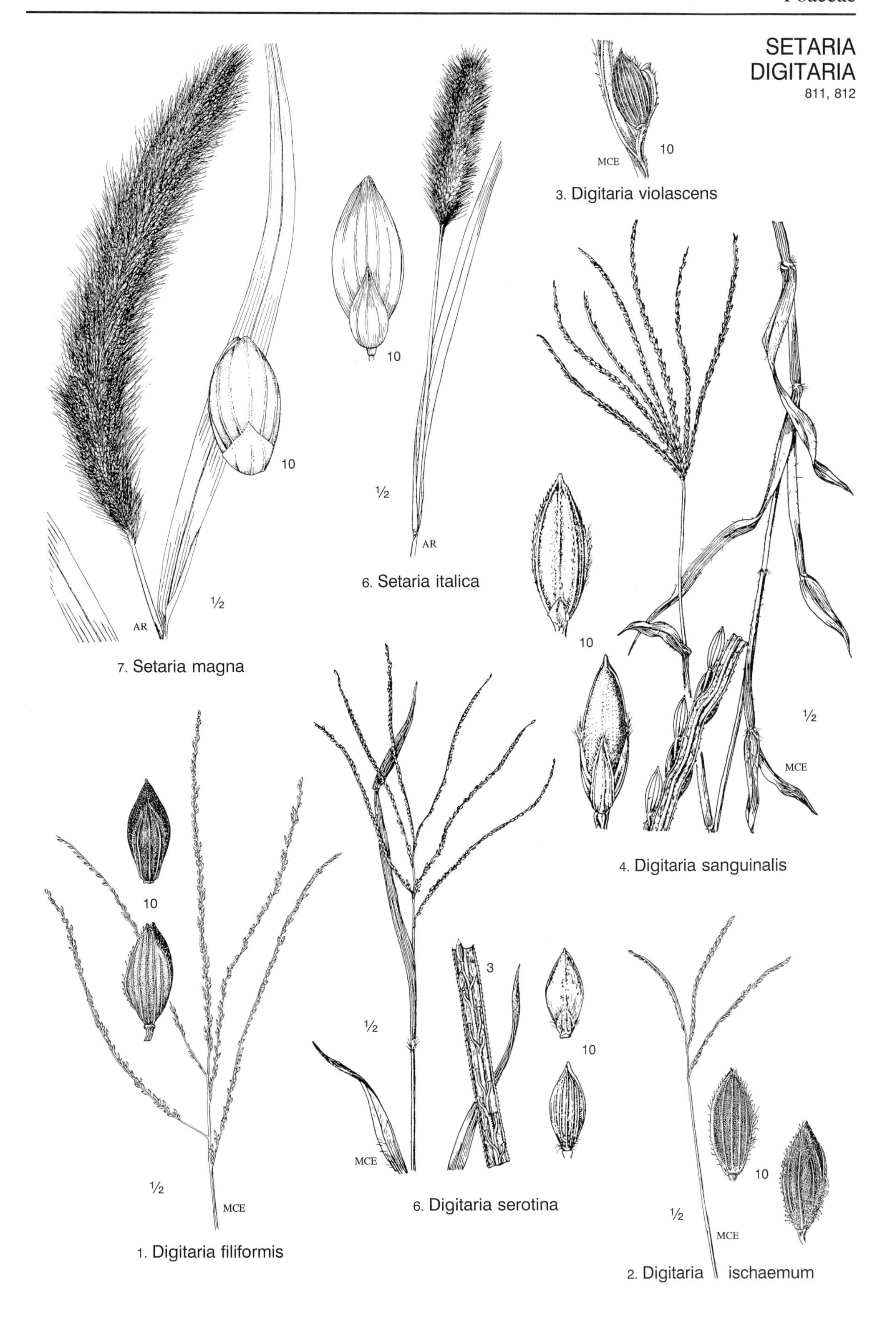

3. Digitaria violascens

7. Setaria magna

6. Setaria italica

4. Digitaria sanguinalis

1. Digitaria filiformis

6. Digitaria serotina

2. Digitaria ischaemum

DIGITARIA
LEPTOLOMA
TRAGUS
812, 813

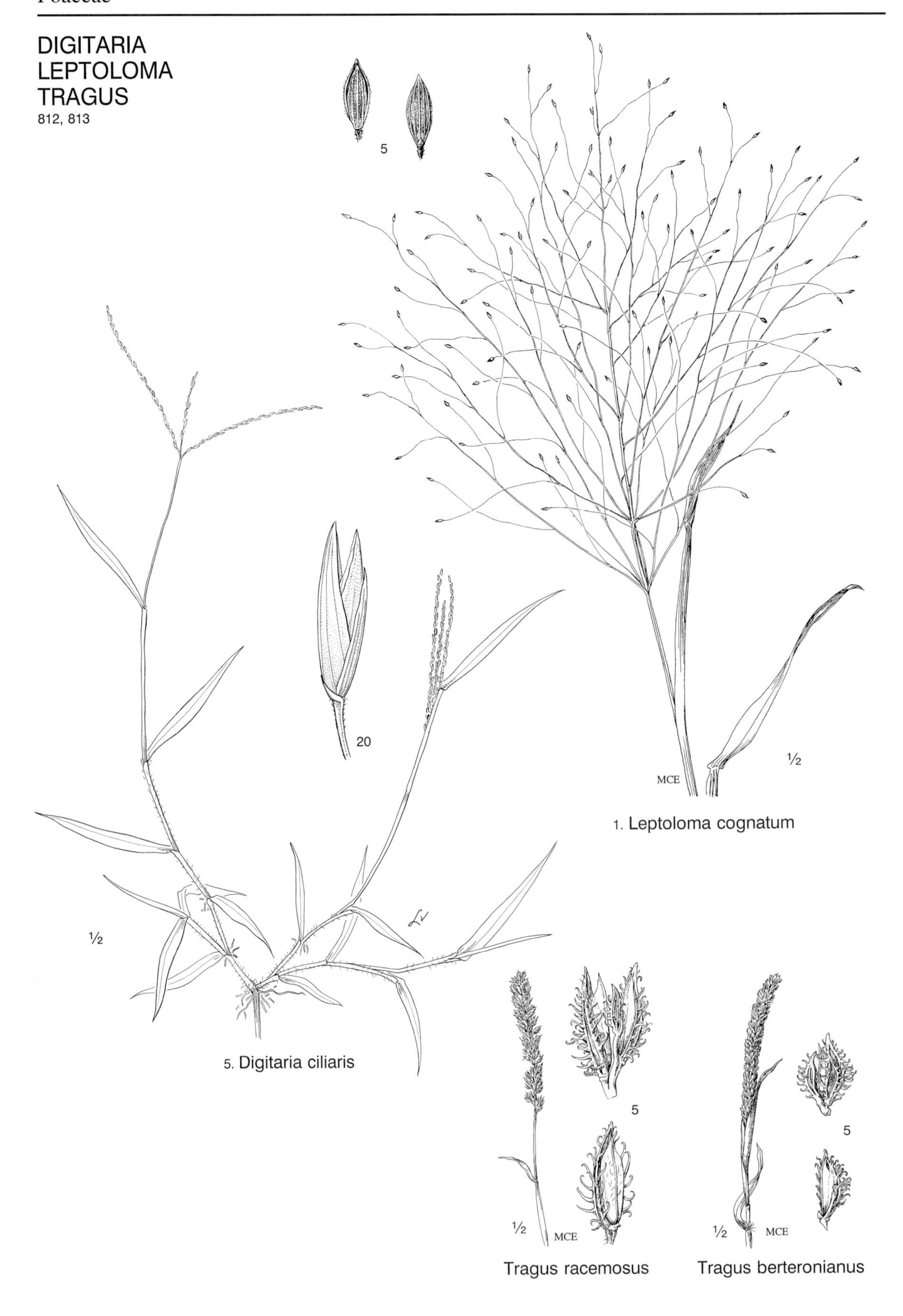

5

20

½

MCE

1. Leptoloma cognatum

½

5. Digitaria ciliaris

5

5

½   MCE   ½   MCE

Tragus racemosus   Tragus berteronianus

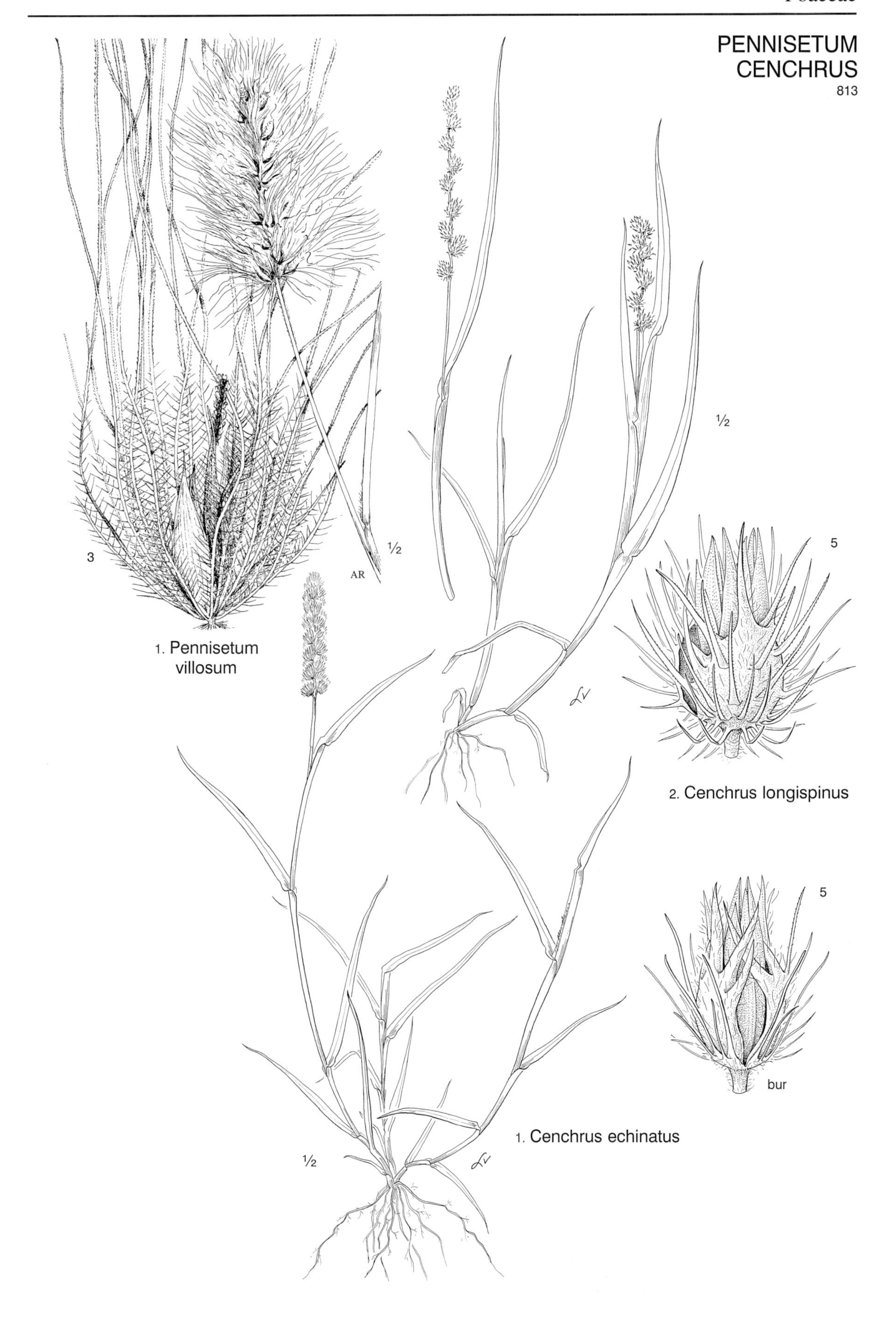

1. Pennisetum
villosum

2. Cenchrus longispinus

1. Cenchrus echinatus

CENCHRUS
ERIANTHUS
813, 814

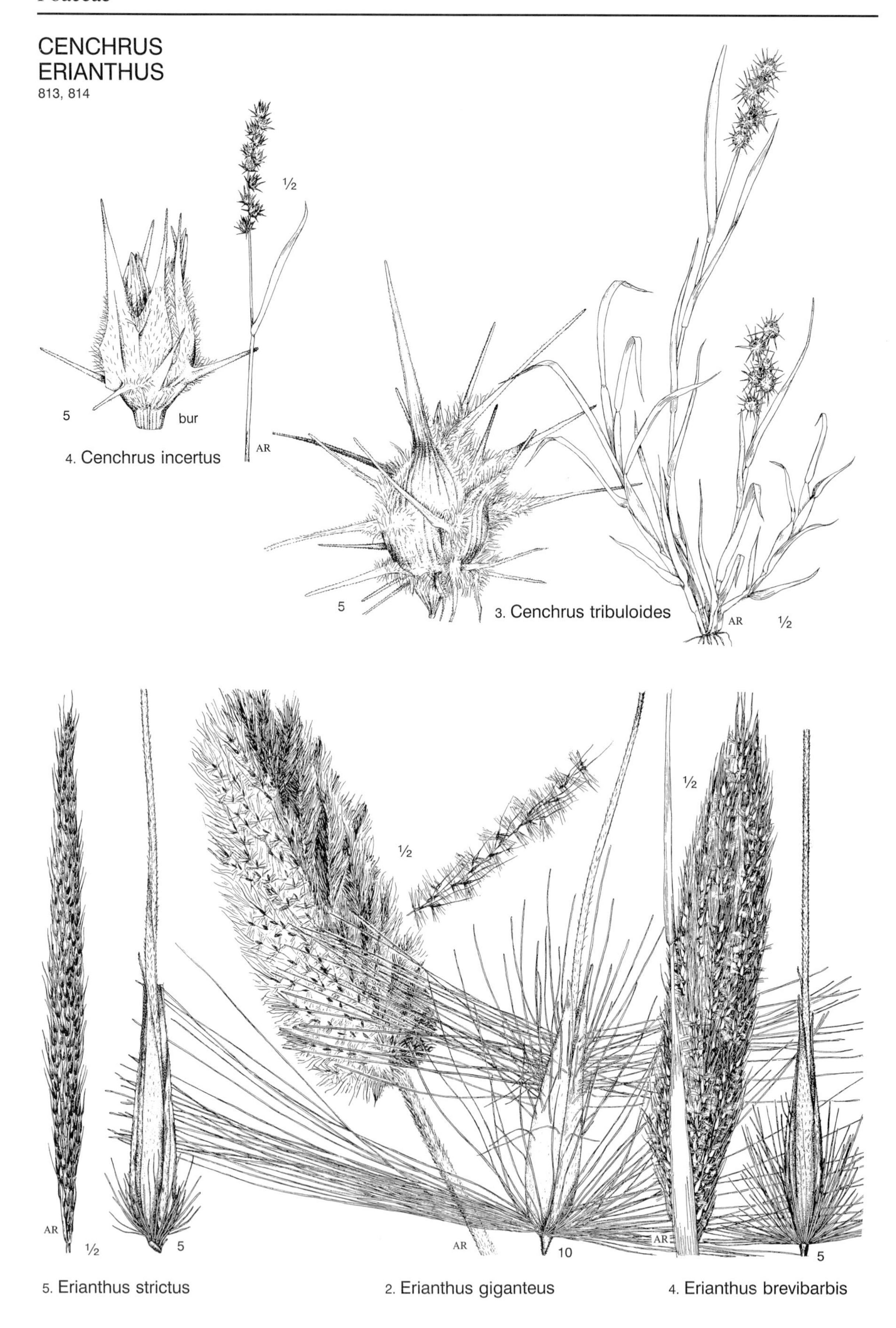

½

5   bur
AR

4. Cenchrus incertus

5   3. Cenchrus tribuloides   AR   ½

½

½

AR   ½   5   AR   10   AR   5

5. Erianthus strictus   2. Erianthus giganteus   4. Erianthus brevibarbis

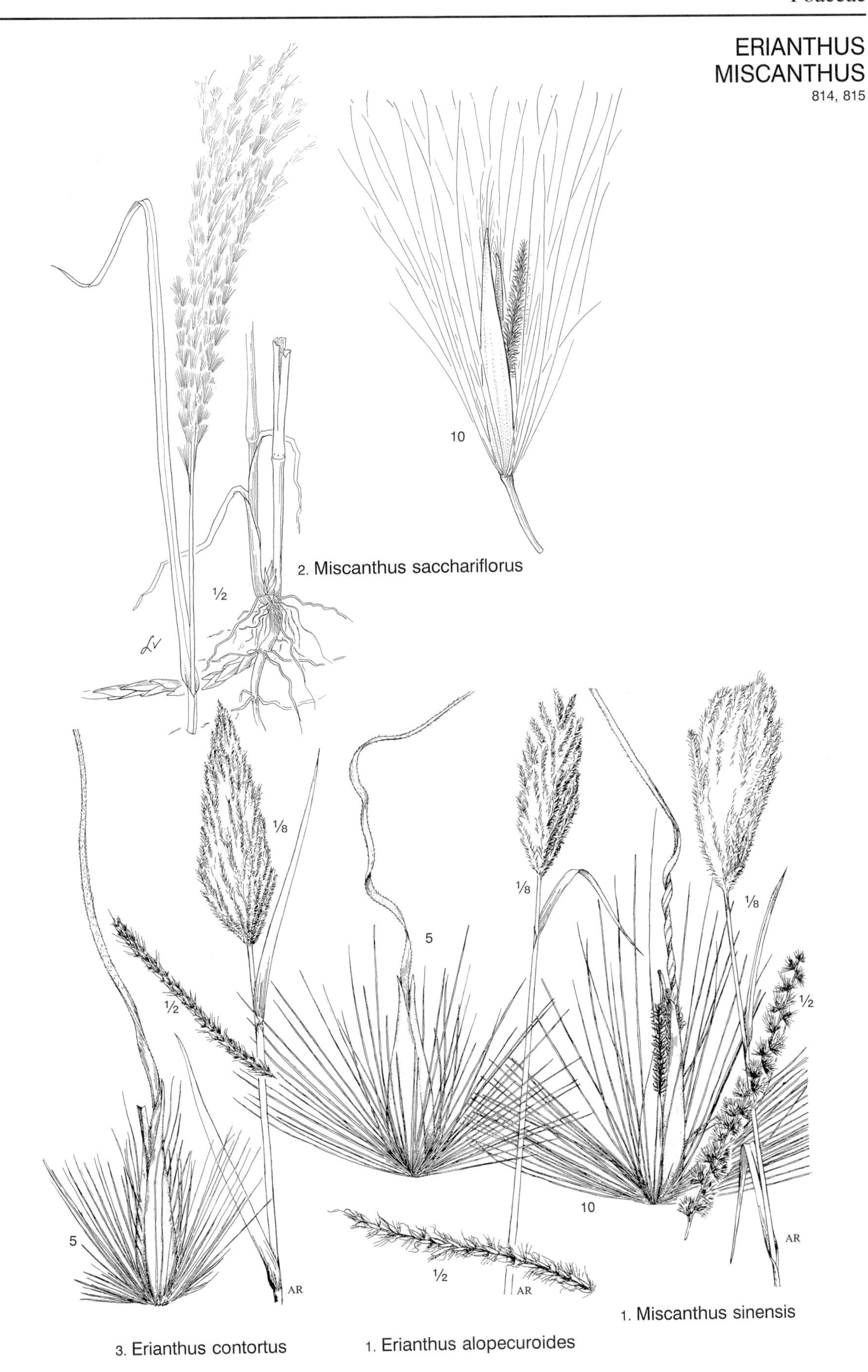

10

2. Miscanthus sacchariflorus

½

½

⅛

5

½

⅛

5

½

⅛

10

AR

½

AR

½

AR

3. Erianthus contortus

1. Erianthus alopecuroides

1. Miscanthus sinensis

MICROSTEGIUM
SORGHUM
SORGHASTRUM
815, 816

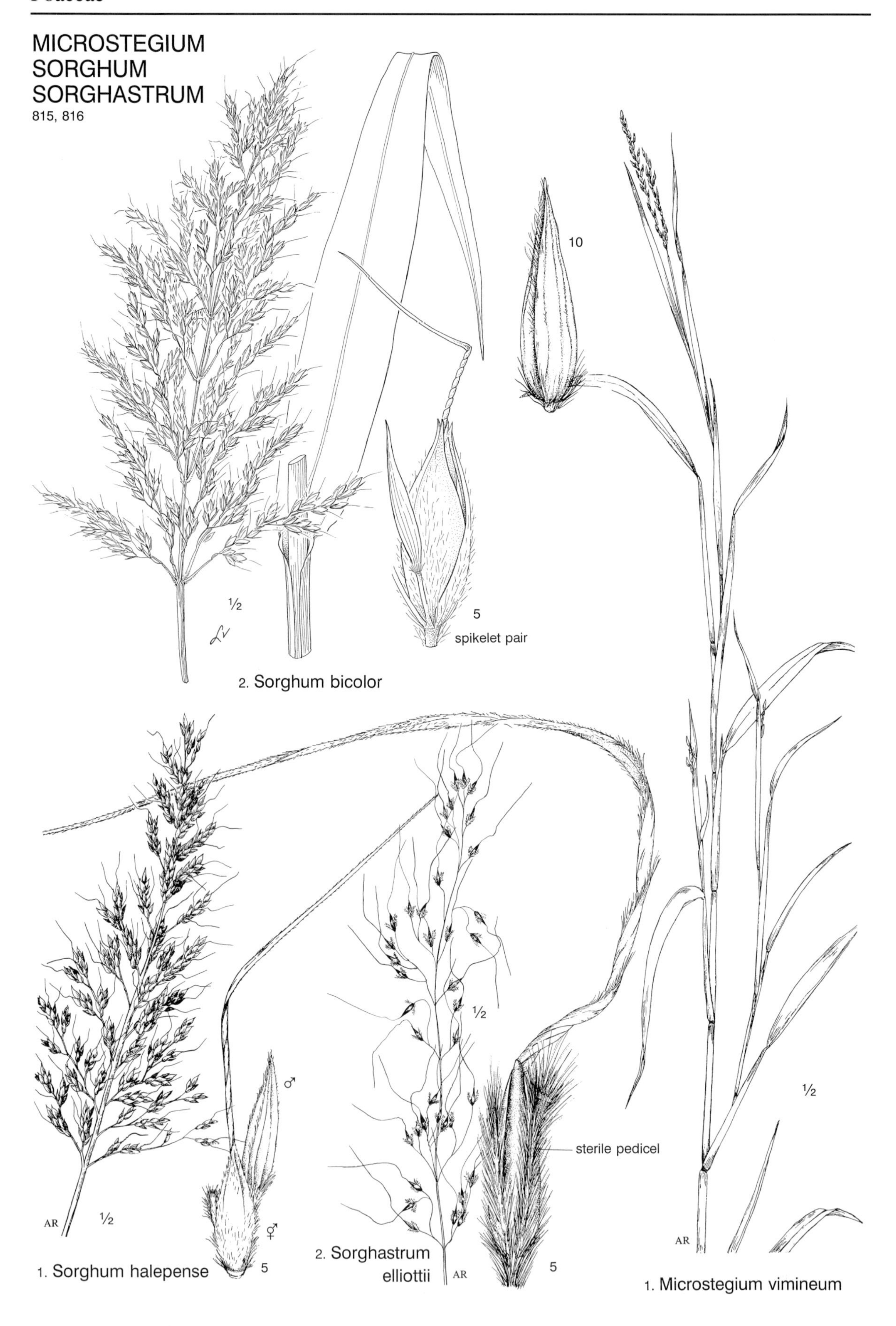

2. Sorghum bicolor

10

5
spikelet pair

1. Sorghum halepense

2. Sorghastrum elliottii

sterile pedicel

1. Microstegium vimineum

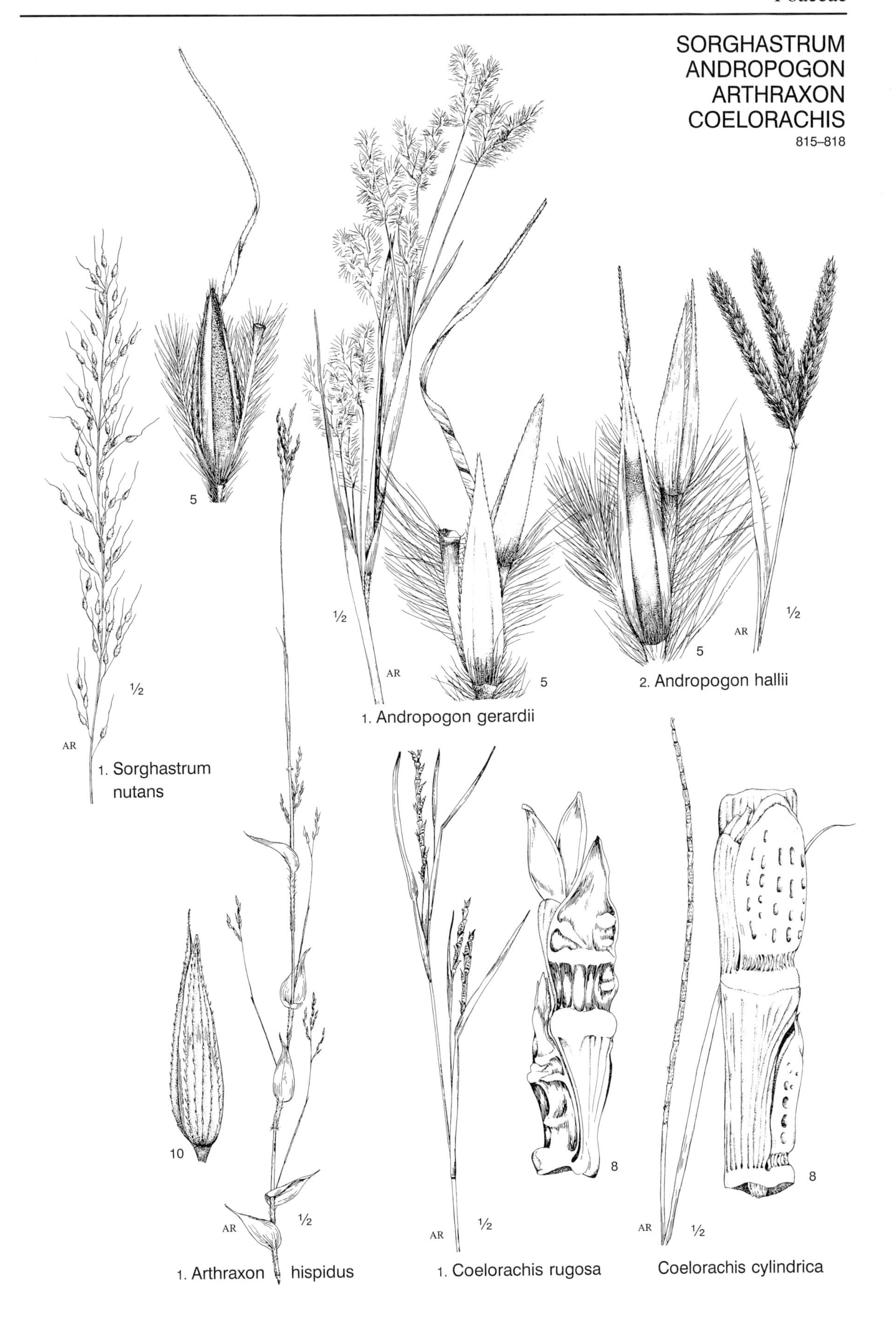

5

½

AR

1. Sorghastrum
nutans

½

AR

5

1. Andropogon gerardii

2. Andropogon hallii

½

AR

5

10

AR

½

1. Arthraxon hispidus

½

AR

8

1. Coelorachis rugosa

½

AR

8

Coelorachis cylindrica

ANDROPOGON
TRIPSACUM
816–818

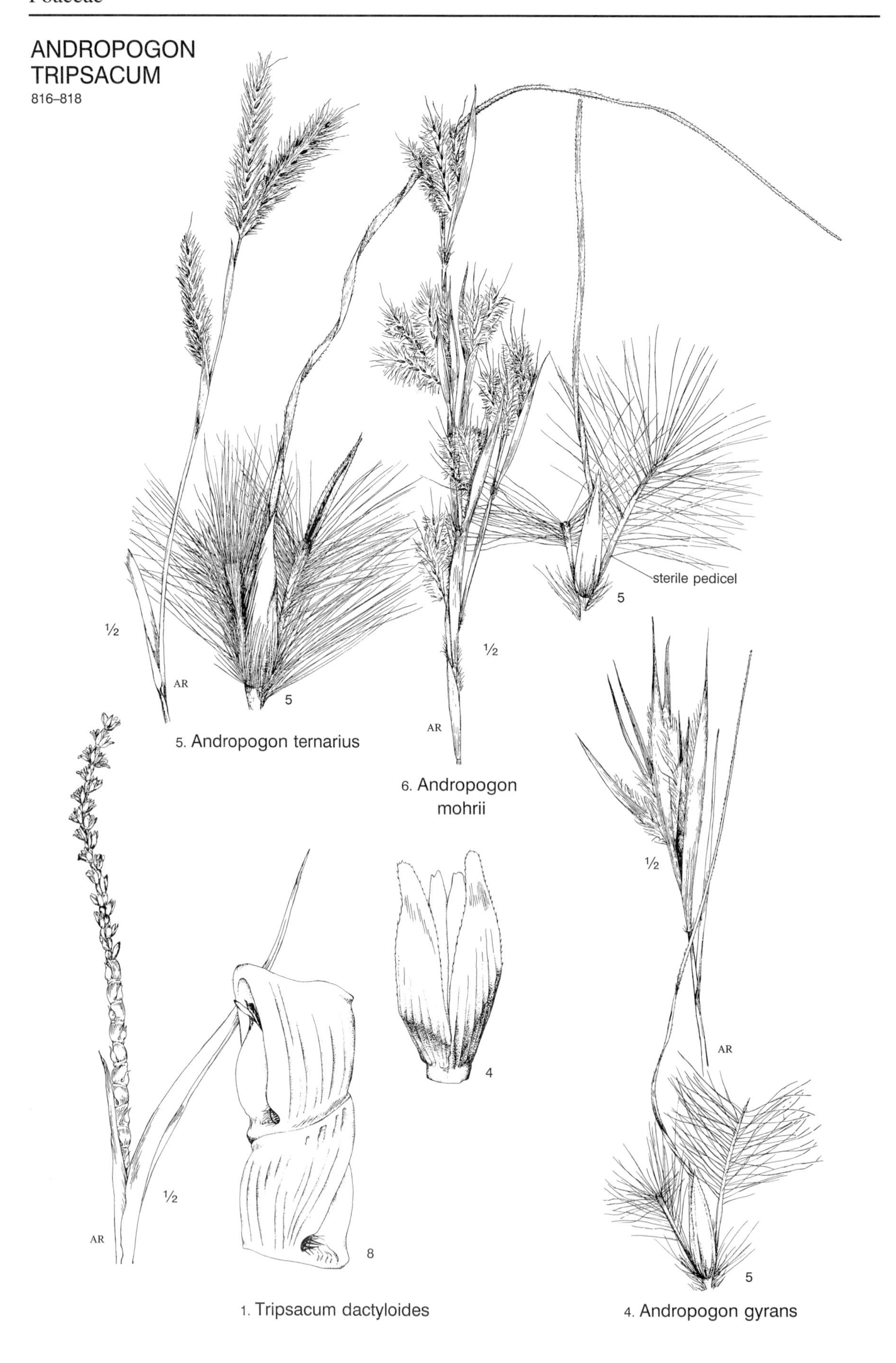

5. Andropogon ternarius

sterile pedicel

6. Andropogon mohrii

1. Tripsacum dactyloides

4. Andropogon gyrans

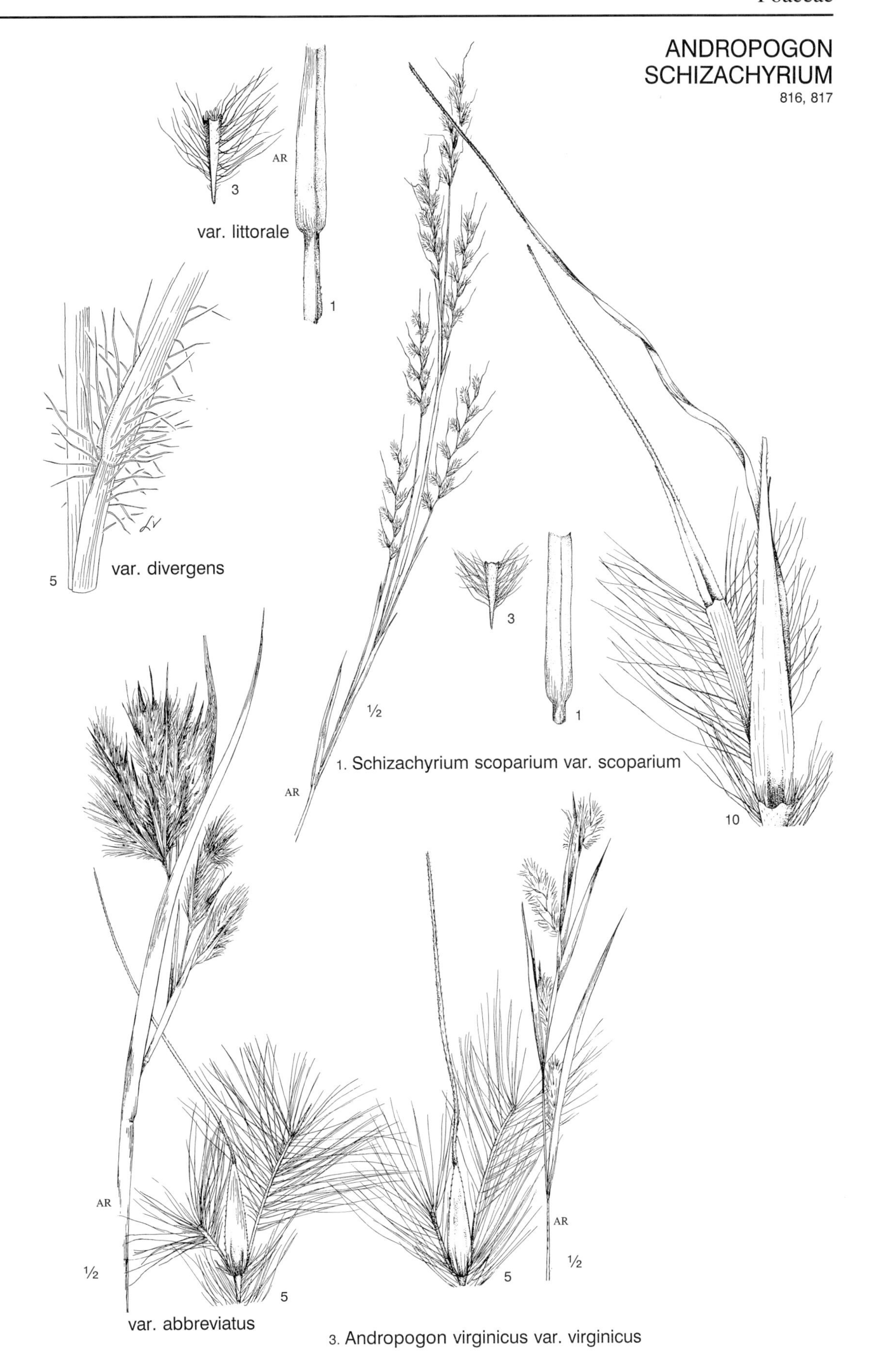

var. littorale

var. divergens

1. Schizachyrium scoparium var. scoparium

var. abbreviatus

3. Andropogon virginicus var. virginicus

SPARGANIUM
818, 819

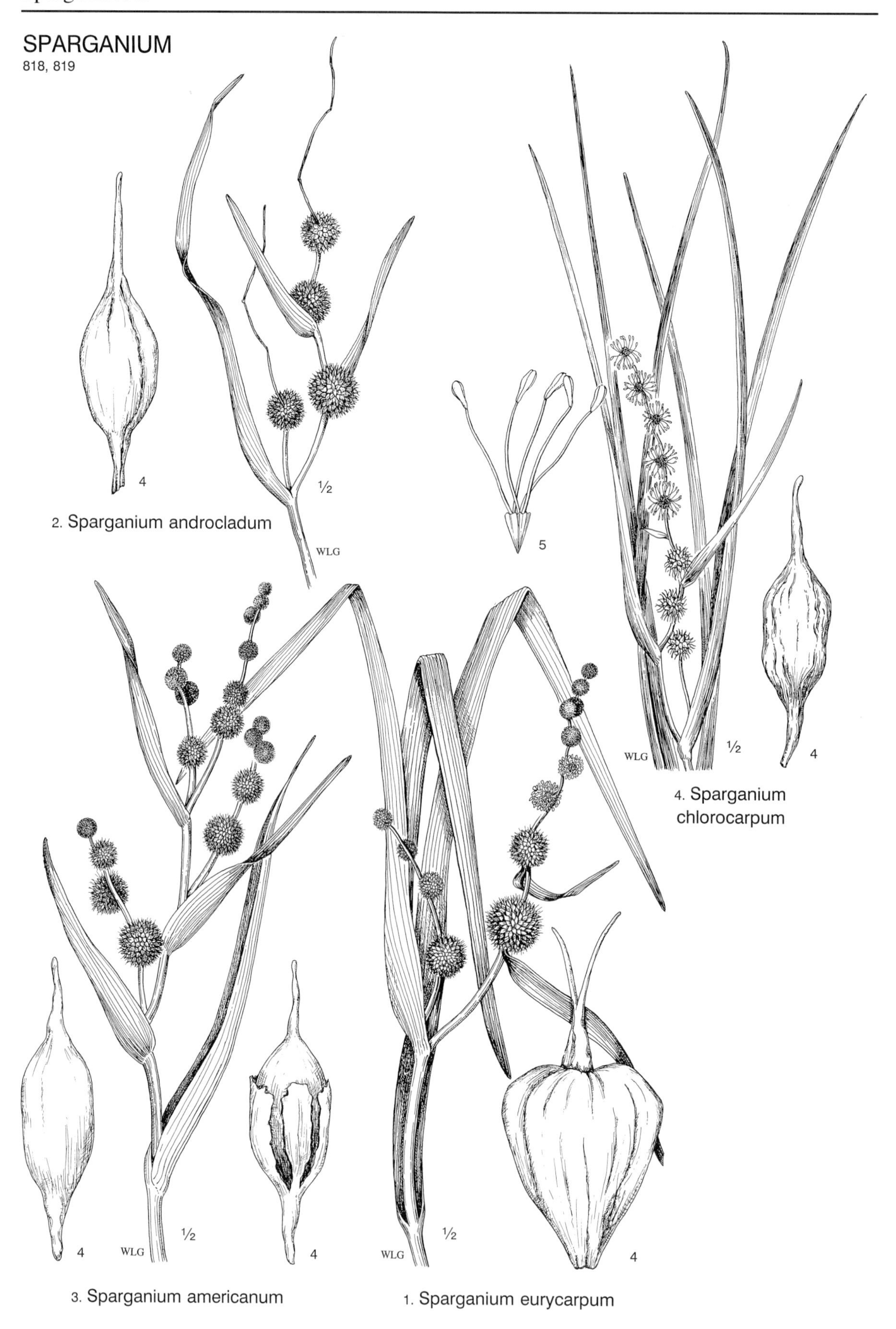

2. Sparganium androcladum

4. Sparganium
   chlorocarpum

3. Sparganium americanum

1. Sparganium eurycarpum

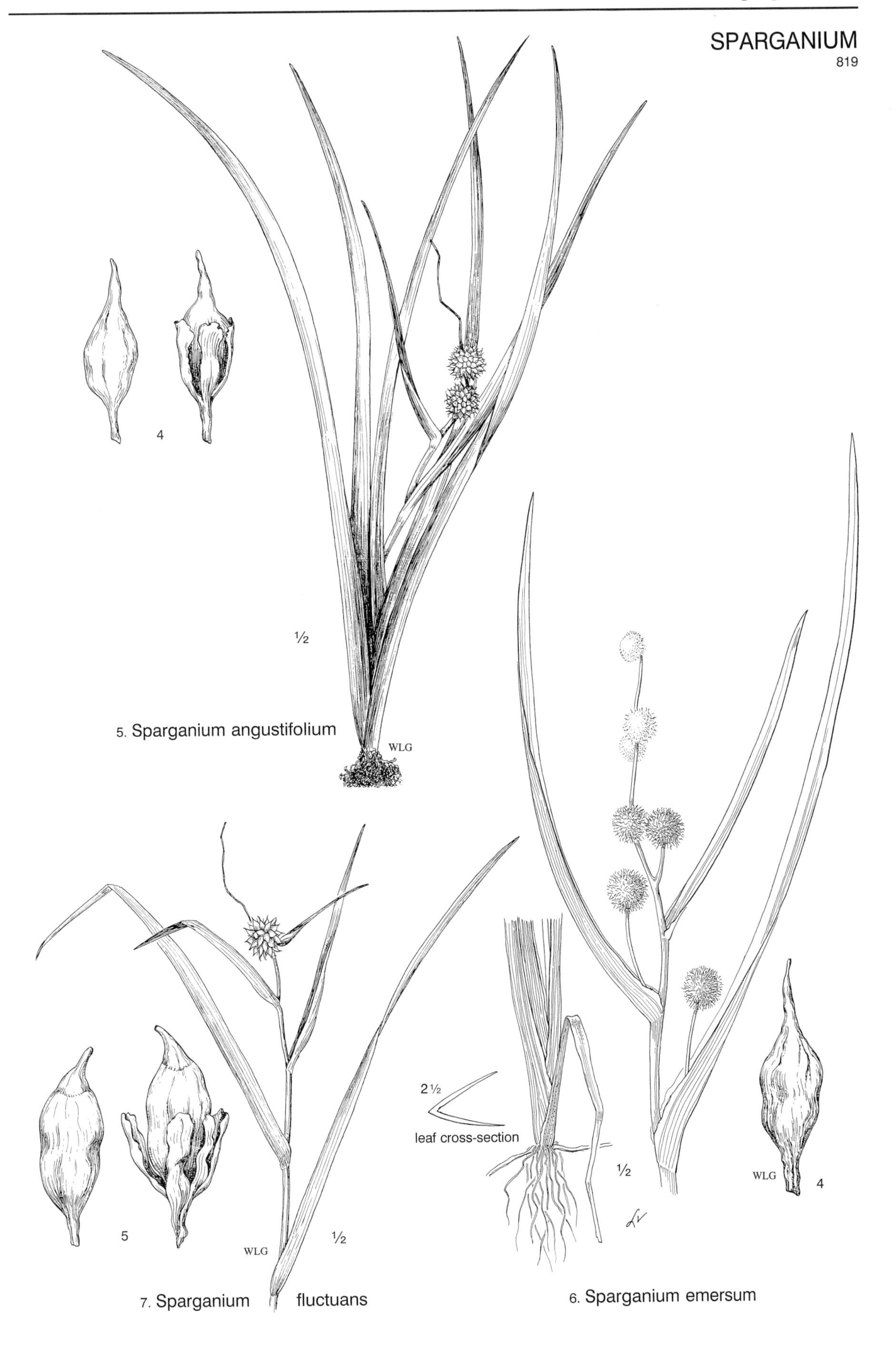

4

½

5. Sparganium angustifolium

WLG

2½
leaf cross-section

7. Sparganium   fluctuans

5

WLG   ½

½

6. Sparganium emersum

WLG   4

SPARGANIUM
TYPHA
819, 820

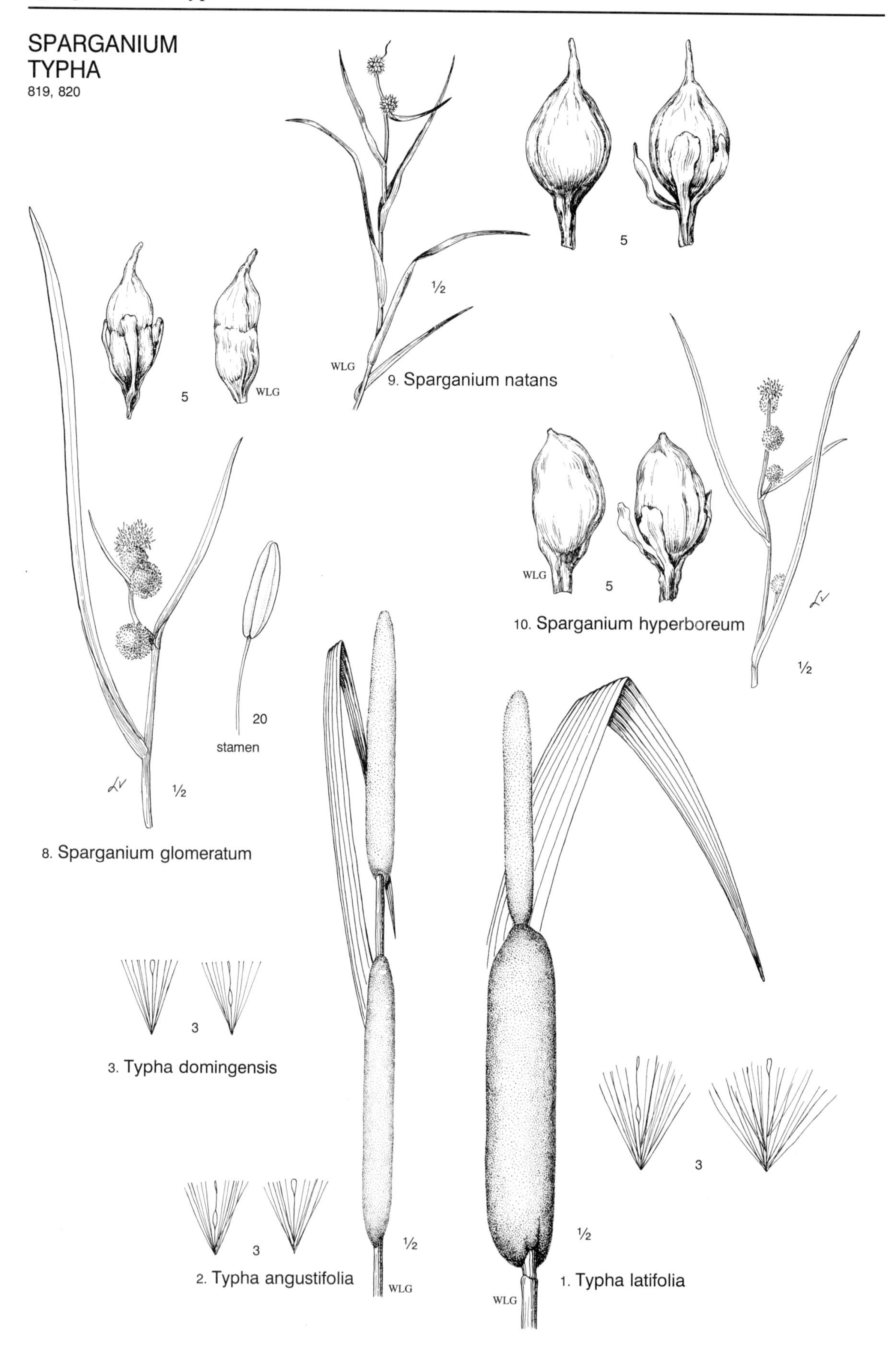

9. Sparganium natans

5

10. Sparganium hyperboreum

1/2

WLG

5

WLG

20
stamen

8. Sparganium glomeratum

3

3. Typha domingensis

3

3

2. Typha angustifolia

WLG

1/2

1/2

WLG

1. Typha latifolia

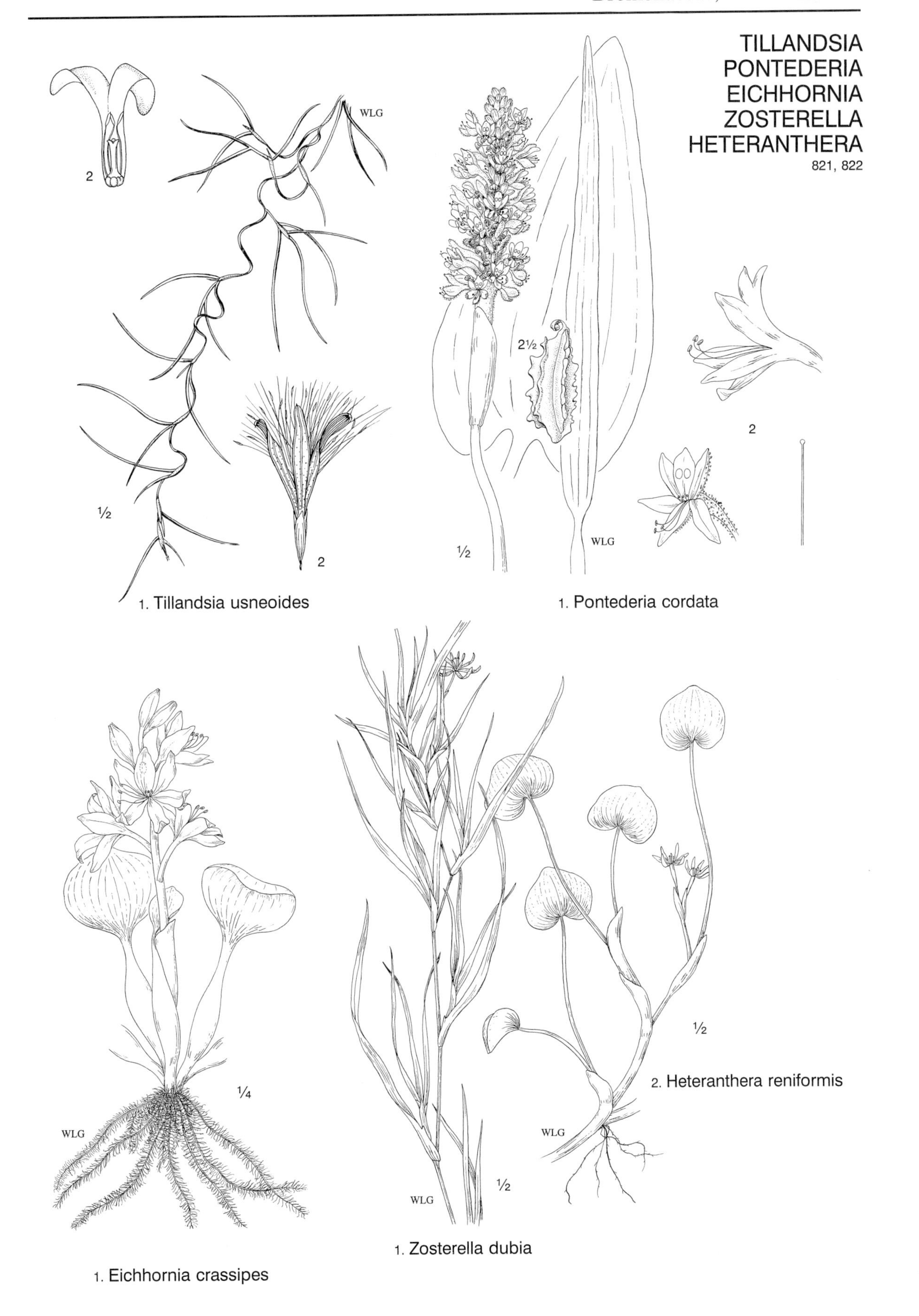

TILLANDSIA
PONTEDERIA
EICHHORNIA
ZOSTERELLA
HETERANTHERA
821, 822

1. Tillandsia usneoides

1. Pontederia cordata

2. Heteranthera reniformis

1. Zosterella dubia

1. Eichhornia crassipes

HETERANTHERA
XEROPHYLLUM
822–825

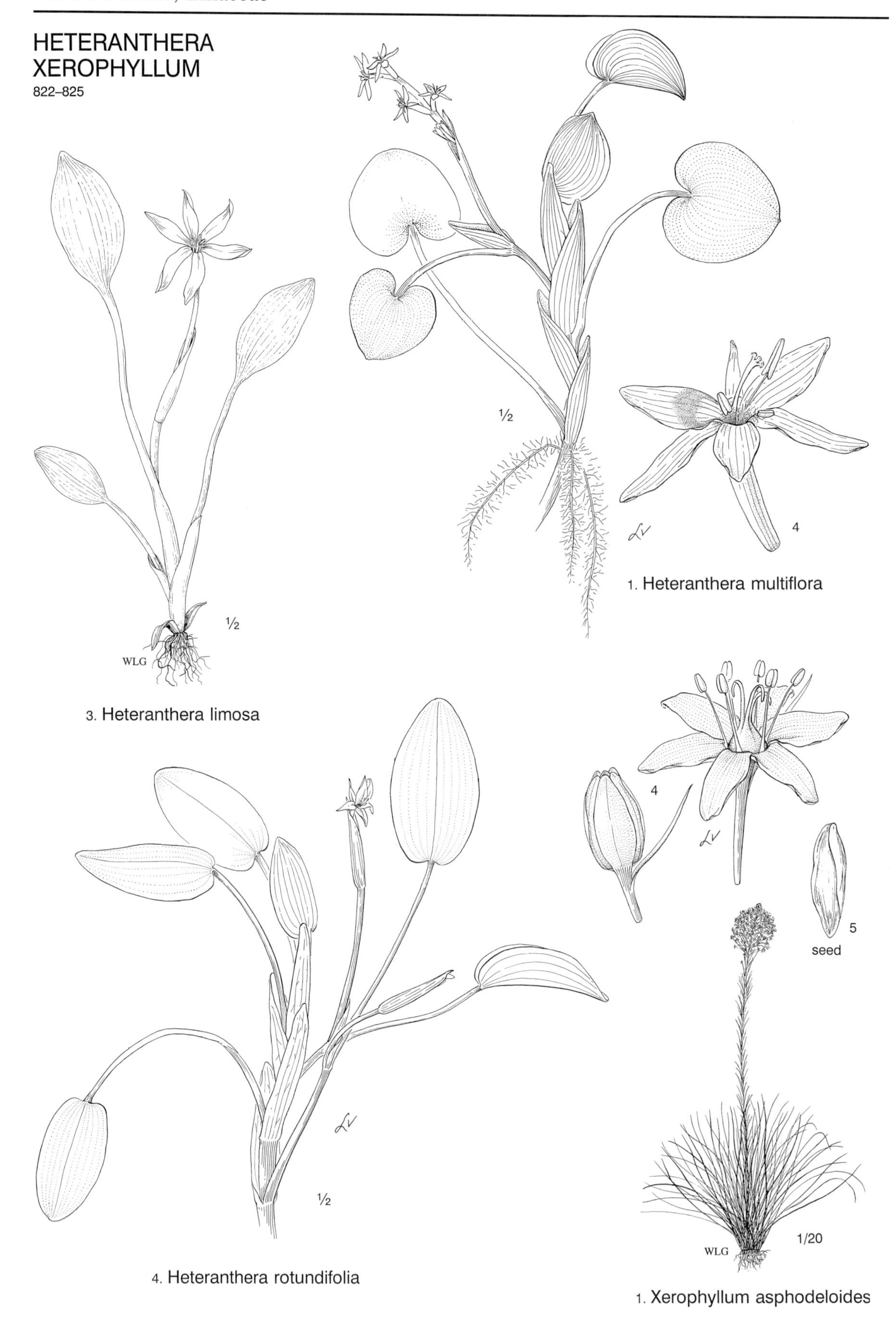

1. Heteranthera multiflora

3. Heteranthera limosa

4. Heteranthera rotundifolia

seed

1. Xerophyllum asphodeloides

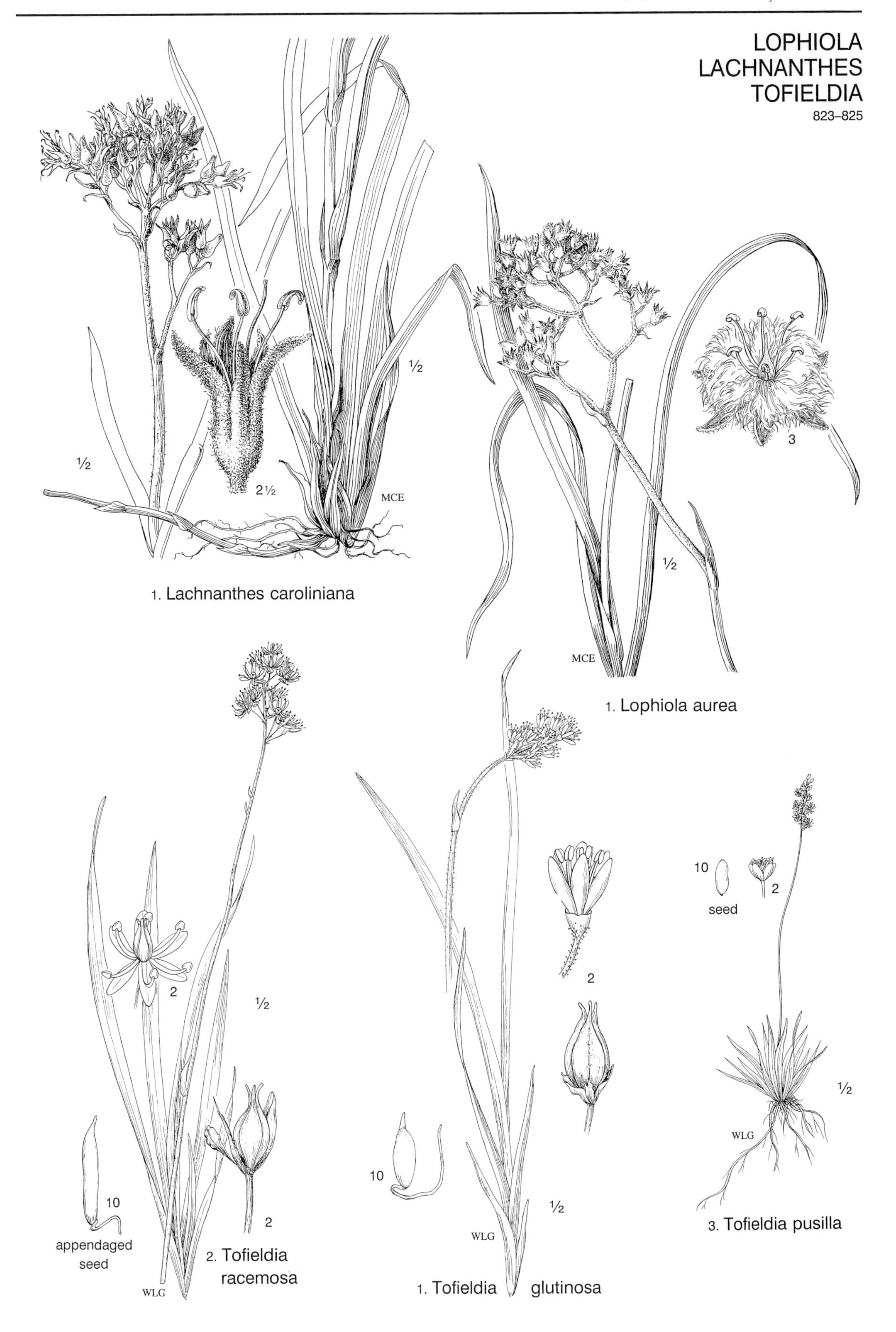

1. Lachnanthes caroliniana

1. Lophiola aurea

3. Tofieldia pusilla

2. Tofieldia racemosa

appendaged seed

1. Tofieldia glutinosa

AMIANTHIUM
STENANTHIUM
ZIGADENUS
826

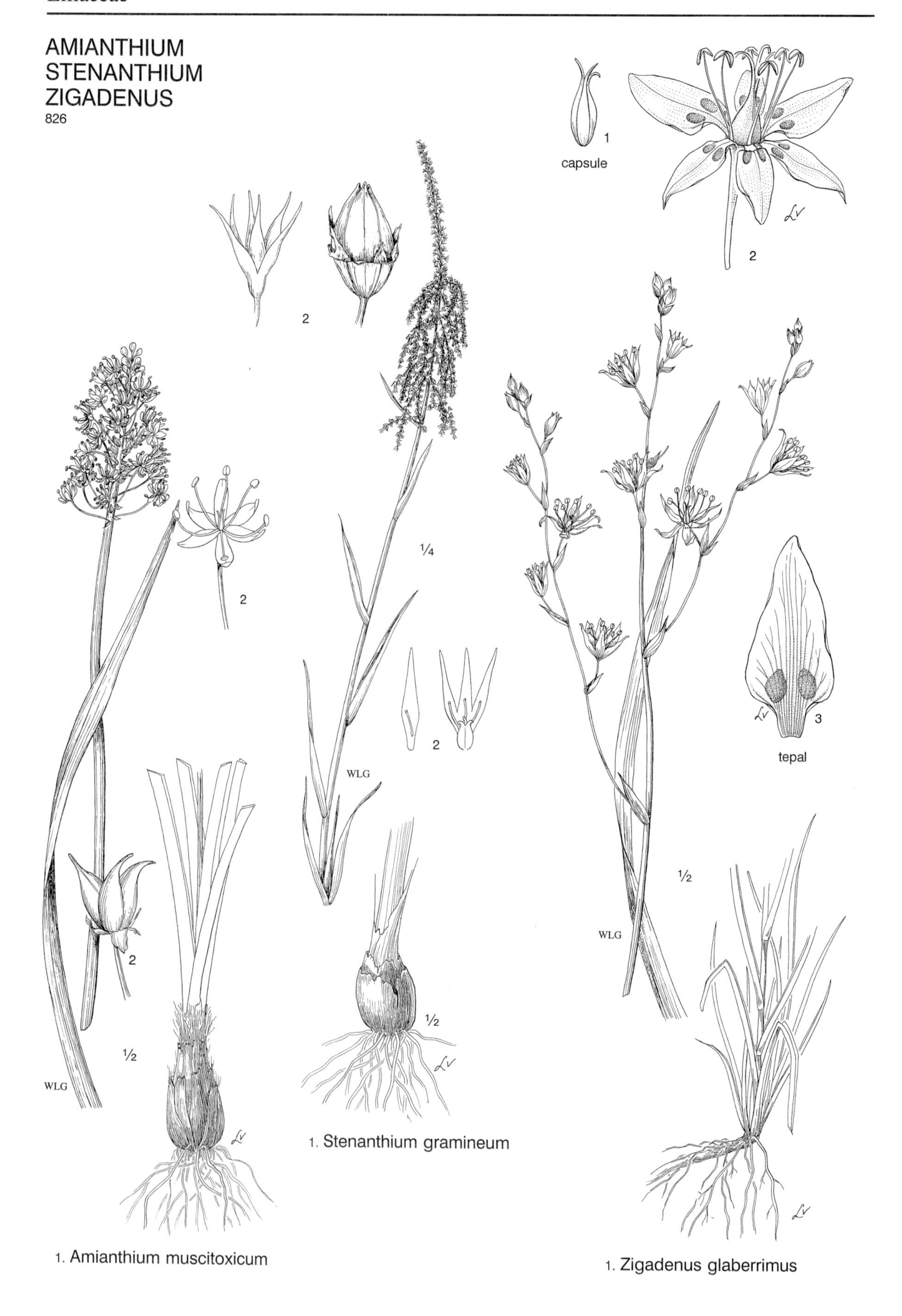

capsule

1

2

2

2

2

1/4

WLG

2

WLG

1/2

WLG

1. Amianthium muscitoxicum

1. Stenanthium gramineum

tepal

3

1/2

WLG

1. Zigadenus glaberrimus

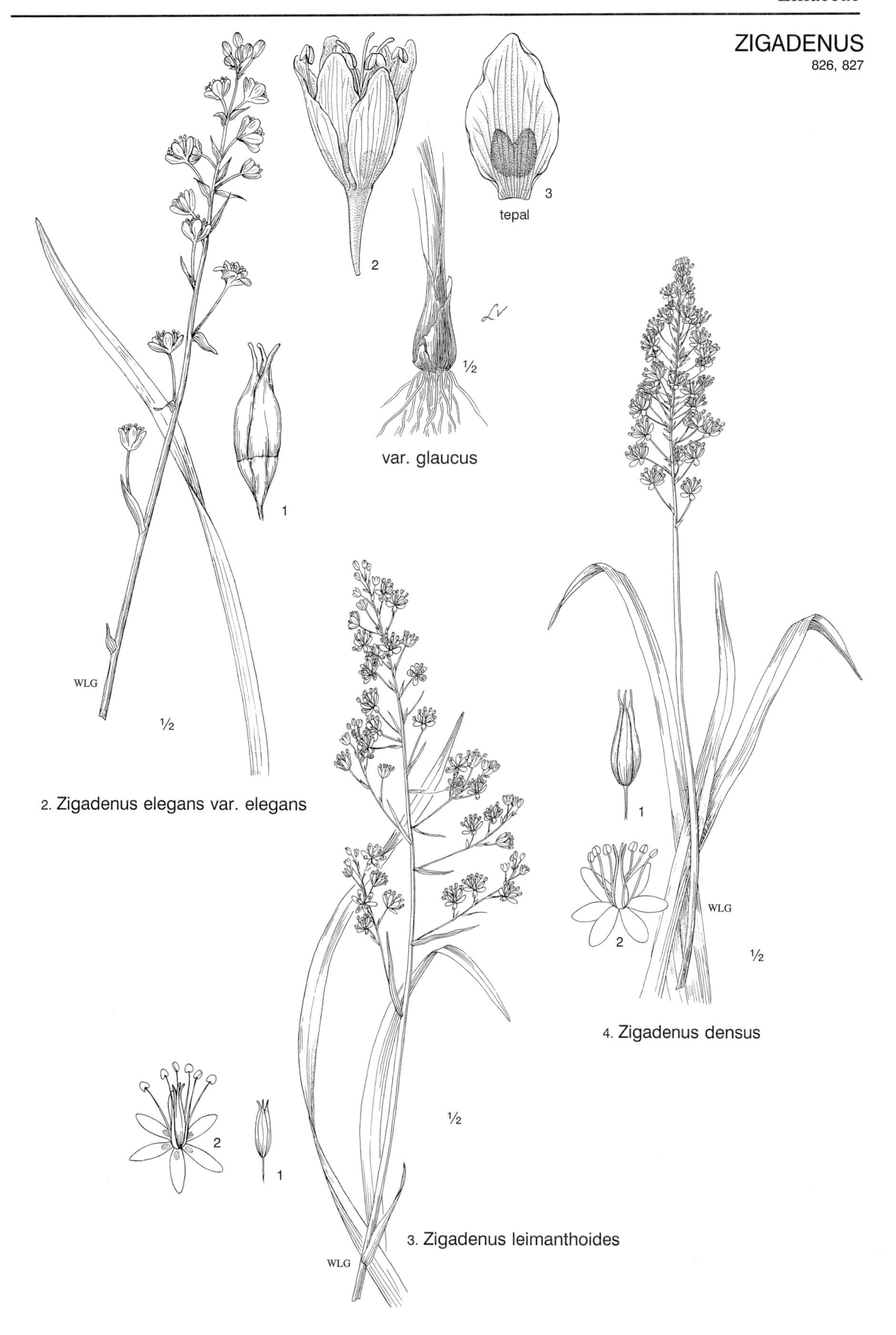

var. glaucus

1

WLG

½

2. Zigadenus elegans var. elegans

2

3

tepal

2

1

WLG

½

4. Zigadenus densus

½

2

1

WLG

3. Zigadenus leimanthoides

MELANTHIUM
VERATRUM
827

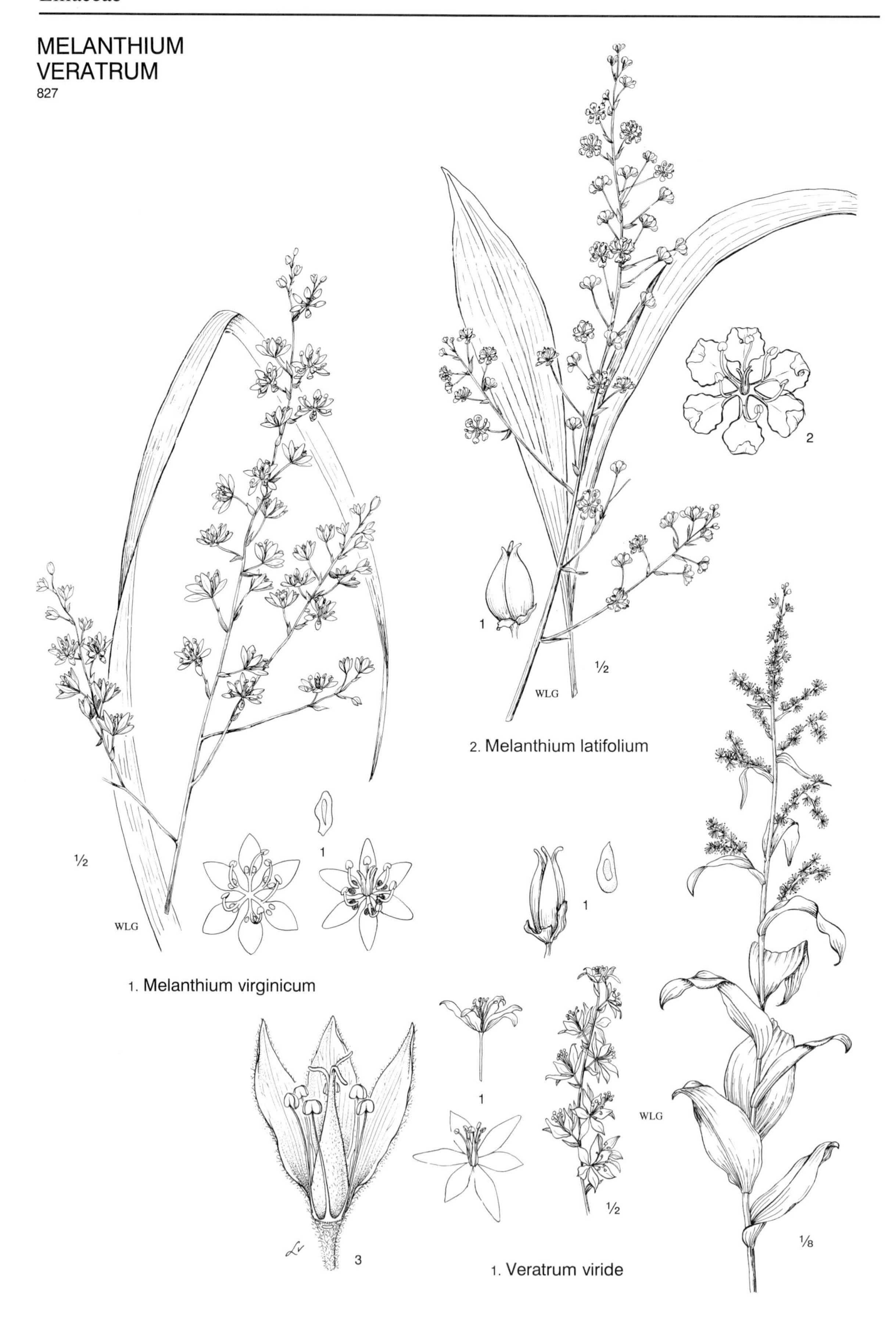

2. Melanthium latifolium

1. Melanthium virginicum

1. Veratrum viride

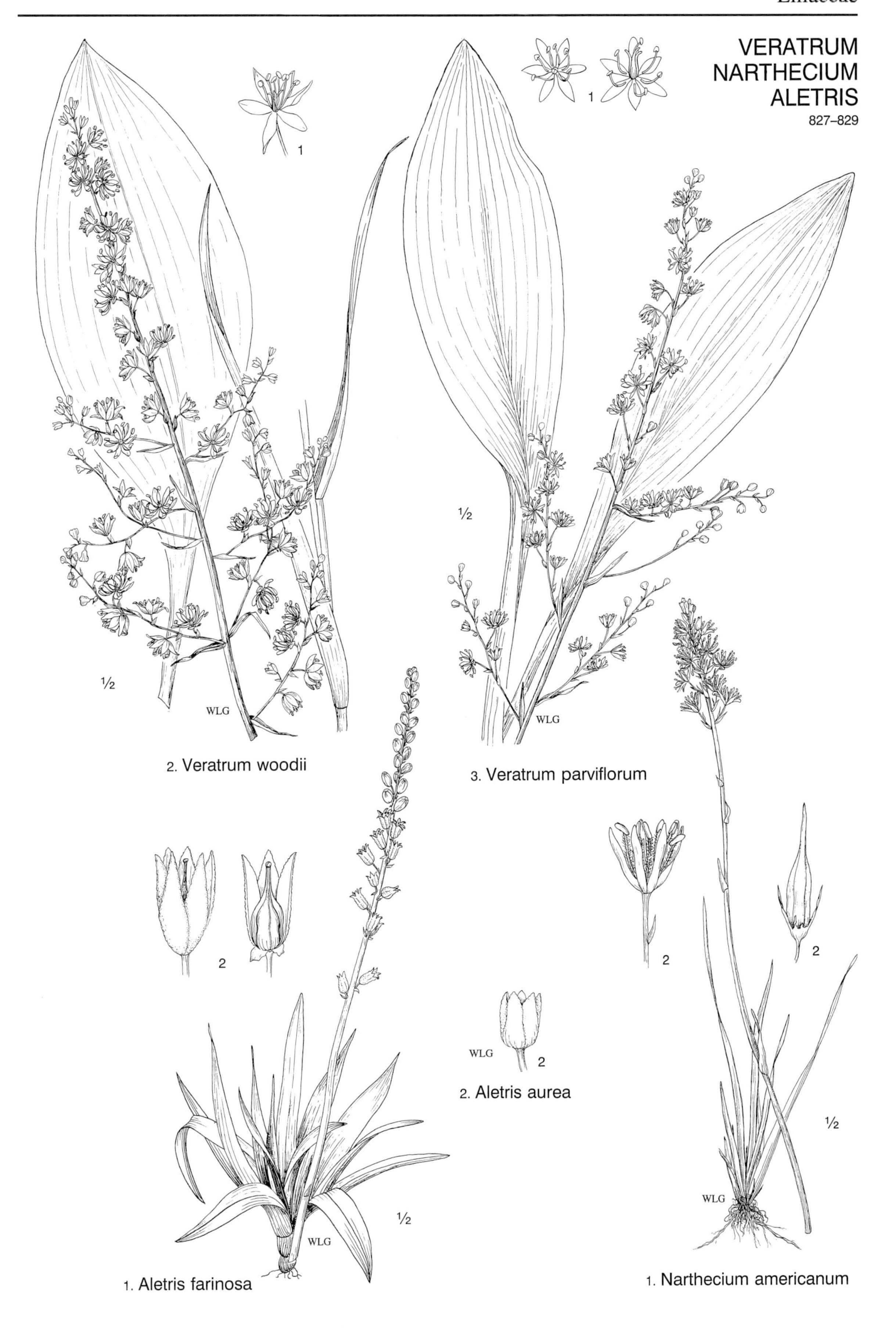

1

1

2. Veratrum woodii

WLG

½

3. Veratrum parviflorum

WLG

½

2

WLG   2

2. Aletris aurea

1. Aletris farinosa

WLG

½

2

2

WLG

½

1. Narthecium americanum

HELONIAS
CAMASSIA
SCILLA
828, 829

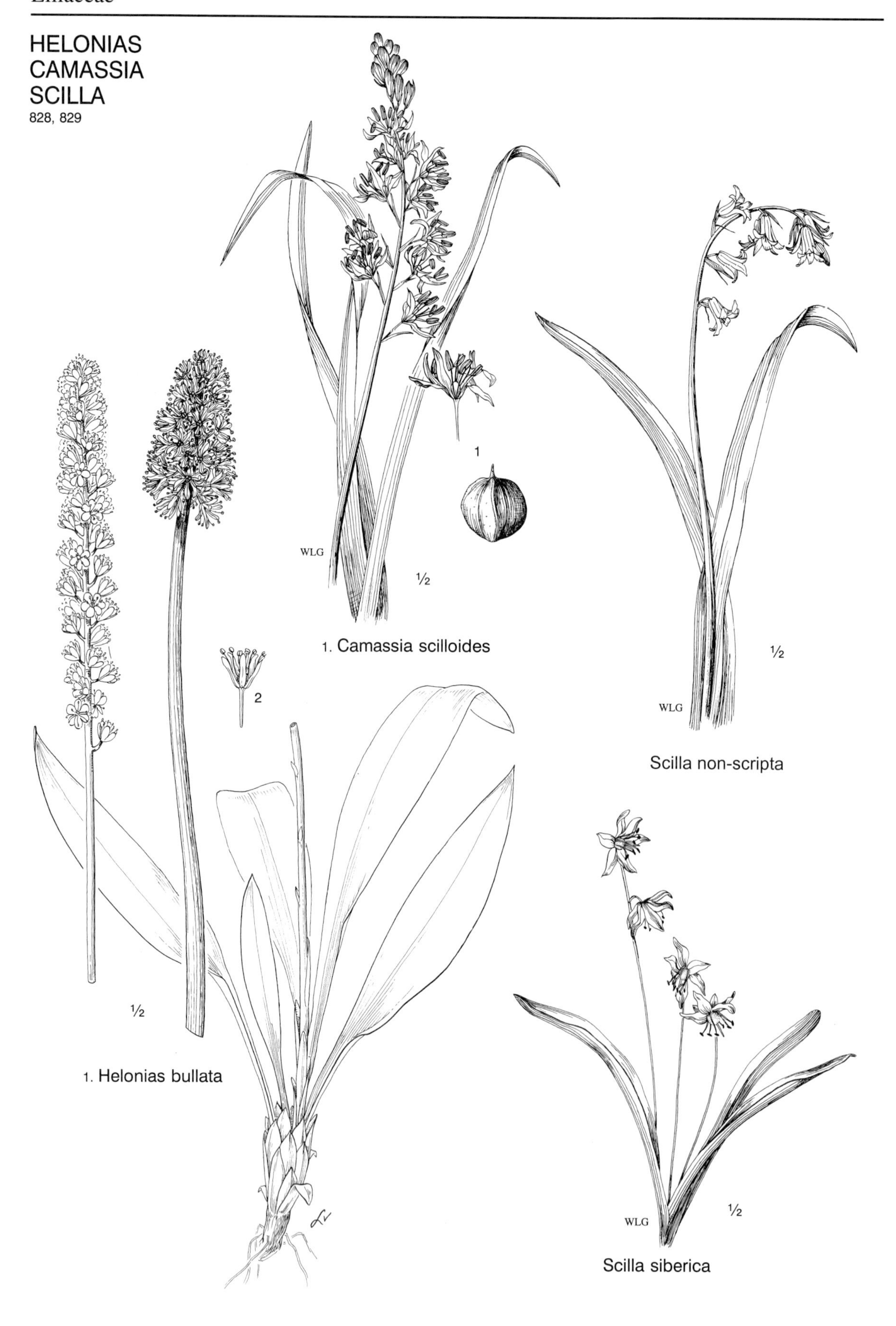

1. Camassia scilloides

WLG

½

2

Scilla non-scripta

WLG

½

1. Helonias bullata

½

Scilla siberica

WLG

½

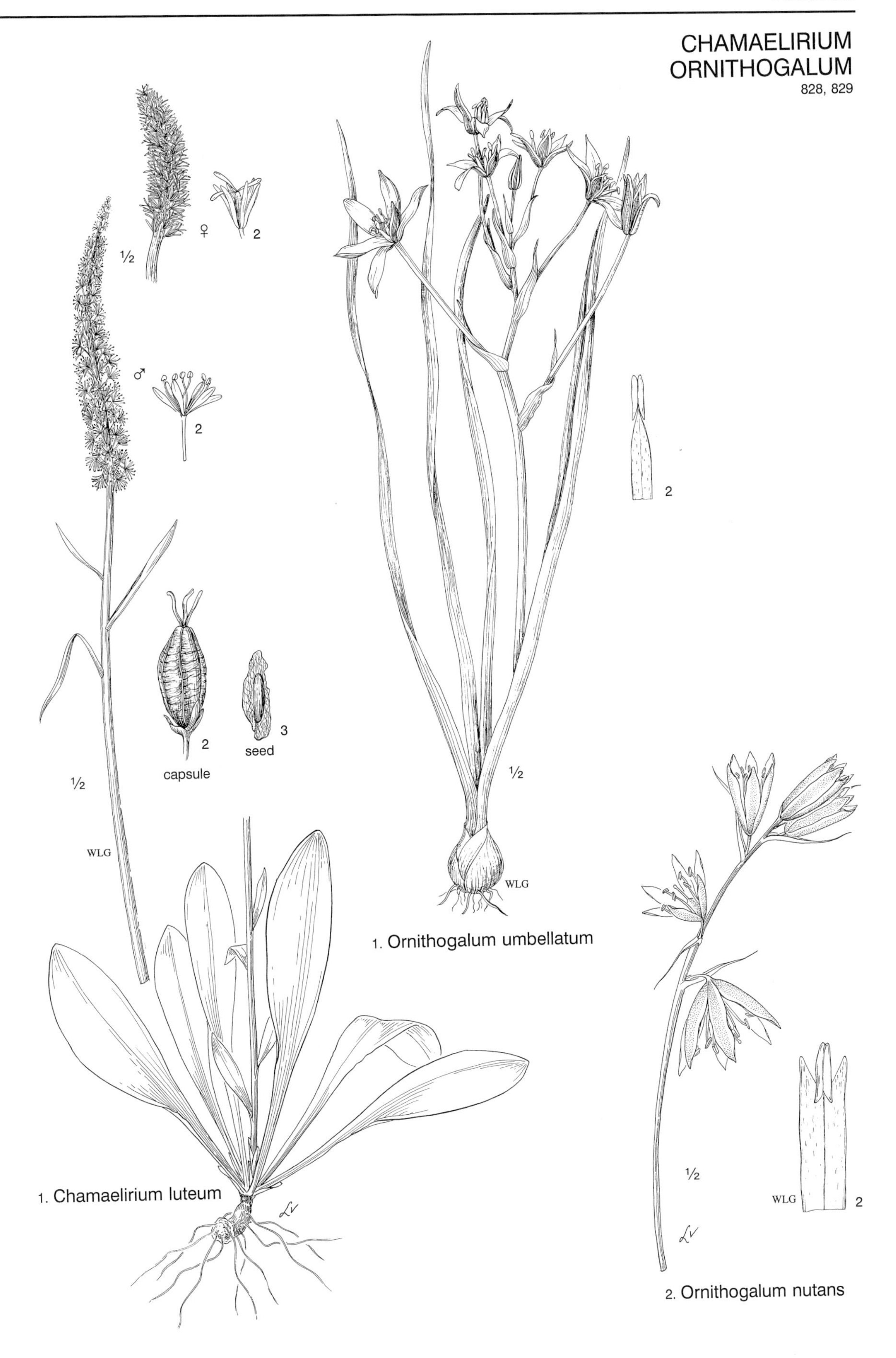

½

♀   2

♂

2

2

capsule   2   3   seed

½

WLG

½

WLG

1. Ornithogalum umbellatum

1. Chamaelirium luteum

½

WLG   2

2. Ornithogalum nutans

MUSCARI
HOSTA
829, 830

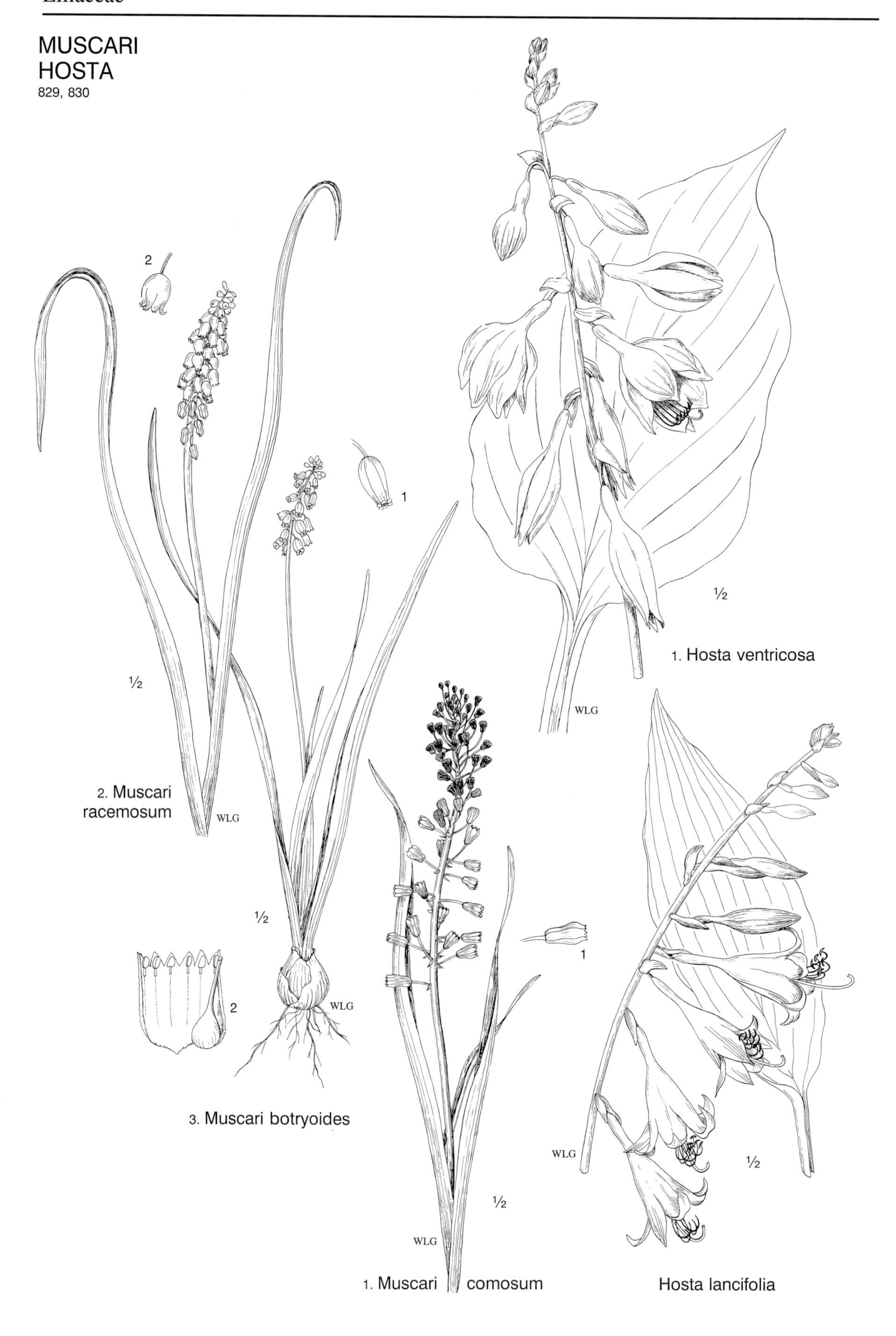

1. Hosta ventricosa

2. Muscari racemosum

3. Muscari botryoides

1. Muscari comosum

Hosta lancifolia

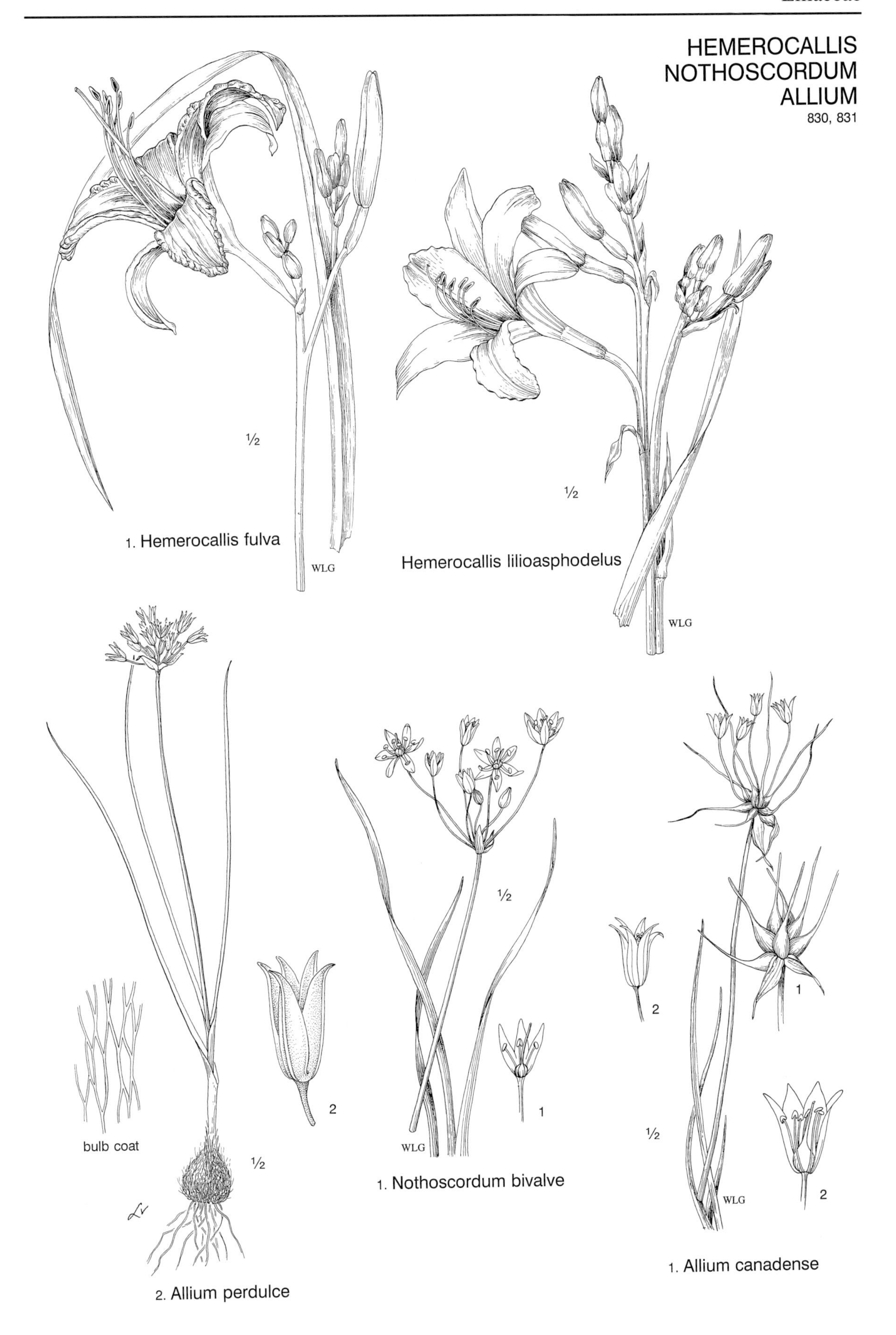

1. Hemerocallis fulva

WLG

Hemerocallis lilioasphodelus

WLG

bulb coat

2. Allium perdulce

½

2

1. Nothoscordum bivalve

WLG

1

2

1

½

WLG

2

1. Allium canadense

ALLIUM
831, 832

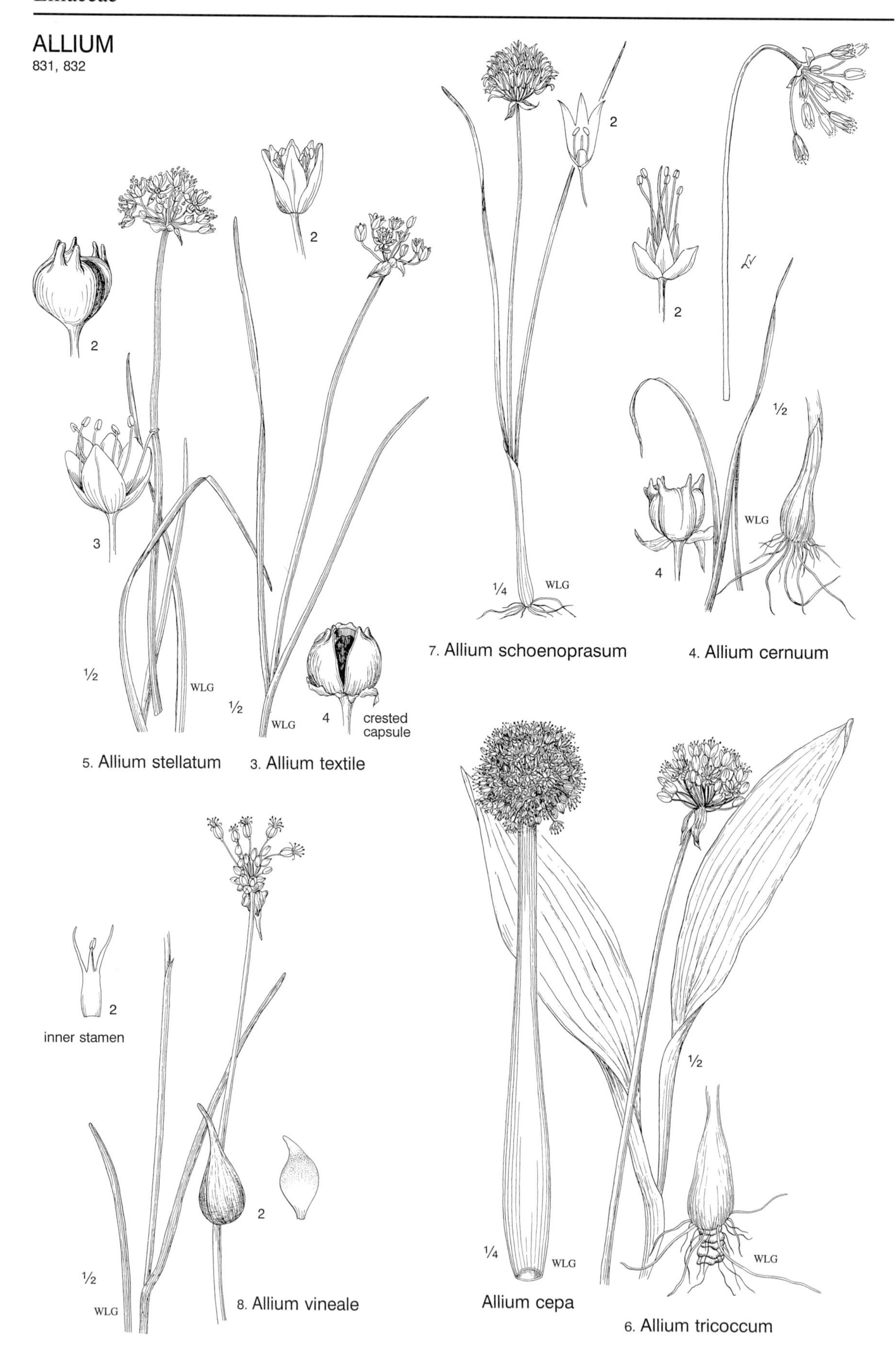

2

2

2

3

2

½

WLG

½

WLG

2

½

WLG

4 crested capsule

5. Allium stellatum   3. Allium textile

7. Allium schoenoprasum

¼   WLG

½

WLG

4

4. Allium cernuum

2

inner stamen

2

½

WLG

8. Allium vineale

¼   WLG

Allium cepa

½

WLG

6. Allium tricoccum

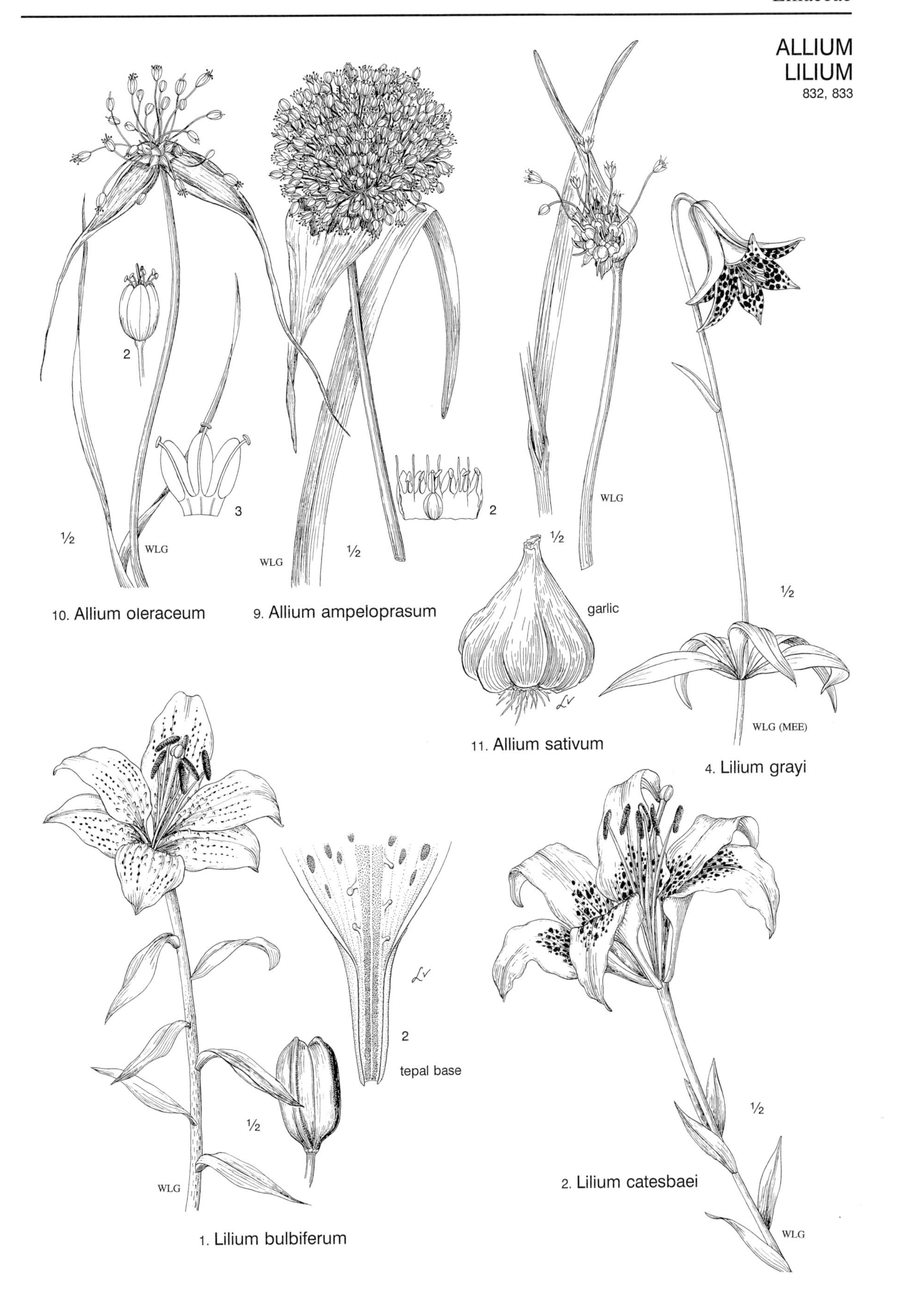

10. Allium oleraceum

9. Allium ampeloprasum

11. Allium sativum

garlic

4. Lilium grayi

WLG (MEE)

1. Lilium bulbiferum

tepal base

2. Lilium catesbaei

LILIUM
833

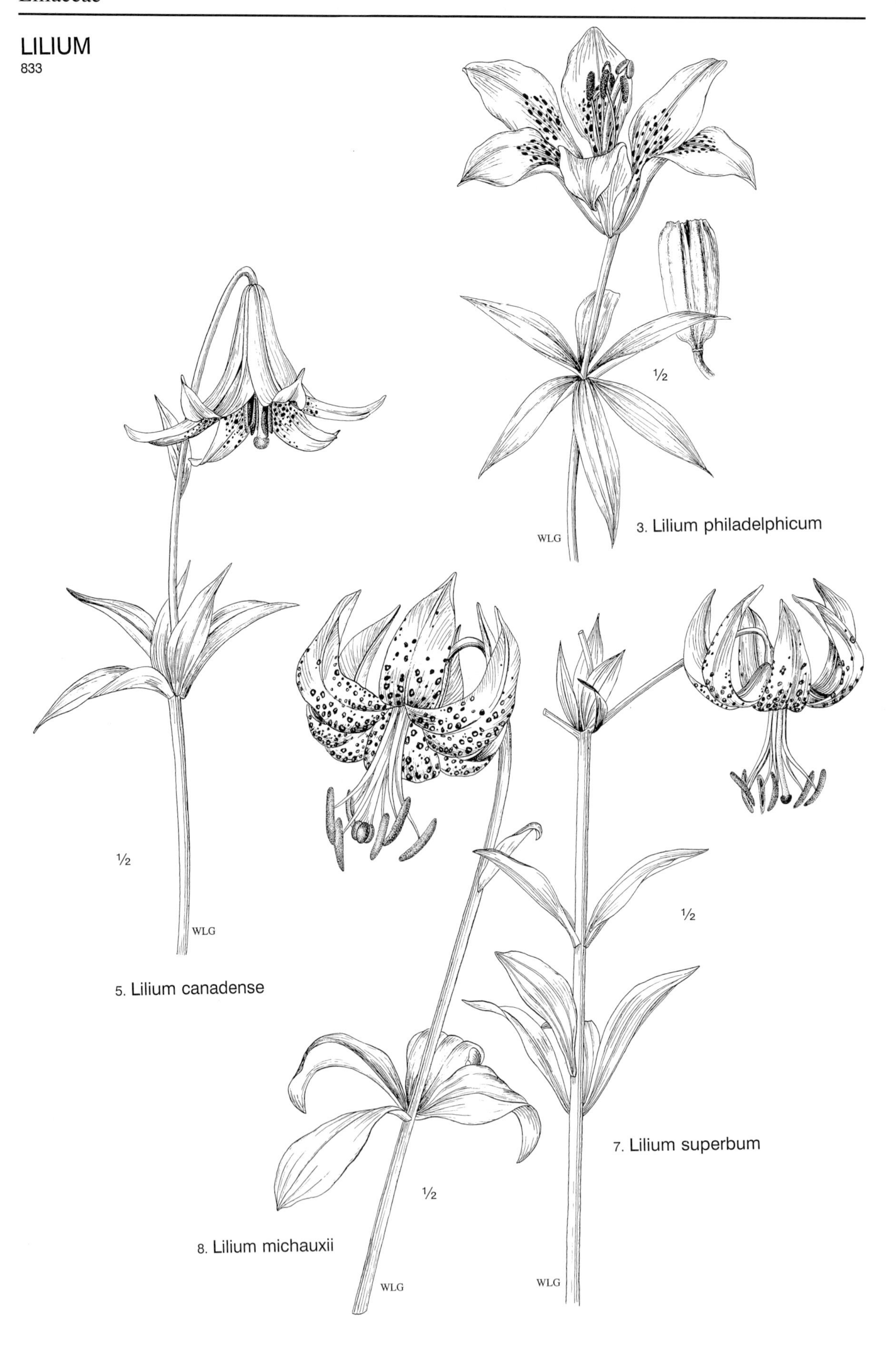

3. Lilium philadelphicum

WLG

½

5. Lilium canadense

WLG

½

7. Lilium superbum

½

8. Lilium michauxii

½

WLG

WLG

ILLUSTRATED COMPANION TO

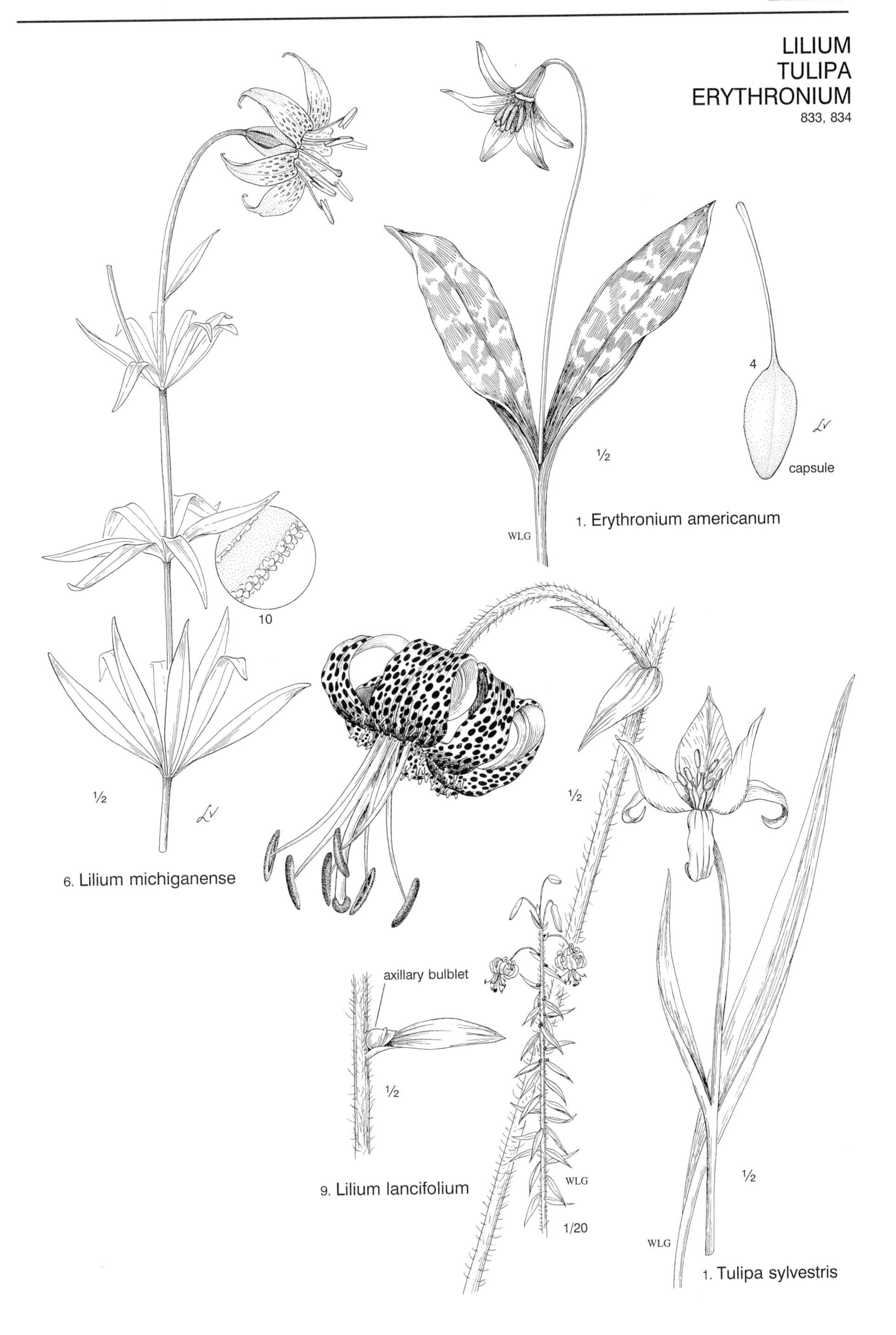

1. Erythronium americanum

capsule

6. Lilium michiganense

10

9. Lilium lancifolium

axillary bulblet

1. Tulipa sylvestris

ERYTHRONIUM
COLCHICUM
834

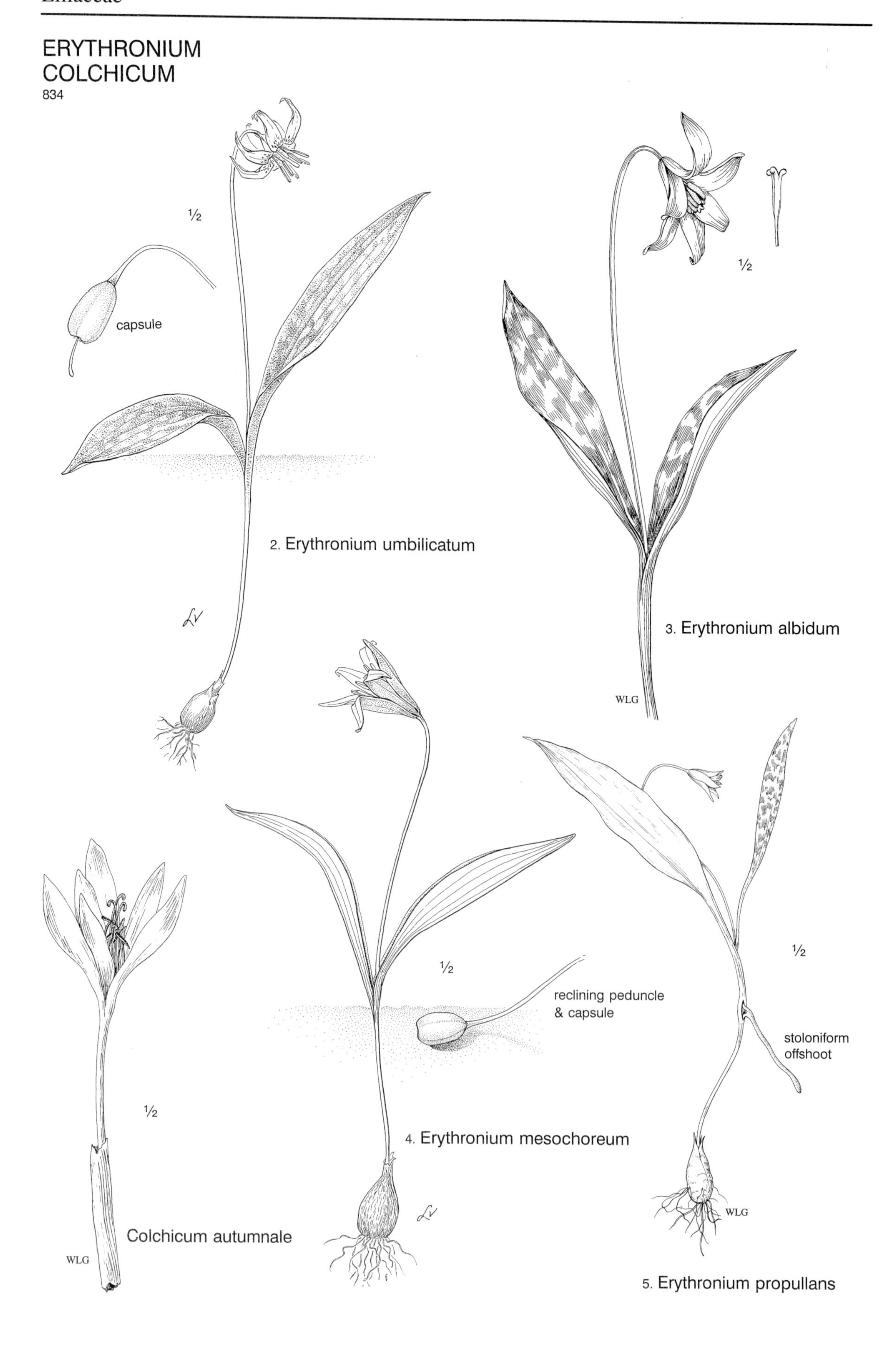

½

capsule

2. Erythronium umbilicatum

½

3. Erythronium albidum

WLG

½

reclining peduncle
& capsule

½

stoloniform
offshoot

4. Erythronium mesochoreum

Colchicum autumnale

½

WLG

WLG

5. Erythronium propullans

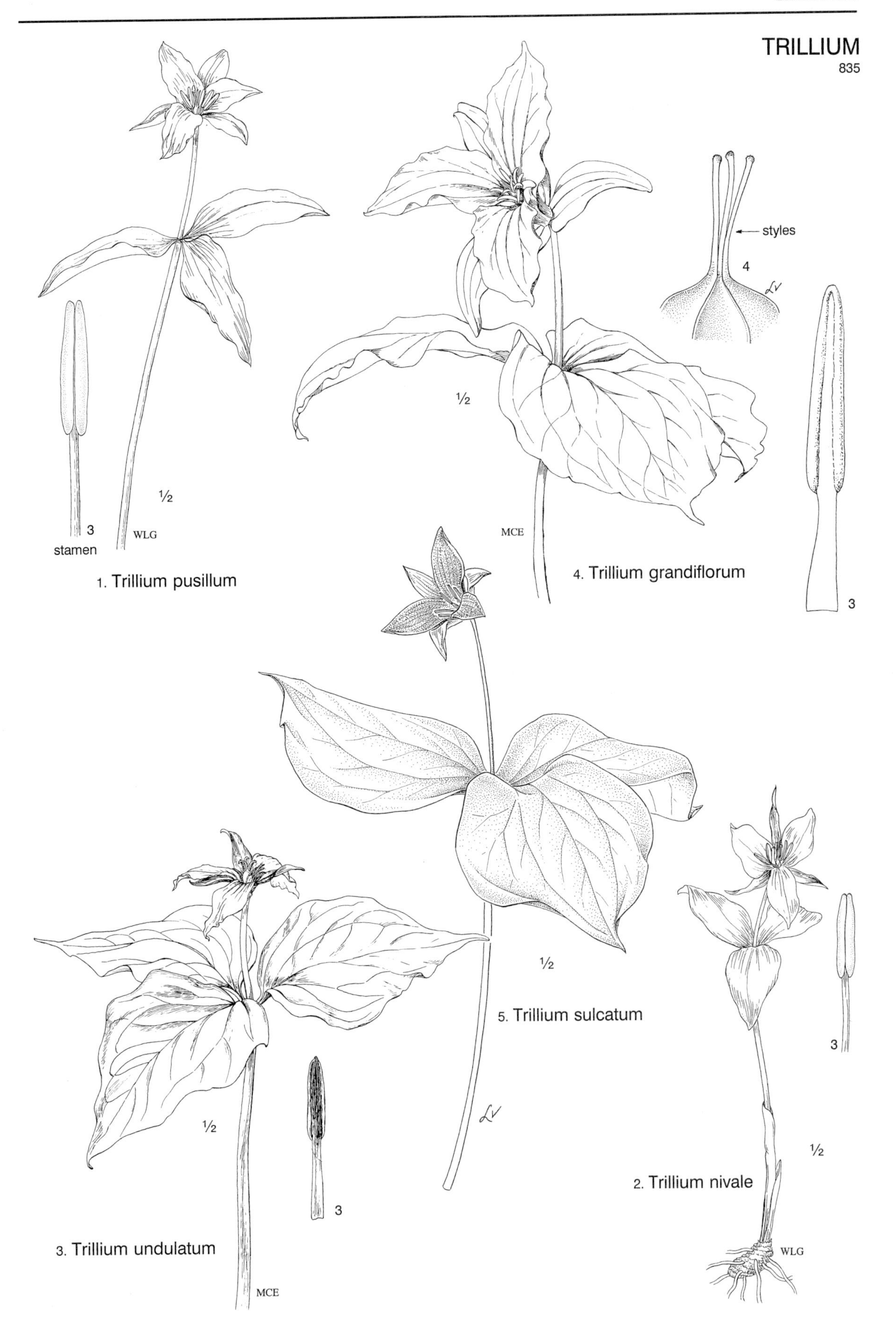

3
stamen
½
WLG

1. Trillium pusillum

½
MCE

4. Trillium grandiflorum

styles
4

3

½

5. Trillium sulcatum

½

3. Trillium undulatum

3

MCE

2. Trillium nivale
½
3
WLG

TRILLIUM
836

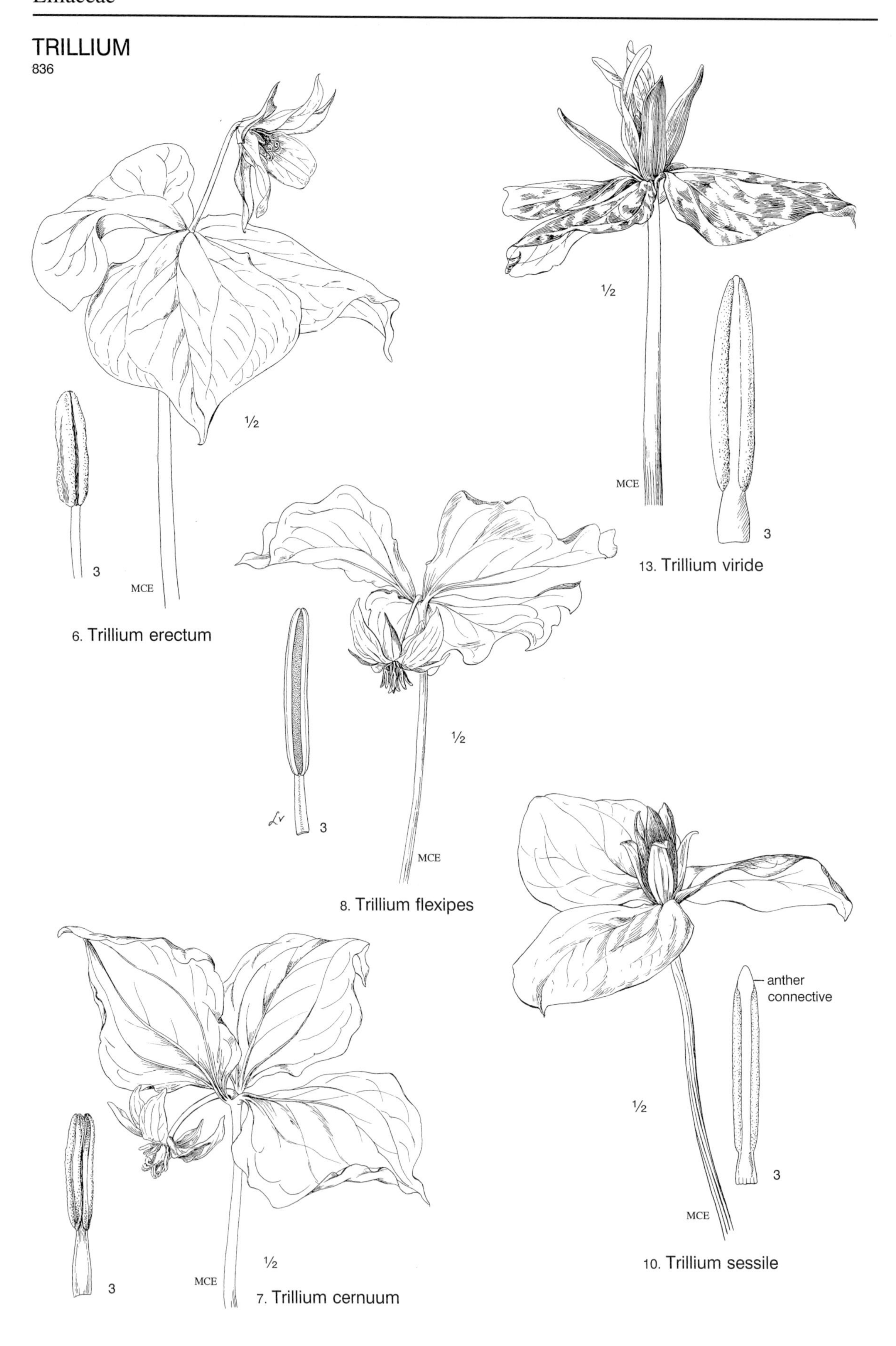

6. Trillium erectum

13. Trillium viride

8. Trillium flexipes

anther connective

10. Trillium sessile

7. Trillium cernuum

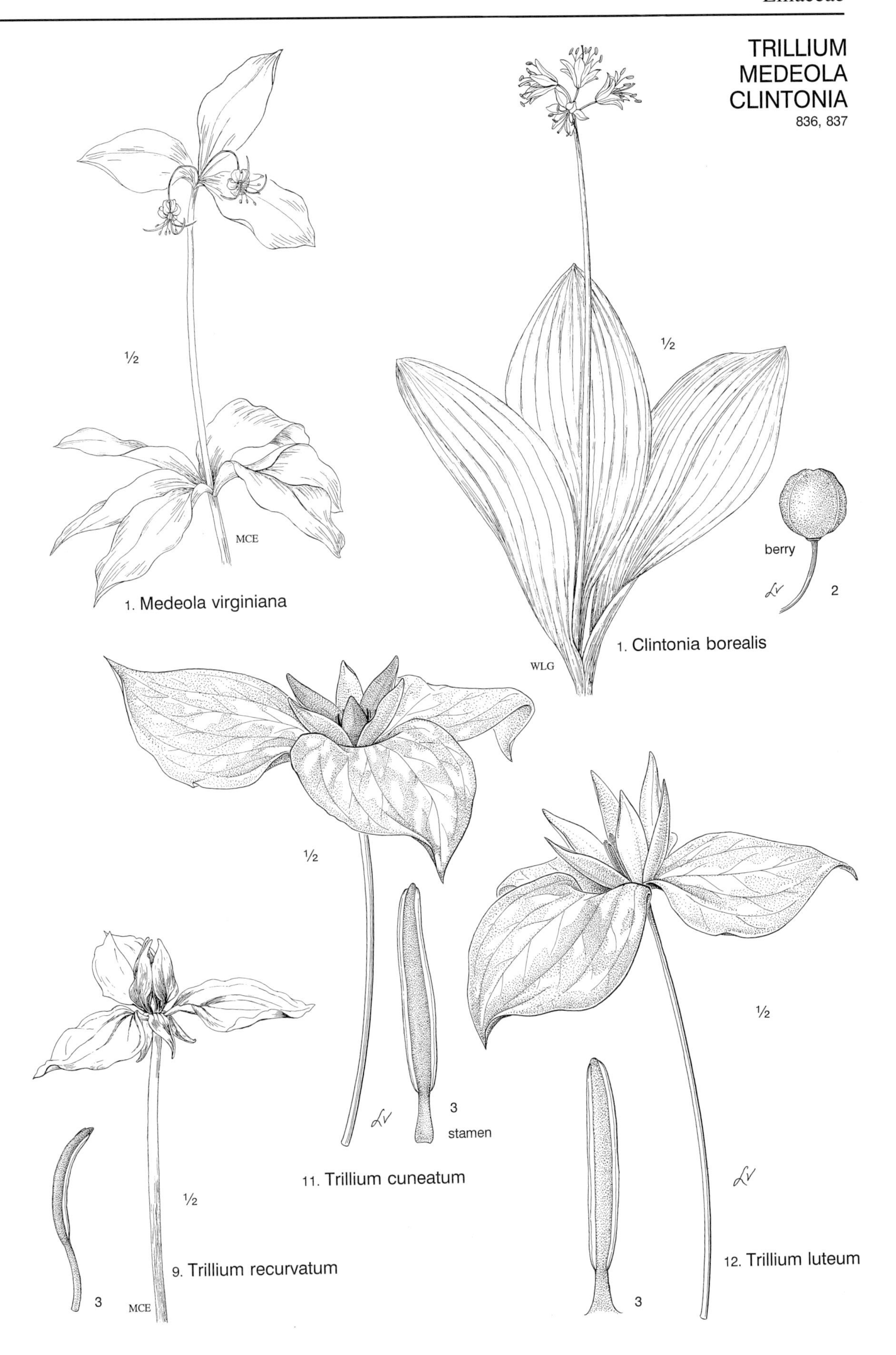

1. Medeola virginiana

MCE

1. Clintonia borealis

WLG

berry

2

9. Trillium recurvatum

3

MCE

11. Trillium cuneatum

stamen

3

12. Trillium luteum

3

CLINTONIA
UVULARIA
837, 838

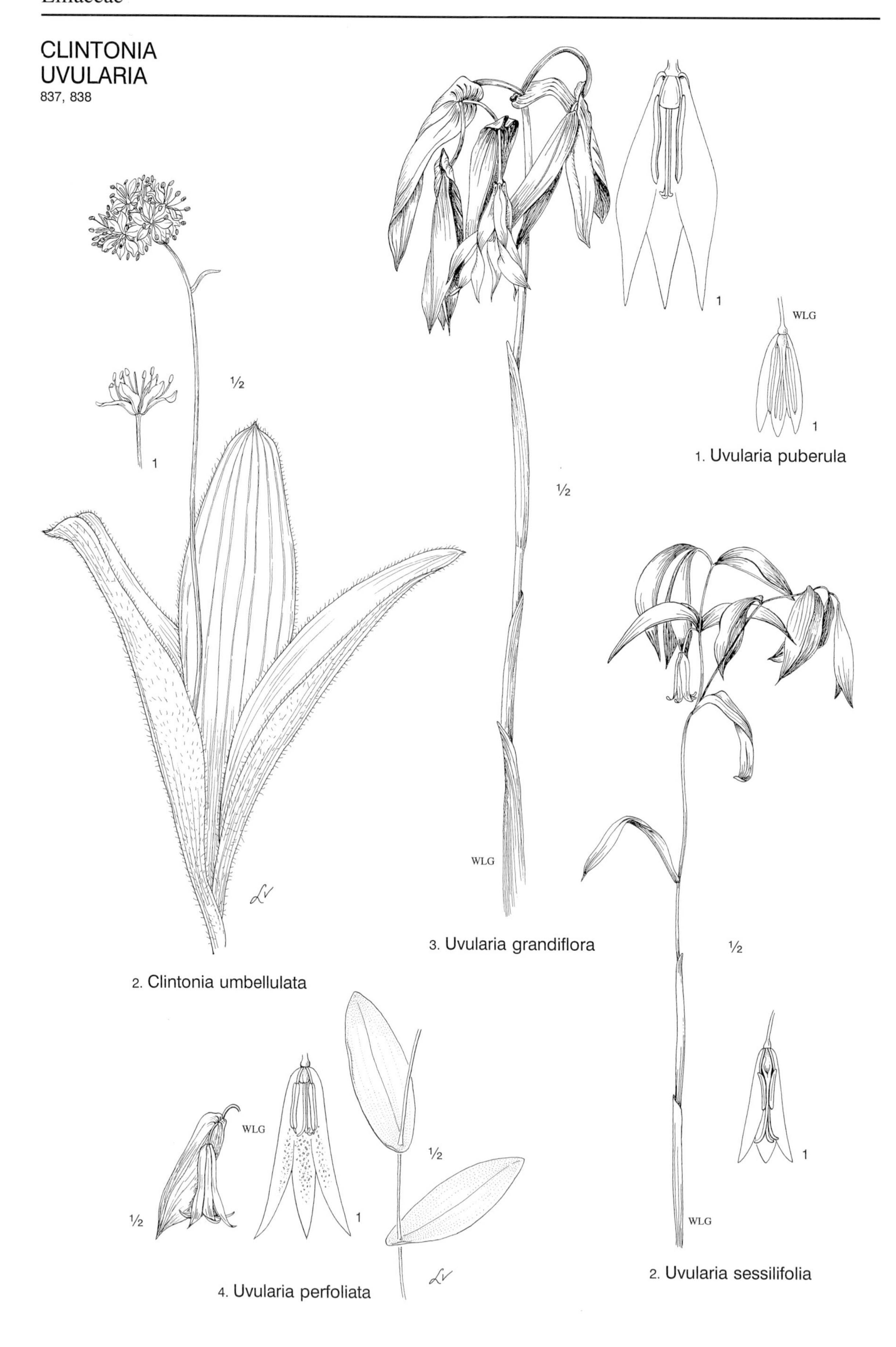

1. Uvularia puberula

2. Clintonia umbellulata

3. Uvularia grandiflora

4. Uvularia perfoliata

2. Uvularia sessilifolia

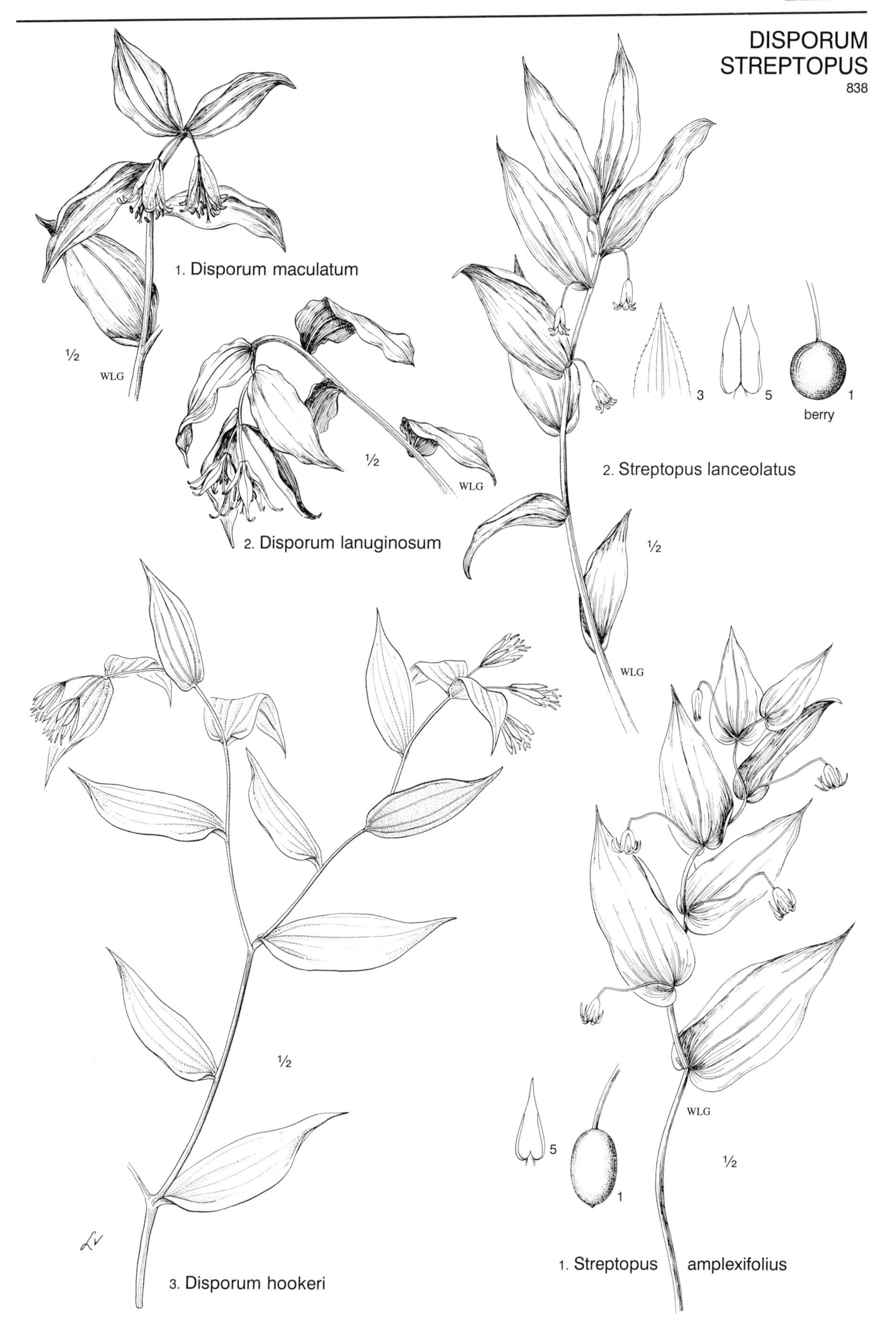

1. Disporum maculatum

½ WLG

2. Disporum lanuginosum

½ WLG

3 5 1
berry

2. Streptopus lanceolatus

½

WLG

3. Disporum hookeri

½

5 1

1. Streptopus   amplexifolius

WLG

½

SMILACINA
MAIANTHEMUM
CONVALLARIA
839

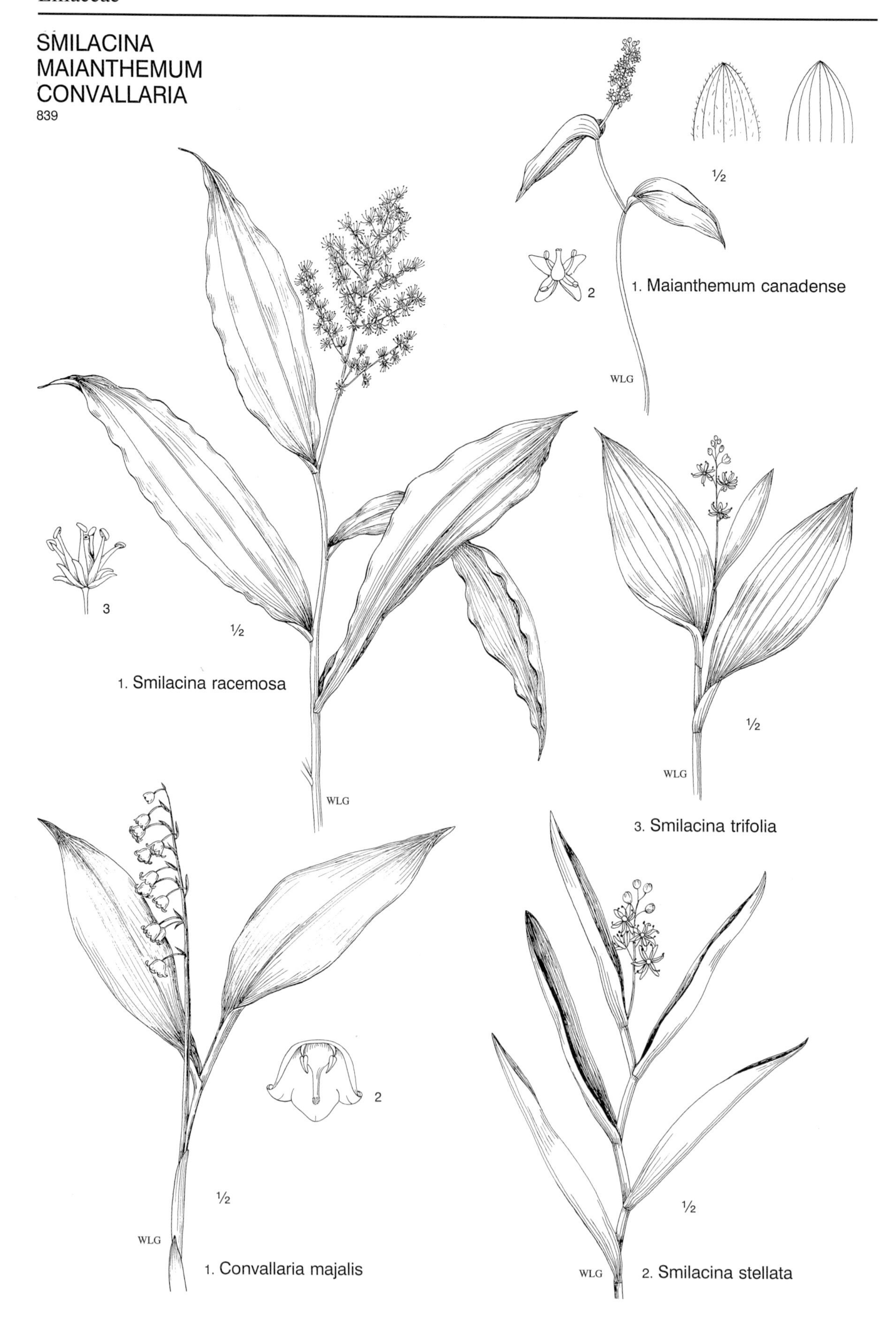

½

1. Maianthemum canadense

2

WLG

3

½

1. Smilacina racemosa

½

WLG

WLG

3. Smilacina trifolia

2

½

WLG

1. Convallaria majalis

½

WLG   2. Smilacina stellata

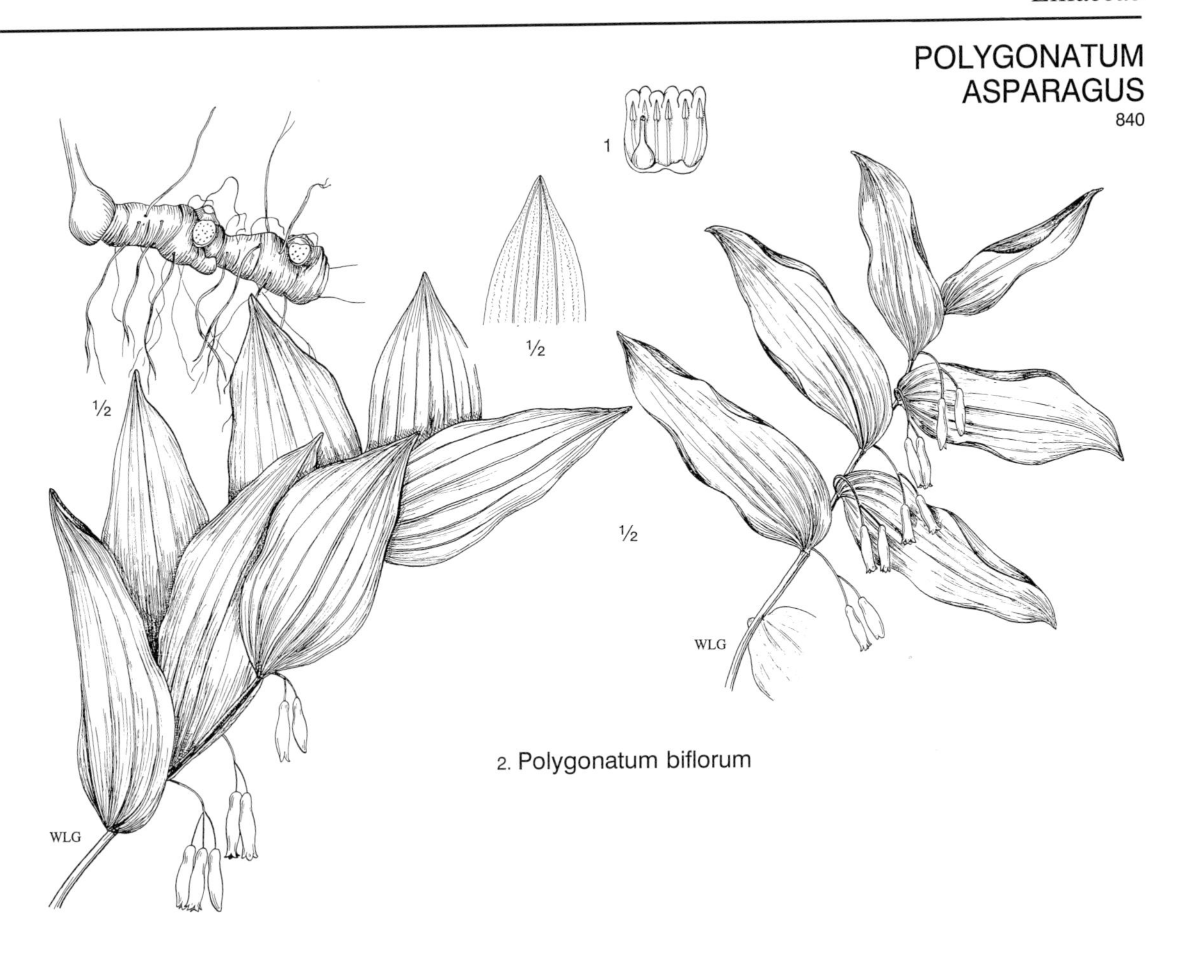

2. Polygonatum biflorum

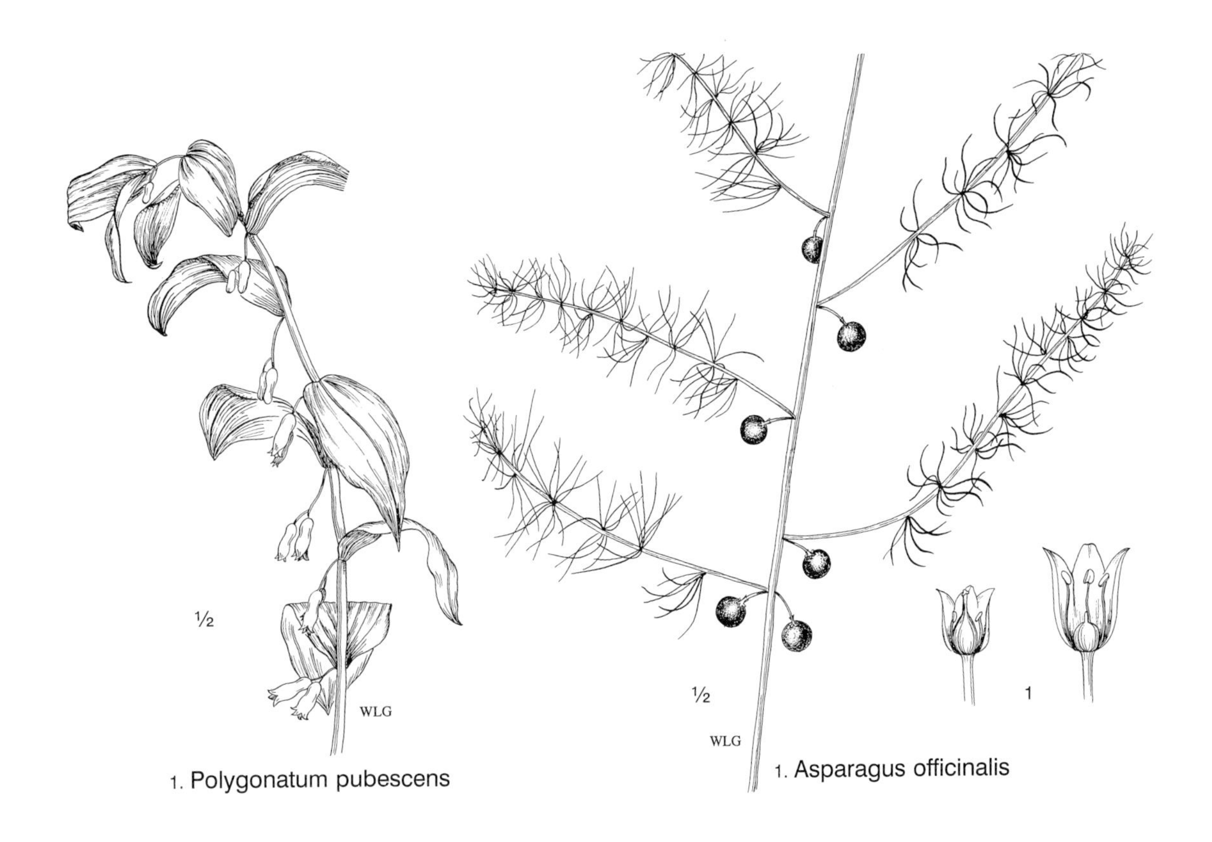

1. Polygonatum pubescens

1. Asparagus officinalis

HYPOXIS
840, 841

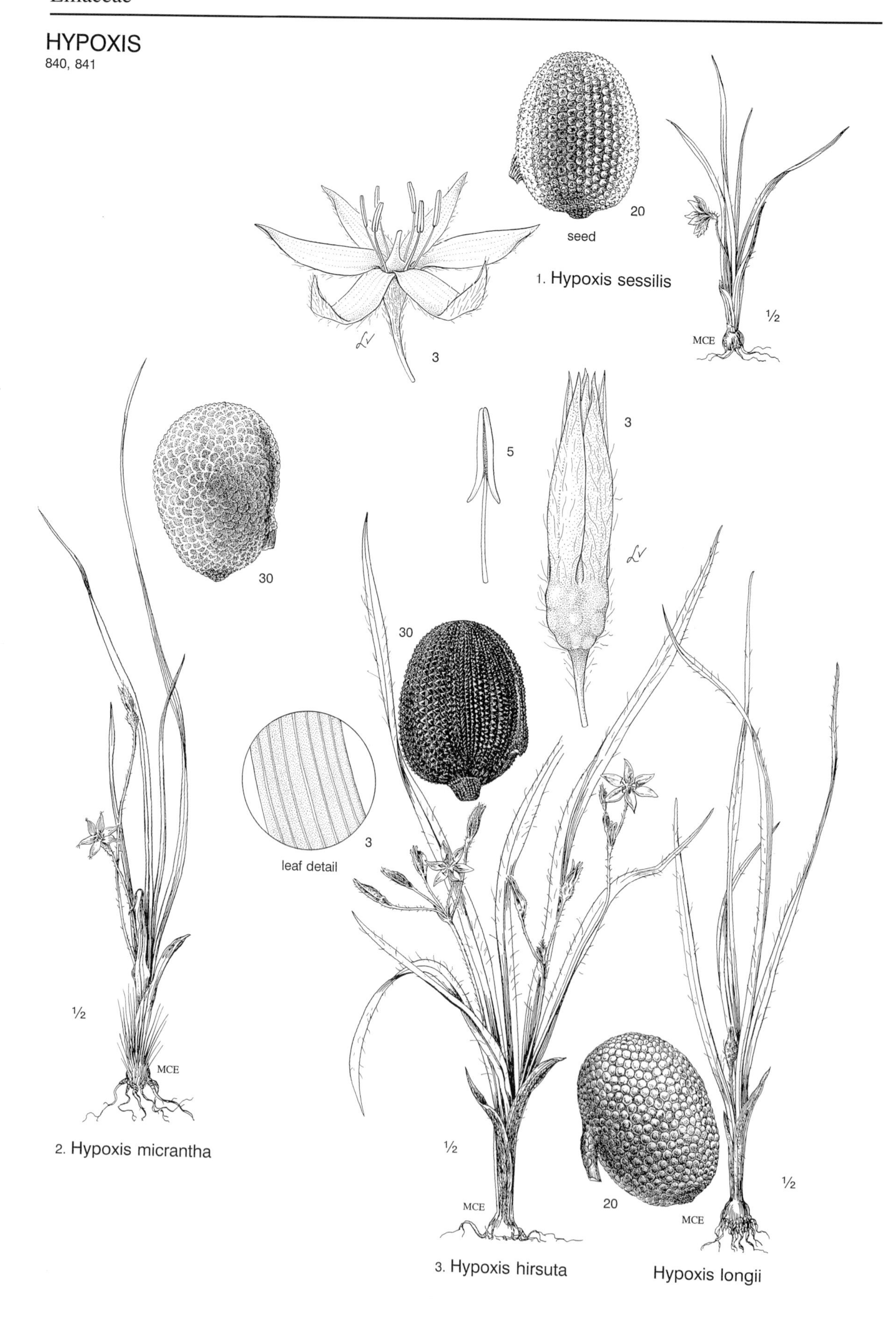

seed

20

1. Hypoxis sessilis

½

MCE

30

5

3

30

leaf detail

3

2. Hypoxis micrantha

½

MCE

3. Hypoxis hirsuta

½

MCE

Hypoxis longii

½

MCE

20

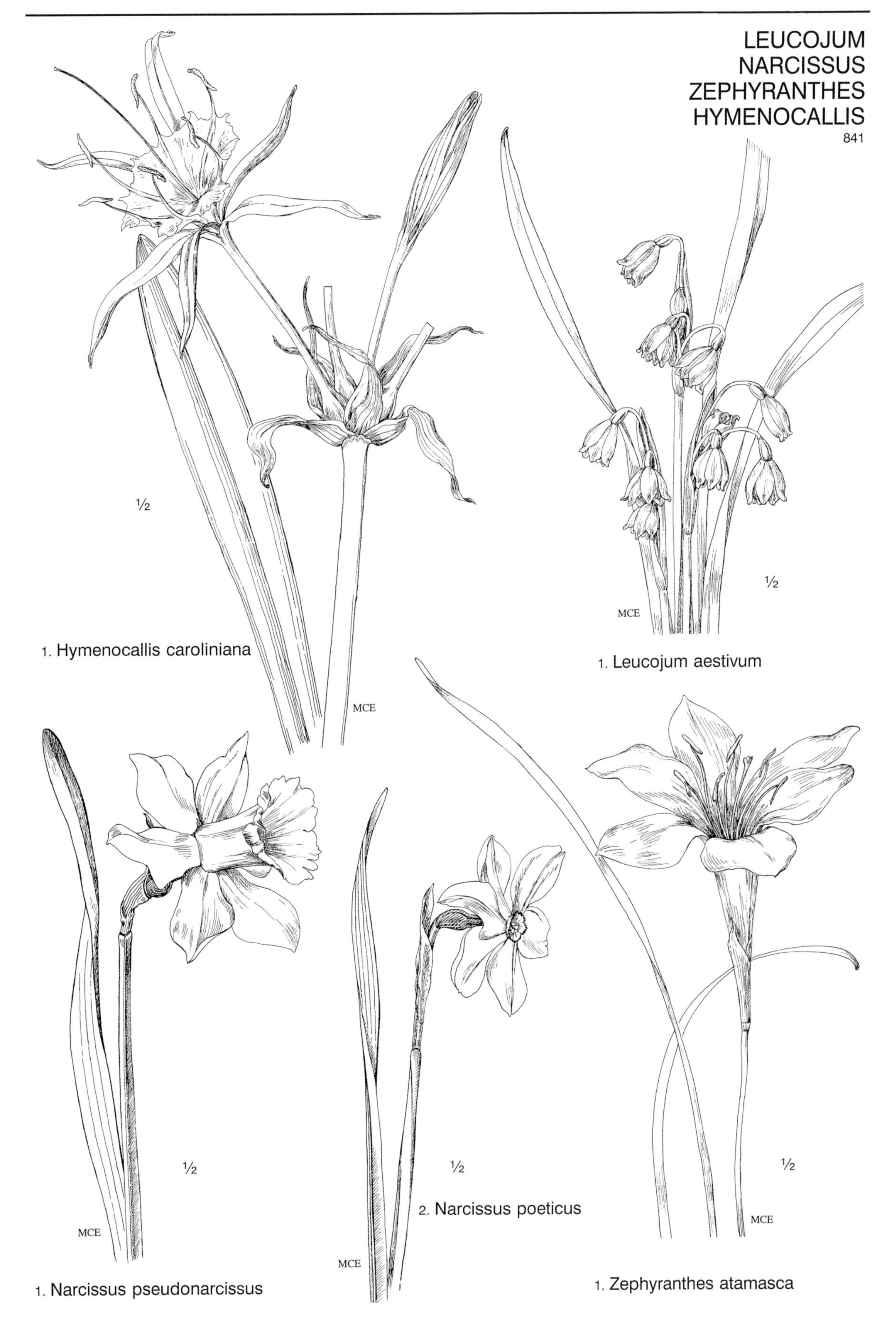

½

1. Hymenocallis caroliniana

MCE

½

MCE

1. Leucojum aestivum

½

MCE

1. Narcissus pseudonarcissus

MCE

½

2. Narcissus poeticus

½

MCE

1. Zephyranthes atamasca

YUCCA
AGAVE
SMILAX
842, 843

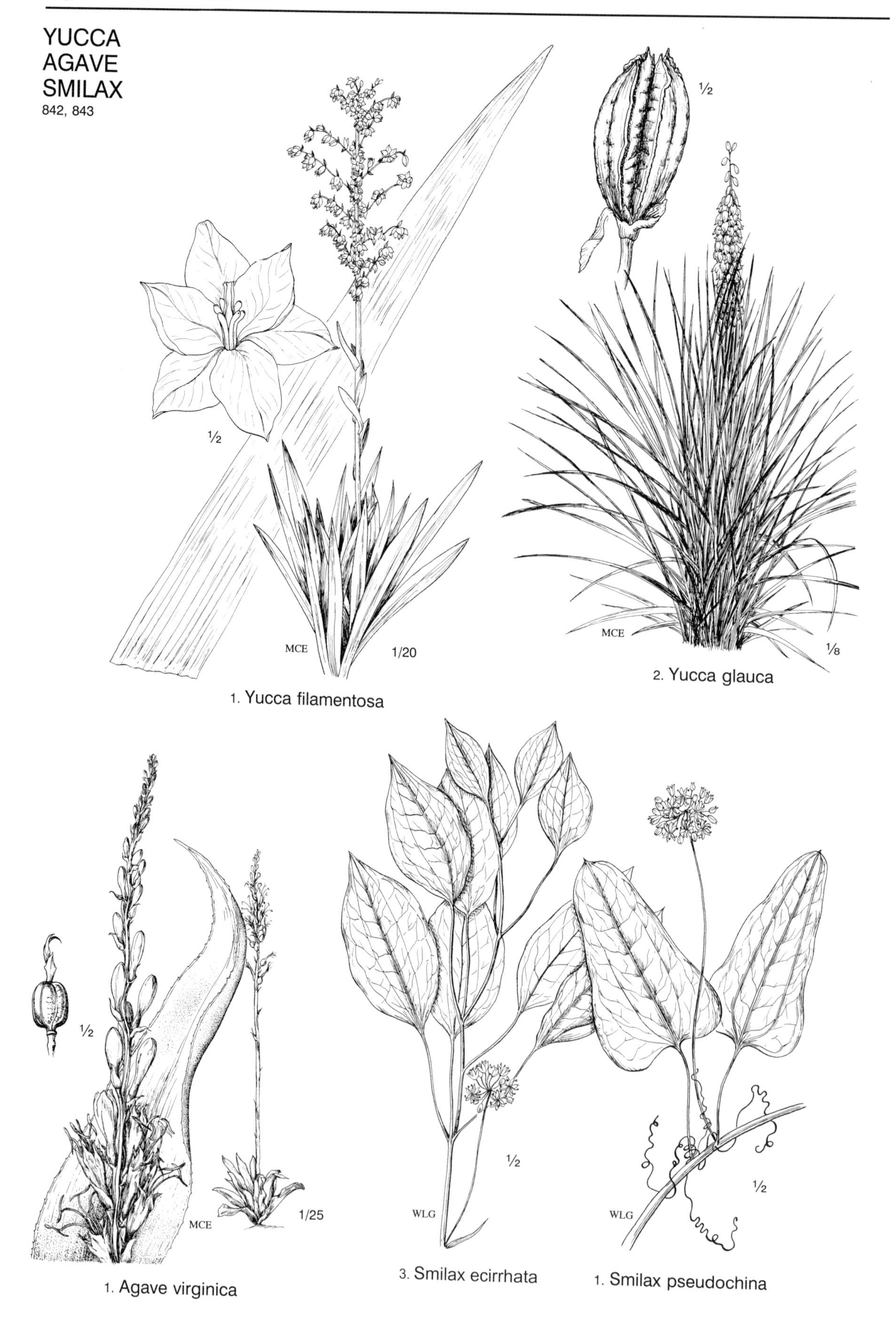

½

½

1. Yucca filamentosa

MCE   1/20

2. Yucca glauca

MCE   ⅛

½

1. Agave virginica

MCE   1/25

3. Smilax ecirrhata

WLG   ½

1. Smilax pseudochina

WLG   ½

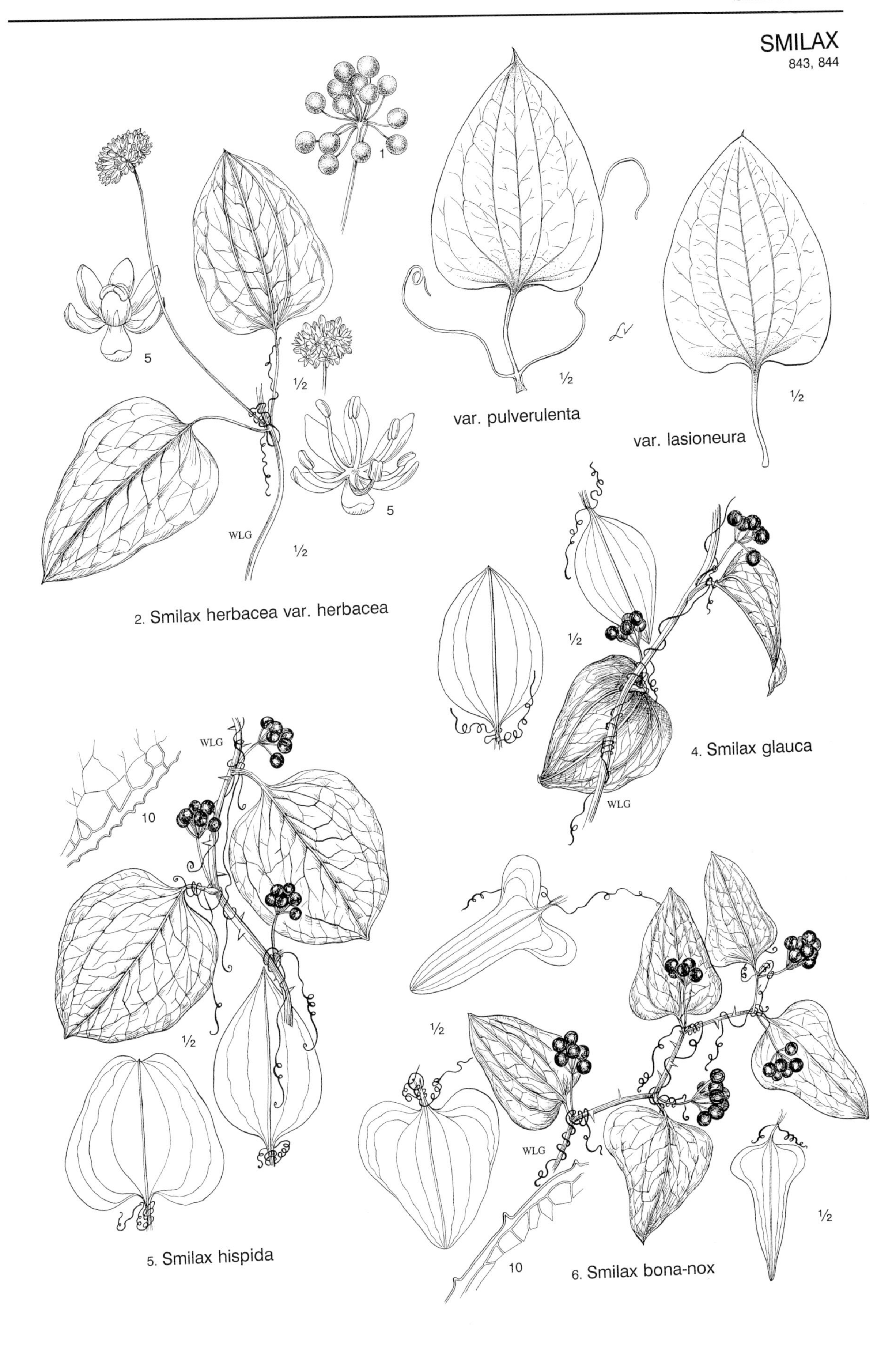

var. pulverulenta

var. lasioneura

2. Smilax herbacea var. herbacea

4. Smilax glauca

5. Smilax hispida

6. Smilax bona-nox

WLG

SMILAX
DIOSCOREA
844, 845

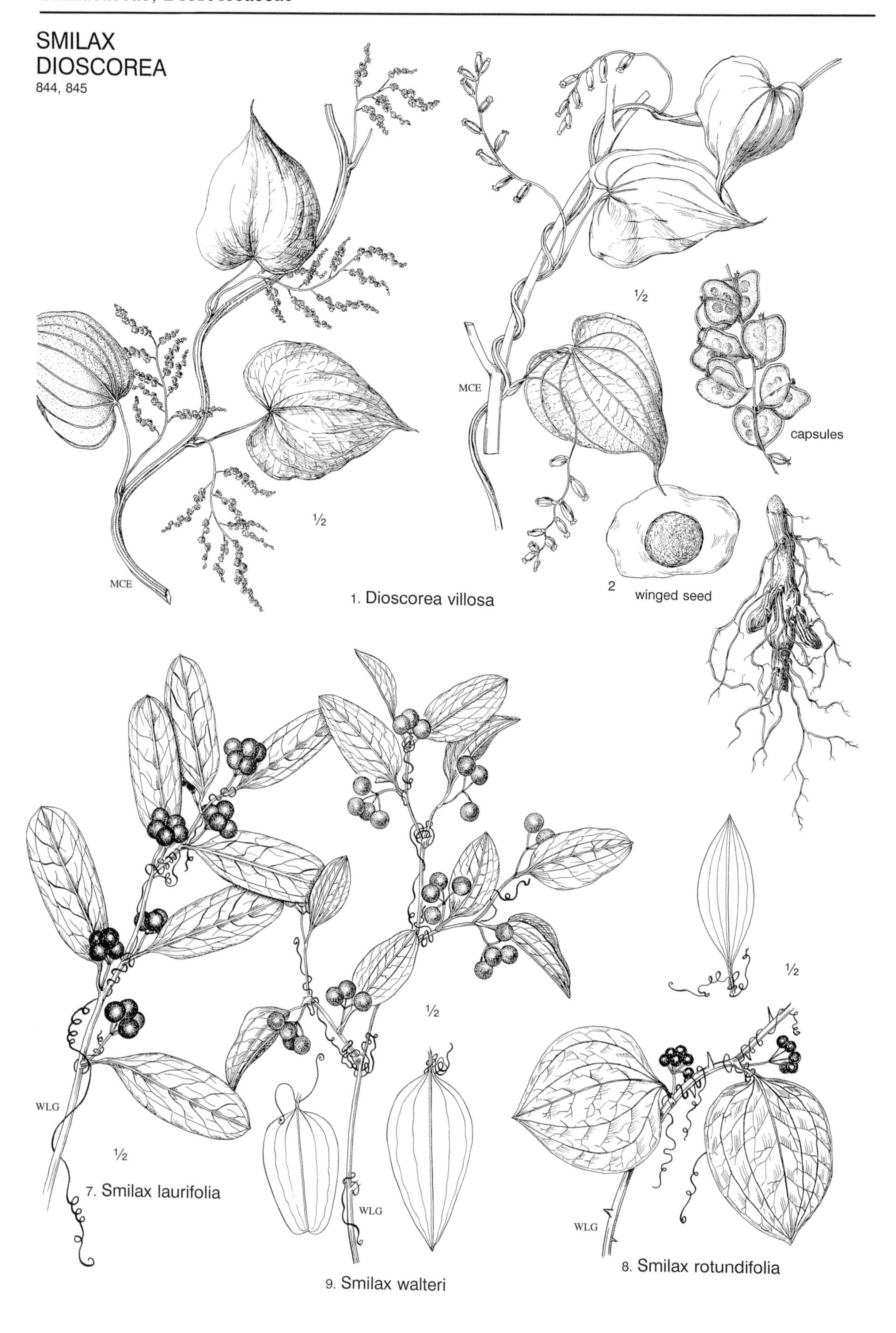

1. Dioscorea villosa

2 winged seed

capsules

7. Smilax laurifolia

9. Smilax walteri

8. Smilax rotundifolia

MCE

WLG

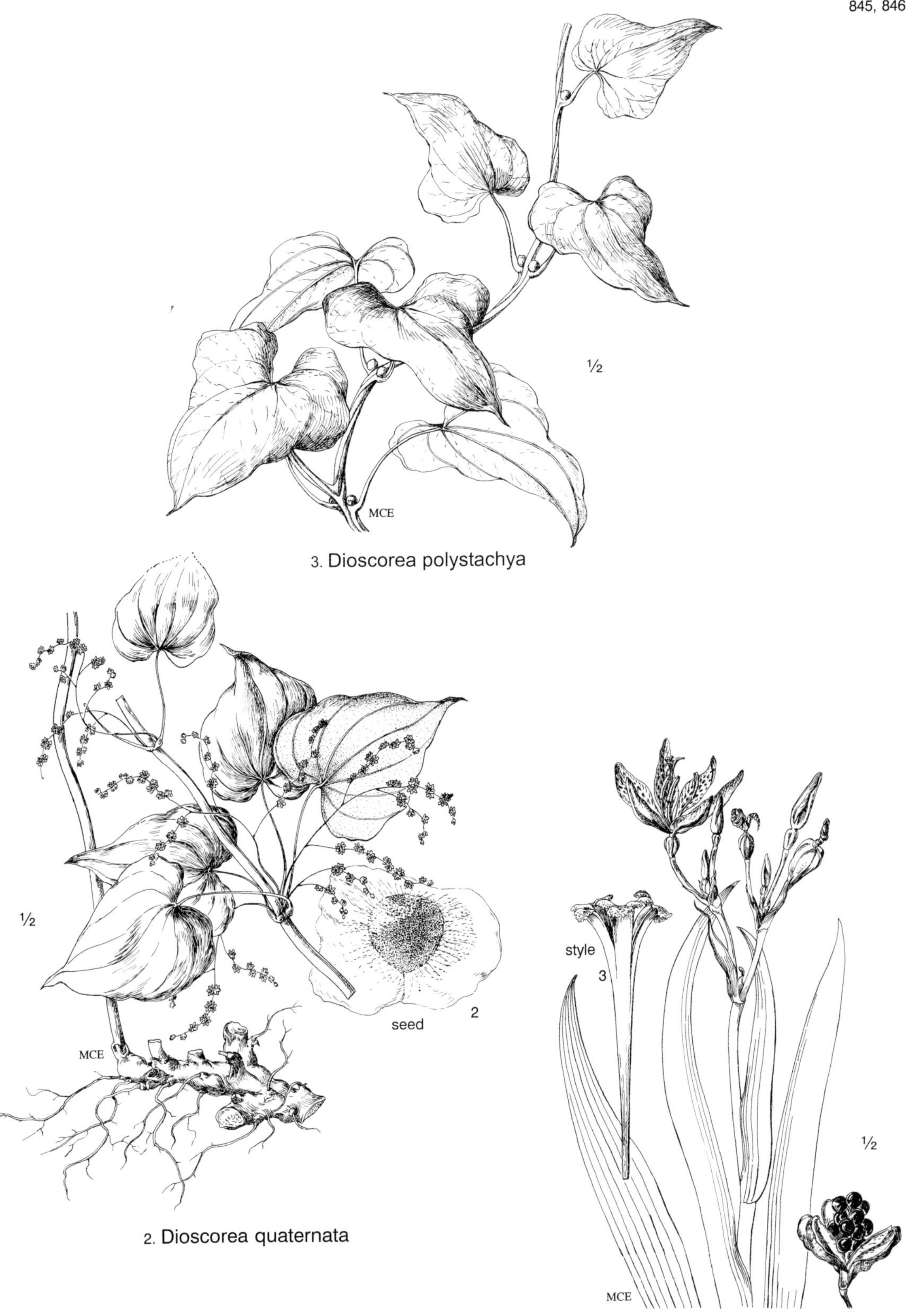

3. Dioscorea polystachya

½

seed 2

½

2. Dioscorea quaternata

MCE

style
3

½

MCE

1. Belamcanda chinensis

SISYRINCHIUM
846

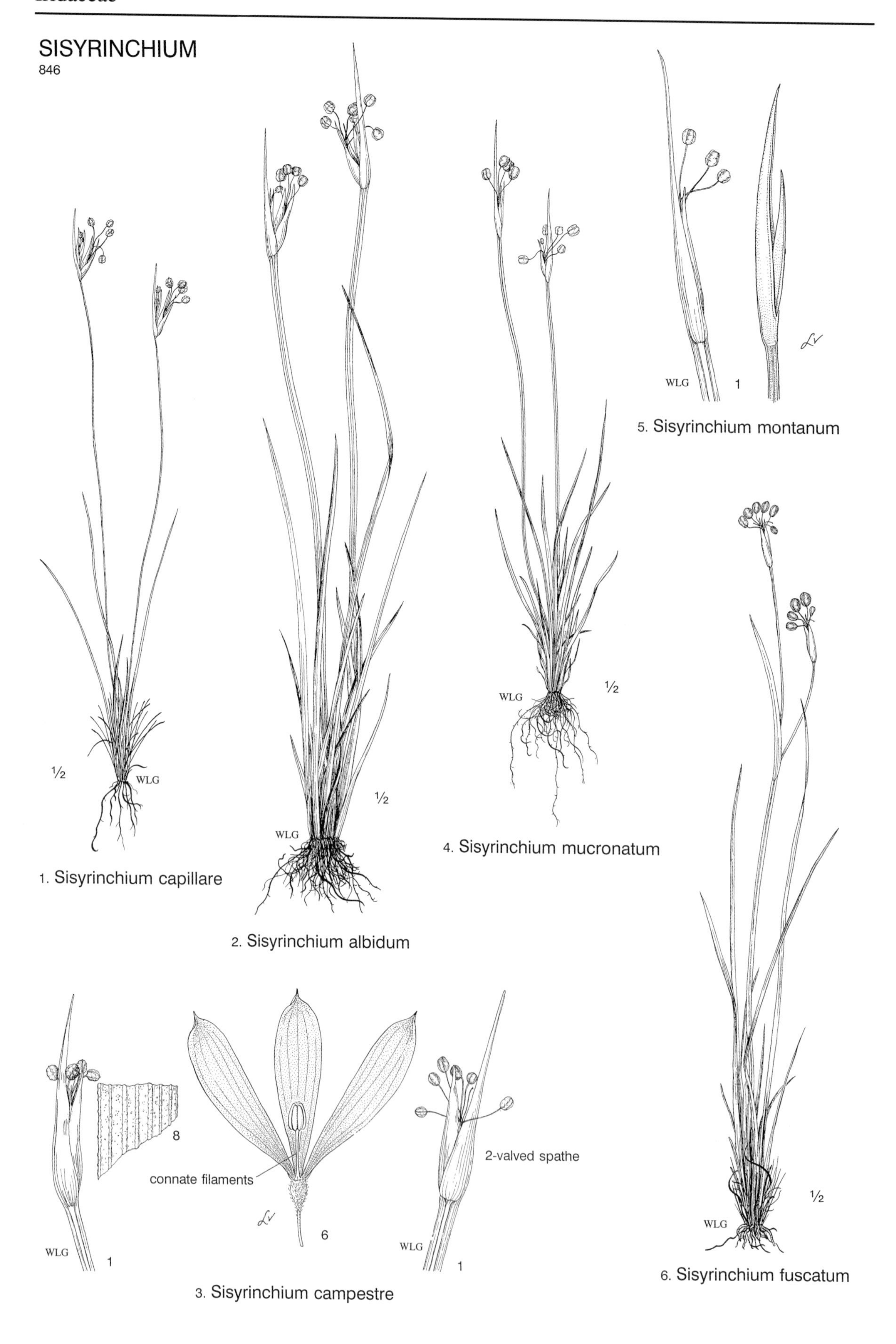

1. Sisyrinchium capillare

2. Sisyrinchium albidum

4. Sisyrinchium mucronatum

5. Sisyrinchium montanum

connate filaments

2-valved spathe

3. Sisyrinchium campestre

6. Sisyrinchium fuscatum

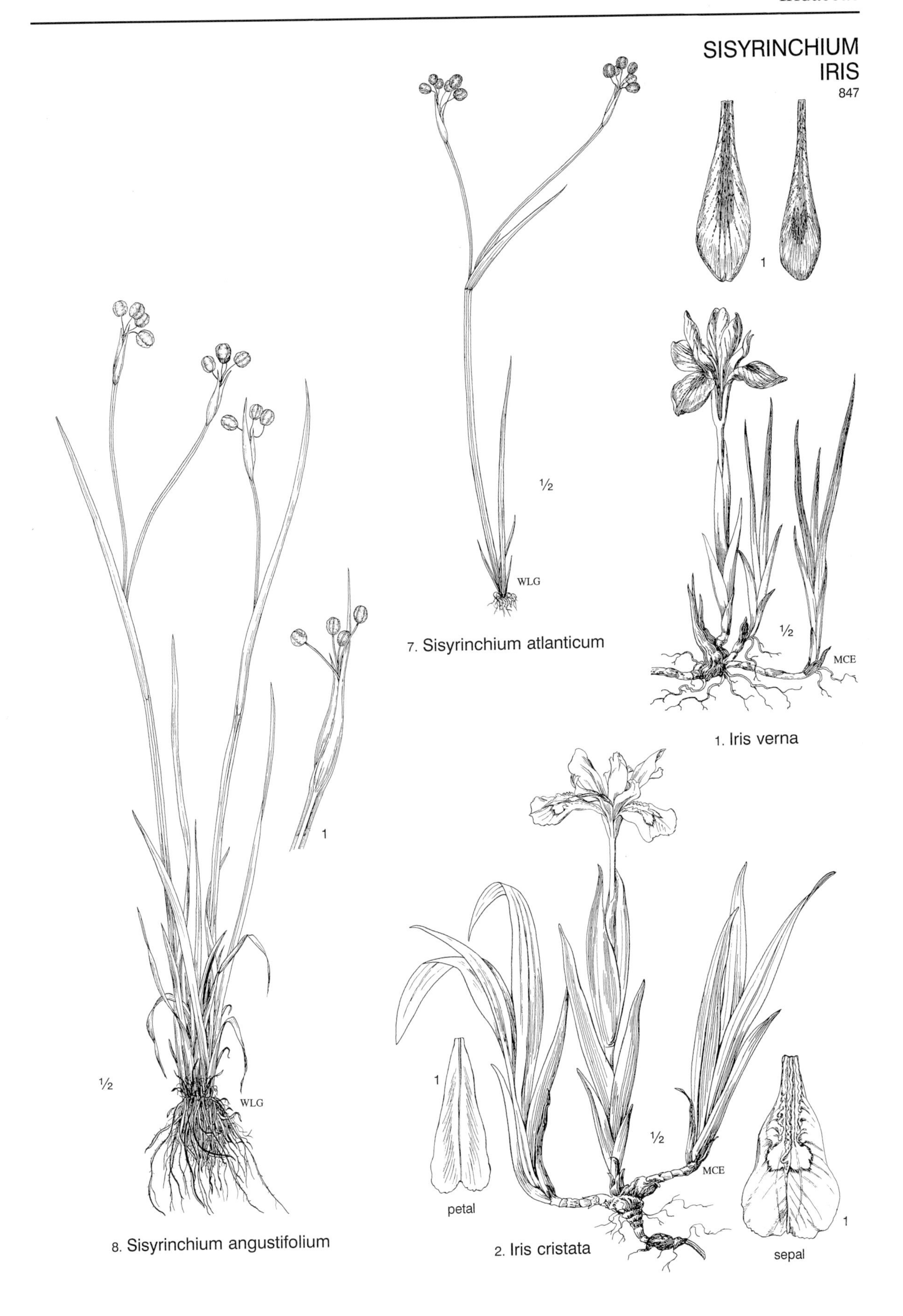

7. Sisyrinchium atlanticum

1. Iris verna

8. Sisyrinchium angustifolium

petal

2. Iris cristata

sepal

IRIS
848

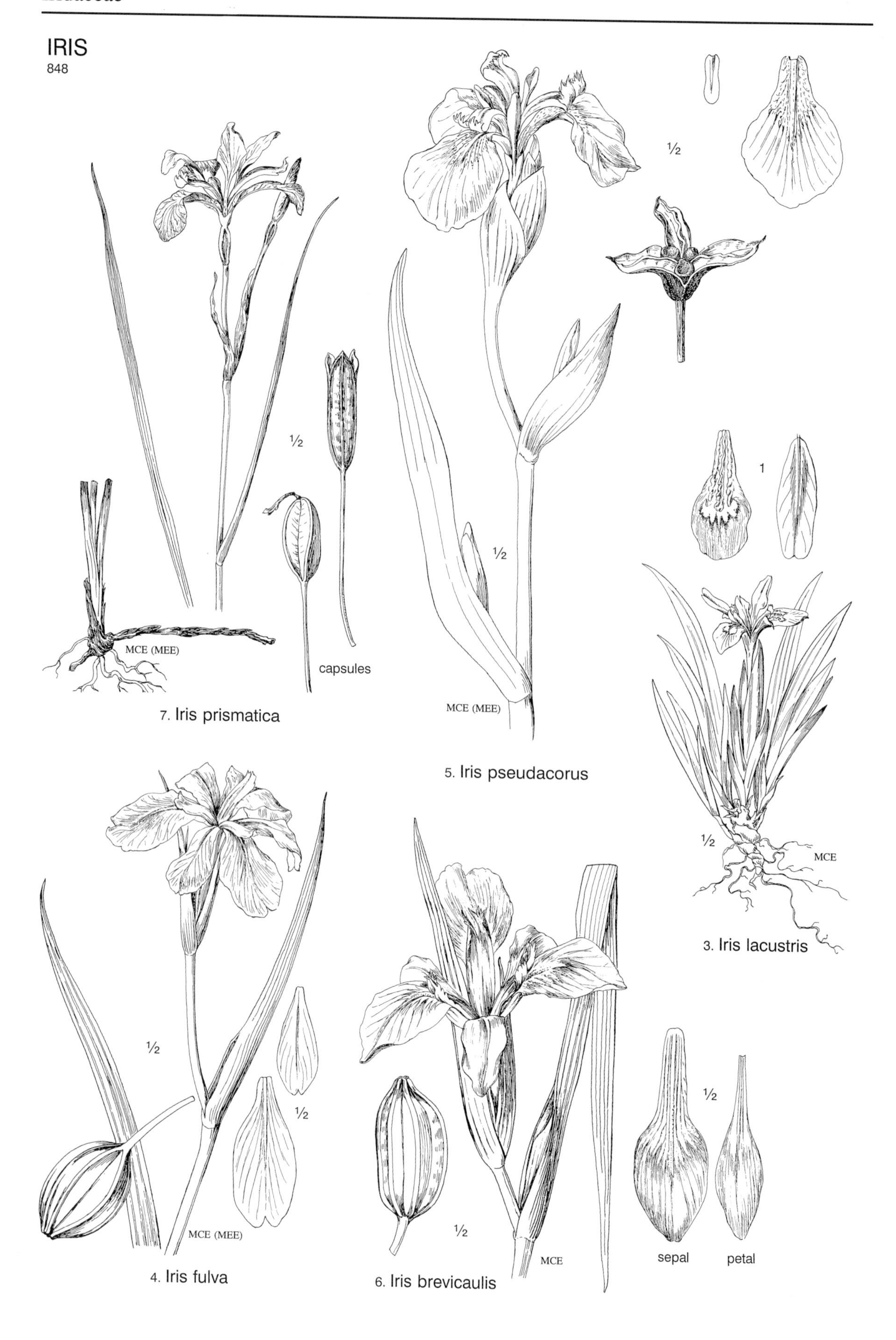

½

1

MCE (MEE)

capsules

7. Iris prismatica

½

½

5. Iris pseudacorus

MCE (MEE)

MCE

3. Iris lacustris

½

½

½

MCE (MEE)

4. Iris fulva

½

MCE

6. Iris brevicaulis

½

sepal   petal

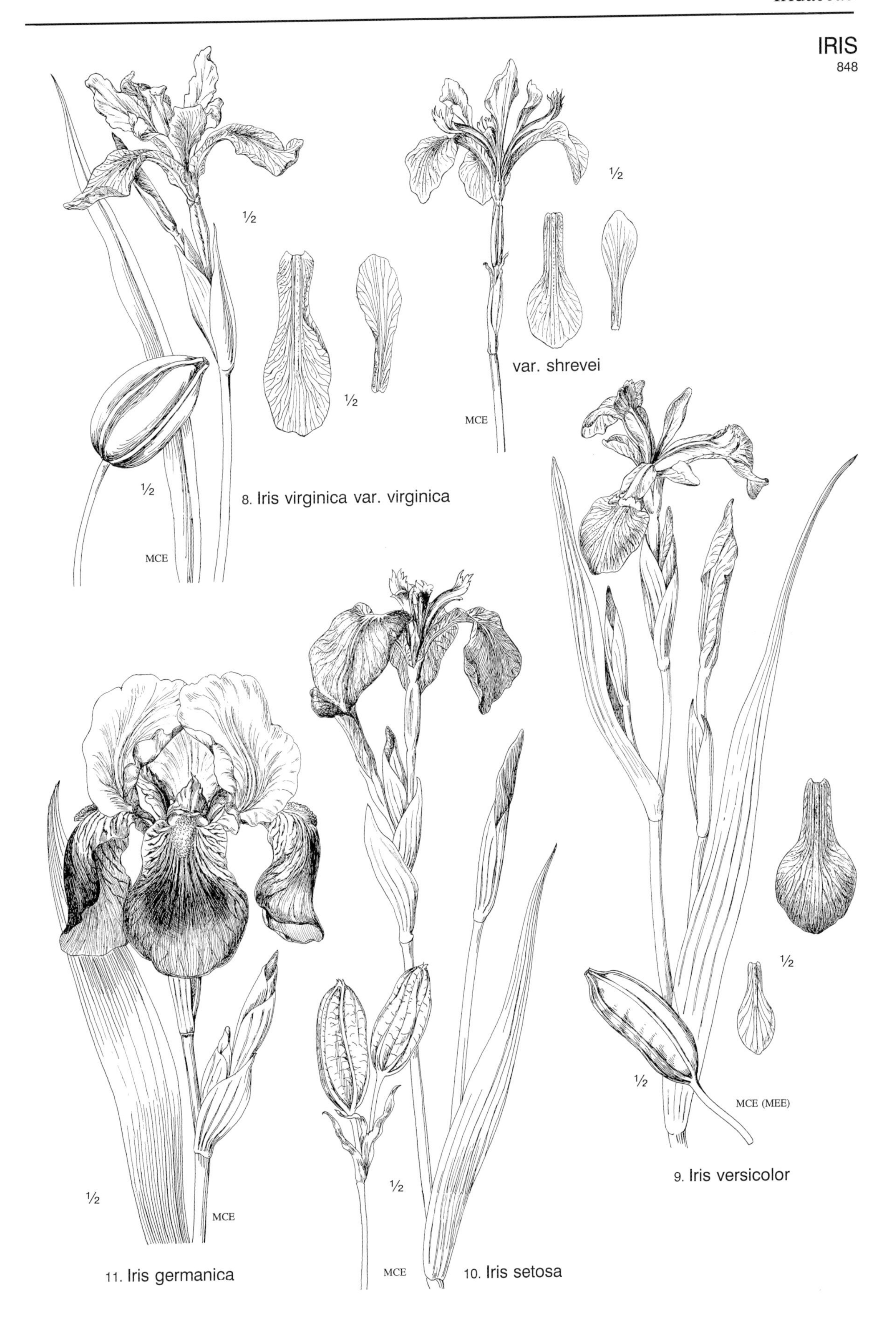

½

½

var. shrevei

MCE

½

½

MCE

8. Iris virginica var. virginica

½

½

MCE (MEE)

9. Iris versicolor

½

MCE

11. Iris germanica

½

MCE

10. Iris setosa

BURMANNIA
THISMIA
CYPRIPEDIUM
849–851

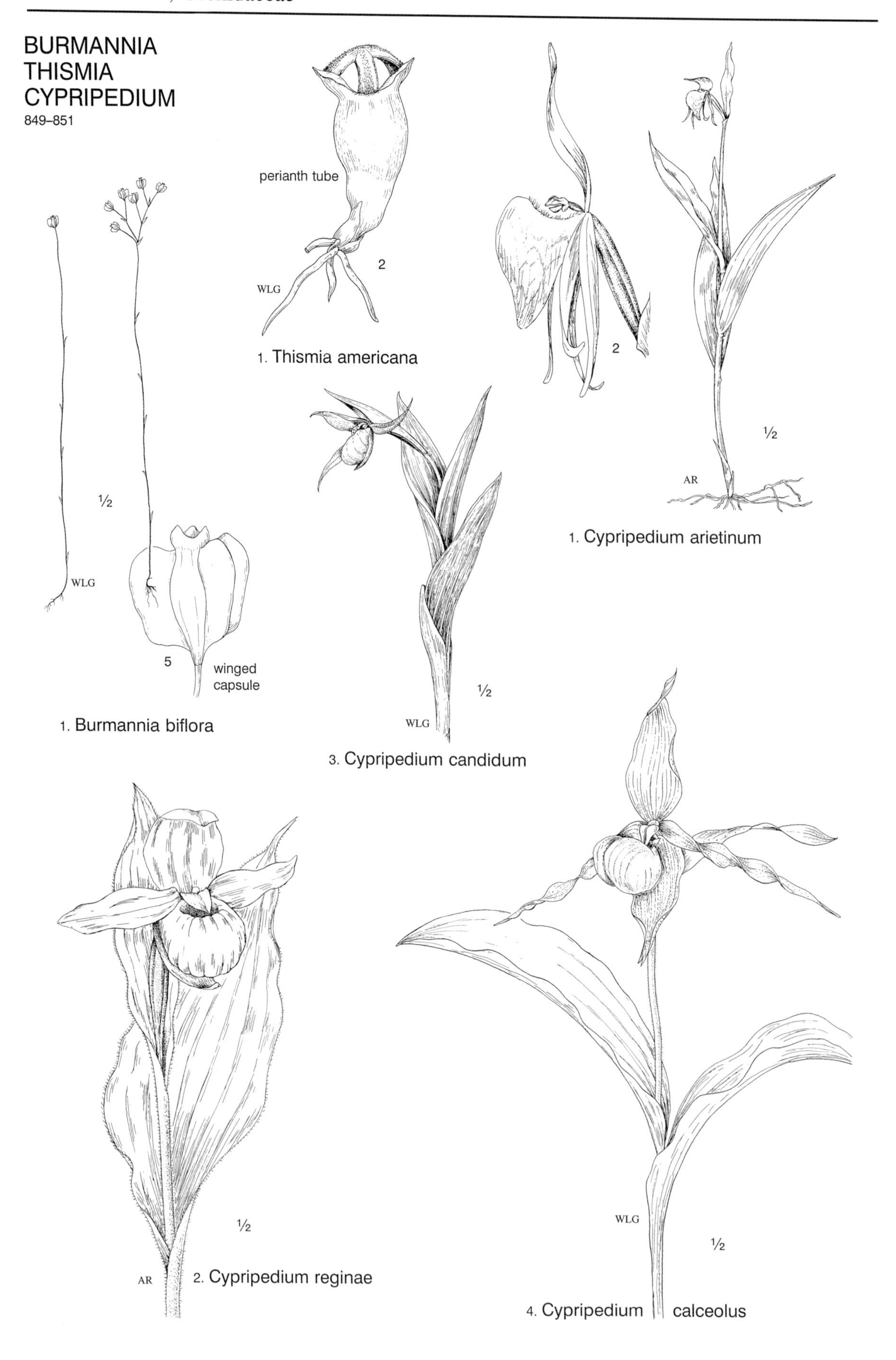

perianth tube

WLG

2

1. Thismia americana

2

½

AR

1. Cypripedium arietinum

½

WLG

5

winged
capsule

1. Burmannia biflora

½

WLG

3. Cypripedium candidum

½

AR

2. Cypripedium reginae

WLG

½

4. Cypripedium   calceolus

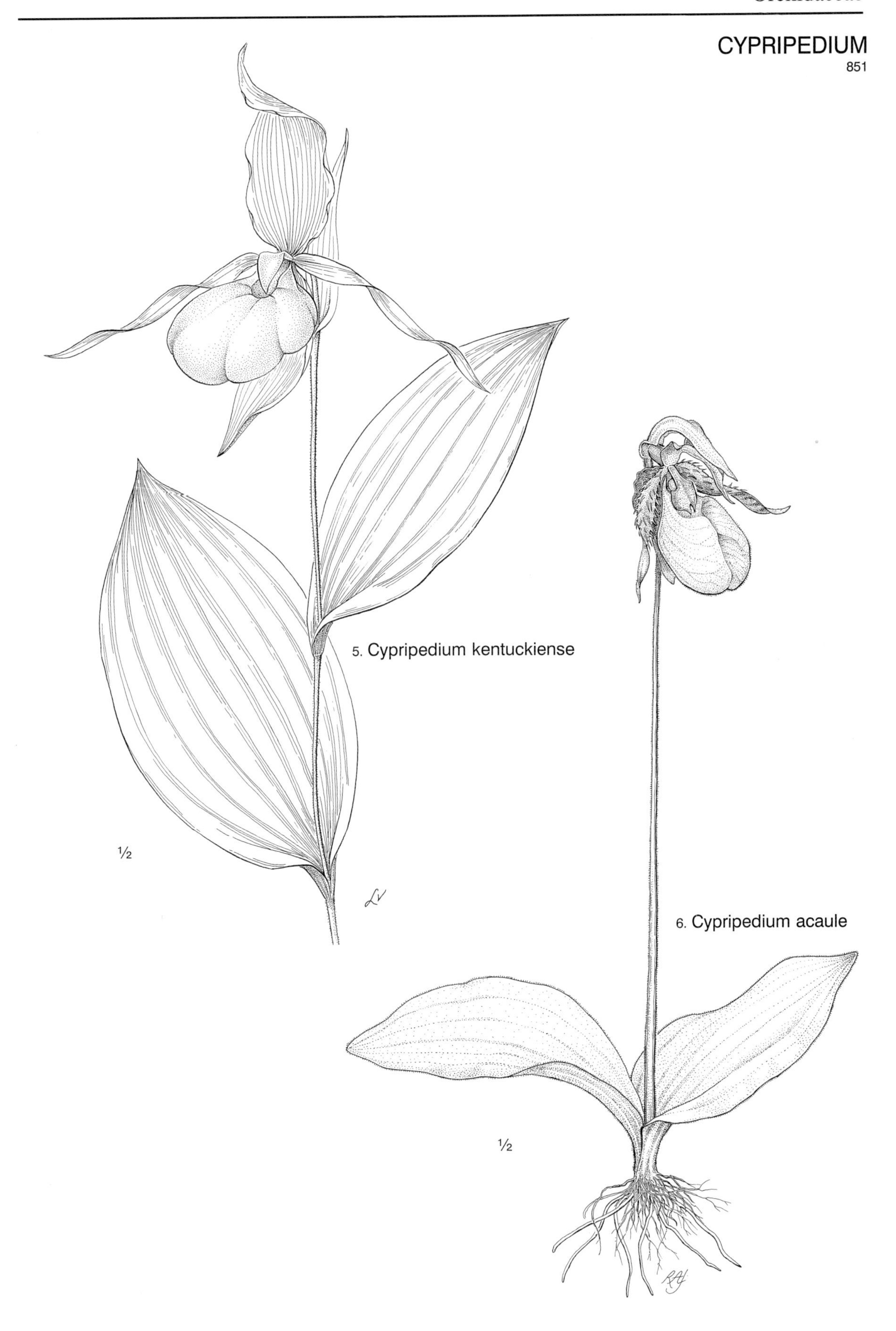

5. Cypripedium kentuckiense

½

6. Cypripedium acaule

½

EPIPACTIS
LISTERA
851, 852

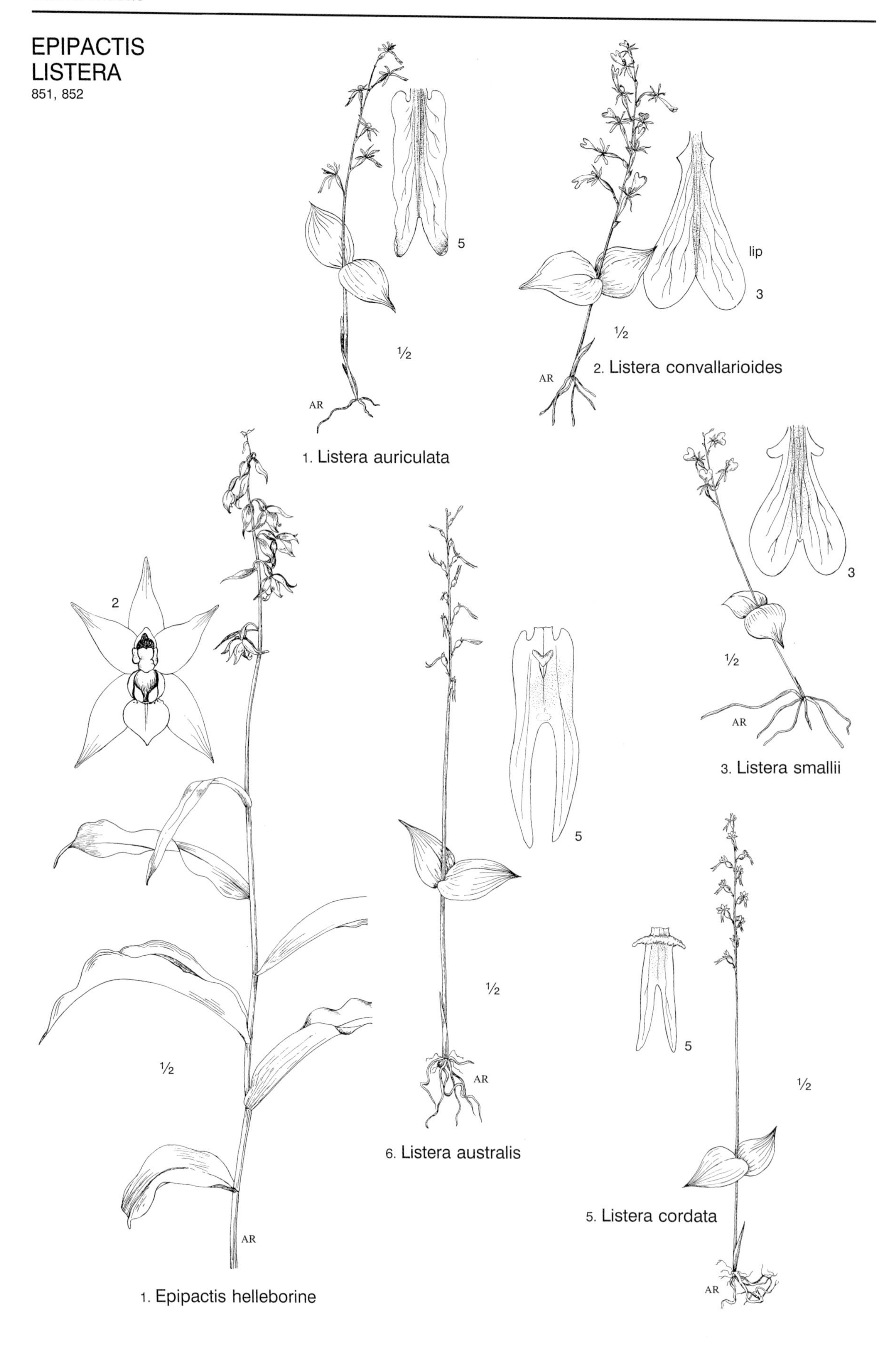

5

½

AR

1. Listera auriculata

lip

3

½

AR

2. Listera convallarioides

2

½

1. Epipactis helleborine

5

½

AR

6. Listera australis

3

½

AR

3. Listera smallii

5

½

AR

5. Listera cordata

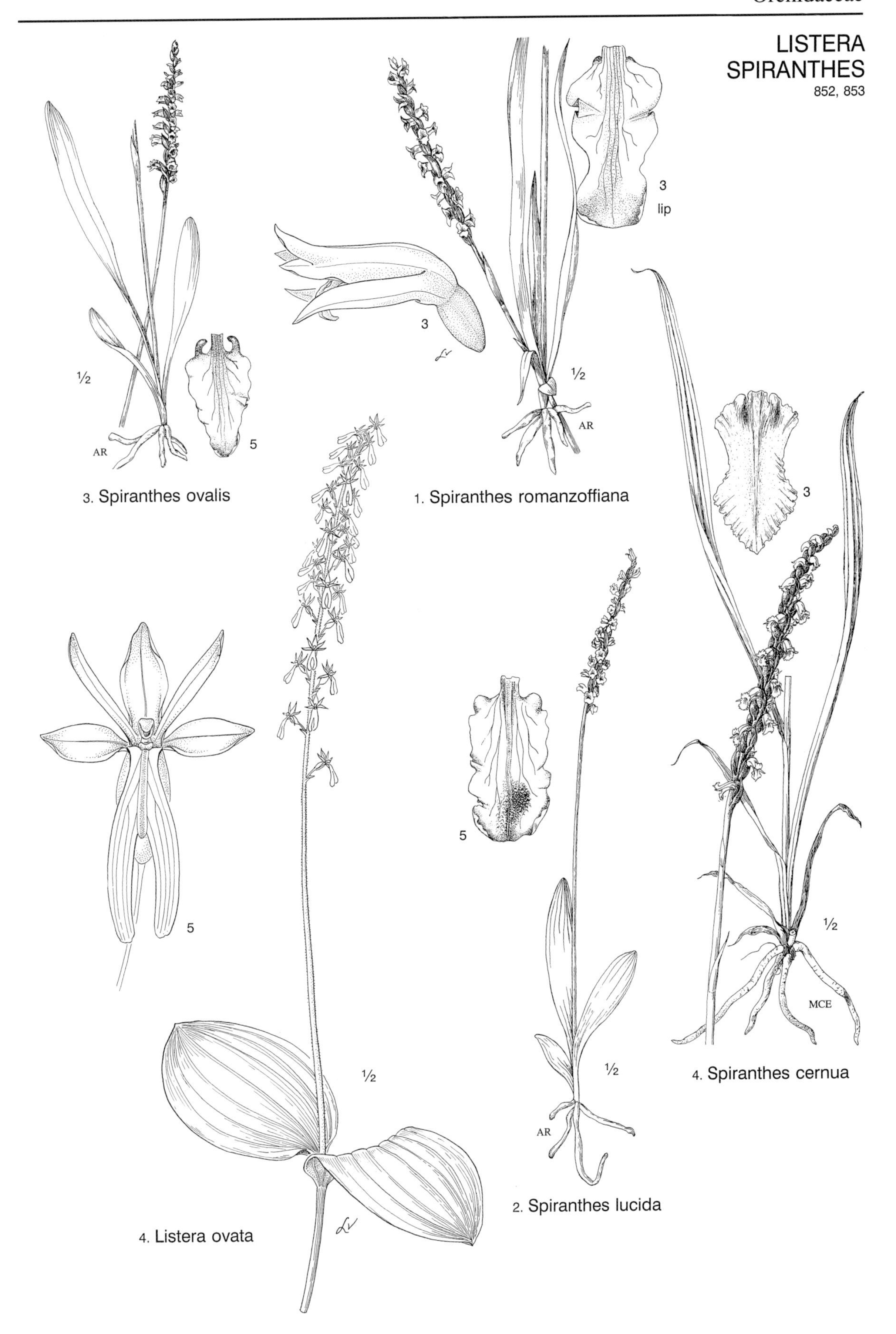

3. Spiranthes ovalis

1. Spiranthes romanzoffiana

3
lip

3

4. Listera ovata

2. Spiranthes lucida

4. Spiranthes cernua

SPIRANTHES
854

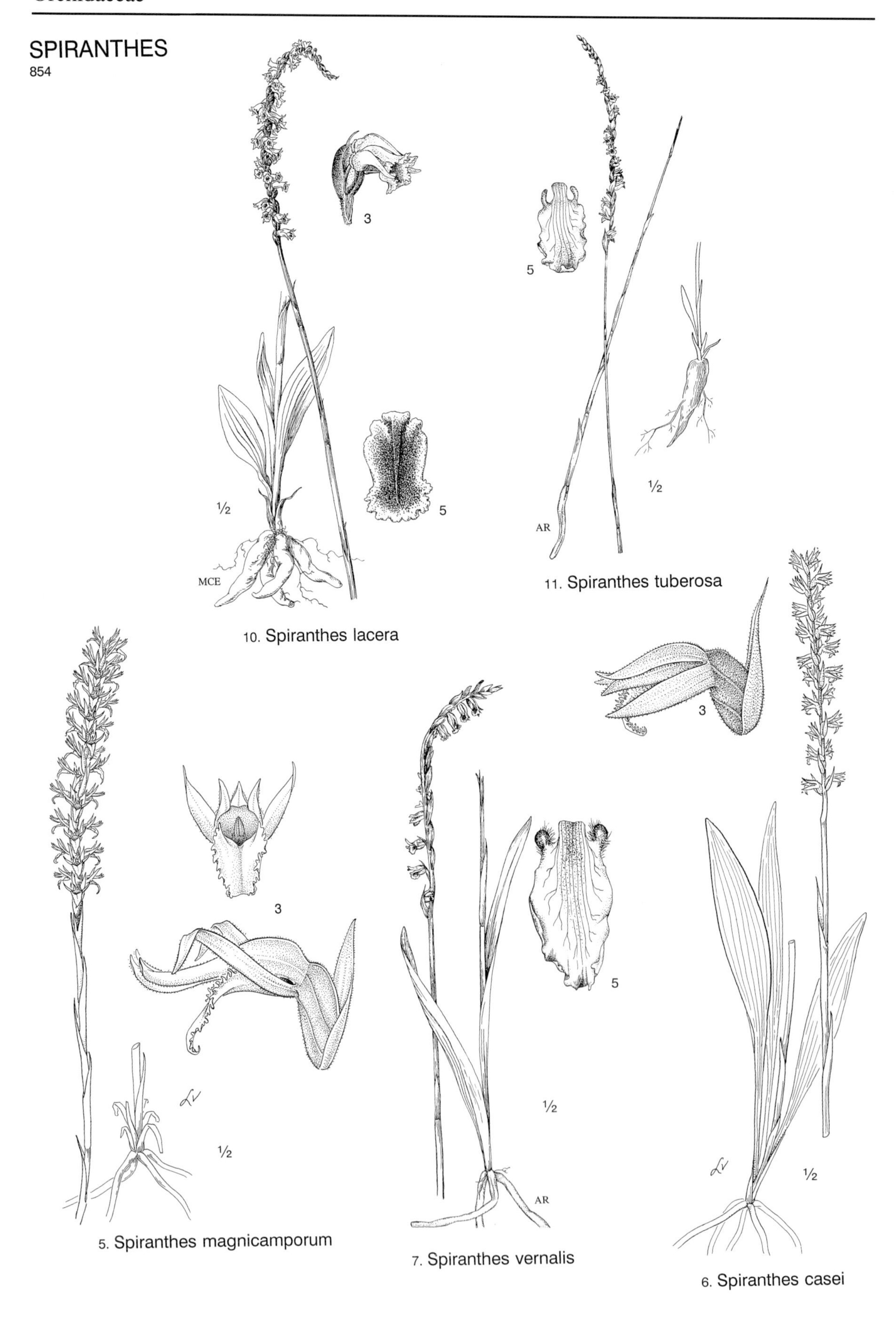

10. Spiranthes lacera

11. Spiranthes tuberosa

5. Spiranthes magnicamporum

7. Spiranthes vernalis

6. Spiranthes casei

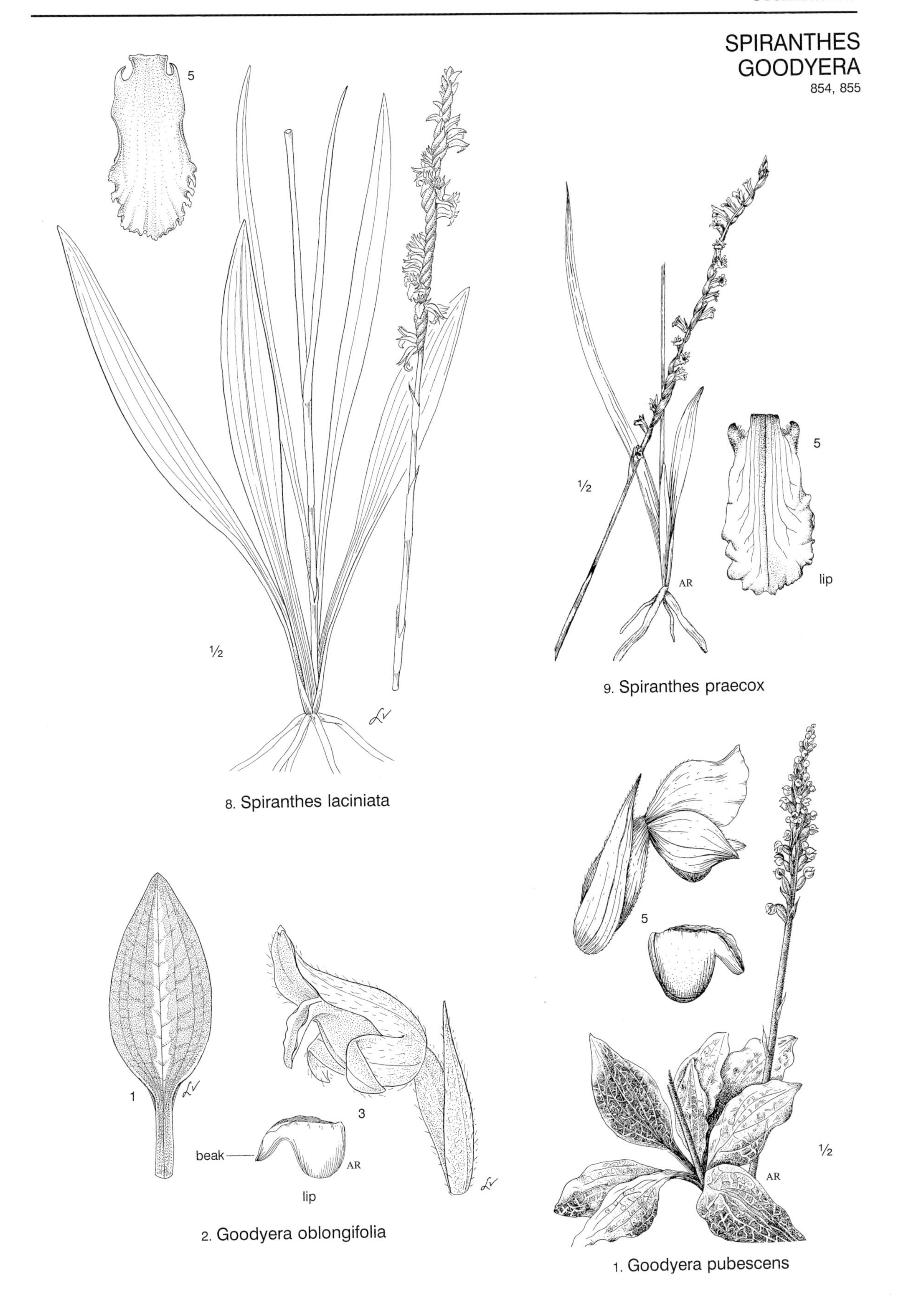

8. Spiranthes laciniata

9. Spiranthes praecox

2. Goodyera oblongifolia

beak

lip

1. Goodyera pubescens

GOODYERA
PONTHIEVA
ORCHIS
855, 856

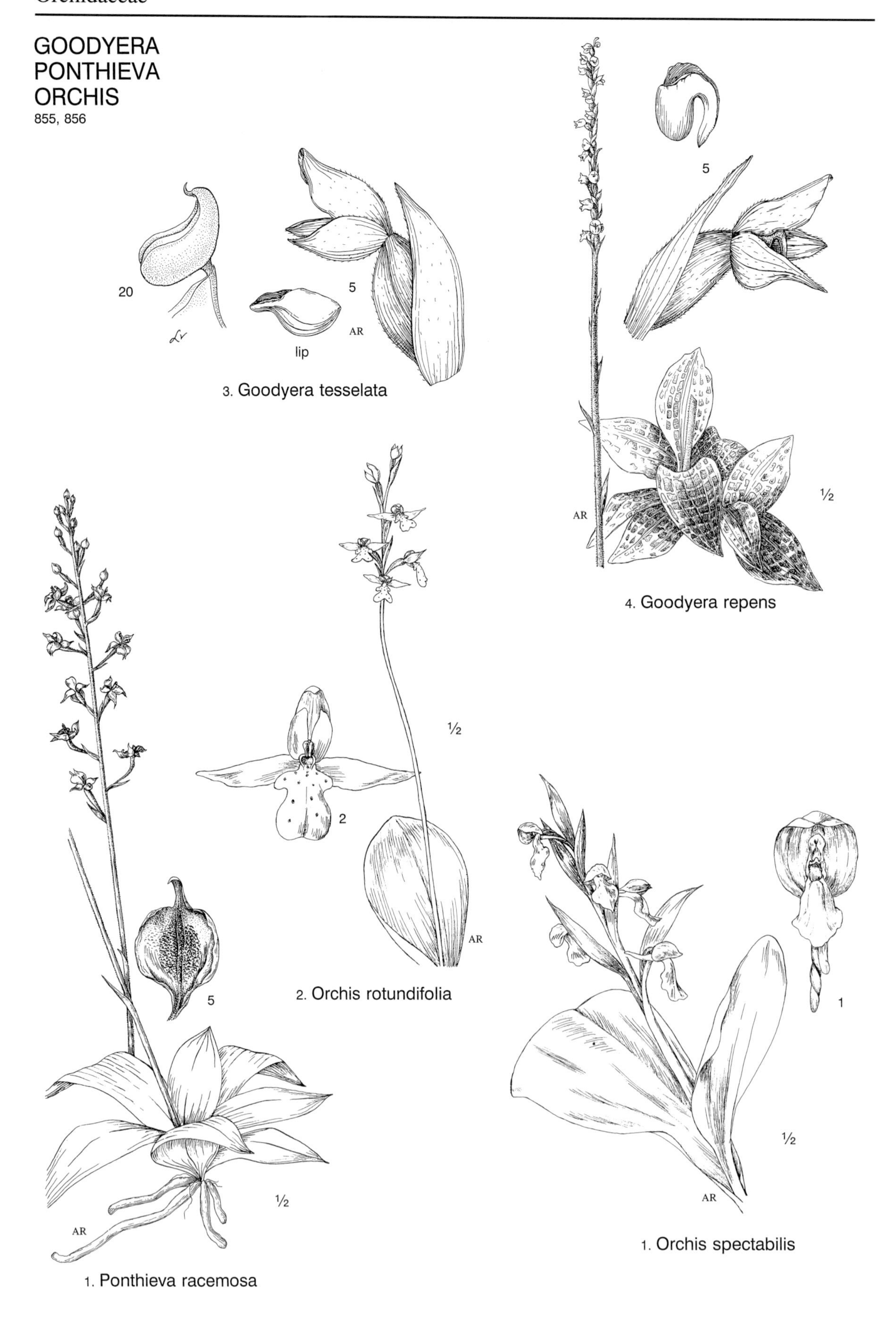

3. Goodyera tesselata

4. Goodyera repens

2. Orchis rotundifolia

1. Ponthieva racemosa

1. Orchis spectabilis

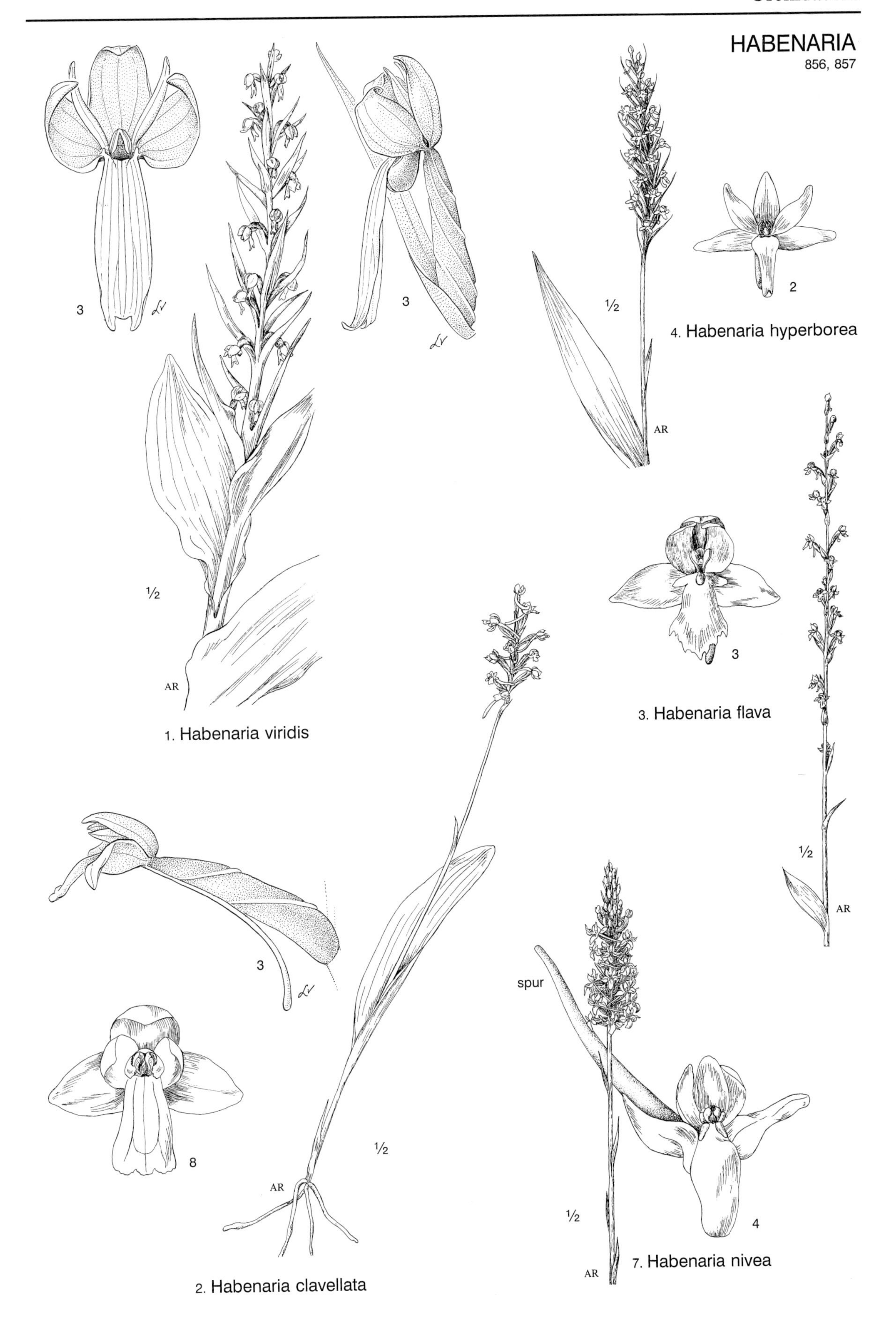

3

½

AR

1. Habenaria viridis

3

3

4. Habenaria hyperborea

2

½

AR

3

3. Habenaria flava

½

AR

3

8

½

AR

2. Habenaria clavellata

spur

½

AR

4

7. Habenaria nivea

HABENARIA
857, 858

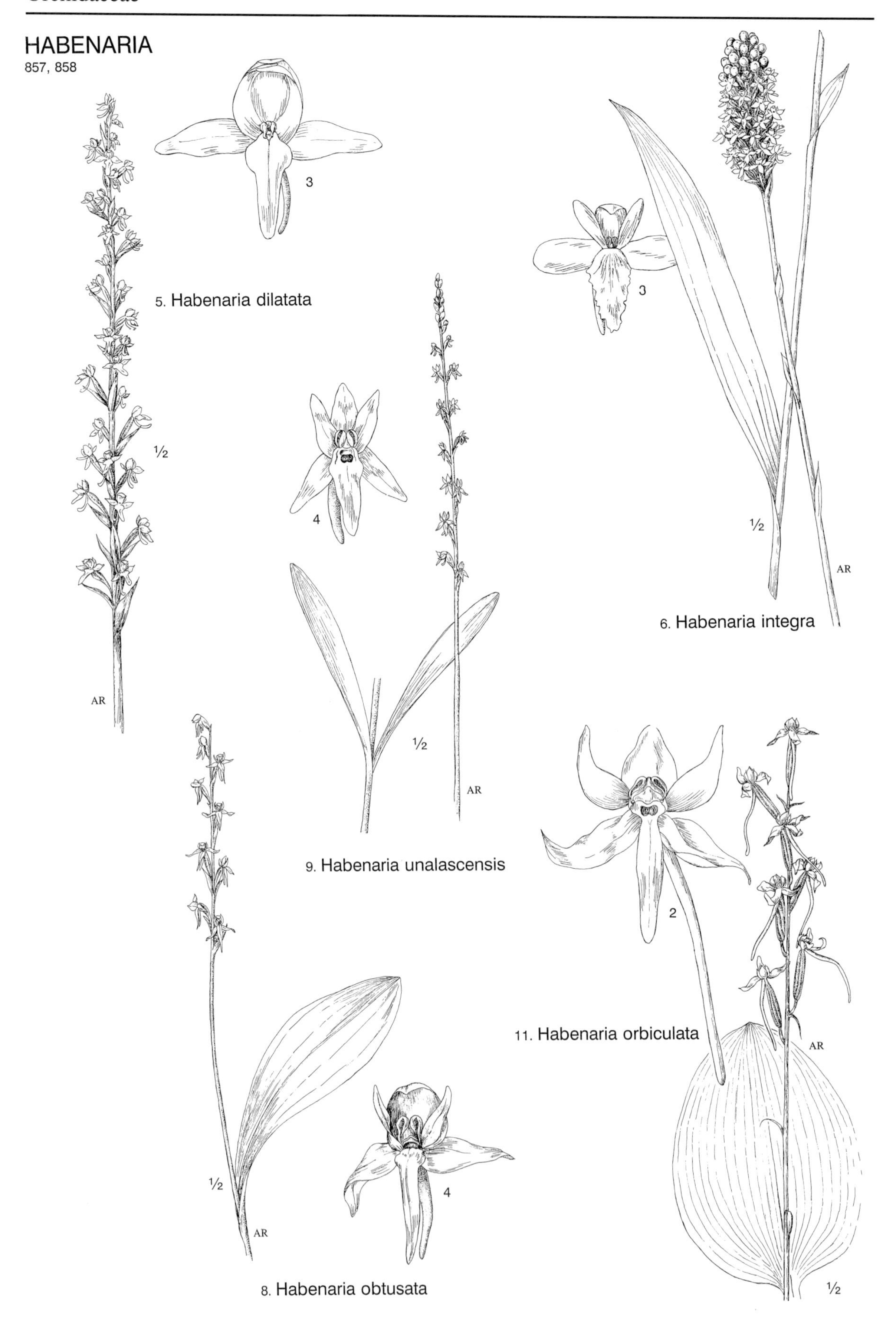

5. Habenaria dilatata

6. Habenaria integra

9. Habenaria unalascensis

11. Habenaria orbiculata

8. Habenaria obtusata

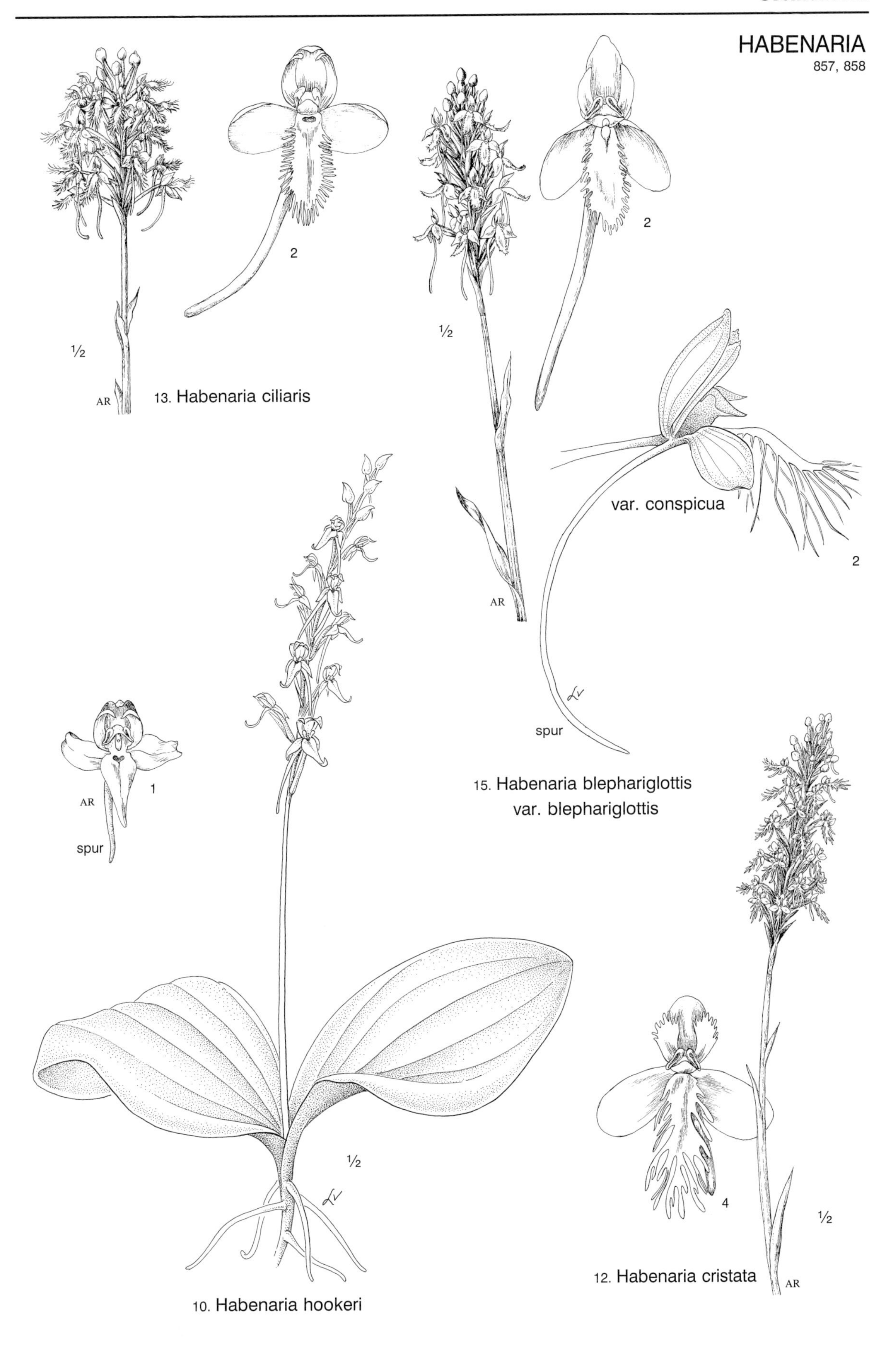

13. Habenaria ciliaris

2

var. conspicua

2

spur

15. Habenaria blephariglottis
var. blephariglottis

AR
1
spur

½

10. Habenaria hookeri

4

½

12. Habenaria cristata

HABENARIA
858, 859

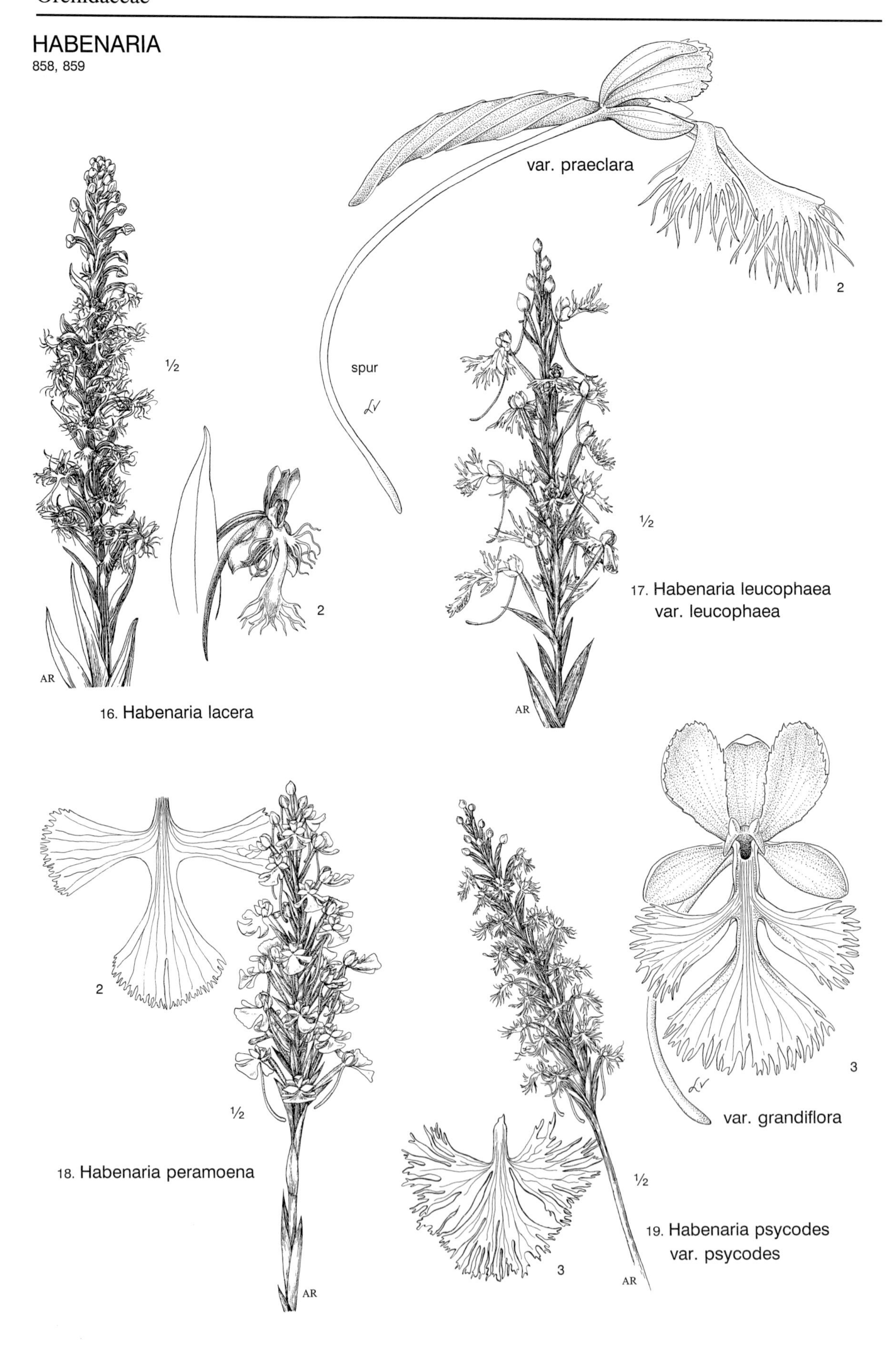

var. praeclara

spur

2

½

2

16. Habenaria lacera

½

17. Habenaria leucophaea
var. leucophaea

2

½

18. Habenaria peramoena

var. grandiflora

3

½

3

19. Habenaria psycodes
var. psycodes

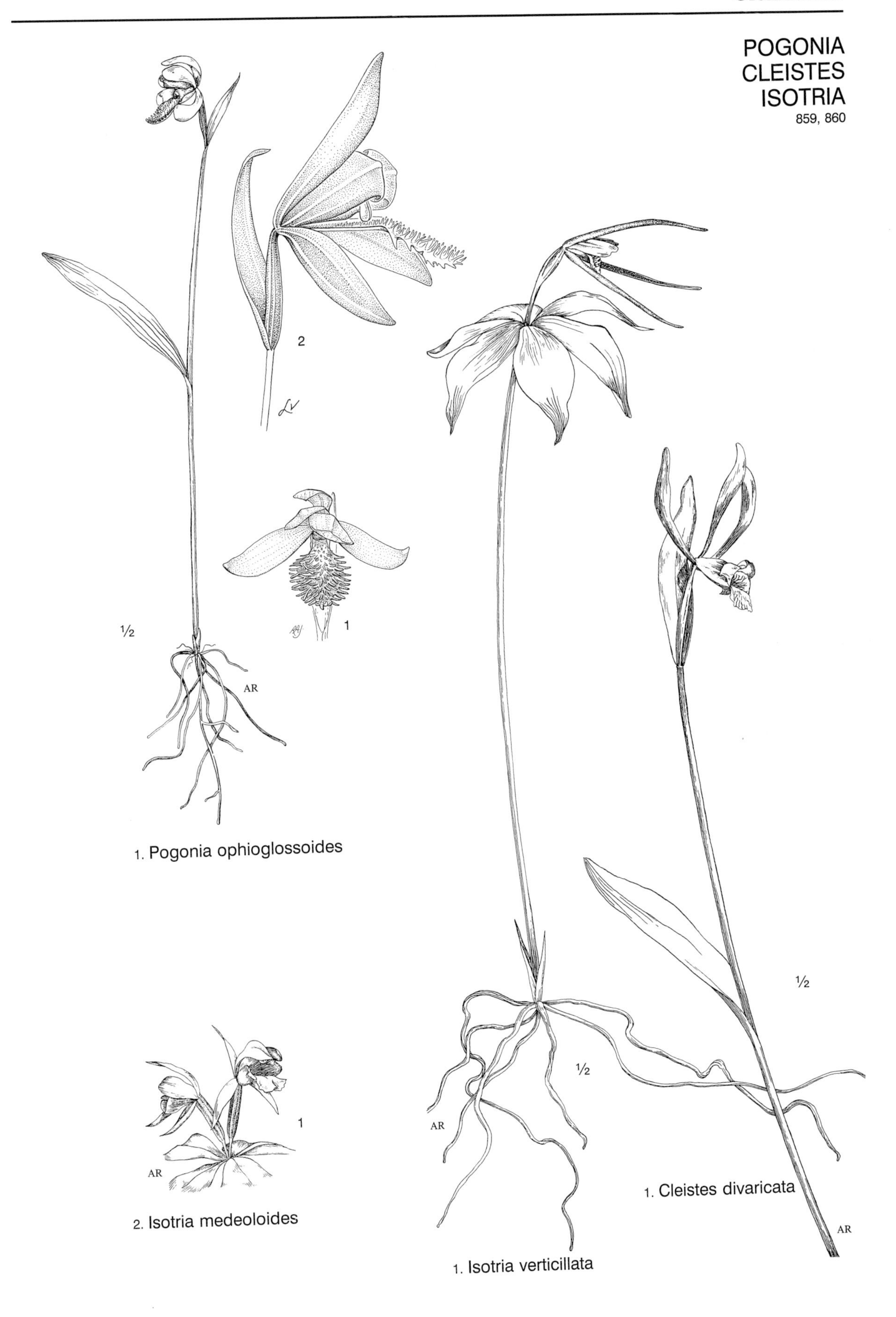

1. Pogonia ophioglossoides

2. Isotria medeoloides

1. Isotria verticillata

1. Cleistes divaricata

TRIPHORA
CALOPOGON
ARETHUSA
HEXALECTRIS
860, 861

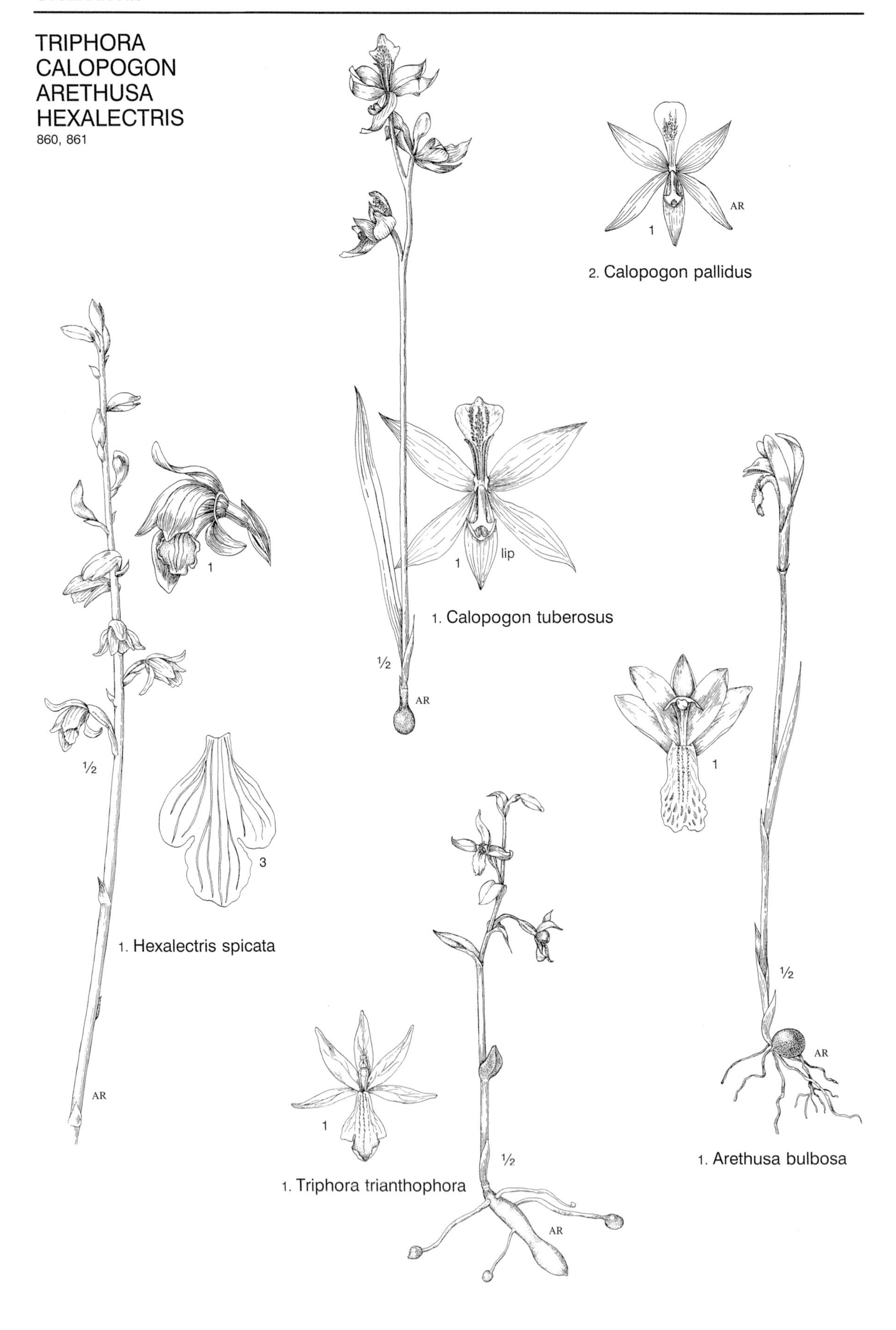

2. Calopogon pallidus

1. Calopogon tuberosus

lip

½

AR

1. Hexalectris spicata

½

AR

3

1. Arethusa bulbosa

½

AR

1. Triphora trianthophora

½

AR

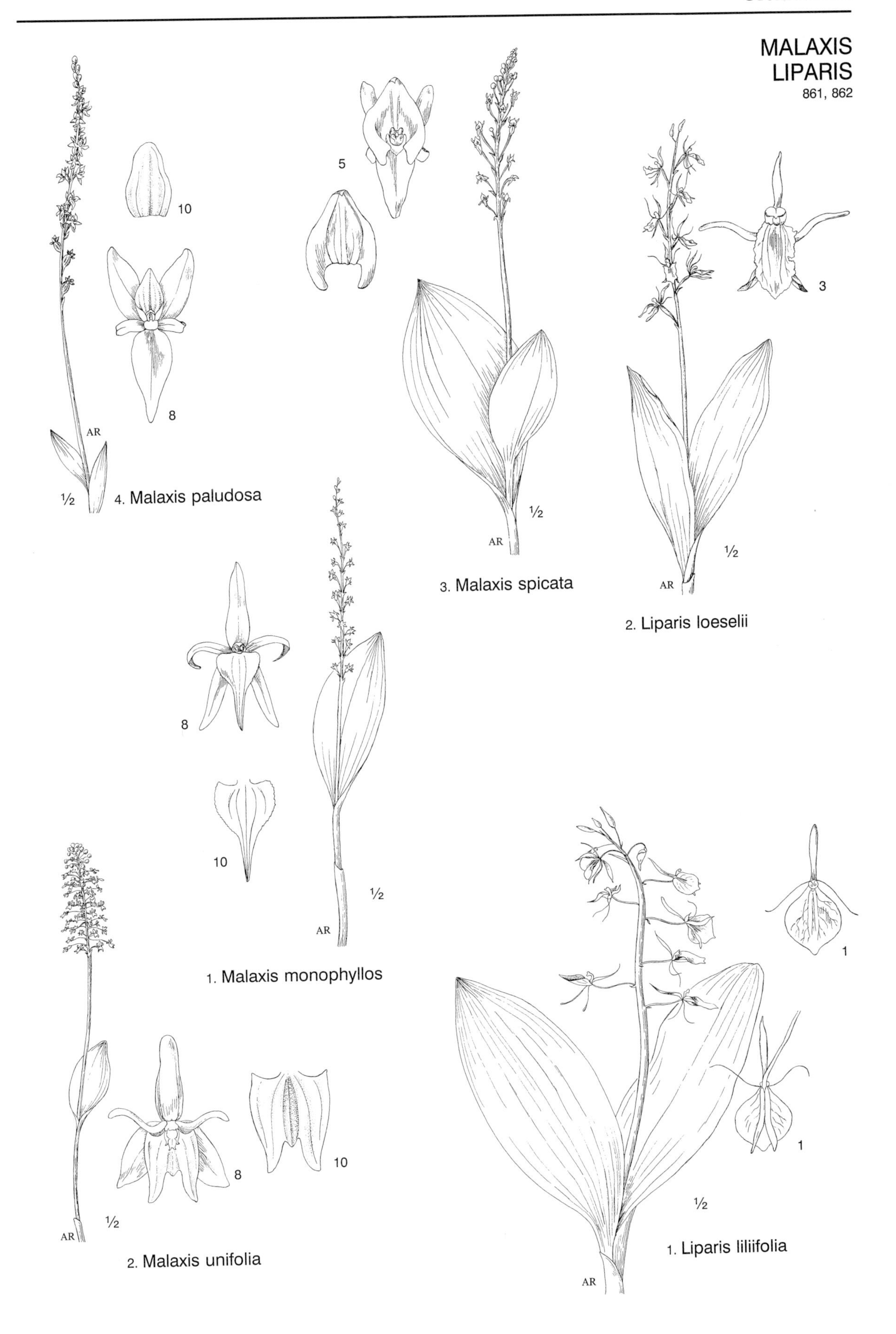

10

5

3

8

AR

½   4. Malaxis paludosa

3. Malaxis spicata

AR

½

2. Liparis loeselii

8

10

AR

½

1. Malaxis monophyllos

1

8   10

AR

½

2. Malaxis unifolia

1

1

½

1. Liparis liliifolia

AR

CALYPSO
APLECTRUM
TIPULARIA
CORALLORHIZA
862, 863

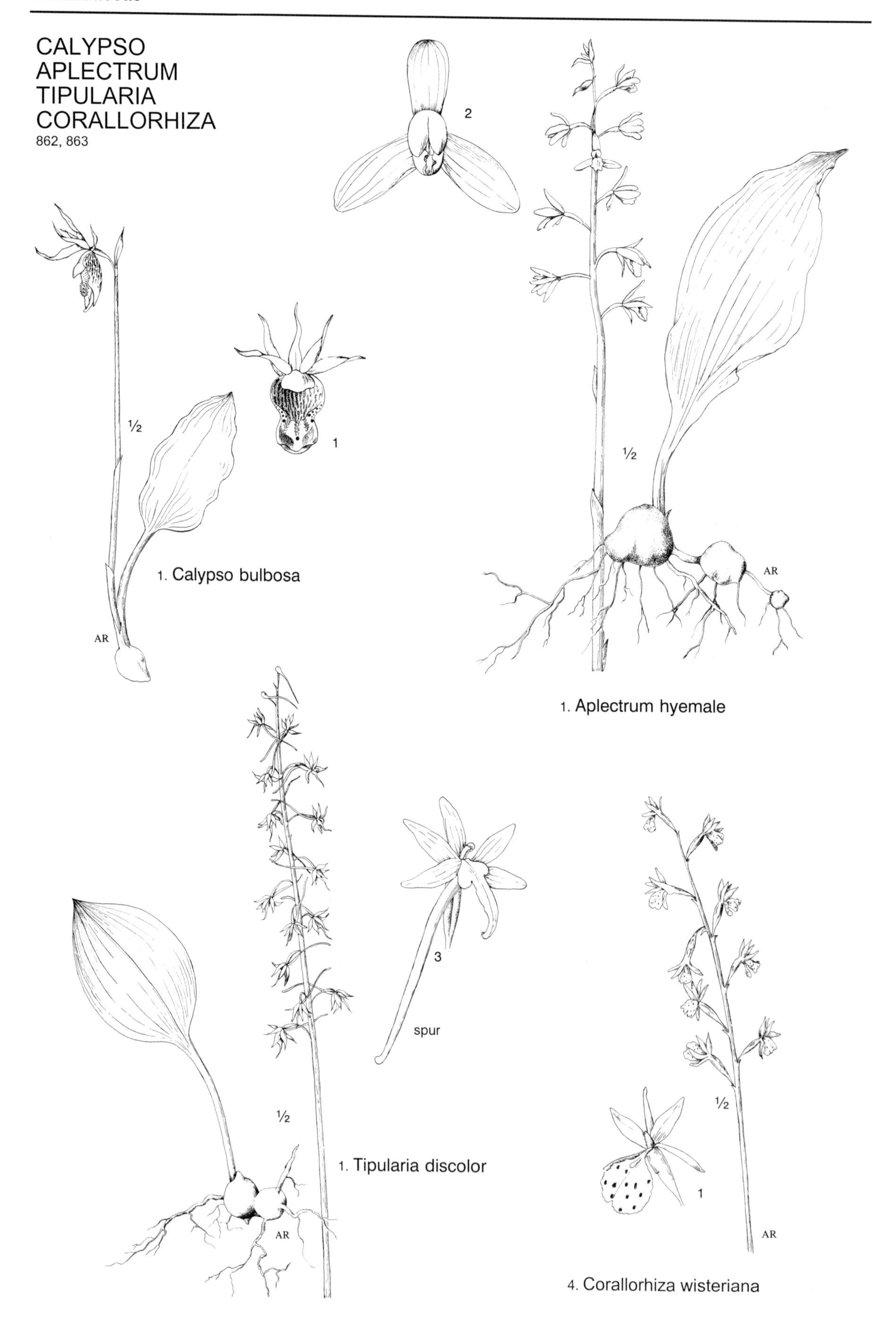

1. Calypso bulbosa

1. Aplectrum hyemale

1. Tipularia discolor

spur

4. Corallorhiza wisteriana

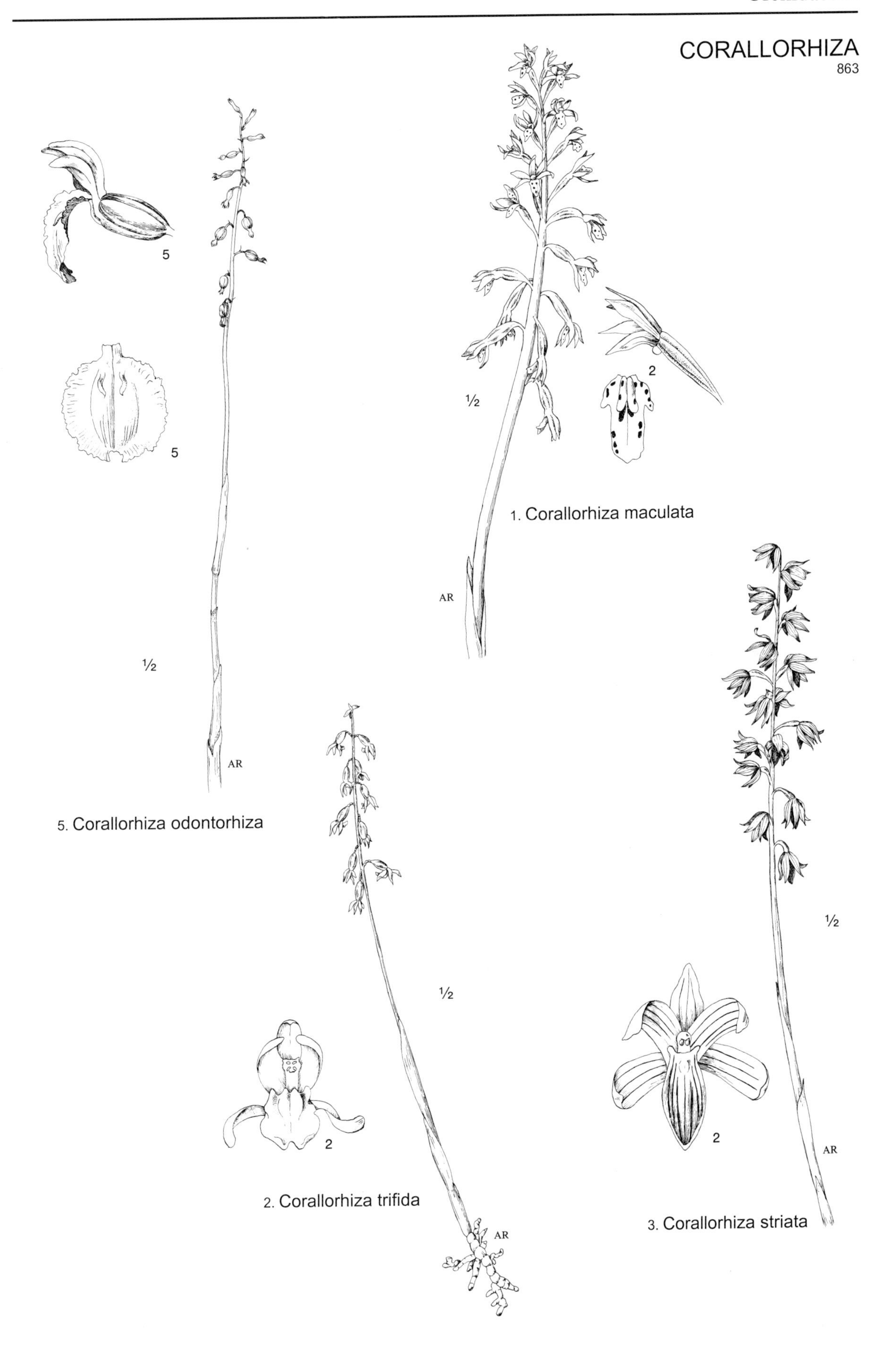

1. Corallorhiza maculata

5. Corallorhiza odontorhiza

2. Corallorhiza trifida

3. Corallorhiza striata

INDEX TO COMMON NAMES

The following index includes all common names used in Gleason & Cronquist's *Manual of Vascular Plants of Northeastern United States and Adjacent Canada, second edition* (1991).

absinthium (*Artemisia absinthium*), 522
acacia, rose- (*Robinia hispida, R. viscosa*), 262
acanthus family (ACANTHACEAE), 459
acid-water arrow-head (*Sagittaria engelmanniana*), 603
aconite (*Aconitum*), 47
  garden- (*Aconitum napellus*), 47
aconite-saxifrage (*Boykinia aconitifolia*), 215
Adam and Eve (*Aplectrum hyemale*), 826
Adam's needle (*Yucca filamentosa*), 804
adder's-mouth (*Malaxis*), 825
  bog (*Malaxis paludosa*), 825
  Florida (*Malaxis spicata*), 825
  green (*Malaxis unifolia*), 825
  white (*Malaxis monophyllos*), 825
adder's-tongue (*Ophioglossum*), 9
  northern (*Ophioglossum vulgatum* var. *pseudopodum*), 9
  southern (*Ophioglossum vulgatum* var. *pycnostichum*), 9
adder's-tongue family (OPHIOGLOSSACEAE), 8
adonis
  fall- (*Adonis annua*), 59
  spring- (*Adonis vernalis*), 59
agalinis (*Agalinis*), 451, 452
  blunt-leaved (*Agalinis obtusifolia*), 452
  common (*Agalinis tenuifolia*), 452
  fascicled (*Agalinis fasciculata*), 451
  midwestern (*Agalinis skinneriana*), 452
  perennial (*Agalinis linifolia*), 451
  rough (*Agalinis aspera*), 451
  salt-marsh (*Agalinis maritima*), 451
  sand-plain (*Agalinis acuta*), 452
  smooth (*Agalinis purpurea*), 451
  thread- (*Agalinis setacea*), 451
agave (*Agave*), 804
agave family (AGAVACEAE), 804
agrimony (*Agrimonia*), 236
  common (*Agrimonia gryposepala*), 236
  downy (*Agrimonia pubescens*), 236
  low (*Agrimonia microcarpa*), 236
  medicinal (*Agrimonia eupatoria*), 236
  roadside- (*Agrimonia striata*), 236
  southern (*Agrimonia parviflora*), 236
  woodland- (*Agrimonia rostellata*), 236
ague-weed (*Gentianella quinquefolia*), 366
Alabama grape-fern (*Botrychium jenmanii*), 9
Alabama lip-fern (*Cheilanthes alabamensis*), 15
Alaska orchid (*Habenaria unalascensis*), 820
alder (*Alnus*), 84
  black (*Alnus glutinosa*), 84
  brook- (*Alnus maritima*), 84
  coast white (*Clethra alnifolia*), 183

alder (*Alnus*) (continued)
  green (*Alnus viridis*), 84
  mountain (*Alnus viridis*), 84
  mountain white (*Clethra acuminata*), 183
  smooth (*Alnus serrulata*), 84
  speckled (*Alnus incana*), 84
  white (*Clethra*), 183
alder-buckthorn
  American (*Rhamnus alnifolia*), 322
  European (*Rhamnus frangula*), 321
alfalfa (*Medicago sativa*), 271
  blue (*Medicago sativa* subsp. *sativa*), 271
  yellow (*Medicago sativa* subsp. *falcata*), 271
alga-pondweed (*Potamogeton confervoides*), 610
alkali-blite (*Chenopodium rubrum*), 87
alkali-bulrush (*Scirpus maritimus*), 634
alkali-grass
  European (*Puccinellia distans*), 700
  Nuttall's (*Puccinellia nuttalliana*), 699
  seaside (*Puccinellia maritima*), 699
alkali milk-vetch (*Astragalus racemosus*), 264
alkali muhly (*Muhlenbergia asperifolia*), 747
alkaline water-nymph (*Najas marina*), 614
alkali-sacaton (*Sporobolus airoides*), 743
alkanet (*Anchusa*), 402
Allegheny-crowfoot (*Ranunculus alleghaniensis*), 56
Allegheny monkey-flower (*Mimulus ringens*), 440
Allegheny-phlox (*Phlox ovata*), 391
Allegheny-plum (*Prunus alleghaniensis*), 242
Allegheny-skullcap (*Scutellaria serrata*), 410
Allegheny-spurge (*Pachysandra procumbens*), 311
Allegheny-vine (*Adlumia*), 66
alligator-weed (*Alternanthera philoxeroides*), 98
alloploid rattlesnake-plantain (*Goodyera tesselata*), 818
allspice, Carolina (*Calycanthus*), 38
aloe, false (*Agave virginica*), 804
alpine azalea (*Loiseleuria*), 187
alpine bearberry (*Arctostaphylos alpina*), 190
alpine bluegrass (*Poa alpina*), 705
alpine-brook saxifrage (*Saxifraga rivularis*), 211
alpine cliff-fern (*Woodsia alpina*), 20
alpine clubmoss (*Lycopodium alpinum*), 3
alpine cudweed (*Gnaphalium supinum*), 564
alpine enchanter's nightshade (*Circaea alpina*), 303
alpine milk-vetch (*Astragalus alpinus*), 265
alpine saxifrage
  purple (*Saxifraga oppositifolia*), 212
  white (*Saxifraga aizoon*), 212
  yellow (*Saxifraga aizoides*), 212
alpine willow-herb (*Epilobium alpinum*), 298

alsike clover (*Trifolium hybridum*), 270
alta-fescue (*Festuca arundinacea*), 695
alum-root (*Heuchera*), 215
    Appalachian (*Heuchera pubescens*), 216
    closed-flowered (*Heuchera longiflora*), 216
    common (*Heuchera americana*), 215
    maple-leaved (*Heuchera villosa*), 215
    prairie (*Heuchera richardsonii*), 216
    small-flowered (*Heuchera parviflora*), 215
alyssum
    hoary (*Berteroa incana*), 168
    sweet (*Lobularia maritima*), 168
amaranth (*Amaranthus*), 94
    globe- (*Gomphrena globosa*), 97
    green (*Amaranthus viridis*), 96
    red (*Amaranthus cruentus*), 95
    sandhill- (*Amaranthus arenicola*), 96
    spiny (*Amaranthus spinosus*), 95
amaranth family (AMARANTHACEAE), 94
American alder-buckthorn (*Rhamnus alnifolia*), 322
American barberry (*Berberis canadensis*), 62
American beautyberry (*Callicarpa americana*), 407
American beech (*Fagus grandifolia*), 75
American bittersweet (*Celastrus scandens*), 308
American bugbane (*Cimicifuga americana*), 46
American burnet (*Sanguisorba canadensis*), 237
American burning bush (*Euonymus americanus*), 308
American Christmas-mistletoe (*Phoradendron serotinum*), 307
American columbo (*Frasera caroliniensis*), 367
American dog-violet (*Viola conspersa*), 148
American dragon-head (*Dracocephalum parviflorum*), 422
American dunegrass (*Leymus mollis*), 728
American elm (*Ulmus americana*), 68
American figwort (*Scrophularia lanceolata*), 446
American frog's-bit (*Limnobium spongia*), 607
American germander (*Teucrium canadense*), 408
American ginseng (*Panax quinquefolius*), 343
American grass-of-Parnassus (*Parnassia glauca*), 216
American gromwell (*Lithospermum latifolium*), 400
American hazel-nut (*Corylus americana*), 82
American holly (*Ilex opaca*), 309
American hydrangea (*Hydrangea arborescens*), 205
American linden (*Tilia americana*), 131
American lotus-lily (*Nelumbo lutea*), 42
American mannagrass (*Glyceria grandis*), 708
American marsh willow-herb (*Epilobium leptophyllum*), 297
American mountain-ash (*Sorbus americana*), 246
American pennyroyal (*Hedeoma pulegioides*), 412
American plantain (*Plantago rugelii*), 432
American sloughgrass (*Beckmannia syzigachne*), 720
American snowbell (*Styrax americanus*), 198
American speedwell (*Veronica americana*), 455
American vetch (*Vicia americana*), 274
American water-horehound (*Lycopus americanus*), 420
American water-willow (*Justicia americana*), 460
American willow-herb (*Epilobium ciliatum*), 298
American yew (*Taxus canadensis*), 31
Amur silver-grass (*Miscanthus sacchariflorus*), 771
androsace
    northern (*Androsace septentrionalis*), 203
    western (*Androsace occidentalis*), 203
anemone (*Anemone*), 49
    Canadian (*Anemone canadensis*), 49
    cut-leaved (*Anemone multifida*), 49
    lance-leaved (*Anemone lancifolia*), 49

anemone (*Anemone*) (continued)
    long-headed (*Anemone cylindrica*), 50
    prairie- (*Anemone caroliniana*), 49
    rue- (*Anemonella*), 59
    small-flowered (*Anemone parviflora*), 49
    tall (*Anemone virginiana*), 50
    wood- (*Anemone quinquefolia*), 49
angelica (*Angelica*), 358
    hairy (*Angelica venenosa*), 359
    mountain- (*Angelica triquinata*), 358
    purplestem- (*Angelica atropurpurea*), 358
    seabeach- (*Angelica lucida*), 359
angelico (*Ligusticum canadense*), 354
angle-pod
    Atlantic (*Matelea suberosa*), 375
    common (*Matelea gonocarpa*), 375
animated oats (*Avena sterilis*), 710
annual buffalo-clover (*Trifolium reflexum*), 270
annual bursage (*Ambrosia acanthicarpa*), 515
annual false foxglove (*Aureolaria pedicularia*), 450
annual fleabane (*Erigeron annuus*), 561
annual knawel (*Scleranthus annuus*), 105
annual loosestrife (*Lythrum hyssopifolia*), 292
annual salt-marsh aster (*Aster subulatus*), 559
annual sea-pink (*Sabatia stellaris*), 363
annual skeleton-weed (*Lygodesmia rostrata*), 589
annual toadflax (*Linaria repens*), 447
annual willow-herb (*Epilobium brachycarpum*), 298
annual woolly bean (*Strophostyles helvula*), 285
annual wormwood (*Artemisia annua*), 521
anoda, spurred (*Anoda cristata*), 138
Appalachian alum-root (*Heuchera pubescens*), 216
Appalachian arrow-head (*Sagittaria australis*), 604
Appalachian Barbara's buttons (*Marshallia grandiflora*), 509
Appalachian beard-tongue (*Penstemon canescens*), 445
Appalachian bellflower (*Campanula divaricata*), 465
Appalachian bitter-cress (*Cardamine clematitis*), 173
Appalachian blazing star (*Liatris turgida*), 575
Appalachian bristle-fern (*Trichomanes boschianum*), 11
Appalachian coneflower (*Echinacea laevigata*), 497
Appalachian false foxglove (*Aureolaria laevigata*), 449
Appalachian fame-flower (*Talinum teretifolium*), 99
Appalachian flat-topped white aster (*Aster infirmus*), 558
Appalachian gentian (*Gentiana decora*), 364
Appalachian gooseberry (*Ribes rotundifolium*), 205
Appalachian groundsel (*Senecio anonymus*), 524
Appalachian ironweed (*Vernonia glauca*), 577
Appalachian Jacob's ladder (*Polemonium vanbruntiae*), 392
Appalachian loosestrife (*Lysimachia tonsa*), 200
Appalachian milkwort (*Polygala curtissii*), 329
Appalachian mock-orange (*Philadelphus inodorus*), 204
Appalachian mountain-holly (*Nemopanthus collinus*), 310
Appalachian parthenium (*Parthenium auriculatum*), 513
Appalachian phacelia (*Phacelia dubia*), 393
Appalachian skullcap (*Scutellaria arguta*), 410
Appalachian spiraea (*Spiraea virginiana*), 218
Appalachian sunflower (*Helianthus atrorubens*), 492
Appalachian toothwort (*Cardamine angustata*), 172
Appalachian twayblade (*Listera smallii*), 814
Appalachian white lettuce (*Prenanthes roanensis*), 588
Appalachian whitlow-wort (*Paronychia virginica*), 106
apple (*Pyrus*), 243; (*P. malus*), 244
    balsam- (*Echinocystis lobata*), 150
    crab- (*Pyrus*), 243
    Indian (*Datura wrightii*), 382
    may- (*Podophyllum*), 62

apple-mint (*Mentha suaveolens*), 418
apple of Peru (*Nicandra*), 376
apricot (*Prunus armeniaca*), 240
arbor vitae (*Thuja*), 35
arbutus, trailing (*Epigaea*), 190
arctic blue flag (*Iris setosa*), 811
arctic bluegrass (*Poa arctica*), 701
arctic eyebright (*Euphrasia disjuncta*), 456
arctic grass-of-Parnassus (*Parnassia palustris*), 217
arnica (*Arnica*), 500
   heart-leaved (*Arnica cordifolia*), 500
   New England (*Arnica lanceolata*), 501
   southeastern (*Arnica acaulis*), 501
aromatic aster (*Aster oblongifolius*), 554
arrow-arum (*Peltandra*), 615
arrowfeather three-awn (*Aristida purpurascens*), 733
arrow-grass (*Triglochin*), 608
   common (*Triglochin maritima*), 608
   slender (*Triglochin palustris*), 608
   southern (*Triglochin striata*), 608
arrow-grass family (JUNCAGINACEAE), 608
arrow-head (*Sagittaria*), 603
   acid-water (*Sagittaria engelmanniana*), 603
   Appalachian (*Sagittaria australis*), 604
   common (*Sagittaria latifolia*), 604
   midwestern (*Sagittaria brevirostra*), 604
   Mississippi (*Sagittaria calycina*), 603
   northern (*Sagittaria cuneata*), 604
   sessile-fruited (*Sagittaria rigida*), 603
arrowhead sweet-coltsfoot (*Petasites sagittatus*), 529
arrowhead-violet (*Viola sagittata*), 143
arrowleaf sida (*Sida rhombifolia*), 137
arrow-leaved aster (*Aster sagittifolius*), 552
arrow-leaved tearthumb (*Polygonum sagittatum*), 122
arrow-wood (*Viburnum dentatum*), 483
   downy (*Viburnum rafinesquianum*), 482
   Missouri (*Viburnum molle*), 482
artichoke, Jerusalem (*Helianthus tuberosus*), 492
arum
   arrow- (*Peltandra*), 615
   water- (*Calla*), 615
arum family (ARACEAE), 615
ash (*Fraxinus*), 436
   American mountain- (*Sorbus americana*), 246
   black (*Fraxinus nigra*), 437
   blue (*Fraxinus quadrangulata*), 437
   Carolina (*Fraxinus caroliniana*), 437
   common prickly (*Zanthoxylum americanum*), 336
   European mountain- (*Sorbus aucuparia*), 246
   mountain- (*Sorbus*), 246
   prickly (*Zanthoxylum*), 336
   pumpkin- (*Fraxinus profunda*), 437
   showy mountain- (*Sorbus decora*), 246
   water- (*Fraxinus caroliniana*), 437
   white (*Fraxinus americana*), 436
ash-leaved maple (*Acer negundo*), 333
ashy sunflower (*Helianthus mollis*), 491
Asiatic hawk's beard (*Youngia japonica*), 596
asparagus (*Asparagus*), 801
aspen (*Populus tremuloides*), 152
   big-toothed (*Populus grandidentata*), 152
   quaking (*Populus tremuloides*), 152
asphodel
   bog- (*Narthecium*), 785
   false (*Tofieldia*), 781

aster
   annual salt-marsh (*Aster subulatus*), 559
   Appalachian flat-topped white (*Aster infirmus*), 558
   aromatic (*Aster oblongifolius*), 554
   arrow-leaved (*Aster sagittifolius*), 552
   awl- (*Aster pilosus*), 550
   big-headed (*Aster grandiflorus*), 554
   big-leaved (*Aster macrophyllus*), 559
   bottomland- (*Aster ontarionis*), 549
   bristly (*Aster puniceus*), 547
   clasping (*Aster patens*), 553
   clasping heart-leaved (*Aster undulatus*), 553
   common blue heart-leaved (*Aster cordifolius*), 552
   common white heart-leaved (*Aster divaricatus*), 555
   eastern lined (*Aster lanceolatus*), 549
   eastern serpentine- (*Aster depauperatus*), 551
   eastern silvery (*Aster concolor*), 551
   falcate golden (*Chrysopsis falcata*), 531
   goblet- (*Aster lateriflorus*), 550
   golden (*Chrysopsis*), 530; (*Heterotheca*), 531
   grass-leaved golden (*Chrysopsis graminifolia*), 531
   hairy heart-leaved (*Aster drummondii*), 552
   leafy bog- (*Aster nemoralis*), 557
   limestone heart-leaved (*Aster anomalus*), 553
   long-stalked (*Aster dumosus*), 550
   low rough (*Aster radula*), 556
   low showy (*Aster spectabilis*), 556
   midwestern blue heart-leaved (*Aster shortii*), 552
   midwestern white heart-leaved (*Aster furcatus*), 555
   narrow-leaved smooth (*Aster concinnus*), 548
   narrow-leaved white topped (*Aster solidagineus*), 558
   New England (*Aster novae-angliae*), 554
   Newfoundland (*Aster crenifolius*), 547
   New York (*Aster novi-belgii*), 547
   northern bog- (*Aster borealis*), 547
   northern heart-leaved (*Aster ciliolatus*), 551
   northwestern flat-topped white (*Aster pubentior*), 557
   northwestern sticky (*Aster modestus*), 554
   Ozark heart-leaved (*Aster texanus*), 553
   perennial salt-marsh (*Aster tenuifolius*), 558
   prairie- (*Aster turbinellus*), 555
   prairie golden (*Chrysopsis camporum*), 530
   prairie heart-leaved (*Aster oolentangiensis*), 551
   shaggy golden (*Chrysopsis mariana*), 530
   shining (*Aster firmus*), 547
   shore- (*Aster tradescantii*), 549
   slender (*Aster gracilis*), 555
   small-headed (*Aster parviceps*), 551; (*A. racemosus*), 550
   smooth (*Aster laevis*), 548
   smooth heart-leaved (*Aster lowrieanus*), 552
   southern swamp- (*Aster elliottii*), 548
   squarrose white (*Aster ericoides*), 550
   stiff (*Aster linariifolius*), 558
   St. Lawrence annual (*Aster laurentianus*), 558
   tall flat-topped white (*Aster umbellatus*), 557
   Tatarian (*Aster tataricus*), 555
   toothed white-topped (*Aster paternus*), 558
   veiny lined (*Aster praealtus*), 549
   western annual (*Aster brachyactis*), 559
   western lined (*Aster hesperius*), 548
   western silvery (*Aster sericeus*), 551
   white prairie- (*Aster falcatus*), 550
   whorled (*Aster acuminatus*), 557
   wild (*Aster*), 547
   woolly-headed golden (*Chrysopsis gossypina*), 530
   zigzag (*Aster prenanthoides*), 548

aster family (ASTERACEAE), 489
atamasco-lily (*Zephyranthes atamasca*), 803
Atlantic angle-pod (*Matelea suberosa*), 375
Atlantic mock bishop-weed (*Ptilimnium capillaceum*), 356
Atlantic mudwort (*Limosella subulata*), 441
Atlantic white cedar (*Chamaecyparis thyoides*), 35
Atlantic wisteria (*Wisteria frutescens*), 261
auricled twayblade (*Listera auriculata*), 814
autumn-bent (*Agrostis perennans*), 718
autumn coral-root (*Corallorhiza odontorhiza*), 827
autumn-crocus (*Colchicum autumnale*), 794
autumn-willow (*Salix serissima*), 155
avens (*Geum*), 227
    big-leaved (*Geum macrophyllum*), 228
    cream-colored (*Geum virginianum*), 227
    rough (*Geum laciniatum*), 227
    spring- (*Geum vernum*), 227
    water- (*Geum rivale*), 228
    white (*Geum canadense*), 227
    White-Mountain (*Geum peckii*), 229
    yellow (*Geum aleppicum*), 228
awl-aster (*Aster pilosus*), 550
awlwort (*Subularia*), 167
awned melic (*Melica smithii*), 708
axillary goldenrod (*Solidago caesia*), 537
azalea
    alpine (*Loiseleuria*), 187
    coastal (*Rhododendron atlanticum*), 185
    Cumberland (*Rhododendron bakeri*), 186
    flame- (*Rhododendron calendulaceum*), 186
    hammock sweet (*Rhododendron serrulatum*), 185
    piedmont- (*Rhododendron canescens*), 185
    rosebud- (*Rhododendron prinophyllum*), 185
    smooth (*Rhododendron arborescens*), 186
    swamp- (*Rhododendron viscosum*), 185
    sweet (*Rhododendron arborescens*), 186

baby's breath (*Gypsophila paniculata*), 111
    cushion (*Gypsophila muralis*), 111
bachelor's button (*Centaurea cyanus*), 584
baked-apple-berry (*Rubus chamaemorus*), 229
bald cypress (*Taxodium distichum*), 34
bald cypress family (TAXODIACEAE), 34
ball-mustard (*Neslia paniculata*), 167
balloon-vine (*Cardiospermum*), 330
balm (*Melissa*), 412
    basil-bee- (*Monarda clinopodia*), 427
    bee- (*Monarda*), 427
    common (*Melissa officinalis*), 412
    horse- (*Collinsonia*), 418
    lemon- (*Melissa officinalis*), 412
    lemon-bee- (*Monarda citriodora*), 427
    northern horse- (*Collinsonia canadensis*), 418
    whorled horse- (*Collinsonia verticillata*), 418
balm of Gilead (*Populus* ×*jackii*), 153
balsam (*Impatiens balsamina*), 342
    she- (*Abies fraseri*), 31
balsam-apple (*Echinocystis lobata*), 150
balsam-fir (*Abies balsamea*), 31
balsam-poplar (*Populus balsamifera*), 152
balsam-willow (*Salix pyrifolia*), 157
bamboo, Mexican (*Polygonum cuspidatum*), 124
baneberry (*Actaea*), 47
    red (*Actaea rubra*), 47
bank-vervain (*Verbena riparia*), 403
Barbara's buttons (*Marshallia*), 509

Barbara's buttons (*Marshallia*) (*continued*)
    Appalachian (*Marshallia grandiflora*), 509
    colonial (*Marshallia trinervia*), 509
barberry (*Berberis*), 62
    American (*Berberis canadensis*), 62
    common (*Berberis vulgaris*), 62
    European (*Berberis vulgaris*), 62
    Japanese (*Berberis thunbergii*), 62
barberry family (BERBERIDACEAE), 62
barberry hawthorn (*Crataegus berberifolia*), 249
Barksdale's trillium (*Trillium sulcatum*), 795
barley (*Hordeum*), 729; (*H. vulgare*), 729
    foxtail- (*Hordeum jubatum*), 729
    little (*Hordeum pusillum*), 729
    meadow- (*Hordeum brachyantherum*), 729
    Mediterranean (*Hordeum geniculatum*), 729
barnyard-grass (*Echinochloa crusgalli*), 763; (*E. muricata*),
    762
barren brome (*Bromus sterilis*), 724
barren strawberry (*Waldsteinia*), 221
basal-leaved rosin-weed (*Silphium terebinthinaceum*), 513
basil (*Ocimum basilicum*), 430
    wild (*Satureja vulgaris*), 413
basil-bee-balm (*Monarda clinopodia*), 427
basil-thyme (*Satureja calamintha*), 413
basin wild rye (*Leymus cinereus*), 728
basket-flower (*Centaurea americana*), 585
basket-willow (*Salix purpurea*), 160
bassia, five-hook (*Bassia hyssopifolia*), 90
basswood (*Tilia*), 131; (*T. americana*), 131
bastard toad-flax (*Comandra*), 306
bay
    bull- (*Magnolia grandiflora*), 36
    Lappland rose- (*Rhododendron lapponicum*), 185
    rose- (*Rhododendron catawbiense, R. maximum*), 184
    swamp- (*Persea palustris*), 38
    sweet (*Magnolia virginiana*), 36
bayberry
    northern (*Myrica pensylvanica*), 75
    southern (*Myrica heterophylla*), 75
bayberry family (MYRICACEAE), 75
bay-leaved willow (*Salix pentandra*), 155
bayonet, Spanish (*Yucca*), 804
beach-dune marsh-elder (*Iva imbricata*), 515
beach-grass (*Ammophila*), 717
beach-pea (*Lathyrus maritimus*), 274
beach-plum (*Prunus maritima*), 242
beach-wormwood (*Artemisia stelleriana*), 520
bead-grass (*Paspalum*), 763
bead-lily (*Clintonia*), 797
beaked dodder (*Cuscuta rostrata*), 387
beaked hawkweed (*Hieracium gronovii*), 594
beaked hazel-nut (*Corylus cornuta*), 82
beaked sanicle (*Sanicula trifoliata*), 346
beaked willow (*Salix bebbiana*), 156
beak-rush (*Rhynchospora*), 645
bean (*Phaseolus*), 285
    annual woolly (*Strophostyles helvula*), 285
    broad (*Vicia faba*), 272
    buck- (*Menyanthes*), 388; (*Thermopsis*), 259
    castor- (*Ricinus communis*), 311
    Egyptian (*Dolichos lablab*), 285
    hyacinth (*Dolichos lablab*), 285
    perennial woolly (*Strophostyles umbellata*), 286
    piedmont-buck- (*Thermopsis mollis*), 259
    small-flowered woolly (*Strophostyles leiosperma*), 286

bean (*Phaseolus*) (*continued*)
  wild (*Phaseolus polystachios*), 285
  woolly (*Strophostyles*), 285
bean family (FABACEAE), 258
bearberry (*Arctostaphylos uva-ursi*), 190
  alpine (*Arctostaphylos alpina*), 190
bearberry-honeysuckle (*Lonicera involucrata*), 478
bearberry-willow (*Salix uva-ursi*), 156
beard
  Asiatic hawk's (*Youngia japonica*), 596
  false goat's- (*Astilbe biternata*), 211
  fistulous goat's- (*Tragopogon dubius*), 601
  goat's- (*Aruncus dioicus*), 219; (*Tragopogon*), 601
  golden crown- (*Verbesina encelioides*), 501
  hawk's (*Crepis*), 595
  showy goat's- (*Tragopogon pratensis*), 601
  turkey- (*Xerophyllum asphodeloides*), 780
bearded meadow-parsnip (*Thaspium barbinode*), 353
beard-grass (*Gymnopogon*), 748
beard-tongue (*Penstemon*), 444
  Appalachian (*Penstemon canescens*), 445
  eastern (*Penstemon laevigatus*), 445
  eastern white (*Penstemon pallidus*), 445
  foxglove (*Penstemon cobaea*), 444
  Kentucky (*Penstemon tenuiflorus*), 444
  large (*Penstemon grandiflorus*), 444
  northeastern (*Penstemon hirsutus*), 444
  prairie white (*Penstemon albidus*), 444
  slender (*Penstemon gracilis*), 446
  southeastern (*Penstemon australis*), 445
  tall white (*Penstemon digitalis*), 445
  tube (*Penstemon tubiflorus*), 444
bear-oak (*Quercus ilicifolia*), 79
beautyberry (*Callicarpa*), 407
  American (*Callicarpa americana*), 407
  Chinese (*Callicarpa dichotoma*), 407
bedstraw (*Galium*), 473
  berry- (*Galium hispidulum*), 476; (*G. uniflorum*), 477
  bluntleaf- (*Galium obtusum*), 475
  forest- (*Galium circaezans*), 476
  Labrador- (*Galium labradoricum*), 475
  marsh- (*Galium palustre*), 474
  northern (*Galium boreale*), 473
  northern three-lobed (*Galium trifidum*), 475
  rough (*Galium asprellum*), 474
  shining (*Galium concinnum*), 475
  southern three-lobed (*Galium tinctorium*), 476
  sweet-scented (*Galium triflorum*), 477
  wide-leaved (*Galium latifolium*), 473
  yellow (*Galium verum*), 474
bee-balm (*Monarda*), 427
  basil- (*Monarda clinopodia*), 427
  lemon- (*Monarda citriodora*), 427
beech (*Fagus*), 75
  American (*Fagus grandifolia*), 75
  blue (*Carpinus caroliniana*), 82
beech-drops (*Epifagus*), 459
beech family (FAGACEAE), 75
beech-fern
  broad (*Thelypteris hexagonoptera*), 23
  narrow (*Thelypteris phegopteris*), 22
  northern (*Thelypteris phegopteris*), 22
  southern (*Thelypteris hexagonoptera*), 23
beggar's lice (*Hackelia*), 398; (*Lappula*), 398
beggar-ticks (*Bidens*), 506
  devil's (*Bidens frondosa*), 507

beggar-ticks (*Bidens*) (*continued*)
  few-bracted (*Bidens discoidea*), 507
  New England estuarine (*Bidens eatonii*), 507
  northern estuarine (*Bidens hyperborea*), 506
  purplestem (*Bidens connata*), 507
  southern estuarine (*Bidens bidentoides*), 507
  strawstem (*Bidens comosa*), 506
  tall (*Bidens vulgata*), 507
  water (*Bidens beckii*), 508
bellflower (*Campanula*), 465
  Appalachian (*Campanula divaricata*), 465
  clustered (*Campanula glomerata*), 465
  marsh- (*Campanula aparinoides*), 465
  rover- (*Campanula rapunculoides*), 465
  tall (*Campanula americana*), 465
bellflower family (CAMPANULACEAE), 465
bell-lily (*Lilium grayi*), 791
bellwort (*Uvularia*), 798
bent (*Agrostis*), 717
  autumn- (*Agrostis perennans*), 718
  creeping (*Agrostis stolonifera*), 718
  Rhode Island (*Agrostis capillaris*), 717
  velvet (*Agrostis canina*), 719
bent-awn plumegrass (*Erianthus contortus*), 771
bent-grass (*Agrostis*), 717
bent trillium (*Trillium flexipes*), 796
bergamot
  purple (*Monarda media*), 428
  wild (*Monarda fistulosa*), 427
Bermuda-grass (*Cynodon dactylon*), 749
berry-bedstraw (*Galium hispidulum*), 476; (*G. uniflorum*), 477
betony, wood- (*Pedicularis canadensis*), 457
Bicknell's yellow flax (*Linum intercursum*), 326
biennial gaura (*Gaura biennis*), 302
biennial waterleaf (*Hydrophyllum appendiculatum*), 394
biennial wormwood (*Artemisia biennis*), 521
big bluestem (*Andropogon gerardii*), 773
big cord-grass (*Spartina cynosuroides*), 749
big floating heart (*Nymphoides aquatica*), 389
bigflower-tickseed (*Coreopsis grandiflora*), 504
bigflower-vetch (*Vicia grandiflora*), 272
bigfruit-dodder (*Cuscuta megalocarpa*), 387
big hawkbit (*Leontodon hispidus*), 601
big-headed aster (*Aster grandiflorus*), 554
bigleaf-holly (*Ilex montana*), 310
bigleaf-magnolia (*Magnolia macrophylla*), 37
bigleaf-pondweed (*Potamogeton amplifolius*), 612
bigleaf-snowbell (*Styrax grandifolius*), 198
big-leaved aster (*Aster macrophyllus*), 559
big-leaved avens (*Geum macrophyllum*), 228
big-leaved goldenrod (*Solidago macrophylla*), 532
big-leaved grass-of-Parnassus (*Parnassia grandifolia*), 217
big marsh-elder (*Iva xanthiifolia*), 514
big mock bishop-weed (*Ptilimnium costatum*), 356
big quaking grass (*Briza maxima*), 700
big-seed scorpion-grass (*Myosotis macrosperma*), 397
bigsheath-pondweed (*Potamogeton vaginatus*), 609
big tick-trefoil (*Desmodium cuspidatum*), 279
big-toothed aspen (*Populus grandidentata*), 152
bigtree-plum (*Prunus mexicana*), 242
big twayblade (*Listera ovata*), 815
big white trillium (*Trillium grandiflorum*), 795
big yellow wood-sorrel (*Oxalis grandis*), 338
bilberry
  bog- (*Vaccinium uliginosum*), 191

bilberry *(continued)*
    dwarf *(Vaccinium cespitosum)*, 191
    mountain- *(Vaccinium membranaceum)*, 191
    tall *(Vaccinium ovalifolium)*, 192
billion-dollar grass *(Echinochloa crusgalli var. frumentacea)*, 763
Biltmore-hawthorn *(Crataegus intricata)*, 250
bindweed
    black *(Polygonum convolvulus)*, 123
    field- *(Convolvulus arvensis)*, 384
    fringed *(Polygonum cilinode)*, 124
    hedge- *(Calystegia sepium)*, 385
    Japanese *(Calystegia hederacea)*, 385
    low *(Calystegia spithamaea)*, 385
birch *(Betula)*, 82
    cherry *(Betula lenta)*, 82
    dwarf *(Betula glandulosa)*, 83
    gray *(Betula populifolia)*, 83
    Newfoundland dwarf *(Betula michauxii)*, 83
    paper- *(Betula papyrifera)*, 83
    red *(Betula nigra)*, 83
    river *(Betula nigra)*, 83
    swamp- *(Betula pumila)*, 83
    sweet *(Betula lenta)*, 82
    white *(Betula papyrifera)*, 83
    yellow *(Betula alleghaniensis)*, 82
birch family (BETULACEAE), 82
birch-leaved spiraea *(Spiraea betulifolia)*, 218
birdseye-speedwell *(Veronica persica)*, 454
birdsfoot-trefoil *(Lotus corniculatus, L. tenuis)*, 268
birdsfoot-violet *(Viola pedata)*, 142
bird-vetch *(Vicia cracca)*, 274
birthwort *(Aristolochia clematitis)*, 41
birthwort family (ARISTOLOCHIACEAE), 40
bishop's cap *(Mitella)*, 214
bishop-weed
    Atlantic mock *(Ptilimnium capillaceum)*, 356
    big mock *(Ptilimnium costatum)*, 356
    mock *(Ptilimnium)*, 356
    Ozark mock *(Ptilimnium nuttallii)*, 356
    riverbank mock *(Ptilimnium nodosum)*, 356
bitter-cress *(Cardamine)*, 171
    Appalachian *(Cardamine clematitis)*, 173
    dry-land *(Cardamine parviflora)*, 173
    hoary *(Cardamine hirsuta)*, 173
    northern *(Cardamine bellidifolia)*, 172
    Pennsylvania *(Cardamine pensylvanica)*, 173
    salt-marsh *(Cardamine longii)*, 172
    trailing *(Cardamine rotundifolia)*, 172
bitter dock *(Rumex obtusifolius)*, 115
bitter milkwort *(Polygala polygama)*, 329
bitternut-hickory *(Carya cordiformis)*, 73
bittersweet *(Celastrus)*, 308; *(Solanum dulcamara)*, 379
    American *(Celastrus scandens)*, 308
    oriental *(Celastrus orbiculatus)*, 308
black alder *(Alnus glutinosa)*, 84
black ash *(Fraxinus nigra)*, 437
blackberry
    bristly *(Rubus setosus)*, 232
    common *(Rubus allegheniensis)*, 233
    evergreen *(Rubus laciniatus)*, 235
    Himalayan *(Rubus discolor)*, 234
    Pennsylvania *(Rubus pensilvanicus)*, 234
    sand- *(Rubus cuneifolius)*, 233
    smooth *(Rubus canadensis)*, 233
    southern *(Rubus argutus)*, 234

blackberry-lily *(Belamcanda)*, 807
black bindweed *(Polygonum convolvulus)*, 123
black bulrush *(Scirpus atrovirens)*, 634
black-cap *(Rubus occidentalis)*, 230
black chokeberry *(Aronia melanocarpa)*, 245
black cohosh *(Cimicifuga racemosa)*, 46
black currant
    eastern *(Ribes americanum)*, 207
    garden *(Ribes nigrum)*, 206
    western *(Ribes hudsonianum)*, 207
black-eyed Susan *(Rudbeckia hirta)*, 495
black-footed quillwort *(Isoetes melanopoda)*, 4
black grass *(Juncus gerardii)*, 622
black gum *(Nyssa sylvatica)*, 306
black haw *(Viburnum prunifolium)*, 484
    southern *(Viburnum rufidulum)*, 484
black hawthorn *(Crataegus douglasii)*, 247
black hickory *(Carya texana)*, 74
black horehound *(Ballota)*, 424
black huckleberry *(Gaylussacia baccata)*, 194
black-jack oak *(Quercus marilandica)*, 79
black knapweed *(Centaurea nigra)*, 585
black locust *(Robinia pseudoacacia)*, 262
black maple *(Acer nigrum)*, 332
black medick *(Medicago lupulina)*, 272
black mustard *(Brassica nigra)*, 162
black nightshade *(Solanum nigrum)*, 379
black oak *(Quercus velutina)*, 80
black poplar *(Populus nigra)*, 153
black raspberry *(Rubus occidentalis)*, 230
blackseed-ricegrass *(Oryzopsis racemosa)*, 693
black snakeroot *(Cimicifuga racemosa)*, 46; *(Sanicula marilandica)*, 346
black spruce *(Picea mariana)*, 31
black-stalked spleenwort *(Asplenium resiliens)*, 17
black swallow-wort *(Vincetoxicum nigrum)*, 374
black tupelo *(Nyssa sylvatica)*, 306
black walnut *(Juglans nigra)*, 72
black willow *(Salix nigra)*, 154
bladder-campion *(Silene vulgaris)*, 109
bladder-fern *(Cystopteris)*, 20
    brittle *(Cystopteris fragilis)*, 20
    bulblet *(Cystopteris bulbifera)*, 20
    lowland *(Cystopteris protrusa)*, 20
bladder-nut *(Staphylea)*, 330
bladder-nut family (STAPHYLEACEAE), 330
bladder-senna *(Colutea arborescens)*, 264
bladderwort *(Utricularia)*, 462
    common *(Utricularia vulgaris)*, 464
    creeping *(Utricularia gibba)*, 464
    floating *(Utricularia radiata)*, 463
    inflated *(Utricularia inflata)*, 463
    lesser *(Utricularia minor)*, 463
    mixed *(Utricularia geminiscapa)*, 463
    naked *(Utricularia cornuta)*, 462
    northern *(Utricularia intermedia)*, 463
    resupinate *(Utricularia resupinata)*, 463
    rushlike *(Utricularia juncea)*, 462
    slender *(Utricularia subulata)*, 463
    spotted *(Utricularia purpurea)*, 463
bladderwort family (LENTIBULARIACEAE), 462
Blanchard's dewberry *(Rubus recurvicaulis)*, 232
bland sweet cicely *(Osmorhiza claytonii)*, 347
blanket-flower *(Gaillardia)*, 499
    common *(Gaillardia aristata)*, 499
    rosering- *(Gaillardia pulchella)*, 498

blazing star (*Liatris*), 574
  Appalachian (*Liatris turgida*), 575
  few-headed (*Liatris cylindracea*), 576
  grass-leaved (*Liatris graminifolia*), 576
  lacerate (*Liatris aspera*), 575
  northern (*Liatris scariosa*), 574
  northern plains (*Liatris ligulistylis*), 575
  piedmont (*Liatris regimontis*), 575
  plains (*Liatris squarrosa*), 576
  sessile (*Liatris spicata*), 576
  small-headed (*Liatris microcephala*), 576
  southern (*Liatris squarrulosa*), 575
  thick-spike (*Liatris pycnostachya*), 576
bleeding heart, wild (*Dicentra eximia*), 66
blessed thistle (*Cnicus*), 586
blinks (*Montia fontana*), 99
blister-plant (*Ranunculus acris*), 54
blite
  alkali- (*Chenopodium rubrum*), 87
  northeastern sea- (*Suaeda americana*), 93
  plains sea- (*Suaeda calceoliformis*), 93
  sea- (*Suaeda*), 93
  southern sea- (*Suaeda linearis*), 93
  strawberry- (*Chenopodium capitatum*), 87
  white sea- (*Suaeda maritima*), 93
bloodleaf (*Iresine*), 98
blood-milkwort (*Polygala sanguinea*), 329
bloodroot (*Sanguinaria*), 64
bloodwort family (HAEMODORACEAE), 781
blue alfalfa (*Medicago sativa* subsp. *sativa*), 271
blue ash (*Fraxinus quadrangulata*), 437
blue beech (*Carpinus caroliniana*), 82
bluebell (*Mertensia*), 396
  eastern (*Mertensia virginica*), 396
  English (*Scilla non-scripta*), 786
  northern (*Mertensia paniculata*), 396
  seaside- (*Mertensia maritima*), 396
blueberry
  common lowbush- (*Vaccinium angustifolium*), 192
  creeping (*Vaccinium crassifolium*), 192
  highbush- (*Vaccinium corymbosum*), 193
  hillside- (*Vaccinium pallidum*), 193
  southern (*Vaccinium tenellum*), 192
  southern highbush- (*Vaccinium elliottii*), 194
  velvet-leaf (*Vaccinium myrtilloides*), 193
blue-bottle (*Muscari racemosum*), 788
bluebunch wheatgrass (*Elytrigia spicata*), 727
blue buttons (*Knautia arvensis*), 488
blue cohosh (*Caulophyllum*), 62
blue curls (*Trichostema*), 409
blue-eyed grass (*Sisyrinchium*), 808
blue-eyed Mary (*Collinsia*), 447
  cordilleran (*Collinsia parviflora*), 447
  eastern (*Collinsia verna*), 447
  Ozarkian (*Collinsia violacea*), 446
blue false indigo (*Baptisia australis*), 258
blue grama (*Bouteloua gracilis*), 750
bluegrass (*Poa*), 701
  alpine (*Poa alpina*), 705
  arctic (*Poa arctica*), 701
  bulbous (*Poa bulbosa*), 702
  Canada (*Poa compressa*), 701
  Canby- (*Poa canbyi*), 704
  forest- (*Poa sylvestris*), 703
  inland- (*Poa interior*), 703
  Kentucky (*Poa pratensis*), 701

bluegrass (*Poa*) (*continued*)
  marsh- (*Poa paludigena*), 703
  plains- (*Poa arida*), 701
  rough (*Poa trivialis*), 702
  tall (*Poa chaixii*), 704
  wood- (*Poa nemoralis*), 704
blue-hearts (*Buchnera*), 452
blue-jack oak (*Quercus cinerea*), 78
bluejoint (*Calamagrostis canadensis*), 716
blueleaf-willow (*Salix myricoides*), 157
blue lettuce (*Lactuca pulchella*), 590
  tall (*Lactuca biennis*), 590
blue marsh-violet (*Viola cucullata*), 142
blue mustard (*Chorispora tenella*), 179
Blue-Ridge gentian (*Gentiana austromontana*), 365
Blue-Ridge phacelia (*Phacelia fimbriata*), 393
blue scorpion-grass (*Myosotis micrantha*), 397
bluestar (*Amsonia*), 368
  common (*Amsonia tabernaemontana*), 368
  Ozark (*Amsonia illustris*), 368
bluestem (*Andropogon*), 773
  big (*Andropogon gerardii*), 773
  little (*Schizachyrium scoparium*), 775
  sand- (*Andropogon hallii*), 773
  splitbeard- (*Andropogon ternarius*), 774
bluets (*Hedyotis caerulea*), 470
  mountain- (*Hedyotis michauxii*), 471
  small (*Hedyotis crassifolia*), 470
blue wax-weed (*Cuphea viscosissima*), 293
blue-weed (*Echium vulgare*), 401
blue wild rye (*Elymus glaucus*), 725
blunt cliff-fern (*Woodsia obtusa*), 21
bluntleaf-bedstraw (*Galium obtusum*), 475
bluntleaf-orchid (*Habenaria obtusata*), 820
bluntleaf-pondweed (*Potamogeton obtusifolius*), 611
blunt-leaved agalinis (*Agalinis obtusifolia*), 452
blunt-lobed grape-fern (*Botrychium oneidense*), 8
bluntscale-bulrush (*Scirpus smithii*), 632
blunt spike-rush (*Eleocharis ovata*), 642
blunt sweet cicely (*Osmorhiza depauperata*), 347
bog adder's-mouth (*Malaxis paludosa*), 825
bog-asphodel (*Narthecium*), 785
bog-bilberry (*Vaccinium uliginosum*), 191
bog-clubmoss (*Lycopodium inundatum*), 2
bog-orchid
  tall northern (*Habenaria hyperborea*), 819
  tall white (*Habenaria dilatata*), 820
bog-rosemary (*Andromeda*), 187
bog-stitchwort (*Stellaria alsine*), 104
bog-willow (*Salix pedicellaris*), 157
boneset (*Eupatorium perfoliatum*), 569
  false (*Kuhnia eupatorioides*), 573
borage (*Borago*), 402
borage family (BORAGINACEAE), 395
Boston-ivy (*Parthenocissus tricuspidata*), 323
bottlebrush-grass (*Elymus hystrix*), 726
bottle-gentian (*Gentiana andrewsii, G. clausa*), 365
bottomland-aster (*Aster ontarionis*), 549
bouncing bet (*Saponaria*), 111
boxelder (*Acer negundo*), 333
box-huckleberry (*Gaylussacia brachycera*), 195
boxwood family (BUXACEAE), 311
bracken family (DENNSTAEDTIACEAE), 12
bracken fern (*Pteridium*), 13
bracted orchid (*Habenaria viridis*), 819
Bradley's spleenwort (*Asplenium bradleyi*), 17

Brainerd-hawthorn (*Crataegus brainerdii*), 251
brake
　　cliff- (*Pellaea*), 16
　　purple cliff- (*Pellaea atropurpurea*), 16
　　rock- (*Cryptogramma*), 14
　　slender rock- (*Cryptogramma stelleri*), 14
　　smooth cliff- (*Pellaea glabella*), 16
bramble (*Rubus*), 229
branching broom-rape (*Orobanche ramosa*), 459
branching centaury (*Centaurium pulchellum*), 363
Braun's holly-fern (*Polystichum braunii*), 28
Brazilian vervain (*Verbena brasiliensis*), 403
Brazilian water-weed (*Egeria densa*), 606
bridal wreath (*Spiraea ×vanhouttei*), 218
brier, sensitive (*Mimosa quadrivalvis*), 255
bristle-fern (*Trichomanes*), 11
　　Appalachian (*Trichomanes boschianum*), 11
bristly aster (*Aster puniceus*), 547
bristly blackberry (*Rubus setosus*), 232
bristly crowfoot (*Ranunculus pensylvanicus*), 54
bristly greenbrier (*Smilax hispida*), 805
bristly locust (*Robinia hispida*), 262
bristly mallow (*Modiola caroliniana*), 134
bristly meadow-pitchers (*Rhexia aristosa*), 303
bristly ox-tongue (*Picris echioides*), 600
bristly rose (*Rosa acicularis*), 240
bristly sarsaparilla (*Aralia hispida*), 342
brittle bladder-fern (*Cystopteris fragilis*), 20
broad bean (*Vicia faba*), 272
broad beech-fern (*Thelypteris hexagonoptera*), 23
broadleaf-hawthorn (*Crataegus dilatata*), 252
broad-leaved toothwort (*Cardamine diphylla*), 171
broad-leaved twayblade (*Listera convallarioides*), 814
broad-leaved wild rye (*Elymus wiegandii*), 724
brome (*Bromus*), 720
　　barren (*Bromus sterilis*), 724
　　California (*Bromus carinatus*), 720
　　fringed (*Bromus ciliatus*), 721
　　smooth (*Bromus inermis*), 721
brome-grass (*Bromus*), 720
bromeliad family (BROMELIACEAE), 779
brook-alder (*Alnus maritima*), 84
brook-five-fingers (*Potentilla rivalis*), 223
brook-grass (*Catabrosa*), 706
brooklime (*Veronica beccabunga*), 455
broom, Scotch (*Cytisus scoparius*), 259
broomcorn millet (*Panicum miliaceum*), 753
broom-crowberry (*Corema*), 184
broom-rape (*Orobanche*), 458
　　branching (*Orobanche ramosa*), 459
　　clustered (*Orobanche fasciculata*), 459
　　lesser (*Orobanche minor*), 459
　　prairie (*Orobanche ludoviciana*), 458
broom-rape family (OROBANCHACEAE), 458
broom-sedge (*Andropogon virginicus*), 775
broomweed (*Amphiachyris dracunculoides*), 546
brown knapweed (*Centaurea jacea*), 584
brown mustard (*Brassica juncea*), 162
brown plumegrass (*Erianthus brevibarbis*), 770
buckbean (*Menyanthes*), 388; (*Thermopsis*), 259
　　piedmont- (*Thermopsis mollis*), 259
buckbean family (MENYANTHACEAE), 388
buckeye (*Aesculus*), 330
　　Georgia (*Aesculus sylvatica*), 331
　　Ohio (*Aesculus glabra*), 331
　　painted (*Aesculus sylvatica*), 331

buckeye (*Aesculus*) (*continued*)
　　red (*Aesculus pavia*), 331
　　sweet (*Aesculus flava*), 331
　　yellow (*Aesculus flava*), 331
buckhorn (*Plantago aristata*), 434
buckthorn (*Rhamnus*), 321
　　American alder- (*Rhamnus alnifolia*), 322
　　Carolina (*Rhamnus caroliniana*), 322
　　common (*Rhamnus cathartica*), 322
　　European alder- (*Rhamnus frangula*), 321
　　lance-leaved (*Rhamnus lanceolata*), 322
　　smooth southern (*Bumelia lycioides*), 198
　　southern (*Bumelia*), 198
　　woolly southern (*Bumelia lanuginosa*), 198
buckthorn family (RHAMNACEAE), 321
buckwheat (*Fagopyrum esculentum*), 124
　　false (*Polygonum scandens*), 124
buckwheat-vine (*Brunnichia cirrhosa*), 124
buffalo-berry (*Shepherdia argentea*), 288
buffalo-bur (*Solanum rostratum*), 380
buffalo-clover
　　annual (*Trifolium reflexum*), 270
　　running (*Trifolium stoloniferum*), 270
buffalo-currant (*Ribes odoratum*), 207
buffalo-grass (*Buchloe*), 751
buffalo-nut (*Pyrularia pubera*), 307
bugbane (*Cimicifuga*), 46
　　American (*Cimicifuga americana*), 46
　　false (*Trautvetteria*), 46
bugle (*Ajuga*), 408
　　carpet- (*Ajuga reptans*), 408
　　standing (*Ajuga genevensis*), 408
bugle-weed (*Lycopus*), 419
bugloss (*Anchusa*), 402
　　common (*Anchusa officinalis*), 402
　　showy (*Anchusa azurea*), 402
　　small (*Anchusa arvensis*), 402
bugseed (*Corispermum*), 90
bulbiliferous water-hemlock (*Cicuta bulbifera*), 355
bulbil-loosestrife (*Lysimachia terrestris*), 202
bulbil-saxifrage
　　leafy (*Saxifraga cernua*), 211
　　naked (*Saxifraga foliolosa*), 212
bulblet bladder-fern (*Cystopteris bulbifera*), 20
bulbous bluegrass (*Poa bulbosa*), 702
bulbous buttercup (*Ranunculus bulbosus*), 54
bull-bay (*Magnolia grandiflora*), 36
bull-nettle (*Cnidoscolus stimulosus*), 312
bull-thistle (*Cirsium vulgare*), 581
bulrush (*Scirpus*), 631
　　alkali- (*Scirpus maritimus*), 634
　　black (*Scirpus atrovirens*). 634
　　bluntscale- (*Scirpus smithii*), 632
　　hardstem- (*Scirpus acutus*), 633
　　river- (*Scirpus fluviatilis*), 634
　　salt-marsh (*Scirpus robustus*), 634
　　seaside- (*Scirpus rufus*), 633
　　sharpscale- (*Scirpus supinus*), 632
　　slender (*Scirpus heterochaetus*), 633
　　softstem- (*Scirpus validus*), 633
　　swamp- (*Scirpus etuberculatus*), 633
　　water- (*Scirpus subterminalis*), 632
bunchberry (*Cornus canadensis*), 305
bunch-flower (*Melanthium*), 784
bundleflower (*Desmanthus illinoensis*), 255

bur
  buffalo- (*Solanum rostratum*), 380
  cockle- (*Xanthium*), 516
  common cockle- (*Xanthium strumarium*), 516
  common sand- (*Cenchrus longispinus*), 769
  dune-sand- (*Cenchrus tribuloides*), 770
  Paraguay- (*Acanthospermum australe*), 511
  sand- (*Cenchrus*), 769
  spiny cockle- (*Xanthium spinosum*), 516
bur-buttercup (*Ranunculus testiculatus*), 58
bur-chervil (*Anthriscus caucalis*), 348
bur-clover
  downy (*Medicago minima*), 272
  smooth (*Medicago polymorpha*), 272
bur-cucumber (*Sicyos angulatus*), 151
burdock (*Arctium*), 579
  common (*Arctium minus*), 579
  cotton- (*Arctium tomentosum*), 579
  great (*Arctium lappa*), 579
bur-foxtail (*Setaria verticillata*), 766
bur-head (*Echinodorus*), 605
  creeping (*Echinodorus cordifolius*), 605
  little (*Echinodorus tenellus*), 605
  tall (*Echinodorus berteroi*), 606
burmannia family (BURMANNIACEAE), 812
bur-marigold (*Bidens cernua*), 506
  showy (*Bidens laevis*), 506
burnet (*Sanguisorba*), 237
  American (*Sanguisorba canadensis*), 237
  great (*Sanguisorba officinalis*), 237
  salad- (*Sanguisorba minor*), 237
burnet-saxifrage (*Pimpinella saxifraga*), 353
burning bush
  American (*Euonymus americanus*), 308
  winged (*Euonymus alatus*), 308
bur-oak (*Quercus macrocarpa*), 77
bur-reed (*Sparganium*), 776
  giant (*Sparganium eurycarpum*), 776
bur-reed family (SPARGANIACEAE), 776
bursage
  annual (*Ambrosia acanthicarpa*), 515
  skeletonleaf- (*Ambrosia tomentosa*), 514
bush-clover (*Lespedeza capitata*), 280
bush-honeysuckle (*Diervilla*), 478
Bush's poppy-mallow (*Callirhoe bushii*), 136
bushy-branched eupatorium (*Eupatorium cuneifolium*), 571
bushy wallflower (*Erysimum repandum*), 179
Butler's quillwort (*Isoetes butleri*), 4
butter-and-eggs (*Linaria vulgaris*), 447
buttercup (*Ranunculus*), 54
  bulbous (*Ranunculus bulbosus*), 54
  bur- (*Ranunculus testiculatus*), 58
  common (*Ranunculus acris*), 54
  creeping (*Ranunculus repens*), 54
  hispid (*Ranunculus hispidus*), 55
  Lappland (*Ranunculus lapponicus*), 58
  meadow- (*Ranunculus acris*), 54
  Ozark (*Ranunculus harveyi*), 56
  prairie- (*Ranunculus rhomboideus*), 56
  thick-root (*Ranunculus fascicularis*), 55
buttercup family (RANUNCULACEAE), 45
buttercup-pennywort (*Hydrocotyle ranunculoides*), 344
butterfly-bush (*Buddleja*), 434
  orange-eye (*Buddleja davidii*), 434
butterfly-bush family (BUDDLEJACEAE), 434
butterfly-dock (*Petasites hybridus*), 528

butterfly-pea (*Clitoria*), 285
butterfly-weed (*Asclepias tuberosa*), 370
butternut (*Juglans cinerea*), 72
butterwort (*Pinguicula*), 462
  violet (*Pinguicula vulgaris*), 462
buttonbush (*Cephalanthus occidentalis*), 471
buttonbush-dodder (*Cuscuta cephalanthi*), 388
buttonweed (*Diodia teres*), 471; (*Spermacoce glabra*), 472
  Virginia (*Diodia virginiana*), 472
buttonwood (*Platanus occidentalis*), 68

cabbage, skunk- (*Symplocarpus*), 615
cacao family (STERCULIACEAE), 131
cactus family (CACTACEAE), 86
caesalpinia family (CAESALPINIACEAE), 256
California brome (*Bromus carinatus*), 720
caltrop (*Kallstroemia*), 337
  small-flowered (*Kallstroemia parviflora*), 337
calypso (*Calypso bulbosa*), 826
camas (*Camassia*), 786
  death- (*Zigadenus*), 782
camphor-weed (*Heterotheca subaxillaris*), 531
campion (*Silene*), 108
  bladder- (*Silene vulgaris*), 109
  moss- (*Silene acaulis*), 110
  red (*Silene dioica*), 108
  starry (*Silene stellata*), 109
  white (*Silene latifolia*), 108; (*S. nivea*), 109
Canada bluegrass (*Poa compressa*), 701
Canada columbine (*Aquilegia canadensis*), 59
Canada hawkweed (*Hieracium kalmii*), 594
Canada mayflower (*Maianthemum canadense*), 800
Canada milk-vetch (*Astragalus canadensis*), 266
Canada plum (*Prunus nigra*), 243
Canada sanicle (*Sanicula canadensis*), 346
Canada thistle (*Cirsium arvense*), 583
Canada wild rye (*Elymus canadensis*), 725
Canadian anemone (*Anemone canadensis*), 49
Canadian phacelia (*Phacelia franklinii*), 392
Canadian tick-trefoil (*Desmodium canadense*), 277
canary-grass (*Phalaris canariensis*), 714
  reed (*Phalaris arundinacea*), 714
Canby-bluegrass (*Poa canbyi*), 704
Canby's water-dropwort (*Oxypolis canbyi*), 356
cancer-root (*Orobanche uniflora*), 458
cancerwort (*Kickxia*), 448
candytuft (*Iberis*), 167
cane (*Arundinaria*), 690
  giant (*Arundinaria gigantea*), 690
  maiden- (*Panicum hemitomon*), 751
cap, bishop's (*Mitella*), 214
caper family (CAPPARACEAE), 161
caper-spurge (*Euphorbia lathyris*), 316
caraway (*Carum carvi*), 348
cardinal-flower (*Lobelia cardinalis*), 469
careless weed (*Amaranthus palmeri*), 96
Carey's saxifrage (*Saxifraga careyana*), 213
Caribbean mitrewort (*Cynoctonum mitreola*), 361
Carolina allspice (*Calycanthus*), 38
Carolina ash (*Fraxinus caroliniana*), 437
Carolina buckthorn (*Rhamnus caroliniana*), 322
Carolina clover (*Trifolium carolinianum*), 270
Carolina crane's-bill (*Geranium carolinianum*), 339
Carolina foxtail (*Alopecurus carolinianus*), 720
Carolina hemlock (*Tsuga caroliniana*), 32
Carolina larkspur (*Delphinium carolinianum*), 48

Carolina lily (*Lilium michauxii*), 792
Carolina lovegrass (*Eragrostis pectinacea*), 739
Carolina moonseed (*Cocculus carolinus*), 63
Carolina saxifrage (*Saxifraga caroliniana*), 212
Carolina shagbark-hickory (*Carya carolinae-septentrionalis*), 74
Carolina whipgrass (*Scleria pauciflora*), 659
Carolina wild indigo (*Baptisia cinerea*), 259
Carolina willow (*Salix caroliniana*), 154
carpet-bugle (*Ajuga reptans*), 408
carpet-weed (*Mollugo*), 100
carpet-weed family (MOLLUGINACEAE), 100
carrot, wild (*Daucus carota*), 350
carrot family (APIACEAE), 344
cashew family (ANACARDIACEAE), 334
castor-bean (*Ricinus communis*), 311
catalpa (*Catalpa*), 461
    northern (*Catalpa speciosa*), 461
    southern (*Catalpa bignonioides*), 462
catbird-grape (*Vitis palmata*), 324
catbrier (*Smilax*), 804
catbrier family (SMILACACEAE), 804
catchfly (*Silene*), 108; (*S. noctiflora*), 108
    German (*Lychnis viscaria*), 107
    sweet william (*Silene armeria*), 110
catchfly-grass (*Leersia lenticularis*), 690
catchweed (*Asperugo*), 396
cat-grape (*Vitis palmata*), 324
catnip (*Nepeta cataria*), 421
catnip giant-hyssop (*Agastache nepetoides*), 421
cat's-ear (*Hypochaeris*), 600
    smooth (*Hypochaeris glabra*), 600
    spotted (*Hypochaeris radicata*), 600
cat-tail (*Typha*), 778
    common (*Typha latifolia*), 778
    narrow-leaved (*Typha angustifolia*), 778
    southern (*Typha domingensis*), 778
cat-tail family (TYPHACEAE), 778
cave-goldenrod (*Solidago albopilosa*), 536
cedar
    Atlantic white (*Chamaecyparis thyoides*), 35
    eastern red (*Juniperus virginiana*), 35
    northern ground- (*Lycopodium complanatum*), 3
    northern white (*Thuja occidentalis*), 35
    southern ground- (*Lycopodium digitatum*), 3
    wiry ground- (*Lycopodium tristachyum*), 3
cedar-glade prairie-clover (*Dalea foliosa*), 282
celandine (*Chelidonium*), 64
    lesser (*Ranunculus ficaria*), 58
celandine-poppy (*Stylophorum*), 64
celery (*Apium graveolens*), 351
    water- (*Vallisneria americana*), 607
centaury (*Centaurium*), 363
    branching (*Centaurium pulchellum*), 363
    forking (*Centaurium erythraea*), 363
    spike- (*Centaurium spicatum*), 363
century-plants (*Agave*), 804
chaffseed (*Schwalbea*), 456
chaffweed (*Centunculus*), 203
chain-fern (*Woodwardia*), 30
    netted (*Woodwardia areolata*), 30
    Virginia (*Woodwardia virginica*), 30
chamomile (*Anthemis*), 516; (*Matricaria recutita*), 519
    corn- (*Anthemis arvensis*), 516
    scentless (*Matricaria maritima*), 519
    yellow (*Anthemis tinctoria*), 516

charlock (*Sinapis arvensis*), 162
    jointed (*Raphanus raphanistrum*), 163
chaste tree (*Vitex agnus-castus*), 407
cheat (*Bromus*), 720; (*B. secalinus*), 723
checkerberry (*Gaultheria procumbens*), 190
cheeses (*Malva neglecta*), 135
cherry
    choke- (*Prunus virginiana*), 241
    clammy ground- (*Physalis heterophylla*), 376
    downy ground- (*Physalis pubescens*), 378
    European bird- (*Prunus padus*), 241
    ground- (*Physalis*), 376; (*Prunus fruticosa*), 242
    longleaf ground- (*Physalis longifolia*), 377
    Mahaleb- (*Prunus mahaleb*), 241
    Ozark ground- (*Physalis missouriensis*), 378
    pie- (*Prunus cerasus*), 241
    pin- (*Prunus pensylvanica*), 241
    plains-sandhill ground- (*Physalis hispida*), 377
    prairie ground- (*Physalis pumila*), 377
    sand- (*Prunus pumila*), 241
    sour (*Prunus cerasus*), 241
    sweet (*Prunus avium*), 241
    Virginia ground- (*Physalis virginiana*), 377
    white-flowered ground- (*Leucophysalis grandiflora*), 376
    wild black (*Prunus serotina*), 240
cherrybark-oak (*Quercus pagoda*), 80
cherry birch (*Betula lenta*), 82
chervil (*Anthriscus cerefolium*), 347
    bur- (*Anthriscus caucalis*), 348
    southern (*Chaerophyllum tainturieri*), 349
    spreading (*Chaerophyllum procumbens*), 348
    wild (*Anthriscus sylvestris*), 347
chess (*Bromus*), 720; (*B. secalinus*), 723
    downy (*Bromus tectorum*), 723
    foxtail- (*Bromus rubens*), 723
    hairy (*Bromus commutatus*), 723
    Japanese (*Bromus japonicus*), 724
    rattlesnake- (*Bromus briziformis*), 722
    soft (*Bromus hordeaceus*), 722
chestnut (*Castanea dentata*), 76
    horse- (*Aesculus hippocastanum*), 330
    water- (*Trapa*), 293
chestnut oak
    rock (*Quercus prinus*), 78
    swamp (*Quercus michauxii*), 77
Chickasaw-plum (*Prunus angustifolia*), 242
chicken-grape (*Vitis vulpina*), 325
chick-pea milk-vetch (*Astragalus cicer*), 266
chickweed (*Stellaria*), 103
    clammy (*Cerastium viscosum*), 102
    common (*Stellaria media*), 103
    field- (*Cerastium arvense*), 102
    forked (*Paronychia canadensis, P. fastigiata*), 106
    giant (*Stellaria aquatica*), 103
    jagged (*Holosteum*), 101
    mouse-ear (*Cerastium*), 102
    star- (*Stellaria pubera*), 103
chicory (*Cichorium intybus*), 599
childing pink (*Petrorhagia prolifera*), 112
Chile-tarweed (*Madia sativa*), 510
chinaberry (*Melia azedarach*), 335
Chinese beautyberry (*Callicarpa dichotoma*), 407
Chinese cup-grass (*Eriochloa villosa*), 762
Chinese-lantern plant (*Physalis alkekengi*), 376
Chinese lespedeza (*Lespedeza cuneata*), 281
Chinese mustard (*Brassica juncea*), 162

Chinese parasol-tree (*Firmiana simplex*), 131
chinquapin (*Castanea pumila*), 76
   water- (*Nelumbo lutea*), 42
chinquapin-oak (*Quercus prinoides*), 78
chokeberry (*Aronia*), 245
   black (*Aronia melanocarpa*), 245
   red (*Aronia arbutifolia*), 245
chokecherry (*Prunus virginiana*), 241
Christmas-fern (*Polystichum acrostichoides*), 28
Christmas-mistletoe, American (*Phoradendron serotinum*), 307
Christmas-mistletoe family (VISCACEAE), 307
chrysanthemum (*Chrysanthemum*), 518
   corn- (*Chrysanthemum segetum*), 518
churchmouse-three-awn (*Aristida dichotoma*), 733
cicely, sweet (*Osmorhiza*), 347
cilantro (*Coriandrum sativum*), 349
cinnamon-fern (*Osmunda cinnamomea*), 10
cinnamon-rose (*Rosa majalis*), 240
cinnamon-vine (*Dioscorea polystachya*), 807
cinquefoil (*Potentilla*), 222
city-goosefoot (*Chenopodium urbicum*), 88
clammy chickweed (*Cerastium viscosum*), 102
clammy cudweed (*Gnaphalium macounii*), 565
clammy ground-cherry (*Physalis heterophylla*), 376
clammy locust (*Robinia viscosa*), 262
clammy-weed (*Polanisia*), 161
clary (*Salvia sclarea*), 430
clasping aster (*Aster patens*), 553
clasping dogbane (*Apocynum sibiricum*), 369
clasping heart-leaved aster (*Aster undulatus*), 553
clasping milkweed (*Asclepias amplexicaulis*), 371
clasping mullein (*Verbascum phlomoides*), 442
clasping pepperweed (*Lepidium perfoliatum*), 164
claw-saxifrage (*Saxifraga michauxii*), 213
clearweed (*Pilea*), 71
cleavers (*Galium*), 473; (*G. aparine*), 477
clematis (*Clematis*), 51
   purple (*Clematis occidentalis*), 51
   yam-leaved (*Clematis terniflora*), 51
clethra family (CLETHRACEAE), 183
cliff-brake (*Pellaea*), 16
   purple (*Pellaea atropurpurea*), 16
   smooth (*Pellaea glabella*), 16
cliff-fern (*Woodsia*), 20
   alpine (*Woodsia alpina*), 20
   blunt (*Woodsia obtusa*), 21
   mountain (*Woodsia scopulina*), 21
   rusty (*Woodsia ilvensis*), 21
   smooth (*Woodsia glabella*), 20
   western (*Woodsia oregana*), 21
climbing fern (*Lygodium*), 11
climbing hempweed (*Mikania*), 573
climbing hydrangea (*Decumaria barbara*), 204
climbing prairie-rose (*Rosa setigera*), 237
Clinton's wood-fern (*Dryopteris clintoniana*), 26
cloak-fern (*Notholaena*), 16
   powdery (*Notholaena dealbata*), 16
closed-flowered alum-root (*Heuchera longiflora*), 216
cloudberry (*Rubus chamaemorus*), 229
clover (*Trifolium*), 269
   alsike (*Trifolium hybridum*), 270
   annual buffalo- (*Trifolium reflexum*), 270
   bush- (*Lespedeza capitata*), 280
   Carolina (*Trifolium carolinianum*), 270
   cedar-glade prairie- (*Dalea foliosa*), 282

clover (*Trifolium*) (continued)
   crimson (*Trifolium incarnatum*), 269
   downy bur- (*Medicago minima*), 272
   downy prairie- (*Dalea villosa*), 282
   European water- (*Marsilea quadrifolia*), 30
   hairy water- (*Marsilea vestita*), 30
   Japanese (*Lespedeza striata*), 281
   Korean (*Lespedeza stipulacea*), 281
   little hop- (*Trifolium dubium*), 271
   owl- (*Orthocarpus*), 458
   palmate hop- (*Trifolium aureum*), 271
   Persian (*Trifolium resupinatum*), 270
   pinnate hop- (*Trifolium campestre*), 271
   purple prairie- (*Dalea purpurea*), 282
   rabbitfoot- (*Trifolium arvense*), 269
   red (*Trifolium pratense*), 269
   rounded-headed prairie- (*Dalea multiflora*), 282
   running buffalo- (*Trifolium stoloniferum*), 270
   shale-barren (*Trifolium virginicum*), 270
   smooth bur- (*Medicago polymorpha*), 272
   Spanish (*Lotus purshianus*), 268
   stinking (*Cleome serrulata*), 161
   strawberry- (*Trifolium fragiferum*), 269
   sweet (*Melilotus*), 271
   water- (*Marsilea*), 30
   white (*Trifolium repens*), 270
   white prairie- (*Dalea candida*), 282
   white sweet (*Melilotus albus*), 271
   yellow sweet (*Melilotus officinalis*), 271
   zigzag (*Trifolium medium*), 269
clubmoss (*Lycopodium*), 1
   alpine (*Lycopodium alpinum*), 3
   bog- (*Lycopodium inundatum*), 2
   featherstem- (*Lycopodium prostratum*), 1
   fir- (*Lycopodium selago*), 1
   foxtail- (*Lycopodium alopecuroides*), 1
   juniper- (*Lycopodium sabinifolium*), 3
   rock- (*Lycopodium porophilum*), 1
   shining (*Lycopodium lucidulum*), 1
   Sitka (*Lycopodium sitchense*), 3
   slender (*Lycopodium carolinianum*), 2
   southern (*Lycopodium appressum*), 2
   stiff (*Lycopodium annotinum*), 2
clubmoss family (LYCOPODIACEAE), 1
club-spur orchid (*Habenaria clavellata*), 819
clustered bellflower (*Campanula glomerata*), 465
clustered broom-rape (*Orobanche fasciculata*), 459
clustered poppy-mallow (*Callirhoe triangulata*), 136
cluster-leaf tick-trefoil (*Desmodium glutinosum*), 276
cluster-sanicle (*Sanicula gregaria*), 346
coarse smartweed (*Polygonum robustius*), 119
coastal azalea (*Rhododendron atlanticum*), 185
coastal mannagrass (*Glyceria obtusa*), 707
coastal plain dewberry (*Rubus trivialis*), 231
coastal plain flat-topped goldenrod (*Euthamia tenuifolia*), 545
coastal plain gentian (*Gentiana catesbaei*), 364
coastal plain mermaid-weed (*Proserpinaca pectinata*), 290
coastal plain rattlebox (*Crotalaria purshii*), 260
coastal plain serviceberry (*Amelanchier obovalis*), 254
coastal plain sida (*Sida elliottii*), 137
coastal plain spearwort (*Ranunculus texensis*), 57
coastal plain tickseed-sunflower (*Bidens mitis*), 508
coastal plain water-purslane (*Ludwigia brevipes*), 294
coastal plain water-willow (*Justicia ovata*), 460
coastal plain yellow flax (*Linum floridanum*), 326

coastal silver-weed (*Potentilla pacifica*), 225
coastal swamp-goldenrod (*Solidago elliottii*), 541
coastal tall thistle (*Cirsium nuttallii*), 583
coast white alder (*Clethra alnifolia*), 183
cockle
    corn- (*Agrostemma*), 107
    red (*Silene dioica*), 108
    sticky (*Silene noctiflora*), 108
    white (*Silene latifolia*), 108
cocklebur (*Xanthium*), 516
    common (*Xanthium strumarium*), 516
    spiny (*Xanthium spinosum*), 516
cockspur-thorn (*Crataegus crus-galli*), 248
coffee-senna (*Senna occidentalis*), 258
coffee-tree, Kentucky (*Gymnocladus dioicus*), 256
cohosh
    black (*Cimicifuga racemosa*), 46
    blue (*Caulophyllum*), 62
    summer- (*Cimicifuga americana*), 46
colic-root (*Aletris*), 785; (*Dioscorea villosa*), 806
collared dodder (*Cuscuta indecora*), 387
colonial Barbara's buttons (*Marshallia trinervia*), 509
colonial dwarf dandelion (*Krigia dandelion*), 598
coltsfoot (*Tussilago*), 528
    arrowhead sweet- (*Petasites sagittatus*), 529
    northern sweet- (*Petasites frigidus*), 528
columbine (*Aquilegia*), 59
    Canada (*Aquilegia canadensis*), 59
    European (*Aquilegia vulgaris*), 59
columbo, American (*Frasera caroliniensis*), 367
columnar coneflower (*Ratibida columnifera*), 498
comb-five-fingers (*Potentilla flabelliformis*), 224
comfrey (*Symphytum*), 402
    common (*Symphytum officinale*), 402
    rough (*Symphytum asperum*), 402
    wild (*Cynoglossum virginianum*), 399
common agalinis (*Agalinis tenuifolia*), 452
common agrimony (*Agrimonia gryposepala*), 236
common alum-root (*Heuchera americana*), 215
common angle-pod (*Matelea gonocarpa*), 375
common arrow-grass (*Triglochin maritima*), 608
common arrow-head (*Sagittaria latifolia*), 604
common balm (*Melissa officinalis*), 412
common barberry (*Berberis vulgaris*), 62
common blackberry (*Rubus allegheniensis*), 233
common bladderwort (*Utricularia vulgaris*), 464
common blanket-flower (*Gaillardia aristata*), 499
common blue heart-leaved aster (*Aster cordifolius*), 552
common bluestar (*Amsonia tabernaemontana*), 368
common buckthorn (*Rhamnus cathartica*), 322
common bugloss (*Anchusa officinalis*), 402
common burdock (*Arctium minus*), 579
common buttercup (*Ranunculus acris*), 54
common cat-tail (*Typha latifolia*), 778
common chickweed (*Stellaria media*), 103
common cocklebur (*Xanthium strumarium*), 516
common comfrey (*Symphytum officinale*), 402
common cosmos (*Cosmos bipinnatus*), 508
common cottonweed (*Froelichia floridana*), 98
common dandelion (*Taraxacum officinale*), 599
common day-flower (*Commelina communis*), 620
common dodder (*Cuscuta gronovii*), 387
common eel-grass (*Zostera marina*), 614
common elder (*Sambucus canadensis*), 485
common enchanter's nightshade (*Circaea lutetiana*), 303
common evening-primrose (*Oenothera biennis*), 299

common flat-topped goldenrod (*Euthamia graminifolia*), 545
common flax (*Linum usitatissimum*), 326
common foxglove (*Digitalis purpurea*), 449
common golden alexanders (*Zizia aurea*), 353
common goldenrod (*Solidago canadensis*), 542
common ground-nut (*Apios americana*), 286
common groundsel (*Senecio vulgaris*), 526
common hop-tree (*Ptelea trifoliata*), 336
common horsetail (*Equisetum arvense*), 7
common juniper (*Juniperus communis*), 35
common lilac (*Syringa vulgaris*), 435
common lowbush-blueberry (*Vaccinium angustifolium*), 192
common mallow (*Malva neglecta*), 135
common marsh-pink (*Sabatia angularis*), 362
common mermaid-weed (*Proserpinaca palustris*), 290
common milkweed (*Asclepias syriaca*), 371
common morning-glory (*Ipomoea purpurea*), 386
common mountain-holly (*Nemopanthus mucronatus*), 310
common mullein (*Verbascum thapsus*), 443
common petunia (*Petunia Xhybrida*), 383
common plantain (*Plantago major*), 432
common poison-ivy (*Toxicodendron radicans*), 335
common polypody (*Polypodium virginianum*), 12
common prickly ash (*Zanthoxylum americanum*), 336
common privet (*Ligustrum vulgare*), 435
common purslane (*Portulaca oleracea*), 98
common quickweed (*Galinsoga quadriradiata*), 509
common ragweed (*Ambrosia artemisiifolia*), 515
common reed (*Phragmites australis*), 732
common rue (*Ruta graveolens*), 336
common sandbur (*Cenchrus longispinus*), 769
common scouring rush (*Equisetum hyemale*), 5
common snailseed-pondweed (*Potamogeton diversifolius*), 609
common snapdragon (*Antirrhinum majus*), 448
common sneezeweed (*Helenium autumnale*), 499
common sow-thistle (*Sonchus oleraceus*), 591
common speedwell (*Veronica officinalis*), 454
common star-grass (*Hypoxis hirsuta*), 802
common stitchwort (*Stellaria graminea*), 104
common St. John's-wort (*Hypericum perforatum*), 128
common sunflower (*Helianthus annuus*), 489
common tansy (*Tanacetum vulgare*), 519
common teasel (*Dipsacus sylvestris*), 487
common threesquare (*Scirpus pungens*), 632
common velvet-grass (*Holcus lanatus*), 712
common vervain (*Verbena hastata*), 404
common vetch (*Vicia sativa*), 272
common water-dropwort (*Oxypolis rigidior*), 356
common water-hemlock (*Cicuta maculata*), 355
common water-milfoil (*Myriophyllum sibiricum*), 290
common water-purslane (*Ludwigia palustris*), 296
common water-weed (*Elodea canadensis*), 606
common wheat (*Triticum aestivum*), 730
common white heart-leaved aster (*Aster divaricatus*), 555
common woodreed (*Cinna arundinacea*), 719
common wormwood (*Artemisia absinthium*), 522
common yarrow (*Achillea millefolium*), 517
common yellow-cress (*Rorippa palustris*), 178
common yellow flax (*Linum medium*), 327
common yellow monkey-flower (*Mimulus guttatus*), 440
common yellow wood-sorrel (*Oxalis stricta*), 337
compass-plant (*Silphium laciniatum*), 512
coneflower (*Echinacea*), 497; (*Ratibida*), 498; (*Rudbeckia*), 495; (*Rudbeckia laciniata*), 497
    Appalachian (*Echinacea laevigata*), 497

coneflower (*continued*)
   columnar (*Ratibida columnifera*), 498
   eastern (*Rudbeckia fulgida*), 495
   globular (*Ratibida pinnata*), 498
   Missouri (*Rudbeckia missouriensis*), 495
   piney-woods (*Rudbeckia heliopsidis*), 496
   prairie- (*Echinacea pallida*), 498
   purple (*Echinacea purpurea*), 497
   sweet (*Rudbeckia subtomentosa*), 496
   three-lobed (*Rudbeckia triloba*), 496
conifer cotton-grass (*Eriophorum tenellum*), 637
coontail (*Ceratophyllum*), 44
Cooper milk-vetch (*Astragalus neglectus*), 264
copper-iris (*Iris fulva*), 810
copperleaf (*Acalypha*), 314
   rhombic (*Acalypha rhomboidea*), 314
   rough-pod (*Acalypha ostryifolia*), 314
   short-stalk (*Acalypha gracilens*), 314
   two-seed (*Acalypha deamii*), 314
   Virginia (*Acalypha virginica*), 314
coppery St. John's-wort (*Hypericum denticulatum*), 129
coral beads (*Cocculus carolinus*), 63
coralberry (*Symphoricarpos orbiculatus*), 481
coral-root (*Corallorhiza*), 827
   autumn (*Corallorhiza odontorhiza*), 827
   crested (*Hexalectris spicata*), 824
   northern (*Corallorhiza trifida*), 827
   spotted (*Corallorhiza maculata*), 827
   spring (*Corallorhiza wisteriana*), 826
   striped (*Corallorhiza striata*), 827
cord-grass (*Spartina*), 749
   big (*Spartina cynosuroides*), 749
   prairie (*Spartina pectinata*), 749
   salt-meadow (*Spartina patens*), 750
   smooth (*Spartina alterniflora*), 749
cordilleran blue-eyed Mary (*Collinsia parviflora*), 447
coriander (*Coriandrum sativum*), 349
corn, squirrel (*Dicentra canadensis*), 66
corn-chamomile (*Anthemis arvensis*), 516
corn-chrysanthemum (*Chrysanthemum segetum*), 518
corn-cockle (*Agrostemma*), 107
corn-crowfoot (*Ranunculus arvensis*), 55
cornel
   dwarf (*Cornus canadensis*), 305
   Swedish (*Cornus suecica*), 306
cornflower (*Centaurea cyanus*), 584
corn-gromwell (*Lithospermum arvense*), 400
Cornish heath (*Erica vagans*), 190
corn-leaved water-dropwort (*Oxypolis canbyi*), 356
corn-poppy (*Papaver rhoeas*), 65
corn-salad (*Valerianella*), 486
   European (*Valerianella locusta*), 486
   Great Lakes (*Valerianella chenopodifolia*), 486
corn-speedwell (*Veronica arvensis*), 453
corydalis (*Corydalis*), 67
   golden (*Corydalis aurea*), 67
   short-spurred (*Corydalis flavula*), 67
   slender (*Corydalis micrantha*), 67
   tall (*Corydalis sempervirens*), 67
cosmos (*Cosmos*), 508
   common (*Cosmos bipinnatus*), 508
   orange (*Cosmos sulphureus*), 509
costmary (*Chrysanthemum balsamita*), 519
cotton (*Gossypium*), 131
   lavender (*Santolina chamaecyparissus*), 518
   upland- (*Gossypium hirsutum*), 131

cotton-burdock (*Arctium tomentosum*), 579
cotton-grass (*Eriophorum*), 637
   conifer (*Eriophorum tenellum*), 637
   dark-scale (*Eriophorum viridicarinatum*), 638
   rusty (*Eriophorum chamissonis*), 637
   short-anthered (*Eriophorum brachyantherum*), 637
   slender (*Eriophorum gracile*), 638
   tawny (*Eriophorum virginicum*), 638
   thin-scale (*Eriophorum angustifolium*), 638
   tussock (*Eriophorum vaginatum*), 637
cotton-gum (*Nyssa aquatica*), 306
cottonweed (*Froelichia*), 98
   common (*Froelichia floridana*), 98
   slender (*Froelichia gracilis*), 98
cottonwood (*Populus deltoides*), 153
   swamp- (*Populus heterophylla*), 152
cow-herb (*Vaccaria*), 111
cow-parsnip (*Heracleum lanatum*), 357
cowslip (*Caltha*), 45
cow-wheat (*Melampyrum*), 457
crab
   prairie- (*Pyrus ioensis*), 245
   southern (*Pyrus angustifolia*), 244
   sweet (*Pyrus coronaria*), 244
   Toringo (*Pyrus sieboldii*), 244
crab-apple (*Pyrus*), 243
crab-grass (*Digitaria*), 767
   dwarf (*Digitaria serotina*), 767
   northern (*Digitaria sanguinalis*), 767
   smooth (*Digitaria ischaemum*), 767
   southern (*Digitaria ciliaris*), 768
crack-willow (*Salix fragilis*), 155
cranberry (*Vaccinium macrocarpon*), 194
   mountain- (*Vaccinium vitis-idaea*), 192
   small (*Vaccinium oxycoccos*), 194
   southern mountain- (*Vaccinium erythrocarpum*), 194
crane-fly orchid (*Tipularia discolor*), 826
crane's-bill (*Geranium*), 338
   Carolina (*Geranium carolinianum*), 339
   dove's-foot (*Geranium molle*), 340
   longstalk (*Geranium columbinum*), 339
   Siberian (*Geranium sibiricum*), 339
   small-flowered (*Geranium pusillum*), 340
crawling phlox (*Phlox stolonifera*), 390
cream-colored avens (*Geum virginianum*), 227
creamy violet (*Viola striata*), 147
creeper
   scarlet (*Ipomoea coccinea*), 386
   trumpet (*Campsis*), 462
   Virginia (*Parthenocissus quinquefolia*), 323
creeping bent (*Agrostis stolonifera*), 718
creeping bladderwort (*Utricularia gibba*), 464
creeping blueberry (*Vaccinium crassifolium*), 192
creeping bur-head (*Echinodorus cordifolius*), 605
creeping buttercup (*Ranunculus repens*), 54
creeping cucumber (*Melothria pendula*), 150
creeping day-flower (*Commelina diffusa*), 620
creeping five-fingers (*Potentilla reptans*), 222
creeping juniper (*Juniperus horizontalis*), 35
creeping lettuce (*Ixeris stolonifera*), 591
creeping rush (*Juncus repens*), 626
creeping snowberry (*Gaultheria hispidula*), 190
creeping spearwort (*Ranunculus flammula*), 57
creeping speedwell (*Veronica filiformis*), 454
creeping water-primrose (*Ludwigia peploides*), 294
creeping yellow-cress (*Rorippa sylvestris*), 177

creeping yellow wood-sorrel (*Oxalis corniculata*), 337

creosote-bush family (*ZYGOPHYLLACEAE*), 337

cress

    Appalachian bitter- (*Cardamine clematitis*), 173

    bitter- (*Cardamine*), 171

    common yellow- (*Rorippa palustris*), 178

    creeping yellow- (*Rorippa sylvestris*), 177

    dry-land bitter- (*Cardamine parviflora*), 173

    early winter- (*Barbarea verna*), 178

    field- (*Lepidium campestre*), 164

    field penny- (*Thlaspi arvense*), 166

    field yellow- (*Rorippa austriaca*), 177

    garden- (*Lepidium sativum*), 165

    globe-podded hoary (*Cardaria pubescens*), 166

    heart-podded hoary (*Cardaria draba*), 166

    hoary (*Cardaria*), 166

    hoary bitter- (*Cardamine hirsuta*), 173

    Indian yellow- (*Rorippa indica*), 177

    lake- (*Armoracia lacustris*), 170

    lens-podded hoary (*Cardaria chalepensis*), 166

    mouse-ear (*Arabidopsis thaliana*), 181

    northern bitter- (*Cardamine bellidifolia*), 172

    northern winter- (*Barbarea orthoceras*), 178

    Pennsylvania bitter- (*Cardamine pensylvanica*), 173

    penny- (*Thlaspi*), 166

    pink spring- (*Cardamine douglassii*), 172

    rock- (*Arabis*), 174

    salt-marsh bitter- (*Cardamine longii*), 172

    shepherd's (*Teesdalia*), 166

    southern yellow- (*Rorippa sessiliflora*), 178

    spring- (*Cardamine rhomboidea*), 172

    swine- (*Coronopus didymus*), 165

    thoroughwort penny- (*Thlaspi perfoliatum*), 166

    trailing bitter- (*Cardamine rotundifolia*), 172

    wart- (*Coronopus*), 165

    water- (*Rorippa nasturtium-aquaticum*), 177

    western yellow- (*Rorippa sinuata*), 177

    winter- (*Barbarea*), 178

crested coral-root (*Hexalectris spicata*), 824

crested fringed orchid (*Habenaria cristata*), 821

crested wheatgrass (*Agropyron*), 730

crested wood-fern (*Dryopteris cristata*), 25

crimson clover (*Trifolium incarnatum*), 269

crocus, autumn- (*Colchicum autumnale*), 794

cross-leaved heath (*Erica tetralix*), 190

cross-vine (*Bignonia*), 462

croton

    one-seed (*Croton monanthogynus*), 312

    Texas (*Croton texensis*), 313

    tooth-leaved (*Croton glandulosus*), 312

    woolly (*Croton capitatus*), 312

crowberry (*Empetrum*), 183

    broom- (*Corema*), 184

    red (*Empetrum rubrum*), 183

crowberry family (EMPETRACEAE), 183

crowfoot (*Ranunculus*), 54

    Allegheny- (*Ranunculus allegheniensis*), 56

    bristly (*Ranunculus pensylvanicus*), 54

    corn- (*Ranunculus arvensis*), 55

    cursed (*Ranunculus sceleratus*), 57

    hooked (*Ranunculus recurvatus*), 54

    ivy-leaved (*Ranunculus hederaceus*), 58

    marsh- (*Ranunculus macounii*), 54

    seaside- (*Ranunculus cymbalaria*), 58

    small-flowered (*Ranunculus abortivus, R. micranthus*), 56

crowfoot (*Ranunculus*) (*continued*)

    stickseed- (*Ranunculus parviflorus*), 56

    white water- (*Ranunculus longirostris, R. subrigidus, R. trichophyllus*), 58

    yellow water- (*Ranunculus flabellaris, R. gmelinii*), 57

crowfoot-grass (*Dactyloctenium aegyptium*), 747

crownbeard, golden (*Verbesina encelioides*), 501

crown-vetch (*Coronilla varia*), 268

crow-poison (*Zigadenus densus*), 783

cuckoo-flower (*Cardamine pratensis*), 173

cucumber

    bur- (*Sicyos angulatus*), 151

    creeping (*Melothria pendula*), 150

    wild (*Echinocystis lobata*), 150

cucumber-root, Indian (*Medeola*), 797

cucumber-tree (*Magnolia acuminata*), 36

cudweed (*Gnaphalium*), 564

    alpine (*Gnaphalium supinum*), 564

    clammy (*Gnaphalium macounii*), 565

    fragrant (*Gnaphalium obtusifolium*), 565

    low (*Gnaphalium uliginosum*), 565

    purple (*Gnaphalium purpureum*), 565

    woodland- (*Gnaphalium sylvaticum*), 564

cultivated onion (*Allium cepa*), 790

culver's root (*Veronicastrum*), 453

Cumberland azalea (*Rhododendron bakeri*), 186

Cumberland false foxglove (*Aureolaria patula*), 450

Cumberland mock-orange (*Philadelphus hirsutus*), 204

Cumberland rosin-weed (*Silphium brachiatum*), 512

Cumberland spurge (*Euphorbia mercurialina*), 318

cup-grass (*Eriochloa*), 762

    Chinese (*Eriochloa villosa*), 762

    prairie (*Eriochloa contracta*), 762

cup-plant (*Silphium perfoliatum*), 512

cupseed (*Calycocarpum*), 63

curly dock (*Rumex crispus*), 114

curly grass (*Schizaea*), 11

curly grass family (SCHIZAEACEAE), 11

curly pondweed (*Potamogeton crispus*), 609

curly-top gum-weed (*Grindelia squarrosa*), 532

currant

    buffalo- (*Ribes odoratum*), 207

    eastern black (*Ribes americanum*), 207

    garden black (*Ribes nigrum*), 206

    garden red (*Ribes sativum*), 207

    skunk- (*Ribes glandulosum*), 206

    spiny swamp- (*Ribes lacustre*), 206

    swamp red (*Ribes triste*), 207

    western black (*Ribes hudsonianum*), 207

cursed crowfoot (*Ranunculus sceleratus*), 57

cushion baby's breath (*Gypsophila muralis*), 111

custard-apple family (ANNONACEAE), 38

cut-grass, rice (*Leersia oryzoides*), 690

cut-leaf evening-primrose (*Oenothera laciniata*), 300

cut-leaf ironplant (*Haplopappus spinulosus*), 529

cut-leaf teasel (*Dipsacus laciniatus*), 487

cut-leaved anemone (*Anemone multifida*), 49

cut-leaved nightshade (*Solanum triflorum*), 380

cypress

    bald (*Taxodium distichum*), 34

    standing (*Gilia rubra*), 392

    summer- (*Kochia scoparia*), 90

cypress family (CUPRESSACEAE), 35

cypress-spurge (*Euphorbia cyparissias*), 317

cypress-vine (*Ipomoea quamoclit*), 386

cyrilla family (CYRILLACEAE), 183

daffodil (*Narcissus pseudonarcissus*), 803
daisy (*Astranthium*), 563; (*Erigeron*), 559
    doll's (*Boltonia diffusa*), 562
    English (*Bellis perennis*), 563
    hyssop- (*Erigeron hyssopifolius*), 561
    lakeside (*Hymenoxys herbacea*), 500
    Michaelmas- (*Aster*), 547
    Montauk (*Chrysanthemum nipponicum*), 518
    Nippon- (*Chrysanthemum nipponicum*), 518
    oak-leaved (*Erigeron quercifolius*), 560
    ox-eye (*Chrysanthemum leucanthemum*), 518
    Philadelphia (*Erigeron philadelphicus*), 560
    southern white (*Erigeron vernus*), 560
    trimorphic (*Erigeron acris*), 561
daisy-leaved grape-fern (*Botrychium matricariifolium*), 8
Dakota vervain (*Verbena bipinnatifida*), 405
dalea
    hare's-foot (*Dalea leporina*), 282
    sailpod (*Dalea enneandra*), 282
Dallis-grass (*Paspalum dilatatum*), 764
dalmatian toadflax (*Linaria dalmatica*), 447
dame's rocket (*Hesperis*), 179
dandelion (*Taraxacum*), 598
    colonial dwarf (*Krigia dandelion*), 598
    common (*Taraxacum officinale*), 599
    dwarf (*Krigia*), 597
    fall- (*Leontodon autumnalis*), 601
    false (*Agoseris*), 596; (*Microseris cuspidata*), 597
    marsh- (*Taraxacum palustre*), 599
    northern (*Taraxacum ceratophorum*), 598
    opposite-leaved dwarf (*Krigia oppositifolia*), 598
    orange dwarf (*Krigia biflora*), 597
    red-seeded (*Taraxacum laevigatum*), 598
    Virginia dwarf (*Krigia virginica*), 598
dangleberry (*Gaylussacia frondosa*), 194
dark-scale cotton-grass (*Eriophorum viridicarinatum*), 638
Darlington-oak (*Quercus hemisphaerica*), 79
darnel (*Lolium temulentum*), 698
    Persian (*Lolium persicum*), 698
day-flower (*Commelina*), 620
    common (*Commelina communis*), 620
    creeping (*Commelina diffusa*), 620
    erect (*Commelina erecta*), 621
    Virginia (*Commelina virginica*), 621
day-lily (*Hemerocallis*), 789
dead nettle (*Lamium*), 423
    red (*Lamium purpureum*), 424
    spotted (*Lamium maculatum*), 423
    white (*Lamium album*), 423
death-camas (*Zigadenus*), 782
deciduous mountain-fetterbush (*Eubotrys recurva*), 189
deciduous swamp-fetterbush (*Eubotrys racemosa*), 188
deep-water quillwort (*Isoetes lacustris*), 4
deerberry (*Vaccinium stamineum*), 191
deer-fern family (*BLECHNACEAE*), 30
Deptford-pink (*Dianthus armeria*), 112
devil's beggar-ticks (*Bidens frondosa*), 507
devil's bit (*Chamaelirium*), 787; (*Succisa pratensis*), 488; (*Succisella inflexa*), 488
devil's claw (*Proboscidea louisianica*), 461
devil's club (*Oplopanax horridus*), 343
devil's lettuce (*Amsinckia tessellata*), 398
devil-wood (*Osmanthus americanus*), 435
dewberry
    Blanchard's (*Rubus recurvicaulis*), 232
    coastal-plain (*Rubus trivialis*), 231

dewberry (*continued*)
    European (*Rubus caesius*), 231
    northern (*Rubus flagellaris*), 232
    southern (*Rubus enslenii*), 232
    swamp- (*Rubus hispidus*), 231
dewdrop (*Dalibarda repens*), 235
diamond-willow (*Salix eriocephala*), 157
diapensia family (DIAPENSIACEAE), 197
diffuse potentilla (*Potentilla paradoxa*), 223
dill (*Anethum graveolens*), 352
dingleberry (*Vaccinium erythrocarpum*), 194
dissected toothwort (*Cardamine dissecta*), 171
ditch-grass (*Ruppia*), 613
ditch-grass family (RUPPIACEAE), 613
ditch-stonecrop (*Penthorum*), 211
dittany (*Cunila*), 419
divaricate sunflower (*Helianthus divaricatus*), 490
dock (*Rumex*), 113
    bitter (*Rumex obtusifolius*), 115
    bur- (*Arctium*), 579
    butterfly- (*Petasites hybridus*), 528
    common bur- (*Arctium minus*), 579
    cotton-bur- (*Arctium tomentosum*), 579
    curly (*Rumex crispus*), 114
    fiddle- (*Rumex pulcher*), 114
    golden (*Rumex maritimus*), 115
    great bur- (*Arctium lappa*), 579
    great water- (*Rumex orbiculatus*), 113
    pale (*Rumex altissimus*), 114
    patience- (*Rumex patientia*), 114
    seabeach- (*Rumex pallidus*), 114
    spatter- (*Nuphar*), 42
    swamp- (*Rumex verticillatus*), 114
    veiny (*Rumex venosus*), 113
    water- (*Rumex verticillatus*), 114
    western (*Rumex occidentalis*), 113
    yard- (*Rumex longifolius*), 113
dock-leaved smartweed (*Polygonum lapathifolium*), 120
dockmackie (*Viburnum acerifolium*), 483
dodder (*Cuscuta*), 386
    beaked (*Cuscuta rostrata*), 387
    bigfruit- (*Cuscuta megalocarpa*), 387
    buttonbush- (*Cuscuta cephalanthi*), 388
    collared (*Cuscuta indecora*), 387
    common (*Cuscuta gronovii*), 387
    field- (*Cuscuta pentagona*), 387
    flax- (*Cuscuta epilinum*), 388
    hazel- (*Cuscuta coryli*), 388
    legume- (*Cuscuta epithymum*), 388
    rope- (*Cuscuta glomerata*), 387
    smartweed- (*Cuscuta polygonorum*), 388
    southern (*Cuscuta obtusiflora*), 387
dodder family (CUSCUTACEAE), 386
dogbane (*Apocynum*), 369
    clasping (*Apocynum sibiricum*), 369
    hemp- (*Apocynum cannabinum*), 369
    spreading (*Apocynum androsaemifolium*), 369
dogbane family (APOCYNACEAE), 368
dogberry (*Ribes cynosbati*), 206
dog-fennel (*Anthemis cotula*), 517; (*Eupatorium capillifolium*), 570
dog-laurel (*Leucothoe*), 188
    mountain (*Leucothoe walteri*), 188
    swamp (*Leucothoe axillaris*), 188
dog-mustard (*Erucastrum gallicum*), 163
dog-rose (*Rosa canina*), 238

dog's-tail grass (*Cynosurus*), 699

dogwood (*Cornus*), 304

   flowering (*Cornus florida*), 306

   knob-styled (*Cornus amomum*), 304

   northern swamp- (*Cornus racemosa*), 305

   pagoda- (*Cornus alternifolia*), 304

   red osier- (*Cornus sericea*), 305

   rough-leaved (*Cornus drummondii*), 305

   round-leaved (*Cornus rugosa*), 304

   southern swamp- (*Cornus stricta*), 305

dogwood family (CORNACEAE), 304

doll's daisy (*Boltonia diffusa*), 562

doll's eyes (*Actaea alba*), 47

dooryard knotweed (*Polygonum arenastrum*), 117

dooryard violet (*Viola sororia*), 143

dotted hawthorn (*Crataegus punctata*), 249

dotted smartweed (*Polygonum punctatum*), 120

dove's-foot crane's-bill (*Geranium molle*), 340

downy agrimony (*Agrimonia pubescens*), 236

downy arrow-wood (*Viburnum rafinesquianum*), 482

downy bur-clover (*Medicago minima*), 272

downy chess (*Bromus tectorum*), 723

downy false foxglove (*Aureolaria virginica*), 451

downy ground-cherry (*Physalis pubescens*), 378

downy hawthorn (*Crataegus mollis*), 251

downy oatgrass (*Danthonia sericea*), 733

downy paintbrush (*Castilleja sessiliflora*), 458

downy prairie-clover (*Dalea villosa*), 282

downy rattlesnake-plantain (*Goodyera pubescens*), 817

downy serviceberry (*Amelanchier arborea*), 254

downy trailing lespedeza (*Lespedeza procumbens*), 279

downy wild rye (*Elymus villosus*), 725

dragon, green (*Arisaema dracontium*), 616

dragon-head (*Dracocephalum*), 422

   American (*Dracocephalum parviflorum*), 422

   thyme-leaved (*Dracocephalum thymiflorum*), 422

dragon-root (*Arisaema dracontium*), 616

dragon's mouth (*Arethusa bulbosa*), 824

driftless area goldenrod (*Solidago sciaphila*), 533

drooping woodreed (*Cinna latifolia*), 719

dropseed (*Sporobolus*), 743

   prairie- (*Sporobolus heterolepis*), 743

   rough (*Sporobolus clandestinus*), 742

   sand- (*Sporobolus cryptandrus*), 741

   tall (*Sporobolus asper*), 741

dropwort (*Filipendula vulgaris*), 226

   Canby's water- (*Oxypolis canbyi*), 356

   common water- (*Oxypolis rigidior*), 356

   corn-leaved water- (*Oxypolis canbyi*), 356

   three-leaved water- (*Oxypolis ternata*), 356

   water- (*Oxypolis*), 356

drum-heads (*Polygala cruciata*), 330

dry-land bitter-cress (*Cardamine parviflora*), 173

duckweed (*Lemna*), 616; (*Spirodela*), 616

   greater (*Spirodela polyrhiza*), 616

   lesser (*Lemna minor*), 616

   star- (*Lemna trisulca*), 616

duckweed family (LEMNACEAE), 616

dull meadow-pitchers (*Rhexia mariana*), 304

dunegrass, American (*Leymus mollis*), 728

dune-sandbur (*Cenchrus tribuloides*), 770

dune-spurge (*Euphorbia geyeri*), 319

dune-thistle (*Cirsium pitcheri*), 582

dune-willow (*Salix cordata*), 158

dusty goldenrod (*Solidago puberula*), 533

dusty miller (*Artemisia stelleriana*), 520

Dutchman's breeches (*Dicentra cucullaria*), 66

Dutchman's pipe (*Aristolochia macrophylla*), 41

dwarf bilberry (*Vaccinium cespitosum*), 191

dwarf birch (*Betula glandulosa*), 83

   Newfoundland (*Betula michauxii*), 83

dwarf cornel (*Cornus canadensis*), 305

dwarf crab-grass (*Digitaria serotina*), 767

dwarf crested iris (*Iris cristata*), 809

dwarf cut-leaf white lettuce (*Prenanthes nana*), 588

dwarf dandelion (*Krigia*), 597

   colonial (*Krigia dandelion*), 598

   opposite-leaved (*Krigia oppositifolia*), 598

   orange (*Krigia biflora*), 597

   Virginia (*Krigia virginica*), 598

dwarf elder (*Sambucus ebulus*), 485

dwarf fleabane (*Conyza ramosissima*), 561

dwarf ginseng (*Panax trifolius*), 343

dwarf glasswort (*Salicornia bigelovii*), 92

dwarf hackberry (*Celtis tenuifolia*), 69

dwarf huckleberry (*Gaylussacia dumosa*), 195

dwarf iris (*Iris verna*), 809

dwarf lake-iris (*Iris lacustris*), 810

dwarf larkspur (*Delphinium tricorne*), 48

dwarf mallow (*Malva rotundifolia*), 135

dwarf milkweed (*Asclepias ovalifolia*), 371

dwarf mistletoe (*Arceuthobium*), 307

   eastern (*Arceuthobium pusillum*), 307

dwarf nipplewort (*Arnoseris minima*), 600

dwarf prairie-rose (*Rosa arkansana*), 239

dwarf raspberry (*Rubus pubescens*), 230

dwarf scouring rush (*Equisetum scirpoides*), 5

dwarf serviceberry (*Amelanchier spicata*), 254

dwarf simple white lettuce (*Prenanthes boottii*), 588

dyer's greenweed (*Genista tinctoria*), 259

dyer's rocket (*Reseda luteola*), 182

early goldenrod (*Solidago juncea*), 538

early meadow-rue (*Thalictrum dioicum*), 60

early saxifrage (*Saxifraga virginiensis*), 214

early scorpion-grass (*Myosotis verna*), 397

early winter-cress (*Barbarea verna*), 178

Easter-bell (*Stellaria holostea*), 103

eastern beard-tongue (*Penstemon laevigatus*), 445

eastern black currant (*Ribes americanum*), 207

eastern bluebell (*Mertensia virginica*), 396

eastern blue-eyed Mary (*Collinsia verna*), 447

eastern coneflower (*Rudbeckia fulgida*), 495

eastern dwarf mistletoe (*Arceuthobium pusillum*), 307

eastern false gromwell (*Onosmodium virginianum*), 401

eastern figwort (*Scrophularia marilandica*), 446

eastern hemlock (*Tsuga canadensis*), 32

eastern lined aster (*Aster lanceolatus*), 549

eastern mannagrass (*Glyceria septentrionalis*), 706

eastern mosquito-fern (*Azolla caroliniana*), 30

eastern parthenium (*Parthenium integrifolium*), 514

eastern prickly pear (*Opuntia humifusa*), 86

eastern red cedar (*Juniperus virginiana*), 35

eastern serpentine-aster (*Aster depauperatus*), 551

eastern serviceberry (*Amelanchier canadensis*), 254

eastern shooting star (*Dodecatheon meadia*), 203

eastern silvery aster (*Aster concolor*), 551

eastern tansy (*Tanacetum huronense*), 519

eastern waterleaf (*Hydrophyllum virginianum*), 393

eastern white beard-tongue (*Penstemon pallidus*), 445

eastern willow-herb (*Epilobium coloratum*), 298

Eaton's lip-fern (*Cheilanthes eatonii*), 16

ebony family (EBENACEAE), 198
ebony-spleenwort (*Asplenium platyneuron*), 17
eel-grass (*Zostera*), 614
    common (*Zostera marina*), 614
eel-grass family (ZOSTERACEAE), 614
Eglantine (*Rosa eglanteria, R. micrantha*), 238
Egyptian bean (*Dolichos lablab*), 285
elder (*Sambucus*), 485
    beach-dune marsh- (*Iva imbricata*), 515
    big marsh- (*Iva xanthiifolia*), 514
    common (*Sambucus canadensis*), 485
    dwarf (*Sambucus ebulus*), 485
    maritime marsh- (*Iva frutescens*), 514
    marsh- (*Iva*), 514
    red-berried (*Sambucus racemosa*), 485
    rough marsh- (*Iva annua*), 515
elecampane (*Inula helenium*), 563
elegant milk-vetch (*Astragalus eucosmus*), 265
elephant's foot (*Elephantopus*), 578
    leafy (*Elephantopus carolinianus*), 578
elk-sedge (*Carex geyeri*), 661
elliptic shinleaf (*Pyrola elliptica*), 196
elm (*Ulmus*), 68
    American (*Ulmus americana*), 68
    red (*Ulmus rubra*), 68
    rock- (*Ulmus thomasii*), 69
    September (*Ulmus serotina*), 69
    slippery (*Ulmus rubra*), 68
    white (*Ulmus americana*), 68
    winged (*Ulmus alata*), 68
elm family (ULMACEAE), 68
elm-leaved goldenrod (*Solidago ulmifolia*), 539
empress-tree (*Paulownia*), 461
enchanter's nightshade (*Circaea*), 303
    alpine (*Circaea alpina*), 303
    common (*Circaea lutetiana*), 303
Engelmann's quillwort (*Isoetes engelmannii*), 4
English bluebell (*Scilla non-scripta*), 786
English daisy (*Bellis perennis*), 563
English ivy (*Hedera*), 343
English plantain (*Plantago lanceolata*), 433
English ryegrass (*Lolium perenne* var. *perenne*), 698
erect day-flower (*Commelina erecta*), 621
eryngo (*Eryngium*), 345
    marsh- (*Eryngium aquaticum*), 345
    spreading (*Eryngium prostratum*), 346
eulalia (*Miscanthus sinensis*), 771
eupatorium
    bushy-branched (*Eupatorium cuneifolium*), 571
    late (*Eupatorium serotinum*), 569
    pine-barren (*Eupatorium resinosum*), 569
    pink (*Eupatorium incarnatum*), 568
    ragged (*Eupatorium pilosum*), 572
    round-leaved (*Eupatorium rotundifolium*), 571
    tall (*Eupatorium altissimum*), 572
    tuberous (*Eupatorium mohrii*), 571
    white-bracted (*Eupatorium album*), 570
European alder-buckthorn (*Rhamnus frangula*), 321
European alkali-grass (*Puccinellia distans*), 700
European barberry (*Berberis vulgaris*), 62
European bird-cherry (*Prunus padus*), 241
European columbine (*Aquilegia vulgaris*), 59
European corn-salad (*Valerianella locusta*), 486
European dewberry (*Rubus caesius*), 231
European eyebright (*Euphrasia officinalis*), 456
European field-pansy (*Viola arvensis*), 149

European fly-honeysuckle (*Lonicera xylosteum*), 479
European frog's-bit (*Hydrocharis morsus-ranae*), 606
European hawkweed (*Hieracium lachenalii*), 593
European heliotrope (*Heliotropium europaeum*), 395
European mock-orange (*Philadelphus coronarius*), 204
European mountain-ash (*Sorbus aucuparia*), 246
European spindle-tree (*Euonymus europaeus*), 308
European vervain (*Verbena officinalis*), 403
European water-clover (*Marsilea quadrifolia*), 30
European water-horehound (*Lycopus europaeus*), 420
European water-milfoil (*Myriophyllum spicatum*), 290
eutrophic water-nymph (*Najas minor*), 614
evening-primrose (*Oenothera brachycarpum*), 299
    common (*Oenothera biennis*), 299
    cut-leaf (*Oenothera laciniata*), 300
    garden (*Oenothera glazioviana*), 299
    longspike (*Oenothera rhombipetala*), 299
    shale-barren (*Oenothera argillicola*), 299
    small-flowered (*Oenothera parviflora*), 299
    spreading (*Oenothera humifusa*), 300
    stemless (*Oenothera triloba*), 300
    white (*Oenothera speciosa*), 300
    white-stemmed (*Oenothera nuttallii*), 300
    wing-fruit (*Oenothera macrocarpa*), 301
evening-primrose family (ONAGRACEAE), 294
evening star (*Mentzelia decapetala*), 151
evergreen blackberry (*Rubus laciniatus*), 235
evergreen mountain-fetterbush (*Pieris floribunda*), 189
evergreen swamp-fetterbush (*Lyonia lucida*), 188
everlasting (*Gnaphalium*), 564
    pearly (*Anaphalis*), 565
    sunflower- (*Heliopsis*), 503
everlasting pea (*Lathyrus latifolius*), 275
eyebane (*Euphorbia nutans*), 320
eyebright (*Euphrasia*), 456; (*Odontites*), 456
    arctic (*Euphrasia disjuncta*), 456
    European (*Euphrasia officinalis*), 456
    Nova Scotian (*Euphrasia randii*), 456
    White-Mountain (*Euphrasia oakesii*), 457

fairy-bells (*Disporum*), 799
falcate golden aster (*Chrysopsis falcata*), 531
fall-adonis (*Adonis annua*), 59
fall-dandelion (*Leontodon autumnalis*), 601
fall witch-grass (*Leptoloma*), 768
false aloe (*Agave virginica*), 804
false asphodel (*Tofieldia*), 781
false boneset (*Kuhnia eupatorioides*), 573
false buckwheat (*Polygonum scandens*), 124
false bugbane (*Trautvetteria*), 46
false dandelion (*Agoseris*), 596; (*Microseris cuspidata*), 597
false flax (*Camelina*), 181
    large-seed (*Camelina sativa*), 181
    small-seed (*Camelina microcarpa*), 181
false foxglove (*Aureolaria*), 449
    annual (*Aureolaria pedicularia*), 450
    Appalachian (*Aureolaria laevigata*), 449
    Cumberland (*Aureolaria patula*), 450
    downy (*Aureolaria virginica*), 451
    smooth (*Aureolaria flava*), 450
    western (*Aureolaria grandiflora*), 450
false goat's-beard (*Astilbe biternata*), 211
false gromwell (*Onosmodium*), 401
    eastern (*Onosmodium virginianum*), 401
    western (*Onosmodium molle*), 401
false heather (*Hudsonia*), 140

false hellebore (*Veratrum*), 784
false indigo (*Amorpha fruticosa*), 283; (*Baptisia*), 258
   blue (*Baptisia australis*), 258
   shining (*Amorpha nitens*), 282
false mermaid (*Floerkea*), 341
false nettle (*Boehmeria cylindrica*), 71
false nutsedge (*Cyperus strigosus*), 653
false pennyroyal (*Isanthus*), 409
false pimpernel (*Lindernia*), 441
false Solomon's seal (*Smilacina*), 800
false water-pepper (*Polygonum hydropiperoides*), 119
fame-flower (*Talinum*), 99
   Appalachian (*Talinum teretifolium*), 99
   prairie (*Talinum parviflorum*), 99
   rockpink (*Talinum calycinum*), 99
   sand (*Talinum rugospermum*), 99
family
   acanthus (ACANTHACEAE), 459
   adder's-tongue (OPHIOGLOSSACEAE), 8
   agave (AGAVACEAE), 804
   amaranth (AMARANTHACEAE), 94
   arrow-grass (JUNCAGINACEAE), 608
   arum (ARACEAE), 615
   aster (ASTERACEAE), 489
   bald cypress (TAXODIACEAE), 34
   barberry (BERBERIDACEAE), 62
   bayberry (MYRICACEAE), 75
   bean (FABACEAE), 258
   beech (FAGACEAE), 75
   bellflower (CAMPANULACEAE), 465
   birch (BETULACEAE), 82
   birthwort (ARISTOLOCHIACEAE), 40
   bladder-nut (STAPHYLEACEAE), 330
   bladderwort (LENTIBULARIACEAE), 462
   bloodwort (HAEMODORACEAE), 781
   borage (BORAGINACEAE), 395
   boxwood (BUXACEAE), 311
   bracken (DENNSTAEDTIACEAE), 12
   bromeliad (BROMELIACEAE), 779
   broom-rape (OROBANCHACEAE), 458
   buckbean (MENYANTHACEAE), 388
   buckthorn (RHAMNACEAE), 321
   burmannia (BURMANNIACEAE), 812
   bur-reed (SPARGANIACEAE), 776
   buttercup (RANUNCULACEAE), 45
   butterfly-bush (BUDDLEJACEAE), 434
   cacao (STERCULIACEAE), 131
   cactus (CACTACEAE), 86
   caesalpinia (CAESALPINIACEAE), 256
   caper (CAPPARACEAE), 161
   carpet-weed (MOLLUGINACEAE), 100
   carrot (APIACEAE), 344
   cashew (ANACARDIACEAE), 334
   catbrier (SMILACACEAE), 804
   cat-tail (TYPHACEAE), 778
   Christmas-mistletoe (VISCACEAE), 307
   clethra (CLETHRACEAE), 183
   clubmoss (LYCOPODIACEAE), 1
   creosote-bush (ZYGOPHYLLACEAE), 337
   crowberry (EMPETRACEAE), 183
   curly grass (SCHIZAEACEAE), 11
   custard-apple (ANNONACEAE), 38
   cypress (CUPRESSACEAE), 35
   cyrilla (CYRILLACEAE), 183
   deer-fern (BLECHNACEAE), 30
   diapensia (DIAPENSIACEAE), 197

family (*continued*)
   ditch-grass (RUPPIACEAE), 613
   dodder (CUSCUTACEAE), 386
   dogbane (APOCYNACEAE), 368
   dogwood (CORNACEAE), 304
   duckweed (LEMNACEAE), 616
   ebony (EBENACEAE), 198
   eel-grass (ZOSTERACEAE), 614
   elm (ULMACEAE), 68
   evening-primrose (ONAGRACEAE), 294
   fig marigold (AIZOACEAE), 85
   figwort (SCROPHULARIACEAE), 438
   filmy fern (HYMENOPHYLLACEAE), 11
   flax (LINACEAE), 326
   floating fern (SALVINIACEAE), 30
   flowering-rush (BUTOMACEAE), 602
   four-o'clock (NYCTAGINACEAE), 85
   frog's-bit (HYDROCHARITACEAE), 606
   fumitory (FUMARIACEAE), 66
   gentian (GENTIANACEAE), 362
   geranium (GERANIACEAE), 338
   ginseng (ARALIACEAE), 342
   gooseberry (GROSSULARIACEAE), 205
   goosefoot (CHENOPODIACEAE), 86
   gourd (CUCURBITACEAE), 150
   grape (VITACEAE), 323
   grass (POACEAE), 690
   heath (ERICACEAE), 184
   holly (AQUIFOLIACEAE), 309
   honeysuckle (CAPRIFOLIACEAE), 478
   horned pondweed (ZANNICHELLIACEAE), 614
   hornwort (CERATOPHYLLACEAE), 44
   horse-chestnut (HIPPOCASTANACEAE), 330
   horsetail (EQUISETACEAE), 5
   hydrangea (HYDRANGEACEAE), 204
   Indian hemp (CANNABACEAE), 70
   Indian pipe (MONOTROPACEAE), 196
   iris (IRIDACEAE), 807
   lardizabala (LARDIZABALACEAE), 63
   laurel (LAURACEAE), 38
   leadwort (PLUMBAGINACEAE), 125
   lily (LILIACEAE), 780
   linden (TILIACEAE), 131
   lizard's tail (SAURURACEAE), 39
   loasa (LOASACEAE), 151
   logania (LOGANIACEAE), 361
   loosestrife (LYTHRACEAE), 291
   lotus-lily (NELUMBONACEAE), 42
   madder (RUBIACEAE), 469
   magnolia (MAGNOLIACEAE), 36
   mahogany (MELIACEAE), 335
   maidenhair fern (ADIANTACEAE), 14
   mallow (MALVACEAE), 131
   mangosteen (CLUSIACEAE), 126
   maple (ACERACEAE), 331
   mare's-tail (HIPPURIDACEAE), 431
   meadow-foam (LIMNANTHACEAE), 341
   melastome (MELASTOMATACEAE), 303
   mezereum (THYMELAEACEAE), 293
   mignonette (RESEDACEAE), 182
   milkweed (ASCLEPIADACEAE), 370
   milkwort (POLYGALACEAE), 328
   mimosa (MIMOSACEAE), 255
   mint (LAMIACEAE), 408
   moonseed (MENISPERMACEAE), 63
   morning-glory (CONVOLVULACEAE), 384

family *(continued)*
  moschatel (ADOXACEAE), 485
  mulberry (MORACEAE), 70
  mustard (BRASSICACEAE), 162
  nettle (URTICACEAE), 71
  nightshade (SOLANACEAE), 376
  oleaster (ELAEAGNACEAE), 288
  olive (OLEACEAE), 435
  orchid (ORCHIDACEAE), 812
  passion-flower (PASSIFLORACEAE), 149
  pea (FABACEAE), 258
  phlox (POLEMONIACEAE), 389
  pine (PINACEAE), 31
  pink (CARYOPHYLLACEAE), 100
  pipewort (ERIOCAULACEAE), 621
  pitcher-plant (SARRACENIACEAE), 138
  plane-tree (PLATANACEAE), 68
  plantain (PLANTAGINACEAE), 431
  pokeweed (PHYTOLACCACEAE), 84
  polypody (POLYPODIACEAE), 12
  pondweed (POTAMOGETONACEAE), 608
  poppy (PAPAVERACEAE), 64
  primrose (PRIMULACEAE), 199
  purslane (PORTULACACEAE), 98
  quassia (SIMAROUBACEAE), 335
  quillwort (ISOETACEAE), 4
  river-weed (PODOSTEMACEAE), 289
  rock-rose (CISTACEAE), 139
  rose (ROSACEAE), 217
  royal fern (OSMUNDACEAE), 10
  rue (RUTACEAE), 336
  rush (JUNCACEAE), 622
  sandalwood (SANTALACEAE), 306
  sapodilla (SAPOTACEAE), 198
  saxifrage (SAXIFRAGACEAE), 211
  scheuchzeria (SCHEUCHZERIACEAE), 607
  sedge (CYPERACEAE), 631
  selaginella (SELAGINELLACEAE), 4
  sensitive fern (ONOCLEACEAE), 29
  sesame (PEDALIACEAE), 460
  shinleaf (PYROLACEAE), 195
  smartweed (POLYGONACEAE), 112
  soapberry (SAPINDACEAE), 330
  spiderwort (COMMELINACEAE), 619
  spleenwort (ASPLENIACEAE), 17
  spurge (EUPHORBIACEAE), 311
  staff-tree (CELASTRACEAE), 308
  stonecrop (CRASSULACEAE), 208
  storax (STYRACACEAE), 198
  strawberry-shrub (CALYCANTHACEAE), 38
  sundew (DROSERACEAE), 138
  sweet flag (ACORACEAE), 615
  sweet-leaf (SYMPLOCACEAE), 199
  tamarisk (TAMARICACEAE), 149
  tea (THEACEAE), 125
  teasel (DIPSACACEAE), 487
  touch-me-not (BALSAMINACEAE), 341
  trumpet-creeper (BIGNONIACEAE), 461
  valerian (VALERIANACEAE), 486
  vervain (VERBENACEAE), 403
  violet (VIOLACEAE), 142
  walnut (JUGLANDACEAE), 72
  water-chestnut (TRAPACEAE), 293
  water-clover (MARSILEACEAE), 30
  water-hyacinth (PONTEDERIACEAE), 779
  waterleaf (HYDROPHYLLACEAE), 392

family *(continued)*
  water-lily (NYMPHAEACEAE), 42
  water-milfoil (HALORAGACEAE), 289
  water-nymph (NAJADACEAE), 614
  water-plantain (ALISMATACEAE), 602
  water-shield (CABOMBACEAE), 44
  water-starwort (CALLITRICHACEAE), 431
  waterwort (ELATINACEAE), 125
  willow (SALICACEAE), 152
  witch hazel (HAMAMELIDACEAE), 68
  wood-sorrel (OXALIDACEAE), 337
  yam (DIOSCOREACEAE), 806
  yellow-eyed grass (XYRIDACEAE), 617
  yew (TAXACEAE), 31
fancy wood-fern (*Dryopteris intermedia*), 26
fanleaf-hawthorn (*Crataegus flabellata*), 253
fanwort (*Cabomba caroliniana*), 44
farkleberry (*Vaccinium arboreum*), 191
fascicled agalinis (*Agalinis fasciculata*), 451
fascicled piney-woods goldenrod (*Solidago pinetorum*), 538
fawn-lily (*Erythronium*), 793
feather
  gay (*Liatris*), 574
  parrot's (*Myriophyllum aquaticum*), 289
  prince's (*Amaranthus hypochondriacus*), 95; (*Polygonum orientale*), 122
feather finger-grass (*Chloris virgata*), 747
featherfoil (*Hottonia*), 203
feather-geranium (*Chenopodium botrys*), 87
featherstem-clubmoss (*Lycopodium prostratum*), 1
feathertop (*Calamagrostis epigejos*), 714; (*Pennisetum villosum*), 769
felwort (*Gentianella amarella*), 366
  marsh- (*Lomatogonium*), 367
fennel (*Foeniculum vulgare*), 351
  dog- (*Anthemis cotula*), 517; (*Eupatorium capillifolium*), 570
  water- (*Oenanthe*), 354
fennel-flower (*Nigella*), 45
fen-orchid (*Liparis loeselii*), 825
fern
  Alabama grape- (*Botrychium jenmanii*), 9
  Alabama lip- (*Cheilanthes alabamensis*), 15
  alpine cliff- (*Woodsia alpina*), 20
  Appalachian bristle- (*Trichomanes boschianum*), 11
  bladder- (*Cystopteris*), 20
  blunt cliff- (*Woodsia obtusa*), 21
  blunt-lobed grape- (*Botrychium oneidense*), 8
  bracken (*Pteridium*), 13
  Braun's holly- (*Polystichum braunii*), 28
  bristle- (*Trichomanes*), 11
  brittle bladder- (*Cystopteris fragilis*), 20
  broad beech- (*Thelypteris hexagonoptera*), 23
  bulblet bladder- (*Cystopteris bulbifera*), 20
  chain- (*Woodwardia*), 30
  Christmas- (*Polystichum acrostichoides*), 28
  cinnamon- (*Osmunda cinnamomea*), 10
  cliff- (*Woodsia*), 20
  climbing (*Lygodium*), 11
  Clinton's wood- (*Dryopteris clintoniana*), 26
  cloak- (*Notholaena*), 16
  crested wood- (*Dryopteris cristata*), 25
  daisy-leaved grape- (*Botrychium matricariifolium*), 8
  eastern mosquito- (*Azolla caroliniana*), 30
  Eaton's lip- (*Cheilanthes eatonii*), 16
  fancy wood- (*Dryopteris intermedia*), 26

fern *(continued)*
   fragrant wood- (*Dryopteris fragrans*), 24
   glade- (*Athyrium pycnocarpon*), 19
   Goldie's wood- (*Dryopteris goldiana*), 25
   grape- (*Botrychium*), 8
   grass- (*Asplenium septentrionale*), 18
   hay-scented (*Dennstaedtia punctilobula*), 12
   interrupted (*Osmunda claytoniana*), 10
   lace-frond grape- (*Botrychium dissectum*), 8
   lady- (*Athyrium filix-femina*), 19
   lance-leaved grape- (*Botrychium lanceolatum*), 8
   leathery grape- (*Botrychium multifidum*), 8
   limestone oak- (*Gymnocarpium robertianum*), 22
   lip- (*Cheilanthes*), 15
   little grape- (*Botrychium simplex*), 8
   log- (*Dryopteris celsa*), 26
   lowland bladder- (*Cystopteris protrusa*), 20
   maidenhair (*Adiantum*), 14
   male (*Dryopteris filix-mas*), 26
   marginal wood- (*Dryopteris marginalis*), 25
   marsh- (*Thelypteris palustris*), 24
   Massachusetts (*Thelypteris simulata*), 23
   Mexican mosquito- (*Azolla mexicana*), 30
   mosquito- (*Azolla*), 30
   mountain cliff- (*Woodsia scopulina*), 21
   mountain wood- (*Dryopteris campyloptera*), 27
   narrow beech- (*Thelypteris phegopteris*), 22
   netted chain- (*Woodwardia areolata*), 30
   New York (*Thelypteris noveboracensis*), 23
   northern beech- (*Thelypteris phegopteris*), 22
   northern holly- (*Polystichum lonchitis*), 28
   northern maidenhair (*Adiantum pedatum*), 14
   northern wood- (*Dryopteris expansa*), 27
   oak- (*Gymnocarpium*), 22; (*G. dryopteris*), 22
   ostrich- (*Matteuccia*), 29
   parsley- (*Cryptogramma crispa*), 14
   pod- (*Aspidotis densa*), 15
   powdery cloak- (*Notholaena dealbata*), 16
   rattlesnake- (*Botrychium virginianum*), 9
   resurrection- (*Polypodium polypodioides*), 12
   royal (*Osmunda regalis*), 10
   rusty cliff- (*Woodsia ilvensis*), 21
   sensitive (*Onoclea*), 29
   shield- (*Dryopteris*), 24
   silvery glade- (*Athyrium thelypterioides*), 19
   slender lip- (*Cheilanthes feei*), 15
   smooth cliff- (*Woodsia glabella*), 20
   southern beech- (*Thelypteris hexagonoptera*), 23
   southern maidenhair (*Adiantum capillus-veneris*), 14
   sparse-lobed grape- (*Botrychium biternatum*), 8
   sweet (*Comptonia*), 75
   toothed wood- (*Dryopteris carthusiana*), 27
   Virginia chain- (*Woodwardia virginica*), 30
   walking (*Asplenium rhizophyllum*), 18
   western cliff- (*Woodsia oregana*), 21
   wood- (*Dryopteris*), 24
   woolly lip- (*Cheilanthes lanosa*), 15; (*C. tomentosa*), 16
Fernald's tick-trefoil (*Desmodium fernaldii*), 277
fern-pondweed (*Potamogeton robbinsii*), 609
fescue (*Festuca*), 695
   alta- (*Festuca arundinacea*), 695
   hard (*Festuca trachyphylla*), 696
   meadow- (*Festuca pratensis*), 695
   nodding (*Festuca subverticillata*), 695
   rat-tail (*Vulpia myuros*), 697
   red (*Festuca rubra*), 697

fescue (*Festuca*) *(continued)*
   rough (*Festuca altaica*), 696
   sheep- (*Festuca ovina*), 696
   shortleaf (*Festuca brachyphylla*), 696
   six-weeks (*Vulpia octoflora*), 697
   squirrel-tail (*Vulpia elliotea*), 697
   tall (*Festuca arundinacea*), 695
   western (*Festuca occidentalis*), 695
fetterbush (*Eubotrys*), 188
   deciduous mountain- (*Eubotrys recurva*), 189
   deciduous swamp- (*Eubotrys racemosa*), 188
   evergreen mountain- (*Pieris floribunda*), 189
   evergreen swamp- (*Lyonia lucida*), 188
feverfew (*Chrysanthemum parthenium*), 518
few-bracted beggar-ticks (*Bidens discoidea*), 507
few-flowered milkweed (*Asclepias lanceolata*), 372
few-flowered tick-trefoil (*Desmodium pauciflorum*), 276
few-headed blazing star (*Liatris cylindracea*), 576
few-headed sneezeweed (*Helenium brevifolium*), 499
fiddle-dock (*Rumex pulcher*), 114
fiddleneck (*Amsinckia*), 398
field-bindweed (*Convolvulus arvensis*), 384
field-chickweed (*Cerastium arvense*), 102
field-cress (*Lepidium campestre*), 164
field-dodder (*Cuscuta pentagona*), 387
field-garlic (*Allium oleraceum*), 791; (*A. vineale*), 790
field hedge-nettle (*Stachys arvensis*), 427
field hedge-parsley (*Torilis arvensis*), 350
field-horsetail (*Equisetum arvense*), 7
field-madder (*Sherardia arvensis*), 473
field milk-vetch (*Astragalus agrestis*), 266
field-mint (*Mentha arvensis*), 416
field penny-cress (*Thlaspi arvense*), 166
field-pussytoes (*Antennaria neglecta*), 566
field scorpion-grass (*Myosotis arvensis*), 397
field-speedwell (*Veronica agrestis*), 454
field-thistle (*Cirsium discolor*), 582
field yellow-cress (*Rorippa austriaca*), 177
fig marigold family (AIZOACEAE), 85
figwort (*Scrophularia*), 446
   American (*Scrophularia lanceolata*), 446
   eastern (*Scrophularia marilandica*), 446
figwort family (SCROPHULARIACEAE), 438
filaree (*Erodium*), 340
   redstem- (*Erodium cicutarium*), 340
   whitestem- (*Erodium moschatum*), 341
filmy fern family (HYMENOPHYLLACEAE), 11
finger-grass (*Chloris*), 747
   feather (*Chloris virgata*), 747
   windmill (*Chloris verticillata*), 747
finger poppy-mallow (*Callirhoe digitata*), 136
finger-tickseed (*Coreopsis palmata*), 505
fir (*Abies*), 31
   balsam- (*Abies balsamea*), 31
   Fraser- (*Abies fraseri*), 31
fir-clubmoss (*Lycopodium selago*), 1
fireberry-hawthorn (*Crataegus chrysocarpa*), 250
fire-on-the-mountain (*Euphorbia cyathophora*), 317
fire-pink (*Silene virginica*), 108
fireweed (*Epilobium angustifolium*), 296; (*Erechtites hieraciifolia*), 527
fistulous goat's-beard (*Tragopogon dubius*), 601
five-fingers (*Potentilla*), 222
   brook- (*Potentilla rivalis*), 223
   comb- (*Potentilla flabelliformis*), 224
   creeping (*Potentilla reptans*), 222

five-fingers (*Potentilla*) *(continued)*
  old-field (*Potentilla simplex*), 222
  running (*Potentilla canadensis*), 222
  shrubby (*Potentilla fruticosa*), 226
  silvery (*Potentilla argentea*), 223
  sulphur- (*Potentilla recta*), 223
  trailing (*Potentilla anglica*), 222
five-hook bassia (*Bassia hyssopifolia*), 90
five-leaves (*Isotria*), 823
  larger (*Isotria verticillata*), 823
  little (*Isotria medeoloides*), 823
five-parted toothwort (*Cardamine concatenata*), 171
flag (*Iris*), 809
  arctic blue (*Iris setosa*), 811
  northern blue (*Iris versicolor*), 811
  slender blue (*Iris prismatica*), 810
  southern blue (*Iris virginica*), 811
  sweet (*Acorus*), 615
  water- (*Iris pseudacorus*), 810
  yellow (*Iris pseudacorus*), 810
flame-azalea (*Rhododendron calendulaceum*), 186
flatsedge (*Cyperus*), 650
  globe- (*Cyperus echinatus*), 654
  redroot (*Cyperus erythrorhizos*), 652
flatseed-sunflower
  Ozark (*Verbesina helianthoides*), 501
  southern (*Verbesina occidentalis*), 501
flatstem-pondweed (*Potamogeton zosteriformis*), 610
flat-topped goldenrod (*Euthamia*), 545
  coastal plain (*Euthamia tenuifolia*), 545
  common (*Euthamia graminifolia*), 545
  Great Plains (*Euthamia gymnospermoides*), 546
  lakes (*Euthamia remota*), 546
  Mississippi-valley (*Euthamia leptocephala*), 546
  Nova Scotia (*Euthamia galetorum*), 545
flax (*Linum*), 326
  annual toad- (*Linaria repens*), 447
  bastard toad- (*Comandra*), 306
  Bicknell's yellow (*Linum intercursum*), 326
  coastal plain yellow (*Linum floridanum*), 326
  common (*Linum usitatissimum*), 326
  common yellow (*Linum medium*), 327
  dalmatian toad- (*Linaria dalmatica*), 447
  false (*Camelina*), 181
  grooved yellow (*Linum sulcatum*), 326
  large-seed false (*Camelina sativa*), 181
  lesser toad- (*Chaenorrhinum minus*), 449
  ridgestem yellow (*Linum striatum*), 327
  small-seed false (*Camelina microcarpa*), 181
  stiffstem yellow (*Linum rigidum*), 327
  toad- (*Linaria*), 447
  Virginia yellow (*Linum virginianum*), 327
  white (*Linum catharticum*), 326
  wild blue (*Linum perenne*), 326
flax-dodder (*Cuscuta epilinum*), 388
flax family (LINACEAE), 326
fleabane (*Erigeron*), 559
  annual (*Erigeron annuus*), 561
  dwarf (*Conyza ramosissima*), 561
  marsh- (*Pluchea*), 564
  rough (*Erigeron strigosus*), 560
fleshy hawthorn (*Crataegus succulenta*), 248
fleshy stitchwort (*Stellaria crassifolia*), 104
fleur-de-lys (*Iris*), 809
floating bladderwort (*Utricularia radiata*), 463
floating fern family (SALVINIACEAE), 30

floating heart (*Nymphoides*), 389
  big (*Nymphoides aquatica*), 389
  little (*Nymphoides cordata*), 389
  yellow (*Nymphoides peltata*), 389
floating pondweed (*Potamogeton natans*), 613
Florida adder's-mouth (*Malaxis spicata*), 825
Florida pusley (*Richardia scabra*), 472
flowering dogwood (*Cornus florida*), 306
flowering maple (*Viburnum acerifolium*), 483
flowering moss (*Pyxidanthera*), 197
flowering raspberry (*Rubus odoratus*), 229
flowering-rush (*Butomus*), 602
flowering-rush family (BUTOMACEAE), 602
flowering spurge (*Euphorbia corollata*), 318
flowering wintergreen (*Polygala paucifolia*), 328
flower of an hour (*Hibiscus trionum*), 133
fly-honeysuckle (*Lonicera canadensis*), 478
  European (*Lonicera xylosteum*), 479
  swamp (*Lonicera oblongifolia*), 479
fly-poison (*Amianthium*), 782
foam-flower (*Tiarella*), 214
fogfruit (*Phyla*), 406
fool's parsley (*Aethusa*), 349
forest-bedstraw (*Galium circaezans*), 476
forest-bluegrass (*Poa sylvestris*), 703
forest-goldenrod (*Solidago arguta*), 539
forest-lousewort (*Pedicularis canadensis*), 457
forest-muhly (*Muhlenbergia sylvatica*), 744
forest-pea (*Lathyrus venosus*), 275
forest-phacelia (*Phacelia bipinnatifida*), 393
forest-phlox (*Phlox divaricata*), 389
forest-skullcap (*Scutellaria ovata*), 409
forest-sunflower (*Helianthus decapetalus*), 490
forest-tickseed (*Coreopsis major*), 504
forget-me-not (*Myosotis*), 396
  garden (*Myosotis sylvatica*), 397
  smaller (*Myosotis laxa*), 397
forked chickweed (*Paronychia canadensis, P. fastigiata*), 106
forked spleenwort (*Asplenium septentrionale*), 18
forking centaury (*Centaurium erythraea*), 363
forktip-three-awn (*Aristida basiramea*), 734
four-leaved milkweed (*Asclepias quadrifolia*), 372
four-o'clock (*Mirabilis jalapa*), 85
four-o'clock family (NYCTAGINACEAE), 85
four-seed vetch (*Vicia tetrasperma*), 273
fowl-mannagrass (*Glyceria striata*), 708
fowl meadow-grass (*Poa palustris*), 704
foxglove (*Digitalis*), 449
  annual false (*Aureolaria pedicularia*), 450
  Appalachian false (*Aureolaria laevigata*), 449
  common (*Digitalis purpurea*), 449
  Cumberland false (*Aureolaria patula*), 450
  downy false (*Aureolaria virginica*), 451
  false (*Aureolaria*), 449
  Grecian (*Digitalis lanata*), 449
  mullein- (*Dasistoma*), 449
  smooth false (*Aureolaria flava*), 450
  western false (*Aureolaria grandiflora*), 450
foxglove beard-tongue (*Penstemon cobaea*), 444
fox-grape (*Vitis labrusca*), 324
foxtail (*Alopecurus*), 720
  bur- (*Setaria verticillata*), 766
  Carolina (*Alopecurus carolinianus*), 720
  giant (*Setaria faberi*), 766
  green (*Setaria viridis*), 766
  knotroot- (*Setaria parviflora*), 766

foxtail (*Alopecurus*) (continued)
  marsh- (*Alopecurus geniculatus*), 720
  meadow- (*Alopecurus pratensis*), 720
  nodding (*Setaria faberi*), 766
  salt-marsh (*Setaria magna*), 767
  short-awn (*Alopecurus aequalis*), 720
  water- (*Alopecurus geniculatus*), 720
  yellow (*Setaria glauca*), 766
foxtail-barley (*Hordeum jubatum*), 729
foxtail-chess (*Bromus rubens*), 723
foxtail-clubmoss (*Lycopodium alopecuroides*), 1
foxtail-grass (*Setaria*), 766
foxtail-millet (*Setaria italica*), 767
fragrant cudweed (*Gnaphalium obtusifolium*), 565
fragrant wood-fern (*Dryopteris fragrans*), 24
Fraser-fir (*Abies fraseri*), 31
free-flowered water-weed (*Elodea nuttallii*), 606
freeway-sedge (*Carex praegracilis*), 662
French rose (*Rosa gallica*), 238
Fries' pondweed (*Potamogeton friesii*), 610
fringed bindweed (*Polygonum cilinode*), 124
fringed brome (*Bromus ciliatus*), 721
fringed gentian (*Gentianopsis*), 366; (*G. crinita*), 366
fringed loosestrife (*Lysimachia ciliata*), 199
fringed orchid
  crested (*Habenaria cristata*), 821
  prairie (*Habenaria leucophaea*), 822
  purple (*Habenaria psycodes*), 822
  ragged (*Habenaria lacera*), 822
  white (*Habenaria blephariglottis*), 821
  yellow (*Habenaria ciliaris*), 821
fringeleaf-ruellia (*Ruellia humilis*), 460
fringe-tree (*Chionanthus*), 435
frog's-bit (*Hydrocharis*), 606; (*Limnobium*), 607
  American (*Limnobium spongia*), 607
  European (*Hydrocharis morsus-ranae*), 606
frog's-bit family (HYDROCHARITACEAE), 606
frosted hawthorn (*Crataegus pruinosa*), 252
frost-grape (*Vitis riparia, V. vulpina*), 325
frostweed (*Helianthemum*), 139; (*Verbesina virginica*), 502
fullers' teasel (*Dipsacus fullonum*), 487
fumitory (*Fumaria*), 66
fumitory family (FUMARIACEAE), 66
Furbish's lousewort (*Pedicularis furbishiae*), 457

gale, sweet (*Myrica gale*), 75
gallberry (*Ilex coriacea*), 309
gall-of-the-earth (*Prenanthes trifoliolata*), 588
gama-grass (*Tripsacum*), 774
garden-aconite (*Aconitum napellus*), 47
garden black currant (*Ribes nigrum*), 206
garden-cress (*Lepidium sativum*), 165
garden evening-primrose (*Oenothera glazioviana*), 299
garden forget-me-not (*Myosotis sylvatica*), 397
garden-gooseberry (*Ribes uva-crispa*), 206
garden-heliotrope (*Valeriana officinalis*), 486
garden-loosestrife (*Lysimachia vulgaris*), 201
garden-petunia (*Petunia ×hybrida*), 383
garden-pink (*Dianthus plumarius*), 111
garden red currant (*Ribes sativum*), 207
garden-rocket (*Eruca sativa*), 163
garden-sage (*Salvia officinalis*), 430
garden-zinnia (*Zinnia elegans*), 502
garlic (*Allium sativum*), 791
  field- (*Allium oleraceum*), 791; (*A. vineale*), 790
garlic-mustard (*Alliaria*), 180

gaura
  biennial (*Gaura biennis*), 302
  scarlet (*Gaura coccinea*), 302
  small-flowered (*Gaura parviflora*), 302
  threadstalk (*Gaura filipes*) 302
gay feather (*Liatris*), 574
gentian (*Gentiana*), 364; (*Gentianella*), 366; (*Gentianopsis*), 366
  Appalachian (*Gentiana decora*), 364
  Blue-Ridge (*Gentiana austromontana*), 365
  bottle- (*Gentiana andrewsii, G. clausa*), 365
  coastal plain (*Gentiana catesbaei*), 364
  fringed (*Gentianopsis*), 366; (*G. crinita*), 366
  Great Lakes (*Gentiana rubricaulis*), 365
  horse- (*Triosteum*), 481
  lesser fringed (*Gentianopsis virgata*), 366
  lesser horse- (*Triosteum angustifolium*), 481
  meadow closed (*Gentiana clausa*), 365
  narrow-leaved (*Gentiana linearis*), 365
  northern (*Gentianella amarella*), 366
  one-flowered (*Gentiana autumnalis*), 364
  pale (*Gentiana flavida*), 366
  perfoliate horse- (*Triosteum perfoliatum*), 481
  pine-barren (*Gentiana autumnalis*), 364
  Plymouth rose (*Sabatia kennedyana*), 363
  prairie- (*Gentiana puberulenta*), 364
  prairie closed (*Gentiana andrewsii*), 365
  soapwort- (*Gentiana saponaria*), 364
  stiff (*Gentianella quinquefolia*), 366
  striped (*Gentiana villosa*), 366
gentian family (GENTIANACEAE), 362
Georgia buckeye (*Aesculus sylvatica*), 331
geranium
  feather- (*Chenopodium botrys*), 87
  meadow- (*Geranium pratense*), 339
  wild (*Geranium*), 338; (*G. maculatum*), 338
geranium family (GERANIACEAE), 338
German catchfly (*Lychnis viscaria*), 107
germander (*Teucrium*), 408
  American (*Teucrium canadense*), 408
  wood- (*Teucrium scorodonia*), 408
germander-speedwell (*Veronica chamaedrys*), 455
German iris (*Iris germanica*), 811
giant bur-reed (*Sparganium eurycarpum*), 776
giant cane (*Arundinaria gigantea*), 690
giant chickweed (*Stellaria aquatica*), 103
giant foxtail (*Setaria faberi*), 766
giant hogweed (*Heracleum mantegazzianum*), 358
giant horsetail (*Equisetum telmateia*), 7
giant-hyssop (*Agastache*), 421
  catnip (*Agastache nepetoides*), 421
  lavender (*Agastache foeniculum*), 421
  purple (*Agastache scrophulariifolia*), 421
giant knotweed (*Polygonum sachalinense*), 124
giant ladies'-tresses (*Spiranthes praecox*), 817
giant ragweed (*Ambrosia trifida*), 515
gill-over-the-ground (*Glechoma hederacea*), 421
ginger, wild (*Asarum*), 41
ginseng
  American (*Panax quinquefolius*), 343
  dwarf (*Panax trifolius*), 343
ginseng family (ARALIACEAE), 342
glade-fern (*Athyrium pycnocarpon*), 19
  silvery (*Athyrium thelypterioides*), 19
glade-mallow (*Napaea*), 136

glasswort (*Salicornia*), 92
  dwarf (*Salicornia bigelovii*), 92
  perennial (*Salicornia virginica*), 92
  western (*Salicornia rubra*), 93
glaucous hawkweed (*Hieracium floribundum*), 592
glaucous king-devil (*Hieracium piloselloides*), 593
glaucous white lettuce (*Prenanthes racemosa*), 586
globe-amaranth (*Gomphrena globosa*), 97
globe-flatsedge (*Cyperus echinatus*), 654
globe-flower (*Trollius*), 45
globe-podded hoary cress (*Cardaria pubescens*), 166
globe-thistle (*Echinops*), 584
globular coneflower (*Ratibida pinnata*), 498
glory-bower (*Clerodendrum japonicum*), 407
goat-grass (*Aegilops*), 730
  jointed (*Aegilops cylindrica*), 730
goat's-beard (*Aruncus dioicus*), 219; (*Tragopogon*), 601
  false (*Astilbe biternata*), 211
  fistulous (*Tragopogon dubius*), 601
  showy (*Tragopogon pratensis*), 601
goat's-rue (*Tephrosia virginiana*), 261
  southern (*Tephrosia spicata*), 261
goblet-aster (*Aster lateriflorus*), 550
gold-crest (*Lophiola aurea*), 781
golden alexanders (*Zizia*), 353
  common (*Zizia aurea*), 353
  heart-leaved (*Zizia aptera*), 353
  mountain (*Zizia trifoliata*), 353
golden aster (*Chrysopsis*), 530; (*Heterotheca*), 531
  falcate (*Chrysopsis falcata*), 531
  grass-leaved (*Chrysopsis graminifolia*), 531
  prairie (*Chrysopsis camporum*), 530
  shaggy (*Chrysopsis mariana*), 530
  woolly-headed (*Chrysopsis gossypina*), 530
golden bell (*Forsythia*), 436
golden carpet (*Sedum acre*), 209
golden club (*Orontium*), 615
golden corydalis (*Corydalis aurea*), 67
golden crownbeard (*Verbesina encelioides*), 501
golden dock (*Rumex maritimus*), 115
golden glow (*Rudbeckia laciniata*), 497
golden-rain tree (*Koelreuteria paniculata*), 330
goldenrod (*Solidago*), 532
  axillary (*Solidago caesia*), 537
  big-leaved (*Solidago macrophylla*), 532
  cave- (*Solidago albopilosa*), 536
  coastal plain flat-topped (*Euthamia tenuifolia*), 545
  coastal swamp- (*Solidago elliottii*), 541
  common (*Solidago canadensis*), 542
  common flat-topped (*Euthamia graminifolia*), 545
  driftless area (*Solidago sciaphila*), 533
  dusty (*Solidago puberula*), 533
  early (*Solidago juncea*), 538
  elm-leaved (*Solidago ulmifolia*), 539
  fascicled piney-woods (*Solidago pinetorum*), 538
  flat-topped (*Euthamia*), 545
  forest- (*Solidago arguta*), 539
  gray (*Solidago nemoralis*), 537
  Great Lakes (*Solidago houghtonii*), 544
  Great Plains flat-topped (*Euthamia gymnospermoides*), 546
  hairy (*Solidago hispida*), 534
  hairy piney-woods (*Solidago fistulosa*), 541
  Kentucky (*Solidago shortii*), 543
  lakes flat-topped (*Euthamia remota*), 546
  leafy piney-woods (*Solidago tortifolia*), 541

goldenrod (*Solidago*) (*continued*)
  licorice- (*Solidago odora*), 541
  Mississippi-valley flat-topped (*Euthamia leptocephala*), 546
  Missouri (*Solidago missouriensis*), 538
  New England (*Solidago cutleri*), 533
  northern bog- (*Solidago uliginosa*), 534
  Nova Scotia flat-topped (*Euthamia galetorum*), 545
  Ohio (*Solidago ohioensis*), 544
  Ozark (*Solidago drummondii*), 539
  riverbank- (*Solidago rupestris*), 542
  Roan Mountain (*Solidago roanensis*), 533
  rough (*Solidago radula*), 543
  rough-leaved (*Solidago patula*), 538
  seaside- (*Solidago sempervirens*), 535
  short-pappus (*Solidago sphacelata*), 540
  showy (*Solidago speciosa*), 534
  smooth (*Solidago gigantea*), 541
  southern bog- (*Solidago gracillima*), 535
  squarrose (*Solidago squarrosa*), 532
  stiff (*Solidago rigida*), 543
  velvety (*Solidago mollis*), 543
  wand- (*Solidago stricta*), 535
  wrinkle-leaved (*Solidago rugosa*), 540
  zigzag (*Solidago flexicaulis*), 536
golden seal (*Hydrastis*), 61
goldenthread (*Coptis*), 61
goldenweed, wax- (*Haplopappus ciliatus*), 529
Goldie's wood-fern (*Dryopteris goldiana*), 25
gold-star (*Chrysogonum*), 513
Good King Henry (*Chenopodium bonus-henricus*), 87
gooseberry
  Appalachian (*Ribes rotundifolium*), 205
  garden- (*Ribes uva-crispa*), 206
  Missouri (*Ribes missouriense*), 205
  northern (*Ribes oxyacanthoides*), 206
gooseberry family (GROSSULARIACEAE), 205
goosefoot (*Chenopodium*), 86
  city- (*Chenopodium urbicum*), 88
  maple-leaved (*Chenopodium simplex*), 88
  nettle-leaved (*Chenopodium murale*), 88
  oak-leaved (*Chenopodium glaucum*), 87
  pitseed (*Chenopodium berlandieri*), 89
  stinking (*Chenopodium vulvaria*), 88
  woodland- (*Chenopodium standleyanum*), 88
goosefoot family (CHENOPODIACEAE), 86
gorse (*Ulex europaeus*), 260
gourd family (CUCURBITACEAE), 150
goutweed (*Aegopodium podagraria*), 347
grama
  blue (*Bouteloua gracilis*), 750
  hairy (*Bouteloua hirsuta*), 750
  side-oats (*Bouteloua curtipendula*), 750
grama-grass (*Bouteloua*), 750
grape (*Vitis*), 324
  cat- (*Vitis palmata*), 324
  catbird- (*Vitis palmata*), 324
  chicken- (*Vitis vulpina*), 325
  fox- (*Vitis labrusca*), 324
  frost- (*Vitis riparia, V. vulpina*), 325
  graybark- (*Vitis cinerea*), 325
  muscadine- (*Vitis rotundifolia*), 324
  New England (*Vitis novae-angliae*), 325
  pigeon- (*Vitis aestivalis*), 324; (*V. cinerea*), 325
  possum- (*Vitis baileyana*), 325
  raccoon- (*Ampelopsis cordata*), 323

grape (*Vitis*) (*continued*)
 red (*Vitis palmata*), 324
 riverbank- (*Vitis riparia*), 325
 sand- (*Vitis rupestris*), 324
 scuppernong- (*Vitis rotundifolia*), 324
 sugar- (*Vitis rupestris*), 324
 summer- (*Vitis aestivalis*), 324
 winter- (*Vitis vulpina*), 325
grape family (VITACEAE), 323
grape-fern (*Botrychium*), 8
 Alabama (*Botrychium jenmanii*), 9
 blunt-lobed (*Botrychium oneidense*), 8
 daisy-leaved (*Botrychium matricariifolium*), 8
 lace-frond (*Botrychium dissectum*), 8
 lance-leaved (*Botrychium lanceolatum*), 8
 leathery (*Botrychium multifidum*), 8
 little (*Botrychium simplex*), 8
 sparse-lobed (*Botrychium biternatum*), 8
grape-honeysuckle (*Lonicera prolifera*), 480
grape-hyacinth (*Muscari*), 788; (*M. botryoides*), 788
grape-woodbine (*Parthenocissus vitacea*), 323
grass
 alpine blue- (*Poa alpina*), 705
 American dune- (*Leymus mollis*), 728
 American manna- (*Glyceria grandis*), 708
 American slough- (*Beckmannia syzigachne*), 720
 Amur silver- (*Miscanthus sacchariflorus*), 771
 arctic blue- (*Poa arctica*), 701
 arrow- (*Triglochin*), 608
 barnyard- (*Echinochloa crusgalli*), 763; (*E. muricata*), 762
 beach- (*Ammophila*), 717
 bead- (*Paspalum*), 763
 beard- (*Gymnopogon*), 748
 bent- (*Agrostis*), 717
 bent-awn plume- (*Erianthus contortus*), 771
 Bermuda- (*Cynodon dactylon*), 749
 big cord- (*Spartina cynosuroides*), 749
 big quaking (*Briza maxima*), 700
 big-seed scorpion- (*Myosotis macrosperma*), 397
 billion-dollar (*Echinochloa crusgalli var. frumentacea*), 763
 black (*Juncus gerardii*), 622
 blackseed-rice- (*Oryzopsis racemosa*), 693
 blue- (*Poa*), 701
 bluebunch wheat- (*Elytrigia spicata*), 727
 blue-eyed (*Sisyrinchium*), 808
 blue scorpion- (*Myosotis micrantha*), 397
 bottlebrush- (*Elymus hystrix*), 726
 brome- (*Bromus*), 720
 brook- (*Catabrosa*), 706
 brown plume- (*Erianthus brevibarbis*), 770
 buffalo- (*Buchloe*), 751
 bulbous blue- (*Poa bulbosa*), 702
 Canada blue- (*Poa compressa*), 701
 canary- (*Phalaris canariensis*), 714
 Canby-blue- (*Poa canbyi*), 704
 Carolina love- (*Eragrostis pectinacea*), 739
 Carolina whip- (*Scleria pauciflora*), 659
 catchfly- (*Leersia lenticularis*), 690
 Chinese cup- (*Eriochloa villosa*), 762
 coastal manna- (*Glyceria obtusa*), 707
 common arrow- (*Triglochin maritima*), 608
 common eel- (*Zostera marina*), 614
 common star- (*Hypoxis hirsuta*), 802
 common velvet- (*Holcus lanatus*), 712
 conifer cotton- (*Eriophorum tenellum*), 637

grass (*continued*)
 cord- (*Spartina*), 749
 cotton- (*Eriophorum*), 637
 crab- (*Digitaria*), 767
 crested wheat- (*Agropyron*), 730
 crowfoot- (*Dactyloctenium aegyptium*), 747
 cup- (*Eriochloa*), 762
 curly (*Schizaea*), 11
 Dallis- (*Paspalum dilatatum*), 764
 dark-scale cotton- (*Eriophorum viridicarinatum*), 638
 ditch- (*Ruppia*), 613
 dog's-tail (*Cynosurus*), 699
 downy oat- (*Danthonia sericea*), 733
 dwarf crab- (*Digitaria serotina*), 767
 early scorpion- (*Myosotis verna*), 397
 eastern manna- (*Glyceria septentrionalis*), 706
 eel- (*Zostera*), 614
 English rye- (*Lolium perenne var. perenne*), 698
 European alkali- (*Puccinellia distans*), 700
 fall witch- (*Leptoloma*), 768
 feather finger- (*Chloris virgata*), 747
 field scorpion- (*Myosotis arvensis*), 397
 finger- (*Chloris*), 747
 forest-blue- (*Poa sylvestris*), 703
 fowl-manna- (*Glyceria striata*), 708
 fowl meadow- (*Poa palustris*), 704
 foxtail- (*Setaria*), 766
 gama- (*Tripsacum*), 774
 goat- (*Aegilops*), 730
 grama- (*Bouteloua*), 750
 gray hair- (*Corynephorus canescens*), 713
 green needle- (*Stipa viridula*), 693
 hair- (*Aira*), 713; (*Deschampsia*), 712; (*Muhlenbergia capillaris*), 746
 heather- (*Danthonia decumbens*), 731
 hedgehog- (*Cenchrus echinatus*), 769
 India-love- (*Eragrostis pilosa*), 739
 Indian (*Sorghastrum nutans*), 773
 Indian rice- (*Oryzopsis hymenoides*), 692
 inland-blue- (*Poa interior*), 703
 Italian rye- (*Lolium perenne var. aristatum*), 698
 Japanese lawn- (*Zoysia japonica*), 750
 Johnson- (*Sorghum halepense*), 772
 jointed goat- (*Aegilops cylindrica*), 730
 June- (*Bromus tectorum*), 723; (*Koeleria pyramidata*), 709
 Kentucky blue- (*Poa pratensis*), 701
 knot- (*Paspalum distichum*), 765
 Korean lawn- (*Zoysia japonica*), 750
 lace- (*Eragrostis capillaris*), 740
 little quaking (*Briza minor*), 700
 love- (*Eragrostis*), 738
 lyme- (*Leymus arenarius*), 728
 manna- (*Glyceria*), 706
 marsh- (*Scolochloa festucacea*), 698
 marsh blue- (*Poa paludigena*), 703
 Merlin's (*Isoetes*), 4
 moor- (*Molinia caerulea*), 732
 moor-mat- (*Nardus stricta*), 692
 narrow plume- (*Erianthus strictus*), 770
 needle- (*Stipa*), 693
 needle-and-thread (*Stipa comata*), 693
 northeastern manna- (*Glyceria melicaria*), 707
 northern crab- (*Digitaria sanguinalis*), 767
 northern manna- (*Glyceria borealis*), 706
 Nuttall's alkali- (*Puccinellia nuttalliana*), 699
 oat- (*Arrhenatherum*), 709

grass *(continued)*
   orange- *(Hypericum gentianoides)*, 130
   orchard- *(Dactylis)*, 705
   panic- *(Panicum)*, 751
   pepper- *(Lepidium)*, 164
   perennial quaking *(Briza media)*, 700
   plains-blue- *(Poa arida)*, 701
   plains reed- *(Calamagrostis montanensis)*, 715
   plume- *(Erianthus)*, 771
   pod- *(Scheuchzeria palustris)*, 607
   porcupine- *(Stipa)*, 693; *(S. spartea)*, 694
   poverty- *(Sporobolus vaginiflorus)*, 743
   poverty-oat- *(Danthonia spicata)*, 731
   prairie cord- *(Spartina pectinata)*, 749
   prairie cup- *(Eriochloa contracta)*, 762
   purple love- *(Eragrostis spectabilis)*, 738
   purple reed- *(Calamagrostis purpurascens)*, 714
   purple sand- *(Triplasis purpurea)*, 736
   quack- *(Elytrigia repens)*, 727
   quaking *(Briza)*, 700
   rabbitfoot- *(Polypogon monspeliensis)*, 719
   rattlesnake-manna- *(Glyceria canadensis)*, 707
   redhead- *(Potamogeton perfoliatus)*, 613
   reed- *(Calamagrostis)*, 714
   reed canary- *(Phalaris arundinacea)*, 714
   rescue- *(Bromus catharticus)*, 721
   rib- *(Plantago lanceolata)*, 433
   rice- *(Oryzopsis)*, 692
   rice cut- *(Leersia oryzoides)*, 690
   ripgut- *(Bromus rigidus)*, 724
   rough blue- *(Poa trivialis)*, 702
   rough-leaved rice- *(Oryzopsis asperifolia)*, 693
   rusty cotton- *(Eriophorum chamissonis)*, 637
   rye- *(Lolium perenne)*, 698
   salt- *(Distichlis)*, 735
   salt-marsh wheat- *(Elytrigia pungens)*, 727
   salt-meadow cord- *(Spartina patens)*, 750
   sand- *(Triplasis)*, 736
   sand-love- *(Eragrostis trichodes)*, 738
   saw- *(Cladium jamaicense)*, 650
   scorpion- *(Myosotis)*, 396
   scratch- *(Muhlenbergia asperifolia)*, 747
   seashore salt- *(Distichlis spicata* var. *spicata)*, 735
   seaside alkali- *(Puccinellia maritima)*, 699
   short-anthered cotton- *(Eriophorum brachyantherum)*, 637
   silver-hair- *(Aira caryophyllea)*, 713
   silver-plume- *(Erianthus alopecuroides)*, 771
   slender arrow- *(Triglochin palustris)*, 608
   slender cotton- *(Eriophorum gracile)*, 638
   slender wheat- *(Elymus trachycaulus)*, 725
   smooth cord- *(Spartina alterniflora)*, 749
   smooth crab- *(Digitaria ischaemum)*, 767
   smut- *(Sporobolus indicus)*, 742
   southern arrow- *(Triglochin striata)*, 608
   southern crab- *(Digitaria ciliaris)*, 768
   spear- *(Poa annua)*, 703
   star- *(Hypoxis)*, 802
   stink- *(Eragrostis cilianensis)*, 739
   Sudan- *(Sorghum bicolor)*, 772
   sugar-cane plume- *(Erianthus giganteus)*, 770
   sweet *(Hierochloe odorata)*, 713
   sweet vernal *(Anthoxanthum)*, 714
   switch- *(Panicum virgatum)*, 753
   tall blue- *(Poa chaixii)*, 704
   tall oat- *(Arrhenatherum elatius)*, 709
   tall wheat- *(Elytrigia elongata)*, 728

grass *(continued)*
   tape- *(Vallisneria)*, 607
   tawny cotton- *(Eriophorum virginicum)*, 638
   thick-spike wheat- *(Elytrigia dasystachya)*, 727
   thin-scale cotton- *(Eriophorum angustifolium)*, 638
   tickle- *(Agrostis hyemalis)*, 717
   timber-oat- *(Danthonia intermedia)*, 732
   toothache- *(Ctenium aromaticum)*, 748
   tufted hair- *(Deschampsia cespitosa)*, 712
   tumble- *(Schedonnardus paniculatus)*, 748
   tussock cotton- *(Eriophorum vaginatum)*, 637
   umbrella- *(Fuirena)*, 658
   velvet- *(Holcus)*, 712
   velvet panic *(Panicum scoparium)*, 759
   water scorpion- *(Myosotis scorpioides)*, 396
   water star- *(Zosterella)*, 779
   wedge- *(Sphenopholis obtusata)*, 711
   weeping love- *(Eragrostis curvula)*, 739
   western wheat- *(Elytrigia smithii)*, 727
   wheat- *(Elytrigia)*, 727
   whip- *(Scleria triglomerata)*, 659
   white *(Leersia virginica)*, 690
   whitlow- *(Draba verna)*, 168
   widgeon- *(Ruppia)*, 613
   wild oat- *(Danthonia)*, 731
   windmill finger- *(Chloris verticillata)*, 747
   wiry witch- *(Panicum flexile)*, 753
   witch- *(Panicum capillare)*, 753
   wood-blue- *(Poa nemoralis)*, 704
   wool- *(Scirpus cyperinus)*, 637
   worm- *(Spigelia)*, 361
   yard- *(Eleusine indica)*, 741
   yellow and blue scorpion- *(Myosotis discolor)*, 397
   yellow-eyed *(Xyris)*, 617
grass family (POACEAE), 690
grass-fern *(Asplenium septentrionale)*, 18
grass-leaved blazing star *(Liatris graminifolia)*, 576
grass-leaved golden aster *(Chrysopsis graminifolia)*, 531
grass-leaved sagittaria *(Sagittaria graminea)*, 605
grass-leaved water-plantain *(Alisma gramineum)*, 602
grass-of-Parnassus *(Parnassia)*, 216
   American *(Parnassia glauca)*, 216
   arctic *(Parnassia palustris)*, 217
   big-leaved *(Parnassia grandifolia)*, 217
   kidney-leaved *(Parnassia asarifolia)*, 216
   small-flowered *(Parnassia parviflora)*, 217
grass-pink *(Calopogon)*, 824
grasswort, starry *(Cerastium arvense)*, 102
graybark-grape *(Vitis cinerea)*, 325
gray birch *(Betula populifolia)*, 83
gray goldenrod *(Solidago nemoralis)*, 537
gray groundsel *(Senecio canus)*, 523
gray hairgrass *(Corynephorus canescens)*, 713
gray scurf-pea *(Psoralidium tenuiflorum)*, 284
gray tansy-mustard *(Descurainia richardsonii)*, 182
great burdock *(Arctium lappa)*, 579
great burnet *(Sanguisorba officinalis)*, 237
greater duckweed *(Spirodela polyrhiza)*, 616
greater periwinkle *(Vinca major)*, 368
great Indian-plantain *(Cacalia muehlenbergii)*, 527
Great Lakes corn-salad *(Valerianella chenopodifolia)*, 486
Great Lakes gentian *(Gentiana rubricaulis)*, 365
Great Lakes goldenrod *(Solidago houghtonii)*, 544
Great Plains flat-topped goldenrod *(Euthamia gymnospermoides)*, 546

Great Plains ladies'-tresses (*Spiranthes magnicamporum*), 816
Great Plains potentilla (*Potentilla effusa*), 225
great-spurred violet (*Viola selkirkii*), 145
great-water-dock (*Rumex orbiculatus*), 113
great willow-herb (*Epilobium angustifolium*), 296
Grecian foxglove (*Digitalis lanata*), 449
green adder's-mouth (*Malaxis unifolia*), 825
green alder (*Alnus viridis*), 84
green amaranth (*Amaranthus viridis*), 96
greenbrier (*Smilax*), 804
  bristly (*Smilax hispida*), 805
green dragon (*Arisaema dracontium*), 616
green foxtail (*Setaria viridis*), 766
green hawthorn (*Crataegus viridis*), 249
green hellebore (*Helleborus viridis*), 45
green milkweed (*Asclepias viridiflora*), 374
green needle-grass (*Stipa viridula*), 693
green sorrel (*Rumex acetosa*), 113
green spleenwort (*Asplenium viride*), 18
green trillium (*Trillium viride*), 796
green violet (*Hybanthus*), 142
greenweed, dyer's (*Genista tinctoria*), 259
gromwell (*Lithospermum*), 399; (*L. officinale*), 400
  American (*Lithospermum latifolium*), 400
  corn- (*Lithospermum arvense*), 400
  eastern false (*Onosmodium virginianum*), 401
  false (*Onosmodium*), 401
  western false (*Onosmodium molle*), 401
grooved yellow flax (*Linum sulcatum*), 326
ground-cedar
  northern (*Lycopodium complanatum*), 3
  southern (*Lycopodium digitatum*), 3
  wiry (*Lycopodium tristachyum*), 3
ground-cherry (*Physalis*), 376; (*Prunus fruticosa*), 242
  clammy (*Physalis heterophylla*), 376
  downy (*Physalis pubescens*), 378
  longleaf (*Physalis longifolia*), 377
  Ozark (*Physalis missouriensis*), 378
  plains-sandhill (*Physalis hispida*), 377
  prairie (*Physalis pumila*), 377
  Virginia (*Physalis virginiana*), 377
  white-flowered (*Leucophysalis grandiflora*), 376
ground-hemlock (*Taxus canadensis*), 31
ground-ivy (*Glechoma*), 421
ground-nut (*Apios*), 286
  common (*Apios americana*), 286
  Kentucky (*Apios priceana*), 286
ground-pine (*Ajuga chamaepitys*), 408; (*Lycopodium obscurum*), 2
ground-plum (*Astragalus crassicarpus*), 265
groundsel (*Senecio*), 523
  Appalachian (*Senecio anonymus*), 524
  common (*Senecio vulgaris*), 526
  gray (*Senecio canus*), 523
  heart-leaved (*Senecio aureus*), 525
  New England (*Senecio schweinitzianus*), 525
  northern meadow- (*Senecio pauperculus*), 524
  northern swamp- (*Senecio congestus*), 523
  Platte (*Senecio plattensis*), 524
  running (*Senecio obovatus*), 524
  seaside- (*Senecio pseudoarnica*), 523
  shale-barren (*Senecio antennariifolius*), 523
  single-stemmed (*Senecio integerrimus*), 523
  southern woolly (*Senecio tomentosus*), 523
  sticky (*Senecio viscosus*), 525
groundsel (*Senecio*) *(continued)*
  taller discoid (*Senecio indecorus*), 524
  western heart-leaved (*Senecio pseudaureus*), 525
  woodland- (*Senecio sylvaticus*), 526
groundsel-tree (*Baccharis halimifolia*), 562
gum
  black (*Nyssa sylvatica*), 306
  cotton- (*Nyssa aquatica*), 306
  red (*Liquidambar*), 68
  sour (*Nyssa*), 306
  sweet (*Liquidambar*), 68
gum-weed (*Grindelia*), 531
  curly-top (*Grindelia squarrosa*), 532
  spiny-tooth (*Grindelia lanceolata*), 531

hackberry (*Celtis*), 69
  dwarf (*Celtis tenuifolia*), 69
  northern (*Celtis occidentalis*), 69
  southern (*Celtis laevigata*), 69
hairgrass (*Aira*), 713; (*Deschampsia*), 712; (*Muhlenbergia capillaris*), 746
  gray (*Corynephorus canescens*), 713
  silver- (*Aira caryophyllea*), 713
  tufted (*Deschampsia cespitosa*), 712
hairy angelica (*Angelica venenosa*), 359
hairy chess (*Bromus commutatus*), 723
hairy goldenrod (*Solidago hispida*), 534
hairy grama (*Bouteloua hirsuta*), 750
hairy heart-leaved aster (*Aster drummondii*), 552
hairy honeysuckle (*Lonicera hirsuta*), 480
hairy lespedeza (*Lespedeza hirta*), 281
hairy milk-pea (*Galactia volubilis*), 287
hairy nightshade (*Solanum sarrachoides*), 379
hairy phlox (*Phlox amoena*), 389
hairy piney-woods goldenrod (*Solidago fistulosa*), 541
hairy spurge (*Euphorbia vermiculata*), 320
hairy sunflower (*Helianthus hirsutus*), 490
hairy tall lettuce (*Lactuca hirsuta*), 589
hairy tickseed (*Coreopsis pubescens*), 503
hairy umbrella-wort (*Mirabilis hirsuta*), 85
hairy vetch (*Vicia villosa*), 273
hairy water-clover (*Marsilea vestita*), 30
hairy waterleaf (*Hydrophyllum macrophyllum*), 393
hairy water-primrose (*Ludwigia pilosa*), 295
hairy willow-herb (*Epilobium hirsutum*), 297
  small-flowered (*Epilobium parviflorum*), 297
halberd-leaved tearthumb (*Polygonum arifolium*), 123
hammock sweet azalea (*Rhododendron serrulatum*), 185
harbinger of spring (*Erigenia*), 361
hard fescue (*Festuca trachyphylla*), 696
hardhack (*Spiraea tomentosa*), 219
hard-heads (*Centaurea scabiosa*), 584
hard maple (*Acer saccharum*), 331
hardstem-bulrush (*Scirpus acutus*), 633
harebell (*Campanula*), 465; (*C. rotundifolia*), 464
hare's-ear mustard (*Conringia*), 164
hare's-foot dalea (*Dalea leporina*), 282
Hart's tongue (*Phyllitis*), 18
harvest-lice (*Agrimonia*), 236
hastate Indian-plantain (*Cacalia suaveolens*), 527
haw
  black (*Viburnum prunifolium*), 484
  possum- (*Viburnum nudum var. nudum*), 484
  southern black (*Viburnum rufidulum*), 484
hawkbit
  big (*Leontodon hispidus*), 601

hawkbit *(continued)*
little (*Leontodon taraxacoides*), 601
hawk's beard (*Crepis*), 595
Asiatic (*Youngia japonica*), 596
hawkweed (*Hieracium*), 592
beaked (*Hieracium gronovii*), 594
Canada (*Hieracium kalmii*), 594
European (*Hieracium lachenalii*), 593
glaucous (*Hieracium floribundum*), 592
long-haired (*Hieracium longipilum*), 594
mouse-ear (*Hieracium pilosella*), 592
northeastern (*Hieracium robinsonii*), 593
northern (*Hieracium umbellatum*), 594
panicled (*Hieracium paniculatum*), 595
shale-barren (*Hieracium traillii*), 593
sticky (*Hieracium scabrum*), 595
veiny (*Hieracium venosum*), 594
wall- (*Hieracium murorum*), 593
whiplash- (*Hieracium flagellare*), 592
hawkweed ox-tongue (*Picris hieracioides*), 600
hawthorn (*Crataegus*), 247
barberry (*Crataegus berberifolia*), 249
Biltmore- (*Crataegus intricata*), 250
black (*Crataegus douglasii*), 247
Brainerd- (*Crataegus brainerdii*), 251
broadleaf- (*Crataegus dilatata*), 252
dotted (*Crataegus punctata*), 249
downy (*Crataegus mollis*), 251
fanleaf- (*Crataegus flabellata*), 253
fireberry- (*Crataegus chrysocarpa*), 250
fleshy (*Crataegus succulenta*), 248
frosted (*Crataegus pruinosa*), 252
green (*Crataegus viridis*), 249
Kansas- (*Crataegus coccinioides*), 251
littlehip- (*Crataegus spathulata*), 247
one-flower (*Crataegus uniflora*), 248
one-seed (*Crataegus monogyna*), 247
parsley- (*Crataegus marshallii*), 247
pear- (*Crataegus calpodendron*), 248
scarlet (*Crataegus coccinea*), 252
yellow (*Crataegus michauxii*), 250
hay-scented fern (*Dennstaedtia punctilobula*), 12
hazel-dodder (*Cuscuta coryli*), 388
hazel-nut (*Corylus*), 82
American (*Corylus americana*), 82
beaked (*Corylus cornuta*), 82
heart(s)
big floating (*Nymphoides aquatica*), 389
blue- (*Buchnera*), 452
floating (*Nymphoides*), 389
little floating (*Nymphoides cordata*), 389
wild bleeding (*Dicentra eximia*), 66
yellow floating (*Nymphoides peltata*), 389
heart-leaf (*Hexastylis*), 40
heart-leaved arnica (*Arnica cordifolia*), 500
heart-leaved golden alexanders (*Zizia aptera*), 353
western (*Senecio pseudaureus*), 525
heart-leaved tragia (*Tragia cordata*), 313
heart-leaved twayblade (*Listera cordata*), 814
heart-leaved umbrella-wort (*Mirabilis nyctaginea*), 85
heart-podded hoary cress (*Cardaria draba*), 166
heath (*Erica*), 190
Cornish (*Erica vagans*), 190
cross-leaved (*Erica tetralix*), 190

heath (*Erica*) *(continued)*
mountain- (*Phyllodoce caerulea*), 187
twisted (*Erica cinerea*), 190
heather (*Calluna*), 190
false (*Hudsonia*), 140
heather-grass (*Danthonia decumbens*), 731
heath family (ERICACEAE), 184
hedge-bindweed (*Calystegia sepium*), 385
hedgehog-grass (*Cenchrus echinatus*), 769
hedge-hyssop (*Gratiola*), 438
yellow (*Gratiola aurea*), 439
hedge-maple (*Acer campestre*), 332
hedge-mustard (*Sisymbrium officinale*), 180
hedge-nettle (*Stachys*), 425; (*S. palustris*), 425
field (*Stachys arvensis*), 427
hispid (*Stachys hispida*), 425
hyssop- (*Stachys hyssopifolia*), 425
smooth (*Stachys tenuifolia*), 426
hedge-parsley (*Torilis*), 350
field (*Torilis arvensis*), 350
Japanese (*Torilis japonica*), 350
hedge-vetch (*Vicia sepium*), 273
heliotrope (*Heliotropium*), 395
European (*Heliotropium europaeum*), 395
garden- (*Valeriana officinalis*), 486
pasture- (*Heliotropium tenellum*), 395
seaside- (*Heliotropium curassavicum*), 395
hellebore
false (*Veratrum*), 784
green (*Helleborus viridis*), 45
helleborine (*Epipactis helleborine*), 814
hemlock (*Tsuga*), 32
bulbiliferous water- (*Cicuta bulbifera*), 355
Carolina (*Tsuga caroliniana*), 32
common water- (*Cicuta maculata*), 355
eastern (*Tsuga canadensis*), 32
ground- (*Taxus canadensis*), 31
poison- (*Conium maculatum*), 351
water- (*Cicuta*), 355
hemlock-parsley (*Conioselinum*), 355
hemp (*Cannabis*), 70
Indian (*Cannabis*), 70
hemp-dogbane (*Apocynum cannabinum*), 369
hemp-nettle (*Galeopsis*), 424
hempweed, climbing (*Mikania*), 573
hen-and-chickens (*Sempervivum tectorum*), 210
henbane (*Hyoscyamus*), 382
henbit (*Lamium amplexicaule*), 424
hepatica (*Hepatica*), 50
round-lobed (*Hepatica americana*), 50
sharp-lobed (*Hepatica acutiloba*), 50
herba impia (*Filago germanica*), 563
herb-like willow (*Salix herbacea*), 156
herb-mercury (*Mercurialis annua*), 315
herb-Robert (*Geranium robertianum*), 339
herb-sophia (*Descurainia sophia*), 182
Hercules' club (*Aralia spinosa*), 342; (*Zanthoxylum clava-herculis*), 336
hickory (*Carya*), 73
bitternut- (*Carya cordiformis*), 73
black (*Carya texana*), 74
Carolina shagbark- (*Carya carolinae-septentrionalis*), 74
mockernut- (*Carya tomentosa*), 73
Ozark (*Carya texana*), 74
pale (*Carya pallida*), 74
pignut- (*Carya glabra*), 74

hickory (*Carya*) *(continued)*
red (*Carya ovalis*), 74
shagbark- (*Carya ovata*), 74
shellbark- (*Carya laciniosa*), 73
sweet pignut- (*Carya ovalis*), 74
water- (*Carya aquatica*), 73
highbush-blueberry (*Vaccinium corymbosum*), 193
southern (*Vaccinium elliottii*), 194
high mallow (*Malva sylvestris*), 135
hillside-blueberry (*Vaccinium pallidum*), 193
Hill's pondweed (*Potamogeton hillii*), 610
Himalayan blackberry (*Rubus discolor*), 234
hispid buttercup (*Ranunculus hispidus*), 55
hispid hedge-nettle (*Stachys hispida*), 425
hoary alyssum (*Berteroa incana*), 168
hoary bitter-cress (*Cardamine hirsuta*), 173
hoary cress (*Cardaria*), 166
globe-podded (*Cardaria pubescens*), 166
heart-podded (*Cardaria draba*), 166
lens-podded (*Cardaria chalepensis*), 166
hoary puccoon (*Lithospermum canescens*), 399
hoary tick-trefoil (*Desmodium canescens*), 277
hoary vervain (*Verbena stricta*), 405
hobble-bush (*Viburnum alnifolium*), 485
hog-peanut (*Amphicarpaea*), 286
hogweed, giant (*Heracleum mantegazzianum*), 358
hollow-rooted thistle (*Cirsium hillii*), 581
hollow-stemmed Joe-Pye weed (*Eupatorium fistulosum*), 568
holly (*Ilex*), 309
American (*Ilex opaca*), 309
Appalachian mountain- (*Nemopanthus collinus*), 310
bigleaf- (*Ilex montana*), 310
common mountain- (*Nemopanthus mucronatus*), 310
mountain- (*Nemopanthus*), 310
Sarvis- (*Ilex amelanchier*), 310
holly family (AQUIFOLIACEAE), 309
holly-fern
Braun's (*Polystichum braunii*), 28
northern (*Polystichum lonchitis*), 28
hollyhock (*Althaea rosea*), 134
honesty (*Lunaria annua*), 168
perennial (*Lunaria rediviva*), 168
honewort (*Cryptotaenia*), 346
honey-locust (*Gleditsia triacanthos*), 257
honeysuckle (*Lonicera*), 478
bearberry- (*Lonicera involucrata*), 478
bush- (*Diervilla*), 478
European fly- (*Lonicera xylosteum*), 479
fly- (*Lonicera canadensis*), 478
grape- (*Lonicera prolifera*), 480
hairy (*Lonicera hirsuta*), 480
Japanese (*Lonicera japonica*), 479
swamp fly- (*Lonicera oblongifolia*), 479
Tatarian (*Lonicera tatarica*), 478
trumpet- (*Lonicera sempervirens*), 480
wild (*Lonicera dioica*), 480
yellow (*Lonicera flava*), 480
honeysuckle family (CAPRIFOLIACEAE), 478
hooded ladies'-tresses (*Spiranthes romanzoffiana*), 815
hooked crowfoot (*Ranunculus recurvatus*), 54
hooked scaleseed (*Spermolepis echinata*), 350
Hooker's orchid (*Habenaria hookeri*), 821
hook-spurred violet (*Viola adunca*), 148
hop-clover
little (*Trifolium dubium*), 271
hop-clover *(continued)*
palmate (*Trifolium aureum*), 271
pinnate (*Trifolium campestre*), 271
hop-hornbeam (*Ostrya*), 82
hops (*Humulus*), 70; (*H. lupulus*), 70
Japanese (*Humulus japonicus*), 70
hop-tree (*Ptelea*), 336
common (*Ptelea trifoliata*), 336
horehound (*Marrubium vulgare*), 412
American water- (*Lycopus americanus*), 420
black (*Ballota*), 424
European water- (*Lycopus europaeus*), 420
narrow-leaved water- (*Lycopus angustifolius*), 420
northern water- (*Lycopus uniflorus*), 419
stalked water- (*Lycopus rubellus*), 420
Virginia water- (*Lycopus virginicus*), 419
water- (*Lycopus*), 419
western water- (*Lycopus asper*), 420
horehound-motherwort (*Leonurus marrubiastrum*), 424
hornbeam (*Carpinus caroliniana*), 82
hop- (*Ostrya*), 82
horned pondweed (*Zannichellia*), 614
horned pondweed family (ZANNICHELLIACEAE), 614
horned poppy (*Glaucium*), 64
Hornemann's willow-herb (*Epilobium hornemannii*), 298
hornwort (*Ceratophyllum*), 44
hornwort family (CERATOPHYLLACEAE), 44
horse-balm (*Collinsonia*), 418
northern (*Collinsonia canadensis*), 418
whorled (*Collinsonia verticillata*), 418
horse-chestnut (*Aesculus hippocastanum*), 330
horse-chestnut family (HIPPOCASTANACEAE), 330
horse-gentian (*Triosteum*), 481
lesser (*Triosteum angustifolium*), 481
perfoliate (*Triosteum perfoliatum*), 481
horse-mint (*Mentha longifolia*), 417; (*Monarda punctata*), 427
horse-nettle (*Solanum carolinense*), 380
horse-radish (*Armoracia rusticana*), 170
horse-sugar (*Symplocos*), 199
horsetail(s) (*Equisetum*), 5
common (*Equisetum arvense*), 7
field- (*Equisetum arvense*), 7
giant (*Equisetum telmateia*), 7
marsh- (*Equisetum palustre*), 6
meadow- (*Equisetum pratense*), 6
water- (*Equisetum fluviatile*), 6
woodland- (*Equisetum sylvaticum*), 7
horsetail family (EQUISETACEAE), 5
horseweed (*Conyza canadensis*), 561
hortulan plum (*Prunus hortulana*), 243
hound's tongue (*Cynoglossum officinale*), 399
huckleberry (*Gaylussacia*), 194
black (*Gaylussacia baccata*), 194
box- (*Gaylussacia brachycera*), 195
dwarf (*Gaylussacia dumosa*), 195
Hudson River water-nymph (*Najas muenscheri*), 614
Hudson sagittaria (*Sagittaria subulata*), 602
hunger-weed (*Ranunculus arvensis*), 55
hyacinth
grape- (*Muscari*), 788; (*M. botryoides*), 788
water- (*Eichhornia crassipes*), 779
wild (*Camassia scilloides*), 786
hyacinth bean (*Dolichos lablab*), 285
hydrangea (*Hydrangea*), 205
American (*Hydrangea arborescens*), 205

hydrangea (*Hydrangea*) (*continued*)
    climbing (*Decumaria barbara*), 204
    oak-leaved (*Hydrangea quercifolia*), 205
hydrangea family (HYDRANGEACEAE), 204
hyssop (*Hyssopus officinalis*), 414
    catnip giant- (*Agastache nepetoides*), 421
    giant- (*Agastache*), 421
    hedge- (*Gratiola*), 438
    lavender giant- (*Agastache foeniculum*), 421
    purple giant- (*Agastache scrophulariifolia*), 421
    water- (*Bacopa*), 438
    yellow hedge- (*Gratiola aurea*), 439
hyssop-daisy (*Erigeron hyssopifolius*), 561
hyssop-hedge-nettle (*Stachys hyssopifolia*), 425

Illinois pondweed (*Potamogeton illinoensis*), 613
India-lovegrass (*Eragrostis pilosa*), 739
Indian apple (*Datura wrightii*), 382
Indian cucumber-root (*Medeola*), 797
Indian grass (*Sorghastrum nutans*), 773
Indian hemp (*Cannabis*), 70
Indian hemp family (CANNABACEAE), 70
Indian mustard (*Brassica juncea*), 162
Indian paintbrush (*Castilleja*), 458
Indian-physic
    midwestern (*Porteranthus stipulatus*), 220
    mountain (*Porteranthus trifoliatus*), 220
Indian pipe (*Monotropa uniflora*), 196
Indian pipe family (MONOTROPACEAE), 196
Indian-plantain (*Cacalia*), 527
    great (*Cacalia muehlenbergii*), 527
    hastate (*Cacalia suaveolens*), 527
    pale (*Cacalia atriplicifolia*), 527
    tuberous (*Cacalia plantaginea*), 528
Indian ricegrass (*Oryzopsis hymenoides*), 692
Indian strawberry (*Duchesnea indica*), 221
Indian tobacco (*Lobelia inflata*), 467
Indian turnip (*Arisaema triphyllum*), 616
Indian yellow-cress (*Rorippa indica*), 177
India wheat (*Fagopyrum tataricum*), 124
indigo
    blue false (*Baptisia australis*), 258
    Carolina wild (*Baptisia cinerea*), 259
    false (*Amorpha fruticosa*), 283; (*Baptisia*), 258
    milky wild (*Baptisia lactea*), 259
    plains wild (*Baptisia bracteata*), 258
    shining false (*Amorpha nitens*), 282
    white wild (*Baptisia alba*), 259
    wild (*Baptisia*), 258
    yellow wild (*Baptisia tinctoria*), 259
inflated bladderwort (*Utricularia inflata*), 463
inkberry (*Ilex glabra*), 309
inland-bluegrass (*Poa interior*), 703
interrupted fern (*Osmunda claytoniana*), 10
Ipecac-spurge (*Euphorbia ipecacuanhae*), 318
iris (*Iris*), 809
    copper- (*Iris fulva*), 810
    dwarf (*Iris verna*), 809
    dwarf crested (*Iris cristata*), 809
    dwarf lake- (*Iris lacustris*), 810
    German (*Iris germanica*), 811
    zigzag (*Iris brevicaulis*), 810
iris family (IRIDACEAE), 807
ironplant, cutleaf- (*Haplopappus spinulosus*), 529
ironweed (*Vernonia*), 577
    Appalachian (*Vernonia glauca*), 577

ironweed (*Vernonia*) (*continued*)
    Missouri (*Vernonia missurica*), 577
    New York (*Vernonia noveboracensis*), 577
    Ozark (*Vernonia arkansana*), 577
    smooth (*Vernonia fasciculata*), 578
    tall (*Vernonia gigantea*), 578
    western (*Vernonia baldwinii*), 578
ironwood (*Carpinus caroliniana*), 82; (*Ostrya*), 82
ironwort (*Sideritis*), 412
Italian ryegrass (*Lolium perenne* var. *aristatum*), 698
ivy (*Hedera*), 343
    Boston- (*Parthenocissus tricuspidata*), 323
    common poison- (*Toxicodendron radicans*), 335
    English (*Hedera*), 343
    ground- (*Glechoma*), 421
    Kenilworth- (*Cymbalaria*), 448
    western poison- (*Toxicodendron rydbergii*), 335
ivy-leaved crowfoot (*Ranunculus hederaceus*), 58
ivy-leaved morning-glory (*Ipomoea hederacea*), 385
ivy-leaved speedwell (*Veronica hederifolia*), 454

Jack-in-the-pulpit (*Arisaema triphyllum*), 616
Jack-oak (*Quercus imbricaria*), 78
Jack-pine (*Pinus banksiana*), 33
Jacob's ladder (*Polemonium*), 392
    Appalachian (*Polemonium vanbruntiae*), 392
    spreading (*Polemonium reptans*), 392
    western (*Polemonium occidentale*), 392
jagged chickweed (*Holosteum*), 101
Japanese barberry (*Berberis thunbergii*), 62
Japanese bindweed (*Calystegia hederacea*), 385
Japanese chess (*Bromus japonicus*), 724
Japanese clover (*Lespedeza striata*), 281
Japanese hedge-parsley (*Torilis japonica*), 350
Japanese honeysuckle (*Lonicera japonica*), 479
Japanese hops (*Humulus japonicus*), 70
Japanese knotweed (*Polygonum cuspidatum*), 124
Japanese lawngrass (*Zoysia japonica*), 750
Japanese rose (*Rosa rugosa*), 238
Japanese spiraea (*Spiraea japonica*), 218
Japanese wisteria (*Wisteria floribunda*), 261
Jerusalem artichoke (*Helianthus tuberosus*), 492
Jerusalem oak (*Chenopodium botrys*), 87
Jerusalem sage (*Phlomis tuberosa*), 423
jessamine, yellow (*Gelsemium*), 361
jewel-weed (*Impatiens capensis, I. pallida*), 341
jim-hill mustard (*Sisymbrium altissimum*), 180
jimson-weed (*Datura stramonium*), 383
Joe-Pye weed
    hollow-stemmed (*Eupatorium fistulosum*), 568
    purple-node (*Eupatorium purpureum*), 568
    spotted (*Eupatorium maculatum*), 567
    Steele's (*Eupatorium steelei*), 567
    three-nerved (*Eupatorium dubium*), 567
Johnny-jump-up (*Viola tricolor*), 149
Johnson-grass (*Sorghum halepense*), 772
jointed charlock (*Raphanus raphanistrum*), 163
jointed goat-grass (*Aegilops cylindrica*), 730
joint-vetch (*Aeschynomene*), 263
    northern (*Aeschynomene virginica*), 263
jointweed (*Polygonella*), 115
jumpseed (*Polygonum virginianum*), 123
Juneberry (*Amelanchier*), 253
Junegrass (*Bromus tectorum*), 723; (*Koeleria pyramidata*), 709
jungle-rice (*Echinochloa colona*), 761

juniper (*Juniperus*), 35
  common (*Juniperus communis*), 35
  creeping (*Juniperus horizontalis*), 35
juniper-clubmoss (*Lycopodium sabinifolium*), 3

Kansas-hawthorn (*Crataegus coccinioides*), 251
katune (*Salsola collina*), 93
Kenilworth-ivy (*Cymbalaria*), 448
Kentucky beard-tongue (*Penstemon tenuiflorus*), 444
Kentucky bluegrass (*Poa pratensis*), 701
Kentucky coffee-tree (*Gymnocladus dioicus*), 256
Kentucky goldenrod (*Solidago shortii*), 543
Kentucky ground-nut (*Apios priceana*), 286
Kentucky lady-slipper (*Cypripedium kentuckiense*), 813
Kentucky pearlwort (*Sagina fontinalis*), 100
kidney-leaved grass-of-Parnassus (*Parnassia asarifolia*), 216
kidney-leaved violet (*Viola renifolia*), 145
kidney-vetch (*Anthyllis vulneraria*), 268
king-devil
  glaucous (*Hieracium piloselloides*), 593
  orange-red (*Hieracium aurantiacum*), 592
  yellow (*Hieracium caespitosum*), 592
king-root (*Plantago cordata*), 432
kitten-tails (*Besseya bullii*), 456
knapweed (*Centaurea*), 585
  black (*Centaurea nigra*), 585
  brown (*Centaurea jacea*), 584
  Russian (*Centaurea repens*), 585
  short-fringed (*Centaurea dubia*), 585
  spotted (*Centaurea maculosa*), 585
  tumble- (*Centaurea diffusa*), 585
knawel (*Scleranthus*), 105
  annual (*Scleranthus annuus*), 105
knob-styled dogwood (*Cornus amomum*), 304
knotgrass (*Paspalum distichum*), 765
knotroot-foxtail (*Setaria parviflora*), 766
knotweed
  dooryard (*Polygonum arenastrum*), 117
  giant (*Polygonum sachalinense*), 124
  Japanese (*Polygonum cuspidatum*), 124
  seaside- (*Polygonum glaucum*), 116
Korean clover (*Lespedeza stipulacea*), 281
Korean lawngrass (*Zoysia japonica*), 750
kudzu-vine (*Pueraria lobata*), 287

Labrador-bedstraw (*Galium labradoricum*), 475
Labrador-tea (*Ledum groenlandicum*), 184
lace-frond grape-fern (*Botrychium dissectum*), 8
lace-grass (*Eragrostis capillaris*), 740
lace-lipped ladies'-tresses (*Spiranthes laciniata*), 817
lacerate blazing star (*Liatris aspera*), 575
ladies'-tresses (*Spiranthes*), 815
  giant (*Spiranthes praecox*), 817
  Great Plains (*Spiranthes magnicamporum*), 816
  hooded (*Spiranthes romanzoffiana*), 815
  lace-lipped (*Spiranthes laciniata*), 817
  little (*Spiranthes tuberosa*), 816
  nodding (*Spiranthes cernua*), 815
  oval (*Spiranthes ovalis*), 815
  shining (*Spiranthes lucida*), 815
  slender (*Spiranthes lacera*), 816
  spring (*Spiranthes vernalis*), 816
lady-fern (*Athyrium filix-femina*), 19
lady-slipper (*Cypripedium*), 812
  Kentucky (*Cypripedium kentuckiense*), 813
  pink (*Cypripedium acaule*), 813

lady-slipper (*Cypripedium*) (*continued*)
  ram's head (*Cypripedium arietinum*), 812
  showy (*Cypripedium reginae*), 812
  white (*Cypripedium candidum*), 812
  yellow (*Cypripedium calceolus*), 812
lady's mantle (*Alchemilla vulgaris*), 235
lady's thumb (*Polygonum persicaria*), 121
lake-cress (*Armoracia lacustris*), 170
lakes flat-topped goldenrod (*Euthamia remota*), 546
lakeside daisy (*Hymenoxys herbacea*), 500
lamb's quarters (*Chenopodium album*), 89
lanceleaf ragweed (*Ambrosia bidentata*), 515
lance-leaved anemone (*Anemone lancifolia*), 49
lance-leaved buckthorn (*Rhamnus lanceolata*), 322
lance-leaved grape-fern (*Botrychium lanceolatum*), 8
lance-leaved loosestrife (*Lysimachia lanceolata*), 200
lance-leaved sagittaria (*Sagittaria lancifolia*), 605
Lappland buttercup (*Ranunculus lapponicus*), 58
Lappland rose-bay (*Rhododendron lapponicum*), 185
larch (*Larix*), 32
lardizabala family (LARDIZABALACEAE), 63
large beard-tongue (*Penstemon grandiflorus*), 444
larger five-leaves (*Isotria verticillata*), 823
larger whorled pogonia (*Isotria verticillata*), 823
large-seed false flax (*Camelina sativa*), 181
large twayblade (*Liparis liliifolia*), 825
larkspur (*Delphinium*), 48
  Carolina (*Delphinium carolinianum*), 48
  dwarf (*Delphinium tricorne*), 48
  prairie- (*Delphinium virescens*), 48
  rocket (*Delphinium ambiguum*), 48
  tall (*Delphinium exaltatum*), 48
late eupatorium (*Eupatorium serotinum*), 569
laurel (*Kalmia*), 187
  dog- (*Leucothoe*), 188
  mountain- (*Kalmia latifolia*), 186
  mountain dog- (*Leucothoe walteri*), 188
  red (*Rhododendron catawbiense*), 184
  sheep- (*Kalmia angustifolia*), 187
  swamp- (*Kalmia polifolia*), 187
  swamp dog- (*Leucothoe axillaris*), 188
  white (*Rhododendron maximum*), 184
laurel family (LAURACEAE), 38
laurel-oak (*Quercus laurifolia*), 79
Laurentide primrose (*Primula laurentiana*), 203
lavender cotton (*Santolina chamaecyparissus*), 518
lavender giant-hyssop (*Agastache foeniculum*), 421
lawngrass
  Japanese (*Zoysia japonica*), 750
  Korean (*Zoysia japonica*), 750
lawn-pennywort (*Hydrocotyle sibthorpioides*), 344
lead-plant (*Amorpha canescens*), 282
  smooth (*Amorpha nana*), 282
leadwort family (PLUMBAGINACEAE), 125
leaf-cup (*Polymnia*), 511
  pale-flowered (*Polymnia canadensis*), 511
  yellow-flowered (*Polymnia uvedalia*), 511
leafy bog-aster (*Aster nemoralis*), 557
leafy bulbil-saxifrage (*Saxifraga cernua*), 211
leafy elephant's foot (*Elephantopus carolinianus*), 578
leafy piney-woods goldenrod (*Solidago tortifolia*), 541
leafy pondweed (*Potamogeton foliosus*), 611
leafy spurge (*Euphorbia esula*), 317
leafy-stemmed plantain (*Plantago psyllium*), 431
least trillium (*Trillium pusillum*), 795

leatherleaf (*Chamaedaphne*), 189
leatherwood (*Cyrilla racemiflora*), 183; (*Dirca*), 293
leathery grape-fern (*Botrychium multifidum*), 8
leek, wild (*Allium ampeloprasum*), 791; (*A. tricoccum*), 790
legume-dodder (*Cuscuta epithymum*), 388
lemon-balm (*Melissa officinalis*), 412
lemon-bee-balm (*Monarda citriodora*), 427
lemon-mint (*Mentha ×citrata*), 417
lens-podded hoary cress (*Cardaria chalepensis*), 166
lespedeza (*Lespedeza*), 279
   Chinese (*Lespedeza cuneata*), 281
   downy trailing (*Lespedeza procumbens*), 279
   hairy (*Lespedeza hirta*), 281
   narrow-leaved (*Lespedeza angustifolia*), 280
   prairie- (*Lespedeza leptostachya*), 281
   smooth trailing (*Lespedeza repens*), 279
   velvety (*Lespedeza stuevei*), 280
   violet (*Lespedeza violacea*), 280
   Virginia (*Lespedeza virginica*), 280
   wand- (*Lespedeza intermedia*), 280
lesser basal-leaved rosin-weed (*Silphium compositum*), 513
lesser bladderwort (*Utricularia minor*), 463
lesser broom-rape (*Orobanche minor*), 459
lesser celandine (*Ranunculus ficaria*), 58
lesser duckweed (*Lemna minor*), 616
lesser fringed gentian (*Gentianopsis virgata*), 366
lesser horse-gentian (*Triosteum angustifolium*), 481
lesser quickweed (*Galinsoga parviflora*), 509
lesser rattlesnake-plantain (*Goodyera repens*), 818
lesser snapdragon (*Antirrhinum orontium*), 448
lesser toadflax (*Chaenorrhinum minus*), 449
lettuce (*Lactuca*), 589
   Appalachian white (*Prenanthes roanensis*), 588
   blue (*Lactuca pulchella*), 590
   creeping (*Ixeris stolonifera*), 591
   devil's (*Amsinckia tessellata*), 398
   dwarf cut-leaf white (*Prenanthes nana*), 588
   dwarf simple white (*Prenanthes boottii*), 588
   glaucous white (*Prenanthes racemosa*), 586
   hairy tall (*Lactuca hirsuta*), 589
   midwestern white (*Prenanthes crepidinea*), 587
   prairie- (*Lactuca ludoviciana*), 589
   prickly (*Lactuca serriola*), 591
   rough white (*Prenanthes aspera*), 587
   tall (*Lactuca canadensis*), 590
   tall blue (*Lactuca biennis*), 590
   tall white (*Prenanthes altissima*), 588
   wall- (*Lactuca muralis*), 589
   white (*Prenanthes*), 586
   willow leaf- (*Lactuca saligna*), 591
   woodland- (*Lactuca floridana*), 590
lettuce-saxifrage (*Saxifraga micranthidifolia*), 212
licorice (*Glycyrrhiza*), 267
   wild (*Galium circaezans, G. lanceolatum*), 476; (*Glycyrrhiza lepidota*), 267
licorice-goldenrod (*Solidago odora*), 541
licorice milk-vetch (*Astragalus glycyphyllos*), 266
lilac (*Syringa*), 435
   common (*Syringa vulgaris*), 435
   wild (*Ceanothus sanguineus*), 321
lily (*Lilium*), 791
   American lotus- (*Nelumbo lutea*), 42
   atamasco- (*Zephyranthes atamasca*), 803
   bead- (*Clintonia*), 797
   bell- (*Lilium grayi*), 791
   blackberry- (*Belamcanda*), 807

lily (*Lilium*) (*continued*)
   Carolina (*Lilium michauxii*), 792
   day- (*Hemerocallis*), 789
   fawn- (*Erythronium*), 793
   lotus- (*Nelumbo*), 42
   Michigan (*Lilium michiganense*), 793
   orange (*Lilium bulbiferum*), 791
   pine- (*Lilium catesbaei*), 791
   plantain- (*Hosta*), 788
   pygmy water- (*Nymphaea tetragona*), 43
   sand- (*Mentzelia decapetala*), 151
   spider- (*Hymenocallis*), 803
   tiger- (*Lilium lancifolium*), 793
   trout- (*Erythronium*), 793
   Turk's-cap (*Lilium superbum*), 792
   water- (*Nymphaea*), 43
   wild yellow (*Lilium canadense*), 792
   wood- (*Lilium philadelphicum*), 792
   yellow water- (*Nuphar*), 42
lily family (LILIACEAE), 780
lily of the valley (*Convallaria*), 800
limestone heart-leaved aster (*Aster anomalus*), 553
limestone oak-fern (*Gymnocarpium robertianum*), 22
linden (*Tilia*), 131
   American (*Tilia americana*), 131
linden family (TILIACEAE), 131
lingonberry (*Vaccinium vitis-idaea*), 192
lion's foot (*Prenanthes serpentaria*), 587
lion's tail (*Leonurus cardiaca*), 424
lip-fern (*Cheilanthes*), 15
   Alabama (*Cheilanthes alabamensis*), 15
   Eaton's (*Cheilanthes eatonii*), 16
   slender (*Cheilanthes feei*), 15
   woolly (*Cheilanthes lanosa*), 15; (*C. tomentosa*), 16
little barley (*Hordeum pusillum*), 729
little bluestem (*Schizachyrium scoparium*), 775
little brown jugs (*Hexastylis*), 40
little bur-head (*Echinodorus tenellus*), 605
little five-leaves (*Isotria medeoloides*), 823
little floating heart (*Nymphoides cordata*), 389
little grape-fern (*Botrychium simplex*), 8
little hawkbit (*Leontodon taraxacoides*), 601
littlehip-hawthorn (*Crataegus spathulata*), 247
little hop-clover (*Trifolium dubium*), 271
little ladies'-tresses (*Spiranthes tuberosa*), 816
little-leaf tick-trefoil (*Desmodium ciliare*), 278
little mallow (*Malva parviflora*), 135
little plantain (*Plantago pusilla*), 433
little prickly pear (*Opuntia fragilis*), 86
little quaking grass (*Briza minor*), 700
little shinleaf (*Pyrola minor*), 195
little skullcap (*Scutellaria parvula*), 411
little sundrops (*Oenothera perennis*), 301
little whorled pogonia (*Isotria medeoloides*), 823
live forever (*Sedum purpureum*), 208
live oak (*Quercus virginiana*), 78
   sand (*Quercus geminata*), 78
lizard's tail (*Saururus*), 39
lizard's tail family (SAURURACEAE), 39
loasa family (LOASACEAE), 151
lobed spleenwort (*Asplenium pinnatifidum*), 18
lobed tickseed (*Coreopsis auriculata*), 504
lobelia (*Lobelia*), 466
   spiked (*Lobelia spicata*), 466
   water- (*Lobelia dortmanna*), 468
loblolly-pine (*Pinus taeda*), 33

loco-weed
  purple (*Oxytropis lambertii*), 267
  showy (*Oxytropis splendens*), 267
locust (*Robinia*), 262
  black (*Robinia pseudoacacia*), 262
  bristly (*Robinia hispida*), 262
  clammy (*Robinia viscosa*), 262
  honey- (*Gleditsia triacanthos*), 257
  water- (*Gleditsia aquatica*), 256
locust-weed (*Chamaecrista fasciculata*), 256
Loesel's twayblade (*Liparis loeselii*), 825
logania family (LOGANIACEAE), 361
log-fern (*Dryopteris celsa*), 26
London-rocket (*Sisymbrium irio*), 181
long-haired hawkweed (*Hieracium longipilum*), 594
long-headed anemone (*Anemone cylindrica*), 50
longleaf ground-cherry (*Physalis longifolia*), 377
longleaf-pine (*Pinus palustris*), 33
longleaf-pondweed (*Potamogeton nodosus*), 612
long-leaved speedwell (*Veronica longifolia*), 453
long-leaved stitchwort (*Stellaria longifolia*), 104
longspike evening-primrose (*Oenothera rhombipetala*), 299
long-spurred violet (*Viola rostrata*), 148
longstalk crane's-bill (*Geranium columbinum*), 339
long-stalked aster (*Aster dumosus*), 550
long-stalked stitchwort (*Stellaria longipes*), 104
longstalk-tickseed (*Coreopsis lanceolata*), 504
long-styled sweet cicely (*Osmorhiza longistylis*), 347
long-tube valerian (*Valeriana pauciflora*), 486
loose milkwort (*Polygala ambigua*), 329
loosestrife (*Lysimachia*), 199; (*Lythrum*), 292
  annual (*Lythrum hyssopifolia*), 292
  Appalachian (*Lysimachia tonsa*), 200
  bulbil- (*Lysimachia terrestris*), 202
  fringed (*Lysimachia ciliata*), 199
  garden- (*Lysimachia vulgaris*), 201
  lance-leaved (*Lysimachia lanceolata*), 200
  Mississippi-valley (*Lysimachia hybrida*), 200
  narrow-leaved (*Lythrum lineare*), 292
  purple (*Lythrum salicaria*), 292
  smooth (*Lysimachia quadriflora*), 200
  spotted (*Lysimachia punctata*), 201
  swamp- (*Lysimachia thyrsiflora*), 201
  trailing (*Lysimachia radicans*), 200
  whorled (*Lysimachia quadrifolia*), 201
  winged (*Lythrum alatum*), 292
loosestrife family (LYTHRACEAE), 291
lopseed (*Phryma*), 407
lotus-lily (*Nelumbo*), 42
  American (*Nelumbo lutea*), 42
lotus-lily family (NELUMBONACEAE), 42
lotus milk-vetch (*Astragalus lotiflorus*), 264
lousewort (*Pedicularis*), 457
  forest- (*Pedicularis canadensis*), 457
  Furbish's (*Pedicularis furbishiae*), 457
  swamp- (*Pedicularis lanceolata*), 457
lovage (*Levisticum*), 359; (*Ligusticum*), 354
  Scotch (*Ligusticum scothicum*), 354
lovegrass (*Eragrostis*), 738
  Carolina (*Eragrostis pectinacea*), 739
  India- (*Eragrostis pilosa*), 739
  purple (*Eragrostis spectabilis*), 738
  sand- (*Eragrostis trichodes*), 738
  weeping (*Eragrostis curvula*), 739
love-in-a-mist (*Nigella damascena*), 45
love-lies-bleeding (*Amaranthus caudatus*), 96

low agrimony (*Agrimonia microcarpa*), 236
low bindweed (*Calystegia spithamaea*), 385
low cudweed (*Gnaphalium uliginosum*), 565
lowland bladder-fern (*Cystopteris protrusa*), 20
low pine-barren milkwort (*Polygala ramosa*), 328
low rattlebox (*Crotalaria rotundifolia*), 260
low rough aster (*Aster radula*), 556
low showy aster (*Aster spectabilis*), 556
low spearwort (*Ranunculus pusillus*), 57
lupine (*Lupinus*), 260
  sundial- (*Lupinus perennis*), 260
lychnis, scarlet (*Lychnis chalcedonica*), 108
lyme-grass (*Leymus arenarius*), 728

Macartney-rose (*Rosa bracteata*), 237
Macree's milk-pea (*Galactia macreei*), 287
madder, field- (*Sherardia arvensis*), 473
madder family (RUBIACEAE), 469
madwort (*Asperugo*), 396
magnolia (*Magnolia*), 36
  bigleaf (*Magnolia macrophylla*), 37
  mountain- (*Magnolia fraseri*), 37
  southern (*Magnolia grandiflora*), 36
magnolia family (MAGNOLIACEAE), 36
Mahaleb-cherry (*Prunus mahaleb*), 241
mahogany family (MELIACEAE), 335
maiden-cane (*Panicum hemitomon*), 751
maidenhair fern (*Adiantum*), 14
  northern (*Adiantum pedatum*), 14
  southern (*Adiantum capillus-veneris*), 14
maidenhair fern family (ADIANTACEAE), 14
maidenhair-spleenwort (*Asplenium trichomanes*), 18
maiden-pink (*Dianthus deltoides*), 112
male-berry (*Lyonia ligustrina*), 187
male fern (*Dryopteris filix-mas*), 26
mallow (*Malva*), 134
  bristly (*Modiola caroliniana*), 134
  Bush's poppy- (*Callirhoe bushii*), 136
  clustered poppy- (*Callirhoe triangulata*), 136
  common (*Malva neglecta*), 135
  dwarf (*Malva rotundifolia*), 135
  finger poppy- (*Callirhoe digitata*), 136
  glade- (*Napaea*), 136
  high (*Malva sylvestris*), 135
  little (*Malva parviflora*), 135
  marsh- (*Althaea officinalis*), 134
  musk- (*Malva moschata*), 134
  pale poppy- (*Callirhoe alcaeoides*), 136
  poppy- (*Callirhoe*), 136
  purple poppy- (*Callirhoe involucrata*), 136
  rose- (*Hibiscus moscheutos*), 132
  scarlet (*Sphaeralcea*), 137
  seashore- (*Kosteletzkya virginica*), 133
  smooth rose- (*Hibiscus laevis*), 132
  vervain- (*Malva alcea*), 134
  whorled (*Malva verticillata*), 135
mallow family (MALVACEAE), 131
Maltese cross (*Lychnis chalcedonica*), 108
Maltese star-thistle (*Centaurea melitensis*), 586
mandrake (*Podophyllum*), 62
mangosteen family (CLUSIACEAE), 126
mannagrass (*Glyceria*), 706
  American (*Glyceria grandis*), 708
  coastal (*Glyceria obtusa*), 707
  eastern (*Glyceria septentrionalis*), 706

mannagrass (*Glyceria*) (*continued*)
    fowl- (*Glyceria striata*), 708
    northeastern (*Glyceria melicaria*), 707
    northern (*Glyceria borealis*), 706
    rattlesnake- (*Glyceria canadensis*), 707
many-seeded plantain (*Plantago heterophylla*), 433
maple (*Acer*), 331
    ash-leaved (*Acer negundo*), 333
    black (*Acer nigrum*), 332
    flowering (*Viburnum acerifolium*), 483
    hard (*Acer saccharum*), 331
    hedge- (*Acer campestre*), 332
    mountain- (*Acer spicatum*), 333
    Norway (*Acer platanoides*), 331
    red (*Acer rubrum*), 333
    silver- (*Acer saccharinum*), 333
    soft (*Acer saccharinum*), 333
    southern sugar- (*Acer barbatum*), 332
    striped (*Acer pensylvanicum*), 333
    sugar- (*Acer saccharum*), 331
    sycamore- (*Acer pseudoplatanus*), 332
maple family (ACERACEAE), 331
maple-leaved alum-root (*Heuchera villosa*), 215
maple-leaved goosefoot (*Chenopodium simplex*), 88
maple-leaved waterleaf (*Hydrophyllum canadense*), 394
mare's-tail (*Hippuris vulgaris*), 431
mare's-tail family (HIPPURIDACEAE), 431
marginal wood-fern (*Dryopteris marginalis*), 25
marigold (*Tagetes*), 510
    bur- (*Bidens cernua*), 506
    marsh- (*Caltha*), 45
    showy bur- (*Bidens laevis*), 506
    stinking (*Dyssodia papposa*), 510
maritime marsh-elder (*Iva frutescens*), 514
marjoram, wild (*Origanum vulgare*), 413
marsh-bedstraw (*Galium palustre*), 474
marsh-bellflower (*Campanula aparinoides*), 465
marsh-bluegrass (*Poa paludigena*), 703
marsh-crowfoot (*Ranunculus macounii*), 54
marsh-dandelion (*Taraxacum palustre*), 599
marsh-elder (*Iva*), 514
    beach-dune (*Iva imbricata*), 515
    big (*Iva xanthiifolia*), 514
    maritime (*Iva frutescens*), 514
    rough (*Iva annua*), 515
marsh-eryngo (*Eryngium aquaticum*), 345
marsh-felwort (*Lomatogonium*), 367
marsh-fern (*Thelypteris palustris*), 24
marsh-fleabane (*Pluchea*), 564
marsh-foxtail (*Alopecurus geniculatus*), 720
marsh-grass (*Scolochloa festucacea*), 698
marsh-horsetail (*Equisetum palustre*), 6
marsh-mallow (*Althaea officinalis*), 134
marsh-marigold (*Caltha*), 45
marsh-muhly (*Muhlenbergia glomerata*), 746
marsh-pea (*Lathyrus palustris*), 275
marsh-pennywort (*Hydrocotyle americana*), 344
marsh-pink (*Sabatia*), 362
    common (*Sabatia angularis*), 362
    slender (*Sabatia campanulata*), 362
    western (*Sabatia campestris*), 362
marsh-potentilla (*Potentilla palustris*), 226
marsh-skullcap (*Scutellaria galericulata*), 411
marsh St. John's-wort (*Triadenum*), 130
marsh-thistle (*Cirsium palustre*), 583

marsh-violet
    blue (*Viola cucullata*), 142
    northern (*Viola palustris*), 145
marsh willow-herb (*Epilobium palustre*), 297
    American (*Epilobium leptophyllum*), 297
Maryland milkwort (*Polygala mariana*), 329
Maryland tick-trefoil (*Desmodium marilandicum*), 278
Massachusetts fern (*Thelypteris simulata*), 23
masterwort (*Peucedanum ostruthium*), 360
matchbrush (*Gutierrezia sarothrae*), 546
matgrass, moor- (*Nardus stricta*), 692
mat-muhly (*Muhlenbergia richardsonis*), 746
matrimony-vine (*Lycium*), 381
matted tick-trefoil (*Desmodium lineatum*), 276
mauve sleekwort (*Liparis liliifolia*), 825
Maximilian-sunflower (*Helianthus maximilianii*), 494
may-apple (*Podophyllum*), 62
mayflower, Canada (*Maianthemum canadense*), 800
maypops (*Passiflora incarnata*), 149
meadow-barley (*Hordeum brachyantherum*), 729
meadow-beauty (*Rhexia*), 303
meadow-buttercup (*Ranunculus acris*), 54
meadow closed gentian (*Gentiana clausa*), 365
meadow-fescue (*Festuca pratensis*), 695
meadow-foam family (LIMNANTHACEAE), 341
meadow-foxtail (*Alopecurus pratensis*), 720
meadow-geranium (*Geranium pratense*), 339
meadow-horsetail (*Equisetum pratense*), 6
meadow-parsnip (*Thaspium*), 353
    bearded (*Thaspium barbinode*), 353
    mountain (*Thaspium pinnatifidum*), 354
    smooth (*Thaspium trifoliatum*), 353
meadow-pea (*Lathyrus pratensis*), 274
meadow-phlox (*Phlox maculata*), 391
meadow-pitchers (*Rhexia*), 303
    bristly (*Rhexia aristosa*), 303
    dull (*Rhexia mariana*), 304
    Nash's (*Rhexia nashii*), 304
    short-stemmed (*Rhexia petiolata*), 303
    showy (*Rhexia interior*), 303
    wingstem (*Rhexia virginica*), 303
meadow-plagiobothrys (*Plagiobothrys scouleri*), 397
meadow-rue (*Thalictrum*), 60
    early (*Thalictrum dioicum*), 60
    mountain (*Thalictrum clavatum*), 60
    northern (*Thalictrum venulosum*), 60
    purple (*Thalictrum dasycarpum*), 61
    skunk (*Thalictrum revolutum*), 61
    small-leaved (*Thalictrum macrostylum*), 60
    tall (*Thalictrum pubescens*), 60
meadow-sage (*Salvia pratensis*), 429
meadow-spikemoss (*Selaginella apoda*), 4
meadow-starwort (*Stellaria palustris*), 103
meadowsweet (*Spiraea alba*), 219
meadow-willow (*Salix petiolaris*), 160
medicinal agrimony (*Agrimonia eupatoria*), 236
medick, black (*Medicago lupulina*), 272
Mediterranean barley (*Hordeum geniculatum*), 729
melastome family (MELASTOMATACEAE), 303
melic (*Melica*), 708
    awned (*Melica smithii*), 708
    three-flower (*Melica nitens*), 708
    two-flower (*Melica mutica*), 708
melonette (*Melothria*), 150
memorial rose (*Rosa wichuraiana*), 237

mercury (*Mercurialis*), 315
  herb- (*Mercurialis annua*), 315
Merlin's grass (*Isoetes*), 4
mermaid, false (*Floerkea*), 341
mermaid-weed (*Proserpinaca*), 290
  coastal plain (*Proserpinaca pectinata*), 290
  common (*Proserpinaca palustris*), 290
merrybells (*Uvularia*), 798
mesquite, vine- (*Panicum obtusum*), 751
Mexican bamboo (*Polygonum cuspidatum*), 124
Mexican mosquito-fern (*Azolla mexicana*), 30
Mexican poppy (*Argemone mexicana*), 65
Mexican tea (*Chenopodium ambrosioides*), 86
mezereum (*Daphne mezereum*), 293
mezereum family (THYMELAEACEAE), 293
Miami-mist (*Phacelia purshii*), 392
Michaelmas-daisy (*Aster*), 547
Michigan lily (*Lilium michiganense*), 793
midwestern agalinis (*Agalinis skinneriana*), 452
midwestern arrow-head (*Sagittaria brevirostra*), 604
midwestern blue heart-leaved aster (*Aster shortii*), 552
midwestern Indian-physic (*Porteranthus stipulatus*), 220
midwestern sundrops (*Oenothera pilosella*), 301
midwestern tickseed-sunflower (*Bidens aristosa*), 508
midwestern white heart-leaved aster (*Aster furcatus*), 555
midwestern white lettuce (*Prenanthes crepidinea*), 587
mignonette (*Reseda odorata*), 183
mignonette family (RESEDACEAE), 182
mile-a-minute (*Polygonum perfoliatum*), 123
milfoil
  common water- (*Myriophyllum sibiricum*), 290
  European water- (*Myriophyllum spicatum*), 290
  water- (*Myriophyllum*), 289
milk-pea (*Galactia*), 287
  hairy (*Galactia volubilis*), 287
  Macree's (*Galactia macreei*), 287
  trailing (*Galactia regularis*), 287
milk-purslane (*Euphorbia maculata*), 320
milk-thistle (*Silybum*), 583
milk-vetch (*Astragalus*), 264
  alkali (*Astragalus racemosus*), 264
  alpine (*Astragalus alpinus*), 265
  Canada (*Astragalus canadensis*), 266
  chick-pea (*Astragalus cicer*), 266
  Cooper (*Astragalus neglectus*), 264
  elegant (*Astragalus eucosmus*), 265
  field (*Astragalus agrestis*), 266
  licorice (*Astragalus glycyphyllos*), 266
  lotus (*Astragalus lotiflorus*), 264
  Nuttall (*Astragalus missouriensis*), 264
  Ozark (*Astragalus distortus*), 264
  pliant (*Astragalus flexuosus*), 265
  pulse (*Astragalus tenellus*), 264
  Robbins (*Astragalus robbinsii*), 265
  standing (*Astragalus adsurgens*), 266
  Tennessee (*Astragalus tennesseensis*), 265
milkweed (*Asclepias*), 370
  clasping (*Asclepias amplexicaulis*), 371
  common (*Asclepias syriaca*), 371
  dwarf (*Asclepias ovalifolia*), 371
  few-flowered (*Asclepias lanceolata*), 372
  four-leaved (*Asclepias quadrifolia*), 372
  green (*Asclepias viridiflora*), 374
  narrow-leaved (*Asclepias stenophylla*), 373
  Ozark (*Asclepias viridis*), 373
  prairie- (*Asclepias hirtella*), 373

milkweed (*Asclepias*) (*continued*)
  purple (*Asclepias purpurascens*), 372
  red (*Asclepias rubra*), 373
  showy (*Asclepias speciosa*), 372
  smooth (*Asclepias sullivantii*), 373
  smoothseed- (*Asclepias perennis*), 370
  swamp- (*Asclepias incarnata*), 370
  tall (*Asclepias exaltata*), 371
  white (*Asclepias variegata*), 372
  whorled (*Asclepias verticillata*), 370
  woolly (*Asclepias lanuginosa*), 374
milkweed family (ASCLEPIADACEAE), 370
milkwort (*Polygala*), 328
  Appalachian (*Polygala curtissii*), 329
  bitter (*Polygala polygama*), 329
  blood- (*Polygala sanguinea*), 329
  loose (*Polygala ambigua*), 329
  low pine-barren (*Polygala ramosa*), 328
  Maryland (*Polygala mariana*), 329
  Nuttall's (*Polygala nuttallii*), 329
  orange (*Polygala lutea*), 328
  pink (*Polygala incarnata*), 328
  sea- (*Glaux*), 202
  shortleaf (*Polygala brevifolia*), 330
  tall pine-barren (*Polygala cymosa*), 328
  white (*Polygala alba*), 328
  whorled (*Polygala verticillata*), 329
milkwort family (POLYGALACEAE), 328
milky wild indigo (*Baptisia lactea*), 259
millet
  broomcorn (*Panicum miliaceum*), 753
  foxtail- (*Setaria italica*), 767
  proso (*Panicum miliaceum*), 753
  Texas (*Panicum texanum*), 751
milo (*Sorghum bicolor*), 772
mimosa (*Albizia julibrissin*), 255
mimosa family (MIMOSACEAE), 255
Minnesota wild rye (*Elymus diversiglumis*), 726
minnie-bush (*Menziesia pilosa*), 184
mint (*Mentha*), 416
  apple- (*Mentha suaveolens*), 418
  field- (*Mentha arvensis*), 416
  horse- (*Mentha longifolia*), 417; (*Monarda punctata*), 427
  lemon- (*Mentha ×citrata*), 417
  mountain- (*Pycnanthemum*), 414
  pepper- (*Mentha ×piperita*), 417
  perilla- (*Perilla frutescens*), 419
  red (*Mentha ×gentilis*), 417
  Scotch (*Mentha ×gentilis*), 417
  spear- (*Mentha spicata*), 418
  stone- (*Cunila*), 419
  water- (*Mentha aquatica*), 417
mint family (LAMIACEAE), 408
Mississippi arrow-head (*Sagittaria calycina*), 603
Mississippi-valley flat-topped goldenrod (*Euthamia leptocephala*), 546
Mississippi-valley loosestrife (*Lysimachia hybrida*), 200
Mississippi wisteria (*Wisteria macrostachya*), 262
Missouri arrow-wood (*Viburnum molle*), 482
Missouri coneflower (*Rudbeckia missouriensis*), 495
Missouri goldenrod (*Solidago missouriensis*), 538
Missouri gooseberry (*Ribes missouriense*), 205
Missouri ironweed (*Vernonia missurica*), 577
Missouri spurge (*Euphorbia missurica*), 319
mist, Scotch (*Galium sylvaticum*), 474
Mistassini primrose (*Primula mistassinica*), 203

mist-flower (*Eupatorium coelestinum*), 568

mistletoe

American Christmas- (*Phoradendron serotinum*), 307

dwarf (*Arceuthobium*), 307

eastern dwarf (*Arceuthobium pusillum*), 307

mitrewort (*Cynoctonum*), 361; (*Mitella*), 214

Caribbean (*Cynoctonum mitreola*), 361

naked (*Mitella nuda*), 214

small-leaved (*Cynoctonum sessilifolium*), 361

two-leaved (*Mitella diphylla*), 214

mixed bladderwort (*Utricularia geminiscapa*), 463

moccasin-flower (*Cypripedium acaule*), 813

mock bishop-weed (*Ptilimnium*), 356

Atlantic (*Ptilimnium capillaceum*), 356

big (*Ptilimnium costatum*), 356

Ozark (*Ptilimnium nuttallii*), 356

riverbank (*Ptilimnium nodosum*), 356

mockernut-hickory (*Carya tomentosa*), 73

mock-orange (*Philadelphus*), 204

Appalachian (*Philadelphus inodorus*), 204

Cumberland (*Philadelphus hirsutus*), 204

European (*Philadelphus coronarius*), 204

Ozark (*Philadelphus pubescens*), 204

money-plant (*Lunaria rediviva*), 168

moneywort (*Lysimachia nummularia*), 201

monkey-flower (*Mimulus*), 440

Allegheny (*Mimulus ringens*), 440

common yellow (*Mimulus guttatus*), 440

musky (*Mimulus moschatus*), 440

round-leaf (*Mimulus glabratus*), 440

sharpwing (*Mimulus alatus*), 440

monkshood (*Aconitum*), 47

New York (*Aconitum noveboracense*), 47

southern (*Aconitum uncinatum*), 47

white (*Aconitum reclinatum*), 47

Montauk daisy (*Chrysanthemum nipponicum*), 518

moonseed (*Menispermum*), 63

Carolina (*Cocculus carolinus*), 63

moonseed family (MENISPERMACEAE), 63

moonwort (*Botrychium lunaria*), 8

moorgrass (*Molinia caerulea*), 732

moor-matgrass (*Nardus stricta*), 692

moosewood (*Acer pensylvanicum*), 333

morning-glory (*Ipomoea*), 385

common (*Ipomoea purpurea*), 386

ivy-leaved (*Ipomoea hederacea*), 385

red (*Ipomoea coccinea*), 386

white (*Ipomoea lacunosa*), 386

morning-glory family (CONVOLVULACEAE), 384

moschatel (*Adoxa*), 485

moschatel family (ADOXACEAE), 485

mosquito-fern (*Azolla*), 30

eastern (*Azolla caroliniana*), 30

Mexican (*Azolla mexicana*), 30

moss

flowering (*Pyxidanthera*), 197

meadow spike- (*Selaginella apoda*), 4

northern spike- (*Selaginella selaginoides*), 4

rock-spike- (*Selaginella rupestris*), 4

Spanish (*Tillandsia usneoides*), 779

spike- (*Selaginella*), 4

moss-campion (*Silene acaulis*), 110

moss-pink (*Phlox subulata*), 390

moss-rose (*Portulaca grandiflora*), 99

mother-of-thyme (*Satureja acinos*), 413

motherwort (*Leonurus cardiaca*), 424

horehound- (*Leonurus marrubiastrum*), 424

moth-mullein (*Verbascum blattaria*), 442

mountain alder (*Alnus viridis*), 84

mountain-angelica (*Angelica triquinata*), 358

mountain-ash (*Sorbus*), 246

American (*Sorbus americana*), 246

European (*Sorbus aucuparia*), 246

showy (*Sorbus decora*), 246

mountain-bilberry (*Vaccinium membranaceum*), 191

mountain-bluets (*Hedyotis michauxii*), 471

mountain cliff-fern (*Woodsia scopulina*), 21

mountain-cranberry (*Vaccinium vitis-idaea*), 192

southern (*Vaccinium erythrocarpum*), 194

mountain dog-laurel (*Leucothoe walteri*), 188

mountain-fetterbush

deciduous (*Eubotrys recurva*), 189

evergreen (*Pieris floribunda*), 189

mountain golden alexanders (*Zizia trifoliata*), 353

mountain-heath (*Phyllodoce caerulea*), 187

mountain-holly (*Nemopanthus*), 310

Appalachian (*Nemopanthus collinus*), 310

common (*Nemopanthus mucronatus*), 310

mountain Indian-physic (*Porteranthus trifoliatus*), 220

mountain-laurel (*Kalmia latifolia*), 186

mountain-lover (*Paxistima*), 309

mountain-magnolia (*Magnolia fraseri*), 37

mountain-maple (*Acer spicatum*), 333

mountain meadow-parsnip (*Thaspium pinnatifidum*), 354

mountain meadow-rue (*Thalictrum clavatum*), 60

mountain-mint (*Pycnanthemum*), 414

mountain serviceberry (*Amelanchier bartramiana*), 253

mountain sorrel (*Oxyria*), 113

mountain-spleenwort (*Asplenium montanum*), 17

mountain stewartia (*Stewartia ovata*), 125

mountain-tarweed (*Madia glomerata*), 510

mountain-timothy (*Phleum alpinum*), 720

mountain white alder (*Clethra acuminata*), 183

mountain white potentilla (*Potentilla tridentata*), 225

mountain wood-fern (*Dryopteris campyloptera*), 27

mouse-ear chickweed (*Cerastium*), 102

mouse-ear cress (*Arabidopsis thaliana*), 181

mouse-ear hawkweed (*Hieracium pilosella*), 592

mouse-tail (*Myosurus*), 59

moxie (*Gaultheria hispidula*), 190

mud-plantain (*Heteranthera*), 780

mudwort (*Limosella*), 441

Atlantic (*Limosella subulata*), 441

northern (*Limosella aquatica*), 441

mugwort (*Artemisia*), 520; (*A. vulgaris*), 520

muhly (*Muhlenbergia*), 744

alkali (*Muhlenbergia asperifolia*), 747

forest- (*Muhlenbergia sylvatica*), 744

marsh- (*Muhlenbergia glomerata*), 746

mat- (*Muhlenbergia richardsonis*), 746

wirestem- (*Muhlenbergia mexicana*), 744

mulberry (*Morus*), 70

paper- (*Broussonetia*), 70

red (*Morus rubra*), 70

white (*Morus alba*), 70

mulberry family (MORACEAE), 70

mullein (*Verbascum*), 442

clasping (*Verbascum phlomoides*), 442

common (*Verbascum thapsus*), 443

moth- (*Verbascum blattaria*), 442

mullein (*Verbascum*) (*continued*)
 purple (*Verbascum phoeniceum*), 442
 white (*Verbascum lychnitis*), 442
mullein-foxglove (*Dasistoma*), 449
mullein-pink (*Lychnis coronaria*), 107
multiflora-rose (*Rosa multiflora*), 237
muscadine-grape (*Vitis rotundifolia*), 324
musclewood (*Carpinus caroliniana*), 82
musk-mallow (*Malva moschata*), 134
musk-thistle (*Carduus nutans*), 580
musky monkey-flower (*Mimulus moschatus*), 440
mustard (*Brassica*), 162
 ball- (*Neslia paniculata*), 167
 black (*Brassica nigra*), 162
 blue (*Chorispora tenella*), 179
 brown (*Brassica juncea*), 162
 Chinese (*Brassica juncea*), 162
 dog- (*Erucastrum gallicum*), 163
 garlic- (*Alliaria*), 180
 gray tansy- (*Descurainia richardsonii*), 182
 hare's-ear (*Conringia*), 164
 hedge- (*Sisymbrium officinale*), 180
 Indian (*Brassica juncea*), 162
 jim-hill (*Sisymbrium altissimum*), 180
 tall hedge- (*Sisymbrium loeselii*), 181
 tall wormseed- (*Erysimum hieracifolium*), 180
 tansy- (*Descurainia pinnata, D. sophia*), 182
 tower- (*Arabis glabra*), 174
 treacle- (*Erysimum repandum*), 179
 tumbling (*Sisymbrium altissimum*), 180
 white (*Sinapis alba*), 162
 wormseed- (*Erysimum cheiranthoides*), 179
mustard family (BRASSICACEAE), 162
myrtle
 sand- (*Leiophyllum*), 184
 sea- (*Baccharis halimifolia*), 562
 wax- (*Myrica cerifera*), 75
myrtle-spurge (*Euphorbia lathyris*), 316

naked bladderwort (*Utricularia cornuta*), 462
naked bulbil-saxifrage (*Saxifraga foliolosa*), 212
naked mitrewort (*Mitella nuda*), 214
naked-stemmed sunflower (*Helianthus occidentalis*), 492
naked tick-trefoil (*Desmodium nudiflorum*), 276
nannyberry (*Viburnum lentago*), 484
Narbonne-vetch (*Vicia narbonensis*), 273
narcissus, poet's (*Narcissus poeticus*), 803
narrow beech-fern (*Thelypteris phegopteris*), 22
narrow-leaved cat-tail (*Typha angustifolia*), 778
narrow-leaved gentian (*Gentiana linearis*), 365
narrow-leaved lespedeza (*Lespedeza angustifolia*), 280
narrow-leaved loosestrife (*Lythrum lineare*), 292
narrow-leaved milkweed (*Asclepias stenophylla*), 373
narrow-leaved puccoon (*Lithospermum incisum*), 400
narrow-leaved smooth aster (*Aster concinnus*), 548
narrow-leaved sneezeweed (*Helenium amarum*), 500
narrow-leaved speedwell (*Veronica scutellata*), 455
narrow-leaved sunflower (*Helianthus angustifolius*), 494
narrow-leaved tick-trefoil (*Desmodium tenuifolium*), 277
narrow-leaved tragia (*Tragia urens*), 313
narrow-leaved triodanis (*Triodanis leptocarpa*), 466
narrow-leaved umbrella-wort (*Mirabilis linearis*), 85
narrow-leaved vervain (*Verbena simplex*), 405
narrow-leaved vetch (*Vicia angustifolia*), 272
narrow-leaved water-horehound (*Lycopus angustifolius*), 420
narrow-leaved water-primrose (*Ludwigia linearis*), 296

narrow-leaved white topped aster (*Aster solidagineus*), 558
narrow plumegrass (*Erianthus strictus*), 770
Nash's meadow-pitchers (*Rhexia nashii*), 304
needle, Adam's (*Yucca filamentosa*), 804
needle-and-thread grass (*Stipa comata*), 693
needle-grass (*Stipa*), 693
 green (*Stipa viridula*), 693
needlerush (*Juncus roemerianus*), 625
netted chain-fern (*Woodwardia areolata*), 30
nettle (*Urtica*), 71
 bull- (*Cnidoscolus stimulosus*), 312
 dead (*Lamium*), 423
 false (*Boehmeria cylindrica*), 71
 field hedge- (*Stachys arvensis*), 427
 hedge- (*Stachys*), 425; (*S. palustris*), 425
 hemp- (*Galeopsis*), 424
 hispid hedge- (*Stachys hispida*), 425
 horse- (*Solanum carolinense*), 380
 hyssop-hedge- (*Stachys hyssopifolia*), 425
 red dead (*Lamium purpureum*), 424
 smooth hedge- (*Stachys tenuifolia*), 426
 spotted dead (*Lamium maculatum*), 423
 stinging (*Urtica*), 71
 white dead (*Lamium album*), 423
nettle family (URTICACEAE), 71
nettle-leaved goosefoot (*Chenopodium murale*), 88
nettle-leaved sage (*Salvia urticifolia*), 429
New England arnica (*Arnica lanceolata*), 501
New England aster (*Aster novae-angliae*), 554
New England estuarine beggar-ticks (*Bidens eatonii*), 507
New England goldenrod (*Solidago cutleri*), 533
New England grape (*Vitis novae-angliae*), 325
New England groundsel (*Senecio schweinitzianus*), 525
New England rose (*Rosa nitida*), 239
New England serviceberry (*Amelanchier sanguinea*), 254
Newfoundland aster (*Aster crenifolius*), 547
Newfoundland dwarf birch (*Betula michauxii*), 83
New Jersey tea (*Ceanothus americanus*), 321
New York aster (*Aster novi-belgii*), 547
New York fern (*Thelypteris noveboracensis*), 23
New York ironweed (*Vernonia noveboracensis*), 577
New York monkshood (*Aconitum noveboracense*), 47
niger-seed (*Guizotia abyssinica*), 503
nightshade (*Solanum*), 379
 alpine enchanter's (*Circaea alpina*), 303
 black (*Solanum nigrum*), 379
 common enchanter's (*Circaea lutetiana*), 303
 cut-leaved (*Solanum triflorum*), 380
 enchanter's (*Circaea*), 303
 hairy (*Solanum sarrachoides*), 379
 silverleaf (*Solanum elaeagnifolium*), 381
 sticky (*Solanum sisymbriifolium*), 380
nightshade family (SOLANACEAE), 376
nimblewill (*Muhlenbergia schreberi*), 746
ninebark (*Physocarpus*), 217
nipplewort (*Lapsana communis*), 599
 dwarf (*Arnoseris minima*), 600
Nippon-daisy (*Chrysanthemum nipponicum*), 518
nits-and-lice (*Hypericum drummondii*), 130
nodding fescue (*Festuca subverticillata*), 695
nodding foxtail (*Setaria faberi*), 766
nodding ladies'-tresses (*Spiranthes cernua*), 815
nodding onion (*Allium cernuum*), 790
nodding trillium (*Trillium cernuum*), 796
nondo (*Ligusticum canadense*), 354
northeastern beard-tongue (*Penstemon hirsutus*), 444

northeastern hawkweed (*Hieracium robinsonii*), 593

northeastern mannagrass (*Glyceria melicaria*), 707

northeastern paintbrush (*Castilleja septentrionalis*), 458

northeastern sea-blite (*Suaeda americana*), 93

northeastern willow-herb (*Epilobium strictum*), 297

northern adder's-tongue (*Ophioglossum vulgatum* var. *pseudopodum*), 9

northern androsace (*Androsace septentrionalis*), 203

northern arrow-head (*Sagittaria cuneata*), 604

northern bayberry (*Myrica pensylvanica*), 75

northern bedstraw (*Galium boreale*), 473

northern beech-fern (*Thelypteris phegopteris*), 22

northern bitter-cress (*Cardamine bellidifolia*), 172

northern bladderwort (*Utricularia intermedia*), 463

northern blazing star (*Liatris scariosa*), 574

northern bluebell (*Mertensia paniculata*), 396

northern blue flag (*Iris versicolor*), 811

northern bog-aster (*Aster borealis*), 547

northern bog-goldenrod (*Solidago uliginosa*), 534

northern catalpa (*Catalpa speciosa*), 461

northern coral-root (*Corallorhiza trifida*), 827

northern crab-grass (*Digitaria sanguinalis*), 767

northern dandelion (*Taraxacum ceratophorum*), 598

northern dewberry (*Rubus flagellaris*), 232

northern estuarine beggar-ticks (*Bidens hyperborea*), 506

northern gentian (*Gentianella amarella*), 366

northern gooseberry (*Ribes oxyacanthoides*), 206

northern ground-cedar (*Lycopodium complanatum*), 3

northern hackberry (*Celtis occidentalis*), 69

northern hawkweed (*Hieracium umbellatum*), 594

northern heart-leaved aster (*Aster ciliolatus*), 551

northern holly-fern (*Polystichum lonchitis*), 28

northern horse-balm (*Collinsonia canadensis*), 418

northern joint-vetch (*Aeschynomene virginica*), 263

northern maidenhair fern (*Adiantum pedatum*), 14

northern mannagrass (*Glyceria borealis*), 706

northern marsh-violet (*Viola palustris*), 145

northern meadow-groundsel (*Senecio pauperculus*), 524

northern meadow-rue (*Thalictrum venulosum*), 60

northern mudwort (*Limosella aquatica*), 441

northern pin-oak (*Quercus ellipsoidalis*), 81

northern plains blazing star (*Liatris ligulistylis*), 575

northern red oak (*Quercus rubra*), 80

northern sand-spurrey (*Spergularia canadensis*), 106

northern snailseed-pondweed (*Potamogeton spirillus*), 608

northern speedwell (*Veronica wormskjoldii*), 453

northern spikemoss (*Selaginella selaginoides*), 4

northern stitchwort (*Stellaria borealis*), 104

northern sundrops (*Oenothera tetragona*), 301

northern swamp-dogwood (*Cornus racemosa*), 305

northern swamp-groundsel (*Senecio congestus*), 523

northern sweet-coltsfoot (*Petasites frigidus*), 528

northern three-lobed bedstraw (*Galium trifidum*), 475

northern tickseed-sunflower (*Bidens coronata*), 508

northern water-horehound (*Lycopus uniflorus*), 419

northern water-nymph (*Najas flexilis*), 614

northern water-plantain (*Alisma triviale*), 602

northern white cedar (*Thuja occidentalis*), 35

northern wild senna (*Senna hebecarpa*), 257

northern willow-herb (*Epilobium glandulosum*), 298

northern winter-cress (*Barbarea orthoceras*), 178

northern wood-fern (*Dryopteris expansa*), 27

northern wood-sorrel (*Oxalis acetosella*), 338

northwestern flat-topped white aster (*Aster pubentior*), 557

northwestern sticky aster (*Aster modestus*), 554

Norway maple (*Acer platanoides*), 331

Norway pine (*Pinus resinosa*), 32

Nova Scotia flat-topped goldenrod (*Euthamia galetorum*), 545

Nova Scotian eyebright (*Euphrasia randii*), 456

nutsedge (*Cyperus rotundus*), 650

   false (*Cyperus strigosus*), 653

   yellow (*Cyperus esculentus*), 651

Nuttall milk-vetch (*Astragalus missouriensis*), 264

Nuttall's alkali-grass (*Puccinellia nuttalliana*), 699

Nuttall's milkwort (*Polygala nuttallii*), 329

oak (*Quercus*), 76

   bear- (*Quercus ilicifolia*), 79

   black (*Quercus velutina*), 80

   black-jack (*Quercus marilandica*), 79

   blue-jack (*Quercus cinerea*), 78

   bur- (*Quercus macrocarpa*), 77

   cherrybark- (*Quercus pagoda*), 80

   chinquapin- (*Quercus prinoides*), 78

   Darlington- (*Quercus hemisphaerica*), 79

   Jack- (*Quercus imbricaria*), 78

   Jerusalem (*Chenopodium botrys*), 87

   laurel- (*Quercus laurifolia*), 79

   live (*Quercus virginiana*), 78

   northern pin- (*Quercus ellipsoidalis*), 81

   northern red (*Quercus rubra*), 80

   overcup- (*Quercus lyrata*), 77

   pin- (*Quercus palustris*), 81

   poison- (*Toxicodendron pubescens*), 335

   post- (*Quercus stellata*), 76

   rock chestnut (*Quercus prinus*), 78

   sand live (*Quercus geminata*), 78

   sand post- (*Quercus margaretta*), 77

   scarlet (*Quercus coccinea*), 81

   shingle- (*Quercus imbricaria*), 78

   Shumard (*Quercus shumardii*), 81

   southern red (*Quercus falcata*), 79

   Spanish (*Quercus falcata*), 79

   swamp chestnut- (*Quercus michauxii*), 77

   swamp white (*Quercus bicolor*), 77

   turkey- (*Quercus laevis*), 80

   water- (*Quercus nigra*), 79

   white (*Quercus alba*), 76

   willow- (*Quercus phellos*), 79

   yellow (*Quercus muehlenbergii*), 78

Oakes' pondweed (*Potamogeton oakesianus*), 613

oak-fern (*Gymnocarpium*), 22; (*G. dryopteris*), 22

   limestone (*Gymnocarpium robertianum*), 22

oak-leaved daisy (*Erigeron quercifolius*), 560

oak-leaved goosefoot (*Chenopodium glaucum*), 87

oak-leaved hydrangea (*Hydrangea quercifolia*), 205

oatgrass (*Arrhenatherum*), 709

   downy (*Danthonia sericea*), 733

   poverty- (*Danthonia spicata*), 731

   tall (*Arrhenatherum elatius*), 709

   timber- (*Danthonia intermedia*), 732

   wild (*Danthonia*), 731

oats (*Avena*), 710; (*A. sativa*), 710

   animated (*Avena sterilis*), 710

   sea- (*Uniola paniculata*), 735

   wild (*Avena fatua*), 710; (*Chasmanthium latifolium*), 731

obedience (*Physostegia virginiana*), 422

   western (*Physostegia parviflora*), 422

Ogden's pondweed (*Potamogeton ogdenii*), 610

Ohio buckeye (*Aesculus glabra*), 331

Ohio goldenrod (*Solidago ohioensis*), 544

oil-nut (*Pyrularia pubera*), 307
old-field five-fingers (*Potentilla simplex*), 222
old plainsman (*Hymenopappus scabiosaeus*), 516
oleaster (*Elaeagnus angustifolia*), 288
oleaster family (ELAEAGNACEAE), 288
olive, Russian (*Elaeagnus angustifolia*), 288
olive family (OLEACEAE), 435
Olney-threesquare (*Scirpus americanus*), 632
one-flowered gentian (*Gentiana autumnalis*), 364
one-flowered shinleaf (*Moneses*), 196
one-flower hawthorn (*Crataegus uniflora*), 248
one-leaf orchis (*Orchis rotundifolia*), 818
one-seed croton (*Croton monanthogynus*), 312
one-seed hawthorn (*Crataegus monogyna*), 247
one-sided shinleaf (*Pyrola secunda*), 195
onion (*Allium*), 789
    cultivated (*Allium cepa*), 790
    nodding (*Allium cernuum*), 790
opium-poppy (*Papaver somniferum*), 65
opposite-leaved dwarf dandelion (*Krigia oppositifolia*), 598
orache (*Atriplex hortensis*), 91
    red (*Atriplex rosea*), 92
    seabeach- (*Atriplex arenaria*), 91
orange
    Appalachian mock- (*Philadelphus inodorus*), 204
    Cumberland mock- (*Philadelphus hirsutus*), 204
    European mock- (*Philadelphus coronarius*), 204
    mock- (*Philadelphus*), 204
    osage- (*Maclura*), 70
    Ozark mock- (*Philadelphus pubescens*), 204
orange cosmos (*Cosmos sulphureus*), 509
orange dwarf dandelion (*Krigia biflora*), 597
orange-eye butterfly-bush (*Buddleja davidii*), 434
orange-grass (*Hypericum gentianoides*), 130
orange lily (*Lilium bulbiferum*), 791
orange milkwort (*Polygala lutea*), 328
orange-red king-devil (*Hieracium aurantiacum*), 592
orange touch-me-not (*Impatiens capensis*), 341
orchard-grass (*Dactylis*), 705
orchid
    Alaska (*Habenaria unalascensis*), 820
    bluntleaf- (*Habenaria obtusata*), 820
    bracted (*Habenaria viridis*), 819
    club-spur (*Habenaria clavellata*), 819
    crane-fly (*Tipularia discolor*), 826
    crested fringed (*Habenaria cristata*), 821
    fen- (*Liparis loeselii*), 825
    Hooker's (*Habenaria hookeri*), 821
    prairie fringed (*Habenaria leucophaea*), 822
    purple fringed (*Habenaria psycodes*), 822
    purple fringeless (*Habenaria peramoena*), 822
    ragged fringed (*Habenaria lacera*), 822
    rein- (*Habenaria*), 819
    round-leaved (*Habenaria orbiculata*), 820
    snowy (*Habenaria nivea*), 819
    tall northern bog- (*Habenaria hyperborea*), 819
    tall white bog- (*Habenaria dilatata*), 820
    three-birds (*Triphora trianthophora*), 824
    tubercled (*Habenaria flava*), 819
    white fringed (*Habenaria blephariglottis*), 821
    yellow (*Habenaria integra*), 820
    yellow fringed (*Habenaria ciliaris*), 821
orchid family (ORCHIDACEAE), 812
orchis
    one-leaf (*Orchis rotundifolia*), 818
    showy (*Orchis spectabilis*), 818

oriental bittersweet (*Celastrus orbiculatus*), 308
osage-orange (*Maclura*), 70
osier
    purple (*Salix purpurea*), 160
    silky (*Salix viminalis*), 160
osier-dogwood, red (*Cornus sericea*), 305
ostrich-fern (*Matteuccia*), 29
Oswego-tea (*Monarda didyma*), 428
oval ladies'-tresses (*Spiranthes ovalis*), 815
overcup-oak (*Quercus lyrata*), 77
owl-clover (*Orthocarpus*), 458
ox-eye, sea (*Borrichia*), 502
ox-eye daisy (*Chrysanthemum leucanthemum*), 518
ox-tongue (*Picris*), 600
    bristly (*Picris echioides*), 600
    hawkweed (*Picris hieracioides*), 600
Ozark bluestar (*Amsonia illustris*), 368
Ozark buttercup (*Ranunculus harveyi*), 56
Ozark flatseed-sunflower (*Verbesina helianthoides*), 501
Ozark goldenrod (*Solidago drummondii*), 539
Ozark ground-cherry (*Physalis missouriensis*), 378
Ozark heart-leaved aster (*Aster texanus*), 553
Ozark hickory (*Carya texana*), 74
Ozarkian blue-eyed Mary (*Collinsia violacea*), 446
Ozark ironweed (*Vernonia arkansana*), 577
Ozark milk-vetch (*Astragalus distortus*), 264
Ozark milkweed (*Asclepias viridis*), 373
Ozark mock bishop-weed (*Ptilimnium nuttallii*), 356
Ozark mock-orange (*Philadelphus pubescens*), 204
Ozark parthenium (*Parthenium hispidum*), 513
Ozark phacelia (*Phacelia gilioides*), 392
Ozark sunflower (*Helianthus silphioides*), 492
Ozark tickseed-sunflower (*Bidens polylepis*), 508

pagoda-dogwood (*Cornus alternifolia*), 304
paintbrush
    downy (*Castilleja sessiliflora*), 458
    Indian (*Castilleja*), 458
    northeastern (*Castilleja septentrionalis*), 458
painted buckeye (*Aesculus sylvatica*), 331
painted cup (*Castilleja coccinea*), 458
painted trillium (*Trillium undulatum*), 795
pale dock (*Rumex altissimus*), 114
pale-flowered leaf-cup (*Polymnia canadensis*), 511
pale gentian (*Gentiana flavida*), 366
pale hickory (*Carya pallida*), 74
pale Indian-plantain (*Cacalia atriplicifolia*), 527
pale poppy-mallow (*Callirhoe alcaeoides*), 136
pale umbrella-wort (*Mirabilis albida*), 85
pale vetch (*Vicia caroliniana*), 274
palmate hop-clover (*Trifolium aureum*), 271
panic-grass (*Panicum*), 751
    velvet (*Panicum scoparium*), 759
panicled hawkweed (*Hieracium paniculatum*), 595
pansy
    European field- (*Viola arvensis*), 149
    wild (*Viola rafinesquii*), 148
paper-birch (*Betula papyrifera*), 83
paper-mulberry (*Broussonetia*), 70
Paraguay-bur (*Acanthospermum australe*), 511
parasol-tree, Chinese (*Firmiana simplex*), 131
parrot's feather (*Myriophyllum aquaticum*), 289
parsley (*Petroselinum crispum*), 351
    field hedge- (*Torilis arvensis*), 350
    fool's (*Aethusa*), 349

parsley *(continued)*
  hedge- *(Torilis)*, 350
  hemlock- *(Conioselinum)*, 355
  Japanese hedge- *(Torilis japonica)*, 350
  prairie- *(Polytaenia nuttallii)*, 360
  white wild *(Lomatium orientale)*, 360
  wild *(Cymopterus acaulis)*, 361
  yellow wild *(Lomatium foeniculaceum)*, 360
parsley-fern *(Cryptogramma crispa)*, 14
parsley-hawthorn *(Crataegus marshallii)*, 247
parsnip *(Pastinaca sativa)*, 357
  bearded meadow- *(Thaspium barbinode)*, 353
  cow- *(Heracleum lanatum)*, 357
  meadow- *(Thaspium)*, 353
  mountain meadow- *(Thaspium pinnatifidum)*, 354
  smooth meadow- *(Thaspium trifoliatum)*, 353
  water- *(Sium)*, 355
parthenium
  Appalachian *(Parthenium auriculatum)*, 513
  eastern *(Parthenium integrifolium)*, 514
  Ozark *(Parthenium hispidum)*, 513
partridge-berry *(Mitchella repens)*, 471; *(Vaccinium vitis-idaea)*, 192
partridge-pea *(Chamaecrista fasciculata)*, 256
pasque-flower *(Anemone patens)*, 50
passion-flower *(Passiflora)*, 149
passion-flower family (PASSIFLORACEAE), 149
pasture-heliotrope *(Heliotropium tenellum)*, 395
pasture-rose *(Rosa carolina)*, 239
pasture-thistle *(Cirsium pumilum)*, 581
path-rush *(Juncus tenuis)*, 623
patience-dock *(Rumex patientia)*, 114
pawpaw *(Asimina)*, 38
pea
  beach- *(Lathyrus maritimus)*, 274
  butterfly- *(Clitoria)*, 285
  everlasting *(Lathyrus latifolius)*, 275
  forest- *(Lathyrus venosus)*, 275
  gray scurf- *(Psoralidium tenuiflorum)*, 284
  hairy milk- *(Galactia volubilis)*, 287
  Macree's milk- *(Galactia macreei)*, 287
  marsh- *(Lathyrus palustris)*, 275
  meadow- *(Lathyrus pratensis)*, 274
  milk- *(Galactia)*, 287
  partridge- *(Chamaecrista fasciculata)*, 256
  silvery scurf- *(Pediomelum argophyllum)*, 284
  trailing milk- *(Galactia regularis)*, 287
  white *(Lathyrus ochroleucus)*, 274
  wild *(Lathyrus)*, 274
peach *(Prunus persica)*, 240
peach-leaf-willow *(Salix amygdaloides)*, 154
pea family (FABACEAE), 258
peanut *(Arachis)*, 263
  hog- *(Amphicarpaea)*, 286
pear *(Pyrus)*, 243; *(P. communis)*, 243
pear-hawthorn *(Crataegus calpodendron)*, 248
pearlwort *(Sagina)*, 100
  Kentucky *(Sagina fontinalis)*, 100
pearly everlasting *(Anaphalis)*, 565
pecan *(Carya illinoinensis)*, 73
pellitory *(Parietaria)*, 72
pencil-flower *(Stylosanthes)*, 263
Pennsylvania bitter-cress *(Cardamine pensylvanica)*, 173
Pennsylvania blackberry *(Rubus pensilvanicus)*, 234
Pennsylvania smartweed *(Polygonum pensylvanicum)*, 120

penny-cress *(Thlaspi)*, 166
  field *(Thlaspi arvense)*, 166
  thoroughwort *(Thlaspi perfoliatum)*, 166
pennyroyal *(Mentha pulegium)*, 417
  American *(Hedeoma pulegioides)*, 412
  false *(Isanthus)*, 409
pennywort *(Hydrocotyle)*, 344; *(Obolaria)*, 367
  buttercup- *(Hydrocotyle ranunculoides)*, 344
  lawn- *(Hydrocotyle sibthorpioides)*, 344
  marsh- *(Hydrocotyle americana)*, 344
  water- *(Hydrocotyle umbellata)*, 344
  whorled *(Hydrocotyle verticillata)*, 344
pepper, poor-man's *(Lepidium virginicum)*, 164
pepper-bush, sweet *(Clethra)*, 183
peppergrass *(Lepidium)*, 164
peppermint *(Mentha ✕piperita)*, 417
pepper-vine *(Ampelopsis arborea)*, 322
pepperweed *(Lepidium)*, 164
  clasping *(Lepidium perfoliatum)*, 164
  perennial *(Lepidium latifolium)*, 165
  prairie- *(Lepidium densiflorum)*, 165
  stinking *(Lepidium ruderale)*, 165
perennial agalinis *(Agalinis linifolia)*, 451
perennial glasswort *(Salicornia virginica)*, 92
perennial honesty *(Lunaria rediviva)*, 168
perennial pepperweed *(Lepidium latifolium)*, 165
perennial quaking grass *(Briza media)*, 700
perennial salt-marsh aster *(Aster tenuifolius)*, 558
perennial sea-pink *(Sabatia dodecandra)*, 363
perennial sow-thistle *(Sonchus arvensis)*, 591
perennial woolly bean *(Strophostyles umbellata)*, 286
perfoliate horse-gentian *(Triosteum perfoliatum)*, 481
perfoliate pondweed *(Potamogeton perfoliatus)*, 613
perilla-mint *(Perilla frutescens)*, 419
periwinkle *(Vinca minor)*, 368
  greater *(Vinca major)*, 368
Persian clover *(Trifolium resupinatum)*, 270
Persian darnel *(Lolium persicum)*, 698
persimmon *(Diospyros)*, 198
petty spurge *(Euphorbia peplus)*, 317
petunia *(Petunia)*, 383
  common *(Petunia ✕hybrida)*, 383
  garden- *(Petunia ✕hybrida)*, 383
  wild *(Petunia parviflora)*, 383
phacelia *(Phacelia)*, 392
  Appalachian *(Phacelia dubia)*, 393
  Blue-Ridge *(Phacelia fimbriata)*, 393
  Canadian *(Phacelia franklinii)*, 392
  forest- *(Phacelia bipinnatifida)*, 393
  Ozark *(Phacelia gilioides)*, 392
pheasant's eye *(Adonis)*, 59
Philadelphia daisy *(Erigeron philadelphicus)*, 560
phlox *(Phlox)*, 389
  Allegheny- *(Phlox ovata)*, 391
  crawling *(Phlox stolonifera)*, 390
  forest- *(Phlox divaricata)*, 389
  hairy *(Phlox amoena)*, 389
  meadow- *(Phlox maculata)*, 391
  piney-woods *(Phlox nivalis)*, 390
  prairie- *(Phlox pilosa)*, 389
  sand- *(Phlox bifida)*, 390
  shale-barren *(Phlox buckleyi)*, 391
  smooth *(Phlox glaberrima)*, 391
  spotted *(Phlox maculata)*, 391
  summer- *(Phlox paniculata)*, 390
  wide-leaved *(Phlox amplifolia)*, 390

phlox family (POLEMONIACEAE), 389
pickerel-weed (*Pontederia*), 779
pie-cherry (*Prunus cerasus*), 241
piedmont-azalea (*Rhododendron canescens*), 185
piedmont blazing star (*Liatris regimontis*), 575
piedmont-buckbean (*Thermopsis mollis*), 259
pie marker (*Abutilon theophrasti*), 133
pigeon-grape (*Vitis aestivalis*), 324; (*V. cinerea*), 325
pignut-hickory (*Carya glabra*), 74
    sweet (*Carya ovalis*), 74
pigweed (*Chenopodium album*), 89
    rough (*Amaranthus retroflexus*), 95
    smooth (*Amaranthus hybridus*), 96
    winged (*Cycloloma*), 90
pimpernel (*Anagallis*), 202
    false (*Lindernia*), 441
    shale-barren (*Taenidia montana*), 352
    water- (*Samolus*), 203
    yellow (*Taenidia integerrima*), 352
pin-cherry (*Prunus pensylvanica*), 241
pincushions (*Scabiosa atropurpurea*), 488
pine (*Pinus*), 32
    ground- (*Ajuga chamaepitys*), 408; (*Lycopodium obscurum*), 2
    Jack- (*Pinus banksiana*), 33
    loblolly- (*Pinus taeda*), 33
    longleaf- (*Pinus palustris*), 33
    Norway (*Pinus resinosa*), 32
    pitch- (*Pinus rigida*), 34
    pond- (*Pinus serotina*), 33
    prince's (*Chimaphila umbellata*), 195
    princess- (*Lycopodium obscurum*), 2
    red (*Pinus resinosa*), 32
    running (*Lycopodium clavatum*), 2
    Scotch (*Pinus sylvestris*), 32
    scrub- (*Pinus virginiana*), 34
    shortleaf (*Pinus echinata*), 34
    table-mountain (*Pinus pungens*), 33
    Virginia (*Pinus virginiana*), 34
    white (*Pinus strobus*), 32
    yellow (*Pinus echinata*), 34
pineapple-weed (*Matricaria matricarioides*), 519
pine-barren eupatorium (*Eupatorium resinosum*), 569
pine-barren gentian (*Gentiana autumnalis*), 364
pine-barren tick-trefoil (*Desmodium strictum*), 277
pine-drops (*Pterospora*), 197
pine family (PINACEAE), 31
pine-lily (*Lilium catesbaei*), 791
pinesap (*Monotropa hypopithys*), 196
    sweet (*Monotropsis*), 196
piney-woods coneflower (*Rudbeckia heliopsidis*), 496
piney-woods phlox (*Phlox nivalis*), 390
pink (*Dianthus*), 112
    annual sea- (*Sabatia stellaris*), 363
    childing (*Petrorhagia prolifera*), 112
    common marsh- (*Sabatia angularis*), 362
    Deptford- (*Dianthus armeria*), 112
    fire- (*Silene virginica*), 108
    garden- (*Dianthus plumarius*), 111
    grass- (*Calopogon*), 824
    maiden- (*Dianthus deltoides*), 112
    marsh- (*Sabatia*), 362
    moss- (*Phlox subulata*), 390
    mullein- (*Lychnis coronaria*), 107
    perennial sea- (*Sabatia dodecandra*), 363
    rush- (*Lygodesmia*), 588

pink (*Dianthus*) (continued)
    sea- (*Sabatia*), 362
    slender marsh- (*Sabatia campanulata*), 362
    swamp- (*Helonias*), 786
    western marsh- (*Sabatia campestris*), 362
    wild (*Silene caroliniana*), 110; (*S. regia*), 109
pink eupatorium (*Eupatorium incarnatum*), 568
pink family (CARYOPHYLLACEAE), 100
pink lady-slipper (*Cypripedium acaule*), 813
pink milkwort (*Polygala incarnata*), 328
pinkroot (*Spigelia*), 361
pink shinleaf (*Pyrola asarifolia*), 196
pink spiderwort (*Tradescantia rosea*), 620
pink spring-cress (*Cardamine douglassii*), 172
pinkster-flower (*Rhododendron periclymenoides*), 185
pink tickseed (*Coreopsis rosea*), 505
pinnate hop-clover (*Trifolium campestre*), 271
pin-oak (*Quercus palustris*), 81
    northern (*Quercus ellipsoidalis*), 81
pinweed (*Lechea*), 141
pipe
    Dutchman's (*Aristolochia macrophylla*), 41
    Indian (*Monotropa uniflora*), 196
pipe-vine (*Aristolochia tomentosa*), 41
pipewort (*Eriocaulon*), 621
pipewort family (ERIOCAULACEAE), 621
pipsissewa (*Chimaphila umbellata*), 195
pitcher-plant (*Sarracenia*), 138
pitcher-plant family (SARRACENIACEAE), 138
pitch-pine (*Pinus rigida*), 34
pitseed goosefoot (*Chenopodium berlandieri*), 89
plagiobothrys, meadow- (*Plagiobothrys scouleri*), 397
plains blazing star (*Liatris squarrosa*), 576
plains-bluegrass (*Poa arida*), 701
plainsman, old (*Hymenopappus scabiosaeus*), 516
plains prickly pear (*Opuntia macrorhiza*), 86
plains-puccoon (*Lithospermum caroliniense*), 399
plains-pussytoes (*Antennaria parvifolia*), 566
plains reed-grass (*Calamagrostis montanensis*), 715
plains-sandhill ground-cherry (*Physalis hispida*), 377
plains sea-blite (*Suaeda calceoliformis*), 93
plains-stoneseed (*Lithospermum incisum*), 400
plains-sunflower (*Helianthus petiolaris*), 489
plains-tickseed (*Coreopsis tinctoria*), 505
plains wild indigo (*Baptisia bracteata*), 258
planer-tree (*Planera*), 69
plane-tree family (PLATANACEAE), 68
plantain (*Plantago*), 431
    alloploid rattlesnake- (*Goodyera tesselata*), 818
    American (*Plantago rugelii*), 432
    common (*Plantago major*), 432
    downy rattlesnake- (*Goodyera pubescens*), 817
    English (*Plantago lanceolata*), 433
    grass-leaved water- (*Alisma gramineum*), 602
    great Indian- (*Cacalia muehlenbergii*), 527
    hastate Indian- (*Cacalia suaveolens*), 527
    Indian- (*Cacalia*), 527
    leafy-stemmed (*Plantago psyllium*), 431
    lesser rattlesnake- (*Goodyera repens*), 818
    little (*Plantago pusilla*), 433
    many-seeded (*Plantago heterophylla*), 433
    mud- (*Heteranthera*), 780
    northern water- (*Alisma triviale*), 602
    pale Indian- (*Cacalia atriplicifolia*), 527
    rattlesnake- (*Goodyera*), 817

plantain (*Plantago*) (*continued*)
   Robin's (*Erigeron pulchellus*), 560
   seaside- (*Plantago maritima*), 432
   southern water- (*Alisma subcordatum*), 603
   tuberous Indian- (*Cacalia plantaginea*), 528
   water- (*Alisma*), 602
   western rattlesnake- (*Goodyera oblongifolia*), 817
   woolly (*Plantago patagonica*), 434
   woolly-crowned (*Plantago eriopoda*), 432
plantain family (PLANTAGINACEAE), 431
plantain-lily (*Hosta*), 788
plantain-pussytoes (*Antennaria plantaginifolia*), 566
Platte groundsel (*Senecio plattensis*), 524
pliant milk-vetch (*Astragalus flexuosus*), 265
plum
   Allegheny- (*Prunus alleghaniensis*), 242
   beach- (*Prunus maritima*), 242
   bigtree- (*Prunus mexicana*), 242
   Canada (*Prunus nigra*), 243
   Chickasaw- (*Prunus angustifolia*), 242
   ground- (*Astragalus crassicarpus*), 265
   hortulan (*Prunus hortulana*), 243
   prairie- (*Astragalus crassicarpus*), 265
   wild (*Prunus americana*), 242
   wildgoose- (*Prunus munsoniana*), 243
plumegrass (*Erianthus*), 771
   bent-awn (*Erianthus contortus*), 771
   brown (*Erianthus brevibarbis*), 770
   narrow (*Erianthus strictus*), 770
   silver- (*Erianthus alopecuroides*), 771
   sugar-cane (*Erianthus giganteus*), 770
plumeless thistle (*Carduus acanthoides*), 580
Plymouth rose gentian (*Sabatia kennedyana*), 363
pod-fern (*Aspidotis densa*), 15
pod-grass (*Scheuchzeria palustris*), 607
poet's narcissus (*Narcissus poeticus*), 803
pogonia
   larger whorled (*Isotria verticillata*), 823
   little whorled (*Isotria medeoloides*), 823
   rose (*Pogonia ophioglossoides*), 823
   spreading (*Cleistes divaricata*), 823
   whorled (*Isotria*), 823
poison
   crow- (*Zigadenus densus*), 783
   fly- (*Amianthium*), 782
poison-hemlock (*Conium maculatum*), 351
poison-ivy
   common (*Toxicodendron radicans*), 335
   western (*Toxicodendron rydbergii*), 335
poison-oak (*Toxicodendron pubescens*), 335
poison-sumac (*Toxicodendron vernix*), 335
pokeberry (*Phytolacca*), 84
pokeweed (*Phytolacca*), 84
pokeweed family (PHYTOLACCACEAE), 84
polypody (*Polypodium*), 12
   common (*Polypodium virginianum*), 12
polypody family (POLYPODIACEAE), 12
pond-pine (*Pinus serotina*), 33
pondweed (*Potamogeton*), 608
   alga- (*Potamogeton confervoides*), 610
   bigleaf- (*Potamogeton amplifolius*), 612
   bigsheath- (*Potamogeton vaginatus*), 609
   bluntleaf- (*Potamogeton obtusifolius*), 611
   common snailseed- (*Potamogeton diversifolius*), 609
   curly (*Potamogeton crispus*), 609

pondweed (*Potamogeton*) (*continued*)
   fern- (*Potamogeton robbinsii*), 609
   flatstem- (*Potamogeton zosteriformis*), 610
   floating (*Potamogeton natans*), 613
   Fries' (*Potamogeton friesii*), 610
   Hill's (*Potamogeton hillii*), 610
   horned (*Zannichellia*), 614
   Illinois (*Potamogeton illinoensis*), 613
   leafy (*Potamogeton foliosus*), 611
   longleaf- (*Potamogeton nodosus*), 612
   northern snailseed- (*Potamogeton spirillus*), 608
   Oakes' (*Potamogeton oakesianus*), 613
   Ogden's (*Potamogeton ogdenii*), 610
   perfoliate (*Potamogeton perfoliatus*), 613
   red (*Potamogeton alpinus*), 612
   ribbonleaf- (*Potamogeton epihydrus*), 611
   sago- (*Potamogeton pectinatus*), 609
   slender (*Potamogeton pusillus*), 611
   spotted (*Potamogeton pulcher*), 612
   straight-leaved (*Potamogeton strictifolius*), 610
   threadleaf- (*Potamogeton filiformis*), 608
   variable (*Potamogeton gramineus*), 612
   Vasey's (*Potamogeton vaseyi*), 611
   whitestem- (*Potamogeton praelongus*), 613
pondweed family (POTAMOGETONACEAE), 608
poorjoe (*Diodia teres*), 471
poor-man's pepper (*Lepidium virginicum*), 164
poplar
   balsam- (*Populus balsamifera*), 152
   black (*Populus nigra*), 153
   silver (*Populus alba*), 152
   white (*Populus alba*), 152
   yellow (*Liriodendron*), 38
poppy (*Papaver*), 65
   celandine- (*Stylophorum*), 64
   corn- (*Papaver rhoeas*), 65
   horned (*Glaucium*), 64
   Mexican (*Argemone mexicana*), 65
   opium- (*Papaver somniferum*), 65
   prickly (*Argemone*), 65
   Shirley- (*Papaver rhoeas*), 65
   white prickly (*Argemone albiflora*), 65
poppy family (PAPAVERACEAE), 64
poppy-mallow (*Callirhoe*), 136
   Bush's (*Callirhoe bushii*), 136
   clustered (*Callirhoe triangulata*), 136
   finger (*Callirhoe digitata*), 136
   pale (*Callirhoe alcaeoides*), 136
   purple (*Callirhoe involucrata*), 136
porcelain-berry (*Ampelopsis brevipedunculata*), 323
porcupine-grass (*Stipa*), 693; (*S. spartea*), 694
possum-grape (*Vitis baileyana*), 325
possum-haw (*Ilex decidua*), 310; (*Viburnum nudum* var. *nudum*), 484
post-oak (*Quercus stellata*), 76
   sand (*Quercus margaretta*), 77
potato (*Solanum tuberosum*), 380
   wild (*Ipomoea pandurata*), 386; (*Solanum jamesii*), 380
potentilla
   diffuse (*Potentilla paradoxa*), 223
   Great Plains (*Potentilla effusa*), 225
   marsh- (*Potentilla palustris*), 226
   mountain white (*Potentilla tridentata*), 225
   strawberry- (*Potentilla sterilis*), 222
   tall (*Potentilla arguta*), 226

potentilla *(continued)*
White-Mountain (*Potentilla robbinsiana*), 224
woolly (*Potentilla hippiana*), 225
poverty-grass (*Sporobolus vaginiflorus*), 743
poverty-oatgrass (*Danthonia spicata*), 731
poverty-weed (*Monolepis nuttalliana*), 90
powdery cloak-fern (*Notholaena dealbata*), 16
prairie alum-root (*Heuchera richardsonii*), 216
prairie-anemone (*Anemone caroliniana*), 49
prairie-aster (*Aster turbinellus*), 555
prairie broom-rape (*Orobanche ludoviciana*), 458
prairie-buttercup (*Ranunculus rhomboideus*), 56
prairie closed gentian (*Gentiana andrewsii*), 365
prairie-clover
cedar-glade (*Dalea foliosa*), 282
downy (*Dalea villosa*), 282
purple (*Dalea purpurea*), 282
rounded-headed (*Dalea multiflora*), 282
white (*Dalea candida*), 282
prairie-coneflower (*Echinacea pallida*), 498
prairie cord-grass (*Spartina pectinata*), 749
prairie-crab (*Pyrus ioensis*), 245
prairie cup-grass (*Eriochloa contracta*), 762
prairie-dropseed (*Sporobolus heterolepis*), 743
prairie fame-flower (*Talinum parviflorum*), 99
prairie fringed orchid (*Habenaria leucophaea*), 822
prairie-gentian (*Gentiana puberulenta*), 364
prairie golden aster (*Chrysopsis camporum*), 530
prairie ground-cherry (*Physalis pumila*), 377
prairie heart-leaved aster (*Aster oolentangiensis*), 551
prairie-larkspur (*Delphinium virescens*), 48
prairie-lespedeza (*Lespedeza leptostachya*), 281
prairie-lettuce (*Lactuca ludoviciana*), 589
prairie-milkweed (*Asclepias hirtella*), 373
prairie-parsley (*Polytaenia nuttallii*), 360
prairie-pepperweed (*Lepidium densiflorum*), 165
prairie-phlox (*Phlox pilosa*), 389
prairie-plum (*Astragalus crassicarpus*), 265
prairie-redroot (*Ceanothus herbaceus*), 321
prairie-rose
climbing (*Rosa setigera*), 237
dwarf (*Rosa arkansana*), 239
prairie rosin-weed (*Silphium integrifolium*), 512
prairie-sagewort (*Artemisia frigida*), 522
prairie-smoke (*Anemone patens*), 50; (*Geum triflorum*), 228
prairie-spiderwort (*Tradescantia occidentalis*), 620
prairie-spurge (*Euphorbia spathulata*), 315
prairie-tea (*Croton monanthogynus*), 312
prairie-thistle (*Cirsium flodmanii*), 582
prairie three-awn (*Aristida oligantha*), 733
prairie tick-trefoil (*Desmodium illinoense*), 277
prairie-turnip
shaggy (*Pediomelum esculentum*), 284
strigose (*Pediomelum cuspidatum*), 284
three-leaflet (*Pediomelum canescens*), 283
prairie white beard-tongue (*Penstemon albidus*), 444
prickly ash (*Zanthoxylum*), 336
common (*Zanthoxylum americanum*), 336
prickly lettuce (*Lactuca serriola*), 591
prickly pear (*Opuntia*), 86
eastern (*Opuntia humifusa*), 86
little (*Opuntia fragilis*), 86
plains (*Opuntia macrorhiza*), 86
prickly poppy (*Argemone*), 65
white (*Argemone albiflora*), 65
prickly sida (*Sida spinosa*), 137
prickly sow-thistle (*Sonchus asper*), 591
primrose (*Primula*), 203
common evening- (*Oenothera biennis*), 299
creeping water- (*Ludwigia peploides*), 294
cut-leaf evening- (*Oenothera laciniata*), 300
garden evening- (*Oenothera glazioviana*), 299
hairy water- (*Ludwigia pilosa*), 295
Laurentide (*Primula laurentiana*), 203
longspike evening- (*Oenothera rhombipetala*), 299
Mistassini (*Primula mistassinica*), 203
narrow-leaved water- (*Ludwigia linearis*), 296
Rafinesque's water- (*Ludwigia hirtella*), 295
round-pod water- (*Ludwigia sphaerocarpa*), 296
shale-barren evening- (*Oenothera argillicola*), 299
showy water- (*Ludwigia uruguayensis*), 294
small-flowered evening- (*Oenothera parviflora*), 299
small-flowered water- (*Ludwigia glandulosa*), 296
spreading evening- (*Oenothera humifusa*), 300
square-pod water- (*Ludwigia alternifolia*), 295
stemless evening- (*Oenothera triloba*), 300
top-pod water- (*Ludwigia polycarpa*), 295
water- (*Ludwigia*), 294
white evening- (*Oenothera speciosa*), 300
white-stemmed evening- (*Oenothera nuttallii*), 300
wing-cup water- (*Ludwigia alata*), 295
wing-fruit evening- (*Oenothera macrocarpa*), 301
wingstem water- (*Ludwigia decurrens*), 294
primrose family (PRIMULACEAE), 199
primrose-leaved violet (*Viola primulifolia*), 145
primrose-willow (*Ludwigia*), 294
prince's feather (*Amaranthus hypochondriacus*), 95; (*Polygonum orientale*), 122
prince's pine (*Chimaphila umbellata*), 195
princess-pine (*Lycopodium obscurum*), 2
privet (*Ligustrum*), 435
common (*Ligustrum vulgare*), 435
swamp- (*Forestiera acuminata*), 436
proso millet (*Panicum miliaceum*), 753
prostrate vervain (*Verbena bracteata*), 405
puccoon (*Lithospermum*), 399
hoary (*Lithospermum canescens*), 399
narrow-leaved (*Lithospermum incisum*), 400
plains- (*Lithospermum caroliniense*), 399
pulse milk-vetch (*Astragalus tenellus*), 264
pumpkin, wild (*Cucurbita foetidissima*), 150
pumpkin-ash (*Fraxinus profunda*), 437
puncture-vine (*Tribulus terrestris*), 337
purple alpine saxifrage (*Saxifraga oppositifolia*), 212
purple bergamot (*Monarda media*), 428
purple clematis (*Clematis occidentalis*), 51
purple cliff-brake (*Pellaea atropurpurea*), 16
purple coneflower (*Echinacea purpurea*), 497
purple cudweed (*Gnaphalium purpureum*), 565
purple fringed orchid (*Habenaria psycodes*), 822
purple fringeless orchid (*Habenaria peramoena*), 822
purple giant-hyssop (*Agastache scrophulariifolia*), 421
purple loco-weed (*Oxytropis lambertii*), 267
purple loosestrife (*Lythrum salicaria*), 292
purple lovegrass (*Eragrostis spectabilis*), 738
purple meadow-rue (*Thalictrum dasycarpum*), 61
purple milkweed (*Asclepias purpurascens*), 372
purple mullein (*Verbascum phoeniceum*), 442
purple-node Joe-Pye weed (*Eupatorium purpureum*), 568
purple osier (*Salix purpurea*), 160

purple poppy-mallow (*Callirhoe involucrata*), 136
purple prairie-clover (*Dalea purpurea*), 282
purple reed-grass (*Calamagrostis purpurascens*), 714
purple rocket (*Iodanthus pinnatifidus*), 179
purple sand-grass (*Triplasis purpurea*), 736
purple spurge (*Euphorbia purpurea*), 315
purple star-thistle (*Centaurea calcitrapa*), 586
purplestem-angelica (*Angelica atropurpurea*), 358
purplestem beggar-ticks (*Bidens connata*), 507
purple three-awn (*Aristida purpurea*), 733
purpletop (*Tridens flavus*), 736
purple trillium (*Trillium erectum*), 796
purple turtlehead (*Chelone obliqua*), 443
purslane
  coastal plain water- (*Ludwigia brevipes*), 294
  common (*Portulaca oleracea*), 98
  common water- (*Ludwigia palustris*), 296
  milk- (*Euphorbia maculata*), 320
  sea- (*Sesuvium*), 85
  water- (*Didiplis*), 292
purslane family (PORTULACACEAE), 98
purslane-speedwell (*Veronica peregrina*), 453
pusley, Florida (*Richardia scabra*), 472
pussytoes (*Antennaria*), 566
  field- (*Antennaria neglecta*), 566
  plains- (*Antennaria parvifolia*), 566
  plantain- (*Antennaria plantaginifolia*), 566
  rosy (*Antennaria microphylla*), 566
  shale-barren (*Antennaria virginica*), 566
  southern single-head (*Antennaria solitaria*), 566
pussy-willow (*Salix discolor*), 159
putty-root (*Aplectrum hyemale*), 826
pygmy water-lily (*Nymphaea tetragona*), 43
pygmy-weed (*Crassula aquatica*), 210
pyxie (*Pyxidanthera*), 197

quack-grass (*Elytrigia repens*), 727
quaking aspen (*Populus tremuloides*), 152
quaking grass (*Briza*), 700
  big (*Briza maxima*), 700
  little (*Briza minor*), 700
  perennial (*Briza media*), 700
quassia family (SIMAROUBACEAE), 335
Queen Anne's lace (*Daucus carota*), 350
queen of the meadow (*Filipendula ulmaria*), 227
queen of the prairie (*Filipendula rubra*), 226
queen's delight (*Stillingia sylvatica*), 313
quickweed (*Galinsoga*), 509
  common (*Galinsoga quadriradiata*), 509
  lesser (*Galinsoga parviflora*), 509
quill-leaved sagittaria (*Sagittaria teres*), 605
quillwort (*Isoetes*), 4
  black-footed (*Isoetes melanopoda*), 4
  Butler's (*Isoetes butleri*), 4
  deep-water (*Isoetes lacustris*), 4
  Engelmann's (*Isoetes engelmannii*), 4
  riverbank- (*Isoetes riparia*), 4
  spiny-spored (*Isoetes echinospora*), 4
  Tuckerman's (*Isoetes tuckermanii*), 4
  Virginia (*Isoetes virginica*), 4
quillwort family (ISOETACEAE), 4

rabbit-berry (*Shepherdia canadensis*), 288
rabbitfoot-clover (*Trifolium arvense*), 269
rabbitfoot-grass (*Polypogon monspeliensis*), 719

raccoon-grape (*Ampelopsis cordata*), 323
radish (*Raphanus sativus*), 163
  horse- (*Armoracia rusticana*), 170
Rafinesque's water-primrose (*Ludwigia hirtella*), 295
ragged eupatorium (*Eupatorium pilosum*), 572
ragged fringed orchid (*Habenaria lacera*), 822
ragged robin (*Lychnis flos-cuculi*), 107
ragweed (*Ambrosia*), 514
  common (*Ambrosia artemisiifolia*), 515
  giant (*Ambrosia trifida*), 515
  lanceleaf (*Ambrosia bidentata*), 515
  western (*Ambrosia psilostachya*), 515
ragwort (*Senecio*), 523
  tansy- (*Senecio jacobaea*), 526
ramps (*Allium tricoccum*), 790
ram's head lady-slipper (*Cypripedium arietinum*), 812
raspberry
  black (*Rubus occidentalis*), 230
  dwarf (*Rubus pubescens*), 230
  flowering (*Rubus odoratus*), 229
  red (*Rubus idaeus*), 230
  strawberry- (*Rubus illecebrosus*), 231
rat-tail fescue (*Vulpia myuros*), 697
rattlebox (*Crotalaria*), 260
  coastal plain (*Crotalaria purshii*), 260
  low (*Crotalaria rotundifolia*), 260
  showy (*Crotalaria spectabilis*), 261
  weedy (*Crotalaria sagittalis*), 260
rattlesnake-chess (*Bromus briziformis*), 722
rattlesnake-fern (*Botrychium virginianum*), 9
rattlesnake-mannagrass (*Glyceria canadensis*), 707
rattlesnake-master (*Eryngium yuccifolium*), 345
rattlesnake-plantain (*Goodyera*), 817
  alloploid (*Goodyera tesselata*), 818
  downy (*Goodyera pubescens*), 817
  lesser (*Goodyera repens*), 818
  western (*Goodyera oblongifolia*), 817
rattlesnake-root (*Prenanthes*), 586; (*P. alba*), 587
  slender (*Prenanthes autumnalis*), 587
red amaranth (*Amaranthus cruentus*), 95
red baneberry (*Actaea rubra*), 47
red-berried elder (*Sambucus racemosa*), 485
red birch (*Betula nigra*), 83
red buckeye (*Aesculus pavia*), 331
redbud (*Cercis canadensis*), 256
red campion (*Silene dioica*), 108
red chokeberry (*Aronia arbutifolia*), 245
red clover (*Trifolium pratense*), 269
red cockle (*Silene dioica*), 108
red crowberry (*Empetrum rubrum*), 183
red currant
  garden (*Ribes sativum*), 207
  swamp (*Ribes triste*), 207
red dead nettle (*Lamium purpureum*), 424
red elm (*Ulmus rubra*), 68
red fescue (*Festuca rubra*), 697
red grape (*Vitis palmata*), 324
red gum (*Liquidambar*), 68
redhead-grass (*Potamogeton perfoliatus*), 613
red hickory (*Carya ovalis*), 74
red laurel (*Rhododendron catawbiense*), 184
red maple (*Acer rubrum*), 333
red milkweed (*Asclepias rubra*), 373
red mint (*Mentha ×gentilis*), 417
red morning-glory (*Ipomoea coccinea*), 386
red mulberry (*Morus rubra*), 70

red oak
    northern (*Quercus rubra*), 80
    southern (*Quercus falcata*), 79
red orache (*Atriplex rosea*), 92
red osier-dogwood (*Cornus sericea*), 305
red pine (*Pinus resinosa*), 32
red pondweed (*Potamogeton alpinus*), 612
red raspberry (*Rubus idaeus*), 230
redroot (*Amaranthus retroflexus*), 95; (*Ceanothus americanus*), 321; (*Lachnanthes caroliniana*), 781
    prairie- (*Ceanothus herbaceus*), 321
redroot flatsedge (*Cyperus erythrorhizos*), 652
red-seeded dandelion (*Taraxacum laevigatum*), 598
red sorrel (*Rumex acetosella*), 113
red sprangletop (*Leptochloa filiformis*), 737
red spruce (*Picea rubens*), 31
redstem (*Ammannia auriculata*), 291
redstem-filaree (*Erodium cicutarium*), 340
redtop (*Agrostis gigantea*), 718
reed
    bur- (*Sparganium*), 776
    common (*Phragmites australis*), 732
    common wood- (*Cinna arundinacea*), 719
    drooping wood- (*Cinna latifolia*), 719
    giant bur- (*Sparganium eurycarpum*), 776
    sand- (*Calamovilfa longifolia*), 745
    wood- (*Cinna*), 719
reed canary-grass (*Phalaris arundinacea*), 714
reed-grass (*Calamagrostis*), 714
    plains (*Calamagrostis montanensis*), 715
    purple (*Calamagrostis purpurascens*), 714
rein-orchid (*Habenaria*), 819
rescue-grass (*Bromus catharticus*), 721
rest-harrow (*Ononis*), 269
resupinate bladderwort (*Utricularia resupinata*), 463
resurrection-fern (*Polypodium polypodioides*), 12
rheumatism root (*Jeffersonia*), 62
Rhode Island bent (*Agrostis capillaris*), 717
rhodora (*Rhododendron canadense*), 186
rhombic copperleaf (*Acalypha rhomboidea*), 314
rhubarb (*Rheum rhabarbarum*), 113
ribbonleaf-pondweed (*Potamogeton epihydrus*), 611
ribgrass (*Plantago lanceolata*), 433
rice (*Oryza sativa*), 690
    jungle- (*Echinochloa colona*), 761
    southern wild (*Zizaniopsis miliacea*), 691
    wild (*Zizania*), 691
rice cut-grass (*Leersia oryzoides*), 690
ricegrass (*Oryzopsis*), 692
    blackseed- (*Oryzopsis racemosa*), 693
    Indian (*Oryzopsis hymenoides*), 692
    rough-leaved (*Oryzopsis asperifolia*), 693
ridge-seed spurge (*Euphorbia glyptosperma*), 320
ridgestem yellow flax (*Linum striatum*), 327
ripgut-grass (*Bromus rigidus*), 724
riverbank-goldenrod (*Solidago rupestris*), 542
riverbank-grape (*Vitis riparia*), 325
riverbank mock bishop-weed (*Ptilimnium nodosum*), 356
riverbank-quillwort (*Isoetes riparia*), 4
river birch (*Betula nigra*), 83
river-bulrush (*Scirpus fluviatilis*), 634
river-weed (*Podostemum*), 289
river-weed family (PODOSTEMACEAE), 289
roadside-agrimony (*Agrimonia striata*), 236
roadside sand-spurrey (*Spergularia rubra*), 105
Roan Mountain goldenrod (*Solidago roanensis*), 533

Robbins milk-vetch (*Astragalus robbinsii*), 265
Robin's plantain (*Erigeron pulchellus*), 560
rock-brake (*Cryptogramma*), 14
    slender (*Cryptogramma stelleri*), 14
rock chestnut oak (*Quercus prinus*), 78
rock-clubmoss (*Lycopodium porophilum*), 1
rock-cress (*Arabis*), 174
rock-elm (*Ulmus thomasii*), 69
rocket
    dame's (*Hesperis*), 179
    dyer's (*Reseda luteola*), 182
    garden- (*Eruca sativa*), 163
    London- (*Sisymbrium irio*), 181
    purple (*Iodanthus pinnatifidus*), 179
    sand- (*Diplotaxis muralis*), 163
    sea- (*Cakile*), 164
    slimleaf wall- (*Diplotaxis tenuifolia*), 163
    stinking wall- (*Diplotaxis muralis*), 163
    yellow (*Barbarea vulgaris*), 178
rocket larkspur (*Delphinium ambiguum*), 48
rocket-salad (*Eruca*), 163
rock-house white snakeroot (*Eupatorium luciae-brauniae*), 569
rockpink fame-flower (*Talinum calycinum*), 99
rock-rose family (CISTACEAE), 139
rock-skullcap (*Scutellaria saxatilis*), 410
rock-spikemoss (*Selaginella rupestris*), 4
rope-dodder (*Cuscuta glomerata*), 387
rose (*Rosa*), 237
    bristly (*Rosa acicularis*), 240
    cinnamon- (*Rosa majalis*), 240
    climbing prairie- (*Rosa setigera*), 237
    dog- (*Rosa canina*), 238
    dwarf prairie- (*Rosa arkansana*), 239
    French (*Rosa gallica*), 238
    Japanese (*Rosa rugosa*), 238
    Macartney- (*Rosa bracteata*), 237
    memorial (*Rosa wichuraiana*), 237
    moss- (*Portulaca grandiflora*), 99
    multiflora- (*Rosa multiflora*), 237
    New England (*Rosa nitida*), 239
    pasture- (*Rosa carolina*), 239
    Scotch (*Rosa pimpinellifolia*), 238
    smooth (*Rosa blanda*), 240
    swamp- (*Rosa palustris*), 239
    Virginia (*Rosa virginiana*), 239
    western (*Rosa woodsii*), 240
rose-acacia (*Robinia hispida, R. viscosa*), 262
rose-bay (*Rhododendron catawbiense, R. maximum*), 184
    Lappland (*Rhododendron lapponicum*), 185
rosebud-azalea (*Rhododendron prinophyllum*), 185
rose family (ROSACEAE), 217
rose-mallow (*Hibiscus moscheutos*), 132
    smooth (*Hibiscus laevis*), 132
rosemary, bog- (*Andromeda*), 187
rose of Sharon (*Hibiscus syriacus*), 132
rose pogonia (*Pogonia ophioglossoides*), 823
rosering-blanket-flower (*Gaillardia pulchella*), 498
roseroot (*Sedum rosea*), 208
rose-vervain (*Verbena canadensis*), 406
rosin-weed (*Silphium*), 512
    basal-leaved (*Silphium terebinthinaceum*), 513
    Cumberland (*Silphium brachiatum*), 512
    lesser basal-leaved (*Silphium compositum*), 513
    prairie (*Silphium integrifolium*), 512
    shaggy (*Silphium mohrii*), 512

rosin-weed (*Silphium*) (*continued*)
    southern (*Silphium asteriscus*), 511
    whorled (*Silphium trifoliatum*), 511
rosy pussytoes (*Antennaria microphylla*), 566
rough agalinis (*Agalinis aspera*), 451
rough avens (*Geum laciniatum*), 227
rough bedstraw (*Galium asprellum*), 474
rough bluegrass (*Poa trivialis*), 702
rough comfrey (*Symphytum asperum*), 402
rough dropseed (*Sporobolus clandestinus*), 742
rough fescue (*Festuca altaica*), 696
rough fleabane (*Erigeron strigosus*), 560
rough goldenrod (*Solidago radula*), 543
rough-leaved dogwood (*Cornus drummondii*), 305
rough-leaved goldenrod (*Solidago patula*), 538
rough-leaved ricegrass (*Oryzopsis asperifolia*), 693
rough-leaved sunflower (*Helianthus strumosus*), 490
rough marsh-elder (*Iva annua*), 515
rough pigweed (*Amaranthus retroflexus*), 95
rough-pod copperleaf (*Acalypha ostryifolia*), 314
rough vervain (*Verbena scabra*), 404
rough white lettuce (*Prenanthes aspera*), 587
rounded-headed prairie-clover (*Dalea multiflora*), 282
rounded shinleaf (*Pyrola rotundifolia*), 196
round-leaf monkey-flower (*Mimulus glabratus*), 440
round-leaved dogwood (*Cornus rugosa*), 304
round-leaved eupatorium (*Eupatorium rotundifolium*), 571
round-leaved orchid (*Habenaria orbiculata*), 820
round-leaved spurge (*Euphorbia serpens*), 319
round-leaved tick-trefoil (*Desmodium rotundifolium*), 276
round-leaved triodanis (*Triodanis perfoliata*), 466
round-leaved yellow violet (*Viola rotundifolia*), 142
round-lobed hepatica (*Hepatica americana*), 50
round-pod water-primrose (*Ludwigia sphaerocarpa*), 296
rover-bellflower (*Campanula rapunculoides*), 465
rowan-tree (*Sorbus aucuparia*), 246
royal fern (*Osmunda regalis*), 10
royal fern family (OSMUNDACEAE), 10
rue (*Ruta*), 336
    common (*Ruta graveolens*), 336
    early meadow- (*Thalictrum dioicum*), 60
    goat's- (*Tephrosia virginiana*), 261
    meadow- (*Thalictrum*), 60
    mountain meadow- (*Thalictrum clavatum*), 60
    northern meadow- (*Thalictrum venulosum*), 60
    purple meadow- (*Thalictrum dasycarpum*), 61
    skunk meadow- (*Thalictrum revolutum*), 61
    small-leaved meadow- (*Thalictrum macrostylum*), 60
    southern goat's- (*Tephrosia spicata*), 261
    tall meadow- (*Thalictrum pubescens*), 60
    wall- (*Asplenium ruta-muraria*), 17
rue-anemone (*Anemonella*), 59
rue family (RUTACEAE), 336
ruellia, fringeleaf- (*Ruellia humilis*), 460
running buffalo-clover (*Trifolium stoloniferum*), 270
running five-fingers (*Potentilla canadensis*), 222
running groundsel (*Senecio obovatus*), 524
running pine (*Lycopodium clavatum*), 2
running strawberry-bush (*Euonymus obovatus*), 308
rush (*Juncus*), 622
    alkali-bul- (*Scirpus maritimus*), 634
    beak- (*Rhynchospora*), 645
    black bul- (*Scirpus atrovirens*), 634
    bluntscale-bul- (*Scirpus smithii*), 632
    blunt spike- (*Eleocharis ovata*), 642
    bul- (*Scirpus*), 631

rush (*Juncus*) (*continued*)
    common scouring (*Equisetum hyemale*), 5
    creeping (*Juncus repens*), 626
    dwarf scouring (*Equisetum scirpoides*), 5
    flowering- (*Butomus*), 602
    hardstem-bul- (*Scirpus acutus*), 633
    needle- (*Juncus roemerianus*), 625
    path- (*Juncus tenuis*), 623
    river-bul- (*Scirpus fluviatilis*), 634
    salt-marsh bul- (*Scirpus robustus*), 634
    scabrous (*Juncus caesariensis*), 626
    seaside-bul- (*Scirpus rufus*), 633
    sharpscale-bul- (*Scirpus supinus*), 632
    slender bul- (*Scirpus heterochaetus*), 633
    slimpod- (*Juncus diffusissimus*), 629
    smooth scouring (*Equisetum laevigatum*), 5
    soft (*Juncus effusus*), 625
    softstem-bul- (*Scirpus validus*), 633
    spike- (*Eleocharis*), 638
    stone- (*Scleria*), 659
    swamp-bul- (*Scirpus etuberculatus*), 633
    toad- (*Juncus bufonius*), 624
    twig- (*Cladium mariscoides*), 650
    variegated scouring (*Equisetum variegatum*), 5
    water-bul- (*Scirpus subterminalis*), 632
    wire- (*Juncus arcticus*), 624
    wood- (*Luzula*), 630
rushes, scouring (*Equisetum*), 5
rush family (JUNCACEAE), 622
rushlike bladderwort (*Utricularia juncea*), 462
rush-pink (*Lygodesmia*), 588
Russian knapweed (*Centaurea repens*), 585
Russian olive (*Elaeagnus angustifolia*), 288
Russian thistle (*Salsola tragus*), 94
rusty cliff-fern (*Woodsia ilvensis*), 21
rusty cotton-grass (*Eriophorum chamissonis*), 637
rye (*Secale*), 730; (*S. cereale*), 730
    basin wild (*Leymus cinereus*), 728
    blue wild (*Elymus glaucus*), 725
    broad-leaved wild (*Elymus wiegandii*), 724
    Canada wild (*Elymus canadensis*), 725
    downy wild (*Elymus villosus*), 725
    Minnesota wild (*Elymus diversiglumis*), 726
    streambank wild (*Elymus riparius*), 726
    Virginia wild (*Elymus virginicus*), 726
    wild (*Elymus*), 725
ryegrass (*Lolium perenne*), 698
    English (*Lolium perenne* var. *perenne*), 698
    Italian (*Lolium perenne* var. *aristatum*), 698

sage (*Artemisia*), 520; (*Salvia*), 429
    annual bur- (*Ambrosia acanthicarpa*), 515
    garden- (*Salvia officinalis*), 430
    Jerusalem (*Phlomis tuberosa*), 423
    meadow- (*Salvia pratensis*), 429
    nettle-leaved (*Salvia urticifolia*), 429
    skeletonleaf bur- (*Ambrosia tomentosa*), 514
    toothed (*Artemisia serrata*), 520
    white (*Artemisia ludoviciana*), 521
    wood- (*Teucrium*), 408; (*T. scorodonia*), 408
sage-leaved willow (*Salix candida*), 160
sagewort, prairie- (*Artemisia frigida*), 522
sagittaria
    grass-leaved (*Sagittaria graminea*), 605
    Hudson (*Sagittaria subulata*), 602

sagittaria *(continued)*
lance-leaved (*Sagittaria lancifolia*), 605
quill-leaved (*Sagittaria teres*), 605
tidal (*Sagittaria spathulata*), 602
sago-pondweed (*Potamogeton pectinatus*), 609
sailpod dalea (*Dalea enneandra*), 282
sainfoin (*Onobrychis viciifolia*), 276
salad-burnet (*Sanguisorba minor*), 237
salsify (*Tragopogon porrifolius*), 601
salt-grass (*Distichlis*), 735
seashore (*Distichlis spicata* var. *spicata*), 735
salt-marsh agalinis (*Agalinis maritima*), 451
salt-marsh bitter-cress (*Cardamine longii*), 172
salt-marsh bulrush (*Scirpus robustus*), 634
salt-marsh foxtail (*Setaria magna*), 767
salt-marsh sand-spurrey (*Spergularia marina*), 105
salt-marsh stitchwort (*Stellaria humifusa*), 104
salt-marsh wheatgrass (*Elytrigia pungens*), 727
salt-meadow cord-grass (*Spartina patens*), 750
saltwort (*Salsola kali*), 94
samphire (*Salicornia*), 92; (*S. europaea*), 93
Sampson's snakeroot (*Orbexilum pedunculatum*), 283
sandalwood family (SANTALACEAE), 306
sandbar-willow (*Salix exigua*), 156
sand-blackberry (*Rubus cuneifolius*), 233
sand-bluestem (*Andropogon hallii*), 773
sandbur (*Cenchrus*), 769
common (*Cenchrus longispinus*), 769
dune- (*Cenchrus tribuloides*), 770
sand-cherry (*Prunus pumila*), 241
sand-dropseed (*Sporobolus cryptandrus*), 741
sand fame-flower (*Talinum rugospermum*), 99
sand-grape (*Vitis rupestris*), 324
sand-grass (*Triplasis*), 736
purple (*Triplasis purpurea*), 736
sandhill-amaranth (*Amaranthus arenicola*), 96
sandhill-thistle (*Cirsium repandum*), 581
sand-lily (*Mentzelia decapetala*), 151
sand live oak (*Quercus geminata*), 78
sand-lovegrass (*Eragrostis trichodes*), 738
sand-myrtle (*Leiophyllum*), 184
sand-phlox (*Phlox bifida*), 390
sand-plain agalinis (*Agalinis acuta*), 452
sand post-oak (*Quercus margaretta*), 77
sand-reed (*Calamovilfa longifolia*), 745
sand-rocket (*Diplotaxis muralis*), 163
sand-spurrey (*Spergularia*), 105
northern (*Spergularia canadensis*), 106
roadside (*Spergularia rubra*), 105
salt-marsh (*Spergularia marina*), 105
sandvine (*Ampelamus*), 375
sandwort (*Arenaria*), 100
thyme-leaf (*Arenaria serpyllifolia*), 101
sanicle (*Sanicula*), 346
beaked (*Sanicula trifoliata*), 346
Canada (*Sanicula canadensis*), 346
cluster- (*Sanicula gregaria*), 346
southern (*Sanicula smallii*), 346
Santa Maria (*Parthenium hysterophorus*), 514
sapodilla family (SAPOTACEAE), 198
sarsaparilla
bristly (*Aralia hispida*), 342
wild (*Aralia nudicaulis*), 342
Sarvis-holly (*Ilex amelanchier*), 310
Saskatoon (*Amelanchier alnifolia*), 253
sassafras (*Sassafras*), 39
satiny willow (*Salix pellita*), 159
savory, summer- (*Satureja hortensis*), 413
saw-grass (*Cladium jamaicense*), 650
sawtooth sunflower (*Helianthus grosseserratus*), 494
saxifrage (*Saxifraga*), 211
aconite- (*Boykinia aconitifolia*), 215
alpine-brook (*Saxifraga rivularis*), 211
burnet- (*Pimpinella saxifraga*), 353
Carey's (*Saxifraga careyana*), 213
Carolina (*Saxifraga caroliniana*), 212
claw- (*Saxifraga michauxii*), 213
early (*Saxifraga virginiensis*), 214
leafy bulbil- (*Saxifraga cernua*), 211
lettuce- (*Saxifraga micranthidifolia*), 212
naked bulbil- (*Saxifraga foliolosa*), 212
purple alpine (*Saxifraga oppositifolia*), 212
swamp- (*Saxifraga pensylvanica*), 213
three-toothed (*Saxifraga tricuspidata*), 211
white alpine (*Saxifraga aizoon*), 212
yellow alpine (*Saxifraga aizoides*), 212
saxifrage family (SAXIFRAGACEAE), 211
scabious (*Scabiosa columbaria*), 488
sweet (*Scabiosa atropurpurea*), 488
scabrous rush (*Juncus caesariensis*), 626
scaleseed (*Spermolepis*), 350
hooked (*Spermolepis echinata*), 350
southern (*Spermolepis divaricata*), 350
western (*Spermolepis inermis*), 350
scallions (*Allium vineale*), 790
scarlet creeper (*Ipomoea coccinea*), 386
scarlet gaura (*Gaura coccinea*), 302
scarlet hawthorn (*Crataegus coccinea*), 252
scarlet lychnis (*Lychnis chalcedonica*), 108
scarlet mallow (*Sphaeralcea*), 137
scarlet oak (*Quercus coccinea*), 81
scentless chamomile (*Matricaria maritima*), 519
scheuchzeria family (SCHEUCHZERIACEAE), 607
scorpion-grass (*Myosotis*), 396
big-seed (*Myosotis macrosperma*), 397
blue (*Myosotis micrantha*), 397
early (*Myosotis verna*), 397
field (*Myosotis arvensis*), 397
water (*Myosotis scorpioides*), 396
yellow and blue (*Myosotis discolor*), 397
Scotch broom (*Cytisus scoparius*), 259
Scotch lovage (*Ligusticum scothicum*), 354
Scotch mint (*Mentha* ×*gentilis*), 417
Scotch mist (*Galium sylvaticum*), 474
Scotch pine (*Pinus sylvestris*), 32
Scotch rose (*Rosa pimpinellifolia*), 238
Scotch thistle (*Onopordum acanthium*), 584
scouring rush(es) (*Equisetum*), 5
common (*Equisetum hyemale*), 5
dwarf (*Equisetum scirpoides*), 5
smooth (*Equisetum laevigatum*), 5
variegated (*Equisetum variegatum*), 5
scratchgrass (*Muhlenbergia asperifolia*), 747
screw-stem (*Bartonia paniculata*), 367
scrub-pine (*Pinus virginiana*), 34
scuppernong-grape (*Vitis rotundifolia*), 324
scurf-pea
gray (*Psoralidium tenuiflorum*), 284
silvery (*Pediomelum argophyllum*), 284
seabeach-angelica (*Angelica lucida*), 359
seabeach-dock (*Rumex pallidus*), 114
seabeach-orache (*Atriplex arenaria*), 91

sea-blite (*Suaeda*), 93
    northeastern (*Suaeda americana*), 93
    plains (*Suaeda calceoliformis*), 93
    southern (*Suaeda linearis*), 93
    white (*Suaeda maritima*), 93
sea-lavender (*Limonium*), 125
sea-milkwort (*Glaux*), 202
sea-myrtle (*Baccharis halimifolia*), 562
sea-oats (*Uniola paniculata*), 735
sea ox-eye (*Borrichia*), 502
sea-pink (*Sabatia*), 362
    annual (*Sabatia stellaris*), 363
    perennial (*Sabatia dodecandra*), 363
sea-purslane (*Sesuvium*), 85
sea-rocket (*Cakile*), 164
seashore-mallow (*Kosteletzkya virginica*), 133
seashore salt-grass (*Distichlis spicata* var. *spicata*), 735
seaside alkali-grass (*Puccinellia maritima*), 699
seaside-bluebell (*Mertensia maritima*), 396
seaside-bulrush (*Scirpus rufus*), 633
seaside-crowfoot (*Ranunculus cymbalaria*), 58
seaside-goldenrod (*Solidago sempervirens*), 535
seaside-groundsel (*Senecio pseudoarnica*), 523
seaside-heliotrope (*Heliotropium curassavicum*), 395
seaside-knotweed (*Polygonum glaucum*), 116
seaside-plantain (*Plantago maritima*), 432
seaside-spurge (*Euphorbia polygonifolia*), 319
    southern (*Euphorbia ammannioides*), 319
sedge (*Carex*), 660
    broom- (*Andropogon virginicus*), 775
    elk- (*Carex geyeri*), 661
    false nut- (*Cyperus strigosus*), 653
    flat- (*Cyperus*), 650
    freeway- (*Carex praegracilis*), 662
    globe-flat- (*Cyperus echinatus*), 654
    redroot flat- (*Cyperus erythrorhizos*), 652
    star- (*Carex echinata*), 668
    white-topped (*Rhynchospora colorata*), 645
    yellow nut- (*Cyperus esculentus*), 651
sedge family (CYPERACEAE), 631
selaginella family (SELAGINELLACEAE), 4
self-heal (*Prunella*), 423
Seneca-snakeroot (*Polygala senega*), 328
senna
    bladder- (*Colutea arborescens*), 264
    coffee- (*Senna occidentalis*), 258
    northern wild (*Senna hebecarpa*), 257
    southern wild (*Senna marilandica*), 257
sensitive brier (*Mimosa quadrivalvis*), 255
sensitive fern (*Onoclea*), 29
sensitive fern family (ONOCLEACEAE), 29
sensitive plant, wild (*Chamaecrista nictitans*), 257
September elm (*Ulmus serotina*), 69
Serradella (*Ornithopus sativus*), 268
serviceberry (*Amelanchier*), 253
    coastal plain (*Amelanchier obovalis*), 254
    downy (*Amelanchier arborea*), 254
    dwarf (*Amelanchier spicata*), 254
    eastern (*Amelanchier canadensis*), 254
    mountain (*Amelanchier bartramiana*), 253
    New England (*Amelanchier sanguinea*), 254
    smooth (*Amelanchier laevis*), 255
    St. Lawrence (*Amelanchier fernaldii*), 254
    western (*Amelanchier alnifolia*), 253
sesame (*Sesamum indicum*), 460
sesame family (PEDALIACEAE), 460

sesban (*Sesbania exaltata*), 262
sessile blazing star (*Liatris spicata*), 576
sessile-fruited arrow-head (*Sagittaria rigida*), 603
sessile-leaved tick-trefoil (*Desmodium sessilifolium*), 277
sessile tooth-cup (*Ammannia robusta*), 292
shadbush (*Amelanchier*), 253; (*A. canadensis*), 254
shadow-witch (*Ponthieva racemosa*), 818
shagbark-hickory (*Carya ovata*), 74
    Carolina (*Carya carolinae-septentrionalis*), 74
shaggy golden aster (*Chrysopsis mariana*), 530
shaggy prairie-turnip (*Pediomelum esculentum*), 284
shaggy rosin-weed (*Silphium mohrii*), 512
shale-barren clover (*Trifolium virginicum*), 270
shale-barren evening-primrose (*Oenothera argillicola*), 299
shale-barren groundsel (*Senecio antennariifolius*), 523
shale-barren hawkweed (*Hieracium traillii*), 593
shale-barren phlox (*Phlox buckleyi*), 391
shale-barren pimpernel (*Taenidia montana*), 352
shale-barren pussytoes (*Antennaria virginica*), 566
shale-barren sunflower (*Helianthus laevigatus*), 491
sharp-lobed hepatica (*Hepatica acutiloba*), 50
sharpscale-bulrush (*Scirpus supinus*), 632
sharpwing monkey-flower (*Mimulus alatus*), 440
she-balsam (*Abies fraseri*), 31
sheepberry (*Viburnum lentago*), 484
sheep-fescue (*Festuca ovina*), 696
sheep-laurel (*Kalmia angustifolia*), 187
sheep's bit (*Jasione montana*), 466
shellbark-hickory (*Carya laciniosa*), 73
shepherd's cress (*Teesdalia*), 166
shepherd's purse (*Capsella*), 167
shield-fern (*Dryopteris*), 24
shingle-oak (*Quercus imbricaria*), 78
shining aster (*Aster firmus*), 547
shining bedstraw (*Galium concinnum*), 475
shining clubmoss (*Lycopodium lucidulum*), 1
shining false indigo (*Amorpha nitens*), 282
shining ladies'-tresses (*Spiranthes lucida*), 815
shining spurge (*Euphorbia lucida*), 317
shining sumac (*Rhus copallina*), 334
shining willow (*Salix lucida*), 154
shinleaf (*Pyrola*), 195
    elliptic (*Pyrola elliptica*), 196
    little (*Pyrola minor*), 195
    one-flowered (*Moneses*), 196
    one-sided (*Pyrola secunda*), 195
    pink (*Pyrola asarifolia*), 196
    rounded (*Pyrola rotundifolia*), 196
shinleaf family (PYROLACEAE), 195
Shirley-poppy (*Papaver rhoeas*), 65
shooting star (*Dodecatheon*), 203
    eastern (*Dodecatheon meadia*), 203
    western (*Dodecatheon radicatum*), 204
shore-aster (*Aster tradescantii*), 549
short-anthered cotton-grass (*Eriophorum brachyantherum*), 637
short-awn foxtail (*Alopecurus aequalis*), 720
short-fringed knapweed (*Centaurea dubia*), 585
shortleaf fescue (*Festuca brachyphylla*), 696
shortleaf milkwort (*Polygala brevifolia*), 330
shortleaf pine (*Pinus echinata*), 34
short-pappus goldenrod (*Solidago sphacelata*), 540
short-spurred corydalis (*Corydalis flavula*), 67
short-stalk copperleaf (*Acalypha gracilens*), 314
short-stemmed meadow-pitchers (*Rhexia petiolata*), 303
showy bugloss (*Anchusa azurea*), 402

showy bur-marigold (*Bidens laevis*), 506
showy goat's-beard (*Tragopogon pratensis*), 601
showy goldenrod (*Solidago speciosa*), 534
showy lady-slipper (*Cypripedium reginae*), 812
showy loco-weed (*Oxytropis splendens*), 267
showy meadow-pitchers (*Rhexia interior*), 303
showy milkweed (*Asclepias speciosa*), 372
showy mountain-ash (*Sorbus decora*), 246
showy orchis (*Orchis spectabilis*), 818
showy rattlebox (*Crotalaria spectabilis*), 261
showy water-primrose (*Ludwigia uruguayensis*), 294
showy yellow wood-sorrel (*Oxalis macrantha*), 338
shrubby five-fingers (*Potentilla fruticosa*), 226
Shumard oak (*Quercus shumardii*), 81
Siberian crane's-bill (*Geranium sibiricum*), 339
sickle-pod (*Arabis canadensis*), 175; (*Senna obtusifolia*), 258
sida
  arrowleaf (*Sida rhombifolia*), 137
  coastal plain (*Sida elliottii*), 137
  prickly (*Sida spinosa*), 137
  Virginia (*Sida hermaphrodita*), 138
side-oats grama (*Bouteloua curtipendula*), 750
silk-tree (*Albizia julibrissin*), 255
silk-vine (*Periploca graeca*), 370
silky osier (*Salix viminalis*), 160
silky stewartia (*Stewartia malacodendron*), 125
silky willow (*Salix sericea*), 158
silver-bell tree (*Halesia*), 199
silver-berry (*Elaeagnus commutata*), 288
silver-grass, Amur (*Miscanthus sacchariflorus*), 771
silver-hairgrass (*Aira caryophyllea*), 713
silverleaf nightshade (*Solanum elaeagnifolium*), 381
silverling (*Paronychia argyrocoma*), 106
silver-maple (*Acer saccharinum*), 333
silver-plumegrass (*Erianthus alopecuroides*), 771
silver poplar (*Populus alba*), 152
silver-rod (*Solidago bicolor*), 534
silverscale (*Atriplex argentea*), 92
silver-weed (*Potentilla anserina*), 225
  coastal (*Potentilla pacifica*), 225
silver whitlow-wort (*Paronychia argyrocoma*), 106
silvery five-fingers (*Potentilla argentea*), 223
silvery glade-fern (*Athyrium thelypterioides*), 19
silvery scurf-pea (*Pediomelum argophyllum*), 284
silvery willow (*Salix argyrocarpa*), 159
single-stemmed groundsel (*Senecio integerrimus*), 523
Sitka clubmoss (*Lycopodium sitchense*), 3
six-angled spurge (*Euphorbia hexagona*), 318
six-weeks fescue (*Vulpia octoflora*), 697
skeletonleaf-bursage (*Ambrosia tomentosa*), 514
skeleton-weed (*Chondrilla juncea*), 599; (*Lygodesmia juncea*), 588
  annual (*Lygodesmia rostrata*), 589
skullcap (*Scutellaria*), 409
  Allegheny- (*Scutellaria serrata*), 410
  Appalachian (*Scutellaria arguta*), 410
  forest- (*Scutellaria ovata*), 409
  little (*Scutellaria parvula*), 411
  marsh- (*Scutellaria galericulata*), 411
  rock- (*Scutellaria saxatilis*), 410
skunk-cabbage (*Symplocarpus*), 615
skunk-currant (*Ribes glandulosum*), 206
skunk meadow-rue (*Thalictrum revolutum*), 61
sleekwort, mauve (*Liparis liliifolia*), 825
slender arrow-grass (*Triglochin palustris*), 608
slender aster (*Aster gracilis*), 555
slender beard-tongue (*Penstemon gracilis*), 446
slender bladderwort (*Utricularia subulata*), 463
slender blue flag (*Iris prismatica*), 810
slender bulrush (*Scirpus heterochaetus*), 633
slender clubmoss (*Lycopodium carolinianum*), 2
slender corydalis (*Corydalis micrantha*), 67
slender cotton-grass (*Eriophorum gracile*), 638
slender cottonweed (*Froelichia gracilis*), 98
slender ladies'-tresses (*Spiranthes lacera*), 816
slender lip-fern (*Cheilanthes feei*), 15
slender marsh-pink (*Sabatia campanulata*), 362
slender pondweed (*Potamogeton pusillus*), 611
slender rattlesnake-root (*Prenanthes autumnalis*), 587
slender rock-brake (*Cryptogramma stelleri*), 14
slender water-nymph (*Najas gracillima*), 614
slender wheatgrass (*Elymus trachycaulus*), 725
slimleaf wall-rocket (*Diplotaxis tenuifolia*), 163
slimpod-rush (*Juncus diffusissimus*), 629
slimspike three-awn (*Aristida longespica*), 734
slippery elm (*Ulmus rubra*), 68
sloughgrass, American (*Beckmannia syzigachne*), 720
small bluets (*Hedyotis crassifolia*), 470
small bugloss (*Anchusa arvensis*), 402
small cranberry (*Vaccinium oxycoccos*), 194
smaller forget-me-not (*Myosotis laxa*), 397
small-flowered alum-root (*Heuchera parviflora*), 215
small-flowered anemone (*Anemone parviflora*), 49
small-flowered caltrop (*Kallstroemia parviflora*), 337
small-flowered crane's-bill (*Geranium pusillum*), 340
small-flowered crowfoot (*Ranunculus abortivus, R. micranthus*), 56
small-flowered evening-primrose (*Oenothera parviflora*), 299
small-flowered gaura (*Gaura parviflora*), 302
small-flowered grass-of-Parnassus (*Parnassia parviflora*), 217
small-flowered hairy willow-herb (*Epilobium parviflorum*), 297
small-flowered water-primrose (*Ludwigia glandulosa*), 296
small-flowered woolly bean (*Strophostyles leiosperma*), 286
small-headed aster (*Aster parviceps*), 551; (*A. racemosus*), 550
small-headed blazing star (*Liatris microcephala*), 576
small-headed sunflower (*Helianthus microcephalus*), 491
small-leaved meadow-rue (*Thalictrum macrostylum*), 60
small-leaved mitrewort (*Cynoctonum sessilifolium*), 361
small-leaved white snakeroot (*Eupatorium aromaticum*), 568
small-seed false flax (*Camelina microcarpa*), 181
smartweed
  coarse (*Polygonum robustius*), 119
  dock-leaved (*Polygonum lapathifolium*), 120
  dotted (*Polygonum punctatum*), 120
  Pennsylvania (*Polygonum pensylvanicum*), 120
  water (*Polygonum amphibium*), 119
smartweed-dodder (*Cuscuta polygonorum*), 388
smartweed family (POLYGONACEAE), 112
smoke, prairie- (*Anemone patens*), 50; (*Geum triflorum*), 228
smooth agalinis (*Agalinis purpurea*), 451
smooth alder (*Alnus serrulata*), 84
smooth aster (*Aster laevis*), 548
  narrow-leaved (*Aster concinnus*), 548
smooth azalea (*Rhododendron arborescens*), 186
smooth blackberry (*Rubus canadensis*), 233
smooth brome (*Bromus inermis*), 721
smooth bur-clover (*Medicago polymorpha*), 272
smooth cat's-ear (*Hypochaeris glabra*), 600

smooth cliff-brake (*Pellaea glabella*), 16
smooth cliff-fern (*Woodsia glabella*), 20
smooth cord-grass (*Spartina alterniflora*), 749
smooth crab-grass (*Digitaria ischaemum*), 767
smooth false foxglove (*Aureolaria flava*), 450
smooth goldenrod (*Solidago gigantea*), 541
smooth heart-leaved aster (*Aster lowrieanus*), 552
smooth hedge-nettle (*Stachys tenuifolia*), 426
smooth ironweed (*Vernonia fasciculata*), 578
smooth lead-plant (*Amorpha nana*), 282
smooth loosestrife (*Lysimachia quadriflora*), 200
smooth meadow-parsnip (*Thaspium trifoliatum*), 353
smooth milkweed (*Asclepias sullivantii*), 373
smooth phlox (*Phlox glaberrima*), 391
smooth pigweed (*Amaranthus hybridus*), 96
smooth rose (*Rosa blanda*), 240
smooth rose-mallow (*Hibiscus laevis*), 132
smooth scouring rush (*Equisetum laevigatum*), 5
smoothseed-milkweed (*Asclepias perennis*), 370
smooth serviceberry (*Amelanchier laevis*), 255
smooth southern buckthorn (*Bumelia lycioides*), 198
smooth spiderwort (*Tradescantia ohiensis*), 620
smooth sumac (*Rhus glabra*), 334
smooth tick-trefoil (*Desmodium laevigatum*), 278
smooth trailing lespedeza (*Lespedeza repens*), 279
smooth winterberry (*Ilex laevigata*), 310
smut-grass (*Sporobolus indicus*), 742
snake-mouth (*Pogonia ophioglossoides*), 823
snakeroot
  black (*Cimicifuga racemosa*), 46; (*Sanicula marilandica*), 346
  rock-house white (*Eupatorium luciae-brauniae*), 569
  Sampson's (*Orbexilum pedunculatum*), 283
  Seneca- (*Polygala senega*), 328
  small-leaved white (*Eupatorium aromaticum*), 568
  Virginia (*Aristolochia serpentaria*), 41
  white (*Eupatorium rugosum*), 569
snapdragon (*Antirrhinum*), 448
  common (*Antirrhinum majus*), 448
  lesser (*Antirrhinum orontium*), 448
sneezeweed (*Achillea ptarmica*), 517; (*Helenium*), 499
  common (*Helenium autumnale*), 499
  few-headed (*Helenium brevifolium*), 499
  narrow-leaved (*Helenium amarum*), 500
  southern (*Helenium flexuosum*), 500
  Virginia (*Helenium virginicum*), 499
snowbell
  American (*Styrax americanus*), 198
  bigleaf- (*Styrax grandifolius*), 198
snowberry (*Symphoricarpos albus*), 481
  creeping (*Gaultheria hispidula*), 190
snowflake (*Leucojum*), 803; (*L. aestivum*), 803
snow-on-the-mountain (*Euphorbia marginata*), 319
snow-trillium (*Trillium nivale*), 795
snowy orchid (*Habenaria nivea*), 819
soapberry family (SAPINDACEAE), 330
soap-plant (*Yucca glauca*), 804
soapwort (*Saponaria*), 111
soapwort-gentian (*Gentiana saponaria*), 364
soft chess (*Bromus hordeaceus*), 722
soft maple (*Acer saccharinum*), 333
soft rush (*Juncus effusus*), 625
softstem-bulrush (*Scirpus validus*), 633
Solomon's seal (*Polygonatum*), 801
  false (*Smilacina*), 800
sorghum (*Sorghum bicolor*), 772
sorrel (*Rumex*), 113
  big yellow wood- (*Oxalis grandis*), 338
  common yellow wood- (*Oxalis stricta*), 337
  creeping yellow wood- (*Oxalis corniculata*), 337
  green (*Rumex acetosa*), 113
  mountain (*Oxyria*), 113
  northern wood- (*Oxalis acetosella*), 338
  red (*Rumex acetosella*), 113
  showy yellow wood- (*Oxalis macrantha*), 338
  southern yellow wood- (*Oxalis dillenii*), 337
  violet wood- (*Oxalis violacea*), 338
  wild (*Rumex hastatulus*), 113
  wood- (*Oxalis*), 337
sour cherry (*Prunus cerasus*), 241
sour gum (*Nyssa*), 306
sourwood (*Oxydendrum*), 189
southeastern arnica (*Arnica acaulis*), 501
southeastern beard-tongue (*Penstemon australis*), 445
southern adder's-tongue (*Ophioglossum vulgatum* var. *pycnostichum*), 9
southern agrimony (*Agrimonia parviflora*), 236
southern arrow-grass (*Triglochin striata*), 608
southern bayberry (*Myrica heterophylla*), 75
southern beech-fern (*Thelypteris hexagonoptera*), 23
southern blackberry (*Rubus argutus*), 234
southern black haw (*Viburnum rufidulum*), 484
southern blazing star (*Liatris squarrulosa*), 575
southern blueberry (*Vaccinium tenellum*), 192
southern blue flag (*Iris virginica*), 811
southern bog-goldenrod (*Solidago gracillima*), 535
southern buckthorn (*Bumelia*), 198
  smooth (*Bumelia lycioides*), 198
  woolly (*Bumelia lanuginosa*), 198
southern catalpa (*Catalpa bignonioides*), 462
southern cat-tail (*Typha domingensis*), 778
southern chervil (*Chaerophyllum tainturieri*), 349
southern clubmoss (*Lycopodium appressum*), 2
southern crab (*Pyrus angustifolia*), 244
southern crab-grass (*Digitaria ciliaris*), 768
southern dewberry (*Rubus enslenii*), 232
southern dodder (*Cuscuta obtusiflora*), 387
southern estuarine beggar-ticks (*Bidens bidentoides*), 507
southern flatseed-sunflower (*Verbesina occidentalis*), 501
southern goat's-rue (*Tephrosia spicata*), 261
southern ground-cedar (*Lycopodium digitatum*), 3
southern hackberry (*Celtis laevigata*), 69
southern highbush-blueberry (*Vaccinium elliottii*), 194
southern magnolia (*Magnolia grandiflora*), 36
southern maidenhair fern (*Adiantum capillus-veneris*), 14
southern monkshood (*Aconitum uncinatum*), 47
southern mountain-cranberry (*Vaccinium erythrocarpum*), 194
southern red oak (*Quercus falcata*), 79
southern rosin-weed (*Silphium asteriscus*), 511
southern sanicle (*Sanicula smallii*), 346
southern scaleseed (*Spermolepis divaricata*), 350
southern sea-blite (*Suaeda linearis*), 93
southern seaside-spurge (*Euphorbia ammannioides*), 319
southern single-head pussytoes (*Antennaria solitaria*), 566
southern sneezeweed (*Helenium flexuosum*), 500
southern stoneseed (*Lithospermum tuberosum*), 400
southern sugar-maple (*Acer barbatum*), 332
southern sundrops (*Oenothera fruticosa*), 302
southern swamp-aster (*Aster elliottii*), 548
southern swamp-dogwood (*Cornus stricta*), 305

southern three-lobed bedstraw (*Galium tinctorium*), 476
southern twayblade (*Listera australis*), 814
southern water-nymph (*Najas guadalupensis*), 614
southern water-plantain (*Alisma subcordatum*), 603
southern white daisy (*Erigeron vernus*), 560
southern whitlow-wort (*Paronychia riparia*), 106
southern wild rice (*Zizaniopsis miliacea*), 691
southern wild senna (*Senna marilandica*), 257
southernwood (*Artemisia abrotanum*), 520
southern woolly groundsel (*Senecio tomentosus*), 523
southern woolly violet (*Viola villosa*), 144
southern yellow-cress (*Rorippa sessiliflora*), 178
southern yellow wood-sorrel (*Oxalis dillenii*), 337
sow-thistle (*Sonchus*), 591
    common (*Sonchus oleraceus*), 591
    perennial (*Sonchus arvensis*), 591
    prickly (*Sonchus asper*), 591
Spanish bayonet (*Yucca*), 804
Spanish clover (*Lotus purshianus*), 268
Spanish moss (*Tillandsia usneoides*), 779
Spanish needles (*Bidens bipinnata*), 508
Spanish oak (*Quercus falcata*), 79
sparse-lobed grape-fern (*Botrychium biternatum*), 8
spatterdock (*Nuphar*), 42
speargrass (*Poa annua*), 703
spear-leaved violet (*Viola hastata*), 146
spearmint (*Mentha spicata*), 418
spearscale (*Atriplex patula*), 91
spearwort (*Ranunculus*), 54
    coastal-plain (*Ranunculus texensis*), 57
    creeping (*Ranunculus flammula*), 57
    low (*Ranunculus pusillus*), 57
    water-plantain (*Ranunculus ambigens*), 57
speckled alder (*Alnus incana*), 84
speedwell (*Veronica*), 453
    American (*Veronica americana*), 455
    birdseye- (*Veronica persica*), 454
    common (*Veronica officinalis*), 454
    corn- (*Veronica arvensis*), 453
    creeping (*Veronica filiformis*), 454
    field- (*Veronica agrestis*), 454
    germander- (*Veronica chamaedrys*), 455
    ivy-leaved (*Veronica hederifolia*), 454
    long-leaved (*Veronica longifolia*), 453
    narrow-leaved (*Veronica scutellata*), 455
    northern (*Veronica wormskjoldii*), 453
    purslane- (*Veronica peregrina*), 453
    thyme-leaved (*Veronica serpyllifolia*), 453
    water- (*Veronica anagallis-aquatica*), 455
spice-bush (*Lindera*), 39
spider-flower (*Cleome hassleriana*), 161
spider-lily (*Hymenocallis*), 803
spiderwort (*Tradescantia*), 619
    pink (*Tradescantia rosea*), 620
    prairie- (*Tradescantia occidentalis*), 620
    smooth (*Tradescantia ohiensis*), 620
    sticky (*Tradescantia bracteata*), 620
    Virginia (*Tradescantia virginiana*), 619
    wide-leaved (*Tradescantia subaspera*), 619
spiderwort family (COMMELINACEAE), 619
spike-centaury (*Centaurium spicatum*), 363
spiked lobelia (*Lobelia spicata*), 466
spike-moss (*Selaginella*), 4
    meadow- (*Selaginella apoda*), 4
    northern (*Selaginella selaginoides*), 4
    rock- (*Selaginella rupestris*), 4

spikenard (*Aralia racemosa*), 343
spike-rush (*Eleocharis*), 638
    blunt (*Eleocharis ovata*), 642
spindle-tree, European (*Euonymus europaeus*), 308
spiny amaranth (*Amaranthus spinosus*), 95
spiny cocklebur (*Xanthium spinosum*), 516
spiny-spored quillwort (*Isoetes echinospora*), 4
spiny swamp-currant (*Ribes lacustre*), 206
spiny-tooth gum-weed (*Grindelia lanceolata*), 531
spiraea (*Spiraea*), 217
    Appalachian (*Spiraea virginiana*), 218
    birch-leaved (*Spiraea betulifolia*), 218
    Japanese (*Spiraea japonica*), 218
spleenwort (*Asplenium*), 17
    black-stalked (*Asplenium resiliens*), 17
    Bradley's (*Asplenium bradleyi*), 17
    ebony- (*Asplenium platyneuron*), 17
    forked (*Asplenium septentrionale*), 18
    green (*Asplenium viride*), 18
    lobed (*Asplenium pinnatifidum*), 18
    maidenhair- (*Asplenium trichomanes*), 18
    mountain- (*Asplenium montanum*), 17
spleenwort family (ASPLENIACEAE), 17
splitbeard-bluestem (*Andropogon ternarius*), 774
spotted bladderwort (*Utricularia purpurea*), 463
spotted cat's-ear (*Hypochaeris radicata*), 600
spotted coral-root (*Corallorhiza maculata*), 827
spotted dead nettle (*Lamium maculatum*), 423
spotted Joe-Pye weed (*Eupatorium maculatum*), 567
spotted knapweed (*Centaurea maculosa*), 585
spotted loosestrife (*Lysimachia punctata*), 201
spotted phlox (*Phlox maculata*), 391
spotted pondweed (*Potamogeton pulcher*), 612
spotted spurge (*Euphorbia maculata*), 320
spotted St. John's-wort (*Hypericum punctatum*), 128
spotted wintergreen (*Chimaphila maculata*), 195
sprangletop (*Leptochloa*), 737; (*Scolochloa festucacea*), 698
    red (*Leptochloa filiformis*), 737
spreading chervil (*Chaerophyllum procumbens*), 348
spreading dogbane (*Apocynum androsaemifolium*), 369
spreading eryngo (*Eryngium prostratum*), 346
spreading evening-primrose (*Oenothera humifusa*), 300
spreading Jacob's ladder (*Polemonium reptans*), 392
spreading pogonia (*Cleistes divaricata*), 823
spring-adonis (*Adonis vernalis*), 59
spring-avens (*Geum vernum*), 227
spring-beauty (*Claytonia*), 99
spring coral-root (*Corallorhiza wisteriana*), 826
spring-cress (*Cardamine rhomboidea*), 172
    pink (*Cardamine douglassii*), 172
spring ladies'-tresses (*Spiranthes vernalis*), 816
spring-thistle (*Cirsium carolinianum*), 581
spring-vetch (*Vicia lathyroides*), 272
spruce (*Picea*), 31
    black (*Picea mariana*), 31
    red (*Picea rubens*), 31
    white (*Picea glauca*), 31
spurge (*Euphorbia*), 315
    Allegheny- (*Pachysandra procumbens*), 311
    caper- (*Euphorbia lathyris*), 316
    Cumberland- (*Euphorbia mercurialina*), 318
    cypress- (*Euphorbia cyparissias*), 317
    dune- (*Euphorbia geyeri*), 319
    flowering (*Euphorbia corollata*), 318
    hairy (*Euphorbia vermiculata*), 320
    Ipecac- (*Euphorbia ipecacuanhae*), 318

spurge *(Euphorbia)* *(continued)*
  leafy *(Euphorbia esula)*, 317
  Missouri *(Euphorbia missurica)*, 319
  myrtle- *(Euphorbia lathyris)*, 316
  petty *(Euphorbia peplus)*, 317
  prairie- *(Euphorbia spathulata)*, 315
  purple *(Euphorbia purpurea)*, 315
  ridge-seed *(Euphorbia glyptosperma)*, 320
  round-leaved *(Euphorbia serpens)*, 319
  seaside- *(Euphorbia polygonifolia)*, 319
  shining *(Euphorbia lucida)*, 317
  six-angled *(Euphorbia hexagona)*, 318
  southern seaside- *(Euphorbia ammannioides)*, 319
  spotted *(Euphorbia maculata)*, 320
  thyme-leaved *(Euphorbia serpyllifolia)*, 320
  tinted *(Euphorbia commutata)*, 316
  toothed *(Euphorbia dentata)*, 317
  woodland- *(Euphorbia obtusata)*, 315
spurge family (EUPHORBIACEAE), 311
spurred anoda *(Anoda cristata)*, 138
spurrey *(Spergula)*, 105
  northern sand- *(Spergularia canadensis)*, 106
  roadside sand- *(Spergularia rubra)*, 105
  salt-marsh sand- *(Spergularia marina)*, 105
  sand- *(Spergularia)*, 105
squarehead *(Tetragonotheca)*, 495
square-pod water-primrose *(Ludwigia alternifolia)*, 295
squarrose goldenrod *(Solidago squarrosa)*, 532
squarrose white aster *(Aster ericoides)*, 550
squashberry *(Viburnum edule)*, 483
squaw-bush *(Rhus aromatica)*, 334
squaw-root *(Conopholis)*, 459
squirrel corn *(Dicentra canadensis)*, 66
squirrel-tail fescue *(Vulpia elliotea)*, 697
staff-tree family (CELASTRACEAE), 308
staggerbush *(Lyonia mariana)*, 188
staghorn-sumac *(Rhus typhina)*, 334
stalked water-horehound *(Lycopus rubellus)*, 420
standing bugle *(Ajuga genevensis)*, 408
standing cypress *(Gilia rubra)*, 392
standing milk-vetch *(Astragalus adsurgens)*, 266
St. Andrew's cross *(Hypericum hypericoides)*, 126
star
  Appalachian blazing *(Liatris turgida)*, 575
  blazing *(Liatris)*, 574
  blue- *(Amsonia)*, 368
  common blue- *(Amsonia tabernaemontana)*, 368
  eastern shooting *(Dodecatheon meadia)*, 203
  evening *(Mentzelia decapetala)*, 151
  few-headed blazing *(Liatris cylindracea)*, 576
  gold- *(Chrysogonum)*, 513
  grass-leaved blazing *(Liatris graminifolia)*, 576
  lacerate blazing *(Liatris aspera)*, 575
  northern blazing *(Liatris scariosa)*, 574
  northern plains blazing *(Liatris ligulistylis)*, 575
  Ozark blue- *(Amsonia illustris)*, 368
  piedmont blazing *(Liatris regimontis)*, 575
  plains blazing *(Liatris squarrosa)*, 576
  sessile blazing *(Liatris spicata)*, 576
  shooting *(Dodecatheon)*, 203
  small-headed blazing *(Liatris microcephala)*, 576
  southern blazing *(Liatris squarrulosa)*, 575
  thick-spike blazing *(Liatris pycnostachya)*, 576
  western shooting *(Dodecatheon radicatum)*, 204
star-chickweed *(Stellaria pubera)*, 103

star-duckweed *(Lemna trisulca)*, 616
starflower *(Trientalis)*, 202
star-grass *(Hypoxis)*, 802
  common *(Hypoxis hirsuta)*, 802
  water *(Zosterella)*, 779
star of Bethlehem *(Ornithogalum umbellatum)*, 787
starry campion *(Silene stellata)*, 109
starry grasswort *(Cerastium arvense)*, 102
star-sedge *(Carex echinata)*, 668
star-thistle *(Centaurea)*, 584
  Maltese *(Centaurea melitensis)*, 586
  purple *(Centaurea calcitrapa)*, 586
  yellow *(Centaurea solstitialis)*, 586
starwort *(Stellaria)*, 103
  meadow- *(Stellaria palustris)*, 103
  water- *(Callitriche)*, 431
Steele's Joe-Pye weed *(Eupatorium steelei)*, 567
stemless evening-primrose *(Oenothera triloba)*, 300
stewartia
  mountain *(Stewartia ovata)*, 125
  silky *(Stewartia malacodendron)*, 125
stickleaf *(Mentzelia oligosperma)*, 151
stickseed *(Hackelia)*, 398; *(Lappula)*, 398
  two-row *(Lappula squarrosa)*, 398
  western *(Lappula redowskii)*, 398
stickseed-crowfoot *(Ranunculus parviflorus)*, 56
sticky cockle *(Silene noctiflora)*, 108
sticky groundsel *(Senecio viscosus)*, 525
sticky hawkweed *(Hieracium scabrum)*, 595
sticky nightshade *(Solanum sisymbriifolium)*, 380
sticky spiderwort *(Tradescantia bracteata)*, 620
stiff aster *(Aster linariifolius)*, 558
stiff clubmoss *(Lycopodium annotinum)*, 2
stiff gentian *(Gentianella quinquefolia)*, 366
stiff goldenrod *(Solidago rigida)*, 543
stiffstem yellow flax *(Linum rigidum)*, 327
stiff sunflower *(Helianthus pauciflorus)*, 493
stiff tick-trefoil *(Desmodium rigidum)*, 278
stinging nettle *(Urtica)*, 71
stink-grass *(Eragrostis cilianensis)*, 739
stinking Benjamin *(Trillium erectum)*, 796
stinking clover *(Cleome serrulata)*, 161
stinking goosefoot *(Chenopodium vulvaria)*, 88
stinking marigold *(Dyssodia papposa)*, 510
stinking pepperweed *(Lepidium ruderale)*, 165
stinking wall-rocket *(Diplotaxis muralis)*, 163
stitchwort *(Stellaria)*, 103
  bog- *(Stellaria alsine)*, 104
  common *(Stellaria graminea)*, 104
  fleshy *(Stellaria crassifolia)*, 104
  long-leaved *(Stellaria longifolia)*, 104
  long-stalked *(Stellaria longipes)*, 104
  northern *(Stellaria borealis)*, 104
  salt-marsh *(Stellaria humifusa)*, 104
St. John's-wort *(Hypericum)*, 126
  common *(Hypericum perforatum)*, 128
  coppery *(Hypericum denticulatum)*, 129
  marsh *(Triadenum)*, 130
  spotted *(Hypericum punctatum)*, 128
St. Lawrence annual aster *(Aster laurentianus)*, 558
St. Lawrence serviceberry *(Amelanchier fernaldii)*, 254
stonecrop *(Sedum)*, 208
  ditch- *(Penthorum)*, 211
stonecrop family (CRASSULACEAE), 208
stone-mint *(Cunila)*, 419
stone-root *(Collinsonia)*, 418

stone-rush (*Scleria*), 659
stoneseed (*Lithospermum*), 399
  plains- (*Lithospermum incisum*), 400
  southern (*Lithospermum tuberosum*), 400
storax (*Styrax*), 198
storax family (STYRACACEAE), 198
stork's-bill (*Erodium*), 340
St. Peter's-wort (*Hypericum stans*), 126
straight-leaved pondweed (*Potamogeton strictifolius*), 610
strap-leaved violet (*Viola lanceolata*), 146
strawberry (*Fragaria*), 220
  barren (*Waldsteinia*), 221
  Indian (*Duchesnea indica*), 221
  thick-leaved wild (*Fragaria virginiana*), 220
  thin-leaved wild (*Fragaria vesca*), 220
strawberry-blite (*Chenopodium capitatum*), 87
strawberry-bush (*Euonymus americanus*), 308
  running (*Euonymus obovatus*), 308
strawberry-clover (*Trifolium fragiferum*), 269
strawberry-potentilla (*Potentilla sterilis*), 222
strawberry-raspberry (*Rubus illecebrosus*), 231
strawberry-shrub (*Calycanthus*), 38
strawberry-shrub family (CALYCANTHACEAE), 38
strawberry-weed (*Potentilla norvegica*), 223
strawstem beggar-ticks (*Bidens comosa*), 506
streambank wild rye (*Elymus riparius*), 726
strigose prairie-turnip (*Pediomelum cuspidatum*), 284
striped coral-root (*Corallorhiza striata*), 827
striped gentian (*Gentiana villosa*), 366
striped maple (*Acer pensylvanicum*), 333
Sudan-grass (*Sorghum bicolor*), 772
sugarberry (*Celtis*), 69
sugar-cane plumegrass (*Erianthus giganteus*), 770
sugar-grape (*Vitis rupestris*), 324
sugar-maple (*Acer saccharum*), 331
  southern (*Acer barbatum*), 332
sulphur-five-fingers (*Potentilla recta*), 223
sumac (*Rhus*), 334
  poison- (*Toxicodendron vernix*), 335
  shining (*Rhus copallina*), 334
  smooth (*Rhus glabra*), 334
  staghorn- (*Rhus typhina*), 334
summer-cohosh (*Cimicifuga americana*), 46
summer-cypress (*Kochia scoparia*), 90
summer-grape (*Vitis aestivalis*), 324
summer-phlox (*Phlox paniculata*), 390
summer-savory (*Satureja hortensis*), 413
sundew (*Drosera*), 138
sundew family (DROSERACEAE), 138
sundial-lupine (*Lupinus perennis*), 260
sundrops
  little (*Oenothera perennis*), 301
  midwestern (*Oenothera pilosella*), 301
  northern (*Oenothera tetragona*), 301
  southern (*Oenothera fruticosa*), 302
  threadleaf- (*Oenothera linifolia*), 301
sunflower (*Helianthus*), 489
  Appalachian (*Helianthus atrorubens*), 492
  ashy (*Helianthus mollis*), 491
  coastal plain tickseed- (*Bidens mitis*), 508
  common (*Helianthus annuus*), 489
  divaricate (*Helianthus divaricatus*), 490
  forest- (*Helianthus decapetalus*), 490
  hairy (*Helianthus hirsutus*), 490
  Maximilian- (*Helianthus maximilianii*), 494

sunflower (*Helianthus*) (*continued*)
  midwestern tickseed- (*Bidens aristosa*), 508
  naked-stemmed (*Helianthus occidentalis*), 492
  narrow-leaved (*Helianthus angustifolius*), 494
  northern tickseed- (*Bidens coronata*), 508
  Ozark (*Helianthus silphioides*), 492
  Ozark flatseed- (*Verbesina helianthoides*), 501
  Ozark tickseed- (*Bidens polylepis*), 508
  plains- (*Helianthus petiolaris*), 489
  rough-leaved (*Helianthus strumosus*), 490
  sawtooth (*Helianthus grosseserratus*), 494
  shale-barren (*Helianthus laevigatus*), 491
  small-headed (*Helianthus microcephalus*), 491
  southern flatseed- (*Verbesina occidentalis*), 501
  stiff (*Helianthus pauciflorus*), 493
  swamp- (*Helianthus giganteus*), 493
  Tennessee (*Helianthus eggertii*), 491
  willow-leaved (*Helianthus salicifolius*), 493
sunflower-everlasting (*Heliopsis*), 503
supple-jack (*Berchemia*), 322
swallow-wort (*Vincetoxicum*), 374
  black (*Vincetoxicum nigrum*), 374
  white (*Vincetoxicum hirundinaria*), 374
swamp-azalea (*Rhododendron viscosum*), 185
swamp-bay (*Persea palustris*), 38
swamp-birch (*Betula pumila*), 83
swamp-bulrush (*Scirpus etuberculatus*), 633
swamp chestnut-oak (*Quercus michauxii*), 77
swamp-cottonwood (*Populus heterophylla*), 152
swamp-dewberry (*Rubus hispidus*), 231
swamp-dock (*Rumex verticillatus*), 114
swamp dog-laurel (*Leucothoe axillaris*), 188
swamp-dogwood
  northern (*Cornus racemosa*), 305
  southern (*Cornus stricta*), 305
swamp-fetterbush
  deciduous (*Eubotrys racemosa*), 188
  evergreen (*Lyonia lucida*), 188
swamp fly-honeysuckle (*Lonicera oblongifolia*), 479
swamp-laurel (*Kalmia polifolia*), 187
swamp-loosestrife (*Lysimachia thyrsiflora*), 201
swamp-lousewort (*Pedicularis lanceolata*), 457
swamp-milkweed (*Asclepias incarnata*), 370
swamp-pink (*Helonias*), 786
swamp-privet (*Forestiera acuminata*), 436
swamp red currant (*Ribes triste*), 207
swamp-rose (*Rosa palustris*), 239
swamp-saxifrage (*Saxifraga pensylvanica*), 213
swamp-sunflower (*Helianthus giganteus*), 493
swamp-thistle (*Cirsium muticum*), 583
swamp-tickseed (*Coreopsis gladiata*), 505
swamp white oak (*Quercus bicolor*), 77
Swedish cornel (*Cornus suecica*), 306
sweet alyssum (*Lobularia maritima*), 168
sweet azalea (*Rhododendron arborescens*), 186
sweet bay (*Magnolia virginiana*), 36
sweet birch (*Betula lenta*), 82
sweetbrier (*Rosa eglanteria, R. micrantha*), 238
sweet buckeye (*Aesculus flava*), 331
sweet cherry (*Prunus avium*), 241
sweet cicely (*Osmorhiza*), 347
  bland (*Osmorhiza claytonii*), 347
  blunt (*Osmorhiza depauperata*), 347
  long-styled (*Osmorhiza longistylis*), 347
  tapering (*Osmorhiza berterii*), 347

sweet clover (*Melilotus*), 271
    white (*Melilotus albus*), 271
    yellow (*Melilotus officinalis*), 271
sweet-coltsfoot
    arrowhead (*Petasites sagittatus*), 529
    northern (*Petasites frigidus*), 528
sweet coneflower (*Rudbeckia subtomentosa*), 496
sweet crab (*Pyrus coronaria*), 244
sweet fern (*Comptonia*), 75
sweet flag (*Acorus*), 615
sweet flag family (ACORACEAE), 615
sweet gale (*Myrica gale*), 75
sweet grass (*Hierochloe odorata*), 713
sweet gum (*Liquidambar*), 68
sweet hurts (*Vaccinium boreale*), 192
sweet-leaf (*Symplocos*), 199
sweet-leaf family (SYMPLOCACEAE), 199
sweet pepper-bush (*Clethra*), 183
sweet pignut-hickory (*Carya ovalis*), 74
sweet pinesap (*Monotropsis*), 196
sweet scabious (*Scabiosa atropurpurea*), 488
sweet-scented bedstraw (*Galium triflorum*), 477
sweet vernal grass (*Anthoxanthum*), 714
sweet violet (*Viola odorata*), 145
sweet white violet (*Viola blanda*), 145
sweet william (*Dianthus barbatus*), 112; (*Phlox*), 389
sweet william catchfly (*Silene armeria*), 110
sweet woodruff (*Galium odoratum*), 477
swine-cress (*Coronopus didymus*), 165
switchgrass (*Panicum virgatum*), 753
sycamore (*Platanus occidentalis*), 68
sycamore-maple (*Acer pseudoplatanus*), 332

table-mountain pine (*Pinus pungens*), 33
tall anemone (*Anemone virginiana*), 50
tall beggar-ticks (*Bidens vulgata*), 507
tall bellflower (*Campanula americana*), 465
tall bilberry (*Vaccinium ovalifolium*), 192
tall bluegrass (*Poa chaixii*), 704
tall blue lettuce (*Lactuca biennis*), 590
tall bur-head (*Echinodorus berteroi*), 606
tall corydalis (*Corydalis sempervirens*), 67
tall dropseed (*Sporobolus asper*), 741
taller discoid groundsel (*Senecio indecorus*), 524
tall eupatorium (*Eupatorium altissimum*), 572
tall fescue (*Festuca arundinacea*), 695
tall flat-topped white aster (*Aster umbellatus*), 557
tall hedge-mustard (*Sisymbrium loeselii*), 181
tall ironweed (*Vernonia gigantea*), 578
tall larkspur (*Delphinium exaltatum*), 48
tall lettuce (*Lactuca canadensis*), 590
    hairy (*Lactuca hirsuta*), 589
tall meadow-rue (*Thalictrum pubescens*), 60
tall milkweed (*Asclepias exaltata*), 371
tall northern bog-orchid (*Habenaria hyperborea*), 819
tall oatgrass (*Arrhenatherum elatius*), 709
tall pine-barren milkwort (*Polygala cymosa*), 328
tall potentilla (*Potentilla arguta*), 226
tall thistle (*Cirsium altissimum*), 581
tall tickseed (*Coreopsis tripteris*), 504
tall wheatgrass (*Elytrigia elongata*), 728
tall white beard-tongue (*Penstemon digitalis*), 445
tall white bog-orchid (*Habenaria dilatata*), 820
tall white lettuce (*Prenanthes altissima*), 588
tall white violet (*Viola canadensis*), 147
tall wormseed-mustard (*Erysimum hieracifolium*), 180

tamarack (*Larix laricina*), 32
tamarisk (*Tamarix*), 149
tamarisk family (TAMARICACEAE), 149
tansy (*Tanacetum*), 519
    common (*Tanacetum vulgare*), 519
    eastern (*Tanacetum huronense*), 519
tansy-mustard (*Descurainia pinnata, D. sophia*), 182
    gray (*Descurainia richardsonii*), 182
tansy-ragwort (*Senecio jacobaea*), 526
tape-grass (*Vallisneria*), 607
taprooted valerian (*Valeriana edulis*), 486
tarweed (*Madia*), 510
    Chile- (*Madia sativa*), 510
    mountain- (*Madia glomerata*), 510
Tatarian aster (*Aster tataricus*), 555
Tatarian honeysuckle (*Lonicera tatarica*), 478
tawny cotton-grass (*Eriophorum virginicum*), 683
tea
    Labrador- (*Ledum groenlandicum*), 184
    Mexican (*Chenopodium ambrosioides*), 86
    New Jersey (*Ceanothus americanus*), 321
    Oswego- (*Monarda didyma*), 428
    prairie- (*Croton monanthogynus*), 312
tea family (THEACEAE), 125
tearthumb
    arrow-leaved (*Polygonum sagittatum*), 122
    halberd-leaved (*Polygonum arifolium*), 123
teasel (*Dipsacus*), 487
    common (*Dipsacus sylvestris*), 487
    cut-leaf (*Dipsacus laciniatus*), 487
    fullers' (*Dipsacus fullonum*), 487
    wild (*Dipsacus sylvestris*), 487
teasel family (DIPSACACEAE), 487
Tennessee milk-vetch (*Astragalus tennesseensis*), 265
Tennessee sunflower (*Helianthus eggertii*), 491
Texas croton (*Croton texensis*), 313
Texas millet (*Panicum texanum*), 751
thick-leaved wild strawberry (*Fragaria virginiana*), 220
thick-root buttercup (*Ranunculus fascicularis*), 55
thick-spike blazing star (*Liatris pycnostachya*), 576
thick-spike wheatgrass (*Elytrigia dasystachya*), 727
thimbleberry (*Rubus parviflorus*), 229
thimbleweed (*Anemone cylindrica, A. virginiana*), 50
thin-leaved wild strawberry (*Fragaria vesca*), 220
thin-scale cotton-grass (*Eriophorum angustifolium*), 638
thistle (*Carduus*), 580; (*Cirsium*), 580
    blessed (*Cnicus*), 586
    bull- (*Cirsium vulgare*), 581
    Canada (*Cirsium arvense*), 583
    coastal tall (*Cirsium nuttallii*), 583
    common sow- (*Sonchus oleraceus*), 591
    dune- (*Cirsium pitcheri*), 582
    field- (*Cirsium discolor*), 582
    globe- (*Echinops*), 584
    hollow-rooted (*Cirsium hillii*), 581
    Maltese star- (*Centaurea melitensis*), 586
    marsh- (*Cirsium palustre*), 583
    milk- (*Silybum*), 583
    musk- (*Carduus nutans*), 580
    pasture- (*Cirsium pumilum*), 581
    perennial sow- (*Sonchus arvensis*), 591
    plumeless (*Carduus acanthoides*), 580
    prairie- (*Cirsium flodmanii*), 582
    prickly sow- (*Sonchus asper*), 591
    purple star- (*Centaurea calcitrapa*), 586

thistle *(continued)*
Russian (*Salsola tragus*), 94
sandhill- (*Cirsium repandum*), 581
Scotch (*Onopordum acanthium*), 584
sow- (*Sonchus*), 591
spring- (*Cirsium carolinianum*), 581
star- (*Centaurea*), 584
swamp- (*Cirsium muticum*), 583
tall (*Cirsium altissimum*), 581
Virginia (*Cirsium virginianum*), 582
wavy-leaved (*Cirsium undulatum*), 582
welted (*Carduus crispus*), 580
yellow (*Cirsium horridulum*), 580
yellow star- (*Centaurea solstitialis*), 586
thorn
cockspur- (*Crataegus crus-galli*), 248
Washington (*Crataegus phaenopyrum*), 247
thoroughwax (*Bupleurum rotundifolium*), 352
thoroughwort penny-cress (*Thlaspi perfoliatum*), 166
thread-agalinis (*Agalinis setacea*), 451
thread-foot (*Podostemum ceratophyllum*), 289
threadleaf-pondweed (*Potamogeton filiformis*), 608
threadleaf-sundrops (*Oenothera linifolia*), 301
threadleaf-tickseed (*Coreopsis verticillata*), 505
threadstalk gaura (*Gaura filipes*), 302
three-awn (*Aristida*), 733
arrowfeather (*Aristida purpurascens*), 733
churchmouse- (*Aristida dichotoma*), 733
forktip- (*Aristida basiramea*), 734
prairie (*Aristida oligantha*), 733
purple (*Aristida purpurea*), 733
slimspike (*Aristida longespica*), 734
woolly (*Aristida lanosa*), 735
three-birds orchid (*Triphora trianthophora*), 824
three-flower melic (*Melica nitens*), 708
three-leaflet prairie-turnip (*Pediomelum canescens*), 283
three-leaved toothwort (*Cardamine ×maxima*), 171
three-leaved water-dropwort (*Oxypolis ternata*), 356
three-lobed coneflower (*Rudbeckia triloba*), 496
three-nerved Joe-Pye weed (*Eupatorium dubium*), 567
three-parted violet (*Viola tripartita*), 146
threesquare
common (*Scirpus pungens*), 632
Olney- (*Scirpus americanus*), 632
Torrey- (*Scirpus torreyi*), 632
three-toothed saxifrage (*Saxifraga tricuspidata*), 211
throatwort (*Campanula trachelium*), 464
thyme (*Thymus*), 414
basil- (*Satureja calamintha*), 413
mother-of- (*Satureja acinos*), 413
wild (*Thymus serpyllum*), 414
thyme-leaf sandwort (*Arenaria serpyllifolia*), 101
thyme-leaved dragon-head (*Dracocephalum thymiflorum*), 422
thyme-leaved speedwell (*Veronica serpyllifolia*), 453
thyme-leaved spurge (*Euphorbia serpyllifolia*), 320
ticklegrass (*Agrostis hyemalis*), 717
ticks
beggar- (*Bidens*), 506
devil's beggar- (*Bidens frondosa*), 507
few-bracted beggar- (*Bidens discoidea*), 507
New England estuarine beggar- (*Bidens eatonii*), 507
northern estuarine beggar- (*Bidens hyperborea*), 506
purplestem beggar- (*Bidens connata*), 507
southern estuarine beggar- (*Bidens bidentoides*), 507
strawstem beggar- (*Bidens comosa*), 506
ticks *(continued)*
tall beggar- (*Bidens vulgata*), 507
water beggar- (*Bidens beckii*), 508
tickseed (*Coreopsis*), 503
bigflower- (*Coreopsis grandiflora*), 504
finger- (*Coreopsis palmata*), 505
forest- (*Coreopsis major*), 504
hairy (*Coreopsis pubescens*), 503
lobed (*Coreopsis auriculata*), 504
longstalk- (*Coreopsis lanceolata*), 504
pink (*Coreopsis rosea*), 505
plains- (*Coreopsis tinctoria*), 505
swamp- (*Coreopsis gladiata*), 505
tall (*Coreopsis tripteris*), 504
threadleaf- (*Coreopsis verticillata*), 505
tickseed-sunflower
coastal plain (*Bidens mitis*), 508
midwestern (*Bidens aristosa*), 508
northern (*Bidens coronata*), 508
Ozark (*Bidens polylepis*), 508
tick-trefoil (*Desmodium*), 276
big (*Desmodium cuspidatum*), 279
Canadian (*Desmodium canadense*), 277
cluster-leaf (*Desmodium glutinosum*), 276
Fernald's (*Desmodium fernaldii*), 277
few-flowered (*Desmodium pauciflorum*), 276
hoary (*Desmodium canescens*), 277
little-leaf (*Desmodium ciliare*), 278
Maryland (*Desmodium marilandicum*), 278
matted (*Desmodium lineatum*), 276
naked (*Desmodium nudiflorum*), 276
narrow-leaved (*Desmodium tenuifolium*), 277
pine-barren (*Desmodium strictum*), 277
prairie (*Desmodium illinoense*), 277
round-leaved (*Desmodium rotundifolium*), 276
sessile-leaved (*Desmodium sessilifolium*), 277
smooth (*Desmodium laevigatum*), 278
stiff (*Desmodium rigidum*), 278
velvety (*Desmodium viridiflorum*), 278
white (*Desmodium ochroleucum*), 276
tidal sagittaria (*Sagittaria spathulata*), 602
tiger-lily (*Lilium lancifolium*), 793
timber-oatgrass (*Danthonia intermedia*), 732
timothy (*Phleum pratense*), 720
mountain- (*Phleum alpinum*), 720
tinted spurge (*Euphorbia commutata*), 316
tiny vetch (*Vicia hirsuta*), 273
ti-ti (*Cyrilla racemiflora*), 183
toadflax (*Linaria*), 447
annual (*Linaria repens*), 447
bastard (*Comandra*), 306
dalmatian (*Linaria dalmatica*), 447
lesser (*Chaenorrhinum minus*), 449
toad-rush (*Juncus bufonius*), 624
toadshade (*Trillium sessile*), 796
tobacco (*Nicotiana*), 382
Indian (*Lobelia inflata*), 467
tomatillo (*Physalis philadelphica*), 379
tomato (*Lycopersicon*), 381
toothache-grass (*Ctenium aromaticum*), 748
tooth-cup (*Ammannia coccinea*), 291; (*Rotala ramosior*), 291
sessile (*Ammannia robusta*), 292
toothed sage (*Artemisia serrata*), 520
toothed spurge (*Euphorbia dentata*), 317

toothed white-topped aster (*Aster paternus*), 558
toothed wood-fern (*Dryopteris carthusiana*), 27
tooth-leaved croton (*Croton glandulosus*), 312
toothwort (*Cardamine*), 171
    Appalachian (*Cardamine angustata*), 172
    broad-leaved (*Cardamine diphylla*), 171
    dissected (*Cardamine dissecta*), 171
    five-parted (*Cardamine concatenata*), 171
    three-leaved (*Cardamine ×maxima*), 171
top-pod water-primrose (*Ludwigia polycarpa*), 295
Toringo crab (*Pyrus sieboldii*), 244
Torrey-threesquare (*Scirpus torreyi*), 632
touch-me-not (*Impatiens*), 341
    orange (*Impatiens capensis*), 341
    yellow (*Impatiens pallida*), 341
touch-me-not family (BALSAMINACEAE), 341
tower-mustard (*Arabis glabra*), 174
tragia
    heart-leaved (*Tragia cordata*), 313
    narrow-leaved (*Tragia urens*), 313
trailing arbutus (*Epigaea*), 190
trailing bitter-cress (*Cardamine rotundifolia*), 172
trailing five-fingers (*Potentilla anglica*), 222
trailing lespedeza
    downy (*Lespedeza procumbens*), 279
    smooth (*Lespedeza repens*), 279
trailing loosestrife (*Lysimachia radicans*), 200
trailing milk-pea (*Galactia regularis*), 287
trail-plant (*Adenocaulon bicolor*), 529
treacle-mustard (*Erysimum repandum*), 179
tree of heaven (*Ailanthus altissima*), 335
trefoil
    big tick- (*Desmodium cuspidatum*), 279
    birdsfoot- (*Lotus corniculatus, L. tenuis*), 268
    Canadian tick- (*Desmodium canadense*), 277
    cluster-leaf tick- (*Desmodium glutinosum*), 276
    Fernald's tick- (*Desmodium fernaldii*), 277
    few-flowered tick- (*Desmodium pauciflorum*), 276
    hoary tick- (*Desmodium canescens*), 277
    little-leaf tick- (*Desmodium ciliare*), 278
    Maryland tick- (*Desmodium marilandicum*), 278
    matted tick- (*Desmodium lineatum*), 276
    naked tick- (*Desmodium nudiflorum*), 276
    narrow-leaved tick- (*Desmodium tenuifolium*), 277
    pine-barren tick- (*Desmodium strictum*), 277
    prairie tick- (*Desmodium illinoense*), 277
    round-leaved tick- (*Desmodium rotundifolium*), 276
    sessile-leaved tick- (*Desmodium sessilifolium*), 277
    smooth tick- (*Desmodium laevigatum*), 278
    stiff tick- (*Desmodium rigidum*), 278
    tick- (*Desmodium*), 276
    velvety tick- (*Desmodium viridiflorum*), 278
    white tick- (*Desmodium ochroleucum*), 276
trillium (*Trillium*), 795
    Barksdale's (*Trillium sulcatum*), 795
    bent (*Trillium flexipes*), 796
    big white (*Trillium grandiflorum*), 795
    green (*Trillium viride*), 796
    least (*Trillium pusillum*), 795
    nodding (*Trillium cernuum*), 796
    painted (*Trillium undulatum*), 795
    purple (*Trillium erectum*), 796
    snow- (*Trillium nivale*), 795
    yellow (*Trillium luteum*), 797
trimorphic daisy (*Erigeron acris*), 561

triodanis
    narrow-leaved (*Triodanis leptocarpa*), 466
    round-leaved (*Triodanis perfoliata*), 466
trout-lily (*Erythronium*), 793
trumpet-creeper (*Campsis*), 462
trumpet-creeper family (BIGNONIACEAE), 461
trumpet-honeysuckle (*Lonicera sempervirens*), 480
tube beard-tongue (*Penstemon tubiflorus*), 444
tubercled orchid (*Habenaria flava*), 819
tuberous eupatorium (*Eupatorium mohrii*), 571
tuberous Indian-plantain (*Cacalia plantaginea*), 528
tuberous vetchling (*Lathyrus tuberosus*), 275
tuckahoe (*Peltandra virginica*), 615
Tuckerman's quillwort (*Isoetes tuckermanii*), 4
tufted hairgrass (*Deschampsia cespitosa*), 712
tulip (*Tulipa*), 793
tulip-tree (*Liriodendron*), 38
tumble-grass (*Schedonnardus paniculatus*), 748
tumble-knapweed (*Centaurea diffusa*), 585
tumbleweed (*Amaranthus albus*), 94
tumbling mustard (*Sisymbrium altissimum*), 180
tupelo
    black (*Nyssa sylvatica*), 306
    water- (*Nyssa aquatica*), 306
turkey-beard (*Xerophyllum asphodeloides*), 780
turkeyfoot (*Andropogon hallii*), 773
turkey-oak (*Quercus laevis*), 80
Turk's-cap lily (*Lilium superbum*), 792
turnip
    Indian (*Arisaema triphyllum*), 616
    shaggy prairie- (*Pediomelum esculentum*), 284
    strigose prairie- (*Pediomelum cuspidatum*), 284
    three-leaflet prairie- (*Pediomelum canescens*), 283
turnsole (*Heliotropium indicum*), 395
turtlehead (*Chelone*), 443
    purple (*Chelone obliqua*), 443
    white (*Chelone glabra*), 443
tussock cotton-grass (*Eriophorum vaginatum*), 637
twayblade (*Liparis*), 825; (*Listera*), 814
    Appalachian (*Listera smallii*), 814
    auricled (*Listera auriculata*), 814
    big (*Listera ovata*), 815
    broad-leaved (*Listera convallarioides*), 814
    heart-leaved (*Listera cordata*), 814
    large (*Liparis liliifolia*), 825
    Loesel's (*Liparis loeselii*), 825
    southern (*Listera australis*), 814
twig-rush (*Cladium mariscoides*), 650
twinflower (*Linnaea borealis*), 482
twinleaf (*Jeffersonia*), 62
twisted heath (*Erica cinerea*), 190
twisted stalk (*Streptopus*), 799
two-flower melic (*Melica mutica*), 708
two-leaved mitrewort (*Mitella diphylla*), 214
two-row stickseed (*Lappula squarrosa*), 398
two-seed copperleaf (*Acalypha deamii*), 314

umbrella-grass (*Fuirena*), 658
umbrella-leaf (*Diphylleia*), 62
umbrella-tree (*Magnolia fraseri, M. macrophylla, M. tripetala*), 37
umbrella-wort (*Mirabilis*), 85
    hairy (*Mirabilis hirsuta*), 85
    heart-leaved (*Mirabilis nyctaginea*), 85
    narrow-leaved (*Mirabilis linearis*), 85
    pale (*Mirabilis albida*), 85

unicorn-plant (*Proboscidea*), 461
upland-cotton (*Gossypium hirsutum*), 131
upland-willow (*Salix humilis*), 159

valerian (*Valeriana*), 486
   long-tube (*Valeriana pauciflora*), 486
   taprooted (*Valeriana edulis*), 486
valerian family (VALERIANACEAE), 486
variable pondweed (*Potamogeton gramineus*), 612
variegated scouring rush (*Equisetum variegatum*), 5
Vasey's pondweed (*Potamogeton vaseyi*), 611
vegetable-oyster (*Tragopogon porrifolius*), 601
veiny dock (*Rumex venosus*), 113
veiny hawkweed (*Hieracium venosum*), 594
veiny lined aster (*Aster praealtus*), 549
velvet bent (*Agrostis canina*), 719
velvet-grass (*Holcus*), 712
   common (*Holcus lanatus*), 712
velvet-leaf (*Abutilon theophrasti*), 133
velvet-leaf blueberry (*Vaccinium myrtilloides*), 193
velvet panic-grass (*Panicum scoparium*), 759
velvety goldenrod (*Solidago mollis*), 543
velvety lespedeza (*Lespedeza stuevei*), 280
velvety tick-trefoil (*Desmodium viridiflorum*), 278
Venus' comb (*Scandix*), 349
Venus' looking-glass (*Specularia speculum-veneris*), 466
vernal grass, sweet (*Anthoxanthum*), 714
vervain (*Verbena*), 403
   bank- (*Verbena riparia*), 403
   Brazilian (*Verbena brasiliensis*), 403
   common (*Verbena hastata*), 404
   Dakota (*Verbena bipinnatifida*), 405
   European (*Verbena officinalis*), 403
   hoary (*Verbena stricta*), 405
   narrow-leaved (*Verbena simplex*), 405
   prostrate (*Verbena bracteata*), 405
   rose- (*Verbena canadensis*), 406
   rough (*Verbena scabra*), 404
   white (*Verbena urticifolia*), 404
vervain family (VERBENACEAE), 403
vervain-mallow (*Malva alcea*), 134
vetch (*Vicia*), 272
   alkali milk- (*Astragalus racemosus*), 264
   alpine milk- (*Astragalus alpinus*), 265
   American (*Vicia americana*), 274
   bigflower- (*Vicia grandiflora*), 272
   bird- (*Vicia cracca*), 274
   Canada milk- (*Astragalus canadensis*), 266
   chick-pea milk- (*Astragalus cicer*), 266
   common (*Vicia sativa*), 272
   Cooper milk- (*Astragalus neglectus*), 264
   crown- (*Coronilla varia*), 268
   elegant milk- (*Astragalus eucosmus*), 265
   field milk- (*Astragalus agrestis*), 266
   four-seed (*Vicia tetrasperma*), 273
   hairy (*Vicia villosa*), 273
   hedge- (*Vicia sepium*), 273
   joint- (*Aeschynomene*), 263
   kidney- (*Anthyllis vulneraria*), 268
   licorice milk- (*Astragalus glycyphyllos*), 266
   lotus milk- (*Astragalus lotiflorus*), 264
   milk- (*Astragalus*), 264
   Narbonne- (*Vicia narbonensis*), 273
   narrow-leaved (*Vicia angustifolia*), 272
   northern joint- (*Aeschynomene virginica*), 263
   Nuttall milk- (*Astragalus missouriensis*), 264

vetch (*Vicia*) (*continued*)
   Ozark milk- (*Astragalus distortus*), 264
   pale (*Vicia caroliniana*), 274
   pliant milk- (*Astragalus flexuosus*), 265
   pulse milk- (*Astragalus tenellus*), 264
   Robbins milk- (*Astragalus robbinsii*), 265
   spring- (*Vicia lathyroides*), 272
   standing milk- (*Astragalus adsurgens*), 266
   Tennessee milk- (*Astragalus tennesseensis*), 265
   tiny (*Vicia hirsuta*), 273
   woolly-pod (*Vicia dasycarpa*), 273
vetchling (*Lathyrus*), 274
   tuberous (*Lathyrus tuberosus*), 275
vine
   Allegheny- (*Adlumia*), 66
   balloon- (*Cardiospermum*), 330
   buckwheat- (*Brunnichia cirrhosa*), 124
   cinnamon- (*Dioscorea polystachya*), 807
   cross- (*Bignonia*), 462
   cypress- (*Ipomoea quamoclit*), 386
   kudzu- (*Pueraria lobata*), 287
   matrimony- (*Lycium*), 381
   pepper- (*Ampelopsis arborea*), 322
   pipe- (*Aristolochia tomentosa*), 41
   puncture- (*Tribulus terrestris*), 337
   sand- (*Ampelamus*), 375
   silk- (*Periploca graeca*), 370
vine-mesquite (*Panicum obtusum*), 751
violet (*Viola*), 142
   American dog- (*Viola conspersa*), 148
   arrowhead- (*Viola sagittata*), 143
   bird's-foot (*Viola pedata*), 142
   blue marsh- (*Viola cucullata*), 142
   creamy (*Viola striata*), 147
   dooryard (*Viola sororia*), 143
   great-spurred (*Viola selkirkii*), 145
   green (*Hybanthus*), 142
   hook-spurred (*Viola adunca*), 148
   kidney-leaved (*Viola renifolia*), 145
   long-spurred (*Viola rostrata*), 148
   northern marsh- (*Viola palustris*), 145
   primrose-leaved (*Viola primulifolia*), 145
   round-leaved yellow (*Viola rotundifolia*), 142
   southern woolly (*Viola villosa*), 144
   spear-leaved (*Viola hastata*), 146
   strap-leaved (*Viola lanceolata*), 146
   sweet (*Viola odorata*), 145
   sweet white (*Viola blanda*), 145
   tall white (*Viola canadensis*), 147
   three-parted (*Viola tripartita*), 146
   Walter's (*Viola walteri*), 148
   wild white (*Viola macloskeyi*), 145
   wood- (*Viola palmata*), 144
   yellow forest- (*Viola pubescens*), 147
   yellow prairie- (*Viola nuttallii*), 146
violet butterwort (*Pinguicula vulgaris*), 462
violet family (VIOLACEAE), 142
violet lespedeza (*Lespedeza violacea*), 280
violet wood-sorrel (*Oxalis violacea*), 338
Virginia buttonweed (*Diodia virginiana*), 472
Virginia chain-fern (*Woodwardia virginica*), 30
Virginia copperleaf (*Acalypha virginica*), 314
Virginia creeper (*Parthenocissus quinquefolia*), 323
Virginia day-flower (*Commelina virginica*), 621
Virginia dwarf dandelion (*Krigia virginica*), 598
Virginia ground-cherry (*Physalis virginiana*), 377

Virginia lespedeza (*Lespedeza virginica*), 280
Virginia pine (*Pinus virginiana*), 34
Virginia quillwort (*Isoetes virginica*), 4
Virginia rose (*Rosa virginiana*), 239
Virginia sida (*Sida hermaphrodita*), 138
Virginia snakeroot (*Aristolochia serpentaria*), 41
Virginia sneezeweed (*Helenium virginicum*), 499
Virginia spiderwort (*Tradescantia virginiana*), 619
Virginia thistle (*Cirsium virginianum*), 582
Virginia water-horehound (*Lycopus virginicus*), 419
Virginia wild rye (*Elymus virginicus*), 726
Virginia willow (*Itea*), 205
Virginia yellow flax (*Linum virginianum*), 327
virgin's bower (*Clematis virginiana*), 51

wahoo (*Euonymus atropurpureus*), 308
wake-robin (*Trillium*), 795
walking fern (*Asplenium rhizophyllum*), 18
wallflower
  bushy (*Erysimum repandum*), 179
  western (*Erysimum asperum*), 179
wall-hawkweed (*Hieracium murorum*), 593
wall-lettuce (*Lactuca muralis*), 589
wall-rocket
  slimleaf (*Diplotaxis tenuifolia*), 163
  stinking (*Diplotaxis muralis*), 163
wall-rue (*Asplenium ruta-muraria*), 17
walnut, black (*Juglans nigra*), 72
walnut family (JUGLANDACEAE), 72
Walter's violet (*Viola walteri*), 148
wand-goldenrod (*Solidago stricta*), 535
wand-lespedeza (*Lespedeza intermedia*), 280
wapato (*Sagittaria cuneata, S. latifolia*), 604
wart-cress (*Coronopus*), 165
wartweed (*Euphorbia helioscopia*), 315
Washington thorn (*Crataegus phaenopyrum*), 247
water-arum (*Calla*), 615
water-ash (*Fraxinus caroliniana*), 437
water-avens (*Geum rivale*), 228
water beggar-ticks (*Bidens beckii*), 508
waterberry (*Lonicera caerulea*), 478
water-bulrush (*Scirpus subterminalis*), 632
water-celery (*Vallisneria americana*), 607
water-chestnut (*Trapa*), 293
water-chestnut family (TRAPACEAE), 293
water-chinquapin (*Nelumbo lutea*), 42
water-clover (*Marsilea*), 30
  European (*Marsilea quadrifolia*), 30
  hairy (*Marsilea vestita*), 30
water-clover family (MARSILEACEAE), 30
water-cress (*Rorippa nasturtium-aquaticum*), 177
water-crowfoot
  white (*Ranunculus longirostris, R. subrigidus, R. trichophyllus*), 58
  yellow (*Ranunculus flabellaris, R. gmelinii*), 57
water-dock (*Rumex verticillatus*), 114
  great (*Rumex orbiculatus*), 113
water-dropwort (*Oxypolis*), 356
  Canby's (*Oxypolis canbyi*), 356
  common (*Oxypolis rigidior*), 356
  corn-leaved (*Oxypolis canbyi*), 356
  three-leaved (*Oxypolis ternata*), 356
water-fennel (*Oenanthe*), 354
water-flag (*Iris pseudacorus*), 810
water-foxtail (*Alopecurus geniculatus*), 720
water-hemlock (*Cicuta*), 355
  bulbiliferous (*Cicuta bulbifera*), 355
  common (*Cicuta maculata*), 355
water-hickory (*Carya aquatica*), 73
water-horehound (*Lycopus*), 419
  American (*Lycopus americanus*), 420
  European (*Lycopus europaeus*), 420
  narrow-leaved (*Lycopus angustifolius*), 420
  northern (*Lycopus uniflorus*), 419
  stalked (*Lycopus rubellus*), 420
  Virginia (*Lycopus virginicus*), 419
  western (*Lycopus asper*), 420
water-horsetail (*Equisetum fluviatile*), 6
water-hyacinth (*Eichhornia crassipes*), 779
water-hyacinth family (PONTEDERIACEAE), 779
water-hyssop (*Bacopa*), 438
waterleaf (*Hydrophyllum*), 393
  biennial (*Hydrophyllum appendiculatum*), 394
  eastern (*Hydrophyllum virginianum*), 393
  hairy (*Hydrophyllum macrophyllum*), 393
  maple-leaved (*Hydrophyllum canadense*), 394
waterleaf family (HYDROPHYLLACEAE), 392
water-lily (*Nymphaea*), 43
  pygmy (*Nymphaea tetragona*), 43
  yellow (*Nuphar*), 42
water-lily family (NYMPHAEACEAE), 42
water-lobelia (*Lobelia dortmanna*), 468
water-locust (*Gleditsia aquatica*), 256
water-meal (*Wolffia*), 616
water-milfoil (*Myriophyllum*), 289
  common (*Myriophyllum sibiricum*), 290
  European (*Myriophyllum spicatum*), 290
water-milfoil family (HALORAGACEAE), 289
water-mint (*Mentha aquatica*), 417
water-nut (*Trapa*), 293
water-nymph (*Najas*), 614
  alkaline (*Najas marina*), 614
  eutrophic (*Najas minor*), 614
  Hudson River (*Najas muenscheri*), 614
  northern (*Najas flexilis*), 614
  slender (*Najas gracillima*), 614
  southern (*Najas guadalupensis*), 614
water-nymph family (NAJADACEAE), 614
water-oak (*Quercus nigra*), 79
water-parsnip (*Sium*), 355
water-pennywort (*Hydrocotyle umbellata*), 344
water-pepper (*Polygonum hydropiper*), 120
  false (*Polygonum hydropiperoides*), 119
water-pimpernel (*Samolus*), 203
water-plantain (*Alisma*), 602
  grass-leaved (*Alisma gramineum*), 602
  northern (*Alisma triviale*), 602
  southern (*Alisma subcordatum*), 603
water-plantain family (ALISMATACEAE), 602
water-plantain spearwort (*Ranunculus ambigens*), 57
water-pod (*Ellisia nyctelea*), 394
water-primrose (*Ludwigia*), 294
  creeping (*Ludwigia peploides*), 294
  hairy (*Ludwigia pilosa*), 295
  narrow-leaved (*Ludwigia linearis*), 296
  Rafinesque's (*Ludwigia hirtella*), 295
  round-pod (*Ludwigia sphaerocarpa*), 296
  showy (*Ludwigia uruguayensis*), 294
  small-flowered (*Ludwigia glandulosa*), 296
  square-pod (*Ludwigia alternifolia*), 295
  top-pod (*Ludwigia polycarpa*), 295

water-primrose (*Ludwigia*) (*continued*)
 wing-cup (*Ludwigia alata*), 295
 wingstem (*Ludwigia decurrens*), 294
water-purslane (*Didiplis*), 292
 coastal plain (*Ludwigia brevipes*), 294
 common (*Ludwigia palustris*), 296
water scorpion-grass (*Myosotis scorpioides*), 396
water-shield (*Brasenia*), 44
water-shield family (CABOMBACEAE), 44
water smartweed (*Polygonum amphibium*), 119
water-speedwell (*Veronica anagallis-aquatica*), 455
water star-grass (*Zosterella*), 779
water-starwort (*Callitriche*), 431
water-starwort family (CALLITRICHACEAE), 431
water-tupelo (*Nyssa aquatica*), 306
water-weed (*Elodea*), 606
 Brazilian (*Egeria densa*), 606
 common (*Elodea canadensis*), 606
 free-flowered (*Elodea nuttallii*), 606
water-willow (*Decodon*), 291; (*Justicia*), 460; (*Ludwigia leptocarpa*), 294
 American (*Justicia americana*), 460
 coastal plain (*Justicia ovata*), 460
waterwort (*Elatine*), 125
waterwort family (ELATINACEAE), 125
wavy-leaved thistle (*Cirsium undulatum*), 582
wax-goldenweed (*Haplopappus ciliatus*), 529
wax-myrtle (*Myrica cerifera*), 75
wax-weed, blue (*Cuphea viscosissima*), 293
wayfaring tree (*Viburnum lantana*), 484
wedge-grass (*Sphenopholis obtusata*), 711
weedy rattlebox (*Crotalaria sagittalis*), 260
weeping lovegrass (*Eragrostis curvula*), 739
weeping willow (*Salix babylonica*), 156
welted thistle (*Carduus crispus*), 580
western androsace (*Androsace occidentalis*), 203
western annual aster (*Aster brachyactis*), 559
western black currant (*Ribes hudsonianum*), 207
western cliff-fern (*Woodsia oregana*), 21
western dock (*Rumex occidentalis*), 113
western false foxglove (*Aureolaria grandiflora*), 450
western false gromwell (*Onosmodium molle*), 401
western fescue (*Festuca occidentalis*), 695
western glasswort (*Salicornia rubra*), 93
western heart-leaved groundsel (*Senecio pseudaureus*), 525
western ironweed (*Vernonia baldwinii*), 578
western Jacob's ladder (*Polemonium occidentale*), 392
western lined aster (*Aster hesperius*), 548
western marsh-pink (*Sabatia campestris*), 362
western obedience (*Physostegia parviflora*), 422
western poison-ivy (*Toxicodendron rydbergii*), 335
western ragweed (*Ambrosia psilostachya*), 515
western rattlesnake-plantain (*Goodyera oblongifolia*), 817
western rose (*Rosa woodsii*), 240
western scaleseed (*Spermolepis inermis*), 350
western serviceberry (*Amelanchier alnifolia*), 253
western shooting star (*Dodecatheon radicatum*), 204
western silvery aster (*Aster sericeus*), 551
western stickseed (*Lappula redowskii*), 398
western wallflower (*Erysimum asperum*), 179
western water-horehound (*Lycopus asper*), 420
western wheatgrass (*Elytrigia smithii*), 727
western yellow-cress (*Rorippa sinuata*), 177
wheat (*Triticum*), 730
 common (*Triticum aestivum*), 730

wheat (*Triticum*) (*continued*)
 cow- (*Melampyrum*), 457
 India (*Fagopyrum tataricum*), 124
wheatgrass (*Elytrigia*), 727
 bluebunch (*Elytrigia spicata*), 727
 crested (*Agropyron*), 730
 salt-marsh (*Elytrigia pungens*), 727
 slender (*Elymus trachycaulus*), 725
 tall (*Elytrigia elongata*), 728
 thick-spike (*Elytrigia dasystachya*), 727
 western (*Elytrigia smithii*), 727
whipgrass (*Scleria triglomerata*), 659
 Carolina (*Scleria pauciflora*), 659
whiplash-hawkweed (*Hieracium flagellare*), 592
white adder's-mouth (*Malaxis monophyllos*), 825
white alder (*Clethra*), 183
 coast (*Clethra alnifolia*), 183
 mountain (*Clethra acuminata*), 183
white alpine saxifrage (*Saxifraga aizoon*), 212
white ash (*Fraxinus americana*), 436
white avens (*Geum canadense*), 227
white birch (*Betula papyrifera*), 83
white-bracted eupatorium (*Eupatorium album*), 570
white campion (*Silene latifolia*), 108; (*S. nivea*), 109
white cedar
 Atlantic (*Chamaecyparis thyoides*), 35
 northern (*Thuja occidentalis*), 35
white clover (*Trifolium repens*), 270
white cockle (*Silene latifolia*), 108
white dead nettle (*Lamium album*), 423
white elm (*Ulmus americana*), 68
white evening-primrose (*Oenothera speciosa*), 300
white flax (*Linum catharticum*), 326
white-flowered ground-cherry (*Leucophysalis grandiflora*), 376
white fringed orchid (*Habenaria blephariglottis*), 821
white grass (*Leersia virginica*), 690
white lady-slipper (*Cypripedium candidum*), 812
white laurel (*Rhododendron maximum*), 184
white lettuce (*Prenanthes*), 586
 Appalachian (*Prenanthes roanensis*), 588
 dwarf cut-leaf (*Prenanthes nana*), 588
 dwarf simple (*Prenanthes boottii*), 588
 glaucous (*Prenanthes racemosa*), 586
 midwestern (*Prenanthes crepidinea*), 587
 rough (*Prenanthes aspera*), 587
 tall (*Prenanthes altissima*), 588
white milkweed (*Asclepias variegata*), 372
white milkwort (*Polygala alba*), 328
white monkshood (*Aconitum reclinatum*), 47
white morning-glory (*Ipomoea lacunosa*), 386
White-Mountain avens (*Geum peckii*), 229
White-Mountain eyebright (*Euphrasia oakesii*), 457
White-Mountain potentilla (*Potentilla robbinsiana*), 224
white mulberry (*Morus alba*), 70
white mullein (*Verbascum lychnitis*), 442
white mustard (*Sinapis alba*), 162
white oak (*Quercus alba*), 76
 swamp (*Quercus bicolor*), 77
white pea (*Lathyrus ochroleucus*), 274
white pine (*Pinus strobus*), 32
white poplar (*Populus alba*), 152
white prairie-aster (*Aster falcatus*), 550
white prairie-clover (*Dalea candida*), 282
white prickly poppy (*Argemone albiflora*), 65
white sage (*Artemisia ludoviciana*), 521

white sea-blite (*Suaeda maritima*), 93
white snakeroot (*Eupatorium rugosum*), 569
   rock-house (*Eupatorium luciae-brauniae*), 569
   small-leaved (*Eupatorium aromaticum*), 568
white spruce (*Picea glauca*), 31
whitestem-filaree (*Erodium moschatum*), 341
white-stemmed evening-primrose (*Oenothera nuttallii*), 300
whitestem-pondweed (*Potamogeton praelongus*), 613
white swallow-wort (*Vincetoxicum hirundinaria*), 374
white sweet clover (*Melilotus albus*), 271
white tick-trefoil (*Desmodium ochroleucum*), 276
white top (*Cardaria*), 166
white-topped sedge (*Rhynchospora colorata*), 645
white turtlehead (*Chelone glabra*), 443
white vervain (*Verbena urticifolia*), 404
white violet
   sweet (*Viola blanda*), 145
   tall (*Viola canadensis*), 147
   wild (*Viola macloskeyi*), 145
white water-crowfoot (*Ranunculus longirostris, R. subrigidus, R. trichophyllus*), 58
white wild indigo (*Baptisia alba*), 259
white wild parsley (*Lomatium orientale*), 360
white willow (*Salix alba*), 155
whitlow-grass (*Draba verna*), 168
whitlow-wort (*Paronychia*), 106
   Appalachian (*Paronychia virginica*), 106
   silver (*Paronychia argyrocoma*), 106
   southern (*Paronychia riparia*), 106
whorled aster (*Aster acuminatus*), 557
whorled horse-balm (*Collinsonia verticillata*), 418
whorled loosestrife (*Lysimachia quadrifolia*), 201
whorled mallow (*Malva verticillata*), 135
whorled milkweed (*Asclepias verticillata*), 370
whorled milkwort (*Polygala verticillata*), 329
whorled pennywort (*Hydrocotyle verticillata*), 344
whorled pogonia (*Isotria*), 823
   larger (*Isotria verticillata*), 823
   little (*Isotria medeoloides*), 823
whorled rosin-weed (*Silphium trifoliatum*), 511
wide-leaved bedstraw (*Galium latifolium*), 473
wide-leaved phlox (*Phlox amplifolia*), 390
wide-leaved spiderwort (*Tradescantia subaspera*), 619
widgeon-grass (*Ruppia*), 613
widow's tears (*Tradescantia virginiana*), 619
wild aster (*Aster*), 547
wild basil (*Satureja vulgaris*), 413
wild bean (*Phaseolus polystachios*), 285
wild bergamot (*Monarda fistulosa*), 427
wild black cherry (*Prunus serotina*), 240
wild bleeding heart (*Dicentra eximia*), 66
wild blue flax (*Linum perenne*), 326
wild carrot (*Daucus carota*), 350
wild chervil (*Anthriscus sylvestris*), 347
wild comfrey (*Cynoglossum virginianum*), 399
wild cucumber (*Echinocystis lobata*), 150
wild geranium (*Geranium*), 338; (*G. maculatum*), 338
wild ginger (*Asarum*), 41
wildgoose-plum (*Prunus munsoniana*), 243
wild honeysuckle (*Lonicera dioica*), 480
wild hyacinth (*Camassia scilloides*), 786
wild indigo (*Baptisia*), 258
   Carolina (*Baptisia cinerea*), 259
   milky (*Baptisia lactea*), 259
   plains (*Baptisia bracteata*), 258

wild indigo (*Baptisia*) (*continued*)
   white (*Baptisia alba*), 259
   yellow (*Baptisia tinctoria*), 259
wild leek (*Allium ampeloprasum*), 791; (*A. tricoccum*), 790
wild licorice (*Galium circaezans, G. lanceolatum*), 476; (*Glycyrrhiza lepidota*), 267
wild lilac (*Ceanothus sanguineus*), 321
wild marjoram (*Origanum vulgare*), 413
wild oatgrass (*Danthonia*), 731
wild oats (*Avena fatua*), 710; (*Chasmanthium latifolium*), 731
wild pansy (*Viola rafinesquii*), 148
wild parsley (*Cymopterus acaulis*), 361
   white (*Lomatium orientale*), 360
   yellow (*Lomatium foeniculaceum*), 360
wild pea (*Lathyrus*), 274
wild petunia (*Petunia parviflora*), 383
wild pink (*Silene caroliniana*), 110; (*S. regia*), 109
wild plum (*Prunus americana*), 242
wild potato (*Ipomoea pandurata*), 386; (*Solanum jamesii*), 380
wild pumpkin (*Cucurbita foetidissima*), 150
wild rice (*Zizania*), 691
   southern (*Zizaniopsis miliacea*), 691
wild rye (*Elymus*), 725
   basin (*Leymus cinereus*), 728
   blue (*Elymus glaucus*), 725
   broad-leaved (*Elymus wiegandii*), 724
   Canada (*Elymus canadensis*), 725
   downy (*Elymus villosus*), 725
   Minnesota (*Elymus diversiglumis*), 726
   streambank (*Elymus riparius*), 726
   Virginia (*Elymus virginicus*), 726
wild sarsaparilla (*Aralia nudicaulis*), 342
wild senna
   northern (*Senna hebecarpa*), 257
   southern (*Senna marilandica*), 257
wild sensitive plant (*Chamaecrista nictitans*), 257
wild sorrel (*Rumex hastatulus*), 113
wild teasel (*Dipsacus sylvestris*), 487
wild thyme (*Thymus serpyllum*), 414
wild white violet (*Viola macloskeyi*), 145
wild yellow lily (*Lilium canadense*), 792
willow (*Salix*), 154
   American water- (*Justicia americana*), 460
   autumn- (*Salix serissima*), 155
   balsam- (*Salix pyrifolia*), 157
   basket- (*Salix purpurea*), 160
   bay-leaved (*Salix pentandra*), 155
   beaked (*Salix bebbiana*), 156
   bearberry- (*Salix uva-ursi*), 156
   black (*Salix nigra*), 154
   blueleaf- (*Salix myricoides*), 157
   bog- (*Salix pedicellaris*), 157
   Carolina (*Salix caroliniana*), 154
   coastal plain water- (*Justicia ovata*), 460
   crack- (*Salix fragilis*), 155
   diamond- (*Salix eriocephala*), 157
   dune- (*Salix cordata*), 158
   herb-like (*Salix herbacea*), 156
   meadow- (*Salix petiolaris*), 160
   peach-leaf- (*Salix amygdaloides*), 154
   primrose- (*Ludwigia*), 294
   pussy- (*Salix discolor*), 159
   sage-leaved (*Salix candida*), 160
   sandbar- (*Salix exigua*), 156

willow (*Salix*) (*continued*)
satiny (*Salix pellita*), 159
shining (*Salix lucida*), 154
silky (*Salix sericea*), 158
silvery (*Salix argyrocarpa*), 159
upland- (*Salix humilis*), 159
Virginia (*Itea*), 205
water- (*Decodon*), 291; (*Justicia*), 460; (*Ludwigia leptocarpa*), 294
weeping (*Salix babylonica*), 156
white (*Salix alba*), 155
willow family (SALICACEAE), 152
willow-herb (*Epilobium*), 296
alpine (*Epilobium alpinum*), 298
American (*Epilobium ciliatum*), 298
American marsh (*Epilobium leptophyllum*), 297
annual (*Epilobium brachycarpum*), 298
eastern (*Epilobium coloratum*), 298
great (*Epilobium angustifolium*), 296
hairy (*Epilobium hirsutum*), 297
Hornemann's (*Epilobium hornemannii*), 298
marsh (*Epilobium palustre*), 297
northeastern (*Epilobium strictum*), 297
northern (*Epilobium glandulosum*), 298
small-flowered hairy (*Epilobium parviflorum*), 297
willow leaf-lettuce (*Lactuca saligna*), 591
willow-leaved sunflower (*Helianthus salicifolius*), 493
willow-oak (*Quercus phellos*), 79
windflower (*Anemone*), 49
windmill finger-grass (*Chloris verticillata*), 747
wineberry (*Rubus phoenicolasius*), 231
wing-cup water-primrose (*Ludwigia alata*), 295
winged burning bush (*Euonymus alatus*), 308
winged elm (*Ulmus alata*), 68
winged loosestrife (*Lythrum alatum*), 292
winged pigweed (*Cycloloma*), 90
wing-fruit evening-primrose (*Oenothera macrocarpa*), 301
wingstem (*Verbesina alternifolia*), 502
wingstem meadow-pitchers (*Rhexia virginica*), 303
wingstem water-primrose (*Ludwigia decurrens*), 294
winterberry (*Ilex verticillata*), 309
smooth (*Ilex laevigata*), 310
winter-cress (*Barbarea*), 178
early (*Barbarea verna*), 178
northern (*Barbarea orthoceras*), 178
winter-grape (*Vitis vulpina*), 325
wintergreen (*Gaultheria procumbens*), 190; (*Pyrola*), 195
flowering (*Polygala paucifolia*), 328
spotted (*Chimaphila maculata*), 195
wire-rush (*Juncus arcticus*), 624
wirestem-muhly (*Muhlenbergia mexicana*), 744
wiry ground-cedar (*Lycopodium tristachyum*), 3
wiry witch-grass (*Panicum flexile*), 753
wisteria (*Wisteria*), 261
Atlantic (*Wisteria frutescens*), 261
Japanese (*Wisteria floribunda*), 261
Mississippi (*Wisteria macrostachya*), 262
witch-grass (*Panicum capillare*), 753
fall (*Leptoloma*), 768
wiry (*Panicum flexile*), 753
witch-hazel (*Hamamelis*), 68
witch hazel family (HAMAMELIDACEAE), 68
withe-rod (*Viburnum nudum var. cassinoides*), 484
woad (*Isatis*), 166
wolfberry (*Symphoricarpos occidentalis*), 481
wood-anemone (*Anemone quinquefolia*), 49
wood-betony (*Pedicularis canadensis*), 457
woodbine (*Lonicera periclymenum*), 479; (*Parthenocissus quinquefolia*), 323
grape- (*Parthenocissus vitacea*), 323
wood-bluegrass (*Poa nemoralis*), 704
wood-fern (*Dryopteris*), 24
Clinton's (*Dryopteris clintoniana*), 26
crested (*Dryopteris cristata*), 25
fancy (*Dryopteris intermedia*), 26
fragrant (*Dryopteris fragrans*), 24
Goldie's (*Dryopteris goldiana*), 25
marginal (*Dryopteris marginalis*), 25
mountain (*Dryopteris campyloptera*), 27
northern (*Dryopteris expansa*), 27
toothed (*Dryopteris carthusiana*), 27
wood-germander (*Teucrium scorodonia*), 408
woodland-agrimony (*Agrimonia rostellata*), 236
woodland-cudweed (*Gnaphalium sylvaticum*), 564
woodland-goosefoot (*Chenopodium standleyanum*), 88
woodland-groundsel (*Senecio sylvaticus*), 526
woodland-horsetail (*Equisetum sylvaticum*), 7
woodland-lettuce (*Lactuca floridana*), 590
woodland-spurge (*Euphorbia obtusata*), 315
wood-lily (*Lilium philadelphicum*), 792
woodreed (*Cinna*), 719
common (*Cinna arundinacea*), 719
drooping (*Cinna latifolia*), 719
woodruff, sweet (*Galium odoratum*), 477
wood-rush (*Luzula*), 630
wood-sage (*Teucrium*), 408; (*T. scorodonia*), 408
wood-sorrel (*Oxalis*), 337
big yellow (*Oxalis grandis*), 338
common yellow (*Oxalis stricta*), 337
creeping yellow (*Oxalis corniculata*), 337
northern (*Oxalis acetosella*), 338
showy yellow (*Oxalis macrantha*), 338
southern yellow (*Oxalis dillenii*), 337
violet (*Oxalis violacea*), 338
wood-sorrel family (OXALIDACEAE), 337
wood-violet (*Viola palmata*), 144
wool-grass (*Scirpus cyperinus*), 637
woolly bean (*Strophostyles*), 285
annual (*Strophostyles helvula*), 285
perennial (*Strophostyles umbellata*), 286
small-flowered (*Strophostyles leiosperma*), 286
woolly croton (*Croton capitatus*), 312
woolly-crowned plantain (*Plantago eriopoda*), 432
woolly-headed golden aster (*Chrysopsis gossypina*), 530
woolly lip-fern (*Cheilanthes lanosa*), 15; (*C. tomentosa*), 16
woolly milkweed (*Asclepias lanuginosa*), 374
woolly plantain (*Plantago patagonica*), 434
woolly-pod vetch (*Vicia dasycarpa*), 273
woolly potentilla (*Potentilla hippiana*), 225
woolly southern buckthorn (*Bumelia lanuginosa*), 198
woolly three-awn (*Aristida lanosa*), 735
wormgrass (*Spigelia*), 361
wormseed (*Chenopodium ambrosioides*), 86
wormseed-mustard (*Erysimum cheiranthoides*), 179
wormwood (*Artemisia*), 520
annual (*Artemisia annua*), 521
beach- (*Artemisia stelleriana*), 520
biennial (*Artemisia biennis*), 521
common (*Artemisia absinthium*), 522
wrinkle-leaved goldenrod (*Solidago rugosa*), 540

yam (*Dioscorea*), 806
yam family (DIOSCOREACEAE), 806
yam-leaved clematis (*Clematis terniflora*), 51
yard-dock (*Rumex longifolius*), 113
yard-grass (*Eleusine indica*), 741
yarrow (*Achillea*), 517
   common (*Achillea millefolium*), 517
yaupon (*Ilex vomitoria*), 309
yellow alfalfa (*Medicago sativa* subsp. *falcata*), 271
yellow alpine saxifrage (*Saxifraga aizoides*), 212
yellow and blue scorpion-grass (*Myosotis discolor*), 397
yellow avens (*Geum aleppicum*), 228
yellow bedstraw (*Galium verum*), 474
yellow birch (*Betula alleghaniensis*), 82
yellow buckeye (*Aesculus flava*), 331
yellow chamomile (*Anthemis tinctoria*), 516
yellow-cress
   common (*Rorippa palustris*), 178
   creeping (*Rorippa sylvestris*), 177
   field (*Rorippa austriaca*), 177
   Indian (*Rorippa indica*), 177
   southern (*Rorippa sessiliflora*), 178
   western (*Rorippa sinuata*), 177
yellow-eyed grass (*Xyris*), 617
yellow-eyed grass family (XYRIDACEAE), 617
yellow flag (*Iris pseudacorus*), 810
yellow flax
   Bicknell's (*Linum intercursum*), 326
   coastal plain (*Linum floridanum*), 326
   common (*Linum medium*), 327
   grooved (*Linum sulcatum*), 326
   ridgestem (*Linum striatum*), 327
   stiffstem (*Linum rigidum*), 327
   Virginia (*Linum virginianum*), 327
yellow floating heart (*Nymphoides peltata*), 389
yellow-flowered leaf-cup (*Polymnia uvedalia*), 511
yellow forest-violet (*Viola pubescens*), 147
yellow foxtail (*Setaria glauca*), 766
yellow fringed orchid (*Habenaria ciliaris*), 821
yellow hawthorn (*Crataegus michauxii*), 250

yellow hedge-hyssop (*Gratiola aurea*), 439
yellow honeysuckle (*Lonicera flava*), 480
yellow jessamine (*Gelsemium*), 361
yellow king-devil (*Hieracium caespitosum*), 592
yellow lady-slipper (*Cypripedium calceolus*), 812
yellow nutsedge (*Cyperus esculentus*), 651
yellow oak (*Quercus muehlenbergii*), 78
yellow orchid (*Habenaria integra*), 820
yellow pimpernel (*Taenidia integerrima*), 352
yellow pine (*Pinus echinata*), 34
yellow poplar (*Liriodendron*), 38
yellow prairie-violet (*Viola nuttallii*), 146
yellow rattle (*Rhinanthus*), 457
yellow rocket (*Barbarea vulgaris*), 178
yellow-root (*Xanthorhiza*), 61
yellow star-thistle (*Centaurea solstitialis*), 586
yellow sweet clover (*Melilotus officinalis*), 271
yellow thistle (*Cirsium horridulum*), 580
yellowtop (*Senecio glabellus*), 526
yellow touch-me-not (*Impatiens pallida*), 341
yellow trillium (*Trillium luteum*), 797
yellow violet
   round-leaved (*Viola rotundifolia*), 142
yellow water-crowfoot (*Ranunculus flabellaris, R. gmelinii*),
   57
yellow water-lily (*Nuphar*), 42
yellow-weed (*Reseda luteola*), 182
yellow wild indigo (*Baptisia tinctoria*), 259
yellow wild parsley (*Lomatium foeniculaceum*), 360
yellow-wood (*Cladrastis*), 258
yerba-de-tajo (*Eclipta prostrata*), 502
yew (*Taxus*), 31
   American (*Taxus canadensis*), 31
yew family (TAXACEAE), 31

zigzag aster (*Aster prenanthoides*), 548
zigzag clover (*Trifolium medium*), 269
zigzag goldenrod (*Solidago flexicaulis*), 536
zigzag iris (*Iris brevicaulis*), 810
zinnia, garden- (*Zinnia elegans*), 502

INDEX TO SCIENTIFIC NAMES

Abies, 31
  balsamea, 31
  fraseri, 31
Abution theophrasti, 133
Acalypha, 314
  deamii, 314
  gracilens, 314
  ostryifolia, 314
  rhomboidea, 314
  virginica, 314
ACANTHACEAE, 459, 460
Acanthospermum australe, 511
Acer, 331–333
  barbatum, 332
  campestre, 332
  negundo, 333
  nigrum, 332
  pensylvanicum, 333
  platanoides, 331
  pseudoplatanus, 332
  rubrum, 333
  saccharinum, 333
  saccharum, 331
  spicatum, 333
ACERACEAE, 331–333
Achillea, 517
  millefolium, 517
  ptarmica, 517
  sibirica, 517
Achyranthes japonica, 97
Aconitum, 47
  napellus, 47
  noveboracense, 47
  reclinatum, 47
  uncinatum, 47
    var. muticum, 47
    var. uncinatum, 47
ACORACEAE, 615
Acorus calamus, 615
Actaea, 47
  alba, 47
  rubra, 47
Adenocaulon bicolor, 529
ADIANTACEAE, 14–16
Adiantum, 14
  capillus-veneris, 14
  pedatum, 14
Adlumia fungosa, 66
Adonis, 59
  annua, 59
  vernalis, 59
ADOXACEAE, 485
Adoxa moschatellina, 485
Aegilops, 730
  cylindrica, 730
  ovata, 730

Aegopodium podagraria, 347
Aeschynomene virginica, 263
Aesculus, 330, 331
  flava, 331
  glabra, 331
  hippocastanum, 330
  pavia, 331
  sylvatica, 331
Aethusa cynapium, 349
Agalinis, 451, 452
  acuta, 452
  aspera, 451
  auriculata, 452
  fasciculata, 451
  gattingeri, 452
  linifolia, 451
  maritima, 451
  obtusifolia, 452
  purpurea, 451
    var. neoscotica, 451
    var. parviflora, 451
    var. purpurea, 451
  setacea, 451
  skinneriana, 452
  tenuifolia, 452
Agastache, 421
  foeniculum, 421
  nepetoides, 421
  scrophulariifolia, 421
AGAVACEAE, 804
Agave virginica, 804
Agoseris glauca, 596
Agrimonia, 236
  eupatoria, 236
  gryposepala, 236
  microcarpa, 236
  parviflora, 236
  pubescens, 236
  rostellata, 236
  striata, 236
Agropyron cristatum, 730
Agrostemma githago, 107
Agrostis, 717–719
  canina, 719
  capillaris, 717
  elliottiana, 719
  gigantea, 718
  hyemalis, 717
    var. hyemalis, 717
    var. scabra, 717
  mertensii, 718
  perennans, 718
  stolonifera, 718
    var. palustris, 718
    var. stolonifera, 718
  viridis, 718

Ailanthus altissima, 335
Aira, 713
  caryophyllea, 713
  elegantissima, 713
  praecox, 713
AIZOACEAE, 85
Ajuga, 408
  chamaepitys, 408
  genevensis, 408
  reptans, 408
Akebia quinata, 63
Albizia julibrissin, 255
Alchemilla vulgaris, 235
Aletris, 785
  aurea, 785
  farinosa, 785
Alisma, 602, 603
  gramineum, 602
  subcordatum, 603
  triviale, 602
ALISMATACEAE, 602–606
Alliaria petiolata, 180
Allium, 789–791
  ampeloprasum, 791
  canadense, 789
  cepa, 790
  cernuum, 790
  oleraceum, 791
  perdulce, 789
  sativum, 791
  schoenoprasum, 790
  stellatum, 790
  textile, 790
  tricoccum, 790
  vineale, 790
Alnus, 84
  glutinosa, 84
  incana var. americana, 84
  maritima, 84
  serrulata, 84
  viridis var. crispa, 84
Alopecurus, 720
  aequalis, 720
  carolinianus, 720
  geniculatus, 720
  myosuroides, 720
  pratensis, 720
Alternanthera philoxeroides, 98
Althaea officinalis, 134
Alyssum alyssoides, 168
AMARANTHACEAE, 94–98
Amaranthus, 94–97
  albus, 94
  arenicola, 96
  blitoides, 94
  blitum, 95

Amaranthus *(continued)*
cannabinus, 97
caudatus, 96
crassipes, 94
crispus, 94
cruentus, 95
deflexus, 96
hybridus, 96
hypochondriacus, 95
palmeri, 96
powellii, 95
pumilus, 95
retroflexus, 95
rudis, 96
spinosus, 95
tuberculatus, 97
viridis, 96
Ambrosia, 514, 515
acanthicarpa, 515
artemisiifolia, 515
bidentata, 515
psilostachya, 515
tomentosa, 514
trifida, 515
Amelanchier, 253–255
alnifolia, 253
arborea, 254
bartramiana, 253
canadensis, 254
fernaldii, 254
laevis, 255
obovalis, 254
sanguinea, 254
spicata, 254
Amianthium muscitoxicum, 782
Ammannia, 291, 292
auriculata, 291
coccinea, 291
latifolia, 291
robusta, 292
Ammophila breviligulata, 717
Amorpha, 282, 283
canescens, 282
fruticosa, 283
nana, 282
nitens, 282
Ampelamus albidus, 375
Ampelopsis, 322, 323
arborea, 322
brevipedunculata, 323
cordata, 323
Amphiachyris dracunculoides, 546
Amphicarpaea bracteata, 286
Amphicarpum amphicarpon, 761
Amsinckia, 398
lycopsoides, 398
retrorsa, 398
tessellata, 398
Amsonia, 368
illustris, 368
tabernaemontana, 368
ANACARDIACEAE, 334, 335
Anagallis arvensis, 202
Anaphalis margaritacea, 565
Anchusa, 402
arvensis, 402

Anchusa *(continued)*
azurea, 402
officinalis, 402
Andromeda glaucophylla, 187
Andropogon, 773–775
gerardii, 773
gyrans, 774
hallii, 773
mohrii, 774
ternarius, 774
virginicus, 775
var. abbreviatus, 775
var. virginicus, 775
Androsace, 203
occidentalis, 203
septentrionalis, 203
Anemone, 49, 50
canadensis, 49
caroliniana, 49
cylindrica, 50
lancifolia, 49
multifida, 49
parviflora, 49
patens, 50
quinquefolia, 49
virginiana, 50
Anemonella thalictroides, 59
Anethum graveolens, 352
Angelica, 358, 359
atropurpurea, 358
lucida, 359
triquinata, 358
venenosa, 359
ANNONACEAE, 38
Anoda cristata, 138
Antennaria, 566
microphylla, 566
neglecta, 566
var. neglecta, 566
var. neodioica, 566
var. petaloidea, 566
parvifolia, 566
plantaginifolia, 566
solitaria, 566
virginica, 566
Anthemis, 516, 517
arvensis, 516
cotula, 517
tinctoria, 516
Anthoxanthum, 714
aristatum, 714
odoratum, 714
Anthriscus, 347, 348
caucalis, 348
cerefolium, 347
sylvestris, 347
Anthyllis vulneraria, 268
Antirrhinum, 448
majus, 448
orontium, 448
Apera, 717
interrupta, 717
spica-venti, 717
Aphanes microcarpa, 235
APIACEAE, 344–361

Apios, 286
americana, 286
priceana, 286
Apium, 351
graveolens, 351
leptophyllum, 351
nodiflorum, 351
repens, 351
Aplectrum hyemale, 826
APOCYNACEAE, 368, 369
Apocynum, 369
androsaemifolium, 369
cannabinum, 369
sibiricum, 369
AQUIFOLIACEAE, 309, 310
Aquilegia, 59
canadensis, 59
vulgaris, 59
Arabidopsis thaliana, 181
Arabis, 174–176
canadensis, 175
divaricarpa, 176
drummondii, 175
glabra, 174
hirsuta, 174
var. adpressipilis, 174
var. glabrata, 174
var. pycnocarpa, 174
holboellii var. retrofracta, 176
laevigata, 175
lyrata, 174
missouriensis, 175
patens, 174
perstellata, 176
serotina, 175
shortii, 176
ARACEAE, 615, 616
Arachis hypogaea, 263
Aralia, 342, 343
hispida, 342
nudicaulis, 342
racemosa, 343
spinosa, 342
ARALIACEAE, 342, 343
Arceuthobium pusillum, 307
Arctium, 579
lappa, 579
minus, 579
tomentosum, 579
Arctostaphylos, 190
alpina, 190
uva-ursi, 190
Arenaria, 100, 101
caroliniana, 101
groenlandica, 101
lanuginosa, 101
lateriflora, 100
macrophylla, 100
patula, 101
rubella, 101
serpyllifolia, 101
stricta, 101
Arethusa bulbosa, 824
Argemone, 65
albiflora, 65
mexicana, 65

Arisaema, 616
  dracontium, 616
  triphyllum, 616
Aristida, 733–735
  adscensionis, 734
  basiramea, 734
    var. basiramea, 734
    var. curtissii, 734
  desmantha, 734
  dichotoma, 733
  lanosa, 735
  longespica, 734
    var. geniculata, 734
    var. longespica, 734
  oligantha, 733
  palustris, 733
  purpurascens, 733
  purpurea, 733
  ramosissima, 734
  tuberculosa, 734
  virgata, 733
Aristolochia, 41
  clematitis, 41
  macrophylla, 41
  serpentaria, 41
  tomentosa, 41
ARISTOLOCHIACEAE, 40, 41
Armoracia, 170
  lacustris, 170
  rusticana, 170
Arnica, 500, 501
  acaulis, 501
  cordifolia, 500
  lanceolata, 501
  lonchophylla, 500
Arnoseris minima, 600
Aronia, 245
  arbutifolia, 245
  melanocarpa, 245
  ×prunifolia, 245
Arrhenatherum elatius, 709
Artemisia, 520–522
  abrotanum, 520
  absinthium, 522
  annua, 521
  biennis, 521
  campestris, 522
  carruthii, 521
  dracunculus, 522
  frigida, 522
  ludoviciana, 521
    var. ludoviciana, 521
    var. mexicana, 521
  pontica, 520
  serrata, 520
  stelleriana, 520
  vulgaris, 520
Arthraxon hispidus, 773
Aruncus dioicus, 219
Arundinaria gigantea, 690
Asarum canadense, 41
    var. canadense, 41
    var. reflexum, 41
ASCLEPIADACEAE, 370–376
Asclepias, 370–374
  amplexicaulis, 371

Asclepias (continued)
  exaltata, 371
  hirtella, 373
  incarnata, 370
  lanceolata, 372
  lanuginosa, 374
  longifolia, 374
  meadii, 373
  ovalifolia, 371
  perennis, 370
  purpurascens, 372
  quadrifolia, 372
  rubra, 373
  speciosa, 372
  stenophylla, 373
  sullivantii, 373
  syriaca, 371
  tuberosa, 370
  variegata, 372
  verticillata, 370
  viridiflora, 374
  viridis, 373
Asimina, 38
  parviflora, 38
  triloba, 38
Asparagus officinalis, 801
Asperugo procumbens, 396
Asperula arvensis, 473
Aspidotis densa, 15
ASPLENIACEAE, 17–28
Asplenium, 17, 18
  bradleyi, 17
  montanum, 17
  pinnatifidum, 18
  platyneuron, 17
  resiliens, 17
  rhizophyllum, 18
  ruta-muraria, 17
  septentrionale, 18
  trichomanes, 18
  viride, 18
Aster, 547–559
  acuminatus, 557
  anomalus, 553
  ×blakei, 557
  borealis, 547
  brachyactis, 559
  ciliolatus, 551
  concinnus, 548
  concolor, 551
  cordifolius, 552
  crenifolius, 547
  depauperatus, 551
  divaricatus, 555
    var. chlorolepis, 555
    var. divaricatus, 555
  drummondii, 552
  dumosus, 550
  elliottii, 548
  ericoides, 550
  falcatus, 550
  firmus, 547
  furcatus, 555
  gracilis, 555
  grandiflorus, 554
  hemisphericus, 556

Aster (continued)
  ×herveyi, 556
  hesperius, 548
  infirmus, 558
  laevis, 548
  lanceolatus, 549
  lateriflorus, 550
  laurentianus, 558
  linariifolius, 558
  ×longulus, 547
  lowrieanus, 552
  macrophyllus, 559
  modestus, 554
  nemoralis, 557
  novae-angliae, 554
  novi-belgii, 547
  oblongifolius, 554
  ontarionis, 549
  oolentangiensis, 551
  parviceps, 551
  patens, 553
    var. patens, 553
    var. phlogifolius, 553
  paternus, 558
  pilosus, 550
  praealtus, 549
  prenanthoides, 548
  pubentior, 557
  puniceus, 547
  racemosus, 550
  radula, 556
  sagittifolius, 552
  saxicastellii, 556
  schreberi, 554
  sericeus, 551
  shortii, 552
  solidagineus, 558
  spectabilis, 556
  subulatus, 559
  surculosus, 556
  tataricus, 555
  tenuifolius, 558
  texanus, 553
  tradescantii, 549
  turbinellus, 555
  umbellatus, 557
  undulatus, 553
ASTERACEAE, 489–601
Astilbe biternata, 211
Astragalus, 264–266
  adsurgens, 266
  agrestis, 266
  alpinus, 265
    var. alpinus, 265
    var. brunetianus, 265
  canadensis, 266
  cicer, 266
  crassicarpus, 265
    var. crassicarpus, 265
    var. trichocalyx, 265
  distortus, 264
  eucosmus, 265
  flexuosus, 265
  glycyphyllos, 266
  lotiflorus, 264
  missouriensis, 264

Astragalus *(continued)*
  neglectus, 264
  racemosus, 264
  robbinsii, 265
    var. jesupii, 265
  tenellus, 264
  tennesseensis, 265
Astranthium integrifolium, 563
Athyrium, 19
  filix-femina, 19
    var. asplenioides, 19
    var. michauxii, 19
  pycnocarpon, 19
  thelypterioides, 19
Atriplex, 91, 92
  arenaria, 91
  argentea, 92
  glabriuscula, 91
  hastata, 92
  hortensis, 91
  laciniata, 92
  littoralis, 92
  patula, 91
  rosea, 92
Aureolaria, 449–451
  flava, 450
  grandiflora, 450
  laevigata, 449
  patula, 450
  pedicularia, 450
    var. ambigens, 450
    var. austromontana, 450
    var. pectinata, 450
    var. pedicularia, 450
  virginica, 451
Avena, 710
  fatua, 710
  sativa, 710
  sterilis, 710
Axonopus furcatus, 766
Axyris amaranthoides, 90
Azolla, 30
  caroliniana, 30
  mexicana, 30

Baccharis halimifolia, 562
Bacopa, 438
  caroliniana, 438
  innominata, 438
  monnieri, 438
  rotundifolia, 438
Ballota nigra, 424
BALSAMINACEAE, 341, 342
Baptisia, 258, 259
  alba, 259
  australis, 258
  bracteata, 258
  cinerea, 259
  lactea, 259
  tinctoria, 259
Barbarea, 178
  orthoceras, 178
  verna, 178
  vulgaris, 178
Bartonia, 367
  paniculata, 367

Bartonia *(continued)*
    var. iodandra, 367
    var. paniculata, 367
  virginica, 367
Bassia, 90
  hirsuta, 90
  hyssopifolia, 90
Beckmannia syzigachne, 720
Belamcanda chinensis, 807
Bellis perennis, 563
BERBERIDACEAE, 62
Berberis, 62
  canadensis, 62
  thunbergii, 62
  vulgaris, 62
Berchemia scandens, 322
Bergia texana, 125
Berteroa incana, 168
Berula erecta, 355
Besseya bullii, 456
Betula, 82, 83
  alleghaniensis, 82
  ×caerulea-grandis, 83
  glandulosa, 83
  lenta, 82
  michauxii, 83
  nigra, 83
  papyrifera, 83
  populifolia, 83
  pumila, 83
BETULACEAE, 82–84
Bidens, 506–508
  aristosa, 508
  beckii, 508
  bidentoides, 507
  bipinnata, 508
  cernua, 506
  comosa, 506
  connata, 507
  coronata, 508
  discoidea, 507
  eatonii, 507
  frondosa, 507
  hyperborea, 506
  laevis, 506
  mitis, 508
  polylepis, 508
  vulgata, 507
Bignonia capreolata, 462
BIGNONIACEAE, 461, 462
BLECHNACEAE, 30
Blephilia, 428
  ciliata, 428
  hirsuta, 428
Boehmeria cylindrica, 71
Boltonia, 562
  asteroides, 562
    var. asteroides, 562
    var. latisquama, 562
    var. recognita, 562
  caroliniana, 562
  decurrens, 562
  diffusa, 562
BORAGINACEAE, 395–402
Borago officinalis, 402
Borrichia frutescens, 502

Botrychium, 8, 9
  biternatum, 8
  dissectum, 8
  jenmanii, 9
  lanceolatum, 8
  lunaria, 8
  matricariifolium, 8
  multifidum, 8
  oneidense, 8
  rugulosum, 8
  simplex, 8
  virginianum, 9
Bouteloua, 750
  curtipendula, 750
  gracilis, 750
  hirsuta, 750
Boykinia aconitifolia, 215
Brachyelytrum erectum, 692
    var. erectum, 692
    var. glabratum, 692
Brasenia schreberi, 44
Brassica, 162
  juncea, 162
  nigra, 162
  rapa, 162
BRASSICACEAE, 162–182
Braya humilis, 181
Brickellia grandiflora, 573
Briza, 700
  maxima, 700
  media, 700
  minor, 700
BROMELIACEAE, 779
Bromus, 720–724
  arvensis, 723
  briziformis, 722
  carinatus, 720
  catharticus, 721
  ciliatus, 721
  commutatus, 723
  erectus, 722
  hordeaceus, 722
  inermis, 721
  japonicus, 724
  kalmii, 721
  latiglumis, 721
  pubescens, 722
  pumpellianus, 721
  racemosus, 722
  rigidus, 724
  rubens, 723
  secalinus, 723
  squarrosus, 722
  sterilis, 724
  tectorum, 723
Broussonetia papyrifera, 70
Brunnichia cirrhosa, 124
Buchloe dactyloides, 751
Buchnera americana, 452
Buckleya distichophylla, 307
BUDDLEJACEAE, 434
Buddleja davidii, 434
Bulbostylis, 644
  capillaris, 644
  ciliatifolia, 644

Bumelia, 198
  lanuginosa, 198
  lycioides, 198
Bunias orientalis 167
Bupleurum, 352
  fontanesii, 352
  lancifolium, 352
  rotundifolium, 352
Burmannia biflora, 812
BURMANNIACEAE, 812
BUTOMACEAE, 602
Butomus umbellatus, 602
BUXACEAE, 311

Cabomba caroliniana, 44
CABOMBACEAE, 44
Cacalia, 527, 528
  atriplicifolia, 527
  muehlenbergii, 527
  plantaginea, 528
  suaveolens, 527
CACTACEAE, 86
CAESALPINIACEAE, 256–258
Cakile, 164
  edentula, 164
  maritima, 164
Calamagrostis, 714–717
  canadensis, 716
  cinnoides, 715
  epigejos, 714
  insperata, 716
  lacustris, 716
  montanensis, 715
  perplexa, 717
  pickeringii, 715
  porteri, 716
  purpurascens, 714
  stricta, 715
Calamovilfa, 745
  brevipilis, 745
  longifolia, 745
Calla palustris, 615
Callicarpa, 407
  americana, 407
  dichotoma, 407
Callirhoe, 136
  alcaeoides, 136
  bushii, 136
  digitata, 136
  involucrata, 136
  triangulata, 136
CALLITRICHACEAE, 431
Callitriche, 431
  anceps, 431
  hermaphroditica, 431
  heterophylla, 431
  palustris, 431
  stagnalis, 431
  terrestris, 431
Calluna vulgaris, 190
Calopogon, 824
  pallidus, 824
  tuberosus, 824
Caltha, 45
  natans, 45
  palustris, 45

CALYCANTHACEAE, 38
Calycanthus floridus, 38
Calycocarpum lyonii, 63
Calylophus serrulatus, 302
Calypso bulbosa, 826
Calystegia, 385
  hederacea, 385
  sepium, 385
  spithamaea, 385
Camassia scilloides, 786
Camelina, 181
  microcarpa, 181
  sativa, 181
Campanula, 464, 465
  americana, 465
  aparinoides, 465
    var. aparinoides, 465
    var. grandiflora, 465
  divaricata, 465
  glomerata, 465
  patula, 464
  rapunculoides, 465
  rotundifolia, 464
  trachelium, 464
CAMPANULACEAE, 464–469
Campsis radicans, 462
CANNABACEAE, 70
Cannabis sativa, 70
CAPPARACEAE, 161
CAPRIFOLIACEAE, 478–485
Capsella bursa-pastoris, 167
Cardamine, 171–173
  angustata, 172
  bellidifolia, 172
  clematitis, 173
  concatenata, 171
  diphylla, 171
  dissecta, 171
  douglassii, 172
  hirsuta, 173
  longii, 172
  ×maxima, 171
  parviflora, 173
  pensylvanica, 173
  pratensis, 173
    var. palustris, 173
    var. pratensis, 173
  rhomboidea, 172
  rotundifolia, 172
Cardaria, 166
  chalepensis, 166
  draba, 166
  pubescens, 166
Cardiospermum halicacabum, 330
Carduus, 580
  acanthoides, 580
  crispus, 580
  nutans, 580
Carex, 660–689
  abscondita, 676
  acutiformis, 686
  adusta, 669
  ×aestivaliformis, 678
  aestivalis, 678
  alata, 671
  albicans, 672

Carex (continued)
  var. albicans, 672
  var. emmonsii, 672
  albolutescens, 671
  albursina, 675
  alopecoidea, 665
  amphibola, 677
  appalachica, 664
  aquatilis, 684
  arcta, 667
  arctata, 679
  arenaria, 662
  argyrantha, 668
  assiniboinensis, 678
  atherodes, 687
  atlantica, 667
    var. atlantica, 667
    var. capillacea, 667
  atratiformis, 683
  aurea, 673
  backii, 672
  baileyi, 689
  barrattii, 682
  bebbii, 670
  bicknellii, 670
  bigelowii, 684
  blanda, 675
  brevior, 670
  bromoides, 668
  brunnescens, 666
  bullata, 688
  bushii, 681
  buxbaumii, 683
  canescens, 667
  capillaris, 679
  capitata, 660
  careyana, 676
  caroliniana, 681
  caryophyllea, 671
  castanea, 678
  cephalophora, 662
    var. cephalophora, 662
    var. mesochorea, 662
  chordorrhiza, 662
  collinsii, 685
  communis, 672
  comosa, 686
  complanata, 681
    var. complanata, 681
    var. hirsuta, 681
  concinna, 673
  conjuncta, 665
  conoidea, 677
  crawei, 676
  crawfordii, 669
  crebriflora, 675
  crinita, 684
  cristatella, 669
  crus-corvi, 665
  cryptolepis, 680
  cumulata, 671
  davisii, 678
  debilis, 679
    var. debilis, 679
    var. pubera, 679
    var. rudgei, 679

Carex (continued)
decomposita, 665
deflexa, 673
deweyana, 668
diandra, 666
digitalis, 676
dioica, 660
disperma, 664
divisa, 661
divulsa, 663
douglasii, 662
eburnea, 674
echinata, 668
emoryi, 685
exilis, 667
extensa, 680
festucacea, 670
filifolia, 661
flacca, 682
flaccosperma, 677
flava, 680
foenea, 668
folliculata, 685
    var. australis, 685
    var. folliculata, 685
formosa, 678
frankii, 687
geyeri, 661
gigantea, 689
glareosa, 666
glaucescens, 682
gracilescens, 675
gracillima, 678
granularis, 676
gravida, 664
    var. gravida, 664
    var. lunelliana, 664
grayi, 689
gynandra, 684
haydenii, 685
heleonastes, 666
hirta, 681
hirtifolia, 674
hitchcockiana, 677
houghtoniana, 681
hyalinolepis, 686
hystericina, 686
interior, 667
intumescens, 689
jamesii, 672
joorii, 682
katahdinensis, 677
kobomugi, 661
lacustris, 686
laeviconica, 687
laevivaginata, 665
lasiocarpa, 682
laxiculmis, 676
laxiflora, 675
leavenworthii, 663
lenticularis, 684
lepidocarpa, 680
leporina, 668
leptalea, 660
leptonervia, 674
limosa, 683

Carex (continued)
livida, 674
longii, 671
louisianica, 689
lucorum, 673
lupuliformis, 689
lupulina, 689
lurida, 688
mackenziei, 666
×mainensis, 688
meadii, 674
melanostachya, 686
michauxiana, 685
mitchelliana, 684
muehlenbergii, 664
    var. australis, 664
    var. muehlenbergii, 664
muricata, 663
muskingumensis, 669
nigra, 685
nigromarginata, 672
normalis, 670
norvegica, 683
novae-angliae, 672
obtusata, 660
oligocarpa, 677
oligosperma, 687
oronensis, 668
oxylepis, 678
paleacea, 684
pallescens, 680
panicea, 674
parryana, 683
pauciflora, 660
paupercula, 683
peckii, 672
pedunculata, 673
pellita, 681
pensylvanica, 672
    var. digyna, 672
    var. pensylvanica, 672
picta, 661
plantaginea, 676
platyphylla, 676
polymorpha, 674
praegracilis, 662
prairea, 666
prasina, 678
praticola, 668
projecta, 669
pseudocyperus, 686
purpurifera, 675
radiata, 663
rariflora, 683
reniformis, 670
retroflexa, 663
    var. retroflexa, 663
    var. texensis, 663
retrorsa, 688
richardsonii, 673
rosea, 664
rossii, 673
rostrata, 688
ruthii, 667
salina, 684
sartwellii, 662

Carex (continued)
saxatilis, 687
scabrata, 682
schweinitzii, 689
scirpoidea, 661
scoparia, 669
seorsa, 668
shortiana, 682
siccata, 662
silicea, 671
socialis, 663
sparganioides, 664
    var. aggregata, 664
    var. cephaloidea, 664
    var. sparganioides, 664
spicata, 663
sprengelii, 679
squarrosa, 687
stenophylla, 662
sterilis, 667
stipata, 665
    var. maxima, 665
    var. stipata, 665
straminea, 671
    var. invisa, 671
    var. straminea, 671
striata, 681
striatula, 675
stricta, 685
styloflexa, 675
suberecta, 670
supina, 671
swanii, 680
sychnocephala, 668
sylvatica, 679
tenera, 670
tenuiflora, 666
tetanica, 674
torreyi, 680
torta, 684
tribuloides, 669
trichocarpa, 687
trisperma, 666
tuckermanii, 688
typhina, 687
umbellata, 673
utriculata, 688
vaginata, 674
venusta, 679
    var. minor, 679
    var. venusta, 679
vesicaria, 688
vestita, 682
virescens, 680
viridula, 679
vulpinoidea, 665
    var. ambigua, 665
    var. vulpinoidea, 665
wiegandii, 667
willdenowii, 672
woodii, 674
xerantica, 669
Carlina vulgaris, 584
Carphephorus, 573, 574
    bellidifolius, 574
    tomentosus, 573

Carpinus caroliniana, 82
Carum carvi, 348
Carya, 73, 74
    aquatica, 73
    carolinae-septentrionalis, 74
    cordiformis, 73
    glabra, 74
    illinoinensis, 73
    laciniosa, 73
    ovalis, 74
    ovata, 74
    pallida, 74
    texana, 74
    tomentosa, 73
CARYOPHYLLACEAE, 100–112
Castanea, 76
    dentata, 76
    pumila, 76
Castilleja, 458
    coccinea, 458
    septentrionalis, 458
    sessiliflora, 458
Catabrosa aquatica, 706
Catalpa, 461, 462
    bignonioides, 462
    speciosa, 461
Caulophyllum thalictroides, 62
Ceanothus, 321
    americanus, 321
        var. americanus, 321
        var. intermedius, 321
    herbaceus, 321
    sanguineus, 321
CELASTRACEAE, 308, 309
Celastrus, 308
    orbiculatus, 308
    scandens, 308
Celosia argentea, 97
Celtis, 69
    laevigata, 69
    occidentalis, 69
    tenuifolia, 69
Cenchrus, 769, 770
    echinatus, 769
    incertus, 770
    longispinus, 769
    tribuloides, 770
Centaurea, 584–586
    americana, 585
    calcitrapa, 586
    cyanus, 584
    diffusa, 585
    dubia, 585
    jacea, 584
    maculosa, 585
    melitensis, 586
    nigra, 585
    repens, 585
    scabiosa, 584
    solstitialis, 586
Centaurium, 363
    erythraea, 363
    pulchellum, 363
    spicatum, 363
Centella erecta, 344
Centrosema virginianum, 285

Centunculus minimus, 203
Cephalanthus occidentalis, 471
Cerastium, 102, 103
    arvense, 102
    brachypetalum, 102
    dubium, 102
    nutans, 103
    pumilum, 102
    semidecandrum, 103
    viscosum, 102
    vulgatum, 102
CERATOPHYLLACEAE, 44
Ceratophyllum, 44
    demersum, 44
    echinatum, 44
Cercis, 256
    canadensis, 256
Chaenorrhinum minus, 449
Chaerophyllum, 348, 349
    bulbosum, 349
    procumbens, 348
    tainturieri, 349
    temulum, 349
Chamaecrista, 256, 257
    fasciculata, 256
    nictitans, 257
Chamaecyparis thyoides, 35
Chamaedaphne calyculata, 189
Chamaelirium luteum, 787
Chamaerhodos erecta, 221
Chasmanthium, 731
    latifolium, 731
    laxum, 731
    sessiliflorum, 731
Cheilanthes, 15, 16
    alabamensis, 15
    eatonii, 16
    feei, 15
    lanosa, 15
    tomentosa, 16
Chelidonium majus, 64
Chelone, 443
    cuthbertii, 443
    glabra, 443
    obliqua, 443
CHENOPODIACEAE, 86–94
Chenopodium, 86–89
    album, 89
    ambrosioides, 86
    aristatum, 88
    berlandieri, 89
    bonus-henricus, 87
    botrys, 87
    capitatum, 87
    glaucum, 87
    graveolens, 87
    leptophyllum, 89
    multifidum, 86
    murale, 88
    polyspermum, 88
    pratericola, 89
    pumilio, 87
    rubrum, 87
    simplex, 88
    standleyanum, 88
    urbicum, 88

Chenopodium (continued)
    vulvaria, 88
Chimaphila, 195
    maculata, 195
    umbellata, 195
Chionanthus virginicus, 435
Chloris, 747
    verticillata, 747
    virgata, 747
Chondrilla juncea, 599
Chorispora tenella, 179
Chrysanthemum, 518, 519
    balsamita, 519
    leucanthemum, 518
    nipponicum, 518
    parthenium, 518
    segetum, 518
Chrysogonum virginianum, 513
Chrysopsis, 530, 531
    camporum, 530
    falcata, 531
    gossypina, 530
    graminifolia, 531
    mariana, 530
    villosa, 530
Chrysosplenium, 217
    americanum, 217
    iowense, 217
Cichorium intybus, 599
Cicuta, 355
    bulbifera, 355
    maculata, 355
Cimicifuga, 46
    americana, 46
    racemosa, 46
Cinna, 719
    arundinacea, 719
    latifolia, 719
Circaea, 303
    alpina, 303
    lutetiana, 303
Cirsium, 580–583
    altissimum, 581
    arvense, 583
    carolinianum, 581
    discolor, 582
    flodmanii, 582
    hillii, 581
    horridulum, 580
    muticum, 583
    nuttallii, 583
    palustre, 583
    pitcheri, 582
    pumilum, 581
    repandum, 581
    undulatum, 582
    virginianum, 582
    vulgare, 581
CISTACEAE, 139–142
Cladium, 650
    jamaicense, 650
    mariscoides, 650
Cladrastis lutea, 258
Claytonia, 99
    caroliniana, 99
    virginica, 99

Cleistes divaricata, 823
Clematis, 51–53
  addisonii, 53
  albicoma, 53
  catesbyana, 51
  coactilis, 53
  crispa, 52
  florida, 51
  fremontii, 53
  glaucophylla, 52
  occidentalis, 51
  ochroleuca, 53
  orientalis, 51
  pitcheri, 52
  terniflora, 51
  versicolor, 52
  viorna, 52
  virginiana, 51
  viticaulis, 53
Cleome, 161
  hassleriana, 161
  serrulata, 161
Clerodendrum japonicum, 407
Clethra, 183
  acuminata, 183
  alnifolia, 183
CLETHRACEAE, 183
Clintonia, 797, 798
  borealis, 797
  umbellulata, 798
Clitoria mariana, 285
CLUSIACEAE, 126–130
Cnicus benedictus, 586
Cnidoscolus stimulosus, 312
Cocculus carolinus, 63
Coelorachis, 773
  cylindrica, 773
  rugosa, 773
Colchicum autumnale, 794
Collinsia, 446, 447
  parviflora, 447
  verna, 447
  violacea, 446
Collinsonia, 418
  canadensis, 418
  verticillata, 418
Collomia linearis, 391
Colutea arborescens, 264
Comandra umbellata, 306
  var. pallida, 306
  var. umbellata, 306
Commelina, 620, 621
  caroliniana, 621
  communis, 620
  diffusa, 620
  erecta, 621
  virginica, 621
COMMELINACEAE, 619–621
Comptonia peregrina, 75
Conioselinum chinense, 355
Conium maculatum, 351
Conopholis americana, 459
Conradina verticillata, 412
Conringia orientalis, 164
Convallaria majalis, 800
CONVOLVULACEAE, 384–386

Convolvulus arvensis, 384
Conyza, 561, 562
  bonariensis, 562
  canadensis, 561
  ramosissima, 561
Coptis trifolia, 61
Corallorhiza, 826, 827
  maculata, 827
  odontorhiza, 827
  striata, 827
  trifida, 827
  wisteriana, 826
Corema conradii, 184
Coreopsis, 503–505
  auriculata, 504
  gladiata, 505
    var. linifolia, 505
  grandiflora, 504
  lanceolata, 504
  major, 504
  palmata, 505
  pubescens, 503
  rosea, 505
  tinctoria, 505
  tripteris, 504
  verticillata, 505
Coriandrum, 349
  sativum, 349
Corispermum, 90
  hyssopifolium, 90
  nitidum, 90
CORNACEAE, 304–306
Cornus, 304–306
  alternifolia, 304
  amomum, 304
    var. amomum, 304
    var. schuetzeana, 304
  canadensis, 305
  drummondii, 305
  florida, 306
  racemosa, 305
  rugosa, 304
  sericea, 305
  stricta, 305
  suecica, 306
Coronilla varia, 268
Coronopus, 165
  didymus, 165
  squamatus, 165
Corydalis, 67
  aurea, 67
  flavula, 67
  micrantha, 67
  sempervirens, 67
Corylus, 82
  americana, 82
  cornuta, 82
Corynephorus canescens, 713
Coryphantha, 86
  missouriensis, 86
  vivipara, 86
Cosmos, 508, 509
  bipinnatus, 508
  sulphureus, 509
Cotula coronopifolia, 520
Crassula aquatica, 210

CRASSULACEAE, 208–210
Crataegus, 247–253
  berberifolia, 249
  brainerdii, 251
  calpodendron, 248
  chrysocarpa, 250
  coccinea, 252
  coccinioides, 251
  crus-galli, 248
  dilatata, 252
  douglasii, 247
  flabellata, 253
  intricata, 250
  marshallii, 247
  michauxii, 250
  mollis, 251
  monogyna, 247
  phaenopyrum, 247
  pruinosa, 252
  punctata, 249
  spathulata, 247
  succulenta, 248
  uniflora, 248
  viridis, 249
Crepis, 595, 596
  capillaris, 595
  nicaeensis, 596
  pulchra, 596
  runcinata, 595
  setosa, 596
  tectorum, 595
  vesicaria, 596
Crotalaria, 260, 261
  purshii, 260
  rotundifolia, 260
  sagittalis, 260
  spectabilis, 261
Croton, 312, 313
  capitatus, 312
  glandulosus, 312
  monanthogynus, 312
  texensis, 313
Crotonopsis, 313
  elliptica, 313
  linearis, 313
Crypsis schoenoides, 744
Cryptogramma, 14
  crispa, 14
  stelleri, 14
Cryptotaenia canadensis, 346
Ctenium aromaticum, 748
CUCURBITACEAE, 150, 151
Cucurbita foetidissima, 150
Cunila origanoides, 419
Cuphea viscosissima, 293
CUPRESSACEAE, 35
Cuscuta, 386–388
  cephalanthi, 388
  compacta, 386
  coryli, 388
  cuspidata, 386
  epilinum, 388
  epithymum, 388
  europaea, 388
  glomerata, 387
  gronovii, 387

Cuscuta (continued)
  indecora, 387
  megalocarpa, 387
  obtusiflora, 387
  pentagona, 387
  polygonorum, 388
  rostrata, 387
CUSCUTACEAE, 386–388
Cycloloma atriplicifolium, 90
Cymbalaria muralis, 448
Cymophyllus fraserianus, 689
Cymopterus acaulis, 361
Cynoctonum, 361
  mitreola, 361
  sessilifolium, 361
Cynodon dactylon, 749
Cynoglossum, 399
  officinale, 399
  virginianum, 399
    var. boreale, 399
    var. virginianum, 399
Cynosciadium digitatum, 357
Cynosurus cristatus, 699
CYPERACEAE, 631–689
Cyperus, 650–657
  acuminatus, 653
  bipartitus, 657
  brevifolioides, 657
  compressus, 651
  croceus, 654
  dentatus, 651
  diandrus, 657
  difformis, 652
  echinatus, 654
  erythrorhizos, 652
  esculentus, 651
  filicinus, 657
  filiculmis, 656
  flavescens, 656
  flavicomus, 656
  flavus, 654
  fuscus, 652
  grayi, 655
  grayioides, 655
  haspan, 651
  houghtonii, 655
  hystricinus, 654
  iria, 652
  lancastriensis, 653
  lupulinus, 655
  microiria, 652
  odoratus, 656
  plukenetii, 653
  polystachyos, 657
    var. texensis, 657
  pseudovegetus, 653
  refractus, 653
  retrofractus, 654
  retrorsus, 654
  rotundus, 650
  schweinitzii, 655
  setigerus, 650
  squarrosus, 652
  strigosus, 653
  tenuifolius, 657

Cypripedium, 812, 813
  acaule, 813
  arietinum, 812
  calceolus, 812
  candidum, 812
  kentuckiense, 813
  reginae, 812
CYRILLACEAE, 183
Cyrilla racemiflora, 183
Cystopteris, 20
  bulbifera, 20
  fragilis, 20
    var. fragilis, 20
    var. mackayi, 20
  protrusa, 20
Cytisus scoparius, 259

Dactylis glomerata, 705
Dactyloctenium aegyptium, 747
Dalea, 282
  candida, 282
  enneandra, 282
  foliosa, 282
  leporina, 282
  multiflora, 282
  purpurea, 282
  villosa, 282
Dalibarda repens, 235
Danthonia, 731–733
  compressa, 732
  decumbens, 731
  intermedia, 732
  sericea, 733
  spicata, 731
Daphne mezereum, 293
Dasistoma macrophylla, 449
Datura, 382, 383
  metel, 382
  stramonium, 383
  wrightii, 382
Daucus, 350
  carota, 350
  pusillus, 350
Decodon verticillatus, 291
Decumaria barbara, 204
Delphinium, 48
  ambiguum, 48
  carolinianum, 48
  exaltatum, 48
  tricorne, 48
  virescens, 48
DENNSTAEDTIACEAE, 12, 13
Dennstaedtia punctilobula, 12
Deschampsia, 712
  atropurpurea, 712
  cespitosa, 712
  flexuosa, 712
Descurainia, 182
  pinnata, 182
  richardsonii, 182
  sophia, 182
Desmanthus illinoensis, 255
Desmodium, 276–279
  canadense, 277
  canescens, 277

Desmodium (continued)
  ciliare, 278
  cuspidatum, 279
  fernaldii, 277
  glabellum, 278
  glutinosum, 276
  humifusum, 278
  illinoense, 277
  laevigatum, 278
  lineatum, 276
  marilandicum, 278
  nudiflorum, 276
  ochroleucum, 276
  paniculatum, 278
  pauciflorum, 276
  rigidum, 278
  rotundifolium, 276
  sessilifolium, 277
  strictum, 277
  tenuifolium, 277
  viridiflorum, 278
Dianthus, 111, 112
  armeria, 112
  barbatus, 112
  deltoides, 112
  plumarius, 111
DIAPENSIACEAE, 197
Diapensia lapponica, 197
Diarrhena americana, 691
    var. americana, 691
    var. obovata, 691
Dicentra, 66
  canadensis, 66
  cucullaria, 66
  eximia, 66
Dichondra carolinensis, 384
Dicliptera brachiata, 460
Didiplis diandra, 292
Diervilla lonicera, 478
Digitalis, 449
  lanata, 449
  lutea, 449
  purpurea, 449
Digitaria, 767, 768
  ciliaris, 768
  filiformis, 767
  ischaemum, 767
  sanguinalis, 767
  serotina, 767
  violascens, 767
Dioclea multiflora, 287
Diodia, 471, 472
  teres, 471
  virginiana, 472
Dioscorea, 806, 807
  polystachya, 807
  quaternata, 807
  villosa, 806
DIOSCOREACEAE, 806, 807
Diospyros virginiana, 198
Diphylleia cymosa, 62
Diplotaxis, 163
  muralis, 163
  tenuifolia, 163
DIPSACACEAE, 487, 488

Dipsacus, 487
fullonum, 487
laciniatus, 487
sylvestris, 487
Dirca palustris, 293
Disporum, 799
hookeri, 799
lanuginosum, 799
maculatum, 799
Distichlis, 735
spicata, 735
var. spicata, 735
var. stricta, 735
Dodecatheon, 203, 204
meadia, 203
var. brachycarpum, 203
var. frenchii, 203
var. meadia, 203
radicatum, 204
Dolichos lablab, 285
Draba, 168, 169
arabisans, 169
brachycarpa, 169
cana, 169
cuneifolia, 169
glabella, 169
incana, 169
nemorosa, 169
norvegica, 169
ramosissima, 169
reptans, 168
verna, 168
Dracocephalum, 422
parviflorum, 422
thymiflorum, 422
Drosera, 138, 139
anglica, 139
brevifolia, 139
capillaris, 138
filiformis, 138
intermedia, 139
linearis, 139
rotundifolia, 138
DROSERACEAE, 138, 139
Dryopteris, 24–27
campyloptera, 27
carthusiana, 27
celsa, 26
clintoniana, 26
cristata, 25
expansa, 27
filix-mas, 26
fragrans, 24
goldiana, 25
intermedia, 26
marginalis, 25
Duchesnea indica, 221
Dulichium arundinaceum, 658
Dyssodia papposa, 510

EBENACEAE, 198
Echinacea, 497, 498
laevigata, 497
pallida, 498
purpurea, 497
Echinochloa, 761–763
colona, 761
crusgalli, 763
var. crusgalli, 763
var. frumentacea, 763
muricata, 762
var. microstachya, 762
var. muricata, 762
walteri, 762
Echinocystis lobata, 150
Echinodorus, 605, 606
berteroi, 606
cordifolius, 605
tenellus, 605
Echinops sphaerocephalus, 584
Echium vulgare, 401
Eclipta prostrata, 502
Egeria densa, 606
Eichhornia crassipes, 779
ELAEAGNACEAE, 288
Elaeagnus, 288
angustifolia, 288
commutata, 288
umbellata, 288
ELATINACEAE, 125
Elatine, 125
minima, 125
triandra, 125
var. americana, 125
var. brachysperma, 125
var. triandra, 125
Eleocharis, 638–642
acicularis, 639
albida, 639
atropurpurea, 642
baldwinii, 640
caribaea, 642
compressa, 640
equisetoides, 639
fallax, 642
flavescens, 642
intermedia, 640
melanocarpa, 639
microcarpa, 641
ovata, 642
palustris, 641
parvula, 638
pauciflora, 638
quadrangulata, 639
radicans, 639
robbinsii, 639
rostellata, 638
tenuis, 640
var. borealis, 640
var. pseudoptera, 640
var. tenuis, 640
tortilis, 641
tricostata, 640
tuberculosa, 641
vivipara, 641
wolfii, 639
Elephantopus, 578, 579
carolinianus, 578
nudatus, 579
tomentosus, 578
Eleusine indica, 741
Ellisia nyctelea, 394
Elodea, 606
canadensis, 606
nuttallii, 606
Elsholtzia ciliata, 420
Elymus, 724–726
canadensis, 725
diversiglumis, 726
glaucus, 725
hystrix, 726
riparius, 726
trachycaulus, 725
villosus, 725
virginicus, 726
wiegandii, 724
Elytrigia, 727, 728
dasystachya, 727
elongata, 728
pungens, 727
repens, 727
smithii, 727
spicata, 727
EMPETRACEAE, 183, 184
Empetrum, 183
nigrum, 183
rubrum, 183
var. atropurpureum, 183
var. eamesii, 183
Epifagus virginiana, 459
Epigaea repens, 190
Epilobium, 296–298
alpinum, 298
angustifolium, 296
brachycarpum, 298
ciliatum, 298
coloratum, 298
glandulosum, 298
hirsutum, 297
hornemannii, 298
leptophyllum, 297
palustre, 297
parviflorum, 297
strictum, 297
Epipactis helleborine, 814
EQUISETACEAE, 5–7
Equisetum, 5–7
arvense, 7
fluviatile, 6
hyemale, 5
laevigatum, 5
palustre, 6
pratense, 6
scirpoides, 5
sylvaticum, 7
telmateia, 7
variegatum, 5
Eragrostis, 738–740
capillaris, 740
cilianensis, 739
curvula, 739
frankii, 739
hirsuta, 738
hypnoides, 740
intermedia, 738
mexicana, 740
minor, 739

Eragrostis (continued)
  pectinacea, 739
  pilosa, 739
  refracta, 739
  reptans, 740
  spectabilis, 738
  trichodes, 738
Erechtites hieraciifolia, 527
    var. hieraciifolia, 527
    var. megalocarpa, 527
Erianthus, 770, 771
  alopecuroides, 771
  brevibarbis, 770
  contortus, 771
  giganteus, 770
  strictus, 770
Erica, 190
  cinerea, 190
  tetralix, 190
  vagans, 190
ERICACEAE, 184–195
Erigenia bulbosa, 361
Erigeron, 559–561
  acris, 561
  annuus, 561
  glabellus, 559
  hyssopifolius, 561
  philadelphicus, 560
  pulchellus, 560
  quercifolius, 560
  strigosus, 560
  vernus, 560
ERIOCAULACEAE, 621, 622
Eriocaulon, 621
  aquaticum, 621
  compressum, 621
  decangulare, 621
  parkeri, 621
Eriochloa, 762
  contracta, 762
  villosa, 762
Eriogonum, 112
  allenii, 112
  longifolium, 112
Eriophorum, 637, 638
  angustifolium, 638
  brachyantherum, 637
  chamissonis, 637
  gracile, 638
  tenellum, 637
  vaginatum, 637
  virginicum, 638
  viridicarinatum, 638
Erodium, 340, 341
  cicutarium, 340
  moschatum, 341
Eruca sativa, 163
Erucastrum gallicum, 163
Eryngium, 345, 346
  aquaticum, 345
  campestre, 345
  maritimum, 345
  planum, 345
  prostratum, 346
  yuccifolium, 345

Erysimum, 179, 180
  asperum, 179
  cheiranthoides, 179
  hieracifolium, 180
  inconspicuum, 180
  repandum, 179
Erythronium, 793, 794
  albidum, 794
  americanum, 793
  mesochoreum, 794
  propullans, 794
  umbilicatum, 794
Eubotrys, 188, 189
  racemosa, 188
  recurva, 189
Euonymus, 308
  alatus, 308
  americanus, 308
  atropurpureus, 308
  europaeus, 308
  obovatus, 308
Eupatorium, 567–572
  album, 570
    var. album, 570
    var. subvenosum, 570
    var. vaseyi, 570
  altissimum, 572
  aromaticum, 568
  capillifolium, 570
  coelestinum, 568
  cuneifolium, 571
  dubium, 567
  fistulosum, 568
  godfreyanum, 571
  hyssopifolium, 572
  incarnatum, 568
  leucolepis, 570
    var. leucolepis, 570
    var. novae-angliae, 570
  luciae-brauniae, 569
  maculatum, 567
    var. foliosum, 567
    var. maculatum, 567
  mohrii, 571
  perfoliatum, 569
  pilosum, 572
  purpureum, 568
  resinosum, 569
  rotundifolium, 571
    var. cordigerum, 571
    var. ovatum, 571
    var. rotundifolium, 571
  rugosum, 569
  semiserratum, 572
  serotinum, 569
  sessilifolium, 570
  steelei, 567
Euphorbia, 315–321
  ammannioides, 319
  commutata, 316
  corollata, 318
  cyathophora, 317
  cyparissias, 317
  dentata, 317
  esula, 317
  exigua, 316

Euphorbia (continued)
  falcata, 316
  geyeri, 319
  glyptosperma, 320
  helioscopia, 315
  hexagona, 318
  hirta, 320
  humistrata, 321
  ipecacuanhae, 318
  lathyris, 316
  lucida, 317
  maculata, 320
  marginata, 319
  marilandica, 318
  mercurialina, 318
  missurica, 319
  nutans, 320
  obtusata, 315
  peplus, 317
  platyphyllos, 315
  polygonifolia, 319
  prostrata, 320
  purpurea, 315
  serpens, 319
  serpyllifolia, 320
  spathulata, 315
  vermiculata, 320
EUPHORBIACEAE, 311–321
Euphrasia, 456, 457
  disjuncta, 456
  oakesii, 457
  officinalis, 456
  randii, 456
Euthamia, 545, 546
  galetorum, 545
  graminifolia, 545
  gymnospermoides, 546
  leptocephala, 546
  remota, 546
  tenuifolia, 545
    var. microcephala, 545
    var. tenuifolia, 545
Evolvulus, 384
  alsinoides, 384
  nuttallianus, 384

FABACEAE, 258–287
FAGACEAE, 75–81
Fagopyrum, 124
  esculentum, 124
  tataricum, 124
Fagus grandifolia, 75
Falcaria vulgaris, 348
Festuca, 695–697
  altaica, 696
  arundinacea, 695
  brachyphylla, 696
  filiformis, 696
  gigantea, 695
  occidentalis, 695
  ovina, 696
  paradoxa, 695
  pratensis, 695
  rubra, 697
  subverticillata, 695
  trachyphylla, 696

Filago, 563
  arvensis, 563
  germanica, 563
Filipendula, 226, 227
  rubra, 226
  ulmaria, 227
  vulgaris, 226
Fimbristylis, 643, 644
  annua, 644
  autumnalis, 643
  caroliniana, 643
  castanea, 643
  miliacea, 643
  perpusilla, 643
  puberula, 644
  vahlii, 643
Firmiana simplex, 131
Flaveria campestris, 510
Floerkea proserpinacoides, 341
Foeniculum vulgare, 351
Forestiera, 436
  acuminata, 436
  ligustrina, 436
Forsythia, 436
  suspensa, 436
  viridissima, 436
Fragaria, 220, 221
  chiloensis, 221
  vesca, 220
    var. americana, 220
    var. vesca, 220
  virginiana, 220
Frasera caroliniensis, 367
Fraxinus, 436, 437
  americana, 436
  caroliniana, 437
  nigra, 437
  pennsylvanica, 436
  profunda, 437
  quadrangulata, 437
Froelichia, 98
  floridana, 98
  gracilis, 98
Fuirena, 658, 659
  breviseta, 659
  pumila, 659
  scirpoidea, 658
  simplex, 658
  squarrosa, 658
FUMARIACEAE, 66, 67
Fumaria officinalis, 66

Gaillardia, 498, 499
  aestivalis, 499
  aristata, 499
  pulchella, 498
Galactia, 287
  macreei, 287
  regularis, 287
  volubilis, 287
Galax aphylla, 197
Galeopsis, 424, 425
  ladanum, 425
  tetrahit, 424
Galinsoga, 509
  parviflora, 509

Galinsoga (continued)
  quadriradiata, 509
Galium, 473–477
  aparine, 477
  asprellum, 474
  boreale, 473
  circaezans, 476
  concinnum, 475
  glaucum, 473
  hispidulum, 476
  kamtschaticum, 476
  labradoricum, 475
  lanceolatum, 476
  latifolium, 473
  mollugo, 474
  obtusum, 475
    var. filifolium, 475
    var. obtusum, 475
  odoratum, 477
  palustre, 474
  parisiense, 477
  pedemontanum, 477
  pilosum, 476
  sylvaticum, 474
  tinctorium, 476
  trifidum, 475
  triflorum, 477
  uniflorum, 477
  verum, 474
    subsp. verum, 474
    subsp. wirtgenii, 474
Gaultheria, 190
  hispidula, 190
  procumbens, 190
Gaura, 302
  biennis, 302
  coccinea, 302
  filipes, 302
  parviflora, 302
Gaylussacia, 194, 195
  baccata, 194
  brachycera, 195
  dumosa, 195
  frondosa, 194
Gelsemium sempervirens, 361
Genista tinctoria, 259
Gentiana, 364–366
  andrewsii, 365
  austromontana, 365
  autumnalis, 364
  catesbaei, 364
  clausa, 365
  decora, 364
  flavida, 366
  linearis, 365
  puberulenta, 364
  rubricaulis, 365
  saponaria, 364
  villosa, 366
GENTIANACEAE, 362–367
Gentianella, 366
  amarella, 366
  quinquefolia, 366
    var. occidentalis, 366
    var. quinquefolia, 366

Gentianopsis, 366
  crinita, 366
  virgata, 366
Geocaulon lividum, 307
GERANIACEAE, 338–341
Geranium, 338–341
  bicknellii, 340
  carolinianum, 339
  columbinum, 339
  dissectum, 339
  maculatum, 338
  molle, 340
  pratense, 339
  pusillum, 340
  pyrenaicum, 341
  robertianum, 339
  rotundifolium, 340
  sanguineum, 338
  sibiricum, 339
Geum, 227–229
  aleppicum, 228
  canadense, 227
  laciniatum, 227
  macrophyllum, 228
  peckii, 229
  rivale, 228
  triflorum, 228
  urbanum, 228
  vernum, 227
  virginianum, 227
Gilia rubra, 392
Glaucium flavum, 64
Glaux maritima, 202
Glechoma hederacea, 421
Gleditsia, 256, 257
  aquatica, 256
  triacanthos, 257
Glinus lotoides, 100
Glyceria, 706–708
  acutiflora, 706
  borealis, 706
  canadensis, 707
  fluitans, 707
  grandis, 708
  ×laxa, 707
  melicaria, 707
  obtusa, 707
  septentrionalis, 706
  striata, 708
Glycyrrhiza lepidota, 267
Gnaphalium, 564, 565
  helleri, 565
  macounii, 565
  obtusifolium, 565
  purpureum, 565
    var. falcatum, 565
    var. purpureum, 565
  supinum, 564
  sylvaticum, 564
  uliginosum, 565
Gomphrena globosa, 97
Goodyera, 817, 818
  oblongifolia, 817
  pubescens, 817
  repens, 818
  tesselata, 818

Gossypium hirsutum, 131
Gratiola, 438, 439
  aurea, 439
  neglecta, 439
  pilosa, 438
  ramosa, 439
  virginiana, 439
  viscidula, 439
Grindelia, 531, 532
  lanceolata, 531
  squarrosa, 532
    var. quasiperennis, 532
    var. serrulata, 532
    var. squarrosa, 532
GROSSULARIACEAE, 205–207
Guizotia abyssinica, 503
Gutierrezia sarothrae, 546
Gymnocarpium, 22
  dryopteris, 22
  robertianum, 22
Gymnocladus dioicus, 256
Gymnopogon, 748
  ambiguus, 748
  brevifolius, 748
Gypsophila, 111
  muralis, 111
  paniculata, 111
  scorzonerifolia, 111

Habenaria, 819–822
  blephariglottis, 821
    var. blephariglottis, 821
    var. conspicua, 821
  ciliaris, 821
  clavellata, 819
  cristata, 821
  dilatata, 820
  flava, 819
  hookeri, 821
  hyperborea, 819
  integra, 820
  lacera, 822
  leucophaea, 822
    var. leucophaea, 822
    var. praeclara, 822
  nivea, 819
  obtusata, 820
  orbiculata, 820
  peramoena, 822
  psycodes, 822
    var. grandiflora, 822
    var. psycodes, 822
  unalascensis, 820
  viridis, 819
Hackelia, 398
  deflexa, 398
  virginiana, 398
HAEMODORACEAE, 781
Halenia deflexa, 367
Halesia tetraptera, 199
HALORAGACEAE, 289–291
HAMAMELIDACEAE, 68
Hamamelis virginiana, 68
Haplopappus, 529, 530
  ciliatus, 529

Haplopappus (continued)
  divaricatus, 530
  spinulosus, 529
Harrimanella hypnoides, 187
Hedeoma
  hispida, 412
  pulegioides, 412
Hedera helix, 343
Hedyotis, 469–471
  boscii, 471
  caerulea, 470
  canadensis, 469
  crassifolia, 470
  longifolia, 469
  michauxii, 471
  nigricans, 470
  nuttalliana, 470
  purpurea, 470
  uniflora, 470
Hedysarum alpinum, 275
Helenium, 499, 500
  amarum, 500
  autumnale, 499
  brevifolium, 499
  flexuosum, 500
  virginicum, 499
Helianthemum, 139, 140
  bicknellii, 140
  canadense, 139
  dumosum, 139
  propinquum, 140
Helianthus, 489–494
  angustifolius, 494
  annuus, 489
  atrorubens, 492
  decapetalus, 490
  divaricatus, 490
  eggertii, 491
  giganteus, 493
  grosseserratus, 494
  hirsutus, 490
  ×laetiflorus, 492
  laevigatus, 491
  maximilianii, 494
  microcephalus, 491
  mollis, 491
  occidentalis, 492
  pauciflorus, 493
    var. pauciflorus, 493
    var. subrhomboideus, 493
  petiolaris, 489
  salicifolius, 493
  silphioides, 492
  strumosus, 490
  tuberosus, 492
Helictotrichon pubescens, 709
Heliopsis helianthoides, 503
Heliotropium, 395
  amplexicaule, 395
  curassavicum, 395
  europaeum, 395
  indicum, 395
  tenellum, 395
Helleborus viridis, 45
Helonias bullata, 786

Hemerocallis, 789
  fulva, 789
  lilioasphodelus, 789
Hemianthus micranthemoides, 441
Hemicarpha micrantha, 658
Hepatica, 50
  acutiloba, 50
  americana, 50
Heracleum, 357, 358
  lanatum, 357
  mantegazzianum, 358
  sphondylium, 358
Herniaria glabra, 107
Hesperis matronalis, 179
Heteranthera, 779, 780
  limosa, 780
  multiflora, 780
  reniformis, 779
  rotundifolia, 780
Heterotheca subaxillaris, 531
Heuchera, 215, 216
  americana, 215
  ×hirsuticaulis, 216
  ×hispida, 215
  longiflora, 216
  parviflora, 215
  pubescens, 216
  richardsonii, 216
  villosa, 215
Hexalectris spicata, 824
Hexastylis, 40
  arifolia, 40
    var. arifolia, 40
    var. ruthii, 40
  shuttleworthii, 40
  virginica, 40
Hibiscus, 132, 133
  laevis, 132
  moscheutos, 132
    var. moscheutos, 132
    var. occidentalis, 132
  syriacus, 132
  trionum, 133
Hieracium, 592–595
  aurantiacum, 592
  caespitosum, 592
  flagellare, 592
  floribundum, 592
  gronovii, 594
  kalmii, 594
    var. fasciculatum, 594
    var. kalmii, 594
  lachenalii, 593
  longipilum, 594
  murorum, 593
  paniculatum, 595
  pilosella, 592
  piloselloides, 593
  robinsonii, 593
  sabaudum, 593
  scabrum, 595
  traillii, 593
  umbellatum, 594
  venosum, 594
Hierochloe, 713
  alpina, 713

Hierochloe *(continued)*
odorata, 713
HIPPOCASTANACEAE, 330, 331
HIPPURIDACEAE, 431
Hippuris vulgaris, 431
Holcus, 712
lanatus, 712
mollis, 712
Holosteum umbellatum, 101
Honckenya peploides, 101
Hordeum, 729
brachyantherum, 729
geniculatum, 729
jubatum, 729
leporinum, 729
murinum, 729
pusillum, 729
vulgare, 729
Hosta, 788
lancifolia, 788
ventricosa, 788
Hottonia inflata, 203
Hudsonia, 140
ericoides, 140
tomentosa, 140
Humulus, 70
japonicus, 70
lupulus, 70
var. lupuloides, 70
var. lupulus, 70
var. pubescens, 70
Hybanthus concolor, 142
Hydrangea, 205
arborescens, 205
paniculata, 205
quercifolia, 205
HYDRANGEACEAE, 204, 205
Hydrastis canadensis, 61
Hydrilla verticillata, 606
Hydrocharis morsus-ranae, 606
HYDROCHARITACEAE, 606, 607
Hydrocotyle, 344
americana, 344
ranunculoides, 344
sibthorpioides, 344
umbellata, 344
verticillata, 344
var. triradiata, 344
var. verticillata, 344
Hydrolea, 394, 395
ovata, 394
quadrivalvis, 395
uniflora, 395
HYDROPHYLLACEAE, 392–395
Hydrophyllum, 393, 394
appendiculatum, 394
canadense, 394
macrophyllum, 393
virginianum, 393
Hymenocallis caroliniana, 803
Hymenopappus scabiosaeus, 516
HYMENOPHYLLACEAE, 11
Hymenoxys herbacea, 500
Hyoscyamus niger, 382
Hypericum, 126–130
adpressum, 127

Hypericum *(continued)*
boreale, 129
canadense, 129
densiflorum, 126
denticulatum, 129
var. acutifolium, 129
var. denticulatum, 129
dolabriforme, 127
drummondii, 130
ellipticum, 127
frondosum, 127
gentianoides, 130
gymnanthum, 129
hypericoides, 126
kalmianum, 126
lobocarpum, 126
majus, 129
mitchellianum, 128
mutilum, 129
nudiflorum, 127
perforatum, 128
prolificum, 126
pseudomaculatum, 128
punctatum, 128
pyramidatum, 128
setosum, 128
sphaerocarpum, 127
stans, 126
stragulum, 126
Hypochaeris, 600
glabra, 600
radicata, 600
Hypoxis, 802
hirsuta, 802
longii, 802
micrantha, 802
sessilis, 802
Hyssopus officinalis, 414

Iberis, 167
amara, 167
umbellata, 167
Ilex, 309, 310
amelanchier, 310
coriacea, 309
decidua, 310
glabra, 309
laevigata, 310
montana, 310
opaca, 309
verticillata, 309
vomitoria, 309
Illiamna remota, 133
Impatiens, 341, 342
balsamina, 342
capensis, 341
glandulifera, 342
pallida, 341
parviflora, 341
Inula, 563
graveolens, 563
helenium, 563
Iodanthus pinnatifidus, 179
Ipomoea, 385, 386
coccinea, 386
hederacea, 385

Ipomoea *(continued)*
lacunosa, 386
pandurata, 386
purpurea, 386
quamoclit, 386
Iresine rhizomatosa, 98
IRIDACEAE, 807–811
Iris, 809–811
brevicaulis, 810
cristata, 809
fulva, 810
germanica, 811
lacustris, 810
prismatica, 810
pseudacorus, 810
setosa, 811
verna, 809
versicolor, 811
virginica, 811
var. shrevei, 811
var. virginica, 811
Isanthus brachiatus, 409
Isatis tinctoria, 166
ISOETACEAE, 4
Isoetes, 4
butleri, 4
echinospora, 4
engelmannii, 4
lacustris, 4
melanopoda, 4
riparia, 4
tuckermanii, 4
virginica, 4
Isopyrum biternatum, 59
Isotria, 823
medeoloides, 823
verticillata, 823
Itea virginica, 205
Iva, 514, 515
annua, 515
frutescens, 514
imbricata, 515
xanthiifolia, 514
Ixeris stolonifera, 591

Jacquemontia tamnifolia, 385
Jasione montana, 466
Jeffersonia diphylla, 62
JUGLANDACEAE, 72–74
Juglans, 72
cinerea, 72
nigra, 72
JUNCACEAE, 622–631
JUNCAGINACEAE, 608
Juncus, 622–629
acuminatus, 629
alpinoarticulatus, 628
arcticus, 624
articulatus, 629
biflorus, 626
brachycarpus, 629
brachycephalus, 627
brachyphyllus, 622
brevicaudatus, 627
bufonius, 624
var. bufonius, 624

Juncus (continued)
  var. halophilus, 624
  bulbosus, 628
  caesariensis, 626
  canadensis, 626
  compressus, 622
  conglomeratus, 624
  coriaceus, 623
  debilis, 629
  diffusissimus, 629
  effusus, 625
  elliottii, 629
  ensifolius, 625
  filiformis, 625
  gerardii, 622
  greenei, 623
  gymnocarpus, 624
  inflexus, 624
  longistylis, 626
  marginatus, 626
  megacephalus, 627
  militaris, 628
  nodatus, 629
  nodosus, 628
  pelocarpus, 628
  pylaei, 625
  repens, 626
  roemerianus, 625
  scirpoides, 627
  secundus, 622
  stygius, 625
  subcaudatus, 626
  subtilis, 628
  tenuis, 623
    var. dichotomus, 623
    var. dudleyi, 623
    var. tenuis, 623
  torreyi, 627
  trifidus, 622
    var. carolinianus, 622
    var. trifidus, 622
  validus, 627
  vaseyi, 623
Juniperus, 35
  communis, 35
  horizontalis, 35
  virginiana, 35
Justicia, 460
  americana, 460
  ovata, 460

Kallstroemia parviflora, 337
Kalmia, 186, 187
  angustifolia, 187
  latifolia, 186
  polifolia, 187
Kickxia, 448
  elatine, 448
  spuria, 448
Knautia arvensis, 488
Kochia scoparia, 90
Koeleria pyramidata, 709
Koelreuteria paniculata, 330
Kosteletzkya virginica, 133
Krigia, 597, 598
  biflora, 597

Krigia (continued)
  dandelion, 598
  oppositifolia, 598
  virginica, 598
Kuhnia eupatorioides, 573
    var. corymbulosa, 573
    var. eupatorioides, 573
    var. texana, 573

Lachnanthes caroliniana, 781
Lachnocaulon anceps, 622
Lactuca, 589–591
  biennis, 590
  canadensis, 590
  floridana, 590
  hirsuta, 589
  ludoviciana, 589
  muralis, 589
  pulchella, 590
  saligna, 591
  serriola, 591
    var. integrata, 591
LAMIACEAE, 408–430
Lamium, 423, 424
  album, 423
  amplexicaule, 424
  maculatum, 423
  purpureum, 424
Laportea canadensis, 71
Lappula, 398
  redowskii, 398
  squarrosa, 398
Lapsana communis, 599
LARDIZABALACEAE, 63
Larix laricina, 32
Lathyrus, 274, 275
  hirsutus, 275
  latifolius, 275
  maritimus, 274
  ochroleucus, 274
  palustris, 275
  pratensis, 274
  tuberosus, 275
  venosus, 275
LAURACEAE, 38, 39
Leavenworthia, 170, 171
  exigua, 170
  torulosa, 171
  uniflora, 170
Lechea, 140–142
  intermedia, 142
  maritima, 141
  minor, 140
  mucronata, 141
  pulchella, 141
  racemulosa, 141
  stricta, 141
  tenuifolia, 141
Ledum groenlandicum, 184
Leersia, 690
  hexandra, 690
  lenticularis, 690
  oryzoides, 690
  virginica, 690
Leiophyllum buxifolium, 184

Lemna, 616
  gibba, 616
  minor, 616
  minuta, 616
  perpusilla, 616
  trisulca, 616
  valdiviana, 616
LEMNACEAE, 616
LENTIBULARIACEAE, 462–464
Leontodon, 601
  autumnalis, 601
  hispidus, 601
  taraxacoides, 601
Leonurus, 424
  cardiaca, 424
  marrubiastrum, 424
  sibiricus, 424
Lepidium, 164, 165
  campestre, 164
  densiflorum, 165
  latifolium, 165
  perfoliatum, 164
  ruderale, 165
  sativum, 165
  virginicum, 164
Leptochloa, 736, 737
  fascicularis, 736
    var. acuminata, 736
    var. fascicularis, 736
    var. maritima, 736
  filiformis, 737
  panicoides, 737
  uninervia, 737
Leptoloma cognatum, 768
Lespedeza, 279–281
  angustifolia, 280
  bicolor, 279
  capitata, 280
  cuneata, 281
  hirta, 281
  intermedia, 280
  leptostachya, 281
  ×nuttallii, 281
  procumbens, 279
  repens, 279
  ×simulata, 281
  stipulacea, 281
  striata, 281
  stuevei, 280
  thunbergii, 279
  violacea, 280
  virginica, 280
Lesquerella, 170
  globosa, 170
  ludoviciana, 170
Leucojum aestivum, 803
Leucophysalis grandiflora, 376
Leucospora multifida, 439
Leucothoe, 188
  axillaris, 188
  walteri, 188
Levisticum officinale, 359
Leymus, 728
  arenarius, 728
  cinereus, 728
  mollis, 728

Liatris, 574–577
  aspera, 575
  cylindracea, 576
  graminifolia, 576
  ligulistylis, 575
  microcephala, 576
  punctata, 577
  pycnostachya, 576
  regimontis, 575
  scariosa, 574
    var. nieuwlandii, 574
    var. novae-angliae, 574
    var. scariosa, 574
  spicata, 576
  squarrosa, 576
    var. hirsuta, 576
    var. squarrosa, 576
  squarrulosa, 575
  turgida, 575
Ligusticum, 354
  canadense, 354
  scothicum, 354
Ligustrum, 435
  amurense, 435
  obtusifolium, 435
  ovalifolium, 435
  sinense, 435
  vulgare, 435
Lilaeopsis, 357
  attenuata, 357
  chinensis, 357
LILIACEAE, 780–803
Lilium, 791–793
  bulbiferum, 791
  canadense, 792
  catesbaei, 791
  grayi, 791
  lancifolium, 793
  michauxii, 792
  michiganense, 793
  philadelphicum, 792
  superbum, 792
LIMNANTHACEAE, 341
Limnobium spongia, 607
Limonium carolinianum, 125
Limosella, 441
  aquatica, 441
  subulata, 441
LINACEAE, 326, 327
Linaria, 447, 448
  canadensis, 448
    var. canadensis, 448
    var. texana, 448
  dalmatica, 447
  genistifolia, 447
  repens, 447
  vulgaris, 447
Lindera, 39
  benzoin, 39
Lindernia, 441
  dubia, 441
    var. anagallidea, 441
    var. dubia, 441
  procumbens, 441
Linnaea borealis, 482

Linum, 326, 327
  catharticum, 326
  floridanum, 326
  intercursum, 326
  medium, 327
  perenne, 326
  rigidum, 327
  striatum, 327
  sulcatum, 326
  usitatissimum, 326
  virginianum, 327
Liparis, 825
  liliifolia, 825
  loeselii, 825
Lipocarpha maculata, 658
Liquidambar styraciflua, 68
Liriodendron tulipifera, 38
Listera, 814, 815
  auriculata, 814
  australis, 814
  convallarioides, 814
  cordata, 814
  ovata, 815
  smallii, 814
Lithospermum, 399, 400
  arvense, 400
  canescens, 399
  caroliniense, 399
  incisum, 400
  latifolium, 400
  officinale, 400
  tuberosum, 400
Littorella uniflora, 434
LOASACEAE, 151
Lobelia, 466–469
  boykinii, 467
  canbyi, 467
  cardinalis, 469
  dortmanna, 468
  elongata, 468
  georgiana, 468
  glandulosa, 468
  inflata, 467
  kalmii, 467
  nuttallii, 467
  puberula, 468
  siphilitica, 469
  spicata, 466
    var. leptostachys, 466
    var. scaposa, 466
    var. spicata, 466
Lobularia maritima, 168
LOGANIACEAE, 361
Loiseleuria procumbens, 187
Lolium, 698
  perenne, 698
    var. aristatum, 698
    var. perenne, 698
  persicum, 698
  temulentum, 698
Lomatium, 360
  foeniculaceum, 360
  orientale, 360
Lomatogonium rotatum, 367
Lonicera, 478–480
  ×bella, 478

Lonicera (continued)
  caerulea, 478
  canadensis, 478
  caprifolium, 480
  dioica, 480
  flava, 480
  hirsuta, 480
  involucrata, 478
  japonica, 479
  maackii, 479
  morrowii, 479
  oblongifolia, 479
  periclymenum, 479
  prolifera, 480
  sempervirens, 480
  tatarica, 478
  xylosteum, 479
Lophiola aurea, 781
Lotus, 268
  corniculatus, 268
  purshianus, 268
  tenuis, 268
Ludwigia, 294–296
  alata, 295
  alternifolia, 295
  brevipes, 294
  decurrens, 294
  glandulosa, 296
  hirtella, 295
  leptocarpa, 294
  linearis, 296
  palustris, 296
  peploides, 294
  pilosa, 295
  polycarpa, 295
  sphaerocarpa, 296
  uruguayensis, 294
Lunaria, 168
  annua, 168
  rediviva, 168
Lupinus perennis, 260
Luzula, 630, 631
  acuminata, 630
    var. acuminata, 630
    var. carolinae, 630
  bulbosa, 630
  campestris, 630
  confusa, 630
  echinata, 631
  luzuloides, 631
  multiflora, 630
  pallidula, 630
  parviflora, 630
  spicata, 631
Lychnis, 107, 108
  chalcedonica, 108
  coronaria, 107
  flos-cuculi, 107
  viscaria, 107
Lycium barbarum, 381
Lycopersicon esculentum, 381
LYCOPODIACEAE, 1–3
Lycopodium, 1–3
  alopecuroides, 1
  alpinum, 3
  annotinum, 2

Lycopodium *(continued)*
  appressum, 2
  carolinianum, 2
  clavatum, 2
  complanatum, 3
  digitatum, 3
  inundatum, 2
  lucidulum, 1
  obscurum, 2
  porophilum, 1
  prostratum, 1
  sabinifolium, 3
  selago, 1
  sitchense, 3
  tristachyum, 3
Lycopus, 419, 420
  americanus, 420
  amplectens, 420
  angustifolius, 420
  asper, 420
  europaeus, 420
  rubellus, 420
  uniflorus, 419
  virginicus, 419
Lygodesmia, 588, 589
  juncea, 588
  rostrata, 589
Lygodium palmatum, 11
Lyonia, 187, 188
  ligustrina, 187
  lucida, 188
  mariana, 188
Lysimachia, 199–202
  ciliata, 199
  hybrida, 200
  lanceolata, 200
  nummularia, 201
  ×producta, 202
  punctata, 201
  quadriflora, 200
  quadrifolia, 201
  radicans, 200
  terrestris, 202
  thyrsiflora, 201
  tonsa, 200
  vulgaris, 201
LYTHRACEAE, 291–293
  Lythrum, 292
  alatum, 292
  hyssopifolia, 292
  lineare, 292
  salicaria, 292
  virgatum, 292

Maclura pomifera, 70
Madia, 510
  glomerata, 510
  sativa, 510
Magnolia, 36, 37
  acuminata, 36
  fraseri, 37
  grandiflora, 36
  macrophylla, 37
  tripetala, 37
  virginiana, 36
MAGNOLIACEAE, 36–38

Maianthemum canadense, 800
Malaxis, 825
  monophyllos, 825
  paludosa, 825
  spicata, 825
  unifolia, 825
Malcolmia maritima, 179
Malva, 134, 135
  alcea, 134
  moschata, 134
  neglecta, 135
  parviflora, 135
  rotundifolia, 135
  sylvestris, 135
  verticillata, 135
MALVACEAE, 131–138
Malvastrum hispidum, 137
Marrubium vulgare, 412
Marshallia, 509
  grandiflora, 509
  obovata, 509
  trinervia, 509
Marsilea, 30
  quadrifolia, 30
  vestita, 30
MARSILEACEAE, 30
Matelea, 375, 376
  carolinensis, 375
  decipiens, 375
  gonocarpa, 375
  obliqua, 376
  suberosa, 375
Matricaria, 519
  maritima, 519
  matricarioides, 519
  recutita, 519
Matteuccia struthiopteris, 29
Mazus, 441
  miquelii, 441
  pumilus, 441
Mecardonia acuminata, 438
Medeola virginiana, 797
Medicago, 271, 272
  lupulina, 272
  minima, 272
  polymorpha, 272
  sativa, 271
    subsp. falcata, 271
    subsp. sativa, 271
Meehania cordata, 421
Melampyrum lineare, 457
    var. americanum, 457
    var. latifolium, 457
    var. lineare, 457
    var. pectinatum, 457
Melanthium, 784
  latifolium, 784
  virginicum, 784
MELASTOMATACEAE, 303, 304
Melia azedarach, 335
MELIACEAE, 335
Melica, 708
  mutica, 708
  nitens, 708
  smithii, 708

Melilotus, 271
  albus, 271
  altissimus, 271
  indicus, 271
  officinalis, 271
Melissa officinalis, 412
Melothria pendula, 150
MENISPERMACEAE, 63
Menispermum canadense, 63
Mentha, 416–418
  aquatica, 417
  arvensis, 416
  ×citrata, 417
  ×gentilis, 417
  longifolia, 417
  ×piperita, 417
  pulegium, 417
  ×rotundifolia, 418
  spicata, 418
  suaveolens, 418
  ×verticillata, 417
Mentzelia, 151
  decapetala, 151
  oligosperma, 151
MENYANTHACEAE, 388, 389
Menyanthes trifoliata, 388
Menziesia pilosa, 184
Mercurialis annua, 315
Mertensia, 396
  maritima, 396
  paniculata, 396
  virginica, 396
Metaplexis japonica, 374
Micranthemum umbrosum, 441
Microseris cuspidata, 597
Microstegium vimineum, 772
Mikania scandens, 573
Milium effusum, 694
MIMOSACEAE, 255
Mimosa quadrivalvis, 255
    var. angustata, 255
    var. nuttallii, 255
Mimulus, 440
  alatus, 440
  glabratus, 440
  guttatus, 440
  moschatus, 440
  ringens, 440
Mirabilis, 85
  albida, 85
  hirsuta, 85
  jalapa, 85
  linearis, 85
  nyctaginea, 85
Miscanthus, 771
  sacchariflorus, 771
  sinensis, 771
Mitchella repens, 471
Mitella, 214
  diphylla, 214
  nuda, 214
Modiola caroliniana, 134
Molinia caerulea, 732
MOLLUGINACEAE, 100
Mollugo verticillata, 100

Monarda, 427, 428
  citriodora, 427
  clinopodia, 427
  didyma, 428
  fistulosa, 427
  media, 428
  punctata, 427
  russeliana, 428
  virgata, 428
Moneses uniflora, 196
Monolepis nuttalliana, 90
Monotropa, 196
  hypopithys, 196
  uniflora, 196
MONOTROPACEAE, 196,197
Monotropsis odorata, 196
Montia, 99
  chamissoi, 99
  fontana, 99
MORACEAE, 70
Morus, 70
  alba, 70
  rubra, 70
Mosla dianthera, 419
Muhlenbergia, 744–747
  asperifolia, 747
  bushii, 746
  capillaris, 746
  cuspidata, 747
  frondosa, 745
  glabrifloris, 745
  glomerata, 746
  mexicana, 744
  racemosa, 745
  richardsonis, 746
  schreberi, 746
  sobolifera, 744
  sylvatica, 744
  tenuiflora, 744
  torreyana, 747
  uniflora, 747
Murdannia keisak, 620
Muscari, 788
  botryoides, 788
  comosum, 788
  racemosum, 788
Myagrum perfoliatum, 167
Myosotis, 396, 397
  arvensis, 397
  discolor, 397
  laxa, 397
  macrosperma, 397
  micrantha, 397
  scorpioides, 396
  sylvatica, 397
  verna, 397
Myosurus minimus, 59
Myrica, 75
  cerifera, 75
  gale, 75
  heterophylla, 75
  pensylvanica, 75
MYRICACEAE, 75
Myriophyllum, 289, 290
  alterniflorum, 289
  aquaticum, 289

Myriophyllum (continued)
  farwellii, 289
  heterophyllum, 290
  humile, 289
  pinnatum, 289
  sibiricum, 290
  spicatum, 290
  tenellum, 289
  verticillatum, 290

NAJADACEAE, 614
Najas, 614
  flexilis, 614
  gracillima, 614
  guadalupensis, 614
  marina, 614
  minor, 614
  muenscheri, 614
Napaea dioica, 136
Narcissus, 803
  poeticus, 803
  pseudonarcissus, 803
Nardus stricta, 692
Narthecium americanum, 785
Nelumbo lutea, 42
NELUMBONACEAE, 42
Nemopanthus, 310
  collinus, 310
  mucronatus, 310
Nemophila triloba, 394
Nepeta cataria, 421
Neslia paniculata, 167
Nestronia umbellula, 307
Nicandra physalodes, 376
Nicotiana, 382, 383
  rustica, 383
  tabacum, 382
Nigella damascena, 45
Notholaena dealbata, 16
Nothoscordum bivalve, 789
Nuphar, 42, 43
  advena, 43
  microphylla, 42
  sagittifolia, 43
  variegata, 42
NYCTAGINACEAE, 85
Nymphaea, 43
  odorata, 43
  tetragona, 43
NYMPHAEACEAE, 42, 43
Nymphoides, 389
  aquatica, 389
  cordata, 389
  peltata, 389
Nyssa, 306
  aquatica, 306
  sylvatica, 306
    var. biflora, 306
    var. sylvatica, 306

Obolaria virginica, 367
Ocimum basilicum, 430
Odontites serotina, 456
Oenanthe aquatica, 354
Oenothera, 299–302
  argillicola, 299

Oenothera (continued)
  biennis, 299
    var. biennis, 299
    var. canescens, 299
  clelandii, 300
  fruticosa, 302
  glazioviana, 299
  humifusa, 300
  laciniata, 300
  linifolia, 301
  macrocarpa, 301
  nuttallii, 300
  parviflora, 299
  perennis, 301
  pilosella, 301
  rhombipetala, 299
  speciosa, 300
  tetragona, 301
  triloba, 300
OLEACEAE, 435–437
ONAGRACEAE, 294–303
Onobrychis viciifolia, 276
ONOCLEACEAE, 29
Onoclea sensibilis, 29
Ononis campestris, 269
Onopordum acanthium, 584
Onosmodium, 401
  molle, 401
    var. hispidissimum, 401
    var. molle, 401
    var. occidentale, 401
  virginianum, 401
OPHIOGLOSSACEAE, 8, 9
Ophioglossum, 9
  engelmannii, 9
  vulgatum, 9
    var. pseudopodum, 9
    var. pycnostichum, 9
Oplopanax horridus, 343
Opuntia, 86
  fragilis, 86
  humifusa, 86
  macrorhiza, 86
Orbexilum, 283
  onobrychis, 283
  pedunculatum, 283
    var. gracile, 283
    var. pedunculatum, 283
ORCHIDACEAE, 812–827
Orchis, 818
  rotundifolia, 818
  spectabilis, 818
Origanum vulgare, 413
Ornithogalum, 787
  nutans, 787
  umbellatum, 787
Ornithopus sativus, 268
OROBANCHACEAE, 458, 459
Orobanche, 458, 459
  fasciculata, 459
  ludoviciana, 458
  minor, 459
  ramosa, 459
  uniflora, 458
Orontium aquaticum, 615
Orthocarpus luteus, 458

Oryza sativa, 690
Oryzopsis, 692, 693
  asperifolia, 693
  canadensis, 692
  hymenoides, 692
  pungens, 692
  racemosa, 693
Osmanthus americanus, 435
Osmorhiza, 347
  berterii, 347
  claytonii, 347
  depauperata, 347
  longistylis, 347
Osmunda, 10
  cinnamomea, 10
  claytoniana, 10
  regalis, 10
OSMUNDACEAE, 10
Ostrya virginiana, 82
OXALIDACEAE, 337, 338
Oxalis, 337, 338
  acetosella, 338
  corniculata, 337
  dillenii, 337
  grandis, 338
  macrantha, 338
  stricta, 337
  violacea, 338
Oxydendrum arboreum, 189
Oxypolis, 356
  canbyi, 356
  rigidior, 356
  ternata, 356
Oxyria digyna, 113
Oxytropis, 267
  campestris, 267
    var. chartacea, 267
    var. johannensis, 267
  lambertii, 267
  splendens, 267
  viscida, 267

Pachysandra procumbens, 311
Panax, 343
  quinquefolius, 343
  trifolius, 343
Panicum, 751–761
  aciculare, 759
  amarulum, 752
  amarum, 752
  anceps, 751
  boreale, 757
  boscii, 760
  capillare, 753
  clandestinum, 759
  columbianum, 755
  commonsianum, 755
  commutatum, 759
  consanguineum, 760
  depauperatum, 754
  dichotomiflorum, 753
  dichotomum, 757
  ensifolium, 758
  flexile, 753
  hemitomon, 751
  hians, 752

Panicum (continued)
  lancearium, 755
  lanuginosum, 756
    var. fasciculatum, 756
    var. implicatum, 756
    var. lanuginosum, 756
    var. lindheimeri, 756
    var. tennesseense, 756
  latifolium, 760
  laxiflorum, 754
  leibergii, 761
  leucothrix, 756
  linearifolium, 754
  longifolium, 752
  miliaceum, 753
  obtusum, 751
  oligosanthes, 761
  philadelphicum, 754
  polyanthes, 760
  ravenelii, 760
  rigidulum, 752
  scabriusculum, 758
  scoparium, 759
  sphaerocarpon, 760
  spretum, 755
  strigosum, 754
  tenue, 758
  texanum, 751
  verrucosum, 752
  villosissimum, 757
  virgatum, 753
  wilcoxianum, 760
  wrightianum, 756
  xanthophysum, 761
  yadkinense, 758
Papaver, 65
  dubium, 65
  rhoeas, 65
  somniferum, 65
PAPAVERACEAE, 64, 65
Parapholis incurva, 706
Parietaria pensylvanica, 72
Parnassia, 216, 217
  asarifolia, 216
  glauca, 216
  grandifolia, 217
  palustris, 217
  parviflora, 217
Paronychia, 106
  argyrocoma, 106
  canadensis, 106
  fastigiata, 106
  riparia, 106
  virginica, 106
Parthenium, 513, 514
  auriculatum, 513
  hispidum, 513
  hysterophorus, 514
  integrifolium, 514
Parthenocissus, 323
  quinquefolia, 323
  tricuspidata, 323
  vitacea, 323
Paspalum, 763–765
  bifidum, 765
  boscianum, 765

Paspalum (continued)
  dilatatum, 764
  dissectum, 763
  distichum, 765
  floridanum, 764
  fluitans, 763
  laeve, 764
  praecox, 765
  pubiflorum, 764
  setaceum, 765
    var. ciliatifolium, 765
    var. muhlenbergii, 765
    var. psammophilum, 765
    var. setaceum, 765
    var. stramineum, 765
  urvillei, 764
Passiflora, 149
  incarnata, 149
  lutea, 149
PASSIFLORACEAE, 149
Pastinaca sativa, 357
Paulownia tomentosa, 461
Paxistima canbyi, 309
PEDALIACEAE, 460, 461
Pedicularis, 457
  canadensis, 457
  furbishiae, 457
  lanceolata, 457
  palustris, 457
Pediomelum, 283, 284
  argophyllum, 284
  canescens, 283
  cuspidatum, 284
  esculentum, 284
Pellaea, 16
  atropurpurea, 16
  glabella, 16
Peltandra virginica, 615
Pennisetum villosum, 769
Penstemon, 444–446
  albidus, 444
  australis, 445
  canescens, 445
  cobaea, 444
  digitalis, 445
  gracilis, 446
  grandiflorus, 444
  hirsutus, 444
  laevigatus, 445
  pallidus, 445
  tenuiflorus, 444
  tubiflorus, 444
Penthorum sedoides, 211
Perideridia americana, 354
Perilla frutescens, 419
Periploca graeca, 370
Persea palustris, 38
Petasites, 528, 529
  frigidus, 528
  hybridus, 528
  sagittatus, 529
Petrorhagia, 112
  prolifera, 112
  saxifraga, 112
Petroselinum crispum, 351

Petunia, 383
  ×hybrida, 383
  parviflora, 383
Peucedanum ostruthium, 360
Phacelia, 392, 393
  bipinnatifida, 393
  dubia, 393
  fimbriata, 393
  franklinii, 392
  gilioides, 392
  purshii, 392
  ranunculacea, 393
Phalaris, 714
  arundinacea, 714
  canariensis, 714
  caroliniana, 714
Phaseolus polystachios, 285
Philadelphus, 204
  coronarius, 204
  hirsutus, 204
  inodorus, 204
  pubescens, 204
Phleum, 720
  alpinum, 720
  pratense, 720
Phlomis tuberosa, 423
Phlox, 389–391
  amoena, 389
  amplifolia, 390
  bifida, 390
  buckleyi, 391
  divaricata, 389
  glaberrima, 391
  maculata, 391
  nivalis, 390
  ovata, 391
  paniculata, 390
  pilosa, 389
  stolonifera, 390
  subulata, 390
Phoradendron serotinum, 307
Phragmites australis, 732
Phryma leptostachya, 407
Phyla, 406
  lanceolata, 406
  nodiflora, 406
Phyllanthus caroliniensis, 311
Phyllitis scolopendrium, 18
Phyllodoce caerulea, 187
Physalis, 376–379
  alkekengi, 376
  angulata, 378
  heterophylla, 376
  hispida, 377
  longifolia, 377
  missouriensis, 378
  peruviana, 377
  philadelphica, 379
  pubescens, 378
    var. grisea, 378
    var. integrifolia, 378
    var. pubescens, 378
  pumila, 377
  virginiana, 377
  walteri, 377
Physocarpus opulifolius, 217

Physostegia, 422
  leptophylla, 422
  parviflora, 422
  virginiana, 422
Phytolacca americana, 84
PHYTOLACCACEAE, 84
Picea, 31
  glauca, 31
  mariana, 31
  rubens, 31
Picris, 600
  echioides, 600
  hieracioides, 600
Pieris floribunda, 189
Pilea, 71, 72
  fontana, 72
  pumila, 71
Pimpinella saxifraga, 353
PINACEAE, 31–34
Pinellia ternata, 616
Pinguicula vulgaris, 462
Pinus, 32–34
  banksiana, 33
  echinata, 34
  palustris, 33
  pungens, 33
  resinosa, 32
  rigida, 34
  serotina, 33
  strobus, 32
  sylvestris, 32
  taeda, 33
  virginiana, 34
Piptochaetium avenaceum, 694
Plagiobothrys scouleri, 397
Planera aquatica, 69
PLANTAGINACEAE, 431–434
Plantago, 431–434
  aristata, 434
  cordata, 432
  coronopus, 434
  elongata, 433
  eriopoda, 432
  heterophylla, 433
  lanceolata, 433
  major, 432
  maritima, 432
  media, 432
  patagonica, 434
  psyllium, 431
  pusilla, 433
  rhodosperma, 433
  rugelii, 432
  virginica, 433
PLATANACEAE, 68
Platanus occidentalis, 68
Pluchea, 564
  camphorata, 564
  foetida, 564
  odorata, 564
PLUMBAGINACEAE, 125
Poa, 701–705
  alpina, 705
  alsodes, 702
  annua, 703
  arctica, 701

Poa (continued)
  arida, 701
  autumnalis, 705
  bulbosa, 702
  canbyi, 704
  chaixii, 704
  chapmaniana, 705
  compressa, 701
  cuspidata, 701
  fernaldiana, 704
  glauca, 703
  interior, 703
  languida, 702
  nemoralis, 704
  paludigena, 703
  palustris, 704
  pratensis, 701
  saltuensis, 702
  sylvestris, 703
  trivialis, 702
  wolfii, 703
POACEAE, 690–775
Podophyllum peltatum, 62
PODOSTEMACEAE, 289
Podostemum ceratophyllum, 289
Pogonia ophioglossoides, 823
Polanisia, 161
  dodecandra, 161
    var. dodecandra, 161
    var. trachysperma, 161
  jamesii, 161
POLEMONIACEAE, 389–392
Polemonium, 392
  occidentale, 392
  reptans, 392
  vanbruntiae, 392
Polygala, 328–330
  alba, 328
  ambigua, 329
  brevifolia, 330
  cruciata, 330
  curtissii, 329
  cymosa, 328
  incarnata, 328
  lutea, 328
  mariana, 329
  nuttallii, 329
  paucifolia, 328
  polygama, 329
  ramosa, 328
  sanguinea, 329
  senega, 328
  verticillata, 329
POLYGALACEAE, 328–330
POLYGONACEAE, 112–124
Polygonatum, 801
  biflorum, 801
  pubescens, 801
Polygonella, 115
  articulata, 115
  polygama, 115
Polygonum, 116–124
  achoreum, 116
  amphibium, 119
  arenastrum, 117
  arifolium, 123

Polygonum *(continued)*
aviculare, 117
bistorta, 118
bungeanum, 122
buxiforme, 118
careyi, 121
cespitosum, 121
cilinode, 124
convolvulus, 123
cuspidatum, 124
densiflorum, 119
douglasii, 116
erectum, 116
fowleri, 117
glaucum, 116
hydropiper, 120
hydropiperoides, 119
var. hydropiperoides, 119
var. opelousanum, 119
var. setaceum, 119
lapathifolium, 120
neglectum, 117
nepalense, 122
orientale, 122
oxyspermum, 117
pensylvanicum, 120
perfoliatum, 123
persicaria, 121
prolificum, 118
punctatum, 120
var. confertiflorum, 120
var. punctatum, 120
ramosissimum, 118
robustius, 119
sachalinense, 124
sagittatum, 122
scandens, 124
var. cristatum, 124
var. dumetorum, 124
var. scandens, 124
tenue, 116
virginianum, 123
viviparum, 118
Polymnia, 511
canadensis, 511
uvedalia, 511
POLYPODIACEAE, 12
Polypodium, 12
polypodioides, 12
virginianum, 12
Polypogon monspeliensis, 719
Polypremum procumbens, 434
Polystichum, 28
acrostichoides, 28
braunii, 28
lonchitis, 28
Polytaenia nuttallii, 360
PONTEDERIACEAE, 779, 780
Pontederia cordata, 779
Ponthieva racemosa, 818
Populus, 152, 153
alba, 152
balsamifera, 152
deltoides, 153
var. deltoides, 153
var. occidentalis, 153
Populus *(continued)*
grandidentata, 152
heterophylla, 152
×jackii, 153
nigra, 153
tremuloides, 152
Porteranthus, 220
stipulatus, 220
trifoliatus, 220
Portulaca, 98, 99
grandiflora, 99
oleracea, 98
retusa, 98
PORTULACACEAE, 98, 99
Potamogeton, 608–613
alpinus, 612
amplifolius, 612
confervoides, 610
crispus, 609
diversifolius, 609
epihydrus, 611
filiformis, 608
foliosus, 611
friesii, 610
gramineus, 612
hillii, 610
illinoensis, 613
natans, 613
nodosus, 612
oakesianus, 613
obtusifolius, 611
ogdenii, 610
pectinatus, 609
perfoliatus, 613
praelongus, 613
pulcher, 612
pusillus, 611
richardsonii, 613
robbinsii, 609
spirillus, 608
strictifolius, 610
vaginatus, 609
vaseyi, 611
zosteriformis, 610
POTAMOGETONACEAE, 608–613
Potentilla, 222–226
anglica, 222
anserina, 225
argentea, 223
arguta, 226
canadensis, 222
effusa, 225
flabelliformis, 224
fruticosa, 226
hippiana, 225
intermedia, 224
norvegica, 223
pacifica, 225
palustris, 226
paradoxa, 223
pensylvanica, 225
var. arida, 225
var. bipinnatifida, 225
var. pensylvanica, 225
pulcherrima, 224
recta, 223
Potentilla *(continued)*
reptans, 222
rivalis, 223
robbinsiana, 224
simplex, 222
sterilis, 222
tridentata, 225
Prenanthes, 586–588
alba, 587
altissima, 588
aspera, 587
autumnalis, 587
boottii, 588
crepidinea, 587
nana, 588
racemosa, 586
var. multiflora, 586
var. racemosa, 586
roanensis, 588
serpentaria, 587
trifoliolata, 588
Primula, 203
laurentiana, 203
mistassinica, 203
PRIMULACEAE, 199–204
Proboscidea louisianica, 461
Proserpinaca, 290, 291
intermedia, 291
palustris, 290
var. ambylogona, 290
var. crebra, 290
var. palustris, 290
pectinata, 290
Prunella, 423
laciniata, 423
vulgaris, 423
Prunus, 240–243
alleghaniensis, 242
americana, 242
angustifolia, 242
armeniaca, 240
avium, 241
cerasus, 241
domestica, 243
fruticosa, 242
hortulana, 243
mahaleb, 241
maritima, 242
mexicana, 242
munsoniana, 243
nigra, 243
padus, 241
pensylvanica, 241
persica, 240
pumila, 241
serotina, 240
virginiana, 241
Psoralidium, 284
lanceolatum, 284
tenuiflorum, 284
Ptelea trifoliata, 336
Pteridium aquilinum, 13
var. latiusculum, 13
var. pseudocaudatum, 13
var. pubescens, 13
Pterospora andromedea, 197

Ptilimnium, 356
  capillaceum, 356
  costatum, 356
  nodosum, 356
  nuttallii, 356
Puccinellia, 699, 700
  distans, 700
  fasciculata, 699
  maritima, 699
  nuttalliana, 699
  pallida, 699
  pumila, 700
Pueraria lobata, 287
Pycnanthemum, 414–416
  albescens, 415
  beadlei, 415
  clinopodioides, 414
  flexuosum, 414
  incanum, 414
  loomisii, 415
  montanum, 415
  muticum, 416
  pycnanthemoides, 415
  setosum, 414
  tenuifolium, 416
  torrei, 416
  verticillatum, 416
  virginianum, 416
Pyrola, 195, 196
  asarifolia, 196
  chlorantha, 195
  elliptica, 196
  minor, 195
  rotundifolia, 196
  secunda, 195
PYROLACEAE, 195, 196
Pyrrhopappus carolinianus, 597
Pyrularia pubera, 307
Pyrus, 243–245
  angustifolia, 244
  communis, 243
  coronaria, 244
  ioensis, 245
  malus, 244
  sieboldii, 244
Pyxidanthera barbulata, 197

Quercus, 76–81
  alba, 76
  bicolor, 77
  cinerea, 78
  coccinea, 81
  ellipsoidalis, 81
  falcata, 79
  geminata, 78
  hemisphaerica, 79
  ilicifolia, 79
  imbricaria, 78
  laevis, 80
  laurifolia, 79
  lyrata, 77
  macrocarpa, 77
  margaretta, 77
  marilandica, 79
  michauxii, 77
  muehlenbergii, 78

Quercus (continued)
  nigra, 79
  pagoda, 80
  palustris, 81
  phellos, 79
  prinoides, 78
  prinus, 78
  rubra, 80
  shumardii, 81
  stellata, 76
  velutina, 80
  virginiana, 78

Radiola linoides, 327
RANUNCULACEAE, 45–61
Ranunculus, 54–58
  abortivus, 56
  acris, 54
  allegheniensis, 56
  ambigens, 57
  arvensis, 55
  bulbosus, 54
  cymbalaria, 58
  fascicularis, 55
  ficaria, 58
  flabellaris, 57
  flammula, 57
    var. filiformis, 57
    var. flammula, 57
  gmelinii, 57
  harveyi, 56
  hederaceus, 58
  hispidus, 55
    var. caricetorum, 55
    var. hispidus, 55
    var. nitidus, 55
  lapponicus, 58
  longirostris, 58
  macounii, 54
  micranthus, 56
  parviflorus, 56
  pensylvanicus, 54
  pusillus, 57
  recurvatus, 54
  repens, 54
  rhomboideus, 56
  sardous, 55
  sceleratus, 57
  subrigidus, 58
  testiculatus, 58
  texensis, 57
  trichophyllus, 58
Raphanus, 163
  raphanistrum, 163
  sativus, 163
Rapistrum rugosum, 164
Ratibida, 498
  columnifera, 498
  pinnata, 498
Reseda, 182, 183
  alba, 182
  lutea, 182
  luteola, 182
  odorata, 183
RESEDACEAE, 182, 183
RHAMNACEAE, 321, 322

Rhamnus, 321, 322
  alnifolia, 322
  caroliniana, 322
  cathartica, 322
  davurica, 322
  frangula, 321
  lanceolata, 322
Rheum rhabarbarum, 113
Rhexia, 303, 304
  aristosa, 303
  interior, 303
  mariana, 304
  nashii, 304
  petiolata, 303
  virginica, 303
Rhinanthus crista-galli, 457
Rhododendron, 184–186
  arborescens, 186
  atlanticum, 185
  bakeri, 186
  calendulaceum, 186
  canadense, 186
  canescens, 185
  catawbiense, 184
  lapponicum, 185
  maximum, 184
  periclymenoides, 185
  prinophyllum, 185
  serrulatum, 185
  viscosum, 185
Rhus, 334
  aromatica, 334
  copallina, 334
  glabra, 334
  typhina, 334
Rhynchosia, 286
  difformis, 286
  tomentosa, 286
Rhynchospora, 645–650
  alba, 646
  caduca, 649
  capillacea, 647
  capitellata, 647
  cephalantha, 646
  chalarocephala, 646
  colorata, 645
  corniculata, 646
  debilis, 649
  fascicularis, 648
  filifolia, 648
  fusca, 647
  globularis, 650
    var. globularis, 650
    var. recognita, 650
  glomerata, 647
  gracilenta, 648
  grayi, 648
  harveyi, 648
  inexpansa, 649
  inundata, 645
  knieskernii, 647
  macrostachya, 645
  microcephala, 646
  miliacea, 649
  nitens, 645
  oligantha, 646

Rhynchospora *(continued)*
  pallida, 647
  perplexa, 649
  rariflora, 649
  scirpoides, 645
  torreyana, 649
Ribes, 205–207
  americanum, 207
  cynosbati, 206
  glandulosum, 206
  hirtellum, 206
  hudsonianum, 207
  lacustre, 206
  missouriense, 205
  nigrum, 206
  odoratum, 207
  oxyacanthoides, 206
  rotundifolium, 205
  sativum, 207
  triste, 207
  uva-crispa, 206
Richardia, 472
  brasiliensis, 472
  scabra, 472
Ricinus communis, 311
Robinia, 262
  hispida, 262
  pseudoacacia, 262
  viscosa, 262
Rorippa, 177, 178
  amphibia, 177
  austriaca, 177
  indica, 177
  nasturtium-aquaticum, 177
  palustris, 178
    var. fernaldiana, 178
    var. hispida, 178
    var. palustris, 178
  sessiliflora, 178
  sinuata, 177
  sylvestris, 177
  tenerrima, 178
Rosa, 237–240
  acicularis, 240
  arkansana, 239
  blanda, 240
  bracteata, 237
  canina, 238
  carolina, 239
  eglanteria, 238
  gallica, 238
  majalis, 240
  micrantha, 238
  multiflora, 237
  nitida, 239
  palustris, 239
  pimpinellifolia, 238
  rugosa, 238
  setigera, 237
  virginiana, 239
  wichuraiana, 237
  woodsii, 240
ROSACEAE, 217–255
Rotala ramosior, 291
RUBIACEAE, 469–477

Rubus, 229–235
  acaulis, 230
  allegheniensis, 233
  argutus, 234
  bifrons, 235
  caesius, 231
  canadensis, 233
  chamaemorus, 229
  cuneifolius, 233
  discolor, 234
  enslenii, 232
  flagellaris, 232
  hispidus, 231
  idaeus, 230
  illecebrosus, 231
  laciniatus, 235
  occidentalis, 230
  odoratus, 229
  orarius, 233
  parviflorus, 229
  pensilvanicus, 234
  phoenicolasius, 231
  pubescens, 230
  recurvicaulis, 232
  setosus, 232
  trivialis, 231
Rudbeckia, 495–497
  fulgida, 495
    var. fulgida, 495
    var. speciosa, 495
    var. umbrosa, 495
  grandiflora, 496
  heliopsidis, 496
  hirta, 495
  laciniata, 497
    var. digitata, 497
    var. humilis, 497
    var. laciniata, 497
  maxima, 496
  missouriensis, 495
  subtomentosa, 496
  triloba, 496
Ruellia, 459, 460
  caroliniensis, 460
  humilis, 460
  pedunculata, 459
  strepens, 459
Rumex, 113–115
  acetosa, 113
  acetosella, 113
  alpinus, 113
  altissimus, 114
  conglomeratus, 115
  crispus, 114
  hastatulus, 113
  longifolius, 113
  maritimus, 115
    var. fueginus, 115
    var. persicarioides, 115
  obtusifolius, 115
  occidentalis, 113
  orbiculatus, 113
  pallidus, 114
  patientia, 114
  pulcher, 114
  salicifolius, 114

Rumex *(continued)*
  venosus, 113
  verticillatus, 114
Ruppia, 613
RUPPIACEAE, 613
Ruta, 336
RUTACEAE, 336

Sabatia, 362, 363
  angularis, 362
  brachiata, 362
  calycina, 363
  campanulata, 362
  campestris, 362
  difformis, 362
  dodecandra, 363
  kennedyana, 363
  quadrangula, 362
  stellaris, 363
Sacciolepis striata, 761
Sagina, 100
  decumbens, 100
  fontinalis, 100
  nodosa, 100
  procumbens, 100
Sagittaria, 602–605
  australis, 604
  brevirostra, 604
  calycina, 603
  cuneata, 604
  engelmanniana, 603
  graminea, 605
  lancifolia, 605
  latifolia, 604
  rigida, 603
  spathulata, 602
  subulata, 602
  teres, 605
SALICACEAE, 152–160
Salicornia, 92, 93
  bigelovii, 92
  europaea, 93
  rubra, 93
  virginica, 92
Salix, 154–160
  alba, 155
  amygdaloides, 154
  arctophila, 157
  argyrocarpa, 159
  babylonica, 156
  bebbiana, 156
  candida, 160
  caroliniana, 154
  cordata, 158
  discolor, 159
  eriocephala, 157
  exigua, 156
  fragilis, 155
  herbacea, 156
  humilis, 159
  lucida, 154
  myricoides, 157
  nigra, 154
  occidentalis, 159
  pedicellaris, 157
  pellita, 159

Salix *(continued)*
  pentandra, 155
  petiolaris, 160
  planifolia, 158
  purpurea, 160
  pyrifolia, 157
  sericea, 158
  serissima, 155
  uva-ursi, 156
  viminalis, 160
Salsola, 93, 94
  collina, 93
  kali, 94
  tragus, 94
Salvia, 429, 430
  lyrata, 430
  nemorosa, 429
  officinalis, 430
  pitcheri, 429
  pratensis, 429
  reflexa, 429
  sclarea, 430
  urticifolia, 429
  verbenacea, 429
  verticillata, 430
SALVINIACEAE, 30
Sambucus, 485
  canadensis, 485
  ebulus, 485
  racemosa, 485
    var. pubens, 485
Samolus floribundus, 203
Sanguinaria canadensis, 64
Sanguisorba, 237
  canadensis, 237
  minor, 237
  officinalis, 237
Sanicula, 346
  canadensis, 346
  gregaria, 346
  marilandica, 346
  smallii, 346
  trifoliata, 346
SANTALACEAE, 306, 307
Santolina chamaecyparissus, 518
SAPINDACEAE, 330
Saponaria officinalis, 111
SAPOTACEAE, 198
Sarracenia, 138
  flava, 138
  purpurea, 138
SARRACENIACEAE, 138
Sassafras albidum, 39
Satureja, 413
  acinos, 413
  calamintha, 413
  glabella, 413
    var. angustifolia, 413
    var. glabella, 413
  hortensis, 413
  vulgaris, 413
SAURURACEAE, 39
Saururus cernuus, 39
Saxifraga, 211–214
  aizoides, 212
  aizoon, 212

Saxifraga *(continued)*
  careyana, 213
  caroliniana, 212
  cernua, 211
  foliolosa, 212
  michauxii, 213
  micranthidifolia, 212
  oppositifolia, 212
  pensylvanica, 213
  rivularis, 211
  tricuspidata, 211
  virginiensis, 214
SAXIFRAGACEAE, 211–217
Scabiosa, 488
  atropurpurea, 488
  columbaria, 488
Scandix pecten-veneris, 349
Schedonnardus paniculatus, 748
SCHEUCHZERIACEAE, 607
Scheuchzeria palustris, 607
Schizachne purpurascens, 709
Schizachyrium scoparium, 775
    var. divergens, 775
    var. littorale, 775
    var. scoparium, 775
SCHIZAEACEAE, 11
Schizaea pusilla, 11
Schwalbea americana, 456
Scilla, 786
  non-scripta, 786
  siberica, 786
Scirpus, 631-637
  acutus, 633
  americanus, 632
  atrovirens, 634
    var. atrovirens, 634
    var. pallidus, 634
  cespitosus, 631
  clintonii, 631
  cylindricus, 634
  cyperinus, 637
  divaricatus, 636
  etuberculatus, 633
  expansus, 635
  fluviatilis, 634
  heterochaetus, 633
  hudsonianus, 631
  koilolepis, 633
  lineatus, 636
  longii, 635
  maritimus, 634
    var. maritimus, 634
    var. paludosus, 634
  microcarpus, 636
  pendulus, 636
  polyphyllus, 635
  pungens, 632
  robustus, 634
  rufus, 633
  smithii, 632
  subterminalis, 632
  supinus, 632
  torreyi, 632
  validus, 633
  verecundus, 631
Scleranthus annuus, 105

Scleria, 659, 660
  ciliata, 659
  minor, 659
  oligantha, 659
  pauciflora, 659
  reticularis, 660
  triglomerata, 659
  verticillata, 660
Sclerolepis uniflora, 573
Scolochloa festucacea, 698
Scrophularia, 446
  lanceolata, 446
  marilandica, 446
  nodosa, 446
SCROPHULARIACEAE, 438–458
Scutellaria, 409–412
  arguta, 410
  ×churchilliana, 411
  elliptica, 410
  galericulata, 411
  incana, 410
  integrifolia, 410
  lateriflora, 411
  leonardii, 411
  nervosa, 412
  ovata, 409
    var. calcarea, 409
    var. ovata, 409
    var. versicolor, 409
  parvula, 411
    var. australis, 411
    var. parvula, 411
  saxatilis, 410
  serrata, 410
Secale, 730
Sedum, 208–210
  acre, 209
  album, 208
  ×erythrostictum, 208
  glaucophyllum, 209
  hispanicum, 209
  pulchellum, 209
  purpureum, 208
  reflexum, 210
  rosea, 208
  sarmentosum, 209
  sexangulare, 210
  spurium, 208
  telephioides, 208
  ternatum, 209
Selaginella, 4
  apoda, 4
  rupestris, 4
  selaginoides, 4
SELAGINELLACEAE, 4
Sempervivum tectorum, 210
Senecio, 523–526
  anonymus, 524
  antennariifolius, 523
  aureus, 525
  canus, 523
  congestus, 523
  glabellus, 526
  indecorus, 524
  integerrimus, 523
  jacobaea, 526

Senecio *(continued)*
  obovatus, 524
  pauperculus, 524
  plattensis, 524
  pseudaureus, 525
  pseudoarnica, 523
  schweinitzianus, 525
  sylvaticus, 526
  tomentosus, 523
  viscosus, 525
  vulgaris, 526
Senna, 257, 258
  hebecarpa, 257
  marilandica, 257
  obtusifolia, 258
  occidentalis, 258
Sesamum indicum, 460
Sesbania exaltata, 262
Seseli libanotis, 349
Sesuvium maritimum, 85
Setaria, 766, 767
  faberi, 766
  glauca, 766
  italica, 767
  magna, 767
  parviflora, 766
  verticillata, 766
    var. ambigua, 766
    var. verticillata, 766
  viridis, 766
Seymeria cassioides, 449
Shepherdia, 288
  argentea, 288
  canadensis, 288
Sherardia arvensis, 473
Shortia galacifolia, 197
Sibara virginica, 174
Sibbaldia procumbens, 221
Sicyos angulatus, 151
Sida, 137, 138
  elliottii, 137
  hermaphrodita, 138
  rhombifolia, 137
  spinosa, 137
Sideritis romana, 412
Sigesbeckia orientalis, 503
Silene, 108–110
  acaulis, 110
  antirrhina, 110
  armeria, 110
  caroliniana, 110
  conica, 110
  conoidea, 110
  csereii, 109
  dichotoma, 110
  dioica, 108
  drummondii, 108
  gallica, 110
  latifolia, 108
  nivea, 109
  noctiflora, 108
  nutans, 110
  ovata, 109
  regia, 109
  rotundifolia, 109

Silene *(continued)*
  stellata, 109
  virginica, 108
  vulgaris, 109
Silphium, 511–513
  asteriscus, 511
  brachiatum, 512
  compositum, 513
  integrifolium, 512
  laciniatum, 512
  mohrii, 512
  perfoliatum, 512
  terebinthinaceum, 513
  trifoliatum, 511
Silybum marianum, 583
SIMAROUBACEAE, 335
Sinapis, 162
  alba, 162
  arvensis, 162
Sisymbrium, 180, 181
  altissimum, 180
  irio, 181
  loeselii, 181
  officinale, 180
Sisyrinchium, 808, 809
  albidum, 808
  angustifolium, 809
  atlanticum, 809
  campestre, 808
  capillare, 808
  fuscatum, 808
  montanum, 808
  mucronatum, 808
Sium suave, 355
SMILACACEAE, 804–806
Smilacina, 800
  racemosa, 800
  stellata, 800
  trifolia, 800
Smilax, 804–806
  bona-nox, 805
  ecirrhata, 804
  glauca, 805
  herbacea, 805
    var. herbacea, 805
    var. lasioneura, 805
    var. pulverulenta, 805
  hispida, 805
  laurifolia, 806
  pseudochina, 804
  rotundifolia, 806
  walteri, 806
SOLANACEAE, 376–383
Solanum, 379–381
  carolinense, 380
  citrullifolium, 380
  dulcamara, 379
  elaeagnifolium, 381
  jamesii, 380
  nigrum, 379
  rostratum, 380
  sarrachoides, 379
  sisymbriifolium, 380
  triflorum, 380
  tuberosum, 380

Solidago, 532–544
  albopilosa, 536
  arguta, 539
    var. arguta, 539
    var. caroliniana, 539
    var. harrisii, 539
  bicolor, 534
  caesia, 537
  calcicola, 541
  canadensis, 542
    var. canadensis, 542
    var. gilvocanescens, 542
    var. salebrosa, 542
    var. scabra, 542
  curtisii, 536
  cutleri, 533
  drummondii, 539
  elliottii, 541
  erecta, 534
  fistulosa, 541
  flaccidifolia, 535
  flexicaulis, 536
  gigantea, 541
  gracillima, 535
  hispida, 534
  houghtonii, 544
  juncea, 538
  lancifolia, 536
  macrophylla, 532
  missouriensis, 538
  mollis, 543
  nemoralis, 537
    var. haleana, 537
    var. longipetiolata, 537
    var. nemoralis, 537
  odora, 541
  ohioensis, 544
  patula, 538
  petiolaris, 537
  pinetorum, 538
  ptarmicoides, 544
  puberula, 533
    var. puberula, 533
    var. pulverulenta, 533
  radula, 543
  riddellii, 544
  rigida, 543
  roanensis, 533
  rugosa, 540
    subsp. aspera, 540
    var. rugosa, 540
    var. sphagnophila, 540
    var. villosa, 540
  rupestris, 542
  sciaphila, 533
  sempervirens, 535
  shortii, 543
  simplex, 533
    var. gillmanii, 533
    var. monticola, 533
    var. racemosa, 533
  speciosa, 534
    var. jejunifolia, 534
    var. rigidiuscula, 534
    var. speciosa, 534
  sphacelata, 540

Solidago *(continued)*
squarrosa, 532
stricta, 535
tarda, 540
tortifolia, 541
uliginosa, 534
ulmifolia, 539
Sonchus, 591
arvensis, 591
var. arvensis, 591
var. glabrescens, 591
asper, 591
oleraceus, 591
Sorbaria sorbifolia, 219
Sorbus, 246
americana, 246
aucuparia, 246
decora, 246
Sorghastrum, 772, 773
elliottii, 772
nutans, 773
Sorghum, 772
bicolor, 772
halepense, 772
SPARGANIACEAE, 776–778
Sparganium, 776–778
americanum, 776
androcladum, 776
angustifolium, 777
chlorocarpum, 776
emersum, 777
eurycarpum, 776
fluctuans, 777
glomeratum, 778
hyperboreum, 778
natans, 778
Spartina, 749, 750
alterniflora, 749
cynosuroides, 749
patens, 750
pectinata, 749
Specularia speculum-veneris, 466
Spergula, 105
arvensis, 105
morisonii, 105
pentandra, 105
Spergularia, 105, 106
canadensis, 106
marina, 105
media, 105
rubra, 105
Spermacoce glabra, 472
Spermolepis, 350
divaricata, 350
echinata, 350
inermis, 350
Sphaeralcea coccinea, 137
Sphenopholis, 711, 712
filiformis, 712
nitida, 711
obtusata, 711
var. major, 711
var. obtusata, 711
pensylvanica, 711
Spigelia marilandica, 361
Spilanthes americana, 502

Spiraea, 217–219
alba, 219
var. alba, 219
var. latifolia, 219
betulifolia, 218
chamaedryfolia, 218
japonica, 218
prunifolia, 217
thunbergia, 218
tomentosa, 219
×vanhouttei, 218
virginiana, 218
Spiranthes, 815–817
casei, 816
cernua, 815
lacera, 816
laciniata, 817
lucida, 815
magnicamporum, 816
ovalis, 815
praecox, 817
romanzoffiana, 815
tuberosa, 816
vernalis, 816
Spirodela, 616
polyrhiza, 616
punctata, 616
Sporobolus, 741–743
airoides, 743
asper, 741
clandestinus, 742
contractus, 742
cryptandrus, 741
heterolepis, 743
indicus, 742
junceus, 741
neglectus, 743
ozarkanus, 742
pyramidatus, 741
vaginiflorus, 743
virginicus, 742
Stachys, 425–427
annua, 427
arvensis, 427
aspera, 425
byzantina, 427
clingmanii, 426
cordata, 426
eplingii, 426
germanica, 427
hispida, 425
hyssopifolia, 425
latidens, 426
palustris, 425
var. homotricha, 425
var. palustris, 425
var. pilosa, 425
subcordata, 426
tenuifolia, 426
STAPHYLEACEAE, 330
Staphylea trifolia, 330
Stellaria, 103, 104
alsine, 104
aquatica, 103
borealis, 104
crassifolia, 104

Stellaria *(continued)*
graminea, 104
holostea, 103
humifusa, 104
longifolia, 104
longipes, 104
media, 103
palustris, 103
pubera, 103
Stenanthium gramineum, 782
STERCULIACEAE, 131
Stewartia, 125
malacodendron, 125
ovata, 125
Stillingia sylvatica, 313
Stipa, 693, 694
comata, 693
spartea, 694
viridula, 693
Stipulicida setacea, 102
Streptopus, 799
amplexifolius, 799
lanceolatus, 799
Strophostyles, 285, 286
helvula, 285
leiosperma, 286
umbellata, 286
Stylisma, 384
humistrata, 384
pickeringii, 384
Stylophorum diphyllum, 64
Stylosanthes biflora, 263
STYRACACEAE, 198, 199
Styrax, 198
americanus, 198
grandifolius, 198
Suaeda, 93
americana, 93
calceoliformis, 93
linearis, 93
maritima, 93
richii, 93
Subularia aquatica, 167
Succisa pratensis, 488
Succisella inflexa, 488
Sullivantia sullivantii, 214
Symphoricarpos, 481
albus, 481
occidentalis, 481
orbiculatus, 481
Symphytum, 402
asperum, 402
officinale, 402
tuberosum, 402
SYMPLOCACEAE, 199
Symplocarpus foetidus, 615
Symplocos tinctoria, 199
Synandra hispidula, 422
Syringa vulgaris, 435

Taenidia, 352
integerrima, 352
montana, 352
Tagetes minuta, 510
Talinum, 99
calycinum, 99

Talinum *(continued)*
   parviflorum, 99
   rugospermum, 99
   teretifolium, 99
TAMARICACEAE, 149
Tamarix chinensis, 149
Tanacetum, 519
   huronense, 519
   vulgare, 519
Taraxacum, 598, 599
   ceratophorum, 598
   laevigatum, 598
   officinale, 599
   palustre, 599
   spectabile, 599
TAXACEAE, 31
TAXODIACEAE, 34
Taxodium distichum, 34
Taxus canadensis, 31
Teesdalia nudicaulis, 166
Tephrosia, 261
   spicata, 261
   virginiana, 261
Tetragonotheca helianthoides, 495
Teucrium, 408, 409
   botrys, 409
   canadense, 408
     var. canadense, 408
     var. occidentale, 408
     var. virginicum, 408
   scorodonia, 408
Thalictrum, 60, 61
   clavatum, 60
   coriaceum, 60
   dasycarpum, 61
   dioicum, 60
   macrostylum, 60
   mirabile, 60
   pubescens, 60
   revolutum, 61
   steeleanum, 60
   venulosum, 60
     var. confine, 60
Thaspium, 353, 354
   barbinode, 353
   pinnatifidum, 354
   trifoliatum, 353
THEACEAE, 125
Thelesperma megapotamicum, 506
Thelypteris, 22–24
   hexagonoptera, 23
   noveboracensis, 23
   palustris, 24
   phegopteris, 22
   simulata, 23
Thermopsis mollis, 259
Thismia americana, 812
Thlaspi, 166
   arvense, 166
   perfoliatum, 166
Thuja occidentalis, 35
THYMELAEACEAE, 293
Thymelaea passerina, 293
Thymus serpyllum, 414
Tiarella cordifolia, 214
Tidestromia lanuginosa, 97

Tilia americana, 131
TILIACEAE, 131
Tillandsia usneoides, 779
Tipularia discolor, 826
Tofieldia, 781
   glutinosa, 781
   pusilla, 781
   racemosa, 781
Torilis, 350
   arvensis, 350
   japonica, 350
Toxicodendron, 335
   pubescens, 335
   radicans, 335
   rydbergii, 335
   vernix, 335
Trachelospermum difforme, 368
Tradescantia, 619, 620
   bracteata, 620
   occidentalis, 620
   ohiensis, 620
   rosea, 620
   subaspera, 619
   virginiana, 619
Tragia, 313
   cordata, 313
   urens, 313
Tragopogon, 601
   dubius, 601
   porrifolius, 601
   pratensis, 601
Tragus, 768
   berteronianus, 768
   racemosus, 768
TRAPACEAE, 293
Trapa natans, 293
Trautvetteria caroliniensis, 46
Trepocarpus aethusae, 348
Triadenum, 130
   fraseri, 130
   tubulosum, 130
   virginicum, 130
   walteri, 130
Trianthema portulacastrum, 85
Tribulus terrestris, 337
Trichomanes boschianum, 11
Trichostema, 409
   dichotomum, 409
   setaceum, 409
Tridens flavus, 736
Trientalis borealis, 202
Trifolium, 269–271
   arvense, 269
   aureum, 271
   campestre, 271
   carolinianum, 270
   dubium, 271
   fragiferum, 269
   hirtum, 269
   hybridum, 270
   incarnatum, 269
   medium, 269
   pratense, 269
   reflexum, 270
   repens, 270

Trifolium *(continued)*
   resupinatum, 270
   stoloniferum, 270
   striatum, 269
   virginicum, 270
Triglochin, 608
   maritima, 608
   palustris, 608
   striata, 608
Trillium, 795–797
   cernuum, 796
   cuneatum, 797
   erectum, 796
   flexipes, 796
   grandiflorum, 795
   luteum, 797
   nivale, 795
   pusillum, 795
   recurvatum, 797
   sessile, 796
   sulcatum, 795
   undulatum, 795
   viride, 796
Triodanis, 466
   biflora, 466
   leptocarpa, 466
   perfoliata, 466
Triosteum, 481, 482
   angustifolium, 481
   aurantiacum, 482
   perfoliatum, 481
Triphora trianthophora, 824
Triplasis purpurea, 736
Tripsacum dactyloides, 774
Trisetum, 709–711
   flavescens, 711
   melicoides, 710
   spicatum, 709
Triticum aestivum, 730
Trollius laxus, 45
Tsuga, 32
   canadensis, 32
   caroliniana, 32
Tulipa sylvestris, 793
Tussilago farfara, 528
Typha, 778
   angustifolia, 778
   domingensis, 778
   latifolia, 778
TYPHACEAE, 778

Ulex, 260
   europaeus, 260
ULMACEAE, 68, 69
Ulmus, 68, 69
   alata, 68
   americana, 68
   rubra, 68
   serotina, 69
   thomasii, 69
Uniola paniculata, 735
Urtica, 71
   chamaedryoides, 71
   dioica, 71
   urens, 71

URTICACEAE, 71, 72
Utricularia, 462–464
  biflora, 464
  cornuta, 462
  fibrosa, 464
  geminiscapa, 463
  gibba, 464
  inflata, 463
  intermedia, 463
  juncea, 462
  minor, 463
  purpurea, 463
  radiata, 463
  resupinata, 463
  subulata, 463
  vulgaris, 464
Uvularia, 798
  grandiflora, 798
  perfoliata, 798
  puberula, 798
  sessilifolia, 798

Vaccaria hispanica, 111
Vaccinium, 191–194
  angustifolium, 192
  arboreum, 191
  boreale, 192
  cespitosum, 191
  corymbosum, 193
  crassifolium, 192
  elliottii, 194
  erythrocarpum, 194
  macrocarpon, 194
  membranaceum, 191
  myrtilloides, 193
  ovalifolium, 192
  oxycoccos, 194
  pallidum, 193
  stamineum, 191
  tenellum, 192
  uliginosum, 191
  vitis-idaea, 192
Valeriana, 486
  edulis, 486
  officinalis, 486
  pauciflora, 486
  uliginosa, 486
VALERIANACEAE, 486, 487
Valerianella, 486, 487
  chenopodifolia, 486
  locusta, 486
  radiata, 486
  umbilicata, 487
Vallisneria americana, 607
Veratrum, 784, 785
  parviflorum, 785
  viride, 784
  woodii, 785
Verbascum, 442, 443
  blattaria, 442
  densiflorum, 443
  lychnitis, 442
  phlomoides, 442
  phoeniceum, 442
  thapsus, 443

Verbena, 403–406
  bipinnatifida, 405
  bonariensis, 403
  bracteata, 405
  brasiliensis, 403
  canadensis, 406
  hastata, 404
  hybrida, 406
  lasiostachys, 405
  officinalis, 403
  riparia, 403
  scabra, 404
  simplex, 405
  stricta, 405
  tenuisecta, 406
  urticifolia, 404
    var. leiocarpa, 404
    var. urticifolia, 404
VERBENACEAE, 403–407
Verbesina, 501, 502
  alternifolia, 502
  encelioides, 501
  helianthoides, 501
  occidentalis, 501
  virginica, 502
Vernonia, 577, 578
  arkansana, 577
  baldwinii, 578
    var. baldwinii, 578
    var. interior, 578
  fasciculata, 578
    var. fasciculata, 578
    var. nebraskensis, 578
  gigantea, 578
  glauca, 577
  missurica, 577
  noveboracensis, 577
Veronica, 453–455
  agrestis, 454
  americana, 455
  anagallis-aquatica, 455
  arvensis, 453
  beccabunga, 455
  catenata, 455
  chamaedrys, 455
  dillenii, 453
  filiformis, 454
  hederifolia, 454
  longifolia, 453
  officinalis, 454
  peregrina, 453
  persica, 454
  polita, 454
  scutellata, 455
  serpyllifolia, 453
    var. humifusa, 453
    var. serpyllifolia, 453
  teucrium, 455
  verna, 453
  wormskjoldii, 453
Veronicastrum virginicum, 453
Viburnum, 482–485
  acerifolium, 483
  alnifolium, 485
  dentatum, 483
    var. deamii, 483

Viburnum (continued)
    var. dentatum, 483
    var. indianense, 483
    var. lucidum, 483
    var. scabrellum, 483
    var. venosum, 483
  edule, 483
  lantana, 484
  lentago, 484
  molle, 482
  nudum, 484
    var. cassinoides, 484
    var. nudum, 484
  opulus, 483
  plicatum, 485
  prunifolium, 484
  rafinesquianum, 482
  rufidulum, 484
  sieboldii, 482
Vicia, 272–274
  americana, 274
    var. americana, 274
    var. minor, 274
  angustifolia, 272
  caroliniana, 274
  cracca, 274
  dasycarpa, 273
  faba, 272
  grandiflora, 272
  hirsuta, 273
  lathyroides, 272
  narbonensis, 273
  sativa, 272
  sepium, 273
  tenuifolia, 274
  tetrasperma, 273
  villosa, 273
Vinca, 368
  major, 368
  minor, 368
Vincetoxicum, 374
  hirundinaria, 374
  nigrum, 374
Viola, 142–149
  adunca, 148
    var. adunca, 148
    var. glabra, 148
  arvensis, 149
  blanda, 145
  canadensis, 147
  conspersa, 148
  cucullata, 142
  hastata, 146
  lanceolata, 146
  macloskeyi, 145
  nephrophylla, 144
  nuttallii, 146
  odorata, 145
  palmata, 144
    var. palmata, 144
    var. pedatifida, 144
  palustris, 145
  pedata, 142
  primulifolia, 145
  pubescens, 147
  rafinesquii, 148

Viola *(continued)*
  renifolia, 145
  rostrata, 148
  rotundifolia, 142
  sagittata, 143
  selkirkii, 145
  sororia, 143
  striata, 147
  tricolor, 149
  tripartita, 146
  villosa, 144
  walteri, 148
VIOLACEAE, 142–149
VISCACEAE, 307
VITACEAE, 322–325
Vitex, 407
  agnus-castus, 407
  negundo, 407
Vitis, 324, 325
  aestivalis, 324
  baileyana, 325
  cinerea, 325
  labrusca, 324
  novae-angliae, 325
  palmata, 324
  riparia, 325
  rotundifolia, 324
  rupestris, 324
  vulpina, 325
Vulpia, 697
  bromoides, 697
  elliotea, 697
  myuros, 697
  octoflora, 697

Waldsteinia fragarioides, 221
  var. fragarioides, 221
  var. parviflora, 221

Wisteria, 261, 262
  floribunda, 261
  frutescens, 261
  macrostachya, 262
Wolffia, 616
  borealis, 616
  brasiliensis, 616
  columbiana, 616
Wolffiella gladiata, 616
Woodsia, 20, 21
  alpina, 20
  glabella, 20
  ilvensis, 21
  obtusa, 21
  oregana, 21
  scopulina, 21
Woodwardia, 30
  areolata, 30
  virginica, 30

Xanthium, 516
  spinosum, 516
  strumarium, 516
Xanthorhiza simplicissima, 61
Xerophyllum asphodeloides, 780
XYRIDACEAE, 617–619
Xyris, 617–619
  ambigua, 619
  caroliniana, 618
  difformis, 618
    var. curtissii, 618
    var. difformis, 618
  fimbriata, 618
  iridifolia, 617
  jupicai, 617
  montana, 617
  platylepis, 617

Xyris *(continued)*
  smalliana, 617
  torta, 618

Youngia japonica, 596
Yucca, 804
  filamentosa, 804
  glauca, 804

ZANNICHELLIACEAE, 614
Zannichellia palustris, 614
Zanthoxylum, 336
  americanum, 336
  clava-herculis, 336
Zenobia pulverulenta, 189
Zephyranthes atamasca, 803
Zigadenus, 782, 783
  densus, 783
  elegans, 783
    var. elegans, 783
    var. glaucus, 783
  glaberrimus, 782
  leimanthoides, 783
Zinnia elegans, 502
Zizania, 691
  aquatica, 691
  palustris, 691
Zizaniopsis miliacea, 691
Zizia, 353
  aptera, 353
  aurea, 353
  trifoliata, 353
Zornia bracteata, 263
ZOSTERACEAE, 614
Zostera marina, 614
Zosterella dubia, 779
Zoysia japonica, 750
ZYGOPHYLLACEAE, 337